620 Wild Plants of North America

—Fully Illustrated

Financial Sponsors

Anonymous

Robert Bateman

S.M. Blair Family Foundation

The DeFehr Foundation

Ducks Unlimited (Canada)

Inco Canada (Manitoba Division)

Louisiana-Pacific Canada

Manitoba Culture, Heritage and Tourism

Manitoba Department of Conservation

Manitoba Hydro

Manitoba Sustainable Co-ordination Unit

Monsanto Canada Inc.

Nature Manitoba

Thomas Reaume

Thomas Sill Foundation

620 Wild Plants of North America

—Fully Illustrated

Tom Reaume

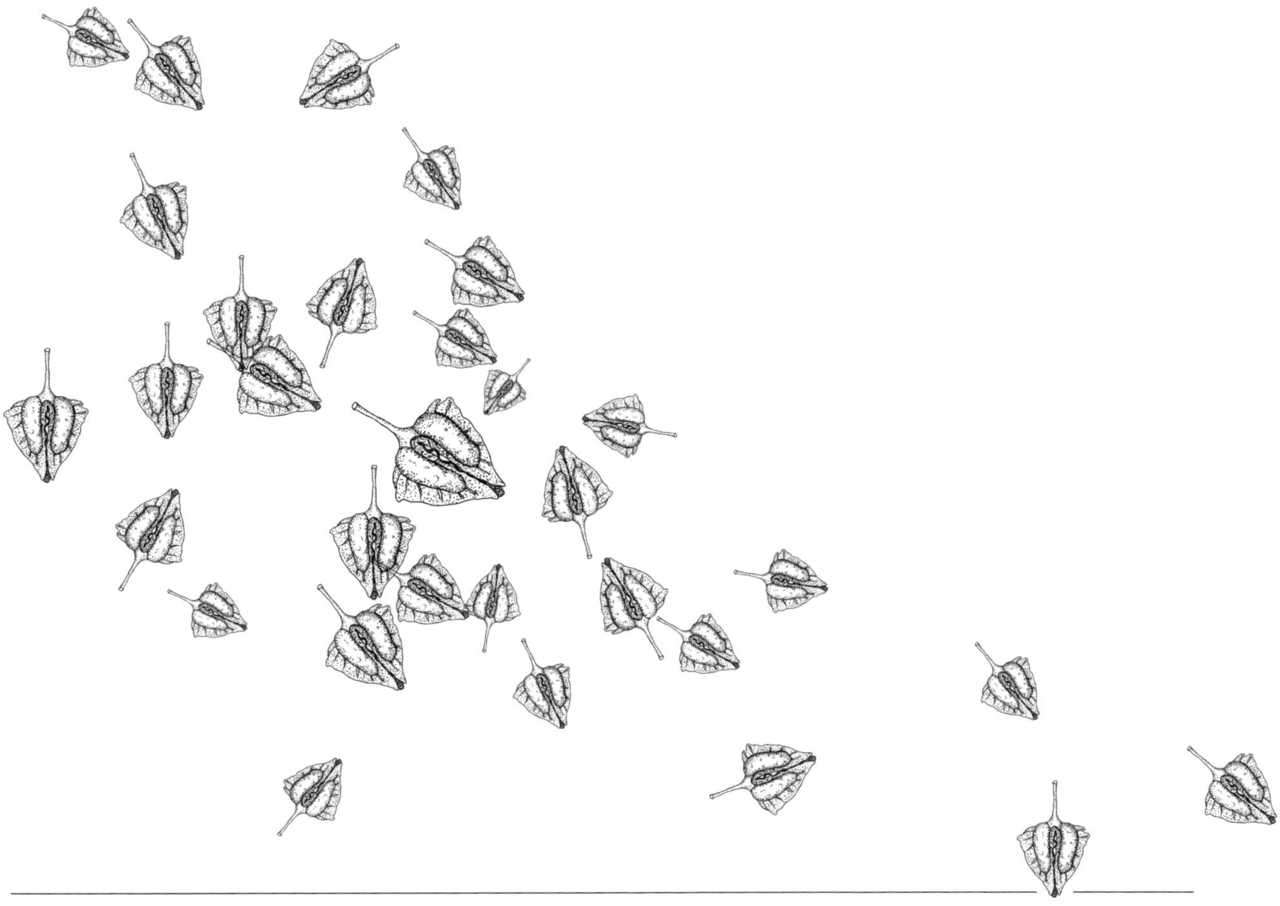

UNIVERSITY OF REGINA

2009

Canadian Plains Research Center

Copyright © 2009 Nature Manitoba

COPYRIGHT NOTICE All rights reserved. No part of this work covered by the copyrights hereon may be reproduced or used in any form or by any means—graphic, electronic, or mechanical—without the prior written permission of the copyright holder. Any request for photocopying, recording, taping or placement in information storage and retrieval systems of any sort shall be directed in writing to the Canadian Reprographic Collective.

401 - 63 Albert Street
Winnipeg, Manitoba R3B 1G4
204-943-9029
mns1@mts.net
www.naturemanitoba.ca

Library and Archives Canada Cataloguing in Publication

Reaume, Tom, 1944-
    620 wild plants of North America : fully illustrated / Tom Reaume.

(Canadian plains studies 0317-6290 57)
Includes bibliographical references and indexes.
ISBN 978-0-88977-214-4

    1. Plants—North America—Identification. 2. Plants—North America—Pictorial works. I. University of Regina. Canadian Plains Research Center II. Title. III. Title: Six hundred twenty wild plants of North America. IV. Series: Canadian plains studies ; 57

QK110.R436 2009   581.97   C2008-907957-4

Text, maps, illustrations, research, design and layout by Tom Reaume.
Contact Tom at tj333reaume@yahoo.com

Cover Design: Duncan Campbell, Canadian Plains Research Center

Printed in Canada.

Published by
Canadian Plains Research Center
University of Regina
Regina, Saskatchewan, Canada S4S 0A2
canadian.plains@uregina.ca
www.cprcpress.com

Distributed by
University of Toronto Press
5201 Dufferin Street
Toronto, Ontario, Canada M3H 5T8
utpbooks@utpress.utoronto.ca

The publication of this book was made possible through the financial support of the George Ira Hanson Trust.

We acknowledge the financial support of the Government of Canada through the Book Publishing Industry Development Program (BPIDP) for our publishing activities.

We acknowledge the support of the Canada Council for the Arts for our publishing program.

    Canada Council for the Arts   Conseil des Arts du Canada

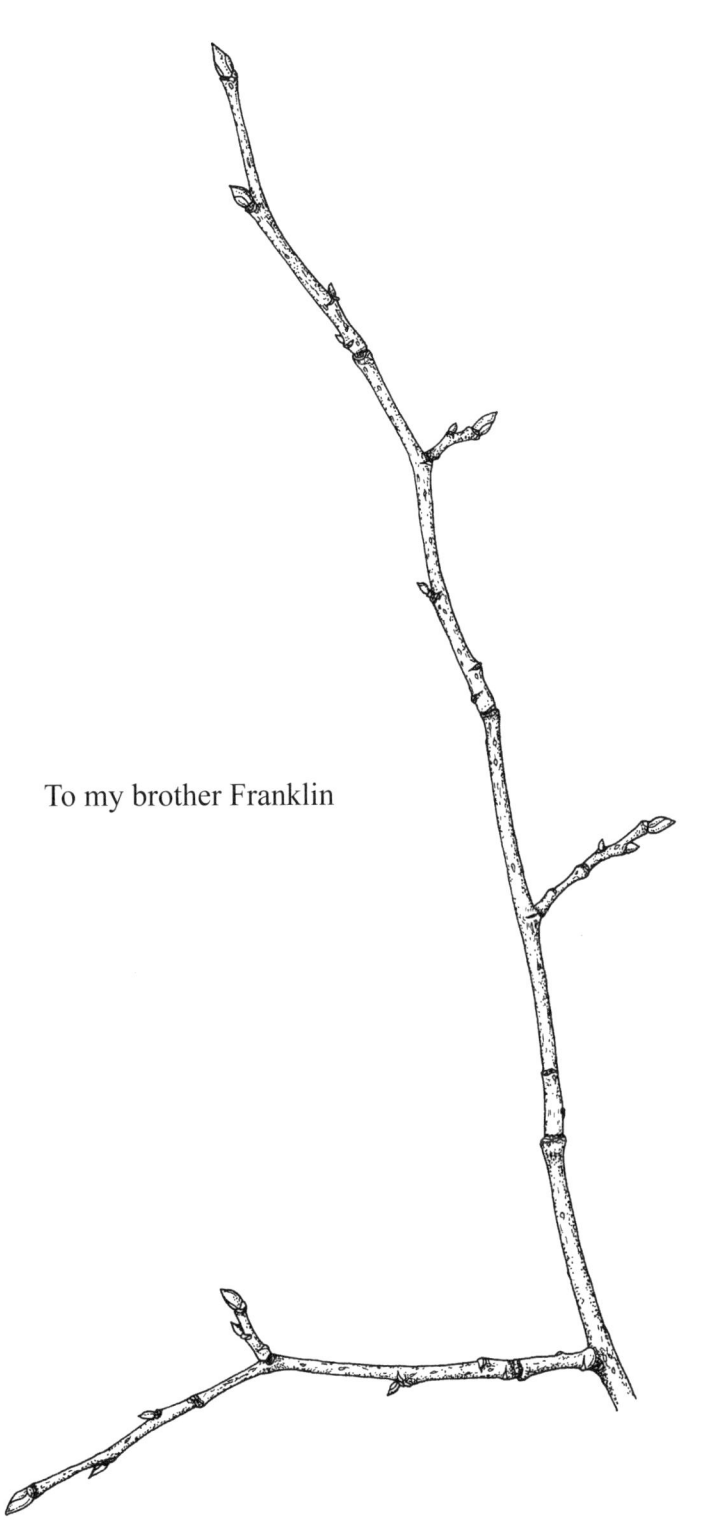

To my brother Franklin

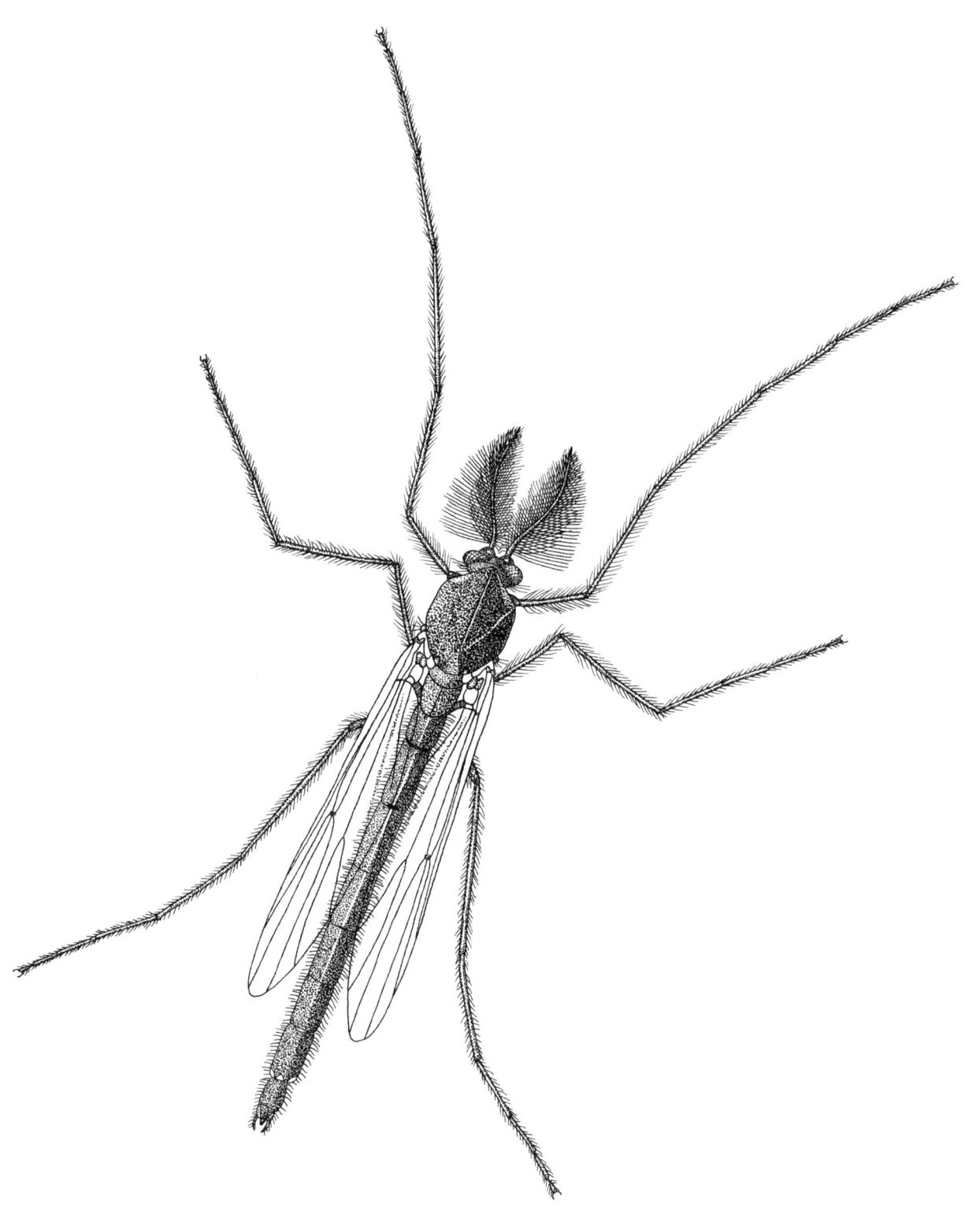

Contents

Financial Sponsors	2
Foreword	8
Preface	9
Acknowledgments	10
Introduction	13
Fieldplay	14
Arrangement of Species	14
Names	14
Illustrations	15
Species' Descriptive text	19
Measurements	19
Wildlife use (w)	20
Native prairie plant restoration (p)	20
Maps of plant ranges	21

MAPS	
1	24
2	25
Family—Species List	27
PTERIDOPHYTES	47
GYMNOSPERMS	59
ANGIOSPERMS	
Dicotyledons	71
Monocotyledons	521
Glossary	701
Abbreviations	724
References	725
Index	747
Short Index	inside back cover

Foreword

What is the best way to learn about plants? I have been asked this question, and have asked it of myself, many times over the decades I have been teaching plant taxonomy at Luther College at the University of Regina. Learning about plants is approached in so many ways. Some learn to recognize plants by location, habitat, flower color, and other characteristics visible to the naked eye. Others are intrigued by the minute differences among species, poring over plants with a hand lens. For many, knowing plants is essential to their jobs in forestry, resource industries, conservation, environmental management, and inventories. For all of these jobs it is essential to understand the morphology of plants: what the different parts look like, how they are described in floras or in wildflower guides, how to interpret what you see as you look through a hand lens or a microscope, and what to look for in distinguishing among similar species using a key. Often, the greatest challenge is in translating word descriptions, which typically use very technical language, into an image of what the feature looks like in a living plant. Tom Reaume's *620 Wild Plants of North America* is a valuable, new resource in addressing this challenge.

This book makes it possible for all of us—skilled and amateur botanists and naturalists alike—to understand plants better. It includes exquisitely detailed and skillfully rendered drawings of species from the northern Great Plains, each labeled and showing the parts for learning morphology and identifying species.

The usefulness of these drawings, however, is not limited to the study of those species shown; the drawings are, in fact, also helpful in understanding related plants. For example, in keys to the species of club-mosses it is necessary to compare the sporophylls to the sterile leaves and to determine the placement of the cone (strobilus). Reaume's drawings for Ground-pine (*Lycopodium dendroideum*), Stiff Club-moss (*Lycopodium annotinum*) and Ground-cedar (*Diphasiastrum complanatum*) illustrate and label the cone, the sporophylls, and many more structures. The reader can readily transfer this expertise to the identification of all club-mosses. Similarly, in looking at sedges, it is necessary to determine the characteristics of the perigynium, the sac-like bract enclosing the pistil. Reaume has drawings for 34 different species of sedges (*Carex*), clearly illustrating all of the key features of perigynia in general and of these species in particular.

The value of this book, therefore, goes far beyond its illustrations of 620 species. Its greater value is in opening the door to understanding the unique morphological features of plants in all of the major families represented in the flora of central North America and using this information for accurate identifications. In short, one can learn technical botany from this book.

620 Wild Plants of North America should be used as an illustrated companion to floras and wildflower guides. Cross-referencing among sources is facilitated by the inclusion of common names and synonyms for Latin names. You will also want to read this book, however, simply for its beauty—the beauty of the drawings, the beauty of the plants shown, the beauty of the poetry and ecological insights, and the beauty of this part of the world.

Mary A. Vetter
Academic Dean and Professor of Biology
Luther College at the University of Regina

Preface

Overhead the geese are a line,
a moving scar. Wavering
like a strand of pollen on the surface of a pond.
— Anne Michaels

To increase our awareness of this green-and-blue planet that sustains us, it is our privilege to rise at dawn, breakfast, and travel under June's sapphire sky to the nearest natural area. There we can revel in the freshness of the morning, touch a flower's yellow petal, absorb the clear melody of a Western Meadowlark, track the flight of a White Admiral, watch Richardson's Ground Squirrels play on a nearby knoll, sit at the edge of an animated slough and wonder how plants came to be this way.

This botanical book is presented from a naturalist's viewpoint. It is a picture book. By Ernest Rutherford's measure, "All science is either physics or stamp collecting." In the spirit of stamp collecting, this publication will serve as a reference for those wishing to learn how certain plants look, their names, parts and general range.

Although 620 species of wild vascular plants are included here, they represent only a thin slice of the total species inhabiting the area within the partial North American map on which the range of each species is presented. Nevertheless, artists and aristocrats, botanists and bureaucrats, developers and divas, farmers and financiers, nurses and naturalists, politicians and plumbers, wildlife managers and window makers, and many others should find the descriptions useful.

Over the years I have grown weary of botanical books and field guides filled with tiny unlabeled pictures. As a counterbalance, each of my plant species benefits from the luxury of a letter-size page with readable print, a map of its range and large, labelled drawings in ink.

In a book on the flora of a state or province, all the species are described. In such a book, dichotomous keys are the best way to reach the identity of a plant. In this book, which is thoroughly illustrated but not comprehensive with regards to the number of species, paired keys are unnecessary. An alphabetical arrangement by family and species is more suitable. Looking through the book over the winter will help you to become familiar with some of the individual species and family characteristics. It should be used in combination with several other field guides and floras.

I shunned the familiar botanical concepts of "weed" and "waste places." From a naturalist's perspective, all plants are native to this planet and were drawn with little thought to their place among undulating human attitudes.

Waste places were not apparent. Bogs and brush, ditches and drylands, farms and fields, riverbanks and railways, sidewalks and sloughs, each full of magic, made the summers pass much too quickly.

Through a blend of imagery and text, along with poetry and decorative drawings, I have tried to add some visual liveliness to the pages.

Although not formally trained in botany or as an illustrator, my interest in plants along with my drawing skills have grown over the years.

Acknowledgments

Modern man, the world eater, respects no space and no thing green or furred as sacred. The march of the machines has entered his blood.

— Loren Eiseley

This book is one more building block depicting the botanical fabric that dresses our planet. It could not have been produced without the previous work of botanists, technicians in herbaria, curators, illustrators, naturalists, collectors, scientists, managers, researchers, taxonomists and biologists who have walked and climbed over much of North America and published and illustrated their findings on plants in books, monographs and journals.

A crystal vase of rushes and lilies to my mother Elizabeth for giving me the patience and ability to draw and for quietly and slowly revealing to me how pleasurable it is to keep the artistic spirit alive in daily undertakings.

Wild prairie roses to my contemporaries for their assistance in many large and small, but always important, ways. Terry Bellhouse, Debbie Friesen, Herta Gudauskas and Susan McLarty at Nature Manitoba (formerly the Manitoba Naturalists Society) helped throughout the years with computer operations, fund-raising and financial administration.

Patrick O'Connell (1944–2005) was commissioned to write poems that provide my readers with a different style of imagery on plants. He is also responsible for my portrait on the back cover.

From the University of Winnipeg, Richard Staniforth reviewed and commented on my work in progress and provided a letter of support to enliven my grant applications. Anne Adkins, always generous, gave constructive comments, drove me one sunny morning to a field replete with dazzling, fringed white orchids, one of which I drew, and provided a dissecting microscope that allowed me to work effectively over long days at home.

Red Windflowers to Karen Johnson and Diana Robson at the Manitoba Museum who provided access to herbarium specimens, maps and books, and wrote letters of support for my grant applications. Kevin Szwaluk assisted in a variety of ways.

From the University of Manitoba, Gloria Keleher provided additional range map information and reviewed some species. Bruce Ford assisted with plant names, text and maps, a letter of support, and in the clarification of some *Carex* (sedge) species.

Jackie Krindle, Janet Moore, Marilyn Latta, Elizabeth Punter, Kim Ottenbreit, Diana Robson, Catherine Foster and Al Rogosin were additional reviewers of text and images. Jackie also took the time to check flower colors. Mr. Rogosin, with enthusiasm, wit and generosity, also showed me some botanically interesting sites in and around Brandon, Manitoba. His ability to identify plants from my over-the-phone descriptions still amazes me.

Naturalists and educators at the Living Prairie Museum in Winnipeg directed me to certain species and gave names to unfamiliar plants.

An azure jar of Golden Alexanders to Lisette Ross and Pat Caldwell at the Institute of Wetland and Waterfowl Research, Oak Hammock Marsh Conservation Center, who made my work more efficient and enjoyable. Chicory to Bob Laidler of Ducks Unlimited who arranged transportation for me to and from the Center each day.

Kimberly Caldwell and Pat Caldwell of Wildlife Landscapes developed the wildlife use rating for plant species of particular value.

Acknowledgments

Bonnie, a librarian at the Centennial Library in Winnipeg, became very familiar through her voice by leaving numerous messages on my phone about the latest botanical book arrival, which she obtained for me through the interlibrary loan system.

Brian Mlazgar, the Publications Coordinator at the Canadian Plains Research Center, University of Regina, responded quickly and favorably to my initial contact. Working with Donna Grant (CPRC), we were able to coax this visual book into being. Her vision matched mine and gave the book enough sunlight to grow. She created the index and was very helpful throughout the year it took to prepare the manuscript for printing.

Dean Nernberg and Eva Pip helped with the identity of a few species.

At Wellman Lake, Nebrazka Williamson harvested fruit of Yellow Pondlilies for a drawing and helped to count the seeds.

An appreciative nod to David Scott and Jackie Krindle for comments on the opening prose.

Pink Wintergreen to Peggy Ann Ryan and Vernon Harms at the University of Saskatchewan for allowing me to spend several months working at the W. P. Fraser Herbarium, where I slowly completed dot-distributional maps from the specimens' labels. They found time to answer my questions and made me feel comfortable.

Through the good offices of Anna Leighton I was able to complete the Wood Lily. She arrived at the W. P. Fraser Herbarium one morning with fruit. She did not bring a tasty pome, but rather some dry, dehiscent lily capsules and seeds eager to be models.

C. C. Chinnappa graciously permitted me several days of quiet time to add information on distribution from specimens at the University of Calgary's herbarium.

Gentians and sedges to several individuals who put together dot-distributional maps or provided computer printouts of plant locations for their area: Erika North at Lakehead University in northwestern Ontario; John Hudson for identifying a *Carex* (sedge) or two, and subjecting himself to the drudgery of converting information on herbarium species' labels to maps for about fifty species in Saskatchewan; Catherine Seibert at Montana State University for adding recent information to her maps and arranging the pages by family and genera; Joshua Brokaw at Washington State University for maps on distribution for some species; Lawrence Magrath at the University of Science and Arts in Oklahoma for maps; Roberta Mason-Gamer at the University of Idaho for a computer printout on the distribution of grasses; Mark Widrlechner of the University of Iowa (Ames) for his publication on the *Rubus* species; Robert Kaul of the University of Nebraska at Lincoln for providing a very usable computer printout of the distribution of vascular plants by county and for sending two recent papers on the Cyperaceae—Sedges in Nebraska which added to my pleasure in creating this book and David Sutherland of the University of Nebraska at Omaha for mailing about 80 maps with species' ranges. He, Robert Kaul and Steven Rolfsmeier published *The Flora of Nebraska* in 2007. John Pearson at the Iowa Department of Natural Resources sent me maps showing the occurrences of 21 rare species in Iowa.

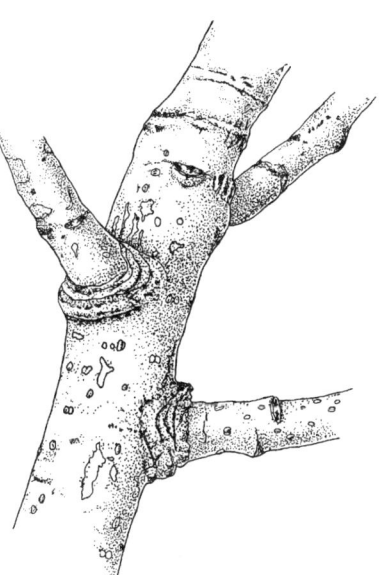

Willows and bedstraws to the many publishers and individuals who gave copyright permission to adapt maps from their publications and web sites to my style of expression. Included are: the University of Toronto Press for maps in E. Moss's book, *Flora of Alberta*, 1959, revised by J. Packer, 1994; *Pacific Northwest Ferns and Their Allies*, 1970, by T. Taylor; the Canadian Museum of Nature for maps in *Rare Vascular Plants in the Canadian Arctic*, 1993, Syllo-

Acknowledgments

geus 72, by C. McJannet, G. Argus, S. Edlund and J. Cayouette; the Minister of Supply and Services, Printing and Publishing Division, Ottawa, for maps in *Trees In Canada*, 1995, by J. Farrar, and *Ferns and Fern Allies of Canada*, 1989, by W. Cody and D. Britton; the Southern Illinois University Press for maps in the 1978 *Distribution of Illinois Vascular Plants*, by R. Mohlenbrock and D. Ladd; Amy McPherson at the Missouri Botanical Garden Press for maps from G. Yatskievych's 1999 *Steyermark's Flora of Missouri*, vol. 1, and J. Steyermark's 1963 *Flora of Missouri*, vols. 1 & 2, along with some information from the *Annals of the Missouri Botanical Garden*; the University of Nebraska Press for maps in *Common Legumes of the Great Plains: An Illustrated Guide*, 1989, by J. Stubbendieck, E. Conrad and B. Jansen, and *North American Range Plants*, 1997, 5th edition, by J. Stubbendieck, S. Hatch and C. Butterfield; the University of Iowa Press for map information in *The Vascular Plants of Iowa: An Annotated Checklist and Natural History*, 1994, by L. Eilers and D. Roosa; Edwin Smith for maps in his 1988 book, *An Atlas and Annotated List of the Vascular Plants of Arkansas*; the Utah Museum of Natural History for maps in *Atlas of the Vascular Plants of Utah*, 1988, by B. Albee, L. Shultz, and S. Goodrich; the University of Texas Press for maps in *The Legumes of Texas*, 1959, by B. Turner; Koeltz Scientific Books for maps in *Atlas of Northern European Vascular Plants North of the Tropic of Cancer*, 1986, by E. Hulten and M. Fries; Peter Rice at the University of Montana for e-maps of the weedy species at (http://invader.dbs.umt.edu/) and Ronald Hartman for e-maps of the state of Wyoming at (www.rmh.uwyo.edu).

A blue bottle of Graceful Goldenrod to Amy Farstad of the Biota of North America Program at the University of North Carolina, Chapel Hill, who explained some recent name changes on the BONAP website and provided a few range maps to accompany the new names.

John and Constance Taylor of Southeastern Oklahoma State University at Durant kindly sent me a copy of their 1994 book, *An Annotated List of the Ferns, Fern Allies, Gymnosperms and Flowering Plants of Oklahoma*.

Meadow Blazingstars to publishers and editors of scientific journals for copyright permission to adapt some of their maps and other bits of information: Anna and Ted Leighton for the *Blue Jay*; Mr. Zimdahl for *Weed Science*; the National Research Council Research Press in Ottawa for *The Canadian Journal of Botany*; Robert McIntosh of *The American Midland Naturalist*; Karl Niklas at *The American Journal of Botany*, and Barney Lipscomb for *Sida*.

Finally, a basket of Showy Sunflowers (think of a Van Gogh painting) to the uncommon individuals who felt this project was beneficial to the people of Manitoba and beyond. Their financial support allowed me to spend glorious days in the field and at the computer during the creation of this book. My apology if I have forgotten to mention someone.

The British philosopher Robin Collingwood expressed my feelings during this twelve-year project quite accurately when he wrote, "Perfect freedom is reserved for the man who lives by his own work, and in that work does what he wants to do." I have been very fortunate.

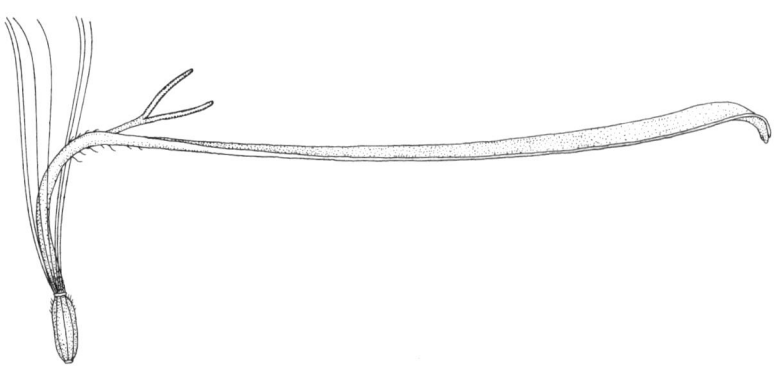

Introduction

It is, perchance,
that sweet scent of the earth
of which the ancients speak.
— H. D. Thoreau

Beauty, as much as science, is the driving force behind the creation of this book. Throughout the centuries, botany and beauty have been the eternal twins courted by all cultures. Whether I was looking intently at a whole plant while surrounded by field and sky, or at a part of it through a dissecting microscope in the seclusion of my home, beauty remained a constant companion. The subtle, the inconspicuous, the flagrant, all contributed to my visual pleasure and learning. David Gross reminds us that, "At the fundamental level nature, for whatever reason, prefers beauty."

Prairies and woodlands have a beauty which should be savored up close. One has to sustain a genuflection to appreciate and show respect to the specific beauty of a grass or sedge. In contrast, the western view of the projecting mountains, grand and majestic like an abstract painting by Robert Motherwell, is best appreciated from a distance, to encompass the jagged sweep of the whole.

Prairie mythology is as rich and deep as the soil created by the plants. Whitcomb (1994) reflects, "The plains are best discovered by following the sun. As one goes from east to west, they seem always to be rising. Some Native Americans believe that there is a place where the plains rise so high that they meet the sky, and that in that place the blue prairie flowers were created from the sky and the red and yellow flowers were painted by the sun."

Botany can lead us in many directions: in rangeland we can watch the breadth of a warm morning's light stretch lazily across a tall grass prairie; at a marsh we might pretend to ride the wind-induced waves flowing through a wild stand of cattails; at a bog we can inhale the aromatic soft scent after a summery rain.

A plant, any plant, is a remarkable concentration of place, spirit, movement, culture, beauty, relationships and lifeforce. Although fragile, plants should not be taken lightly. Those of us accessible to the spirit of a living plant realize this spirit is not found in the herbaria of the world, or in a painting, however imitative. Nor do we need a deep wilderness experience to appreciate the essence of plants. An old fertile ditch replete with species is a wonderful place to spend a summery morning. There the plants, tangled and lush, consummate an eternal light, transforming it, giving more than taking.

Concerned with primary producers, botany provides a warm atmosphere for interspecies fraternizing. Through quiet observation and a little study, we can begin to appreciate the connectedness of plants to other plants, or to microbes, fish, insects, mammals, reptiles, invertebrates and even birds.

Like the grand passage of constellations overhead in the night's sky, earthbound plants contribute to the many natural rhythms in our lives: daily rhythms through the opening of a flower bud; weekly rhythms with the appearance of an inflorescence where only stem and leaves grew before; and seasonal rhythms

Introduction

where plants in the landscape change from the many tough tones of brown in winter, reminiscent of a tempera by Andrew Wyeth, to summer's promotion of green, red, yellow and blue as exaggerated in an oil by Vincent Van Gogh.

Fieldplay

In Manitoba, plants were located, drawn and identified in Winnipeg, at Oak Hammock Marsh, in Brandon, near Pine Falls and at the Goldeneye Field Station on the shore of Wellman Lake in the Duck Mountain Provincial Park. One glorious summer of fieldplay took place in and around Saskatoon, Saskatchewan.

This book's 620 species of trees, shrubs, vines, wildflowers, grasses, sedges, rushes, clubmosses and horsetails were drawn from living plants in the field.

Arrangement of Species

It is impossible to organize a botanical book that is user-friendly to all of its readers. Everyone who uses this publication carries a different level of botanical awareness, understanding and criticism.

It is no exaggeration to remind my readers that this book illustrates only a handful of the species growing in the state or province where they live (*see* map 1, page 24). Seventeen of the largest families in this book have their characteristics described at the

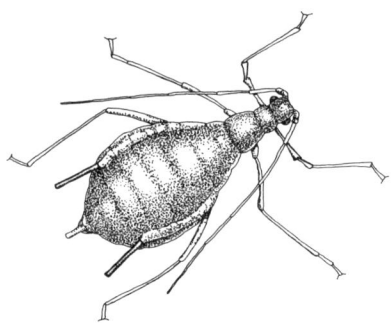

start of the family. In North America there are an estimated 34,500 plant species (Govaerts, 2001).

The families and species are arranged alphabetically within four broad categories according to Harms' (2003) *Checklist of Saskatchewan's Vascular Plants*. I begin with fern allies, horsetails and a few club-mosses, followed by coniferous trees and shrubs, then finish with flowering plants, the dicotyledons (dicots) and the monocotyledons (monocots). Within each broad category, plants are listed alphabetically, first by family, then by genus and species.

Names

In the top left corner of each species' page, the family's common and scientific names are listed; in the top right corner you will find a general description of the type of plant—for example, wildflower, tree, shrub—followed by the color of its flower and the type and number of floral parts. Just above the line near the top of each species' page are the plant's common name and its scientific name followed by the nomenclatural authority. The common names follow those in *Budd's Flora of the Canadian Prairie Provinces* (last reprint 1994). The scientific name and authority used are the accepted names according to ITIS (Integrated Taxonomic Information System; http://www.itis.gov/). Latin synonyms are included at the end of the descriptive text on each page. Additional common names are listed at the base of each species' page. One should realize that botanical common names are not regulated by a committee as are the common names of birds. Consequently, you may call a plant by any common name you like or are familiar with. Using a different name certainly will not produce a ruction from me. Nor will the Botany Police break down your door at 3:00 a.m. and do odd things to your involucral bracts.

My only wish is that more common names could be as singular as Madonna in my species, or Chicory among plants, rather than Great Plains White Fringed Orchid, if only to fit more easily within the narrow columns we assign to a book's index. On the other hand, some names are full of pleasing imagery—Russian Blue Lettuce, Two-grooved Milk-vetch, Common Milkweed, Fall Sneezeweed, Rough Cocklebur and Field Mouse-ear Chickweed. Suggesting place, color, texture, abundance, season, sound and

Introduction

interspecies association, the names reflect the ties we humans are fond of making when describing our natural world.

There are no taxonomic breakthroughs. I simply tried to keep up with recent name changes, of which there are more than a few.

Species in this book that were under the genus Aster now fall under the genus *Symphyotrichum*. Most of those species in the genus *Scirpus* are now included within the genus *Schoenoplectus*.

These are the two obvious genera changes. Other recent name changes are included.

My plant identifications are based on texts I have read, along with illustrations and photographs viewed. I did not compare my plants to any type specimens. I am solely responsible for any misidentified species or improper labels.

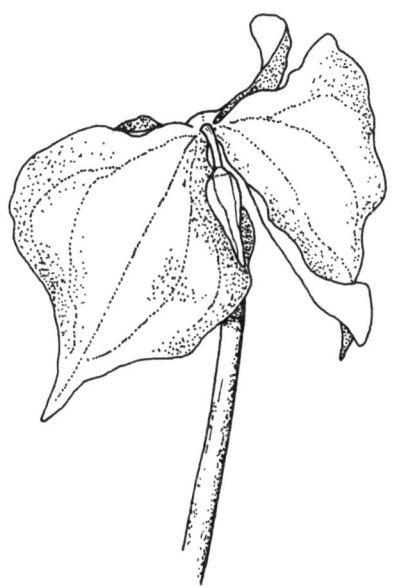

Botany is a nascent science and research is the cultural force enlightening us about the relationships among plants at various taxonomic levels. Research methods employing a scanning electron microscope (SEM) to picture an achene's surface, chromosome counts from cells in a root's tip, DNA sequencing, etc., can reveal differences between plants. It is the researcher, however, who determines at what level the differences are significant. She decides whether new varieties, species, genera or families are warranted. It is a practiced art to interpret the results and is therefore open to misinterpretation and disagreement.

Research uncovers new identities and scientists provide new names. But even now synonyms (historical names) play a strategic role in lowering my botanical morale. Due to the stagy number trailing behind some species, synonyms make me pant trying to keep up with them when reading botanical books and journals. They linger in the literature like old shoes in a closet. The synonymic race, although not an Olympic event, is fueled, at least in part, by botanical rivalries between countries, universities and taxonomists.

Illustrations

"Art is the universal language, and careful drawings after life are made in the image of nature itself, universally accessible to all. Those who would study the book of nature do not require Latin to read the characters, to determine the essence of the living thing; it can be read in line and paint" (Dickenson, 1998).

Plant and animal drawings, as well as abstract and geometric shapes, have been produced by humans for thousands of years. In France, Spain, Australia and elsewhere, people used a ritualistic code between 23,000 and 12,000 years ago to produce images on walls. Ephemeral drawings in mud or sand surely go back much further in time as women and men communicated their ideas about flora and fauna pictorially. Historically, a twig made a wonderful drawing tool, as it does today.

In comparison, abstract text relying on the creation of many different alphabets developed much later, between 5,000 and 4,000 years ago, often from existing pictographs.

Today our ritualistic code begins with a blank piece of paper. Upon this paper are placed curved lines and dots. This is how we represent plants in two-dimensional black-and-white imagery. To do

Introduction

this it is necessary to teach the hand to see, to work on its own to give visual pleasure.

My initial pencil drawing of a plant is light and with mistakes, a simple outline. Later, when transformed into a pen-and-ink rendition, it somehow appears more precise or real. The use of texture and shading allows a flat figure to achieve a modest success in the third dimension. Fruit, or a plant's seed, is a favorite botanical object of mine to draw, especially a seed with texture.

Think of a seed. It is a compression of the entire history and future of a species. With its biological clock set to nature's time, a seed waits days, months, years or decades in the soil before heat, light, cold, moisture and microbes kickstart its germination. While drawing a fruit or seed I naturally think of the still life tradition by artists like Paul Cézanne. However, Anne Michaels (1997) explains it this way: "A still life isn't about fruit, but about time."

The images in this book should not be considered fine art. They are simple, rather crude drawings of plants and their parts. Although fun to do, they involve no creativity or leap in the imagination, surely two

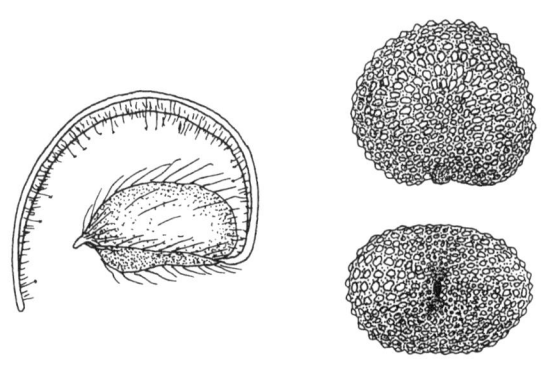

important prerequisites for the sense of wonder and excitement one feels while viewing fine art. Georges Braque, the French painter, wrote, "Art is meant to disturb, science reassures." I believe my illustrations are much more reassuring than disturbing.

Although botanists require a finely-tuned visual acuity to distinguish species in the field and laboratory, few of them can readily draw a plant. Historically, even before the Swedish naturalist Carolus Linnaeus (1707–78), they relied more on textual descriptions than on a visual presentation. This is partly because scientific illustrations, although easily read, may include a little imaginative embellishment to fill the gaps when actual plant material is incomplete or dry and flat in a herbarium. For this reason there has developed an unnecessary split between text and images in botanical literature. Unlike birders, some botanists consider looking at images almost cheating. The use of dichotomous artificial keys, they insist, is the proper means to identification. However, "examining the way in which images and texts are interwoven offers new and clearer perspectives on the use of imagery and its role in thinking about the natural world." Furthermore, "the ability to counterfeit the appearance of a living thing meant that not only could the image explain a text, but it could also stand for the object itself" (Dickenson, 1998).

In botanical literature, a pen-and-ink drawing is still the preferred medium for clearly depicting the salient characteristics of a species and offering them up to the scientific gaze. In contrast, color photographs are, by definition, full of color. Contrary to general opinion, they often lack full clarity and are unable to reveal many details, especially among the finely flowered grasses and sedges.

The separation of a living plant into its smaller parts, and the subsequent measuring, drawing and labelling is performed out of curiosity. Each part has an aesthetic which should be appreciated as part of the whole.

Because there is an abundance of labels and directional lines on the pages devoted to each species, I have placed decorative drawings in other parts of the book. The drawings permit us to appreciate the form and beauty of plants and other species without attaching a name or label. Science fills in the details, often unnecessary details, which most of us can do without. Our green alphabet, used to recognize a plant's pattern, remains a visual whisper in our imagination long after we have forgotten the many details.

Introduction

It was not my intention to describe and draw each species in a thorough, standardized manner. The complexity of a plant's form decided what was shown and how on a page. Plants completed in the first few months of this project were not as thoroughly described as plants in the final years. The amount of detail noticed and described grew each year.

Nor was it my intention to provide all the species' pages with an identical layout. Not only would this prove uninteresting to a reader, it would also be an insult to the community of plants. I allowed the shape of some plants to suggest different ways to design a page, rather than forcing a plant to conform to a preconceived layout. A rigid design would indicate an artificial system of control which we still believe we must exert over nature. When the diversity and mixture of plants in the field are severely organized into a book, plants are presented the way some of us would like them to be in nature—neat and gardenesque. A sense of paternalistic domination rather than a relational dependence is fostered and maintained.

Claude Monet's impressionistic vision of aquatic plants (water lilies) at his garden at Giverny reveals more about the plant and its environment and how he related to it with a brush, canvas and oils than does my limited view using a technical pen with a 0.25 mm nib, white paper and black ink. All things considered, on most days I greatly prefer Monet's view of nature to mine.

Art and the simple botanical illustrations in this book can only be viewed through the cultural monocle one holds between the eye and the image. Our vision is continually being adjusted by our personal experiences and the culture in which we find ourselves inadvertently placed. Previous eyes have interpreted the natural world differently. As an example, compare a bird painted by John James Audubon in the early 1800s to one by James Fenwick Lansdowne in the late 1900s.

Like most naturalists I am a wanderer, moving easily about the fields, marshes, urban paths, and woods. Looking for a plant to draw teaches one to move with the speed of a prairie river in July. Walking slowly makes it easy to let the world pass through you.

Drawing plants in a city has its moments. Almost no one in a city has witnessed a man sitting on the ground next to a plant for an hour while drawing and measuring it. One day I did a drawing of Lesser Burdock. Finished with my field work, I retired to my home to convert the pencil drawing to one in ink. Unfortunately, I forgot to measure one part of the plant. The next morning I returned to find the burdock I was working with and all the burdocks in the surrounding half a block cut down and removed.

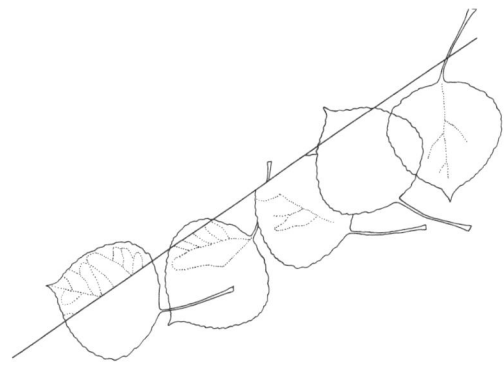

Obviously, someone was furtively watching me from their window as I completed a delightful drawing of this large plant. Possibly he thought I was a weed inspector. When I left he ventured out of his cave and solved his "perceived problem" with skill, stealth and a knife.

In the field a visually satisfying plant is selected. The colored form becomes the model for a drawing. Plants covered with dozens of open complicated flowers, as in some Asteraceae—Aster, are passed over in favor of a plant of the same species with a few open flowering heads and the remainder in bud. The shape of an inflorescence is easier to draw and appreciate from a drawing when not obscured by too many open flowers. My drawings are meant to reveal and clarify the structure of plants, not to impress with an astonishing number of dots. Detail was not sought

Introduction

for its own sake. After all, I had to locate, draw and identify almost one species per day to complete a reasonable number of plants each summer.

The initial pencil sketches were executed in the field while sitting on the ground next to the plant-of-the-day. This was a relaxing one to two hours. Sitting while drawing allowed me to touch the earth near a plant, to feel the earth's texture, warmth and moisture. I was privileged to mingle with the flora and momentarily become part of a fold in the earth's green skirt.

Drawing a plant in the field provided insight into the day's playful string of contingencies: shapes and textures of white-and-gray cumulus clouds, a magpie's black-and-white song, a bumblebee which nestled among yellow flowers for a nap, and the bounce of a tan coyote moving surely along the bottom of a dry ditch.

Once a reasonable pencil sketch was completed, the model plant was usually uprooted and carried home. The roots were washed and studied. If they supported a tuber or rhizome, and room on the page permitted, the lower part of the plant was drawn. Finally, the smaller parts of the flower were teased apart and drawn before the next sleep whenever possible. Viewing them through a dissecting microscope allowed me to describe and illustrate their details.

Plants were revisited a few times in the field until the fruit was ripe. The mature fruit and seeds were picked and drawn since their form and color may be useful in identifying a species. Also, the visual presence of fruit and a seed reaffirms the cyclicity of the species.

A leaf's outline and veins were traced using a light box—a useful shortcut, especially when a blade is compound. When a leaf of average size was drawn, and then scanned at 70 percent to fit comfortably on a page, its subtitle, **leaf x 0.7**, refers only to the leaf illustrated. In some cases the shape of a leaf is a more useful taxonomic indicator than size, although in Lesser Burdock, size does matter.

The pappus of an achene is often labelled "pappus *partial*." This means the full length is not shown since it could result in a pappus almost filling the page at the magnification indicated. As well, only several fine bristles of the pappus are drawn to preserve visual clarity (*see below*).

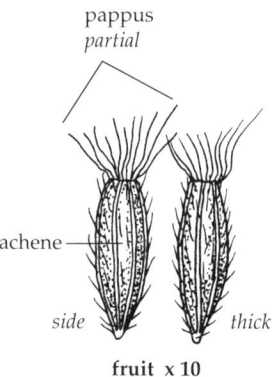

Some plant parts are presented in more than one view in order to appreciate their spacial properties. Italicized words such as *side*, *thick*, *dorsal*, *ventral* or *above*, etc., appear near the part to help orient the reader to a particular aspect.

In the vigorous pursuit of a simple life, all my original drawings, their copyright, and the book's royalties are donated to Nature Manitoba.

Introduction

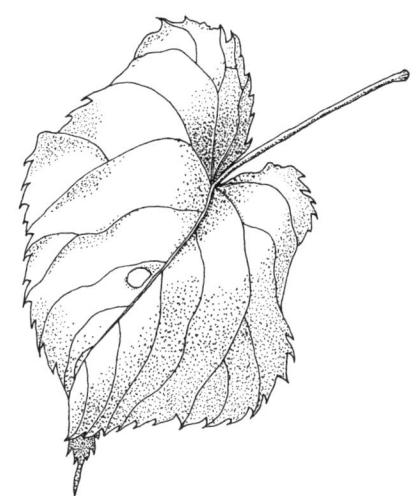

Species' Descriptive Text

The text describing each species is divided into four or five unequal parts:

■ **SKETCH** Briefly deals with plant type, height, lower and underground plant parts, habitat, and if the plant was introduced and is now naturalized. Monoecious or dioecious refers to the sexual lifestyle of the plant. A wildlife use rating (w) and native prairie restoration rating (p) are given for some species.

■ **FLOWERS** Provides information on color, blooming period, inflorescence and branching. Most of the parts of a flower are described. **FRUIT** is included at the end of the flowers to emphasize the continuity of a plant into its next generation. Fruit type and seeds are described. For the Brassicaceae—Mustard species I have followed Rollins' (1993) concept and used only silique as a name for the family's fruit. A silicle, previously applied to a short, wide fruit, is at times arbitrarily separated from the silique. Rollins limits us to silique, plus a description, drawing, or photograph as support.

■ **LEAVES** Describes arrangement, style and type, then the parts.

■ **STEM** Explains the type, shape, buds, color, solidity, node description in grasses, and thickness near the base.

■ **SYN** Lists one to a few recent synonyms. I noted synonyms as far back as H. J. Scoggan's 1957 book, *Flora of Manitoba*. Because the map covers such a large area, there is insufficient space or interest on my part to mention all the different subspecies, varieties and forms in the various populations. These can be found in more scholarly publications. The avid collector of synonyms should consult Kartesz's (1994) volume 1. For example, the Dandelion, *Taraxacum officinale*, has a history of at least 57 synonyms within three subspecies. Its present scientific name represents a roomful of research money. Other than botanists, very few people are interested in taxonomic revelations below the specific level.

Measurements

The millimeter does matter. Even for naturalists, tenths of a millimeter should not be too fine-grained. In Jonathan Weiner's book, *The Beak of the Finch*, based on the fieldwork of Rosemary and Peter Grant, the millimeter comes to life. Studying the relationships of Darwin's finches on the Galapagos Islands, equatorially clustered off the west coast of South America, the Grants noticed that finches with minutely different sizes of bills fed on food of different sizes which enabled some finches to survive periods of environmental change.

In the field, one to a few plants had their larger parts measured to obtain ranges (e.g., leaf blades 6–12 cm long). At home, the smaller parts were measured while viewing them through a microscope. Sometimes these measurements introduced new information for a species. The numbers were compared to those in books and journals and adjusted accordingly.

The larger, obvious parts (e.g., leaf length and width, inflorescence length, etc.) of pressed specimens in the herbarium at the University of Saskatchewan were measured to improve the size ranges. Even so, a range of measurements is rarely complete since there was no opportunity to measure specimens throughout a plant's range.

Where a measurement is difficult to obtain or limited to a single number (e.g., *c.* 3 mm long), the abbreviation *c.* for circa (approximately) is used before that number.

Introduction

"But measurements have meaning only to the extent that we respect their various limits. No measurements, when you get right down to it, are straightforward. All involve disentangling things that can't be separated, or quantifying things that can't be counted, or defining things you can't quite put your finger on" (Cole, 1998).

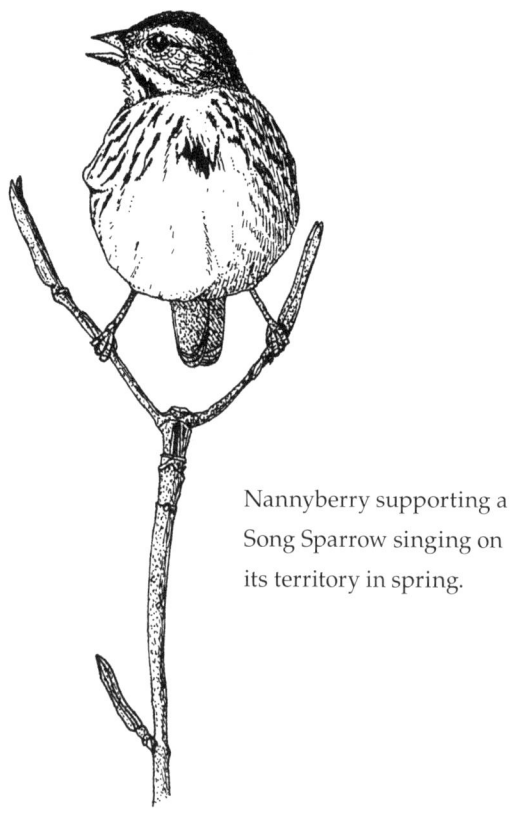

Nannyberry supporting a Song Sparrow singing on its territory in spring.

Working with plants, you quickly realize that botany is not so much about "getting it perfect," whether a plant's name, range or description, but about improving our relationship with our flora. As scientists enter their deep green phase with probes reaching beneath the skin of plants to help them re-work relationships, it is often a relief to abandon the lab and literature and spend a few moments under the big sky. Stepping into a place without walls or a roof, our senses settle on the arch of a leaf's blade, an aster dancing along the top of a ridge, or the labor of a bee on a blue corolla.

Wildlife Use (w)

All plants are used directly or indirectly by wildlife. Multidimensionally useful, plants produce oxygen, inhale carbon dioxide, create and bind soil with the help of microbes and provide food and shelter for all animals including people. The destruction of plant life in an area reduces the diversity of its wildlife.

For this book, wildlife is limited to birds and mammals. Wildlife use, as a rough initial measure, includes the utilization of a plant for food, cover or a place to nest. To rate a plant for wildlife use, two categories were used at the end of the SKETCH part of the text: **w1** indicates the plant was important to one or more species (e.g., *c.* 25% of their diet at certain times of the year), and **w2** indicates moderate use by several wildlife species or good use by one or two species.

Not every plant species can be given a wildlife use rating. Our knowledge of the interactions between plants and animals is limited. During training in the natural sciences, one generally specializes (due to personal interest) in plants or animals, rarely both. Even so, botanists and ecologists must find it odd when the captions of published photographs of birds and mammals, even in nature magazines, rarely name or even mention vegetation in the picture with the identified vertebrate.

Native Prairie Plant Restoration (p)

At the end of the SKETCH part in the descriptive text there may be a **p1** or a **p2**. A **p1** rating is for a species with high priority of use in restoring native prairie, whereas **p2** indicates a species with less priority in the restorative process. The rating is mainly applicable to the three Canadian prairie provinces of Manitoba, Saskatchewan and Alberta, and the northern states immediately below them. It is adapted, with permission, from *Restoring Canada's Native Prairies: A Practical Manual* (Morgan et al., 1995).

In Morgan's book each of the three prairie provinces is divided into four or five native prairie restoration zones. The number appearing after **p** is an average of the ratings provided for each species

Introduction

The Prairie-crocus blooms early in the spring.

in the restoration zones of the three provinces. Some zones in Morgan's book are left blank if the species is not found in that prairie type or the distribution of the species is unknown.

W1 and **p1** is the highest rating given to a plant for use both by wildlife and in the restoration of a native prairie.

Maps of Plant Ranges

Botanical maps come in two main styles. First there is the dot-distributional map, common in journals, monographs and some books. A dot can show where a species, now in a herbarium, was collected. Dots are used in three locational levels: **1)** a dot on a map may represent the "exact" location where one plant or several were collected; **2)** with less precision a dot is often placed in the middle of a state's county to indicate the species' presence somewhere in that county. This works well where the counties are small and numerous (eastern American states) and is less effective in the west where the counties are generally larger; **3)** at the extreme end, one black dot appears in a state or province, as in recent volumes of the *Flora of North America*, to indicate presence only.

Canadian prairie provinces are not divided into counties. The dots on our distributional maps indicate where a plant or a group of plants was collected.

The second style is a map showing the range of a species. Derived from dot-distributional maps, the range uses a continuous gray or colored area to show where a species usually occurs. Visually, it is easier to grasp.

I use a gray area to indicate where a species might be found. Not having the botanical resources of, say, the New York Botanical Garden at my disposal, my maps provide only a rough approximation. When converting dot-distributional maps to a gray area, it is sometimes difficult to decide how many sinuses or fingers to draw. How wavy should the boundary be? How much power to give a dot located near the edge of a plant's range is always a question in the mind of the map maker. The boundary becomes especially difficult to locate when the dots are few and widely scattered.

The hard-edge boundary of the gray area should not be considered a force of science or nature. Rather, view it as an indistinct boundary, like the edge of a cloud's shadow passing over the face of the prairie. Another person would probably draw the boundaries somewhat differently using the same information. Some of my maps are more accurate than others, depending on how much recent work has been published on a species and whether or not I was able to locate that information.

Keep in mind that within the gray area on a map there may be half a state or a large section of a province without the species in question due to mountains, forest, a lake, an expanse of desert, or simply a lack of botanical collecting.

The distribution of a species in one state or province can influence how a map is drawn in an adjacent state or province.

After spending more than one year compiling the information to produce these maps, including five months on dot-distributional maps for Saskatchewan,

Introduction

I am still not certain of the correctness of any map. Eventually the law of diminishing returns takes over (not to mention the law of diminishing grant money) and the search for more information on a species' range comes to a halt.

Behind the scenes, as they say on the set of a Hollywood movie, there exists one glumey floral secret which I found surprising. There are only a few Canadian and American herbaria with current dot-distributional maps on file for all species found in their province or state. Botanists and collectors associated with herbaria, except for a precious few, apparently lack the will, money, or the gene for mapping.

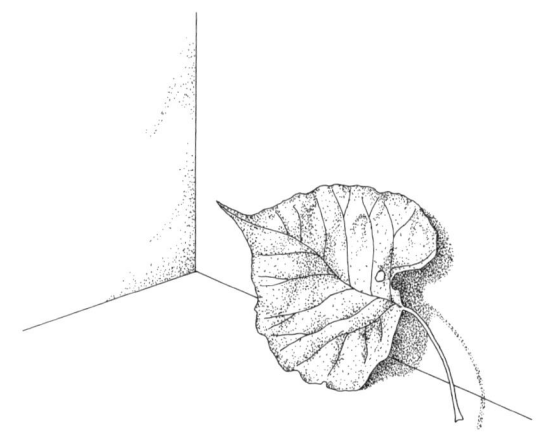

Common sense should dictate that once a plant has been collected and processed for placement in a herbarium's cabinet, a dot representing where the plant was collected would be entered on a provincial or state map for each species in a herbarium. Throughout the twentieth century, this mapping procedure has rarely taken place. I could not send a curator of a herbarium my list of species and expect photocopies of current dot-distributional maps a few weeks later.

Occasionally, a botanist is very protective of information on distribution and refuses to share it. Thankfully, most botanists, both in Canada and the United States, are quite willing to have the information utilized.

Criticism is easy; solutions to the map problem are more difficult. With herbaria in North America closing or losing their staff due to financial cutbacks, two possible solutions are: 1) botanists need more splash-and-dash to get a lot more money pouring into their departments; 2) where a herbarium is part of a University, make it mandatory for each student working on an advanced degree in the natural sciences to visit the herbarium and maintain dot maps for 30 species during their years of research. That way, someone with an interest in birds, beetles or bats might actually get to know a few plants by the time they finish their academic training.

Editors of botanical journals should place the word map in the index and list any species with a map of its range in their journal. This simple addition would make maps much easier to locate in the numerous journals. Presently, the scientific method of flipping pages is the only way to locate maps.

The maps in this book were puzzled together from many sources. Dickenson (1998) informs us that the first herbaria (dried gardens) refer to botanical gardens on the grounds of medical institutions. The earliest of these were established in Europe in the beginning of the 1500s, about the time Leonardo da Vinci painted the Mona Lisa. In those early years, plants, as well as illustrations of plants, were eagerly collected by personnel in herbaria.

I used recently published maps representing specimens in herbaria from the 1800s to the present. Information was also slowly mapped (a very boring job) from labels of specimens to which I was given access in herbaria in Saskatchewan and Alberta. Fortunately, the two largest herbaria in Manitoba have dot-distributional maps available for all their species. The curators placed these maps at my disposal. Also, in a wonderful spirit of scientific cooperation, unpublished/published dot-distributional maps arrived in the mail from botanists, curators and technicians at several Canadian and American herbaria. Some even took the time to produce local maps for many of my species.

Introduction

Obviously a dot representing a plant collected in 1860 may now represent the site of a natural history museum or a corn field. I wonder if any personnel at a herbarium made a fresh start on 1 January 2000 to separately map the collections that will be made in the twenty-first century? Considering the large number of plant species that will disappear in this ecologically dysfunctional century, it probably was a good idea.

Occasionally my maps contain one to many squarish dots outside the gray area. These are extralimital locations of a plant, which I decided not to include within the gray area. A dot may represent one to several plants, two collecting locations close together, or simply that the species is present in, say, the eastern part of a province, state, or county. In other words, check with local botanists, naturalists, personnel in herbaria, or local floras if detailed information is necessary.

The internet is a valuable resource (H. D. Wilson, 2001). Some state dot-distributional maps online are based on those from published books. As the *Flora of North America* volumes are published (two per year), their maps become available online at (www.fna.org).

Finally, when adapting published dot-distributional maps or even textual descriptions of ranges in books and journals to establish the ranges for this publication, it was impossible to recheck the original identifications or make adjustments for some recent taxonomic changes.

However, the transformation of those maps into gray areas on my maps may hide or lessen the impact of certain errors.

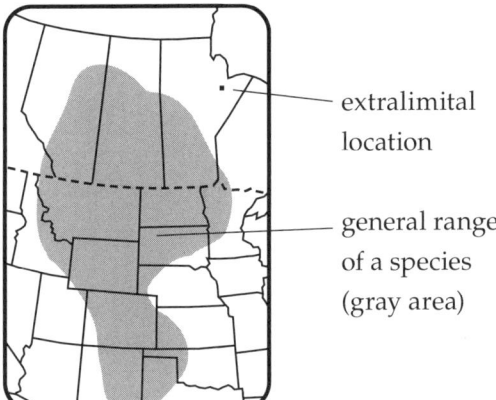

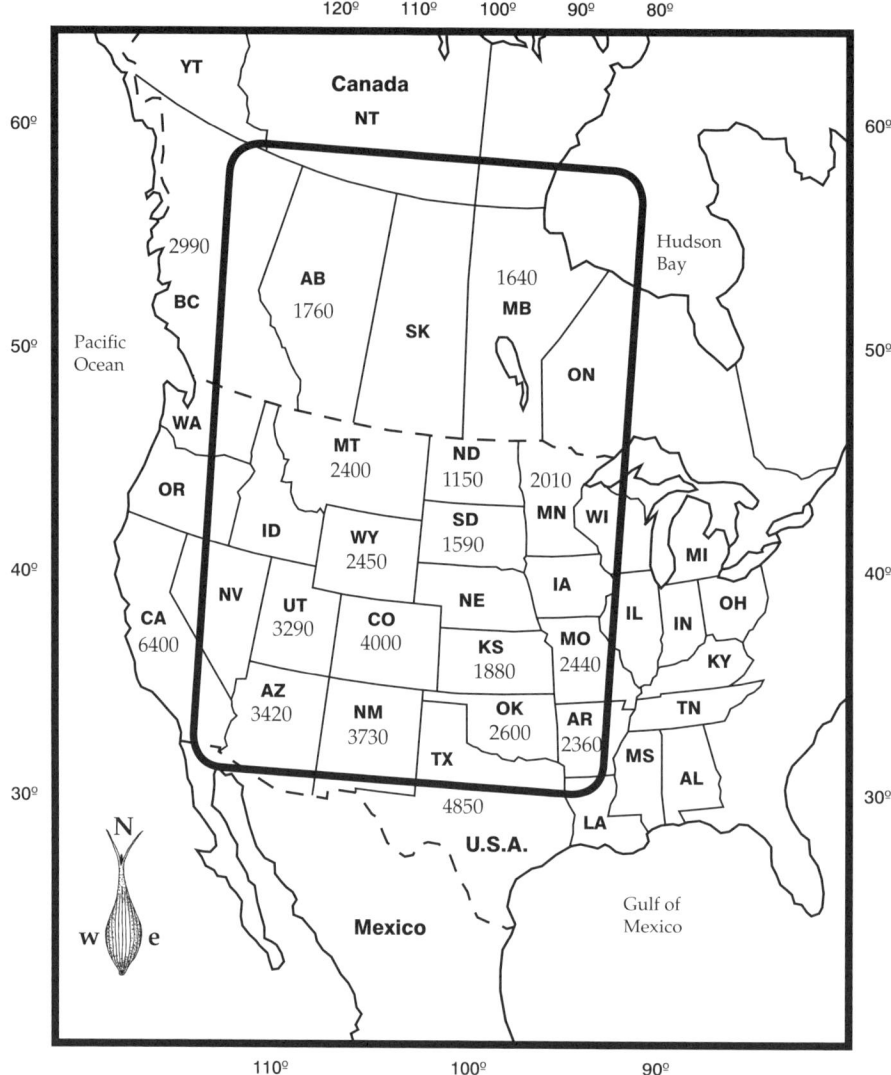

Map 1. Part of North America depicting Canadian provincial and American state boundaries. The rectangular insert, with rounded corners, is the area covered by the map for the range of each species. The numbers are the approximate number of vascular plant species in some provinces and states.

AB	Alberta	MB	Manitoba	OK	Oklahoma
AL	Alabama	MI	Michigan	ON	Ontario
AR	Arkansas	MN	Minnesota	OR	Oregon
AZ	Arizona	MO	Missouri	SD	South Dakota
BC	British Columbia	MS	Mississippi	SK	Saskatchewan
CA	California	MT	Montana	TN	Tennessee
CO	Colorado	ND	North Dakota	TX	Texas
IA	Iowa	NE	Nebraska	UT	Utah
ID	Idaho	NM	New Mexico	WA	Washington
IL	Illinois	NT	Northwest Territories*	WI	Wisconsin
IN	Indiana	NV	Nevada	WY	Wyoming
KS	Kansas	OH	Ohio	YT	Yukon Territory
KY	Kentucky				
LA	Louisiana				

* Includes the new Canadian territory of Nunavut.

24

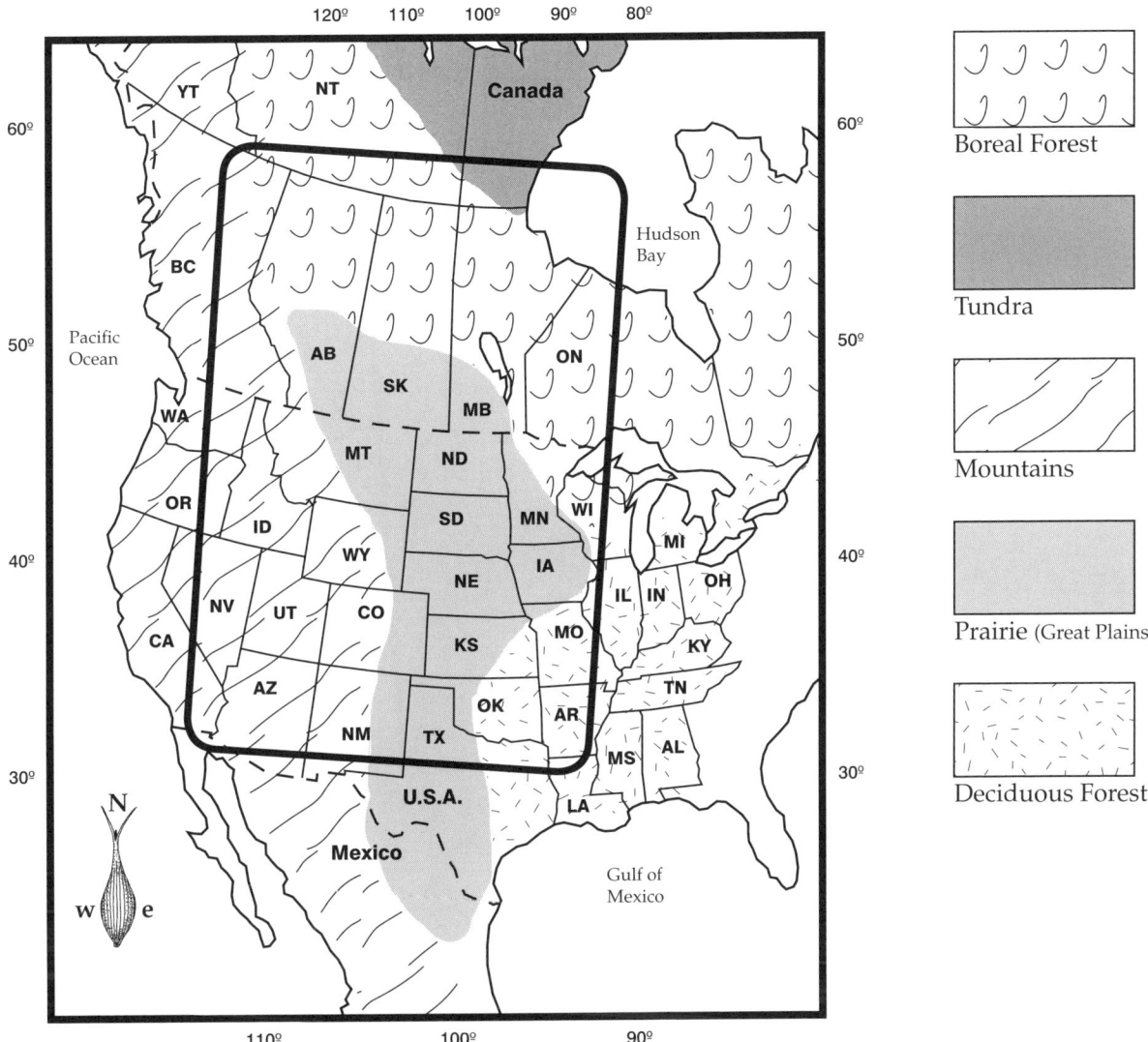

Map 2. Part of North America depicting Canadian provincial and American state boundaries. The rectangular insert, with rounded corners, is the area covered by the map for the range of each species. The general location of mountains and four vegetative types are presented with five patterns.

AB	Alberta	MB	Manitoba	OK	Oklahoma
AL	Alabama	MI	Michigan	ON	Ontario
AR	Arkansas	MN	Minnesota	OR	Oregon
AZ	Arizona	MO	Missouri	SD	South Dakota
BC	British Columbia	MS	Mississippi	SK	Saskatchewan
CA	California	MT	Montana	TN	Tennessee
CO	Colorado	ND	North Dakota	TX	Texas
IA	Iowa	NE	Nebraska	UT	Utah
ID	Idaho	NM	New Mexico	WA	Washington
IL	Illinois	NT	Northwest Territories*	WI	Wisconsin
IN	Indiana	NV	Nevada	WY	Wyoming
KS	Kansas	OH	Ohio	YT	Yukon Territory
KY	Kentucky				
LA	Louisiana				

* Includes the new Canadian territory of Nunavut.

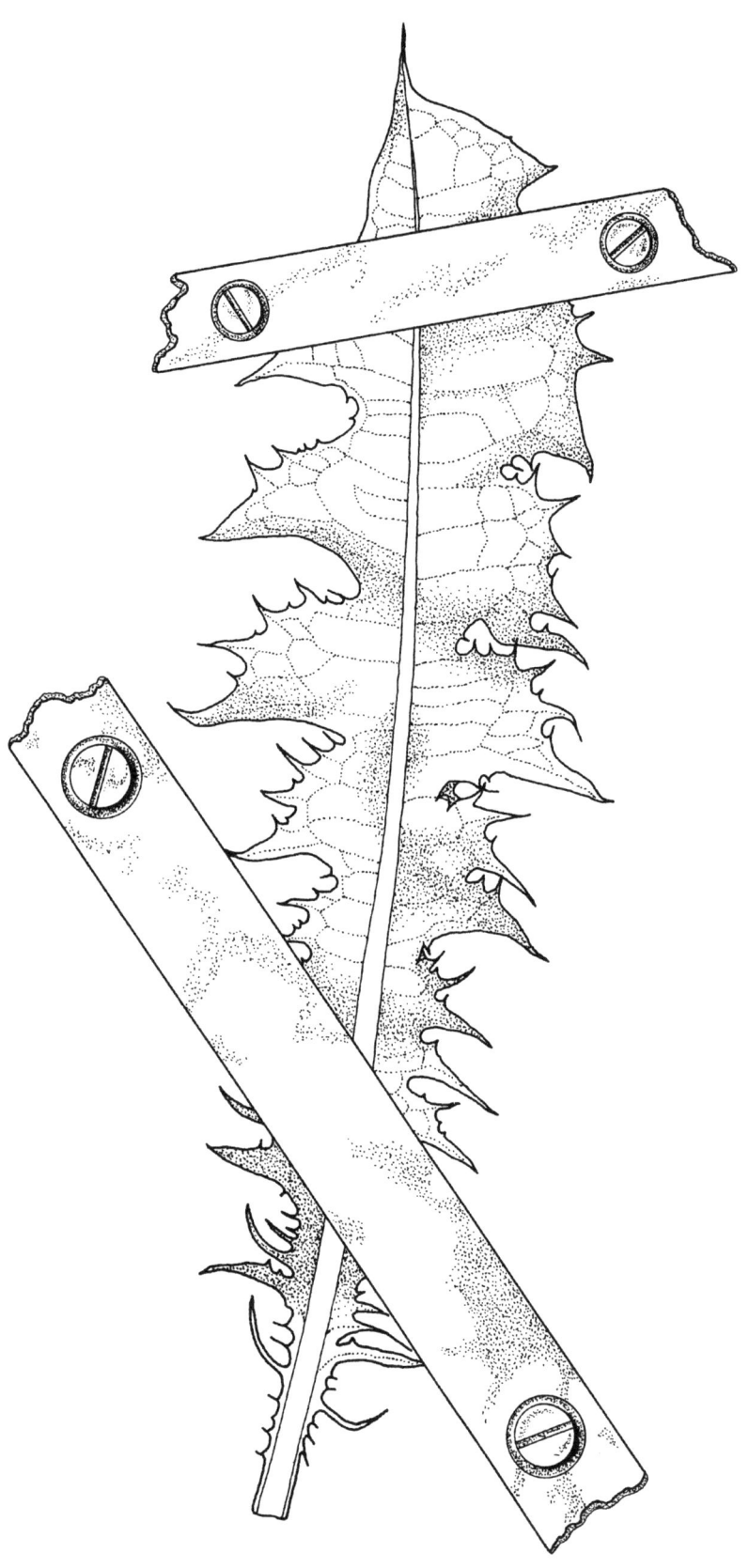

89 Families   620 *Species*

In a sense, all the patterns of nature, from flowering trees to ocean swells, from mountains to koala bears, are the emergent properties of simple interactions between subatomic particles that over time add up to far more than the sum of their parts.

— K. C. Cole

Scientific family and species' names are arranged alphabetically. Common names follow those in *Budd's Flora of the Canadian Prairie Provinces*. The Latin names submit to ITIS (Integrated Taxonomic Information System; http://www.itis.gov/). Accepted names change and vary among treatments; therefore, Latin synonyms in common use are on the species' pages.

ACERACEAE ⚘ Maple
Manitoba Maple *Acer negundo* L. 72
Silver Maple *Acer saccharinum* L. 73
Mountain Maple *Acer spicatum* Lam. 74

ACORACEAE ⚘ Calamus
Sweet Flag *Acorus americanus* (Raf.) Rat. 522

ALISMATACEAE ⚘ Water-plantain (Alismaceae)
Common Water-plantain *Alisma triviale* Pursh 523
Arum-leaved Arrowhead *Sagittaria cuneata* Sheldon 524
Broad-leaved Arrowhead *Sagittaria latifolia* Willd. 525

AMARANTHACEAE ⚘ Amaranth
Prostrate Amaranth *Amaranthus blitoides* S. Wats. 75
Red-root Pigweed *Amaranthus retroflexus* L. 76

ANACARDIACEAE ⚘ Sumac
Smooth Sumac *Rhus glabra* L. 77
Poison-ivy *Toxicodendron rydbergii* (Small ex Rydb.) Greene 78

APIACEAE ⚘ Carrot (Umbelliferae) 79
Caraway *Carum carvi* L. 80
Bulb-bearing Water-hemlock *Cicuta bulbifera* L. 81
Water-hemlock *Cicuta maculata* L. 82
Cow-parsnip *Heracleum maximum* Bartr. 83

Family—Species

[APIACEAE 🔲 Carrot]
Smooth Sweet Cicely *Osmorhiza longistylis* (Torr.) DC. **84**
Wild Parsnip *Pastinaca sativa* L. **85**
Snakeroot *Sanicula marilandica* L. **86**
Water-parsnip *Sium suave* Walt. **87**
Heart-leaved Alexanders *Zizia aptera* (Gray) Fern. **88**
Golden Alexanders *Zizia aurea* (L.) W.D.J. Koch **89**

APOCYNACEAE 🔲 Dogbane
Spreading Dogbane *Apocynum androsaemifolium* L. **90**
Indian-hemp *Apocynum cannabinum* L. **91**

ARACEAE 🔲 Arum
Water Calla *Calla palustris* L. **526**

ARALIACEAE 🔲 Ginseng
Wild Sarsaparilla *Aralia nudicaulis* L. **92**

ASCLEPIADACEAE 🔲 Milkweed
Swamp Milkweed *Asclepias incarnata* L. **93**
Dwarf Milkweed *Asclepias ovalifolia* Dcne. **94**
Showy Milkweed *Asclepias speciosa* Torr. **95**
Silky Milkweed *Asclepias syriaca* L. **96**
Whorled Milkweed *Asclepias verticillata* L. **97**

ASTERACEAE 🔲 Aster (Compositae) 98–99
Yarrow *Achillea millefolium* L. **100**
Many-flowered Yarrow *Achillea sibirica* Ledeb. **101**
False Dandelion *Agoseris glauca* (Pursh) Raf. **102**
Common Ragweed *Ambrosia artemisiifolia* L. **103**
Perennial Ragweed *Ambrosia psilostachya* DC. **104**
Great Ragweed *Ambrosia trifida* L. **105**
Pearly Everlasting *Anaphalis margaritacea* (L.) Benth. **106**
Small-leaf Pussytoes *Antennaria microphylla* Rydb. **107**
Field Pussytoes *Antennaria neglecta* Greene **108**
Lesser Burdock *Arctium minus* Bernh. **109**
Cotton Burdock *Arctium tomentosum* P. Mill. **110**

Absinthe *Artemisia absinthium* L. **111**
Biennial Wormwood *Artemisia biennis* Willd. **112**
Plains Wormwood *Artemisia campestris* L. **113**
Pasture Sage *Artemisia frigida* Willd. **114**
Prairie Sage *Artemisia ludoviciana* Nutt. **115**

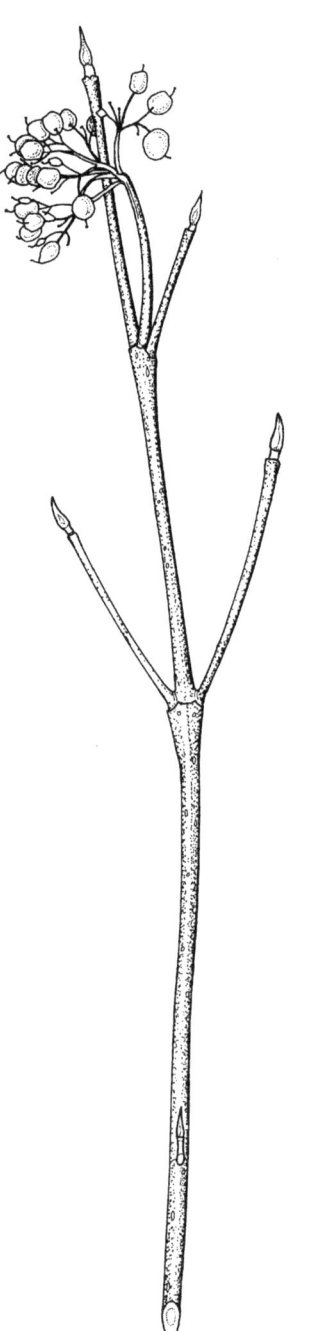

[ASTERACEAE Aster]
    Smooth Beggarticks *Bidens cernua* L. **116**
    Common Beggarticks *Bidens frondosa* L. **117**
    Nodding Thistle *Carduus nutans* L. **118**
    Diffuse Knapweed *Centaurea diffusa* Lam. **119**
    Chicory *Cichorium intybus* L. **120**
    Canada Thistle *Cirsium arvense* (L.) Scop. **121**
    Flodman's Thistle *Cirsium flodmanii* (Rydb.) Arthur **122**
    Swamp Thistle *Cirsium muticum* Michx. **123**
    Bull Thistle *Cirsium vulgare* (Savi) Ten. **124**

    Canada Fleabane *Conyza canadensis* (L.) Cronq. **125**
    Common Tickseed *Coreopsis tinctoria* Nutt. **126**
    Scapose Hawk's-beard *Crepis runcinata* (James) Torr. & Gray **127**
    Narrow-leaved Hawk's-beard *Crepis tectorum* L. **128**
    Flat-topped White Aster *Doellingeria umbellata* (P. Mill.) Nees **129**
    Purple Coneflower *Echinacea angustifolia* DC. **130**
    Northern Daisy Fleabane *Erigeron acris* L. **131**
    Tufted Fleabane *Erigeron caespitosus* Nutt. **132**
    Smooth Fleabane *Erigeron glabellus* Nutt. **133**
    Hirsute Fleabane *Erigeron lonchophyllus* Hook. **134**
    Philadelphia Fleabane *Erigeron philadelphicus* L. **135**
    Whitetop *Erigeron strigosus* Muhl. ex Willd. **136**

    Spotted Joe-pye Weed *Eupatorium maculatum* L. **137**
    Flat-topped Goldenrod *Euthamia graminifolia* (L.) Nutt. **138**
    Great-flowered Gaillardia *Gaillardia aristata* Pursh **139**
    Galinsoga *Galinsoga quadriradiata* Cav. **140**
    Gumweed *Grindelia squarrosa* (Pursh) Dunal **141**
    Common Broomweed *Gutierrezia sarothrae* (Pursh) Britt. & Rusby **142**
    Mountain Sneezeweed *Helenium autumnale* L. **143**

    Showy Sunflower *Helianthus annuus* L. **144**
    Narrow-leaved Sunflower *Helianthus maximiliani* Schrad. **145**
    Tuberous-rooted Sunflower *Helianthus nuttallii* Torr. & Gray **146**
    Beautiful Sunflower *Helianthus pauciflorus* Nutt. **147**
    Jerusalem Artichoke *Helianthus tuberosus* L. **148**
    Rough False Sunflower *Heliopsis helianthoides* (L.) Sweet **149**
    Hairy Golden-aster *Heterotheca villosa* (Pursh) Shinners **150**
    Canada Hawkweed *Hieracium umbellatum* L. **151**
    Colorado Rubberweed *Hymenoxys richardsonii* (Hook.) Cockerell **152**
    False Ragweed *Iva xanthifolia* Nutt. **153**
    Lobed Prickly Lettuce *Lactuca serriola* L. **154**

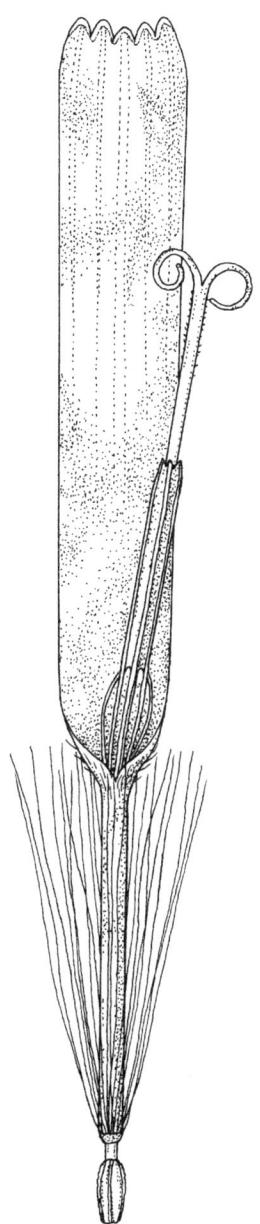

Family—Species

[ASTERACEAE Aster]

Blue Lettuce *Lactuca tatarica* (L.) C. A. Mey. **155**
Ox-eye Daisy *Leucanthemum vulgare* Lam. **156**
Meadow Blazingstar *Liatris ligulistylis* (A. Nels.) K. Schum. **157**
Dotted Blazingstar *Liatris punctata* Hook. **158**
Skeletonweed *Lygodesmia juncea* (Pursh) D. Don ex Hook. **159**
Spiny Ironplant *Machaeranthera pinnatifida* (Hook.) Shinners **160**
Pineappleweed *Matricaria discoidea* DC. **161**
Upland White Goldenrod *Oligoneuron album* (Nutt.) Nesom **162**
Stiff Goldenrod *Oligoneuron rigidum* (L.) Small **163**

Golden Ragwort *Packera aurea* (L.) A. & D. Löve **164**
Silvery Groundsel *Packera cana* (Hook.) W. A. Weber & A. Löve **165**
Palmate-leaved Colt's-foot *Petasites frigidus* (L.) Fries var. *palmatus* (Ait.) Cronq. **166**
Arrow-leaved Colt's-foot *Petasites frigidus* (L.) Fr. var. *sagittatus* (Banks ex Pursh) Cherniawsky **167**
White Lettuce *Prenanthes alba* L. **168**
Glaucous White Lettuce *Prenanthes racemosa* Michx. **169**
Long-headed Coneflower *Ratibida columnifera* (Nutt.) Woot. & Standl. **170**
Black-eyed Susan *Rudbeckia hirta* L. **171**
Tall Coneflower *Rudbeckia laciniata* L. **172**
Marsh Ragwort *Senecio congestus* (R. Br.) DC. **173**
Common Groundsel *Senecio vulgaris* L. **174**

Graceful Goldenrod *Solidago canadensis* L. **175**
Late Goldenrod *Solidago gigantea* Ait. **176**
Low Goldenrod *Solidago missouriensis* Nutt. **177**
Velvety Goldenrod *Solidago mollis* Bartl. **179**
Showy Goldenrod *Solidago nemoralis* Ait. **180**
Perennial Sow-thistle *Sonchus arvensis* L. **181**
Annual Sow-thistle *Sonchus oleraceus* L. **182**

Rush Aster *Symphyotrichum boreale* (Torr. & Gray) A. & D. Löve **183**
Rayless Aster *Symphyotrichum ciliatum* (Ledeb.) Nesom **184**
Lindley's Aster *Symphyotrichum ciliolatum* (Lindl.) A. & D. Löve **185**
Many-flowered Aster *Symphyotrichum ericoides* (L.) Nesom **186**
White Prairie Aster *Symphyotrichum falcatum* (Lindl.) Nesom **187**
Smooth Aster *Symphyotrichum laeve* (L.) A. & D. Löve **188**
Small Blue Aster *Symphyotrichum lanceolatum* (Willd.) Nesom **189**
Wood Aster *Symphyotrichum lateriflorum* (L.) A. & D. Löve **190**
New England Aster *Symphyotrichum novae-angliae* (L.) Nesom. **191**
Purple-stemmed Aster *Symphyotrichum puniceum* (L.) A. & D. Löve **192**
Tansy *Tanacetum vulgare* L. **193**
Red-seeded Dandelion *Taraxacum leavigatum* (Willd.) DC. **194**

[ASTERACEAE 🌼 Aster]
    Dandelion *Taraxacum officinale* G. H. Weber ex Wiggers **195**
    Yellow Goat's-beard *Tragopogon dubius* Scop. **196**
    Scentless Chamomile *Tripleurospermum perforata* (Mérat) M. Lainz **197**
    Cocklebur *Xanthium strumarium* L. **198**

BALSAMINACEAE 🌼 Touch-me-not
    Spotted Touch-me-not *Impatiens capensis* Meerb. **199**

BETULACEAE 🌼 Birch
    Speckled Alder *Alnus incana* (Linnaeus) Moench. **201**
    Green Alder *Alnus viridis* (Vill.) Lam. & DC. **202**
    Scrub Birch *Betula nana* L. **203**
    River Birch *Betula occidentalis* Hook. **204**
    White Birch *Betula papyrifera* Marsh. **205**
    Swamp Birch *Betula pumila* L. **206**
    American Hazelnut *Corylus americana* Walt. **207**
    Beaked Hazelnut *Corylus cornuta* Marsh. **208**

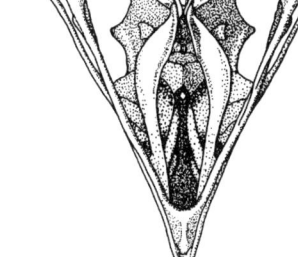

BORAGINACEAE 🌼 Borage
    Nodding Stickseed *Hackelia deflexa* (Wahlenb.) Opiz **209**
    Bluebur *Lappula squarrosa* (Retz.) Dumort. **210**
    Hoary Puccoon *Lithospermum canescens* (Michx.) Lehm. **211**
    Tall Lungwort *Mertensia paniculata* (Ait.) G. Don **212**

BRASSICACEAE 🌼 Mustard (Cruciferae) **213**
    Purple Rock Cress *Arabis* X *divaricarpa* A. Nels. (pro sp.) **214**
    Reflexed Rock Cress *Arabis holboellii* Hornem. **215**
    Shepherd's-purse *Capsella bursa-pastoris* (L.) Medik. **216**
    Bitter Cress *Cardamine pensylvanica* Muhl. ex Willd. **217**
    Gray Tansy Mustard *Descurainia incana* (Bernh. ex Fisch. & C. A. Mey.) Dorn **218**
    Flixweed *Descurainia sophia* (L.) Webb ex Prantl **219**
    Yellow Whitlow-grass *Draba nemorosa* L. **220**
    Dog Mustard *Erucastrum gallicum* (Willd.) O. E. Schulz **221**
    Wormseed Mustard *Erysimum cheiranthoides* L. **222**
    Dame's-rocket *Hesperis matronalis* L. **223**
    Common Pepper-grass *Lepidium densiflorum* Schrad. **224**
    Branched Pepper-grass *Lepidium ramosissimum* A. Nels. **225**
    Marsh Yellow Cress *Rorippa palustris* (L.) Bess. **226**
    Wild Mustard *Sinapis arvensis* L. **227**
    Tumbling Mustard *Sisymbrium altissimum* L. **228**
    Tall Hedge Mustard *Sisymbrium loeselii* L. **229**
    Stinkweed *Thlaspi arvense* L. **230**

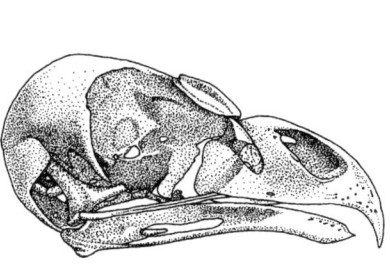

Family—Species

BUTOMACEAE 🝆 Flowering Rush
Flowering Rush *Butomus umbellatus* L. 527

CALLITRICHACEAE 🝆 Water-starwort
Vernal Water-starwort *Callitriche palustris* L. 231

CAMPANULACEAE 🝆 Bellflower
Marsh Bellflower *Campanula aparinoides* Pursh 232
Creeping Bluebell *Campanula rapunculoides* L. 233
Harebell *Campanula rotundifolia* L. 234

CANNABACEAE 🝆 Hemp (Cannabinaceae)
Common Hop *Humulus lupulus* L. 235

CAPRIFOLIACEAE 🝆 Honeysuckle 236
Bush-honeysuckle *Diervilla lonicera* P. Mill. 237
Twinflower *Linnaea borealis* L. 238
Twining Honeysuckle *Lonicera dioica* L. 239
Tatarian Honeysuckle *Lonicera tatarica* L. 240
Blue Fly Honeysuckle *Lonicera villosa* (Michx.) J. A. Schultes 241
Red Elderberry *Sambucus racemosa* L. 242
Western Snowberry *Symphoricarpos occidentalis* Hook. 243
Low Bush-cranberry *Viburnum edule* (Michx.) Raf. 244
Nannyberry *Viburnum lentago* L. 245
High Bush-cranberry *Viburnum opulus* L. 246
Downy Arrowwood *Viburnum rafinesquianum* J. A. Schultes 247

CARYOPHYLLACEAE 🝆 Pink 248
Field Chickweed *Cerastium arvense* L. 249
Mouse-ear Chickweed *Cerastium fontanum* Baumg. 250
Baby's-breath *Gypsophila paniculata* L. 251
Blunt-leaved Sandwort *Moehringia lateriflora* (L.) Fenzl 252
Bouncing Bet *Saponaria officinalis* L. 253
Smooth Catchfly *Silene csereii* Baumg. 254
White Cockle *Silene latifolia* Poir. 255
Night-flowering Catchfly *Silene noctiflora* L. 256
Long-leaved Stitchwort *Stellaria longifolia* Muhl. ex Willd. 257
Common Chickweed *Stellaria media* (L.) Vill. 258

CHENOPODIACEAE 🝆 Goosefoot 259
Nuttall's Atriplex *Atriplex gardneri* (Moq.) D. Dietr. 260
Garden Atriplex *Atriplex hortensis* L. 261
Orache *Atriplex prostrata* Bouchér ex DC. 262

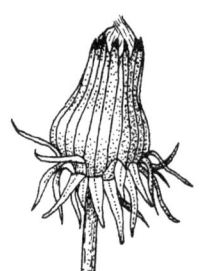

[CHENOPODIACEAE 🙂 Goosefoot]
- Russian Pigweed *Axyris amaranthoides* L. 263
- Lamb's-quarters *Chenopodium album* L. 264
- Pit-seed Goosefoot *Chenopodium berlandieri* Moq. 265
- Strawberry Blite *Chenopodium capitatum* (L.) Ambrosi 266
- Fremont's Goosefoot *Chenopodium fremontii* S. Wats. 267
- Saline Goosefoot *Chenopodium glaucum* L. 268
- Red Goosefoot *Chenopodium rubrum* L. 269
- Maple-leaved Goosefoot *Chenopodium simplex* (Torr.) Raf. 270
- Summer-cypress *Kochia scoparia* (L.) Schrad. 271
- Winterfat *Krascheninnikovia lanata* (Pursh) A.D.J. Meeuse & Smit 272
- Spear-leaved Goosefoot *Monolepis nuttalliana* (J. A. Schultes) Greene 273
- Western Sea-blite *Suaeda calceoliformis* (Hook.) Moq. 274

CONVOLVULACEAE 🙂 Morning-glory
- Hedge Bindweed *Calystegia sepium* (L.) R. Br. 275
- Field Bindweed *Convolvulus arvensis* L. 276

CORNACEAE 🙂 Dogwood
- Bunchberry *Cornus canadensis* L. 277
- Red-osier Dogwood *Cornus sericea* L. 278

CUCURBITACEAE 🙂 Cucumber
- Wild Cucumber *Echinocystis lobata* (Michx.) Torr. & Gray 279

CUPRESSACEAE 🙂 Cypress
- Low Juniper *Juniperus communis* L. 61
- Creeping Juniper *Juniperus horizontalis* Moench 62
- White Cedar *Thuja occidentalis* L. 63

CYPERACEAE 🙂 Sedge 529–530
- Water Sedge *Carex aquatilis* Wahlenb. 531
- Awned Sedge *Carex atherodes* Spreng. 532
- Golden Sedge *Carex aurea* Nutt. 533
- Bebb's Sedge *Carex bebbii* Olney ex Fern. 534
- Broad-fruited Sedge *Carex brevior* (Dewey) Mackenzie 535
- Hair-like Sedge *Carex capillaris* L. 536
- Dewey's Sedge *Carex deweyana* Schwein. 537
- Two-stamened Sedge *Carex diandra* Schrank 538
- Two-seeded Sedge *Carex disperma* Dewey 539
- Thread-leaved Sedge *Carex filifolia* Nutt. 540
- Hay Sedge *Carex foenea* var. *foenea* Willd. 541
- Slender Sedge *Carex gracillima* Schwein. 542

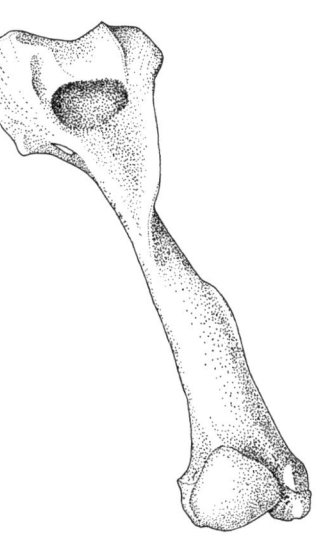

Family—Species

[CYPERACEAE Sedge]
  Granular Sedge *Carex granularis* Muhl. ex Willd. 543
  Northern Bog Sedge *Carex gynocrates* Wormsk. ex Drej. 544
  Sand Sedge *Carex houghtoniana* Torr. ex Dewey 545
  Swollen Sedge *Carex intumescens* Rudge 546
  Bristle-stalked Sedge *Carex leptalea* Wahlenb. 547
  Bog Sedge *Carex magellanica* Lam. 548
  Parry's Sedge *Carex parryana* Dewey 549
  Peck's Sedge *Carex peckii* Howe 550
  Woolly Sedge *Carex pellita* Muhl ex Willd. 551
  Sun-loving Sedge *Carex pensylvanica* Lam. 552
  Graceful Sedge *Carex praegracilis* W. Boott 553
  Cyperus-like Sedge *Carex pseudocyperus* L. 554
  Eastern Star Sedge *Carex radiata* (Wahlenb.) Small 555
  Turned Sedge *Carex retrorsa* Schwein. 556
  Sartwell's Sedge *Carex sartwellii* Dew. 557
  Sprengel's Sedge *Carex sprengelii* Dewey ex Spreng. 558
  Awl-fruited Sedge *Carex stipata* Muhl. ex Willd. 559
  Quill Sedge *Carex tenera* Dewey 560
  Thin-flowered Sedge *Carex tenuiflora* Wahl. 561
  Northern Beaked Sedge *Carex utriculata* Boott 562
  Green Sedge *Carex viridula* Michx. 563
  White-scaled Sedge *Carex xerantica* Bailey 564

  Awned Nut-grass *Cyperus squarrosus* L. 565
  Needle Spike-rush *Eleocharis acicularis* (L.) Roemer & J.A. Schultes 566
  Creeping Spike-rush *Eleocharis palustris* (L.) Roemer & J.A. Schultes 567
  Slender Cotton-grass *Eriophorum gracile* W.D.J. Koch 568
  Viscid Great Bulrush *Schoenoplectus acutus* (Muhl. ex Bigelow) A. & D. Löve 569
  River Bulrush *Schoenoplectus fluviatilis* (Torr.) M.T. Strong 570
  Prairie Bulrush *Schoenoplectus maritimus* (L.) Lye 571
  Three-square Bulrush *Schoenoplectus pungens* (Vahl) Palla 572
  Great Bulrush *Schoenoplectus tabernaemontani* (K.C. Gmel.) Palla 573
  Wool-grass *Scirpus cyperinus* (L.) Kunth 574
  Small-fruited Bulrush *Scirpus microcarpus* J. & K. Presl 575
  Pale-green Bulrush *Scirpus pallidus* (Britt.) Fern. 576
  Alpine Cotton-grass *Trichophorum alpinum* (L.) Pers. 577

ELAEAGNACEAE  Oleaster
  Russian Olive *Elaeagnus angustifolia* L. 280
  Wolf-willow *Elaeagnus commutata* Bernh. ex Rydb. 281
  Buffaloberry *Shepherdia argentea* (Pursh) Nutt. 282
  Canada Buffaloberry *Shepherdia canadensis* (L.) Nutt. 283

Family—Species

EQUISETACEAE Horsetail
Common Horsetail *Equisetum arvense* L. **49**
Swamp Horsetail *Equisetum fluviatile* L. **50**
Common Scouring-rush *Equisetum hyemale* L. **51**
Smooth Scouring-rush *Equisetum laevigatum* A. Braun **52**
Dwarf Scouring-rush *Equisetum scirpoides* Michx. **53**
Woodland Horsetail *Equisetum sylvaticum* L. **54**
Variegated Horsetail *Equisetum variegatum* Schleich. ex F. Weber & D.M.H. Mohr **55**

ERICACEAE Heath
Bearberry *Arctostaphylos uva-ursi* (L.) Spreng. **284**
Creeping Snowberry *Gaultheria hispidula* (L.) Muhl. ex Bigelow **285**
Labrador-tea *Ledum groenlandicum* Oeder **286**
Blueberry *Vaccinium angustifolium* Ait. **287**
Velvet-leaved Blueberry *Vaccinium myrtilloides* Michx. **288**
Dry-ground Cranberry *Vaccinium vitis-idaea* L. **289**

EUPHORBIACEAE Spurge
Rib-seed Sandmat *Chamaesyce glyptosperma* (Engelm.) Small **290**
Thyme-leaved Spurge *Chamaesyce serpyllifolia* (Pers.) Small **291**
Leafy Spurge *Euphorbia esula* L. **292**

FABACEAE Pea (Leguminosae) **294**
Dwarf False Indigo *Amorpha nana* Nutt. **295**
Hog-peanut *Amphicarpaea bracteata* (L.) Fern. **296**
Purple Milk-vetch *Astragalus agrestis* Dougl. ex G. Don **297**
Two-grooved Milk-vetch *Astragalus bisulcatus* (Hook.) Gray **298**
Canadian Milk-vetch *Astragalus canadensis* L. **299**
Ground-plum *Astragalus crassicarpus* Nutt. **300**
Missouri Milk-vetch *Astragalus missouriensis* Nutt. **301**
Narrow-leaved Milk-vetch *Astragalus pectinatus* (Hook.) Dougl. ex G. Don **302**

Common Caragana *Caragana arborescens* Lam. **303**
Field Crown-vetch *Coronilla varia* L. **304**
White Prairie-clover *Dalea candida* Michx. ex Willd. **305**
Purple Prairie-clover *Dalea purpurea* Vent. **306**
Wild Licorice *Glycyrrhiza lepidota* Pursh **307**
American Hedysarum *Hedysarum alpinum* L. **308**
Cream-colored Vetchling *Lathyrus ochroleucus* Hook. **309**
Marsh Vetchling *Lathyrus palustris* L. **310**
Wild Peavine *Lathyrus venosus* Muhl. ex Willd. **311**
Bird's-foot Trefoil *Lotus corniculatus* L. **312**
Black Medick *Medicago lupulina* L. **313**

Family—Species

[FABACEAE Pea]
   Alfalfa *Medicago sativa* L. **314**
   White Sweet-clover *Melilotus alba* Medikus **315**
   Yellow Sweet-clover *Melilotus officinalis* (L.) Lam **316**
   Late Yellow Locoweed *Oxytropis campestris* (L.) DC. **317**
   Silverleaf Psoralea *Pediomelum argophyllum* (Pursh) J. Grimes **318**
   Indian Breadroot *Pediomelum esculentum* (Pursh) Rydb. **319**

   Golden-bean *Thermopsis rhombifolia* (Nutt. ex Pursh) Nutt. ex Richards. **320**
   Yellow Field Clover *Trifolium campestre* Schreb. **321**
   Alsike Clover *Trifolium hybridum* L. **322**
   Red Clover *Trifolium pratense* L. **323**
   White Clover *Trifolium repens* L. **324**
   American Vetch *Vicia americana* Muhl. ex Willd. **325**
   Tufted Vetch *Vicia cracca* L. **326**

FAGACEAE Beech
   Bur Oak *Quercus macrocarpa* Michx. **327**

FUMARIACEAE Fumitory
   Golden Corydalis *Corydalis aurea* Willd. **328**
   Pink Corydalis *Corydalis sempervirens* (L.) Pers. **329**

GENTIANACEAE Gentian
   Oblong-leaved Gentian *Gentiana affinis* Griseb. **330**
   Closed Gentian *Gentiana andrewsii* Griseb. **331**
   Northern Gentian *Gentianella amarella* (L.) Boerner **332**
   Fringed Gentian *Gentianopsis virgata* (Raf.) Holub **333**
   Marsh Felwort *Lomatogonium rotatum* (L.) Fries ex Fern. **334**

GERANIACEAE Geranium
   Stork's-bill *Erodium cicutarium* (L.) L'Her. ex Ait. **335**
   Bicknell's Geranium *Geranium bicknellii* Britt. **336**

GROSSULARIACEAE Currant
   Wild Black Currant *Ribes americanum* P. Mill. **337**
   Skunkberry *Ribes glandulosum* Grauer **338**
   Northern Black Currant *Ribes hudsonianum* Richards. **339**
   Northern Gooseberry *Ribes oxyacanthoides* L. **340**
   Swamp Red Currant *Ribes triste* Pallas **341**

HALORAGACEAE Water-milfoil
   Spiked Water-milfoil *Myriophyllum sibiricum* Komarov **342**

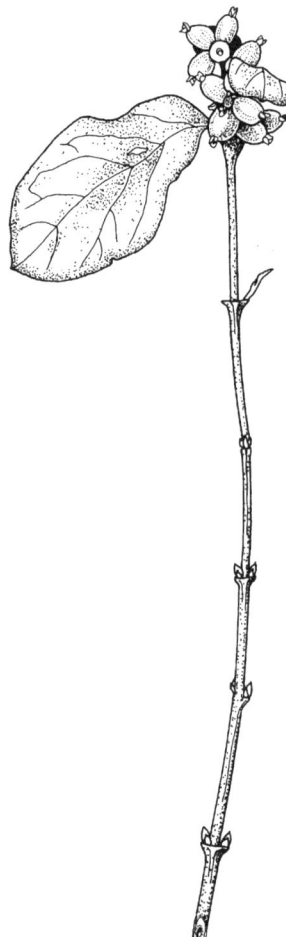

HIPPURIDACEAE 🔲 Mare's-tail
Mare's-tail *Hippuris vulgaris* L. 343

IRIDACEAE 🔲 Iris
Water Flag *Iris pseudacorus* L. 578
Blue Flag *Iris versicolor* L. 579
Common Blue-eyed Grass *Sisyrinchium montanum* Greene 580

JUNCACEAE 🔲 Rush
Alpine Rush *Juncus alpinoarticulatus* Chaix 581
Baltic Rush *Juncus balticus* Willd. 582
Toad Rush *Juncus bufonius* L. 583
Flattened Rush *Juncus compressus* Jacq. 584
Long-styled Rush *Juncus longistylis* Torr. 585
Knotted Rush *Juncus nodosus* L. 586
Torrey's Rush *Juncus torreyi* Coville 588
Hairy Wood-rush *Luzula acuminata* Raf. 589
Small-flowered Wood-rush *Luzula parviflora* (Ehrh.) Desv. 590

JUNCAGINACEAE 🔲 Arrow-grass
Seaside Arrow-grass *Triglochin maritima* L. 591
Marsh Arrow-grass *Triglochin palustre* L. 592

LAMIACEAE 🔲 Mint (Labiatae) 344
Giant-hyssop *Agastache foeniculum* (Pursh) Kuntze 345
American Dragonhead *Dracocephalum parviflorum* Nutt. 346
Hemp-nettle *Galeopsis bifida* Boenn. 347
Ground-ivy *Glechoma hederacea* L. 348
Motherwort *Leonurus cardiaca* L. 349
Water-horehound *Lycopus americanus* Muhl. ex W. Bart. 350
Western Water-horehound *Lycopus asper* Greene 351
Northern Water-horehound *Lycopus uniflorus* Michx. 352
Field Mint *Mentha arvensis* L. 353
Wild Bergamot *Monarda fistulosa* L. 354
False Dragonhead *Physostegia virginiana* (L.) Benth. 355
Marsh Skullcap *Scutellaria galericulata* L. 356
Blue Skullcap *Scutellaria lateriflora* L. 357
Marsh Hedge-nettle *Stachys pilosa* Nutt. 358

LEMNACEAE 🔲 Duckweed
Ivy-leaved Duckweed *Lemna trisulca* L. 593
Lesser Duckweed *Lemna turionifera* Landolt 594
Larger Duckweed *Spirodela polyrrhiza* (L.) Schleid 595

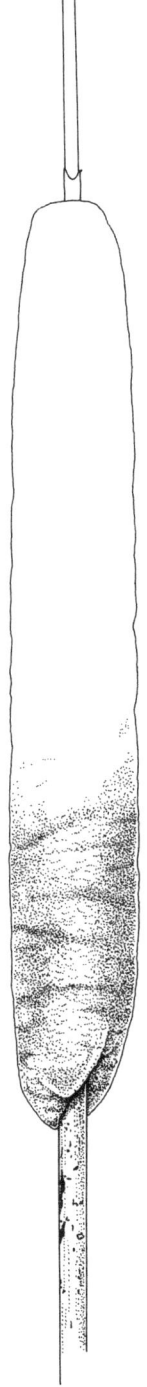

Family—Species

LENTIBULARIACEAE 🔲 Bladderwort
Flat-leaved Bladderwort *Utricularia intermedia* Hayne 359
Greater Bladderwort *Utricularia macrorhiza* Le Conte 360
Lesser Bladderwort *Utricularia minor* L. 361

LILIACEAE 🔲 Lily 596
Wild Chives *Allium schoenoprasum* L. 597
Pink-flowered Onion *Allium stellatum* Nutt. ex Ker-Gawl. 598
Asparagus *Asparagus officinalis* L. 599
Yellow Star-grass *Hypoxis hirsuta* (L.) Coville 600
Wood Lily *Lilium philadelphicum* L. 601
Two-leaved Solomon's-seal *Maianthemum canadense* Desf. 602
Star-flowered Solomon's-seal *Maianthemum stellatum* (L.) Link 603
Three-leaved Solomon's-seal *Maianthemum trifolium* (L.) Sloboda 604
Common Solomon's-seal *Polygonatum biflorum* (Walt.) Ell. 605
Nodding Wakerobin *Trillium cernuum* L. 606
Smooth Camas *Zigadenus elegans* Pursh 607

LINACEAE 🔲 Flax
Lewis Wild Flax *Linum lewisii* Pursh 362
Large-flowered Yellow Flax *Linum rigidum* Pursh 363

LOBELIACEAE 🔲 Lobelia
Kalm's Lobelia *Lobelia kalmii* L. 364
Spiked Lobelia *Lobelia spicata* Lam. 365

LYCOPODIACEAE 🔲 Club-moss
Trailing Club-moss *Diphasiastrum complanatum* (L.) Holub 56
Stiff Club-moss *Lycopodium annotinum* L. 57
Ground-pine *Lycopodium dendroideum* Michx. 58

LYTHRACEAE 🔲 Loosestrife
Purple Loosestrife *Lythrum salicaria* L. 366

MALVACEAE 🔲 Mallow
Small-flowered Mallow *Malva parviflora* L. 367
Round-leaved Mallow *Malva rotundifolia* L. 368
Scarlet Mallow *Sphaeralcea coccinea* (Nutt.) Rydb. 369

MENISPERMACEAE 🔲 Moonseed
Yellow Parilla *Menispermum canadense* L. 370

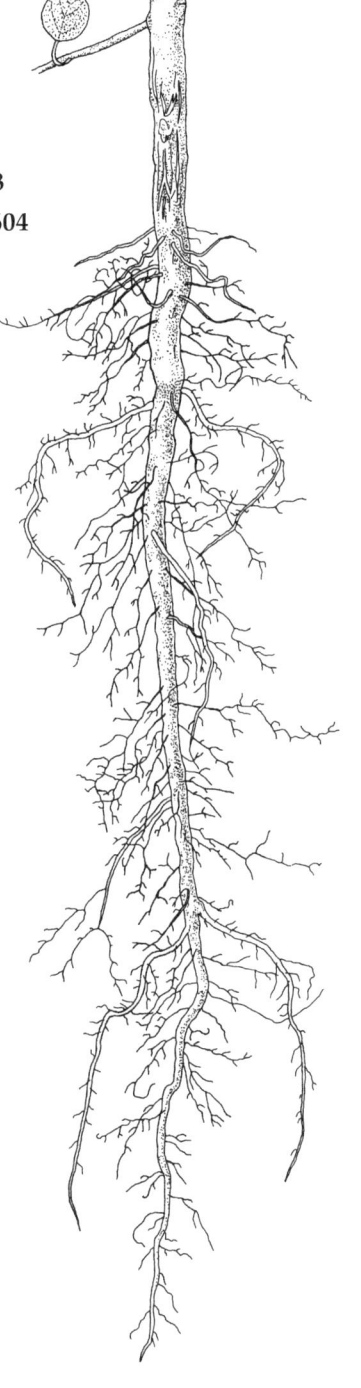

Family—Species

MENYANTHACEAE 🌼 Buck-bean
Buck-bean *Menyanthes trifoliata* L. **371**

MONOTROPACEAE 🌼 Indian Pipe
Indian-pipe *Monotropa uniflora* L. **372**

NYCTAGINACEAE 🌼 Four-o'clock
Hairy Umbrellawort *Mirabilis hirsuta* (Pursh) MacM. **373**

NYMPHAEACEAE 🌼 Water-lily
Yellow Pond-lily *Nuphar lutea* ssp. *variegata* (Dur.) E.O. Beal **374**

OLEACEAE 🌼 Olive
Black Ash *Fraxinus nigra* Marsh. **375**
Green Ash *Fraxinus pennsylvanica* Marsh. **376**

ONAGRACEAE 🌼 Evening-primrose (Oenotheraceae)
Shrubby Evening-primrose *Calylophus serrulatus* (Nutt.) Raven **377**
Fireweed *Chamerion angustifolium* (L.) Holub. **378**
Northern Willowherb *Epilobium ciliatum* Raf. **379**
Scarlet Gaura *Gaura coccinea* Nutt. ex Pursh **380**
Yellow Evening-primrose *Oenothera biennis* L. **381**
White Evening-primrose *Oenothera nuttallii* Sweet **382**

ORCHIDACEAE 🌼 Orchid **608**
Round-leaved Orchid *Amerorchis rotundifolia* (Banks ex Pursh) Hultén **609**
Long-bracted Orchid *Coeloglossum viride* (L.) Hartman **610**
Spotted Coralroot *Corallorhiza maculata* (Raf.) Raf. **611**
Striped Coralroot *Corallorhiza striata* Lindl. **612**
Early Coralroot *Corallorhiza trifida* Châtelain **613**
Small White Lady's-slipper *Cypripedium candidum* Muhl. ex Willd. **614**
Yellow Lady's-slipper *Cypripedium parviflorum* Salisb. **615**
Lesser Rattlesnake-plantain *Goodyera repens* (L.) R. Br. ex Ait. f. **616**
Heart-leaved Twayblade *Listera cordata* (L.) R. Br. ex Ait. f. **617**
Green-flowered Bog Orchid *Platanthera aquilonis* Sheviak **618**
Great Plains White Fringed Orchid *Platanthera praeclara* Sheviak & Bowles **619**

OXALIDACEAE 🌼 Wood-sorrel
Yellow Wood-sorrel *Oxalis stricta* L. **383**

PINACEAE 🌼 Pine
Balsam Fir *Abies balsamea* (L.) P. Mill. **65**

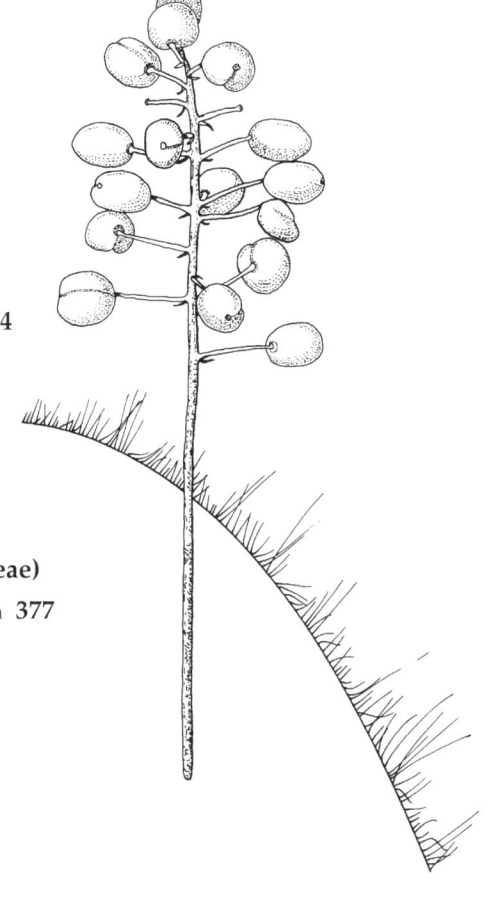

Family—Species

[PINACEAE 🌼 Pine]
    Tamarack *Larix laricina* (Du Roi) K. Koch **66**
    White Spruce *Picea glauca* (Moench) Voss **67**
    Black Spruce *Picea mariana* (P. Mill.) B. S. P. **68**
    Jack Pine *Pinus banksiana* Lamb. **69**

PLANTAGINACEAE 🌼 Plantain
    Saline Plantain *Plantago eriopoda* Torr. **384**
    Common Plantain *Plantago major* L. **385**

POACEAE 🌼 Grass (Gramineae) **621**
    Crested Wheatgrass *Agropyron cristatum* (L.) Gaertn. **622**
    Redtop *Agrostis stolonifera* L. **623**
    Short-awned Foxtail *Alopecurus aequalis* Sobol. **624**
    Big Bluestem *Andropogon gerardii* Vitman **625**
    Wild Oat *Avena fatua* L. **626**
    Slough Grass *Beckmannia syzigachne* (Steud.) Fern. **627**
    Side-oats Grama *Bouteloua curtipendula* (Michx.) Torr. **628**
    Blue Grama *Bouteloua gracilis* (Willd. ex Kunth) Lag. ex Griffiths **629**
    Fringed Brome *Bromus ciliatus* L. **630**
    Smooth Brome *Bromus inermis* Leyss. **631**
    Nodding Brome *Bromus porteri* (Coult.) Nash **632**

    Marsh Reed Grass *Calamagrostis canadensis* (Michx.) Beauv. **633**
    Northern Reed Grass *Calamagrostis stricta* (Timm) Koel. **634**
    Sand Grass *Calamovilfa longifolia* (Hook.) Scribn. **635**
    Slender Wood Grass *Cinna latifolia* (Trev. ex Göpp.) Griseb. **636**
    Orchard Grass *Dactylis glomerata* L. **637**
    Tufted Hair Grass *Deschampsia caespitosa* (L.) Beauv. **638**
    Leiberg's Rosette Grass *Dichanthelium leibergii* (Vasey) Freckmann **639**
    Barnyard Grass *Echinochloa crus-galli* (L.) Beauv. **640**
    Canada Wild Rye *Elymus canadensis* L. **641**
    Quack Grass *Elymus repens* (L.) Gould **642**
    Virginia Wild Rye *Elymus virginicus* L. **643**
    Stinkgrass *Eragrostis cilianensis* (All.) Vign. ex Janchen **644**
    Creeping Love Grass *Eragrostis hypnoides* (Lam.) B. S. P. **645**

    Tall Manna Grass *Glyceria grandis* S. Wats. **646**
    Fowl Manna Grass *Glyceria striata* (Lam.) A. S. Hitchc. **647**
    Spear Grass *Hesperostipa comata* (Trin. & Rupr.) Barkworth **648**
    Porcupine Grass *Hesperostipa spartea* (Trin.) Barkworth **649**
    Sweet Grass *Hierochloë odorata* (L.) Beauv. **650**
    Wild Barley *Hordeum jubatum* L. **651**

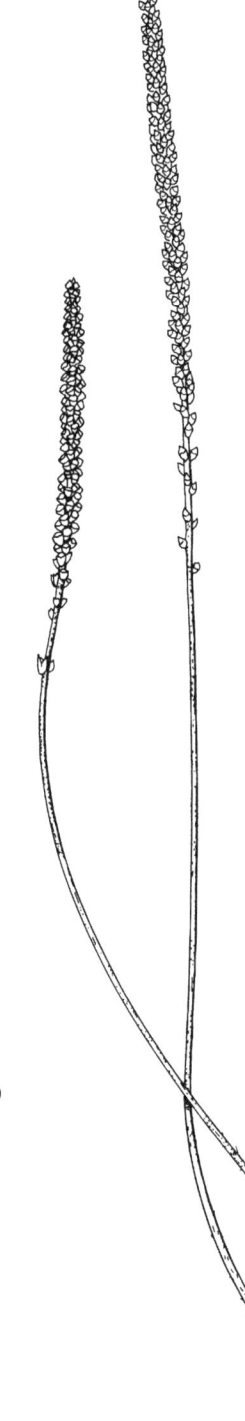

[POACEAE 🌿 Grass]

June Grass *Koeleria macrantha* (Ledeb.) J.A. Schultes 652
Rice Cut Grass *Leersia oryzoides* (L.) Sw. 653
Hairy Wild Rye *Leymus innovatus* (Beal) Pilger 654
Perennial Rye Grass *Lolium perenne* L. 655
Scratch Grass *Muhlenbergia asperifolia* (Nees & Meyen ex Trin.) Parodi 656
Prairie Muhly *Muhlenbergia cuspidata* (Torr. ex Hook.) Rydb. 657
Mat Muhly *Muhlenbergia richardsonis* (Trin.) Rydb. 658
Green Needle Grass *Nassella viridula* (Trin.) Barkworth 659
White-grained Mountain Rice Grass *Oryzopsis asperifolia* Michx. 660

Witch Grass *Panicum capillare* L. 661
Broomcorn Millet *Panicum miliaceum* L. 662
Switch Grass *Panicum virgatum* L. 663
Western Wheatgrass *Pascopyrum smithii* (Rydb.) A. Löve 664
Reed Canary Grass *Phalaris arundinacea* L. 665
Canary Grass *Phalaris canariensis* L. 666
Timothy *Phleum pratense* L. 667
Common Reed Grass *Phragmites australis* (Cav.) Trin. ex Steud. 668
Canada Blue Grass *Poa compressa* L. 669
Fowl Blue Grass *Poa palustris* L. 670
Kentucky Blue Grass *Poa pratensis* L. 671
Russian Wild Rye *Psathyrostachys juncea* (Fisch.) Nevski 672
Slender Salt-meadow Grass *Puccinellia distans* (Jacq.) Parl. 673
Nuttall's Salt-meadow Grass *Puccinellia nuttalliana* (J.A. Schultes) A. S. Hitchc. 674

Purple Oat Grass *Schizachne purpurascens* (Torr.) Swallen 675
Little Bluestem *Schizachyrium scoparium* (Michx.) Nash 676
Sprangletop *Scolochloa festucacea* (Willd.) Link 677
Yellow Foxtail *Setaria pumila* (Poir.) Roemer & J.A. Schultes 678
Green Foxtail *Setaria viridis* (L.) Beauv. 679
Indian Grass *Sorghastrum nutans* (L.) Nash 680
Alkali Cord Grass *Spartina gracilis* Trin. 681
Prairie Cord Grass *Spartina pectinata* Bosc ex Link 682
Rough Dropseed *Sporobolus compositus* (Poir.) Merr. 683
Sand Dropseed *Sporobolus cryptandrus* (Torr.) Gray 684
Prairie Dropseed *Sporobolus heterolepis* (Gray) Gray 685
Annual Dropseed *Sporobolus neglectus* Nash 686
Annual Wild-rice *Zizania aquatica* L. 687

POLEMONIACEAE 🌿 Phlox

Moss Phlox *Phlox hoodii* Richards. 386

Family—Species

POLYGALACEAE ▫ Milkwort
Seneca Snakeroot *Polygala senega* L. 387

POLYGONACEAE ▫ Buckwheat 388
Yellow Umbrellaplant *Eriogonum flavum* Nutt. 389
Garden Buckwheat *Fagopyrum esculentum* Moench 390
Striate Knotweed *Polygonum achoreum* Blake 391
Swamp Persicaria *Polygonum amphibium* var. *emersum* Michx. 392
Doorweed *Polygonum aviculare* L. 393
Bindweed *Polygonum cilinode* Michx. 394
Wild Buckwheat *Polygonum convolvulus* var. *convolvulus* L. 395
Water-pepper *Polygonum hydropiper* L. 396
Pale Persicaria *Polygonum lapathifolium* L. 397
Lady's-thumb *Polygonum persicaria* L. 398
Bushy Knotweed *Polygonum ramosissimum* Michx. 399
Arrow-leaf Tear-thumb *Polygonum sagittatum* L. 400
False Buckwheat *Polygonum scandens* var. *scandens* L. 401

Sour Dock *Rumex acetosa* L. 402
Western Dock *Rumex aquaticus* var. *fenestratus* (Greene) Dorn 403
Curled Dock *Rumex crispus* L. 404
Golden Dock *Rumex maritimus* L. 405
Triangular-valved Dock *Rumex salicifolius* var. *mexicanus* (Meisn.) C.L. Hitchc. 406

PORTULACACEAE ▫ Purslane
Purslane *Portulaca oleracea* L. 407

POTAMOGETONACEAE ▫ Pondweed (Najadaceae)
Northern Pondweed *Potamogeton alpinus* Balbis 688
Leafy Pondweed *Potamogeton foliosus* Raf. 689
Floating-leaf Pondweed *Potamogeton natans* L. 690
Richardson's Pondweed *Potamogeton richardsonii* (Benn.) Rydb. 691
Slender Pondweed *Stuckenia filiformis* (Pers.) Börner 692
Sago Pondweed *Stuckenia pectinata* (L.) Börner 693

PRIMULACEAE ▫ Primrose
Western Pygmyflower *Androsace occidentalis* Pursh 408
Pygmyflower *Androsace septentrionalis* L. 409
Sea-milkwort *Glaux maritima* L. 410
Fringed Loosestrife *Lysimachia ciliata* L. 411
Tufted Loosestrife *Lysimachia thyrsiflora* L. 412
Dwarf Primrose *Primula mistassinica* Michx. 413
Northern Starflower *Trientalis borealis* Raf. 414

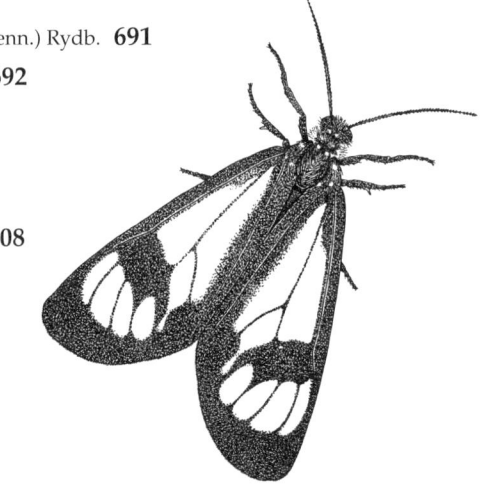

Family—Species

PYROLACEAE 🌼 Wintergreen
One-flowered Wintergreen *Moneses uniflora* (L.) Gray **415**
One-sided Wintergreen *Orthilia secunda* (L.) House **416**
Pink Wintergreen *Pyrola asarifolia* Michx. **417**
Greenish-flowered Wintergreen *Pyrola chlorantha* Sw. **418**

RANUNCULACEAE 🌼 Buttercup **419**
Red Baneberry *Actaea rubra* (Ait.) Willd. **420**
Canada Anemone *Anemone canadensis* L. **421**
Long-fruited Anemone *Anemone cylindrica* Gray **422**
Cut-leaved Anemone *Anemone multifida* Poir. **423**
Wood Anemone *Anemone quinquefolia* L. **424**
Small-flowered Columbine *Aquilegia brevistyla* Hook. **425**
Wild Columbine *Aquilegia canadensis* L. **426**
Marsh-marigold *Caltha palustris* L. **427**
Western Virgin's-bower *Clematis ligusticifolia* Nutt. **428**
Clematis *Clematis tangutica* (Maxim.) Korsh. **429**
Goldthread *Coptis trifolia* (L.) Salisb. **430**
Crocus Anemone *Pulsatilla patens* (L.) P. Mill **431**

Smooth-leaved Buttercup *Ranunculus abortivus* L. **432**
Tall Buttercup *Ranunculus acris* L. **433**
Large-leaved Watercrowfoot *Ranunculus aquatilis* L. **434**
Seaside Buttercup *Ranunculus cymbalaria* Pursh **435**
Creeping Spearwort *Ranunculus flammula* L. **436**
Bristly Buttercup *Ranunculus pensylvanicus* L. f. **437**
Prairie Buttercup *Ranunculus rhomboideus* Goldie **438**
Celery-leaved Buttercup *Ranunculus sceleratus* L. **439**
Tall Meadow-rue *Thalictrum dasycarpum* Fisch. & Avé-Lall. **440**
Veiny Meadow-rue *Thalictrum venulosum* Trel. **441**

RHAMNACEAE 🌼 Buckthorn
Buckthorn *Rhamnus cathartica* L. **442**

ROSACEAE 🌼 Rose **443**
Hooked Agrimony *Agrimonia gryposepala* Wallr. **444**
Saskatoon *Amelanchier alnifolia* (Nutt.) Nutt. ex M. Roemer **445**
Silverweed *Argentina anserina* (L.) Rydb. **446**
Chamaerhodos *Chamaerhodos erecta* (L.) Bunge **447**
Marsh Cinquefoil *Comarum palustre* L. **448**
Round-leaved Hawthorn *Crataegus chrysocarpa* Ashe **449**
Shrubby Cinquefoil *Dasiphora floribunda* (Pursh) Kartesz, comb. nov. ined. **450**
Smooth Wild Strawberry *Fragaria virginiana* Duchesne **451**

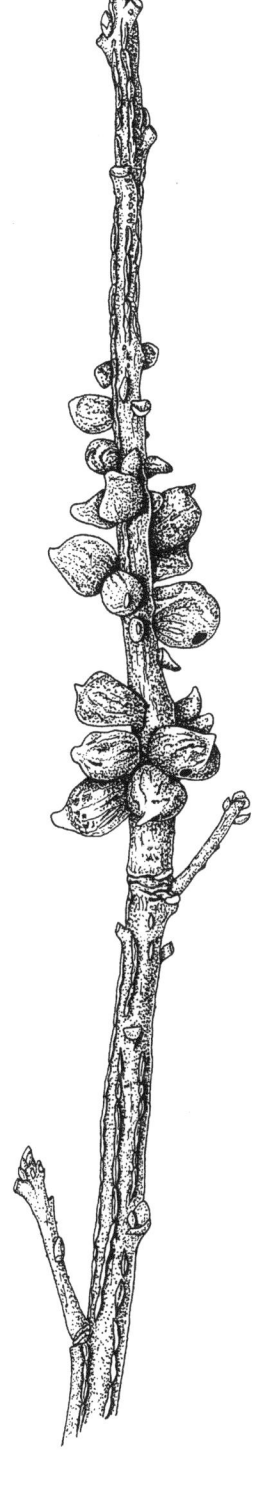

Family—Species

[ROSACEAE 🌹 Rose]
  Yellow Avens *Geum aleppicum* Jacq. **452**
  Large-leaved Avens *Geum macrophyllum* Willd. **453**
  Purple Avens *Geum rivale* L. **454**
  Three-flowered Avens *Geum triflorum* Pursh **455**
  Ninebark *Physocarpus opulifolius* (L.) Maxim. **456**
  White Cinquefoil *Potentilla arguta* Pursh **457**
  Graceful Cinquefoil *Potentilla gracilis* Dougl. ex Hook. **458**
  Rough Cinquefoil *Potentilla norvegica* L. **459**
  Canada Plum *Prunus nigra* Ait. **460**
  Pin Cherry *Prunus pensylvanica* L. f. **461**
  Red-fruited Choke Cherry *Prunus virginiana* L. **462**

  Prickly Rose *Rosa acicularis* Lindl. **463**
  Low Prairie Rose *Rosa arkansana* Porter **464**
  Wood's Rose *Rosa woodsii* Lindl. **465**
  Stemless Raspberry *Rubus arcticus* L. **466**
  Wild Red Raspberry *Rubus idaeus* L. **467**
  Dewberry *Rubus pubescens* Raf. **468**
  Three-toothed Cinquefoil *Sibbaldiopsis tridentata* (Ait.) Rydb. **469**
  Narrow-leaved Meadowsweet *Spiraea alba* Du Roi **470**

RUBIACEAE 🌿 Madder
  Northern Bedstraw *Galium boreale* L. **471**
  Small Bedstraw *Galium trifidum* L. **472**
  Sweet-scented Bedstraw *Galium triflorum* Michx. **473**
  Long-leaved Bluets *Houstonia longifolia* Gaertn. **474**

SALICACEAE 🌿 Willow **475**
  Balsam Poplar *Populus balsamifera* L. **476**
  Cottonwood *Populus deltoides* Bartr. ex Marsh. **477**
  Aspen Poplar *Populus tremuloides* Michx. **478**
  Beaked Willow *Salix bebbiana* Sarg. **479**
  Hoary Willow *Salix candida* Flueggé ex Willd. **480**
  Pussy Willow *Salix discolor* Muhl. **481**
  Yellow Willow *Salix eriocephala* Michx. **482**
  Sandbar Willow *Salix interior* Rowlee **483**
  Velvet-fruited Willow *Salix maccalliana* Rowlee **484**
  Basket Willow *Salix petiolaris* Sm. **485**

SANTALACEAE 🌿 Sandalwood
  Bastard Toadflax *Comandra umbellata* (L.) Nutt. **486**

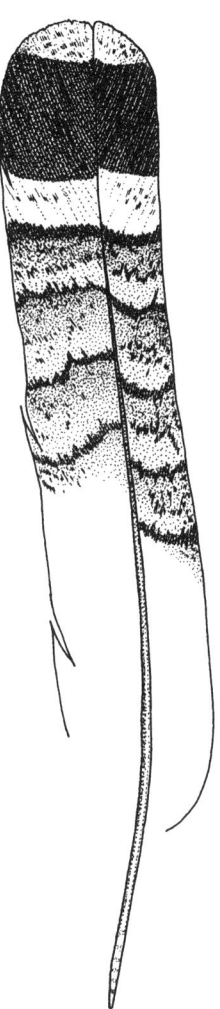

SAXIFRAGACEAE 🌼 Saxifrage
Golden Saxifrage *Chrysosplenium tetrandrum* (Lund ex Malmgr.) Th. Fries **487**
Alumroot *Heuchera richardsonii* R. Br. **488**
Bishop's-cap *Mitella nuda* L. **489**
Glaucous Grass-of-parnassus *Parnassia glauca* Raf. **490**
Northern Grass-of-parnassus *Parnassia palustris* L. **491**
Early Saxifrage *Saxifraga virginiensis* Michx. **492**

SCROPHULARIACEAE 🌼 Figwort 493
Slender Agalinis *Agalinis tenuifolia* (Vahl) Raf. **494**
Red Indian Paintbrush *Castilleja miniata* Dougl. ex Hook. **495**
Small-snapdragon *Chaenorhinum minus* (L.) Lange **496**
Mudwort *Limosella aquatica* L. **497**
Butter-and-eggs *Linaria vulgaris* P. Mill. **498**
Blue Monkeyflower *Mimulus ringens* L. **499**
Swamp Lousewort *Pedicularis lanceolata* Michx. **500**
Lilac-flowered Beardtongue *Penstemon gracilis* Nutt. **501**
Smooth Blue Beardtongue *Penstemon nitidus* Dougl. ex Benth. **502**
Common Mullein *Verbascum thapsus* L. **503**
American Speedwell *Veronica americana* Schwein. ex Benth. **504**
Hairy Speedwell *Veronica peregrina* L. **505**
Culver's-root *Veronicastrum virginicum* (L.) Farw. **506**

SMILACACEAE 🌼 Greenbrier
Carrionflower *Smilax lasioneura* Hook. **694**

SOLANACEAE 🌼 Nightshade
Bittersweet *Solanum dulcamara* L. **507**

SPARGANIACEAE 🌼 Bur-reed
Stemless Bur-reed *Sparganium emersum* Rehmann **695**
Broad-fruited Bur-reed *Sparganium eurycarpum* Engelm. ex Gray **696**
Clustered Bur-reed *Sparganium glomeratum* (Laestad.) L. Neum. **697**
Small Bur-reed *Sparganium natans* L. **698**

TILIACEAE 🌼 Linden
Basswood *Tilia americana* L. **508**

TYPHACEAE 🌼 Cattail
Narrow-leaved Cattail *Typha angustifolia* L. **699**
Common Cattail *Typha latifolia* L. **700**

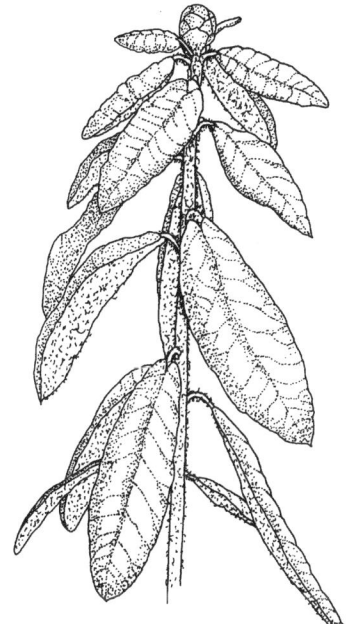

Family—Species

ULMACEAE 🌿 Elm
American Elm *Ulmus americana* L. **509**

URTICACEAE 🌿 Nettle
Wood Nettle *Laportea canadensis* (L.) Weddell **510**
Stinging Nettle *Urtica dioica* L. **511**

VALERIANACEAE 🌿 Valerian
Northern Valerian *Valeriana dioica* L. **512**

VERBENACEAE 🌿 Vervain
Bracted Vervain *Verbena bracteata* Lag. & Rodr. **513**

VIOLACEAE 🌿 Violet
Early Blue Violet *Viola adunca* Sm. **514**
Western Canada Violet *Viola canadensis* L. **515**
Northern Bog Violet *Viola nephrophylla* Greene **516**
Crowfoot Violet *Viola pedatifida* G. Don **517**
Downy Yellow Violet *Viola pubescens* Ait. **518**

VITACEAE 🌿 Grape
Large-toothed Virginia Creeper *Parthenocissus vitacea* (Knerr) A. S. Hitchc. **519**

Pteridophytes

IN THIS FIRST small section is a group of plants that produces a soft cone (strobilus) at the tip of a stem or branch. The plants, mostly herbs, comprise only about 3% of the vascular plants in North America (Canada and the United States). Reproduction is by spores, not pollen. The cones are composed of numerous sporophylls that contain spore sacs (sporangia) under an outer shield (cover).

SINCE THERE ARE no flowers, there are no stamens, pistils, sepals and petals. There is no ovary to mature into a fruit and seeds.

I DID NOT concern myself with describing the true ferns (for example: Ostrich Fern, Crested Shield Fern or the Purple Cliff Brake, etc.).

INSTEAD, I LOCATED and described some of the perennial fern allies: a few club-mosses in the Lycopodiaceae family and several horsetails in the Equisetaceae.

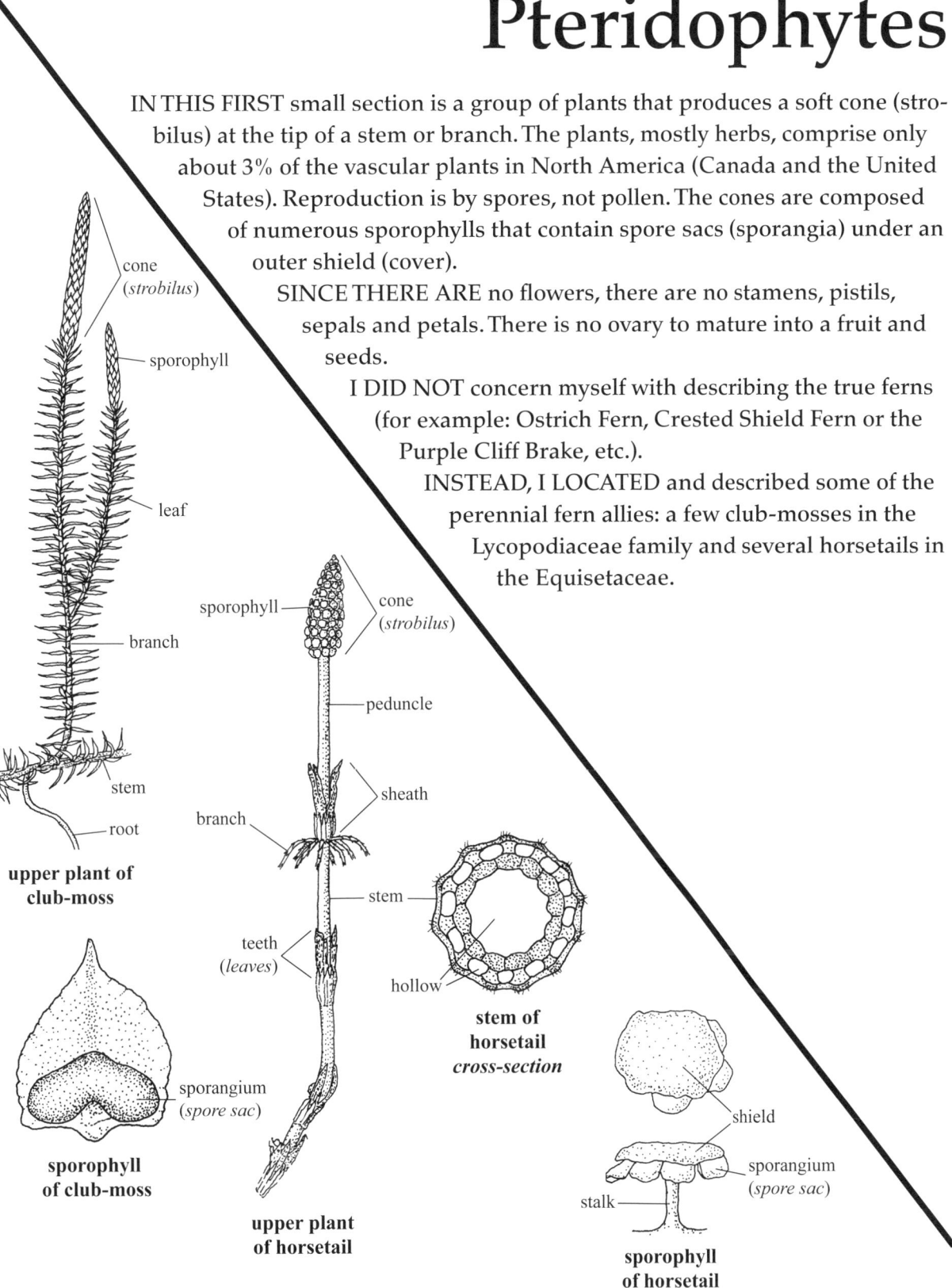

47

Horsetail

It's like living inside a pearl
where the light is so fantastic
and all the beauty of the world
I'm talking about the Horsetail
truly a gracious thing in all creation
humble too
like some blind man with a tin cup
whispering in the rain

Horsetail—*Equisetaceae*

Cones pink to tan, 1–3.5 cm long, blunt

COMMON HORSETAIL *Equisetum arvense* L.

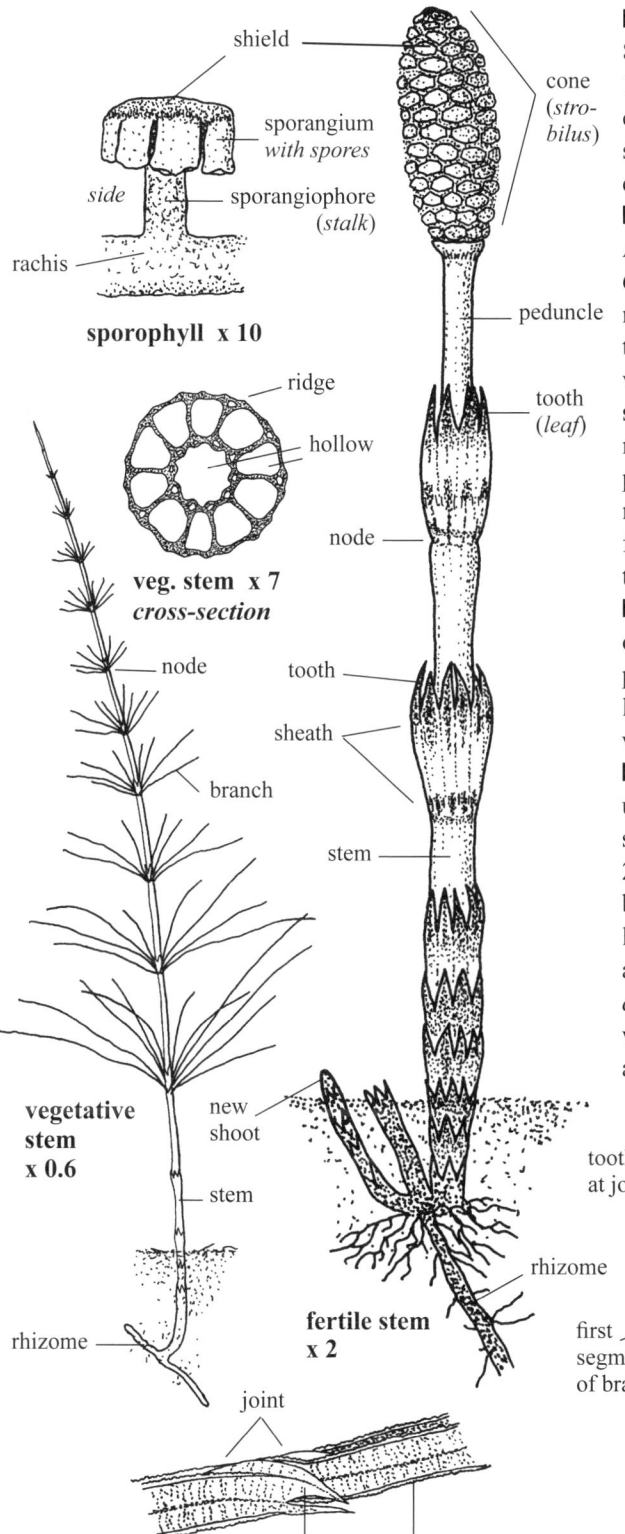

sporophyll x 10

veg. stem x 7 *cross-section*

vegetative stem x 0.6

fertile stem x 2

branch section x 8

lowest node of vegetative stem x 4

■ **SKETCH** A variable **perennial herb** with two forms, 8–100 cm tall with dark brown, tuber-bearing **rhizomes** 1–4 mm thick by 10–150 cm long, solitary or in colonies covering several m^2; in moist soils along ditches, pond and stream banks, grain fields, orchards, tundra, open woodland edges and pastures. **w1**

■ **REPRODUCTIVE PARTS Cones** (strobili) active April–July, pinkish to tan, terminal, blunt, 1–3.5 cm long by 6–10 mm wide, oval; **peduncles** 1–6.2 cm long; **sporophylls** numerous, *c.* 2 mm high; **sporangiophores** (stalks) *c.* 1 mm tall, pale green; **shields** tan, 1.7–2 mm long by *c.* 1.5 mm wide, hairless, smooth, usually 6-sided, edges rounded; **sporangia** (spore sacs) six or seven, attached to the outer margin of the shield, 0.6–0.7 mm long by *c.* 0.5 mm wide, pale green, hairless, the opening facing inward; **spores** medium green, microscopic, numerous, round, each with four spirally wound bands (elaters) with slightly enlarged tips, the bands unravelling as the spores dry in air.

■ **LEAVES** Whorls fused to a nodal sheath; **1) sheaths** on fertile stems inflated, 5–20 mm long at each node, thin, papery, with 8–12 pointed, dark brown, erect teeth 3–10 mm long; **2) sheaths** on vegetative stems tight, 2–10 mm long, with 4–14 erect, dark teeth 1.5–3.5 mm long.

■ **STEM** Two annual erect types: **1) fertile stems** tan, unbranched, first to appear in spring, 7–30 cm tall, fleshy, simple, thick-walled, hollow, short-lived; base 4–8 mm thick; **2) vegetative stems** and branches green, pinkish to dark brown near the 2–5 mm thick base, 20–100 cm tall, hollow, long-lasting, with 7–14 longitudinal ridges; **branches** mostly ascending, simple, in whorls of 7–14, each 2–20 cm long by *c.* 1 mm thick, 3-ridged (4-) with tiny silica spicules; **joints** with 3 or 4 teeth; **first branch segment** longer than sheath and teeth at lower nodes.

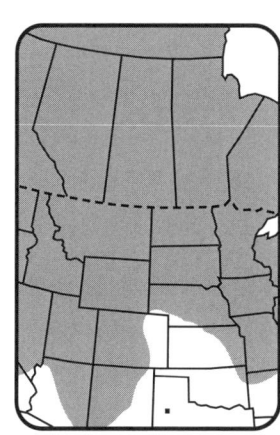

Field Horsetail

Horsetail—*Equisetaceae* Cones tan, 1–3.5 cm long, blunt

Swamp Horsetail   *Equisetum fluviatile* L.

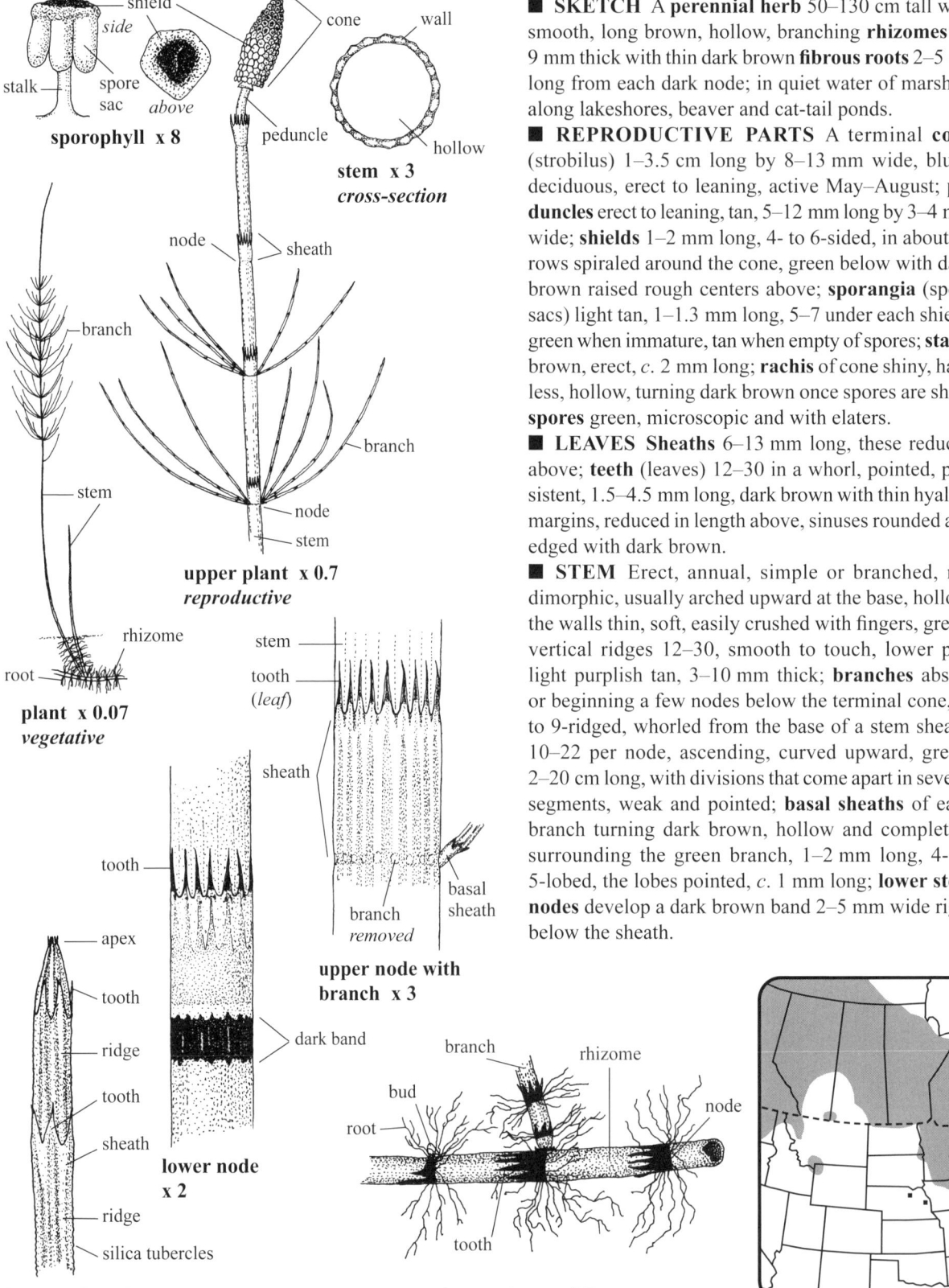

- **SKETCH** A **perennial herb** 50–130 cm tall with smooth, long brown, hollow, branching **rhizomes** 5–9 mm thick with thin dark brown **fibrous roots** 2–5 cm long from each dark node; in quiet water of marshes, along lakeshores, beaver and cat-tail ponds.
- **REPRODUCTIVE PARTS** A terminal **cone** (strobilus) 1–3.5 cm long by 8–13 mm wide, blunt, deciduous, erect to leaning, active May–August; **peduncles** erect to leaning, tan, 5–12 mm long by 3–4 mm wide; **shields** 1–2 mm long, 4- to 6-sided, in about 15 rows spiraled around the cone, green below with dark brown raised rough centers above; **sporangia** (spore sacs) light tan, 1–1.3 mm long, 5–7 under each shield, green when immature, tan when empty of spores; **stalks** brown, erect, *c.* 2 mm long; **rachis** of cone shiny, hairless, hollow, turning dark brown once spores are shed; **spores** green, microscopic and with elaters.
- **LEAVES** Sheaths 6–13 mm long, these reduced above; **teeth** (leaves) 12–30 in a whorl, pointed, persistent, 1.5–4.5 mm long, dark brown with thin hyaline margins, reduced in length above, sinuses rounded and edged with dark brown.
- **STEM** Erect, annual, simple or branched, not dimorphic, usually arched upward at the base, hollow, the walls thin, soft, easily crushed with fingers, green, vertical ridges 12–30, smooth to touch, lower part light purplish tan, 3–10 mm thick; **branches** absent or beginning a few nodes below the terminal cone, 4- to 9-ridged, whorled from the base of a stem sheath, 10–22 per node, ascending, curved upward, green, 2–20 cm long, with divisions that come apart in several segments, weak and pointed; **basal sheaths** of each branch turning dark brown, hollow and completely surrounding the green branch, 1–2 mm long, 4- or 5-lobed, the lobes pointed, *c.* 1 mm long; **lower stem nodes** develop a dark brown band 2–5 mm wide right below the sheath.

Water Horsetail

Horsetail—*Equisetaceae*   Cones tan, 0.5–2.5 cm long, pointed

COMMON SCOURING-RUSH *Equisetum hyemale* L.

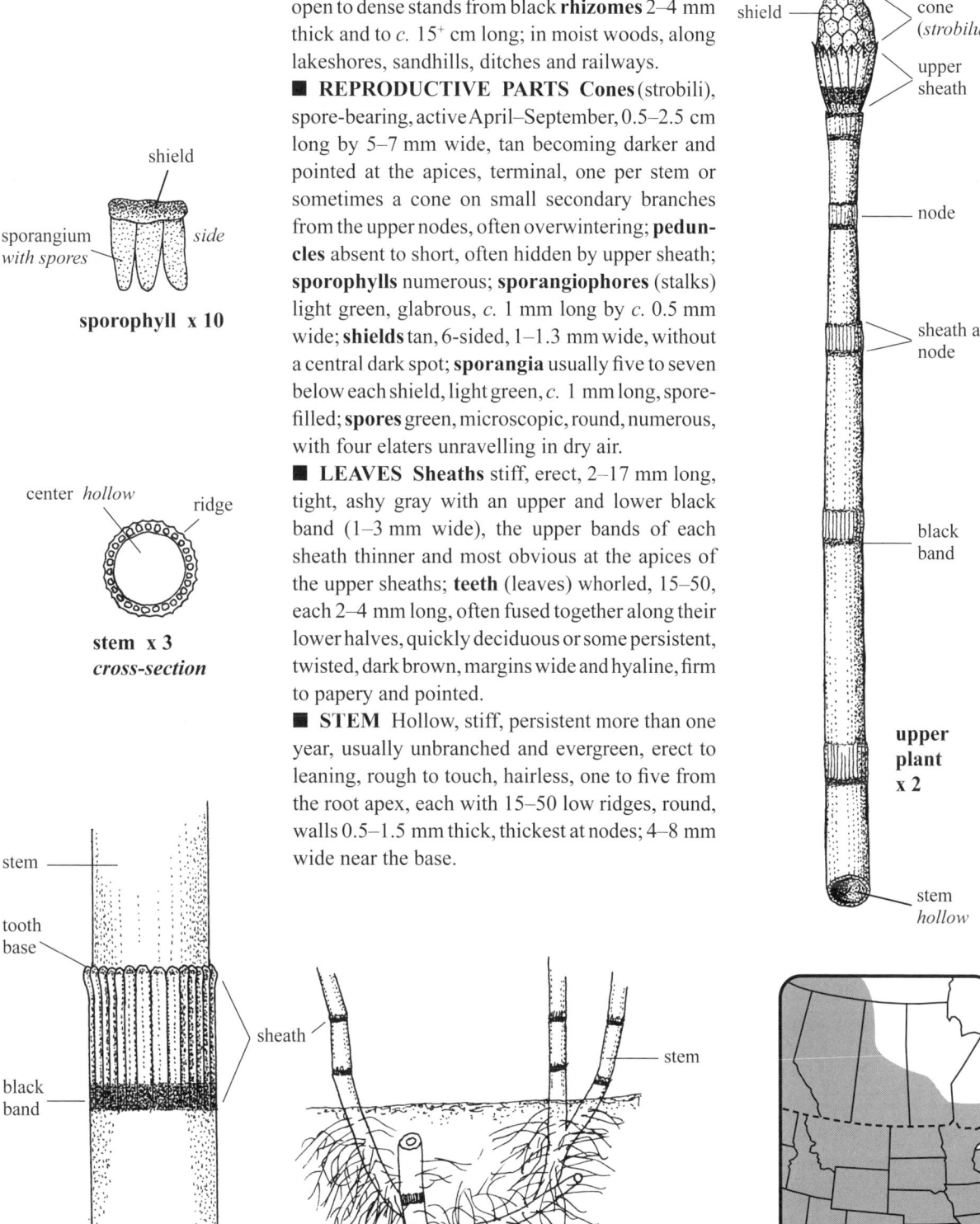

■ **SKETCH** A **perennial herb** 20–220 cm tall, in open to dense stands from black **rhizomes** 2–4 mm thick and to *c.* 15⁺ cm long; in moist woods, along lakeshores, sandhills, ditches and railways.

■ **REPRODUCTIVE PARTS Cones** (strobili), spore-bearing, active April–September, 0.5–2.5 cm long by 5–7 mm wide, tan becoming darker and pointed at the apices, terminal, one per stem or sometimes a cone on small secondary branches from the upper nodes, often overwintering; **peduncles** absent to short, often hidden by upper sheath; **sporophylls** numerous; **sporangiophores** (stalks) light green, glabrous, *c.* 1 mm long by *c.* 0.5 mm wide; **shields** tan, 6-sided, 1–1.3 mm wide, without a central dark spot; **sporangia** usually five to seven below each shield, light green, *c.* 1 mm long, spore-filled; **spores** green, microscopic, round, numerous, with four elaters unravelling in dry air.

■ **LEAVES Sheaths** stiff, erect, 2–17 mm long, tight, ashy gray with an upper and lower black band (1–3 mm wide), the upper bands of each sheath thinner and most obvious at the apices of the upper sheaths; **teeth** (leaves) whorled, 15–50, each 2–4 mm long, often fused together along their lower halves, quickly deciduous or some persistent, twisted, dark brown, margins wide and hyaline, firm to papery and pointed.

■ **STEM** Hollow, stiff, persistent more than one year, usually unbranched and evergreen, erect to leaning, rough to touch, hairless, one to five from the root apex, each with 15–50 low ridges, round, walls 0.5–1.5 mm thick, thickest at nodes; 4–8 mm wide near the base.

Tall Scouring-rush, Scouringrush Horsetail

Horsetail—*Equisetaceae* Cones green or tan, 0.5–2 cm long, pointed to blunt

SMOOTH SCOURING-RUSH   *Equisetum laevigatum* A. Braun

■ **SKETCH** A **perennial herb** 20–150 cm tall, single or tufted, sometimes in rows, with a vertical, dark brown **rhizome** 3–4 mm wide and branching at the apex; along sandy river banks, dry gravelly railways, meadows, pastures, prairies and roadsides.

■ **REPRODUCTIVE PARTS** Green ripening to tan; **cones** (strobili) spore-bearing, active May–August, terminal, one per stem, 0.5–2 cm long by 4–8 mm wide, pointed to blunt, apiculum often dark brown and 0.5–1 mm long; **peduncles** tan, hollow, 0–10 mm long by 3–5 mm wide, ridged with a medium brown ring around its wide apex; **sporophylls** 5- or 6-angled, flat; **sporangiophore** (stalk) green, erect; **shields** flat, edges blunt, 1–1.5 mm wide and long by *c.* 0.5 mm thick, upper ones may have a black, irregular, central "splash"; **sporangia** (spore sacs) whitish, *c.* 1 mm long, usually six per sporophyll, attached to underside of shield; **spores** green, each with four elaters which unravel when exposed to dry air.

■ **LEAVES Sheaths** at each node 5–15 mm long, slightly expanded upper margin with a microscopic black dot visible at the tip of each ridge once the deciduous teeth have dropped, lower sheath apices with a black band *c.* 0.2 mm wide or most of sheath black; **teeth** (reduced leaves) whorled, early deciduous and wavy, 15–32, each 1.5–3 mm long, often united, tapered to the apices, hyaline margins white, central vein thin and dark brown.

■ **STEM** Green, annual, soft, erect, hollow, walls 0.5–0.8 mm thick with 15–32 longitudinal ridges, slightly scabrous to smooth with a dry feel, mostly unbranched, easily pulled apart at the nodes, sometimes with two or three short, ascending branches from a slight indent at the base of the upper nodes; 2–8 mm wide near the base.

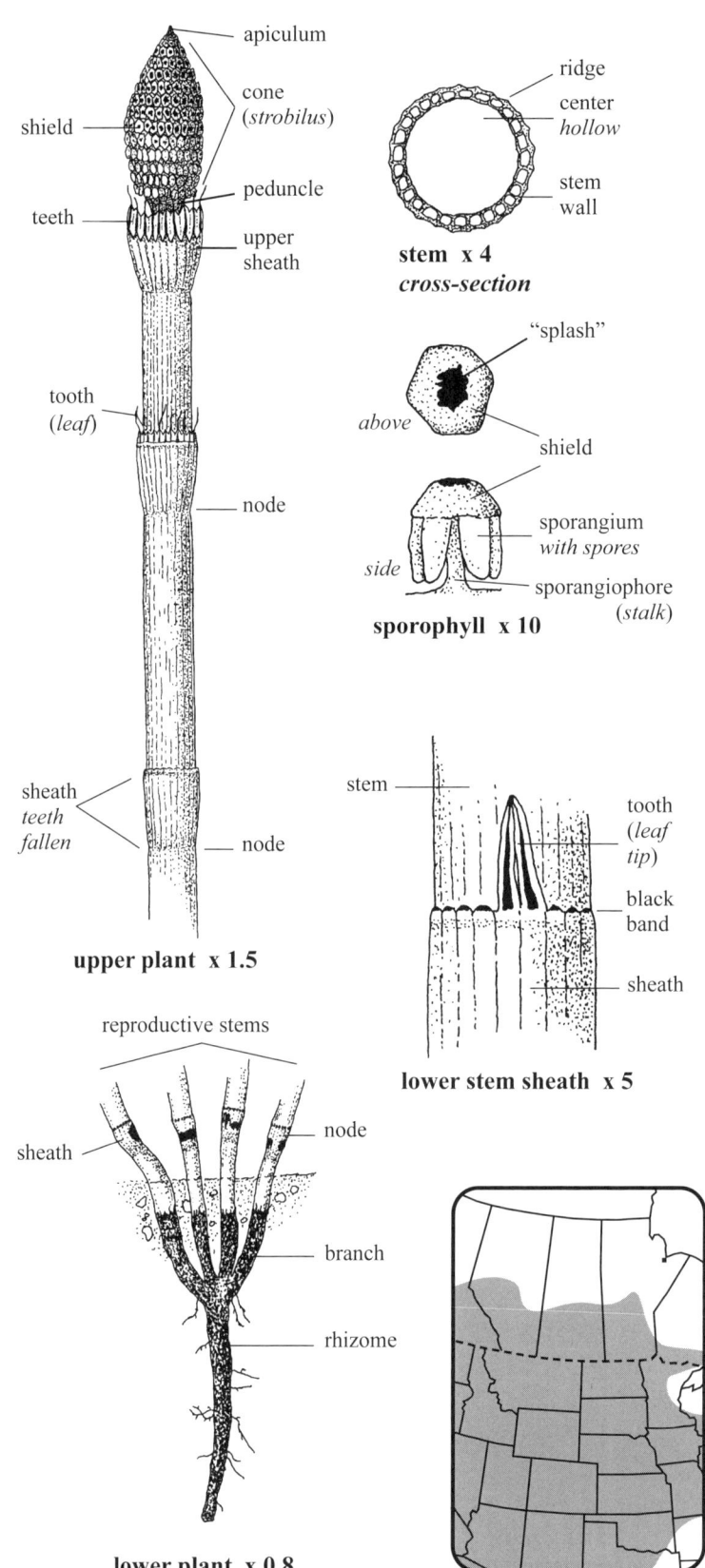

Dwarf Scouring-rush *Equisetum scirpoides* Michx.

- **SKETCH** A **perennial herb** 5–25 cm tall in tangled clusters with dark brown to black **rhizomes** *c.* 0.6 mm thick and wrinkled; in moist coniferous woods, swamps, tundra and mossy woods.
- **REPRODUCTIVE PARTS** A terminal **cone** (strobilus) 2–4 mm long by *c.* 1.5 mm wide, active May–August, often overwintering, partially covered with three upper teeth from the node; **apex** blunt to pointed, black, with spore sacs; **peduncles** *c.* 0.5 mm long, usually hidden by the three upper teeth; **shields** black when exposed above the three teeth, *c.* 1.5 mm wide by *c.* 1 mm tall, 6-sided, interlocking, lower ones slightly transparent and green; **sporophylls** 6–9 per cone, three in a row around the cone and two or three deep (columns); **sporangia** (spore sacs) 4–6 under each shield, pale green, 0.3–0.6 mm long; **stalks** greenish yellow, transparent, winged; **spores** microscopic, light green, with elaters.
- **LEAVES** Sheaths 1–2.5 mm long; **teeth** (reduced leaves) whorled, three at each node, each 1–1.2 mm long, erect to ascending, often broken off on older stems, margins white hyaline, central nerve dark brown, a black band 1–1.5 mm below the teeth, apices thin, 0.5–1 mm long, dark brown and deciduous.
- **STEM** Erect to prostrate, evergreen, a mixture of fertile and vegetative, simple, persistent, not branched, curvy, mostly 5- or 6-ridged, the ridges rough with rows of silica tubercles (microscopic), the top of some stems arched downward, center solid but side cavities quite large; 0.6–0.8 mm wide near the bare base.

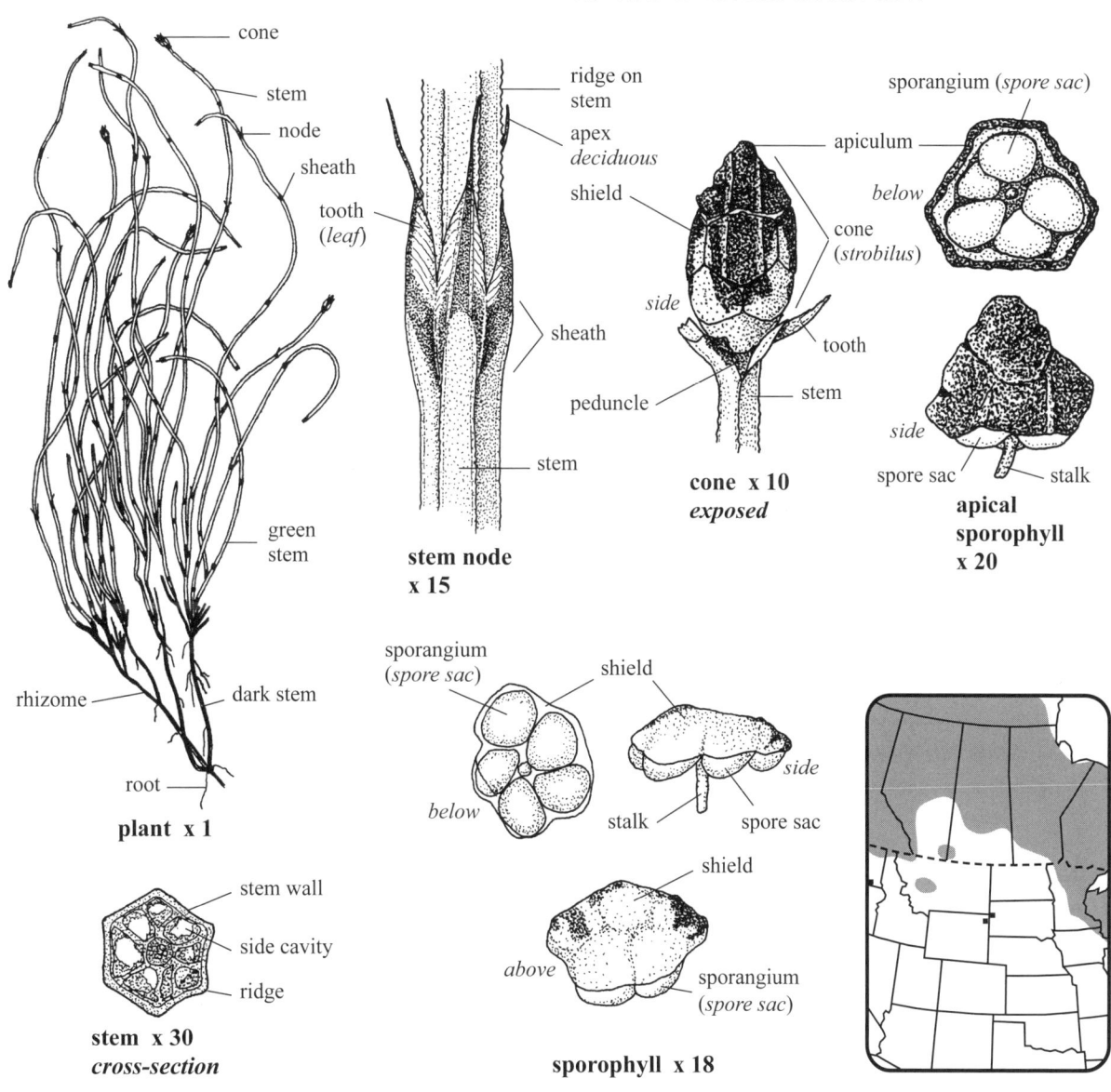

Horsetail—*Equisetaceae* Cones tan, 1–4 cm long, blunt

WOODLAND HORSETAIL   *Equisetum sylvaticum* L.

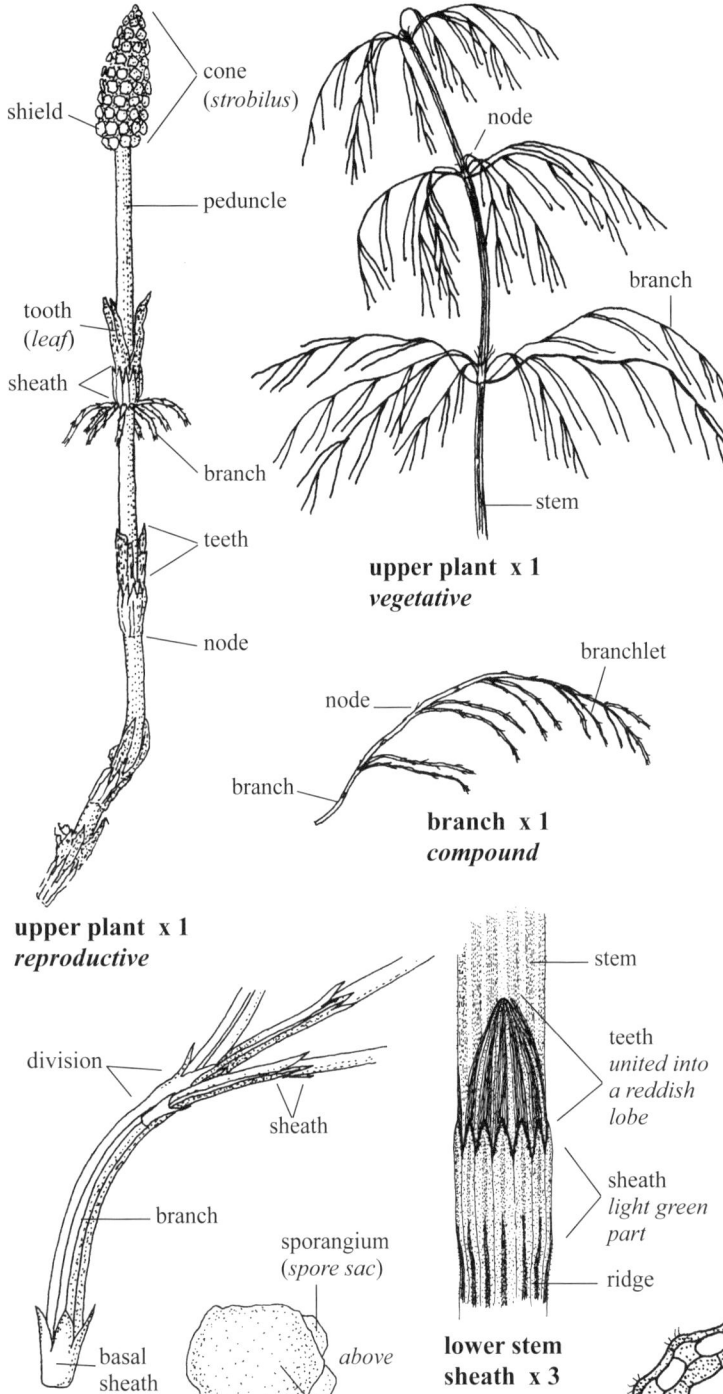

- **SKETCH** A variable **perennial herb** 10–80 cm tall with a deep, smooth, branched **rhizome**; in open or shady sites along lakeshores, marshes and boggy woods.
- **REPRODUCTIVE PARTS** A tan **cone** (strobilus) 1–4 cm long by 5–10 mm wide, active May–August, terminal, tapered to a blunt tip; **peduncles** 2–4 cm long by 1.5–2.5 mm wide, erect, pinkish tan; **sporophylls** in 8–14 rows per cone; **shields** often slightly reddish near the apex, 1–2 mm wide, 6-sided, each with 6–8 thin-walled white sporangia (spore sacs) below; **sporangiophores** (stalks of sporophylls) 1–3 mm long, erect, light tan; **spores** microscopic and with elaters.
- **LEAVES Sheaths** 3–6 mm long at nodes, green in lower half; **teeth** (leaves) whorled, 8–18, reddish brown, united to form 3 or 4 wide lobes 6–13 mm long.
- **STEM** Erect, annual, hollow, of 2 types: **1) vegetative stems** green, branched, round, walls 0.6–1 mm thick, central cavity up to two-thirds of the diameter, 10- to 18-ridged, slightly scabrous from microscopic spreading bristles along each ridge; **branches** in whorls of 5–14 per node, compound, 4- or 5-ridged, light green, ascending to arching and descending, 1–15 cm long, reduced above; **basal sheaths** (of branches) 4- or 5-pointed, apices pale green to reddish brown; **sheaths** (at branch joints) 3-pointed; **nodes** on stems 2–6 cm apart, reduced above, lower two or three nodes without branches but sheaths and teeth present; **2) fertile stems** 10–40 cm tall, flesh-colored at first and unbranched, becoming green and branched after spore dispersal; 1.5–4 mm thick near base.

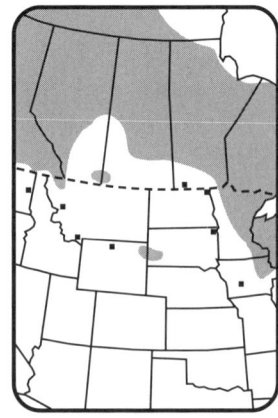

54   Wood Horsetail, Sylvan's Horsetail

Horsetail—*Equisetaceae* | Cones tan and black, 4–10 mm long, pointed

VARIEGATED HORSETAIL   *Equisetum variegatum* Schleich. ex F. Weber & D.M.H. Mohr

■ **SKETCH** A tufted **perennial herb** 5–55 cm tall with reddish brown smooth **rhizomes** 2–5 cm long by *c.* 1 mm thick, and thin **roots** from near the nodes, sometimes covering several m^2; in sandy moist soil, ditches, shores, wet thickets, tundra and bogs.

■ **REPRODUCTIVE PARTS** A terminal **cone** (strobilus) 4–10 mm long by 2.5–4 mm wide, active May–July, often overwintering and dark gray; **apicula** black, 0.5–2 mm long, triangular; **peduncles** 3–5 mm long, sometimes bent/leaning to the side; **shields** flat, 5- or 6-sided, *c.* 0.4 mm thick, tan and black in the depressed middle, drying to dark gray; **sporophylls** 1.5–2 mm wide by 1–1.5 mm tall, in 3–6 rows in a complete turn of the cone and in 3–6 columns; **stalks** *c.* 1 mm long; **sporangia** (spore sacs) 5 or 6 under each shield, each *c.* 0.6 mm long; **spores** microscopic, tan, each with four elaters which unravel when exposed to dry air.

■ **LEAVES Sheaths** green, 3–6 mm long with a black band 0.5–2 mm wide extending into the center of each tooth; **teeth** six (3–14), 0.5–2 mm long, erect, usually persistent, united near their bases, whitish with the centers dark brown, the ridge sometimes reaching the tooth's tip, apiculate, the filiform appendage 0.5–2.5 mm long, deciduous and white to dark brown, erect to slightly spreading.

■ **STEM** Erect, stiff, persistent, curved upward, hollow, the central opening 0.2–1.5 mm wide with 6–12 side cavities, unbranched, 5- or 6-ridged (3–12), the ridges with a low narrow mid-groove and two rows of silica tubercles along the groove's edges; **internodal length** 1–4.5 cm; 0.7–2.5 mm thick near the bare green base.

■ **SYN.** *Hippochaete variegata* (Schleich. ex F. Weber & D.M.H. Mohr) Bruhin = *Equisetum variegatum* var. *variegatum* Schleich. ex F. Weber & D.M.H. Mohr.

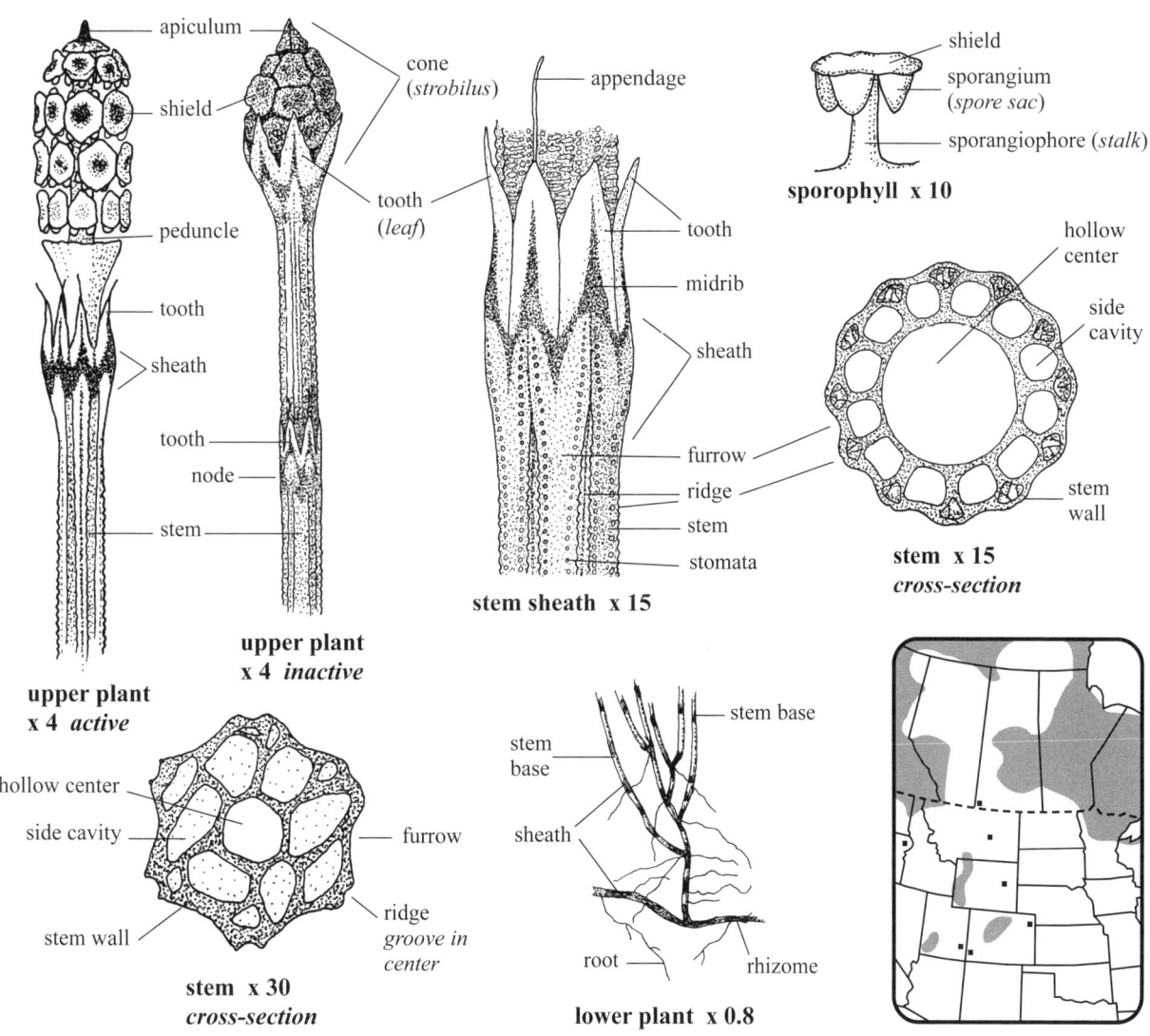

Variegated Scouring-rush, Northern Scouring-rush

Club-moss—*Lycopodiaceae*

TRAILING CLUB-MOSS   *Diphasiastrum complanatum* (L.) Holub

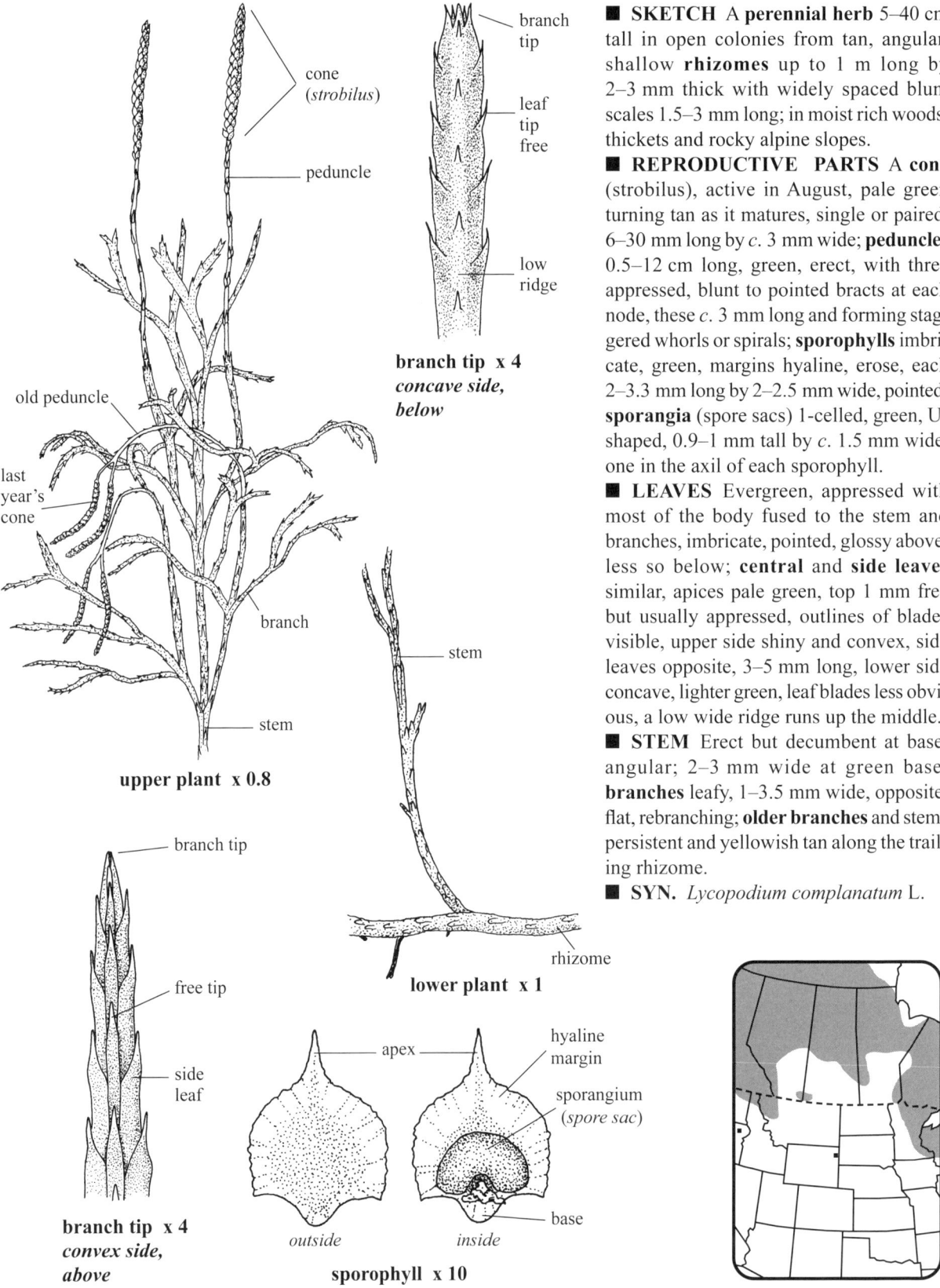

- **SKETCH** A **perennial herb** 5–40 cm tall in open colonies from tan, angular, shallow **rhizomes** up to 1 m long by 2–3 mm thick with widely spaced blunt scales 1.5–3 mm long; in moist rich woods, thickets and rocky alpine slopes.
- **REPRODUCTIVE PARTS** A **cone** (strobilus), active in August, pale green turning tan as it matures, single or paired, 6–30 mm long by *c.* 3 mm wide; **peduncles** 0.5–12 cm long, green, erect, with three appressed, blunt to pointed bracts at each node, these *c.* 3 mm long and forming staggered whorls or spirals; **sporophylls** imbricate, green, margins hyaline, erose, each 2–3.3 mm long by 2–2.5 mm wide, pointed; **sporangia** (spore sacs) 1-celled, green, U-shaped, 0.9–1 mm tall by *c.* 1.5 mm wide, one in the axil of each sporophyll.
- **LEAVES** Evergreen, appressed with most of the body fused to the stem and branches, imbricate, pointed, glossy above, less so below; **central** and **side leaves** similar, apices pale green, top 1 mm free but usually appressed, outlines of blades visible, upper side shiny and convex, side leaves opposite, 3–5 mm long, lower side concave, lighter green, leaf blades less obvious, a low wide ridge runs up the middle.
- **STEM** Erect but decumbent at base, angular; 2–3 mm wide at green base; **branches** leafy, 1–3.5 mm wide, opposite, flat, rebranching; **older branches** and stems persistent and yellowish tan along the trailing rhizome.
- **SYN.** *Lycopodium complanatum* L.

Groundcedar, Trailing Ground-pine, Northern Running-pine, Christmas-green

Club-moss—*Lycopodiaceae* Cones sessile, 12–35 mm long

STIFF CLUB-MOSS *Lycopodium annotinum* L.

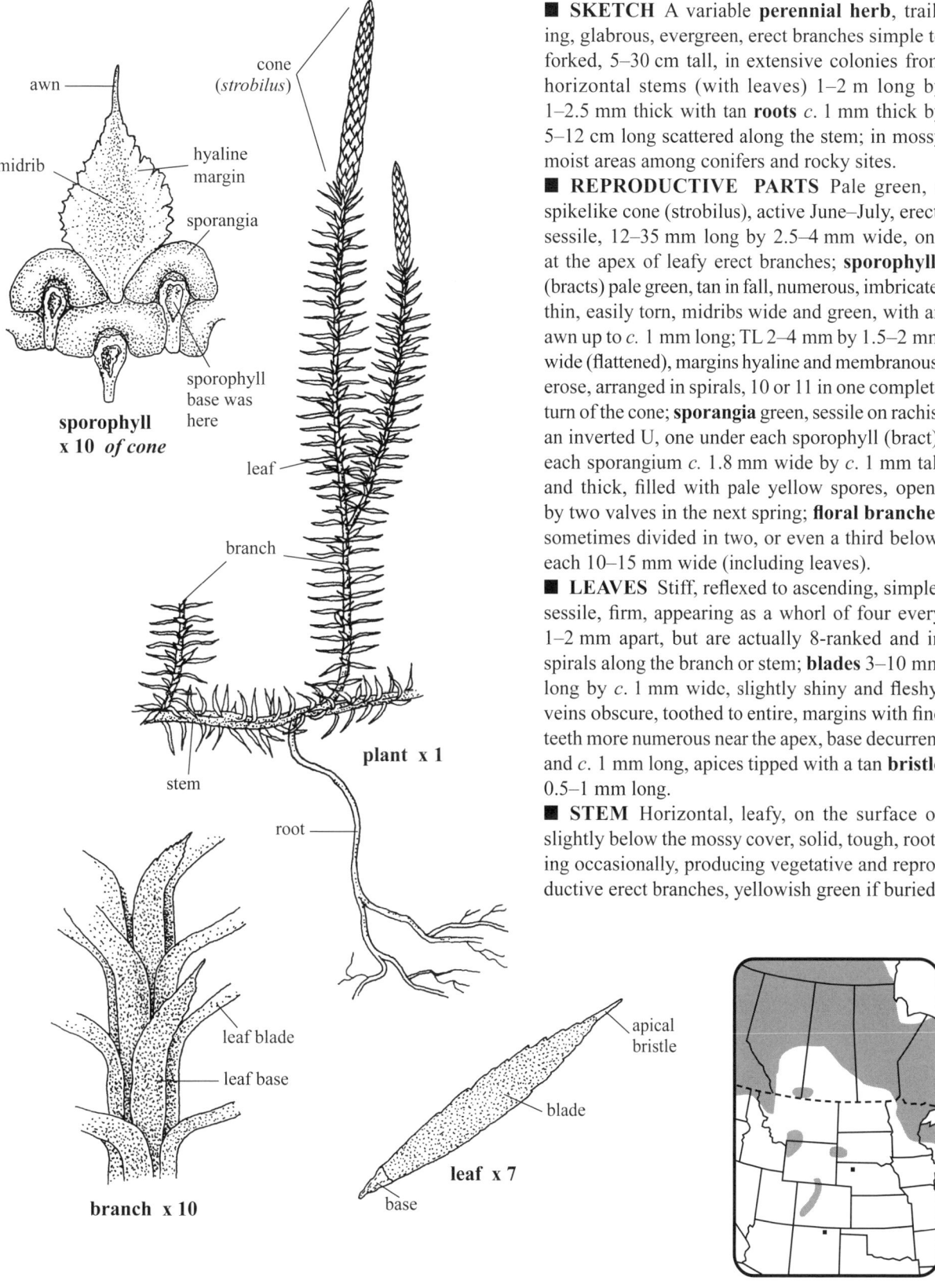

- **SKETCH** A variable **perennial herb**, trailing, glabrous, evergreen, erect branches simple to forked, 5–30 cm tall, in extensive colonies from horizontal stems (with leaves) 1–2 m long by 1–2.5 mm thick with tan **roots** *c.* 1 mm thick by 5–12 cm long scattered along the stem; in mossy moist areas among conifers and rocky sites.
- **REPRODUCTIVE PARTS** Pale green, a spikelike cone (strobilus), active June–July, erect, sessile, 12–35 mm long by 2.5–4 mm wide, one at the apex of leafy erect branches; **sporophylls** (bracts) pale green, tan in fall, numerous, imbricate, thin, easily torn, midribs wide and green, with an awn up to *c.* 1 mm long; TL 2–4 mm by 1.5–2 mm wide (flattened), margins hyaline and membranous, erose, arranged in spirals, 10 or 11 in one complete turn of the cone; **sporangia** green, sessile on rachis, an inverted U, one under each sporophyll (bract), each sporangium *c.* 1.8 mm wide by *c.* 1 mm tall and thick, filled with pale yellow spores, opens by two valves in the next spring; **floral branches** sometimes divided in two, or even a third below, each 10–15 mm wide (including leaves).
- **LEAVES** Stiff, reflexed to ascending, simple, sessile, firm, appearing as a whorl of four every 1–2 mm apart, but are actually 8-ranked and in spirals along the branch or stem; **blades** 3–10 mm long by *c.* 1 mm wide, slightly shiny and fleshy, veins obscure, toothed to entire, margins with fine teeth more numerous near the apex, base decurrent and *c.* 1 mm long, apices tipped with a tan **bristle** 0.5–1 mm long.
- **STEM** Horizontal, leafy, on the surface or slightly below the mossy cover, solid, tough, rooting occasionally, producing vegetative and reproductive erect branches, yellowish green if buried.

Bristly Club-moss, Stiff Ground-pine

Club-moss—*Lycopodiaceae*

GROUND-PINE   *Lycopodium dendroideum* Michx.

■ **SKETCH** An evergreen **perennial herb** forming dense colonies 10–30 cm tall from a deep horizontal orangish brown **rhizome** to about 10^+ cm long by 2–3 mm wide; in moist woods along streams, ravines and bog edges.

■ **REPRODUCTIVE PARTS** A **cone** (strobilus), active in July–October, 1–11 per aerial stem, sessile, 1–6.5 cm long by 3–5 mm wide, from upper branches only, opens in late fall or spring of following year, green turning tan with maturity, persistent for about a year; **sporophylls** imbricate, 3–3.5 mm long and wide, margins erose and white hyaline, pale green in center, tan when mature; **sporangium** (a spore sac), U-shaped, 2–2.8 mm wide by 1–1.4 mm tall, one at base of each sporophyll, pale yellowish green, its two valves opening 90° to each other to release spores.

■ **LEAVES** Stem and branch; **main stem leaves** 6-ranked, two dorsal, two side and two ventral, spreading with tips ascending; **blades** shiny, with free part 3–6 mm long by *c.* 1 mm wide, pointed, widely spaced, glabrous, entire, not prickly; **bases** as long or longer than free part; **branch leaves** similar to stem leaves but shorter and more crowded, especially on the terminal segments, in alternating whorls of three; **blades** 3–4 mm long by 0.7–1 mm wide, shiny, entire, with a short apical bristle, but not prickly to touch, the bases shorter than the blade.

■ **STEM** Erect, covered with leaf bases, green, shiny, glabrous, branched throughout; **branches** ascending, 3–10 cm long, reduced above, dividing into alternate branchlets, these then forking into 4–5 opposite terminal pairs of horizontally spreading segments; **terminal segments** 4–6 mm wide (including leaves); *c.* 2 mm wide at the orangish tan, leafy base.

■ **SYN.** *Lycopodium obscurum* var. *dendroideum* (Michx.) D.C. Eat.

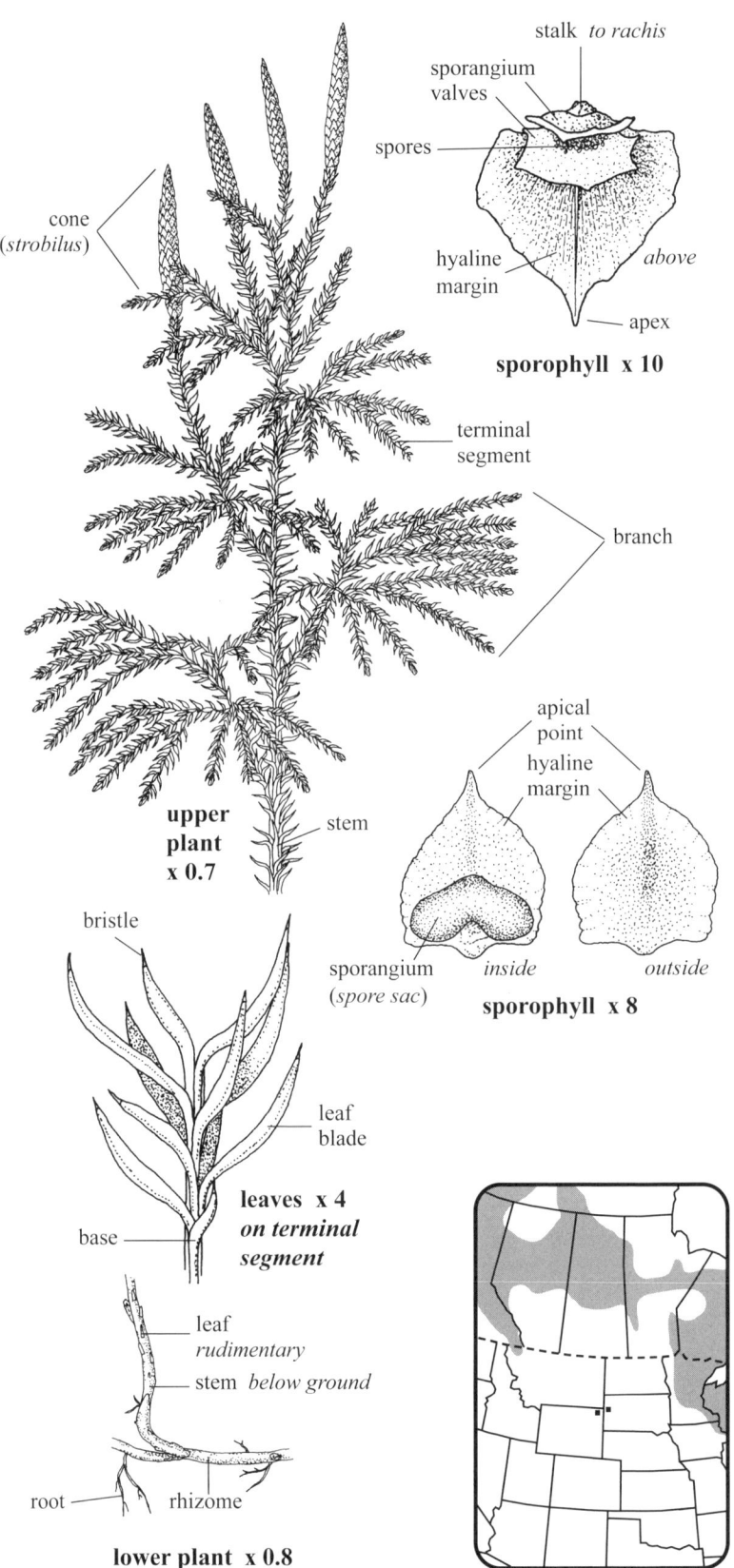

Prickly-tree Club-moss, Tree Groundpine, Ground-cedar, Tree Club-moss, Princess-pine

Gymnosperms

THIS SECOND SMALL section presents some coniferous trees and shrubs. Naked seeds (not included in fruit such as a capsule or an achene) are one obvious characteristic. For instance, White Spruce and Jack Pine, etc., produce pairs of winged seeds that are exposed on the hard woody scales in a ripe cone. In this group the trees are monoecious, with female and male cones on different parts of the same plant. Leaves are needle-like, in tufts (Tamarack) or in pairs (Jack Pine) or single in White and Black Spruce.

THE CONIFEROUS SHRUBS (junipers) in the Cupressaceae family along with one tree, the White Cedar, have spreading to mostly appressed stiff leaves. Cones of the cedar are small with naked winged seeds. Cones of the junipers are roundish and berrylike with a few large, unwinged seeds inside. These shrubs are dioecious, with some plants female and others male. The cedar tree, like those above, is monoecious, with separated male and female structures on each plant.

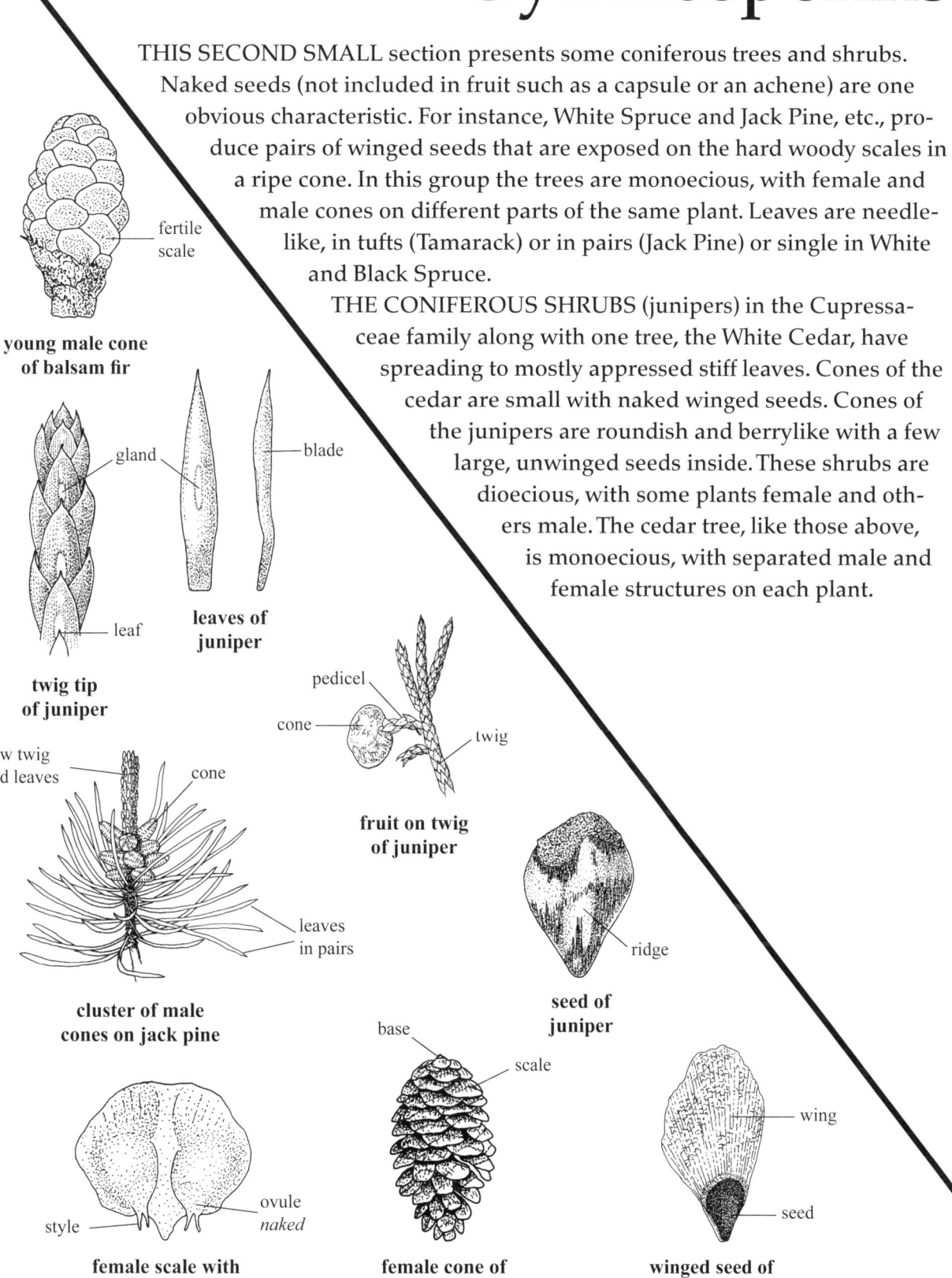

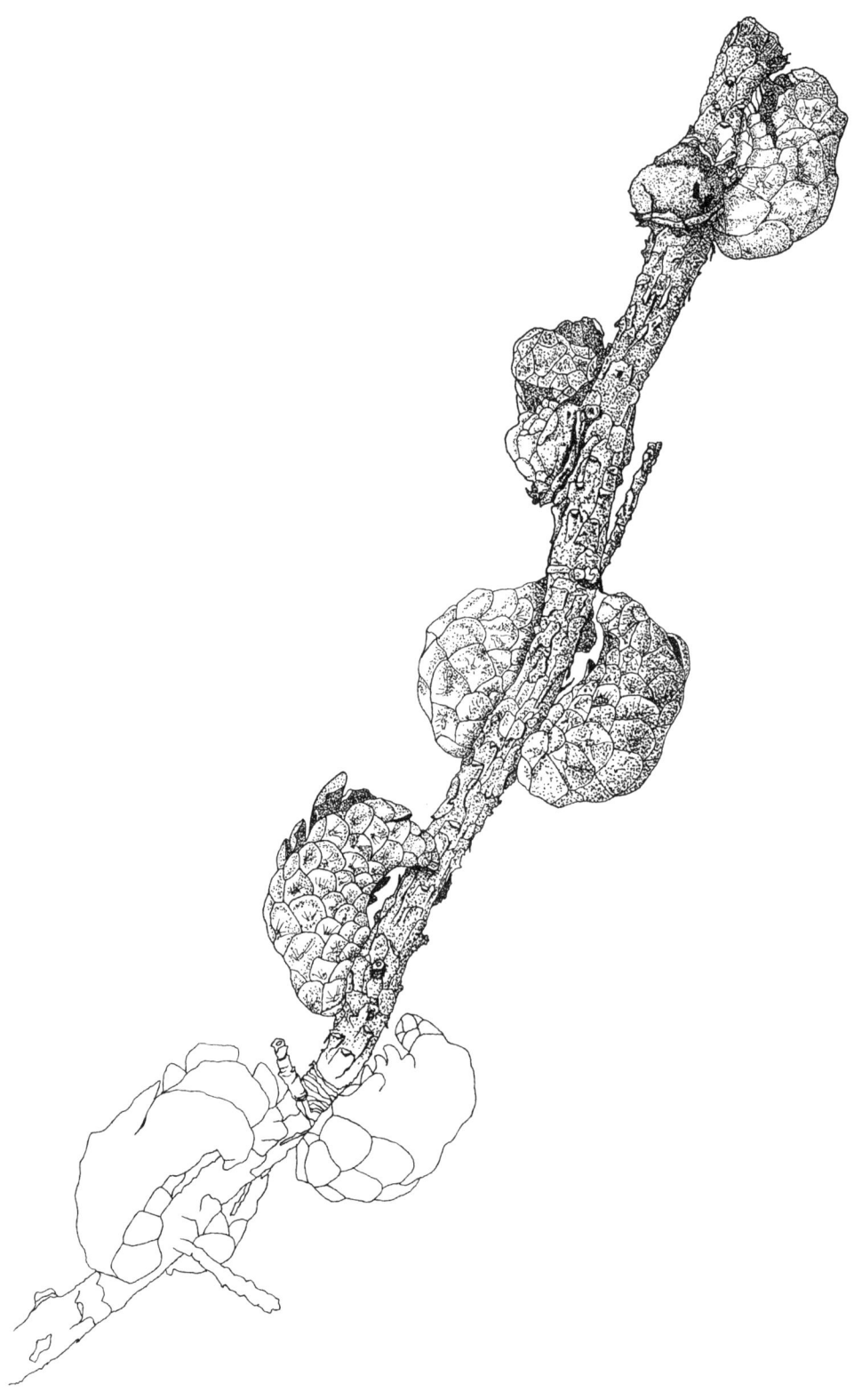

Cypress—*Cupressaceae* **Shrub** Evergreen Cones berrylike

Low Juniper *Juniperus communis* L.

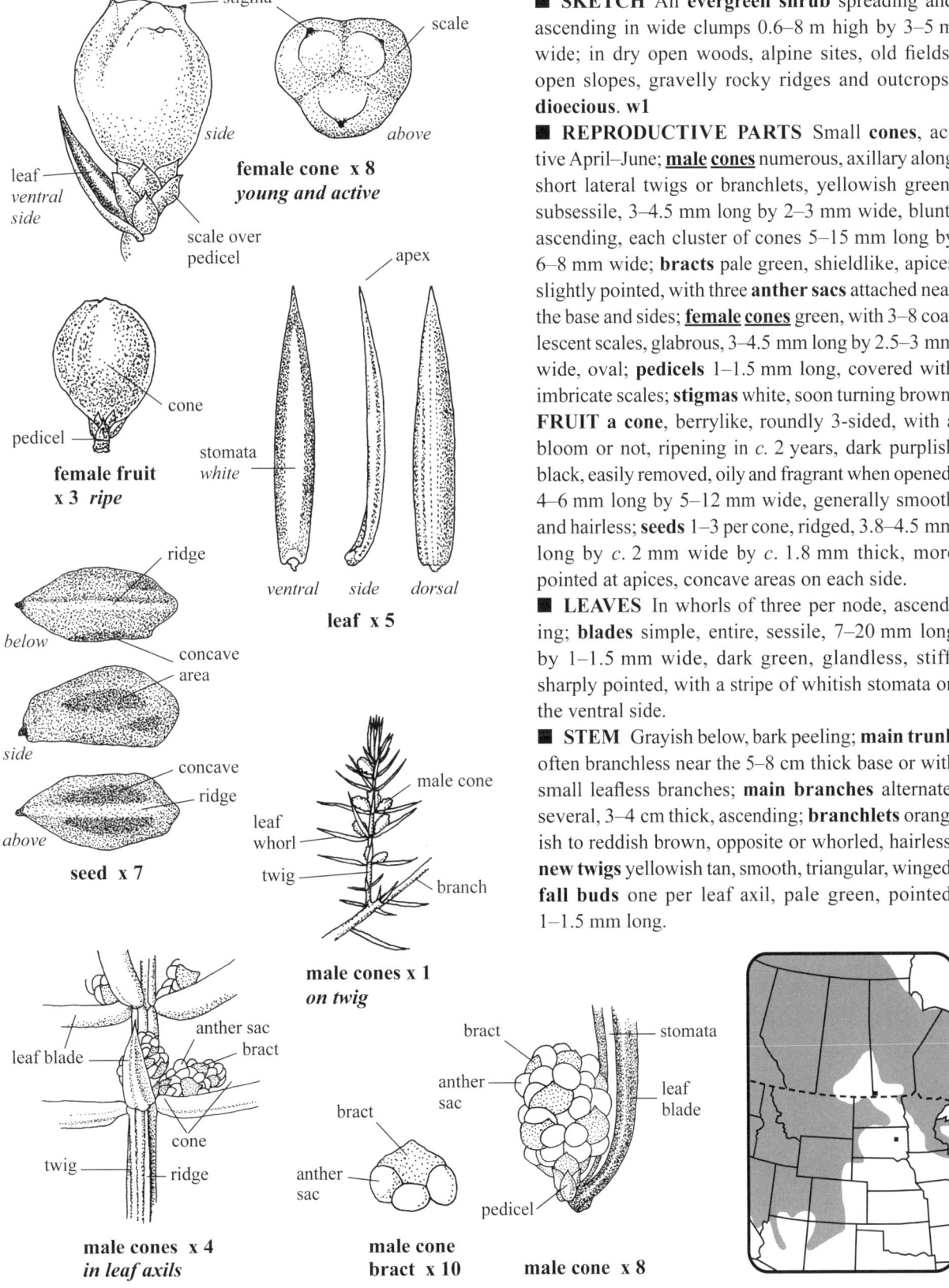

- **SKETCH** An **evergreen shrub** spreading and ascending in wide clumps 0.6–8 m high by 3–5 m wide; in dry open woods, alpine sites, old fields, open slopes, gravelly rocky ridges and outcrops; **dioecious**. w1
- **REPRODUCTIVE PARTS** Small **cones**, active April–June; **male cones** numerous, axillary along short lateral twigs or branchlets, yellowish green, subsessile, 3–4.5 mm long by 2–3 mm wide, blunt, ascending, each cluster of cones 5–15 mm long by 6–8 mm wide; **bracts** pale green, shieldlike, apices slightly pointed, with three **anther sacs** attached near the base and sides; **female cones** green, with 3–8 coalescent scales, glabrous, 3–4.5 mm long by 2.5–3 mm wide, oval; **pedicels** 1–1.5 mm long, covered with imbricate scales; **stigmas** white, soon turning brown. **FRUIT** a cone, berrylike, roundly 3-sided, with a bloom or not, ripening in *c.* 2 years, dark purplish black, easily removed, oily and fragrant when opened, 4–6 mm long by 5–12 mm wide, generally smooth and hairless; **seeds** 1–3 per cone, ridged, 3.8–4.5 mm long by *c.* 2 mm wide by *c.* 1.8 mm thick, more pointed at apices, concave areas on each side.
- **LEAVES** In whorls of three per node, ascending; **blades** simple, entire, sessile, 7–20 mm long by 1–1.5 mm wide, dark green, glandless, stiff, sharply pointed, with a stripe of whitish stomata on the ventral side.
- **STEM** Grayish below, bark peeling; **main trunk** often branchless near the 5–8 cm thick base or with small leafless branches; **main branches** alternate, several, 3–4 cm thick, ascending; **branchlets** orangish to reddish brown, opposite or whorled, hairless; **new twigs** yellowish tan, smooth, triangular, winged; **fall buds** one per leaf axil, pale green, pointed, 1–1.5 mm long.

Common Juniper, Ground Juniper, Dwarf Juniper, Mountain Juniper

Cypress—*Cupressaceae* **Shrub** Evergreen Cones berrylike

CREEPING JUNIPER   *Juniperus horizontalis* Moench

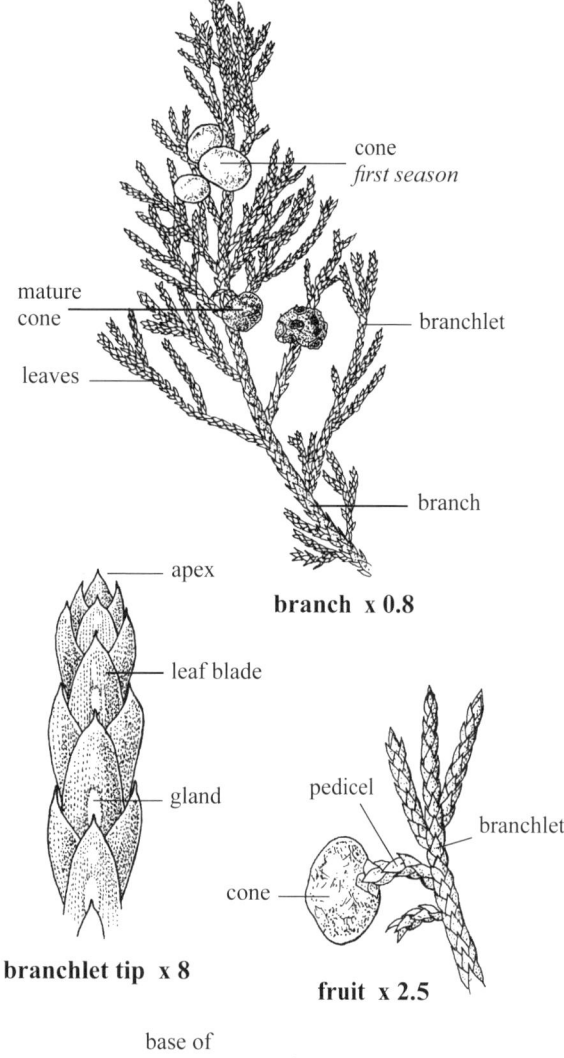

branch x 0.8

branchlet tip x 8

fruit x 2.5

stem section x 1

■ **SKETCH** A mat forming, glabrous, **evergreen shrub** rooting along its prostrate branches, covering several m², and rarely taller than *c.* 20 cm, fragrant; in dry open woods, rocky outcrops, prairies, stream banks, sand-hills and dry clearings; **dioecious. p1**

■ **REPRODUCTIVE PARTS** Green, blooming May–June; **inflorescence** conelike; **male parts** (not shown) at tips of branchlets, cylindrical, yellowish becoming purplish, 3.5–5 mm long by 1–2 mm across, sessile; **pollen sacs** four, sessile; **pedicels** 2–6 mm long, leafy, recurved; **female parts** at tips of often reflexed branchlets, of two to six ovules *c.* 0.3 mm long, on soft scales. **FRUIT a cone**, berrylike, matures in 15–20 months, fragrant, scales fused, 5–10 mm wide by 5–6 mm long, glaucous, blue (first year), usually one to ten, often in a cluster on 5–8 cm long branchlets; **mature fruit** darker blue, slightly glaucous, mealy inside, in persistent clusters; **seeds** one to five per cone, each 3–5 mm long by 2.5–3 mm wide by 2.3–3 mm thick, reddish brown, shiny, not hairy, oval, with a rough (microscopic), dull brown area at one end, streaks on shiny part.

■ **LEAVES** Scalelike, imbricate, appressed, pointed, yellowish green to bluish green and glaucous; **blades** sessile, entire, 1.1–7 mm long by 0.4–1.3 mm wide by 0.2–0.4 mm thick, longest on the new shoots, pointed, convex above, flat to slightly concave below; **gland** dorsal, 0.3–0.5 mm long by 0.2–0.3 mm wide, oval, slightly raised, but in a slight depression near the base of the blade.

■ **STEM** Horizontal, 1–5⁺ m long by 3–6 mm thick, woody, tan inside, with a slightly darker center; **branches** leafy, alternate and ascending, often partially buried under a layer of old leaves and grass; **new vigorous branches** tan to slightly reddish; **branchlets** round, 1–2.5 mm wide with appressed leaves (scales); **bark** gray to brown, peeling.

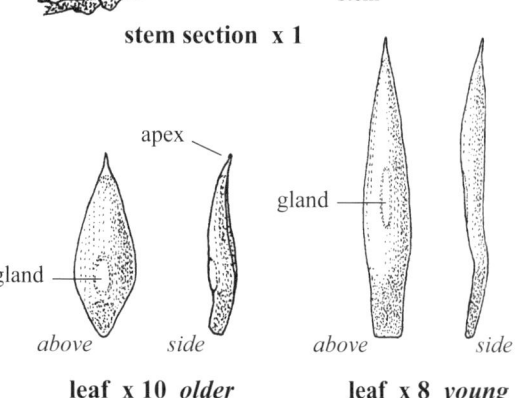

leaf x 10 *older*   leaf x 8 *young*

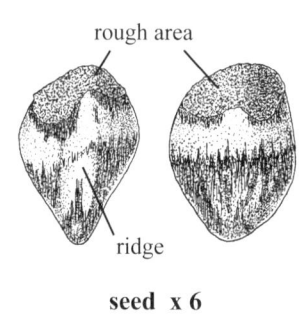

seed x 6

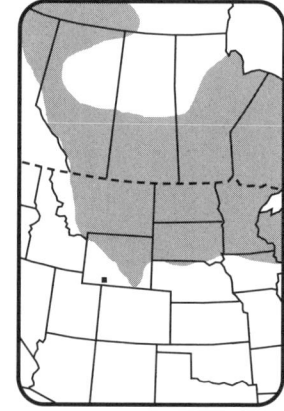

Trailing Juniper, Creeping Savin, Creeping Cedar

Cypress—*Cupressaceae* **Tree** Evergreen Cones 7–14 mm long

WHITE CEDAR   *Thuja occidentalis* L.

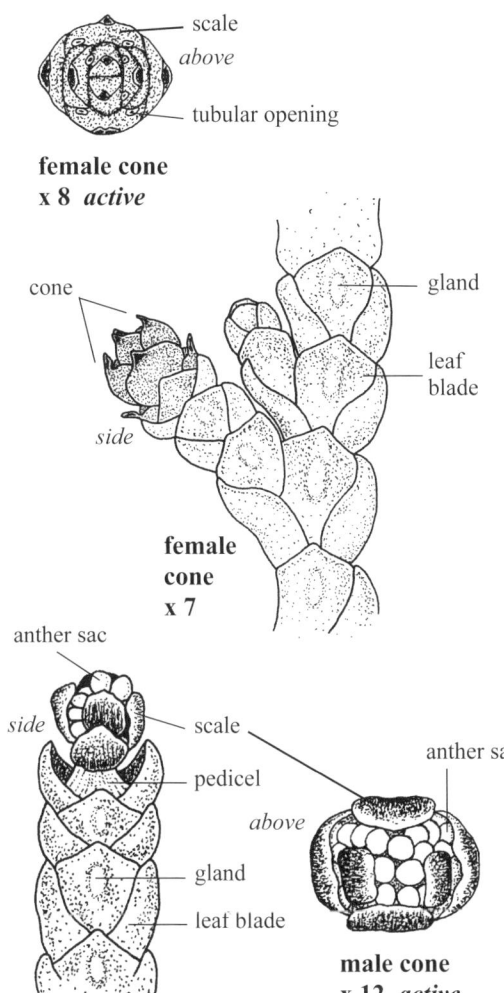

- **SKETCH** A **coniferous tree** 8–30 m tall, cone-shaped, singly or in small groups, branches spreading; in damp woodlands and swamps; **monoecious**. w1
- **REPRODUCTIVE PARTS** Brown, active April–June; **inflorescence** of unisexual cones; **pedicels** brown, wider at the base, shiny and hairless; <u>male cones</u> numerous, *c.* 1.5 mm long by *c.* 2 mm wide by *c.* 1.5 mm thick, appearing at the tips of new, 4–15 mm long branchlets; **scales** six (4 + 2), hairless, dry, crisp, dark brown below, C-shaped, with a hyaline margin, the four outer lower scales to *c.* 1 mm long and wide, the upper two scales *c.* 0.8 mm long by *c.* 0.5 mm wide, all scales attached to a central brown stalk *c.* 1.3 mm wide; **anther** sacs 24, in six groups of four each, included, beneath each scale but not attached to it, sessile, roundish, beige, *c.* 0.5 mm long by *c.* 0.4 mm wide; <u>female cones</u> (in small clusters of three or four on branchlets 2–5 mm long), brown, *c.* 2 mm long by 2–2.4 mm wide by 1.7–2 mm thick, with eight fleshy brown scales (some sterile), each with a dark brown to black pointed tip with eight tubular openings at the base of the inner scales; **pistils** obscure. **FRUIT a cone**, brown, woody, overwintering, clustered, usually 7- to 12-scaled, 7–14 mm long, when open *c.* 10 mm wide, round and long, slightly rough and resinous; **seeds** winged, flat, slightly darker than the wing, 5.5–6.5 mm long by 2.5–2.7 mm wide by *c.* 0.5 mm thick, usually seven or eight per cone, mostly two under four fertile scales; **wings** thin, light tan, the apices notched.
- **LEAVES** Scalelike, opposite, 4-ranked, glabrous, stiff, paired, appressed, imbricate, medium green, lasting three to four years, turning brown before falling, 1–7 mm long by 1–2 mm wide by *c.* 1 mm thick; **gland** small, raised near the tip of larger leaves.
- **TRUNK** Erect, tapered; **branches** spreading and ascending near the top; **bark** thin and shredding in elongated fibrous strips, furrowed, dull, gray to reddish brown; **twigs** alternate, fan-shaped, flattened, covered with scalelike leaves, glabrous, rough from leaf tips when shedding; **lenticels** obscure; **winter buds** hidden beneath leaves; **dbh** 30–100 cm.

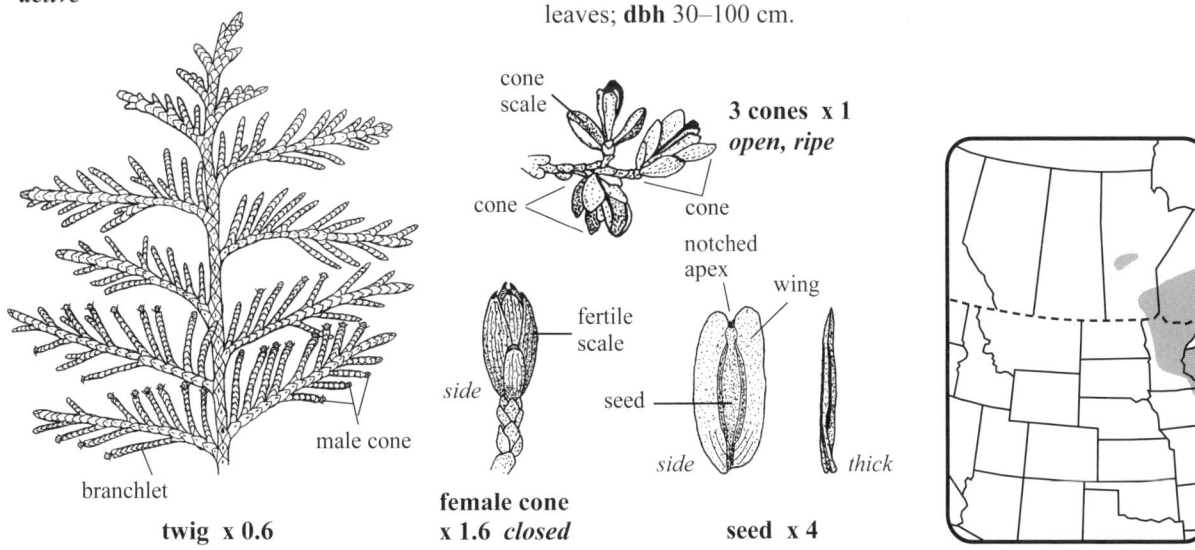

Arborvitae, Eastern Arborvitae, Northern White Cedar, Eastern White Cedar

Tree

Trees that know the way of rain
and how to provide an awning
when we come begging shade

So picture this
a tree in late March
full of hoar frost
with a cawing crow

Some splendid Elm or holy Oak
where birds come one by one

Branches the moon gets caught in
leaves that rustle like harps
in the breeze

Pine—*Pinaceae*   **Tree** Evergreen   Cones erect, 1.5–7 cm long

Balsam Fir   *Abies balsamea* (L.) P. Mill.

■ **SKETCH** An **evergreen tree** 8–25 m tall, symmetrical, narrowly tapered from bottom to top; in moist sites along rivers and streams in the Boreal forest; **monoecious**. **w2**

■ **REPRODUCTIVE PARTS** Cones, unisexual, reddish, active May–June; **male cones** yellow and red, situated at leaf bases along numerous new twigs, easily visible from ground, each subsessile, 5–7 mm long by 3–4 mm wide, glaucous, blunt, each cluster of cones 2–6 cm long, throughout the middle region of a tree; **young female cones** erect, tips of subtending bracts project past the scales. FRUIT **a cone**, erect on branches at the top of the tree, 1.5–7 cm long by 1.5–3 cm wide, resinous with fruit; **subtending bracts** one per scale, 4–5 mm long by 3.8–4.3 mm wide, sometimes exserted, on lower side, light brown, apex awnlike and to *c*. 1 mm long; **scales** 8–13 mm long by 9–16 mm wide, ciliate, dark brown when ripe, margins slightly irregular, deciduous in first fall, leaving the central rough axis exposed for several years; **seeds** two per scale on upper side, each 11–12 mm long (including wing) by 5–6 mm wide; **seed body** 4–6 mm long by 2.5–3 mm wide, shiny, *c*. 2 mm thick, with long dark striations; **wing** thin, dark and shiny at the 5–6 mm wide end.

■ **LEAVES** Needlelike, flattened, spreading; **blades** single, 10–25 mm long by 1–1.2 mm wide, oval in cross-section, shiny green above, below 4–8 rows of tiny white dots (stomata) on both sides of a green midrib, margins entire, apices green and blunt, some with a tiny notch (microscopic), subsessile, twisted near the base, ascending to horizontal, fragrant when crushed; **leaf scars** oval, barely raised and the center slightly concave.

■ **TRUNK** Erect, bark grayish, smooth to scaly with resinous blisters; **dbh** 30–70 cm; **branchlets** in whorls, symmetrical; **new twigs** tan, slightly hairy, the hairs spreading; **older twigs** gray and hairless; **terminal buds** blunt, flanked with two lateral buds, each 4–5 mm long.

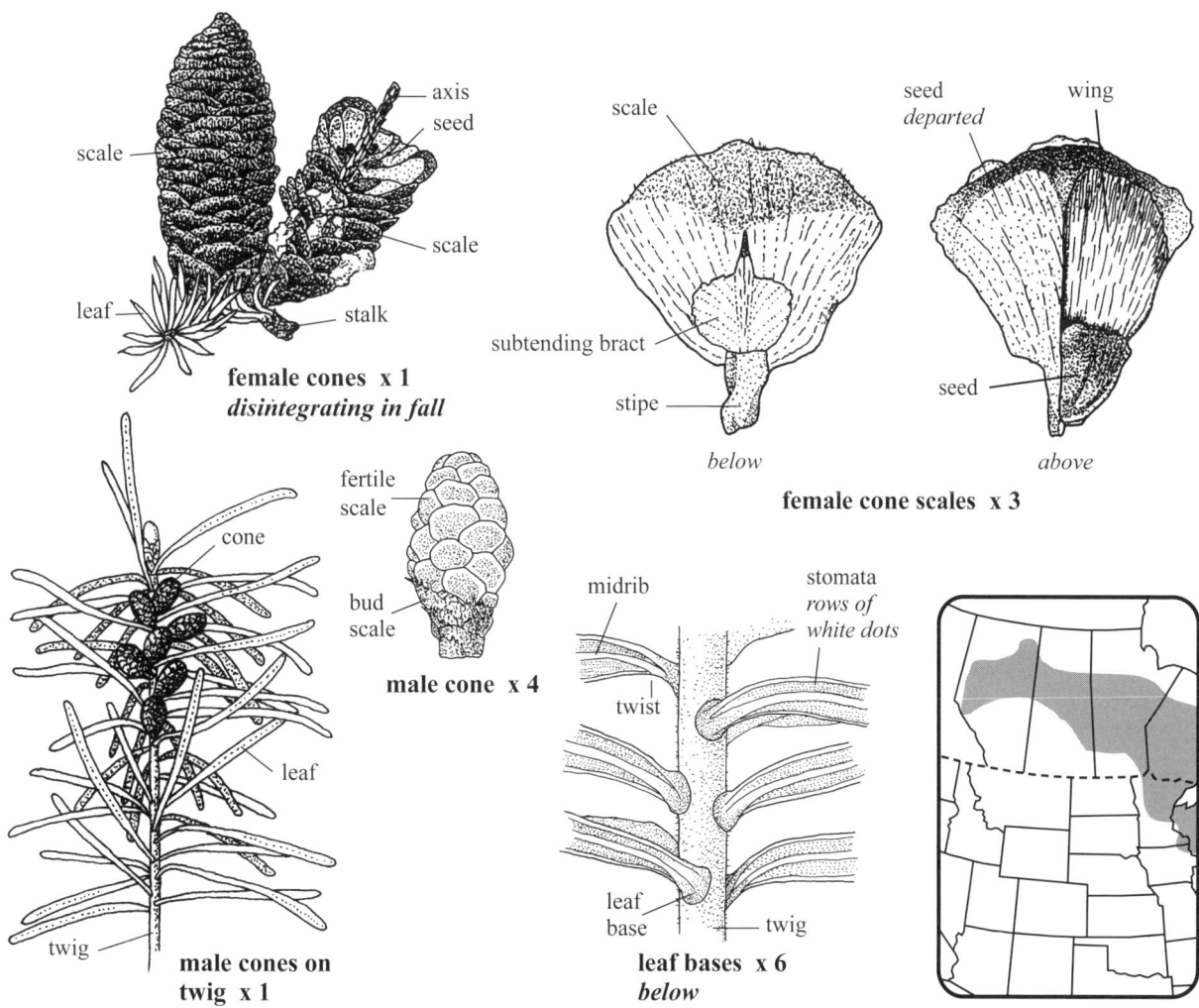

Canada Balsam

Pine—*Pinaceae*　　　**Tree**　Coniferous　Cones 15–25 mm long

TAMARACK　*Larix laricina* (Du Roi) K. Koch

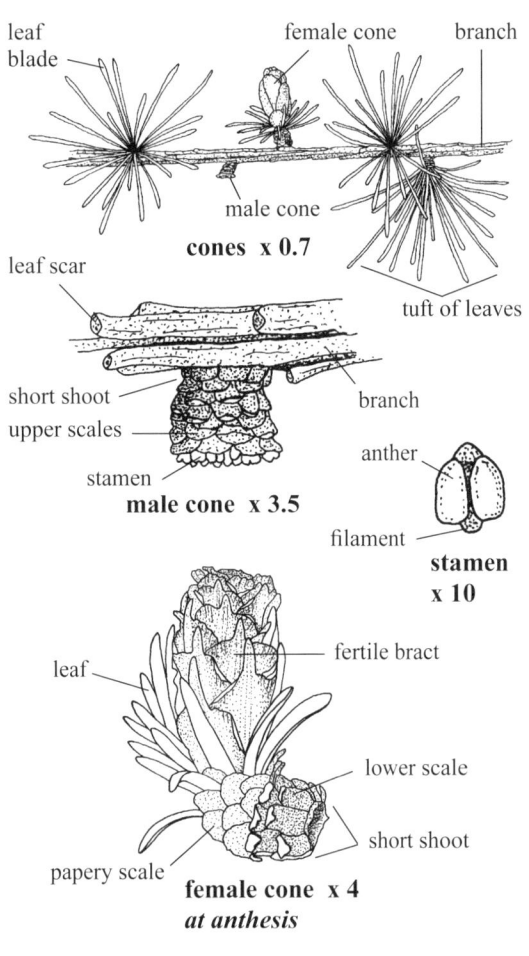

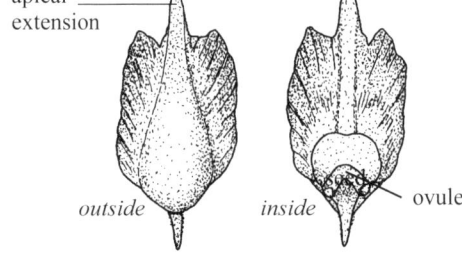

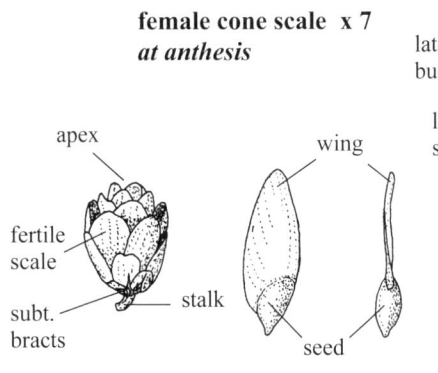

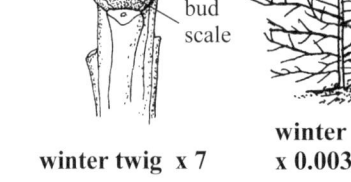

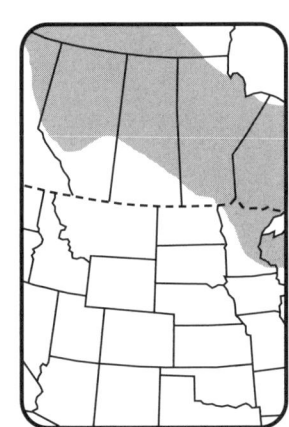

■ **SKETCH** A **coniferous tree** with deciduous leaves (needles) falling each autumn, 10–20 m tall; in wet sphagnum bogs, muskeg and swamps to well-drained sites, in open stands in the Boreal forest; **monoecious**. w2

■ **REPRODUCTIVE PARTS** Unisexual cones active in May–June; **male cones** yellowish tan, numerous, round, alternate, mostly directed down or at a lower angle than the leaves and female cones, usually *c.* 4 mm long and wide (including short, scaly, basal shoot); **stamens** tan, *c.* 1 mm long and wide by *c.* 0.5 mm thick; **female cone basal shoots** scaly, 4–7 mm long, curved upward, composed of two types of scales: **1)** lower ones dark brown, stiff with apices rolled back and whitish; **2)** upper papery scales at base of leaf rosette, apices tan and wide, *c.* 3 mm long by *c.* 1.5 mm wide, base whitish and narrow, glabrous; **female cones** bright red at anthesis, 7–10 mm long by *c.* 3 mm wide, erect, less numerous than male cones, on main branches and twigs; **scales** in a spiral, sides and margins reddish; **ovules** paired, winged, at base of fertile scales, cones set in a rosette of 20–40 modified leaves. **FRUIT** a cone, tan, scattered, with 20–30 rounded scales, erect on 2–8 mm long shoots, overwintering, mature open cones 15–25 mm long by *c.* 10 mm wide; **subtending bracts** (at base of cones) several; **seeds** two per fertile scale, 1–2 mm long, winged, TL 7–10 mm (seed plus wing), tannish red, glabrous; **wing** 2.2–4 mm wide and tan.

■ **LEAVES** Needlelike, deciduous, simple, green, on short lateral shoots, turning yellow in autumn before falling; **blades** (needles) in tufts of 12–45 in a circular groove in the upper edge of short shoots, or singly and spirally on first year shoots, mostly sessile, soft, entire, 12–25 mm long by 0.7–0.8 mm wide, glaucous, with two white lines of stomata near the margins, smooth above, hairy at base, flattened, not shiny.

■ **TRUNK** Erect, smooth when young, rough with age, gray, not shiny; **branches** gray, alternate, hairless, rough with scattered persistent bases of shoots; **lateral winter buds** on new twigs round, alternate, reddish 1–1.5 mm long and wide, scales imbricate with edges irregular and whitish; **dbh** 30–60 cm.

Pine—*Pinaceae* **Tree** Evergreen Cones 2.5–8 cm long, hanging

WHITE SPRUCE *Picea glauca* (Moench) Voss

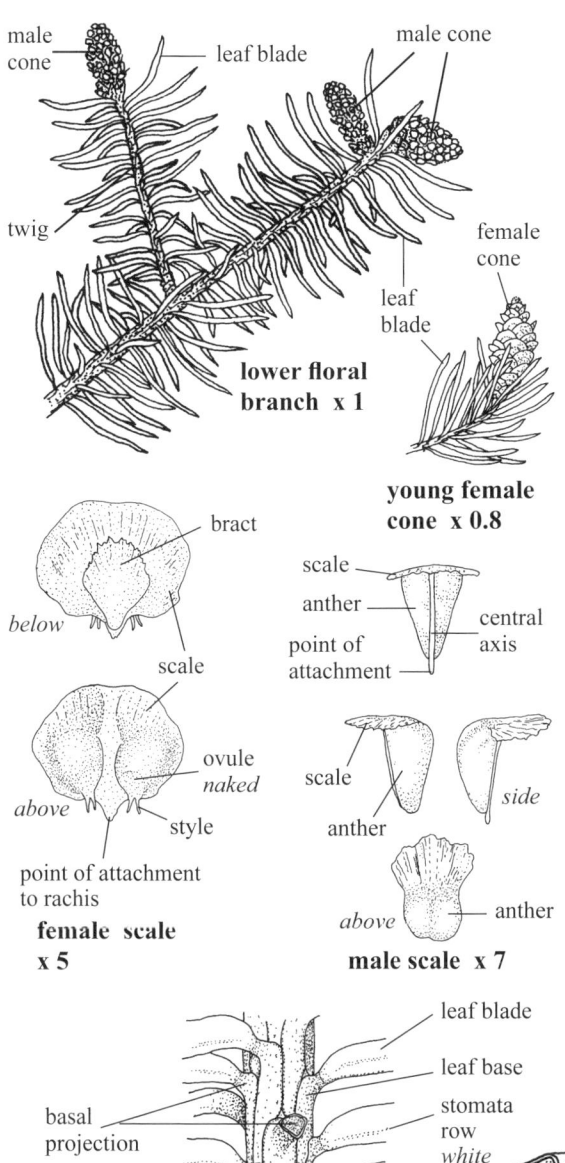

■ **SKETCH** An **evergreen tree** 20–35 m tall, narrow pyramidal in shape; in well-drained areas or in bogs, muskeg, along streams and lake margins, on rocky slopes in northern woods; **monoecious**. w2

■ **REPRODUCTIVE PARTS** Cones reddish, active May–June; **bud scales** (of male cones) medium brown, deciduous; **male cones** oval, one to four in terminal clusters at the ends of year-old twigs, 10–23 mm long by 5–8 mm wide; **male scales** 0.9–1.8 mm wide by 1.3–1.8 mm long, reddish, often with a tan apex, margins irregular, attached at their base, more or less sessile; **anthers** two per scale, yellow; **female cones** sessile, *c.* 1 cm long at anthesis, oval, solitary at the ends of year-old twigs in tree's crown, hanging when mature; **young scales** *c.* 4 mm wide by *c.* 3.5 mm long, thin, smooth-margined, subtended by a small bract, each with two basal naked ovules; **styles** paired, projecting from base. **FRUIT a cone**, woody, 2.5–8 cm long by 2–2.5 cm wide (open), hanging and falling whole the first year, scales woody, thin, margins smooth; **seeds** two per scale, winged, 2–3 mm long by 1.2–1.5 mm wide by 1–1.2 mm thick, the coat dull, rough and dark brown, shed in fall; **wing** tan, papery, 5–7 mm long by 3–3.5 mm wide.

■ **LEAVES** Needlelike, single, spirally arranged on branches and twigs, fewer on the lower side of twigs; **blades** linear, 4-sided, tips pointed, 10–24 mm long by *c.* 1 mm wide, glabrous, medium green, each side with two or three rows of white **stomata**, each blade jointed to a short basal projection.

■ **TRUNK** Erect, tapered, branched; **bark** thin, gray, scaly; **twigs** yellowish tan when young, hairless, slightly shiny, turning gray with age; **branches** horizontal to ascending above, descending with upturned tips below, rough from persistent basal projections of leaves; **dbh** 40–100 cm.

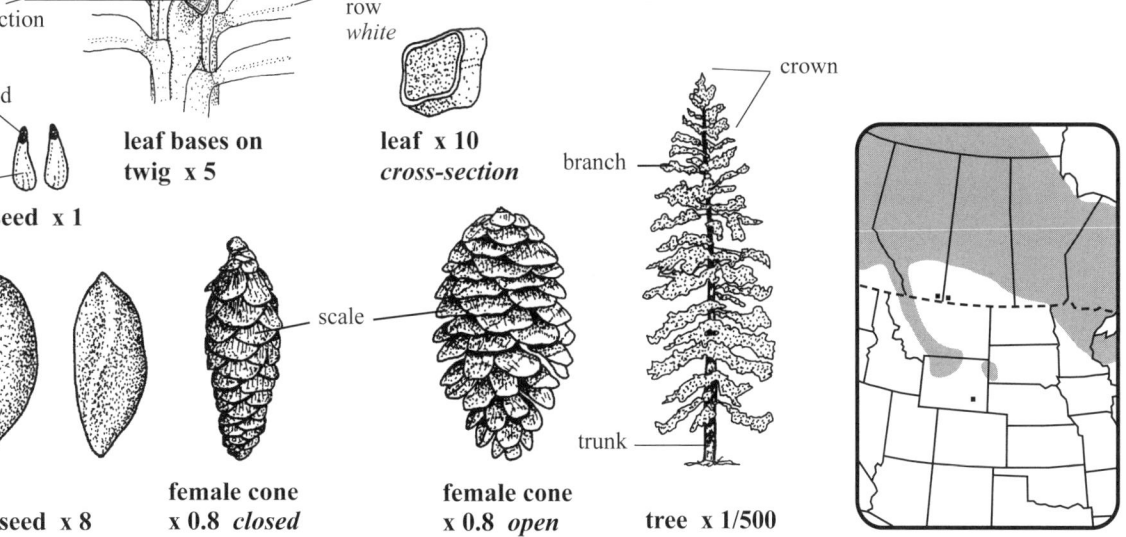

Black Hills Spruce

Pine—*Pinaceae* **Tree** Evergreen Cones 1.5–3.5 cm long, hanging

Black Spruce   *Picea mariana* (P. Mill.) B.S.P.

■ **SKETCH** An **evergreen tree** 7–25 m tall, apex erect and often with an obvious, dense clump of twigs; in poorly drained boggy sites, in swamps, near lakes, streams and beaver ponds in the Boreal forest; **monoecious**. w2

■ **REPRODUCTIVE PARTS** Cones, oval, active in June–July; male **cones** reddish purple, 1–4 terminal and 1–4 side cones often in a whorl, blunt, 5–13 mm long by 4–7 mm wide at anthesis; **peduncles** 2–5 mm long; **fertile scales** imbricate, in spirals, 1.7–2 mm long by 1–2 mm wide, margins erose; **stamens** paired under each scale; **anthers** yellow, *c.* 1.6 mm long, tapered to base; **rachis** green, round, *c.* 1 mm wide; female **cones** purple and hanging at anthesis; **subtending** (primary) **bract** (of fertile scale) *c.* 3 mm long by *c.* 1.5 mm wide, apex dark purple and erose; **fertile scales** in young cones 8–12 mm long and wide, glaucous, tight, greenish brown; **ovules** white, *c.* 2.2 mm long by *c.* 1.3 mm wide. **FRUIT a cone**, 1.5–3.5 cm long by 13–21 mm wide (open), hanging on upper branches in clusters of 1–3, maturing to reddish brown, remaining on the tree for several years; **seeds** two per fertile scale, 2.2–2.5 mm long by *c.* 1.5 mm wide, dark brown, rough, dull, pointed at one end, bare and raised above wing on one side only; **wing** tan, 5–7 mm long by *c.* 4 mm wide, easily torn, erose, striations red.

■ **LEAVES** Evergreen, needlelike, single, stiff, spirally arranged, green; **blades** linear, 4-sided, 0.8–2 cm long by 1–2 mm wide, slightly curved, each side with 2–6 rows of white stomata (dots), apex with a tiny point but not prickly, deciduous from a brown, 4-sided, hairy basal projection.

■ **TRUNK** Erect, simple, bark scaly, scales thin and gray to brown, inner bark greenish yellow; **dbh** 10–50 cm; **branches** not symmetrical in older trees, spreading above to strongly descending in lower half but with the tips ascending; **lower branches** often missing, or if present may root where touching the ground, hairless and dark brown; **new twigs** tan with numerous white to reddish hairs.

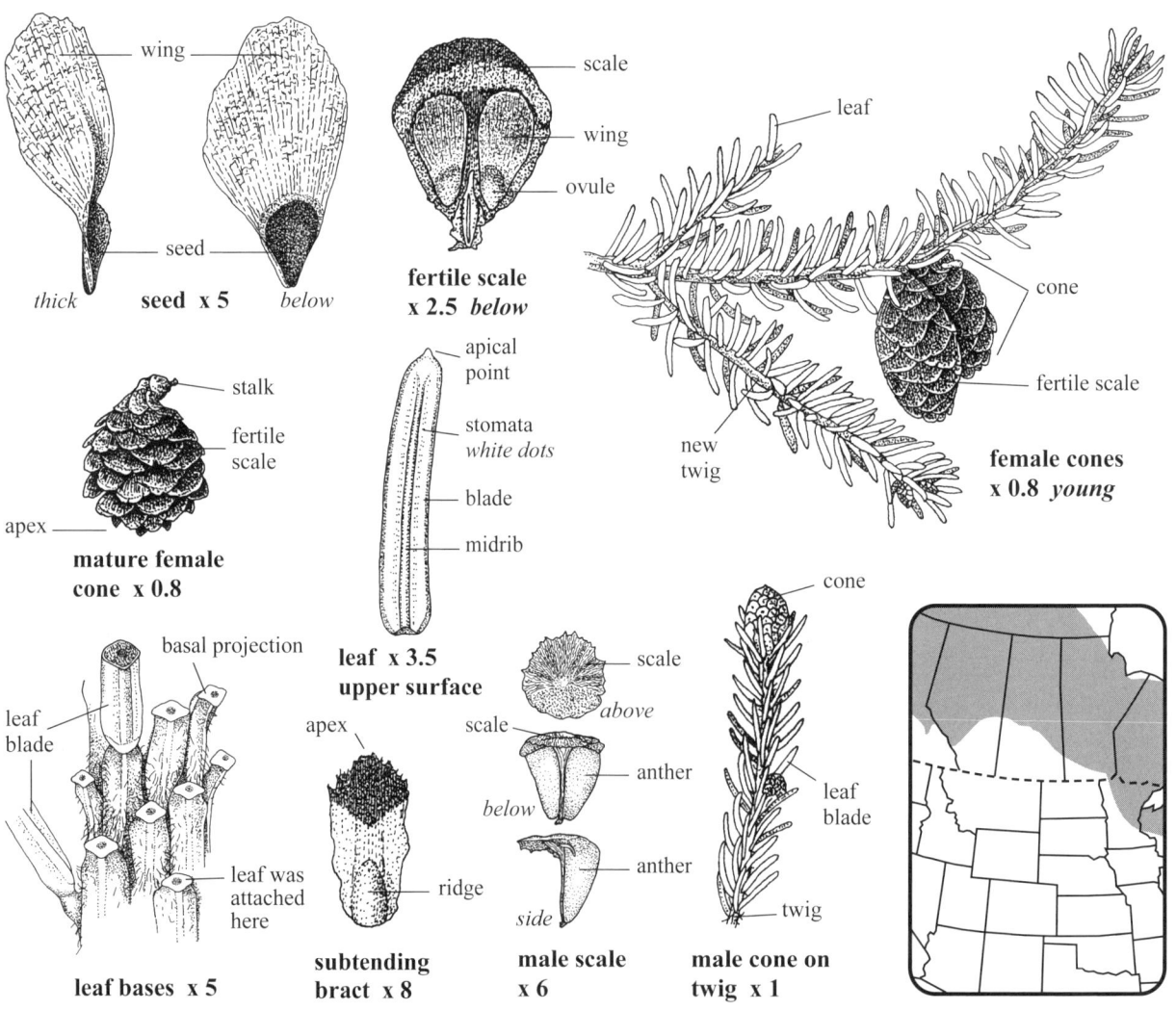

Bog Spruce, Swamp Spruce

Pine—*Pinaceae* **Tree** Coniferous Cones 2.8–6.8 cm long, curved

JACK PINE *Pinus banksiana* Lamb.

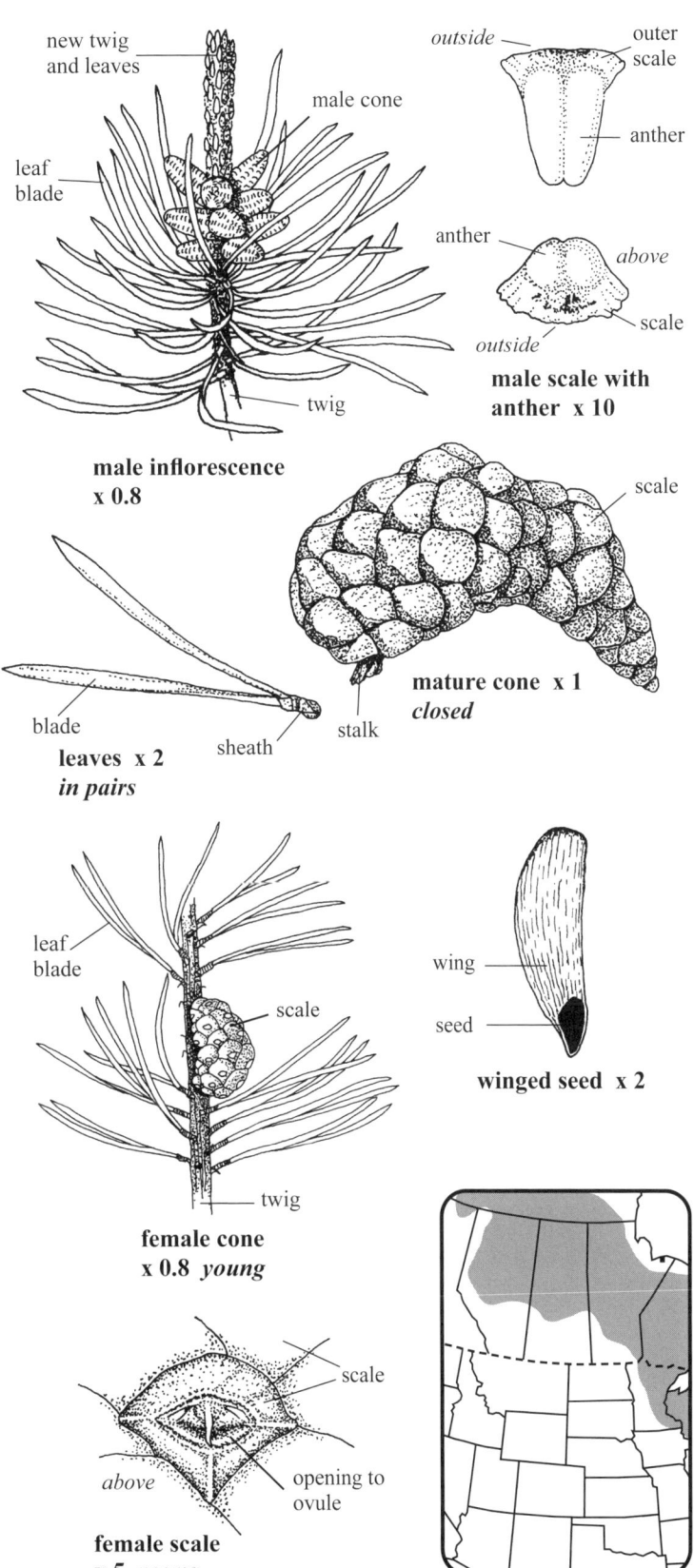

- **SKETCH** A variable, fragrant **coniferous tree** 5–25 m tall; on sandy gravelly ridges, rocky outcrops and burned-over sites in Boreal forest and Rocky mountains; **monoecious**. w2

- **REPRODUCTIVE PARTS** Unisexual, active May–June; **male cones** 7–15 mm long by 2.5–5 mm wide, reddish yellow, in small clusters of 4–20, together 1–3 cm long by 1.5–2.5 cm wide, at the base of new twig tips; **scales** hairless, *c.* 1 mm tall by *c.* 2 mm wide by *c.* 0.8 mm thick, with reddish brown spots and streaks, in 10 long rows (lengthways) around the pollen cone, scales slightly imbricate and margins slightly erose; **anthers** yellow, *c.* 1.5 mm long, 2-lobed; **young female cones** green, 18–25 mm long by 10–15 mm wide, often at base of twigs along branches, curved inward toward apex of the twig, enlarging and turning gray when mature in two years. **FRUIT a cone**, 2.8–6.8 cm long by 1.7–3.2 cm wide, tapered to a point, usually in pairs (1–3), subsessile, persistent on trees for years, expanding and releasing seeds when heated by a forest fire; **scales** slightly shiny, not armed, with round tips; **seeds** two per cone scale, winged, dark brown, 4–5 mm long by *c.* 2 mm wide by *c.* 1 mm thick, smooth, dull; **wing** membranous, tan, 10–14 mm long by 3–5 mm wide, slightly shiny.

- **LEAVES** Ascending, in pairs, base enclosed in a persistent sheath 4–5 mm long with a reddish brown subtending bract; **blades** pointed, scabrous along margins, flat, curved, slightly twisted and spreading, 1.5–4.5 cm long by 0.9–1.6 mm wide, soft, both sides with several parallel lines of white dots (stomata), outer surface convex and shiny, inner side flat and dull.

- **TRUNK** Grayish brown, rough, straight to crooked; **dbh** 15–60 cm; **bark scales** short and with orange margins, thick, irregular, often elevated, brownish and smooth below; **main branches** ascending to descending; **young branches**, reddish brown, fragrant when broken; **twigs** not hairy, orangish brown, ridged from leaf bases; **older twigs** gray and rough from persistent leaf bases *c.* 2 mm long; **autumnal buds** 6–14 mm long by 4–5 mm wide, oval, brown, blunt, resinous.

Scrub Pine, Banksian Pine, Gray Pine

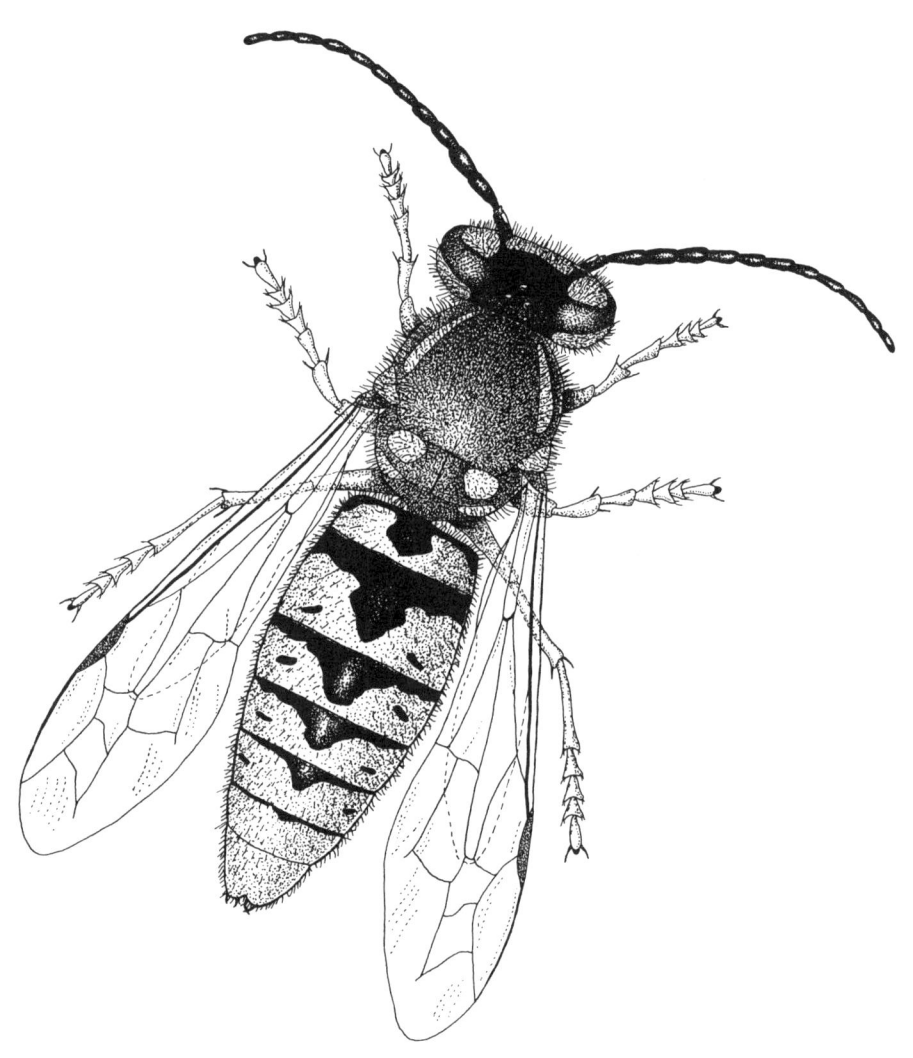

Dicotyledons

THE REMAINING TWO large sections of this book constitute the Angiosperms—the flowering vascular plants. Within the Angiosperms are two major groups: Dicotyledons and Monocotyledons. Among the dicots are many of the colorful wildflowers of prairie and woodland. Some trees, shrubs and vines also belong to this diverse groups of annuals, biennials and perennials.

TWO MAIN CHARACTERISTICS of the dicots are: **1)** flower parts in 4s or 5s. However, 4-merous flowers are not as common as 5-merous flowers. In mustards (Brassicaceae) the 4, often yellow petals are obvious. Wild roses (Rosaceae), peas (Fabaceae) and buttercups (Ranunculaceae) are three readily recognizable dicot families whose members have 5 petals; and **2)** leaves are net-veined.

THE LARGEST DICOT family, the asters (Asteraceae), includes asters, goldenrods, thistles and dandelions, along with many others. Easily reconizable, asters often have numerous small flowers forming three types of flowering heads. Ligules of individual flowers often end with 5, or sometimes 4 teeth.

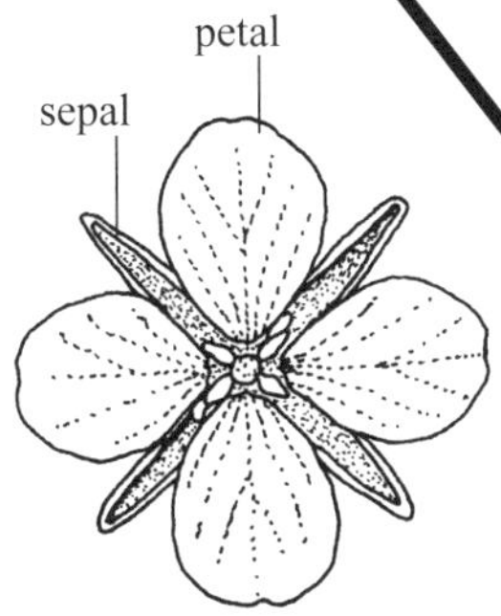

flower of wild mustard
4-merous

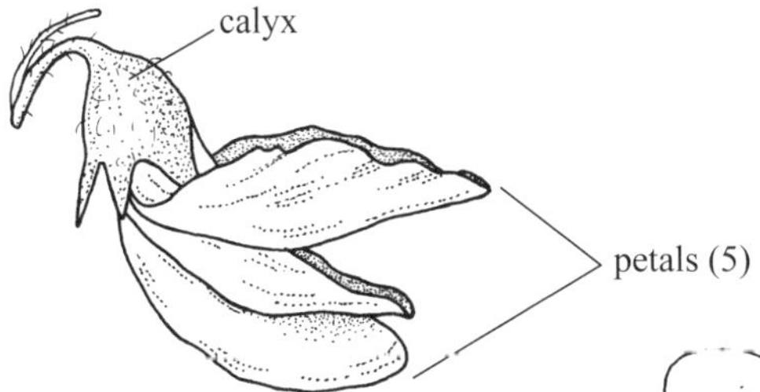

flower of clover
5-merous

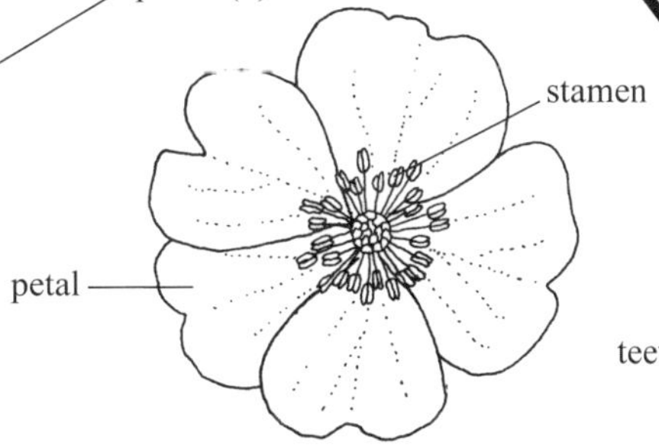

flower of wild rose
5-merous

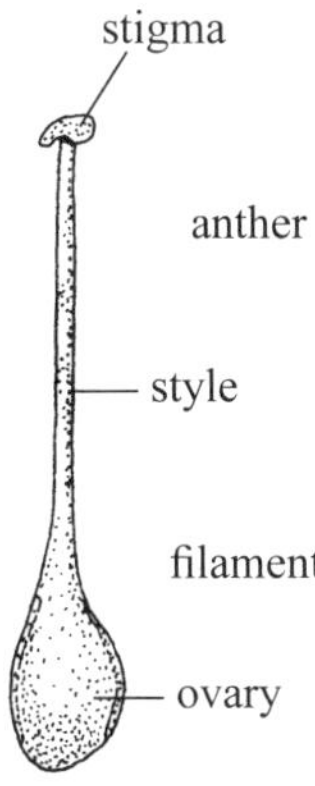

pistil of pin cherry

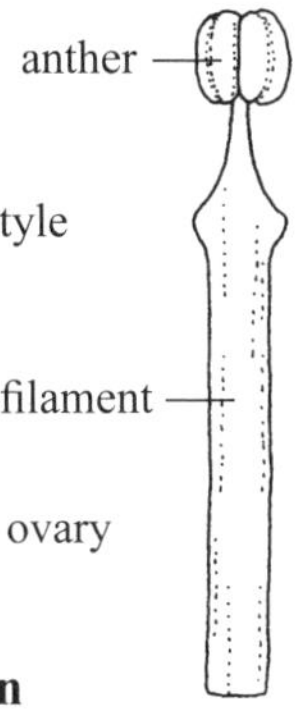

stamen of raspberry

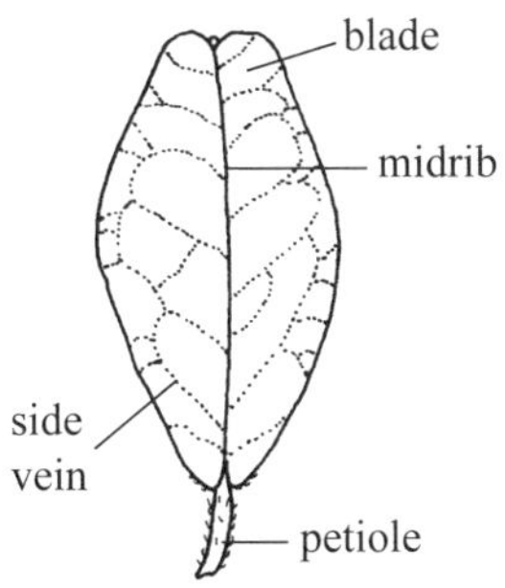

leaf
net-veined

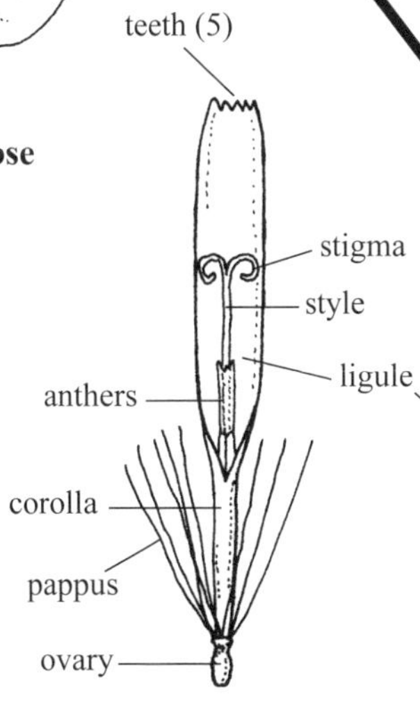

ligulate floret of dandelion

flower head of dandelion
ligulate

71

Maple—*Aceraceae*

Tree Calyx 4- or 5-lobed

Manitoba Maple *Acer negundo* L.

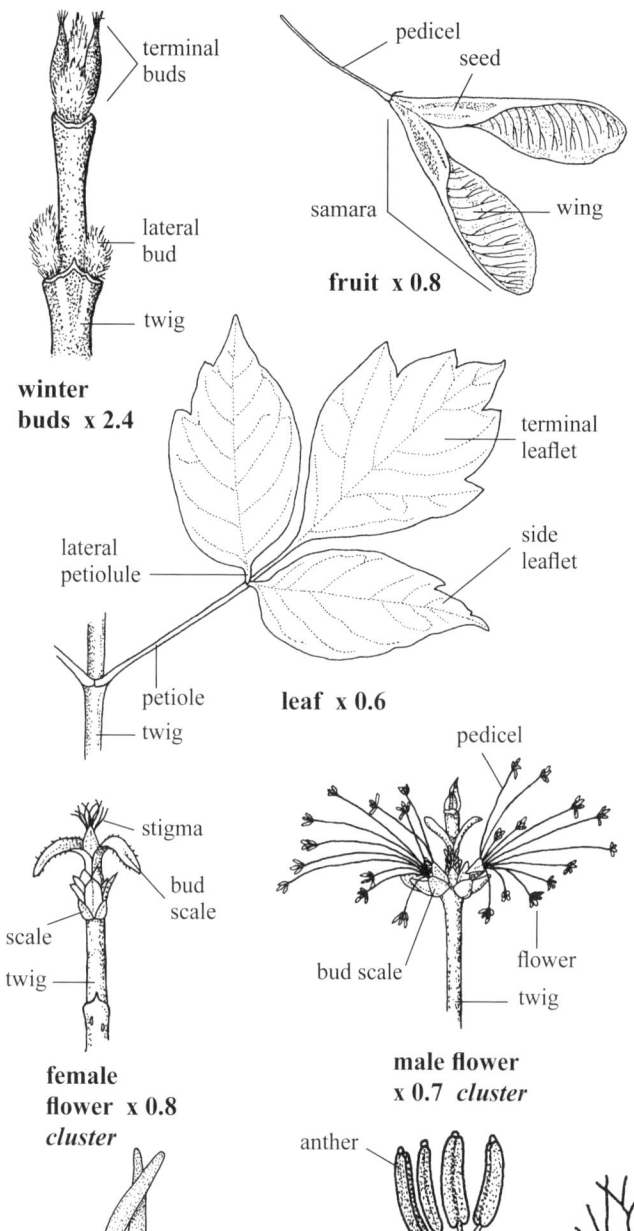

winter buds x 2.4
fruit x 0.8
leaf x 0.6
female flower x 0.8 cluster
male flower x 0.7 cluster
female flower x 5
male flower x 4
tree x 0.005

■ **SKETCH** A **deciduous tree** 5–18 m tall with several thick branches and a spreading crown; in cities and towns, along river and stream banks and ravines, often planted; **dioecious**. w2
■ **FLOWERS** Green and red, blooming March–July; **inflorescence** unisexual, rarely perfect, appears before the leaves develop; **pedicels** reddish, spreading, to *c.* 6 cm long, more hairy near the anthers; <u>male flowers</u> reddish, in umbel-like clusters from twig ends; **calyx** 4- or 5-lobed, *c.* 1.5 mm wide and long, green and hairy; **stamens** usually four (five) per flower, erect to drooping; **filaments** 1–3 mm long, whitish green; **anthers** red, 2.8–3.2 mm long; **pedicels** *c.* 4 mm long and hairy, 20–35 mm long with fruit; <u>female flowers</u> greenish, small, in racemes, clustered, four to nine; **stigmas** paired, 5–6 mm long by *c.* 0.5 mm thick. **FRUIT** a schizocarp of two samaras, each 1-seeded and united basally, each 2.5–5 cm long by *c.* 1 cm wide, winged, persistent overwinter, usually several[+] pairs per cluster, glabrous to hairy.
■ **LEAVES** Opposite, compound (odd-pinnate); **leaflets** toothed and lobed, three to seven, each 5–12 cm long by 3–5 cm wide, hairy, becoming glabrous with age; **petioles** 3–9 cm long, smooth; **petiolules** 0–22 mm long.
■ **TRUNK** Brown to dark gray; **bark** with irregular furrows; **twigs** opposite, gray and velvety in the spring from a dense cover of white hairs *c.* 0.1 mm long to glabrous; **lenticels** narrow, vertical and tan; **winter buds** opposite, roundish, blunt, 2.5–7 mm long by 2.5–4.5 mm wide, covered with white hairs; **scales** usually four are visible; **terminal buds** usually subtended by a pair of opposite hairy buds; **leaf scars** tan, narrow, 0.5–1 mm long; **dbh** 30–60 cm.

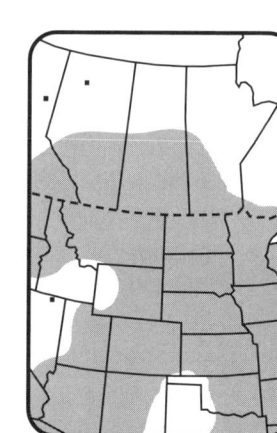

Box Elder, Ashleaf Maple

Maple—*Aceraceae* **Tree** Calyx 5- or 6-lobed

SILVER MAPLE *Acer saccharinum* L.

■ **SKETCH** A **deciduous tree** 10–30 m tall with branches spreading to pendulous; along streams and lakes, in damp woods, around older farms, planted as a shade tree to the north and west, native in east; usually **dioecious**, or **monoecious**. w1

■ **FLOWERS** Pink and green, blooming February–May before the leaves appear; **inflorescence** of clusters of whorled lateral flowers; **flower clusters** 2–3 cm wide, each bud *c.* 6 mm long by 4–7 mm wide and usually up to six in a whorl around a twig; **pedicel** *c.* 1 mm long, erect; <u>male flowers</u> in groups of five (4 or 6) attached to a green basal disc *c.* 3 mm wide with tangled reddish hairs; **calyx tube** 2–3 mm long and wide, pale yellowish green with faint red streaks on the 5 or 6 ciliate lobes; **petals** absent; **stamens** five (4 or 6), exserted to *c.* 8 mm past the bud scales; **filaments** white, erect, glabrous to hairy, 5–7 mm long; **anthers** yellow, 1–1.5 mm long; **pedicels** green, *c.* 1 mm long; <u>female flowers</u> five per bud; **ovaries** paired, hairy, *c.* 2.5 mm wide, flattened; **calyx** tubular, green, slightly hairy, 2.5–3 mm long by 3–4 mm wide; **petals** absent; **style** obscure; **stigmas** 2, exserted, 4–6 mm long, rough; **stamens** *c.* 2 mm long, five, vestigial, attached to base of superior ovary; **anthers** project slightly above calyx lobes; **pedicels** with fruit 2.5–3.5 cm long, turning brown as fruit ripens. **FRUIT** a schizocarp of two 1-seeded samaras (keys); **samaras** 1–3 pairs per flower, ripening to tan and dropping as leaves reach full size, 4–7 cm long by 1.5–2 cm wide (wing), slightly hairy or glabrous, often one samara is not fully formed and only 1.5–2 cm long by *c.* 1 cm wide with an undeveloped seed, shed individually in early summer as the leaves mature; **seeds** *c.* 13 mm long by 6–7 mm wide by 5–5.7 mm thick, tan, smooth, veiny inside, germinating quickly the first year.

■ **LEAVES** Opposite, simple, 3- or 5-lobed, coarsely toothed, sinuses narrow, deep and acute, smaller lobes near the base, in fascicles of 2–4 from new shoot ends; **leaf blades** 6–20 cm long and wide, dark green, glabrous and slightly shiny above, silvery white, dull, and finely hairy below mainly on raised veins, yellowish in the fall; **petioles** 2.5–13.5 cm long, round, slightly flattened near the base, slightly hairy near the blade; **leaf buds** opposite, 6-scaled, 5–7 mm long by *c.* 3 mm wide; **bud scales** red, ciliate, lower scales have dark brown pointed tips.

■ **TRUNK** Gray, bark in narrow strips attached in the middle, ends free, cut stumps produce many new shoots as do branch cuttings; **branches** light gray, smooth, lower branches with tips ascending; **twigs** gray to reddish, slightly shiny, glabrous; **lenticels** vertical, gray; **winter buds** 2–25 per cluster, unisexual, sessile, reddish brown, *c.* 4 mm long by *c.* 3 mm wide by *c.* 2 mm thick, with two obvious pairs of opposite scales, margins ciliate with tan hairs, apices of scales dark brown, slightly shiny; **dbh** 60–100 cm.

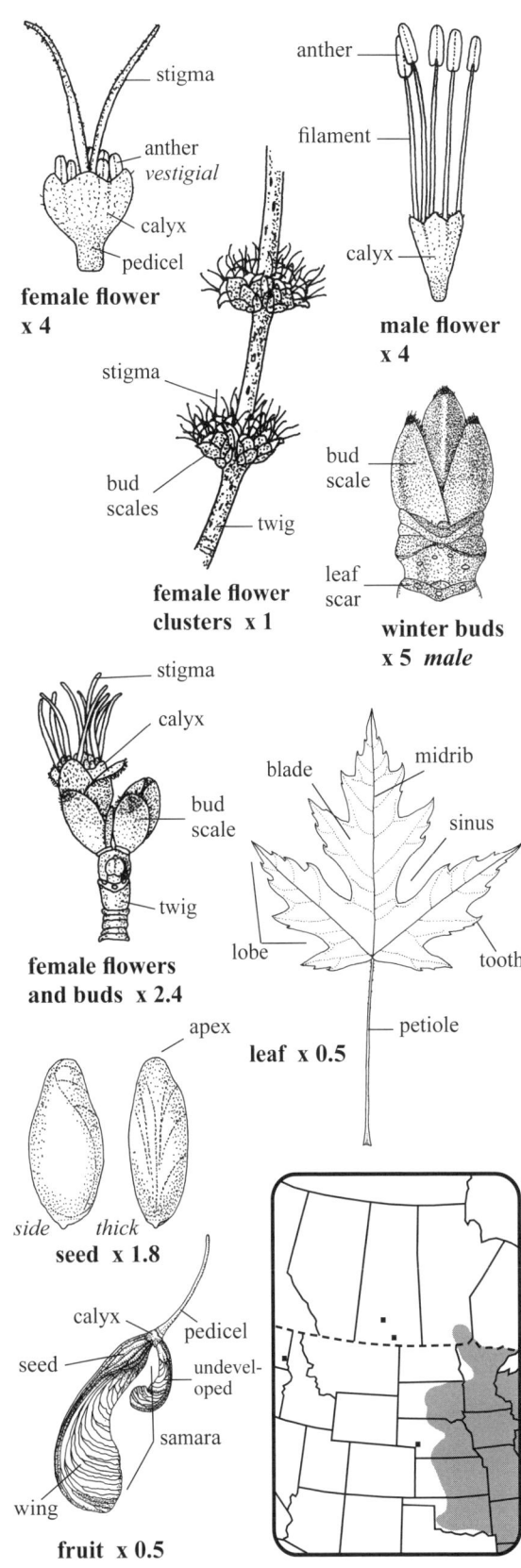

Sugar Maple, Hard Maple, River Maple, Soft Maple

Maple—*Aceraceae* **Shrub** Yellowish green to creamy white Petals 5

Mountain Maple *Acer spicatum* Lam.

■ **SKETCH** A **deciduous shrub** 3–5 (–10) m tall; in moist woods, swamps and thickets, along ridges and in cutovers in the Boreal forest and deciduous woods. **w1**

■ **FLOWERS** Yellowish green to creamy white, blooming May–July; **inflorescence** a panicle (or raceme) 6–10 cm long by 3–4 cm wide of 70–230 flowers, ascending from between paired leaves at the ends of side branches; **peduncles** 3–5.5 cm long, reddish and mostly glabrous with fruit; **pedicels** 3–7 mm long, reflexed with fruit, often dotted with red glands; **flowers** perfect or male, opening well after the leaves; <u>**male flowers**</u> 9–11 mm wide by 3–4 mm tall; **stamens** often eight, exserted; **filaments** *c.* 5 mm long; **anthers** *c.* 1 mm long, yellow; **calyx** hairy, *c.* 2 mm long, 5-lobed, these *c.* 1 mm long; **petals** five, 2.5–4 mm long by *c.* 0.5 mm wide; **pistil** vestigial; **glands** eight, *c.* 0.6 mm long, oval, between the filaments; <u>**perfect flowers**</u> 5–7 mm wide by 5–6 mm tall, often near the center of the panicle; **pistil** one; **ovaries** two, united, winged, *c.* 2 mm long by *c.* 3.5 mm wide by *c.* 1 mm thick, hairs curly, white; **style** one, *c.* 2 mm long, green, hairy near the base; **stigmas** two, curled; **sepals** five, *c.* 1.5 mm long by *c.* 1 mm wide, hairy; **petals** five, white, *c.* 3 mm long by *c.* 0.5 mm wide; **stamens** eight, in pairs; **filaments** white, *c.* 1 mm long; **anthers** yellow, *c.* 1 mm long. **FRUIT a samara**, 18–25 mm long, the wing 6–9 mm wide, developing fruit often hairy at least on the seeds near their joint, less hairy at maturity, white glandular-dotted; **samaras** not widely divergent, reticulately veined on the seed portion, turning red in late summer and remaining on trees into winter; **seeds** reddish tan, hairless, *c.* 6 mm long by *c.* 3.5 mm wide by *c.* 1.1 mm thick, surface somewhat ridged with an indentation on one side.

■ **LEAVES** Opposite, simple, usually 3- or 5-lobed, finely and often doubly toothed, lobes less obvious on young leaves; **blades** 4–16 cm long by 2.5–14 cm wide, dull, lighter green below and usually with white hairs along the midrib with some reddish glands, hairy above mostly near the base of the midrib, with red glandular dots sometimes common along the midrib, lower lateral ribs and central part of blade, turning yellow or red in autumn; **petioles** 1.8–11 cm long, usually hairless, often reddish above and widely grooved, dotted with reddish glands.

■ **TRUNK** Dark grayish brown, without stripes, often several in a clump, 5–15 cm wide near the base; **small branches** purplish gray, minutely hairy; **lenticels** not obvious, but reddish brown on new green shoots; **winter buds** 2–3 mm long by 1.5–2.0 mm wide, pointed, opposite, dark reddish brown, external scales 2 or 3, margins ciliate, especially near the apex, scales slightly shiny and V-shaped.

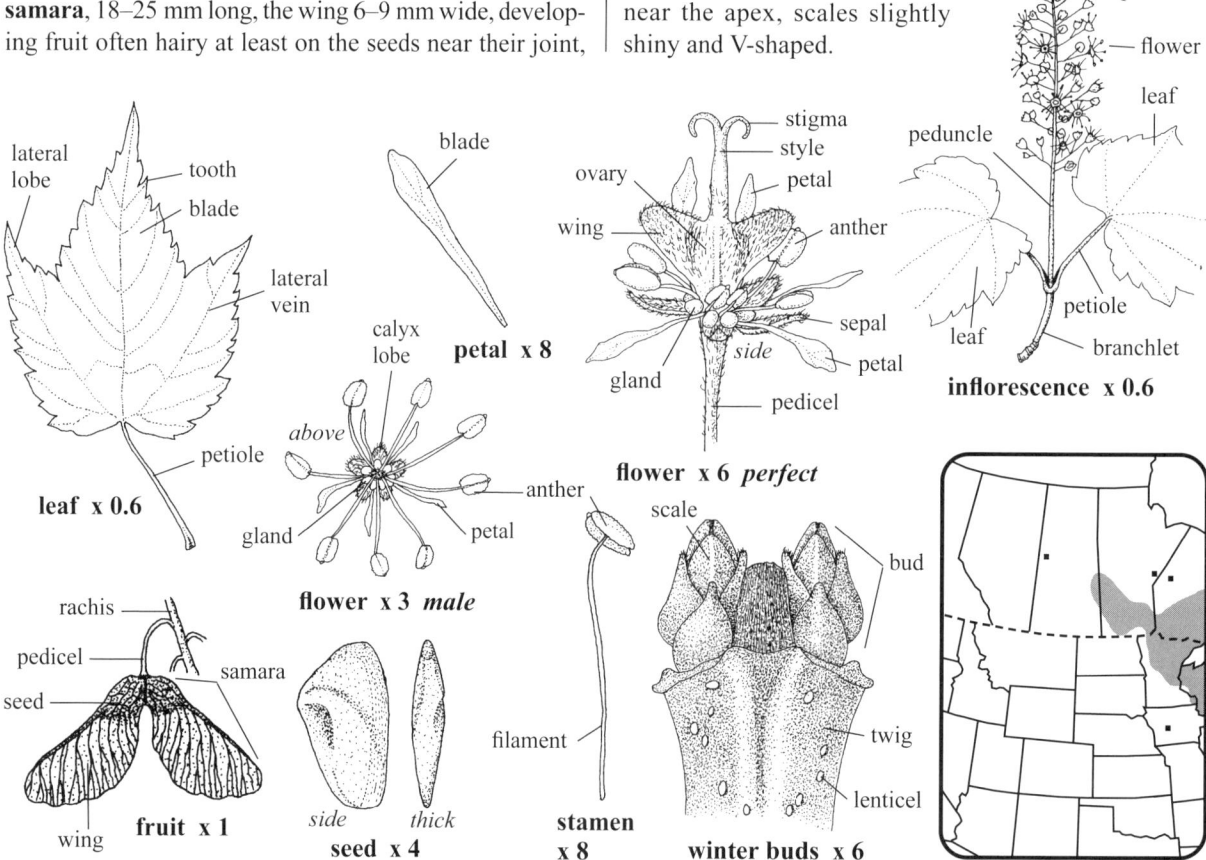

74 White Maple

Amaranth—*Amaranthaceae*   **Wildflower**   Green   Sepals 3–5

PROSTRATE AMARANTH   *Amaranthus blitoides* S. Wats.

■ **SKETCH** A glabrous **annual herb** with several prostrate, branching, pink fleshy stems forming mats 10–70⁺ cm wide from a **taproot** 1–3 mm wide and 4–15⁺ mm long; in gardens, along sidewalks, dry prairies, mesas, pastures and roadsides; **monoecious**. Native and *naturalized* w1

■ **FLOWERS** Pale green, blooming May–November; **inflorescence** of unisexual flowers clustered together in leaf axils along the stem and branches throughout the plant, both sexes quite similar, sessile; **subtending bracts** (of flowers) one to three, pointed, V-shaped, with a central green nerve, 1.5–5 mm long by 0.5–0.9 mm wide (flattened); <u>male</u> **flowers** with three or five whitish **sepals** 1.6–3 mm long by 0.6–1 mm wide, pointed, with a green midvein; **petals** absent; **stamens** three to five, 2–3 mm long, eventually exserted; **filaments** white, to *c.* 2 mm long when exserted; **anthers** pale yellow, 0.8–1 mm long; <u>female</u> <u>flowers</u> usually more abundant than male flowers, membranous, 1–1.3 mm wide by 1.8–2.1 mm long; **sepals** three to five, thin, flat, each 1–3 mm long by *c.* 0.9 mm wide, persistent, pointed, usually not exceeding the flower, each with a faint green midvein, erect; **petals** absent; **style** 3-parted, these spreading and 0.5–0.8 mm long. **FRUIT** a **utricle**, 1-seeded, thin, tan, 2–3 mm long by 1.4–1.9 mm wide, smooth, some tinged with red to dark brown, suture around the middle; **seeds** glossy to dull black, 1.5–1.7 mm long by 1.3–1.6 mm wide by *c.* 0.8 mm thick, smooth, lenticular with a thin marginal ridge.

■ **LEAVES** Alternate, simple, entire, glabrous, numerous; **blades** flat, with a thin white marginal line, 5–40 mm long by 4–20 mm wide, the tips blunt, some with a mucro, veins few, shiny below, less so above; **petioles** whitish green, grooved above, round below, 2–25 mm long, glabrous; **stipules** absent.

■ **STEM** Prostrate, usually hairless, solid, round, light pink to pale green, branched; 1–5 mm thick near the base.

■ **SYN.** *Amaranthus graecizans* auct. non L.

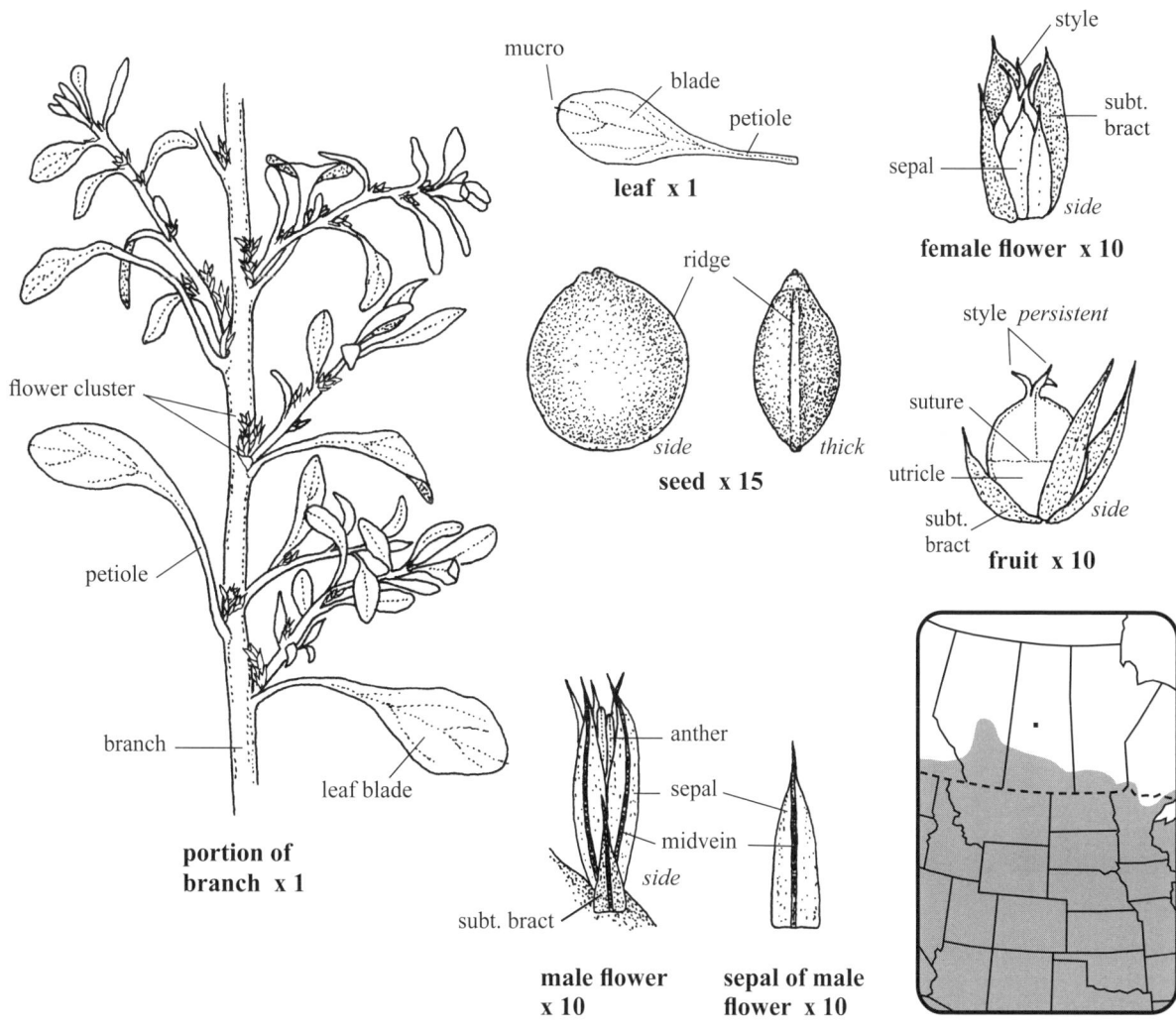

Matweed, Mat Amaranth, Mat Pigweed, Prostrate Pigweed, Tumbleweed

Amaranth—*Amaranthaceae*

Wildflower Green Sepals 5

Red-root Pigweed *Amaranthus retroflexus* L.

■ **SKETCH** A variable **annual herb** 20–300 cm tall from a reddish tan **taproot** 1.5–13 mm wide by 5–15+ cm long; along stream and river banks, roadsides, sidewalks, in cultivated and fallow fields, dry ravines and city lots; **monoecious**. Native to *naturalized* **w1**

■ **FLOWERS** Green, blooming June–October; **inflorescence** a panicle (dense clusters of spikes), terminal and axillary, TL 3–80 cm, each spike 5–25 mm wide with small axillary spikes among the upper leaves; **flowers** unisexual, crowded, the sexes usually adjacent; **petals** absent; **subtending bracts** (of flowers) one to three, glabrous, 3–7 mm long by 1–1.3 mm wide, longer than the sepals, with a flat base and green midrib tapering into a long thin point, the margins white and thin; **male flowers** 2–4 mm wide by 2–3 mm long; **sepals** five, soft, white, hyaline, midvein green; **stamens** five (3 or 4), TL 2–3 mm, spreading, exserted; **filaments** white and filiform; **anthers** yellow, *c.* 1 mm long; **female flowers** usually more numerous than male flowers; **sepals** five, as above, 2–4 mm long by *c.* 1 mm wide, the midvein extending as a short point, persistent around the fruit; **pistil** one, *c.* 1.5 mm long; **style** 3-parted (4-) directly from the ovary, persistent. **FRUIT** a utricle, separating around the middle; **cap** 1.8–2.2 mm long by 1–1.4 mm wide, its base wrinkled; **seeds** one per utricle, smooth, black, shiny, lenticular, 0.9–1.3 mm long and wide by *c.* 0.6 mm thick.

■ **LEAVES** Alternate, simple, toothed; **blades** dull, hairy, especially below on the veins, 1.5–15 cm long by 1–10 cm wide, widest below the middle, margins slightly wavy and sinuate, tips rounded; **petioles** scabrous, 0.7–10 cm long.

■ **STEM** Stout, erect, round to slightly angular, usually branched, sometimes reddish green, covered with short rough hairs; 2–15 mm thick near the base.

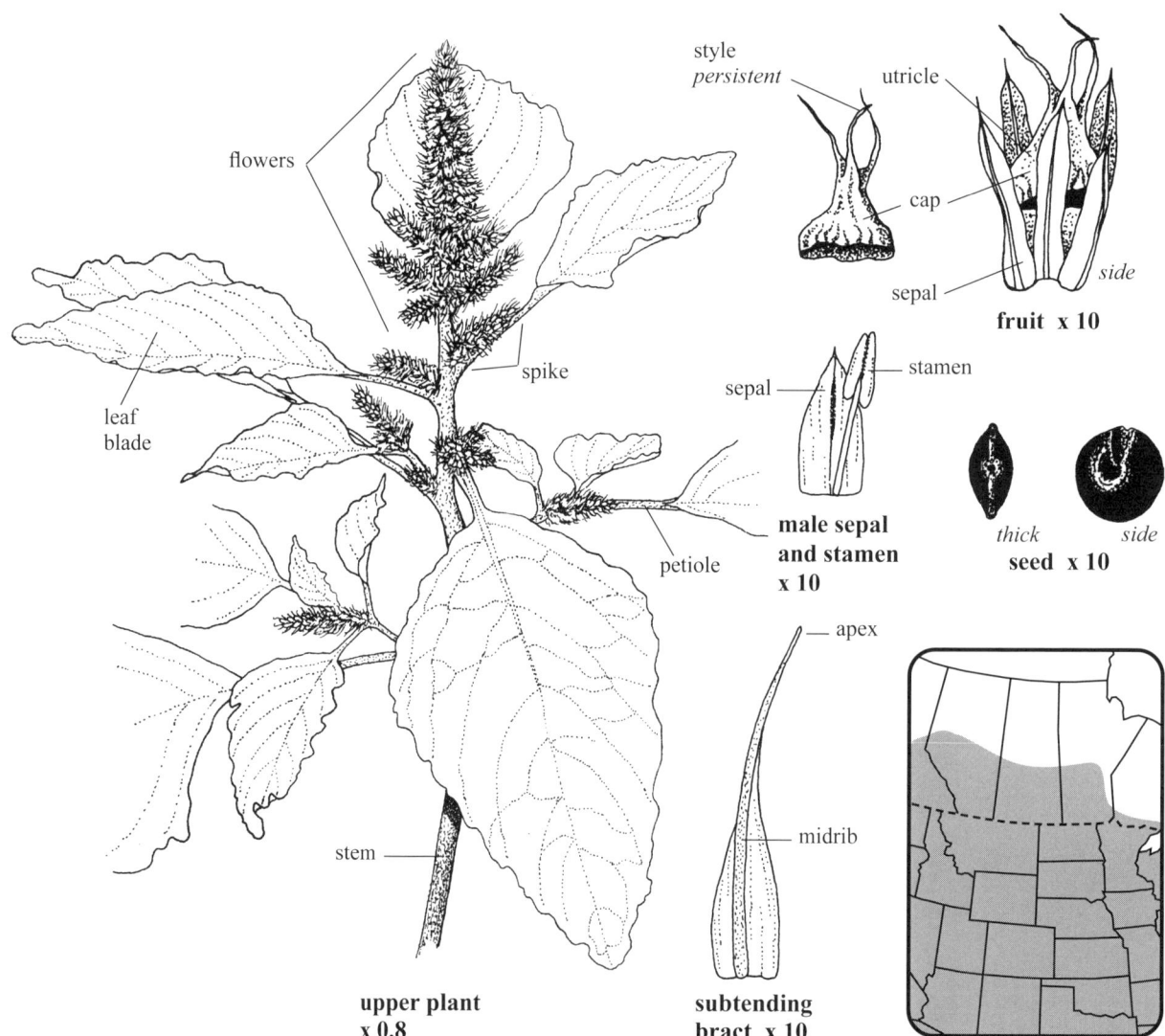

Common Amaranth, Redroot Amaranth, Rough Pigweed, Green Amaranth

Sumac—*Anacardiaceae* **Shrub** Light yellowish green Petals 5 (6)

Smooth Sumac *Rhus glabra* L.

■ **SKETCH** A **deciduous shrub** 0.4–4 m tall by 20–200 cm wide, forming open colonies or thickets from **root suckers**; in open woodlands, along roadsides and trails, on hillsides, rocky outcrops and dry sandy soil; **juice milky polygamo-dioecious.**

■ **FLOWERS** Light yellowish green; blooming May–July; **inflorescence** a panicle, dense, 2–25 cm long by 2–20 cm wide, at ends of upper branches; **floral branches** 1–10 cm long, reduced above, ascending, hairy, reddish in fall, thin gray bark peeling on older lower branches; **peduncles** erect, hairy, 5–13 mm long by 2–4 mm thick, oval; **pedicels** 1–2 mm long, hairy; **subtending bracts** (of pedicels) two or three, hairy on margins, each 1.2–3 mm long by *c.* 0.3 mm wide, involute on drying; **flowers** some perfect, 5–6 mm wide by 2–2.5 mm long; **rachis** with curled hairs; **hypanthium disc** orange, *c.* 1.6 mm wide, moist, flat, around the pistil's base; **sepals** five, erect, *c.* 2 mm long by 0.8–1 mm wide, darker green than petals, persistent; **petals** five (6), *c.* 2.5 mm long by *c.* 1 mm wide, hairy on inner lower surface, glabrous outside, persistent; **stamens** five, *c.* 0.6 mm long, included, inserted on sides of disc, vestigial ones present; **filaments** white, *c.* 0.2 mm long, erect, tapered above; **anthers** pale yellow, *c.* 0.4 mm long; **pistil** one, included, *c.* 1 mm long, from center of orange disc; **ovary** green, hairy, *c.* 0.8 mm long and wide, tapered above; **style** obscure, *c.* 0.1 mm long, 3-parted; **stigmas** three, yellowish, triangular, wider than style; **fruiting heads** erect, dark reddish brown, 5–20 cm long by 2–10 cm wide and thick, overwintering. **FRUIT a drupe**, 1-stoned, reddish brown, 3.4–4.5 mm long by *c.* 4 mm wide by *c.* 2.5 mm thick, covered with spreading reddish hairs, these sometimes glandular, some veins obvious; **stones** one-seeded, smooth, light tan, 3–3.5 mm long by 2.2–2.7 mm wide by *c.* 1.8 mm thick.

■ **LEAVES** Alternate, compound, 5–11 clustered below each panicle, branches and stem otherwise naked; **leaf blades** 5–40 cm long by 5–15 cm wide; **leaflets** 9–31, subsessile, 2–9 cm long by 5–35 mm wide, toothed, shiny above, dull, lighter green and glaucous below, spreading to drooping, red in fall; **petioles** 1.5–7 cm long, slightly hairy, reddish; **petiolules** 0–10 mm long; **rachis** reddish and hairy above, light green and glabrous below.

■ **STEM** Smooth, hairless, gray, slightly shiny, 0.5–4 cm thick hear the base; **winter buds** *c.* 3 mm long and wide, blunt, dense hairs tangled and clumped, outer hairs gray, inner ones tan, no bud scales visible.

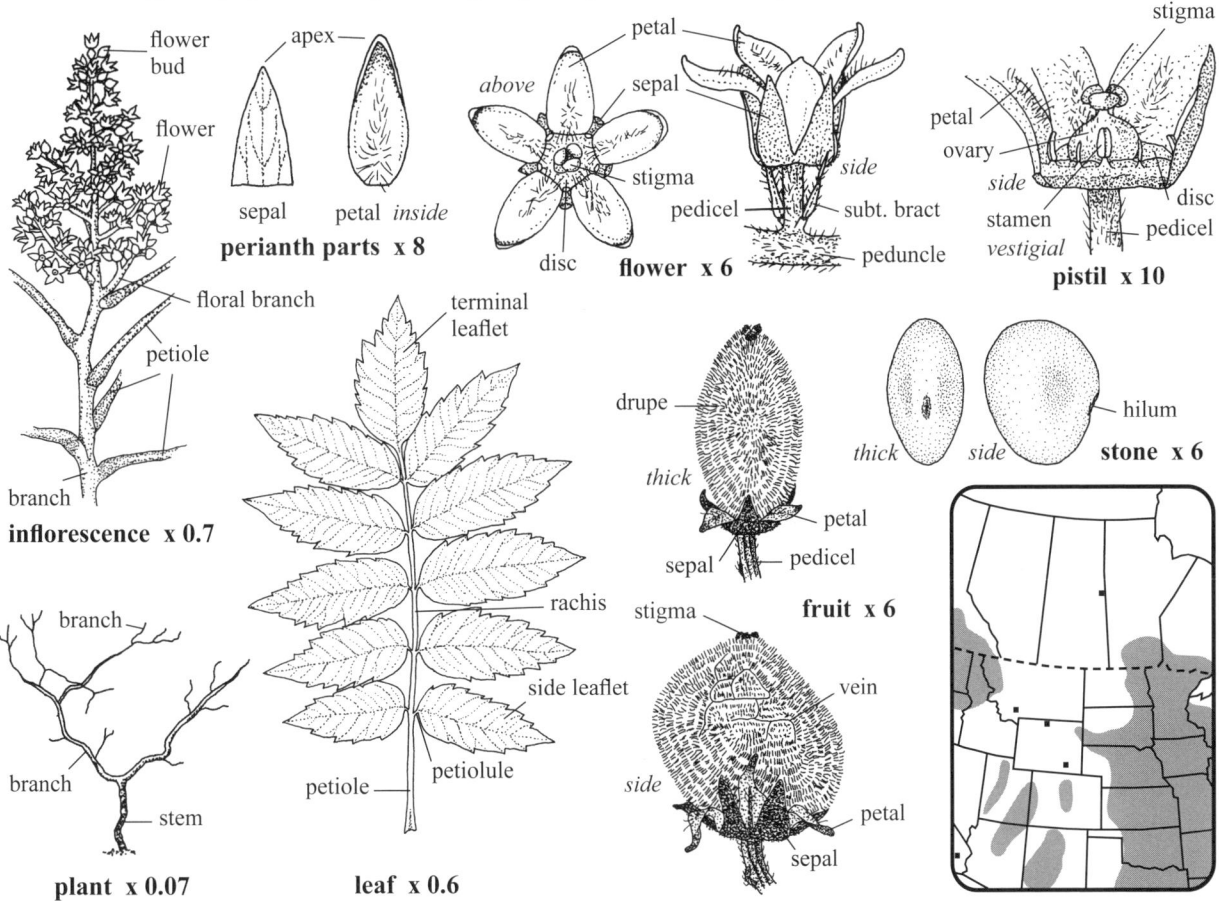

Scarlet Sumac

Sumac—*Anacardiaceae*

Shrub   White   Petals 5

POISON-IVY   *Toxicodendron rydbergii* (Small ex Rydb.) Greene

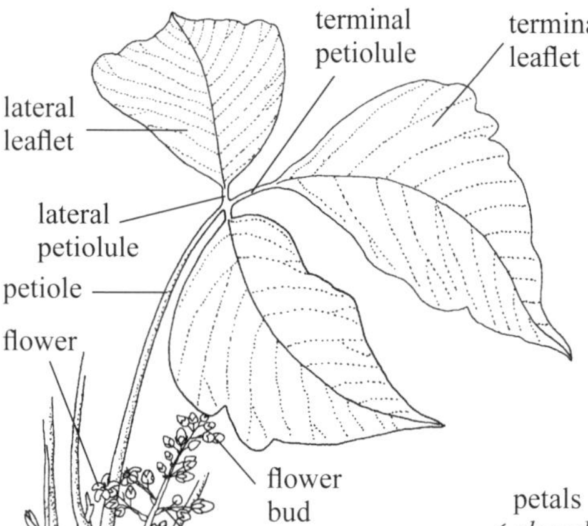

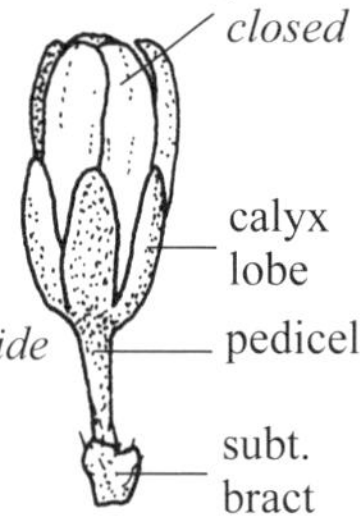

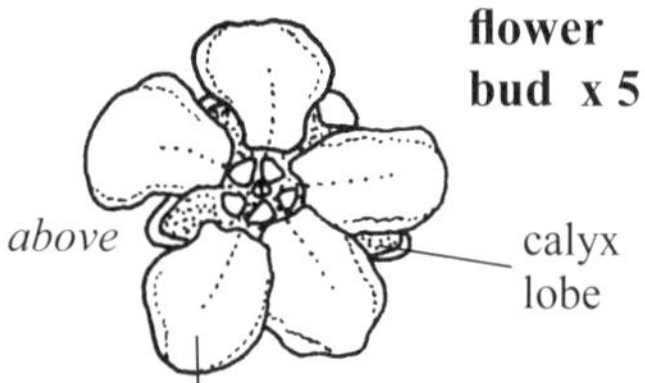

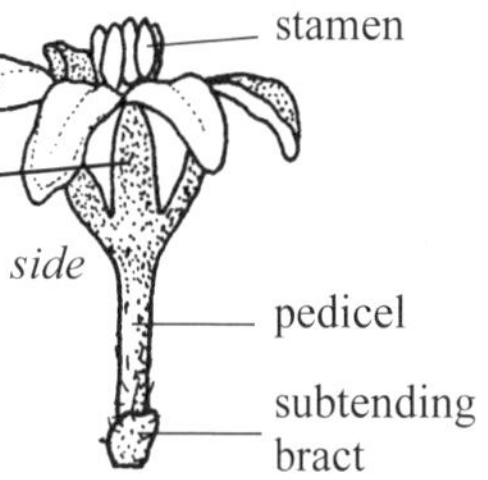

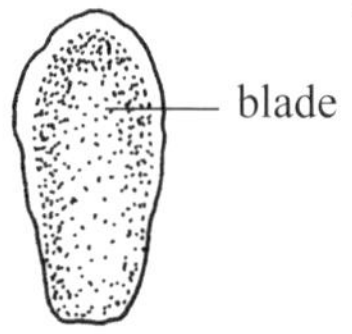

■ **SKETCH** A variable **deciduous shrub** 20–100⁺ cm tall with branched **rhizomes** 3–7 mm thick, aerial roots lacking; in woodland edges and clearings, pond and river edges, canyons, ravines, along railways, roadsides and rocky prairie hillsides. **w1**

■ **FLOWERS** White, blooming May–September; **inflorescence** a panicle 5–20 cm long by 5–15 cm wide, axillary; **peduncles** erect to spreading, 1–2 cm long by 1–2 mm wide; **subtending bracts** (of peduncles) one, 2–3 mm long, pointed; **pedicels** green, 2–5 mm long, glabrous above to slightly hairy below; **subtending bracts** (of pedicels) one, light green, slightly hairy, *c.* 1 mm long by 0.5–1 mm wide, clasping; **flowers** perfect, 4–5 mm wide by *c.* 3 mm long, hidden below the leaves; **calyx** *c.* 2.3 mm long, united below; **calyx lobes** five, erect to ascending, 1.8–2 mm long by *c.* 1 mm wide, glabrous, margins whitish; **petals** five, spreading, tips descending, *c.* 3 mm long by 1.6–2 mm wide, veined, glabrous; **stamens** five, slightly exserted, attached to the edge of the thin, yellowish green, flat, glabrous, central disc; **filaments** white, *c.* 1 mm long, stout and widest at their bases; **anthers** erect, *c.* 1.8 mm long by *c.* 1 mm wide; **pistil** stout, *c.* 1 mm long, included; **stigmas** of two or three blunt, short, lobes, not obvious. **FRUIT a drupe**, 1-stoned, white to yellowish, smooth, veiny, roundish, 4–6 mm wide and long, overwintering; **stones** (endocarp) 1-seeded, medium brown, *c.* 5 mm wide by *c.* 4 mm long by 2.5–3 mm thick, smooth once the waxy mesocarp is scraped away, grooved and notched.

■ **LEAVES** Alternate, compound (trifoliate), crowded at the summit of each stem, turning yellow to red in autumn; **leaflets** 3–14 cm long by 1.5–11 cm wide, pointed, lighter green below, entire or with a few teeth, drooping, shiny when young, dull with age, veins distinct, hairy below, glabrous above; **petioles** erect, 3.5–32 cm long, glabrous to finely hairy; **lateral petiolules** 1–6 mm long, hairy; **terminal petiolules** 10–50 mm long, hairy.

■ **STEM** Erect, simple to branched, rough from leaf scars, light grayish brown; the 5–12 mm wide woody base overwintering.

■ **SYN.** *Rhus radicans* var. *rydbergii* (Small ex Rydb.) Rehd., *Rhus radicans* var. *vulgaris* (Michx.) DC., and *Toxicodendron radicans* var. *rydbergii* (Small ex Rydb.) Erskine.

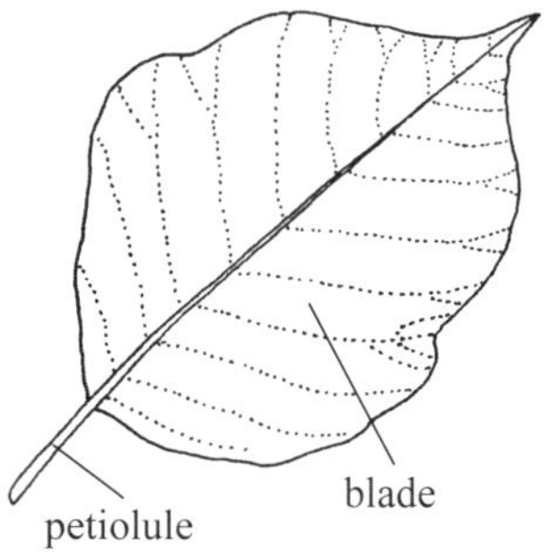

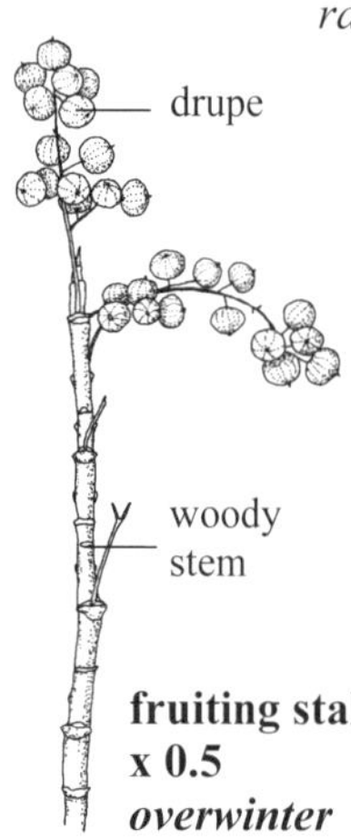

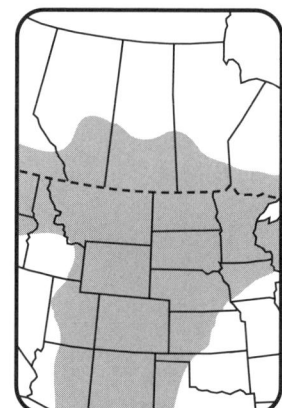

Common Poison-ivy, Western Poison-ivy

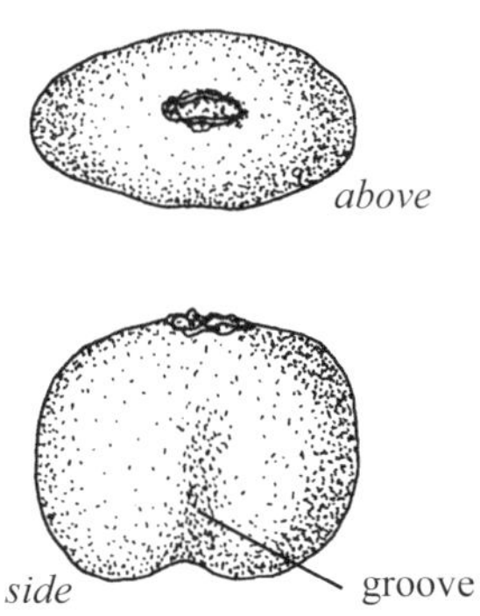

Family Characteristics

Carrot—Apiaceae — Flower parts 5

SKETCH Mostly **dicotyledonous herbs** (wildflowers), annual, biennial or perennial, often aromatic. About 3,500 species worldwide, mostly in temperate regions or tropical mountains.

FLOWERS Mostly perfect, often white or yellow, small. **Inflorescence** an umbel, simple or compound into smaller umbellets at ends of rays; some flowers on edges of umbellets are male only. **Involucre** often present. **Calyx** tube fused to ovary, teeth 5, tiny. **Petals** 5, tips incurved. **Stamens** 5, inserted on edge of central epigynous disc. **Pistil** 1, of 2 united carpels. **Ovary** 1, inferior, with 2-locules. **Styles** 2, often short.

FRUIT A **schizocarp** of 2 mericarps. **Mericarps** eventually splitting apart, each 1-seeded and often attached to a slender carpophore (stalk); mericarp body 5-ribbed, often with dark oil-tubes between the ribs.

LEAVES Alternate or basal. **Leaf blades** usually compound, some simple. **Leaflets** often lobed or toothed. **Petioles** with obvious sheaths at base where they meet the nodes of the stem.

STEM Erect, green, often hollow and with ridges.

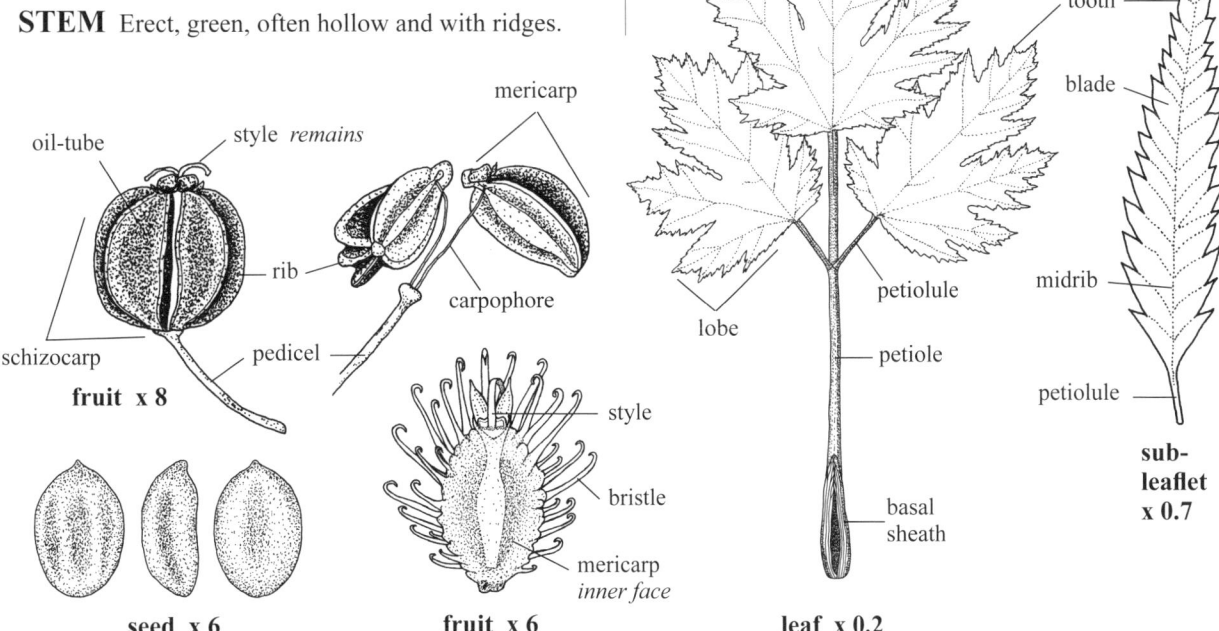

79

Carrot—*Apiaceae* **Wildflower** White Petals 5

CARAWAY *Carum carvi* L.

- **SKETCH** A glabrous **biennial herb** 30–100 cm tall with a light tan **taproot** 8–12 cm long by 1–1.5 cm wide; along gravelly roadsides, in meadows and shallow ditches. *Naturalized*
- **FLOWERS** White, blooming May–July; **inflorescence** an umbel, compound, 11–24 per plant, usually 2–4 umbels at the end of each floral branch, each umbel 2–12 cm wide by 1.5–2.5 cm tall (including rays); **umbellets** 7–14 per umbel, each 5–17 mm across by 3–4 mm tall, (including pedicels) of 7–21 flowers; **floral branches** 3–45 cm long, ascending; **peduncles** 2–14 cm long, grooved; **involucral bracts** (of rays) none to four, green, linear, 6–10 mm long by *c.* 0.2 mm wide, margins at flaring base whitish; **rays** ridged, 5–80 mm long, ascending; **pedicels** 0.5–20 mm long; **involucels** (at base of pedicels) usually absent; **flowers** perfect, 3–4 mm wide by *c.* 2 mm tall; **calyx** fused to top of ovary, appearing as short green teeth between the petals; **petals** white, apex notched and curled inward, 1–1.2 mm long by *c.* 1 mm wide; **stamens** five, exserted, spreading between the petals; **filaments** white, *c.* 1.5 mm long, filiform; **anthers** *c.* 0.5 mm long, yellow with a pink flush, 2-lobed; **ovary** green, hairless, *c.* 0.8 mm long by *c.* 1 mm wide by *c.* 0.5 mm thick, ribbed. **FRUIT a schizocarp**, 3–4 mm long by *c.* 2.3 mm wide by *c.* 1 mm thick; **mericarps** two, 1-seeded, with five long tan ribs and dark brown oil-tubes between and wider than the ribs, attached until ripe at their inner face (commissure); **carpophore** thin, brown.
- **LEAVES** Basal and stem, alternate, compound; **1st year leaves** several, erect, blades 6.5–14.5 cm long by 2.5–4 cm wide; **leaflets** 4–11 opposite pairs, divided, each 1–3.5 cm long by 10–25 mm wide, reduced near tip of leaf blade, dull, slightly lighter green below; **petioles** 3–18 cm long; **sheaths** 1.5–3.5 cm long; **basal blades** ascending, 6–12 cm long by 2–4 cm wide; **petioles** 2–5 cm long, deeply grooved; **sheaths** pale green, 1.5–3.5 cm long by 7–8 mm wide, clasping above; **stem leaves** few, to *c.* 30 cm long by 3–8 cm wide, reduced above; **sheaths** to *c.* 4 cm long, clasping; **petioles** 0–6 cm long.
- **STEM** Glabrous and glaucous, erect, ridged, solid, branched; 3–7 mm wide at base.

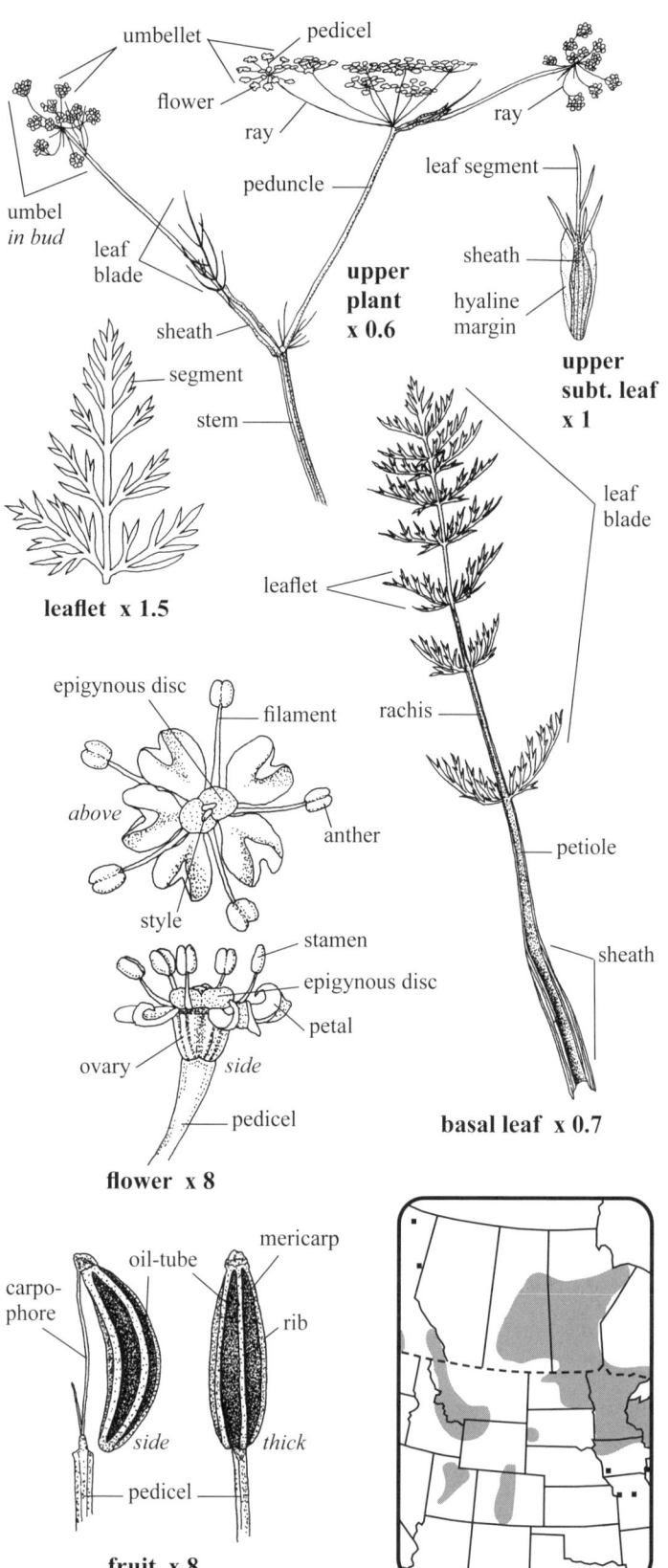

Anis

Carrot—*Apiaceae* **Wildflower White Petals 5**

BULB-BEARING WATER-HEMLOCK *Cicuta bulbifera* L.

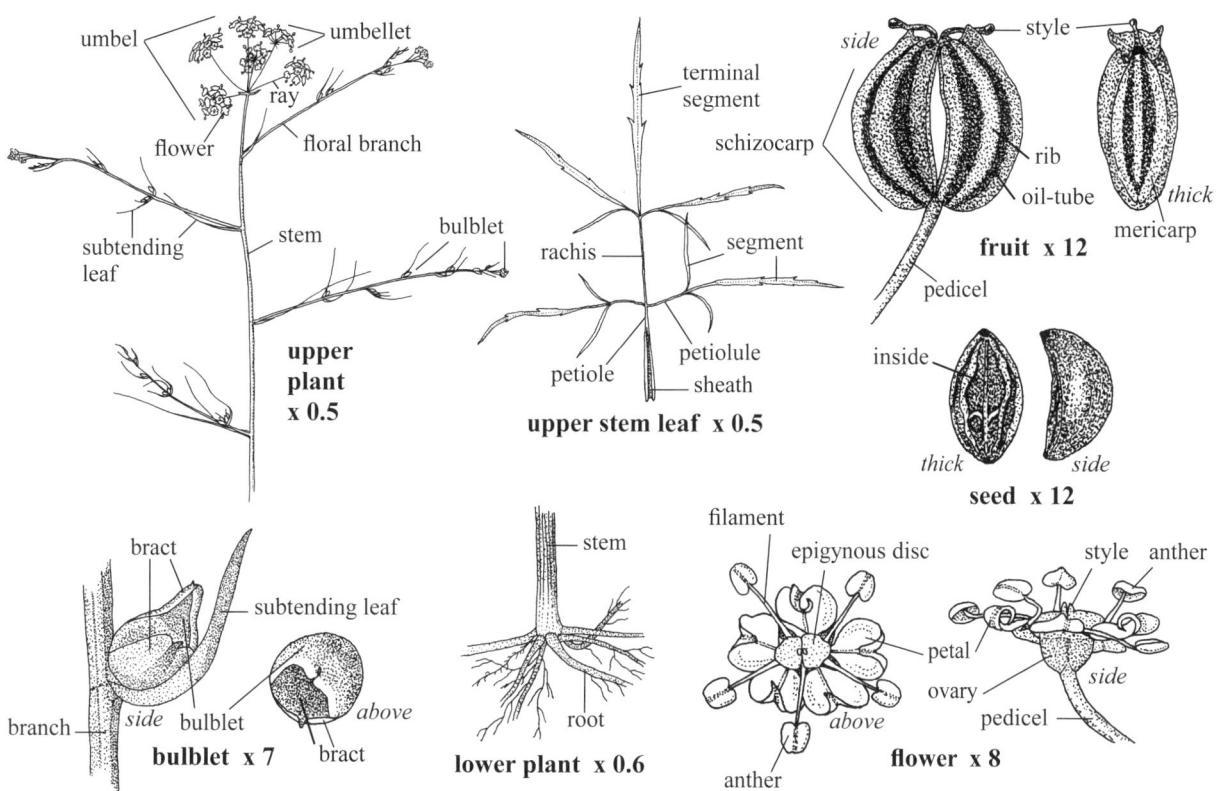

■ **SKETCH** A glabrous **biennial** or **perennial** herb 0.3–1 m tall with several tan **roots** 3–7 cm long by 1–2.5 mm thick; in wet fields, ditches, edges of marshes, ponds and lakes.

■ **FLOWERS** White, blooming July–September; **inflorescence** an umbel, compound, 4–30 umbels per plant, together 10–50 cm long by 15–60 cm wide; **umbels** 2–4.5 cm wide; **floral branches** 10–60 cm long, reduced above, ascending to spreading; **rays** 4–9 per umbel, each 5–25 mm long; **involucral bracts** (of rays) 0–2, each 1–4 mm long, pointed; **umbellets** four to nine per umbel, each 9–12 mm wide; **pedicels** 2–3 mm long; **involucel bractlets** (of pedicels) 5–7, each 1–2 mm long by 0.3–0.6 mm wide, united at their bases, midnerve obvious; **flowers** perfect, 3–4 mm wide (including stamens), 13–20 per umbellet, often infertile and few producing fruit; **epigynous disc** 1–1.3 mm wide, pale green, dull, bisected; **sepals** five, tiny, pointed, triangular; **petals** five, c. 1 mm long and wide, tip pointed and incurved; **stamens** five, exserted, c. 1.5 mm long, from edge of epigynous disc; **filaments** c. 0.9 mm long, filiform, white, spreading to erect; **anthers** white, 2-lobed, c. 0.6 mm long by c. 0.4 mm wide; **styles** paired at center, c. 0.1 mm long, pale green, persistent on fruit. **FRUIT a schizocarp**, dark brown, 1.8–2.2 mm long by 1.8–2.5 mm wide by c. 1 mm thick, dull, glabrous; **mericarps** two, 1-seeded, usually 5-ribbed, ribs lighter brown and wider than the oil-tubes; **seeds** one per mericarp, dark brown, 1.4–1.6 mm long by 0.8–1 mm wide by 0.7–0.9 mm thick, apices pointed and black, dull, sides convex and concave; **bulblets** sessile, one per leaf axil, numerous on upper branches, 2–2.8 mm long by 1.3–1.8 mm wide, green, turning brown and dropping; **beak** c. 1 mm long, TL c. 4 mm by c. 3 mm wide when ripe and shiny, filled with green, soft tissue.

■ **LEAVES** Alternate, stem only, ascending to spreading, compound, dull, base clasping; **blades** 6–18 cm long by 5–15 cm wide, glabrous, twice pinnate, reduced and simple above; **segments** 1–11 cm long by 1–10 mm wide, entire to toothed, teeth 2–13 mm long, narrow and curved; **petioles** 0–15 cm long with the lower 1–6 cm of the petiole widening as a sheath but clasping the stem only at the very base; **sheaths** strongly veined; **petiolules** 0–2 cm long.

■ **STEM** Erect, glaucous and glabrous, hollow, easily bent above; 1.5–8 mm thick near the stiff reddish base.

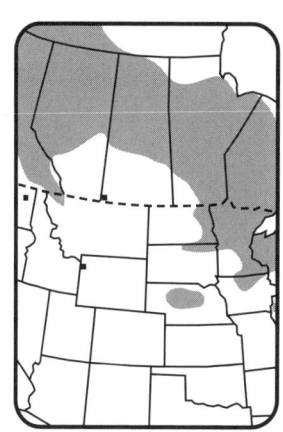

Carrot—*Apiaceae*   **Wildflower** White   Petals 5

WATER-HEMLOCK   *Cicuta maculata* L.

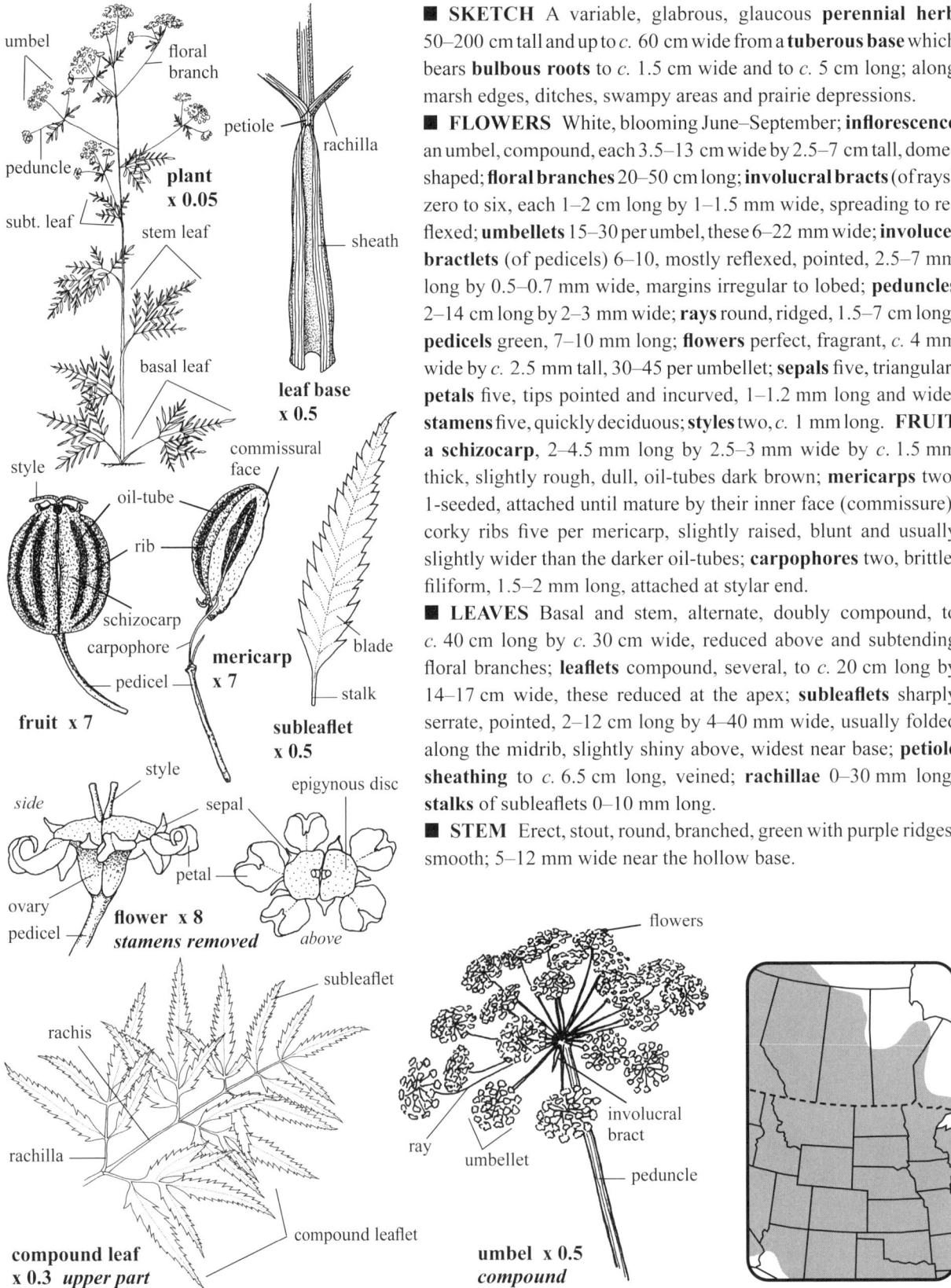

- **SKETCH** A variable, glabrous, glaucous **perennial herb** 50–200 cm tall and up to *c.* 60 cm wide from a **tuberous base** which bears **bulbous roots** to *c.* 1.5 cm wide and to *c.* 5 cm long; along marsh edges, ditches, swampy areas and prairie depressions.
- **FLOWERS** White, blooming June–September; **inflorescence** an umbel, compound, each 3.5–13 cm wide by 2.5–7 cm tall, dome-shaped; **floral branches** 20–50 cm long; **involucral bracts** (of rays) zero to six, each 1–2 cm long by 1–1.5 mm wide, spreading to reflexed; **umbellets** 15–30 per umbel, these 6–22 mm wide; **involucel bractlets** (of pedicels) 6–10, mostly reflexed, pointed, 2.5–7 mm long by 0.5–0.7 mm wide, margins irregular to lobed; **peduncles** 2–14 cm long by 2–3 mm wide; **rays** round, ridged, 1.5–7 cm long; **pedicels** green, 7–10 mm long; **flowers** perfect, fragrant, *c.* 4 mm wide by *c.* 2.5 mm tall, 30–45 per umbellet; **sepals** five, triangular; **petals** five, tips pointed and incurved, 1–1.2 mm long and wide; **stamens** five, quickly deciduous; **styles** two, *c.* 1 mm long. **FRUIT** a **schizocarp**, 2–4.5 mm long by 2.5–3 mm wide by *c.* 1.5 mm thick, slightly rough, dull, oil-tubes dark brown; **mericarps** two, 1-seeded, attached until mature by their inner face (commissure), corky ribs five per mericarp, slightly raised, blunt and usually slightly wider than the darker oil-tubes; **carpophores** two, brittle, filiform, 1.5–2 mm long, attached at stylar end.
- **LEAVES** Basal and stem, alternate, doubly compound, to *c.* 40 cm long by *c.* 30 cm wide, reduced above and subtending floral branches; **leaflets** compound, several, to *c.* 20 cm long by 14–17 cm wide, these reduced at the apex; **subleaflets** sharply serrate, pointed, 2–12 cm long by 4–40 mm wide, usually folded along the midrib, slightly shiny above, widest near base; **petiole sheathing** to *c.* 6.5 cm long, veined; **rachillae** 0–30 mm long; **stalks** of subleaflets 0–10 mm long.
- **STEM** Erect, stout, round, branched, green with purple ridges, smooth; 5–12 mm wide near the hollow base.

Spotted Water-hemlock, Common Water Hemlock, Spotted Cowbane

Carrot—*Apiaceae* **Wildflower**   White to purplish   Petals 5

COW-PARSNIP   *Heracleum maximum* Bartr.

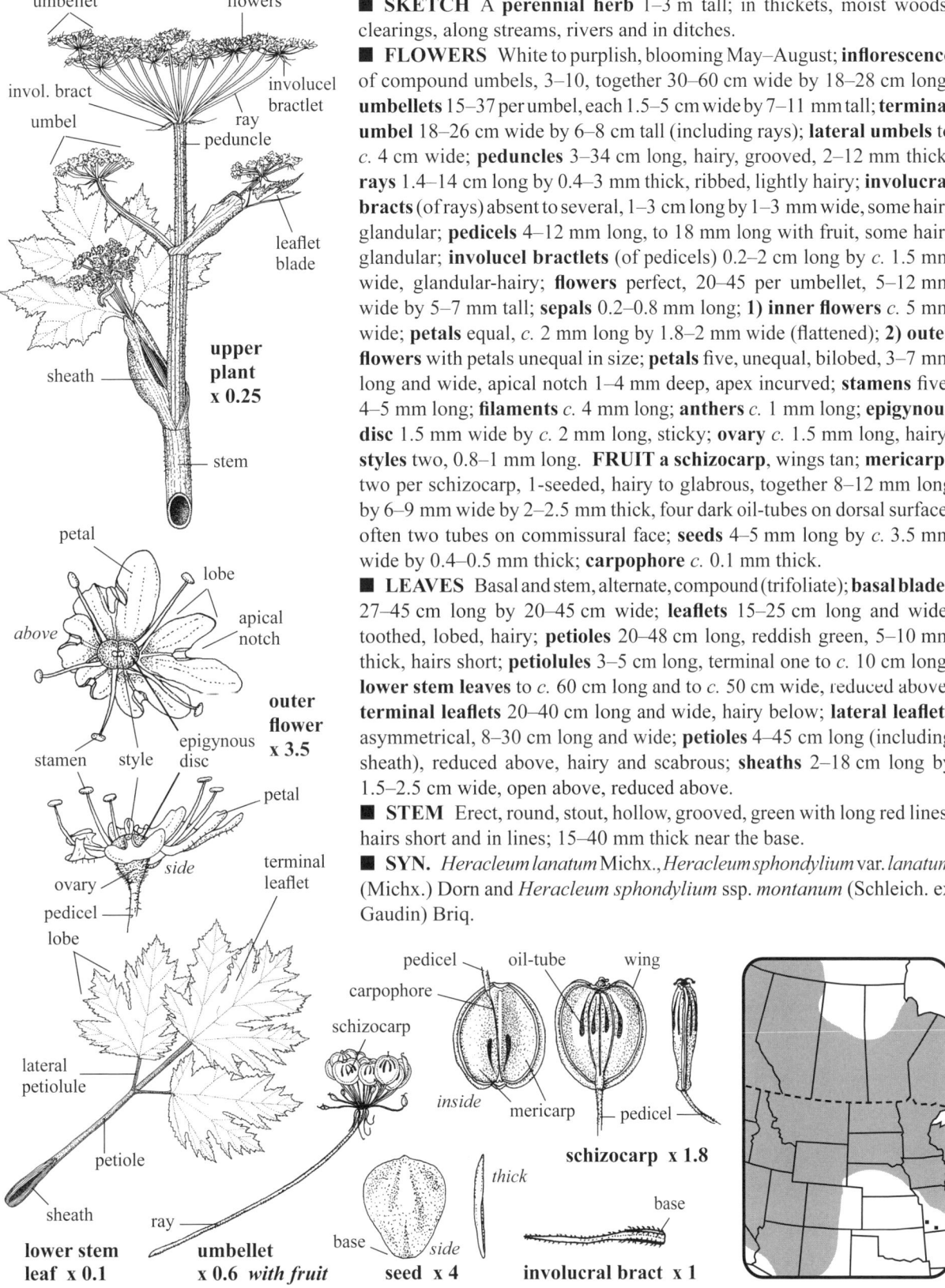

- **SKETCH** A **perennial herb** 1–3 m tall; in thickets, moist woods, clearings, along streams, rivers and in ditches.
- **FLOWERS** White to purplish, blooming May–August; **inflorescence** of compound umbels, 3–10, together 30–60 cm wide by 18–28 cm long; **umbellets** 15–37 per umbel, each 1.5–5 cm wide by 7–11 mm tall; **terminal umbel** 18–26 cm wide by 6–8 cm tall (including rays); **lateral umbels** to *c.* 4 cm wide; **peduncles** 3–34 cm long, hairy, grooved, 2–12 mm thick; **rays** 1.4–14 cm long by 0.4–3 mm thick, ribbed, lightly hairy; **involucral bracts** (of rays) absent to several, 1–3 cm long by 1–3 mm wide, some hairs glandular; **pedicels** 4–12 mm long, to 18 mm long with fruit, some hairs glandular; **involucel bractlets** (of pedicels) 0.2–2 cm long by *c.* 1.5 mm wide, glandular-hairy; **flowers** perfect, 20–45 per umbellet, 5–12 mm wide by 5–7 mm tall; **sepals** 0.2–0.8 mm long; **1) inner flowers** *c.* 5 mm wide; **petals** equal, *c.* 2 mm long by 1.8–2 mm wide (flattened); **2) outer flowers** with petals unequal in size; **petals** five, unequal, bilobed, 3–7 mm long and wide, apical notch 1–4 mm deep, apex incurved; **stamens** five, 4–5 mm long; **filaments** *c.* 4 mm long; **anthers** *c.* 1 mm long; **epigynous disc** 1.5 mm wide by *c.* 2 mm long, sticky; **ovary** *c.* 1.5 mm long, hairy; **styles** two, 0.8–1 mm long. **FRUIT a schizocarp**, wings tan; **mericarps** two per schizocarp, 1-seeded, hairy to glabrous, together 8–12 mm long by 6–9 mm wide by 2–2.5 mm thick, four dark oil-tubes on dorsal surface, often two tubes on commissural face; **seeds** 4–5 mm long by *c.* 3.5 mm wide by 0.4–0.5 mm thick; **carpophore** *c.* 0.1 mm thick.
- **LEAVES** Basal and stem, alternate, compound (trifoliate); **basal blades** 27–45 cm long by 20–45 cm wide; **leaflets** 15–25 cm long and wide, toothed, lobed, hairy; **petioles** 20–48 cm long, reddish green, 5–10 mm thick, hairs short; **petiolules** 3–5 cm long, terminal one to *c.* 10 cm long; **lower stem leaves** to *c.* 60 cm long and to *c.* 50 cm wide, reduced above; **terminal leaflets** 20–40 cm long and wide, hairy below; **lateral leaflets** asymmetrical, 8–30 cm long and wide; **petioles** 4–45 cm long (including sheath), reduced above, hairy and scabrous; **sheaths** 2–18 cm long by 1.5–2.5 cm wide, open above, reduced above.
- **STEM** Erect, round, stout, hollow, grooved, green with long red lines, hairs short and in lines; 15–40 mm thick near the base.
- **SYN.** *Heracleum lanatum* Michx., *Heracleum sphondylium* var. *lanatum* (Michx.) Dorn and *Heracleum sphondylium* ssp. *montanum* (Schleich. ex Gaudin) Briq.

Common Cowparsnip, American Cow-parsnip

Carrot—*Apiaceae* Wildflower   White   Petals 5

Smooth Sweet Cicely   *Osmorhiza longistylis* (Torr.) DC.

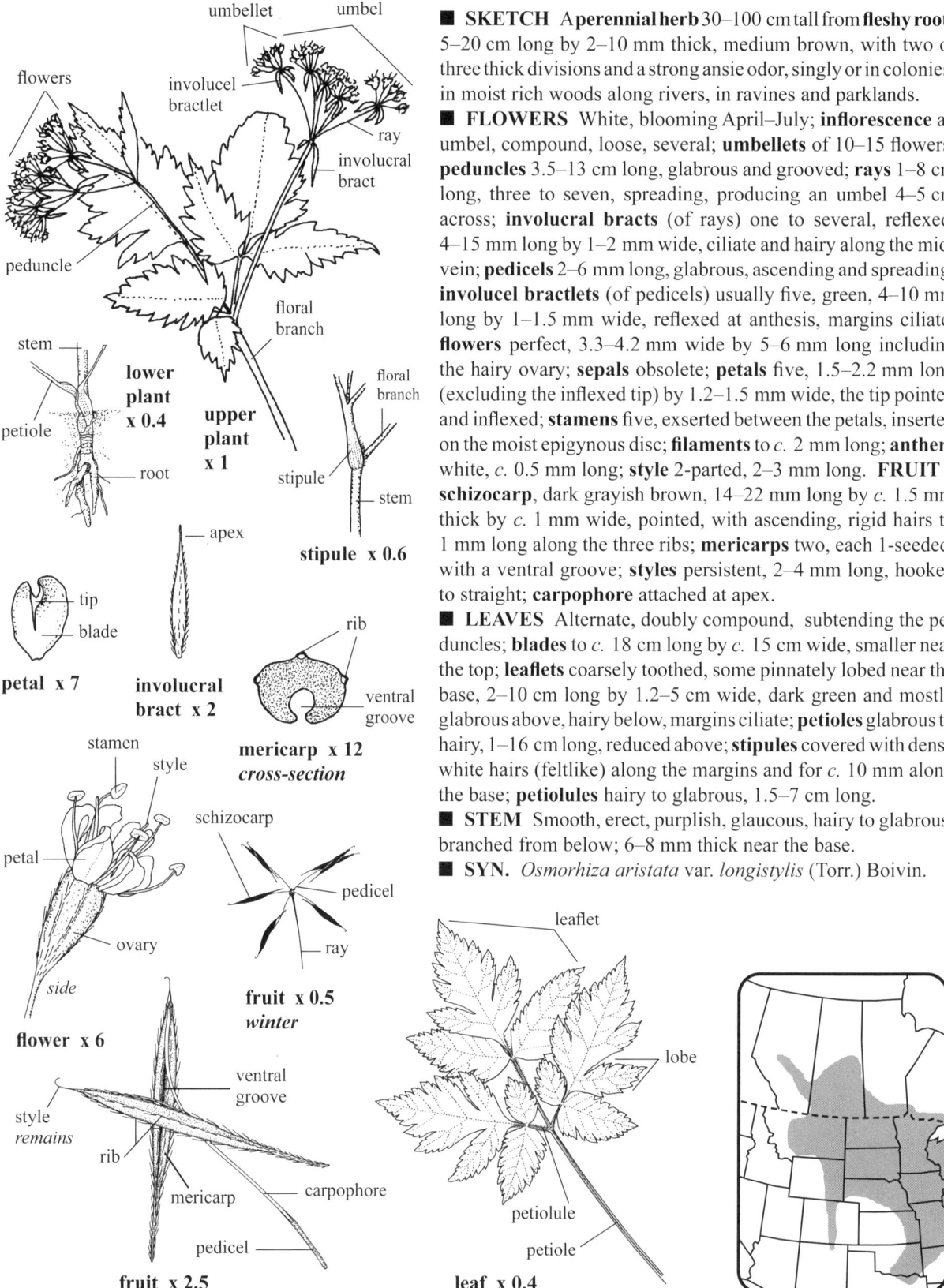

- **SKETCH** A **perennial herb** 30–100 cm tall from **fleshy roots** 5–20 cm long by 2–10 mm thick, medium brown, with two or three thick divisions and a strong ansie odor, singly or in colonies; in moist rich woods along rivers, in ravines and parklands.
- **FLOWERS** White, blooming April–July; **inflorescence** an umbel, compound, loose, several; **umbellets** of 10–15 flowers; **peduncles** 3.5–13 cm long, glabrous and grooved; **rays** 1–8 cm long, three to seven, spreading, producing an umbel 4–5 cm across; **involucral bracts** (of rays) one to several, reflexed, 4–15 mm long by 1–2 mm wide, ciliate and hairy along the midvein; **pedicels** 2–6 mm long, glabrous, ascending and spreading; **involucel bractlets** (of pedicels) usually five, green, 4–10 mm long by 1–1.5 mm wide, reflexed at anthesis, margins ciliate; **flowers** perfect, 3.3–4.2 mm wide by 5–6 mm long including the hairy ovary; **sepals** obsolete; **petals** five, 1.5–2.2 mm long (excluding the inflexed tip) by 1.2–1.5 mm wide, the tip pointed and inflexed; **stamens** five, exserted between the petals, inserted on the moist epigynous disc; **filaments** to *c.* 2 mm long; **anthers** white, *c.* 0.5 mm long; **style** 2-parted, 2–3 mm long. **FRUIT** a **schizocarp**, dark grayish brown, 14–22 mm long by *c.* 1.5 mm thick by *c.* 1 mm wide, pointed, with ascending, rigid hairs to 1 mm long along the three ribs; **mericarps** two, each 1-seeded, with a ventral groove; **styles** persistent, 2–4 mm long, hooked to straight; **carpophore** attached at apex.
- **LEAVES** Alternate, doubly compound, subtending the peduncles; **blades** to *c.* 18 cm long by *c.* 15 cm wide, smaller near the top; **leaflets** coarsely toothed, some pinnately lobed near the base, 2–10 cm long by 1.2–5 cm wide, dark green and mostly glabrous above, hairy below, margins ciliate; **petioles** glabrous to hairy, 1–16 cm long, reduced above; **stipules** covered with dense white hairs (feltlike) along the margins and for *c.* 10 mm along the base; **petiolules** hairy to glabrous, 1.5–7 cm long.
- **STEM** Smooth, erect, purplish, glaucous, hairy to glabrous, branched from below; 6–8 mm thick near the base.
- **SYN.** *Osmorhiza aristata* var. *longistylis* (Torr.) Boivin.

84   Anise-root

Carrot—*Apiaceae* **Wildflower** Yellow to reddish Petals 5

Wild Parsnip *Pastinaca sativa* L.

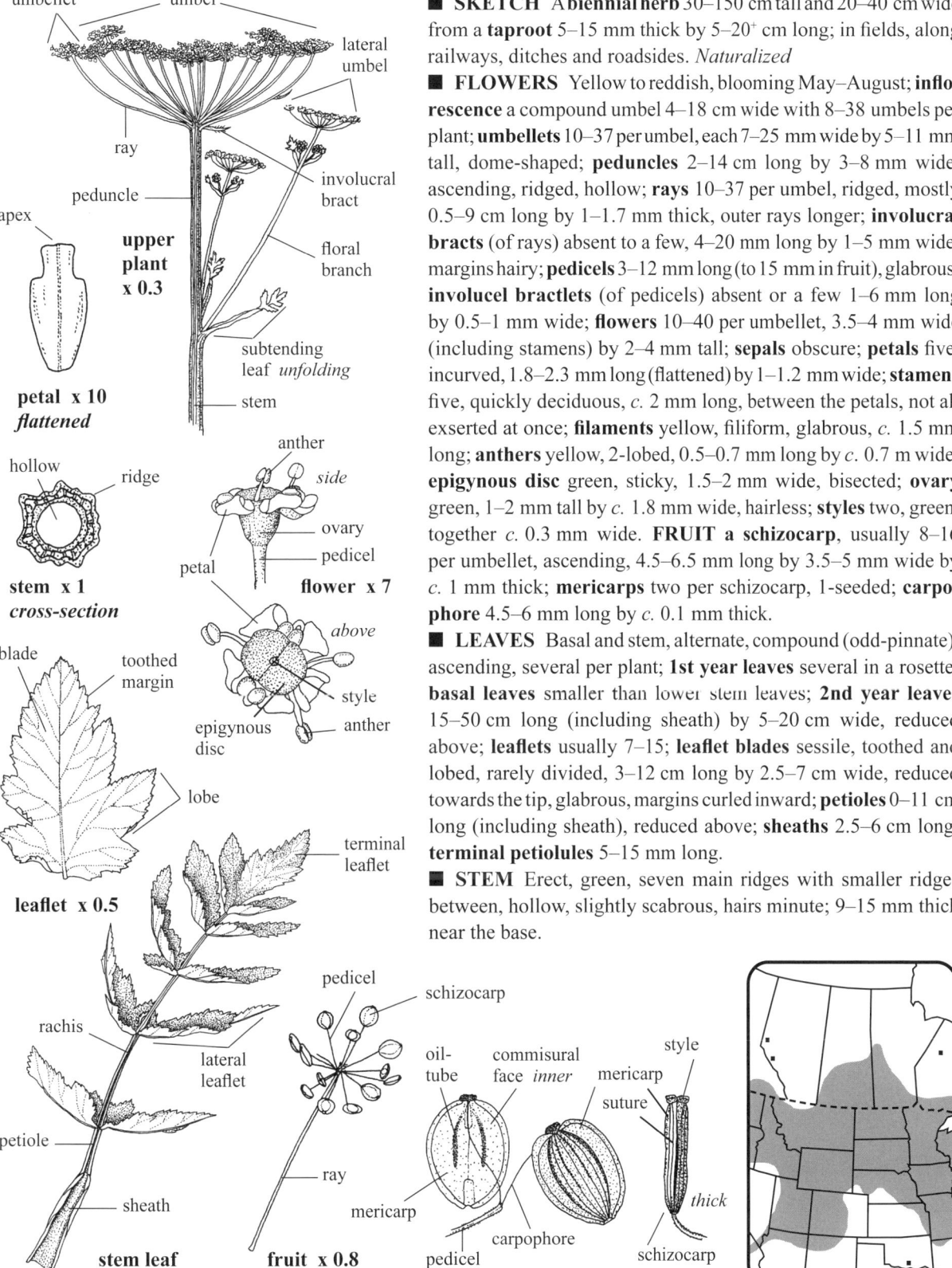

- **SKETCH** A **biennial herb** 30–150 cm tall and 20–40 cm wide from a **taproot** 5–15 mm thick by 5–20⁺ cm long; in fields, along railways, ditches and roadsides. *Naturalized*
- **FLOWERS** Yellow to reddish, blooming May–August; **inflorescence** a compound umbel 4–18 cm wide with 8–38 umbels per plant; **umbellets** 10–37 per umbel, each 7–25 mm wide by 5–11 mm tall, dome-shaped; **peduncles** 2–14 cm long by 3–8 mm wide, ascending, ridged, hollow; **rays** 10–37 per umbel, ridged, mostly 0.5–9 cm long by 1–1.7 mm thick, outer rays longer; **involucral bracts** (of rays) absent to a few, 4–20 mm long by 1–5 mm wide, margins hairy; **pedicels** 3–12 mm long (to 15 mm in fruit), glabrous; **involucel bractlets** (of pedicels) absent or a few 1–6 mm long by 0.5–1 mm wide; **flowers** 10–40 per umbellet, 3.5–4 mm wide (including stamens) by 2–4 mm tall; **sepals** obscure; **petals** five, incurved, 1.8–2.3 mm long (flattened) by 1–1.2 mm wide; **stamens** five, quickly deciduous, *c.* 2 mm long, between the petals, not all exserted at once; **filaments** yellow, filiform, glabrous, *c.* 1.5 mm long; **anthers** yellow, 2-lobed, 0.5–0.7 mm long by *c.* 0.7 m wide; **epigynous disc** green, sticky, 1.5–2 mm wide, bisected; **ovary** green, 1–2 mm tall by *c.* 1.8 mm wide, hairless; **styles** two, green, together *c.* 0.3 mm wide. **FRUIT** a schizocarp, usually 8–16 per umbellet, ascending, 4.5–6.5 mm long by 3.5–5 mm wide by *c.* 1 mm thick; **mericarps** two per schizocarp, 1-seeded; **carpophore** 4.5–6 mm long by *c.* 0.1 mm thick.
- **LEAVES** Basal and stem, alternate, compound (odd-pinnate), ascending, several per plant; **1st year leaves** several in a rosette; **basal leaves** smaller than lower stem leaves; **2nd year leaves** 15–50 cm long (including sheath) by 5–20 cm wide, reduced above; **leaflets** usually 7–15; **leaflet blades** sessile, toothed and lobed, rarely divided, 3–12 cm long by 2.5–7 cm wide, reduced towards the tip, glabrous, margins curled inward; **petioles** 0–11 cm long (including sheath), reduced above; **sheaths** 2.5–6 cm long; **terminal petiolules** 5–15 mm long.
- **STEM** Erect, green, seven main ridges with smaller ridges between, hollow, slightly scabrous, hairs minute; 9–15 mm thick near the base.

Carrot—*Apiaceae* — Wildflower   White (greenish)   Petals 5

SNAKEROOT   *Sanicula marilandica* L.

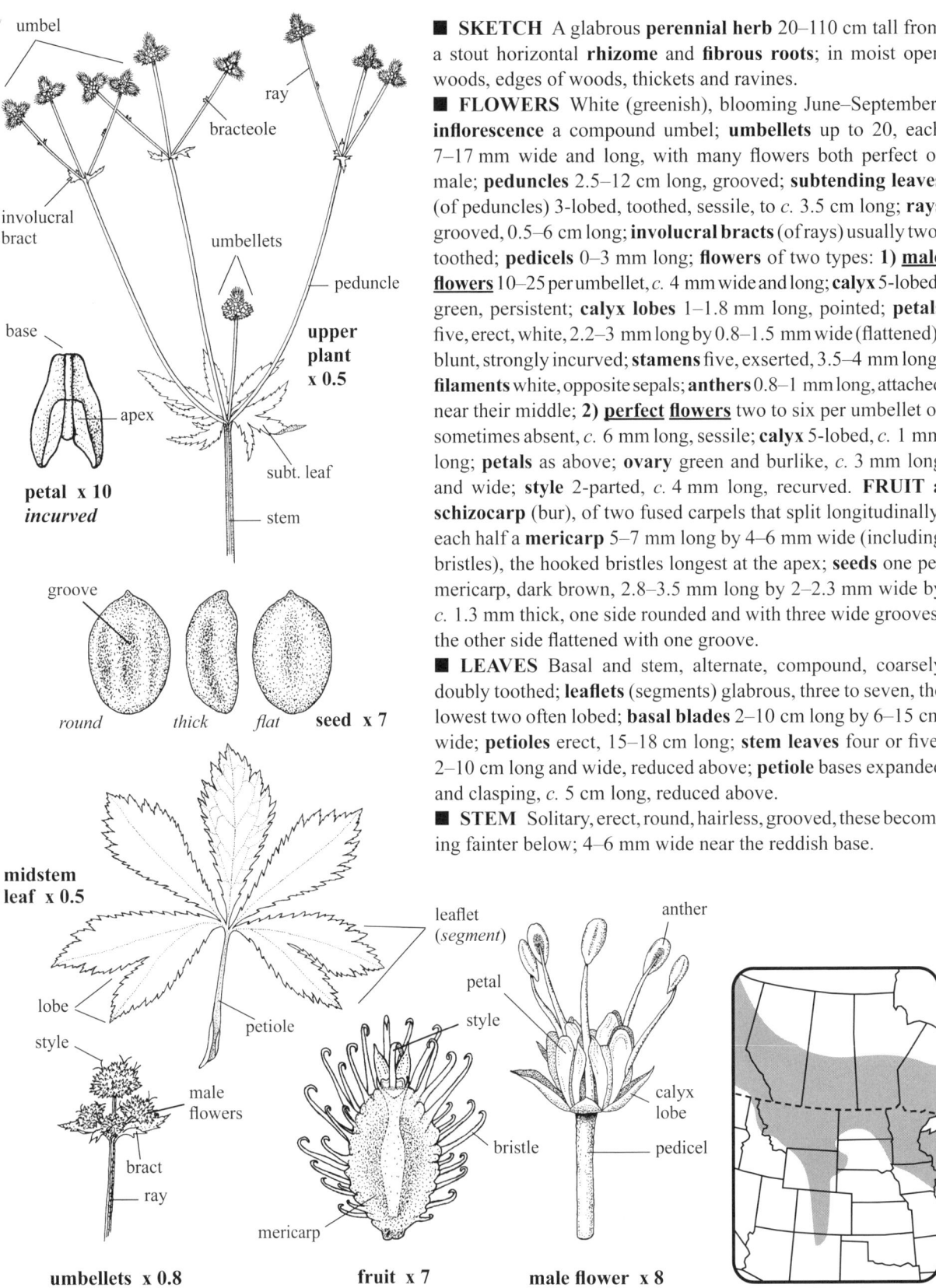

- **SKETCH** A glabrous **perennial herb** 20–110 cm tall from a stout horizontal **rhizome** and **fibrous roots**; in moist open woods, edges of woods, thickets and ravines.
- **FLOWERS** White (greenish), blooming June–September; **inflorescence** a compound umbel; **umbellets** up to 20, each 7–17 mm wide and long, with many flowers both perfect or male; **peduncles** 2.5–12 cm long, grooved; **subtending leaves** (of peduncles) 3-lobed, toothed, sessile, to *c.* 3.5 cm long; **rays** grooved, 0.5–6 cm long; **involucral bracts** (of rays) usually two, toothed; **pedicels** 0–3 mm long; **flowers** of two types: **1) male flowers** 10–25 per umbellet, *c.* 4 mm wide and long; **calyx** 5-lobed, green, persistent; **calyx lobes** 1–1.8 mm long, pointed; **petals** five, erect, white, 2.2–3 mm long by 0.8–1.5 mm wide (flattened), blunt, strongly incurved; **stamens** five, exserted, 3.5–4 mm long; **filaments** white, opposite sepals; **anthers** 0.8–1 mm long, attached near their middle; **2) perfect flowers** two to six per umbellet or sometimes absent, *c.* 6 mm long, sessile; **calyx** 5-lobed, *c.* 1 mm long; **petals** as above; **ovary** green and burlike, *c.* 3 mm long and wide; **style** 2-parted, *c.* 4 mm long, recurved. **FRUIT** a **schizocarp** (bur), of two fused carpels that split longitudinally, each half a **mericarp** 5–7 mm long by 4–6 mm wide (including bristles), the hooked bristles longest at the apex; **seeds** one per mericarp, dark brown, 2.8–3.5 mm long by 2–2.3 mm wide by *c.* 1.3 mm thick, one side rounded and with three wide grooves, the other side flattened with one groove.
- **LEAVES** Basal and stem, alternate, compound, coarsely doubly toothed; **leaflets** (segments) glabrous, three to seven, the lowest two often lobed; **basal blades** 2–10 cm long by 6–15 cm wide; **petioles** erect, 15–18 cm long; **stem leaves** four or five, 2–10 cm long and wide, reduced above; **petiole** bases expanded and clasping, *c.* 5 cm long, reduced above.
- **STEM** Solitary, erect, round, hairless, grooved, these becoming fainter below; 4–6 mm wide near the reddish base.

Sanicle, Black Snakeroot, Maryland Sanicle, Maryland Black Snakeroot

Carrot—*Apiaceae* **Wildflower** White Sepals often absent Petals 5

WATER-PARSNIP *Sium suave* Walt.

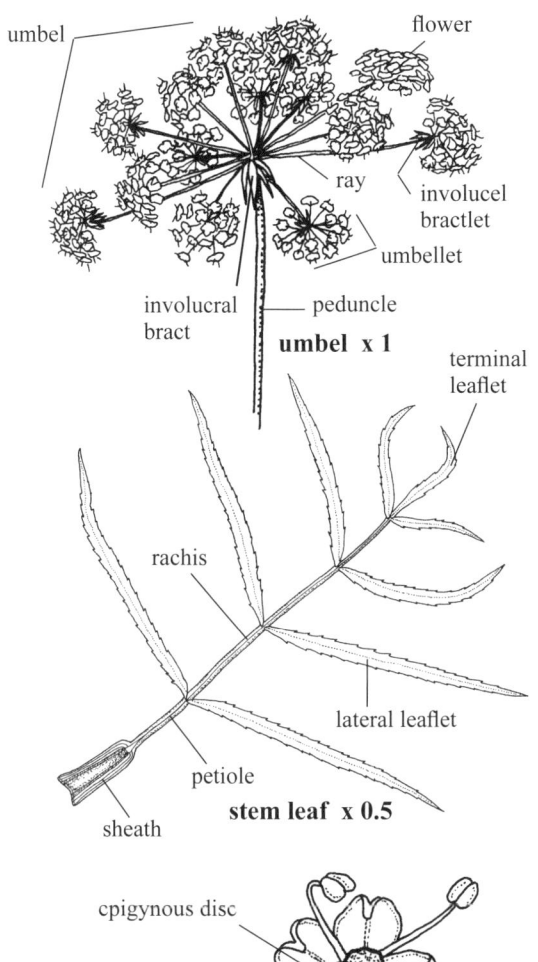

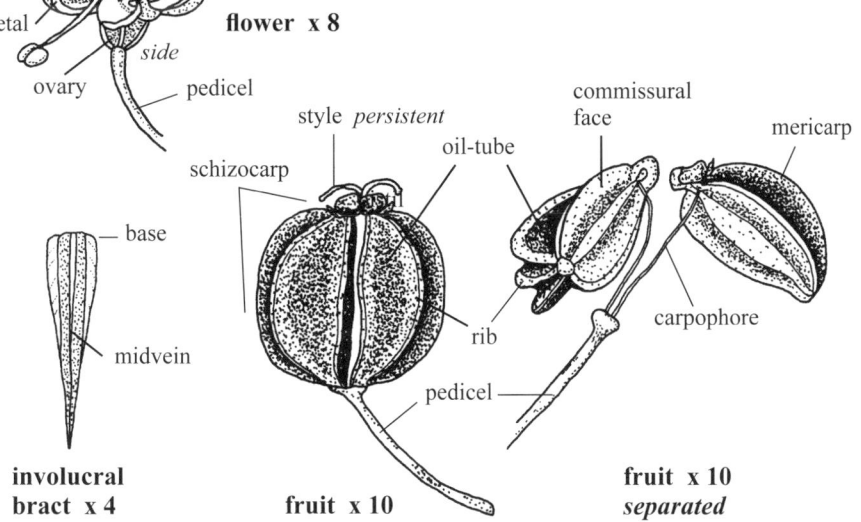

■ **SKETCH** A glabrous **perennial herb** 50–200 cm tall from **fascicled roots**; in wetland edges and roadside ditches.
■ **FLOWERS** White, blooming July–September; **inflorescence** of compound umbels, 1–11, each 5–11 cm wide by 2.5–8 cm tall, terminal and axillary; **floral branches** to *c*. 20 cm long, hollow; **peduncles** 2–13 cm long, with smooth longitudinal ridges; **rays** 8–32, each 1–4 cm long, grooved; **pedicels** 20–40 per umbellet, each 2–8 mm long; **umbellets** 8–32 per umbel, each 10–15 mm wide by 5–6 mm tall with 20–40 flowers; **involucral bracts** (of rays) 6–10, reflexed, 3–15 mm long by 0.7–2 mm wide; **involucel bractlets** (of pedicels) four to eight, reflexed, 1–3 mm long by 0.5–1 mm wide; **flowers** perfect, 3–5 mm wide by *c*. 2 mm tall; **sepals** tiny to absent; **petals** five, *c*. 1 mm long and wide, tips incurved; **stamens** five, between the petals, *c*. 2 mm long, exserted, quickly deciduous; **filaments** white; **anthers** pink and white, *c*. 0.5 mm long by *c*. 0.4 mm wide; **epigynous disc** green, *c*. 1 mm long, 2-parted, oval, edges irregular; **ovary** hairless, ridged, *c*. 0.7 mm long by *c*. 1 mm wide; **style** 2-parted, *c*. 0.1 mm tall to *c*. 0.7 mm on fruit. **FRUIT** a schizocarp, 2.2–2.8 mm long and wide by 1–1.5 mm thick, splitting; **mericarps** two, each 1-seeded, oil-tubes dark brown between the ribs and along the commissural face; **ribs** five per mericarp, corky; **carpophore** 2–2.5 mm long; **commissural faces** flat to slightly concave.
■ **LEAVES** Alternate, compound (odd-pinnate), two types: **1) early underwater leaves** two or three times pinnate with threadlike leaflets; **2) stem leaves above water** (shown), widely spaced; **blades** 5–25 cm long by 6–20 cm wide; **leaflets** sessile, toothed, usually 7–17, each 3–17 cm long by 3–15 mm wide; **petioles** 0–32 cm long, reduced above, hollow, form basal **sheaths** 1–7.5 cm long by 6–10 mm wide, veined, tips not free; **rachis** with a narrow groove above, flat below.
■ **STEM** Stout, erect, hollow, glabrous, several-sided, the sides flat, edges sharp, branched; 5–10 mm thick near the base.

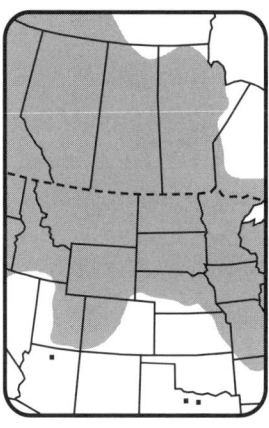

Hemlock Waterparsnip, Common Water-parsnip

Carrot—*Apiaceae*   **Wildflower**   Yellow   Petals 5 (6)

HEART-LEAVED ALEXANDERS   *Zizia aptera* (Gray) Fern.

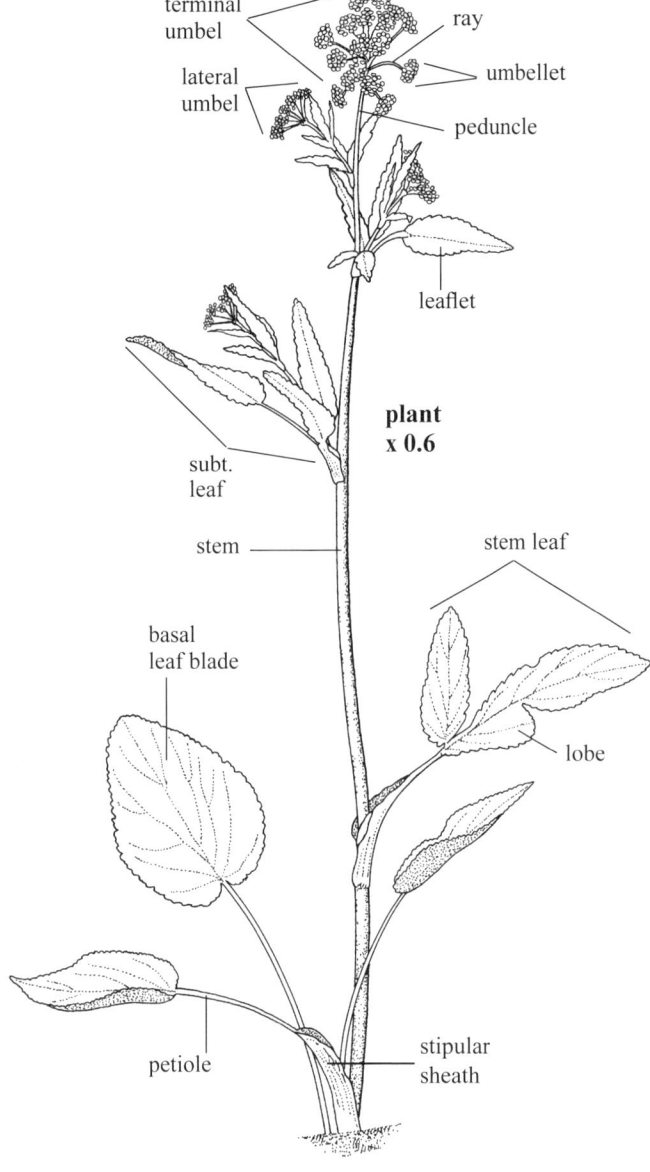

plant x 0.6

■ **SKETCH** A glabrous **perennial herb** 20–70 cm tall from a fascicle of **fleshy roots** each 2–8 mm wide and to *c.* 10⁺ cm long; in moist prairies, parklands, streambanks, dry slopes, meadows and thickets. **p1**

■ **FLOWERS** Yellow, blooming April–August; **inflorescence** an umbel, terminal and lateral, 1–12, each compounded into umbellets, terminal umbel the largest and blooming first; **umbels** 2–4.5 cm wide; **umbellets** up to *c.* 20 in the terminal umbel, each 6–9 mm wide; **peduncles** 2.5–11 cm long, axillary; **rays** 10–20 per umbel, each 6–35 mm long, to 50 mm with fruit; **involucral bracts** (of rays) absent; **pedicels** 1–5 mm long; **involucel bractlets** (of pedicels) ascending, linear, 1–2 mm long; **flowers** perfect, 15–20 per umbellet, 1–2 mm wide, central flower usually sessile; **calyx tube** ribbed and fused around the ovary; **calyx teeth** five, persistent and minute; **petals** five (six), curled inward; **stamens** five (not shown); **styles** two, arising from an epigynous disc. **FRUIT** a **schizocarp**, 2-seeded, 2.5–3.5 mm long by 2–2.5 mm wide by *c.* 1 mm thick, flattened, hairless, erect, dark brown, splitting longitudinally; **mericarps** two, each 1-seeded and with four to six tan ribs, wingless or narrowly winged.

■ **LEAVES** Alternate, simple to compound above; **basal leaves** simple, heart-shaped, finely toothed; **blades** 1.5–8 cm long by 3–5 cm wide, apices blunt; **petioles** 5–10 cm long; **stem leaves** compound, 3.3–8 cm long by 3–6 cm wide; **leaflets** three (five), toothed, 2–5 cm long by 1–2.5 cm wide, lighter green below; **petioles** 2–6 cm long, reduced above; **stipules** as sheathing 0.5–5.5 cm long, clasping the stem, united to petiole, reduced above.

■ **STEM** Round, erect, branched above.

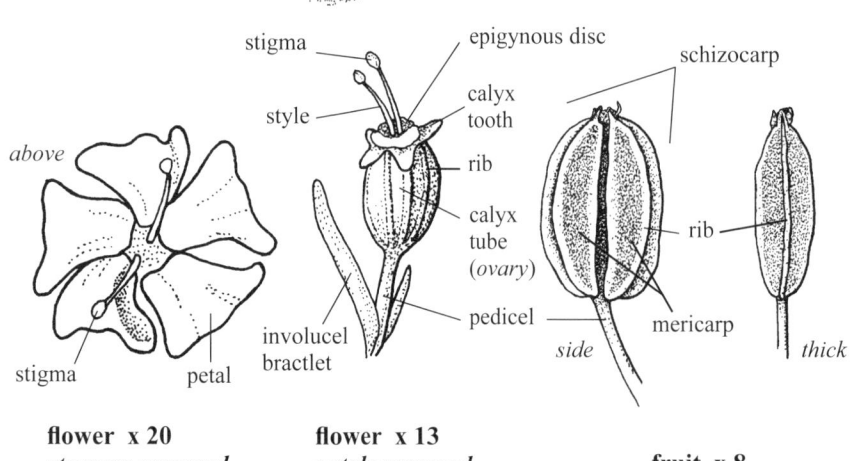

flower x 20
stamens removed

flower x 13
petals removed

fruit x 8

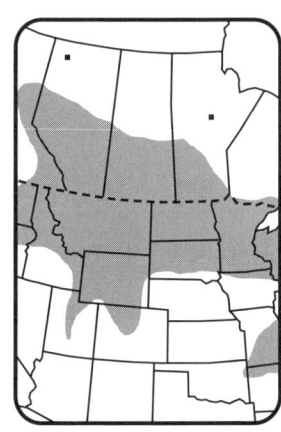

Meadow-parsnip, Meadow Zizia, Heart-leaf Alexanders

Carrot—*Apiaceae*   Wildflower Yellow Petals 5

GOLDEN ALEXANDERS *Zizia aurea* (L.) W.D.J. Koch

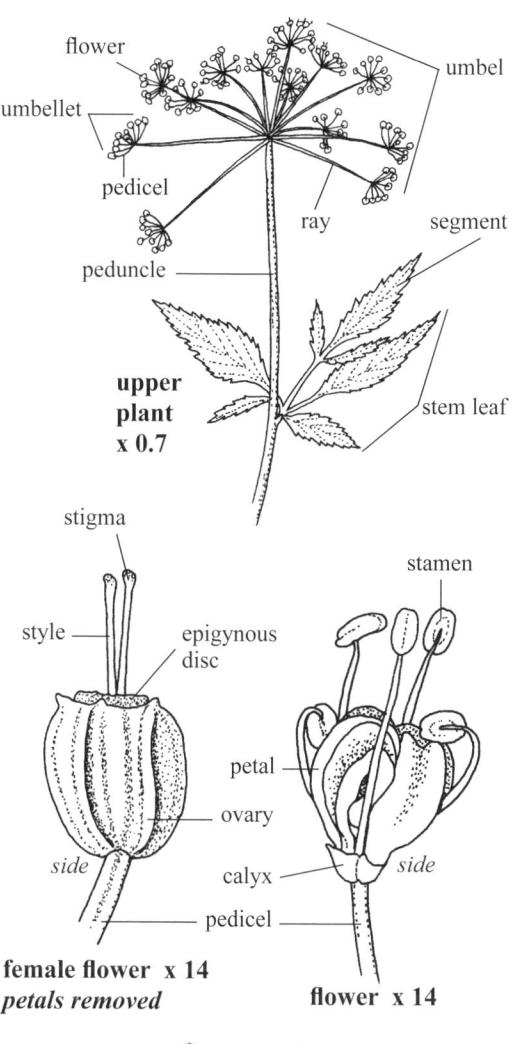

■ **SKETCH** A glabrous **perennial herb** 30–100 cm tall from a **fascicle** of thick **roots**; in open woodlands, ditches, pond margins and meadows. **p2**

■ **FLOWERS** Yellow, blooming April–July; **inflorescence** of two to six compound umbels, terminal or lateral along two main branches; **umbels** 5–6 cm wide by *c.* 3 cm deep; **umbellets** 10–18 per umbel, 8–13 mm wide by 4–5 mm deep; **peduncles** 3–15 cm long, ridged; **rays** unequal, 1.5–4.5 cm long, grooved; **involucral bracts** absent from base of rays; **pedicels** unequal, 1–5 mm long, grooved; **involucel bractlets** (of pedicels) *c.* five, green, spreading, pointed, 1–3 mm long; **flowers** perfect or unisexual, 13–26 per umbellet, *c.* 1.5 mm wide by *c.* 2.5 mm long, the central flower in an umbellet usually sessile, the outer flowers tending to be female (without stamens); **calyx** green, united, fleshy, the small, pointed teeth distinctive; **petals** five, *c.* 1.3 mm long by 0.6–1 mm wide, the tip inflexed; **stamens** zero to five; **filaments** yellow, between the petals, 1.5–2 mm long, erect, exserted; **anthers** yellow, 0.5–0.7 mm long; **epigynous disc** fleshy, *c.* 0.7 mm long and light green; **ovary** ribbed, green; **styles** two, distinct to their bases, *c.* 1.3 mm long, visible as fruit develops. **FRUIT** a schizocarp, hairless, brown with tan ribs, 3–4 mm long by 1.5–2 mm wide by *c.* 1 mm thick, bisected; **mericarps** two, 1-seeded, 5-ribbed with oil-tubes between the ribs.

■ **LEAVES** Basal and stem, alternate, compound; **basal leaves** one or two, twice ternate, glabrous, ascending; **blades** 6–10 cm long and about as wide; **segments** sharply and irregularly toothed, 1.5–6.5 cm long by 1–3.5 cm wide; **petioles** 10–25 cm long, pink at the base; **stipules** (sheath) clasping, 5–20 mm long, light green; **stem leaves** not numerous, compound, reduced upward.

■ **STEM** Erect, the two main branches enveloped in a **sheath** *c.* 2 cm long; 2–5 mm thick near the base.

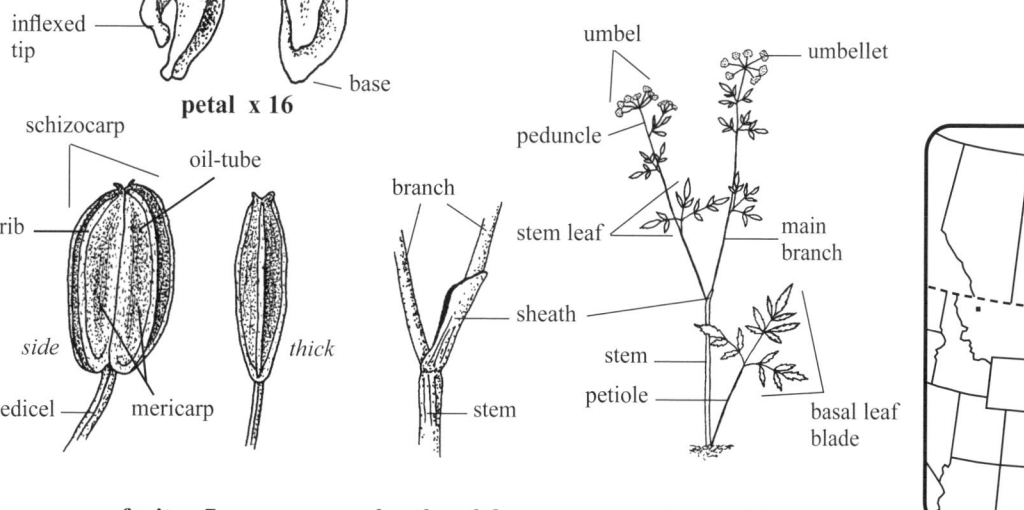

Golden Zizia

Dogbane—*Apocynaceae*   **Wildflower** White and pink   Corolla tubular, 5-lobed

SPREADING DOGBANE   *Apocynum androsaemifolium* L.

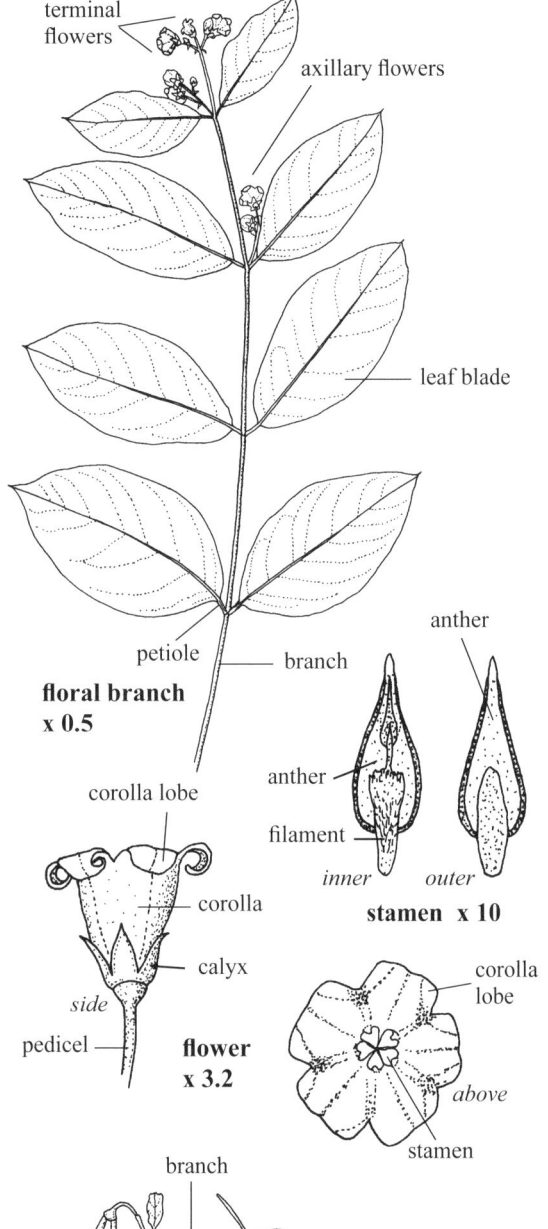

floral branch x 0.5

flower x 3.2

stamen x 10

■ **SKETCH** A glabrous **perennial herb** 30–150 cm tall from a **rhizome** 2–4 mm wide and to *c.* 12+ cm long; along dry woodland edges and clearings, roadsides, lakeshores, sandy areas, in thickets and prairies; **juice milky**. p2
■ **FLOWERS** White and pink, blooming May–September; **inflorescence** of cymes, terminal and axillary from the first one or two pairs of leaves below the terminal flowers; **branches** mostly in the upper half, alternate, to *c.* 35 cm long, usually ascending, reddish green; **pedicels** 2–5 mm long; **subtending bracts** shorter than pedicels; **flowers** perfect, fragrant, erect to drooping; **calyx** 5-lobed, united near the base, erect at anthesis, lobes 1–3 mm long with faint pink margins; **corolla** tubular, 5–8.5 mm long and wide, 5-lobed, these spreading to recurved, pink veined within; **stamens** five, included; **anthers** triangular, *c.* 3.5 mm long, leaning inward over pistil, fertile for 1 mm at tips; **filaments** hairy, *c.* 2 mm long, between five glands; **ovary** *c.* 1 mm long, bisected; **style** tapered, base straddles two apices of ovary; **stigma** glandular, *c.* 0.8 mm wide. **FRUIT a follicle**, usually straight and hanging, often in pairs, 6–15 cm long by 2–3 mm wide, opening along one suture, round, smooth, pointed, reddish brown when ripe; **seeds** 1.8–2.8 mm long by 0.5–0.6 mm wide by *c.* 0.4 mm thick, slightly rough with three to five obscure and sometimes incomplete longitudinal ribs along the slightly curved golden brown body; **coma** hairs white to tawny, numerous, 15–20 mm long, together to *c.* 3.5 cm wide when spreading, readily deciduous as a unit.
■ **LEAVES** Opposite, simple, entire, horizontal or sometimes drooping, flat, usually two to six pairs per branch with a pair at the base of each branch; **blades** 1–10.5 cm long by 0.5–5.2 cm wide with a tiny apical point (mucro), glabrous and darker green above, glabrous to hairy below especially along the raised veins; **petioles** 2–8 mm long, grooved above, round and hairy below, reduced above.
■ **STEM** Round, glabrous, often glaucous, branched, one to five stems; base of stem 3–6 mm thick.

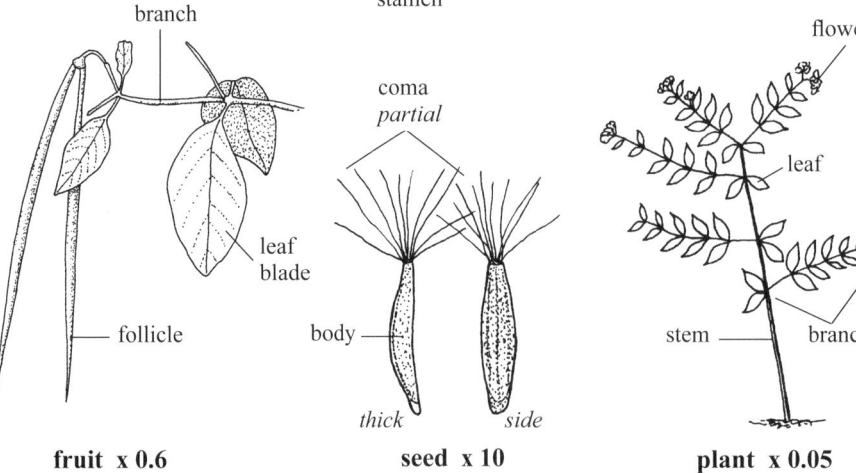

fruit x 0.6   seed x 10   plant x 0.05

Dogbane—*Apocynaceae*   **Wildflower**   White   Corolla tubular, 5-lobed

INDIAN-HEMP *Apocynum cannabinum* L.

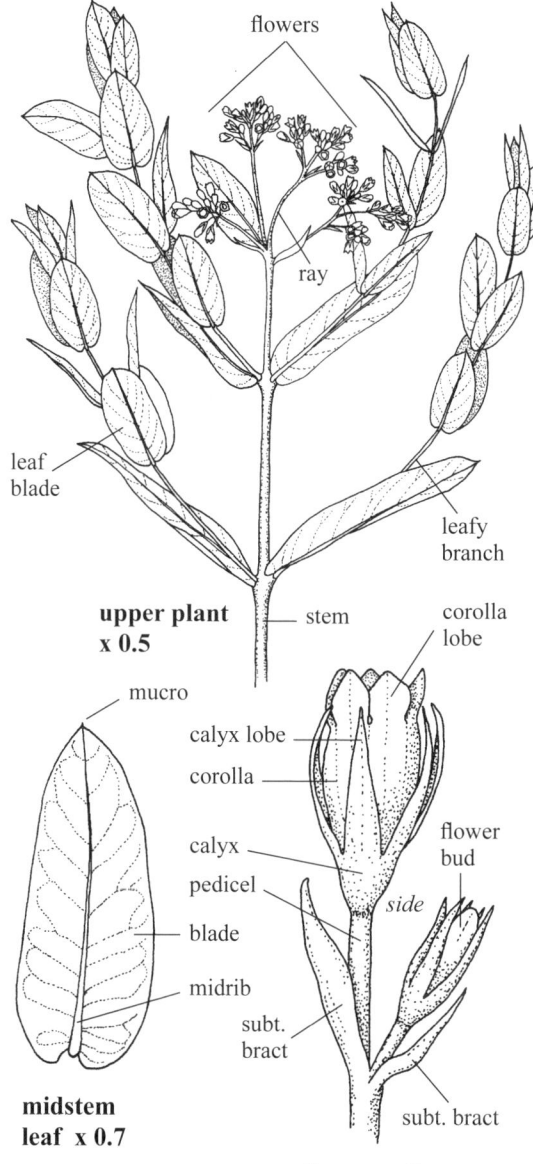

upper plant x 0.5

midstem leaf x 0.7

flower x 7

■ **SKETCH** A variable **perennial herb** 30–150 cm tall by 20–40 cm wide, single or in open colonies with vertical **tuberous roots** 5–8 mm thick and to *c.* 16⁺ cm long; in thickets, prairies, disturbed roadsides and moist fields; **juice milky. p2**

■ **FLOWERS** White, blooming May–September; **inflorescence** a cyme, mostly terminal, 4–6 cm tall by 5–12 cm wide; **upper branches** leafy, opposite, or alternate where a leaf replaces a branch in a pair; **rays** reddish green, 2–4 cm long, spreading, glabrous, subtended by a narrow bract; **pedicels** *c.* 3 mm long, green, glabrous; **subtending bracts** (of pedicels) one, glabrous, 3–4 mm long by *c.* 0.7 mm wide, keeled at the base; **flowers** perfect, fragrant, numerous, glabrous, erect to drooping, 2.5–3.5 mm wide by 2–4.5 mm long; **calyx** tubular, 5-lobed, light green, 2.5–4 mm long, the lobes pointed, 1.2–3.2 mm long by 0.8–1 mm wide, glabrous, with whitish tips; **corolla** tubular, 5-lobed, 2.5–4.5 mm long by 2–3 mm wide, glabrous, each lobe erect, *c.* 1 mm long and wide. **FRUIT a follicle**, tan to reddish, usually in pairs, 7–22 cm long by 3–4 mm wide, glabrous, round in cross-section, slightly curved to straight, suture not obvious; **seeds** numerous, orangish brown, imbricate, with two basal, narrow lateral ridges (usually incomplete) and a narrow whitish line along one side only, 3–6.2 mm long by 0.7–0.9 mm wide and thick, a slight bulge in the middle; **coma** of numerous, white silky hairs 1.5–4 cm long, deciduous as a unit; **septum** in follicle, tan, *c.* 2 mm wide.

■ **LEAVES** Opposite, simple, entire, spreading to ascending above, glabrous to slightly hairy and pale green below, apices point (mucro) *c.* 1 mm long; **blades** 1.5–14 cm long by 0.3–5 cm wide, lower ones reduced, not shiny, turning yellow by late summer; **petioles** 0–10 mm long; **stipules** absent.

■ **STEM** Erect to ascending, round to oval above, glabrous to glaucous to hairy, simple or branched above, solid, turning reddish green in late summer; 5–10 mm thick near the base.

■ **SYN.** *Apocynum sibiricum* Jacq.

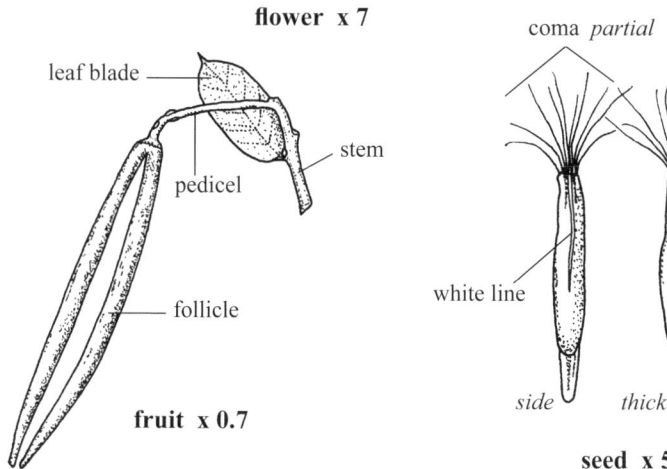

fruit x 0.7

seed x 5

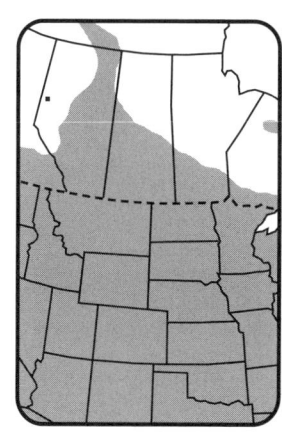

Indian Hemp Dogbane, Prairie Dogbane

Ginseng—*Araliaceae* **Wildflower** Whitish green Petals 5

WILD SARSAPARILLA   *Aralia nudicaulis* L.

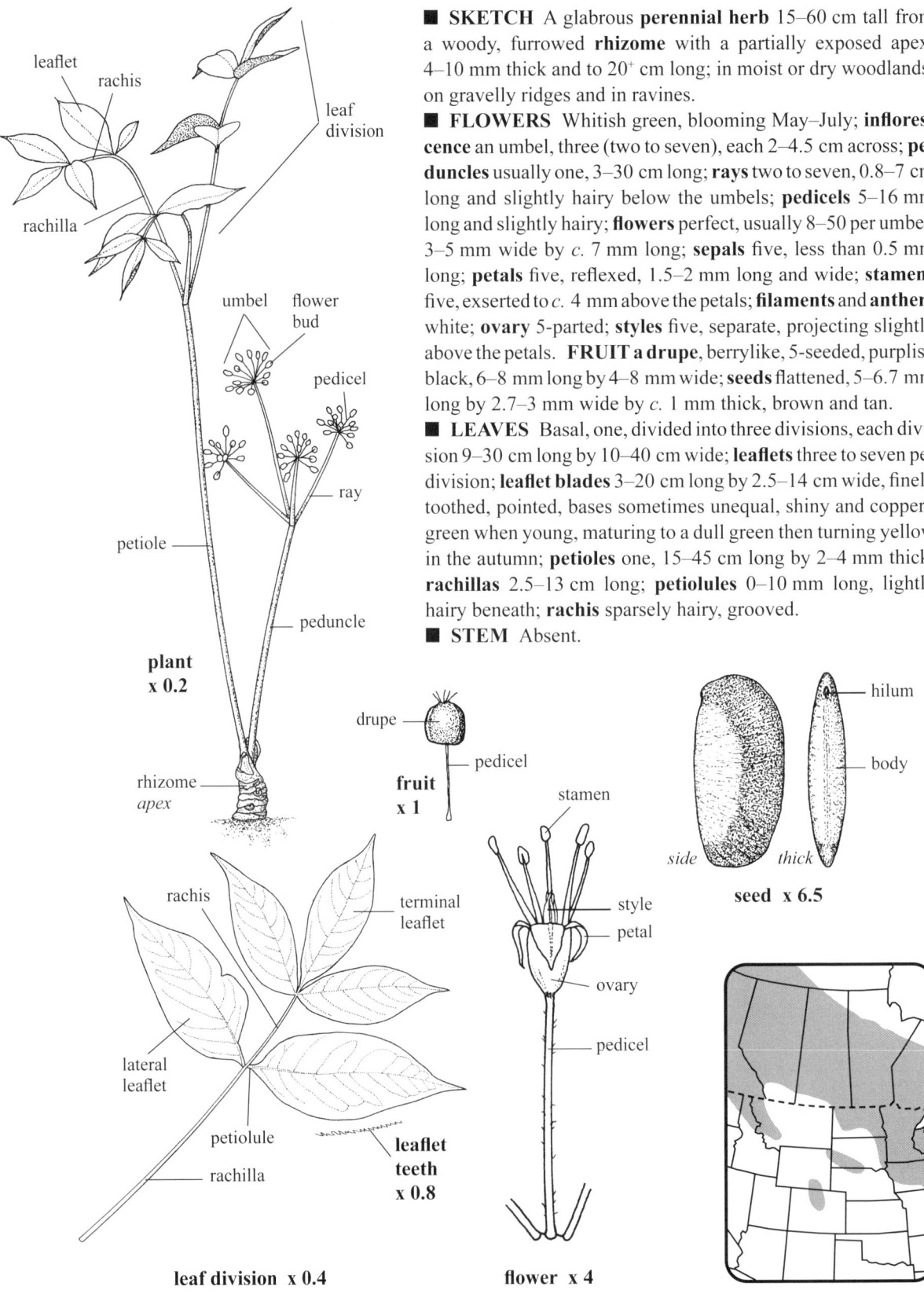

■ **SKETCH** A glabrous **perennial herb** 15–60 cm tall from a woody, furrowed **rhizome** with a partially exposed apex, 4–10 mm thick and to 20⁺ cm long; in moist or dry woodlands, on gravelly ridges and in ravines.

■ **FLOWERS** Whitish green, blooming May–July; **inflorescence** an umbel, three (two to seven), each 2–4.5 cm across; **peduncles** usually one, 3–30 cm long; **rays** two to seven, 0.8–7 cm long and slightly hairy below the umbels; **pedicels** 5–16 mm long and slightly hairy; **flowers** perfect, usually 8–50 per umbel, 3–5 mm wide by *c.* 7 mm long; **sepals** five, less than 0.5 mm long; **petals** five, reflexed, 1.5–2 mm long and wide; **stamens** five, exserted to *c.* 4 mm above the petals; **filaments** and **anthers** white; **ovary** 5-parted; **styles** five, separate, projecting slightly above the petals. FRUIT **a drupe**, berrylike, 5-seeded, purplish black, 6–8 mm long by 4–8 mm wide; **seeds** flattened, 5–6.7 mm long by 2.7–3 mm wide by *c.* 1 mm thick, brown and tan.

■ **LEAVES** Basal, one, divided into three divisions, each division 9–30 cm long by 10–40 cm wide; **leaflets** three to seven per division; **leaflet blades** 3–20 cm long by 2.5–14 cm wide, finely toothed, pointed, bases sometimes unequal, shiny and coppery green when young, maturing to a dull green then turning yellow in the autumn; **petioles** one, 15–45 cm long by 2–4 mm thick; **rachillas** 2.5–13 cm long; **petiolules** 0–10 mm long, lightly hairy beneath; **rachis** sparsely hairy, grooved.

■ **STEM** Absent.

Milkweed—*Asclepiadaceae* **Wildflower** Dark pink Corolla lobes 5

Swamp Milkweed *Asclepias incarnata* L.

■ **SKETCH** A **perennial herb** 50–250 cm tall from a **rhizome** 1–2 mm long by *c.* 7 mm thick with numerous white **fibrous roots** 2–20 cm long by 1–3.3 mm thick, unbranched and smooth; along edges of marshes, lakeshores and ponds, ditches and flood plains; **juice milky**.

■ **FLOWERS** Dark pink, blooming June–September; **inflorescence** an umbel, usually 3–8, each *c.* 2 cm long by *c.* 2.5–4 cm wide, terminal and axillary from upper leaves; **peduncles** light green, 0.2–7 cm long by 1–2 mm thick, erect, hairy; **pedicels** pink, ascending, 10–27 mm long by 0.5–0.6 mm thick, widest at base, erect with fruit and 1.5–2 mm thick, curved, reddish purple, hairs white, ascending, appressed; **subtending bracts** (of pedicels) 5–10 mm long by *c.* 0.5 mm wide, spreading, several at bases of pedicels; **flower buds** reddish purple, 3–4 mm wide by 4–5 mm tall; **flowers** perfect, fragrant, 10–40 per umbel, each 6–9 mm wide by 7–11 mm long, ascending; **calyx lobes** five, green, pointed, reddish near apices, descending, 1.3–2.8 mm long by *c.* 1 mm wide, glabrous to slightly hairy on back; **corolla lobes** five, reddish purple, descending, apices curved upward, 3–7 mm long by *c.* 2.3 mm wide, glabrous, not veiny; **column** green, 1.2–1.8 mm tall by 0.8–1.5 mm wide; **hoods** five, oblong, light pink, attached at base, 2–3.4 mm long by *c.* 1.6 mm wide, not fleshy, rounded at back, blunt, apices about level with the anther heads, lower sides incurved; **horns** five, attached near base of hoods, arching up and over the anther head, not quite reaching the center; **anther head** 2.1–2.5 mm tall by 1.5–2 mm wide. **FRUIT a follicle**, yellowish tan, erect, beaked, 5–9 cm long by 0.8–1.2 cm wide, smooth, hairless to slightly hairy, opening along one inner full length suture, thin and papery with several long low ridges; **seeds** 6.5–9 mm long by 5–6 mm wide by 1.3–2 mm thick, winged, 37–50+ per follicle, slightly shiny, dark brown with a fine ridge on one side from the apex to about the center of the seed; **seed wing** 1–1.5 mm wide, reddish brown, completely encircling the seed; **coma** of smooth white hairs 1.5–2.7 mm long, arched, deciduous as a unit.

■ **LEAVES** Opposite, entire, simple, dull, slighter lighter green below with raised veins, ascending to spreading; **blades** 4–18 cm long by 5–45 mm wide, reduced below, margins revolute and hairy, slightly hairy along veins on both sides; **petioles** 3–17 mm long, ascending; **stipules** absent.

■ **STEM** Green, erect, thick-walled, hollow, smooth, round below, simple to branched above, oval beneath flowers and slightly hairy above in lines, with some vegetative leafy branches to *c.* 30 cm long ascending from upper leaf axils; 5–12 mm thick near the naked base.

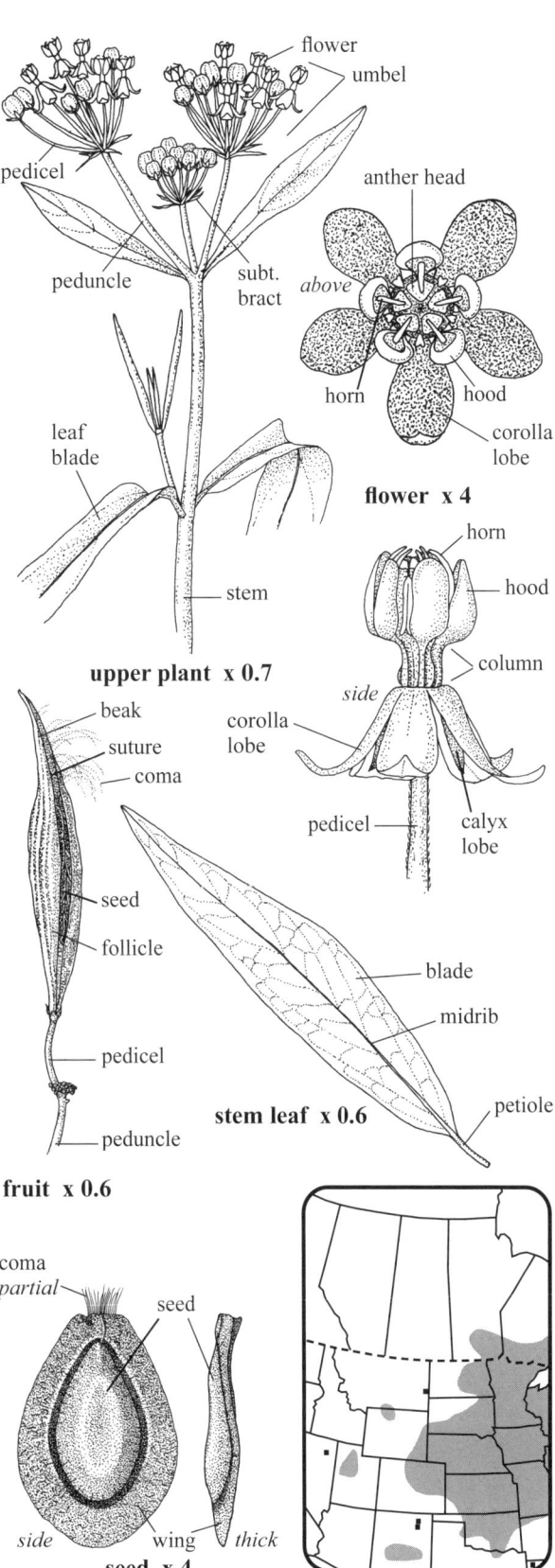

93

Milkweed—*Asclepiadaceae* **Wildflower** White to pale green Corolla 5-lobed

Dwarf Milkweed *Asclepias ovalifolia* Dcne.

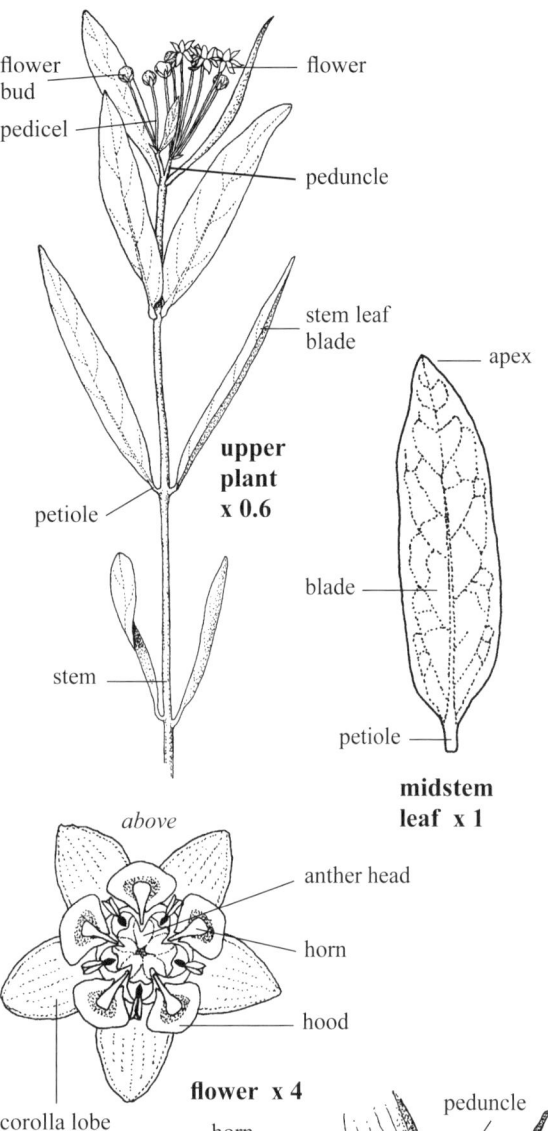

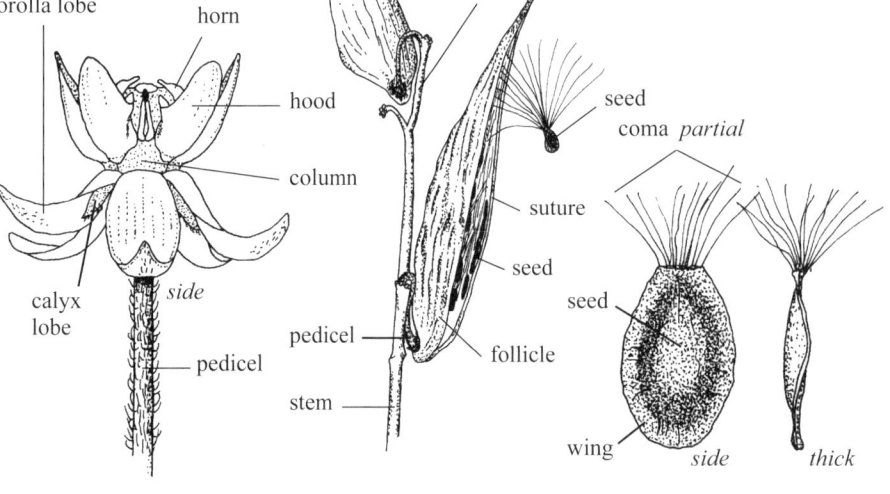

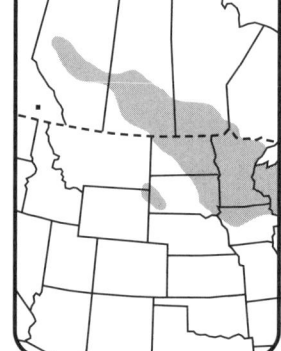

■ **SKETCH** A **perennial herb** 20–50 cm tall from a shallow, slender **rhizome**; in sandy prairies, slopes, along railways and in open woods; **juice milky. p2**

■ **FLOWERS** White to pale green, blooming May–August; **inflorescence** an umbel, one to four, terminal or from upper leaf axils; **peduncles** 0.3–4 cm long, hairs white and mostly ascending; **pedicels** ascending, 1.5–2.5 cm long by 0.5–0.6 mm wide and to *c.* 1 mm thick in fruit, hairy; **flowers** perfect, 3–25 per umbel, *c.* 10 mm wide by 7.5–10 mm tall; **calyx lobes** five, 2.3–3.5 mm long by 1–1.3 mm wide, pointed, hairy on outside, glabrous above; **corolla** 5-lobed, recurved, the tips ascending, 4–7 mm long by 2.5–3 mm wide, glabrous above, pinkish green below and slightly hairy especially near the tip; **column** *c.* 0.5 mm tall by 1.2–2 mm wide; **hoods** five, 3.8–5 mm long, each with a white, incurved horn; **horns** five, arching over the tip of the 3 mm wide anther head. **FRUIT a follicle**, tan to medium brown, usually one or two per plant, 6–9 cm long by 0.8–1.5 cm wide, round in cross-section, downy, wrinkled, without tubercles, erect, one full-length suture opposite the deflexed pedicel; **seeds** numerous, imbricate, brown, hairless, 5–7 mm long by 3.5–4.5 mm wide by 1–1.5 mm thick, winged; **coma** white to tan, 2–3.5 cm long, of numerous silky hairs.

■ **LEAVES** Opposite, simple, entire, few; **blades** 2–8.5 cm long by 0.7–5.2 cm wide, largest at midstem, hairs bent and numerous below and on midrib, less hairy above, margins with hairs arched forward; **petioles** 2–10 mm long, hairs ascending and curved.

■ **STEM** Erect, round, hollow, green to reddish below, solitary or paired, usually simple, hairy above; hairs few near the 2–4 mm thick base.

94 Oval-leaved Milkweed

Milkweed—*Asclepiadaceae*   **Wildflower**   Pink   Corolla 5-lobed

Showy Milkweed   *Asclepias speciosa* Torr.

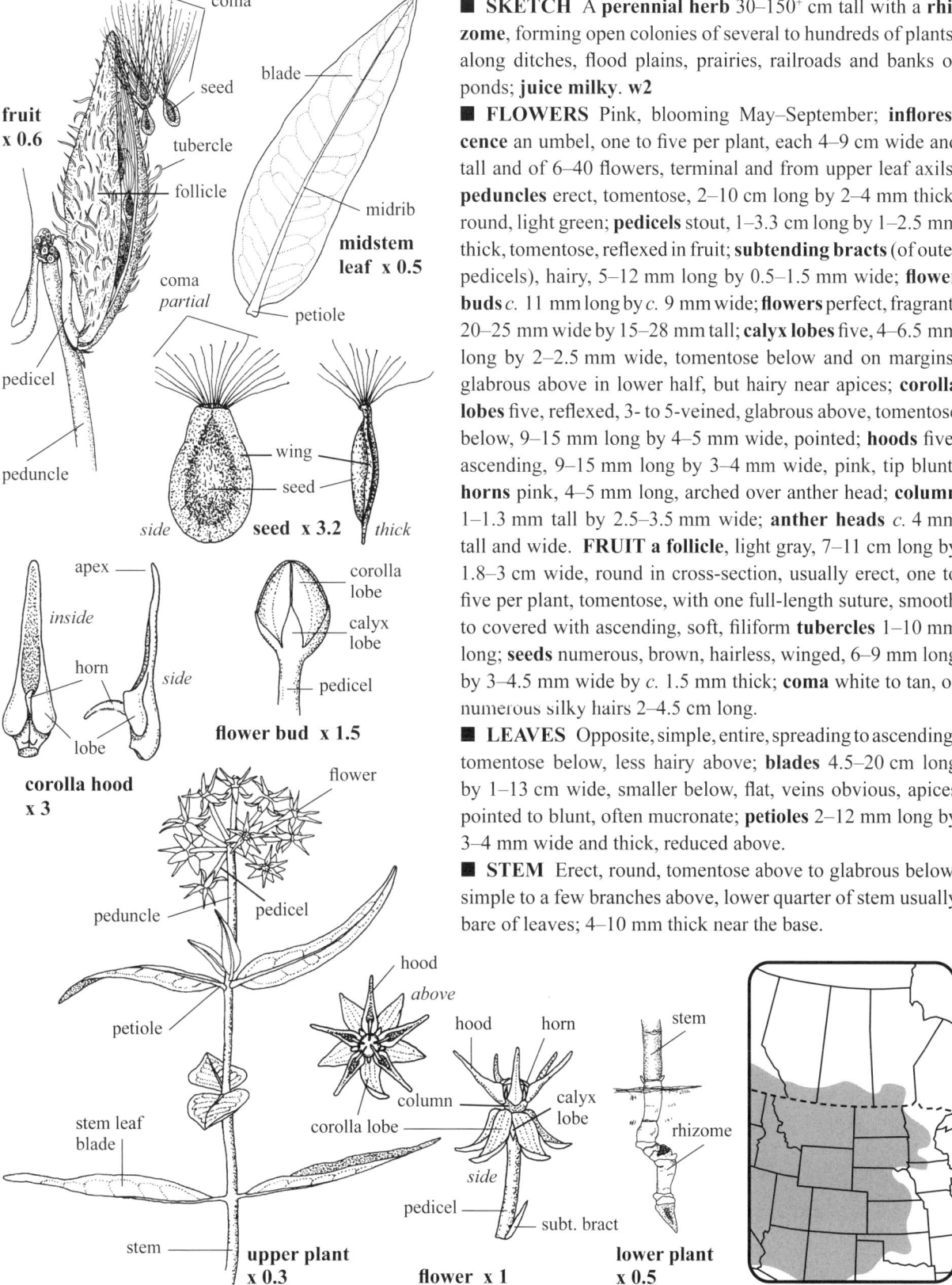

■ **SKETCH** A **perennial herb** 30–150+ cm tall with a **rhizome**, forming open colonies of several to hundreds of plants; along ditches, flood plains, prairies, railroads and banks of ponds; **juice milky**. **w2**

■ **FLOWERS** Pink, blooming May–September; **inflorescence** an umbel, one to five per plant, each 4–9 cm wide and tall and of 6–40 flowers, terminal and from upper leaf axils; **peduncles** erect, tomentose, 2–10 cm long by 2–4 mm thick, round, light green; **pedicels** stout, 1–3.3 cm long by 1–2.5 mm thick, tomentose, reflexed in fruit; **subtending bracts** (of outer pedicels), hairy, 5–12 mm long by 0.5–1.5 mm wide; **flower buds** *c.* 11 mm long by *c.* 9 mm wide; **flowers** perfect, fragrant, 20–25 mm wide by 15–28 mm tall; **calyx lobes** five, 4–6.5 mm long by 2–2.5 mm wide, tomentose below and on margins, glabrous above in lower half, but hairy near apices; **corolla lobes** five, reflexed, 3- to 5-veined, glabrous above, tomentose below, 9–15 mm long by 4–5 mm wide, pointed; **hoods** five, ascending, 9–15 mm long by 3–4 mm wide, pink, tip blunt; **horns** pink, 4–5 mm long, arched over anther head; **column** 1–1.3 mm tall by 2.5–3.5 mm wide; **anther heads** *c.* 4 mm tall and wide. **FRUIT a follicle**, light gray, 7–11 cm long by 1.8–3 cm wide, round in cross-section, usually erect, one to five per plant, tomentose, with one full-length suture, smooth to covered with ascending, soft, filiform **tubercles** 1–10 mm long; **seeds** numerous, brown, hairless, winged, 6–9 mm long by 3–4.5 mm wide by *c.* 1.5 mm thick; **coma** white to tan, of numerous silky hairs 2–4.5 cm long.

■ **LEAVES** Opposite, simple, entire, spreading to ascending, tomentose below, less hairy above; **blades** 4.5–20 cm long by 1–13 cm wide, smaller below, flat, veins obvious, apices pointed to blunt, often mucronate; **petioles** 2–12 mm long by 3–4 mm wide and thick, reduced above.

■ **STEM** Erect, round, tomentose above to glabrous below, simple to a few branches above, lower quarter of stem usually bare of leaves; 4–10 mm thick near the base.

Milkweed—*Asclepiadaceae* **Wildflower** Pink to purplish green Corolla 5- (6-) lobed

SILKY MILKWEED *Asclepias syriaca* L.

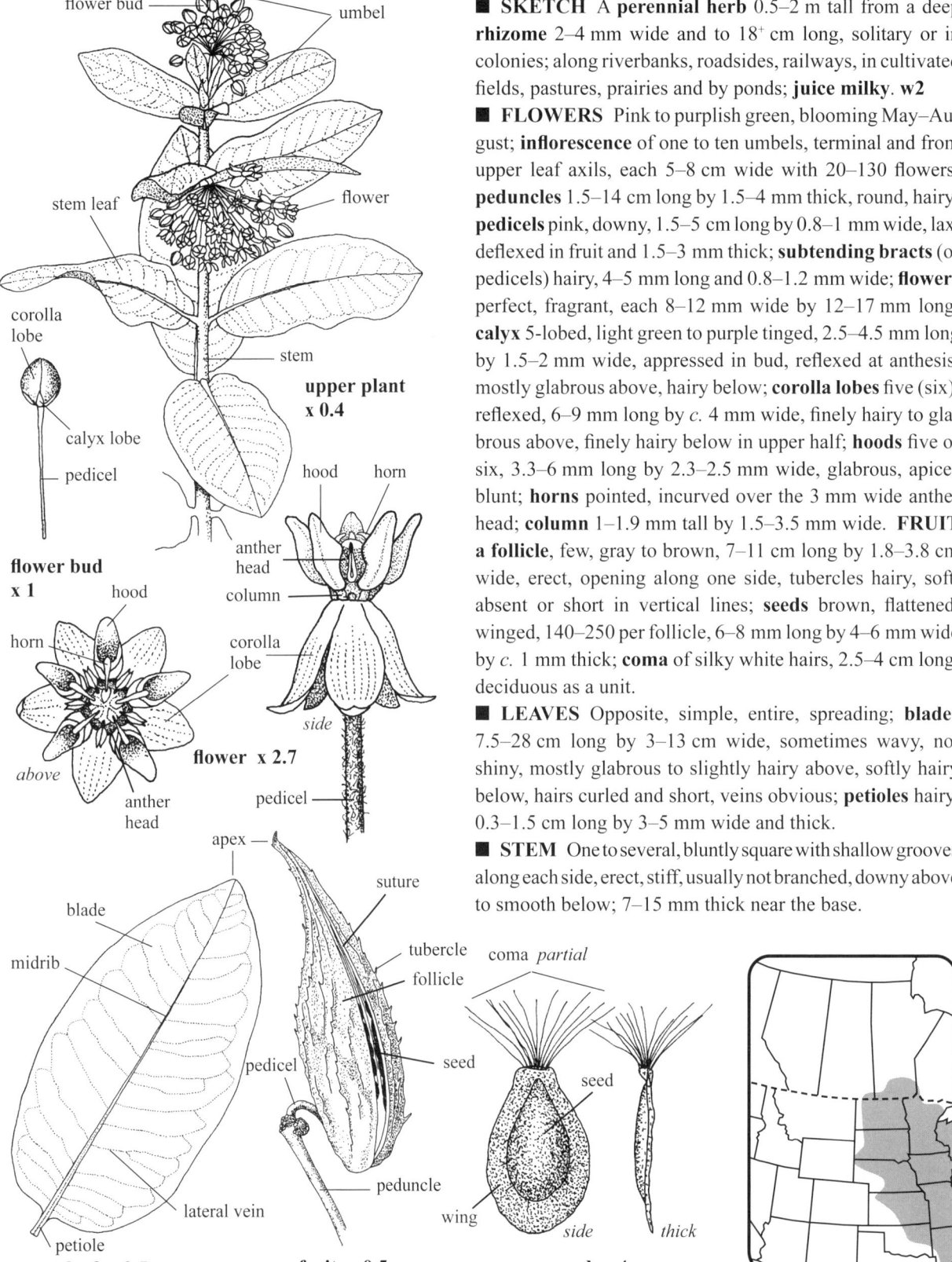

- **SKETCH** A **perennial herb** 0.5–2 m tall from a deep **rhizome** 2–4 mm wide and to 18+ cm long, solitary or in colonies; along riverbanks, roadsides, railways, in cultivated fields, pastures, prairies and by ponds; **juice milky**. w2
- **FLOWERS** Pink to purplish green, blooming May–August; **inflorescence** of one to ten umbels, terminal and from upper leaf axils, each 5–8 cm wide with 20–130 flowers; **peduncles** 1.5–14 cm long by 1.5–4 mm thick, round, hairy; **pedicels** pink, downy, 1.5–5 cm long by 0.8–1 mm wide, lax, deflexed in fruit and 1.5–3 mm thick; **subtending bracts** (of pedicels) hairy, 4–5 mm long and 0.8–1.2 mm wide; **flowers** perfect, fragrant, each 8–12 mm wide by 12–17 mm long; **calyx** 5-lobed, light green to purple tinged, 2.5–4.5 mm long by 1.5–2 mm wide, appressed in bud, reflexed at anthesis, mostly glabrous above, hairy below; **corolla lobes** five (six), reflexed, 6–9 mm long by *c.* 4 mm wide, finely hairy to glabrous above, finely hairy below in upper half; **hoods** five or six, 3.3–6 mm long by 2.3–2.5 mm wide, glabrous, apices blunt; **horns** pointed, incurved over the 3 mm wide anther head; **column** 1–1.9 mm tall by 1.5–3.5 mm wide. **FRUIT a follicle**, few, gray to brown, 7–11 cm long by 1.8–3.8 cm wide, erect, opening along one side, tubercles hairy, soft, absent or short in vertical lines; **seeds** brown, flattened, winged, 140–250 per follicle, 6–8 mm long by 4–6 mm wide by *c.* 1 mm thick; **coma** of silky white hairs, 2.5–4 cm long, deciduous as a unit.
- **LEAVES** Opposite, simple, entire, spreading; **blades** 7.5–28 cm long by 3–13 cm wide, sometimes wavy, not shiny, mostly glabrous to slightly hairy above, softly hairy below, hairs curled and short, veins obvious; **petioles** hairy, 0.3–1.5 cm long by 3–5 mm wide and thick.
- **STEM** One to several, bluntly square with shallow grooves along each side, erect, stiff, usually not branched, downy above to smooth below; 7–15 mm thick near the base.

96 Common Milkweed

Milkweed—*Asclepiadaceae* **Wildflower**   Greenish white   Corolla 5-lobed

WHORLED MILKWEED   *Asclepias verticillata* L.

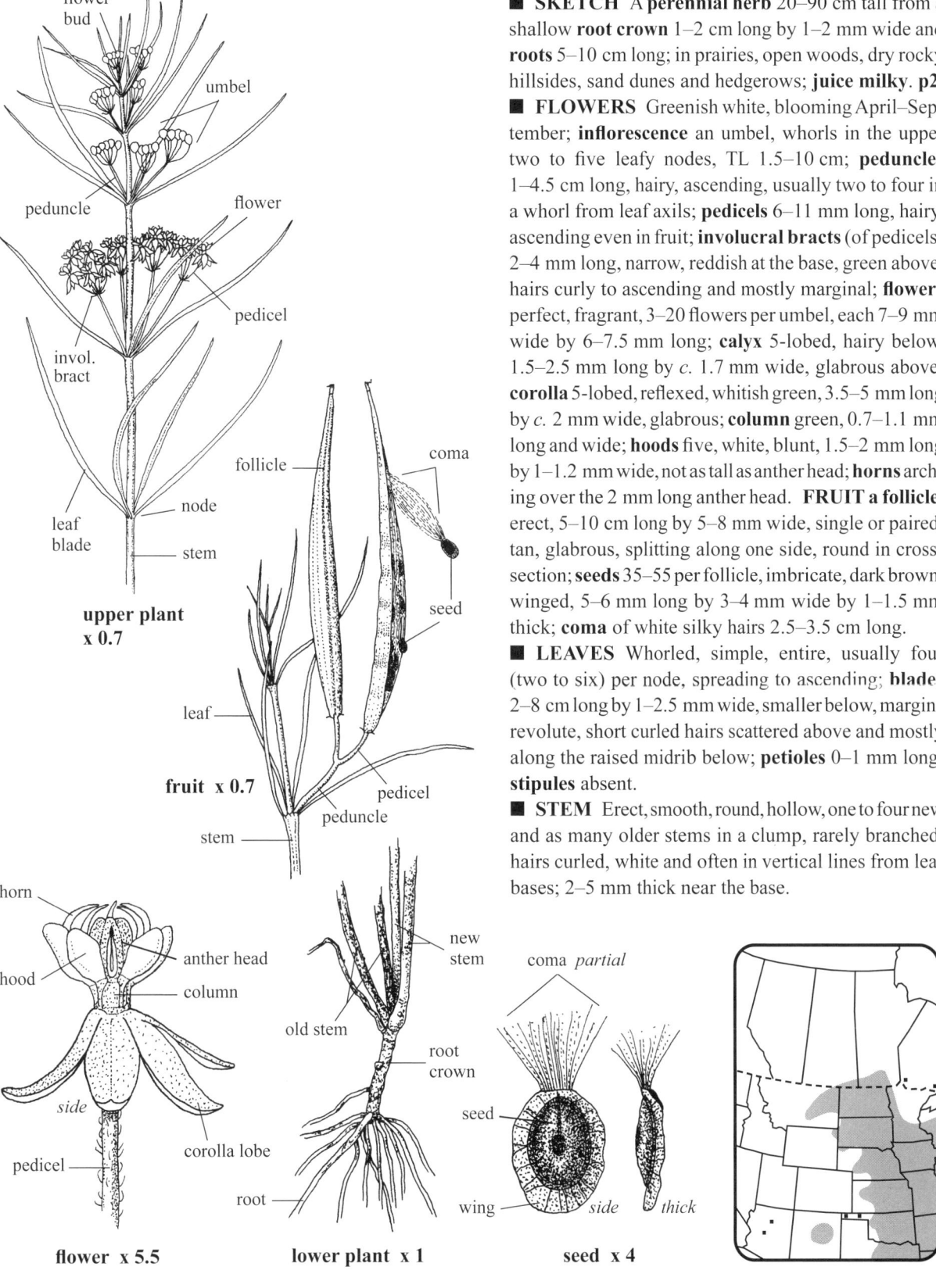

■ **SKETCH** A **perennial herb** 20–90 cm tall from a shallow **root crown** 1–2 cm long by 1–2 mm wide and **roots** 5–10 cm long; in prairies, open woods, dry rocky hillsides, sand dunes and hedgerows; **juice milky**. p2

■ **FLOWERS** Greenish white, blooming April–September; **inflorescence** an umbel, whorls in the upper two to five leafy nodes, TL 1.5–10 cm; **peduncles** 1–4.5 cm long, hairy, ascending, usually two to four in a whorl from leaf axils; **pedicels** 6–11 mm long, hairy, ascending even in fruit; **involucral bracts** (of pedicels) 2–4 mm long, narrow, reddish at the base, green above, hairs curly to ascending and mostly marginal; **flowers** perfect, fragrant, 3–20 flowers per umbel, each 7–9 mm wide by 6–7.5 mm long; **calyx** 5-lobed, hairy below, 1.5–2.5 mm long by *c.* 1.7 mm wide, glabrous above; **corolla** 5-lobed, reflexed, whitish green, 3.5–5 mm long by *c.* 2 mm wide, glabrous; **column** green, 0.7–1.1 mm long and wide; **hoods** five, white, blunt, 1.5–2 mm long by 1–1.2 mm wide, not as tall as anther head; **horns** arching over the 2 mm long anther head. **FRUIT a follicle**, erect, 5–10 cm long by 5–8 mm wide, single or paired, tan, glabrous, splitting along one side, round in cross-section; **seeds** 35–55 per follicle, imbricate, dark brown, winged, 5–6 mm long by 3–4 mm wide by 1–1.5 mm thick; **coma** of white silky hairs 2.5–3.5 cm long.

■ **LEAVES** Whorled, simple, entire, usually four (two to six) per node, spreading to ascending; **blades** 2–8 cm long by 1–2.5 mm wide, smaller below, margins revolute, short curled hairs scattered above and mostly along the raised midrib below; **petioles** 0–1 mm long; **stipules** absent.

■ **STEM** Erect, smooth, round, hollow, one to four new and as many older stems in a clump, rarely branched, hairs curled, white and often in vertical lines from leaf bases; 2–5 mm thick near the base.

Family Characteristics

Aster—Asteraceae — Flower heads of 3 types

(two types on this page)

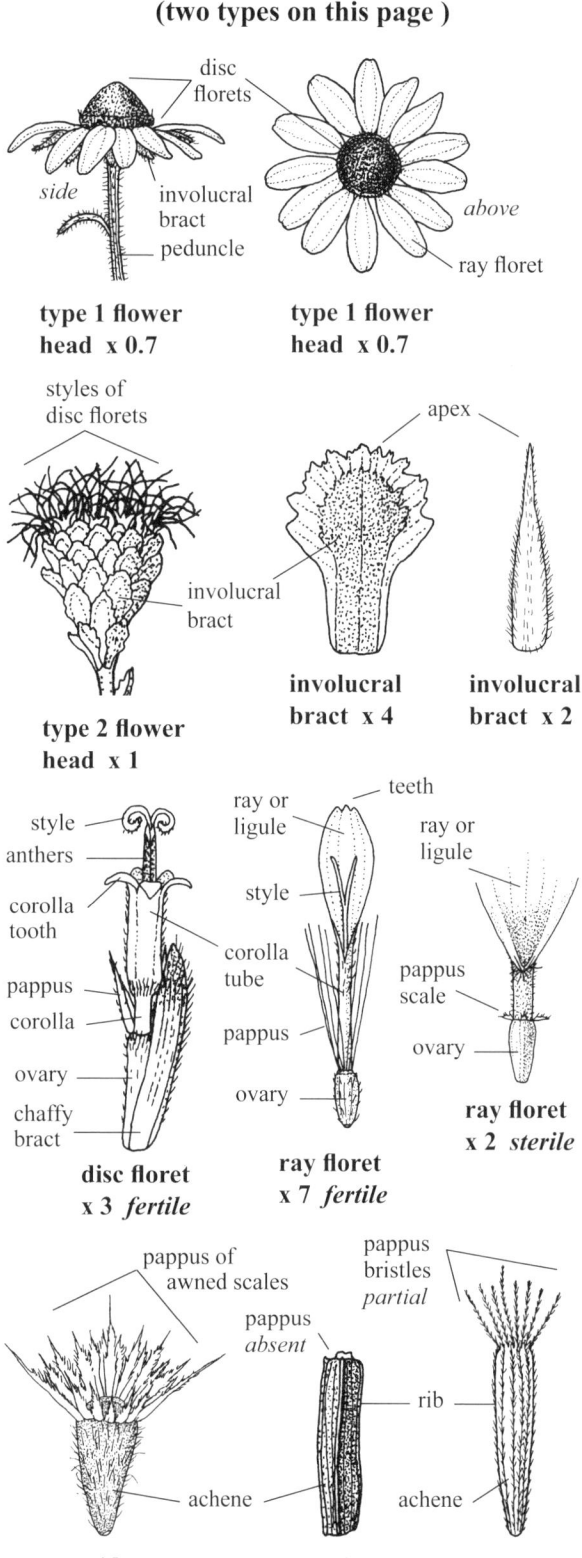

SKETCH Dicotyledonous herbs (wildflowers), annual, biennial, or perennial. About 23,000 species worldwide; adapted for cooler temperate regions.

FLOWER HEADS Composed of sessile florets (2–1,000 small flowers) on a common receptacle with involucral bracts below. **Inflorescence** of various colors, (white and yellow are common) with one to many heads (variously arranged) per plant. **Heads** of three types (two on this page): **1)** with both central disc florets and outer ray florets (Sunflower, *Helianthus*); **2)** with disc florets only, the outer ray florets missing (Meadow Blazingstar, *Lygodesmia*).

Disc florets have a regular, tubular corolla usually with 5 (4) small teeth or lobes, normally perfect and fertile. Some genera have a **chaffy bract** for each ovary and fruit. **Stamens** mostly 5, inserted on corolla tube; filaments usually with upper portion free; anthers united into a tube through which the style/stigmas are exserted.

Ray florets with a corolla tube at their base and then often expanded into a large flat ray or ligule with 3–5 teeth, usually female (ovary and style present, stamens absent), or else sterile with stamens and style absent.

For both florets, **pistil** 1, of 2 united carpels. **Ovary** of 1 locule and 1 ovule, erect, forming a single seed when mature. **Style** 2-branched, these stigmatic. **Pappus** (calyx) atop the fruit (achene), modified into hairy to plumose bristles, or awns, scales, a low crown, or rarely absent; persistent, or deciduous as a unit or separately.

FRUIT An achene, 1-seeded, usually with a pappus, glabrous or hairy, often brown and ribbed.

LEAVES Mostly alternate, some opposite, sometimes opposite and alternate on same plant, rarely whorled; simple to dissected, rarely compound; entire to variously toothed and/or lobed. **Stipules** absent.

STEM Without milky juice, usually erect and solid.

Family Characteristics

Aster—Asteraceae — Flower heads of 3 types

(third type)

FLOWER HEADS Composed of sessile florets (small flowers) on a common receptacle with involucral bracts below. **Inflorescence** of various colors (mostly yellow) with one to many heads per plant, variously arranged. **Heads** of type three ligulate or ray, composed entirely of ligulate florets (Dandelion, *Taraxacum*) with disc florets absent. The ligulate florets, several to many, are perfect and fertile.

Ligulate florets have a short corolla tube above the ovary which then expands on one side to form the wide ligule. Three to five apical teeth are usually present. **Stamens** 5, inserted on corolla tube; filaments usually free above; anthers united into a tube through which the style/stigmas are exserted. **Pistil** 1, of 2 united carpels. **Ovary** 1-loculed with 1 ovule, erect, maturing into a fruit. **Style** 2-branched. **Pappus** (modified calyx) atop the fruit (achene) consists of hairy to plumose bristles, or awns, scales, a low crown, or rarely absent; persistent, or deciduous as a unit or individual parts.

FRUIT An achene, 1-seeded, beaked or not, usually with a pappus, glabrous to hairy, mostly brown and with ribs.

LEAVES Variable, often basal, or basal and stem, alternate, simple to dissected, rarely compound, entire to variously toothed and/or lobed. **Stipules** absent.

STEM A scape, usually erect, with milky juice, often hollow, smooth or with ridges.

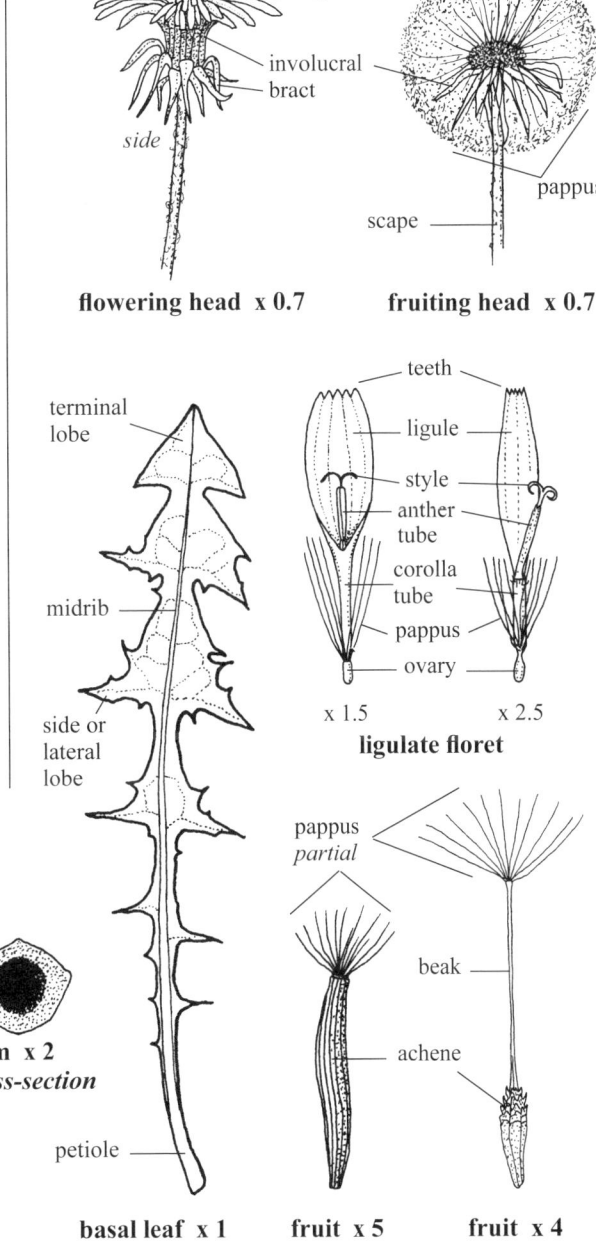
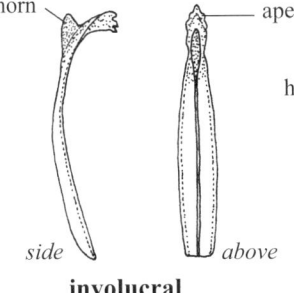
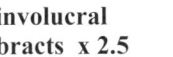

Aster—*Asteraceae* **Wildflower** White to pinkish Ray florets 5

YARROW *Achillea millefolium* L.

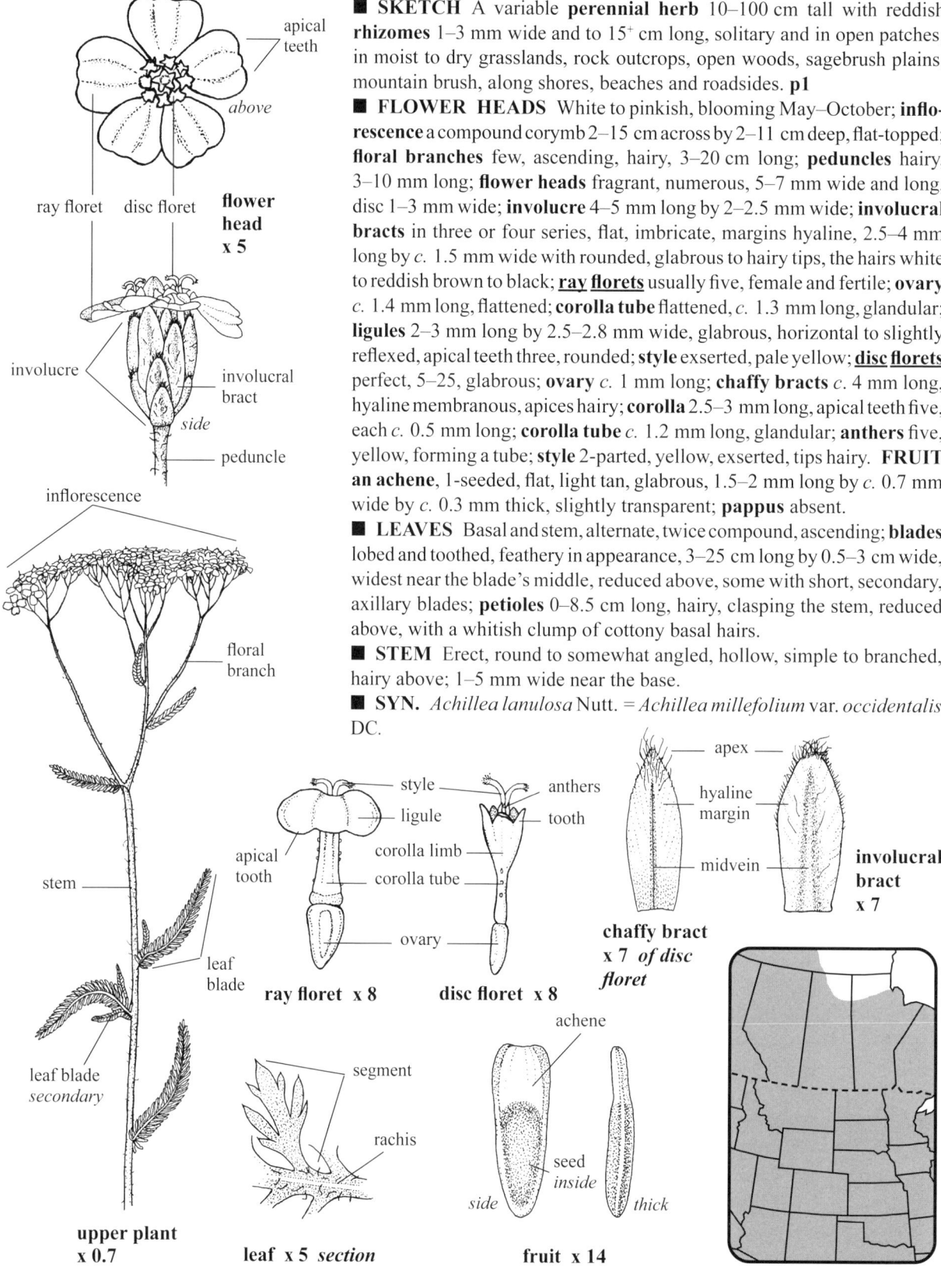

■ **SKETCH** A variable **perennial herb** 10–100 cm tall with reddish **rhizomes** 1–3 mm wide and to 15⁺ cm long, solitary and in open patches; in moist to dry grasslands, rock outcrops, open woods, sagebrush plains, mountain brush, along shores, beaches and roadsides. **p1**

■ **FLOWER HEADS** White to pinkish, blooming May–October; **inflorescence** a compound corymb 2–15 cm across by 2–11 cm deep, flat-topped; **floral branches** few, ascending, hairy, 3–20 cm long; **peduncles** hairy, 3–10 mm long; **flower heads** fragrant, numerous, 5–7 mm wide and long, disc 1–3 mm wide; **involucre** 4–5 mm long by 2–2.5 mm wide; **involucral bracts** in three or four series, flat, imbricate, margins hyaline, 2.5–4 mm long by *c.* 1.5 mm wide with rounded, glabrous to hairy tips, the hairs white to reddish brown to black; **ray florets** usually five, female and fertile; **ovary** *c.* 1.4 mm long, flattened; **corolla tube** flattened, *c.* 1.3 mm long, glandular; **ligules** 2–3 mm long by 2.5–2.8 mm wide, glabrous, horizontal to slightly reflexed, apical teeth three, rounded; **style** exserted, pale yellow; **disc florets** perfect, 5–25, glabrous; **ovary** *c.* 1 mm long; **chaffy bracts** *c.* 4 mm long, hyaline membranous, apices hairy; **corolla** 2.5–3 mm long, apical teeth five, each *c.* 0.5 mm long; **corolla tube** *c.* 1.2 mm long, glandular; **anthers** five, yellow, forming a tube; **style** 2-parted, yellow, exserted, tips hairy. **FRUIT** an achene, 1-seeded, flat, light tan, glabrous, 1.5–2 mm long by *c.* 0.7 mm wide by *c.* 0.3 mm thick, slightly transparent; **pappus** absent.

■ **LEAVES** Basal and stem, alternate, twice compound, ascending; **blades** lobed and toothed, feathery in appearance, 3–25 cm long by 0.5–3 cm wide, widest near the blade's middle, reduced above, some with short, secondary, axillary blades; **petioles** 0–8.5 cm long, hairy, clasping the stem, reduced above, with a whitish clump of cottony basal hairs.

■ **STEM** Erect, round to somewhat angled, hollow, simple to branched, hairy above; 1–5 mm wide near the base.

■ **SYN.** *Achillea lanulosa* Nutt. = *Achillea millefolium* var. *occidentalis* DC.

Milfoil, Common Yarrow, Western Yarrow

Aster—*Asteraceae* **Wildflower** White, center pale green Ray florets 7–12

MANY-FLOWERED YARROW *Achillea sibirica* Ledeb.

■ **SKETCH** A **perennial herb** 30–115 cm tall with a dark brown rough **rhizome** 2–2.5 cm long by *c.* 1.5 cm thick with tan **roots** 2–15 cm long by 0.5–1 mm thick; in open woodlands, disturbed sites, in moist to dry thickets and stream banks in forests.

■ **FLOWER HEADS** White, the disc pale green, blooming July–August; **inflorescence** corymbiform, 9–18 cm wide by 5–15 cm tall; **floral branches** ascending, 2–20 cm long; **flower heads** 4–5 mm wide by 5–8 mm tall, numerous, in clusters of 8–19 at ends of branches, hairy; **peduncles** 1–16 mm long, hairy; **involucre** 4–5 mm long by 3–6 mm wide, hairy, pale green; **involucral bracts** in 1 or 2 series, imbricate, hairy, hyaline whitish sides, midrib raised, 2.5–4 mm long by 1–2 mm wide (flattened), apices and margins light brown; **ray florets** 7–12, female, 4–5 mm long, hairless; **ovary** green, flattened, 2–2.3 mm long by *c.* 1 mm wide, shiny, winged; **pappus** absent; **corolla tube** 1–1.5 mm long, flat, greenish, glandular, translucent; **ligules** white, 1–2 mm long by 1.2–1.7 mm wide, 3-lobed, center lobe the shortest, 2-grooved; **style** one, 2-parted, pale green, *c.* 1.5 mm long; **chaffy bracts** one per ray floret, mostly hyaline; **disc florets** perfect, 25–35 per head, TL *c.* 5 mm, glabrous, glandular; **ovary** *c.* 2 mm long by *c.* 1 mm wide, flat, winged, translucent; **pappus** absent; **corolla tube** flattened, *c.* 1.3 mm long, glandular at apex; **corolla limb** white, glandular, *c.* 1 mm long; **corolla teeth** five, each 0.2–0.3 mm long; **anthers** yellow, five, united into a tube *c.* 1 mm long; **style** *c.* 1.5 mm long, 2-parted, exserted through anthers; **chaffy bracts** one per floret, each *c.* 4 mm long by *c.* 1.3 mm wide (flattened), margins brownish hyaline, apex hairy, midrib green and raised; **fruiting heads** brown, 4–5 mm wide by 5–7 mm long, involucral bracts erect. **FRUIT an achene**, 1-seeded, light tan, hairless, dull, 2.3–3 mm long by 0.9–1.1 mm wide by *c.* 0.3 mm thick, blunt at ends, transparent; **seeds** dark brown, shiny, pointed at base, 1.8–2 mm long by 0.4–0.5 mm wide by 0.2–0.3 mm thick.

■ **LEAVES** Alternate, stem only, simple, sessile, dull, ascending to arched, often with 1–4 smaller leaves from each leaf axil; **blades** 2.5–12 cm long by 4–14 mm wide, reduced above, the lobes toothed, lobes and apices with a clear bristle 0.1–0.3 mm long, more hairy below, glandular-punctate on both sides, midrib obvious, raised and hairy below, side veins obscure.

■ **STEM** Green, erect, stiff, ridged, hairy below; 3–6 mm thick near the base.

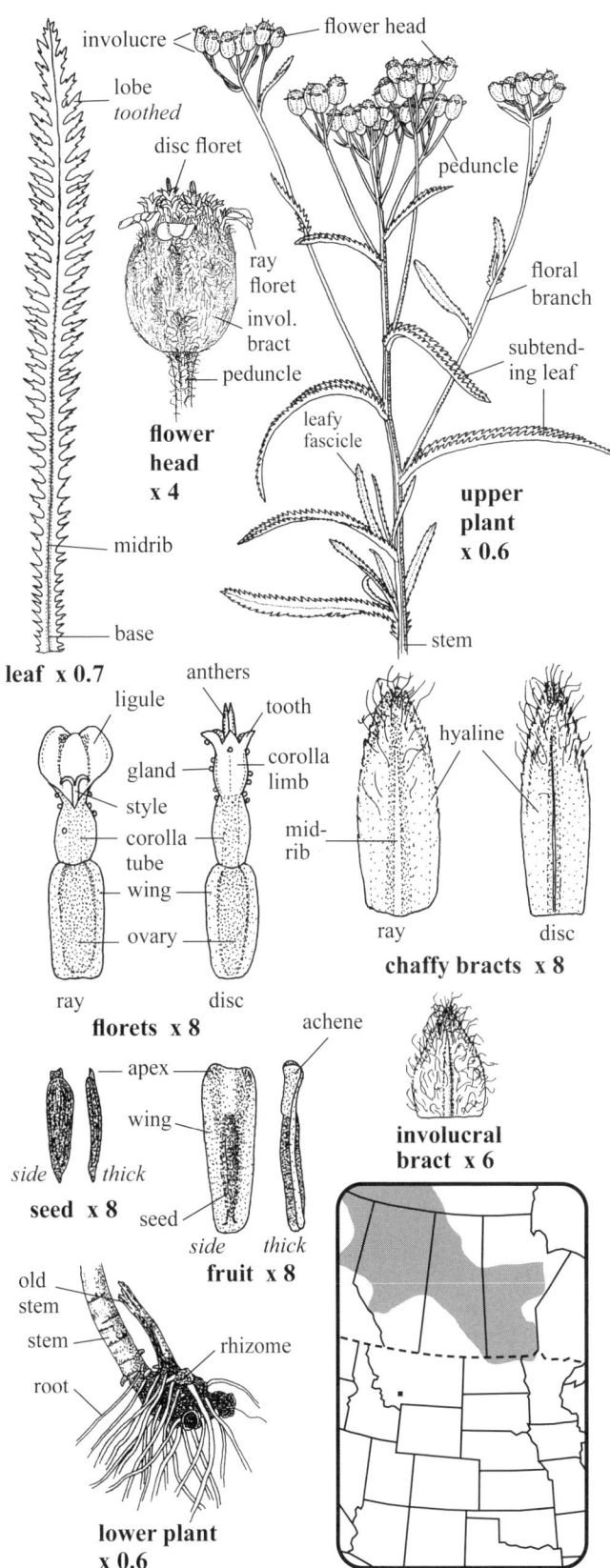

Siberian Yarrow

Aster—*Asteraceae* **Wildflower** Yellow to orange Ligulate florets 23–50

FALSE DANDELION *Agoseris glauca* (Pursh) Raf.

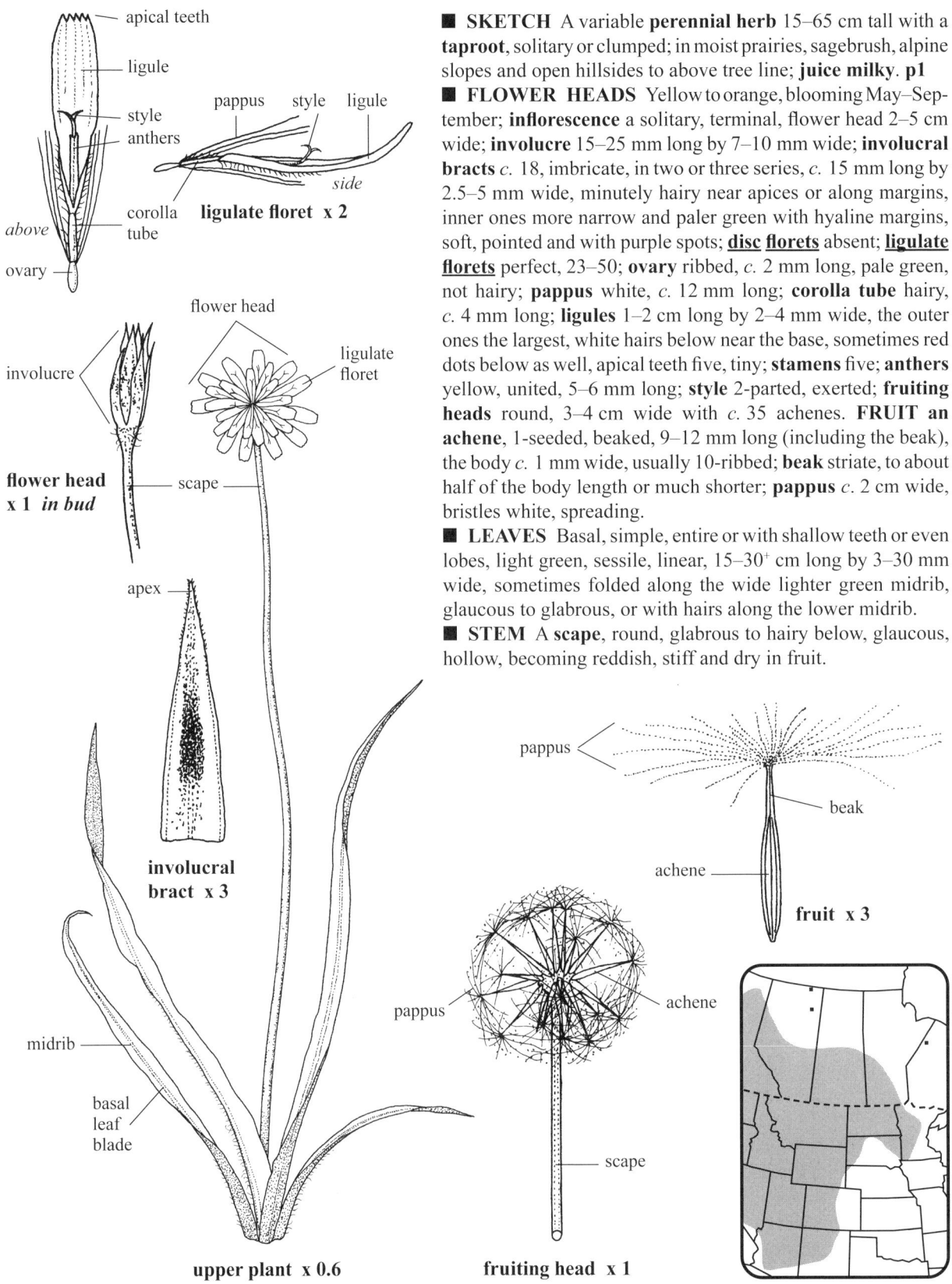

■ **SKETCH** A variable **perennial herb** 15–65 cm tall with a **taproot**, solitary or clumped; in moist prairies, sagebrush, alpine slopes and open hillsides to above tree line; **juice milky**. p1
■ **FLOWER HEADS** Yellow to orange, blooming May–September; **inflorescence** a solitary, terminal, flower head 2–5 cm wide; **involucre** 15–25 mm long by 7–10 mm wide; **involucral bracts** *c.* 18, imbricate, in two or three series, *c.* 15 mm long by 2.5–5 mm wide, minutely hairy near apices or along margins, inner ones more narrow and paler green with hyaline margins, soft, pointed and with purple spots; **disc florets** absent; **ligulate florets** perfect, 23–50; **ovary** ribbed, *c.* 2 mm long, pale green, not hairy; **pappus** white, *c.* 12 mm long; **corolla tube** hairy, *c.* 4 mm long; **ligules** 1–2 cm long by 2–4 mm wide, the outer ones the largest, white hairs below near the base, sometimes red dots below as well, apical teeth five, tiny; **stamens** five; **anthers** yellow, united, 5–6 mm long; **style** 2-parted, exerted; **fruiting heads** round, 3–4 cm wide with *c.* 35 achenes. FRUIT an **achene**, 1-seeded, beaked, 9–12 mm long (including the beak), the body *c.* 1 mm wide, usually 10-ribbed; **beak** striate, to about half of the body length or much shorter; **pappus** *c.* 2 cm wide, bristles white, spreading.
■ **LEAVES** Basal, simple, entire or with shallow teeth or even lobes, light green, sessile, linear, 15–30⁺ cm long by 3–30 mm wide, sometimes folded along the wide lighter green midrib, glaucous to glabrous, or with hairs along the lower midrib.
■ **STEM** A **scape**, round, glabrous to hairy below, glaucous, hollow, becoming reddish, stiff and dry in fruit.

Large-flowered False Dandelion, Pale Agoseris, Pale Goat-chicory

Aster—*Asteraceae* **Wildflower** Green Disc florets only

Common Ragweed *Ambrosia artemisiifolia* L.

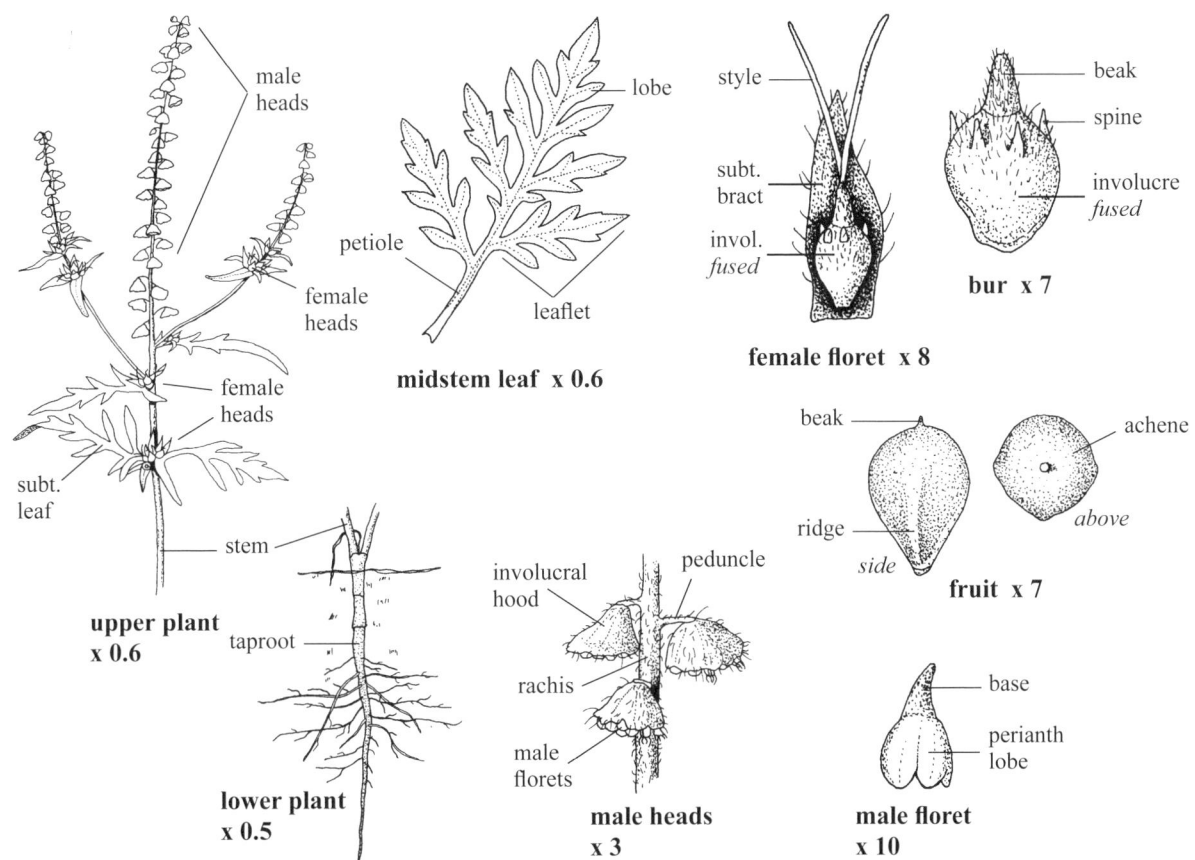

■ **SKETCH** An **annual herb** 30–110⁺ cm tall from a **taproot** 1–5 mm thick by 10–15⁺ cm long; in moist sites along roadsides and in prairies; **monoecious**. w1
■ **FLOWER HEADS** Green, blooming July–October; **inflorescence** of numerous unisexual heads, male above the female; **floral branches** alternate, 5–25⁺ cm long, subtended by a leaf (bract); **peduncles** 0–2 mm long; **male heads** in a raceme 2–15 cm long by *c.* 7 mm wide; **rachis** green, slightly hairy, erect, *c.* 1 mm thick; **involucral bracts** forming a cuplike circular hood 2–4 mm wide by 2–2.5 mm high, nodding, margins crenate, glabrous or white hairs spreading, not nerved; **male florets** usually 13–20 per (head) hood, 1.5–2 mm long by *c.* 1 mm wide; **perianth** 5-lobed, united below, lobes glabrous, pale yellow, membranous, with dark green margins, base tapered and pale green; **stamens** five; **anthers** yellow, *c.* 0.8 mm long, tip deflexed then erect when yellow pollen is released; **pistil** vestigial, *c.* 0.7 mm long; **female heads** in clusters of 2–15 florets in upper leaf axils; **subtending bracts** green, of various lengths; **female florets** 1-flowered, erect, *c.* 1 mm wide (not including style) by *c.* 2 mm long, body reddish green; **involucral bracts** fused, with pointed spines around the hairy upper body and *c.* 0.6 mm long hairy beak; **style** 2-parted, *c.* 2.7 mm long, pale yellowish green; **bur** dark brown, 3–4 mm long, with several hard, pointed spines 0.2–0.8 mm long around the base of the 1–1.8 mm long hairy beak. **FRUIT** an **achene**, 1-seeded, 2.5–3 mm long by 1.7–2 mm wide and thick, ridged at base, otherwise smooth; **beak** 0.1–0.2 mm long.
■ **LEAVES** Opposite below, alternate above, compound; **blades** 2–10 cm long by 1.5–7 cm wide, lobed and toothed, less divided above, hairy, especially below; **petioles** 0–4 cm long, reduced above, with marginal hairs.
■ **STEM** Stout, erect, stiff, hollow, branched, reddish green, scabrous, hairs scattered and minute below, longer hairs above especially on floral branches; 1–6 mm thick near the almost woody base.
■ **SYN**. *Ambrosia elatior* L. = *Ambrosia artemisiifolia* var. *elatior* (L.) Descourtils.

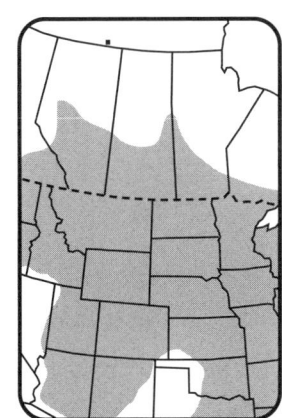

Annual Ragweed, Short Ragweed

Aster—*Asteraceae* **Wildflower** Green Disc florets only

PERENNIAL RAGWEED *Ambrosia psilostachya* DC.

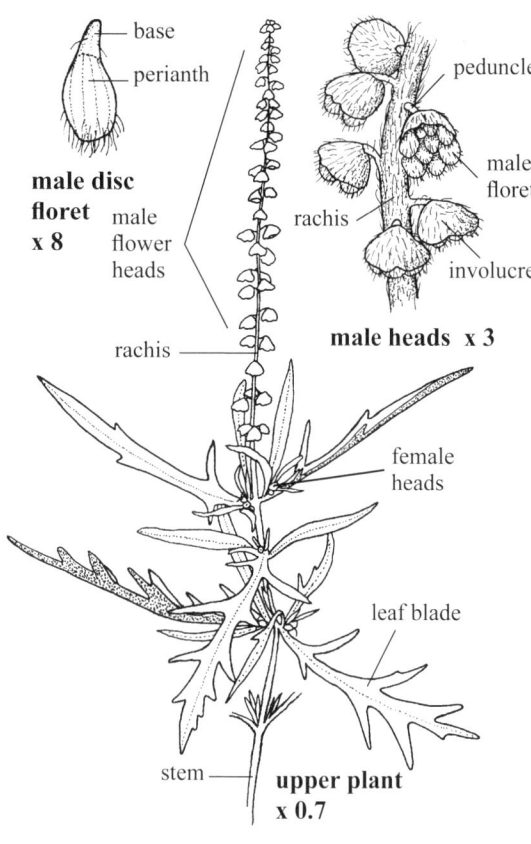

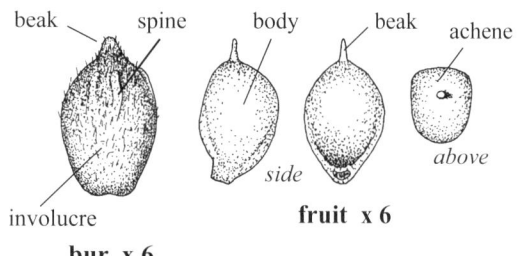

■ **SKETCH** An erect, hairy **perennial herb** 30–100⁺ cm tall with a deep, horizontal **rhizome** 1–3 mm thick; often in extensive colonies along banks of streams, meadows, in dry prairies, desert sagebrush, mountain brush and disturbed sites; **monoecious**. w1

■ **FLOWER HEADS** Green, blooming July–October; **inflorescence** of unisexual flower heads; **floral branches** one to several pairs, opposite, leafy, ascending, 2–40 cm long; **peduncles** hairy, *c.* 1 mm long; **male heads** in racemes 2–11 cm long by 5–11 mm wide, terminating a central stalk and floral branches; **involucral bracts** fused, nodding, hairy, 2.5–3 mm wide with irregular margins, smallest near the top; **male florets** 15–30 per head; **perianth** 5-lobed, transparent, *c.* 1.5 mm long by *c.* 1 mm wide with a hairy apex; **anthers** yellow, *c.* 1 mm long; **female heads** green, sessile, in axillary clusters of one to six florets, below the male flowers; **female florets** 1-flowered, sessile; **corolla** and **pappus** absent; **involucral bracts** fused, the body 1.3–1.5 mm wide by 2–2.5 mm long, covered with ascending hairs, becoming burlike, erect, beak constricted, *c.* 1 mm long; **style** 2-parted, each half *c.* 2 mm long, glabrous and squarish, recurved; **bur** indehiscent, 3–4 mm long by 2–2.5 mm wide and thick, with the tops of 0–4 bracts visible as blunt spines or tubercles to *c.* 0.5 mm long, slightly hairy, falling entire; **beak** thick, 0.5–0.8 mm long. FRUIT an achene, 1-seeded, shiny, dark brown, glabrous, 2.8–3.8 mm long by 1.5–2.2 mm wide and thick, sides blunt; **beak** *c.* 0.5 mm long.

■ **LEAVES** Opposite below, with one or two alternate leaves near the base of the male inflorescence, compound; **blades** hairy on both sides, 2.4–10 cm long by 1–5 cm wide, the lobes entire or toothed; **petioles** 0–2 cm long.

■ **STEM** Erect, branched, angular, hairs ascending; 1–6 mm thick near the base.

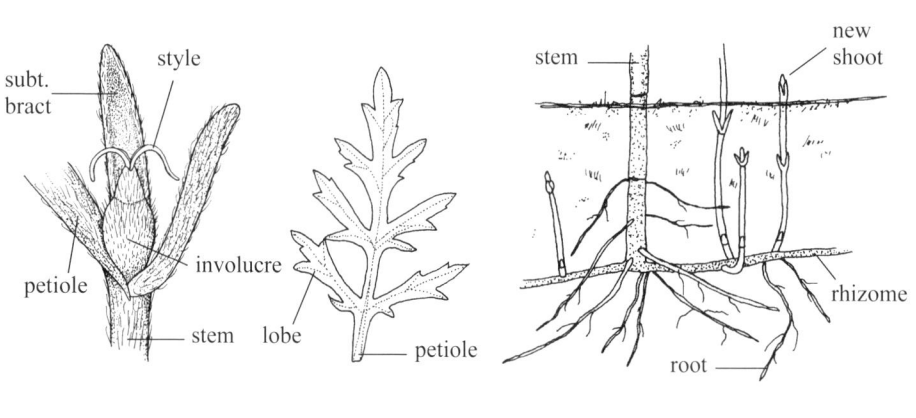

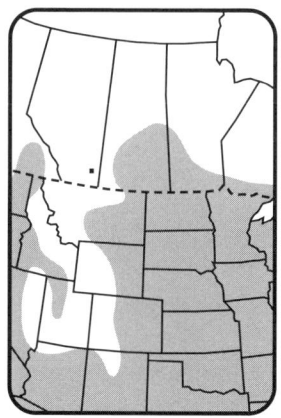

104   Western Ragweed, Cuman Ragweed

Aster—*Asteraceae* **Wildflower** Green Disc florets only

GREAT RAGWEED *Ambrosia trifida* L.

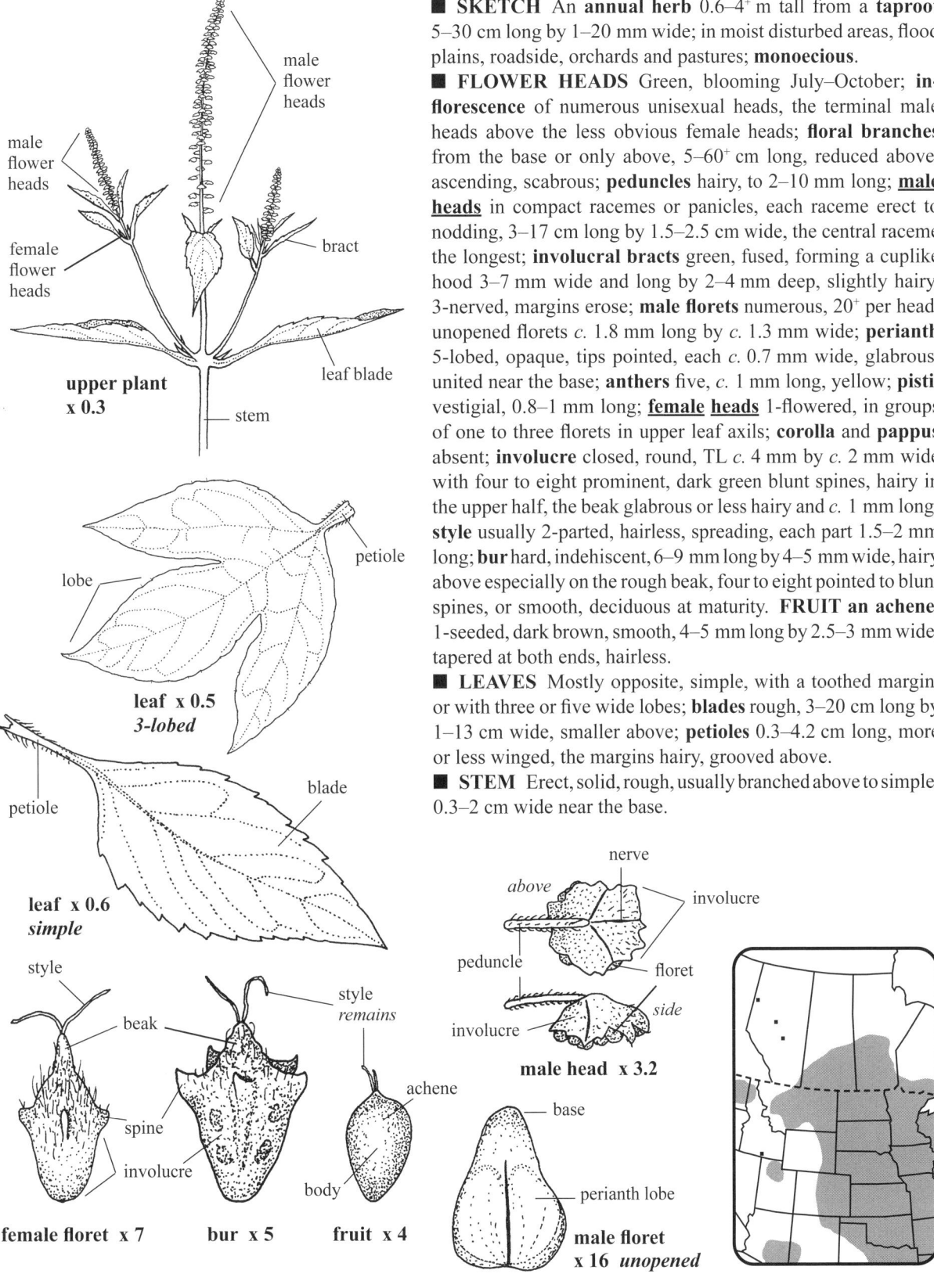

■ **SKETCH** An **annual herb** 0.6–4⁺ m tall from a **taproot** 5–30 cm long by 1–20 mm wide; in moist disturbed areas, flood plains, roadside, orchards and pastures; **monoecious**.

■ **FLOWER HEADS** Green, blooming July–October; **inflorescence** of numerous unisexual heads, the terminal male heads above the less obvious female heads; **floral branches** from the base or only above, 5–60⁺ cm long, reduced above, ascending, scabrous; **peduncles** hairy, to 2–10 mm long; **male heads** in compact racemes or panicles, each raceme erect to nodding, 3–17 cm long by 1.5–2.5 cm wide, the central raceme the longest; **involucral bracts** green, fused, forming a cuplike hood 3–7 mm wide and long by 2–4 mm deep, slightly hairy, 3-nerved, margins erose; **male florets** numerous, 20⁺ per head, unopened florets *c.* 1.8 mm long by *c.* 1.3 mm wide; **perianth** 5-lobed, opaque, tips pointed, each *c.* 0.7 mm wide, glabrous, united near the base; **anthers** five, *c.* 1 mm long, yellow; **pistil** vestigial, 0.8–1 mm long; **female heads** 1-flowered, in groups of one to three florets in upper leaf axils; **corolla** and **pappus** absent; **involucre** closed, round, TL *c.* 4 mm by *c.* 2 mm wide with four to eight prominent, dark green blunt spines, hairy in the upper half, the beak glabrous or less hairy and *c.* 1 mm long; **style** usually 2-parted, hairless, spreading, each part 1.5–2 mm long; **bur** hard, indehiscent, 6–9 mm long by 4–5 mm wide, hairy above especially on the rough beak, four to eight pointed to blunt spines, or smooth, deciduous at maturity. **FRUIT an achene**, 1-seeded, dark brown, smooth, 4–5 mm long by 2.5–3 mm wide, tapered at both ends, hairless.

■ **LEAVES** Mostly opposite, simple, with a toothed margin, or with three or five wide lobes; **blades** rough, 3–20 cm long by 1–13 cm wide, smaller above; **petioles** 0.3–4.2 cm long, more or less winged, the margins hairy, grooved above.

■ **STEM** Erect, solid, rough, usually branched above to simple; 0.3–2 cm wide near the base.

Giant Ragweed   105

Aster—*Asteraceae* **Wildflower** White, center pale yellow Disc florets only

Pearly Everlasting *Anaphalis margaritacea* (L.) Benth.

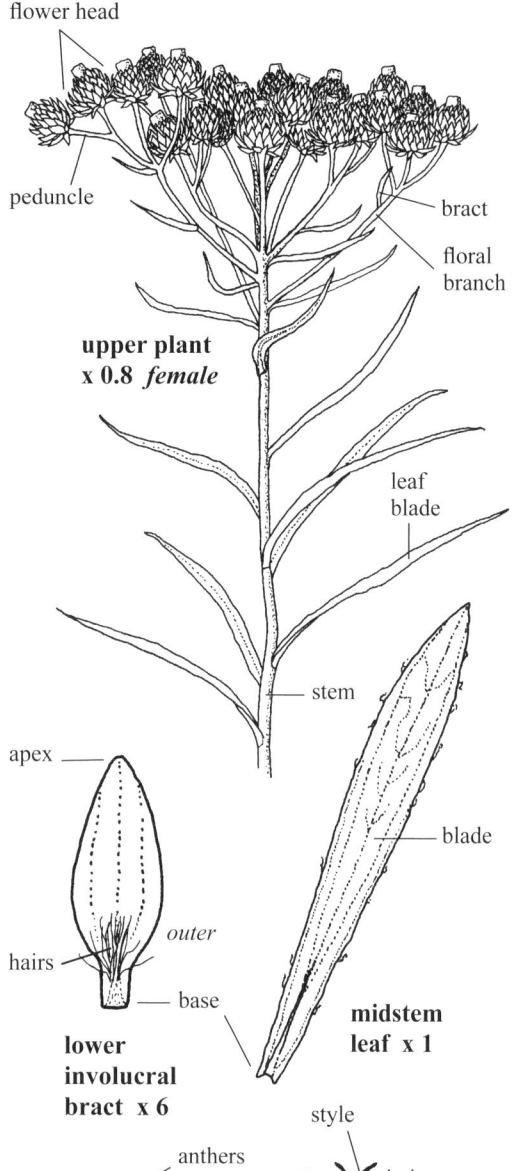

■ **SKETCH** A variable **perennial herb** 30–80 cm tall, solitary or clumped, with **rhizomes**; in open woodlands, dry meadows, along roadsides, valleys and streamsides; mostly **dioecious**.

■ **FLOWER HEADS** White, center pale yellow, blooming July–September; **inflorescence** a cymose panicle, generally flat-topped, 2–15 cm wide by 1.5–5 cm tall with 10–80 heads; **peduncles** 5–15 mm long, hairy; **flower heads** mostly unisexual, (some female heads with a few central male flowers), 8–10 mm wide by 7–9 mm tall; **involucral bracts** (of either sex), white, papery, dry, in several series, 5–5.5 mm long by 1.5–2 mm wide, spreading as fruit is released, cottony hairs on basal outer surface, inner side glabrous, base hyaline with the central area green or sometimes a dark spot near the base; **ray florets** absent; male heads with a central yellow tuft of disc florets 2–2.5 mm wide by 1.5–2 mm tall; **disc florets** *c.* 5 mm long; **ovary** vestigial; **pappus** of *c.* 12 white bristles *c.* 4 mm long; **corolla** tubular, *c.* 3 mm long by 0.2–0.3 mm wide; **upper limb** *c.* 1 mm long and yellow; **teeth** five, spreading, *c.* 0.6 mm long; **anthers** five, yellow, forming a tube 1.6–1.8 mm long; **style** 2-parted, yellow, the parts *c.* 0.3 mm long, included in the anthers; female heads with a central pale yellow tuft of disc florets 3–3.5 mm wide by 1.5–2 mm tall; **disc florets** TL *c.* 5.5 mm; **ovary** *c.* 0.7 mm long; **pappus** white, bristles *c.* 20 and *c.* 4.5 mm long; **corolla tube** *c.* 4 mm long by *c.* 0.2 mm wide; **teeth** three, filiform, *c.* 0.5 mm long; **style** 2-parted, tan, eventually exserted. **FRUIT an achene**, 1-seeded, dark brown, shiny, 0.6–1 mm long by 0.2–0.3 mm wide and thick, papillate; **pappus** of *c.* 20 white bristles 4.5–5 mm long, spreading; **receptacle** hairless, *c.* 4 mm wide, convex.

■ **LEAVES** Alternate, simple, entire, sessile; **midstem leaves** the largest, spreading; **blades**, 2.5–12 cm long by 2–20 mm wide, 3-nerved, reduced above, soft, margins revolute, tomentose below, less hairy and green above.

■ **STEM** Erect, whitish green, mostly white-woolly throughout, simple or branched, round, solid; 3–5 mm thick near the base.

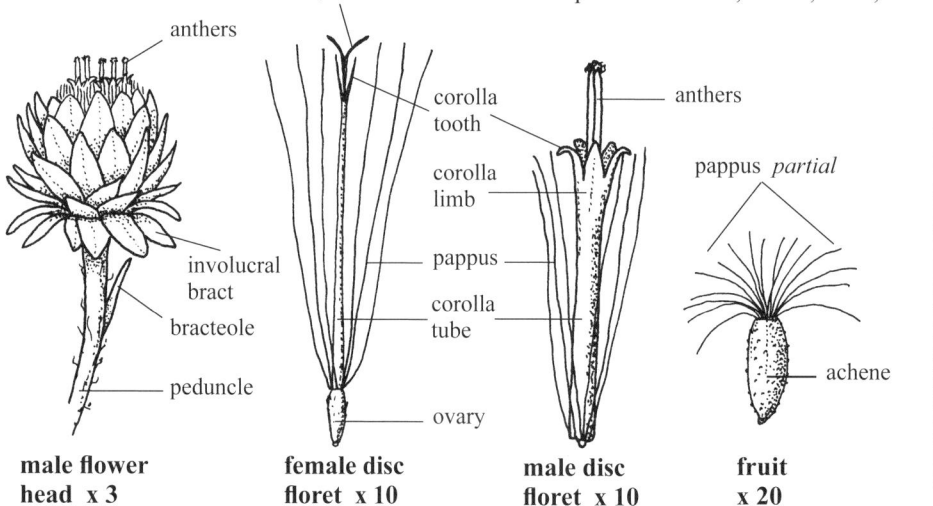

Aster—*Asteraceae*   **Wildflower**   White to pink   Disc florets only

SMALL-LEAF PUSSYTOES   *Antennaria microphylla* Rydb.

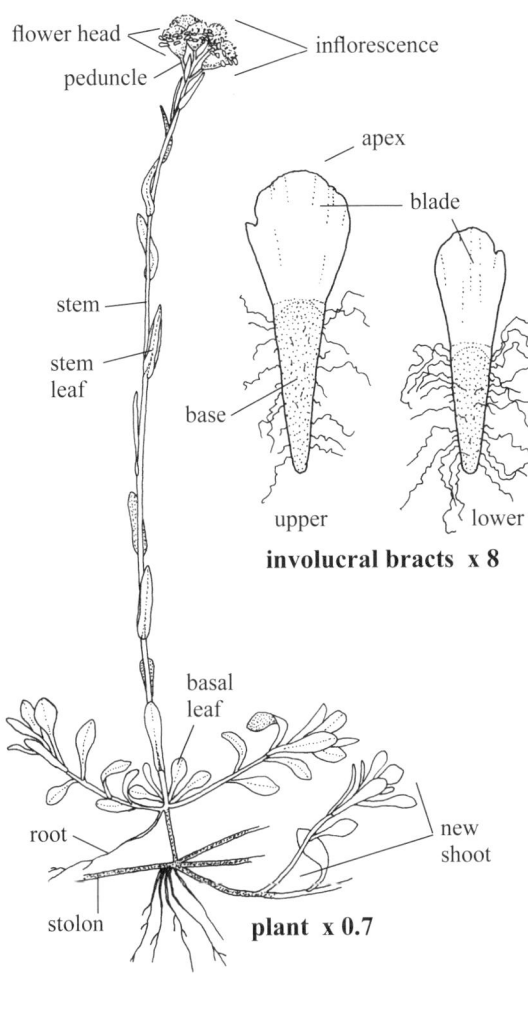

■ **SKETCH** A very hairy, mat-forming, variable **perennial herb** 5–30 cm tall from **stolons** *c.* 1 mm thick and **fibrous roots** at nodes; in prairies, open woodlands, moist to dry hillsides and roadsides to alpine meadows; **dioecious**. p1

■ **FLOWER HEADS** White to pink, blooming May–July; **inflorescence** a cyme, unisexual, compact to loose, of 3–12+ heads 10–20 mm wide by 8–12 mm tall (including peduncles); **peduncles** 3–6 mm long, hairy; **subtending bracts** (of peduncles) hairy, green, usually longer than the peduncles, often absent in upper peduncles; **flower heads** (male) 5–7 mm wide and tall, erect; **involucre** 4–7 mm long; **involucral bracts** imbricate in 3 or 4 series, white to pink, base narrow and green, 2–3 mm long with long marginal hairs, upper half white, petal-like, spreading, 1.5–2 mm long by 1–2 mm wide, margins slightly erose; ray florets absent; disc florets (male) *c.* 4.5 mm long (to top of exserted anthers); **ovary** green, vestigial, *c.* 0.5 mm long; **pappus** of 15–20 white, 2.5–3 mm long fine hairy bristles with a white expanded tip; **corolla** pale green, tubular, *c.* 3 mm long, glabrous, slightly widened above, corolla teeth five, recurved; **stamens** five; **anthers** *c.* 1 mm long, yellow, exserted, tailed, united around the stigma and style; **style** filiform, *c.* 2 mm long, eventually exserted, yellowish; **stigma** flat, slightly wider than the style, green, one in male florets, two in females. **FRUIT an achene**, 1-seeded, 0.8–1.2 mm long, olivaceous to brownish, smooth to slightly papillate.

■ **LEAVES** Basal and stem, alternate, simple, entire, hairy on both sides, ascending to erect, margins often revolute; **basal blades** one-nerved, 10–25 mm long by 3–7 mm wide, pointed to blunt; **petioles** 3–10 mm long, tapered, hairy; **stem blades** sessile, 10–20 mm long by 2–4 mm wide, reduced above, usually 7–15 per stem.

■ **STEM** Erect, weak, hairy, whitish green, some hairs glandular above; 1–2 mm thick near the base.

■ **SYN.** *Antennaria nitida* Greene, *Antennaria parvifolia* sensu Greene, non Nutt. and *Antennaria rosea* var. *nitida* (Greene) Breitung.

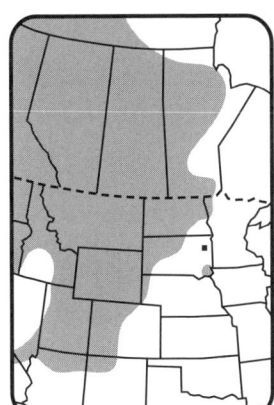

Neat Pussytoes, Small-leaved Everlasting, Littleleaf Pussytoes

Aster—*Asteraceae* **Wildflower** White Disc florets only

FIELD PUSSYTOES *Antennaria neglecta* Greene

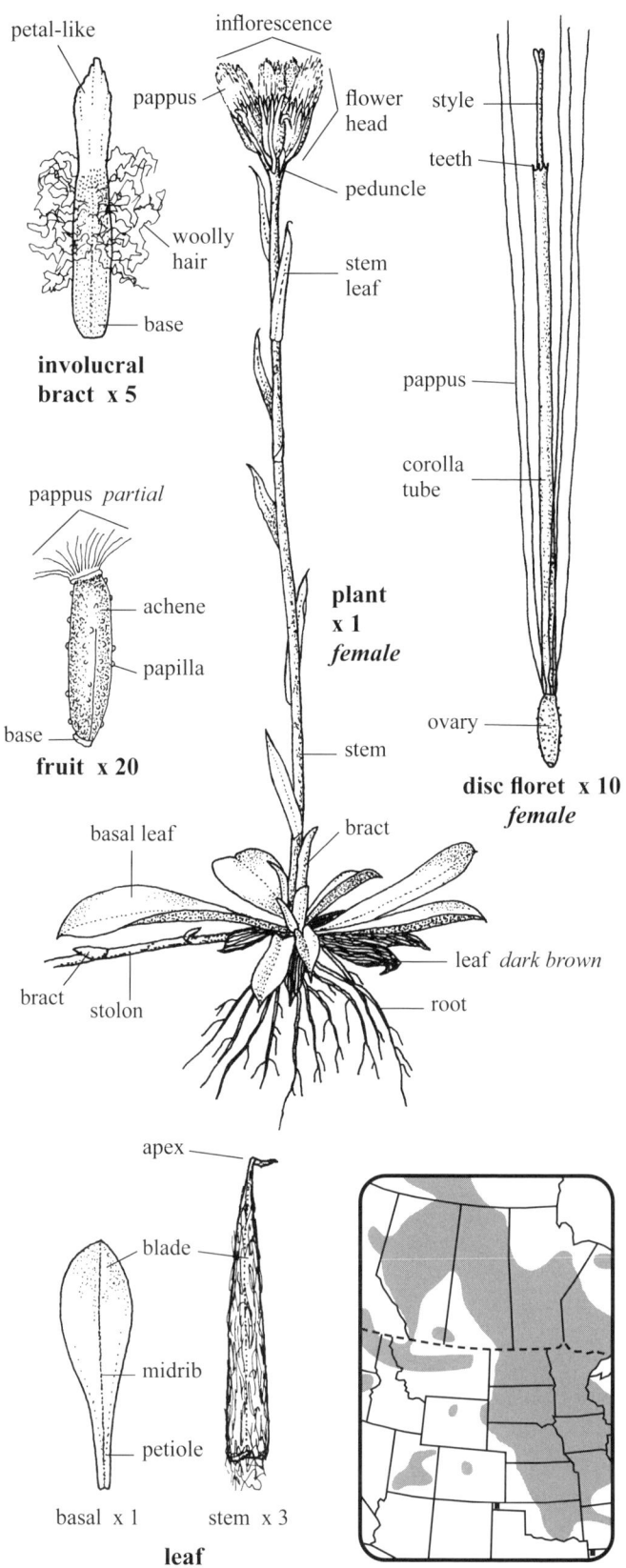

- **SKETCH** A variable, woolly **perennial herb** 15–40 cm tall with **fibrous roots** 2–6 cm long, and white **stolons** 2–6 cm long; in open forests, parklands and grasslands; **dioecious**. p1
- **FLOWER HEADS** White, blooming March–July; **inflorescence** a cyme, open, appearing more racemose when mature, 15–18 mm tall (including peduncles) by 12–15 mm wide, of 3–6 erect heads; **peduncles** 2–4 mm long (to 15 mm and erect with fruit), very hairy; **subtending bracts** (of peduncles) one, hairy, to *c.* 1 cm long, margins with short, red, spreading hairs; **flower heads** unisexual, 7–8 mm wide by *c.* 12 mm tall; **involucre** 7–10 mm long by 7–8 mm wide, cottony hairy; **involucral bracts** imbricate, in 3 or 4 series, 7–8 mm long by 1–1.5 mm wide, lower half yellowish green with pinkish tips, the upper half white, petal-like and *c.* 3.5 mm long by 0.9–1.5 mm wide, apex minutely toothed; <u>ray</u> **florets** absent; <u>disc</u> **florets** pistillate, fertile, numerous, TL *c.* 10 mm; **ovary** pale green, papillose; **pappus** white, bristles 9–10 mm long; **corolla tube** *c.* 7.5 mm long by *c.* 0.2 mm wide, pale greenish yellow, apex with four or five minute teeth; **style** filiform, green turning brown, exserted 1–2 mm past the tip of the corolla tube, apex minutely 2-parted; **fruiting heads** erect, fluffy white, together 2–3 cm long by 3–5 cm wide. **FRUIT** an achene, 1-seeded, minutely papillose, olive to brown, base white, 1.1–1.5 mm long by *c.* 0.3 mm wide and thick, 4-sided; **pappus** of *c.* 25 white bristles with fine hairs ascending, 9–10 mm long, deciduous as a unit; **receptacle** bumpy, hairless.
- **LEAVES** Basal and stem, entire; **basal leaves** in a rosette, green with dark brown ones below; **blades** TL 13–50 mm by 5–20 mm wide, spreading, densely white tomentose below, dull green and thinly hairy to glabrous above, midrib obvious, two side nerves obscure, tapered to base, apical mucro often deflexed; **petioles** 2–10 mm long; **stem leaves** alternate, several, each 15–25 mm long by 1.4–3 mm wide, erect to ascending, sessile; **blades** soft, tomentose above, glabrous below, tip with a hard scarious, red to pale yellow point, straight or twisted and hairless, margins, midribs and bases pinkish, midrib obvious, side veins obscure.
- **STEM** Erect, simple, weak, woolly, pinkish, solid; 1–2 mm wide base is enclosed by several short, leaf-like green bracts.
- **SYN.** *Antennaria athabascensis* Greene, *Antennaria campestris* Rydb., *Antennaria howellii* var. *athabascensis* (Greene) Boivin, and *Antennaria howellii* var. *campestris* (Rydb.) Boivin.

Broad-leaved Pussytoes

Aster—*Asteraceae* **Wildflower** Pinkish purple Disc florets only

LESSER BURDOCK *Arctium minus* Bernh.

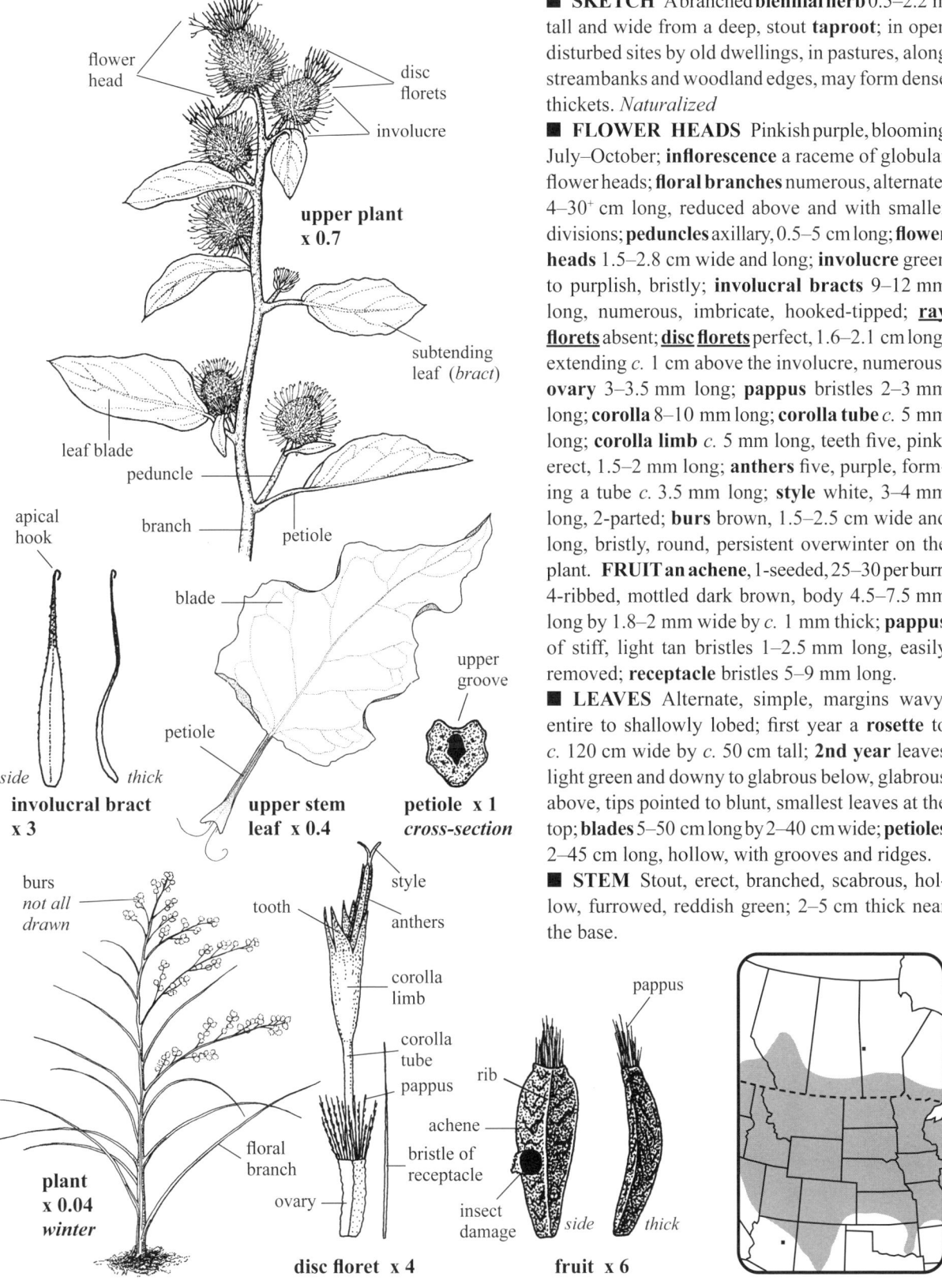

■ **SKETCH** A branched **biennial herb** 0.5–2.2 m tall and wide from a deep, stout **taproot**; in open disturbed sites by old dwellings, in pastures, along streambanks and woodland edges, may form dense thickets. *Naturalized*

■ **FLOWER HEADS** Pinkish purple, blooming July–October; **inflorescence** a raceme of globular flower heads; **floral branches** numerous, alternate, 4–30+ cm long, reduced above and with smaller divisions; **peduncles** axillary, 0.5–5 cm long; **flower heads** 1.5–2.8 cm wide and long; **involucre** green to purplish, bristly; **involucral bracts** 9–12 mm long, numerous, imbricate, hooked-tipped; <u>**ray florets**</u> absent; <u>**disc florets**</u> perfect, 1.6–2.1 cm long, extending *c.* 1 cm above the involucre, numerous; **ovary** 3–3.5 mm long; **pappus** bristles 2–3 mm long; **corolla** 8–10 mm long; **corolla tube** *c.* 5 mm long; **corolla limb** *c.* 5 mm long, teeth five, pink, erect, 1.5–2 mm long; **anthers** five, purple, forming a tube *c.* 3.5 mm long; **style** white, 3–4 mm long, 2-parted; **burs** brown, 1.5–2.5 cm wide and long, bristly, round, persistent overwinter on the plant. **FRUIT an achene**, 1-seeded, 25–30 per burr, 4-ribbed, mottled dark brown, body 4.5–7.5 mm long by 1.8–2 mm wide by *c.* 1 mm thick; **pappus** of stiff, light tan bristles 1–2.5 mm long, easily removed; **receptacle** bristles 5–9 mm long.

■ **LEAVES** Alternate, simple, margins wavy, entire to shallowly lobed; first year a **rosette** to *c.* 120 cm wide by *c.* 50 cm tall; **2nd year** leaves light green and downy to glabrous below, glabrous above, tips pointed to blunt, smallest leaves at the top; **blades** 5–50 cm long by 2–40 cm wide; **petioles** 2–45 cm long, hollow, with grooves and ridges.

■ **STEM** Stout, erect, branched, scabrous, hollow, furrowed, reddish green; 2–5 cm thick near the base.

Common Burdock

Aster—*Asteraceae* **Wildflower** Pinkish purple Disc florets only

COTTON BURDOCK *Arctium tomentosum* P. Mill.

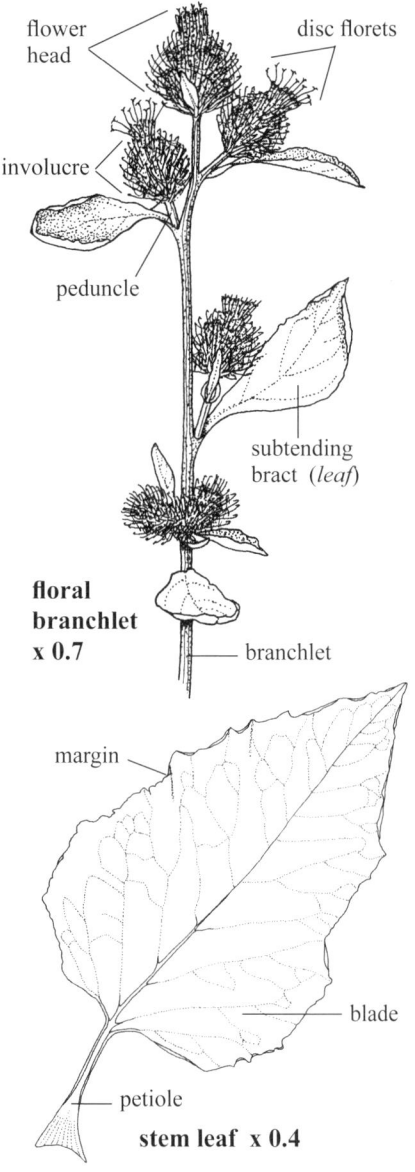

floral branchlet x 0.7

stem leaf x 0.4

■ **SKETCH** A **biennial herb** 0.8–2 m tall and wide from a **taproot**; scattered in moist sites along riverbanks, and in dry gravelly sites along roads and railways. *Naturalized*

■ **FLOWER HEADS** Pinkish purple, blooming July–September; **inflorescence** corymbiform; **floral branches** alternate, 10–110 cm long, ascending, reduced above; **branchlets** with one to eight heads; **peduncles** erect, 2–50 mm long, finely hairy; **subtending bracts** (of peduncles) leaflike, with blades 10–40 mm long by 3–20 mm wide, often asymmetrical; **flower heads** numerous, 15–30 mm wide by 19–22 mm long, throughout the plant; **involucre** green, cottony, 12–15 mm long by 15–30 mm wide; **involucral bracts** imbricate, thin, mostly ascending, 8–10 series, cottony hairy on outside, 4–12 mm long by 0.5–1.1 mm wide, apices hooked, inner bracts the longest; <u>ray florets</u> absent; <u>disc florets</u> perfect, exserted to *c.* 10 mm above the top of the involucre; **ovary** white, *c.* 3 mm long by *c.* 1 mm wide, hairless, rough; **pappus** bristles 0.6–1.8 mm long and white; **receptacle bristles** numerous, 2–7 mm long, erect, several at the base of each floret; **corolla** TL 10–11 mm, glabrous or with glandular hairs; **corolla tube** white, *c.* 5 mm long; **corolla limb** 5–6 mm long, pinkish above; **teeth** *c.* 1 mm long, thickened at apices; **anthers** five, purplish pink, erect, 3–3.5 mm long, united; **styles** pale pink, exserted, 2-parted, not hairy; **burs** brown, bristly, *c.* 20 mm tall, wide, and thick, persisting on the plant overwinter. **FRUIT** an achene, 1-seeded, about 40 per bur, each 5–5.5 mm long by 2–2.3 mm wide by 1–1.3 mm thick, tan with dark, irregular striations, base flat, 4- or 5-ribbed; **pappus** tan, *c.* 50 bristles, 0.5–2 mm long.

■ **LEAVES** Alternate, simple, with margins slightly inrolled, entire to shallowly lobed and toothed; **first year** a large rosette; **second year** leaves basal and stem; **blades** 3.5–40 cm long by 3–34 cm wide, veins raised below, lighter green below due to woolly hairs, dull green and soft to slightly scabrous above; **stem leaves** subtending each large floral branch and branchlet, reduced above; **petioles** 2–35 cm long by 5–10 mm wide and thick, hollow, longitudinally ridged, hairy, widening at whitish base.

■ **STEM** Erect, stout, hollow, reddish green, grooved, branched throughout, softly hairy, hairs mostly appressed, some glandular; 2–4 cm wide near the base.

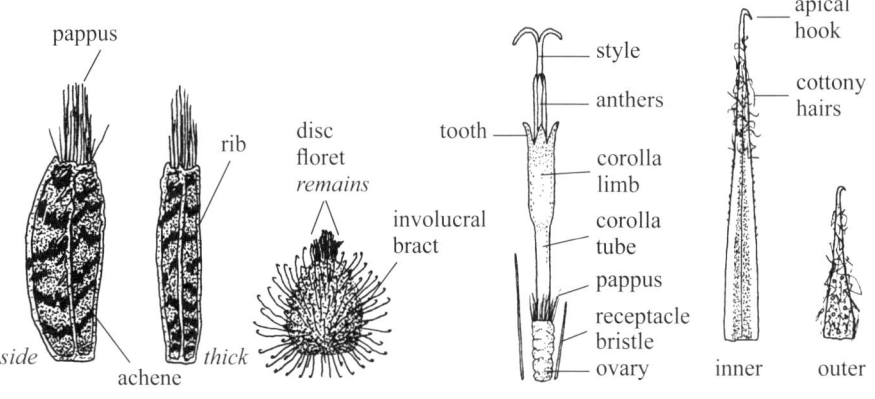

fruit x 5   bur x 1   disc floret x 2.5   involucral bract x 4

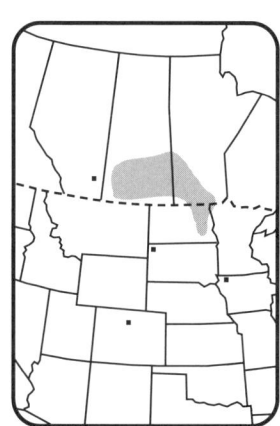

Aster—*Asteraceae* **Wildflower** Green Disc florets only

ABSINTHE *Artemisia absinthium* L.

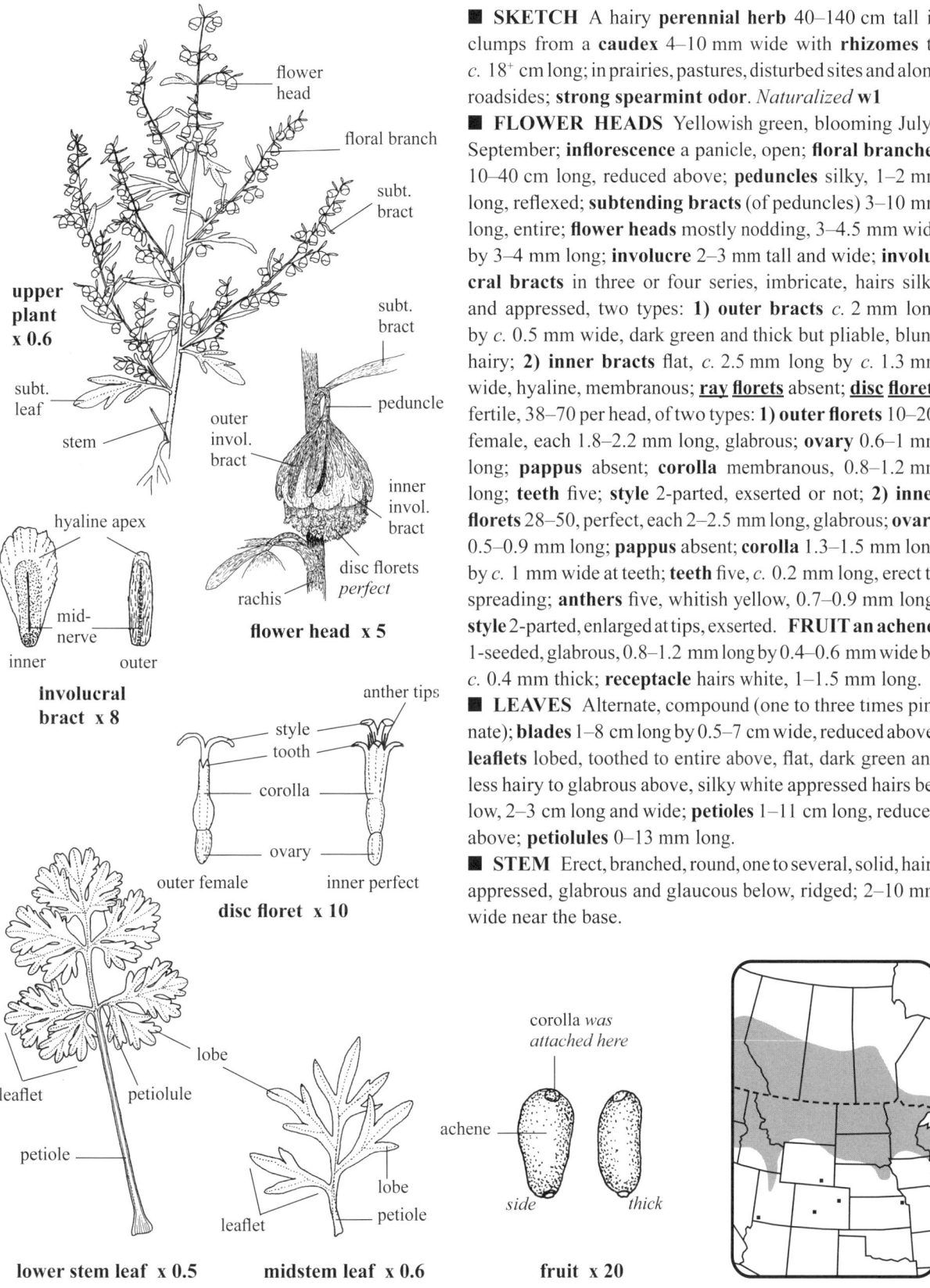

■ **SKETCH** A hairy **perennial herb** 40–140 cm tall in clumps from a **caudex** 4–10 mm wide with **rhizomes** to *c.* 18+ cm long; in prairies, pastures, disturbed sites and along roadsides; **strong spearmint odor**. *Naturalized* **w1**

■ **FLOWER HEADS** Yellowish green, blooming July–September; **inflorescence** a panicle, open; **floral branches** 10–40 cm long, reduced above; **peduncles** silky, 1–2 mm long, reflexed; **subtending bracts** (of peduncles) 3–10 mm long, entire; **flower heads** mostly nodding, 3–4.5 mm wide by 3–4 mm long; **involucre** 2–3 mm tall and wide; **involucral bracts** in three or four series, imbricate, hairs silky and appressed, two types: **1) outer bracts** *c.* 2 mm long by *c.* 0.5 mm wide, dark green and thick but pliable, blunt, hairy; **2) inner bracts** flat, *c.* 2.5 mm long by *c.* 1.3 mm wide, hyaline, membranous; <u>ray florets</u> absent; <u>disc florets</u> fertile, 38–70 per head, of two types: **1) outer florets** 10–20, female, each 1.8–2.2 mm long, glabrous; **ovary** 0.6–1 mm long; **pappus** absent; **corolla** membranous, 0.8–1.2 mm long; **teeth** five; **style** 2-parted, exserted or not; **2) inner florets** 28–50, perfect, each 2–2.5 mm long, glabrous; **ovary** 0.5–0.9 mm long; **pappus** absent; **corolla** 1.3–1.5 mm long by *c.* 1 mm wide at teeth; **teeth** five, *c.* 0.2 mm long, erect to spreading; **anthers** five, whitish yellow, 0.7–0.9 mm long; **style** 2-parted, enlarged at tips, exserted. **FRUIT an achene**, 1-seeded, glabrous, 0.8–1.2 mm long by 0.4–0.6 mm wide by *c.* 0.4 mm thick; **receptacle** hairs white, 1–1.5 mm long.

■ **LEAVES** Alternate, compound (one to three times pinnate); **blades** 1–8 cm long by 0.5–7 cm wide, reduced above; **leaflets** lobed, toothed to entire above, flat, dark green and less hairy to glabrous above, silky white appressed hairs below, 2–3 cm long and wide; **petioles** 1–11 cm long, reduced above; **petiolules** 0–13 mm long.

■ **STEM** Erect, branched, round, one to several, solid, hairs appressed, glabrous and glaucous below, ridged; 2–10 mm wide near the base.

Oldman, Absinthium, Absinthe Wormwood

Aster—*Asteraceae* **Wildflower** Green Disc florets only

BIENNIAL WORMWOOD   *Artemisia biennis* Willd.

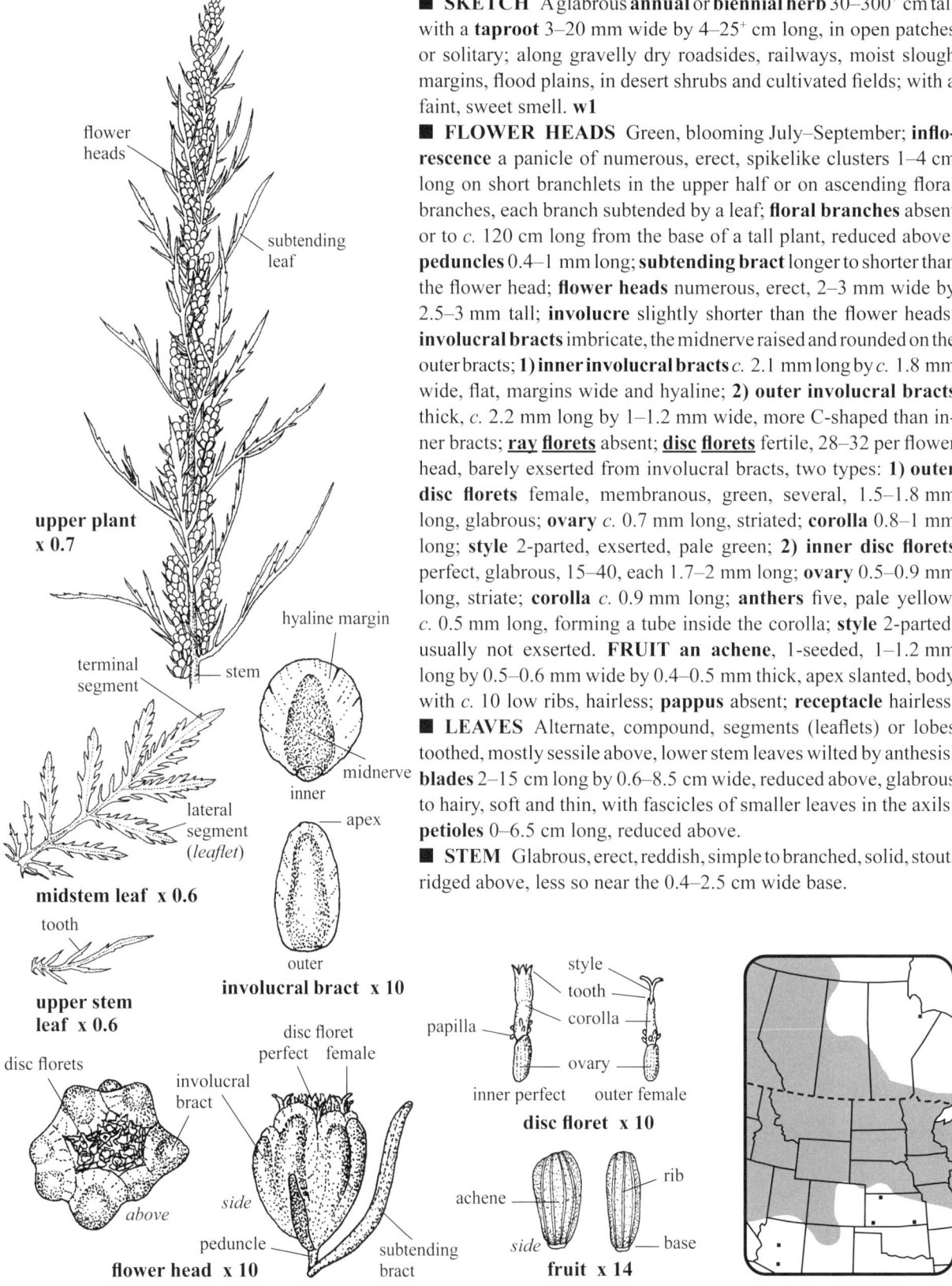

■ **SKETCH** A glabrous **annual** or **biennial herb** 30–300⁺ cm tall with a **taproot** 3–20 mm wide by 4–25⁺ cm long, in open patches or solitary; along gravelly dry roadsides, railways, moist slough margins, flood plains, in desert shrubs and cultivated fields; with a faint, sweet smell. **w1**

■ **FLOWER HEADS** Green, blooming July–September; **inflorescence** a panicle of numerous, erect, spikelike clusters 1–4 cm long on short branchlets in the upper half or on ascending floral branches, each branch subtended by a leaf; **floral branches** absent or to *c.* 120 cm long from the base of a tall plant, reduced above; **peduncles** 0.4–1 mm long; **subtending bract** longer to shorter than the flower head; **flower heads** numerous, erect, 2–3 mm wide by 2.5–3 mm tall; **involucre** slightly shorter than the flower heads; **involucral bracts** imbricate, the midnerve raised and rounded on the outer bracts; **1) inner involucral bracts** *c.* 2.1 mm long by *c.* 1.8 mm wide, flat, margins wide and hyaline; **2) outer involucral bracts** thick, *c.* 2.2 mm long by 1–1.2 mm wide, more C-shaped than inner bracts; **ray florets** absent; **disc florets** fertile, 28–32 per flower head, barely exserted from involucral bracts, two types: **1) outer disc florets** female, membranous, green, several, 1.5–1.8 mm long, glabrous; **ovary** *c.* 0.7 mm long, striated; **corolla** 0.8–1 mm long; **style** 2-parted, exserted, pale green; **2) inner disc florets** perfect, glabrous, 15–40, each 1.7–2 mm long; **ovary** 0.5–0.9 mm long, striate; **corolla** *c.* 0.9 mm long; **anthers** five, pale yellow, *c.* 0.5 mm long, forming a tube inside the corolla; **style** 2-parted, usually not exserted. **FRUIT** an achene, 1-seeded, 1–1.2 mm long by 0.5–0.6 mm wide by 0.4–0.5 mm thick, apex slanted, body with *c.* 10 low ribs, hairless; **pappus** absent; **receptacle** hairless.

■ **LEAVES** Alternate, compound, segments (leaflets) or lobes toothed, mostly sessile above, lower stem leaves wilted by anthesis; **blades** 2–15 cm long by 0.6–8.5 cm wide, reduced above, glabrous to hairy, soft and thin, with fascicles of smaller leaves in the axils; **petioles** 0–6.5 cm long, reduced above.

■ **STEM** Glabrous, erect, reddish, simple to branched, solid, stout, ridged above, less so near the 0.4–2.5 cm wide base.

112   Biennial Sagewort

Aster—*Asteraceae* **Wildflower** Green Disc florets only

Plains Wormwood *Artemisia campestris* L.

■ **SKETCH** A variable **biennial** or **perennial** herb 10–100 cm tall from a thick, woody **taproot** 2–12 mm wide by 9–15+ cm long, some with a woody **caudex** 0.8–2.2 cm wide; in mountain brush, along roadsides, gravel beaches, disturbed dry sites, pastures, woods and shores. **w1, p2**

■ **FLOWER HEADS** Green, blooming July–September; **inflorescence** a panicle, narrow, 10–30+ cm long; **floral branches** numerous, alternate, 2–15 cm long, the longer ones leafy near the base; **peduncles** recurved, 1–3 mm long, glabrous; **subtending bracts** (of peduncles) one, shorter to longer than peduncle, entire, glabrous, pointed; **flower heads** mostly drooping, 2–3 mm wide by 2–4 mm long; **involucre** equal to flower heads in size, glabrous to hairy; **involucral bracts** in three or four series, imbricate, soft, midnerve wide and green: **1) outer bracts** darker green, thick, C-shaped, margins hyaline, 1.3–2 mm long by 1–1.5 mm wide (flattened); **2) inner involucral bracts** more membranous, lighter green, 2.2–2.5 mm long by *c.* 1.5 mm wide, margins and apices hyaline; ray florets absent; disc florets of two types: **1) inner florets** sterile, *c.* 15 per head, 1.7–2 mm long, glabrous; **ovary** pale green, *c.* 0.8 mm long; **corolla** *c.* 1 mm long, body striate with long low ridges, teeth five, membranous; **style** *c.* 0.5 mm long; **2) outer florets** female, fertile, 15–19 per flower head, 1.8–2 mm long; **ovary** *c.* 0.5 mm long, glabrous, pale green, striate; **corolla** 0.7–0.8 mm long, membranous, apex bidentate; **style** 2-parted, the parts *c.* 0.5 mm long, exserted. **FRUIT an achene**, 1-seeded, 0.7–0.8 mm long by 0.3–0.4 mm wide by *c.* 0.2 mm thick, hairless, several-ribbed, whitish at both ends, apex slanted; **pappus** absent; **receptacle** hairless.

■ **LEAVES** Basal and stem, alternate, mostly compound two or three times, lobed; **basal leaves** numerous, TL (including petioles) 20–40 cm by 1–4 cm wide, the ultimate segments linear and up to 2 mm wide; **stem leaves** glabrous, light grayish green, 4–6 cm long by 2–3 cm wide, reduced above especially in inflorescence, usually sessile, ascending, segments mostly 0.5–1 mm wide; **petioles** 0–6.5 cm long, reduced above.

■ **STEM** Erect, stiff, solid, reddish especially below, hairy below to glabrous above, ridged longitudinally, branched above, solitary to several+ stems from caudex; 1.5–5 mm thick near the base.

■ **SYN.** *Artemisia caudata* Michx. = *Artemisia campestris* ssp. *caudata* (Michx.) Hall & Clements, *Artemisia canadensis* Michx. = *Artemisia campestris* var. *borealis* (Pallas) M.E. Peck and *Artemisia camporum* Rydb. = *Artemisia campestris* var. *scouleriana* (Hook.) Cronq.

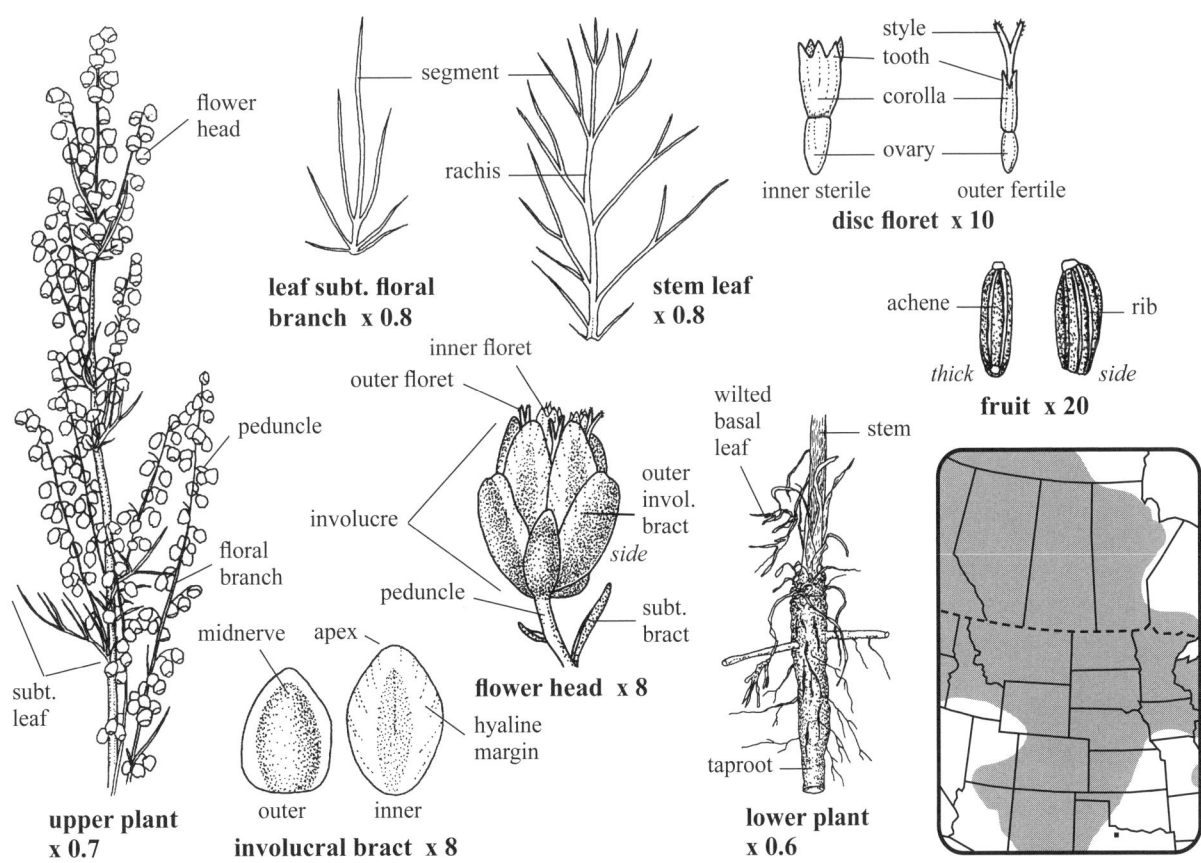

Field Sagewort, Pacific Wormwood, Western Sagewort, Tall Wormwood

Aster—*Asteraceae* **Wildflower Green Disc florets only**

Pasture Sage   *Artemisia frigida* Willd.

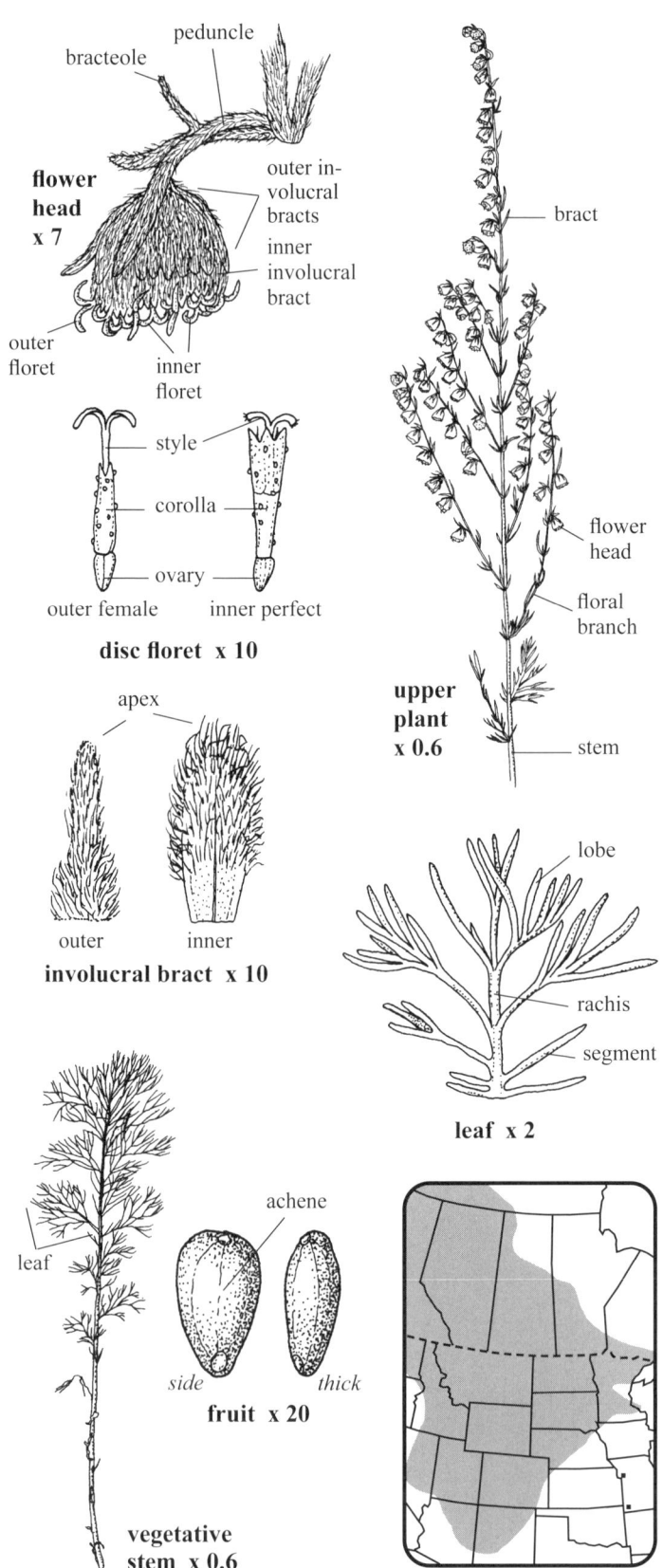

■ **SKETCH** A hairy **perennial herb** (or subshrub), whitish green, in scattered clumps 15–60 cm tall from a woody **caudex** and branching **taproot** 2–3 mm wide and to *c.* 12⁺ cm long; in dry uplands, mountain brush, ridges, overgrazed pastures and sandy soils. *Naturalized* **w1, p1**

■ **FLOWER HEADS** Green, blooming July–September; **inflorescence** a panicle or narrow raceme on some branches, 10–25 cm long by 4–6 cm wide; **floral branches** several, 1–15⁺ cm long, ascending, leafy; **peduncles** bracteolate, 1–3 mm long, reflexed; **flower heads** drooping, 2–4 mm wide by 3–4 mm long, mostly along one side of the rachis or terminal on a short branch; **involucral bracts** in two or three series, 2.5–3 mm long by 1–1.5 mm wide (including hairs), outside hairs hiding veins; <u>ray</u> <u>florets</u> absent; <u>disc</u> <u>florets</u> fertile, of two types: **1) outer florets** female, 10–16 per head, in one outer series, TL *c.* 2.5 mm; **ovary** glabrous, *c.* 0.6 mm long; **pappus** absent; **corolla** 1–1.2 mm long by *c.* 0.3 mm wide, slightly flattened, glandular, transparent, apical teeth two and erect; **style** 2-parted, exserted, yellow, not hairy; **2) inner florets** perfect, 20–40 per head, TL 2–2.5 mm; **ovary** glabrous, *c.* 0.5 mm long, ribbed; **pappus** absent; **corolla** *c.* 1.8 mm long, tapered and greenish below, yellowish above, glandular, apical teeth five, each *c.* 0.3 mm long and erect; **anthers** yellow, united; **style** 2-parted, *c.* 0.8 mm long, yellow, exserted, apex of each part blunt and hairy. **FRUIT an achene**, 1-seeded, tan, glabrous, tapered to the base, round to oval in cross-section, *c.* 1 mm long by *c.* 0.5 mm wide by *c.* 0.4 mm thick, barely ridged longitudinally; **pappus** absent; **receptacle** with white hairs *c.* 1 mm long between the florets.

■ **LEAVES** Basal and stem, alternate, compound (two or three times divided), lobed, pale green, fragrant when crushed, soft white hairs appressed; **basal blades** 1–2.5 cm long by 1–2 cm wide, ascending, tomentose on both sides, segments filiform, soft, 0.4–1 mm wide, flat, apices pointed; **lowest stem leaves** wilted by anthesis; **petioles** 0–5 mm long, reduced above.

■ **STEM** Several to 50⁺ per clump, erect, branched; woody near the 1–3 mm wide base.

Aster—*Asteraceae* **Wildflower** Green Disc florets only

PRAIRIE SAGE *Artemisia ludoviciana* Nutt.

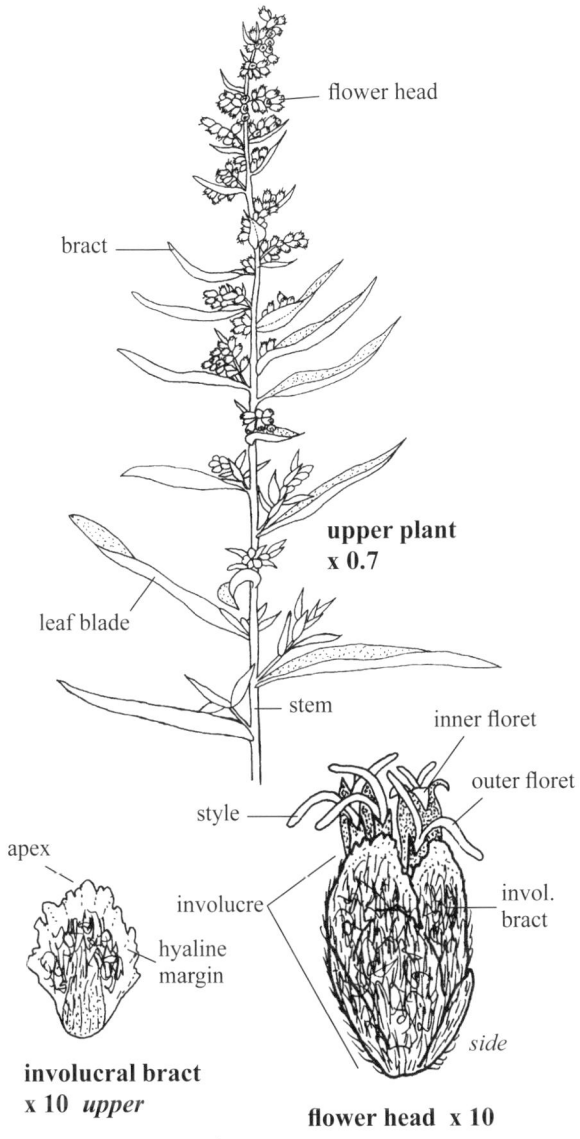

■ **SKETCH** A hairy, variable **perennial herb** 15–110 cm tall, one to four from **rhizomes**; in open moist prairies, sandhills, foothills, roadsides and slough margins. **w2, p1**

■ **FLOWER HEADS** Green, blooming July–October; **inflorescence** a panicle with axillary spikes, narrow to open; **floral branches** to *c.* 10 cm long, axillary in upper half or throughout, reduced above; **flower heads** numerous, 2–4 mm wide by 3–4.5 mm long, mostly sessile; **involucre** woolly hairy, appears whitish green, 2.5–4 mm long by 2–3 mm wide; **involucral bracts** in two or three series, each 1.5–2 mm long by 1.2–1.5 mm wide, imbricate, their outlines hidden by dense white hairs, upper bracts with ragged, hyaline tips and margins; <u>ray florets</u> absent; <u>disc florets</u> 6–13, fertile, of two kinds: **1) outer florets** four to eight, pistillate, TL 2.5–3 mm; **ovary** glabrous, slightly flattened, 0.7–0.9 mm long; **pappus** absent; **corolla** narrow, 1–1.2 mm long, reddish in the upper half, glabrous, the two apical teeth erect; **style** 2-parted, the parts yellowish brown, 0.7–1 mm long; **2) inner florets** two to five, perfect, TL 2–2.5 mm; **ovary** *c.* 0.6 mm long, glabrous; **pappus** absent; **corolla** *c.* 1.5 mm long, glabrous, the upper half reddish purple, apical teeth five, erect to spreading; **anthers** five, yellow, forming a tube *c.* 1 mm long, mostly included; **style** 2-parted, *c.* 1.3 mm long, included or not, the tips reddish. **FRUIT an achene**, 1-seeded, cylindrical, 5-ribbed, 1–1.4 mm long by *c.* 0.4 mm wide by *c.* 0.3 mm thick, hairless; **pappus** absent; **receptacle** hairless.

■ **LEAVES** Alternate, simple, entire to irregularly toothed to deeply lobed, white woolly to glabrous, more hairy below, sessile; **blades** 2–11 cm long by 5–14 mm wide, often partially folded along the midrib, with a downy feel, sometimes twisted, aromatic when crushed.

■ **STEM** Erect to leaning, solid, often branched, covered with white appressed hairs; 2–4 mm thick near the base.

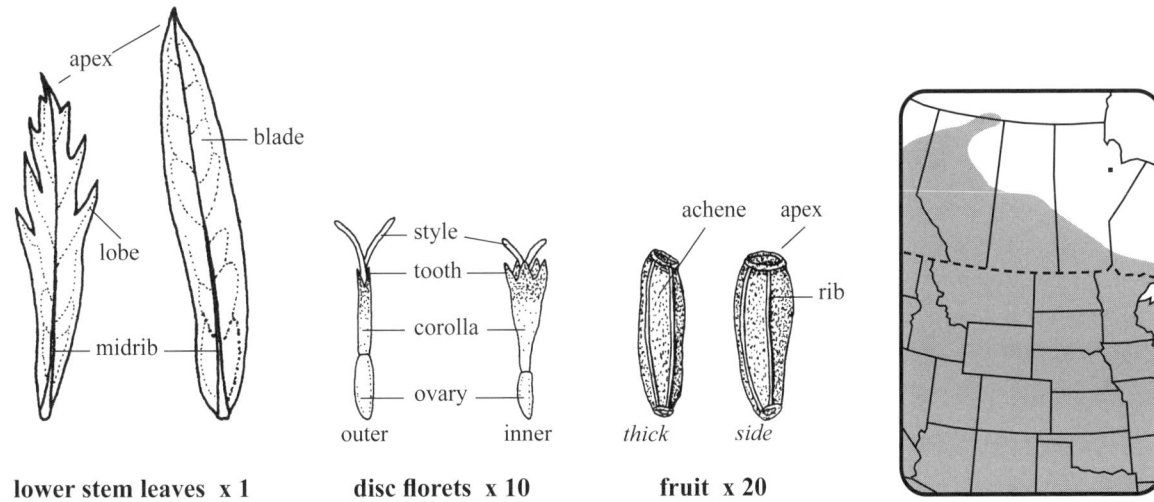

White Sage, White Sagebrush, Western Mugwort

Aster—*Asteraceae* **Wildflower** Yellow Ray florets 6–9

Smooth Beggarticks *Bidens cernua* L.

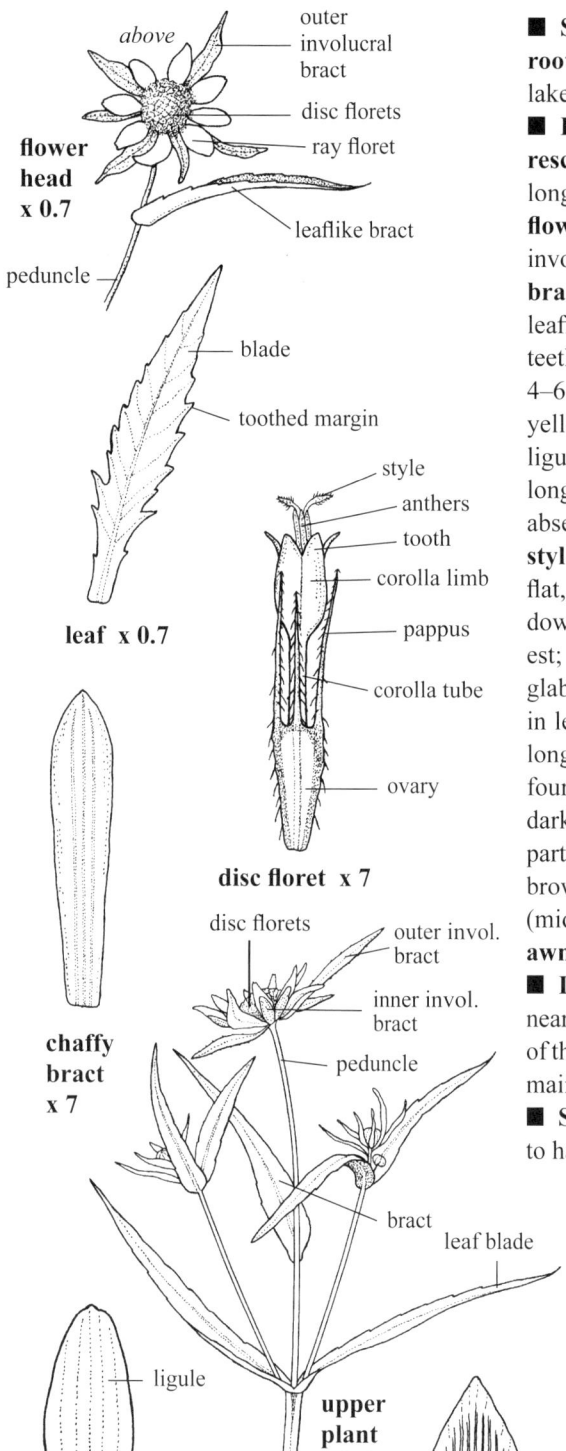

■ **SKETCH** An erect **annual herb** 20–120 cm tall with **fibrous roots** 3–20 cm long by 1–2 mm wide; in moist sites along rivers, lakes, streams, sloughs, in bogs and upper meadows. **w2**

■ **FLOWER HEADS** Yellow, blooming June–October; **inflorescence** of flower heads, often nodding; **floral branches** 9–33 cm long, ascending; **peduncles** smooth to hairy, axillary, 1.5–8 cm long; **flower heads** 18–38 mm wide by 10–12 mm tall (not including outer involucral bracts); **central disc** 9–23 mm wide and convex; **involucre bracts** of two types: **1) outer involucral bracts** five to eight per head, leaflike, 10–50 mm long by 3–7 mm wide, spreading, with minute teeth; **2) inner involucral bracts** membranous, 6–15 mm long by 4–6.5 mm wide, mostly ascending, glabrous with striations, apices yellow, blunt to pointed, beneath and shorter than each yellow ray ligule; <u>ray florets</u> sterile, six to nine; **ovary** vestigial, hairy, *c.* 1 mm long; **pappus** absent; **corolla tube** *c.* 1.3 mm long; **ligules** (rarely absent) 8–15 mm long by 4–8 mm wide with three blunt apical teeth; **style** absent; <u>disc florets</u> perfect, numerous; **ovary** *c.* 2.3 mm long, flat, 4-ridged with marginal retrorse barbs; **pappus** of four awns, downwardly barbed, 2–3 mm long, the two lateral awns the longest; **chaffy bracts** 6–8 mm long by 1–1.4 mm wide, pointed, flat, glabrous, with three central veins and hyaline margins, about equal in length to the disc floret and the mature fruit; **corolla** *c.* 3.7 mm long; **corolla tube** *c.* 2 mm long; **corolla limb** *c.* 1.6 mm long with four or five yellow apical teeth; **filaments** separate; **anthers** five, dark brown, *c.* 1 mm long and forming a tube; **style** 2-parted, the parts yellow, hairy, exserted. **FRUIT an achene**, 1-seeded, flat, dark brown with four ribs, TL 5–8 mm by 1.5–2 mm wide by 0.5 mm thick (midbody); **body** 4–5 mm long, a convex bump at apex; **pappus awns** four (two), tan, 2–3 mm long with retrorse barbs.

■ **LEAVES** Opposite, simple, margins toothed, becoming entire near the blade's tip, sometimes a singular leaflike bract on the peduncle of the terminal flower head; **blades** 3–20 cm long by 0.7–4 cm wide, mainly sessile, clasping, united at bases, glabrous.

■ **STEM** Angular, solid to hollow, simple to branched, glabrous to hairy, erect and green; base 2–10 mm wide.

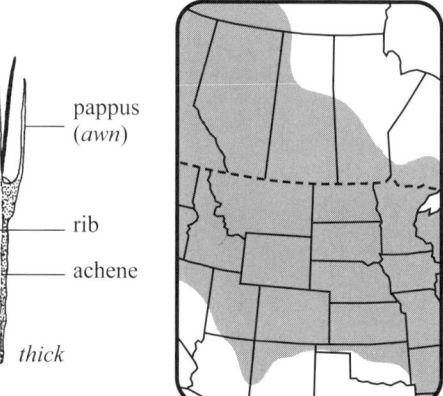

Nodding Beggarticks, Nodding Burr-marigold

Aster—*Asteraceae*   **Wildflower** Yellow Disc florets only Ray florets absent to small

COMMON BEGGARTICKS *Bidens frondosa* L.

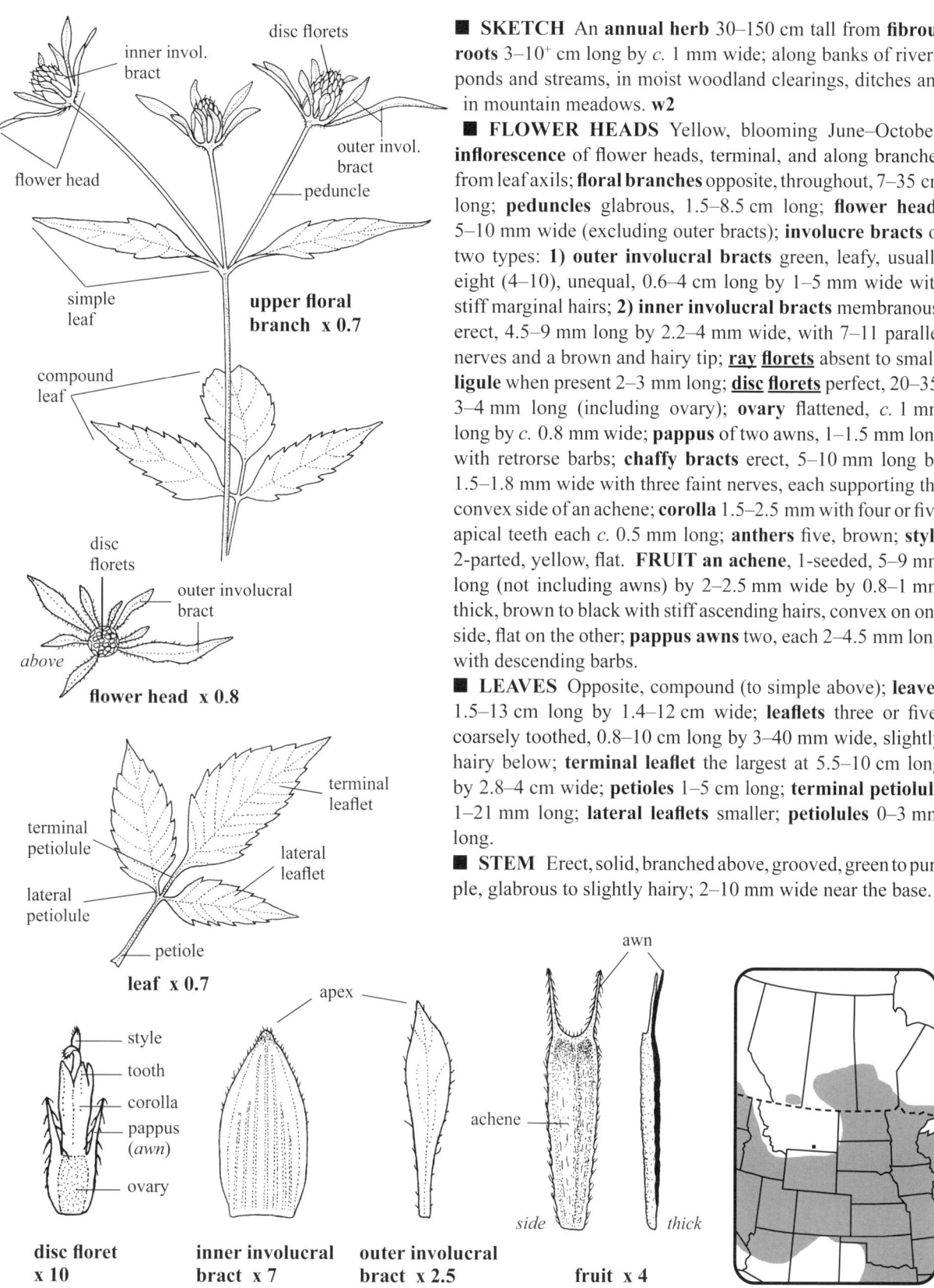

■ **SKETCH** An **annual herb** 30–150 cm tall from **fibrous roots** 3–10⁺ cm long by *c.* 1 mm wide; along banks of rivers, ponds and streams, in moist woodland clearings, ditches and in mountain meadows. **w2**

■ **FLOWER HEADS** Yellow, blooming June–October; **inflorescence** of flower heads, terminal, and along branches from leaf axils; **floral branches** opposite, throughout, 7–35 cm long; **peduncles** glabrous, 1.5–8.5 cm long; **flower heads** 5–10 mm wide (excluding outer bracts); **involucre bracts** of two types: **1) outer involucral bracts** green, leafy, usually eight (4–10), unequal, 0.6–4 cm long by 1–5 mm wide with stiff marginal hairs; **2) inner involucral bracts** membranous, erect, 4.5–9 mm long by 2.2–4 mm wide, with 7–11 parallel nerves and a brown and hairy tip; <u>ray florets</u> absent to small; **ligule** when present 2–3 mm long; <u>disc florets</u> perfect, 20–35, 3–4 mm long (including ovary); **ovary** flattened, *c.* 1 mm long by *c.* 0.8 mm wide; **pappus** of two awns, 1–1.5 mm long with retrorse barbs; **chaffy bracts** erect, 5–10 mm long by 1.5–1.8 mm wide with three faint nerves, each supporting the convex side of an achene; **corolla** 1.5–2.5 mm with four or five apical teeth each *c.* 0.5 mm long; **anthers** five, brown; **style** 2-parted, yellow, flat. **FRUIT an achene**, 1-seeded, 5–9 mm long (not including awns) by 2–2.5 mm wide by 0.8–1 mm thick, brown to black with stiff ascending hairs, convex on one side, flat on the other; **pappus awns** two, each 2–4.5 mm long with descending barbs.

■ **LEAVES** Opposite, compound (to simple above); **leaves** 1.5–13 cm long by 1.4–12 cm wide; **leaflets** three or five, coarsely toothed, 0.8–10 cm long by 3–40 mm wide, slightly hairy below; **terminal leaflet** the largest at 5.5–10 cm long by 2.8–4 cm wide; **petioles** 1–5 cm long; **terminal petiolule** 1–21 mm long; **lateral leaflets** smaller; **petiolules** 0–3 mm long.

■ **STEM** Erect, solid, branched above, grooved, green to purple, glabrous to slightly hairy; 2–10 mm wide near the base.

Devil's Pitchfork, Devil's Beggarticks

Aster—*Asteraceae* **Wildflower** Pink to purplish Disc florets only

NODDING THISTLE   *Carduus nutans* L.

■ **SKETCH** An armed, variable **winter annual** to **biennial herb** 50–300 cm tall from a stout, fleshy **taproot**; in fields, pastures, along roadsides, disturbed slopes, railways and open wooded areas. *Naturalized*

■ **FLOWER HEADS** Pink to purplish, blooming May–September; **inflorescence** of 1–60+ heads, nodding to almost erect, mostly solitary and terminating the main stem and branches; **floral branches** ascending, up to *c*. 100+ cm long from a subtending leaf, with one to five flower heads; **peduncles** 1–7 cm long by 4–7 mm thick, the central peduncle the longest, ridged, cottony hairy, mostly naked except for leaflike bracts; **flower heads** 3–7.5 cm across by 3–6.5 cm tall, fragrant; **disc** 12–25 mm tall, convex; **involucre** 3–7 cm wide by 2–4 cm high, spiny; **involucral bracts** imbricate, green to purplish, 10–26 mm long by 2–8 mm wide, glabrous to hairy, ascending above to descending below, constricted near the middle; **upper bracts** narrow, pointed; **lower** and **middle bracts** green, wider and stiffer; ray florets absent; disc florets perfect, numerous, pink, *c*. 37 mm long; **ovary** white, 2–2.5 mm long by *c*. 1.2 mm wide, with ribs; **pappus** white, 18–20 mm long; **corolla tube** white, 14–15 mm long by *c*. 0.3 mm wide, wrinkled above; **corolla limb** and lobes 10–13 mm long, limb expanded at base, glabrous, *c*. 1.5 mm wide and 5-veined; **corolla lobes** five, 6–7.5 mm long, ascending; **stamens** five, *c*. 10 mm long, included; **filaments** white, hairy, distinct, 3–4 mm long, attached to expanded base of corolla limb; **anthers** 6–8 mm long, erect; **style** white, *c*. 0.2 mm thick; **stigma** erect, rough, medium purple, *c*. 3 mm long by *c*. 0.4 mm wide, stigmatic groove sticky along each side; **fruiting heads** tawny. **FRUIT an achene**, 1-seeded, medium brown with darker crossbands, apex light, slightly shiny, 3.5–4 mm long by *c*. 1.5 mm wide by *c*. 1 mm thick; **pappus bristles** numerous, 12–17 mm long, fine hairs ascending, not plumose.

■ **LEAVES** First year a rosette; **second year** alternate, simple, lobed, toothed, spiny, lower stem leaves descending, ascending above; **blades** 5–40 cm long by 3–15 cm wide, sessile, hairy or not, lobes often twisted, midrib not prickly; **spines** 1–6 mm long, light tan, sharp.

■ **STEM** Erect, tough, branched, hollow, soft whitish lining inside, outside with numerous vertical lines (wings) of spiny lobes 7–25 mm long, width variable, lobes reduced in size above; 2–4 cm thick near the stem base, 4–6 cm wide, including the spiny lobes.

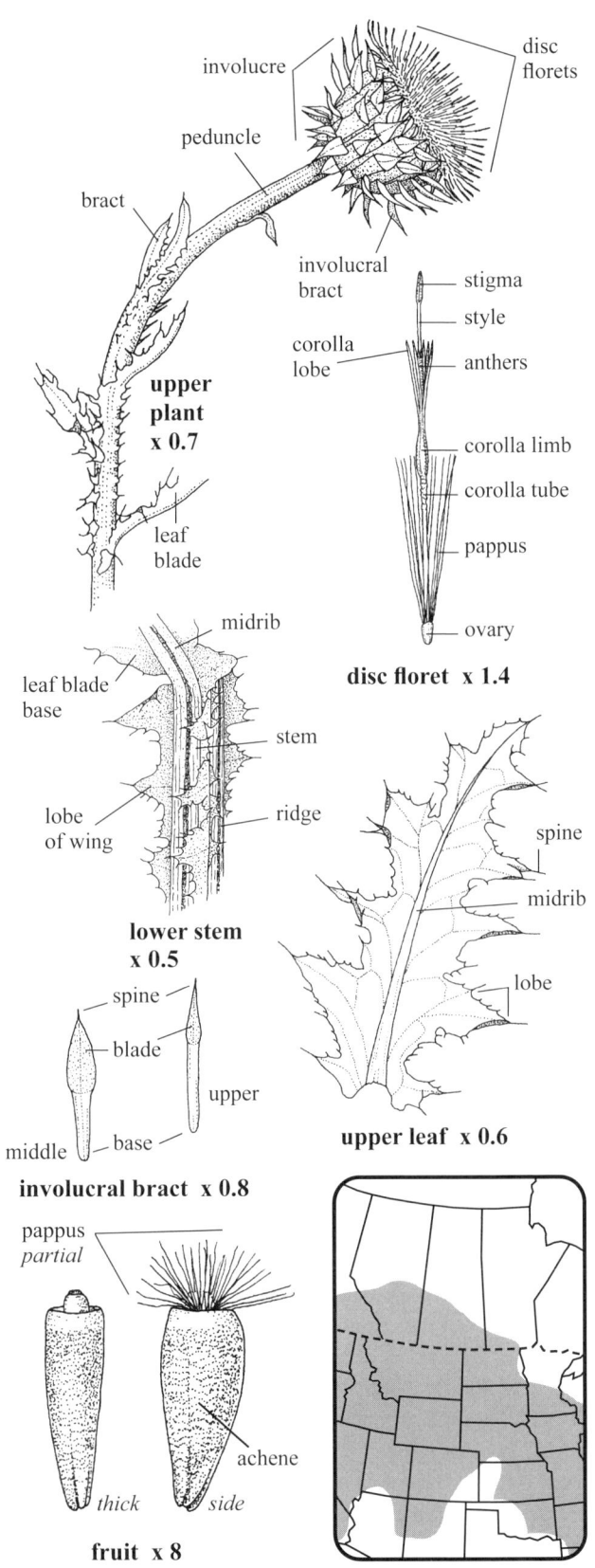

Nodding Plumeless-thistle, Musk Thistle

Aster—*Asteraceae* **Wildflower** White to blue to yellow Disc florets only

Diffuse Knapweed *Centaurea diffusa* Lam.

■ **SKETCH** A variable **annual or biennial herb** 20–70 cm tall and 10–40 cm wide from a **taproot** 5–20 cm long by 3–8 mm wide, singly or in dense colonies over several m²; along railways, roadsides and on dry hills. *Naturalized*

■ **FLOWER HEADS** White to pale blue to yellow, blooming July–September; **inflorescence** of numerous, solitary heads each terminating an ascending branch; **peduncles** stiff, 12–45 mm long, from a leaf axil, with one to three leaflike, entire bracts 6–10 mm long by 0.7–2 mm wide, spine-tipped; **flower heads** 17–23 mm wide and tall; **involucre** armed, tan, 8–11 mm long by 8–10 mm wide (including spines); **involucral bracts** in six or seven series, imbricate, erect around fruit, pliable, 2–10.5 mm long by 0.8–3 mm wide (including spines), three central nerves dark and usually two to four faint incomplete lateral nerves, apices spine-tipped (except for upper bracts), with one to four lateral smaller spines, exposed parts hairy, decreasing in size below; <u>ray</u> **florets** absent; <u>disc</u> **florets** of two types: **1) inner florets** perfect, white, TL 17–18 mm (base of ovary to tip of style); **ovary** *c.* 1.1 mm long, slightly hairy, white; **pappus** usually absent; **corolla tube** 9–10 mm long by *c.* 1 mm wide, wrinkled above ovary, curled hairs along the upper half; **corolla lobes** four or five, white, 3.5–4.5 mm long by *c.* 0.6 mm wide; **stamens** four; **filaments** white, hairy and free in upper half, attached to corolla tube below; **anthers** white, 4–5 mm long, glabrous, united, with basal appendages *c.* 0.2 mm long; **style** with microscopic fringe of hairs 1.3–1.5 mm from the apex, grooved above the hairs; **2) outer florets** 11–12, sterile, spreading to descending, white, TL 18–19 mm; **ovary** white, flat, glabrous, *c.* 1 mm long; **pappus** absent; **corolla tube** glabrous, 7–10 mm long by 1–1.2 mm wide above, ridged, punctate; **lobes** five, 5–8 mm long by 0.4–0.9 mm wide. **FRUIT** an achene, 1-seeded, 2–3 mm long by 1–1.3 mm wide by *c.* 0.8 mm thick, finely hairy, brown and shiny, whitish near the base, with longitudinal light and dark stripes, hilum notched; **pappus** absent, or bristles *c.* 1 mm long.

■ **LEAVES** Basal and stem, alternate, compound (once or twice pinnate), lobed and toothed; **first year rosette** of 6–20 leaves; **blades** 5–15 cm long by 1.5–4.5 cm wide, cottony hairy, segments 1–3 mm wide; **petioles** 2–7 cm long, hairy; **lower stem leaves** punctate; **blades** 7–11 cm long by 3–5 cm wide, reduced above; **petioles** 0–25 mm long; **upper leaves** along floral branches reduced in size, entire, cottony hairy, punctate, marginal hairs spreading, apical spine minute.

■ **STEM** Stiff, punctate, erect, tough, branched, solid, angled but not winged, cottony hairs appressed; 2–10 mm thick near the base; **branches** alternate, leafy, hairy, ridged, dividing into shorter branchlets then peduncles supporting the flower heads.

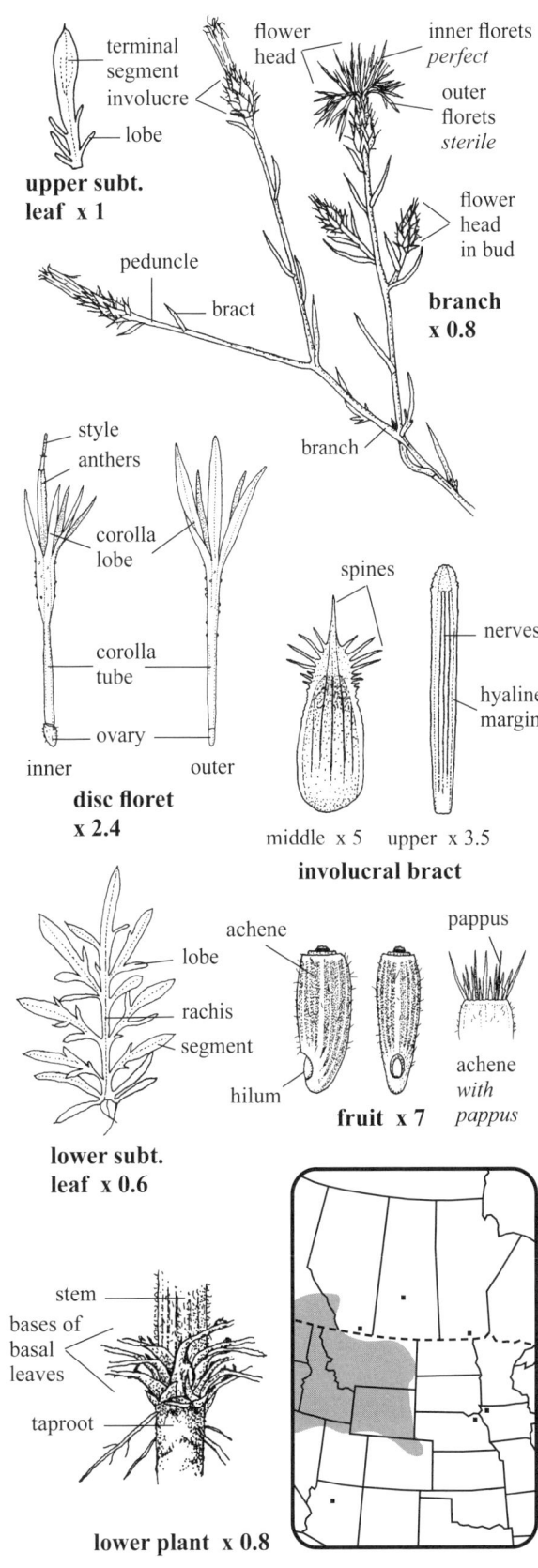

White Knapweed

Aster—*Asteraceae* **Wildflower** Blue Ligulate florets 11–21

CHICORY *Cichorium intybus* L.

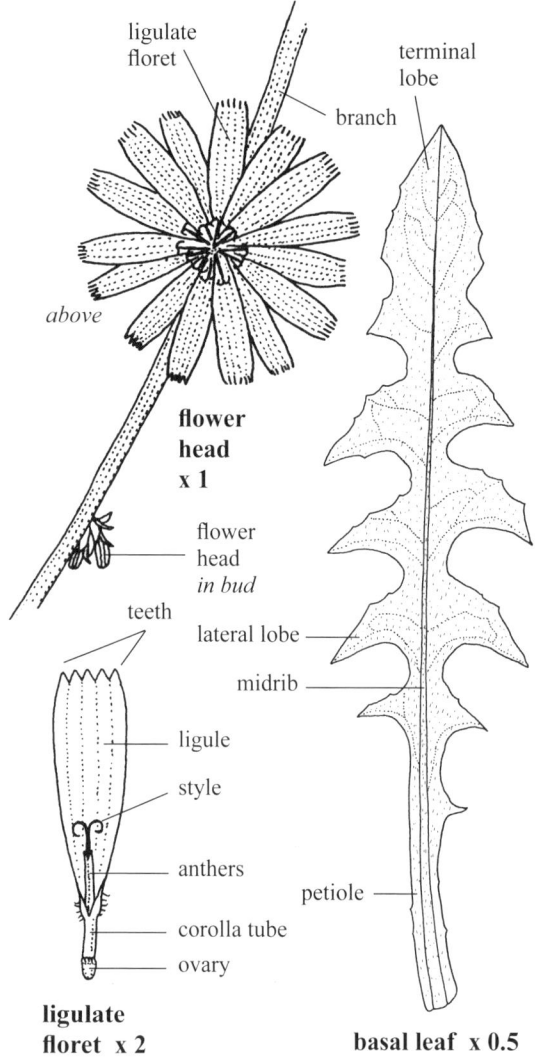

■ **SKETCH** A **perennial herb** 47–208 cm tall from a stout branched **taproot**; along roadsides, railways, hedgerows, field edges and other disturbed sites. *Naturalized*

■ **FLOWER HEADS** Blue, rarely white, blooming June–October; **inflorescence** a raceme with separate clusters of one to nine almost sessile heads alternating along the branches and upper stem; **floral branches** ascending, alternate, usually undivided, 1.5–120 cm long, ridged, subtended by sessile leaves 1.2–16 cm long with auricles; **peduncles** 0–10 cm long, occasionally a flower will terminate a bare branch; **subtending bracts** (of peduncles), 3–8 mm long, lightly hairy, entire to finely toothed; **flowering heads** opening once in the morning, 2.3–4.6 cm wide by 10–15 mm long; **involucral bracts** greenish red, hairy, in two series: **1) outer bracts** five, each 6–7 mm long by 2.5–3 mm wide, slightly imbricate, ascending, some hairs glandular, bases thick; **2) inner involucral bracts** usually eight, each 9–10 mm long by *c.* 1.5 mm wide, hairy at the pointed tips; <u>disc florets</u> absent; <u>ligulate florets</u> perfect, 15 (11–21), spreading; **ovary** white, glabrous, *c.* 1.5 mm long by *c.* 1 mm wide, flattened; **pappus** reduced to scales; **corolla tube** white, hairy where it expands, 2–2.5 mm long by *c.* 0.8 mm wide; **ligules** 16–18 mm long by 4–6 mm wide, with five apical teeth; **anthers** five, blue, forming a tube 3–4 mm long; **style** 2-parted, blue. **FRUIT** an achene, 1-seeded, 2.2–2.8 mm long by 1–1.5 mm wide by 0.7–1 mm thick, with four or five ridges near the base; **pappus** of ragged overlapping erect scales 0.2–0.3 mm long.

■ **LEAVES** Basal and stem, alternate, simple, lobed, toothed; **basal blades** in a rosette, ascending, 7–40 cm long by 2–13 cm wide, rough hairy on both surfaces; **petioles** winged, 1.5–3 cm long; **stem blades** (below branches) 3–15, petiolate to auriculate and clasping above, 3–33 cm long by 0.9–11 cm wide, hairy, reduced above.

■ **STEM** Erect, tough, stiff, branched, hollow; 5–18 mm wide and very hairy near the reddish base.

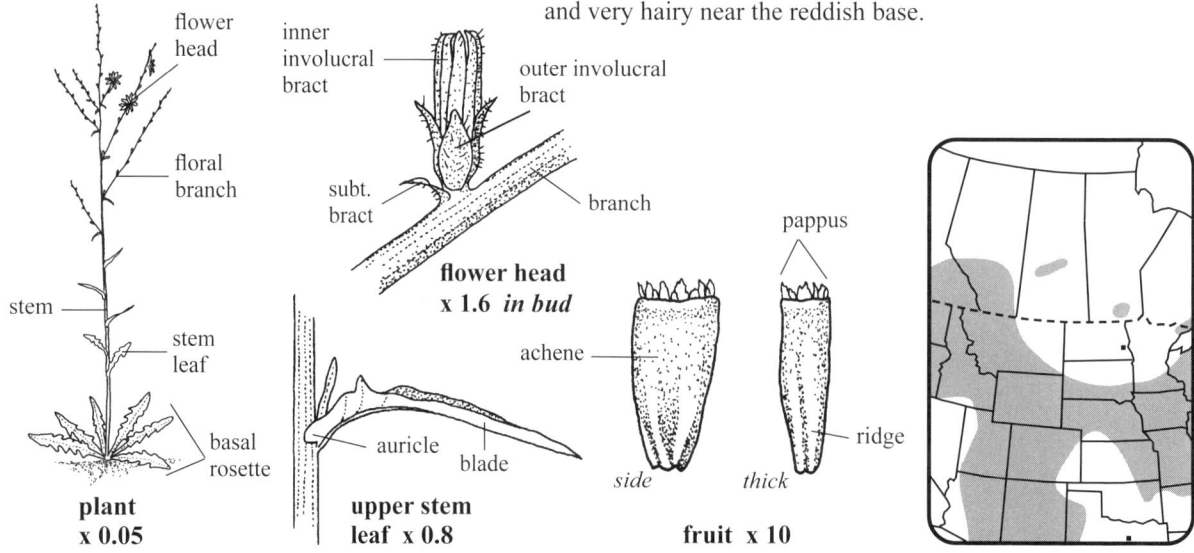

Aster—*Asteraceae* | **Wildflower** Pink Disc florets only

Canada Thistle *Cirsium arvense* (L.) Scop.

■ **SKETCH** An armed, variable **perennial herb** 30–180 cm tall, mostly in large colonies from deep **rhizomes** 1–2 mm wide and to 50$^+$ cm long with adventitious shoots; in pastures, ditches, city lots, prairies, along roadsides and river banks; mostly **dioecious**. *Naturalized* **w2**

■ **FLOWER HEADS** Pink, rarely white, blooming in April–October; **inflorescence** a loose corymb 5–30 cm tall by 3–42 cm wide of unisexual heads at floral branch ends; **floral branches** ascending, 2–52 cm long; **peduncles** 0.5–10 cm long; **flower heads** fragrant, 15–25 mm wide; **involucre** rough, 10–20 mm long by 5–10 mm wide, lengthening in fruit; **involucral bracts** in five or six series, imbricate, the reddish tips bent outward, bracts of two types: **1) lower outer bracts** thick, 2–5 mm long by 1–2 mm wide with a short apical ridge and gland, the upper margin hairy; **2) upper inner bracts** thin, 7–11 mm long by *c.* 1 mm wide, curved, slightly hairy at the reddish purple tips, central ridge dark green; <u>ray florets</u> absent; <u>male disc florets</u> sterile; **ovary** vestigial; **pappus** bristles 10–15 mm long; **corolla** 12–17 mm long, 5-lobed, the lobes linear, flat and thick, 3.5–5.5 mm long; **anthers** five, tubular, 4–5 mm long; **style** pink, 2-parted, usually not spreading, the **stigmatic joint** *c.* 2 mm from the tip; <u>female disc florets</u> perfect; **ovary** white, *c.* 1 mm long; **pappus** bristles as above; **corolla** 15–24 mm long, 5-lobed, these linear, 2–5 mm long; **anthers** five, vestigial at the base of the style; **style** as above, the **stigmatic joint** 1.5–2 mm from the tip; **fruiting heads** with a tan pappus 1–4 cm wide. **FRUIT** an achene, 1-seeded, 20–60 per head, 3-sided, light brown, shiny, curved, 2–4 mm long by 1–1.4 mm wide by 0.7–0.9 mm thick; **pappus** whitish tan, 25–30 mm long.

■ **LEAVES** Alternate, simple, lobed and toothed or not, mostly sessile, clasping above; **blades** with margins wavy and prickly, sharp spines to 5 mm long, 2–26 cm long by 1.3–8 cm wide, reduced above, surfaces glabrous or sometimes white woolly beneath, lobes often erect; **petioles** 0–3 cm long, reduced above; **auricles** clasping or not.

■ **STEM** Erect, stout, branched, ridged and angular, not spiny, hairless, solid; 3–15 mm thick near the base.

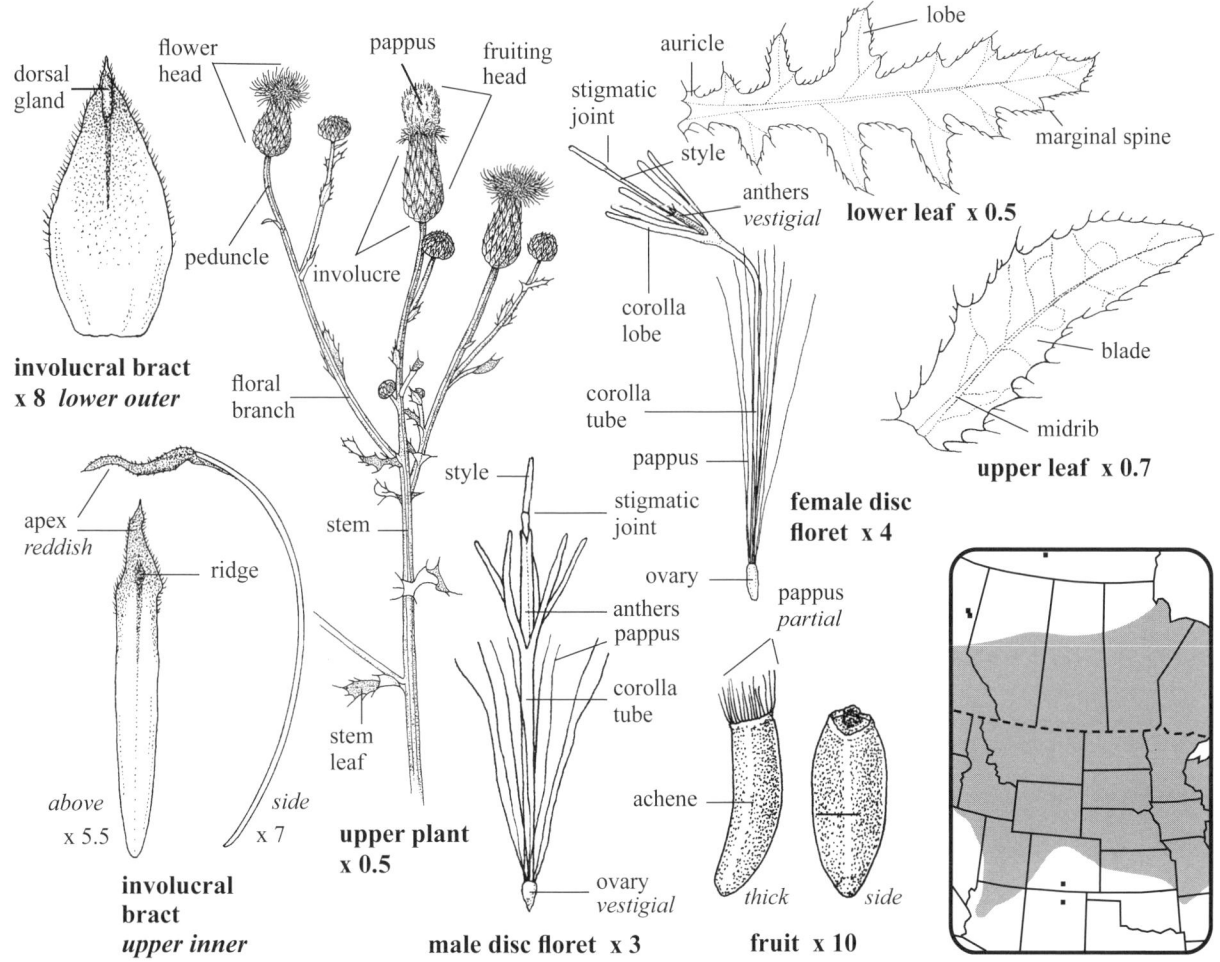

Canadian Thistle

Aster—*Asteraceae* **Wildflower** Pink to purplish Disc florets only

FLODMAN'S THISTLE *Cirsium flodmanii* (Rydb.) Arthur

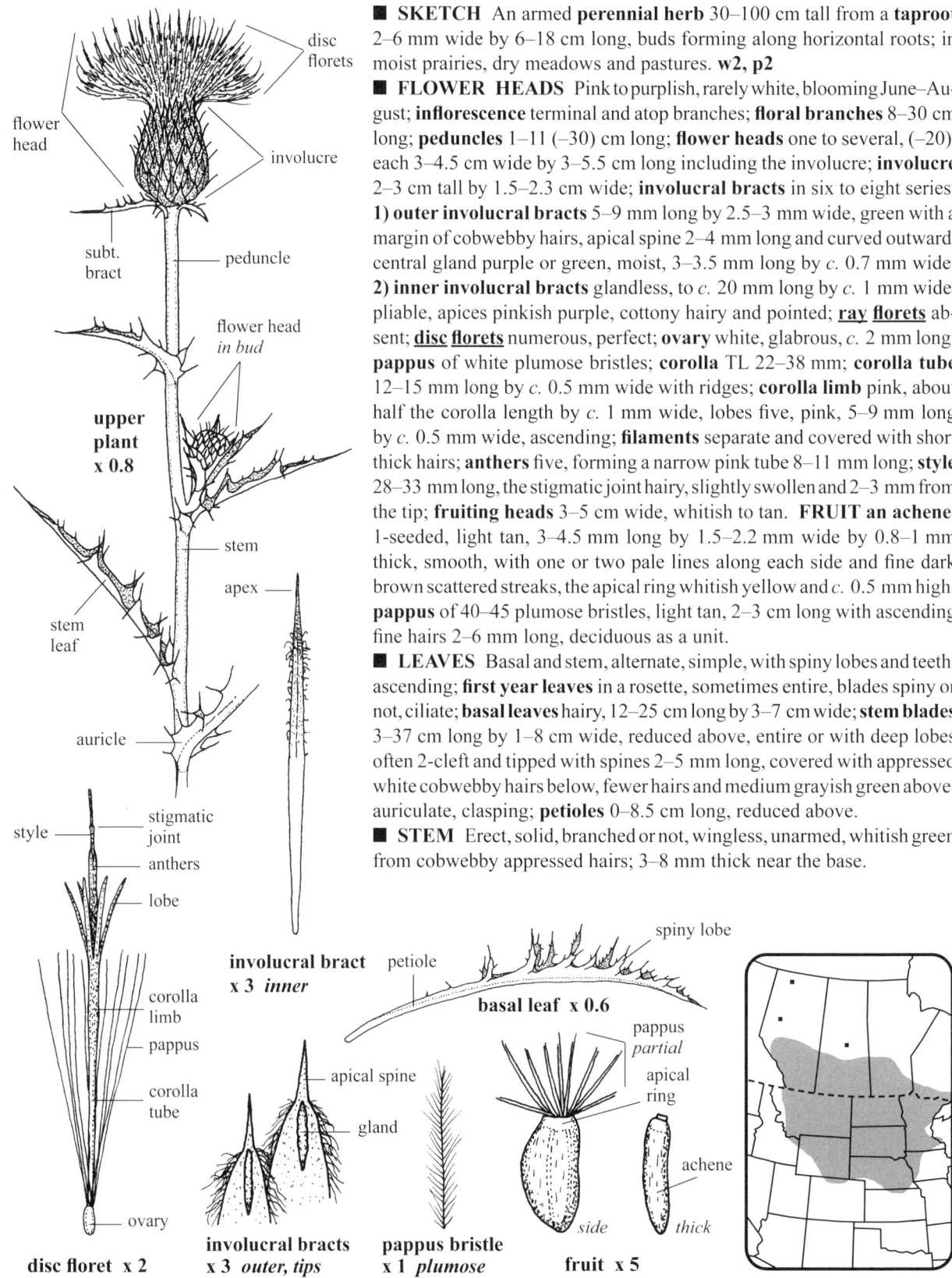

■ **SKETCH** An armed **perennial herb** 30–100 cm tall from a **taproot** 2–6 mm wide by 6–18 cm long, buds forming along horizontal roots; in moist prairies, dry meadows and pastures. **w2, p2**

■ **FLOWER HEADS** Pink to purplish, rarely white, blooming June–August; **inflorescence** terminal and atop branches; **floral branches** 8–30 cm long; **peduncles** 1–11 (–30) cm long; **flower heads** one to several, (–20), each 3–4.5 cm wide by 3–5.5 cm long including the involucre; **involucre** 2–3 cm tall by 1.5–2.3 cm wide; **involucral bracts** in six to eight series: **1) outer involucral bracts** 5–9 mm long by 2.5–3 mm wide, green with a margin of cobwebby hairs, apical spine 2–4 mm long and curved outward, central gland purple or green, moist, 3–3.5 mm long by *c.* 0.7 mm wide; **2) inner involucral bracts** glandless, to *c.* 20 mm long by *c.* 1 mm wide, pliable, apices pinkish purple, cottony hairy and pointed; **ray florets** absent; **disc florets** numerous, perfect; **ovary** white, glabrous, *c.* 2 mm long; **pappus** of white plumose bristles; **corolla** TL 22–38 mm; **corolla tube** 12–15 mm long by *c.* 0.5 mm wide with ridges; **corolla limb** pink, about half the corolla length by *c.* 1 mm wide, lobes five, pink, 5–9 mm long by *c.* 0.5 mm wide, ascending; **filaments** separate and covered with short thick hairs; **anthers** five, forming a narrow pink tube 8–11 mm long; **style** 28–33 mm long, the stigmatic joint hairy, slightly swollen and 2–3 mm from the tip; **fruiting heads** 3–5 cm wide, whitish to tan. **FRUIT** an achene, 1-seeded, light tan, 3–4.5 mm long by 1.5–2.2 mm wide by 0.8–1 mm thick, smooth, with one or two pale lines along each side and fine dark brown scattered streaks, the apical ring whitish yellow and *c.* 0.5 mm high; **pappus** of 40–45 plumose bristles, light tan, 2–3 cm long with ascending fine hairs 2–6 mm long, deciduous as a unit.

■ **LEAVES** Basal and stem, alternate, simple, with spiny lobes and teeth, ascending; **first year leaves** in a rosette, sometimes entire, blades spiny or not, ciliate; **basal leaves** hairy, 12–25 cm long by 3–7 cm wide; **stem blades** 3–37 cm long by 1–8 cm wide, reduced above, entire or with deep lobes often 2-cleft and tipped with spines 2–5 mm long, covered with appressed white cobwebby hairs below, fewer hairs and medium grayish green above, auriculate, clasping; **petioles** 0–8.5 cm long, reduced above.

■ **STEM** Erect, solid, branched or not, wingless, unarmed, whitish green from cobwebby appressed hairs; 3–8 mm thick near the base.

Aster—*Asteraceae* **Wildflower**   Pinkish purple   Disc florets only

Swamp Thistle   *Cirsium muticum* Michx.

■ **SKETCH** An armed **biennial herb** 60–180 cm tall with a tan, fleshy multibranched **taproot** 3–10 cm long by 1–4 mm thick and spreading; solitary or in open colonies along wet woodland edges, ditches, roadsides, riverbanks, meadows and edges of beaver ponds.

■ **FLOWER HEADS** Pinkish purple, blooming July–September; **inflorescence** terminal and branched, with 2–21 heads, together 4–25 cm long by 6–30 cm wide; **floral branches** ascending, 5–35 cm long; **peduncles** 0.5–11 cm long, ascending, ridged, with scattered hairs, sometimes with a leaf *c.* 2 cm long; **subtending bracts** leaf-like, 0–2 below each head, 1.5–4 cm long, toothed and armed; **flower heads** sometimes touching each other at right angles, each head 3–5 cm wide by 3–4.5 cm long and of disc florets only; **involucre** grayish green, 1.7–2.6 cm tall by 1.3–2.2 cm wide, not prickly; **involucral bracts** in 7–9 rows, spirally arranged, imbricate, white hairy on margins, dorsal ridge glandular; **lower bracts** stiff, 4.5–5 mm long by 1.9–2.2 mm wide, hairy apex with a tan spine *c.* 0.3 mm long; **upper bracts** glandless, 17–22 mm long by 1.6–2 mm wide, purplish above, greenish below, apical spine *c.* 0.7 mm long and bent outward; **disc florets** TL 30–35 mm; **ovary** white, 2.3–2.5 mm long, glabrous; **pappus** 13–20 mm long, white near base, tan above, hairy, apices of some bristles slightly flattened and bent; **corolla tube** white, 11–12 mm long by *c.* 0.6 mm wide; **corolla limb** pale purple, 7–8 mm long; **corolla lobes** five, 4–7 mm long by *c.* 0.6 mm wide, pointed, flat; **styles** 23–30 mm long, tip 2-parted, finely hairy; **stigmatic joint** 4–5 mm from tip; **stamens** five; **anthers** 7–9 mm long, white, tips free and pointed. **FRUIT** an achene, 1-seeded, purplish brown, 4–6 mm long by 1.5–2 mm wide by *c.* 1 mm thick with four low tan ridges, one per side, tan apical ring *c.* 0.2 mm wide, smooth and shiny; **pappus** bristles 30–35, plumose, tan, deciduous.

■ **LEAVES** Alternate, spiny, hairy, lobed and toothed, simple; **1st year blades** 2–5 in a rosette, each 5–30 cm long by 3–15 cm wide, prickly, lobes mostly entire; **petioles** 2–15 cm long, hairy, purplish at base; **stem blades** 4–55 cm long by 2–20 cm wide, reduced above, yellowish green, lighter below from appressed cottony hairs, petiolate below, sessile above; **lobes** toothed, 5–8 main ones, sinuses wide and almost reaching the wide midrib, spines along margins 1–3 mm long; **lower stem blades** often brown by anthesis.

■ **STEM** Erect, stiff, ridged, hollow, unarmed, branched, not winged, glabrous above, lower hairs white and spreading to slightly descending; 5–15 mm wide near the purplish, hairy base.

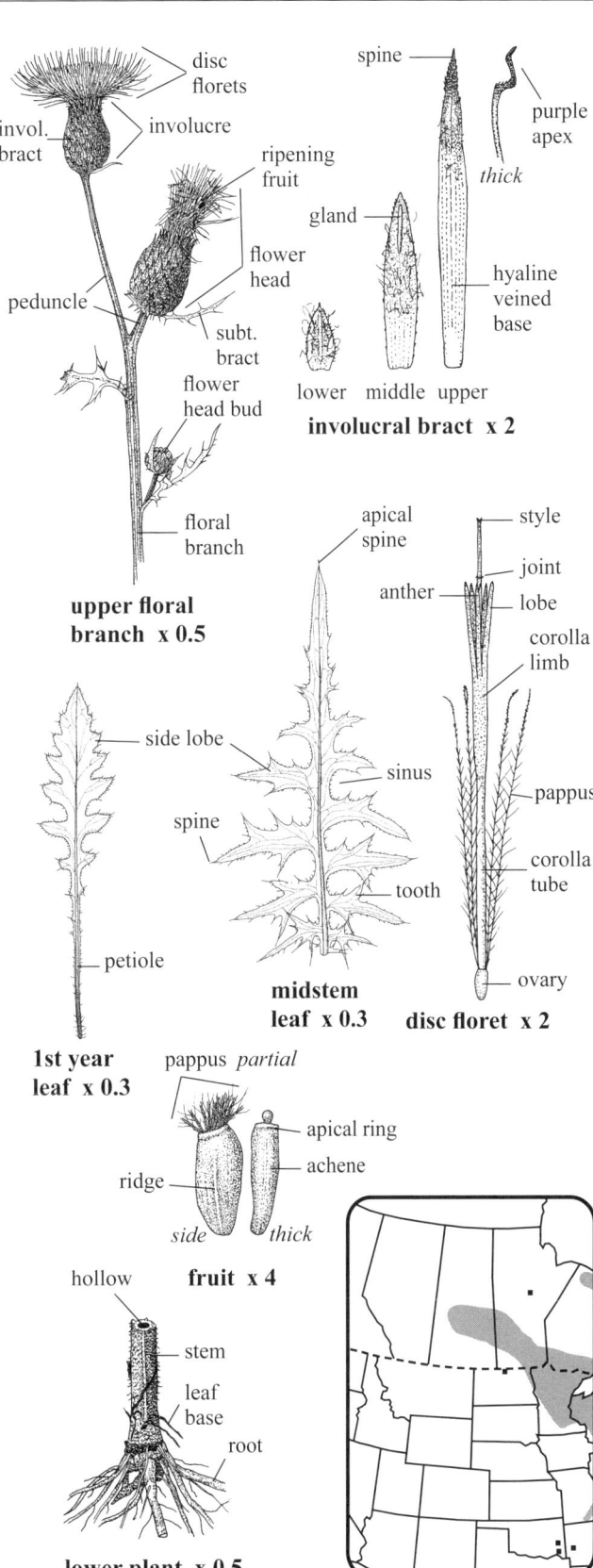

123

Aster—*Asteraceae* **Wildflower** Purplish pink Disc florets only

BULL THISTLE *Cirsium vulgare* (Savi) Ten.

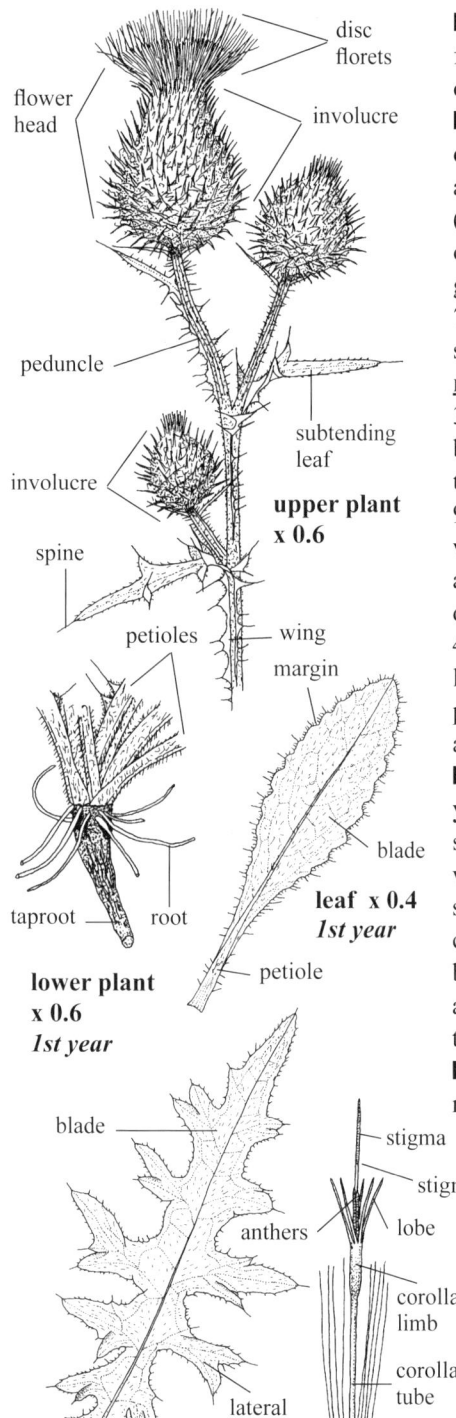

■ **SKETCH** An armed **biennial herb** 50–200 cm tall by 30–150 cm wide from a short, fleshy **taproot**; along field edges, roadsides, in pastures and city lots. *Naturalized*

■ **FLOWER HEADS** Purplish pink, blooming June–October; **inflorescence** terminal and axillary; **floral branches** alternate, ascending, leafy, axillary, 15–60+ cm long; **peduncles** spiny, 2–20 cm long by 2–3 mm thick (not including spines), terminal peduncle the longest; **flower heads** numerous, 2.5–4.5 cm wide by 4–6 cm long, erect to slightly nodding; **involucre** green, cobwebby, 2–4.5 cm wide by 2–4 cm tall; **involucral bracts** hairy, 7–30 mm long by 1–2 mm wide, imbricate in 8–12 series, not glandular, spine-tipped, ascending to descending below; **ray florets** absent; **disc florets** numerous, perfect, exserted 15–20 mm above the involucre, together 3–3.5 cm wide; **ovary** white, glabrous, 2–2.2 mm long, striate; **pappus** bristles white, 20–27 mm long, numerous; **corolla** TL 27–35 mm; **corolla tube** white, 18–23 mm long by *c.* 0.4 mm thick; **corolla limb** pink, TL 9–12 mm by *c.* 1 mm wide, lobes five, filiform, 5–8 mm long by *c.* 0.4 mm wide, flattened, pointed; **anthers** five, pinkish purple, 6–8 mm long, forming a tube; **style** 28–37 mm long, filiform, pink above; **stigma** not spreading, darker pink than the style, *c.* 0.2 mm thick; **stigmatic joint** microscopic, 4–5.5 mm from the tip. **FRUIT an achene**, 1-seeded, whitish tan, 3–5 mm long by 1.3–2 mm wide by 0.9–1.1 mm thick, shiny, finely streaked, long pale lines four; **pappus** tawny, bristles plumose, 20–28 mm long with ascending hairs.

■ **LEAVES** Basal and stem, alternate, simple, toothed, lobed, spiny; **1st year leaves** usually 8–10 in a rosette 23–50 cm wide, prostrate, hairy both sides; **blades** 6–25 cm long by 3–10 cm wide, lobes shallow, to about halfway to the midrib, dull; **petioles** 2–8 cm long, hairy; **2nd year leaves** hairy, spines 3–15 mm long; **blades** 5–40 cm long by 2–15 cm wide, lobes deeply cut and ascending to slightly descending, appressed prickles above, hairy below especially along veins; **petioles** of lower stem leaves 2–4 cm long and not continuing as spiny wings along the stem; **upper leaf** margins and the midribs continue along the stem as three, spiny, vertical lobed wings.

■ **STEM** Round, erect, tough, solid, branched, hairy, with seven to eight low ridges, spiny, winged above; 1–2 cm thick near the brownish, hairy base.

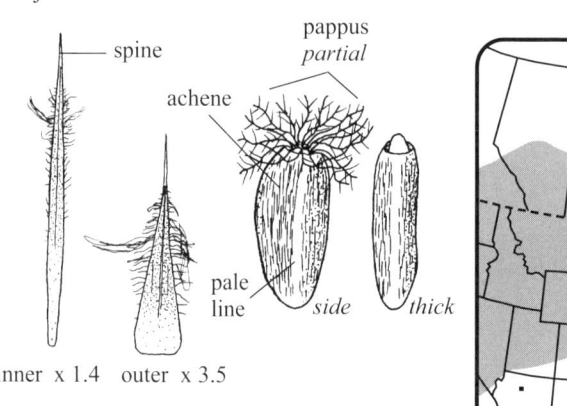

Aster—*Asteraceae* **Wildflower** White, center yellow Ray florets 20–40

Canada Fleabane *Conyza canadensis* (L.) Cronq.

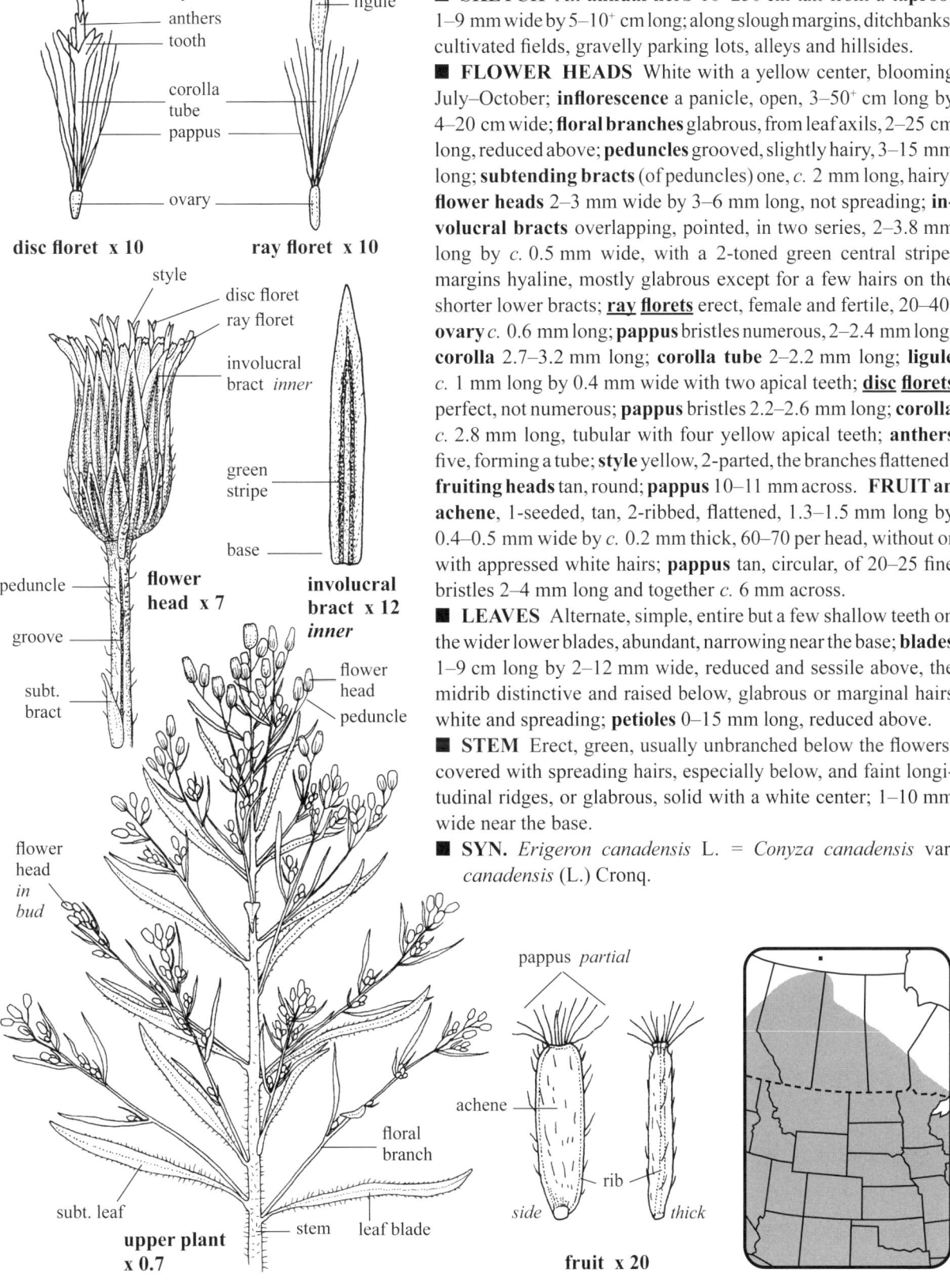

■ **SKETCH** An **annual herb** 10–250 cm tall from a **taproot** 1–9 mm wide by 5–10+ cm long; along slough margins, ditchbanks, cultivated fields, gravelly parking lots, alleys and hillsides.
■ **FLOWER HEADS** White with a yellow center, blooming July–October; **inflorescence** a panicle, open, 3–50+ cm long by 4–20 cm wide; **floral branches** glabrous, from leaf axils, 2–25 cm long, reduced above; **peduncles** grooved, slightly hairy, 3–15 mm long; **subtending bracts** (of peduncles) one, *c.* 2 mm long, hairy; **flower heads** 2–3 mm wide by 3–6 mm long, not spreading; **involucral bracts** overlapping, pointed, in two series, 2–3.8 mm long by *c.* 0.5 mm wide, with a 2-toned green central stripe, margins hyaline, mostly glabrous except for a few hairs on the shorter lower bracts; <u>**ray florets**</u> erect, female and fertile, 20–40; **ovary** *c.* 0.6 mm long; **pappus** bristles numerous, 2–2.4 mm long; **corolla** 2.7–3.2 mm long; **corolla tube** 2–2.2 mm long; **ligule** *c.* 1 mm long by 0.4 mm wide with two apical teeth; <u>**disc florets**</u> perfect, not numerous; **pappus** bristles 2.2–2.6 mm long; **corolla** *c.* 2.8 mm long, tubular with four yellow apical teeth; **anthers** five, forming a tube; **style** yellow, 2-parted, the branches flattened; **fruiting heads** tan, round; **pappus** 10–11 mm across. **FRUIT** an achene, 1-seeded, tan, 2-ribbed, flattened, 1.3–1.5 mm long by 0.4–0.5 mm wide by *c.* 0.2 mm thick, 60–70 per head, without or with appressed white hairs; **pappus** tan, circular, of 20–25 fine bristles 2–4 mm long and together *c.* 6 mm across.
■ **LEAVES** Alternate, simple, entire but a few shallow teeth on the wider lower blades, abundant, narrowing near the base; **blades** 1–9 cm long by 2–12 mm wide, reduced and sessile above, the midrib distinctive and raised below, glabrous or marginal hairs white and spreading; **petioles** 0–15 mm long, reduced above.
■ **STEM** Erect, green, usually unbranched below the flowers, covered with spreading hairs, especially below, and faint longitudinal ridges, or glabrous, solid with a white center; 1–10 mm wide near the base.
■ **SYN.** *Erigeron canadensis* L. = *Conyza canadensis* var. *canadensis* (L.) Cronq.

Aster—*Asteraceae*   **Wildflower**   Yellow, center dark reddish brown   Ray florets 8 (6–10)

COMMON TICKSEED   *Coreopsis tinctoria* Nutt.

■ **SKETCH** A glabrous **annual** or short-lived **perennial** 30–120 cm tall and wide; along slough margins, clay flats, ditches and low sandy sites.

■ **FLOWER HEADS** Yellow with a dark reddish brown center, blooming June–September; **inflorescence** of numerous heads terminating upper stalks; **branches** opposite, ascending to leaning with age; **peduncles** 1.2–7.2 cm long by 0.5–1 mm thick, glabrous, sometimes with a leaflike, entire, bracteole to *c.* 5 mm long; **flower heads** 14–25 mm wide by 6–8 mm tall; **disc** 5–10 mm wide by 4–5 mm tall; **involucre** green, glabrous, 3–4 mm tall by 11–14 mm wide; **involucral bracts** pliable, thin, in two series: 1) **lower outer bracts** about eight, each 2–2.2 mm long by 1–1.3 mm wide, slightly imbricate, margins hyaline; 2) **upper inner bracts** eight, imbricate, spreading at anthesis, 5–8 mm long by 3–3.2 mm wide, upper margins turning reddish brown; <u>**ray florets**</u> eight (6–10), sterile, each 7–15 mm long by 5–7 mm wide, 3- or 4-lobed, spreading, imbricate; **ovary** vestigial, *c.* 1.2 mm long by *c.* 0.8 mm wide, flat, 3-ribbed, glabrous, margins transparent; **pappus** absent; **corolla tube** *c.* 1 mm long, flattened, with two lateral nerves; **ligule** yellow, glabrous except for a few scattered hairs along the raised veins below, sometimes with a dark reddish area at base; **disc florets** numerous, perfect; **ovary** glabrous, 1–1.7 mm long by *c.* 0.6 mm wide; **pappus** absent; **chaffy bracts** scattered, more than one for each floret; **corolla** 2.7–3.2 mm long, glabrous; **corolla tube** 1.4–1.6 mm long, pale yellow; **corolla limb** 1.3–1.6 mm long, yellow, becoming reddish at the apex; **corolla teeth** four, reddish, descending, *c.* 0.3 mm long; **stamens** four, exserted; **anthers** dark brown, united, 1.2–1.3 mm long, apices pointed; **style** yellow, 2-parted, exserted. **FRUIT an achene**, 1-seeded, *c.* 2.5 mm long by *c.* 1.1 mm wide, dark brown, curved, flattened, slightly shiny, papillate, winged or not, wing tan; **pappus** of minute awns or a simple crown.

■ **LEAVES** Opposite (mostly), compound to simple, divided below to entire above, soft, sometimes with paired leafy fascicles in lower leaf axils; **blades** 0.7–15 cm long by 0.6–12 cm wide, segments usually entire, linear and sessile, terminal one the widest at 2–7 mm, midrib obvious, lateral veins obscure; **petioles** 0–8.5 cm long, widely grooved above, lower ones with a few scattered hairs; **stipules** absent.

■ **STEM** Green, oval, erect, glabrous, branched throughout; 5–10 mm thick near the base.

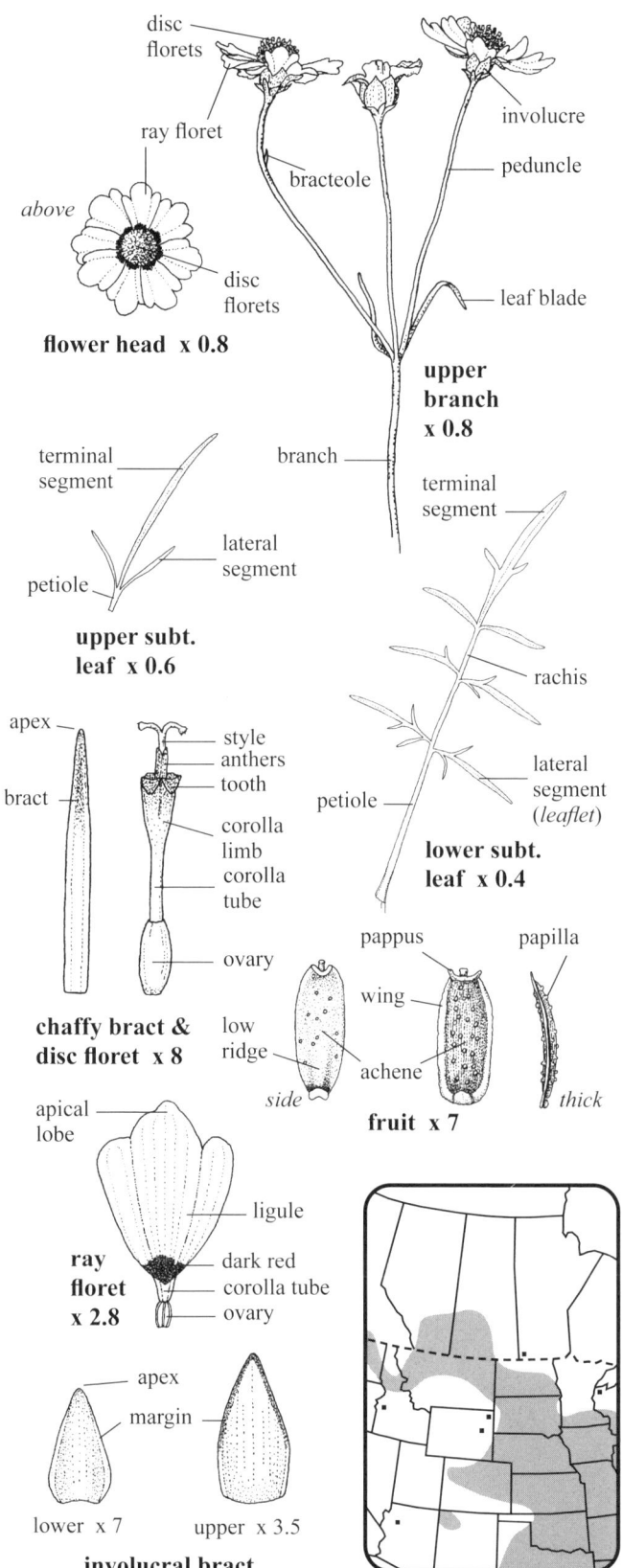

Golden Tickweed, Golden Tickseed, Plains Coreopsis, Garden Coreopsis

Aster—*Asteraceae* **Wildflower Yellow Ligulate florets 20–50**

Scapose Hawk's-beard *Crepis runcinata* (James) Torr. & Gray

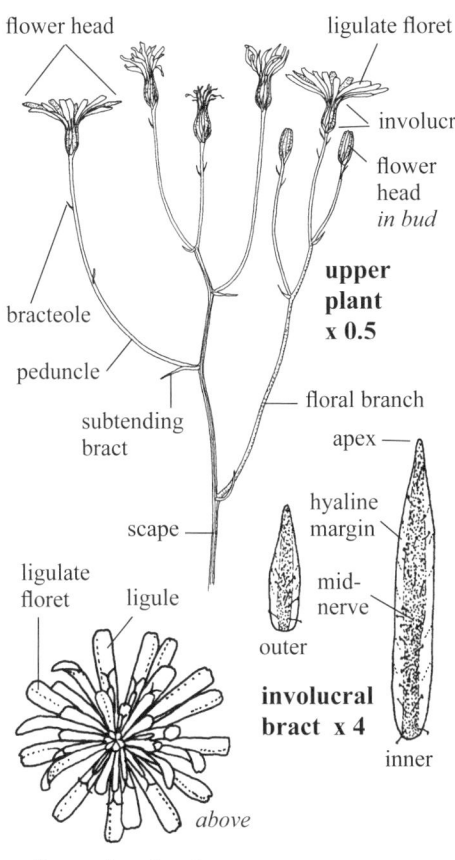

upper plant x 0.5

flower head x 1

basal leaf x 0.4
three shapes

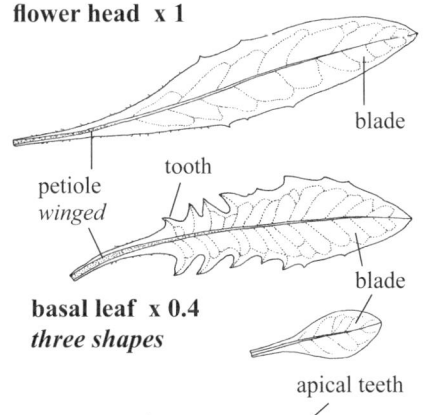

lower plant x 0.5

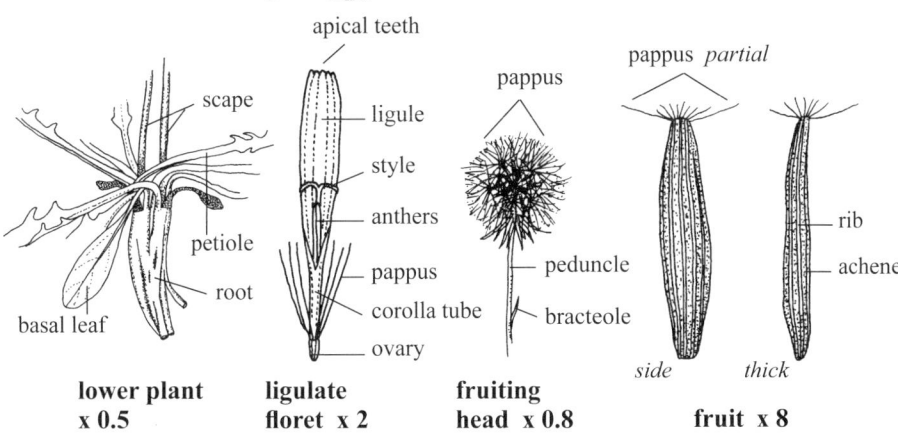

ligulate floret x 2

fruiting head x 0.8

fruit x 8

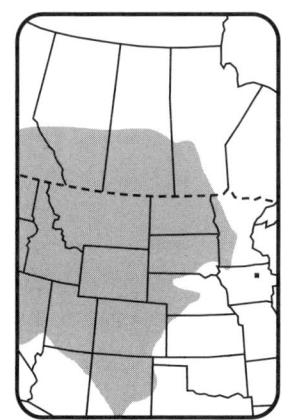

■ **SKETCH** A variable **perennial herb** 10–80 cm tall with tan **fascicled root**s 7–14 mm wide and to 15⁺ cm long, with a short **caudex**; in meadows, prairies, open woods, lakeshores, near springs and damp to drying sites; **juice milky**.

■ **FLOWER HEADS** Yellow, blooming May–August; **inflorescence** an open corymbiform pattern, simple to branching; **peduncles** 0.5–11 cm long; **bracteoles** usually one or two, 2–3 mm long; **subtending bracts** (of peduncles) one, 5–20 mm long by 1–1.5 mm wide; **flower heads** 22–35 mm wide, mostly 3–7 (1–17) per scape; **involucre** 7–16 mm long by 4–7 mm wide; **involucral bracts** green, reflexed in fruit, imbricate, glabrous to glandular-hairy, thin, soft, apices sometimes dark brown, in two series: **1) outer bracts** shorter, unequal and narrow, flat; **2) inner bracts** 10–16, each 4–10 mm long by 1–1.3 mm wide, margins hyaline, central green midnerve slightly raised and rounded, tips spreading under the florets; <u>**disc florets**</u> absent; <u>**ligulate florets**</u> perfect, 20–50 and 10–19 mm long; **ovary** green, hairless, ribbed, 1.3–1.5 mm long; **pappus** of numerous, fine white bristles 6.5–7 mm long; **corolla tube** 4–5.5 mm long; **ligule** 10–13 mm long by 2–3 mm wide, 4-nerved, apical teeth five, blunt and thickened, *c.* 0.2 mm long; **stamens** five; **anthers** yellow, 3.5–4 mm long, forming a tube; **style** yellow, exserted, 2-parted, each part *c.* 1.6 mm long, hairy below the parts; **fruiting heads** white, round, 15–17 mm wide and tall. **FRUIT** an achene, 1-seeded, brown, 3.5–7 mm long by 0.5–0.8 mm wide by *c.* 0.5 mm thick, 9- or 10-ribbed; **pappus** bristles *c.* 5 mm long, 60–70, together *c.* 10 mm wide for each achene.

■ **LEAVES** Mostly basal, simple, toothed to entire, forming a **rosette** 12–30 cm wide, persisting; **blades** 2–15 cm long by 0.7–5.5 cm wide, lightly hairy above and below, especially on the raised midrib below, hairs spreading and usually not glandular; **petioles** hairy, reddish, winged or not, 1–7 cm long; **stem leaves** few to absent, alternate, 2–8 cm long by 2–30 mm wide, reduced above to bracts subtending the peduncles and floral branches; **petioles** 0–3 cm long, reduced above.

■ **STEM** A **scape**, one to three, erect, tough, hollow, reddish green with shallow, longitudinal grooves, hairless, usually leafless; 1.5–3 mm thick near the base.

Smooth Hawk's-beard, Fiddleleaf Hawksbeard

Aster—*Asteraceae*   **Wildflower** Yellow Ligulate florets 30–70

NARROW-LEAVED HAWK'S-BEARD *Crepis tectorum* L.

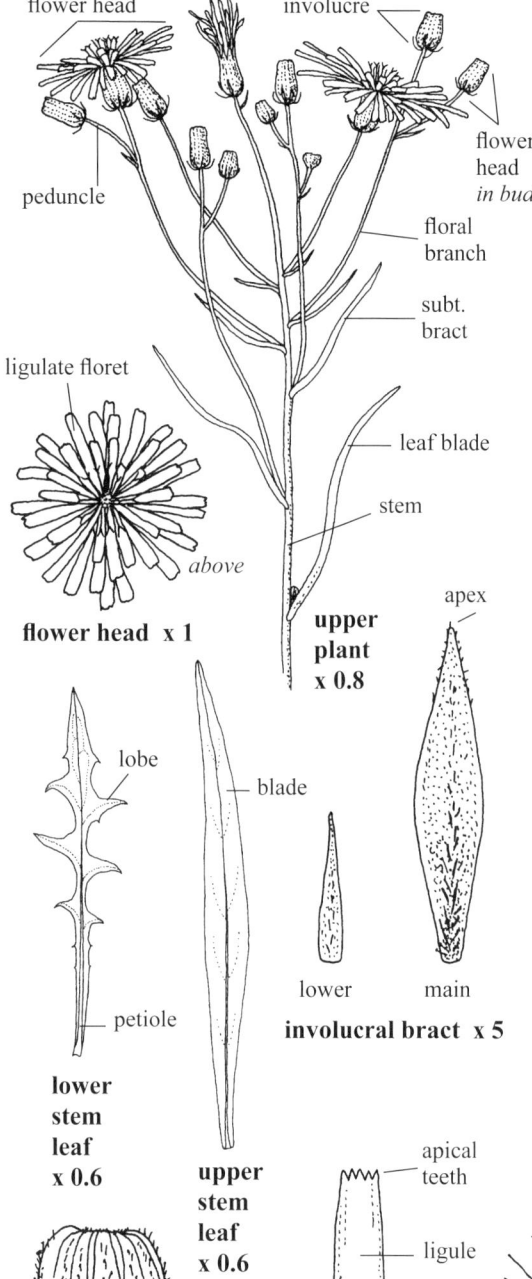

■ **SKETCH** An **annual** or **winter annual herb** 30–100 cm tall from a beige **taproot** 1–5 mm wide and 3–10⁺ cm long; along roadsides, in gravelled parking lots, forage crops, lawns and in gardens; **juice milky**. *Naturalized*

■ **FLOWER HEADS** Yellow, blooming June–August; **inflorescence** an open corymb of 2–60⁺ heads, together to *c.* 30 cm wide by *c.* 15 cm long; **floral branches** 4–30⁺ cm long, ridged and slightly scabrous, with two to several flower heads; **subtending bracts** (of floral branches and peduncles) linear, 1–4 cm long; **peduncles** 0.4–8 cm long, slightly hairy, some hairs gland-tipped; **flower heads** 10–30 mm wide by 10–15 mm tall; **involucre** green, hairy, 6–9 mm tall by 4–5 mm wide, some hairs gland-tipped; **1) main involucral bracts** 12–15, in one series, lightly hairy on both sides, 8–9 mm long, imbricate at the bases, tips hyaline with white hairs, margins hyaline, central nerve low, wide and hairy; **2) lower involucral bracts** of various sizes, about eight, 3.5–5 mm long, filiform; <u>disc florets</u> absent; <u>ligulate florets</u> perfect, 30–70; **ovary** glabrous, *c.* 1 mm long; **pappus** of *c.* 60 white bristles 4–5 mm long; **corolla tube** 2.5–3 mm long; **ligule** 7–10 mm long by 1.5–2.5 mm wide, glabrous, with five apical teeth; **anthers** five, forming a tube *c.* 2.5 mm long, yellow; **style** yellow, 2-parted; **fruiting heads** white, *c.* 15 mm wide and long. FRUIT **an achene**, 1-seeded, dark brown, 2.5–4.5 mm long by *c.* 0.5 mm wide, with 10–12 ribs, mostly round in cross-section, short-beaked; **pappus** as a whole 8–9 mm wide.

■ **LEAVES** Basal and stem, alternate, simple, lobed or toothed to entire; **basal rosette** 15–30 cm wide, of several leaves; **blades** 10–15 cm long by 0.7–4 cm wide, wilting early; **petioles** winged, 0.3–3 cm long; **stem leaves** lightly hairy on both sides, sessile, with narrow auricles, clasping or not; **lower blades** widely toothed or lobed, 5–17 cm long by 0.2–3.5 cm wide; **upper blades** linear, entire, bractlike, to *c.* 7 cm long by 4–10 mm wide, some with flower buds in axils; **petioles** 0–2 cm long.

■ **STEM** Erect, solitary, hollow, ridged above, slightly scabrous; 1–8 mm thick near the pinkish base.

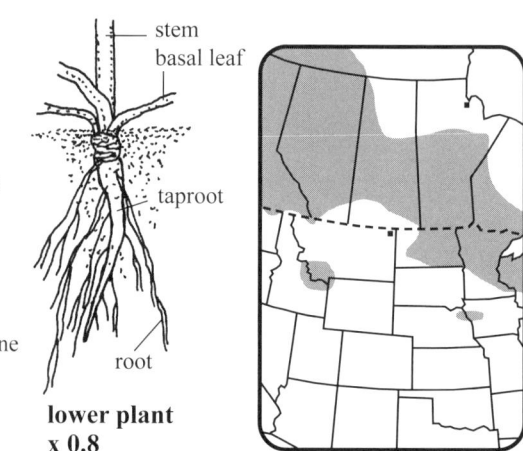

128   Annual Hawk's-beard

Aster—*Asteraceae* **Wildflower** White, center yellow Ray florets 4–8

FLAT-TOPPED WHITE ASTER *Doellingeria umbellata* (P. Mill.) Nees

■ **SKETCH** A **perennial herb** 30–200 cm tall with a woody brown **rhizome** to *c.* 18⁺ cm long by 3–4 mm thick with brown **fibrous roots** 2–10 cm long by *c.* 1 mm thick; in damp woods, thickets, edges of lakes and marshes, ditches, prairies, parklands and forest.

■ **FLOWER HEADS** White, with a yellow center, blooming July–September; **inflorescence** corymbiform, open, generally flat-topped, 3–50 cm tall by 15–30 cm wide; **floral branches** ascending, stiff, 2–50 cm long, reduced above, reddish purple, hairy, naked below; **peduncles** 5–18 mm long, hairs numerous and arched upward; **flower heads** 12–20 mm wide by 9–11 mm tall, numerous; **involucre** pale green, 4–6 mm long by 3–4 mm wide, hairy; **involucral bracts** imbricate, in 2 or 3 series, roundly keeled, 1.8–5 mm long, outer bracts short, thick, hairy, inner bracts longer, hairy at exposed tip, flat, margins pale green, spreading with fruit; <u>ray florets</u> female, fertile, usually 4–8 (–14); **ovary** pale green, *c.* 2 mm long, slightly hairy; **pappus** white, double, 5–5.3 mm long, short bristles *c.* 1 mm long (outer), not thickened at apex like some of the long inner bristles; **corolla tube** pale green, *c.* 3 mm long; **ligule** spreading to slightly descending, 6–8 mm long by 2–2.5 mm wide with 2–3 tiny apical teeth, hairless; **style** pale green, 2-parted, the parts flat, hairless, exserted; <u>disc florets</u> yellow, perfect, 15–30, together 5–8 mm wide; **ovary** and **pappus** as above; **corolla tube** *c.* 1.5 mm long, pale green, hairy; **corolla limb** *c.* 1.5 mm long; **corolla teeth** five, pale yellowish green, *c.* 1 mm long, hairless; **stamens** five; **anthers** united into a yellow tube *c.* 2 mm long; **style** 2-parted, pale green, exserted; **fruiting heads** 13–16 mm wide by 10–11 mm tall, dirty tan. **FRUIT an achene**, 1-seeded, brown, 2.3–3 mm long by *c.* 1 mm wide by 0.6–0.7 mm thick, slightly hairy, 5- to 7-ribbed.

■ **LEAVES** Alternate, simple, entire, stem only, dull, numerous, green to purplish, ascending to spreading; **blades** 2.5–15 cm long by 4–46 mm wide, reduced above and below, margins hairy and scabrous, hairy and lighter green below, hairs white and arched forward, much less hairy above; **petioles** 2–3 mm long, slightly hairy, pale green.

■ **STEM** Erect, reddish, solid, stiff, tough, hairy to glabrous above, hairs curled and ascending, with several long red ridges, branched above; 3–9 mm thick near the reddish, slightly hairy base.

■ **SYN.** *Aster umbellatus* P. Mill. = *Doellingeria umbellata* var. *umbellata* (P. Mill.) Nees and *Aster pubentior* Cronq. = *Doellingeria umbellata* var. *pubens* (Gray) Britt.

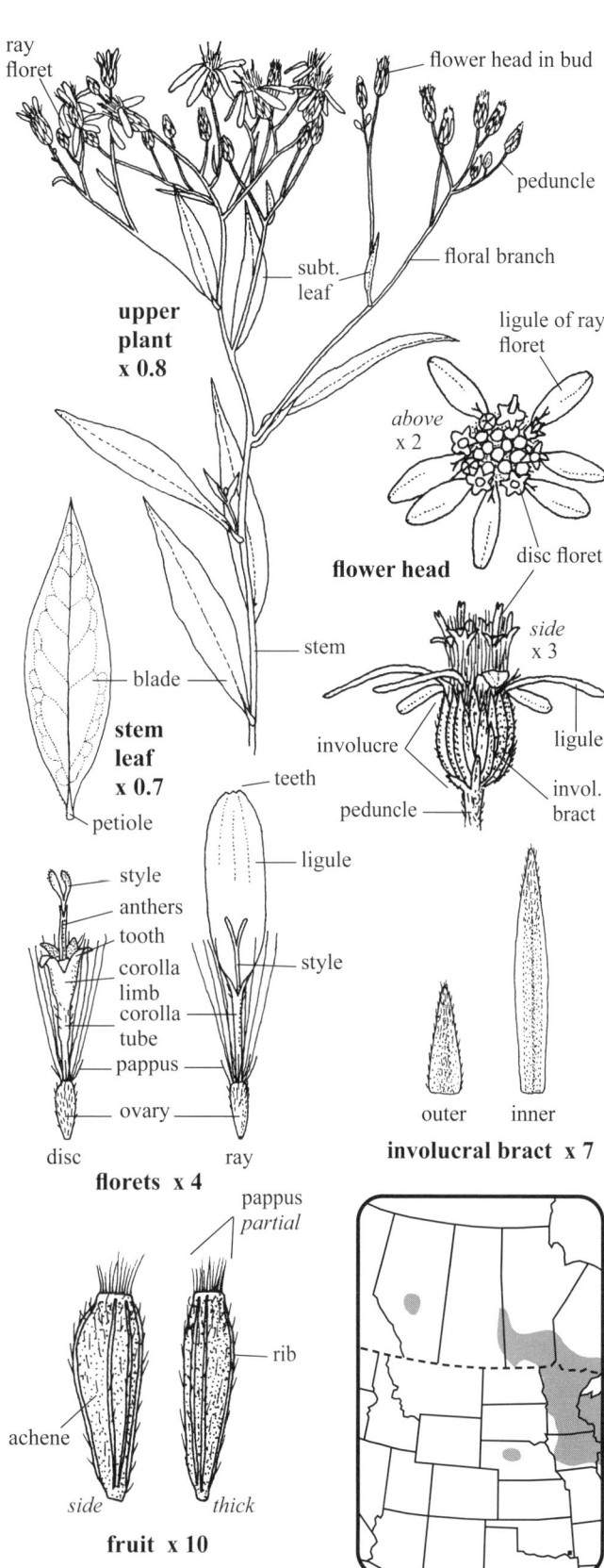

Umbellate Aster, Parasol Whitetop, Flat-top Aster, Tall Flat-top White Aster

Aster—*Asteraceae*   **Wildflower** Pink to pale purple, center reddish brown   Ray florets 16–19

Purple Coneflower *Echinacea angustifolia* DC.

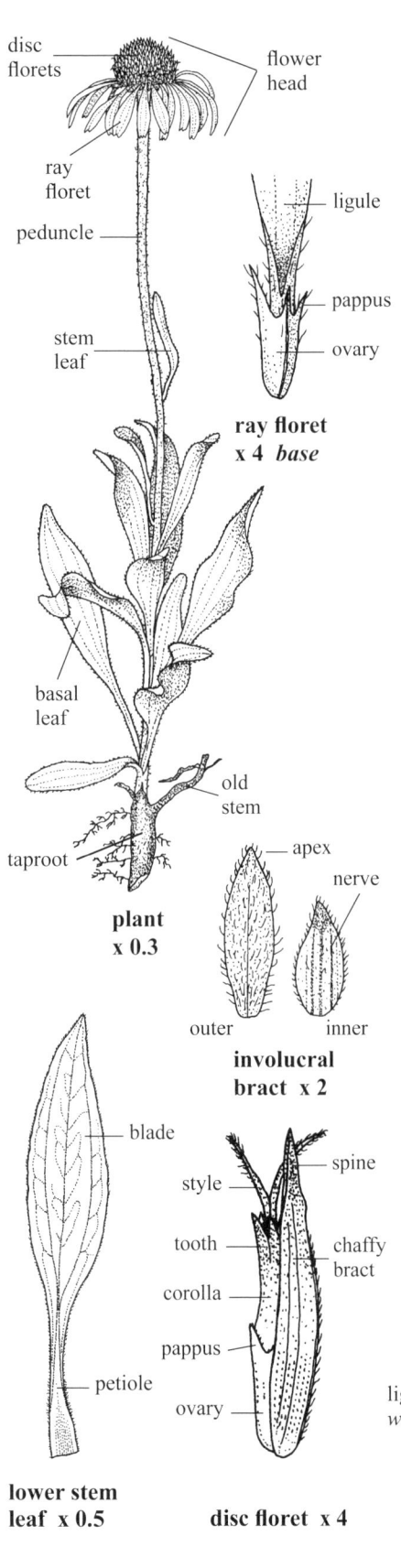

plant x 0.3

ray floret x 4 *base*

involucral bract x 2

lower stem leaf x 0.5

disc floret x 4

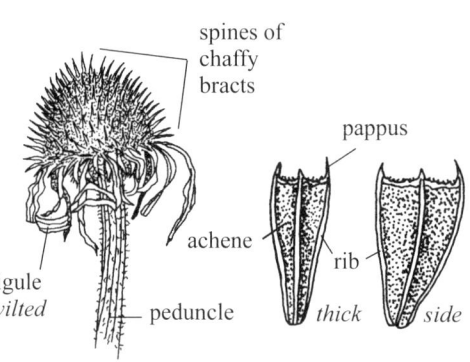

fruiting head x 0.8

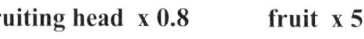

fruit x 5

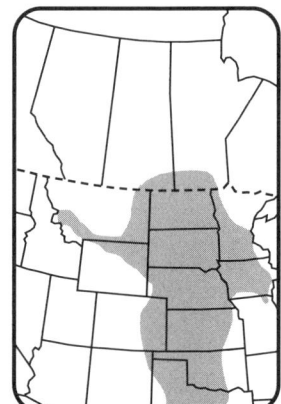

■ **SKETCH** A **perennial herb** 20–60 cm tall in open colonies from a woody **taproot** 2–15 mm thick and to *c.* 10⁺ cm long and a **caudex** to *c.* 3 cm long with one to several branches; in dry gravelly prairies, pastures, and dry hillsides.

■ **FLOWER HEADS** Pink to pale purple, center dark reddish brown, blooming June–August; **inflorescence** a terminal flowering head; **peduncles** 10–15 cm long, hairy; **flower heads** 3.5–8 cm wide by 4–5 cm long; **disc** 1.5–3 cm wide by 1–2 cm tall; **involucral bracts** in three or four series, imbricate, 7–13 mm long by 3–4 mm wide, thick, stiff, spreading, stiff hairs spreading to ascending from thick raised bases; <u>ray florets</u> 16–19, sterile, TL 2.5–4.5 cm, spreading then drooping; **ovary** and **pappus** 4–5.2 mm long; **ligules** 20–40 mm long by 5–8 mm wide, bright pink near the base, apical teeth two or three, each 1–2 mm long; <u>disc florets</u> perfect, blooming from outside to middle, numerous; **ovary** whitish, 3–3.2 mm long, 3-sided, *c.* 1.5 mm wide; **chaffy bracts** stiff, 10–13 mm long by *c.* 1.5 mm wide, greenish below, hairy along keel, 2- or 3-nerved along each side, narrowing to a thick, dark red spine 2–3 mm long; **corolla** 5–8 mm long (including teeth) by *c.* 1.5 mm wide; **apical teeth** five, each 1.5–1.8 mm long, erect, pink to dark red; **filaments** white, *c.* 4 mm long; **anthers** dark brown, 3–3.2 mm long, included to exserted; **pollen** yellow; **style** 2-parted and hairy, exserted; **fruiting heads** bristly from spines of chaffy bracts, 15–22 mm tall (not including involucral bracts) by 15–30 mm wide. **FRUIT an achene**, 1-seeded, 4–5 mm long by 2–2.5 mm wide by 1.5–1.8 mm thick, 4-angled, ribbed, dark brown, striated; **pappus** a short, toothed crown.

■ **LEAVES** Basal and stem, alternate, simple, entire, 10–17 per plant, spreading to ascending, mostly in lower half of stem; **basal blades** 7–25 cm long by 0.8–5 cm wide; **petioles** 3–11 cm long; **stem leaf blades** 3.5–10 cm long by 0.5–2 cm wide, reduced and more erect above, main veins three, margins undulate, tips pointed, sometimes twisted, scabrous, hairs simple with a green raised base, hairs *c.* 1 mm long, curved upward toward the apex; **petioles** 0–6 cm long, hairy, reduced above.

■ **STEM** Erect, stiff, hairy, some branched, green; **hairs** simple on raised bases, white and somewhat clustered; 2–5 mm thick near the base, and 5–7 mm thick below the flower head.

■ **SYN.** *Echinacea pallida* var. *angustifolia* (DC.) Cronq. = *Echinacea angustifolia* var. *angustifolia* DC.

Aster—*Asteraceae*   **Wildflower** Pale purple, center pale yellow   Ray florets 50–60

Northern Daisy Fleabane   *Erigeron acris* L.

■ **SKETCH** A variable **biennial** or **perennial herb** 30–80 cm tall with a simple or branched **caudex** *c.* 15 mm long by *c.* 5 mm wide; **fibrous roots** tan, 2–6 cm long by 0.3–1 mm thick; in open, rocky, gravelly woodland sites, road edges in woods, along lakeshores and streambanks.

■ **FLOWER HEADS** Pale purple, center pale yellow, blooming June–September; **inflorescence** corymbiform, 6–20 cm long by 2–8 cm wide of 7–31 flower heads; **lower floral branches** to *c.* 5 cm long, with two to four heads; **peduncles** 15–30 mm long, hairy, some glandular; **subtending bracts** (of peduncles) leaflike, 9–25 mm long by 3–6 mm wide, entire, ciliate along lower half; **flower heads** 8–12 mm wide by 10–12 mm long, with three types of florets; **involucre** hairy, some glandular; **outer involucral bracts** *c.* 5 mm long by *c.* 0.7 mm wide, *c.* 11, hairs gland-tipped; **inner involucral bracts** mostly green, 6.5–7 mm long by *c.* 0.7 mm wide, long hairs and short cone-shaped hairs, some hairs reddish purple, margins hyaline with scattered ascending hairs; **1) ray florets** 50–60, pistillate, TL 9–10 mm; **ovary** 1.5–1.7 mm long, pale green, slightly hairy at apex; **pappus** *c.* 6 mm long; **corolla tube** *c.* 4 mm long; **ligules** 2–4.5 mm long by *c.* 0.3 mm wide, erect to arched outward, hairy below near base; **2) middle florets** pistillate, forming a distinctive ring; TL *c.* 7.5 mm; **ovary** *c.* 1.5 mm long, as above; **pappus** tan, 6–7.5 mm long; **corolla tube** *c.* 4 mm long by *c.* 0.2 mm wide, apex hairy; **style** 2-parted, exserted *c.* 2 mm past corolla tip; **3) disc florets** perfect, 16–29, together 2.5–3.5 mm wide in center; TL 6–6.5 mm; **ovary** *c.* 1.6 mm long, as above; **pappus** tan, 6–7 mm long, *c.* 30 bristles; **corolla tube** *c.* 3 mm long, whitish; **corolla limb** *c.* 3 mm long, wrinkled; **corolla teeth** five, pale yellow, *c.* 0.5 mm long; **stamens** five; **anthers** yellow, *c.* 1 mm long, forming a tube; **style** 2-parted, exserted; **fruiting heads** *c.* 2 cm wide by 15–18 mm tall; **pappus** pale yellowish tan to white. **FRUIT** an achene, 1-seeded, 2-ribbed, 1.5–1.7 mm long by *c.* 0.3 mm wide by *c.* 0.2 mm thick, apices slightly hairy, base light brown.

■ **LEAVES** Basal and stem, alternate, simple, entire; **basal leaves** spreading, TL 3.5–10 cm; **blades** light green, 2–5.5 cm long by 7–10 mm wide, reduced above, slightly hairy, margins ciliate especially near base of blade, tapering gradually to a petiole; **petioles** 1.5–4.5 cm long, margins ciliate; **stem leaves** arching downward near stem's base, TL 2–10 cm; **blades** hairy, dull, 2–5 cm long by 5–11 mm wide, reduced and sessile above, tapered to petioles below; **petioles** 0–5 cm long, hairy.

■ **STEM** Erect, solitary (or two), slightly scabrous throughout; base hairy, dark brown and 2–3 mm wide.

■ **SYN.** *Trimorpha acris* auct. non = *Erigeron acris* ssp. *debilis* (Gray) Piper (L.) S.F. Gray p.p., *Trimorpha acris* auct. non (L.) S.F. Gray p.p. = *Erigeron acris* ssp. *politus* (Fries) Schinz & R. Keller.

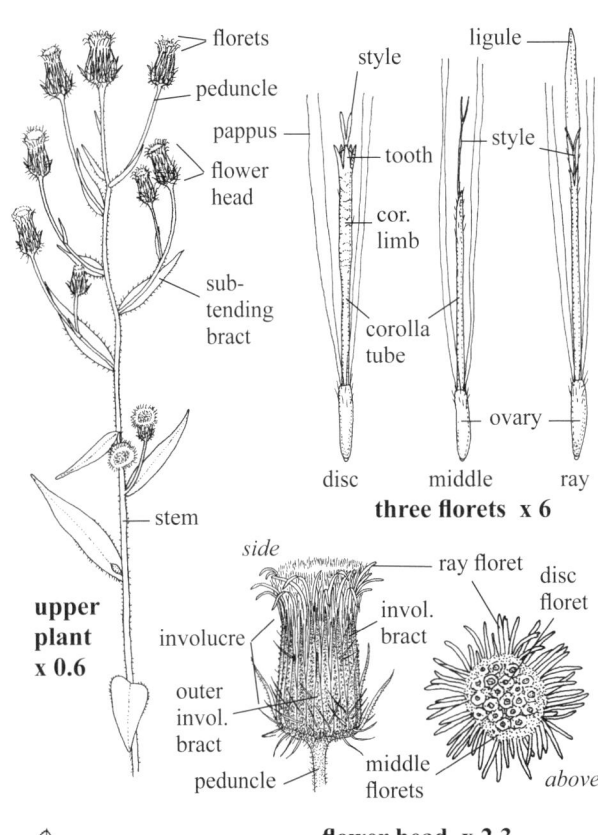

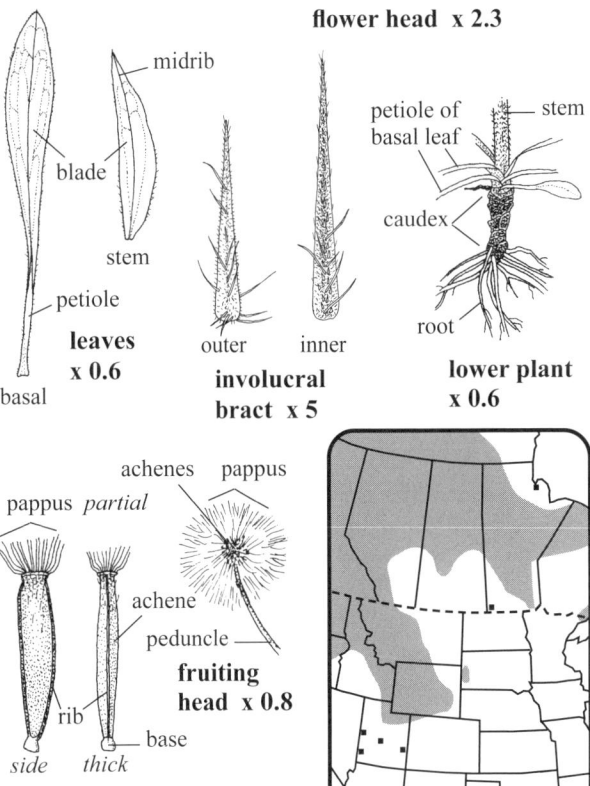

Aster—*Asteraceae*   **Wildflower**   White to pale blue, center yellow   Ray florets 30–100

TUFTED FLEABANE   *Erigeron caespitosus* Nutt.

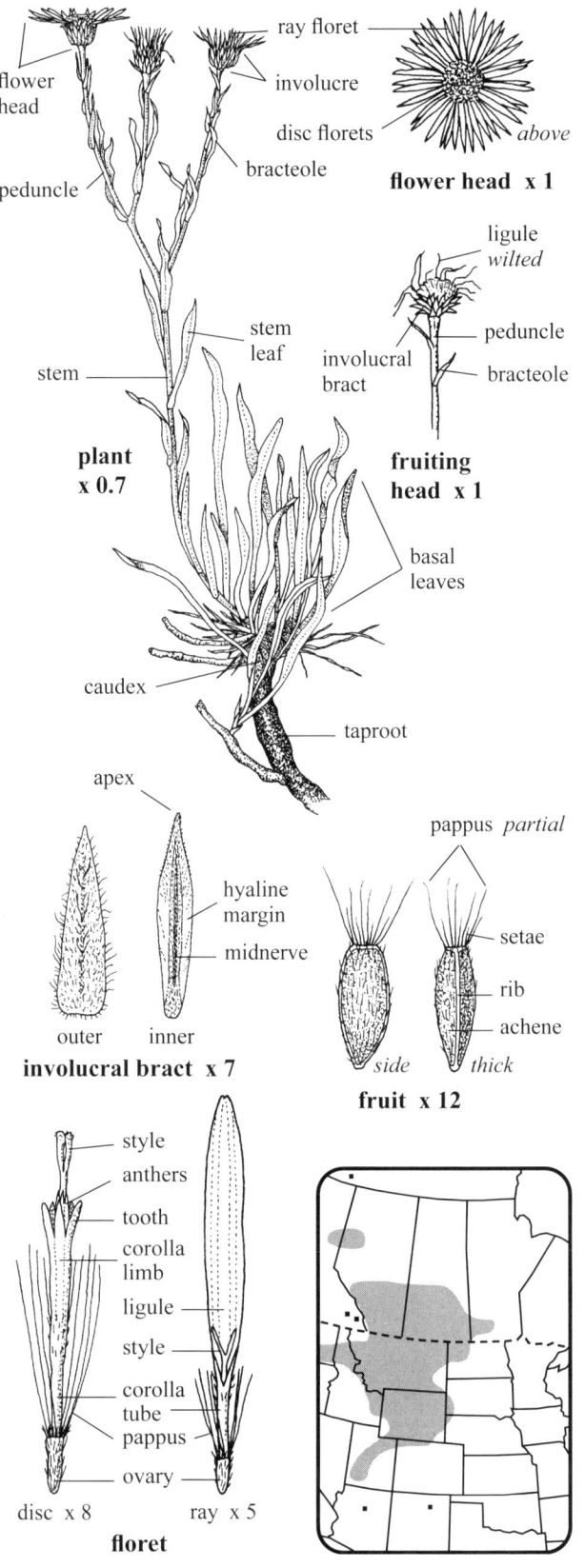

- **SKETCH** A tufted **perennial herb** 5–30 cm tall with a dark brown **taproot** 5–10 mm thick and to *c.* 10⁺ cm long with a branching **caudex**; on dry open hillsides, sagebrush and prairies, to above tree line. **p2**
- **FLOWER HEADS** White to pale blue, center yellow, blooming May–August; **inflorescence** a corymb, usually one to four heads (–9); **peduncles** erect to ascending, 3–4.5 cm long, white hairs declining, bracteoles usually 5–10 mm long by 1–2 mm wide, entire, sessile, hairy, reduced above, some twisted near the apex; **flower heads** 10–20 mm wide by 8–10 mm tall; **disc** 5–7 mm wide; **involucre** green, hairy, 5–7 mm tall by 6–7.5 mm wide, sparsely to densely glandular, hairs white and mostly spreading; **involucral bracts** in three or four series, imbricate, 3.8–5 mm long by 0.7–1 mm wide; **outer lower bracts** hairy on outside and margins, glabrous and thickened below, *c.* 0.5 mm thick at the base, midnerve slightly raised; **inner upper bracts** flat, pliable, glabrous except for short hairs on apical margins, base hairless, margins hyaline; **ray florets** 30–100, female, TL 8–18 mm; **ovary** pale green, *c.* 1 mm long, hairs ascending; **pappus** in two series: 18–25 inner bristles 2–2.8 mm long, and short bristles (setae) *c.* 0.3 mm long; **corolla tube** 2–2.2 mm long, pale green below to darker green above, ascending hairs at apex; **ligule** 5–15 mm long by 0.9–2.5 mm wide, veined, apex pointed to bidentate; **style** exserted, 2-parted, parts *c.* 0.8 mm long; **disc florets** perfect, numerous, yellow; **ovary** hairy and *c.* 1 mm long; **pappus** white, in two series: long inner bristles *c.* 20, each 3–3.5 mm long; short outer bristles (setae) 0.2–0.3 mm long; **corolla tube** 0.8–1.5 mm long, glabrous; **corolla limb** 2.3–3 mm long, pale green, hairs scattered and ascending at base, yellow and glabrous near teeth; **teeth** five, rather blunt, *c.* 0.7 mm long; **anthers** five, pale yellow, 1.5–1.8 mm long; **style** exserted, pale yellow, 2-parted, the parts flat, *c.* 0.8 mm long. **FRUIT an achene**, 1-seeded, brown, sparsely hairy, 2- or 3-ribbed, *c.* 1.5 mm long by *c.* 0.7 mm wide by *c.* 0.5 mm thick; **pappus** tawny, in two series: 15–25 inner bristles, each 2–2.5 mm long; outer bristles (setae) *c.* 0.5 mm long.
- **LEAVES** Basal and stem, alternate, simple, entire, hairs spreading, 3-nerved, midrib pale green and raised below; **basal leaves** crowded, ascending, 1.7–8 cm long by 3–10⁺ mm wide; **petioles** 5–20 mm long, hairy; **stem leaves** sessile, usually 2–3 cm long by 2–3 mm wide, reduced above, apices often twisted.
- **STEM** Erect, decumbent below, ridged from the continuation of midribs of leaves, hairs white and declining; 1.5–2.5 mm thick near the reddish green base.

Aster—*Asteraceae*   **Wildflower**   Light purple to pink or white, center yellow   Ray florets 125–175

Smooth Fleabane   *Erigeron glabellus* Nutt.

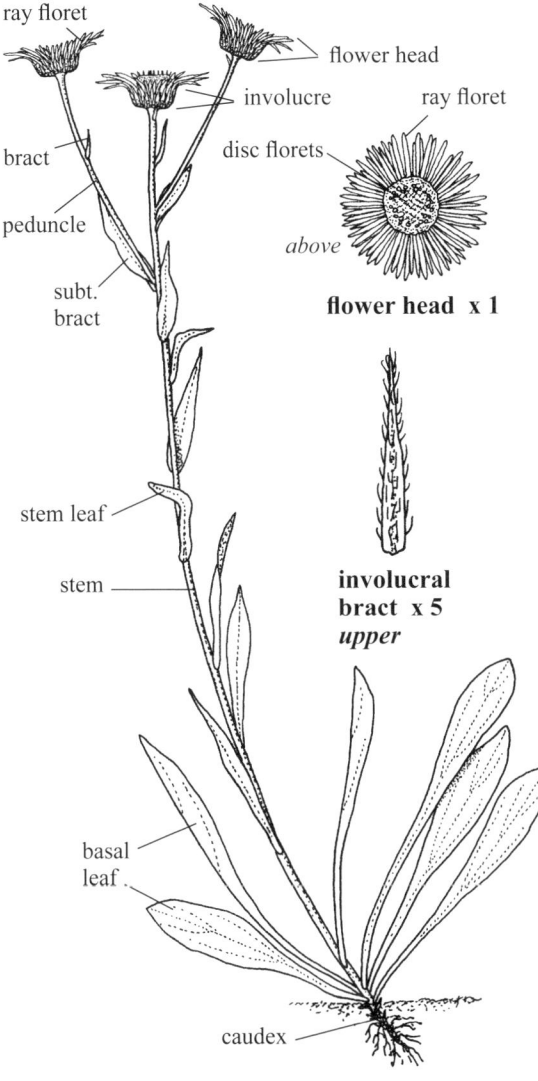

■ **SKETCH** A **biennial** or weakly **perennial herb** 15–70 cm tall with a **caudex** 1–2 cm long and **fibrous roots**; in moist to dry prairies, along streamsides, in open woodlands and meadows. **p1**

■ **FLOWER HEADS** Light purple to pink to white, with a yellow center, blooming May–August; **inflorescence** corymbose, of one to four (–15) flowering heads; **peduncles** scabrous, ridged, 1.5–14 cm long, with numerous white appressed to spreading hairs; **subtending bracts** (of peduncles) one, 1.5–4 cm long by 2–5 mm wide, hairy; **flower heads in bud** drooping; **flower heads** 15–30 mm wide by 6–10 mm long; **disc** flat, yellow, 8–20 mm wide; **involucre** green and very hairy; **involucral bracts** imbricate, in three or four series, 3.5–9 mm long by 0.6–1 mm wide, thick, slightly V-shaped; <u>ray florets</u> pistillate, 125–175, TL 9–19 mm; **ovary** white, hairy, *c.* 1.3 mm long, 2-ribbed; **pappus** bristles 12–15, each 2.5–3 mm long; **corolla tube** 1.5–2.5 mm long; **ligule** 6–15 mm long by 0.7–0.8 mm wide; **style** 2-parted, yellow; <u>disc florets</u> perfect, numerous, TL 6–7 mm; **ovary** as above; **pappus** bristles in two series: outer short setae, and *c.* 16 inner bristles to *c.* 3.3 mm long; **corolla** glabrous, 4–5 mm long, with five apical yellow teeth *c.* 0.5 mm long; **anthers** five, forming a yellow tube; **style** 2-parted, yellow; **fruiting head** light tan, *c.* 13 mm wide by *c.* 10 mm tall. FRUIT an achene, 1-seeded, 2-ribbed, 1.2–1.5 mm long by 0.5–0.7 mm wide by *c.* 0.2 mm thick, hairy; **pappus** double, white to tan, outer short setae and 12–18 inner long bristles 2.5–3 mm.

■ **LEAVES** Basal and stem, alternate, simple, entire to barely toothed; **basal blades** 1.5–15 cm long by 0.5–2 cm wide, glabrous; **petioles** 1–9 cm long, tapered; **stem blades** mostly sessile, entire, usually ascending, 1–8 cm long by 2–11 mm wide, reduced above, midrib obvious, lateral veins faint, hairy on margins and slightly so above, hairy to glabrous below.

■ **STEM** Erect to arching from the base, solid, one to four, pith white, sparsely hairy below, more hairy above and the hairs appressed to spreading; 1–5 mm thick near the base.

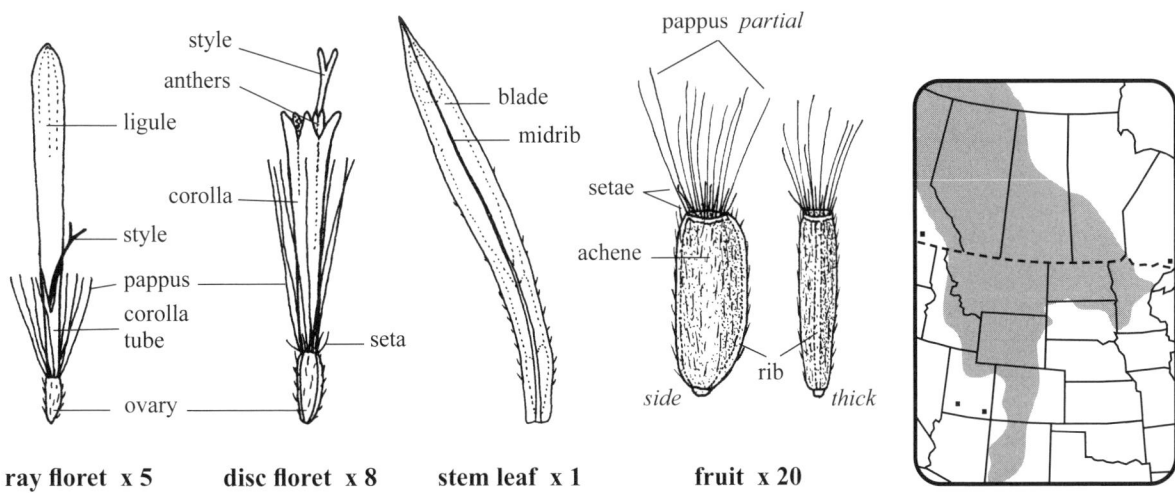

Streamside Fleabane   133

Aster—*Asteraceae*   **Wildflower**   White to pink, center yellow   Ray florets numerous

Hirsute Fleabane   *Erigeron lonchophyllus* Hook.

■ **SKETCH** A hairy **biennial** or **short-lived perennial** 5–60 cm tall from a hard **caudex** 7–10 mm long by *c.* 5 mm wide with whitish tan **fibrous roots** *c.* 5 cm long near its base; in wet to drying prairies, meadows, and sloughs, often in open colonies.

■ **FLOWER HEADS** White to pink, center yellow, blooming July–August; **inflorescence** a raceme, 1.5–23 cm long by 1–11 cm wide of 2–18 heads, terminal and axillary; **lower floral branches** up to *c.* 24 cm long; **peduncles** 1–10 cm long with one to several bracts 5–20 mm long by 0.5–1 mm wide, axillary, ascending, hairy, ridged vertically with dark green; **flower heads** 7–15 mm wide by 8–10 mm tall, erect to nodding; **disc** 2–11 mm wide; **involucre** green, hairy, 5–9 mm long by 3–12 mm wide, not glandular; **involucral bracts** imbricate, pale green, in 2 or 3 series, each 4–6 mm long by 0.5–0.7 mm wide, pointed, margins thinly hyaline, midvein darker green, upper inner tips brown to purplish and curved outward from among the ligule bases, hairs *c.* 1 mm long along the central vein of outer, lower bracts and on the sides near the base; **inner bracts** with small hairs only on the upper exposed half; **rachis** hairy; **ray florets** numerous, female; **ovary** *c.* 1 mm long, slightly hairy; **pappus** of 20–30 fine bristles 4–5 mm long; **corolla tube** pale green, 4–5 mm long by *c.* 0.1 mm wide; **ligules** white to pink, spreading to erect, 2–3 mm long by 0.1–0.5 mm wide, apices entire or with two blunt teeth; **style** 2-parted, these *c.* 0.3 mm long and white; **disc florets** perfect, 20–30, white, 3.5–5 mm long (including ovary); **ovary** *c.* 0.8 mm long, slightly hairy, 2-ribbed; **pappus** as above; **corolla tube** 2–4 mm long; **corolla teeth** five, blunt, *c.* 0.4 mm long; **anthers** yellow, *c.* 0.9 mm long, tips apiculate, united around the style; **style** 2-parted, parts *c.* 0.5 mm long, filiform, slightly exserted; **fruiting heads** light tan, 12–18 mm wide by 10–15 mm tall. **FRUIT** an **achene**, 1-seeded, tan, flattened, 1.2–1.5 mm long by *c.* 0.3 mm wide by *c.* 0.1 mm thick, lightly hairy, with two dark lateral ribs; **pappus** as above.

■ **LEAVES** Alternate, entire, linear, lower margins ciliate with long white hairs arched forward; **basal blades** 0.8–8 cm long by 3–15 mm wide, spatulate or more linear; **petioles** (of basal leaves) 1–8 cm long, hairy along their lower margins, tapered gradually into the blade; **stem blades** erect to ascending, pointed, slightly V-shaped in cross-section, 3–14 cm long by 2–5 mm wide, glabrous, with leafy fascicles to *c.* 5 cm long around the lower blades.

■ **STEM** Erect, light green, simple to branched throughout, stiff in lower half, lax above, spreading scattered hairs to *c.* 1.5 mm long and smaller hairs with enlarged bases; 1–4 mm thick near the base (not including hairs).

■ **SYN.** *Trimorpha lonchophylla* (Hook.) Nesom.

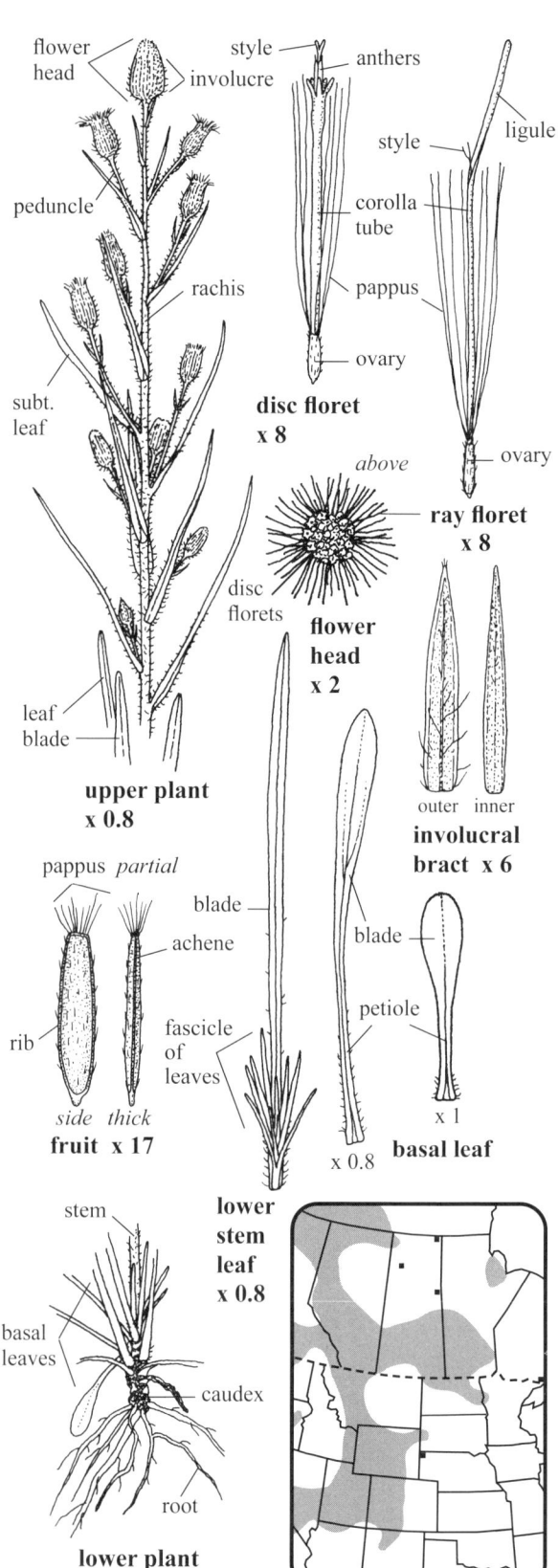

Short-ray Fleabane, Spear-leaved Fleabane

Aster—*Asteraceae* **Wildflower** Pink to pale purple to white, center yellow Ray florets *c*. 150

PHILADELPHIA FLEABANE *Erigeron philadelphicus* L.

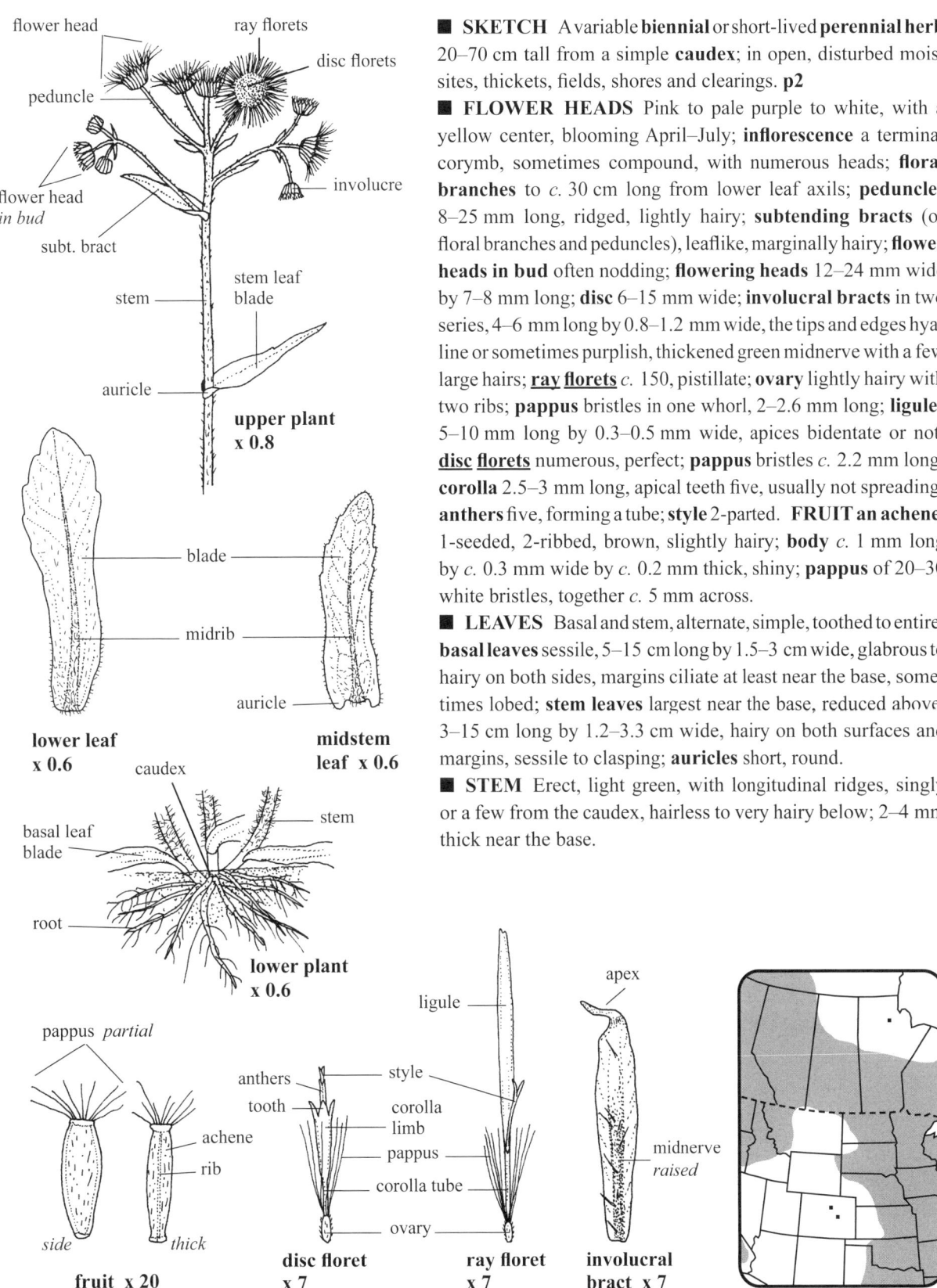

■ **SKETCH** A variable **biennial** or short-lived **perennial herb** 20–70 cm tall from a simple **caudex**; in open, disturbed moist sites, thickets, fields, shores and clearings. **p2**

■ **FLOWER HEADS** Pink to pale purple to white, with a yellow center, blooming April–July; **inflorescence** a terminal corymb, sometimes compound, with numerous heads; **floral branches** to *c*. 30 cm long from lower leaf axils; **peduncles** 8–25 mm long, ridged, lightly hairy; **subtending bracts** (of floral branches and peduncles), leaflike, marginally hairy; **flower heads in bud** often nodding; **flowering heads** 12–24 mm wide by 7–8 mm long; **disc** 6–15 mm wide; **involucral bracts** in two series, 4–6 mm long by 0.8–1.2 mm wide, the tips and edges hyaline or sometimes purplish, thickened green midnerve with a few large hairs; **ray florets** *c*. 150, pistillate; **ovary** lightly hairy with two ribs; **pappus** bristles in one whorl, 2–2.6 mm long; **ligules** 5–10 mm long by 0.3–0.5 mm wide, apices bidentate or not; **disc florets** numerous, perfect; **pappus** bristles *c*. 2.2 mm long; **corolla** 2.5–3 mm long, apical teeth five, usually not spreading; **anthers** five, forming a tube; **style** 2-parted. **FRUIT an achene**, 1-seeded, 2-ribbed, brown, slightly hairy; **body** *c*. 1 mm long by *c*. 0.3 mm wide by *c*. 0.2 mm thick, shiny; **pappus** of 20–30 white bristles, together *c*. 5 mm across.

■ **LEAVES** Basal and stem, alternate, simple, toothed to entire; **basal leaves** sessile, 5–15 cm long by 1.5–3 cm wide, glabrous to hairy on both sides, margins ciliate at least near the base, sometimes lobed; **stem leaves** largest near the base, reduced above, 3–15 cm long by 1.2–3.3 cm wide, hairy on both surfaces and margins, sessile to clasping; **auricles** short, round.

■ **STEM** Erect, light green, with longitudinal ridges, singly or a few from the caudex, hairless to very hairy below; 2–4 mm thick near the base.

Common Fleabane 135

Aster—*Asteraceae* **Wildflower** White to blue, center yellow Ray florets 50–100

WHITETOP *Erigeron strigosus* Muhl. ex Willd.

■ **SKETCH** An **annual** or rarely **biennial herb** 30–75 cm tall from **fibrous roots**; in moist to drying prairies and disturbed sites. **p2**

■ **FLOWER HEADS** White to pink to blue with a yellow central disc, blooming May–August; **inflorescence** corymblike, 4–50⁺ heads; **floral branches** one to several, 12–30 cm long, hairy; **peduncles** 1–7 cm long, drooping when flower head in bud, erect at anthesis, hairs ascending and appressed but not glandular; **flower heads** 8–12 mm wide by 8–9 mm tall, disc 5–10 mm wide; **involucre** hairy, yellowish green, 3–5 mm long by *c.* 5 mm wide, not glandular, ridged due to the raised hairy midnerves of the inner bracts; **involucral bracts** imbricate, in three or four series, 2–3 mm long by 0.4–0.8 mm wide (not including hairs), erect at anthesis, becoming reflexed as fruit ripens, midnerve green, convex, thickened but pliable, margins hyaline; **ray florets** female, 50–100, TL 5.5–8 mm; **ovary** *c.* 0.9 mm long, hairy; **pappus** white, double, fine bristles to *c.* 1 mm long with shorter setae; **corolla** pale green, *c.* 1 mm long, hairy where it begins to widen into a ligule; **ligule** erect at anthesis, 4–6 mm long by 0.5–1 mm wide, glabrous, apex minutely bidentate; **disc florets** perfect, numerous, yellow, TL 2.5–3.5 mm; **ovary** 0.8–1 mm long, hairy; **pappus** white, double, 10–15 fine bristles 1.5–2 mm long with shorter setae; **corolla** 1.5–2.5 mm long, pale green below, yellowish near the five, ascending, *c.* 0.4 mm long apical teeth; **anthers** five, yellow, *c.* 0.9 mm long, included, united; **style** 2-parted, these flattened and *c.* 0.4 mm long, included but not hairy; **fruiting heads** white, 6–8 mm wide by 6–7 mm tall. **FRUIT** an achene, 1-seeded, yellowish tan, *c.* 1 mm long by *c.* 0.3 mm wide by *c.* 0.2 mm thick, 2-ribbed laterally, scattered hairs ascending; **pappus** double, of 10–15 bristles 1.8–2 mm long with shorter setae; **receptacle** naked.

■ **LEAVES** Basal and stem, alternate, simple, usually entire; **basal blades** wilting early, TL 1–15 cm by 0.5–2.5 cm wide; **petioles** 2–5 cm long, hairs ascending on both sides; **stem blades** sparse, flat, scabrous, linear, spreading below to ascending above, 1–8 cm long by 2–20 mm wide; **petioles** hairy, 0–5.5 cm long, reduced above.

■ **STEM** Green to reddish, round, solid, erect, scabrous, hairy throughout, these mostly appressed and ascending; 1–4 mm thick near the base.

■ **SYN.** *Erigeron ramosus* (Walt.) B.S.P. = *Erigeron strigosus* var. *strigosus* Muhl. ex Willd. and *Erigeron annuus* ssp. *strigosus* (Muhl. ex Willd.) Wagenitz = *Erigeron strigosus* var. *strigosus* Muhl. ex Willd.

Prairie Fleabane, Daisy Fleabane

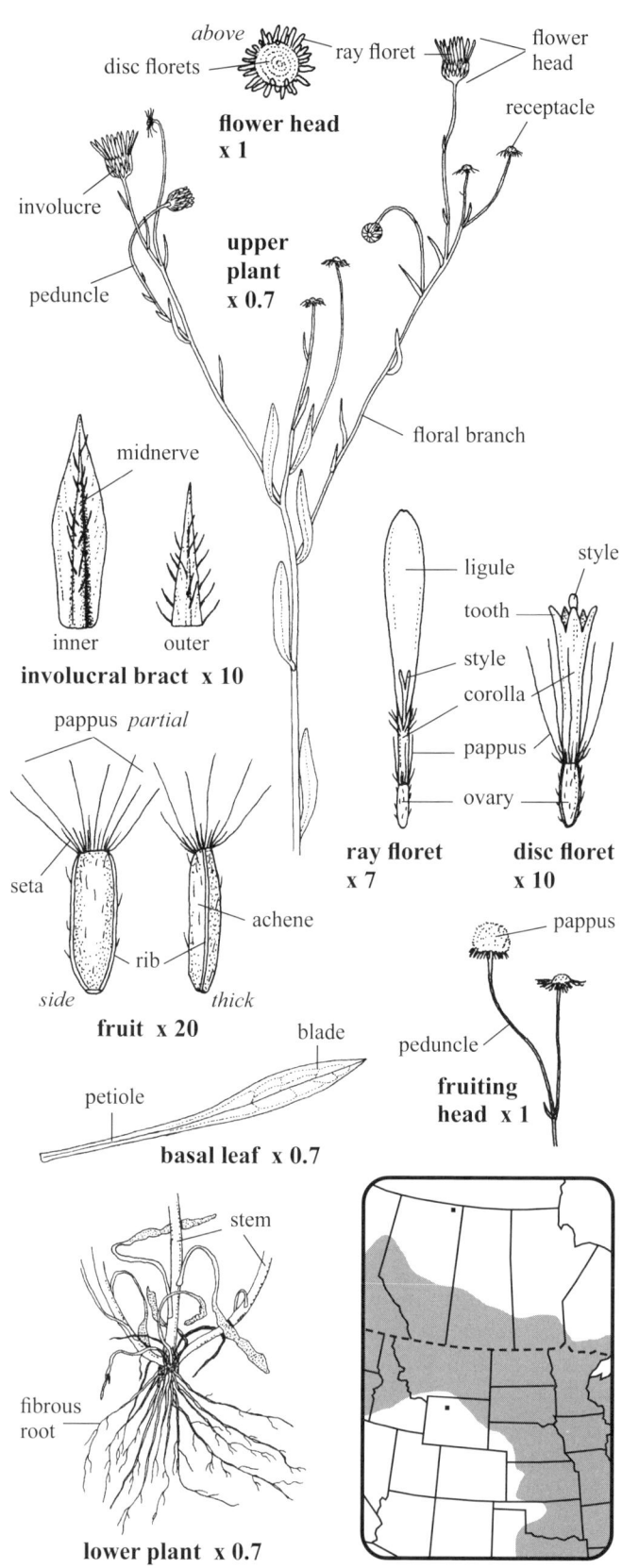

Aster—*Asteraceae* **Wildflower** Pinkish purple Disc florets only, 9–22

Spotted Joe-pye Weed *Eupatorium maculatum* L.

■ **SKETCH** A **perennial herb** 50–200 cm tall with a stout **rhizome** to *c.* 10 mm thick and numerous lateral fibrous **roots** 10–20 cm long by *c.* 1 mm thick; in parklands and Boreal forest, ditches, slough edges, bogs, marshes, moist sites in open wooded areas and along canal banks.

■ **FLOWER HEADS** Pinkish purple, blooming July–September; **inflorescence** a corymbiform cyme forming distinct clusters of numerous heads, 4–15 cm long by 5–25 cm wide, clusters terminal and on floral branches; **floral branches** in whorls of four from upper leaf axils, each 1–11 cm long, white hairs bent forward; **rays** 3–20 mm long, hairy; **stalks** 2–12 mm long; **peduncles** 2–7 mm long, hairy; **flower heads** 10–13 mm wide by 10–15 mm long; **involucre** pink and green, 7–9 mm long by 3.5–5 mm wide, hairy near the base, ascending around fruit; **involucral bracts** imbricate in 3 or 4 series, margins ciliate, apices blunt to pointed above, white to pink hairs short and curled forward on body; **lower bracts** 3-nerved, 2.3–4 mm long by 1–1.8 mm wide, green at base, pinkish at apex, hyaline margins near base, lateral nerves covered by pink hairs; **mid** and **upper involucral bracts** 6–9 mm long by 0.4–1.6 mm wide, tips pink, more narrow and longer above; <u>ray</u> **florets** absent; <u>disc</u> **florets** 9–22 per head, TL 10–12 mm (including styles); **ovary** green, 3–3.5 mm long by *c.* 0.6 mm wide, glandular, 5-ribbed; **corolla** 5–5.5 mm long by *c.* 0.9 mm wide, pale pink, glandular below; **corolla teeth** five, erect to spreading, *c.* 0.8 mm long, apical hairs few but golden glandular dots on the outside of the upper half of each tooth; **pappus** of numerous bristles, light tan, 5–5.5 mm long; **stamens** five; **filaments** white, glabrous, free and separate; **anthers** pinkish purple, *c.* 2 mm long, tips exserted; **pollen** white; **style** one, *c.* 8 mm long, 2-parted above, the parts *c.* 4 mm long, pale pink and mostly exserted; **fruiting heads** *c.* 10 mm long by 8–12 mm wide with the involucral bracts ascending to spreading. **FRUIT an achene**, 1-seeded, gray, golden glandular, 5-ribbed, cottony hairy on ribs in upper half; **body** 3–4 mm long by 0.6–0.7 mm wide and thick, bent at base; **pappus** tan, 5–7 mm long.

■ **LEAVES** Four (3–6) in whorls, simple, coarsely toothed, spreading; **stem blades** 5–20 cm long by 1–8 cm wide, dull, lighter green below, pointed, teeth point forward, rough to touch above, more hairy below, glandular golden dots numerous below, short hairs white and arched forward; **petioles** 3–15 mm long, hairs as on blades.

■ **STEM** Erect, solitary, unbranched, vertically streaked with purple, continuous pinkish purple above, sometimes ridged above, hollow, stout, white hairs arched forward, more hairy below; 3–10 mm wide at the green base.

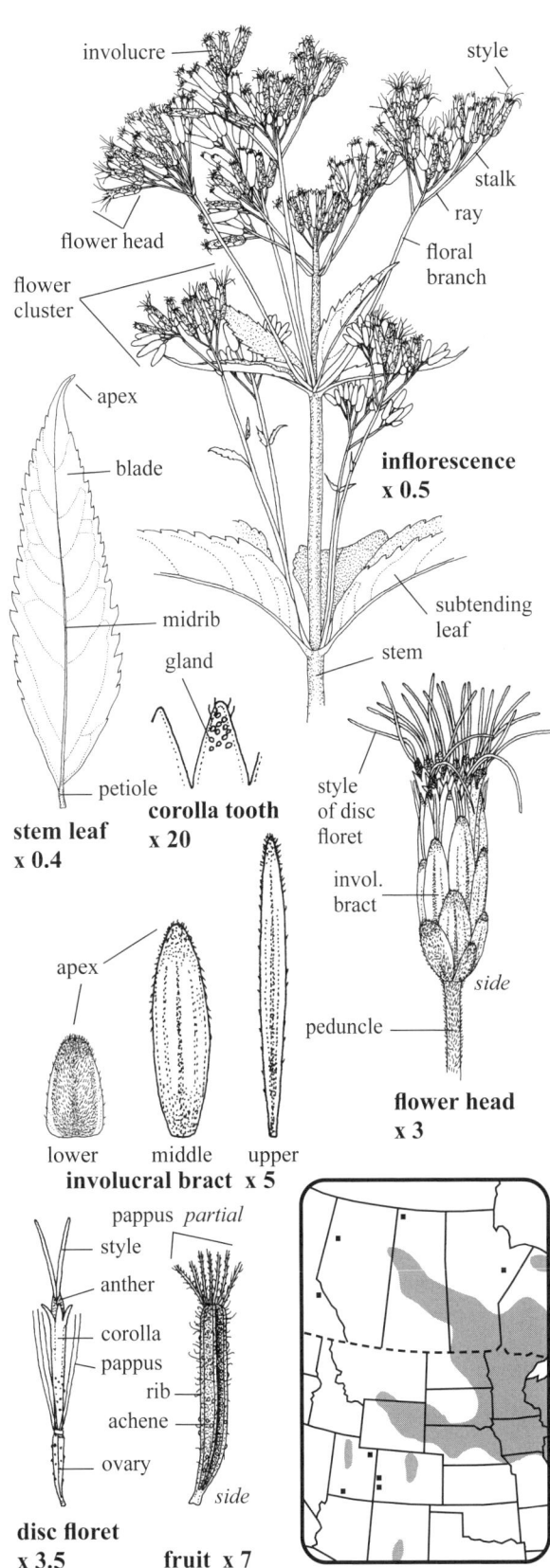

Tall Boneset, Bruner's Trumpetweed, Joe-pye Weed, Purple Boneset

Flat-topped Goldenrod *Euthamia graminifolia* (L.) Nutt.

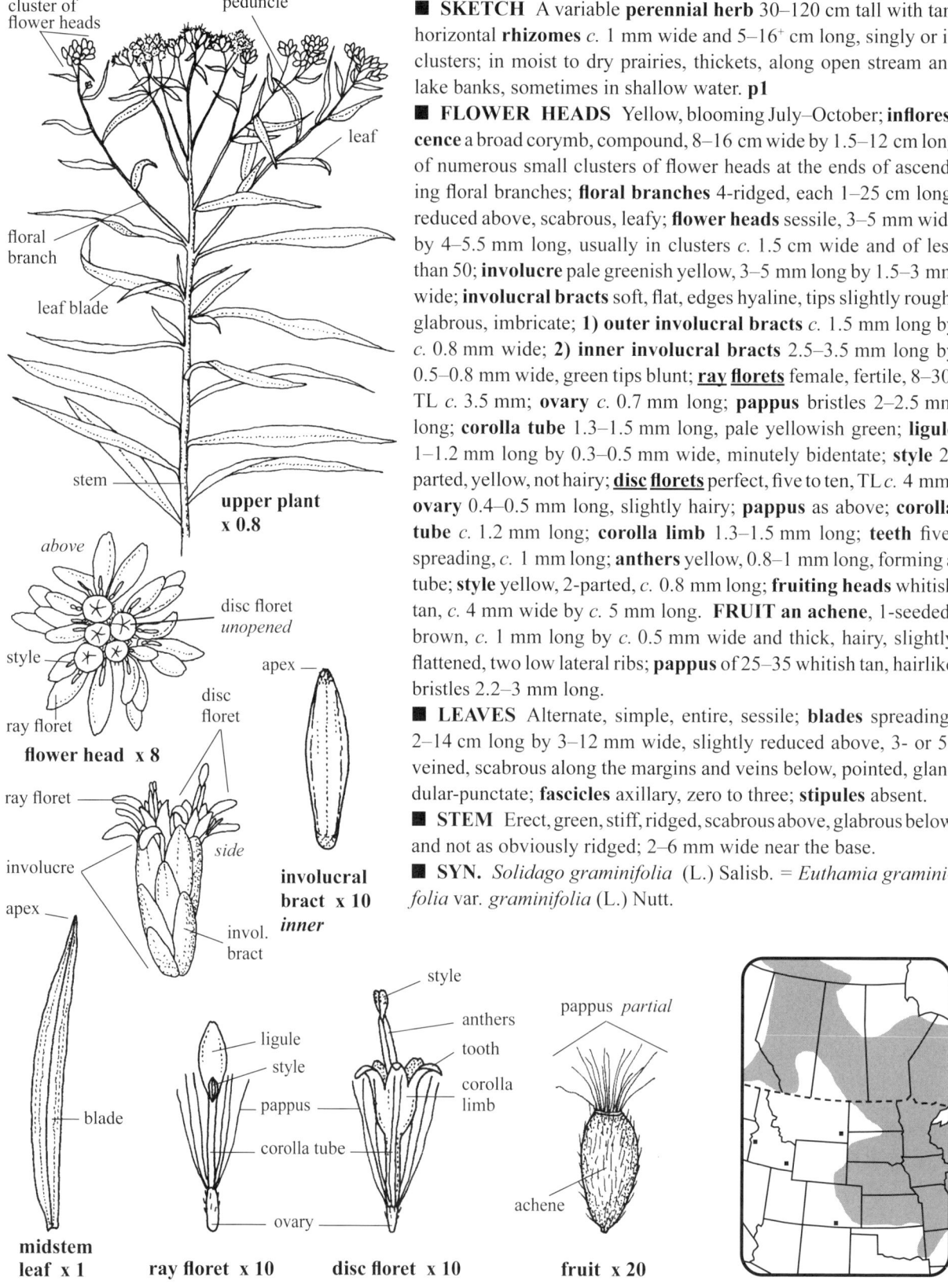

- **SKETCH** A variable **perennial herb** 30–120 cm tall with tan, horizontal **rhizomes** *c.* 1 mm wide and 5–16⁺ cm long, singly or in clusters; in moist to dry prairies, thickets, along open stream and lake banks, sometimes in shallow water. **p1**
- **FLOWER HEADS** Yellow, blooming July–October; **inflorescence** a broad corymb, compound, 8–16 cm wide by 1.5–12 cm long of numerous small clusters of flower heads at the ends of ascending floral branches; **floral branches** 4-ridged, each 1–25 cm long, reduced above, scabrous, leafy; **flower heads** sessile, 3–5 mm wide by 4–5.5 mm long, usually in clusters *c.* 1.5 cm wide and of less than 50; **involucre** pale greenish yellow, 3–5 mm long by 1.5–3 mm wide; **involucral bracts** soft, flat, edges hyaline, tips slightly rough, glabrous, imbricate; **1) outer involucral bracts** *c.* 1.5 mm long by *c.* 0.8 mm wide; **2) inner involucral bracts** 2.5–3.5 mm long by 0.5–0.8 mm wide, green tips blunt; **ray florets** female, fertile, 8–30, TL *c.* 3.5 mm; **ovary** *c.* 0.7 mm long; **pappus** bristles 2–2.5 mm long; **corolla tube** 1.3–1.5 mm long, pale yellowish green; **ligule** 1–1.2 mm long by 0.3–0.5 mm wide, minutely bidentate; **style** 2-parted, yellow, not hairy; **disc florets** perfect, five to ten, TL *c.* 4 mm; **ovary** 0.4–0.5 mm long, slightly hairy; **pappus** as above; **corolla tube** *c.* 1.2 mm long; **corolla limb** 1.3–1.5 mm long; **teeth** five, spreading, *c.* 1 mm long; **anthers** yellow, 0.8–1 mm long, forming a tube; **style** yellow, 2-parted, *c.* 0.8 mm long; **fruiting heads** whitish tan, *c.* 4 mm wide by *c.* 5 mm long. **FRUIT an achene**, 1-seeded, brown, *c.* 1 mm long by *c.* 0.5 mm wide and thick, hairy, slightly flattened, two low lateral ribs; **pappus** of 25–35 whitish tan, hairlike bristles 2.2–3 mm long.
- **LEAVES** Alternate, simple, entire, sessile; **blades** spreading, 2–14 cm long by 3–12 mm wide, slightly reduced above, 3- or 5-veined, scabrous along the margins and veins below, pointed, glandular-punctate; **fascicles** axillary, zero to three; **stipules** absent.
- **STEM** Erect, green, stiff, ridged, scabrous above, glabrous below and not as obviously ridged; 2–6 mm wide near the base.
- **SYN.** *Solidago graminifolia* (L.) Salisb. = *Euthamia graminifolia* var. *graminifolia* (L.) Nutt.

Aster—*Asteraceae* **Wildflower** Yellow, center reddish brown Ray florets 10–18

GREAT-FLOWERED GAILLARDIA *Gaillardia aristata* Pursh

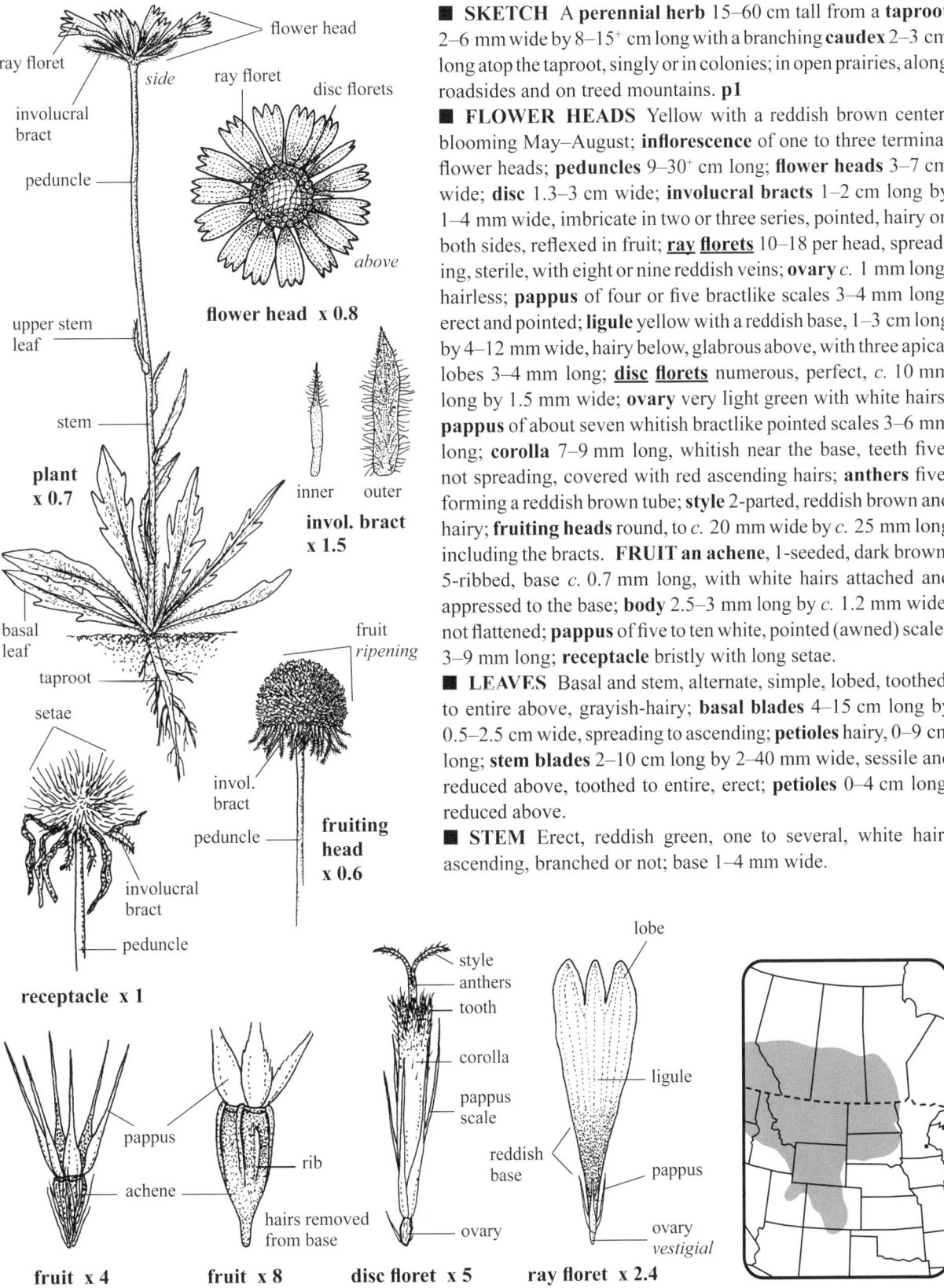

■ **SKETCH** A **perennial herb** 15–60 cm tall from a **taproot** 2–6 mm wide by 8–15⁺ cm long with a branching **caudex** 2–3 cm long atop the taproot, singly or in colonies; in open prairies, along roadsides and on treed mountains. **p1**

■ **FLOWER HEADS** Yellow with a reddish brown center, blooming May–August; **inflorescence** of one to three terminal flower heads; **peduncles** 9–30⁺ cm long; **flower heads** 3–7 cm wide; **disc** 1.3–3 cm wide; **involucral bracts** 1–2 cm long by 1–4 mm wide, imbricate in two or three series, pointed, hairy on both sides, reflexed in fruit; **ray florets** 10–18 per head, spreading, sterile, with eight or nine reddish veins; **ovary** c. 1 mm long, hairless; **pappus** of four or five bractlike scales 3–4 mm long, erect and pointed; **ligule** yellow with a reddish base, 1–3 cm long by 4–12 mm wide, hairy below, glabrous above, with three apical lobes 3–4 mm long; **disc florets** numerous, perfect, c. 10 mm long by 1.5 mm wide; **ovary** very light green with white hairs; **pappus** of about seven whitish bractlike pointed scales 3–6 mm long; **corolla** 7–9 mm long, whitish near the base, teeth five, not spreading, covered with red ascending hairs; **anthers** five, forming a reddish brown tube; **style** 2-parted, reddish brown and hairy; **fruiting heads** round, to c. 20 mm wide by c. 25 mm long including the bracts. **FRUIT an achene**, 1-seeded, dark brown, 5-ribbed, base c. 0.7 mm long, with white hairs attached and appressed to the base; **body** 2.5–3 mm long by c. 1.2 mm wide, not flattened; **pappus** of five to ten white, pointed (awned) scales 3–9 mm long; **receptacle** bristly with long setae.

■ **LEAVES** Basal and stem, alternate, simple, lobed, toothed, to entire above, grayish-hairy; **basal blades** 4–15 cm long by 0.5–2.5 cm wide, spreading to ascending; **petioles** hairy, 0–9 cm long; **stem blades** 2–10 cm long by 2–40 mm wide, sessile and reduced above, toothed to entire, erect; **petioles** 0–4 cm long, reduced above.

■ **STEM** Erect, reddish green, one to several, white hairs ascending, branched or not; base 1–4 mm wide.

Great Blanketflower, Common Gaillardia, Blanket Flower, Gaillardia

Aster—*Asteraceae* Wildflower White to reddish purple, center yellow Ray florets 5 (4–8)

GALINSOGA *Galinsoga quadriradiata* Cav.

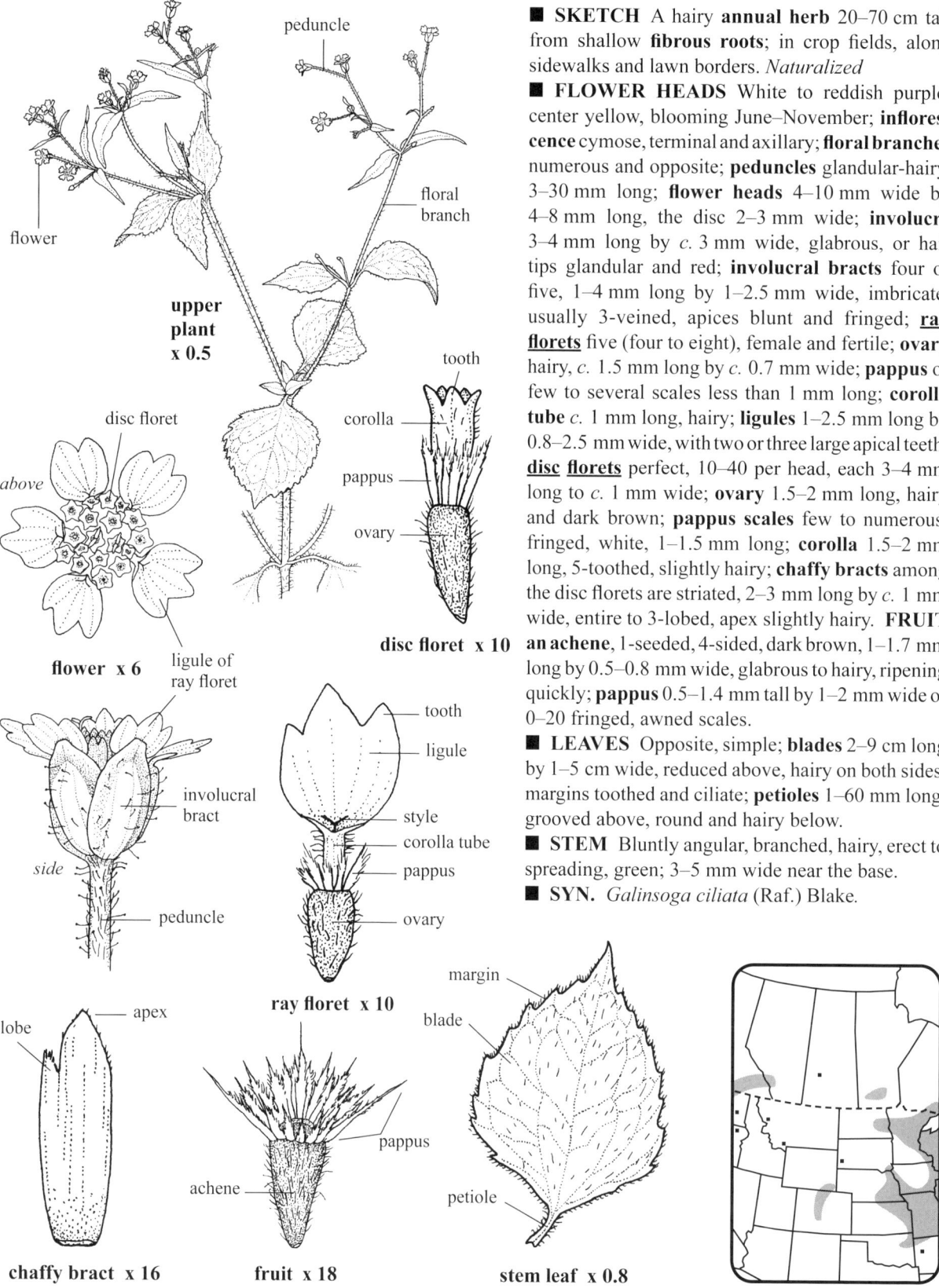

- **SKETCH** A hairy **annual herb** 20–70 cm tall from shallow **fibrous roots**; in crop fields, along sidewalks and lawn borders. *Naturalized*
- **FLOWER HEADS** White to reddish purple, center yellow, blooming June–November; **inflorescence** cymose, terminal and axillary; **floral branches** numerous and opposite; **peduncles** glandular-hairy, 3–30 mm long; **flower heads** 4–10 mm wide by 4–8 mm long, the disc 2–3 mm wide; **involucre** 3–4 mm long by *c.* 3 mm wide, glabrous, or hair tips glandular and red; **involucral bracts** four or five, 1–4 mm long by 1–2.5 mm wide, imbricate, usually 3-veined, apices blunt and fringed; **ray florets** five (four to eight), female and fertile; **ovary** hairy, *c.* 1.5 mm long by *c.* 0.7 mm wide; **pappus** of few to several scales less than 1 mm long; **corolla tube** *c.* 1 mm long, hairy; **ligules** 1–2.5 mm long by 0.8–2.5 mm wide, with two or three large apical teeth; **disc florets** perfect, 10–40 per head, each 3–4 mm long to *c.* 1 mm wide; **ovary** 1.5–2 mm long, hairy and dark brown; **pappus scales** few to numerous, fringed, white, 1–1.5 mm long; **corolla** 1.5–2 mm long, 5-toothed, slightly hairy; **chaffy bracts** among the disc florets are striated, 2–3 mm long by *c.* 1 mm wide, entire to 3-lobed, apex slightly hairy. **FRUIT** an achene, 1-seeded, 4-sided, dark brown, 1–1.7 mm long by 0.5–0.8 mm wide, glabrous to hairy, ripening quickly; **pappus** 0.5–1.4 mm tall by 1–2 mm wide of 0–20 fringed, awned scales.
- **LEAVES** Opposite, simple; **blades** 2–9 cm long by 1–5 cm wide, reduced above, hairy on both sides, margins toothed and ciliate; **petioles** 1–60 mm long, grooved above, round and hairy below.
- **STEM** Bluntly angular, branched, hairy, erect to spreading, green; 3–5 mm wide near the base.
- **SYN.** *Galinsoga ciliata* (Raf.) Blake.

140   Shaggy-soldier, Quickweed, Fringed Quickweed

Aster—*Asteraceae* **Wildflower** Yellow Ray florets 12–51 (rarely none)

GUMWEED *Grindelia squarrosa* (Pursh) Dunal

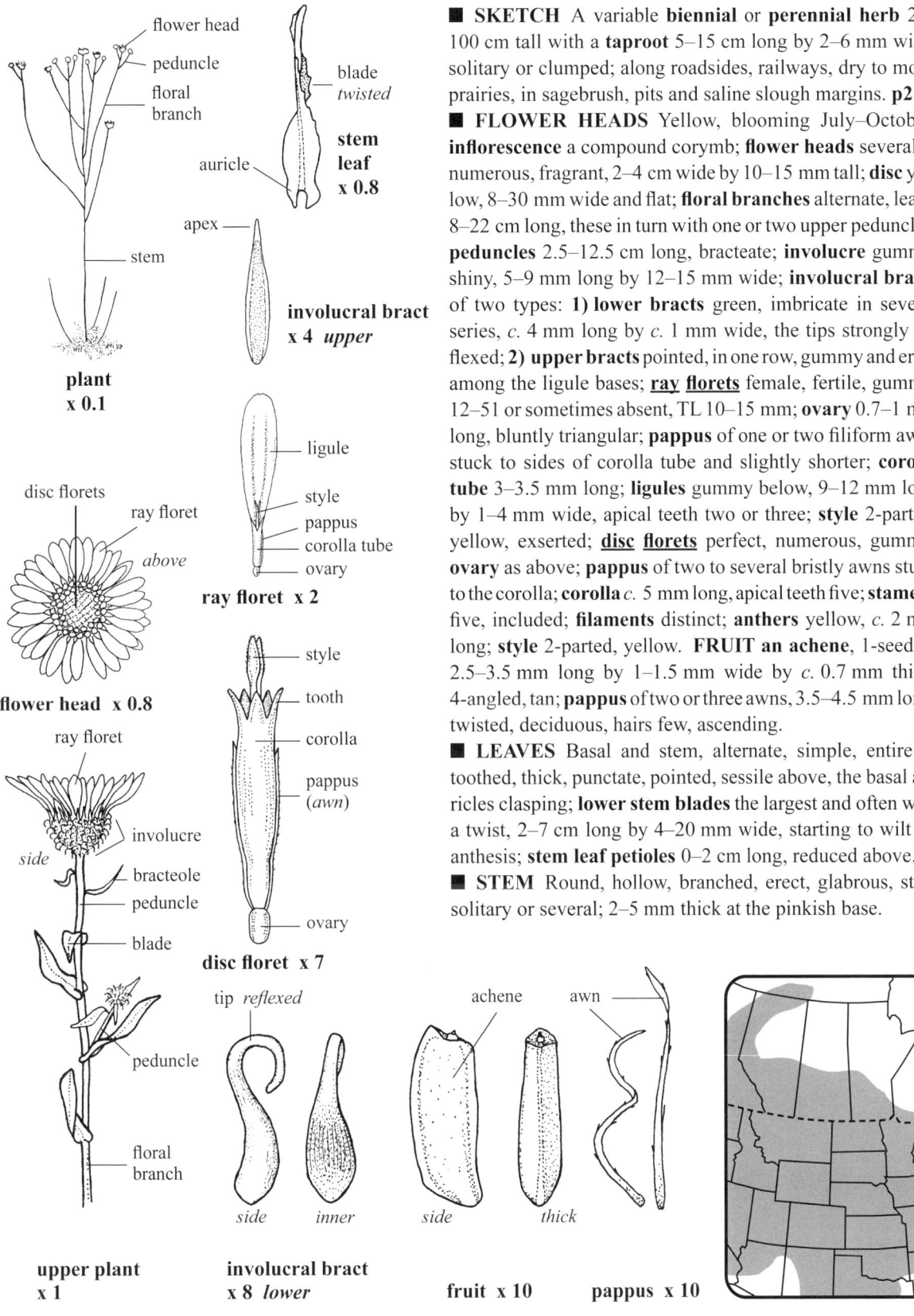

■ **SKETCH** A variable **biennial** or **perennial herb** 20–100 cm tall with a **taproot** 5–15 cm long by 2–6 mm wide, solitary or clumped; along roadsides, railways, dry to moist prairies, in sagebrush, pits and saline slough margins. **p2**

■ **FLOWER HEADS** Yellow, blooming July–October; **inflorescence** a compound corymb; **flower heads** several to numerous, fragrant, 2–4 cm wide by 10–15 mm tall; **disc** yellow, 8–30 mm wide and flat; **floral branches** alternate, leafy, 8–22 cm long, these in turn with one or two upper peduncles; **peduncles** 2.5–12.5 cm long, bracteate; **involucre** gummy, shiny, 5–9 mm long by 12–15 mm wide; **involucral bracts** of two types: 1) **lower bracts** green, imbricate in several series, *c*. 4 mm long by *c*. 1 mm wide, the tips strongly reflexed; 2) **upper bracts** pointed, in one row, gummy and erect among the ligule bases; **ray florets** female, fertile, gummy, 12–51 or sometimes absent, TL 10–15 mm; **ovary** 0.7–1 mm long, bluntly triangular; **pappus** of one or two filiform awns stuck to sides of corolla tube and slightly shorter; **corolla tube** 3–3.5 mm long; **ligules** gummy below, 9–12 mm long by 1–4 mm wide, apical teeth two or three; **style** 2-parted, yellow, exserted; **disc florets** perfect, numerous, gummy; **ovary** as above; **pappus** of two to several bristly awns stuck to the corolla; **corolla** *c*. 5 mm long, apical teeth five; **stamens** five, included; **filaments** distinct; **anthers** yellow, *c*. 2 mm long; **style** 2-parted, yellow. **FRUIT an achene**, 1-seeded, 2.5–3.5 mm long by 1–1.5 mm wide by *c*. 0.7 mm thick, 4-angled, tan; **pappus** of two or three awns, 3.5–4.5 mm long, twisted, deciduous, hairs few, ascending.

■ **LEAVES** Basal and stem, alternate, simple, entire to toothed, thick, punctate, pointed, sessile above, the basal auricles clasping; **lower stem blades** the largest and often with a twist, 2–7 cm long by 4–20 mm wide, starting to wilt by anthesis; **stem leaf petioles** 0–2 cm long, reduced above.

■ **STEM** Round, hollow, branched, erect, glabrous, stiff, solitary or several; 2–5 mm thick at the pinkish base.

Curly-top Gumweed, Curly-cup Gumweed, Curly-cup

Aster—*Asteraceae* **Wildflower** Yellow Ray florets 3–8

Common Broomweed *Gutierrezia sarothrae* (Pursh) Britt. & Rusby

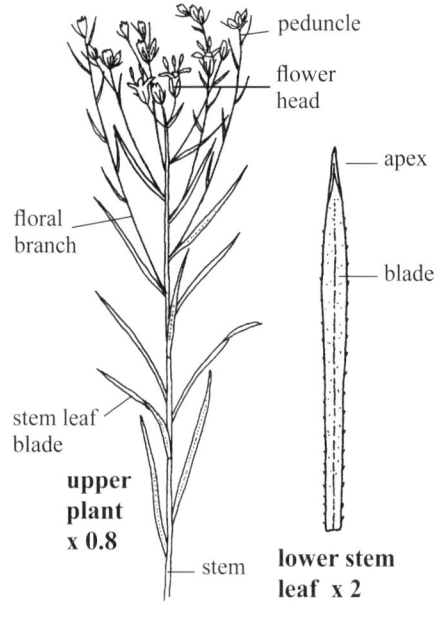

upper plant x 0.8

lower stem leaf x 2

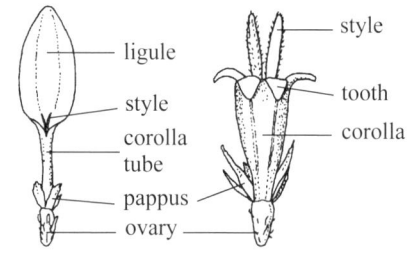

ray floret x 6.5   disc floret x 8

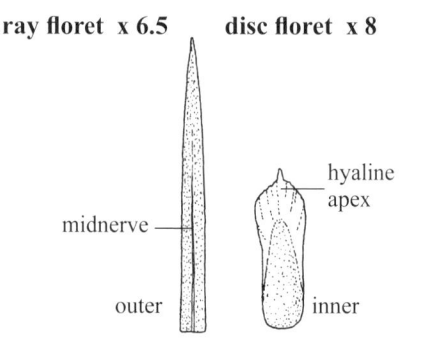

involucral bract x 7

■ **SKETCH** A tufted **perennial herb** 10–70 cm tall from a branching **caudex** 1.5–5 cm long and **rhizomes**; in rocky slopes, overgrazed pastures, dry washes, open plains, uplands and dry hillsides. **w1, p1**
■ **FLOWER HEADS** Yellow, blooming July–November; **inflorescence** corymbose; **floral branches** alternate, usually 2–8 cm long, often rebranched; **peduncles** 1–60 mm long, erect, glabrous or hairs scattered; **flower heads** numerous, some sticky, 5–7.5 mm wide by 6–10 mm tall (to top of style); **involucre** 3.5–6 mm long by 1.5–2.5 mm wide, glabrous to glandular; **involucral bracts** usually in three series, imbricate, 2.5–6 mm long by 0.3–1.5 mm wide, thick, slightly shiny; **1) outer bracts** greenish, thick, pale midnerve obvious near the base, *c.* 0.4 mm wide; **2) inner bracts** more yellowish green, soft, flat, *c.* 3 mm long by *c.* 1 mm wide, opaque to translucent and widest below the short apicular point; <u>ray florets</u> three to eight, female and fertile; TL *c.* 5 mm; **ovary** *c.* 0.8 mm long, pale green, slightly hairy; **pappus** of several scales *c.* 0.5 mm long; **corolla tube** yellow, *c.* 1.5 mm long and glabrous; **ligule** 1–3 mm long by 1–1.3 mm wide, usually 2-veined with no apical teeth; <u>disc florets</u> two to six, perfect, TL 3–4 mm; **ovary** as above; **pappus** of several acute scales 0.5–1.2 mm long, imbricate; **corolla** TL 2.8–3 mm, yellow, glabrous; **teeth** five, yellow, 0.5–0.7 mm long; **stamens** five; **anthers** *c.* 1.5 mm long, yellow, united; **styles** exserted, *c.* 1 mm long, yellow, hairy. **FRUIT an achene**, 1-seeded, of two types: **1) ray fruit** brown, almost completely covered with short, white ascending hairs except on ribs; **body** 1.3–1.5 mm long by *c.* 0.6 mm wide by *c.* 0.5 mm thick; **pappus scales** 8–13, acute, whitish 0.4–0.7 mm long; **2) disc fruit** *c.* 0.7 mm long by *c.* 0.4 mm wide, hairy as above, brown; **pappus scales** 8–10, erect, acute, 0.9–1.4 mm long.
■ **LEAVES** Alternate, simple, ascending; **stem leaves** flat, dull, pointed, sessile; **blades** 5–65 mm long by 1–3 mm wide, reduced above, widest above the middle, sometimes resinous, entire, margins scabrous, apices involute and pointed, midrib scabrous near the base.
■ **STEM** Erect, stiff, branched above or from the woody 1–2 mm wide base, previous year's brown stems usually persisting.
■ **SYN.** *Gutierrezia diversifolia* Greene.

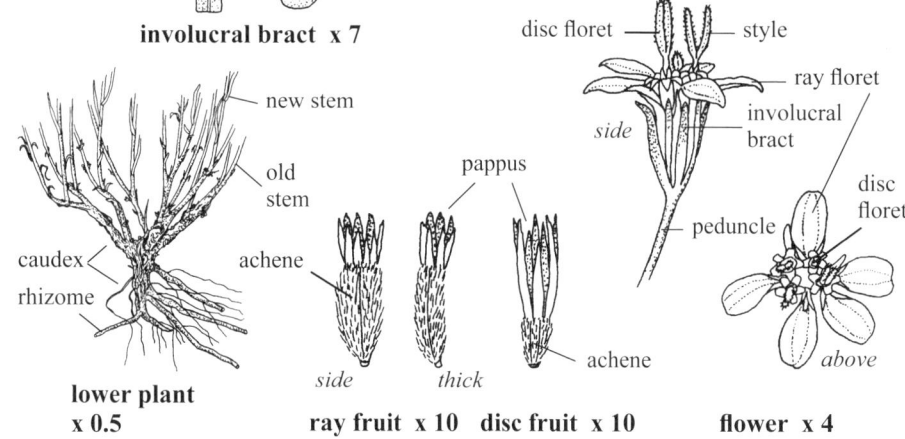

lower plant x 0.5   ray fruit x 10   disc fruit x 10   flower x 4

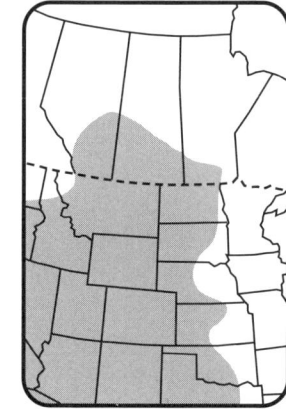

Kindlingweed, Broom Snakeweed, Snakeweed, Broomweed

Aster—*Asteraceae* **Wildflower** Yellow Ray florets 10–20

Mountain Sneezeweed *Helenium autumnale* L.

■ **SKETCH** A variable **perennial herb** 30–130 cm tall with **fibrous roots** and a **caudex**; in wet meadows, dried sloughs, marshes and river banks. **p2**

■ **FLOWER HEADS** Yellow, blooming July–September; **inflorescence** of numerous heads, each 3–4 cm wide by 1.5–3 cm tall; **disc** convex, yellow, 8–22 mm wide and tall; **branches** ascending, to *c.* 30⁺ cm long, winged, wing margins hairy; **peduncles** ridged, 3–7 cm long by *c.* 2 mm thick, hairs white and ascending; **involucre** *c.* 2 cm wide, green; **involucral bracts** pliable, hairy, in two series, inner bracts united at the white bases for *c.* 2 mm, free part 6–13 mm long by 1–5 mm wide (flattened), entire, margins revolute, twisted, tapered to a fine point, arched downward but tips ascending; <u>ray **florets**</u> 10–20, female, fertile, TL 11–15 (–25) mm; **ovary** with a flat base, *c.* 1 mm long and wide, pale green, hairs ascending; **pappus** of several white scales *c.* 1 mm long, toothed, pointed; **ligules** 5–13 mm wide, hairy below, glabrous above, spreading, sometimes imbricate, apical lobes three (two to four), each 2–4 mm wide, punctate below; **style** glabrous, 3–3.5 mm long, pale yellow, 2-parted; <u>disc **florets**</u> numerous, perfect, 4.2–5 mm long (base of ovary to tip of corolla teeth); **ovary** 1–1.2 mm long by *c.* 0.7 mm wide, hairs ascending; **pappus** of several pointed membranous scales 0.6–1.5 mm long; **corolla** *c.* 3.2 mm long, punctate; **corolla tube** *c.* 1.5 mm long; **corolla limb** 1.5–1.7 mm long by *c.* 1 mm wide, lightly hairy, yellow, veined between teeth; **corolla teeth** five, *c.* 0.5 mm long, yellow, erect, hairy; **anthers** *c.* 1 mm long, pale reddish brown, united into a tube; **style** exserted, 2-parted, yellow, tips flattened and hairy; **fruiting heads** brown, 10–15 mm wide, roundish. **FRUIT an achene**, 1-seeded, hairy, 1.4–1.7 mm long by *c.* 0.8 mm wide by *c.* 0.5 mm thick, 6- or 7-ribbed; **pappus** of five or six awned, membranous scales 0.7–1.5 mm long; **receptacle** naked, convex, 4–6 mm wide by 3–4.5 mm tall.

■ **LEAVES** Basal and stem, alternate, simple, entire to slightly toothed, thin, flat; **basal leaves** entire, brown and withered by anthesis, TL 2.5–5 cm and to *c.* 1 cm wide; **stem blades** 4–15 cm long by 1.5–5 cm wide, mostly sessile, dull, punctate and hairy on both surfaces, midrib obvious, margins continuous along stem forming soft wings 1–2 mm tall and darker green than the stem.

■ **STEM** Erect, branched, pale green, walls *c.* 0.7 mm thick, glabrous or hairs arched upward; 7–20 mm wide near the base.

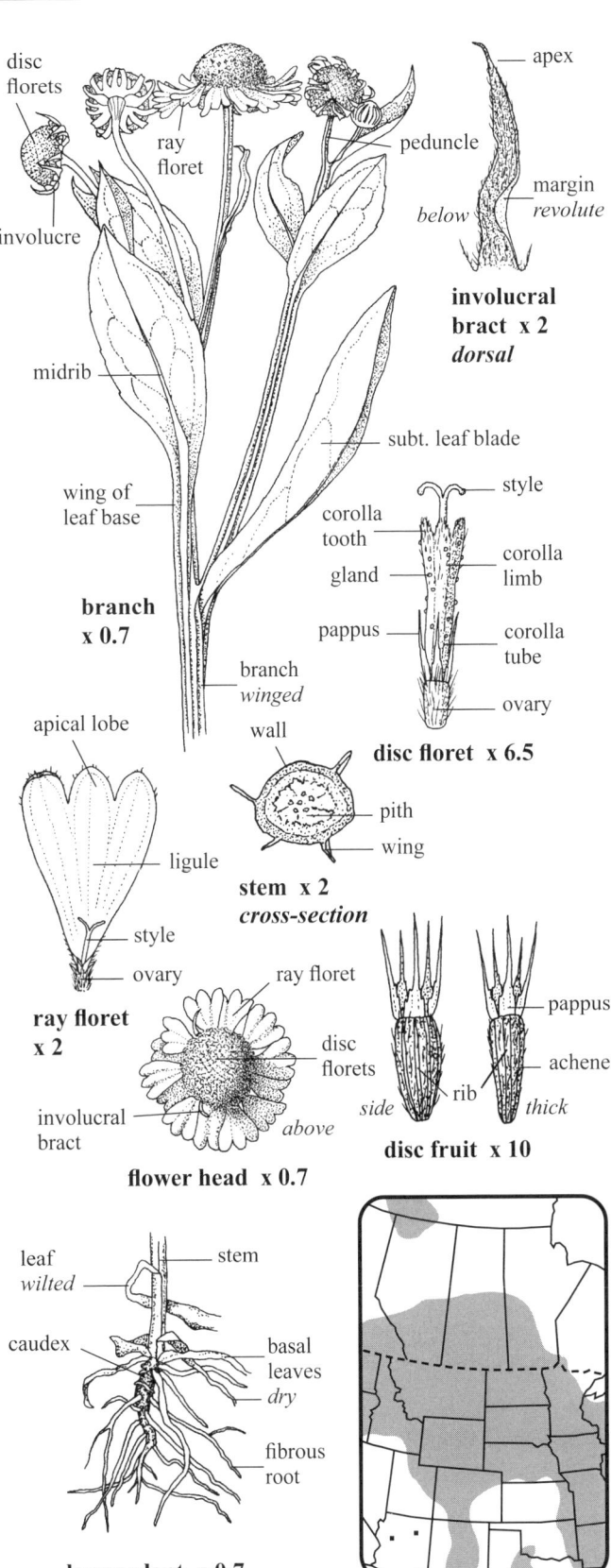

Fall Sneezeweed, Common Sneezeweed, Sneezeweed

Aster—*Asteraceae* **Wildflower** Yellow, center reddish brown Ray florets 12–35

Showy Sunflower *Helianthus annuus* L.

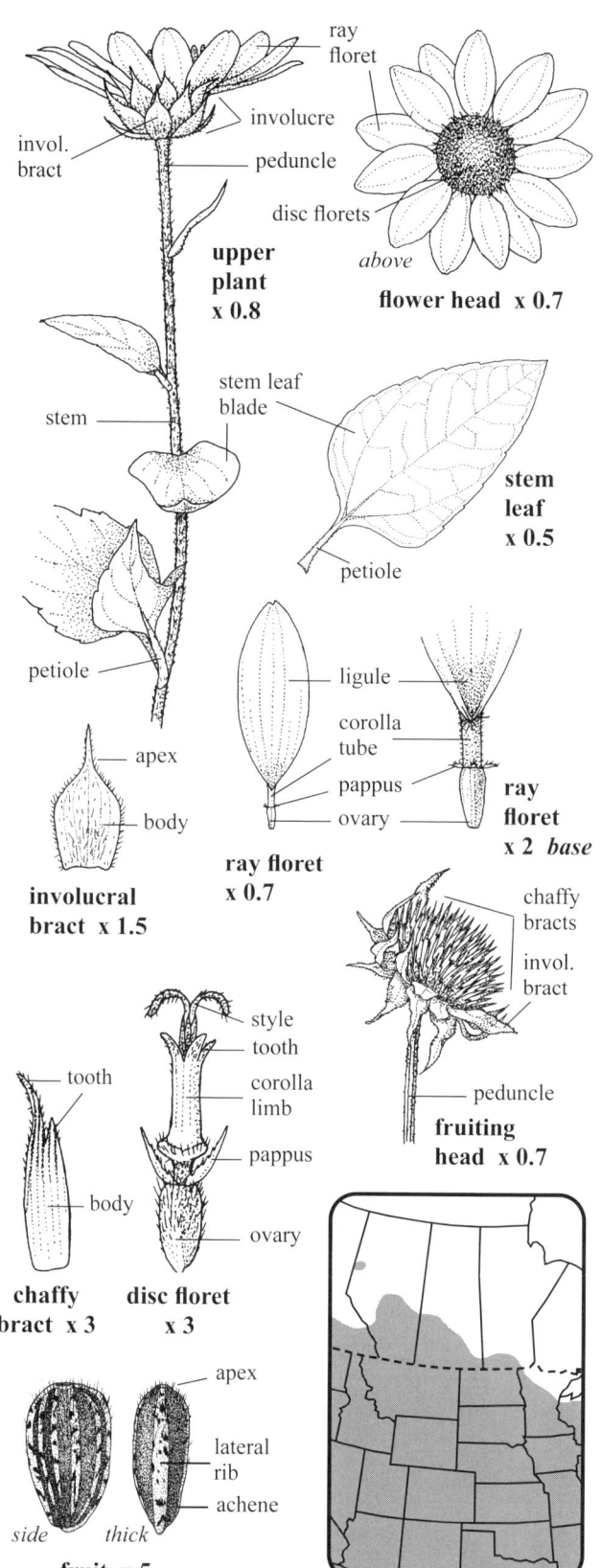

- **SKETCH** A highly variable **annual herb** 40–250 cm tall with a **taproot** 10–20 cm long; in disturbed areas, prairies, abandoned fields and roadsides. w2, p1
- **FLOWER HEADS** Yellow with a reddish brown center, blooming May–October; **inflorescence** a terminal head on the main stalk or branches, 1–25 heads per plant; **lower branches** opposite, those above alternate and rebranched with ascending peduncles and flower heads; **peduncles** hairy, rough from thickened and raised bases of hairs, 7–45 cm long by 2–5 mm thick; **flower heads** erect to slightly nodding, 5–15 cm wide by 1.5–4 cm tall; **central disc** 15–45 mm across, flat; **involucre** green, glabrous to hairy, 2.5–5 cm wide by 1.5–3 cm tall; **involucral bracts** in two or three series, imbricate, pointed, outer surface scabrous, margins ciliate, inside shiny and whitish near the base, 12–22 mm long by 3–12 mm wide, 1- to 5-nerved, sometimes obscurely so, lower bracts spreading to recurved; <u>ray florets</u> 12–35, sterile; **ovary** 3.5–4.5 mm long by 1.5–1.8 mm wide, pale green, 3-sided, glabrous or with scattered hairs along the sides and at the apex, shiny; **pappus** of two main awns, membranous, soft, *c.* 1 mm long, toothed, spreading to ascending, sometimes with one or two smaller awns; **corolla tube** yellow, *c.* 3 mm long, hairy; **ligule** 2.5–5 cm long by 0.8–1.5 cm wide, apex bidentate, slightly hairy below at the base, glabrous above; **style** absent; <u>disc florets</u> numerous, perfect, TL 13–15 mm; **ovary** hairy, 4–4.5 mm long by *c.* 2 mm wide; **chaffy bracts** one per floret, persistent, 9–11 mm long, V-shaped, 3-toothed, shiny, white and glabrous below, reddish purple above, hairy on midrib and margins of teeth; **pappus** of awns *c.* 3 mm long, toothed; **corolla limb** yellow, 5–5.5 mm long by 1.3–1.5 mm wide; **corolla teeth** five, *c.* 1.5 mm long, reddish to purple, hairy; **stamens** five; **filaments** distinct, yellow; **anthers** dark red or purple, *c.* 4 mm long; **style** 2-parted, exserted, dark purplish red, parts reflexed, hairy. **FRUIT** an achene, 1-seeded, 3–6.5 mm long by 2–3.5 mm wide by 1.5–2 mm thick, apex hairy or not, grayish to dark brown with longitudinal stripes and two lateral tan ribs; **pappus** early deciduous.
- **LEAVES** Alternate above, opposite below, simple, widely toothed to entire, spreading, scabrous; **blades** 4–30+ cm long by 1.5–20 cm wide, dull, hairy, pointed, glandular-punctate below (dorsally), less so above; **petioles** 0.5–16 cm long, hairy, deeply grooved above, scabrous, thick, green; **stipules** absent.
- **STEM** Round, erect, scabrous, solid, pith white, simple to branched; 5–30 mm thick near the smooth base.

Common Sunflower, Annual Sunflower

Aster—*Asteraceae* **Wildflower Yellow Ray florets 10–25**

Narrow-leaved Sunflower   *Helianthus maximiliani* Schrad.

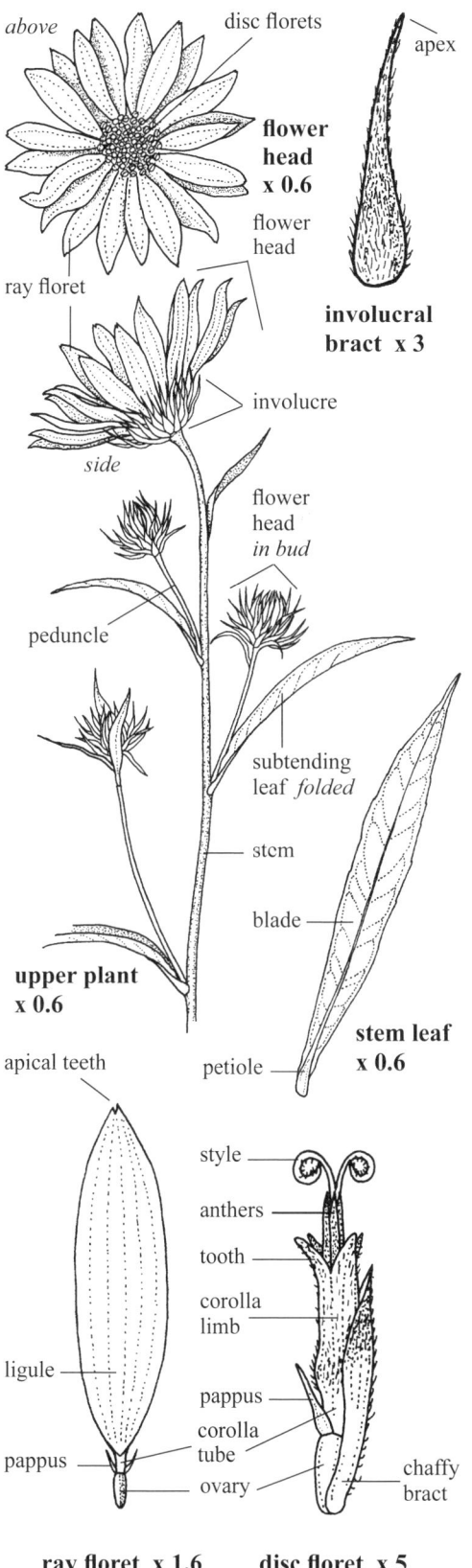

ray floret x 1.6   disc floret x 5

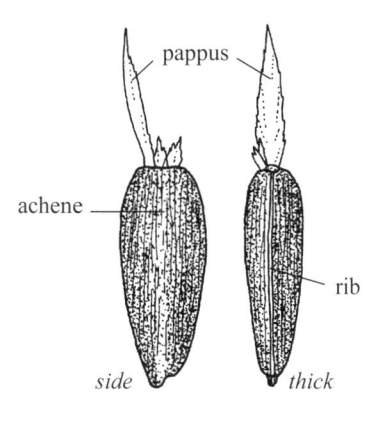

fruit x 8

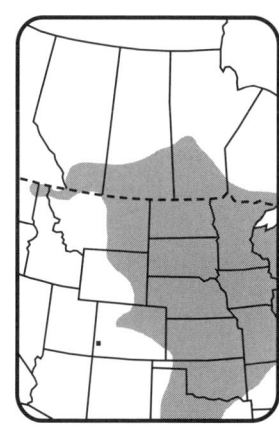

■ **SKETCH** A **perennial herb** 50–250⁺ cm tall, single or clustered, with **rhizomes** 3–10 mm wide by 8–10 cm long; in dry to moist prairies, along roadsides, thickets and in sandy sites. **w2, p2**

■ **FLOWER HEADS** Yellow, blooming July–October; **inflorescence** a raceme of 1–50 flower heads, some slightly nodding; **floral branches** to *c.* 15 cm long; **flower heads** 5–8 cm wide by 2–4 cm tall; **disc** 1.5–2.5 cm wide, yellow; **peduncles** 2–12 cm long, scabrous; **involucre** 25–30 mm wide by 10–12 mm tall; **involucral bracts** numerous, imbricate, 10–14 mm long by *c.* 2.5 mm wide, stiff, with short appressed hairs; **ray florets** 10–25, sterile, glabrous, TL 3–4 cm; **ovary** white, 2.5–3 mm long by *c.* 1 mm wide, triangular; **pappus** absent or reduced to two white awns 1–2 mm long; **corolla tube** *c.* 1.5 mm long, pale yellow; **ligules** 2.3–3.7 cm long by 7–9 mm wide, ascending to spreading, slightly imbricate, glabrous, several-veined, apices sometimes slightly twisted, apical teeth often two, each 1–2 mm long; **disc florets** perfect, TL 10–12 mm; **ovary** white, 2.5–3 mm long, glabrous, flattened; **chaffy bracts** 6–7 mm long by *c.* 1 mm wide across one side, to *c.* 8 mm long in fruit, V-shaped, several nerved, hairy apices pointed, dark brown to whitish green, midvein hairy; **pappus** of two hyaline awns *c.* 2 mm long; **corolla tube** pale yellow, 0.8–1 mm long, glabrous; **corolla limb** 4–4.5 mm long, yellow, slightly hairy; **teeth** usually four (five), ascending, yellow, *c.* 1 mm long; **style** 2-parted, yellow, tips curled, hairy and *c.* 2 mm long; **fruiting heads** erect, 2–3 cm wide by 1.5–2 cm tall. **FRUIT an achene**, 1-seeded, dark gray to brown, 3–4 mm long by *c.* 1.5 mm wide by *c.* 1 mm thick, hairless, striated; **pappus** of two white awns *c.* 2.5 mm long, quickly deciduous, sometimes short awns between the main pair.

■ **LEAVES** Alternate above to opposite below, simple, entire to slightly toothed, grayish green, stiff, scabrous on both sides, hairs appressed and ascending; **blades** arched, usually folded in along midrib, 5–25 cm long by 0.6–4 cm wide, reduced above, lower leaves withered by anthesis; **petioles** 1–20 mm long.

■ **STEM** Round, stiff, erect, solid, reddish green, scabrous from appressed hairs, simple or branched above; 2–12 mm thick near base.

Aster—*Asteraceae* **Wildflower** Yellow, center yellow to brownish Ray florets 8–17

TUBEROUS-ROOTED SUNFLOWER *Helianthus nuttallii* Torr. & Gray

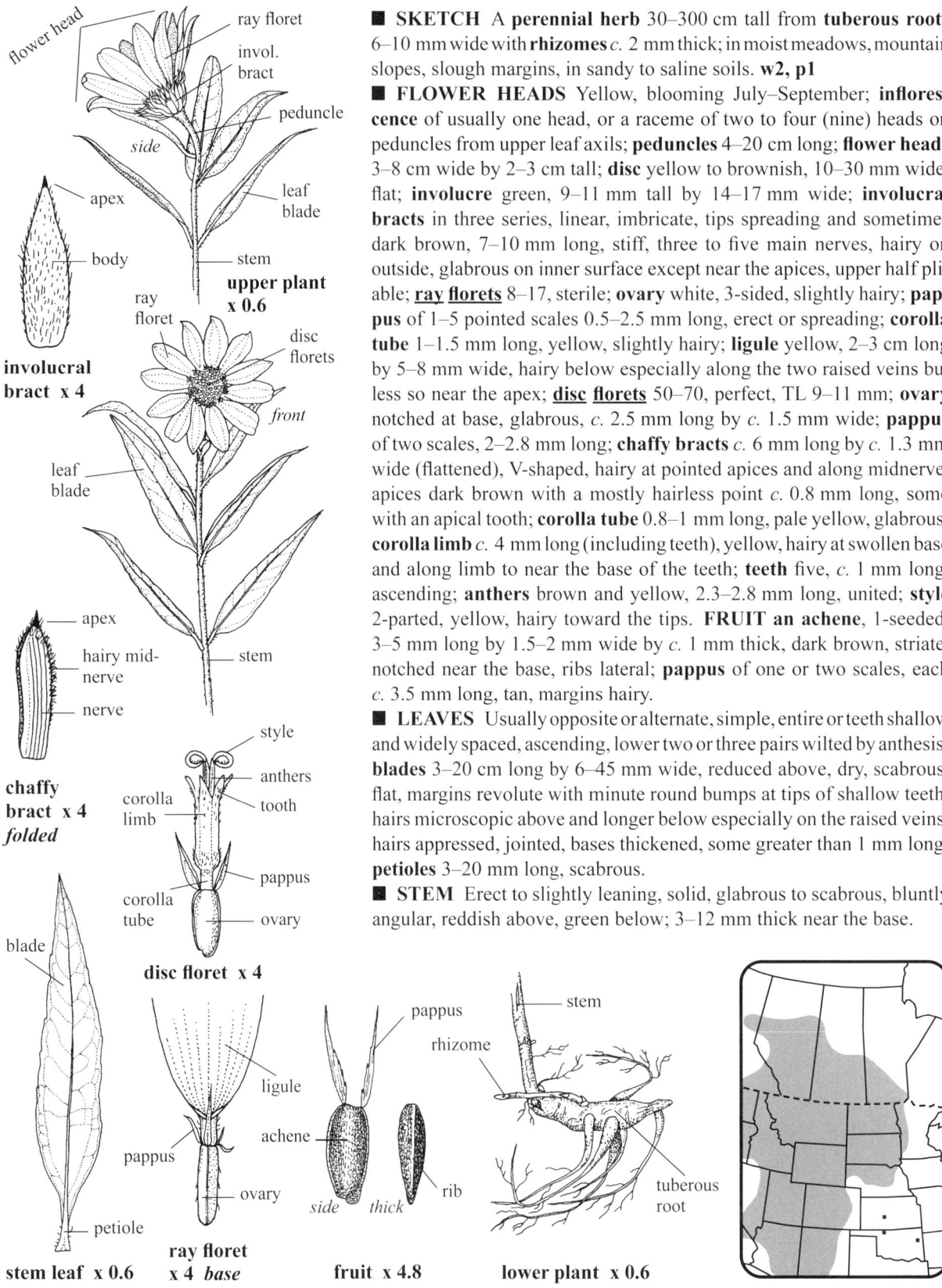

■ **SKETCH** A **perennial herb** 30–300 cm tall from **tuberous roots** 6–10 mm wide with **rhizomes** *c.* 2 mm thick; in moist meadows, mountain slopes, slough margins, in sandy to saline soils. **w2, p1**

■ **FLOWER HEADS** Yellow, blooming July–September; **inflorescence** of usually one head, or a raceme of two to four (nine) heads on peduncles from upper leaf axils; **peduncles** 4–20 cm long; **flower heads** 3–8 cm wide by 2–3 cm tall; **disc** yellow to brownish, 10–30 mm wide, flat; **involucre** green, 9–11 mm tall by 14–17 mm wide; **involucral bracts** in three series, linear, imbricate, tips spreading and sometimes dark brown, 7–10 mm long, stiff, three to five main nerves, hairy on outside, glabrous on inner surface except near the apices, upper half pliable; <u>**ray florets**</u> 8–17, sterile; **ovary** white, 3-sided, slightly hairy; **pappus** of 1–5 pointed scales 0.5–2.5 mm long, erect or spreading; **corolla tube** 1–1.5 mm long, yellow, slightly hairy; **ligule** yellow, 2–3 cm long by 5–8 mm wide, hairy below especially along the two raised veins but less so near the apex; <u>**disc florets**</u> 50–70, perfect, TL 9–11 mm; **ovary** notched at base, glabrous, *c.* 2.5 mm long by *c.* 1.5 mm wide; **pappus** of two scales, 2–2.8 mm long; **chaffy bracts** *c.* 6 mm long by *c.* 1.3 mm wide (flattened), V-shaped, hairy at pointed apices and along midnerve, apices dark brown with a mostly hairless point *c.* 0.8 mm long, some with an apical tooth; **corolla tube** 0.8–1 mm long, pale yellow, glabrous; **corolla limb** *c.* 4 mm long (including teeth), yellow, hairy at swollen base and along limb to near the base of the teeth; **teeth** five, *c.* 1 mm long, ascending; **anthers** brown and yellow, 2.3–2.8 mm long, united; **style** 2-parted, yellow, hairy toward the tips. **FRUIT** an achene, 1-seeded, 3–5 mm long by 1.5–2 mm wide by *c.* 1 mm thick, dark brown, striate, notched near the base, ribs lateral; **pappus** of one or two scales, each *c.* 3.5 mm long, tan, margins hairy.

■ **LEAVES** Usually opposite or alternate, simple, entire or teeth shallow and widely spaced, ascending, lower two or three pairs wilted by anthesis; **blades** 3–20 cm long by 6–45 mm wide, reduced above, dry, scabrous, flat, margins revolute with minute round bumps at tips of shallow teeth, hairs microscopic above and longer below especially on the raised veins, hairs appressed, jointed, bases thickened, some greater than 1 mm long; **petioles** 3–20 mm long, scabrous.

■ **STEM** Erect to slightly leaning, solid, glabrous to scabrous, bluntly angular, reddish above, green below; 3–12 mm thick near the base.

146 Nuttall's Sunflower, Clustered Sunflower

Aster—*Asteraceae* **Wildflower** Yellow, center dark brown Ray florets 10–18

BEAUTIFUL SUNFLOWER *Helianthus pauciflorus* Nutt.

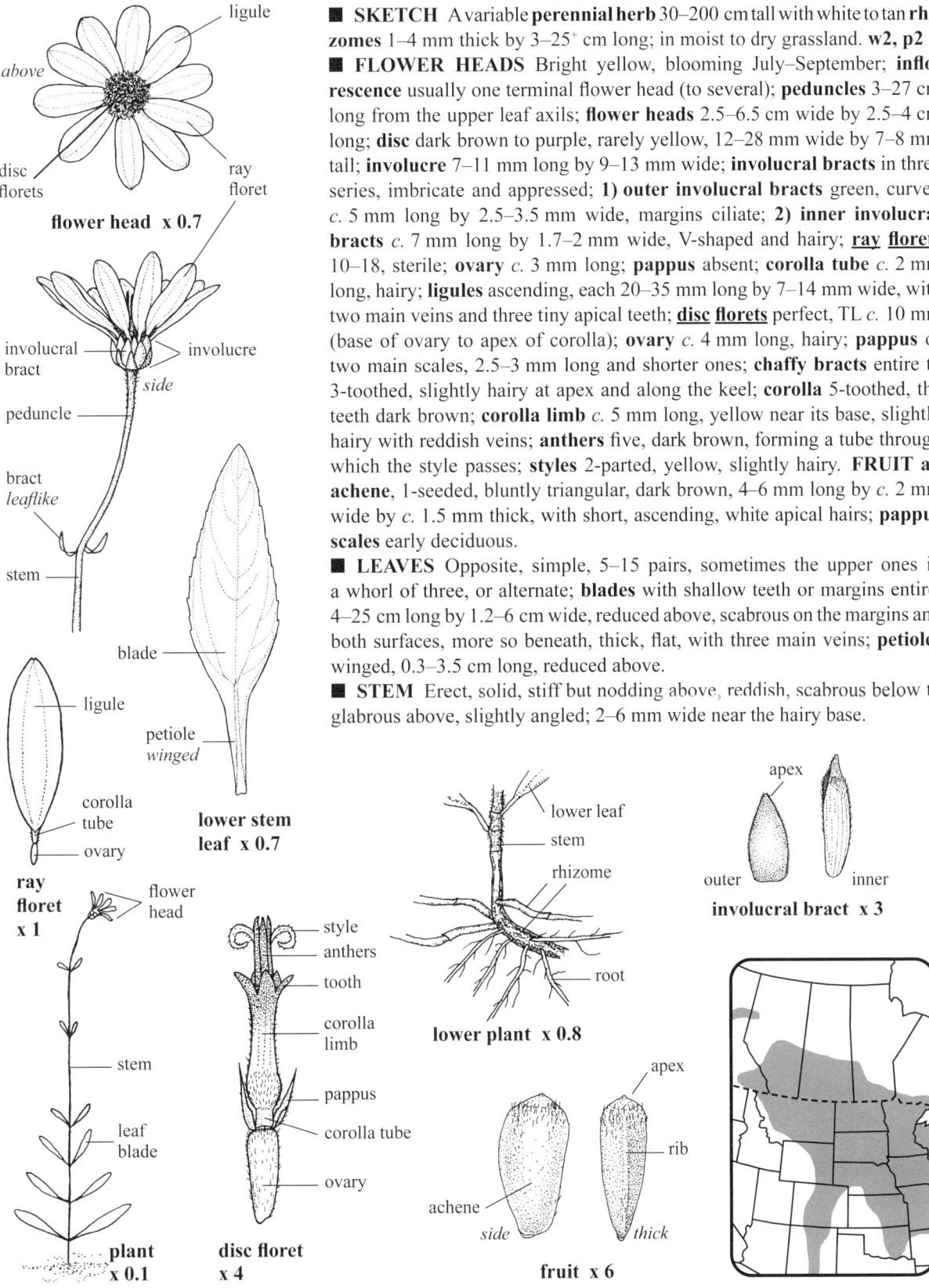

- **SKETCH** A variable **perennial herb** 30–200 cm tall with white to tan **rhizomes** 1–4 mm thick by 3–25+ cm long; in moist to dry grassland. w2, p2
- **FLOWER HEADS** Bright yellow, blooming July–September; **inflorescence** usually one terminal flower head (to several); **peduncles** 3–27 cm long from the upper leaf axils; **flower heads** 2.5–6.5 cm wide by 2.5–4 cm long; **disc** dark brown to purple, rarely yellow, 12–28 mm wide by 7–8 mm tall; **involucre** 7–11 mm long by 9–13 mm wide; **involucral bracts** in three series, imbricate and appressed; **1) outer involucral bracts** green, curved *c.* 5 mm long by 2.5–3.5 mm wide, margins ciliate; **2) inner involucral bracts** *c.* 7 mm long by 1.7–2 mm wide, V-shaped and hairy; **ray florets** 10–18, sterile; **ovary** *c.* 3 mm long; **pappus** absent; **corolla tube** *c.* 2 mm long, hairy; **ligules** ascending, each 20–35 mm long by 7–14 mm wide, with two main veins and three tiny apical teeth; **disc florets** perfect, TL *c.* 10 mm (base of ovary to apex of corolla); **ovary** *c.* 4 mm long, hairy; **pappus** of two main scales, 2.5–3 mm long and shorter ones; **chaffy bracts** entire to 3-toothed, slightly hairy at apex and along the keel; **corolla** 5-toothed, the teeth dark brown; **corolla limb** *c.* 5 mm long, yellow near its base, slightly hairy with reddish veins; **anthers** five, dark brown, forming a tube through which the style passes; **styles** 2-parted, yellow, slightly hairy. FRUIT **an achene**, 1-seeded, bluntly triangular, dark brown, 4–6 mm long by *c.* 2 mm wide by *c.* 1.5 mm thick, with short, ascending, white apical hairs; **pappus scales** early deciduous.
- **LEAVES** Opposite, simple, 5–15 pairs, sometimes the upper ones in a whorl of three, or alternate; **blades** with shallow teeth or margins entire, 4–25 cm long by 1.2–6 cm wide, reduced above, scabrous on the margins and both surfaces, more so beneath, thick, flat, with three main veins; **petioles** winged, 0.3–3.5 cm long, reduced above.
- **STEM** Erect, solid, stiff but nodding above, reddish, scabrous below to glabrous above, slightly angled; 2–6 mm wide near the hairy base.

Rhombic-leaved Sunflower, Stiff Sunflower

Aster—Asteraceae   **Wildflower** Yellow Ray florets 12–20

JERUSALEM ARTICHOKE   *Helianthus tuberosus* L.

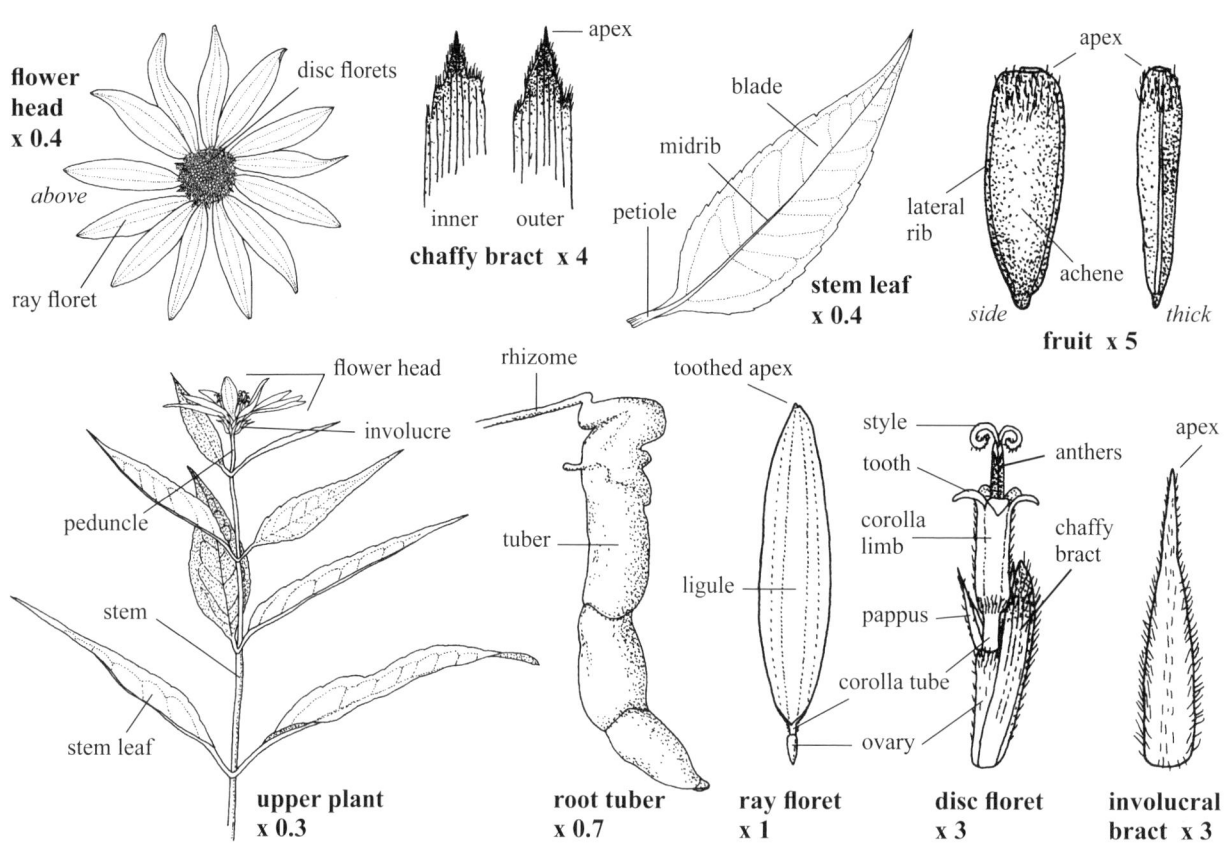

- **SKETCH** A variable **perennial herb** 0.8–3 m tall from fleshy **tubers** 1–20 cm long by 0.8–3 cm thick, attached to **rhizomes** 5–50+ cm long by 1–2 mm thick; along river banks, woodland edges, in crops and moist to drying sites. **w2, p2**
- **FLOWER HEADS** Yellow, blooming August–October; **inflorescence** of 1–20 heads, terminal and atop ascending branches; **peduncles** 2.5–10 cm long, scabrous, with 0–2 bracteoles; **flowers heads** 3–10 cm wide by 2–4 cm tall; **disc** yellow, 13–25 mm wide by 10–13 mm tall; **involucre** 20–27 mm wide by 10–12 mm tall; **involucral bracts** imbricate, 11–16 mm long by 2–3 mm wide, flexible, greenish black on the outside, midnerve slightly raised and rounded with short ascending hairs, margins hairy, tips spreading; **ray florets** 12–20, female, sterile; **ovary** white, glabrous, triangular, 2.5–4 mm long by 1–1.2 mm wide; **corolla tube** yellow, hairy below, 1.2–1.4 mm long; **ligules** yellow, 28–46 mm long by 8–12 mm wide, slightly hairy below, glabrous above, apices pointed with two or three minute teeth; **disc florets** numerous, perfect; **ovary** 4.5–6 mm long, white, hairy at apex and along one side; **pappus awns** two, 3–4 mm long, hyaline, hairy; **chaffy bracts** veined, 7.5–9.5 mm long, V-shaped, apical teeth one to three, hairy; **corolla** 6–7.5 mm long; **corolla tube** 1.3–1.9 mm long, glabrous; **corolla limb** yellow, 4.8–5.5 mm long, hairs in vertical lines; **teeth** five, yellow, each *c*. 1.5 mm long; **anthers** five, dark brown, forming a tube 3–4 mm long; **style** 2-parted, yellow, tips strongly curled and hairy. **FRUIT** an achene, 1-seeded, 5–8 mm long by 2–2.3 mm wide by 1–1.2 mm thick, dark gray, mottled, glabrous to hairy near the blunt apex, lateral ribs low.
- **LEAVES** Opposite, or sometimes alternate above to mostly alternate, simple, toothed; **blades** 5–25 cm long by 4–15 cm wide, dark green, dull, thin, spreading, scabrous above, lighter green, downy and glandular below, margins with ascending hairs; **petioles** 0.6–8 cm long, winged or not, scabrous.
- **STEM** Erect to often leaning, solid, round, simple to much branched, scabrous from hairs above; glabrous near the 5–25 mm thick base.

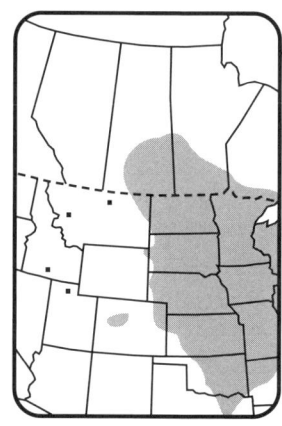

148

Aster—*Asteraceae* **Wildflower** Yellow Ray florets 10–17

ROUGH FALSE SUNFLOWER   *Heliopsis helianthoides* (L.) Sweet

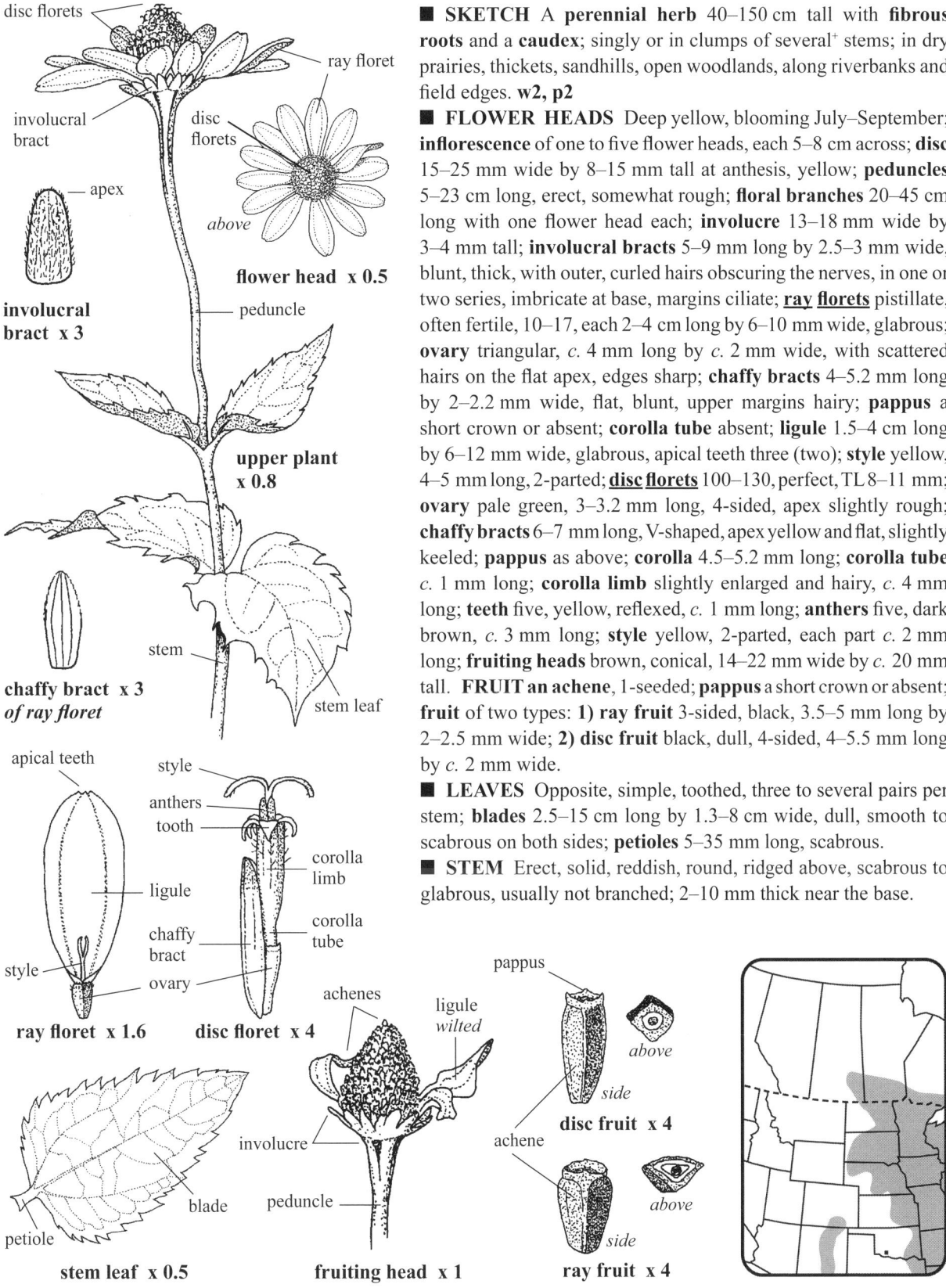

- **SKETCH** A **perennial herb** 40–150 cm tall with **fibrous roots** and a **caudex**; singly or in clumps of several⁺ stems; in dry prairies, thickets, sandhills, open woodlands, along riverbanks and field edges. w2, p2
- **FLOWER HEADS** Deep yellow, blooming July–September; **inflorescence** of one to five flower heads, each 5–8 cm across; **disc** 15–25 mm wide by 8–15 mm tall at anthesis, yellow; **peduncles** 5–23 cm long, erect, somewhat rough; **floral branches** 20–45 cm long with one flower head each; **involucre** 13–18 mm wide by 3–4 mm tall; **involucral bracts** 5–9 mm long by 2.5–3 mm wide, blunt, thick, with outer, curled hairs obscuring the nerves, in one or two series, imbricate at base, margins ciliate; **ray florets** pistillate, often fertile, 10–17, each 2–4 cm long by 6–10 mm wide, glabrous; **ovary** triangular, *c.* 4 mm long by *c.* 2 mm wide, with scattered hairs on the flat apex, edges sharp; **chaffy bracts** 4–5.2 mm long by 2–2.2 mm wide, flat, blunt, upper margins hairy; **pappus** a short crown or absent; **corolla tube** absent; **ligule** 1.5–4 cm long by 6–12 mm wide, glabrous, apical teeth three (two); **style** yellow, 4–5 mm long, 2-parted; **disc florets** 100–130, perfect, TL 8–11 mm; **ovary** pale green, 3–3.2 mm long, 4-sided, apex slightly rough; **chaffy bracts** 6–7 mm long, V-shaped, apex yellow and flat, slightly keeled; **pappus** as above; **corolla** 4.5–5.2 mm long; **corolla tube** *c.* 1 mm long; **corolla limb** slightly enlarged and hairy, *c.* 4 mm long; **teeth** five, yellow, reflexed, *c.* 1 mm long; **anthers** five, dark brown, *c.* 3 mm long; **style** yellow, 2-parted, each part *c.* 2 mm long; **fruiting heads** brown, conical, 14–22 mm wide by *c.* 20 mm tall. FRUIT an achene, 1-seeded; **pappus** a short crown or absent; **fruit** of two types: **1) ray fruit** 3-sided, black, 3.5–5 mm long by 2–2.5 mm wide; **2) disc fruit** black, dull, 4-sided, 4–5.5 mm long by *c.* 2 mm wide.
- **LEAVES** Opposite, simple, toothed, three to several pairs per stem; **blades** 2.5–15 cm long by 1.3–8 cm wide, dull, smooth to scabrous on both sides; **petioles** 5–35 mm long, scabrous.
- **STEM** Erect, solid, reddish, round, ridged above, scabrous to glabrous, usually not branched; 2–10 mm thick near the base.

Rough Oxeye, Smooth Oxeye, Sweet Oxeye, False Sunflower, Ox-eye

Aster—*Asteraceae* **Wildflower   Yellow   Ray florets 15–33**

HAIRY GOLDEN-ASTER   *Heterotheca villosa* (Pursh) Shinners

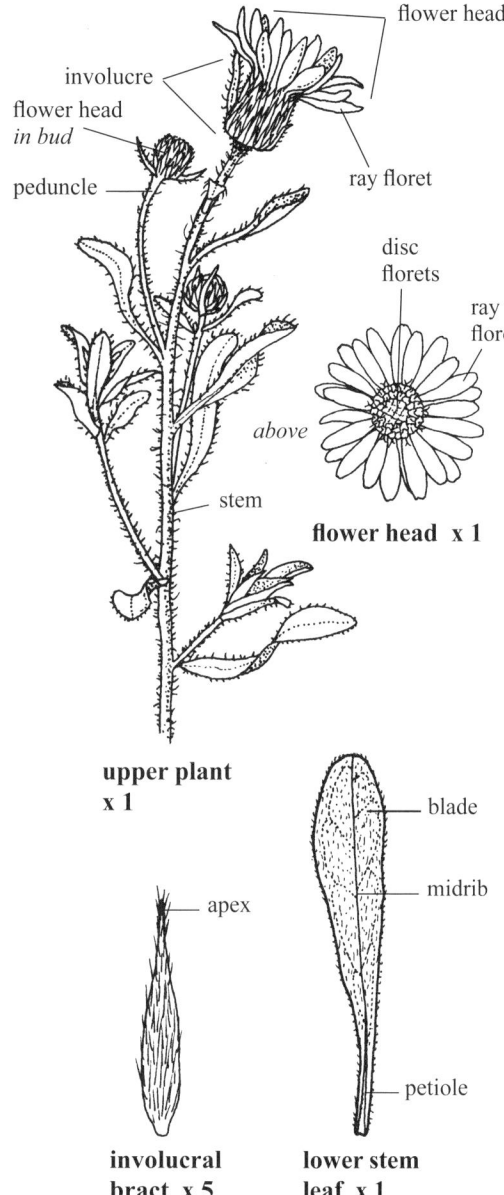

upper plant x 1

involucral bract x 5

lower stem leaf x 1

■ **SKETCH** A hairy, variable **perennial herb** 15–90⁺ cm tall from a thick, woody, **caudex** atop a branching **taproot**, singly or in scattered clumps; in dry sandy prairies and rocky slopes. **p1**
■ **FLOWER HEADS** Yellow, blooming June–September; **inflorescence** a corymb; **peduncles** 0.8–15 cm long, hairy, the outer ones often with two or three flower heads; **flower heads** 20–30 mm wide, the disc 7–10 mm wide; **involucre** green, 6–12 mm long and wide, some glandular; **involucral bracts** in several series, 3–7 mm long by 0.6–1 mm wide, imbricate, not thick, some with a medium brown tip, hairy; <u>**ray florets**</u> 15–33, female, fertile, 11–15 mm long; **ovary** *c.* 0.7 mm long, hairy; **pappus** white, the inner bristles 4.5–5.5 mm long, short bristles *c.* 1 mm long; **corolla tube** 3–4 mm long; **ligules** yellow, 7–12 mm long by 2.5–3.5 mm wide, glabrous; <u>**disc florets**</u> perfect, numerous, TL 9–12 mm; **ovary** as above; **pappus** as above; **corolla** glabrous, 5–8 mm long; **corolla tube** white, 2–3 mm long; **corolla limb** slightly wider and 3–5 mm long with five yellow apical teeth; **anthers** five, yellow, forming a tube *c.* 2 mm long; **style** yellow, 2-parted; **fruiting heads** *c.* 2 cm wide by 1.5 cm long. **FRUIT an achene**, 1-seeded, *c.* 2.5 mm long by *c.* 1 mm wide by *c.* 0.5 mm thick, with four or five light tan ribs, covered with silvery white hairs; **pappus** *c.* 8 mm wide, of numerous tan bristles in two series: **1)** inner bristles 5–6 mm long; **2)** the outer bristles or fringed scales *c.* 1 mm long.
■ **LEAVES** Alternate, simple, entire to toothed; **lowest stem leaves** deciduous by anthesis; **upper blades** 1–6 cm long by 2–13 mm wide, reduced above, firm, flat to slightly undulate, sometimes glandular, hairs appressed to spreading, with a downy feel; **petioles** 0–12 mm long, winged below, reduced above.
■ **STEM** Erect to decumbent, 1–18 per clump, simple to much branched, solid, round, light green; **hairs** stiff, appressed to spreading, of two types (also on peduncles and some leaves): **1)** more spreading and to *c.* 2 mm long with several microscopic joints; **2)** shorter, clumped, numerous, with two or three joints; **stem** 2–5 mm thick near the brown base.
■ **SYN.** *Chrysopsis villosa* (Pursh) Nutt. ex DC. = *Heterotheca villosa* var. *villosa* (Pursh) Shinners and *Chrysopsis hispida* (Hook.) DC. = *Heterotheca villosa* var. *minor* (Hook.) Semple.

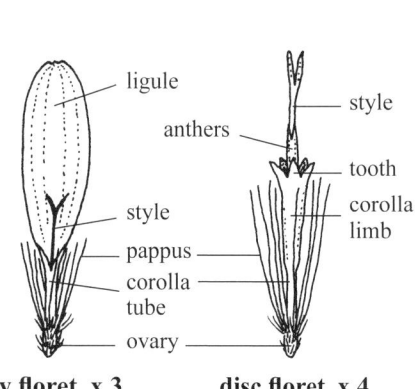

ray floret x 3   **disc floret x 4**

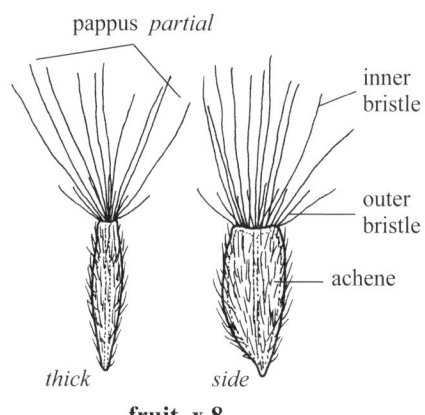

fruit x 8

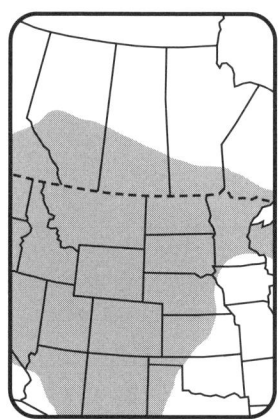

Hairy False Goldenaster

Aster—*Asteraceae* **Wildflower** Yellow Ligulate florets many

CANADA HAWKWEED   *Hieracium umbellatum* L.

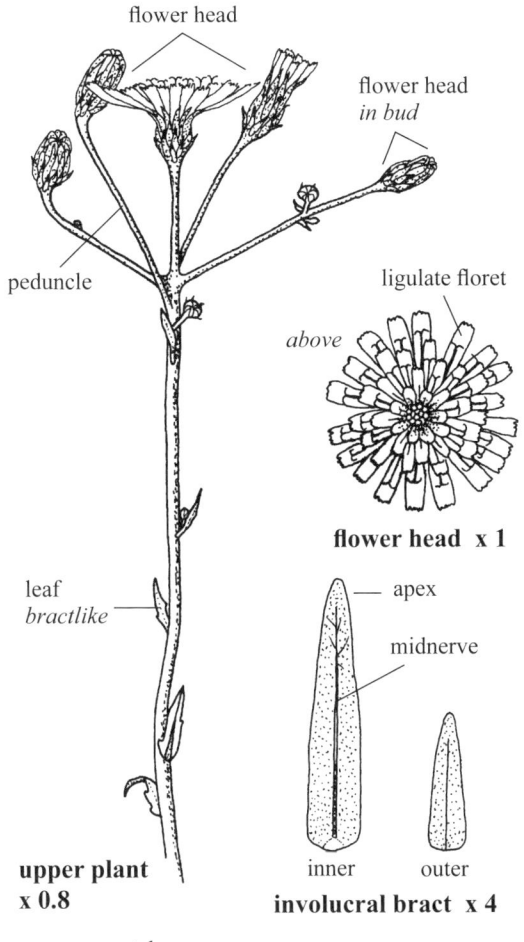

upper plant x 0.8

flower head x 1

involucral bract x 4

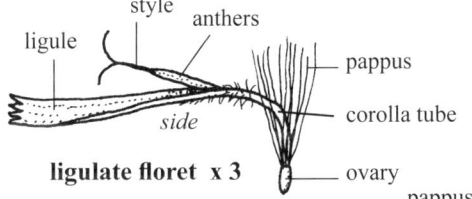

ligulate floret x 3

■ **SKETCH** A perennial herb 30–120 cm tall from **fibrous roots** 5–12 cm long with a short **caudex**; in dry to moist sites along woodland edges or openings and meadows, sometimes in sandy soils or on rocky outcrops; **juice milky**.

■ **FLOWER HEADS** Bright yellow, blooming July–September; **inflorescence** a corymb of 1–50⁺ flower heads; **peduncles** rough, 1.5–15 cm long, with stellate hairs and short black, cone-shaped hairs; **flower heads** 2–3 cm wide; **involucre** 6–12 mm tall by 6–8 wide; **involucral bracts** in three or four series, soft; **1) outer bracts** 3–5 mm long by 1.3–1.5 mm wide, with the tips bent outward, some with white or black gland-tipped hairs along the midnerve; **2) inner bracts** 8–9 mm long by 1.5–1.8 mm wide, some with a few scattered hairs near the apices, spreading beneath the ligules; disc florets absent; ligulate florets perfect, numerous; **ovary** glabrous, the apex flat; **pappus** 4–6 mm long, bristles tawny; **corolla** 9.5–16 mm long; **corolla tube** 3.5–5.5 mm long, white; **ligule** yellow, hairy below, 6–10 mm long by 2–3 mm wide, the largest to the outside, apical teeth five; **anthers** five, yellow, united, *c.* 4 mm long; **style** 2-parted, hairless; **fruiting heads** tan, *c.* 2 cm wide and tall. **FRUIT** an achene, 1-seeded, *c.* 3 mm long by *c.* 0.5 mm wide, dark brown, apex flat, main ribs five, hairless; **pappus** tan, with *c.* 40 bristles, unequal and to *c.* 6 mm long, together 7–8 mm wide.

■ **LEAVES** Basal and stem, alternate, simple, teeth pointing forward or entire; **basal blades** generally wilting by anthesis; **stem leaves** sessile, numerous; **blades** flat, horizontal to ascending, sometimes undulate, 2–11 cm long by 0.7–2 cm wide, reduced above, with a rough dry feel, the margins with numerous cone-shaped and sometimes black hairs.

■ **STEM** Erect to leaning, solitary or in small clusters, the hairs scattered, appressed below, more numerous above; reddish near the 1–5 mm thick, hairy to hairless base.

■ **NOTE** Part of a complex including *H. canadense* to the east (Great Plains Flora Association, 1991).

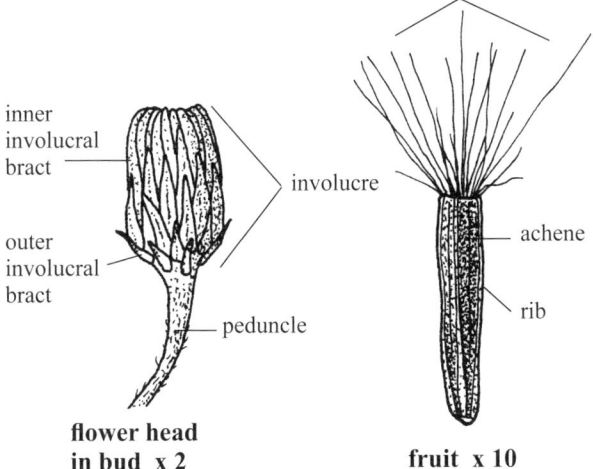

flower head in bud x 2

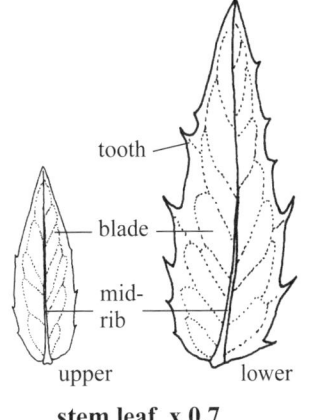

fruit x 10

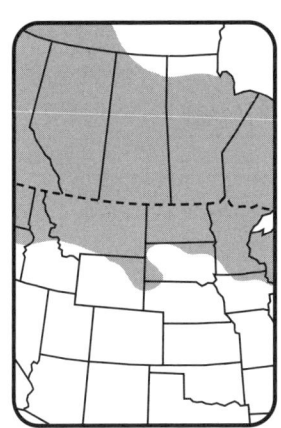

stem leaf x 0.7

Narrow-leaved Hawkweed

Aster—*Asteraceae* **Wildflower** Yellow Ray florets 7–14

COLORADO RUBBERWEED *Hymenoxys richardsonii* (Hook.) Cockerell

■ **SKETCH** A tufted, punctate **perennial herb** 10–30 cm tall with a coarse **taproot** 3–12 mm thick with a branching woody **caudex** covered with persistent hairy leaf bases; in open plains, dry rocky hillsides and dry to moist rangeland, in sandy soils. **p2**

■ **FLOWER HEADS** Yellow, blooming in May–September; **inflorescence** a flat-topped cyme of one to several heads; **peduncles** 2–4 cm long, ridged, punctate, slightly hairy; **flower heads** 2–4 cm across by 8–12 mm tall; **disc** 7–20 mm across; **involucre** 5–8 mm tall by 5–10 mm wide, hairy-punctate; **outer involucral bracts** 8–16, lower half united, stiff, persistent and erect, pale green, thick, upper 3–4 mm distinct, keels wide; **inner involucral bracts** separate, between the outer bracts, pointed, soft, each 5–6 mm long by 2–2.5 mm wide, lower covered parts glabrous, upper exposed parts hairy, apices pointed; **ray florets** 7–14, female and fertile; **ovary** 1–1.5 mm long, very hairy; **pappus** of several scales, imbricate, 2–2.5 mm long; **corolla tube** 1.5–1.8 mm long, pale green; **ligule** yellow, glabrous, 3–20 mm long by 2–5 mm wide, teeth three (four), blunt, 0.5–1.3 mm long; **disc florets** 40–60, perfect, each 5–6 mm long; **ovary** as above; **pappus** of five to seven scales, hyaline 1.5–3 mm long by 0.7–0.8 mm wide, pointed, toothed; **corolla** 3.5–4 mm long; **corolla tube** *c.* 1 mm long; **corolla limb** yellow, papillate, 2.4–3 mm long by *c.* 1.2 mm wide, teeth five, blunt, 0.5–0.7 mm long; **stamens** five, yellow; **anthers** united, *c.* 1.5 mm long, apices exserted; **style** 2-parted, slightly exserted, spreading, apices hairy; **fruiting heads** erect, 7–9 mm wide and long. FRUIT **an achene**, 1-seeded, brown, 2–4 mm long by 1–1.3 mm wide by 0.8–1 mm thick, hairy, 10- to 12-ribbed; **pappus** scales 2–4 mm long; **receptacle** hairless, convex.

■ **LEAVES** Basal and stem, alternate, compound; **basal leaves** few to several, 10–15 cm long; **stem blades** usually with three to seven linear, entire segments 2–5 cm long by 0.9–1.3 mm wide, punctate, thick, soft, widely grooved on the inner (ventral side), pointed; **petioles** 0–4 cm long, reduced above, base hairy.

■ **STEM** Stiff, erect, round, pale green, often branched above, 1.5–3 mm thick near the base; **old stems** sometimes among the new ones.

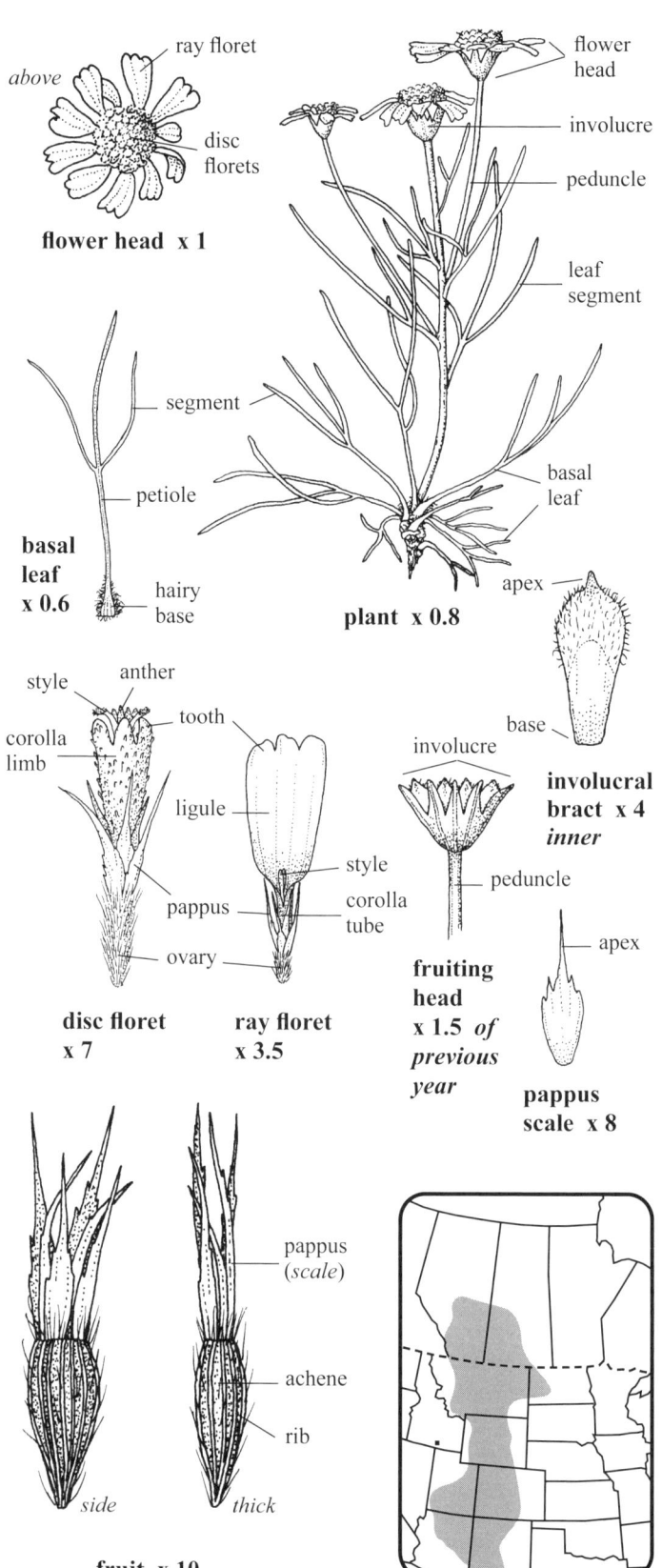

152 Colorado Rubber-plant

Aster—*Asteraceae* **Wildflower** Green Disc florets only

FALSE RAGWEED *Iva xanthifolia* Nutt.

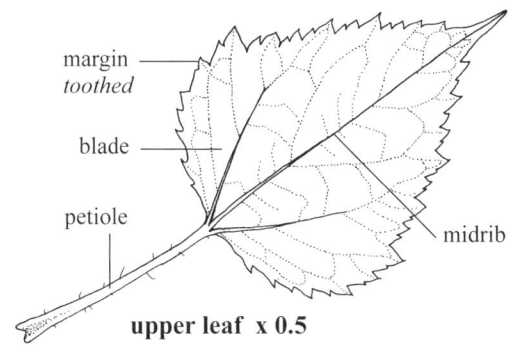
upper leaf x 0.5

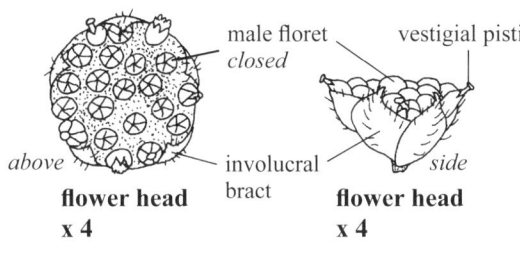
flower head x 4   flower head x 4

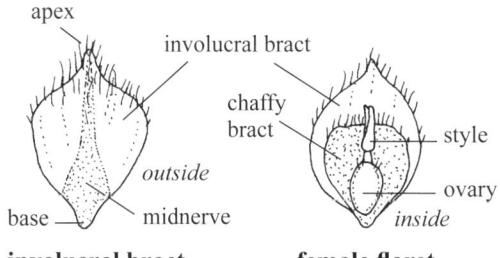
involucral bract x 8   female floret x 8

■ **SKETCH** An **annual herb** 50–200 cm tall with a **taproot** 6–15+ cm long by 2–13 mm wide; in disturbed gravelly to sandy sites, along streams, irrigation canals, fields, roadsides and flood plains, in cities and towns. Native and *naturalized*

■ **FLOWER HEADS** Green, blooming July–October; **inflorescence** paniculate with clusters of congested glomerules, terminal and axillary; **floral branches** leafy, ascending, 4–40 cm long, axillary; **flower heads** 4–5 mm wide by 2–3 mm tall; **involucral bracts** five, green, 2–3 mm long by *c.* 2 mm wide, midnerve thickened and slightly raised, pointed, apices dark green, hairy on outside, glabrous inside; <u>ray florets</u> absent; <u>disc florets</u> of two types: **1) male florets** 8–21 per flower head, 2–2.3 mm long by 0.8–1 mm wide, mostly central in flower heads, some marginal, outer ones bloom first; **corolla** tubular, with five apical teeth, hyaline, opaque; **stamens** five; **filaments** distinct, 0.2–0.4 mm long, green; **anthers** yellow, *c.* 1 mm long, slightly exserted; **pistil** vestigial, *c.* 2 mm long, hyaline, thick, apex flat and expanded, usually hidden among the anthers; **2) female florets** usually five, marginal, green, *c.* 1.8 mm long, included, opposite the involucral bracts; **ovary** *c.* 0.8 mm long, with a few scattered apical hairs; **chaffy bracts** usually five, marginal, *c.* 1.7 mm long, flat apices and hairy, between the involucral bract and ovary; **corolla** absent to tiny; **style** light green, 2-parted, *c.* 0.7 mm long. **FRUIT an achene**, 1-seeded, 2–3 mm long by 1.5–2 mm wide by 0.8–1 mm thick, dark brown, striated; **pappus** absent.

■ **LEAVES** Opposite below to alternate above, simple, coarsely toothed, some lobed; **blades** downy on both sides, 4–22 cm long by 1.5–20 cm wide, flat, spreading, three main veins; **petioles** reddish or green, 1–15 cm long by 1–5 mm wide and almost as thick, hairs scattered, spreading and to *c.* 3 mm long.

■ **STEM** Erect, glabrous to downy, round, with long reddish vertical stripes, branched, ridges blunt; 2–20 mm thick near the base.

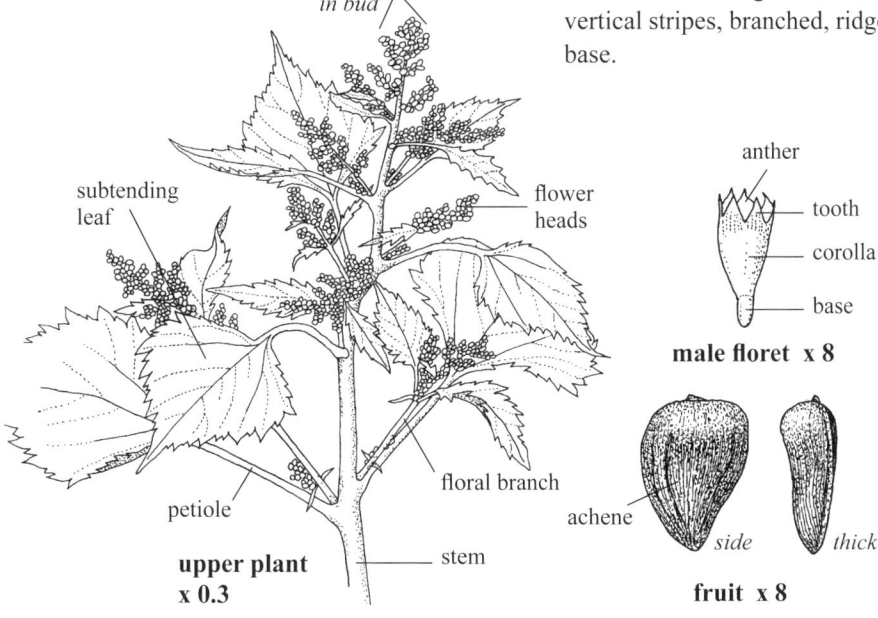
upper plant x 0.3

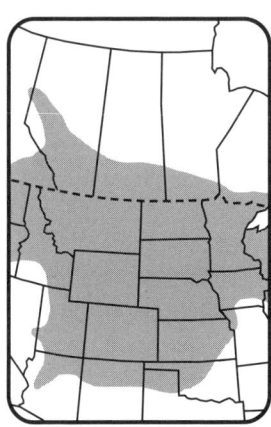
male floret x 8

fruit x 8

Careless-weed, Giant Sumpweed   153

Aster—*Asteraceae* **Wildflower** Yellow Ligulate florets 16–28

LOBED PRICKLY LETTUCE *Lactuca serriola* L.

■ **SKETCH** A **winter annual herb** 20–200 cm tall from a **taproot** 2–15 mm thick by 5–15⁺ cm long; along slough margins, edges of cultivated fields, roadsides, orchards and other disturbed sites; **juice milky**. *Naturalized*

■ **FLOWER HEADS** Pale yellow, blooming May–October; **inflorescence** an open panicle up to *c.* 90 cm long by *c.* 40 cm wide of 50–100⁺ flower heads; **floral branches** ascending but lax at the tips, 2–45 cm long, reduced above; **peduncles** 2–30 mm long; **flower heads** 9–15 mm wide by 10–13 mm long; **involucral bracts** 17, in three or four series, imbricate, green with reddish tips, inner bracts linear and the longest at 7–10 mm; <u>disc florets</u> absent; <u>ligulate florets</u> perfect, 16–28; **ovary** *c.* 1 mm long, glabrous; **pappus** bristles 3–3.5 mm long; **corolla tube** white, 3–3.5 mm long with a tuft of short white curly hairs at its summit; **ligules** 4.5–6.5 mm long by 1.5–2.2 mm wide, striped below, apical teeth five; **anthers** five, yellow, forming a tube *c.* 2 mm long; **style** 2-parted, yellow, exserted; **fruiting heads** white, 15–20 mm across, with *c.* 18 achenes per head. **FRUIT an achene**, 1-seeded, the body tan to dark brown, 3–3.8 mm long by *c.* 1 mm wide by *c.* 0.4 mm thick, with several ribs converging at the hairy apex; **beak** longer than the body, *c.* 4 mm, very thin, tan; **pappus** bristles white, together 10–11 mm across.

■ **LEAVES** Alternate, simple, deeply lobed to toothed to entire, clasping, sessile, 3.5–25 cm long by 1–20 cm wide, larger below; **lower blades** with a partial twist near the stem, the midrib thick and with one row of erect prickles 2–4 mm long below, glabrous above, the margins irregular with short prickles; **upper blades** mostly entire and with fewer prickles; **auricles** pointed.

■ **STEM** Erect, stout, round, solid, tan to reddish green, glabrous or sometimes with prickles (not shown) in the lower quarter; 6–20 mm thick at the base.

■ **SYN.** *Lactuca scariola* L.

154 Prickly Lettuce

Aster—*Asteraceae* **Wildflower** Blue Ligulate florets 16–21

BLUE LETTUCE   *Lactuca tatarica* (L.) C.A. Mey.

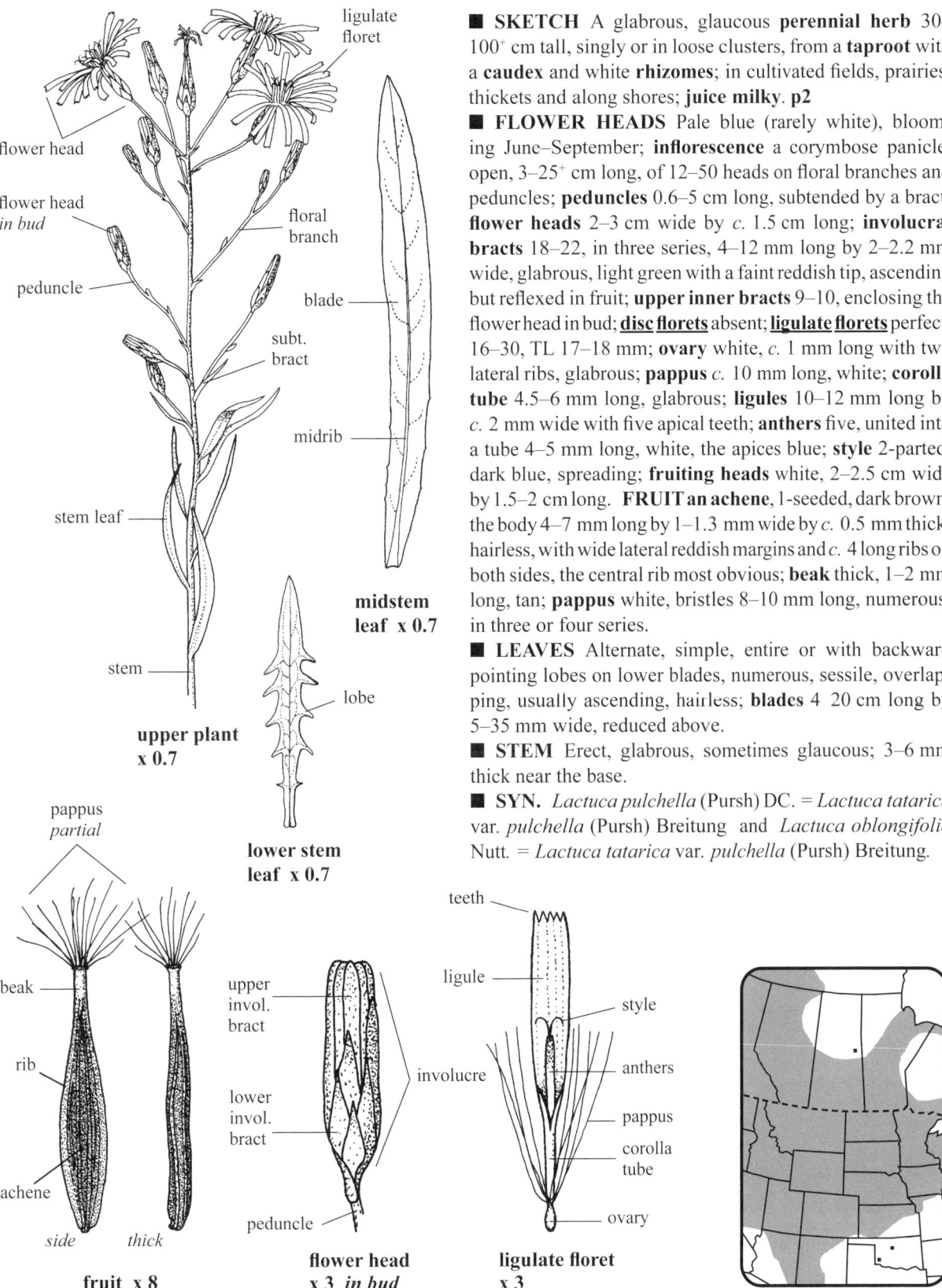

■ **SKETCH** A glabrous, glaucous **perennial herb** 30–100⁺ cm tall, singly or in loose clusters, from a **taproot** with a **caudex** and white **rhizomes**; in cultivated fields, prairies, thickets and along shores; **juice milky. p2**

■ **FLOWER HEADS** Pale blue (rarely white), blooming June–September; **inflorescence** a corymbose panicle, open, 3–25⁺ cm long, of 12–50 heads on floral branches and peduncles; **peduncles** 0.6–5 cm long, subtended by a bract; **flower heads** 2–3 cm wide by *c.* 1.5 cm long; **involucral bracts** 18–22, in three series, 4–12 mm long by 2–2.2 mm wide, glabrous, light green with a faint reddish tip, ascending but reflexed in fruit; **upper inner bracts** 9–10, enclosing the flower head in bud; <u>**disc florets**</u> absent; <u>**ligulate florets**</u> perfect, 16–30, TL 17–18 mm; **ovary** white, *c.* 1 mm long with two lateral ribs, glabrous; **pappus** *c.* 10 mm long, white; **corolla tube** 4.5–6 mm long, glabrous; **ligules** 10–12 mm long by *c.* 2 mm wide with five apical teeth; **anthers** five, united into a tube 4–5 mm long, white, the apices blue; **style** 2-parted, dark blue, spreading; **fruiting heads** white, 2–2.5 cm wide by 1.5–2 cm long. **FRUIT an achene**, 1-seeded, dark brown, the body 4–7 mm long by 1–1.3 mm wide by *c.* 0.5 mm thick, hairless, with wide lateral reddish margins and *c.* 4 long ribs on both sides, the central rib most obvious; **beak** thick, 1–2 mm long, tan; **pappus** white, bristles 8–10 mm long, numerous, in three or four series.

■ **LEAVES** Alternate, simple, entire or with backward pointing lobes on lower blades, numerous, sessile, overlapping, usually ascending, hairless; **blades** 4–20 cm long by 5–35 mm wide, reduced above.

■ **STEM** Erect, glabrous, sometimes glaucous; 3–6 mm thick near the base.

■ **SYN.** *Lactuca pulchella* (Pursh) DC. = *Lactuca tatarica* var. *pulchella* (Pursh) Breitung and *Lactuca oblongifolia* Nutt. = *Lactuca tatarica* var. *pulchella* (Pursh) Breitung.

Russian Blue Lettuce, Common Blue Lettuce

Aster—*Asteraceae* **Wildflower** White, center yellow Ray florets 15–30

Ox-eye Daisy *Leucanthemum vulgare* Lam.

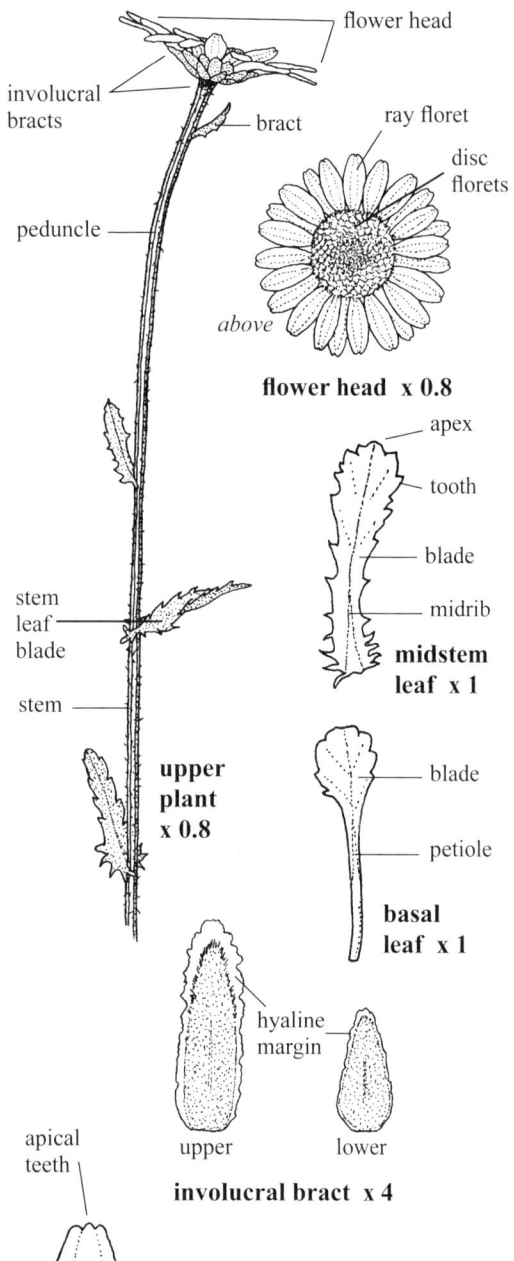

■ **SKETCH** A **perennial herb** 20–100 cm tall with brown **rhizomes** 2–4 mm thick, in open patches; in grassy fields, roadsides and along trail edges. *Naturalized*

■ **FLOWER HEADS** White with a yellow center, blooming May–October; **inflorescence** solitary or few-flowered, terminal; **peduncles** 9–19 cm long with one to three leaves or bracts; **flower heads** 2–5 cm wide by 8–10 mm tall; **disc** 10–20 mm wide, yellow, usually slightly concave in the middle at the start of blooming; **involucre** 12–20 mm wide by 3–4 mm tall, green, spreading, glabrous; **involucral bracts** imbricate, in three or four series, flat, midnerve not thick, 4–7.5 mm long by 2–2.5 mm wide, margins dark brown to hyaline and erose, especially in larger bracts, ascending around the receptacle after fruit has fallen; **ray florets** fertile, 15–35 in one whorl, female, TL 15–25 mm, slightly imbricate; **ovary** 10-ribbed, hairless, flat at both ends, *c.* 1.5 mm long; **pappus** absent; **corolla tube** *c.* 1.5 mm long by *c.* 1 mm wide, green, glabrous, slightly flattened; **ligules** flat, 10–25 mm long by 2–6 mm wide, widest in the middle, 2-veined, **apical teeth** two or three, blunt, minute, glabrous; **style** 2-parted, exserted, tips yellow; **disc florets** perfect, numerous, glabrous; **ovary** as above; **pappus** absent; **corolla tube** squarish, 1.3–1.5 mm long, green; **corolla limb** yellow, *c.* 1.5 mm long including the erect teeth; **teeth** five, *c.* 0.5 mm long. **FRUIT an achene**, 1-seeded, usually 10-ribbed, 1.8–2.3 mm long by 0.8–1 mm wide and thick, round in cross-section, shiny, body dark brown, ribs raised and white hyaline; **pappus** absent.

■ **LEAVES** Basal and stem, alternate, simple, toothed, several per stem, petiolate below; **basal blades** 7–30 mm long and wide, glabrous; **petioles** 1.5–10 cm long, slightly hairy below; **stem leaves** 1–10 cm long by 2–26 mm wide, reduced and sessile above, ascending, clasping, mostly glabrous, some with a few hairs below near the base.

■ **STEM** Erect, simple to few-branched, hairy below, less so above, ridged, round, lighter green than leaves; base 1–4 mm wide.

■ **SYN.** *Chrysanthemum leucanthemum* L.

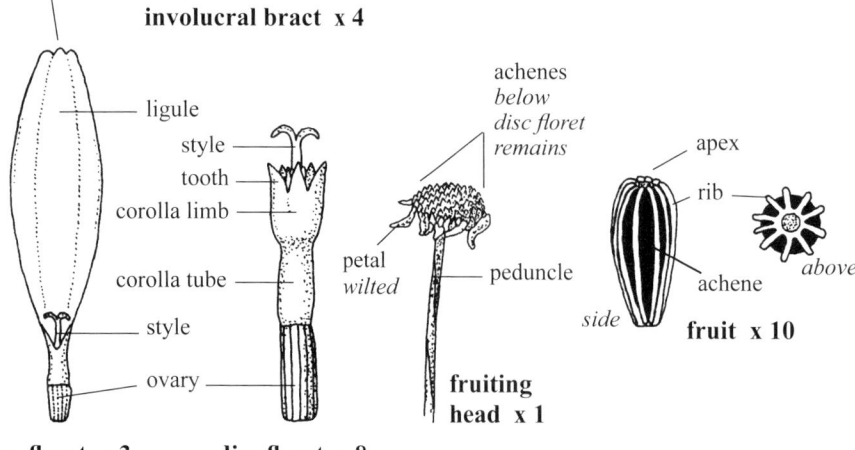

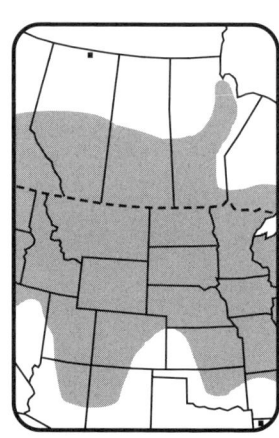

Aster—*Asteraceae* **Wildflower** Pinkish purple Disc florets 30–70

Meadow Blazingstar *Liatris ligulistylis* (A. Nels.) K. Schum.

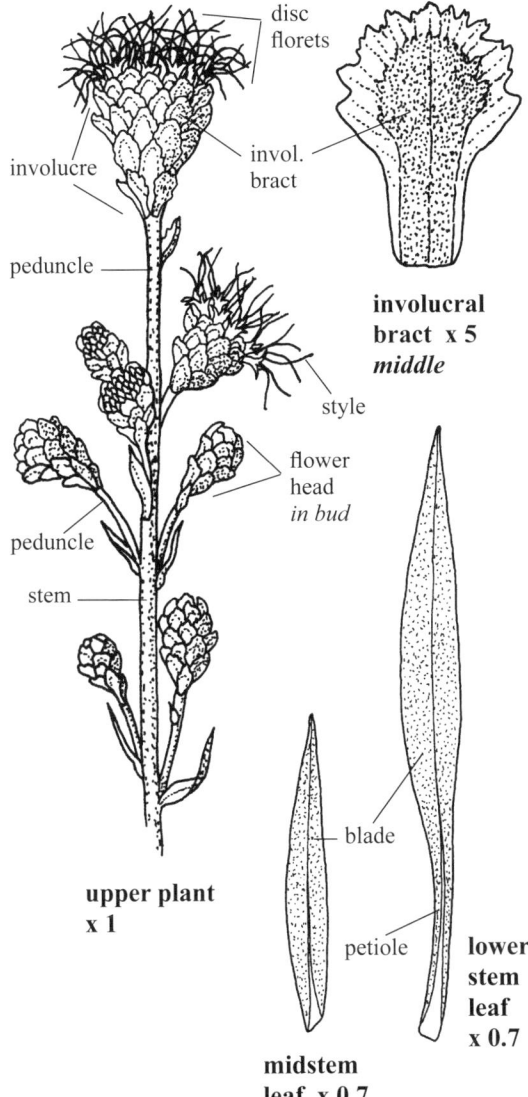

upper plant x 1

midstem leaf x 0.7

involucral bract x 5 *middle*

lower stem leaf x 0.7

■ **SKETCH** A **perennial herb** 20–100 cm tall from a **corm** 1–4 cm wide, in scattered patches; in dry to moist prairies, along slough margins, sandhills and in forest openings. **p1**

■ **FLOWER HEADS** Pinkish purple, blooming July–September; **inflorescence** a raceme 4–45 cm long by 2–7 cm wide, of 3–37 heads; **peduncles** rough hairy, 0.3–6 cm long, ascending; **flower heads** 2–3.3 cm wide by 1.5–2.5 cm long; **involucral bracts** green to reddish purple, glabrous, in four to six series, imbricate, *c.* 7 mm long by 5–7 mm wide, with an irregularly toothed scarious margin, the raised central area gives a bumpy appearance to the involucre, spreading in fruit; <u>ray florets</u> absent; <u>disc florets</u> perfect, 30–70, TL 17–20 mm (base of ovary to tip of style); **ovary** with ascending hairs, white, ribbed, 4–4.5 mm long by *c.* 1 mm wide; **pappus** bristles 6–7 mm long, plumose, with at least the tips of the hairs reddish; **corolla** 8–11 mm long, white; **corolla limb** light purplish pink, apical teeth five, spreading, each tooth to *c.* 2.5 mm long by *c.* 1 mm wide; **anthers** five, forming a tube 2.5–3 mm long; **style** 2-parted, the parts often crossing and extending 5–8 mm past the anthers, purplish pink, hairless; **fruiting heads** of *c.* 60 achenes, 20–25 mm wide by *c.* 15 mm long. **FRUIT an achene**, 1-seeded, 10-ribbed, the ribs with ascending hairs; **body** medium gray, 4–6 mm long by 0.7–1 mm wide by *c.* 0.7 mm thick; **pappus** slightly reddish tan, plumose bristles 20–22, each 6–10 mm long, the short ascending white hairs with a reddish tip.

■ **LEAVES** Basal and stem, alternate, simple, entire, mostly ascending, imbricate; **basal leaves** 4–13 cm long by 8–15 mm wide, reduced above; **petioles** 2–6 cm long; **stem blades** 1–25 cm long by 1–40 mm wide, reduced above, margins scabrous, lateral veins obscure; **petioles** 0–6 cm long, reduced above.

■ **STEM** Erect, solid, one to three, reddish green, unbranched; **hairs** *c.* 0.5 mm long, appressed upward and more numerous above; 2–7 mm thick near the base.

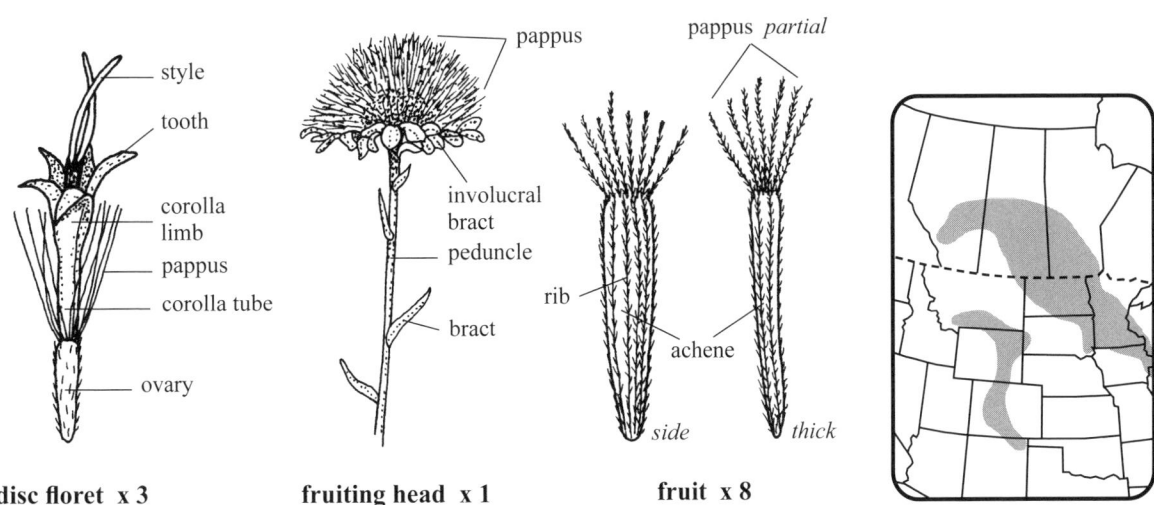

disc floret x 3

fruiting head x 1

fruit x 8

Blazing Star, Strap-style Gayfeather, Rocky Mountain Blazing Star 157

Aster—*Asteraceae* **Wildflower** Pinkish purple Disc florets 4–8

Dotted Blazingstar *Liatris punctata* Hook.

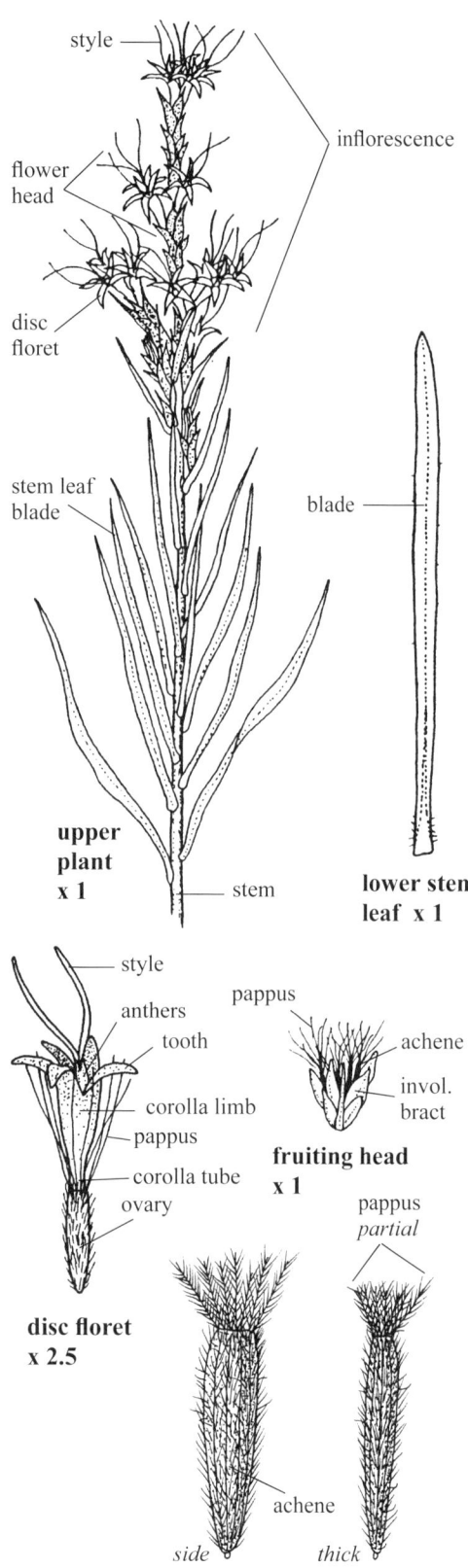

upper plant x 1

lower stem leaf x 1

disc floret x 2.5

fruiting head x 1

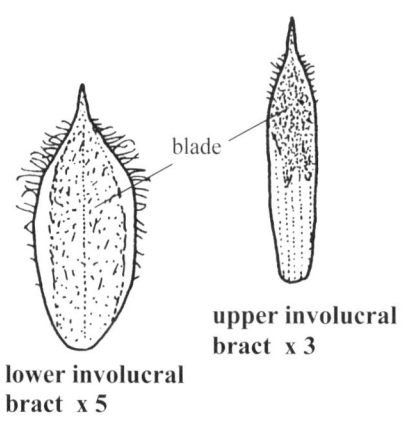

fruit x 4

lower involucral bract x 5

upper involucral bract x 3

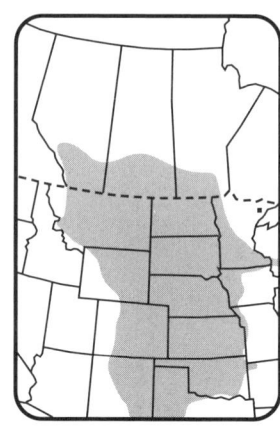

■ **SKETCH** A **perennial herb** 10–80 cm tall with a vertical **corm** 0.7–2.5 cm thick by 3–10⁺ cm long, sometimes topped with a branching **caudex** 1–2 cm long; in dry prairies, hills and uplands, usually in sandy soils. **p1**

■ **FLOWER HEADS** Pinkish purple, rarely white, blooming July–October; **inflorescence** spikelike, 2–30 cm long by 2–4.5 cm wide; **flower heads** 1–30⁺, sessile, each 10–15 mm wide by 1.5–2.5 cm long, subtended by a hairy leaflike bract; **involucre** 13–18 mm long by 4–5 mm wide; **involucral bracts** in three or four series, imbricate, pointed; **1) lower involucral bracts** green, 5–7 mm long by 2–4 mm wide, with scarious ciliate margins, slightly hairy on the outside; **2) upper involucral bracts** 11–14 mm long by 2–3 mm wide, reddish purple on the upper exposed half, green below where covered, 5-nerved, margins ciliate above; <u>ray florets</u> absent; <u>disc florets</u> perfect, four to eight, each *c.* 2 cm long by 7–8 mm wide; **ovary** 6–6.5 mm long by *c.* 1.3 mm wide, hairy, slightly flattened, ribbed; **pappus** 7–10 mm long of 17–22 plumose red bristles with ascending white hairs, tan in fruit; **corolla tube** white, 2–3 mm long; **corolla limb** pink, 7–8 mm long, not hairy inside, with five spreading teeth 3–3.5 mm long by 1.3–1.5 mm wide; **filaments** distinct, white, hairy; **anthers** brown, united, *c.* 3 mm long; **style** 2-parted, filiform, pink, 7–10 mm long, exserted, with two darker pink lines near the base; **fruiting heads** sessile, each 15–20 mm long and wide, the involucral bracts erect to spreading. **FRUIT** an achene, 1-seeded, 10-ribbed, medium gray, 6–8 mm long by 1.3–1.8 mm wide by 0.7–1.2 mm thick, ribs with ascending hairs; **pappus** as above.

■ **LEAVES** Basal and stem, alternate, simple, entire, flat, numerous, overlapping, mostly ascending; **blades** resinous, punctate on both surfaces, stiff, sessile, linear, 2.5–15 cm long by 1–8 mm wide, reduced above, minutely hairy along the scabrous margins; **lower blades** sometimes with a twist.

■ **STEM** Erect, rarely branched, solid, stiff, light green, round, singly or clusters up to several stems, glabrous or sparsely hairy below, more hairy above with white dishevelled hairs especially between the flower heads; 2–5 mm thick near the base.

Aster—*Asteraceae* **Wildflower** Pale blue to pink Ligulate florets 5 (3–6)

SKELETONWEED   *Lygodesmia juncea* (Pursh) D. Don ex Hook.

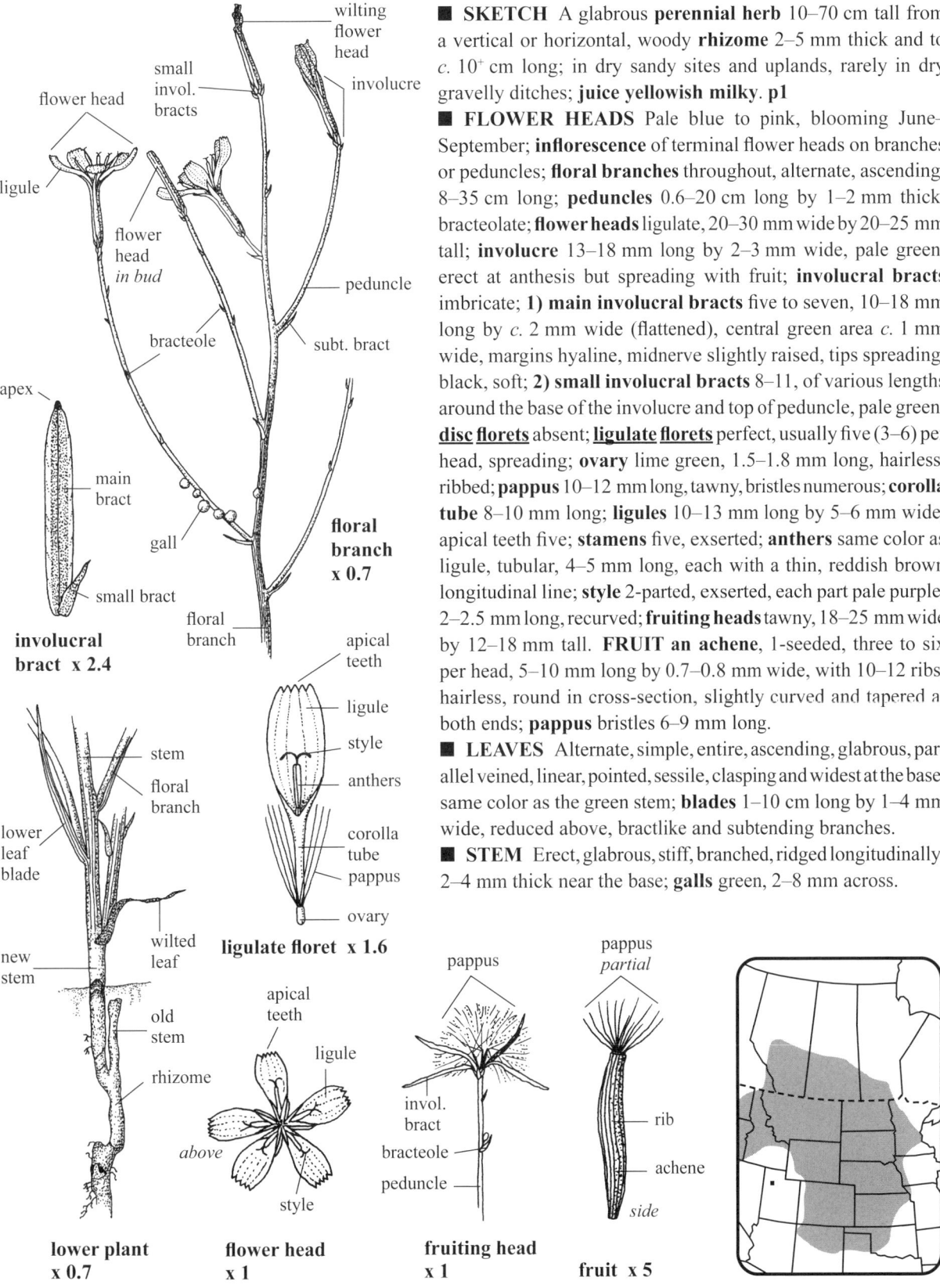

- **SKETCH** A glabrous **perennial herb** 10–70 cm tall from a vertical or horizontal, woody **rhizome** 2–5 mm thick and to *c.* 10⁺ cm long; in dry sandy sites and uplands, rarely in dry gravelly ditches; **juice yellowish milky. p1**
- **FLOWER HEADS** Pale blue to pink, blooming June–September; **inflorescence** of terminal flower heads on branches or peduncles; **floral branches** throughout, alternate, ascending, 8–35 cm long; **peduncles** 0.6–20 cm long by 1–2 mm thick, bracteolate; **flower heads** ligulate, 20–30 mm wide by 20–25 mm tall; **involucre** 13–18 mm long by 2–3 mm wide, pale green, erect at anthesis but spreading with fruit; **involucral bracts** imbricate; **1) main involucral bracts** five to seven, 10–18 mm long by *c.* 2 mm wide (flattened), central green area *c.* 1 mm wide, margins hyaline, midnerve slightly raised, tips spreading, black, soft; **2) small involucral bracts** 8–11, of various lengths around the base of the involucre and top of peduncle, pale green; <u>disc florets</u> absent; <u>ligulate florets</u> perfect, usually five (3–6) per head, spreading; **ovary** lime green, 1.5–1.8 mm long, hairless, ribbed; **pappus** 10–12 mm long, tawny, bristles numerous; **corolla tube** 8–10 mm long; **ligules** 10–13 mm long by 5–6 mm wide, apical teeth five; **stamens** five, exserted; **anthers** same color as ligule, tubular, 4–5 mm long, each with a thin, reddish brown longitudinal line; **style** 2-parted, exserted, each part pale purple, 2–2.5 mm long, recurved; **fruiting heads** tawny, 18–25 mm wide by 12–18 mm tall. **FRUIT an achene**, 1-seeded, three to six per head, 5–10 mm long by 0.7–0.8 mm wide, with 10–12 ribs, hairless, round in cross-section, slightly curved and tapered at both ends; **pappus** bristles 6–9 mm long.
- **LEAVES** Alternate, simple, entire, ascending, glabrous, parallel veined, linear, pointed, sessile, clasping and widest at the base, same color as the green stem; **blades** 1–10 cm long by 1–4 mm wide, reduced above, bractlike and subtending branches.
- **STEM** Erect, glabrous, stiff, branched, ridged longitudinally; 2–4 mm thick near the base; **galls** green, 2–8 mm across.

Rush Skeleton-weed, Rush Skeletonplant

Aster—*Asteraceae* **Wildflower** Yellow Ray florets 15–50

SPINY IRONPLANT *Machaeranthera pinnatifida* (Hook.) Shinners

■ **SKETCH** A tufted, variable **perennial herb** 15–80 cm tall and 20–40 cm wide in scattered clumps from a thick, woody **taproot** and **caudex**; in sandy prairies, along roadsides and in dry foothills. **p1**

■ **FLOWER HEADS** Yellow, blooming May–September; **inflorescence** corymblike, with heads on leafy branches; **flower heads** 1.3–2.2 cm wide by 12–15 mm tall; **disc** yellow, 5–10 mm wide, flat; **involucre** hairy, 5–8 mm tall by 5–12 mm wide, pale green; **involucral bracts** green with whitish bases, tips pointed with a whitish spine 0.5–0.7 mm long and often bent, bracts spreading to ascending, 3–6 mm long by 0.5–1 mm wide, reduced below, in four to six series, imbricate, hairs cottony and tangled on outer dorsal side, upper inner bracts with hyaline margins, green midnerve, hairy near apices; <u>ray florets</u> female, 15–50, TL *c.* 13 mm (including ovary); **ovary** 1–1.2 mm long, covered with ascending white hairs; **pappus** of fine unequal bristles 2–4.5 mm long; **corolla tube** pale yellow, 3–4 mm long, scattered hairs on underside near the top of the tube where it begins to widen; **ligule** 8–10 mm long by 1–1.7 mm wide, ascending to spreading, apical teeth three; **style** pale yellow, exserted, 2-parted, the parts *c.* 1 mm long; <u>disc florets</u> perfect, numerous, 6.5–7.5 mm long (base of ovary to tip of corolla teeth); **ovary** *c.* 1.5 mm long, covered by ascending hairs; **pappus** of unequal bristles to *c.* 5 mm long; **corolla tube** 1.2–2 mm long, green; **corolla limb** 3.5–4 mm long by 0.8–1 mm wide, yellow near the apex; **corolla teeth** yellow, slightly spreading, 0.3–0.4 mm long with a few scattered hairs; **anthers** five, yellow, *c.* 2 mm long, united into a tube around the style; **style** 2-parted, exserted, parts *c.* 1 mm long, flattened, hairy, usually not widely spreading; **fruiting heads** tawny to white, 1–1.5 cm wide and tall, erect. **FRUIT an achene**, 1-seeded, brown, 2–2.8 mm long by 0.7–0.9 mm wide (excluding hairs) by 0.5–0.6 mm thick, lateral ribs two, tan, with a few flat ribs and striations on each side, ascending hairs throughout; **pappus** of *c.* 50 tawny unequal bristles 2–6 mm long.

■ **LEAVES** Alternate, usually compound, sessile; **leaf blades** 0.7–6 cm long by 0.8–10 mm wide, bluish green to gray, lobes and leaflets bristle-tipped, reduced above, upper leaves toothed to entire and more hairy than lower ones, lateral veins obscure.

■ **STEM** Pale green, rarely solitary, usually several from the caudex, horizontal with the apex and branches ascending, somewhat fleshy, 2–3 mm thick near the base.

■ **SYN.** *Haplopappus spinulosus* (Pursh) DC. = *Machaeranthera pinnatifida* var. *pinnatifida* (Hook.) Shinners.

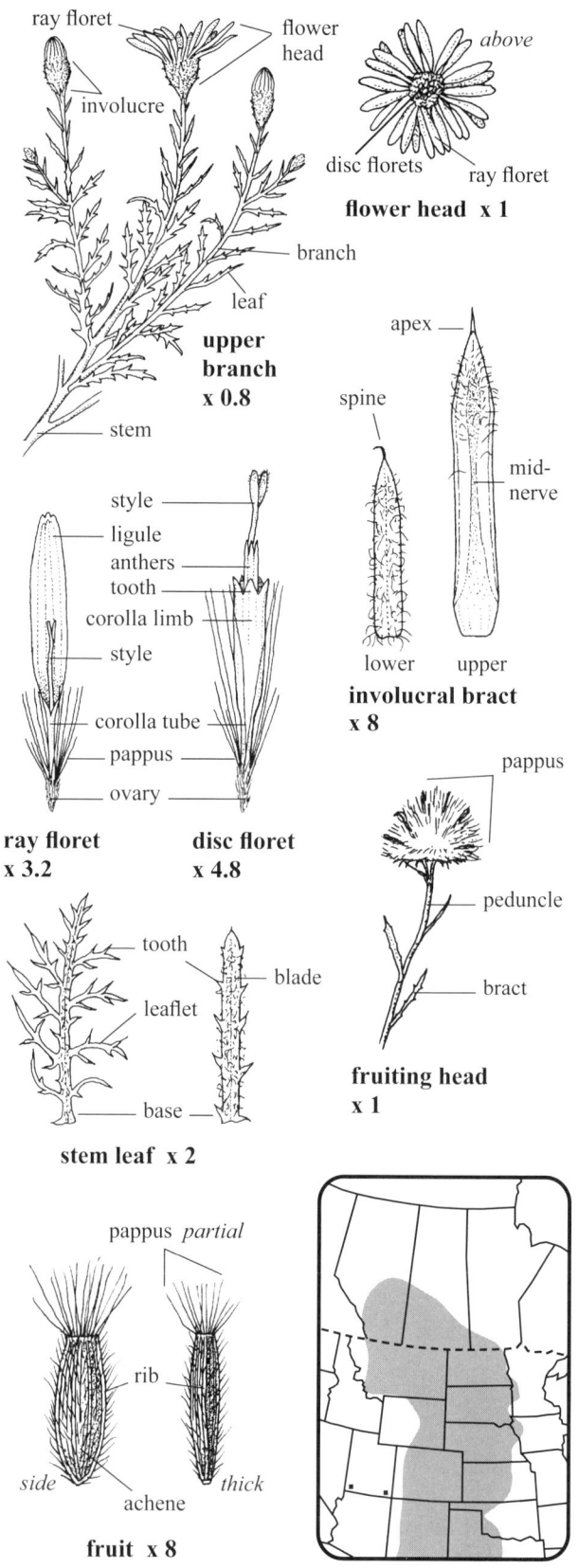

Lacy Tansyaster, Cutleaf Ironplant

Aster—*Asteraceae* **Wildflower** Greenish yellow Disc florets only

PINEAPPLEWEED *Matricaria discoidea* DC.

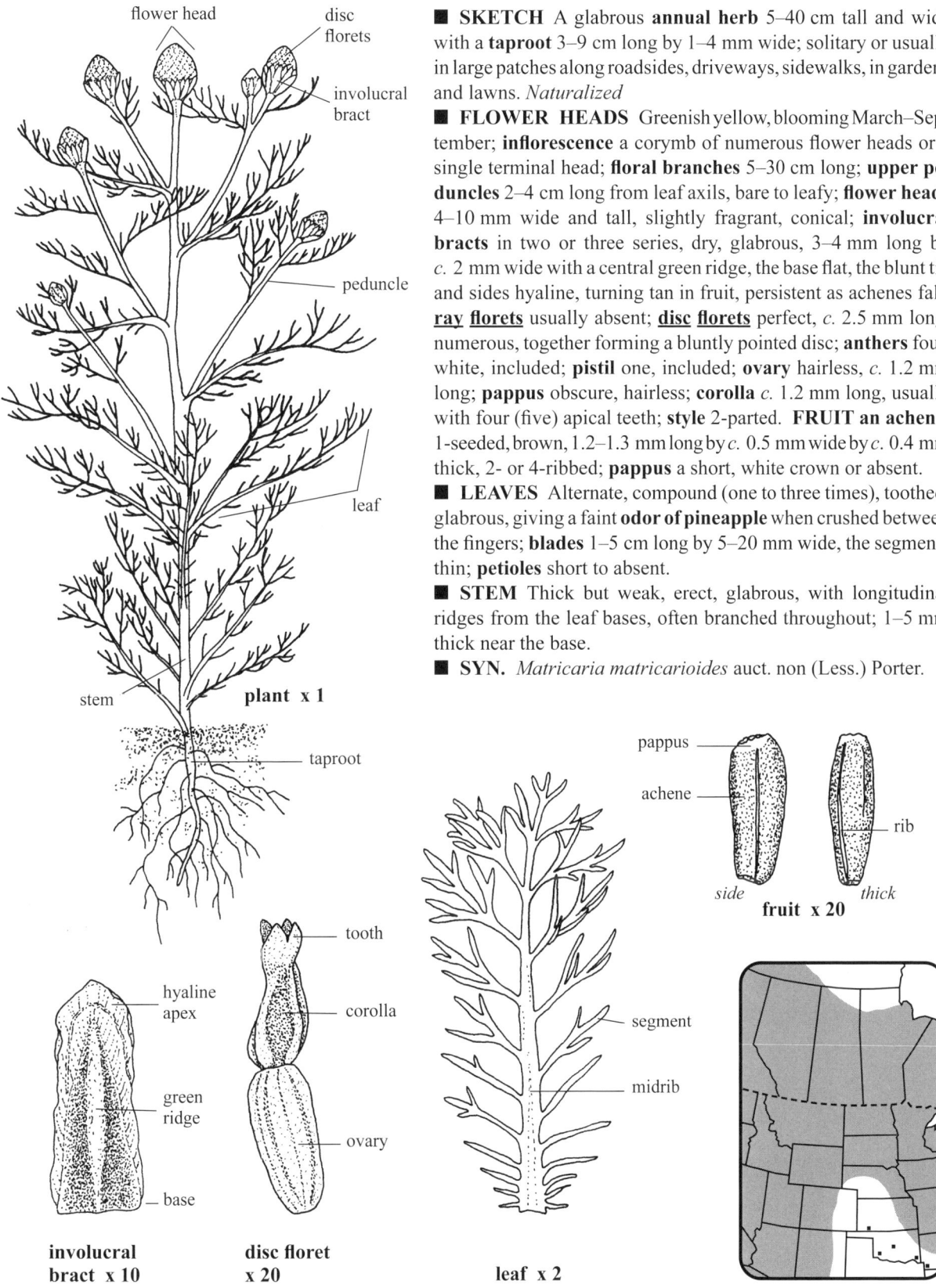

■ **SKETCH** A glabrous **annual herb** 5–40 cm tall and wide with a **taproot** 3–9 cm long by 1–4 mm wide; solitary or usually in large patches along roadsides, driveways, sidewalks, in gardens and lawns. *Naturalized*

■ **FLOWER HEADS** Greenish yellow, blooming March–September; **inflorescence** a corymb of numerous flower heads or a single terminal head; **floral branches** 5–30 cm long; **upper peduncles** 2–4 cm long from leaf axils, bare to leafy; **flower heads** 4–10 mm wide and tall, slightly fragrant, conical; **involucral bracts** in two or three series, dry, glabrous, 3–4 mm long by *c.* 2 mm wide with a central green ridge, the base flat, the blunt tip and sides hyaline, turning tan in fruit, persistent as achenes fall; <u>ray</u> **florets** usually absent; <u>disc</u> **florets** perfect, *c.* 2.5 mm long, numerous, together forming a bluntly pointed disc; **anthers** four, white, included; **pistil** one, included; **ovary** hairless, *c.* 1.2 mm long; **pappus** obscure, hairless; **corolla** *c.* 1.2 mm long, usually with four (five) apical teeth; **style** 2-parted. **FRUIT** an achene, 1-seeded, brown, 1.2–1.3 mm long by *c.* 0.5 mm wide by *c.* 0.4 mm thick, 2- or 4-ribbed; **pappus** a short, white crown or absent.

■ **LEAVES** Alternate, compound (one to three times), toothed, glabrous, giving a faint **odor of pineapple** when crushed between the fingers; **blades** 1–5 cm long by 5–20 mm wide, the segments thin; **petioles** short to absent.

■ **STEM** Thick but weak, erect, glabrous, with longitudinal ridges from the leaf bases, often branched throughout; 1–5 mm thick near the base.

■ **SYN.** *Matricaria matricarioides* auct. non (Less.) Porter.

Disc Mayweed

Aster—*Asteraceae* Wildflower White Ray florets 10–25

UPLAND WHITE GOLDENROD   *Oligoneuron album* (Nutt.) Nesom

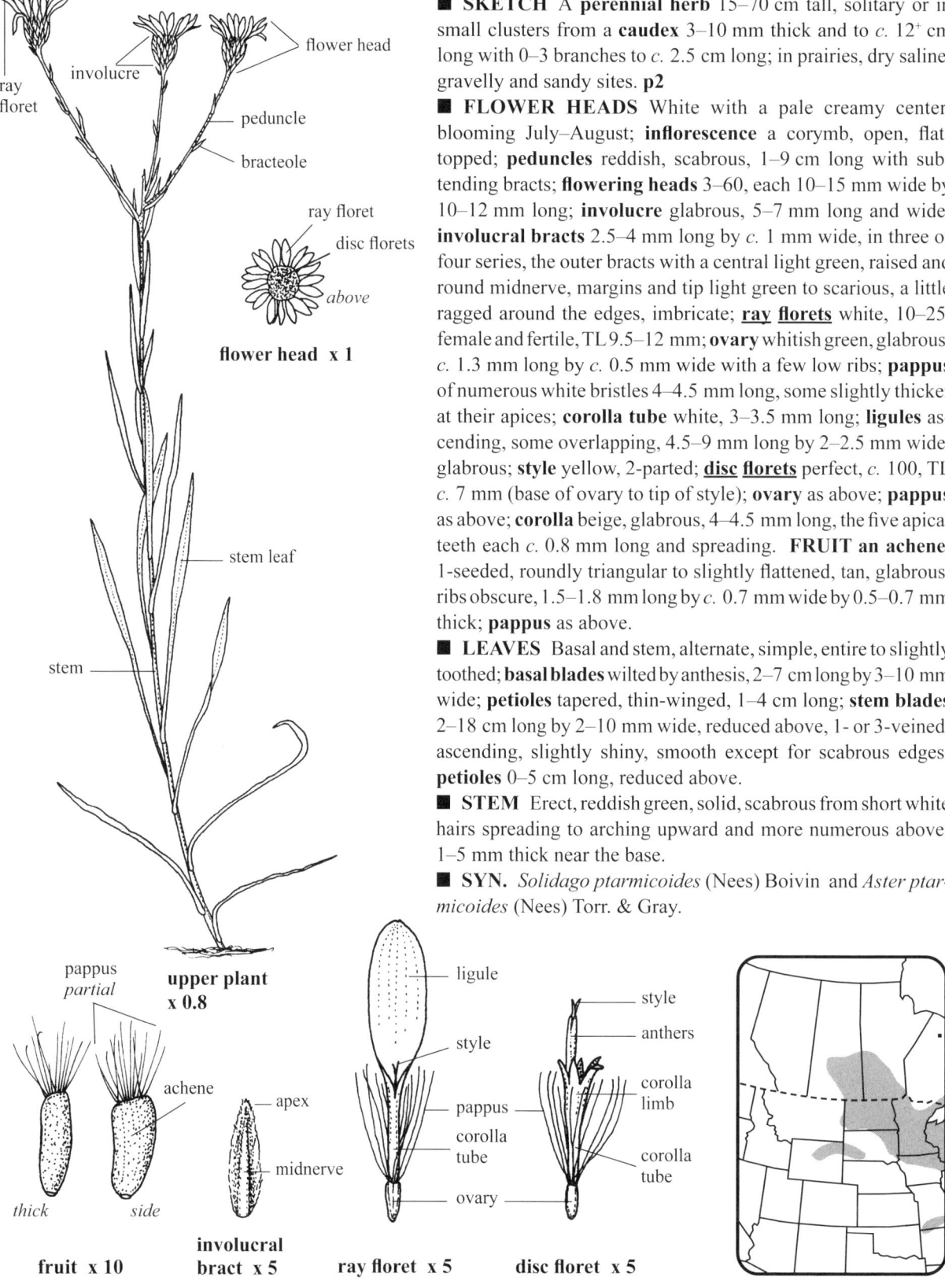

■ **SKETCH** A **perennial herb** 15–70 cm tall, solitary or in small clusters from a **caudex** 3–10 mm thick and to *c.* 12+ cm long with 0–3 branches to *c.* 2.5 cm long; in prairies, dry saline, gravelly and sandy sites. **p2**

■ **FLOWER HEADS** White with a pale creamy center, blooming July–August; **inflorescence** a corymb, open, flat-topped; **peduncles** reddish, scabrous, 1–9 cm long with sub-tending bracts; **flowering heads** 3–60, each 10–15 mm wide by 10–12 mm long; **involucre** glabrous, 5–7 mm long and wide; **involucral bracts** 2.5–4 mm long by *c.* 1 mm wide, in three or four series, the outer bracts with a central light green, raised and round midnerve, margins and tip light green to scarious, a little ragged around the edges, imbricate; <u>ray florets</u> white, 10–25, female and fertile, TL 9.5–12 mm; **ovary** whitish green, glabrous, *c.* 1.3 mm long by *c.* 0.5 mm wide with a few low ribs; **pappus** of numerous white bristles 4–4.5 mm long, some slightly thicker at their apices; **corolla tube** white, 3–3.5 mm long; **ligules** ascending, some overlapping, 4.5–9 mm long by 2–2.5 mm wide, glabrous; **style** yellow, 2-parted; <u>disc florets</u> perfect, *c.* 100, TL *c.* 7 mm (base of ovary to tip of style); **ovary** as above; **pappus** as above; **corolla** beige, glabrous, 4–4.5 mm long, the five apical teeth each *c.* 0.8 mm long and spreading. **FRUIT an achene**, 1-seeded, roundly triangular to slightly flattened, tan, glabrous, ribs obscure, 1.5–1.8 mm long by *c.* 0.7 mm wide by 0.5–0.7 mm thick; **pappus** as above.

■ **LEAVES** Basal and stem, alternate, simple, entire to slightly toothed; **basal blades** wilted by anthesis, 2–7 cm long by 3–10 mm wide; **petioles** tapered, thin-winged, 1–4 cm long; **stem blades** 2–18 cm long by 2–10 mm wide, reduced above, 1- or 3-veined, ascending, slightly shiny, smooth except for scabrous edges; **petioles** 0–5 cm long, reduced above.

■ **STEM** Erect, reddish green, solid, scabrous from short white hairs spreading to arching upward and more numerous above; 1–5 mm thick near the base.

■ **SYN.** *Solidago ptarmicoides* (Nees) Boivin and *Aster ptarmicoides* (Nees) Torr. & Gray.

Prairie Flat-top-goldenrod, White Upland Aster, Sneezewort Aster

Stiff Goldenrod *Oligoneuron rigidum* (L.) Small

■ **SKETCH** A variable **perennial herb** 10–160 cm tall with a thick, branched **caudex**; on dry sites in grasslands, sandhills, open rocky sites and woodland openings. **p1**

■ **FLOWER HEADS** Yellow, blooming July–October; **inflorescence** a compound corymb of crowded heads 2.5–10 cm wide by 2–4 cm deep, flat-topped; **floral branches** 1.5–11 cm long; **peduncles** 1–11 mm long and hairy; **flower heads** 8–13 mm wide by *c.* 12 mm long, the disc 4–8 mm wide; **involucre** 5–9 mm long by 4–5 mm wide, densely hairy on the exposed areas; **involucral bracts** bluntly pointed, 3.3–5 mm long by 1–1.6 mm wide, imbricate, firm, the three nerves often hidden by hairs; **ray florets** female and fertile, 7–14, irregularly spaced; **ovary** glabrous, *c.* 2 mm long; **pappus** 3.5–5.5 mm long, bristles numerous, white with ascending hairs; **corolla tube** *c.* 4 mm long; **ligules** 3–5 mm long by 1.5–1.8 mm wide; **style** 2-parted; **disc florets** perfect, 15–35, each to *c.* 3 mm wide; **pappus** as above; **corolla tube** 2–2.2 mm long; **corolla limb** 5-toothed, the teeth spreading; **anthers** five, yellow, *c.* 1.8 mm long; **style** 2-parted, flat and hairy; **fruiting heads** white, 10–12 mm wide. **FRUIT an achene**, 1-seeded, brown, 1.5–3.5 mm long by *c.* 1 mm wide by *c.* 0.8 mm thick with scattered hairs near the apex or glabrous, the three to five sides bluntly angled and with 10–24 light tan, flat ribs; **pappus** of 70–75 white bristles 2–4.5 mm long and together 8–9 mm wide.

■ **LEAVES** Basal and stem, alternate, simple, entire or finely toothed, firm; **basal leaves** scattered around the stem bases; **blades** downy, 3–23 cm long by 1–10 cm wide; **petioles** downy, 6–15 cm long and folded inward; **stem leaves** downy, ascending, larger and horizontal near the base; **blades** 1.5–11 cm long by 0.6–4 cm wide, reduced and sessile above; **petioles** 0–9 cm long, clasping.

■ **STEM** Downy, erect, reddish green, one to several in a cluster among the basal leaves, solid, round, grooved; 2–6 mm thick near the base.

■ **SYN.** *Solidago rigida* L.

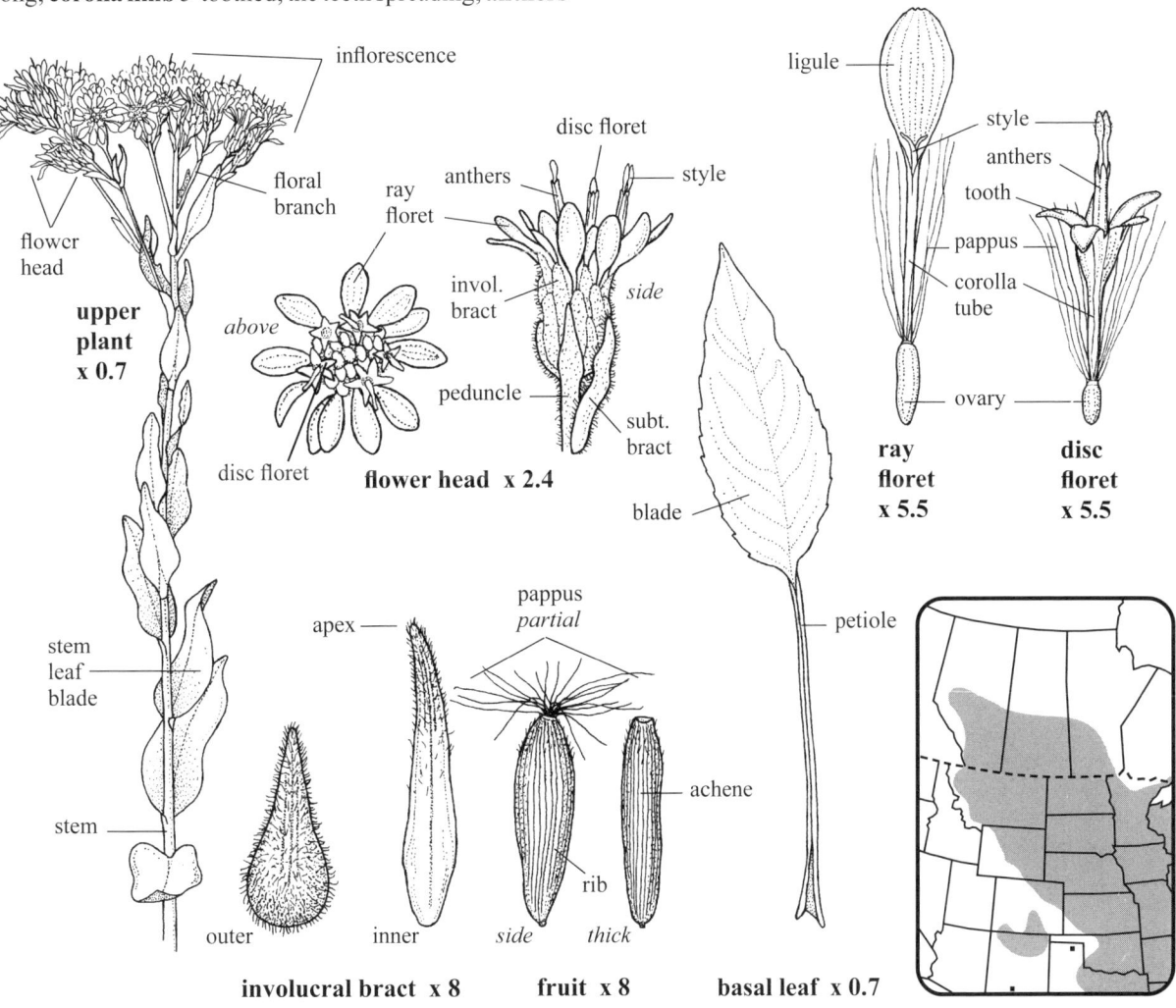

Rigid Goldenrod, Hardleaf Flattop Goldenrod

Aster—*Asteraceae* **Wildflower** Yellow Ray florets 8–12

Golden Ragwort *Packera aurea* (L.) A. & D. Löve

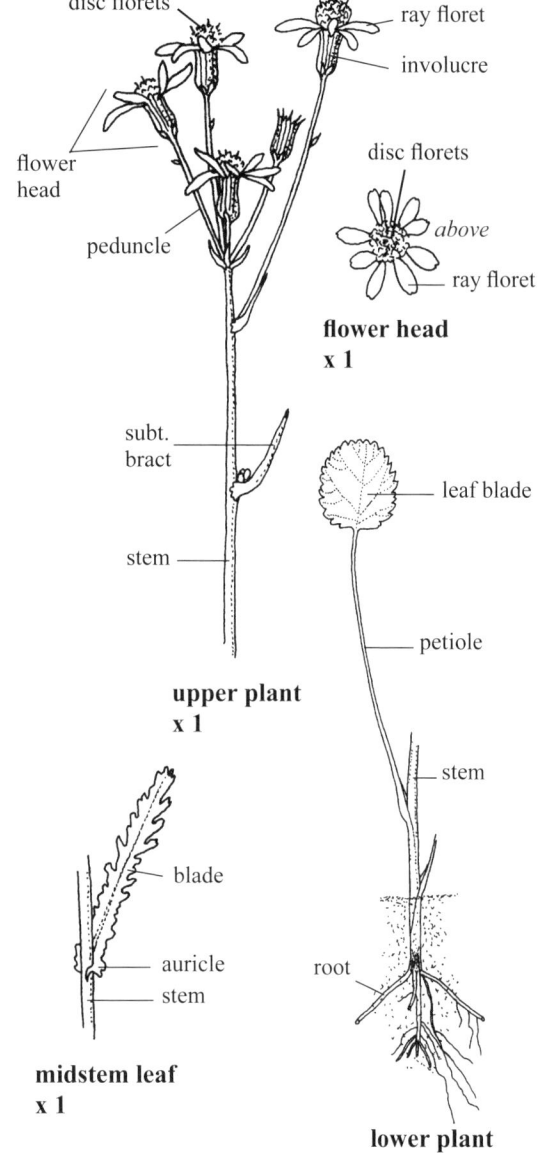

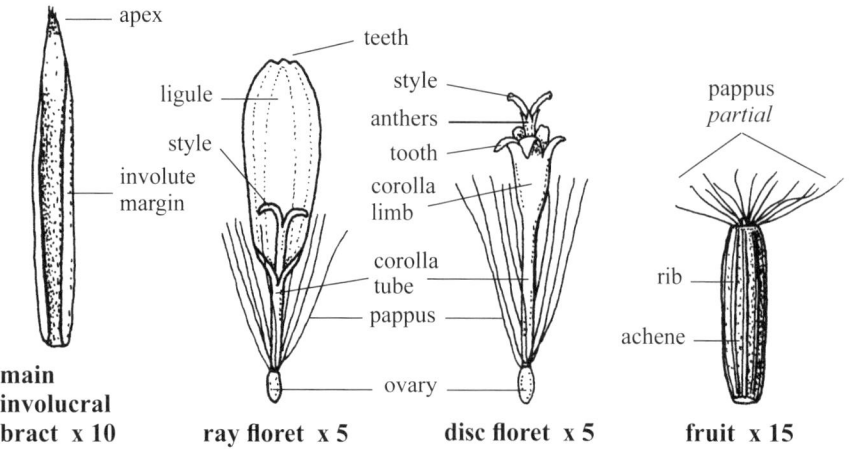

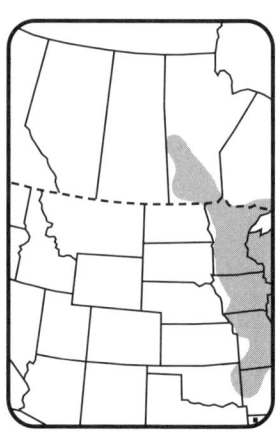

■ **SKETCH** A **perennial herb** 20–60 cm tall, singly or in scattered open colonies from **rhizomes**; in moist meadows, open woodland edges, swamps and other damp areas.
■ **FLOWER HEADS** Deep yellow, blooming April–July; **inflorescence** a cyme, 4–10 cm wide; **peduncles** 7–75 mm long, erect, the center one often the shortest, lightly hairy, ridged, some dividing and with a second flower head; **subtending bracts** (of peduncles) 2–10 mm long with a red tip; **flower heads** 4–18, each 12–18 mm wide by 8–12 mm long; **disc** 5–10 mm wide; **involucre** 5–8 mm long by 3–5 mm wide, hairy near the base; **involucral bracts** in two or three series, some imbricate; **1) main involucral bracts** 4.5–5 mm long by 0.6–0.8 mm wide, pointed, apices slightly reddish and minutely hairy, fleshy, the sides involute; **2) short involucral bracts** few, 1–2 mm long with reddish tips, at the base of the involucre; **ray florets** female, 8–12, unevenly spaced; **ovary** *c.* 0.8 mm long, glabrous; **pappus** of *c.* 60 white, fine bristles 4–4.3 mm long; **corolla tube** *c.* 2.5 mm long, pale yellow; **ligules** 4–8 mm long by 1.7–2.2 mm wide; **style** 2-parted; **disc florets** perfect, numerous; **ovary** *c.* 1 mm long and glabrous; **pappus** 5–6 mm long, of numerous white bristles; **corolla** 5.5–7 mm long; **corolla tube** 3.5–4.5 mm long; **corolla limb** 2–2.5 mm long, the five teeth yellow; **anthers** five, forming a tube; **style** 2-parted, yellow; **fruiting heads** white, round, *c.* 15 mm wide. FRUIT **an achene**, 1-seeded, 10-ribbed, *c.* 1.6 mm long by *c.* 0.4 mm wide and thick, glabrous, round in cross-section, dark brown; **pappus** of many fine white bristles.
■ **LEAVES** Basal and stem, alternate, simple, toothed; **basal blades** 10–40[+] mm long by 15–25[+] mm wide, glabrous, lighter green below; **petioles** 5–13 cm long, pinkish near the base and clasping; **stem blades** 1–4 cm long by 3–8 mm wide, reduced and lightly hairy above, glabrous below, deeply toothed, petiolate to sessile above, auricles toothed and clasping.
■ **STEM** Light green, erect, longitudinally grooved; 2–3 mm thick near the pinkish base.
■ **SYN.** *Senecio aureus* L.

164 Golden Groundsel

Aster—*Asteraceae* **Wildflower Yellow Ray florets 6–13⁺ (0)**

SILVERY GROUNDSEL   *Packera cana* (Hook.) W.A. Weber & A. Löve

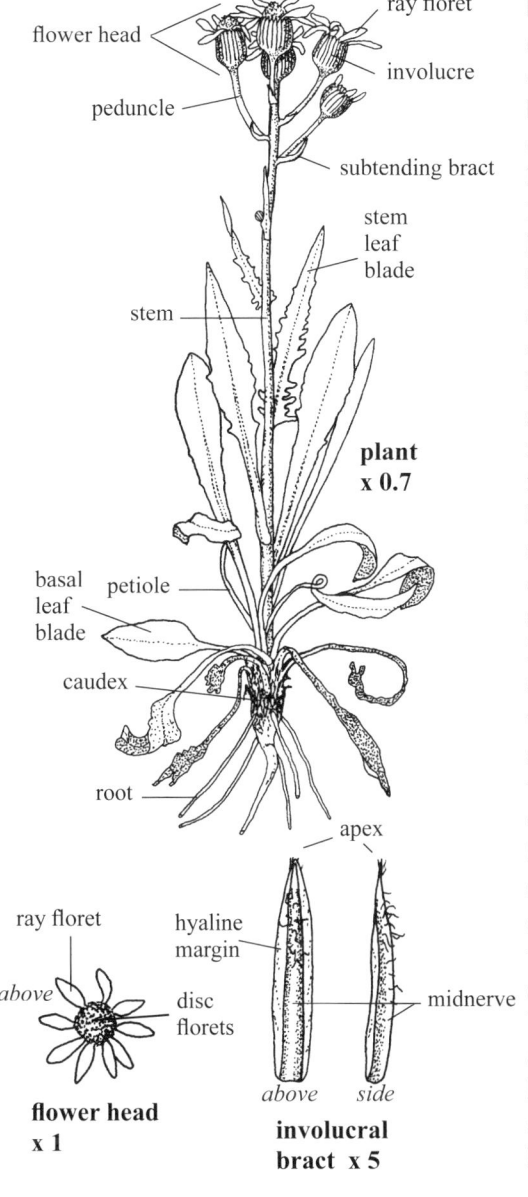

- **SKETCH** A silvery-hairy **perennial herb** 10–40 cm tall from a **caudex** 3–5 mm thick by 1–3 cm long, or a **rhizome**; in dry, rocky or gravelly hills and slopes to open plains and to above tree line.
- **FLOWER HEADS** Yellow, blooming May–August; **inflorescence** a corymb, sometimes compound, loose to congested, of 2–16 heads; **floral branches** 2–13 cm long; **peduncles** hairy, 0.5–7 cm long, ascending; **subtending bracts** (of peduncles) 3–8 mm long, hairy, pointed, soft, ascending; **flower heads** 12–25 mm wide by 9–11 mm tall; **involucre** green, 5–8 mm long by 5–7 mm wide, slightly hairy; **involucral bracts** in one series, usually 20–21 or *c.* 13, each 6–7 mm long by *c.* 1 mm wide, stiff, margins imbricate, hyaline, midnerve thick, rounded and *c.* 0.5 mm wide, hairy above, shiny near the base, reflexed in fruit; <u>ray florets</u> 6–13⁺ or absent, female; **ovary** 1.5–2 mm long, green, glabrous; **pappus** of white, fine bristles 3–3.5 mm long; **corolla tube** *c.* 3 mm long, glabrous; **ligules** 6–10 mm long by *c.* 2 mm wide, 4-veined, apical teeth three and tiny; <u>disc florets</u> perfect, numerous, together 4–7 mm wide; **ovary** 1.5–1.8 mm long, glabrous; **pappus** white, *c.* 3 mm long; **corolla** *c.* 6 mm long; **corolla tube** *c.* 3 mm long; **corolla limb** yellow, *c.* 3 mm long, glabrous, apical teeth five and *c.* 0.7 mm long, apices thickened on the outside; **fruiting heads** 12–16 mm wide and tall, white. **FRUIT an achene**, 1-seeded, dark brown, *c.* 8-ribbed, some ribs incomplete, hairless, 1.9–2.1 mm long by 0.5–0.6 mm wide and thick, round in cross-section, slightly curved; **pappus** of 40–50 fine white bristles 3–4.5 mm long.
- **LEAVES** Basal and stem, alternate, simple, entire to toothed, silvery gray, more hairy below, midrib raised below; **basal leaves** spreading, usually brown by anthesis; **basal blades** entire to weakly toothed, 1–8 cm long by 0.4–3 cm wide; **petioles** 2–7 cm long; **stem leaves** ascending, lower ones crowded; **blades** 2–8 cm long by 6–20 mm wide, pointed, irregularly toothed, reduced, bractlike and more remote above; **petioles** 0–4 cm long.
- **STEM** Erect, hairy, green, solid, round, simple, one to three stems per cluster; 1–4 mm thick near the base.
- **SYN.** *Senecio canus* Hook.

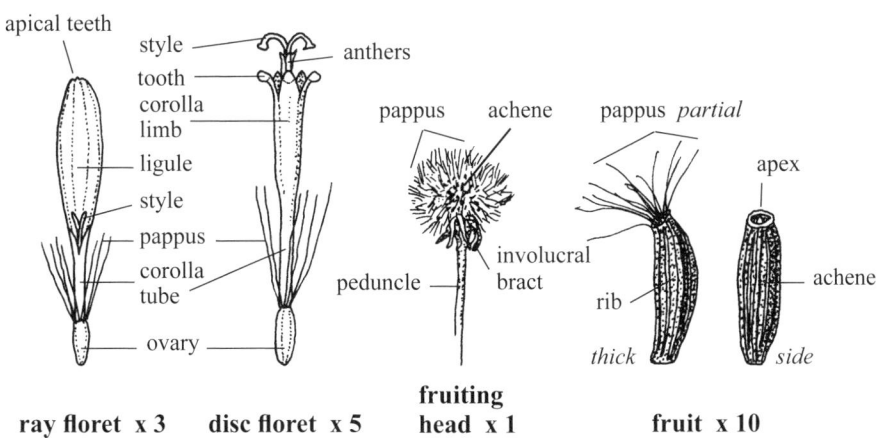

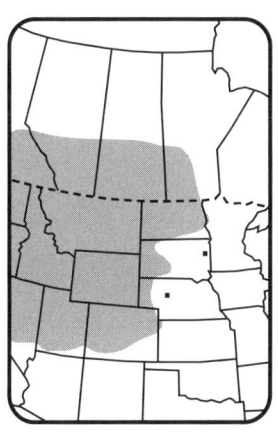

Gray Ragwort, Silver-woolly Groundsel

Palmate-leaved Colt's-foot   *Petasites frigidus* (L.) Fries var. *palmatus* (Ait.) Cronq.

■ **SKETCH** A variable, hairy **perennial herb**, 15–50 cm tall from white **rhizomes** 5–18+ cm long by 3–4 mm thick with **nodal bracts** 4–6 mm long and dark brown; **roots** white, 2–5 cm long; wet woods, ditches, fens, swamps and thickets; **imperfectly dioecious**.

■ **FLOWER HEADS** White, blooming April–June; **inflorescence** a raceme or corymb 3–20 cm long by 5–10 cm wide of fragrant mostly unisexual flower heads; **peduncles** veined, 8–40 mm long, reduced above, bracteate, some white hairs gland-tipped; **subtending bracts** (of peduncles) 7–18 mm long, pointed, hairy on outside and margins; **male heads** 4–35, each 4–5 mm wide by 10–12 mm long, ascending; **involucre** 6–10 mm long by 4–5 mm wide; **involucral bracts** 10–13, somewhat imbricate, glandular hairy near base, 5- or 6-veined (obscure), pink at pointed tip, 6.3–10 mm long by 1–3 mm wide (flattened), thin, margins hyaline; **female heads** 8–42; florets of two types: **1) ray florets** female, fertile, 8–9 mm long, 50–125; **ovary** glabrous, 1.3–1.7 mm long; **pappus** bristles 7–12 mm long; **corolla tube** 5.5–6 mm long by *c.* 0.2 mm wide; **ligule** absent to 6.3 mm long by 0.5–0.8 mm wide, teeth two; **style** exserted, 2-parted, these 0.06–0.8 mm long; **2) disc florets** male (inner in female heads), 1–5, *c.* 7 mm long; **ovary** 1–1.2 mm long, glabrous; **pappus** of white bristles 4–5 mm long; **corolla tube** *c.* 2.2 mm long; **corolla limb** *c.* 2.5 mm long (including teeth); **teeth** five, whitish green, 0.6–2.9 mm long; **stamens** five; **anthers** included, yellow, *c.* 1.2 mm long; **style** white, exserted, apex slightly hairy (microscopic), TL 2–8 mm; **fruiting heads** white, each *c.* 3 cm wide and tall. **FRUIT an achene**, 1-seeded, brown, 2–2.5 mm long by 0.3–0.4 mm wide and thick, 5- or 6-ribbed, *c.* 70 per head; **pappus** of *c.* 70 bristles *c.* 12 mm long.

■ **LEAVES** Stem and basal; **basal leaves** 1–3, erect, appearing later than flowering stems and often one to several cm from them; sometimes only colonies of leaves without flowering stems; **blades** dull green above, 5–16 cm long 5–21 cm wide, veins hairy and raised below, cottony tomentose below, hairs few and scattered above, palmately 5- to 11-lobed, these 1.5–6 cm wide, coarsely toothed; **petioles** 8–33 cm long by 3–5 mm wide; **stem leaves** alternate, 2–5 cm long by 5–23 mm wide, reduced above and below, sessile, ascending, entire or apices sometimes toothed, short hairs gland-tipped, margins densely cottony hairy, reduced to 2 or 3 dark brown bracts below ground.

■ **STEM** Erect, hairy, ridged from leaf bases, hollow, pale green, tomentose; 5–10 mm wide near the green base.

SYN. *Petasites palmatus* (Ait.) Gray.

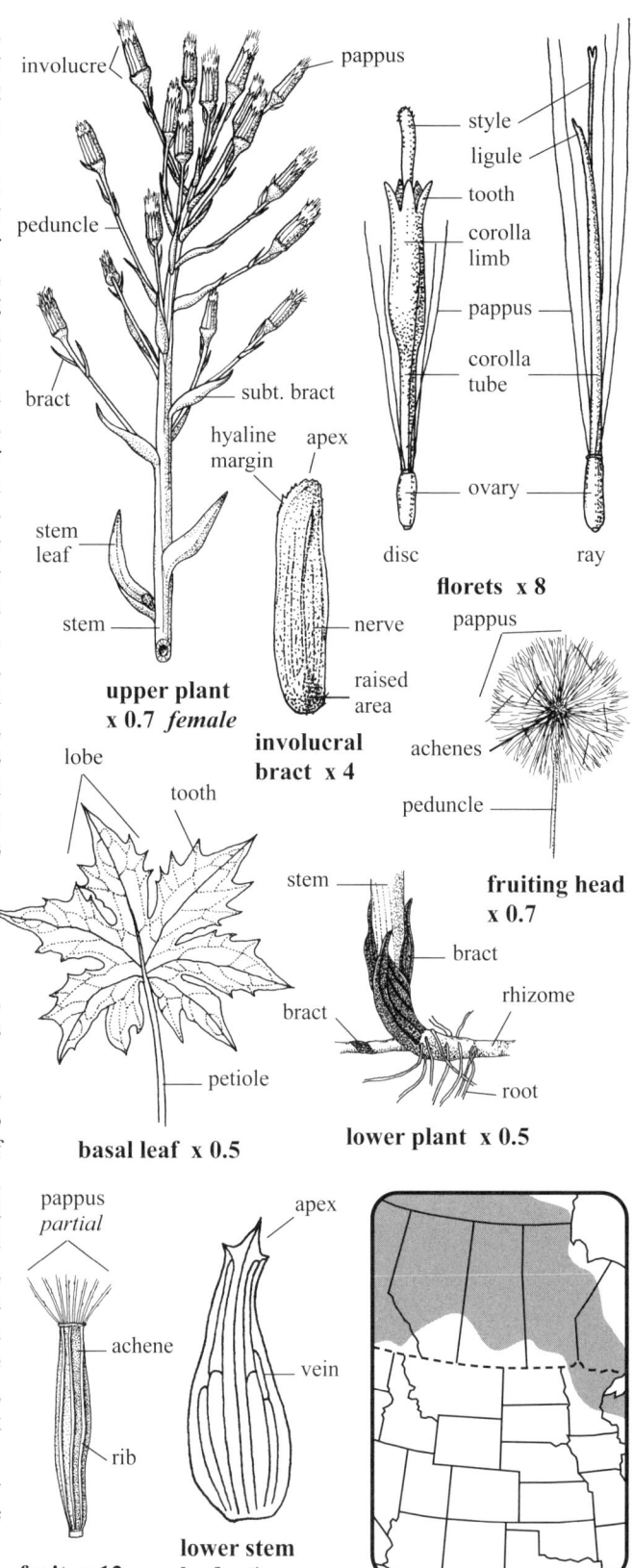

Aster—*Asteraceae* **Wildflower** White Female ray florets 20–140

Arrow-leaved Colt's-foot   *Petasites frigidus* (L.) Fr. **var.** *sagittatus* (Banks ex Pursh) Cherniawsky

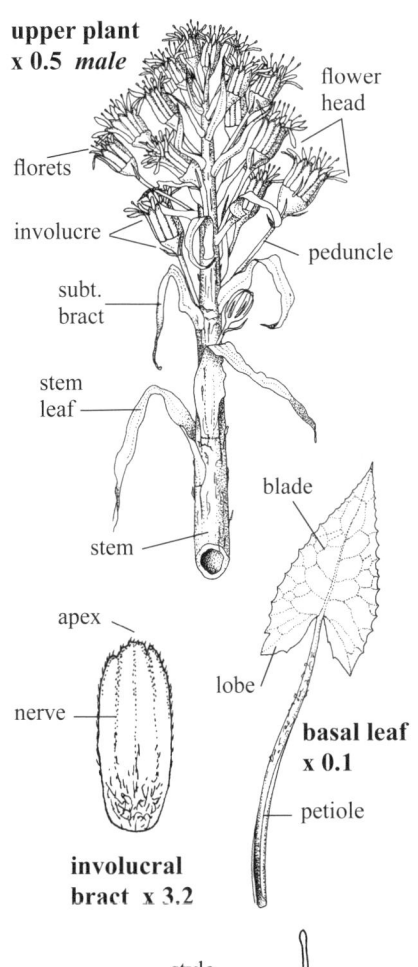

- upper plant x 0.5 *male*
- flower head
- florets
- involucre
- peduncle
- subt. bract
- stem leaf
- blade
- stem
- apex
- nerve
- lobe
- **basal leaf x 0.1**
- petiole
- involucral bract x 3.2

■ **SKETCH** A **perennial herb** growing 20–70 cm tall in open colonies from tan **rhizomes** 2–6 mm wide by 5–80+ cm long; in open wet areas of woods and thickets, slough margins, bogs and fens; **imperfectly dioecious**.

■ **FLOWER HEADS** White, blooming April–July; **inflorescence** a corymb or raceme 3–20 cm long by 6–15 cm wide, of fragrant, mostly unisexual flower heads; **peduncles** hairy, 0.5–8 cm long in flower, reduced above; **subtending bracts** (of peduncles) leaflike, reduced above; <u>**male flower heads**</u> 8–35, 1.5–2 cm wide by *c.* 15 mm tall; **involucre** 7–10 mm long by 6–8 mm wide; **involucral bracts** 15–16, imbricate, in one series, 6–10 mm long by 1.2–3.3 mm wide, apices pointed to blunt and ciliate, margins entire or toothed, body mostly green, hairy at base; **male florets** of two types: **1) outer ray florets** 4–19, sometimes female, TL 12–13 mm; **ovary** green, *c.* 1.4 mm long; **pappus** bristles numerous, 3.5–6.3 mm long; **corolla tube** *c.* 4.5 mm long; **ligules** 1.1–7.7 mm long by 1–4.4 mm wide, glabrous, apices with two tiny teeth; **2) inner disc florets** 22–56, appear perfect but are male, TL 15–16 mm; **ovary** *c.* 2 mm long; **pappus** as above; **corolla** 0.5–9 mm long (including teeth); **teeth** five, 1.2–2.2 mm long; **anthers** 1.5–3.8 mm long; <u>**female flower heads**</u> 6–34 per plant, 13–15 mm wide by *c.* 1 cm long; **involucre** 6–7 mm tall and wide; <u>**female florets**</u> fertile, of two types: **1) outer ray florets** 20–140, TL 9–10 mm, glabrous; **ovary** *c.* 1 mm long; **pappus** white, numerous fine bristles 3–17 mm long; **corolla tube** 3–4 mm long; **ligule** 0.6–5.4 mm long by 0.1–1 mm wide; **2) inner disc florets** 1–5, sometimes male, TL *c.* 9 mm; **pappus** white, 3.5–11 mm long; **anthers** 0.7–1.3 mm long; **style** 2-parted, parts 0.3–1.3 mm long, hairy. **FRUIT** an achene, 1-seeded, brown, 2.3–3.5 mm long by 0.7–0.9 mm wide by *c.* 0.6 mm thick, hairless, 5- to 10-ribbed; **pappus** 15–20 mm long.

■ **LEAVES** Basal and stem, variable, alternate, simple; **basal leaves** one to three in a cluster in large open colonies, erect, prominent all summer; **blades** arrowhead-shaped, 2–45 cm long by 2–40 cm wide, white-woolly below, entire or teeth 5–15 mm long, basal lobes two, each 5–20 cm long; **petioles** 10–33 cm long by *c.* 5 mm wide and thick, hollow, hairy below and less so above; **stem leaves** (bracts above) mostly entire, clasping, sessile, 3–20 cm long by 0.5–3 cm wide, reduced above, spreading to reflexed, cottony hairs below.

■ **STEM** Erect, hollow, round, slightly ridged, hairy; 1–2 cm thick near the downy base.

■ **SYN.** *Petasites sagittatus* (Banks ex Pursh) A. Gray.

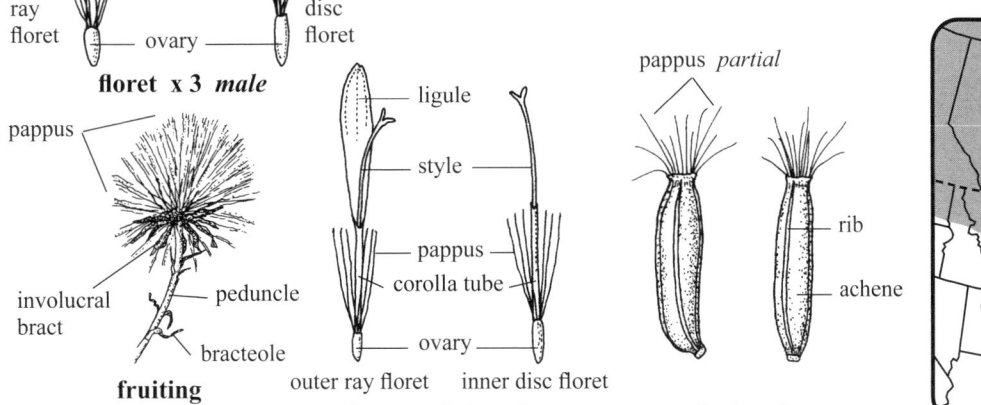

- style
- anthers
- ligule tooth
- style
- corolla limb
- corolla tube
- outer ray floret
- pappus
- ovary
- inner disc floret
- **floret x 3** *male*
- pappus
- involucral bract
- peduncle
- bracteole
- **fruiting head x 0.7**
- ligule
- style
- pappus
- corolla tube
- ovary
- outer ray floret
- inner disc floret
- **floret x 4** *female*
- pappus *partial*
- rib
- achene
- **fruit x 8**

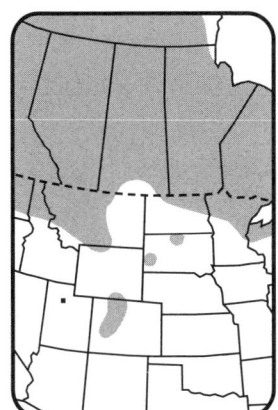

Sweet Colt's-foot, Arrowleaf Sweet Coltsfoot

Aster—*Asteraceae* **Wildflower** White to pale pink Ligulate florets 8–13

White Lettuce *Prenanthes alba* L.

■ **SKETCH** A glaucous **perennial herb** 50–180 cm tall with woody, ridged, orangish brown branched **roots** 2–13 cm long by 10–13 mm wide with knobby projections; **fibrous roots** 1–4 cm long; in open woods, gravelly disturbed sites and along roadsides; **juice milky. p2**

■ **FLOWER HEADS** White to pale pink, blooming August–September; **inflorescence** a panicle, 18–26 cm long by 9–11 cm wide; **floral branches** purplish brown, 1–7 cm long, ascending; **subtending bracts** (of floral branches) leaflike, 0.5–4.5 cm long by 2–37 mm wide, petiolate to sessile above; **peduncles** 5–10 mm long, slightly hairy, reddish brown; **flower heads** nodding, in clusters, ligulate, 15–20 mm wide by 20–22 mm long (including exserted styles); **involucre** 12–16 mm long by 4–5 mm wide; **main involucral bracts** erect, pinkish purple, usually eight, thin, pliable, midnerve dark brown, raised, margins white hyaline, 12–14 mm long by 2–2.5 mm wide, imbricate, apices fuzzy from curled hairs (microscopic), white and blunt; **short involucral bracts** 6–9 at base, 2–4.5 mm long, pointed, margins ciliate, paler than main bracts; **ligulate florets** 8–13, perfect; **ovary** pale green, ribbed, hairless, 2–2.3 mm long; **pappus** 7–9 mm long, light reddish brown, *c.* 80 bristles with fine ascending hairs; **corolla tube** 5–6 mm long by *c.* 1.2 mm wide, white, hairless; **ligules** spreading, 7–8 mm long by 3–3.5 mm wide, apical teeth five (4 or 6), each 0.5–0.7 mm long, thickened and bumpy at apex; **stamens** five; **anthers** brown, *c.* 5 mm long, united into a tube, exserted; **style** dark green, hairy, 2-parted, parts *c.* 2 mm long, curled and yellow, exserted 5–9 mm past anther tips; **fruiting heads** *c.* 2 cm wide, brown. **FRUIT an achene**, 1-seeded, 8–13 per flower, 6-ribbed, each 3.5–4.2 mm long by 1–1.2 mm wide by *c.* 0.8 mm thick, dull, reddish brown; **pappus** 8–9.5 mm long; **receptacle** hairless.

■ **LEAVES** Stem only, alternate, simple, lobed and toothed; **blades** purplish green, 5–15 cm long and wide, reduced above, widely spaced, variable in shape, ascending to spreading, hairy and lighter green below, hairy above along veins; **lower blades** wilted by anthesis; **petioles** 2–11 cm long, reduced and winged above, purplish, arched upward to spreading.

■ **STEM** Erect, glaucous and glabrous, purplish, usually unbranched below, stiff, round, hollow; 4–7 mm thick near the reddish, naked base.

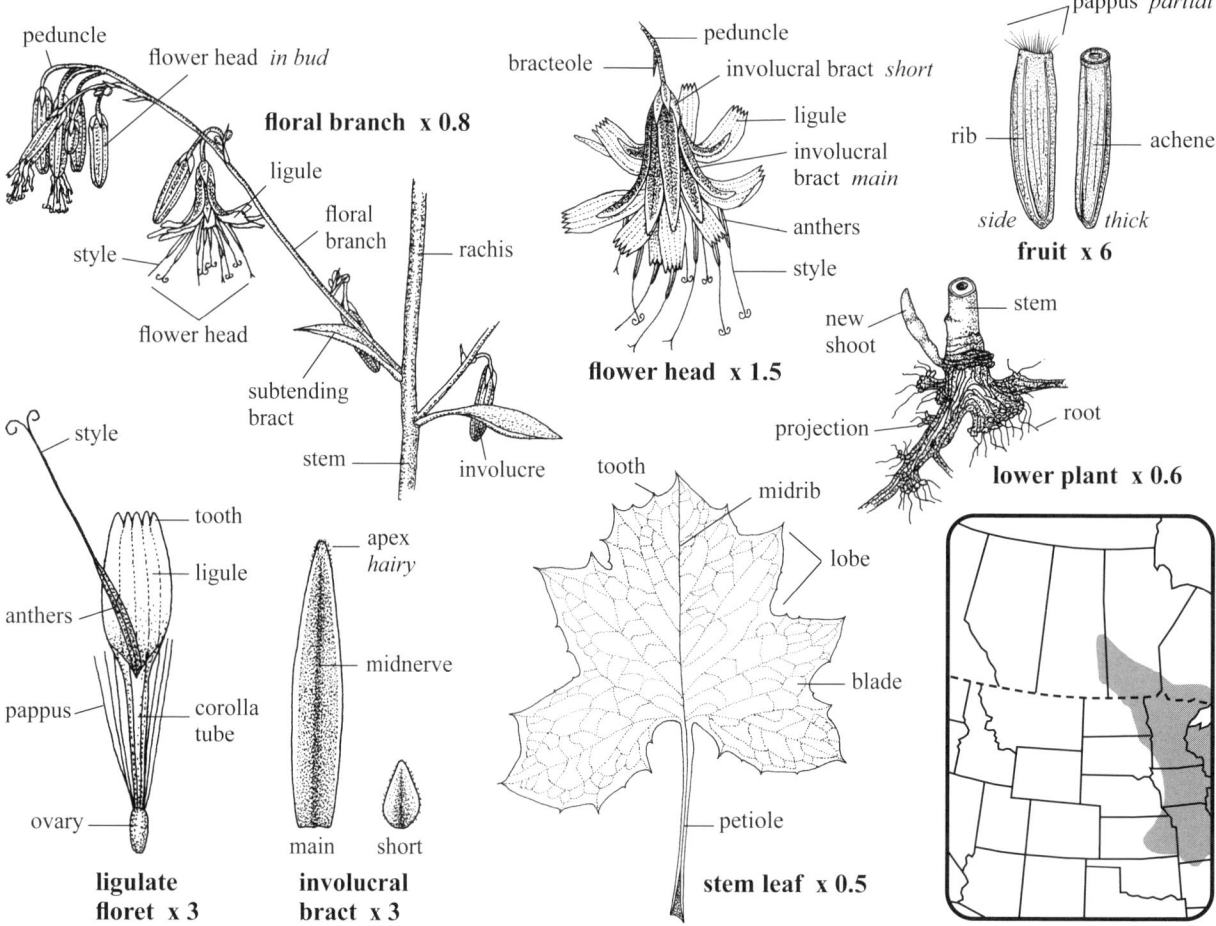

Aster—*Asteraceae* **Wildflower** Pink to pale purple Ligulate florets 10–26

Glaucous White Lettuce   *Prenanthes racemosa* Michx.

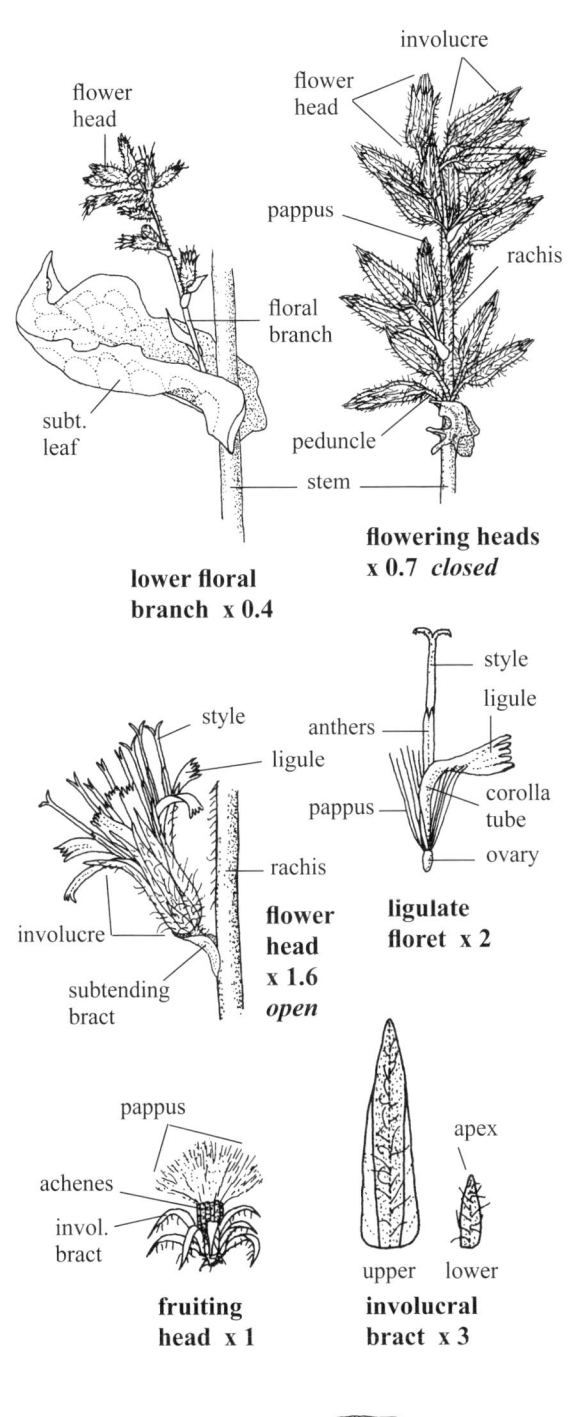

lower floral branch x 0.4

flowering heads x 0.7 *closed*

flower head x 1.6 *open*

ligulate floret x 2

fruiting head x 1

upper   lower
involucral bract x 3

lower stem leaf x 0.2

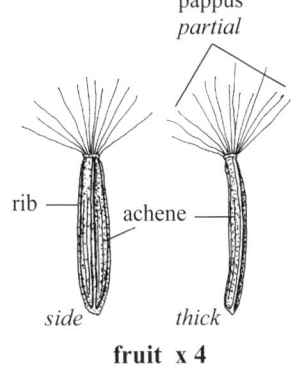

fruit x 4

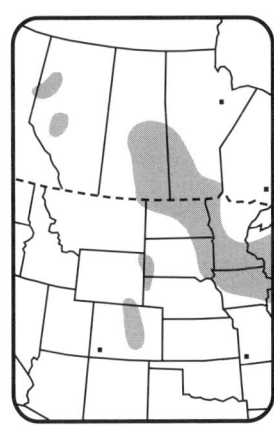

■ **SKETCH** A **perennial herb** 40–150 cm tall with **fibrous roots**; scattered in moist prairies, open woods, gravelly ridges, wet meadows and along stream banks; **juice milky. p2**

■ **FLOWER HEADS** Pink to pale purple, blooming August–September; **inflorescence** paniculate, 5–60 cm long by 4–10 cm wide, terminal and on branches, hairy; **floral branches** 3–16 cm long, hairy, reduced above; **peduncles** 1–10 mm long; **subtending bracts** (of peduncles) as long or longer than peduncles; **flowers heads** spreading to erect, crowded, 15–18 mm wide by 15–20 mm long; **involucre** brownish purple to blackish, 10–15 mm long by 5–6 mm wide; **involucral bracts** hairy along midnerve, upper bracts usually 13 (9–16), each 10–11 mm long by 1–2.5 mm wide, lower bracts 3–5 mm long by *c*. 1 mm wide, pliable, persistent, spreading in fruit; **rachis** hairy; <u>disc **florets**</u> absent; <u>ligulate **florets**</u> perfect, 10–26, exserted, light pink; **ovary** whitish green, *c*. 1.5 mm long, hairless, faintly ribbed; **pappus** of numerous tawny bristles 7–7.5 mm long; **corolla** TL 10–11 mm; **corolla tube** *c*. 4 mm long; **ligules** *c*. 7 mm long by 2.5–2.7 mm wide, apices twisting and recurved quickly along the outside, apical teeth five, each 0.5–1.4 mm long, outer teeth longer, tips of teeth expanded; **anthers** *c*. 4.8 mm long, yellow to pink; **styles** 2-parted, spreading; **fruiting heads** are 17–22 mm long; **pappus** tannish yellow, 10–15 mm wide by 7–8 mm long. **FRUIT an achene**, 1-seeded, 12–25 per head, reddish brown, 4.5–6.5 mm long by 0.9–1.3 mm wide by 0.7–1 mm thick, hairless, 5-ribbed with one or two lesser ribs; **pappus** tannish yellow, *c*. 8 mm long of *c*. 70 bristles, deciduous; **receptacle** *c*. 3 mm wide, round, flat, hairless.

■ **LEAVES** Basal and stem, alternate, simple, shallowly toothed to entire, glabrous; **stem blades** 2–40 cm long by 0.5–10 cm wide, sessile and clasping above, margins undulate, tips often twisted, lower blades folded up along midrib; **petioles** 0–14 cm long and to *c*. 12 mm wide, winged, reduced above.

■ **STEM** Erect, glabrous, glaucous, round, stout, hairy only in inflorescence, green to reddish; 5–10 mm thick near the base.

Purple Rattlesnakeroot   169

Aster—*Asteraceae* **Wildflower** Yellow, center reddish brown Ray florets 6 (4–11)

LONG-HEADED CONEFLOWER *Ratibida columnifera* (Nutt.) Woot. & Standl.

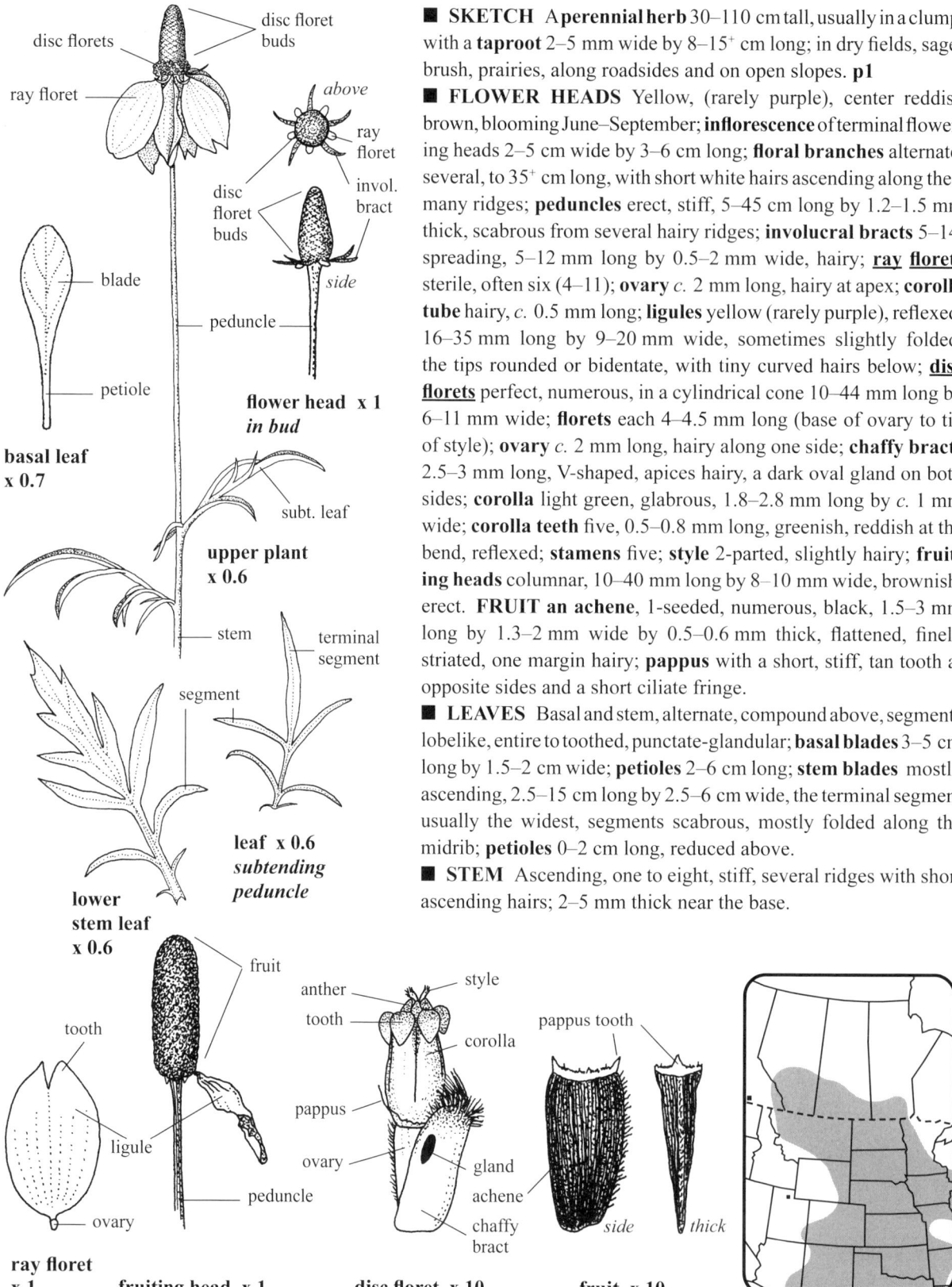

- **SKETCH** A perennial herb 30–110 cm tall, usually in a clump, with a **taproot** 2–5 mm wide by 8–15+ cm long; in dry fields, sagebrush, prairies, along roadsides and on open slopes. p1
- **FLOWER HEADS** Yellow, (rarely purple), center reddish brown, blooming June–September; **inflorescence** of terminal flowering heads 2–5 cm wide by 3–6 cm long; **floral branches** alternate, several, to 35+ cm long, with short white hairs ascending along their many ridges; **peduncles** erect, stiff, 5–45 cm long by 1.2–1.5 mm thick, scabrous from several hairy ridges; **involucral bracts** 5–14, spreading, 5–12 mm long by 0.5–2 mm wide, hairy; **ray florets** sterile, often six (4–11); **ovary** *c.* 2 mm long, hairy at apex; **corolla tube** hairy, *c.* 0.5 mm long; **ligules** yellow (rarely purple), reflexed, 16–35 mm long by 9–20 mm wide, sometimes slightly folded, the tips rounded or bidentate, with tiny curved hairs below; **disc florets** perfect, numerous, in a cylindrical cone 10–44 mm long by 6–11 mm wide; **florets** each 4–4.5 mm long (base of ovary to tip of style); **ovary** *c.* 2 mm long, hairy along one side; **chaffy bracts** 2.5–3 mm long, V-shaped, apices hairy, a dark oval gland on both sides; **corolla** light green, glabrous, 1.8–2.8 mm long by *c.* 1 mm wide; **corolla teeth** five, 0.5–0.8 mm long, greenish, reddish at the bend, reflexed; **stamens** five; **style** 2-parted, slightly hairy; **fruiting heads** columnar, 10–40 mm long by 8–10 mm wide, brownish, erect. **FRUIT** an achene, 1-seeded, numerous, black, 1.5–3 mm long by 1.3–2 mm wide by 0.5–0.6 mm thick, flattened, finely striated, one margin hairy; **pappus** with a short, stiff, tan tooth at opposite sides and a short ciliate fringe.
- **LEAVES** Basal and stem, alternate, compound above, segments lobelike, entire to toothed, punctate-glandular; **basal blades** 3–5 cm long by 1.5–2 cm wide; **petioles** 2–6 cm long; **stem blades** mostly ascending, 2.5–15 cm long by 2.5–6 cm wide, the terminal segment usually the widest, segments scabrous, mostly folded along the midrib; **petioles** 0–2 cm long, reduced above.
- **STEM** Ascending, one to eight, stiff, several ridges with short ascending hairs; 2–5 mm thick near the base.

Prairie Coneflower, Upright Prairie Coneflower, Red-spike Mexican-hat

BLACK-EYED SUSAN *Rudbeckia hirta* L.

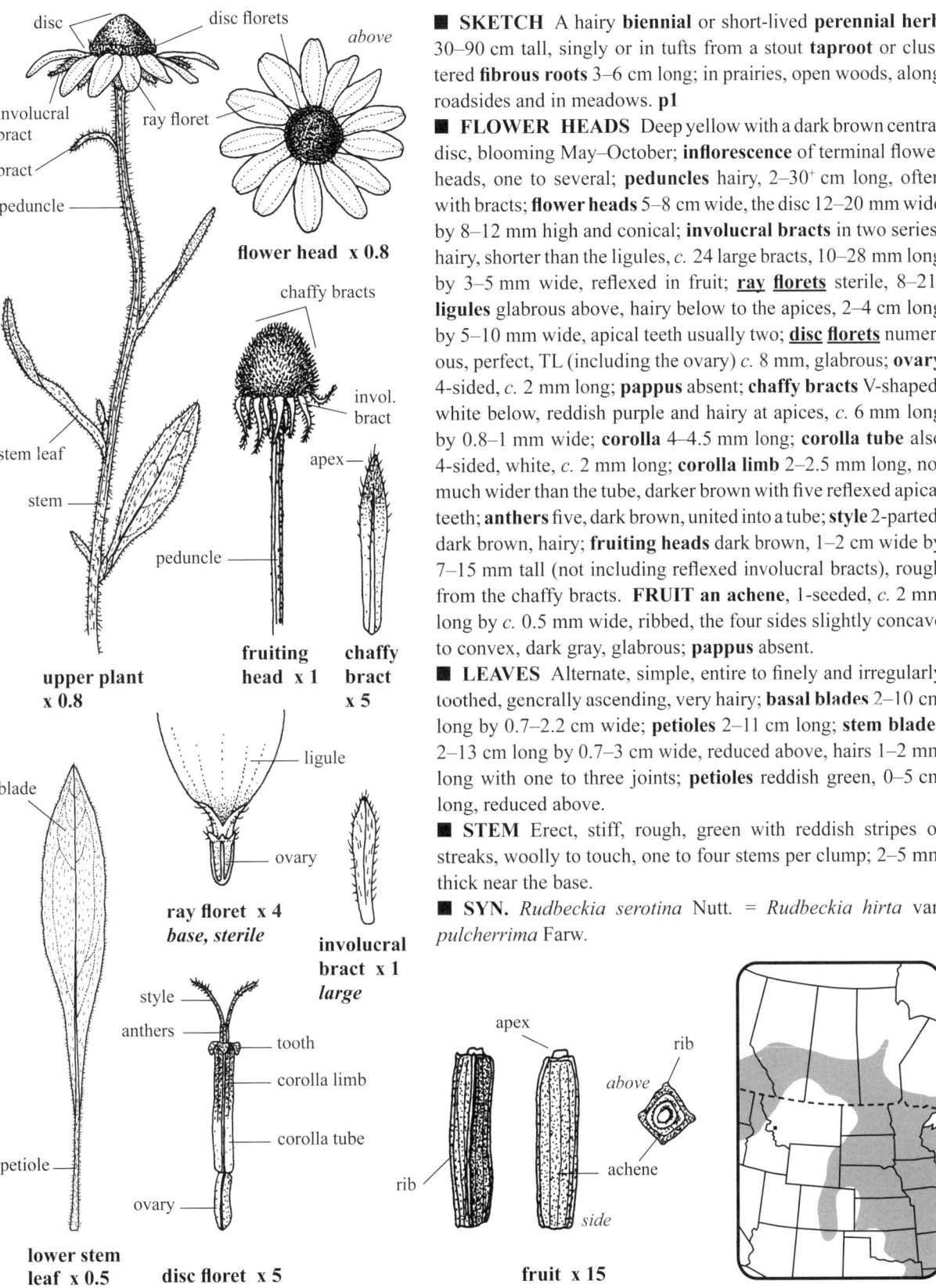

- **SKETCH** A hairy **biennial** or short-lived **perennial herb** 30–90 cm tall, singly or in tufts from a stout **taproot** or clustered **fibrous roots** 3–6 cm long; in prairies, open woods, along roadsides and in meadows. p1
- **FLOWER HEADS** Deep yellow with a dark brown central disc, blooming May–October; **inflorescence** of terminal flower heads, one to several; **peduncles** hairy, 2–30+ cm long, often with bracts; **flower heads** 5–8 cm wide, the disc 12–20 mm wide by 8–12 mm high and conical; **involucral bracts** in two series, hairy, shorter than the ligules, *c.* 24 large bracts, 10–28 mm long by 3–5 mm wide, reflexed in fruit; **ray florets** sterile, 8–21; **ligules** glabrous above, hairy below to the apices, 2–4 cm long by 5–10 mm wide, apical teeth usually two; **disc florets** numerous, perfect, TL (including the ovary) *c.* 8 mm, glabrous; **ovary** 4-sided, *c.* 2 mm long; **pappus** absent; **chaffy bracts** V-shaped, white below, reddish purple and hairy at apices, *c.* 6 mm long by 0.8–1 mm wide; **corolla** 4–4.5 mm long; **corolla tube** also 4-sided, white, *c.* 2 mm long; **corolla limb** 2–2.5 mm long, not much wider than the tube, darker brown with five reflexed apical teeth; **anthers** five, dark brown, united into a tube; **style** 2-parted, dark brown, hairy; **fruiting heads** dark brown, 1–2 cm wide by 7–15 mm tall (not including reflexed involucral bracts), rough from the chaffy bracts. FRUIT an achene, 1-seeded, *c.* 2 mm long by *c.* 0.5 mm wide, ribbed, the four sides slightly concave to convex, dark gray, glabrous; **pappus** absent.
- **LEAVES** Alternate, simple, entire to finely and irregularly toothed, generally ascending, very hairy; **basal blades** 2–10 cm long by 0.7–2.2 cm wide; **petioles** 2–11 cm long; **stem blades** 2–13 cm long by 0.7–3 cm wide, reduced above, hairs 1–2 mm long with one to three joints; **petioles** reddish green, 0–5 cm long, reduced above.
- **STEM** Erect, stiff, rough, green with reddish stripes or streaks, woolly to touch, one to four stems per clump; 2–5 mm thick near the base.
- **SYN.** *Rudbeckia serotina* Nutt. = *Rudbeckia hirta* var. *pulcherrima* Farw.

Aster—*Asteraceae* **Wildflower** Yellow, center yellowish green Ray florets 6–16

Tall Coneflower *Rudbeckia laciniata* L.

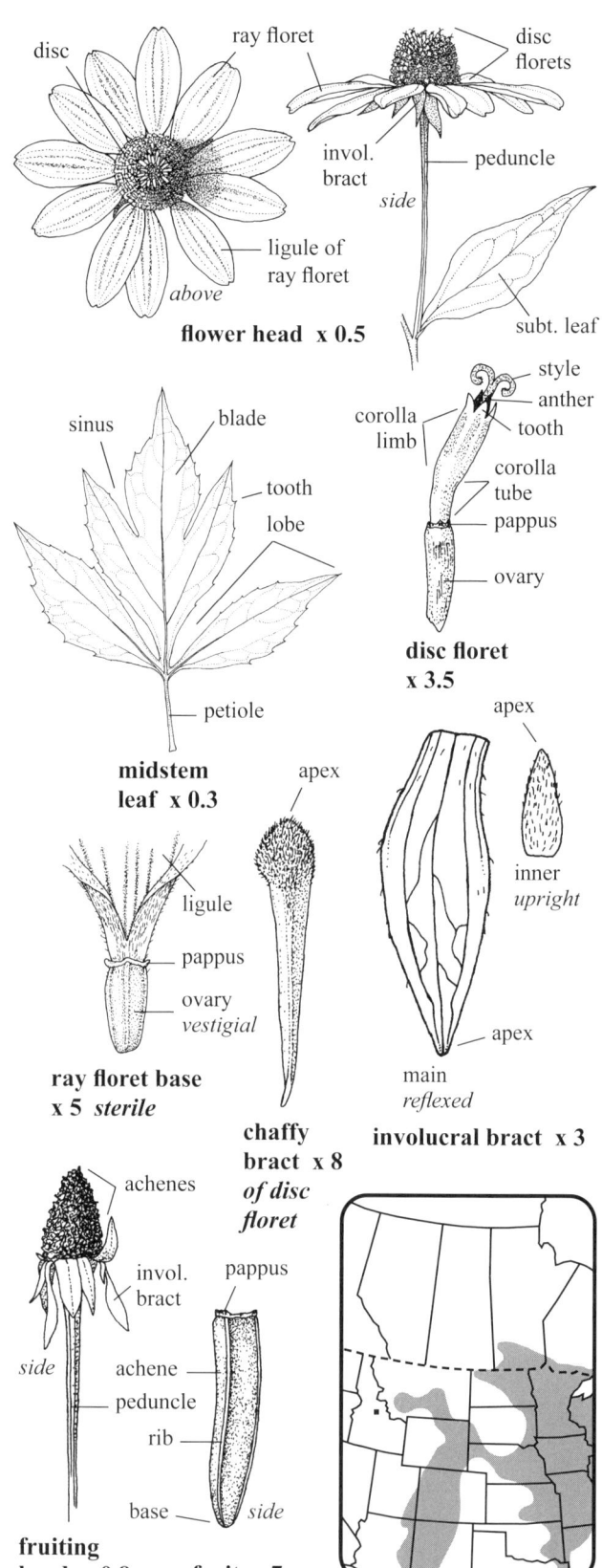

flower head x 0.5
disc floret x 3.5
midstem leaf x 0.3
ray floret base x 5 *sterile*
chaffy bract x 8 *of disc floret*
involucral bract x 3
fruiting head x 0.8 fruit x 7

■ **SKETCH** A **perennial herb** 50–250 cm tall with a thick **rhizome** and numerous **roots** less than 1 mm thick and **pink shoots**; in moist thickets, meadows, open deciduous woods and along woodland trails in parklands.

■ **FLOWER HEADS** Yellow, central disc yellowish green, blooming July–September; **inflorescence** open, to *c.* 20⁺ cm tall by *c.* 25 cm wide, of 1–10⁺ heads, terminal and on floral branches; **floral branches** 12–40⁺ cm long, axillary; **peduncles** 7–12 cm long, erect, stiff, grooved, flattened to oval and finely hairy at apices; **flower heads** 4.5–10 cm wide by *c.* 3 cm tall, central disc 15–22 mm wide by 10–16 mm tall; **involucre** 10–15 mm tall by 1.5–3 cm wide; **involucral bracts** green, 6–9 main ones, 10–15 mm long by 4–5 mm wide, spreading to reflexed, pointed, thin, in 2 or 3 series, imbricate; **short inner bracts** erect, 5–6 mm long by 1.5–2.5 mm wide, pointed; ray florets sterile, 6–16; **chaffy bracts** *c.* 3 mm long, apices hairy; **ovary** white, 2.3–3 mm long; **ligules** yellow, 23–60 mm long by 7–14 mm wide, declining; disc florets perfect, 9–11 mm long, hairless; **chaffy bracts** one per floret, V-shaped, pale green, *c.* 5 mm long by *c.* 1.5 mm wide (flattened), keeled, blunt apex hairy, tapered; **ovary** pinkish purple, 4–5 mm long and 4-angled; **pappus** *c.* 0.3 mm long, crownlike, streaked with purple, erose; **corolla** yellow, 3–5 mm long; **corolla tube** 1–1.5 mm long; **corolla limb** 2–3.5 mm long, teeth five, erect, 0.5–0.7 mm long; **anthers** five, black, *c.* 2 mm long, barely exserted, united into a tube; **style** one, exserted, 2-parted; **fruiting heads** 10–16 mm tall by 10–12 mm wide (not including bracts), tapered, dark brown. **FRUIT an achene** (of disc floret), 1-seeded, purplish black, 4–4.8 mm long by *c.* 1 mm wide, 4-sided; **pappus** hairless, *c.* 0.3 mm long.

■ **LEAVES** Alternate, deeply lobed to compound, simple above, mostly hairless or hairy below, dull, toothed to entire; **1st year blades** to *c.* 30 cm long and wide, compound, the leaflets 2- or 3-lobed, toothed; **petioles** to *c.* 48 cm long and to *c.* 8 mm wide at the reddish purple base, 1–3 per cluster, arched; **petiolules** 10–30 mm long; **basal blades** to *c.* 15 cm long and to *c.* 17 cm wide; **petioles** to *c.* 23 cm long; **lateral petiolules** 5–20 mm long; **main stem blades** 3- to 7-lobed, 10–25 cm long by 12–30 cm wide, reduced above, some lower leaves with small leaves in their axils; **petioles** 0–16 cm long, reduced above; **upper stem leaves** entire, 4.5–7 cm long by 2–3.5 cm wide, subtending peduncles.

■ **STEM** Glabrous and glaucous, erect, one to several, ridged, angled, hollow, branched or simple; 5–12⁺ mm wide near the base.

■ **SYN.** *Rudbeckia ampla* A. Nels. = *Rudbeckia laciniata* var. *ampla* (A. Nels.) Cronq.

Green-headed Coneflower, Golden Glow, Cutleaf Coneflower

Marsh Ragwort *Senecio congestus* (R. Br.) DC.

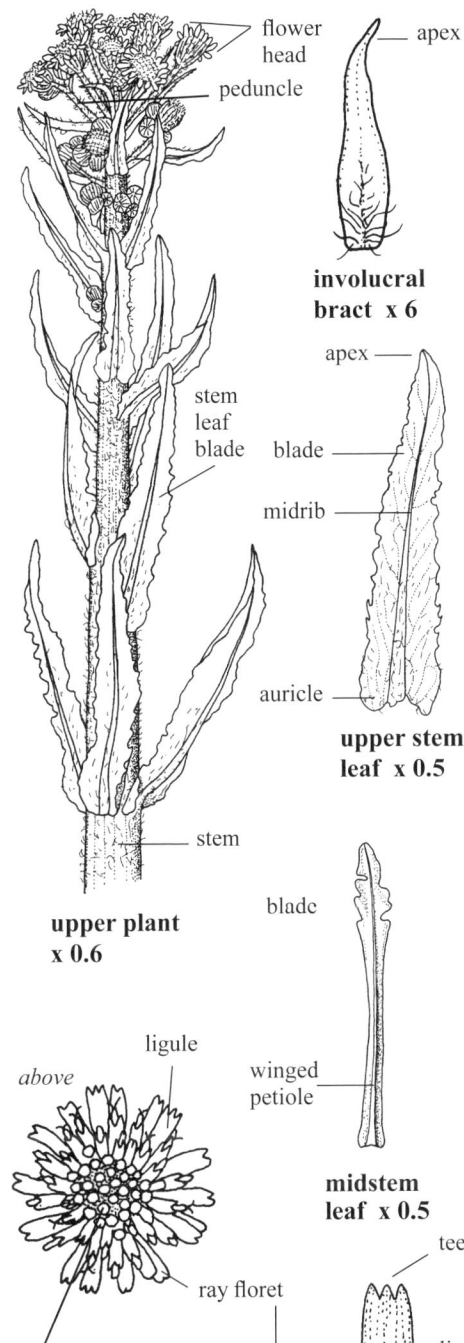

■ **SKETCH** A hairy, variable **annual herb** (biennial?) 15–90 cm tall from a fleshy wide **caudex** with numerous fleshy **fibrous roots**; along pond margins and lakeshores, solitary or forming a shoreline belt, often in shallow water.

■ **FLOWER HEADS** Yellow, blooming May–July; **inflorescence** a corymbiform cyme, crowded, 6–60 heads with a TL of 3.5–30 cm by 4–30 cm wide; **floral branches** ascending, 3–27 cm long; **peduncles** hairy, 10–40 mm long; **subtending bracts** (of peduncles) 5–22 mm long by *c*. 0.7 mm wide, pointed, white hairs spreading below; **flower heads** 10–20 mm wide by 7–12 mm long; **involucre** 6–9 mm long by 10–11 mm wide, fuzzy from spreading white hairs on the outside of the lower half; **involucral bracts** in one series, not imbricate, *c*. 21, each 5–10 mm long by 1.3–1.8 mm wide, thin, tips brown to pink, curled, midnerve obscure; <u>ray florets</u> 17–22, pistillate, yellow, 6–10 mm long; **ovary** *c*. 1 mm long; **pappus** of numerous bristles 2–3 mm long; **corolla tube** 2.5–3.5 mm long; **ligules** 3–5 (–8) mm long by 1.2–2 mm wide, glabrous, apical teeth two or three; <u>disc florets</u> perfect, numerous, 6–7 mm long (base of ovary to top of corolla); **ovary** *c*. 1 mm long; **pappus** bristles *c*. 45, each *c*. 2.5 mm long; **corolla tube** 2.5–3.5 mm long; **corolla limb** yellow, 1.7–2.5 mm long by 1–1.2 mm wide, teeth five, spreading or not, *c*. 0.5 mm long; **anthers** five, united into a tube *c*. 1 mm long; **style** 2-parted, exserted, spreading; **fruiting heads** white, *c*. 2 cm wide and long. **FRUIT an achene**, 1-seeded, brown, 10-ribbed, hairless, *c*. 2 mm long by *c*. 0.5 mm wide and thick; **pappus** bristles 10–11 mm long.

■ **LEAVES** Basal and stem, alternate, simple, toothed to entire and wavy; **basal blades** 3–9 cm long by 6–30 mm wide; **petioles** 0–6 cm long; **stem blades** fleshy, ascending, some tips twisted, TL 2–18 cm by 3–40 mm wide, reduced above, covered with fine cobwebby hairs when young, glabrous with age, lateral veins obscure above, raised below; **upper blades** with **auricles** clasping; **petioles** 0–8 cm long, reduced above, winged below, fleshy, 4–7 mm wide at their bases.

■ **STEM** Erect, hollow, simple to branched above, fleshy, round, ridged, cobwebby when young; 0.6–5 cm thick near the base.

■ **SYN.** *Senecio palustris* (L.) Hook.

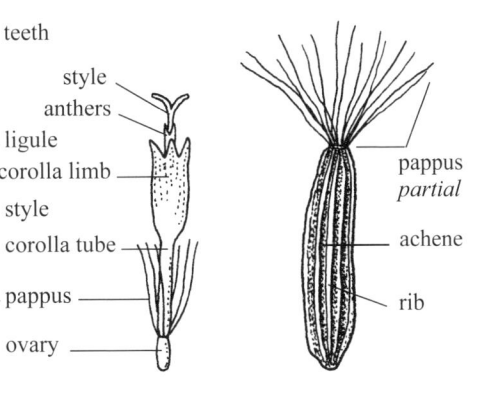

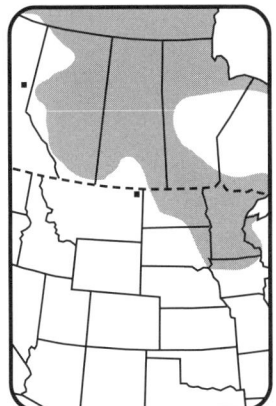

Marsh Fleabane, Clustered Marsh Ragwort, Swamp Ragwort

Aster—*Asteraceae* **Wildflower** Yellow Disc florets only

Common Groundsel *Senecio vulgaris* L.

■ **SKETCH** An **annual herb** 10–50 cm tall from a white **taproot** 1–8+ cm long by 1–3 mm wide, singly or in large patches; in gardens, along sidewalks and around settlements. *Naturalized*

■ **FLOWER HEADS** Yellow, blooming March–November; **inflorescence** of loosely clustered flower heads on branches; **floral branches** 5–25 cm long, from leaf axils, ascending to slightly nodding; **peduncles** 2–30 mm long, upright to lax, grooved and slightly hairy; **subtending bracts** (of peduncles) black-tipped; **flower heads** 8–50+ per plant, not spreading, 2–5 mm wide by 6–11 mm long; **involucre** green, 5–8 mm long by 3–6 mm wide; **involucral bracts** glabrous, of two types: **1) upper involucral bracts** *c*. 21, each 4–7 mm long by 0.7–0.8 mm wide, some black-tipped; **2) basal involucral bracts** *c*. 1.5 mm long by *c*. 0.4 mm wide, green, glabrous, with distinctive black tips, reflexed during fruit dispersal; **ray florets** absent; **disc florets** perfect, numerous; **ovary** smooth and *c*. 2 mm long; **pappus** reaching to the top or beyond the corolla; **corolla** 5.1–5.5 mm long including a 3 mm long tube; **corolla limb** 5-toothed; **style** yellow, 2-parted; **fruiting heads** 10–18 mm wide, white, dandelion-like. **FRUIT an achene**, 1-seeded, 2–2.5 mm long by *c*. 0.5 mm wide and thick, with tan ribs and short ascending hairs to almost hairless; **pappus** 8–10 mm across, the fine bristles with ascending hairs.

■ **LEAVES** Basal and stem, alternate, simple, toothed to lobed, variable in shape, mostly glabrous; **basal leaf blades** 1.5–3 cm long by 6–10 mm wide; **petioles** 1–2 cm long; **stem leaf blades** 2–9 cm long by 5–35 mm wide; **lower blades** spatulate, lobed or not, teeth shallow; **petioles** 0–2 cm long, reduced above, winged.

■ **STEM** Erect, round, simple or extensively branched, hollow, weak, grooved, glabrous; 1–4 mm wide near the base.

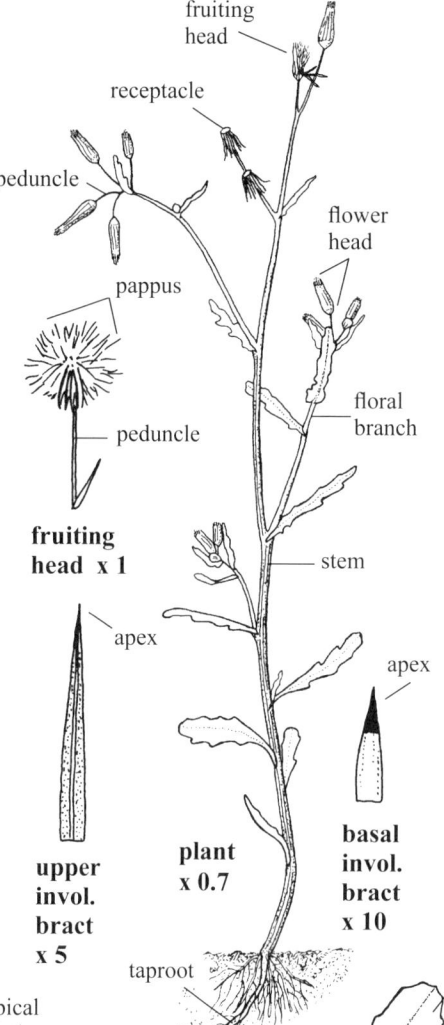

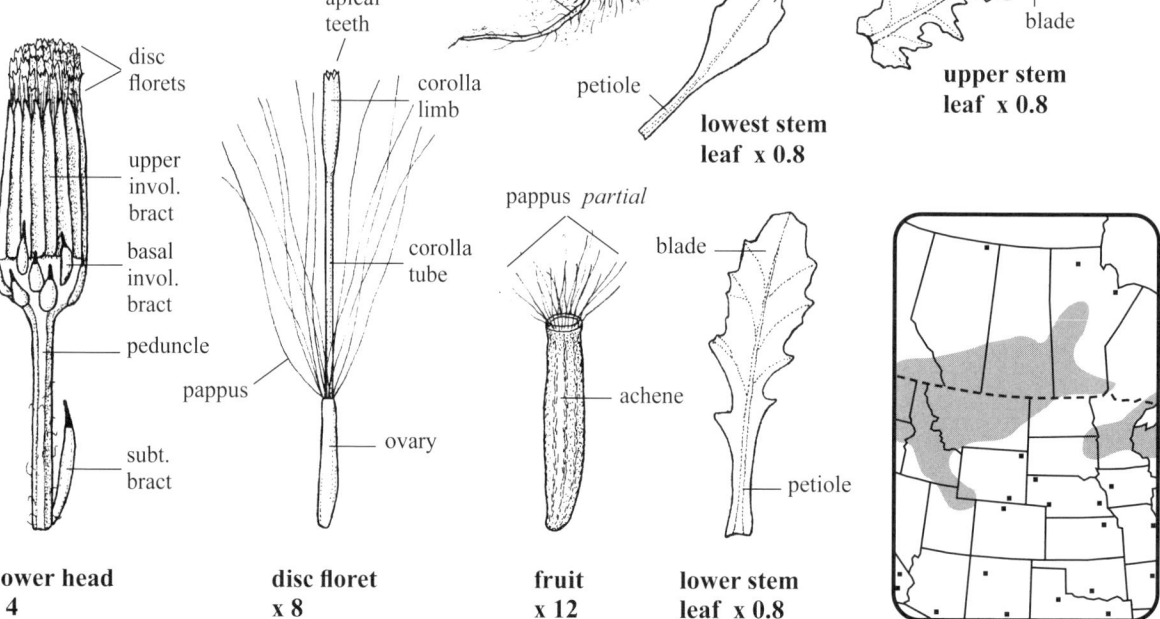

174   Old-man-in-the-spring, Groundsel

Aster—*Asteraceae* **Wildflower** Yellow Ray florets 8–18

GRACEFUL GOLDENROD *Solidago canadensis* L.

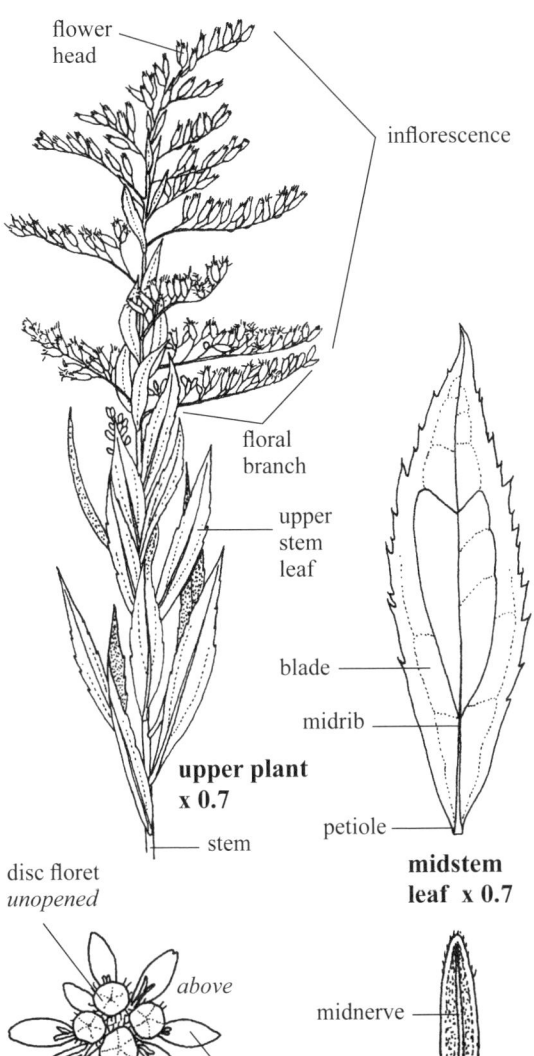

■ **SKETCH** A variable **perennial herb** 30–200 cm tall, singly or in open patches from horizontal, reddish purple **rhizomes** 2–3 mm wide and to *c.* 25⁺ cm long; in woodland edges and abandoned farmlands, in moist to dry sites. **w2, p1**

■ **FLOWER HEADS** Yellow, blooming July–October; **inflorescence** a panicle, 6–10 cm long by 5–9 cm wide with numerous flower heads situated mostly along the upper side of spreading, hairy, floral branches; **subtending bracts** (of floral branches) one; **peduncles** 1–5 mm long, hairy, subtended by a bract; **flower heads** 2.5–3.5 mm wide by 4.5–6.5 mm long; **involucre** pale green, 2–5 mm long by 1–2 mm wide, glabrous; **involucral bracts** in two or three series, thin, 1–3 mm long by *c.* 0.5 mm wide, with a narrow central nerve, pale green sides and ciliate upper margins; **ray florets** female, fertile, (5–) 8–18, TL 4–4.5 mm, irregularly spaced; **ovary** slightly hairy, *c.* 1 mm long; **pappus** of *c.* 40 white bristles 2–2.3 mm long; **corolla tube** whitish yellow, 1.2–1.6 mm long; **ligules** yellow, 1–3 mm long by *c.* 0.5 mm wide, glabrous; **style** yellow, 2-parted; **disc florets** two to eight, perfect, TL 5–6 mm; **ovary** as above; **pappus** as above; **corolla tube** 1.3–1.5 mm long; **corolla limb** yellow, *c.* 1.7 mm long, the four or five apical teeth *c.* 1 mm long; **anthers** five, *c.* 1.3 mm long; **style** yellow, 2-parted. **FRUIT** an achene, 1-seeded, medium brown, with five tan ribs; **body** *c.* 1.3 mm long by *c.* 0.3 mm wide and thick, hairs short, scattered; **pappus** of 25–37 white bristles *c.* 2 mm long.

■ **LEAVES** Alternate, simple, toothed to almost entire, numerous, mostly sessile, the lower ones soon deciduous; **blades** 2.5–14 cm long by 0.5–3.5 cm wide, thick to thin, pointed, with three main veins, flat, glabrous or with a downy feel from the grayish hairy surfaces, more or less hairy below; **petioles** 1–2 mm long.

■ **STEM** Erect, solid, round, pale yellowish green, covered with short, woolly hairs in the upper half giving a downy feel and look; more glabrous and reddish brown near the 2–6 mm wide base.

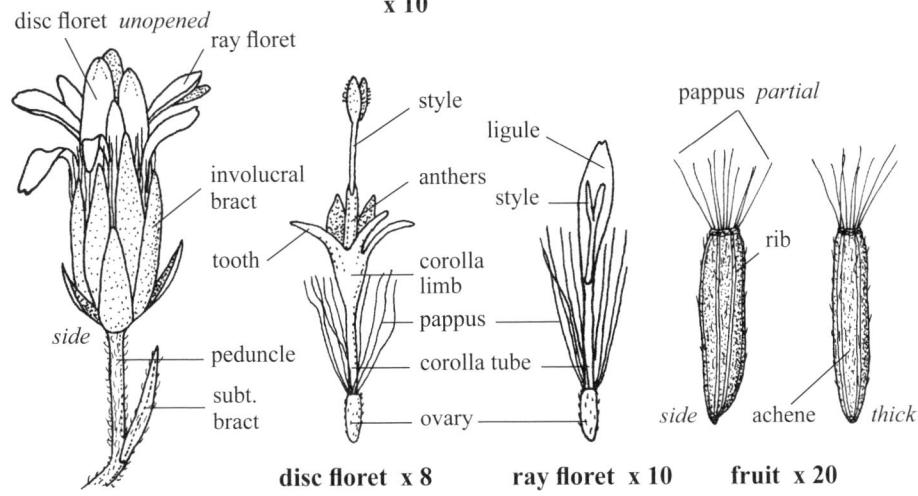

Canada Goldenrod, Canadian Goldenrod 175

Aster—*Asteraceae* **Wildflower** Yellow Ray florets 8–18

Late Goldenrod *Solidago gigantea* Ait.

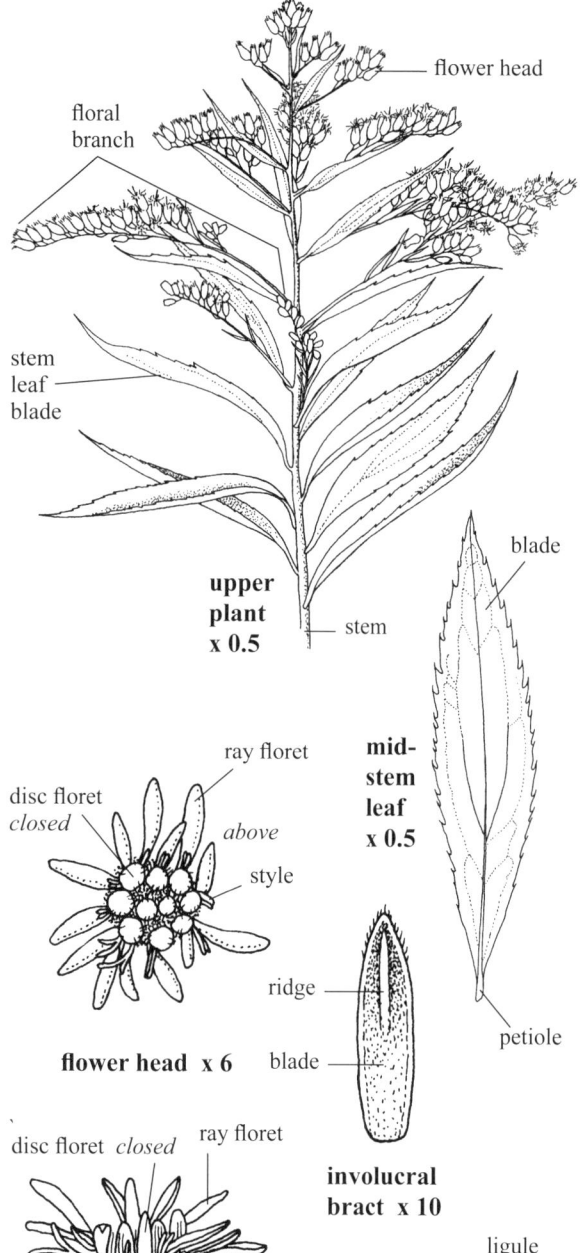

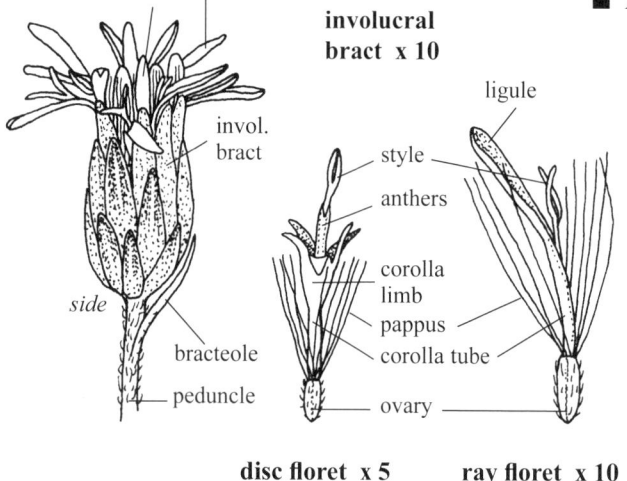

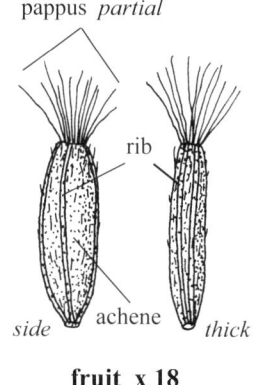

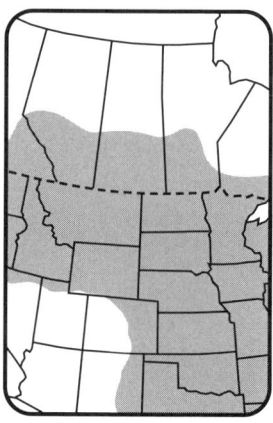

■ **SKETCH** A **perennial herb** 50–200 cm tall from a **rhizome** 1–3 mm wide and to *c.* 25⁺ cm long, singly or in loose clusters; in open sites along woodland edges, streams and thickets.

■ **FLOWER HEADS** Yellow, blooming July–October; **inflorescence** a panicle, 5–15 cm long by 4–25 cm wide, with numerous, mostly upright flower heads; **floral branches** 2–18 cm long, spreading, recurved, hairy; **peduncles** hairy and 0.5–10 mm long; **flower heads** 5–7 mm wide by 6–8 mm long; **involucre** light green, 3–5 mm long by 2.2–3 mm wide, glabrous; **involucral bracts** in three or four series, 2–4 mm long by 0.5–0.8 mm wide, with a central light green low ridge, the edges scarious, apices green and ciliate; <u>ray florets</u> female and fertile, 8–18, TL 4–4.5 mm; **ovary** *c.* 0.8 mm long, short hairs ascending; **pappus** white, bristles numerous, *c.* 3 mm long; **corolla tube** creamy white, 1.8–2 mm long; **ligules** glabrous, yellow, 1.7–2.2 mm long by *c.* 0.5 mm wide; **style** yellow, 2-parted; <u>disc florets</u> 6–10, perfect, TL 7–8 mm; **ovary** *c.* 1 mm long, hairy, ribbed; **pappus** as above; **corolla** glabrous, *c.* 4 mm long; **corolla tube** *c.* 1.8 mm long; **corolla limb** 1.8–2.2 mm long with five spreading yellow apical teeth each *c.* 1 mm long; **anthers** five, yellow, forming a tube; **style** 2-parted, yellow. **FRUIT an achene**, 1-seeded, medium to dark brown, 1.3–1.5 mm long by *c.* 0.5 mm wide by 0.3–0.5 mm thick, glabrous or with scattered ascending short hairs, five or six low ribs about the same color as the body; **pappus** tan, of 30–35 bristles 2–3 mm long.

■ **LEAVES** Alternate, simple, toothed, numerous, glabrous to slightly hairy, mostly sessile; **blades** three-nerved, thin, dull, 4–17 cm long by 0.5–4.5 cm wide with forward pointing teeth, and short ascending hairs along the margins and on main veins beneath or glabrous below; **petioles** 0–8 mm long.

■ **STEM** Erect, glabrous, glaucous, solid, round, not ridged; 3–10 mm thick near the green to purplish red base.

■ **NOTE** Similar to *S. canadensis* but less hairy.

Giant Goldenrod

Aster—*Asteraceae* **Wildflower** Yellow Ray florets 7–13

LOW GOLDENROD *Solidago missouriensis* Nutt.

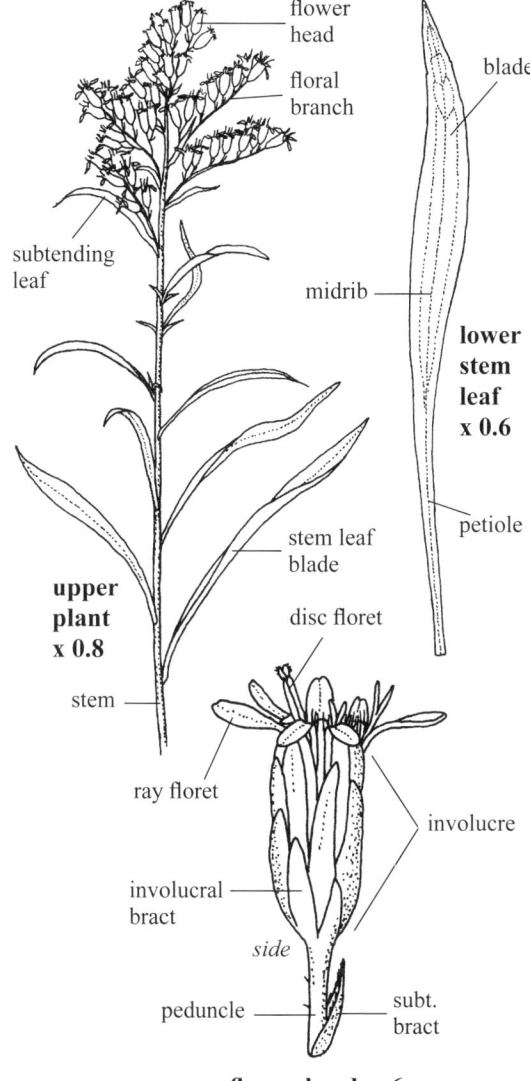

upper plant x 0.8

flower head x 6

■ **SKETCH** A glabrous **perennial herb** 15–50 (–90) cm tall, single or tufted from a **caudex**; **rhizomes** 1–2 mm wide and to *c.* 18+ cm long; in dry prairies, along roadsides and in open woodlands. **p1**

■ **FLOWER HEADS** Yellow, blooming June–October; **inflorescence** a panicle 3–17 cm tall by 3–15 cm wide; **floral branches** several, erect to spreading, 1–11 cm long; **peduncles** 1–7 mm long, each subtended by a bract; **flower heads** mostly erect along the branches, 6–7 mm wide by 5–6 mm long; **involucre** glabrous, yellowish green, 3.5–5 mm long by 2–2.2 mm wide; **involucral bracts** slightly stiff, imbricate, 1.5–4.5 mm long by 0.5–1 mm wide, the tips hairy (microscopic), margins hyaline, midnerve round and thick; **ray florets** 7–13, female, TL *c.* 5 mm; **ovary** slightly hairy, 0.8–1 mm long; **pappus** bristles white, *c.* 3 mm long; **corolla tube** pale yellow, 1.5–2 mm long; **ligules** 1.8–2.5 mm long by *c.* 0.8 mm wide, flat, apical teeth minute; **style** yellow, 2-parted, parts *c.* 1 mm long; **disc florets** perfect, 8–13; **ovary** as above; **corolla** 3.2–3.8 mm long; **corolla tube** *c.* 1.8 mm long, pale green; **corolla limb** 1.8–2 mm long; **teeth** five, yellow, *c.* 0.6 mm long and ascending; **anthers** five, yellow, *c.* 1 mm long, forming a tube around the style; **style** 2-parted, yellow, branches *c.* 1 mm long, tips hairy; **fruiting heads** 5–6 mm wide by *c.* 5 mm tall. **FRUIT an achene**, 1-seeded, 4-sided, *c.* 10 low ribs, tan, 1.5–2.3 mm long by *c.* 0.5 mm wide by *c.* 0.4 mm thick, sparsely hairy or glabrous; **pappus** white, bristles numerous, 2.8–3.2 mm long.

■ **LEAVES** Basal and stem, alternate, simple, entire to sometimes finely serrate below; **basal blades** withering early, 2–11 cm long by 0.5–3 cm wide; **petioles** 1–7 cm long; **stem blades** ascending, 2–13 cm long by 1–3 cm wide, reduced and some with leafy fascicles above, some folded along the midrib, 3-veined, margins slightly scabrous; **petioles** winged, 1–4 cm long.

■ **STEM** Erect, reddish, one to three, smooth, rarely branched, solid, slightly angled, 4- or 5-ridged, these narrow and reddish; 1–3 mm thick near the base.

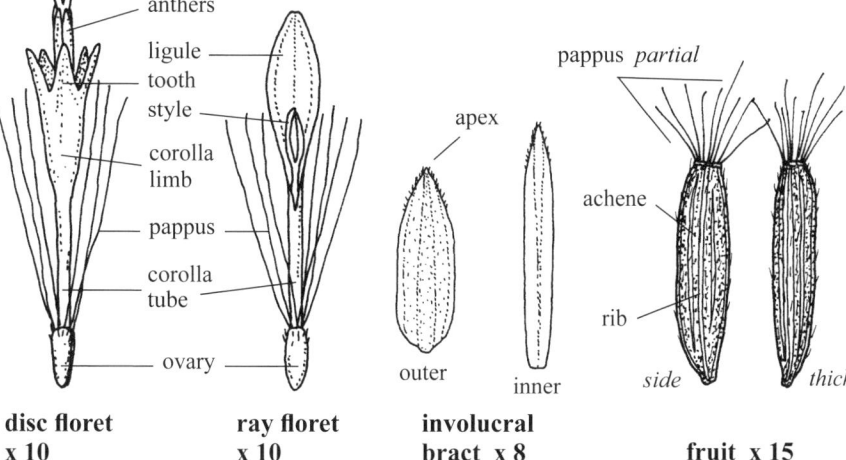

disc floret x 10   ray floret x 10   involucral bract x 8   fruit x 15

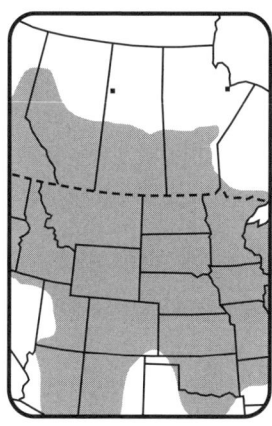

Prairie Goldenrod, Missouri Goldenrod

Wild Flower

O the Missouri Goldenrod is right you know

all things pass quickly away

yet appear again in late April

when the rain on the roof

is only the hounds of heaven

and wildflowers open

like a crowd of blue angels

blossoming in the wind

Velvety Goldenrod *Solidago mollis* Bartl.

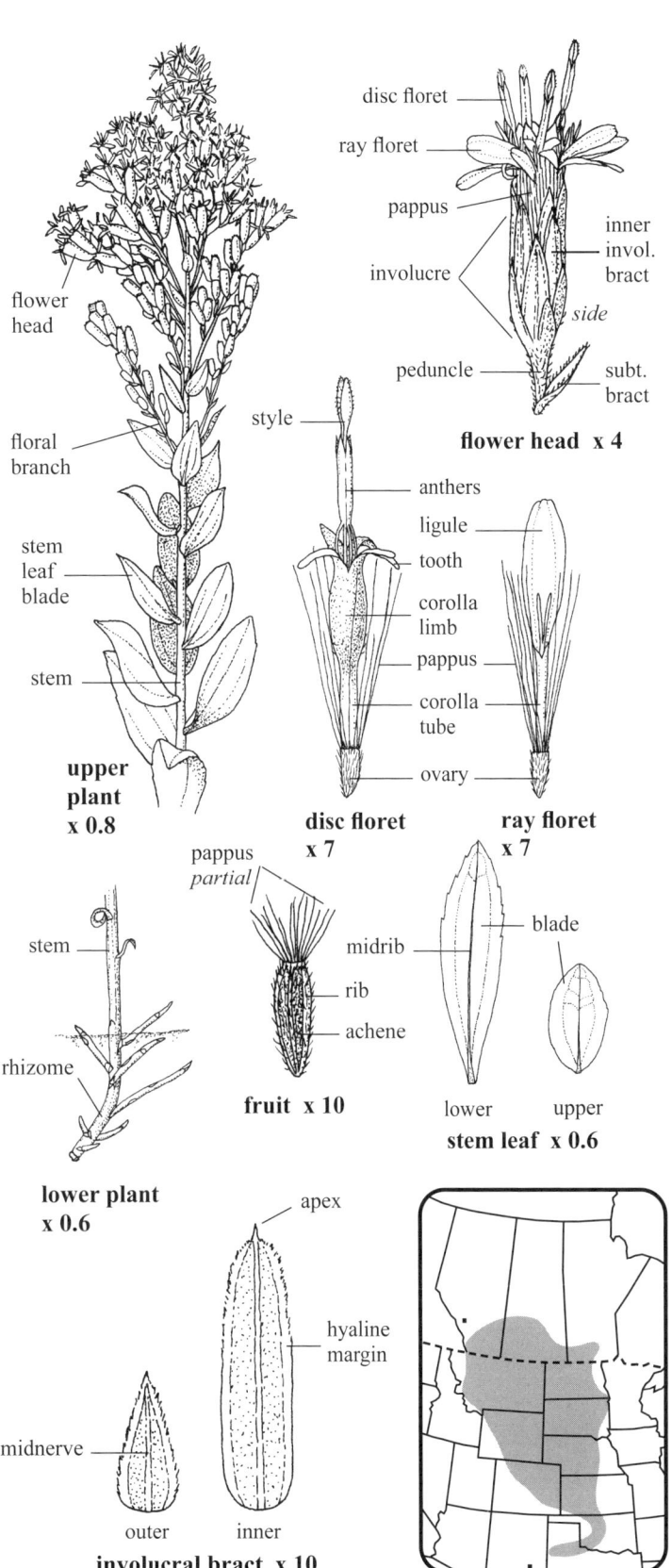

- **SKETCH** A variable, hairy **perennial herb** 10–60 cm tall in open colonies from horizontal, tan **rhizomes** 1–2.5 mm thick and to *c.* 7⁺ cm long; in sandy soils of dry prairies, open woods, roadsides and fencerows. **p2**
- **FLOWER HEADS** Yellow, blooming July–October; **inflorescence** a panicle, usually 2.5–9 cm long by 2–6 cm wide; **floral branches** ascending to recurved and 1–5 cm long; **peduncles** 0.5–2.5 mm long, hairy; **subtending bracts** (of peduncles) hairy, entire, about as long as peduncle; **flower heads** 5–8 mm wide by 9–11 mm long; **involucre** mostly glabrous, pale yellowish green, 3–6 mm long by 1.8–2.2 mm wide; **involucral bracts** imbricate, acute to mucronate, in three or four series, outer bracts slightly hairy, inner ones glabrous, 2–4.2 mm long by 0.8–1.2 mm wide, bases thickened, 1-nerved, margins hyaline and ciliate, spreading to release the fruit; **ray florets** 6–10, pistillate, TL 5–6 mm; **ovary** *c.* 1 mm long, hairy; **pappus** of numerous white bristles 2–4 mm long; **corolla tube** pale yellow, *c.* 2 mm long; **ligules** yellow, 2.5–4 mm long by 0.8–1 mm wide, apical teeth microscopic, two or three, blunt; **style** yellow, 2-parted, each part *c.* 1 mm long; **disc florets** perfect, 3–8; **ovary** as above; **pappus** as above; **corolla** yellow, 2.5–4.2 mm long, glabrous; **corolla tube** pale yellow, 1–2 mm long; **corolla limb** *c.* 2 mm long, teeth five, 1–1.5 mm long, spreading, yellow; **anthers** united, yellow, *c.* 1.8 mm long; **style** 2-parted; **fruiting heads** tawny. **FRUIT an achene**, 1-seeded, medium brown, 1.5–1.8 mm long by *c.* 0.5 mm wide and thick, with several tan ribs partially obscured by short ascending hairs; **pappus** tawny, comprised of *c.* 30 bristles 2.8–3.5 mm long with minute ascending hairs.
- **LEAVES** Alternate, simple, firm, entire to toothed, grayish green, ascending, imbricate, hairy on both sides and velvety to touch; **blades** pointed to blunt, 3-veined, 1–10 cm long by 4–40 mm wide, reduced above, lowest blades early deciduous; **petioles** are 0–2 mm long.
- **STEM** Erect, curved at base, hairy, tough, stiff; 2–4 mm thick near the usually bare and less hairy base.

Aster—*Asteraceae* **Wildflower** Yellow Ray florets 5–10

SHOWY GOLDENROD *Solidago nemoralis* Ait.

■ **SKETCH** A hairy **perennial herb** 20–100+ cm tall, singly or clumped, with a branching **caudex** 3–5 mm thick by 5–10+ cm long, the branches to *c.* 3 cm long; in dry to moist prairies, dry roadsides and in sandy to gravelly soils. **p1**

■ **FLOWER HEADS** Yellow, blooming August–October; **inflorescence** a panicle, nodding, narrow to wide, 3–30 cm long by 2–10 cm wide; **floral branches** 1–10 cm long, longer ones recurved, reduced above; **peduncles** hairy, 1–4 mm long; **flower heads** 6–8 mm wide by *c.* 8 mm long, ascending, mostly along one side of floral branches; **involucre** light yellowish green, mostly glabrous, 3–6 mm long by 2–2.5 mm wide; **involucral bracts** imbricate in three or four series, pliable, 2.5–4.5 mm long by 1–1.5 mm wide, tips blunt to pointed, ciliate on upper hyaline margins; **ray florets** 5–10, female; **ovary** *c.* 0.8 mm long, hairs silky; **pappus** of white bristles *c.* 3.2 mm long; **corolla tube** 2–2.2 mm long; **ligules** yellow, 2.3–5 mm long by *c.* 1 mm wide, glabrous, apical teeth two; **style** yellow, 2-parted; **disc florets** perfect, four to nine; **ovary** and **pappus** as above; **corolla tube** pale yellow, *c.* 1.7 mm long; **corolla limb** 1–1.5 mm long by *c.* 0.8 mm thick, yellow, glabrous, teeth five, 0.6–1 mm long, spreading; **anthers** five, *c.* 1.7 mm long, yellow; **style** 2-parted, apex hairy; **fruiting heads** white, 6–8 mm wide and long. **FRUIT an achene** (of disc floret), 1-seeded, tan, 1–2 mm long by 0.5–0.7 mm wide by 0.4–0.5 mm thick, 6- to 8-ribbed, these covered by white ascending hairs; **pappus** white, the 35–50 bristles 2.5–3.5 mm long.

■ **LEAVES** Basal and stem, alternate, simple, entire to widely toothed, scabrous, light gray; **basal blades** wilting early, 1–22 cm long by 5–40 mm wide, apices pointed, hairs more numerous below and descending along midrib; **petioles** tapered, 1.5–6 cm long, hairs descending; **stem blades** 1–6.5 cm long by 2–3 mm wide, often entire, reduced above and more erect; **petioles** 0–4 cm long.

■ **STEM** Erect, solid, simple, one to four, some slightly decumbent below, white hairy throughout, 11 or 12 low, reddish ridges at midstem; 1–4 mm thick near the base.

Gray Goldenrod

Aster—*Asteraceae* **Wildflower** Yellow Ligulate florets only, numerous

PERENNIAL SOW-THISTLE *Sonchus arvensis* L.

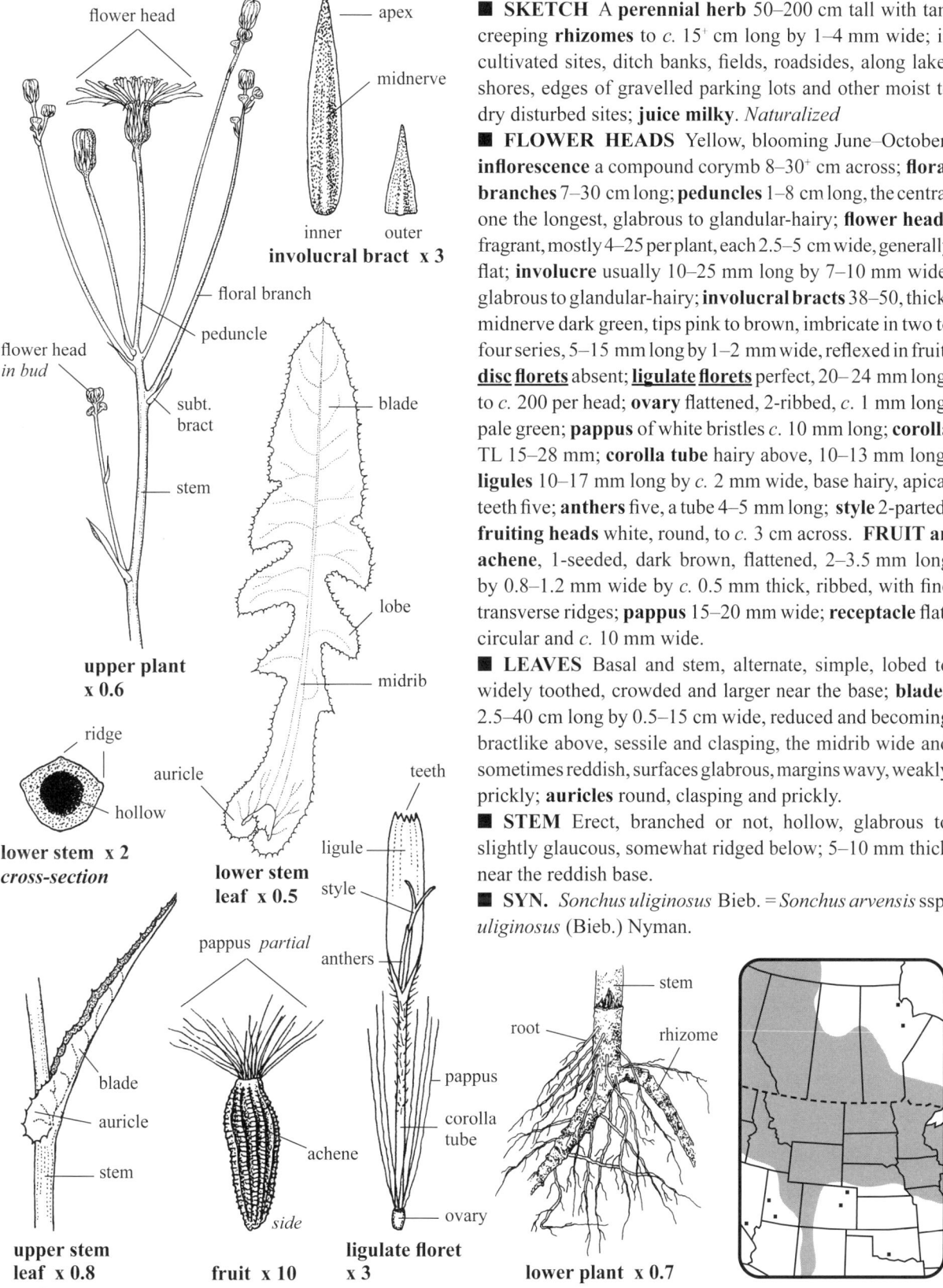

■ **SKETCH** A **perennial herb** 50–200 cm tall with tan, creeping **rhizomes** to *c.* 15⁺ cm long by 1–4 mm wide; in cultivated sites, ditch banks, fields, roadsides, along lakeshores, edges of gravelled parking lots and other moist to dry disturbed sites; **juice milky**. *Naturalized*

■ **FLOWER HEADS** Yellow, blooming June–October; **inflorescence** a compound corymb 8–30⁺ cm across; **floral branches** 7–30 cm long; **peduncles** 1–8 cm long, the central one the longest, glabrous to glandular-hairy; **flower heads** fragrant, mostly 4–25 per plant, each 2.5–5 cm wide, generally flat; **involucre** usually 10–25 mm long by 7–10 mm wide, glabrous to glandular-hairy; **involucral bracts** 38–50, thick, midnerve dark green, tips pink to brown, imbricate in two to four series, 5–15 mm long by 1–2 mm wide, reflexed in fruit; <u>disc florets</u> absent; <u>ligulate florets</u> perfect, 20– 24 mm long, to *c.* 200 per head; **ovary** flattened, 2-ribbed, *c.* 1 mm long, pale green; **pappus** of white bristles *c.* 10 mm long; **corolla** TL 15–28 mm; **corolla tube** hairy above, 10–13 mm long; **ligules** 10–17 mm long by *c.* 2 mm wide, base hairy, apical teeth five; **anthers** five, a tube 4–5 mm long; **style** 2-parted; **fruiting heads** white, round, to *c.* 3 cm across. **FRUIT an achene**, 1-seeded, dark brown, flattened, 2–3.5 mm long by 0.8–1.2 mm wide by *c.* 0.5 mm thick, ribbed, with fine transverse ridges; **pappus** 15–20 mm wide; **receptacle** flat, circular and *c.* 10 mm wide.

■ **LEAVES** Basal and stem, alternate, simple, lobed to widely toothed, crowded and larger near the base; **blades** 2.5–40 cm long by 0.5–15 cm wide, reduced and becoming bractlike above, sessile and clasping, the midrib wide and sometimes reddish, surfaces glabrous, margins wavy, weakly prickly; **auricles** round, clasping and prickly.

■ **STEM** Erect, branched or not, hollow, glabrous to slightly glaucous, somewhat ridged below; 5–10 mm thick near the reddish base.

■ **SYN.** *Sonchus uliginosus* Bieb. = *Sonchus arvensis* ssp. *uliginosus* (Bieb.) Nyman.

Field Sow Thistle   181

Aster—*Asteraceae* **Wildflower** Pale yellow Ligulate florets many

ANNUAL SOW-THISTLE *Sonchus oleraceus* L.

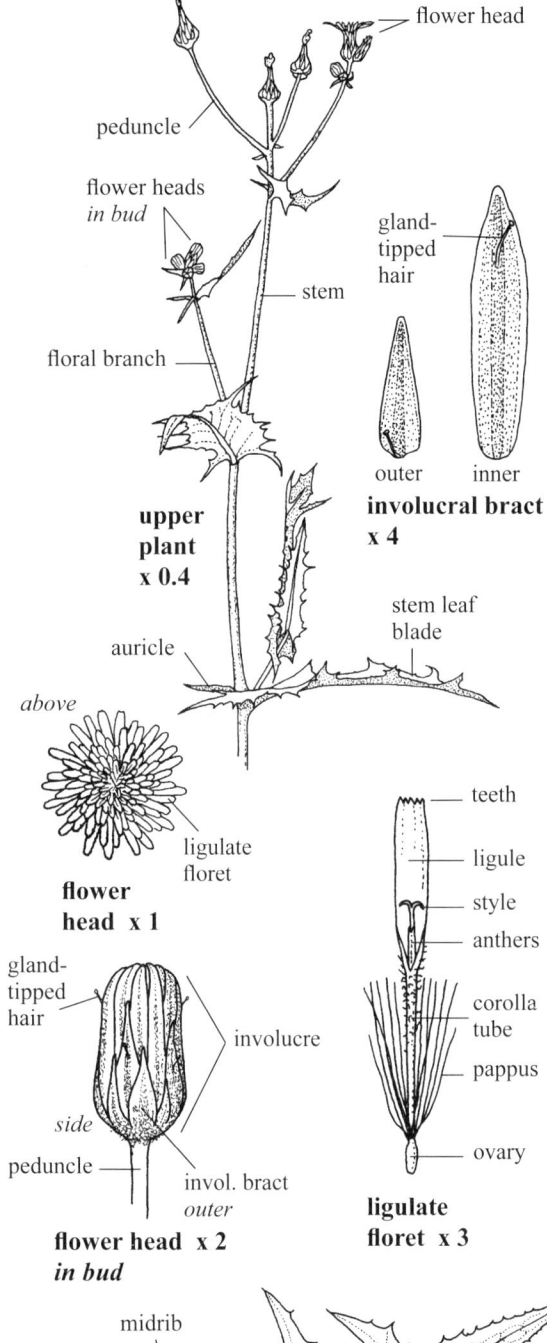

upper plant x 0.4

involucral bract x 4

flower head x 1

flower head x 2 in bud

ligulate floret x 3

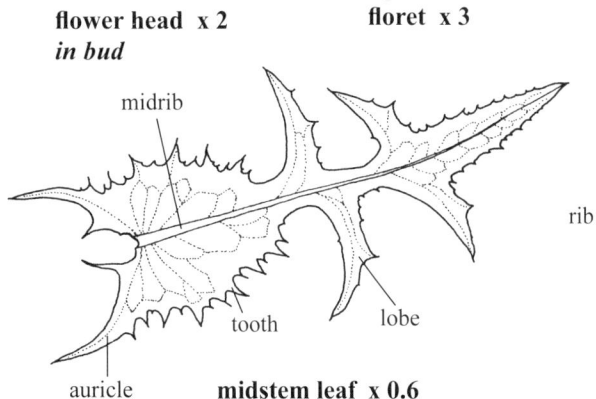

midstem leaf x 0.6

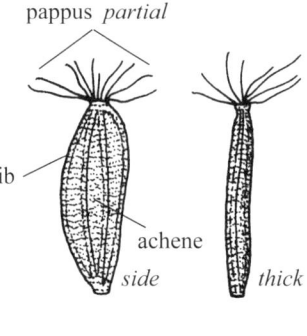

fruit x 10

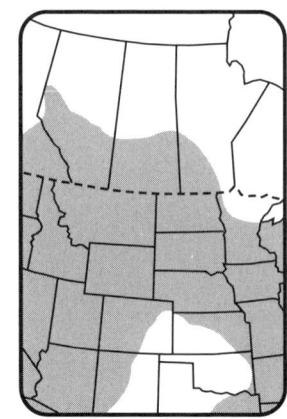

■ **SKETCH** An **annual herb** 20–120 cm tall, singly or in patches from a **taproot** 2–10 mm thick; in cultivated and disturbed sites, ditch banks, city lots and bottomlands; **juice milky**. *Naturalized*

■ **FLOWER HEADS** Pale yellow, blooming in February–October; **inflorescence** a corymb of several+ heads, terminal and on branches; **floral branches** 4–30+ cm long; **peduncles** 0.5–11 cm long; **flower heads** 1.5–2.8 cm across by 10–15 mm long; **involucre** 9–15 mm long, the flower head buds often with a mass of congested white basal hairs resembling frost; **involucral bracts** in three or four series, imbricate, 3–10 mm long by 1.5–2.2 mm wide, margins hyaline, tips pointed, with a few scattered, often gland-tipped hairs, the central nerve thickest at the base of the lower outer bracts; <u>**disc florets**</u> absent; <u>**ligulate florets**</u> numerous, perfect; **ovary** white, hairless, flattened, faintly ribbed, *c.* 1.3 mm long; **pappus** white, of fine bristles 5–7 mm long; **corolla tube** white, *c.* 8 mm long with curly hairs along its upper half; **ligules** yellow, 5–7 mm long by 1.2–1.5 mm wide, with five tiny apical teeth, the outer ligules the longest; **anthers** five, yellow, forming a tube; **fruiting head** white, round, *c.* 2 cm across. **FRUIT an achene**, 1-seeded, medium brown, dull, flattened, 2.5–3 mm long by *c.* 1 mm wide by *c.* 0.3 mm thick, with shallow transverse wrinkles, the faces 3- to 5-ribbed, not beaked; **pappus** of numerous white bristles 5–8 mm long.

■ **LEAVES** Basal and stem, alternate, simple, lobed and toothed, glabrous; **basal leaves** petiolate, auricles lacking or smaller than those above, early deciduous; **petioles** 0–6.5 cm long, reduced above; **stem leaves** with one to several lobes, teeth ending with short weak prickles; **blades** 4–30 cm long by 1.5–15 cm wide; **auricles** pointed and to *c.* 3 cm long, extending well past the stem, usually not clasping.

■ **STEM** Erect, hollow, glabrous, round, usually branched; 3–10 mm thick near the base.

Aster—*Asteraceae* **Wildflower** White, center yellow Ray florets 25–40

Rush Aster *Symphyotrichum boreale* (Torr. & Gray) A. & D. Löve

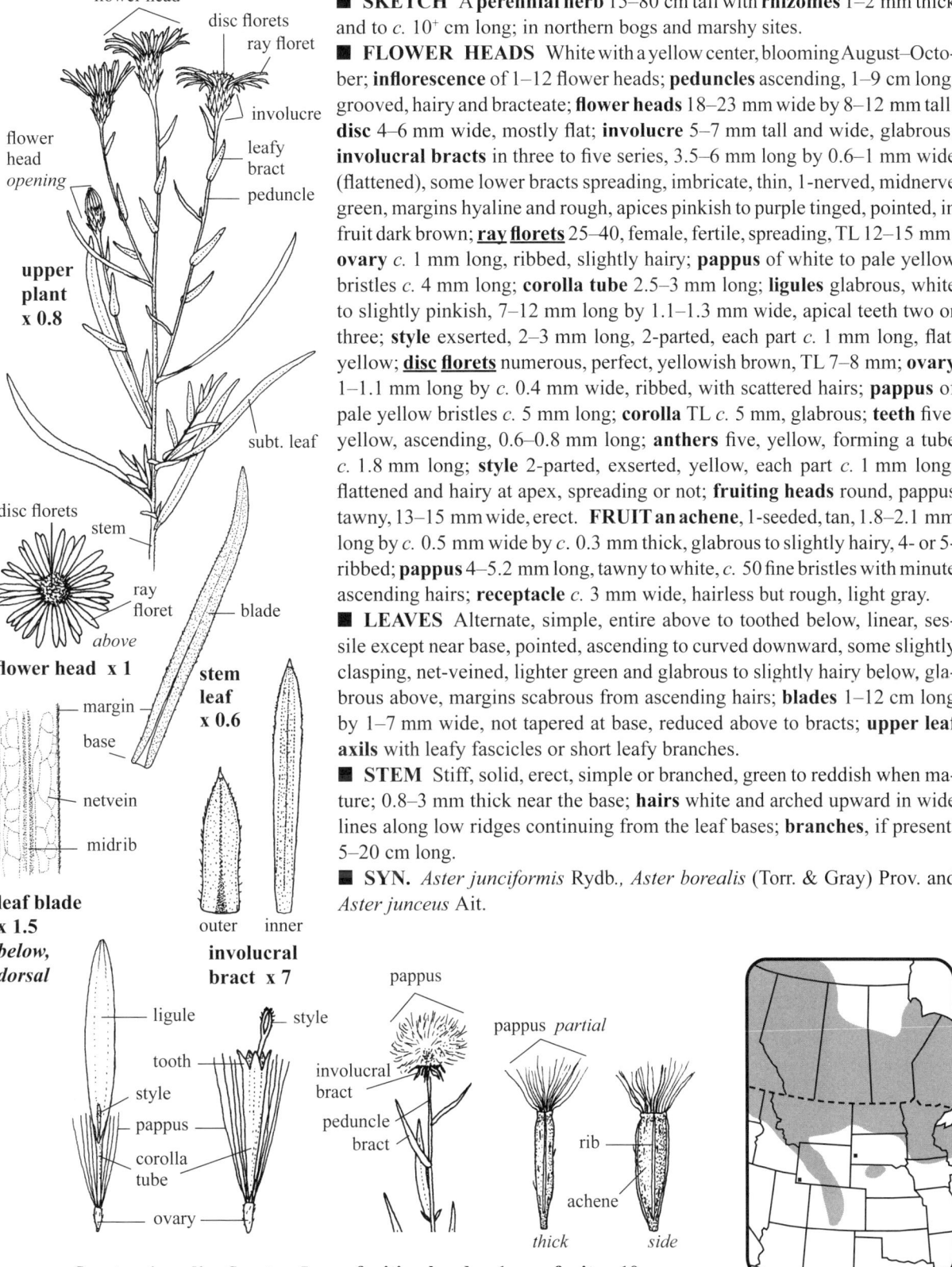

- **SKETCH** A **perennial herb** 15–80 cm tall with **rhizomes** 1–2 mm thick and to *c.* 10⁺ cm long; in northern bogs and marshy sites.
- **FLOWER HEADS** White with a yellow center, blooming August–October; **inflorescence** of 1–12 flower heads; **peduncles** ascending, 1–9 cm long, grooved, hairy and bracteate; **flower heads** 18–23 mm wide by 8–12 mm tall; **disc** 4–6 mm wide, mostly flat; **involucre** 5–7 mm tall and wide, glabrous; **involucral bracts** in three to five series, 3.5–6 mm long by 0.6–1 mm wide (flattened), some lower bracts spreading, imbricate, thin, 1-nerved, midnerve green, margins hyaline and rough, apices pinkish to purple tinged, pointed, in fruit dark brown; <u>ray florets</u> 25–40, female, fertile, spreading, TL 12–15 mm; **ovary** *c.* 1 mm long, ribbed, slightly hairy; **pappus** of white to pale yellow bristles *c.* 4 mm long; **corolla tube** 2.5–3 mm long; **ligules** glabrous, white to slightly pinkish, 7–12 mm long by 1.1–1.3 mm wide, apical teeth two or three; **style** exserted, 2–3 mm long, 2-parted, each part *c.* 1 mm long, flat, yellow; <u>disc florets</u> numerous, perfect, yellowish brown, TL 7–8 mm; **ovary** 1–1.1 mm long by *c.* 0.4 mm wide, ribbed, with scattered hairs; **pappus** of pale yellow bristles *c.* 5 mm long; **corolla** TL *c.* 5 mm, glabrous; **teeth** five, yellow, ascending, 0.6–0.8 mm long; **anthers** five, yellow, forming a tube *c.* 1.8 mm long; **style** 2-parted, exserted, yellow, each part *c.* 1 mm long, flattened and hairy at apex, spreading or not; **fruiting heads** round, pappus tawny, 13–15 mm wide, erect. **FRUIT** an achene, 1-seeded, tan, 1.8–2.1 mm long by *c.* 0.5 mm wide by *c.* 0.3 mm thick, glabrous to slightly hairy, 4- or 5-ribbed; **pappus** 4–5.2 mm long, tawny to white, *c.* 50 fine bristles with minute ascending hairs; **receptacle** *c.* 3 mm wide, hairless but rough, light gray.
- **LEAVES** Alternate, simple, entire above to toothed below, linear, sessile except near base, pointed, ascending to curved downward, some slightly clasping, net-veined, lighter green and glabrous to slightly hairy below, glabrous above, margins scabrous from ascending hairs; **blades** 1–12 cm long by 1–7 mm wide, not tapered at base, reduced above to bracts; **upper leaf axils** with leafy fascicles or short leafy branches.
- **STEM** Stiff, solid, erect, simple or branched, green to reddish when mature; 0.8–3 mm thick near the base; **hairs** white and arched upward in wide lines along low ridges continuing from the leaf bases; **branches**, if present, 5–20 cm long.
- **SYN.** *Aster junciformis* Rydb., *Aster borealis* (Torr. & Gray) Prov. and *Aster junceus* Ait.

Boreal Aster, Northern Bog Aster, Boreal American-aster, Marsh Aster

Aster—*Asteraceae* **Wildflower** White   Ray floret ligules vestigial

Rayless Aster   *Symphyotrichum ciliatum* (Ledeb.) Nesom

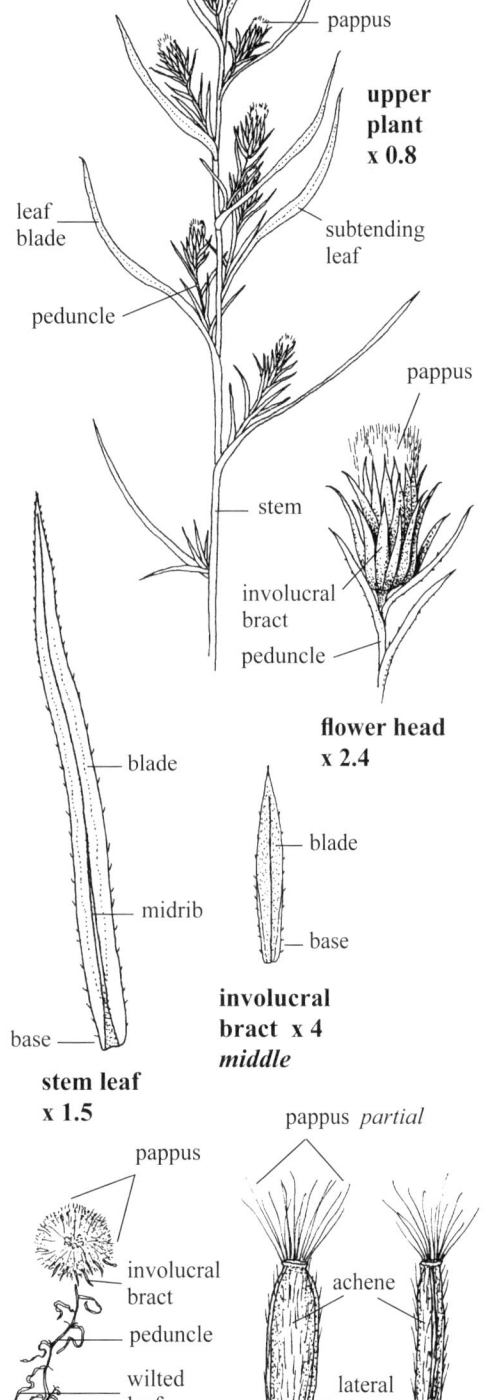

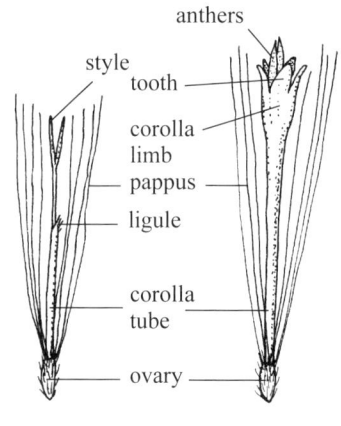

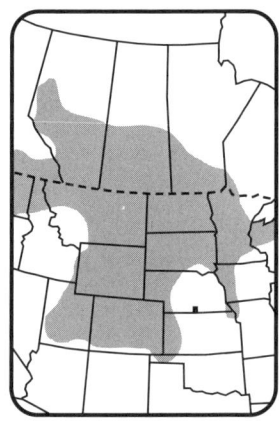

■ **SKETCH** An **annual herb** 15–60 cm tall with a **taproot** 5–10 cm long by 1–4 mm wide; in moist, slightly saline sites in wetlands and ditches, often covering several m².

■ **FLOWER HEADS** White, blooming August–October; **inflorescence** a raceme or panicle, 3–33 cm long by 1.5–24 cm wide; **floral branches** leafy, ascending, 4–25+ cm long, reduced above; **peduncles** leafy, 3–50 mm long; **flower heads** several to several dozen, 6–10 mm wide by 7–10 mm long; **involucre** green to purple, topped by white pappus; **involucral bracts** in two or three series, *c.* 33, imbricate, 3.5–7 mm long by 0.7–0.9 mm wide, flat, pliable, with a mucro, short scattered hairs on both sides; **inner bracts** more narrow than outer ones, margins hyaline, less hairy; **florets** of two types: **1) ray florets** outer, female, not showy, more numerous than inner disc florets; **ovary** *c.* 0.6 mm long, with hairs ascending; **pappus** of numerous white bristles 4–4.5 mm long; **corolla tube** 2–2.3 mm long; **ligules** vestigial; **style** 2-parted, exserted 1.5–2 mm beyond the corolla tube apex; **2) disc florets** perfect, inner; **ovary** *c.* 0.6–0.8 mm long, hairy; **pappus** of 40–50 white bristles 4.5–5.5 mm long; **corolla** hyaline to violet, 3–5 mm long, apical teeth five, each *c.* 0.4 mm long; **corolla tube** 2–3.8 mm long; **corolla limb** 1–1.2 mm long by 0.8 mm wide; **anthers** five, yellow, *c.* 0.9 mm long; **style** 2-parted, exserted; **fruiting heads** white, roundish, 8–18 mm long and wide. **FRUIT** an **achene**, 1-seeded, tan, 1.5–1.8 mm long by *c.* 0.5 mm wide by 0.3–0.4 mm thick, with two lateral ribs and white appressed hairs; **pappus** 5–6 mm long.

■ **LEAVES** Alternate, simple, entire, mostly subtending floral branches or peduncles; **blades** thin, flexible, 1–12 cm long by 1–8 mm wide, reduced above, sessile, glabrous except for short marginal hairs, midrib obvious, side veins two and obscure.

■ **STEM** Erect, solid, tough, reddish brown below, round, grooved above, slightly hairy to glabrous, simple or branches beginning near the base; 1–5 mm wide near the base.

■ **SYN.** *Aster brachyactis* Blake.

Rayless Alkali Aster, Alkali American-aster

Aster—*Asteraceae* **Wildflower** Bluish purple, center pale green to pink   Ray florets 12–25

LINDLEY'S ASTER   *Symphyotrichum ciliolatum* (Lindl.) A. & D. Löve

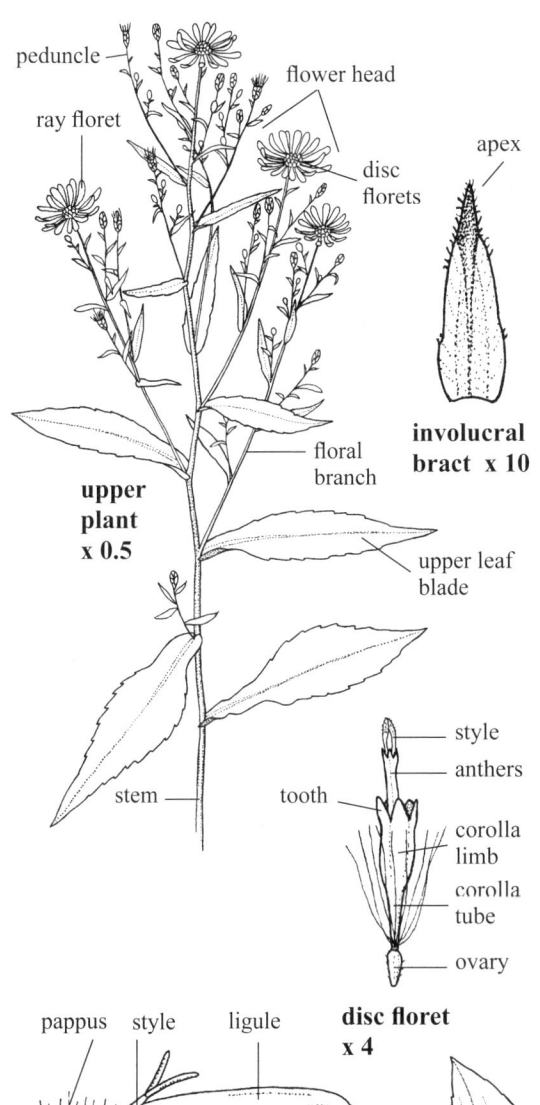
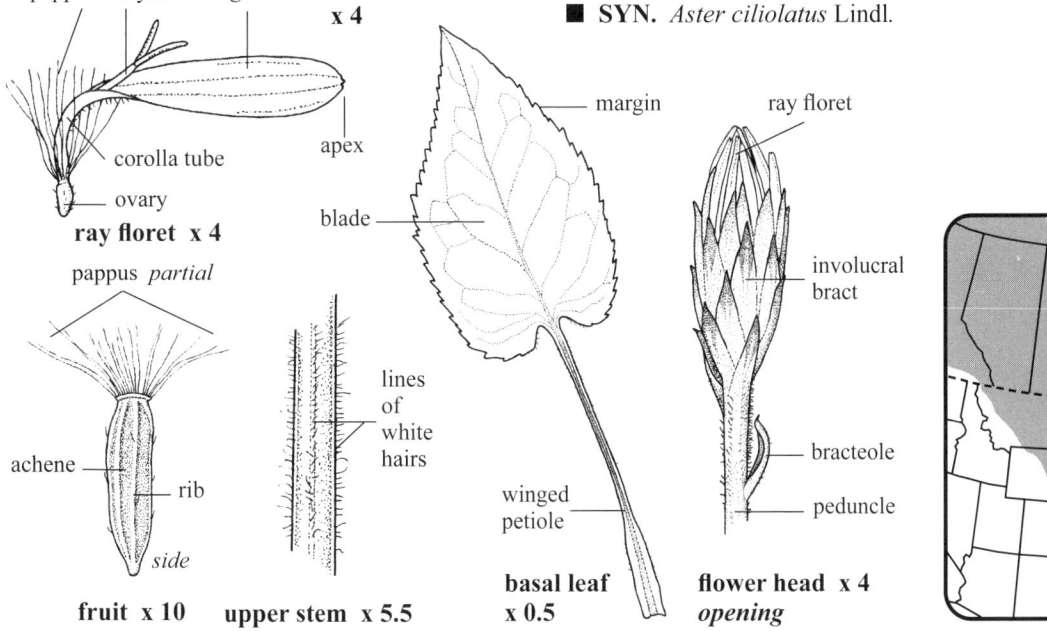
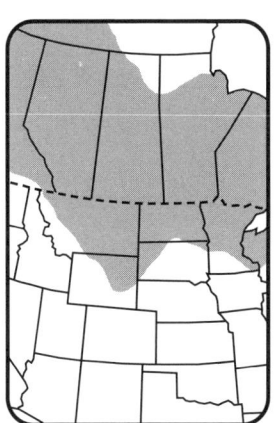

■ **SKETCH** A **perennial herb** 20–100 cm tall with **rhizomes** 1–2 mm wide by 5–12⁺ cm long, or a **caudex** 1.2–3.5 cm long and 4–6 mm thick; in moist openings of woodlands, thickets and rocky sites. **p2**

■ **FLOWER HEADS** Light bluish purple with a pale green or pinkish disc, blooming July–October; **inflorescence** paniculate, 4–40 cm long; **peduncles** 3–25 mm long, some bracteolate; **floral branches** 4–22 cm long; **flower heads** numerous, 15–28 mm wide, the disc 4–6 mm wide; **involucre** 5–8 mm long by 3–6 mm wide; **involucral bracts** 3–6 mm long by *c.* 1 mm wide, imbricate, with a dark green hairy tip; <u>ray florets</u> 10–25 per head, female and fertile; **ovary** slightly hairy; **pappus** *c.* 4 mm long, of fine bristles; **corolla tube** 2.5–3 mm long, white, slightly hairy; **ligules** lightly hairy at base, 5–12 mm long by 1.8–2.5 mm wide, with three blunt apical teeth; **style** 2-parted; <u>disc florets</u> perfect, 12–25 per head; **ovary** slightly hairy; **pappus** as above; **corolla** 5-lobed, 4.5–5 mm long; **corolla limb** *c.* 2.5 mm long; **anthers** five, yellow, forming a tube; **style** 2-parted, yellow; **fruiting heads** tan, 10–12 mm wide. **FRUIT** an achene, 1-seeded, flat, brown, 2- to 5-ribbed, 1.5–3 mm long by *c.* 0.6 mm wide by *c.* 0.4 mm thick, lightly hairy to glabrous; **pappus** 3.5–6 mm long by 7–9 mm wide, tan, of *c.* 40 bristles.

■ **LEAVES** Alternate, simple, toothed to entire; **basal blades** toothed, 1.8–15 cm long by 1–6.5 cm wide, glabrous above, slightly hairy below especially on veins; **petioles** winged, 1–17 cm long and to 15 mm wide, hairy; **stem blades** 4–12 cm long by 1–4 cm wide, reduced above, pointed, glabrous, becoming sessile and clasping; **petioles** winged, reduced above.

■ **STEM** Round, erect, solid, dark red to green, branched above, hairs in vertical lines above, the lower part glabrous; 1–5 mm wide near the reddish brown base.

■ **SYN.** *Aster ciliolatus* Lindl.

Lindley's Blue Aster, Fringed American-aster, Lindley's American-aster, Fringed Aster

Aster—*Asteraceae* **Wildflower** White, center yellow to pinkish Ray florets 10–20

MANY-FLOWERED ASTER   *Symphyotrichum ericoides* (L.) Nesom

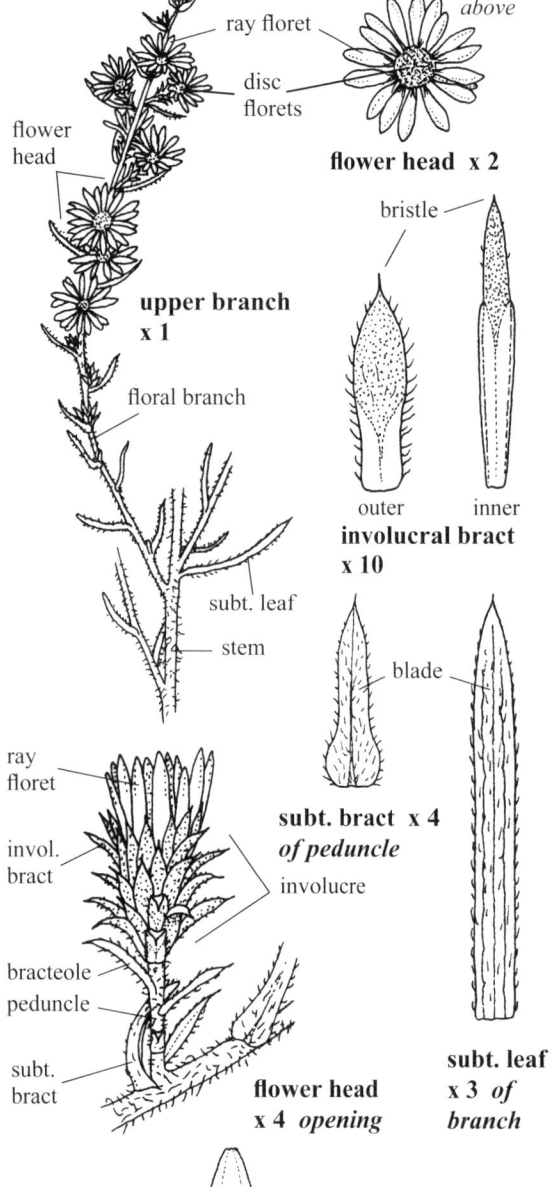

■ **SKETCH** A variable, colonial or tufted **perennial herb** 20–100 cm tall from a thick to thin **rhizome**, some with **stolons**; in dry prairies to alkali flats. **p1**
■ **FLOWER HEADS** White with a yellow to pinkish center, blooming August–October; **inflorescence** racemose, of numerous small heads mostly along one side and towards the ends of the brown hairy floral branches; **floral branches** 2–21 cm long, curved; **peduncles** 2–15 mm long, hairy, often with 2–8 hairy, bristle-tipped bracteoles 1.5–5 mm long; **flower heads** 8–12 mm wide, the disc 2–3 mm wide; **involucre** 3–5.5 mm long and wide; **involucral bracts** 2–4 mm long by 0.5–1 mm wide, in three to five series, imbricate, the tips medium green, spreading, slightly hairy with a tiny bristle, the bases white and the midnerve green; <u>ray florets</u> female and fertile, 10–20; **ovary** hairy, *c.* 0.7 mm long; **pappus** of *c.* 30 white bristles *c.* 3 mm long; **corolla tube** glabrous, 1.6–2 mm long; **ligules** 3–5.5 mm long by 1–1.2 mm wide; **style** 2-parted, exserted; <u>disc florets</u> perfect, 5–25; **ovary** and **pappus** as above; **corolla tube** 1–1.3 mm long; **corolla limb** *c.* 2 mm long, teeth five, yellow, *c.* 0.5 mm long; **anthers** yellow, *c.* 1 mm long; **style** yellow, 2-parted. **FRUIT an achene**, 1-seeded, 1.4–2 mm long by *c.* 0.5 mm wide and thick, medium grayish purple with six or seven whitish tan low ribs, hairs scattered and ascending; **pappus** whitish tan, of *c.* 30 bristles 2–2.5 mm long.
■ **LEAVES** Alternate, simple, entire, hairy, more so above including the margins, all hairs short and ascending; **blades** sessile, not clasping, 1–6 cm long by 1–5 mm wide, reduced above, becoming bractlike and ending with a white apical bristle; **lower blades** deciduous by anthesis.
■ **STEM** Ascending to erect, tan, wiry, branched above, pith white, with numerous white appressed short hairs; 2–3 mm thick near the often bare, glabrous base.
■ **SYN.** *Aster ericoides* L. = *Symphyotrichum ericoides* var. *ericoides* (L.) Nesom and *Aster pansus* (Blake) Cronq. = *Symphyotrichum ericoides* var. *pansum* (Blake) Nesom.

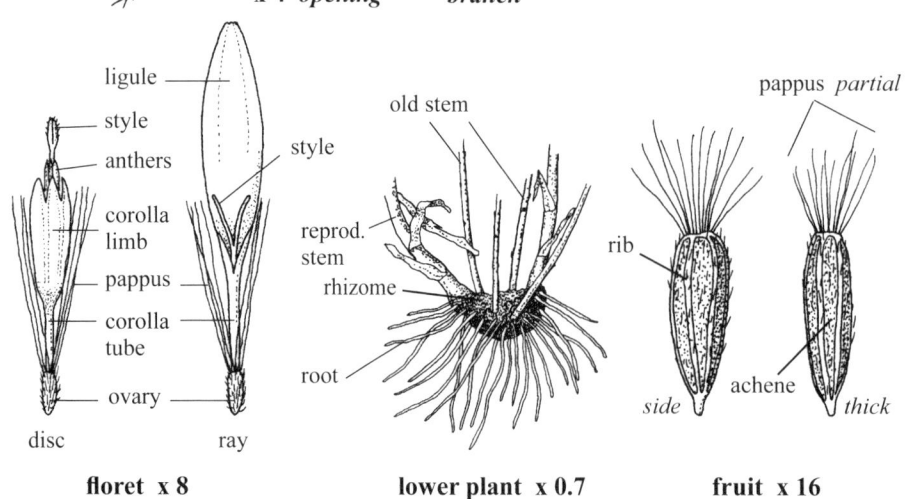

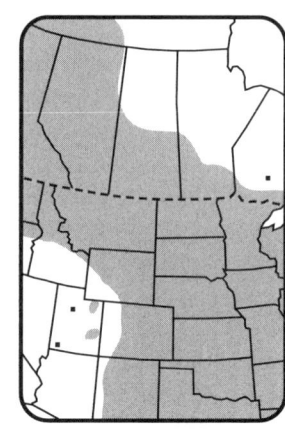

186   Heath Aster, White Heath Aster, White Heath American-aster, White Aster, Tufted White Prairie Aster

Aster—*Asteraceae* **Wildflower** White, center yellow Ray florets 20–35

White Prairie Aster *Symphyotrichum falcatum* (Lindl.) Nesom

■ **SKETCH** A **perennial herb** 30–90 cm tall, tufted or in open colonies from brown **rhizomes** 2–5⁺ mm thick by 2–8⁺ cm long and with pink-tipped, whitish **stolons** 1–2 mm thick; along roadsides, dry prairies and sandy fields. **p1**

■ **FLOWER HEADS** White with a yellow disc which turns reddish with age, blooming July–October; **inflorescence** a panicle (or raceme), flower heads not crowded; **floral branches** 2–30 cm long; **peduncles** ascending, hairy, 3–45 mm long, reduced above, leafy bracteoles 2–9 mm long by 0.7–1.5 mm wide, sessile, hairy, usually spreading and often bristle-tipped; **flower heads** 13–25 mm wide; **central disc** 4–5 mm wide; **involucre** green and white, 5–9 mm long and wide; **involucral bracts** imbricate, in four or five series, apices often bristle-tipped and spreading, hairy on green midnerve; **1) outer bracts** 3–4.5 mm long by 0.7–1.2 mm wide (excluding marginal hairs); **2) inner bracts** 4.3–6 mm long by 0.7–1.2 mm wide, margins hyaline, thicker near the white bases, a few marginal hairs at the green tips; **ray florets** female, 20–35 per head, TL 7–12 mm; **ovary** 0.8–1 mm long, covered with ascending white hairs; **pappus** single, white to tawny, 40–50 fine bristles 3.5–5.5 mm long; **corolla tube** glabrous, white, 1.8–3 mm long; **ligules** slightly imbricate, 4.5–8 mm long by 1–1.8 mm wide, with three to five apical teeth; **style** white, 2-parted, parts 1–1.3 mm long, glabrous; **disc florets** perfect, 14–33 per head, TL 6–9 mm; **ovary** as above; **pappus** as above; **corolla** TL 3.5–5 mm, glabrous, veined; **corolla tube** pale green, 1.3–2 mm long; **corolla limb** 2.2–3 mm long, pale yellow, turning pinkish with age; **teeth** five, *c.* 1 mm long; **anthers** five, united, 1.5–2 mm long, exserted, yellow to pink; **style** 2-parted, exserted, parts hairy, slightly flattened; **fruiting heads** tan, 15–18 mm wide by *c.* 10 mm tall. **FRUIT an achene**, 1-seeded, 1.5–2.4 mm long by 0.5–0.7 mm wide by 0.4–0.6 mm thick, 8- to 10-ribbed, tan to pale purple, hairs ascending; **pappus** single, of 32–50 fine tan bristles 4–6 mm long; **receptacle** flat, rough from projections.

■ **LEAVES** Alternate, simple, entire, sessile, absent below by anthesis; **stem blades** linear, flat, thin, 0.5–5.5 cm long by 1.5–5 mm wide, spreading, midrib obvious, hairs simple and appressed on sides and margins, a short white bristle at apices.

■ **STEM** Ascending to erect, one to eight per clump, solid, stiff, branched below or not, green and hairy above, hairs appressed and ascending; 1–5 mm thick near the hairless, woody base.

■ **SYN.** *Aster falcatus* Lindl. = *Symphyotrichum falcatum* var. *falcatum* (Lindl.) Nesom and *Aster commutatus* (Torr. & Gray) Gray = *Symphyotrichum falcatum* var. *commutatum* (Torr. & Gray) Nesom.

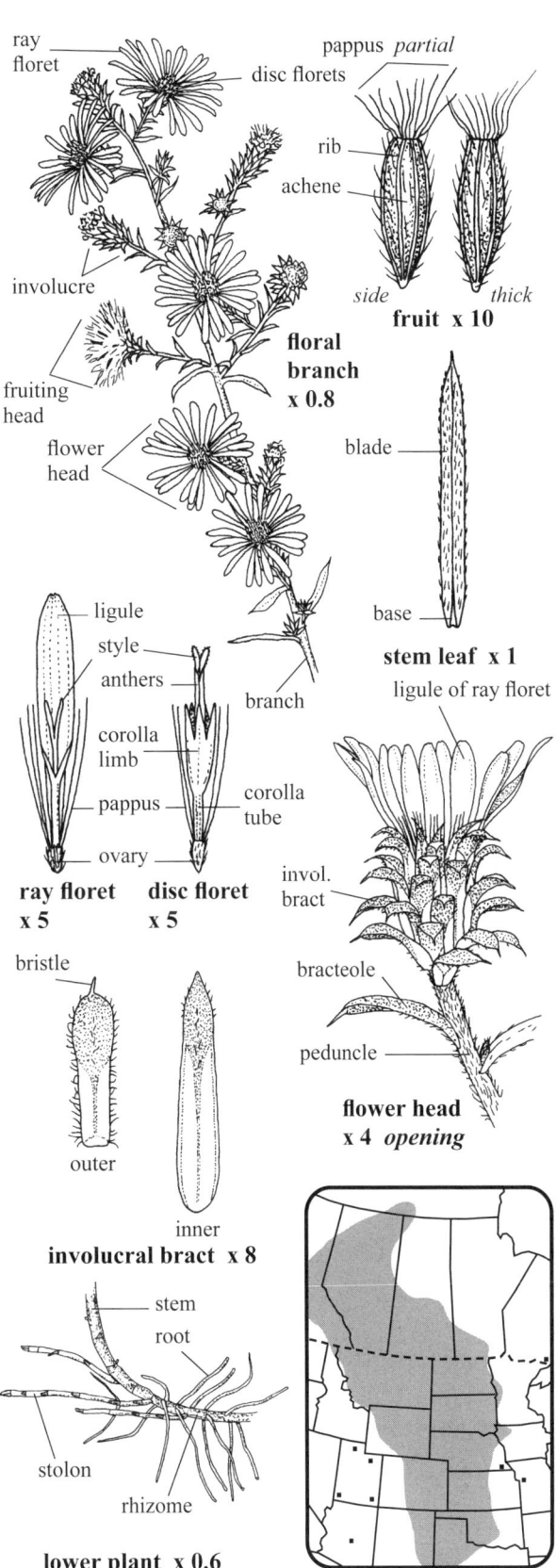

Aster—*Asteraceae* **Wildflower** Blue to light purple, center yellow Ray florets 15–25

SMOOTH ASTER *Symphyotrichum laeve* (L.) A. & D. Löve

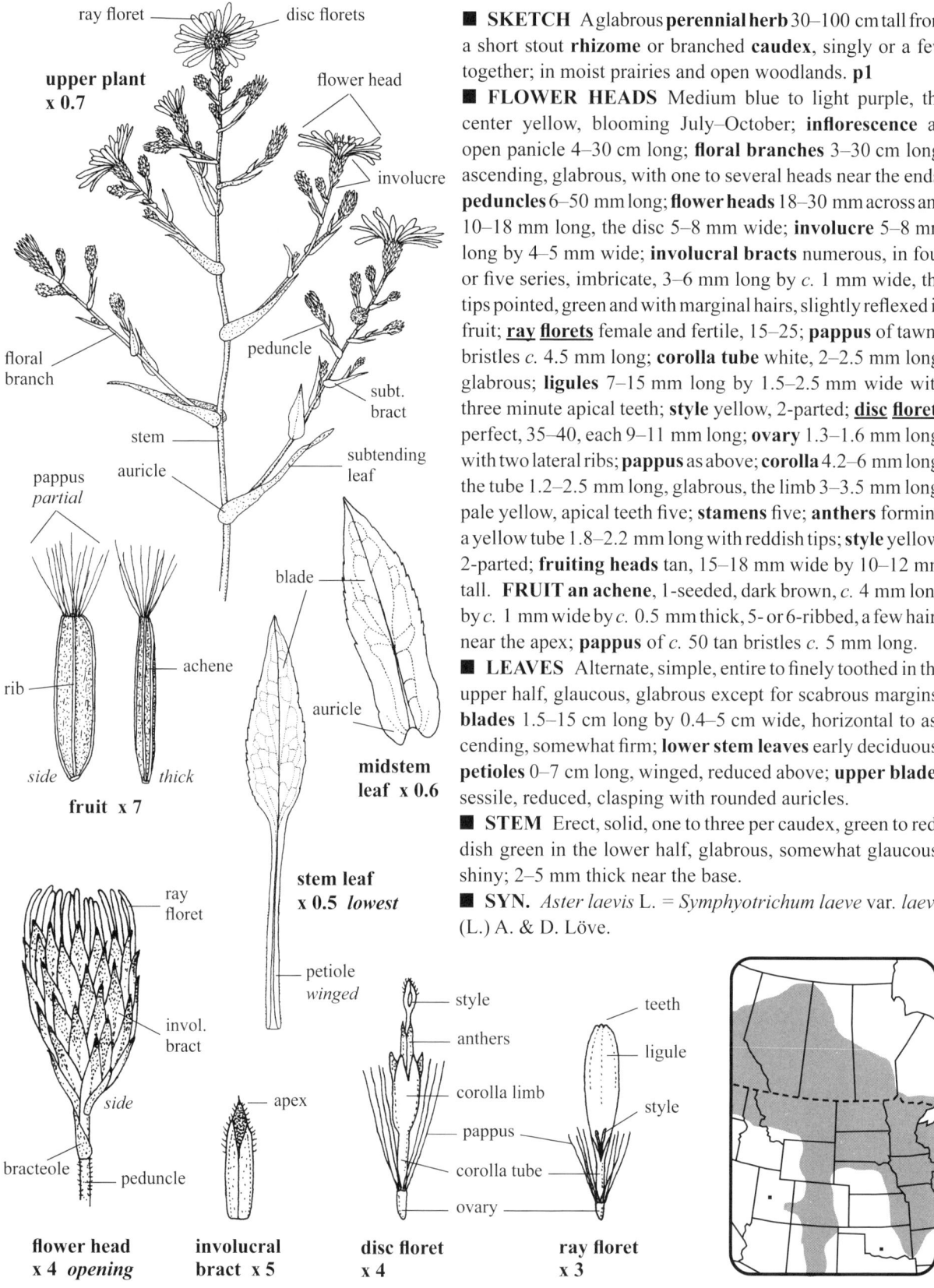

- **SKETCH** A glabrous **perennial herb** 30–100 cm tall from a short stout **rhizome** or branched **caudex**, singly or a few together; in moist prairies and open woodlands. **p1**
- **FLOWER HEADS** Medium blue to light purple, the center yellow, blooming July–October; **inflorescence** an open panicle 4–30 cm long; **floral branches** 3–30 cm long, ascending, glabrous, with one to several heads near the ends; **peduncles** 6–50 mm long; **flower heads** 18–30 mm across and 10–18 mm long, the disc 5–8 mm wide; **involucre** 5–8 mm long by 4–5 mm wide; **involucral bracts** numerous, in four or five series, imbricate, 3–6 mm long by *c.* 1 mm wide, the tips pointed, green and with marginal hairs, slightly reflexed in fruit; <u>**ray florets**</u> female and fertile, 15–25; **pappus** of tawny bristles *c.* 4.5 mm long; **corolla tube** white, 2–2.5 mm long, glabrous; **ligules** 7–15 mm long by 1.5–2.5 mm wide with three minute apical teeth; **style** yellow, 2-parted; <u>**disc florets**</u> perfect, 35–40, each 9–11 mm long; **ovary** 1.3–1.6 mm long, with two lateral ribs; **pappus** as above; **corolla** 4.2–6 mm long, the tube 1.2–2.5 mm long, glabrous, the limb 3–3.5 mm long, pale yellow, apical teeth five; **stamens** five; **anthers** forming a yellow tube 1.8–2.2 mm long with reddish tips; **style** yellow, 2-parted; **fruiting heads** tan, 15–18 mm wide by 10–12 mm tall. **FRUIT** an achene, 1-seeded, dark brown, *c.* 4 mm long by *c.* 1 mm wide by *c.* 0.5 mm thick, 5- or 6-ribbed, a few hairs near the apex; **pappus** of *c.* 50 tan bristles *c.* 5 mm long.
- **LEAVES** Alternate, simple, entire to finely toothed in the upper half, glaucous, glabrous except for scabrous margins; **blades** 1.5–15 cm long by 0.4–5 cm wide, horizontal to ascending, somewhat firm; **lower stem leaves** early deciduous; **petioles** 0–7 cm long, winged, reduced above; **upper blades** sessile, reduced, clasping with rounded auricles.
- **STEM** Erect, solid, one to three per caudex, green to reddish green in the lower half, glabrous, somewhat glaucous, shiny; 2–5 mm thick near the base.
- **SYN.** *Aster laevis* L. = *Symphyotrichum laeve* var. *laeve* (L.) A. & D. Löve.

Smooth Blue Aster, Smooth Blue American-aster

Aster—*Asteraceae* **Wildflower** White to violet, center yellow to purple Ray florets 16–47

SMALL BLUE ASTER *Symphyotrichum lanceolatum* (Willd.) Nesom

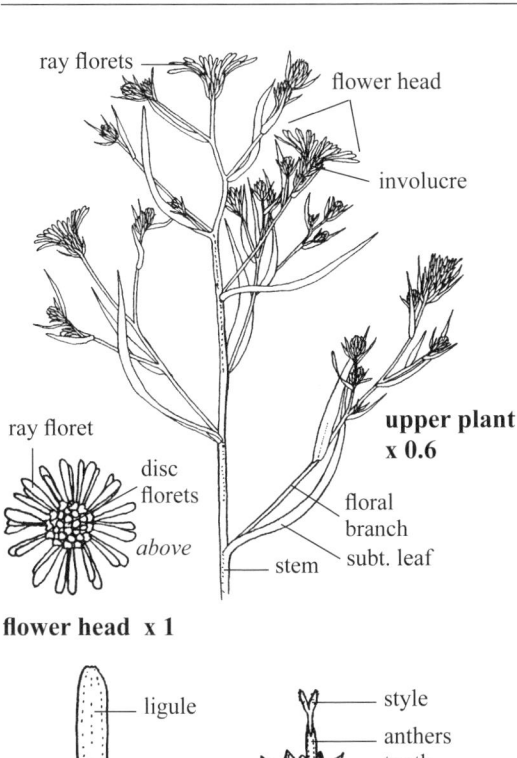

flower head x 1

ray floret x 3.2

involucral bract x 6

fruit x 10

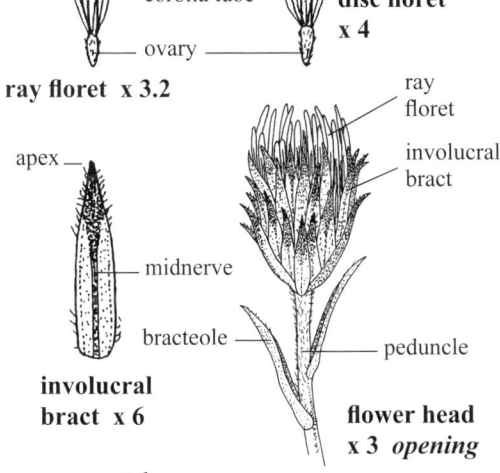

lower plant x 0.6

■ **SKETCH** A variable **perennial herb** 20–200 cm tall from a stout **root** and tan **rhizomes** 1–5 mm thick by 8–18+ cm long; in damp meadows, ditches, along stream and river banks. **p2**

■ **FLOWER HEADS** White to violet or pink, center yellow to purple, blooming August–October; **inflorescence** a panicle, open, usually in the upper half; **floral branches** alternate, leafy, ascending, 4–30 cm long with 3–10 flower heads near the ends; **flower heads** numerous, 14–22 mm wide by 3–9 mm tall, the disc 5–6 mm wide; **involucre** 3–6 mm long by 3–4 mm wide; **involucral bracts** in three or four series, imbricate, 3–4.5 mm long by 0.6–0.8 mm wide, thin, somewhat hairy, the tips appressed or spreading, green to reddish, the base whitish, reflexed in fruit; **ray florets** 16–47, female and fertile, often overlapping, TL 5–12 mm; **pappus** of white bristles *c.* 4 mm long; **corolla tube** 2–2.5 mm long; **ligules** glabrous, 4–10 mm long by 1–1.3 mm wide with two or three minute apical teeth; **disc florets** 13–52, perfect, yellow to purplish; **ovary** 0.7–1.3 mm long and lightly hairy; **pappus** of white to tawny bristles 3.5–6 mm long; **corolla tube** as above; **corolla limb** 2–2.5 mm long, the five apical teeth yellow to purple, 0.4–1.2 mm long, eventually spreading; **anthers** five, yellow, forming a tube 1.5–2 mm long; **style** 2-parted, flattened, tips hairy; **fruiting heads** 13–14 mm wide by 10–12 mm tall, round. **FRUIT an achene**, 1-seeded, 4-ribbed, *c.* 2 mm long by *c.* 0.6 mm wide by *c.* 0.3 mm thick, hairs scattered; **pappus** of white to tawny bristles 3.2–6.4 mm long.

■ **LEAVES** Alternate, simple, entire or with short teeth in the upper half; **stem blades** flat, 2.5–15 cm long by 0.5–3 cm wide, larger below, glabrous but scabrous along the margins, sessile, weakly clasping, thin and weak, lateral veins obscure; **auricles** absent or obscure; **lower blades** deciduous by anthesis.

■ **STEM** Erect, solid, round, branched above, with vertical lines of white hairs descending from the upper leaf bases at nodes, otherwise mostly glabrous below; 2.5–6 mm thick near the reddish green base.

■ **SYN.** *Aster hesperius* Gray = *Symphyotrichum lanceolatum* var. *hesperium* (Gray) Nesom, *Aster lanceolatus* Willd. = *Symphyotrichum lanceolatum* var. *lanceolatum* (Willd.) Nesom, *Aster paniculatus* Lam., p.p. non P. Mill. = *Symphyotrichum lanceolatum* var. *lanceolatum* (Willd.) Nesom and *Aster simplex* Willd. = *Symphyotrichum lanceolatum* var. *lanceolatum* (Willd.) Nesom.

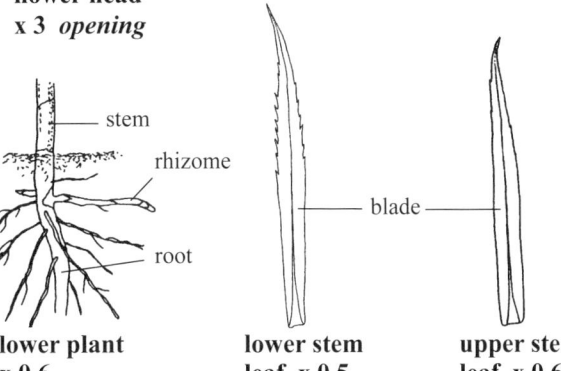

lower stem leaf x 0.5

upper stem leaf x 0.6

Eastern Willow Aster, White Panicle Aster, White Panicled American-aster, Panicled Aster

Aster—*Asteraceae* **Wildflower** White to pale purple Ray florets 9–15

WOOD ASTER   *Symphyotrichum lateriflorum* (L.) A. & D. Löve

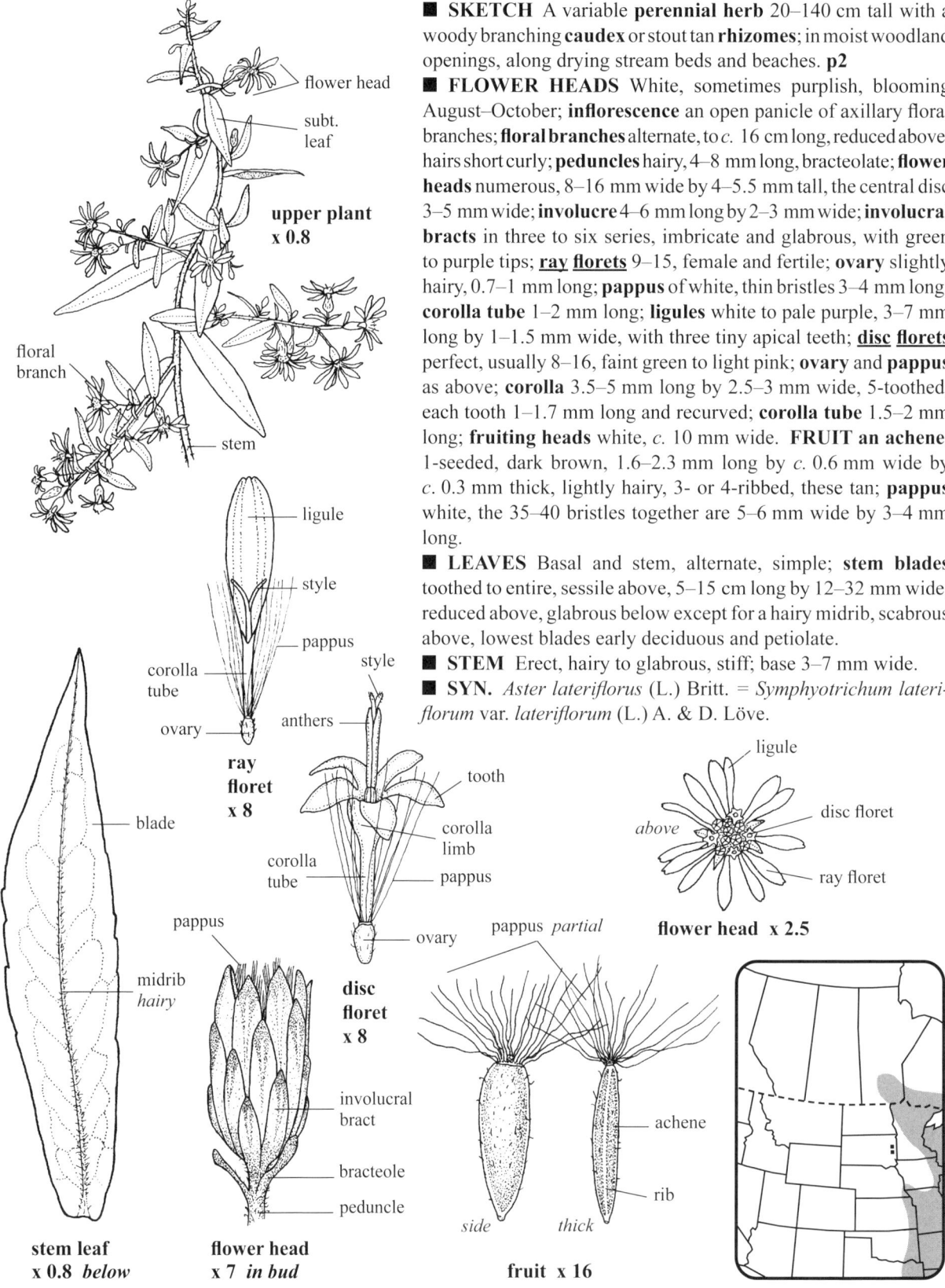

■ **SKETCH** A variable **perennial herb** 20–140 cm tall with a woody branching **caudex** or stout tan **rhizomes**; in moist woodland openings, along drying stream beds and beaches. **p2**

■ **FLOWER HEADS** White, sometimes purplish, blooming August–October; **inflorescence** an open panicle of axillary floral branches; **floral branches** alternate, to *c.* 16 cm long, reduced above, hairs short curly; **peduncles** hairy, 4–8 mm long, bracteolate; **flower heads** numerous, 8–16 mm wide by 4–5.5 mm tall, the central disc 3–5 mm wide; **involucre** 4–6 mm long by 2–3 mm wide; **involucral bracts** in three to six series, imbricate and glabrous, with green to purple tips; ray florets 9–15, female and fertile; **ovary** slightly hairy, 0.7–1 mm long; **pappus** of white, thin bristles 3–4 mm long; **corolla tube** 1–2 mm long; **ligules** white to pale purple, 3–7 mm long by 1–1.5 mm wide, with three tiny apical teeth; disc florets perfect, usually 8–16, faint green to light pink; **ovary** and **pappus** as above; **corolla** 3.5–5 mm long by 2.5–3 mm wide, 5-toothed, each tooth 1–1.7 mm long and recurved; **corolla tube** 1.5–2 mm long; **fruiting heads** white, *c.* 10 mm wide. **FRUIT** an achene, 1-seeded, dark brown, 1.6–2.3 mm long by *c.* 0.6 mm wide by *c.* 0.3 mm thick, lightly hairy, 3- or 4-ribbed, these tan; **pappus** white, the 35–40 bristles together are 5–6 mm wide by 3–4 mm long.

■ **LEAVES** Basal and stem, alternate, simple; **stem blades** toothed to entire, sessile above, 5–15 cm long by 12–32 mm wide, reduced above, glabrous below except for a hairy midrib, scabrous above, lowest blades early deciduous and petiolate.

■ **STEM** Erect, hairy to glabrous, stiff; base 3–7 mm wide.

■ **SYN.** *Aster lateriflorus* (L.) Britt. = *Symphyotrichum lateriflorum* var. *lateriflorum* (L.) A. & D. Löve.

Calico Aster, Farewell Summer, One-sided Aster, White Woodland Aster

Aster—*Asteraceae* **Wildflower** Reddish purple, center yellow Ray florets 40–100

New England Aster *Symphyotrichum novae-angliae* (L.) Nesom

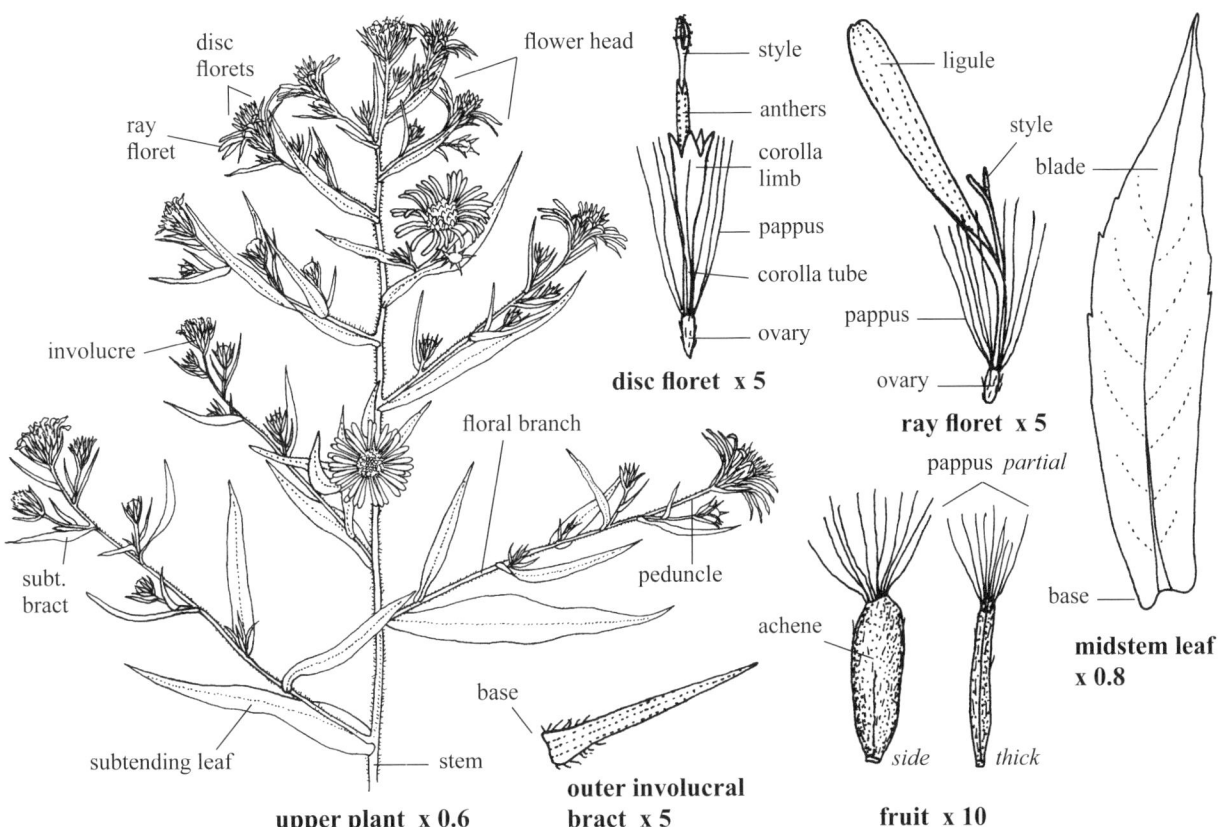

- **SKETCH** A variable **perennial herb** 50–150 cm tall, singly or in small clusters from a branching **caudex** or stout **rhizome**; in moist woodlands, thickets, along roadsides, streambanks and in low fields. p1
- **FLOWER HEADS** Reddish purple (rarely white), center yellow, blooming August–October; **inflorescence** an open panicle with a few to several heads alternating along the upper branches; **flower heads** 2–4 cm wide by 10–12 mm long; **involucre** glandular, 6–10 mm tall; **bracts** in two or three series, thin, 5–7 mm long by 0.6–2 mm wide, of two types: **1) outer bracts** medium green, slightly expanded at the base with hairy margins, one to three long nerves, some tips spreading and purplish; **2) inner bracts** whitish green and glabrous, the central nerve dark and the tips often pale green; **ray florets** female, 40–100; **ovary** *c.* 1 mm long with short ascending hairs; **pappus** bristles *c.* 4 mm long; **corolla tube** 2–3 mm long; **ligules** 6–20 mm long by 1–1.5 mm wide, overlapping, glabrous, with two tiny apical teeth; **style** 2-parted; **disc florets** 70–85, perfect, yellow, 8–10 mm long; **ovary** as above; **pappus** bristles *c.* 5 mm long; **corolla** 5–5.3 mm long; **corolla tube** 2–2.5 mm long, white; **corolla limb** 2.3–2.5 mm long with five yellow apical teeth not widely spreading and *c.* 0.7 mm long; **anthers** five, united into a tube; **style** yellow, 2-parted; **fruiting heads** 7–8 mm long and wide. **FRUIT an achene**, 1-seeded, 2.2–2.5 mm long by 0.7–0.9 mm wide by *c.* 0.5 mm thick with scattered, appressed hairs, dark reddish brown with obscure lateral ribs; **pappus** white to tan, of 55–60 fine bristles.
- **LEAVES** Alternate, simple, mostly entire to slightly toothed, flat, sessile to slightly clasping, scabrous, 3–12 cm long by 1–2 cm wide, reduced above, short hairs above especially near the edges, less or more hairy below; **lower stem blades** early deciduous.
- **STEM** Erect, solid, often clustered, scabrous, white hairs in vertical lines near leaf bases, less hairy below, short hairs glandular or not, longer hairs not glandular; 5–8 mm wide near the base.
- **SYN.** *Aster novae-angliae* L.

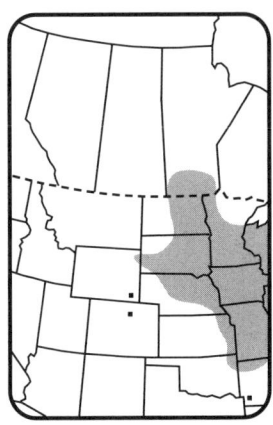

New England American-aster 191

Aster—*Asteraceae* **Wildflower** Pale purple to pink, center yellow Ray florets 30–60

Purple-stemmed Aster *Symphyotrichum puniceum* (L.) A. & D. Löve

■ **SKETCH** A **perennial herb** 40–250 cm tall with pinkish tan, smooth **rhizomes** 5–10+ cm long by 2–3 mm thick and tan **fibrous roots** 1–10 cm long by 0.5–1 mm wide, some with short **stolons**; in marshes, swampy shores, ditches, wet thickets and along stream banks.

■ **FLOWER HEADS** Light purple to blue to pink, blooming August–October; **inflorescence** paniculate, open, 10–15 cm long by 10–20 cm wide; **floral branches** leafy, 4–20 cm long, ascending, hairs spreading; **peduncles** 5–30 mm long, hairy; **flower heads** 30–50+, each 15–40 mm wide by 10–13 mm tall; **disc** 4–10 mm wide and yellow; **involucre** pale green, 5–12 mm long; **involucral bracts** in 2 or 3 series, spreading, tips dark green; **outer ones** 5–15 mm long by 0.8–3 mm wide, margins hairy below or throughout; **inner** (upper) **bracts** 5–8 mm long by 0.9–1 mm wide, hyaline and hairy in upper third; <u>**ray florets**</u> female, 30–60+, spreading to slightly descending, TL 13–21 mm; **ovary** 1–2 mm long, ribbed, slightly hairy near apex; **pappus** 4–4.5 mm long; **corolla tube** 2–3.5 mm long, pale green, hairless; **ligule** 8–16 mm long by 1.2–2 mm wide with 2 or 3 tiny apical teeth; **style** exserted, yellow, 2-parted, *c*. 1.3 mm long; <u>**disc florets**</u> perfect, TL 8–10 mm, 40–62 per head, yellow turning reddish brown; **ovary** as above; **pappus** of whitish bristles 5–6 mm long; **corolla tube** pale green, 2–2.8 mm long; **corolla limb** pale yellow, 3–3.5 mm long (including teeth), teeth five, yellow, *c*. 1 mm long; **anthers** yellow, *c*. 2 mm long, included to exserted. **FRUIT** an achene, 1-seeded, of disc florets 1.5–2.5 mm long by *c*. 0.8 mm wide by *c*. 0.3 mm thick, reddish purple ripening to dark brown, 4- or 5-ribbed, glabrous to slightly hairy, white hairs ascending; **pappus** of white bristles 5–7 mm long.

■ **LEAVES** Alternate, simple, sessile, entire to toothed, thin, dull, auricles clasping; **blades** spreading, 2–22 cm long by 0.6–4 cm wide, glabrous or with appressed hairs, midrib raised and very hairy below, the hairs hyaline, jointed and spreading, margins short ciliate; **lower blades** scabrous with short bristles, often wilted.

■ **STEM** Erect, reddish purple, branching above, solid, stiff, white hairs spreading on floral branches and upper stem, fewer hairs below; 4–10 mm thick near the bare base.

■ **SYN.** *Aster puniceus* L. = *Symphyotrichum puniceum* var. *puniceum* (L.) A. & D. Löve.

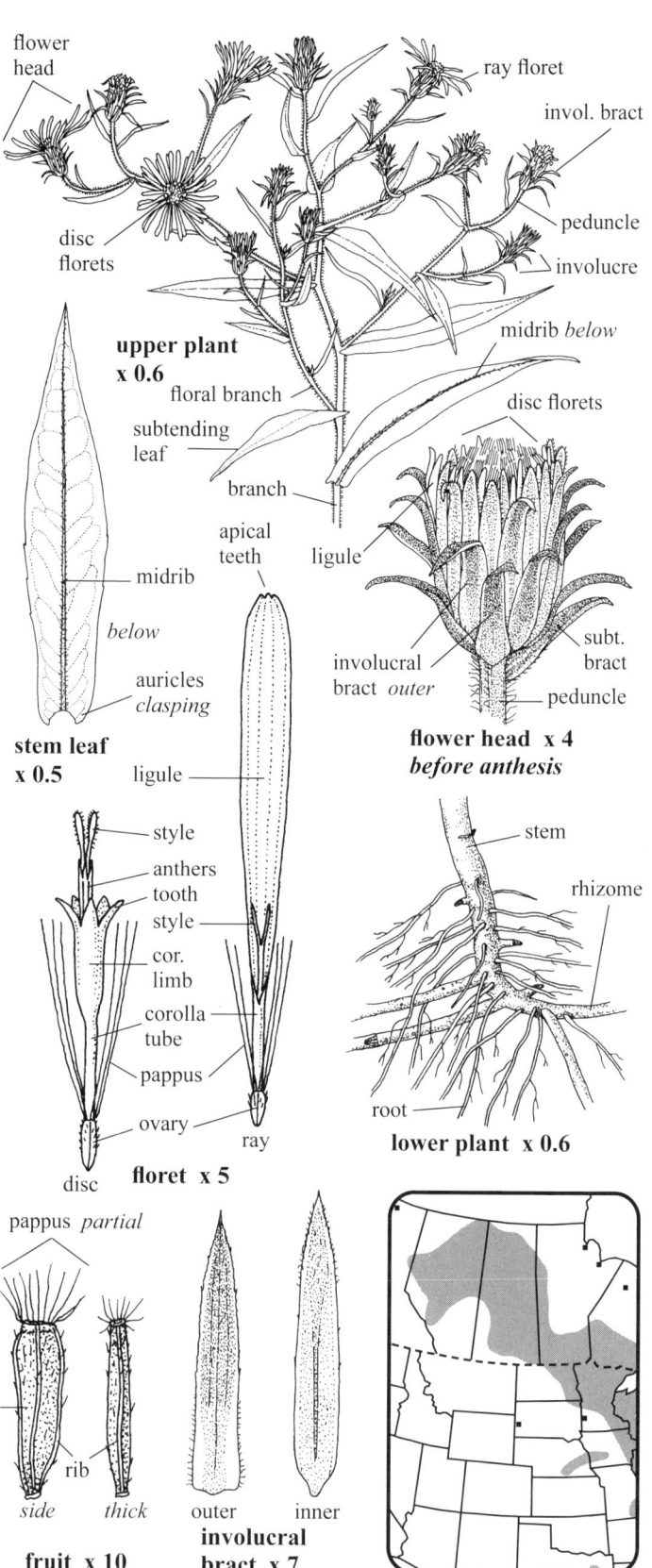

Bog Aster, Swamp Aster, Bristly Aster

Aster—*Asteraceae* **Wildflower** Yellow Disc florets only

Tansy *Tanacetum vulgare* L.

■ **SKETCH** A **perennial herb** 30–150+ cm tall from woody branching **rhizomes** 2–4 mm thick, with one to several stems (old and new) in scattered clumps; in dry sandy fields and along roadsides. *Naturalized*

■ **FLOWER HEADS** Yellow, blooming in July–October; **inflorescence** a compound corymb, flat-topped, 4–12+ cm wide by 3–15 cm tall; **floral branches** alternate, from below the middle, axillary, ascending, shorter to longer than terminal flower heads, 2–30+ cm long, leafless and bractless; **peduncles** green, glabrous, erect to ascending, *c.* 7-ridged, 5–40 mm long by *c.* 1 mm thick, usually three at the apex of a floral branch; **bracteoles** (of peduncles) usually one, 2–3 mm long; **flower heads** fragrant, 8–60 (–200) per stem, each 8–12 mm wide by 5–7.5 mm tall; **involucre** 3–4 mm tall by 6–10 mm wide; **involucral bracts** pale yellowish green, imbricate in three or four series, stiff, 2.5–4.5 mm long by 1–2 mm wide, apices and upper margins hyaline, variable, brownish, ragged, midnerve obvious, central area pale green; **ray florets** absent; **disc florets** numerous, yellow, of two types: **1) outer disc florets** female, 3–4 mm long, usually only one row around the disc margin; **ovary** pale green, glabrous, 1.2–1.4 mm long, 3- or 4-ribbed; **pappus** of several scales, membranous, 0.2–0.4 mm long; **corolla** flattened, *c.* 2.3 mm long by 0.5–0.7 mm wide, glandular; **corolla teeth** three, *c.* 0.5 mm long, usually erect, blunt, yellow; **style** 2-parted, included to exserted, the parts *c.* 0.4 mm long, flattened; **2) inner disc florets** perfect, numerous, crowded, forming most of the disc; TL 3.8–4.5 mm; **ovary** green, usually 5-ribbed, 1–1.4 mm long, sometimes glandular; **pappus** scales as above; **corolla tube** pale green, round, glandular, 1–1.2 mm long; **corolla limb** *c.* 1.5 mm long including teeth, glandular; **teeth** five, yellow, erect, blunt, *c.* 0.3 mm long, thick; **stamens** five; **filaments** *c.* 1 mm long, free above, united in lower half to corolla tube; **anthers** pale yellow, forming a tube *c.* 1 mm long; **style** 2-parted, exserted, recurved, flat, yellow. **FRUIT** an achene, 1-seeded, tan, hairless, 5-ribbed, *c.* 1.4 mm long (including pappus) by *c.* 0.5 mm wide by *c.* 0.4 mm thick, glands between or on some ribs; **pappus** a crown of scales *c.* 0.2 mm long; **receptacle** naked.

■ **LEAVES** Alternate, compound (twice), ascending, sessile, yellowish green, dull; **blades** 5–25 cm long by 2–12 cm wide (flattened), reduced above, lower blades wilted by anthesis, glandular, midrib raised below, fragrant when crushed; **leaflets** toothed, sessile, 1–7 cm long by 4–20 mm wide.

■ **STEM** Solid, erect, reddish brown below, green near the flowers, low narrow ridges longitudinally from the midrib of leaves and leaf edges along the stem; 3–6 mm thick near the base.

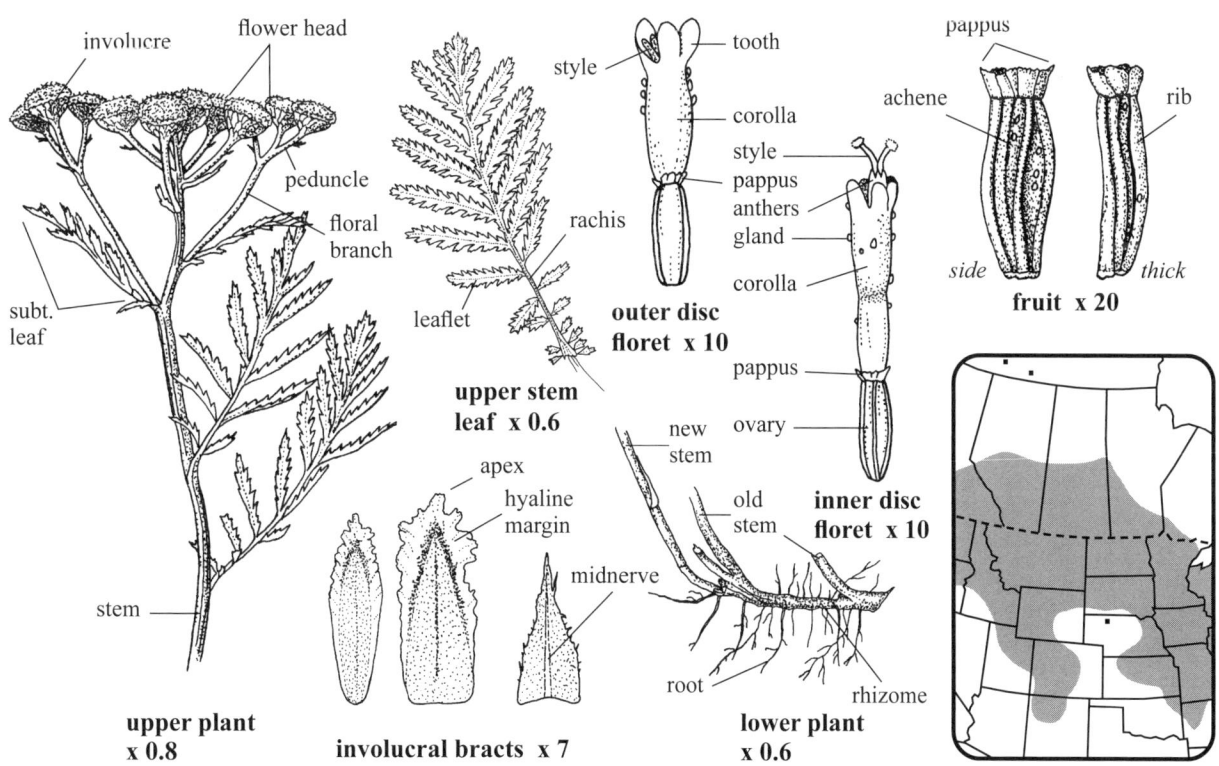

Common Tansy

Aster—*Asteraceae* **Wildflower** Yellow Ligulate florets many

RED-SEEDED DANDELION *Taraxacum laevigatum* (Willd.) DC.

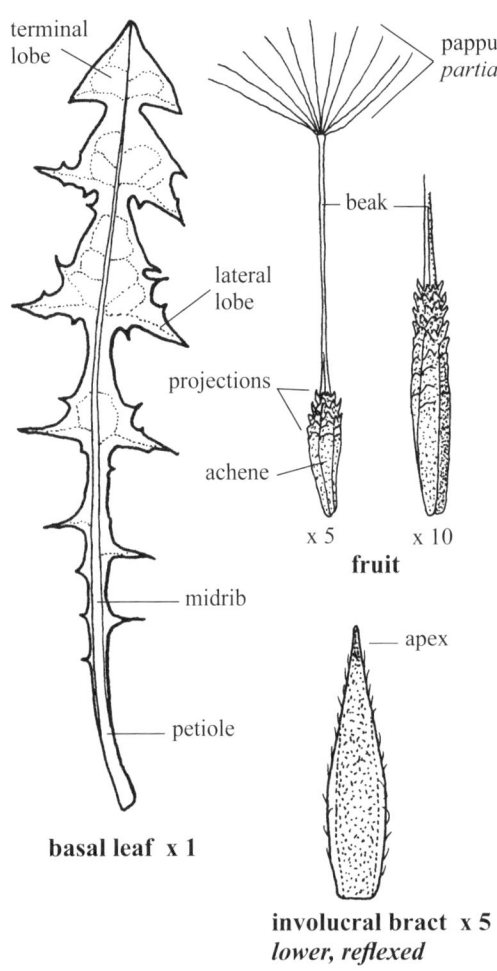

basal leaf x 1

■ **SKETCH** A **perennial herb** 10–40 cm tall from a **taproot** 4–8 mm wide and to *c.* 10⁺ cm long with a fleshy **crown**; in lawns, along sidewalks and roadsides, meadows, pastures and in disturbed areas; **juice milky**. *Naturalized* **w1**

■ **FLOWER HEADS** Yellow, blooming March–September; **inflorescence** a single flower head atop a scape; **flower heads** 20–35 mm wide by 13–15 mm tall; **involucre** glabrous, with two types of bracts: **1) lower involucral bracts** reflexed, 14–16, imbricate, 6–8 mm long by 1.5–3 mm wide, not thick, some with ciliate margins; **2) upper involucral bracts** erect, 11–15, the lower margins slightly imbricate, 10–17 mm long by 1.3–2.5 mm wide, tips usually brownish, horned, the edges scarious; **disc florets** absent; **ligulate florets** perfect, numerous, outer ones the largest, TL *c.* 11 mm; **ovary** glabrous, *c.* 0.6 mm long; **pappus** of numerous white bristles 4–4.5 mm long; **corolla tube** 2.5–3.5 mm long, white, glabrous; **ligules** 5–8 mm long by 1.5–2 mm wide, yellow, with five apical teeth and a few hairs below where it narrows, outer ligules with a wide brown stripe below; **stamens** five; **anthers** yellow to slightly reddish, forming a tube 1–1.5 mm long; **style** greenish yellow, 2-parted, the curled parts *c.* 1.5 mm long; **fruiting heads** whitish, round and *c.* 3 cm wide. FRUIT an achene, 1-seeded, reddish brown to red, the body *c.* 3 mm long by *c.* 0.8 mm wide by *c.* 0.5 mm thick, the top roughened with projections, two grooves along each wide side; **beak** tan, 2–7 mm long by *c.* 0.1 mm thick; **pappus** of whitish bristles *c.* 5 mm long.

■ **LEAVES** Basal, simple, lobed and toothed, a rosette of a dozen or more leaves; **blades** petiolate, soft, not thick, usually glabrous, 2.5–20 cm long by 10–50 mm wide, the lobes small and cut almost to the midrib; **terminal lobe** usually no larger than the upper lateral lobes, midrib raised below, side veins faint; **petioles** 1–7 cm long, winged.

■ **STEM** A **scape** 10–40 cm tall, usually glabrous, hollow, one to several; reddish near the 2–3 mm thick base.

■ **SYN.** *Taraxacum erythrospermum* Andrz. ex Bess.

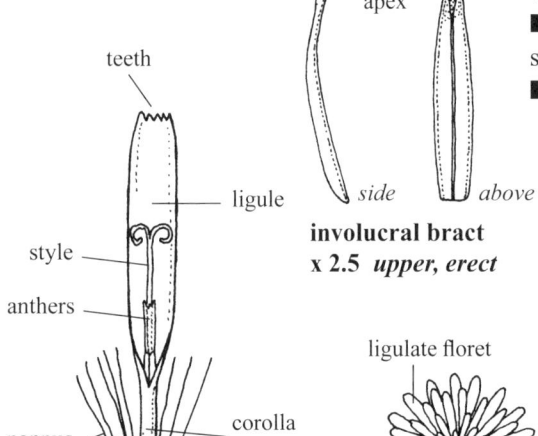

ligulate floret x 5

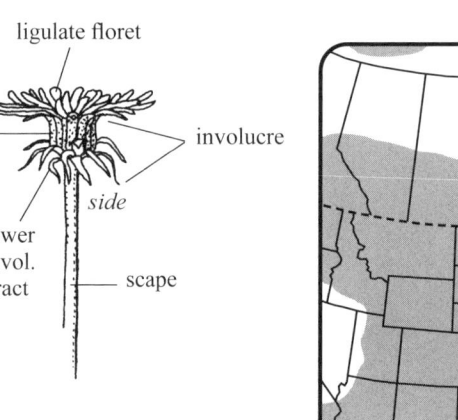

Aster—*Asteraceae* **Wildflower** Yellow Ligulate florets 100–250

DANDELION *Taraxacum officinale* G.H. Weber ex Wiggers

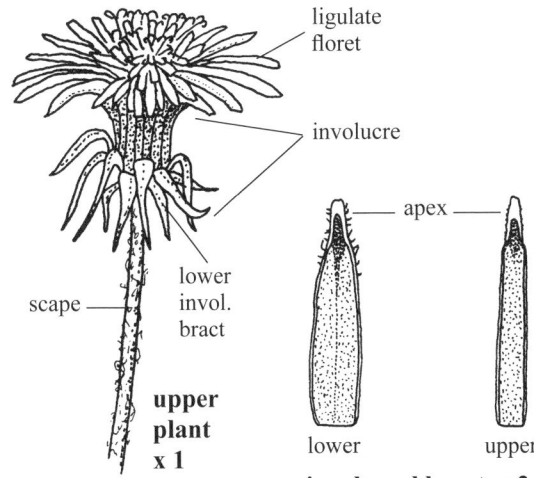

upper plant x 1

involucral bract x 2

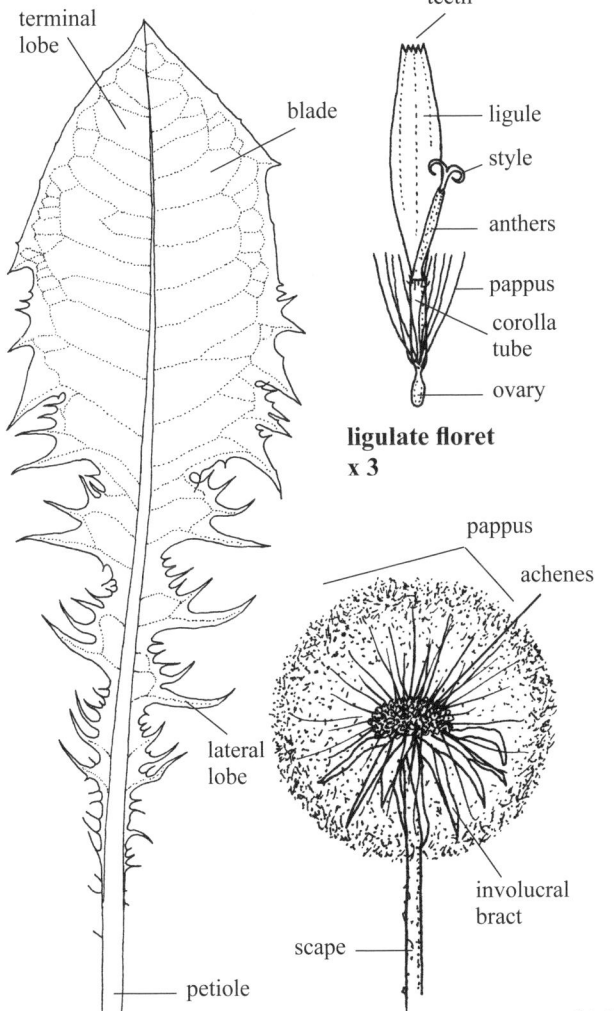

ligulate floret x 3

leaf x 0.7

fruiting head x 1

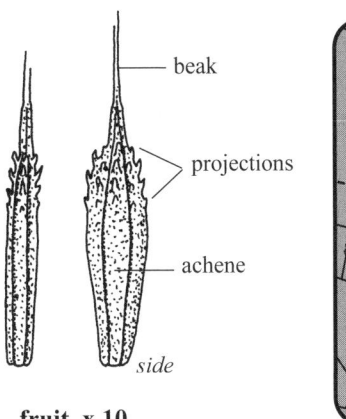

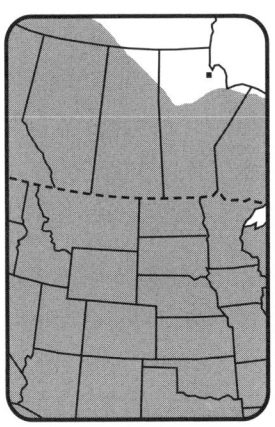

fruit x 10

■ **SKETCH** A **perennial herb** 5–81 cm tall from a deep **taproot** 3–30 mm thick and 10–20 cm long; in lawns, gardens, rangelands, mountain meadows, roadsides and fields; **juice milky**. *Naturalized* **w1**

■ **FLOWER HEADS** Yellow, blooming March–December; **inflorescence** of one to ten flower heads, each 3–5.5 cm wide by 2–4 cm long, one head atop each scape; **involucre** 1.5–2.5 cm tall, glabrous; 1) **lower involucral bracts** reflexed, slightly hairy at blunt pale yellow apices, body light green, to *c.* 15 mm long by 2–4 mm wide; 2) **upper involucral bracts** 15–21, dark green, erect around the flower head when in bud, usually 12–18 mm long by *c.* 2 mm wide, often united at the base, eventually reflexed in fruit, apices winglike and pale yellow; <u>disc florets</u> absent; <u>ligulate florets</u> perfect, 100–250; **ovary** glabrous, *c.* 1.2 mm long; **pappus** of numerous white bristles 4–7 mm long; **corolla tube** white, 3–4 mm long, lightly hairy where it widens; **ligules** 10–15 mm long by 1.5–2.7 mm wide, yellow, with five tiny apical teeth, the outer ligules the largest and with a reddish tan stripe below; **anthers** five, forming a tube 4–5 mm long; **style** 2-parted, slightly hairy; **fruiting heads** round, whitish, 3–5 cm wide, deciduous in the wind. **FRUIT** an achene, 1-seeded, brown, 2.7–4 mm long by 0.8–1 mm wide by 0.4–0.9 mm thick, with short, curved, apical projections, grooved, squarish near the base; **beak** 5.5–12 mm long; **pappus** white, *c.* 10 mm wide for each achene.

■ **LEAVES** Basal, simple, lobed and toothed below, a rosette of 4–25 leaves; **blades** hairy to glabrous, spreading to ascending, 5–40 cm long by 1–15 cm wide, terminal lobe the largest; **petioles** 2–6 cm long and to *c.* 8 mm wide and to *c.* 5 mm thick, some reddish near the base, glabrous or with matted patches of white hairs.

■ **STEM** A **scape**, erect, hollow, light green to reddish, glabrous or with scattered matted hairs especially above; 2–6 mm thick near the base.

Common Dandelion

Aster—*Asteraceae* **Wildflower** Yellow Ligulate florets many

YELLOW GOAT'S-BEARD *Tragopogon dubius* Scop.

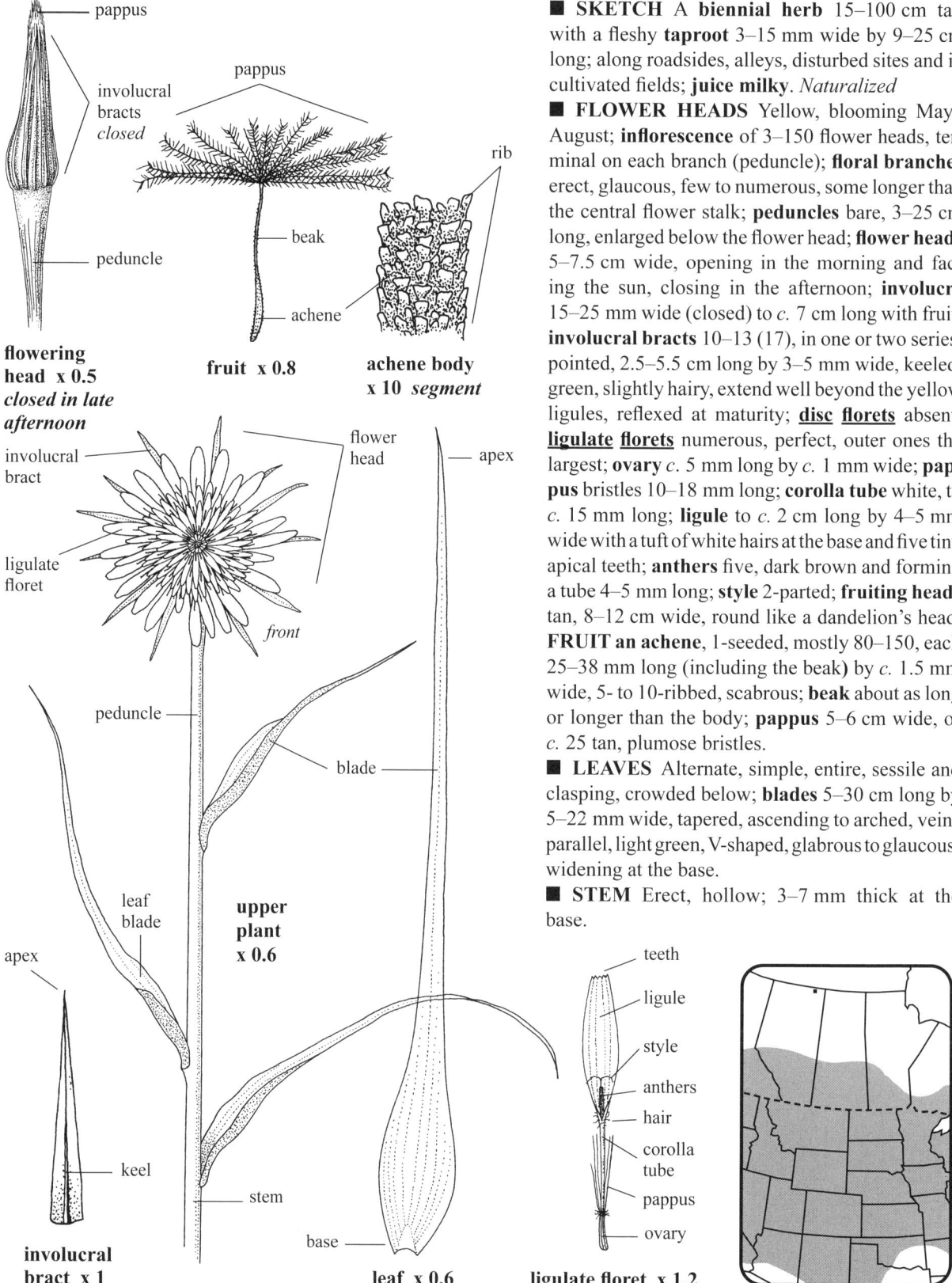

- **SKETCH** A **biennial herb** 15–100 cm tall with a fleshy **taproot** 3–15 mm wide by 9–25 cm long; along roadsides, alleys, disturbed sites and in cultivated fields; **juice milky**. *Naturalized*
- **FLOWER HEADS** Yellow, blooming May–August; **inflorescence** of 3–150 flower heads, terminal on each branch (peduncle); **floral branches** erect, glaucous, few to numerous, some longer than the central flower stalk; **peduncles** bare, 3–25 cm long, enlarged below the flower head; **flower heads** 5–7.5 cm wide, opening in the morning and facing the sun, closing in the afternoon; **involucre** 15–25 mm wide (closed) to *c.* 7 cm long with fruit; **involucral bracts** 10–13 (17), in one or two series, pointed, 2.5–5.5 cm long by 3–5 mm wide, keeled, green, slightly hairy, extend well beyond the yellow ligules, reflexed at maturity; <u>disc</u> **florets** absent; <u>ligulate</u> **florets** numerous, perfect, outer ones the largest; **ovary** *c.* 5 mm long by *c.* 1 mm wide; **pappus** bristles 10–18 mm long; **corolla tube** white, to *c.* 15 mm long; **ligule** to *c.* 2 cm long by 4–5 mm wide with a tuft of white hairs at the base and five tiny apical teeth; **anthers** five, dark brown and forming a tube 4–5 mm long; **style** 2-parted; **fruiting heads** tan, 8–12 cm wide, round like a dandelion's head. FRUIT an **achene**, 1-seeded, mostly 80–150, each 25–38 mm long (including the beak) by *c.* 1.5 mm wide, 5- to 10-ribbed, scabrous; **beak** about as long or longer than the body; **pappus** 5–6 cm wide, of *c.* 25 tan, plumose bristles.
- **LEAVES** Alternate, simple, entire, sessile and clasping, crowded below; **blades** 5–30 cm long by 5–22 mm wide, tapered, ascending to arched, veins parallel, light green, V-shaped, glabrous to glaucous, widening at the base.
- **STEM** Erect, hollow; 3–7 mm thick at the base.

Goat's Beard, Western Salsify, Common Goat's-beard

Aster—*Asteraceae* **Wildflower** White, center yellow Ray florets 12–35

SCENTLESS CHAMOMILE *Tripleurospermum perforata* (Mérat) M. Lainz

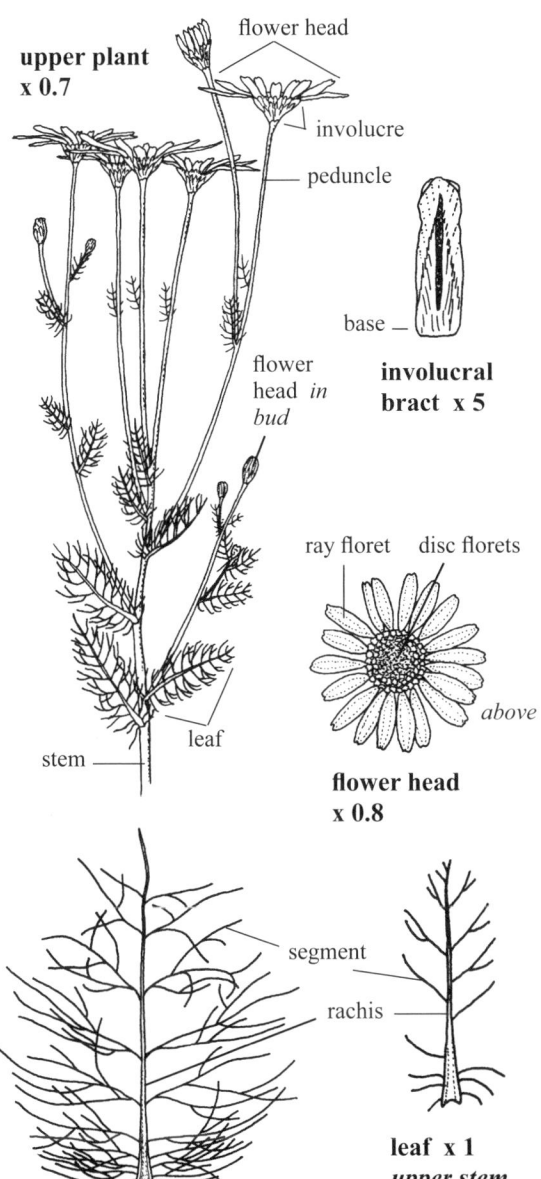

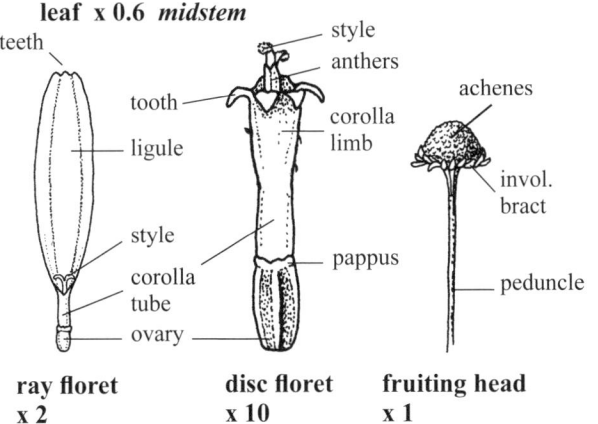

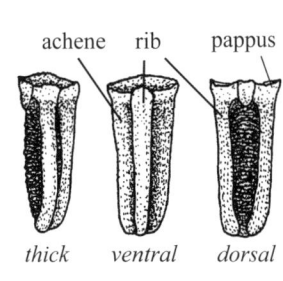

■ **SKETCH** A glabrous **annual** to **perennial herb** 20–100 cm tall from **fibrous roots** or a **taproot** 2–7 mm wide and 5–10+ cm long, often in dense stands; along stream, pond or slough banks, roadsides, fence lines, pastures, in crops and dry disturbed areas in cities and towns. *Naturalized*

■ **FLOWER HEADS** White with a yellow center, blooming June–September; **inflorescence** a corymb; **peduncles** erect, glabrous, 0.7–13 cm long; **flower heads** 5–60+, each 2–4.5 cm wide by *c*. 7 mm long; **disc** yellow, flat to raised, 7–15 mm wide; **involucre** glabrous, 7–10 mm wide by *c*. 5 mm long; **involucral bracts** in three series, 3–4 mm long by 1–1.3 mm wide, the margins and blunt tip hyaline, the central ridge low and dark green with a whitish green border, the base flat; **ray florets** 12–35, female and usually fertile; **ovary** *c*. 1.4 mm long; **pappus** reduced to a short crown of a few scales *c*. 0.3 mm long; **corolla tube** *c*. 2 mm long and flattened; **ligules** white, 6–14 mm long by 2.5–3.5 mm wide, with three apical teeth; **disc florets** 200+, perfect; **ovary** with lateral ribs and a central rib on the ventral (inner) face, the dorsal face is smooth; **corolla** 2–2.5 mm long with the tube *c*. 1.2 mm long, the limb *c*. 1.3 mm long, sometimes lightly hairy, with five apical teeth; **anthers** five, yellow, forming a tube; **style** 2-parted, flattened; **fruiting heads** are *c*. 10 mm wide by *c*. 7 mm tall including the tan papery involucral bracts. **FRUIT an achene**, 1-seeded, dark brown, 1.3–2.6 mm long by 0.8–1.2 mm wide by *c*. 1 mm thick, with one ventral and two lateral corky ribs, the dorsal surface glandular, rough and dark; **pappus** a scaly crown 0.2–0.3 mm long.

■ **LEAVES** Alternate, compound, finely divided into numerous, thin segments less than 1 mm wide; **blades** 1.2–11 cm long by 0.6–6.5 cm wide, soft, glabrous, smaller above, sessile; **rachis** spreading toward its base.

■ **STEM** Erect, dark green, one to several, glabrous, simple to branched, leafy; 2–7 mm thick near the reddish ridged base.

■ **SYN.** *Matricaria maritima* var. *agrestis* (Knaf) Wilmott, *Matricaria maritima* ssp. *inodora* (L.) Clapham, and *Matricaria perforata* Mérat.

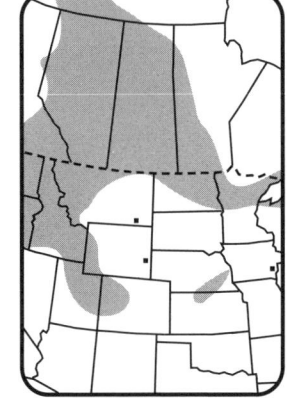

Scentless False Mayweed, Wild Chamomile   197

Aster—*Asteraceae* **Wildflower Green Disc florets only**

COCKLEBUR *Xanthium strumarium* L.

■ **SKETCH** An **annual herb** growing 20–200 cm tall from a **taproot** 2–10 mm wide by 10–15+ cm long; in moist to drying areas along river shores, slough margins, crop fields, water holes and saline sites; **monoecious**.

■ **FLOWER HEADS** Green, blooming June–October; **inflorescence** of axillary clusters in upper leaves; **flower heads** unisexual, sessile or nearly so; **outer subtending bracts** 3–10 mm long and hairy along their entire length; **floral branches** to *c.* 11 cm long from leaf axils, these branches terminating in a pair of leaves with flower heads in their axils; **subtending bracts** (of male florets) *c.* 3.5 mm long by *c.* 1 mm wide, hairy over the upper outer surface; <u>male</u> **heads** a round cone 5–6 mm wide and long; **florets** numerous; **corolla** *c.* 2.5 mm long by 1.2–1.5 mm wide, tubular with five apical teeth; **filaments** *c.* 2.3 mm long, united, white; **anthers** five, distinct, yellow, *c.* 1 mm long; <u>female</u> **heads** of two florets in an involucre; **corolla** absent; **involucre** 4–5 mm wide by 7–10 mm long, closed, covered with hooked prickles and two stout, inwardly curved apical beaks (styles) *c.* 5 mm long, persistent; **bur** (mature) brownish red, variable, with two incurved beaks, 2-chambered, hard, clustered, 2–40+ per plant, each 12–35 mm long by 12–25 mm wide, sessile or nearly so, covered with hairy hooked prickles, these 4–6 mm long by *c.* 1 mm wide (not including hairs). **FRUIT an achene**, 1-seeded, two per bur, one per chamber, beaked, striated, slate gray, glabrous, thin, 12–15 mm long by *c.* 5 mm wide by 1.5–2 mm thick; **pappus** absent.

■ **LEAVES** Alternate, simple, coarsely toothed, with wide shallow lobes, undulate; **blades** 5–20 cm long and wide, scabrous; **petioles** reddish, scabrous, 1.5–11 cm long by 3–4 mm thick, grooved above; **stipules** absent.

■ **STEM** Round, scabrous, branched, erect, solid but weak, with vertical reddish lines, changing to a reddish purple by autumn; 3–12 mm wide near the base.

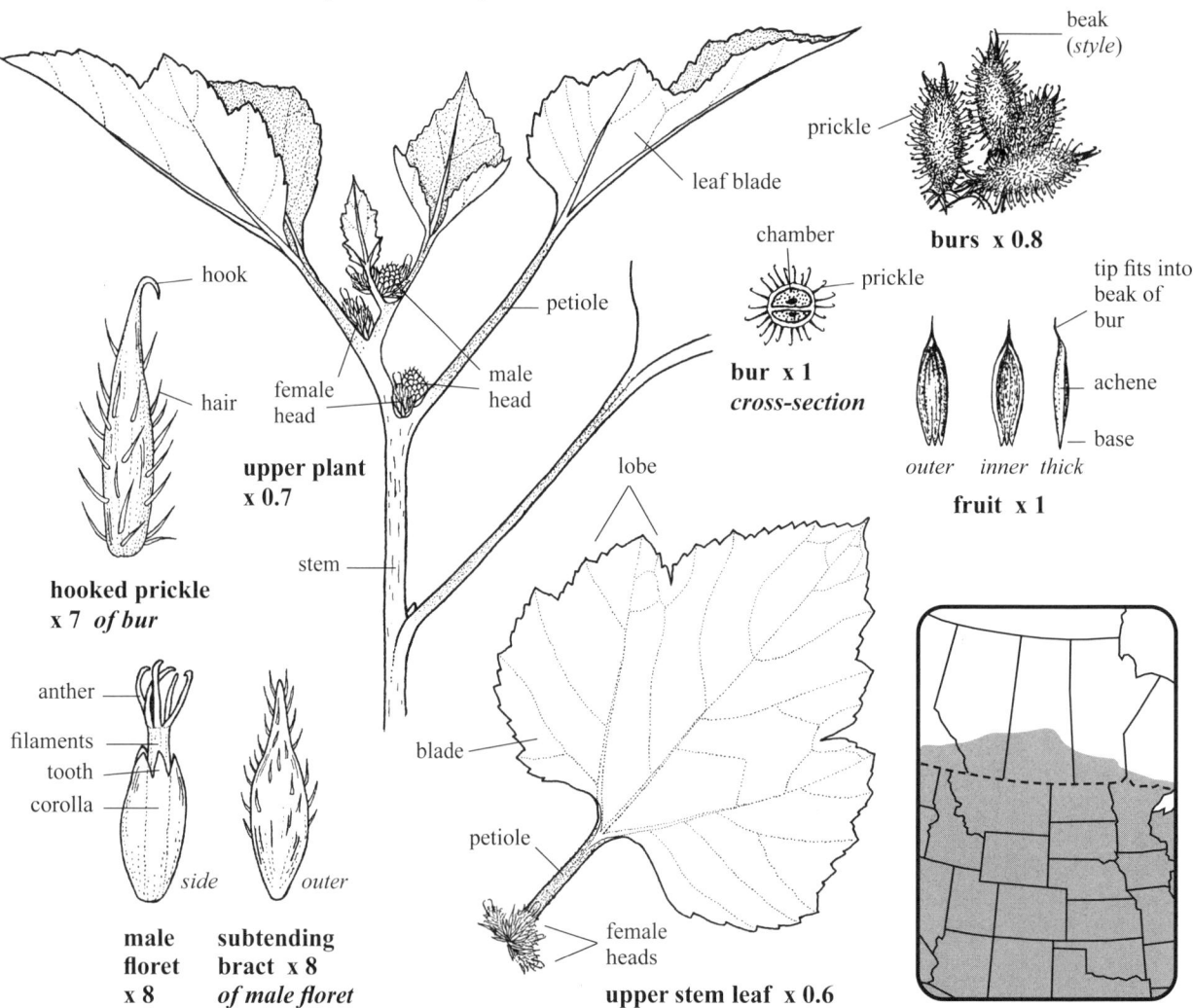

Touch-me-not—*Balsaminaceae* **Wildflower** Yellow and orange, spotted with red
Corolla tubular, 3-lobed

SPOTTED TOUCH-ME-NOT   *Impatiens capensis* Meerb.

- **SKETCH** A glabrous and glaucous **annual herb** 30–170 cm tall from a thick, short horizontal **rhizome** with **fibrous roots** to *c.* 5 mm thick, and some **adventitious roots** from the lowest node of the stem; in low moist woods, slough margins, stream banks, thickets and other wet places. **w2**
- **FLOWERS** Yellow and orange, spotted or marked with red, blooming May–October; **inflorescence** of numerous racemes of 1–5 flowers from leaf axils; **branches** ascending, lower pairs may be opposite, naked below and to *c.* 1 m long, upper branches alternate; juicy when cut; **peduncles** axillary, spreading, 1–2 cm long, branching 2–4 times; **pedicels** hanging, 1–2.5 cm long; **flowers** perfect, showy, 10–15 mm wide by 14–22 mm long and tall, the cleistogamous ones produce fruit; **sepals** three, two side ones asymmetrical, 5–6 mm long by *c.* 5 mm wide, entire, some spotted with red, midvein obvious, lateral veins faint, apices with a short point; **lower middle sepal** with a yellow to orange **spur** 6–9 mm long, slightly expanded at end, curved forward under flower; **corolla** tubular, 3-lobed, 9–12 mm long by 5–7 mm wide, the red spots inside visible through thin walls; **lower lobes** two, apices curved downward, *c.* 8 mm long by *c.* 6 mm wide, reddish from many veins; **upper lobe** one, slightly notched, 6–7 mm wide, hooded, keeled with a white low ridge above, numerous orangish red spots inside; **stamens** five, white, united around the stigma; **filaments** curved, 1–1.5 mm long, white; **anthers** pale yellow, *c.* 2.5 mm long; **pistil** one, *c.* 1.6 mm long by *c.* 0.7 mm wide and thick, cylindrical; **style** obscure; **stigma** flat, circular. **FRUIT a capsule** (pod), glabrous, 12–20 mm long by 3.5–4.5 mm wide, pointed, round in cross-section, 5-veined, green to brownish, hanging on pedicels, the valves curling rapidly from pedicel end when gently squeezed expelling the ripe seeds, capsule falls soon after opening; **valves** five, *c.* 1.5 mm wide; **seeds** 4–8 per capsule, 3.8–5.5 mm long by 2.2–3.5 mm wide by 1.4–2 mm thick with four low ridges tapered to a squarish end, green to dark grey with pale spots.
- **LEAVES** Alternate, simple, toothed, dull, lighter green below, glabrous; **blades** 2–9 cm long by 1.2–5 cm wide, spreading; **petioles** 3–30 mm long, subtending a branch or peduncle; **stipules** absent.
- **STEM** Erect, much branched throughout, stiff, pinkish green to pale green, smooth, almost translucent, hollow, ridges blunt, pale green near the 1–2.5 cm thick base.
- **SYN.** *Impatiens biflora* Walt.

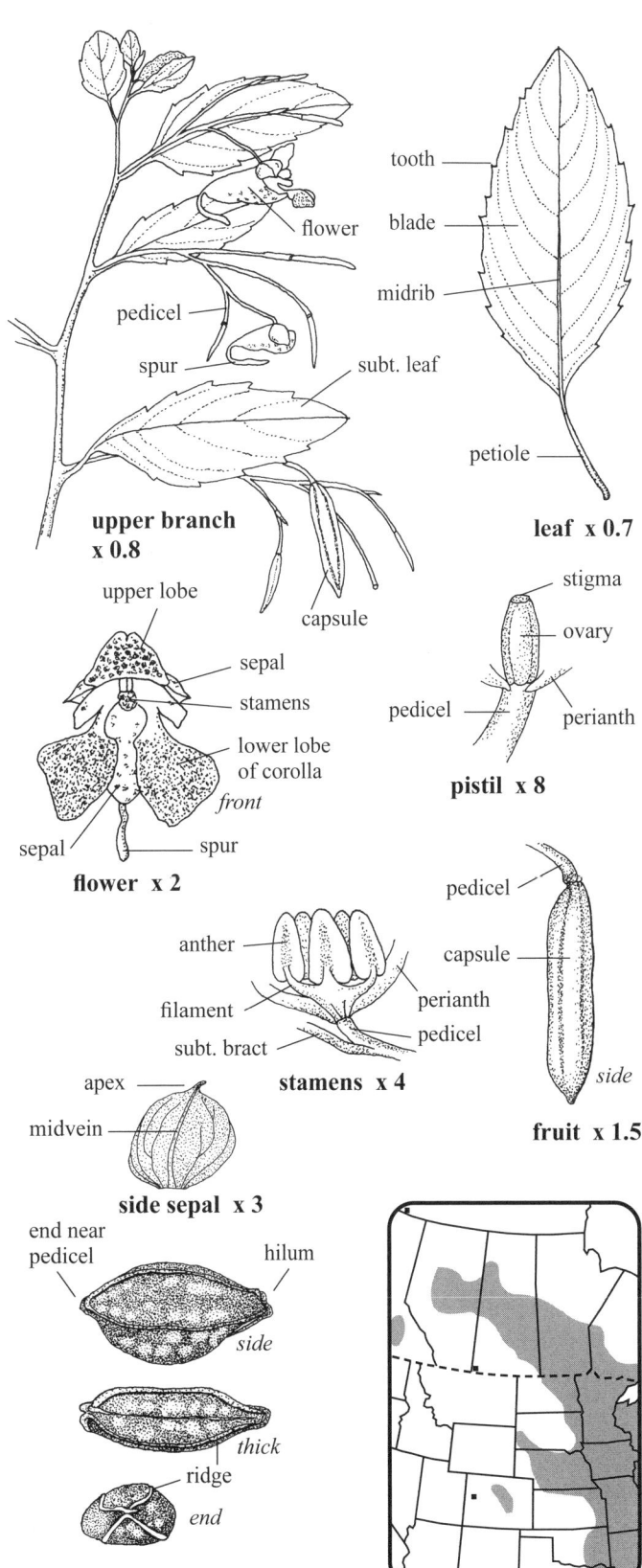

Jewelweed, Spotted Snapweed, Snapweed, Orange Touch-me-not

The Shrub

This might be the twilight shrub
the field speaks about

Just some tawny seed
that settled there
for better or for worse
the thing we tie our ribbons to
in springtime
O here we go round the Nanny-berry Bush
all the way into May

Wild shrub that was there
that bloomed in the night
when high winds rattled the sky

Birch—*Betulaceae* **Shrub** Reddish brown Catkins

SPECKLED ALDER *Alnus incana* (Linnaeus) Moench.

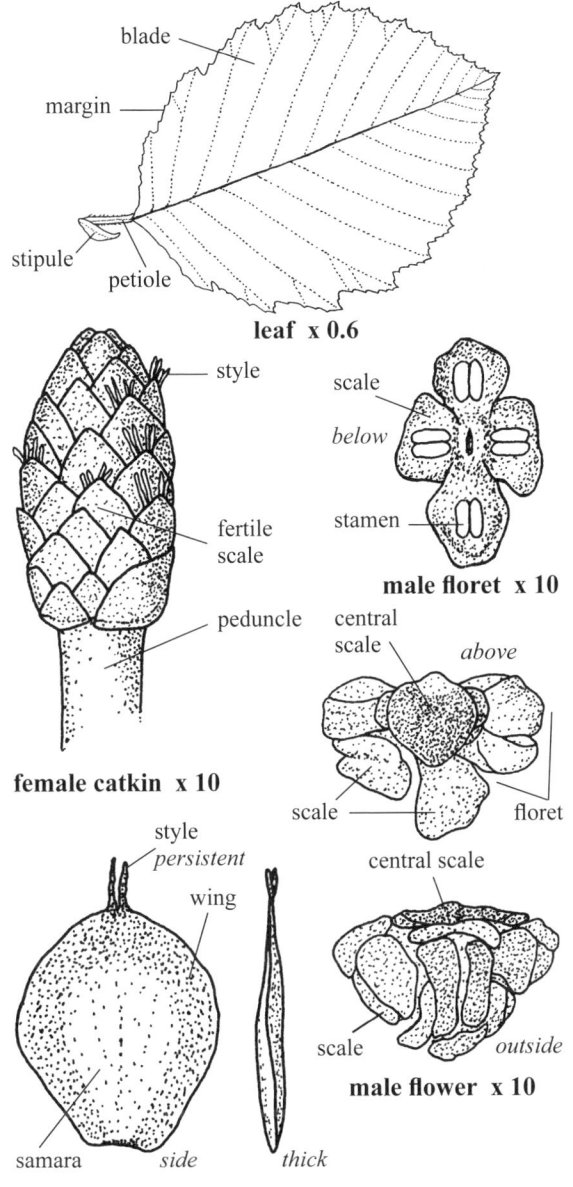

■ **SKETCH** A **deciduous shrub** 5–25 m tall in clumps with nodular roots; along ponds, rivers, in bog borders, muskeg, parklands and mountainous forest edges; **monoecious**. w2

■ **REPRODUCTIVE PARTS** Reddish brown, active in April–June before the leaves expand; **inflorescence** of catkins, unisexual; **peduncles** 0.5–18 mm long, reddish brown; **male catkins** formed by the previous autumn, two to five in a group, overwintering, 17–26 mm long by *c.* 5 mm wide; at anthesis they expand to 4–10 cm long by 5–7 mm wide, hanging; **rachis** tan, hairless, *c.* 1 mm thick; **central scale** with three florets attached, the whole structure 2.8–3 mm wide by 1.5–2.2 mm high as seen from the outside; **male florets** 1.5–2 mm long by 1–1.2 mm wide; **scales** 4-parted, brown, hairless, 1–1.4 mm long by 1–1.2 mm wide; **stamens** mostly four; **anthers** sessile, *c.* 0.6 mm long, light yellow; **peduncles** 0.5–7 mm long; **female catkins** in groups of two to six, dark reddish brown, 3.5–5 mm long by *c.* 2 mm wide; **fertile scales** numerous, imbricate, hairless, hardening with inner bracteoles to form a woody, conelike, dark brown structure 10–20 mm long by 8–13 mm wide, oval, persistent; **female florets** paired with two pairs of red styles *c.* 1 mm long exserted from each scale. **FRUIT** a samara, two per scale, 1-seeded, winged, 2.5–3.5 mm long and wide by *c.* 0.5 mm thick, golden, persistent styles 0.4–1 mm long.

■ **LEAVES** Alternate, simple, doubly toothed, not sticky, secondary teeth rounded to pointed; **blades** 3–14 cm long by 2–11 cm wide, apices pointed to blunt, glabrous above, hairy and glandular to glabrous below, thick to thin; **petioles** 10–28 mm long, hairy, glandular; **stipules** 6–12 mm long by 2–4 mm wide, entire, pointed, margins ciliate near the tip.

■ **TRUNKS** Several in a clump; 6–25 cm wide near the base; **bark** smooth to rough, and gray to reddish; **branches** alternate; **lenticels** tan, round, 0.5–1 mm wide; **winter buds** alternate, short stalked, reddish brown, 3–8 mm long, 2- or 3-scaled, slightly hairy.

■ **SYN.** *Alnus rugosa* (Du Roi) Spreng. = *Alnus incana* ssp. *rugosa* (Du Roi) Clausen and *Alnus tenuifolia* Nutt. = *Alnus incana* ssp. *tenuifolia* (Nutt.) Breitung.

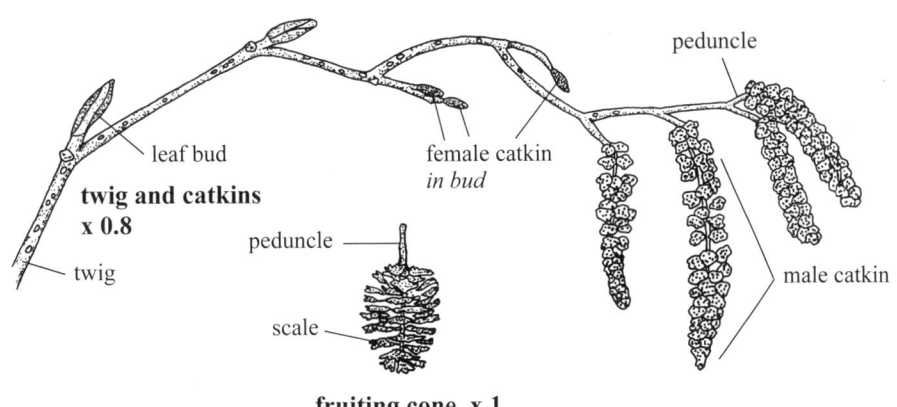
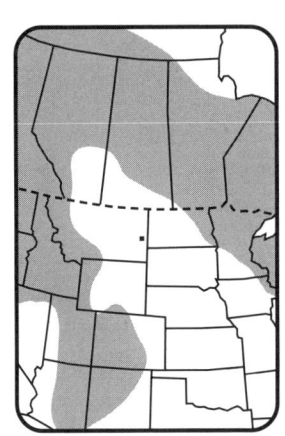

Gray Alder

Birch—*Betulaceae* **Shrub** Reddish brown or green Catkins

GREEN ALDER *Alnus viridis* (Vill.) Lam. & DC.

- **SKETCH** A deciduous shrub 1–10 m tall in clumps; in open woods, well-drained edges of streams and wetlands, roadsides and in sandy sites; **monoecious. w2**
- **REPRODUCTIVE PARTS** Reddish brown or green, active May–June; **inflorescence** of catkins; **peduncles** green, *c.* 0.3 mm long; **male catkins** terminal, 1–4 at ends of new twigs, formed the previous autumn, expanding and hanging in the spring, 2.5–9.5 cm long by 6–11 mm wide at anthesis; **fall catkins** 1.5–2 cm long by 4–5 mm wide, reddish purple, gummy; **florets** in clusters of threes, 4–5.5 mm wide by *c.* 3.5 mm long by 2.5 mm deep; **rachis** pale green, glabrous; **scales** three, brown, together 2.5–3 mm wide, central scale the largest, margins slightly hairy; **sepals** four per scale (12 in total), pale green, soft, apices with 3 or 4 small glands, glabrous, *c.* 1 mm long by 0.7–1 mm wide, blunt, one at the base of each anther; **stamens** four per scale (12 in total), crowded in three rows beneath each scale; **filaments** 0.4–1 mm long; **anthers** greenish yellow, crowded, 4-lobed, *c.* 1.2 mm long by *c.* 1 mm wide; **subtending bracts** (of peduncles) *c.* 1.5 mm long, blunt to toothed, sticky, glandular; **peduncles** 1–2 mm long (to 10 mm in fruit), glandular, hairy; **female catkins** gummy, ascending, 4–7 mm long by *c.* 2 mm wide, green and red; **pistil** (at anthesis) *c.* 0.5 mm long, two per scale; **styles** paired, reddish, exserted; **mature catkins** woody, dark brown, 0.7–1.4 cm long by 7–10 mm wide, in clusters of 1–7 overwintering into the next summer; **cone scales** *c.* 4 mm long by *c.* 3 mm wide, mostly 3-toothed, slightly glandular at apex, arranged in six rows around the rough indented rachis. **FRUIT a samara**, 1-seeded, brown, winged, 2–3.2 mm long (including styles) by 1.8–3 mm wide; **wings** thin, 0.5–1 mm wide, notched at base; **seed** 1.7–2 mm long by *c.* 1 mm wide, flat, thin.
- **LEAVES** Alternate, simple, toothed; **blades** 3–11 cm long by 2–8 cm wide, dull and glabrous above, slightly shiny and sticky below, veins hairy and raised below, teeth 0.4–1 mm long, curved downward; **young leaves** sticky when emerging from spring buds; **petioles** 5–10 mm long, sticky, hairy above along groove; **stipules** green, turning brown and deciduous, 6–9 mm long by 3–4 mm wide, margins entire and revolute, hairy near apices.
- **STEM** Erect, ascending to spreading, smooth, bark gray on older twigs and not peeling, **lenticels** tan and raised; **branches** alternate; **young twigs** reddish brown, slightly shiny; **fall buds** dark reddish purple, shiny, pointed, 8–11 mm long by 2.5–3.5 mm wide and thick, with 2–3 scales exposed, their margins ciliate.
- **SYN.** *Alnus crispa* (Ait.) Pursh = *Alnus viridis* ssp. *crispa* (Ait.) Turrill.

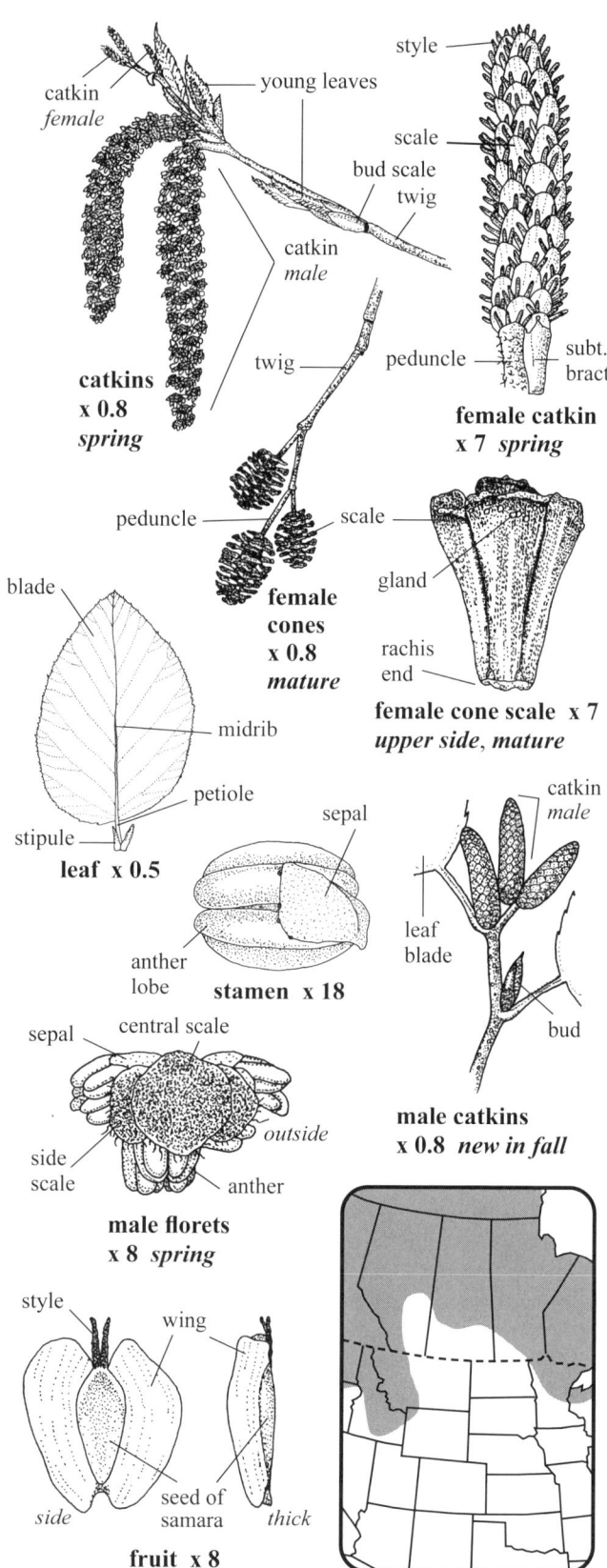

Sitka Alder, Mountain Alder

Birch—*Betulaceae* **Shrub** Green to brown Catkins

Scrub Birch *Betula nana* L.

■ **SKETCH** A variable, glandular **deciduous shrub** 0.5–3 m tall, sometimes forming thickets; in bogs, slough margins, alpine slopes, arctic tundra and streamsides of the Boreal forest; **monoecious**. w1

■ **REPRODUCTIVE PARTS** Green to brown, blooming April–June; **inflorescence** of catkins, unisexual, formed in the fall, blooming in the spring as the leaves appear; <u>**male catkins**</u> solitary, 4–6 mm long by *c.* 2 mm wide in the fall, blunt, 7–25 mm long by 3–5 mm wide at anthesis, drooping; **male florets** in triads under each scale, together *c.* 2 mm long by 2–2.3 mm wide; **filaments** *c.* 0.5 mm long; **anthers** reddish yellow, 0.8–1 mm long, 2-lobed, four anthers per sepal; **sepals** three per scale, hairy or not, each *c.* 0.7 mm long; **peduncles** 4–7 mm long, ascending; <u>**female catkins**</u> erect, pink and green, 12–17 mm long by 3–4 mm wide at anthesis, ascending, 10–20 mm long by 4–5 mm wide with fruit, stiff, ascending; **scales** 3-lobed; **female florets** (three) together *c.* 2.5 mm long by 3–4 mm wide; **pistils** *c.* 2 mm long; **ovary** flat, *c.* 0.6 mm long and wide, slightly winged; **styles** exserted, pink, each 2-parted to base, widely spreading in fruit; **mature scale** *c.* 4 mm long by *c.* 3 mm wide, slightly hairy in sinuses of the lobes. **FRUIT a samara**, 1-seeded, tan, dull, three per scale, 2–2.2 mm long (not including persistent styles) by 2–3.6 mm wide by 0.6–0.7 mm thick, wings more narrow than the seed's 1.7 mm wide body.

■ **LEAVES** Alternate, simple, toothed, slightly leathery, stiff, ascending to arched; **blades** 1–3 cm long by 1–2.5 cm wide, 2–4 side veins and 6–10 blunt teeth on each side of the midrib, shiny above, dull and glabrous to hairy below, especially along the veins, green both sides and turning reddish brown in the autumn, glands resinous; **petioles** 2–5 mm long, finely hairy; **stipules** 2–4 mm long, slightly hairy.

■ **STEM** Erect to leaning, bark dark brown to blackish, smooth, 5–15 mm thick near the base; **lenticels** vertical, tan, slightly raised; **branches** alternate; **twigs** glabrous to finely hairy, reddish, resinous raised glands red, black to white, give the twigs a rough feel; **winter side buds** *c.* 2 mm long with ciliate scales.

■ **SYN.** *Betula glandulosa* Michx.

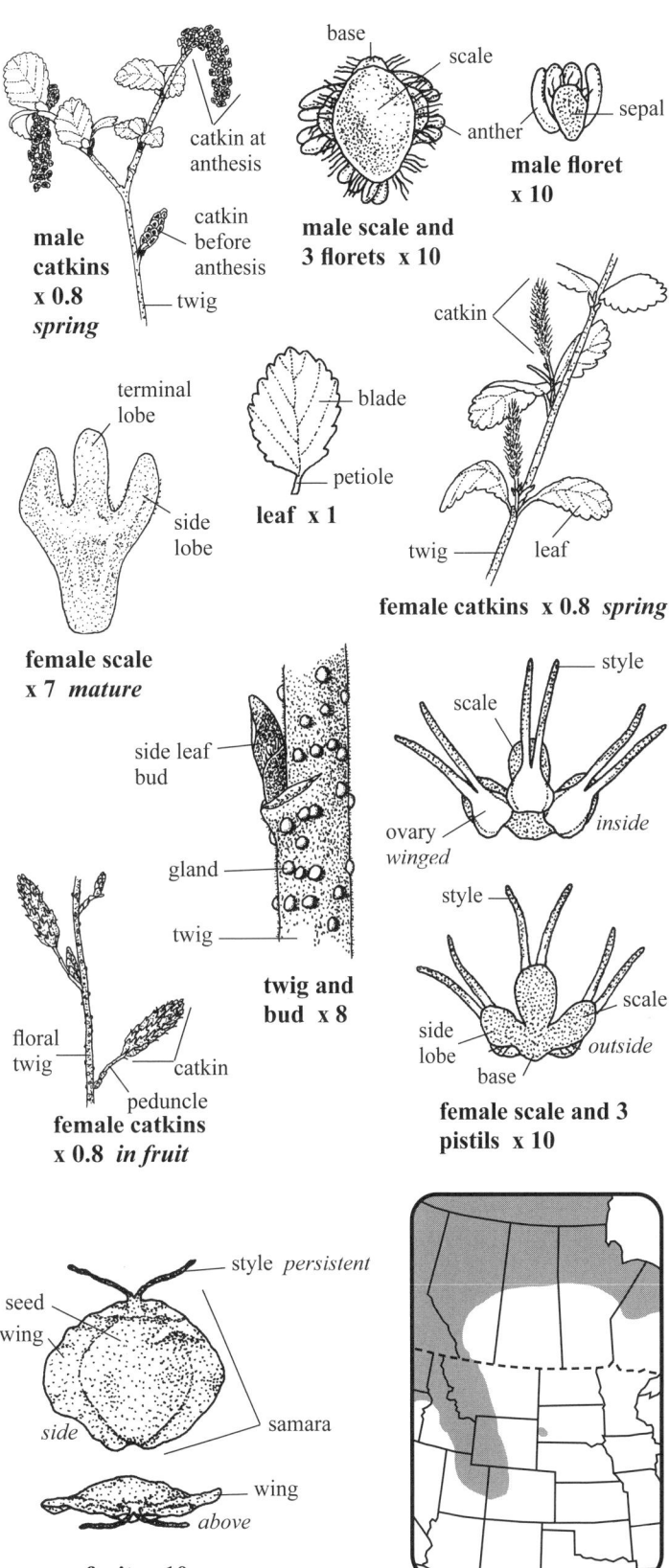

Glandular Birch, Tundra Dwarf Birch, Resin Birch

Birch—*Betulaceae*   **Shrub** Brown and green Catkins

River Birch   *Betula occidentalis* Hook.

- **SKETCH** A **deciduous shrub** 2–10 m tall in clumps of several trunks; in open woods, sandy hills, mountain canyons, along open river banks, lakes, marshes and bog edges; **monoecious**.
- **REPRODUCTIVE PARTS** Brown and green, active April–June; **inflorescence** of catkins, unisexual, active as leaves expand; **peduncles** *c.* 0.5 mm long; **male catkins** terminal on branches, one to three, each 3.5–7.5 cm long by 4–5 mm wide at anthesis, hanging; **floral scales** three, together 1.5–2 mm wide and long, brown, imbricate, margins hairy; **male florets** alternate along the glabrous rachis, in threes, post anthesis 2.5–3 mm wide, composed of a sepal and two stamens; **sepals** three, 1.3–1.5 mm long by *c.* 1 mm wide, pale yellow, apices brown with scattered hairs; **peduncles** hairy and gummy, 2–20 mm long with small leaves or bracts; **female catkins** 1–4 cm long in flower by 1.5–2.5 mm wide, not hanging but on lateral leafy branchlets; **female florets** in threes; **central scales** green, 1.5–2 mm long by 0.8–1 mm wide, apices pointed, slightly spreading, hairy on margins near base; **lateral scales** two, *c.* 1 mm long by *c.* 0.7 mm wide, margins hairy, partially concealed and attached to the central green scale; **pistils** three, *c.* 1.3 mm long by *c.* 0.7 mm wide, one under each scale; **ovary** glabrous, *c.* 0.7 mm long; **styles** two, *c.* 0.6 mm long, exserted, tips reddish, glabrous; **cones** of previous year 10–30 mm long by 7–12 mm wide, usually persistent for at least one year; **cone scale**s brown, 3-veined, 4–4.8 mm long by *c.* 3.5 mm wide, 3-lobed, lateral lobes pointed and ciliate. **FRUIT a samara**, 1-seeded, tan, winged, TW 3–5 mm by 2–3 mm long by 0.3–0.5 mm thick, wing wider than seed.
- **LEAVES** Alternate, may appear opposite, simple, toothed, fascicled; **blades** 2–6 cm long by 1–4 cm wide, sticky and shiny when young, glandular-dotted below, yellow in the fall; **petioles** 5–23 mm long, glandular-dotted; **stipules** ciliate.
- **TRUNK** Erect, branched, several, each 10–35 cm wide near the base; **bark** gray to reddish brown, shiny, smooth, not peeling; **lenticels** 5–12 mm long by 1–2 mm wide and slightly raised, horizontal, light gray; **branches** arched to ascending; **twigs** glandular, some short hairy.
- **SYN.** *Betula fontinalis* Sarg.

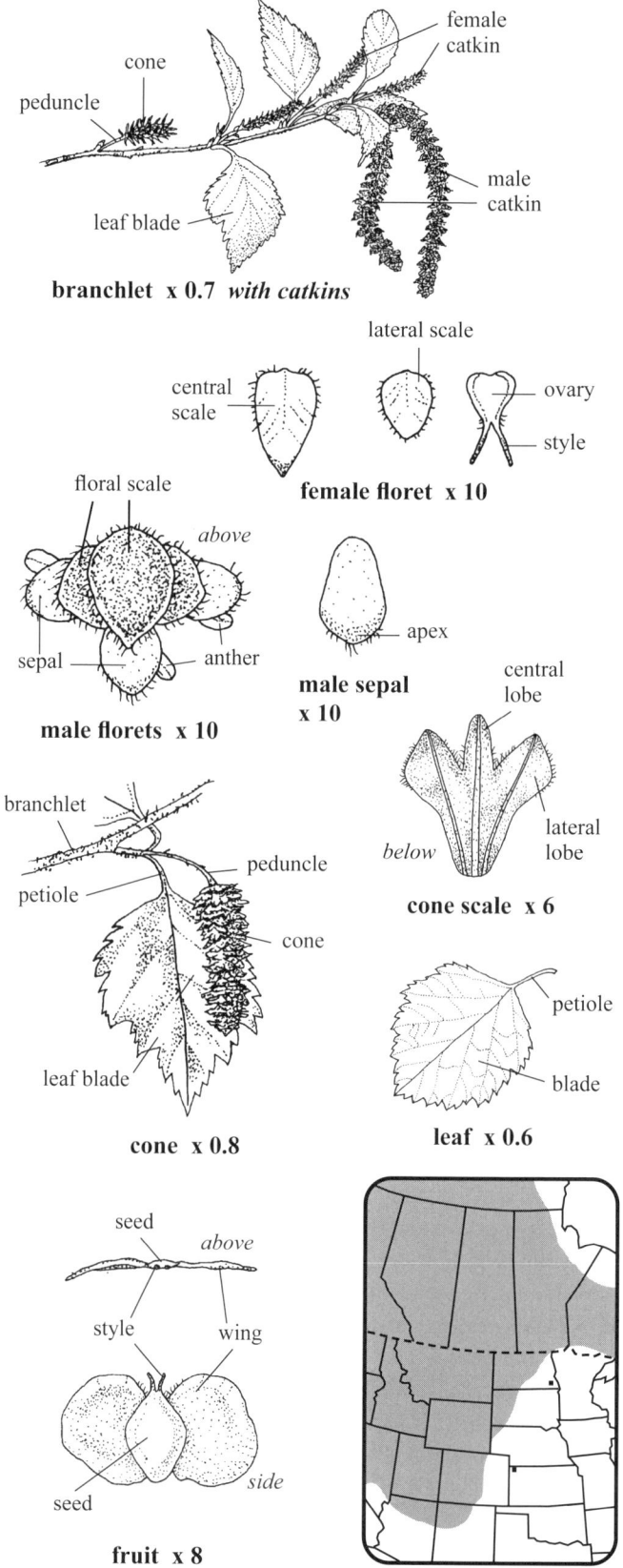

204   Mountain Birch, Water Birch, Black Birch

Birch—*Betulaceae* **Tree** Deciduous Catkins brownish green to pink

WHITE BIRCH *Betula papyrifera* Marsh.

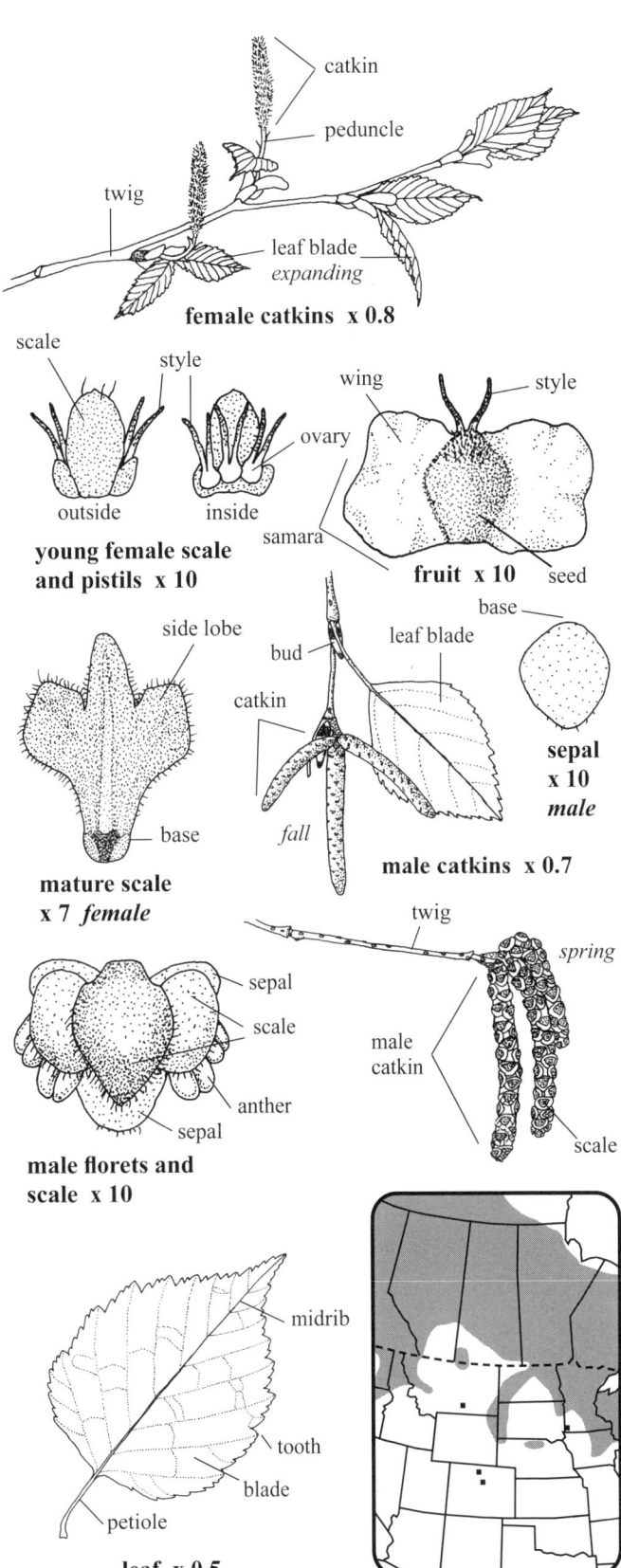

■ **SKETCH** A variable **deciduous tree** with a white to brown trunk, one to several in a clump, 5–30 m tall; on woodland slopes, in canyons, bottomlands, lakeshores, rocky slopes, peat-covered islands and cutover areas; **monoecious. w1**

■ **REPRODUCTIVE PARTS** Unisexual, active May–June; **male catkins** brownish green, hanging in groups of 1–3, formed in the fall, then 2–3 cm long and overwintering, expanding at anthesis to 3–9 cm long by 5–7 mm wide; **male florets** sessile, under a triad of three scales and three sepals *c.* 3 mm wide by *c.* 2 mm long, these ciliate on the margins; **rachis** hairless, *c.* 0.5 mm wide; **stamens** two under each sepal, the two stamens appear as four anthers for a total of 12; **sepals** three, pale green, the front one the largest at 1–1.5 mm long by 1–1.3 mm wide, the lateral sepals *c.* 1 mm long and wide; **anthers** yellow, *c.* 1.3 mm long in the fall; **peduncles** green, ascending, glandular, *c.* 8 mm long, soon descending as fruit ripens; **female catkins** active slightly later than male catkins, as leaf buds are opening, 15–20 mm long (to *c.* 50 mm with fruit) by *c.* 3 mm wide, green below, reddish above, erect, soon drooping after anthesis and with fruit, one or two per bud; **perianth** absent; **scales** three, together *c.* 1.5 mm long by *c.* 1.2 mm wide, maturing into a 3-lobed, ciliate scale 4–4.8 mm long by 3–3.5 mm wide with three fruit; **pistils** three, each *c.* 1 mm long; **style** 2-parted, the upper half reddish purple, persistent. **FRUIT a samara**, 1-seeded, 2–2.5 mm long (including styles) by 3.3–4 mm wide, shed in autumn; **wings** tan but paler than seed; **seed** tan, *c.* 0.3 mm thick.

■ **LEAVES** Alternate in clumps when young, simple, toothed; **young blades** shiny and slightly hairy above; **mature blades** dull, 4–10 cm long by 3–6.5 cm wide, teeth not glandular, glabrous above, lighter green below with long hairs and rusty hairs at junctions of midrib and veins, blades turn yellow in fall; **petioles** 1–4 cm long, glandular and hairy, hairs short and long, ascending to spreading near apices.

■ **TRUNK** Erect, **dbh** 10–60 cm; **bark** white to brown, thin and peeling horizontally in 5–15 cm wide curled sheets exposing smooth tan bark; **branches** and **twigs** reddish brown, ascending to spreading, densely glandular or not, hairy or not; **lenticels** horizontal, 3–12 mm long by 0.6–1 mm wide, gray; **fall buds** 5–7 mm long by 2–2.5 mm wide, sticky.

Paper Birch, Canoe Birch, American White Birch

Birch—*Betulaceae* **Shrub** Green to brown Catkins

Swamp Birch *Betula pumila* L.

■ **SKETCH** A variable, somewhat glandular **deciduous shrub** 1–4 m tall; in clumps along bog, slough and stream sides, wet fields and lake margins, in parklands and Boreal forest; **monoecious**. w1

■ **REPRODUCTIVE PARTS** Green to brown, blooming April–June; **inflorescence** of catkins, unisexual, three florets per scale, one under each of the scale's three lobes, formed in autumn, blooming as leaves appear in spring; <u>male</u> **catkins** solitary and 5–10 mm long over the winter; **scales** dark brown, hairy on margins, catkins become 1.5–3 cm long by 4–5 mm wide at anthesis, drooping; **peduncles** 5–10 mm long, the glands brown and turning white on drying; <u>female</u> **catkins** 1–2 cm long by 3–4 mm wide at anthesis; **scales** 3-lobed, in a spiral around the central axis, 3.5–4 mm long by 2.5–3 mm wide, blunt, the side lobes 1–1.5 mm long, all lobes ciliate and turning brown; **catkins** with fruit, 1–2.5 cm long by 6–10 mm wide, ascending, brown and persisting. **FRUIT** a samara, 1-seeded, 2–2.7 mm long (including styles) by 2.5–3 mm wide by *c*. 0.6 mm thick, three imbricate under each scale; **seed** 1.5–2.5 mm long (not including styles) by 1.1–2 mm wide, concave below, convex above against the scale, wings about as wide as seed; **styles** persistent, 0.5–0.7 mm long.

■ **LEAVES** Alternate, simple, toothed, spreading to slightly ascending; **blades** 15–50⁺ mm long by 15–40 mm wide, medium green and shiny above, some very hairy and pale green below, widest near the middle, tapered to the base, 10–20 teeth per side, slightly glandular on both sides, 4–6 veins on each side of midrib, teeth coarse to fine, glands small, green, round, raised, with a pale margin; **petioles** 2–8 mm long, slightly glandular, slightly hairy, pale greenish red; **stipules** ovate, acute, 3–5 mm long, ciliate to scarious.

■ **STEM** Erect, branches alternate, ascending, in loose clumps of several to over 100 stems, reddish brown with vertical tan lenticels near the 5–10 mm thick bases; **new twigs** gray to reddish brown, glabrous to hairy, ascending, coating easily scraped off, bark doesn't peel, glands brown with a thin pale margin; **winter buds** 2–3 mm long, pointed, resinous.

■ **SYN.** *Betula glandulifera* (Regel) Butler = *Betula pumila* var. *glandulifera* Regel and *Betula glandulosa* var. *glandulifera* (Regel) Gleason = *Betula pumila* var. *glandulifera* Regel.

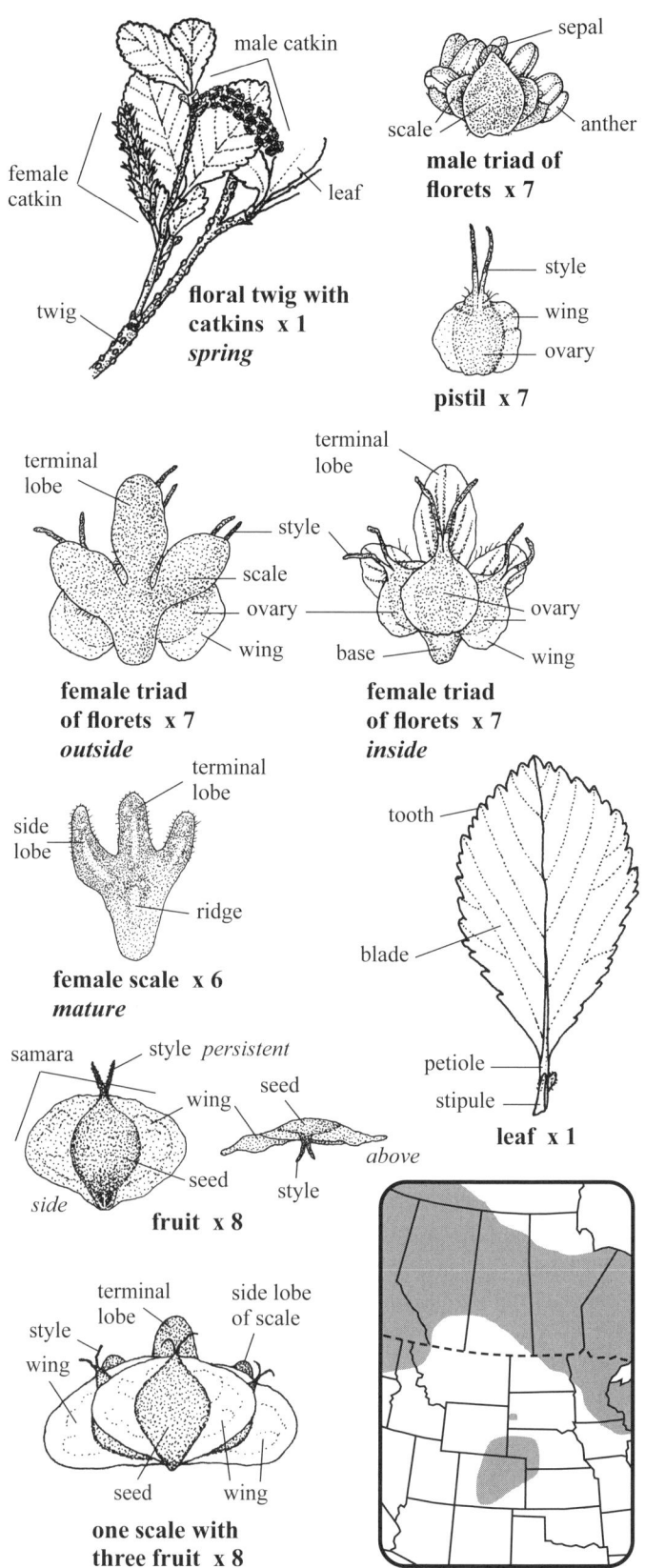

Bog Birch, Dwarf Birch, Yellow Birch

American Hazelnut   *Corylus americana* Walt.

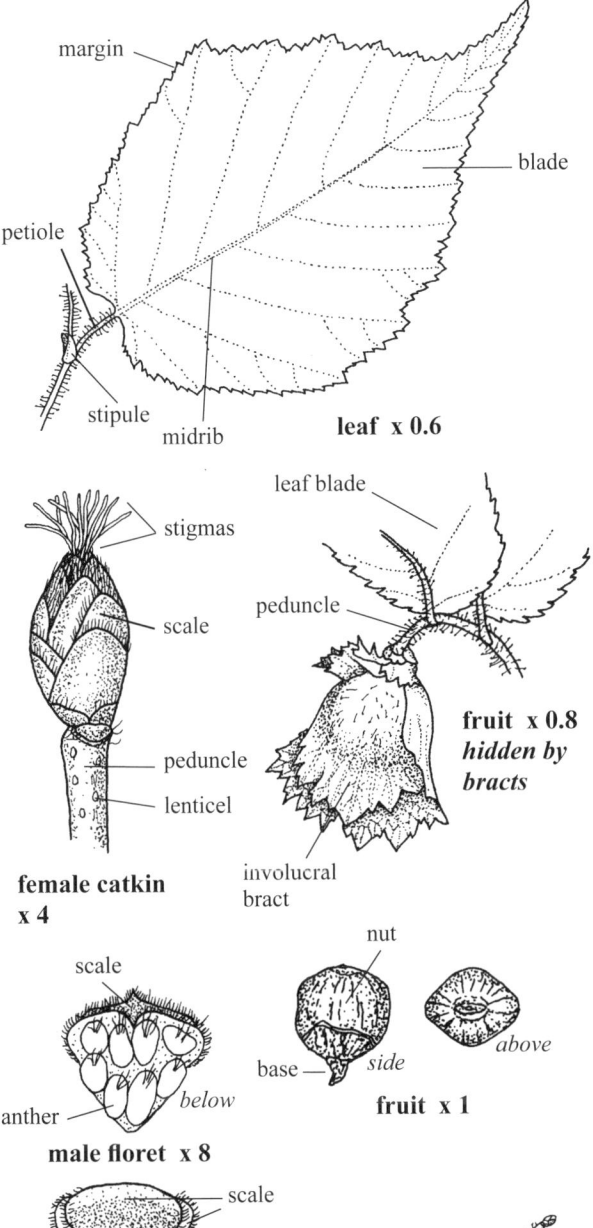

- **SKETCH** A **deciduous shrub** 1–3 (–5) m tall, singly or in small clusters; along forest edges, fencerows, hillsides and in moist thickets; **monoecious**. w2
- **REPRODUCTIVE PARTS** Greenish brown to reddish, active March–May before the leaves appear; **inflorescence** of catkins, unisexual; **peduncles** 2–6 mm long; **male catkins** hanging, brown, 13–53 mm long by 4–6 mm wide, one to three along a hairy reddish brown twig, developing in the fall and overwintering; **scales** three, fringed with white hairs, the central scale reddish brown especially at its short, pointed tip; **male florets** three (two) per scale (bract), *c.* 2 mm long including the tip, by 1.9–2.2 mm wide; **stamens** four, sessile, 0.5–0.7 mm long by 0.5 mm wide; **anthers** divided, appear as eight, with a few hairs; **female catkins** budlike, one to five, oval, terminal or lateral along a peduncle, 5–6 mm long by 3–3.5 mm wide; **scales** numerous, sticky, imbricate and with a hairy fringe; **female florets** two per scale; **stigmas** thin, 2–2.3 mm long, exserted, bright reddish purple, several pairs, joined at their base; **involucral bracts** two, toothed, veined, enclosing fruit, apices sometimes opening, overlapping at one end, some hairs gland-tipped, the whole structure 2–3 cm long and wide by *c.* 1.5 cm thick and usually hidden beneath the leaves. **FRUIT a nut**, 1-seeded, brown, slightly flattened, rough, slightly grooved above, 12–15 mm long by 10–15 mm wide by *c.* 10 mm thick.
- **LEAVES** Alternate, simple, irregularly toothed, reddish or coppery green when young, veins obvious, margins lightly hairy, more hairy beneath, hairs gland-tipped; **blades** 5–16 cm long by 4–12 cm wide, pointed; **petioles** *c.* 1 cm long with spreading gland-tipped hairs; **stipules** light green, entire, 5–6 mm long with short glandular hairs.
- **STEM** Branches gray, bark smooth; **twigs** reddish brown, alternate, usually covered with spreading, reddish purple, gland-tipped hairs and short white hairs; **buds** alternate and terminal, 2–3 mm long by *c.* 2 mm wide, blunt, with several scales.

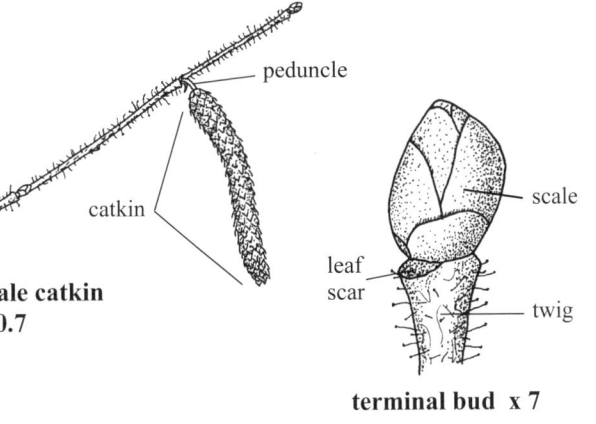

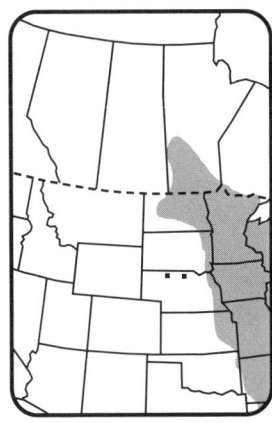

Birch—*Betulaceae* **Shrub** Green to brown Catkins

BEAKED HAZELNUT *Corylus cornuta* Marsh.

■ **SKETCH** A nonglandular **deciduous shrub** 1–10 m tall, in clumps; in parklands and along edges of forests; **monoecious**. w2

■ **REPRODUCTIVE PARTS** Green to brown, formed in the autumn, blooming April–May before the leaves appear; **inflorescence** of catkins, unisexual, near twig or branch ends; **male catkins** often sessile, pale green, 5–8 mm long in fall, 1 to 3 catkins per leaf axil, expanding in spring to 8–25 mm long by 3–4.5 mm wide; **scales** 1.5 mm long and wide; **stamens** four; **filaments** 2-parted; **anthers** eight, pale yellow, 0.5–0.7 mm long, crowded, apices slightly hairy; **subtending bracts** (of male catkins) two or three, hairy, 1.5–3 mm long, pointed; **female catkins** sessile, few-flowered, oval, 3.4–5 mm long by 2–3 mm wide (fall), pointed, tan, hairy, scales 5 or 6, margins ciliate, short scattered white hairs on the body in the fall, catkins single at base of upper leaf petioles; **outer scales** dark brown to reddish brown, upper and inner scales light green, inside numerous white hairs covering green flowers in autumn; **stigmas** reddish, exerted in spring; **involucral bracts** two, hairy below, brown, 4–6 cm long, beaked, enclosing one nut; **beak** narrow, flattened, grooved, 2–4 cm long by 5–8 mm wide with several reticulate-veined teeth 1–5 mm long at the apex, fewer hairs along the beak and teeth. **FRUIT a nut**, 1-seeded, clustered, 1–3, 10–12 mm long and thick by 11–15 mm wide, hairless, tan with a wide, pale, slightly rough base, apical point minute, numerous shallow vertical grooves surrounding the base.

■ **LEAVES** Alternate, simple, double-toothed, with a few shallow lobes on each side, pointed, yellow in the fall; **blades** 4.5–11.5 cm long by 3.5–10.5 cm wide, dull, nonglandular, glabrous but soft above, lighter green and slightly hairy below mostly along the raised yellowish green veins, more hairy when young; **petioles** 7–20 mm long, hairy at first, mostly glabrous by fall, nonglandular.

■ **STEM** Brown, nonglandular; **twigs** hairy when young, hairless with age, bark smooth, a dozen or more stems loosely clumped, erect to leaning, well branched; 5–15 mm wide near the base.

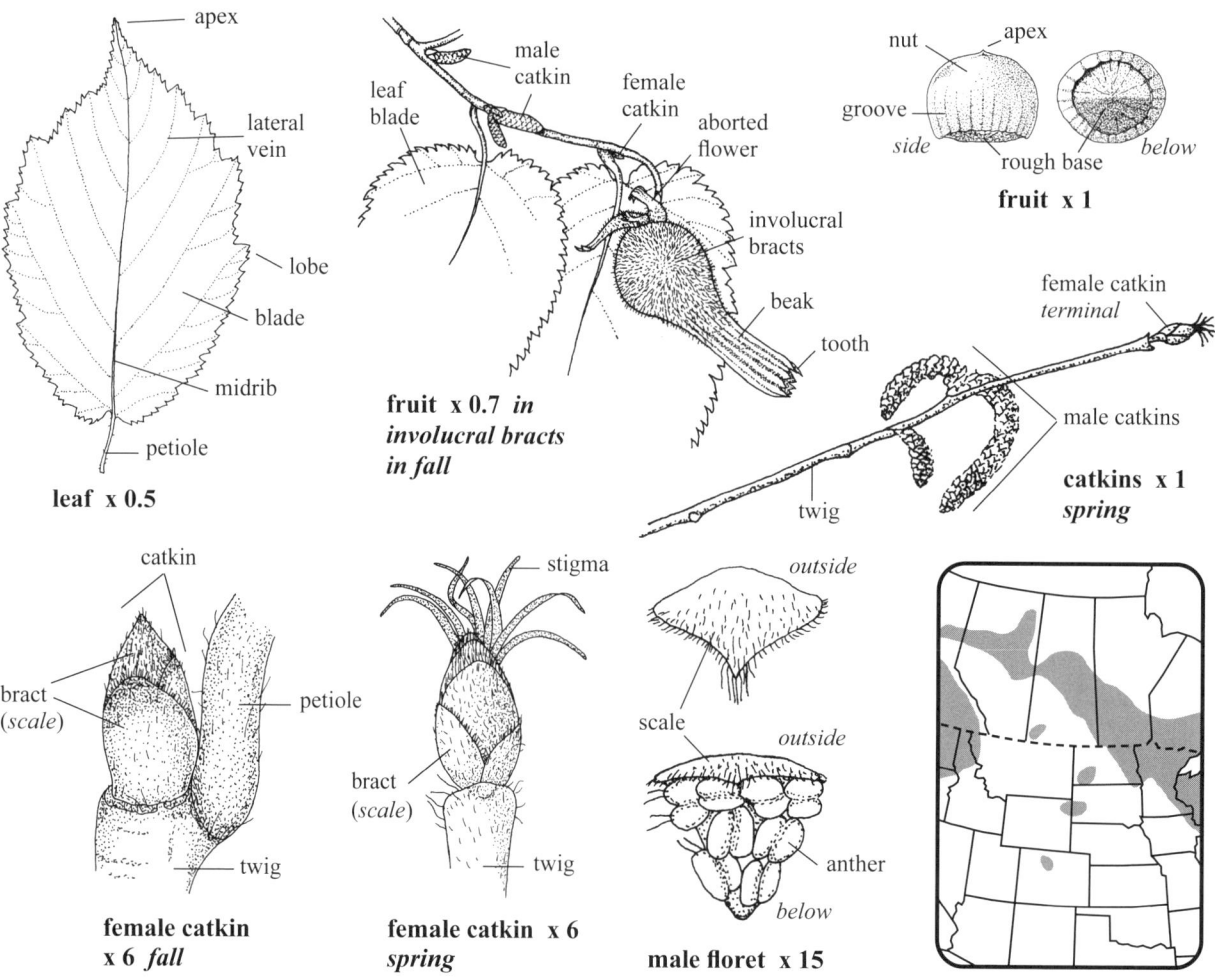

208

Borage—*Boraginaceae* Wildflower   Blue to white   Corolla tubular, 5-lobed

NODDING STICKSEED   *Hackelia deflexa* (Wahlenb.) Opiz

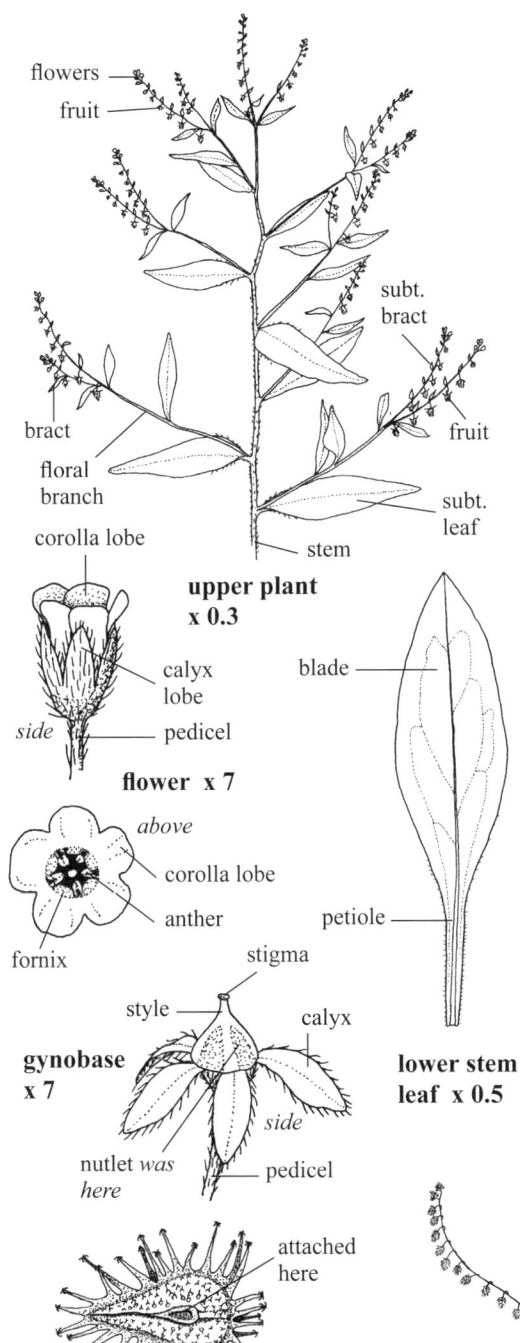

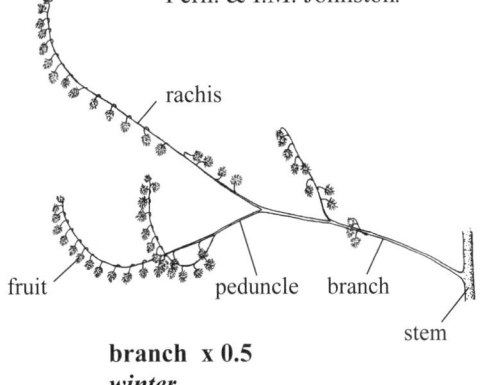

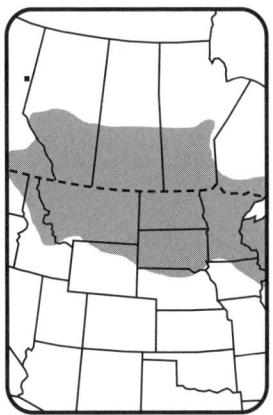

■ **SKETCH** A **biennial** or **perennial herb** 30–90 cm tall from a thick **taproot** 2–13+ cm long, usually scattered; along woodland edges, hillsides and thickets in calcareous soils.

■ **FLOWERS** Pale blue to white, blooming June–August; **inflorescence** of paired cymes (racemelike), at floral branch tips, these lengthen to 5–10 cm with tiny flowers at the apices and the ripening fruit behind; **floral branches** 3–20 cm long, axillary; **pedicels** 1–6 mm long, hairy, mostly recurved with fruit; **subtending bracts** (of pedicels) one, leafy, hairy, 3–7 mm long in fruit; **flowers** perfect; **calyx** 5-lobed, green, united below, shorter than the corolla, ascending in flower, 1–2 mm long by 0.6–0.8 mm wide, with long stiff hairs on the outside, mostly glabrous inside, spreading in fruit; **corolla** tubular, 5-lobed, 1.5–3 mm wide by 2–2.5 mm long, the lobes *c.* 1 mm long and wide, glabrous, slightly spreading; **fornices** five, white, prominent in the throat and opposite each corolla lobe; **stamens** five, included; **anthers** pale yellow, visible between the fornices; **style** one, distinct, short; **stigma** round, turning dark brown in fruit; **gynobase** broadly pyramidal with four nutlets attached. **FRUIT a nutlet**, vertical, four per flower, 3–4.3 mm long by *c.* 3.5 mm wide, usually with one marginal row of hooked bristles distinct to the base, other bristles of various lengths, outer (dorsal) surface with scattered, short bristles.

■ **LEAVES** Basal and stem, alternate, simple, mostly entire; **first year rosette** of several, hairy, wide blades; **basal blades** early deciduous; **stem blades** usually flat, smaller or larger above, 2–12 cm long by 1–4.5 cm wide, hairy and scabrous on both surfaces; **petioles** 0–5 cm long, narrowly winged, hairy, reduced above.

■ **STEM** Erect, single or in clusters to four, ridged, branched, scabrous from short stiff ascending hairs, hollow, sometimes nodding; 1–8 mm wide near the base.

■ **SYN.** *Hackelia americana* (Gray) Fern. = *Hackelia deflexa* var. *americana* (Gray) Fern. & I.M. Johnston and *Lappula americana* (Gray) Rydb. = *Hackelia deflexa* var. *americana* (Gray) Fern. & I.M. Johnston.

Borage—*Boraginaceae*   **Wildflower**   Pale blue, center yellow   Corolla tubular, 5-lobed

BLUEBUR   *Lappula squarrosa* (Retz.) Dumort.

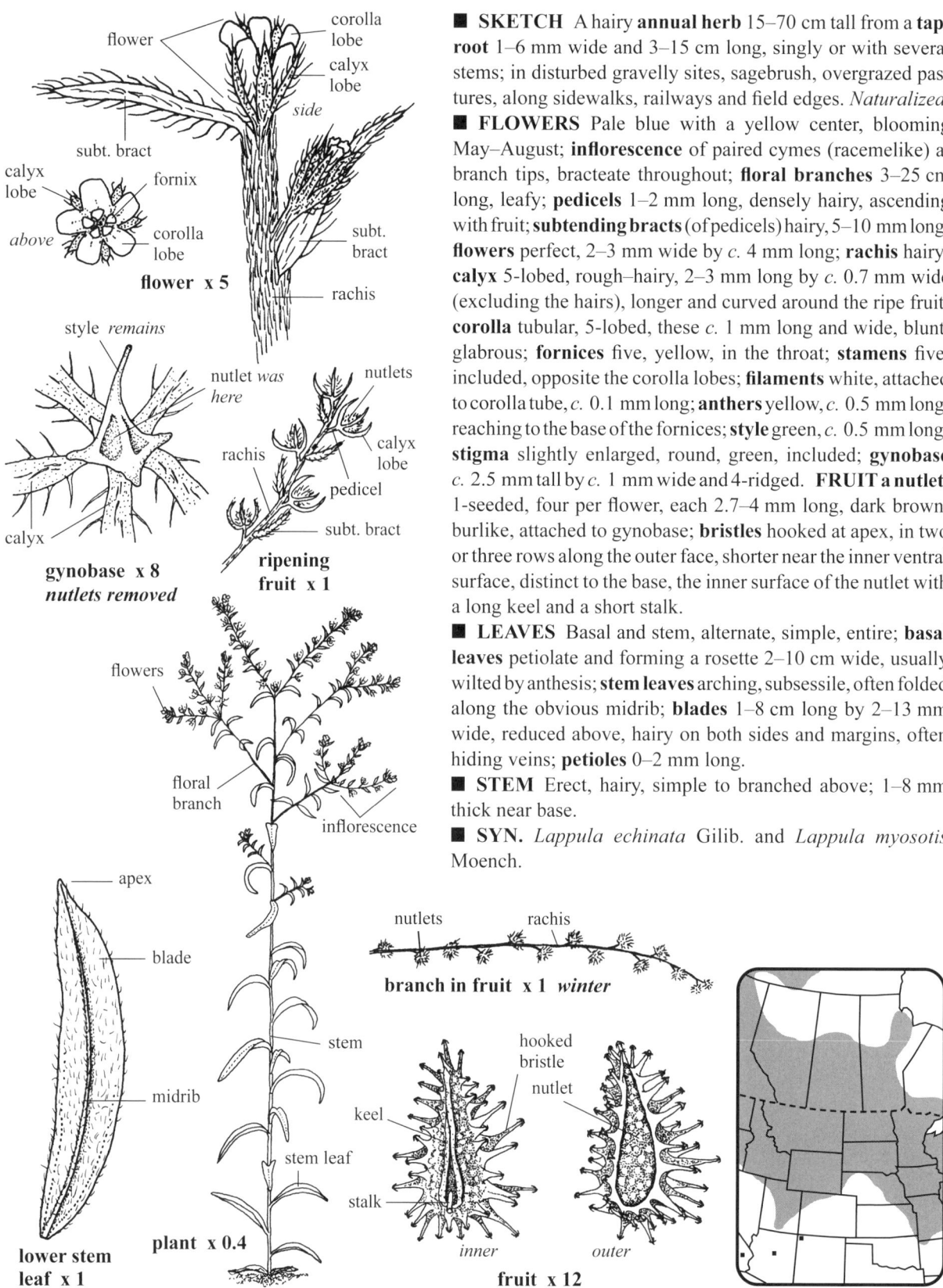

- **SKETCH** A hairy **annual herb** 15–70 cm tall from a **taproot** 1–6 mm wide and 3–15 cm long, singly or with several stems; in disturbed gravelly sites, sagebrush, overgrazed pastures, along sidewalks, railways and field edges. *Naturalized*
- **FLOWERS** Pale blue with a yellow center, blooming May–August; **inflorescence** of paired cymes (racemelike) at branch tips, bracteate throughout; **floral branches** 3–25 cm long, leafy; **pedicels** 1–2 mm long, densely hairy, ascending with fruit; **subtending bracts** (of pedicels) hairy, 5–10 mm long; **flowers** perfect, 2–3 mm wide by *c.* 4 mm long; **rachis** hairy; **calyx** 5-lobed, rough–hairy, 2–3 mm long by *c.* 0.7 mm wide (excluding the hairs), longer and curved around the ripe fruit; **corolla** tubular, 5-lobed, these *c.* 1 mm long and wide, blunt, glabrous; **fornices** five, yellow, in the throat; **stamens** five, included, opposite the corolla lobes; **filaments** white, attached to corolla tube, *c.* 0.1 mm long; **anthers** yellow, *c.* 0.5 mm long, reaching to the base of the fornices; **style** green, *c.* 0.5 mm long; **stigma** slightly enlarged, round, green, included; **gynobase** *c.* 2.5 mm tall by *c.* 1 mm wide and 4-ridged. **FRUIT a nutlet**, 1-seeded, four per flower, each 2.7–4 mm long, dark brown, burlike, attached to gynobase; **bristles** hooked at apex, in two or three rows along the outer face, shorter near the inner ventral surface, distinct to the base, the inner surface of the nutlet with a long keel and a short stalk.
- **LEAVES** Basal and stem, alternate, simple, entire; **basal leaves** petiolate and forming a rosette 2–10 cm wide, usually wilted by anthesis; **stem leaves** arching, subsessile, often folded along the obvious midrib; **blades** 1–8 cm long by 2–13 mm wide, reduced above, hairy on both sides and margins, often hiding veins; **petioles** 0–2 mm long.
- **STEM** Erect, hairy, simple to branched above; 1–8 mm thick near base.
- **SYN.** *Lappula echinata* Gilib. and *Lappula myosotis* Moench.

European Stickseed, Bristly Sheepbur, Blue Stickseed

Borage—*Boraginaceae*   **Wildflower** Deep yellow Corolla tubular, 5-lobed

HOARY PUCCOON   *Lithospermum canescens* (Michx.) Lehm.

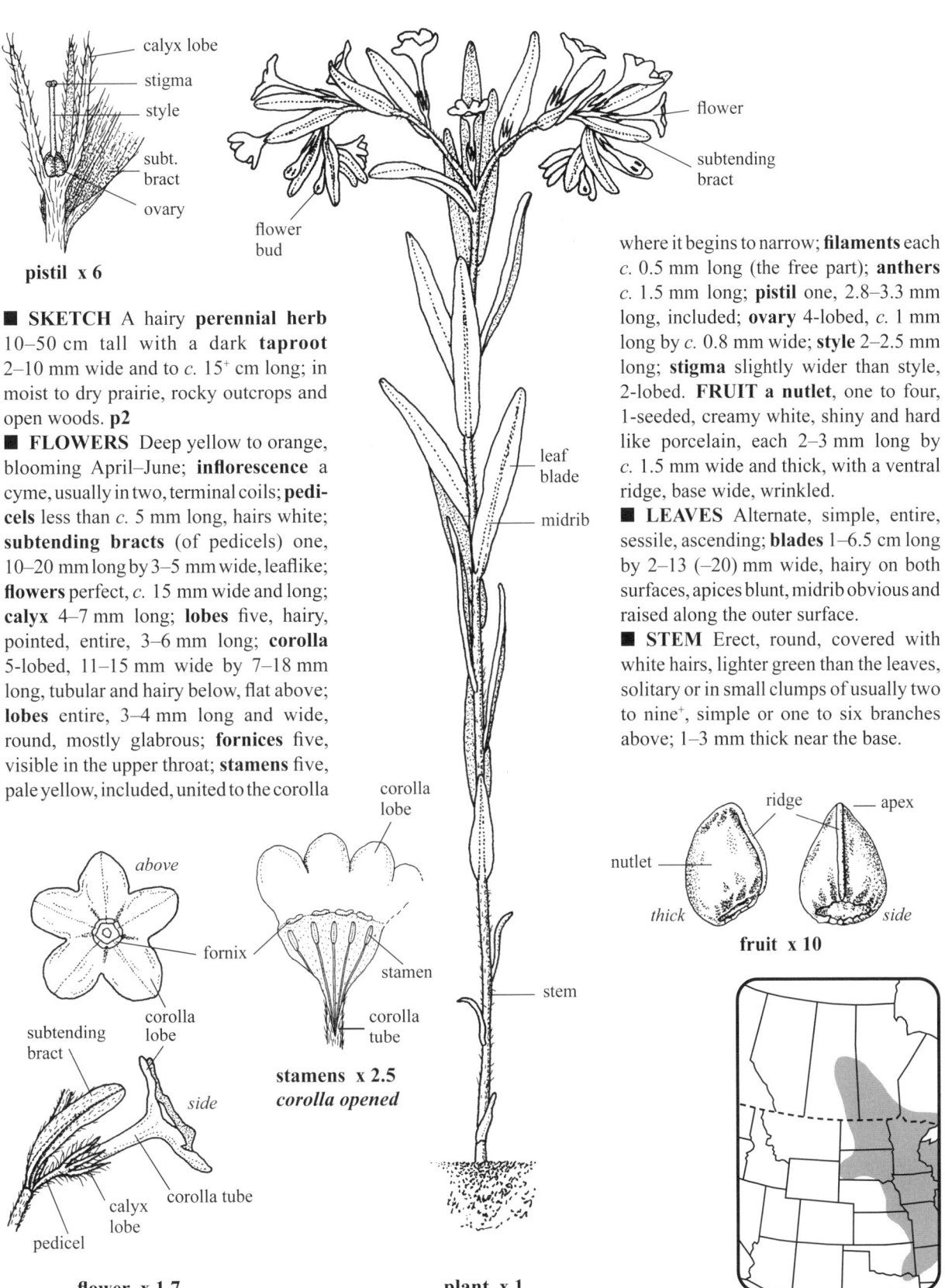

■ **SKETCH** A hairy **perennial herb** 10–50 cm tall with a dark **taproot** 2–10 mm wide and to *c.* 15⁺ cm long; in moist to dry prairie, rocky outcrops and open woods. **p2**

■ **FLOWERS** Deep yellow to orange, blooming April–June; **inflorescence** a cyme, usually in two, terminal coils; **pedicels** less than *c.* 5 mm long, hairs white; **subtending bracts** (of pedicels) one, 10–20 mm long by 3–5 mm wide, leaflike; **flowers** perfect, *c.* 15 mm wide and long; **calyx** 4–7 mm long; **lobes** five, hairy, pointed, entire, 3–6 mm long; **corolla** 5-lobed, 11–15 mm wide by 7–18 mm long, tubular and hairy below, flat above; **lobes** entire, 3–4 mm long and wide, round, mostly glabrous; **fornices** five, visible in the upper throat; **stamens** five, pale yellow, included, united to the corolla where it begins to narrow; **filaments** each *c.* 0.5 mm long (the free part); **anthers** *c.* 1.5 mm long; **pistil** one, 2.8–3.3 mm long, included; **ovary** 4-lobed, *c.* 1 mm long by *c.* 0.8 mm wide; **style** 2–2.5 mm long; **stigma** slightly wider than style, 2-lobed. **FRUIT a nutlet**, one to four, 1-seeded, creamy white, shiny and hard like porcelain, each 2–3 mm long by *c.* 1.5 mm wide and thick, with a ventral ridge, base wide, wrinkled.

■ **LEAVES** Alternate, simple, entire, sessile, ascending; **blades** 1–6.5 cm long by 2–13 (–20) mm wide, hairy on both surfaces, apices blunt, midrib obvious and raised along the outer surface.

■ **STEM** Erect, round, covered with white hairs, lighter green than the leaves, solitary or in small clumps of usually two to nine⁺, simple or one to six branches above; 1–3 mm thick near the base.

211

Borage—*Boraginaceae* **Wildflower** Blue to pink Corolla tubular, 5-lobed

TALL LUNGWORT *Mertensia paniculata* (Ait.) G. Don

■ **SKETCH** A hairy **perennial herb** 20–80 cm tall from a woody **rhizome** 5–20 cm long by 4–7 mm thick with a short **caudex** with one or two stems; in parklands, meadows and streambanks in moist woodlands.

■ **FLOWERS** Blue to pink, blooming May–August; **inflorescence** a cyme with 6–20 flowers often in small nodding clusters, terminal and from upper leaf axils; **pedicels** 1–2 cm long, hairs ascending, green, purplish near the calyx; **flower buds** pink; **flowers** perfect, often drooping; **calyx** tubular, deeply 5-lobed, these 3–6 mm long, hairy; **corolla** tubular, 5-lobed, TL 10–16 mm by 6–8 mm wide, glabrous outside, hairy inside below the fornices, the tube 5–7 mm long, the upper wider limb 5–8 mm long (including the lobes), lobes 1.5–2.5 mm long by 4–4.5 mm wide, spreading to slightly recurved, blunt; **fornices** yellow, *c.* 0.5 mm wide, at the top of the corolla tube between filaments; **stamens** five, included; **filaments** free part *c.* 2 mm long, white, lower part attached to corolla tube; **anthers** 2.5–4 mm long by *c.* 1 mm wide, yellow; **pistil** 12–16 mm long, glabrous; **ovaries** four, green, glabrous, *c.* 0.6 mm tall and wide, positioned around the base of the style; **style** one, filiform, barely exserted, white, tapered; **stigma** pale green, slightly wider than the tip of the style, smooth. **FRUIT a nutlet**, 1-seeded, black, one to four per flower, 3.5–5 mm long by 2.3–3 mm wide by *c.* 1.8 mm thick, attached laterally to the gynobase, triangular, pointed at apex, outside rounded and irregularly ridged, the two inner surfaces flat to slightly concave, surrounded by erect calyx lobes; **gynobase** *c.* 2 mm long and wide, tapered to style above.

■ **LEAVES** Basal and stem, alternate, simple, entire, softly hairy, more hairy below especially on raised veins; **basal leaves** few, wilting early; **petioles** to *c.* 15 cm long; **stem leaf blades** 3–15 cm long by 1–4 cm wide, reduced above, hairs white and long, margins ciliate; **petioles** winged, hairy, 0–6 cm long, reduced above.

■ **STEM** Erect, mostly solitary, nodding, slightly hairy above, green, with narrow grooves from bases of petioles; 3–7 mm thick near the pink, often bent base.

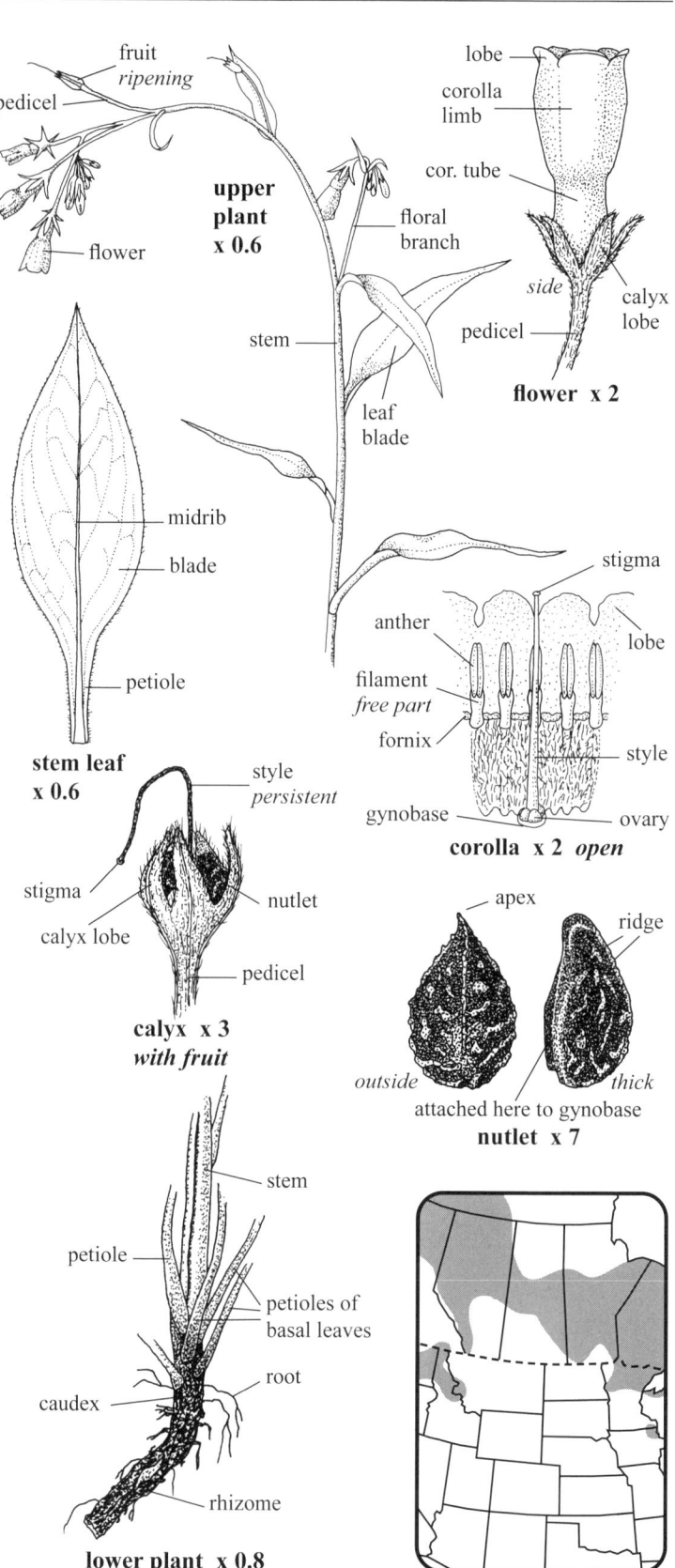

212 Tall Bluebells, Tall Mertensia, Northern Bluebells

Family Characteristics

Mustard—Brassicaceae — Flower parts 4

SKETCH **Dicotyledonous herb** (wildflowers), annual, biennial or perennial. About 3,200 species worldwide; mostly in temperate to cold regions from sea level to alpine habitats.

FLOWERS Perfect, regular, yellow or white, small to showy. **Inflorescence** often a raceme, simple to branched, one to many per plant. **Sepals** 4, in two opposite pairs. **Petals** 4, usually longer than sepals, with a claw. **Stamens** mostly 6, the outer two shorter than the inner four, arising from the base of the pistil. **Pistil** 1. **Ovary** long and cylindrical with 2 carpels and 2 locules, each with two rows of one to many ovules. **Style** simple, often very short. **Stigma** capitate (a round head-like cluster) or 2-lobed.

FRUIT A silique, long and narrow, or flattened, wide and short, 2-valved, these usually spreading when mature to reveal the persistent septum (a partition between the two valves); sometimes opening from base at pedicel end.

LEAVES Usually alternate, often lobed and toothed, simple to pinnately compound, rarely with distinct leaflets. **Stipules** absent.

STEM Erect, green, branched or not, round, most with pungent watery juice.

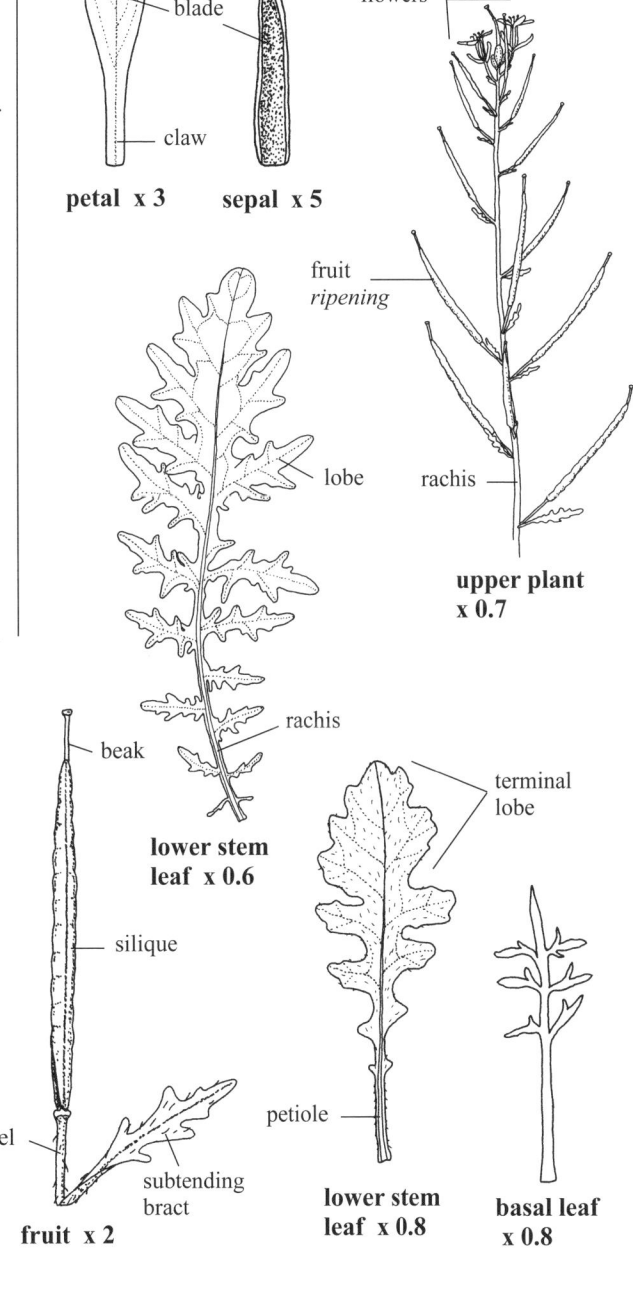

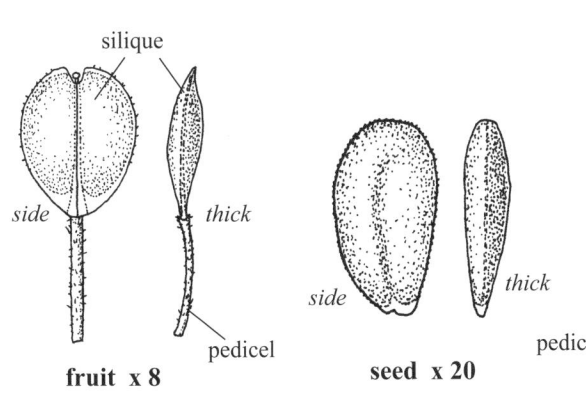

213

Mustard—*Brassicaceae* **Wildflower** White to pale purple to pink Petals 4

PURPLE ROCK CRESS *Arabis* X ***divaricarpa*** A. Nels. (pro sp.)

■ **SKETCH** A variable **biennial herb** 20–90 cm tall from a **taproot** 10–15 cm long by 2–8 mm wide with side roots to *c*. 10 cm long by *c*. 1 mm wide; on rocky outcrops, sandy soils, dry slopes, along railways and gravelly trails.

■ **FLOWERS** White to pale purple to pink, blooming May–July; **inflorescence** a raceme 2–3 cm long and wide, elongating greatly with ripening fruit below; **floral branches** 2–25 cm long, ascending with two or more leaves; **pedicels** 5–7 mm long, ascending to spreading, glabrous, reddish and to *c*. 15 mm long with fruit, reduced above; **flowers** perfect, 3–5 mm wide by 6–8 mm long; **sepals** four, 2.5–4.5 mm long by 1–1.7 mm wide, apices and margins white, glabrous or with a few branching hairs near the apices, soon deciduous; **petals** four, 5–10 mm long by 1–2 mm wide, tapered to a long claw, glabrous, veins obvious; **stamens** six, included, four long and two short, short ones *c*. 3.7 mm long, long ones 5–5.5 mm long; **filaments** white, glabrous; **anthers** yellow, 1.1–1.3 mm long, curved away from the pistil; **pistil** one, included, cylindrical, 3.5–4.5 mm long by *c*. 0.6 mm wide, yellowish green, glabrous; **style** obscure; **stigma** slightly wider than the style, whitish green; **racemes in fruit** each 13–18 cm wide and to *c*. 23 cm long. FRUIT a silique, mostly spreading to erect, 2.5–9 cm long by 1.2–2.5 mm wide, 1-nerved for most of its length except near its apex; **valves** two, opening from the base (at pedicel), glabrous, falling when ripe leaving the septum without seeds; **seeds** more or less in one row when ripe, 1.3–2 mm long by 1–1.3 mm wide by 0.3–0.5 mm thick, golden tan, winged edges hyaline, one-grooved on both sides; **septum** tan, unnerved.

■ **LEAVES** Basal and stem, alternate, simple; **basal leaves** wilting early, hairy on both sides with white 3-parted (stellate) hairs, forming a rosette 6–8 cm wide of *c*. 15 leaves; **blades** 1.3–4 cm long by 4–10 mm wide, entire to toothed; **petioles** winged, 1–2.5 cm long; **stem leaves** sessile, erect, imbricate, entire, clasping, pointed; **blades** 2–7 cm long by 3–15 mm wide, reduced above, glabrous or hairy below on midrib and margins, hairs simple or stellate.

■ **STEM** Erect, simple to branched throughout or at least near the top, round, solid, stiff, green or purplish, glabrous or basal hairs 2- or 3-branched; 1–7 mm wide near the base.

■ **SYN.** *Arabis confinis* var. *interposita* (Greene) Welsh & Reveal.

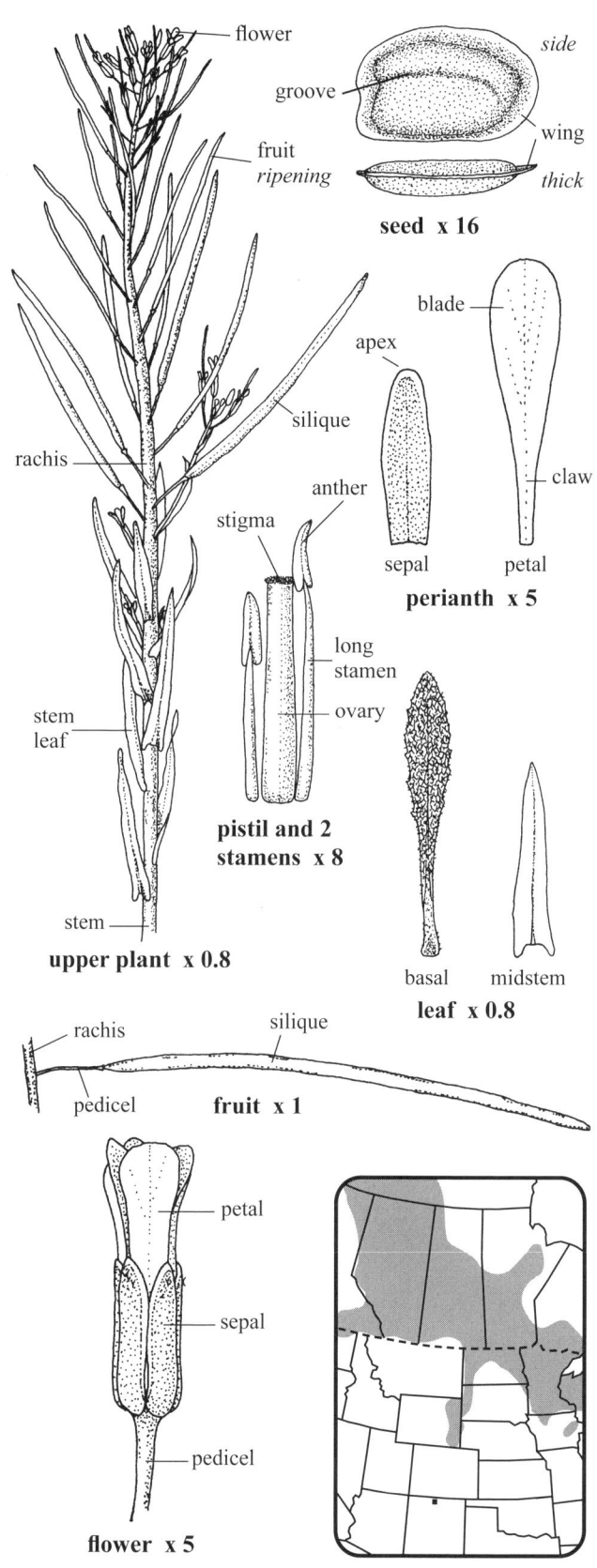

Mustard—*Brassicaceae* **Wildflower** White to pink to purple Petals 4

REFLEXED ROCK CRESS   *Arabis holboellii* Hornem.

■ **SKETCH** A variable **biennial** or **perennial herb** 10–90 cm tall from a tan **taproot** to *c.* 16+ cm long and a simple to branched **caudex** 2–10 mm wide; in sandy prairies, gravelly and rocky sites and mountain meadows.

■ **FLOWERS** White to pink to purple, blooming April–July; **inflorescence** a raceme 12–20 mm wide (flowers only); **pedicels** 3–5 mm long, purplish green, glabrous to hairy, bent downward with fruit and 6–16 mm long; **subtending bracts** absent; **flowers** perfect, 1.5–2.5 mm wide by 4–5 mm long, spreading; **sepals** four, green, 2–4 mm long by 0.5–2 mm wide, erect, blunt, margins and tips hyaline white, hairy on the outer surface, more so near the tips, some hairs branched; **petals** four, white to pink, glabrous, imbricate, exserted 1–1.5 mm beyond the sepals, slightly spreading at the tips, TL 4–10 mm by 1.2–3.5 mm wide, tapered to a claw; **stamens** six, four long and two short, 2–3 mm long, glabrous; **filaments** pale green, shorter or longer than the pistil, round; **anthers** yellow, 0.6–0.7 mm long, oval; **pistil** 2–2.5 mm long, included, glabrous; **ovary** *c.* 0.5 mm wide, cylindrical; **style** obscure; **stigma** flat, round, about as wide as the ovary. **FRUIT a silique** (pod), hanging, slightly curved to straight, 3–7 cm long by 1–2.5 mm wide by 0.7–1 mm thick, glabrous to slightly hairy, flattened, 1-nerved on each valve in lower half, 2-valved, opening from the base; **beak** absent, pod tip is flat and without seeds; **seeds** golden, winged, glabrous, 1–1.7 mm long by 0.6–1.2 mm wide by *c.* 0.3 mm thick, single file to slightly staggered on each side of the septum; **septum** not veined, complete.

■ **LEAVES** Basal and stem, alternate, simple, entire to few-toothed; **basal rosette** 3–9 cm wide, reddish pink at anthesis; **lowest stem blades** 1–5 cm long by 2–8 mm wide, congested, pointed, tapered at the base, white hairs simple to branched throughout; **petioles** 7–15 mm long by 1–2 mm wide; **upper stem leaves** erect, entire, glaucous, auriculate and clasping, sessile, 2–4 cm long by 2–8 mm wide, glabrous above to hairy below, reduced and more distant above; **auricles** 2–3 mm long, rounded, slightly spreading, entire, sometimes absent.

■ **STEM** Erect, 1–15+ in a clump, simple to branched above, hairy throughout or glaucous and glabrous above and hairy below, the stellate hairs with two or three branches; 2–7 mm thick near the reddish to green base.

■ **SYN.** *Arabis retrofracta* Graham = *Arabis holboellii* var. *retrofracta* (Graham) Rydb.

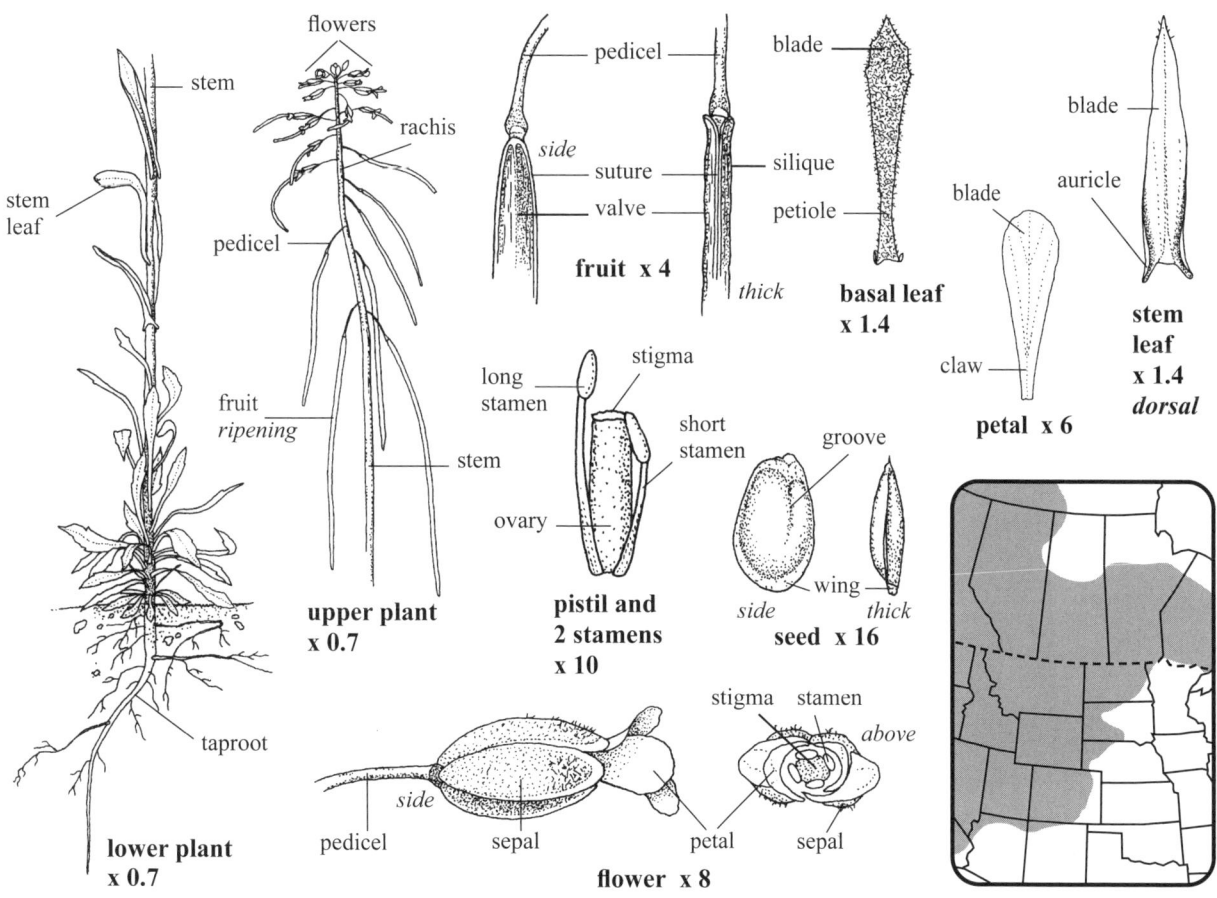

Holboell's Rockcress

Mustard—*Brassicaceae*   Wildflower   White   Petals 4

SHEPHERD'S-PURSE   *Capsella bursa-pastoris* (L.) Medik.

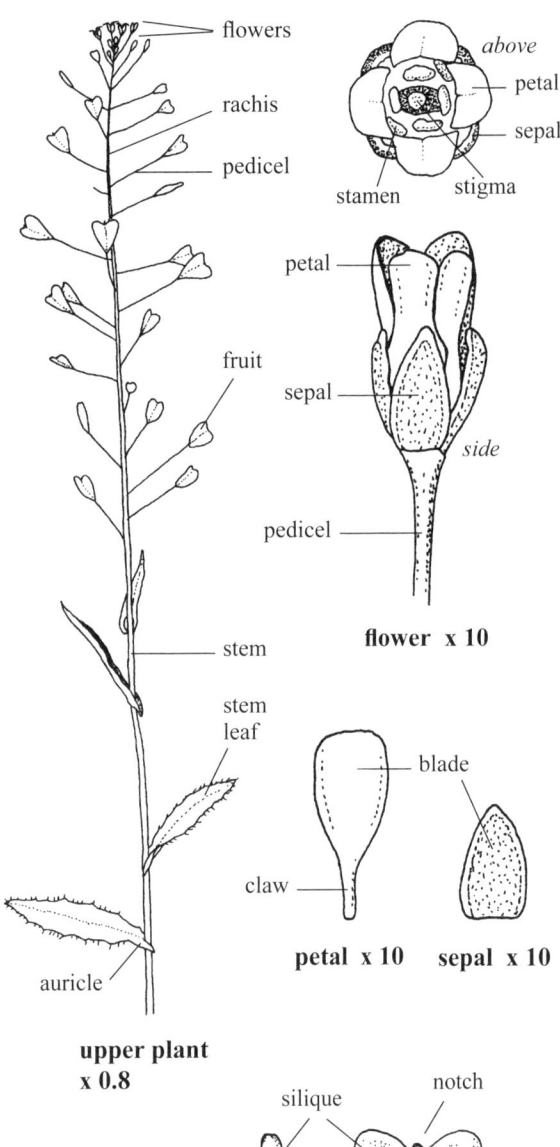

■ **SKETCH** An **annual** or **biennial herb** 15–50 cm tall from a **taproot** 2–3 mm wide and to *c.* 15 cm long; in gardens, orchards, clearings, fields, on open slopes and along roadsides. *Naturalized*

■ **FLOWERS** White, blooming March–December; **inflorescence** of racemes, elongating greatly with fruit; **floral branches** not as long as the central floral stem; **pedicels** glabrous, 3–6 mm long in flower, 10–20 mm long and spreading with fruit; **flowers** perfect, squarish, 1.5–2 mm wide by 2–3 mm long, crowded; **sepals** four, light green to pinkish, pointed, 1.5–2 mm long by *c.* 1 mm wide, glabrous or with a few hairs, margins hyaline; **petals** four, each 1.8–3.8 mm long by 1–1.5 mm wide, clawed; **stamens** six, included, 1–1.5 mm long, four long and two slightly shorter, the shorter outer pair not reaching the stigma as do the four inner ones; **anthers** yellow, *c.* 0.5 mm long; **pistil** flattened, *c.* 1.4 mm long and *c.* 0.8 mm wide, glabrous; **style** and **stigma** 0.3–0.4 mm long; **fruiting stalks** to *c.* 40 cm long by 1.2–4.5 cm wide. **FRUIT a silique**, flattened and notched, dehiscent, 5–8 mm long by 4–5 mm wide by 1.5–2 mm thick, glabrous; **seeds** 14–20 per silique, 0.8–1 mm long by 0.4–0.6 mm wide by *c.* 0.5 mm thick, golden, grooved on each side.

■ **LEAVES** Basal and stem, alternate, simple, variable, toothed to lobed; **basal leaves** forming a rosette of 10–30 leaves, together 8–20 cm across; **blades** flat, 3–10 cm long by 6–30 mm wide, unevenly toothed and sometimes lobed almost to the midrib, slightly hairy along the margins or not, blades covered with appressed stellate hairs on both surfaces, side veins faint; **petioles** 0.6–5 cm long; **stem leaves** sessile, clasping with auricles; **blades** lightly hairy on both surfaces and margins, 1.5–2.5 cm long by 3–6 mm wide, larger near the base of the stem, toothed to entire.

■ **STEM** Erect to decumbent, one to several, glabrous or with stellate hairs below, the main stem sometimes with alternate branching in the lower half; 1–3 mm thick near the base.

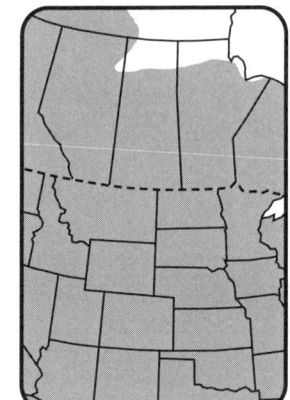

216

Mustard—*Brassicaceae*   **Wildflower White Petals 4**

BITTER CRESS   *Cardamine pensylvanica* Muhl. ex Willd.

■ **SKETCH** An **annual** or **biennial herb** 10–70 cm tall from white **fibrous roots** 3–6 cm long; in moist ditches, wet woodlands, along streams and springs in woodlands, mountains and prairies.

■ **FLOWERS** White, blooming April–July; **inflorescence** a raceme 1–3 cm long by 1–2 cm wide (flowers only), terminal and at ends of upper branches; **pedicels** glabrous, 3–6 mm long with flower, lengthening to 8–10 mm and mostly ascending to erect with fruit; **flowers** perfect, glabrous, 4–6 mm wide by 3–4 mm long, erect; **sepals** four, 1.8–2.2 mm long by *c.* 1 mm wide, erect, upper half slightly reddish green, apices white; **petals** four, white, 1.5–4 mm long by 1–1.5 mm wide, blunt; **stamens** six (didynamous), four long and two short, long ones *c.* 2.5 mm and exserted, short ones *c.* 2 mm long; **filaments** white, glabrous; **anthers** yellow, *c.* 0.3 mm long; **pistil** one, cylindrical, slightly exserted, 2–2.3 mm long by *c.* 0.3 mm wide, glabrous, reddish green; **style** obscure; **stigma** flat, hairy, whitish green, about as wide as the ovary. **FRUIT a silique**, reddish brown, tan and speckled with black when ripe, glabrous, 1–3 cm long by *c.* 1 mm wide, ascending and mostly straight, 2-valved, opening from the pedicel end; **beak** (persistent style) 0.5–2 mm long, seedless; **septum** complete, nerveless, tan; **seeds** *c.* 30 in a 2 cm long silique, in single file on each side of septum, variable in shape, 0.8–1.5 mm long by 0.6–0.7 mm wide by 0.2–0.3 mm thick, smooth and dull, flattened, oblong to almost round, tan but darker brown at hilum and along margins.

■ **LEAVES** Basal and stem, alternate, glabrous, compound, ascending; **basal leaves** in a rosette, pinkish, similar to stem leaves in shape; **petioles** 10–20 mm long; **stem leaves** 1–10 cm long by 0.6–4 cm wide, reduced above; **lateral leaflets** 2–11 pairs, apiculate, 4–15 mm long by 1–12 mm wide, 1-veined, margins usually entire, bases continuous with the rachis; **terminal leaflet** the largest at 5–30 mm long by 2–20 mm wide, usually with two blunt teeth or apiculate side lobes; **petioles** 1–10 mm long by *c.* 1 mm wide; **petiolules** 1–4 mm long; **rachis** ridged on margins.

■ **STEM** Erect, solitary, branched above, pinkish green, with four low ridges, slightly hairy above, more hairy below or glabrous, hairs short and spreading; 1.5–4 mm thick near the hairy base.

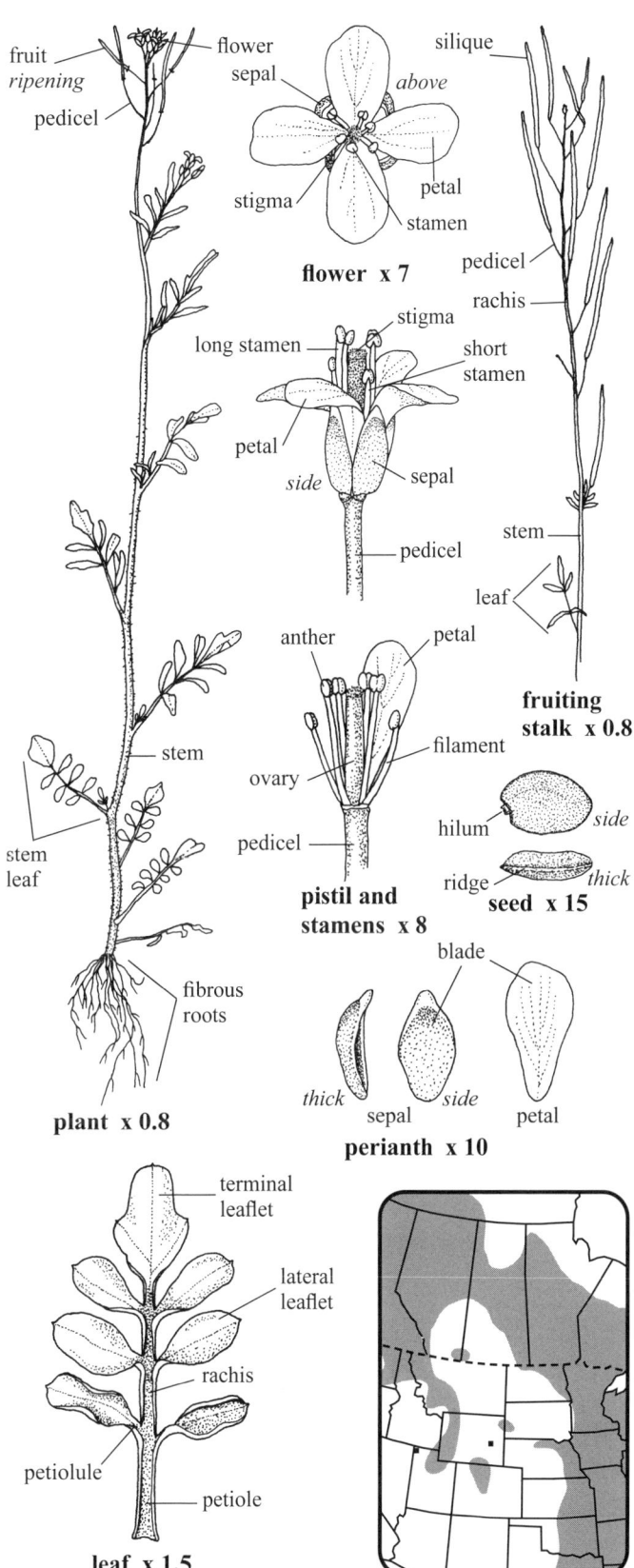

Quaker Bitter-cress, Pennsylvania Bittercress

Mustard—*Brassicaceae* **Wildflower** Yellow Petals 4

GRAY TANSY MUSTARD *Descurainia incana* (Bernh. ex Fisch. & C.A. Mey.) Dorn

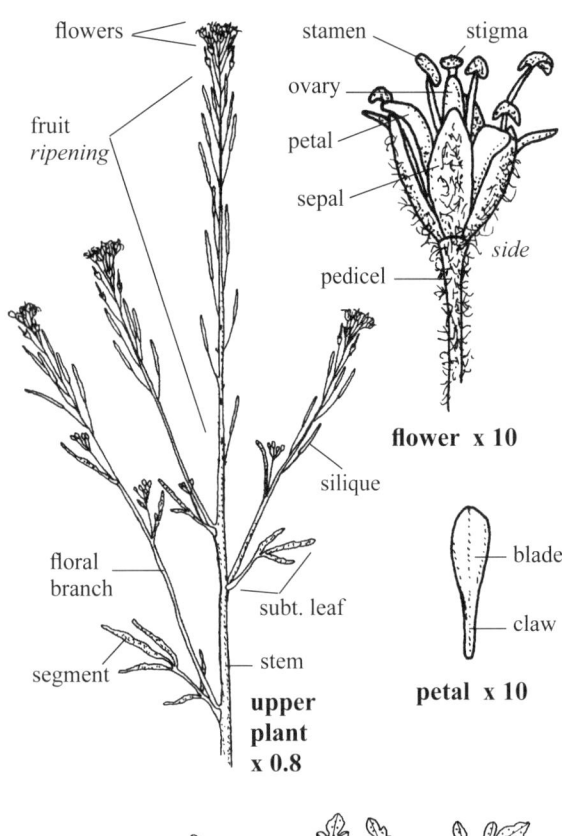

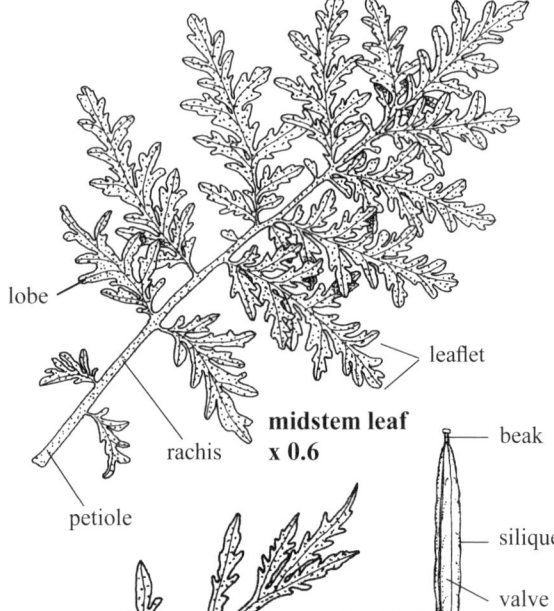

■ **SKETCH** A variable **biennial herb** 30–130 cm tall and to *c.* 40 cm wide with a hairy grayish appearance; in gravelly areas, sagebrush, mountain meadows, open river banks, open woods and along roadsides.

■ **FLOWERS** Yellow, blooming April–August; **inflorescence** a raceme, terminal and on upper floral branches; **floral branches** up to about 25 per plant, reaching to *c.* 45 cm in length, most with leaves below the racemes; **pedicels** hairy, 3–10 mm long, ascending; **flowers** perfect, 2.5–3 mm wide and long; **sepals** four, erect, light yellow, 1–2 mm long by 0.6–0.8 mm wide, hairy on the outer surface, glabrous inside; **petals** four, glabrous, *c.* 2 mm long by *c.* 0.5 mm wide, clawed; **stamens** six, four long and two short; **filaments** 1.5–1.8 mm long; **anthers** yellow, *c.* 0.3 mm long; **ovary** glabrous, exserted; **style** constricted and quite short; **stigma** round and flat; **central fruiting stalk** the longest at 3–25+ cm by *c.* 1 cm wide. **FRUIT a silique**, 2-valved, opening from the base when ripe, close to the rachis, erect, 5–12 mm long by 0.5–1.2 mm wide, glabrous, round in cross-section, midnerve narrow and slightly raised on valves; **beak** 0.4–0.8 mm long, seedless; **seeds** 4–15 per silique, in one row, golden brown, *c.* 1 mm long by 0.5–0.7 mm wide by *c.* 0.5 mm thick, lightly striated, with one or two grooves.

■ **LEAVES** Basal and stem, alternate, compound (once or twice divided), lobed or toothed, grayish green; **basal rosette** to *c.* 18 cm wide; **basal blades** to *c.* 8 cm long and *c.* 3.5 cm wide, often withered by anthesis; **petioles** 2–4 cm long; **stem blades** 2–15 cm long by 1–9 cm wide, smaller and less divided above, downy, some hairs branched, mostly eglandular; **petioles** 0–2.5 cm long, hairy.

■ **STEM** Erect, tough, solid, round, branched above, more hairy and downy above, hairs branched, not glandular; 5–10 mm wide near the smooth base.

■ **SYN.** *Descurainia richardsonii* O.E. Schulz = *Descurainia incana* ssp. *incana* (Bernh. ex Fisch. & C.A. Mey.) Dorn.

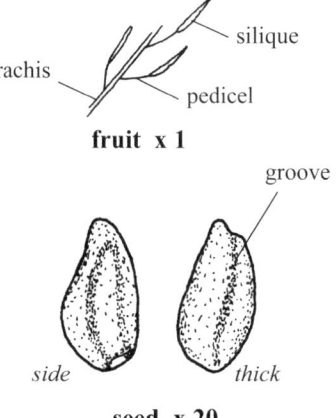

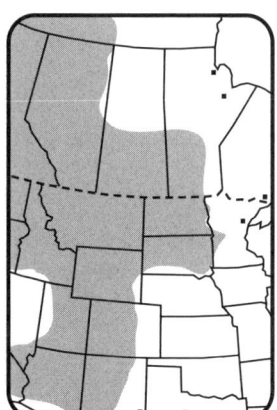

Mustard—*Brassicaceae* Wildflower Yellow Petals 4

FLIXWEED *Descurainia sophia* (L.) Webb ex Prantl

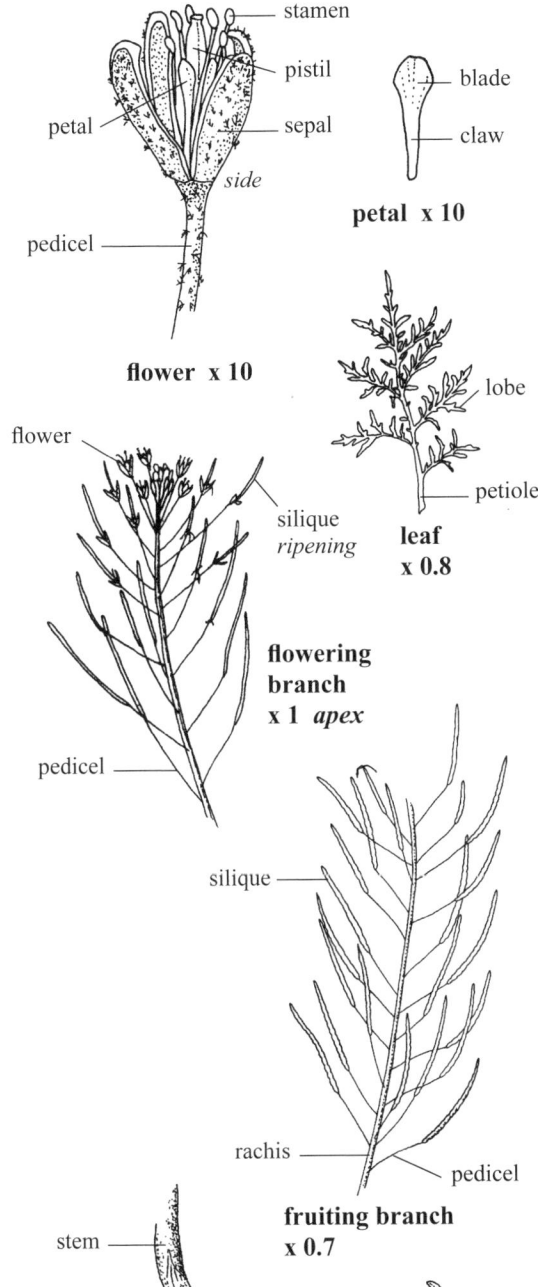

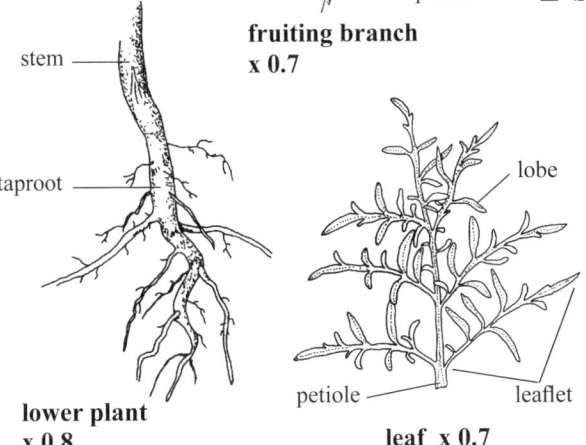

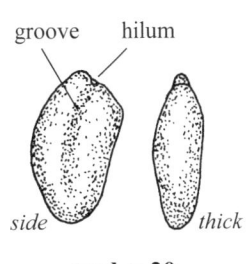

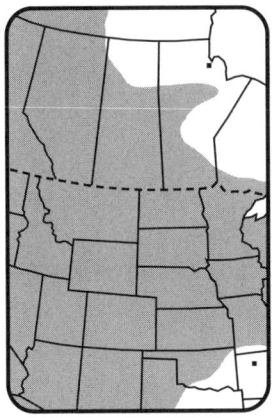

■ **SKETCH** A hairy **annual** or **biennial herb** 30–100 cm tall with a tan **taproot** 7–15 cm long; in lawn edges, deserts, rocky shores, prairies, grassy slopes, canyon bottoms and cultivated fields. *Naturalized*

■ **FLOWERS** Yellow, blooming March–September; **inflorescence** a raceme; **pedicels** green, stellate hairy, 3–15 mm long, ascending in fruit; **flowers** perfect, lightly fragrant, 2–3 mm wide and long, clustered at apices of floral branches with fruit ripening below; **sepals** four, ascending, tips hooded, 2–2.6 mm long by 0.5–0.8 mm wide, hairy to glabrous on outside, margins hyaline, usually obscurely 3-veined, quickly deciduous; **petals** four, pale yellow, between and shorter than the sepals, 1.5–1.8 mm long, glabrous, clawed, apices 0.3–0.4 mm wide; **stamens** six, two short (outer) and four long, 1.5–2 mm long; **filaments** whitish green; **anthers** pale yellow, 0.2–0.5 mm long; **pistil** *c.* 2 mm long by *c.* 0.3 mm thick near the base, pale yellow, apex narrowing to a flat stigma, slightly exserted; **fruiting branches** several per plant, becoming 15–45 cm long by 3.5–5 cm wide with up to *c.* 80 siliques, bases leafless. **FRUIT a silique** (pod), 1–3 cm long by 0.6–1 mm wide, ascending, slightly arched to straight, reddish green, glabrous, 2-valved, round in cross-section, opening from the base, a low midvein along the length of each valve; **beaks** 0.4–1.5 mm long, seedless; **seeds** golden brown, 20–40 per silique, in a single row under each valve, glabrous, 0.8–1.3 mm long by 0.4–0.6 mm wide by 0.3–0.4 mm thick, grooved on both wide surfaces, darker brown at the hilum end; **septum** tan, 2- or 3-nerved.

■ **LEAVES** Alternate, compound (two or three times divided); **blades** 0.8–9 cm long by 1–5 cm wide, scented, grayish green, hairs branched, axils often with leafy fascicles or short flowering clusters; **leaflets** lobed and toothed; **petioles** 0–15 mm long.

■ **STEM** Erect, tough, solid, alternately branched above, round, reddish purple when in fruit, hairs branched to simple; 2–8 mm thick near the base.

■ **SYN.** *Sisymbrium sophia* L.

Herb Sophia, Tansy Mustard   219

Mustard—*Brassicaceae*   **Wildflower**   Yellow   Petals 4

Yellow Whitlow-grass   *Draba nemorosa* L.

■ **SKETCH** An **annual** or **winter annual herb** 8–35 cm tall with a tan **taproot** 3–7 cm long by 1–2 mm thick; in dry sandy soil of hillsides, prairies, dry flats, foothills and along sidewalks.

■ **FLOWERS** Yellow, sometimes fading to pink or white; blooming April–June; **inflorescence** a congested raceme, 10- to 50-flowered, often slightly nodding, elongating with fruit; **pedicels** glabrous, ascending to spreading in flower and fruit, 5–30 mm long; **flowers** perfect, *c.* 3 mm wide by *c.* 2 mm long; **rachis** green, glabrous; **sepals** four, margins lighter green, 1.5–2.4 mm long by 0.8–1 mm wide, body curved with a few outer simple hairs near the blunt apices; **petals** four, pale yellow, glabrous, notched at apices, 2–4 mm long by 1–1.5 mm wide, tapered to a short claw; **stamens** six, four long at *c.* 1.8 mm, and two short at *c.* 1.5 mm; **filaments** thick, green, yellowish near apices; **anthers** yellow, *c.* 0.3 mm long; **pistil** glabrous, *c.* 1.8 mm long by *c.* 0.6 mm wide; **style** obscure; **stigma** not as wide as ovary; **fruiting stalks** 10–25 cm long by 2–5 cm wide. **FRUIT a silique** (pod), 2-valved, glabrous to hairy, 6–13 mm long by 1.2–2 mm wide by 0.8–0.9 mm thick, slightly curved, ascending, opening along two sutures from the base at the pedicel; **beak** minute; **seeds** 40–75 per silique, dull brown, *c.* 0.6 mm long by *c.* 0.4 mm wide *c.* 0. 2 mm thick, in two rows along the margins on both sides of the septum, round at both ends, one groove on each side; **septum** opaque, not nerved.

■ **LEAVES** Basal and stem, alternate, simple, toothed, hairy on both sides; **basal blades** in a rosette, each 3–45 mm long by 3–20 mm wide, larger basal leaves subtending branches, apices blunt, stellate hairs with two to four branches; **petioles** 0–7 mm long, reduced above; **stem leaves** few, in lower half of stems and branches, sessile.

■ **STEM** Main stem erect; 1–2 mm thick near the reddish base; **basal lateral branches** shorter and ascending, with a few leaves, glabrous above, with hairs simple to stellate near the base.

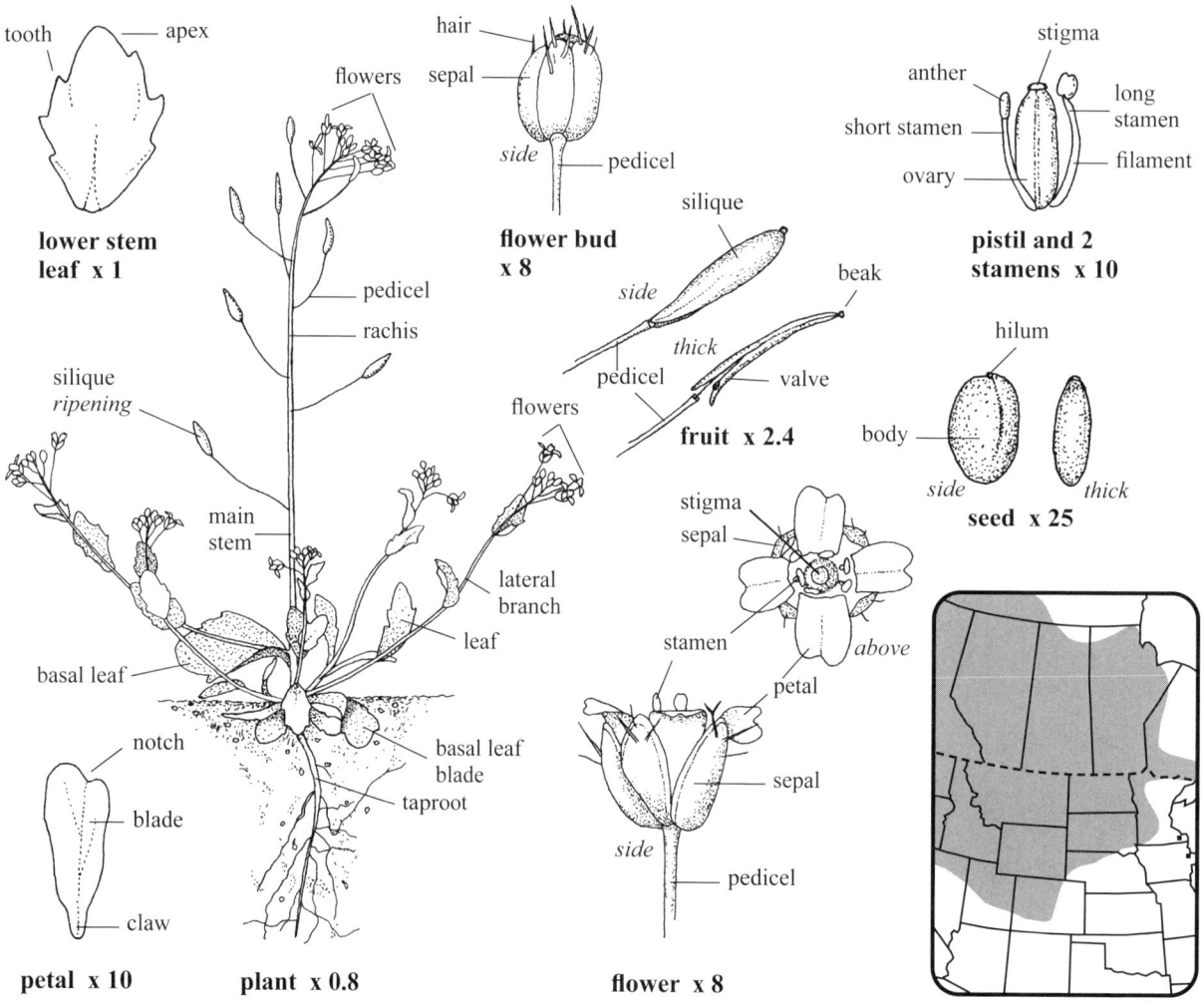

220   Woodland Draba, Woodland Whitlow-grass

Mustard—*Brassicaceae* **Wildflower** Pale yellow Petals 4

DOG MUSTARD *Erucastrum gallicum* (Willd.) O.E. Schulz

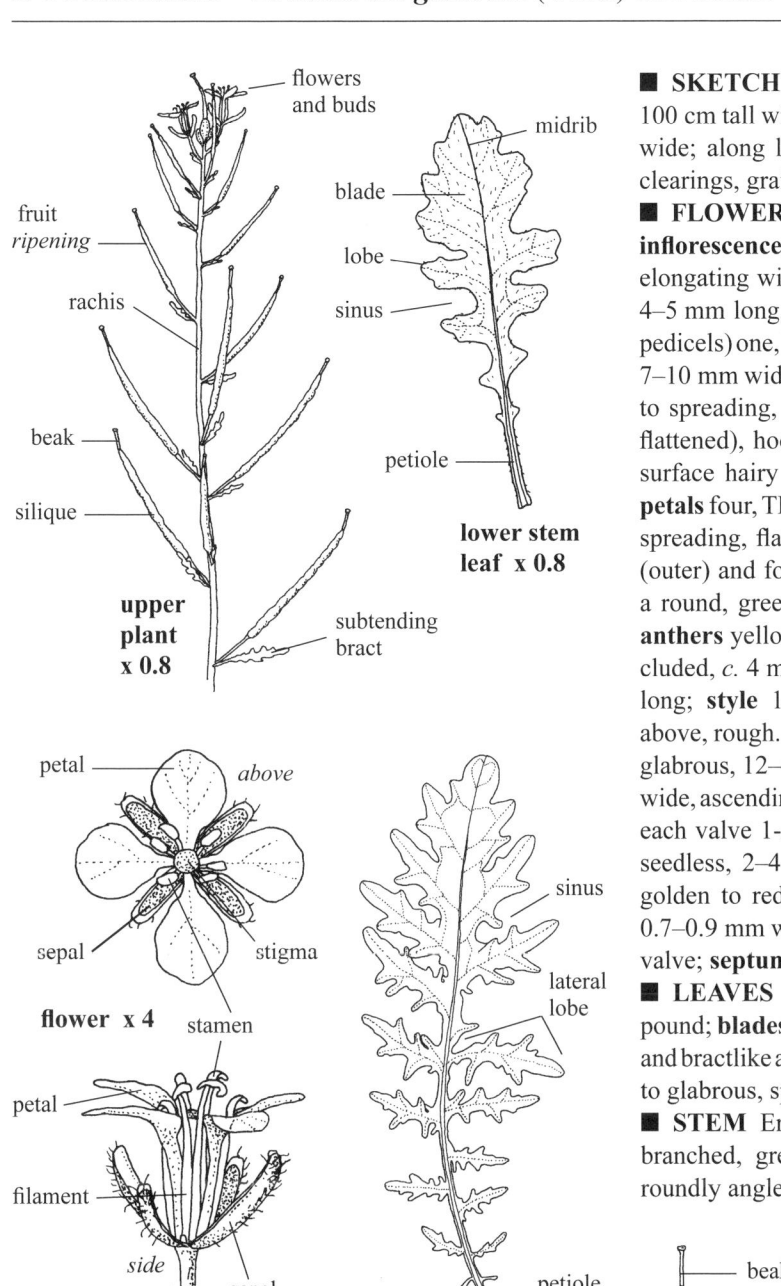

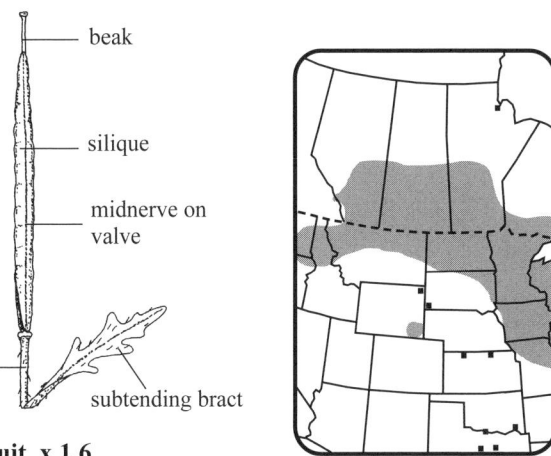

■ **SKETCH** An **annual** or **winter annual herb** 20–100 cm tall with a tan **taproot** 5–11⁺ cm long by 2–4 mm wide; along lawn and field edges, railroads, roadsides, clearings, grainfields and orchards. *Naturalized*

■ **FLOWERS** Pale yellow, blooming March–December; **inflorescence** a raceme, terminal and on floral branches, elongating with fruit; **pedicels** ascending, slightly hairy, 4–5 mm long, 5–18 mm in fruit; **subtending bracts** (of pedicels) one, leaflike, toothed, persistent; **flowers** perfect, 7–10 mm wide by 5–7 mm long; **sepals** four, green, erect to spreading, 2.5–4.5 mm long by *c.* 1.4 mm wide (not flattened), hooded, hyaline and rough at apices, outside surface hairy especially in upper half, glabrous inside; **petals** four, TL 4–8 mm by 2.5–3 mm wide, clawed, blade spreading, flat, entire, glabrous; **stamens** six, two short (outer) and four long (inner), 4–5.5 mm long, each with a round, green gland at its base; **filaments** pale green; **anthers** yellow, *c.* 1 mm long; **pistil** green, glabrous, included, *c.* 4 mm long by *c.* 0.8 mm thick; **ovary** *c.* 3 mm long; **style** 1–2 mm long; **stigma** green, round from above, rough. FRUIT a silique (pod), roundly 4-angled, glabrous, 12–48 mm long (including beak) by 1–2.5 mm wide, ascending to spreading, slightly upcurved, 2-valved, each valve 1-nerved, opening from the base; **beak** thin, seedless, 2–4 mm long; **seeds** oval, 12–30 per silique, golden to reddish brown, glabrous, 1–1.5 mm long by 0.7–0.9 mm wide and thick, one row of seeds under each valve; **septum** entire, not nerved.

■ **LEAVES** Basal and stem, alternate, simple to compound; **blades** 1.5–26 cm long by 1–10 cm wide, reduced and bractlike above, lobed, toothed, flat, thin, slightly hairy to glabrous, spreading; **petioles** hairy, 0–4 cm long.

■ **STEM** Erect to decumbent, solid, simple to few-branched, green, white hairs deflexed and appressed; roundly angled near the 1–5 mm thick base.

Common Dogmustard

Mustard—*Brassicaceae* **Wildflower Yellow Petals 4**

WORMSEED MUSTARD   *Erysimum cheiranthoides* L.

■ **SKETCH** An **annual** or **biennial herb** 20–110 cm tall from a **taproot** 2–15 cm long by 2–5 mm wide in patches several m^2; in meadows, along rivers, roadsides, open hillsides and in cultivated fields. *Naturalized*

■ **FLOWERS** Yellow, blooming May–August; **inflorescence** a raceme, terminal and lateral, central one to *c.* 25$^+$ cm long by 2–2.5 cm wide; **floral branches** up to 20, with small leaves along the lower half, central stalk without leaves at the base; **pedicels** glabrous to lightly hairy, 1–8 mm long in flower, erect, 4–15 mm long but mostly shorter than fruit; **flowers** perfect, each 4–5 mm wide and long, glabrous; **rachis** green with several longitudinal ridges, hairs attached in middle; **sepals** four, erect, light green, glabrous, 2–3.5 mm long by 0.8–1 mm wide; **petals** four, 3–6 mm long by 1.3–1.7 mm wide, with a claw; **stamens** six, four long and two short, included; **long filaments** 2.5–3 mm, free to the base; **anthers** *c.* 0.5 mm long, yellow, erect around the capitate stigma; **ovary** 4-sided, 2.5–3 mm long by *c.* 0.7 mm wide; **style** included; **central fruiting stalk** to *c.* 36 cm long by 3–6 cm wide. FRUIT a **silique**, 2-valved, opening from the base, ascending to erect, 1.2–3 cm long by 0.8–1.2 mm wide, green, glabrous or with stellate hairs, 4-ridged, diamond-shaped in cross-section; **beak** seedless, 0.5–1 mm long; **seeds** 20–40 per silique, 1–1.4 mm long by 0.4–0.6 mm wide by 0.4–0.5 mm thick, golden brown, grooved; **septum** unveined.

■ **LEAVES** Alternate, simple, mostly entire or toothed, reduced above; **lower blades** 2–10 cm long by 1.5–2.3 cm wide, more hairy toward the base, hairs appressed, stellate with two or three points, lateral veins faint; **petioles** 0–5 mm long.

■ **STEM** Erect, tough, solid, simple or branched, with longitudinal ridges, stellate hairs as above; 2–16 mm thick near the green to purplish base.

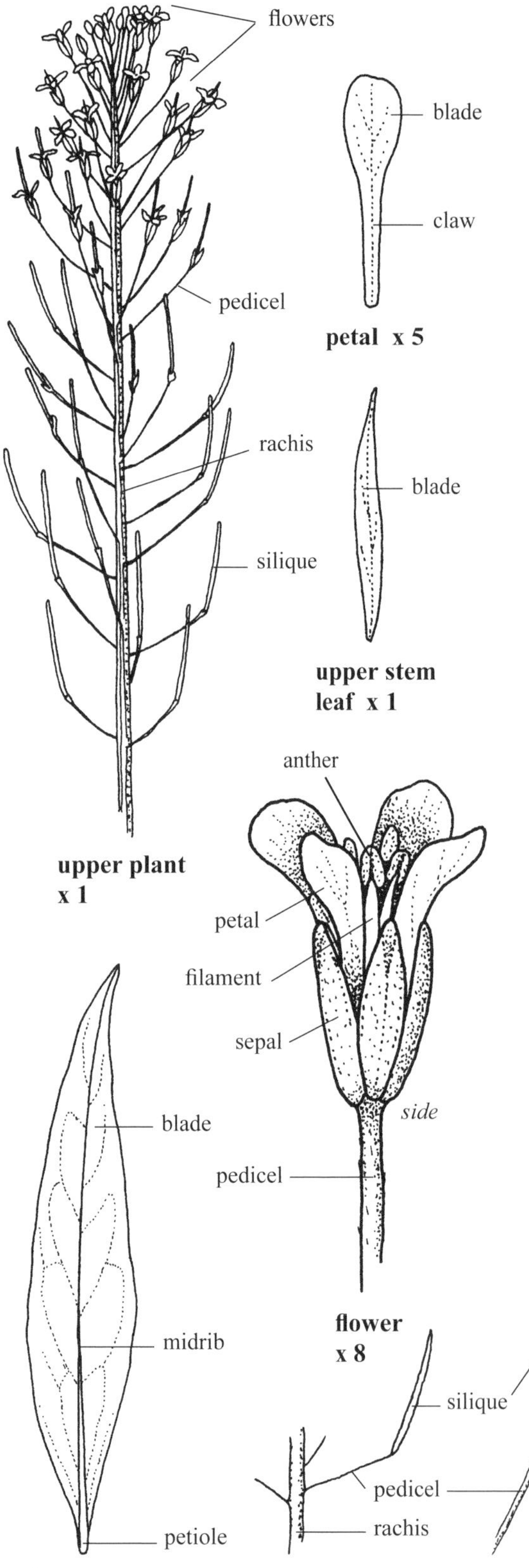

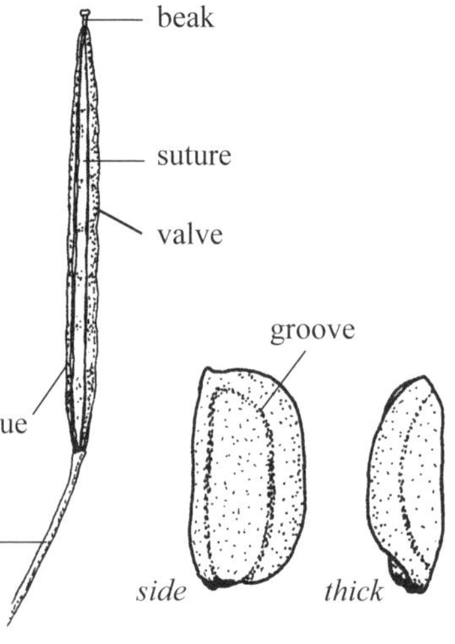

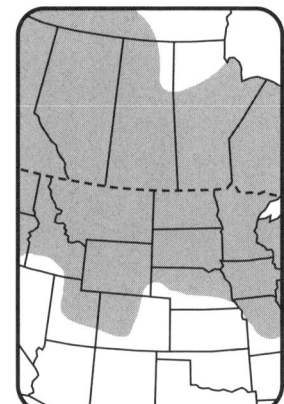

Wormseed Wallflower, Treacle Mustard

Mustard—*Brassicaceae* **Wildflower** Bluish purple Petals 4

DAME'S-ROCKET *Hesperis matronalis* L.

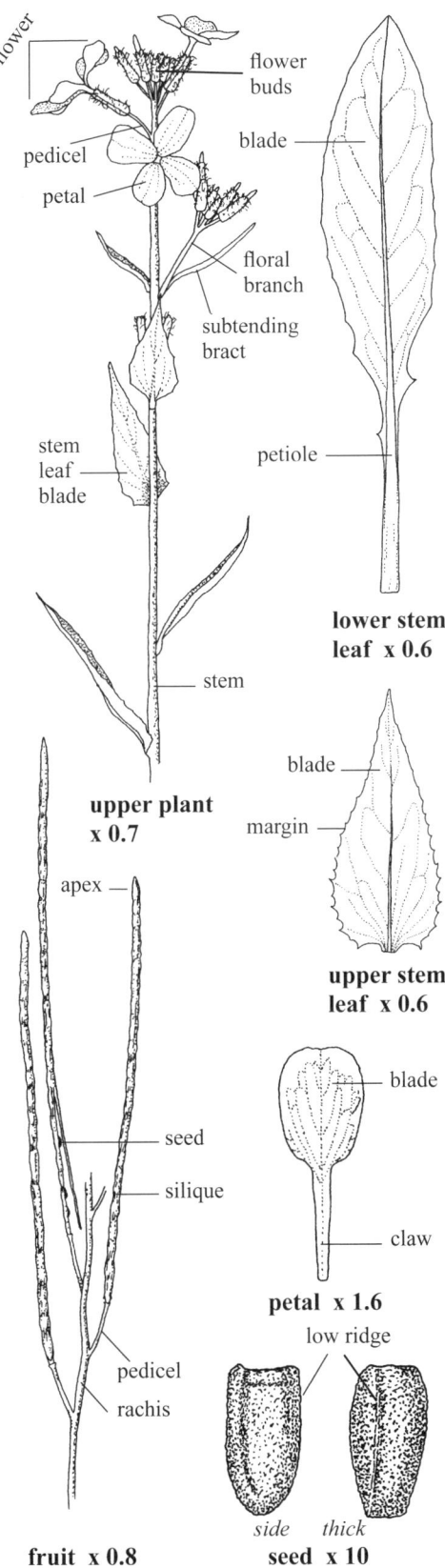

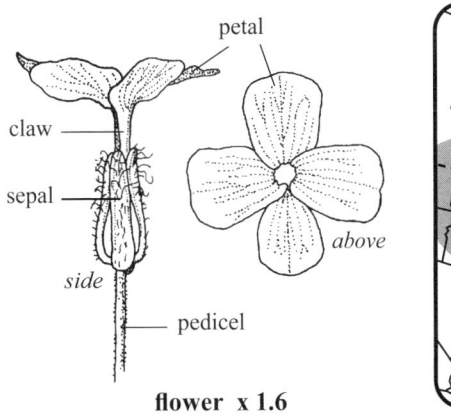

■ **SKETCH** A **biennial** or **perennial herb** 30–120 cm tall in clumps from **rhizomes** 5–10+ cm long; along woodland edges, roadsides and thickets. *Naturalized*

■ **FLOWERS** Bluish purple, blooming April–August; **inflorescence** a raceme 12–20 cm long; **floral branches** ascending from leaf axils, 2–5 cm long, hairs simple to branched; **pedicels** 3–10 mm long, 7–13 mm long in fruit, stout, slightly hairy, hairs simple to branched; **flowers** perfect, fragrant, 15–25 mm across by 15–18 mm long; **sepals** four, erect, 8–11 mm long by 1–2.5 mm wide, apices blunt, bases subequal and reddish, upper half green and hairy, margins pink; **petals** four, 1.5–2.5 cm long; **blades** 10–13 mm long by 6–8 mm wide, claw flat, *c.* 12 mm long by *c.* 1.3 mm wide, longer than sepals; **stamens** six, included, four long and two short, 7–10 mm long; **filaments** pale green, flattened, widening to 0.7–0.9 mm at the base, glabrous, free; **anthers** thin, erect, 3.5–4 mm long; **pistil** included, medium green, erect, 4-angled, hairless, 5–6 mm long by *c.* 0.7 mm thick near the base; **stigma** 2-lobed, descending. **FRUIT a silique**, erect to spreading, with dark streaks and dots, opening from base, 2-valved, several low ridges along each valve, glabrous, slightly shiny, 4–14 cm long by 1–1.5 mm wide, roundish in cross-section, beakless, slightly constricted between seeds; **seeds** in a single row per valve but alternating, about four seeds per cm of silique, medium brown, finely cottony hairy, 2–4 mm long by 1–1.3 mm wide and thick, outer surface convex, 2-ridged, groove shallow; **septum** tan, slightly rough, hairless.

■ **LEAVES** Basal and stem, alternate, simple, toothed, pointed, sessile to petiolate, hairy on both sides; **basal blades** 3–18 cm long by 1–4 cm wide, brown by anthesis; **petioles** 4–11 cm long; **stem blades** 3–13 cm long by 0.8–3 cm wide, ascending to spreading, midrib obvious, lateral veins obscure, soft to touch, hairs simple to branched and more branched below; **petioles** 0–5 cm long, soft and hairy, reduced above.

■ **STEM** Erect, simple to branched above, 1–16 stems per clump, finely hairy throughout, hairs simple to branched; 3–7 mm wide near the base.

Mother-of-the-evening, Dame's Violet, Sweet Rocket

Mustard—*Brassicaceae* **Wildflower** Pale green Petals absent or vestigial Sepals 4

COMMON PEPPER-GRASS *Lepidium densiflorum* Schrad.

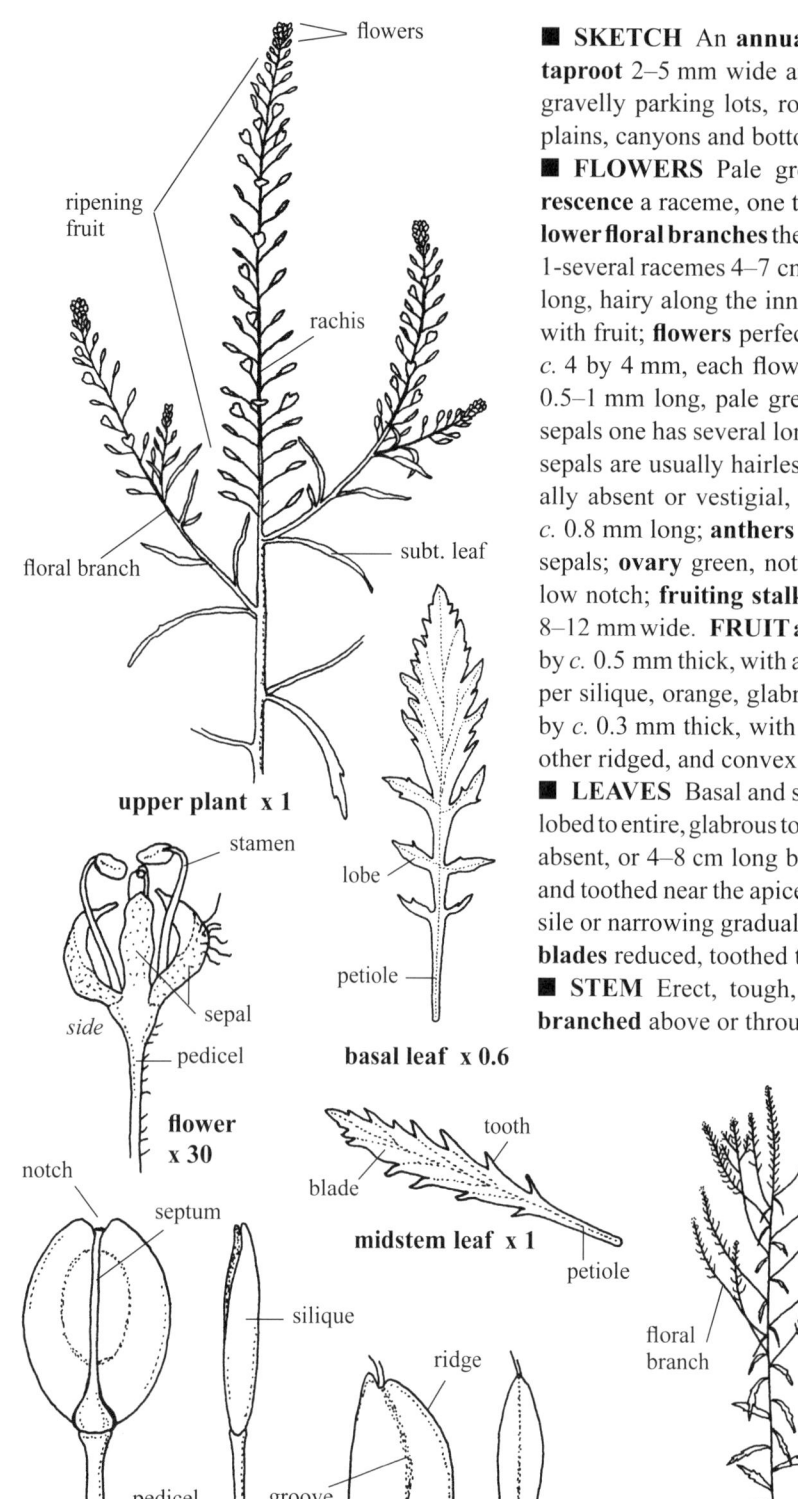

upper plant x 1
flower x 30
basal leaf x 0.6
midstem leaf x 1
fruit x 10
seed x 20
plant x 0.2

■ **SKETCH** An **annual** or **biennial herb** 10–60 cm tall from a **taproot** 2–5 mm wide and 3–20 cm long; in dry or moist sites in gravelly parking lots, roadsides, gardens, open woods, sagebrush plains, canyons and bottom lands.
■ **FLOWERS** Pale green, blooming March–September; **inflorescence** a raceme, one to several[+], ascending, each 4–22 cm long; **lower floral branches** the longest to *c.* 33 cm, often rebranching with 1-several racemes 4–7 cm long; **pedicels** slightly flattened, *c.* 2 mm long, hairy along the inner surface, ascending and to *c.* 4 mm long with fruit; **flowers** perfect, tiny and crowded at the top in a cluster *c.* 4 by 4 mm, each flower 0.8–1 mm long and wide; **sepals** four, 0.5–1 mm long, pale green, quickly deciduous, of the two longer sepals one has several long hairs and a narrow white tip, the shorter sepals are usually hairless, mostly green and incurved; **petals** usually absent or vestigial, white; **stamens** two, exserted; **filaments** *c.* 0.8 mm long; **anthers** *c.* 0.2 mm long, yellow, opposite the long sepals; **ovary** green, notched; **stigma** filling the base of the shallow notch; **fruiting stalks** 3–50[+] per plant, each 5–30 cm long by 8–12 mm wide. FRUIT a **silique**, 2.5–3.5 mm long by 2–3 mm wide by *c.* 0.5 mm thick, with an apical notch, mostly glabrous; **seeds** two per silique, orange, glabrous, 1.3–1.5 mm long by 0.7–1 mm wide by *c.* 0.3 mm thick, with a shallow groove, one edge flattened, the other ridged, and convex.
■ **LEAVES** Basal and stem, alternate, simple, coarsely toothed or lobed to entire, glabrous to very slightly hairy; **basal leaves** sometimes absent, or 4–8 cm long by 1–2 cm wide; **lower stem blades** lobed and toothed near the apices, to *c.* 10 cm long by 2–2.5 cm wide, sessile or narrowing gradually to a petiole, flat, soft, spreading; **upper blades** reduced, toothed to entire.
■ **STEM** Erect, tough, glaucous to glabrous, or hairs minute; **branched** above or throughout; 1–5 mm thick near the base.

Pepper Grass, Miner's Peppergrass, Prairie Peppergrass

Mustard—*Brassicaceae*   **Wildflower**   White and green   Petals 4

Branched Pepper-grass   *Lepidium ramosissimum* A. Nels.

- **SKETCH** An **annual** or **biennial herb** 15–50 cm tall by 15–40 cm wide from a **taproot** 3–12 mm thick by 7–20 cm long; in disturbed sites, roadsides, dry plains and slopes, often in sandy soil.
- **FLOWERS** White and green, blooming April–October; **inflorescence** a raceme, terminal and axillary, each 2–20 cm long by 4–8 mm wide; **pedicels** *c.* 1.2 mm long, to 2 mm long in fruit, green, hairy; **flowers** perfect, *c.* 1.4 mm wide by *c.* 1.2 mm long, numerous; **sepals** four, 0.8–1 mm long by *c.* 0.6 mm wide, C-shaped, margins white, hairs few and scattered on outer surface, apices irregularly blunt; **petals** four, sometimes absent, white, blunt, ascending, glabrous, *c.* 0.6 mm long by *c.* 0.2 mm wide; **stamens** two, exserted, incurved, opposite two sepals, attached at base of ovary, *c.* 1.1 mm long, hairless; **filaments** white, curved, *c.* 0.9 mm long; **anthers** yellow, *c.* 0.2 mm long, roundish; **ovary** green, *c.* 0.8 mm long and wide, short hairs on margins; **style** and **stigma** obscure, set in a notch at apex of ovary. **FRUIT** a silique, tan, 2.2–3.4 mm long by 1.6–2 mm wide by *c.* 0.6 mm thick, short hairs along margin and sparingly on body; **seeds** two per silique, golden brown, 1–1.3 mm long by *c.* 0.8 mm wide by *c.* 0.3 mm thick, hairless, a thin groove on each side, margins rough (microscopic); **septum** complete, pale tan, with a thin midnerve its full length.
- **LEAVES** Basal and stem, alternate, simple, deeply lobed, toothed to entire above; **basal leaves** TL 2–5 cm by 6–15 mm wide, lobed, quickly wilting; **petioles** 10–25 mm long, flat; **stem leaves** subtending branches and along floral branches, numerous, 6–30 mm long by 1–10 mm wide, sessile, thin, clasping, slightly hairy especially along the margins, midrib faint, lateral veins obscure.
- **STEM** Main stem with numerous alternate leafy branches 2–20 cm long, starting at ground level, spreading to ascending, green, with short white descending hairs; 2–8 mm thick near the base.

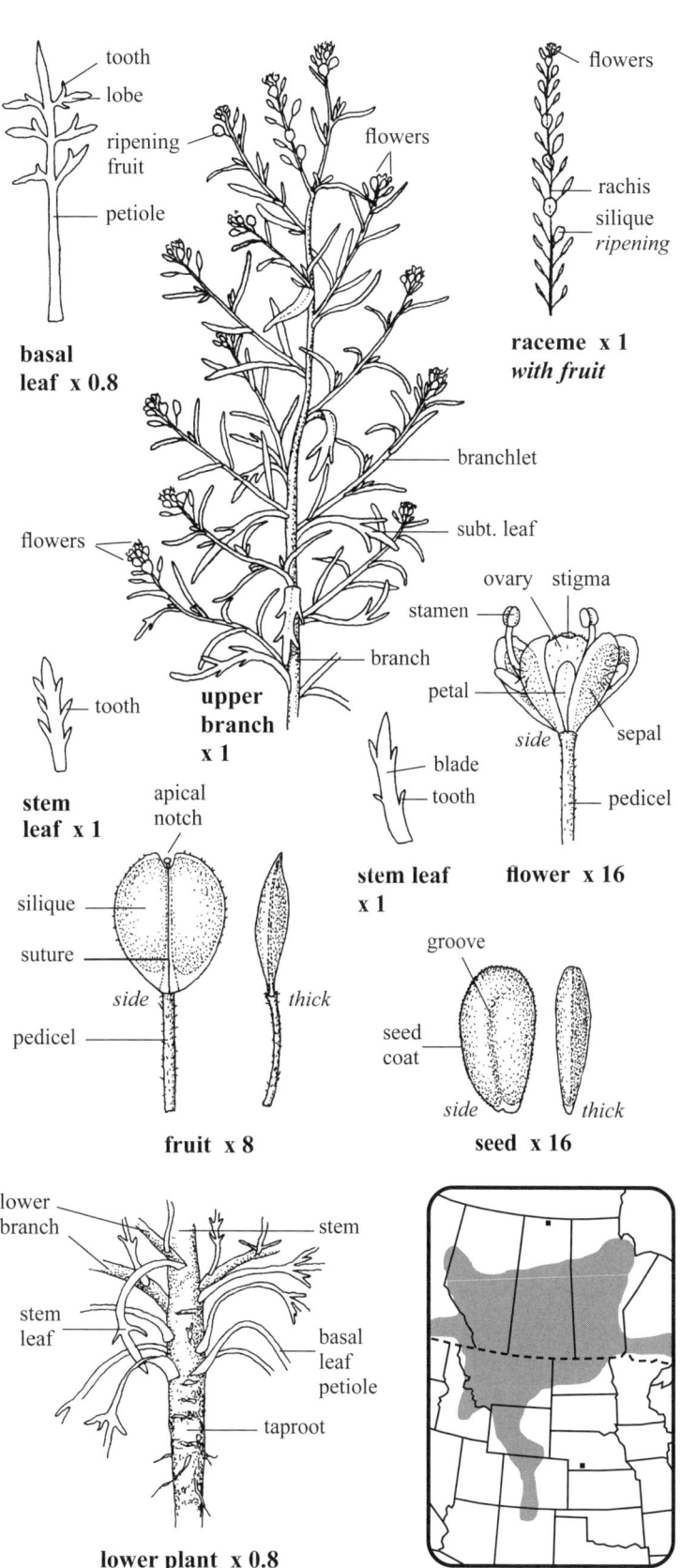

Bushy Peppergrass, Manybranched Pepperweed, Branched Pepperwort

Mustard—*Brassicaceae*   **Wildflower** Yellow Petals 4

Marsh Yellow Cress   *Rorippa palustris* (L.) Bess.

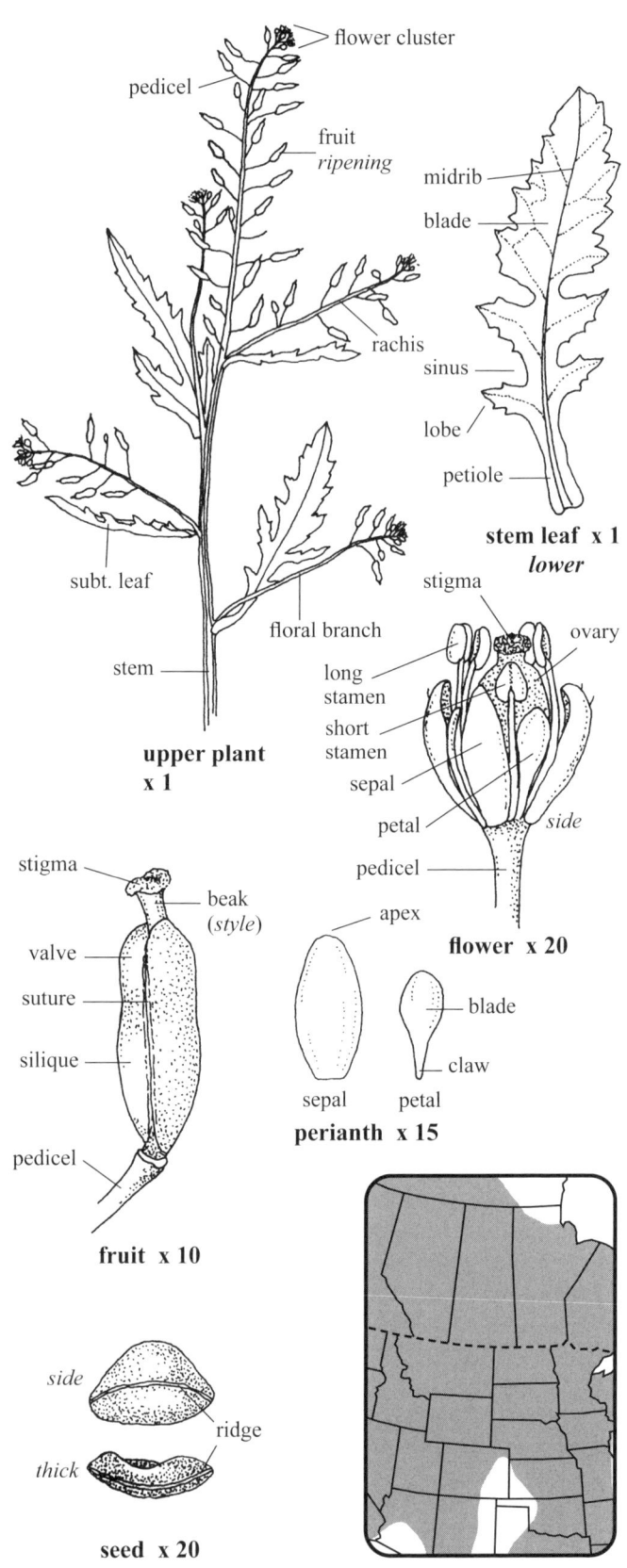

upper plant x 1

stem leaf x 1 *lower*

fruit x 10

flower x 20

perianth x 15

seed x 20

- **SKETCH** A variable, glabrous to hairy **annual** or **biennial herb** 20–100 cm tall with a **taproot**; along edges of rivers, streams, lakes and ponds, wet roadsides and pastures.
- **FLOWERS** Yellow, blooming May–October; **inflorescence** a raceme, terminal and axillary, the tiny flowers clustered at tips of the racemes with developing fruit below; **pedicels** green, 3–7.5 mm long, to *c.* 10 mm long with fruit, spreading to ascending with fruit, glabrous, filiform; **flowers** perfect, 2–3 mm wide by 1–1.5 mm long; **sepals** four, yellow, 1.4–2.5 mm long by *c.* 0.8 mm wide (flattened) usually C-shaped, slightly hooded, not persistent; **petals** four, yellow, 1–3.3 mm long by 0.3–0.5 mm wide, glabrous, tapered to a claw; **stamens** six, yellow, four *c.* 1.2 mm long and two *c.* 1 mm long; **filaments** filiform, glabrous; **anthers** yellow, *c.* 0.3 mm long, 2-lobed, pressed against the stigma at anthesis; **pistil** one, green, glabrous, *c.* 1.2 mm long, exserted; **ovary** slightly oval, 0.7–0.8 mm wide; **style** short, thick, green, not as wide as the stigma, 0.3–1 mm long in fruit, seedless; **stigma** oval, *c.* 0.4 mm wide but not as wide as the ovary, pale in fruit. **FRUIT a silique**, roundish to cylindrical, 3–8 mm long by 1–3 mm wide, glabrous, 2-valved, opening from the base by the pedicel, beaked, straight to slightly curved upward toward the rachis, slightly constricted in the center or not, slightly tapered at the apex or not; **beak** (persistent style) 0.2–1 mm long; **seeds** 20–80 per silique, brown, 0.5–0.9 mm long by 0.3–0.5 mm wide, dull, with a faint low ridge.
- **LEAVES** Alternate, simple, lobed and toothed to pinnatifid, clasping or not, subtending lateral branches; **blades** 2.5–17 cm long by 1–5 cm wide, reduced above, very hairy to almost hairless above and below, 2- to 4-lobed below the large terminal lobe; **petioles** 0–2.5 cm long, reduced above, winged, sometimes clasping, hairs simple.
- **STEM** Light green, erect, rarely prostrate, one to several from a taproot's apex, branches to *c.* 10 cm long and sometimes rebranching, lower half of branches bare, stems bristly hairy to hairless, several-grooved above; 2–4 mm wide near the base.
- **SYN.** *Rorippa hispida* (Desv.) Britt. = *Rorippa palustris* ssp. *hispida* (Desv.) Jonsell.

226   Bog Marsh Cress, Bog Yellowcress, Yellow Cress, Marsh Cress

Mustard—*Brassicaceae* **Wildflower** Yellow Petals 4

WILD MUSTARD *Sinapis arvensis* L.

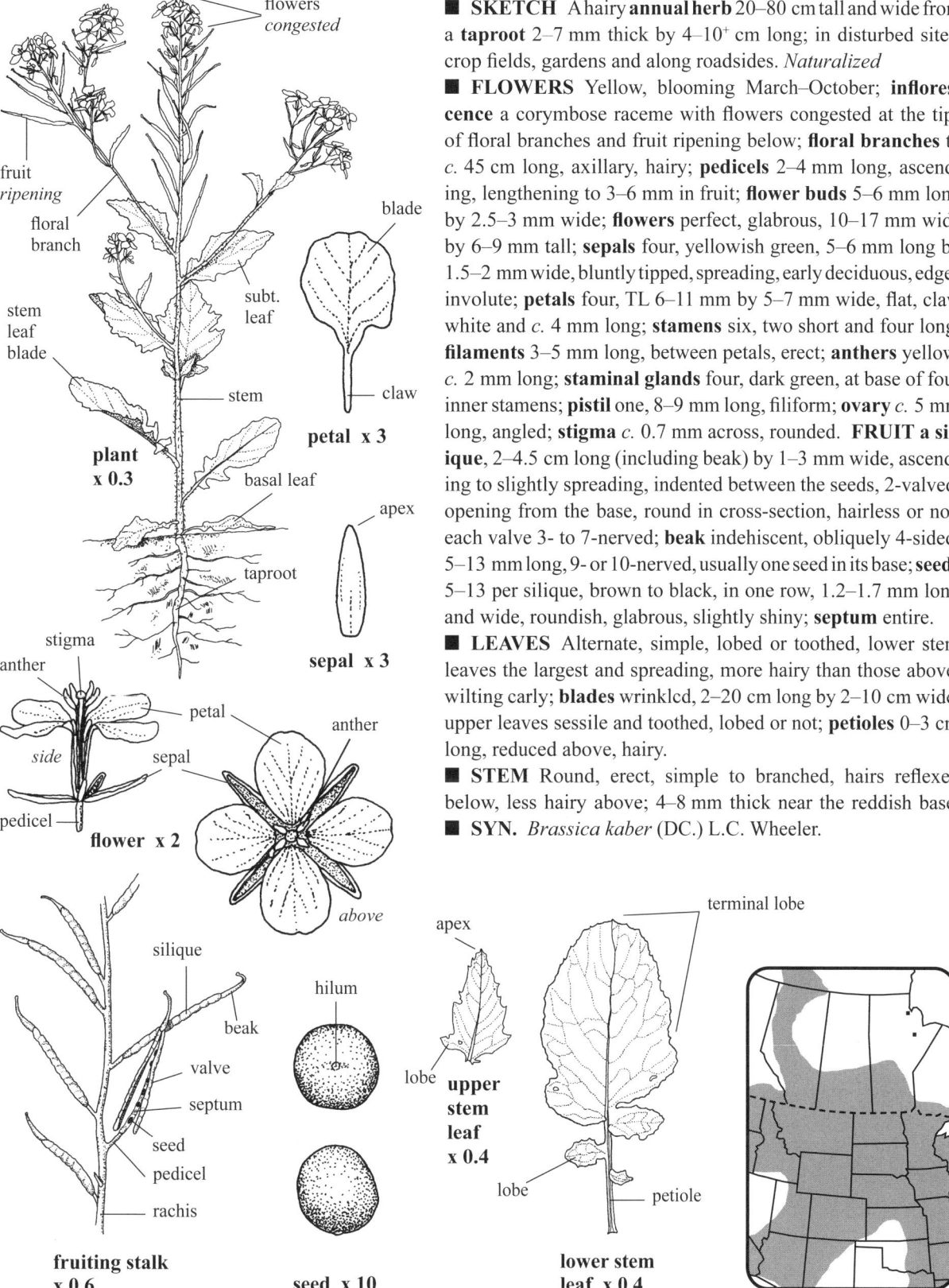

■ **SKETCH** A hairy **annual herb** 20–80 cm tall and wide from a **taproot** 2–7 mm thick by 4–10⁺ cm long; in disturbed sites, crop fields, gardens and along roadsides. *Naturalized*

■ **FLOWERS** Yellow, blooming March–October; **inflorescence** a corymbose raceme with flowers congested at the tips of floral branches and fruit ripening below; **floral branches** to *c.* 45 cm long, axillary, hairy; **pedicels** 2–4 mm long, ascending, lengthening to 3–6 mm in fruit; **flower buds** 5–6 mm long by 2.5–3 mm wide; **flowers** perfect, glabrous, 10–17 mm wide by 6–9 mm tall; **sepals** four, yellowish green, 5–6 mm long by 1.5–2 mm wide, bluntly tipped, spreading, early deciduous, edges involute; **petals** four, TL 6–11 mm by 5–7 mm wide, flat, claw white and *c.* 4 mm long; **stamens** six, two short and four long; **filaments** 3–5 mm long, between petals, erect; **anthers** yellow, *c.* 2 mm long; **staminal glands** four, dark green, at base of four inner stamens; **pistil** one, 8–9 mm long, filiform; **ovary** *c.* 5 mm long, angled; **stigma** *c.* 0.7 mm across, rounded. **FRUIT a silique**, 2–4.5 cm long (including beak) by 1–3 mm wide, ascending to slightly spreading, indented between the seeds, 2-valved, opening from the base, round in cross-section, hairless or not, each valve 3- to 7-nerved; **beak** indehiscent, obliquely 4-sided, 5–13 mm long, 9- or 10-nerved, usually one seed in its base; **seeds** 5–13 per silique, brown to black, in one row, 1.2–1.7 mm long and wide, roundish, glabrous, slightly shiny; **septum** entire.

■ **LEAVES** Alternate, simple, lobed or toothed, lower stem leaves the largest and spreading, more hairy than those above, wilting early; **blades** wrinkled, 2–20 cm long by 2–10 cm wide, upper leaves sessile and toothed, lobed or not; **petioles** 0–3 cm long, reduced above, hairy.

■ **STEM** Round, erect, simple to branched, hairs reflexed below, less hairy above; 4–8 mm thick near the reddish base.

■ **SYN.** *Brassica kaber* (DC.) L.C. Wheeler.

Charlock Mustard, Corn-mustard, Charlock, Crunch-weed

Mustard—*Brassicaceae* **Wildflower** Pale yellow Petals 4

TUMBLING MUSTARD *Sisymbrium altissimum* L.

■ **SKETCH** An **annual herb** 30–150 cm tall from a **taproot** 4–20 cm long by 1–6 mm thick; in dry sandy soil of high deserts, along roadsides, railways, in disturbed prairies and fields. *Naturalized*

■ **FLOWERS** Pale yellow, blooming May–August; **inflorescence** a raceme, loose; **floral branches** ascending, central branch 13–20⁺ cm long, leafless, lateral floral branches rebranched, subtended by a compound leaf 4–9 cm long, reduced above; **pedicels** glabrous, *c.* 5 mm long by *c.* 0.6 mm thick at anthesis, 5–10 mm long and to *c.* 0.7 mm thick in fruit; **flowers** perfect, glabrous, each 9–11 mm wide by 6–7 mm tall; **sepals** four, erect to ascending, 4–6 mm long by *c.* 1 mm wide, pale green, C-shaped, two slightly longer with their apices hooded, the two shorter sepals are less hooded; **petals** four, spreading, 6–9 mm long by 2.3–2.7 mm wide, blades veined, claw white to pale yellow, 3–4 mm long by *c.* 0.9 mm wide; **stamens** six, four long and two short, 4–5.5 mm long, the four long stamens opposite the petals, two short stamens between the petals; **filaments** thick, round, 3.5–4.5 mm long, light green; **anthers** yellow, 1–1.5 mm long; **pistil** green, 5–5.5 mm long by *c.* 0.6 mm thick; **stigma** *c.* 1 mm wide, flat, slightly 2-lobed. **FRUIT** a **silique**, golden tan, glabrous, 5–10 cm long by 0.8–1.3 mm wide and thick, roundly 4-angled, 2-valved, each valve with a raised midnerve and small lateral nerves, opening from base; **beak** (persistent style) 1–2 mm long and seedless; **seeds** single file, one row under each valve, 0.9–1.1 mm long by *c.* 0.6 mm wide by 0.4–0.5 mm thick, one curved longitudinal groove on each side, cream to yellow, slightly shiny to dull; **septum** complete, nerveless, raised between seeds.

■ **LEAVES** Basal and stem, alternate, compound, ascending; **basal blades** usually wilted by anthesis or to *c.* 10 cm long by *c.* 3 cm wide; **lower stem blades** 10–16 cm long by 4–5 cm wide, lobed, toothed, hairs simple, margins hairy, midrib raised below; **petioles** 0–3 cm long, hairy, reduced and glabrous above; **upper stem blades** with the segments linear and 0.5–2 mm wide, entire, soft, glabrous.

■ **STEM** Erect, branched, white hairs pointing in all directions below, glabrous above; 2–8 mm thick near the break-away base, a tumbleweed in the wind.

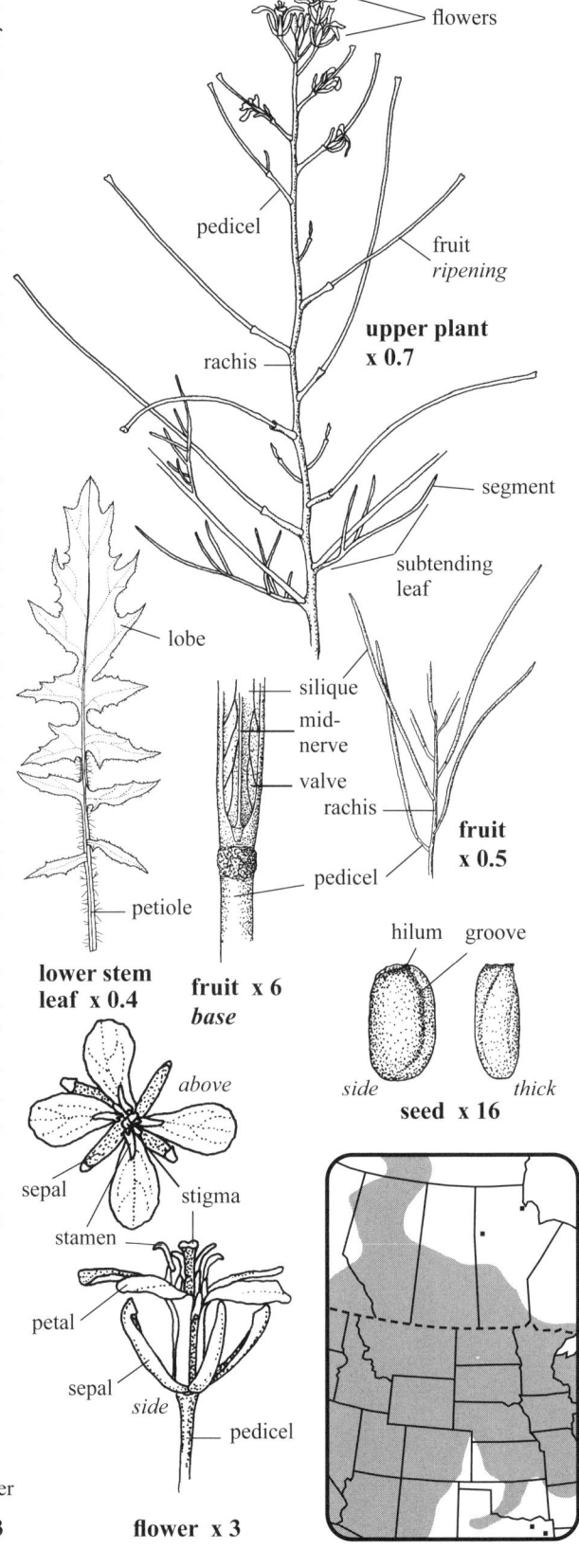

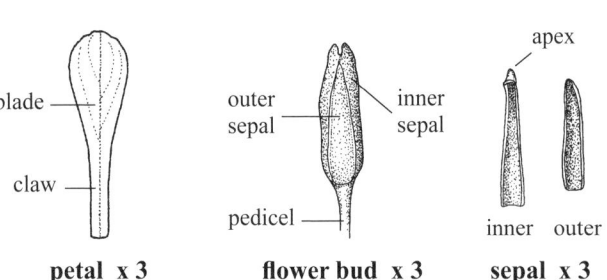

228 Tumbleweed Mustard

Mustard—*Brassicaceae* | **Wildflower** Yellow Petals 4

TALL HEDGE MUSTARD *Sisymbrium loeselii* L.

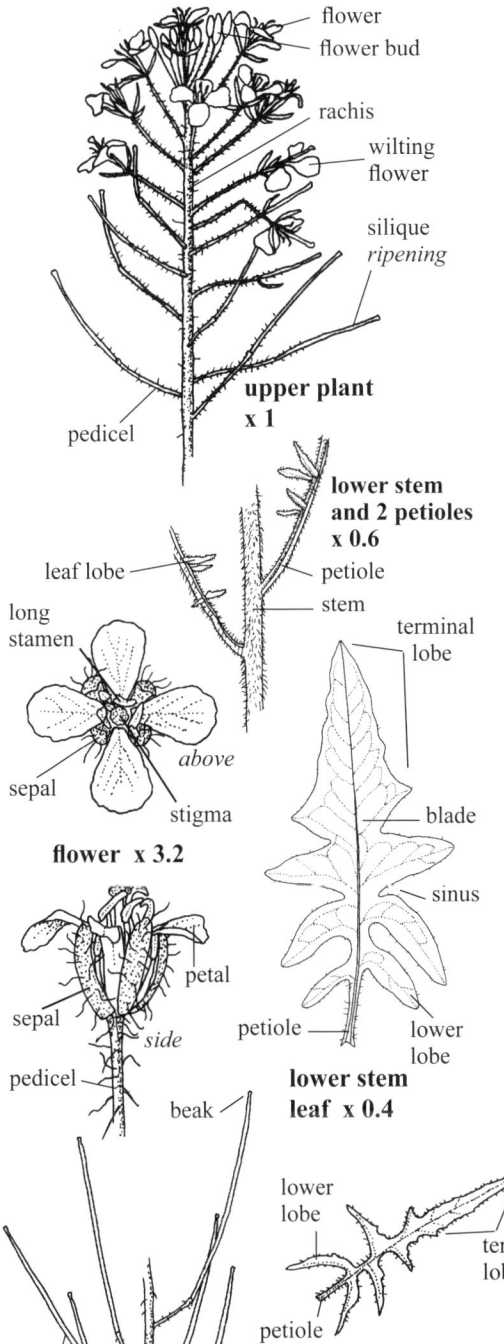

■ **SKETCH** A hairy **annual herb** 40–110 cm tall with a **taproot** 3–10 mm thick and to *c.* 18⁺ cm long; in disturbed sites, canyon bottoms, roadsides and ditches. *Naturalized*

■ **FLOWERS** Yellow, blooming May–August; **inflorescence** a raceme, usually 2–3.5 cm long by 2.5–3.5 cm wide at anthesis (flower cluster only), lengthening in fruit, terminal and on branches; **upper floral branches** stiff, ascending; **subtending bracts** (of branches) leaflike, hairy and lobed; **pedicels** 6–18 mm long by *c.* 0.4 mm wide, hairs to 1 mm long and spreading; **flowers** perfect, 6–10 mm wide by 4–5 mm long; **rachis** slightly hairy; **sepals** four, erect, 3- or 4-veined, 3–4.8 mm long by 1–1.4 mm wide, tips hooded, pale green, with several long hairs spreading, glabrous inside; **petals** four, yellow, 4–8 mm long by 2.5–4 mm wide, blades spreading, margins slightly erose, glabrous, claw *c.* 0.7 mm wide, flat, green; **stamens** six, four long and two short, exserted; **filaments** 2.2–3.5 mm long by 0.4–0.5 mm thick and wider at their base; **anthers** yellow, *c.* 2 mm long, attached near the base; **pistil** glabrous, 4–4.2 mm long by *c.* 0.8 mm wide by *c.* 0.4 mm thick; **ovary** 2-sutured, 3.2–3.5 mm long at anthesis; **style** *c.* 0.2 mm long, glabrous, narrower than ovary and stigma; **stigma** oval, capitate; **fruiting stalks** to *c.* 50 cm long by 3–5 cm wide. **FRUIT a silique**, 2–4 cm long by *c.* 0.7 mm wide, reddish brown, beak obscure, three low reddish nerves along each valve, opening at base; **seeds** 50–70 per silique, orange, one row on each side of septum, 0.8–1 mm long by 0.5–0.6 mm wide by *c.* 0.4 mm thick; **septum** unnerved.

■ **LEAVES** Basal and stem, alternate, simple, lobes more deeply cut and pointing slightly backward at the blade's base, entire to toothed, hairy, especially on midrib below, hairs fewer and more scattered on upper leaves; **basal leaves** 10–20 cm long by 3–5 cm wide, mostly wilted by anthesis; **petioles** to *c.* 4 cm long, hairy; **stem leaves** spreading below to erect, TL 1.5–20 cm by 0.3–5 cm wide, reduced above; **petioles** 0–4 cm long, hairs deflexed to spreading.

■ **STEM** Hollow, erect, stiff, single or tufted with several stems in a clump, white hairs descending; 3–12 mm thick near the hairy base.

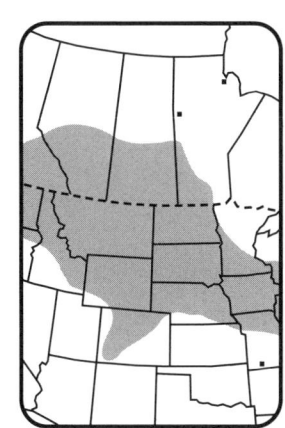

Small Tumbleweed Mustard, False London Rocket

Mustard—*Brassicaceae*   **Wildflower**   White   Petals 4

STINKWEED   *Thlaspi arvense* L.

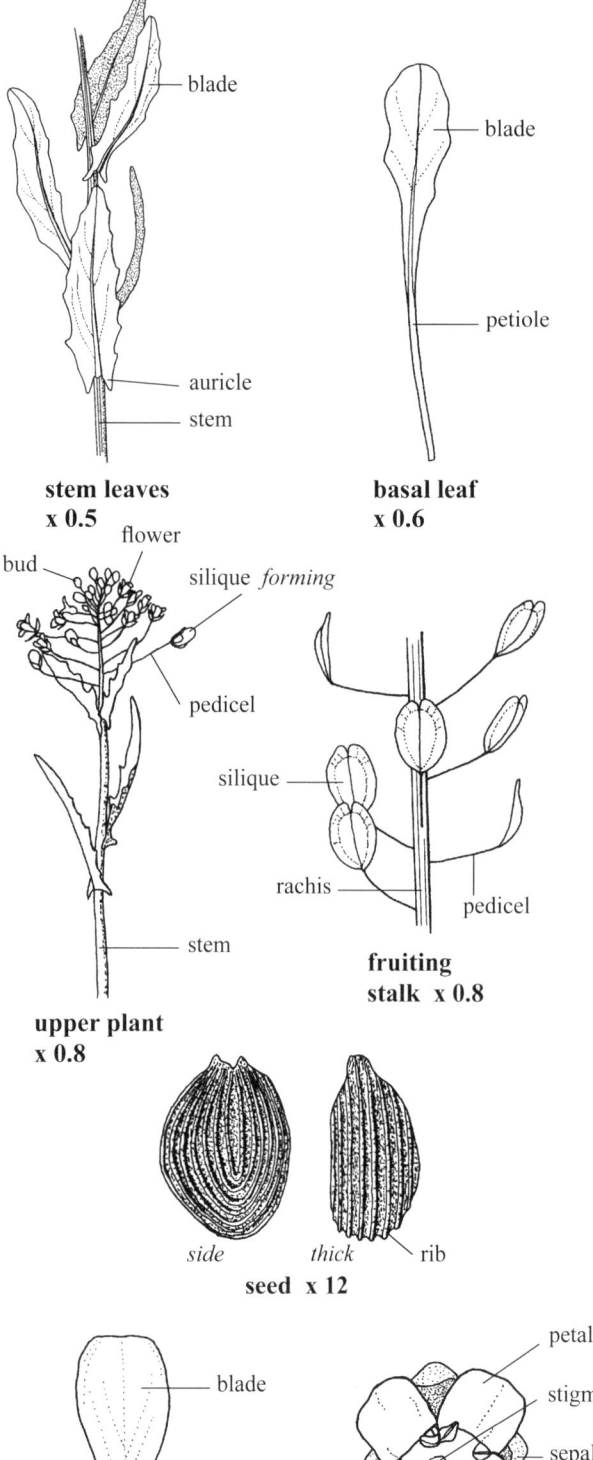

■ **SKETCH** A glabrous **annual** or **winter-annual herb** 3–70 cm tall with a **taproot** 4–11 cm long by 1–3 mm thick; whole plant turning yellowish tan after flowering; in gardens, fields, mountain slopes, along shores of streams, roadsides, alleys and other disturbed moist to dry sites. *Naturalized*

■ **FLOWERS** White with a green interior, blooming April–August; **inflorescence** a raceme, a terminal congestion of flowers and buds *c.* 2 cm wide by 1–1.5 cm high, elongating with fruit; **floral branches** near the top, ascending, each supporting one or two racemes; **pedicels** 2–5 mm long, lengthening to 1–2 cm in fruit; **flowers** perfect, pungent, *c.* 3 mm wide by *c.* 3.5 mm long; **sepals** four, green with white margins, 1.5–2.5 mm long by 1.2–1.5 mm wide, bluntly pointed, curved to slightly hooded; **petals** four, white, 2–4 mm long by 1.2–1.5 mm wide, spatulate, clawed; **stamens** six, four long and two short, 1.8–2 mm long, included; **anthers** *c.* 0.4 mm long, green and basifixed; **stigma** green, included; **fruiting stalks** 4–30 cm long by 2–4.5 cm wide. FRUIT a **silique** (pod), flattened, upright, winged, yellowish tan, oval, 8–18 mm long by 8–12 mm wide by 1–2 mm thick with an apical notch 1–2.5 mm deep; **seeds** 6–16 per silique, purplish brown, 1.5–2.3 mm long by 1.2–1.5 mm wide by *c.* 1 mm thick, shiny, slightly flattened, with concentric ribs and furrows.

■ **LEAVES** Basal and stem, alternate, simple, toothed, hairless, glaucous, thin, larger near the base, midrib obvious; **basal leaves** entire or irregularly toothed, TL 2–8 cm and 5–25 mm wide, tips rounded, early deciduous; **petioles** 1–4.5 cm long and winged; **stem leaves** sessile, 2–9 cm long by 5–20 mm wide, reduced above, sinuately toothed to entire; **auricles** blunt to pointed, 4–7 mm long.

■ **STEM** Simple to branched, 1–15 per clump, central stem usually the tallest, round, hairless, ridged; reddish at the 4–7 mm wide base.

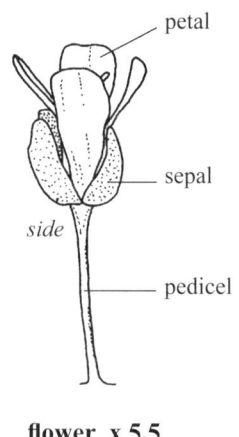

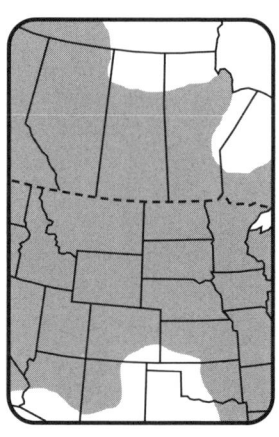

Field Pennycress

Water-starwort—*Callitrichaceae* **Wildflower** Pale green Subtending bracts of flowers 2

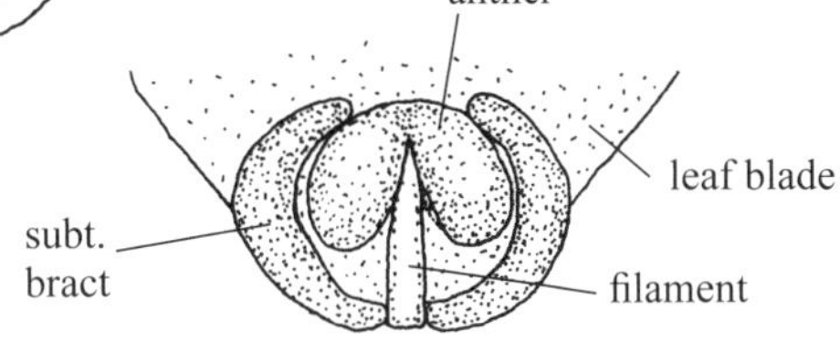

VERNAL WATER-STARWORT   *Callitriche palustris* L.

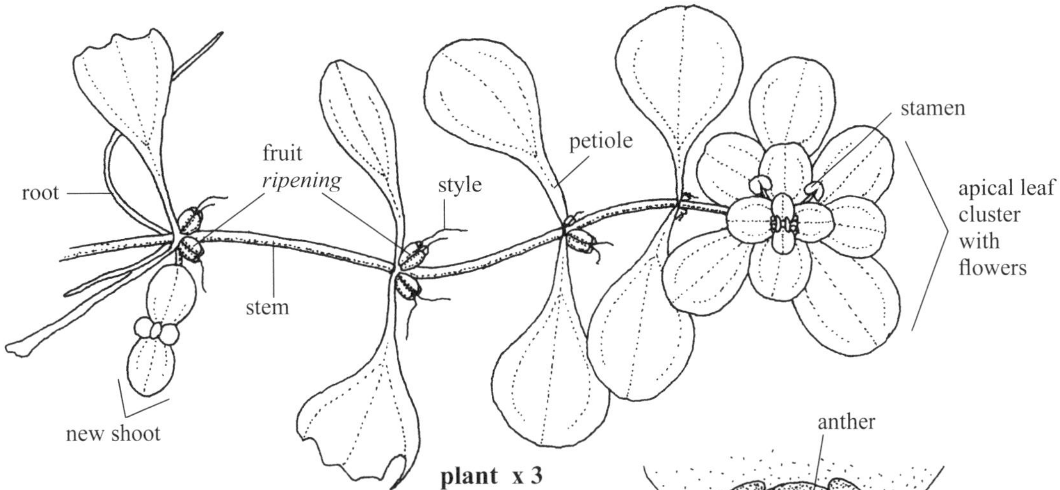

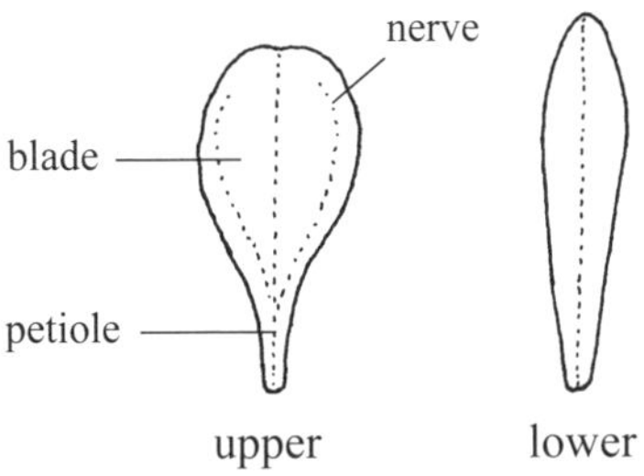

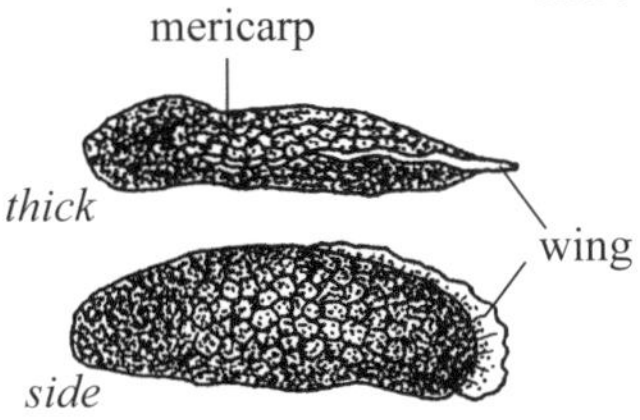

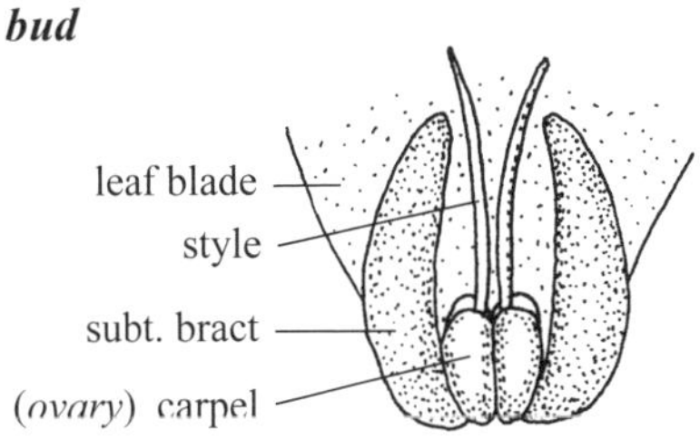

- **SKETCH** An aquatic, glabrous **perennial herb** 5–30 cm long with reddish brown **rhizomes** 3–8 cm long by *c.* 0.3 mm thick with white filiform **fibrous roots** 5–15 mm long from some nodes; in shallow quiet water of small ponds, ditches, on exposed mud, by streams and sloughs; **monoecious**.
- **FLOWERS** Pale green, blooming May–September; **inflorescence** unisexual, sessile, in leaf axils; **male flowers** of one stamen, soon exserted, usually one per leaf axil; **filament** *c.* 0.5 mm long, filiform, extending to *c.* 1 mm or more to protrude above the water's surface; **anthers** pale yellow, *c.* 0.3 mm long by *c.* 0.5 mm wide; **subtending bracts** two, each 0.8–1 mm long by *c.* 0.4 mm wide, fleshy, entire and blunt; **female flowers** of one pistil with two carpels and four ovules; **pistil** green, *c.* 1 mm long, one or two per leaf axil; **styles** two, filiform, whitish hyaline; **subtending bracts** two, as above. **FRUIT** a mericarp, four in an axillary cluster, sessile, each 1-seeded, 0.8–1.7 mm long by 0.4–0.6 mm wide and 0.2–0.3 mm thick, dark brown, winged near the apex, finely net-reticulate, slightly shiny; **wing** light tan, veined, wavy and erose.
- **LEAVES** Opposite, simple, entire, 6–20 mm long by 2–8 mm wide, slightly lighter green below, two forms: 1) sessile narrow submerged blades near the base of the stem, 1-nerved; 2) wider floating blades above, 3-nerved, a cluster of three or four pairs at the stem's apex, the internodal distance lengthens below as do the petioles; **petioles** 1–4 mm long, reduced above, longer away from stem apex as fruit matures, bases connected by a thin membranous wing across the stem; **stipules** absent.
- **STEM** Light green, ascending in water, prostrate on mud when exposed to air, round, smooth, solid; 0.3–0.7 mm thick near the base.
- **SYN.** *Callitriche verna* L.

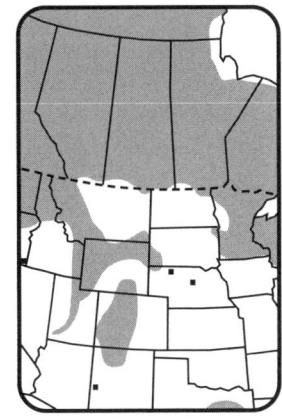

Common Water-starwort, Spring Water-starwort, Water-starwort

Bellflower—*Campanulaceae* **Wildflower** White to pale blue Corolla tubular, 5-lobed

Marsh Bellflower *Campanula aparinoides* Pursh

■ **SKETCH** A **perennial herb** from 15–50 cm tall or long with a tan **rhizome** 0.8–1 mm thick by *c.* 7⁺ cm long and pale green **stolons** *c.* 0.5 mm thick and to *c.* 20 cm long; in wet areas on edges of marshes and lakes, meadows and wooded hillsides or shallow streams.

■ **FLOWERS** White to a pale blue, blooming June–August; **inflorescence** a raceme, open; **pedicels** thin, 3–4.5 cm long, naked, weak, ascending; **flowers** perfect, 10–15 mm wide by 9–12 mm long; **calyx** tubular, 5-lobed, green, the lobes 1–3.5 mm long by 0.8–1.5 mm wide, glabrous, spreading, sometimes with a shallow tooth on each side, more often entire; **corolla** tubular, 5-lobed, glabrous outside, finely hairy inside, hairs short and spreading; **corolla lobes** arched, 6–8 mm long by 3–4 mm wide with three obvious blue veins; **stamens** five, included, attached to ovary apex; **filaments** hyaline, flat and wide at base, tapered above, *c.* 2 mm long by *c.* 0.8 mm wide, hairy along lower inside margins, hairless above; **anthers** white, turning pale yellow, 2.8–3 mm long, 2-lobed, erect, slightly hooded at apex; **pistil** one, included; **ovary** green, *c.* 1.7 mm long and wide, glabrous or with thick stiff hairs along the five ridges; **style** one, white, *c.* 5 mm long, glabrous; **stigmas** three, erect at first, then arched and spreading, short hairy, each *c.* 1.8 mm long. **FRUIT** a **capsule**, tan, hairless, slightly shiny, ribbed, 2–3 mm long and wide, apex flat, usually hidden by dry perianth, opening by pores at base between the ribs; **seeds** smooth, *c.* 20 per capsule, slightly shiny, 0.8–1.2 mm long by 0.6–0.8 mm wide by *c.* 0.5 mm thick, light chestnut brown.

■ **LEAVES** Alternate, simple, entire, margins with stiff, backward pointing hairs; **blades** pointed, sessile, dull, 1-veined, generally descending, 1–7 cm long by 1–7 mm wide, reduced above, hairs pointing backward on midrib below.

■ **STEM** Erect to leaning, sinuous, weak, usually tangled, branched above, hollow, triangular, 3-ridged, the ridges with short stiff descending hairs or hairless at the angles; 0.5–1.4 mm thick near the base.

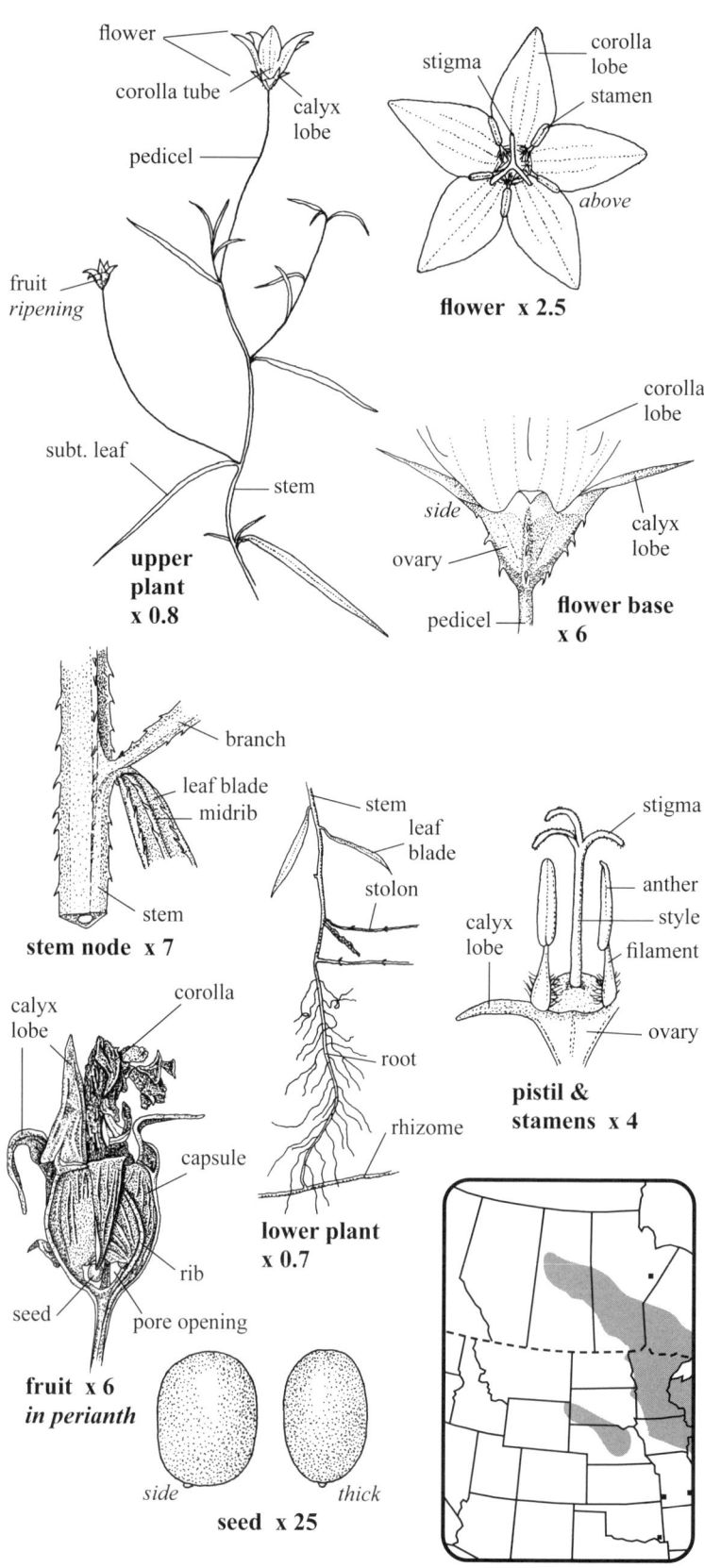

Bellflower—*Campanulaceae* **Wildflower** Blue to purple Corolla tubular, 5-lobed

CREEPING BLUEBELL   *Campanula rapunculoides* L.

■ **SKETCH** A **perennial herb** 40–150 cm tall with woody, tan **rhizomes** 2–5 mm thick; in disturbed sites, along roadsides, exposed streams and river banks. *Naturalized*
■ **FLOWERS** Blue to purplish pink, blooming June–October; **inflorescence** a raceme 30–50+ cm long by *c.* 7 cm wide; **pedicels** arched, glabrous to slightly hairy, 1–10 mm long, longer in fruit; **subtending bracts** (of pedicels), one, leaflike, sessile, margins hairy and finely toothed, veins obvious, 1.5–6 cm long by 4–22 mm wide, reduced and glabrous above; **flowers** perfect, nodding, 2.5–3.5 cm wide by 2–4 cm long, up to three flowers per node, the two lateral ones blooming after the terminal one has wilted; **calyx** tubular, 5-lobed; **lobes** green, scabrous, entire, pointed, strongly reflexed, 5–10 mm long by 1.8–2.3 mm wide, enlarging in fruit, united at the base, glabrous on outside, hairy on inner hidden side, the hairs descending near a lobe's base but ascending above near a lobe's apex and along the margins; **corolla** TL 2–3 cm; **corolla tube** glabrous, 13–15 mm long by *c.* 1 cm wide at the base of the lobes; **corolla lobes** five, 12–14 mm long by 6–7 mm wide, spreading, midnerve obvious, ciliate along the margins, hairs 1–2 mm long; **stamens** five, 12–14 mm long, included; **filaments** white, *c.* 4.5 mm long by *c.* 2 mm wide at the base, the base abruptly widening and hairy on inner surface, glabrous on the outside; **anthers** pale yellow, 9–10 mm long; **pistil** 3–4.5 cm long; **ovary** green, *c.* 3 mm long and wide, with numerous short descending hairs; **style** exserted, 1–1.2 mm thick, round, with slightly ascending hairs; **stigma** 3-lobed, each lobe 3–3.5 mm long, recurved. **FRUIT a capsule**, brown, 6–8 mm long by 5–8 mm wide, 10-ribbed, nodding, opening by basal pores equal in number to the locules, scabrous, veined between the ribs, roundish near the blunt apex, tapered below; **seeds** numerous, shiny, yellowish tan, 1.2–1.4 mm long by *c.* 1 mm wide by *c.* 0.5 mm thick, with a narrow winglike marginal ridge, plano-convex, glabrous, striate longitudinally with a few slightly darker streaks.
■ **LEAVES** Basal and stem, alternate, simple, toothed, pointed, dull green above, slightly lighter green below; **blades** glabrous above, scabrous on margins, slightly hairy below on nerves, 1.5–15 cm long by 1–5 cm wide, reduced upward; **petioles** 0–15 cm long, reduced above, deeply grooved above, hairy mostly on the upper margins.
■ **STEM** Erect, solid, usually unbranched, glabrous below, slightly scabrous among the more angular rachis; 3–8 mm thick near the tough base.

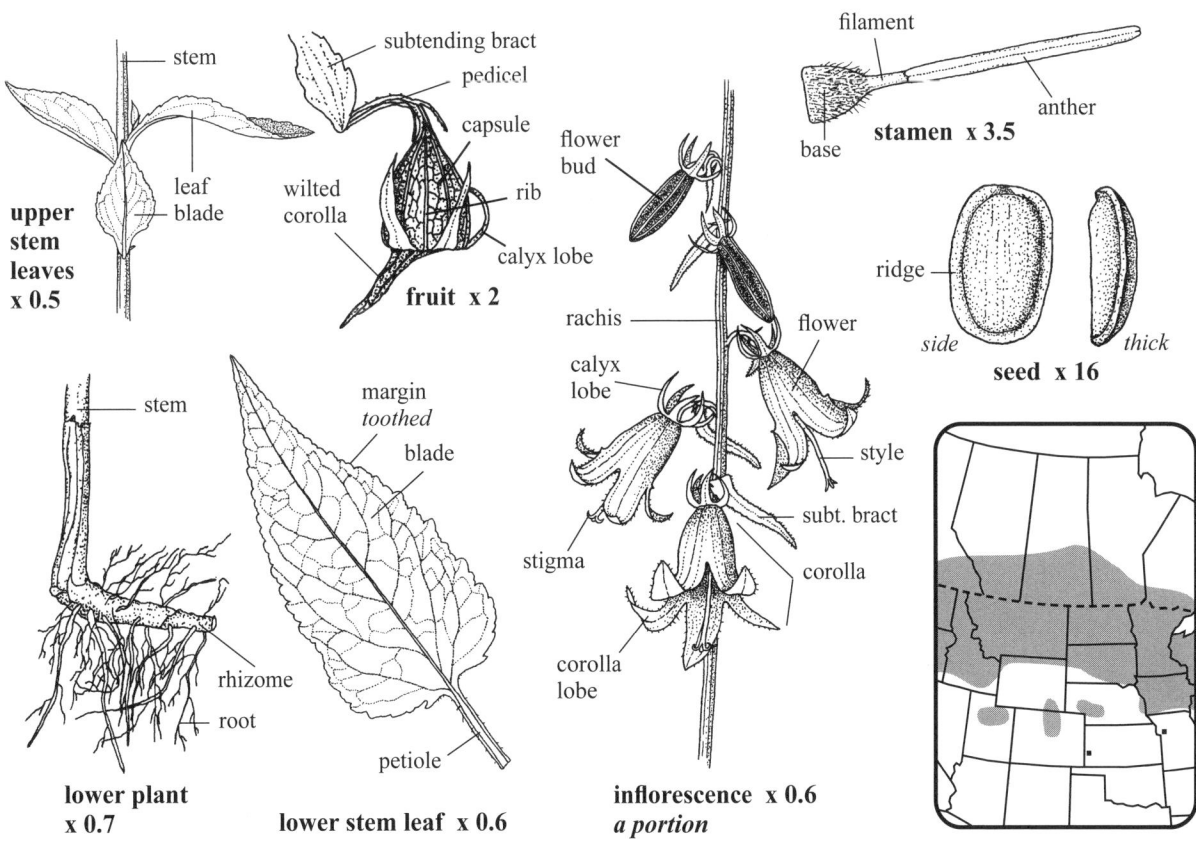

Rover Bellflower, Creeping Bellflower

Bellflower—*Campanulaceae* **Wildflower** Blue Corolla tubular, 5-lobed

HAREBELL *Campanula rotundifolia* L.

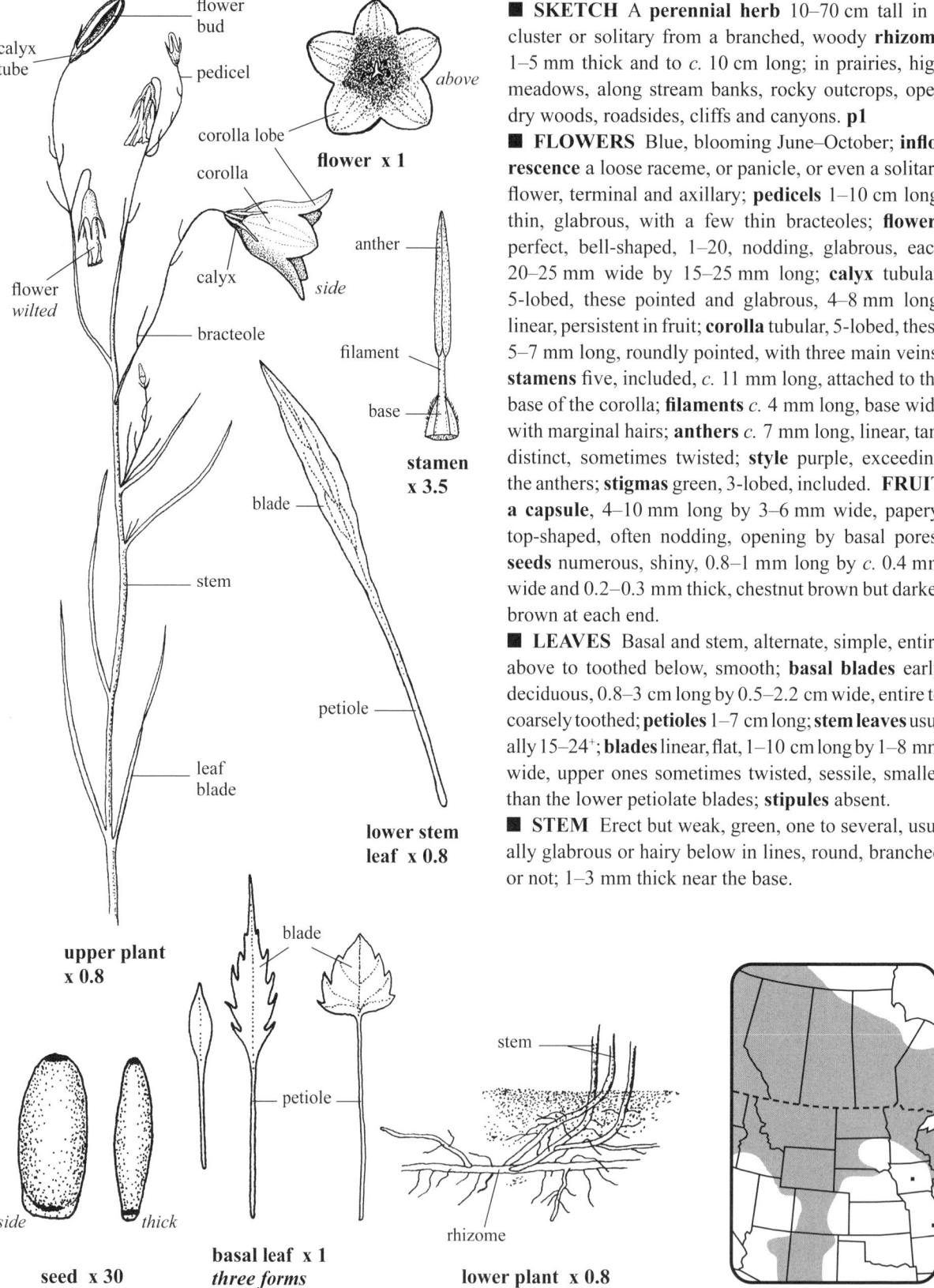

- **SKETCH** A **perennial herb** 10–70 cm tall in a cluster or solitary from a branched, woody **rhizome** 1–5 mm thick and to *c.* 10 cm long; in prairies, high meadows, along stream banks, rocky outcrops, open dry woods, roadsides, cliffs and canyons. **p1**
- **FLOWERS** Blue, blooming June–October; **inflorescence** a loose raceme, or panicle, or even a solitary flower, terminal and axillary; **pedicels** 1–10 cm long, thin, glabrous, with a few thin bracteoles; **flowers** perfect, bell-shaped, 1–20, nodding, glabrous, each 20–25 mm wide by 15–25 mm long; **calyx** tubular, 5-lobed, these pointed and glabrous, 4–8 mm long, linear, persistent in fruit; **corolla** tubular, 5-lobed, these 5–7 mm long, roundly pointed, with three main veins; **stamens** five, included, *c.* 11 mm long, attached to the base of the corolla; **filaments** *c.* 4 mm long, base wide with marginal hairs; **anthers** *c.* 7 mm long, linear, tan, distinct, sometimes twisted; **style** purple, exceeding the anthers; **stigmas** green, 3-lobed, included. **FRUIT** a capsule, 4–10 mm long by 3–6 mm wide, papery, top-shaped, often nodding, opening by basal pores; **seeds** numerous, shiny, 0.8–1 mm long by *c.* 0.4 mm wide and 0.2–0.3 mm thick, chestnut brown but darker brown at each end.
- **LEAVES** Basal and stem, alternate, simple, entire above to toothed below, smooth; **basal blades** early deciduous, 0.8–3 cm long by 0.5–2.2 cm wide, entire to coarsely toothed; **petioles** 1–7 cm long; **stem leaves** usually 15–24[+]; **blades** linear, flat, 1–10 cm long by 1–8 mm wide, upper ones sometimes twisted, sessile, smaller than the lower petiolate blades; **stipules** absent.
- **STEM** Erect but weak, green, one to several, usually glabrous or hairy below in lines, round, branched or not; 1–3 mm thick near the base.

234   Bluebell Bellflower, Bluebell-of-Scotland, Bluebell, Common Harebell

Hemp—*Cannabaceae*

Vine Green Sepals 5 Male petals absent
Female bract 1, forming a cone

Common Hop *Humulus lupulus* L.

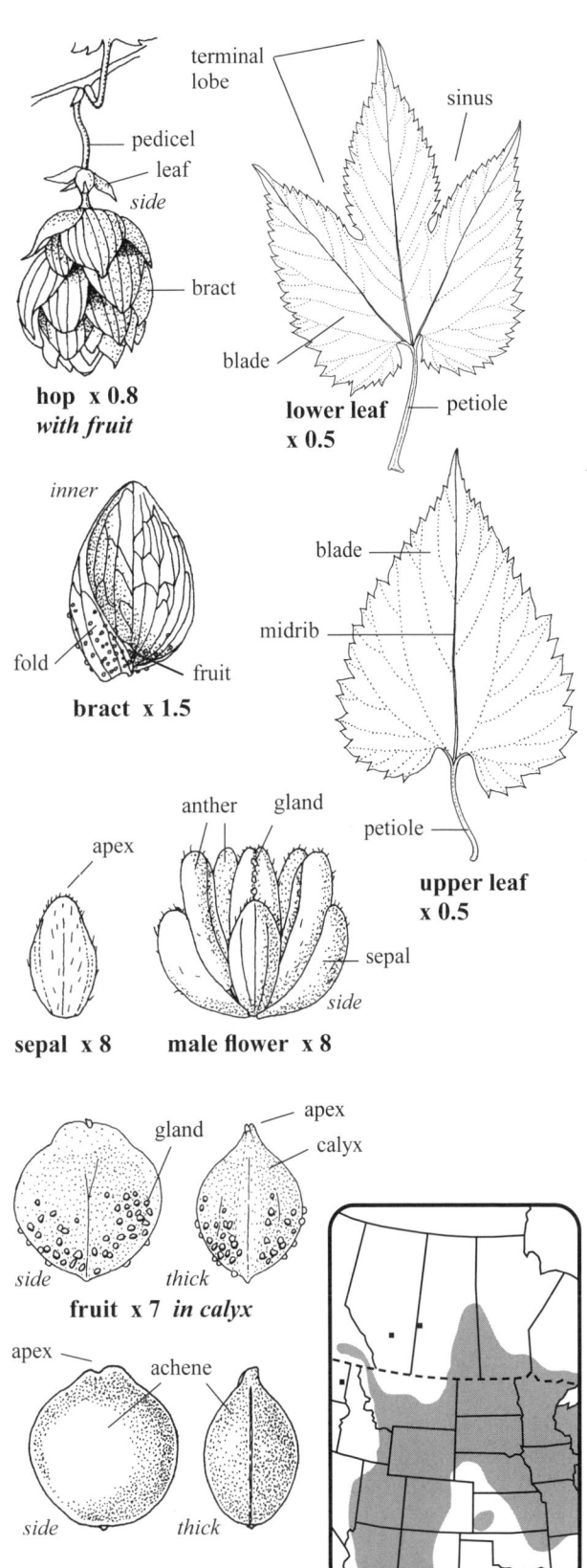

hop x 0.8 with fruit

lower leaf x 0.5

bract x 1.5

upper leaf x 0.5

sepal x 8   male flower x 8

fruit x 7 *in calyx*

fruit x 7

■ **SKETCH** A variable, deciduous **perennial vine** climbing without tendrils and growing 2–12+ m long up and over supportive vegetation from a brown, woody **rhizome** 5–10 mm thick; **lateral roots** 1–5 mm thick; in moist sites of thickets, ditches, stream banks, fencerows, open woodlands in prairies and parklands; **dioecious**. Native and *naturalized*

■ **FLOWERS** Unisexual, blooming June–September; male **inflorescence** an open panicle 7–15 cm long by 1.5–5 cm wide, axillary; **pedicels** 0–3.5 mm long; **male flowers** whitish yellow, 2–3.5 mm wide and tall; **sepals** five, 1.5–2.8 mm long by 1–1.5 mm wide, blunt, 1-veined, C-shaped, glabrous or short hairs outside; **petals** absent; **stamens** five, about as long as or longer than the sepals; **filaments** 0.5–0.8 mm long; **anthers** *c.* 2.5 mm long, 4-lobed, curved, slightly hairy, glandular in outer groove; female **inflorescence** a short spike (catkinlike), drooping, axillary, 5–10 mm long at anthesis, expanding in fruit; **pedicels** 18–40 mm long in fruit with a whorl of 3 or 4 leaves near the hop; **female flowers** in pairs and subtended by a foliaceous bract; **calyx** entire, membranous, unlobed, enclosing the ovary and persisting, glandular; **stigmas** 2-parted, 3–4 mm long, hairy, united near the base, exserted; **ripe hop** yellowish tan, conelike, hanging, 1–3 per leaf axil, 1.8–5 cm long by 1.5–3 cm wide, oval to roundish in cross-section, fragrant with bracts removed; **bracts** 7–20 mm long by 5–13 mm wide, glabrous to hairy, glandular, veined, entire, one side folded over at the base and enclosing the calyx with fruit, reduced above and below on each hop, margins ciliate or not, turning tan and dropping in the autumn; **rachis** of hop very hairy, slightly glandular, zigzagging. **FRUIT** an **ache**ne, 1-seeded, 2–3.3 mm long by 2.2–3 mm wide by 1.7–2 mm thick, 3–12 per hop, smooth, apex blunt, sides convex, fruit easily removed from bract, brown and glandless, enclosed in persistent calyx bearing yellowish green, raised glands in lower half.

■ **LEAVES** Opposite, simple, unlobed or 3- to 7-lobed, toothed, dull, dark green above, lighter below, glandular dotted (golden to green) on both sides, hairless above, glabrous or scattered hairs below on or between the veins, margins slightly hairy and revolute; **lower blades** often 3-lobed, 4–16 cm long by 2.8–16 cm wide; **upper blades** unlobed, 4.5–8 cm long by 3.5–5.5 cm wide; **petioles** 2–15 cm long, slightly hairy, hairs sharply bent and ascending, scabrous from raised hair bases and short, two-branched hairs; **stipules** persistent.

■ **STEM** Solid, branched, slightly juicy, 5- or 6-ridged, angled, rough to the touch from short stiff hairs; 1–5 mm thick near the herbaceous base.

Bine, Hop, European Hop, Hops

Family Characteristics

Honeysuckle—Caprifoliaceae Flowers tubular, 5-lobed

SKETCH Mostly **deciduous shrubs**. Worldwide about 400 species in the northern temperate regions.

FLOWERS Perfect (rarely sterile), regular or irregular, white to yellow, some pinkish. **Inflorescence** often a cyme. **Calyx** tubular, 3- or 5-lobed. **Corolla** tubular, often 2-lipped and 5-lobed. **Stamens** often 5 (4), filaments attached to tube of corolla and visible between the lobes. **Pistil** 1, each of 2–5 united carpels. **Ovary** of 2- to 5-locules, ovules 1 to several per locule. **Styles** distinct or united. **Stigma** unlobed or 3-lobed, wider than the style.

FRUIT Usually a berry, drupe, capsule or schizocarp with seeds, stones or nutlets.

LEAVES Opposite, mostly simple, some compound, blades entire to lobed, often toothed, deciduous. **Stipules** absent, or present and then quite small.

STEM Woody, erect to leaning to trailing, branches opposite.

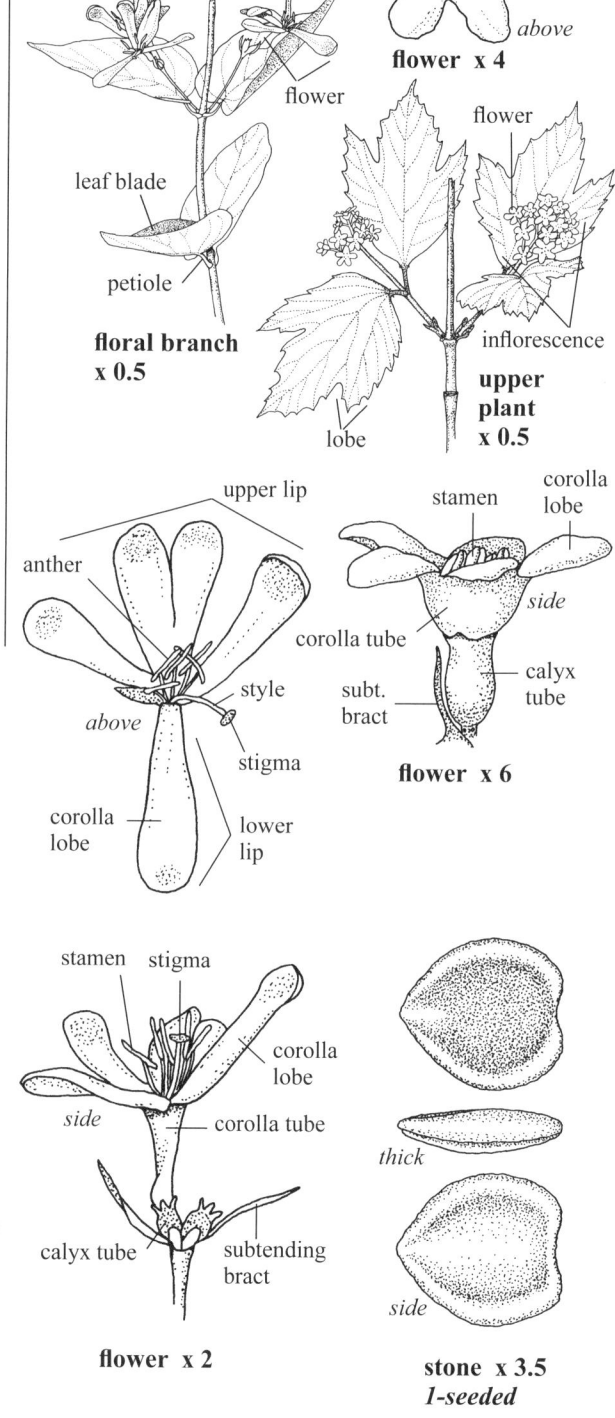

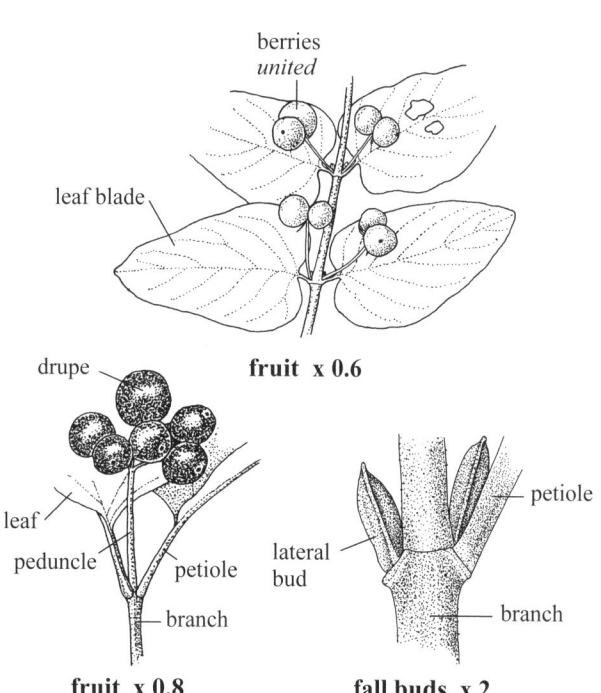

236

Honeysuckle—*Caprifoliaceae*   **Shrub** Yellow   Corolla tubular, 5-lobed

BUSH-HONEYSUCKLE   *Diervilla lonicera* P. Mill.

■ **SKETCH** A **deciduous shrub** 20–130 cm tall, the branches often arched from a long woody **rhizome** 5–8 mm thick and to *c.* 20⁺ cm long with pale tan **roots**; along woodland trails, in thickets, clearings in dry to moist woods and pastures, on cliffs and ridges.

■ **FLOWERS** Yellow, blooming June–August; **inflorescence** a cyme, terminal and from upper leaf axils, 1–10 flowers per axil; **peduncles** 5–15 mm long; **pedicels** 6–23 mm long; **subtending bracts** (of pedicels) 5–15 mm long, filiform; **flowers** perfect, 10–12 mm wide by 15–19 mm long, drooping and usually partially hidden beneath the leaves; **subtending bracts** (of flowers) filiform, *c.* 3 mm long, often paired, erect, green; **calyx** 5-lobed, lobes filiform, 4–7 mm long, erect to spreading and persistent at apex of beak on fruit; **corolla** tubular, 5-lobed, four lobes greenish yellow, 7–9 mm long by 2–2.5 mm wide, often arched, one lower yellow lobe 6–7 mm long by 3–3.5 mm wide, hairy above, red on fading; **corolla tube** hairy on outside (microscopic), glabrous inside, 7–8 mm long by *c.* 2 mm wide at base, wider above, soon deciduous; **stamens** five, pale orange, alternate with the corolla lobes, exserted to *c.* 7 mm; **filaments** pale yellow, very hairy below, hairless above, attached to corolla tube; **anthers** 4–4.5 mm long, pale yellow with reddish dots and streaks; **pistil** one, exserted, 22–25 mm long; **ovary** *c.* 4 mm long by *c.* 1.5 mm wide; **style** filiform, *c.* 18 mm long, hairs declining below, spreading above, hairless near stigma; **stigma** green, *c.* 1.5 mm wide, dome-shaped. **FRUIT** a capsule, brown, 6.7–9 mm long (not including beak) by 2–3 mm wide, persistent; **beak** filiform, 3–4 mm long; **seeds** golden brown, slightly shiny, finely reticulate, 16–43 per capsule, each 0.9–1.1 mm long by 0.7–0.9 mm wide by 0.5–0.7 mm thick.

■ **LEAVES** Opposite, simple, toothed, dull above, lighter green and slightly shiny below, tapered to a point; **blades** 5–15 cm long by 1.5–7.5 cm wide, glabrous to hairy, curled marginal hairs and hairy midrib below; **petioles** 3–11 mm long, ascending, finely hairy along margins of upper reddish groove; **stipules** absent.

■ **STEM** Several, branched, older ones light brown, new leafy branches reddish to green, smooth, round, solid, persistent, two low ridges along its length; 3–5 mm thick near the base.

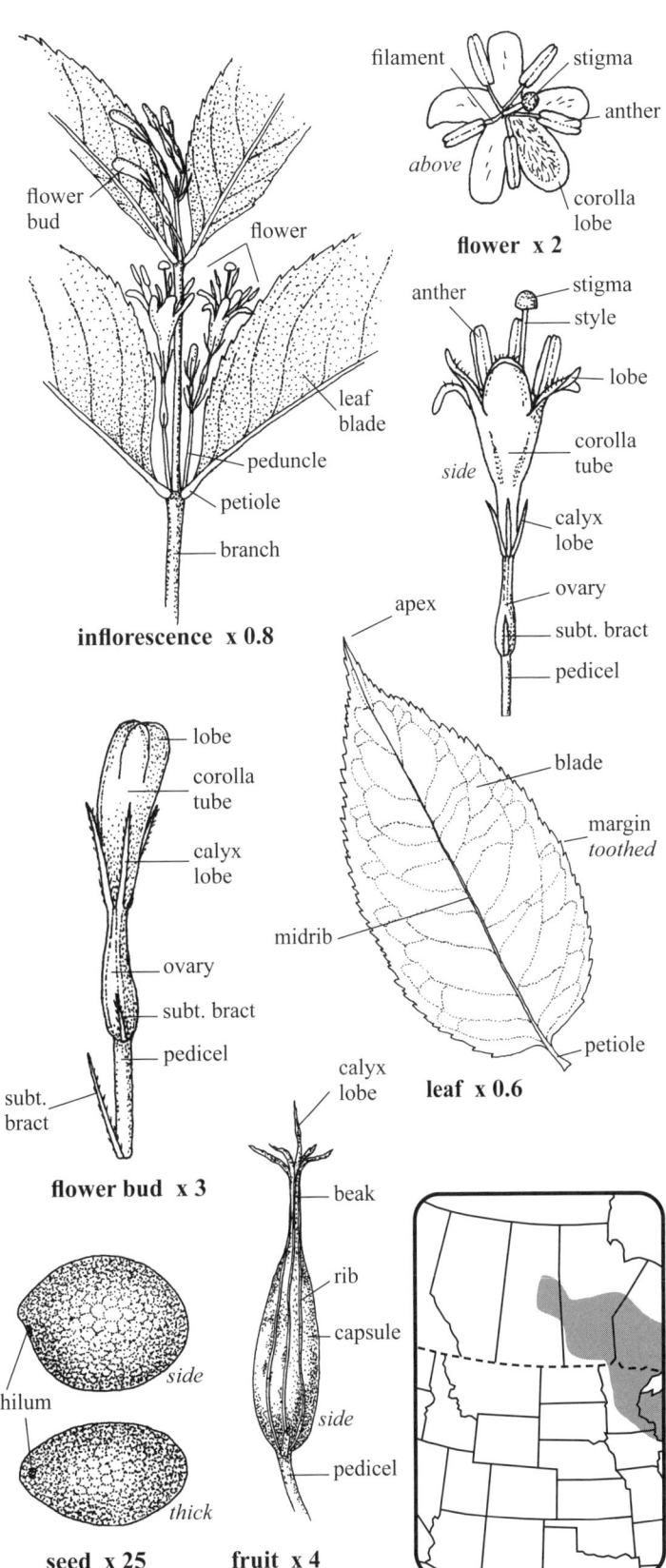

Northern Bush-honeysuckle

Honeysuckle—*Caprifoliaceae* **Wildflower** Pink Evergreen Corolla tubular, 5-lobed

TWINFLOWER *Linnaea borealis* L.

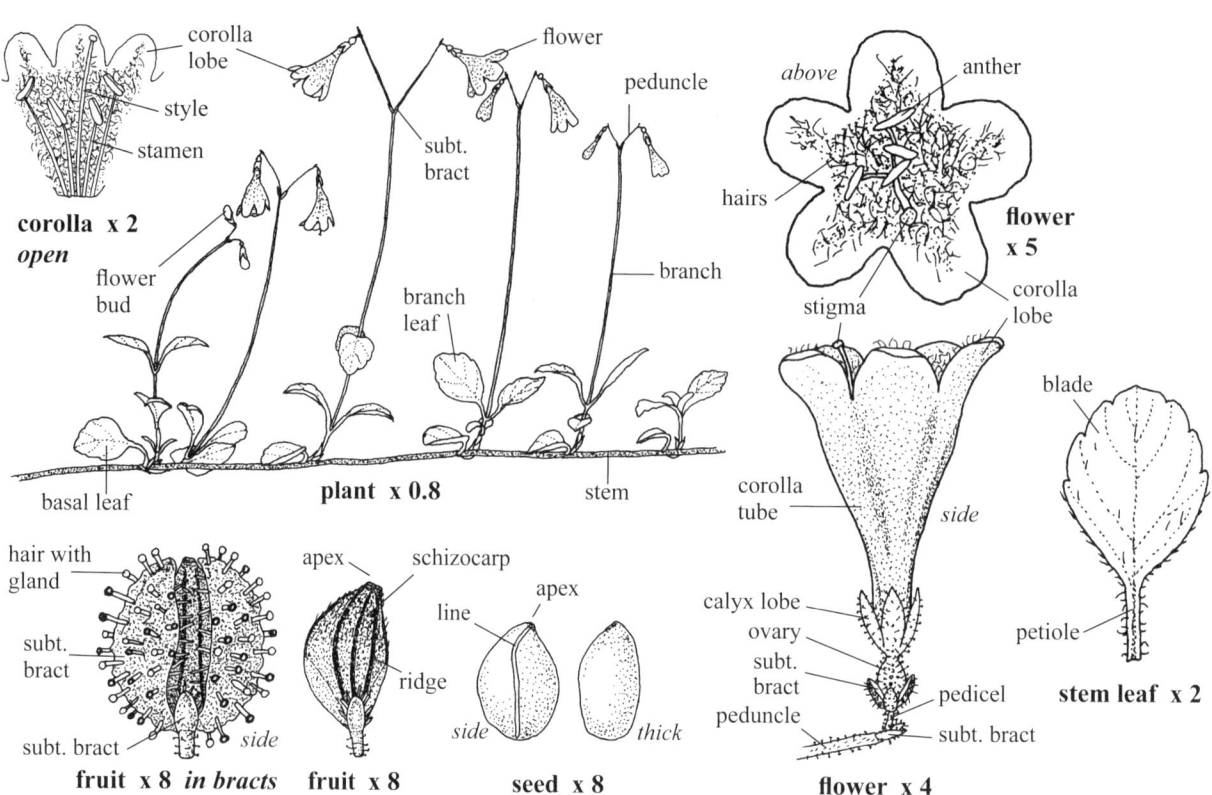

■ **SKETCH** A **perennial herb** with erect flowering branches from green, woody, horizontal **stems** 7–120 cm long by *c.* 1 mm thick, hairy to glabrous; **fibrous roots** white, 2–4 cm long at some nodes; in rocky woodlands and moist depressions on rocky outcrops in Boreal forest.
■ **FLOWERS** Pink, blooming June–August; **inflorescence** an umbel of two flowers; **peduncles** 5–30 mm long, reddish green, ascending and diverging, long hairs spreading, short hairs gland-tipped and deflexed; **subtending bracts** (of peduncles) two, 2.5–4 mm long by 0.8–1.6 mm wide, glandular-hairy; **pedicels** 1–5 mm long, glandular-hairy, short hairs descending; **subtending bracts** (of pedicels) paired, 1–2 mm long by *c.* 0.5 mm wide, erect; **flowers** paired, perfect, fragrant, 8–11 mm wide by 9–14 mm long, descending; **calyx** tubular, 3- or 5-lobed, TL 2–5 mm, glandular-hairy, reddish green, lobes 1.5–2.5 mm long by 0.5–1 mm wide, pointed; **corolla** tubular with white, eglandular hairs inside, 5-lobed, these blunt, 2–3 mm long by 2–3.5 mm wide; **stamens** four, didynamous (two long and two short), included; **filaments** 5–8 mm long, white, lower 2–3 mm attached near base of corolla tube; **anthers** pale yellow to white, 2-lobed, *c.* 2 mm long; **subtending bracts** (of ovaries) four, *c.* 0.8 mm long, reddish glandular-hairy outside, green and glabrous inside, two of them expanding to cover the fruit. **FRUIT a schizocarp**, dry, indehiscent, reddish brown, 1.5–3 mm long by 1.3–1.5 mm wide with dark brown longitudinal ridges and short ascending hairs in the upper half; **seeds** two per schizocarp, pale yellowish brown, 1.5–2 mm long by 0.8–1.5 mm wide by *c.* 1 mm thick, smooth, hairless, with a single pale line on one side only.
■ **LEAVES** Opposite, simple, evergreen, slightly toothed near apices but entire in lower half, lighter green below; **branch leaves** two or three pairs near the base; **blades** 5–20 mm long by 5–15 mm wide, reduced below, spreading to ascending, slightly hairy on lower margins, upper surface and veins below; **petioles** ciliate, 1–5 mm long; **stipules** absent; **basal leaves** one or two pairs, wilting early.
■ **BRANCHES** Erect, 5–15 cm long by 0.8–1 mm wide, reddish, round, hairs descending along the branches, hairs ascending from nodes along the stem.
■ **SYN.** *Linnaea americana* Forbes = *Linnaea borealis* ssp. *americana* (Forbes) Hultén ex Clausen.

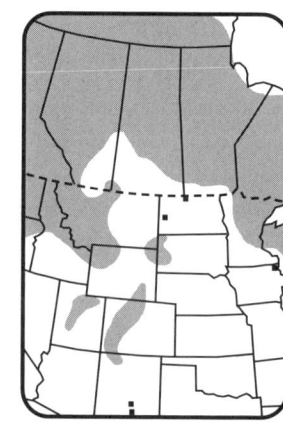

238   American Twinflower

Honeysuckle—*Caprifoliaceae* **Vine** Yellow to orange to red   Corolla tubular, 2-lipped, 4- or 5-lobed

TWINING HONEYSUCKLE *Lonicera dioica* L.

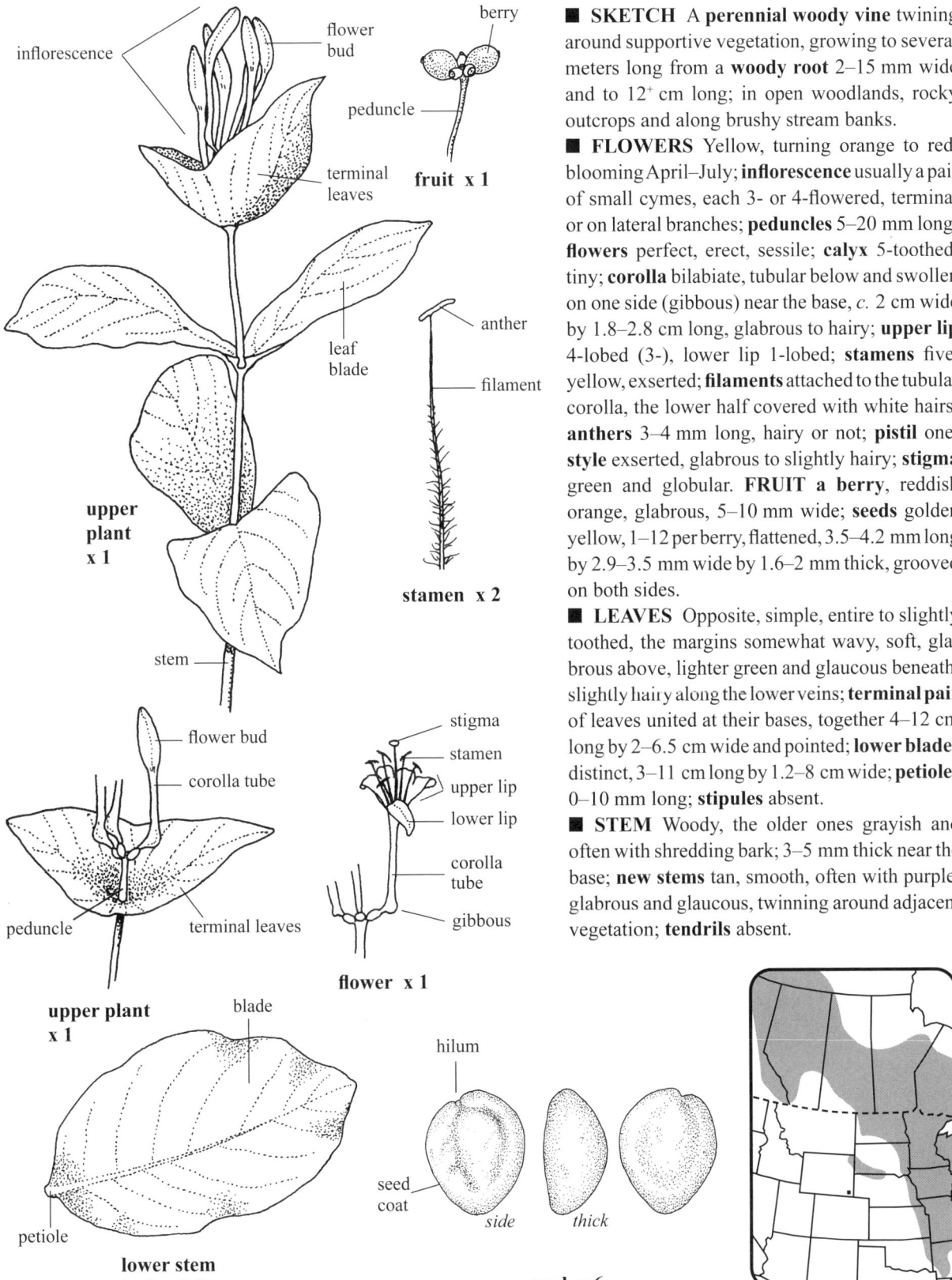

- **SKETCH** A **perennial woody vine** twining around supportive vegetation, growing to several meters long from a **woody root** 2–15 mm wide and to 12$^+$ cm long; in open woodlands, rocky outcrops and along brushy stream banks.
- **FLOWERS** Yellow, turning orange to red, blooming April–July; **inflorescence** usually a pair of small cymes, each 3- or 4-flowered, terminal or on lateral branches; **peduncles** 5–20 mm long; **flowers** perfect, erect, sessile; **calyx** 5-toothed, tiny; **corolla** bilabiate, tubular below and swollen on one side (gibbous) near the base, *c.* 2 cm wide by 1.8–2.8 cm long, glabrous to hairy; **upper lip** 4-lobed (3-), lower lip 1-lobed; **stamens** five, yellow, exserted; **filaments** attached to the tubular corolla, the lower half covered with white hairs; **anthers** 3–4 mm long, hairy or not; **pistil** one; **style** exserted, glabrous to slightly hairy; **stigma** green and globular. **FRUIT** a berry, reddish orange, glabrous, 5–10 mm wide; **seeds** golden yellow, 1–12 per berry, flattened, 3.5–4.2 mm long by 2.9–3.5 mm wide by 1.6–2 mm thick, grooved on both sides.
- **LEAVES** Opposite, simple, entire to slightly toothed, the margins somewhat wavy, soft, glabrous above, lighter green and glaucous beneath, slightly hairy along the lower veins; **terminal pair** of leaves united at their bases, together 4–12 cm long by 2–6.5 cm wide and pointed; **lower blades** distinct, 3–11 cm long by 1.2–8 cm wide; **petioles** 0–10 mm long; **stipules** absent.
- **STEM** Woody, the older ones grayish and often with shredding bark; 3–5 mm thick near the base; **new stems** tan, smooth, often with purple, glabrous and glaucous, twinning around adjacent vegetation; **tendrils** absent.

Limber Honeysuckle, Wild Honeysuckle, Glaucous Honeysuckle, Mountain Honeysuckle

Honeysuckle—*Caprifoliaceae* **Shrub** Pink to white Corolla tubular, 2-lipped, 5-lobed

TATARIAN HONEYSUCKLE *Lonicera tatarica* L.

■ **SKETCH** A **deciduous shrub** 1–3 m tall; along forest trails, woodland edges, stream banks and shrubby pastures. *Naturalized*

■ **FLOWERS** Pink to white, blooming May–July; **inflorescence** in pairs from leaf axils; **peduncles** 5–23 mm long, glabrous, ascending; **subtending bracts** (at apices of peduncles) 4–10 mm long by 0.8–2 mm wide, linear, veined, usually glabrous; **bractlets** four (two below each flower), light green, 1.3–1.5 mm long by *c.* 1 mm wide, glabrous; **flowers** perfect, numerous, 23–25 mm wide by 15–20 mm tall; **calyx** green, glabrous, oval, TL 2–3 mm by 1.3–1.5 mm wide; **calyx lobes** five, lighter green, blunt, erect to ascending, glabrous, 0.5–1 mm long by *c.* 0.4 mm wide near their base; **corolla** tubular, bilabiate, glabrous, upper lip 4-lobed, lower lip 1-lobed, all lobes about equal; **corolla tube** 6–7 mm long, with a slight bulge at the base on one side; **upper lip** with two middle lobes united below, 8–9 mm wide; **lower lip** 10–13 mm long by 5–6 mm wide; **stamens** five, exserted; **filaments** white, the upper 2.5–4 mm are free, green and glabrous, lower half hairy and attached to corolla tube; **anthers** yellow, 2.5–3 mm long, narrow, attached near one end to filament; **style** filiform, white, slightly hairy, 7–8 mm long, exserted *c.* 5 mm, equal to or slightly longer than the stamens; **stigma** green, capitate, rough, *c.* 1.2 mm wide. **FRUIT a berry**, yellow to reddish orange, glabrous, round, slightly shiny, in united pairs at ends of peduncles, each 5–8 mm wide and long; **seeds** two to five per berry, light yellow, flattened with a low ridge on each side, 2.3–3 mm long by 1.9–2.5 mm wide by 1–1.3 mm thick, hairless, surface minutely rough.

■ **LEAVES** Opposite, simple, entire, glabrous; **blades** 1–8 cm long by 1–4.2 cm wide, flat, apices bluntly pointed, lighter green below; **petioles** green, glabrous, 2–8 mm long; **stipules** absent.

■ **STEM** Erect, branches opposite, ascending, light gray to whitish gray above, smooth; **older stems** brownish gray with bark shredding, hollow; 1–2 cm wide near the base.

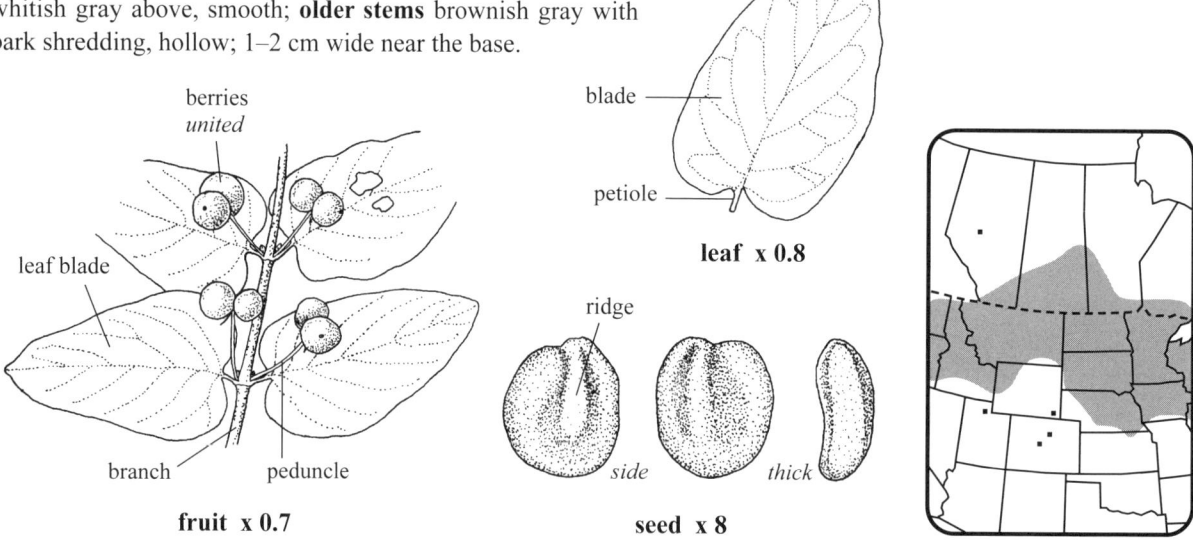

Twinsisters

Honeysuckle—*Caprifoliaceae*　　**Shrub**　Pale yellow　Corolla tubular, (4-) 5-lobed

BLUE FLY HONEYSUCKLE　*Lonicera villosa* (Michx.) J.A. Schultes

■ **SKETCH** A variable **deciduous shrub** 22–100 cm tall with horizontal **roots** to *c.* 30 cm long by *c.* 1 mm thick; in wet woods, bogs and swamps.

■ **FLOWERS** Pale yellow (creamy), blooming May–July; **inflorescence** a raceme of 1 to 3 pairs of flowers in fascicles of leaves along the upper branches; **pedicels** hairy, axillary, 6–8 mm long, curved at anthesis; **flowers** perfect, slightly fragrant, 10–12 mm wide by 9–15 mm long, ascending to drooping; **calyx** green, two atop ovary, *c.* 1 mm long and wide, 5-lobed, lobes blunt, 0.2–0.3 mm long and wide; **corolla** tubular, (4-) 5-lobed; **tube** 5–7 mm long by *c.* 1.5 mm wide and thick, gibbous and slightly hairy at base, ridged, very hairy inside; **corolla lobes** similar, ascending, 4–5 mm long by 2–2.5 mm wide, blunt, hairless; **stamens** five, exserted; **filaments** pale yellow, *c.* 11 mm long, upper free part *c.* 6 mm long, hairless, lower 5 mm hairy and attached to corolla tube; **pistil** one, exserted; **ovary** green, glabrous, *c.* 3.5 mm long and each *c.* 1.2 mm wide, separate to base, both within a smooth, glaucous outer cup; **subtending bracts** (of ovary cup) 5–7.5 mm long, hairy, paired, ascending, persistent at base of fruit; **style** filiform, *c.* 11 mm long, glabrous; **stigma** green rough, wider than style, slightly above anthers. **FRUIT a berry**, dark blue, glaucous, dull, 8–11 mm long by 4.5–8 mm wide, ovoid, tapered to base, 1–4 in a cluster; **seeds** 7–20 per berry, tan, smooth, *c.* 2 mm long by 1.4–1.6 mm wide by 0.7–0.9 mm thick, not ridged, stained red.

■ **LEAVES** Opposite, entire, simple, ascending to spreading; **blades** 1.5–8 cm long by 7–20 mm wide, lighter green and slightly hairy below and along margins, hairless above, soft, dull, blunt; **petioles** 1–2.5 mm long, hairy; **stipules** absent.

■ **STEM** Woody, gray, branching, bark peeling in long thin vertical strips, brown and smooth under the bark; 3–5 mm wide near the base; **new twigs** pinkish green with scattered white hairs to *c.* 1 mm long and a fine covering of shorter hairs; **terminal fall buds** reddish, 6–9 mm long on a stalk *c.* 1 mm long, with two or three pairs of shorter side buds.

■ **SYN.** *Lonicera caerulea* var. *villosa* (Michx.) Torr. & Gray = *Lonicera villosa* var. *villosa* (Michx.) J.A. Schultes.

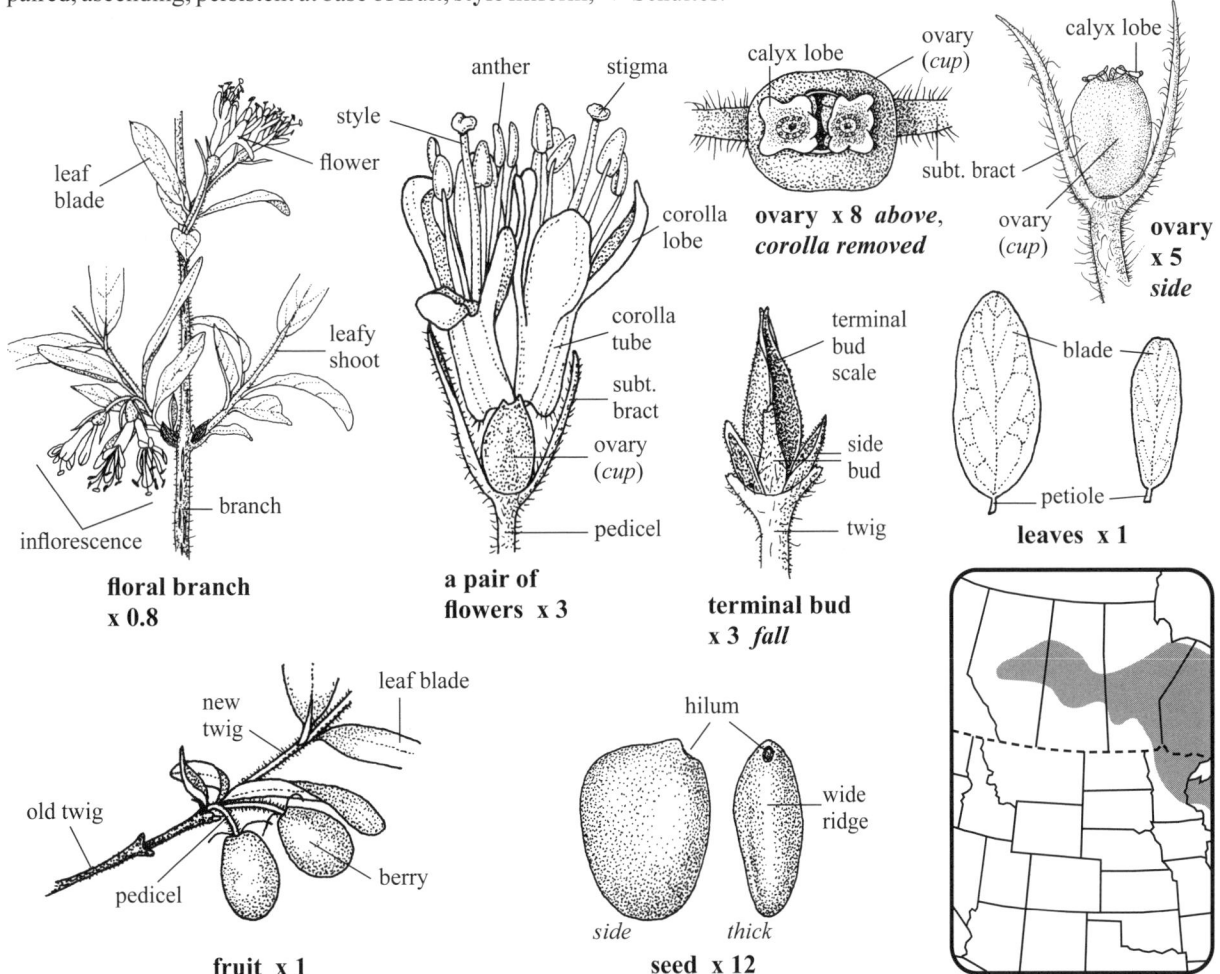

Honeysuckle—*Caprifoliaceae* **Shrub   White   Corolla 5-lobed**

RED ELDERBERRY   *Sambucus racemosa* L.

■ **SKETCH** A **deciduous shrub** 1–6 m tall with several clustered stems; in moist woods, meadows, along river and stream banks. **w1**

■ **FLOWERS** White to slightly green, blooming April–July; **inflorescence** of cymules on numerous small rays forming an oval terminal cluster 4–10 cm long by 3–7 cm wide; **peduncles** 2.5–5.5 cm long by 1–2 mm thick, lightly hairy and pale green; **rays** several, 3–11 mm long, spreading, ridged; **pedicels** 1–3 mm long, pale yellow, lightly hairy; **flowers** perfect, fragrant, 3–5 mm wide, numerous; **calyx** 5-lobed, visible when petals raised, glabrous, widely triangular, the lobes *c.* 0.5 mm long and united into a pale green ring; **corolla** 5-lobed, these strongly reflexed, 2.8–3 mm long by 1.5–2.5 mm wide, glabrous, apex slightly erose, reddish on outside, united at their base and with the filaments; **stamens** five, *c.* 2.5 mm long, spreading between the corolla lobes; **filaments** thick, pale, 1–2 mm long; **anthers** pale yellow, 2-lobed, *c.* 1.5 mm long and wide; **pollen** yellow; **ovary** medium green, glabrous; **stigma** 3-lobed, slightly exserted. **FRUIT a drupe**, berrylike, bright red to black, shiny, juicy, 4–7 mm wide and long, in a cluster to *c.* 8 cm wide and about as long; **nutlets** two to four per drupe, 1-seeded, 2.5–3.1 mm long by 1.5–1.8 mm wide by *c.* 1 mm thick, hard, rough, golden yellow.

■ **LEAVES** Opposite, compound (odd-pinnate), each 5–18 cm long by 6–19 cm wide; **leaflets** toothed, three to nine, dark green, usually 4–17 cm long by 1.5–6 cm wide, pointed, bases often asymmetrical, hairy below, less so above, midribs reddish above; **petioles** hairy, 1.5–5 cm long; **petiolules** 1–18 mm long, terminal one the longest; **stipules** 1–5 mm long by *c.* 1 mm wide, soon deciduous; **rachis** hairy, hairs spreading and purplish.

■ **STEM** Erect, woody, medium gray, smooth, central pith yellowish brown; **branches** mostly opposite and with an undulating appearance; **twigs** glabrous to lightly hairy; **lenticels** oval, 1–2 mm long and raised; **winter buds** opposite, 10–15 mm long by 5–7 mm wide and thick, 6-scaled, glabrous, medium to reddish brown, pointed; **leaf scars** wide, tan, crescent-shaped; **leaf bundles** three.

■ **SYN.** *Sambucus pubens* Michx. = *Sambucus racemosa* var. *racemosa* L.

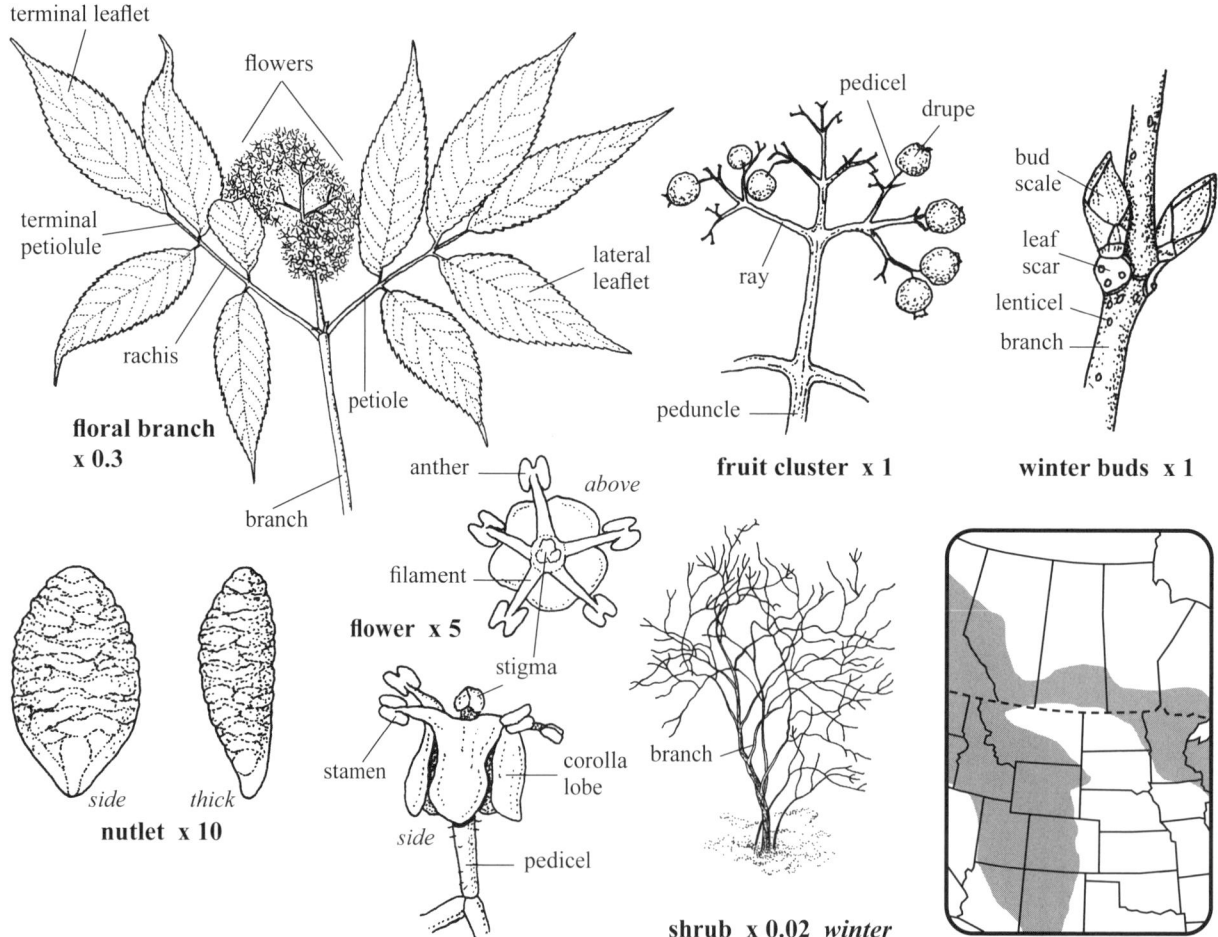

Scarlet Elder, Red-berried Elder, Red Elder, Stinking Elderberry

Honeysuckle—*Caprifoliaceae* **Shrub** White and pink Corolla tubular, 5-lobed

WESTERN SNOWBERRY *Symphoricarpos occidentalis* Hook.

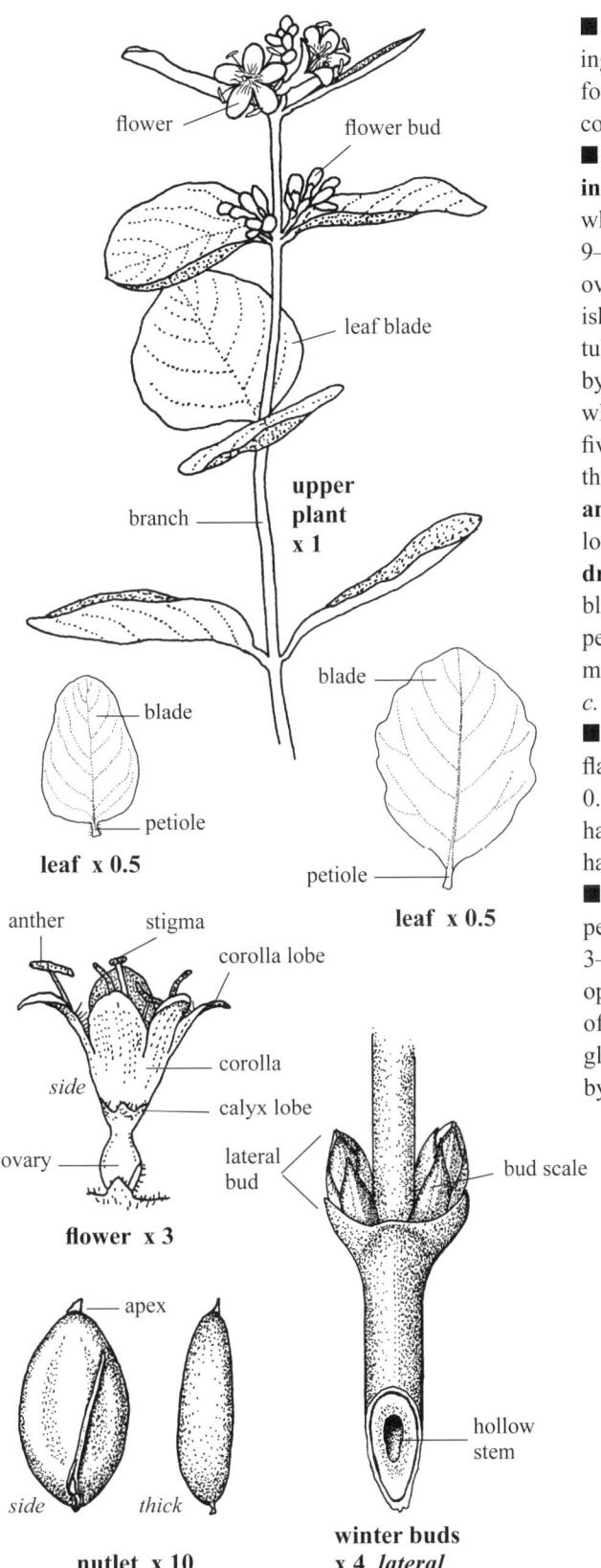

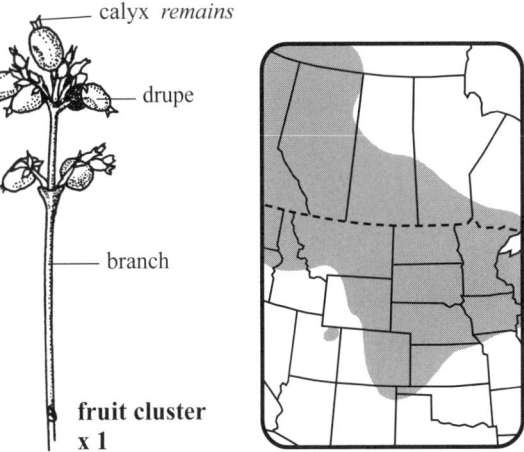

■ **SKETCH** A **deciduous shrub** 0.5–1.5 m tall, spreading by woody **rhizomes** 3–10 mm thick by 25⁺ cm long, forming large colonies; in wet or dry grasslands, ravines, coulees, rocky hillsides and open woodlands. **w1, p2**

■ **FLOWERS** White and pink, blooming May–August; **inflorescence** of axillary clusters in upper leaves; **buds** whitish pink; **flowers** perfect, sessile, 4–12 in a spike, each 9–12 mm wide by 10–13 mm long (including the green ovary); **calyx** usually 5-lobed, these *c.* 0.5 mm long, pinkish green, the margins ciliate, persistent on fruit; **corolla** tubular, 5-lobed (4-), these spreading and 4–5 mm long by 3–3.5 mm wide, pinkish and glabrous on the outside, whitish inside and densely hairy in the throat; **stamens** five, exserted; **filaments** with the lower third attached to the corolla tube, the free part between the corolla lobes; **anthers** 1.8–2.2 mm long, pale yellow; **style** white, 4–7 mm long, exserted; **stigma** round, *c.* 0.8 mm wide. **FRUIT a drupe**, berrylike, white, 6–9 mm wide, fleshy, drying to bluish black overwinter, persistent; **nutlets** 1-seeded, two per drupe, yellowish, flattened, smooth, one side slightly more convex, 2.5–3.5 mm long by 1.5–2 mm wide by *c.* 0.8 mm thick.

■ **LEAVES** Opposite, simple, entire or with blunt teeth, flat, lighter green beneath; **blades** 1.2–7.5 cm long by 0.5–5.5 cm wide, dull, thick, glabrous above and lightly hairy below, more hairy near the base, veins faint; **petioles** hairy or not, 1–7 mm long; **stipules** absent.

■ **STEM** Woody, erect, persistent overwinter, one to four per clump, round, hollow, smooth, light to medium brown; 3–12 mm thick near the base; **bark** peeling; **branches** opposite at the upper nodes, pale green to light tan, one of a pair often longer than the other, ascending, downy to glabrous; **winter buds** opposite, glabrous, 2–3.2 mm long by *c.* 1.3 mm wide, pointed, 5- or 6-scaled.

Wolfberry, Buckbrush

Honeysuckle—*Caprifoliaceae* **Shrub** White Corolla tubular, (4-) 5-lobed

LOW BUSH-CRANBERRY *Viburnum edule* (Michx.) Raf.

■ **SKETCH** A **deciduous shrub**, erect to leaning, 15–200 cm tall, forming thickets with long, woody dark **rhizomes** 5–6 mm thick with branched **fibrous roots** to *c.* 8 cm long; in open woods, ditch banks, gravel ridges, wooded hillsides and ravines.

■ **FLOWERS** White, blooming June–July; **inflorescence** a cyme (umbel-like), compound, each 1.5–4 cm across by 7–11 mm tall on new floral branches with one pair of leaves at base of peduncle; **peduncles** 0.8–2.3 cm long, green to reddish with fruit, glandular-dotted; **subtending bracts** (of floral branches) entire, linear, reddish, 3–5 mm long by *c.* 1 mm wide, pointed; **pedicels** green, glandular, 0.1–0.8 mm long, 2–7 mm long with fruit, ascending to spreading; **subtending bracts** (of pedicels) one, reddish, 2–2.5 mm long by *c.* 0.5 mm wide, linear, reaching almost to calyx lobes; **flowers** perfect, fragrant, 6–7 mm wide by 4–5 mm long, similar, 10–35 per inflorescence; **calyx** tubular, green, toothed, glabrous, 1.5–1.9 mm long by *c.* 1 mm wide; **corolla** tubular, (4-) 5-lobed; **tube** 1.8–2 mm long and wide, glabrous; **corolla lobes** 1.2–2 mm long and wide; **stamens** five, included to almost level with the corolla lobes; **filaments** white, *c.* 1 mm long, attached to top of corolla tube; **anthers** yellow, 2-lobed, *c.* 0.4 mm long; **pistils** mostly enclosed in calyx tube, glabrous; **ovary** white; **style** obscure; **stigmas** 3-lobed; **fruiting heads** 15–25 mm wide by 10–15 mm tall, usually of 2–9 drupes. **FRUIT a drupe**, 1-stoned, red, shiny, smooth, 6–10 mm long and wide, juicy, dropping quickly when ripe; **stone** 1-seeded, light brown, 4–8 mm long by 5–7 mm wide by 1.5–2 mm thick, rough, with a wide marginal ridge and low ridge in the middle of one side.

■ **LEAVES** Opposite, 3-lobed, simple, toothed, pointed, turning reddish purple in autumn; **blades** 3–13 cm long and wide, slightly glandular (green dots) and lighter green below with white hairs on veins and in axils, glabrous above, margins ciliate especially in lower half; **projections** paired, 0.5–1 mm long near petiole; **petioles** bright reddish purple, hairless, slightly glandular in the deep upper groove, 0.4–3 cm long; **stipules** absent.

■ **STEM** Dark gray, smooth or glandular; **branches** opposite, ascending; **new twigs** 4-sided, edges blunt, hairless, brown; **lenticels** tan; **axillary buds** in fall, dark red, 4–7 mm long by 2–2.3 mm wide by 1.7–2.8 mm thick, blunt, scales one, 2-ridged, slightly shiny, hairless; lower 15–20 cm of stem bare; base of stem 4–20 mm wide.

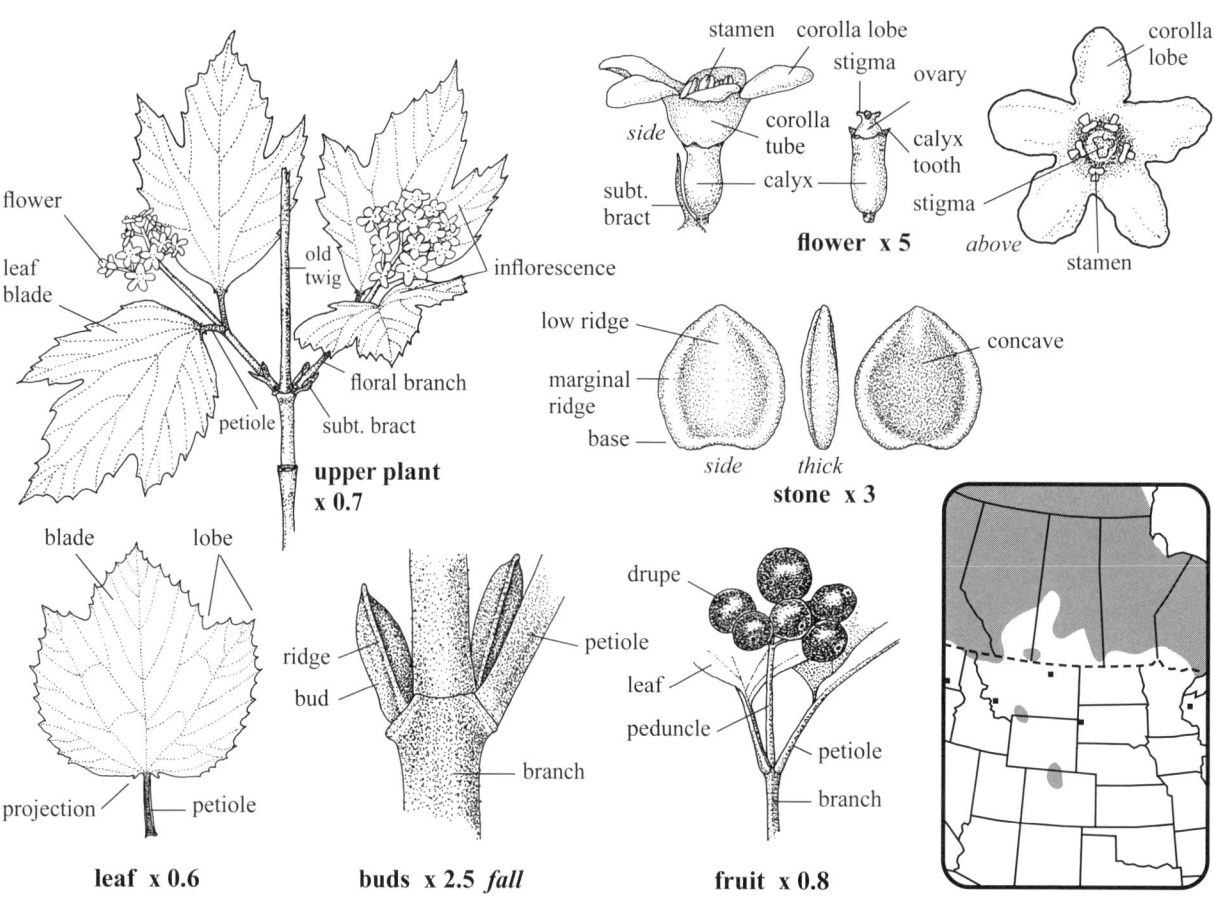

Squashberry, Mooseberry

Honeysuckle—*Caprifoliaceae* **Shrub** White Corolla tubular, 5-lobed

NANNYBERRY *Viburnum lentago* L.

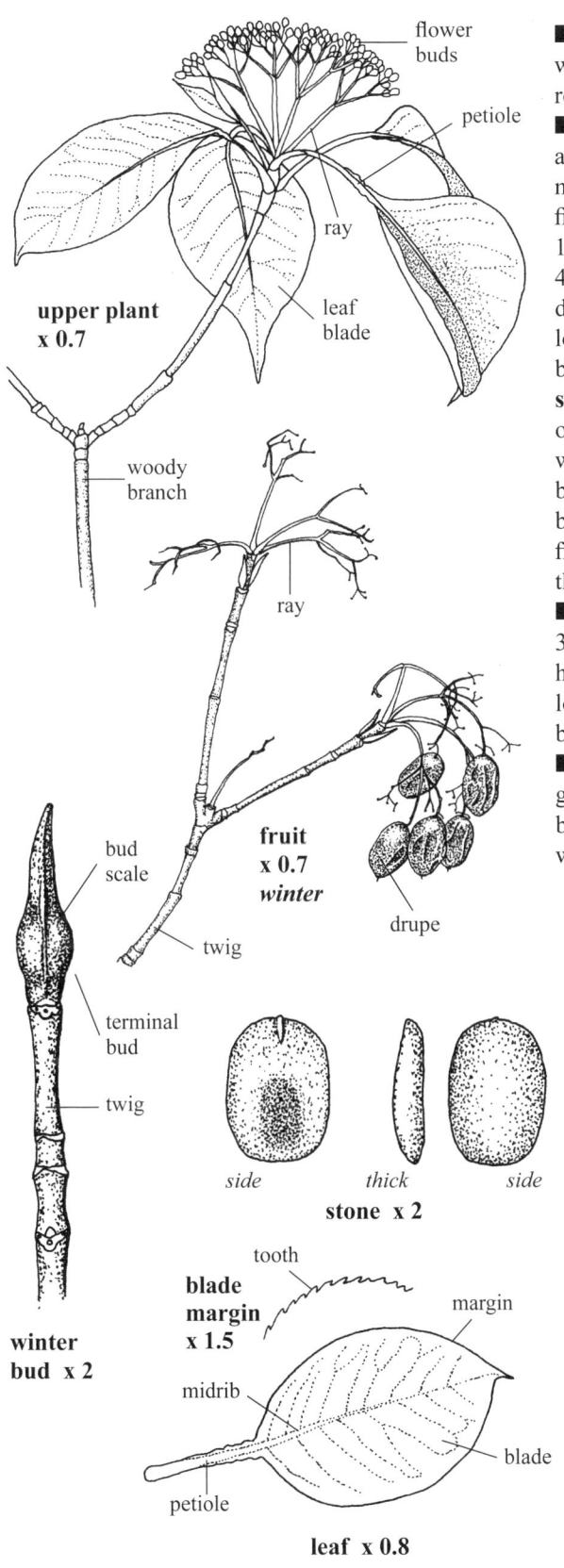

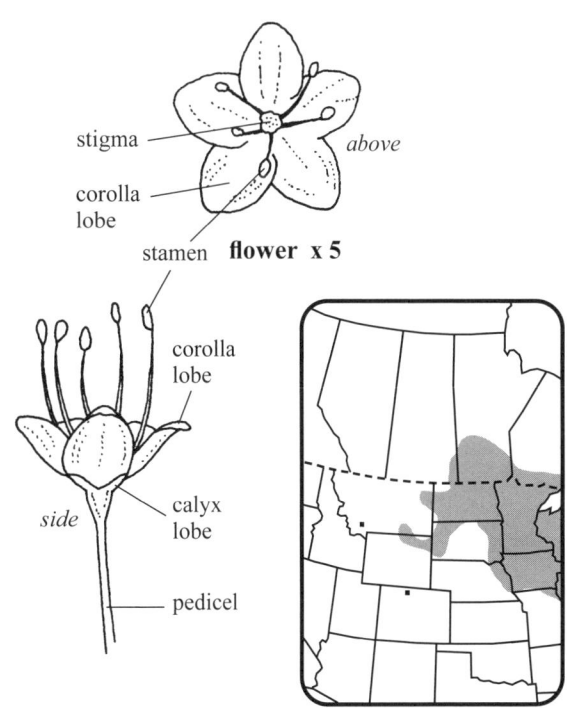

- **SKETCH** A **deciduous shrub** 3–6 m tall; in open moist woodlands, on rocky slopes, along streams, swamps and roadsides. w2
- **FLOWERS** White, blooming May–July; **inflorescence** a cyme, sessile (no peduncle), umbel-like, compound, terminal, flat to rounded, 5–11 cm across; **rays** usually three to five (seven), 1.5–2.5 cm long, turning red in fruit; **pedicels** 1–5 mm long; **flowers** perfect, fragrant, 5–7 mm wide by 4–5 mm long; **calyx** 5-lobed, white to light green, glabrous, deciduous, the lobes 0.5–1 mm long; **corolla** tubular, 5-lobed, subrotate, 2.5–3.5 mm long, the lobes 1–2 mm long by *c.* 2 mm wide, horizontal to slightly ascending, glabrous; **stamens** five, exserted; **filaments** 3–4 mm long; **stigma** green, obscurely 3-lobed, included; **fruiting clusters** to *c.* 10 cm wide by 4–5 cm long. **FRUIT a drupe**, 1-stoned, fleshy, bluish black with a bloom, 8–14 mm long by 6–10 mm wide by 6–8 mm thick; **stone** 1-seeded, tan to yellowish, rough, flattened, 7–11 mm long by 7–9 mm wide by 1.5–2.5 mm thick, slightly concave on one side.
- **LEAVES** Opposite, simple, sharply toothed; **blades** 3.5–10 cm long by 1.8–6 cm wide, glabrous above, thinly hairy below, pointed or with a short bent tip; **petioles** 1–3 cm long, narrowly winged, the wing wavy near the base of the blade, hairs reddish brown and stellate; **stipules** absent.
- **STEM** Erect, 3–20 cm wide near the base; **branches** gray to reddish brown, smooth; **winter terminal bud** reddish brown with two scales 12–20 mm long, glabrous, pointed and with an obvious bulge 4–5 mm wide near its base.

Sheepberry

Honeysuckle—*Caprifoliaceae* **Shrub** White Corolla tubular, 5-lobed

HIGH BUSH-CRANBERRY *Viburnum opulus* L.

- **SKETCH** A **deciduous shrub** 1–5 m tall; in open woodlands, wooded hillsides and along riverbanks. **w2**
- **FLOWERS** White, blooming May–June; **inflorescence** an umbel-like, compound cyme, terminal, 4–12 cm across; **peduncles** 2–9 cm long; **rays** four to six, each 15–20 mm long with the central one the shortest; **pedicels** 1–5 mm long; **flowers** of two types: **1) outer flowers** showy but sterile, blooming first, each 15–25 mm wide; **calyx** 5-lobed and tiny; **corolla** flat with five rotate lobes; **2) inner flowers** perfect, subrotate, numerous, 3–4 mm wide by 4–6 mm long; **corolla** tubular, 5-lobed, these recurved; **stamens** five, white, exserted; **filaments** *c.* 4 mm long; **style** included; **stigma** obscurely 3-lobed. **FRUIT a drupe**, 1-stoned, fleshy, reddish orange, smooth and shiny, 9–12 mm wide and long, persistent; **stone** (endocarp) 1-seeded, rough, flattened, yellowish, 7–10 mm long by 6–8 mm wide by 1.5–2.5 mm thick and pointed at one end.
- **LEAVES** Opposite, simple, mostly 3-lobed, margins entire to toothed, pointed, reddish in autumn; **blades** 4–15 cm long by 4–13 cm wide, glabrous above, tufts of white hairs in axils of veins below; **petioles** 1–3.3 cm long, reddish or green, glabrous with 2–6 glands at their apices; **stipules** 2–6 mm long and hairlike.
- **STEM** Smooth, light gray to brown; **branches** erect to nodding, glabrous; **winter buds** shiny, opposite, reddish brown, scale outlines obscure, 5–8 mm long by 3–5 mm wide; **leaf scars** with three dark bundle scars.
- **SYN.** *Viburnum trilobum* Marsh. = *Viburnum opulus* var. *americanum* Ait.

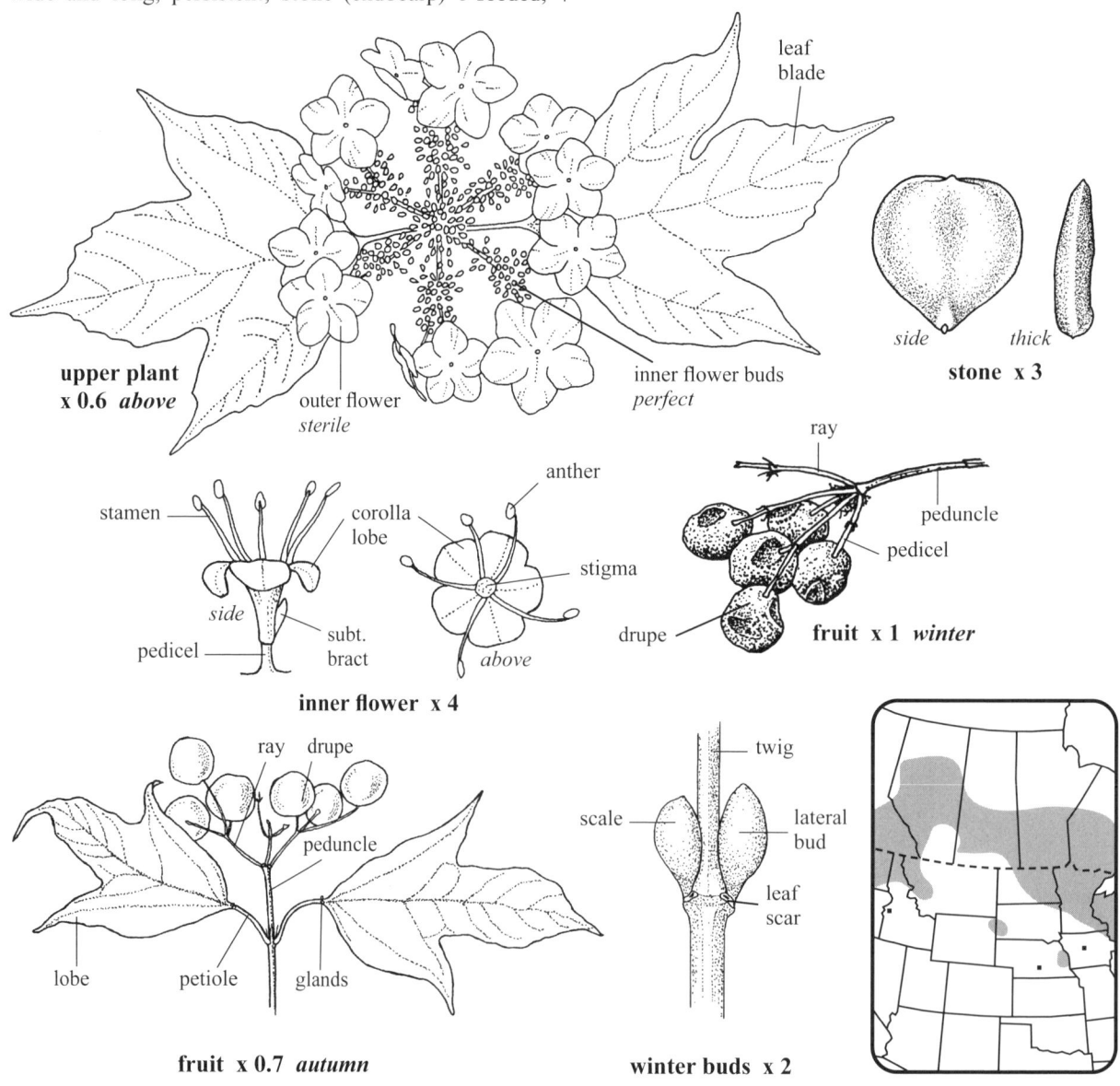

European Cranberrybush, American Bush-cranberry

Honeysuckle—*Caprifoliaceae* **Shrub** White Corolla tubular, 5-lobed

Downy Arrowwood *Viburnum rafinesquianum* J.A. Schultes

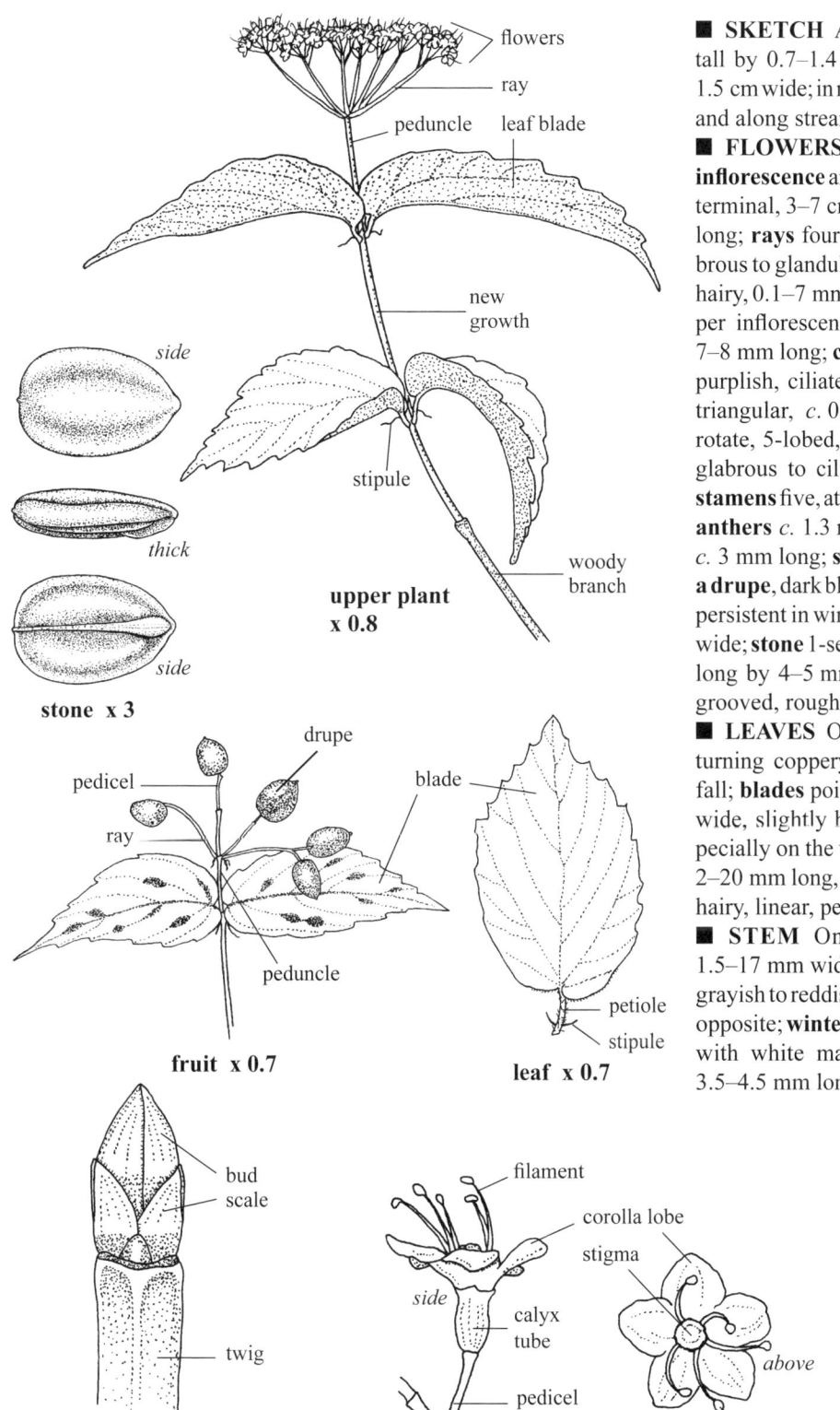

■ **SKETCH** A **deciduous shrub** 0.2–2.4 m tall by 0.7–1.4 m wide with **rhizomes** up to 1.5 cm wide; in moist to dry woodlands, thickets and along streams. **w2**

■ **FLOWERS** White, blooming May–June; **inflorescence** an umbel-like, compound cyme, terminal, 3–7 cm across; **peduncles** 5–50 mm long; **rays** four to seven, 2–20 mm long, glabrous to glandular, reddish; **pedicels** glandular hairy, 0.1–7 mm long; **flowers** perfect, 13–122 per inflorescence, similar, 5–7 mm wide by 7–8 mm long; **calyx** tubular, 5-lobed, green to purplish, ciliate and slightly glandular, lobes triangular, *c.* 0.5 mm long; **corolla** tubular, rotate, 5-lobed, the lobes bluntly pointed and glabrous to ciliate, 1–2 mm long and wide; **stamens** five, attached to corolla tube, exserted; **anthers** *c.* 1.3 mm long; **pistil** one, included, *c.* 3 mm long; **stigma** round and flat. **FRUIT a drupe**, dark blue, each 1-stoned, shiny, pulpy, persistent in winter, 7–11 mm long by 5–7 mm wide; **stone** 1-seeded, yellowish tan, 6–8.5 mm long by 4–5 mm wide by 2.2–2.8 mm thick, grooved, rough, apex pointed.

■ **LEAVES** Opposite, simple, widely toothed, turning coppery red with purple blotches in fall; **blades** pointed, 3–11 cm long by 2–7 cm wide, slightly hairy above, downy below especially on the veins, margins ciliate; **petioles** 2–20 mm long, hairy; **stipules** 1–13 mm long, hairy, linear, persistent.

■ **STEM** One to several in a cluster, 1.5–17 mm wide near the base; **bark** smooth, grayish to reddish brown; **branches** numerous, opposite; **winter buds** reddish brown, 4-scaled with white marginal hairs; **terminal buds** 3.5–4.5 mm long by 2–2.5 mm wide.

Shortstalk Arrowwood

Family Characteristics

Pink—Caryophyllaceae — Flower parts 4 or 5

SKETCH Dicotyledonous herbs (wildflowers), annual or perennial. Worldwide about 3,000 species, mostly in the cool or northern temperate regions.

FLOWERS Perfect or unisexual, regular. **Inflorescence** often a cyme or a solitary flower. **Sepals** 5 (4), separate or slightly united, persistent around fruit. **Petals** 5 (4), separate, often notched and 2-lobed, sometimes clawed, sometimes absent. **Stamens** 5 or 10, distinct. **Filaments** filiform, free, sometimes hairy. **Pistil** 1, of 2–5 united locules. **Ovary** superior with one to many ovules. **Styles** and stigmas 2–5, distinct or united.

FRUIT Often a capsule opening by valves or at the apical teeth. **Seeds** often papillose and few to numerous.

LEAVES Mostly opposite, simple, entire. **Stipules** usually absent.

STEM Erect, weak, often slightly swollen at nodes, often in clusters or colonies, dying back each year.

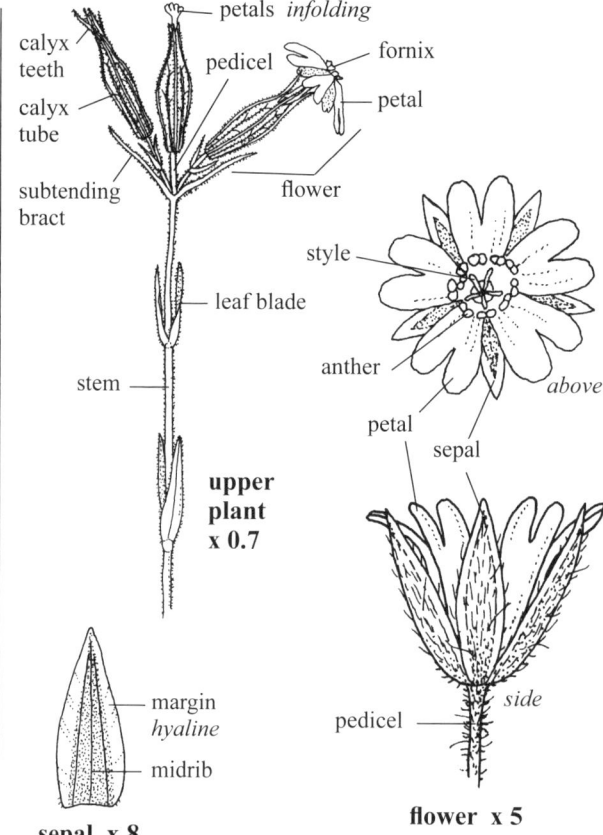

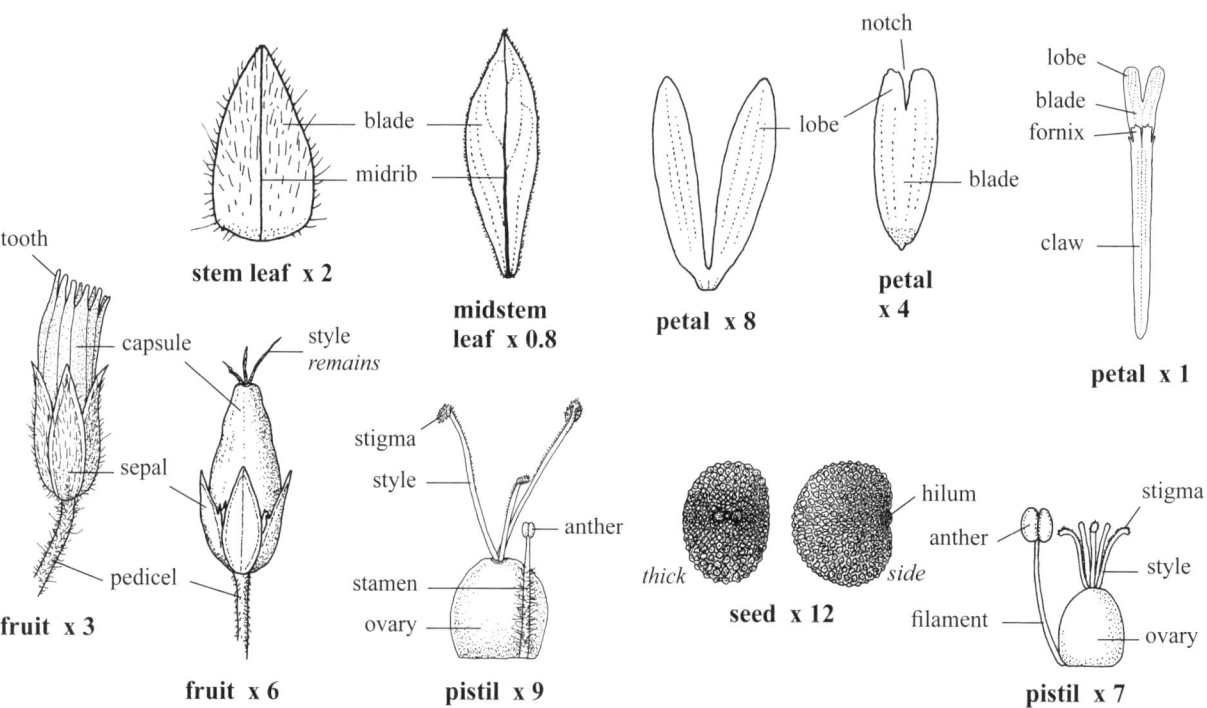

248

Pink—*Caryophyllaceae* **Wildflower** White Petals 5

Field Chickweed *Cerastium arvense* L.

■ **SKETCH** A highly variable **perennial herb** with flowering stems 5–40 cm tall from trailing, tan, horizontal **basal offshoots**; in prairies, pastures, dry roadside ditches, alpine meadows, rocky banks and gravelly eroded sites. *Naturalized*

■ **FLOWERS** White, blooming April–August; **inflorescence** a cyme, 1–11 cm tall by 1–6 cm wide; **peduncles** a few, hairy, 1–5 cm long; **pedicels** 3–42 mm long, densely glandular-hairy; **subtending bracts** (of pedicels) paired, 4–5 mm long by 1.8–3 mm wide, glandular-hairy on outside, glabrous inside, apices white hyaline; **flowers** perfect, 7–15 mm wide by 5–9 mm tall, 1–20 per stem; **sepals** five, 4–7 mm long by 1.8–2 mm wide, erect, midvein faint, hairs glandular, glabrous inside, margins white hyaline; **petals** five, white, ascending to spreading, veiny in lower tapered half, glabrous, 6–12 mm long by 3–4 mm wide, 2-lobed, these rounded and *c.* 2 mm long; **stamens** 10, included, unequal, 3.5–5.5 mm long, glabrous; **filaments** white, 3–4 mm long; **anthers** yellow, 0.8–1 mm long, oval; **pistil** included, 2–2.5 mm long; **ovary** *c.* 1.2 mm tall by 1–1.2 mm wide and thick, round, glabrous, green; **styles** and **stigmas** five, intertwined, filiform, 1–1.3 mm long, persistent on fruit before it opens. **FRUIT a capsule**, membranous, glabrous, green to brown, cylindrical, often curved, 5–11 mm long by 2–4 mm wide, 10-toothed, each tooth *c.* 1 mm long and slightly exserted from the erect, dried sepals; **seeds** *c.* 15 per capsule, reddish to golden brown, 0.7–1.2 mm long by 0.6–0.7 mm wide by *c.* 0.5 mm thick, rough with tubercules.

■ **LEAVES** Opposite, simple, entire, usually four to six pairs, sessile, erect to ascending, some with leafy fascicles in axils, small dried leaves persistent at stem's base; **blades** linear, pointed, more hairy on the outside, sometimes glandular-hairy above, 1–7 cm long by 1–15 mm wide, 1-nerved; **stipules** absent.

■ **STEM** Erect, flowering stem's hairs non-glandular or glandular above and spreading, hairs more tangled and reflexed below, especially on older, thinner, brown stems; 1–2 mm thick near the base.

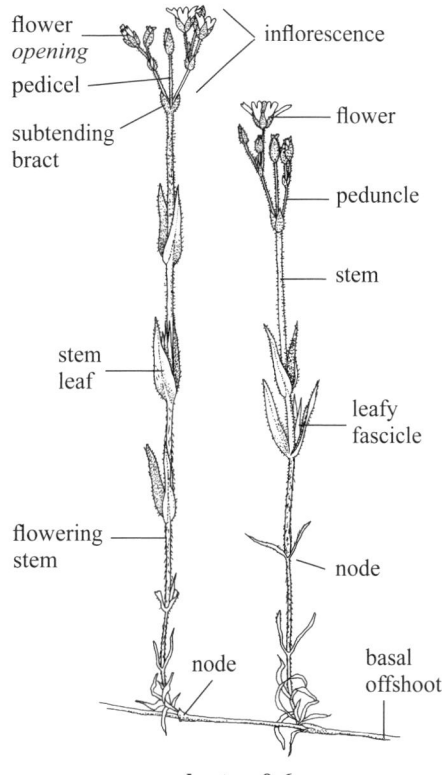

plant x 0.6

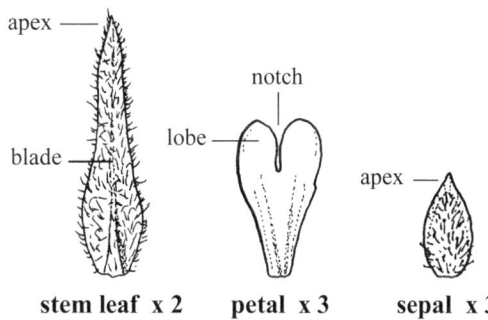

stem leaf x 2   petal x 3   sepal x 3

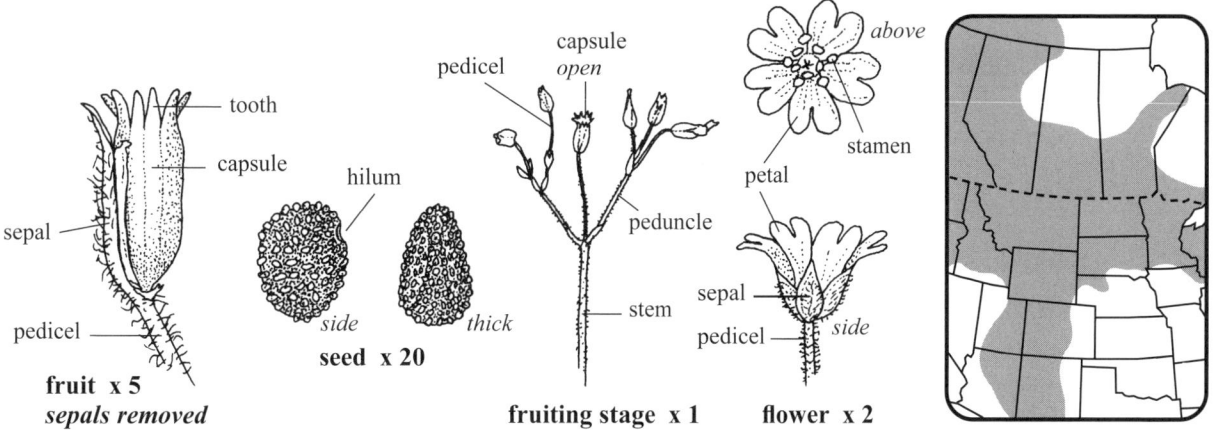

fruit x 5
sepals removed   seed x 20   fruiting stage x 1   flower x 2

Prairie Mouse-ear Chickweed, Prairie Chickweed

Pink—*Caryophyllaceae* **Wildflower** White to pink Petals 5

MOUSE-EAR CHICKWEED *Cerastium fontanum* Baumg.

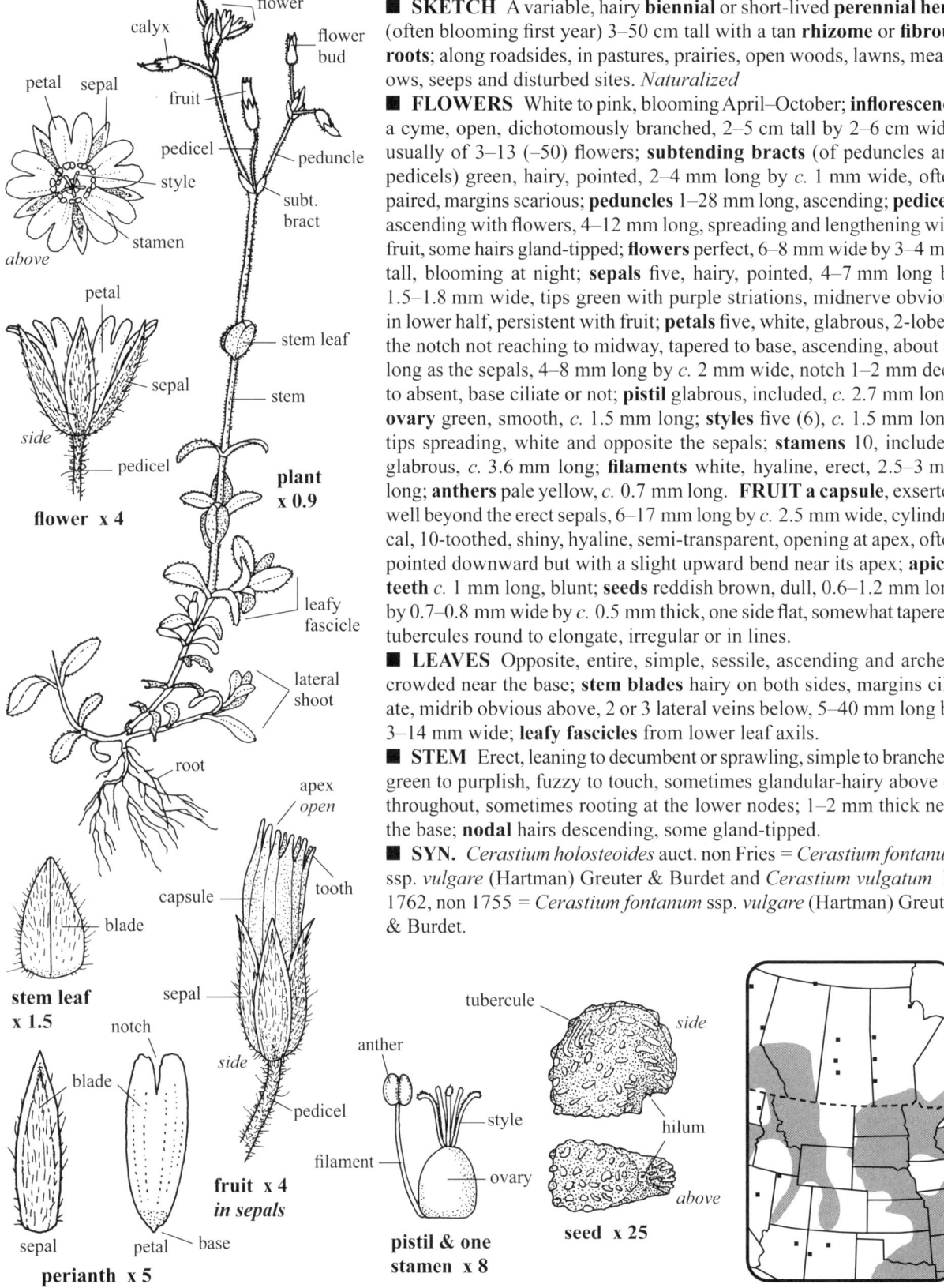

■ **SKETCH** A variable, hairy **biennial** or short-lived **perennial herb** (often blooming first year) 3–50 cm tall with a tan **rhizome** or **fibrous roots**; along roadsides, in pastures, prairies, open woods, lawns, meadows, seeps and disturbed sites. *Naturalized*

■ **FLOWERS** White to pink, blooming April–October; **inflorescence** a cyme, open, dichotomously branched, 2–5 cm tall by 2–6 cm wide, usually of 3–13 (–50) flowers; **subtending bracts** (of peduncles and pedicels) green, hairy, pointed, 2–4 mm long by *c.* 1 mm wide, often paired, margins scarious; **peduncles** 1–28 mm long, ascending; **pedicels** ascending with flowers, 4–12 mm long, spreading and lengthening with fruit, some hairs gland-tipped; **flowers** perfect, 6–8 mm wide by 3–4 mm tall, blooming at night; **sepals** five, hairy, pointed, 4–7 mm long by 1.5–1.8 mm wide, tips green with purple striations, midnerve obvious in lower half, persistent with fruit; **petals** five, white, glabrous, 2-lobed, the notch not reaching to midway, tapered to base, ascending, about as long as the sepals, 4–8 mm long by *c.* 2 mm wide, notch 1–2 mm deep to absent, base ciliate or not; **pistil** glabrous, included, *c.* 2.7 mm long; **ovary** green, smooth, *c.* 1.5 mm long; **styles** five (6), *c.* 1.5 mm long, tips spreading, white and opposite the sepals; **stamens** 10, included, glabrous, *c.* 3.6 mm long; **filaments** white, hyaline, erect, 2.5–3 mm long; **anthers** pale yellow, *c.* 0.7 mm long. **FRUIT a capsule**, exserted well beyond the erect sepals, 6–17 mm long by *c.* 2.5 mm wide, cylindrical, 10-toothed, shiny, hyaline, semi-transparent, opening at apex, often pointed downward but with a slight upward bend near its apex; **apical teeth** *c.* 1 mm long, blunt; **seeds** reddish brown, dull, 0.6–1.2 mm long by 0.7–0.8 mm wide by *c.* 0.5 mm thick, one side flat, somewhat tapered, tubercules round to elongate, irregular or in lines.

■ **LEAVES** Opposite, entire, simple, sessile, ascending and arched, crowded near the base; **stem blades** hairy on both sides, margins ciliate, midrib obvious above, 2 or 3 lateral veins below, 5–40 mm long by 3–14 mm wide; **leafy fascicles** from lower leaf axils.

■ **STEM** Erect, leaning to decumbent or sprawling, simple to branched, green to purplish, fuzzy to touch, sometimes glandular-hairy above or throughout, sometimes rooting at the lower nodes; 1–2 mm thick near the base; **nodal** hairs descending, some gland-tipped.

■ **SYN.** *Cerastium holosteoides* auct. non Fries = *Cerastium fontanum* ssp. *vulgare* (Hartman) Greuter & Burdet and *Cerastium vulgatum* L. 1762, non 1755 = *Cerastium fontanum* ssp. *vulgare* (Hartman) Greuter & Burdet.

250 Common Mouse-ear Chickweed

Pink—*Caryophyllaceae*   **Wildflower** White Petals 5

B ABY'S-BREATH   *Gypsophila paniculata* L.

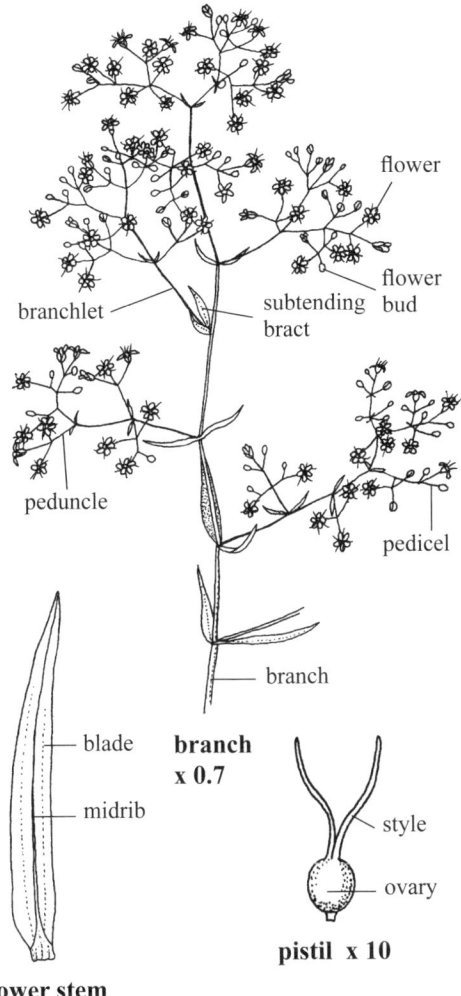

■ **SKETCH** A mostly glabrous and glaucous **perennial herb** 40–120 cm tall and wide from a woody **taproot** 1–6 mm thick and to *c.* 10⁺ cm long; in dry, sandy soils, pastures, hayfields, beaches, roadsides and gravel pits. *Naturalized*

■ **FLOWERS** White, blooming June–September; **inflorescence** cymose-paniculate, delicately branched throughout; **floral branches** alternate below, opposite above, to *c.* 40 cm long below, reduced above, one branch and two leaves per node; **lower branches** leafy, flowers near the ends, upper branches less leafy; **pedicels** stiff, 3–20 mm long; **flowers** perfect, fragrant, numerous, 4–6 mm wide by 4–4.5 mm tall, usually in threes at the ends of peduncles; **calyx** 1.7–2 mm long, united below, 5-lobed, each lobe 1–1.3 mm long by *c.* 1 mm wide, erect, margins white and erose, central wide nerve of each lobe reddish brown, apices slightly hooded; **petals** five, white, 2.4–3.3 mm long by 0.8–1 mm wide, soft, thin, tapered at base, spreading, apices flat to slightly notched; **stamens** 10, exserted, 3–3.5 mm long, equal, reflexed after anthesis, five opposite the petals and five opposite the sepals; **anthers** white, 0.3–0.4 mm long and wide; **pistil** 2–2.5 mm long; **ovary** *c.* 0.7 mm long by *c.* 0.6 mm wide and thick, slightly flattened, reddish green, not lobed; **styles** two, 1.5–2 mm long, white, filiform, spreading, exserted. **FRUIT a capsule**, 2–2.5 mm long by 1.3–2.2 mm wide, roundish, slightly longer than the erect persistent calyx, apical teeth four, broad and blunt, erect to slightly spreading when ripe; **seeds** black, one to three per capsule, tuberculate, 1.1–1.8 mm long by *c.* 1.2 mm wide by *c.* 0.8 mm thick.

■ **LEAVES** Opposite, simple, entire, green, pointed, sessile, glabrous, lower ones wilted by anthesis, subtending branches; **blades** 0.7–8.3 cm long by 1–9 mm wide, reduced above and becoming bractlike, midrib obvious, raised below, usually two lateral parallel veins; **stipules** absent.

■ **STEM** Solid, widely branched, erect to ascending, several in a clump, stiff, glabrous to somewhat (glandular-) hairy below, greenish red, in a zigzagging pattern from branch to branch; 3–6 mm thick near the base; a tumbleweed.

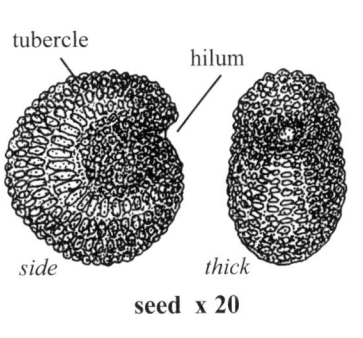

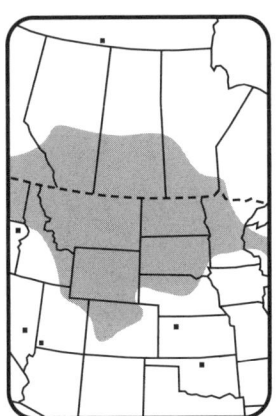

Tall Baby's-breath, Babysbreath Gypsophila   251

Pink—*Caryophyllaceae* **Wildflower White Petals 5 (4)**

BLUNT-LEAVED SANDWORT *Moehringia lateriflora* (L.) Fenzl

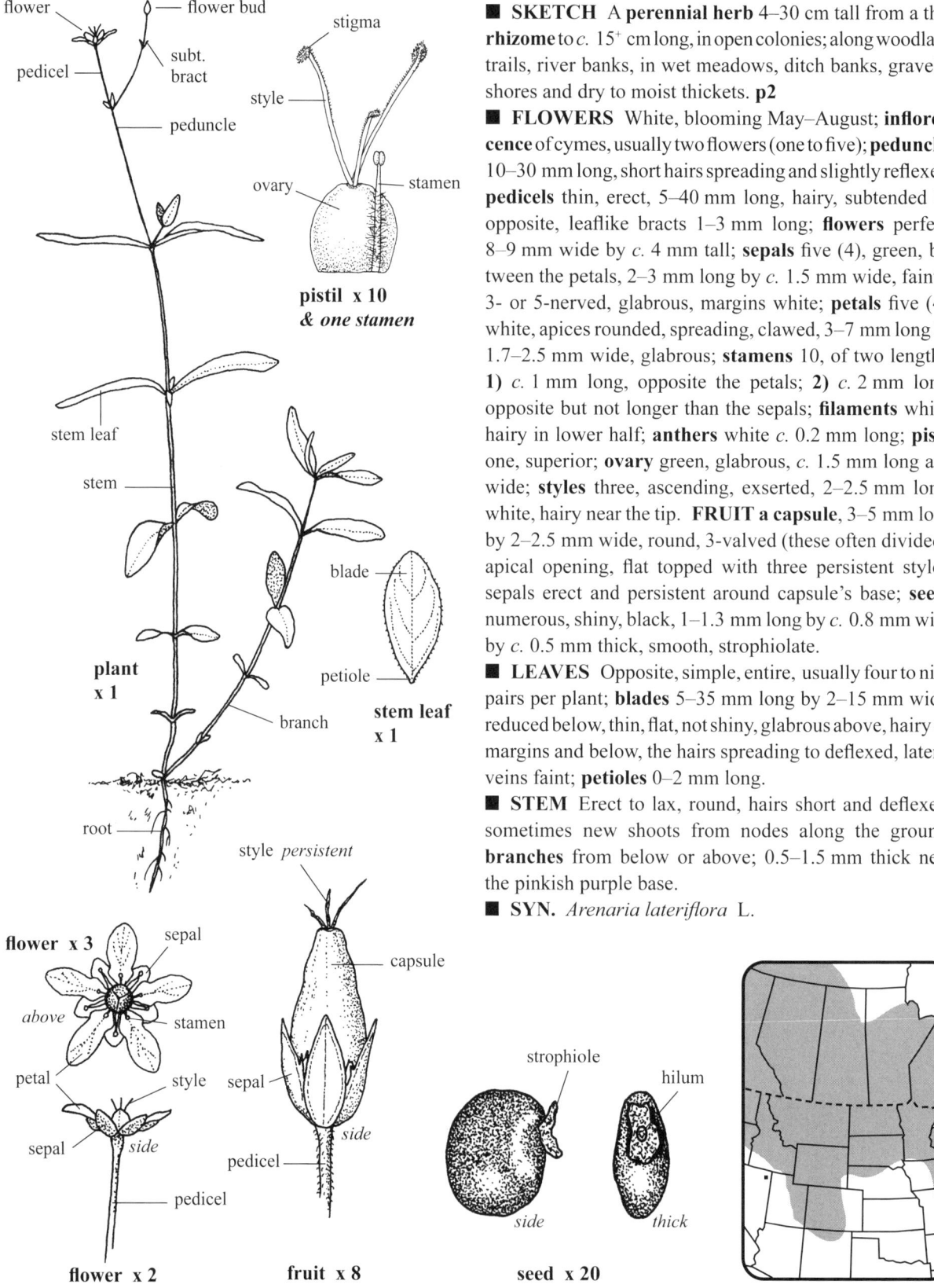

■ **SKETCH** A **perennial herb** 4–30 cm tall from a thin **rhizome** to *c.* 15⁺ cm long, in open colonies; along woodland trails, river banks, in wet meadows, ditch banks, gravelly shores and dry to moist thickets. **p2**

■ **FLOWERS** White, blooming May–August; **inflorescence** of cymes, usually two flowers (one to five); **peduncles** 10–30 mm long, short hairs spreading and slightly reflexed; **pedicels** thin, erect, 5–40 mm long, hairy, subtended by opposite, leaflike bracts 1–3 mm long; **flowers** perfect, 8–9 mm wide by *c.* 4 mm tall; **sepals** five (4), green, between the petals, 2–3 mm long by *c.* 1.5 mm wide, faintly 3- or 5-nerved, glabrous, margins white; **petals** five (4), white, apices rounded, spreading, clawed, 3–7 mm long by 1.7–2.5 mm wide, glabrous; **stamens** 10, of two lengths: 1) *c.* 1 mm long, opposite the petals; 2) *c.* 2 mm long, opposite but not longer than the sepals; **filaments** white, hairy in lower half; **anthers** white *c.* 0.2 mm long; **pistil** one, superior; **ovary** green, glabrous, *c.* 1.5 mm long and wide; **styles** three, ascending, exserted, 2–2.5 mm long, white, hairy near the tip. **FRUIT a capsule**, 3–5 mm long by 2–2.5 mm wide, round, 3-valved (these often divided), apical opening, flat topped with three persistent styles, sepals erect and persistent around capsule's base; **seeds** numerous, shiny, black, 1–1.3 mm long by *c.* 0.8 mm wide by *c.* 0.5 mm thick, smooth, strophiolate.

■ **LEAVES** Opposite, simple, entire, usually four to nine pairs per plant; **blades** 5–35 mm long by 2–15 mm wide, reduced below, thin, flat, not shiny, glabrous above, hairy on margins and below, the hairs spreading to deflexed, lateral veins faint; **petioles** 0–2 mm long.

■ **STEM** Erect to lax, round, hairs short and deflexed; sometimes new shoots from nodes along the ground; **branches** from below or above; 0.5–1.5 mm thick near the pinkish purple base.

■ **SYN.** *Arenaria lateriflora* L.

Pink—*Caryophyllaceae* **Wildflower** White to pink Petals 5

Bouncing Bet *Saponaria officinalis* L.

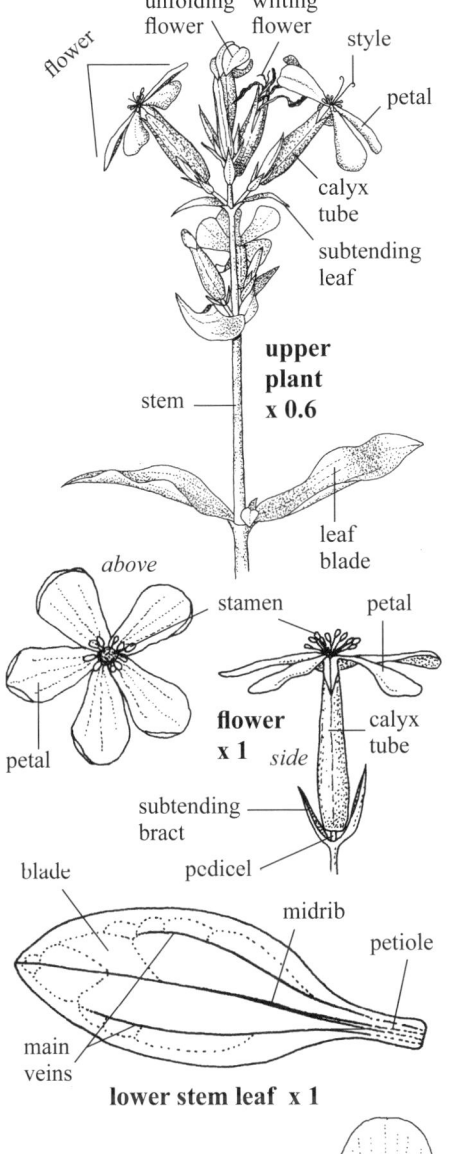

■ **SKETCH** A **perennial herb** 30–90 cm tall from a reddish brown, woody, horizontal **rhizome** 4–6 mm thick; in lawns, along roadsides, railways and shelterbelts. Escaped and *naturalized*

■ **FLOWERS** White to pink, blooming June–September; **inflorescence** a terminal cyme of opposite flowers to *c.* 15 cm long by *c.* 8 cm wide, crowded at the stem's apex, or flowers in upper leaf axils; **pedicels** *c.* 2 mm long, green; **subtending bracts** (of pedicels) one or two, green, leaflike, pointed, usually longer than the pedicel; **flowers** perfect, fragrant, 2.5–3 cm wide by 2.5–3.4 cm long; **calyx** tubular, light green to slightly reddish, 20-nerved (best seen inside), TL 15–24 mm by 4–6 mm wide, rough with microscopic bumps or papillae; **calyx teeth** five, unequal, 1.5–5.5 mm long, pointed, one or two pairs united except for their very tips, one deep sinus between the teeth, thin and persistent around the fruit; **petals** five, glabrous, slightly descending to spreading, blades 8–16 mm long by 7–9 mm wide, rounded, two white fingerlike appendages *c.* 1.5 mm long at the base of the blade, claws pale green, 20–22 mm long by *c.* 2 mm wide, ridged and grooved in the middle; **stamens** 10, exserted, free and whorled at the base; **filaments** white, filiform, 20–23 mm long by *c.* 0.6 mm wide, tapered at both ends, glabrous, winged and 3-sided; **anthers** light gray, 1.2–1.8 mm long; **pistil** 25–27 mm long by *c.* 1.3 mm wide, glabrous; **ovary** *c.* 12 mm long by *c.* 1.3 mm wide, cylindrical, green; **styles** two, rarely three, exserted, white, glabrous, *c.* 14 mm long, separate to base, slightly spreading; **stigmas** two, a slightly roughened line *c.* 6 mm long at the style's apex. **FRUIT a capsule**, tan, erect, 4-valved, 13–20 mm long by 4–6 mm wide, glabrous, thin-walled, opening at the apex; **seeds** 10–45 per capsule, dark brown, dull, 1.5–2.1 mm long by 1.6–1.9 mm wide by 0.8–0.9 mm thick, covered with rounded, low papillae in concentric rows.

■ **LEAVES** Opposite, simple, entire, glabrous to hairy, pointed to blunt, ascending to spreading, flat or with undulate margins; **blades** 4–13 cm long by 1–4 cm wide, with three main veins, lateral veins becoming obscure near the apex and margins; **petioles** winged, 0–1 cm long, reduced above, often a low ridge across the stem where the petioles meet.

■ **STEM** Erect, simple or branched above, solid, slightly scabrous, green with a slight redness above the lower nodes; 3–6 mm thick near the base.

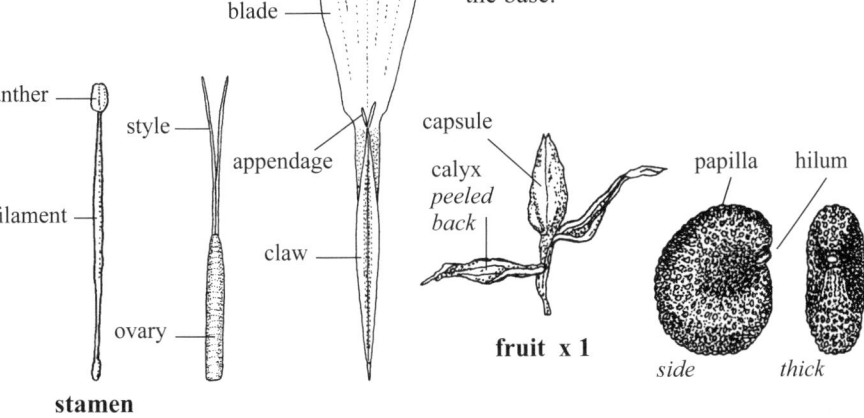

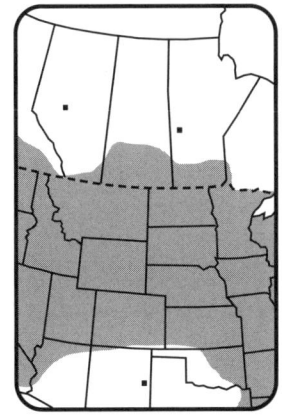

Soapwort

Pink—*Caryophyllaceae* Wildflower White Petals 5

Smooth Catchfly *Silene csereii* Baumg.

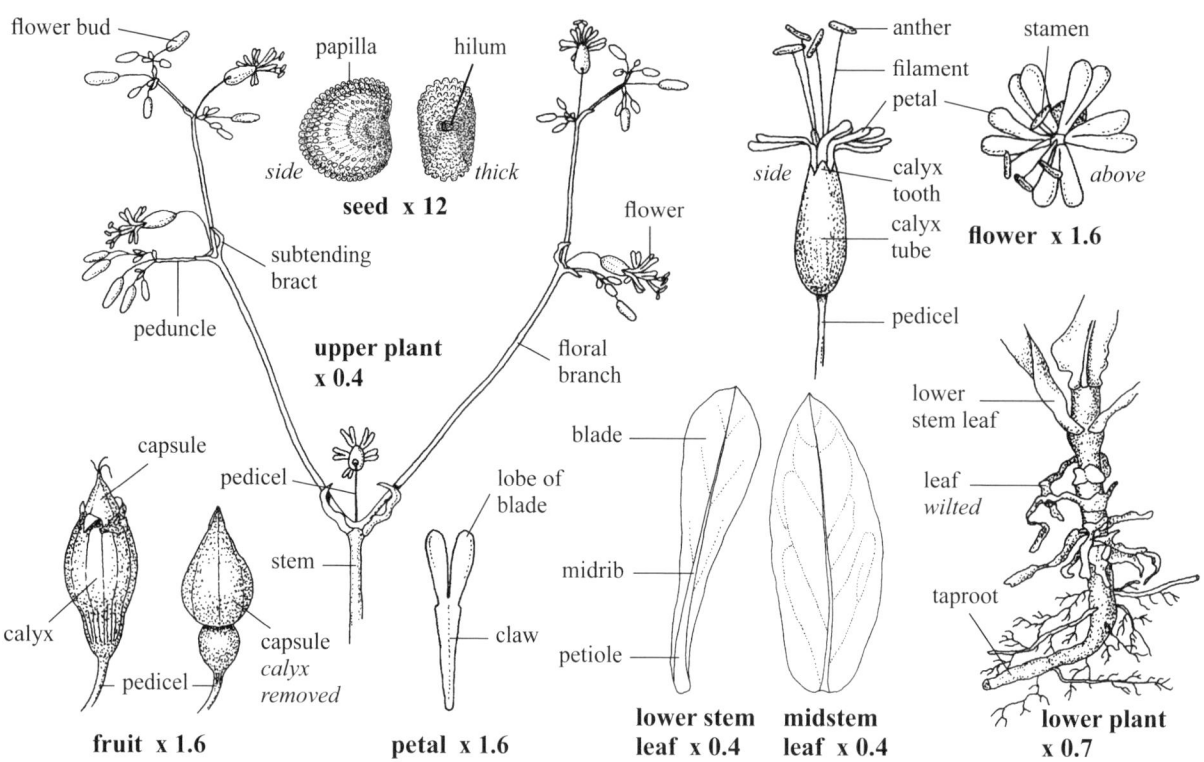

- **SKETCH** A glabrous, glaucous **annual** or **biennial herb**, 10–100 cm tall with a woody **taproot** 2–7 mm wide and 4–16 cm long, single or tufted; in disturbed sites, prairies, along roadsides and railways. *Naturalized*
- **FLOWERS** White, blooming at night and mornings in May–September; **inflorescence** a paniculate cyme 10–30 cm long and wide, from two main divergent floral branches; **floral branches** 15–20 cm long; **subtending bracts** paired, 1–2.5 cm long by 2–7 mm wide, pointed, ascending and often with a twist, reduced above; **peduncles** glabrous, 1–5 cm long, ascending to nodding; **pedicels** 8–40 mm long; **flowers** perfect, clustered, closing by midday, 13–15 mm wide by 20–25 mm long (including exserted stamens); **calyx** tubular, 20-veined, veins obscure, 8–12 mm long by 3–7 mm wide, widest below middle; **calyx teeth** five, erect, 1.5–2 mm long, margins whitish; **petals** five, lobed, spreading, TL 14–15 mm, curved inward during the day; **blades** 2-lobed, each lobe 4–5.5 mm long by 1.7–2 mm wide, claw greenish white, *c.* 9 mm long, tapered, slightly hairy near base; **stamens** 10, exserted 8–11 mm past the petals, TL *c.* 20 mm; **filaments** *c.* 18 mm long, filiform, white, pinkish purple above, bases minutely hairy; **anthers** reddish purple, 2.5–3 mm long; **pistils** *c.* 2 cm long; **ovary** green, elongate, *c.* 4 mm long by 1.5–1.7 mm wide, glabrous, tapered above, round; **styles** three (four), exserted eventually, filiform, whitish green, 12–16 mm long; **stigmas** obscure. **FRUIT** a **capsule**, 5-toothed (6-), *c.* 10 mm long by *c.* 6 mm wide, widest near the base, shiny, round, exserted 2–3 mm past the tight fitting, persistent, papery calyx tube; **seeds** brown *c.* 100 per capsule, 0.8–1.2 mm long by 0.9–1 mm wide by 0.6–0.7 mm thick, papillae in concentric rings.
- **LEAVES** Opposite, simple, entire, ascending, clasping, thick, some with a partial apical twist, pointed; **basal blades** 8–10 cm long by 2–2.5 cm wide; **petioles** 0–3.5 cm long, reduced above, wide and tapered gradually from blades; **stem leaf blades** flat or the margins curled downwards, 1.5–10 cm long by 0.5–4 cm wide, usually 6–9 pairs, more crowded below, top pair of leaves usually 10–25 cm below the two main floral branches.
- **STEM** Light green, erect to ascending, branching; 2–8 mm thick near the often decumbent base.

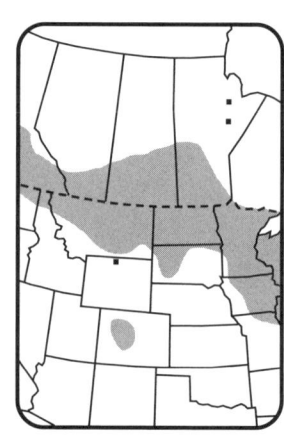

254 Biennial Campion, Balkan Catchfly

Pink—*Caryophyllaceae* Wildflower White Petals 5

WHITE COCKLE *Silene latifolia* Poir.

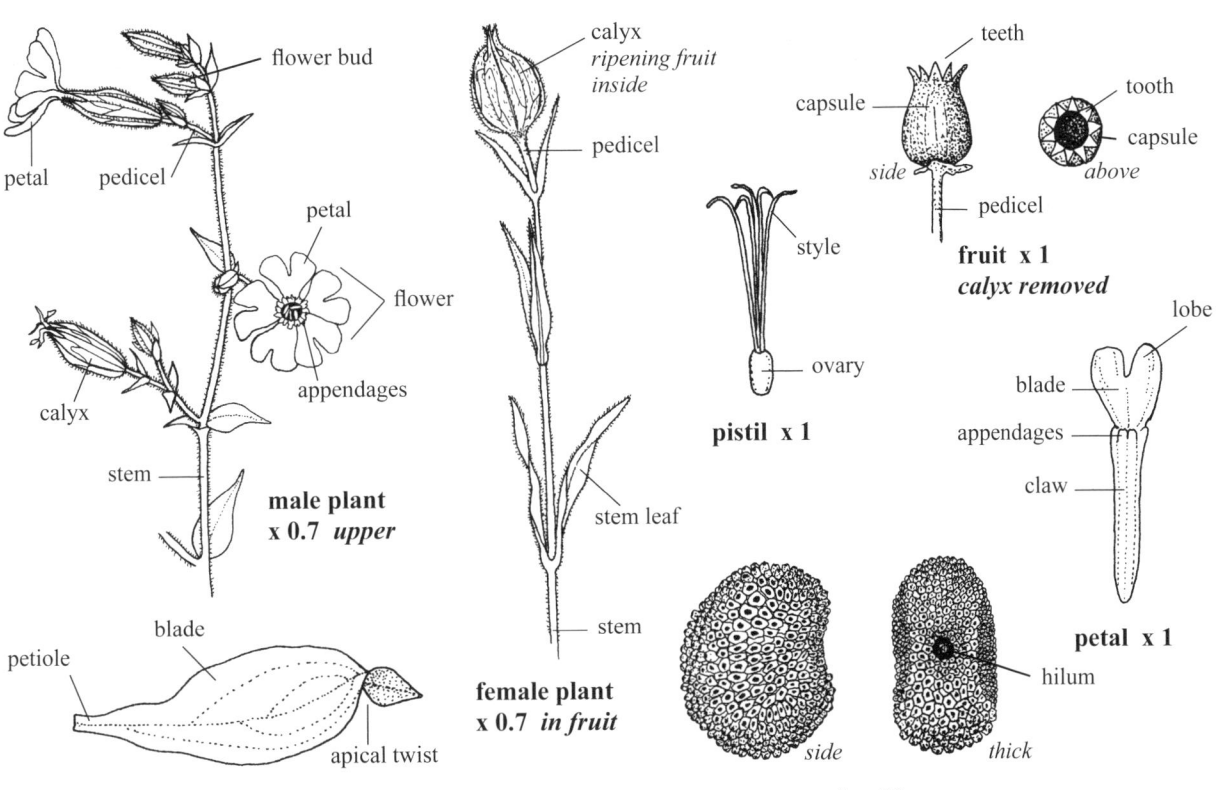

male plant x 0.7 *upper*

lower stem leaf x 0.8

female plant x 0.7 *in fruit*

pistil x 1

fruit x 1 *calyx removed*

seed x 20

petal x 1

■ **SKETCH** A hairy **annual** or short-lived **perennial herb** 25–120 cm tall, singly or in small clusters with a well-branched woody **taproot** 3–8 mm thick; in fields, lawns, disturbed sites and along roadsides; **dioecious**. *Naturalized*

■ **FLOWERS** White, blooming June–October; **inflorescence** a cyme, with one to many flowers, freely branched; **flowers** unisexual, fragrant, blooming at night, closing by noon; **pedicels** 5–25 mm long (–50 female) with gland-tipped hairs; **male flowers** 25–30 mm wide and long; **calyx** tubular, inflated, green, 15–20 mm long by 6–10 mm wide, constricted above, with 10 reddish branching veins, hairs gland-tipped, teeth five, erect, 3–6 mm long; **petals** five, 2.5–3.5 cm long, blade 2-lobed or not, white, glabrous, *c*. 1 cm long and wide, the claw light green, membranous, 1.5–2.3 cm long by 2–4 mm wide and glabrous, exserted to *c*. 4 mm beyond the calyx teeth; **appendages** ragged, 1–1.5 mm long, visible at the center; **stamens** 10, slightly exserted or not; **filaments** white, *c*. 15 mm long, united at their hairy bases to the petals; **anthers** yellow, *c*. 1.8 mm long; **female flowers** white, similar to the male flowers; **ovary** green, glabrous, 5–8 mm long by 3–4 mm wide; **styles** mostly five (4–6), white, 20–22 mm long, filiform, glabrous, distinct to the base, the top 3–5 mm are exserted; **calyx** persistent around the fruit, inflated to 20–30 mm long by 9–15 mm wide, 20-veined. **FRUIT a capsule**, erect, smooth, tan, 11–15 mm long by 9–11 mm wide, the five bifid teeth (10 teeth) *c*. 2 mm long and slightly spreading around the 4–5 mm wide circular apical opening; **seeds** *c*. 200 per capsule, grayish black, 0.8–1.5 mm long by 0.9–1.2 mm wide by 0.6–1 mm thick, bluntly tuberculate.

■ **LEAVES** Opposite, simple, entire, sessile or not, larger below, bractlike above; **blades** 3–15 cm long by 6–40 mm wide, often with an apical twist, some hairs jointed, some gland-tipped; **petioles** 0–3 cm long, reduced above.

■ **STEM** Erect, round, hollow, sticky, branched, densely hairy, some hairs gland-tipped, reddish and stiff with fruit; 2–6 mm thick near the base.

■ **SYN.** *Lychnis alba* P. Mill. = *Silene latifolia* ssp. *alba* (P. Mill.) Greuter & Burdet, *Silene alba* (P. Mill.) Krause = *Silene latifolia* ssp. *alba* (P. Mill.) Greuter & Burdet and *Silene pratensis* (Rafn) Godr. & Gren. = *Silene latifolia* ssp. *alba* (P. Mill.) Greuter & Burdet.

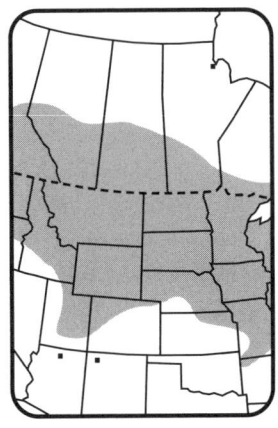

White Campion, Bladder Campion

Pink—*Caryophyllaceae* **Wildflower** White to pale pink Petals 5

NIGHT-FLOWERING CATCHFLY *Silene noctiflora* L.

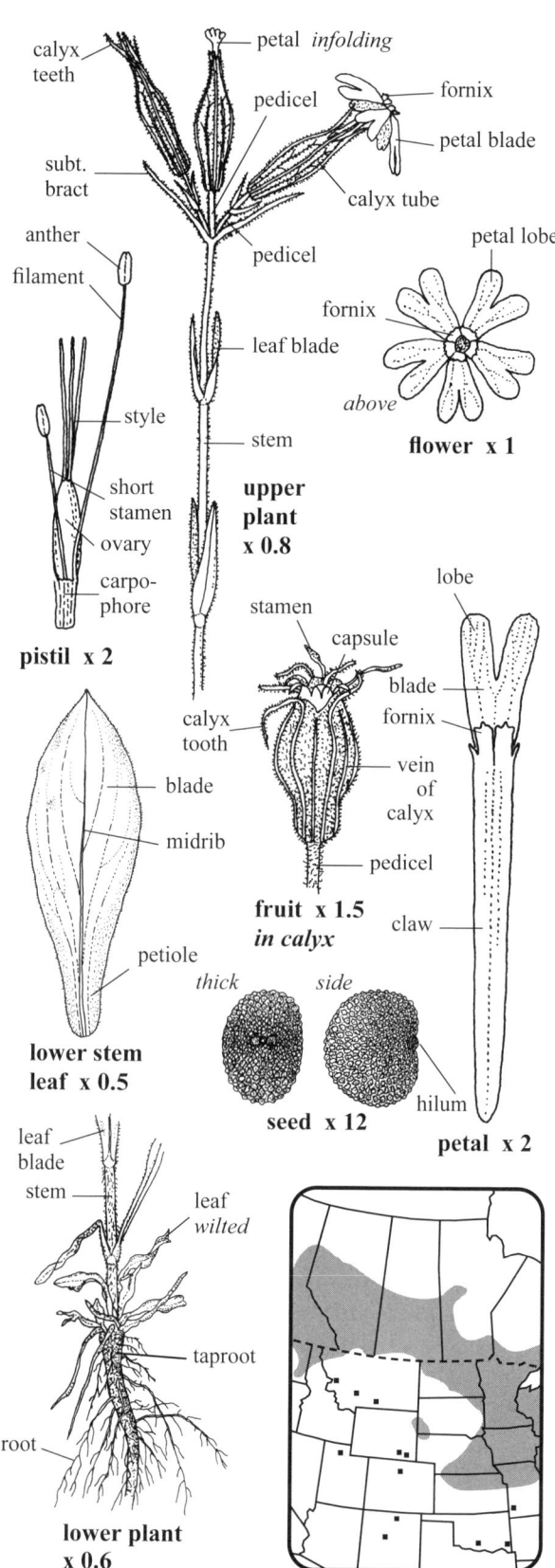

- **SKETCH** A hairy and sticky **annual herb** 20–90 cm tall with a **taproot** 7–15 cm long by 3–5 mm wide; in open colonies in disturbed fields, grainfields and along roadsides; **sometimes dioecious.** *Naturalized*
- **FLOWERS** White to pale pink, blooming at night in June–September; **inflorescence** a cyme, loosely branched, open, of 2–15 flowers, 4–25 cm long by 4–9 cm wide, terminal and a few flowers from the upper leaf axils, or 1–3 dichotomous divisions with two or three flowers at each division and an erect flower at the base of the V; **subtending bracts** (of branches) 6–45 mm long by 1.5–5 mm wide, pointed, ascending, brownish green; **pedicels** 5–30 mm long, green, sticky, subtending bracts paired; **flowers** fragrant, mostly perfect, 2–2.3 cm wide by 30–33 mm long; **calyx** tubular, green, glandular-hairy, TL 13–26 mm by 5–6 mm wide, longer with fruit; **calyx teeth** narrow, erect, 4–12 mm long, veins green to brown and continue as five main veins to base of calyx tube, secondary veins dividing at tip; **petals** five, exserted blades 8–10 mm long by 5–8 mm wide, slightly descending, curled inward by mid-morning, unfolding at dusk, each 2-lobed, these *c.* 5 mm long by 2–2.5 mm wide; **petal claw** light green, *c.* 24 mm long by 2–3 mm wide, about equal to the calyx in length, membranous; **fornices** 2–3 mm long, white, lobed and toothed, exserted in a circle 5–6 mm wide in throat of flower; **stamens** 10, usually included or one or two exserted *c.* 2 mm; **short stamens** five, 11–12 mm long; **long stamens** *c.* 20 mm and attached between the petals; **filaments** white, filiform; **anthers** *c.* 2 mm long, pale green; **carpophore** green, 2–3 mm long, below the ovary; **pistil** one, included, 13–27 mm long, glabrous; **ovary** green, cylindrical, 5–7 mm long by 1.3–2 mm wide; **styles** three, 8–10 mm long, filiform, white, minutely hairy along one side. **FRUIT a capsule**, tan, 12–22 mm long by 7–10 mm wide, apical opening *c.* 3 mm wide with six curved teeth, enclosed in calyx; **seeds** *c.* 150 per capsule, grayish brown, 1–1.4 mm long by 1–1.2 mm wide by *c.* 1 mm thick, tuberculate.
- **LEAVES** Basal and stem, opposite, simple, entire, ascending, dull, hairy on both sides and soft to touch; **basal leaves** several, 4–14 cm long by 2–4.5 cm wide, mostly sessile, usually wilting by flowering, tapered at the base; **stem blades** erect, some stuck to the stem, sessile above, 1.5–12 cm long by 4–40 mm wide, reduced above, pointed, glandular-hairy above, less so below; **petioles** 0–3 mm long and to *c.* 10 mm wide, reduced above.
- **STEM** Erect, simple or branched, long hairs non-glandular and descending below, spreading short hairs gland-tipped above; reddish at the 2–7 mm wide stiff base.

256 Sticky Cockle, Nightflowering Silene, Clammy Cockle

Pink—*Caryophyllaceae* **Wildflower   White   Petals 5**

LONG-LEAVED STITCHWORT   *Stellaria longifolia* Muhl. ex Willd.

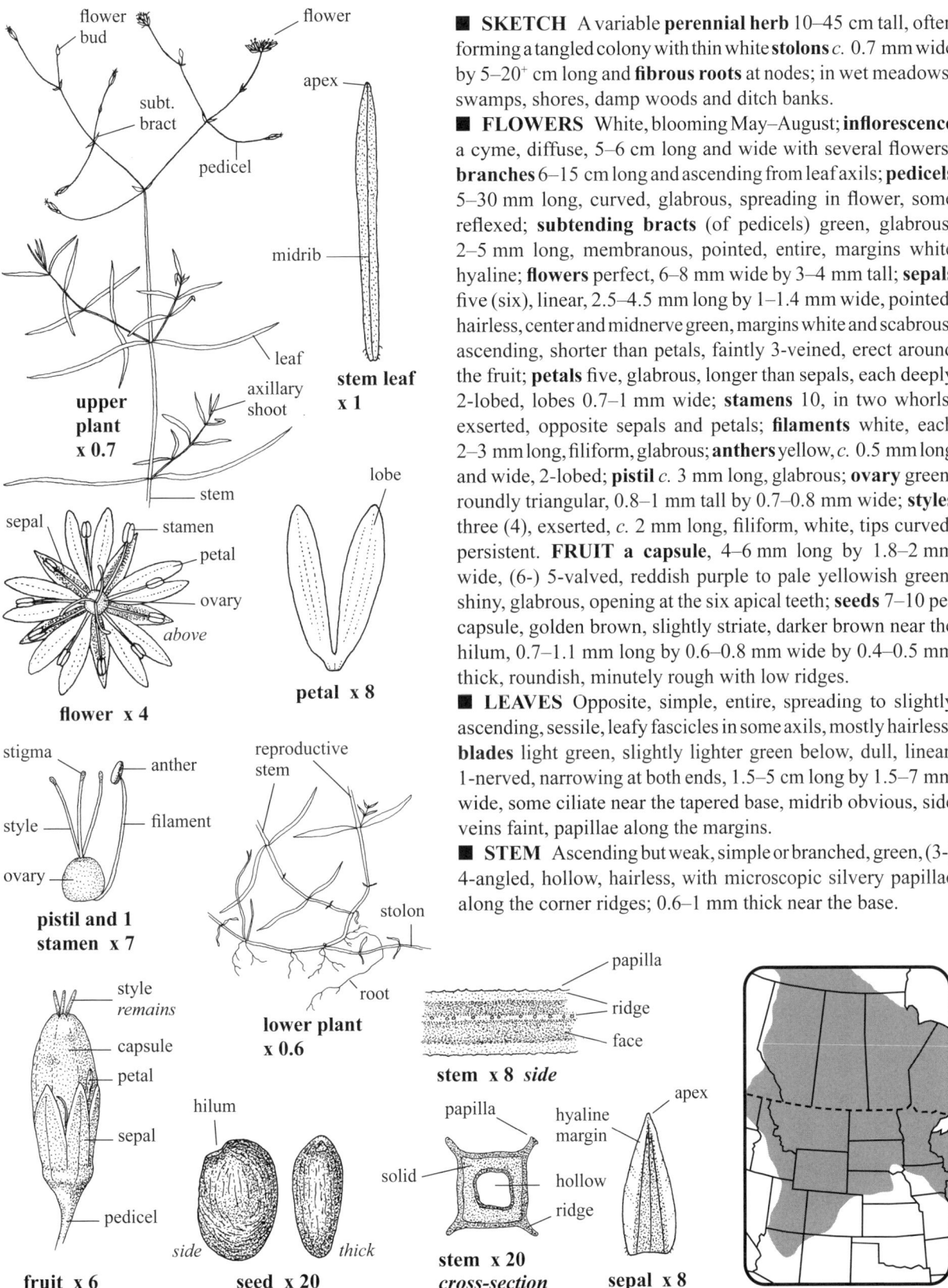

■ **SKETCH** A variable **perennial herb** 10–45 cm tall, often forming a tangled colony with thin white **stolons** *c.* 0.7 mm wide by 5–20⁺ cm long and **fibrous roots** at nodes; in wet meadows, swamps, shores, damp woods and ditch banks.

■ **FLOWERS** White, blooming May–August; **inflorescence** a cyme, diffuse, 5–6 cm long and wide with several flowers; **branches** 6–15 cm long and ascending from leaf axils; **pedicels** 5–30 mm long, curved, glabrous, spreading in flower, some reflexed; **subtending bracts** (of pedicels) green, glabrous, 2–5 mm long, membranous, pointed, entire, margins white hyaline; **flowers** perfect, 6–8 mm wide by 3–4 mm tall; **sepals** five (six), linear, 2.5–4.5 mm long by 1–1.4 mm wide, pointed, hairless, center and midnerve green, margins white and scabrous, ascending, shorter than petals, faintly 3-veined, erect around the fruit; **petals** five, glabrous, longer than sepals, each deeply 2-lobed, lobes 0.7–1 mm wide; **stamens** 10, in two whorls, exserted, opposite sepals and petals; **filaments** white, each 2–3 mm long, filiform, glabrous; **anthers** yellow, *c.* 0.5 mm long and wide, 2-lobed; **pistil** *c.* 3 mm long, glabrous; **ovary** green, roundly triangular, 0.8–1 mm tall by 0.7–0.8 mm wide; **styles** three (4), exserted, *c.* 2 mm long, filiform, white, tips curved, persistent. **FRUIT a capsule**, 4–6 mm long by 1.8–2 mm wide, (6-) 5-valved, reddish purple to pale yellowish green, shiny, glabrous, opening at the six apical teeth; **seeds** 7–10 per capsule, golden brown, slightly striate, darker brown near the hilum, 0.7–1.1 mm long by 0.6–0.8 mm wide by 0.4–0.5 mm thick, roundish, minutely rough with low ridges.

■ **LEAVES** Opposite, simple, entire, spreading to slightly ascending, sessile, leafy fascicles in some axils, mostly hairless; **blades** light green, slightly lighter green below, dull, linear, 1-nerved, narrowing at both ends, 1.5–5 cm long by 1.5–7 mm wide, some ciliate near the tapered base, midrib obvious, side veins faint, papillae along the margins.

■ **STEM** Ascending but weak, simple or branched, green, (3-) 4-angled, hollow, hairless, with microscopic silvery papillae along the corner ridges; 0.6–1 mm thick near the base.

Longleaf Starwort

Pink—*Caryophyllaceae* **Wildflower** White Petals 5

Common Chickweed *Stellaria media* (L.) Vill.

■ **SKETCH** A variable **annual** or **winter herb** 8–50 cm tall or long, matted, usually with a thin **taproot** and **rooting** at the **lowest nodes**; in lawns, irrigated areas, meadows, woods and gardens. *Naturalized* w1

■ **FLOWERS** White, blooming March–October; **inflorescence** a cyme of three to many flowers, terminal and axillary; **pedicels** erect, green, 4–40 mm long, reflexed in fruit, hairs spreading and in a wide line along one side only; **flowers** perfect, 4–6 mm wide by *c.* 3 mm tall; **sepals** five, ascending, pointed, 2.8–5.5 mm long by 1–1.8 mm wide, margins scarious, glabrous inside, veins obscure, hairs on outside spreading, simple and jointed, gland-tipped or not; **petals** five to absent, white, 2-lobed, divide almost to their base, 2–4 mm long by 1–1.5 mm wide, lobes *c.* 0.5 mm wide; **stamens** three to five (–8), included, *c.* 1.5 mm long; **filaments** filiform, a bulge at the base, glabrous; **anthers** pinkish, *c.* 0.2 mm long; **pistil** glabrous, 1.5–1.8 mm long, included; **ovary** green, *c.* 1 mm long by 0.8–0.9 mm wide, bluntly triangular; **styles** white, 3-parted, the parts spreading, filiform. **FRUIT a capsule**, 6-valved, extending past the persistent sepals, TL 3–5 mm by 2–2.5 mm thick, roundish in cross-section, tan, glabrous, thin, papery, six apical teeth often opening while still yellowish green; **seeds** dark brown, one to ten per capsule, dull, 1–1.2 mm long by 1–1.1 mm wide by 0.5–0.7 mm thick, papillate.

■ **LEAVES** Opposite, simple, entire, spreading, sessile above, petiolate below; **blades** flat, pointed, 1–4 cm long by 3–20 mm wide, glabrous, slightly lighter green below, veins faint; **petioles** 0–2.5 cm long, reduced above, slightly hairy along upper margins; **stipules** absent.

■ **STEM** Green, erect to decumbent, round to 4-sided, solid but weak, simple to branched, one or two longitudinal lines of jointed, white hairs along one side of the stem above, glabrous below; 0.5–2 mm thick near the base.

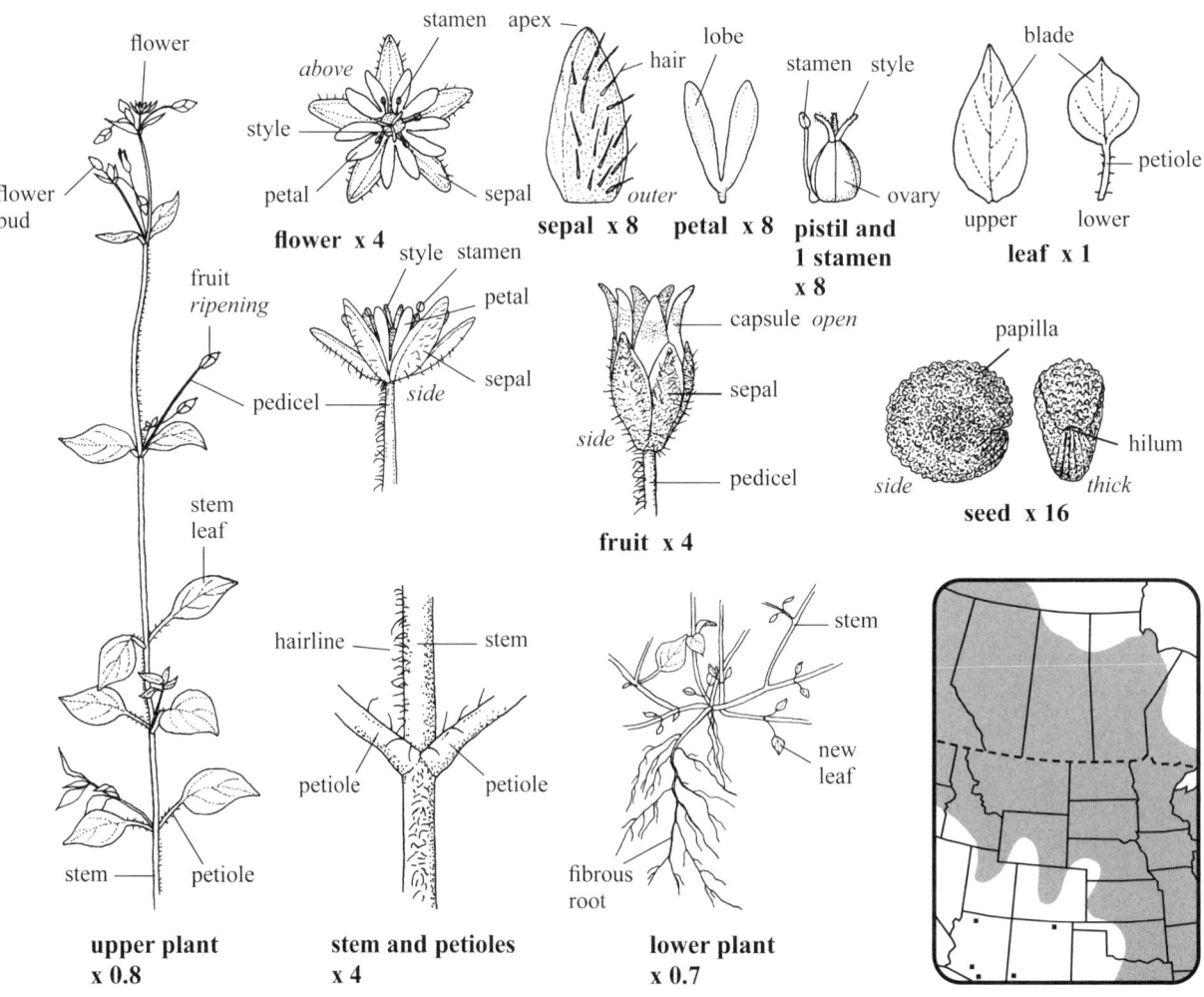

258 Chickweed, Common Starwort

Family Characteristics

Goosefoot—Chenopodiaceae Flower parts 1–5

SKETCH Dicotyledonous herb (wildflowers) or shrubs, annual or perennial, often mealy or scurfy. Dioecious or monoecious. About 1,800 species worldwide.

FLOWERS Perfect or unisexual, regular, greenish, small, solitary or in tight clusters called glomerules. **Inflorescence** a spike or panicles, often numerous on a plant. **Perianth** (calyx) 1- to 5-lobed, these united at their bases, often somewhat fleshy, blunt, persistent around fruit. **Corolla** absent. **Stamens** 1–5, opposite the lobes. **Pistil** 1, green. **Ovary** 1-loculed with one ovule. **Styles** 1–5, short. **Stigmas** mostly 2-parted.

FRUIT A **utricle**, 1-seeded, vertical or horizontal, circumscissile (opening around the middle), pericarp thin, easily removed on some. **Seeds** one per fruit, lenticular. **Embryo** curved.

LEAVES Mostly alternate and simple, entire to lobed to toothed, glabrous to hairy, often fleshy, some mealy. **Stipules** absent.

STEM Erect to prostrate, green to reddish, round, or angular and with ridges, solid, often branched at least in the inflorescence, usually dying back each year.

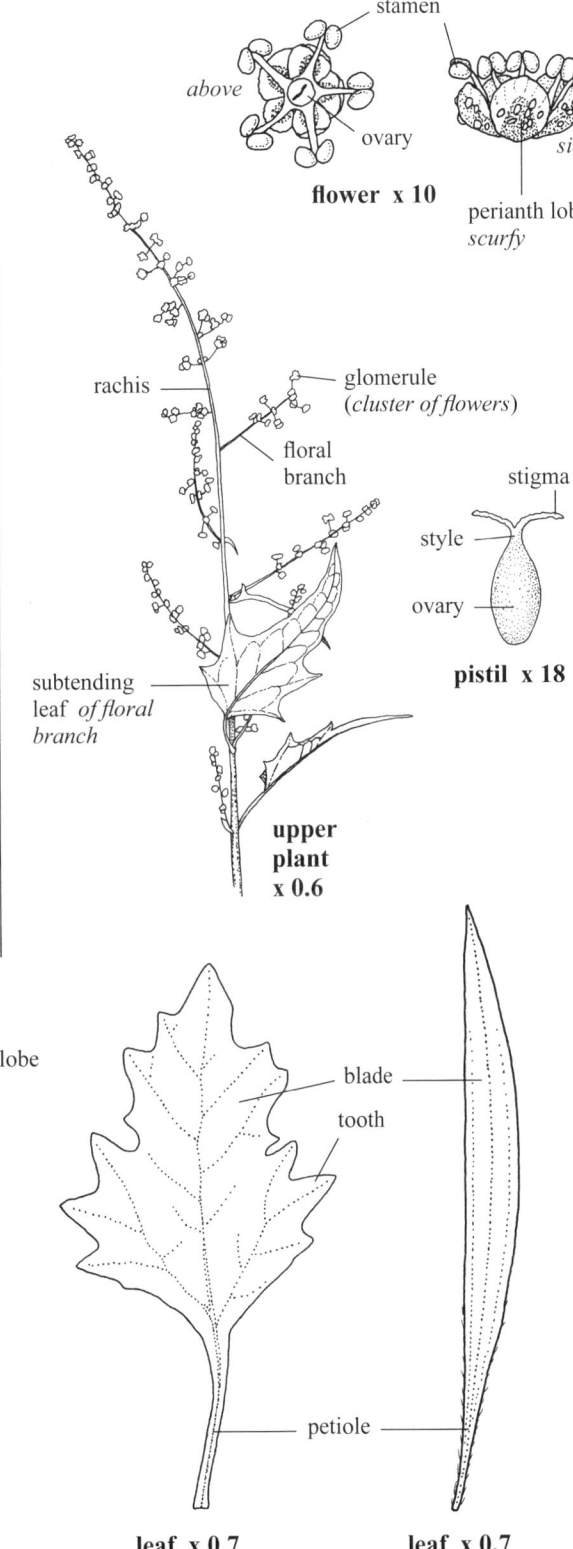
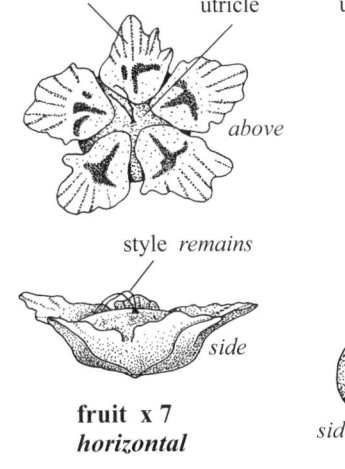
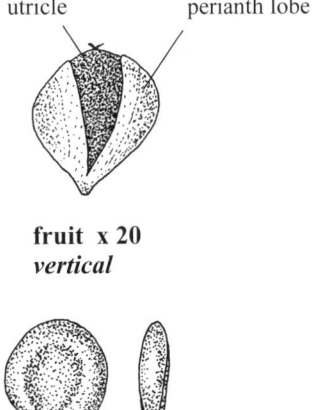

259

Goosefoot—*Chenopodiaceae*   **Wildflower**   Green to brown   Perianth lobes 5 (male), bracteoles 2 (female)

Nuttall's Atriplex   *Atriplex gardneri* (Moq.) D. Dietr.

■ **SKETCH** A variable, shrublike **perennial herb** with many ascending, new herbaceous stems 5–100 cm tall along older prostrate woody stems, with deep **roots**; in badlands, saline sites, sandy prairies and valleys; **dioecious** or **monoecious**. w2, p1

■ **FLOWERS** Green to brown, blooming June–August; **inflorescence** spikelike, 2–30 cm long, of unisexual flowers along new branches, glomerules in upper leaf axils; <u>**male flowers**</u> sessile, 2.5–3 mm wide by *c.* 2 mm tall, yellowish green; **perianth lobes** five, green, 1.3–1.6 mm long by *c.* 1 mm wide, glabrous inside, apices irregular with a few hairs and pointed, united in lower half, margins hyaline, scurfy on outside, a darker green area near the inside apex of each lobe; **stamens** five, exserted, glabrous; **filaments** pale green, 1.3–1.5 mm long, from the middle of the flower, ascending; **anthers** pale reddish yellow, 2-lobed, 0.6–0.7 mm long and wide; <u>**female flowers**</u> sessile, 1.5–2.5 mm wide by *c.* 4 mm long (including style), green, soft; **bracteoles** two, variable, scurfy, bluntly pointed, slightly spreading, mostly united, 2.3–3 mm long by 1.5–2.5 mm wide, thickish, tan, margins light green, smooth or with several fingerlike teeth or appendages; **pistil** exserted, *c.* 3.5 mm long; **ovary** glabrous, round, *c.* 0.5 mm long; **styles** two, erect to spreading, *c.* 3 mm long, hairy, reddish green, united at their glabrous base; **bracteoles** with fruit 3–9 mm long by 2.5–9 mm wide. **FRUIT** a utricle, 1-seeded, inside the two bracteoles; **pericarp** thin; **seeds** reddish brown, *c.* 2 mm long by 1.3–1.8 mm wide by *c.* 0.7 mm thick.

■ **LEAVES** Alternate (some opposite), simple, entire to slightly toothed, sessile to short petiolate, pale green, scurfy, hairless; **blades** flat to slightly V-shaped, blunt, 1–5.4 cm long by 2–25 mm wide, midrib obvious, lateral veins obscure, leafy fascicles in some axils.

■ **STEM** Brown, woody, warty, not stiff; 2–10 mm thick near the horizontal base.

■ **SYN.** *Atriplex nuttallii* ssp. *gardneri* (Moq.) Hall & Clements.

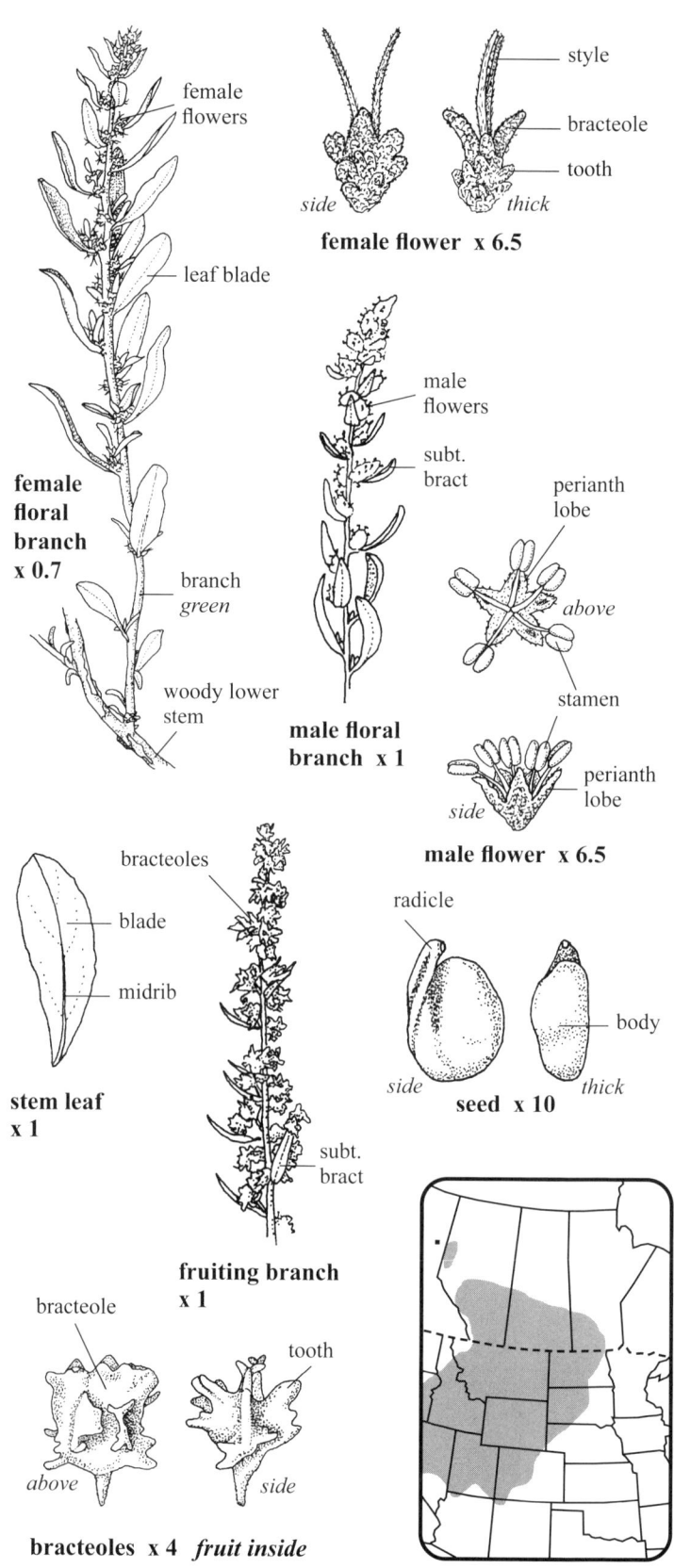

Gardner's Saltbush, Moundscale

Goosefoot—*Chenopodiaceae* Wildflower Green Perianth 5-lobed (male and female), bracteoles 2 (female)

GARDEN ATRIPLEX *Atriplex hortensis* L.

■ **SKETCH** A glabrous **annual herb** 60–250 cm tall from a **taproot** 4–20 mm wide by 6–20⁺ cm long; in disturbed drier sites around towns, cities, lakeshores, stream banks, gardens and roadsides; **monoecious. w1**

■ **FLOWERS** Green, blooming July–September; **inflorescence** of spikes in an extended panicle to *c.* 120 cm long and to *c.* 50 cm wide; **floral branches** 7–70 cm long, redivided, erect; **flowers** unisexual, sessile, separate or clustered, three female types mixed together but type one female flowers less numerous than the other two types; male **flowers** 2–2.5 mm wide by *c.* 1 mm tall, sessile, congested; **perianth** 5-lobed, incurved, parts *c.* 0.6 mm long by *c.* 0.4 mm wide, apices hyaline; **stamens** five, exserted; **filaments** *c.* 1 mm long, spreading; **anthers** *c.* 0.3 mm long by *c.* 0.7 mm wide, pale yellow; female **flowers** of three types: **1)** flower horizontal, *c.* 1 mm wide; **perianth** 5-parted, incurved, 0.2–0.4 mm wide and slightly longer; **pistil** green, *c.* 0.2 mm long (not shown); **2 & 3)** flowers vertical, 0.5–1 mm wide, enclosed by two **bracteoles**, these *c.* 1 mm long and wide by *c.* 0.7 mm thick, hairless, slightly scurfy, pointed, fleshy, expanding to 8–19 mm long as fruit matures inside; **pistil** green situated in middle of bracts, *c.* 0.4 mm wide; **style** 2-parted, *c.* 0.2 mm long. **FRUIT a utricle**, 1-seeded, each enclosed by a thin, easily removed pericarp; three types: **1)** horizontal, sessile, perianth 5-parted, parts 1–1.5 mm wide, pointed, appressed to spreading; **seeds** horizontal, black, biconvex, *c.* 2 mm wide by *c.* 1 mm thick, shiny, with striations; **2)** vertical, enclosed by a pair of small bracteoles *c.* 8 mm long by *c.* 7 mm wide by *c.* 2 mm thick (together), opening from apex, united at base; **seeds** vertical, black, shiny round, biconvex, 2.2–2.5 mm long and wide by *c.* 1 mm thick; **3)** vertical, enclosed by two large bracteoles 11–19 mm long by 9–10 mm wide by *c.* 2.5 mm thick (together) opening from apex; **seeds** vertical, yellowish brown, round, flattened, 3.5–4.5 mm long and wide by *c.* 1.3 mm thick.

■ **LEAVES** Alternate above to opposite near the base, simple, entire to toothed, sometimes lobed at their base, slightly scurfy below, glabrous with age; **blades** ascending to descending, 3–25 cm long by 1–18 cm wide, veins raised below, obvious above, dull, lower part often folded inward; **petioles** 5–55 mm long, glabrous; **stipules** absent.

■ **STEM** Erect to decumbent, usually branched, ridged, apex nodding, shiny, wavy, green to reddish; 0.7–2 cm wide near the base.

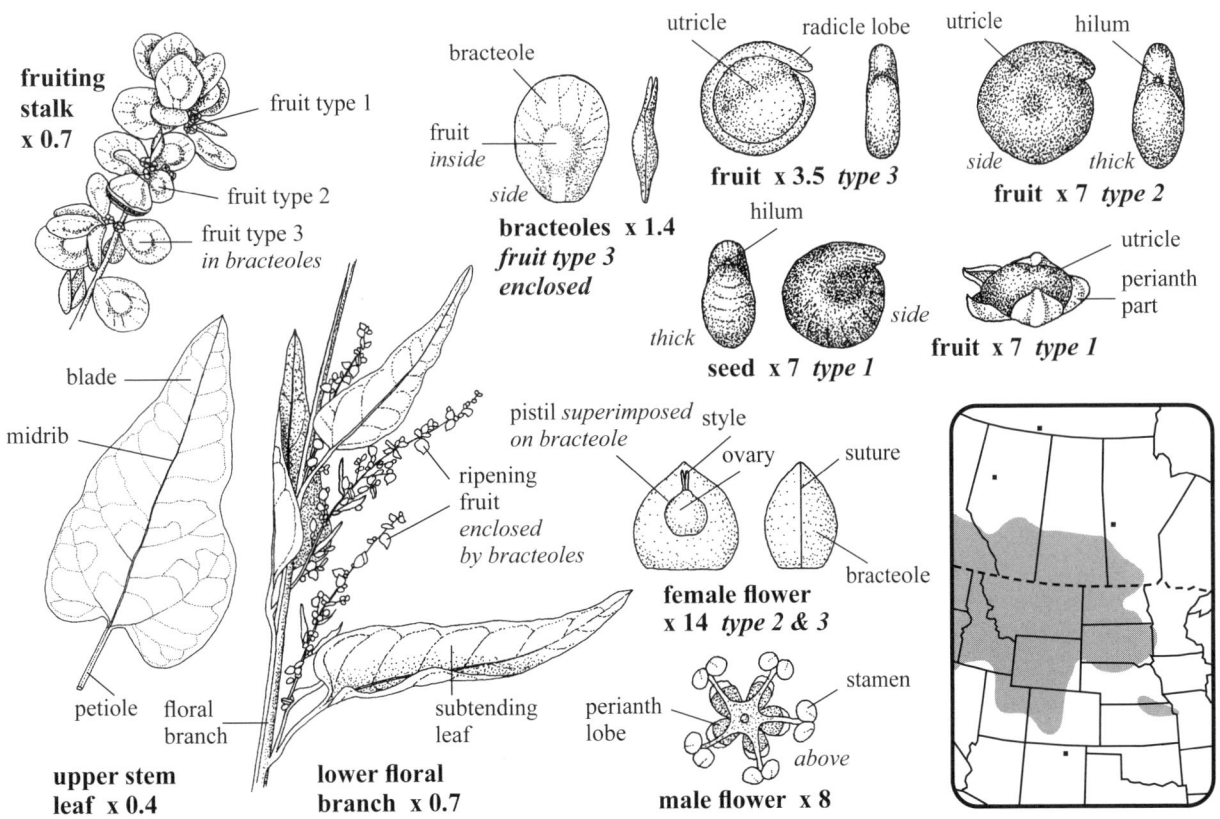

Garden Orache

Goosefoot—*Chenopodiaceae* **Wildflower** Reddish green Perianth lobes 5 (male), bracteoles 2 (female)

ORACHE *Atriplex prostrata* Bouchér ex DC.

■ **SKETCH** An **annual herb** 10–100 cm tall or long with a tan **taproot** 10–20 cm long; in saline sites in fields, gravelly edges of roads and pathways, in dry sloughs and valleys; **monoecious**. w1

■ **FLOWERS** Reddish green, blooming June–September; **inflorescence** a spike, of unisexual, mostly separate glomerules; **spikes** 2–13 cm long, alternating from leaf axils; <u>male flowers</u> (not shown) pink, sessile, *c.* 1 mm wide and long, scattered among the glomerules; **perianth lobes** five, slightly scurfy, united at base, the lobes 0.5–1 mm wide, slightly fleshy, apices pink and strongly curved into the middle; **stamens** five, exserted; **anthers** yellow, *c.* 0.5 mm long; **female glomerules** 3–12 mm wide, of several flowers and developing fruit pointing in all directions; <u>female flowers</u> red and green, 0.8 mm wide by 1–1.2 mm long, slightly scurfy, perianth absent, replaced by succulent bracteoles with tubercles and pointed tips; **ovary** green and glabrous; **style** 2-parted, *c.* 0.4 mm long, included to exserted; **bracteoles** two, enclosing the fruit, sessile, united only near their bases, each pair 3–10 mm long by 4–5.5 mm wide by 3.5–4.5 mm thick, pointed, margins slightly irregular to entire, with one or two tubercles on each side, turning reddish to dark brown. **FRUIT** a utricle, 1-seeded, in two bracteoles; **pericarp** membranous, easily removed; **seeds** of two types: 1) **small black** shiny, ripening early, 1–2 mm across the radicle lobe by *c.* 0.5 mm thick, radicle lobe lateral to ascending; 2) **large brown** 1.5–2.5 mm across the radicle by 0.8–1 mm thick, edges dark brown and slightly shiny, radicle lobe lateral.

■ **LEAVES** Alternate along most upper branches, opposite below, simple, entire to shallowly toothed, often with two spreading lobes near base; **blades** thin, 2–10 cm long by 1–6.5 cm wide, covered with scattered scurfy scales, more numerous below or glabrous and giving a faint grayish green color; **petioles** 3–40 mm long, scurfy.

■ **STEM** Bluntly angular, solid, crisp, mostly erect, branched, often with an obvious red area where the large floral branches meet the main stem, scurfy, 9-ridged, ridges green or pink and green between them; 2–10 mm thick near the almost woody base of large stems.

■ **SYN.** *Atriplex triangularis* Willd., *Atriplex patula* ssp. *hastata sensu* Hall & Clements 1923, non (L.) Hall & Clements, *Atriplex patula* var. *hastata* auct. non (L.) Gray, and *Atriplex patula* var. *triangularis* (Willd.) Thorne & Welsh.

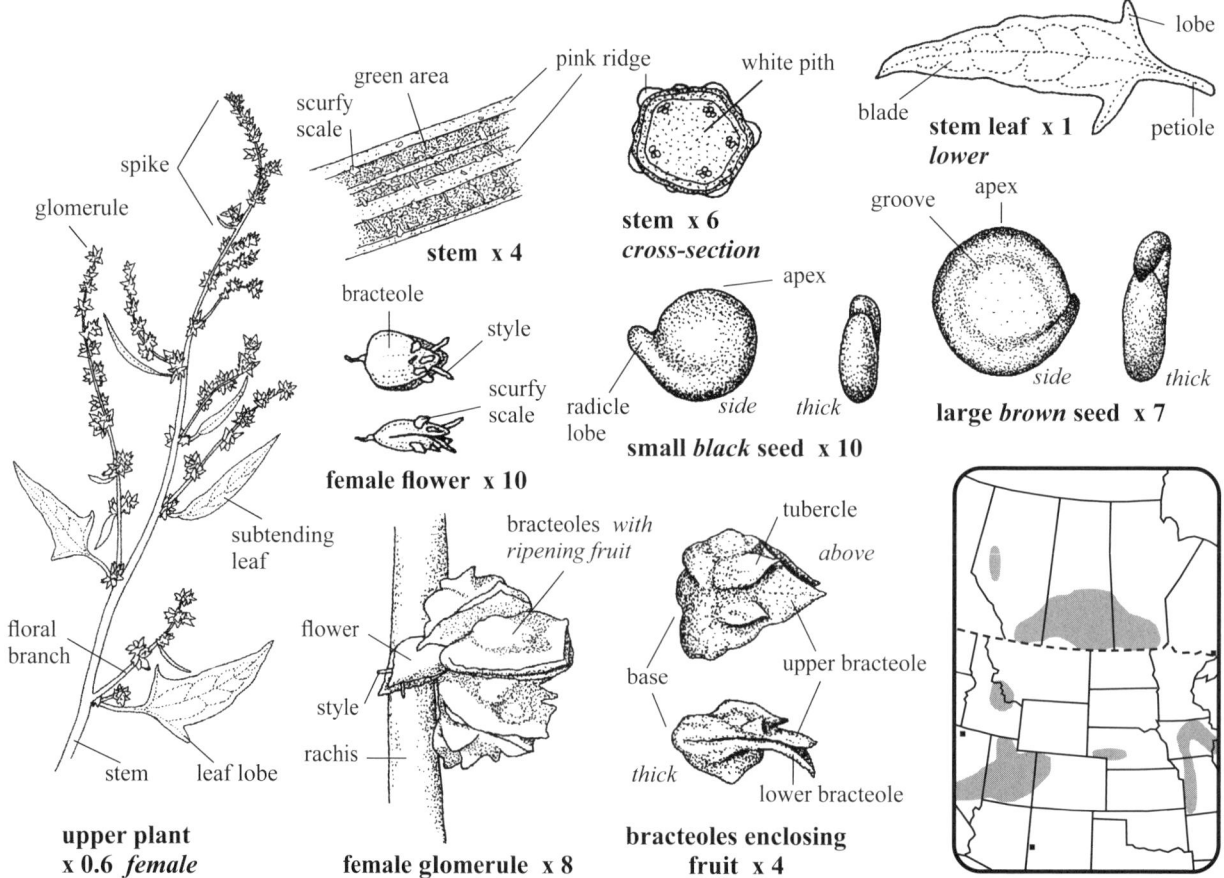

Halberd-leaved Saltbush, Thinleaf Orach, Fat-hen

Goosefoot—*Chenopodiaceae* **Wildflower** Green Perianth 1- or 3-lobed (male), bracteoles 3 (female)

Russian Pigweed *Axyris amaranthoides* L.

■ **SKETCH** An **annual herb** 40–100+ cm tall and almost as wide with a **taproot** 8–12+ cm long by 2–6 mm thick, the entire plant turning straw-colored in fall; along shelterbelts, disturbed dry prairies, cultivated fields, lakeshores and streams; **monoecious**. *Naturalized*

■ **FLOWERS** Green, blooming July–September; **inflorescence** of panicles and terminal spikes, in upper parts of branches and branchlets; **main branches** ascending, to *c.* 70 cm long, leafless in lower third; **branchlets** 2–20 cm long, reduced above, naked in lower half; **flowers** unisexual, minute, sessile; **male spikes** 2–65 mm long by 2–3 mm wide, yellowish tan, at ends of upper branchlets; <u>male flowers</u> congested, each flower *c.* 1 mm wide and tall; **perianth** 3-lobed (mostly), each lobe *c.* 0.7 mm long by *c.* 0.3 mm wide (unflattened), slightly shorter than the stamens, pale green, stellate-hairy on back; **stamens** three, exserted; **filaments** filiform, *c.* 0.8 mm long; **anthers** pale yellow, *c.* 0.3 mm long and wide; <u>female flowers</u> in upper axils of bracts, a few flowers in a cluster surrounded by three main bracteoles 2–4 mm long, these stellate-hairy; **perianth** 3-lobed (4-), lobes 1–1.2 mm long, white hairy outside, glabrous inside, erect around fruit, papery and *c.* 2 mm long by *c.* 1 mm wide; **pistil** *c.* 2 mm long; **ovary** green, glabrous, *c.* 0.2 mm long, lens-shaped; **styles** two, filiform, exserted, ascending, whitish, *c.* 1.8 mm long. **FRUIT** a **utricle**, 1-seeded, of two types: **1)** dark brown, 1.6–3 mm long by 1.1–1.3 mm wide by *c.* 0.8 mm thick, dull, usually irregularly streaked, apical tan wing to *c.* 0.4 mm long, bilobed; **2)** body unstreaked, little or no apical wing, slightly shorter and more symmetrical; **pericarp** thin.

■ **LEAVES** Alternate, simple, entire to slightly wavy; **blades** thin, dull, 2–12 cm long by 5–32 mm wide, reduced above, drooping along stem, glabrous or with scattered stellate hairs especially below; **petioles** 2–25 mm long, stellate-hairy.

■ **STEM** Solid, erect, light green, stiff, much-branched, densely hairy above to glabrous below, apex of plant sometimes nodding; 2–10 mm thick near the base.

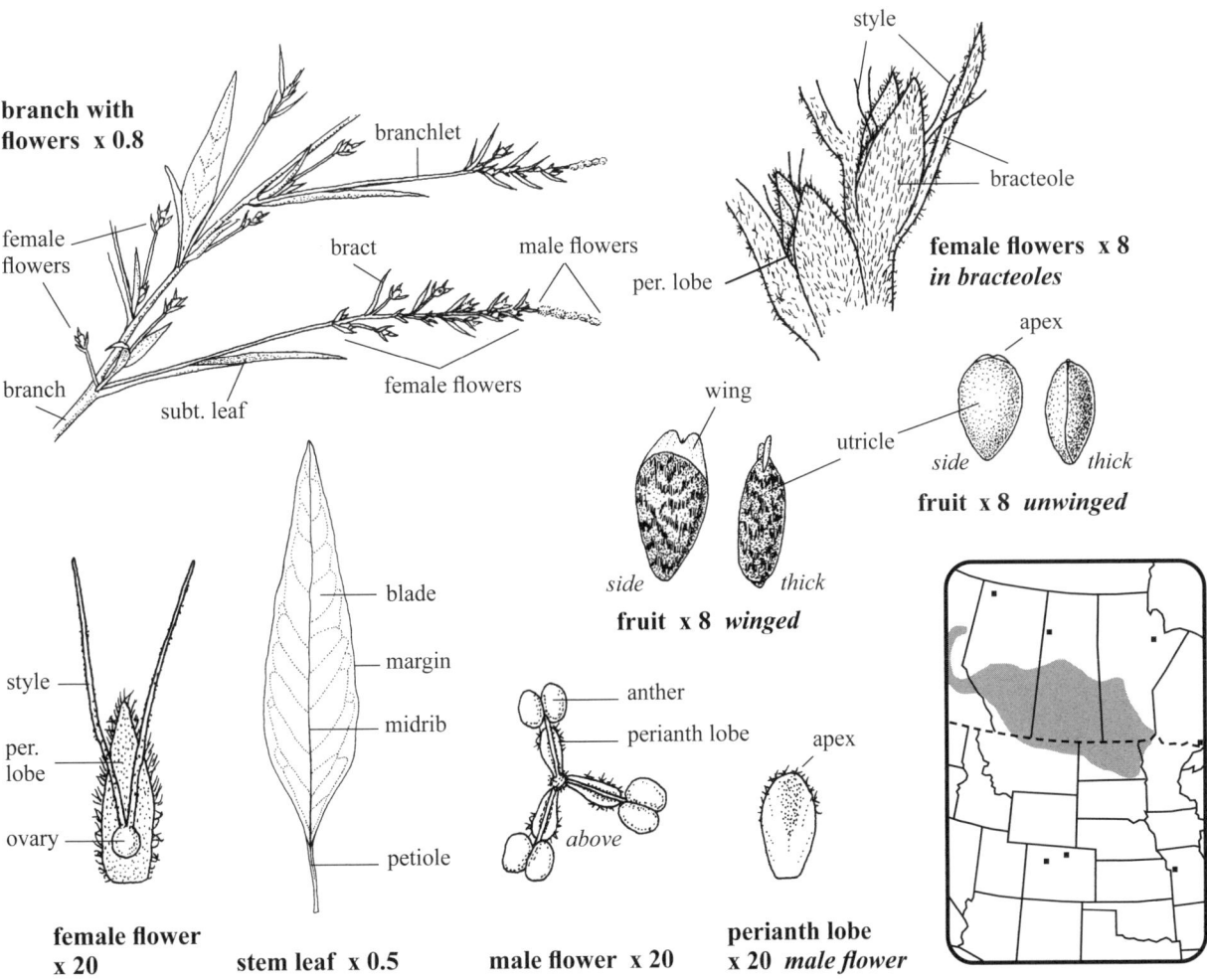

Goosefoot—*Chenopodiaceae*   **Wildflower** Green Perianth tubular, 5-lobed

Lamb's-quarters *Chenopodium album* L.

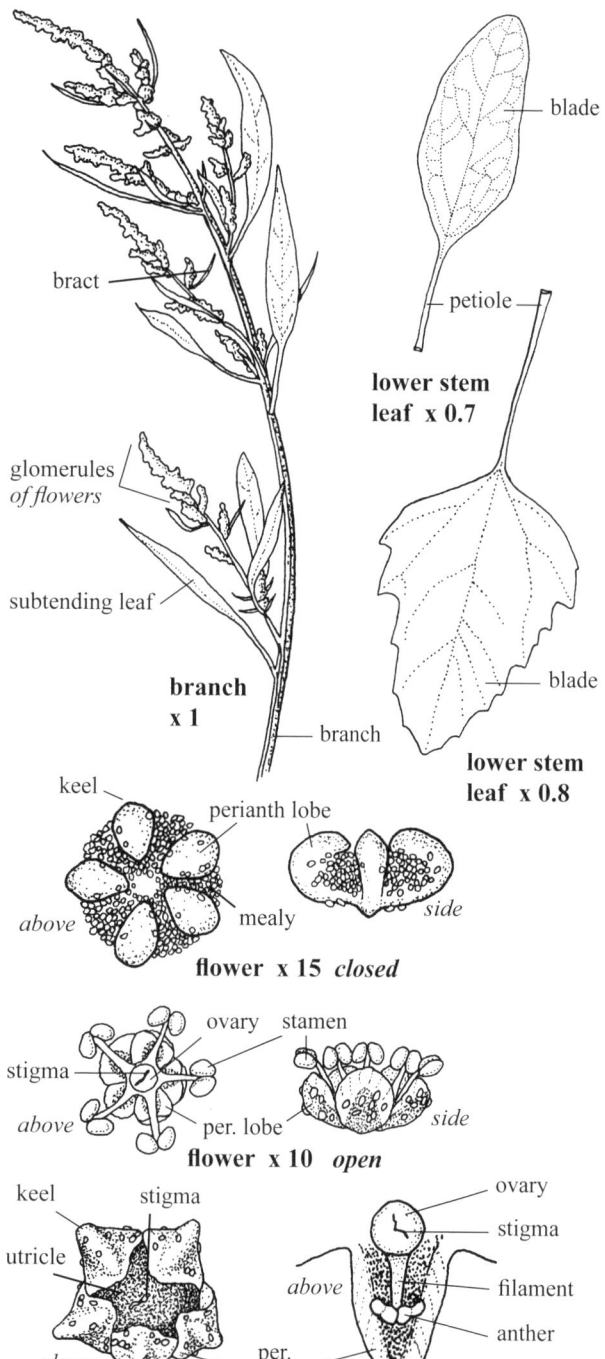

■ **SKETCH** A highly variable **annual herb** 50–200+ cm tall from a **taproot** 1–12 mm wide by 4–20 cm long; in disturbed sites along roadsides, alleys, edges of parking lots, desert grasslands and cultivated fields. w2

■ **FLOWERS** Green, blooming May–October; **inflorescence** a panicle of glomerules in dense elongated interrupted spikes in the upper two-thirds of the plant; **lower branches** to 45+ cm long, these sometimes rebranching; **flowers** perfect, sessile, crowded, 1.3–2 mm wide by *c*. 0.8 mm long; **perianth** tubular, 5-lobed, lobes moderately to densely mealy between the wide, medium green central keels, margins hyaline, 1.5–2 mm long and wide, usually partially covering the fruit, briefly spreading at anthesis, tan and bluntly keeled in fruit; **stamens** five, one opposite each part; **filaments** green, to *c*. 0.5 mm long; **anthers** exserted, yellow, *c*. 0.5 mm wide; **ovary** green, *c*. 0.4 mm wide; **stigmas** two, exserted after the stamens, persistent. **FRUIT** a **utricle**, 1-seeded, horizontal, enclosed, one per flower; **pericarp** thin, smooth to lightly roughened, dull, light brown bands radiating from the center, translucent to opaque, attached to seed but can be scraped away; **seeds** horizontal, lenticular, shiny, smooth, dark brown, round, 1–1.6 mm wide by *c*. 0.7 mm thick, slightly notched, edges rounded.

■ **LEAVES** Alternate, simple, entire above to irregularly toothed below, subtending branches; **blades** glabrous above, mealy below giving a lighter green appearance, 1–10 cm long by 0.5–4 (–8) cm wide, pointed to blunt; **petioles** 0.5–5 cm long, glabrous, with light green longitudinal lines below; **stipules** absent.

■ **STEM** Erect to sprawling, smooth, simple to branched throughout, with long green lines or sometimes with reddish streaks; 1–20 mm wide near the base.

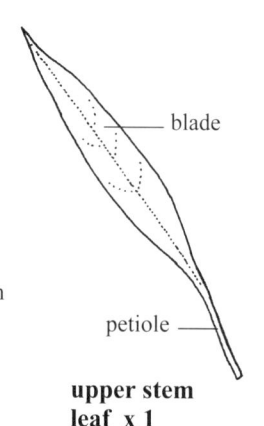

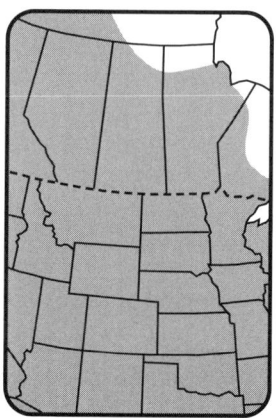

264   Pigweed

Goosefoot—Chenopodiaceae **Wildflower** Green to pale pink Perianth tubular, 5-lobed

PIT-SEED GOOSEFOOT *Chenopodium berlandieri* Moq.

■ **SKETCH** A variable, erect **annual herb** 40–150 cm tall from a **taproot** 1–14 mm wide by 5–20 cm long; in grasslands, dry sloughs and disturbed open ground.

■ **FLOWERS** Green to pale pink, blooming June–September; **inflorescence** of spikes, paniculate, TL 10–110 cm by 1.5–42 cm wide; **glomerules** (clustered flowers) 3–6 mm wide, in continuous to slightly interrupted spikes; **floral branches** alternate, 3–50 cm long, from leaf axils, reduced above; **flowers** perfect, sessile, scurfy; **perianth** tubular, 5-lobed, keeled, 0.8–1.3 mm long and wide, scurfy, persistent, united below, green midvein mostly glabrous, apices incurved over the stamens, margins hyaline; **stamens** five, included; **pistil** 1–1.4 mm long; **ovary** c. 0.8 mm long by c. 1.2 mm wide; **stigmas** two, exserted, filiform, 0.3–0.6 mm long. **FRUIT** a utricle, 1-seeded, horizontal, 1.2–1.5 mm wide by c. 0.7 mm thick; **pericarp** (fruit wall) adherent, thin, appears papillate at low magnification, slightly striated, some with a yellow spot at base of stigmas; **seeds** round from above, dark brown, 1.2–2 mm wide by c. 0.7 mm thick, shiny, minutely roughened from the pericarp, sides convex, groove shallow.

■ **LEAVES** Alternate, simple, toothed, lobed to entire, yellowish green, dull, slightly lighter below; **blades** 1.5–15 cm long by 0.4–9 cm wide, flat, linear above, scurfy below (dorsally), especially when young; **petioles** 0.5–9 cm long, flattened near the leaf blade, scurfy, often twisted 90° showing the full blade.

■ **STEM** Erect, stiff below, simple to branched, solid, ridged, round; 1–14 mm thick near the reddish and sometimes glaucous base.

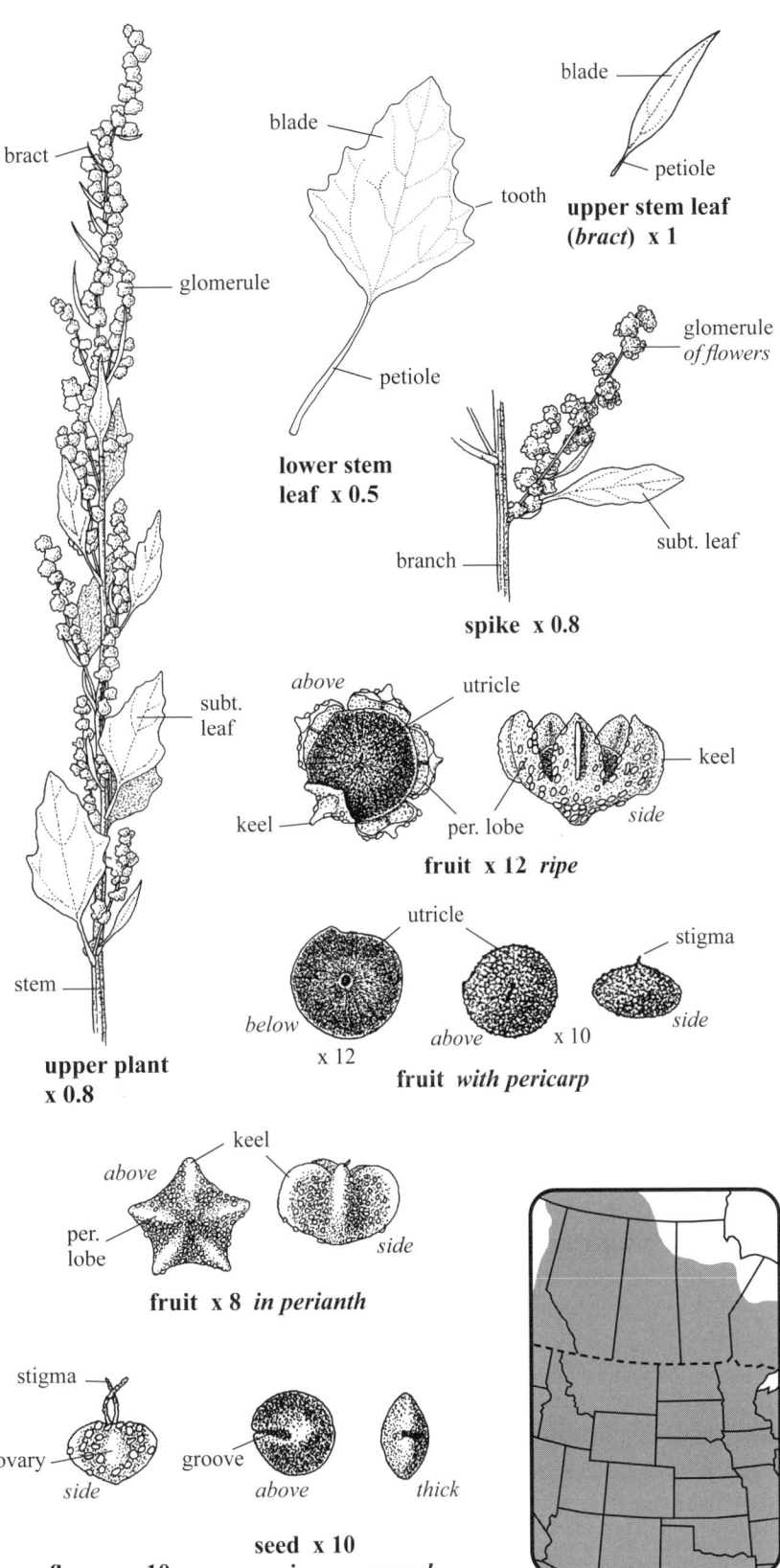

265

Goosefoot—*Chenopodiaceae* **Wildflower** Green Perianth tubular, 3- to 5-lobed

STRAWBERRY BLITE *Chenopodium capitatum* (L.) Ambrosi

■ **SKETCH** A glabrous **annual herb**, 20–100 cm tall with a whitish **taproot** 4–10 cm long by 3–6 mm wide; in disturbed sites, stony soils, river bars, lakeshores, cultivated soil and forest clearings. w1

■ **FLOWERS** Green, blooming June–August; **inflorescence** a spike of glomerules 3–5 mm wide and long at anthesis, sessile above, on floral branches below, overall 5–20 cm long; **flowers** perfect, 0.5–0.6 mm wide by *c.* 1 mm long (including stigmas), crowded; **rachis**, green, 7- or 8-ridged; **perianth** tubular 3- to 5-lobed, erect, 0.8–1 mm long by *c.* 0.4 mm wide at anthesis, becoming *c.* 1 mm long by 0.8–1.4 mm wide and fleshy with fruit, on a **pedicel** up to *c.* 2 mm long; **stamens** *c.* 0.8 mm long, usually 3 or 4, included to eventually exserted; **filaments** clear, *c.* 0.5 mm long; **anthers** yellowish green, 2-lobed, *c.* 0.3 mm long; **pistil** green, 0.4–0.9 mm long; **ovary** oval; **stigmas** two, clear and spreading, turning brown, exserted before the stamens; **fruiting heads** bright red, each glomerule 7–12 mm wide and tall, lower ones in bract axils and upper ones without obvious subtending bracts. **FRUIT a utricle**, 1-seeded, vertical, shiny, one per flower, 0.8–1.2 mm long by 0.8–0.9 mm wide by *c.* 0.5 mm thick, dark brown, smooth; **pericarp** tight.

■ **LEAVES** Basal and stem, alternate, simple, dull, entire to toothed, teeth 2–5 per side, pointed; **basal leaves** several; **blades** 1.5–4 cm long by 1–3 cm wide; **stem blades** 3–10 cm long by 2.5–9 cm wide, reduced above to narrow subtending bracts; **petioles** 1–10 cm long, reduced above; **stipules** absent.

■ **STEM** Erect to decumbent, green, simple to widely branched from the base; **branches** alternate, each subtended by a leaf-like bract; reddish green near the 2–5 mm wide base.

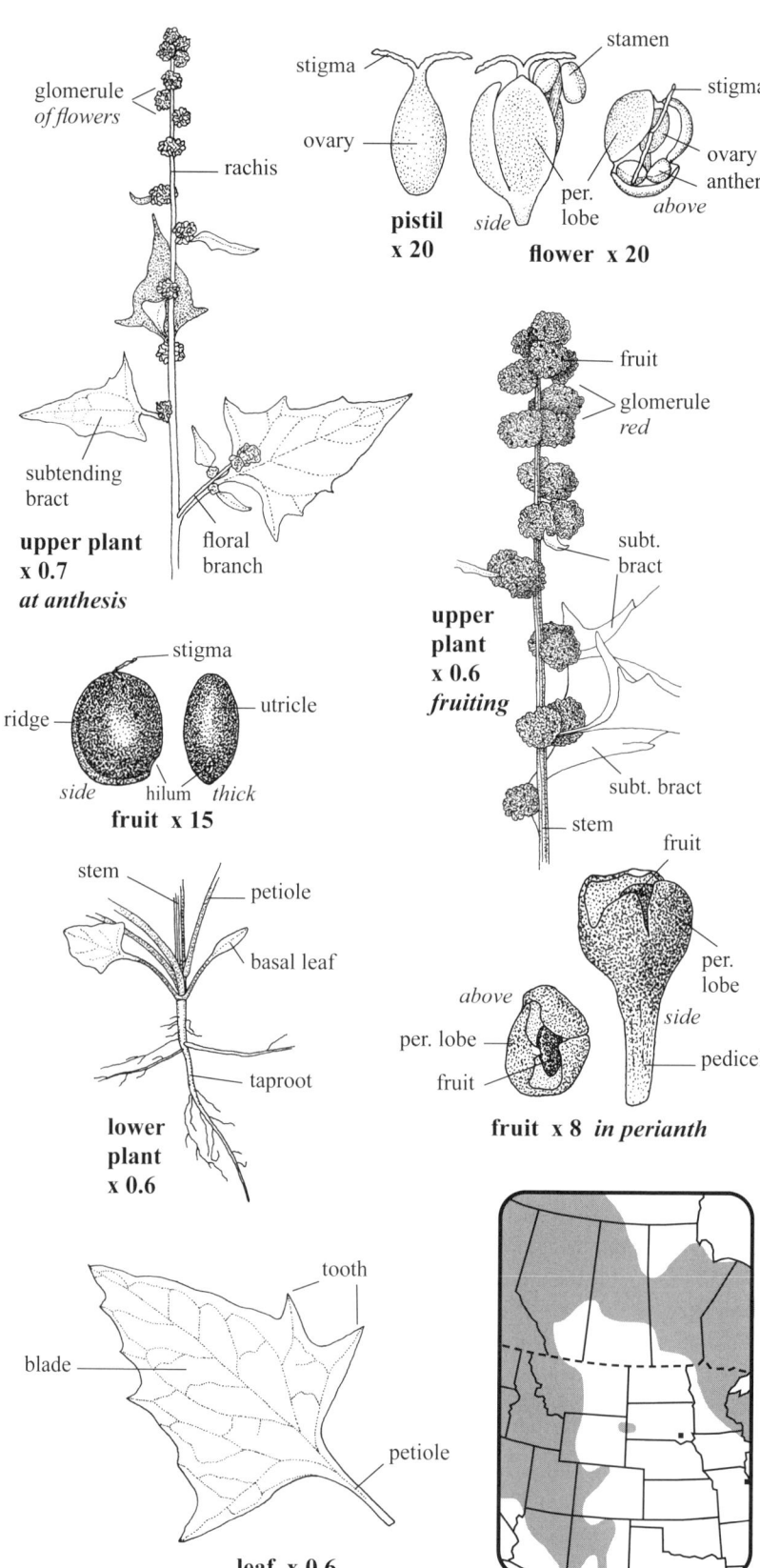

266 Blite Goosefoot, Indianpaint

Goosefoot—*Chenopodiaceae*   **Wildflower**   Green   Perianth tubular, 5-lobed (4-)

FREMONT'S GOOSEFOOT   *Chenopodium fremontii* S. Wats.

■ **SKETCH** A variable **annual herb** 10–80 cm tall with a **taproot** 5–10 cm long by 1–2 mm wide; in moist and often shaded areas among shrubs, along slough margins, woodland sites, deserts and bluffs.

■ **FLOWERS** Green, blooming July–September; **inflorescence** a panicle of well separated spikelike glomerules, terminal and axillary; **floral branches** ascending to spreading, 1–10 cm long, reduced above and below, throughout most of the plant except for the naked base; **flowers** perfect, *c.* 0.7 mm wide and tall, sessile, crowded into glomerules 2–5 mm wide; **perianth** tubular, 5-lobed (4-), lobes scurfy, *c.* 0.8 mm long by *c.* 0.5 mm wide, slightly or not spreading at anthesis, barely keeled, margins hyaline, ascending around fruit or reflexed; **stamens** five, included to exserted, *c.* 0.8 mm long; **filaments** flat, about as long as the perianth, wider at the base; **pistil** *c.* 1 mm long by *c.* 0.4 mm wide; **ovary** *c.* 0.3 mm tall, pale green; **stigmas** two, 0.3–0.5 mm long, exserted, persistent on fruit. **FRUIT a utricle**, 1-seeded, horizontal; **pericarp** (fruit wall) smooth, thin, transparent, tight but removable; **seeds** 1.1–1.4 mm wide by *c.* 0.6 mm thick, lenticular, medium to dark brown, shallowly grooved on one side.

■ **LEAVES** Alternate, simple, light yellowish green, mucronate, glabrous to slightly scurfy above and below, often more scurfy below and on upper young blades; **blades** 1.5–6.5 cm long by 0.8–6 cm wide, reduced above and often subtending the floral branches, two lateral lobes near the base, otherwise entire; **petioles** glabrous to slightly scurfy, 3–25 mm long, grooved on sides near blade, reduced above; **stipules** absent.

■ **STEM** Erect to leaning on nearby vegetation, solid, pith white, simple below to branched above, glabrous, more angular above with seven or eight ridges; 1–3 mm thick near the stiff, pinkish base.

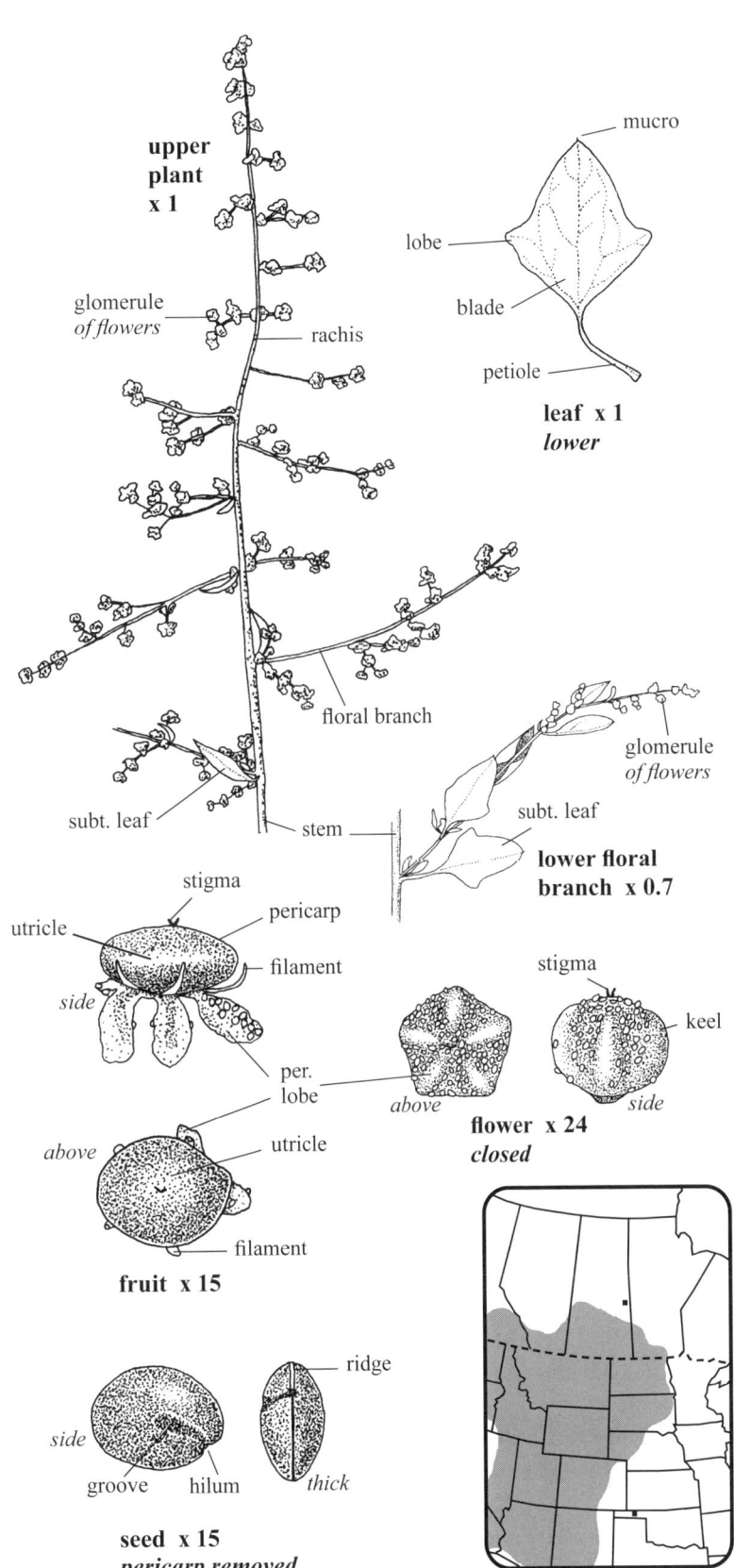

267

Goosefoot—*Chenopodiaceae* **Wildflower** Green Perianth tubular, 3- to 5-lobed

SALINE GOOSEFOOT *Chenopodium glaucum* L.

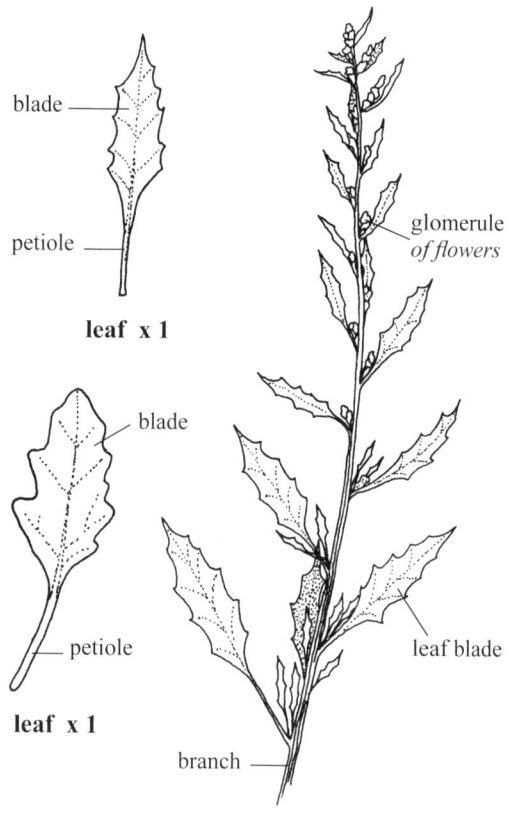

- **SKETCH** A variable **annual herb** 10–40 cm high by 20–50 cm wide with one to several prostrate branches with ascending tips and a **taproot** 1–4 mm wide by 4–8 cm long; in alkaline habitats, shores of marshes, gardens, moist to dry gravelly sites. *Naturalized*
- **FLOWERS** Green, blooming June–August; **inflorescence** of glomerules on short, axillary spikelike branches, bracteate; **subtending bracts** (of glomerules) leaflike, 2–10 mm long by 1–5 mm wide, sometimes absent from terminal glomerules; **flowers** perfect, each c. 0.5 mm wide and long, round, glabrous, not farinose; **perianth** tubular, 3- to 5-lobed, these C-shaped, minute, ascending, persistent; **stamens** one or more, one per perianth part; **filaments** tiny; **anthers** pale yellow, wider than the perianth part; **style** 2-parted, minute. **FRUIT a utricle**, 1-seeded, round, each c. 1 mm wide and long, glabrous, short stalked; **pericarp** smooth, paper-thin, easily removed; **seeds** mostly vertical, some horizontal, dark shiny brown, slightly concave in the middle, 0.6–1.1 mm wide by c. 0.4 mm thick, with a faintly honeycombed surface, not ribbed.
- **LEAVES** Alternate, simple, toothed to entire; **blades** 1–4 cm long by 0.5–2 cm wide, densely farinose below and appearing almost white except for a thin green margin and green at the base of the midrib, glabrous and not farinose above; **petioles** 1–2 cm long, flat above, not grooved.
- **STEM** Erect to prostrate, solitary to much branched from the base, rubbery, fleshy, with red longitudinal stripes; 1–5 mm thick near the base.

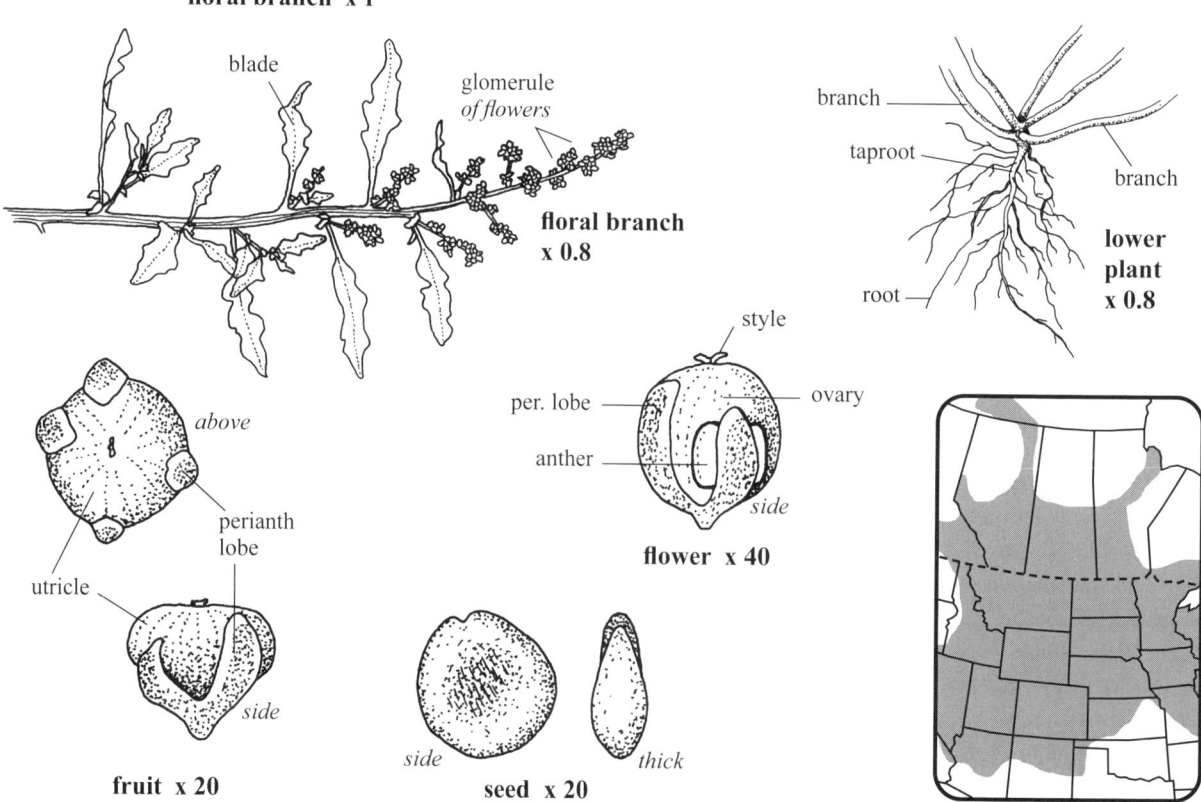

Goosefoot—*Chenopodiaceae* **Wildflower** Green Perianth tubular, 3- to 5-lobed

RED GOOSEFOOT *Chenopodium rubrum* L.

■ **SKETCH** A glabrous **annual herb** 20–100 cm tall from a tan, branched **taproot** to *c.* 10 mm thick and to 20+ cm long; along lake shores and sloughs, mostly in saline, moist areas.

■ **FLOWERS** Green, blooming June–September; **inflorescence** of short spikes of dense, axillary, bracteate glomerules, TL 1.5–70 cm by 3–24 cm wide, spikes shorter to longer than subtending leaf; **floral branches** ascending, 2–30 cm long, alternate and reduced above; **glomerules** 1.5–2 mm wide by 2–6 mm long, yellowish green; **flowers** perfect, sessile, 0.8–1 mm wide by 0.6–0.8 mm tall; **perianth** tubular, united at base, 3- to 5-lobed, slightly fleshy, lobes 0.4–0.7 mm wide, glabrous to slightly farinose, margins hyaline, apices blunt, incurved, turning red in fruit; **stamens** one to three, eventually exserted; **anthers** *c.* 0.3 mm long by *c.* 0.4 mm wide, pale yellow; **ovary** *c.* 0.5 mm tall, glabrous, flattened; **style** 2-parted, persistent, *c.* 0.1 mm long. **FRUIT a utricle**, 1-seeded, 1–1.2 mm long by *c.* 1 mm wide, mostly enclosed by erect perianth, vertical but pointing in all directions on a plant; **pericarp** thin, transparent, dull, smooth, easily removed; **seeds** shiny, dark brown, glabrous, mostly vertical, each 0.6–1 mm long by 0.6–0.8 mm wide, slightly concave on sides.

■ **LEAVES** Alternate and simple, coarsely toothed, sometimes entire, sinuses blunt; **blades** dull, lighter green and glabrous below (dorsally), 2–14 cm long by 0.5–6 cm wide, reduced above, spreading to ascending; **petioles** 5–45 mm long, thick, rounded below; **stipules** absent.

■ **STEM** Pale green to reddish, stiff, solid, erect to ascending, simple to branched throughout; 3–12 mm thick near the often naked base.

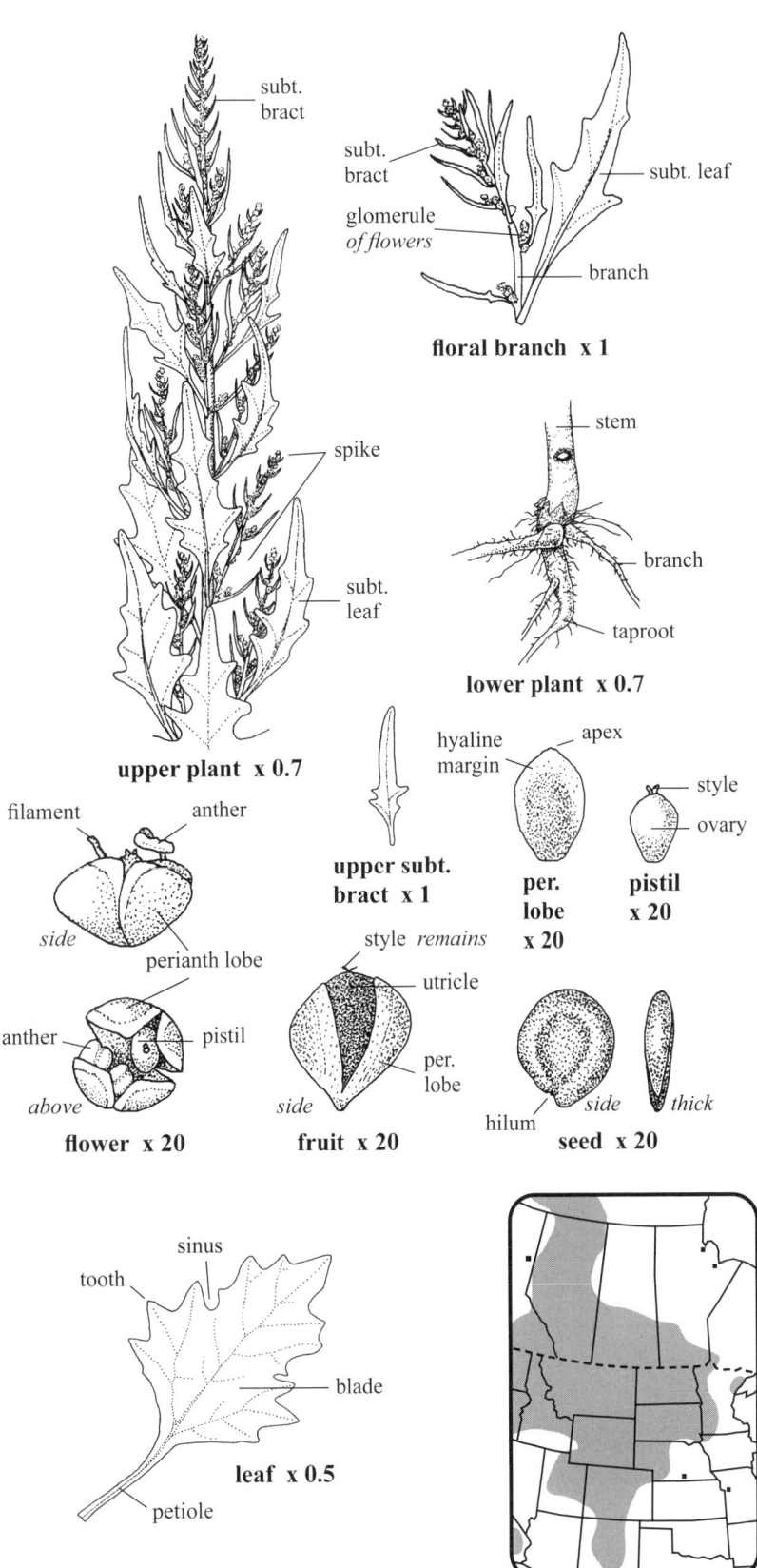

Red Pigweed, Alkali Blite, Coast Blite

Goosefoot—Chenopodiaceae **Wildflower** Green Perianth tubular, 5-lobed

MAPLE-LEAVED GOOSEFOOT *Chenopodium simplex* (Torr.) Raf.

■ **SKETCH** An **annual herb** 30–200 cm tall with a tan **taproot** 5–15 cm long; in disturbed sites often in shade, edges of cultivation, along roadsides, in gardens or in sagebrush and aspen communities. **w1**

■ **FLOWERS** Green, blooming July–September; **inflorescence** a panicle, terminal and axillary, open, 20–90 cm long by 6–40 cm wide; **floral branches** 1.5–30 cm long, reduced above, sometimes re-branching, ascending to spreading, often with one or two small axillary leaves; **short paired stalks** along the branches and upper part of stem form a V with one stalk longer than its pair, upper part of inflorescence without leaves; **flowers** perfect, sessile, in small glomerules of 2–7 flowers, glabrous to slightly farinose, each flower *c*. 0.3 mm wide by *c*. 0.5 mm long, usually a single flower at the base of the V-shaped paired stalks; **rachis** green, slightly farinose; **perianth** tubular, 5-lobed, lobes green with margins whitish hyaline, hooded, in fruit 1–1.3 mm long by 0.7–0.9 mm wide, united at base, the keel broadly rounded, exposing most of the fruit from above; **stamens** five, opposite the lobes, exserted or not; **filaments** white, *c*. 1 mm long; **styles** two, exserted, spreading, shorter than the flower. **FRUIT a utricle**, 1-seeded, horizontal, 1.5–2.7 mm wide, round from above, lenticular with both sides convex; **pericarp** (fruit wall) very thin, easily scraped off the seed, semi-transparent, rough; **seeds** 1.5–2.5 mm wide by 0.8–1.1 mm thick, dark brown to black, generally round from above with a low blunt ridge and indentation on one side, a wide groove on the ventral surface from the indentation to the center.

■ **LEAVES** Alternate, the lower may be opposite, simple, with 1–5 large pointed teeth on each side, apices pointed; **blades** dull to slightly shiny, thin, smooth, slightly lighter green below, 2–20 cm long by 1.7–15 cm wide, ascending, glabrous to slightly farinose, hairless, reduced in size above and below; **petioles** 0.3–6.5 cm long, reddish green, often subtending a short floral branch.

■ **STEM** Erect, simple to branched, light green, glabrous, 5-ridged, solid, slides slightly concave between the light green ridges; 3–7 mm thick near the base.

■ **SYN.** *Chenopodium gigantospermum* Aellen, *Chenopodium hybridum* auct. non L., *Chenopodium hybridum* ssp. *gigantospermum* (Aellen) Hultén, and *Chenopodium hybridum* var. *gigantospermum* (Aellen) Rouleau.

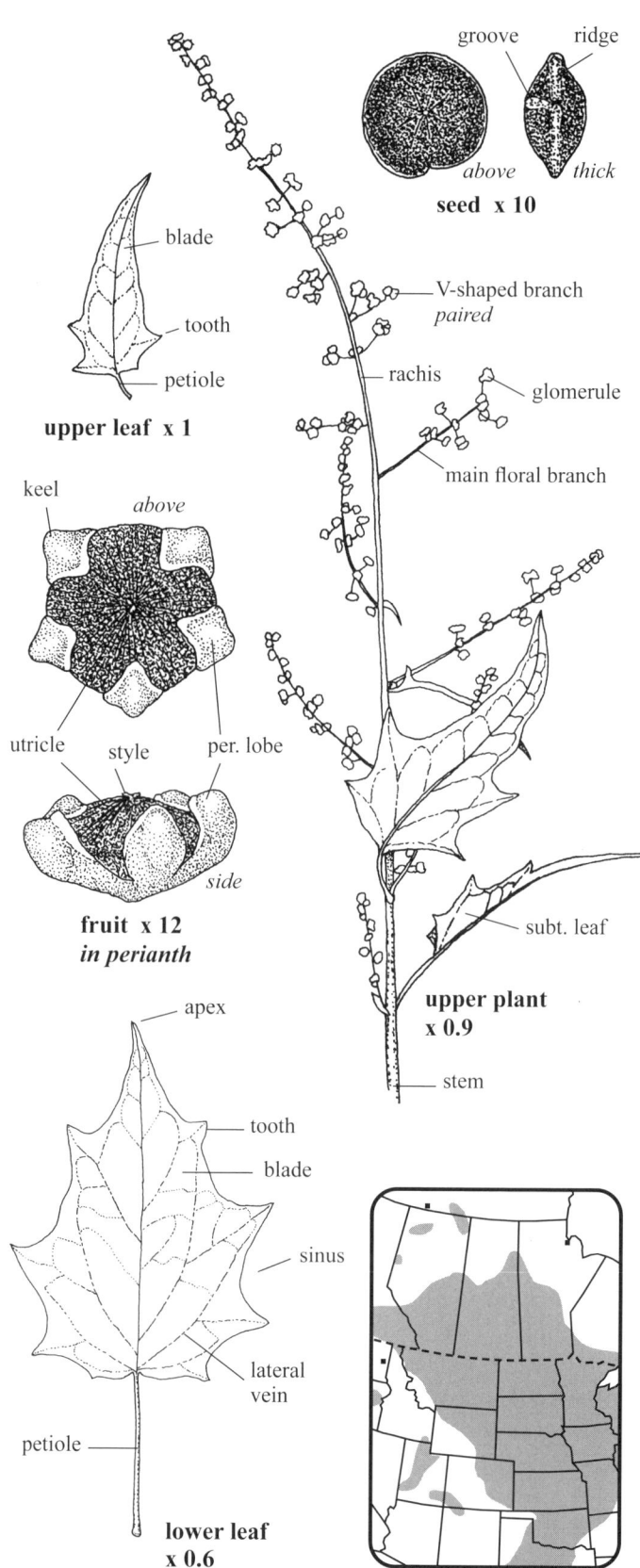

270 Giant-seed Goosefoot, Big-seed Goosefoot, Sowbane

Goosefoot—*Chenopodiaceae* **Wildflower Green Perianth tubular, 5-lobed**

SUMMER-CYPRESS *Kochia scoparia* (L.) Schrad.

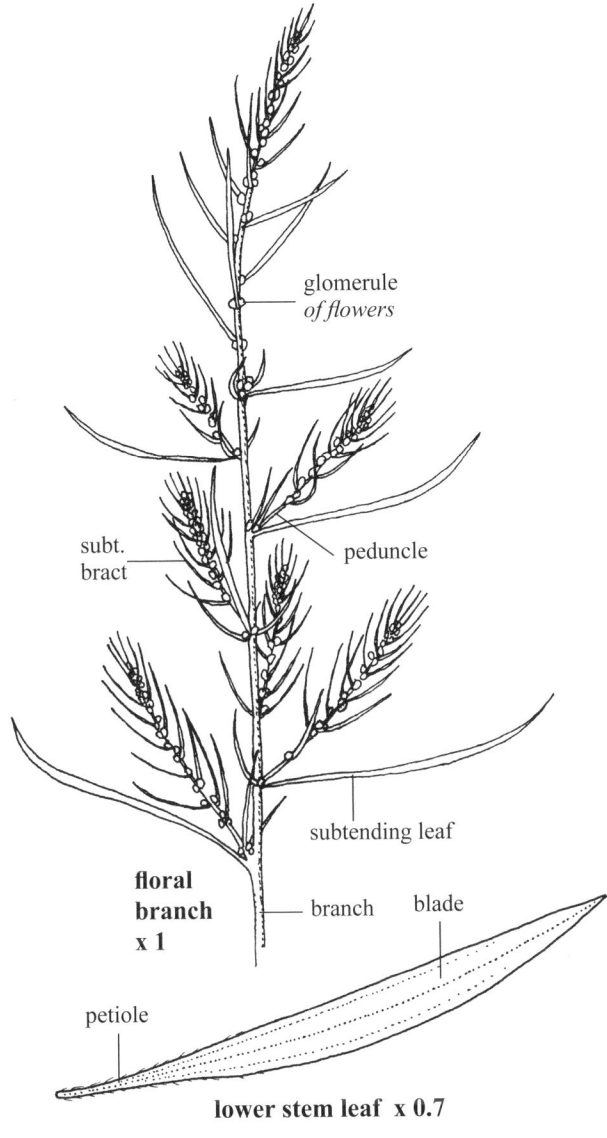

floral branch x 1

lower stem leaf x 0.7

- **SKETCH** An **annual herb** 30–200 cm high and wide from a **taproot** 1–10 mm wide by 5–20+ cm long; deserts, steppes, roadsides, in city lots, cultivated fields, alleys, pastures and other disturbed sites. *Naturalized* **w2**
- **FLOWERS** Green, blooming July–October; **inflorescence** spikelike, of glomerules arranged terminally and on branches; **floral branches** leafy, alternate, 3–40 cm long, ascending, their bases with red streaks, not scabrous but with short curved hairs; **peduncles** to *c.* 2 cm long from upper leaf axils; **subtending bracts** (of glomerules) 3–20 mm long with marginal hairs to *c.* 2 mm long and a tuft of apical hairs; **flowers** perfect, or unisexual along the lateral branches, sessile, each 1.8–2 mm wide and almost as high at the start of anthesis, one to three (usually two) at the base of each subtending bract; **perianth** tubular, 5-lobed, *c.* 1.5 mm long, united except for the five short ciliate lobes, opaque with a medium green central ridge near the apices, in fruit the lobes *c.* 1 mm wide with a tan, patterned horizontal wing *c.* 1 mm long; **anthers** five, light yellow, 4-lobed, 1.3–1.5 mm long by *c.* 1 mm wide, included to exserted when spent; **stigma** 2-parted, filiform. **FRUIT** a utricle, 1-seeded, sessile, *c.* 2 mm wide (not including wings) by 1–1.5 mm thick, pericarp thin, free of seed; **seeds** dark brown to black, horizontal, dull, smooth to granular, 1.3–1.5 mm long by 1–1.3 mm wide by *c.* 0.5 mm thick, concave and grooved.
- **LEAVES** Alternate, simple, entire, smaller above; **lower blades** 1–10 cm long by 1–15 mm wide, flat and pointed, 3-veined, generally glabrous above, hairy on the margins and below especially on smaller blades, reddish in fall; **petioles** 1–2 cm long, reduced above, hairy.
- **STEM** Erect but not rigid, round, somewhat ridged, green turning red, simple to much branched; 1–11 mm thick near the base.
- **SYN.** *Bassia scoparia* (L.) A.J. Scott.

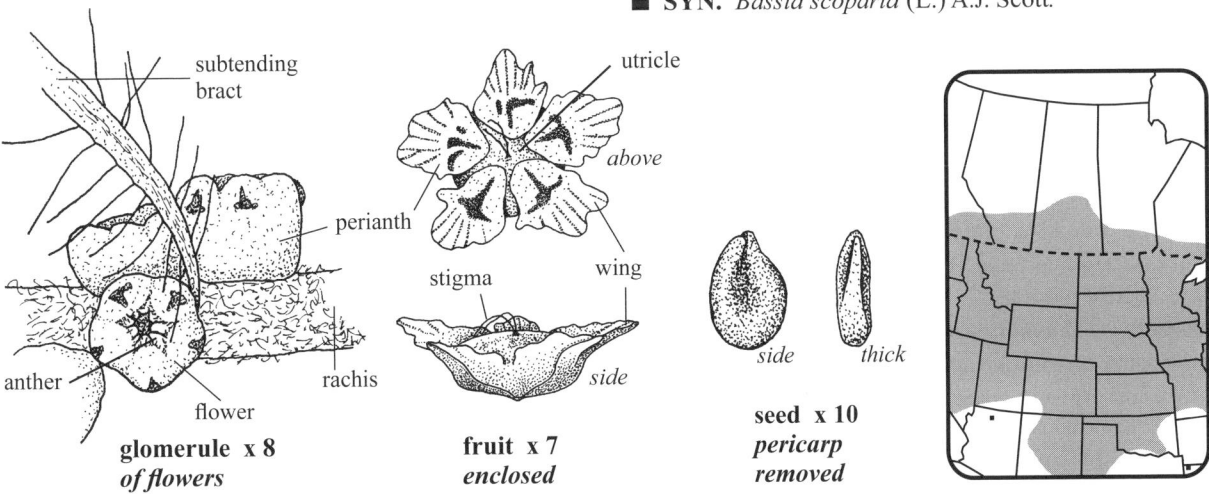

glomerule x 8
of flowers

fruit x 7
enclosed

seed x 10
pericarp removed

Kochia, Mexican Fireweed, Fireweed

Goosefoot—*Chenopodiaceae* **Wildflower** Pale yellow to light green Male perianth 4-lobed, female bracteoles 2

WINTERFAT   *Krascheninnikovia lanata* (Pursh) A.D.J. Meeuse & Smit

■ **SKETCH** A hairy, tufted, shrublike **perennial herb** 15–80 cm tall from a thick **woody root**; in dry plains, sagebrush, foothills, rocky mesas, often in alkaline soils; **monoecious** to **dioecious**. w2

■ **FLOWERS** Pale yellow to pale green, blooming April–September; **inflorescence** spikelike, of unisexual flowers in short axillary clusters; **male flowers** above female flowers, hairy, sessile, mostly six to eight in axillary glomerules 1.5–2.5 mm wide and long; **perianth lobes** four, united at base or free, each lobe 1.1–1.5 mm long by 0.8–1 mm wide (not flattened), glabrous inside, hairy outside, midvein reddish brown; **petals** absent; **stamens** four, exserted; **filaments** glabrous, 1–1.5 mm long; **anthers** reddish yellow, *c.* 1 mm long and wide; **female flowers** one to four per leaf axil, sometimes a few among the lowest male flowers, each 2–4 mm wide by 3–4.5 mm long, sessile, very hairy; **bracteoles** two, fleshy, blunt, erect, *c.* 0.5 mm wide, united in lower half, large tufts of hair below; **pistil** *c.* 3 mm long, exserted; **ovary** green, flattened, *c.* 0.6 mm long by *c.* 0.5 mm wide (not including hairs), hairy on margin; **style** glabrous, *c.* 0.5 mm long; **stigmas** two, brown, 1.8–2 mm long, hairy, filiform, persistent; **fruiting stalks** 5–12 cm long by 10–15 mm wide; **bracteoles** (with fruit) two, united, 5–9 mm long and wide by 1–1.3 mm thick (including hairs), the two thick spreading hornlike tips 1.5–2 mm long, easily opened, the hairs white to reddish tan in two to four ascending tufts 3–5 mm long. **FRUIT a utricle**, 1-seeded, *c.* 5 mm long by *c.* 2 mm wide, beaked, hairy; **seeds** brown, flat, 3.5–4 mm long by 1.5–1.8 mm wide by *c.* 1 mm thick, ridged, hairless.

■ **LEAVES** Alternate, simple, entire with margins revolute, ascending, midrib raised below, hairs stellate to simple, some to *c.* 3 mm long; **blades** 1–4 cm long by 1.7–4 mm wide, reduced above, apices blunt, main leaves with axillary leafy fascicles; **petioles** 0–4 mm long, hairy.

■ **STEM** Woody and herbaceous (green), horizontal to erect; **woody stems** 2–3 mm thick with tan bark peeling; **new stems** 1–2 mm thick near their base, hairs gray and turning reddish.

■ **SYN.** *Eurotia lanata* (Pursh) Moq. and *Ceratoides lanata* (Pursh) J.T. Howell.

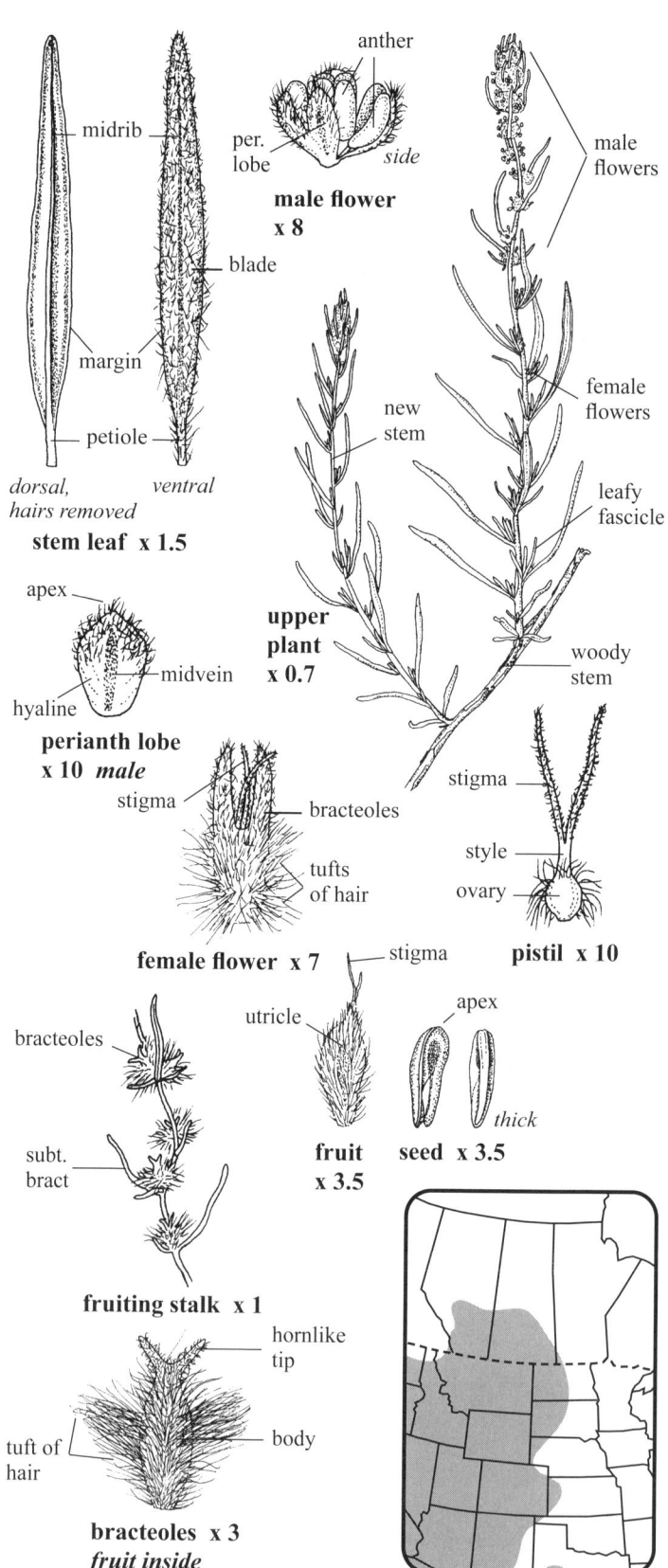

272   White Sage

Goosefoot—*Chenopodiaceae* **Wildflower** Green Perianth one

SPEAR-LEAVED GOOSEFOOT *Monolepis nuttalliana* (J.A. Schultes) Greene

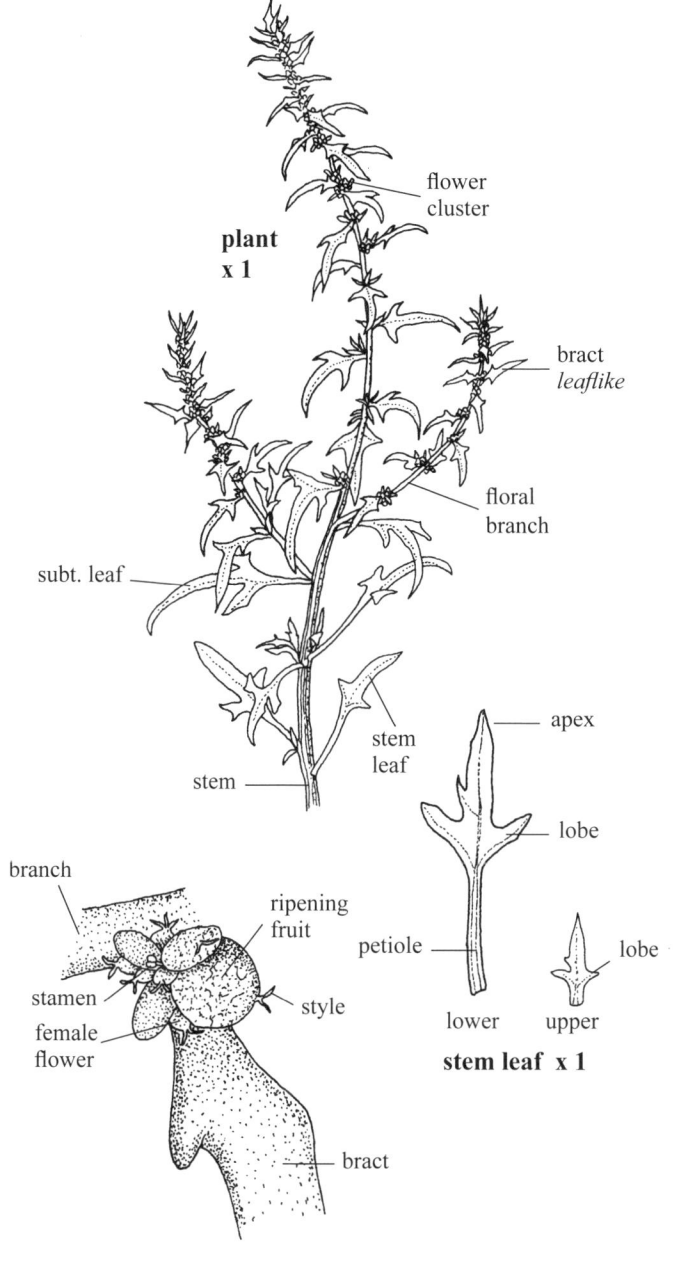

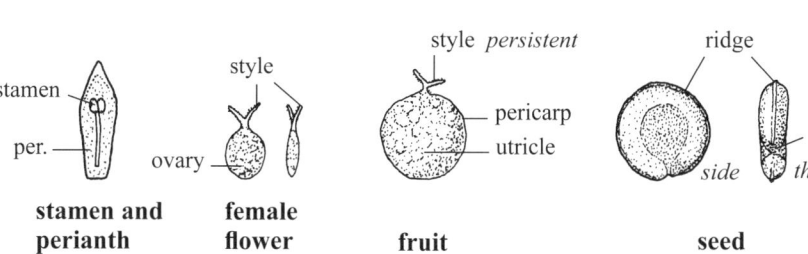

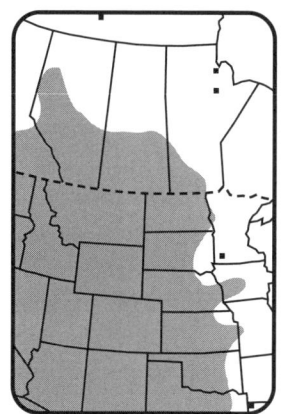

■ **SKETCH** A hairless and glandular-dotted, **annual herb** forming a spreading mat 15–50 cm wide and to *c.* 20 cm tall from a **taproot** 3–6 mm wide by 5–10 cm long; in cultivated fields, along roadsides, railways, salt marshes, deserts and mesas, sometimes in saline prairies; **monoecious**.

■ **FLOWERS** Green, blooming January–September; **inflorescence** axillary, in bracteate clusters along the upper branches; **floral branches** throughout, alternate; **flowers** male or female, minute, sessile, congested, more female than male flowers in each cluster; <u>male flowers</u> of one perianth part 1–2 mm long by *c.* 0.5 mm wide; **stamens** one, attached near base of perianth, included, *c.* 0.7 mm long; **filaments** white, filiform, *c.* 0.5 mm long; **anthers** pale yellow, *c.* 0.2 mm long; <u>female flowers</u> *c.* 0.5 mm wide by 0.8–1 mm long (including styles); **ovary** lenticular; **styles** two (three), spreading, reddish green, *c.* 0.3 mm long. **FRUIT** a utricle, 1-seeded, indehiscent, pericarp adherent to seed, finely pitted at maturity, with translucent dots (microscopic) when developing; **seeds** vertical, 0.8–1.4 mm wide by *c.* 0.4 mm thick, brown to black, dull, thinly ridged, with a small notch.

■ **LEAVES** Alternate, simple, mostly 2-lobed at the blade's base, sometimes with a few teeth, bright green, spreading to ascending; **blades** thick, glabrous, 0.8–5 cm long by 3–25 mm wide, reduced above, punctate, glands slightly raised; **petioles** 0–3 (–6) cm long, reduced above.

■ **STEM** Several, decumbent to erect, branched, green with reddish low ridges, slightly fleshy, lightly mealy to glabrous; 1.5–5 mm thick near the base.

Nuttall's Povertyweed, Poverty Weed

Goosefoot—*Chenopodiaceae* — **Wildflower** Green to reddish Perianth 5-lobed

Western Sea-blite *Suaeda calceoliformis* (Hook.) Moq.

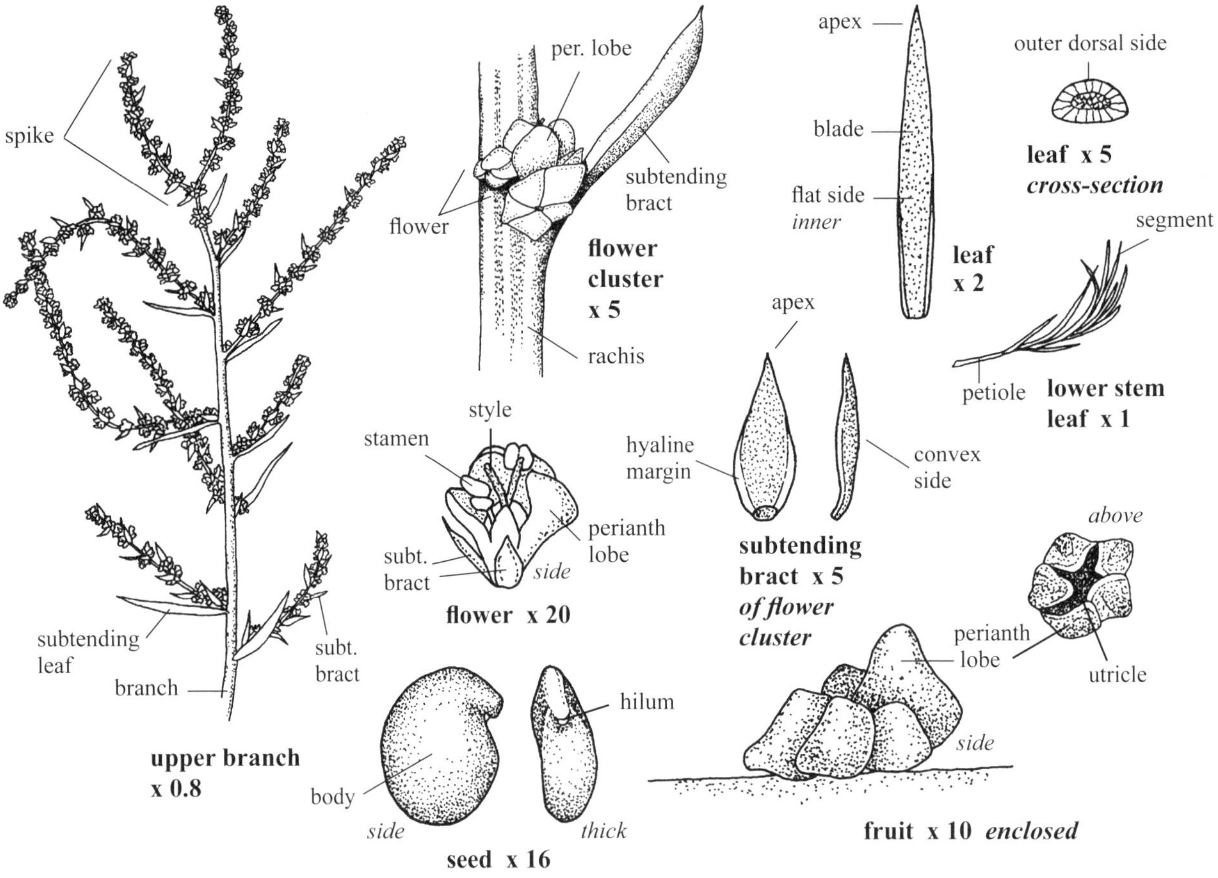

- **SKETCH** A glabrous, glaucous **annual herb** 7–100 cm tall or long, in patches, with a **taproot** 1–7 mm wide and 3–15 cm long; in moist sites, pond or slough edges, among desert shrubs and often in saline soils, the whole plant turning blackish as it matures.
- **FLOWERS** Green to reddish, blooming July–September; **inflorescence** of spikes 1–15 cm long by 2–5 mm wide, axillary, the flower clusteres separated or almost continuous; **subtending bracts** (of flower clusters), 3–7 mm long by 1–1.5 mm wide; **subtending bracts** (of flowers) one to three, membranous, pointed, shorter than perianth; **flowers** perfect, sessile, three to seven per cluster, each *c.* 1 mm wide by 0.8–1 mm tall, base reddish; **perianth** 5-lobed, united below, 1–3 lobes obviously larger, more so in fruit, fleshy, reddish green, hooded, with a honeycomb surface; **petals** absent; **stamens** five or less, *c.* 1 mm long, included to exserted; **filaments** hyaline; **anthers** *c.* 0.2 mm long, pale yellow; **styles** two to five, 0.8–1 mm long, base reddish. **FRUIT a utricle**, 1-seeded, sessile, enclosed by unequal perianth lobes; **seeds** dimorphic; **1)** black, shiny, 0.8–1.7 mm long by 0.5–0.7 mm thick, horizontal; **2)** dull brown, 1–1.5 mm long and wide; **pericarp** thin, gray, easily removed.
- **LEAVES** Alternate, simple, entire; **blades** linear, fleshy, 1–4 cm long by 1–5 mm wide by 0.5–2 mm thick, reduced above, sessile, flat above, edges blunt, dorsal side rounded, tip pointed and often with a short mucro; **lower blades** turning grayish brown early, divided, 2–9 cm long; **segments** 7–25, alternate, filiform, each 0.5–1.7 cm long.
- **STEM** Prostrate to erect, solid, simple to branched, glaucous, round, with *c.* 10 low, wide, green and pinkish ridges; 2–10 mm thick near the base.
- **SYN.** *Suaeda depressa* auct. non (Pursh) S. Wats. and *Suaeda maritima* var. *americana* (Pers.) Boivin.

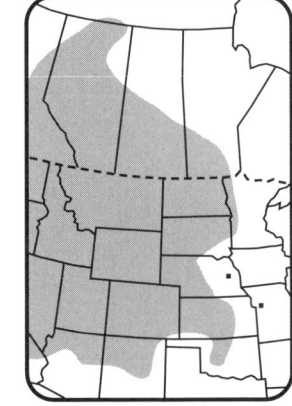

Western Blite, Pursh Seepweed, Bob, Paiuteweed, Sea Blite, Low Sea Blite

Morning-glory—*Convolvulaceae* **Wildflower** White to pink   Corolla funnel-like with 5 blunt corners

Hedge Bindweed   *Calystegia sepium* (L.) R. Br.

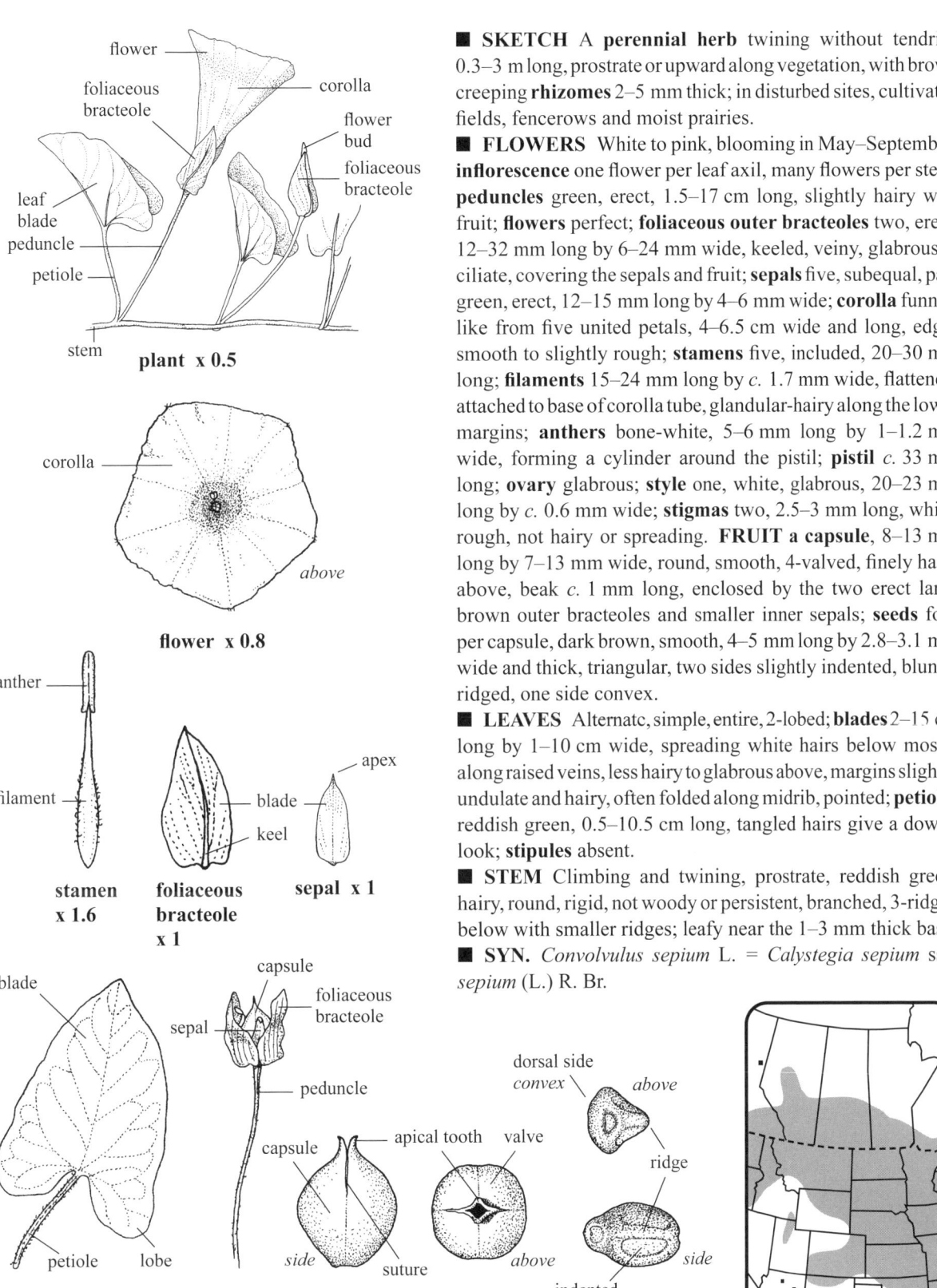

- **SKETCH** A **perennial herb** twining without tendrils, 0.3–3 m long, prostrate or upward along vegetation, with brown creeping **rhizomes** 2–5 mm thick; in disturbed sites, cultivated fields, fencerows and moist prairies.
- **FLOWERS** White to pink, blooming in May–September; **inflorescence** one flower per leaf axil, many flowers per stem; **peduncles** green, erect, 1.5–17 cm long, slightly hairy with fruit; **flowers** perfect; **foliaceous outer bracteoles** two, erect, 12–32 mm long by 6–24 mm wide, keeled, veiny, glabrous to ciliate, covering the sepals and fruit; **sepals** five, subequal, pale green, erect, 12–15 mm long by 4–6 mm wide; **corolla** funnel-like from five united petals, 4–6.5 cm wide and long, edges smooth to slightly rough; **stamens** five, included, 20–30 mm long; **filaments** 15–24 mm long by *c.* 1.7 mm wide, flattened, attached to base of corolla tube, glandular-hairy along the lower margins; **anthers** bone-white, 5–6 mm long by 1–1.2 mm wide, forming a cylinder around the pistil; **pistil** *c.* 33 mm long; **ovary** glabrous; **style** one, white, glabrous, 20–23 mm long by *c.* 0.6 mm wide; **stigmas** two, 2.5–3 mm long, white, rough, not hairy or spreading. **FRUIT a capsule**, 8–13 mm long by 7–13 mm wide, round, smooth, 4-valved, finely hairy above, beak *c.* 1 mm long, enclosed by the two erect large brown outer bracteoles and smaller inner sepals; **seeds** four per capsule, dark brown, smooth, 4–5 mm long by 2.8–3.1 mm wide and thick, triangular, two sides slightly indented, bluntly ridged, one side convex.
- **LEAVES** Alternate, simple, entire, 2-lobed; **blades** 2–15 cm long by 1–10 cm wide, spreading white hairs below mostly along raised veins, less hairy to glabrous above, margins slightly undulate and hairy, often folded along midrib, pointed; **petioles** reddish green, 0.5–10.5 cm long, tangled hairs give a downy look; **stipules** absent.
- **STEM** Climbing and twining, prostrate, reddish green, hairy, round, rigid, not woody or persistent, branched, 3-ridged below with smaller ridges; leafy near the 1–3 mm thick base.
- **SYN.** *Convolvulus sepium* L. = *Calystegia sepium* ssp. *sepium* (L.) R. Br.

Hedge False Bindweed, Wild Morning-glory

Morning-glory—*Convolvulaceae*   **Wildflower**   White to pink   Corolla funnel-like with 5 corners

Field Bindweed *Convolvulus arvensis* L.

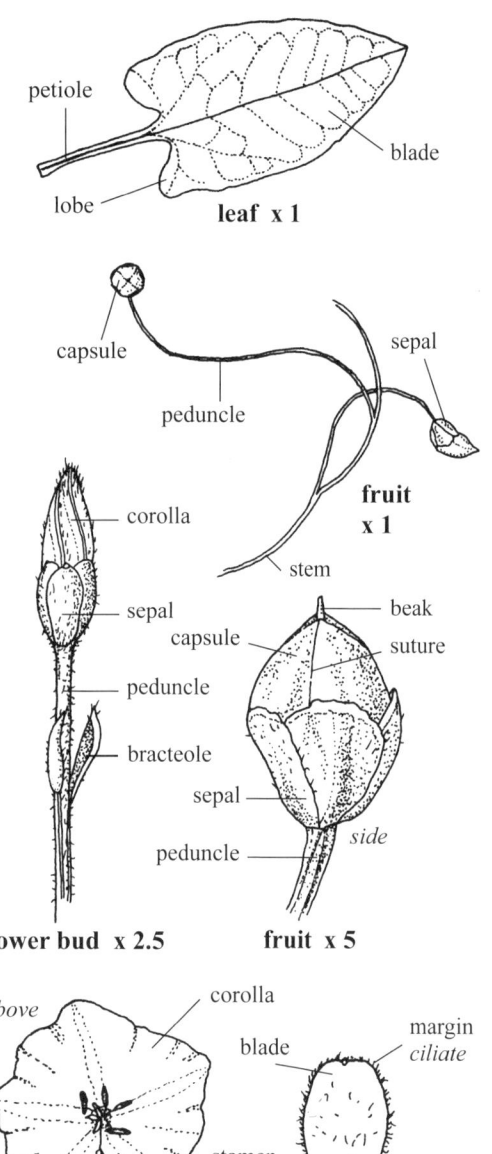

■ **SKETCH** A prostrate **perennial herb** 30–130 cm long with a central, white, fleshy **rhizome** 1–5 mm wide by 1–4 m long, forming large patches; along sidewalks, fences, in pastures, farmyards, along roadsides, and in cultivated fields. *Naturalized*

■ **FLOWERS** White to pink, blooming March–November; **inflorescence** in cymes of one to four flowers; **peduncles** 1.7–8.5 cm long, erect, axillary, slightly hairy, curved in fruit; **bracteoles** (on peduncles) paired, 2–10 mm long by *c.* 0.7 mm wide, slightly hairy on the margins; **flowers** perfect, apices bluntly 5-sided; **sepals** five, blunt, 3–5.5 mm long by 2–4 mm wide, persistent and appressed around the lower half of the fruit, subequal; **1) outer two sepals** distinct to base, margins white, minutely hairy, outer surface green; **2) inner three sepals** united, convex on back, not veined, no obvious midvein, glabrous, shiny; **corolla** funnel-like, 2.5–3.5 cm wide by 1.2–2.5 cm long, glabrous, five pinkish bands radiating to each marginal point; **stamens** five, white, spreading, separate, 8–13 mm long, included; **filaments** erect, filiform, *c.* 7 mm long, attached to corolla tube near the base; **anthers** pinkish purple, 2–3.5 mm long; **pistils** included; **ovary** orange, glabrous, 1.8–2 mm wide; **style** one, 7–10 mm long, white, glabrous, filiform; **stigmas** two, threadlike. FRUIT a capsule, 5–7 mm tall by 4–7 mm wide tan, dull, pointed, glabrous, 4-valved; **seeds** one to four per capsule, dull, dark brown, 3-sided, 3–5 mm long by 2–2.3 mm wide (across the convex side) by *c.* 2 mm thick (across one flat side) hairless and finely rough (tuberculate).

■ **LEAVES** Alternate, simple, entire, basal lobes two, blunt to pointed, each lobe 1–1.3 cm long; **blades** dark green above, 2–10 cm long by 1.5–6 cm wide, glabrous on both sides or slightly hairy below near the base, veins raised below and recessed above; **petioles** 1–4 cm long, glabrous or with a few microscopic hairs; **stipules** and **tendrils** absent.

■ **STEM** Prostrate, 1–3 mm thick, glabrous to hairy, twining to trailing, branched, often tangled, several from one root, green, 5-ridged, angular, flat between the ridges, often twisted; **internodal** distances 1–6 cm.

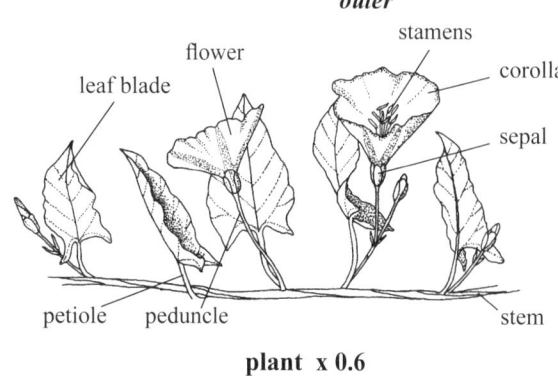

plant x 0.6

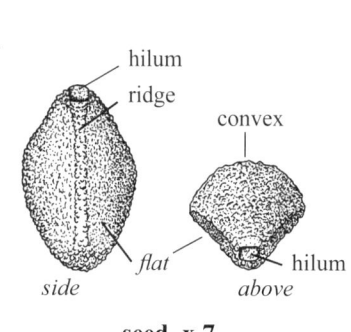

seed x 7

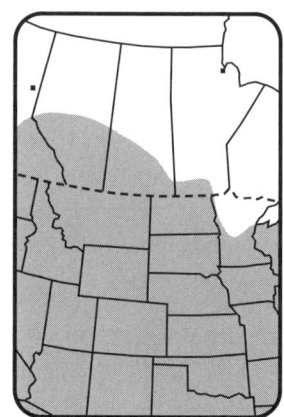

Common Bindweed

Dogwood—*Cornaceae* **Wildflower** White   Petals 4, tiny   Involucral bracts 4, showy

Bunchberry *Cornus canadensis* L.

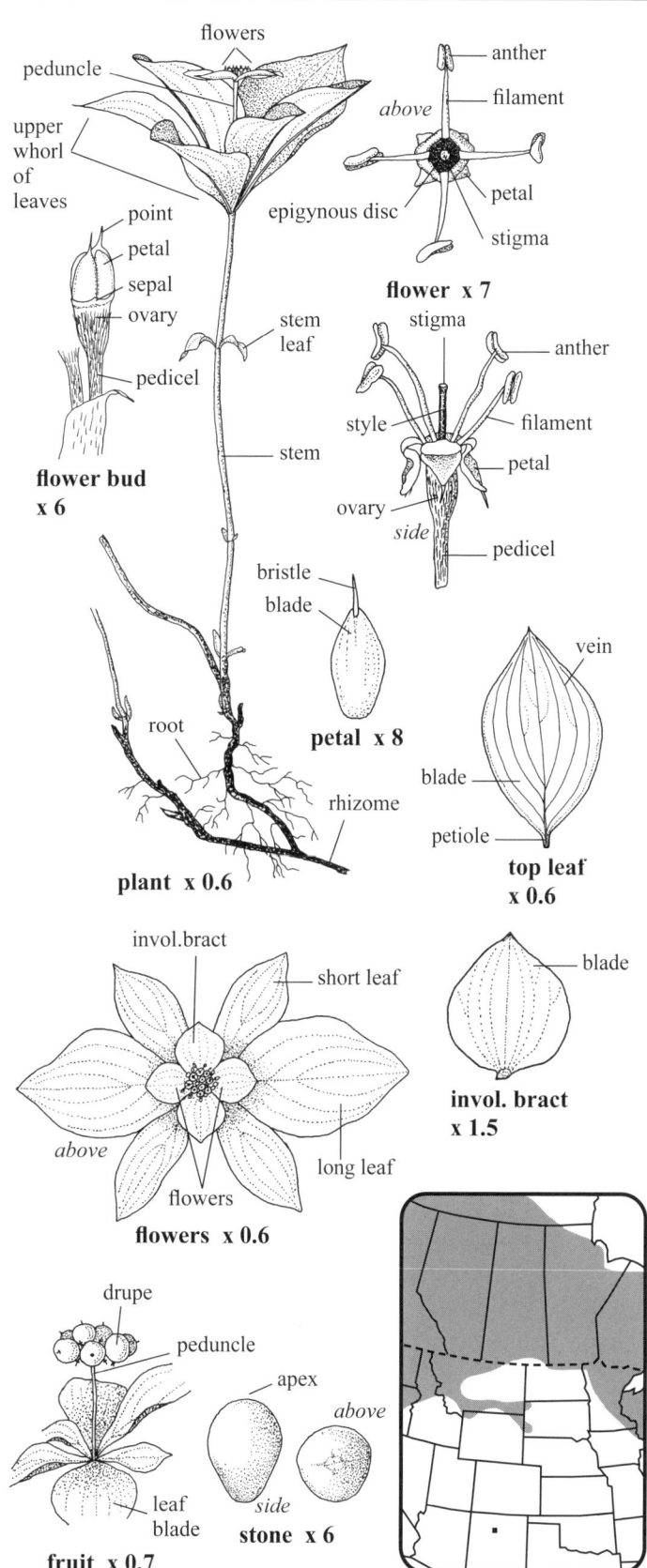

- **SKETCH** A **perennial herb** 5–23 cm tall in loose colonies with creeping, dark brown **rhizomes** 1–1.5 mm wide, smooth to rough; in moist woodlands, shady canyons of parklands and prairies. **w2**
- **FLOWERS** White, blooming May–July; **inflorescence** a cyme of 10–25 flowers clustered and appearing as one, subtended by white involucral bracts; **peduncles** 2–3.5 cm long, square, widely grooved on two sides, white dolabriform hairs on ridges; **pedicels** 1–2 mm long, hairs as on peduncles; **flowers** perfect, 4–4.5 mm wide by 2.5–3.5 mm long, together 7–10 mm wide, exserted above bracts; **involucral bracts** (of flowers) four, white, each 1–2 cm long by 8–15 mm wide, two large and two small, petal-like, spreading, slightly hairy below near base and on lower margins, lightly veined, groups of several flowers attached to base of each bract; **sepals** four, light green, points *c.* 0.2 mm long; **petals** four, reflexed, white, 1–2 mm long and *c.* 1 mm wide, some with a pointed deciduous bristle attached near the top on the lower side; **stamens** four, white, exserted, spreading, glabrous, *c.* 2 mm long, attached to epigynous disc; **filaments** filiform, *c.* 1.5 mm long, attached to middle of anther; **anthers** pale yellow, *c.* 1 mm long; **ovary** inferior, green, *c.* 1 mm long by *c.* 0.8 mm wide, covered with white, dolabriform hairs; **epigynous disc** dark brown, 8-sided; **style** dark brown, one, erect, filiform, exserted; **stigma** slightly wider than the style. **FRUIT a drupe**, 1-stoned, reddish orange, slightly shiny, smooth, 4–8 mm long and wide, usually 2–14 per plant; **stone** 1-seeded, orangish tan, 2.4–2.7 mm long by 1.8–2 mm wide and thick, tapered.
- **LEAVES** Opposite, entire, arched and pointed, simple; **stem blades** 10–16 mm long by 7–10 mm wide, 1–3 pairs, dolabriform hairs appressed to blade below (dorsal side), fewer hairs above and on margins; **petioles** pinkish, 1–2 mm long, hairy below; **top leaves** in a false whorl of six (4–7), subsessile, mostly two long and four short; **long blades** 4–8 cm long by 2.5–5 cm wide, ascending to spreading; **short blades** 3–5.6 cm long by 14–30 mm wide, pointed.
- **STEM** Erect, solitary, unbranched, 2-grooved, the grooves on opposite sides and deeper above; base pink and 1–2 mm wide.

Canadian Bunchberry, Dwarf Dogwood, Dwarf Cornel, Bunchberry Dogwood, Puddingberry

Dogwood—*Cornaceae*   **Shrub**   White   Petals 4 (5)

RED-OSIER DOGWOOD   *Cornus sericea* L.

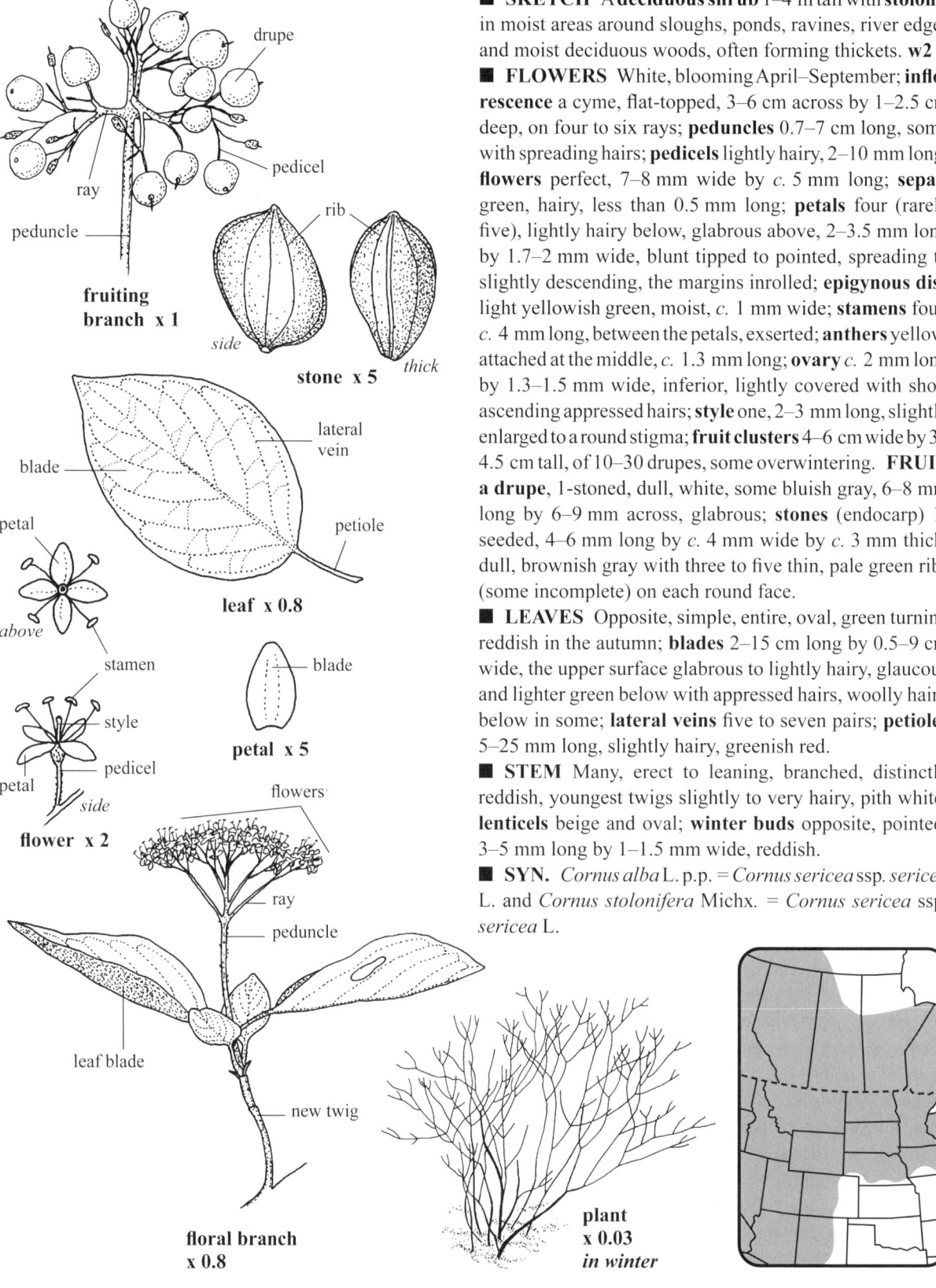

- **SKETCH** A **deciduous shrub** 1–4 m tall with **stolons**; in moist areas around sloughs, ponds, ravines, river edges and moist deciduous woods, often forming thickets. **w2**
- **FLOWERS** White, blooming April–September; **inflorescence** a cyme, flat-topped, 3–6 cm across by 1–2.5 cm deep, on four to six rays; **peduncles** 0.7–7 cm long, some with spreading hairs; **pedicels** lightly hairy, 2–10 mm long; **flowers** perfect, 7–8 mm wide by *c.* 5 mm long; **sepals** green, hairy, less than 0.5 mm long; **petals** four (rarely five), lightly hairy below, glabrous above, 2–3.5 mm long by 1.7–2 mm wide, blunt tipped to pointed, spreading to slightly descending, the margins inrolled; **epigynous disc** light yellowish green, moist, *c.* 1 mm wide; **stamens** four, *c.* 4 mm long, between the petals, exserted; **anthers** yellow, attached at the middle, *c.* 1.3 mm long; **ovary** *c.* 2 mm long by 1.3–1.5 mm wide, inferior, lightly covered with short ascending appressed hairs; **style** one, 2–3 mm long, slightly enlarged to a round stigma; **fruit clusters** 4–6 cm wide by 3–4.5 cm tall, of 10–30 drupes, some overwintering. **FRUIT a drupe**, 1-stoned, dull, white, some bluish gray, 6–8 mm long by 6–9 mm across, glabrous; **stones** (endocarp) 1-seeded, 4–6 mm long by *c.* 4 mm wide by *c.* 3 mm thick, dull, brownish gray with three to five thin, pale green ribs (some incomplete) on each round face.
- **LEAVES** Opposite, simple, entire, oval, green turning reddish in the autumn; **blades** 2–15 cm long by 0.5–9 cm wide, the upper surface glabrous to lightly hairy, glaucous and lighter green below with appressed hairs, woolly hairy below in some; **lateral veins** five to seven pairs; **petioles** 5–25 mm long, slightly hairy, greenish red.
- **STEM** Many, erect to leaning, branched, distinctly reddish, youngest twigs slightly to very hairy, pith white; **lenticels** beige and oval; **winter buds** opposite, pointed, 3–5 mm long by 1–1.5 mm wide, reddish.
- **SYN.** *Cornus alba* L. p.p. = *Cornus sericea* ssp. *sericea* L. and *Cornus stolonifera* Michx. = *Cornus sericea* ssp. *sericea* L.

Redosier, Red Willow

Cucumber—*Cucurbitaceae*   **Vine**   White to pale green   Corolla tubular, 6-lobed

WILD CUCUMBER   *Echinocystis lobata* (Michx.) Torr. & Gray

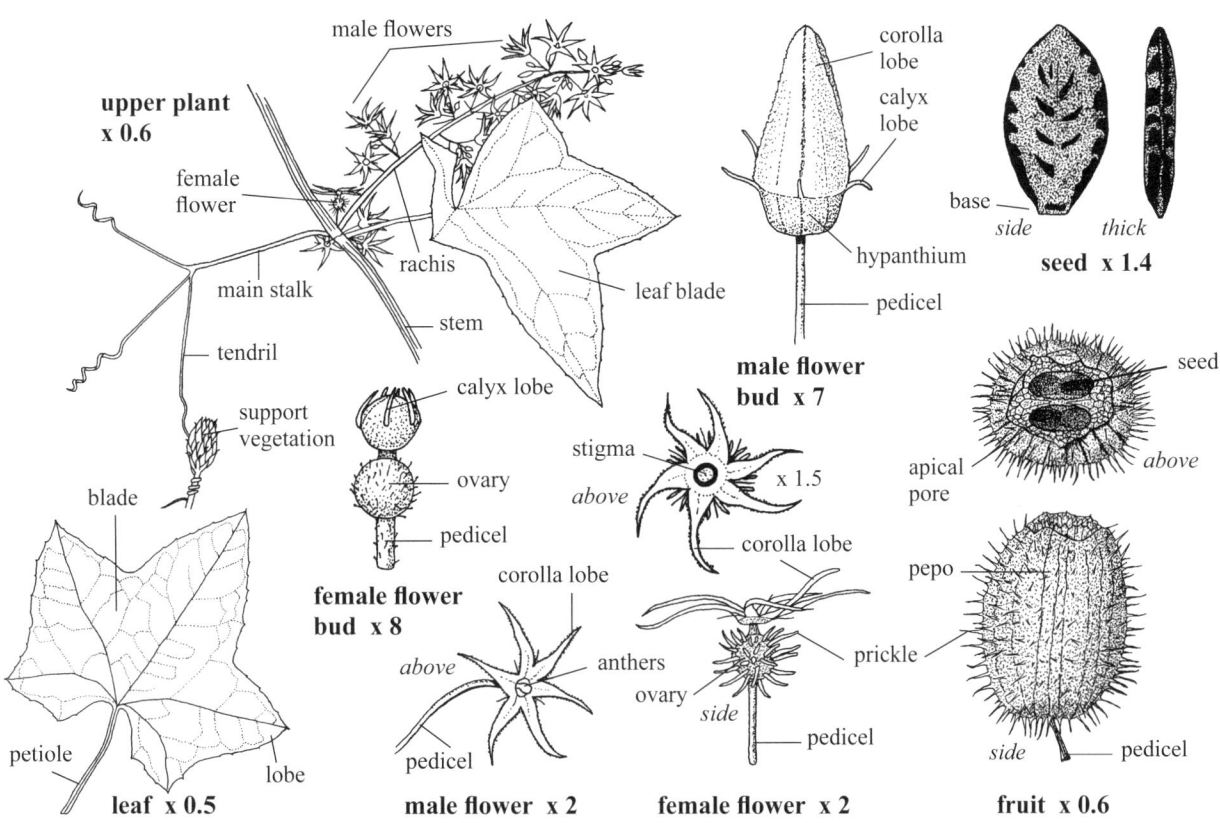

- **SKETCH** An **annual herbaceous vine** 2–6+ m long or tall with a **taproot**, climbing with **tendrils** over vegetation; in moist thickets, ditch banks, fencerows and river shores; **monoecious**. Native and *naturalized*
- **FLOWERS** White to pale green, blooming June–October; **inflorescence** a panicle (raceme), each 7–30 cm long by 2–8 cm wide, axillary from all but the lowest leaves; **pedicels** 4–13 mm long, spreading; **hypanthium** veined, c. 1.5 mm wide; **flowers** unisexual; **calyx lobes** six, spreading, filiform, each 1.5–2 mm long; male flowers numerous, 8–12 mm wide by 4–5 mm tall; **corolla** tubular, 6-lobed, the lobes pointed, spreading, rotate, 4–9 mm long by 1–1.2 mm wide, with short gland-tipped hairs; **stamens** 3 (2), united, hairy; **filaments** c. 0.5 mm long; **anthers** pale green, c. 1 mm long by 0.8 mm wide; female flowers white, 15–25 mm wide by 5–7 mm tall, one to four at base of each panicle; **corolla** tubular, 6-lobed, each lobe 7–12 mm long by 1.2–2 mm wide, united near the base, hairs short and gland-tipped; **ovary** green, prickly, 5–6 mm wide by 3–4 mm tall; **stigmas** blunt, included and c. 1.2 mm wide by c. 1 mm tall. **FRUIT a pepo**, light green to tan, with about 18 thin, long veins, bladderlike, inflated, thin-walled and dry when ripe, prickly, 3–5 cm long by 2.5–3.5 cm wide by 2–2.8 cm thick (not including prickles); **prickles** soft, 2–8 mm long, mostly spreading; **apical pores** two, pointing downward when ripe, each opening 12–13 mm long by 5–6 mm wide, inner lining tan and net-veined; **seeds** brown, four per pepo, smooth, flattened, 12–20 mm long by 7–10 mm wide by 2.5–3.8 mm thick with dark brown markings.
- **LEAVES** Alternate, simple, usually 5-lobed (3- or 7-), barely toothed; **blades** 3–16 cm long and wide, short hairy along margins, scabrous above, glabrous below, some veins ending with thick marginal hairs; **petioles** 1–11.5 cm long, glabrous, ridged; **stipules** absent; **tendrils** axillary, one per node, mostly 3-parted, coiled, main stalk 2.5–15+ cm long.
- **STEM** Climbing to trailing, erect to horizontal, hollow, simple to branched, hairless, ridged; 2–5 mm thick throughout.

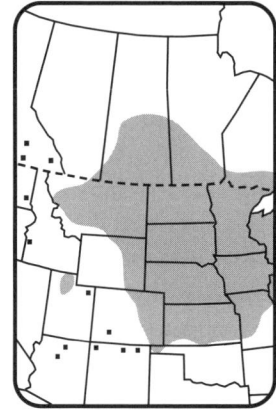

Wild Balsam Apple, Balsam Apple

Oleaster—*Elaeagnaceae* **Shrub** Pale yellow Calyx tubular, 4-lobed

Russian Olive *Elaeagnus angustifolia* L.

■ **SKETCH** A spiny **deciduous shrub** 3–7 m tall and wide, **roots** with **nodules**; in dry to moist sites along woodland edges and sandy flood plains, near sloughs and streams; often planted to attract wildlife. *Naturalized* **w2**

■ **FLOWERS** Yellow, blooming May–July; **inflorescence** one to three flowers in leaf axils, mostly in upper two-thirds of shrub; **pedicels** scurfy, pale green, 3–6 mm long, spreading to drooping with fruit; **flowers** mostly perfect (or unisexual), fragrant, 8–12 mm wide by 7–11 mm long; **calyx** tubular, 4-lobed, scurfy on outside, glabrous inside, the tube 4–6 mm long by 2.5–3 mm wide; **calyx lobes** yellow on upper side, pointed, 3–5 mm long by 2–3 mm wide, thickish with a few white stellate hairs, ascending to spreading; **petals** absent; **stamens** four, included to exserted; **filaments** *c.* 0.3 mm long, white, thickish, attached to calyx where the lobes meet; **anthers** yellow, 2–3 mm long by *c.* 1.3 mm wide; **ovary** *c.* 1 mm long, glabrous, green, enclosed in hypanthium at the bulge above the pedicel and below the constriction of the calyx; **style** yellow, 7–9 mm long, filiform, apex curved, widest and hairy at base; **stigma** slightly enlarged; **mature hypanthium** drupelike, 9–15 mm long by 9–10 mm wide, yellowish green, ovoid, mealy dry inside. **FRUIT an achene**, 1-seeded, dull, 9–11 mm long by 4.5–5.5 mm across, slightly oval, with eight longitudinal dark brown stripes *c.* 0.7 mm wide and eight pale green stripes *c.* 1 mm wide.

■ **LEAVES** Alternate, simple, entire, pale green, dry, thin; **blades** 3–9 cm long by 5–22 mm wide, slightly shiny, pale gray below, veins visible and slightly raised below, scurfy, pointed, tapered at base; **petioles** scurfy, 4–8 mm long; **stipules** absent.

■ **TRUNK** Dark gray, older branches brown and slightly shiny, alternate, ascending to spreading; **new twigs** 3–15 cm long, leafy, those along lower branches vegetative; **upper branches** slightly shiny, dark reddish brown; **spines** scattered, 7–20 mm long from the base of some new twigs; 5–15 cm wide near the furrowed base.

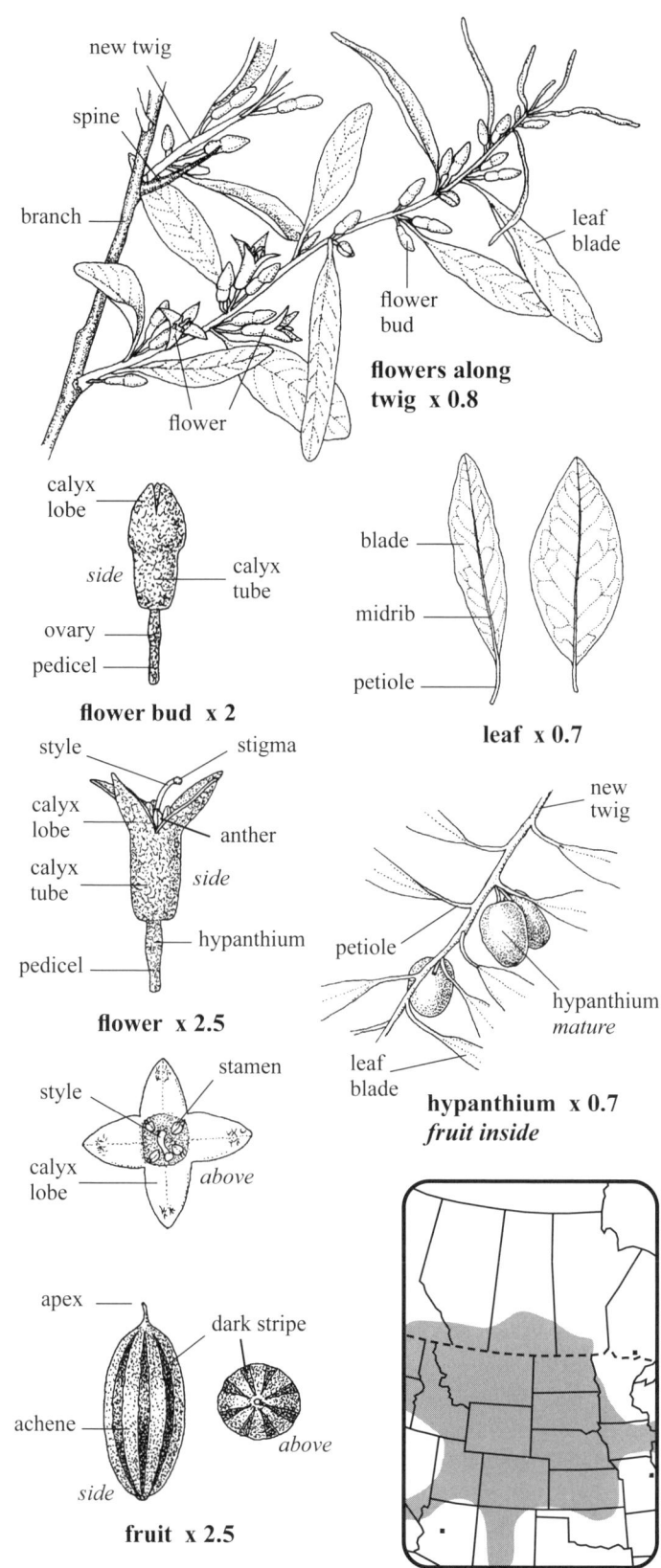

Oleaster—*Elaeagnaceae*　　　　　　　　　　　　　**Shrub**　Yellow　Calyx tubular, 4-lobed

WOLF-WILLOW　*Elaeagnus commutata* Bernh. ex Rydb.

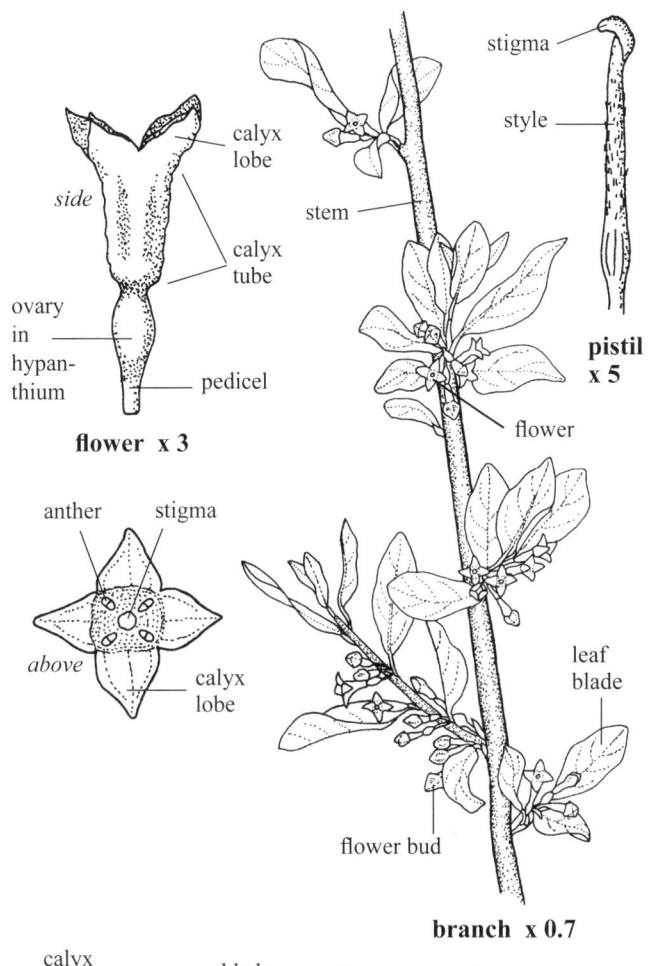

flower x 3

flower bud x 3

leaf x 0.7

branch x 0.7

pistil x 5

■ **SKETCH** An unarmed **deciduous shrub** 1–5 m tall from **woody rhizomes** 4–9 mm thick and 15–30+ cm long with nitrogen-fixing **nodules**; scattered colonies in prairies, along edges of forests, eroded slopes, ravines and along waterways. **w2, p2**

■ **FLOWERS** Pale yellow, blooming May–July; **inflorescence** usually one to four flowers per axillary cluster; **pedicels** 2–3 mm long; **flowers** perfect or unisexual, fragrant, each 6–9 mm wide by 9–14 mm long; **hypanthium** enclosing ovary, expanding around fruit; **calyx** tubular, 4-lobed, the tube 4–7 mm long, the lobes 2–3 mm long and wide, ascending at anthesis, yellow on the inside but silvery grayish green on the outside; **petals** absent; **stamens** four, between the calyx lobes and included; **filaments** attached to the calyx, the top 1–2 mm free; **anthers** 1–1.5 mm long; **pistil** one, 6–7 mm long, included; **ovary** enclosed in hypanthium, glabrous, *c.* 1.5 mm long, surrounded by hairs at base of calyx tube; **style** 3–4 mm long; **stigma** slanted; **mature hypanthium** silvery green, round, 8–15 mm long and wide, dry and mealy, enclosing one fruit. **FRUIT** an achene, 1-seeded, 7–14 mm long by 4–6 mm wide and thick, with eight wide dark brown ridges and eight, thin, tan ribs.

■ **LEAVES** Alternate, simple, entire, light silvery grayish green, along lateral shoots; **blades** glabrous, 1–10 cm long by 0.5–4 cm wide, scurfy on both sides, brown scales below; **petioles** 3–5 mm long, with tiny rusty-colored scales.

■ **STEM** Erect, not thorny, compactly and alternately branched; **older sections** medium gray with a rusty color beneath when scratched; **new branches** and shoots covered with rusty-colored scales; base of stem 1–5 cm wide; **lateral winter buds** alternate, 3.5–5 mm long by 3–4 mm wide by *c.* 1.5 mm thick.

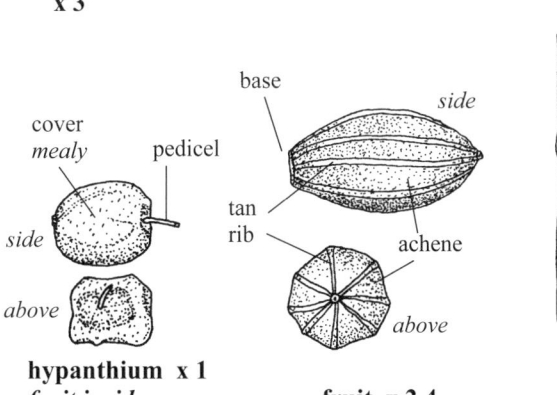

hypanthium x 1 *fruit inside*　　　fruit x 2.4

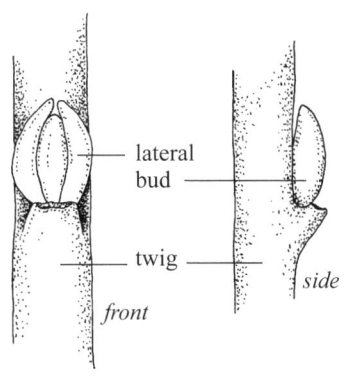

winter bud x 3.5

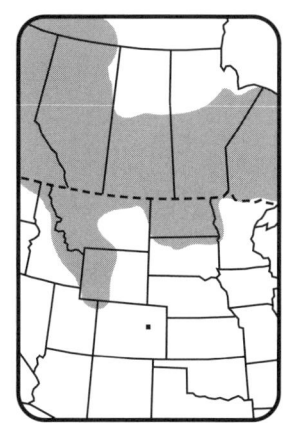

Oleaster—*Elaeagnaceae* **Shrub** Yellow Calyx 4-lobed

BUFFALOBERRY   *Shepherdia argentea* (Pursh) Nutt.

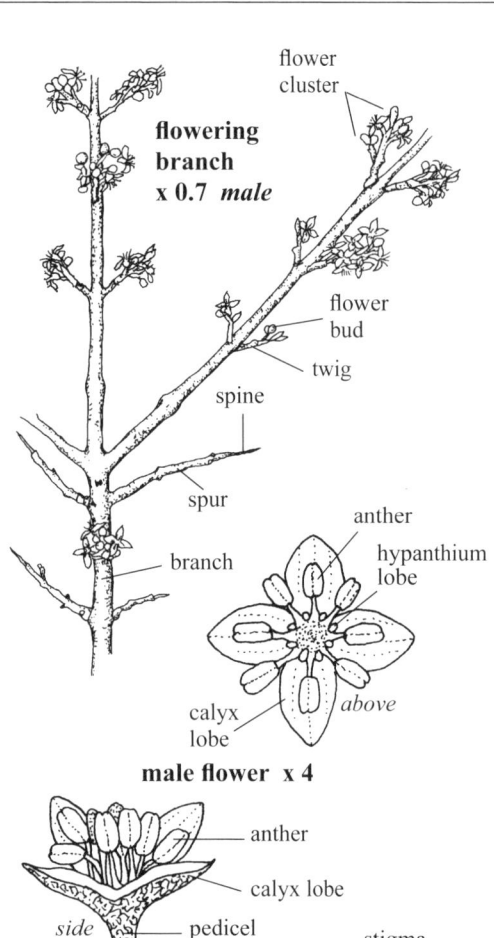

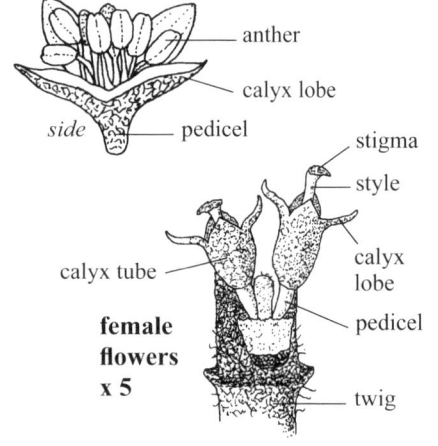

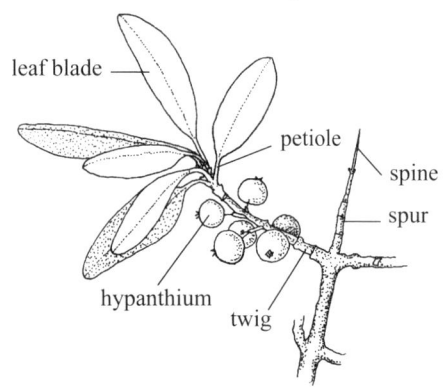

■ **SKETCH** An armed **deciduous shrub** 1–6 m high from a crown on a **taproot**, singly or in shrub communities; in prairies, along riverbanks, canyons, hedgerows and shelterbelts; **dioecious. p1**
■ **FLOWERS** Pale yellow, blooming April–June; **inflorescence** of unisexual flowers in small clusters on young twigs; **spur branches** often ending in a sharp spine 3–10 mm long; **pedicels** *c.* 0.5 mm long, scurfy, 1–2 mm long with fruit; **male flower buds** *c.* 2.8 mm long by *c.* 2 mm wide, light reddish brown; male flowers 5–7 mm wide by 3–5 mm long, two or three per cluster, opening before leaves, mostly in upper half of twigs; **calyx** 4-lobed, each 2–2.8 mm long by 1.5–2 mm wide, bluntly pointed, spreading to descending, pale yellow above to scurfy below, midrib and two lateral veins visible from above; **petals** absent; **stamens** eight, exserted, *c.* 2 mm long; **filaments** about as along as the anthers; **anthers** pale yellow, 0.8–1.2 mm long; **hypanthium disc** 8-lobed, orangish yellow between the filaments; female flowers 1.5–2.3 mm wide by 2.5–3 mm long, often in clusters of six on new twigs; **calyx** 4-lobed, each 1–1.5 mm long, erect to incurved, united below around the pistil; **calyx mouth** nearly covered with dense short hairs; **petals** absent; **style** 1–2 mm long, exserted; **mature hypanthium** fleshy, reddish orange, 4–8 mm long by 5–7 mm wide, roundish, glabrous, often in pairs, persistent. **FRUIT** an achene, 1-seeded, shiny, 3–5 mm long by 2.3–3 mm wide by 1.8–2.5 mm thick, medium brown, one end asymmetrical, a narrow shallow groove on each side.
■ **LEAVES** Opposite, simple, entire, appearing after the flowers, ascending to spreading, crowded; **blades** rounded at apices, tapered to the petioles, lighter green below, scurfy above and below, 1.5–5.5 cm long by 6–21 mm wide, midrib obvious, especially below, lateral veins obscure; **petioles** 3–10 mm long, scurfy, light green.
■ **STEM** Furrowed near the 4–15 cm wide base; **larger lower branches** alternate, small branches and twigs opposite, one of the pair shorter than the other, bark medium gray; **twigs** scurfy gray, reddish below when scrapped; **spines** at ends of spurs; **hairs** on young branches matted and microscopic; **winter buds** 2–4 mm long.

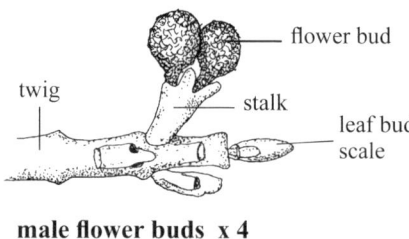

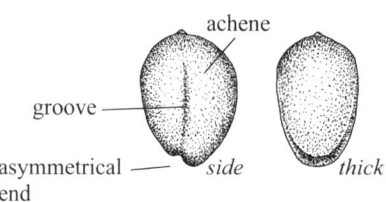

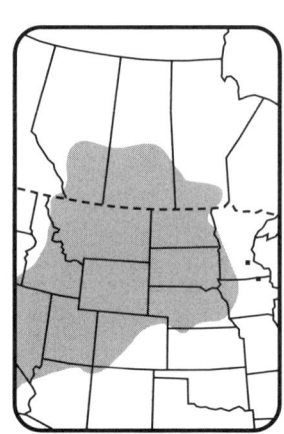

282   Thorny Buffaloberry, Silver Buffaloberry

Oleaster—*Elaeagnaceae* **Shrub** Greenish yellow Calyx 4-lobed

Canada Buffaloberry *Shepherdia canadensis* (L.) Nutt.

■ **SKETCH** An unarmed **deciduous shrub** 0.5–3 m tall, forming thickets; on rocky hillsides, openings in woodlands, on moist slopes and riverbanks; **dioecious. w2**

■ **FLOWERS** Greenish yellow, blooming April–June; **inflorescence** of unisexual racemes, numerous, each one *c.* 10 mm long, 1–6 flowers in opposite pairs; **pedicels** *c.* 1 mm long, scurfy brown; **flowers** with scurfy granules 0.5–0.7 mm wide, flat, with a dark brown center and a white to reddish halo; **petals** absent; <u>male **flowers**</u> 3–6 mm wide by 2.5–3 mm long; **calyx lobes** four, pointed, 1–2.2 mm long by 1–1.7 mm wide, reflexed to spreading, 3-veined, scurfy and brown below; **stamens** eight, erect, outside of and between the hypanthium lobes; **filaments** pale yellow, *c.* 1 mm long, widest at base, curved inward; **anthers** yellow, *c.* 0.5 mm long, 2-lobed; **hypanthium** moist, 8-lobed, *c.* 2 mm wide, pale green; <u>female **flowers**</u> sessile, 2–3.5 mm wide by 2.5–3 mm long; **calyx lobes** 1–1.5 mm long by *c.* 1 mm wide, 3-veined, reflexed to spreading; **stamens** four, vestigial; **pistil** *c.* 3 mm long, hairless, green; **ovary** *c.* 1.5 mm long by *c.* 0.6 mm wide; **style** one, together with the curved stigma *c.* 1.5 mm long; **hypanthium** 8-lobed, 1.3–1.5 mm across, white hairs in throat; becoming drupelike, red, juicy, shiny, smooth, not scaly, 5–8 mm long by 4–6.5 mm wide, subsessile. **FRUIT an achene**, 1-seeded, 4–4.3 mm long by *c.* 3 mm wide by *c.* 2 mm thick, covered with an easily removed dull papery coat, grooved lengthways, shiny; **beak** tiny.

■ **LEAVES** Opposite, simple, entire; **blades** dull, very hairy below from stellate hairs with 8–12 spreading branches, mostly glabrous above, 2–6.7 cm long by 7–30 mm wide, soft and thin, apices blunt, veins visible above, less so below; **petioles** scurfy, reddish brown, very hairy, 4–7 mm long, ascending; **stipules** absent.

■ **STEM** Several in a cluster, grayish brown; **fall buds** in leaf axils, reddish brown, 5–7 mm long by 1.8–2 mm wide by 1–1.3 mm thick, scurfy, blunt; **branches** reddish tan when scurfyness is scraped away; **twigs** opposite.

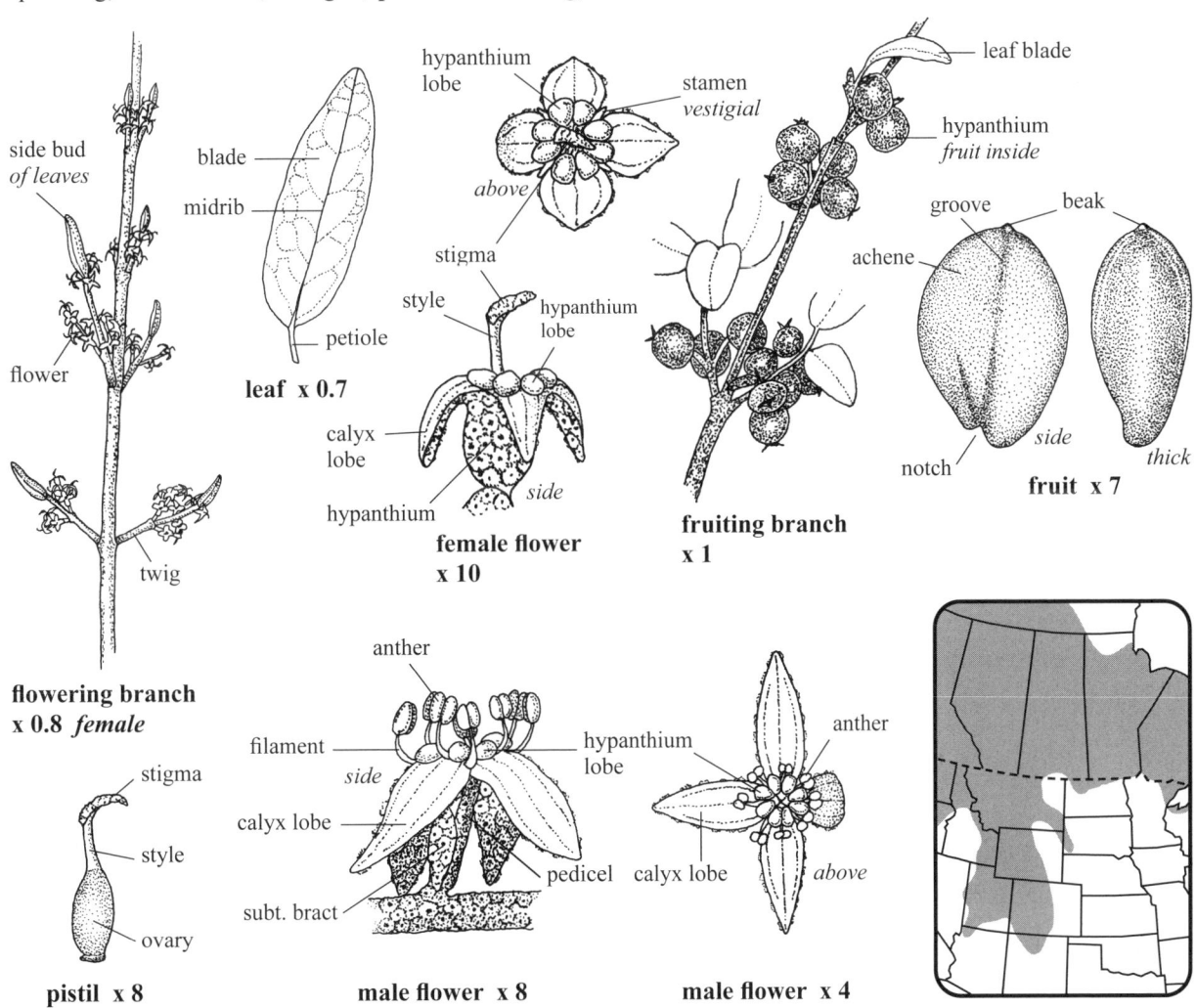

Russet Buffaloberry, Rabbitberry, Soapberry

Heath—*Ericaceae* **Shrub** Evergreen White to pinkish Corolla tubular, (4-) 5-lobed

BEARBERRY *Arctostaphylos uva-ursi* (L.) Spreng.

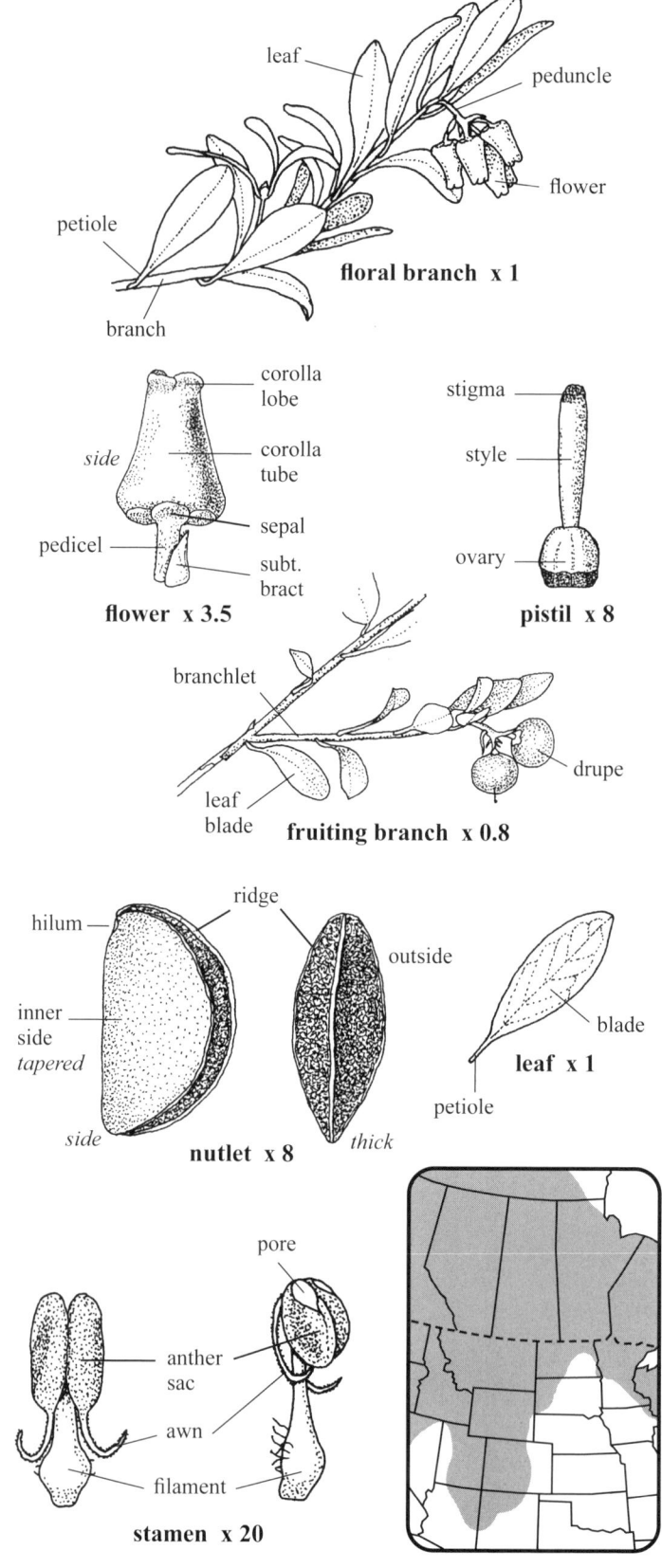

■ **SKETCH** A variable, prostrate **evergreen shrub** forming mats 0.5–2 m wide and often covering several m²; on sandy hillsides, rocky outcrops, river banks, gravelly ridges, shores and open woods. **w2, p1**

■ **FLOWERS** White with pinkish tips, blooming April–July; **inflorescence** a raceme (or panicle) of 3–10 clustered flowers hanging from the more upright tips of leafy branches; **peduncles** deflexed, 3–5 mm long, reddish green, finely hairy; **pedicels** 2–3 mm long, glabrous; **subtending bracts** (of pedicels) pointed, firm, veined outside, green with some reddish spots, 2–6 mm long, margins entire and hairy, persistent; **flowers** perfect, 4–5 mm wide by 4–8 mm long, tapered; **sepals** five, imbricate, 1–1.3 mm long and wide, pinkish, blunt, persistent; **corolla** tubular, (4-) 5-lobed, these blunt, 1.3–1.5 mm wide by *c.* 0.5 mm long, erect to spreading, imbricate, hairy inside, glabrous outside, opaque at the wide base; **stamens** ten (eight), awned, included; **filaments** distinct, bases swollen, tapered above, slightly hairy below, white; **anthers** reddish brown, deflexed when ripe, *c.* 0.8 mm long, each sac opening by an apical pore; **awns** brown, 1–1.3 mm long, curved; **pistil** 3–4 mm long, included, glabrous; **ovary** *c.* 1 mm long and wide; **style** 2–3 mm long by *c.* 0.5 mm thick, pale green; **stigma** darker than style. **FRUIT a drupe**, berrylike, red, slightly shiny, glabrous, 5–8 mm long by 4–10 mm wide, mealy to fleshy, pale yellow inside; **nutlets** four or five per drupe, 1-seeded, united into a central ball 4–5 mm long and wide, each nutlet 3.2–4.2 mm long by 1.6–2.5 mm wide by 1.3–1.8 mm thick, hairless, wedge-shaped, inner side smooth, outside rough with a narrow white ridge and usually two lateral ridges.

■ **LEAVES** Alternate, simple, entire, persistent, evergreen; **blades** leathery, slightly hairy and shiny above, lighter green below, apices blunt, 1–3 cm long by 5–10 mm wide, lateral veins obscure, midrib slightly raised and hairy below; **petioles** 3–5 mm long, ascending, hairs glandular or not; **stipules** absent.

■ **STEM** Flexible, mostly prostrate to decumbent, branched, young leafy branches yellowish green, hairs glandular or not; **bark** gray, peeling, ragged, pale yellow inside stem.

Red Bearberry, Common Bearberry, Kinnikinnick, Hog Cranberry, Bearberry Manzanita

Heath—*Ericaceae* **Shrub** Evergreen White Corolla tubular, 4-lobed

CREEPING SNOWBERRY *Gaultheria hispidula* (L.) Muhl. ex Bigelow

■ **SKETCH** A low **evergreen shrub** with creeping stems 8–40 cm long with dark brown **roots** 5–10 cm long by *c.* 0.6 mm thick; in damp woods, on mossy hummocks in bogs and muskeg. **w2**

■ **FLOWERS** White, blooming May–June; **inflorescence** a raceme; **pedicels** pale green, 1.5–2.5 mm long with ascending red to white hairs (red-tipped); **subtending bracts** (of flowers) two, green, each *c.* 2 mm long and wide, persistent, pointed, hairy; **flowers** perfect, odorless, *c.* 3 mm wide and long, one per leaf axil beneath the leaves, usually a few near the stem's apex; **calyx** tubular, TL 1.8–2 mm, 4-lobed, lobes *c.* 1 mm long and wide, thin, soft, slightly hairy, persistent as green points on the fleshy fruit, body very pale green, whitish along entire margin; **corolla** tubular, 4-lobed, TL *c.* 2 mm, each lobe *c.* 1 mm long, pointed; **stamens** eight, erect, *c.* 1 mm long, included; **filaments** white, wide, flat, *c.* 0.5 mm long from the base of the ovary; **anthers** yellow, paired, 2-lobed, each lobe with a golden soft appendage at apex, these often spreading, with a small apical pore on the pistil side (inside); **pistil** green, glabrous, *c.* 1.5 mm long, included; **ovary** roundish, *c.* 1 mm wide; **style** one, erect, persistent; **stigma** green, obscure. **FRUIT** a capsule, berrylike, enclosed in a fleshy white calyx, 4–7 mm long and wide, mealy, speckled with short appressed reddish hairs; **seeds** *c.* 35 per fruit, 0.8–1 mm long by 0.6–0.7 mm wide by *c.* 0.4 mm thick, white, smooth, glossy, sides slightly concave, corners rounded.

■ **LEAVES** Alternate, entire, simple, evergreen; **blades** 5–10 mm long by 3–6 mm wide, glabrous and shiny above, the midrib recessed, margins slightly revolute with reddish appressed hairs giving a scabrous appearance, pale green and slightly shiny below with several scattered appressed reddish hairs pointing toward the apex, midrib obscure; **petioles** pale green, *c.* 1 mm long with a few scattered reddish hairs; **stipules** absent.

■ **STEM** Prostrate, light green when young, brown when mature, branching, reddish hairs pointing forward giving the stem a rough appearance; 0.7–1 mm thick near the tan, hairy base.

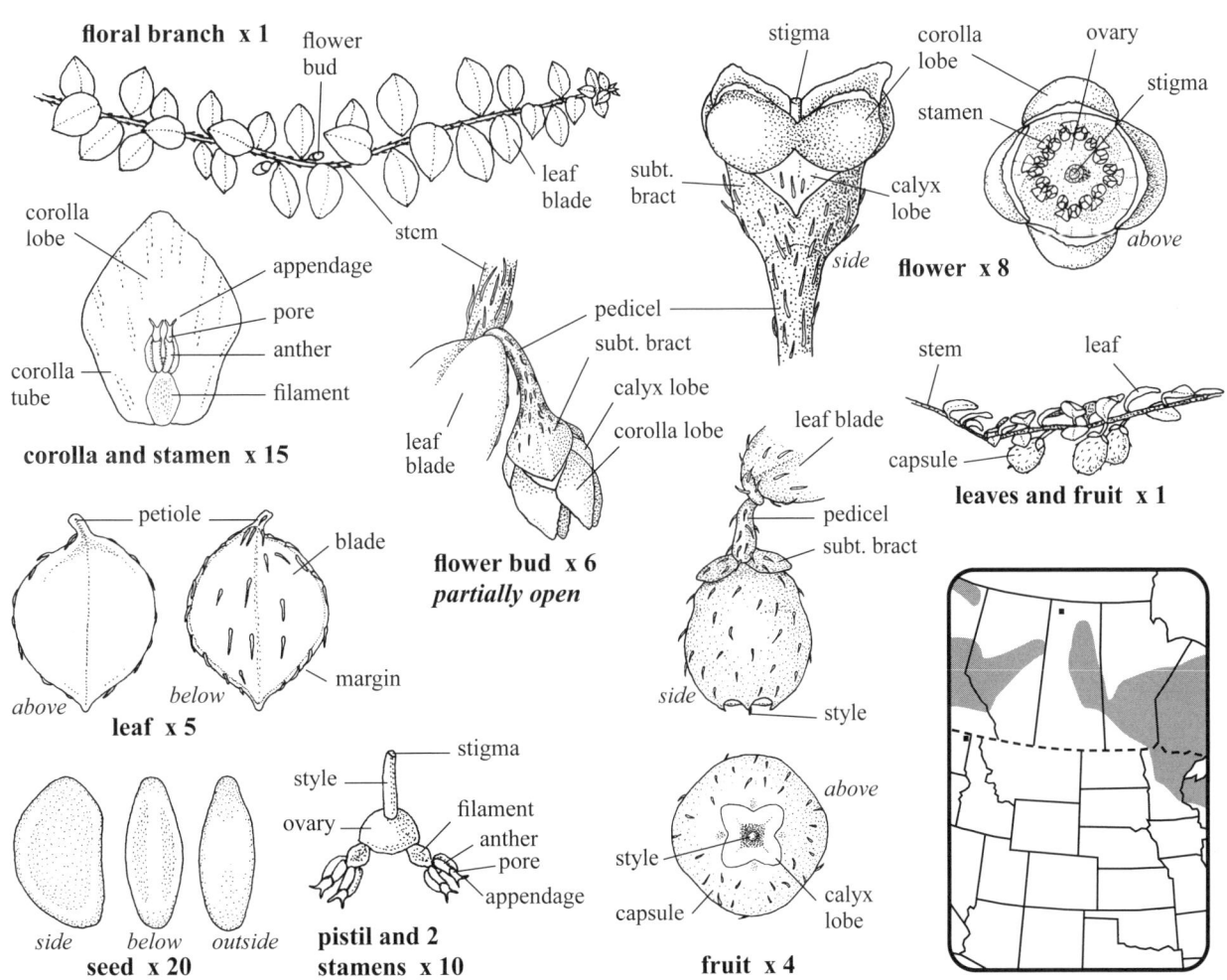

285

Heath—*Ericaceae* **Shrub** Evergreen White Petals 5

LABRADOR-TEA *Ledum groenlandicum* Oeder

■ **SKETCH** A evergreen shrub 15–80 cm tall with reddish brown, woody, **rhizomes** to *c.* 50 cm long by 4–13 mm wide with **fibrous roots**; in bogs.

■ **FLOWERS** White, blooming June–July; **inflorescence** a cyme (umbel-like) 3–3.5 cm long by 2.5–4.5 cm wide, terminal on new stems; **pedicels** 1–2 cm long, arched and ascending with flowers, drooping with fruit, green to brown; **subtending scales** (of pedicels) golden brown, glandular-dotted, finely hairy, 4–7 mm long by 2–3 mm wide, loose; **flowers** perfect, 10–11 mm wide by 6–8 mm tall (including stamens); **calyx** green, tubular, glandular, *c.* 2 mm wide; **lobes** five, blunt, 0.3–0.7 mm long by *c.* 1 mm wide; **petals** five, white, ascending, 5–8 mm long by *c.* 3 mm wide, apices blunt and often hooded, hairy at base on inside only; **stamens** exserted, usually 5–7; **filaments** whitish green, 7–9 mm long, slightly hairy at base; **anthers** white, 2-lobed, *c.* 1 mm long, opening by apical pores; **pistil** one, green, exserted, 6–7 mm long; **ovary** 5-sided, *c.* 2 mm long by 1.5 mm wide, corners blunt, densely glandular; **style** glabrous, 4–5 mm long, tapered, persistent on fruit; **stigma** obscure, rough, no wider than style. **FRUIT** a capsule, reddish brown, hairy, 4–7 mm long (not including persistent style) by 1.8–2.8 mm wide and thick, glandular-dotted, 5-valved, hanging, opening from base; **seeds** numerous, reddish tan, winged, 2–2.2 mm long by *c.* 0.3 mm wide by *c.* 0.15 mm thick, ends tapered, finely striated.

■ **LEAVES** Alternate, simple, entire, spreading to reflexed; **blades** 1.5–7.8 cm long by 7–20 mm wide (flat), green and dull above, turning yellow in autumn, margins strongly revolute, hairs dense and rusty below, midrib raised but often obscure below; **petioles** 3–8 mm long, hairy, recurved; **stipules** absent.

■ **STEM** Erect, reddish brown and glabrous, peeling bark reveals reticulation below; 2–3 mm wide near the bare base; **new twigs** yellowish green and hairy.

■ **SYN.** *Rhododendron groenlandicum* (Oeder) Kron & Judd.

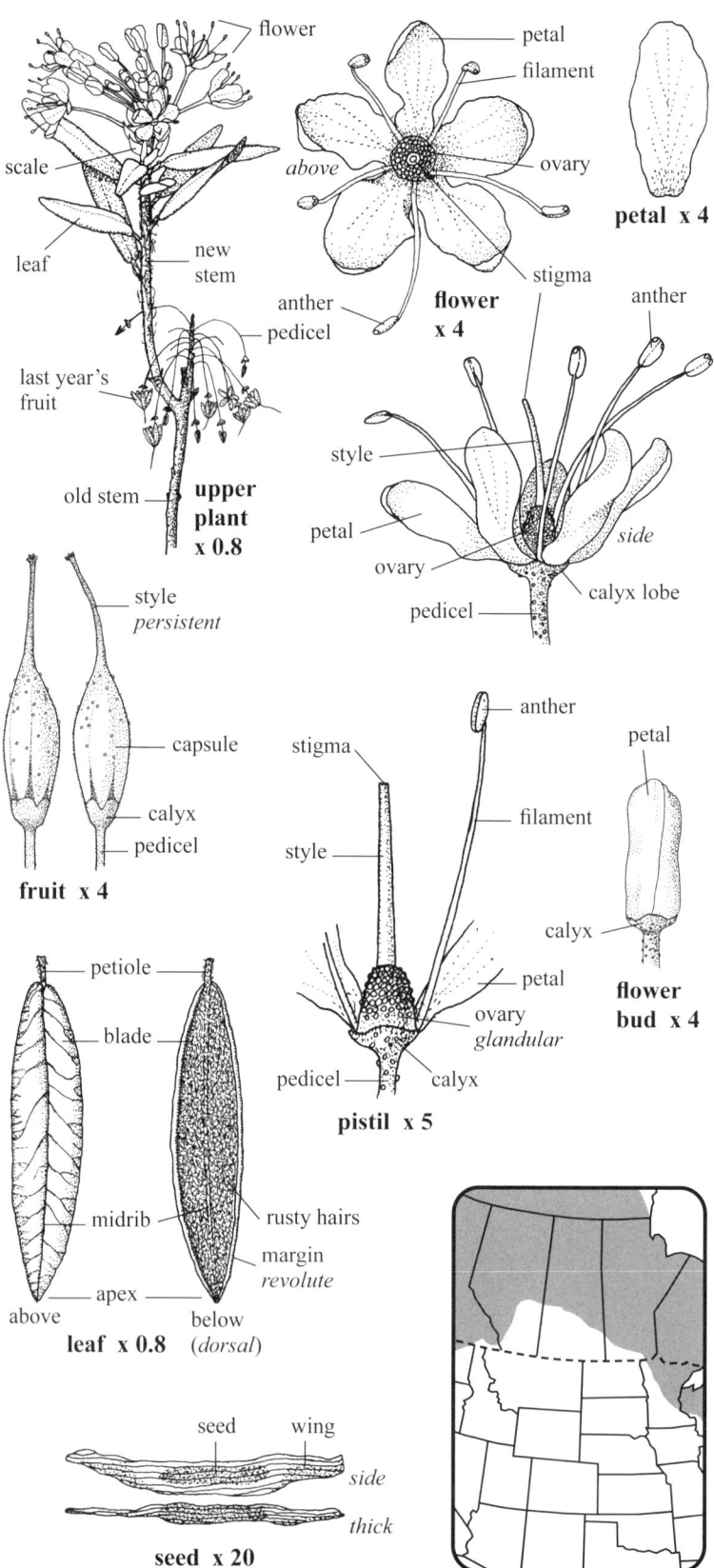

Greenland Labrador-tea, Rusty Labrador-tea, Common Labrador Tea, Bog Labrador Tea

Heath—*Ericaceae* **Shrub** White to pale green Corolla tubular, 5-lobed

BLUEBERRY *Vaccinium angustifolium* Ait.

- **SKETCH** A **deciduous shrub** 10–40 cm tall in large patches from a **woody rhizome** 2–30+ cm long by 2–6 mm thick with fine roots; in sphagnum bogs and on rocky outcrops in the Boreal forest. **w1**
- **FLOWERS** White to pale green, blooming May–June; **inflorescence** a raceme 1–1.5 cm long and wide, of 3–7 flowers, terminal on new leafy branches; **pedicels** green, recurved, 2–4 mm long; **flowers** perfect, 3–4.5 mm wide by 4–6 mm long, drooping, crowded; **subtending bracts** (of flowers) 2.5–3 mm long by *c.* 1 mm wide, glabrous, mostly entire; **hypanthium** a low cup *c.* 2 mm wide, oval; **calyx** green, glabrous, 2–2.8 mm long by 3–4 mm wide, persistent; **calyx** lobes five, green, *c.* 1 mm long by *c.* 1.8 mm wide; **corolla** tubular, 5-lobed, 4–5 mm long, hairless, lobes pointed, recurved, *c.* 2 mm wide; **stamens** 8 or 10, included, *c.* 3.5 mm long; **filaments** pale green, 1.8–2 mm long by *c.* 0.7 mm wide, curved slightly inward, margins hairy; **anthers** two-lobed, golden brown, *c.* 2.2 mm long by *c.* 0.9 mm wide; **tubes** with an apical pore, paired, *c.* 1.5 mm long; **style** one, green, filiform, *c.* 4 mm long, included; **stigma** pale green, about level with the apex of the corolla. **FRUIT a berry**, blue with a glaucous bloom, 4–12 mm wide and long with the dry calyx lobes at the apex, these spreading to erect; **seeds** a few to several per berry, golden brown, slightly shiny, 1.2–1.3 mm long by 0.8–1 mm wide by 0.4–0.7 mm thick, variable in shape, rough with elongated papillae.
- **LEAVES** Alternate, in clusters, simple, pointed, thin, reddish along margins and tips when unfolding, finely toothed, margins hairy to glabrous; **blades** slightly shiny on both sides, lighter green below, slightly hairy on midrib below, 1–4 cm long by 4–15 mm wide, veins obvious, apices with a thick tip *c.* 0.3 mm long; **petioles** 1–2 mm long, pale green, slightly hairy on margins.
- **STEM** Erect, woody, bark peeling; 2–4 mm wide near the base, glabrous to slightly hairy; **new twigs** with low, numerous, pale green glands, hairs in narrow lines with two lines from each alternate bud, the short hairs arched forward, the lines of hairs running the length of the twig.

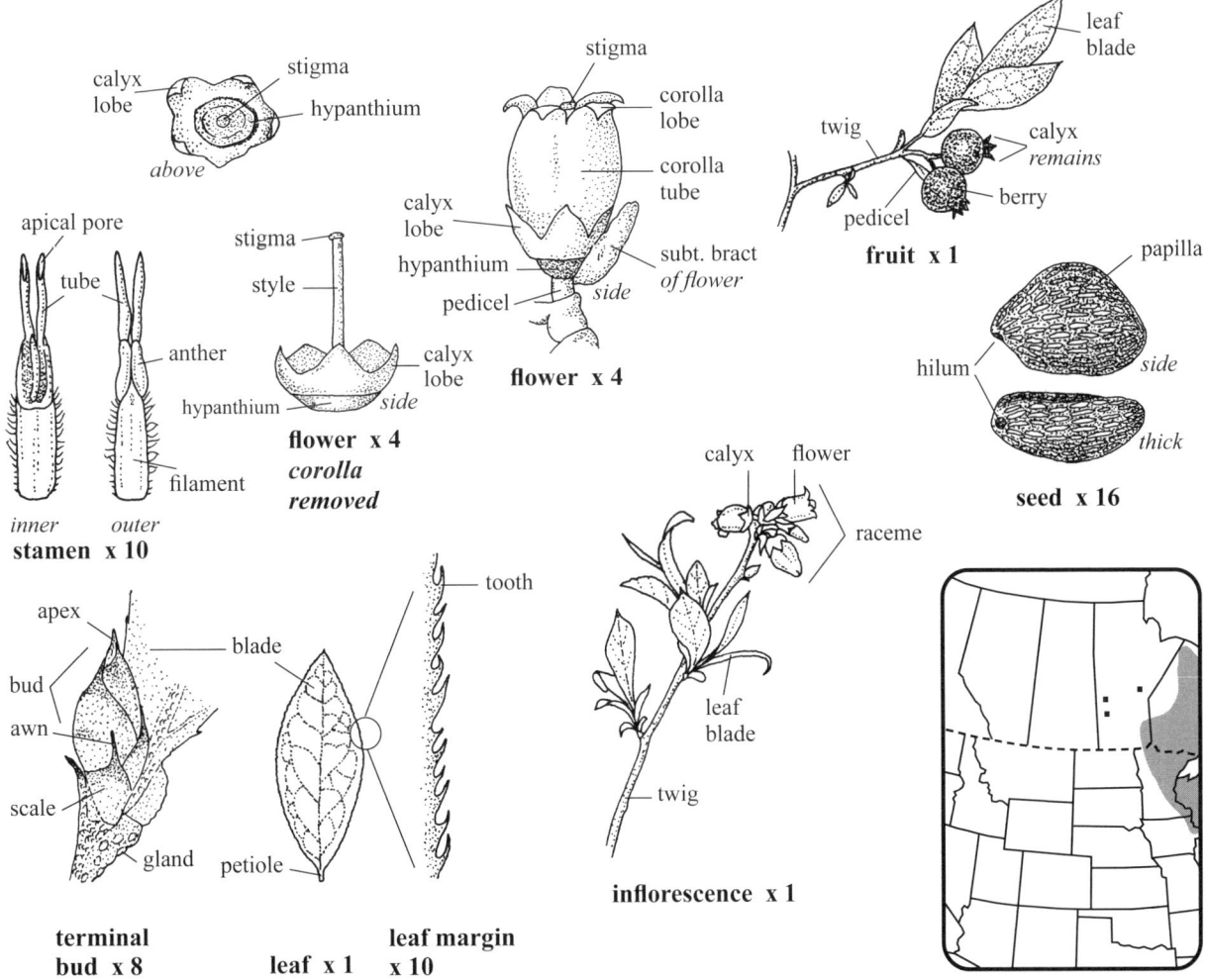

Late Low Blueberry, Low Sweet Blueberry, Lowbush Blueberry

Heath—*Ericaceae* **Shrub** White to greenish pink Corolla tubular, 5-lobed

VELVET-LEAVED BLUEBERRY *Vaccinium myrtilloides* Michx.

■ **SKETCH** A **deciduous shrub** 10–40 cm tall in colonies from dark brown, woody **rhizomes** to 20⁺ cm long by 4–5 mm thick with fine, short **fibrous roots**; in open woods, muskeg, rocky outcrops, in gravelly, coarse to light soils. w1

■ **FLOWERS** White to greenish pink, blooming June–July; **inflorescence** a raceme 8–15 mm long and wide at the ends of short new twigs; **floral branches** ascending, 1–4 cm long, hairy, leafy; **pedicels** hairy, 2–3 mm long, glands green; **subtending bracts** (of pedicels) *c.* 2 mm long by *c.* 1 mm wide, apices reddish and bent; **flowers** perfect, 4–5 mm wide by 6–7 mm long, hanging; **hypanthium** dark green, moist, a several-sided ring atop the ovary and surrounding the base of the style; **calyx tube** 0.5–1 mm long, green, glabrous; **calyx lobes** five, *c.* 2 mm long and wide, pointed, tips often reddish, erect atop the fruit; **corolla tube** 4–5 mm long and wide, glabrous, faintly veined; **corolla lobes** five, 1–1.5 mm long by 2–2.5 mm wide, reflexed; **stamens** 10, included, 4–4.5 mm long; **filaments** pale green, flat, *c.* 3 mm long by *c.* 0.8 mm wide, margins hairy except at base; **anthers** 0.8–1 mm long, 2-lobed, paired, tubes 1–1.2 mm long with an apical pore; **pistil** one, green and glabrous, *c.* 5 mm long, slightly exserted; **ovary** enclosed by calyx; **style** *c.* 5 mm long; **stigma** one, flat, slightly wider than style. **FRUIT** a **berry**, 5–8 mm long by 5–9 mm wide, usually 1–9 in a cluster, dark blue with a bloom; **seeds** golden brown, 2–25 per berry, angular, 1–1.1 mm long by 0.7–1 mm wide by 0.4–0.6 mm thick.

■ **LEAVES** Alternate, simple, entire, reddish green, turning reddish in the autumn; **blades** 1–4 cm long by 5–15 mm wide, apiculate, margins ciliate, some hairs curved, both sides of blades hairy, but less hairy above; **petioles** green, *c.* 1 mm long, hairy; **stipules** absent.

■ **STEM** Erect, branched, solid, hairy, glandular-dotted; the greenish base 1.5–2.5 mm thick; **terminal fall buds** reddish brown, pointed, 2.5–5 mm long by *c.* 2 mm wide and thick, hairless; upper bud scales several and pointed, imbricate, midnerve raised, margins white or not, lowest four bracts with awns 1–1.5 mm long.

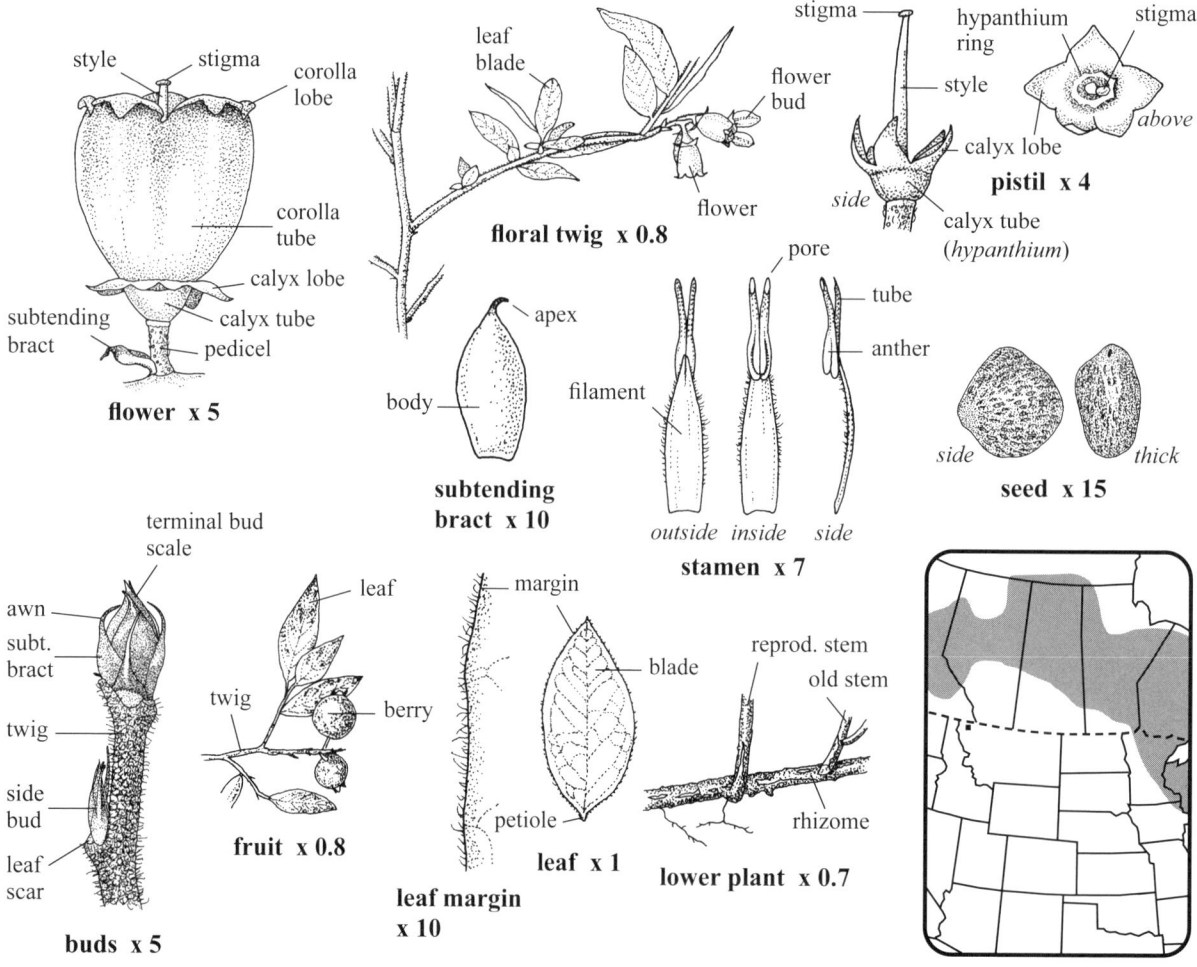

Sour-top Blueberry

Heath—*Ericaceae*   **Shrub** Evergreen White to pink Corolla tubular, 4-lobed

DRY-GROUND CRANBERRY   *Vaccinium vitis-idaea* L.

■ **SKETCH** An evergreen dwarf **shrub** 8–20 cm tall, in colonies with trailing stems or woody **rhizomes** to *c.* 30+ cm long by 0.5–1 mm thick with peeling bark and **fibrous roots**; in mossy areas and swamps, muskeg, bogs, rocky outcrops and sandy woodlots. **w1**

■ **FLOWERS** White to pink, blooming May–July; **inflorescence** a raceme, terminal, 1–2 cm long by 0.7–1 cm wide, of 3–6 flowers at the ends of short new leafy branches; **pedicels** 1–2 mm long, wrinkled, hairs white and curled upward; **subtending bracts** (of pedicels) one; **bracteoles** (on pedicels) two, 2–2.7 mm long by *c.* 2 mm wide, C-shaped, veined, green to red; **flowers** perfect, 4–5 mm wide by 6–8 mm long (including style), nodding to hanging; **hypanthium ring** green to dark red, *c.* 1.5 mm wide by *c.* 1 mm long, 8-lobed, recessed in center; **calyx** tubular, TL *c.* 1.7 mm, 4-lobed, lobes 1–1.3 mm long by 1.5–1.7 mm wide, pale green turning red, red glands along the margins; **corolla** tubular, TL 4–6 mm, 4-lobed, lobes 1.7–2.5 mm wide, apices recurved; **stamens** eight, 3.5–4 mm long, shorter than style; **filaments** *c.* 1.7 mm long, white, hairy on outside and margins, pink base attached to outside of hypanthium ring; **anthers** reddish purple, *c.* 2 mm long (including tubes), turning golden brown after anthesis; **tubes** *c.* 1 mm long with an apical pore; **pistil** one; **style** filiform, green, 5–6 mm long, exserted *c.* 1 mm past top of corolla; **stigma** flat, round, *c.* 0.3 mm wide, slightly wider than style. **FRUIT a berry**, dark red, slightly shiny, 5–10 mm long and wide, hanging; **seeds** tan, 3–22 per berry, each 1.4–1.7 mm long by 0.7–0.9 mm wide by 0.4–0.6 mm thick, clear at both ends, striated lengthwise, with a thin ridge end to end.

■ **LEAVES** Alternate, simple, evergreen, entire, blunt, stiff and shiny above with the midrib obviously recessed; **blades** dark green above (new summer leaves lighter green), margins revolute, 6–21 mm long by 3–10 mm wide, spreading, leathery, notched, scattered white hairs along the upper midrib near the petiole, dull and lighter green below with brown dots (microscopic gland-tipped hairs); **petioles** green, 1–3.5 mm long, arched white hairs ascending; **stipules** absent.

■ **STEM** Erect, branched, solid, reddish brown with white, curled, short hairs; **new twigs** light green with white curled hairs; 1–2 mm thick near the woody base.

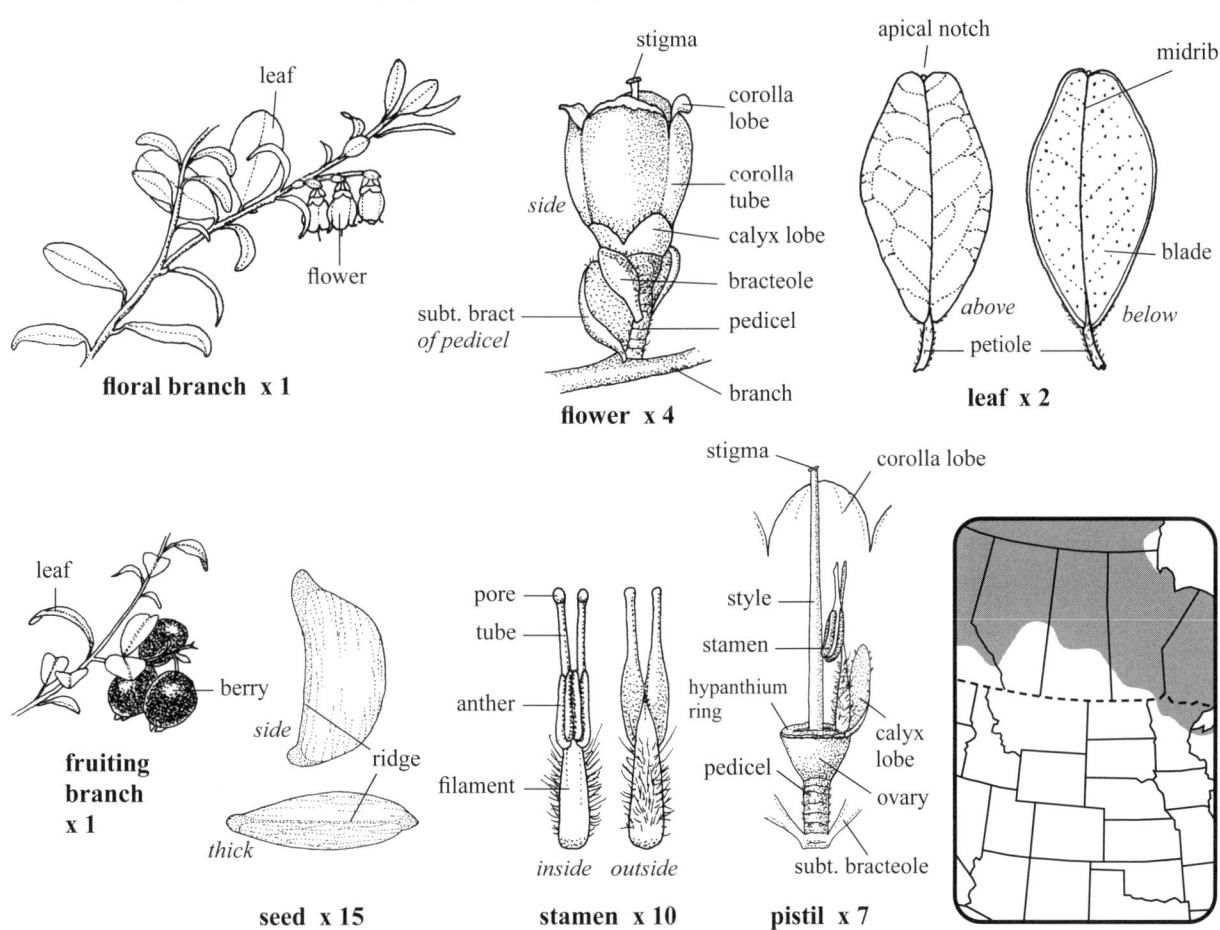

Lingonberry, Northern Mountain-cranberry, Cowberry

Spurge—*Euphorbiaceae* **Wildflower** White and pink Involucre tubular, 4-lobed

RIB-SEED SANDMAT *Chamaesyce glyptosperma* (Engelm.) Small

■ **SKETCH** A glabrous **annual herb**, often prostrate, mats 5–50 cm across with a tan **taproot** 1–2 mm thick by 5–10 cm long; in disturbed, dry soils in yards, along railways, sidewalks and in mountain brush; **juice** milky.

■ **FLOWERS** White and pink, blooming May–October; **inflorescence** of unisexual flowers in a cyathium, each flower solitary; **involucre** tubular, 4-lobed, *c.* 0.5 mm wide by 0.7–1.2 mm long; **flowers** unisexual but appear perfect (due to closeness) in the involucre; **glands** four, pink; **lobes** four, white, petal-like 1- or 2-notched, about as long and wide as the glands, becoming larger as fruit develops; male flowers usually four (1–7) per involucre; **stamens** (each one a flower) clustered and in various stages of maturity, to *c.* 1 mm long and partially exserted; female flowers included, green, glabrous; **pistil** 0.7–1 mm long (including 0.5 mm long stipe), one per involucre; **ovary** enlarges into a capsule which becomes reflexed; **style** 3-parted, each part notched at tip, hyaline. **FRUIT a capsule**, glabrous, 1.2–1.8 mm long by 1.5–1.8 mm wide and thick, sharply triangular, usually visible on underside of branchlets; **stipe** 1.5–2 mm long, reflexed; **seeds** three per capsule, 1–1.3 mm long by *c.* 0.6 mm wide across a side, 4-sided and 4-ridged, sides with four or five shallow transverse grooves, ripe seeds are ejected; **coat** reddish to tan, thin; **septum** with a long narrow opening in the middle of each valve.

■ **LEAVES** Opposite, simple, margins or apices with shallow teeth, dull, flat, lighter green below, midrib dark green, side veins obscure; **blades** 3–15 mm long by 1.5–4 mm wide, base asymmetrical; **petioles** pink to pale green, to *c.* 1 mm long, round, fleshy; **stipules** white to pink, paired, 0.5–2 mm long, entire or with one to four filiform segments erect to curved.

■ **STEM** Prostrate to ascending, tan to pinkish, glabrous, slightly flattened centrally, 0.5–1 mm thick near the base, dividing into several branches, these rebranching alternately; **branchlets** 5–35 mm long with involucre and capsules near their apices among crowded leaves; **lower nodes** bare and slightly swollen.

■ **SYN.** *Euphorbia glyptosperma* Engelm.

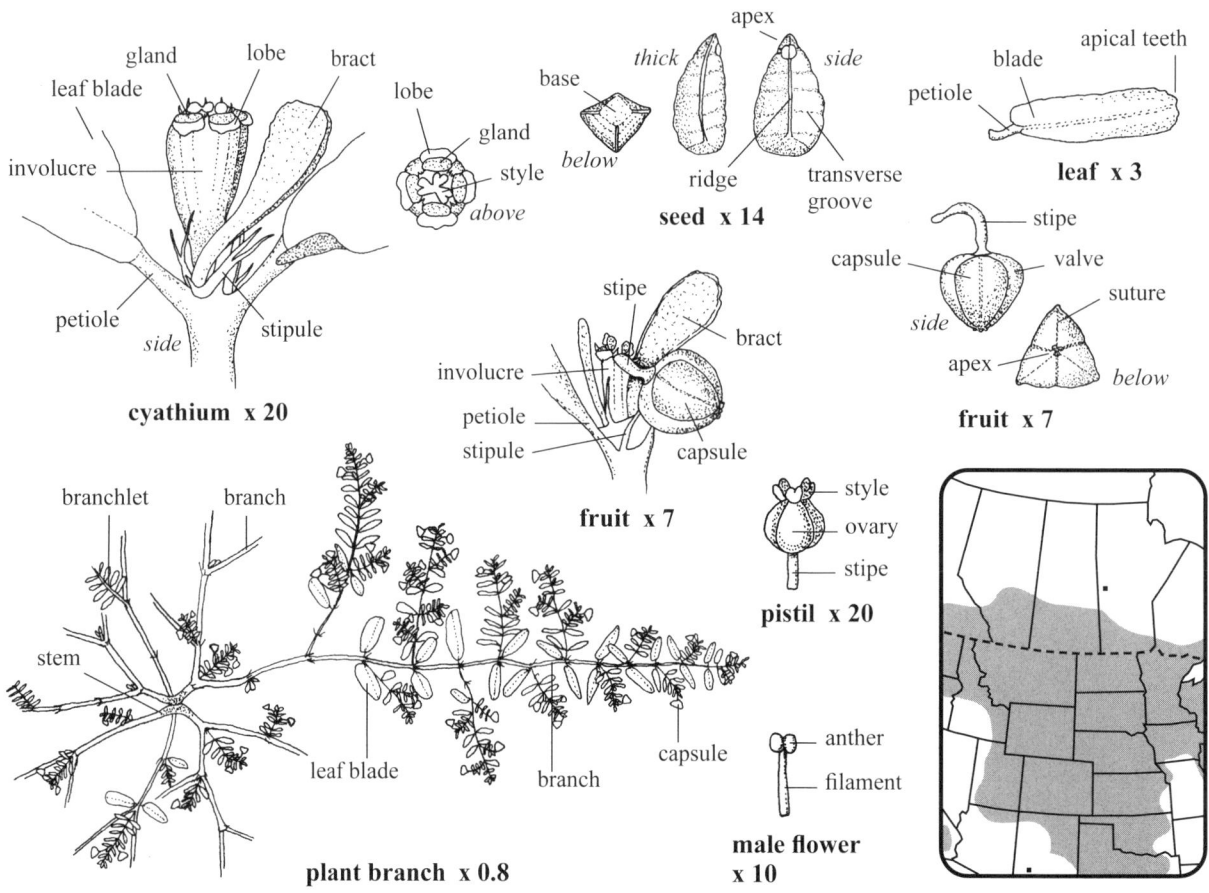

Spurge—*Euphorbiaceae*   **Wildflower**   White and green   Involucre tubular, 4-lobed

THYME-LEAVED SPURGE   *Chamaesyce serpyllifolia* (Pers.) Small

■ **SKETCH** A glabrous **annual herb** 5–50 cm tall or long from a tan **taproot** 0.4–2 mm thick by 4–10 cm long; in disturbed sites, dry prairies, dry sloughs, fields, railways, roadsides and open woods; **juice milky**.

■ **FLOWERS** White and green, blooming June–September; **inflorescence** of unisexual flowers in a cyathium, solitary between two leaves at each upper node; **involucre** tubular, 4-lobed, 1–1.2 mm wide at apex by *c.* 1.5 mm long (including a 0.5 mm long stipe); **glands** four, green; **lobes** four, white, 1- or 2-notched, 0.3–0.6 mm long, narrow at the base where attached to the oval gland; <u>male flowers</u> usually 5–18, mostly included, various sizes due to stage of maturity; **stamens** each a separate male flower; **anthers** green, turning reddish when eventually exserted; <u>female flowers</u> a pistil only; **style** 3-parted, filiform, spreading, each part notched. **FRUIT a capsule**, 3-sided, 1.5–2 mm long by *c.* 2 mm wide, the sides slightly convex; **seeds** three per capsule, dark brown after thin, whitish membrane is scraped away, 1–1.6 mm long by *c.* 0.7 mm wide and thick, 3- or 4-sided, low ridged at corners, sides smooth to slightly wrinkled, expelled several cm when capsule snaps open; **septum** pale green, incomplete in middle and exposing the seed inside.

■ **LEAVES** Opposite, simple, margin entire below but with several tiny teeth across the blunt, often pink-edged apices, lower ones early deciduous; **blades** glabrous, 3–16 mm long by 2–7 mm wide, dull, lighter green below, often with a dark red streak along midrib of upper surface, base usually asymmetrical; **petioles** green to pink, 1–2 mm long, slightly flattened above and round below; **stipules** pointed, mostly ascending, *c.* 2 mm long, entire to separated into two or more tapered divisions, one division usually longer.

■ **STEM** Ascending to prostrate, reddish at least near the base, usually branched close to the base; **branches** alternate, slightly flattened, oval, with two lateral, longitudinal ridges in line with the petioles at each node.

■ **SYN.** *Euphorbia serpyllifolia* Pers. = *Chamaesyce serpyllifolia* ssp. *serpyllifolia* (Pers.) Small.

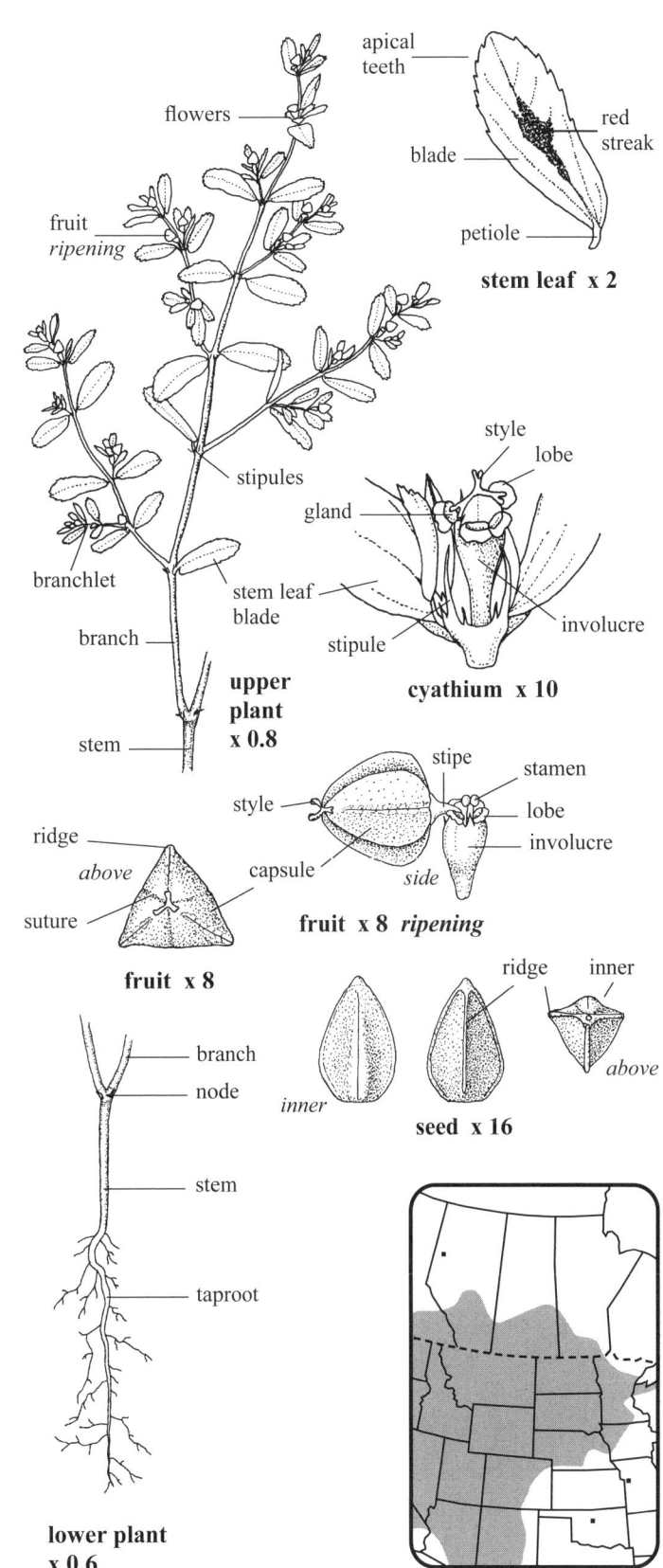

Thyme-leaved Sand-mat

Spurge—*Euphorbiaceae* **Wildflower** Green Involucre 4-lobed

LEAFY SPURGE   *Euphorbia esula* L.

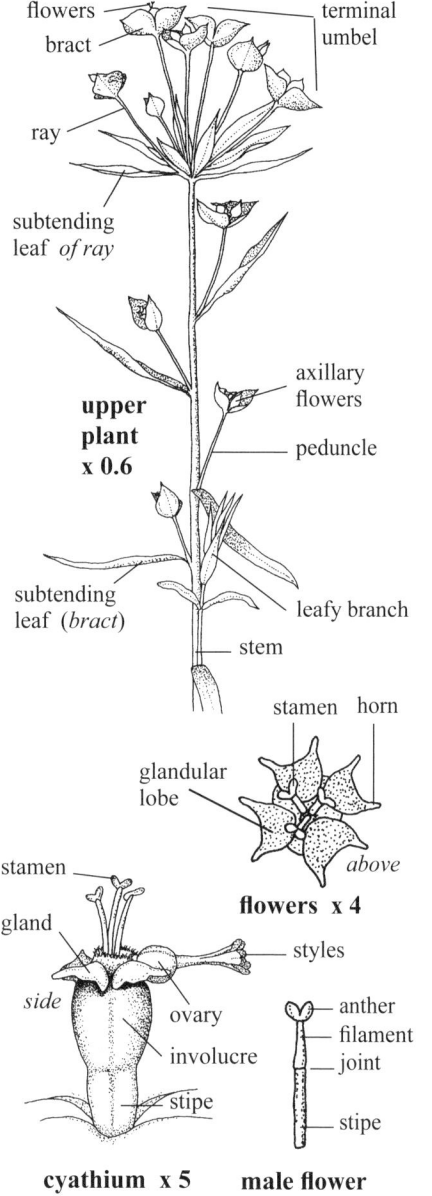

■ **SKETCH** A glabrous variable **perennial herb** 30–90 cm tall in open colonies from a branching woody **rhizome** 1–5 mm thick and a stout **caudex** 6–8 mm thick and deep **roots**; in disturbed open fields, along roadsides and stream valleys; **monoecious**; **juice milky**. *Naturalized*

■ **FLOWERS** Green, blooming May–October; **inflorescence** an umbel, TL 5–30 cm by 4–17 cm wide, terminal and axillary; **floral branches** to *c*. 15 cm long from upper leaf axils; **peduncles** from upper leaf axils 3–8 cm long; **rays** of terminal umbel 5–17, ascending, 2–6 cm long; **bracts** (atop rays) two, green with a thin yellowish green margin, apices pointed, together 14–21 mm long by 8–17 mm wide, forming a cuplike structure in which the flowers and more bracts with flowers develop; **flowers** unisexual, (appearing perfect with male and female parts exserted), together 4–5 mm wide by 5–6 mm long; **stipe** 1.5–2 mm long and wide, flattened; **involucre** tubular, 4-lobed, 2–3 mm wide and long; **glands** four, green, a fleshy body *c*. 2.8 mm long by *c*. 1.5 mm wide with two horns; **male flowers** clustered, 10–25 per involucre, each flower 1.5–4 mm long; **stamens** green; **filaments** erect, with a joint between the stipe and filament; **anthers** *c*. 0.5 mm long by *c*. 0.8 mm wide; **female flowers** one, horizontal, leaning out the top of the involucre; **ovary** exserted, 3-lobed, *c*. 1.2 mm long and wide, roundly triangular; **styles** three, 2–3 mm long, each 2-lobed. **FRUIT** a **capsule**, 2.5–3.5 mm long by 3–4 mm wide, 3-lobed, tan, exserted from involucre, snaps open expelling the seeds several cm; **stipe** *c*. 2 mm long; **seeds** three per capsule, light gray to brownish, spotted, glabrous, 2–3 mm long by 1.2–1.9 mm wide by 1.5–2 mm thick; **caruncle** pale yellow and *c*. 0.8 mm long and wide, rough.

■ **LEAVES** Alternate, simple, entire, sessile to short petiolate, linear; **blades** flat, 1–10 cm long by 3–17 mm wide, tapered at both ends, largest about midstem, apices rounded to pointed, midrib obvious, side veins faint; **subtending leaves** (of rays) at terminal umbels 5–17, each 1–3.5 cm long by 3–8 mm wide, spreading to erect.

■ **STEM** Tough, erect, light yellowish green, one to three from the caudex, glabrous, often branched above, ridged; 2–5 mm wide near the base.

■ **SYN.** *Euphorbia virgata* Waldst. & Kit., non Desf. = *Euphorbia esula* var. *uralensis* (Fisch. ex Link) Dorn.

■ **NOTE** The species *E.* x *pseudovirgata* (a presumed hybrid of *Ee* with *E. virgata*) has pointed leaves and is common (Great Plains Flora Association, 1986, p. 549).

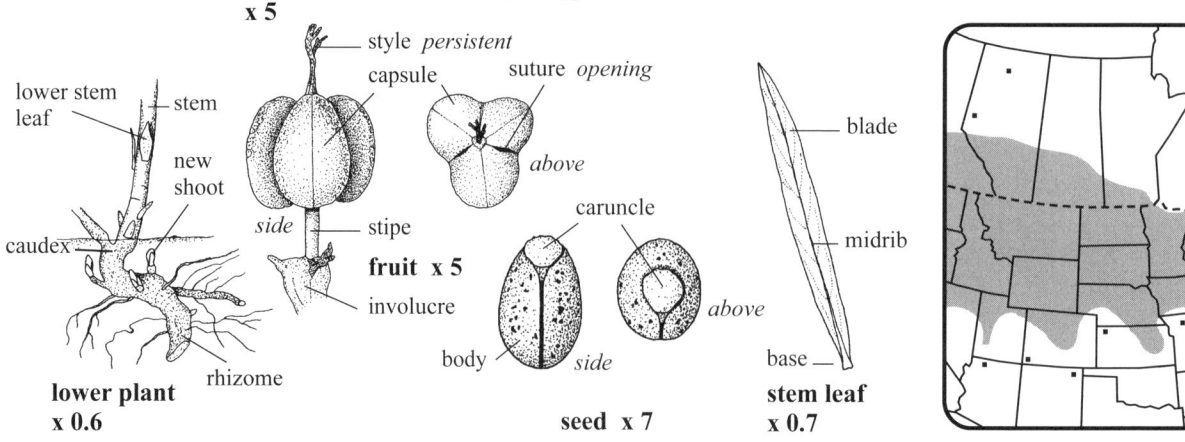

292   Wolf's-milk

Wild Vine

The vine that keeps on winding

and winding

up the old post of a fence

like the one that turned

the curling stem

and grew the flower

the thing the mind knew once

on the edge of the wild

the vines entangled

where those twisting tendrils

join the earth to the sky

Family Characteristics

Pea—Fabaceae — Flower parts 5

SKETCH Dicotyledonous herbs (wildflowers) and shrubs, annual or perennial. **Roots** often with nitrogen-fixing bacteria in small nodules. About 18,000 species worldwide.

FLOWERS Usually perfect, often blue to pale purple, pinkish to yellow, regular to irregular, mostly 5-merous. **Inflorescence** a spike or raceme. **Calyx** mostly tubular, usually 5-toothed or 5-lobed, these equal or unequal in length. **Petals** 5 (rarely 1); the banner (1 upper petal), wings (2 side petals), keel (2 lower petals) enclosing the stamens and pistil; wings and keels often with a narrow claw. **Stamens** mostly 10, diadelphous (9+1). **Filaments** often united into a tube that is open along one side, one filament mostly free (only its base united to the other nine). **Pistil** 1, simple, 1-loculed with one to several ovules. **Style** 1. **Stigma** 1.

FRUIT A **legume** (pod), opening along 2 sutures, 1- to several-seeded, or a **loment** (each segment 1-seeded).

LEAVES Mostly alternate and compound. **Leaflets** often numerous, entire, the terminal one sometimes replaced with a tendril. **Stipules** present or absent.

STEM Erect to prostrate, simple to branched.

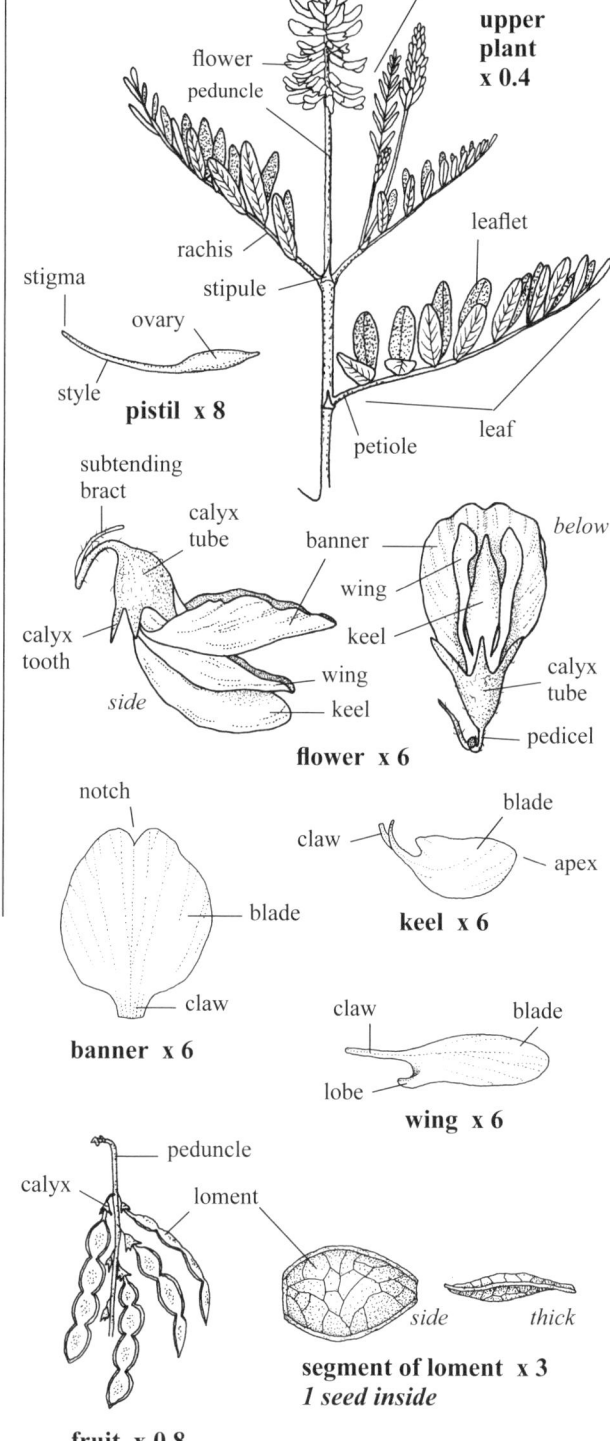

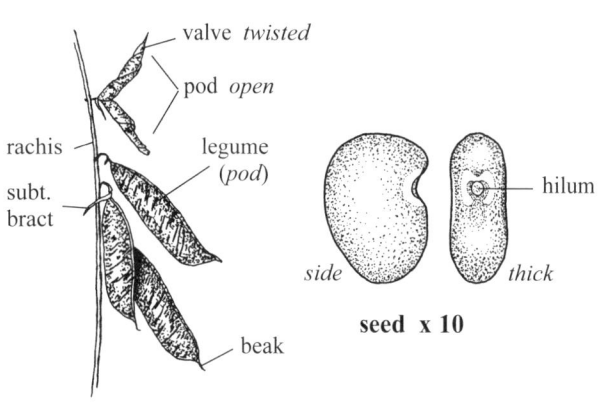

Pea—*Fabaceae* **Shrub** Purple Banner only

Dwarf False Indigo *Amorpha nana* Nutt.

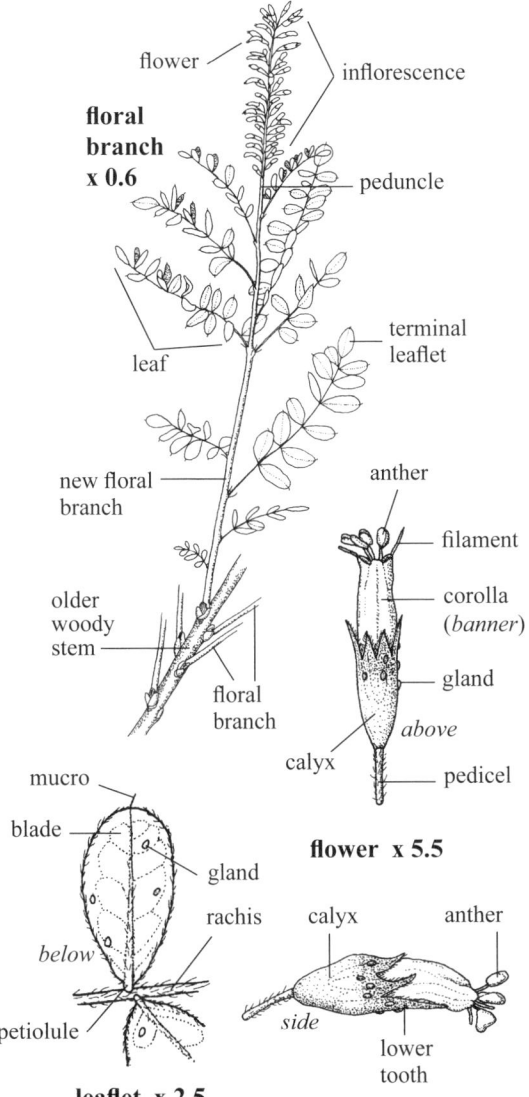

■ **SKETCH** A glandular-punctate, mostly hairless **deciduous shrub** 20–100 cm tall, some from **rhizomes**, singly or clustered, scattered; in dry prairies, hillsides and buttes.

■ **FLOWERS** Purple, blooming May–July; **inflorescence** a raceme, usually solitary on new branches, each 2.5–9 cm long by 10–15 mm wide, erect; **floral branches** glabrous to slightly hairy, to *c.* 13 cm long; **peduncles** 5–15 mm long, axillary; **pedicels** reddish purple, 1.5–2.5 mm long, hairs mostly ascending; **subtending bracts** (of pedicels) 3–4 mm long by *c.* 0.3 mm wide; **flowers** perfect, fragrant, spreading, *c.* 2 mm wide by 5–6 mm long by *c.* 1 mm thick; **calyx tube** *c.* 2 mm long by *c.* 1 mm wide by *c.* 1.5 mm thick, reddish purple and glandular-punctate in upper half, persistent; **calyx teeth** five, 0.8–2 mm long, the lower one the longest, tooth margins and inside hairy; **corolla** (banner only), purple, glabrous, 3–6 mm long, open along the bottom, claw *c.* 0.8 mm long; **stamens** 10, exserted, *c.* 4 mm long; **filaments** purple pink, glabrous, united below; **anthers** reddish purple, 0.5–0.7 mm long; **pistil** 4–6 mm long; **ovary** glabrous; **style** one, purplish pink, 3.5–5 mm long, thin, hairy; **stigma** filiform, hairless. **FRUIT a legume** (pod), 4.5–5.5 mm long by 2.2–3 mm wide by *c.* 1.7 mm thick, erect, glandular-punctate, beaked; **seeds** one per pod, olive brown, 2.5–3 mm long by 1.2–1.5 mm wide by *c.* 0.8 mm thick.

■ **LEAVES** Alternate, compound (odd-pinnate), spreading to ascending, 4–11 on new floral branches; **blades** 2.5–10 cm long by 5–50 mm wide; **leaflets** 4–20 pairs, plus one, per leaf, 3–18 mm long by 1.3–8 mm wide, entire, flat, slightly shiny above, stiff, glandular-punctate with hairs white below, margins hairy, apical mucro 0.5–1.5 mm long; **petioles** 3–15 mm long; **stipules** filiform, hairy, early deciduous, 3–5 mm long and slightly enlarged at their apices; **lateral petiolules** 0.3–1.5 mm long, hairy above; **rachis** slightly hairy, 0.3–0.7 mm wide.

■ **STEM** Woody, gray, ascending to erect, rough from previous annual floral branch dieback; 2–5 mm thick near the base.

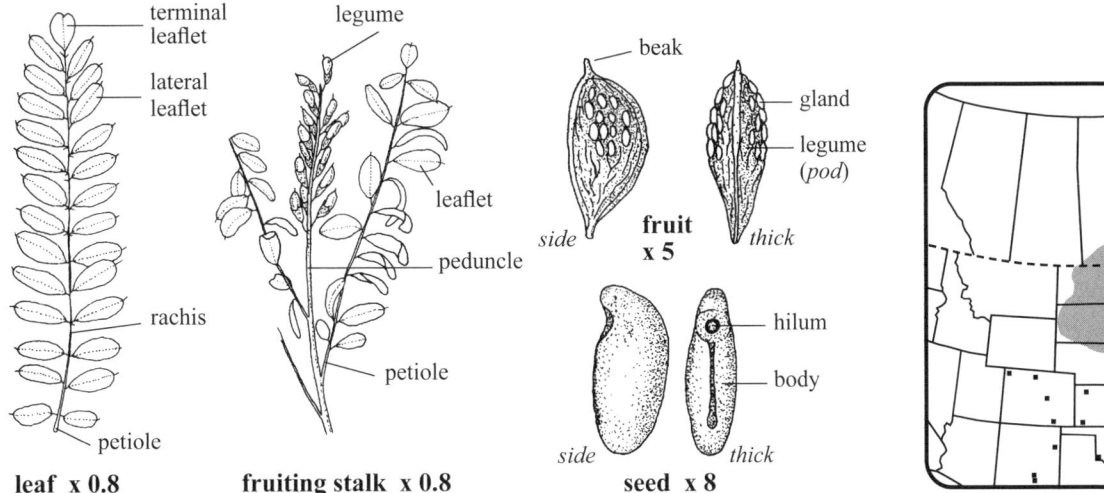

Fragrant False Indigo, Dwarf Wild Indigo, Dwarf Indigo-bush Amorpha

Pea—*Fabaceae* **Wildflower** Pale purple to white Petals 5

Hog-peanut *Amphicarpaea bracteata* (L.) Fern.

■ **SKETCH** A variable, twining **annual herb** from 30–200 cm long with a thin, brown **taproot** and small nitrogen-fixing, tan **nodules**; in moist riverbottoms, woodlands, along roadsides, in dry to moist prairie thickets, solitary or with several plants tangled together. **w2**

■ **FLOWERS** Pale purple to white, blooming July–October; **inflorescence** a raceme of 2–20 flowers; **pedicels** hairy, 2–6 mm long; **subtending bracts** (of pedicels) paired, 2–4 mm long; **flowers** perfect, often in pairs, of three types: **1)** chasmogamous aerial flowers are *c.* 5 mm wide by 5–16 mm long by 5–6 mm tall; **calyx** pale yellowish green, tubular, 6–9 mm long (including teeth) by *c.* 3 mm tall by 2–2.5 mm wide; **calyx teeth** five, 1–2.5 mm long, hairy, the upper two fused; **petals** five; **banner** (one petal) pale purplish, 12–14 mm long by 5–6 mm wide, glabrous; **wings** (two petals) white, *c.* 11 mm long, blade *c.* 5 mm long by *c.* 2 mm wide, claw 6–7 mm long; **keels** (two petals) white, apex with a pale orange spot on both sides, united near the tip, blade 4–4.5 mm long by *c.* 2.8 mm wide, claw *c.* 6 mm long; **stamens** 10, diadelphous (9 + 1), included; **filaments** white, glabrous; **anthers** yellow; **pistil** 8–9 mm long; **ovary** laterally flattened, 3–4 mm long by *c.* 1 mm wide, hairy; **style** filiform, 4–5 mm long, hairy near the base, glabrous above; **stigma** slightly wider than the style; **2 & 3)** cleistogamous flowers near the stem base or below ground; **petals** poorly developed. **FRUIT a legume** (pod), three types: **1)** above ground pods are 1.5–4 cm long by 6–8 mm wide by 3–4 mm thick, tan, hairy; **beak** 0–2 mm long; **seeds** two or three per pod, smooth, grayish brown with black markings, 3.5–6 mm long by 2.7–4 mm wide by 2–3 mm thick; **2 & 3)** underground pods fleshy, 1-seeded, tan, 6–12 mm long and wide, indehiscent, ripe in autumn.

■ **LEAVES** Mostly alternate, compound; **leaflets** three, entire, lighter green below, 2–10 cm long by 2–7 cm wide, glabrous to hairy, especially below; **petioles** 2–10 cm long, hairy; **stipules** two, 2–8 mm long by 2–3 mm wide, striate, partially united about half their length, white hairs at the base; **lateral petiolules** hairy, 1–2.5 mm long; **terminal petiolules** 5–40 mm long, hairy; **bracteoles** (at base of petiolules) paired, *c.* 1 mm long.

■ **STEM** Twining, branched, glabrous or with descending to spreading hairs 1–1.5 mm long, *c.* 1 mm thick near the base; **tendrils** absent.

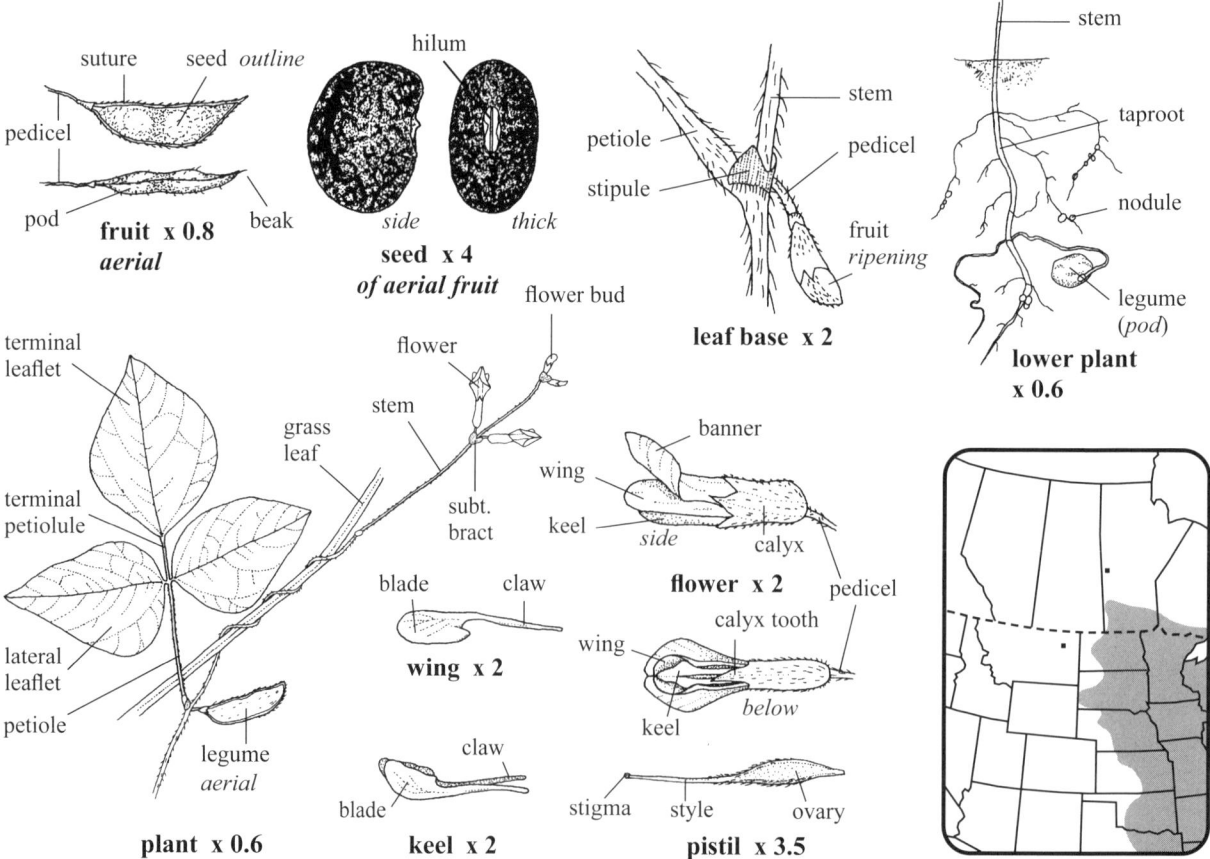

296 American Hog-peanut

Pea—*Fabaceae* **Wildflower** Purple to blue, rarely white Petals 5

PURPLE MILK-VETCH *Astragalus agrestis* Dougl. ex G. Don

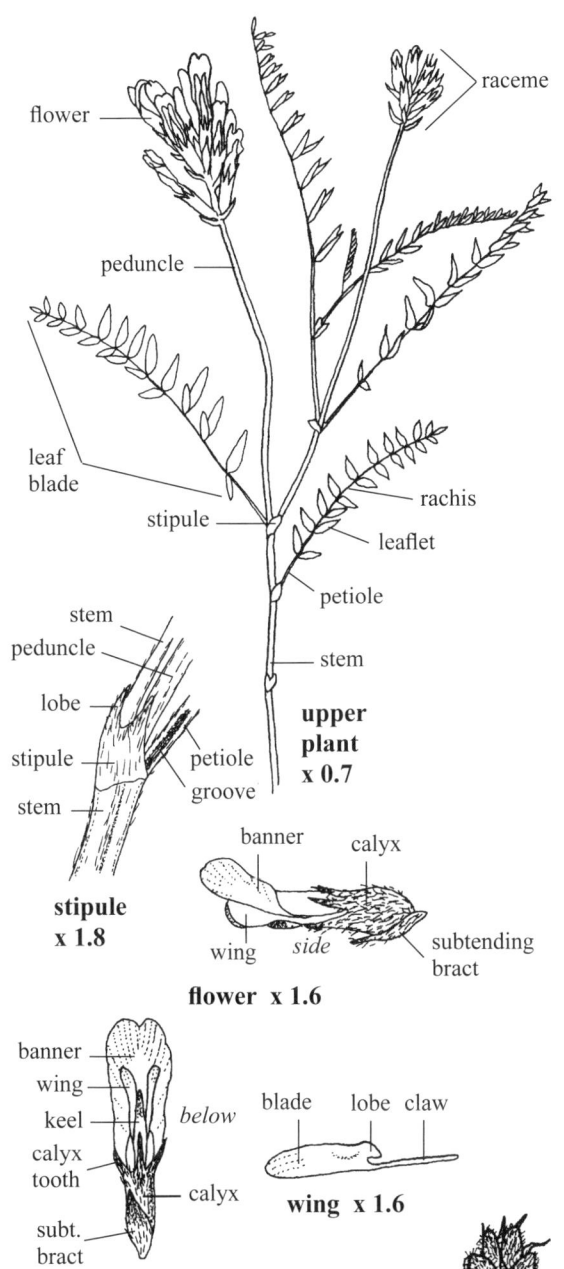

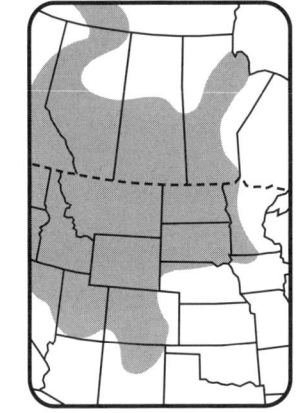

■ **SKETCH** A loosely tufted, variable **perennial herb** 5–35 cm tall with a **taproot** 2–10 mm thick and to 8+ cm long and a branching **caudex**; in moist prairies, roadsides, open sagebrush, open brushy hillsides and moist mountain meadows. **p2**

■ **FLOWERS** Purple to blue, rarely white, blooming April–August; **inflorescence** a raceme, 1–10 per plant, each with 5–15 ascending flowers in clusters 1.5–5 cm long by 1.5–3 cm wide; **peduncles** 2–18 cm long, ridged, hairy; **pedicels** 1–1.5 mm long, to *c.* 2 mm long in fruit, hairy; **subtending bracts** (of pedicels) one, 4–7 mm long, sparsely hairy below; **flowers** perfect, usually 5–8 mm wide by 14–20 mm long; **calyx** tubular, 5-toothed, tube 5–8 mm long by *c.* 4 mm wide by 2.5–3 mm thick with black or white basifixed hairs; **calyx teeth** five, subequal, 3–5 mm long, hairy; **petals** five; **banner** (one petal) 16–22 mm long by 5–8 mm wide, notched; **wings** (two petals) 15–18 mm long by 2–2.5 mm wide, lobed, claw white, 6–8 mm long, distinct; **keels** (two petals) united above, 11–14 mm long (including claws), claws 6–8 mm long; **stamens** 10, diadelphous (9 + 1), included; **anthers** yellow; **fruiting heads** usually of 7–12 pods, hairy, 1–2.5 cm long by 1–1.4 cm wide. **FRUIT a legume** (pod), 8–12 mm long by 3–5 mm wide by 2–3 mm thick, opening from apex, ascending, brown, very hairy with white hairs; **style** persistent and 1–2 mm long; **seeds** tan to black, 10–15 per pod, smooth, 1.5–2 mm long by 1.3–1.6 mm wide by *c.* 0.8 mm thick.

■ **LEAVES** Alternate, compound (odd-pinnate); **blades** 3–11 cm long by 10–36 mm wide; **leaflets** 11–25, entire, each 4–20 mm long by 1–9 mm wide, sessile, opposite, darker green above, smaller towards the leaf tip, very hairy, more so below with midrib obvious; **petioles** 8–33 mm long; **stipules** 2–10 mm long, 2-lobed, hairy especially on outside and margins; **petiolules** *c.* 0.4 mm long, hairy, glandular at base and in the rachis groove; **rachis** grooved above, hairs white.

■ **STEM** Erect to decumbent, weak, grooved, glabrous or hairs appressed and basifixed; reddish near the 1–2 mm thick base.

■ **SYN.** *Astragalus danicus* var. *dasyglottis* (Fisch. ex DC.) Boivin, *Astragalus dasyglottis* Fisch. ex DC., *Astragalus goniatus* Nutt. and *Astragalus hypoglottis* Hook.

Cock's-head, Field Milk-vetch 297

Pea—*Fabaceae* **Wildflower** Pinkish purple to white Petals 5

Two-grooved Milk-vetch *Astragalus bisulcatus* (Hook.) Gray

■ **SKETCH** A tufted **perennial herb** 20–100 cm tall with a **taproot** and branching **caudex**; along river banks, in foothills, sagebrush, badlands, prairies, along roadsides and canyon floors. **p2**

■ **FLOWERS** Pinkish purple to white, blooming April–August; **inflorescence** a raceme, 5–18 cm long by 2.5–3 cm wide, 25- to 75-flowered; **peduncles** erect, axillary, 3–14 cm long, with 10–12 longitudinal low ridges, hairs basifixed; **pedicels** 1–3 mm long, hairy, reflexed in flower and fruit; **subtending bracts** (of pedicels) one, 2.8–4 mm long, margins hairy; **flowers** perfect, drooping, fragrant, 5–9 mm wide by 12–18 mm long; **rachis** ridged, hairy; **calyx** 5-veined, TL 4.5–9.5 mm by 2–2.5 mm wide by *c.* 2.5 mm tall, pinkish, hairs white or black, teeth five, these 1–3.5 mm long, narrow; **petals** five; **banner** (one petal) TL 10–17 mm by 4–9 mm wide, glabrous, patterned, claw tapered; **wings** (two petals) distinct, TL 8–15 mm, blade 2–3 mm wide, spreading, claw 3–5 mm long, white; **keels** (two petals) TL 7–14 mm, blade 2.7–3.2 mm wide (unflattened), united below, open above, claw 3.8–5 mm long, united except for base; **stamens** 10, diadelphous (9 + 1), 11–13 mm long; **filaments** white, glabrous, mostly united into a split tube; **anthers** orange, 0.5–0.8 mm long; **pistil** glabrous, 11–12 mm long; **stipe** *c.* 3 mm long; **ovary** green, 3–4 mm long by *c.* 0.8 mm wide; **style** *c.* 5 mm long, reddish near the apex; **stigma** obscure, flat. **FRUIT a legume** (pod), glabrous to slightly hairy, 7–22 mm long by 3–4 mm wide by 2–2.5 mm thick, widely 2-grooved, hanging, 2-valved, beak (persistent style) 2–5 mm long; **seeds** mostly two to ten, each 2.5–3.5 mm long by 2–2.4 mm wide by *c.* 1 mm thick, glabrous, slightly shiny, brown.

■ **LEAVES** Alternate, compound (odd-pinnate), TL 4–12 cm by 20–60 mm wide; **leaflets** 13–35, each 5–35 mm long by 2–12 mm wide, entire, dull, hairy to glabrous, margins hairy, apices slightly notched to mucronate; **petioles** 0–15 mm long; **stipules** erect to spreading, united, 3–12 mm long by 3–5 mm wide, pale green, hairy; **petiolules** 0.5–1.5 mm long, hairy; **rachis** hairy.

■ **STEM** Ascending to erect, several[+], reddish green, hollow, glabrous to slightly hairy, with low reddish ridges; 3–8 mm thick near the base.

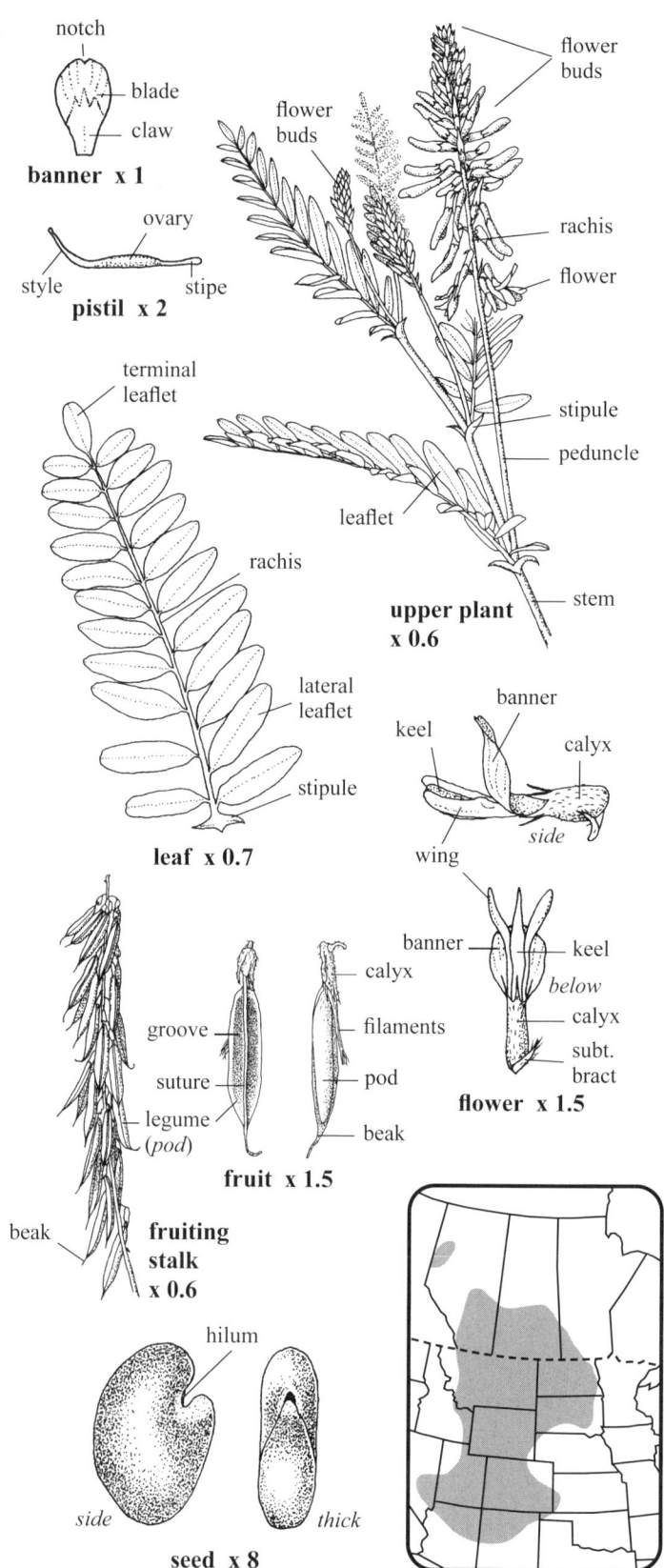

Pea—*Fabaceae* **Wildflower** Greenish white to yellowish Petals 5

Canadian Milk-vetch *Astragalus canadensis* L.

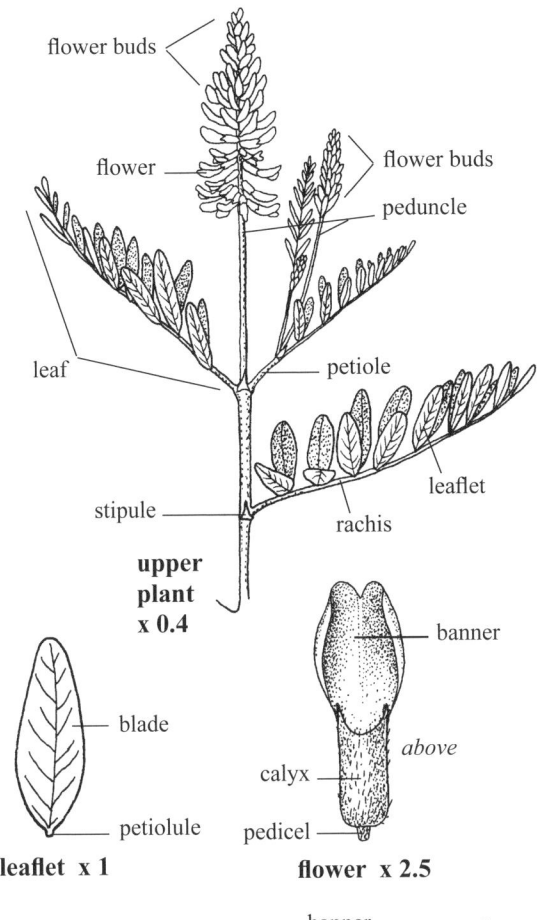

leaflet x 1

flower x 2.5

keel x 2

fruit cluster x 0.8 *winter*

seed x 12

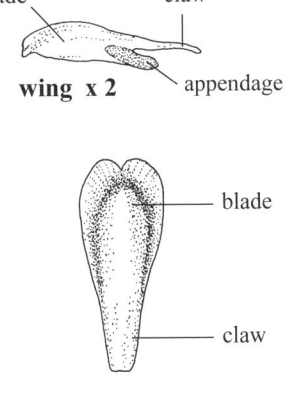

wing x 2

banner x 2

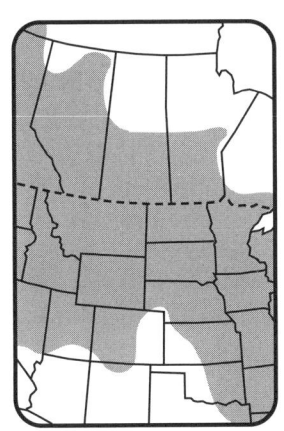

■ **SKETCH** A variable **perennial herb** 20–100⁺ cm tall from **rhizomes** 4–6 mm wide; in moist prairies, along river and lake shores, ditches, open wooded hills and ravines. **p1**

■ **FLOWERS** Greenish white to yellowish, blooming April–September; **inflorescence** a raceme (spikelike), dense, 5–18 cm long by 3–5 cm wide, blooming sequence bottom to top; **peduncle** erect, straight, usually one to six per plant, each 4–22 cm long from leaf axils; **pedicels** 1–3 mm long; **subtending bracts** (of flowers) one, 4–9 mm long by *c.* 0.5 mm wide, pointed, very hairy on the outer surface, V-shaped, reflexed in fruit; **flowers** perfect, spreading to declining, 5–8 mm wide by 12–15 mm long by *c.* 10 mm high; **calyx** tubular, glabrous to hairy, hairs black or white, appressed, the tube 4–9 mm long; **calyx teeth** five, 1.5–3 mm long; **petals** five, glabrous, clawed; **banner** (one petal) 12–17 mm long by 4.5–8 mm wide; **wings** (two petals) 10–15 mm long, with an appendage; **keels** (two petals) 9–13 mm long, partly united; **stamens** 10, diadelphous (9 + 1), included in the keel; **fruit clusters** erect, 4–13.5 cm long by 1.8–3.5 cm wide, with 25–125 pods per cluster. **FRUIT** a legume (pod), glabrous or rarely hairy, 10–20 mm long by 3–5 mm wide, clustered, erect, splitting lengthways from apex, dark brown, mottled, beak 2–5 mm long; **seeds** *c.* 20, each 1.7–2.3 mm long by 1.5–1.8 mm wide by 0.8–1 mm thick, tan to grayish brown.

■ **LEAVES** Alternate, compound (odd-pinnate); **blades** 5–32 cm long by 2.5–7.3 cm wide (flattened); **leaflets** generally ascending, 9–35, mostly 1–4.7 cm long by 6–18 mm wide, tips blunt, smaller towards the leaf tip, entire, glabrous to hairy above with hairs attached at the base or near the middle; **petioles** 10–52 mm long; **petiolules** *c.* 1 mm long; **stipules** pointed, 3–18 mm long and as wide as the stem, greenish tan, entire, erect, the lower ones united on the side opposite the leaf.

■ **STEM** Erect, solid, persistent for about a year, single to a few, reddish with a woody feel, glabrous to lightly hairy; 2–8 mm thick near the base.

Milk-vetch 299

Pea—*Fabaceae* **Wildflower** Purple to pink, geen, blue or white Petals 5

GROUND-PLUM *Astragalus crassicarpus* Nutt.

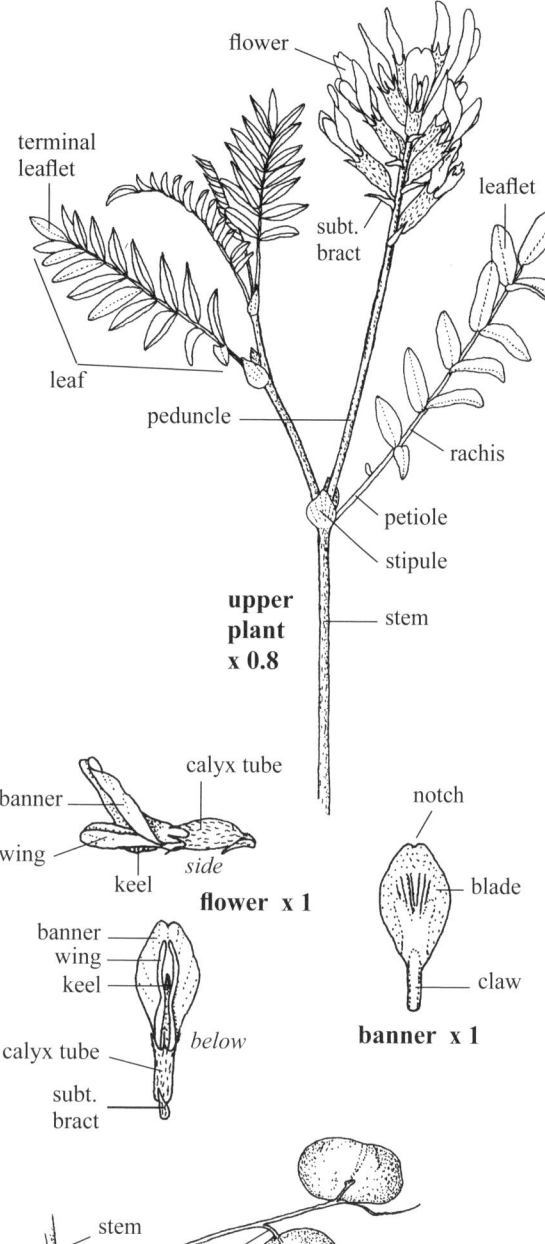

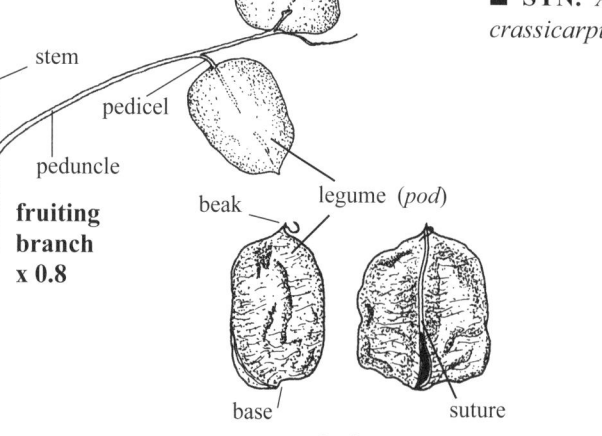

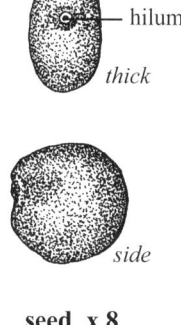

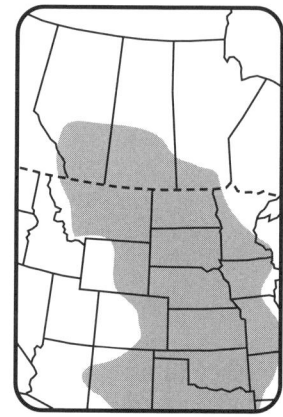

■ **SKETCH** A variable **perennial herb** 10–60 cm long or tall from a woody **taproot** 6–20 mm thick and a branched **caudex**; in prairies, sagebrush, along roadsides and in parklands. **w2, p1**

■ **FLOWERS** Purple to pink, green, fading to blue or white, blooming March–June; **inflorescence** a raceme 2.2–6.5 cm long by 3–5 cm wide, of 5–30 flowers ascending to spreading; **peduncles** erect, 2.5–13 cm long, grooved, hairs white; **pedicels** 1–2.5 mm long, to 3–6 mm long in fruit, hairs black or white; **subtending bracts** (of pedicels) one, 3–4 mm long by *c.* 1.7 mm wide, midrib obvious, pointed, hairs black or white on outside and margins; **flowers** perfect, 7–10 mm wide by 15–25 mm long; **calyx tube** pale reddish green, 5–9.5 mm long, hairs appressed and black or white; **calyx teeth** five, 2–4.5 mm long; **petals** five, glabrous; **banner** (one petal) slightly notched, 16–25 mm long by 10–12 mm wide (flattened), claw white, tapered, *c.* 2 mm wide; **wings** (two petals) 16–20 mm long by *c.* 3 mm wide (side), claw white, *c.* 7 mm long by *c.* 1 mm wide; **keels** (two petals) 11–18 mm long, apices deep purple pink, united, claws white, *c.* 7 mm long by *c.* 1 mm wide; **stamens** 10, diadelphous (9 + 1), included; **pistil** included and extending slightly past the longest stamens; **stigma** obscure. **FRUIT a legume** (pod), one to ten per raceme, glabrous, tan to reddish purple, 1–3.5 cm long by 15–18 mm wide by 12–15 mm thick, fleshy then dry, opening along basal suture; **beak** 1–3 mm long; **seeds** black, 23–30 per pod, 2–4 mm long by *c.* 2 mm wide by 1.2–1.7 mm thick, smooth, rattle when the pod is shaken, not stirred.

■ **LEAVES** Alternate, compound (odd-pinnate); **blades** 3–15 cm long by 1.5–4 cm wide; **leaflets** 11–33, each 5–24 mm long by 3–8 mm wide, entire, glabrous to hairy above with a thin white margin of hairs, white hairy below; **petioles** 5–30 mm long, hairy; **stipules** separate, pointed, hairy on margins and tips, 3–10 mm long and to *c.* 6 mm wide; **petiolules** *c.* 0.5 mm long.

■ **STEM** Several, prostrate to ascending, round, forming a clump 20–80 cm wide, light to reddish green, slightly hairy; 2–4 mm thick near the hollow base.

■ **SYN.** *Astragalus caryocarpus* Ker-Gawl. = *Astragalus crassicarpus* var. *crassicarpus* Nutt.

Ground Milkvetch, Buffalo-bean

Pea—*Fabaceae* **Wildflower** Pinkish purple Petals 5

MISSOURI MILK-VETCH *Astragalus missouriensis* Nutt.

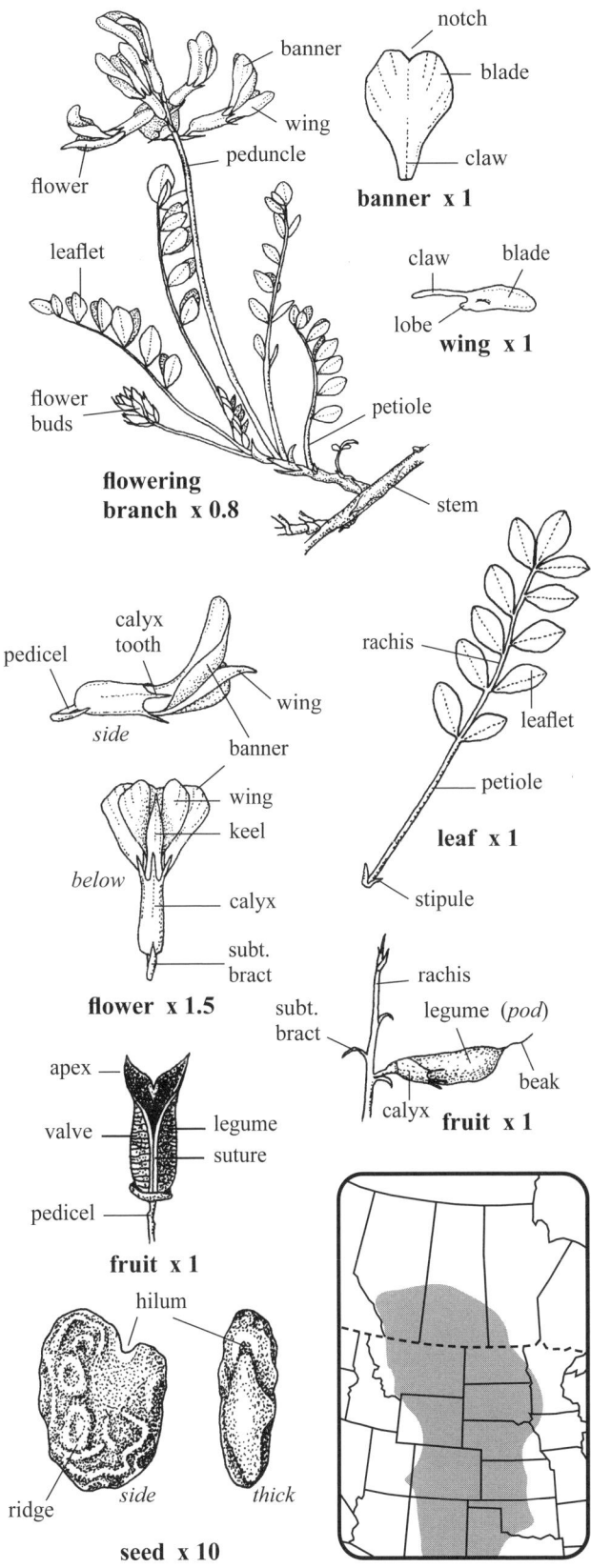

■ **SKETCH** A tufted **perennial herb** 3–20 cm tall or long with a **taproot** and branched **caudex**; in dry prairies, gravel flats, stony washes, slopes, clay banks, rocky bluffs, roadsides and stream valleys. **p1**

■ **FLOWERS** Pinkish purple, becoming blue on fading, blooming March–July; **inflorescence** a raceme, 2–5 cm long by 3–4.5 cm wide; **peduncles** 2–14 cm long, ridges low and pink, covered with white, basifixed, appressed hairs; **pedicels** 1–2 mm long, hairy, to *c.* 3 mm long in fruit; **subtending bracts** (of pedicels) 3–7 mm long, entire, pointed, hairs mostly on outside; **flowers** perfect, 3–12 per raceme, not crowded, each 8–11 mm wide by 15–22 mm long, spreading to ascending; **rachis** green with low red ridges as it ages, hairs white, appressed and ascending; **calyx** tube 5-veined, 6–9.5 mm long by *c.* 3 mm tall by *c.* 2 mm wide, persistent and splitting as fruit develops, hairs mostly white, some black, calyx teeth five, narrow, green, hairy, 1–4 mm long; **petals** five; **banner** (one petal) TL 10–24 mm by 10–13 mm wide, claw 4–6 mm long; **wings** (two petals) TL 11–19 mm, blade *c.* 3 mm wide, lobe *c.* 1.5 mm long, claw white, 7–10 mm long by *c.* 0.7 mm wide; **keels** (two petals) united above, separated for 3–4 mm at base, lobes *c.* 1 mm long, TL 10–18 mm, blade 5–8 mm long by 6–7 mm wide (open and flat), the claw 5–10 mm long; **stamens** 10, diadelphous (9 + 1), 14–15 mm long, glabrous, included in the keel; **anthers** orange, 0.4–0.7 mm long; **pistil** glabrous; **ovary** green, round, *c.* 4 mm long; **style** *c.* 12 mm long, filiform; **stigma** obscure, slightly wider than the style. **FRUIT** a **legume** (pod), often persistent overwinter, dark brown to gray, 1.5–3 cm long by 6–8 mm wide by 4–6 mm tall, ascending to spreading, white hairs appressed, 2-valved, beaked, firm, apical opening, ripening pod green and red, less hairy but more veins visible as it ripens; **seeds** shiny, 2–3 mm long by 1.8–2.1 mm wide by 0.8–1 mm thick, medium brown, hairless, tan at hilum, irregularly ridged.

■ **LEAVES** Alternate, compound (odd-pinnate); **blades** 3–10 cm long by 12–26 mm wide; **leaflets** 9–21, each 5–15 mm long by 3.5–7 mm wide, entire, silvery white, midrib usually visible, hairs appressed and aligned toward the margin; **petioles** hairy, 2–5 cm long; **petiolules** *c.* 0.5 mm long; **stipules** hairy, pointed, 3–8 mm long by *c.* 1 mm wide.

■ **STEM** Tufted, mostly prostrate, with silvery-hairy rubbery stems 3–4 mm thick, older stems less hairy.

301

Pea—*Fabaceae* **Wildflower** White to creamy yellow Petals 5

Narrow-leaved Milk-vetch *Astragalus pectinatus* (Hook.) Dougl. ex G. Don

■ **SKETCH** A **perennial herb** from 20–70 cm tall in clumps 10–60 cm wide, with a woody **taproot** and long branched **caudex**; on dry, gravelly hillsides, sandy flats and prairies. **p1**

■ **FLOWERS** White to creamy yellow, blooming May–June; **inflorescence** a raceme 3–15 cm long by 3–5 cm wide, one to three per stem from upper leaf axils; **peduncles** slightly hairy, curved, flattened, 2–11 cm long, with about eight low ridges; **pedicels** 2–4 mm long, with white or black hairs, deflexed in fruit; **subtending bracts** (of pedicels) one, entire, 3–7 mm long by 1–1.5 mm wide, pointed, persistent in fruit, hairs white or black; **flowers** perfect, fragrant, 3–30 per raceme, crowded to loose, 8–12 mm wide by 20–25 mm long; **rachis** ridged, with scattered white hairs; **calyx** tubular, 7–13 mm long by *c.* 2.5 mm wide by *c.* 3 mm high, hairs white or black, or mixed, less hairy on banner side; **calyx teeth** five, each 1.5–3.5 mm long, veined; **petals** five, glabrous; **banner** (one petal) 20–25 mm long by 9–12 mm wide (flattened), apically notched, claw stiff, C-shaped, 5–9 mm long; **wings** (two petals) distinct, each 17–20 mm long by 2.5–3 mm wide, blade veined, spreading, a lobe and appendage near the base of the blade, claw 7–9 mm long; **keels** (two petals) each 14–16 mm long, united along one side, blade *c.* 7 mm long by *c.* 3 mm wide, claw 8–9 mm long, distinct near the base; **stamens** 10, diadelphous (9 + 1), 13–15 mm long, glabrous, included in keel; **filaments** pale green; **anthers** orange, *c.* 1 mm long; **pistil** 13–15 mm long, glabrous; **ovary** 5–6 mm long; **style** 8–9 mm long, filiform; **stigma** not hairy. **FRUIT** a legume (pod), descending, hairless, tan, 2-valved, apical opening, TL (including beak) 1.3–2.5 cm by 5–8 mm wide by 4.8–7 mm thick, hard when dry, sutures raised; **seeds** 12–20 per legume, brown, glabrous, shiny, 2.8–3.8 mm long by 2–2.5 mm wide by 1–1.3 mm thick.

■ **LEAVES** Alternate, compound (odd-pinnate); **blades** 4–11 cm long by 1.5–7 cm wide; **leaflets** 7–21, opposite to alternate, 1–6.7 cm long by 1–2.5 mm wide, revolute, linear, soft, entire, sessile, glabrous above and hairy below; **petioles** 1–5 mm long; **stipules** entire, 2–10 mm long, united at base, the lobes 2–4 mm long.

■ **STEM** Ascending to prostrate, branched, reddish green; 3–5 mm thick near the base.

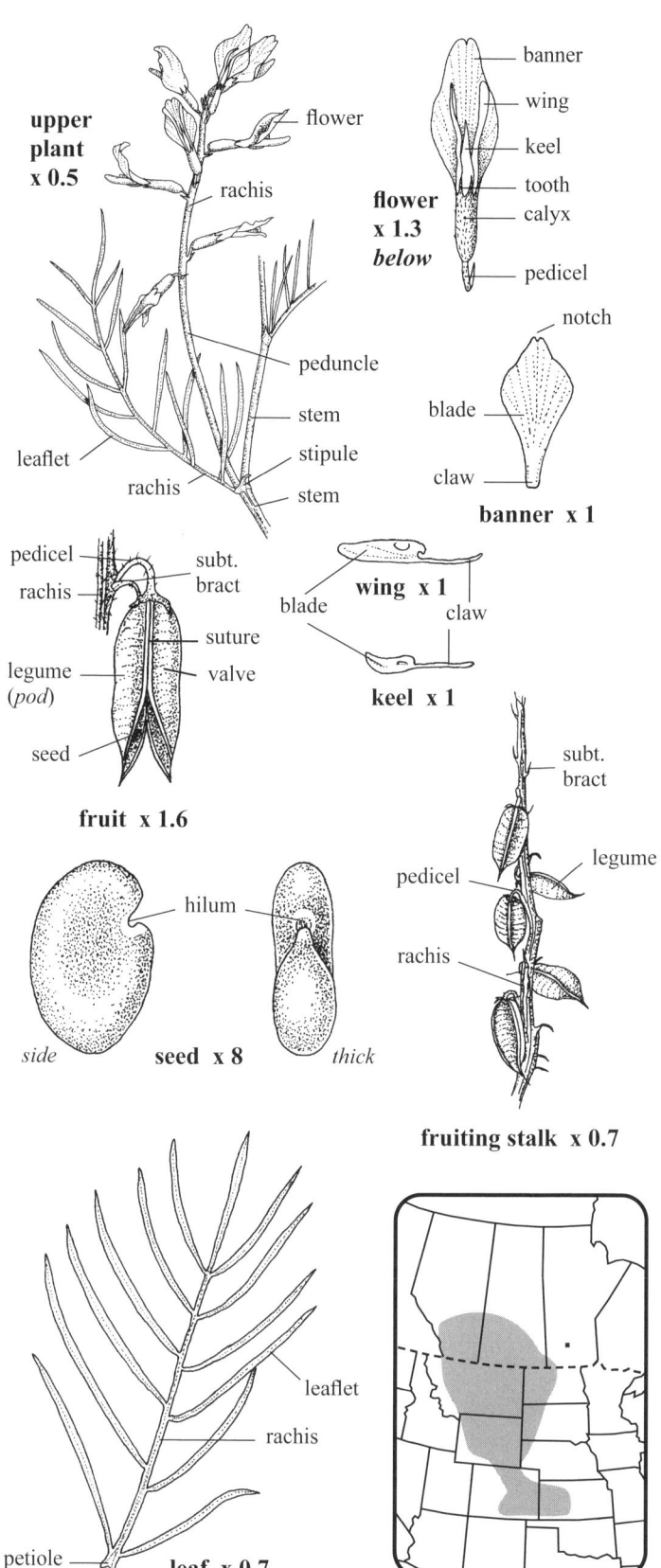

Tine-leaved Milk-vetch, Narrow-leaved Poison-vetch

Pea—*Fabaceae* **Shrub Yellow Petals 5**

COMMON CARAGANA   *Caragana arborescens* Lam.

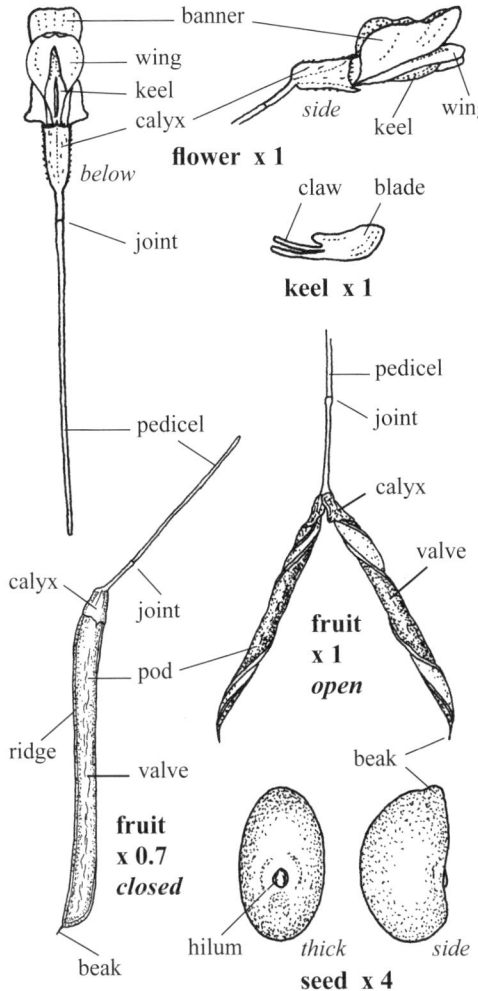

■ **SKETCH** A **deciduous shrub** 2–6 m tall; along shelterbelts, hedges and trail edges, often planted. *Naturalized*

■ **FLOWERS** Yellow, blooming May–June; **inflorescence** of clusters of 1–15 flowers among the alternating leaf clusters; **pedicels** 2–5.7 cm long, hairy, with a joint 5–7 mm below the calyx; **flowers** perfect, numerous, 9–10 mm wide by 15–24 mm long; **calyx** light green, hairy, 7–9 mm long by *c.* 5 mm wide by *c.* 4 mm thick with five faint green veins, persistent in fruit; **calyx teeth** five, 1–2 mm long, the upper two touching; **petals** five, irregular, glabrous; **banner** (one petal) 18–21 mm long and wide (flat), claws whitish, 4–6 mm long; **wings** (two petals) 18–21 mm long by 4–5 mm wide, claws 7–8 mm long, distinct; **keels** (two petals) 15–17 mm long by *c.* 4.5 mm wide, united at apices, claws 6–7 mm long, distinct; **stamens** 10, diadelphous (9 + 1), included in keel; **anthers** *c.* 0.5 mm long; **pistil** thin, included, *c.* 2 mm beyond the longest stamens; **stigma** not wider than style.   **FRUIT a legume** (pod), 3–6.5 cm long by 4–6 mm wide and thick, 2-valved, 2-ridged, reddish brown, veiny, opening at apex, beak short; **seeds** five to nine per pod, smooth, medium brown, 4–5 mm long by 2.5–3 mm wide and thick, plump, some marbled with dark brown streaks, slightly shiny, beak tiny.

■ **LEAVES** Compound (even-pinnate), in alternating clusters of five to eight blades; **blades** to *c.* 9 cm long by 1.8–5 cm wide; **leaflets** 4–12, each 7–37 mm long by 7–17 mm wide, entire, lateral veins often branching, flat, thin, with a mucro, hairy when young, especially near the base, glabrous with age, darker green above; **petioles** 9–38 mm long, hairy; **stipules** 5–8 mm long, yellowish green, thin, stiff, persistent, each attached on opposite sides of the oval leaf scar; **petiolules** *c.* 1 mm long, hairy; **rachis** apex a thin, soft point 1–2 mm long.

■ **STEM** Branched, grayish to green, young branches pubescent, alternate, ascending, with four or five low ridges or gray lines, several stems in a clump, 2–8 cm thick near the base; **lenticels** horizontal, 3–8 mm long; **lateral buds** tan, 3.5–6 mm long by 2–2.5 mm wide by *c.* 1 mm thick, 3- or 4-scaled, these slightly hairy on margins; **leaf scars** *c.* 1.5 mm wide, almost round.

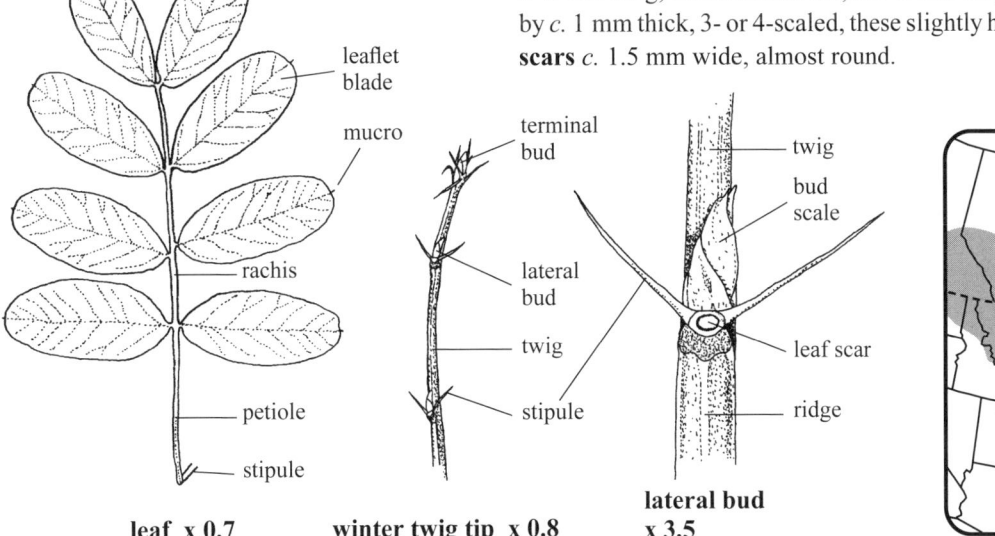

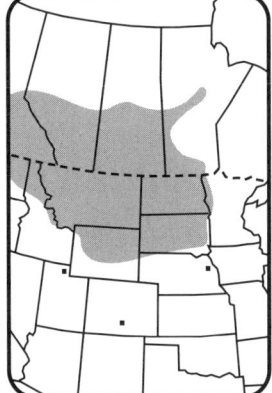

Siberian Peashrub

Pea—*Fabaceae* **Wildflower** Pink and white Petals 5

Field Crown-vetch *Coronilla varia* L.

■ **SKETCH** A variable, glabrous **perennial herb** 20–120 cm tall with a **taproot** 3–10⁺ cm long by 1–5 mm thick and a short branching **caudex**; along roadsides (planted as cover), railways, ditches and open areas. *Naturalized*

■ **FLOWERS** Pink and white, blooming May–September; **inflorescence** an umbel 3–3.5 cm wide by 1.5–2.5 cm tall, axillary; **branches** alternate, ascending, each subtended by a leaf; **peduncles** axillary, 6- or 7-ridged, 5–15 cm long by *c.* 1.5 mm thick; **pedicels** 3–7 mm long; **subtending bracts** (of pedicels) narrow, usually less than 1 mm long, apices reddish and bidentate; **flowers** perfect, fragrant, 10–20 per umbel, each 5–7 mm wide by 10–15 mm long by 9–12 mm tall; **calyx** TL 2.5–3 mm by 3–3.5 mm tall by 1–1.3 mm thick; **calyx tube** 1–2 mm long, minutely hairy on margins between teeth, upper two teeth united except for the tips, lower three calyx teeth 0.5–1 mm long, distinct, glabrous; **petals** five; **banner** (one petal) TL 10–13 mm by 7–8 mm wide (flattened), claw 1–2 mm long, pale green; **wings** (two petals) distinct, TL 11–15 mm by 4.5–5 mm wide, with a shallow groove along most of its length, folded over at apex to support long beak of keel, claws 2–3 mm long, thin, white; **keels** (two petals) TL 10–13 mm, united except for the upper part of blade, tapered to a dark red beak *c.* 4 mm long, blade *c.* 3 mm wide, claws 2.8–3 mm long, white, distinct; **stamens** 10, diadelphous (9 + 1), included in keel; **filaments** pale green, nine united into an open tube except for the apical 4 mm, one distinct; **anthers** golden, *c.* 0.5 mm long; **pistil** 12–13 mm long; **ovary** green, glabrous, *c.* 6 mm long by *c.* 0.5 mm wide, apex tapered; **style** pale green, filiform, extends *c.* 2 mm past the anthers; **stigma** pale green, slightly wider than the style, *c.* 0.5 mm long, slanted. **FRUIT a loment**, 2–7 cm long with 1–12 segments, each separating at maturity, apical beak tan, curved, *c.* 5 mm long; **loment segments** 1-seeded, veined, each 5–7 mm long by *c.* 2 mm wide and thick, 4-angled, 2-valved; **septum** nerveless, whitish, surrounding each seed; **seeds** reddish brown, dull, 3.5–4 mm long by 1–1.4 mm wide by 1–1.3 mm thick, glabrous.

■ **LEAVES** Alternate, compound (odd-pinnate); **leaf blades** 4–14 cm long by 1.5–4 cm wide; **leaflets** 9–25, each 5–25 mm long by 4–12 mm wide, largest at middle of leaf blade, entire, glabrous, midrib obvious, slightly paler green below, apices blunt, mucronate; **petioles** 0–3 cm long, reduced above; **stipules** linear, 2–3 mm long, persistent; **petiolules** *c.* 1 mm long; **rachis** grooved.

■ **STEM** Green, hollow, 8- to 10-ridged, branched, roundish; 1–4 mm thick near the slightly reddish base.

■ **SYN.** *Securigera varia* (L.) Lassen.

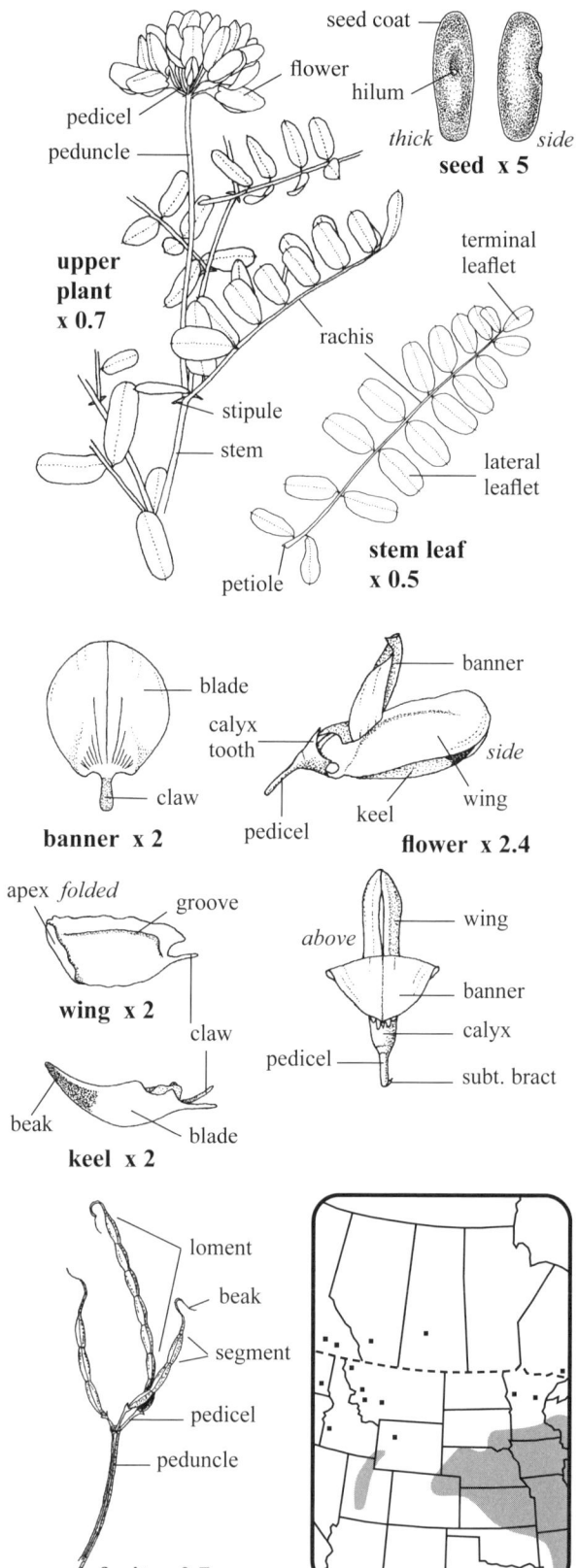

304 Crown Vetch, Purple Crownvetch, Axseed

WHITE PRAIRIE-CLOVER   *Dalea candida* Michx. ex Willd.

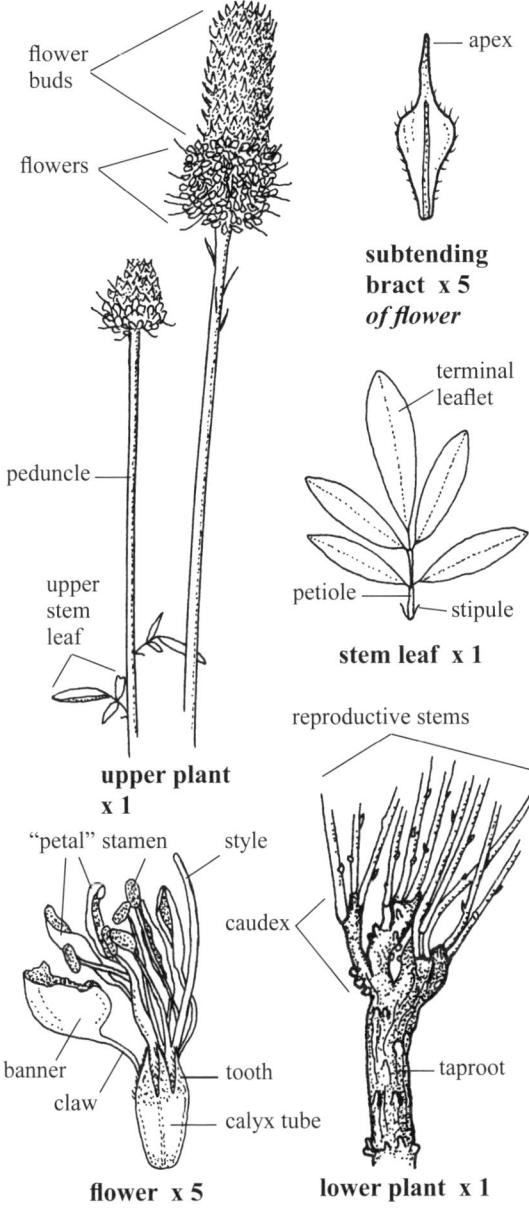

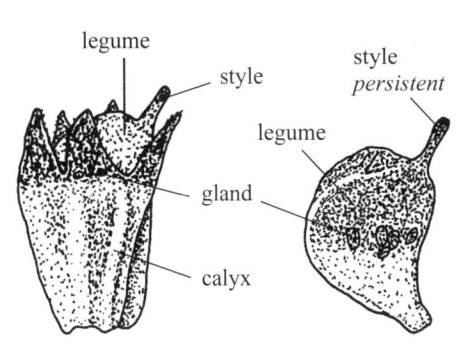

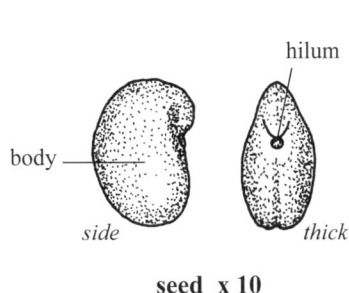

■ **SKETCH** A **perennial herb** 20–100 cm tall from a woody **taproot** 6–11 mm thick and to *c.* 15⁺ cm long with a branched **caudex**; in dry prairies, open woodlands, desert shrubs, along roadsides, knolls and gravelly hills. **p1**

■ **FLOWERS** White, blooming June–September; **inflorescence** a spike, one to three, each 1–8.5 cm long by 12–20 mm wide when in bloom; **subtending bracts** (of flowers) one, 4–5 mm long by *c.* 1.5 mm wide, pointed, V-shaped, ciliate, some hairs glandular; **flowers** perfect, sessile, *c.* 4 mm wide by 6–8 mm long; **rachis** glabrous; **calyx** TL 2–4 mm by *c.* 2 mm wide, base pinkish, glandular above, body glabrous to hairy, 10-veined; **calyx teeth** five, 0.6–1.8 mm long, hairy on the margins; **petals** five (1 + 4), clawed; **banner** (one petal) 4–6 mm long, claws 2–3.8 mm long, C-shaped; **wings** and **keels** (two "petals" staminodia each) with blades 1.5–2 mm long by 1–2 mm wide, with claws extending beyond the calyx; **stamens** five, fertile, slightly shorter than the "petals"; **filaments** filiform, distinct for 2–4.5 mm, united below into a staminal tube; **anthers** yellow, *c.* 0.7 mm long; **style** white, exserted, curved, persistent in fruit; **fruiting heads** to *c.* 6 cm long by 7–8 mm wide, dark brown due to the teeth of the calyces. **FRUIT** a **legume** (pod), 2.5–4.5 mm long by 1.7–2 mm wide by 1–1.5 mm thick, glandular, light brown and glabrous below, lightly hairy at the apex; **seeds** brown, one per pod, 1.6–2.5 mm long by 1.3–1.5 mm wide by 0.8–1 mm thick.

■ **LEAVES** Alternate, compound (odd-pinnate), numerous below; **blades** 2–6 cm long by 2–3 cm wide, glabrous; **leaflets** three to seven (nine), glandular-punctate below, 0.7–3.5 cm long by 1–8 mm wide, entire, the terminal leaflet the largest; **petioles** usually less than 5 mm long; **stipules** hairlike, 2–5 mm long.

■ **STEM** Erect to decumbent, tough, with longitudinal ridges below the inflorescence, smooth near the base, glabrous, simple or branched above, one to twelve in scattered clumps; 1–2 mm thick near the base.

■ **SYN.** *Petalostemon candidus* Michx. = *Dalea candida* var. *candida* Michx. ex Willd. and *Petalostemon candidum* (Michx. ex Willd.) Michx. = *Dalea candida* var. *candida* Michx. ex Willd. *Petalostemum* is an alternate spelling.

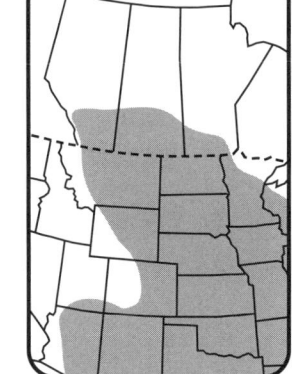

Pea—*Fabaceae* **Wildflower** Purple to pink Petals 5

Purple Prairie-clover   *Dalea purpurea* Vent.

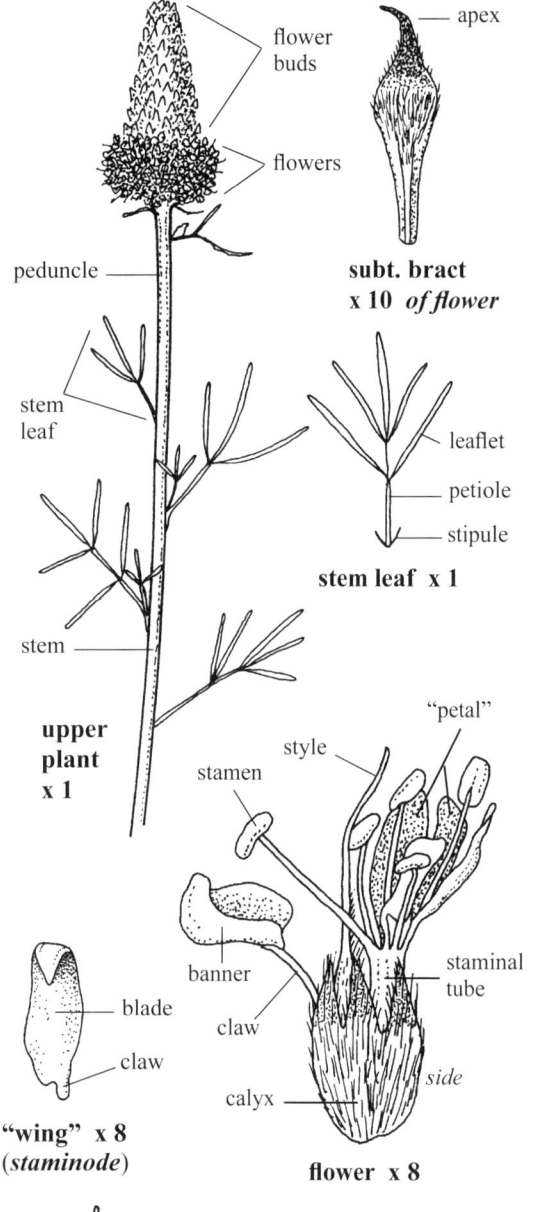

■ **SKETCH** A **perennial herb** 20–90 cm tall in small open clumps from a **taproot** 3–15 mm wide and to 30⁺ cm long with a branched **caudex** 1–2 cm long with lateral branches producing above ground shoots intermittently; in open prairies, hillsides, sandy barrens and open woodlands. **p1**

■ **FLOWERS** Purple to pink, blooming June–August; **inflorescence** a terminal spike, dense, usually 1–11, each cylindrical and 1–6.5 cm long by 12–25 mm wide in flower; **peduncles** 1–15 cm long; **subtending bracts** (of flowers) one, hairy, 3–6 mm long by 1–2 mm wide, keeled; **flowers** perfect, *c.* 6 mm long, fragrant; **rachis** hairy; **calyx** TL 3.3–4 mm, punctate, very hairy, calyx body tubular, slightly compressed to *c.* 2 mm wide, calyx teeth five, erect, dark green, 1–2 mm long; **petals** five (1 + 4), clawed; **banner** (one petal) 4–7 mm long, hooded, with a claw 2.5–4.5 mm long; **wings** and **keels** each of two "petals" (staminodia), blades 2–5 mm long by *c.* 1 mm wide with a short claw; **stamens** five, the free part of the filaments purple, 2.5–3 mm long, united below into a white staminal tube *c.* 3 mm long; **anthers** yellowish orange, *c.* 0.8 mm long; **ovary** white, *c.* 1 mm long; **style** *c.* 5 mm long, exserted, white hairs below. **FRUIT a legume** (pod), tan, indehiscent, 2–2.5 mm long by 1.8–2 mm wide by *c.* 1 mm thick, hairy above and punctate, the base membranous, transparent, revealing the seed inside, hairy style persistent; **seeds** one per pod, smooth, 1.6–2 mm long by 1.3–1.7 mm wide by 1–1.4 mm thick, light to medium brown.

■ **LEAVES** Alternate, compound, 1.5–4.5 cm long, along the stem and peduncle almost to the inflorescence; **blades** larger near the base; **leaflets** three to seven, entire, linear, 5–28 mm long by 0.5–2 mm wide with slightly inrolled margins, glabrous to hairy with dark glandular dots below; **petioles** to *c.* 10 mm long; **stipules** hairlike, to *c.* 3 mm long, lightly hairy at base.

■ **STEM** Erect to prostrate, ridged, tough, 2–15 in a clump, simple or branched above, mostly glabrous, some very hairy, green to purple near the top; 1–3 mm thick near the base.

■ **SYN.** *Petalostemon purpureus* (Vent.) Rydb. = *Dalea purpurea* var. *purpurea* Vent and *Petalostemon purpureum* (Vent.) Rydb. = *Dalea purpurea* var. *purpurea* Vent. *Petalostemum* is an alternate spelling.

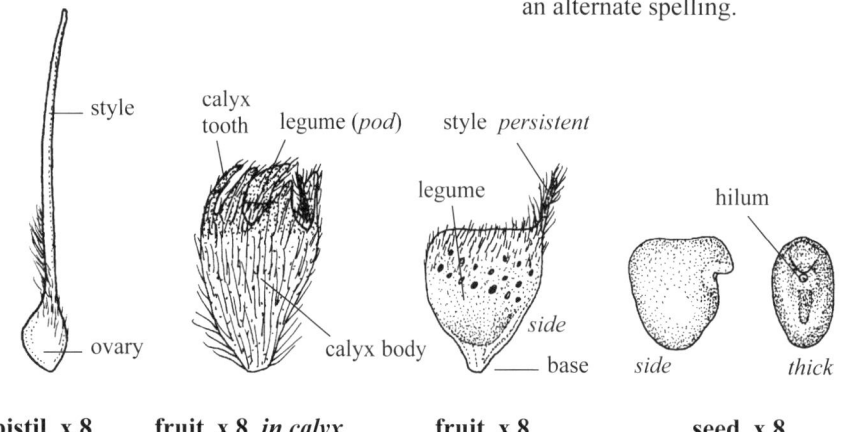

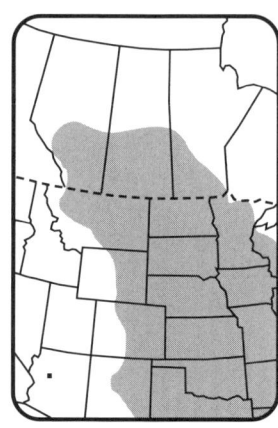

306   Violet Prairie-clover

Pea—*Fabaceae* **Wildflower** Pale yellow Petals 5

Wild Licorice *Glycyrrhiza lepidota* Pursh

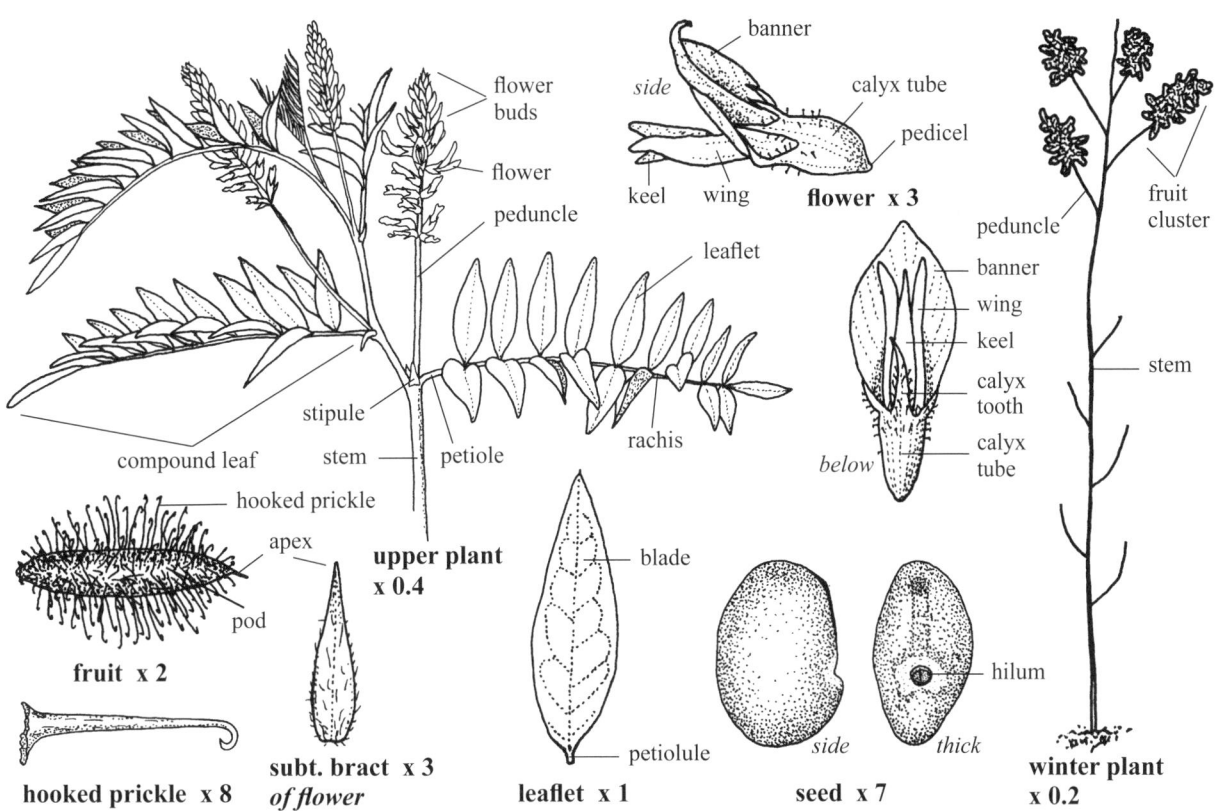

- **SKETCH** A **perennial herb** 30–110 cm tall from extensive woody **rhizomes** 2–15 mm thick and to 90+ cm long with a slight licorice flavor, in large patches to several m² or singly; in prairies, along railroads, stream valleys, slough banks, roadsides and ravines. p1
- **FLOWERS** Pale yellow to slightly pinkish purple, blooming May–August; **inflorescence** a raceme (spike-like), vertical, each 1.5–8.2 cm long by 2–3 cm wide of 10–60 flowers, usually with one to eight (–36) racemes per plant; **peduncles** axillary, glabrous, 1–9.5 cm long; **pedicels** less than 1 mm long in flower, 0.5–3 mm long in fruit; **subtending bracts** (of flowers), 5–8 mm long by *c.* 2 mm wide with scattered hairs; **flowers** perfect, 5–7 mm wide by 9–15 mm long; **rachis** hairy; **calyx** TL 5–8 mm, tubular, with scattered short-stalked glands, teeth five, 2.5–4 mm long; **petals** five, glabrous; **banner** (one petal) 10–14 mm long, pointed; **wings** (two petals) slightly shorter and clawed; **keels** (two petals) partially united, clawed; **stamens** 10, diadelphous (9 + 1), enclosed in keel; **pistil** included in keel; **fruit clusters** 3–9 cm long by 2.5–3.5 cm wide, overwintering. **FRUIT a legume** (pod), 15–48 per cluster, reddish brown, 10–20 mm long by 8–12 mm wide (including prickles), with an apical projection *c.* 2 mm long, pods eventually splitting lengthways, each half *c.* 4 mm wide (not including prickles); **hooked prickles** 2–3.5 mm long; **seeds** two to five per pod, each 2.5–3.5 mm long by 2–2.5 mm wide by *c.* 2 mm thick, smooth, olive green to gray, producing a soft rattle when a cluster of pods is shaken, not stirred.
- **LEAVES** Alternate, compound (odd-pinnate); **blades** 3–20 cm long by 2–13 cm wide; **leaflets** 7–21 per leaf, not widely spreading, entire, mucronate, 1.5–6.5 cm long by 0.3–2.5 cm wide, more hairy below on raised midrib, glandular-punctate below; **petioles** 1–5 cm long and hairy; **stipules** 3–9 mm long by 1–2 mm wide, pointed, deciduous; **petiolules** 1–2 mm long; **rachis** hairy, with tiny scales.
- **STEM** Erect, stiff, one to several, ridged, hairy, with tiny white scales, stalked glands or not, by late summer turning a medium to dark brown, striate and mottled, branched or not, overwintering; 2–6 mm wide near the base.

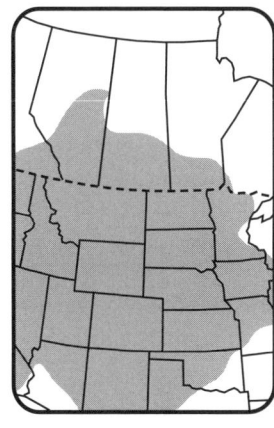

American Licorice 307

Pea—*Fabaceae* **Wildflower** Pink to reddish purple Petals 5

American Hedysarum *Hedysarum alpinum* L.

■ **SKETCH** A variable **perennial herb** 15–80 cm tall in small clusters from a **taproot** and short branching **caudex**; in moist meadows, open wooded hillsides, alpine sites and along riverbanks. **p1**

■ **FLOWERS** Pink to reddish purple, rarely white, blooming June–August; **inflorescence** a raceme 10–16 cm long by 2–3.5 cm wide, usually three to six racemes from upper leaf axils; **peduncles** axillary, ascending, 4–18 cm long by 1–2 mm wide, ridged, hairy; **pedicels** hairy, arched, 2–3 mm long; **subtending bracts** (of pedicels) 3–4 mm long, linear, hairy, entire, early deciduous; **subtending bracteoles** (of flowers) two, each *c.* 1 mm long; **flowers** perfect, mostly nodding to spreading, 4–5 mm wide by 12–18 mm long; **calyx** light green, hairy, TL 3–4.5 mm by *c.* 3 mm wide, reddish green, persistent, teeth unequal, 1–2 mm long; **corolla** of five petals; **banner** (one petal) barely notched, 10–15 mm long by *c.* 6 mm wide; **wings** (two petals) 10–13 mm long, lobe about as long as the 2–3 mm long claw; **keels** (two petals) 13–18 mm long by *c.* 5 mm wide (across one side), eventually opening above, claw 3–4 mm long and lobe *c.* 1 mm long, both white; **stamens** 10, diadelphous (9 + 1), included in the keel; **anthers** yellow, 0.4–0.5 mm long; **pistil** included to eventually exserted, *c.* 16 mm long; **ovary** *c.* 5 mm long, glabrous except for white hairs along the upper and lower sutures; **style** filiform, glabrous, *c.* 6 mm long; **stigma** capitate, slightly wider than the style. **FRUIT a loment**, brown, glabrous to hairy, veined, of two to five segments, each segment 5–10 mm long by 3.5–6.5 mm wide by *c.* 1.4 mm thick, winged, sutures 0.2–0.5 mm long; **stipe** 3–5 mm long, flattened; **seeds** brown, smooth, one per segment, dull, 3–4 mm long by 2–2.8 mm wide by *c.* 1 mm thick.

■ **LEAVES** Alternate, compound (odd-pinnate), arched, four or five; **blades** 10–15 cm long by 2.5–6 cm wide; **leaflets** 11–25, entire, mostly opposite, 10–38 mm long by 5–10 mm wide, reduced near leaf apex, veins obvious, apices mucronate, glabrous above, hairy below especially on veins; **petioles** 0–4 cm long, reduced above, reddish green, hairy; **stipules** two, united, 5–20 mm long by 7–10 mm wide, entire, clasping, hairy on outside, glabrous inside, many-veined, tips lobed; **petiolules** 0.2–1 mm long.

■ **STEM** Erect, branched above, hollow, ridged longitudinally, stiff, scattered ascending hairs below, more hairy above, hairs white and appressed; 3–5 mm thick near the reddish green base.

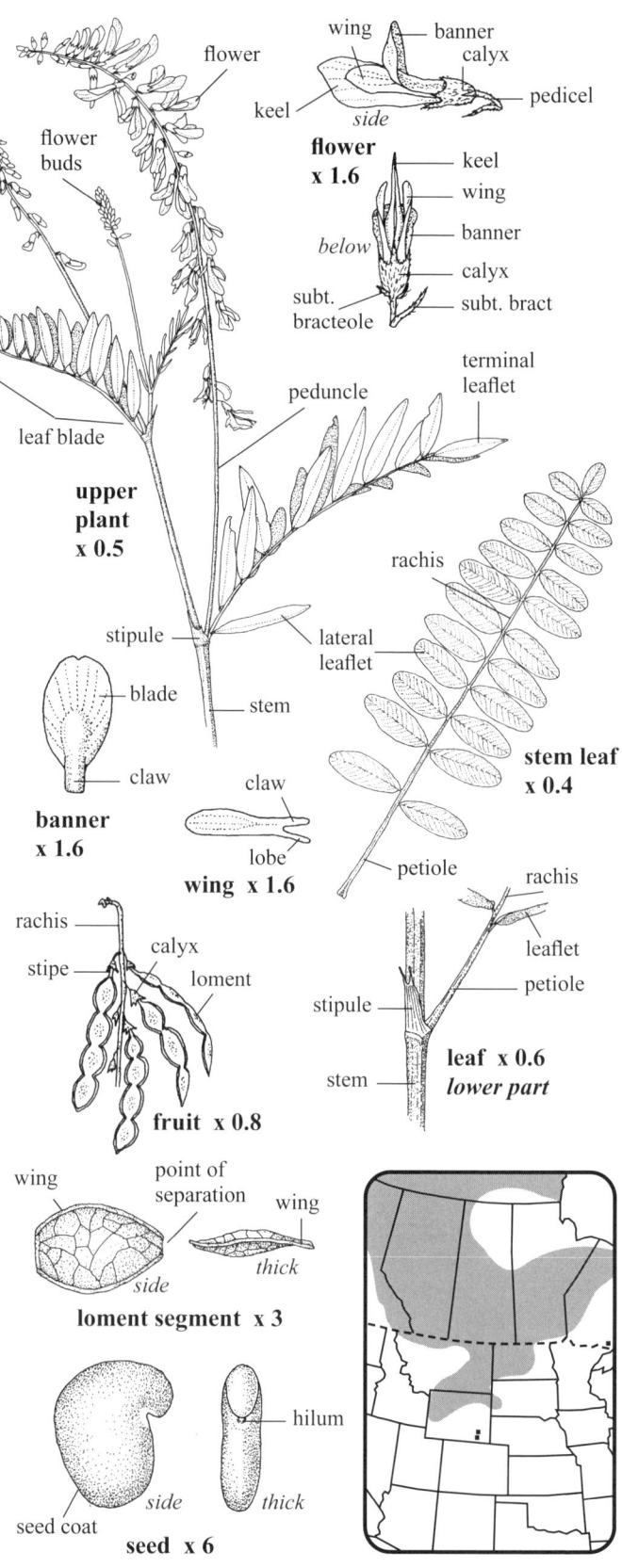

Pea—*Fabaceae*   **Vine**   Creamy yellow   Petals 5

CREAM-COLORED VETCHLING   *Lathyrus ochroleucus* Hook.

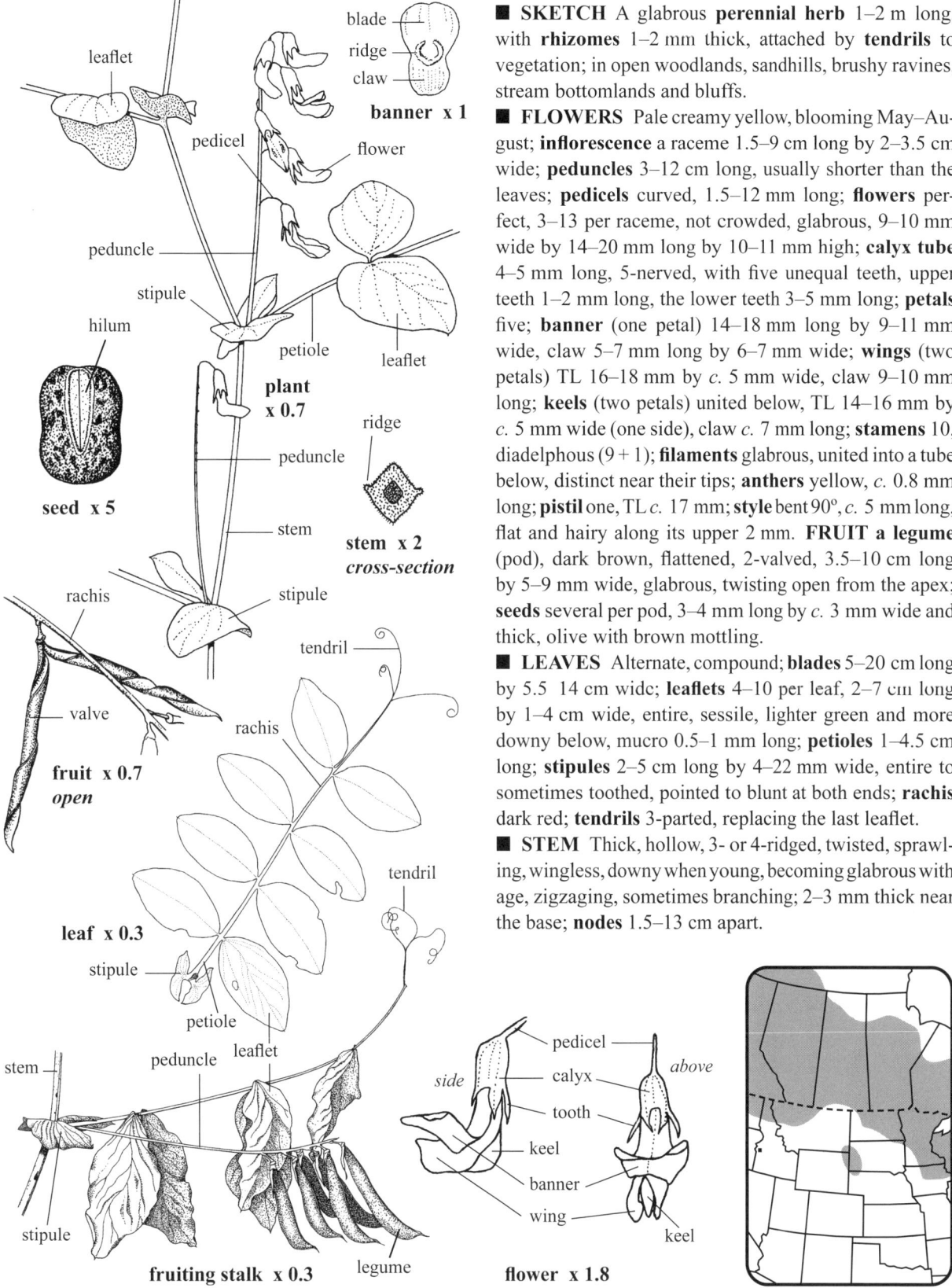

■ **SKETCH** A glabrous **perennial herb** 1–2 m long, with **rhizomes** 1–2 mm thick, attached by **tendrils** to vegetation; in open woodlands, sandhills, brushy ravines, stream bottomlands and bluffs.

■ **FLOWERS** Pale creamy yellow, blooming May–August; **inflorescence** a raceme 1.5–9 cm long by 2–3.5 cm wide; **peduncles** 3–12 cm long, usually shorter than the leaves; **pedicels** curved, 1.5–12 mm long; **flowers** perfect, 3–13 per raceme, not crowded, glabrous, 9–10 mm wide by 14–20 mm long by 10–11 mm high; **calyx tube** 4–5 mm long, 5-nerved, with five unequal teeth, upper teeth 1–2 mm long, the lower teeth 3–5 mm long; **petals** five; **banner** (one petal) 14–18 mm long by 9–11 mm wide, claw 5–7 mm long by 6–7 mm wide; **wings** (two petals) TL 16–18 mm by *c.* 5 mm wide, claw 9–10 mm long; **keels** (two petals) united below, TL 14–16 mm by *c.* 5 mm wide (one side), claw *c.* 7 mm long; **stamens** 10, diadelphous (9 + 1); **filaments** glabrous, united into a tube below, distinct near their tips; **anthers** yellow, *c.* 0.8 mm long; **pistil** one, TL *c.* 17 mm; **style** bent 90°, *c.* 5 mm long, flat and hairy along its upper 2 mm. **FRUIT** a legume (pod), dark brown, flattened, 2-valved, 3.5–10 cm long by 5–9 mm wide, glabrous, twisting open from the apex; **seeds** several per pod, 3–4 mm long by *c.* 3 mm wide and thick, olive with brown mottling.

■ **LEAVES** Alternate, compound; **blades** 5–20 cm long by 5.5–14 cm wide; **leaflets** 4–10 per leaf, 2–7 cm long by 1–4 cm wide, entire, sessile, lighter green and more downy below, mucro 0.5–1 mm long; **petioles** 1–4.5 cm long; **stipules** 2–5 cm long by 4–22 mm wide, entire to sometimes toothed, pointed to blunt at both ends; **rachis** dark red; **tendrils** 3-parted, replacing the last leaflet.

■ **STEM** Thick, hollow, 3- or 4-ridged, twisted, sprawling, wingless, downy when young, becoming glabrous with age, zigzaging, sometimes branching; 2–3 mm thick near the base; **nodes** 1.5–13 cm apart.

Yellow Vetchling, Pale Vetchling, Creamy Peavine, Cream Pea

Pea—*Fabaceae* — **Vine** Reddish purple to blue Petals 5

MARSH VETCHLING *Lathyrus palustris* L.

- **SKETCH** A variable **perennial vine** with **tendrils**, 30–100 cm long or tall from tan **rhizomes** 1–2 mm thick, roots with round **nodules** 1–2.5 mm wide; in wet prairies, woodlands, along lakeshores and stream valleys. **p2**
- **FLOWERS** Reddish purple to blue on fading, blooming June–August; **inflorescence** a raceme, each 2–6 cm long by 3–5 cm wide; **peduncles** axillary, horizontal to ascending, 4–10 cm long, stiff, slightly hairy to glabrous; **pedicels** 2–6 mm long, glabrous to slightly hairy; **flowers** perfect, two to eight per raceme, 12–16 mm wide by 11–22 mm long, mostly drooping; **calyx** TL 7–10 mm by 3.5–4 mm wide and high; **calyx tube** 5-veined, 2.5–4 mm long, one red vein per tooth, glabrous to slightly hairy especially below; **calyx teeth** five, unequal, upper two teeth curved inward, 1–3 mm long by *c.* 3 mm wide, middle lower tooth 4–6 mm long by *c.* 1 mm wide, tapered, two lateral teeth 3–5 mm long by *c.* 2 mm wide; **petals** five, glabrous; **banner** (one petal) 12–20 mm long by 12–16 mm wide, notched at apex, tapered to the base; **wings** (two petals) spreading, TL *c.* 18 mm by 5–6 mm wide, blade 10–11 mm long, claw white, 7–8 mm long by 1–1.2 mm wide; **keels** (two petals) easily separated, blade 7–8 mm long by 6–7 mm wide, claw 7–8 mm long; **stamens** 10, diadelphous (9 + 1), included; **filaments** are white, TL *c.* 16 mm, distinct filament *c.* 1 mm wide near the base; **anthers** yellow, 1–1.2 mm long; **pistil** green, glabrous; **style** *c.* 5 mm long by *c.* 0.5 mm wide, flattened, hairy in the upper 2–3 mm; **stigma** hairless, beaklike. **FRUIT** a **legume** (pod), 3.5–6 cm long by 4–6.5 mm wide by *c.* 3 mm thick, brown, beaked, veiny, 2-valved, glabrous or with scattered, short, red glandular hairs; **seeds** several[+] per pod, greenish brown, dull, mottled with dark brown, 3–3.8 mm long by 2.8–3.5 mm wide by 2–3 mm thick.
- **LEAVES** Alternate, compound (even pinnate); **blades** 5–14 cm long by 7–16 cm wide; **leaflets** linear, four or six (eight) per leaf, 1–8 cm long by 3–20 mm wide, entire, glabrous, pointed, spreading to ascending; **tendrils** terminal, branched once or twice, encircling nearby vegetation; **petioles** 5–30 mm long; **stipules** 0.7–3 cm long by 1.3–8 mm wide, 2-lobed, the lobes unequal, entire to toothed, pointed; **rachis** glabrous to slightly hairy.
- **STEM** Vinelike, trailing to ascending, glaucous and glabrous to slightly hairy, twisted, winged or not, 2.5–5 mm wide (including wing) by 1–1.2 mm thick; wingless and slightly reddish near the 1–2 mm thick base.

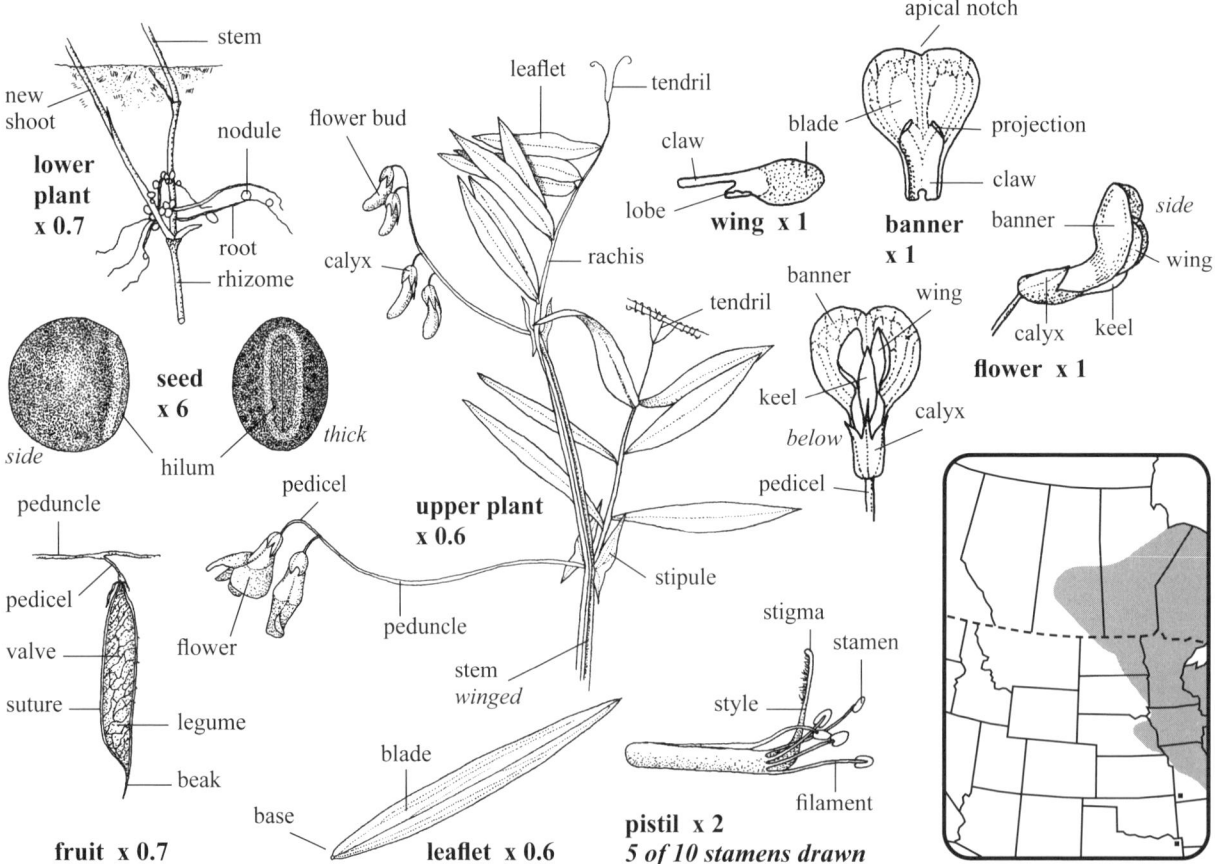

310 Marsh Pea, Vetchling

Pea—*Fabaceae* Vine   Pinkish purple   Petals 5

WILD PEAVINE   *Lathyrus venosus* Muhl. ex Willd.

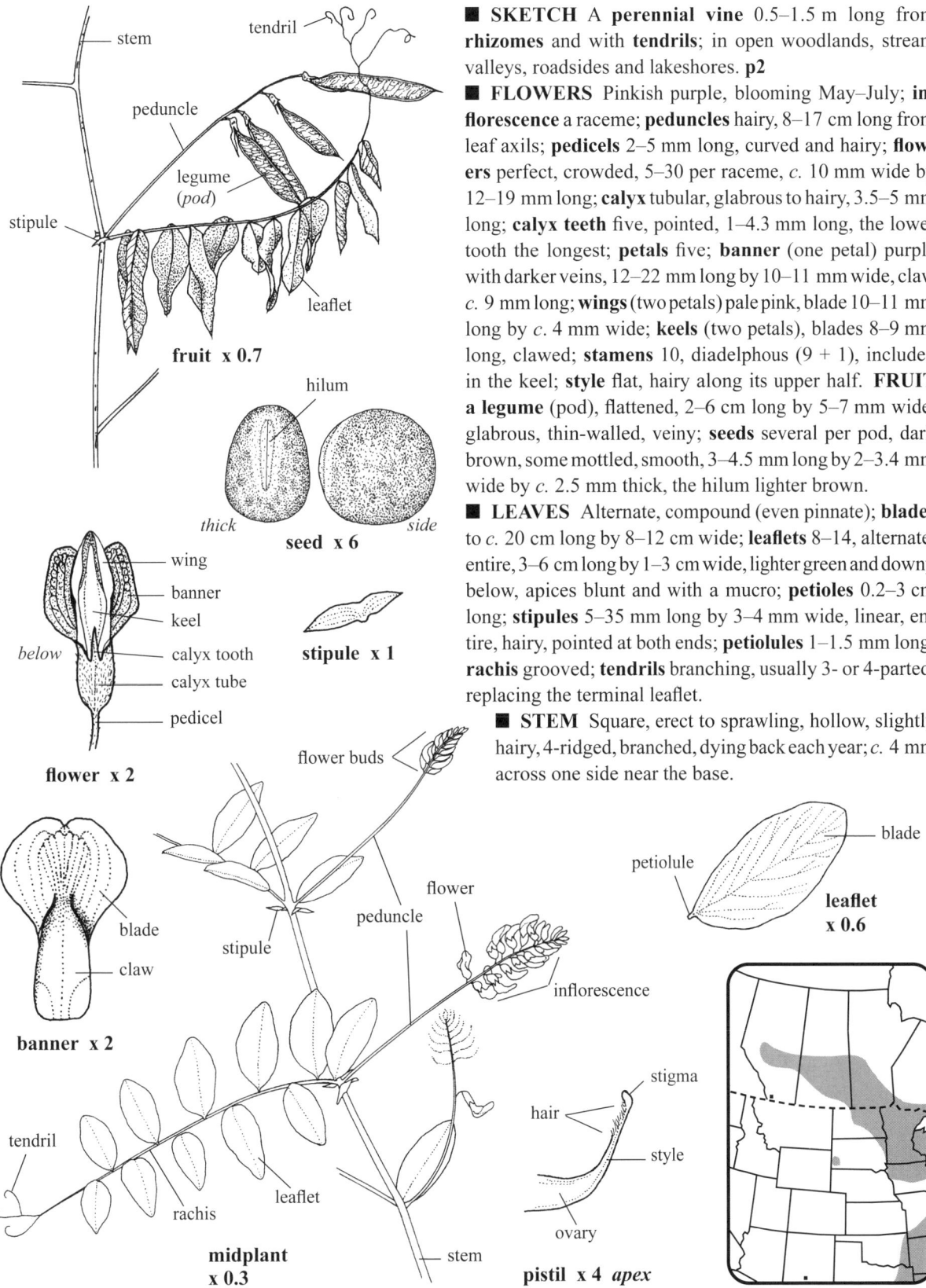

- **SKETCH** A **perennial vine** 0.5–1.5 m long from **rhizomes** and with **tendrils**; in open woodlands, stream valleys, roadsides and lakeshores. p2
- **FLOWERS** Pinkish purple, blooming May–July; **inflorescence** a raceme; **peduncles** hairy, 8–17 cm long from leaf axils; **pedicels** 2–5 mm long, curved and hairy; **flowers** perfect, crowded, 5–30 per raceme, *c.* 10 mm wide by 12–19 mm long; **calyx** tubular, glabrous to hairy, 3.5–5 mm long; **calyx teeth** five, pointed, 1–4.3 mm long, the lower tooth the longest; **petals** five; **banner** (one petal) purple with darker veins, 12–22 mm long by 10–11 mm wide, claw *c.* 9 mm long; **wings** (two petals) pale pink, blade 10–11 mm long by *c.* 4 mm wide; **keels** (two petals), blades 8–9 mm long, clawed; **stamens** 10, diadelphous (9 + 1), included in the keel; **style** flat, hairy along its upper half. **FRUIT** a legume (pod), flattened, 2–6 cm long by 5–7 mm wide, glabrous, thin-walled, veiny; **seeds** several per pod, dark brown, some mottled, smooth, 3–4.5 mm long by 2–3.4 mm wide by *c.* 2.5 mm thick, the hilum lighter brown.
- **LEAVES** Alternate, compound (even pinnate); **blades** to *c.* 20 cm long by 8–12 cm wide; **leaflets** 8–14, alternate, entire, 3–6 cm long by 1–3 cm wide, lighter green and downy below, apices blunt and with a mucro; **petioles** 0.2–3 cm long; **stipules** 5–35 mm long by 3–4 mm wide, linear, entire, hairy, pointed at both ends; **petiolules** 1–1.5 mm long; **rachis** grooved; **tendrils** branching, usually 3- or 4-parted, replacing the terminal leaflet.
- **STEM** Square, erect to sprawling, hollow, slightly hairy, 4-ridged, branched, dying back each year; *c.* 4 mm across one side near the base.

Bushy Vetchling, Veiny Vetchling, Veiny Pea, Purple Peavine

Pea—*Fabaceae* **Wildflower** Yellow to orange Petals 5

BIRD'S-FOOT TREFOIL *Lotus corniculatus* L.

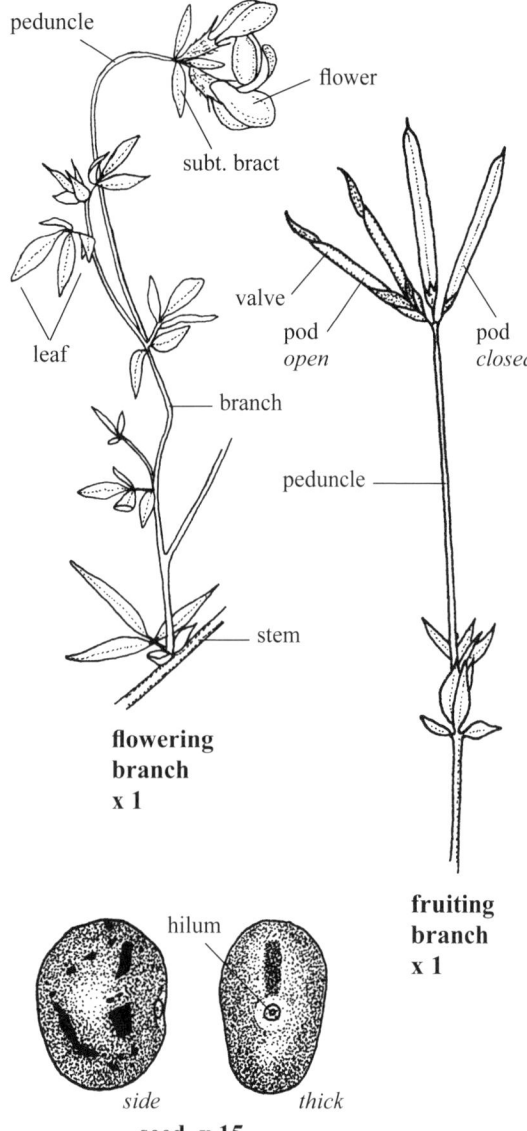

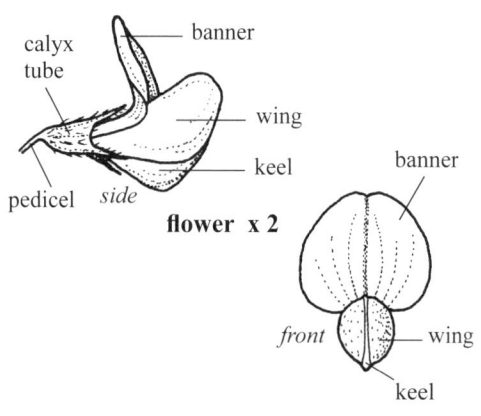

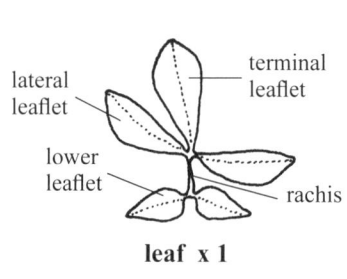

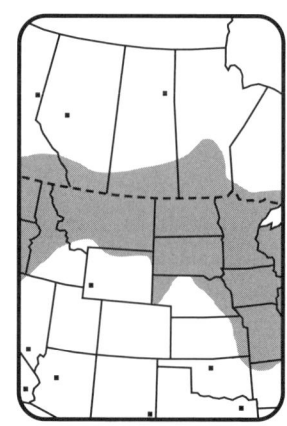

■ **SKETCH** A variable **perennial herb** 10–60 cm tall from a long woody **taproot** 1–6 mm thick with a branching **caudex**; along roadsides, field edges and streambanks, where it often forms tangles by late summer. *Naturalized*

■ **FLOWERS** Bright yellow to orange, blooming June–September; **inflorescence** of umbellate clusters, numerous, 19–30 mm wide by 15–20 mm tall, each 1- to 10-flowered; **branches** to *c.* 20 cm long, alternate and rebranched; **subtending bracts** (of flowers) leaflike, 3–8 mm long; **peduncles** nodding, slightly hairy to glabrous, thin, 3–9 cm long from leaf axils; **pedicels** hairy, 1–3 mm long; **flowers** perfect, 8–9 mm wide by 8–16 mm long by *c.* 11 mm high; **calyx** pale green, 6–7 mm long, the tube 2–4 mm long, lightly hairy; **calyx teeth** five, subequal, 1–3 mm long, each with a dark green nerve; **corolla** of five petals; **banner** (one petal) 12–16 mm long by 8–9 mm wide, with reddish veins; **wings** (two petals) oblong, 10–14 mm long by *c.* 5 mm wide, claws 3–4 mm long; **keels** (two petals) 12–14 mm long, claws 2–4 mm long; **stamens** ten, diadelphous (9 + 1), included in keel; **filaments** white, subequal, sometimes enlarged right below the anthers, nine of them forming a tube around the glabrous ovary; **anthers** pale yellow, *c.* 0.5 mm long; **style** thin, whitish green, *c.* 6 mm long, included; **stigma** barely enlarged. **FRUIT** a **legume** (pod), 2-valved, dark brown, glabrous, 1.2–4 cm long by 1.5–3 mm wide, the two valves tightly twisted after opening; **seeds** *c.* 14 in a two cm long pod, smooth, each 1.3–1.8 mm long by 1–1.3 mm wide by *c.* 1 mm thick, medium to dark brown, often with brown blotches (mottled), slightly shiny.

■ **LEAVES** Alternate, pinnately compound, sessile; **blades** 2–3 cm long by 2–2.5 cm wide; **leaflets** five, each 6–20 mm long and 3–9 mm wide, entire to finely toothed, sessile, lightly hairy when young, flat, thin, the midrib obvious, apices pointed or blunt, the lower pair reduced and appearing like stipules.

■ **STEM** Erect to lax, round, solid, one to several, glabrous to hairy, much branched; 1–2 mm thick near the base.

312 Birdfoot Deervetch, Garden Bird's-foot Trefoil

Black Medick *Medicago lupulina* L.

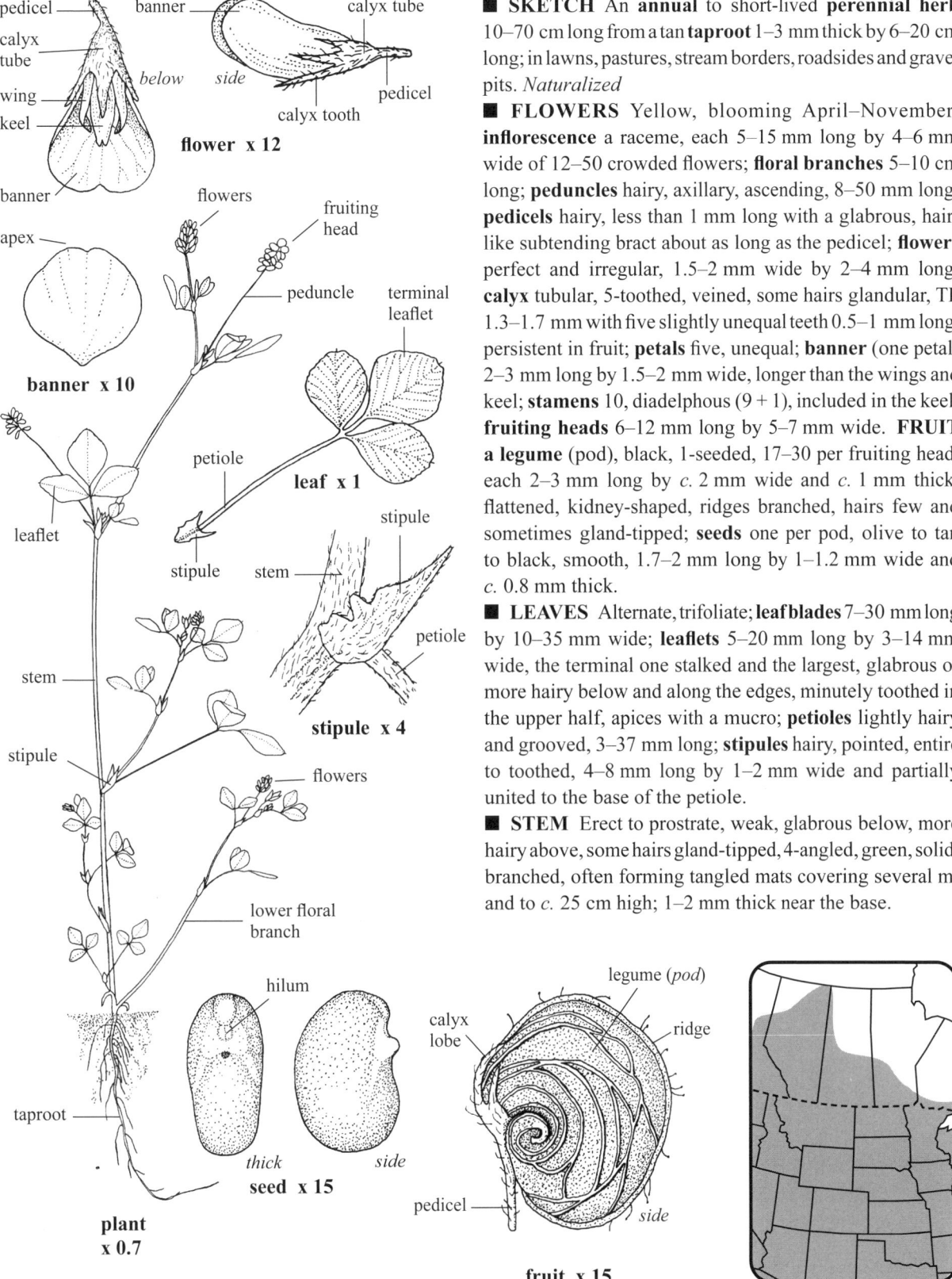

■ **SKETCH** An **annual** to short-lived **perennial herb** 10–70 cm long from a tan **taproot** 1–3 mm thick by 6–20 cm long; in lawns, pastures, stream borders, roadsides and gravel pits. *Naturalized*

■ **FLOWERS** Yellow, blooming April–November; **inflorescence** a raceme, each 5–15 mm long by 4–6 mm wide of 12–50 crowded flowers; **floral branches** 5–10 cm long; **peduncles** hairy, axillary, ascending, 8–50 mm long; **pedicels** hairy, less than 1 mm long with a glabrous, hair-like subtending bract about as long as the pedicel; **flowers** perfect and irregular, 1.5–2 mm wide by 2–4 mm long; **calyx** tubular, 5-toothed, veined, some hairs glandular, TL 1.3–1.7 mm with five slightly unequal teeth 0.5–1 mm long, persistent in fruit; **petals** five, unequal; **banner** (one petal) 2–3 mm long by 1.5–2 mm wide, longer than the wings and keel; **stamens** 10, diadelphous (9 + 1), included in the keel; **fruiting heads** 6–12 mm long by 5–7 mm wide. **FRUIT** a legume (pod), black, 1-seeded, 17–30 per fruiting head, each 2–3 mm long by *c.* 2 mm wide and *c.* 1 mm thick, flattened, kidney-shaped, ridges branched, hairs few and sometimes gland-tipped; **seeds** one per pod, olive to tan to black, smooth, 1.7–2 mm long by 1–1.2 mm wide and *c.* 0.8 mm thick.

■ **LEAVES** Alternate, trifoliate; **leaf blades** 7–30 mm long by 10–35 mm wide; **leaflets** 5–20 mm long by 3–14 mm wide, the terminal one stalked and the largest, glabrous or more hairy below and along the edges, minutely toothed in the upper half, apices with a mucro; **petioles** lightly hairy and grooved, 3–37 mm long; **stipules** hairy, pointed, entire to toothed, 4–8 mm long by 1–2 mm wide and partially united to the base of the petiole.

■ **STEM** Erect to prostrate, weak, glabrous below, more hairy above, some hairs gland-tipped, 4-angled, green, solid, branched, often forming tangled mats covering several m^2 and to *c.* 25 cm high; 1–2 mm thick near the base.

Pea—*Fabaceae*   **Wildflower**   Blue to yellow   Petals 5

ALFALFA   *Medicago sativa* L.

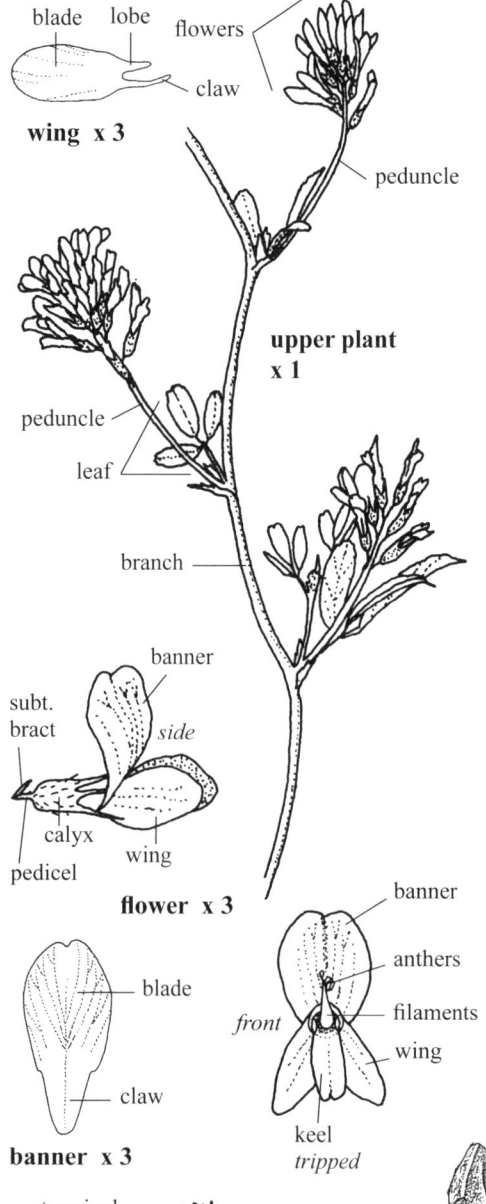

■ **SKETCH** A **perennial herb** 30–100 cm tall from a deep **taproot** 3–11 mm thick and a branching **caudex**; in disturbed sites along riverbanks, roadsides, fields and prairies, widely grown as a crop for livestock. *Naturalized* or *introduced*

■ **FLOWERS** Blue to yellow, blooming May–September; **inflorescence** of numerous racemes, compact (headlike), each 10–45 mm long by 10–25 mm wide of 5–40 flowers; **peduncles** 10–45 mm long, ascending from leaf axils, lightly hairy; **pedicels** 1–2 mm long, hairy; **subtending bracts** (of pedicels) one, narrow, pale, about as long as the pedicel; **flowers** perfect, 4–6 mm wide by 5–11 mm long; **calyx** tubular, slightly hairy, TL 4.5–6 mm including five pointed teeth 2.5–3.5 mm long, persistent in fruit; **petals** five, hairless; **banner** (one petal) notched, with dark purple veins, 7–10 mm long by 4–5 mm wide, claw 2.5–3.5 mm long, tapered; **wings** (two petals) *c.* 7 mm long by 2–2.5 mm wide, claw 2–2.5 mm long; **keels** (two petals) 5–6 mm long, when tripped exposes the reproductive parts; **stamens** 10, diadelphous, nine of the ten **filaments** united into a curved, green tube *c.* 4 mm long by *c.* 0.8 mm wide, split along its length, the apex with nine distinctive filaments *c.* 1 mm long; **anthers** pale yellow; **pistil** 4–5 mm long; **style** *c.* 1 mm long; **stigma** slightly enlarged from the style, green, round. **FRUIT a legume** (pod), splitting along both sutures, hairy or not, straight (from yellow flowers), slightly curved or with one to three coils, *c.* 5 mm long and wide (from blue flowers); **seeds** several per pod, smooth, *c.* 2.3 mm long by *c.* 1.4 mm wide by *c.* 1 mm thick, tan to yellowish.

■ **LEAVES** Alternate, compound; **blades** 0.7–4 cm long by 0.8–4.5 cm wide; **leaflets** three, 10–35 mm long by 2–15 mm wide, hairy below, glabrous to slightly hairy above, the upper half toothed; **petioles** 2–50 mm long and hairy; **stipules** united at the base to the petiole, lightly hairy on both sides or glabrous above, pointed and toothed, 5–20 mm long; **side petiolules** *c.* 1 mm long.

■ **STEM** Tough, usually glabrous, branched, erect to leaning, forming scattered clumps; 2–5 mm thick near the base.

■ **SYN.** *Medicago falcata* L. = *Medicago sativa* ssp. *falcata* (L.) Arcang. (yellow flowers).

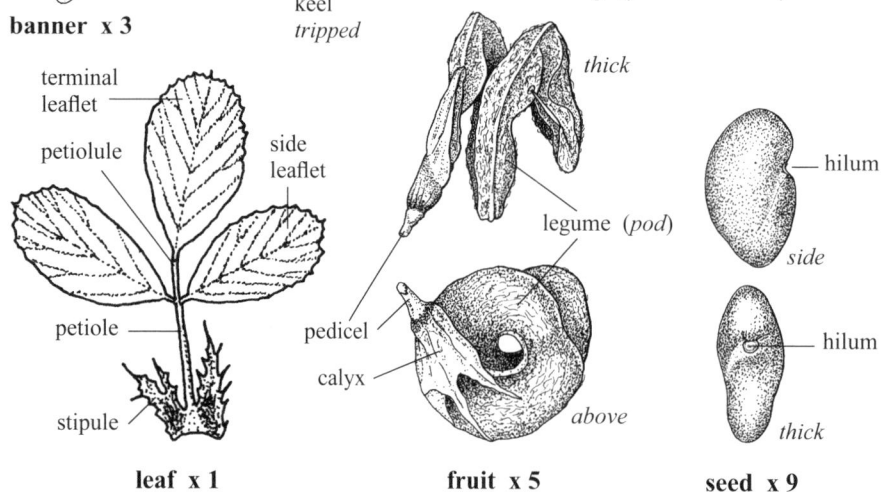
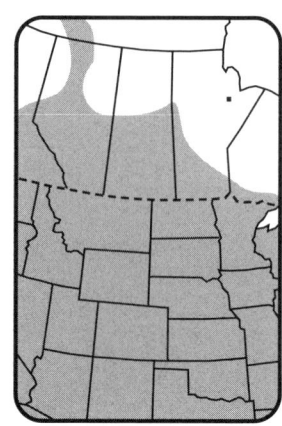

314   Lucerne

Pea—*Fabaceae* **Wildflower White Petals 5**

WHITE SWEET-CLOVER *Melilotus alba* Medikus

■ **SKETCH** An **annual** or **biennial herb** 30–280 cm tall from a stout **taproot** 3–13 mm wide by 10–30 cm long, with side roots 2–3 mm thick, often with an apical caudex supporting several stems, forming large patches; along roadsides, railways and in prairies. *Naturalized* **w2**

■ **FLOWERS** White, blooming May–October; **inflorescence** a raceme, numerous, 4–8 cm long by 8–12 mm wide (to 17 cm by 4–6 mm with fruit), mostly ascending, 20–90 flowers per raceme; **floral branches** numerous, alternate 10–80 cm long throughout the plant; **peduncles** axillary, glabrous, grooved, 1–4 cm long; **pedicels** 1–1.5 mm long, slightly hairy, reddish, curved downward; **subtending bracts** (of pedicels) one, about as long as the pedicel; **flowers** perfect, fragrant, average about eight per cm of raceme, each 2.5–3.5 mm wide by 4–5 mm long by *c.* 3.5 mm high; **rachis** glabrous; **calyx** green, tubular, tube slightly hairy, *c.* 1 mm long by *c.* 1 mm tall, teeth five, subequal, 0.5–0.7 mm long, pointed, persistent; **corolla** of five glabrous petals; **banner** (one petal) 4–4.2 mm long by 3.5–3.7 mm wide, with a shallow apical notch, claw *c.* 0.5 mm long; **wings** (two petals) *c.* 4.5 mm long by *c.* 1 mm wide, claw *c.* 1.5 mm long, lobe thick, 0.5–0.7 mm long; **keels** (two petals) 2.3–2.5 mm long by 1.3–1.5 mm wide, claw 1–1.3 mm long, separate most of their length; **stamens** 10, diadelphous (9 + 1), 2.5–3.5 mm long; **filaments** white, mostly united to form a split tube *c.* 2 mm long around the ovary, distinct at their tips; **anthers** yellow, *c.* 0.3 mm long; **pistil** one, *c.* 3.5 mm long, hairless; **stipe** *c.* 0.2 mm long; **ovary** *c.* 1.2 mm long, dark green; **style** curved, *c.* 2.3 mm long, light green, persistent; **stigma** obscure. **FRUIT** a **legume** (pod), compressed, with low reticulate veins, 3.5–4.3 mm long by 2–2.5 mm wide, 2-valved, many dropping early before ripe; **seeds** mostly one (two) per pod, smooth, tan to olive green, some with purple spots, 1.7–2.2 mm long by 1.3–1.5 mm wide by *c.* 1 mm thick.

■ **LEAVES** Alternate, compound, at the bases of branches or peduncles; **leaf blades** 1–5.5 cm long by 2–6.4 cm wide, glabrous; **leaflets** three, 1–3.7 cm long by 0.3–1.8 cm wide, flat, toothed, slightly hairy below, glabrous above, the terminal stalked leaflet the largest; **petioles** glabrous, 3–26 mm long; **petiolules** lighter green, hairs white, 1–7 mm long; **stipules** 3–6 mm long, paired, marginally hairy, free their entire length.

■ **STEM** Tough, erect to leaning, 1–10 per cluster, branched throughout, glabrous to slightly hairy above, pith white; 2–12 mm thick near the reddish base.

Yellow Sweet-clover   *Melilotus officinalis* (L.) Lam.

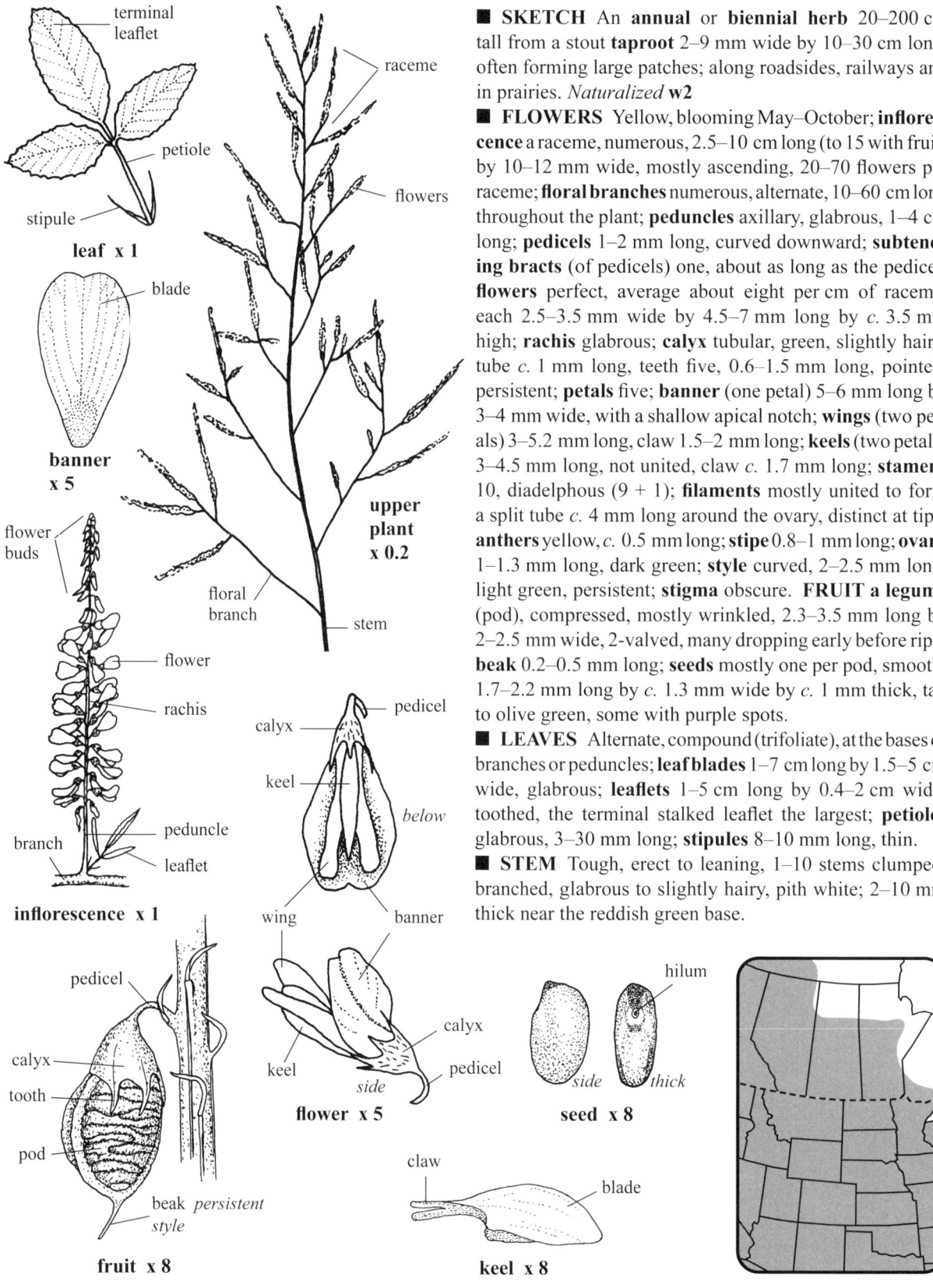

- **SKETCH** An **annual** or **biennial herb** 20–200 cm tall from a stout **taproot** 2–9 mm wide by 10–30 cm long, often forming large patches; along roadsides, railways and in prairies. *Naturalized* w2
- **FLOWERS** Yellow, blooming May–October; **inflorescence** a raceme, numerous, 2.5–10 cm long (to 15 with fruit) by 10–12 mm wide, mostly ascending, 20–70 flowers per raceme; **floral branches** numerous, alternate, 10–60 cm long throughout the plant; **peduncles** axillary, glabrous, 1–4 cm long; **pedicels** 1–2 mm long, curved downward; **subtending bracts** (of pedicels) one, about as long as the pedicel; **flowers** perfect, average about eight per cm of raceme, each 2.5–3.5 mm wide by 4.5–7 mm long by *c.* 3.5 mm high; **rachis** glabrous; **calyx** tubular, green, slightly hairy, tube *c.* 1 mm long, teeth five, 0.6–1.5 mm long, pointed, persistent; **petals** five; **banner** (one petal) 5–6 mm long by 3–4 mm wide, with a shallow apical notch; **wings** (two petals) 3–5.2 mm long, claw 1.5–2 mm long; **keels** (two petals) 3–4.5 mm long, not united, claw *c.* 1.7 mm long; **stamens** 10, diadelphous (9 + 1); **filaments** mostly united to form a split tube *c.* 4 mm long around the ovary, distinct at tips; **anthers** yellow, *c.* 0.5 mm long; **stipe** 0.8–1 mm long; **ovary** 1–1.3 mm long, dark green; **style** curved, 2–2.5 mm long, light green, persistent; **stigma** obscure. **FRUIT a legume** (pod), compressed, mostly wrinkled, 2.3–3.5 mm long by 2–2.5 mm wide, 2-valved, many dropping early before ripe; **beak** 0.2–0.5 mm long; **seeds** mostly one per pod, smooth, 1.7–2.2 mm long by *c.* 1.3 mm wide by *c.* 1 mm thick, tan to olive green, some with purple spots.
- **LEAVES** Alternate, compound (trifoliate), at the bases of branches or peduncles; **leaf blades** 1–7 cm long by 1.5–5 cm wide, glabrous; **leaflets** 1–5 cm long by 0.4–2 cm wide, toothed, the terminal stalked leaflet the largest; **petioles** glabrous, 3–30 mm long; **stipules** 8–10 mm long, thin.
- **STEM** Tough, erect to leaning, 1–10 stems clumped, branched, glabrous to slightly hairy, pith white; 2–10 mm thick near the reddish green base.

Pea—*Fabaceae* **Wildflower** Pale yellow to purple to pink Petals 5

LATE YELLOW LOCOWEED *Oxytropis campestris* (L.) DC.

■ **SKETCH** A hairy, variable **perennial herb**, tufted, 15–45 cm tall from a stout **taproot** and branched **caudex**; in prairies, open woodlots, moist banks, mountains and brushy ravines. **p1**

■ **FLOWERS** Pale yellow to purple to pink, blooming May–July; **inflorescence** a raceme, erect, 3–8 cm long by 18–25 mm wide; **pedicels** 0.5–1 mm long, to 2.5 mm long in fruit, hairy; **subtending bracts** (of pedicels) 7–10 mm long by 1.2–1.5 mm wide, entire, pointed, very hairy, hairs white or brown; **flowers** perfect, 3–20 per raceme, crowded, ascending, 5–7 mm wide by 12–15 mm long by 6–7 mm tall; **calyx** tube glabrous inside, TL 5–9 mm by 3–3.5 mm wide and thick, very hairy outside, some black hairs among the white ones, calyx teeth five, 1.5–3 mm long, about equal, hairy on both sides; **petals** five; **banner** (one petal) TL 12–18 mm by 6–9 mm wide, veins obscure, apical notch slight, claw 5–7 mm long by 1.5–2 mm wide; **wings** (two petals) TL 10–16 mm by 2.5–6 mm wide, lobe 1.5–2 mm long, claw 5–6 mm long; **keels** (two petals) TL 10–14 mm by 2.5–3 mm wide (across flat side), some blades with purple blotches, apices pointed, united below, claw 5–6 mm long; **stamens** 10, diadelphous (9 + 1), included in keel; **filaments** glabrous, 8–10 mm long; **anthers** brown, *c.* 0.4 mm long; **pistil** *c.* 11 mm long; **ovary** pale green, white hairs on the lower 5–6 mm; **stigma** *c.* 0.3 mm wide atop a glabrous persistent style; **fruiting stalks** erect, 4–10 cm long by 1.5–2.3 cm wide (including beaks). **FRUIT a legume** (pod), ascending, reddish tan, 2-valved, papery, 10–20 mm long (including beak) by 2.5–4 mm wide and thick, some hairs black; **seeds** dark brown, dull, 1.3–2.4 mm long by 1.1–1.3 mm wide by *c.* 0.9 mm thick.

■ **LEAVES** Basal, compound (odd-pinnate), erect to ascending, TL 6–25 cm by 2–4 cm wide; **leaflets** 7–33, pointed, entire, sessile, 6–25 mm long by 2–5 mm wide, reduced above, white hairs appressed and basifixed; **petioles** hairy, 3.5–5.5 cm long; **stipules** glabrous to hairy dorsally, 4–16 mm long, tapered to a fine point.

■ **STEM** A **scape**, erect, several to 48 in a clump, stiff, with 10–12 low longitudinal ridges, hairy, usually 10–35 cm long by 1.5–2.5 mm wide.

■ **SYN.** *Oxytropis varians* (Rydb.) K. Schum. = *Oxytropis campestris* var. *varians* (Rydb.) Barneby.

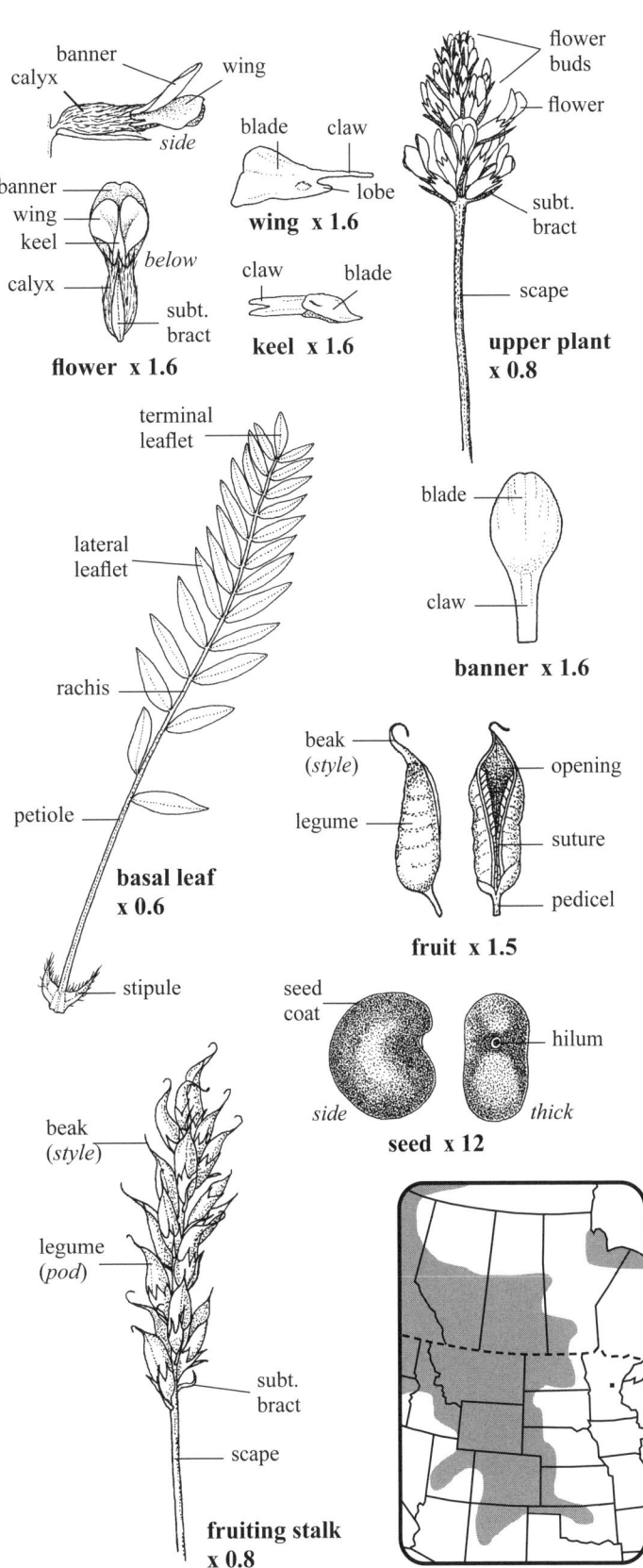

Field Locoweed, Northern Yellow Locoweed, Early Yellow Locoweed

Pea—*Fabaceae* Wildflower Purple Petals 5

Silverleaf Psoralea *Pediomelum argophyllum* (Pursh) J. Grimes

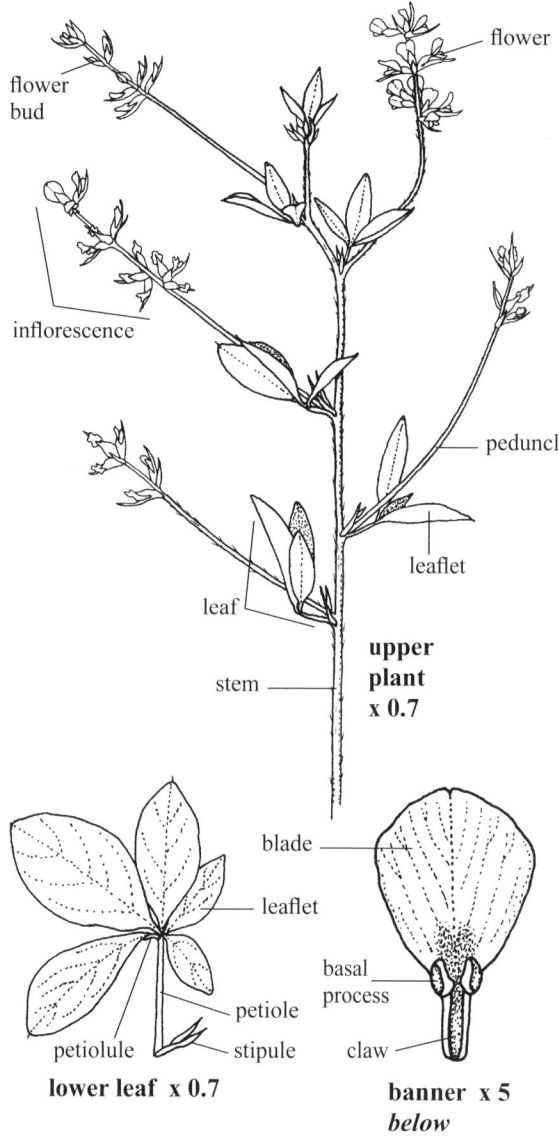

upper plant x 0.7

lower leaf x 0.7

banner x 5 below

■ **SKETCH** A silvery-hairy **perennial herb** 30–80 cm tall from a **taproot** 4–14 mm thick and to *c.* 10⁺ cm long with a **branching caudex**, often in loose colonies; in dry to moist prairie, open woodlands, dunes and along stream valleys. **p1**

■ **FLOWERS** Purple, quickly fading to light brown, blooming June–September; **inflorescence** a spike, numerous, open, each 1.5–7 cm long by 1–1.5 cm wide of two to six distinct whorls, each whorl with two to six (usually three) sessile flowers; **lowest branches** to *c.* 35 cm long and ascending; **peduncles** ascending, axillary, 2–9 cm long; **subtending bracts** (of flowers) one, narrow, hairy, 5–6 mm long; **flowers** perfect, sessile, 3–4 mm wide by 6–9 mm long; **calyx** hairy, bell-shaped, tube 3–5 mm long, the upper four teeth 2.5–3 mm long, the lower tooth 5–10 mm long and extending beyond the bract and petals, persistent in fruit; **petals** five, glabrous; **banner** (one petal) 5–7 mm long by 4–5 mm wide, its white claw 1–2 mm long with two raised basal processes *c.* 1 mm long; **wings** (two petals) 4–6 mm long; **keels** (two petals) 4–5 mm long; **stamens** 10, diadelphous (9 + 1), included; **anthers** yellow. FRUIT a legume (pod), 7–9 mm long by 3–3.5 mm wide by *c.* 1.7 mm thick, hairy, beaked; **seeds** one per pod, 3–4.5 mm long by 2–3 mm wide by 1–1.5 mm thick, dark brown to olivaceous, smooth.

■ **LEAVES** Alternate, compound; **leaf blades** 1.8–7.2 cm long by 2.5–7.5 cm wide; **leaflets** entire, 2–4.5 cm long by 5–22 mm wide, usually three or five leaflets below, reduced to three above, more hairy below producing a whitish-silvery look; **petioles** 0.5–5 cm long; **stipules** 0.5–1.8 cm long, ascending; **petiolules** 1–2.5 mm long.

■ **STEM** Erect, with a silvery look due to white hairs; branched and rebranched; 2–5 mm wide at the base, breaking off at ground level in late summer and tumbling with the wind aids in seed dispersal.

■ **SYN.** *Psoralea argophylla* Pursh.

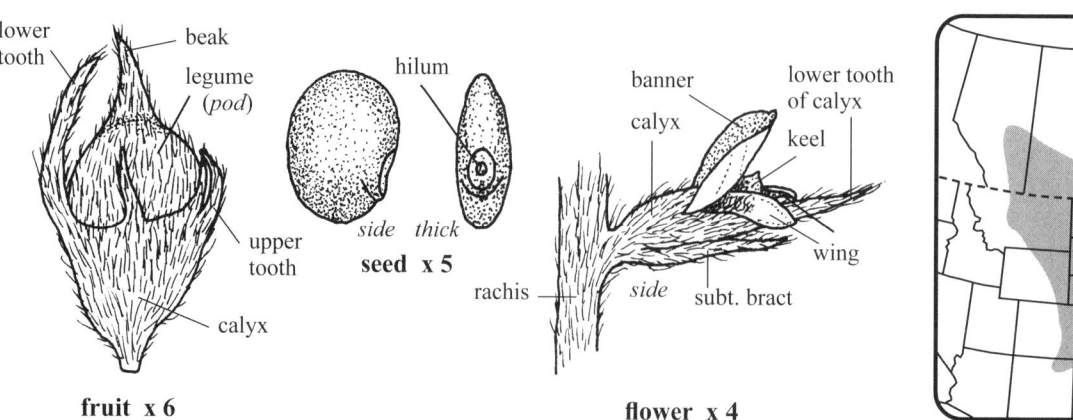

fruit x 6

seed x 5

flower x 4

Silverleaf Indian Breadroot, Silver-leaf Scurf-pea

Pea—*Fabaceae* **Wildflower** Pale bluish purple Petals 5

Indian Breadroot *Pediomelum esculentum* (Pursh) Rydb.

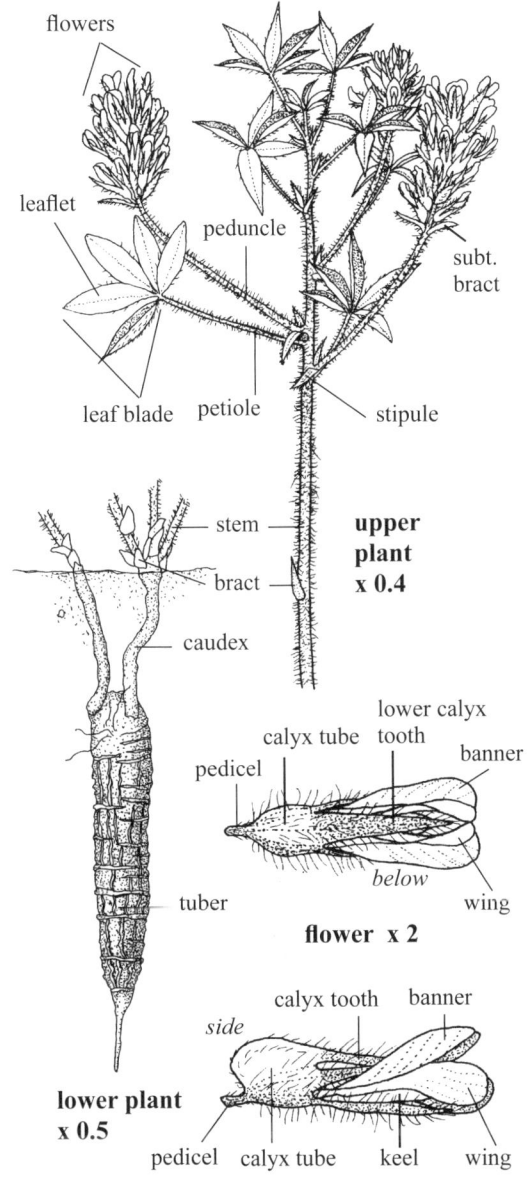

- **SKETCH** A hairy, variable **perennial herb** 10–50 cm tall with dark brown **tuberous roots** 4–20+ cm long by 1.3–8 cm wide; **caudices** one to three, 3–5 cm long by *c.* 5 mm wide; in moist to dry prairies, valleys, open woods, roadsides and sandy banks. p2
- **FLOWERS** Light bluish purple, quickly fading to pale yellow, blooming May–July; **inflorescence** a raceme (spikelike), three to eight per plant, each 2–9 cm long by 2–4.5 cm wide (including subt. bracts); **peduncles** hairy, 2–15 cm long, purplish, axillary; **pedicels** 0–1.5 mm long; **subtending bracts** (of pedicels) hairy outside, glabrous inside, entire, 10–15 mm long by 4–5 mm wide; **flowers** perfect, in whorls of three to seven, 5–6 mm wide by 15–20 mm long; **calyx tube** hairy outside but glabrous inside, 4–7 mm long by *c.* 3 mm thick by *c.* 4.5 mm wide (side), persistent, calyx teeth five, unequal, hairy, some hairs inside of teeth, lowest tooth 6–9 mm long by *c.* 2 mm wide, side and upper teeth 5–7 mm long; **petals** five; **banner** (one petal) TL 13–16 mm and to *c.* 7 mm wide (flat), with two basal processes, claw 4–6 mm long; **wings** (two petals) bluish purple, TL 14–18 mm by *c.* 4 mm wide, claw 5–7 mm long; **keels** (two petals) TL *c.* 11 mm by *c.* 3 mm wide (one side), united at the purple tips, claws two, *c.* 5.5 mm long; **stamens** 10, diadelphous (9 + 1), included; **filaments** mostly united, apices distinct; **anthers** *c.* 0.5 mm long, yellow, attached to expanded tips of filaments; **ovary** glabrous, *c.* 2 mm long; **style** filiform, hairy below for about half its length. FRUIT a legume (pod), body 5–7 mm long, enclosed in calyx, beak (style) and cap 14–20 mm long, deciduous; **seeds** one (two) per pod, 4–5 mm long by 3.2–3.5 mm wide by 2–2.2 mm thick, dark brown, some spotted, glabrous.
- **LEAVES** Alternate, palmately compound, glandless; **blades** 2–5 cm long by 2.2–6.8 cm wide; **leaflets** five (three), 1.5–5 cm long by 7–18 mm wide, slightly folded, entire, white hairy below especially along margins, slightly hairy above and mostly along midribs and some side veins; **petioles** 1.6–13 cm long, hairy; **stipules** entire, 10–20 mm long by 3–8 mm wide, pointed, glabrous inside, hairy outside; **petiolules** 1–2 mm long.
- **STEM** Erect, round, one to three, lower part of stem leafless except for a few bracts; 2–10 mm thick near the base (including white hairs); **tumbles** in wind to disperse seeds.
- **SYN.** *Psoralea esculenta* Pursh.

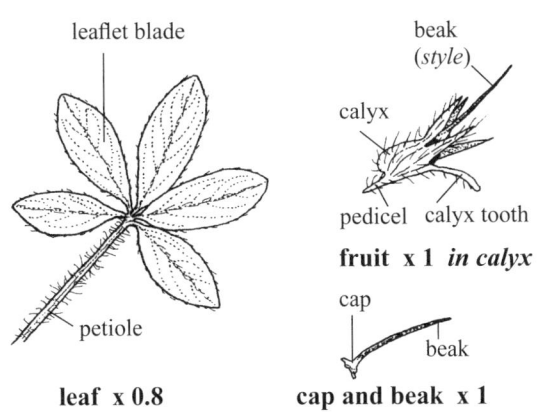
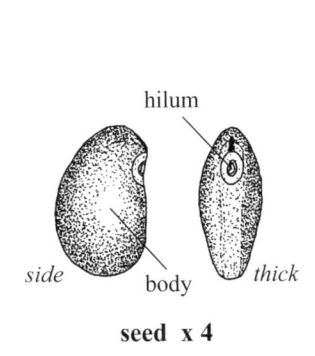
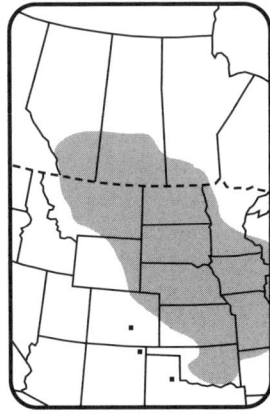

Prairie Turnip, Large Indian Breadroot

Pea—*Fabaceae* **Wildflower** Yellow Petals 5

GOLDEN-BEAN   *Thermopsis rhombifolia* (Nutt. ex Pursh) Nutt. ex Richards.

■ **SKETCH** A **perennial herb** 15–60 cm tall with brown woody **rhizomes** 3–5 mm thick, forming small clumps or large colonies; in moist to dry sites on plains and hillsides, along roads and railways, in badlands, canyons and meadows. **p1**

■ **FLOWERS** Yellow, slightly fragrant, blooming April–July; **inflorescence** a raceme, terminal and axillary, 5–20 cm long by 4–5 cm wide, of 8–25 flowers, blooming sequence bottom to top; **pedicels** 4–10 mm long, hairy; **subtending bracts** (of pedicels) one, 5–11 mm long by 3–6 mm wide, hairy, veiny, pointed; **flowers** perfect, 14–18 mm wide by 15–23 mm long; **calyx** TL 7–10 mm, tubular, persistent, hairy on the outside, calyx tube 5-veined, 4–5 mm long, glabrous inside, calyx teeth five, 2.5–5 mm long, hairy, upper two teeth mostly united, three lower teeth distinct and pointed; **petals** five, glabrous; **banner** (one petal) TL 15–20 mm by 14–18 mm wide, notched, often with purplish dots or streaks inside near the base of the blade, claw 2–5 mm long, tapered; **wings** (two petals) TL 15–22 mm, blade 12–16 mm long by 5–8 mm wide, basal lobe *c.* 2 mm long, claw 3–6 mm long, pale yellow; **keels** (two petals) TL 14–19 mm by 6–7 mm wide on one side, united below the middle of the blade, notched or open at apex; **stamens** 10, included, distinct, 10–15 mm long; **filaments** pale green, glabrous; **anthers** yellow, 1.4–1.8 mm long; **pistil** included, TL 16–19 mm; **stipe** *c.* 2 mm long, glabrous; **ovary** hairy, *c.* 1 cm long by *c.* 1 mm wide by *c.* 0.6 mm thick; **style** tapered, 5–7 mm long, hairy only at base; **stigma** obscure. **FRUIT a legume** (pod), dark brown, flattened, curved to straight, 4–10 cm long by 6–7 mm wide by *c.* 3 mm thick, constricted between seeds, 2-valved, glabrous to hairy, slightly veiny, pointed; **seeds** 4–15 per pod, each 3.4–5 mm long by 3–3.5 mm wide by *c.* 2 mm thick, dark brown to yellowish, smooth.

■ **LEAVES** Alternate, compound; **lowest stem leaves** bractlike, mostly united and clasping, 10–15 mm long and wide; **leaflets** three, sessile, entire, 1.5–4 cm long by 1–2.5 cm wide, more hairy below, usually glabrous above, pointed; **petioles** 1–3.8 cm long, hairy; **stipules** paired, more hairy below, entire, 10–28 mm long by 10–18 mm wide.

■ **STEM** Erect, green, angular, simple to few-branched, hairs short, appressed and white; 3–5 mm thick near the base.

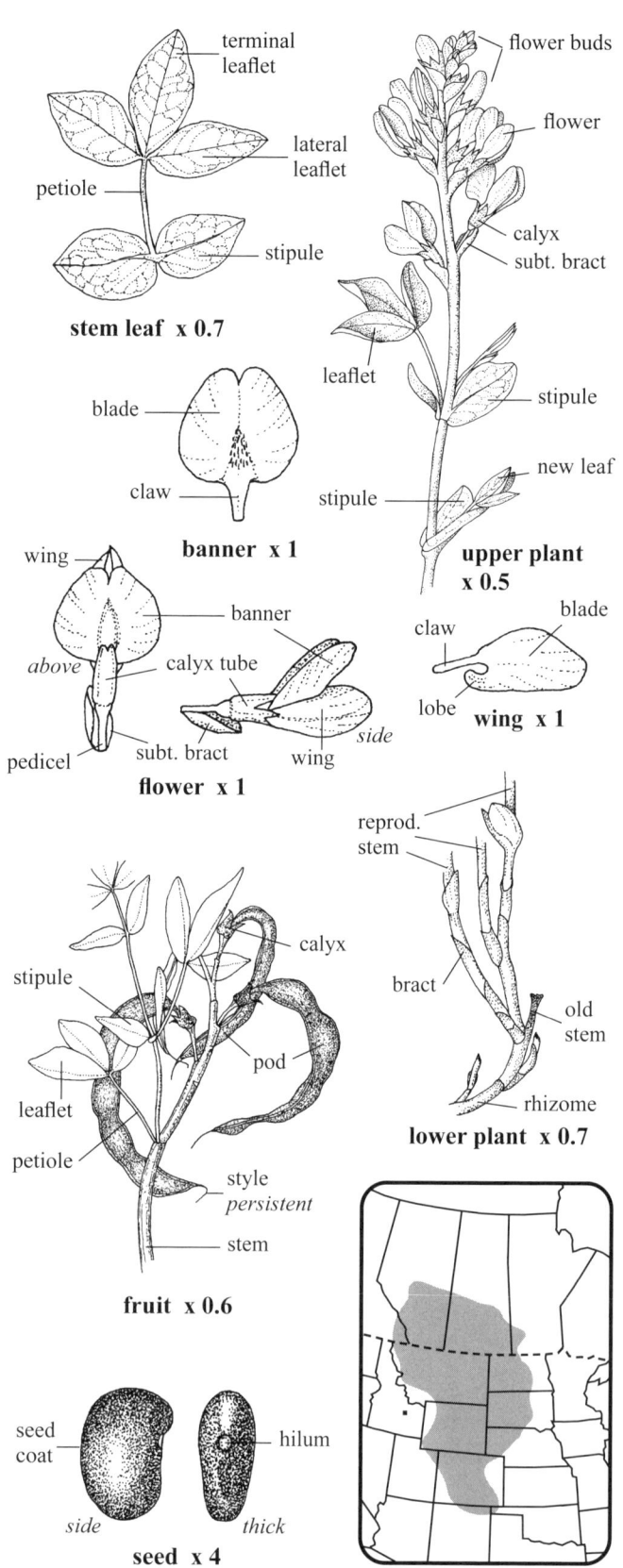

320   Prairie Golden-banner, Prairie Thermopsis, Prairie Buck Bean, Yellow Pea

Yellow Field Clover   *Trifolium campestre* Schreb.

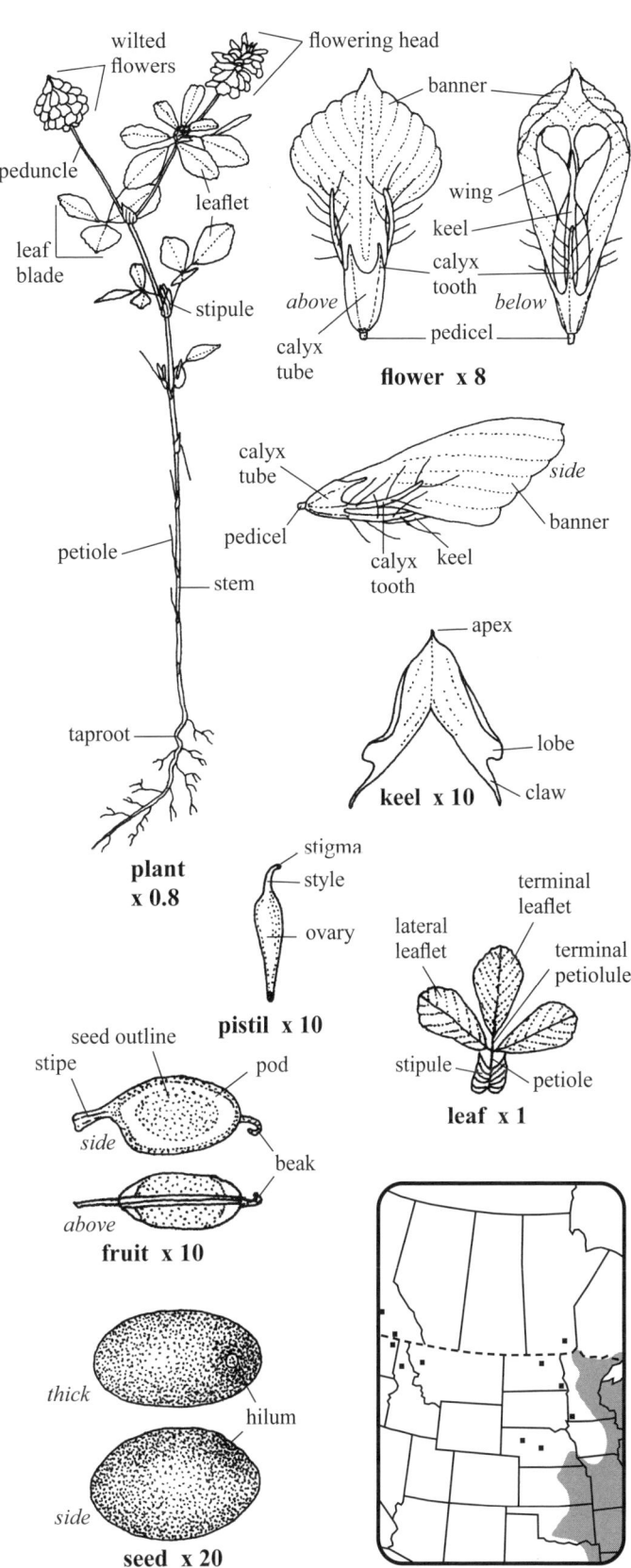

■ **SKETCH** An **annual herb** 10–40 cm tall with a tan stiff **taproot** 3–10 cm long by 0.5–2 mm thick; in open rocky or sandy pastures, dry gravelly roadsides, lawns and open woodlands. *Naturalized* **w2**

■ **FLOWERS** Yellow, blooming May–September; **inflorescence** a raceme, forming a roundish head 10–15 mm long by 9–10 mm wide of 30–45 flowers; **floral branches** 2–15 cm long, ascending, alternate, from upper half of stem; **peduncles** erect, 15–38 mm long, axillary, white hairs ascending and appressed; **pedicels** 0.2–0.6 mm long, glabrous, green to tan with fruit; **flowers** 2.5–3.5 mm wide by 3.5–4.5 mm long, drooping with fruit and reddish brown; **rachis** tan, slightly hairy; **calyx** tubular, membranous, tube 0.7–1 mm long, 5-nerved; **calyx teeth** five, upper two 0.1–0.5 mm long, side and lower teeth filiform and 0.8–1.5 mm long, sinuses rounded, the lower three teeth with long spreading hairs; **petals** five; **banner** (one petal) veiny, *c.* 5 mm long by *c.* 4 mm wide, hairless; **wings** (two petals) slightly longer than keels, apices spreading, *c.* 3.8 mm long by *c.* 1.5 mm wide, clawed; **keels** (two petals) united near tip below, *c.* 2.8 mm long by *c.* 1 mm wide across one side, hairless; **stamens** 10, TL 2–2.3 mm; **filaments** form a membranous tube around pistil, free part of filaments 0.7–1 mm long; **anthers** yellow, *c.* 0.1 mm long; **pistil** 1.5–2 mm long, glabrous; **ovary** tapered, green; **style** white, bent; **stigma** green, not obvious. **FRUIT a pod**, 1-seeded, enclosed in perianth parts, tan, hairless, slightly transparent, body 1.8–2 mm long by 1–1.2 mm wide by *c.* 0.8 mm thick; **stipe** 0.7–1 mm long, tan; **beak** 0.3–0.5 mm long; **seeds** smooth, shiny, medium brown, 0.8–1.2 mm long by 0.8–0.9 mm wide by *c.* 0.7 mm thick.

■ **LEAVES** Alternate, compound; **leaf blades** 1–1.5 cm long by 1.5–2.5 cm wide; **leaflets** three, widest above their middle, each 8–15 mm long by 5–10 mm wide, with shallow teeth in the upper half, hairy on midrib below, glabrous above; **petioles** 1–30 mm long, reduced above, hairy, bare ones below; **petiolules** hairy, side ones 0.5–1 mm long, terminal ones 2–3 mm long; **stipules** pointed, attached to petiole, 3–8 mm long by *c.* 2.5 mm wide, strongly veined, hairy on margins.

■ **STEM** Erect to prostrate, stiff, roundish, not ridged, solitary, branched above, hairs simple, appressed and ascending; 1–2 mm thick near the bare, reddish brown base.

■ **SYN.** *Trifolium procumbens* L. 1755, non 1753.

Pea—*Fabaceae* **Wildflower** White to pink Petals 5

ALSIKE CLOVER *Trifolium hybridum* L.

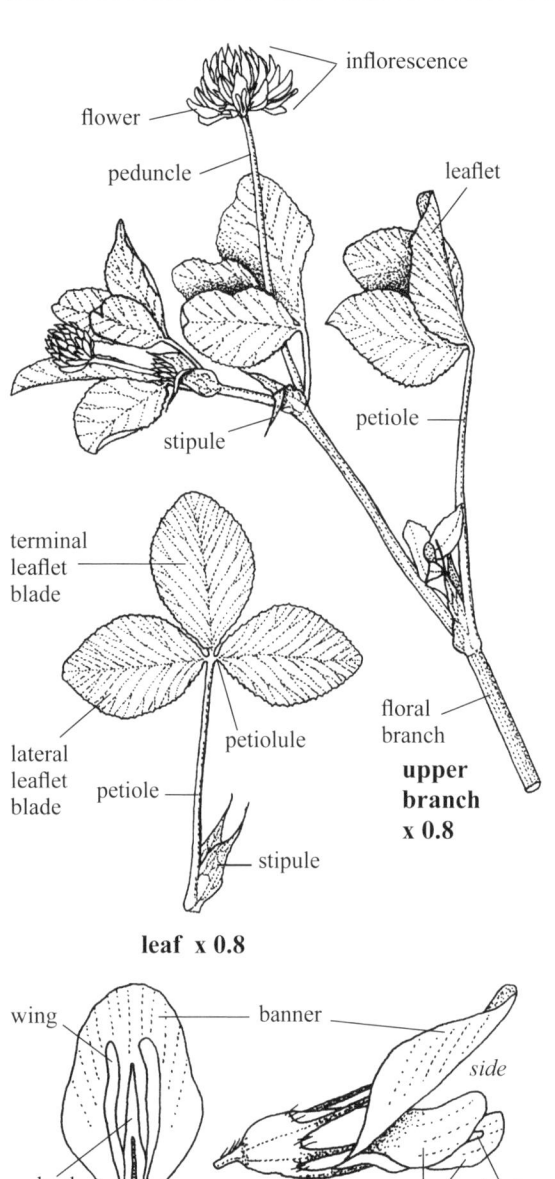

leaf x 0.8

upper branch x 0.8

flower x 5

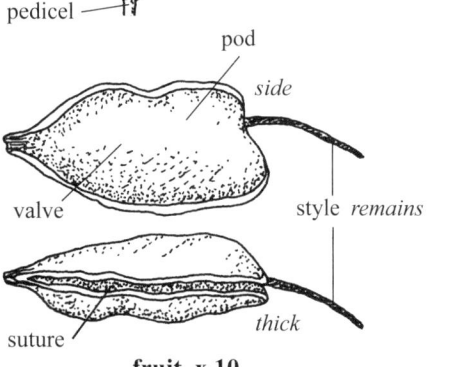

fruit x 10

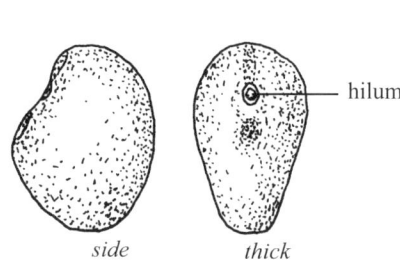

seed x 20

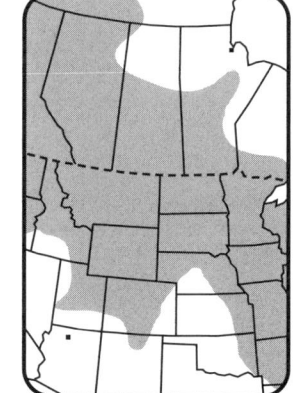

■ **SKETCH** A **perennial herb** 20–80 cm tall and wide from a **taproot** to *c.* 15⁺ cm long; along ditches, streams, roadsides, in moist meadows and pastures. *Naturalized* **w2**

■ **FLOWERS** White to pink, blooming May–October; **inflorescence** a raceme (headlike), each 13–35 mm wide by 11–28 mm tall, of 30–60 flowers; **branches** ascending from several radiating stems; **peduncles** 3–12 cm long, erect from upper leaf axils, rigid, with scattered white hairs; **pedicels** 1–5 mm long, deflexed in fruit, white hairs ascending; **flowers** perfect, *c.* 4 mm wide by 8–12 mm long; **calyx** 2.5–4.8 mm long (including teeth) by *c.* 1.2 mm thick; **calyx tube**, light green, 1.5–2.3 mm long, 5-veined; **teeth** five, tapered, thin, subequal to equal, 1.2–2.5 mm long, some with hairs at base of lobes; **petals** five; **banner** (one petal) 8–12 mm long (including the 2 mm long claw) by *c.* 4 mm wide, light pink, veined; **wings** (two petals) *c.* 7 mm long by *c.* 2 mm wide, claw *c.* 3 mm long; **keels** (two petals) *c.* 6 mm long by *c.* 1.5 mm wide, claws white, *c.* 3 mm long; **stamens** 10, diadelphous (9 + 1), included in keel; **anthers** yellow, *c.* 0.3 mm long; **pistil** slightly shorter than longest stamens; **ovary** glabrous; **style** 1.5–2.3 mm long, persistent; **fruiting heads** reddish brown, flowers deflexed. **FRUIT a legume** (pod), pale green to black, flattened, 3–4 mm long by 1.5–1.8 mm wide by 1–1.5 mm thick, 2-valved, glabrous, enclosed by persistent calyx and withered petals; **seeds** one to four per pod, green to black, some mottled, glabrous, each 1.1–1.4 mm long by 1–1.3 mm wide by 0.5–0.8 mm thick.

■ **LEAVES** Alternate, compound, ascending; **blades** 2–4.5 cm long by 2–7 cm wide; **leaflets** three, 7–35 mm long by 7–30 mm wide, glabrous, finely toothed, lateral veins often divided, chevron absent; **petioles** 1.3–15 cm long, glabrous; **stipules** paired, pointed, 10–30 mm long by 3–7 mm wide, veined, partially united to petioles along the lower third, tips free; **petiolules** *c.* 1 mm long or less, glabrous to lightly hairy.

■ **STEM** Several, erect to arching, oval, pith white, grooved above, glabrous to slightly hairy, simple to branched, reddish green; 2–5 mm thick near the base.

Pea—*Fabaceae*　　　**Wildflower**　Pink (red to white)　Petals 5

Red Clover　*Trifolium pratense* L.

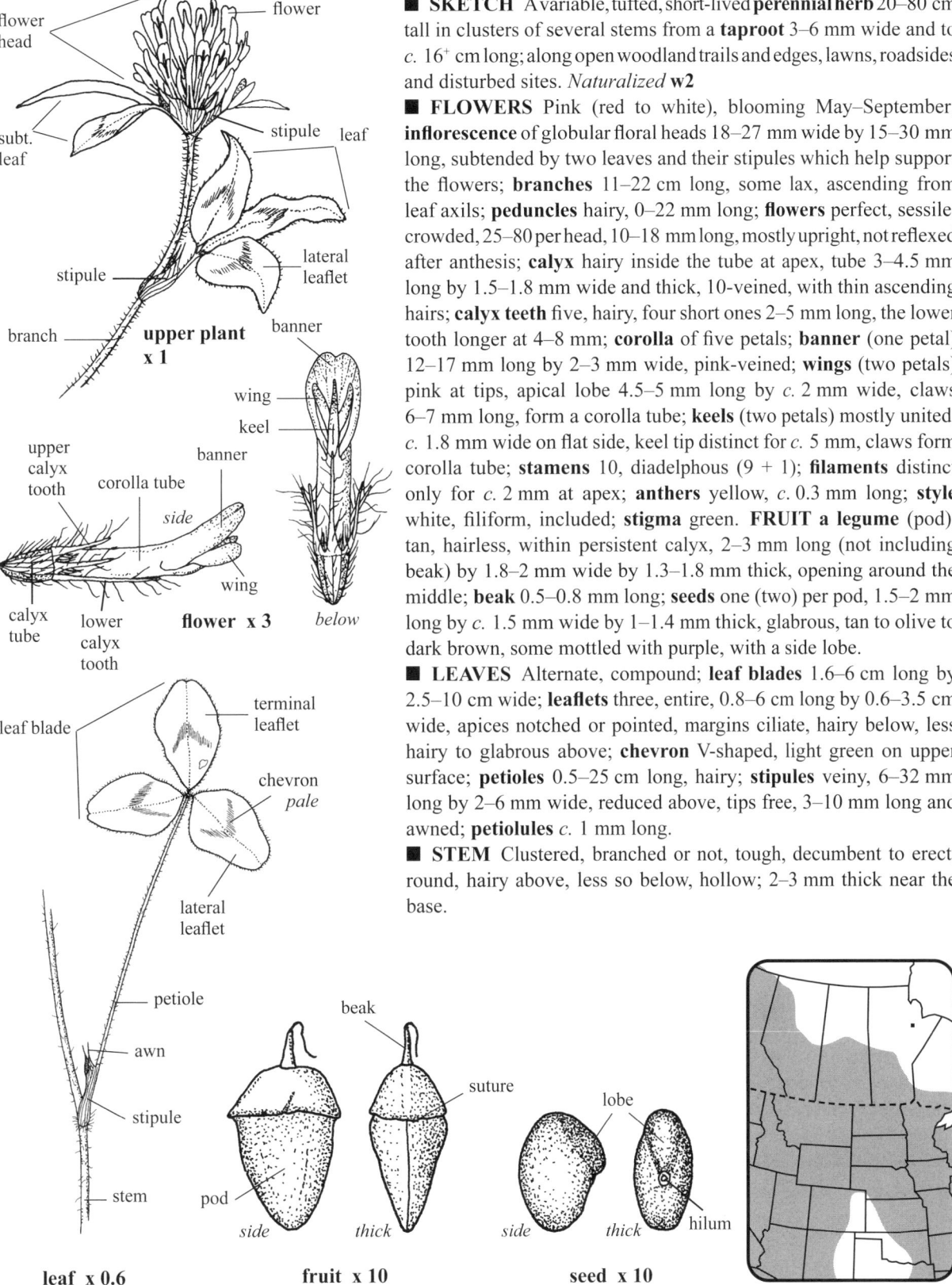

■ **SKETCH** A variable, tufted, short-lived **perennial herb** 20–80 cm tall in clusters of several stems from a **taproot** 3–6 mm wide and to *c.* 16⁺ cm long; along open woodland trails and edges, lawns, roadsides and disturbed sites. *Naturalized* **w2**

■ **FLOWERS** Pink (red to white), blooming May–September; **inflorescence** of globular floral heads 18–27 mm wide by 15–30 mm long, subtended by two leaves and their stipules which help support the flowers; **branches** 11–22 cm long, some lax, ascending from leaf axils; **peduncles** hairy, 0–22 mm long; **flowers** perfect, sessile, crowded, 25–80 per head, 10–18 mm long, mostly upright, not reflexed after anthesis; **calyx** hairy inside the tube at apex, tube 3–4.5 mm long by 1.5–1.8 mm wide and thick, 10-veined, with thin ascending hairs; **calyx teeth** five, hairy, four short ones 2–5 mm long, the lower tooth longer at 4–8 mm; **corolla** of five petals; **banner** (one petal) 12–17 mm long by 2–3 mm wide, pink-veined; **wings** (two petals) pink at tips, apical lobe 4.5–5 mm long by *c.* 2 mm wide, claws 6–7 mm long, form a corolla tube; **keels** (two petals) mostly united, *c.* 1.8 mm wide on flat side, keel tip distinct for *c.* 5 mm, claws form corolla tube; **stamens** 10, diadelphous (9 + 1); **filaments** distinct only for *c.* 2 mm at apex; **anthers** yellow, *c.* 0.3 mm long; **style** white, filiform, included; **stigma** green. **FRUIT a legume** (pod), tan, hairless, within persistent calyx, 2–3 mm long (not including beak) by 1.8–2 mm wide by 1.3–1.8 mm thick, opening around the middle; **beak** 0.5–0.8 mm long; **seeds** one (two) per pod, 1.5–2 mm long by *c.* 1.5 mm wide by 1–1.4 mm thick, glabrous, tan to olive to dark brown, some mottled with purple, with a side lobe.

■ **LEAVES** Alternate, compound; **leaf blades** 1.6–6 cm long by 2.5–10 cm wide; **leaflets** three, entire, 0.8–6 cm long by 0.6–3.5 cm wide, apices notched or pointed, margins ciliate, hairy below, less hairy to glabrous above; **chevron** V-shaped, light green on upper surface; **petioles** 0.5–25 cm long, hairy; **stipules** veiny, 6–32 mm long by 2–6 mm wide, reduced above, tips free, 3–10 mm long and awned; **petiolules** *c.* 1 mm long.

■ **STEM** Clustered, branched or not, tough, decumbent to erect, round, hairy above, less so below, hollow; 2–3 mm thick near the base.

323

Pea—*Fabaceae* **Wildflower** White to pinkish Petals 5

WHITE CLOVER *Trifolium repens* L.

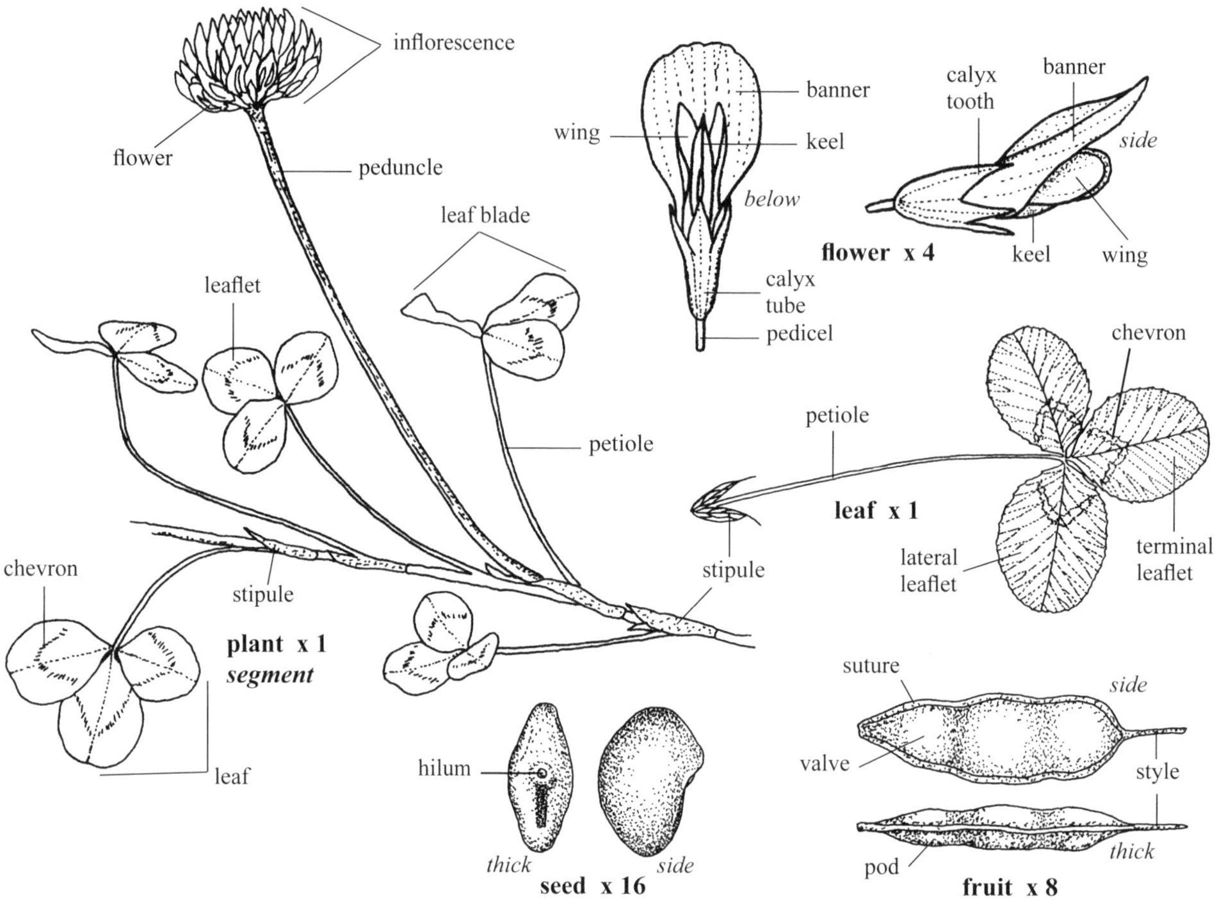

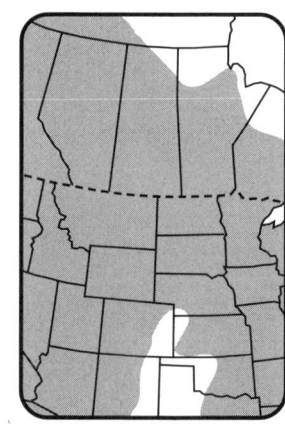

■ **SKETCH** A variable **perennial herb** 15–35 cm long from a **taproot** 1–2 mm thick and to *c.* 15⁺ cm long; in disturbed sites, roadsides, lawns, moist meadows, pastures and clearings. *Naturalized* **w2**

■ **FLOWERS** White to pinkish, blooming May–October; **inflorescence** a raceme of 40–80 flowers in round headlike clusters 14–35 mm wide by 10–22 mm tall; **peduncles** grooved, axillary, 2.5–27 cm long by 1–2 mm thick, ascending; **pedicels** 1–4 mm long, glabrous to slightly hairy; **flowers** perfect, fragrant, glabrous, 8–11 mm long; **calyx** tubular, 10-nerved, the body 2–4 mm long by 1.3–1.5 mm wide; **calyx** teeth five, 2–3 mm long, three short and two slightly longer, not spreading; **petals** five; **banner** (one petal) 6–11 mm long by *c.* 4 mm wide; **wings** (two petals) *c.* 2 mm wide, united by their claws to the staminal tube; **keels** (two petals) slightly shorter than the wings, also united by their claws to the staminal tube; **stamens** 10, diadelphous (9 + 1), included within the keel; **anthers** yellow; **pistil** green, glabrous, *c.* 5 mm long. **FRUIT** a **legume** (pod), linear, 4–6 mm long by 1.5–1.8 mm wide by 0.7–1 mm thick, 2-valved, smooth, slightly indented around the seeds, often with a persistent style; **seeds** usually three (two to four) per pod, yellow to golden brown, smooth, dull, 1–1.7 mm long by *c.* 1 mm wide by 0.6–1 mm thick, usually not mottled.

■ **LEAVES** Alternate, compound; **leaflets** three (four), 1–3 cm long and 7–28 mm wide, finely toothed, mostly glabrous, slightly notched at apices, an inverted V-shaped chevron of light green streaks on the upper surface; **petioles** 2–20 cm long; **stipules** partially united, veined, membranous, 3–11 mm long with tips spreading; **petiolules** *c.* 1 mm long.

■ **STEM** Creeping, reddish green, rooting at some nodes, glabrous to slightly hairy, several, branched and often mat-forming; 1–2 mm thick near the base.

324 Dutch Clover, Ladino Clover

Pea—*Fabaceae* **Vine** Bluish to pinkish purple Petals 5

AMERICAN VETCH   *Vicia americana* Muhl. ex Willd.

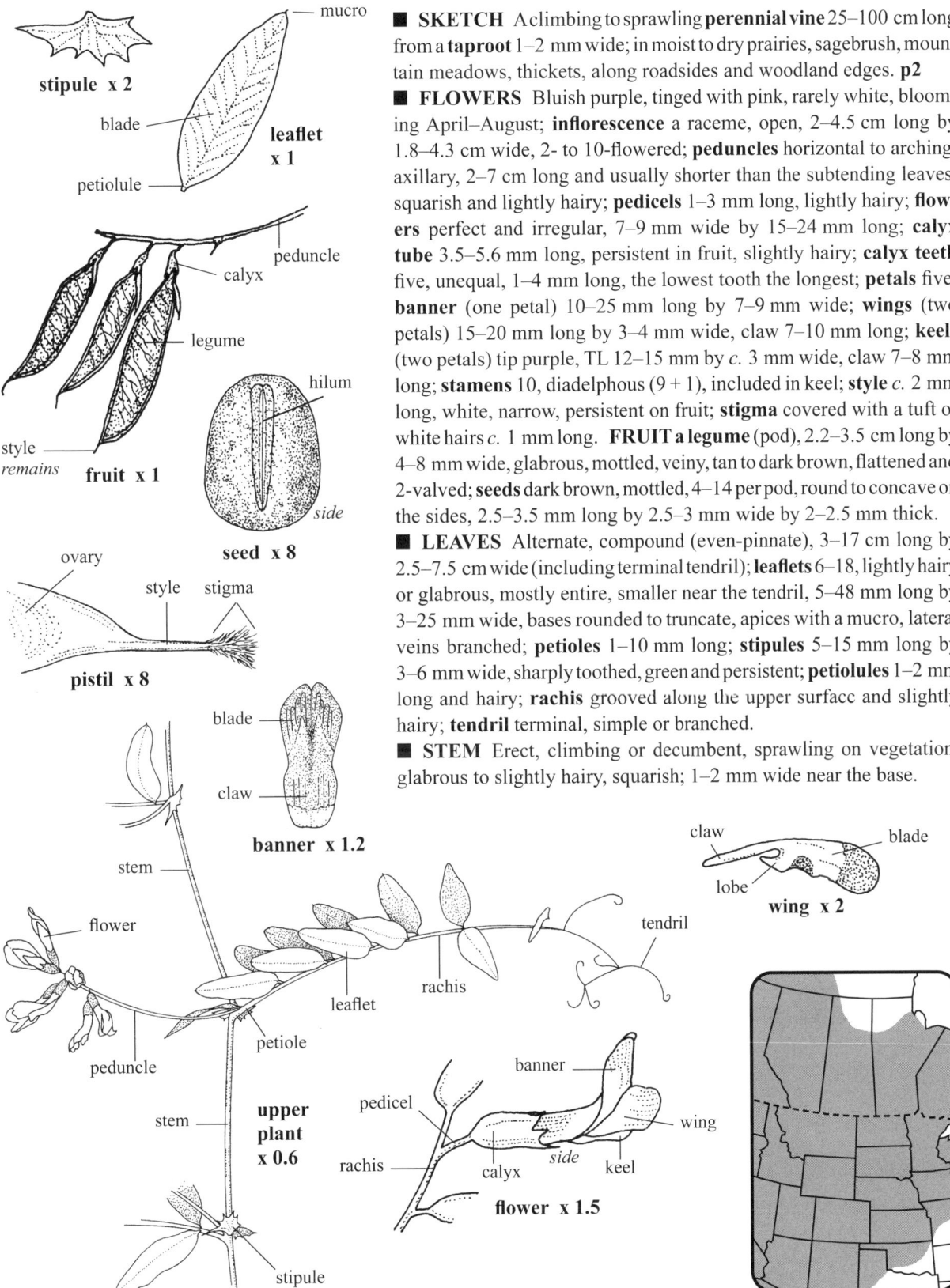

- **SKETCH** A climbing to sprawling **perennial vine** 25–100 cm long from a **taproot** 1–2 mm wide; in moist to dry prairies, sagebrush, mountain meadows, thickets, along roadsides and woodland edges. p2
- **FLOWERS** Bluish purple, tinged with pink, rarely white, blooming April–August; **inflorescence** a raceme, open, 2–4.5 cm long by 1.8–4.3 cm wide, 2- to 10-flowered; **peduncles** horizontal to arching, axillary, 2–7 cm long and usually shorter than the subtending leaves, squarish and lightly hairy; **pedicels** 1–3 mm long, lightly hairy; **flowers** perfect and irregular, 7–9 mm wide by 15–24 mm long; **calyx tube** 3.5–5.6 mm long, persistent in fruit, slightly hairy; **calyx teeth** five, unequal, 1–4 mm long, the lowest tooth the longest; **petals** five; **banner** (one petal) 10–25 mm long by 7–9 mm wide; **wings** (two petals) 15–20 mm long by 3–4 mm wide, claw 7–10 mm long; **keels** (two petals) tip purple, TL 12–15 mm by *c.* 3 mm wide, claw 7–8 mm long; **stamens** 10, diadelphous (9 + 1), included in keel; **style** *c.* 2 mm long, white, narrow, persistent on fruit; **stigma** covered with a tuft of white hairs *c.* 1 mm long. **FRUIT** a **legume** (pod), 2.2–3.5 cm long by 4–8 mm wide, glabrous, mottled, veiny, tan to dark brown, flattened and 2-valved; **seeds** dark brown, mottled, 4–14 per pod, round to concave on the sides, 2.5–3.5 mm long by 2.5–3 mm wide by 2–2.5 mm thick.
- **LEAVES** Alternate, compound (even-pinnate), 3–17 cm long by 2.5–7.5 cm wide (including terminal tendril); **leaflets** 6–18, lightly hairy or glabrous, mostly entire, smaller near the tendril, 5–48 mm long by 3–25 mm wide, bases rounded to truncate, apices with a mucro, lateral veins branched; **petioles** 1–10 mm long; **stipules** 5–15 mm long by 3–6 mm wide, sharply toothed, green and persistent; **petiolules** 1–2 mm long and hairy; **rachis** grooved along the upper surface and slightly hairy; **tendril** terminal, simple or branched.
- **STEM** Erect, climbing or decumbent, sprawling on vegetation, glabrous to slightly hairy, squarish; 1–2 mm wide near the base.

Purple Vetch, American Purple Vetch, Wild Vetch

Pea—*Fabaceae* **Vine** Pinkish blue Petals 5

TUFTED VETCH *Vicia cracca* L.

■ **SKETCH** A **perennial vine** 50–125 cm tall or long, climbing, often covering several m²; in prairies and roadside ditches. Native and/or *naturalized*

■ **FLOWERS** Pinkish blue, blooming May–August; **inflorescence** a raceme, several, 2–9 cm long by 15–25 mm wide, one-sided, of 18–40 reflexed flowers; blooming sequence bottom to top; **peduncles** 2–11 cm long, slightly ridged, from leaf axils; **pedicels** 1–2 mm long, reflexed as flowers fade, hairy; **flowers** perfect, 8–13 mm long (excluding pedicel); **calyx tube** pinkish, hairy, 2–2.8 mm long by *c.* 2.2 mm wide by *c.* 1.5 mm thick; **calyx teeth** five, two teeth behind banner *c.* 0.5 mm long, curved inward, hairy and pink, lower teeth 1.3–2.5 mm long; **petals** five; **banner** (one petal) 4–5 mm wide, base *c.* 9 mm long by *c.* 5 mm wide, glabrous, not clawed; **wings** (two petals) 8–9 mm long by *c.* 2 mm wide, claw 3–4 mm long; **keels** (two petals) apices dark purple, united below, claw *c.* 3.5 mm long by *c.* 1.8 mm wide with two narrow projections *c.* 1 mm long; **stamens** 10, diadelphous (9 + 1), included; **anthers** *c.* 0.3 mm long; **ovary** glabrous; **style** included in keel, *c.* 1.5 mm long, hairy near apex; **stigma** oval, green, glabrous. **FRUIT a legume** (pod), brown with dark streaks, usually one to six per peduncle, descending, 1.5–3 cm long by 4–7 mm wide by 3–4 mm thick; **valves** two, twisting open from the apex; **seeds** four to eight per pod, each 2.5–3.5 mm long, wide and thick, dull, glabrous, brown, some mottled.

■ **LEAVES** Alternate, sessile, compound (even-pinnate), 4.5–12 cm long by 1.3–6 cm wide, those lower on the stem with secondary axillary leaves or a branch; **leaflets** 8–30, each 5–30 mm long by 1–6 mm wide, entire, apices pointed to rounded, mucronate, lower blades the largest, mostly ascending to erect along rachis, hairy on both sides, appearing frosty below, lateral veins obscure, hairs forming a thin white marginal line; **tendrils** green, replacing the terminal leaflet, twisting around nearby vegetation; **stipules** 5–10 mm long by 3–4 mm wide, hairy on outside, entire or with one or two pointed narrow lobes near the base; **petiolules** hairy, *c.* 1 mm long or less.

■ **STEM** Erect to horizontal along the top of vegetation, hollow, 4-ridged below, hairs dishevelled below but mostly appressed above; *c.* 3 mm thick near the base, not woody.

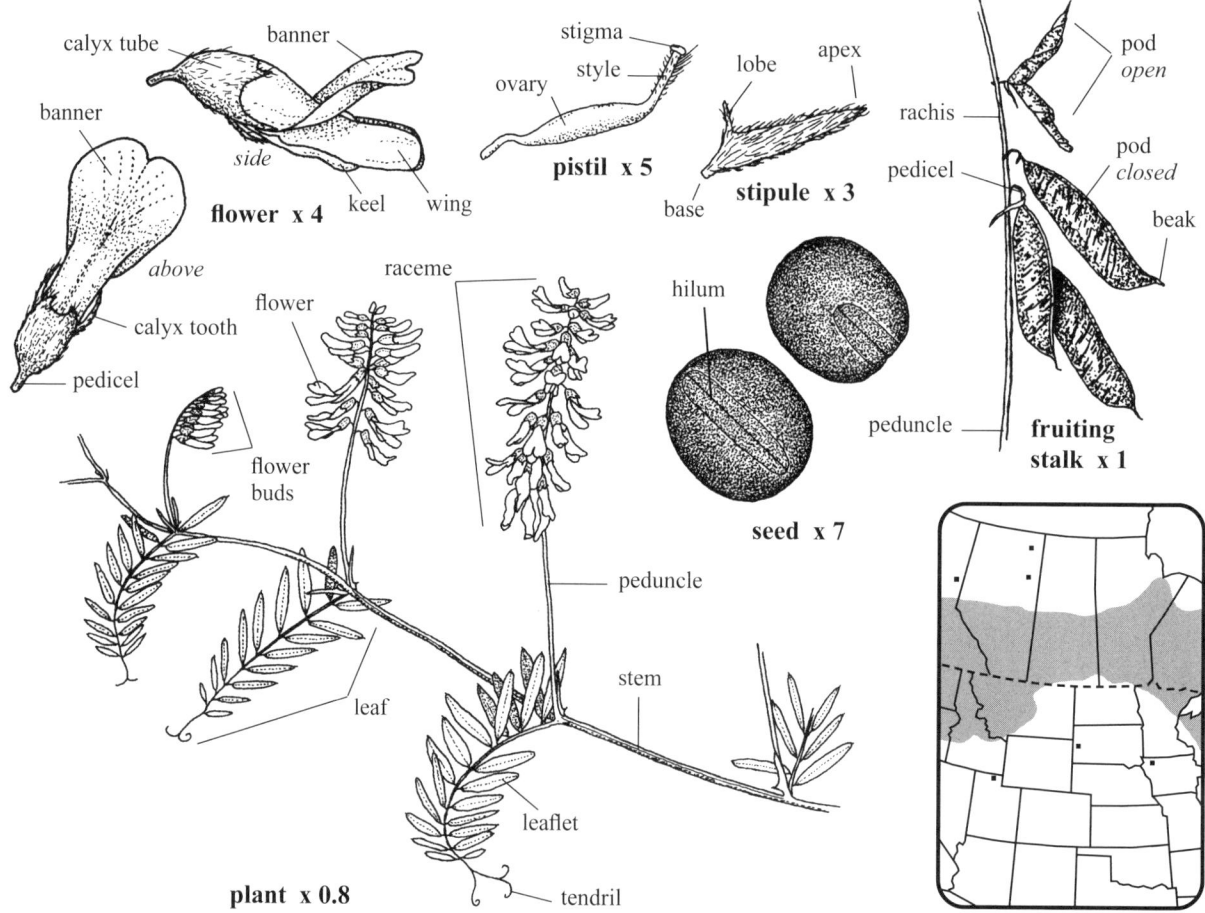

Bird Vetch, Purple-white Tufted Vetch, Cow Vetch

Beech—*Fagaceae*   **Tree**   Green to greenish yellow   Catkins

Bur Oak   *Quercus macrocarpa* Michx.

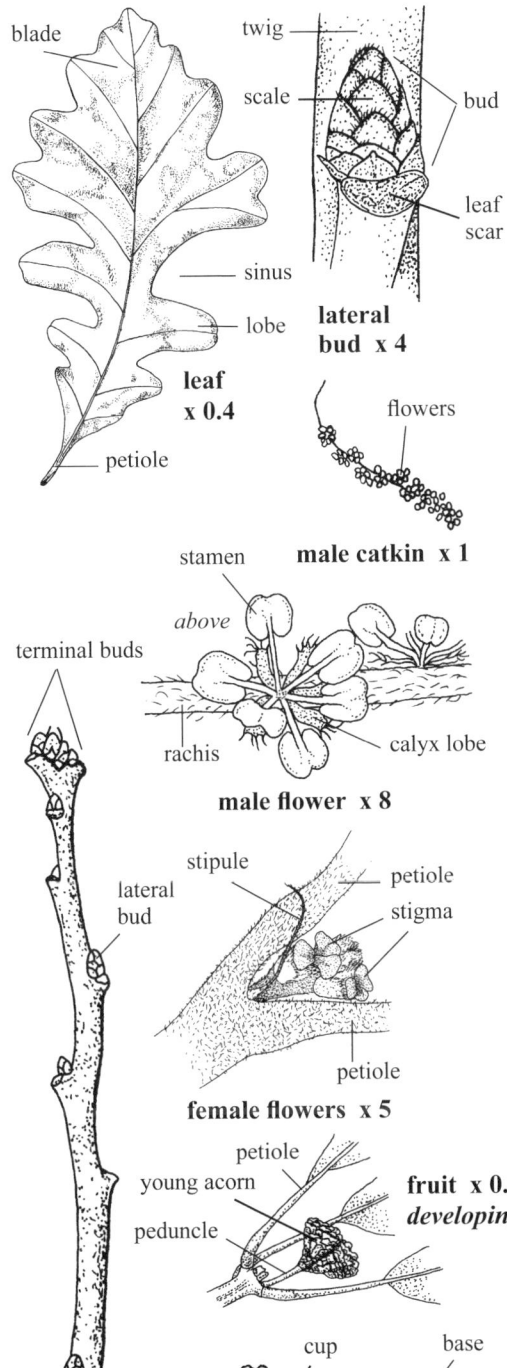

- **SKETCH** A **deciduous tree** 4–50 m tall; in gravelly to moist woodlots, parklands, prairies, bottomlands and dry slopes; wood is very hard; **monoecious**. w1
- **FLOWERS** Green to greenish yellow, blooming April–June; **inflorescence** of unisexual catkins; **male catkins** to *c*. 6 cm long by *c*. 5 mm wide, hanging, one to four per bud; **rachis** hairy, *c*. 0.5 mm wide, base naked; **male flowers** mostly sessile, to *c*. 3 mm wide by 1–1.5 mm long, alternate; **calyx** 3- to 5-lobed, these *c*. 1 mm long with a fringed, often dark tip; **stamens** usually five to eight, erect to spreading; **filaments** 1–2 mm long; **anthers** 1–1.5 mm long, light green; **pollen** yellow; **peduncles** erect, 1–2 mm long at anthesis, lengthening to 6–20 mm in fruit; **female flowers** in clusters of one to five; **flowers** 1–2 mm wide and long, sessile; **calyx** 5- or 6-lobed, green with reddish tips, hairy; **styles** distinct, greenish yellow; **stigmas** thick, 3-lobed, together 1–1.5 mm wide. **FRUIT a nut** (acorn), one to three per peduncle, 1-seeded, maturing the first year; **cup** deciduous, covering one half or more of the acorn, light brown to gray, 1–5 cm long by 1–6 cm wide, rough from scaly involucre and a fringed margin with 5–10 mm long awns; **acorn** smooth, brown, 1.5–5 cm long by 1–5 cm wide, hard and round, the base whitish tan, blunt at the apex; **seed** round, not shiny, medium brown with shallow furrows, 11–20 mm long by 9–18 mm wide, enclosed in a papery brown cover.
- **LEAVES** Alternate, simple, shallowly to deeply lobed; **blades** 5–30 cm long by 3.5–15 cm wide, the upper surface shiny and glabrous, lighter green below and lightly hairy on the blade and raised veins; **petioles** 6–30 mm long, round, lightly hairy; **stipules** paired, filiform, 5–7 mm long, tapered, turning brown and falling early.
- **TRUNK** Bark dark gray, deeply furrowed, scaly; **new twigs** tan at the treetop, finely hairy; **older twigs** brownish gray, often with corky ridges; **lenticels** oval, often obscured; **winter buds** alternate, 2.5–6 mm long by *c*. 3 mm wide, blunt, the terminal cluster light gray to reddish brown, finely hairy, scales numerous, margins ciliate; **leaf scars** wide, oval; **bundle scars** obscured by hairs; **dbh** 20–150 cm.

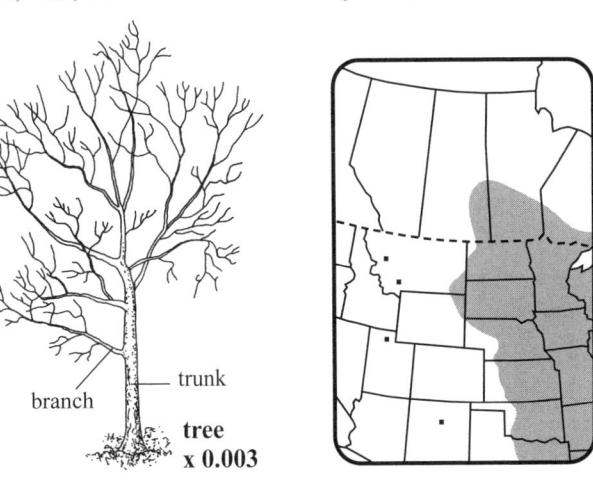

Mossycup Oak   327

Fumitory—*Fumariaceae* **Wildflower** Yellow Petals 4

Golden Corydalis   *Corydalis aurea* Willd.

■ **SKETCH** A glabrous, glaucous **winter annual** or **biennial herb** 10–50 cm tall or long and 30–45 cm wide from a **taproot**; in open woods, moist prairies, sagebrush, along railways, shores, rocky slopes and disturbed sites.

■ **FLOWERS** Yellow, blooming March–August; **inflorescence** a raceme, each 1.5–7 cm long by 1.5–3 cm wide, rising above the leaves or not; **floral branches** several per plant, to *c.* 20 cm long, fleshy, pinkish, erect to prostrate with age, hollow, 7- or 8-angled, each with one to four racemes; **pedicels** 2–7 mm long, ascending, recurved and to *c.* 9 mm long with fruit; **subtending bracts** (of pedicels) one, 4–10 mm long by 1.5–2.5 mm wide, V-shaped, erose near the pointed tip; **flowers** perfect, 10–30 per raceme, 3–4 mm wide by 11–15 mm long by 8–10 mm tall; **sepals** two, erose, early deciduous, 1–3 mm long by 1–2 mm wide; **petals** four, crests green and *c.* 2.5 mm long by *c.* 0.8 mm high, toothed or not; **lower petal** 8–13 mm long, spurless; **inner petals** two, united at the tip, 7–11 mm long by 2–2.2 mm wide, the crest fitting into the upper petal's crest, claw *c.* 4 mm long and narrow; **upper petal** one, 13–18 mm long, its spur yellow, 4–9 mm long with a slight curve; **stamens** six (3 + 3), diadelphous, included, 7–8 mm long, the spurs 2–6 mm long; **filaments** 6–7 mm long, widening at their base to 1–1.5 mm, wrapped around the pistil; **anthers** dimorphic, yellow, *c.* 0.7 mm long, two outer anthers in each group with one pollen sac, the inner or middle anthers with two sacs; **pistil** 10–12 mm long (including stipe), glabrous; **style** narrow, 3–4 mm long; **stigma** flattened, lobed and green. **FRUIT a capsule**, glabrous, 15–30 mm long by 2.5–4.2 mm wide by 2–2.5 mm thick, erect to descending, straight to curved, 2-valved, opening from base, persistent style *c.* 2 mm long; **seeds** 8–20 per capsule, shiny black, in two rows, 1.8–2.1 mm long by 1.7–1.9 mm wide by 1–1.2 mm thick, a whitish, opaque, parallel-veined appendage attached near the hilum and wrapped around about half of the seed's body (not illustrated).

■ **LEAVES** Basal and stem, alternate, compound (twice divided); **basal blades** several, light bluish green, glaucous below, 2–14 cm long by 1–8 cm wide, not stiff; **leaflets** usually 7–13, alternate to opposite, lobed and toothed, ascending, 2–3 cm long by 1–2.5 cm wide (flattened); **petioles** 1–16 cm long by *c.* 2 mm wide and thick, reduced above, weak, fleshy; **stipules** absent; **petiolules** 1–6 mm long, reduced above; **stem leaves** similar and reduced above.

■ **STEM** Several, hollow, weak, simple to branched, erect to prostrate; 5–10 mm thick near the base.

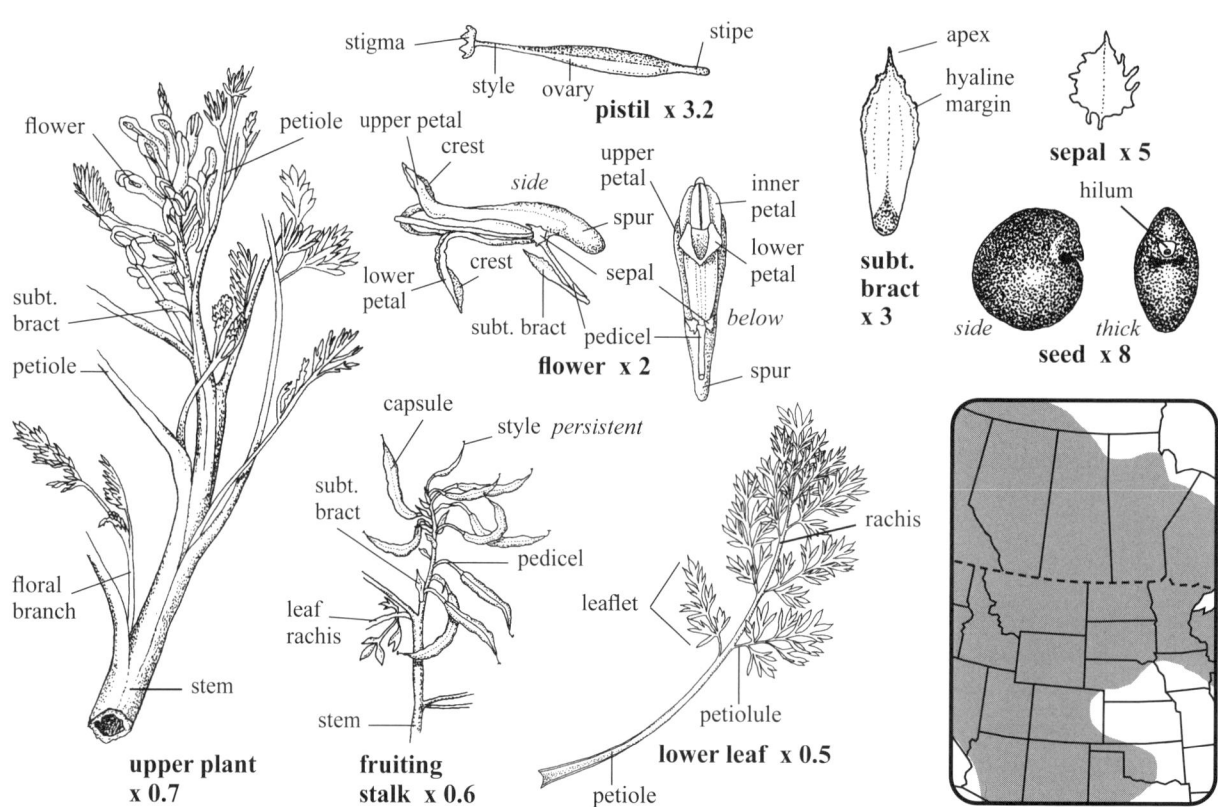

328   Scrambled Eggs

Fumitory—*Fumariaceae* **Wildflower** Pink, tips yellow Petals 4

Pink Corydalis *Corydalis sempervirens* (L.) Pers.

■ **SKETCH** A glaucous and glabrous **biennial herb** 10–80 cm tall with a **taproot** 3–20 cm long; in open woods, talus and depressions on rocky outcrops in forests.

■ **FLOWERS** Light pink with yellow tips, blooming May–September; **inflorescence** a raceme or panicle 3–15 cm long by 3–10 cm wide, terminal on main stem and branches; **floral branches** few, 2–35 cm long, ascending to arched; **pedicels** 3–20 mm long, ascending, reddish with fruit; **subtending bracts** (of pedicels) ascending, shorter than the pedicel and 1–1.5 mm wide, mostly entire, pointed; **flowers** perfect, 3–4 mm wide by 10–16 mm long by 5–6 mm tall, 3–15 per raceme; **sepals** two, pale green to pink with reddish margins, erose, 3–4 mm long by 1.3–2 mm wide, appressed, early deciduous; **petals** four, spur rounded, 3–5 mm long; **stamens** six, diadelphous, lower three in lower petal, 8–9 mm long; **filaments** united, whitish, flattened, membranous, *c.* 1.2 mm wide, attached to the petal along the lowest 2–3 mm, apices divided; **anthers** yellow, *c.* 0.5 mm long, the two lateral ones with only one anther sac, the other four with two pollen sacs; **pistils** hairless, 10–12 mm long by *c.* 0.5 mm thick, reddish green, pale green at apex; **stigmas** 3- or 4-lobed, persistent on fruit. **FRUIT a capsule**, tan, erect to ascending, glabrous, 1.5–4.7 cm long by 1–1.5 mm thick, valves two, opening from base along two sutures, beak 1–3 mm long, seedless; **seeds** 20–28 per capsule, 1.2–1.4 mm long by *c.* 1.1 mm wide by *c.* 0.7 mm thick, in one row, slightly overlapping, black, shiny, with low concentric papillose rows; **aril** curved, 0.3–1 mm long, ridged, tan near hilum, whitish near apex.

■ **LEAVES** Basal and stem, alternate, doubly compound, pale green; **basal leaves** few to several, ascending; **petioles** 2–7.5 cm long; **petiolules** as below; **stem leaves** ascending, 2–10 cm long by 2–9 cm wide, reduced above; **side leaflets** 1–3.5 cm long and wide, lobed and toothed; **terminal leaflets** 2–3.5 cm long by 1.5–5 cm wide; **petioles** 0.4–8.5 cm long, reduced above; **petiolules** (of side leaflets) 0.2–2 cm long.

■ **STEM** Erect, often one, weak, ridged, glabrous and glaucous, green and simple to branched above; pinkish near the 2–7 mm thick base.

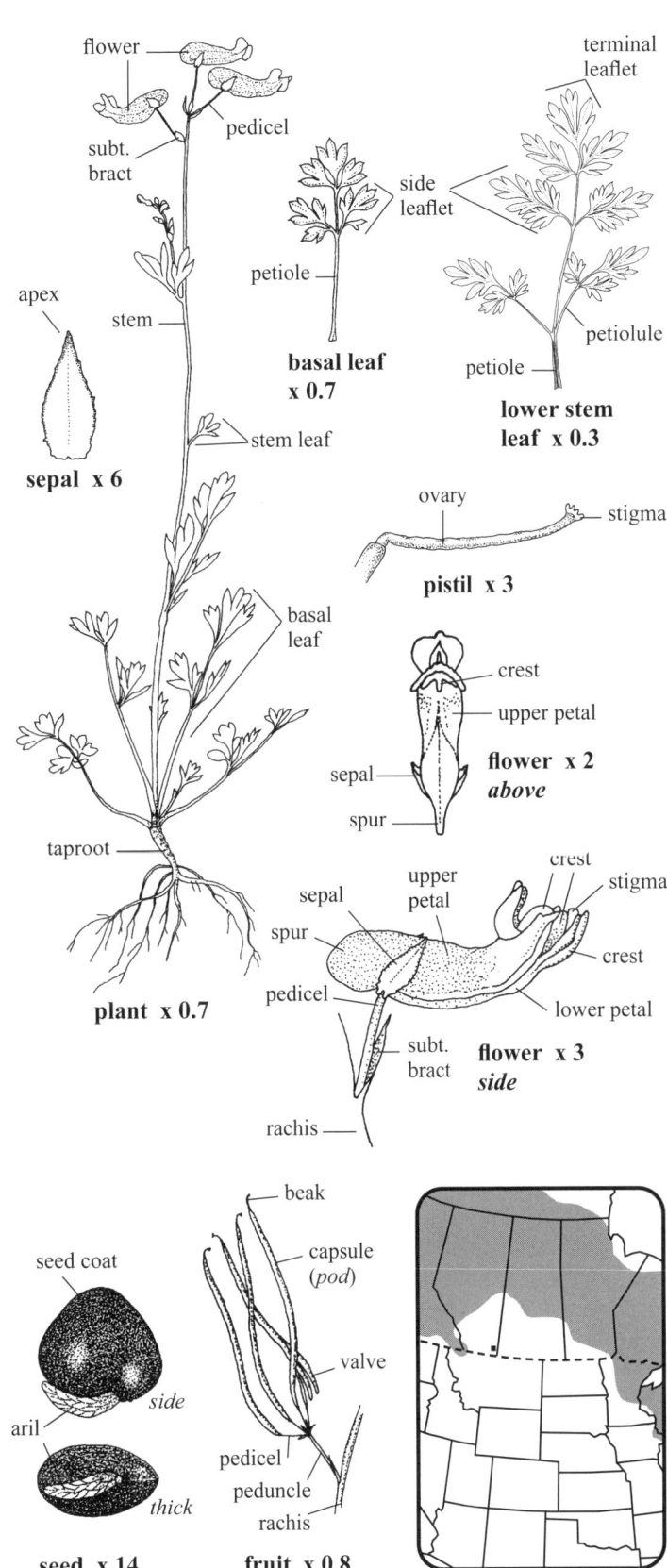

Rock-harlequin, Pale Corydalis, Harlequin-flower

Gentian—*Gentianaceae* **Wildflower** Blue Corolla tubular, 5-lobed

OBLONG-LEAVED GENTIAN   *Gentiana affinis* Griseb.

■ **SKETCH** A glabrous **perennial herb** 10–50 cm tall from a white, branching **taproot** with faint narrow horizontal grooves or rings in the upper 10 mm or so and on some of the larger side roots near the main root; in moist areas of meadows, hilly areas, lakeshores, in sandy to gravelly sites in open forests and mountains. **p2**

■ **FLOWERS** Blue, pale blue interior, blooming July–August; **inflorescence** racemose, a terminal cluster of 1–5 flowers *c.* 3 cm long by *c.* 2.2 cm wide, in axils of upper leaves; **main subtending bracts** (of flower clusters) two, V-shaped, *c.* 12 mm long; **secondary subtending bracts** (of each flower) *c.* 7 mm long; **flowers** perfect, *c.* 15 mm wide by 23–30 mm long; **calyx** reddish green, 8–14 mm long by 3–4 mm wide; **calyx tube** 4–9 mm long; **calyx lobes** five, pointed, widest above the base, 3–7 mm long by *c.* 1 mm wide, erect, a membrane *c.* 1 mm wide between the lobes; **corolla** tubular, 5-lobed, the main lobes pointed, 3–6 mm long by *c.* 3 mm wide, not fringed, greenish on the outside, with 15–25 pale green spots each *c.* 0.3 mm wide on the upper surface; **pleat lobes** light blue on the outside, smaller, 2-toothed, between each main lobe; **stamens** five, included, glabrous; **filaments** pale green, the upper free part 3–4.5 mm long, lower part attached to corolla walls; **anthers** yellow, *c.* 2.3 mm long by *c.* 1 mm wide, 2-lobed, erect, straight; **pistil** one, green, glabrous, *c.* 15 mm long; **stigma** 2-lobed, pale green, spreading, together *c.* 2 mm wide. **FRUIT a capsule**, tan, hairless, body 2.5–3 cm long by 4–8 mm wide by 2–3.5 mm thick, usually wrinkled in lower half, opening at apex, 2-valved, erect, enclosed by dry corolla and calyx; **stipe** 4–10 mm long, smooth; **seeds** more than 100 per capsule, dark brown, 0.8–1.1 mm long by 0.5–0.8 mm wide by *c.* 0.5 mm thick, covered with elongated papillae.

■ **LEAVES** Opposite, simple, entire, sessile, pointed, ascending; **basal leaves** absent; **stem blades** usually 6–13 pairs, each 1–3.5 cm long by 0.3–1.5 cm wide, margins microscopically rough, reduced below to bractlike clasping blades, midrib obvious, two lateral veins obscure.

■ **STEM** Erect or sometimes prostrate, one to several, stiff, reddish green to dark reddish brown, round, solid, with paired low vertical rough ridges from the paired leaf bases; base 1.5–3 mm wide.

■ **SYN.** *Pneumonanthe affinis* (Griseb.) W.A. Weber.

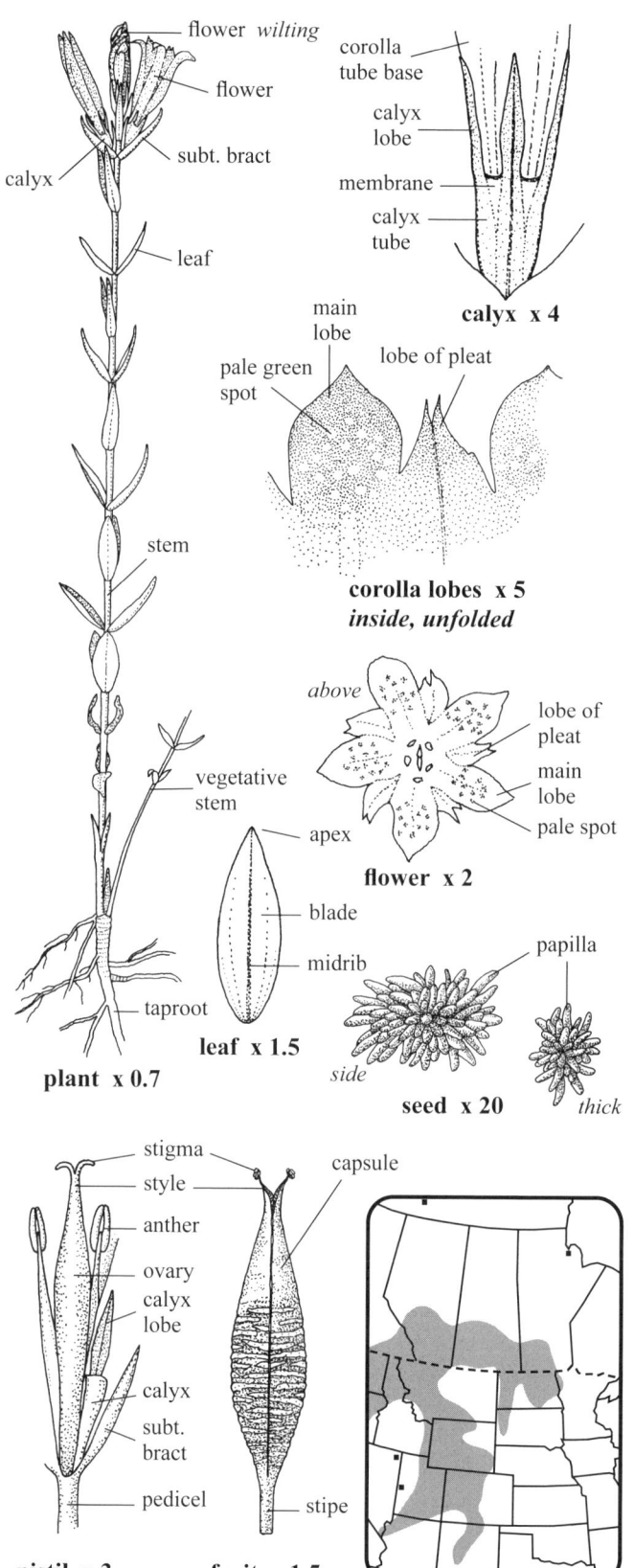

Gentian—*Gentianaceae* **Wildflower** Blue Corolla closed Petals 5, united, pleated

CLOSED GENTIAN *Gentiana andrewsii* Griseb.

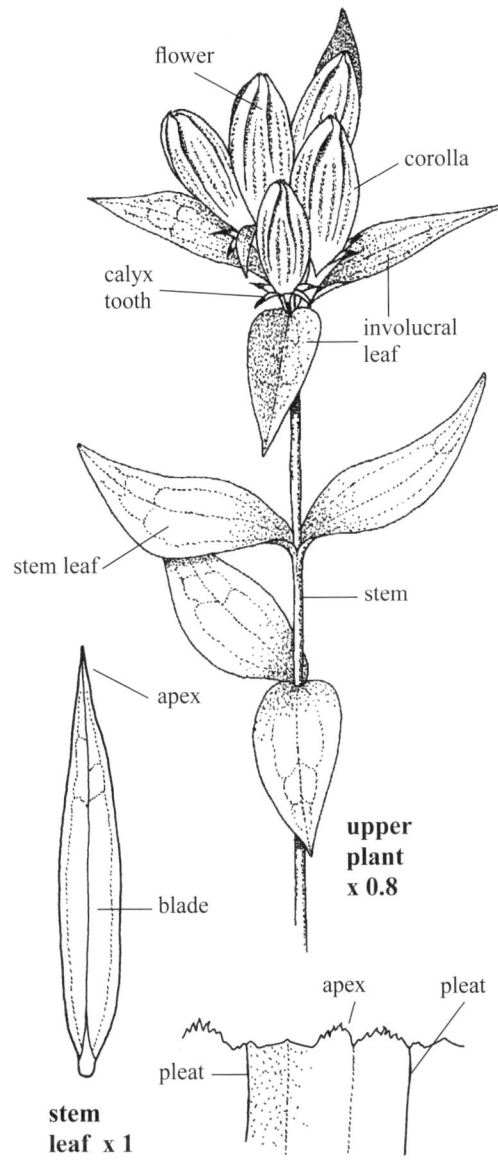

■ **SKETCH** A **perennial herb** 20–100 cm tall with thickened **fibrous roots**, singly or in open patches; in moist prairies, swamps, roadside ditches and open wet woodlands. **p1**

■ **FLOWERS** Blue, quickly fading to reddish purple, rarely white; blooming July–October; **inflorescence** a crowded terminal cluster and from upper leaf axils; **flowers** perfect, one to twenty, each 10–12 mm wide by 30–45 mm long, sessile and erect; **subtending bracts** (of flowers) two, *c.* 7 mm long, pointed, keeled, glabrous; **calyx** tubular, 12–27 mm long, green, glabrous, tube 8–13 mm long by 3–4 mm wide, round, 5-ridged; **calyx teeth** five, unequal, each 4–14 mm long by 1–4.5 mm wide, tips purple inside, ascending to spreading, a tiny membrane between the base of each tooth; **corolla** tubular, closed; **petals** five, united, 2.8–4.5 cm long (unfolded one petal is 7–8 mm wide with apical teeth 0.5–1.5 mm long), forming an oval tube, slightly papery, apices incurved and irregularly toothed, pale blue in the five pleats; **stamens** five, *c.* 22 mm long; **filaments** with upper free part *c.* 10 mm long, lower 10 mm adnate to corolla tube; **anthers** pale yellow, *c.* 3 mm long, united; **stipe** *c.* 10 mm long by *c.* 2 mm wide, hollow, glabrous, lengthening to 13–20 mm in fruit; **ovaries** glabrous, *c.* 15 mm long by *c.* 3 mm wide, tapered at the ridged base; **style** obscure; **stigma** of two pale green flat lobes, recurved, *c.* 2 mm long. **FRUIT a capsule**, tan to reddish, included in the corolla, 2-valved, erect, 18–25 mm long (not including stipe) by 5–9 mm wide by 3.5–7 mm thick, grooved, hairless, opening at apex; **seeds** numerous, flat, winged, in two columns per valve, hairless, 2–2.5 mm long by *c.* 1 mm wide by *c.* 0.3 mm thick, wing often wavy, pointed.

■ **LEAVES** Opposite, simple, entire, 8–18 pairs of main stem leaves, spreading to ascending, sessile; **blades** 3- or 5-nerved, slightly shiny, reduced below, 1.5–14 cm long by 5–45 mm wide, firm, flat or upper blades slightly V-shaped along midrib, glabrous with margins slightly scabrous; **stem base** with 2–35 small leaves, these not showy and usually less than 1 cm long.

■ **STEM** Erect, simple, glabrous, one to several, green to reddish above, slightly ridged, solid; 1–5 mm thick near the base.

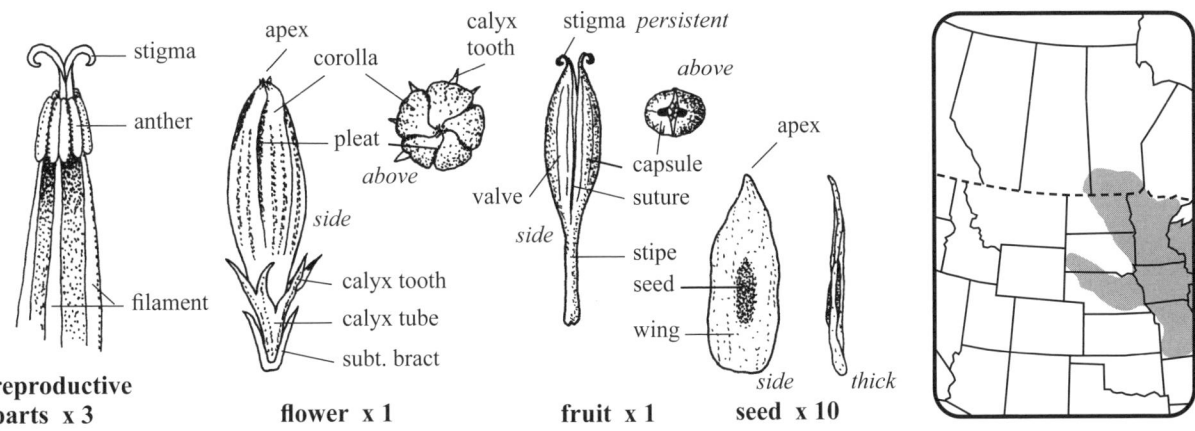

Bottle Gentian, Closed Bottle Gentian

Gentian—*Gentianaceae* **Wildflower** Violet to pale blue to pale yellow
Corolla tubular, 4- or 5-lobed

Northern Gentian *Gentianella amarella* (L.) Boerner

■ **SKETCH** A variable, glabrous **annual herb** 10–50 cm tall with a shallow, light yellow **taproot** 3–7 cm long by 2–3 mm wide, with yellowish **side roots** 2–5 cm long; in open woods, low fields, disturbed sites, gravelly ravines and alpine meadows. **p1**

■ **FLOWERS** Violet to blue to purplish pink to yellowish white, blooming July–September; **inflorescence** terminal and axillary from all leaf axils, 3–8 cm wide and almost as long as the plant; **floral branches** 2–12 cm long, ascending to slightly lax; **pedicels** 4–31 mm long, 4-sided, ridged; **flowers** perfect, a few per axil, 6–143 per plant, *c.* 10 mm wide by 10–15 mm long, throat white and frilly; **calyx** tubular, 4- or 5-lobed, tube 1.5–2 mm long, 8- to 10-ridged, lobes 4–10 mm long by 1–2 mm wide, erect, persistent; **corolla** tubular, TL 8–15 mm by 3.5–4 mm wide, lobes 4 or 5, each 3–5 mm long by 1.5–2.5 mm wide, faintly veined on outside, erect to spreading at anthesis, each fringed with 7–10 white, erect, filiform fingers 1.5–3 mm long with pale purple tips; **stamens** five, included, TL *c.* 8 mm; **filaments** white, flat, attached near base of corolla tube, upper free part 4–5 mm long by *c.* 0.8 mm wide; **anthers** dark purple, 2-lobed, *c.* 1 mm long; **pistil** one, green, cylindrical, included, 8–9.5 mm long by *c.* 1.3 mm wide, sessile; **style** *c.* 0.5 mm long, tapered; **stigmas** two, lobes flat, erect to spreading, *c.* 1 mm long, blunt. **FRUIT a capsule**, erect, brown, 12–15 mm long by 2–2.5 mm wide by 1.4–1.8 mm thick, 2-valved, the apices spreading and exserted from the persistent perianth; **seeds** roundish, 43–77 per capsule, light greenish brown, 0.6–0.7 mm long by *c.* 0.5 mm wide and thick, smooth and slightly shiny.

■ **LEAVES** Opposite, entire, simple, dull, 5–8 pairs per stem; **basal leaves** one or two pairs, spatulate, blunt; **blades** 4–15 mm long by 4–7 mm wide, 3-veined; **petioles** (of basal leaves) 3–7 mm long; **stem blades** 1.2–4.3 cm long by 0.2–1.7 cm wide, reduced above and below, sessile, pointed.

■ **STEM** Erect, simple to branched throughout, square, edges ridged, hollow, tough; reddish near the 1–3 mm wide base.

■ **SYN.** *Gentiana amarella* auct. p.p. non L. and *Gentiana acuta* Michx. = *Gentianella amarella* ssp. *acuta* (Michx.) J. Gillett.

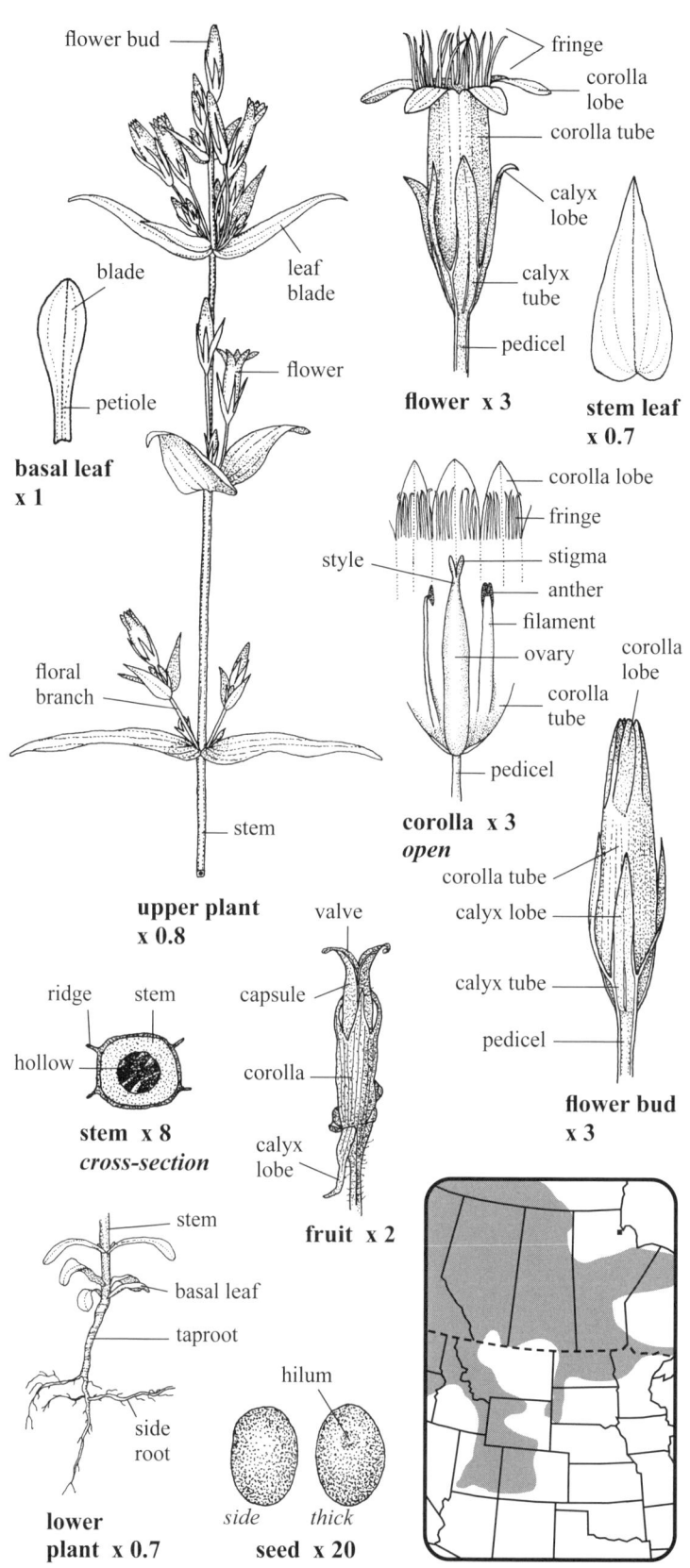

Felwort, Autumn Dwarf Gentian

Gentian—*Gentianaceae*   **Wildflower** Blue   Corolla tubular, 4-lobed

FRINGED GENTIAN *Gentianopsis virgata* (Raf.) Holub

■ **SKETCH** A variable **annual** or **biennial herb** 15–60 cm tall with branched white **roots** 1–7 cm long; in bogs, wet meadows and thickets, especially on calcareous soil. **p2**

■ **FLOWERS** Blue, blooming August–October; **inflorescence** of 1–10 flowers per stem, terminal and axillary; **peduncles** 2–22 cm long by 0.7–1.1 mm wide, ridges most obvious near apices; **calyx** tubular, 4-lobed, green, TL 12–37 mm by 5–10 mm wide, persistent around fruit, tube 7–15 mm long, hairless; **calyx lobes** with a raised midvein rough with microscopic papillae, erect between the corolla lobes, pointed, margins hyaline, of two types: **1) narrow lobes** 4–22 mm long by 1–4 mm wide; **2) wide lobes** 3.5–20 mm long by 2–8 mm wide; **corolla** tubular, 4-lobed, glabrous, TL 20–37 mm; **corolla tube** 13–22 mm long, pale green below where covered by calyx; **corolla lobes** 4–13 mm long by 4–8 mm wide, spreading to ascending, variously toothed and fringed on sides and apices with hairs 0.2–2.5 mm long; **stamens** four, included, alternate with the corolla lobes; **filaments** white hyaline, flat, inserted about midway on corolla tube, free part of filament 7–9 mm long by 0.8–2 mm wide, glabrous to slightly hairy near the middle; **anthers** yellow, 1–2.6 mm long, nodding; **nectaries** four, dark green, c. 0.5 mm long by 0.2–0.6 mm wide at bases of filaments; **pistil** TL 14–20 mm, glabrous, included; **stipe** pale green, c. 5 mm long, round; **ovary** c. 12 mm long by c. 3 mm wide by c. 2 mm thick, flattened, diamond-shaped, lateral sutures two; **style** c. 1 mm long; **stigmas** 2-lobed, included, 1.2–2.2 mm long and wide, sides folded inward on drying. **FRUIT a capsule**, tan, erect, TL 28–35 mm (including the 5 mm stipe) by 4–6 mm wide by 3–4 mm thick, 2-valved, opening from apex, wrinkled, opaque, stigma persistent; **seeds** numerous, 0.8–1.1 mm long by 0.6–0.8 mm wide, dark brown, papillae covering the seed are tan, transparent, 0.9–1.1 mm long, shiny and bubblelike.

■ **LEAVES** Opposite, simple, entire, linear, glabrous; **blades** 1–7 cm long by 1–2 cm wide, reduced below, midrib obvious, lateral veins obscure, pointed, sessile above and somewhat clasping, ascending; **petioles** 0–4 mm long, reduced above.

■ **STEM** Erect, round, stiff, slightly ridged, hollow, glabrous; usually 0.5–1.5 mm thick near the base.

■ **SYN.** *Gentiana crinita* var. *browniana* (Hook.) Boivin, *Gentianella crinita* ssp. *procera* (Holm) J. Gillett, *Gentiana procera* Holm, *Gentianopsis procera* (Holm) Ma.

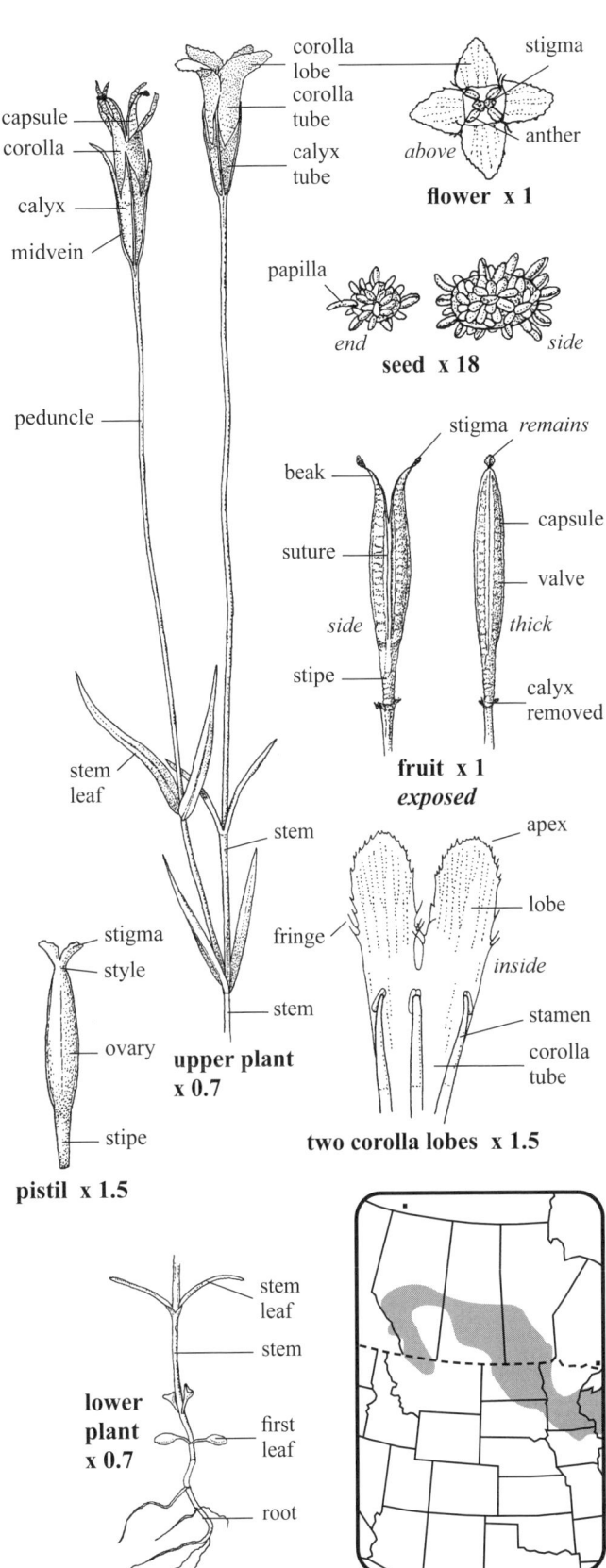

Narrow-leaved Fringed Gentian, Small Fringed Gentian, Lesser Fringed Gentian

Gentian—*Gentianaceae* **Wildflower** White to blue Corolla tubular, 5-lobed (4-)

Marsh Felwort *Lomatogonium rotatum* (L.) Fries ex Fern.

■ **SKETCH** An **annual herb** 10–45 cm tall from a tan **taproot** 3–5 cm long by *c.* 1 mm thick; in bogs, wet meadows, edges of saline flats and drainage channels.

■ **FLOWERS** White to blue, blooming August–September; **inflorescence** a raceme or panicle 5–20 cm long by 2–5 cm wide; **floral branches** strongly ascending above or throughout, the lower branches with one to several flowers; **pedicels** ascending, 5–45 mm long, axillary, 4-ridged, squarish, pinkish green; **flowers** perfect, 2–30+ per stem, 12–16 mm wide by 5–7 mm tall, often three terminating the top pair of stem leaves; **calyx** tubular, 5-lobed, green, united at base for 1–1.5 mm, lobes linear, pointed, entire, 8–18 mm long by 0.8–2 mm wide, midvein obscure, spreading between the corolla lobes; **corolla** tubular, 5-lobed (4-), united below for a short length, each lobe 10–13 mm long by 3–5.5 mm wide, slightly asymmetrical, 3- or 5-veined, pointed, with filiform white appendages 0.5–1 mm long from the apices of two lateral glands; **stamens** four or five, 5–8 mm long, glabrous, between the corolla lobes; **filaments** flat, light purple, 4–6.3 mm long by *c.* 0.5 mm wide; **anthers** yellow, 1.5–1.8 mm long, 2-lobed; **pistil** glabrous, pale yellow, exserted, 8–12 mm long including a **stipe** *c.* 1 mm long; **style** obscure; **stigmas** are two rough lines along the middle of each wide side of the ovary. **FRUIT a capsule**, dark brown, 2-valved, hairless, 8–13 mm long by *c.* 2 mm wide by 1.3–1.5 mm thick, striated along the valves, opening at apex, teeth spreading; **seeds** 70–90 per capsule, shiny, medium to dark brown, 0.3–0.6 mm long by 0.2–0.4 mm wide by *c.* 0.3 mm thick, sides flat to oval, ends blunt, appear wrinkled but are honeycombed (microscopic); **beak** minute.

■ **LEAVES** Basal and stem, opposite, simple, entire, glabrous, sessile, 1-nerved; **basal blades** spatulate, wilting early; **stem blades** linear, green, 12–45 mm long by 1–4.5 mm wide, strongly ascending above, margins slightly revolute.

■ **STEM** Erect, glabrous, squarish, 4-ridged, simple or branched above; 0.5–1.5 mm thick near the reddish brown base.

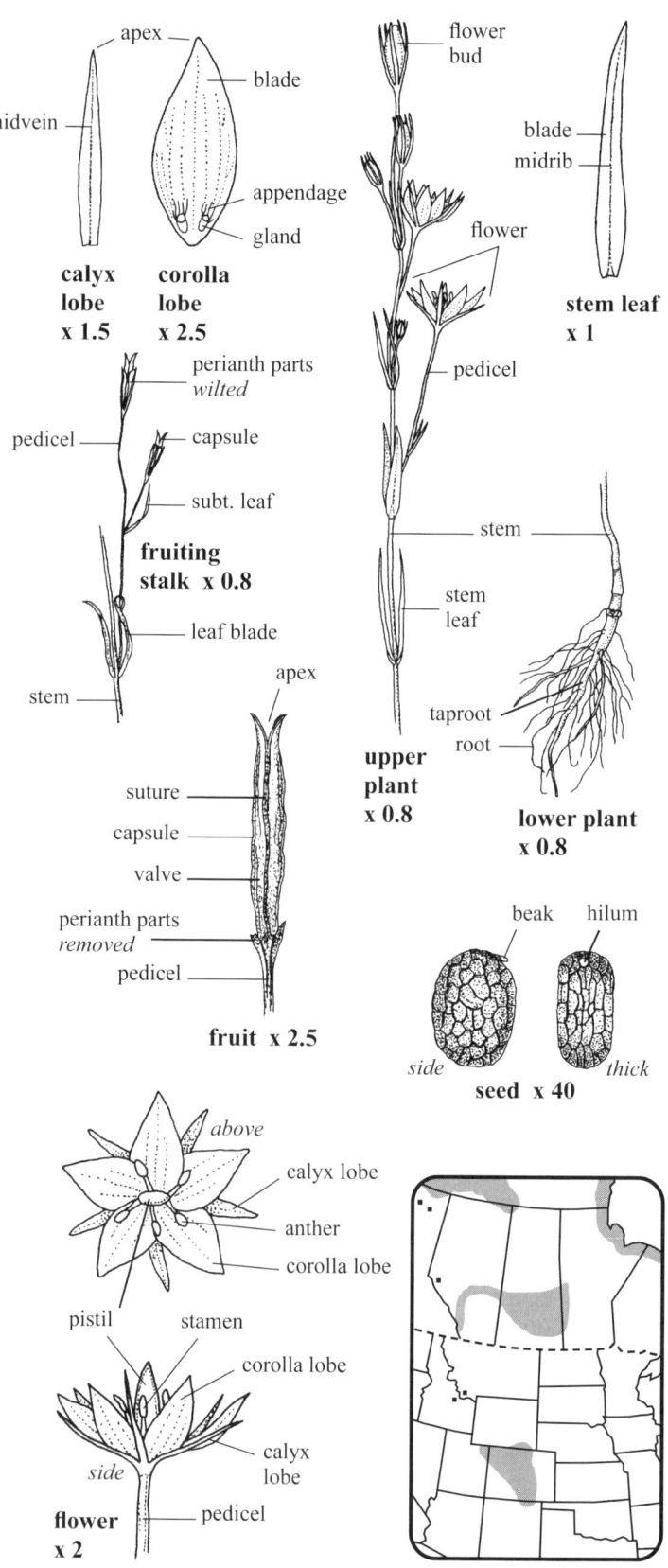

334

Geranium—*Geraniaceae* **Wildflower   Pink to purple   Petals 5**

Stork's-bill *Erodium cicutarium* (L.) L'Hér. ex Ait.

■ **SKETCH** A hairy **winter annual** or **biennial** 8–50 cm tall and wide from a white **taproot** 2–4 mm thick and to *c.* 8⁺ cm long with a **caudex**; in fields, mesas, slopes, lawns, gardens and along roadsides. *Naturalized*

■ **FLOWERS** Pink to purple, blooming February–October; **inflorescence** an umbel with 2–12 flowers; **peduncles** hairy, reddish green, erect to ascending, stiff, to *c.* 5⁺ cm long by 1–2 mm thick, slightly thicker and oval at the base, one or two from each stem node; **pedicels** reddish green, slightly swollen at both ends, 2–9 mm long, increasing to *c.* 25 mm long in fruit, hairs white, scattered and short; **involucre** (at base of pedicels) of several pointed lobes united at their bases, 1-nerved, hairy, *c.* 2 mm long by *c.* 1 mm wide; **flowers** perfect, 8–12 mm wide by 3–5 mm tall, closing in late afternoon; **sepals** five, green, hairy outside, glabrous inside, 2–6 mm long by 1.3–2 mm wide, 3-veined, spreading, with one or two spines to *c.* 1 mm long at apices, margins hyaline and involute; **petals** five, each 4–6.5 mm long by 2–3 mm wide, 3-veined, blade glabrous, entire, spreading, tapered to a short hairy claw *c.* 0.6 mm long; **stamens** 10, exserted, in two series: **1) inner series** five, fertile and opposite the sepals; **filaments** *c.* 2.5 mm long, erect, widest at base, pinkish purple, glabrous with a basal gland *c.* 0.5 mm wide, oval and purplish brown; **anthers** pink to purple, *c.* 0.9 mm long; **2) outer series** five, sterile, *c.* 1.2 mm long and filiform; **pistil** of five united carpels *c.* 2.5 mm long by *c.* 1.5 mm wide; **styles** five, united, hairy, *c.* 1.4 mm long at anthesis; **stigmas** *c.* 0.5 mm long, not hairy, reddish purple, fingerlike, spreading among the anthers. **FRUIT** of carpels, five, brown, hairy, body *c.* 4.5 mm long by *c.* 1 mm wide and thick, with a concave depression on each side at the hairless apex, base a pointed hairy callus; **styles** 2–5 cm long, straight, then looped twice when separated and dry, hairs 1–4 mm long; **seeds** one per carpel, each 2–4 mm long by *c.* 1 mm wide and thick, glabrous, dark brown, dull, apex blunt, base tapered with a short ridge.

■ **LEAVES** Basal and stem, opposite, compound (once or twice); **basal leaves** hairy, in a rosette and usually dry by anthesis; **stem** and **basal blades** 4–9 cm long by 1.5–4 cm wide; **leaflets** 7–13, sessile, 7–24 mm long by 6–12 mm wide, lobed and toothed, crowded near the apex, hairs white and glandular; **petioles** 1–30 mm long, reduced above, the groove with descending white hairs; **stipules** membranous, 1.5–5 mm long by 1–2 mm wide, margins entire and hairy.

■ **STEM** Ascending, hairy, roundish above, several from the caudex; 2–5 mm thick near the oval base.

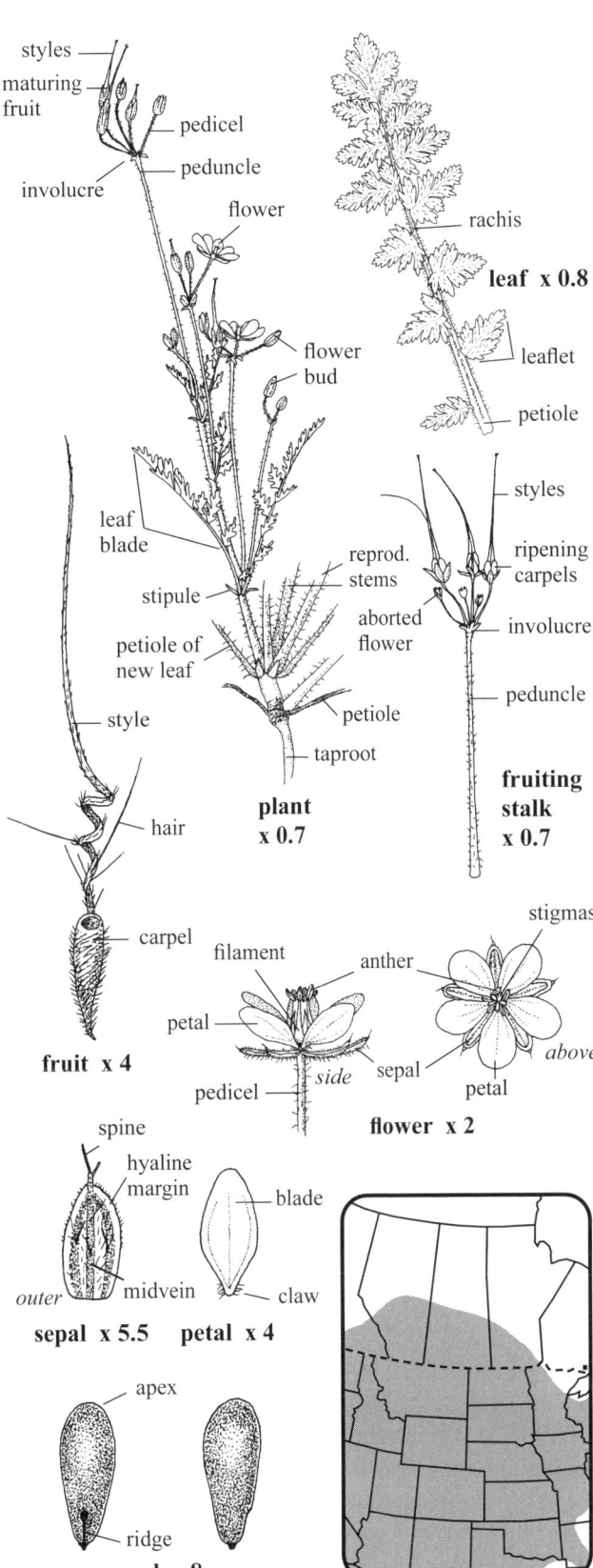

plant x 0.7

leaf x 0.8

fruiting stalk x 0.7

fruit x 4

flower x 2

sepal x 5.5   petal x 4

seed x 8

Redstem Stork's-bill, Filaria, Alfilaria

Geranium—*Geraniaceae*   **Wildflower**   Pink   Petals 5

BICKNELL'S GERANIUM   *Geranium bicknellii* Britt.

■ **SKETCH** A glandular-hairy **annual** or **biennial herb** 15–60 cm tall from a short **taproot**; in upland woods, disturbed sites along open woodland trails and sunny roadsides, common after fires in parklands and the Boreal forest. w2

■ **FLOWERS** Pink, blooming May–September; **inflorescence** of flowers mostly in pairs on axillary peduncles, open, loose, terminating axillary branches; **peduncles** spreading to ascending; **pedicels** glandular-hairy, 12–22 mm long, bending upward with fruit, short hairs numerous, long hairs spreading, scattered, gland-tipped; **flowers** perfect, 5–8 mm wide by 4–5 mm tall; **sepals** five, 5.5–9 mm long by 2–3 mm wide, glabrous inside, glandular-hairy mostly along the three obvious veins and the hyaline margins which are often revolute, apices awnlike, reddish, *c.* 1 mm long, hairless, persistent and brown with fruit; **petals** five, 5–8 mm long by *c.* 3 mm wide, the notch 0.7–1 mm deep, blade 3-veined, slightly hairy on margins near the whitish base; **stamens** 10, included, 2–3 mm long, of two lengths, erect, free to base; **filaments** white, hairy along the lower wide basal margins; **anthers** purplish pink, *c.* 0.5 mm long, clustered around the stigmas; **pistils** of five carpels, TL 3–3.2 mm, green, included, lengthening as they ripen; **carpels** covered with ascending white hairs; **styles** five, persistent, elongating in fruit to *c.* 16 mm, glandular-hairy on the outside, curled around the carpel; **stigmas** five, persistent in fruit, *c.* 1 mm long in flower, hairy near their base; **beak** 3–6 mm long. **FRUIT a carpel**, five, brown, hairy, 2–3 mm long by *c.* 1.5 mm wide, opening along one side, attached at one end of the style; **seeds** one per carpel, five per flower, brown, 1.4–1.6 mm long by 1.3–1.4 mm wide by *c.* 1 mm thick, papillose, catapulted a few meters.

■ **LEAVES** Opposite, simple, deeply lobed and toothed, 3–5 segments cleft almost to base, margins finely ciliate; **stem blades** 5–30 mm long by 5–70 mm wide, scattered white hairs on veins especially below, fewer on blades above; **petioles** 0–6 cm long, reduced above, glandular-hairy.

■ **STEM** Erect, dichotomously branched, flattened, white hairs of various lengths spreading to descending below, some gland-tipped; **long hairs** scattered, gland-tipped and fewer than the more numerous short hairs above; 1–4 mm wide near the reddish green base.

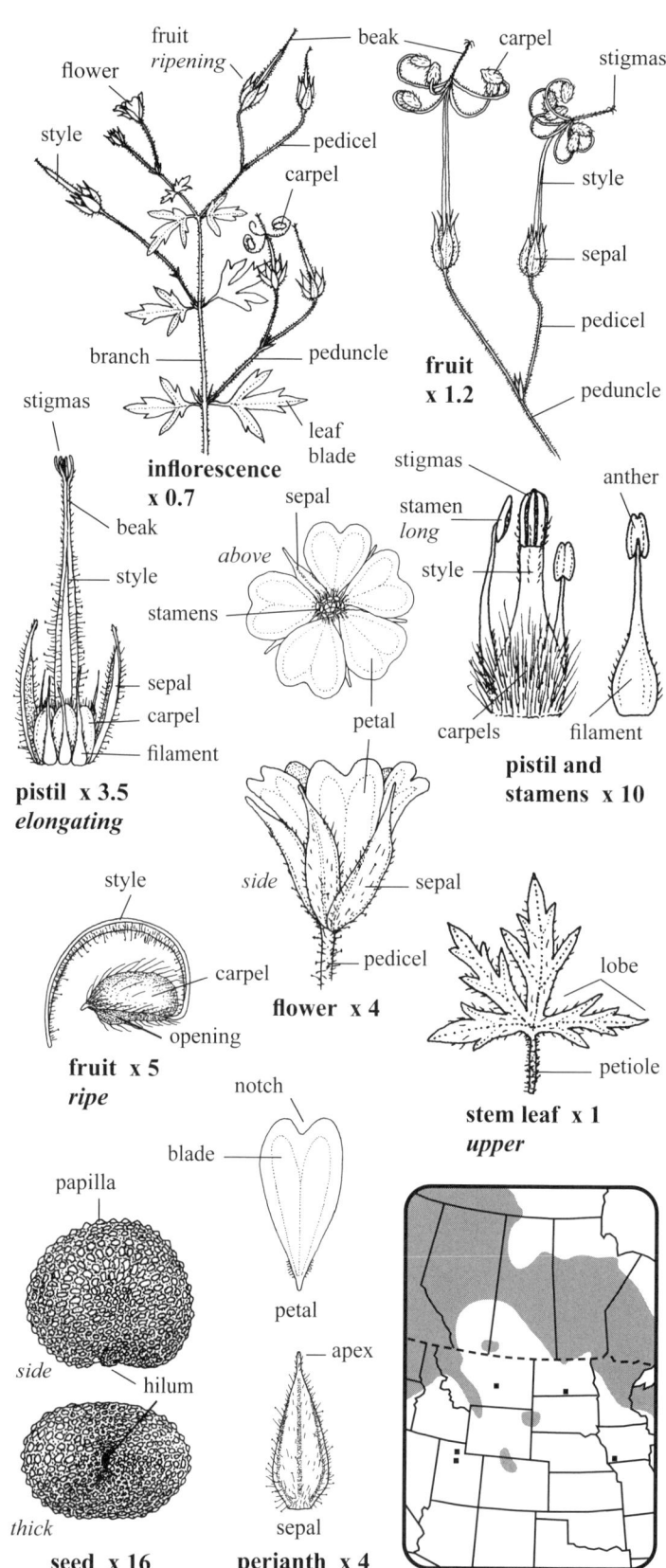

Currant—*Grossulariaceae* **Shrub** Pale green to creamy white Petals 5

Wild Black Currant *Ribes americanum* P. Mill.

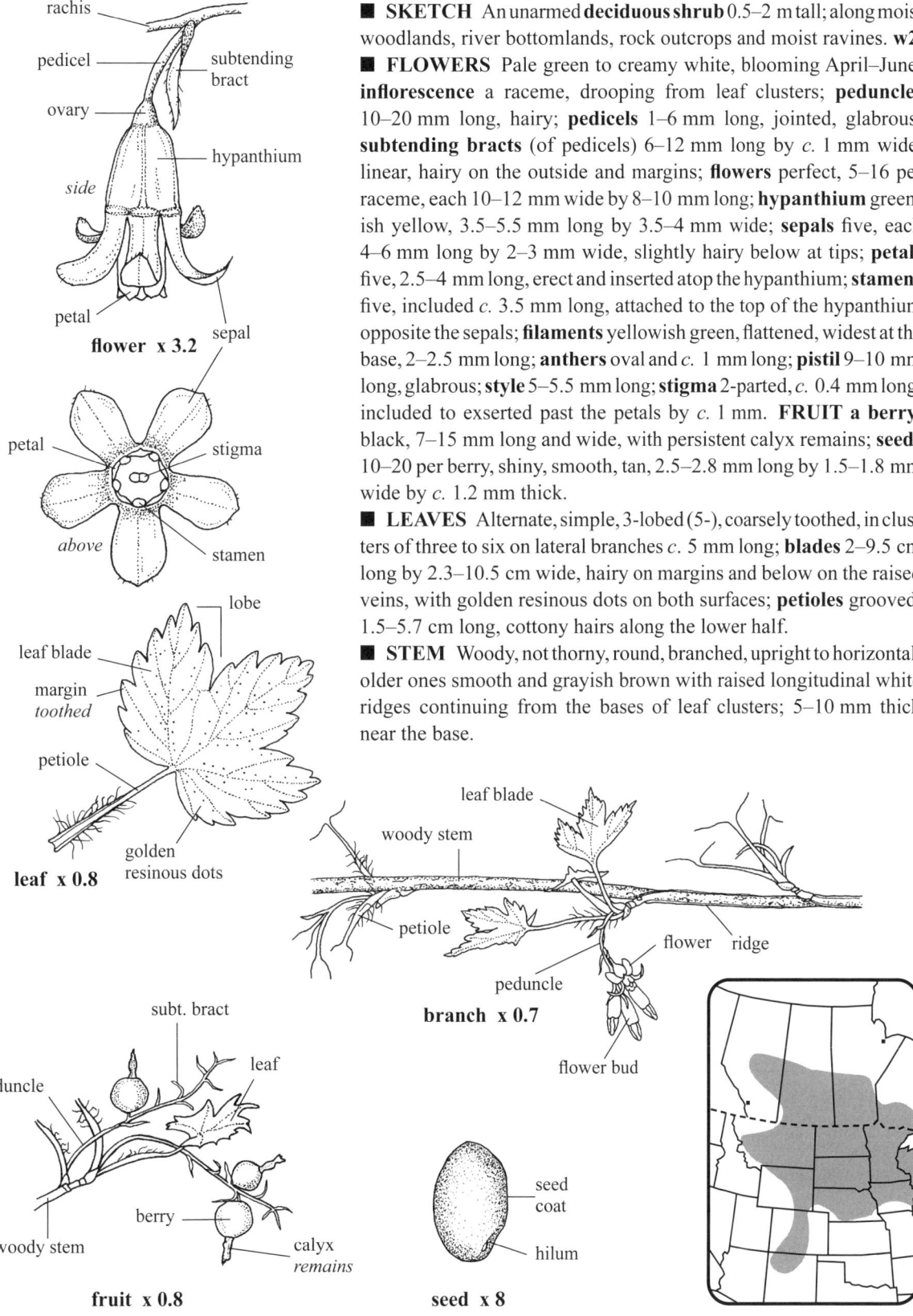

- **SKETCH** An unarmed **deciduous shrub** 0.5–2 m tall; along moist woodlands, river bottomlands, rock outcrops and moist ravines. **w2**
- **FLOWERS** Pale green to creamy white, blooming April–June; **inflorescence** a raceme, drooping from leaf clusters; **peduncles** 10–20 mm long, hairy; **pedicels** 1–6 mm long, jointed, glabrous; **subtending bracts** (of pedicels) 6–12 mm long by *c.* 1 mm wide, linear, hairy on the outside and margins; **flowers** perfect, 5–16 per raceme, each 10–12 mm wide by 8–10 mm long; **hypanthium** greenish yellow, 3.5–5.5 mm long by 3.5–4 mm wide; **sepals** five, each 4–6 mm long by 2–3 mm wide, slightly hairy below at tips; **petals** five, 2.5–4 mm long, erect and inserted atop the hypanthium; **stamens** five, included *c.* 3.5 mm long, attached to the top of the hypanthium opposite the sepals; **filaments** yellowish green, flattened, widest at the base, 2–2.5 mm long; **anthers** oval and *c.* 1 mm long; **pistil** 9–10 mm long, glabrous; **style** 5–5.5 mm long; **stigma** 2-parted, *c.* 0.4 mm long, included to exserted past the petals by *c.* 1 mm. **FRUIT** a berry, black, 7–15 mm long and wide, with persistent calyx remains; **seeds** 10–20 per berry, shiny, smooth, tan, 2.5–2.8 mm long by 1.5–1.8 mm wide by *c.* 1.2 mm thick.
- **LEAVES** Alternate, simple, 3-lobed (5-), coarsely toothed, in clusters of three to six on lateral branches *c.* 5 mm long; **blades** 2–9.5 cm long by 2.3–10.5 cm wide, hairy on margins and below on the raised veins, with golden resinous dots on both surfaces; **petioles** grooved, 1.5–5.7 cm long, cottony hairs along the lower half.
- **STEM** Woody, not thorny, round, branched, upright to horizontal, older ones smooth and grayish brown with raised longitudinal white ridges continuing from the bases of leaf clusters; 5–10 mm thick near the base.

American Black Currant

Currant—*Grossulariaceae*   **Shrub**   White and pink   Petals 5, smaller than sepals

Skunkberry   *Ribes glandulosum* Grauer

■ **SKETCH** An unarmed **deciduous shrub** 40–100 cm tall or long; in moist woods, thickets, clearings, on rocky slopes, swamps, shorelines and rocky outcrops of Boreal forest. **w1**

■ **FLOWERS** White and pink, blooming May–June; **inflorescence** a raceme, erect from leafy fascicles, 3–8 cm long by 1.5–3 cm wide, with two to six leaves at base of peduncle; **peduncles** green, erect, 2–2.5 cm long, slightly glandular-hairy, hanging with fruit; **pedicels** 3–8 mm long, green, white hairs tipped with red glands, jointed at apices, ascending; **subtending bracts** (of pedicels) 2–3 mm long by 1–2 mm wide, finely toothed with red glands at the tip of each tooth, pointed, not hairy, green, turning pink with fruit; **flowers** perfect, 5–15 per raceme, each 5–6 mm wide by 3–4 mm long, sometimes one flower arising from the base of the peduncle; **rachis** green, pink with fruit, glandular-hairy, the hairs spreading and short; **hypanthium** 1–1.5 mm long and wide, glabrous; **sepals** five (4), ascending to spreading, white, 2–2.5 mm long and wide, tapered at base; **petals** five, 0.5–1 mm long by *c.* 1 mm wide, white fading to pink; **stamens** five, erect, included, one opposite each sepal, 1.2–2 mm long, hairless; **filaments** red, *c.* 1 mm long; **anthers** pink, *c.* 0.5 mm long; **ovary** glandular-hairy; **style** included, 2-parted almost to base, *c.* 1.3 mm long. **FRUIT a berry**, red, glandular-hairy, shiny, 5–12 mm long (not including the hairs) by 5–9 mm wide, perianth parts persistent at apex, hairs and glands red, skin slightly transparent and smooth; **seeds** roundly triangular in cross-section, 10–45 per berry, medium reddish brown, smooth, enveloped in a translucent, wrinkled membrane, bare seeds 1.5–2 mm long by *c.* 1 mm wide and thick.

■ **LEAVES** Alternate, simple, 3- to 7-lobed, toothed, often in small clusters; **blades** shiny on both sides, 2–8 cm long and wide, non-glandular, with a few marginal hairs, glabrous above, slightly hairy below on veins or near the petiole, torn blades have a slight skunkish odor; **petioles** 2–5 cm long, slightly glandular-hairy, expanded at base; **stipules** attached to petiole base, *c.* 4 mm long, margins glandular-hairy.

■ **STEM** Erect to ascending or sprawling, unarmed, gray near apex, two thin, low ridges running along the length of the brown stem from node to node; reddish brown near the 3–5 mm thick base; **nodes** 1–3 cm apart.

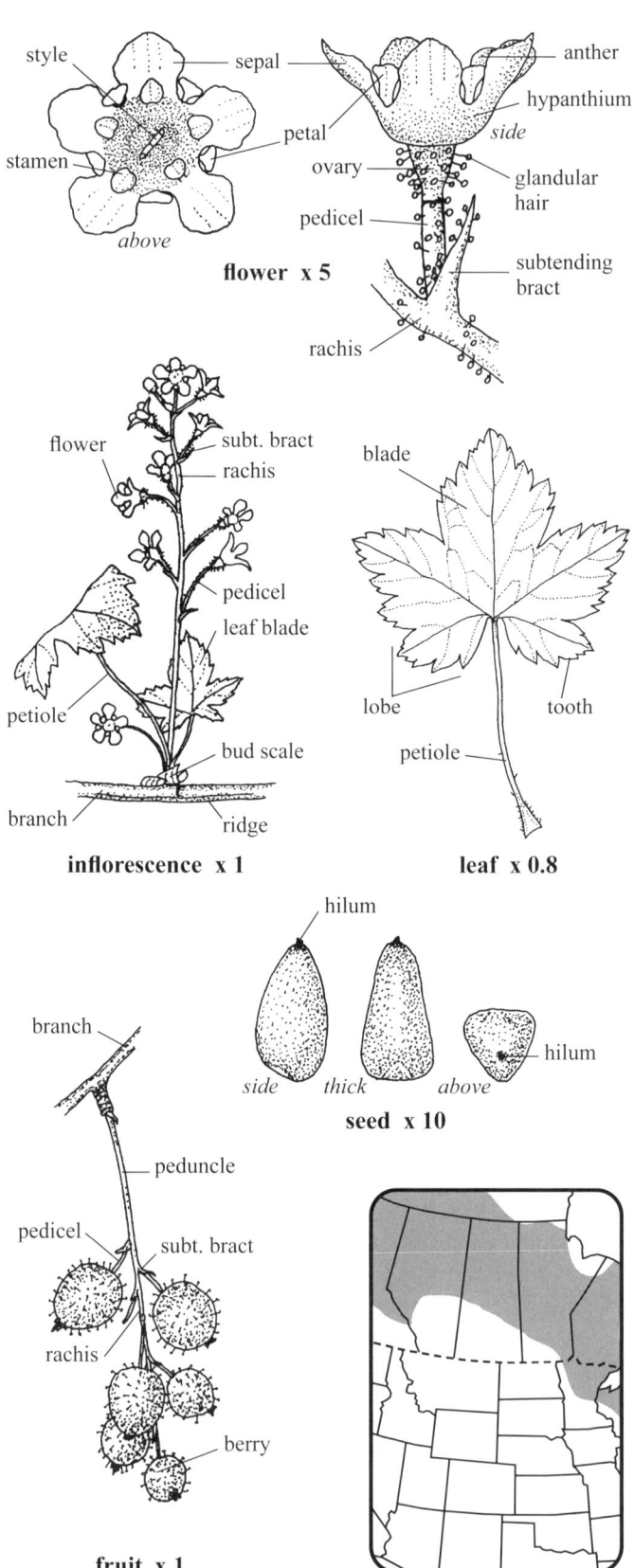

Skunk Currant

Currant—*Grossulariaceae* **Shrub** White Petals 5, smaller than sepals

Northern Black Currant *Ribes hudsonianum* Richards.

■ **SKETCH** An unarmed, odorous **deciduous shrub** 80–200 cm tall or long, erect to leaning; in wet woods, along ditches and streams. **w1**

■ **FLOWERS** White, blooming May–July; **inflorescence** a raceme, each 6- to 12-flowered, many, ascending, 2–3 cm long by 1.8–2.5 cm wide near branch tips, among alternating leaf clusters; **pedicels** 7–12 mm long, hairs short, often with a few yellow resin dots; **subtending bracts** (of pedicels) 1.5–3 mm long, narrow, 1-veined, pointed; **flowers** perfect, 7–9 mm wide by *c.* 5 mm long, fragrant, appear as leaves are expanding; **hypanthium** (around ovary) green, hairy, *c.* 2.8 mm wide, 5-ridged, glandular-dotted; **sepals** five, white, 3–5 mm long by 1.5–1.8 mm wide, 3-veined, very hairy on outside, spreading; **petals** five, white, erect between the sepals, 1.5–1.8 mm long by *c.* 1 mm wide, tapered, apices erose to slightly pointed, hairless; **stamens** five, *c.* 1.8 mm long, white, hairless, attached to top of hypanthium; **filaments** *c.* 1 mm long; **anthers** white, 2-lobed, *c.* 0.8 mm long; **styles** *c.* 1.5 mm long, green, two-parted; **stigmas** two, facing outward; **fruit clusters** spreading to ascending. **FRUIT a berry**, black, slightly shiny, 5–10 mm long by 6–8 mm wide, smooth, roundish; **seeds** 10–20 per berry, 1.5–2.5 mm long by 1–1.2 mm wide by 0.6–0.9 mm thick, smooth, reddish brown, oval to 3-sided in cross-section, each enclosed in a clear gelatinous sac which is easily removed.

■ **LEAVES** In alternating clusters of three to six on short spurs, simple, 3- or 5-lobed, toothed, golden resin-dotted below with white hairs on raised veins, short white hairs scattered above; **blades** 4–14 cm long by 5–15 cm wide, dull to slightly shiny, lighter green below; **petioles** 4–7 cm long, with a few short scattered hairs to *c.* 2 mm long, deeply grooved and spreading near base, yellow resin dots in the lower half; **fall axillary buds** pale green, 6–7 mm long by 2–3 mm wide, golden glandular-dotted, pointed, margins ciliate.

■ **STEM** Reddish gray; **newer twigs** on branch tips light tannish gray, yellow resin-dotted, short hairs recurved; **branches** alternate; **lenticels** black dots; 5–12 mm thick near the bare brown base.

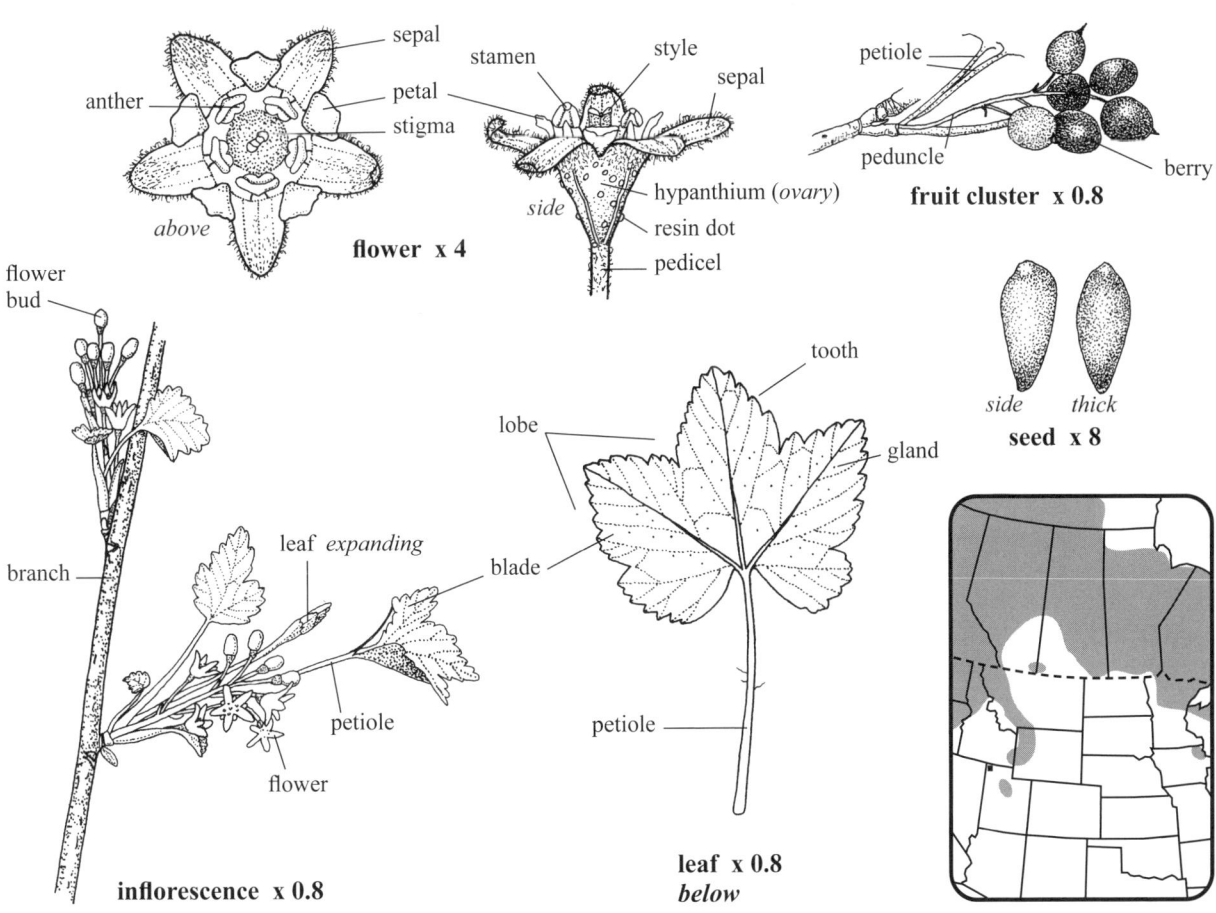

Hudson Bay Currant, Wild Black Currant, Canadian Black Currant

Currant—*Grossulariaceae* **Shrub** Pale green to white Petals 5

Northern Gooseberry *Ribes oxyacanthoides* L.

■ **SKETCH** A variable, armed **deciduous shrub** 40–150 cm tall or long, ascending to sprawling; in moist woods, along river banks, thickets and rocky sites. **w1**

■ **FLOWERS** Pale green to white, blooming April–June; **inflorescence** a corymb with one to five (often two) flowers per axillary cluster; **peduncles** 1–10 mm long, glabrous to glandular; **pedicels** glabrous, unjointed, 2–5 mm long; **subtending bracts** (of pedicels) 1–2 mm long, clasping, margins and apices with gland-tipped hairs; **flowers** perfect, 5–10 mm wide by 8–10 mm long; **hypanthium** glabrous, fleshy, 2–5 mm long by 2–4.5 mm wide, veined, pale green, hairy inside; **sepals** five, green to white, blunt, 3–5 mm long by 1.5–2.2 mm wide, spreading to reflexed; **petals** five, erect, glabrous, 1.5–3 mm long by *c.* 2 mm wide, tapered to the base, apices slightly erose; **stamens** five, 2–2.5 mm long, usually included and attached to apex of hypanthium; **filaments** oval, 1.5–2 mm long by *c.* 0.5 mm wide, pink, glabrous; **anthers** *c.* 1 mm long, green; **ovary** glabrous, shiny, *c.* 2 mm long by 1.5–2 mm wide, inferior; **style** slightly exserted, 6–8 mm long, united below, 2-parted at apex, slightly spreading, apices generally not hairy, very hairy below, hairs white and spreading; **stigmas** capitate, slightly wider than the style. **FRUIT a berry**, round, glabrous, 6–10 mm long by 7–14 mm wide, deep blue to purple, 10-veined, topped with persistent, erect perianth parts 5–8 mm long; **seeds** 5–25 per berry, glabrous, each 1.5–2.2 mm long by 0.8–1.1 mm wide by 0.7–1 mm thick, reddish to dark brown, 3-sided, one side roundish, two sides flat and tapered to a thin edge.

■ **LEAVES** Alternate, simple, in fascicles, three or four leaves per spur shoot with fruit; **blades** dull, 1–3.5 cm long and wide, 3- or 5-lobed, toothed, palmately veined, hairy below especially on veins, punctate (brown dots) below especially near lobe tips, less punctate and less hairy above; **petioles** glabrous to glandular-hairy, 5–30 mm long, short hairs curled to arched; **subtending bud scales** several, around base of fascicled petioles, imbricate, 1–5 mm long by 2–2.5 mm wide, pointed, margins ciliate, dry, midvein obvious, brown to slightly reddish brown.

■ **STEM** Erect to nodding, irregularly branched, finely short hairy, spiny and prickly; 5–8 mm thick near the base; **prickles** 1–2 mm long, stiff, reddish tan, slightly wider at their base, numerous to absent on older stems, also slightly declining; **older stems** with gray and brown bark, few if any prickles, new branches greenish tan; **nodes** with 1–4 larger, tan to reddish brown to gray spines 5–14 mm long, subtending the spur shoots.

■ **SYN.** *Ribes setosum* Lindl. = *Ribes oxyacanthoides* ssp. *setosum* (Lindl.) Sinnott.

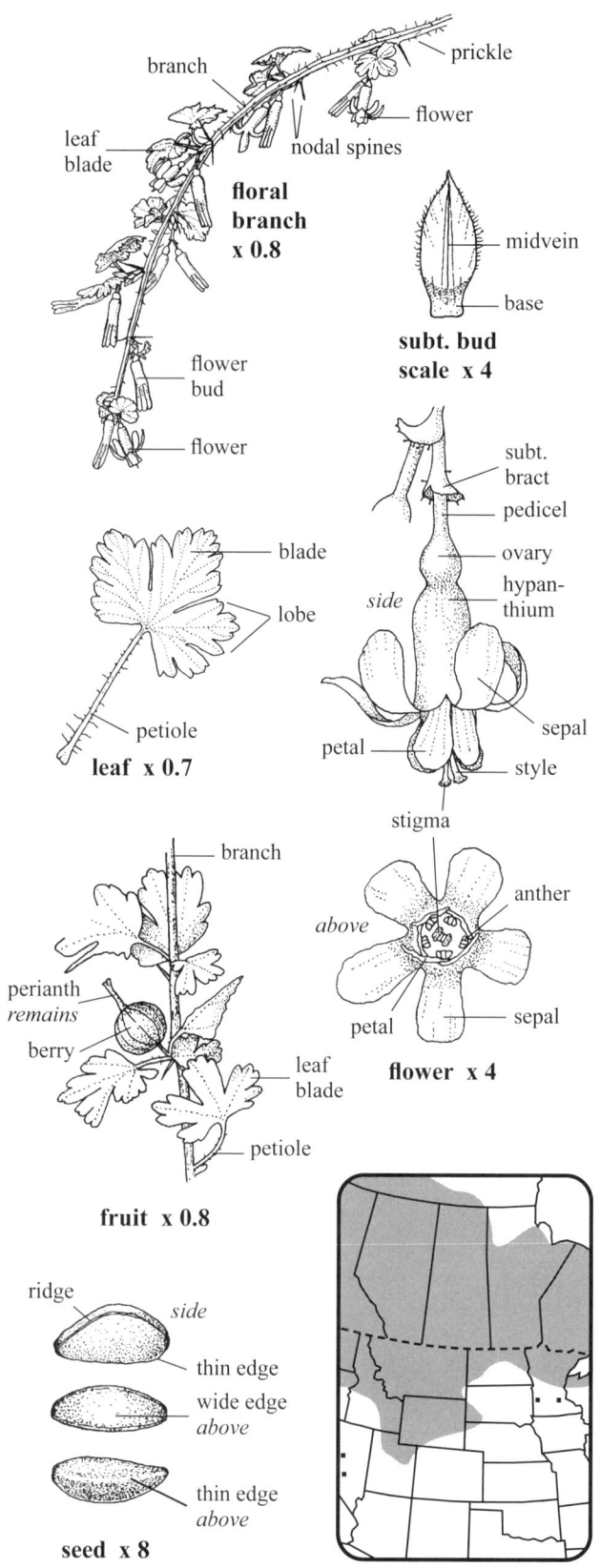

Canada Gooseberry, Canadian Gooseberry, Wild Gooseberry

Currant—*Grossulariaceae*   **Shrub**   Pale green to reddish   Petals 5

Swamp Red Currant *Ribes triste* Pallas

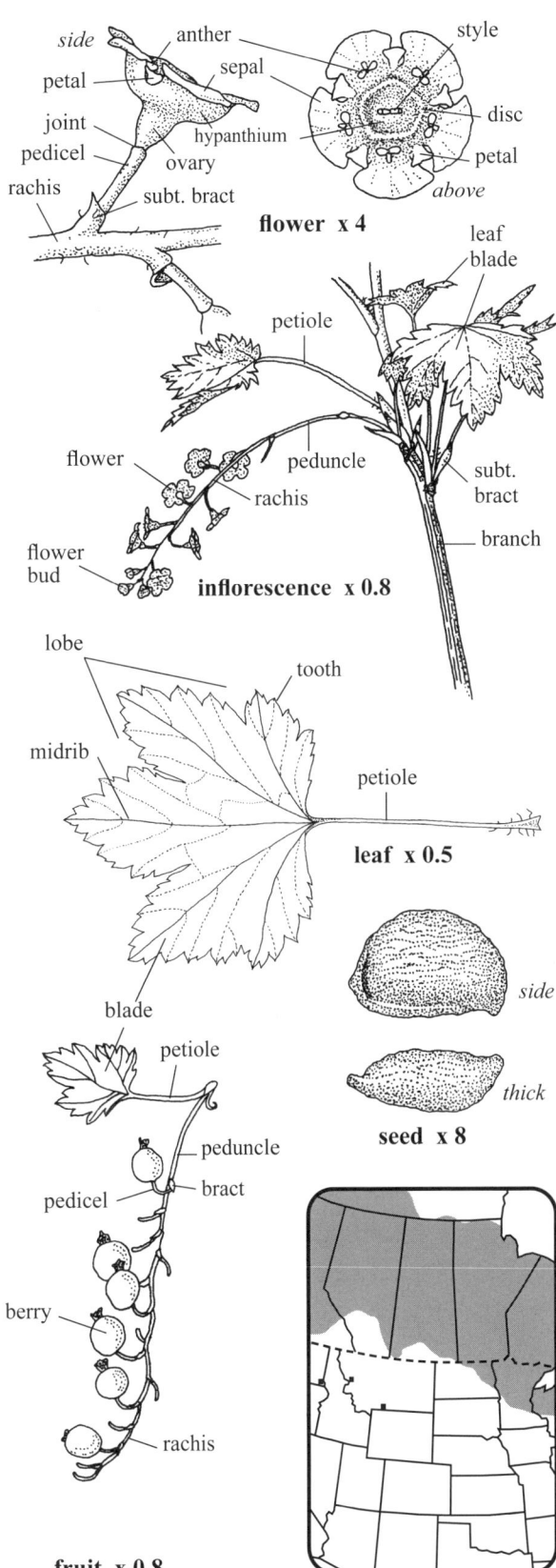

flower x 4

inflorescence x 0.8

leaf x 0.5

seed x 8

fruit x 0.8

- **SKETCH** An unarmed **deciduous shrub** 0.2–1.2 m tall or long, branches erect to prostrate and rooting; in moist woods, bogs and swamps. **w1**
- **FLOWERS** Pale green, some with reddish tinges, odorless, blooming May–June; **inflorescence** a raceme 2–7 cm long by 10–20 mm wide, drooping, with 6–14 flowers; **peduncles** 1.5–3.5 cm long, arched, glabrous to hairy; **subtending bracts** (of peduncles) 5–8 mm long, pointed; **pedicels** 2–4 mm long, glabrous to glandular-hairy, curved upward with fruit; **subtending bracts** (of pedicels) *c.* 1 mm long, to *c.* 2 mm long in fruit, pointed, with a few hairs, wrapped around the base of a pedicel; **flowers** perfect, regular, 6–7 mm wide by 2–3 mm tall; **rachis** glabrous or glandular-hairy; **hypanthia** green, moist, *c.* 3 mm wide by *c.* 1 mm deep, 5-lobed, forming a disc atop the ovary; **sepals** five, green or purple tinged, 1.5–2.2 mm long by 2–3.5 mm wide, glabrous, apices slightly erose; **petals** five, stiff, *c.* 1 mm long and wide, green to purplish, erect, tapered to a narrow base; **stamens** five, included, *c.* 1 mm long, leaning inward; **filaments** green, erect; **anthers** pale yellow, 2-lobed, 0.7–1 mm wide; **pistil** with two carpels; **style** green, glabrous, united most of its length, *c.* 1 mm long, 2-parted near the enlarged stigmatic tip. **FRUIT a berry**, red, glabrous, shiny, 4–8 mm long and wide, ascending on curved pedicels; **seeds** 2–4 per berry, golden tan but stained pink, 2.5–3 mm long by 1.8–2.3 mm wide by 1–1.3 mm thick, angular, rough with low irregular ridges, each enclosed in a reticulate sac as it ripens.
- **LEAVES** Alternate, simple, 3- or 5-lobed, toothed, in clusters of 2–5 on short spurs *c.* 10 mm long; **blades** 3.5–10.7 cm long by 4–12 cm wide, dull, glandless above with a few scattered hairs, margins hairy, raised veins hairy below with some scattered glands, frost-hardy; **subtending bracts** 2 or 3, whitish, at base of leafy fascicles, each 8–10 mm long by 2–3 mm wide, pointed, some tips brown, body hairless, marginal hairs often glandular; **petioles** 1.3–7.2 cm long, green, ascending, with a few hairs at the base or throughout, glands scattered.
- **STEM** Solitary or a few clustered, branched near the base or throughout, smooth, dark brown, bark peeling below on older stems, some branches whitish gray, new terminal growth pale green, hairy and glandular, two ridges from bases of petioles run along the new growth to the next leaf; bases 3–9 mm wide; **winter buds** dark brown, 3–4 mm long by 1.5–2 mm wide, pointed, hairy on margins of scales.
- **SYN.** *Ribes rubrum* var. *propinquum* (Turcz.) Trautv. & C.A. Mey.

Wild Red Currant, Red Currant, Swamp Currant, Red Swamp Currant

Water-milfoil—*Haloragaceae* **Wildflower** Green and red Petals 4

SPIKED WATER-MILFOIL   *Myriophyllum sibiricum* Komarov

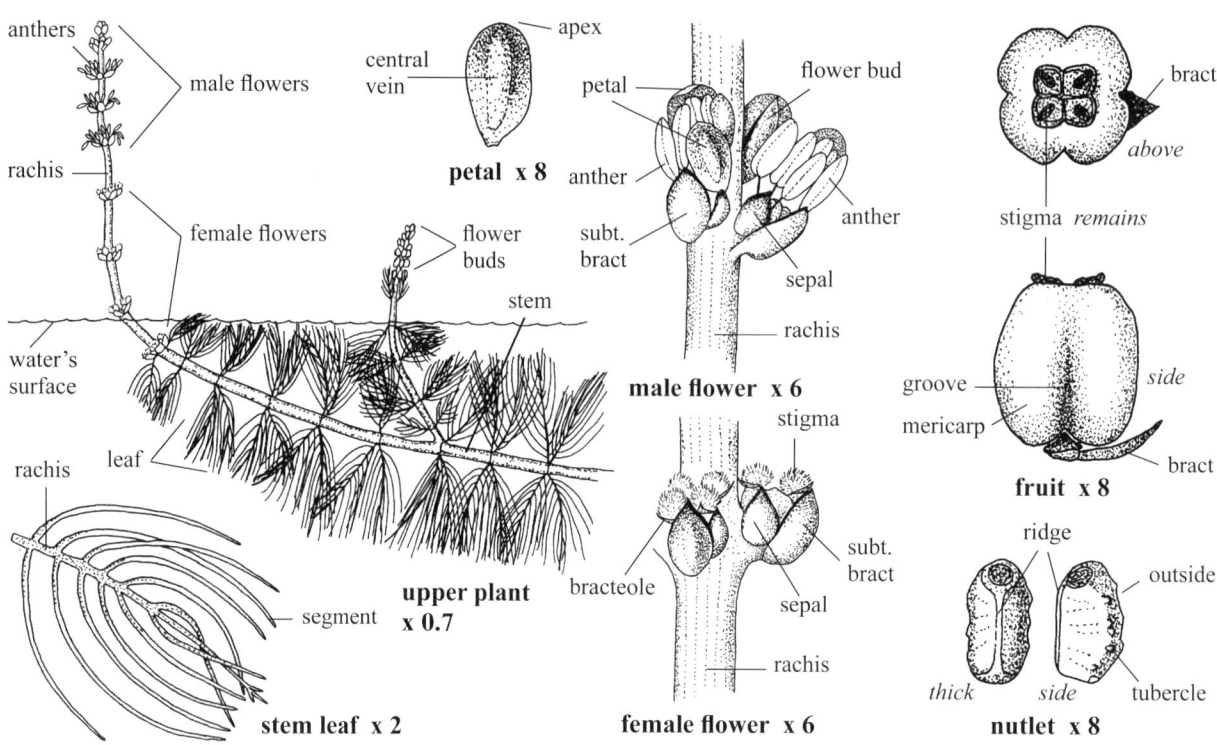

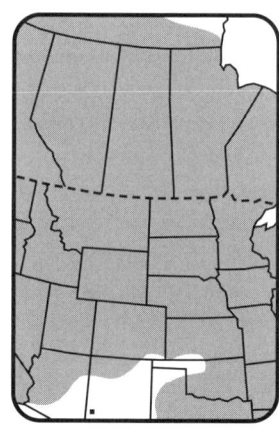

- **SKETCH** An **aquatic perennial herb** 30–150 cm long from **rhizomes**; in ponds, marshes and sloughs, often free-floating or rooted, in water to *c.* 13 m deep; **monoecious**. w1
- **FLOWERS** Green and red, blooming April–September; **inflorescence** a spike with 3–13 whorls, TL 2–10 cm by 5–9 mm wide; **flowers** usually four per whorl, mainly unisexual, sessile; **subtending bracts** (of flowers) one, 1.5 mm long by *c.* 1 mm wide, C-shaped, apices and margins dark brown, glabrous, stiff; **rachis** glabrous, erect, red; **male flowers** usually in three to five whorls at the top, the whorls 5–9 mm wide by *c.* 3 mm tall; **sepals** four, entire, *c.* 1 mm long and wide; **petals** four, soft and membranous in bud, quickly deciduous, entire, glabrous, C-shaped, 1.8–2 mm long by 1–1.2 mm wide; **ovaries** four, vestigial, green; **stamens** eight per flower; **filaments** 1–1.5 mm long, white; **anthers** yellowish green, 1.4–2.5 mm long; **female flowers** pink, each *c.* 1.7 mm wide by 1.8–2 mm tall, three to five whorls below the male flowers, the lowest female flowers may be at or below the water's surface; **bracts** 0.8–1 mm long by 0.5–0.7 mm wide, tips triangular, the pointed apex with a brown dried margin; **sepals** and **petals** as above; **stigmas** four, feathery; **fruiting stalks** submerged. **FRUIT** a mericarp, of four nutlets, the mericarp 2.5–3.5 mm long by 3–3.3 mm wide and thick, glabrous, sessile, apex rough, thin-walled; **nutlets** 1-seeded, 1-ridged, brown, triangular, apices slanted, outside rounded and tuberculate, *c.* 2 mm long by *c.* 1.2 mm thick by *c.* 1.1 mm wide.
- **LEAVES** In whorls of four per node, compound, submerged, divided leaves end abruptly where female flowers begin; **internodal distances** usually 8–12 mm; **stem blades** spreading, flaccid, 0.7–4 cm long by 4–25 mm wide, flat; **segments** 6–14 pairs, entire, glabrous, slightly oval; **rachis** 0.4–0.6 mm thick.
- **STEM** Pinkish, submerged, branching, horizontal to vertical, smooth and hairless, filled with air pockets; 2–5 mm thick.
- **SYN.** *Myriophyllum exalbescens* Fern., *Myriophyllum spicatum* var. *capillaceum* Lange, *Myriophyllum spicatum* var. *exalbescens* (Fern.) Jepson, *Myriophyllum spicatum* ssp. *exalbescens* (Fern.) Hultén, *Myriophyllum spicatum* ssp. *squamosum* Laestad. ex Hartman and *Myriophyllum spicatum* var. *squamosum* (Laestad. ex Hartman) Hartman.

American Milfoil, Shortspike Watermilfoil, Siberian Water-milfoil

Mare's-tail—*Hippuridaceae* **Wildflower** Reddish purple Calyx tubular, not lobed
Petals absent

MARE'S-TAIL *Hippuris vulgaris* L.

■ **SKETCH** A glabrous, aquatic **perennial herb** 4–60 cm tall with solid, whitish **rhizomes** 2–4 mm thick, rooting at nodes; in shallow fresh water of ditches, ponds, marshes, lakes and slow streams; sometimes **monoecious**, or **dioecious**. w2

■ **FLOWERS** Reddish purple, blooming June–August; **inflorescence** axillary, with one flower per exposed leaf axil; **flowering stems** 4–10 cm tall above water; **flowers** mostly perfect, or unisexual, *c.* 0.7 mm wide by 2–2.5 mm long, sessile in leaf axils; **calyx** tubular, 0.5–0.7 mm long, fused to ovary, persistent in fruit; **petals** absent; **stamens** one, exserted, *c.* 1 mm long; **filaments** filiform, longer than the calyx; **anthers** reddish green, *c.* 1 mm long and wide, 2-lobed; **style** one, erect, white, 1–1.5 mm long, stigmatic along its length, persistent. **FRUIT nutlike**, 1-seeded, one per flower, hard, 1.5–3 mm long by 1–1.3 mm wide and thick, includes the hardened calyx, apex concave where style was attached; **seeds** one per fruit, glabrous, 1.7–2.5 mm long by 1–1.3 mm wide, apex concave.

■ **LEAVES** Whorled, simple, entire, 6–13 blades at nodes 4–17 mm apart; **blades** (of vegetative stems) 5–50 mm long by 1–2.2 mm wide by 0.2–0.7 mm thick, reduced above and below water, sessile, pointed, slightly shiny, midrib obvious near apex, spreading to ascending, margins slightly paler green; **submerged blades** scalelike, erect, 4–7 mm long, bases flat; **blades** (of reproductive stems) above water 5–12 mm long, those below water 2–3 cm long by *c.* 1 mm wide, soft, thin, spreading, flaccid in air.

■ **STEM** Erect, simple, finely ridged, lighter green than leaves; **vegetative stems** 5–20 mm thick near the base.

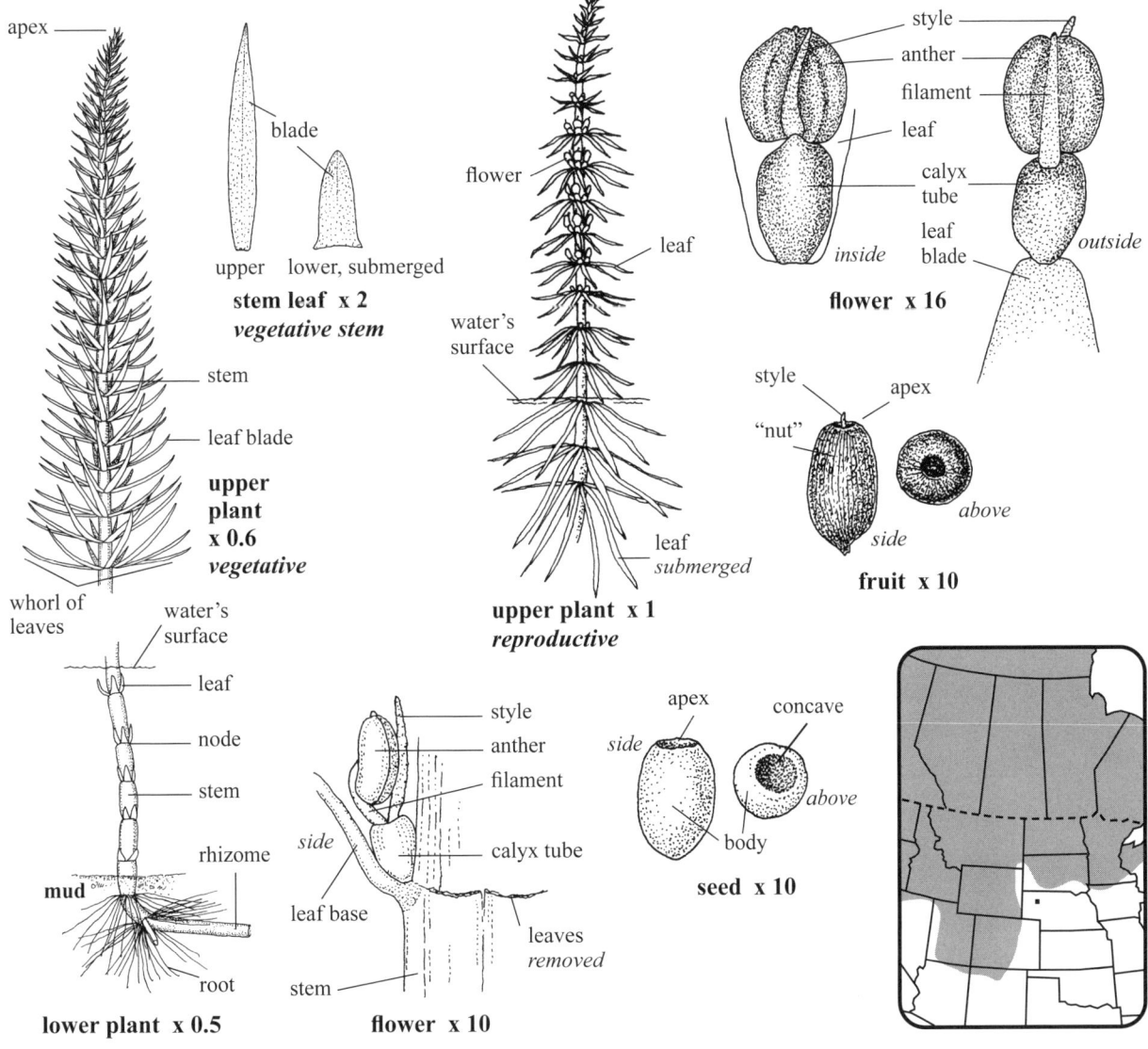

Common Mare's-tail

Family Characteristics

Mint—Lamiaceae — Flower parts 5

SKETCH Dicotyledonous herbs (wildflowers) and shrubs; annual, biennial or perennial; some fragrant. About 6,500 species worldwide.

FLOWERS Perfect, 5-merous, white to blue, violet to pink. **Inflorescence** in cymose clusters (verticillasters) in leaf axils. **Calyx** tubular, regular or 2-lipped (bilabiate), teeth or lobes 5, some form a scutellum, persistent with fruit. **Corolla** tubular, irregular or 2-lipped (bilabiate), each lip 2- or 3-lobed. **Stamens** 4 (2), didynamous (2 long and 2 short), or reduced to two, inserted on corolla tube and alternate with the lobes of the corolla; the upper 5th stamen lacking or present as a staminode. **Pistil** 1, of 4 locules. **Ovary** 4-lobed with 1 ovule per lobe. **Style** single, terminal, often branched at apex, branches (stigmas) unequal.

FRUIT A **nutlet**, each 1-seeded, hard, four located at bases of persistent calyces.

LEAVES Opposite, simple, often toothed. **Stipules** absent.

STEM Usually erect to leaning, often squarish.

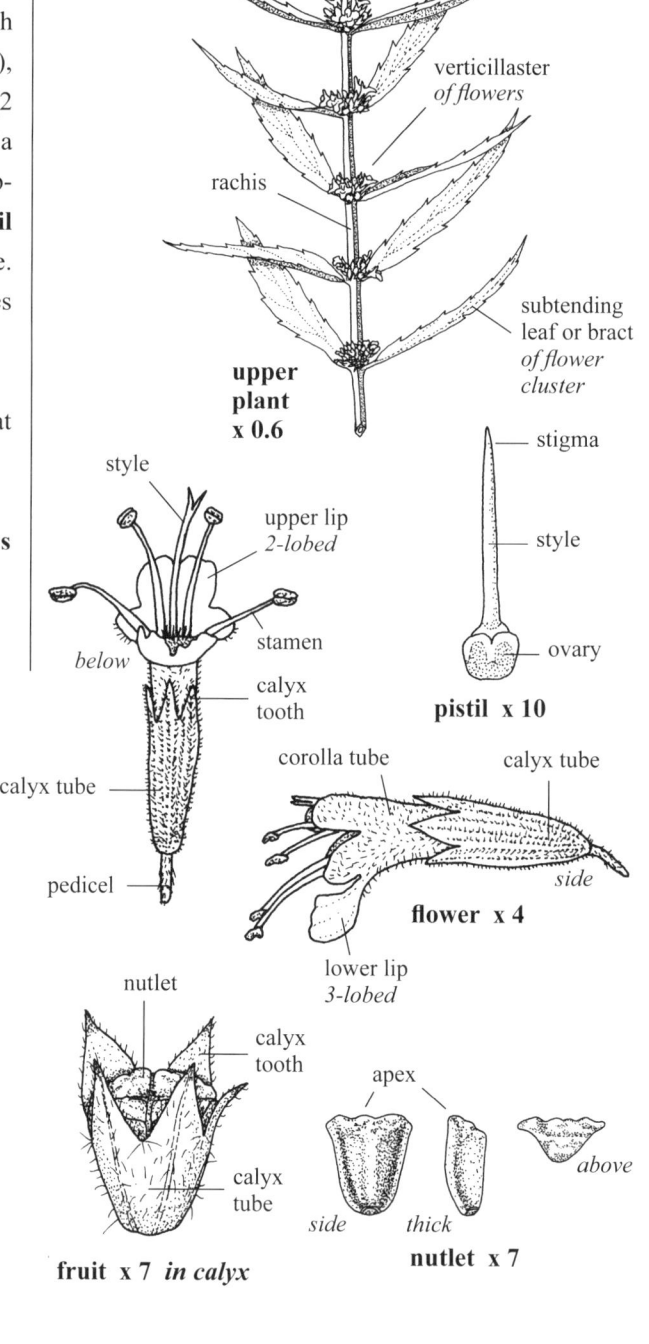

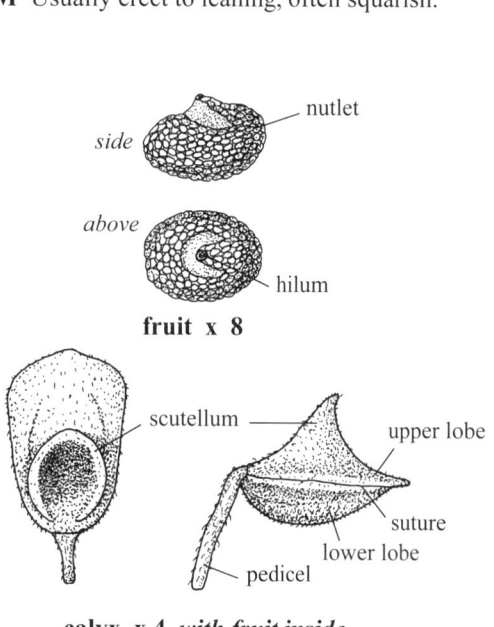

Mint—*Lamiaceae*   **Wildflower**   Pale blue to pale purple   Corolla tubular, 2-lipped, 4- or 5-lobed

GIANT-HYSSOP   *Agastache foeniculum* (Pursh) Kuntze

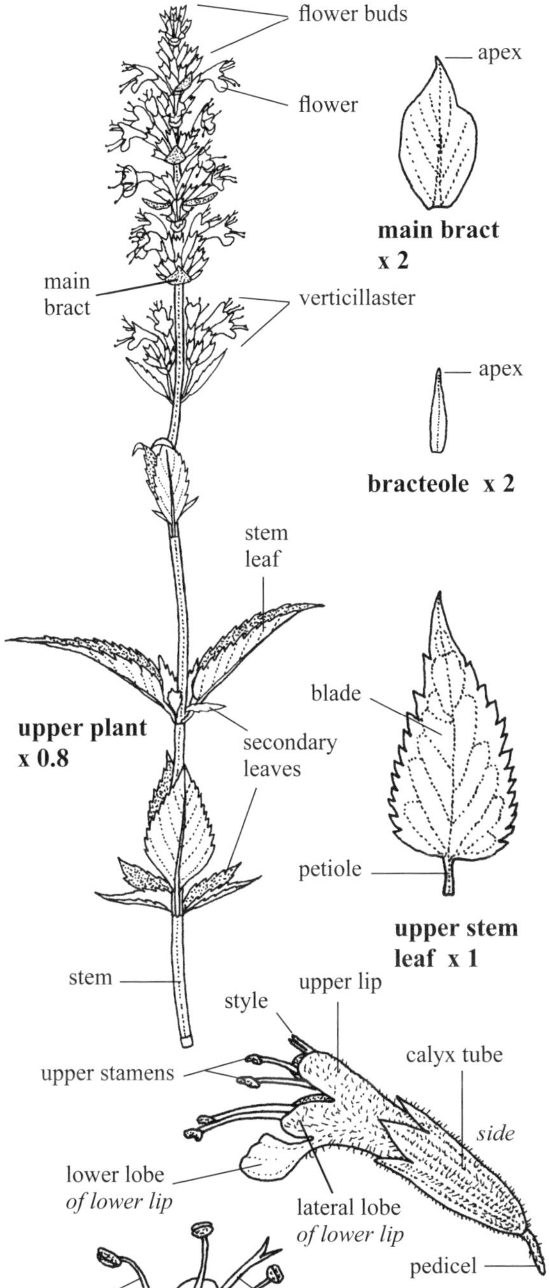

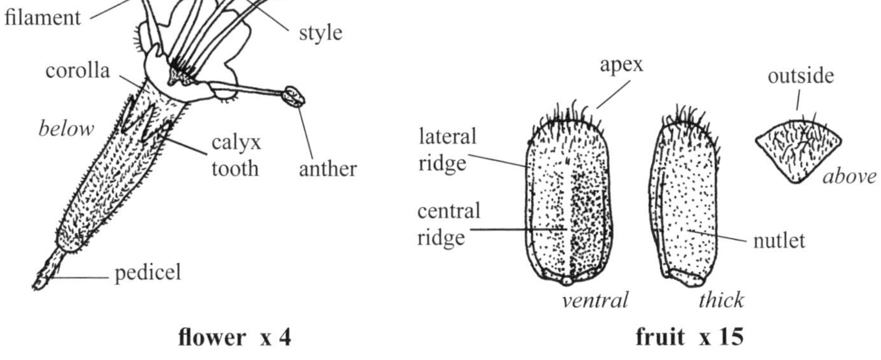

■ **SKETCH** A **perennial herb** 30–120 cm tall, singly or in clumps of 2–30 stems with a dark brown **rhizome** 1.5–3 mm thick; in moist open woodlands, along lakeshores, ditches, in dry thickets and prairies; **licorice odor**.

■ **FLOWERS** Pale blue to pale purple, blooming June–September; **inflorescence** of cymes (spikelike), terminal and on branches, TL 2–15 cm by 1.5–3.5 cm wide, the lower verticillasters sometimes remote; **floral branches** two to several, erect and opposite, 10–33 cm long, mostly from upper leaf axils; **pedicels** *c.* 2 mm long, hairs deflexed; **flowers** perfect, TL 11–13 mm, spreading, 15–20 per verticillaster, congested; **bracts** of two types subtending each cyme: **1) main bracts** *c.* 10 mm long by *c.* 5 mm wide, woolly below, glabrous above; **2) bracteoles** *c.* 10, in pairs, 5–6 mm long by *c.* 1 mm wide; **calyx** tubular, 4–6 mm long, 15-veined, hairs spreading and short; **calyx teeth** five, medium purplish pink, each 1–1.8 mm long, lower tooth the shortest; **corolla** tubular, bilabiate, 7–10 mm long by 3.5–4 mm wide by 5–5.5 mm high, hairy outside except on the lower lip; **upper lip** 2-lobed, *c.* 2 mm long by *c.* 2.5 mm wide, apex barely notched, glabrous inside; **lower lip** 3-lobed, hairy in throat, lateral lobes *c.* 1.3 mm long; **stamens** four, two pairs, exserted; **filaments** *c.* 7 mm long, glabrous, filiform, upper pair attached at base of upper lip; **anthers** dark purplish pink, 0.7–0.9 mm long; **ovary** hairy at apex; **style** pale purple, 2-parted, exserted. **FRUIT a nutlet**, 1-seeded, one to four at base of persistent calyx, dark brown, *c.* 1.5 mm long by *c.* 0.8 mm wide by *c.* 0.7 mm thick, 3-sided, outside rounded, central ridge blunt, apex hairy.

■ **LEAVES** Opposite, simple, toothed, slightly shiny; **blades** 2–11.5 cm long by 1–6 cm wide, reduced above, whitish below from appressed microscopic hairs, glabrous and green above; **petioles** 2–35 mm long, reduced above, slightly hairy, often with a pair of secondary leaves in axils.

■ **STEM** Erect, bluntly square, stout, simple to branched, hollow, mostly hairless to slightly hairy above or throughout; 2–5 mm thick near the base.

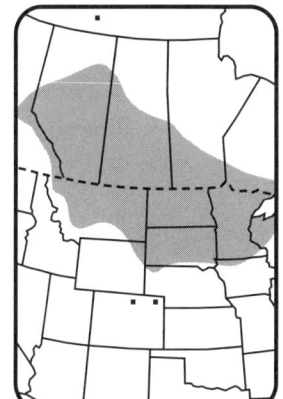

Blue Giant Hyssop, Lavender Hyssop   345

Mint—*Lamiaceae* **Wildflower** Pink to pale blue Corolla tubular, 2-lipped, 5-lobed

AMERICAN DRAGONHEAD *Dracocephalum parviflorum* Nutt.

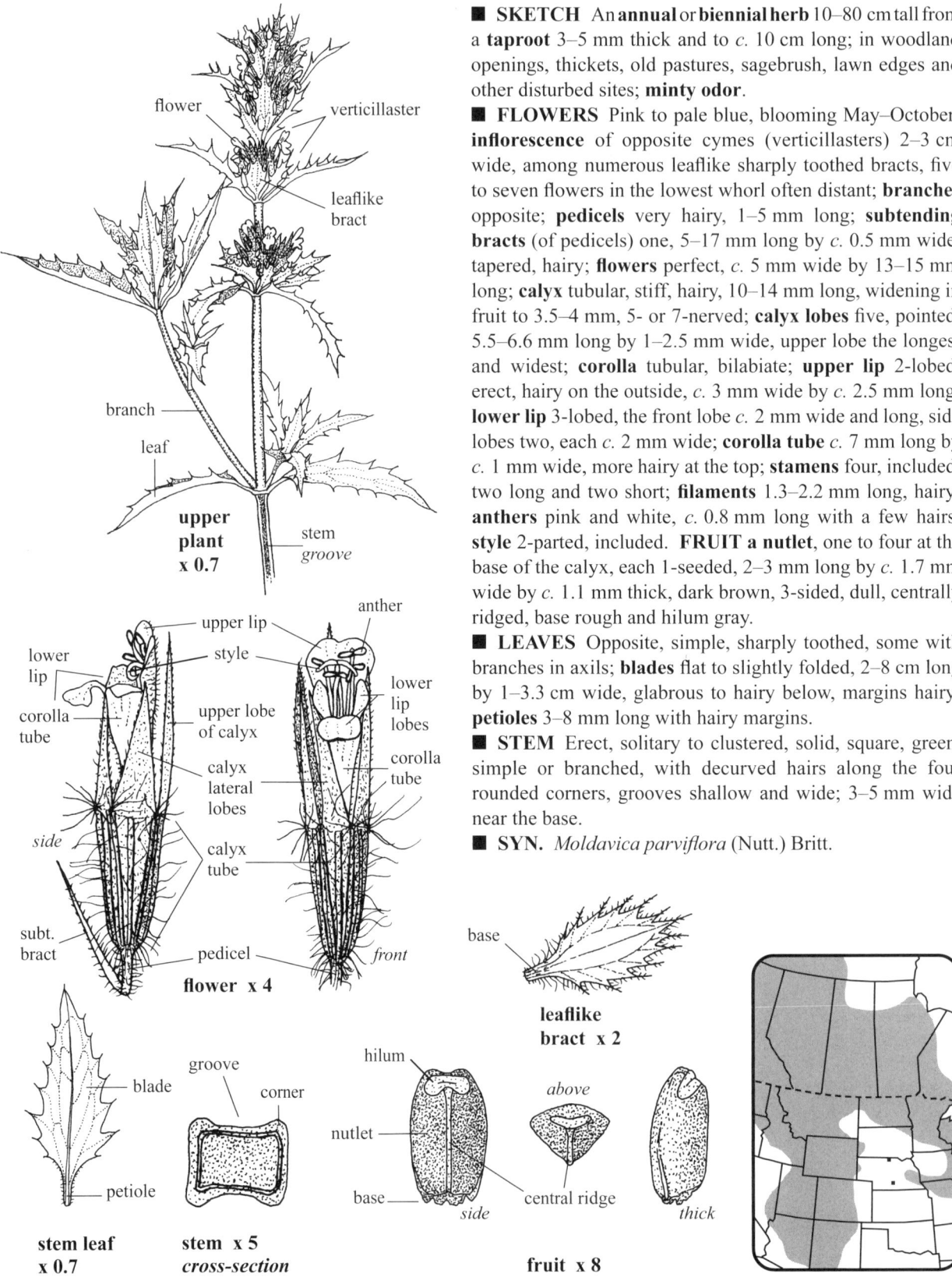

- **SKETCH** An **annual** or **biennial herb** 10–80 cm tall from a **taproot** 3–5 mm thick and to *c.* 10 cm long; in woodland openings, thickets, old pastures, sagebrush, lawn edges and other disturbed sites; **minty odor**.
- **FLOWERS** Pink to pale blue, blooming May–October; **inflorescence** of opposite cymes (verticillasters) 2–3 cm wide, among numerous leaflike sharply toothed bracts, five to seven flowers in the lowest whorl often distant; **branches** opposite; **pedicels** very hairy, 1–5 mm long; **subtending bracts** (of pedicels) one, 5–17 mm long by *c.* 0.5 mm wide, tapered, hairy; **flowers** perfect, *c.* 5 mm wide by 13–15 mm long; **calyx** tubular, stiff, hairy, 10–14 mm long, widening in fruit to 3.5–4 mm, 5- or 7-nerved; **calyx lobes** five, pointed, 5.5–6.6 mm long by 1–2.5 mm wide, upper lobe the longest and widest; **corolla** tubular, bilabiate; **upper lip** 2-lobed, erect, hairy on the outside, *c.* 3 mm wide by *c.* 2.5 mm long; **lower lip** 3-lobed, the front lobe *c.* 2 mm wide and long, side lobes two, each *c.* 2 mm wide; **corolla tube** *c.* 7 mm long by *c.* 1 mm wide, more hairy at the top; **stamens** four, included, two long and two short; **filaments** 1.3–2.2 mm long, hairy; **anthers** pink and white, *c.* 0.8 mm long with a few hairs; **style** 2-parted, included. **FRUIT a nutlet**, one to four at the base of the calyx, each 1-seeded, 2–3 mm long by *c.* 1.7 mm wide by *c.* 1.1 mm thick, dark brown, 3-sided, dull, centrally ridged, base rough and hilum gray.
- **LEAVES** Opposite, simple, sharply toothed, some with branches in axils; **blades** flat to slightly folded, 2–8 cm long by 1–3.3 cm wide, glabrous to hairy below, margins hairy; **petioles** 3–8 mm long with hairy margins.
- **STEM** Erect, solitary to clustered, solid, square, green, simple or branched, with decurved hairs along the four rounded corners, grooves shallow and wide; 3–5 mm wide near the base.
- **SYN.** *Moldavica parviflora* (Nutt.) Britt.

Mint—*Lamiaceae* **Wildflower** Pinkish pale purple to white Corolla tubular, 2-lipped, lobed

HEMP-NETTLE   *Galeopsis bifida* Boenn.

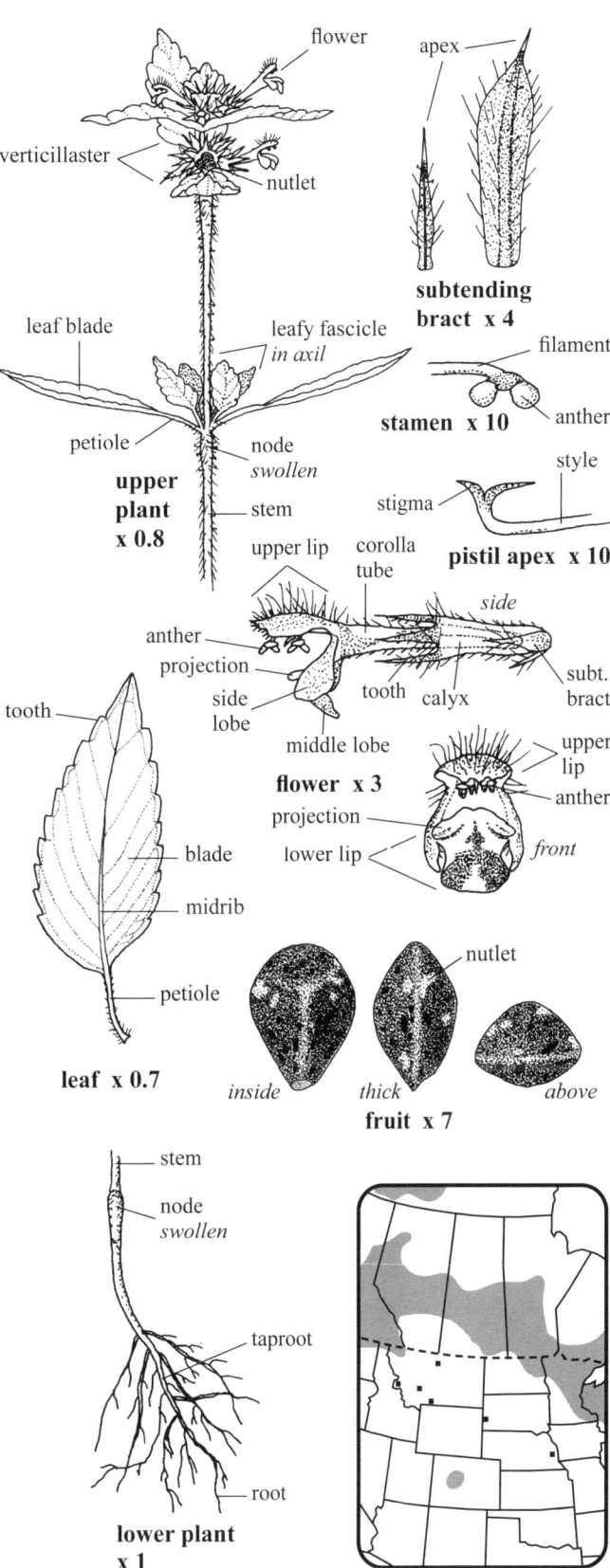

■ **SKETCH** A variable, hairy **annual herb** 30–100 cm tall with a tan **taproot** 3–10 cm long by 1–5 mm thick; along lakeshores, in moist fields, clearings, roadsides and open woods. *Naturalized* **w2**

■ **FLOWERS** Pinkish pale purple to white, blooming July–September; **inflorescence** terminal and in upper leaf axils; **lower floral branches** to *c.* 35 cm long, ascending; **verticillasters** of 3–9 flowers per axil; **subtending bracts** (of flowers) usually two, 5–8 mm long by 0.5–2 mm wide, hairy with a few gland-tipped hairs; **flowers** perfect, sessile, TL 10–22 mm by 3–5 mm wide; **calyx** tubular, green, TL 7–11 mm (including teeth); **calyx tube** 4–5.5 mm long, hairy inside, 10-nerved, teeth five, these 3–5.5 mm long (–11 mm in fruit), glandular-hairy; **corolla** bilabiate, tube *c.* 10 mm long, white below, pinkish near lobes, outside hairs ascending, shorter gland-tipped outer hairs spreading; **upper lip** 3–4 mm long and wide, with 3 or 4 blunt teeth, arched, hairy, short hairs gland-tipped, long clear hairs 1–1.5 mm long; **lower lip** 3-lobed, hairy below, side lobes 2–3.8 mm long by *c.* 2 mm wide with a projection, middle lobe recurved; **stamens** four, didynamous (2 + 2), two long and two short, included under the upper lip; **filaments** attached in corolla throat, upper free part 2–4 mm long, hairy below; **anthers** two, yellow, hairy or not, together *c.* 1 mm wide; **ovary** green, glabrous, 4-lobed, 0.7–0.9 mm long; **style** filiform, white, 9–14 mm long, pinkish at apex; **stigma** 2-parted *c.* 1 mm long and pointed downward, included under upper lip. **FRUIT a nutlet**, 1-seeded, 1–4 in base of calyx tube, dark brown with black and tan spots, smooth, shiny, 2.8–3.8 mm long by 2.3–2.6 mm wide by 1.8–2 mm thick.

■ **LEAVES** Opposite, simple, usually toothed, stem only; **blades** 2–12 cm long by 1–5 cm wide, spreading, hairy below along side veins, margins and midrib, slightly hairy on blade, slightly glandular spotted, more hairy above, hairs appressed and parallel with side veins; **lower leaf axils** often with small leaves or a floral branch; **petioles** 3–25 mm long, hairy inside groove, the hairs facing the blade.

■ **STEM** Erect, solid, square, edges blunt, simple to branched throughout, hairs descending and 1–2 mm long, less hairy below; base 1.5–6 mm wide and hairy, the hairs transparent, long hairs with 2–4 swollen joints, upper short hairs gland-tipped and spreading; **nodes** hairy, swollen for 1–2 cm below the petioles, some nodal hairs gland-tipped.

■ **SYN.** *Galeopsis tetrahit* var. *bifida* (Boenn.) Lej. & Court.

Split-lip Hemp-nettle, Common Hemp-nettle, Brittle-stem Hemp-nettle, Dog Nettle, Bee Nettle

Mint—*Lamiaceae* **Wildflower** Blue to purple Corolla tubular, 2-lipped, 5-lobed

GROUND-IVY *Glechoma hederacea* L.

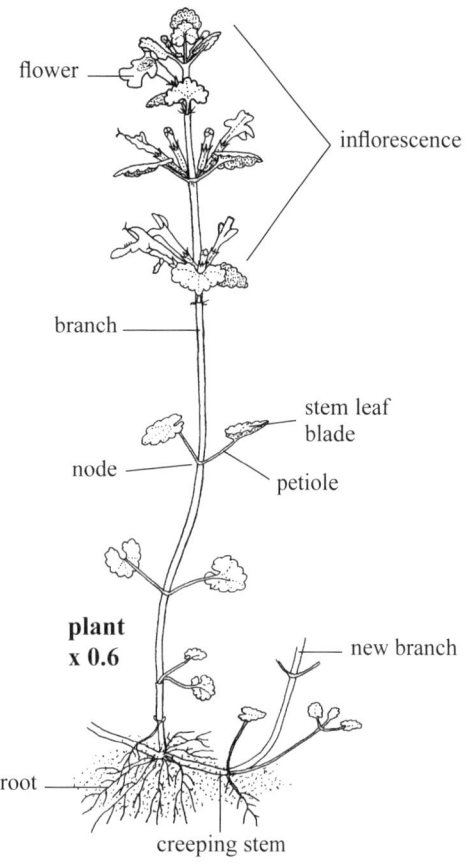

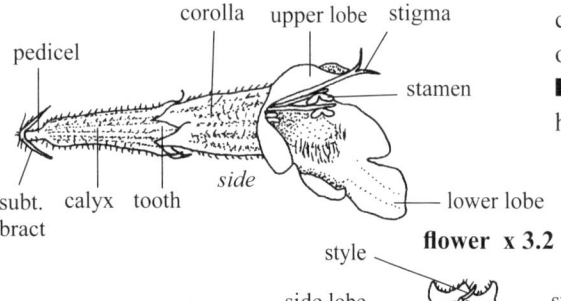

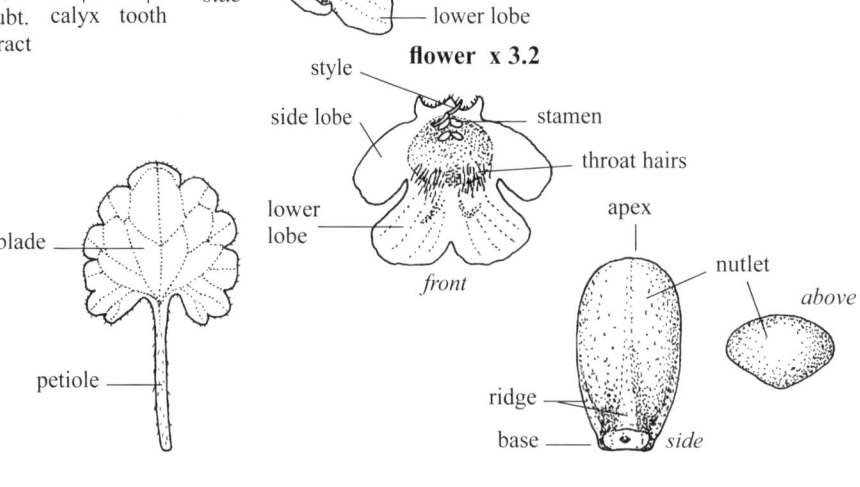

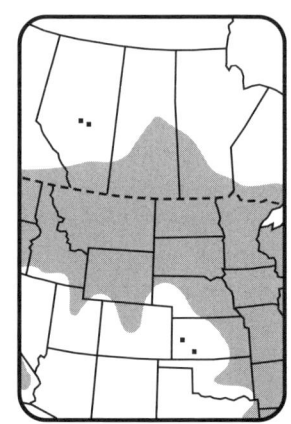

■ **SKETCH** A **perennial herb** of creeping stems up to *c.* 1 m long with **fibrous roots** developing at the nodes and the upright flowering branches 6–45 cm long; in woodland edges, thickets, lawns, roadsides and pastures. *Naturalized*

■ **FLOWERS** Blue to light purple, blooming April–June; **inflorescence** of ascending flowers in a whorl (cymule) from each upper leaf axil; **pedicels** 1–3 mm long, lengthening to 2–4 mm with fruit, hairy; **subtending bracts** (of pedicels) two, 1–1.5 mm long, thin, reddish, hairy; **flowers** perfect, 10–22 mm long, up to six at opposite leaf axils, absent from lower leaf axils; **calyx** tubular, 15-nerved, hairy, 4–8 mm long by *c.* 2 mm wide, apical teeth five, pointed, reddish and *c.* 1.5 mm long, three nerves per tooth, persistent in fruit; **corolla** tubular, bilabiate, 5-lobed, 8–12 mm wide by 10–22 mm long, hairy outside, especially along the upper two lobes, dark purple stripes inside the corolla tube, these usually extending as two to four veins on the lower two lobes, white hairs upright and clustered on either side of the lower throat with a gap between them; **stamens** four, variable in length, didynamous (2 + 2); **anthers** pink and white, *c.* 1 mm long, one pair near the base of the upper two lobes, the second pair shorter; **pollen** white; **stigma** 2-parted, glabrous, barely exserted, close to the upper lobes. **FRUIT a nutlet**, 1-seeded, one to four at base of calyx tube, dark brown, roundly triangular, smooth to slightly rough, hairless, 1.5–2 mm long by *c.* 1 mm wide by 0.6–0.7 mm thick, the three ridges converging at the whitish base.

■ **LEAVES** Opposite, simple, roundly toothed; **blades** 10–40 mm long by 8–45 mm wide, mostly glabrous above, margins ciliate, velvety hairy below; **petioles** 0.4–4.5(–10) cm long, hairy, often with a visible tuft of white hairs at the base.

■ **STEM** Square, the four edges with short downward pointing hairs or almost glabrous; 1–2 mm thick near the reddish base.

348 Creeping Charlie, Gill-over-the-ground

Mint—*Lamiaceae* **Wildflower** White to pale pink Corolla tubular, 2-lipped, 4-lobed

MOTHERWORT *Leonurus cardiaca* L.

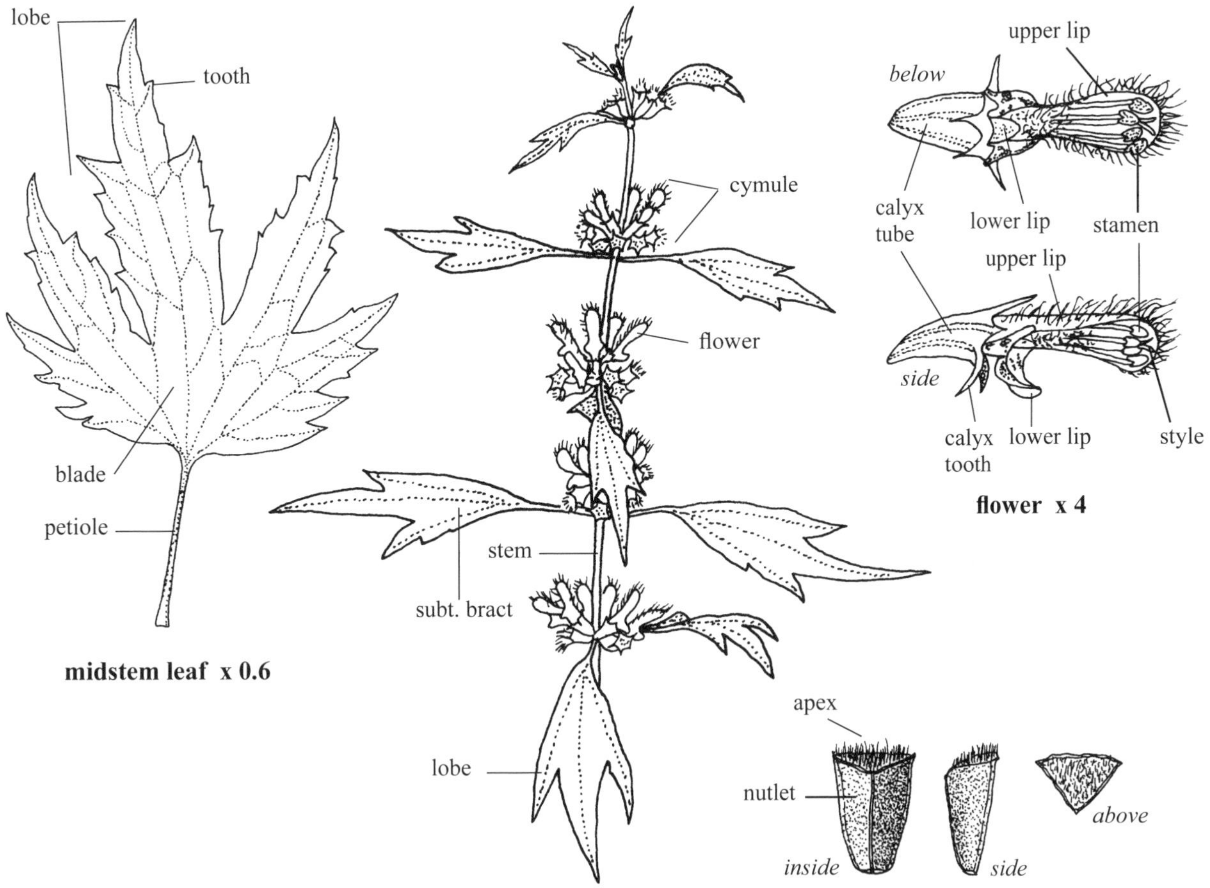

■ **SKETCH** A perennial herb 50–200 cm tall with short **rhizomes** 3–4 mm thick; in shelterbelts, mountain brush, along streams and riverbanks, roadsides and in ditches. *Naturalized*

■ **FLOWERS** White to pale pink, blooming June–August; **inflorescence** of 20–30 crowded cymules or whorls 15–20 mm wide in the upper leaf axils; **subtending bracts** leaflike, with three large teeth; **floral branches** 3–50 cm long, opposite, ascending; **flowers** perfect, 10–15 per whorl, each *c.* 3 mm wide by 8–12 mm long by 5–7 mm high, blooming sequence bottom to top; **calyx** 5-nerved and angled, 4–8 mm long by *c.* 2 mm wide, usually glabrous, teeth five, pointed, 1.5–2 mm long, deflexed below, persistent around the fruit; **corolla** tubular, 2-lipped, 8–12 mm long with pink to purple spots; **upper lip** hairy on the outside, glabrous and light pink inside; **lower lip** 3-lobed, glabrous, with dark pink spots; **stamens** four, included, two slightly shorter; **filaments** pink and white with white hairs; **pistil** included, white, glabrous; **style** 2-parted at the tip, situated between the stamens and the upper lip. **FRUIT** a **nutlet**, 1-seeded, one to four per flower, 1.7–2.2 mm long by 1.3–1.5 mm wide by *c.* 1 mm thick, triangular, dark brown, apex flat with white hairs.

■ **LEAVES** Opposite, simple, 3- to 7-lobed, toothed, and spreading; **blades** glabrous above, 3–12 cm long by 1.5–12 cm wide, lightly hairy below especially along the raised veins, slightly darker green above; **upper blades** (bracts) smaller, 3-lobed or only toothed; **petioles** with a groove above, 1–8.2 cm long, white hairs where petioles merge with the stem; **stipules** absent.

■ **STEM** Square, erect, and branched, 4-ridged, slightly hairy to downy, hollow; 3–10 mm wide near the hairless base.

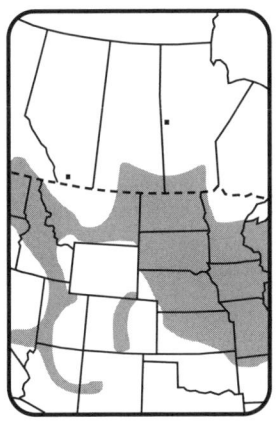

Common Motherwort   349

Mint—*Lamiaceae* **Wildflower** White Corolla tubular, 4-lobed

WATER-HOREHOUND *Lycopus americanus* Muhl. ex W. Bart.

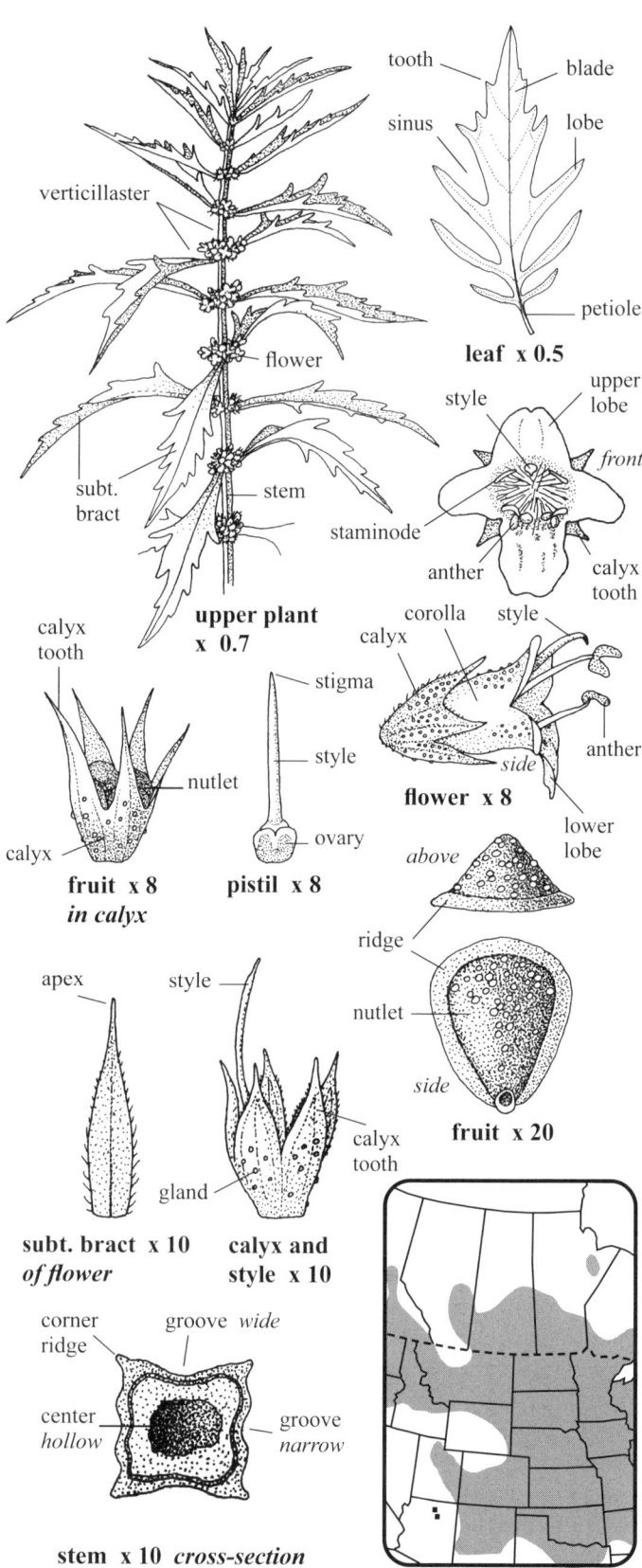

■ **SKETCH** A **perennial herb** 20–90 cm tall with **rhizomes** to *c.* 20 cm long and 1–2 mm wide, **without tubers**; in moist sites along stream banks, sloughs, ditches, forest edges, lakeshores and marshes.

■ **FLOWERS** White, blooming July–September; **inflorescence** 20–40 cm long of **verticillasters** 7- to 26-flowered and 7–10 mm wide by 5–7 mm tall; **subtending bracts** (of flowers) one, green, pointed, entire, 1–3 mm long by 0.2–0.6 mm wide, 1-veined, ciliate; **flowers** sessile, perfect, 2.5–3.5 mm wide by 3–4 mm long; **calyx** green, glandular-punctate, tubular, the tube finely hairy, 2–3 mm long, 5-toothed, the teeth pointed and mostly equal, 1–1.5 mm long, each tooth 3-veined, finely ciliate, erect around and surpassing the fruit; **corolla** tubular, 4-lobed, 2.5–3.5 mm long, easily separated from the calyx; **corolla lobes** 1–1.4 mm long and wide, side lobes rounded, upper lobe slightly notched, white hairs spreading in the throat, lower lobe curved downward with a few irregular pinkish purple markings; **staminodia** two; **stamens** two, exserted; **filaments** white, *c.* 1.5 mm long, attached to the corolla tube on opposite lower sides; **anthers** pink, turning brown, divergent, 2-lobed, *c.* 0.4 mm long by *c.* 0.5 mm wide; **pistil** *c.* 3.3 mm long; **ovary** 4-parted, green with an orange base, *c.* 0.6 mm long by *c.* 0.8 mm wide; **style** one, white, *c.* 2.5 mm long, exserted. **FRUIT** a **nutlet**, 1-seeded, one to four per flower, 1–1.3 mm long by 0.6–1 mm wide by *c.* 0.6 mm thick, dark brown, dull, apex corky and punctate, lateral ridge winglike; **body** tapered to base, enclosed by persistent calyx.

■ **LEAVES** Opposite, simple, narrowly lobed, with two or three main lobes per side, toothed; **blades** 1.5–15 cm long by 0.5–9 cm wide, reduced above, glabrous above, slightly hairy below on the veins, hairs white, appressed and pointing forward, glandular-punctate on both sides, dull, lighter green below, spreading; **lower blades** lobed, upper blades less lobed, some merely toothed to entire; **petioles** 2–10 mm long, hairy below, tapered.

■ **STEM** Erect, green, square, stiff, hollow, simple or branched, punctate, glabrous to densely hairy especially along the four ridges, white hairs ascending and appressed, a wide groove on one pair of opposite sides and a narrow groove on the other pair of sides, white hairy at base of verticillasters between the petioles; 2–5 mm thick at the base.

350   Cut-leaf Water-horehound, American Water-horehound, American Bugleweed

Mint—*Lamiaceae* **Wildflower** White Corolla tubular, 4-lobed

WESTERN WATER-HOREHOUND *Lycopus asper* Greene

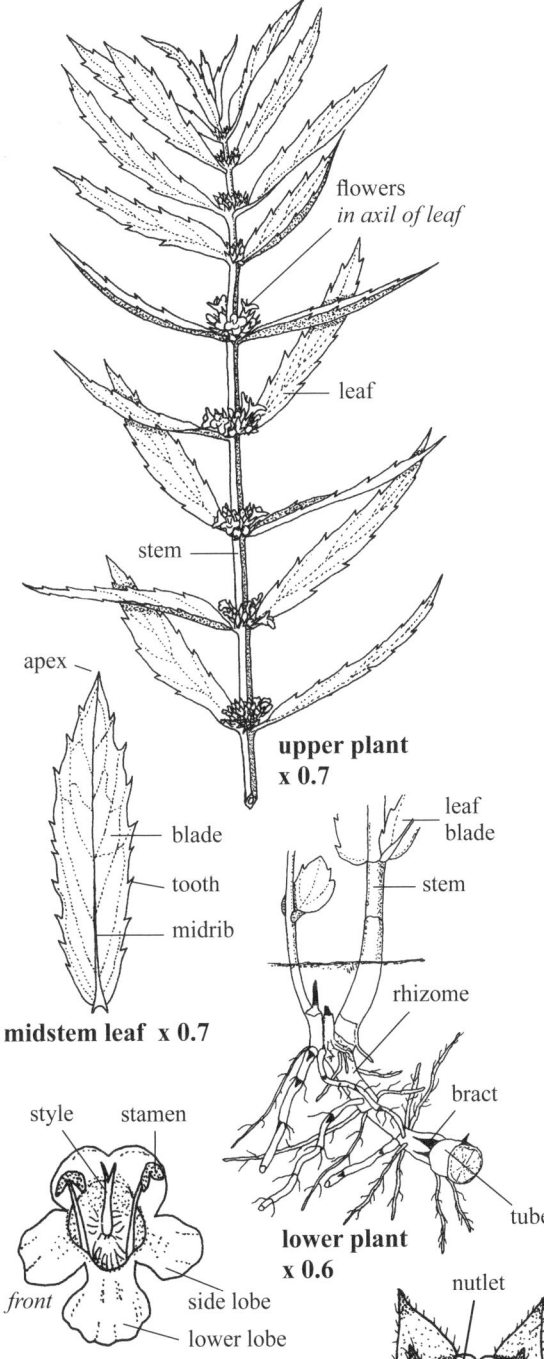

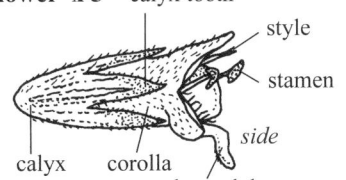

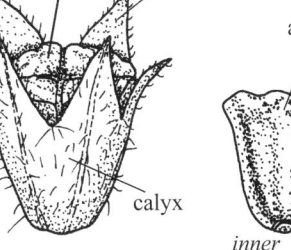

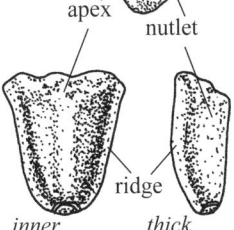

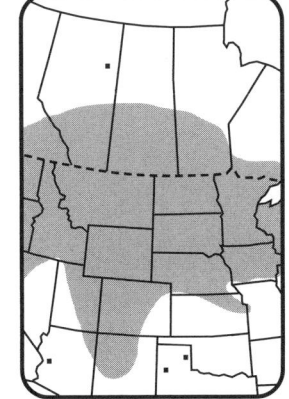

■ **SKETCH** A **perennial herb** 20–130 cm tall from **rhizomes** 1–2 mm thick and to *c.* 30⁺ cm long with paired bracts 1–4 mm long at nodes, and with white **tubers** 3–10 mm thick, single or in crowded colonies; in wet meadows and marshes; no minty odor.

■ **FLOWERS** White, blooming in July–September; **inflorescence** 7–60 cm long on main stem and branches, with whorls of 6–24 flowers in leaf axils; **verticillasters** (in axils) 10–15 mm wide by 5–8 mm tall; **subtending bracts** (of flowers) one, 1.5–3 mm long to *c.* 1 mm wide, hairy, pointed; **flowers** perfect, usually ascending to spreading, sessile, 4.5–5 mm wide by 4–5 mm tall; **calyx** tubular, 2.5–4.5 mm long by *c.* 2 mm wide, 15-nerved, persistent, tan, some hairs glandular; **calyx teeth** five, 2–2.5 mm long, reddish green, margins hairy, erect above the fruit; **corolla** tubular, 4-lobed, white, 4–6 mm long, tube mostly glabrous, 2–3 mm long by *c.* 1.6 mm wide, exposed area is hairy outside and in the throat; **lower corolla lobe** 2–2.3 mm long by *c.* 2 mm wide, with faint pink spots; **side lobes** two, blushed with pink; **stamens** two, fertile, included, sometimes two are nonfunctional; **filaments**, glabrous, 2.5–3 mm long, inserted at the top of the corolla tube; **anthers** pink and white, *c.* 1 mm long and wide, glabrous; **ovary** 4-parted, green, with a rough apex *c.* 1 mm wide; **style** included, glabrous, 2-parted, 4–4.5 mm long, filiform. **FRUIT a nutlet**, 1-seeded, up to four per calyx, dark brown, 1.7–2.2 mm long by 1.3–2 mm wide by 0.7–0.9 mm thick, triangular, outer surface smooth and slightly convex, side ridges two, the apex rough, inside face bluntly V-shaped.

■ **LEAVES** Opposite, simple, toothed, flat, greenish yellow, spreading to ascending, sessile, glabrous to hairy, glandular-punctate above; **blades** 2.4–11 cm long by 7–35 mm wide, reduced below.

■ **STEM** Erect, green, square, hollow, simple, one to three clustered stems, corners blunt and hairy, the faces slightly concave between the corners and glabrous; 2–6 mm thick near the base.

Rough Water-horehound, Rough Bugleweed   351

Mint—*Lamiaceae* **Wildflower** White to pale blue Corolla tubular, 4- or 5-lobed

NORTHERN WATER-HOREHOUND *Lycopus uniflorus* Michx.

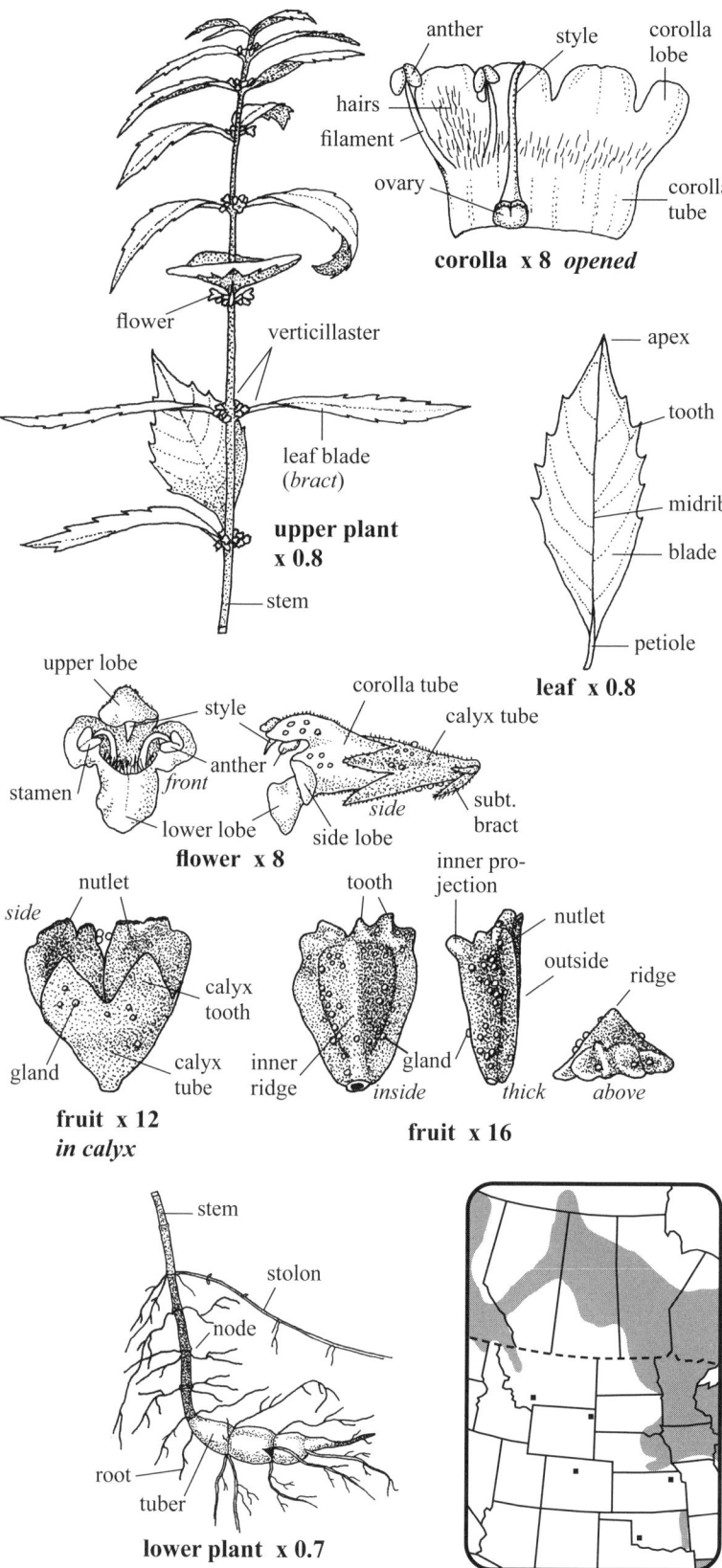

- **SKETCH** A **perennial herb** 15–70 cm tall with a whitish **tuber** 8–35 mm long by 4–7 mm thick; **roots** tan, 10–30 mm long and **stolons** to 15+ cm long by *c.* 0.5 mm thick; in wooded areas along marshes, bogs, creeks, springs and streams.
- **FLOWERS** White to pale blue, blooming July–September; **inflorescence** axillary, 10–50 cm long with verticillasters in all but the lowest leaf axils; **floral branches** opposite, to *c.* 25 cm long; **flowers** perfect, 2–2.5 mm wide by 3–4 mm long, sessile, spreading to ascending; **subtending bracts** two or three, 0.5–1 mm long, hairy; **calyx** tubular, green, 2–2.5 mm long, short hairy outside and slightly glandular, 4- or 5-toothed, the teeth pointed to blunt, about equal, erect, slightly shorter than the fruit; **corolla** mostly regular, tubular with 4- or 5-lobes, short hairs outside, longer hairs inside in a band at the base of the filaments, glandular, lobes flaring, lower lobe *c.* 1 mm long and wide, notched or not, upper lobe arched; **stamens** two (3), included to slightly exserted, *c.* 1.2 mm long, white; **filaments** *c.* 1 mm long, hairy near base; **anthers** 2-lobed, the lobes divergent; **pistil** one, 2.5–3 mm long; **ovary** green, 4-lobed, of four carpels, *c.* 0.5 mm long by *c.* 0.7 mm wide; **style** filiform, pointing down from below the upper lobe; **stigma** obscure. FRUIT a nutlet, 1-seeded, 1–4 per flower, brown, apex toothed, glandular, each 1–1.6 mm long by 0.8–1.1 mm wide by 0.6–0.8 mm thick, exserted past calyx teeth, triangular, inner two sides glandular with a blunt inward projection.
- **LEAVES** Opposite, stem only, simple, toothed, dull, spreading, lighter green below; **blades** 2–10 cm long by 0.6–4 cm wide, pointed, veins raised below with short and long hairs bent forward; **petioles** 2–6 mm long, hairy.
- **STEM** Erect, hollow, square, simple to branched, corners blunt to sharp, glabrous or white hairs appressed and arched forward, hairy on ridges; 1–1.5 mm wide near the reddish brown base.
- **SYN.** *Lycopus virginicus* var. *pauciflorus* Benth.

Mint—*Lamiaceae* **Wildflower** Pale bluish purple to pink Corolla tubular, 2-lipped, 4-lobed (5-)

FIELD MINT *Mentha arvensis* L.

■ **SKETCH** A variable **perennial herb** 15–90 cm tall or long from tan **rhizomes** 1–5 mm thick by 5–20+ cm long; in damp meadows or woods, ditches, edges of sloughs, shores of streams and lakes; **minty odor. p2**

■ **FLOWERS** Pale bluish purple to pink, blooming July–September; **inflorescence** of verticillasters in leaf axils, TL 3–30 cm; **flower clusters** 1.5–2.5 cm wide by 7–15 mm tall, each cymule 8- to 30-flowered; **floral branches** opposite, to *c.* 35 cm long, squarish, hairy; **peduncles** green, hairy, mostly 1–4 mm long, one per leaf axil; **bracteoles** paired at top of peduncles, green, 4–12 mm long by 1–5 mm wide, hairy below and on margins, glabrous above, spreading; **foliaceous bracts** spreading to ascending, reduced above, glandular dots (microscopic) more obvious below on lighter green surface, blades glabrous with margins ciliate, petioles and veins hairy; **pedicels** reddish, 1–3 mm long, glabrous; **flowers** perfect, weakly bilabiate, 4–5 mm wide by 6–8 mm long, 4- or 5-lobed; **calyx** tubular, 2.5–3.3 mm long by *c.* 2 mm wide, 10-veined, these easily seen on inside, punctate, hairy, erect to leaning with fruit, 5-toothed, these subequal, 0.6–1 mm long and spreading with fruit; **corolla** tubular, 2-lipped, hairy in throat, 4.5–6.5 mm long, hairy outside on lobes, tubular part glabrous outside and 3–5 mm long; **lower lobes** three, each *c.* 1.5 mm long by 1–1.2 mm wide; **upper lobe** 1.5–2 mm long by 1.8–2 mm wide, barely notched; **stamens** four, *c.* 6 mm long, exserted, spreading between the corolla's lobes; **filaments** free above for *c.* 4 mm, then attached below to the corolla tube for *c.* 2 mm; **anthers** *c.* 0.8 mm long, pinkish lavender; **ovary** 4-lobed, glabrous, 0.7–0.8 mm long; **style** *c.* 6 mm long, exserted, pale lavender, 2-parted. **FRUIT a nutlet**, 1-seeded, one to four per calyx, yellowish brown, glabrous, apices blunt, base whitish, 3-sided, 0.7–1.3 mm long by 0.6–0.8 mm wide by *c.* 0.6 mm thick.

■ **LEAVES** Opposite, simple, toothed, scabrous, some glandular, pointed, fragrant when crushed; **blades** 3–12 cm long by 5–45 mm wide, reduced above to foliaceous bracts, the base of the blade usually entire; **petioles** 2–20 mm long, the hairs arched forward; **stipules** absent.

■ **STEM** Erect to decumbent, simple or branched, hollow, square with the sides slightly concave and glandular-punctate, hairs deflexed on corners; 1–4 mm thick near the glabrous base; **nodes** hairy.

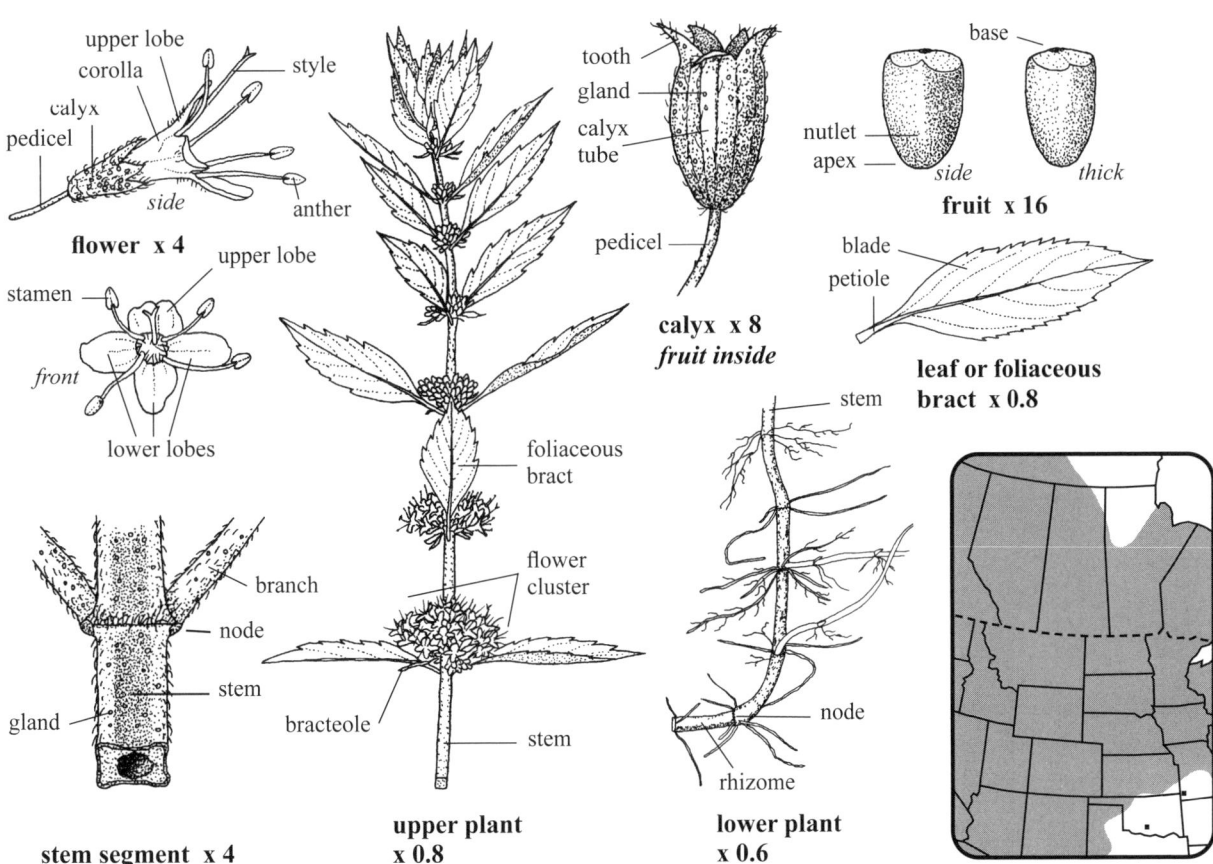

American Wild Mint, Wild Mint

Mint—*Lamiaceae* **Wildflower** Pink to purple to white Corolla tubular, 2-lipped, lobed

Wild Bergamot *Monarda fistulosa* L.

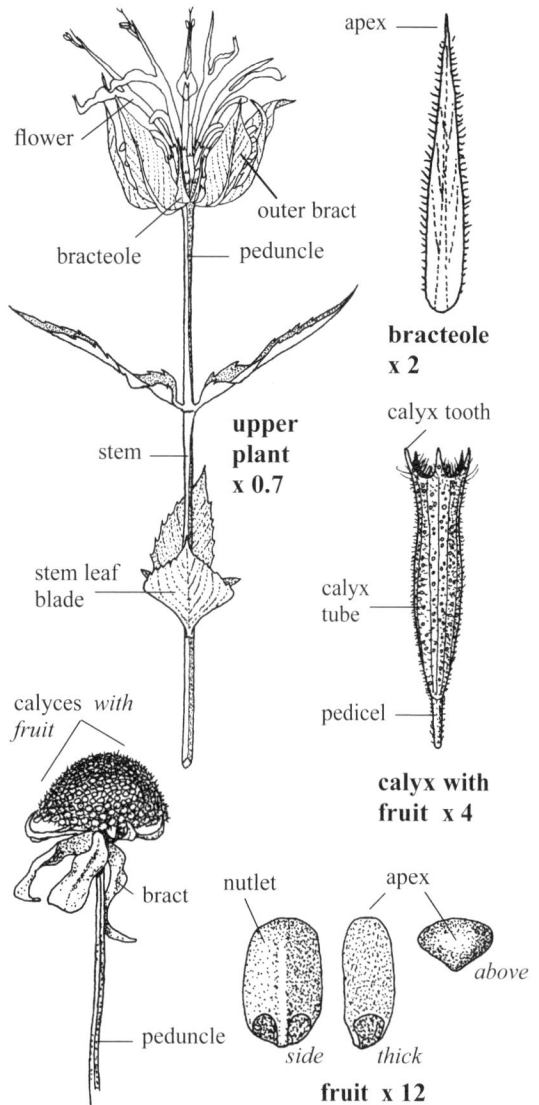

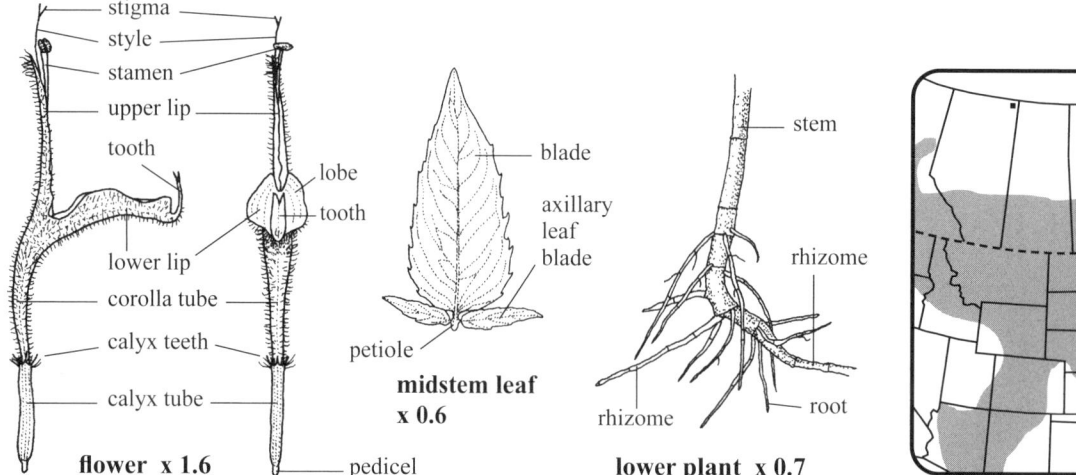

■ **SKETCH** A variable **perennial herb** 30–120 cm tall in clumps to open colonies from slender creeping **rhizomes** 1–2 mm thick by 9–18+ cm long; in prairies, thickets, sagebrush, pastures, along roadsides and streamsides. p1

■ **FLOWERS** Pink to purple to white, blooming June–September; **inflorescence** of one to six compact heads, each head 3–8 cm wide by 3–5 cm tall, terminal on a peduncle; **peduncles** in fruit 3–5 cm long, hairy, square; **pedicels** 1–2 mm long; **outer bracts** four to six, leaflike, green, erect to spreading, punctate, 1.8–3 cm long by 7–15 mm wide; **inner floral bracts** several, erect, 10–15 mm long by 2–3 mm wide; **bracteoles** 0–13, punctate, 6–20 mm long by 1–2.5 mm wide, margins hairy; **flowers** perfect, 40–50 per head, irregular, fragrant, 5–6 mm wide and to *c.* 33 mm long; **calyx** tubular, 6–11 mm long by 1–1.3 mm wide, persistent, punctate, hairs purple, 10- to 13-veined, glabrous and veined inside, throat hairy; **calyx teeth** five, purple, 1–1.5 mm long; **corolla** tubular, bilabiate, 20–33 mm long, punctate, glabrous inside, hairy above, tubular part 16–25 mm long; **1) upper lip** erect, hairy outside, 8–17 mm long, split along entire length; **2) lower lip** 10–14 mm long by 5–6 mm wide, 2-lobed with an erect bidentate tooth *c.* 4 mm long; **stamens** two; **filaments** white, attached near base of upper lip, 12–22 mm long, glabrous, slightly exserted; **anthers** *c.* 2 mm long, medium purple; **style** *c.* 15 mm long by *c.* 0.4 mm wide, exserted, hairy in lower two-thirds, white, filiform; **stigma** 2-parted; **fruiting heads** 19–25 mm wide by 12–15 mm tall. **FRUIT a nutlet**, 1-seeded, up to four per calyx, each 1.4–2 mm long by *c.* 0.9 mm wide by *c.* 0.7 mm thick, glabrous, 3-sided, outside slightly rounded.

■ **LEAVES** Opposite, simple, toothed, spreading to ascending, punctate; **blades** 2.3–11 cm long by 0.7–4.5 cm wide, larger below, glabrous to hairy above and below and on margins, axillary leaves to *c.* 15 mm long; **petioles** 2–28 mm long, hairy, grooved above; **stipules** absent.

■ **STEM** Square, erect, simple or branched, upper stem with numerous reflexed or spreading hairs, smooth below; geniculate or not at the 2–5 mm thick base.

Oswego-tea, Horse-mint, Western Wild Bergamot

Mint—*Lamiaceae* **Wildflower** Pinkish purple to white Corolla tubular, 2-lipped, lobed

False Dragonhead *Physostegia virginiana* (L.) Benth.

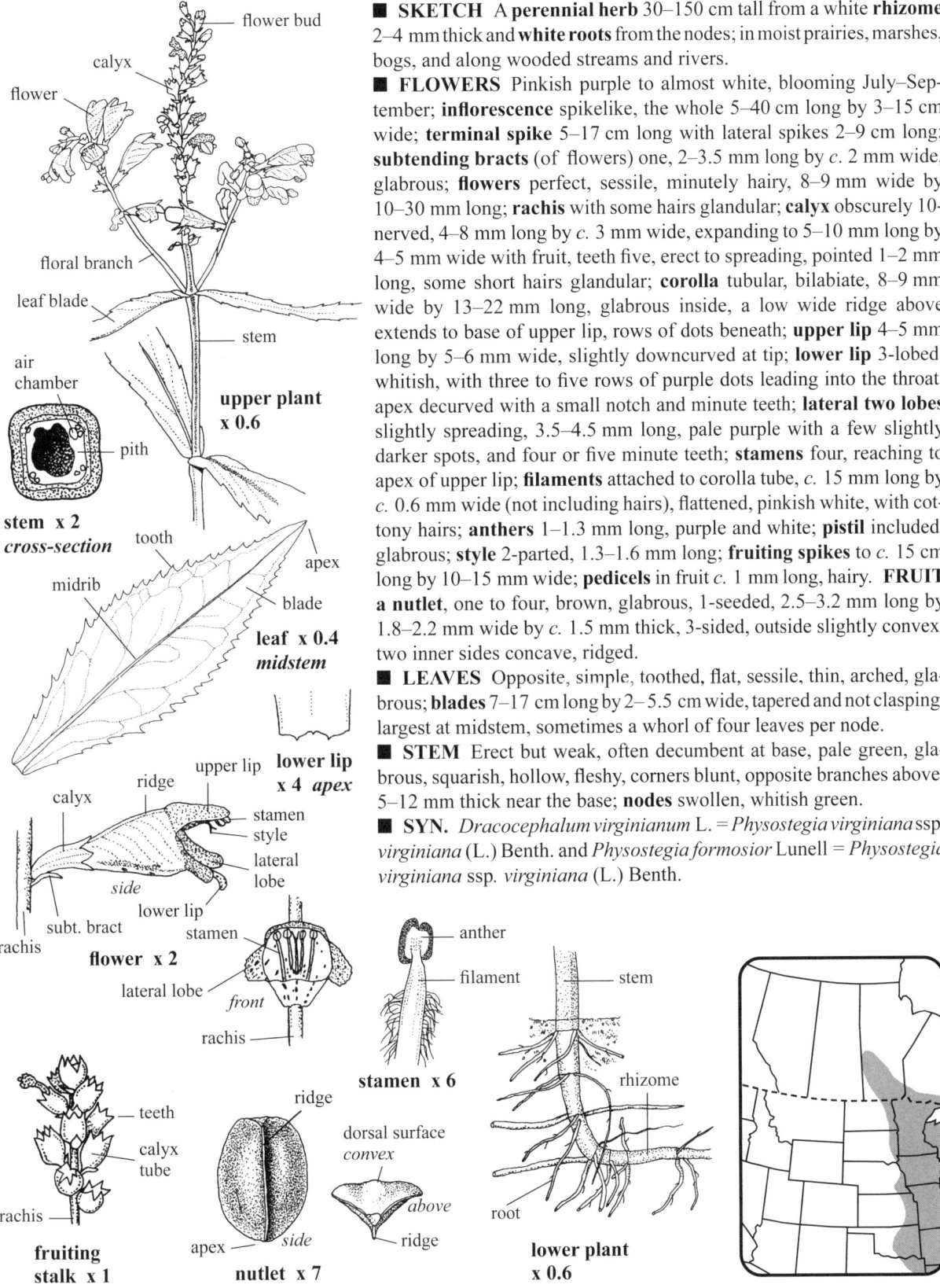

■ **SKETCH** A **perennial herb** 30–150 cm tall from a white **rhizome** 2–4 mm thick and **white roots** from the nodes; in moist prairies, marshes, bogs, and along wooded streams and rivers.
■ **FLOWERS** Pinkish purple to almost white, blooming July–September; **inflorescence** spikelike, the whole 5–40 cm long by 3–15 cm wide; **terminal spike** 5–17 cm long with lateral spikes 2–9 cm long; **subtending bracts** (of flowers) one, 2–3.5 mm long by *c*. 2 mm wide, glabrous; **flowers** perfect, sessile, minutely hairy, 8–9 mm wide by 10–30 mm long; **rachis** with some hairs glandular; **calyx** obscurely 10-nerved, 4–8 mm long by *c*. 3 mm wide, expanding to 5–10 mm long by 4–5 mm wide with fruit, teeth five, erect to spreading, pointed 1–2 mm long, some short hairs glandular; **corolla** tubular, bilabiate, 8–9 mm wide by 13–22 mm long, glabrous inside, a low wide ridge above extends to base of upper lip, rows of dots beneath; **upper lip** 4–5 mm long by 5–6 mm wide, slightly downcurved at tip; **lower lip** 3-lobed, whitish, with three to five rows of purple dots leading into the throat, apex decurved with a small notch and minute teeth; **lateral two lobes** slightly spreading, 3.5–4.5 mm long, pale purple with a few slightly darker spots, and four or five minute teeth; **stamens** four, reaching to apex of upper lip; **filaments** attached to corolla tube, *c*. 15 mm long by *c*. 0.6 mm wide (not including hairs), flattened, pinkish white, with cottony hairs; **anthers** 1–1.3 mm long, purple and white; **pistil** included, glabrous; **style** 2-parted, 1.3–1.6 mm long; **fruiting spikes** to *c*. 15 cm long by 10–15 mm wide; **pedicels** in fruit *c*. 1 mm long, hairy. **FRUIT** a nutlet, one to four, brown, glabrous, 1-seeded, 2.5–3.2 mm long by 1.8–2.2 mm wide by *c*. 1.5 mm thick, 3-sided, outside slightly convex, two inner sides concave, ridged.
■ **LEAVES** Opposite, simple, toothed, flat, sessile, thin, arched, glabrous; **blades** 7–17 cm long by 2–5.5 cm wide, tapered and not clasping, largest at midstem, sometimes a whorl of four leaves per node.
■ **STEM** Erect but weak, often decumbent at base, pale green, glabrous, squarish, hollow, fleshy, corners blunt, opposite branches above; 5–12 mm thick near the base; **nodes** swollen, whitish green.
■ **SYN.** *Dracocephalum virginianum* L. = *Physostegia virginiana* ssp. *virginiana* (L.) Benth. and *Physostegia formosior* Lunell = *Physostegia virginiana* ssp. *virginiana* (L.) Benth.

Virginia Lionsheart

Mint—*Lamiaceae* **Wildflower** Blue Corolla tubular, 2-lipped

Marsh Skullcap *Scutellaria galericulata* L.

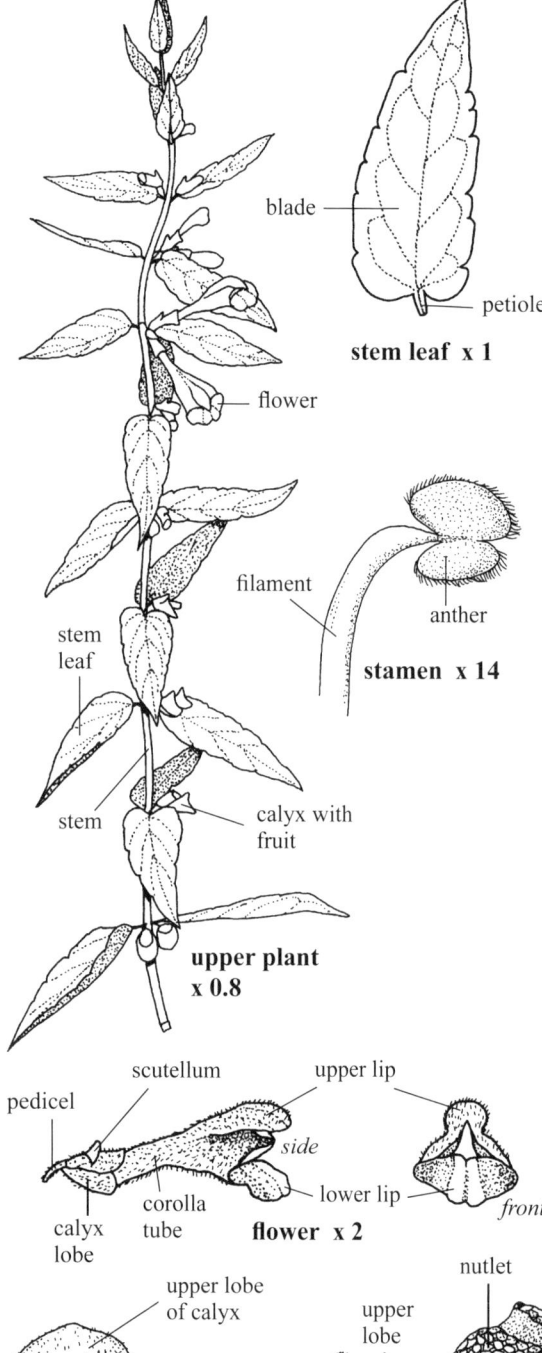

upper plant x 0.8

stem leaf x 1

stamen x 14

■ **SKETCH** A **perennial herb** 20–80+ cm tall with white **rhizomes** 1–2 mm wide by 5–30+ cm long; in marshy areas, along shores and in wet meadows; no minty odor.
■ **FLOWERS** Blue, blooming June–September; **inflorescence** axillary, in pairs or singly in leaf axils, TL 5–40 cm by *c.* 3.5 cm wide (each branch); **pedicels** 2–3 mm long, to *c.* 4 mm long in fruit, dark brown, hairs reflexed; **flowers** perfect, 7–8 mm wide by 15–25 mm long by *c.* 7 mm tall, finely hairy; **calyx** 2-lobed, 3–4 mm long by *c.* 3 mm tall by *c.* 2 mm thick (across the scutellum), enlarged in fruit to 5–6 mm long by 4–5 mm tall by *c.* 4 mm wide, glandular, hairs deflexed; **upper lobe** with a **scutellum** 2–3 mm tall and persistent in fruit; **lower lobe** slightly shorter than upper lobe, with four faint veins in fruit; **corolla** tubular, 2-lipped, hairy on outside; **corolla tube** light blue to whitish, tapered to 1.5–2 mm thick near the calyx, a few white hairs scattered inside; **stamens** four, included; **filaments** attached to corolla tube, two free above for *c.* 7 mm and two free for *c.* 3 mm; **anthers** purple and white, hairy, *c.* 1 mm long and wide; **style** filiform, glabrous, white, as long as flower, curving at the blunt tip. **FRUIT a nutlet**, 1-seeded, four per calyx, *c.* 2 mm long by *c.* 1.5 mm wide by 1.4–1.5 mm thick, glandular, covered with short blunt papillae except near the hilum.
■ **LEAVES** Opposite, simple, shallowly toothed, soft, thin, mostly spreading; **blades** 2–8.6 cm long by 1–3.5 cm wide, finely hairy above or glabrous, lighter green and more hairy below especially on raised veins, hairs deflexed; **petioles** 0–4 mm long, reduced above, hairs white and reflexed; **stipules** absent.
■ **STEM** Erect to trailing, hollow, simple or with opposite branches, square, 4-ridged, hairs on ridges recurved, scattered hairs on flat to slightly concave surfaces between the ridges; 1–3 mm thick near the base.
■ **SYN.** *Scutellaria epilobiifolia* A. Hamilton.

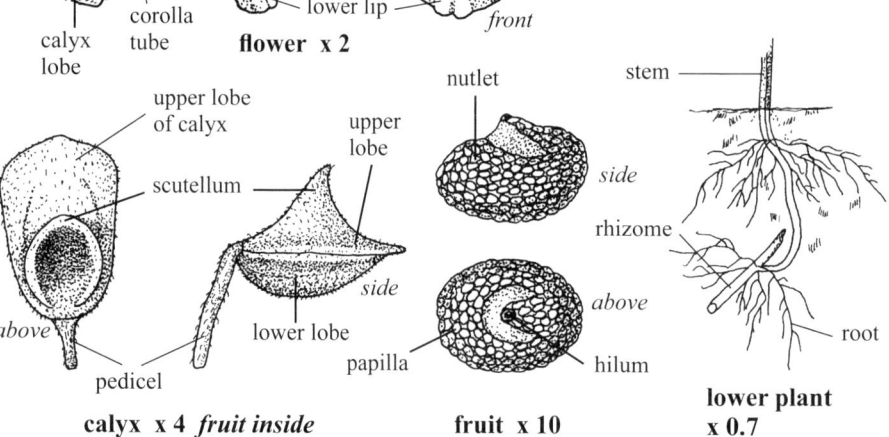

flower x 2

calyx x 4 *fruit inside*

fruit x 10

lower plant x 0.7

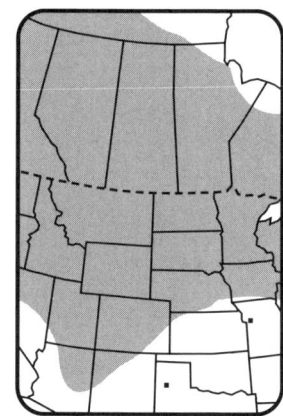

Common Skullcap, Hooded Skullcap

Mint—*Lamiaceae* **Wildflower** Blue to violet Corolla tubular, 2-lipped

Blue Skullcap *Scutellaria lateriflora* L.

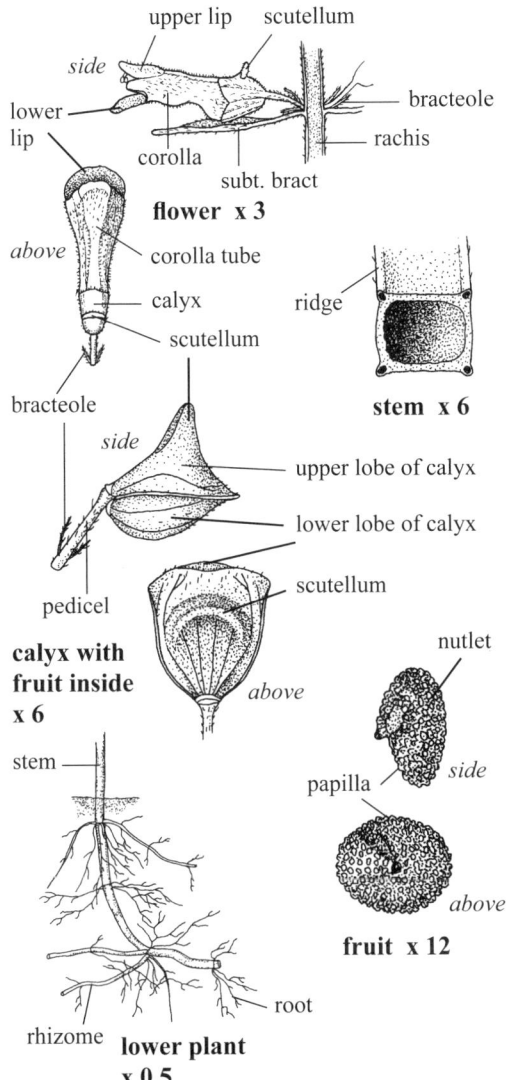

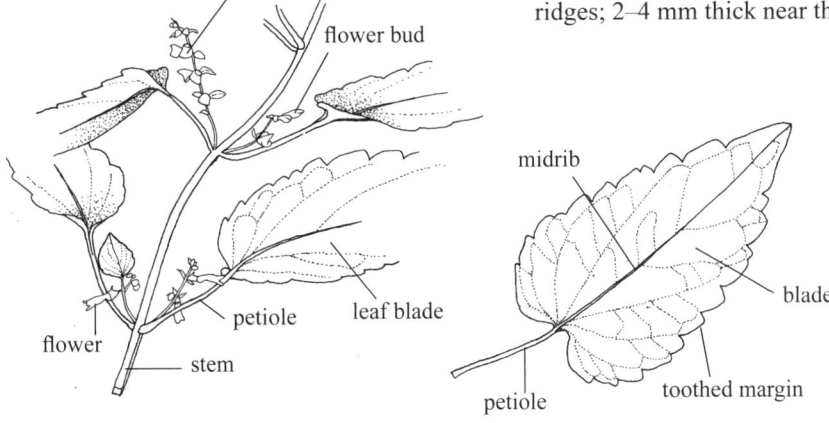

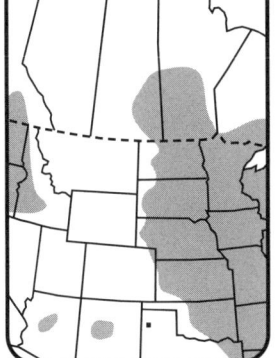

■ **SKETCH** A **perennial herb** 20–100 cm tall or long from tan **rhizomes** 0.5–2 mm thick and to *c.* 30 cm long; along woodland stream and river banks, in thickets and swamps.

■ **FLOWERS** Blue to violet, blooming July–September; **inflorescence** of axillary (sometimes terminal) racemes with 2–24 flowers in pairs along a stalk 0.5–16 cm long; **pedicels** 0.2–1.5 mm long by *c.* 0.3 mm wide at anthesis, reduced above, green, hairy; **bracteoles** (of pedicels) paired, 0.8–1 mm long, filiform, hairy, green, between subtending bract and pedicel; **subtending bracts** (of pedicels) leaflike, 3–13 mm long by 3–5 mm wide, reduced and glabrous above, hairy below on the veins; **flowers** perfect, 5–9 mm long; **calyx** 2–3.3 mm long by 2.5–2.8 mm wide by 2–3.3 mm tall (including scutellum), veins obscure, short hairs ascending on the outside, hairless inside; **scutellum** 0.6–1 mm tall and wide at anthesis, hairy on outside, to *c.* 2 mm wide in fruit; **corolla** tubular, 2-lipped, hairy on the outside, hairy in the tube but not inside at the lips, lobes, or throat, no spots on the corolla lobes; **corolla tube** 4.5–6 mm long with a bulge below near the calyx and widening toward the lobes; **upper corolla lip** 0.8–1 mm long by *c.* 1.5 mm wide, apical margin slightly irregular; **lower lip** 3-lobed, *c.* 2 mm long by *c.* 3 mm wide, lateral two lobes of lower lip *c.* 1 mm long and wide; **stamens** four, didynamous (2 + 2), mostly included; **filaments** white, filiform, hairy; **anthers** pale yellow, slightly hairy, minute; **ovary** 4-lobed, glabrous, together the lobes are 0.5–0.6 mm across. **FRUIT a nutlet**, 1-seeded, up to four per flower, light tan, papillate, *c.* 1.3 mm long by *c.* 1 mm wide by *c.* 0.8 mm thick, enclosed in calyx.

■ **LEAVES** Opposite, simple, widely toothed, thin, pointed, mostly hairless, spreading to slightly ascending, stem leaves only; **blades** slightly lighter green below, 2–11 cm long by 1–5.5 cm wide, reduced above; **petioles** 0.7–5 cm long, reduced above; **stipules** absent.

■ **STEM** Green, hollow, squarish, 4-ridged, ridges hollow, simple to branched above, glabrous or with tiny ascending hairs along the ridges; 2–4 mm thick near the reddish base.

Mad-dog Skullcap

Mint—*Lamiaceae* **Wildflower** Purple to pink Corolla tubular, 2-lipped, 4-lobed

MARSH HEDGE-NETTLE *Stachys pilosa* Nutt.

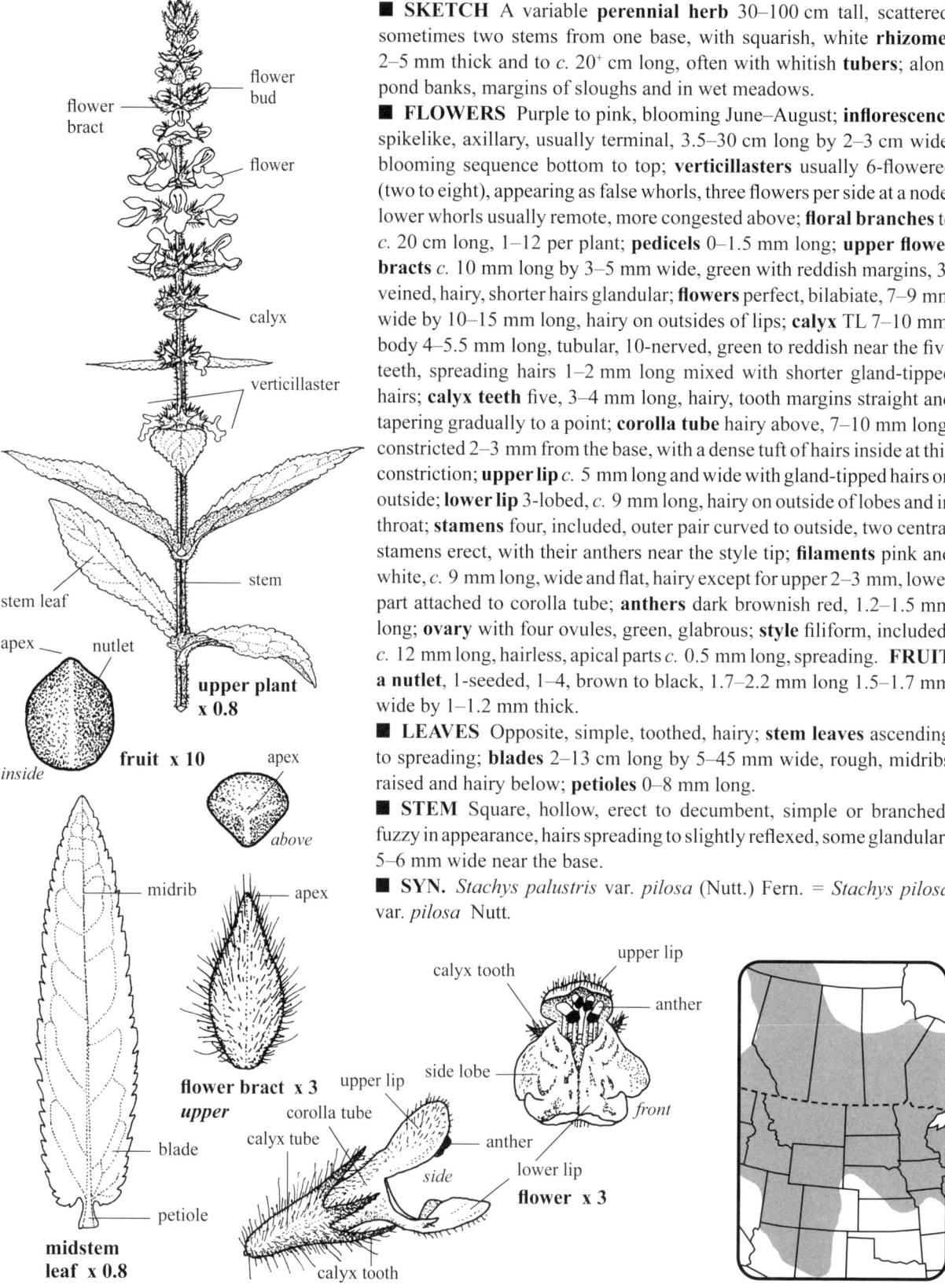

- **SKETCH** A variable **perennial herb** 30–100 cm tall, scattered, sometimes two stems from one base, with squarish, white **rhizomes** 2–5 mm thick and to *c.* 20+ cm long, often with whitish **tubers**; along pond banks, margins of sloughs and in wet meadows.
- **FLOWERS** Purple to pink, blooming June–August; **inflorescence** spikelike, axillary, usually terminal, 3.5–30 cm long by 2–3 cm wide, blooming sequence bottom to top; **verticillasters** usually 6-flowered (two to eight), appearing as false whorls, three flowers per side at a node, lower whorls usually remote, more congested above; **floral branches** to *c.* 20 cm long, 1–12 per plant; **pedicels** 0–1.5 mm long; **upper flower bracts** *c.* 10 mm long by 3–5 mm wide, green with reddish margins, 3-veined, hairy, shorter hairs glandular; **flowers** perfect, bilabiate, 7–9 mm wide by 10–15 mm long, hairy on outsides of lips; **calyx** TL 7–10 mm, body 4–5.5 mm long, tubular, 10-nerved, green to reddish near the five teeth, spreading hairs 1–2 mm long mixed with shorter gland-tipped hairs; **calyx teeth** five, 3–4 mm long, hairy, tooth margins straight and tapering gradually to a point; **corolla tube** hairy above, 7–10 mm long, constricted 2–3 mm from the base, with a dense tuft of hairs inside at this constriction; **upper lip** *c.* 5 mm long and wide with gland-tipped hairs on outside; **lower lip** 3-lobed, *c.* 9 mm long, hairy on outside of lobes and in throat; **stamens** four, included, outer pair curved to outside, two central stamens erect, with their anthers near the style tip; **filaments** pink and white, *c.* 9 mm long, wide and flat, hairy except for upper 2–3 mm, lower part attached to corolla tube; **anthers** dark brownish red, 1.2–1.5 mm long; **ovary** with four ovules, green, glabrous; **style** filiform, included, *c.* 12 mm long, hairless, apical parts *c.* 0.5 mm long, spreading. **FRUIT** a nutlet, 1-seeded, 1–4, brown to black, 1.7–2.2 mm long 1.5–1.7 mm wide by 1–1.2 mm thick.
- **LEAVES** Opposite, simple, toothed, hairy; **stem leaves** ascending to spreading; **blades** 2–13 cm long by 5–45 mm wide, rough, midribs raised and hairy below; **petioles** 0–8 mm long.
- **STEM** Square, hollow, erect to decumbent, simple or branched, fuzzy in appearance, hairs spreading to slightly reflexed, some glandular; 5–6 mm wide near the base.
- **SYN.** *Stachys palustris* var. *pilosa* (Nutt.) Fern. = *Stachys pilosa* var. *pilosa* Nutt.

Hairy Hedgenettle, Hedgenettle, Marsh Betony, Swamp Hedgenettle, Woundwort

Bladderwort—*Lentibulariaceae* **Wildflower** Yellow Corolla 2-lipped

FLAT-LEAVED BLADDERWORT   *Utricularia intermedia* Hayne

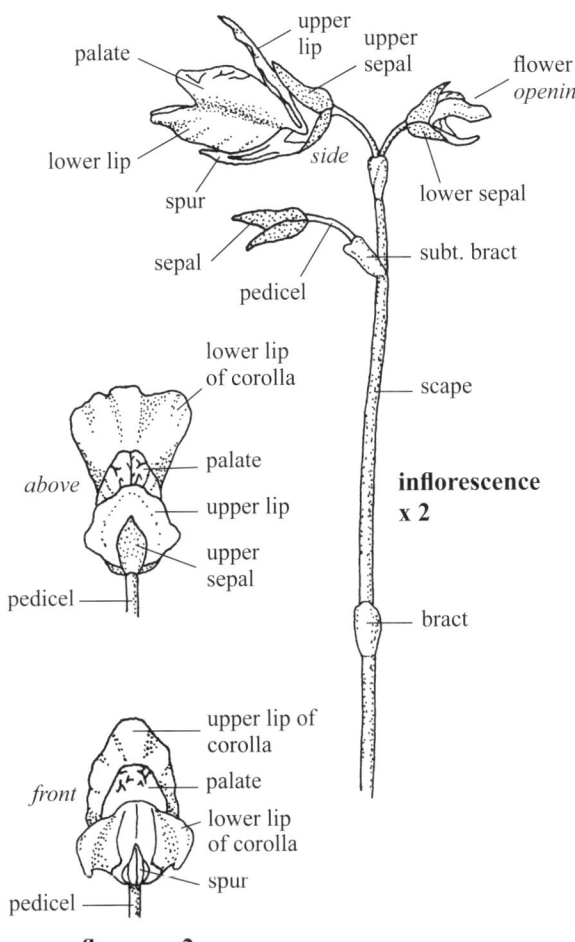

flower x 2

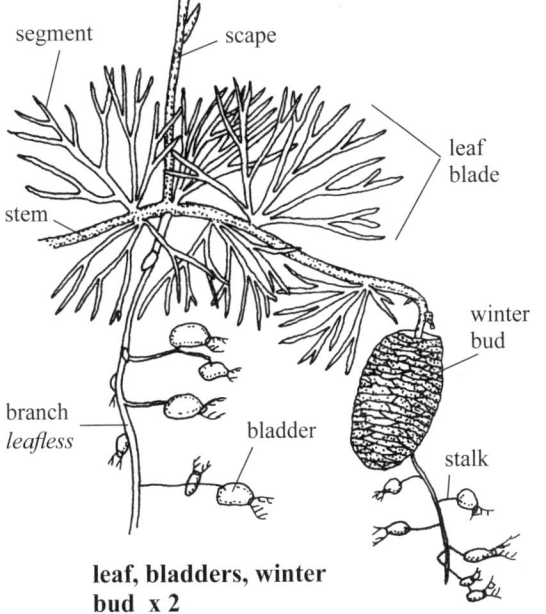

leaf, bladders, winter bud x 2

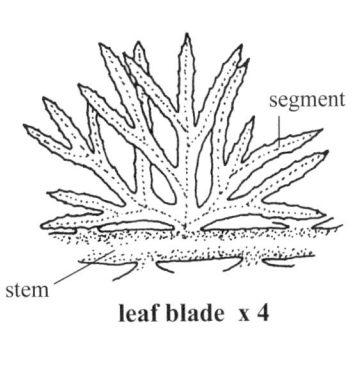

leaf blade x 4

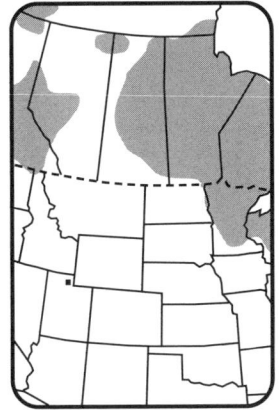

- **SKETCH** A carnivorous, aquatic **perennial herb** 6–20 cm tall with **rootless stems**; in quiet shallow water or mud in sloughs, bogs, ponds, lakes and springs.
- **FLOWERS** Yellow, blooming July–August; **inflorescence** a raceme, lax, 2–5 cm long of 2–5 flowers, terminal and above water; **scapes**, erect, 1–1.3 mm wide, rising above the water, whitish to green; **bracts** on scape few, 1.5–4 mm long, entire, erect, clasping; **pedicels** ascending, glabrous, 3–15 mm long, arched; **subtending bracts** (of pedicels) 3-lobed, *c.* 4 mm long, wrapped around stem and pedicel; **flowers** perfect, 7–9 mm wide by 8–12 (–18) mm long by 10–12 mm tall; **sepals** two, upper one 2.5–5 mm long by 2–3 mm wide, pointed, entire; **lower sepal** 2.5–5 mm long by *c.* 4 mm wide, descending, persistent around the base of the fruit; **corolla** 2-lipped; **upper lip** ascending; **lower lip** 6–18 mm long by 7–20 mm wide, palate *c.* 5 mm wide by *c.* 4 mm high with several fine brownish red lines; **spur** an extension of the lower lip and more than half its length, 4–12 mm long, cylindrical, straight; **stamens** two, included, inserted near base of corolla tube; **style** short, included. **FRUIT a capsule**, roundish, *c.* 3 mm wide, partially enclosed by calyx, glabrous; **seeds** numerous, angled, ridged, attached to central round placenta.
- **LEAVES** Alternate, compound, 4–30 mm long by 10–17 mm wide, ultimate divisions pointed, the segments 2 or 3, flat, narrow, minutely toothed and each tooth with an apical hair; **blades** sessile, quickly divided; **internodal length** 2–6 mm; **bladders** numerous, 2–7 mm long by *c.* 2 mm wide by *c.* 1 mm thick on slender stalks 0.5–4 mm long on separate leafless branches, branched hairs adorn the opening and help to trap and direct small invertebrates (food) into the bladder for digestion; **winter buds** 5–15 mm long by 5–10 mm wide, filled with numerous tiny leaves bordered with dense tufts of hair.
- **STEM** Horizontal, leafy, submerged to resting on mud, *c.* 1 mm wide, glabrous.

Bladderwort—*Lentibulariaceae* **Wildflower** Yellow Corolla 2-lipped

GREATER BLADDERWORT *Utricularia macrorhiza* Le Conte

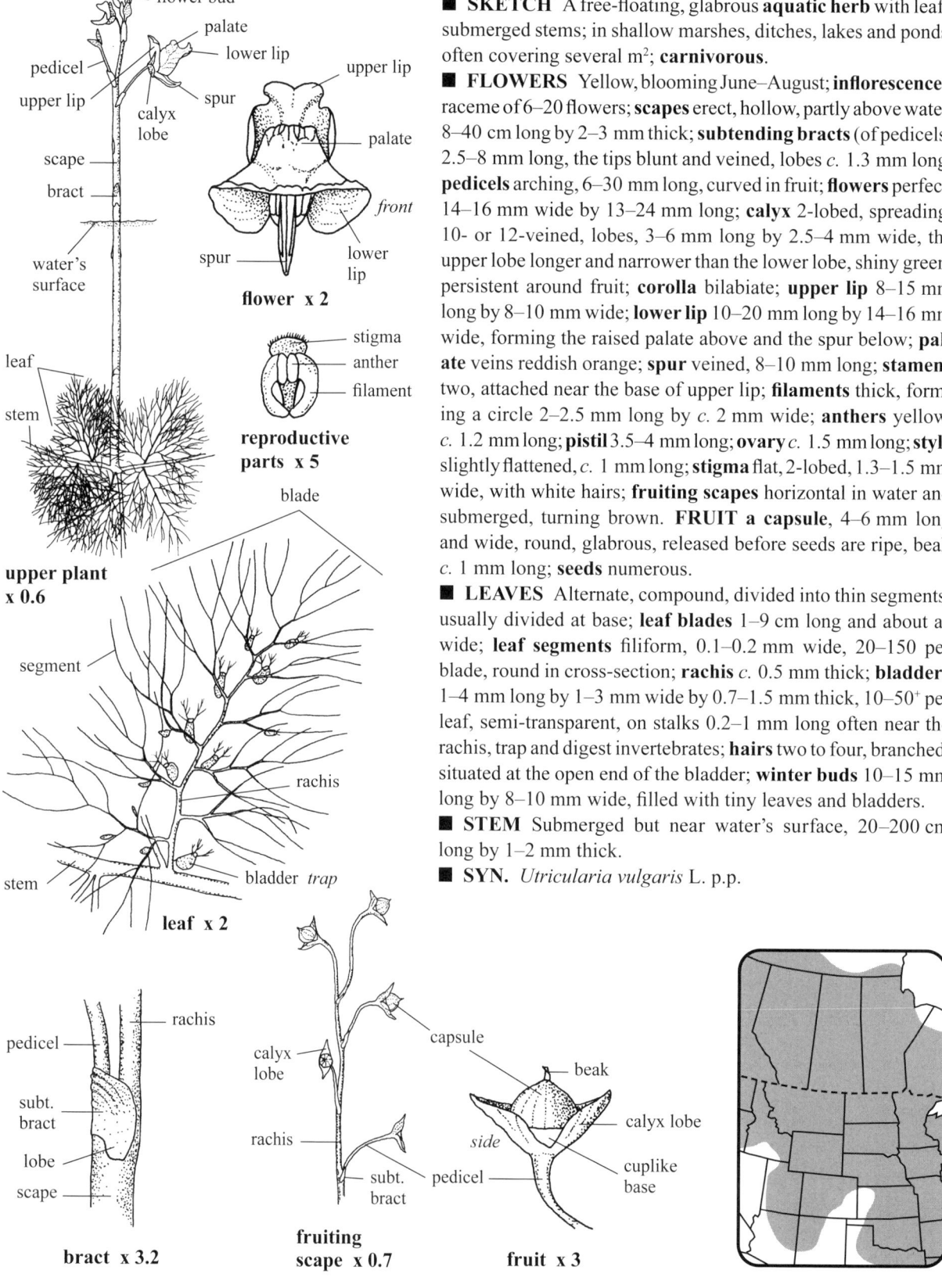

■ **SKETCH** A free-floating, glabrous **aquatic herb** with leafy submerged stems; in shallow marshes, ditches, lakes and ponds, often covering several m^2; **carnivorous**.

■ **FLOWERS** Yellow, blooming June–August; **inflorescence** a raceme of 6–20 flowers; **scapes** erect, hollow, partly above water, 8–40 cm long by 2–3 mm thick; **subtending bracts** (of pedicels) 2.5–8 mm long, the tips blunt and veined, lobes *c.* 1.3 mm long; **pedicels** arching, 6–30 mm long, curved in fruit; **flowers** perfect, 14–16 mm wide by 13–24 mm long; **calyx** 2-lobed, spreading, 10- or 12-veined, lobes, 3–6 mm long by 2.5–4 mm wide, the upper lobe longer and narrower than the lower lobe, shiny green, persistent around fruit; **corolla** bilabiate; **upper lip** 8–15 mm long by 8–10 mm wide; **lower lip** 10–20 mm long by 14–16 mm wide, forming the raised palate above and the spur below; **palate** veins reddish orange; **spur** veined, 8–10 mm long; **stamens** two, attached near the base of upper lip; **filaments** thick, forming a circle 2–2.5 mm long by *c.* 2 mm wide; **anthers** yellow, *c.* 1.2 mm long; **pistil** 3.5–4 mm long; **ovary** *c.* 1.5 mm long; **style** slightly flattened, *c.* 1 mm long; **stigma** flat, 2-lobed, 1.3–1.5 mm wide, with white hairs; **fruiting scapes** horizontal in water and submerged, turning brown. **FRUIT** a capsule, 4–6 mm long and wide, round, glabrous, released before seeds are ripe, beak *c.* 1 mm long; **seeds** numerous.

■ **LEAVES** Alternate, compound, divided into thin segments, usually divided at base; **leaf blades** 1–9 cm long and about as wide; **leaf segments** filiform, 0.1–0.2 mm wide, 20–150 per blade, round in cross-section; **rachis** *c.* 0.5 mm thick; **bladders** 1–4 mm long by 1–3 mm wide by 0.7–1.5 mm thick, 10–50$^+$ per leaf, semi-transparent, on stalks 0.2–1 mm long often near the rachis, trap and digest invertebrates; **hairs** two to four, branched, situated at the open end of the bladder; **winter buds** 10–15 mm long by 8–10 mm wide, filled with tiny leaves and bladders.

■ **STEM** Submerged but near water's surface, 20–200 cm long by 1–2 mm thick.

■ **SYN.** *Utricularia vulgaris* L. p.p.

Common Bladderwort

Bladderwort—*Lentibulariaceae* **Wildflower** Yellow Corolla 2-lipped

LESSER BLADDERWORT *Utricularia minor* L.

■ **SKETCH** A carnivorous, aquatic **perennial herb** 5–15 cm tall; in shallow quiet water of sloughs, fens and bogs, along the bottom or floating in ponds and lakes.

■ **FLOWERS** Yellow, blooming July–August; **inflorescence** a raceme of 2–10 flowers, 2–8 cm long by 2–3 cm wide, flowers above water; **scapes** 4–15 cm long by *c.* 0.7 mm wide, erect, straight, rising above the water, reddish above, greenish near the base, leafless but with a few ascending widely spaced bracts 1–2 mm long; **pedicels** 2–10 mm long by *c.* 0.4 mm thick, arched, glabrous, green to reddish; **subtending bracts** (of pedicels) purplish, 1.5–2 mm long, blunt; **flowers** perfect, 4–5.5 mm wide by 4–8 mm long by 3–4 mm tall; **sepals** two, entire, pointed, reddish, 1–2.5 mm long by *c.* 1.5 mm wide, spreading, upper sepal with a raised area at its base, persistent and enclosing the base of the fruit; **corolla** bilabiate; **upper lip** *c.* 3 mm long by 2.5–3 mm wide; **lower corolla lip** 4–8 mm long, notched at apex, palate with red lines, recessed in the throat and not well developed; **spur** elevated, reddish yellow, about half the length of the lower lip, not obvious since no pointed tube is formed, with three red parallel lines along the raised area; **stamens** two, included; **pistil** included. **FRUIT a capsule**, round, 2–2.5 mm long and wide, lower half enclosed by sepals, body reddish with short persistent style; **placenta** parietal, a globular disc *c.* 1.5 mm wide with seeds attached to its indented surface; **seeds** *c.* 20 per capsule, *c.* 1 mm long by *c.* 0.7 mm wide by *c.* 0.4 mm thick, papillate, 4- or 5-sided, ridged.

■ **LEAVES** Alternate, compound, sessile, divided into filiform flat segments; **blades** 3–10 mm long by 4–12 mm wide, usually with three divisions at their bases, main segments 0.4–1 mm wide, ultimate segments pointed, often curved, entire or with a few microscopic teeth or marginal bumps, central veins not obvious; **small leaves** pinkish with wide segments; **bladders** 1.5–2 mm long by 0.7–1.5 mm wide by 0.7–1 mm thick, common on leaves but also on separate leafless branches (in mud); **subtending bracts** of side branches *c.* 0.5 mm long and with 3–5 teeth; **side branches** *c.* 3 mm long, often with two bladders near their ends or bladders singly on stalks less than 1 mm long; **winter buds** 2–5 mm wide, reddish green, tiny leaves not bordered with hairs.

■ **STEM** Horizontal, leafy; 0.4–0.5 mm wide.

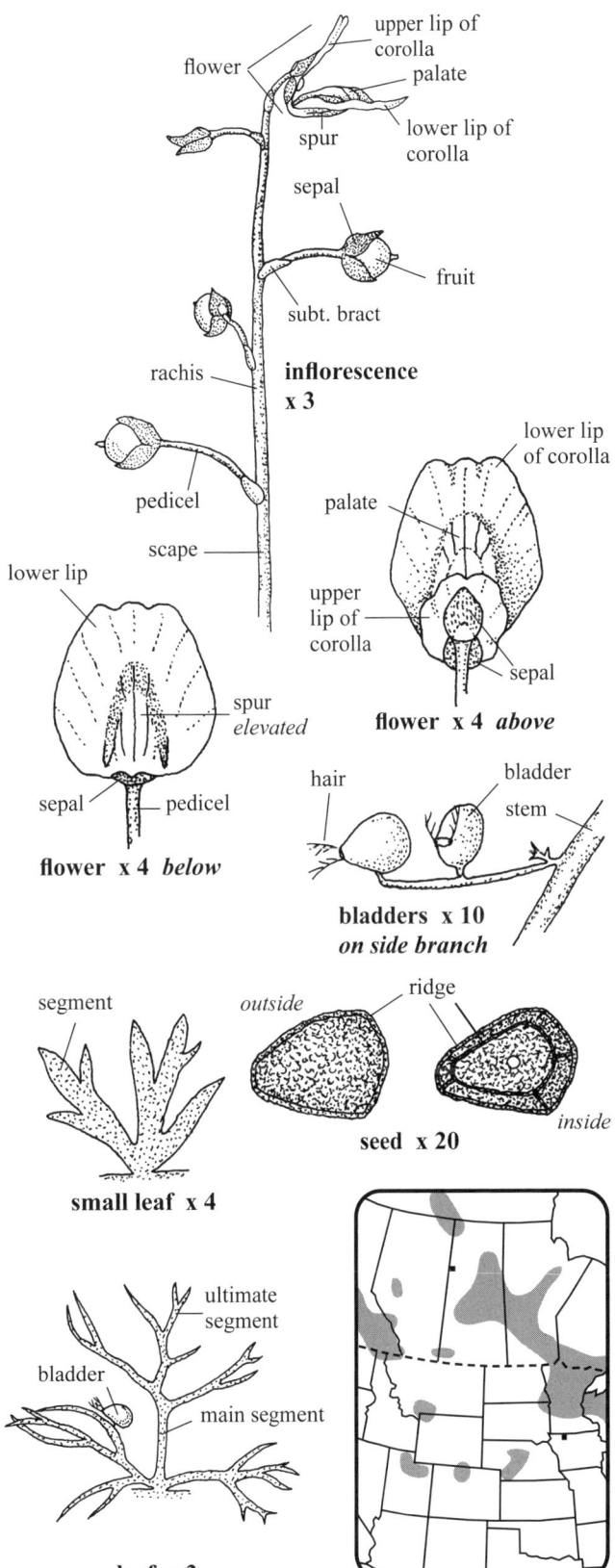

Small Bladderwort

Flax—*Linaceae* **Wildflower** Blue, center yellow Petals 5

Lewis Wild Flax *Linum lewisii* Pursh

■ **SKETCH** A glabrous **perennial herb** 20–80 cm tall from a creamy white, woody **taproot** 2–8 mm thick by 5–15 cm long, sometimes with a **caudex**; in moist to dry prairies, disturbed sites along river banks, on rocky wooded hillsides and among desert shrubs. **p1**

■ **FLOWERS** Pale blue with a pale yellow center, a white form in the north, blooming May–August; **inflorescence** an open panicle 5–20 cm long with floral branches 5–25+ cm long; **pedicels** thin, erect to arching, 0.5–4 cm long; **flowers** perfect, erect to nodding, 20–30 mm wide by *c.* 10 mm long; **sepals** five, pointed, 3.5–6 mm long by 2–3 mm wide, entire, glabrous, imbricate, green with white glandless margins, 3-nerved, persistent in fruit and more than half as long as the capsule; **petals** five, 10–18 mm long by 9–12 mm wide, thin, with a finely erose apex and numerous thin, dark blue veins on both surfaces, yellow near the hairy base, lasting one to two days but falling within minutes when a flower is picked; **stamens** five, projecting slightly above the petals; **filaments** white, 6–7 mm long, twisted into a column; **anthers** pale yellow, 1–1.5 mm long; **ovary** glabrous, green, *c.* 2 mm long by *c.* 1.5 mm wide; **styles** five, distinct, 3–9 mm long, blue at the base, slightly shorter to longer than the stamens; **stigmas** blue to green, capitate. **FRUIT a capsule**, tan, erect, 4.5–8 mm long and wide, five false septa incomplete and long ciliate, separating into 10 segments; **seeds** 10 per capsule, flattened, 3.5–5 mm long by 1.8–2.5 mm wide by *c.* 1 mm thick with a lateral ridge, smooth and shiny, medium to dark brown.

■ **LEAVES** Alternate, simple, entire, crowded, overlapping; **blades** linear, 0.8–3 cm long by 1–4 mm wide, pointed, ascending to erect, sessile, 3-veined, the two lateral veins faint; **stipular glands** absent.

■ **STEM** Erect, weak, glabrous, unbranched to branching above or from the base to form a clump of 1–24 stems; 1–2 mm thick near the base.

■ **SYN.** *Linum perenne* var. *lewisii* (Pursh) Eat. & J. Wright = *Linum lewisii* var. *lewisii* Pursh.

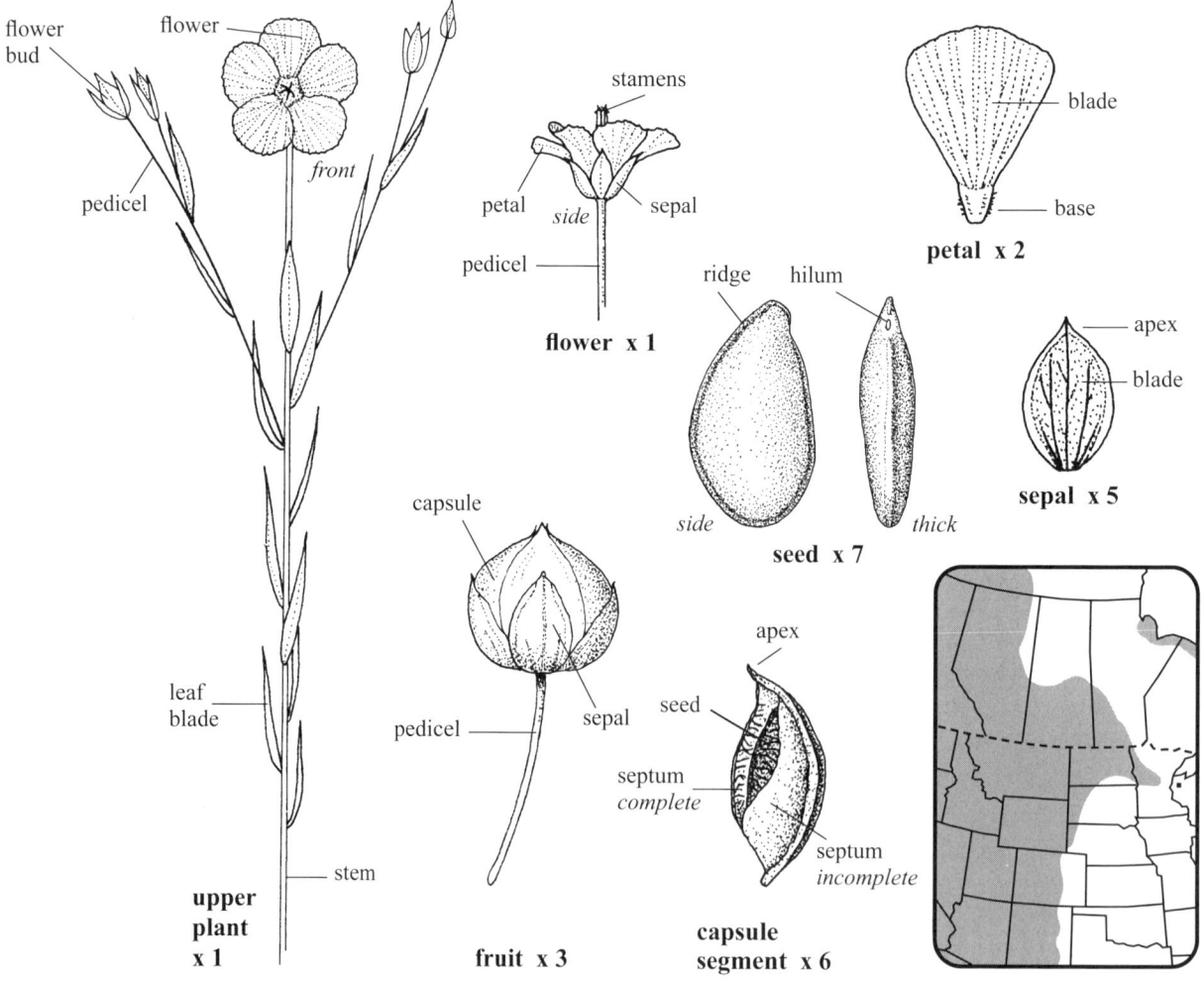

362 Prairie Flax, Blue Flax, Wild Blue Flax

Flax—*Linaceae* **Wildflower** Yellow Petals 5

LARGE-FLOWERED YELLOW FLAX *Linum rigidum* Pursh

■ **SKETCH** A variable **annual herb** 5–50 cm tall from a thin brown **taproot** 5–10 cm long by 1–3 mm thick; in sandy soils, hillsides and prairies. **p2**

■ **FLOWERS** Yellow, blooming May–October; **inflorescence** a panicle, open to flat-topped; **pedicels** ascending, 4-ridged, hairless, axillary, 4–10 mm long; **flowers** perfect, erect, 13–26 mm wide; **sepals** five, opposite petals, dropping as fruit ripens, pointed, 3.5–10 mm long by 1.8–2.2 mm wide, 1- or 3-nerved, stalked white glands along the narrow hyaline margins; **petals** five, finely veined, 7–16+ mm long by 4–6 mm wide, bases hairy and pointed, apices slightly erose, dropping quickly when flowers are picked; **stamens** five, 5–9 mm long, included; **filaments** yellow, 4–5 mm long, filiform, erect, widening and united at the slightly hairy base around the superior ovary; **anthers** yellow, 1–1.5 mm long by *c.* 0.8 mm wide; **pistil** 5–12 mm long, included, glabrous; **ovary** green, roundish, 10-nerved, 1.6–2.2 mm long by 1.3–1.5 mm wide; **styles** five, 2.5–10 mm long, yellowish, united most of their length, the upper 0.5 mm distinct; **stigmas** capitate, green, obvious atop the filiform styles. FRUIT a **capsule**, glabrous, 10-nerved, erect, 3.5–4.5 mm tall by 2.5–4 mm wide, divided into five, 2-seeded segments, the segments 1.8–2 mm wide with a thickened base; **false septa** partially complete and transparent; **seeds** 10 per capsule, grooved, medium to reddish brown, 2.5–3.5 mm long by 1–1.5 mm wide by *c.* 0.5 mm thick, rounded on one side.

■ **LEAVES** Alternate, simple, entire, sessile, ascending to spreading, 5–30 mm long by 1–4 mm wide, reduced above, 1- or 3-nerved, lower ones falling early.

■ **STEM** Erect, angled and branched above; round near the glabrous to hairy 0.7–3 mm thick base; **leaf scars** tan along the reddish base.

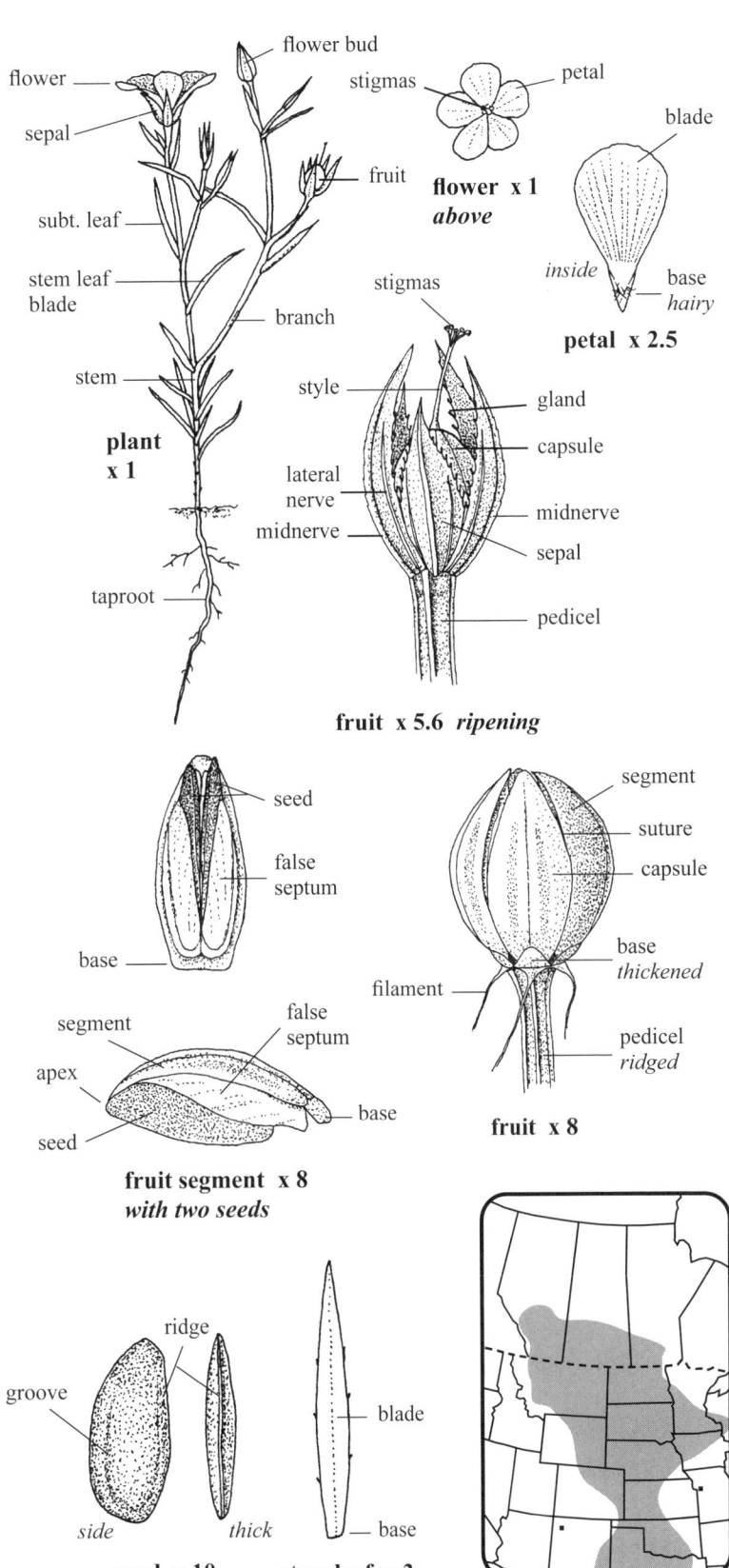

Stiff Flax, Stiffstem Flax, Yellow Flax

Lobelia—*Lobeliaceae* **Wildflower** Light blue, center white Corolla tubular, 2-lipped, 5-lobed

Kalm's Lobelia *Lobelia kalmii* L.

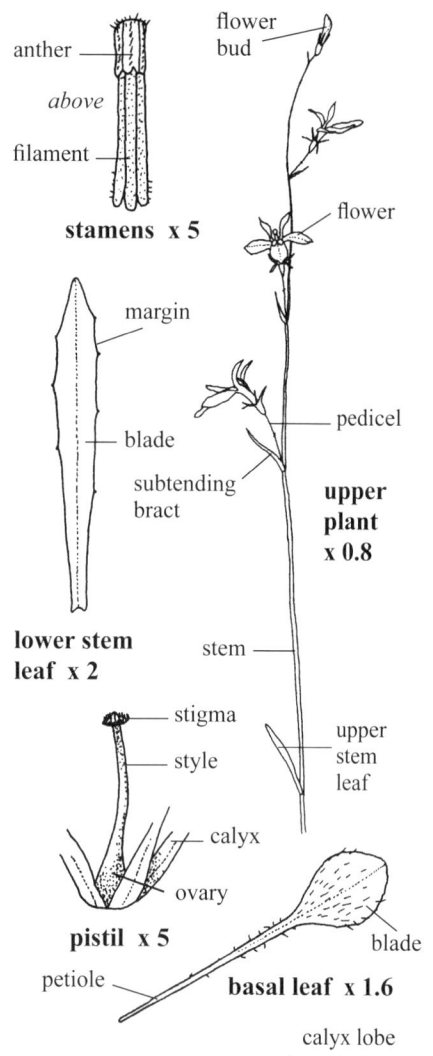

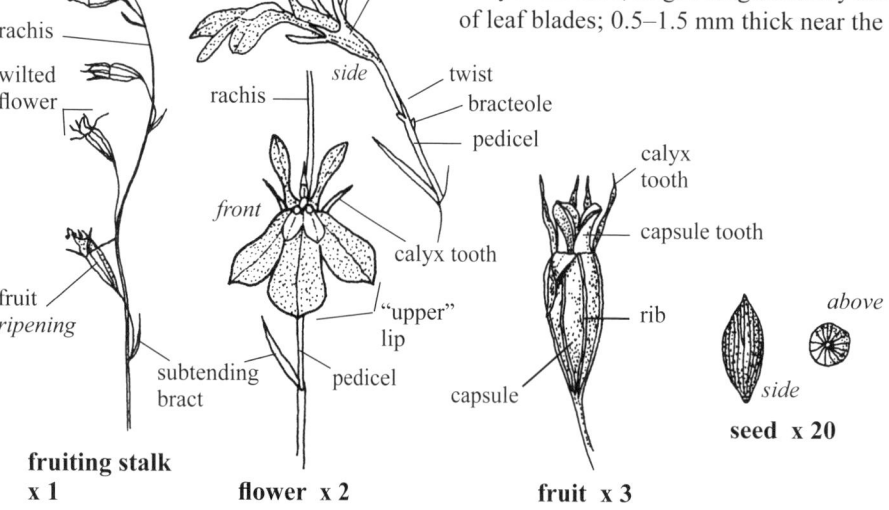

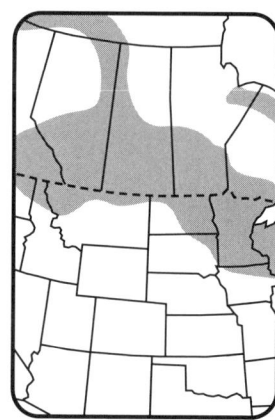

■ **SKETCH** A **perennial** or **biennial herb** 10–50 cm tall from a **rhizome** *c.* 0.3 mm thick and to *c.* 10 cm long and **fibrous roots** 1–3 cm long; in ditches, moist fields, bogs, pond margins and around willow thickets. p2

■ **FLOWERS** Light blue with white near the throat, blooming July–August; **inflorescence** an open raceme, 1–15 cm long by 8–20 mm wide, of 2–14 flowers, mostly terminal, sometimes on short lateral floral branches 3–13 cm long from upper leaf axils; **pedicels** glabrous, 3–17 mm long, erect in fruit, twisted 180°; **bracteoles** two, *c.* 0.4 mm long near the middle of the pedicel where it is twisted; **subtending bracts** (of pedicels), narrow, 5–25 mm long, entire; **flowers** perfect, inverted, 10–14 mm wide by 8–13 mm long by 7–9 mm tall, some nodding; **calyx** 3–6 mm long by 1–2 mm wide, glabrous, reddish green, expanded in fruit; **calyx tube** *c.* 2 mm long by *c.* 1 mm wide, 5-ribbed; **calyx teeth** five, pointed, 2.5–4 mm long by 0.6–1 mm wide, entire; **corolla** tubular, bilabiate, 7–10 mm long; "**upper**" **lip** 3-lobed, the lobes declining, 3–3.5 mm wide, the central lobe *c.* 4 mm long, midvein obvious, white near the throat; "**lower**" **lip** 2-lobed, ascending to reflexed, each lobe pointed, spatulate, widest near the apex, glabrous, midvein obvious; **stamens** five, 4–5 mm long; **filaments** 2.7–3.3 mm long, hairy at bases; **anthers** gray, hairy, 1.3–1.7 mm long, united into a tube around the stigma; **style** 4–5 mm long, obscured by the stamens; **stigma** reddish, 2-lobed, with a circle of white hairs around its base. **FRUIT** a **capsule**, brown, 2-valved, ribbed, erect, 4–9 mm long by 2.5–6 mm wide, the two apical teeth *c.* 2 mm long and wide with a middle ridge, obscured by persistent floral parts; **seeds** numerous, medium brown, ribbed, *c.* 0.7 mm long by *c.* 0.3 mm wide and thick, ends pointed.

■ **LEAVES** Alternate, simple, entire or with a few wide teeth, basal and stem; **basal leaves** an early rosette 1–5 cm wide; **basal leaf blades** oval, 8–15 mm long by 1–6 mm wide, hairs above simple and erect, glabrous below, midrib obvious; **petioles** 5–18 mm long, slightly hairy; **stem leaf blades** linear, 5–50 mm long by 0.5–5 mm wide, reduced above, apices tapered, slightly hairy above, at least on the lowest leaves, glabrous below; **petioles** 0–10 mm long, reduced above.

■ **STEM** Erect but weak, green, simple to branched, glabrous to slightly hairy at the base, ridged longitudinally as a continuation from the margins of leaf blades; 0.5–1.5 mm thick near the base.

Lobelia—*Lobeliaceae* **Wildflower** Pale blue to white Corolla tubular, 2-lipped, 5-lobed

Spiked Lobelia *Lobelia spicata* Lam.

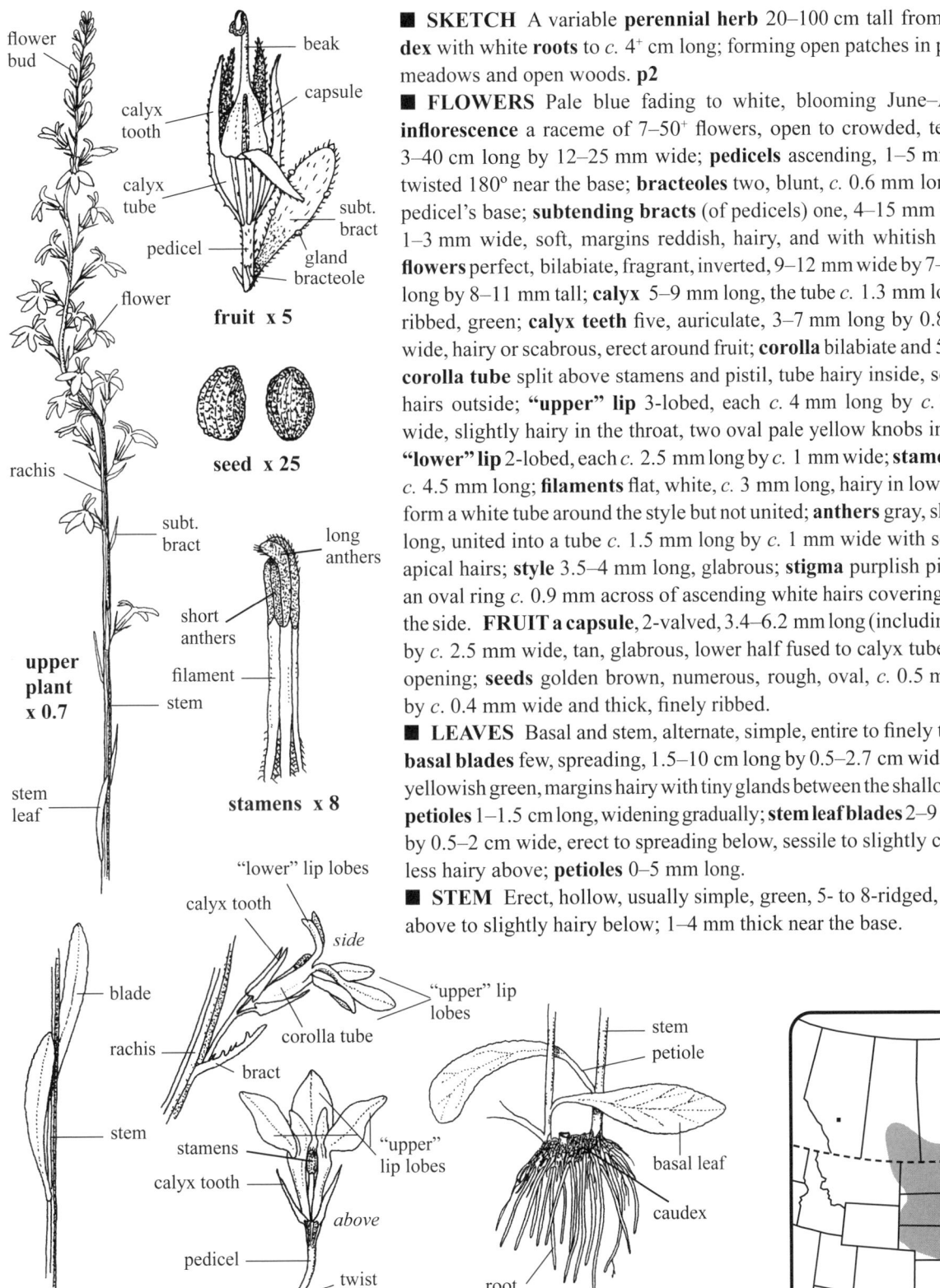

■ **SKETCH** A variable **perennial herb** 20–100 cm tall from a **caudex** with white **roots** to *c.* 4⁺ cm long; forming open patches in prairies, meadows and open woods. p2

■ **FLOWERS** Pale blue fading to white, blooming June–August; **inflorescence** a raceme of 7–50⁺ flowers, open to crowded, terminal, 3–40 cm long by 12–25 mm wide; **pedicels** ascending, 1–5 mm long, twisted 180° near the base; **bracteoles** two, blunt, *c.* 0.6 mm long, near pedicel's base; **subtending bracts** (of pedicels) one, 4–15 mm long by 1–3 mm wide, soft, margins reddish, hairy, and with whitish glands; **flowers** perfect, bilabiate, fragrant, inverted, 9–12 mm wide by 7–13 mm long by 8–11 mm tall; **calyx** 5–9 mm long, the tube *c.* 1.3 mm long, 10-ribbed, green; **calyx teeth** five, auriculate, 3–7 mm long by 0.8–1 mm wide, hairy or scabrous, erect around fruit; **corolla** bilabiate and 5-lobed; **corolla tube** split above stamens and pistil, tube hairy inside, scattered hairs outside; "**upper**" **lip** 3-lobed, each *c.* 4 mm long by *c.* 2.5 mm wide, slightly hairy in the throat, two oval pale yellow knobs in throat; "**lower**" **lip** 2-lobed, each *c.* 2.5 mm long by *c.* 1 mm wide; **stamens** five, *c.* 4.5 mm long; **filaments** flat, white, *c.* 3 mm long, hairy in lower third, form a white tube around the style but not united; **anthers** gray, short and long, united into a tube *c.* 1.5 mm long by *c.* 1 mm wide with scattered apical hairs; **style** 3.5–4 mm long, glabrous; **stigma** purplish pink with an oval ring *c.* 0.9 mm across of ascending white hairs covering it from the side. **FRUIT a capsule**, 2-valved, 3.4–6.2 mm long (including beak) by *c.* 2.5 mm wide, tan, glabrous, lower half fused to calyx tube, apical opening; **seeds** golden brown, numerous, rough, oval, *c.* 0.5 mm long by *c.* 0.4 mm wide and thick, finely ribbed.

■ **LEAVES** Basal and stem, alternate, simple, entire to finely toothed; **basal blades** few, spreading, 1.5–10 cm long by 0.5–2.7 cm wide, blunt, yellowish green, margins hairy with tiny glands between the shallow teeth; **petioles** 1–1.5 cm long, widening gradually; **stem leaf blades** 2–9 cm long by 0.5–2 cm wide, erect to spreading below, sessile to slightly clasping, less hairy above; **petioles** 0–5 mm long.

■ **STEM** Erect, hollow, usually simple, green, 5- to 8-ridged, hairless above to slightly hairy below; 1–4 mm thick near the base.

Palespike Lobelia 365

Loosestrife—*Lythraceae* **Wildflower** Pinkish purple Petals 6 (5 or 7)

PURPLE LOOSESTRIFE *Lythrum salicaria* L.

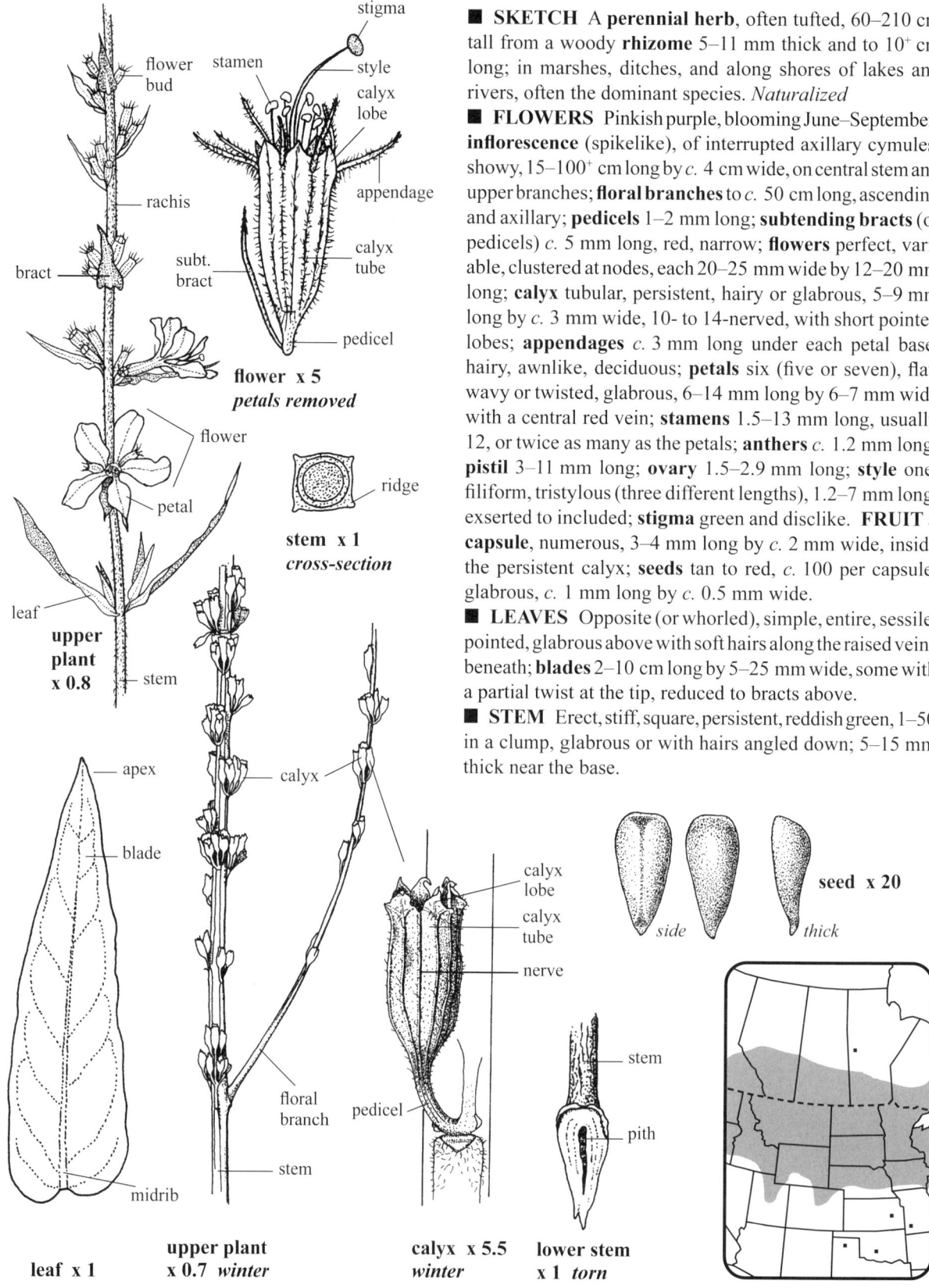

■ **SKETCH** A **perennial herb**, often tufted, 60–210 cm tall from a woody **rhizome** 5–11 mm thick and to 10⁺ cm long; in marshes, ditches, and along shores of lakes and rivers, often the dominant species. *Naturalized*

■ **FLOWERS** Pinkish purple, blooming June–September; **inflorescence** (spikelike), of interrupted axillary cymules, showy, 15–100⁺ cm long by *c.* 4 cm wide, on central stem and upper branches; **floral branches** to *c.* 50 cm long, ascending and axillary; **pedicels** 1–2 mm long; **subtending bracts** (of pedicels) *c.* 5 mm long, red, narrow; **flowers** perfect, variable, clustered at nodes, each 20–25 mm wide by 12–20 mm long; **calyx** tubular, persistent, hairy or glabrous, 5–9 mm long by *c.* 3 mm wide, 10- to 14-nerved, with short pointed lobes; **appendages** *c.* 3 mm long under each petal base, hairy, awnlike, deciduous; **petals** six (five or seven), flat, wavy or twisted, glabrous, 6–14 mm long by 6–7 mm wide with a central red vein; **stamens** 1.5–13 mm long, usually 12, or twice as many as the petals; **anthers** *c.* 1.2 mm long; **pistil** 3–11 mm long; **ovary** 1.5–2.9 mm long; **style** one, filiform, tristylous (three different lengths), 1.2–7 mm long, exserted to included; **stigma** green and disclike. **FRUIT a capsule**, numerous, 3–4 mm long by *c.* 2 mm wide, inside the persistent calyx; **seeds** tan to red, *c.* 100 per capsule, glabrous, *c.* 1 mm long by *c.* 0.5 mm wide.

■ **LEAVES** Opposite (or whorled), simple, entire, sessile, pointed, glabrous above with soft hairs along the raised veins beneath; **blades** 2–10 cm long by 5–25 mm wide, some with a partial twist at the tip, reduced to bracts above.

■ **STEM** Erect, stiff, square, persistent, reddish green, 1–50 in a clump, glabrous or with hairs angled down; 5–15 mm thick near the base.

366

Mallow—*Malvaceae* **Wildflower** White to pink to pale blue Petals 5

SMALL-FLOWERED MALLOW *Malva parviflora* L.

■ **SKETCH** An **annual herb** 40–120 cm tall from a **taproot** 3–10+ mm thick and to *c.* 16+ cm long; in gardens, roadsides, ditches, thickets, cultivated, irrigated land and along city streets. *Naturalized*

■ **FLOWERS** White with pinkish tips, to pale blue, blooming February–January; **inflorescence** of fascicles in leaf axils throughout the plant; **floral branches** alternate, lower branches horizontal to ascending at apices, reduced above, not densely hairy, subtended by leaves; **pedicels** hairy, 2–20 mm long; **flowers** perfect, 5–6 mm wide by 4–5 mm long; **involucels** three, hairy, linear, 2–2.5 mm long by *c.* 0.5 mm wide, flat, ascending; **calyx** 3.5–4.5 mm long, green, hairy on outside, hairs simple to stellate, embracing the ripe carpels; **calyx tube** 2–2.5 mm long; **calyx lobes** five, *c.* 2 mm long by *c.* 2.5 mm wide, pointed, ascending in flower, slightly shorter than the petals; **petals** five, 3.5–6 mm long by 1.8–2.5 mm wide, margins entire, irregular to wrinkled, veins microscopic, apical lobes two, subequal, glabrous to minutely hairy at the tapered base; **stamens** several, forming an erect, glabrous, loose column around the styles; **anthers** pale yellow, *c.* 0.4 mm long; **styles** light green, as many as there are carpels, 1–1.3 mm wide by *c.* 1 mm long, filiform, often curled and tangled. **FRUIT a schizocarp**, usually 10 (8–11) carpels united into a disclike ring 5–7 mm wide around a central, round depression 1.5–2 mm wide; **carpels** 2–2.5 mm thick by 2–2.2 mm wide by *c.* 2.5 mm tall, one radial line per marginal tooth, hairy on outer convex side, transverse ridges or net-veins between the marginal teeth, separating and falling at maturity; **seeds** one per carpel, each 1.8–2 mm long by 1.6–1.7 mm wide by *c.* 1.2 mm thick, exposed at base of carpel, roundish with an indent at one end.

■ **LEAVES** Alternate, simple, with five or seven shallow lobes, bluntly toothed, dull, palmately-veined; **blades** soft, thin, 2–8 cm long by 2–11 cm wide, lighter green below, hairs simple to stellate, a red spot at the base where blade and petiole meet; **petioles** 1–21 cm long, shallowly grooved above, hairs simple to stellate and more numerous near the blade; **stipules** linear to triangular, deciduous, 4–5 mm long by 1.5–2.5 mm wide, flat, separated, hairy only on entire margins, hairs simple and to *c.* 1 mm long.

■ **STEM** Erect to ascending, solid, reddish green, branched from base, hairs simple to 5-pointed (stellate) on a slightly raised base; 3–10 mm thick near the base.

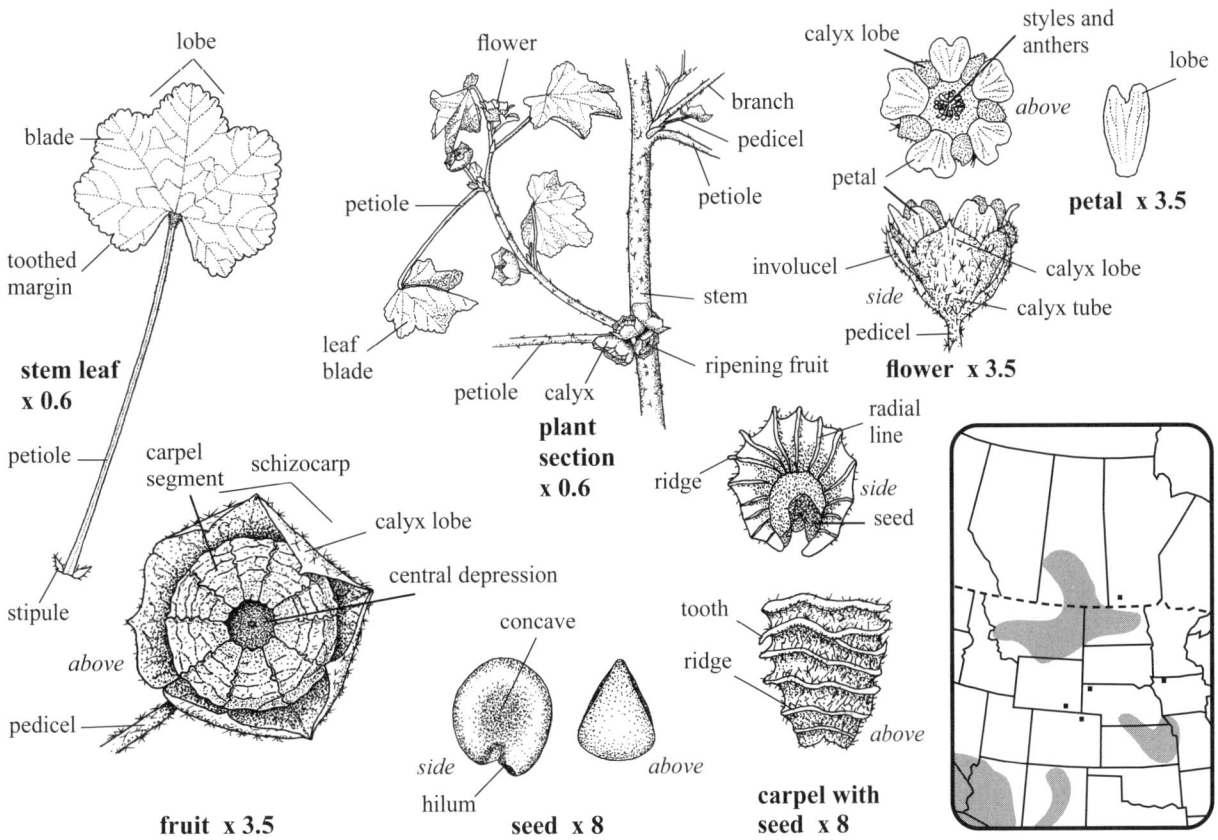

Small-fruited Mallow, Cheeseweed Mallow, Small-whorl Mallow

Mallow—*Malvaceae*

Wildflower White to lilac Petals 5

Round-leaved Mallow *Malva rotundifolia* L.

- **SKETCH** An **annual** or **biennial herb** with prostrate to decumbent branches 10–110 cm long; **taproot** 2–10 mm thick and to *c.* 15⁺ cm long; in disturbed sites, gardens, cities, towns and cultivated fields. *Naturalized* **w2**
- **FLOWERS** White to lilac, blooming May–September; **inflorescence** a cluster of one to ten (–20⁺) flowers per leaf axil; **pedicels** 5–30 mm long, glabrous or hairy, reflexed with fruit; **flowers** perfect, 3–4 mm wide by 4–6 mm long; **involucels** three, with a few hairs, 2.5–3 mm long by *c.* 0.7 mm wide; **calyx** 3–5 mm long by *c.* 3 mm wide, body hairs stellate with two to five parts; **calyx lobes** five, triangular, 2–2.3 mm long by *c.* 2 mm wide, with wavy margins and silvery white marginal hairs to 1.5 mm long, some with white pustular bases; **petals** five, erect, 3–5.2 mm long by 2–2.5 mm wide, slightly longer than the calyx, not showy, base *c.* 1 mm wide and hairy on margins; **stamens** 10–15, included; **filaments** *c.* 0.2 mm long, the central column *c.* 0.5 mm wide; **anthers** white, *c.* 0.5 mm long, in a ring; **pistil** one; **styles** usually 10, filiform, each *c.* 1.5 mm long, erect. **FRUIT** a **schizocarp** surrounded by the calyx, the brown disc 5–7 mm wide by *c.* 2 mm deep, with a central depression 1–1.2 mm wide; **carpels** 8–14, rough, reticulated, more or less hairy, each 2–2.2 mm long by 1.6–2 mm wide by 0.8–1.8 mm thick (outer surface), the edges slightly wavy and sharp, the lateral tan surface with *c.* 12 radial lines, the inner area usually exposing the seed; **seeds** one per carpel, 325–5,260 per plant, each 1.8–2 mm long by 1.5–2 mm wide by *c.* 1 mm thick, slightly concave on sides, smooth, medium brown to black.
- **LEAVES** Alternate, simple, mostly 5- or 7-lobed, these shallow and toothed; **blades** 1–8 cm long by 1.5–9 cm wide, with wavy margins, stellate hairs on both surfaces, five main veins more hairy below; **petioles** 2–24 cm long, with a shallow hairy groove above and round below; **stipules** triangular, with hairy margins and apices, to *c.* 7 mm long by *c.* 3 mm wide at the base, recurved.
- **STEM** Prostrate to ascending, solid, round, glabrous to slightly hairy, radiating branches several, these rebranched, green to reddish; 2–6 mm thick near the base.
- **SYN.** *Malva pusilla* Sm.

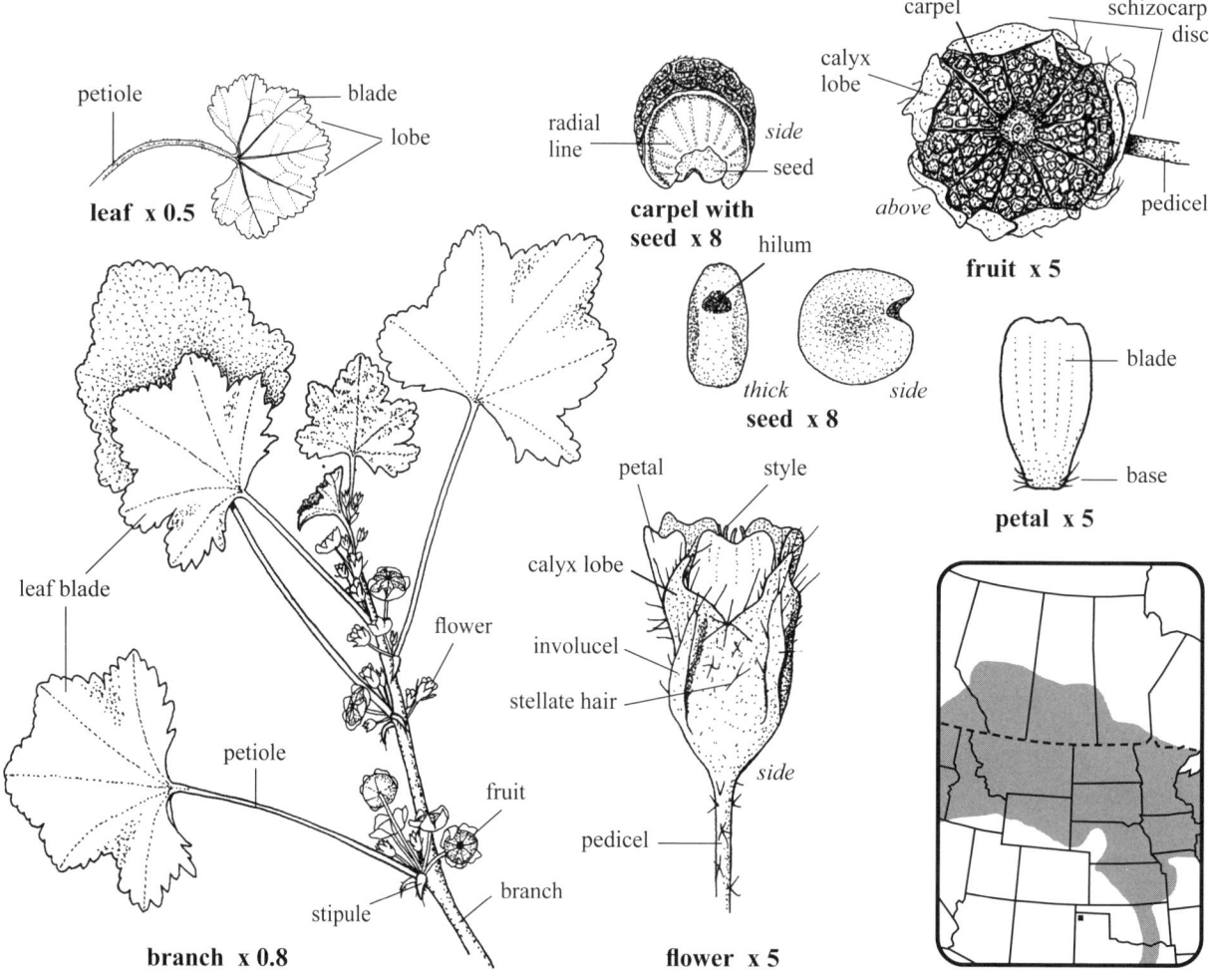

leaf x 0.5

carpel with seed x 8

seed x 8

fruit x 5

petal x 5

branch x 0.8

flower x 5

Low Mallow

Mallow—*Malvaceae* Wildflower Orange to red Petals 5

Scarlet Mallow *Sphaeralcea coccinea* (Nutt.) Rydb.

■ **SKETCH** A stellate-hairy **perennial herb** 10–50 cm tall or long, in patches, with a woody **caudex**; in dry prairies, sagebrush, along open trails and roadsides. **w2, p1**

■ **FLOWERS** Orange to red, blooming April–August; **inflorescence** a raceme, terminal, 2–12 cm long by 3–3.5 cm wide; **pedicels** 1–5 mm long, hairy; **subtending bracts** (of pedicels) reddish tan, early deciduous, stellate-hairy, 8–9 mm long by 0.3–0.5 mm wide, apices pointed, usually 3-veined; **flowers** perfect, 1–2.7 cm wide by 8–15 mm tall, 5–25 per raceme; **calyx** tubular, 5-lobed, pale green, TL 5–10 mm, tube 3–5 mm long and wide, calyx lobes 3.5–6 mm long by 2–2.5 mm wide, pointed, spreading, densely stellate-hairy; **petals** five, blunt to notched, 10–20 mm long by 7–11 mm wide, pale green near the hairy base; **stamens** numerous, included to exserted, erect in a central column 4–8 mm long; **filaments** pale green, in a united column *c.* 1 mm wide by 3–4 mm long with a few stellate hairs, distinct above for 1–1.5 mm; **anthers** yellow, *c.* 0.5 mm long; **pistils** equal to the number of carpels, forming a central column 3–5 mm wide by 8–10 mm tall, yellowish; **ovary** *c.* 3 mm wide by *c.* 1.5 mm tall, round, hairy at apex; **styles** 10+, each *c.* 8 mm long by *c.* 0.2 mm thick, distinct above, glabrous, each filiform, united in lower 2 mm above the ovary; **stigmas** slightly wider than styles, round. **FRUIT** a schizocarp of 10+ **mericarps** (carpel segments), together 3–5.5 mm wide by 2.5–3.5 mm tall, gray, hairy, disc slightly depressed in the center; **mericarps** reticulate on sides, *c.* 2.5 mm long by *c.* 3 mm wide by *c.* 2 mm thick; **seeds** *c.* 2 mm long by *c.* 1.5 mm wide by 1–1.2 mm thick, one per mericarp, dark brown, rounded on back, slightly hairy on inner concave sides.

■ **LEAVES** Alternate, compound (palmately), lobed and toothed, three or five divisions, flat to slightly involute; **blades** 1–6 cm long and wide, stellate-hairy on upper surface, whitish green below; **petioles** hairy, 1–3.5 cm long; **stipules** deciduous.

■ **STEM** Several in a clump, densely hairy, whitish green; 1.5–3 mm thick near the base.

■ **SYN.** *Malvastrum coccineum* (Nutt.) Gray = *Sphaeralcea coccinea* ssp. *coccinea* (Nutt.) Rydb.

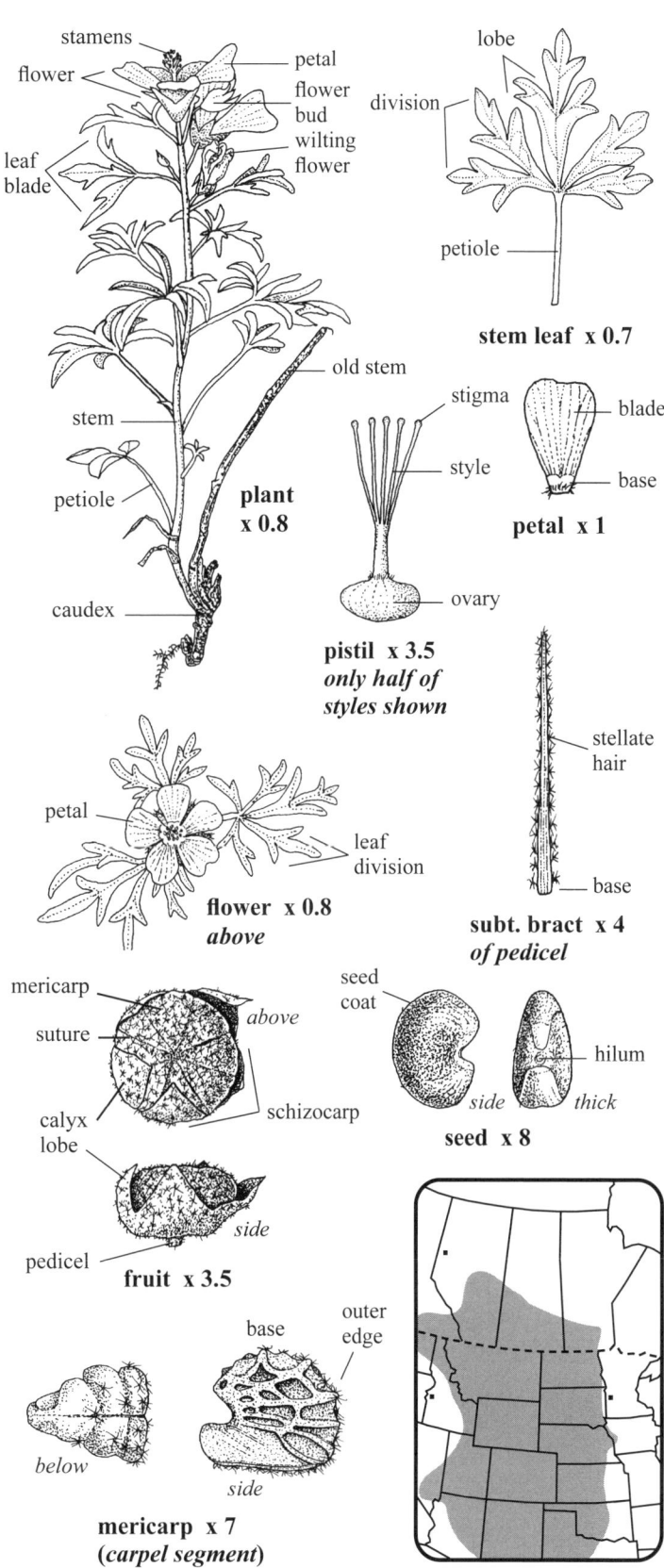

Red False Mallow, Scarlet Globemallow

Moonseed—*Menispermaceae* **Vine** White to pale green Perianth parts six (8–20)

Yellow Parilla *Menispermum canadense* L.

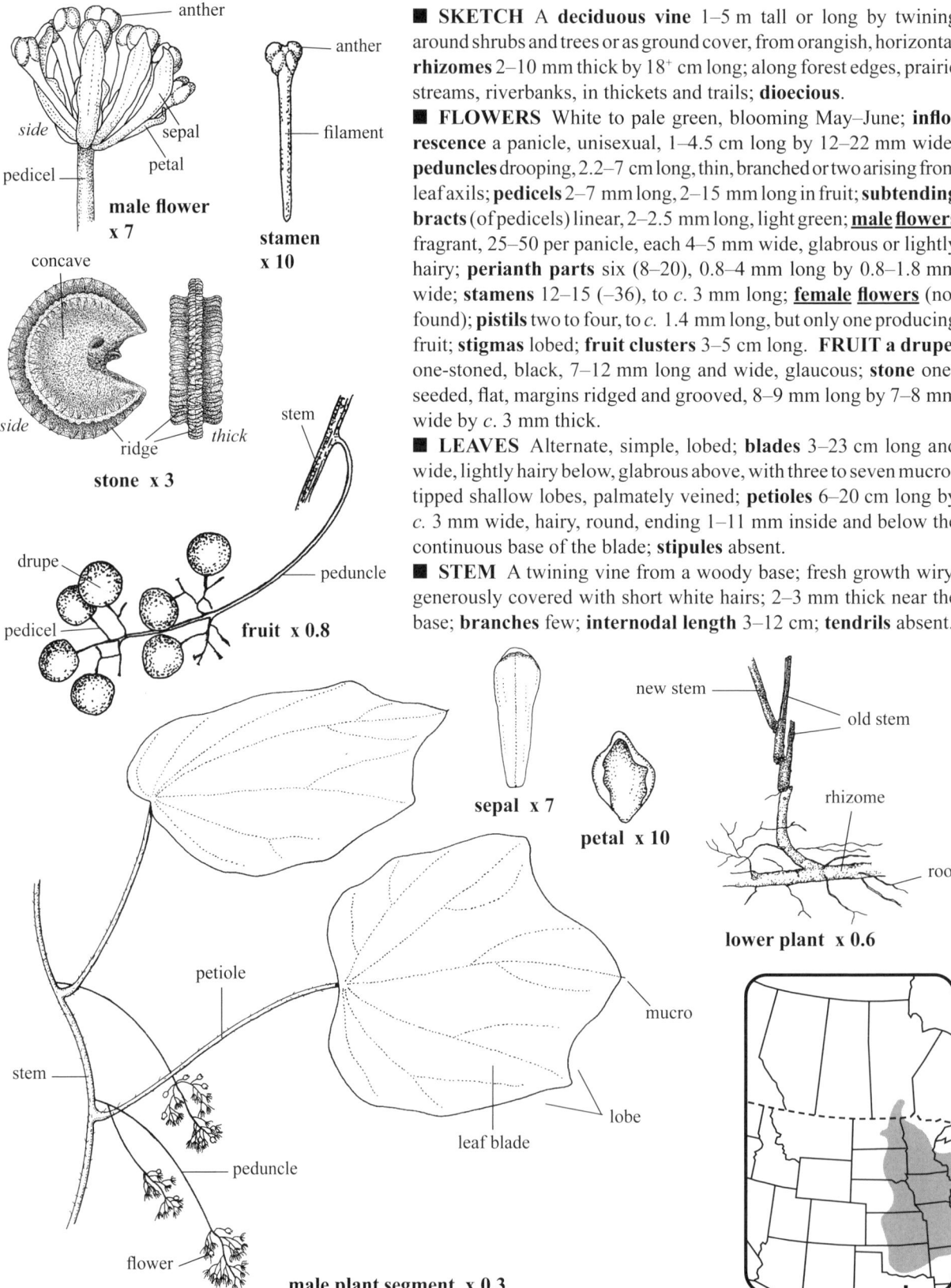

■ **SKETCH** A **deciduous vine** 1–5 m tall or long by twining around shrubs and trees or as ground cover, from orangish, horizontal **rhizomes** 2–10 mm thick by 18+ cm long; along forest edges, prairie streams, riverbanks, in thickets and trails; **dioecious**.

■ **FLOWERS** White to pale green, blooming May–June; **inflorescence** a panicle, unisexual, 1–4.5 cm long by 12–22 mm wide; **peduncles** drooping, 2.2–7 cm long, thin, branched or two arising from leaf axils; **pedicels** 2–7 mm long, 2–15 mm long in fruit; **subtending bracts** (of pedicels) linear, 2–2.5 mm long, light green; male flowers fragrant, 25–50 per panicle, each 4–5 mm wide, glabrous or lightly hairy; **perianth parts** six (8–20), 0.8–4 mm long by 0.8–1.8 mm wide; **stamens** 12–15 (–36), to *c.* 3 mm long; female flowers (not found); **pistils** two to four, to *c.* 1.4 mm long, but only one producing fruit; **stigmas** lobed; **fruit clusters** 3–5 cm long. **FRUIT a drupe**, one-stoned, black, 7–12 mm long and wide, glaucous; **stone** one-seeded, flat, margins ridged and grooved, 8–9 mm long by 7–8 mm wide by *c.* 3 mm thick.

■ **LEAVES** Alternate, simple, lobed; **blades** 3–23 cm long and wide, lightly hairy below, glabrous above, with three to seven mucro-tipped shallow lobes, palmately veined; **petioles** 6–20 cm long by *c.* 3 mm wide, hairy, round, ending 1–11 mm inside and below the continuous base of the blade; **stipules** absent.

■ **STEM** A twining vine from a woody base; fresh growth wiry, generously covered with short white hairs; 2–3 mm thick near the base; **branches** few; **internodal length** 3–12 cm; **tendrils** absent.

Canadian Moonseed, Common Moonseed, Moonseed

Buck-bean—*Menyanthaceae* **Wildflower** White to pink Corolla tubular, 5-lobed (4 or 6)

BUCK-BEAN *Menyanthes trifoliata* L.

■ **SKETCH** An aquatic **perennial herb** 10–35 cm tall with a whitish green, solid **rhizome** 20–50+ cm long by 4–11 mm thick; **internodes** 0.7–3.5 cm long; **roots** 3–52 cm long by 1–3 mm wide, white, some with wrinkled bases; in marshes, bogs, fens, beaver ponds and margins of shallow lakes. w2

■ **FLOWERS** White to pink, blooming May–August; **inflorescence** a raceme 4–8 cm long by 3–4 cm wide; **pedicels** ascending, 5–20 mm long; **subtending bracts** (of pedicels) 4–7 mm long, entire; **flowers** perfect, 15–23 per raceme, each 12–17 mm wide by 12–14 mm long; **calyx** tubular, 5-lobed; **tube** *c.* 1 mm long, hairless; **lobes** 2–3.5 mm long by 1–2 mm wide, 3-veined, erect, persistent; **corolla** tubular, usually 5-lobed (4- or 6-); **tube** glabrous, *c.* 5 mm long by *c.* 4 mm wide; **lobes** recurved, 5–8 mm long by 2–2.8 mm wide, white hairs numerous on inner surface, 2–3 mm long and tangled; **stamens** five, exserted or included, 6.7–12.2 mm long; **filaments** white, united below to corolla tube, 8–11 mm long, glabrous; **anthers** dark reddish brown, 2.5–2.8 mm long; **pistil** one, glabrous, 6–10 mm long, persistent; **ovary** 3–5 mm long by *c.* 2.5 mm wide, bluntly triangular; **style** one, erect, 3–5 mm long, included to exserted; **stigma** 2-lobed, *c.* 1 mm wide, green. **FRUIT a capsule**, 3-valved, ascending, brown, corky, 5–10 mm long by 3–7 mm wide by 2.8–5 mm thick, opening from apex; **seeds** 1–5 per capsule, shiny, smooth, golden tan, 2.6–3 mm long by 2.3–2.5 mm wide by 1.2–1.5 mm thick.

■ **LEAVES** Basal, alternate, arising from nodes of rhizome near stem, compound; **leaflets** three, glabrous, each 3–11.3 cm long by 1–5.3 cm wide, ascending, dull, entire or with wide shallow teeth; **petioles** 8–25 cm long by 3–11 mm wide, reddish green, solid, smooth, oval; **petiolules** 2–5 mm long; **stipular sheathing** 9–15 cm long, the base wrapped around the rhizome, veined, open to base, membranous, attached to petioles, apex pointed and with *c.* 1 mm free, persistent.

■ **STEM** A **scape**, oval in cross-section, smooth, solid, green to reddish; 4–6 mm wide near the base.

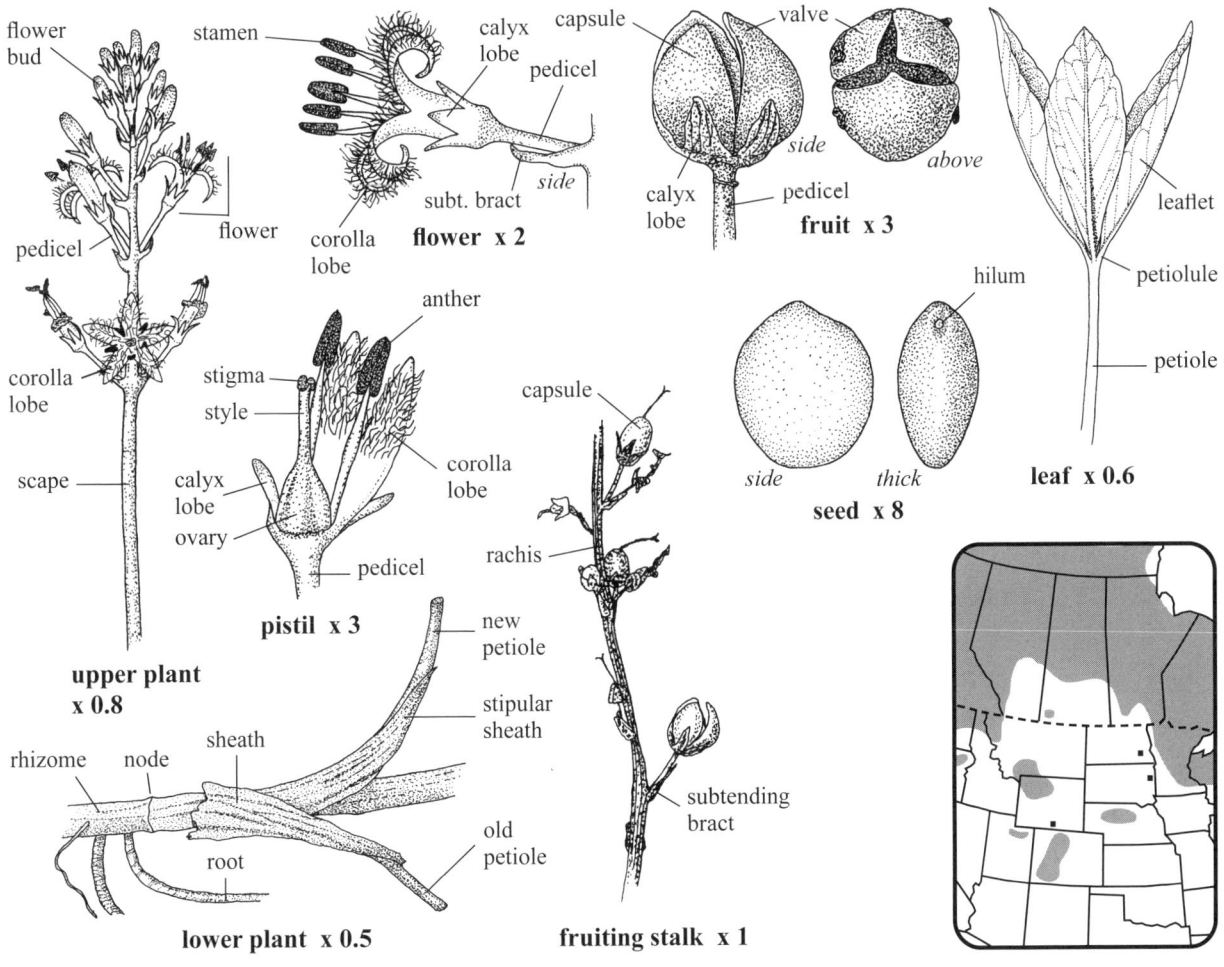

Bogbean, Marsh-trefoil

Indian Pipe—*Monotropaceae* **Wildflower** White Petals 5

INDIAN-PIPE *Monotropa uniflora* L.

■ **SKETCH** A white **perennial herb** lacking chlorophyll, 4–30 cm tall, singly or in clusters up to *c.* 50 stems; **coral-like roots** branching, brown, *c.* 0.7 mm thick and easily broken, with thin, intertwined **white roots**; in decaying vegetation of rich woods, along trails, in semi-open areas.

■ **FLOWERS** White, blooming July–September; **inflorescence** a terminal, nodding flower; **flowers** one per stem, 10–15 mm wide by 1.4–3 cm long (including subtending scales), odorless; **sepals** similar in texture to scales, 0–4, each *c.* 13 mm long by 2–2.2 mm wide, pointed, apices erose, 1-nerved; **petals** five, white, 0.8–2 cm long by *c.* 7 mm wide, white hairy on inside and lower margins, hairless outside, saclike hairless base 1.5–2 mm deep by *c.* 4 mm wide; **stamens** 10, included, attached to base of ovary; **filaments** white, ascending, 7–8 mm long by *c.* 1 mm wide, hairy, convex on outside, V-shaped inside to fit grooves in the ovary; **anthers** pale yellow, *c.* 1.4 mm long by 1.8–2 mm wide, opening by apical transverse slits; **pistil** 10–11 mm long, included; **ovary** white to pale pink, 7–8 mm long by *c.* 7 mm wide, with narrow white appendages 1.5–2 mm long projecting downward between the filament bases; **style** *c.* 1.5 mm long, flaring outward; **stigma** a moist, circular disc atop the round style, 4–5 mm wide by *c.* 1 mm high, concave in middle, 5-lobed, persistent. **FRUIT a capsule**, dull, dark brown, erect, persistent overwinter, 9–13 mm long (including style) by 9–10 mm wide and thick, partially covered by persistent perianth and filaments, 5-lobed, each lobe bisected into two rounded lobes, apex a dark ring *c.* 3 mm wide; **seeds** hundreds per capsule, each 0.6–1.1 mm long by 0.1–0.2 mm wide, reddish brown, wing parallel to net-veined, transparent, brown.

■ **LEAVES** Alternate, entire to slightly toothed, sessile; **blades** thin, semi-transparent, scalelike, appressed, 6–12 mm long by 4–7 mm wide, ascending, with three faint veins, midrib the most obvious, covering the upper stem, widely spaced below; **scales** at flower's base similar in texture but 15–17 mm long by 4.5–7.5 mm wide.

■ **STEM** White, solid, crisp, unbranched, with low ridges *c.* 2 mm wide from the base of each scale, brown and erect with fruit; bare base 4–6 mm wide.

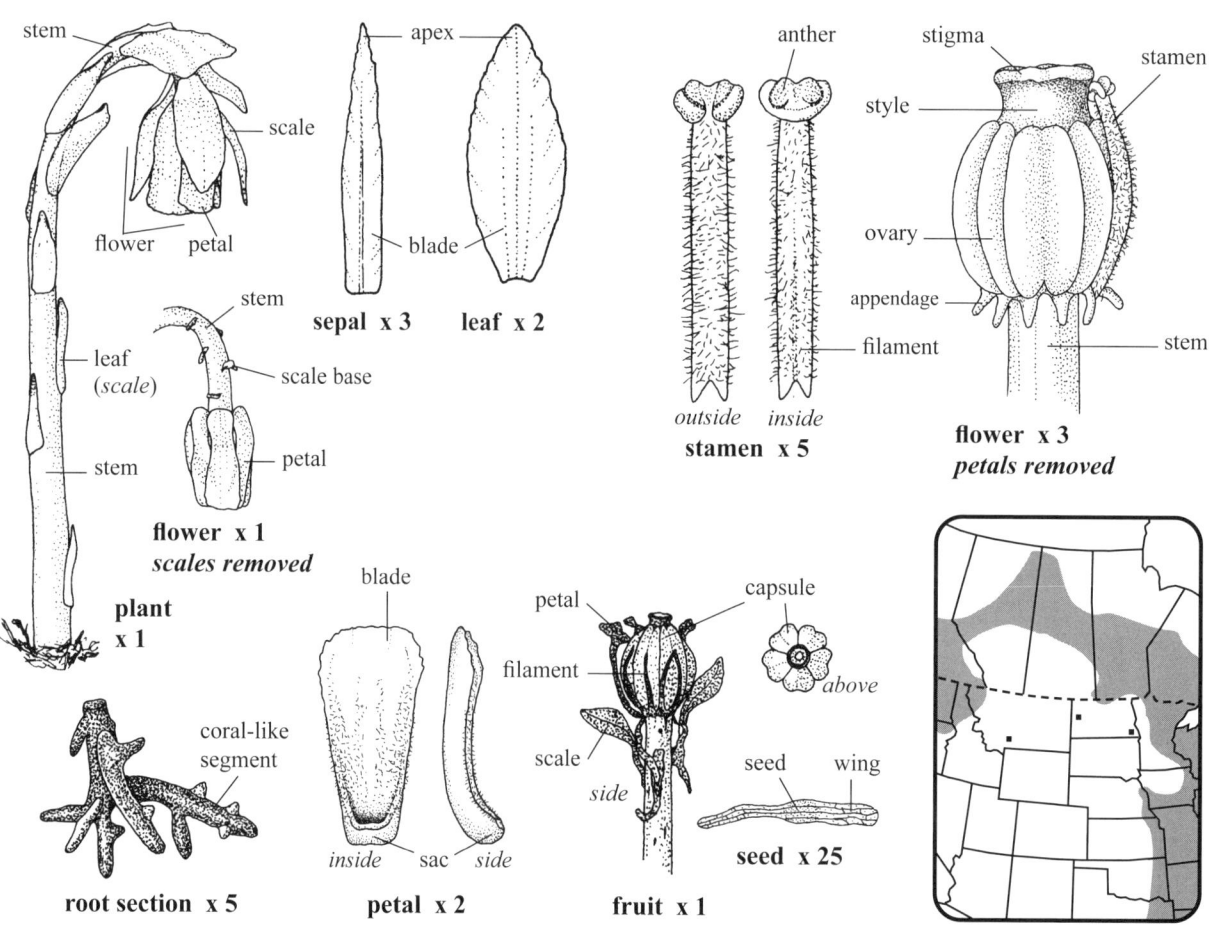

One-flower Indianpipe

Four-o'clock—*Nyctaginaceae* **Wildflower** Pink to purplish red   Corolla absent
Calyx tubular, 5-lobed

Hairy Umbrellawort *Mirabilis hirsuta* (Pursh) MacM.

■ **SKETCH** A hairy, variable **perennial herb** 20–100 cm tall, solitary or in clusters with a woody **taproot** 3–8 mm thick and to *c.* 15⁺ cm long crowned with a branching **caudex**; in sandy soils of stream valleys, prairies, roadsides, pasture and rocky hillsides.

■ **FLOWERS** Pink to purplish red, blooming May–October; **inflorescence** a panicle, cymose, loose, terminal and near the ends of branches; **floral branches** axillary, 4–25 cm long, opposite with one branch of the pair usually longer, bracteate, ascending, reduced above, very hairy, often glandular, with short hairs ascending and numerous between spreading hairs; **pedicels** hairy, 3–6 mm long at anthesis, 15–21 mm long and descending in fruit; **subtending bracts** (of pedicels) leaflike, *c.* 2 mm long; **flowers** perfect, not fragrant, usually three per involucre, 10–17 mm wide and long, blooming in the evening, sometimes cleistogamous; **involucre** 5-lobed, pale green, veiny, hairy, more so below and along margins, each lobe 4–7 mm long by 2–3 mm wide, pointed and often with reddish margins at anthesis, folded inward in the mornings hiding the pink calyx, spreading in fruit to 10–33 mm wide; **calyx** tubular, 5-lobed, tubes 6–7 mm long, mostly green, persisting, hard, ribbed in fruit; **calyx lobes** pink, *c.* 7 mm wide, each with a notch 2–3 mm deep, spreading with a star-shaped pattern in the middle; **corolla** absent; **stamens** three to five, exserted; **filaments** pale pink, filiform; **anthers** yellow, *c.* 1 mm long; **pistil** exserted; **ovary** green, hairy, superior, *c.* 2 mm long by *c.* 1.3 mm wide. **FRUIT an anthocarp**, 1-seeded, usually three per involucre, greenish gray, often falling early, 4–5 mm long by 2.5–3 mm wide, 5-ribbed, base and apex flat, ridges and body tuberculate, hairs spreading and mostly along the ridges and smaller tubercles; **seeds** glabrous, yellowish tan, 2.9–3.5 mm long by *c.* 2 mm wide and thick, with a low wide ridge, hilum (base) slightly pointed, apices blunt.

■ **LEAVES** Opposite, simple, entire, ascending; **blades** 2–7(–12) cm long by 1–4.5 cm wide, dull, scattered hairs on margins, less so on blades, hairs simple, arched forward toward apex or spreading, slightly lighter green below, margins undulate, apices blunt to pointed; **petioles** 0–9 mm long, reduced above, hairy like the stem.

■ **STEM** Ascending, stiff, solid, oval to rectangular, often with two longitudinal grooves, white hairs 1–3 mm long and spreading, sometimes glandular; 2–5 mm thick near the usually hairy base; **nodes** always hairy.

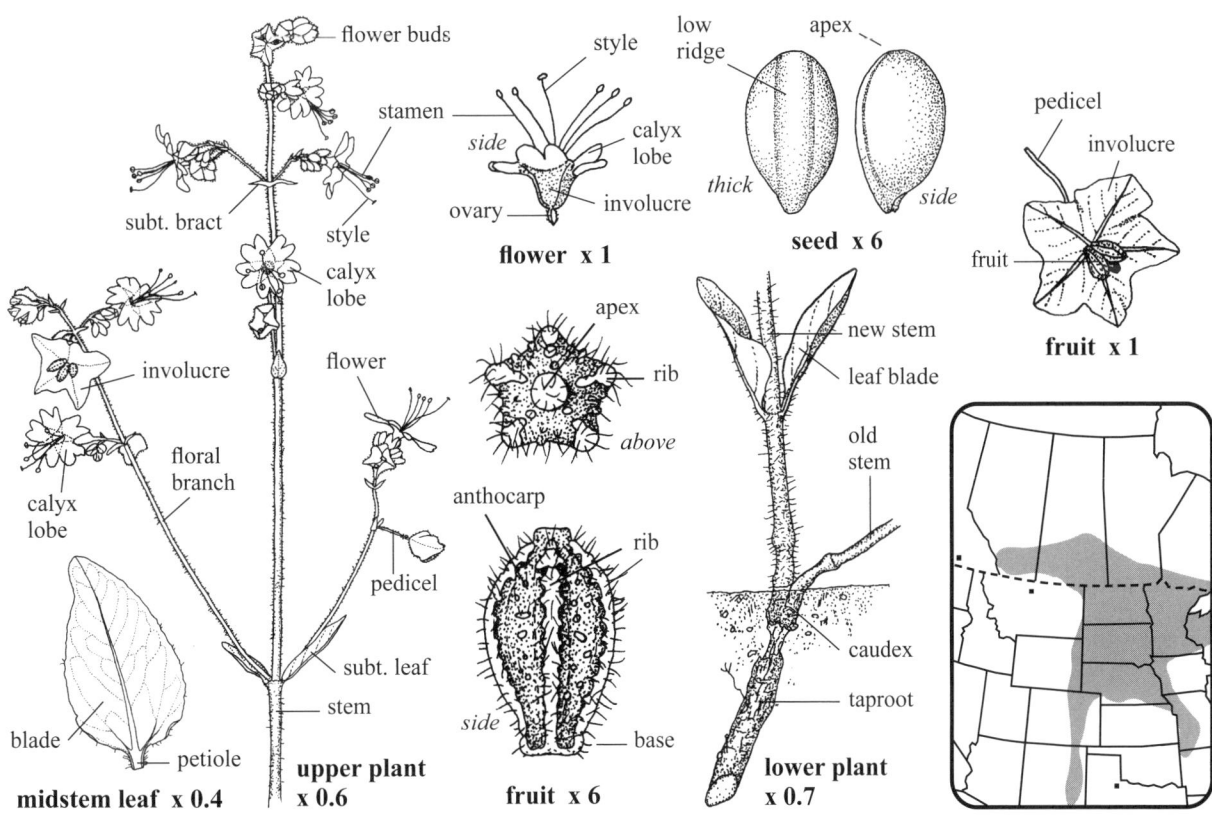

Water-lily—*Nymphaeaceae* **Wildflower** Yellow Sepals 6

YELLOW POND-LILY *Nuphar lutea* **ssp.** *variegata* (Dur.) E.O. Beal

■ **SKETCH** An aquatic, glabrous **perennial herb** 0.5–3 m long or tall (depending on water's depth) in colonies from **rhizomes** 5–20 cm long by 2–10 cm thick, some branching, covered with scars in spirals; in small lakes, beaver ponds and slowly moving water. **w1**

■ **FLOWERS** Yellow, blooming June–October; **inflorescence** a terminal flower at or held several cm above the water's surface; **peduncles** equal to water's depth, round, 8–15 mm thick, mostly smooth; **flower** erect, 4–7 cm wide by 2–2.5 cm tall, perfect, not fragrant, base recessed; **sepals** six, yellow with a dark reddish purple central area on inside; **small (outer) ones** greenish on outside, 23–35 mm long by 15–20 mm wide; **large (inner) sepals** 28–40 mm long by 28–35 mm wide, C-shaped, green on outside near base, with the stigmatic rays imprinted on their yellow apices; **stamens** yellow, 160–220, in five layers, each 9–11 mm long by *c.* 2 mm wide, arched downward forming a circle *c.* 3 cm wide, imbricate, apices thickened, attached at base of pistil; **filaments** flat, 3–4 mm long by *c.* 2 mm wide, lower side with a ridge; **anthers** two per stamen, each 2-lobed, the lobes 3–6 mm long by *c.* 0.7 mm wide; **staminodia** petaloid, *c.* 18 in a single whorl below the stamens, 7–9 mm long by 3.5–7.5 mm wide, yellow, apices thickened and glandular below (sticky); **pistils** 7–25, united, 10–12 mm long by 9–10 mm wide at base; **locules** *c.* 5 mm long; **ovules** numerous; **styles** fused, obscure; **stigmatic disc** yellow to greenish yellow, slightly concave, 13–16 mm wide, roundish from above, recessed in center with a tiny opening surrounded by 7–25 narrow **stigmatic rays** *c.* 4 mm long radiating to the lobed edges. **FRUIT a berry**, leathery outside with numerous long ridges, 3–4 cm long by 2–2.8 cm wide, apex concave and 10–20 mm wide, body shiny green, ripening to purplish red, white and spongy inside, the inner tissue becoming mucilaginous as it decays; **seeds** oval, smooth, 55–90 per berry, 3–5 mm long by 3.8–4 mm wide by 3.5–3.8 mm thick, shiny, yellowish brown.

■ **LEAVES** Alternate, entire, simple, floating flatly on, to erect at the water's surface, several; **blades** 8–35 cm long by 7–15 cm wide, veins visible below, more obscure above; **basal lobes** two, rounded, 4–8 cm long, sinus narrow; **petioles** winged or not, oval, solid, buoyant, length depends on the water's depth.

■ **STEM** Absent.

■ **SYN.** *Nuphar variegata* Dur.

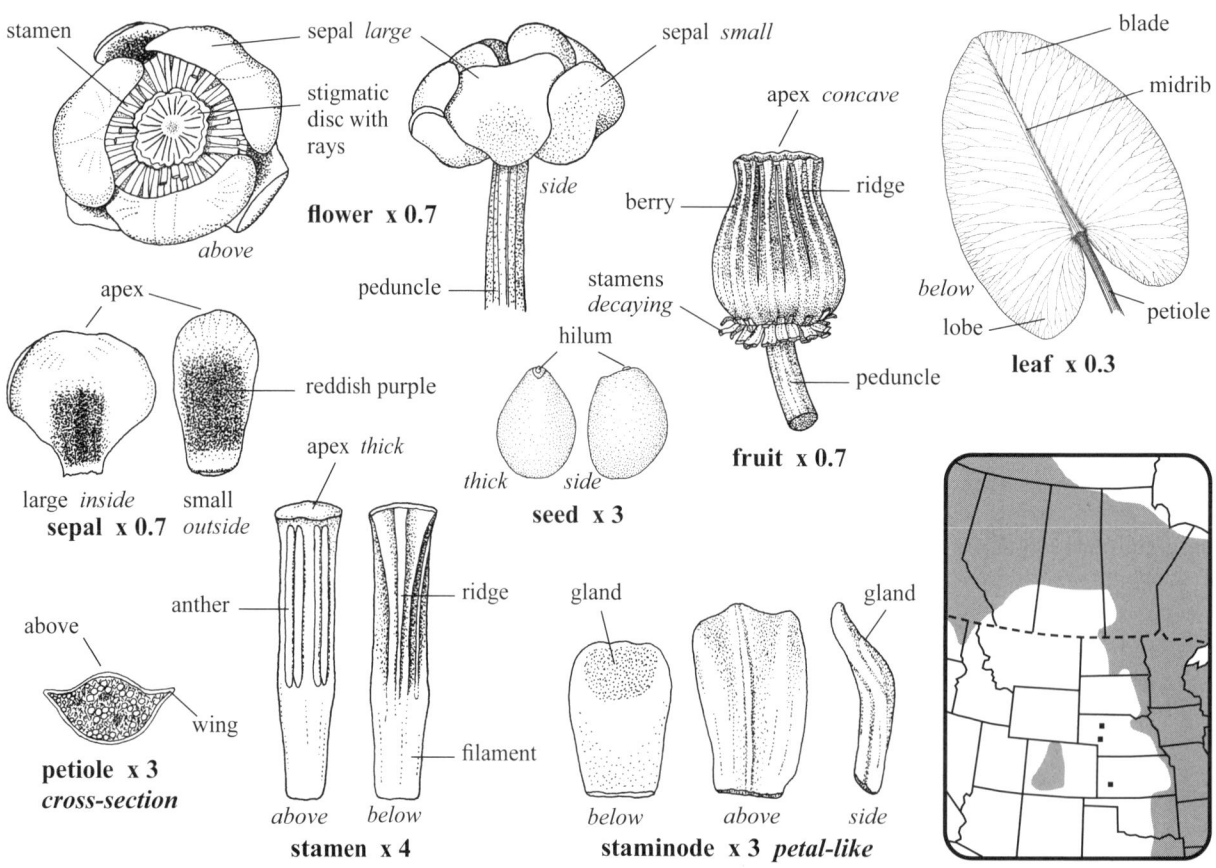

Bullhead-lily, Spatterdock, Cow Lily, Small Yellow Pond-lily

Olive—*Oleaceae* **Tree** Green to reddish brown

Black Ash *Fraxinus nigra* Marsh.

■ **SKETCH** A **deciduous tree** 7–25 m tall; in wet sites in bogs, swamps, streams and lakeshores of woodlands; **dioecious** to **monoecious**; a hardwood. **w2**

■ **FLOWERS** Green to reddish brown, blooming May–June; **inflorescence** unisexual to perfect at tips of twigs; <u>male flowers</u> in a raceme, clustered, together 2–5 cm long and wide; **perianth parts** absent; **stamens** in exposed bundles; **filaments** green, filiform, very short; **anthers** reddish purple, 1–1.8 mm long, 2-lobed, apices blunt; **subtending bracts** 2–4 mm long by *c.* 0.5 mm wide, green and covered with tangled tan hairs, the stalk slightly hairy; <u>female flowers</u> (not shown) 5–6 mm long, clustered; **stipe** tan, 1–2 mm long; **calyx cup** pale green, *c.* 1 mm long, apex toothed, enclosing the ovary; **style** pinkish tan, exserted, filiform, *c.* 3 mm long, 2-parted at the apex; **subtending bracts** (of several flowers) 2–3 mm long and wide, covered with reddish tan tangled hairs; **pedicels** (of fruit) 4–6 mm long. **FRUIT a samara**, 1-seeded, tan, flattened, 2–4.5 cm long by 0.7–1 cm wide; **wing** tan, usually with a tiny notch, sometimes twisted; **seed** not shiny, flattened, darker brown than the wing, 14–18 mm long by 4–5 mm wide by *c.* 1 mm thick, tapered at both ends.

■ **LEAVES** Opposite, compound, in clusters near tip; **blades** 20–40 cm long by 17–26 cm wide, falling as a unit; **leaflets** 7–11, turn yellow in fall, lateral ones sessile, finely and shallowly toothed, 5–15 cm long by 1.5–5 cm wide, reddish matted hairs at base of leaflets and where the midrib and side veins meet below; **petioles** 4–12 cm long, slightly glandular below; **rachis** and midrib with reddish brown glands; **petiolule** of terminal leaflet 15–25 mm long; **bracts** 2–4, yellowish green at the base of petioles, these 1–2 cm long by 7–12 mm wide, pointed, glandular, especially below.

■ **TRUNK** Straight, slender; **dbh** 20–50 cm; **bark** light gray with corky ridges when young, mature bark scaly, not deeply furrowed; **branches** coarse, mostly ascending, tips also ascending at least on lower branches; **twigs** coarse, dull, hairless; **winter buds** dark brown, opposite; **terminal bud** 5–10 mm long by 3.5–6.5 mm wide by 3.2–5.2 mm thick, pointed, often slightly bent, 2- to 4-scaled, dull; **lateral buds** blunt, 1.5–3.5 mm long by 2.5–4.5 mm wide; **leaf scars** round and obvious.

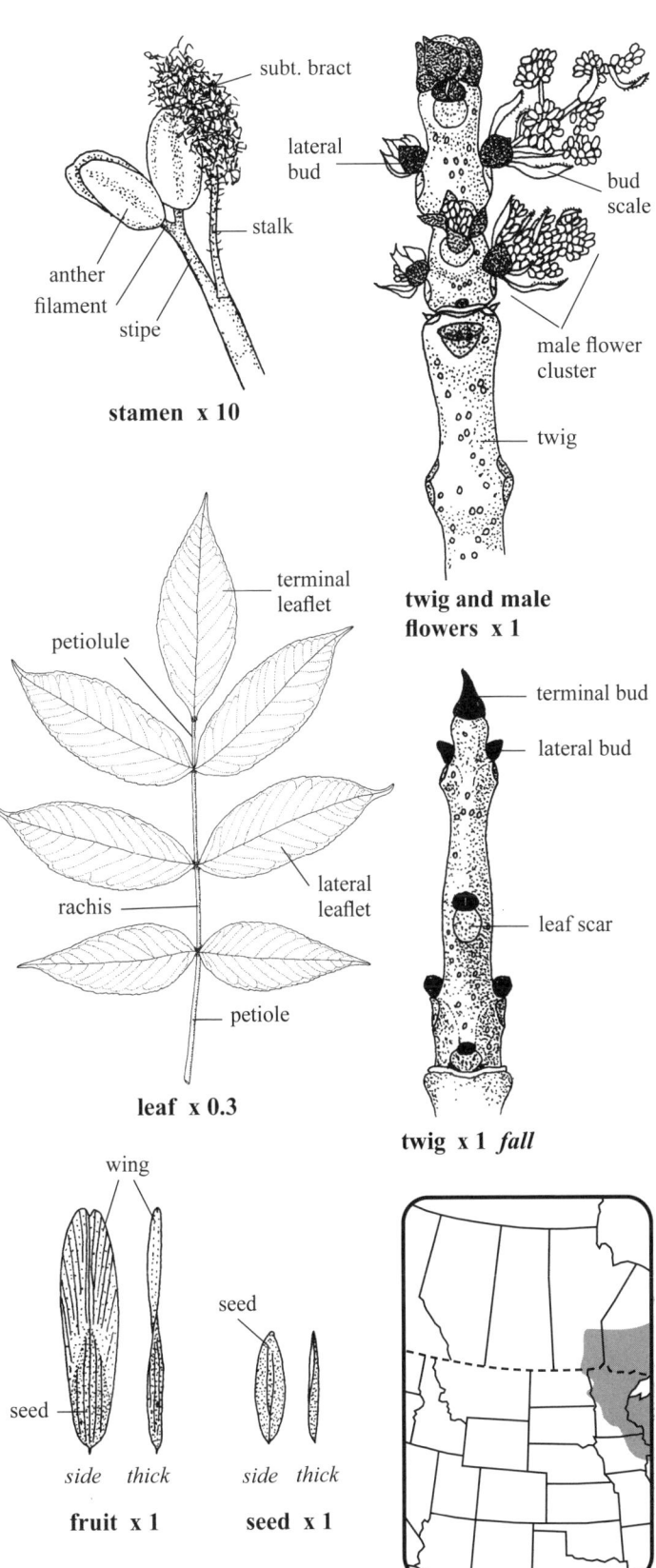

Swamp Ash

Olive—*Oleaceae*   **Tree**   Green to reddish   Male calyx cup 4-toothed

GREEN ASH   *Fraxinus pennsylvanica* Marsh.

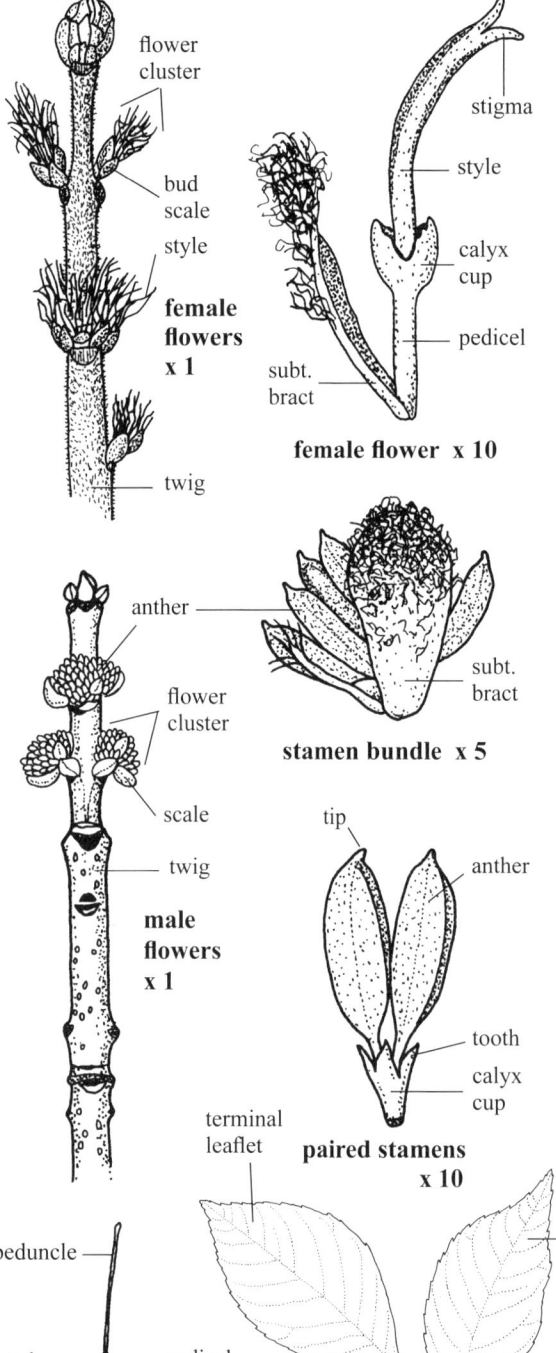

■ **SKETCH** A **deciduous tree** 8–25 m tall; in woodlands, along streams or lake margins, ravines, prairies and planted in towns; a **dioecious** hardwood. **w2**

■ **FLOWERS** Green to reddish, blooming April–May; **inflorescence** unisexual; **subtending bracts** 4–6 mm long, hairy in upper half; <u>**male flowers**</u> reddish brown, in compact clusters 7–10 mm wide by 8–9 mm high by *c.* 7 mm thick; **calyx cup** usually with four teeth, glabrous, *c.* 1 mm long; **petals** absent; **stamen bundles** under each large, hairy bract; **filaments** green, glabrous *c.* 0.5 mm long, partially hidden by the calyx cup; **anthers** reddish purple, 2.5–4 mm long by *c.* 1 mm thick, with a short flexible tip; **female flower clusters** green and pink, 10–20 mm long by *c.* 10 mm wide, ascending; **pedicels** green, 1–2 mm long in flower, 4–10 mm long with fruit; **subtending bracts** 1–4 mm long (including hairs) by 0.1–0.7 mm wide, beige to light green; <u>**female flowers**</u> hairless, *c.* 6 mm long, several bunches of 3–10 flowers in a cluster; **styles** 3–3.3 mm long, curved, thickened along the side, reddish green; **stigma** 2-lobed, each lobe *c.* 0.8 mm long, spreading, rough but not hairy. **FRUIT a samara**, 1-seeded, tan, 2–5 cm long by 4–7 mm wide, the wing symmetrical, apex blunt, extending about halfway along the sides of the elongated, round, 1–2 mm thick seed.

■ **LEAVES** Opposite, compound (odd-pinnate); **leaf blades** 10–30 cm long by 8–22 cm wide; **leaflets** five to nine, dark green, glabrous and slightly shiny above, toothed, 3.5–17 cm long by 1.5–8.2 cm wide, veins raised and woolly hairy to glabrous below; **petioles** glabrous to hairy, 3–7 cm long; **side petiolules** 2–7 mm long; **stipules** absent.

■ **TRUNK** Brownish gray, furrowed, branched above; **dbh** 25–50 cm; **branches** with many ascending twigs at their ends; **twigs** glabrous or covered with fine whitish hairs giving them a gray appearance; **leaf scars** half-moon-shaped; **bundle scars** not obvious; **terminal buds** reddish brown.

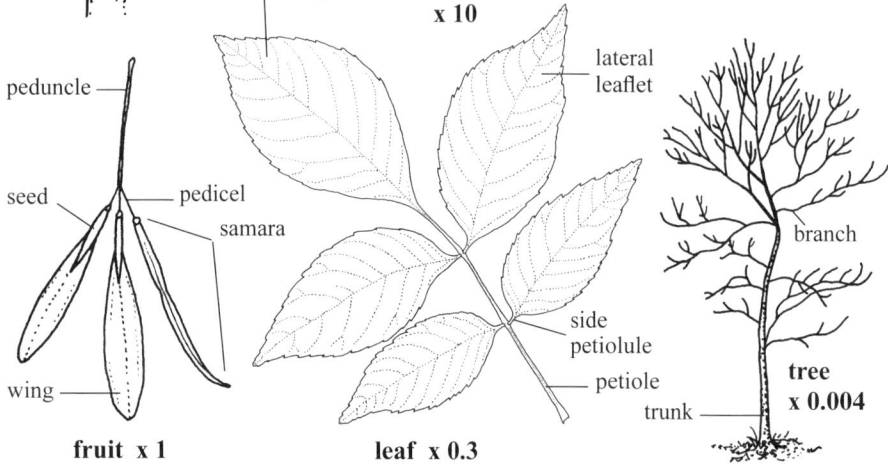

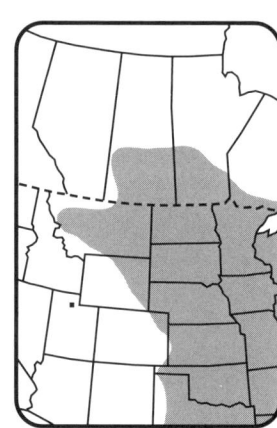

Red Ash

Evening-primrose—*Onagraceae* **Wildflower** Yellow Petals 4

Shrubby Evening-primrose *Calylophus serrulatus* (Nutt.) Raven

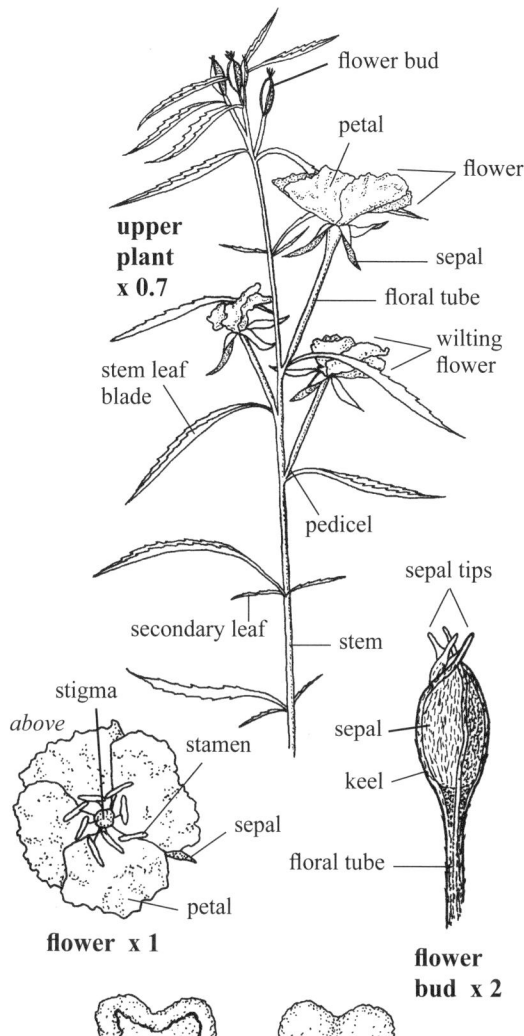

- **SKETCH** A variable **perennial herb** 20–60 cm tall from a woody, often branched **caudex**, in open colonies; in dry to moist prairies, sandhills, open woodlands and roadsides. **p2**
- **FLOWERS** Yellow, blooming March–September; **inflorescence** a raceme of several flowers, each from an upper leaf axil; **pedicels** 2–4 mm long, hairy; **flower buds** *c.* 13 mm long by *c.* 5 mm wide, 4-sided, the sepal tips twisted at the apices; **flowers** perfect, 2–3 cm wide by 10–16 mm tall; **sepals** four, united into two pairs or distinct, keeled, whitish green, 2–10 mm long by 2.5–5 mm wide, hairy outside, glabrous inside, tips green and distinct for 1–3 mm; **petals** four, ascending, glabrous, 5–16 mm long by 7–20 mm wide, wrinkled, apices erose; **floral tube** 5–20 mm long by 1.5–2 mm wide, 4-ribbed, squarish, covered with short white ascending hairs; **stamens** eight, of two lengths, the four from the sepals 3.5–4 mm long and the four from the petals 1.5–3 mm long; **filaments** glabrous, attached about one-third from the base; **anthers** 2–4 mm long, pale yellow; **ovary** 4–12 mm long; **style** light green, 5–15 mm long, glabrous; **stigma** included, 4-lobed, club-shaped, 1–2 mm long and wide, greenish yellow. **FRUIT a capsule**, hairy, 10–30 mm long by 1–3 mm wide, squarish, erect, brown, 4-valved, opening at blunt apex; **seeds** in two rows, brown, 1–2 mm long by 0.6–0.8 mm wide by 0.3–0.5 mm thick, smooth, dull, angular, apices truncate to pointed.
- **LEAVES** Alternate, simple, toothed to entire, narrow, spreading to descending below; **blades** 2–10 cm long by 1–12 mm wide, larger above, apices pointed to rounded, V-shaped along midrib, narrowing gradually to the petiole, more hairy near the base and below, often with a pair of tiny secondary blades in axils; **petioles** hairy, 1–10 mm long.
- **STEM** Erect, branched or not, leafless below, more hairy among the flowers; 1–2 mm thick near the slightly woody base.
- **SYN.** *Oenothera serrulata* Nutt.

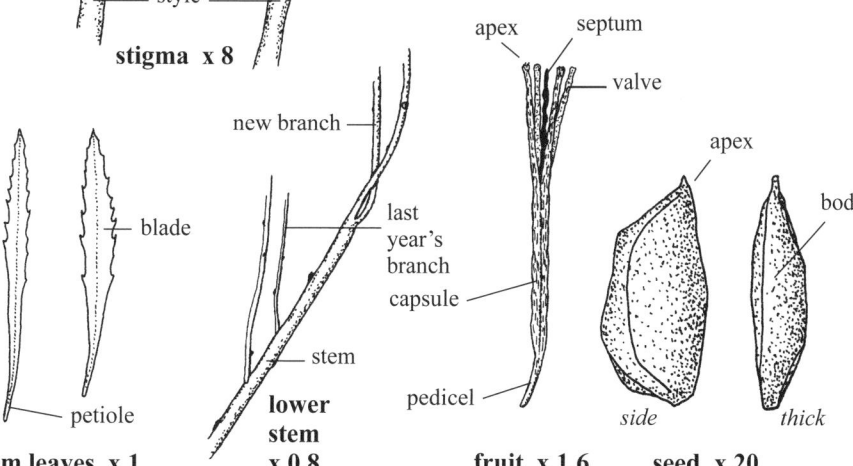

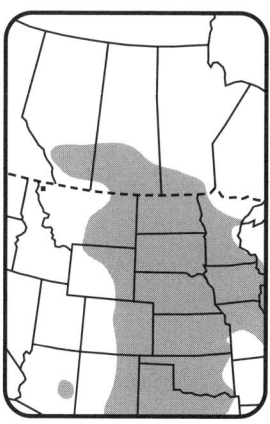

Yellow Evening-primrose, Plains Yellow Primrose, Yellow Sundrops   377

Evening-primrose—*Onagraceae*   **Wildflower**   Pink to pale purple   Petals 4

FIREWEED   *Chamerion angustifolium* (L.) Holub.

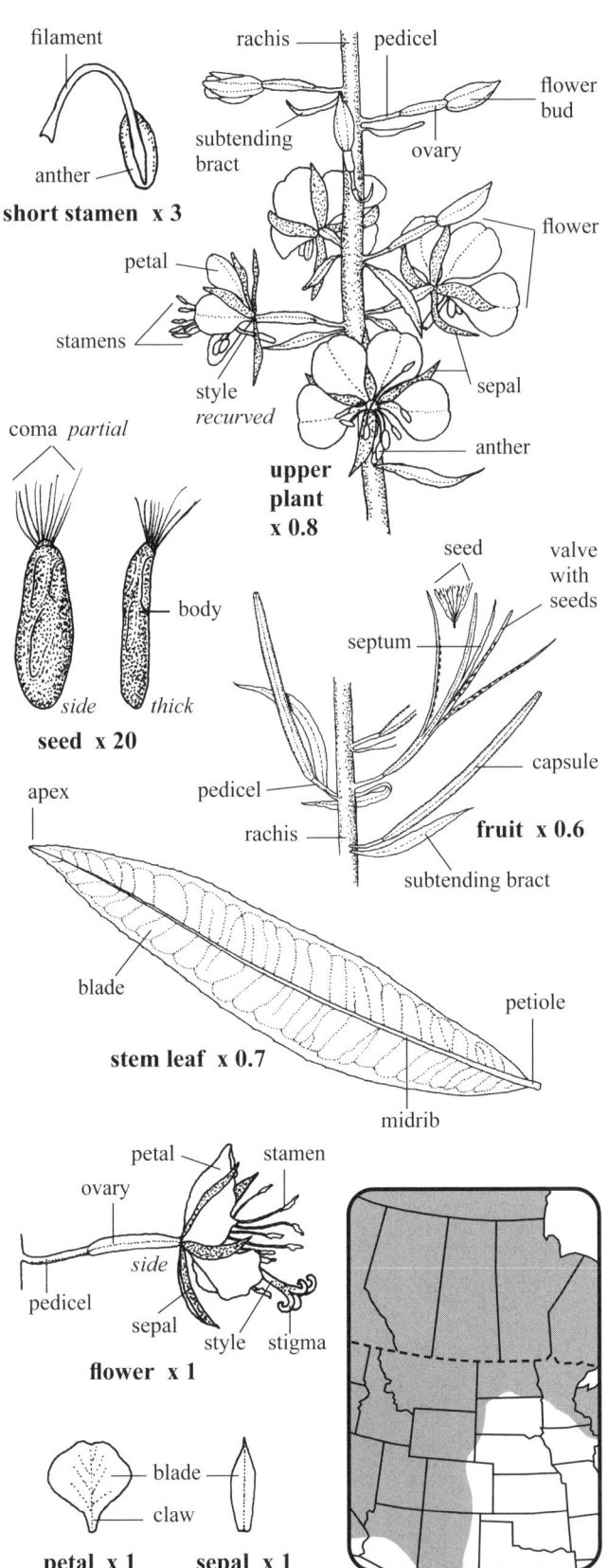

■ **SKETCH** A variable **perennial herb** 30–250⁺ cm tall in colonies from a woody **rootstock** 3–6 mm thick, with spreading **rhizomes** to 1⁺ m long; in burned, disturbed and cleared, moist to dry forest sites, along roadsides, rocky outcrops and moist prairies. **p2**

■ **FLOWERS** Pink to pale purple, rarely white, blooming June–September; **inflorescence** a raceme 20–50⁺ cm long by 6–9 cm wide, terminal and lateral, apex in bud, erect to nodding; **pedicels** reddish green, glabrous, 5–12 mm long, spreading at anthesis, 6–20 mm long with fruit; **subtending bracts** (of pedicels) leaflike, reduced above; **flowers** perfect, spreading to slightly ascending, 2.5–3 cm wide by 3–4 cm long (including stigmas when exserted); **rachis** glabrous to hairy, glaucous, reddish green; **sepals** four, glabrous, green to red, 7–16 mm long by 1.6–3 mm wide, pointed, midrib obvious, spreading between and projecting to about the tips of the petals; **petals** four, glabrous, 10–20 mm long by 7–15 mm wide, claw 2–4 mm long; **stamens** eight, to *c.* 15 mm long, included to exserted, subequal; **filaments** white, free and glabrous, 4–12 mm long, forming a circle *c.* 2 mm wide at their bases; **anthers** pinkish tan, 1.5–3.5 mm long by 1–2 mm wide; **ovary** 8–18 mm long by *c.* 2 mm wide and thick, hairy, grooved; **style** hairy at base, exserted, 10–20 mm long, strongly recurved, then elevated as the anthers and petals wilt; **stigma** 4-lobed, white, the lobes recurved and together *c.* 6 mm wide. **FRUIT a capsule**, 3–10 cm long by 2.5–3 mm wide, 4-valved, with a shallow groove along each valve, opening at the blunt apex, hairs appressed and ascending; **seeds** numerous, brown, 1–1.4 mm long by 0.3–0.4 mm wide by *c.* 0.2 mm thick, in two rows under each valve, each pair of rows separated by a longitudinal ridge; **coma** of white hairs 9–13 mm long, easily removed.

■ **LEAVES** Alternate to almost in whorls of four, simple, entire to finely toothed, spreading; **blades** pointed, 3.5–20 cm long by 0.5–4 cm wide, reduced and bractlike above and below, flat, dull green above, lighter green below, margins slightly undulate, veins raised and sometimes hairy below; **upper blades** with short, leafy branches to *c.* 8 cm long from axils; **petioles** 0–4 mm long.

■ **STEM** Erect, glabrous to hairy above, glaucous, round, simple to branched above, solid; naked near the 2–8 mm thick base.

■ **SYN.** *Epilobium angustifolium* L. = *Chamerion angustifolium* ssp. *angustifolium* (L.) Holub.

Evening-primrose—*Onagraceae* **Wildflower** Pink to white Corolla tubular, 4-lobed

NORTHERN WILLOWHERB *Epilobium ciliatum* Raf.

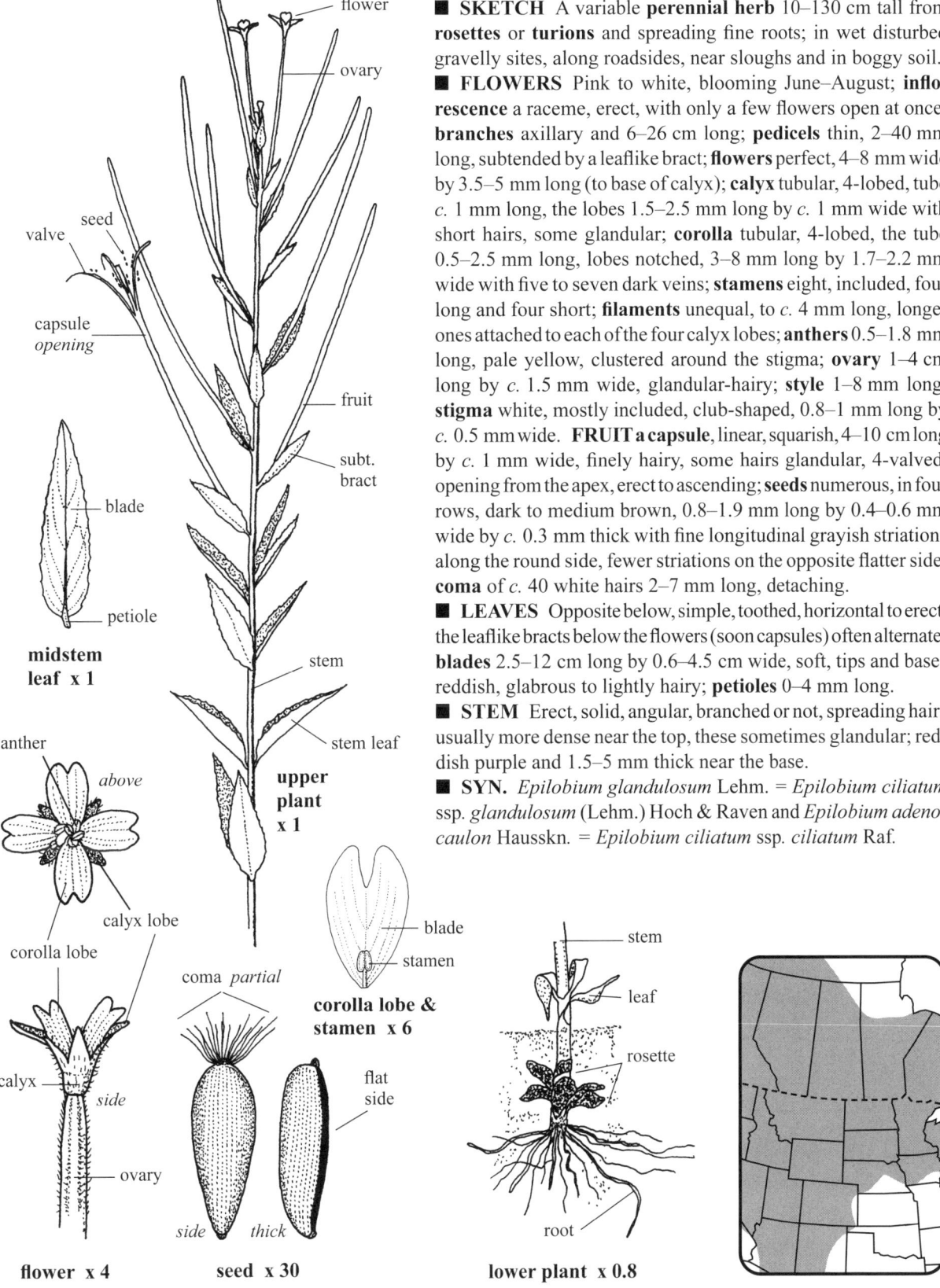

■ **SKETCH** A variable **perennial herb** 10–130 cm tall from **rosettes** or **turions** and spreading fine roots; in wet disturbed gravelly sites, along roadsides, near sloughs and in boggy soil.
■ **FLOWERS** Pink to white, blooming June–August; **inflorescence** a raceme, erect, with only a few flowers open at once; **branches** axillary and 6–26 cm long; **pedicels** thin, 2–40 mm long, subtended by a leaflike bract; **flowers** perfect, 4–8 mm wide by 3.5–5 mm long (to base of calyx); **calyx** tubular, 4-lobed, tube *c.* 1 mm long, the lobes 1.5–2.5 mm long by *c.* 1 mm wide with short hairs, some glandular; **corolla** tubular, 4-lobed, the tube 0.5–2.5 mm long, lobes notched, 3–8 mm long by 1.7–2.2 mm wide with five to seven dark veins; **stamens** eight, included, four long and four short; **filaments** unequal, to *c.* 4 mm long, longer ones attached to each of the four calyx lobes; **anthers** 0.5–1.8 mm long, pale yellow, clustered around the stigma; **ovary** 1–4 cm long by *c.* 1.5 mm wide, glandular-hairy; **style** 1–8 mm long; **stigma** white, mostly included, club-shaped, 0.8–1 mm long by *c.* 0.5 mm wide. **FRUIT a capsule**, linear, squarish, 4–10 cm long by *c.* 1 mm wide, finely hairy, some hairs glandular, 4-valved, opening from the apex, erect to ascending; **seeds** numerous, in four rows, dark to medium brown, 0.8–1.9 mm long by 0.4–0.6 mm wide by *c.* 0.3 mm thick with fine longitudinal grayish striations along the round side, fewer striations on the opposite flatter side; **coma** of *c.* 40 white hairs 2–7 mm long, detaching.
■ **LEAVES** Opposite below, simple, toothed, horizontal to erect, the leaflike bracts below the flowers (soon capsules) often alternate; **blades** 2.5–12 cm long by 0.6–4.5 cm wide, soft, tips and bases reddish, glabrous to lightly hairy; **petioles** 0–4 mm long.
■ **STEM** Erect, solid, angular, branched or not, spreading hairs usually more dense near the top, these sometimes glandular; reddish purple and 1.5–5 mm thick near the base.
■ **SYN.** *Epilobium glandulosum* Lehm. = *Epilobium ciliatum* ssp. *glandulosum* (Lehm.) Hoch & Raven and *Epilobium adenocaulon* Hausskn. = *Epilobium ciliatum* ssp. *ciliatum* Raf.

Fringed Willowherb, Purple-leaved Willowherb

Evening-primrose—*Onagraceae* **Wildflower** White, turning pink Petals 4

Scarlet Gaura *Gaura coccinea* Nutt. ex Pursh

■ **SKETCH** A hairy to almost hairless, tufted **perennial herb** 15–60 cm tall from a thick **taproot** and branched **caudex**, in colonies; on dry sandy prairies and slopes, open wooded hillsides, along roadsides, trail edges, stream valleys and slightly disturbed sites. **p2**

■ **FLOWERS** White, turning pink to red when fading, blooming April–September; **inflorescence** a spike, 5–50 cm long by 2.5–5 cm wide; **branches** alternate, hairy and leafy, each branch subtended by a leaf; **subtending bracts** (of flowers) one, leaflike, entire, 3–7 mm long by 0.5–2 mm wide, pointed, enlarging with fruit; **flowers** perfect, sessile, spreading, 15–17 mm wide (including stamens) by 15–25 mm long; **rachis** hairy, ridged; **sepals** four, reflexed, 5–10 mm long by *c.* 1 mm wide, involute, entire, united at first into two parts, glabrous inside, hairy on outside; **floral tube** 4–12 mm long by 1–1.5 mm wide, veined, hairy inside and sometimes outside; **petals** four, 3–8 mm long by 2–5 mm wide (flattened), oval to roundish, claw 2–3 mm long; **stamens** eight, exserted, opposite petals and sepals; **filaments** pale green, 3–7 mm long, filiform, hairy on the inner side; **anthers** reddish, 2.5–5 mm long, slightly curved, attached near their base; **ovary** hairy, 4–5.5 mm long at anthesis, 4-ridged, square in cross-section, *c.* 1 mm wide per side; **style** exserted, 7–20 mm long, glabrous above; **stigma** 4-lobed, light green, sticky, 1.5–2 mm wide by *c.* 1 mm long. **FRUIT a capsule**, 4-angled, thick-walled, nutlike, indehiscent, medium brown, TL 4–8 mm by 3–4 mm wide and thick, constricted into a thick stipe about half as long as the body, deciduous before ripe, hairs short, ascending and white; **seeds** one to four per capsule, each 1.8–2.2 mm long by 1–1.3 mm wide by *c.* 1 mm thick, glabrous, tan to reddish brown, extending partly into wide stipe.

■ **LEAVES** Alternate, simple, entire to slightly toothed, ascending, sessile, pointed; **blades** 5–50 mm long by 1–10 mm wide, margins flat to undulate, hairy on both sides, midrib pale, lateral veins obscure.

■ **STEM** Erect to decumbent, pale green, several in a clump, sometimes branched; 3–5 mm thick near the glabrous to hairy base.

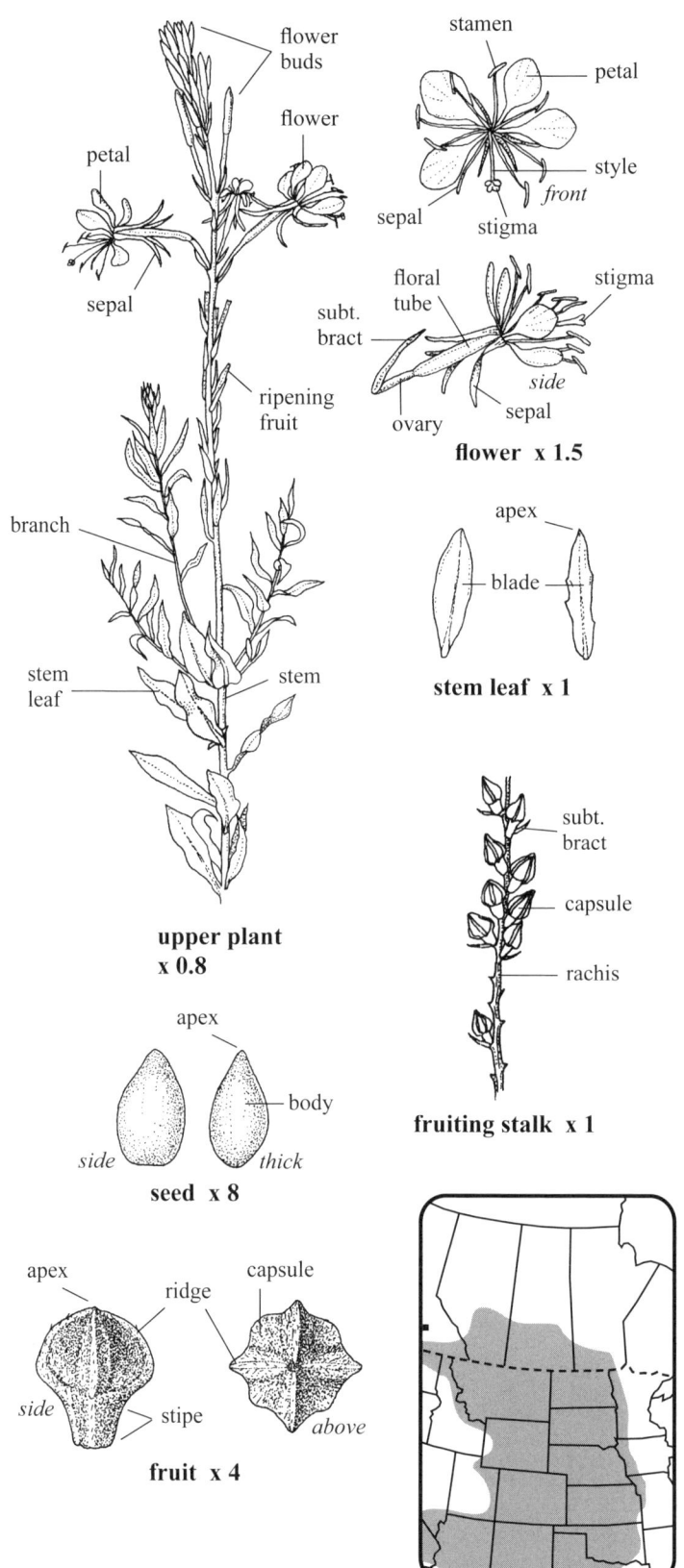

Scarlet Beeblossom, Scarlet Butterflyweed

Evening-primrose—*Onagraceae* Wildflower Yellow Petals 4

Yellow Evening-primrose   *Oenothera biennis* L.

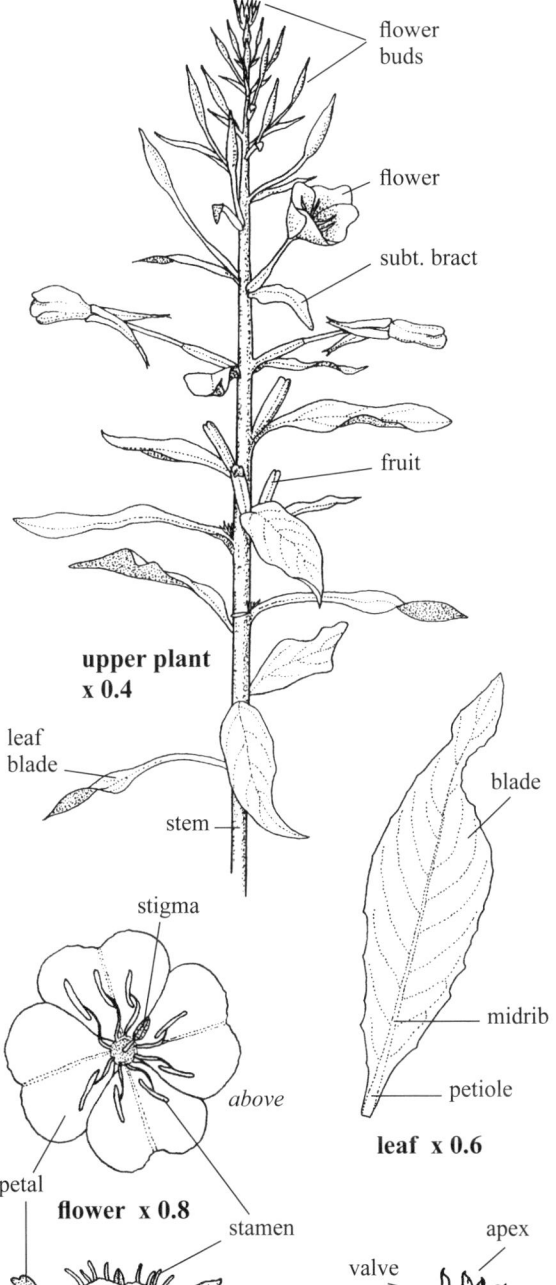

upper plant x 0.4
leaf blade
stem
stigma
above
petal
flower x 0.8
stamen
side
floral tube
ovary
paired sepals
subt. bract
flower buds
flower
subt. bract
fruit
blade
midrib
petiole
leaf x 0.6

■ **SKETCH** A **biennial herb**, usually solitary, 50–200 cm tall from a fleshy **taproot** 4–25 mm thick and to *c.* 20⁺ cm long; in open woodlands, along the shores of rivers, streams and lakes, along roadsides, open slopes and railways, in sandy to gravelly sites. *Naturalized* **p1**
■ **FLOWERS** Yellow, blooming June–October; **inflorescence** a spike, terminal, 20–70 cm long, bracteate; blooming sequence bottom to top; **subtending bracts** (of flowers) one, spreading, entire, lanceolate, 1–5 cm long, reduced above; **flowers** perfect, sessile, crowded above, opening late in the day, 3–4.5 cm wide by 3–6 cm long; **floral tube** 2–5 cm long, lightly to densely hairy, some hairs glandular; **sepals** four, light green, in two pairs, each pair united except for 1–4 mm at the pointed tips, reflexed or twisted, margins hairy, each 10–25 mm long by 4–5 mm wide; **petals** four, 15–25 mm long by 20–25 mm wide, roundish, glabrous and overlapping, slightly notched; **stamens** eight, yellow; **filaments** 9–12 mm long; **anthers** 9–10 mm long; **ovary** hairy, *c.* 15 mm long by *c.* 3 mm wide, cylindrical; **stigma** light green, 5–8 mm long, 4-cleft. **FRUIT a capsule**, tan, cylindrical, 1.5–4 cm long by 4–7 mm wide, 4-valved, opening at apex, glabrous to hairy; **seeds** numerous, in two rows in the four compartments, medium brown, irregularly shaped and sized, 1.3–2 mm long by *c.* 1 mm wide and thick.
■ **LEAVES** A rosette the first year; **second year stem leaves** alternate, simple, entire or teeth shallow; **blades** 5–30 cm long by 0.7–7 cm wide, larger below, often twisted or wavy, glabrous above, edges and lower surface downy, light green below, midrib slightly reddish; **petioles** 0–20 mm long, reduced above; **stipules** absent.
■ **STEM** Erect, round, stout, hollow, green to reddish with appressed reddish hairs; 2–25 mm thick near the base; **branches** if present, alternate and ascending, 10–70 cm tall, not as long as the central stalk.

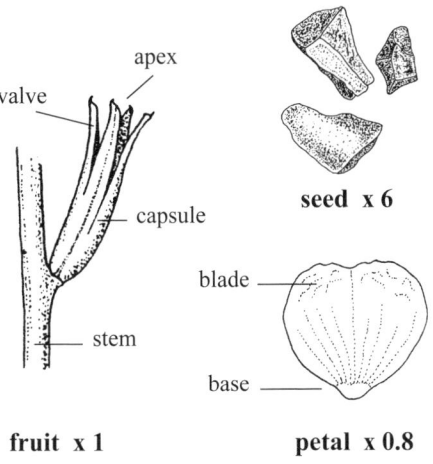

apex
valve
capsule
stem
fruit x 1

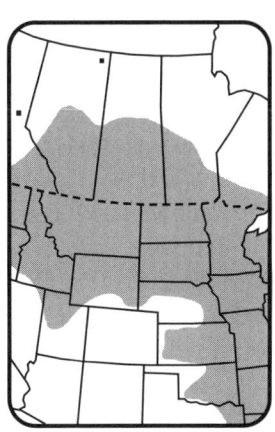

seed x 6
blade
base
petal x 0.8

Common Evening Primrose, Evening Primrose, King's-cureall

Evening-primrose—*Onagraceae*   **Wildflower** White Petals 4

WHITE EVENING-PRIMROSE   *Oenothera nuttallii* Sweet

■ **SKETCH** A **perennial herb** 30–100 cm tall with white fleshy **rootstocks** and a **caudex**, often spreading by adventitious shoots from **lateral roots**; along roadsides, field edges, sandy prairies, open wooded hillsides and stream valleys. **p2**

■ **FLOWERS** White, blooming June–September; **inflorescence** spikelike, open, axillary from upper leaves of branches; **flower buds** drooping, 2–3 cm long by 4–5 mm wide, shiny, pale green to pinkish; **flowers** perfect, smelly, sessile, 3.5–4.5 cm wide by 2.5–3 cm tall (to base of sepals), turning pink when fading, opening in evening for the night; **floral tube** pale green, 1.5–3.8 cm long by *c*. 1.5 mm wide, 4-ridged with many short spreading hairs, some gland-tipped; **sepals** four, entire, bent at 90° from the tube, pale green, each 2–3 cm long by *c*. 4 mm wide, united into two pairs, the tips split and distinct for 1–4 mm, glabrous inside, short spreading hairs, some gland-tipped on the outside (dorsal); **petals** four, soft, 1.5–2.7 cm long by 1.5–2.3 cm wide, glabrous, light green and shiny at the base, veins obvious, apices slightly erose; **stamens** eight, erect, exserted; **filaments** white, filiform, 10–14 mm long, glabrous, attached at the base of the petals atop the floral tube; **anthers** pale yellow, 8–11 mm long by *c*. 1 mm wide, 2-lobed; **ovary** green, shiny, 16–20 mm long by *c*. 2 mm wide, 4-ridged, not twisted, some hairs gland-tipped; **style** one, 18–25 mm long by *c*. 0.5 mm wide, exserted between the stamens; **stigmas** four, rough, hairless, spreading, 4–9 mm long by *c*. 1 mm wide. **FRUIT a capsule**, ascending, 2–3 cm long by 2.5–3 mm thick, 4-ridged, sessile, opening from apex, tan, slightly hairy, one per leaf axil, septum dark brown; **seeds** in one row in each of the four locules, reddish brown, 1.9–2.2 mm long by *c*. 0.8 mm wide by *c*. 0.5 mm thick, ridged along one side, surface marked with tiny purplish dots, slightly shiny, one end winglike.

■ **LEAVES** Alternate, simple, entire to slightly toothed at apices, linear and tapered to a narrow base, dull, spreading to ascending, glabrous above, slightly hairy and lighter below, midrib pale green and raised below, lateral veins obscure; **blades** 2–10 cm long by 2–10 mm wide, small leafy fascicles from some leaf axils; **stipules** absent.

■ **STEM** Erect, solid, one to several from the 5–10 mm thick woody caudex; **branches** alternate; **bark** tan and shed in long strips to reveal the white stem below which becomes pale green near the flowers.

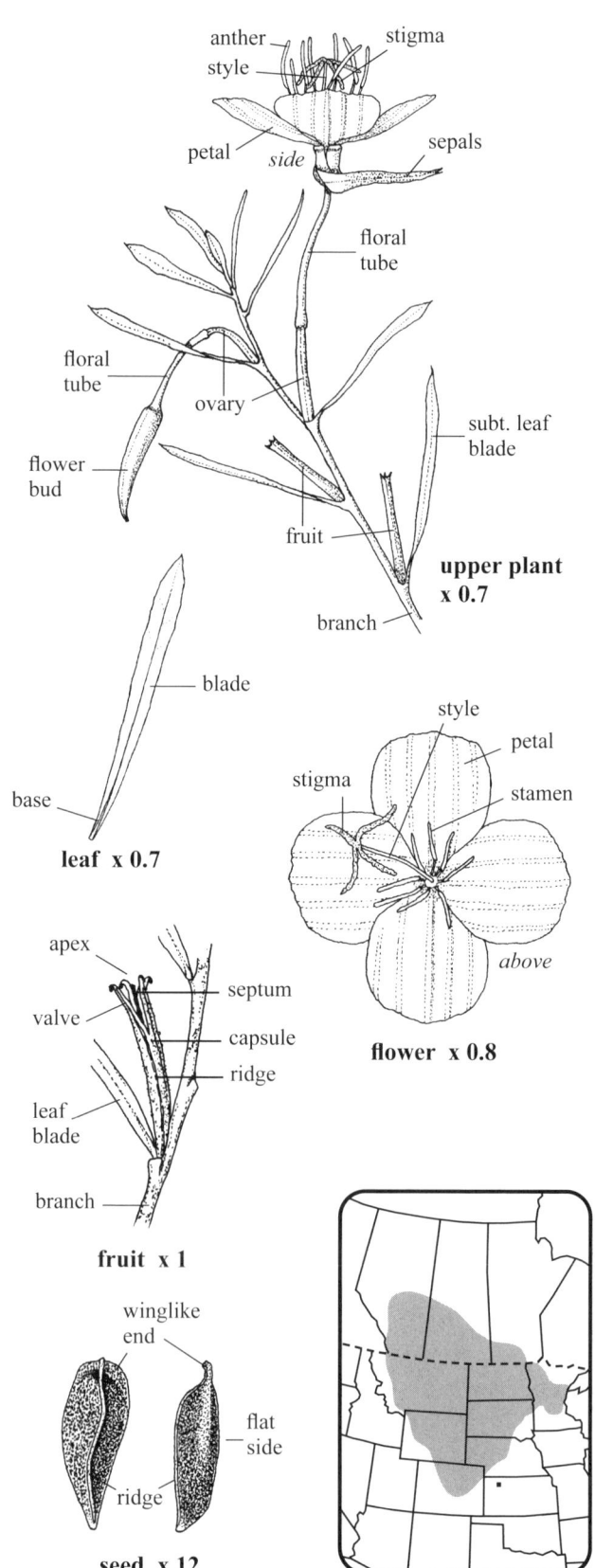

382   Nuttall's Evening-primrose, White-stemmed Evening Primrose

Wood-sorrel—*Oxalidaceae* **Wildflower Yellow Petals 5**

YELLOW WOOD-SORREL *Oxalis stricta* L.

■ **SKETCH** A variable **annual** or **perennial herb** 15–78 cm tall with a narrow white to pink **taproot** 5–12 cm long; in open woods, ravines, along roadsides, in shallow ditches, gardens and prairies. **w1**

■ **FLOWERS** Yellow, blooming April–October; **inflorescence** cymose, umbel-like, 1- to 7-flowered; **peduncles** green, axillary from upper leaves, 4–6.5 cm long, with short scattered ascending appressed hairs, erect with flower to spreading with erect fruit at their ends; **subtending bracts** (of pedicels) short, pointed; **pedicels** hairy, 5–22 mm long by 0.5–0.7 mm thick, curved with fruit, often with a pair of green bracteoles 1–2 mm long; **flowers** perfect, erect, 10–15 mm wide by *c.* 5 mm tall; **sepals** five, hairy, 4–7 mm long by 1–1.5 mm wide, ascending, apices hairy, some with long scattered hairs, appressed hairs at base, persistent with fruit; **petals** five, yellow, 5–11 mm long by 4–5 mm wide, glabrous, apices rounded; **pistils** of five carpels, 5–6 mm long, green, united in lower 2 mm, free above and microscopically hairy in upper half; **ovary** *c.* 2 mm long by *c.* 1 mm wide; **styles** five, filiform, 1–5 mm long, free, curved at apices, hairs appressed; **stigmas** green, no wider than styles; **stamens** 10, five long and five short, shorter than styles; **filaments** green, united below around the carpels, free part 1 and 2 mm long for the short and long stamens, hairless; **anthers** yellow, *c.* 0.4 mm long. **FRUIT** a **capsule**, erect, 8–20 mm long by 3–4 mm wide, hairless, finely hairy or with several long spreading hairs at the base, 5 sutured, valves splitting open while still green to reveal the seeds; **seeds** reddish brown, 25–35 per capsule, 1.2–1.4 mm long by *c.* 1 mm wide by 0.4–0.5 mm thick, lenticular with eight or nine transverse ridges on each side and a groove along the edges, body pointed at one end, enclosed in an aril; **aril** semi-transparent, 1.5–1.9 mm long by *c.* 1 mm wide, smooth outside, snaps open as it dries to expel the seed, open along one side only, rough textured when the aril turns inside out to eject the seed.

■ **LEAVES** Alternate, compound, trifoliate, green to purplish; **leaf blades** 3–4.5 cm long and wide; **leaflets** three, entire, 1–3.5 cm long and wide, usually reflexed to spreading, both sides dull, thin, notched, tapered to a narrow base, lighter green below, thinly veined, hairless but margins with hairs arched upward; **petioles** 3–6 cm long, filiform, with long scattered hairs near the base, short appressed hairs above and ascending; **petiolules** *c.* 0.5 mm long, hairy; **stipules** absent.

■ **STEM** Erect to leaning, tangled, weak, green, hairy above with short white appressed hairs arched upward, stem often branched below, not rooting at lower nodes; reddish near the 1–3 mm wide base.

■ **SYN.** *Oxalis europaea* Jord.

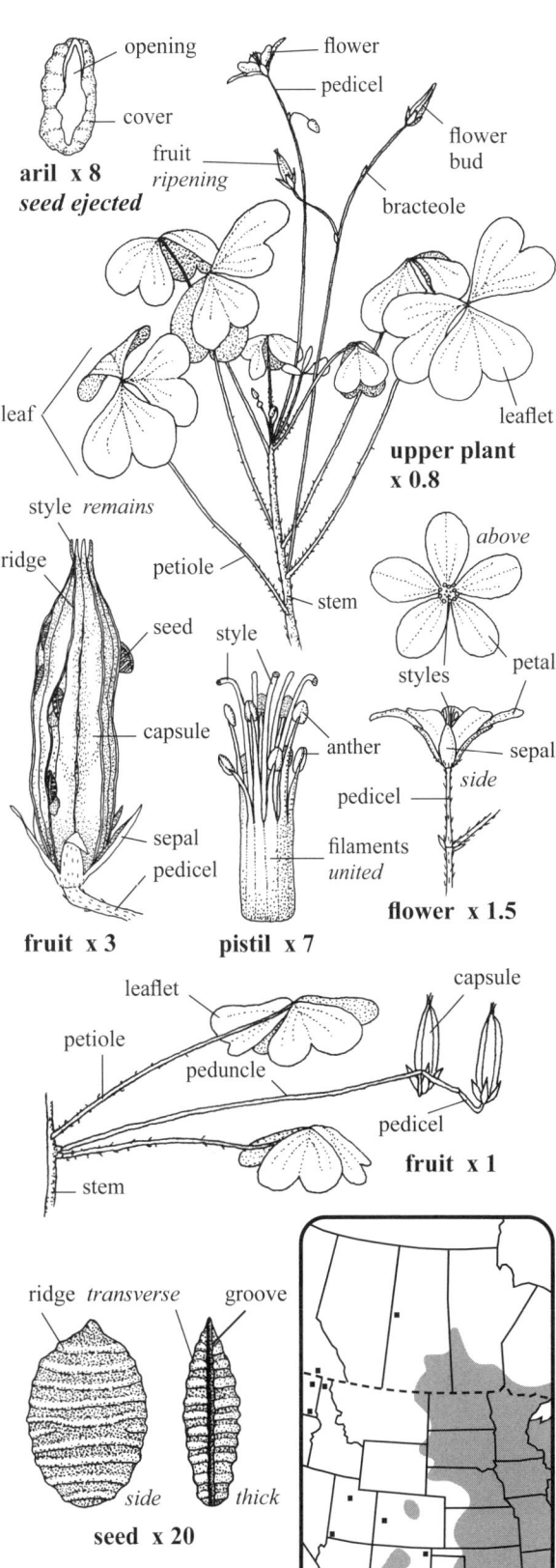

Upright Yellow Wood-sorrel, Common Yellow Wood-sorrel

Plantain—*Plantaginaceae* **Wildflower** Green Corolla tubular, 4-lobed

SALINE PLANTAIN *Plantago eriopoda* Torr.

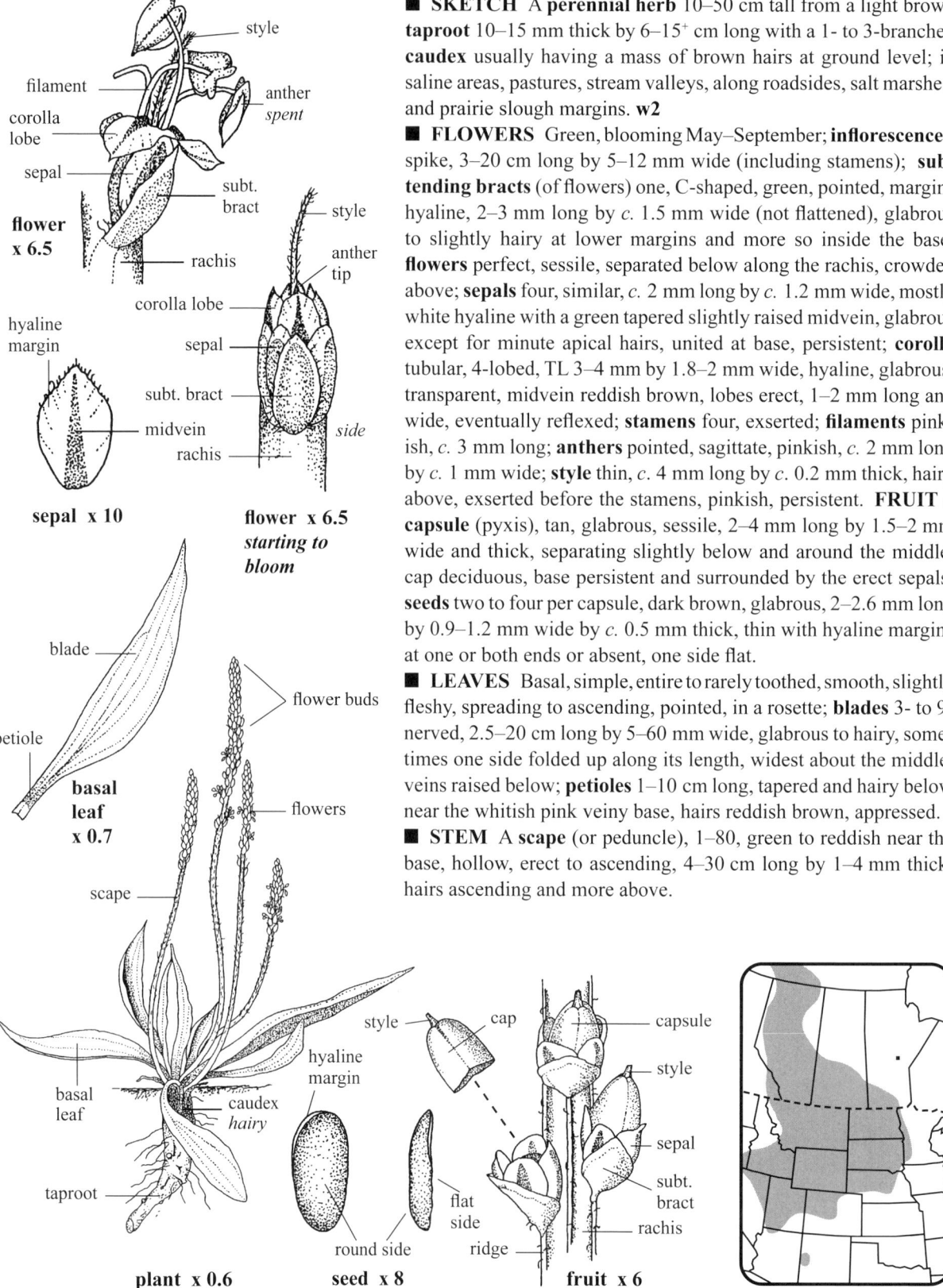

■ **SKETCH** A **perennial herb** 10–50 cm tall from a light brown **taproot** 10–15 mm thick by 6–15⁺ cm long with a 1- to 3-branched **caudex** usually having a mass of brown hairs at ground level; in saline areas, pastures, stream valleys, along roadsides, salt marshes, and prairie slough margins. **w2**

■ **FLOWERS** Green, blooming May–September; **inflorescence** a spike, 3–20 cm long by 5–12 mm wide (including stamens); **subtending bracts** (of flowers) one, C-shaped, green, pointed, margins hyaline, 2–3 mm long by *c.* 1.5 mm wide (not flattened), glabrous to slightly hairy at lower margins and more so inside the base; **flowers** perfect, sessile, separated below along the rachis, crowded above; **sepals** four, similar, *c.* 2 mm long by *c.* 1.2 mm wide, mostly white hyaline with a green tapered slightly raised midvein, glabrous except for minute apical hairs, united at base, persistent; **corolla** tubular, 4-lobed, TL 3–4 mm by 1.8–2 mm wide, hyaline, glabrous, transparent, midvein reddish brown, lobes erect, 1–2 mm long and wide, eventually reflexed; **stamens** four, exserted; **filaments** pinkish, *c.* 3 mm long; **anthers** pointed, sagittate, pinkish, *c.* 2 mm long by *c.* 1 mm wide; **style** thin, *c.* 4 mm long by *c.* 0.2 mm thick, hairy above, exserted before the stamens, pinkish, persistent. **FRUIT** a **capsule** (pyxis), tan, glabrous, sessile, 2–4 mm long by 1.5–2 mm wide and thick, separating slightly below and around the middle, cap deciduous, base persistent and surrounded by the erect sepals; **seeds** two to four per capsule, dark brown, glabrous, 2–2.6 mm long by 0.9–1.2 mm wide by *c.* 0.5 mm thick, thin with hyaline margins at one or both ends or absent, one side flat.

■ **LEAVES** Basal, simple, entire to rarely toothed, smooth, slightly fleshy, spreading to ascending, pointed, in a rosette; **blades** 3- to 9-nerved, 2.5–20 cm long by 5–60 mm wide, glabrous to hairy, sometimes one side folded up along its length, widest about the middle, veins raised below; **petioles** 1–10 cm long, tapered and hairy below near the whitish pink veiny base, hairs reddish brown, appressed.

■ **STEM** A **scape** (or peduncle), 1–80, green to reddish near the base, hollow, erect to ascending, 4–30 cm long by 1–4 mm thick, hairs ascending and more above.

384 Alkali Plantain, Redwool Plantain

Plantain—*Plantaginaceae* **Wildflower** Green Corolla tubular, 4-lobed

Common Plantain *Plantago major* L.

■ **SKETCH** A **perennial herb** (some blooming first year) 10–55 cm tall with a short, hairless **caudex** 0.7–2 cm thick with numerous, tan, **fibrous roots** 5–15 cm long by *c.* 1 mm thick; in lawns, pastures, along roadsides, farmyards, ditches and fields. *Naturalized* **w2**

■ **FLOWERS** Green, blooming March–November; **inflorescence** a spike, dense but open near the base, 8–33 cm long by 2–5 mm wide (to 10 mm wide in fruit); **peduncles** 1–25+, tough, 4–28 cm long by 1–4 mm thick, glabrous to slightly hairy; **subtending bracts** (of flowers) one, 1.5–2.3 mm long, pointed, similar to but usually shorter than the sepals; **flowers** numerous, perfect, sessile, 2–3 mm wide by 4.5–5.6 mm long; **sepals** four, 1.6–2.5 mm long with a wide green midvein (keel) and white borders, glabrous, persistent; **corolla** tubular, 4-lobed, pointed, these ascending at first around the base of the style, then reflexed with the appearance of the stamens, the lobes *c.* 1 mm long, tan, and membranous; **stamens** four, exserted; **filaments** white, *c.* 3 mm long; **anthers** purple and yellow, *c.* 1 mm long, horned; **style** one, thin, hairy, exserted. **FRUIT** a **capsule** (pyxis), 2–4 mm long by 1–2 mm wide, the cap separating near the middle; **seeds** 8–25 per capsule, golden to dark brown, shiny, irregular in shape, 0.8–1.6 mm long by 0.8–1 mm wide by *c.* 0.5 mm thick, finely ridged, sticky when wet.

■ **LEAVES** Basal, simple, entire to slightly toothed, a **rosette** 20–60+ cm across; **blades** ascending to spreading, glabrous to slightly hairy, 3.5–20 cm long by 2–15 cm wide, abruptly tapered to the petiole, margins wavy, five to nine main parallel veins, these raised below and continue along the rounded underside of the petiole; **petioles** tough and stringy, 1.5–20 cm long by 4–8 mm wide and thick, deeply grooved above, green to reddish.

■ **STEM** Absent, or a peduncle.

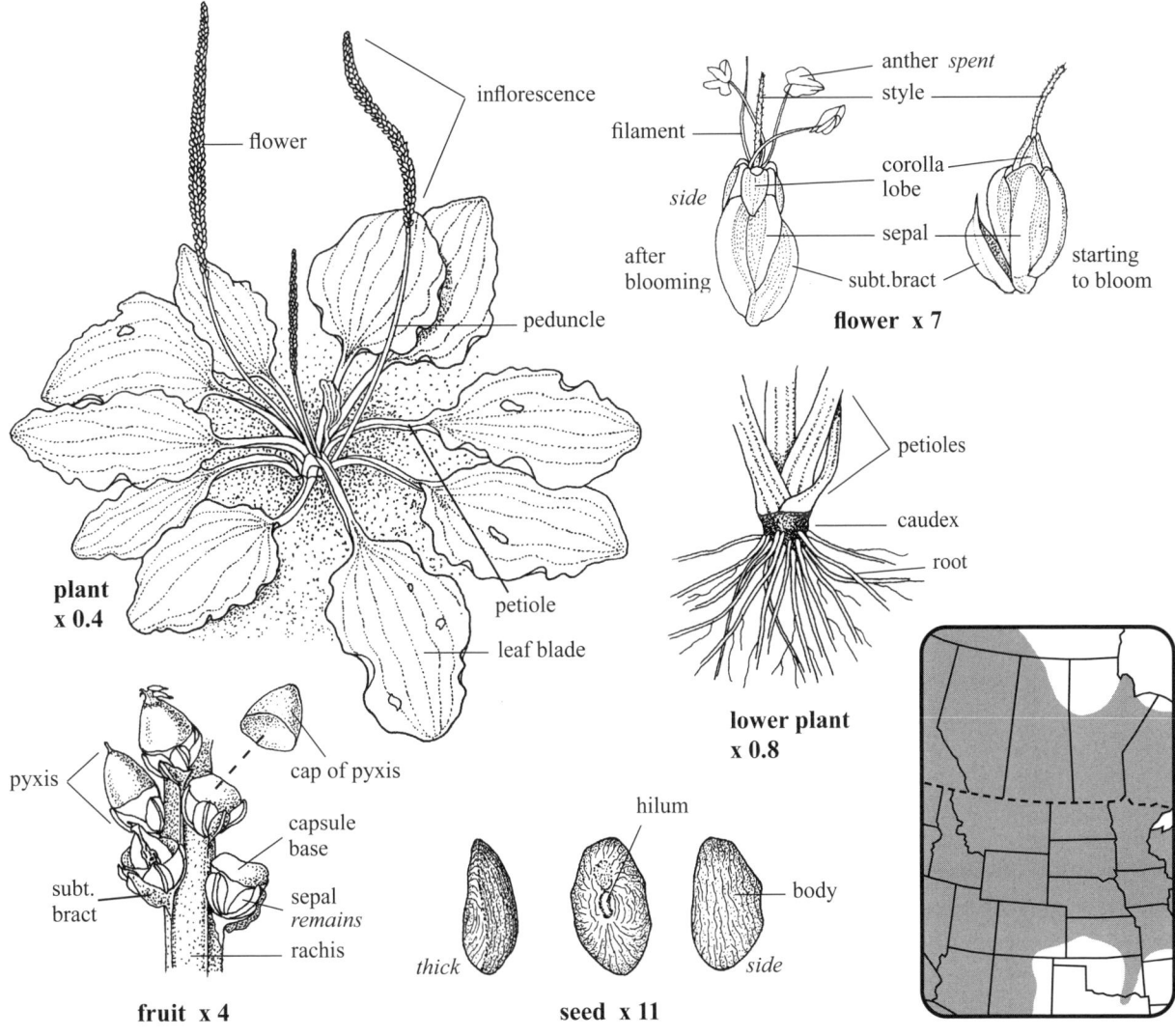

385

Phlox—*Polemoniaceae* **Wildflower** White Corolla tubular, 5-lobed

Moss Phlox *Phlox hoodii* Richards.

■ **SKETCH** A tufted, compact **perennial herb** 2–7 cm tall from a **woody taproot** 1–5 mm thick and 3–15+ cm long; covering the ground like moss, in open prairies, woods, on sandy to rocky slopes, in desert shrub and often eroded dry sites.

■ **FLOWERS** White, some with a yellow throat, blooming April–July; **inflorescence** terminal and solitary, one flower per reproductive branch; **branches** and leafy shoots usually opposite, 1–4 cm long, somewhat hairy to glabrous above; **pedicels** 0–4 mm long; **subtending bracts** (of flowers) leaflike, pointed, margins hairy; **flowers** perfect, 10–12 mm wide; **calyx** TL 4–7 mm by *c.* 2 mm wide at the tube, some glandular-hairy; **calyx lobes** five, 2.5–3 mm long by *c.* 0.8 mm wide at the base, tapered to a point, slightly spreading and darker green than the tube, cottony tangled hairs in the lower inside, upper half roundish in cross-section, tips brown; **corolla tube** 3.5–8 mm long by *c.* 1 mm wide, glabrous, widest below the lobes; **corolla lobes** five, white, entire, spreading to slightly ascending, 3–5 mm long by 1.5–4 mm wide, glabrous, apices mucronate; **stamens** five, included, unequal in length; **filaments** pale yellow, attached to the corolla tube, upper 0.5–1 mm free; **anthers** yellow, *c.* 1 mm long, between the corolla lobes; **pistil** pale yellow, included, 3–8 mm long, glabrous; **ovary** oval, *c.* 1 mm long; **style** one, filiform, 1.5–6 mm long; **stigmas** three, each part *c.* 1 mm long, filiform, usually not spreading. **FRUIT a capsule**, oval, yellowish tan, hairless, 3–3.8 mm long by *c.* 2 mm wide, 3-valved, thick-walled, opening from apex, pointed from style remains; **seeds** three to six per capsule.

■ **LEAVES** Mostly opposite, simple, entire, linear, rigid, sessile, thickish, pointed, prickly at apices; **blades** 4–10 mm long by 0.7–1 mm wide, cottony hairs along the lower margins, flat near the base, becoming roundish near the apices, ascending to erect, crowded, imbricate, often with smaller leaves in axils of larger leaves, sometimes glandular-hairy.

■ **STEM** Brown, slightly scaly from the old leaves; *c.* 1 mm thick near the base.

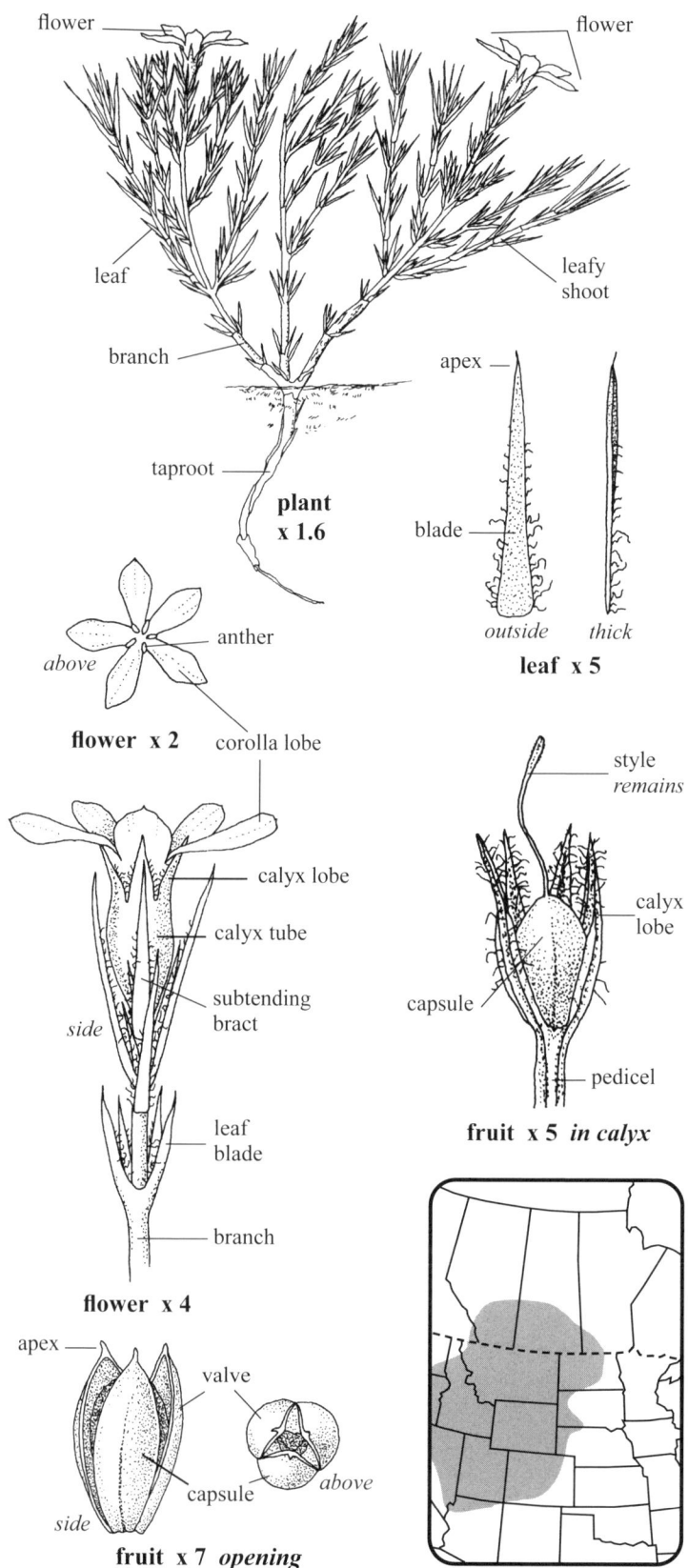

Carpet Phlox, Spiny Phlox, Hood's Phlox

Milkwort—*Polygalaceae* **Wildflower** White Corolla tubular, 3-lobed

Seneca Snakeroot *Polygala senega* L.

■ **SKETCH** A **perennial herb** 10–50 cm tall, singly or in a cluster of arched stems from a thick **crown** and woody, thick, branching **root** system; in prairies, edges of bluffs and open woodlands. p2

■ **FLOWERS** White, blooming May–July; **inflorescence** a raceme (spikelike), solitary, 1.5–6 cm long by 6–12 mm wide, compact and cylindrical; **pedicels** 0.2–0.5 mm long, ascending, glabrous; **flowers** perfect, 4.5–6.5 mm wide by 3.5–4 mm long, most closed at any one time; blooming sequence bottom to top; **sepals** five, distinct, glabrous, the lower two *c.* 1.3 mm long by *c.* 0.5 mm wide, whitish green; the lateral two "wings" *c.* 3 mm long by *c.* 2 mm wide, colored like the corolla, the sepal in back of the flower is pointed and *c.* 2 mm long by *c.* 1 mm wide, not extending past the flower, persistent around the fruit; **corolla** 3-lobed, tubular, the two back lobes erect, imbricate and clawed, 2.5–3 mm long by *c.* 1.2 mm wide, glabrous, pointed, entire to toothed, the tips folded forward and incurved; **middle lobe** often crested on the back and with finger-like appendages which spread at anthesis; **stamens** *c.* 0.6 mm long, usually eight, united to the corolla tube in two rows; **anthers** yellow, about half the stamen length; **pistil** green, included, glabrous, *c.* 1.5 mm long; **ovary** dark green, *c.* 0.6 mm wide and *c.* 0.2 mm thick, flattened; **stigma** *c.* 0.6 mm long and slightly thicker than the ovary. **FRUIT a capsule**, plump, rounded, 2.5–4.2 mm long by *c.* 3.3 mm wide by *c.* 2 mm thick, reddish brown, thin-walled, soft, easily opened, with a complete septum between the seeds; **seeds** two per capsule, 2.5–3.5 mm long by *c.* 1.3 mm wide by *c.* 1.4 mm thick, hairy, black, shiny, pitted; **aril** creamy white, thick, almost as long as the seed's body.

■ **LEAVES** Alternate, simple, entire, ascending; **blades** 0.4–8 cm long by 4–30 mm wide, reduced below, glabrous or with microscopic hairs; **petioles** 1–2 mm long; **stipules** absent.

■ **STEM** Round, green to reddish, arched below and ascending, unbranched, one to *c.* 30 in a cluster, glabrous to minutely hairy below; 1–2 mm thick near the base with its tiny, reddish purple bracts.

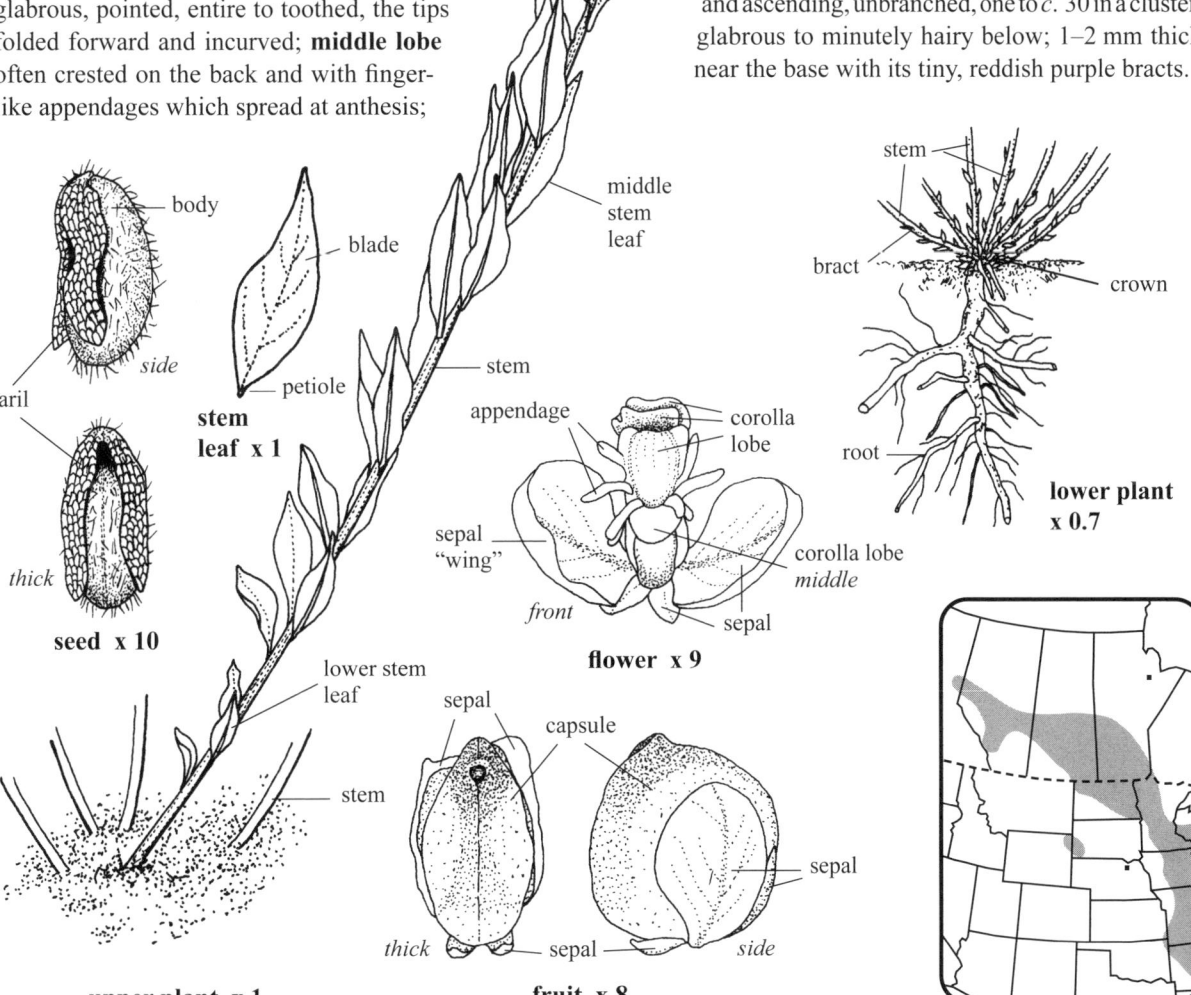

Seneca Root

Family Characteristics

Buckwheat—Polygonaceae   Flowers 3- to 6-lobed

SKETCH Dicotyledonous herbs (wildflowers or vines), annual or perennial, sometimes woody at base. Worldwide about 1,200 species.

FLOWERS Mostly perfect, or unisexual, regular, small, greenish to slightly reddish, numerous, loose to crowded into obvious clusters or whorls; the bases of buds and jointed pedicels enclosed in membranous bracts (ocreolae). **Inflorescence** a panicle or raceme, rarely umbel-like. **Perianth** sepal-like, united below, 3- to 6-lobed, these persistent and generally enclosing all or most of the fruit, sometimes in two whorls, enlarging and forming veiny valves with or without a tubercle. **Stamens** 4–9, free or inserted on base of perianth, usually include, sometimes exserted. **Pistil** 1. **Ovary** with 1 locule and 1 ovule. **Styles** 2 or 3, separate or united at base.

FRUIT An **achene**, 1-seeded, triangular or lenticular.

LEAVES Mostly alternate, sometimes all basal, simple, entire, some with two basal lobes, petiolate to sessile. **Stipules** forming a membranous tubular sheath (ocrea), often with apical hairs, around the stem above the node, or sometimes absent.

STEM Erect or even climbing, solid to hollow, glabrous to hairy. **Nodes** often swollen.

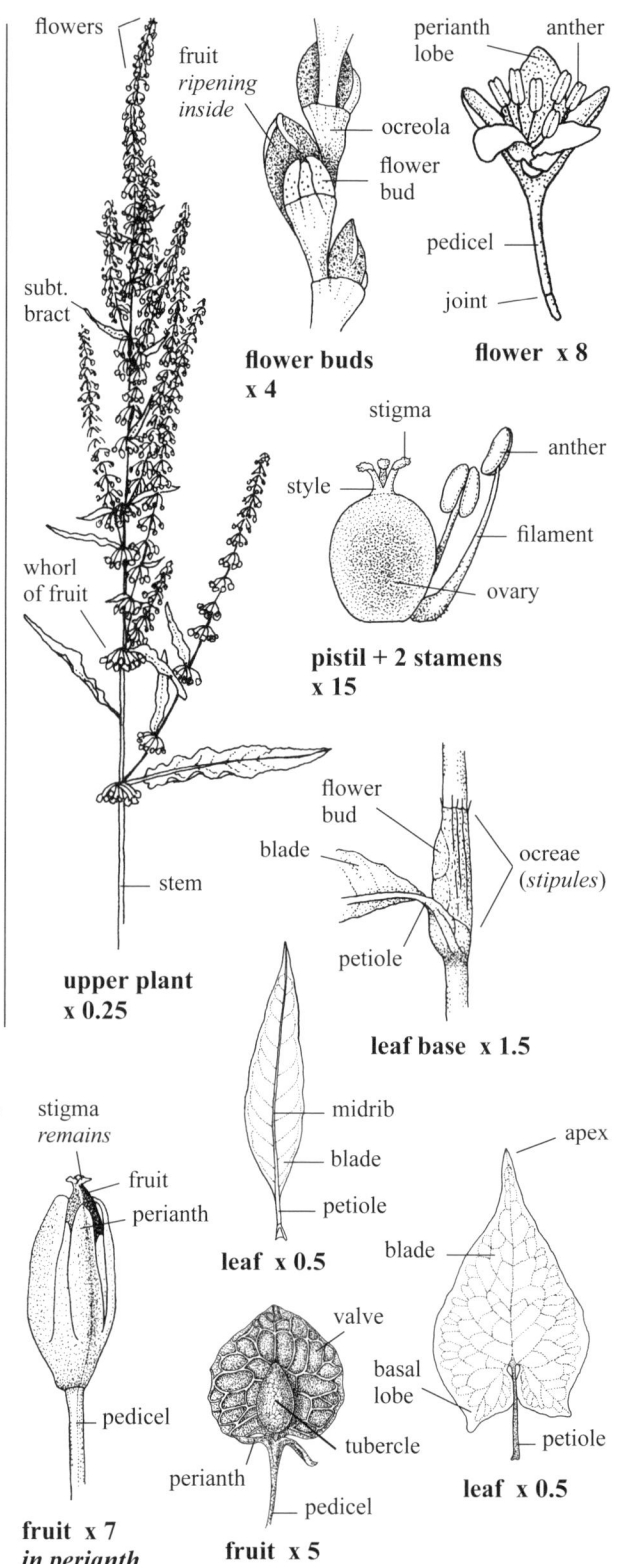

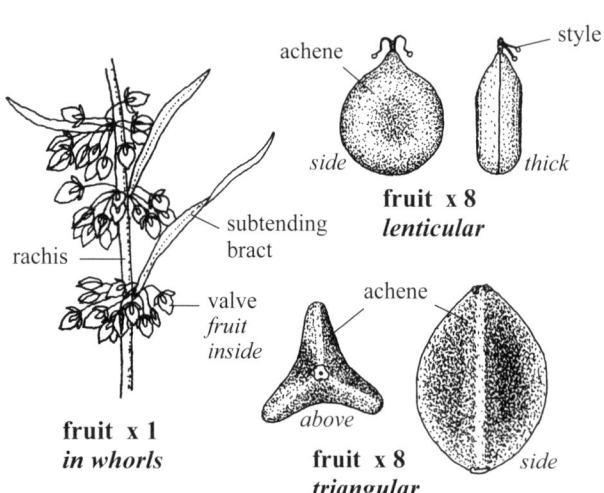

388

Buckwheat—*Polygonaceae* Wildflower Yellow Perianth tubular, 6-lobed

YELLOW UMBRELLAPLANT *Eriogonum flavum* Nutt.

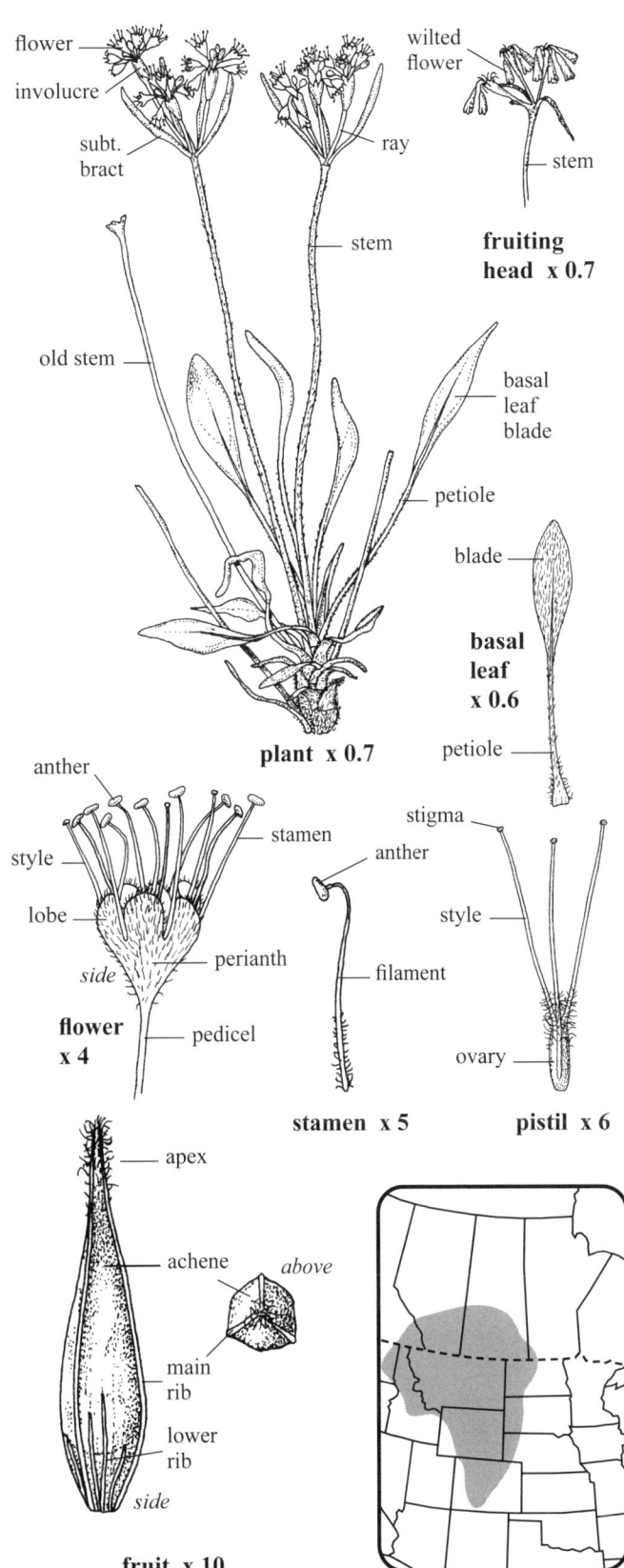

fruiting head x 0.7

basal leaf x 0.6

plant x 0.7

flower x 4

stamen x 5 pistil x 6

fruit x 10

■ **SKETCH** A variable **perennial herb**, silvery gray, softly hairy throughout, 5–30 cm tall in small clumps from a scaly **caudex** *c.* 10 mm thick; on dry eroded hillsides, badlands, alpine slopes, dry plains, canyons and cliff edges. **p2**

■ **FLOWERS** Yellow, blooming May–September; **inflorescence** umbellate with three to six rays; **umbels** 2–3.5 cm wide; **rays** 5–30 mm long, ascending, hairy; **subtending bracts** (of rays) one to six, 1–2.5 cm long by 2–7 mm wide, leaflike, hairy, entire, pointed, tapered to base, midrib visible, margins revolute; **involucre** hairy, erect, atop rays, tubular, 4–9 mm long by 3–5 mm wide, apical teeth five to eight or absent, *c.* 1 mm long, easily torn, bluntly pointed, keeled to base; **flowers** perfect, fragrant, 10–15 from each involucre, each flower 7–8 mm wide and tall; **perianth** tubular, 6-lobed, these *c.* 1.5 mm long and wide, incurved, TL 3.5–4.5 mm, 5–6 mm long in fruit, yellowish green, hairy on outside, glabrous and green inside; **stamens** nine, exserted, attached near base of perianth; **filaments** green, 6–7 mm long, hairy near base; **anthers** *c.* 0.7 mm long, yellowish green; **pistil** green, exserted; **ovary** green, bluntly triangular, 1.3–1.5 mm long by *c.* 0.5 mm wide, hairy at apex; **styles** three, coiled at first, then ascending, *c.* 5 mm long, filiform, hairy at base; **stigma** glabrous, green, slightly wider than style.

FRUIT an achene, 1-seeded, shiny, golden tan to brown, each 3–5.5 mm long by 1–1.3 mm wide and thick, tapered above, 3-sided, three main ribs with shorter ribs below, sides flat to convex, apex hairy.

■ **LEAVES** Basal, (rarely a pair of opposite leaves along the lower part of the stem), simple, entire, erect to ascending, often facing in one direction, persistent; **blades** 2–8 cm long by 3–15 mm wide, margins often revolute, white tomentose below (dorsal), dark green and slightly hairy above (ventral), apices pointed to blunt, bases tapered, midrib pale green and faint near apex; **petioles** hairy, 1–4 cm long, widest at base; **stipules** absent.

■ **STEM** Pale green, simple, erect, hairy, old stems persistent; 1–2 mm thick near the base.

Alpine Golden Wild Buckwheat, Yellow Wild Buckwheat

Buckwheat—*Polygonaceae*

Wildflower White Perianth tubular, 5-lobed

Garden Buckwheat *Fagopyrum esculentum* Moench

■ **SKETCH** An **annual herb** 10–90 cm tall with a white **taproot** several centimeters long; in cultivated fields, disturbed sites, roadsides and along railway tracks. *Naturalized* **w2**

■ **FLOWERS** White, blooming June–September; **inflorescence** paniclelike, terminal and axillary from upper leaves, TL 3–15 cm by 3–6 cm wide; **peduncles** ascending, 8–38 mm long, ciliate along the inner side, thinner and usually longer than the petioles; **pedicels** pale green, glabrous, 2–4 mm long, filiform, recurved in fruit; **ocreolae** membranous, 1.5–2 mm long, around the bases of pedicels, beneath the subtending bracts, whitish, very thin and easily torn; **flowers** perfect, congested, 5–6 mm wide by *c.* 3 mm tall; **subtending bracts** (of flowers) green, pointed, *c.* 3 mm long, enveloping about three flower buds; **perianth** tubular, 5-lobed, petal-like, spreading, greenish near the base, 2–5 mm long by 2–2.7 mm wide, two lobes with a raised keel below and *c.* 2 mm wide compared to the other three lobes which are flat and *c.* 2.5 mm wide; **stamens** eight, exserted, *c.* 2 mm long; **filaments** white, filiform, five outer ones between the perianth lobes, inner three opposite a lobe; **anthers** pink, 2-lobed, *c.* 0.7 mm long by *c.* 0.5 mm wide (in flower buds); **disc** yellowish green, *c.* 1 mm wide by *c.* 1.2 mm long with eight round, yellowish glands along its margin; **pistil** greenish yellow, exserted; **ovary** *c.* 0.5 mm long and wide, tapered, 3-ridged, hairless; **styles** three, erect, filiform, separate, 1–2.5 mm long, persistent in fruit; **stigmas** pale green, capitate, slightly wider than the styles. **FRUIT an achene**, 1-seeded, triangular, 5–7 mm long by 4–6 mm wide, sides slightly concave, shiny and dark brown, exserted from perianth.

■ **LEAVES** Alternate, entire, simple, margins slightly hairy, pointed, base 2-lobed, dull, lighter green below, veins reddish at base of blade; **blades** glabrous, 3–8 cm long by 2–8 cm wide, reduced above, horizontal to slightly descending below; **petioles** green to pink, 0–9 cm long, reduced above, ascending to arched, ciliate along the margins of the upper groove; **stipules** membranous, forming a tube around the stem, 2–8 mm long, apices slightly pointed.

■ **STEM** Erect but weak, unbranched or branched, glabrous to hairy, reddish throughout or at least below, some glaucous below; 2–5 mm thick near the base.

■ **SYN.** *Fagopyrum sagittatum* Gilib. and *Polygonum fagopyrum* L.

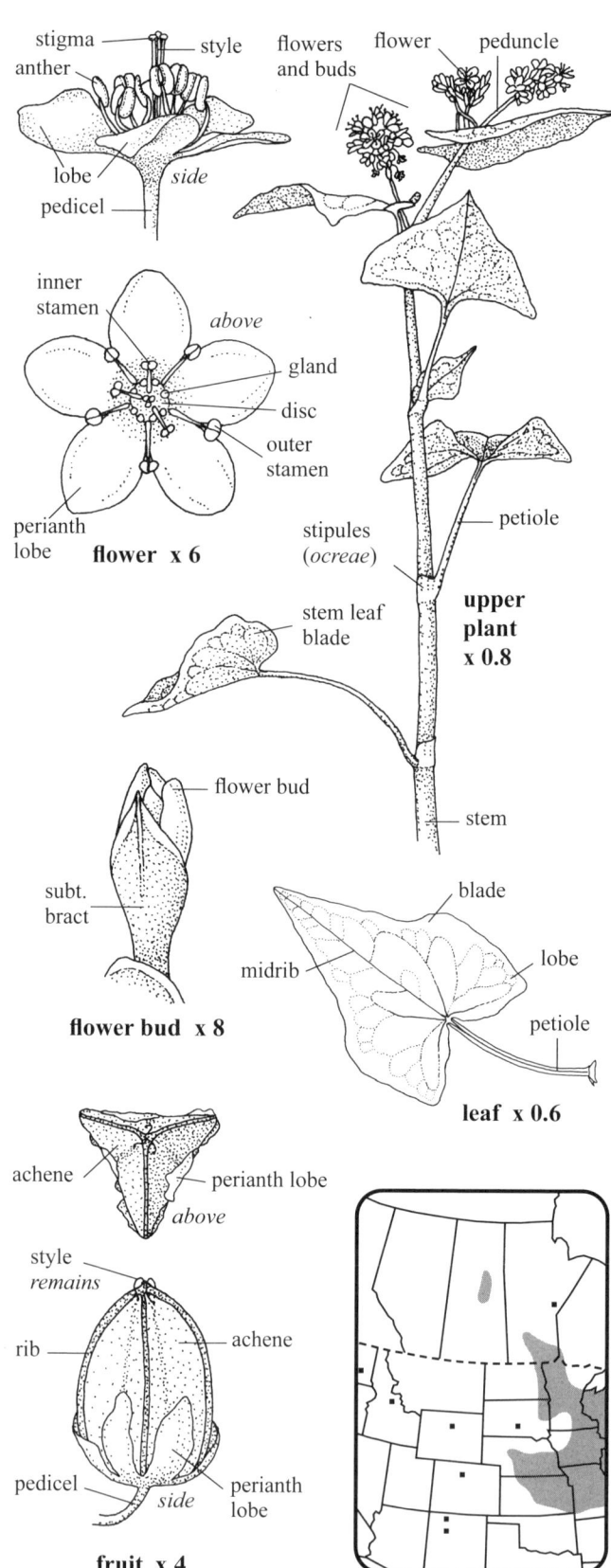

390 Buckwheat

Buckwheat—*Polygonaceae* **Wildflower** Green to pinkish Perianth tubular, 5-lobed (4-)

STRIATE KNOTWEED *Polygonum achoreum* Blake

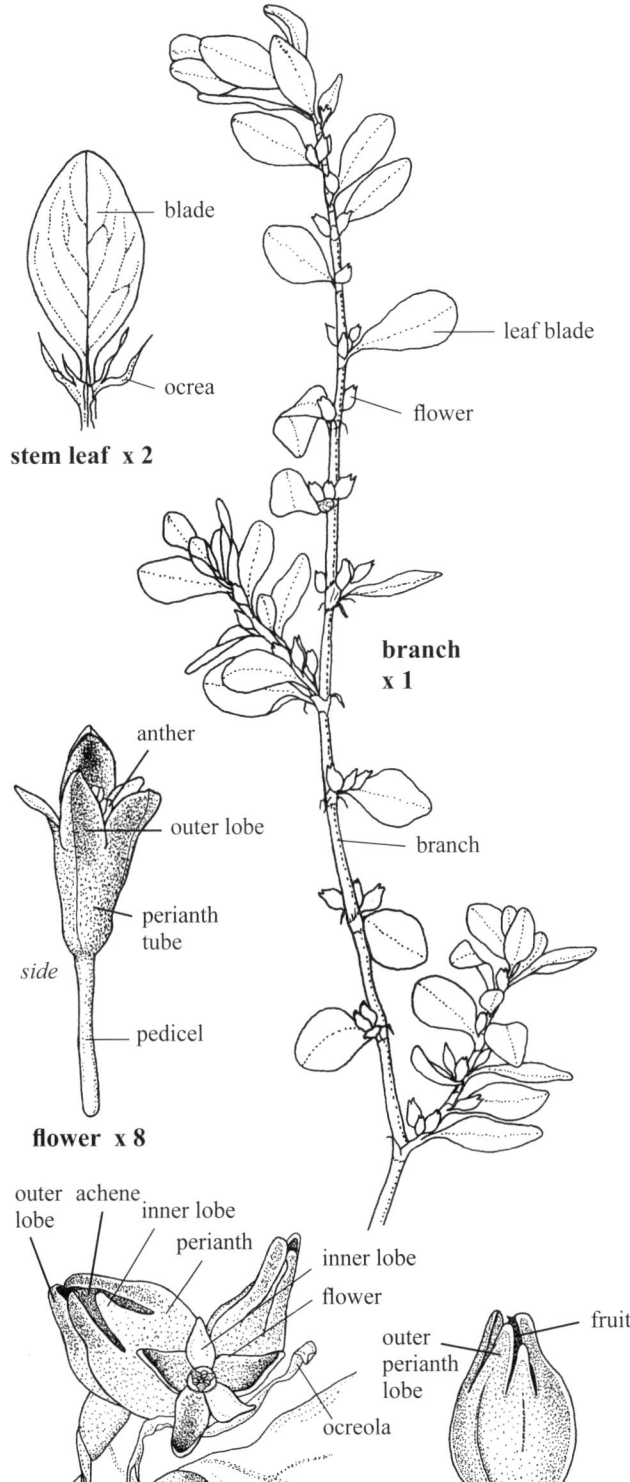

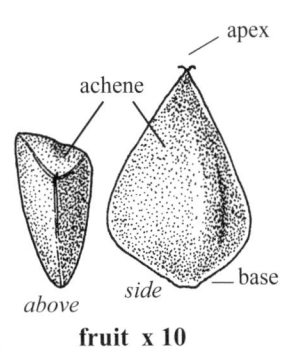

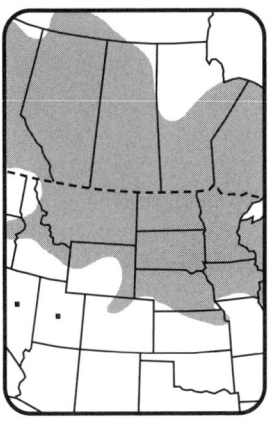

■ **SKETCH** An **annual herb** with several radiating branches to *c.* 70 cm long by *c.* 25 cm tall with the ends upturned, from a brown **taproot** 0.2–1 cm thick and to *c.* 10 cm long; in gardens, lawns, along walkways, in packed soil of cities, towns and farmyards. **w1**

■ **FLOWERS** Green to somewhat pinkish, blooming July–September; **inflorescence** of one to five flowers in axillary clusters throughout the plant; **pedicels** erect, glabrous, 1.3–2.5 mm long, enclosed in ocreae and erect in fruit; **bracts** (ocreolae) among the flowers, torn and stringy; **flowers** perfect, 2.5–3 mm wide by 2.5–3.5 mm long, not showy; **perianth** tubular, 5-lobed (4-), thick, persistent, erect and 3–4 mm long around the fruit, glabrous, united from the base to above the middle; **outer three lobes** slightly keeled and hooded, long, green with greenish white borders; **inner two lobes** short, flat and whitish; **stamens** three to eight, included; **filaments** whitish, wider at the base; **anthers** yellow; **stigmas** 2-lobed, included, about as high as the anthers. **FRUIT an achene**, 1-seeded, 3-sided, medium golden brown, glabrous, 2.5–3.5 mm long by 1.8–2 mm wide by *c.* 1 mm thick, the sides unequal and the narrow side barely concave, the other two sides convex.

■ **LEAVES** Alternate, simple, entire, numerous, glabrous, spreading, overlapping or not, little reduced above; **blades** 7–35 mm long by 4–15 mm wide, the tips mostly round, often covered with a grayish fungal bloom in late summer; **petioles** 0.3–5 mm long; **stipular sheathing** (ocrea) membranous and tubular, 5–12 mm long and often torn.

■ **STEM** Tough, solid, erect to decumbent, branched, round to slightly triangular, glabrous, with alternating light and medium green striations along its length; 2–4 mm thick near the taproot.

■ **SYN.** *Polygonum erectum* ssp. *achoreum* (S.F. Blake) A. & D. Löve.

Leathery Knotweed

Buckwheat—*Polygonaceae* **Wildflower** Pink Perianth tubular, 5-lobed

SWAMP PERSICARIA *Polygonum amphibium* var. *emersum* Michx.

■ **SKETCH** A highly variable (depending on moisture) **perennial herb** 20–120 cm tall or long from branching **rhizomes** *c.* 5 mm wide and to 25+ cm long; in quiet water (floating), in mud along the shore (decumbent) and in dry uplands (erect). A shoreline plant growing in mud is illustrated. **w1**

■ **FLOWERS** Pink, blooming June–September; **inflorescence** a raceme (spikelike), terminal, 1.3–14 cm long by 10–20 mm wide; **secondary racemes** (one or two) may develop from upper leaf axils; **peduncles** 1–10 cm long, glabrous to hairy, sometimes glandular-hairy; **flowers** perfect to unisexual, 4–7 mm wide and long, congested; **subtending bracts** (ocreolae) persist at the base of each flower; **perianth** 5-lobed, the lobes 3.3–5 mm long, C-shaped, glabrous, united at the base; **stamens** five to eight, pink, included to exserted; **styles** 2-parted, included to exserted, united below. **FRUIT an achene**, 1-seeded, dark reddish brown to black, lens-shaped, glossy to dull, 2.5–3 mm long by 1.5–2.5 mm wide by *c.* 1.5 mm thick; **beak** short.

■ **LEAVES** Alternate, simple, entire, glabrous to hairy; **blades** 2.5–20 cm long by 1–8 cm wide, reduced above; **petioles** 0.1–6 cm long, round, glabrous or hairy; **stipules** (ocreae) membranous, tubular, 5–50 mm long, often torn; **land form** may have a broad, flared, ciliate collar.

■ **STEM** Horizontal to erect, hollow, round, simple or branched, glabrous to hairy, singly or in a group, rooting at the **nodes** from adventitious buds, then ascending and erect to 20–50 cm tall; base 4–6 mm thick.

■ **SYN.** *Persicaria amphibia* (L.) S.F. Gray p.p. and *Polygonum coccineum* Muhl. ex Willd.

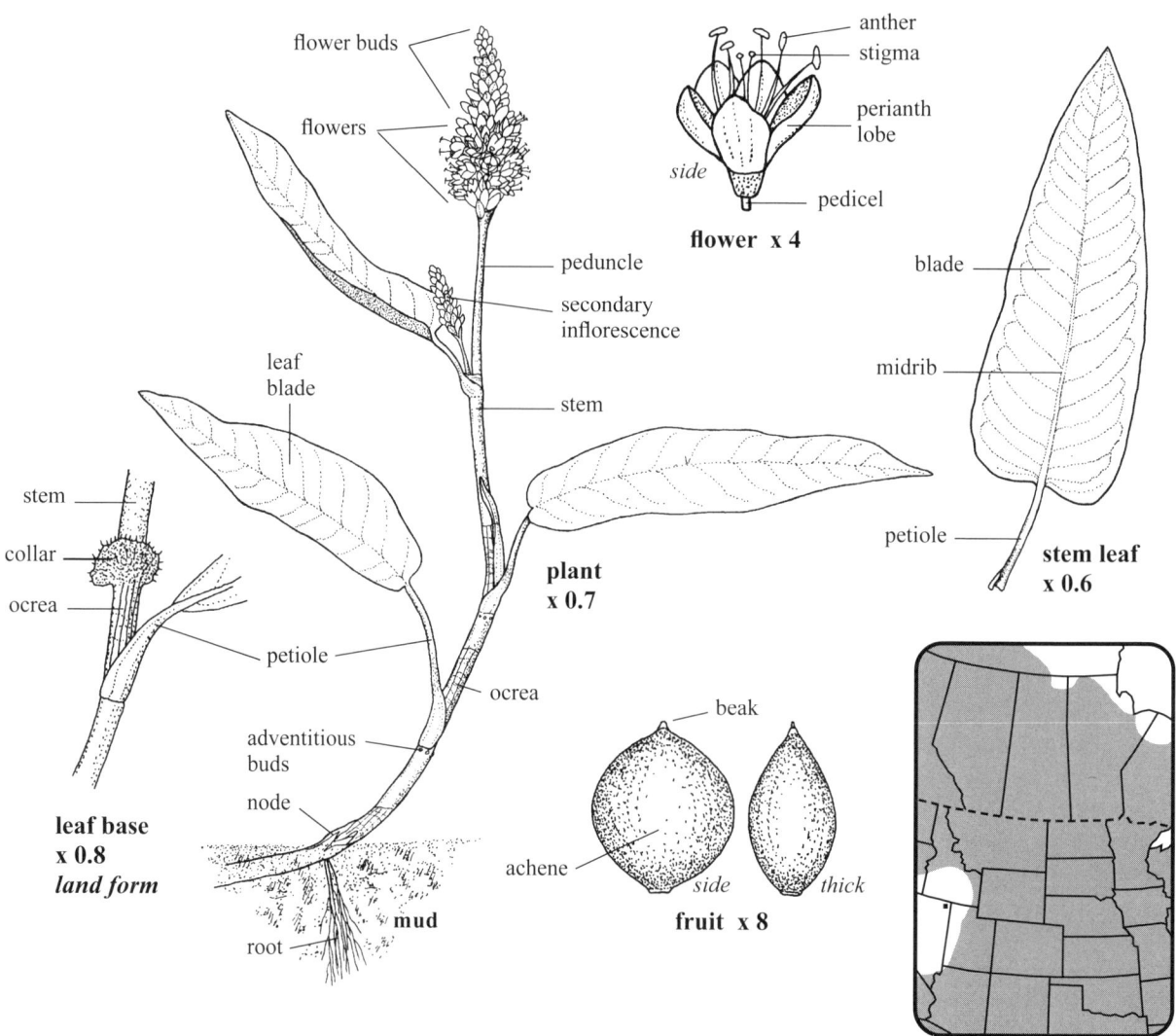

Water Smartweed, Water Knotweed

Buckwheat—*Polygonaceae* **Wildflower** White to pink Perianth tubular, 5-lobed

DOORWEED *Polygonum aviculare* L.

■ **SKETCH** A variable, glabrous **annual herb** 10–200 cm tall or long from a **taproot** 1–3 mm thick and to *c*. 10⁺ cm long; in cultivated fields, lawns, roadsides and dry sloughs. *Naturalized*

■ **FLOWERS** White to pink, blooming March–October; **inflorescence** of cymes, axillary, 1–8 flowers per axil along most of the stem and branches; **branches** alternate, 5–60⁺ cm long, reduced above; **pedicels** 1.5–5 mm long; **flowers** perfect, erect, 1.8–2.3 mm wide by 2.5–5 mm long; **perianth** tubular, 5-lobed, the lobes about equal or the inner two slightly longer than the outer three, beginning at or below the flower's middle, margins white to pink, erect around the fruit; **outer perianth lobes** three, slightly keeled, *c*. 1.6 mm long by *c*. 1 mm wide at anthesis; **inner perianth lobes** two, flat, 0.9–1.2 mm wide; **ocreolae** membranous, subtending flowers and covering pedicels and lower part of perianths, 2–5 mm long, easily torn, lobes pointed and usually imbricate; **stamens** five to eight, included, *c*. 1 mm long, one or two attached at the base of perianth lobes; **filaments** filiform, widening at base, pale green; **anthers** yellow, *c*. 0.2 mm long; **pistil** included, *c*. 1 mm long; **ovary** triangular, green, *c*. 0.7 mm long by *c*. 0.5 mm wide; **stigmas** three, yellowish green. **FRUIT** an achene, 1-seeded, light to dark reddish brown, dull, some tuberculate, included to slightly exserted, 3-sided, erect to arched on pedicel, 1.5–5 mm long by 1.5–2 mm wide, triangular, corners blunt, one narrow side concave, two wider sides convex, or two sides concave and one convex (both types shown).

■ **LEAVES** Alternate, simple, entire, green to yellowish or bluish green, dull, pointed to rounded, veins obscure; **blades** 1–3 cm long by 1–10 mm wide, flat, reduced above; **petioles** jointed at base, 0.5–9 mm long, reduced above; **ocreae** (stipules) veined longitudinally, 3–15 mm long, membranous, tubular around the stem, silvery, usually torn or ragged at the apices.

■ **STEM** Sprawling to mat-forming to erect, branched, stiff below, lax above, ridges numerous and low above; 1–3 mm thick near the smooth, round base.

■ **SYN.** *Polygonum arenastrum* Jord. ex Boreau = *Polygonum aviculare* var. *arenastrum* (Jord. ex Boreau) Rouy.

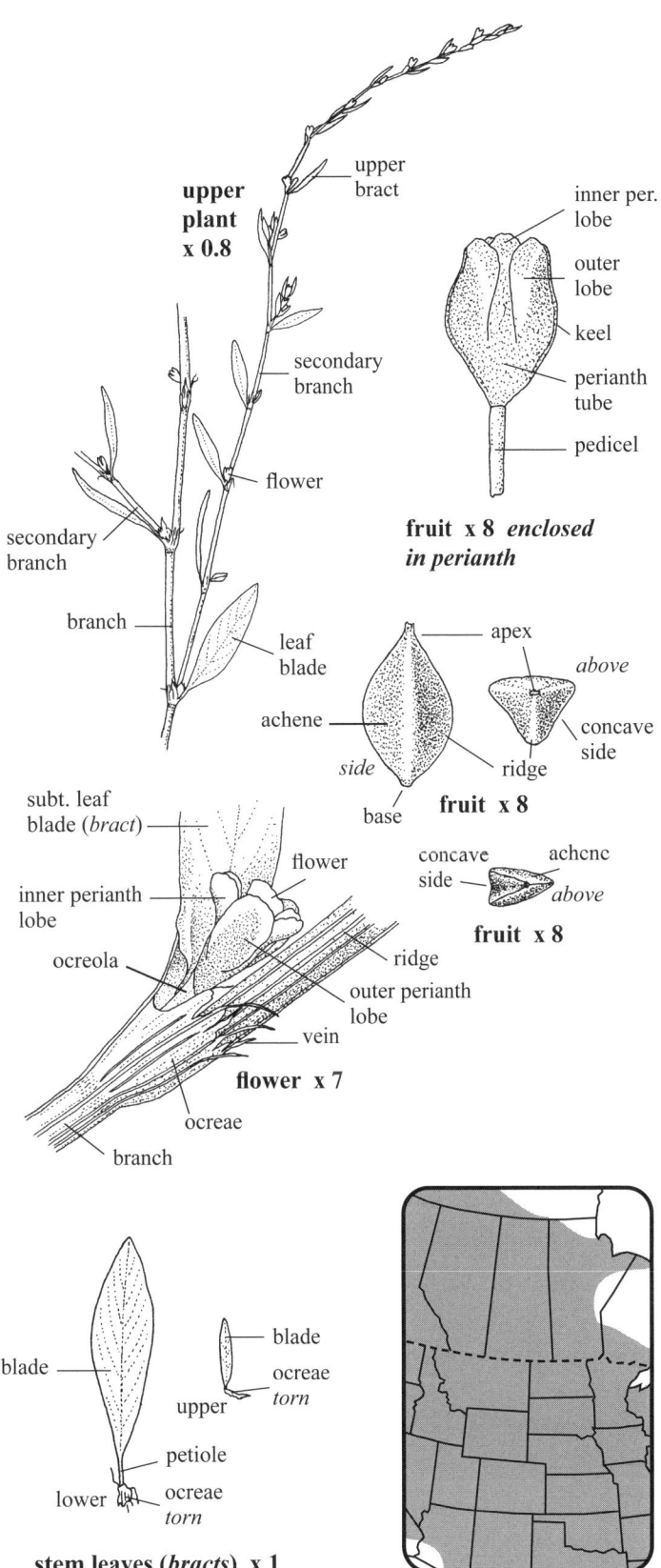

Common Knotweed, Oval-leaf Knotweed

Buckwheat—*Polygonaceae*   **Vine**   White   Perianth tubular, 4- to 6-lobed   Tendrils absent

BINDWEED   *Polygonum cilinode* Michx.

■ **SKETCH** A **perennial vine** 1.5–2 m tall or long without **tendrils**, twining around vegetation, erect to trailing; in forest clearing, on ridges and rocky outcrops. w1

■ **FLOWERS** White, blooming July–September; **inflorescence** a panicle, 5–15 cm long by 3–8 cm wide; **floral branches** one per node, 3–25 cm long with few to several side branches, these sometimes rebranching, reddish, hairs descending; **pedicels** pale green, arched, joint faint and near the flower, 2–3 mm long, recurved with fruit; **ocreolae** (of pedicels) reddish green, pointed, 0.6–1.5 mm long, hairless, often with two tiny teeth at the apices; **flowers** perfect, 4–5 mm wide by *c.* 3 mm long, ascending to slightly drooping; **perianth** tubular, 4- to 6-lobed, white, lobes spreading; **outer lobes** *c.* 1.6 mm long by *c.* 1.1 mm wide; **inner perianth lobes** *c.* 1.5 mm long by *c.* 1.3 mm wide, persistent and erect around fruit, lobes then 3–4 mm long; **stamens** eight, exserted, 4 long and 4 short; **filaments** white, 1–1.8 mm long, swollen and minutely hairy at base; **anthers** pale yellow, *c.* 0.5 mm long and wide; **pistil** exserted, hairless, *c.* 1.5 mm tall by *c.* 1 mm wide; **ovary** green, 3-angled, sides concave; **styles** three, white, each *c.* 0.2 mm long; **stigmas** three, white. **FRUIT an achene**, 1-seeded, black, shiny, 3–3.8 mm long by 2–2.5 mm wide, 3-angled, enclosed in slightly winged perianth lobes, hanging, tip of achene may be visible.

■ **LEAVES** Alternate, simple, entire to toothed, 2-lobed basally, one to three leaves at lower nodes; **blades** dull, 2–11 cm long by 1.5–9 cm wide, short hairy above, hairy on raised veins below, lobes pointed to round on large leaves and 0.2–3 cm long; **stipules** (ocreae) 5–7.5 mm long, tubular, tan, apices split into two teeth, each half with three to five veins, open above, enveloping the stem below, base a ridge with declining, clear hairs usually forming a ring around the stem giving the nodes a fringed appearance; **petioles** 0.4–9 cm long, short hairs reflexed.

■ **STEM** Twining, reddish purple, smooth, hollow, 3- or 4-sided with blunt corners, branching, tapered toward the tip, short white hairs descending; 3–4 mm thick near the brown bare base.

■ **SYN.** *Fallopia cilinodis* (Michx.) Holub.

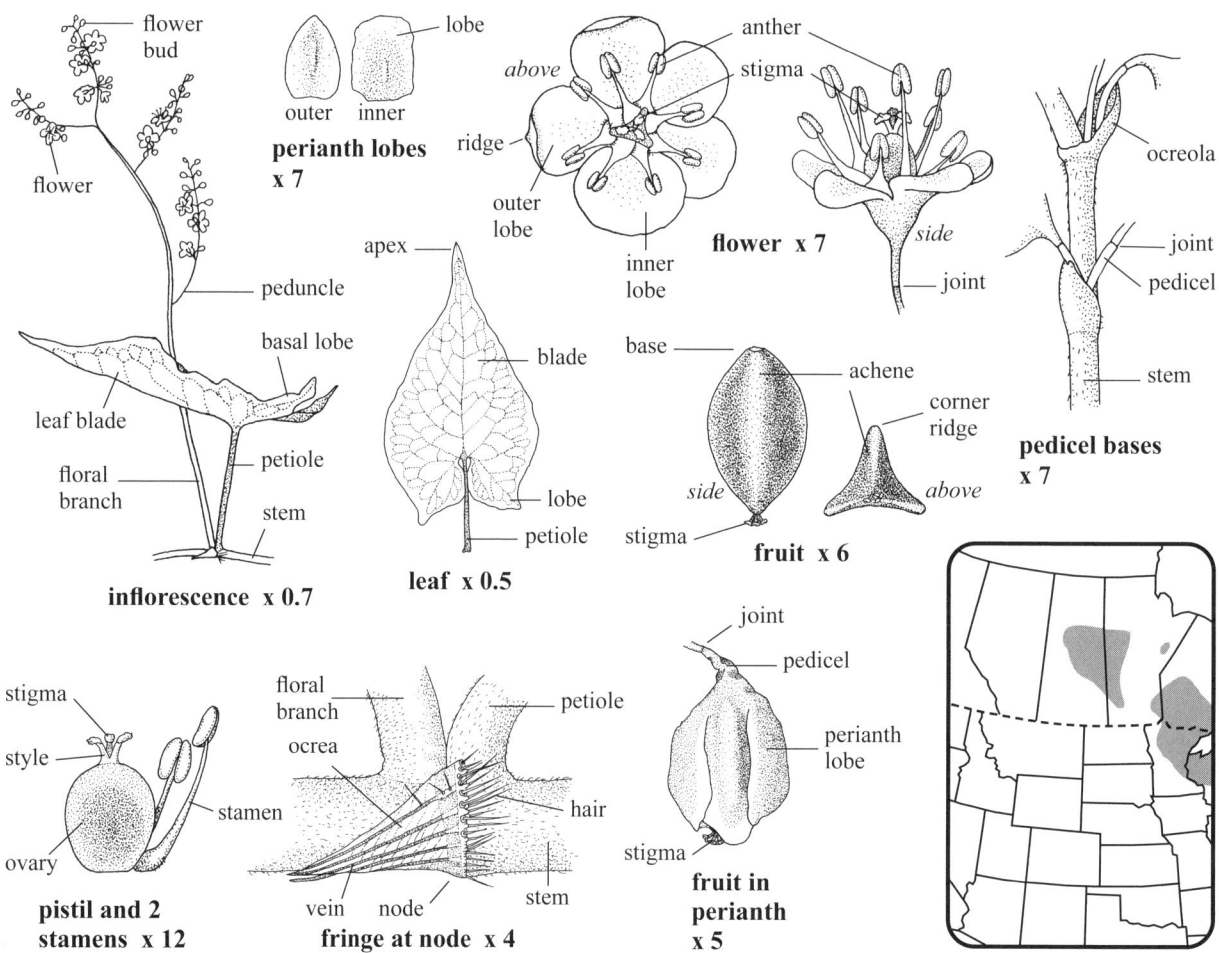

394   Fringed Black Bindweed

Buckwheat—*Polygonaceae* **Wildflower** Green Perianth tubular, 5-lobed

WILD BUCKWHEAT *Polygonum convolvulus* var. *convolvulus* L.

■ **SKETCH** An **annual herb**, twining or trailing 0.5–2 m long from a **taproot** 2–6 mm wide and to *c.* 10+ cm long; in disturbed sites, gardens, roadsides and fields. **w1**

■ **FLOWERS** Green, blooming May–October; **inflorescence** a raceme, interrupted and axillary; **peduncles** one (two) per node, 3–12 cm long by 0.5–1 mm thick, erect, reddish green, 7-ridged, scabrous from bumps and minute hairs; **pedicels** glabrous, 1–3 mm long, erect to recurved in fruit, pink or green with a joint near apices right below the flower or fruit; **ocreolae** membranous, pointed, 1–1.5 mm long, not torn, partially covering the pedicels; **flowers** perfect, in whorls of 3–6, each 1.3–2 mm wide by 3–5 mm long, ascending to drooping; **perianth** tubular, 5-lobed, 1.5–2 mm long by *c.* 1.8 mm wide, increasing to 4–5 mm long by *c.* 3 mm wide with fruit; **1) outer perianth parts** three, scabrous on margins and body, keeled, hooded, margins white; **2) inner parts** two, whitish pink near the tapered base, glabrous, flatter, entire, *c.* 1.5 mm long by *c.* 1.3 mm wide; **stamens** five to eight, *c.* 1 mm long; **filaments** white, *c.* 0.8 mm long; **anthers** pink to dark purple, included, *c.* 0.3 mm long by 0.4 mm wide, 2-lobed, surrounding the three styles; **pistil** *c.* 1.2 mm long, green, glabrous; **ovary** *c.* 0.9 mm long, with three bluntly rounded edges, the sides slightly concave. **FRUIT an achene**, 1-seeded, hanging, included, black, 3–5 mm long by 1.8–2.5 mm wide and thick, sides concave, ridges blunt, granular and dull.

■ **LEAVES** Opposite, simple, entire, base 2-lobed; **blades** 1–10 cm long by 0.6–8 cm wide; **petioles** 1–6.5 cm long, slightly scabrous, pink to green, one to a few flowers at their bases; **stipules** (ocreae) 2–4 mm long, membranous, hairless, entire.

■ **STEM** Trailing or twining, 7-ridged, red and green, slightly scabrous, branched; 1–3 mm thick near the base.

■ **SYN.** *Fallopia convolvulus* (L.) A. Löve.

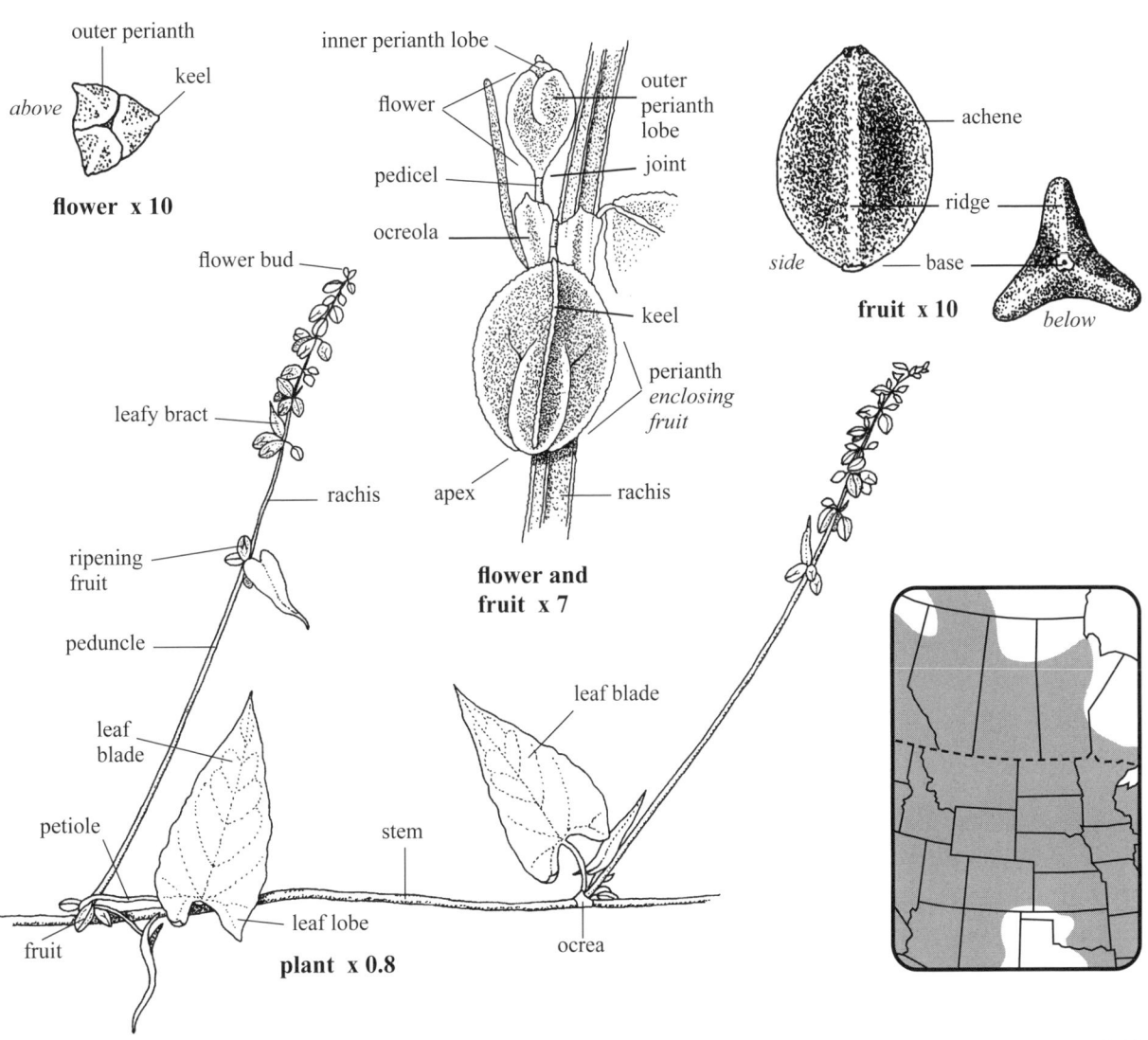

Climbing Buckwheat, Black Bindweed

Buckwheat—*Polygonaceae* **Wildflower** Whitish green Perianth tubular, 4- or 5-lobed

WATER-PEPPER *Polygonum hydropiper* L.

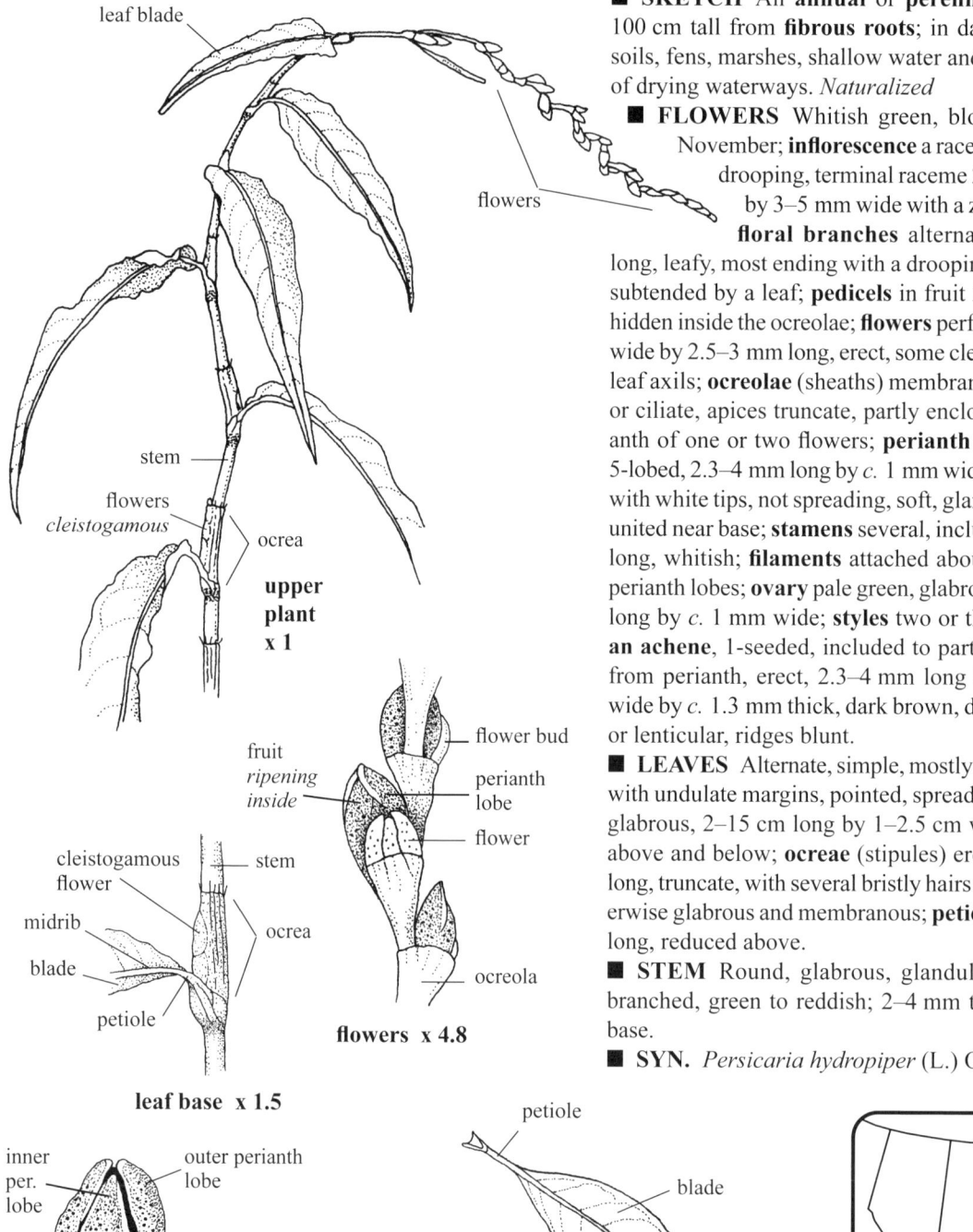

- **SKETCH** An **annual** or **perennial herb** 10–100 cm tall from **fibrous roots**; in damp disturbed soils, fens, marshes, shallow water and along shores of drying waterways. *Naturalized*
- **FLOWERS** Whitish green, blooming May–November; **inflorescence** a raceme, arching to drooping, terminal raceme 2.5–6 cm long by 3–5 mm wide with a zigzag pattern; **floral branches** alternate, 10–40 cm long, leafy, most ending with a drooping raceme and subtended by a leaf; **pedicels** in fruit 2–3 mm long, hidden inside the ocreolae; **flowers** perfect, 1–1.5 mm wide by 2.5–3 mm long, erect, some cleistogamous in leaf axils; **ocreolae** (sheaths) membranous, glabrous or ciliate, apices truncate, partly enclosing the perianth of one or two flowers; **perianth** tubular, 4- or 5-lobed, 2.3–4 mm long by *c.* 1 mm wide, erect, green with white tips, not spreading, soft, glandular-dotted, united near base; **stamens** several, included, *c.* 1 mm long, whitish; **filaments** attached about midway on perianth lobes; **ovary** pale green, glabrous, *c.* 1.5 mm long by *c.* 1 mm wide; **styles** two or three. **FRUIT** an achene, 1-seeded, included to partially exserted from perianth, erect, 2.3–4 mm long by 1.5–2 mm wide by *c.* 1.3 mm thick, dark brown, dull, triangular or lenticular, ridges blunt.
- **LEAVES** Alternate, simple, mostly entire; **blades** with undulate margins, pointed, spreading to arched, glabrous, 2–15 cm long by 1–2.5 cm wide, reduced above and below; **ocreae** (stipules) erect, 5–15 mm long, truncate, with several bristly hairs at apices, otherwise glabrous and membranous; **petioles** 1–12 mm long, reduced above.
- **STEM** Round, glabrous, glandular, simple to branched, green to reddish; 2–4 mm thick near the base.
- **SYN.** *Persicaria hydropiper* (L.) Opiz.

Marsh-pepper Smartweed, Mild Water-pepper, Common Smartweed

Buckwheat—*Polygonaceae* **Wildflower** Greenish white to pink Perianth tubular, 4-lobed

PALE PERSICARIA *Polygonum lapathifolium* L.

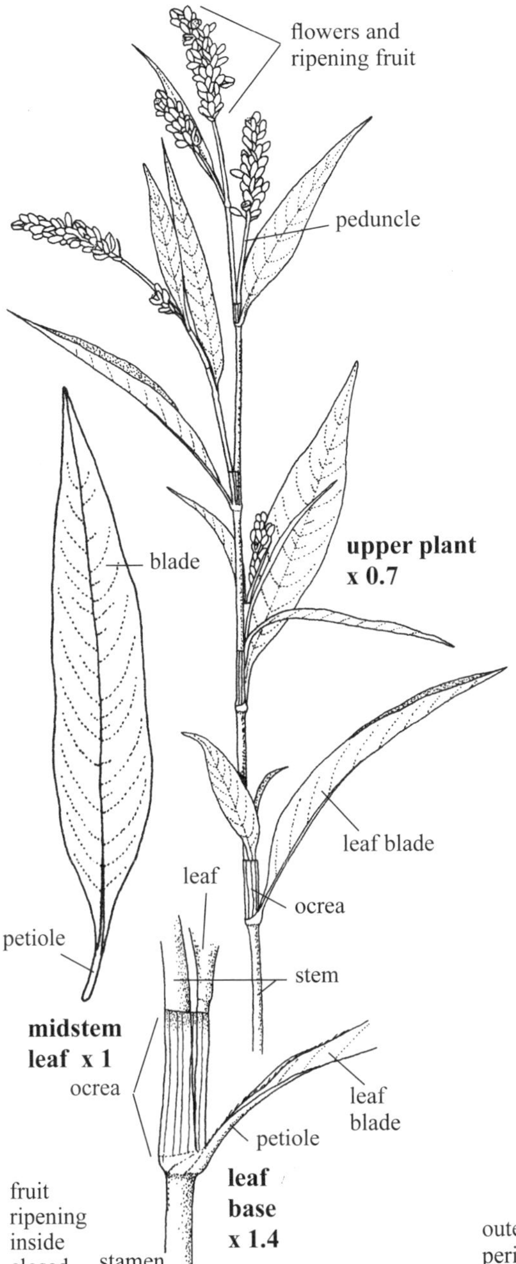

■ **SKETCH** A variable **annual herb** 10–130 cm tall from a **taproot** 2–6 mm thick and to *c.* 10+ cm long; in ditches, drying pond banks, disturbed sites, marsh edges and wet meadows. **w1**

■ **FLOWERS** Greenish white to pink, blooming June–November; **inflorescence** a raceme, nodding to ascending, each 0.5–8 cm long by 6–10 mm wide; **floral branches** 7–17 cm long, two to six per plant from upper leaf axils; **peduncles** 3–26 mm long, slightly scabrous to glabrous, some glandular; **ocreolae** (of flowers) pointed, 1- or 3-nerved, membranous, whitish green, 2–4 mm long; **pedicels** glabrous, ascending, *c.* 1.5 mm long; **flowers** perfect, 2–2.5 mm wide by 2–3 mm long by 1.5–1.8 mm thick; **perianth** tubular, 4-lobed (or 5-) ascending, glabrous, often glandular, lobes C-shaped; **1) outer perianth lobes** two, 2–3.5 mm long by *c.* 2 mm wide (flattened), apices blunt, with three distinct veins, each vein with two recurved branches at its apex, these most obvious at fruiting time; **2) inner perianth lobes** two, flat, 2–2.5 mm long by *c.* 1 mm wide, apices blunt, two green veins as above; **stamens** six (5), attached to base of perianth lobes, 1–1.5 mm long, included at anthesis to barely exserted, slightly longer than the styles; **filaments** filiform, white; **anthers** white and pink, 0.3–0.4 mm long; **styles** two (3), white, *c.* 1 mm long, spreading, included; **stigmas** round, whitish. **FRUIT** an achene, 1-seeded, one per flower, enclosed in perianth, mostly lenticular, 1.5–3.2 mm long by 1.8–2.5 mm wide by 0.7–0.8 mm thick, glabrous, slightly shiny, with a low, thin, marginal ridge, sides with concave centers.

■ **LEAVES** Alternate, simple, entire, pointed, lower ones wilting by anthesis, lower surface glandular dotted (microscopic); **blades** yellowish green, usually 3–20 cm long by 5–60 mm wide, scabrous below on raised midrib, long hairs scattered, slightly hairy to glabrous above with hairs appressed; **petioles** 2–15 mm long, scabrous on margins; **ocreae** (stipules) tubular, membranous, longitudinally veined, surrounding the stem, 5–35 mm long, apices flat, rarely hairy, some glandular-dotted, often torn.

■ **STEM** Usually erect, hollow, usually branched, glabrous, slightly yellowish green; 1–7 mm thick near the base.

■ **SYN.** *Persicaria lapathifolia* (L.) S.F. Gray.

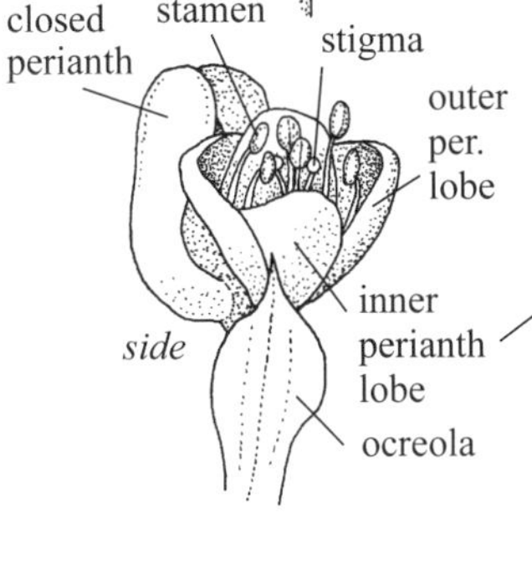

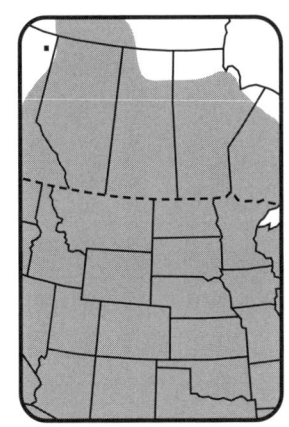

Dockleaf Smartweed, Curlytop Knotweed, Pale Smartweed, Nodding Smartweed

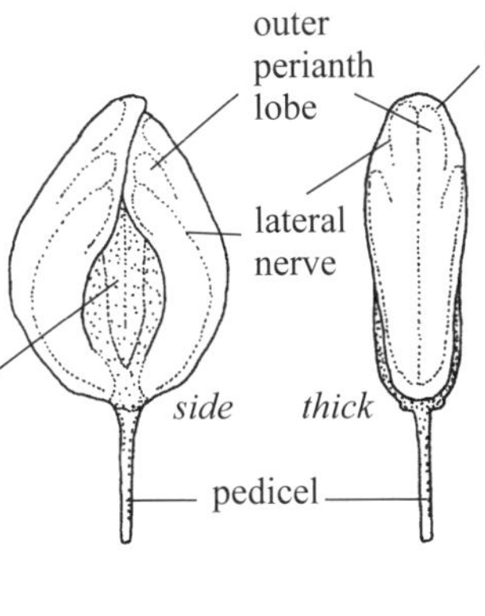

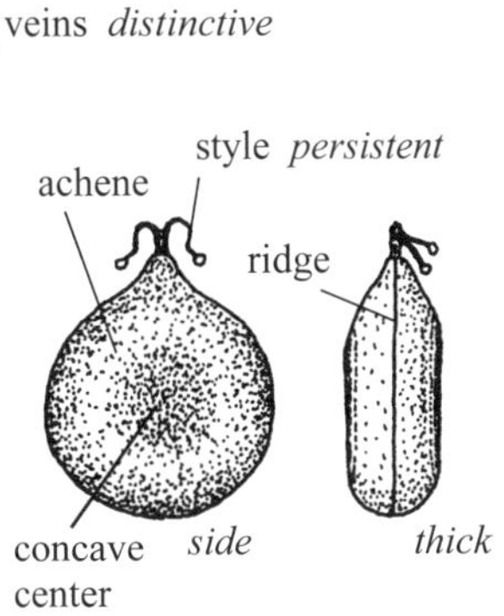

Buckwheat—*Polygonaceae* **Wildflower** White to greenish red Perianth tubular, lobes 4 or 5

LADY'S-THUMB *Polygonum persicaria* L.

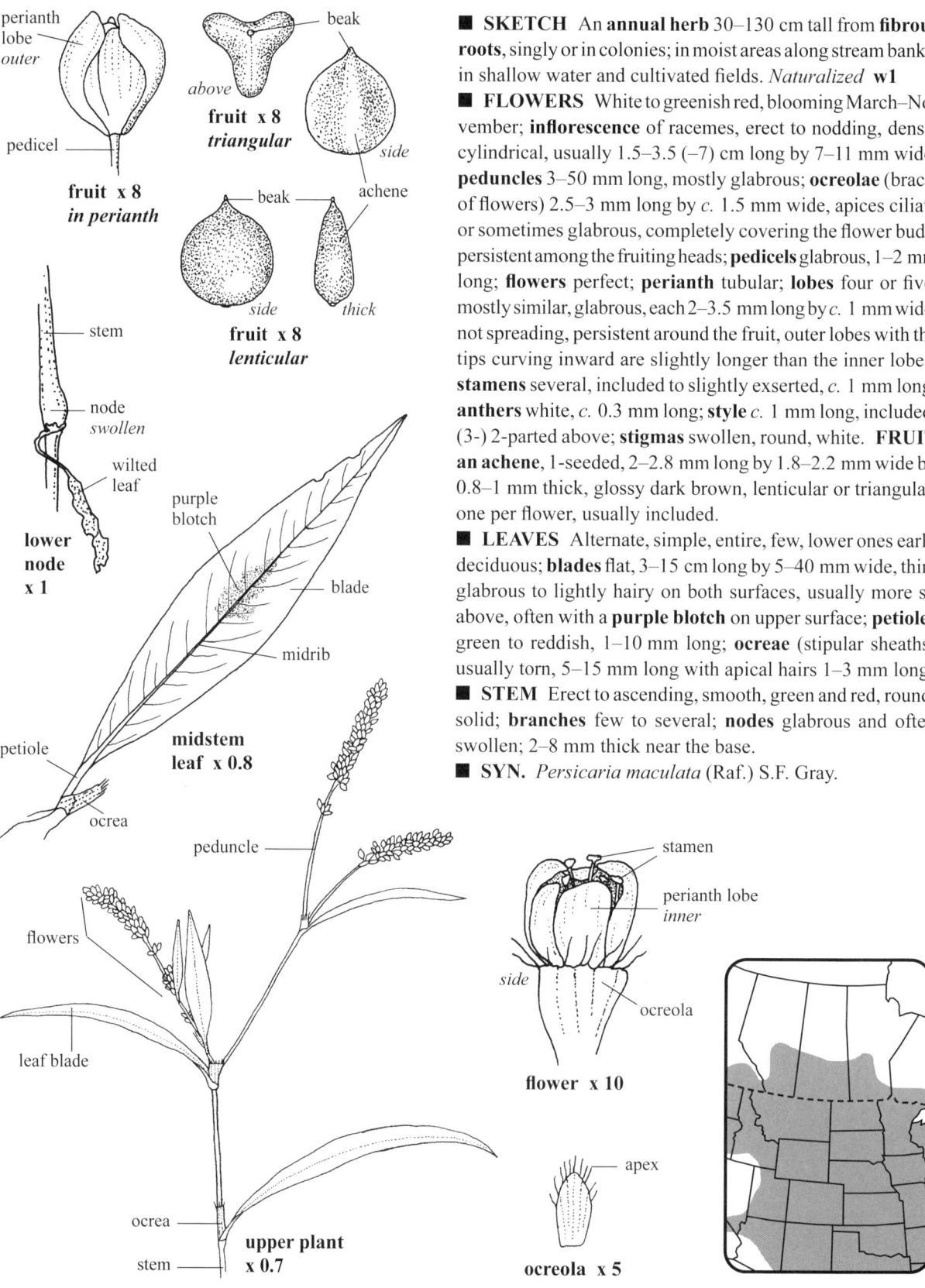

- **SKETCH** An **annual herb** 30–130 cm tall from **fibrous roots**, singly or in colonies; in moist areas along stream banks, in shallow water and cultivated fields. *Naturalized* **w1**
- **FLOWERS** White to greenish red, blooming March–November; **inflorescence** of racemes, erect to nodding, dense, cylindrical, usually 1.5–3.5 (–7) cm long by 7–11 mm wide; **peduncles** 3–50 mm long, mostly glabrous; **ocreolae** (bracts of flowers) 2.5–3 mm long by *c.* 1.5 mm wide, apices ciliate or sometimes glabrous, completely covering the flower buds, persistent among the fruiting heads; **pedicels** glabrous, 1–2 mm long; **flowers** perfect; **perianth** tubular; **lobes** four or five, mostly similar, glabrous, each 2–3.5 mm long by *c.* 1 mm wide, not spreading, persistent around the fruit, outer lobes with the tips curving inward are slightly longer than the inner lobes; **stamens** several, included to slightly exserted, *c.* 1 mm long; **anthers** white, *c.* 0.3 mm long; **style** *c.* 1 mm long, included, (3-) 2-parted above; **stigmas** swollen, round, white. **FRUIT** an achene, 1-seeded, 2–2.8 mm long by 1.8–2.2 mm wide by 0.8–1 mm thick, glossy dark brown, lenticular or triangular, one per flower, usually included.
- **LEAVES** Alternate, simple, entire, few, lower ones early deciduous; **blades** flat, 3–15 cm long by 5–40 mm wide, thin, glabrous to lightly hairy on both surfaces, usually more so above, often with a **purple blotch** on upper surface; **petioles** green to reddish, 1–10 mm long; **ocreae** (stipular sheaths) usually torn, 5–15 mm long with apical hairs 1–3 mm long.
- **STEM** Erect to ascending, smooth, green and red, round, solid; **branches** few to several; **nodes** glabrous and often swollen; 2–8 mm thick near the base.
- **SYN.** *Persicaria maculata* (Raf.) S.F. Gray.

Spotted Ladysthumb, Redleg, Redshank

Buckwheat—*Polygonaceae* **Wildflower** Green and white Perianth tubular, 5-lobed (6-)

BUSHY KNOTWEED *Polygonum ramosissimum* Michx.

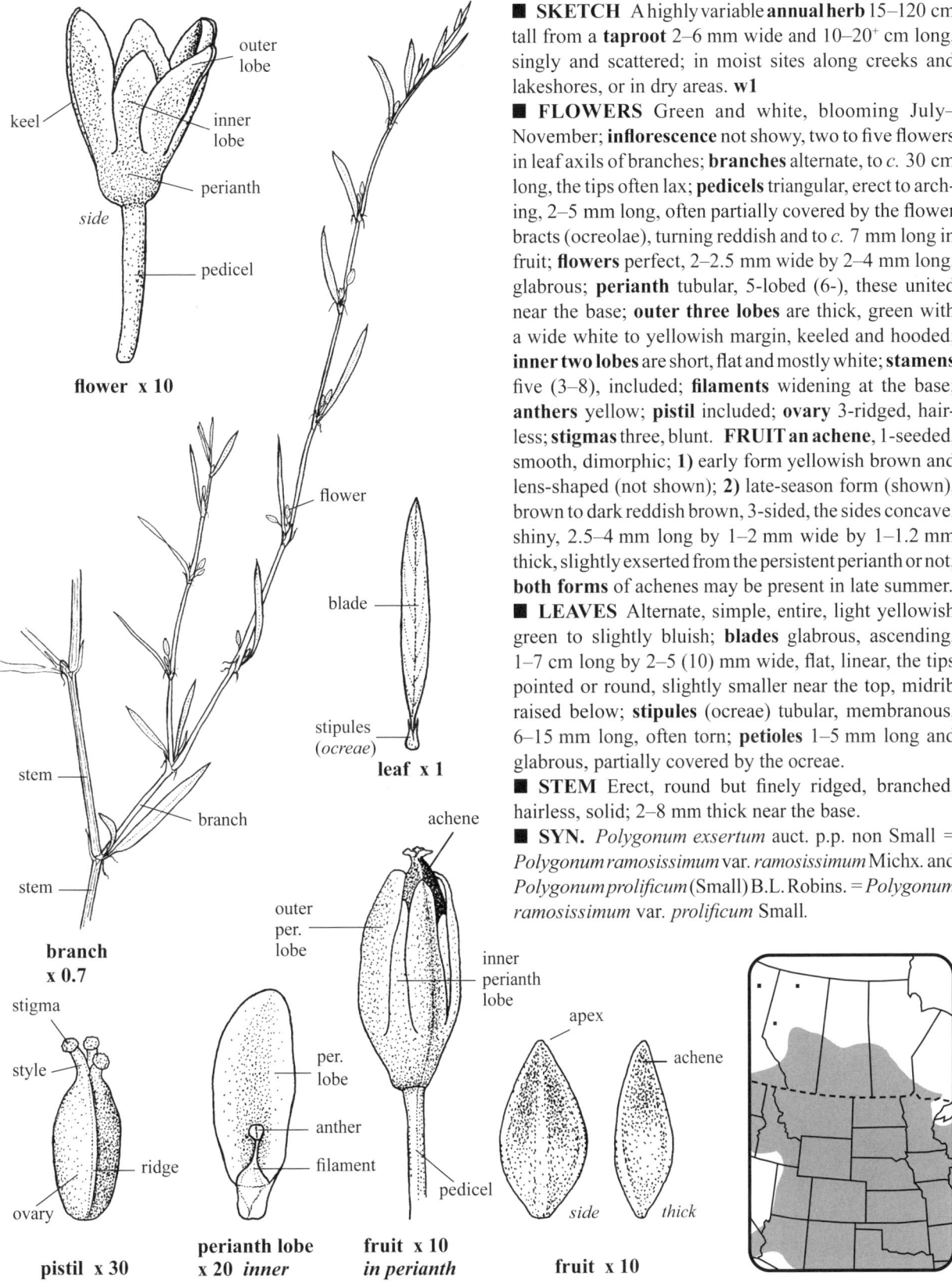

■ **SKETCH** A highly variable **annual herb** 15–120 cm tall from a **taproot** 2–6 mm wide and 10–20+ cm long, singly and scattered; in moist sites along creeks and lakeshores, or in dry areas. w1

■ **FLOWERS** Green and white, blooming July–November; **inflorescence** not showy, two to five flowers in leaf axils of branches; **branches** alternate, to *c.* 30 cm long, the tips often lax; **pedicels** triangular, erect to arching, 2–5 mm long, often partially covered by the flower bracts (ocreolae), turning reddish and to *c.* 7 mm long in fruit; **flowers** perfect, 2–2.5 mm wide by 2–4 mm long, glabrous; **perianth** tubular, 5-lobed (6-), these united near the base; **outer three lobes** are thick, green with a wide white to yellowish margin, keeled and hooded; **inner two lobes** are short, flat and mostly white; **stamens** five (3–8), included; **filaments** widening at the base; **anthers** yellow; **pistil** included; **ovary** 3-ridged, hairless; **stigmas** three, blunt. **FRUIT an achene**, 1-seeded, smooth, dimorphic; **1)** early form yellowish brown and lens-shaped (not shown); **2)** late-season form (shown), brown to dark reddish brown, 3-sided, the sides concave, shiny, 2.5–4 mm long by 1–2 mm wide by 1–1.2 mm thick, slightly exserted from the persistent perianth or not; **both forms** of achenes may be present in late summer.

■ **LEAVES** Alternate, simple, entire, light yellowish green to slightly bluish; **blades** glabrous, ascending, 1–7 cm long by 2–5 (10) mm wide, flat, linear, the tips pointed or round, slightly smaller near the top, midrib raised below; **stipules** (ocreae) tubular, membranous, 6–15 mm long, often torn; **petioles** 1–5 mm long and glabrous, partially covered by the ocreae.

■ **STEM** Erect, round but finely ridged, branched, hairless, solid; 2–8 mm thick near the base.

■ **SYN.** *Polygonum exsertum* auct. p.p. non Small = *Polygonum ramosissimum* var. *ramosissimum* Michx. and *Polygonum prolificum* (Small) B.L. Robins. = *Polygonum ramosissimum* var. *prolificum* Small.

Buckwheat—*Polygonaceae*   **Wildflower**   White to pink   Perianth tubular, 5-lobed

Arrow-leaf Tear-thumb   *Polygonum sagittatum* L.

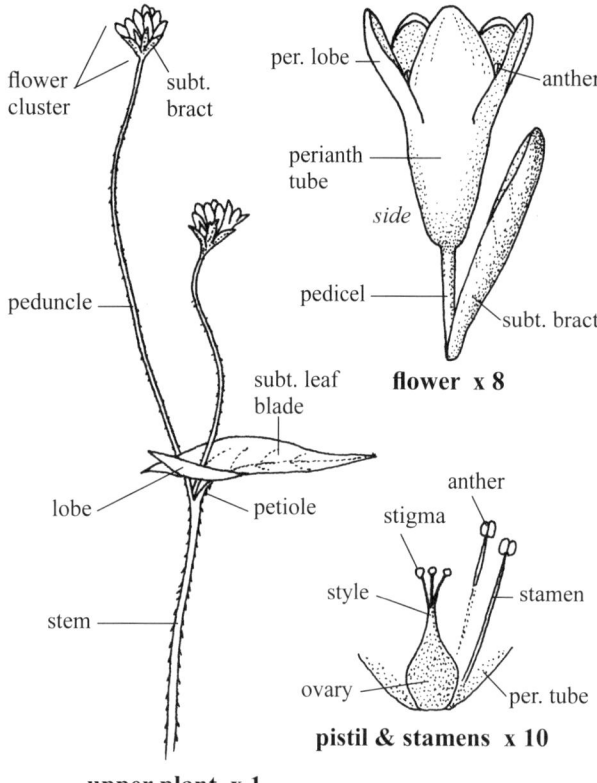

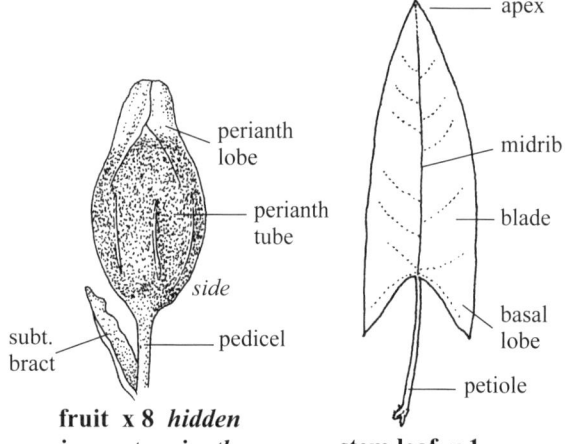

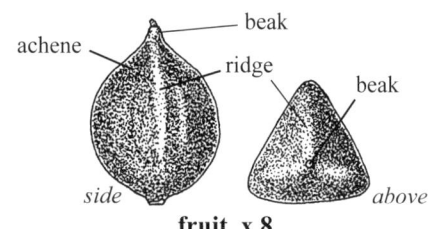

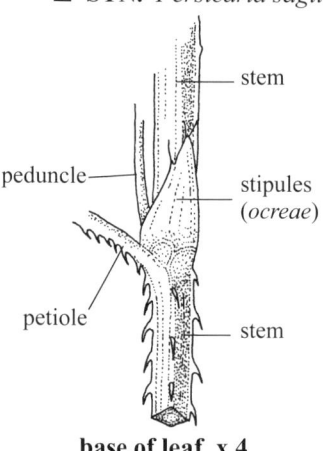

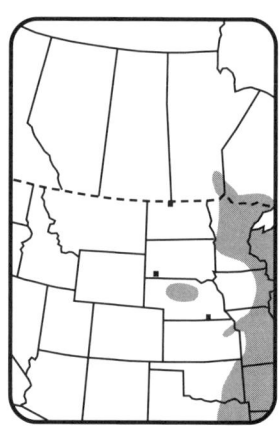

■ **SKETCH** An **annual herb** 30–200 cm tall or long, often sprawling over vegetation; **fibrous roots** pinkish tan, 8–15 cm long by 0.5–0.8 mm thick; in marshy areas, damp meadows, fens and shorelines. **w1**

■ **FLOWERS** White to pink, blooming June–October; **inflorescence** a raceme (compact), each 5–8 mm long by 4–15 mm wide of 2–8 flowers (mostly buds or fruit) at ends of peduncles; **peduncles** 0.7–12 cm long, slightly scabrous; **pedicels** 1–3 mm long, erect with flowers and fruit, glabrous, pale green; **ocreolae** (subt. bracts) of flowers membranous, 4–5 mm long, green, apices often torn, brown and pointed; **flowers** perfect, *c.* 3 mm wide by *c.* 4 mm long; **perianth** tubular, 5-lobed, these 2.3–2.5 mm long by 2.5–2.8 mm wide, blunt; **outer lobes** three, **inner lobes** two, slightly shorter than the outer lobes, apices hooded; **stamens** 8 or 9, white, included, three stamens *c.* 2 mm long around the pistil, one stamen attached at the inside edge of each lobe, these *c.* 1 mm long; **filaments** filiform; **anthers** white, 2-lobed, *c.* 0.3 mm long; **pistil** one, included, *c.* 2 mm long; **style** 3-parted. **FRUIT** an achene, 1-seeded, triangular, 2.3–4 mm long by 1.8–2.5 mm wide, sides slightly convex to straight, shiny to dull, dark brown, hidden in perianth.

■ **LEAVES** Alternate, entire, simple, subtending a branch or peduncle; **blades** dull, thin, 2–12 cm long by 8–30 mm wide, spreading, 2-lobed, glabrous above, midribs raised below and with retrorse barbs near base, side veins obscure; **lobes** not spreading, lobe's outer margins with retrorse hairs; **petioles** 3–40 mm long, grooved above, retrorse stiff hairs below; **stipules** (ocreae) tubular, 3–13 mm long, a few hairs along apical margin.

■ **STEM** Green, square, 4-ridged with thick prickles along each ridge, faces flat, much branched; pinkish near the 2–4 mm thick base; **nodes** swollen, some up to 5 mm long, the lowest 3–5 nodes pinkish tan, some plants with a few roots from the lowest node.

■ **SYN.** *Persicaria sagittata* (L.) H. Gross.

Buckwheat—*Polygonaceae* **Wildflower** Whitish green Perianth tubular, 5-lobed

FALSE BUCKWHEAT *Polygonum scandens* var. *scandens* L.

■ **SKETCH** A variable **perennial herb**, twining or trailing, 1–5 m long with fine tan **roots** 5–20 cm long by *c.* 0.5 mm thick; in disturbed sites, moist woodlands, roadsides, cultivated fields and shorelines. Native and *naturalized* **w1**

■ **FLOWERS** Whitish green, blooming June–November; **inflorescence** a raceme, 2–28 cm long by 1–2 cm wide, from upper leaf axils, one (two) racemes per node, flowers and fruit crowded above, more remote below with one or two small leaves along the rachis (peduncle); **pedicels** 4–8 mm long, green to pinkish with fruit, recurved, joint near the middle, wider above the joint; **flowers** perfect, 3–3.5 mm wide by 2–3 mm long, erect, drooping as fruit ripens; **ocreolae** 1.5–2.5 mm long, partially covering the base of the pedicels of two or three flowers, glabrous, with a subtending green narrow bract; **perianth** tubular, 5-lobed, three long (outer lobes) and two short (inner), 8–15 mm long by 3–8 mm wide in fruit, ascending, green, margins winged, pale green, slightly erose; **stamens** 7 or 8, included, attached to bases of perianth lobes; TL 0.5–0.6 mm; **anthers** pale yellow, *c.* 0.2 mm long by *c.* 0.3 mm wide, 2-lobed; **pistil** one, green, *c.* 1 mm long by *c.* 0.7 mm wide, glabrous, 3-sided. **FRUIT an achene**, 1-seeded, ridged, enclosed and hidden by perianth, dark brown to black, smooth, glossy, 3-sided, one side more concave than the other two, 3–6 mm long by 1.5–3.5 mm wide.

■ **LEAVES** Opposite, simple, entire, 2-lobed at base; **blades** 2.5–12 cm long by 1.5–8 cm wide, margins hyaline, papillose, midrib papillose below; **petioles** 0.5–8 cm long, upper margins papillose; **stipules** (ocreae) a membranous, rusty sheath 1.5–6 mm long, hairless, tight, truncate.

■ **STEM** Stiff, twining, solid, angular, branched, glabrous with a waxy feel; 0.8–2 mm thick and reddish brown, stiff and naked; **nodes** 7–12 cm apart.

■ **SYN.** *Fallopia scandens* (L.) Holub.

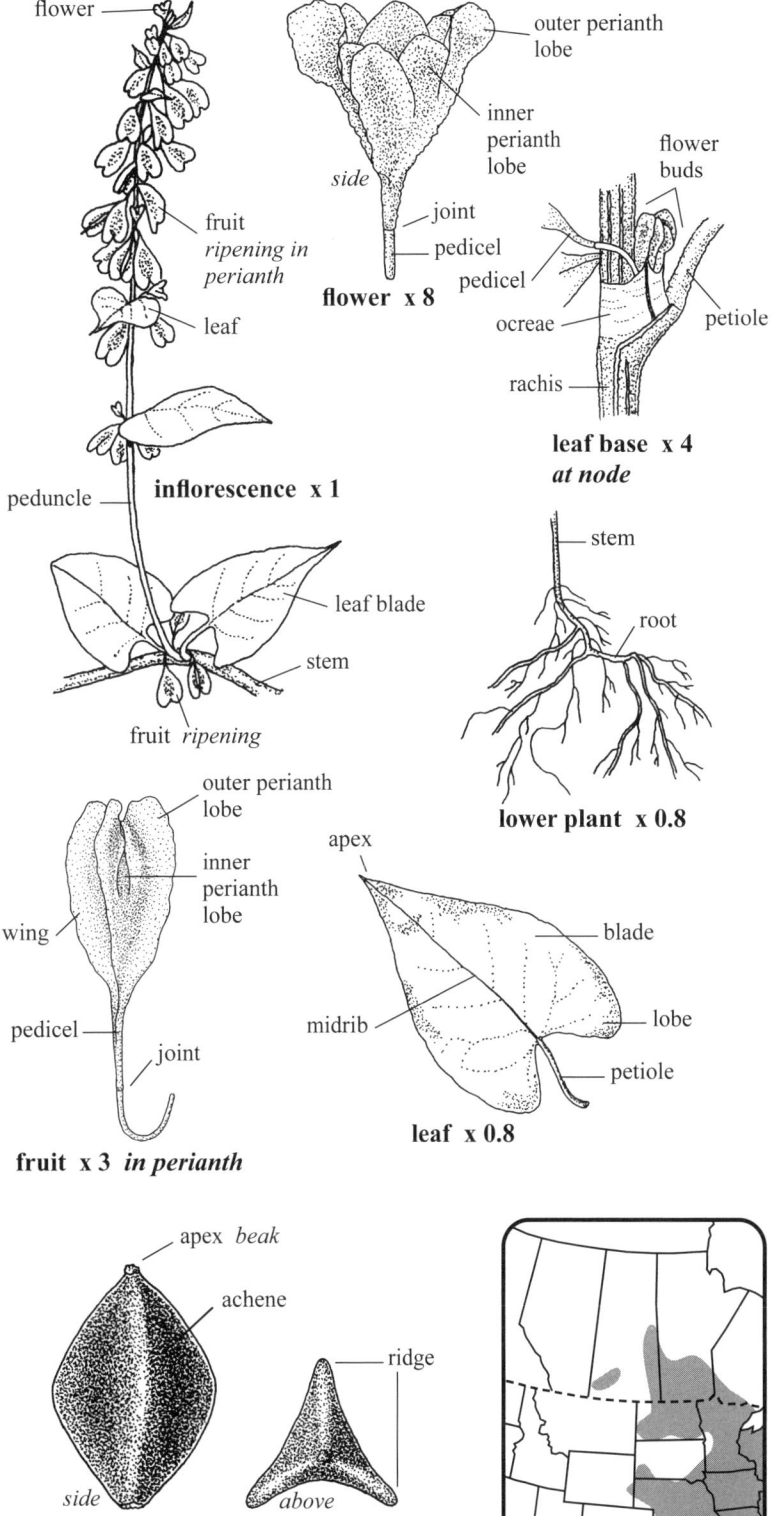

Climbing False Buckwheat

Buckwheat—*Polygonaceae* **Wildflower** Green Perianth tubular, 5- or 6-lobed

Sour Dock *Rumex acetosa* L.

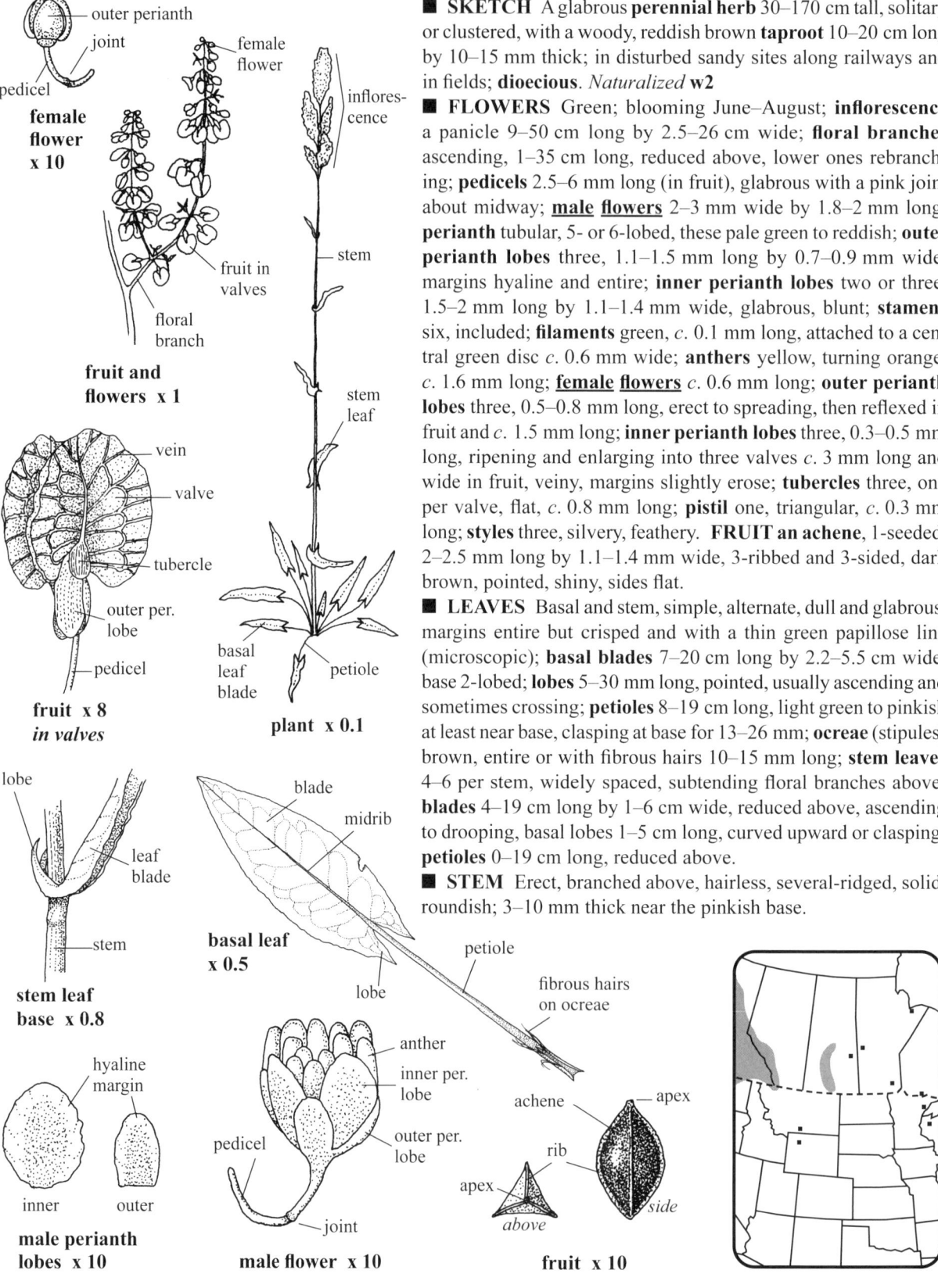

■ **SKETCH** A glabrous **perennial herb** 30–170 cm tall, solitary or clustered, with a woody, reddish brown **taproot** 10–20 cm long by 10–15 mm thick; in disturbed sandy sites along railways and in fields; **dioecious**. *Naturalized* w2

■ **FLOWERS** Green; blooming June–August; **inflorescence** a panicle 9–50 cm long by 2.5–26 cm wide; **floral branches** ascending, 1–35 cm long, reduced above, lower ones rebranching; **pedicels** 2.5–6 mm long (in fruit), glabrous with a pink joint about midway; male flowers 2–3 mm wide by 1.8–2 mm long; **perianth** tubular, 5- or 6-lobed, these pale green to reddish; **outer perianth lobes** three, 1.1–1.5 mm long by 0.7–0.9 mm wide, margins hyaline and entire; **inner perianth lobes** two or three, 1.5–2 mm long by 1.1–1.4 mm wide, glabrous, blunt; **stamens** six, included; **filaments** green, *c.* 0.1 mm long, attached to a central green disc *c.* 0.6 mm wide; **anthers** yellow, turning orange, *c.* 1.6 mm long; female flowers *c.* 0.6 mm long; **outer perianth lobes** three, 0.5–0.8 mm long, erect to spreading, then reflexed in fruit and *c.* 1.5 mm long; **inner perianth lobes** three, 0.3–0.5 mm long, ripening and enlarging into three valves *c.* 3 mm long and wide in fruit, veiny, margins slightly erose; **tubercles** three, one per valve, flat, *c.* 0.8 mm long; **pistil** one, triangular, *c.* 0.3 mm long; **styles** three, silvery, feathery. FRUIT an achene, 1-seeded, 2–2.5 mm long by 1.1–1.4 mm wide, 3-ribbed and 3-sided, dark brown, pointed, shiny, sides flat.

■ **LEAVES** Basal and stem, simple, alternate, dull and glabrous, margins entire but crisped and with a thin green papillose line (microscopic); **basal blades** 7–20 cm long by 2.2–5.5 cm wide, base 2-lobed; **lobes** 5–30 mm long, pointed, usually ascending and sometimes crossing; **petioles** 8–19 cm long, light green to pinkish at least near base, clasping at base for 13–26 mm; **ocreae** (stipules) brown, entire or with fibrous hairs 10–15 mm long; **stem leaves** 4–6 per stem, widely spaced, subtending floral branches above; **blades** 4–19 cm long by 1–6 cm wide, reduced above, ascending to drooping, basal lobes 1–5 cm long, curved upward or clasping; **petioles** 0–19 cm long, reduced above.

■ **STEM** Erect, branched above, hairless, several-ridged, solid, roundish; 3–10 mm thick near the pinkish base.

402 Garden Sorrel, Green Sorrel

Buckwheat—Polygonaceae   Wildflower   Green   Perianth tubular, 6-lobed

WESTERN DOCK *Rumex aquaticus* var. *fenestratus* (Greene) Dorn

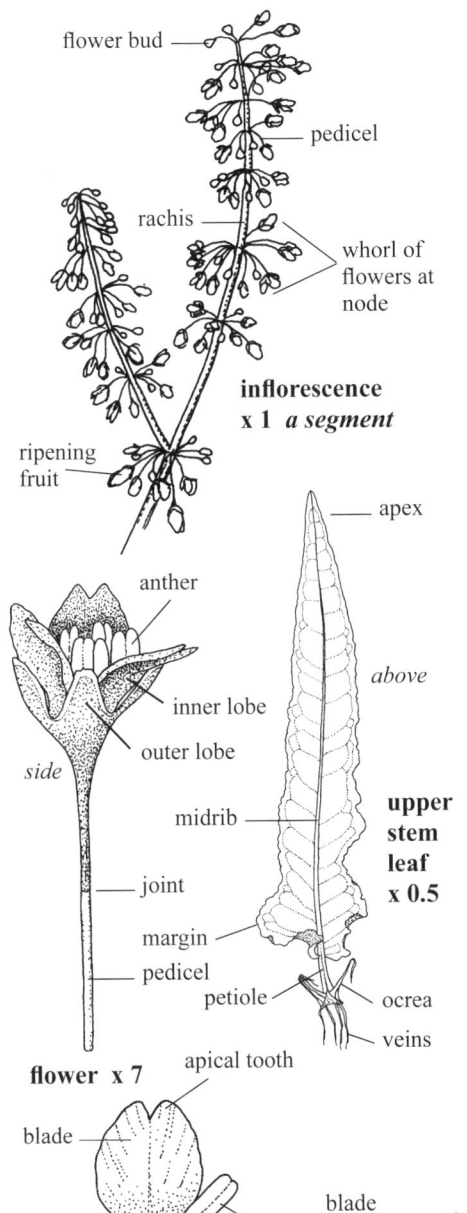

■ **SKETCH** A glabrous **perennial herb** 50–200 cm tall from a dark brown **taproot** 1–2 cm thick; in ditches, bogs, moist meadows, willow thickets, slough margins and along roadsides. w2

■ **FLOWERS** Green, blooming May–August; **inflorescence** paniculate, 30–60⁺ cm long, of numerous whorls of a dozen or more drooping flowers at each node; **floral branches** glabrous, ascending from upper leaf axils; **pedicels** reflexed, 5–14 mm long, much longer than the perianth, joint obscure and near the middle; **flowers** perfect, glabrous, *c.* 3 mm wide and long; **perianth** tubular, 6-lobed; 1) **outer lobes** three, truncate, *c.* 1.3 mm long by *c.* 1 mm wide near base, ascending, united below, green with margins whitish hyaline, slightly rough at apices, spreading in fruit; 2) **inner perianth lobes** three, margins slightly irregular, bidentate or blunt apices, developing into valves as they mature, thin, *c.* 2.5 mm long by 1.8–2 mm wide; **stamens** six, 2–2.5 mm long; **anthers** pale yellow, 1.1–1.3 mm long; **pistil** included, *c.* 2 mm long; **ovary** triangular, glabrous, *c.* 1 mm long; **style** obscure; **stigmas** three, each branch *c.* 1 mm long, apices feathery, spreading to reflexed; **valves** three, without tubercles, dark brown, 4–7 mm long and wide, net-veined, margins entire to toothed, enclosing the achene. **FRUIT an achene**, 1-seeded, one per flower, brown, 3–4.5 mm long by 1.6–2.5 mm wide, 3-sided, glabrous, sides slightly concave.

■ **LEAVES** Basal and stem, alternate, simple, margins undulate and toothed, the teeth blunt and shallow, ascending to spreading below; **basal leaves** mostly entire and petiolate; **blades** 10–25 cm long, margins slightly undulate; **stem blades** crisped, dull, 7–45 cm long by 2–16 cm wide, pointed, midrib round and raised below, lateral veins barely raised below; **petioles** 1–30 cm long, reduced above, base surrounds the stem and swollen node, glabrous; **stipules** (ocreae) membranous, partly enclosing the stem above the petiole, tips free or torn.

■ **STEM** Erect, hairless, ridged longitudinally at the top, reddish green, branched above; smooth near the 7–20 mm thick base; **nodes** enlarged and reddish along the top.

■ **SYN.** *Rumex occidentalis* S. Wats.

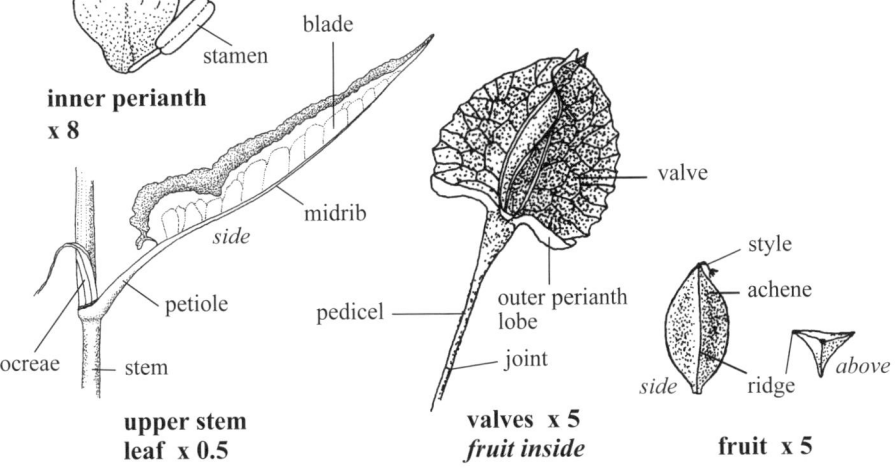

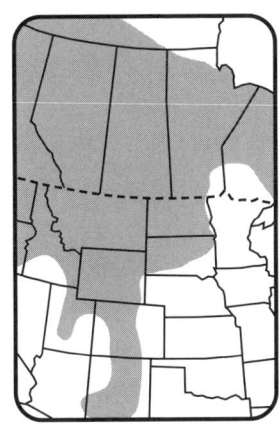

403

Buckwheat—*Polygonaceae* **Wildflower** Green Perianth tubular, 6-lobed

CURLED DOCK *Rumex crispus* L.

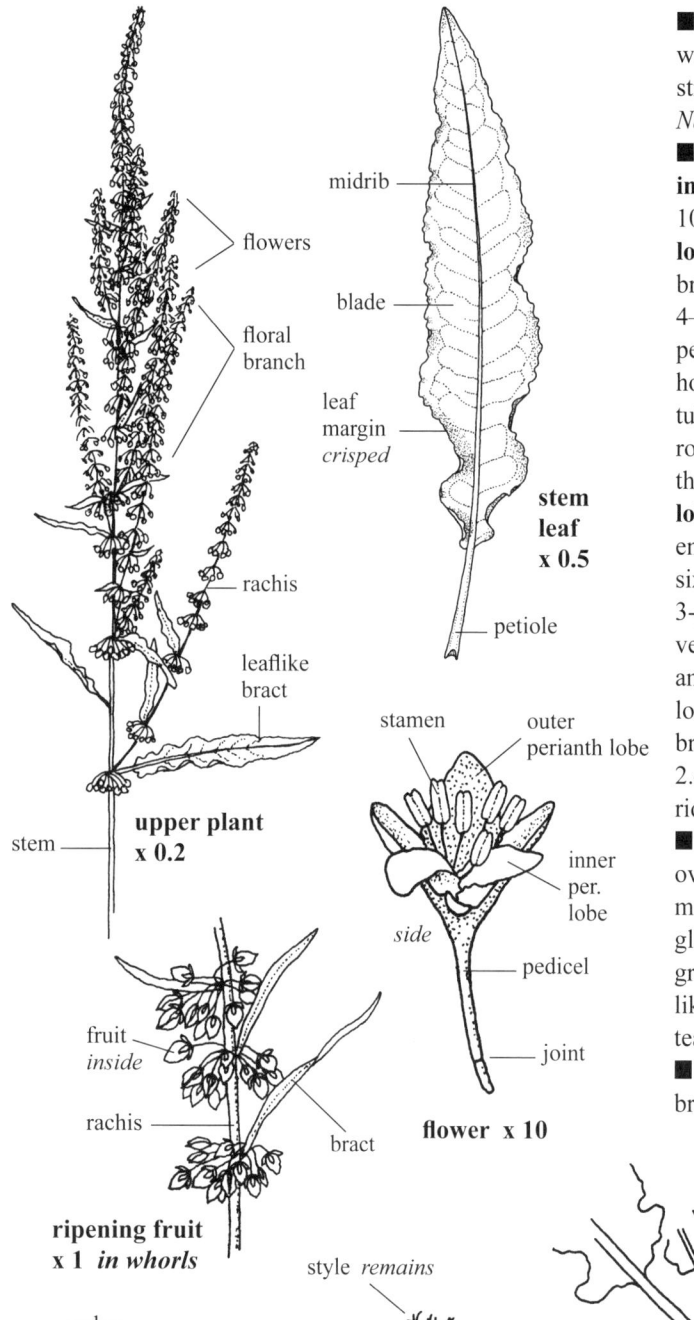

■ **SKETCH** A **perennial herb** 40–150 cm tall with a thick deep **taproot**; in disturbed sites along streams, rivers, ditches, parking lots and sidewalks. *Naturalized*

■ **FLOWERS** Green, blooming May–October; **inflorescence** a panicle of numerous erect branches 10–30 cm long, each subtended by a leaflike bract; **lower floral branches** to 80 cm long, turning reddish brown in fruit; **pedicels** *c.* 2.5 mm long, glabrous, 4–8 mm long in fruit with a joint near the base; **flowers** perfect, glabrous, 2–2.5 mm wide by 2–3 mm long, horizontal to nodding, in whorls of 10–25; **perianth** tubular, 6-lobed, in two series: **1) outer lobes** three, narrow and C-shaped, brown in fruit and usually cupping the bases of the valves or sometimes reflexed; **2) inner lobes** three, flat, wider, more spreading, enlarging and enveloping the fruit and then called valves; **stamens** six, included; **anthers** pale yellow, *c.* 1 mm long; **styles** 3-parted, included; **stigmas** white and feathery; **valves** veined, reddish brown, entire to toothed, 3–6 mm long and wide, each usually with a brown **tubercle** 1–1.5 mm long by *c.* 1 mm wide. **FRUIT an achene**, 1-seeded, brown, 3-sided, one enclosed by the three valves, shiny, 2.4–3 mm long by 1.3–2 mm wide with three narrow ridges, the sides slightly convex.

■ **LEAVES** Alternate, simple, toothed, first year overwintering as a rosette; **blades** with undulating margins (crisped), 10–30 cm long by 2–7 cm wide, glabrous, reduced above; **petioles** 1–12 cm long, not grooved above; **stipules** (ocreae) membranous, sheathlike, veined, to *c.* 10 mm long, enveloping the stem, tearing with age.

■ **STEM** Erect, glabrous, solid, round, usually branched; 5–15 mm wide near the base.

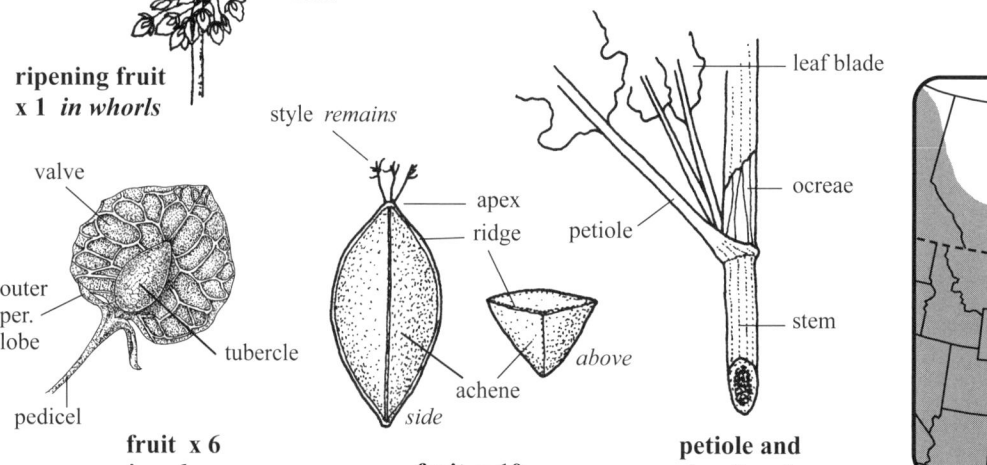

Yellow Dock, Curly Dock

Buckwheat—*Polygonaceae* — Wildflower   Green   Perianth tubular, 6-lobed

Golden Dock   *Rumex maritimus* L.

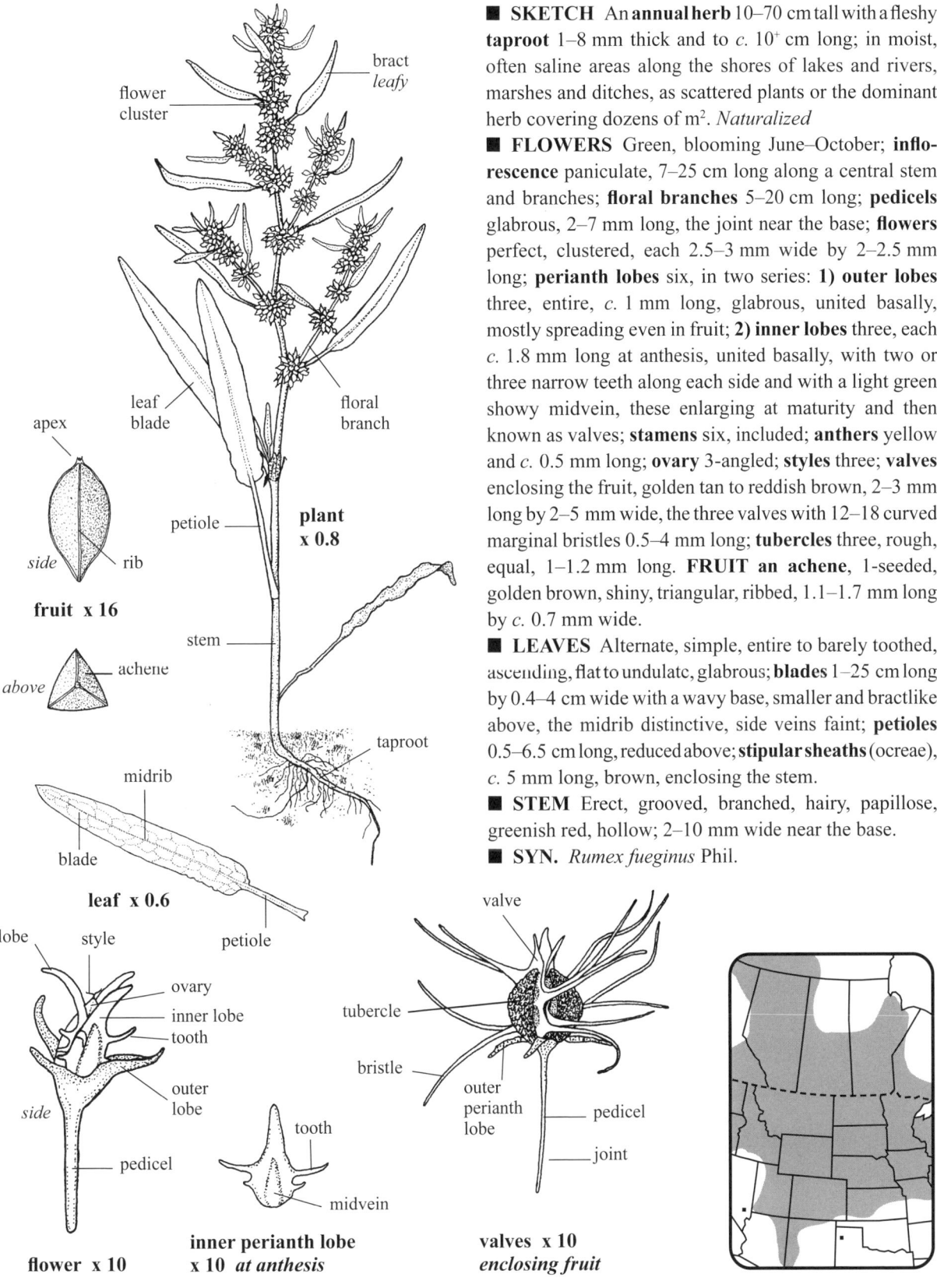

■ **SKETCH** An **annual herb** 10–70 cm tall with a fleshy **taproot** 1–8 mm thick and to *c.* 10⁺ cm long; in moist, often saline areas along the shores of lakes and rivers, marshes and ditches, as scattered plants or the dominant herb covering dozens of m². *Naturalized*

■ **FLOWERS** Green, blooming June–October; **inflorescence** paniculate, 7–25 cm long along a central stem and branches; **floral branches** 5–20 cm long; **pedicels** glabrous, 2–7 mm long, the joint near the base; **flowers** perfect, clustered, each 2.5–3 mm wide by 2–2.5 mm long; **perianth lobes** six, in two series: **1) outer lobes** three, entire, *c.* 1 mm long, glabrous, united basally, mostly spreading even in fruit; **2) inner lobes** three, each *c.* 1.8 mm long at anthesis, united basally, with two or three narrow teeth along each side and with a light green showy midvein, these enlarging at maturity and then known as valves; **stamens** six, included; **anthers** yellow and *c.* 0.5 mm long; **ovary** 3-angled; **styles** three; **valves** enclosing the fruit, golden tan to reddish brown, 2–3 mm long by 2–5 mm wide, the three valves with 12–18 curved marginal bristles 0.5–4 mm long; **tubercles** three, rough, equal, 1–1.2 mm long. **FRUIT an achene**, 1-seeded, golden brown, shiny, triangular, ribbed, 1.1–1.7 mm long by *c.* 0.7 mm wide.

■ **LEAVES** Alternate, simple, entire to barely toothed, ascending, flat to undulate, glabrous; **blades** 1–25 cm long by 0.4–4 cm wide with a wavy base, smaller and bractlike above, the midrib distinctive, side veins faint; **petioles** 0.5–6.5 cm long, reduced above; **stipular sheaths** (ocreae), *c.* 5 mm long, brown, enclosing the stem.

■ **STEM** Erect, grooved, branched, hairy, papillose, greenish red, hollow; 2–10 mm wide near the base.

■ **SYN.** *Rumex fueginus* Phil.

405

Triangular-valved Dock *Rumex salicifolius* var. *mexicanus* (Meisn.) C.L. Hitchc.

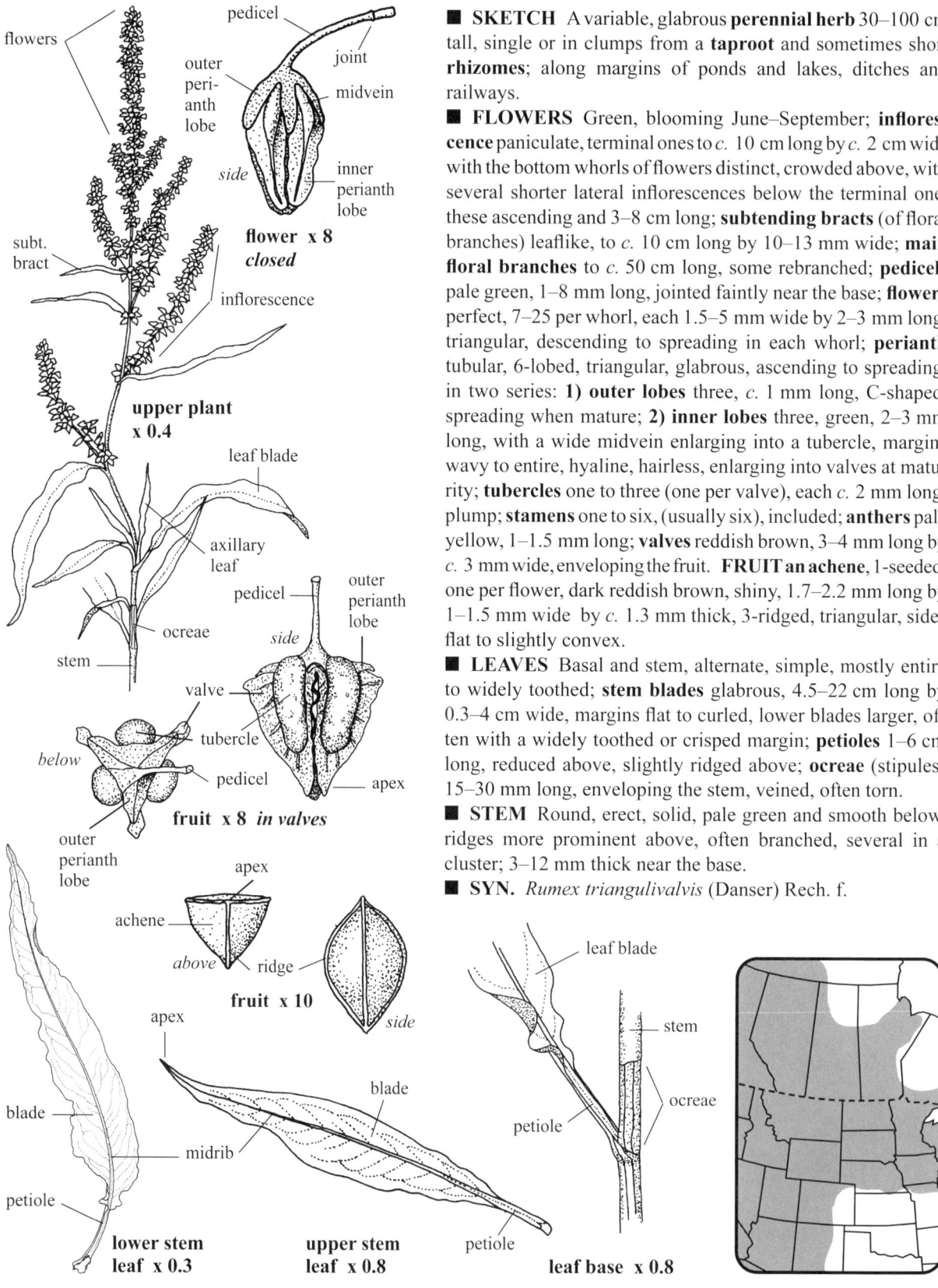

■ **SKETCH** A variable, glabrous **perennial herb** 30–100 cm tall, single or in clumps from a **taproot** and sometimes short **rhizomes**; along margins of ponds and lakes, ditches and railways.

■ **FLOWERS** Green, blooming June–September; **inflorescence** paniculate, terminal ones to *c.* 10 cm long by *c.* 2 cm wide with the bottom whorls of flowers distinct, crowded above, with several shorter lateral inflorescences below the terminal one, these ascending and 3–8 cm long; **subtending bracts** (of floral branches) leaflike, to *c.* 10 cm long by 10–13 mm wide; **main floral branches** to *c.* 50 cm long, some rebranched; **pedicels** pale green, 1–8 mm long, jointed faintly near the base; **flowers** perfect, 7–25 per whorl, each 1.5–5 mm wide by 2–3 mm long, triangular, descending to spreading in each whorl; **perianth** tubular, 6-lobed, triangular, glabrous, ascending to spreading, in two series: **1) outer lobes** three, *c.* 1 mm long, C-shaped, spreading when mature; **2) inner lobes** three, green, 2–3 mm long, with a wide midvein enlarging into a tubercle, margins wavy to entire, hyaline, hairless, enlarging into valves at maturity; **tubercles** one to three (one per valve), each *c.* 2 mm long, plump; **stamens** one to six, (usually six), included; **anthers** pale yellow, 1–1.5 mm long; **valves** reddish brown, 3–4 mm long by *c.* 3 mm wide, enveloping the fruit. **FRUIT** an achene, 1-seeded, one per flower, dark reddish brown, shiny, 1.7–2.2 mm long by 1–1.5 mm wide by *c.* 1.3 mm thick, 3-ridged, triangular, sides flat to slightly convex.

■ **LEAVES** Basal and stem, alternate, simple, mostly entire to widely toothed; **stem blades** glabrous, 4.5–22 cm long by 0.3–4 cm wide, margins flat to curled, lower blades larger, often with a widely toothed or crisped margin; **petioles** 1–6 cm long, reduced above, slightly ridged above; **ocreae** (stipules) 15–30 mm long, enveloping the stem, veined, often torn.

■ **STEM** Round, erect, solid, pale green and smooth below, ridges more prominent above, often branched, several in a cluster; 3–12 mm thick near the base.

■ **SYN.** *Rumex triangulivalvis* (Danser) Rech. f.

Purslane—*Portulacaceae* **Wildflower** Yellow Petals 5 (4 or 6)

PURSLANE *Portulaca oleracea* L.

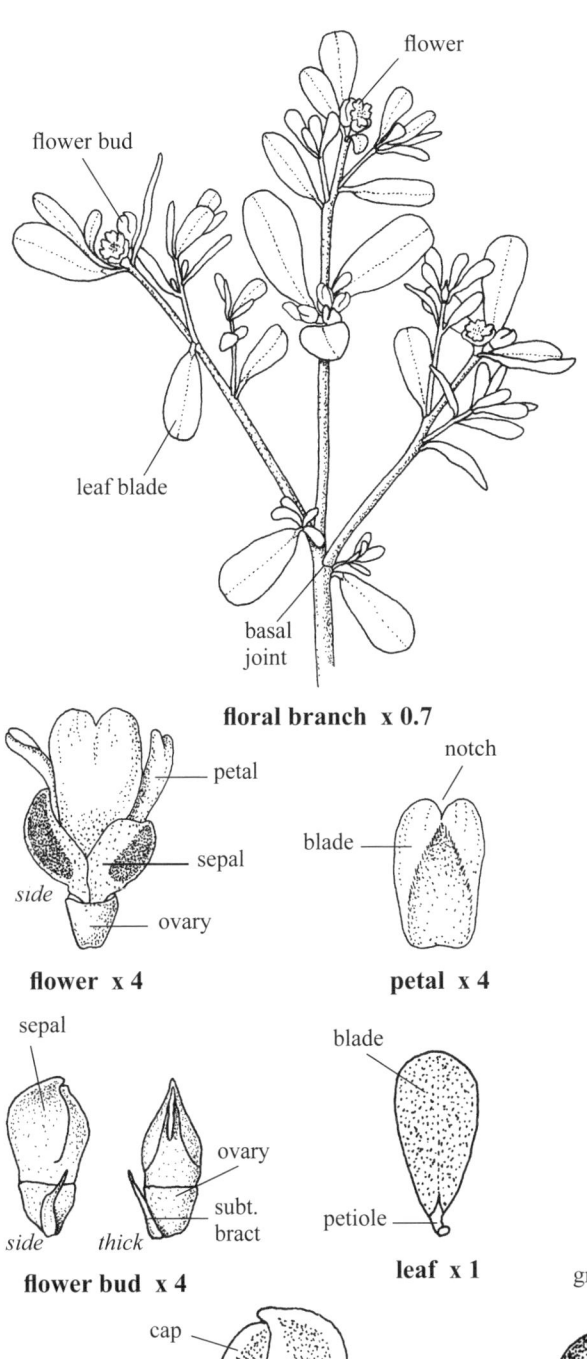

floral branch x 0.7
flower x 4
petal x 4
flower bud x 4
leaf x 1
fruit x 5

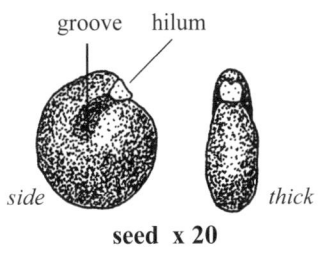

seed x 20

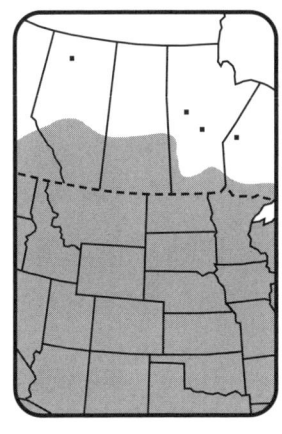

■ **SKETCH** A fleshy, glabrous, variable **annual herb** forming dense prostrate mats 20–80 cm across from a thick **taproot** 2–11 cm long; in disturbed areas, cultivated fields, city lots, gardens, mountain slopes and sidewalks. *Naturalized* **w2**

■ **FLOWERS** Yellow, blooming April–October; **inflorescence** of one to five sessile flowers at the base of a leafy rosette along the many branches, usually only one flower per cluster in bloom at a time; **subtending bracts** (of flowers) one, *c.* 2 mm long; **flower buds** pale yellow, *c.* 5 mm long by 2.8–3 mm wide by *c.* 2 mm thick; **flowers** perfect, 5–10 mm wide by 4–6 mm long, sessile, open in bright light about noon; **sepals** two, green, keeled, 2.5–4.5 mm long by 2.8–3.8 mm wide, the shorter one fitting into the slightly larger one in buds, fused to ovary, spreading but closing by early afternoon, persistent in fruit; **petals** five (four or six), each 3–6 mm long by 2–4 mm wide, with two rounded apical lobes; **stamens** 6–15[+], yellow, included; **styles** yellow, filiform, included, *c.* 2 mm long, 4- to 6-parted. **FRUIT a capsule**, tan inside the base, 2.8–3 mm across by 3–8 mm long, round, greenish red, persistent; **cap** *c.* 4 mm long, deciduous, flattened above; **seeds** 40–50 per capsule, black, slightly rough, circular and flattened, 0.7–1 mm long by *c.* 0.7 mm wide by *c.* 0.4 mm thick.

■ **LEAVES** Alternate to subopposite, simple, entire, medium green above, lighter green below, with only the faint midrib visible on either side; **blades** flat, fleshy, glabrous, 1–5 cm long by 5–20 mm wide, subtending the branches; **petioles** 1–3 mm long, jointed; **stipules** reduced to bristles.

■ **STEM** Round, fleshy, solid, glabrous, shiny, green to pink, prostrate; **branches** easily snapped off by hand at their basal joints; 5–6 mm thick near the taproot.

Common Purslane, Little Hogweed, Pusley

Primrose—*Primulaceae*   **Wildflower**   White to pinkish   Corolla tubular, 5-lobed

WESTERN PYGMYFLOWER   *Androsace occidentalis* Pursh

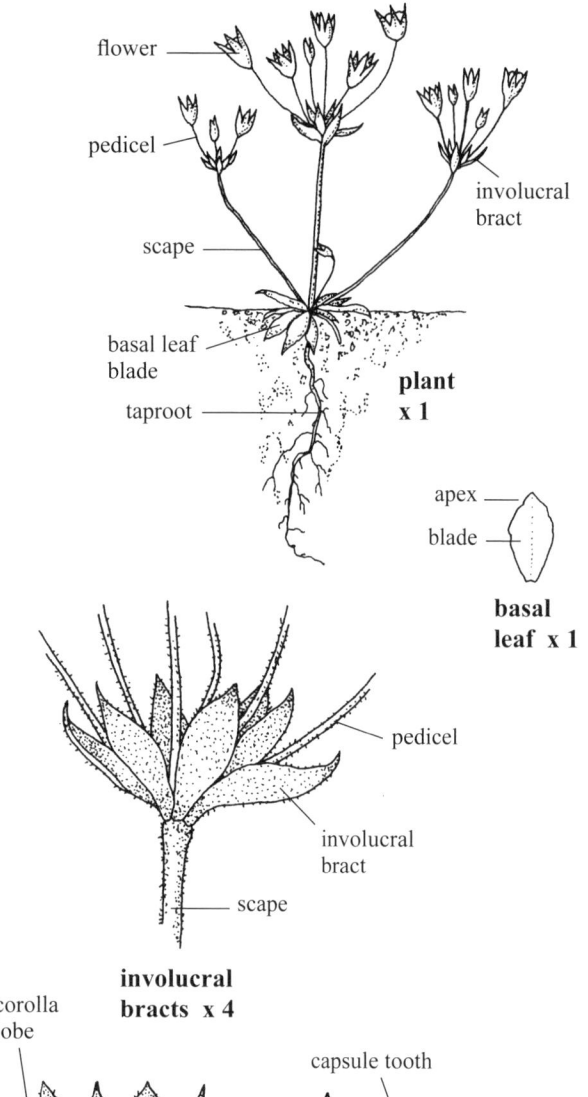

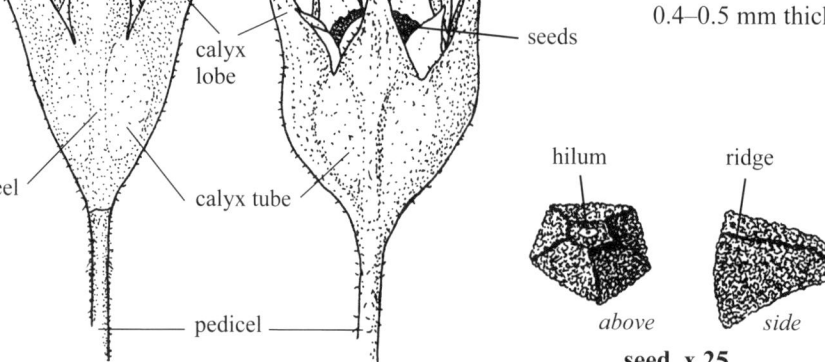

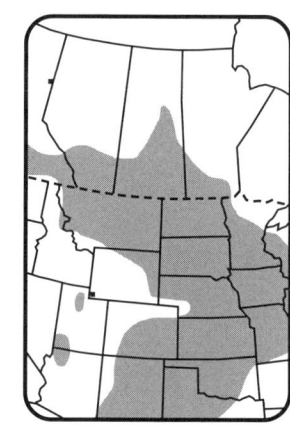

■ **SKETCH** An **annual herb** 2–13 cm tall from a tan **taproot** 2–5 cm long by 0.4–0.6 mm thick; in dry sandy sites in cities, along sidewalks and parking lots, in open woods and dry hillsides.

■ **FLOWERS** White to pinkish, blooming March–June; **inflorescence** an umbel, one to several per plant, terminal and lateral; **pedicels** 2–17 per umbel, unequal, each 3–42 mm long, hairs microscopic and stellate; **involucral bracts** (of pedicels) erect to spreading, usually 3.5–6.5 mm long by 0.8–4.2 mm wide, largest ones below outer pedicels on central scape; **flowers** perfect, 3–4 mm wide by 4–5 mm long; **calyx** tubular, 5-lobed, green, keeled, tube 1.8–3.5 mm long, hairy; **calyx lobes** pointed, 2–3.5 mm long, erect, tips reddish, slightly longer than the tube on most flowers; **corolla** tubular, 5-lobed, usually shorter than or equal to calyx length, constricted at base of lobes then expanding around the ovary, lobes *c.* 1 mm long; **stamens** five, included; **pistil** included. **FRUIT** a **capsule**, 5-valved, opening above, TL *c.* 3 mm by 2–2.5 mm wide, enclosed within calyx, glabrous, somewhat transparent, 5-toothed, these *c.* 1.4 mm wide; **seeds** dark brown, several per capsule, pyramidal, 5- or 6-sided, 0.7–0.9 mm long, with ridges, bases slightly curved, apices tapered.

■ **LEAVES** Basal, simple, slightly toothed near apices, the rosette 1–6 cm wide; **blades** mostly sessile, numerous, tapered to the base, 6–30 mm long by 2–7 mm wide, minute hairs above, these spreading and obvious along the margins, glabrous below, veins obscure.

■ **STEM** A **scape**, one to several, erect to ascending, hairs stellate, green to reddish at the base, usually 2–10 cm long; the central scape erect and 0.6–0.7 mm thick compared to the ascending lateral scapes at 0.4–0.5 mm thick.

Western Rock Jasmine

Primrose—*Primulaceae*   **Wildflower** White, center yellow Corolla tubular, 5-lobed

PYGMYFLOWER   *Androsace septentrionalis* L.

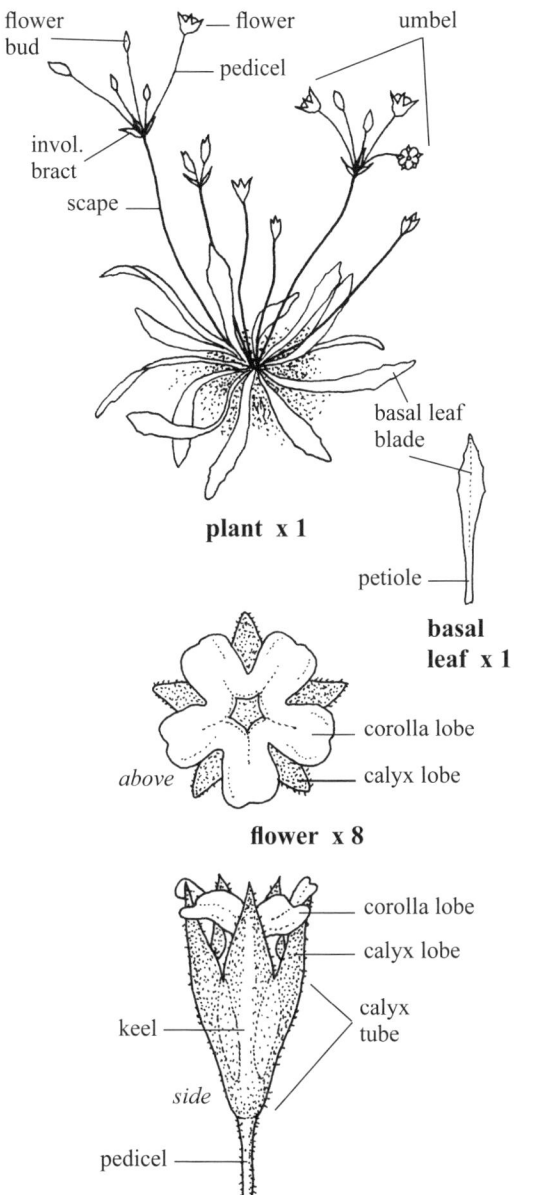

■ **SKETCH** A (winter) **annual herb** 2–30 cm tall from a thin **taproot**; in dry sandy prairies, stubble fields, open woods, ravines, rocky outcrops, meadows, lawns and edges of parking lots.

■ **FLOWERS** White to slightly tinged with pink, center yellow, blooming March–August; **inflorescence** an umbel on a scape, one to several; **pedicels** 2–20, green, ascending, 0.3–8.5 cm long, short hairy; **involucral bracts** (of pedicels) one, ascending, slightly hairy, 2.5–5 mm long by 0.5–1 mm wide, pointed, slightly reddish; **flowers** perfect, 2–4 mm wide by 3–4 mm long, 2–20 per umbel; **calyx** TL 2–5 mm; **calyx tube** whitish green, usually 2–3 mm long, concave between the keels; **calyx lobes** five, triangular, slightly keeled, tips reddish and pointed, glabrous to hairy on outside, 1–2 mm long by 1–1.3 mm wide, erect in fruit; **corolla** tubular, 5-lobed, TL 2–3.3 mm; **corolla tube** green, 1.8–2 mm long by 0.8–1 mm wide, constricted slightly below the lobes; **lobes** glabrous, 1–1.3 mm long by 0.8–1 mm wide, apices blunt and with a shallow notch; **stamens** five, included, *c.* 0.5 mm long, attached to base of corolla tube, opposite corolla lobes; **filaments** slightly shorter than anthers; **anthers** yellow, 0.3–0.4 mm long; **ovary** green, glabrous, *c.* 1 mm long and wide. **FRUIT a capsule**, erect, tan, 5-valved, 2.4–2.8 mm long by *c.* 2 mm wide, opening at apex, seeds partially visible inside; **teeth** five, *c.* 1.5 mm long by 1–1.2 mm wide; **seeds** dark brown, 10–25 per capsule, irregular in shape, usually 5- or 6-sided, tapered to a blunt apex, *c.* 1 mm long by 0.6–0.7 mm wide and thick, edges or low ridges between facets are slightly rough, hilum at narrow tip.

■ **LEAVES** Basal, simple, the rosette 1–8 cm wide; **blades** flat, entire to toothed near the apex, 6–45 mm long by 1–9 mm wide, thin, dull, pointed, gradually tapered, hairy above and on margins, glabrous below, midrib somewhat obvious above, more visible below, lateral veins obscure; **petioles** 0–1.5 cm long.

■ **STEM** A scape, 1–40+ per plant, erect, green to reddish, 2–20 cm long, central one the thickest at 0.5–0.7 mm, lateral ones slightly thinner; **hairs** short, stellate, white, some glandular.

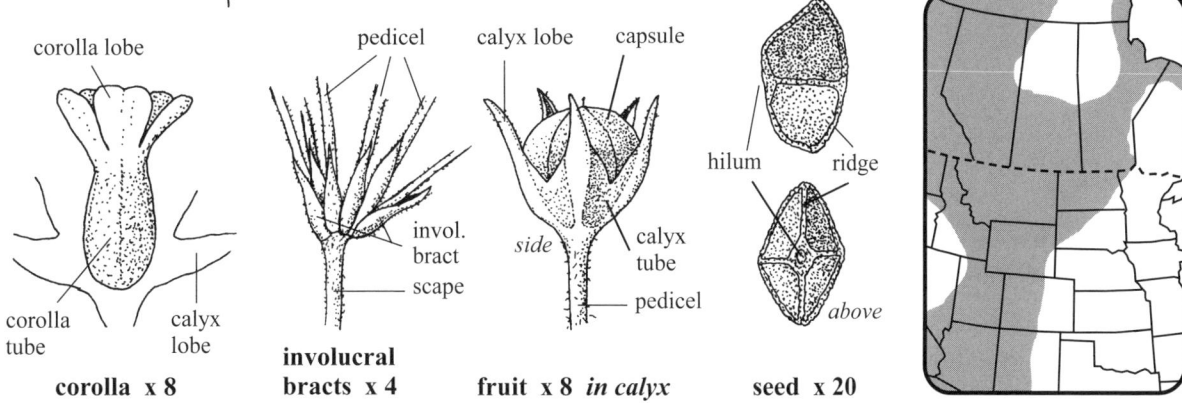

Northern Pygmy-flower, Pygmyflower Rockjasmine, Northern Rock Jasmine

Primrose—*Primulaceae*   **Wildflower**   White to pinkish   Calyx tubular, 5-lobed

SEA-MILKWORT   *Glaux maritima* L.

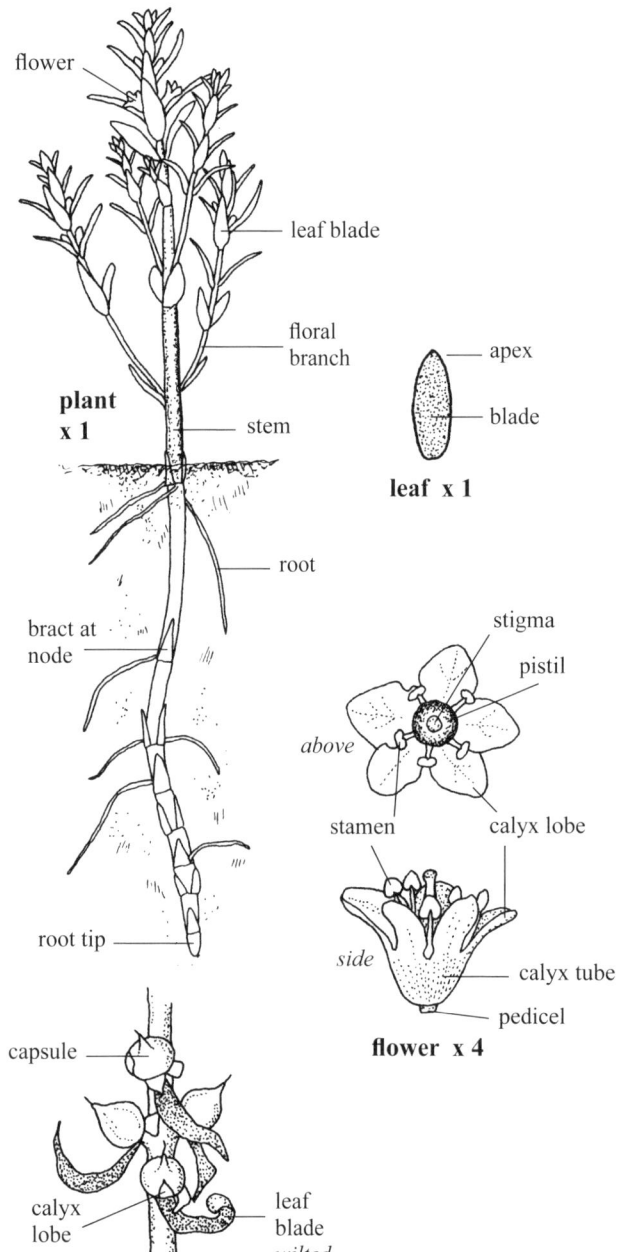

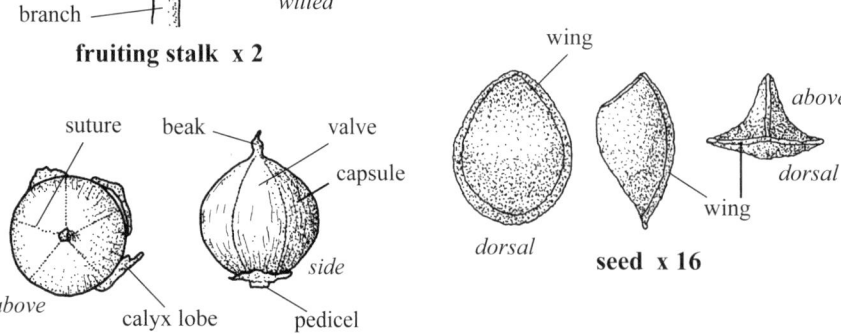

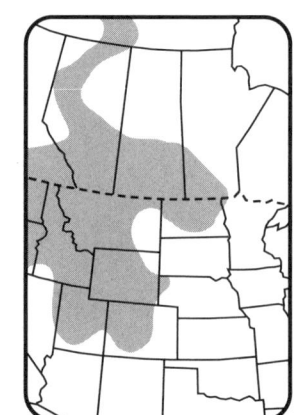

■ **SKETCH** A glabrous **perennial herb** 3.5–20 cm tall, often in patches, from a long whitish tan **rootstock** with opposite **bracts** and white roots and horizontal slender **rhizomes**; along saline pond edges, beaches, salt flats and in wet meadows.

■ **FLOWERS** White to pinkish, blooming May–July; **inflorescence** a raceme (spikelike), axillary, one flower per leaf axil, blooming sequence bottom to top; **flowers** perfect, 5–6.5 mm wide by 3–4 mm long, subsessile; **pedicels** very short; **calyx** tubular, 5-lobed, 3–4 mm long, tube *c.* 2.5 mm wide; **calyx lobes** round, 2–2.2 mm long by 1.7–1.8 mm wide, hairless, tips white; **petals** absent; **stamens** five, included, between the calyx lobes; **filaments** pinkish, 2–2.2 mm long; **anthers** yellow with pink spots, *c.* 0.8 mm long by *c.* 0.7 mm wide; **pistil** one, *c.* 3.2 mm long, exserted, expanded green base *c.* 1.5 mm long by *c.* 1.2 mm wide; **stigma** slightly enlarged. **FRUIT a capsule**, 2.2–3 mm wide and tall, 5-valved, reddish brown, slightly rough, subsessile, ascending; **beak** (persistent style) to *c.* 1 mm long; **seeds** black, six or seven, minutely rough, triangular, 1–1.5 mm long by *c.* 0.7 mm thick by *c.* 1 mm wide across the winged, outer convex surface, shed with placenta as a unit out the top.

■ **LEAVES** Opposite below, becoming alternate above, simple, entire, fleshy, pliable, sessile, glabrous; **blades** 3–20 mm long by 1–5 mm wide, veins obscure above, midrib slightly raised below.

■ **STEM** Erect to leaning, smooth, leafy, alternatively grooved on sides opposite leaf pairs, reddish especially below, branches opposite; 1–3 mm thick near the fleshy base.

410

Primrose—*Primulaceae* **Wildflower** Yellow Corolla tubular, 5-lobed

FRINGED LOOSESTRIFE *Lysimachia ciliata* L.

■ **SKETCH** A **perennial herb** 10–120 cm tall in small groups from **rhizomes** 1–3 mm wide and to *c.* 10⁺ cm long; in moist open woodlands, wet meadows and stream banks, marshes and ditches.

■ **FLOWERS** Yellow, blooming May–September; **inflorescence** a terminal to axillary raceme; **pedicels** thin, arched, glabrous, 1–7.5 cm long, erect in fruit; **flowers** perfect, inverted, 15–25 mm wide, flat, solitary from the upper one to six leaf axils; **calyx** tubular, 3- or 5-veined, glabrous, 5-lobed, the lobes 2–9 mm long by 1–7 mm wide, pointed and persistent; **corolla** tubular, deeply cut, 5-lobed, the lobes 5–12 mm long by 3–9 mm wide, reddish near the inside base, the apices flat with a short soft mucro; **stamens** five, 4–4.5 mm long, exserted; **filaments** 1–2 mm long, with gland-tipped hairs; **anthers** 2.5–4 mm long, pale yellow; **staminodia** five, sterile, 1–2 mm long, glandular, erect between the filaments; **style** 3.5–5 mm long, thin, slightly enlarged at its tip, persistent. **FRUIT a capsule**, tan, glabrous, 4–5 mm long by 4–8 mm wide, 5-valved, partially covered by the brown calyx lobes; **seeds** 25–40 per capsule, dark brown with a tan netlike pattern and tan ridges along the face edges, hairless, angular, *c.* 2 mm long by 1–1.5 mm wide.

■ **LEAVES** Opposite, simple, entire, widely spaced, four to ten pairs with two or three pairs crowded at the top; **blades** 1.5–15 cm long by 1–7.5 cm wide, lower blades rounder, veins distinctive, glabrous on both surfaces, lighter green below, pointed, margins downy; **petioles** 0.7–7 cm long, reduced above, white hairs along both sides of the upper groove; **stipules** absent.

■ **STEM** Erect but weak, glabrous, usually unbranched, bluntly square in the lower half; reddish near the 1–5 mm thick base; **nodes** above often glandular hairy.

■ **SYN.** *Steironema ciliatum* (L.) Baudo.

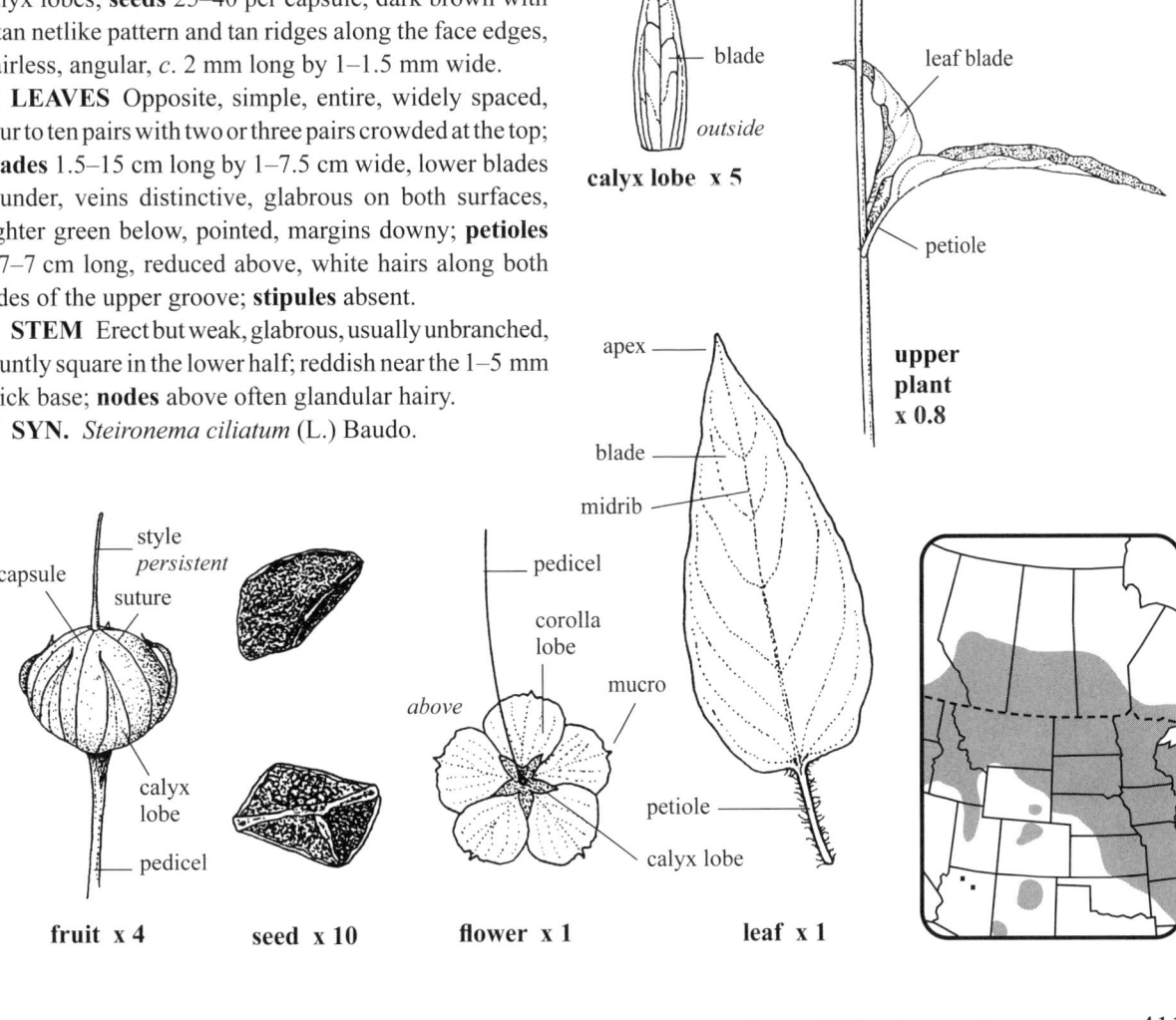

Fringed Yellow Loosestrife

Primrose—*Primulaceae*   Wildflower Yellow Corolla lobes 5–8

Tufted Loosestrife   *Lysimachia thyrsiflora* L.

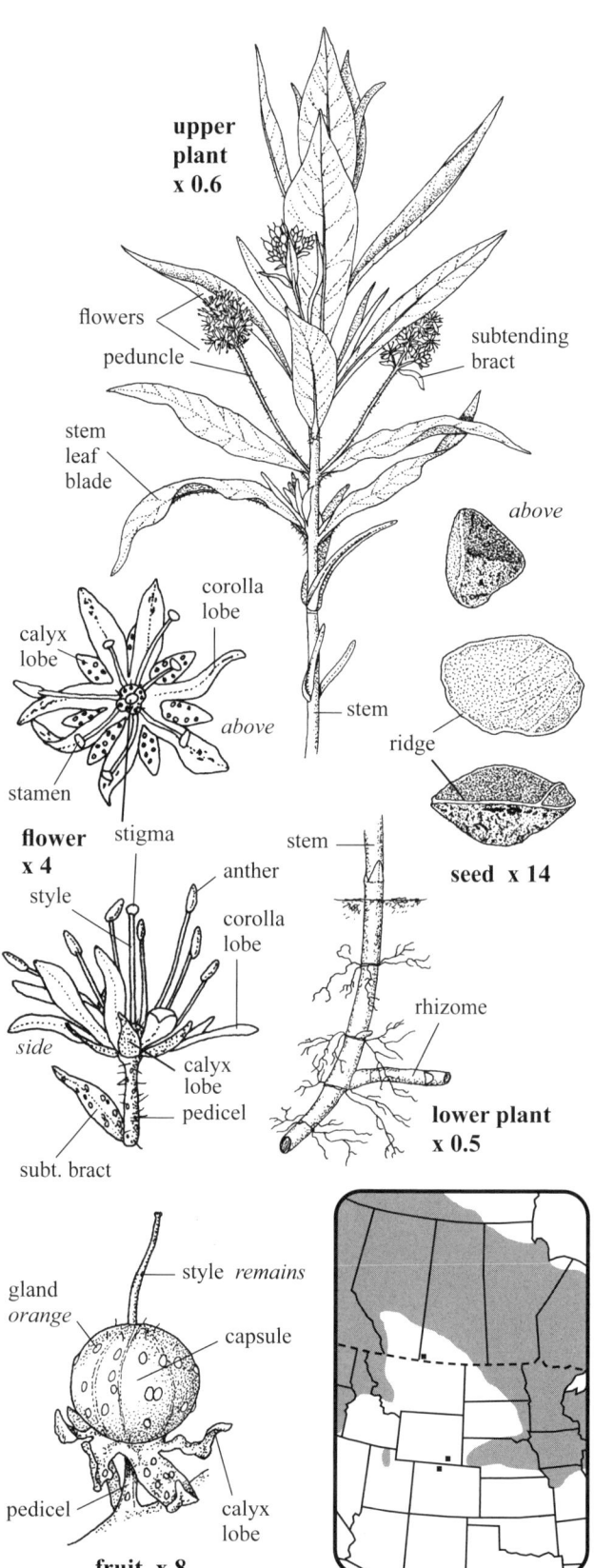

- **SKETCH** A **perennial herb** 20–80 cm tall with white creeping **rhizomes** 1–5 mm thick and to *c.* 30⁺ cm long; in fens, bogs, marshes and wet meadows, sometimes in shallow water.
- **FLOWERS** Yellow, blooming May–August; **inflorescence** a raceme, 2–13 per stem, each 0.8–2.8 cm long by 1.5–2.5 cm wide; **peduncles** axillary, ascending to spreading, 1.4–5.5 cm long by 0.7–1 mm wide, hairy; **pedicels** 0.5–4 mm long, hairy to glabrous to glandular, hairs straight and spreading; **subtending bracts** (of pedicels) 3–12 mm long, orange glandular-dotted; **flowers** 6–12 mm wide by 5–7 mm tall; **calyx** 5- to 7-lobed, glabrous, lobes 2–3 mm long by 0.6–1 mm wide, green, pointed, spreading, with reddish orange scattered glands; **corolla lobes** five to eight, linear, yellow, united at base, 3–5 mm long by 1–2 mm wide, glabrous, orange dotted above, midvein obvious; **stamens** usually five to seven, yellow, exserted, glabrous, 4–6 mm long, opposite the corolla lobes; **filaments** 3–5 mm long, filiform; **anthers** yellow, 0.7–1.2 mm long by *c.* 0.5 mm wide; **pistil** exserted, 5–7 mm long; **ovary** *c.* 0.8 mm high by *c.* 1.2 mm wide, green, roundish, slightly lobed, red gland-dotted, slightly hairy; **style** one, filiform, glabrous, 4–6 mm long, persistent; **stigma** round, *c.* 0.4 mm across. **FRUIT** a **capsule**, 2–2.5 mm tall (not including style) by 2–3.5 mm wide, globular, erect, slightly hairy near the apex, 5-valved, with numerous orange glands; **style** persistent, 2–3 mm long but often broken near the capsule; **seeds** six to eight per capsule, tan, pitted, 3-angled, slightly convex on two sides, 1.2–1.5 mm long by *c.* 1 mm wide by *c.* 0.8 mm thick, central ridge blunt.
- **LEAVES** Opposite, simple, entire, ascending to spreading, in upper half of stem; **upper stem blades** pointed, 2–16 cm long by 0.4–5 cm wide, reduced below, margins with orangish brown glands (dots) and some glands scattered below, midrib raised and hairy below, glandular and hairless above, lateral veins obvious; **petioles** 0–5 mm long, hairy and reduced below; **lower stem leaves** bractlike, *c.* 5 mm long by 5–6 mm wide, scabrous, appressed, light tannish pink.
- **STEM** Erect, green and hairy above especially at leaf bases, reddish and glabrous below to glandular-punctate throughout, sometimes branched from lower nodes, roundish below to bluntly squarish above; 3–7 mm thick near the base.
- **SYN.** *Naumburgia thyrsiflora* (L.) Duby.

412   Tufted Yellow-loosestrife

Primrose—*Primulaceae*   **Wildflower**   Pale pinkish purple, center yellow   Corolla tubular, 5-lobed

DWARF PRIMROSE   *Primula mistassinica* Michx.

■ **SKETCH** A variable **perennial herb** 2–15 cm tall with white **fibrous roots** 2–12 cm long, in small open colonies; in wet fields, marshy areas, cliffs, disturbed sites and along shores in forests.

■ **FLOWERS** Pale pinkish purple, center yellow, blooming May–June; **inflorescence** an umbel of 1–10 flowers, together 10–30 mm wide by 1–2 cm tall; **pedicels** green, filiform, 4–23 mm long (to 25 mm and erect with fruit); **involucral bracts** (of pedicels) mealy below, pointed, 3.5–7 mm long; **flowers** perfect, 10–18 mm wide by 6–9 mm long; **calyx** tubular, 5-lobed, erect, glabrous, TL 3–5 mm, lobes 1.5–2 mm long, hairless, pointed, pale yellow and mealy inside; **corolla** tubular, 5-lobed (3- or 4-); **corolla tube** slightly longer than calyx, yellow, 5–8 mm long, apical opening *c.* 1.2 mm wide, persistent until fruit is ripe, then falling off as a unit exposing the lid; **corolla lobes** 1.5–2 mm long by *c.* 4 mm wide, notched, bases yellow, hairless and not mealy; **stamens** five, *c.* 1 mm long, included; **filaments** white, *c.* 0.2 mm long, attached near top of corolla tube; **anthers** 0.8–1.3 mm long, yellow, leaning over the top of the stigma; **pistil** 3–4 mm long, glabrous, included, green; **ovary** 1–1.8 mm long by *c.* 1.2 mm wide; **style** one, 2–3.7 mm long, persistent, deciduous with the lid; **stigma** capitate, light green, slightly wider than the style. **FRUIT a capsule**, erect, smooth, 2–5 clustered atop scape, tan, apical lid *c.* 1 mm wide, capsule body 5–7 mm long by 1.3–2 mm wide, light tan to reddish brown; **seeds** brown, *c.* 100 per capsule, 4-sided, winged along one end, dull, 0.3–0.7 mm long by *c.* 0.5 mm wide by *c.* 0.3 mm thick.

■ **LEAVES** Basal only, a rosette of 8–15 simple leaves, ascending to spreading; **blades** 1–5 cm long by 4–16 mm wide, apices shallowly toothed, slightly mealy (yellowish) below; **petiole** 0.3–3 cm long, winged, mealy.

■ **STEM** A **scape**, erect, round and solid, not mealy, smooth, reddish green; 0.6–1.7 mm wide near the base.

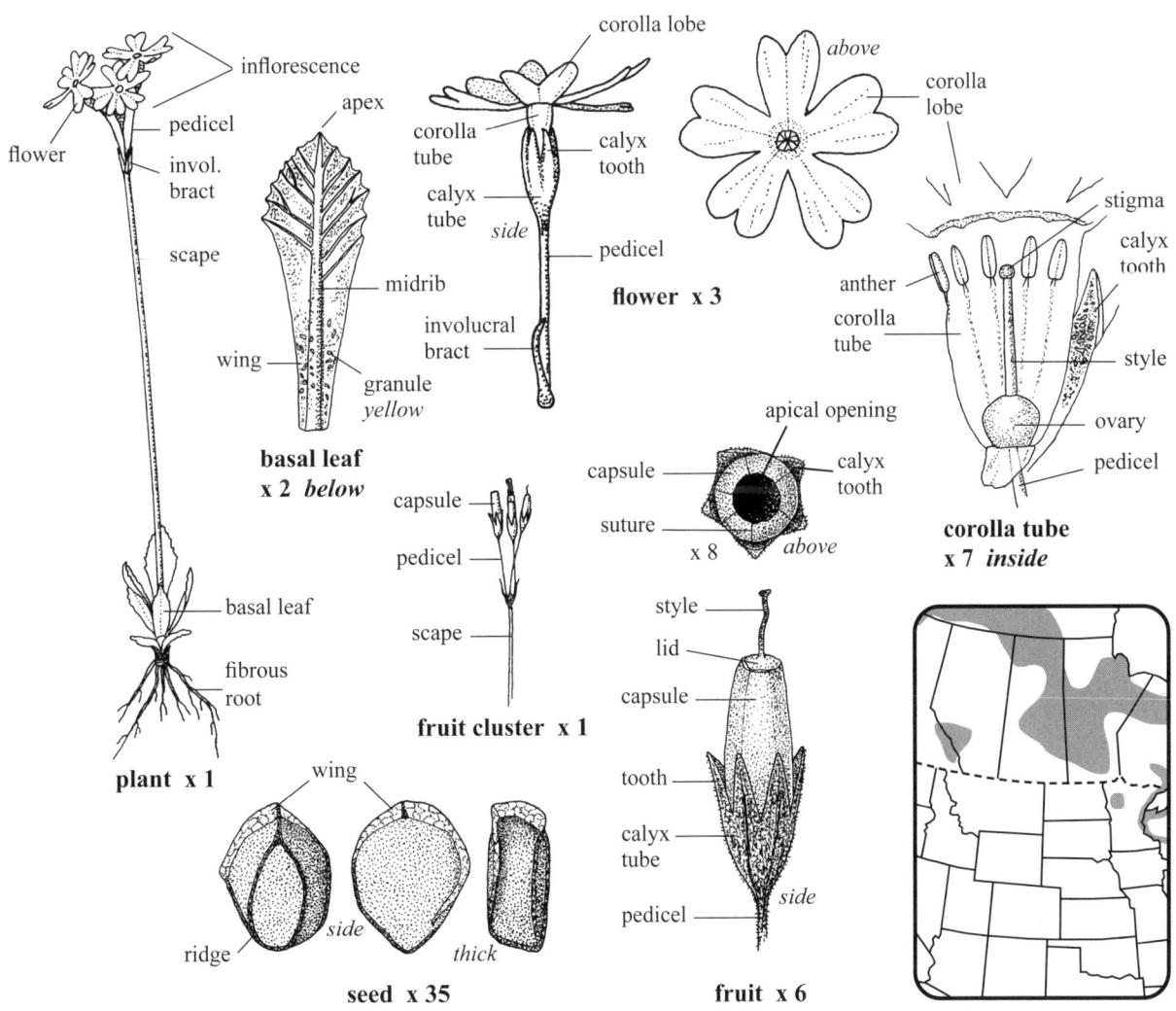

Dwarf Canadian Primrose, Mistassini Primrose, Lake Mistassini Primrose

Primrose—*Primulaceae* **Wildflower** White Petals 6 (5–9)

Northern Starflower *Trientalis borealis* Raf.

■ **SKETCH** A **perennial herb** 6–18 (–30) cm tall with white, smooth **rhizomes** 2–4⁺ cm long by *c.* 1 mm thick with white **fibrous roots** to *c.* 7 cm long; in damp open northern woods, often forming large open colonies.

■ **FLOWERS** White, blooming May–July; **inflorescence** terminal, 1–4 flowers; **pedicels** green, ascending, 15–45 mm long by *c.* 0.2 mm thick, axillary, slightly glandular and hairy; **flowers** perfect, 10–21 mm wide by 5–8 mm tall (including stamens); **sepals** green, equal to the number of petals, 3–7 mm long by *c.* 1 mm wide, thin; **petals** six (5–9), white, pointed, 4–10 mm long by 2–4.5 mm wide, glabrous, veins faint, spreading, slightly erose near tip; **stamens** six or seven, exserted, 3.5–4.5 mm long, glabrous, attached to top of light green glandular ring; **filaments** white, filiform, ascending, 3–4 mm long; **anthers** yellow, curled inward when spent, *c.* 1.5 mm long, 4-lobed; **pistil** one, green, erect, 4.5–5.5 mm long; **ovary** *c.* 1.4 mm tall by *c.* 0.8 mm wide, lower half enclosed by calyx; **style** pale green, filiform, persistent on fruit and *c.* 4 mm long; **stigma** obscure, rough, flat, slightly wider than style. **FRUIT a capsule**, tan, hairless, pointed, 2.3–2.5 mm long by 2–2.5 mm wide, erect, 5- or 6-valved, sutures opening from base, shiny and smooth inside; **seeds** 9–11 per capsule, dark brown, *c.* 1.4 mm long by 1–1.2 mm wide by 0.7–0.8 mm thick, bluntly angular, enclosed in a white, net-veined, easily removed material.

■ **LEAVES** Stem, and in a terminal whorl; **stem blades** bractlike, two or three, ascending, sessile, 2.5–8 mm long by 1–3 mm wide, red-streaked, with red marginal glands, simple to three leaflets, each with an axillary reddish bud; **terminal leaves** whorled, 5–10, spreading, glabrous, thin, dull, imbricate; **blades** 2–10 cm long by 6–35 mm wide, margins toothed (microscopic) the teeth up to *c.* 0.2 mm long; **petioles** 1–5 mm long.

■ **STEM** Erect, weak, round, solid, simple, light green with red streaks; reddish near the 0.7–1.5 mm wide base.

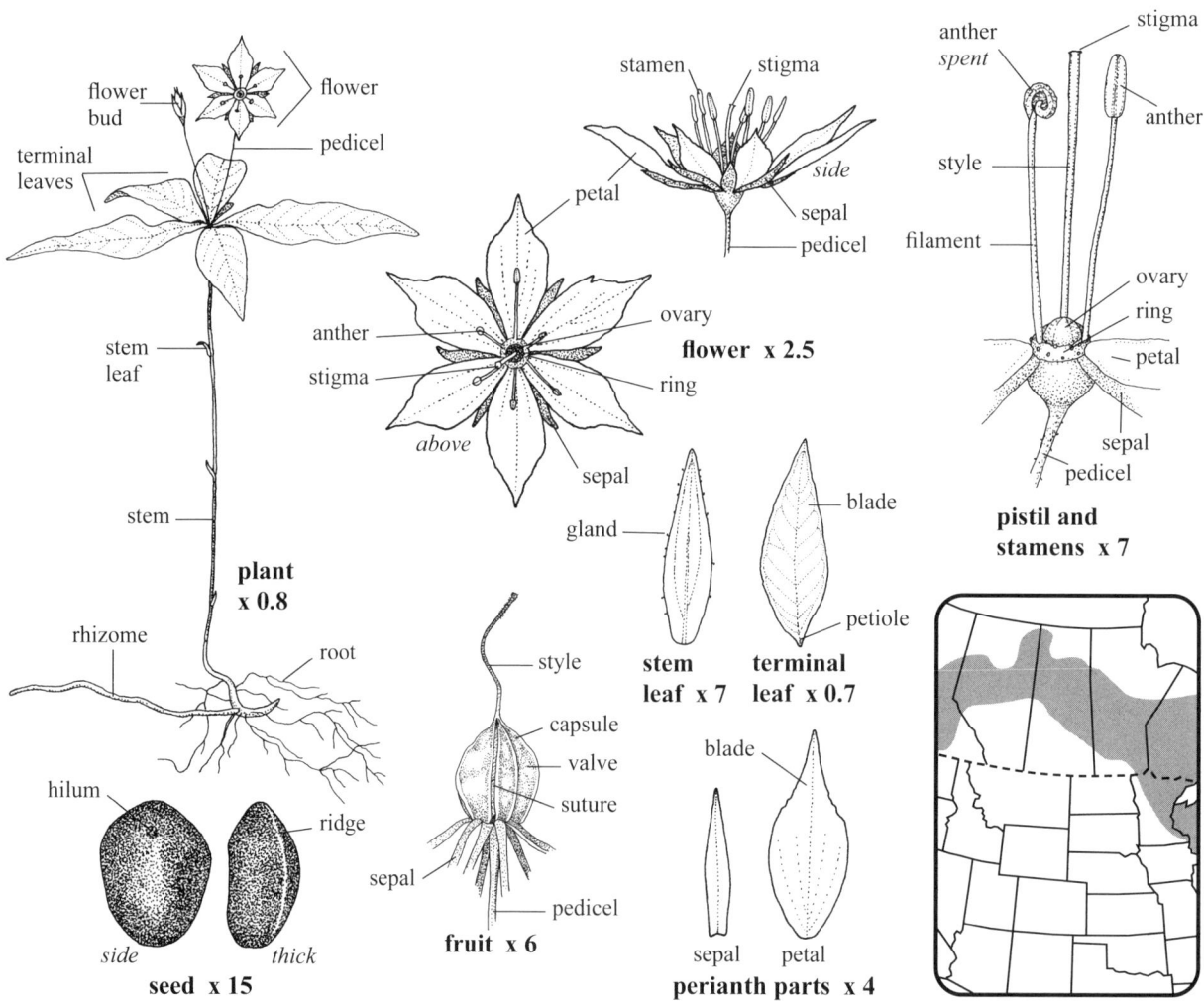

American Starflower, Starflower, Maystar

Wintergreen—*Pyrolaceae*

Wildflower White Petals 5 (4)

ONE-FLOWERED WINTERGREEN *Moneses uniflora* (L.) Gray

■ **SKETCH** A **perennial herb** 3–10 (–15) cm tall in small groups or singly with a whitish tan **rhizome** 5–20 cm long by *c.* 0.2 mm thick and thin **roots**; in mossy, boggy areas of forest and parklands.

■ **FLOWERS** White, blooming July–August; **inflorescence** a single, terminal, nodding flower; **flowers** perfect, fragrant, 1.5–2.5 cm wide by 10–12 mm long (including stigma); **sepals** five (four), 2–2.5 mm long by *c.* 2 mm wide, greenish yellow, margins ciliate and white hyaline, persistent, spreading and appressed to the petals, eventually reflexed, apices blunt, body slightly ridged or wrinkled; **petals** five (four), 8–12 mm long by 7–8 mm wide, spreading, margins undulate, veins faint; **stamens** 10, included, attached at base of ovary; **filaments** 5–6 mm long, incurved, wider below, whitish green; **anthers** pale yellowish green, 2–3.5 mm long, 4-lobed, apical tubes golden brown and pointing outward; **pistil** one, green, exserted, glabrous, TL *c.* 8 mm; **ovary** 5-sided, *c.* 4.2 mm wide, 5-grooved; **style** 3–6 mm long, persistent on fruit; **stigma** wider than style, usually 5-lobed, these green and bent inward, then opening and spreading to *c.* 2 mm wide. **FRUIT** a capsule, reddish brown, dull, hairless, 5-lobed, erect or nodding, 4–5 mm long by 4–7 mm wide, opening from depressed apex; **style** (beak) erect, 4–5 mm long; **seeds** numerous, brown, 0.6–0.9 mm long by *c.* 0.1 mm wide and thick, wings net-veined with one end blunt.

■ **LEAVES** Evergreen, three to ten, usually in whorls of three crowded near the base, the lower pair may be opposite; **bracts** small, C-shaped, among the petiole bases; **blades** 0.8–2.5 cm long by 5–12 mm wide, finely toothed to entire, glabrous, dull and darker green above, veins not raised below, easily seen above; **petioles** 3–10 mm long, ascending; **stipules** absent.

■ **STEM** A **scape** (peduncle), pale green, erect but nodding at apex, round, weak, solid, with 0–2 **bracts** 2–3 mm long below the upper curved portion; *c.* 1 mm thick near the bare base.

■ **SYN.** *Pyrola uniflora* L.

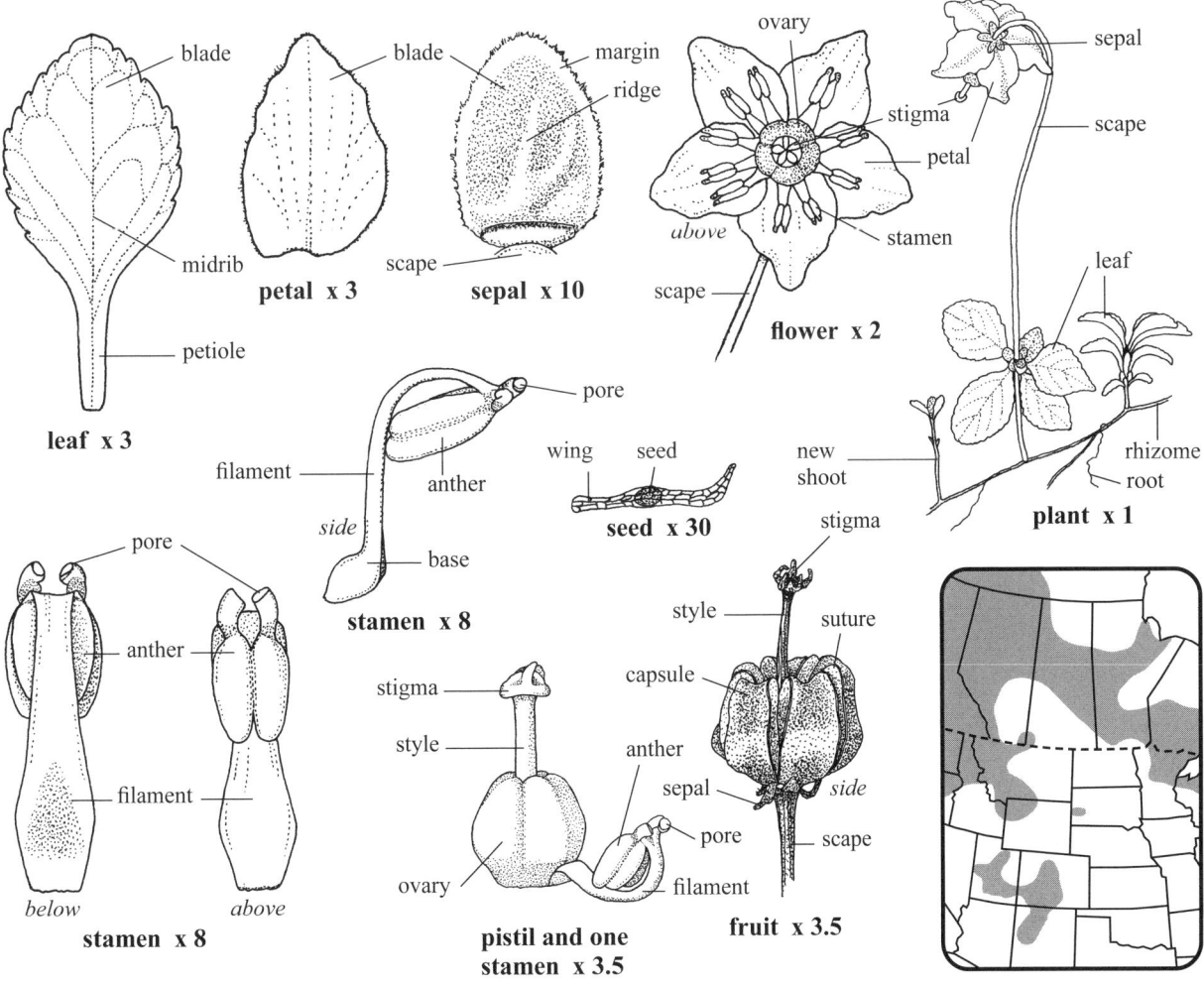

One-flowered Pyrola, Single Delight, Wood-nymph

Wintergreen—*Pyrolaceae*   **Wildflower** Evergreen Pale green Petals 5 (6)

ONE-SIDED WINTERGREEN *Orthilia secunda* (L.) House

■ **SKETCH** An evergreen **perennial herb** 5–20 cm tall in dense colonies or solitary from a stiff, branching **rhizome** 2–5⁺ cm long by *c.* 1 mm wide with a few thin scattered **fibrous roots**; in woods, bluffs and thickets. **w2**

■ **FLOWERS** Pale green, blooming June–August; **inflorescence** a raceme, one-sided, at first erect then horizontal at anthesis, 1–3 cm long by 1–1.5 cm wide; **pedicels** 3–7 mm long, pale green, with lines of microscopic silicate tubercles; **subtending bracts** (of pedicels) 2–4.3 mm long by 1–2 mm wide, pale green, pointed, margins finely erose; **flowers** perfect, 6–20, hanging, 4–5 mm wide by 8–9 mm long (including style), 2-ranked along the rachis; **rachis** pale green, with lines of silicate tubercles; **calyx** 4- or 5-lobed, appressed against petal bases, lobes 0.5–1.2 mm long by *c.* 1 mm wide, margins erose, membranous; **petals** five (6), C-shaped, pale green, 4–6 mm long by 2–4 mm wide (flat), veins obscure, with a glandular ridge near the base and sometimes two tiny basal lobes; **stamens** 10, barely exserted, *c.* 5 mm long; **filaments** pale green, *c.* 5 mm long and bent near the anthers, flattened, attached to base of petals; **anthers** pale yellow, *c.* 2 mm long, pores at end near filaments, becoming inverted; **pistil** one, exserted, green; **ovary** 5-lobed, *c.* 2 mm long by *c.* 3 mm wide, base covered by calyx; **style** pale green, exserted, protruding *c.* 4 mm, straight to bent, thicker than pedicel; **stigma** 5-lobed, *c.* 1.5 mm wide. **FRUIT a capsule**, hanging from erect rachis, 5-lobed, brown, dull, 2.2–2.8 mm tall (excluding style) by 3.5–4.5 mm wide; **seeds** numerous, golden brown, shiny, 0.4–0.6 mm long by *c.* 0.1 mm wide and thick, wing transparent, ridged lengthways, straight to slightly curved.

■ **LEAVES** Basal and stem, alternate, simple, toothed, crowded near stem's base, ascending to spreading; **basal blades** 2–10, dull to shiny, 1–6 cm long by 5–20 mm wide, teeth blunt projections 0.2–0.3 mm long; **petioles** 1–2 cm long, pinkish green; **stem** bracts, sessile, 4–5 mm long by *c.* 2 mm wide, several near base, appressed to ascending, pointed, veins raised below, margins white hyaline, erose to minutely hairy; **stipules** absent.

■ **STEM** Erect, pinkish green, solid, 3- to 5-ridged, roundish; 1.5–2 mm wide near the reddish purple base.

■ **SYN.** *Pyrola secunda* L.

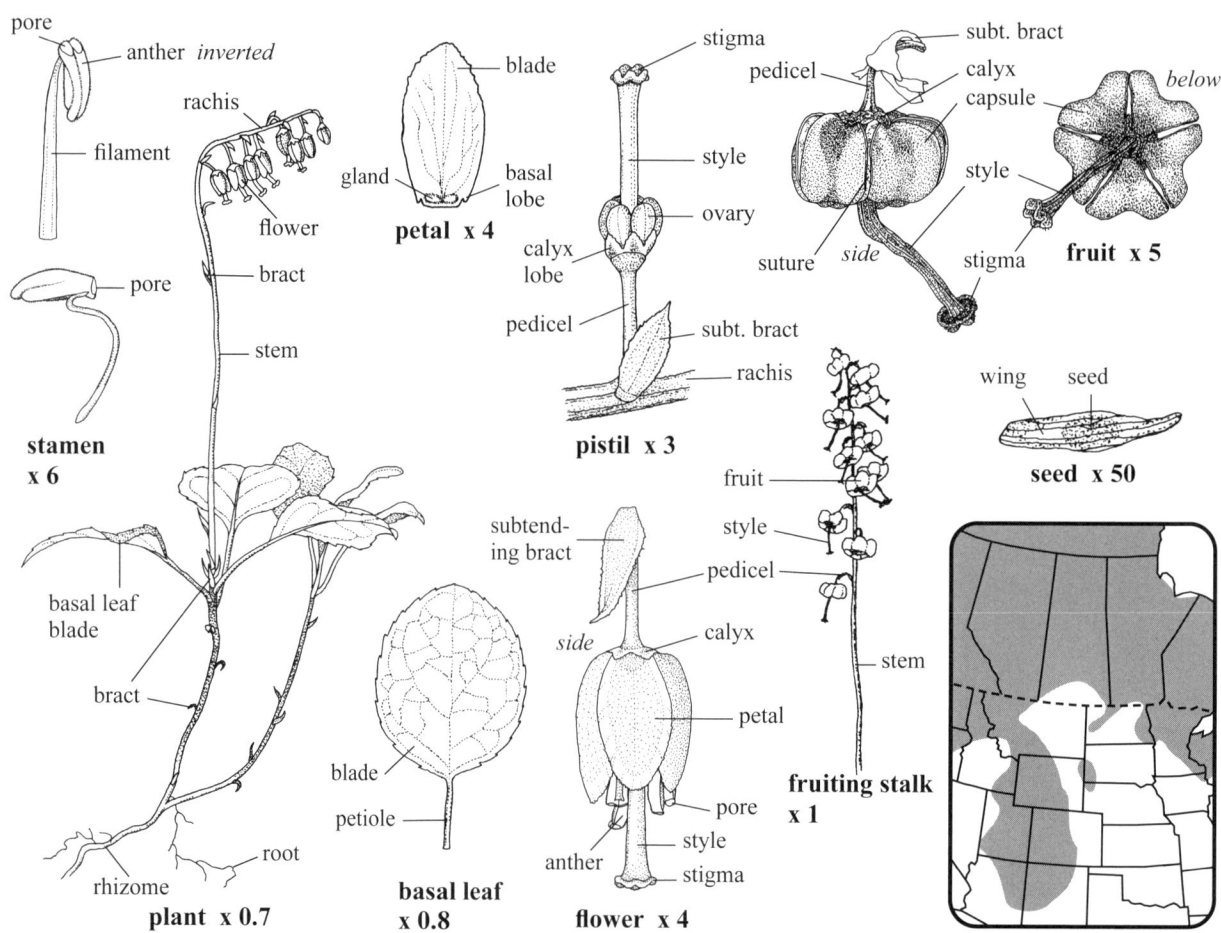

One-sided Pyrola, Sidebells, Sidebells Wintergreen

Wintergreen—*Pyrolaceae* **Wildflower   Pink   Petals 5**

PINK WINTERGREEN   *Pyrola asarifolia* Michx.

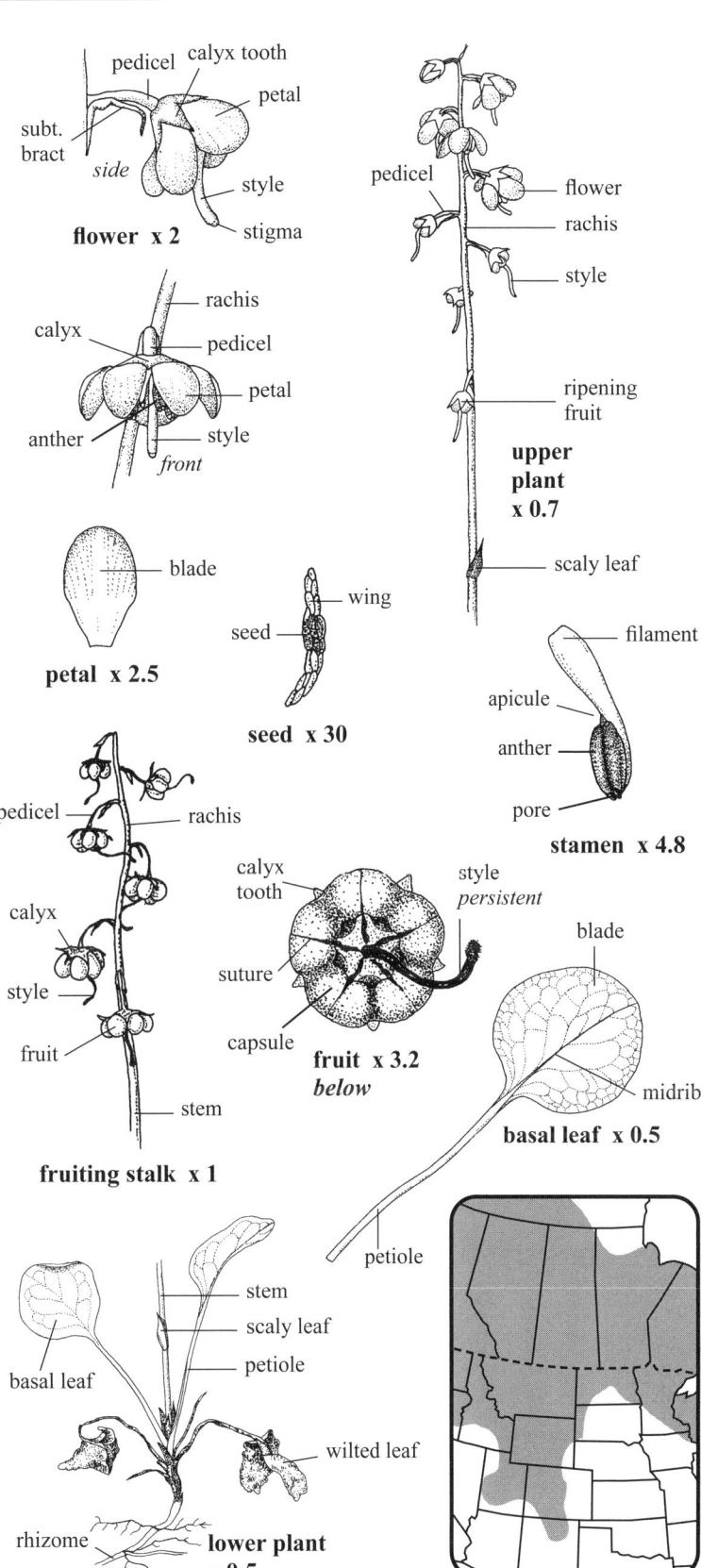

■ **SKETCH** A glabrous, evergreen **perennial herb** 10–30 cm tall from a white **rhizome** *c.* 2 mm thick; in moist woods, bogs, thickets and along treed river banks.
■ **FLOWERS** Pink, blooming July–August; **inflorescence** a raceme 6–22 cm long by 2–3 cm wide; **pedicels** 4–8 mm long by *c.* 0.7 mm thick, curved downward; **subtending bracts** (of pedicels) 4–6 mm long by 1–1.5 mm wide, pointed, entire; **flowers** perfect, nodding, 7–20 per raceme, 8–14 mm wide by 6–9 mm long (including style); **calyx** pale green to slightly pinkish, TL 3–4 mm by 5–6 mm wide, tubular base 1.3–1.5 mm long; **calyx teeth** five, *c.* 2.5 mm long by 1.6–2 mm wide, entire, spreading between the petals, persistent in fruit; **petals** five, 6–8 mm long by 4–5 mm wide, tapered to base, C-shaped, entire, veins faint; **stamens** 10 (9), included, 4–5 mm long; **filaments** pink, curved, *c.* 3 mm long, widening to *c.* 1 mm and flat near the base; **anthers** pink in flower buds, turning brown, 2–2.5 mm long, apicule *c.* 0.3 mm long at base, two pores at other end; **ovary** green, round, 2–2.3 mm wide by 1–1.3 mm tall, 5-parted; **style** one, 5–7 mm long by *c.* 0.8 mm thick, pale green, exserted, curved, persistent; **stigma** obscure.
FRUIT a capsule, dark brown, 4–8 mm wide by 3–4 mm tall (not including style), 5-parted, center concave; **seeds** numerous, golden brown, winged; TL 0.5–0.7 mm by *c.* 0.1 mm wide; **seed** *c.* 0.2 mm long.
■ **LEAVES** Basal and stem, simple, entire, two to several, persistent, ascending to spreading, glabrous; **basal blades** 3–6 cm long and wide, thin, apices round, slightly to very shiny above, lighter green and dull below, veins obvious; **petioles** reddish, 4–7 cm long by 1.5–2 mm wide; **stipules** absent; **scaly stem leaves** 5–12 mm long by 2–6 mm wide, pointed, V-shaped, one to three along the scape, sessile and clasping.
■ **STEM** Simple, solid, stiff, angular, reddish to pale green; 1.5–2.5 mm thick near the base.
■ **SYN.** *Pyrola uliginosa* Torr. & Gray ex Torr. = *Pyrola asarifolia* ssp. *asarifolia* Michx.

Liverleaf Wintergreen, Pink Pyrola, Bog Wintergreen, Common Pink Wintergreen

Wintergreen—*Pyrolaceae*   **Wildflower** Evergreen Pale greenish white Petals 5

GREENISH-FLOWERED WINTERGREEN   *Pyrola chlorantha* Sw.

■ **SKETCH** A glabrous **perennial herb** 6–25 cm tall in clusters from spreading **rhizomes** *c.* 1 mm wide with short bracts; in moist to dry upland woods and thickets.

■ **FLOWERS** Pale greenish white, blooming June–August; **inflorescence** a raceme, 4.5–10 cm long by 2–2.5 cm wide; **pedicels** 2–6 mm long, green, arched, with warty ridges; **subtending bracts** (of pedicels) 3–7 mm long; **flowers** 3–10 per stalk, each 7–12 mm wide by 6–8 mm long (including style), drooping; **calyx** 5-lobed, lobes with a thin, white, entire to slightly erose margin, *c.* 1 mm long by *c.* 1.5 mm wide, persistent, roundly pointed; **petals** five, greenish white, blunt, 4–7 mm long by 3–4.2 mm wide, veins obscure, apical margin thin and white; **stamens** 10, each 4–5 mm long, included to slightly exserted at anthesis; **filaments** white, thick, curved, bisected at apices and attached to bases of apical tubes; **anthers** pale yellow, *c.* 2 mm long, lobed, with two apical pores on two yellowish tan apical tubes, pores facing stigma as flower opens; **pistil** one, exserted, green, hairless; **ovary** 5-lobed, *c.* 3 mm long and wide, each lobe with one outside suture; **style** 6–7 mm long, light green, eventually exserted, curved away from rachis; **stigma** green, *c.* 0.8 mm long, 5-lobed above a narrow collar. **FRUIT a capsule**, brown, 2–4 mm long by 4–8 mm wide, the center depressed, 5-lobed, splitting open along the five sutures from below; **seeds** numerous, brown, winged, *c.* 0.5 mm long by *c.* 0.1 mm thick.

■ **LEAVES** Alternate (appear basal), slightly toothed, evergreen; **basal blades** 4–11, each 1–3 cm long and wide, dull, veins obvious above, hairless; **petioles** pinkish green, ascending, 2–4 cm long, usually longer than the blades; **stem blades** none to two, sessile, linear, 0.8–1 mm long.

■ **STEM** Erect, stiff, one or two, pinkish green, solid, hairless, 4-sided, ridges green, slightly twisted, unbranched; 1–2 mm wide with bracts around the base.

■ **SYN.** *Pyrola virens* Schreb.

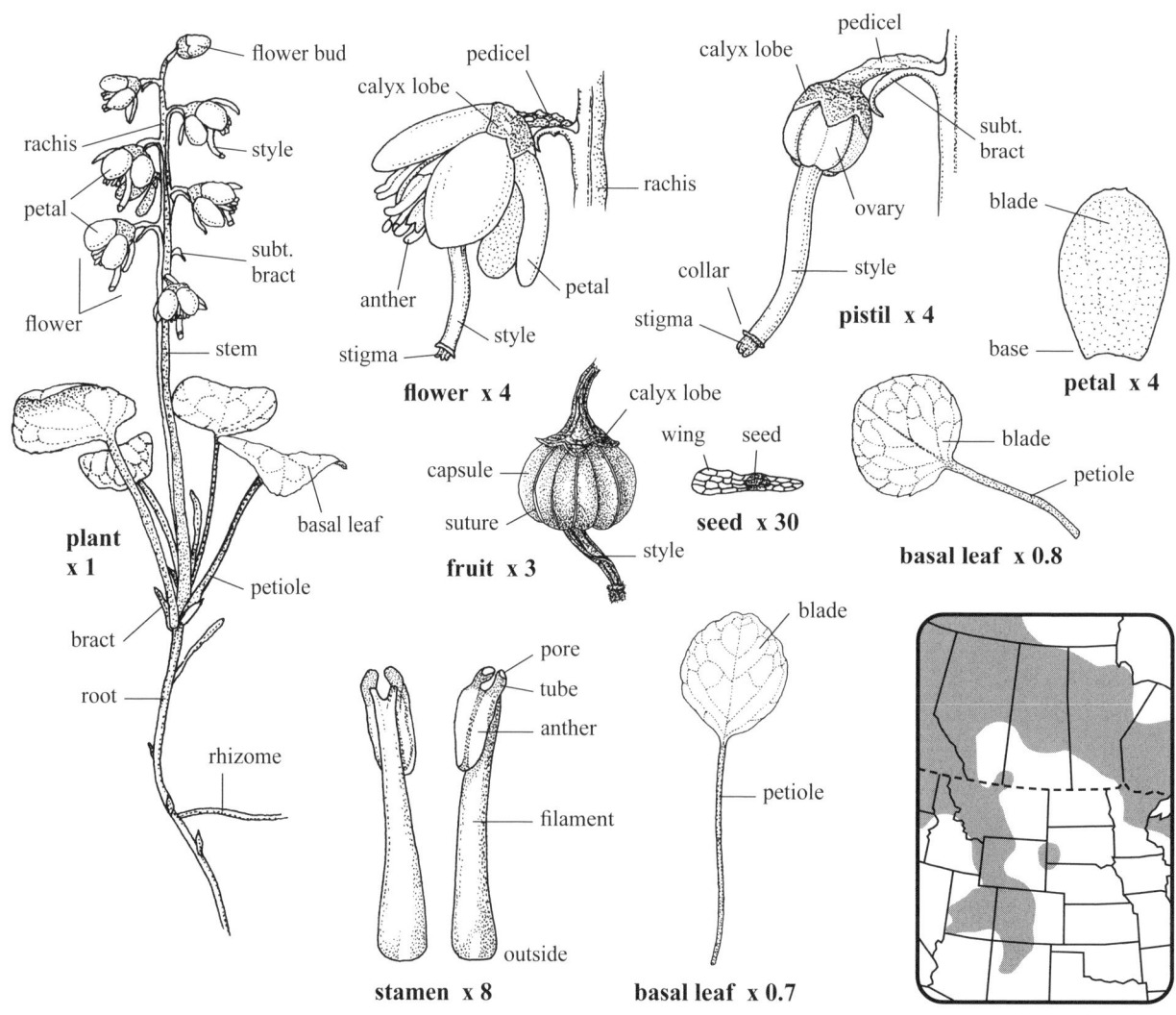

Greenish-flowered Pyrola, Green Wintergreen, Greenflowered Wintergreen

Family Characteristics

Buttercup—Ranunculaceae Flower parts 5 (3–8)

SKETCH Dicotyledonous herbs (wildflowers), perennial, some woody climbers. Many have an acrid juice. About 2,500 species worldwide in cooler regions.

FLOWERS Mostly perfect, regular to irregular, the parts free and distinct, some with a spur, variable, 5-merous or not. **Inflorescence** a cyme, panicle or solitary. **Sepals** usually 5 (3–8), sometime petal-like, green to yellowish to purplish. **Petals** absent or a few (5) to many, yellow to white or blue, often with a nectariferous gland at the base, some with a long spur. **Stamens** numerous, in spirals or whorls. **Receptacle** globular to cylindrical. **Pistils** simple, separate, few to many. **Ovary** 1, each 1-loculed, ovules one to many. **Style** 1, erect to bent. **Stigma** 1.

FRUIT Usually an achene, 1-seeded, or a follicle, usually not fleshy, with many seeds, rarely a berry.

LEAVES Mostly alternate, rarely opposite or basal, simple or compound (often divided into three parts or even filiform segments), entire to lobed and toothed, sessile or with petioles. **Stipules** mostly absent, sometimes present, free or attached to petiole.

STEM Erect to climbing to floating in aquatic plants, simple to branched, some fleshy.

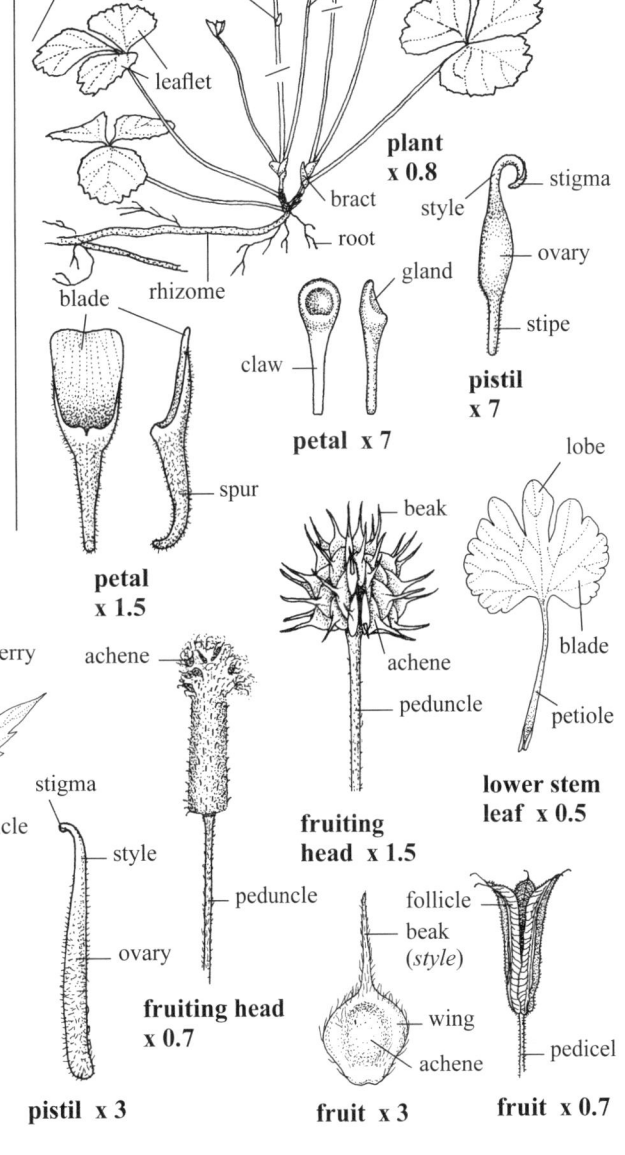

419

Buttercup—*Ranunculaceae*   **Wildflower**   White   Petals 3–10

Red Baneberry   *Actaea rubra* (Ait.) Willd.

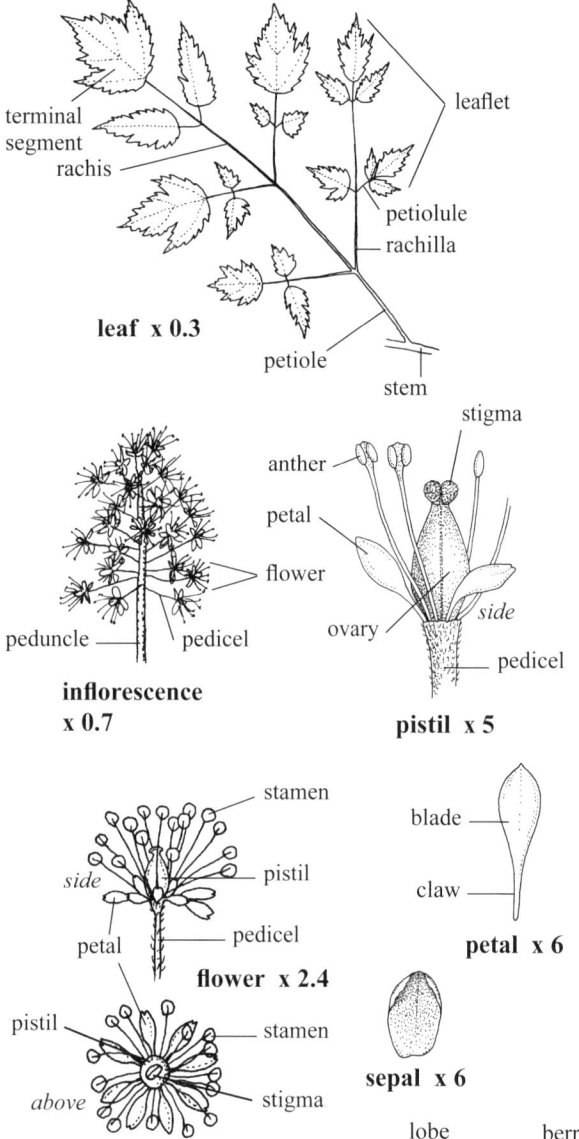

- **SKETCH** A **perennial herb** 30–100 cm tall, singly or in clusters of several stems; in moist woods, mountain forests, edges along rivers and wooded ravines.
- **FLOWERS** White, blooming May–July; **inflorescence** a raceme, terminal, 3–5.5 cm long and wide, of 30–40 flowers; **peduncles** 2–16 cm long by 1–2.5 mm thick; **pedicels** thin, hairy, 8–10 mm long, spreading, to 25 mm long in fruit; **flowers** perfect, fragrant, 7–9 mm wide by *c.* 7 mm long; **sepals** three to five, hooded to mostly C-shaped, green, glabrous, enclosing the bud, 2–3.7 mm long by 1–2 mm wide, deciduous as the flowers unfold; **petals** usually 3–10, spreading, 2–4 mm long by *c.* 1 mm wide, narrow at the base, glabrous, tip pointed to bidentate; **stamens** 14–19, each 5–6 mm long, from base of ovary, exserted; **anthers** white, *c.* 0.7 mm long and wide; **pistil** one, thick; **ovary** light green, glabrous, *c.* 3 mm long by 1.6 mm thick with a shallow groove; **stigma** rough, cleft lengthways, sessile, more narrow than the ovary. **FRUIT a berry**, fleshy, red (or white), shiny, grooved on one side, 7–13 mm long by 6–11 mm wide; **seeds** 9–16 per berry, dark brown, wedge-shaped, 2.9–3.6 mm long by 1.5–2 mm wide by 1–1.5 mm thick, rough with a pitted surface, bumpy edges, hairless, 2-ridged along the round side.
- **LEAVES** Alternate, stem only, compound (twice), usually two (one or three) per plant; **leaf blades** 10–35 cm long by 8–16 cm wide with three compound leaflets; **segments** 2- or 3-lobed and toothed; **terminal segments** the largest at 2.2–8 cm long by 1.8–9 cm wide, hairy to glabrous below, glabrous and darker green above; **petioles** 1–16 cm long, glabrous; **petiolules** 5–20 mm long, hairy when young.
- **STEM** One to several in a cluster, erect, glabrous and glaucous; 2–9 mm thick near the base.
- **NOTE** Plants produce either all red or all white berries (Great Plains Flora Association, 1986).

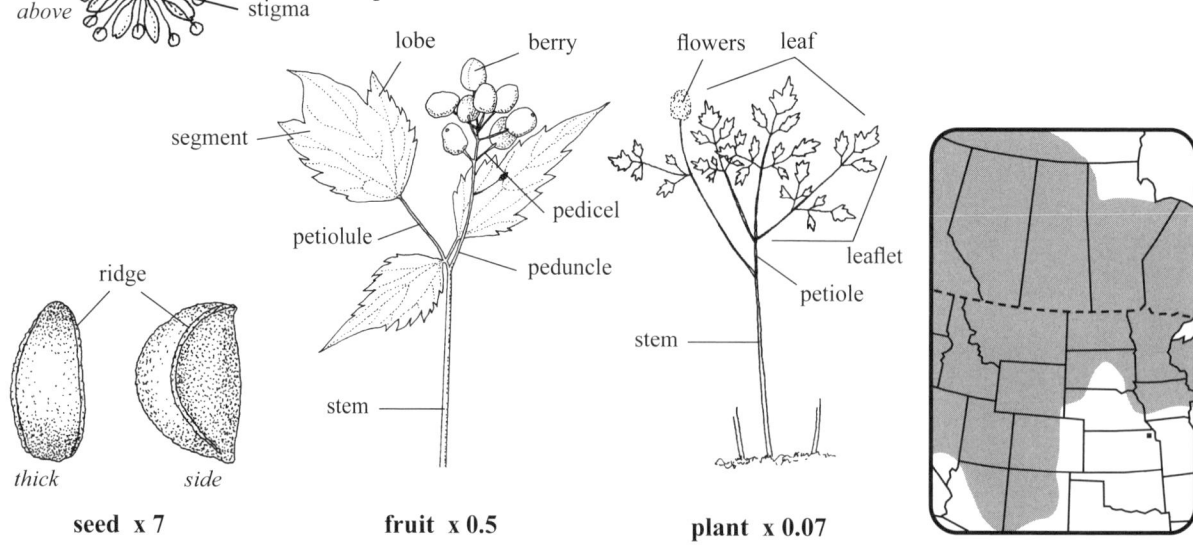

Baneberry, Red and White Baneberry

Buttercup—*Ranunculaceae*   **Wildflower** White Sepals 5 (4 or 6)

Canada Anemone   *Anemone canadensis* L.

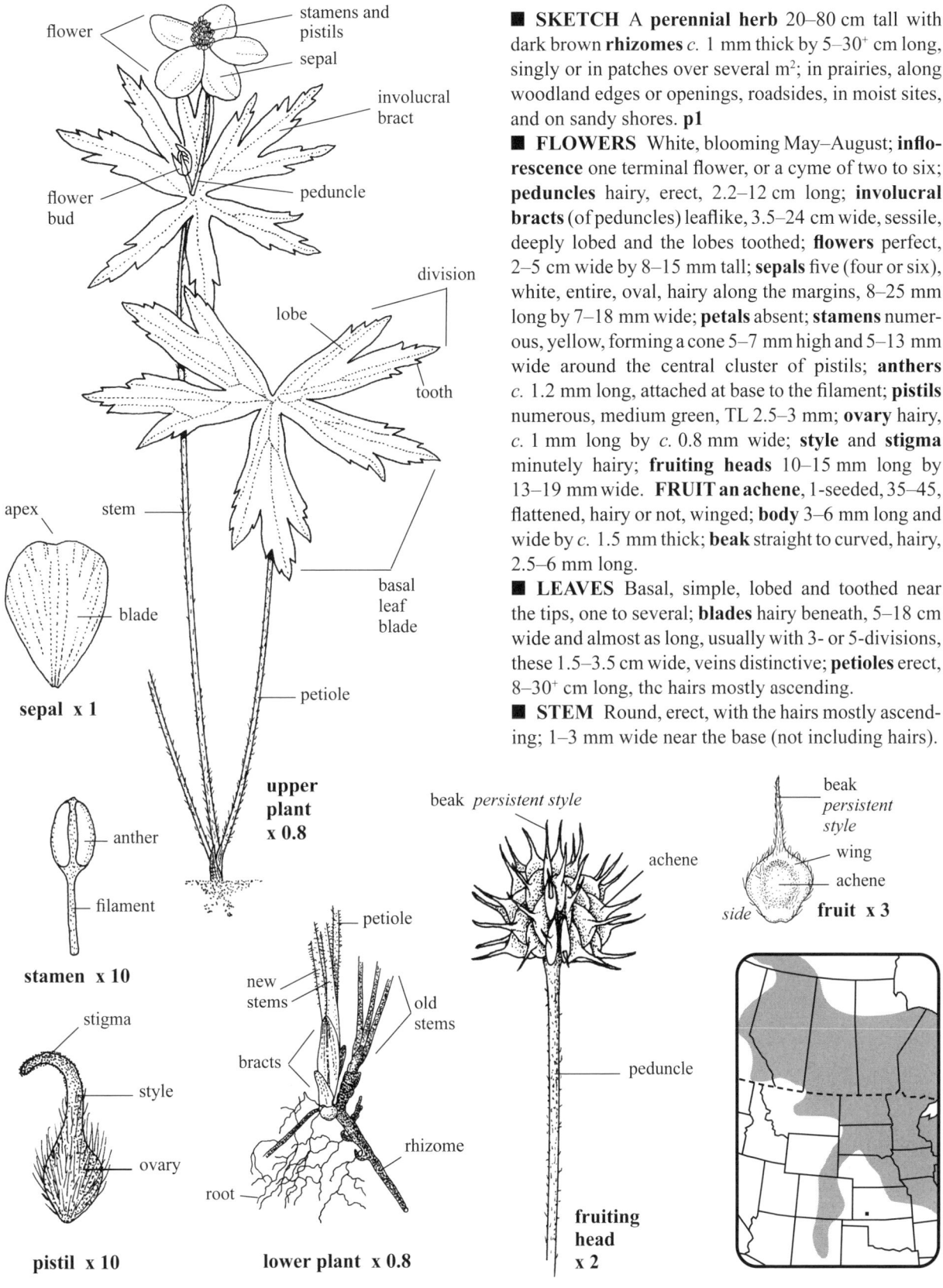

- **SKETCH** A **perennial herb** 20–80 cm tall with dark brown **rhizomes** *c.* 1 mm thick by 5–30⁺ cm long, singly or in patches over several m²; in prairies, along woodland edges or openings, roadsides, in moist sites, and on sandy shores. **p1**
- **FLOWERS** White, blooming May–August; **inflorescence** one terminal flower, or a cyme of two to six; **peduncles** hairy, erect, 2.2–12 cm long; **involucral bracts** (of peduncles) leaflike, 3.5–24 cm wide, sessile, deeply lobed and the lobes toothed; **flowers** perfect, 2–5 cm wide by 8–15 mm tall; **sepals** five (four or six), white, entire, oval, hairy along the margins, 8–25 mm long by 7–18 mm wide; **petals** absent; **stamens** numerous, yellow, forming a cone 5–7 mm high and 5–13 mm wide around the central cluster of pistils; **anthers** *c.* 1.2 mm long, attached at base to the filament; **pistils** numerous, medium green, TL 2.5–3 mm; **ovary** hairy, *c.* 1 mm long by *c.* 0.8 mm wide; **style** and **stigma** minutely hairy; **fruiting heads** 10–15 mm long by 13–19 mm wide. FRUIT **an achene**, 1-seeded, 35–45, flattened, hairy or not, winged; **body** 3–6 mm long and wide by *c.* 1.5 mm thick; **beak** straight to curved, hairy, 2.5–6 mm long.
- **LEAVES** Basal, simple, lobed and toothed near the tips, one to several; **blades** hairy beneath, 5–18 cm wide and almost as long, usually with 3- or 5-divisions, these 1.5–3.5 cm wide, veins distinctive; **petioles** erect, 8–30⁺ cm long, the hairs mostly ascending.
- **STEM** Round, erect, with the hairs mostly ascending; 1–3 mm wide near the base (not including hairs).

Round-leaf Thimbleweed, Meadow Anemone

Buttercup—*Ranunculaceae*   **Wildflower**   White to whitish green   Sepals 5 (4 or 6)

LONG-FRUITED ANEMONE   *Anemone cylindrica* Gray

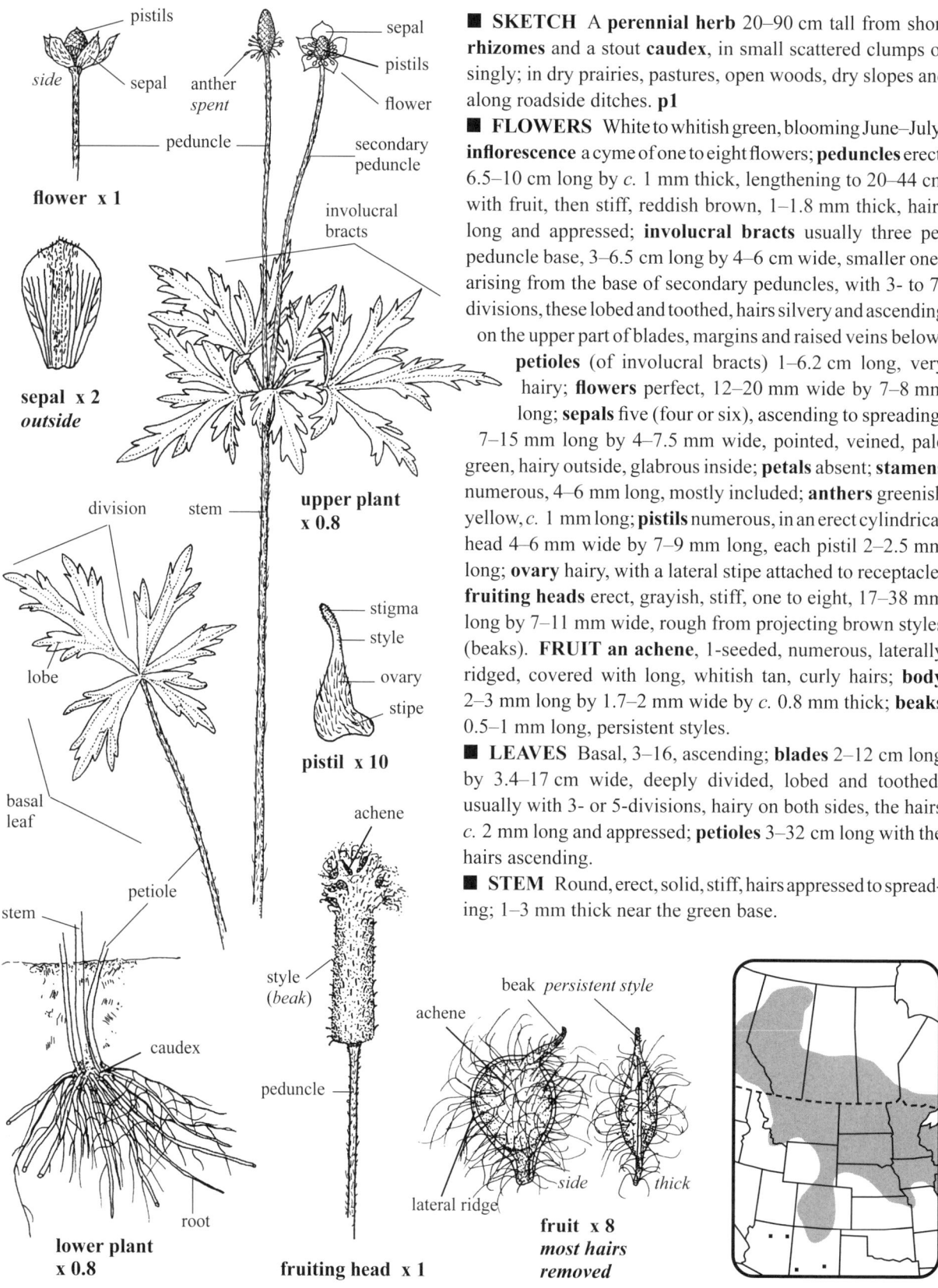

- **SKETCH** A **perennial herb** 20–90 cm tall from short **rhizomes** and a stout **caudex**, in small scattered clumps or singly; in dry prairies, pastures, open woods, dry slopes and along roadside ditches. p1
- **FLOWERS** White to whitish green, blooming June–July; **inflorescence** a cyme of one to eight flowers; **peduncles** erect, 6.5–10 cm long by *c.* 1 mm thick, lengthening to 20–44 cm with fruit, then stiff, reddish brown, 1–1.8 mm thick, hairs long and appressed; **involucral bracts** usually three per peduncle base, 3–6.5 cm long by 4–6 cm wide, smaller ones arising from the base of secondary peduncles, with 3- to 7- divisions, these lobed and toothed, hairs silvery and ascending on the upper part of blades, margins and raised veins below; **petioles** (of involucral bracts) 1–6.2 cm long, very hairy; **flowers** perfect, 12–20 mm wide by 7–8 mm long; **sepals** five (four or six), ascending to spreading, 7–15 mm long by 4–7.5 mm wide, pointed, veined, pale green, hairy outside, glabrous inside; **petals** absent; **stamens** numerous, 4–6 mm long, mostly included; **anthers** greenish yellow, *c.* 1 mm long; **pistils** numerous, in an erect cylindrical head 4–6 mm wide by 7–9 mm long, each pistil 2–2.5 mm long; **ovary** hairy, with a lateral stipe attached to receptacle; **fruiting heads** erect, grayish, stiff, one to eight, 17–38 mm long by 7–11 mm wide, rough from projecting brown styles (beaks). **FRUIT** an achene, 1-seeded, numerous, laterally ridged, covered with long, whitish tan, curly hairs; **body** 2–3 mm long by 1.7–2 mm wide by *c.* 0.8 mm thick; **beaks** 0.5–1 mm long, persistent styles.
- **LEAVES** Basal, 3–16, ascending; **blades** 2–12 cm long by 3.4–17 cm wide, deeply divided, lobed and toothed, usually with 3- or 5-divisions, hairy on both sides, the hairs *c.* 2 mm long and appressed; **petioles** 3–32 cm long with the hairs ascending.
- **STEM** Round, erect, solid, stiff, hairs appressed to spreading; 1–3 mm thick near the green base.

Long-headed Anemone, Candle Anemone, Long-headed Thimbleweed

Buttercup—*Ranunculaceae*   **Wildflower**   Purplish pink, red, green, yellow to white
Sepals 5 (4–9)

CUT-LEAVED ANEMONE   *Anemone multifida* Poir.

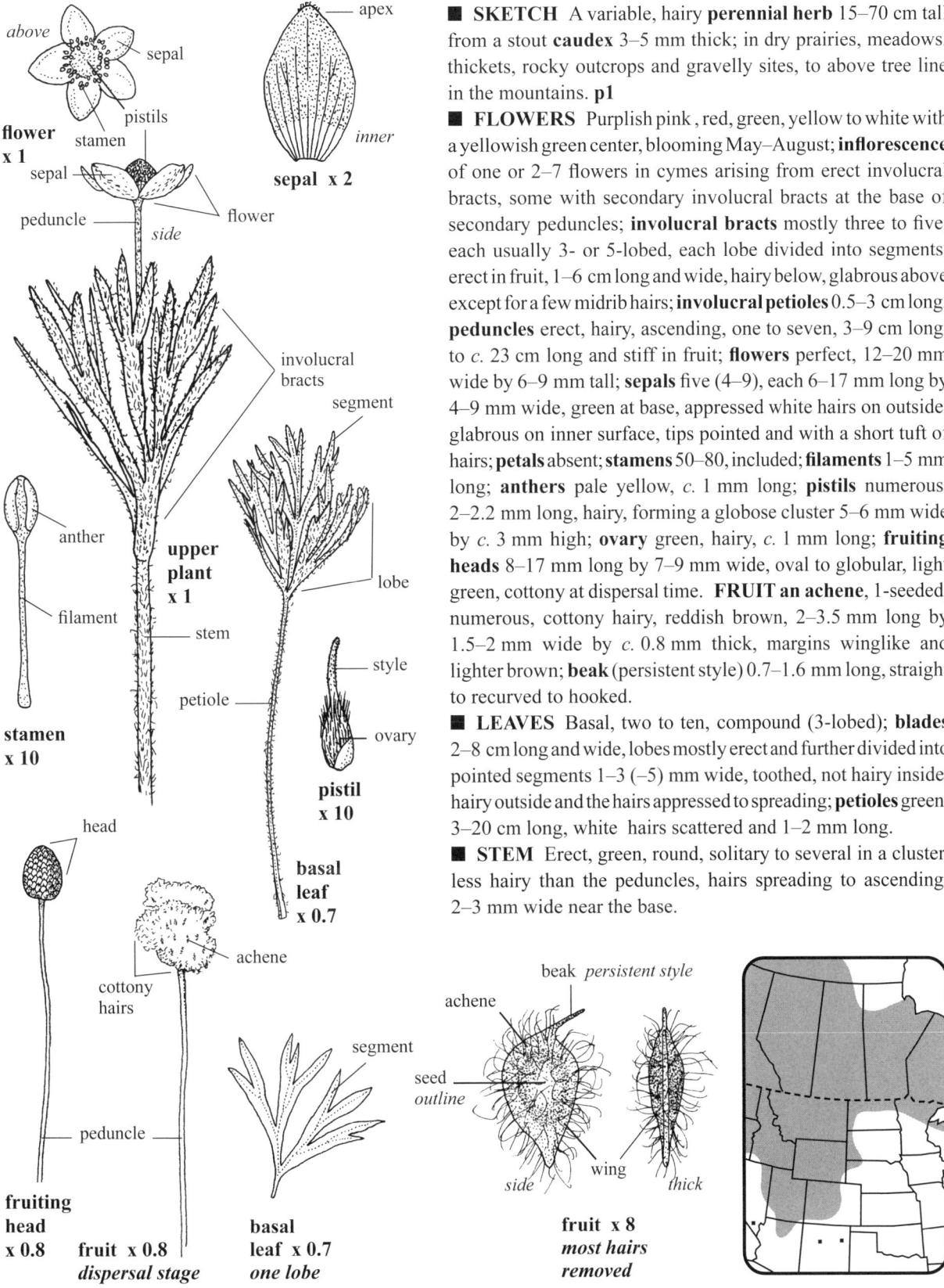

- **SKETCH** A variable, hairy **perennial herb** 15–70 cm tall from a stout **caudex** 3–5 mm thick; in dry prairies, meadows, thickets, rocky outcrops and gravelly sites, to above tree line in the mountains. **p1**
- **FLOWERS** Purplish pink, red, green, yellow to white with a yellowish green center, blooming May–August; **inflorescence** of one or 2–7 flowers in cymes arising from erect involucral bracts, some with secondary involucral bracts at the base of secondary peduncles; **involucral bracts** mostly three to five, each usually 3- or 5-lobed, each lobe divided into segments, erect in fruit, 1–6 cm long and wide, hairy below, glabrous above except for a few midrib hairs; **involucral petioles** 0.5–3 cm long; **peduncles** erect, hairy, ascending, one to seven, 3–9 cm long, to *c.* 23 cm long and stiff in fruit; **flowers** perfect, 12–20 mm wide by 6–9 mm tall; **sepals** five (4–9), each 6–17 mm long by 4–9 mm wide, green at base, appressed white hairs on outside, glabrous on inner surface, tips pointed and with a short tuft of hairs; **petals** absent; **stamens** 50–80, included; **filaments** 1–5 mm long; **anthers** pale yellow, *c.* 1 mm long; **pistils** numerous, 2–2.2 mm long, hairy, forming a globose cluster 5–6 mm wide by *c.* 3 mm high; **ovary** green, hairy, *c.* 1 mm long; **fruiting heads** 8–17 mm long by 7–9 mm wide, oval to globular, light green, cottony at dispersal time. **FRUIT an achene**, 1-seeded, numerous, cottony hairy, reddish brown, 2–3.5 mm long by 1.5–2 mm wide by *c.* 0.8 mm thick, margins winglike and lighter brown; **beak** (persistent style) 0.7–1.6 mm long, straight to recurved to hooked.
- **LEAVES** Basal, two to ten, compound (3-lobed); **blades** 2–8 cm long and wide, lobes mostly erect and further divided into pointed segments 1–3 (–5) mm wide, toothed, not hairy inside, hairy outside and the hairs appressed to spreading; **petioles** green, 3–20 cm long, white hairs scattered and 1–2 mm long.
- **STEM** Erect, green, round, solitary to several in a cluster, less hairy than the peduncles, hairs spreading to ascending; 2–3 mm wide near the base.

Red Anemone, Red Windflower, Pacific Anemone

Buttercup—*Ranunculaceae*   **Wildflower**   White   Sepals 5 (4–9)

WOOD ANEMONE   *Anemone quinquefolia* L.

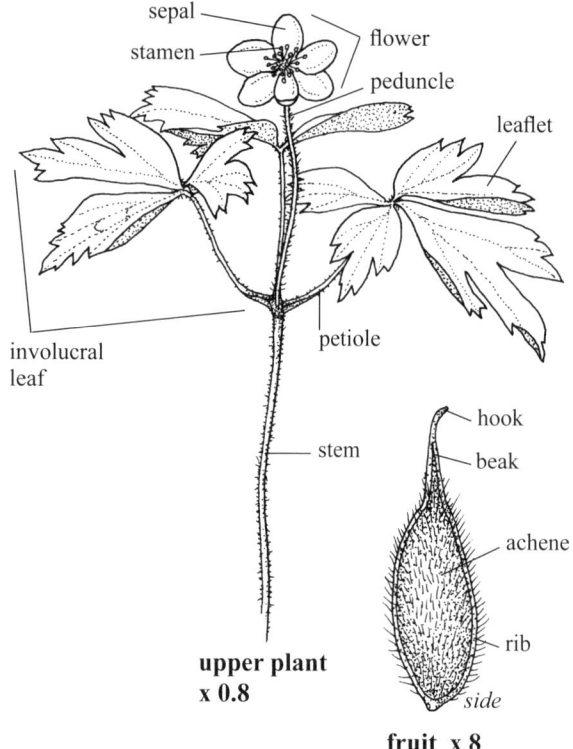

upper plant x 0.8

fruit x 8

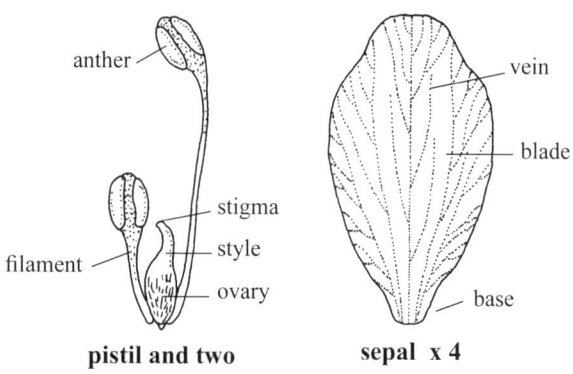

pistil and two stamens x 10

sepal x 4

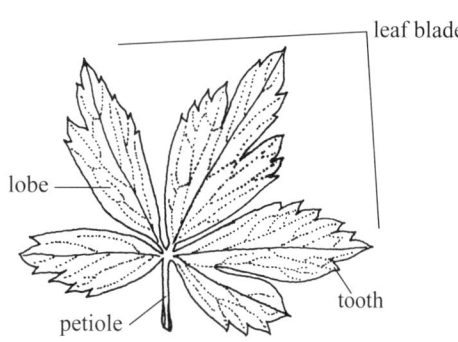

involucral leaf x 1

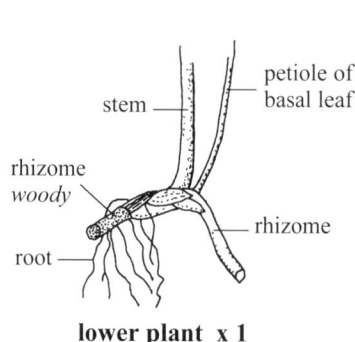

lower plant x 1

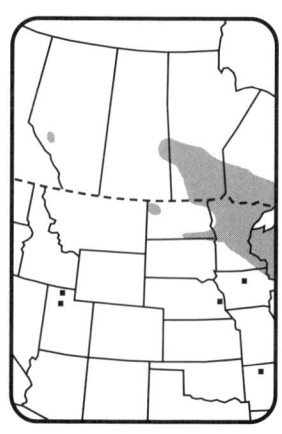

■ **SKETCH** A **perennial herb** 10–50 cm tall, solitary or in patches with a brown **rhizome** several centimeters long by *c.* 2 mm wide; in gravelly, rocky areas of forests, thickets and ditches.

■ **FLOWERS** White, blooming May–June; **inflorescence** a solitary flower; **peduncles** erect, one, 2.3–5 cm long, often nodding or arched at apex, green, white hairs ascending; **involucral leaves** three, whorled, appear compound, each 3–5 cm long by 4–7 cm wide, ascending from base of peduncle; **leaflets** 3–5, each 2.5–4 cm long by 1.3–2.8 cm wide, some lobed, toothed, slightly hairy on both sides and margins, hairs appressed and ascending mostly along the veins; **petioles** 0.5–3 cm long, hairs appressed; **petiolules** *c.* 1 mm long, purplish, hairy; **flowers** one, perfect, 1.7–3.5 cm wide by 5–7 mm long, often nodding, center cluster of pistils and stamens 5–8 mm wide; **sepals** five (4–9), petal-like, white, ascending to spreading, 8–22 mm long by 4–8 mm wide, veined on inside, apices blunt, outside slightly pinkish, glabrous on both sides; **petals** absent; **stamens** numerous, glabrous, up to *c.* 4 mm long, exserted or not; **filaments** white, tapered above; **anthers** white, 0.7–0.8 mm long, curved inward over stigmas; **pistils** green, 15–20 per flower, included, crowded, each *c.* 1.3 mm long by *c.* 0.4 mm wide; **fruiting heads** 8–10 mm wide, roundish. **FRUIT an achene**, 1-seeded, brown, 3.5–5 mm long by *c.* 1.5 mm wide, 2-ribbed, hairy; **beak** (style) 1–2 mm long with an apical hook.

■ **LEAVES** Basal; **stem leaves** absent; **basal leaf** one or more, compound, appear late, dull above, erect; **leaflets** three to five, 2.5–3.8 cm long by 1.6–2.5 cm wide, lobed, toothed above near apices, base entire; **petioles** 5–8 cm long by *c.* 1 mm wide, reddish purple near base, slightly hairy, hairs white and spreading.

■ **STEM** Erect, unbranched, white hairs above mostly spreading, glabrous below; 1–2 mm thick near the reddish brown base.

Nightcaps, American Wood Anemone, Windflower

Buttercup—*Ranunculaceae* Wildflower Blue and white Petals 5

SMALL-FLOWERED COLUMBINE *Aquilegia brevistyla* Hook.

■ **SKETCH** A **perennial herb** 20–80 cm tall from a brown **caudex** to *c.* 10 cm long by *c.* 10 mm wide; in open woods, roadside ditches and banks.

■ **FLOWERS** Blue and white, blooming May–August; **inflorescence** a terminal flower with 1–4 (–8) lateral flowers from upper leaf axils; **floral branches** reddish, hairy, 1–15 cm long, reduced above; **pedicels** 5–15 mm long, hairy, reddish; **flowers** perfect, nodding, 2.6–3.5 cm wide by 15–22 mm long; **subtending bracts** (of flowers) 7–8 mm long by *c.* 1 mm wide, hairless, white membranous, margins undulate; **sepals** five, blue with white tips, spreading, 1–2 cm long by 5–7 mm wide, hairy on outside and margins, hairless inside, claw 2–3 mm long, hairy on margins; **petals** five, blue with white to pale yellow apices, 15–21 mm long by 6–8 mm wide, hairy on outside; **spur** blue, 5–10 mm long, hooked toward pedicel; **stamens** *c.* 20, included to barely showing, 9–11 mm long, glabrous; **filaments** white, wide and wrinkled near base; **anthers** yellow, 1.5–2 mm long; **carpels** (pistils) five to seven, *c.* 11 mm long by *c.* 1 mm thick, barely exserted, green, hairy; **ovary** triangular in cross-section; **styles** curved and hairless above, persistent on fruit; **stigmas** green, curved outward. **FRUIT a follicle**, brown, heavily veined, 14–27 mm long by 2.5–3 mm wide and thick, some hairs gland-tipped, opening along one inner suture, beaks spreading outward, 2–5 mm long; **seeds** 26–110 per follicle, black, shiny, ridged on one side, 1.8–2 mm long by *c.* 1.2 mm wide by *c.* 1 mm thick (ridge in middle).

■ **LEAVES** Basal and stem, trifoliate; **basal leaves** 2–9, ascending; **leaf blades** 2–8 cm long by 3–15 cm wide, twice compound; **leaflets** 1.5–4.8 cm long by 0.7–3.3 cm wide, lobed and toothed, hairy below, less so on margins and slightly hairy above; **petioles** reddish, hairy, 3–22 cm long; **petiolules** 0.5–6 cm long, reddish and hairy; **stem leaves** three to five, alternate, well spaced, leaflets 0.8–5 cm long by 0.3–4.5 cm wide, lobed and toothed, entire and reduced above; **petioles** 0–8.5 cm long, reduced above; **petiolules** 0.5–3.8 cm long.

■ **STEM** Erect, often branched, nodding at apex; 1–3 mm wide at hairy, reddish base.

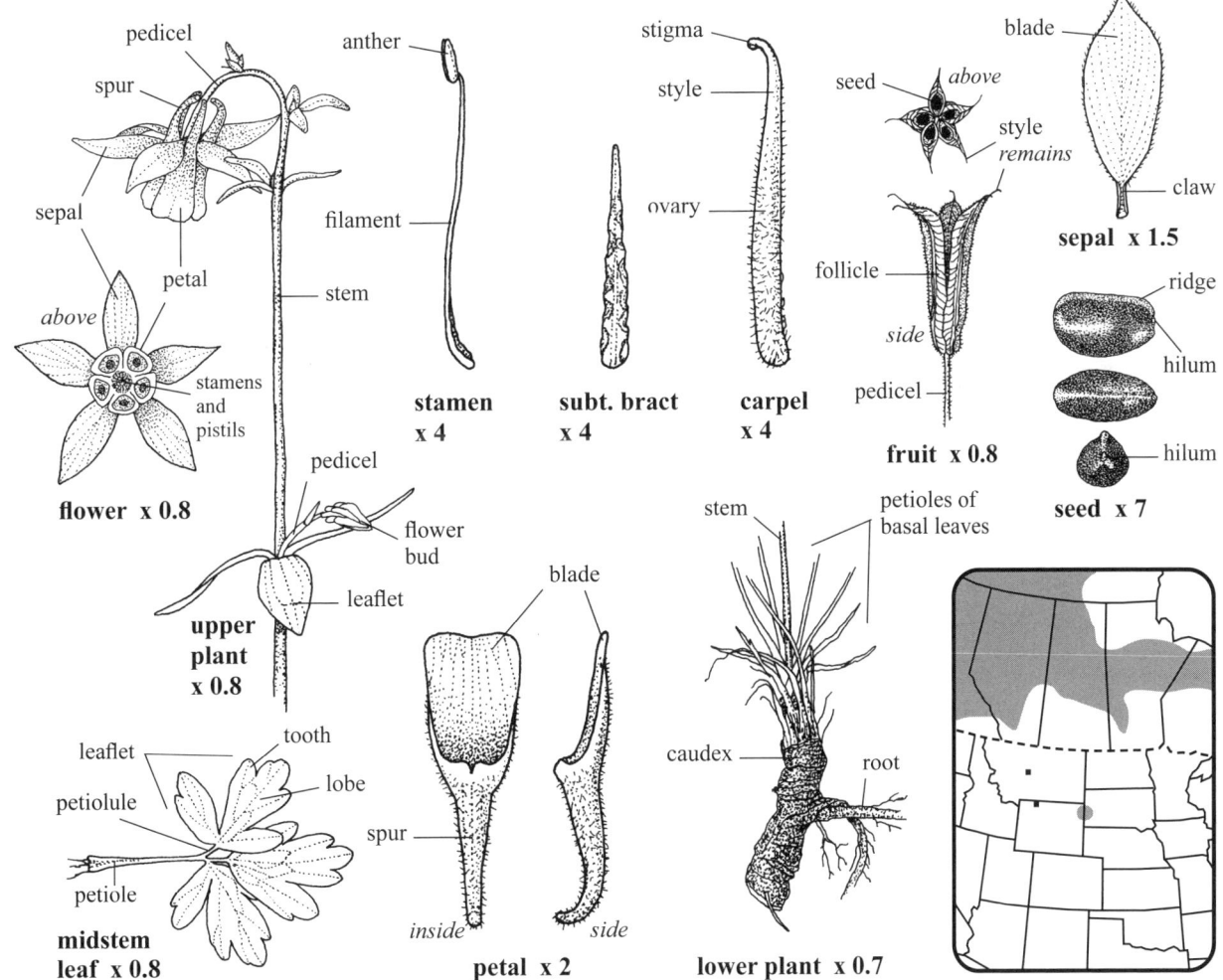

Blue Columbine

Buttercup—*Ranunculaceae* **Wildflower** Orange to red, center yellow Petals 5

Wild Columbine   *Aquilegia canadensis* L.

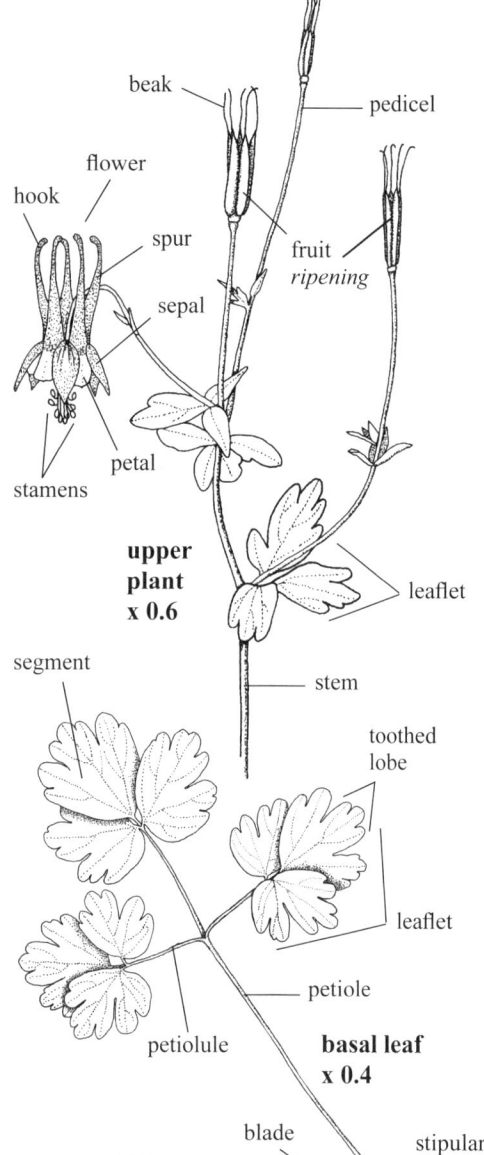

■ **SKETCH** A **perennial herb** 20–100+ cm tall from a **caudex** to c. 2.5 cm long with one or two branches; in open moist woodlands, rocky woods, slopes, thickets, bogs and clearings.

■ **FLOWERS** Orange to red, center yellow, blooming March–July; **inflorescence** a panicle, open, of four to eight flowers per branch; **pedicels** 3–11 cm long, hairy, arched in flower, erect in fruit; **flowers** perfect, nodding, regular, 14–35 mm wide by 35–45 mm long; **sepals** five, reddish, 1–2 cm long by 5–8 mm wide, pointed, claw 3–4 mm long, glandular hairs near the base, margins and blade hairy, glabrous below; **petals** five, yellow at apices, red basally, blade 6–9 mm long by 5–8 mm wide, glabrous; **spur** tubular, 18–32 mm long, hairy, the hook at tip is reddish purple; **stamens** c. 25, yellow, exserted to c. 1 cm past petal tips, forming a column 3–5 mm wide; **filaments** yellow, glabrous, recurved or straight, 9–20 mm long, hairless; **anthers** yellow, 1.2–2 mm long; **staminodia** between filaments and carpels, c. 8, each 6–7 mm long by 1–1.2 mm wide, hyaline, edges wavy, hairless; **pistils** compound, of five carpels, c. 15 mm long; **carpels** 5–6 mm long, green, hairs fine, spreading; **styles** five, filiform, hairless, 8–18 mm long, exserted well past the yellow petal tips, persistent as beaks on fruit; **stigmas** obscure. **FRUIT a follicle**, five, hairy, erect, 2–3.5 cm long by 2–3 mm wide, opening from apex, veined, hairs scattered, beaks 1–2 cm long, mostly ascending; **seeds** c. 10 per follicle, black, shiny, 1.8–2 mm long by c. 1.2 mm wide by c. 1 mm thick, ridged, round on one side.

■ **LEAVES** Basal and stem, alternate, compound (twice 3-parted); **leaflets** 2.3–6 cm wide and long; **segments** three per leaflet, each 2–4.5 cm long and wide, usually 3-lobed, teeth blunt, glabrous and glaucous above, whitish green and lightly hairy below; **petioles** 0–36 cm long, reduced above, purplish green, glabrous to slightly hairy, expanded at base; **stipular wings** to c. 2 cm long, glaucous, upper margins whitish; **petiolules** 0.2–9.5 cm long, hairy or not.

■ **STEM** Erect, purplish green, hollow, round, glabrous to glandular-hairy above, glaucous; 4–5 mm thick and ridged near the base.

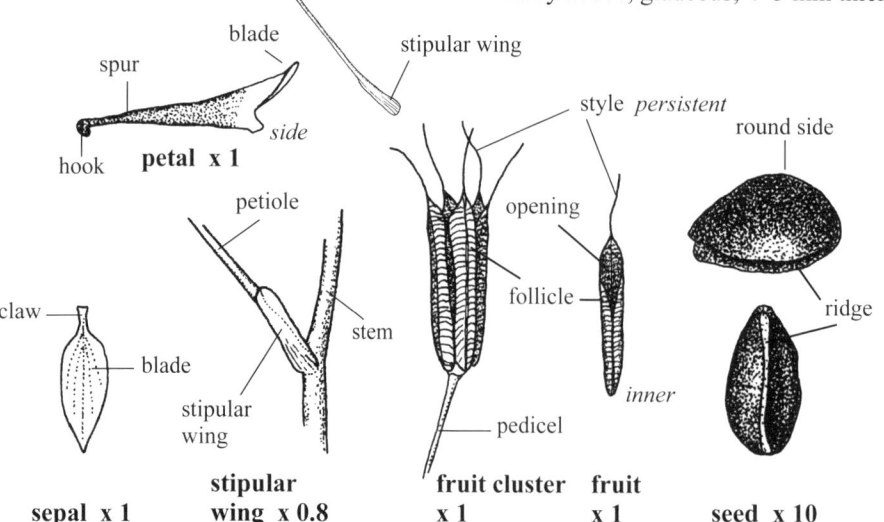

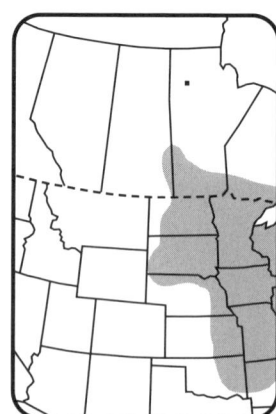

426   Red Columbine, Canadian Columbine

Buttercup—*Ranunculaceae* Wildflower   Yellow   Sepals 5 or 6 (–9)

Marsh-marigold *Caltha palustris* L.

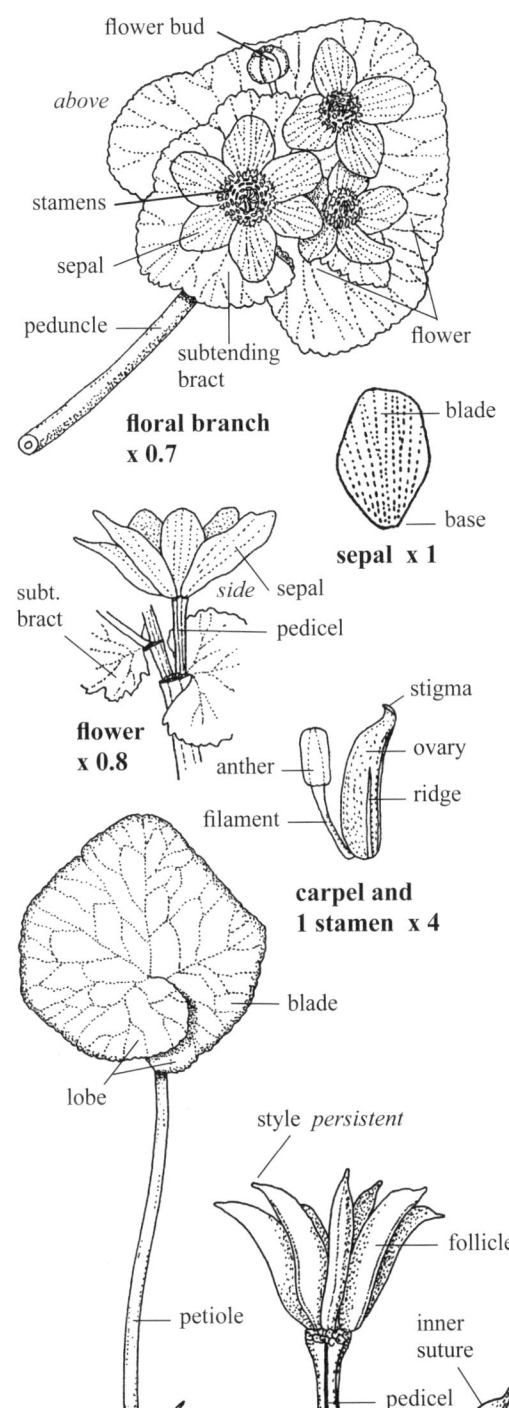

floral branch x 0.7

sepal x 1

flower x 0.8

carpel and 1 stamen x 4

basal leaf x 0.6

fruit x 2 clustered

fruit x 2

seed x 10

■ **SKETCH** A glabrous **perennial herb**, fleshy, 15–60 cm tall and wide from fleshy **fibrous roots** 1–2 mm wide and to 30$^+$ cm long; in wet woods, marshes, pastures and bogs, often in shallow, slowly moving water.

■ **FLOWERS** Yellow, blooming April–July; **inflorescence** of clusters of two to seven flowers, terminal and axillary; **floral branches** few to 20$^+$ per plant, ascending, hollow, 10–20 cm long by *c.* 5 mm thick; **peduncles** erect, 5–8 cm long by 2–4 mm thick, ridged; **pedicels** 1–8 cm long, to 18 cm long in fruit, ridged; **subtending bracts** (of pedicels) three to five; **bract blades** *c.* 2 cm wide and long, reduced above, toothed, veins purplish below, forming a cup around the flower bud clusters; **flowers** perfect, 2–5 cm wide by 10–15 mm tall, four to six per peduncle; **sepals** yellow, usually five or six (nine), each 1–2.7 cm long by 6–18 mm wide, dull, more greenish beneath, veined above, ascending; **petals** absent; **stamens** yellow, numerous, the central cluster *c.* 5 mm tall by 10–15 mm wide, included; **filaments** 3–4 mm long, glabrous; **anthers** yellow, 1.5–1.8 mm long; **carpels** included, 5–25, green, tapered, sessile, yellow near the apices, *c.* 5 mm long by 1.3–1.5 mm wide by *c.* 1.2 mm thick; **stigma** angled; **fruiting heads** green, erect, often hidden by leaves, usually of seven (4–12) follicles in an ascending cluster 11–17 mm long and wide. **FRUIT a follicle**, opening along the inner full-length suture, 8–17 mm long by 3–4 mm wide by 2–2.5 mm thick; **beak** (persistent style) 1–2 mm long; **seeds** 9–12 per follicle, greenish, 1.7–2.4 mm long by 0.8–0.9 mm wide by *c.* 0.7 mm thick, finely pitted, wider near the base.

■ **LEAVES** Basal and stem, alternate; **basal leaves** round; **blades** 3–19 cm long by 3–21 cm wide, simple, 2-lobed basally, thin, dull, medium green above, margins roundly toothed and purplish, veins smooth and purplish below; **lobes** two, basal, round, imbricate, each 2–5 cm long; **petioles** 1–44 cm long by 3–5 mm wide and thick, hollow; **stipules** 30–40 mm long, united except for spreading brown tips, stiff and membranous; **stem leaves** mostly sessile, roundish, flat, base lobed, reduced and bractlike above.

■ **STEM** Ascending, hollow, fleshy, with a few low ridges, branched above; 5–6 mm thick near the base.

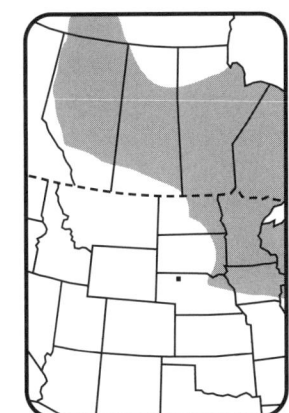

Yellow Marsh-marigold, Cowslip

Buttercup—*Ranunculaceae* Vine White Sepals 4 (5)

Western Virgin's-bower *Clematis ligusticifolia* Nutt.

■ **SKETCH** A **perennial vine** 1–20 m long, horizontal to vertical, **without tendrils**; along streambanks, moist slopes, ravines, on bushes, trees and fences; **dioecious**.

■ **FLOWERS** Creamy white, blooming March–September; **inflorescence** a compound cyme, from most leaf axils, each 4–6 cm long and wide; **peduncles** 2.5–10 cm long, reddish green; **pedicels** green, 5–30 mm long, slightly hairy, rigid; **flowers** unisexual, fragrant, usually 7–20+ per cyme; <u>male flowers</u> with *c.* 40+ white stamens; **pistils** absent; <u>female flowers</u> 16–24 mm wide by 9–11 mm tall; **sepals** white, four (five), petal-like, each 5–12 mm long by 3–6 mm wide, spreading, tips often reflexed, 5- or 7-nerved, slightly hairy; **petals** absent; **staminodia** *c.* 25, sterile, ascending to eventually spreading, 6–8 mm long; **filaments** white, glabrous, 5–6 mm long; **anthers** creamy white, 1.5–1.8 mm long; **pistils** 20–65, each 7–9 mm long, slightly twisted at the tip, forming a central column 2.5–3.3 mm wide; **ovary** green, *c.* 0.7 mm long, obscured by silky white hairs; **styles** filiform, 5–6 mm long, silky hairy, persistent; **stigma** obscure; **fruiting heads** are usually of 20–45 achenes, together 8–12 mm wide by 6–8 mm tall (excluding the styles). **FRUIT an achene**, 1-seeded, smooth, 2–3.5 mm long by 2–2.2 mm wide by 1–1.5 mm thick with a lateral ridge, hairs silky and ascending; **styles** 2–6 cm long and tapered to a fine point, hairs *c.* 4 mm long.

■ **LEAVES** Opposite, compound, deciduous; **leaf blades** 10–15 cm long by 9–13 cm wide; **leaflets** five (three or seven), lobed, mostly coarsely toothed; **blades** 2–9 cm long by 1–7 cm wide, dull, glabrous above, minute appressed hairs below, margins hairy; **petioles** hairy, 2–6.5 cm long, forming a ridge across the stem, twisting around vegetation; **stipules** absent; **petiolules** 2–20 mm long, with a patch of short white hairs where they meet the rachis, and along the reddish margins of the groove.

■ **STEM** Reddish to yellowish brown, slightly angular, branched, solid, glabrous to slightly hairy, horizontal to vertical; 2–10 mm thick near the woody base.

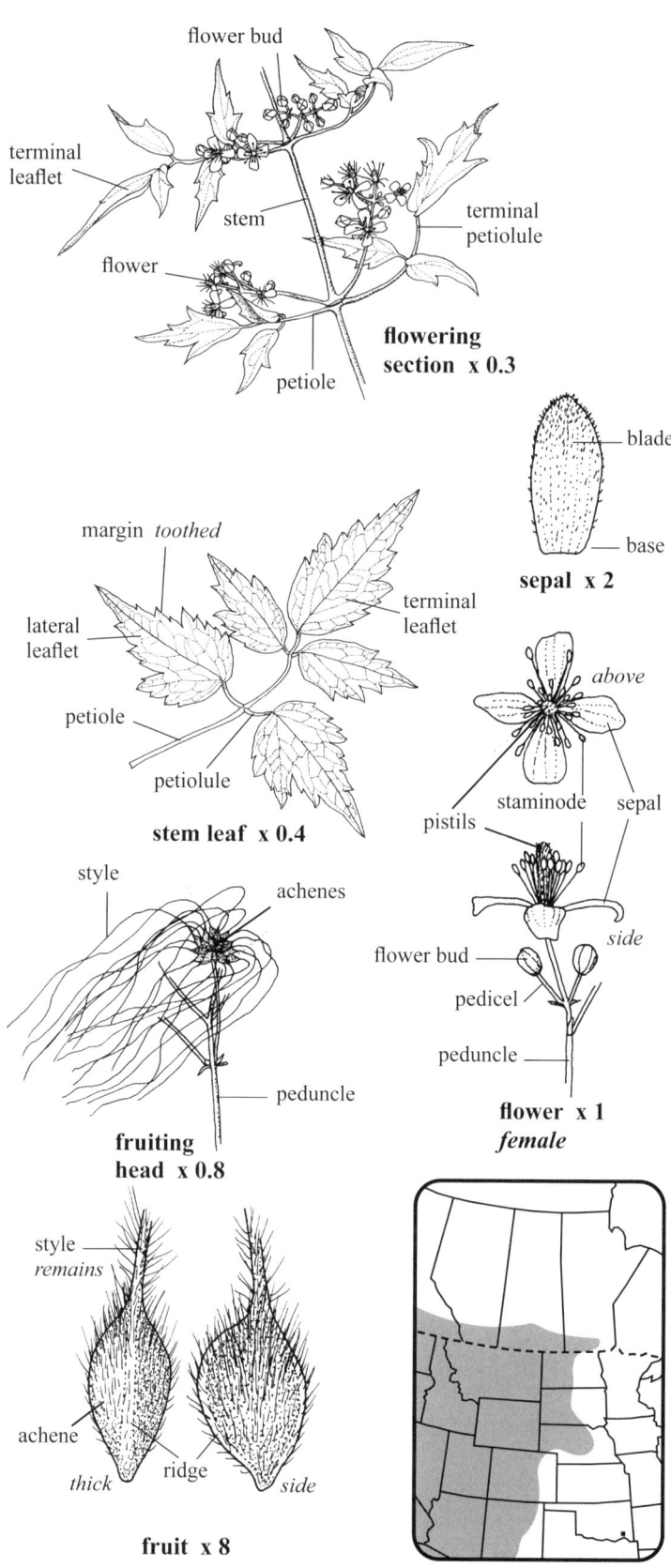

428 *Deciduous Traveler's-joy, Western White Clematis, Western Clematis*

Buttercup—*Ranunculaceae*

Vine   Yellow   Sepals 4

CLEMATIS   *Clematis tangutica* (Maxim.) Korsh.

■ **SKETCH** A **perennial vine** 2–4 m tall or long **without tendrils**; along railways, coulees, among shrubs and sprawling on fences. *Naturalized*

■ **FLOWERS** Yellow, blooming May–September; **inflorescence** one or two flowers from each leaf axil throughout most of the plant; **pedicels** erect, ridged, apices nodding, reddish green, 5–12 cm long, extending to *c*. 15 cm long in fruit; **flowers** perfect, 3–5 cm wide by 3.5–4.5 cm long, nodding, one to four per node, central column of pistils and stamens *c*. 2 cm long by 10–11 mm wide; **sepals** four, yellow, glabrous inside, hairy outside, slightly shiny, veiny, 2–4.5 cm long by 10–15 mm wide, margins entire and involute, tapered to a fine point, turning reddish brown on fading; **petals** absent; **stamens** numerous, included, 10–13 mm long, forming a column around the base of the pistils; **filaments** reddish green, *c*. 10 mm long, hairy in lower half; **anthers** 2–3 mm long, pale yellow, a few scattered hairs at their bases; **pistils** numerous, 20–22 mm long by *c*. 7 mm wide (together), congested in the center, included, each *c*. 11 mm long; **ovary** *c*. 1 mm long, covered by silky white hairs 1–1.5 mm long; **styles** green, filiform, completely hidden by silky white ascending hairs; **stigmas** pale green, filiform, slightly roughened, bent; **fruiting heads** tan, stiff, erect, 7–10 cm long and wide, the styles arched downward. **FRUIT an achene**, 1-seeded, 160 on one head, hairy, pointed, 3–4 mm long by 1.5–2 mm wide by *c*. 0.6 mm thick (not including hairs), reddish brown; **beak** (style) tan, plumose, 5–8 cm long, with ascending hairs 3–4 mm long; **receptacle** tan, oval, 5–7 mm long by 3–4 mm wide, rough.

■ **LEAVES** Opposite, compound, 5–15 cm long by 2–10 cm wide, apices twisted like a tendril around vegetation; **leaflets** three to seven, toothed, lobed, 1–6 cm long by 5–20 mm wide, reduced above, often hairy below; **petioles** reddish green, ridged, 3–6 cm long; **petiolules** 1–2.5 cm long, reddish; **stipules** absent.

■ **STEM** Ridged; 0.5–2 cm thick near the woody base; **bark** gray and shed in long strips; **branches** reddish brown, ridged, a few ascending from the lower nodes.

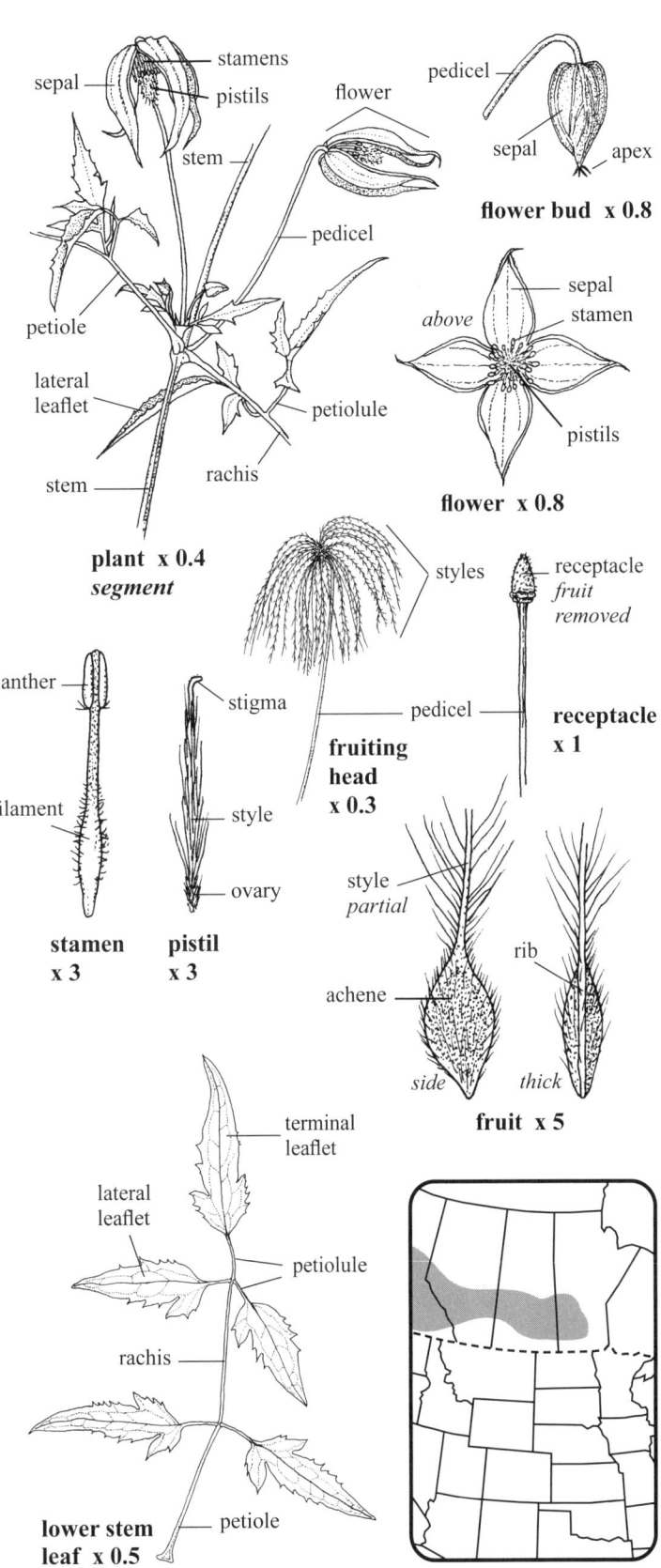

Oriental Virgin's-bower

Buttercup—*Ranunculaceae* Wildflower Evergreen White Sepals 4–7

GOLDTHREAD *Coptis trifolia* (L.) Salisb.

■ **SKETCH** An evergreen **perennial herb** 5–15 cm tall with orangish yellow, smooth **rhizomes** 10–20 cm long by *c.* 1 mm wide; colonies in damp woods and muskeg.

■ **FLOWERS** White, blooming May–August; **inflorescence** of one terminal flower on a scape; **flowers** perfect, 12–20 mm wide by 5–6 mm long; **sepals** 4–7, white, glabrous, 7–11 mm long by 2.5–3.5 mm wide, 3- or 5-veined, spreading, tapered to base; **petals** 4–7, each 2.5–3.5 mm long by *c.* 0.7 mm wide, ascending, expanded into a flattened, yellowish glandular nectary indented on one side, claw white, blade absent; **stamens** 30–55, each 3–4.5 mm long, white, ascending; **filaments** filiform; **anthers**, 2-lobed, *c.* 0.4 mm long and wide; **pistils** (carpels) 3–9, green, ascending, 3.7–5 mm long by *c.* 0.7 mm wide; **stipe** 1.5–2 mm long, slightly hairy, grooved on inside; **ovary** 1.5–2 mm long, tapered above; **style** one, reflexed at apex, 1.5–2 mm long; **stigma** *c.* 0.6 mm long, angled and grooved. **FRUIT a follicle**, 4–7 mm long (not including beak) by 2.3–2.5 mm wide by 1.5–2 mm thick, transparent, 3–9 per scape, ascending, glabrous, encircled by one rib; **stipes** 5–6 mm long, yellowish green, slightly hairy; **beaks** ascending, 2–4 mm long with a tiny apical hook; **seeds** 8–12 per follicle, shiny, dark brown, hilum above the ridge, ridged along one side, convex and rough on the opposite side, 1.2–1.4 mm long by 0.5–0.6 mm wide by 0.4–0.5 mm thick.

■ **LEAVES** Basal only, 2–5 per scape, arising from dark brown bracts at base, evergreen, compound; **leaf blades** 2.3–3.5 cm long by 2.3–4 cm wide; **leaflets** three, shiny on both sides, 12–21 mm long by 11–18 mm wide, glabrous, lobed and toothed; **petioles** (of leaves) 3–6 cm long, thin, glabrous; **petiolules** (of leaflets) 1–2 mm long, with microscopic hairs.

■ **STEM** A **scape**, green, 5–15 cm long, erect with flower and fruit, usually with one **bracteole** *c.* 2 mm long by *c.* 1 mm wide about midway on the scape; 0.7–1.1 mm wide near the bare green base.

■ **SYN.** *Coptis groenlandica* (Oeder) Fern.

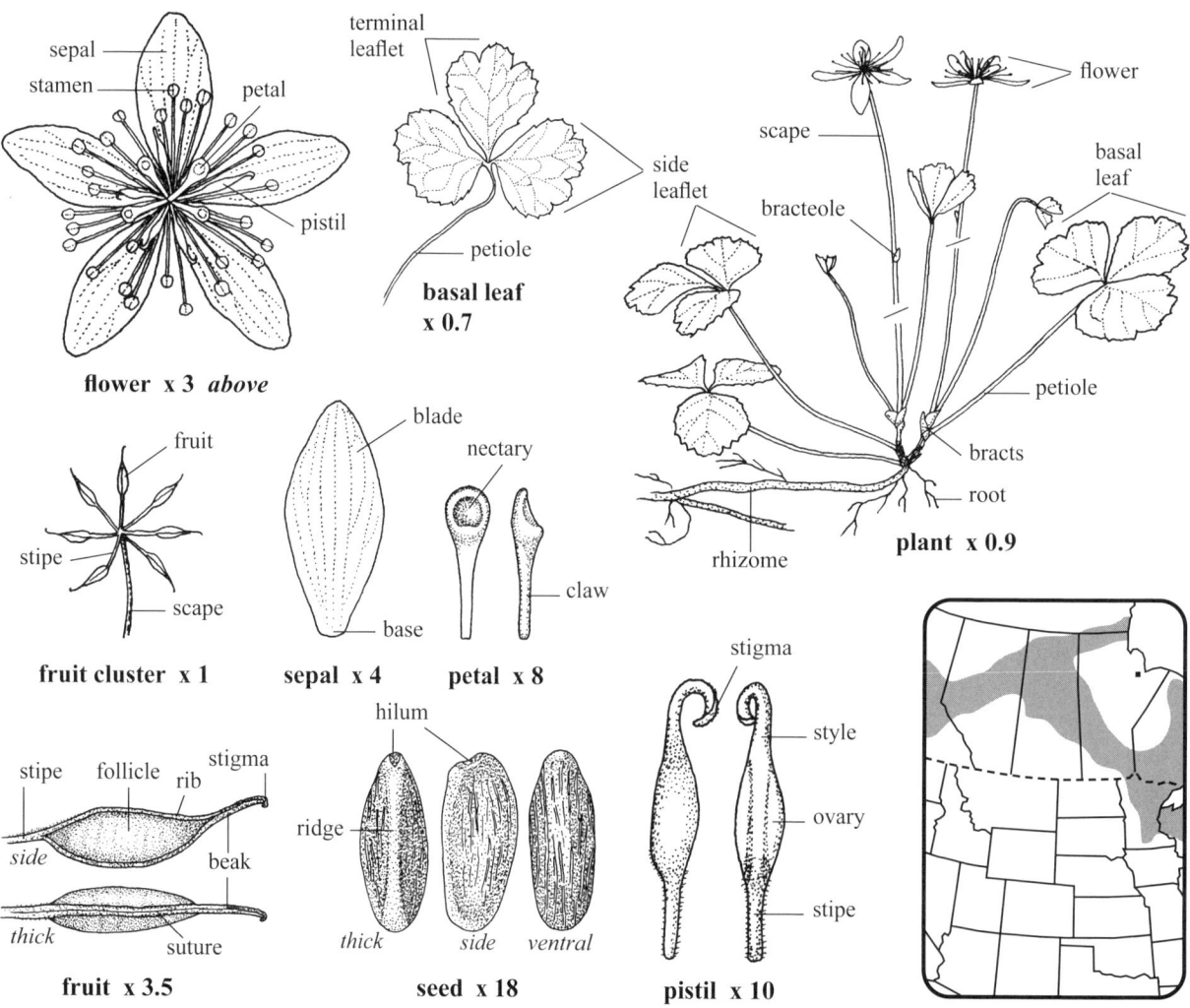

430 Three-leaf Goldthread

Buttercup—*Ranunculaceae* **Wildflower** Pale blue or purple to white, center yellow
Sepals 6 (5–8)

CROCUS ANEMONE *Pulsatilla patens* (L.) P. Mill.

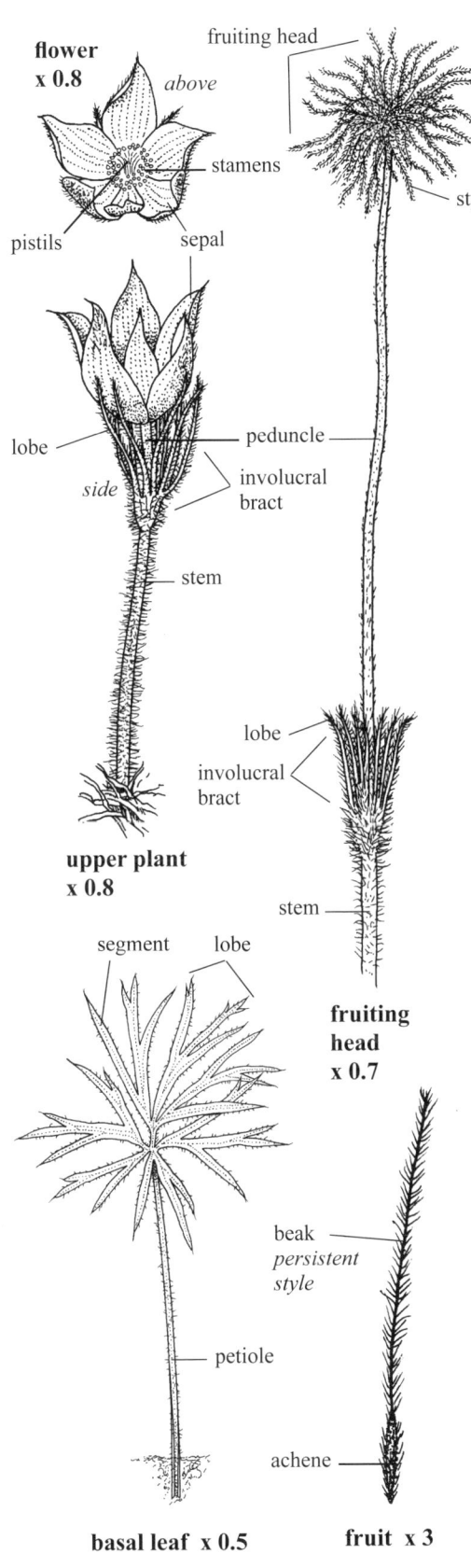

upper plant x 0.8

basal leaf x 0.5

fruiting head x 0.7

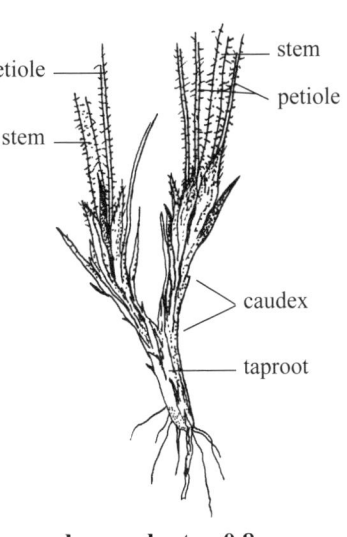

fruit x 3

lower plant x 0.8

■ **SKETCH** A hairy **perennial herb** 5–60 cm tall from a **taproot** 4–10 cm long by 4–10 mm thick, in small bunches from a **caudex** of one to five branches 1–5 cm long and 2–12 mm wide; in open prairies, open woods, rocky soil, on hillsides, along railways and in overgrazed pasture. **p1**

■ **FLOWERS** Pale blue or purple to white, with a yellow center, blooming March–July; **inflorescence** a terminal, solitary flower; **involucral bracts** three, 2–5 cm long, forming a sessile whorl at the base of the peduncle, united below with 14- to 20-lobed (segments) above, the lobes *c.* 2 cm long by 1–2 mm wide, hairy on the outside, erect, a few bisected near the tips, very hairy in fruit with white to beige hairs to *c.* 4 mm long; **peduncles** hairy or not, 1–12 cm long with flower, erect and up to *c.* 24 cm long in fruit; **flowers** perfect, 2.5–7 cm across; **sepals** petal-like, six (five to eight), 15–43 mm long by 7–18 mm wide, lightly hairy only on the outside, whiter inside, veiny, pointed; **petals** glandular, 1–2 mm long; **stamens** numerous, yellow, as a group 10–20 mm wide and *c.* 10 mm tall; **pistils** numerous; **fruiting head** 3–6 cm long and wide, 60–80 achenes per head. **FRUIT** an achene, 1-seeded, the body 3–5 mm long by *c.* 1 mm wide, pointed at one end, covered with ascending white hairs; **beak** (style) persistent, 2–4.5 cm long, pinkish brown, plumose from ascending thick hairs 1–2 mm long.

■ **LEAVES** Basal, simple but deeply lobed and toothed (appearing compound); **blades** 1–15 per clump, appear after anthesis, 6–12 cm long and wide, deeply 3- to 7-lobed, the lobes dissected into linear segments 2–5 mm wide, marginal hairs, the tips with a tuft of white hairs *c.* 2 mm long, hairy below along the veins, mostly glabrous above; **basal leaf petioles** 5–20 cm long, reddish near the base, covered with cottony hairs.

■ **STEM** Erect, unbranched, covered with silky white spreading hairs; 2–3 mm wide near the base.

■ **SYN.** *Anemone patens* L.

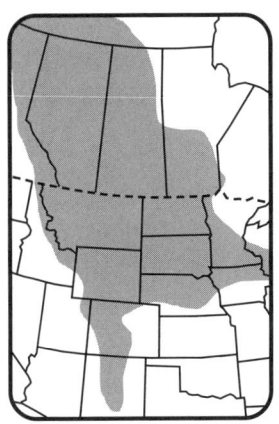

Prairie Crocus, American Pasqueflower, Pasque Flower

Buttercup—*Ranunculaceae* **Wildflower** Yellow Petals 5

SMOOTH-LEAVED BUTTERCUP *Ranunculus abortivus* L.

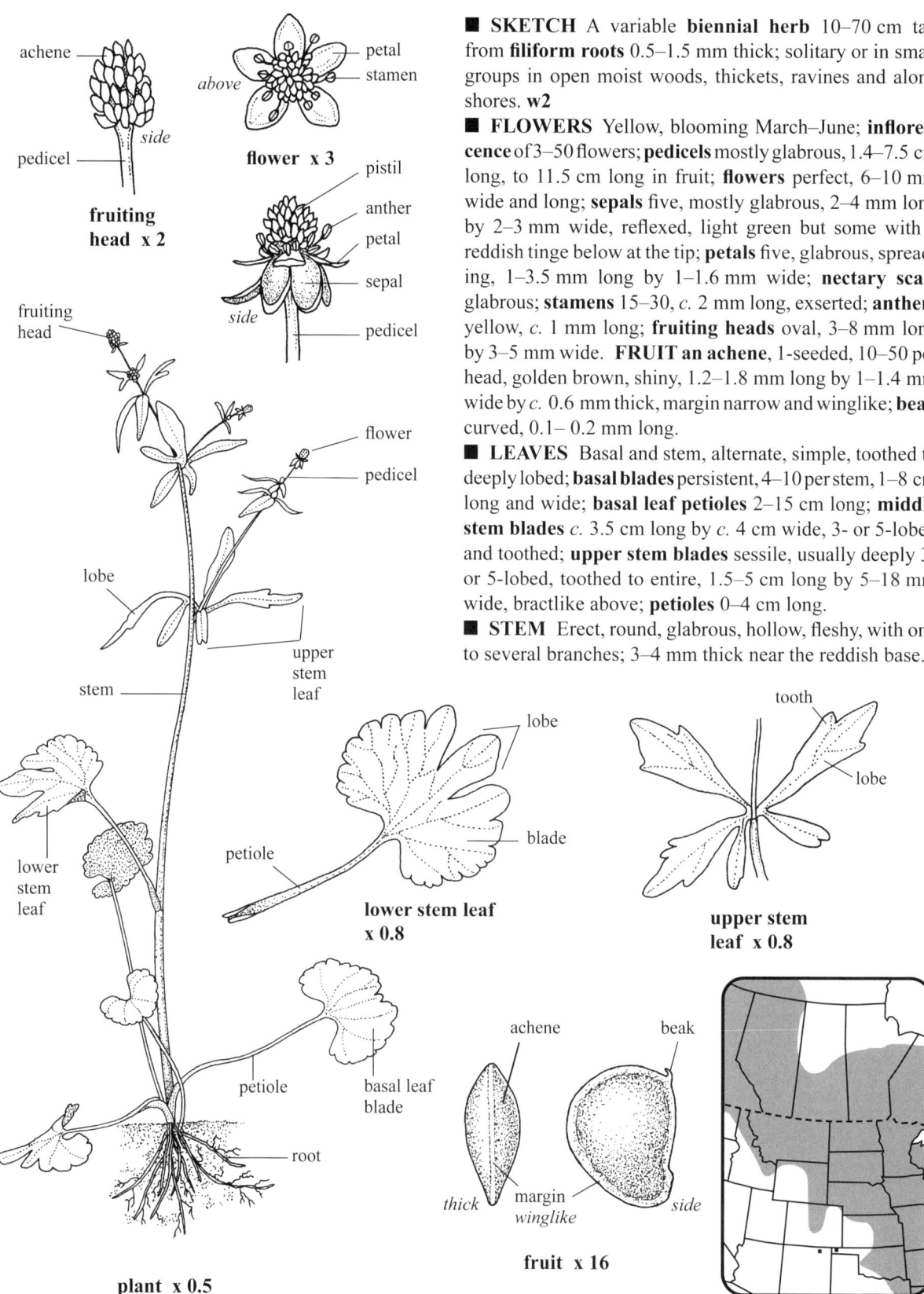

■ **SKETCH** A variable **biennial herb** 10–70 cm tall from **filiform roots** 0.5–1.5 mm thick; solitary or in small groups in open moist woods, thickets, ravines and along shores. **w2**

■ **FLOWERS** Yellow, blooming March–June; **inflorescence** of 3–50 flowers; **pedicels** mostly glabrous, 1.4–7.5 cm long, to 11.5 cm long in fruit; **flowers** perfect, 6–10 mm wide and long; **sepals** five, mostly glabrous, 2–4 mm long by 2–3 mm wide, reflexed, light green but some with a reddish tinge below at the tip; **petals** five, glabrous, spreading, 1–3.5 mm long by 1–1.6 mm wide; **nectary scale** glabrous; **stamens** 15–30, *c.* 2 mm long, exserted; **anthers** yellow, *c.* 1 mm long; **fruiting heads** oval, 3–8 mm long by 3–5 mm wide. **FRUIT an achene**, 1-seeded, 10–50 per head, golden brown, shiny, 1.2–1.8 mm long by 1–1.4 mm wide by *c.* 0.6 mm thick, margin narrow and winglike; **beak** curved, 0.1– 0.2 mm long.

■ **LEAVES** Basal and stem, alternate, simple, toothed to deeply lobed; **basal blades** persistent, 4–10 per stem, 1–8 cm long and wide; **basal leaf petioles** 2–15 cm long; **middle stem blades** *c.* 3.5 cm long by *c.* 4 cm wide, 3- or 5-lobed and toothed; **upper stem blades** sessile, usually deeply 3- or 5-lobed, toothed to entire, 1.5–5 cm long by 5–18 mm wide, bractlike above; **petioles** 0–4 cm long.

■ **STEM** Erect, round, glabrous, hollow, fleshy, with one to several branches; 3–4 mm thick near the reddish base.

Kidney-leaved Buttercup, Littleleaf Buttercup, Small-flowered Buttercup, Early Wood Buttercup

Buttercup—*Ranunculaceae* **Wildflower** Yellow Petals 5

TALL BUTTERCUP *Ranunculus acris* L.

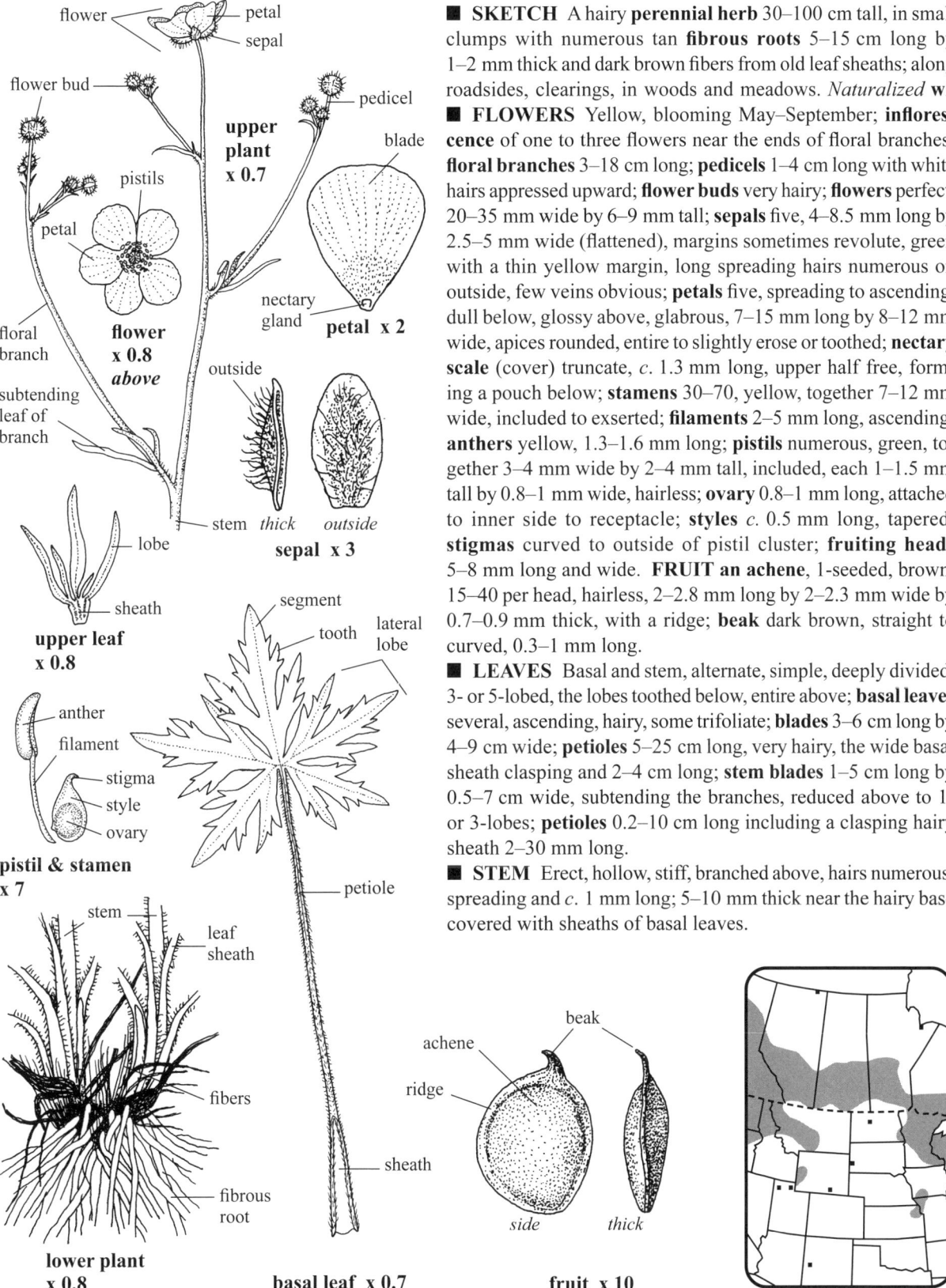

■ **SKETCH** A hairy **perennial herb** 30–100 cm tall, in small clumps with numerous tan **fibrous roots** 5–15 cm long by 1–2 mm thick and dark brown fibers from old leaf sheaths; along roadsides, clearings, in woods and meadows. *Naturalized* **w2**

■ **FLOWERS** Yellow, blooming May–September; **inflorescence** of one to three flowers near the ends of floral branches; **floral branches** 3–18 cm long; **pedicels** 1–4 cm long with white hairs appressed upward; **flower buds** very hairy; **flowers** perfect, 20–35 mm wide by 6–9 mm tall; **sepals** five, 4–8.5 mm long by 2.5–5 mm wide (flattened), margins sometimes revolute, green with a thin yellow margin, long spreading hairs numerous on outside, few veins obvious; **petals** five, spreading to ascending, dull below, glossy above, glabrous, 7–15 mm long by 8–12 mm wide, apices rounded, entire to slightly erose or toothed; **nectary scale** (cover) truncate, *c.* 1.3 mm long, upper half free, forming a pouch below; **stamens** 30–70, yellow, together 7–12 mm wide, included to exserted; **filaments** 2–5 mm long, ascending; **anthers** yellow, 1.3–1.6 mm long; **pistils** numerous, green, together 3–4 mm wide by 2–4 mm tall, included, each 1–1.5 mm tall by 0.8–1 mm wide, hairless; **ovary** 0.8–1 mm long, attached to inner side to receptacle; **styles** *c.* 0.5 mm long, tapered; **stigmas** curved to outside of pistil cluster; **fruiting heads** 5–8 mm long and wide. **FRUIT an achene**, 1-seeded, brown, 15–40 per head, hairless, 2–2.8 mm long by 2–2.3 mm wide by 0.7–0.9 mm thick, with a ridge; **beak** dark brown, straight to curved, 0.3–1 mm long.

■ **LEAVES** Basal and stem, alternate, simple, deeply divided, 3- or 5-lobed, the lobes toothed below, entire above; **basal leaves** several, ascending, hairy, some trifoliate; **blades** 3–6 cm long by 4–9 cm wide; **petioles** 5–25 cm long, very hairy, the wide basal sheath clasping and 2–4 cm long; **stem blades** 1–5 cm long by 0.5–7 cm wide, subtending the branches, reduced above to 1- or 3-lobes; **petioles** 0.2–10 cm long including a clasping hairy sheath 2–30 mm long.

■ **STEM** Erect, hollow, stiff, branched above, hairs numerous, spreading and *c.* 1 mm long; 5–10 mm thick near the hairy base covered with sheaths of basal leaves.

Common Buttercup, Meadow Buttercup

Buttercup—*Ranunculaceae*   **Wildflower**   White   Petals 5 (6)

Large-leaved Watercrowfoot   *Ranunculus aquatilis* L.

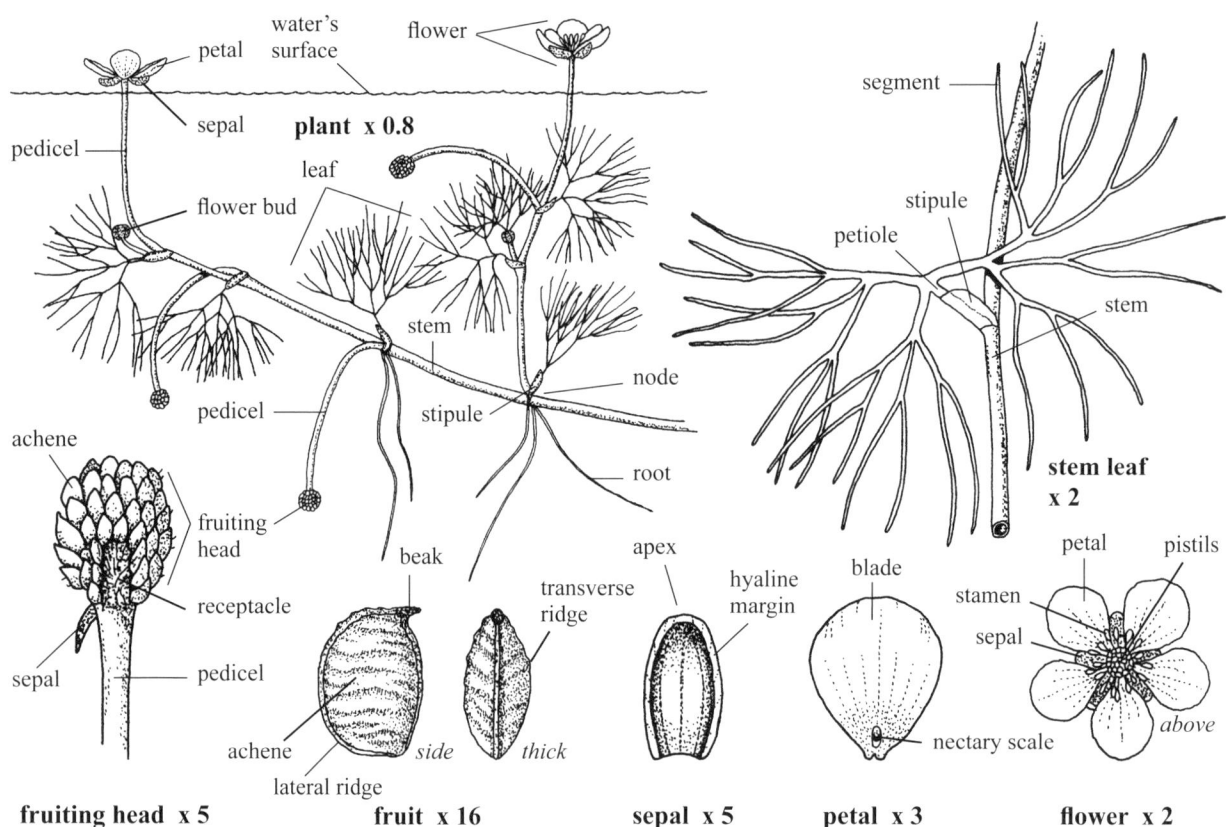

- **SKETCH** A variable, glabrous, aquatic **perennial herb** with a submerged horizontal stem to *c.* 60 cm long; in shallow water of ponds, marshes, ditches and slow streams, sometimes covering several m². **w1**
- **FLOWERS** White, blooming May–September; **inflorescence** an axillary flower; **pedicels** 1.5–12 cm long by 1–1.5 mm thick, hollow, straight to reflexed in fruit; **flowers** perfect, 10–17 mm wide by 5–7 mm tall, 0–5 cm above the water's surface; **sepals** five (six), glabrous, greenish gray, blunt, veins three, 3–4 mm long by 2–2.5 mm wide (not flattened), yellowish gray inside with the tips silvery to almost black, spreading to reflexed, early deciduous, C-shaped, bases truncate, margins hyaline green and rolled outward; **petals** five (six), 6- or 8-veined, 4–8.5 mm long by 3–6.2 mm wide, white with a yellow base, entire, flat; **nectary scale** yellow, indented, to *c.* 1 mm long by *c.* 0.4 mm wide; **stamens** 10–20, from the base of the pistils; **filaments** 1–2 mm long, yellowish green; **anthers** yellow, *c.* 1 mm long; **pistils** 30–80, green, in a globular cluster *c.* 2 mm wide and tall; **fruiting heads** 3–6 mm long and wide, submerged. **FRUIT an achene**, 1-seeded, 20–60+ per head, a few hairs along the dorsal ridge when green, body oval, from 1–2 mm long by 0.7–1.4 mm wide by 0.5–0.7 mm thick with wavy, blunt, transverse ridges, often falling from receptacle before fully ripe; **beak** 0.2–1.2 mm long; **receptacle** hairy, rough.
- **LEAVES** Alternate, compound, green to brown; **blades** 1–4 cm long and wide, divided three times, with entire filiform segments *c.* 0.2 mm thick which gradually taper towards the tips, usually collapsing out of water; **petioles** hidden by stipules or sometimes extending past them by 1–10 mm; **stipules** usually 4–5 mm long by 3–4 mm wide, 3-veined, hairy or not.
- **STEM** Hollow, light green, glabrous, breaking easily, often in short segments 10–60 cm long by 1–3 mm thick; **roots** from nodes, white, simple, one to several, each *c.* 0.4 mm wide.

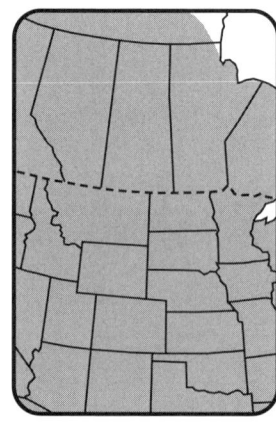

Thread-leaved Buttercup, Whitewater Crowfoot, White Water Buttercup

Buttercup—*Ranunculaceae* Wildflower Yellow Petals 5

Seaside Buttercup *Ranunculus cymbalaria* Pursh

■ **SKETCH** A **perennial herb** 5–25 cm tall with nodal **fibrous roots** and **stolons** *c.* 1 mm thick and to 30⁺ cm long, often forming colonies over wide areas; along marsh shores, bogs, stream banks, ditches, muddy shores and at seepages, including saline sites. **p2**

■ **FLOWERS** Yellow, blooming April–September; **inflorescence** few-flowered; **pedicels** mostly erect, 1.5–15 cm long by *c.* 1 mm thick, slightly hairy; **flowers** perfect, 5–10 mm wide with a conical center *c.* 2 mm wide; **sepals** five, greenish yellow, 3–6 mm long by 1.5–3 mm wide, slightly shorter or longer than the petals, spreading; **petals** five (usually), yellow, shiny, 2–7 mm long by 1–3 mm wide, slightly recurved; **stamens** 10–30, shorter than pistils; **pistils** numerous, forming a conical disc *c.* 3 mm tall; **fruiting heads** cylindrical, 3–15 mm long by 3–7 mm wide. **FRUIT an achene**, 1-seeded, 40–200 per head, tan, slightly hairy, 1.4–2.2 mm long by 0.7–1.2 mm wide by *c.* 0.5 mm thick with two to six ribs on each face; **beak** triangular, *c.* 0.2 mm long; **seeds** pointed, *c.* 0.5 mm long by *c.* 0.3 mm wide by 0.2 mm thick.

■ **LEAVES** Basal, simple, teeth shallow, blunt, apical lobes rounded; **blades** 5–38 mm long by 3–33 mm wide, glabrous to lightly hairy; **petioles** 1–10 cm long, widening at the base, glabrous to lightly hairy.

■ **STEM** A **scape**, one to several, glabrous to slightly hairy, thin, mostly erect, round, simple or few-branched.

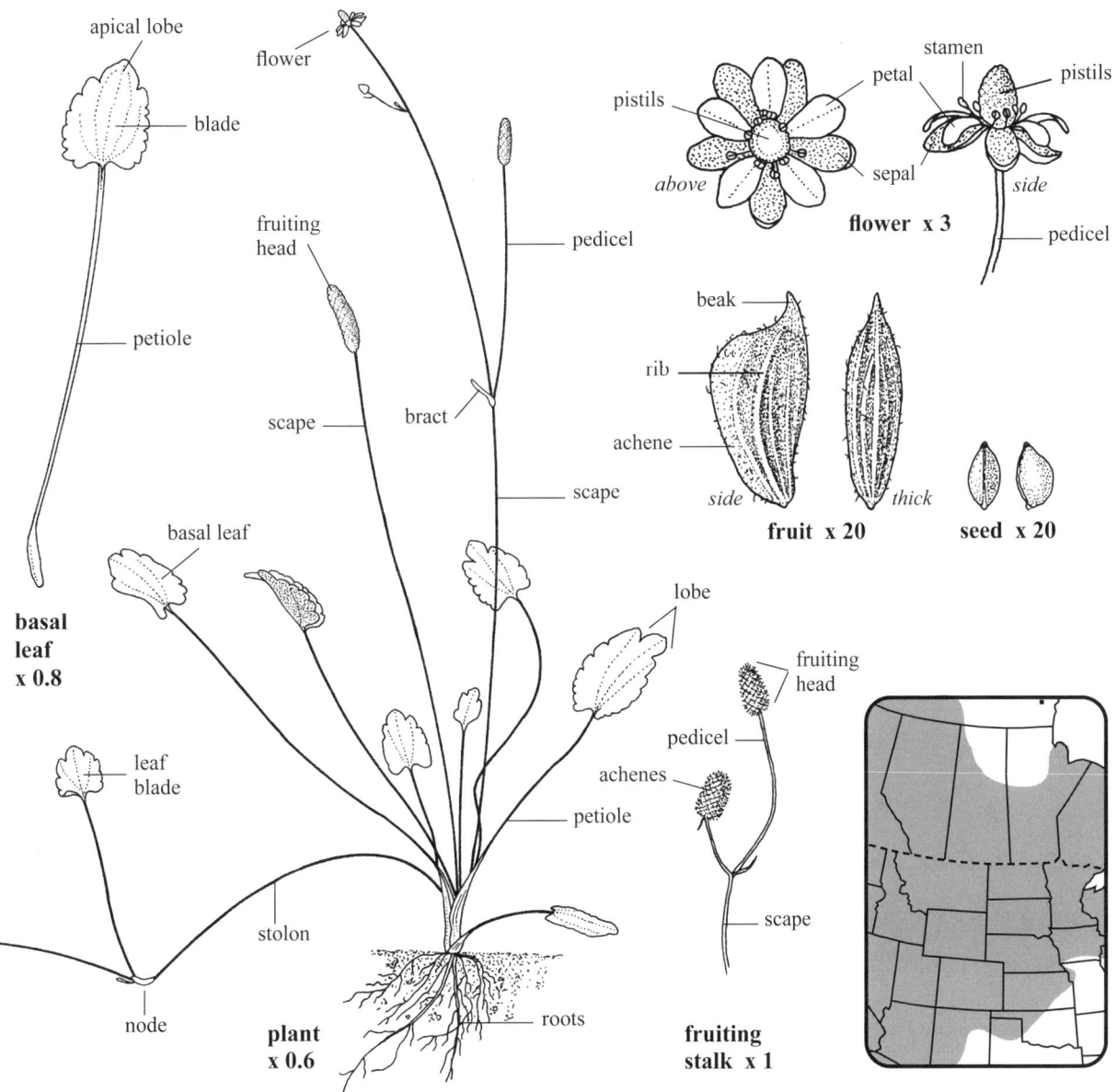

Seaside Crowfoot, Shore Buttercup, Alkali Buttercup

Buttercup—*Ranunculaceae*   **Wildflower** Yellow Petals 5 (6)

CREEPING SPEARWORT *Ranunculus flammula* L.

■ **SKETCH** A variable **perennial herb** with the upright tips 4–15 cm tall from long, prostrate stems with a few **fibrous roots** 1–3 cm long at nodes; in wet ditches, muddy shores and along marshes, rivers, lakes and ponds. **w2**

■ **FLOWERS** Yellow, blooming June–August; **inflorescence** a terminal flower, well-spaced along the stem; **pedicels** slightly hairy, ascending, pale green to pinkish, 0.5–10 cm long by $c.$ 0.5 mm wide; **flowers** perfect, 8–12 mm wide by 4–5 mm long; **sepals** five, C-shaped, yellow, spreading, 2–3.2 (–5) mm long by 1.2–2 mm wide, hairy on outside when young, glabrous with age, some with two or three shallow blunt teeth, margins revolute with age; **petals** five (6), yellow, arched, tips ascending, 2.5–7 mm long by 1.7–4 mm wide, shiny, glabrous, slightly notched at apices, veins obscure, claw white; **nectary scale** microscopic, $c.$ 1 mm from base; **stamens** yellow, included, 16–50, each $c.$ 2 mm long; **filaments** arising from base of pistils, $c.$ 1.4 mm long; **anthers** 2-lobed, $c.$ 0.8 mm long; **pistils** 10–25, included, a central cluster $c.$ 1.5 mm wide, green, each $c.$ 1.5 mm long, glabrous; **ovary** $c.$ 1 mm long, with a faint ridge along the inner aspect; **style** $c.$ 0.3 mm long; **stigma** yellowish green, rough, arching to the outside; **fruiting heads** almost round, 2.5–4 mm long by 3–4.5 mm wide, with 5–20 fruit. **FRUIT an achene**, 1-seeded, 1.3–2 mm long (including beak) by 1.2–1.5 mm wide by $c.$ 0.7 mm thick, greenish yellow, with a low dorsal ridge; **beak** straight to curved, 0.2–0.6 mm long; **receptacle** $c.$ 1 mm long, hairless.

■ **LEAVES** Alternate, one or more at each rooting node along the stem; **upper stem leaves** fleshy, entire, simple, linear, 2–8 cm long by 0.8–1 mm wide by $c.$ 0.5 mm thick, sessile; **lower leaves** spoon-shaped; **blades** linear, 1-nerved, 5–23 mm long by 2–7 (–20) mm wide; **petioles** 5–25 mm long; **stipules** paired, generally transparent, 3–14 mm long, apices hairy, enveloping stem and leaf base but open to base, sides overlapping, becoming hairless with age.

■ **STEM** Round to oval, stoloniferous, prostrate, pale green, hairs scattered and pointed forward, hollow; hairless near the 0.8–2 mm thick base.

■ **SYN.** *Ranunculus reptans* L. = *Ranunculus flammula* var. *filiformis* (Michx.) Hook.

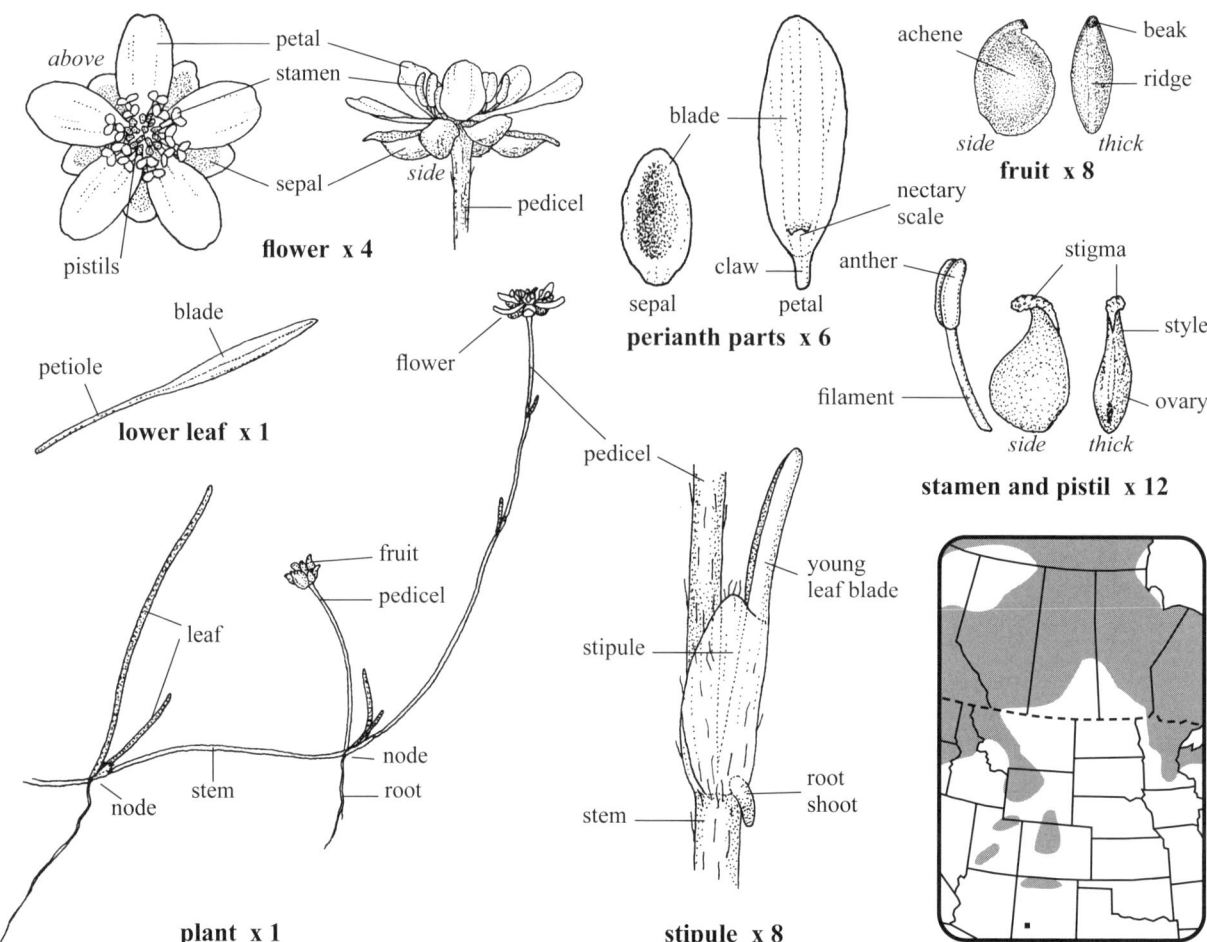

Creeping Buttercup, Spearwort, Lesser Spearwort, Greater Creeping Spearwort

Buttercup—*Ranunculaceae* **Wildflower Yellow Petals 5**

Bristly Buttercup *Ranunculus pensylvanicus* L. f.

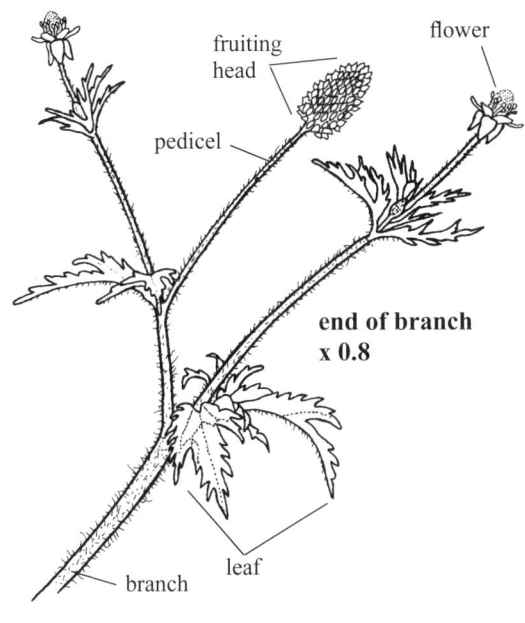

end of branch x 0.8

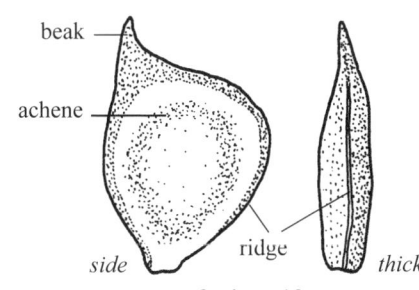

fruit x 10

■ **SKETCH** A hairy **annual** or **short-lived perennial herb**, solitary, 20–100 cm tall from fleshy **fibrous roots**; in wet meadows and forests, along stream and wooded river banks.

■ **FLOWERS** Yellow, blooming June–August; **inflorescence** paniculate, open, with flowers at the ends of several branches; **branches** alternate, to *c.* 35 cm long, ascending, axillary and hairy; **pedicels** hairy, 1–7 cm long, axillary; **flowers** perfect, regular, 6–10 mm wide and long; **sepals** five, greenish yellow, 3.4–7 mm long by 2–3 mm wide, reflexed, with scattered, white, 1 mm long hairs beneath; **petals** five, shiny, glabrous, flat, horizontal to slightly ascending, 2–4 mm long by 1.8–3 mm wide, mostly shorter than the sepals; **stamens** yellow, 15–25; **anthers** *c.* 1 mm long; **pistils** numerous, forming a green oval head; **fruiting heads** cylindrical, 10–18 mm long by 5–10 mm wide. **FRUIT** an achene, 1-seeded, 40–125 per head, each 1.8–3.5 mm long by 1.5–2.3 mm wide by *c.* 0.7 mm thick, with a narrow marginal ridge; **beak** 0.6–1.4 mm long; **receptacle** elongate, slightly hairy, 2–3 mm wide.

■ **LEAVES** Alternate, trifoliate; **leaflets** 2- or 3-lobed, coarsely toothed; **basal blades** withering by anthesis, 2–6.5 cm long by 2.5–9 cm wide, similar to stem blades; **lower stem blades** opposite, one or two pairs; **upper stem blades** 4–16 cm long by 4–21 cm wide, smaller toward the top; **terminal leaflet** the largest at *c.* 10 cm long by *c.* 12 cm wide; **lateral leaflets** smaller, to *c.* 9 cm long and wide; **petioles** 1–25 cm long, hairy, widening and clasping; **petiolules** hairy, 2.5–7.5 cm long.

■ **STEM** Erect, stout, covered with spreading soft hairs 1–3 mm long, these becoming less dense toward the top; 5–20 mm thick near the base.

■ **NOTE** Similar to *R. macounii* ("Flora of North America," www.efloras.org).

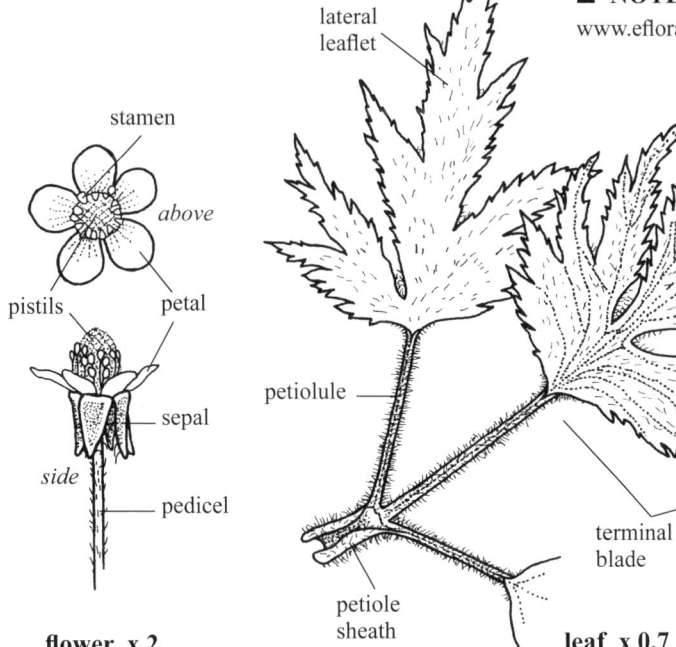

leaf x 0.7

flower x 2

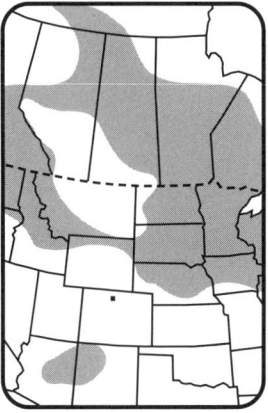

Bristly Crowfoot, Pennsylvania Buttercup

Buttercup—*Ranunculaceae* **Wildflower** Yellow Petals 5 (6)

PRAIRIE BUTTERCUP *Ranunculus rhomboideus* Goldie

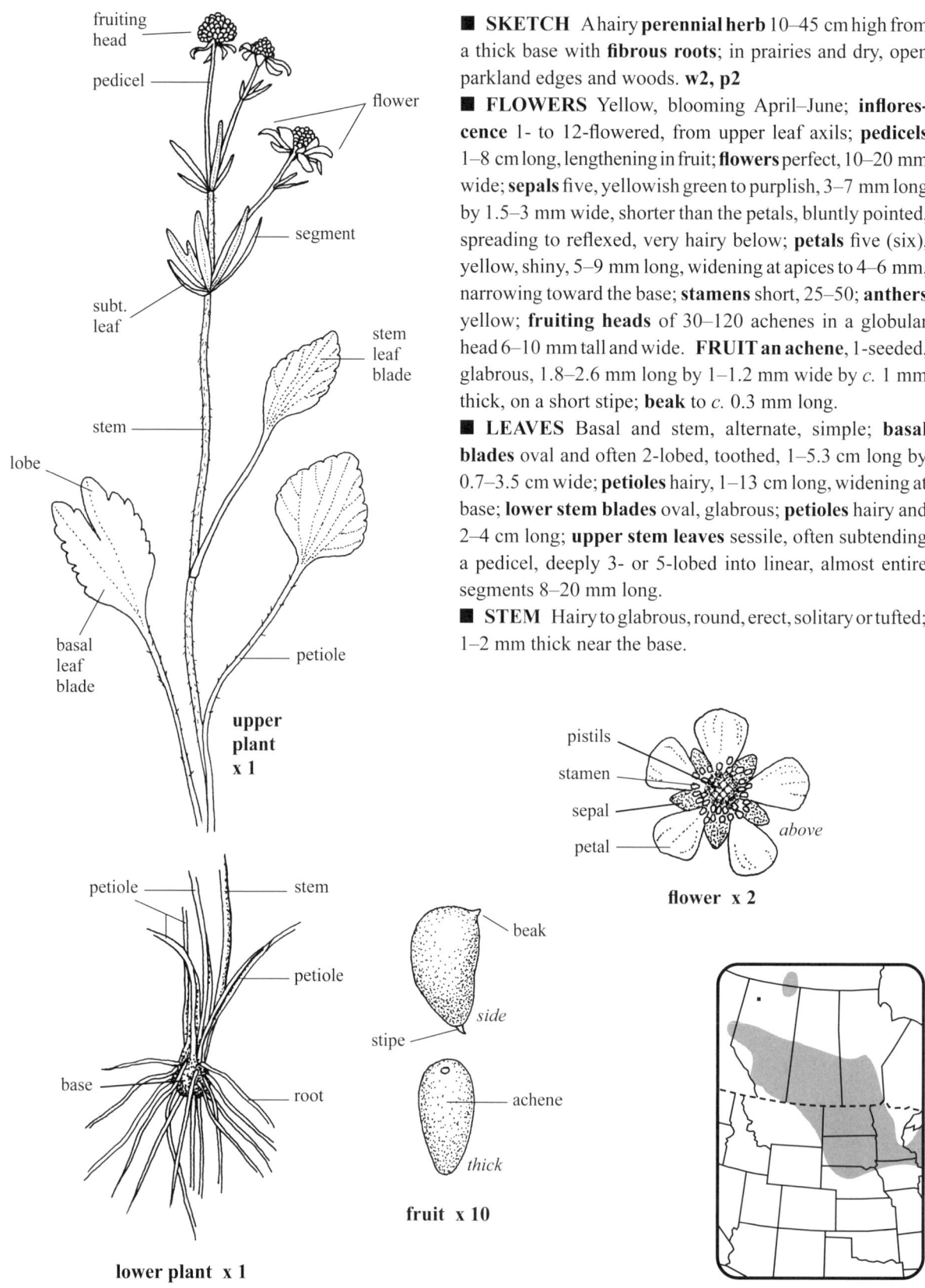

■ **SKETCH** A hairy **perennial herb** 10–45 cm high from a thick base with **fibrous roots**; in prairies and dry, open parkland edges and woods. **w2, p2**

■ **FLOWERS** Yellow, blooming April–June; **inflorescence** 1- to 12-flowered, from upper leaf axils; **pedicels** 1–8 cm long, lengthening in fruit; **flowers** perfect, 10–20 mm wide; **sepals** five, yellowish green to purplish, 3–7 mm long by 1.5–3 mm wide, shorter than the petals, bluntly pointed, spreading to reflexed, very hairy below; **petals** five (six), yellow, shiny, 5–9 mm long, widening at apices to 4–6 mm, narrowing toward the base; **stamens** short, 25–50; **anthers** yellow; **fruiting heads** of 30–120 achenes in a globular head 6–10 mm tall and wide. **FRUIT an achene**, 1-seeded, glabrous, 1.8–2.6 mm long by 1–1.2 mm wide by *c.* 1 mm thick, on a short stipe; **beak** to *c.* 0.3 mm long.

■ **LEAVES** Basal and stem, alternate, simple; **basal blades** oval and often 2-lobed, toothed, 1–5.3 cm long by 0.7–3.5 cm wide; **petioles** hairy, 1–13 cm long, widening at base; **lower stem blades** oval, glabrous; **petioles** hairy and 2–4 cm long; **upper stem leaves** sessile, often subtending a pedicel, deeply 3- or 5-lobed into linear, almost entire segments 8–20 mm long.

■ **STEM** Hairy to glabrous, round, erect, solitary or tufted; 1–2 mm thick near the base.

Buttercup—*Ranunculaceae*   **Wildflower** Yellow Petals 5 (3 or 4)

CELERY-LEAVED BUTTERCUP   *Ranunculus sceleratus* L.

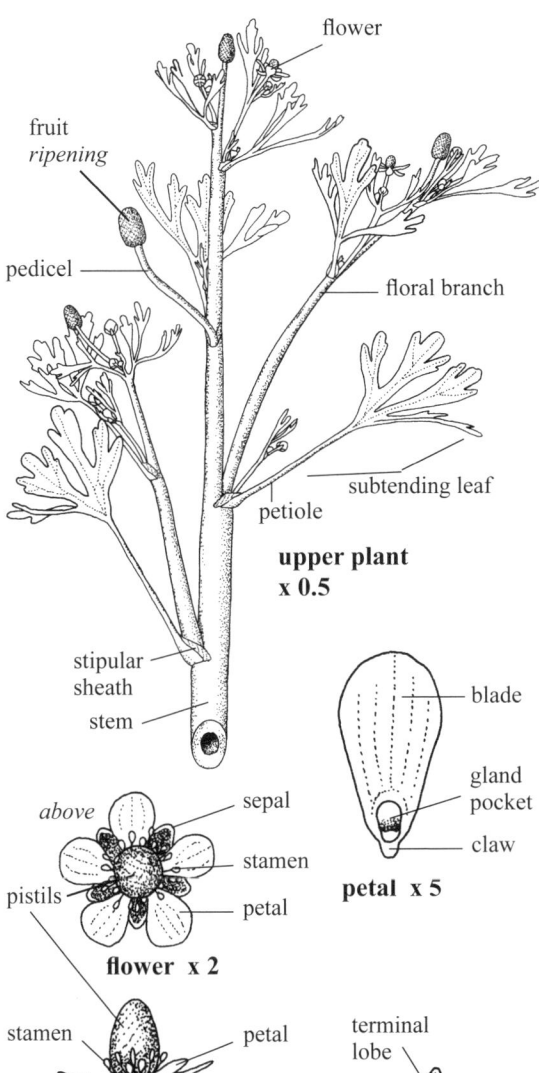

■ **SKETCH** A variable, glabrous, semi-aquatic **annual herb** 5–60 cm tall with white, fleshy **fibrous roots**; in shallow water or drying ditches, wet meadows, seeps and along pond margins.

■ **FLOWERS** Yellow, blooming May–September; **inflorescence** terminal and axillary on branches; **pedicels** 1–4 cm long at anthesis, longer with fruit, ascending, glabrous to slightly hairy, green, 1–2 mm thick, hollow, triangular or two-sided; **flowers** perfect, 6–12 mm wide by 6–8 mm long; **sepals** five (four), green, 2–5 mm long by 2–2.5 mm wide, spreading to slightly reflexed, glabrous, margins yellowish white to hyaline; **petals** five (3 or 4), shiny, spreading, 2–5.5 mm long by 2–4 mm wide, flat, apices round, early deciduous, glabrous; **nectary gland**, glabrous, basal, in a small pocket; **stamens** yellow, 10–25, exserted, ascending, 2–3 mm long; **filaments** 1–2 mm long; **anthers** yellow, *c.* 1 mm long; **pistils** green, numerous, forming an oval *c.* 5 mm tall by *c.* 3.5 mm wide, exserted; **fruiting heads** a cylinder 4–13 mm long by 3.5–7 mm wide. **FRUIT** an achene, 1-seeded, 40–300 per head, 1–1.4 mm long by 0.8–1.2 mm wide by *c.* 0.5 mm thick, not ridged or veiny, slightly rough (microscopic), often falling before fully ripe; **beak** (style) tiny and blunt.

■ **LEAVES** Basal and stem, alternate, simple, lobed, toothed; **basal leaves** may be floating early in season; **blades** shiny, flat, thin, 1–7 cm long by 3–9 cm wide, lighter green and veins slightly raised below; **petioles** 2–30 cm long, base widened into a **stipular sheath** 5–10 mm long; **stem blades** 1–6 cm long by 1.5–6 cm wide, reduced above, 3- or 5-lobed (deeply above) with rounded, shallow blunt teeth; **petioles** similar to those of basal leaves, reduced above.

■ **STEM** Round to oval, hollow, glabrous, one to several, erect to arched, branched; 0.3–1 cm thick near the base.

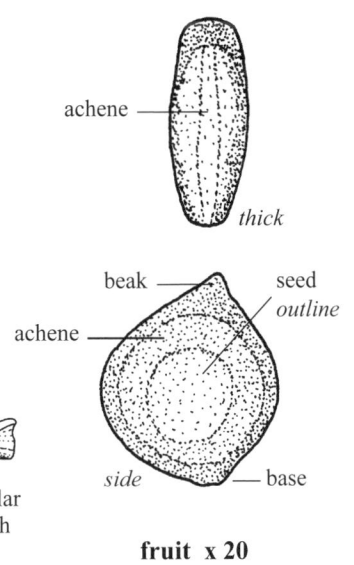

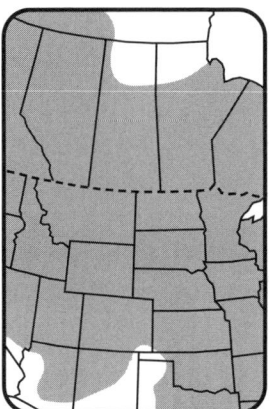

Cursed Buttercup, Cursed Crowfoot   439

Buttercup—*Ranunculaceae*   **Wildflower**   Green   Sepals 4 (5)

TALL MEADOW-RUE   *Thalictrum dasycarpum* Fisch. & Avé-Lall.

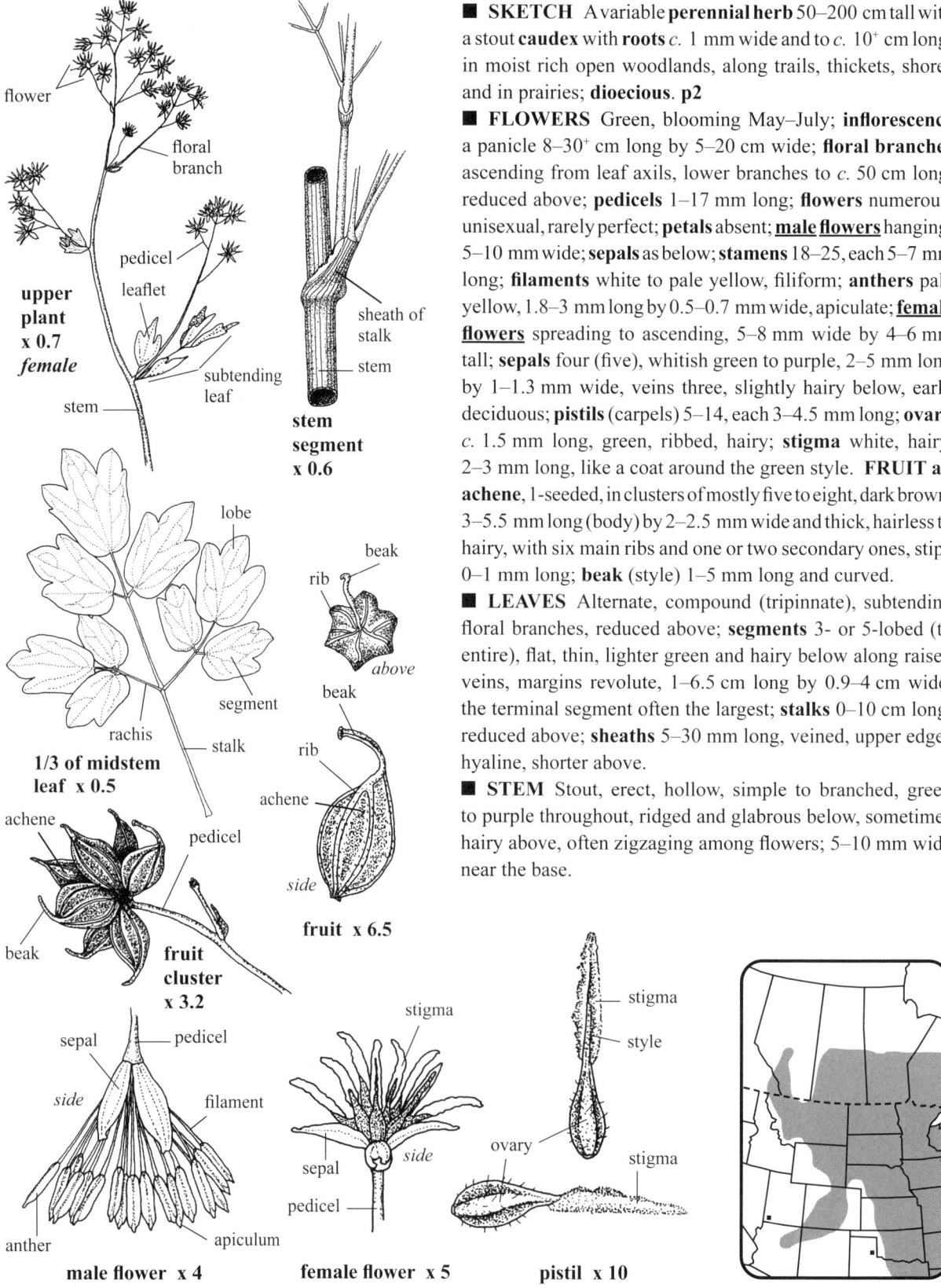

■ **SKETCH** A variable **perennial herb** 50–200 cm tall with a stout **caudex** with **roots** *c.* 1 mm wide and to *c.* 10⁺ cm long; in moist rich open woodlands, along trails, thickets, shores and in prairies; **dioecious**. p2

■ **FLOWERS** Green, blooming May–July; **inflorescence** a panicle 8–30⁺ cm long by 5–20 cm wide; **floral branches** ascending from leaf axils, lower branches to *c.* 50 cm long, reduced above; **pedicels** 1–17 mm long; **flowers** numerous, unisexual, rarely perfect; **petals** absent; **male flowers** hanging, 5–10 mm wide; **sepals** as below; **stamens** 18–25, each 5–7 mm long; **filaments** white to pale yellow, filiform; **anthers** pale yellow, 1.8–3 mm long by 0.5–0.7 mm wide, apiculate; **female flowers** spreading to ascending, 5–8 mm wide by 4–6 mm tall; **sepals** four (five), whitish green to purple, 2–5 mm long by 1–1.3 mm wide, veins three, slightly hairy below, early deciduous; **pistils** (carpels) 5–14, each 3–4.5 mm long; **ovary** *c.* 1.5 mm long, green, ribbed, hairy; **stigma** white, hairy, 2–3 mm long, like a coat around the green style. **FRUIT** an achene, 1-seeded, in clusters of mostly five to eight, dark brown, 3–5.5 mm long (body) by 2–2.5 mm wide and thick, hairless to hairy, with six main ribs and one or two secondary ones, stipe 0–1 mm long; **beak** (style) 1–5 mm long and curved.

■ **LEAVES** Alternate, compound (tripinnate), subtending floral branches, reduced above; **segments** 3- or 5-lobed (to entire), flat, thin, lighter green and hairy below along raised veins, margins revolute, 1–6.5 cm long by 0.9–4 cm wide, the terminal segment often the largest; **stalks** 0–10 cm long, reduced above; **sheaths** 5–30 mm long, veined, upper edges hyaline, shorter above.

■ **STEM** Stout, erect, hollow, simple to branched, green to purple throughout, ridged and glabrous below, sometimes hairy above, often zigzaging among flowers; 5–10 mm wide near the base.

440

Buttercup—*Ranunculaceae*   **Wildflower   Green   Sepals 4**

VEINY MEADOW-RUE   *Thalictrum venulosum* Trel.

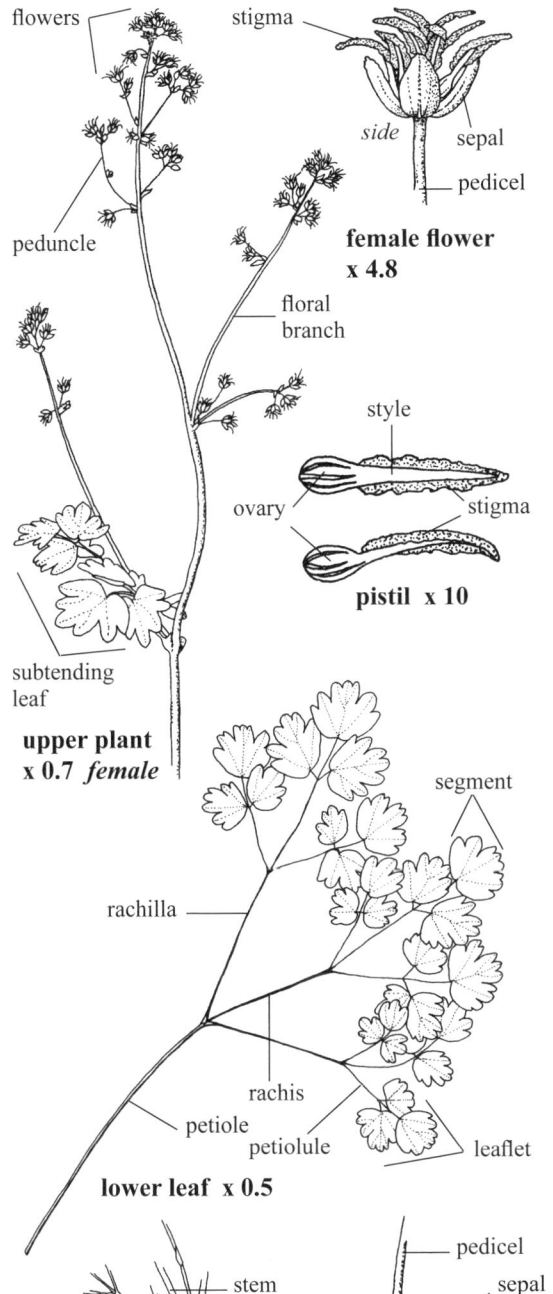

- **SKETCH** A variable, mostly glabrous **perennial herb** 20–150+ cm tall with brown, woody **rhizomes** and numerous brown **roots** 1–1.5 mm wide and to *c.* 10+ cm long; in moist prairies, meadows, clearings and woodland sites; **dioecious**. p2
- **FLOWERS** Green, blooming May–August; **inflorescence** a panicle, open, 5–20+ cm long by 3–15 cm wide; **peduncles** 1.5–6 cm long, ascending to spreading; **pedicels** 1–10 mm long; **subtending bracts** (of pedicels) 2–4 mm long, pointed, clasping, glaucous; **flowers** numerous, unisexual; **petals** absent; <u>male flowers</u> hanging, 5–11 mm wide by 8–10 mm long; **sepals** four, pale green to purplish with white hyaline margins, apices slightly erose, veins faint, 1.5–5 mm long by 2–4 mm wide (flattened), deciduous; **stamens** 10–20, each 5.5–9 mm long, exserted; **filaments** pinkish, 2–5.5 mm long, filiform, slightly thicker at both ends, free; **anthers** yellow, hanging, 2.5–4.5 mm long by *c.* 0.8 mm wide, apiculate; <u>female flowers</u> 4–5 mm wide by 2.5–3 mm long, ascending to descending; **sepals** four, pale green, 3-nerved, 1.5–2 mm long by *c.* 1.2 mm wide; **pistils** 5–17, each 2.5–2.8 mm long; **ovary** green, *c.* 0.8 mm long, ribbed, hairless, oval to round; **style** 1.5–2.5 mm long, persistent as a beak 0.7–2.5 mm long on fruit; **stigma** reddish, feathery, 1.8–2.5 mm long by *c.* 0.4 mm wide, covering the outside surface of the style, tapered toward the tip. **FRUIT an achene**, 1-seeded, beaked, 2–17 per cluster, ascending, 8-ribbed, 3–5 mm long by 1.5–1.9 mm thick, glabrous to glandular, round in cross-section.
- **LEAVES** Basal and stem, alternate, compound, (tripinnate), widely separated, ascending, reduced above; **main blades** 5–15 cm long by 4–13 cm wide, glabrous to thinly hairy below on blades; **leaflets** 3-parted, 4–8 cm long by 4–7 cm wide; **segments** 8–35 mm long by 9–40 mm wide, bluish green, usually 3-lobed, lobes entire to roundly toothed; **petioles** 0.4–9 cm long by 1–2 mm thick, round, expanded into a basal sheath, reduced above; **petiolules** 1–2 cm long.
- **STEM** Erect, branched, hollow, glaucous; base 2–5 mm wide.
- **SYN.** *Thalictrum confine* Fern.

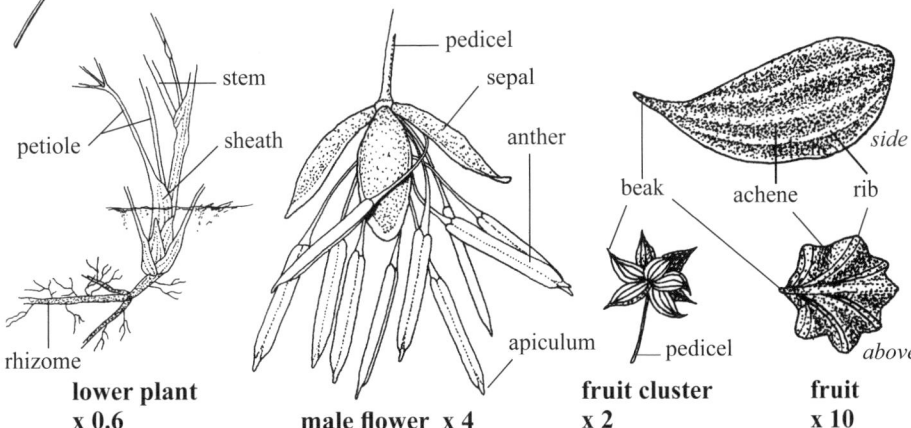

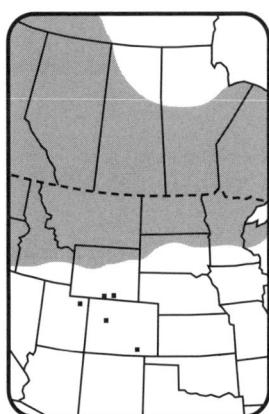

Veiny-leaf Meadow-rue, Early Meadow-rue   441

Buckthorn—*Rhamnaceae* **Shrub** Green Petals 4

BUCKTHORN *Rhamnus cathartica* L.

■ **SKETCH** A spiny, **deciduous shrub** 1–6 m tall; on dry sites in open woods, along fences, coulees and hedgerows; mainly **dioecious**, some **polygamous**. *Naturalized*

■ **FLOWERS** Green, blooming May–June; **inflorescence** of unisexual flower clusters (some may be perfect) among the leaves; **pedicels** glabrous, 3–6 mm long; **male flowers** 5–6 mm wide, in clusters of 5–50 on spur branches; **hypanthium** glabrous, *c.* 2 mm long and wide; **calyx lobes** four, 2–3 mm long; **petals** four, 1–1.6 mm long by 0.6–1 mm wide, notched or not at apices, deciduous; **stamens** four, 1.3–1.5 mm long; **anthers** *c.* 0.4 mm long; **pistil** vestigial; **female flowers** fragrant, 4–6 mm wide by *c.* 4 mm long, in clusters of 2–15 along spur branches; **calyx lobes** four, spreading, *c.* 2 mm long by *c.* 1 mm wide; **petals** four, *c.* 0.5 mm long between the calyx lobes; **styles** 3- or 4-parted, *c.* 2.5 mm long; **stamens** vestigial, *c.* 1 mm long. **FRUIT a drupe**, berrylike, round, black, shiny, 5–9 mm across, persisting into winter; **stones** usually four, 1-seeded, dark brown, 4–5 mm long by 2.5–3.5 mm wide, 3-sided, one side slightly rounded with a shallow groove and the two inner sides divided by a ridge.

■ **LEAVES** Opposite, simple, finely toothed; **blades** 1.5–7.5 cm long by 1–5.5 cm wide, frost hardy, glabrous above, hairy below, the pointed tip often with a slight twist, two to four lateral curved veins; **petioles** 0.5–2 cm long and finely hairy; **stipules** 2–4 mm long, linear.

■ **STEM** Dark gray bark with light gray lenticels; 2–12 cm wide near the rough base; **spines** small, straight, sharp, at branch tips; **winter buds** dull, subopposite, dark brown, several-scaled, 4–6 mm long by 1.5–2.3 mm wide.

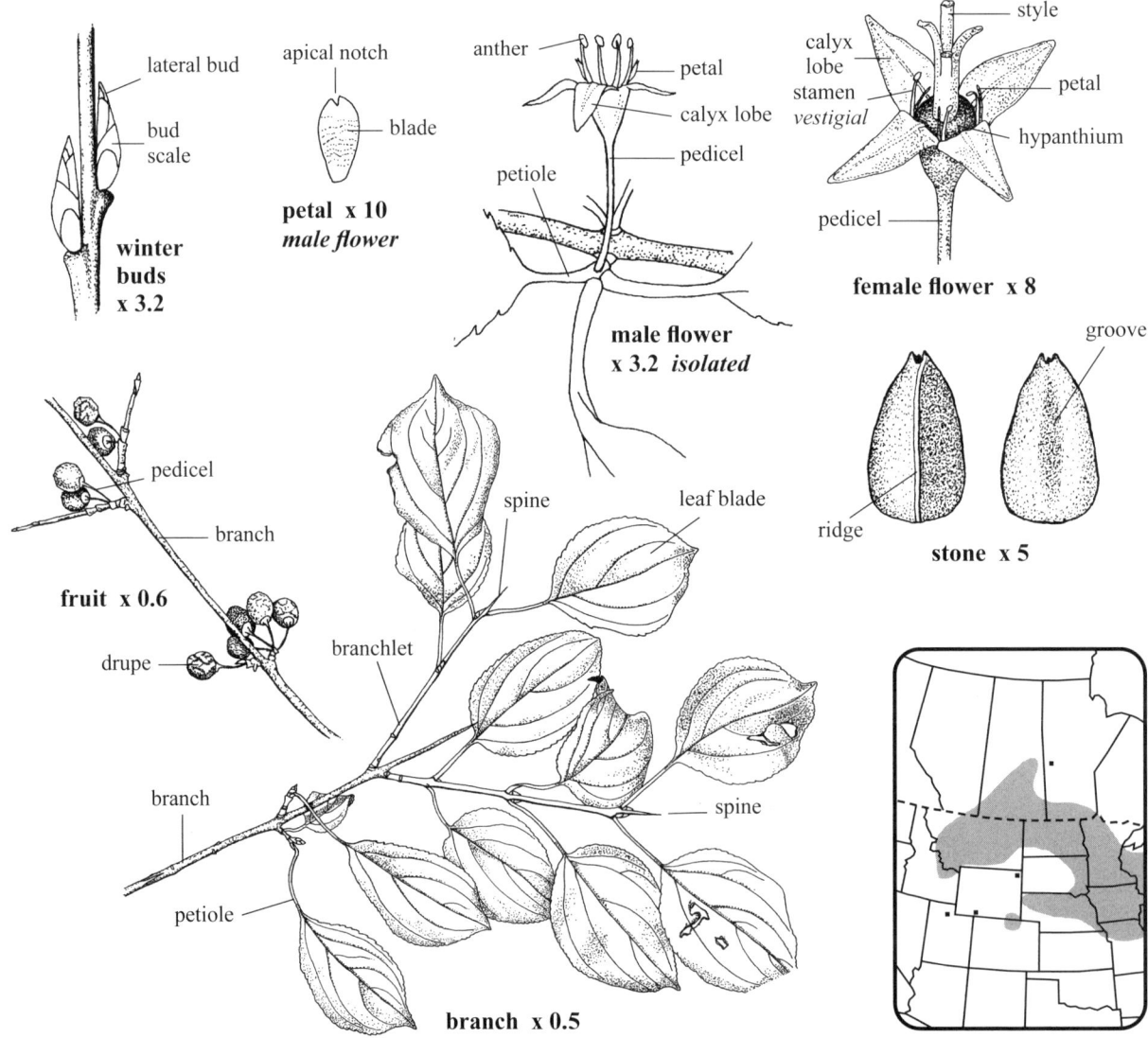

442   Common Buckthorn, European Buckthorn

Family Characteristics

Rose—Rosaceae   Flower parts 5 (4 or 6)

SKETCH Dicotyledonous herbs (wildflowers) and shrubs, sometimes armed. About 3,000 species worldwide, widely distributed.

FLOWERS Perfect and regular, 5-merous, showy, pink to purple to white, some fragrant. **Hypanthium** saucer-shaped or cup-like, free or attached (adnate) to the ovary. **Inflorescence** variable, often a raceme or cyme. **Sepals** 5 (4 or 6), mostly green and not united, often with 5 small bractlets between them. **Petals** 5 (4 or 6), free. **Stamens** often numerous or five, in whorls, not united, often persistent. **Pistil** one to many, with free or united locules. **Ovary** with 1–several ovules. **Styles** usually one per pistil. **Stigmas** often obscure.

FRUIT Variable, a 1-seeded achene, or a follicle, pome, drupe or an aggregate of achenes or of drupelets (raspberry). **Seeds** often lacking endosperm.

LEAVES Mostly alternate or basal, simple or compound, leaves or leaflets often toothed. **Stipules** often present and obvious, free or partially attached to petiole, some deciduous.

STEM Green and dying each year or somewhat woody and persisting for years.

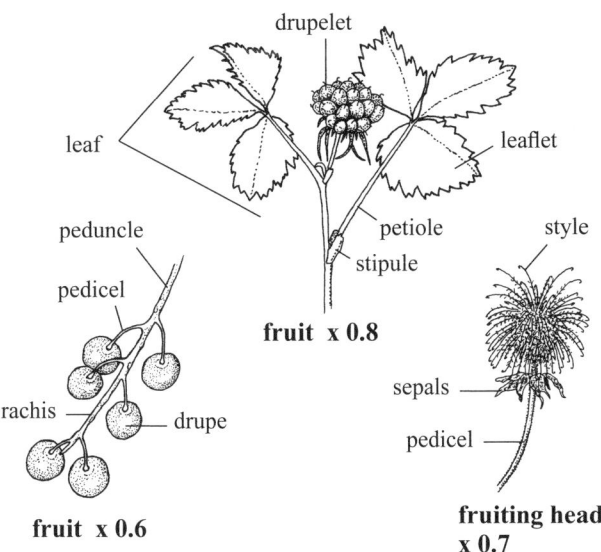
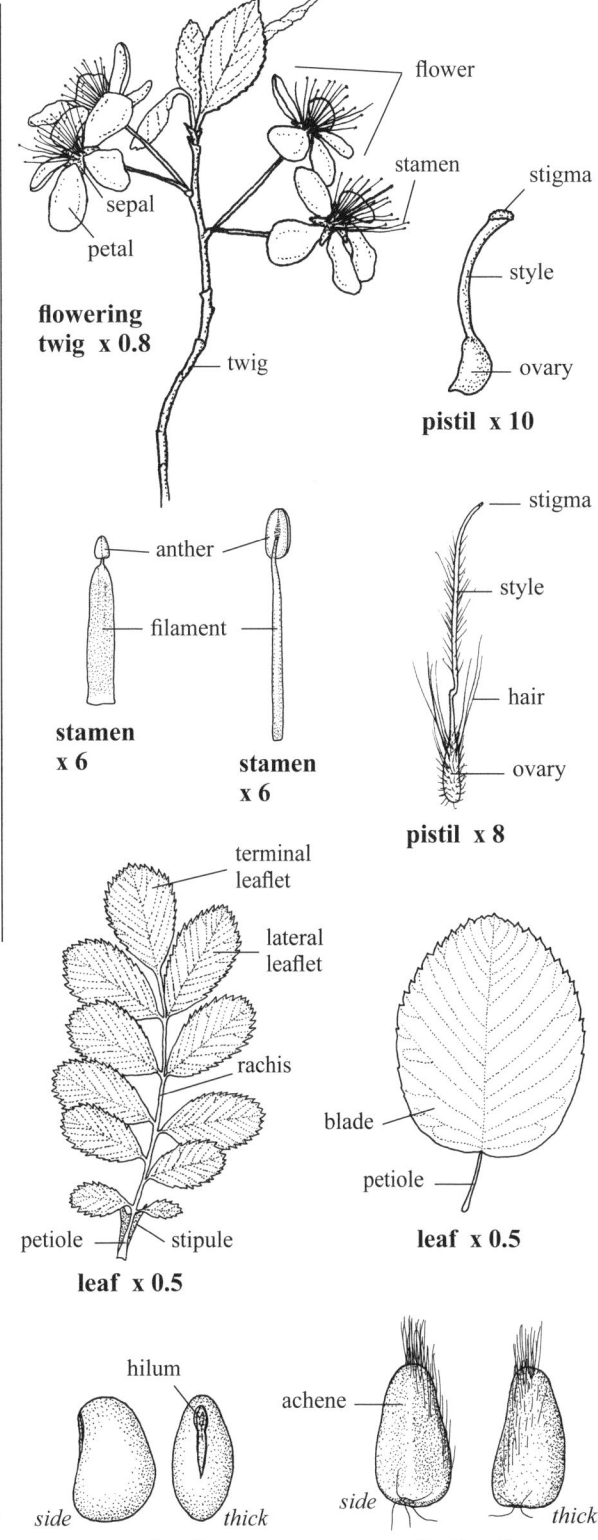

443

Rose—*Rosaceae*   **Wildflower**   Yellow   Petals 5

Hooked Agrimony   *Agrimonia gryposepala* Wallr.

■ **SKETCH** A hairy **perennial herb** 35–120+ tall, singly or in small clumps with a knobby, dark, hard **rhizome** 2–3 cm long by *c.* 1 cm thick, with **fibrous roots** and woody black smooth **rhizomes** 1–2 mm thick; in open woods and thickets.

■ **FLOWERS** Yellow, blooming July–August; **inflorescence** a raceme, blooming bottom to top, 4–7 cm long by *c.* 1.5 cm wide; **subtending bracts** (of pedicels) 3-lobed, 4–8 mm long, glandular, central lobe the longest and lobe margins ciliate; **pedicels** green, ascending in flower, reflexed with fruit, 1–2 mm long; **subtending bracts** (of flowers) 2–3 mm long, paired, margins glandular-hairy; **flowers** perfect, 7–8 mm wide by *c.* 3 mm tall, ascending; **rachis** with short gland-tipped hairs and scattered long tan hairs longer than rachis is wide, spreading to ascending, often bent; **hypanthium** green, partly covered by subtending bracts, with bracts removed silvery and with *c.* 1 mm long glands in the 10 vertical grooves, enclosing ovary, several white straight hairs ascending from its base, hairless above; **hooked bristles** in 4 or 5 rows, forming an erect band *c.* 1 mm wide at the top of the hypanthium, spreading in fruit; **sepals** five, green, spreading, hairy 2.5–3 mm long by 1.2–1.5 mm wide, pointed, 3-veined, minutely hairy on margins and tips; **petals** five, spreading to ascending, 3–5 mm long by *c.* 2 mm wide with 5 or 6 faint veins; **stamens** five, (or 15), ascending, exserted, opposite the sepals, *c.* 2 mm long; **filaments** pale greenish yellow, from central disc; **anthers** pale green, *c.* 0.9 mm wide, orange at both ends; **central disc** pale green, 1.3–1.5 mm wide, smooth; **pistils** two, free but touching, *c.* 2.5 mm long, not connate; **ovaries** two, separate, green, glabrous; **styles** two, pale green, *c.* 1.2 mm long, filiform; **stigmas** flattened, slightly wider than the style. **FRUIT** an achene (or nutlet), 1-seeded, one per flower, roundish, 2.8–3.3 mm wide, enclosed in the hard 3–5 mm long hypanthium.

■ **LEAVES** Alternate, compound, ascending; **leaflets** 5–9, each 2–6 cm long by 10–30 mm wide, the terminal one the largest, toothed, glandular on both sides, margins ciliate, scattered long hairs on blades and veins below, less so to hairless above; **stipules** 1.5–2 cm long by 6–8 mm wide, toothed to entire, spreading, pointed, slightly glandular, united near the base and wrapped around the stem; **petioles** 1–4 cm long, glandular, with scattered tan hairs, grooved above.

■ **STEM** Erect, stiff, long tan hairs scattered and spreading to ascending, short microscopic glandular hairs throughout but more so above especially in the inflorescence; 2–5 mm wide near the pinkish green base.

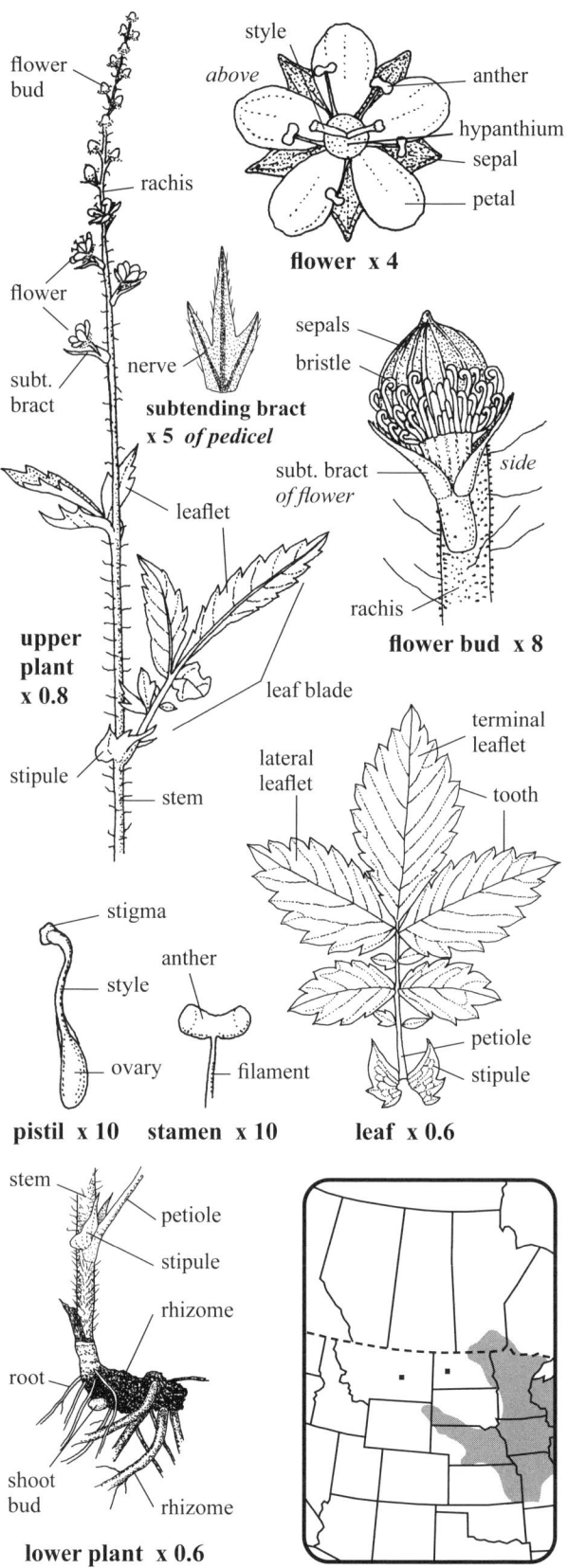

444   Tall Hairy Groovebur, Tall Hairy Agrimony, Feverfew

Rose—*Rosaceae*   **Shrub** White Petals 5

Saskatoon   *Amelanchier alnifolia* (Nutt.) Nutt. ex M. Roemer

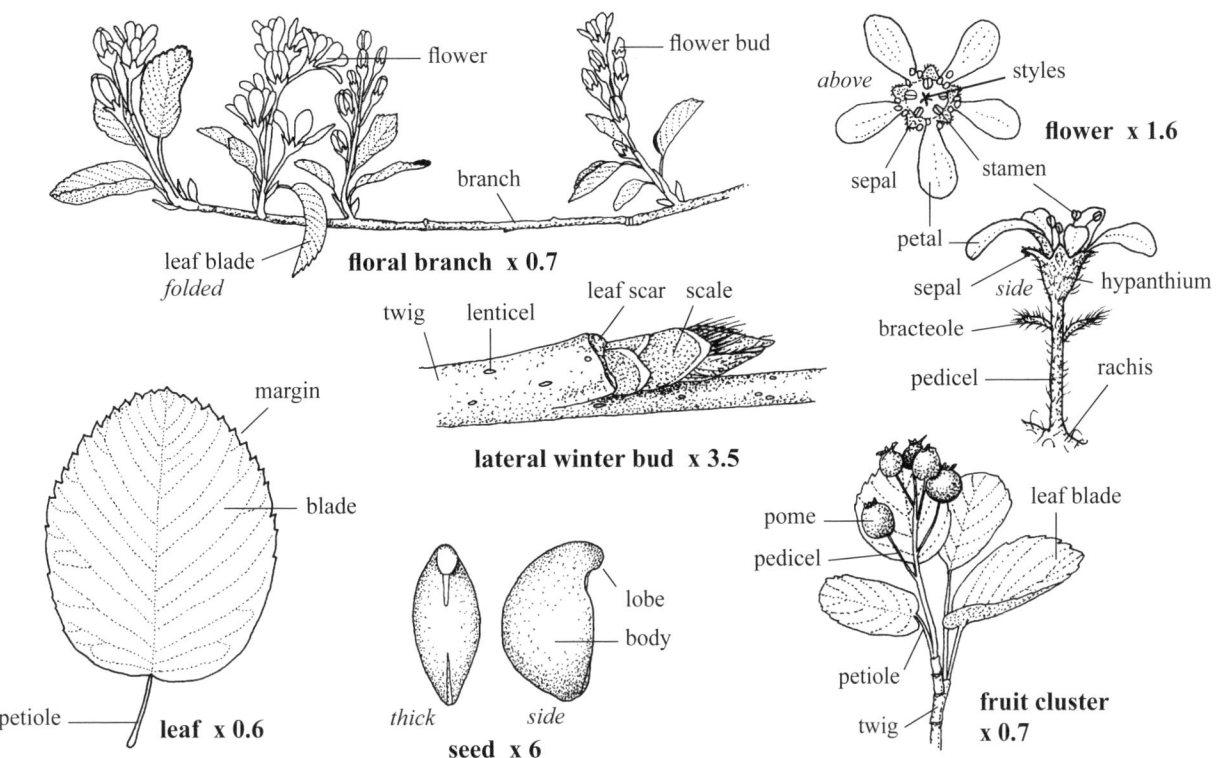

■ **SKETCH** A variable **deciduous shrub** 0.5–7 m tall, single or forming clumps covering several m², spreading by **stolons**; along prairie edges, canyons to dry mountain slopes, woodland edges and lakeshores. w2
■ **FLOWERS** White, blooming April–August; **inflorescence** a raceme, numerous, erect, 1.5–7 cm long by 1.5–3 cm wide, alternating along new or older branches; **pedicels** hairy, 5–20 mm long, with a pair of hairy reddish **bracteoles**, these early deciduous; **flowers** perfect, 2–18 per raceme, each 15–25 mm wide; **rachis** covered with dishevelled white hairs, becoming glabrous; **hypanthium** 3–4 mm wide, green, hairy; **sepals** five, lobes 1.5–3 mm long, hairy on both surfaces and margins, some with a reddish tip, spreading; **petals** five, tapered at the base, 5–13 mm long by 3–5.5 mm wide, hairs microscopic; **stamens** 20, white; **filaments** white, *c.* 2 mm long, slightly taller than the styles; **anthers** *c.* 0.8 mm long; **ovary** apex covered with a tangle of white hairs; **styles** five, *c.* 3 mm long, united or mostly distinct their entire length, green, glabrous; **stigmas** green, round, slightly larger than the styles. **FRUIT a pome**, berrylike, round, 6–10 mm wide by 10–14 mm long, juicy, reddish purple with a faint bloom; **seeds** 1–10 per pome, reddish brown, rarely white, shiny, smooth, 3–4 mm long by 1.8–2.3 mm wide by 1.3–1.5 mm thick with a stout lobe.

■ **LEAVES** Alternate, simple, toothed mostly in the upper half or entire, folded when young; **blades** 7- to 13-veined, 1.5–6.5 cm long and wide, young blades reddish green, more hairy below especially along the midrib, glabrous above to sparsely hairy, apices blunt to pointed, usually two to five leaves along the seasonal stalks below the flowers; **petioles** woolly hairy, 3–22 mm long, round, not grooved; **stipules** very hairy, tan, 2–6 mm long, linear, early deciduous.
■ **STEM** Medium gray bark becoming reddish brown among the flowers; **branches** alternate; **winter twigs** reddish brown, dull; **lenticels** tan, oval, *c.* 0.5 mm long; **lateral buds** alternate, 5–6 mm long and *c.* 2 mm wide, 6- or 7-scaled, the scales dark reddish brown, with lighter tan margins and white hairs projecting out from under the middle scales, the hairs *c.* 4 mm long, tips of buds pointed; **terminal buds** also hairy, *c.* 8 mm long by 2–3 mm wide.

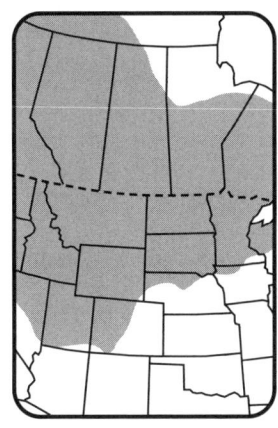

Saskatoon Serviceberry, Juneberry, Alderleaf Juneberry

Rose—Rosaceae   **Wildflower** Yellow Petals 5

SILVERWEED   *Argentina anserina* (L.) Rydb.

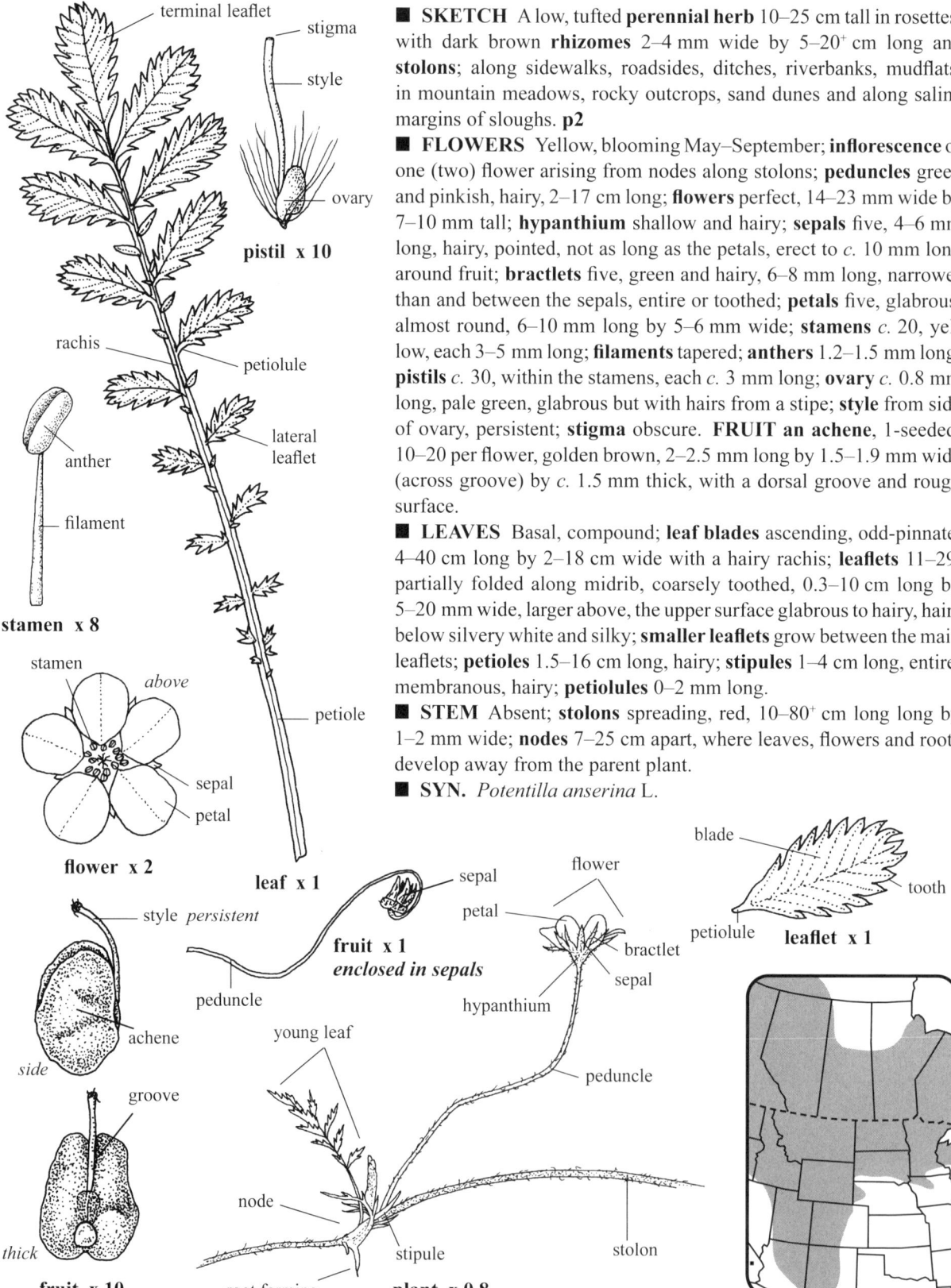

■ **SKETCH** A low, tufted **perennial herb** 10–25 cm tall in rosettes, with dark brown **rhizomes** 2–4 mm wide by 5–20⁺ cm long and **stolons**; along sidewalks, roadsides, ditches, riverbanks, mudflats, in mountain meadows, rocky outcrops, sand dunes and along saline margins of sloughs. p2

■ **FLOWERS** Yellow, blooming May–September; **inflorescence** of one (two) flower arising from nodes along stolons; **peduncles** green and pinkish, hairy, 2–17 cm long; **flowers** perfect, 14–23 mm wide by 7–10 mm tall; **hypanthium** shallow and hairy; **sepals** five, 4–6 mm long, hairy, pointed, not as long as the petals, erect to *c.* 10 mm long around fruit; **bractlets** five, green and hairy, 6–8 mm long, narrower than and between the sepals, entire or toothed; **petals** five, glabrous, almost round, 6–10 mm long by 5–6 mm wide; **stamens** *c.* 20, yellow, each 3–5 mm long; **filaments** tapered; **anthers** 1.2–1.5 mm long; **pistils** *c.* 30, within the stamens, each *c.* 3 mm long; **ovary** *c.* 0.8 mm long, pale green, glabrous but with hairs from a stipe; **style** from side of ovary, persistent; **stigma** obscure. FRUIT an achene, 1-seeded, 10–20 per flower, golden brown, 2–2.5 mm long by 1.5–1.9 mm wide (across groove) by *c.* 1.5 mm thick, with a dorsal groove and rough surface.

■ **LEAVES** Basal, compound; **leaf blades** ascending, odd-pinnate, 4–40 cm long by 2–18 cm wide with a hairy rachis; **leaflets** 11–29, partially folded along midrib, coarsely toothed, 0.3–10 cm long by 5–20 mm wide, larger above, the upper surface glabrous to hairy, hairs below silvery white and silky; **smaller leaflets** grow between the main leaflets; **petioles** 1.5–16 cm long, hairy; **stipules** 1–4 cm long, entire, membranous, hairy; **petiolules** 0–2 mm long.

■ **STEM** Absent; **stolons** spreading, red, 10–80⁺ cm long long by 1–2 mm wide; **nodes** 7–25 cm apart, where leaves, flowers and roots develop away from the parent plant.

■ **SYN.** *Potentilla anserina* L.

446   Common Silverweed, Silverweed Cinquefoil

CHAMAERHODOS *Chamaerhodos erecta* (L.) Bunge

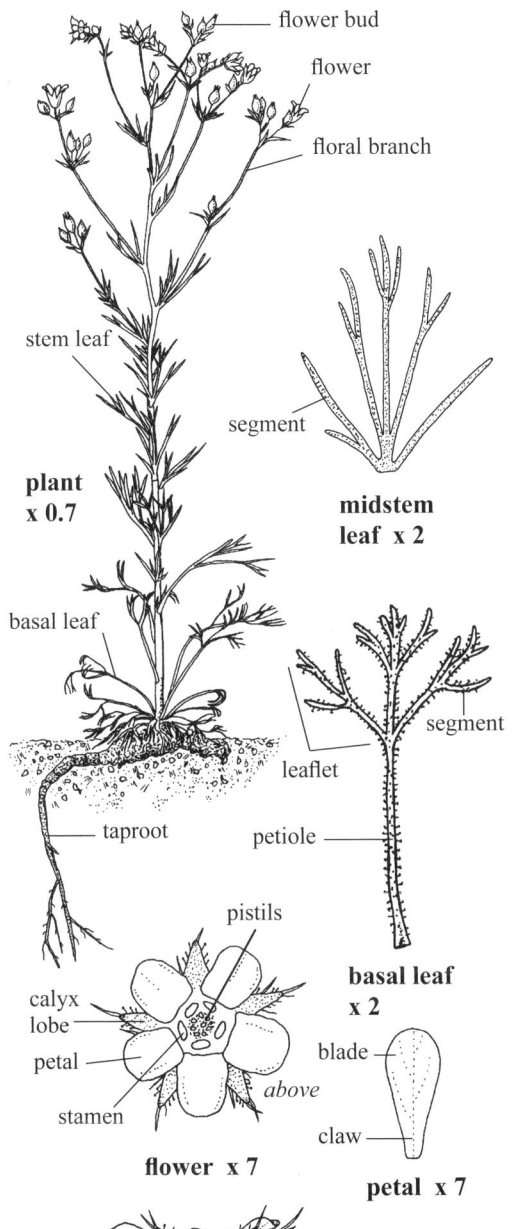

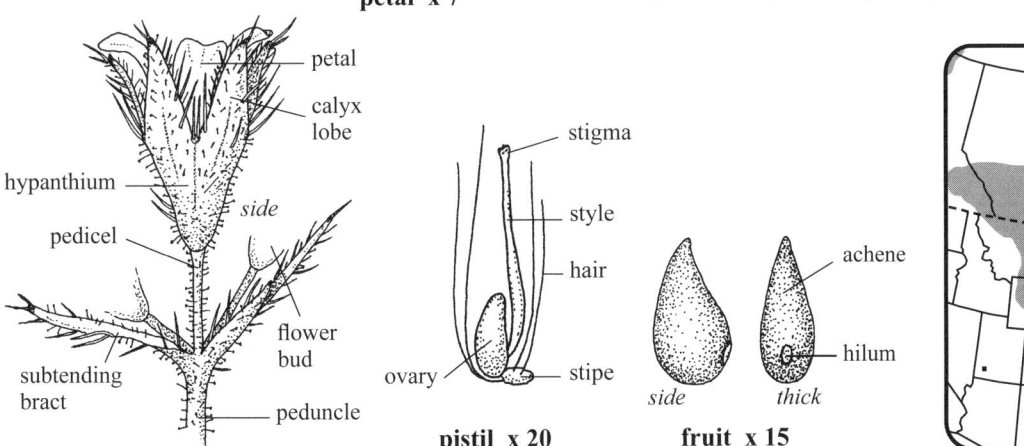

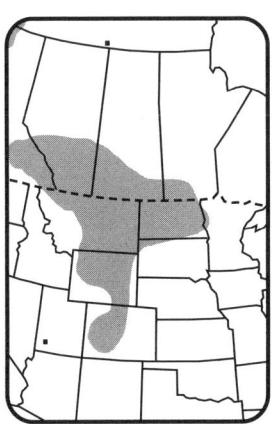

■ **SKETCH** A short-lived, glandular-hairy **perennial herb** 10–30 cm tall from a dark, woody **taproot** 1–4 mm thick and to *c.* 10^+ cm long; in dry gravelly sites, ravines and open woodlands. **p2**

■ **FLOWERS** White to pinkish, blooming June–August; **inflorescence** a cyme, branched, spreading to congested; **floral branches** 3–20 cm long, sometimes rebranching; **peduncles** 1–5 cm long; **pedicels** 1–4 mm long, hairs gland-tipped; **subtending bracts** (of pedicels) one, 1–4 mm long, soft, hairy, entire to toothed; **flowers** perfect, 3–4 mm wide by *c.* 4 mm long; **hypanthium** 1.5–2.4 mm wide by *c.* 2 mm long, green, hairy; **calyx lobes** five, pointed, 1.5–2 mm long by *c.* 1 mm wide, tips reddish and hairy, long hairs ascending and without glands, short hairs (microscopic) are glandular, lobes persistent around fruit, closed at first then slightly opening as fruit matures; **petals** five, white, glabrous, 2–3 mm long by 1–1.2 mm wide, blade rounded and faintly veined, claw *c.* 0.5 mm long, attached to top of hypanthium; **stamens** five, TL 0.8–1 mm, opposite petals, included; **filaments** white, erect, filiform, glabrous; **anthers** yellow, *c.* 0.5 mm long by *c.* 0.6 mm wide; **pistils** erect, included, *c.* 1.5 mm long, $5-10^+$ clustered together among erect hairs; **ovary** glabrous, *c.* 0.5 mm long; **style** filiform, *c.* 1.4 mm long, glabrous, attached to side of ovary. **FRUIT** an achene, 1-seeded, several per flower, raindrop-shaped, olive to grayish brown, glabrous, 1.3–1.5 mm long by 0.6–0.7 mm wide by 0.4–0.5 mm thick, hilum near rounded end.

■ **LEAVES** Basal and stem, alternate, compound (two or three times), ascending, imbricate, often with a fascicle of leaves developing in axils; **basal leaves** numerous, forming a rosette 2–5 cm wide of new and old blades; **blades** 1–2 cm long and wide, thick and fleshy, segments entire, *c.* 0.5 mm wide, veins obscure, glandular-hairy, glands dark red; **petioles** 1–2 cm long by 0.6–0.9 mm thick, hairy; **stem leaves** 1–5 cm long by 8–13 mm wide, less glandular-hairy, some hairs without glands, segments 0.4–0.7 mm wide, all parts grooved longitudinally below; **petioles** 0–15 mm long, reduced above.

■ **STEM** Erect, one to three, branched above or throughout, light green, round, very glandular-hairy, hairs spreading; 1–4 mm thick near the base.

■ **SYN.** *Chamaerhodos nuttallii* Pickering ex Rydb. = *Chamaerhodos erecta* ssp. *nuttallii* (Pickering ex Rydb.) Hultén.

Little Rose, Little Ground Rose

Rose—*Rosaceae*　**Wildflower**　Dark reddish purple　Petals 5 (small)　Sepals 5 (large)

Marsh Cinquefoil *Comarum palustre* L.

■ **SKETCH** A glandular-hairy **perennial herb** 20–60 cm tall or long with dark brown, hollow, woody **rhizomes** 2–4 mm thick by 10–20+ cm long; in bogs and marshes, and in shallow water along lakes, rivers and streams.

■ **FLOWERS** Dark reddish purple, blooming June–August; **inflorescence** a cyme, open, 6–10+ cm tall by 5–11 cm wide, stiff, ascending above the water's surface; **floral branches** 2.5–10 cm long, some hairs gland-tipped and ascending; **pedicels** 12–35 mm long, very hairy, hairs tipped with a dark purple gland; **flowers** perfect, 15–42 mm wide by 6–13 mm tall; **hypanthia** 5–8 mm wide by *c.* 1 mm deep, pale green, glandular-hairy below, hairy above; **sepals** five, pointed, spreading at anthesis, ascending with fruit, 6–15 mm long by 4–8 mm wide, veiny and hairy below, marginal hairs gland-tipped; **bractlets** five (6), between the sepals, 3–6 mm long by 0.8–1.5 mm wide, hairy on both sides, erect around the fruit; **petals** five, 3–6 mm long by 1–2 mm wide, declining, hairless; **stamens** 20, dark purple, exserted; **filaments** *c.* 3 mm long, thick; **anthers** *c.* 1.5 mm long by *c.* 1 mm wide, falling early; **pistils** numerous, together 3–6 mm wide and tall, each 1.5–2 mm long; **ovary** hairless, 0.6–0.8 mm long, slightly flattened, apex reddish; **style** filiform, dark purple, 1.3–1.5 mm long, hairless, from the side of the ovary; **stigma** obscure; **fruiting heads** each *c.* 10 mm long and wide, usually drooping below the leaves. **FRUIT an achene**, 1-seeded, golden brown, numerous, smooth, 1–1.5 mm long by 1–1.2 mm wide by 0.8–1 mm thick, beak short; **receptacle** cone-shaped, white, covered with gray hairs, *c.* 5 mm long with fruit.

■ **LEAVES** Alternate, compound, above water; **leaf blades** 4–15 cm long by 4.5–10 cm wide, spreading; **leaflets** 3–7, usually three above, mostly sessile, toothed, 2–10 cm long by 3–30 mm wide, dull, lighter green below with raised veins, hairy on both sides or less above, hairs appressed, pointed toward the margins; **petioles** 3–7 cm long, hairs appressed, the lower 2–3.5 cm an open unclasping sheath; **stipules** of upper leaves united to petiole, pointed, 5–20 mm long by 5–7 mm wide, hairy on both sides.

■ **STEM** Round, stiff, ascending, reddish green, hollow, hairless or with spreading hairs, dull with shallow grooves, ascending above water; 3–5 mm thick near the pinkish, smooth base.

■ **SYN.** *Potentilla palustris* (L.) Scop.

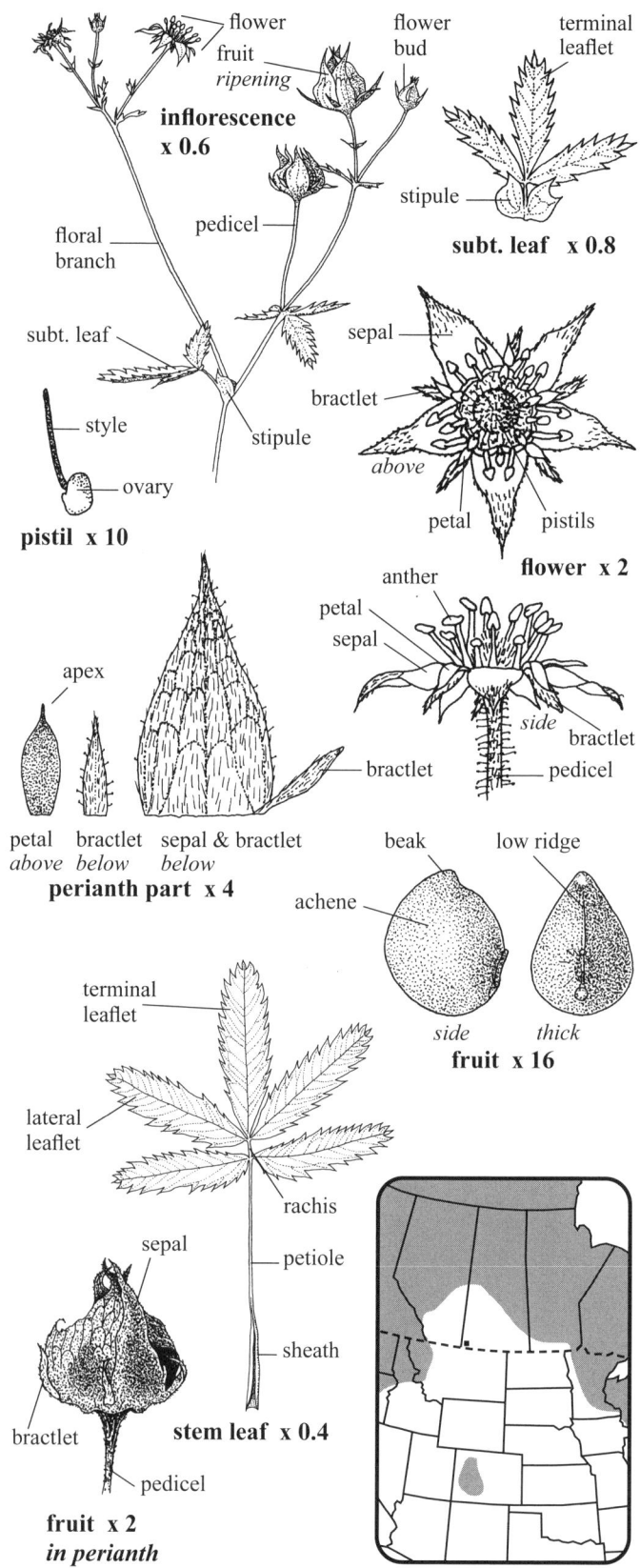

Purple Marshlocks, Marsh Five-finger

Rose—*Rosaceae*   **Shrub White Armed Petals 5**

Round-leaved Hawthorn *Crataegus chrysocarpa* Ashe

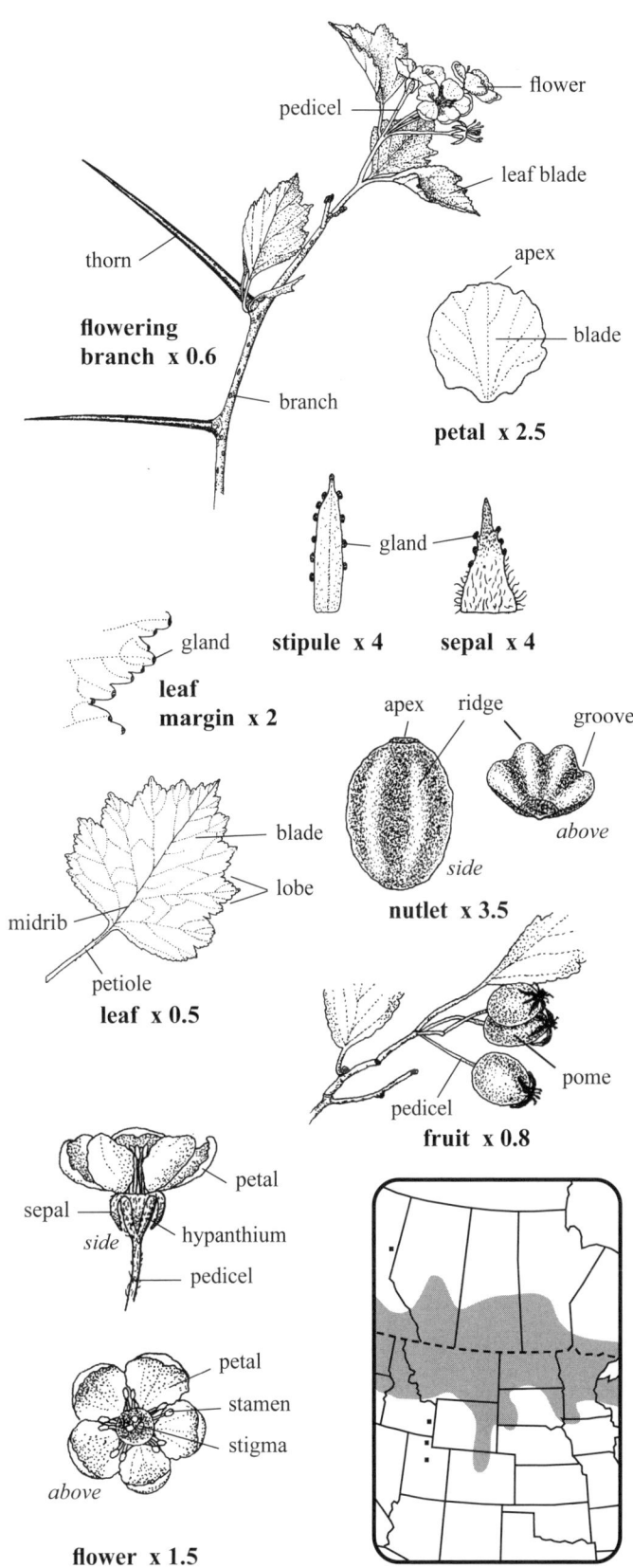

- **SKETCH** A variable, armed, deciduous **shrub** 1.5–5+ m tall, forming thickets; along stream and riverbanks, prairie ravines, coulees and open woods.
- **FLOWERS** White, blooming May–June; **inflorescence** a corymb of 3–15 flowers from leaf axils throughout the plant; **pedicels** hairy, 3–16 mm long; **flowers** perfect, 10–15 mm wide by 7–10 mm long; **hypanthium** 3–4 mm long by 2.5–3 mm wide, scattered hairy outside (or rarely glabrous), and at the inside base around the styles; **sepals** five, reflexed, pointed, hairy on ventral surface and margins, *c.* 4 mm long by *c.* 2 mm wide, slightly darker green and glandular along margins near apices; **petals** five, white, glabrous, 6–8 mm long and wide, roundish, margins erose; **stamens** 10, erect, each *c.* 4 mm long, glabrous; **filaments** white to pale green, from the apex of the hypanthium, paired, opposite the sepals; **anthers** 1–1.5 mm long, creamy white; **styles** 3–5, filiform, glabrous, erect, *c.* 4.5 mm long or about as high as the stamens, distinct; **stigma** greenish brown, *c.* 0.5 mm long, angled, slightly wider than the style. **FRUIT a pome**, mostly glabrous or with scattered hairs, 7–15 mm long and wide, golden to deep red when ripe, slightly shiny, usually from one to three, yellowish pulp inside; **nutlets** three or four per pome, 1-seeded, touching but not fused, triangular, 6–8 mm long by 4–5 mm wide by 3–4 mm thick, outer surface with two rounded ridges, golden brown, darker brown at the apices, not pitted.
- **LEAVES** Alternate, simple, lobed and toothed, slightly darker green above, young leaves light green with hairs along the upper veins and scattered on the blade, less hairy to glabrous below, teeth gland-tipped, glands brown, darkening with age; **blades** flat, 3–7 cm long by 2–5.5 cm wide, thin; **petioles** 1–4 cm long, hairy when young, often with a few glands along the upper groove; **stipules** linear, 4–7 mm long by *c.* 1 mm wide, margins glandular, early deciduous.
- **STEM** Erect and branched; **thorns** 2–7 cm long by 2–3 mm wide near the base, reddish brown, shiny with minute white lenticels, slightly declining, sharp, stout, alternate, persistent; **mature branches** alternate, greenish yellow, glabrous; **new floral branches** green and hairy, with a few leaves and flowers; **bark** gray, furrowed near the base.
- **SYN.** *Crataegus columbiana* var. *chrysocarpa* (Ashe) Dorn and *Crataegus rotundifolia* Moench p.p. non Lamb.

Fire-berry Hawthorn, Northern Hawthorn

Rose—*Rosaceae* **Shrub Yellow Petals 5**

SHRUBBY CINQUEFOIL   *Dasiphora floribunda* (Pursh) Kartesz, comb. nov. ined.

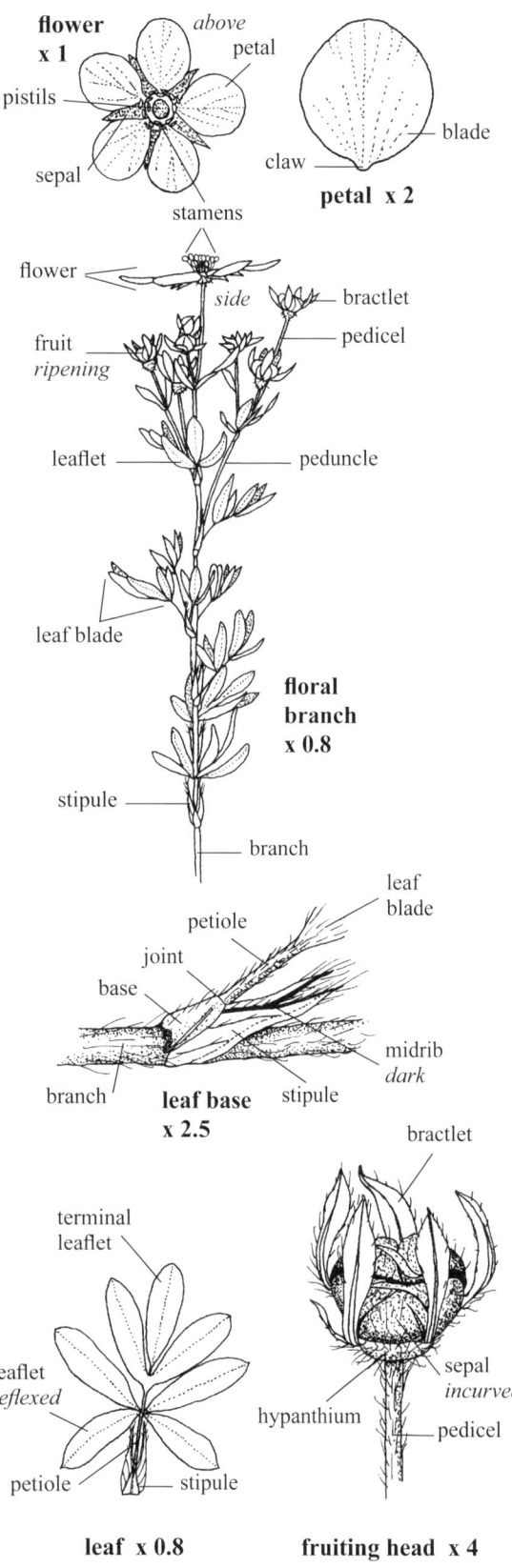

- **SKETCH** A hairy **deciduous shrub** 15–120 cm tall with a thick, orangish tan branching **rootstock** up to *c.* 20+ cm long; in moist ground in alpine meadows, sagebrush, bogs, moist prairies, open woods, rocky outcrops and canyons. **w2, p1**
- **FLOWERS** Yellow, blooming May–October; **inflorescence** of single flowers or to six in terminal cymes; **peduncles** 5–20 mm long, hairy; **pedicels** 5–50 mm long, silky hairs ascending to spreading; **flowers** perfect, 15–30 mm wide by 5–7 mm tall; **hypanthium** green, hairy, 3–5 mm wide by *c.* 2 mm tall, saucer-shaped; **sepals** five, 4–7 mm long by 3–3.8 mm wide, pointed, yellowish green, hairy below and on tips above, persistent in fruit; **bractlets** five+, one to three between the sepals, 0.7–2 mm wide, pointed, green, silky hairy on both sides, margins often revolute, erect in fruit; **petals** five, flat, glabrous, round, 8–12 mm long and wide, veined, claw short; **stamens** exserted, 2–3 mm long, 15–30, staggered from a yellow 5–6 mm wide nectar ring; **filaments** yellow; **anthers** yellow, *c.* 1 mm long, oval, not hairy; **pistils** numerous, yellow, each 1–2 mm long and together *c.* 3 mm across forming the central cluster; **ovary** very hairy; **style** from the side; **fruiting heads** 5–7 mm wide by 6–7 mm tall. **FRUIT an achene**, 1-seeded, brown, oval below, pointed above, 1–2 mm long by 0.6–0.8 mm wide by 0.5–0.6 mm thick, white hairs 0.8–1 mm long and mostly along one side.
- **LEAVES** Alternate, compound (odd-pinnate); **blades** hairy, more so below, 1.3–4.2 cm long by 0.7–4 cm wide; **leaflets** three to seven, 3–26 mm long by 1–8 mm wide, sessile, entire, oval, veins raised and silky hairy below, margins revolute; **petioles** 4–17 mm long, hairy, widening at the jointed clasping base; **stipules** membranous, persistent, scarious, 4–12 mm long by 2–4 mm wide, veins reddish brown, midrib dark and hairy.
- **STEM** Erect to twisted, branched throughout, woody, tan to reddish bark peeling and shredding into strips, young branches hairy, more glabrous with age; 5–30 mm thick near base.
- **SYN.** *Dasiphora fruticosa* auct. non (L.) Rydb., *Potentilla fruticosa* auct. non L., *Pentaphylloides floribunda* (Pursh) A. Löve nom. super. and *Potentilla floribunda* Pursh.

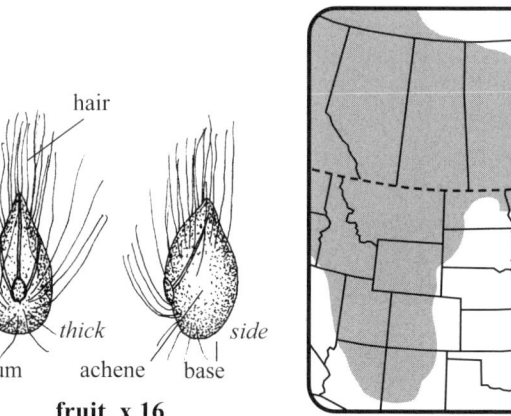

Golden-hardhack, Bush Cinquefoil

Rose—*Rosaceae* **Wildflower** White, center yellow Petals 5

Smooth Wild Strawberry *Fragaria virginiana* Duchesne

■ **SKETCH** A variable **perennial herb** 5–15 cm tall from **rhizomes** 4–5 mm thick by 2–6+ cm long and **stolons** 1–2 mm thick; in prairies, roadsides, valleys, open woodlands to subalpine elevations. **p1**

■ **FLOWERS** White, blooming April–June; **inflorescence** a corymbiform cluster of one to ten flowers; **peduncles** with ascending hairs, one to three per node, 2.5–14 cm long; **pedicels** hairy, 5–40 mm long; **flowers** perfect to unisexual, 12–22 mm wide by 5–6 mm long; **sepals** five, ascending in fruit, pointed, hairy, 5–10 mm long by 2–2.5 mm wide; **foliaceous bracts** between the sepals, ascending in fruit, pointed, hairy, 5–7 mm long by *c.* 1 mm wide; **petals** five, round, 7–14 mm long by 5–9 mm wide, glabrous; **stamens** *c.* 20; **filaments** 1.5–3 mm long, in three whorls along the lip of the hypanthium; **anthers** yellow, *c.* 1 mm long; **styles** numerous, *c.* 1 mm long, green, shorter than the long stamens. **FRUIT an achene**, 1-seeded, numerous, tan to reddish brown, slightly veiny, 1.1–1.5 mm long by 0.9–1.1 mm wide by 0.5–0.7 mm thick, in shallow pits on the surface of the receptacle; **receptacle** (accessory) when mature 5–15 mm long and wide, red, fleshy.

■ **LEAVES** Basal, compound (trifoliate); **leaf blades** 2.5–9 cm long by 3.5–13 cm wide; **leaflets** three, 1.7–8.2 cm long by 0.8–4.2 cm wide, coarsely toothed in the upper two-thirds, terminal tooth smaller than the two adjacent teeth, usually with silky hairs below and along the margins, slightly hairy to glabrous above; **petioles** reddish, 2–26 cm long, hairs ascending to spreading; **stipules** hairy, 1–2 cm long, tan to reddish, persistent, appearing as scales on the root crown; **petiolules** hairy, 1–5 mm long.

■ **STEM** Absent; **stolons** reddish, hairy, rooting at nodes 9–30+ cm apart.

■ **SYN.** *Fragaria glauca* (S. Wats.) Rydb. = *Fragaria virginiana* ssp. *glauca* (S. Wats.) Staudt.

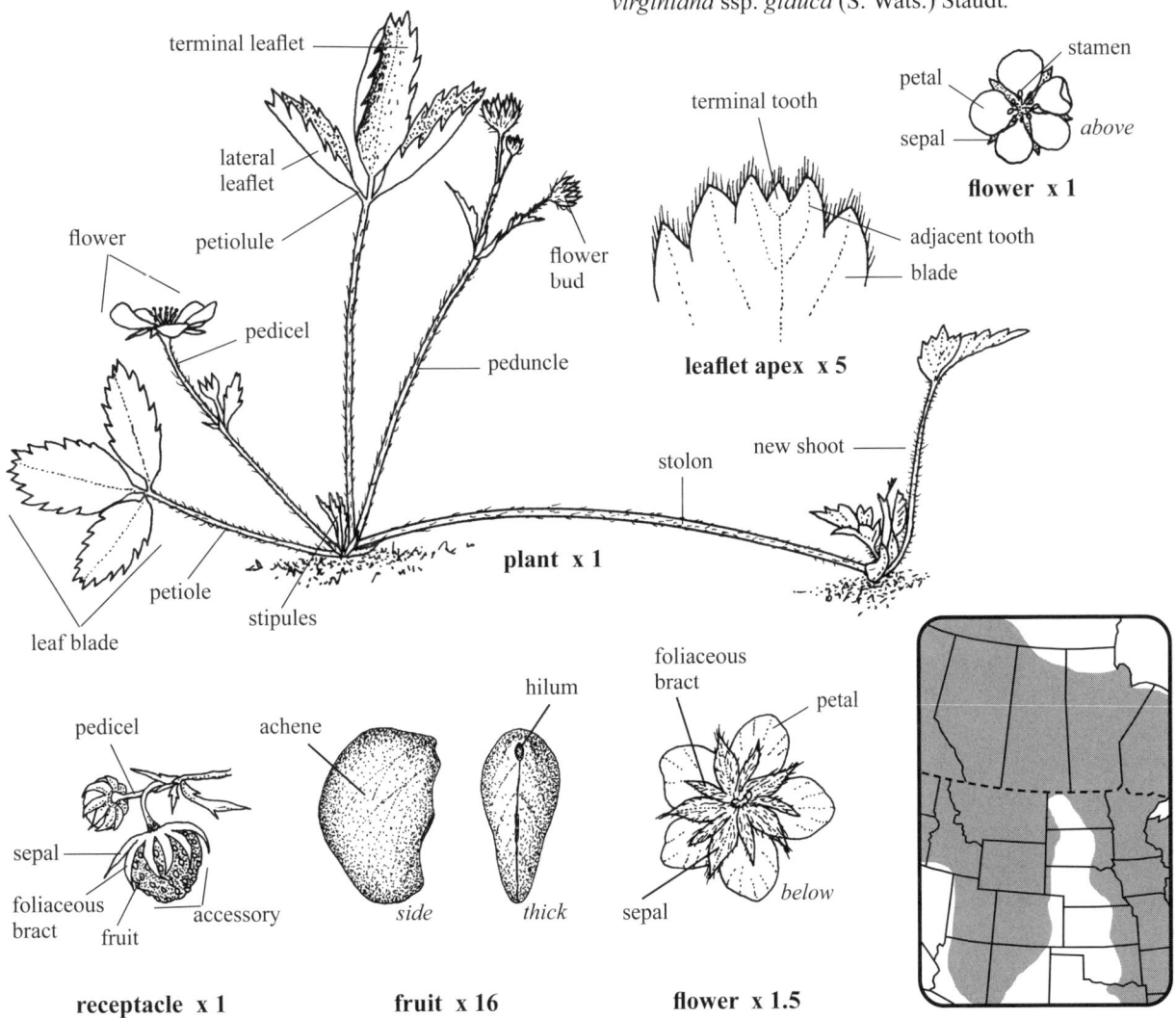

Virginia Strawberry, Wild Strawberry

Rose—*Rosaceae* — Wildflower Yellow Petals 5

YELLOW AVENS *Geum aleppicum* Jacq.

■ **SKETCH** A hairy **perennial herb** 30–120 cm tall with a short **caudex** or horizontal **rhizome**; in open woods, thickets, moist ravines, prairies and wet to dry meadows. **p1**

■ **FLOWERS** Yellow, blooming June–August; **inflorescence** a cyme, open, unsymmetrical; **pedicels** 3.5–10 cm long, spreading, with short hairs and scattered long hairs *c.* 1 mm; **flowers** perfect, 10–25 mm wide by 8–10 mm tall, two to several; **hypanthium** 3–4 mm long; **sepals** five, green, pointed, descended in flower to reflexed in fruit, 5–8 mm long by 3–5 mm wide, margins hairy; **bractlets** five, green, pointed, entire, margins hairy, spreading between the sepals; **petals** five, yellow, 6–10 mm long by 7–9 mm wide, ascending to spreading, glabrous, short clawed, veined, apices sometimes slightly twisted; **stamens** 60+, erect, yellow, 3–4 mm long, shorter to almost as tall as the pistils; **filaments** pale yellow; **anthers** yellow, 2-lobed, *c.* 0.8 mm long and wide; **pistils** green, numerous, forming an oval head 5–6 mm wide and tall, each 2.8–3.5 mm long, hairy; **ovary** green, *c.* 1 mm wide, flattened, short hairy and with white bristles 1.5–2 mm long ascending from near the top; **style** one, bent, near ovary, with ascending short hairs above the bend, lower part glabrous, apex hairless; **lower part** 0.6–0.8 mm long, with an apical hook, non-glandular; **upper part** 1.5–2 mm long, deciduous; **stigma** one, slightly angled, not wider than style; **fruiting heads** 2–6 per stem, each 1.5–2.2 cm long and wide. **FRUIT an achene**, 1-seeded, reddish brown, 200+ per flower, declining to spreading, body flattened, 3–4.5 mm long by 1–1.3 mm wide by *c.* 0.5 mm thick with a dark ridge, hairs to *c.* 2 mm long along the ridge and at the base of the beak; **beak** hairless except at its base, apex hooked; **receptacle** erect, green, 10–14 mm long by *c.* 2 mm wide, flattened, with erect white bristles *c.* 0.5 mm long between the crowded fruit.

■ **LEAVES** Alternate, compound (5–9 leaflets), toothed, some lobed, dull with small leaflets below the main leaflets; **leaf blades** 1–15 cm long by 1–12 cm wide, reduced above, hairy on both sides; **lateral leaflets** 0.7–6.5 cm long by 0.5–4.5 cm wide, reduced above; **terminal leaflets** 4–9 cm long by 4–9.5 cm wide; **basal leaves** few, hairy throughout, soft to touch, terminal leaflet to *c.* 9 cm long and wide, 5-lobed, toothed, lighter green below; **stipules** paired, toothed, 5–15 mm long by 4–12 mm wide, hairy throughout, more so below, hairs *c.* 1 mm long, ascending; **petioles** 0–14 cm long, reduced above, widely grooved above, hairs *c.* 2 mm long and tan to brownish red, spreading and continuing onto the rachis.

■ **STEM** Erect, simple, stiff, reddish tan, one to several, hairs *c.* 1 mm long, spreading and scattered; 2–5 mm thick near the base.

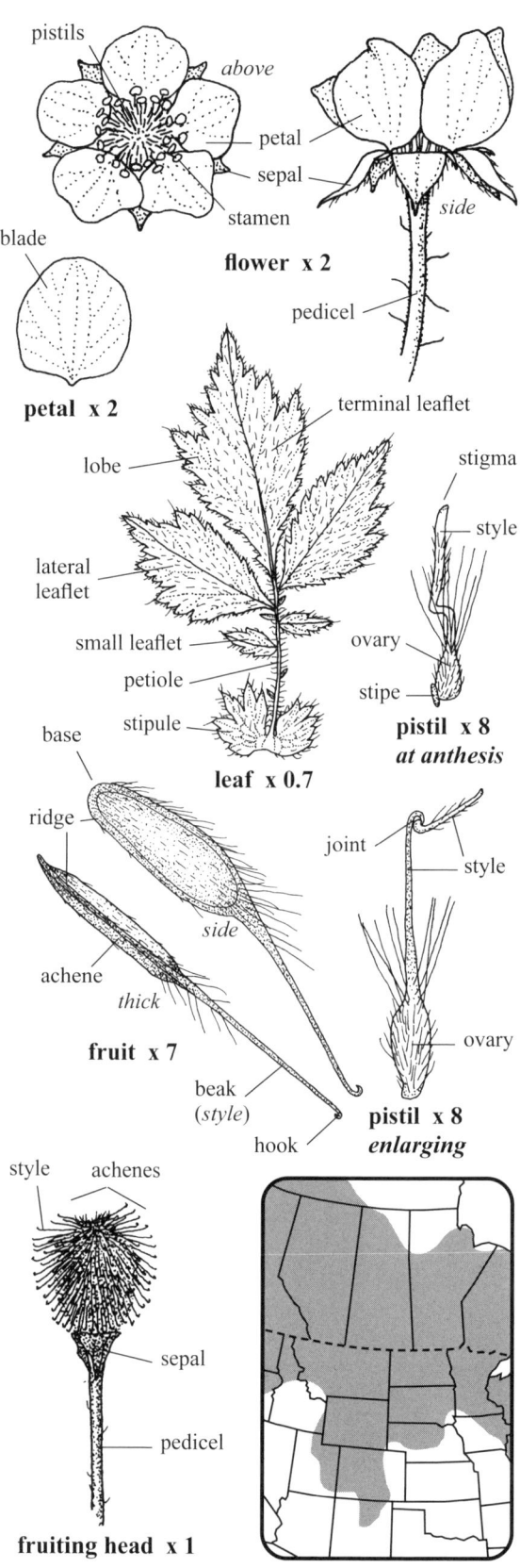

Rose—*Rosaceae* **Wildflower** Yellow Petals 5

LARGE-LEAVED AVENS *Geum macrophyllum* Willd.

■ **SKETCH** A variable **perennial herb** 30–100 cm tall from a **caudex** 1–1.5 cm long by 0.5–2 cm wide, hidden by tan **roots** 1.5–10⁺ cm long by 0.5–2 mm wide; **rhizomes** present or not; in open ditches, brushy sites and stream banks in woodlands.

■ **FLOWERS** Yellow, blooming June–August; **inflorescence** a cyme, uneven, 8–10 cm wide by 3–4 cm tall (not including peduncles); **peduncles** (floral branches) 3–4 cm long, reddish green, some short hairs gland-tipped; **subtending bracts** (of peduncles) 2–2.5 cm long, lobed and toothed; **pedicels** 3–25 mm long, ascending, short hairs often gland-tipped; **flowers** perfect, 8–14 mm wide by 9–10 mm tall, 7–24 per inflorescence; **hypanthium** green and hairy, 4–4.5 mm wide by *c.* 1 mm tall; **sepals** five, green, reflexed, 3–5 mm long by 3–4 mm wide, pointed, marginally hairy, some short hairs gland-tipped; **bractlets** 1–1.2 mm long, slightly hairy; **petals** five, yellow, ascending, 5–7 mm long by 4–6 mm wide, glabrous; **stamens** *c.* 80, yellow, included, attached to hypanthium ring; **filaments** hairless, *c.* 2 mm long; **anthers** yellow, *c.* 0.7 mm long; **pistils** 2.3–2.5 mm long, numerous, included, forming a green cone 3–4 mm tall; **ovary** *c.* 0.8 mm long, hairs at apex very long; **style** jointed at bend, upper part with a few hairs near its base, apex bare; **stigma** obscure; **fruiting heads** brown, 12–15 mm long by 11–15 (–20) mm wide. **FRUIT** an achene, 1-seeded, brown, body 2.5–3 mm long by *c.* 0.8 mm wide by *c.* 0.4 mm thick, slightly shiny, sides convex, hairs to *c.* 1.5 mm long, golden, mostly marginal along the rough ridge; **beak** (persistent style), hooked at apex, 2.5–3 mm long, hairs few, scattered, gland-tipped or not; **receptacle** brown, *c.* 4 mm long by *c.* 1.5 mm wide, hairless or with microscopic hairs *c.* 0.1 mm long.

■ **LEAVES** Basal and stem, compound, lobed and toothed; **basal leaves** ascending, with 5–15 pairs of interrupted leaflets, terminal leaflet the largest at 4–15 cm long by 3–17 cm wide, 3-lobed, hairy both sides, leaflets reduced below; **petioles** 1.5–14 cm long, white hairs to *c.* 3 mm long and spreading; **rachis** hairy; **stem leaves** 2–5 (–7), alternate, simple above, well spaced; **blades** 4–8 cm long by 3–7 cm wide, ascending, hairy; **petioles** 5–10 mm long; **stipules** hairy, 10–15 mm long by 5–7 mm wide, 2- to 4-toothed, ascending.

■ **STEM** Erect, 1–3, hairy or slightly so, green, round, solid to hollow, short hairs numerous, spreading, some gland-tipped; 3–6 mm thick near the base.

■ **SYN.** *Geum perincisum* Rydb. = *Geum macrophyllum* var. *perincisum* (Rydb.) Raup.

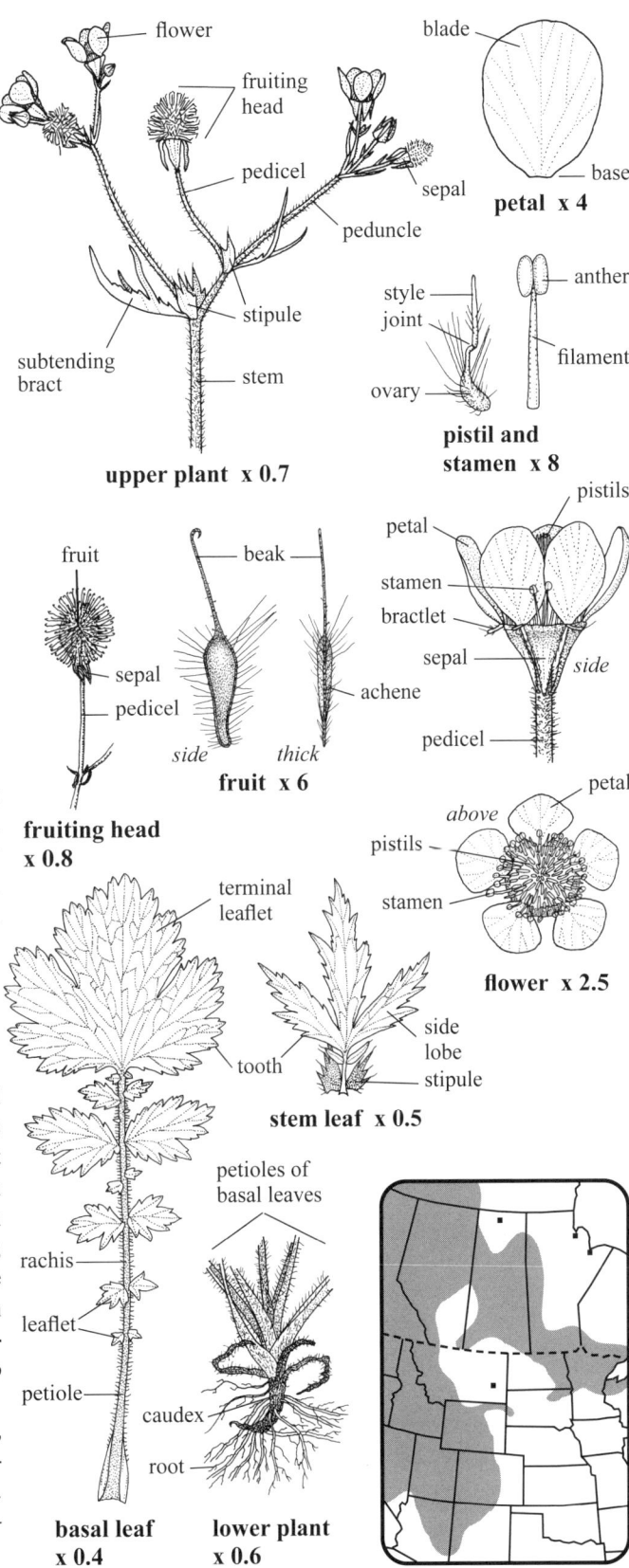

Cut-leaf Avens

Rose—*Rosaceae* **Wildflower** Reddish green Petals 5, pale yellow

PURPLE AVENS *Geum rivale* L.

■ **SKETCH** A **perennial herb** 30–100 cm tall with a **rhizome** 5–10 mm thick covered with dark brown bases of old leaves; **roots** whitish tan, 3–10 cm long by 0.5–1.5 mm thick; in moist areas, stream banks and marshes.

■ **FLOWERS** Reddish green and pale yellow, blooming May–August; **inflorescence** cymose, 3–6 cm long by 10–15 cm wide; **floral branches** 10–30 cm long, hairy, ascending, stiff with fruit; **pedicels** 2–8 cm long, arched with flowers, erect with fruit, hairs long, short hairs gland-tipped; **flowers** perfect, 5 or 6 (3–10), each 12–20 mm wide by 13–16 mm long, nodding or hanging; **hypanthium** *c.* 6 mm wide by *c.* 2 mm deep at anthesis, hairy outside; **sepals** five, reddish, erect, 7–15 mm long by 5–6 mm wide, veiny, some hairs gland-tipped; **bractlets** five, 3–5 mm long by 0.5–0.6 mm wide, ascending; **petals** light yellow, 7–10 mm long by 7–8 mm wide, apices flat, turning orange with age, glabrous, veins faint, claw 2.5–3 mm long; **stamens** included, numerous, shorter than pistils; **filaments** pale green, glabrous, 2.5–4.2 mm long, attached to disc; **anthers** yellow, 1–1.2 mm long, attached in middle, facing inward; **pistils** green, numerous, 4–5 mm long, middle ones slightly exserted; **ovaries** *c.* 1 mm long, with hairs 1–2 mm long near apices; **styles** filiform, long white hairs ascending except near the bare base and tip; **stigma** obscure, brown; **fruiting receptacle** on a short stipe, cylindrical, 9–10 mm long by *c.* 1 mm thick, spreading hairs mostly tan, lower third fruitless and with short gland-tipped hairs; **fruiting heads** *c.* 2 cm long and wide, erect. **FRUIT** an achene, 1-seeded, greenish brown when falling; **body** 3–4 mm long by 1–1.2 mm wide by *c.* 0.9 mm thick, very hairy, short hairs at apex tipped with red glands; **beak** 5.5–8 mm long, hooked at the bare tip, hairy in lower two-thirds, short hairs gland-tipped.

■ **LEAVES** Basal and stem, alternate, toothed, compound below, simple above, hairy both sides; **basal blades** 14–28 cm long by 8–19 cm wide; **petioles** 9–27 cm long, hairy; **stem leaves** two to five, widely spaced, leaflets three or blades simple; **leaf blades** 3.5–7 cm long by 3.5–8 cm wide; **petioles** 0.5–2.5 cm long; **stipules** 10–22 mm long by 5–7 mm wide, hairy, entire or with 1–3 teeth, partially united to base of petiole.

■ **STEM** Green, hairy, simple to somewhat branched, erect to leaning with fruit; base 2–4 mm thick, stiff, bare, ridges low and hairy.

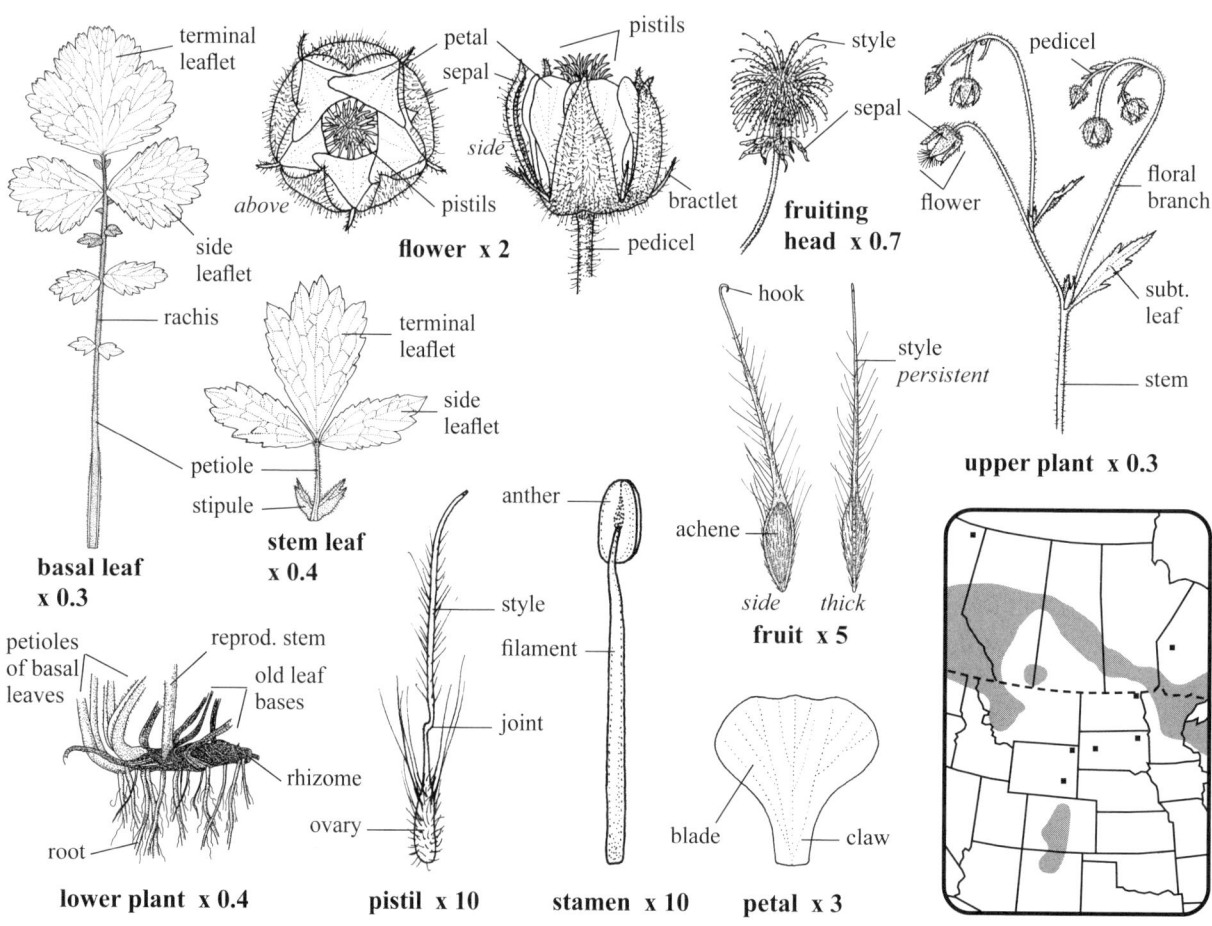

Water Avens, Chocolate-root

Rose—*Rosaceae* **Wildflower** Dark pinkish purple Petals 5, hidden

THREE-FLOWERED AVENS *Geum triflorum* Pursh

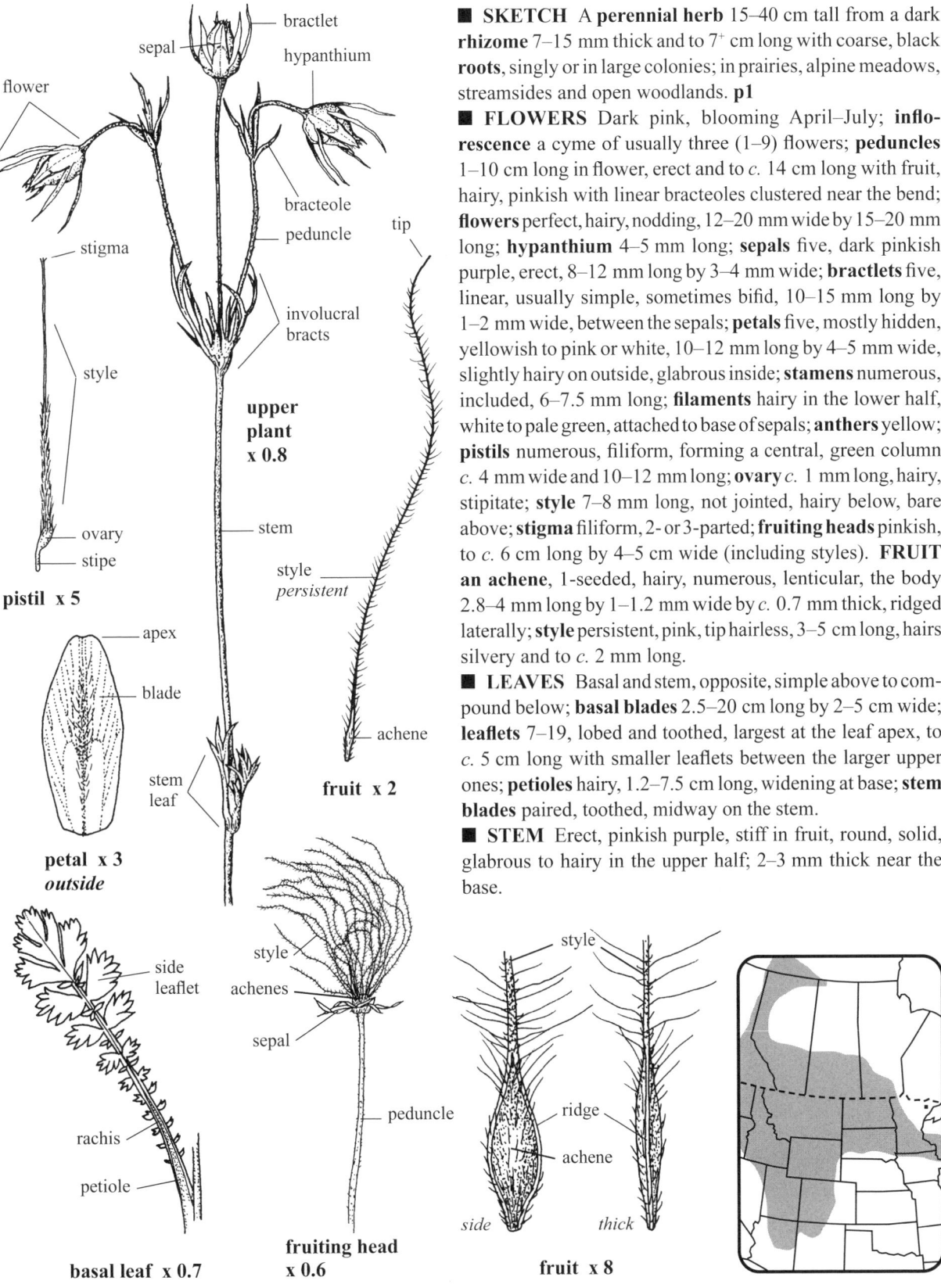

■ **SKETCH** A **perennial herb** 15–40 cm tall from a dark **rhizome** 7–15 mm thick and to 7+ cm long with coarse, black **roots**, singly or in large colonies; in prairies, alpine meadows, streamsides and open woodlands. **p1**

■ **FLOWERS** Dark pink, blooming April–July; **inflorescence** a cyme of usually three (1–9) flowers; **peduncles** 1–10 cm long in flower, erect and to *c.* 14 cm long with fruit, hairy, pinkish with linear bracteoles clustered near the bend; **flowers** perfect, hairy, nodding, 12–20 mm wide by 15–20 mm long; **hypanthium** 4–5 mm long; **sepals** five, dark pinkish purple, erect, 8–12 mm long by 3–4 mm wide; **bractlets** five, linear, usually simple, sometimes bifid, 10–15 mm long by 1–2 mm wide, between the sepals; **petals** five, mostly hidden, yellowish to pink or white, 10–12 mm long by 4–5 mm wide, slightly hairy on outside, glabrous inside; **stamens** numerous, included, 6–7.5 mm long; **filaments** hairy in the lower half, white to pale green, attached to base of sepals; **anthers** yellow; **pistils** numerous, filiform, forming a central, green column *c.* 4 mm wide and 10–12 mm long; **ovary** *c.* 1 mm long, hairy, stipitate; **style** 7–8 mm long, not jointed, hairy below, bare above; **stigma** filiform, 2- or 3-parted; **fruiting heads** pinkish, to *c.* 6 cm long by 4–5 cm wide (including styles). **FRUIT** an **achene**, 1-seeded, hairy, numerous, lenticular, the body 2.8–4 mm long by 1–1.2 mm wide by *c.* 0.7 mm thick, ridged laterally; **style** persistent, pink, tip hairless, 3–5 cm long, hairs silvery and to *c.* 2 mm long.

■ **LEAVES** Basal and stem, opposite, simple above to compound below; **basal blades** 2.5–20 cm long by 2–5 cm wide; **leaflets** 7–19, lobed and toothed, largest at the leaf apex, to *c.* 5 cm long with smaller leaflets between the larger upper ones; **petioles** hairy, 1.2–7.5 cm long, widening at base; **stem blades** paired, toothed, midway on the stem.

■ **STEM** Erect, pinkish purple, stiff in fruit, round, solid, glabrous to hairy in the upper half; 2–3 mm thick near the base.

Old Man's Whiskers, Prairie Smoke, Torch Flower, Maidenhair, Long-plumed Purple Avens

Rose—Rosaceae Shrub White, center yellow Petals 5

NINEBARK *Physocarpus opulifolius* (L.) Maxim.

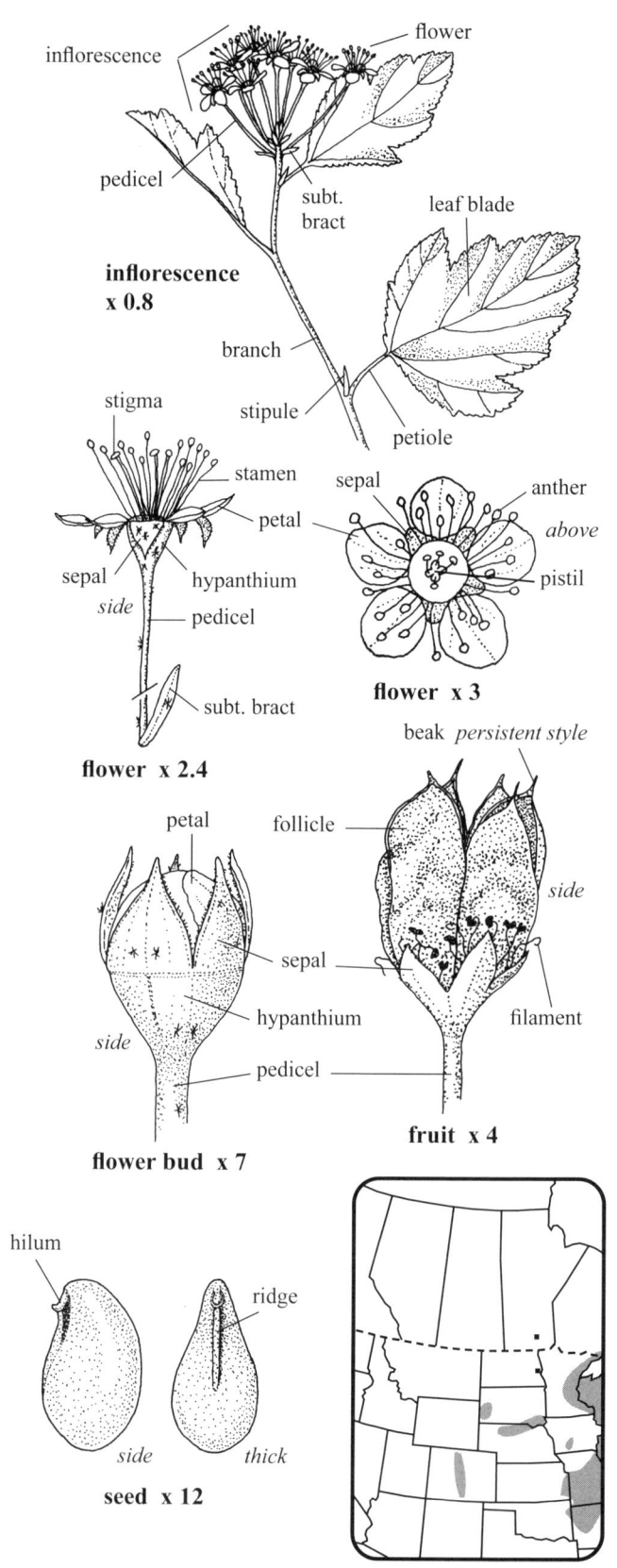

■ **SKETCH** A **deciduous shrub** 1–3 m tall; on rocky banks, streambanks, pond edges and in open woods.

■ **FLOWERS** White, center yellow, blooming May–July; **inflorescence** a corymb (umbel-like) of 5–30+ flowers, 2–4.5 cm wide by 2–3 cm tall, (including pedicels), terminating new seasonal growth; **pedicels** green, ascending, 1.3–2 cm long with scattered stellate hairs; **subtending bracts** (of pedicels) 4–6 mm long by 0.4–1.5 mm wide, reduced above, green, pointed, ascending, usually with a few stellate hairs along the margins, persistent; **flowers** perfect, regular, 8–11 mm wide by 6–7 mm long; **hypanthia** recessed and bright yellow from above, glabrous, 3–4 mm wide by 1.5–2 mm deep, 15-grooved; **sepals** five, light green, 3-nerved, pointed, 2–2.5 mm long and wide, triangular, recurved at anthesis, persistent and erect around the fruit, stellate hairs on both surfaces and margins; **petals** five, white, round, 3–4 mm long by 2–3.5 mm wide, quickly falling, not clawed, base light green and slightly tapered; **stamens** 20–40, exserted, ascending, 4–6 mm long, attached to lip of hypanthium disc; **filaments** white, hairless; **anthers** dark reddish brown, *c.* 0.5 mm long, attached near middle, pollen pale yellow; **pistils** 3 or 4, inserted at center and base of hypanthium cup below the ring of stamens; **ovaries** light yellow, *c.* 1.3 mm long, glabrous, touching along their inner side; **styles** and **stigmas** *c.* 4 mm long, light green, among the stamens and slightly shorter; **stigmas** round, *c.* 0.3 mm wide, flat, slightly wider than the filiform style. **FRUIT** a **follicle** (ripe carpel), opening along both lateral sutures, 3 or 4 per flower, beaked, stellate-hairy to glabrous, 6–12 mm long by *c.* 3 mm wide, turning red then brown when ripe, united near their bases, erect in sepals; **seeds** shiny, 2–4 per follicle, 1.8–2.5 mm long by 1.2 mm wide by *c.* 1 mm thick, smooth, ivory to light brown, pear-shaped, attached along inner suture near the middle of fruit cluster.

■ **LEAVES** Alternate, simple, usually 3- or 5-lobed, finely toothed, stellate-hairy above and below especially along veins, much less on the blade; **blades** 2.5–12 cm long by 2.5–9 cm wide; **petioles** 7–25 mm long, grooved above, usually glabrous; **stipules** 5–7 mm long, caducous.

■ **STEM** Branches brown, glabrous, usually branched above the middle with bark peeling off in long narrow strips, yellowish brown beneath bark.

■ **SYN.** *Physocarpus intermedius* (Rydb.) Schneid. = *Physocarpus opulifolius* var. *intermedius* (Rydb.) B.L. Robins.

Rose—*Rosaceae*   **Wildflower**   White, center yellow   Petals 5

WHITE CINQUEFOIL   *Potentilla arguta* Pursh

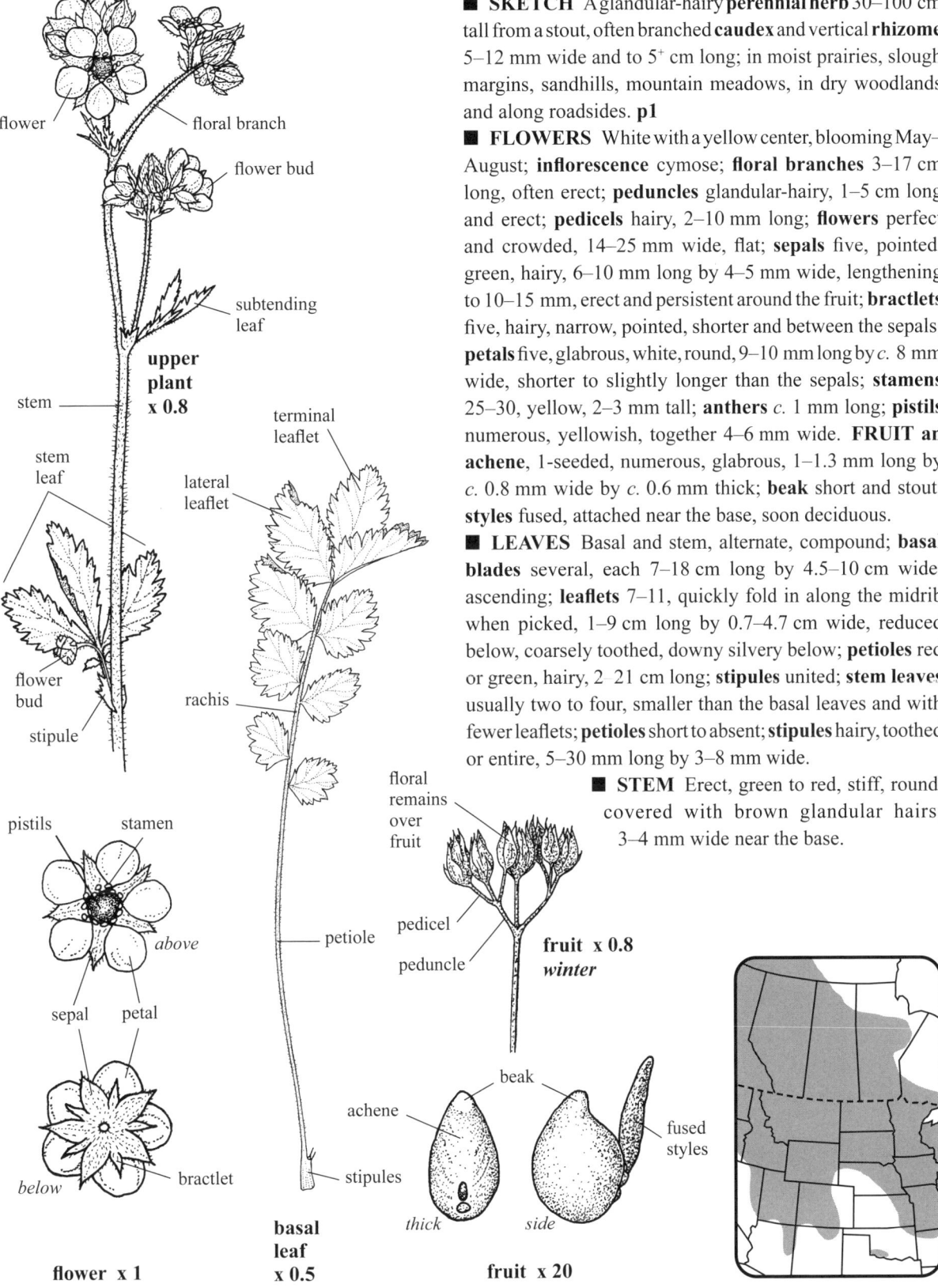

■ **SKETCH** A glandular-hairy **perennial herb** 30–100 cm tall from a stout, often branched **caudex** and vertical **rhizome** 5–12 mm wide and to 5+ cm long; in moist prairies, slough margins, sandhills, mountain meadows, in dry woodlands and along roadsides. **p1**

■ **FLOWERS** White with a yellow center, blooming May–August; **inflorescence** cymose; **floral branches** 3–17 cm long, often erect; **peduncles** glandular-hairy, 1–5 cm long and erect; **pedicels** hairy, 2–10 mm long; **flowers** perfect and crowded, 14–25 mm wide, flat; **sepals** five, pointed, green, hairy, 6–10 mm long by 4–5 mm wide, lengthening to 10–15 mm, erect and persistent around the fruit; **bractlets** five, hairy, narrow, pointed, shorter and between the sepals; **petals** five, glabrous, white, round, 9–10 mm long by *c.* 8 mm wide, shorter to slightly longer than the sepals; **stamens** 25–30, yellow, 2–3 mm tall; **anthers** *c.* 1 mm long; **pistils** numerous, yellowish, together 4–6 mm wide. **FRUIT an achene**, 1-seeded, numerous, glabrous, 1–1.3 mm long by *c.* 0.8 mm wide by *c.* 0.6 mm thick; **beak** short and stout; **styles** fused, attached near the base, soon deciduous.

■ **LEAVES** Basal and stem, alternate, compound; **basal blades** several, each 7–18 cm long by 4.5–10 cm wide, ascending; **leaflets** 7–11, quickly fold in along the midrib when picked, 1–9 cm long by 0.7–4.7 cm wide, reduced below, coarsely toothed, downy silvery below; **petioles** red or green, hairy, 2 21 cm long; **stipules** united; **stem leaves** usually two to four, smaller than the basal leaves and with fewer leaflets; **petioles** short to absent; **stipules** hairy, toothed or entire, 5–30 mm long by 3–8 mm wide.

■ **STEM** Erect, green to red, stiff, round, covered with brown glandular hairs; 3–4 mm wide near the base.

Tall Cinquefoil

Graceful Cinquefoil *Potentilla gracilis* Dougl. ex Hook.

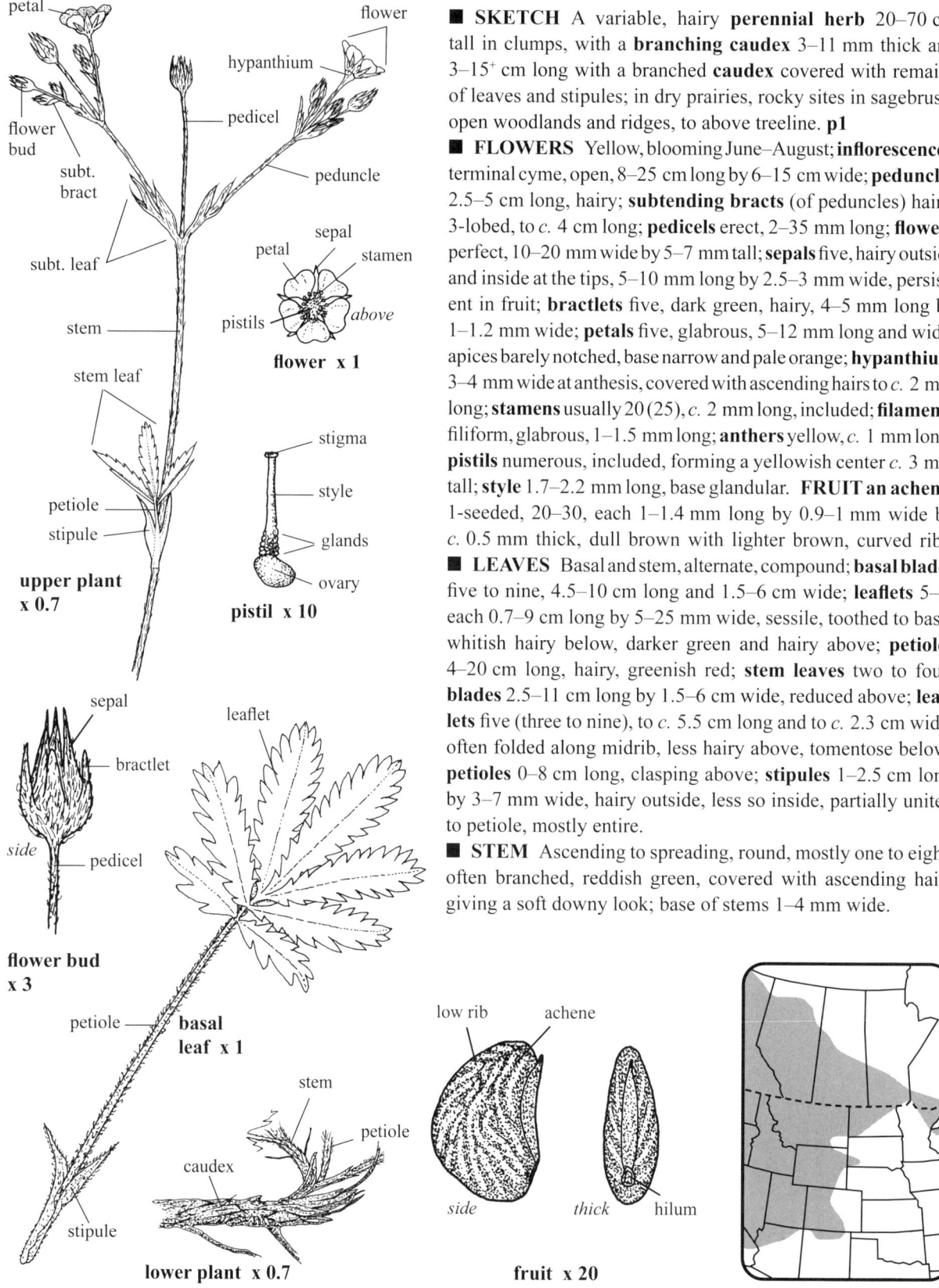

- **SKETCH** A variable, hairy **perennial herb** 20–70 cm tall in clumps, with a **branching caudex** 3–11 mm thick and 3–15+ cm long with a branched **caudex** covered with remains of leaves and stipules; in dry prairies, rocky sites in sagebrush, open woodlands and ridges, to above treeline. **p1**
- **FLOWERS** Yellow, blooming June–August; **inflorescence** a terminal cyme, open, 8–25 cm long by 6–15 cm wide; **peduncles** 2.5–5 cm long, hairy; **subtending bracts** (of peduncles) hairy, 3-lobed, to *c.* 4 cm long; **pedicels** erect, 2–35 mm long; **flowers** perfect, 10–20 mm wide by 5–7 mm tall; **sepals** five, hairy outside and inside at the tips, 5–10 mm long by 2.5–3 mm wide, persistent in fruit; **bractlets** five, dark green, hairy, 4–5 mm long by 1–1.2 mm wide; **petals** five, glabrous, 5–12 mm long and wide, apices barely notched, base narrow and pale orange; **hypanthium** 3–4 mm wide at anthesis, covered with ascending hairs to *c.* 2 mm long; **stamens** usually 20 (25), *c.* 2 mm long, included; **filaments** filiform, glabrous, 1–1.5 mm long; **anthers** yellow, *c.* 1 mm long; **pistils** numerous, included, forming a yellowish center *c.* 3 mm tall; **style** 1.7–2.2 mm long, base glandular. **FRUIT an achene**, 1-seeded, 20–30, each 1–1.4 mm long by 0.9–1 mm wide by *c.* 0.5 mm thick, dull brown with lighter brown, curved ribs.
- **LEAVES** Basal and stem, alternate, compound; **basal blades** five to nine, 4.5–10 cm long and 1.5–6 cm wide; **leaflets** 5–9, each 0.7–9 cm long by 5–25 mm wide, sessile, toothed to base, whitish hairy below, darker green and hairy above; **petioles** 4–20 cm long, hairy, greenish red; **stem leaves** two to four; **blades** 2.5–11 cm long by 1.5–6 cm wide, reduced above; **leaflets** five (three to nine), to *c.* 5.5 cm long and to *c.* 2.3 cm wide, often folded along midrib, less hairy above, tomentose below; **petioles** 0–8 cm long, clasping above; **stipules** 1–2.5 cm long by 3–7 mm wide, hairy outside, less so inside, partially united to petiole, mostly entire.
- **STEM** Ascending to spreading, round, mostly one to eight, often branched, reddish green, covered with ascending hairs giving a soft downy look; base of stems 1–4 mm wide.

Rose—*Rosaceae* Wildflower Yellow Petals 5

ROUGH CINQUEFOIL   *Potentilla norvegica* L.

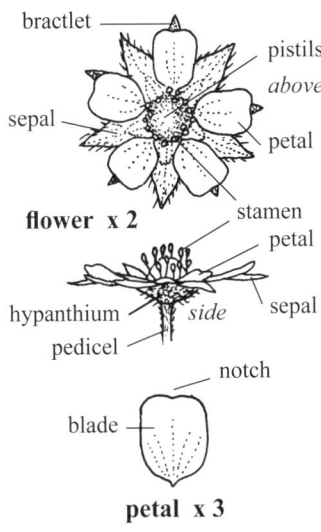

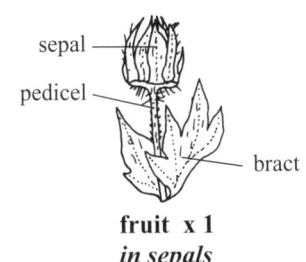

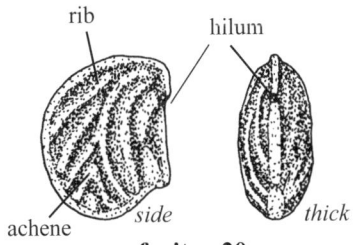

■ **SKETCH** A variable, hairy **annual**, **biennial** or short-lived **perennial herb** 15–80 cm tall from a **taproot** 2–5 mm wide and to *c.* 18+ cm long; along ditches, and streams, lakeshores, in mountain meadows and moist prairies.

■ **FLOWERS** Yellow, blooming May–September; **inflorescence** a series of cymes, leafy-bracteate; **branches** ascending, alternate, hairy, subtended by leaves; **pedicels** hairy, 3–5 mm long, lengthening to 5–15 mm in fruit; **hypanthium** green, shallow, hairy, 4–5 mm wide; **flowers** perfect, 8–14 mm wide by 4–6 mm tall; **sepals** five, 4–6 mm long by 2.5–3 mm wide, usually slightly longer than the petals, pointed, hairy below and in a wide band along margins above, lateral veins obscured by long hairs below, persistent and 9–16 mm long in fruit; **bractlets** five, hairy on both sides, *c.* 5 mm long by *c.* 2 mm wide, 3-veined, situated below each petal and slightly shorter to longer than the petals, darker green than the sepals; **petals** five, yellow, 4–5.5 mm long by 2.8–3.5 mm wide, base pointed, glabrous, apices blunt with a shallow notch, usually shorter than the sepals; **stamens** 15–20, included to exserted, length variable, 1–2.5 mm long, shorter to slightly taller than the central pistils; **filaments** yellow, filiform, erect; **anthers** yellow, 0.3–0.5 mm long, attached in the middle; **pistils** numerous, yellowish green, form a central conical disc *c.* 3 mm wide by *c.* 2.5 mm tall; **fruiting heads** 6–8 mm long and wide, with sepals and bractlets erect and covering the fruit. **FRUIT** an achene, 1-seeded, numerous, 0.9–1.3 mm long by 0.8–0.9 mm wide by 0.4–0.5 mm thick, tan, covered with wavy ribs or rarely smooth, not hairy.

■ **LEAVES** Basal and stem, alternate, compound (mostly trifoliate), toothed; **basal blades** 0.7–3 cm long by 1–3 cm wide; **petioles** 1–12 cm long; **stem blades** 2–9 cm long by 2–11 cm wide, slightly hairy, hairs appressed, lighter green and more scabrous below; **lateral leaflets** sessile, 1–6 cm long by 0.3–4 cm wide, reduced above; **petioles** 0–7 cm long, hairy, reduced above; **petiolule** of terminal leaflets 2–5 mm long, hairy; **stipules** leafletlike, 6–25 mm long by 2–17 mm wide, entire to toothed, hairs simple, 1–2 mm long and appressed.

■ **STEM** Erect to decumbent at base, one to several from the root apex, branched, round, slightly reddish green, more hairy below; 1–7 mm wide near the base.

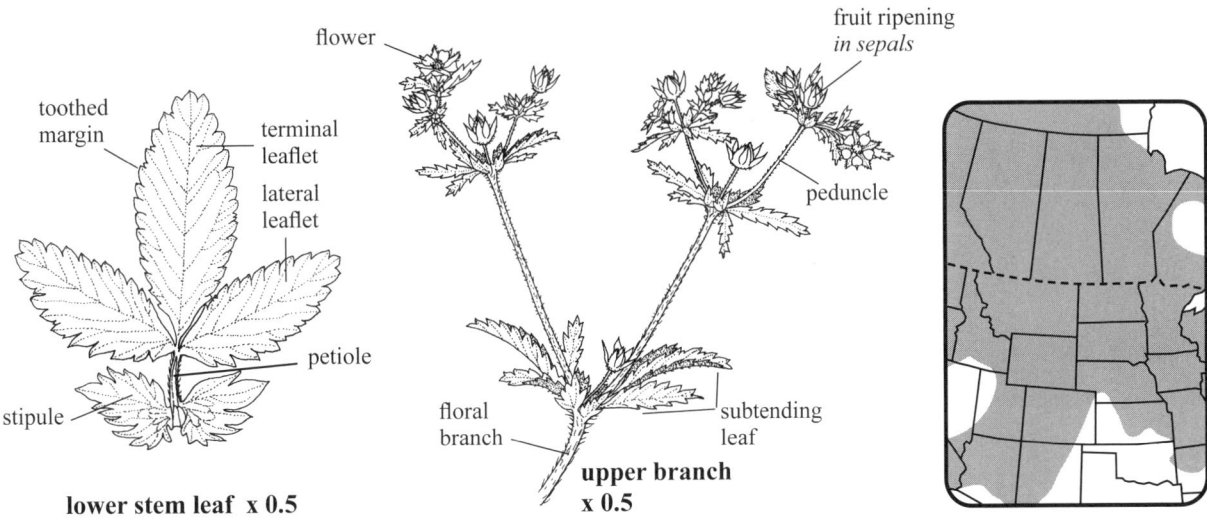

Norwegian Cinquefoil

Rose—*Rosaceae* | **Shrub** White fading to pink Petals 5

Canada Plum *Prunus nigra* Ait.

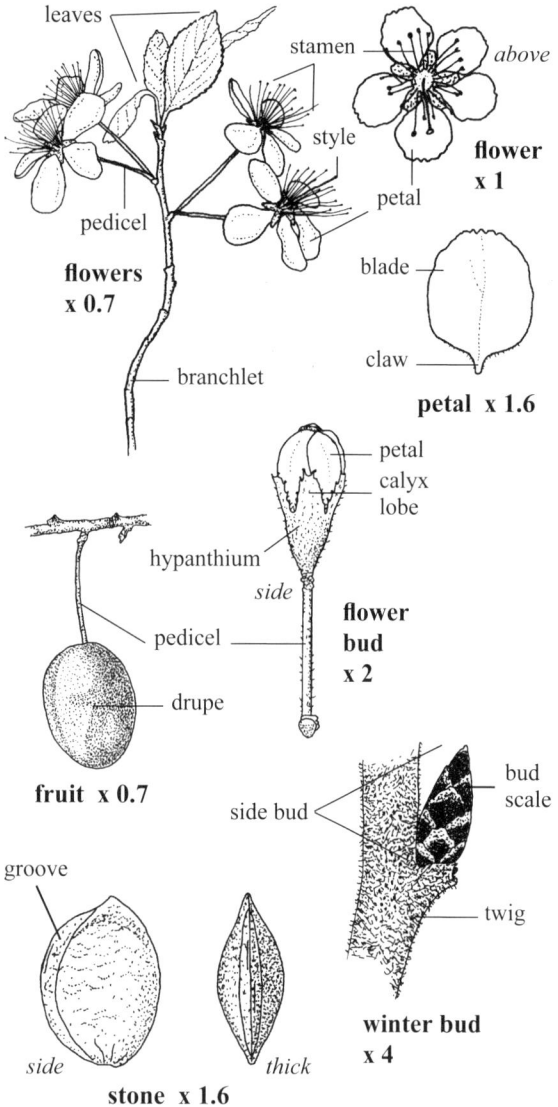

- **SKETCH** A **deciduous shrub** 2–8 m tall with zigzagging branches; along bluffs and woodland edges. **w2**
- **FLOWERS** White, turning pink as they fade, blooming May–June as leaves expand; **inflorescence** a raceme in alternate clusters of (one) two to five flowers near the ends of branchlets; **pedicels** 7–20 mm long, reddish, finely hairy to glabrous, some wrinkled; **flowers** perfect, fragrant, 1.5–3 cm wide by 12–15 mm long; **hypanthium** 3–4 mm long, reddish green, lightly hairy or glabrous; **calyx lobes** five, 3–4 mm long by 1.5–2 mm wide, glandular teeth near blunt apices, reddish green, hairy at inner base; **petals** five, soon deciduous, 9–14 mm long by 7–9 mm wide, spreading, claw short, some ciliate; **stamens** *c.* 30, exserted, 7–11 mm long; **filaments** white, in two series arising from edge of hypanthium; **anthers** yellow, 0.7–0.9 mm long; **style** one, exserted, 9–12 mm long; **stigma** oval, 0.5–0.7 mm wide.
- **FRUIT a drupe**, 1-stoned, fleshy, oval, 15–35 mm long by 13–18 mm wide, glabrous, not glaucous, orangish red when ripe; **stones** 1-seeded, tan, 1.4–1.8 cm long by 1–1.3 cm wide by 5–6 mm thick, slightly wrinkled, with a groove.
- **LEAVES** Alternate, simple, toothed; **blades** 2–12 cm long by 1–7 cm wide, teeth blunt, gland-tipped, white hairs along midrib and side veins on both surfaces; **petioles** 1–2.5 cm long, hairy, with two large glands near the blade; **stipules** with glandular margins, 4–8 mm long, 1- to several-lobed, early deciduous, hairy base united with petiole.
- **STEM** A **trunk**, dark gray with age, scaly, bark splits vertically on trunk; base 5–20 cm thick; **branches** often crooked; **spurs** not sharply pointed, but stiff; **branchlets** glabrous or covered with tiny white hairs; **twigs** dark glandular spotted, hairy, bark peeling to reveal reddish brown glabrous bark below; **winter buds** dark brown, alternate, 3–4 mm long by 1.5–2 mm wide and thick, 7- to 9-scaled, dull, scale margins erose and light gray, not hairy; **leaf scars** oval, dark brown.

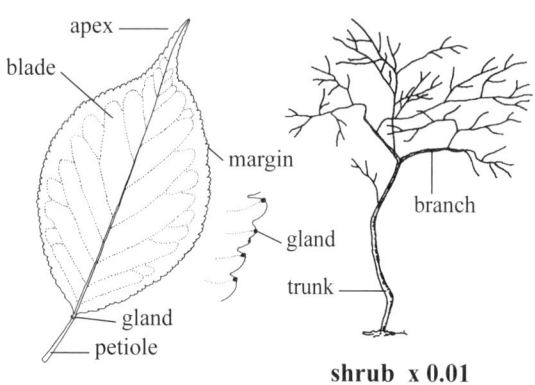

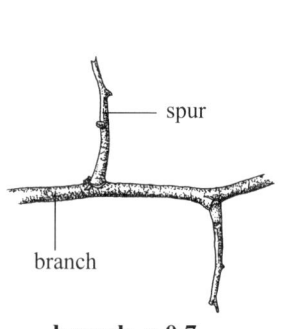

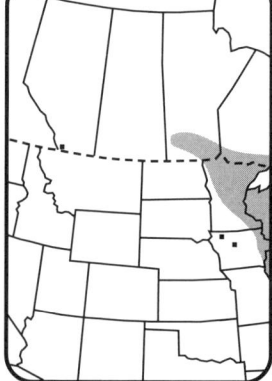

Rose—*Rosaceae* Shrub White Petals 5

Pin Cherry *Prunus pensylvanica* L. f.

■ **SKETCH** A **deciduous shrub** or small tree 1–10 m tall; in aspen woodlands along trails and bluffs, recent clearings, ravines and fencerows in dry to moist sites. **w1**

■ **FLOWERS** White, blooming March–July; **inflorescence** a corymb (umbel-like) of 2–8 flowers per cluster, the clusters 2–4 cm wide, flat-topped at anthesis; **peduncles** absent; **pedicels** green, glabrous, 12–27 mm long with a small toothed bud scale near the base; **flowers** perfect, 10–15 mm wide by 7–9 mm long, fragrant; **hypanthium** glabrous, veined outside, orange inside; **sepals** five, glabrous, reddish green, apices roundish, *c.* 2 mm long and wide; **petals** five, white, spreading, 5–6 mm long by 4–5 mm wide, claw pale pink and not obvious, slightly hairy below; **stamens** 10 or 20, exserted, 7–8 mm long; **filaments** white, filiform; **anthers** yellow, *c.* 1 mm long; **pistil** one, green and glabrous, 7–8 mm long; **ovary** *c.* 1.5 mm long by *c.* 1 mm wide, few ridged; **style** erect, *c.* 0.2 mm thick, round in cross-section, exserted *c.* 5 mm past hypanthium; **stigma** pale green, irregularly shaped, 0.7–0.9 mm wide. **FRUIT** a drupe, 1-stoned, red, fleshy and shiny, smooth, 5–8 mm wide and long, round; **stones** 1-seeded, tan, 5–6 mm long by 4–5 mm wide by 3.5–4 mm thick with a 1 mm wide ridge on the ventral side, coat slightly rough, roundish.

■ **LEAVES** Alternate, simple, finely toothed, several in alternate clusters; **blades** 2–12 cm long by 1–5 cm wide, somewhat drooping, shiny, lighter green below, reddish purple in fall, teeth numerous, blunt and gland-tipped, the glands orangish tan; **petioles** 0.7–3 cm long, hairless, with one or two reddish round glands near the blade; **stipules** paired, 2–5 mm long, pale green, glandular-toothed, glabrous, shiny above, quickly falling.

■ **TRUNK** Gray, 3–25 cm thick; **new twigs** reddish, shiny, hairless; **autumnal buds** lateral (alternate) and terminal, hairless, 2–4.5 mm long by 1–2 mm wide and thick, pointed, slightly shiny, 7- or 8-scaled, these tan and reddish brown; **terminal buds** one or more and larger than side buds.

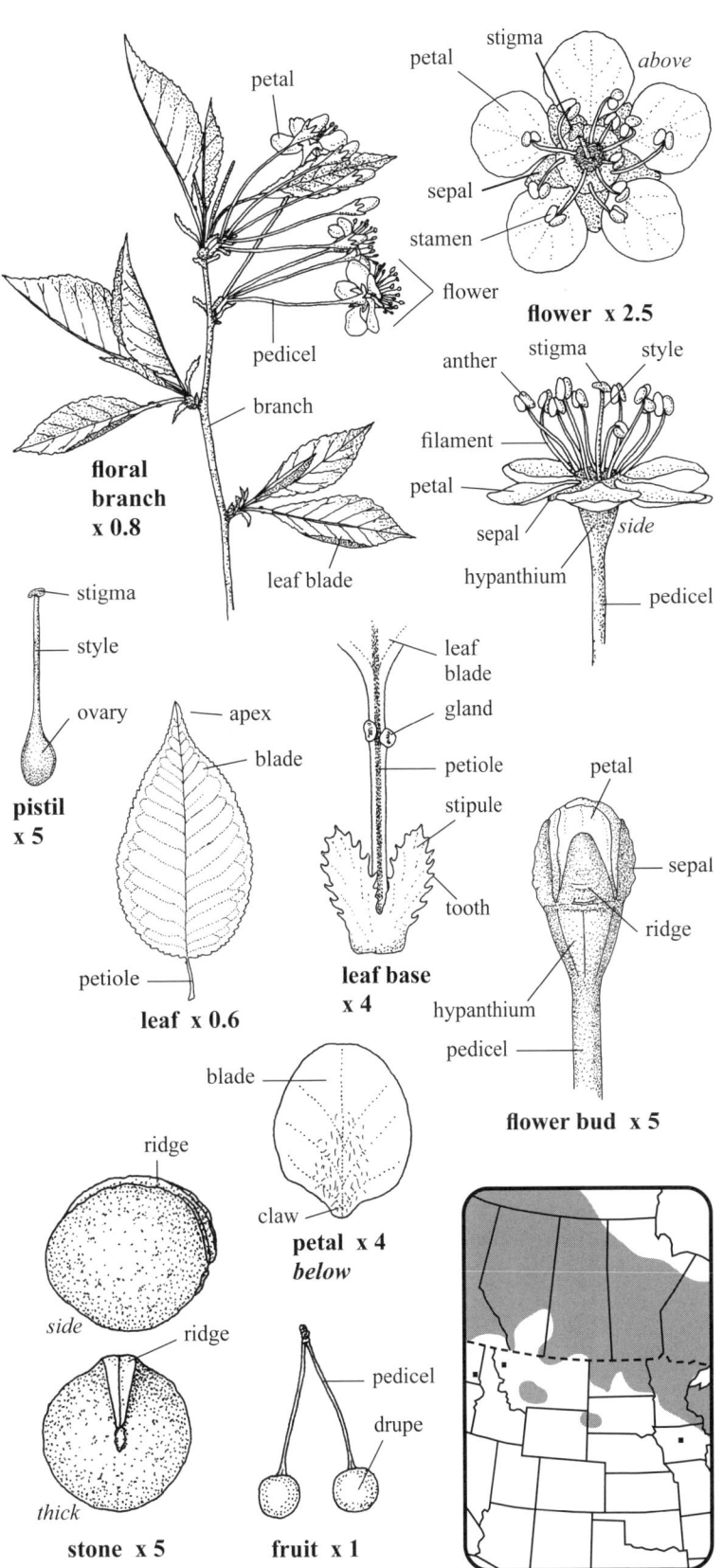

Fire Cherry, Bird Cherry, Wild Red Cherry

Rose—*Rosaceae*   **Shrub** White Petals 5

Red-fruited Choke Cherry *Prunus virginiana* L.

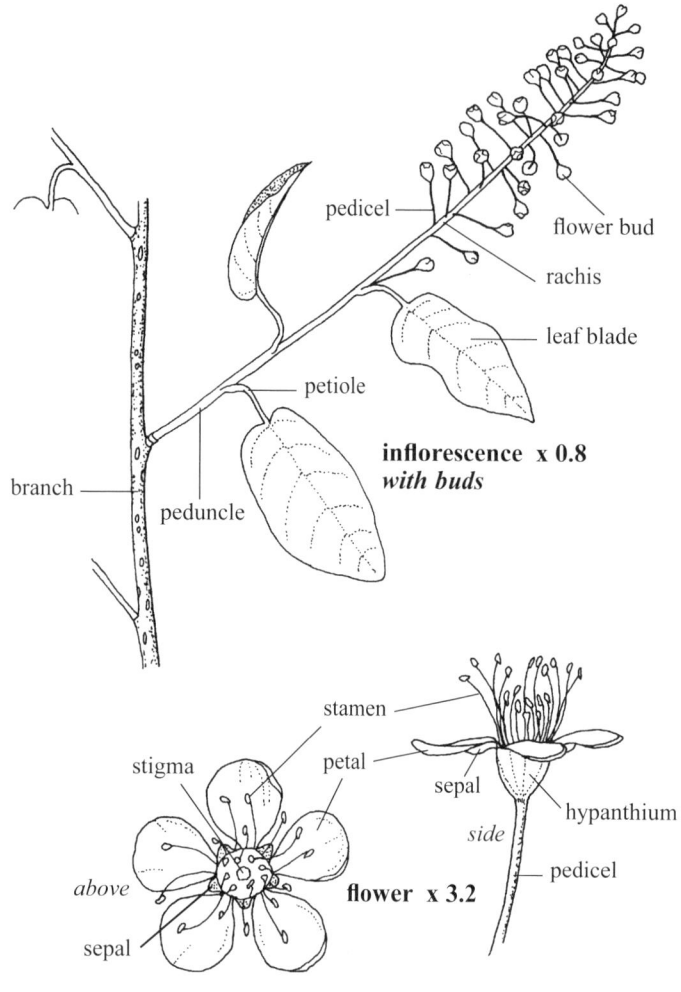

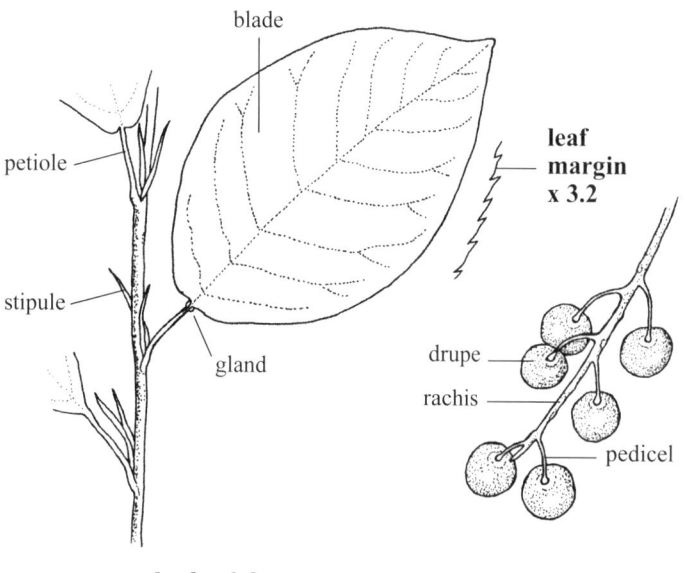

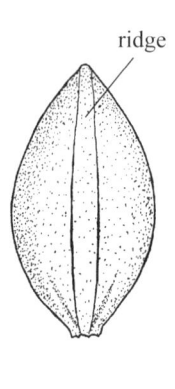

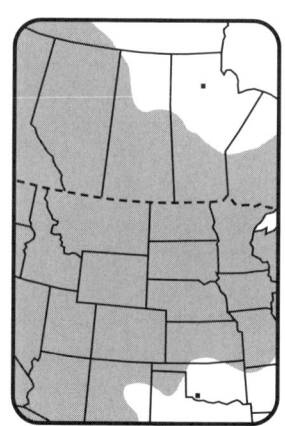

■ **SKETCH** A **deciduous shrub** or **small tree** 1–10 m tall; along riverbanks, woodland edges, hedgerows, thickets, roadsides, sagebrush, rocky outcrops and prairies. **w1**

■ **FLOWERS** White, blooming April–July; **inflorescence** a raceme, each 3–14 cm long by 2–3 cm wide; **peduncles** alternate, 0.5–6 cm long with one to five leaves; **pedicels** glabrous to hairy, 3–8 mm long, to *c.* 10 mm with fruit; **flowers** fragrant, perfect, 10–15 mm wide, 15–50 per raceme; **hypanthium** 1.2–2.5 mm long and wide, glabrous; **sepals** five, 1–1.5 mm long, roundly triangular, hairless, glandular-erose, deciduous soon after anthesis; **petals** five, 3–5 mm long, horizontal and almost round; **stamens** 15–30, in whorls, exserted; **anthers** yellow; **ovary** glabrous, 1.5–1.7 mm long; **style** one, green, about as tall as the shortest stamens. **FRUIT** a drupe, 1-stoned, globular, dark red to black, 6–10 mm long and wide; **stones** 1-seeded, tan, 7–9 mm long by 5–6 mm wide and thick, with a flat ridge on one side.

■ **LEAVES** Alternate, simple, finely toothed, oval, thin, slightly shiny above, glabrous or with tufted hairs in vein axils below, rarely hairy on blade; **blades** 2.5–12 cm long by 1.5–6.2 cm wide; **petioles** green to red, 0.7–3.2 cm long; **glands** usually two at the base of the leaf blade; **stipules** paired, 8–15 mm long by 1–2 mm wide, pointed, entire to finely toothed, white, papery, linear, caducous.

■ **STEM** Woody, round, the bark grayish brown and smooth, with horizontal, elongated, tan lenticels; 5–18 cm thick near the base; **branches** horizontal below to ascending above.

■ **NOTE** The variety *melanocarpa* produces black fruit (Great Plains Flora Association, 1986).

462   Common Chokecherry, Choke Cherry

Rose—*Rosaceae* — **Shrub** Pink Petals 5

PRICKLY ROSE *Rosa acicularis* Lindl.

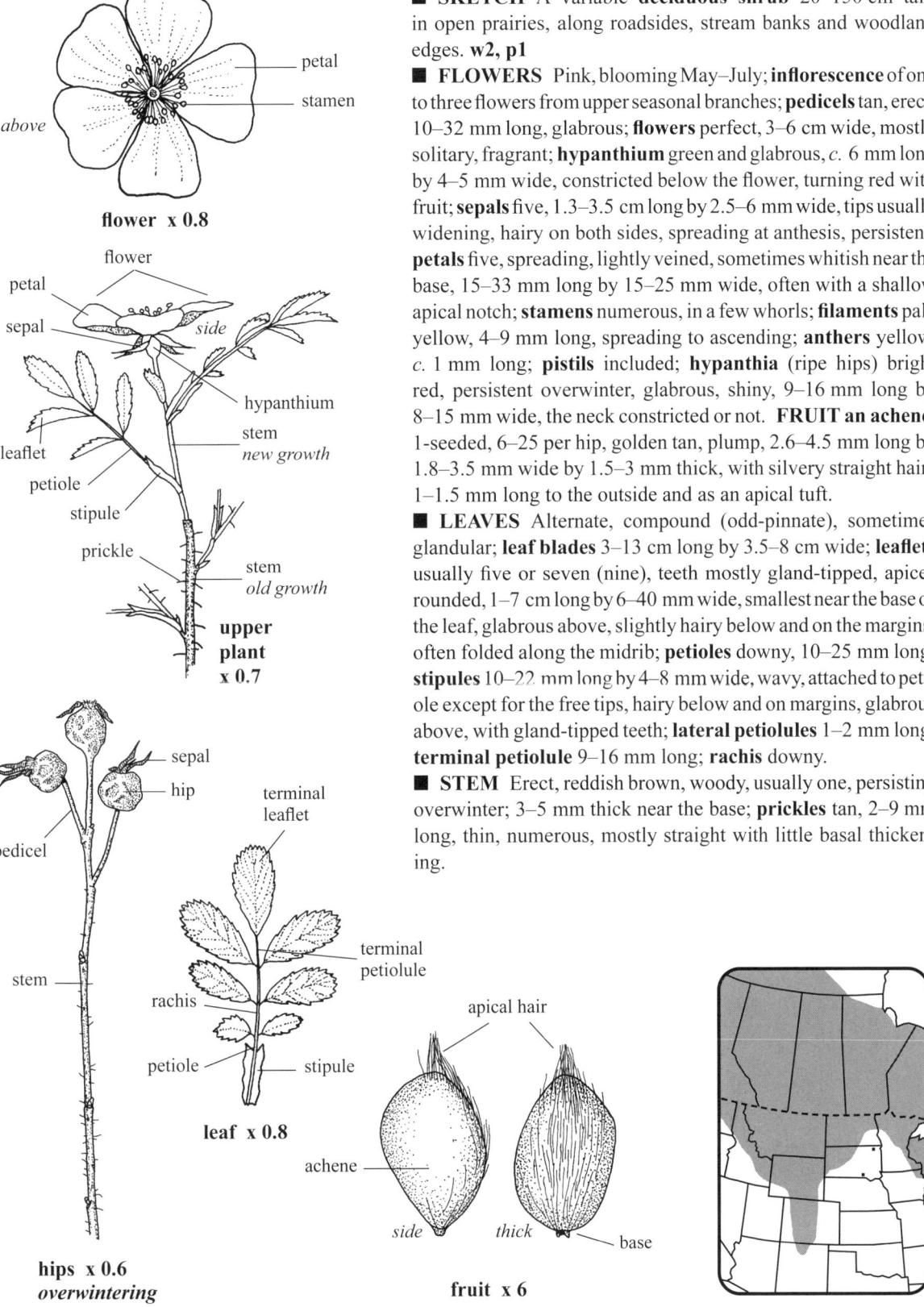

■ **SKETCH** A variable **deciduous shrub** 20–150 cm tall; in open prairies, along roadsides, stream banks and woodland edges. **w2, p1**

■ **FLOWERS** Pink, blooming May–July; **inflorescence** of one to three flowers from upper seasonal branches; **pedicels** tan, erect, 10–32 mm long, glabrous; **flowers** perfect, 3–6 cm wide, mostly solitary, fragrant; **hypanthium** green and glabrous, *c.* 6 mm long by 4–5 mm wide, constricted below the flower, turning red with fruit; **sepals** five, 1.3–3.5 cm long by 2.5–6 mm wide, tips usually widening, hairy on both sides, spreading at anthesis, persistent; **petals** five, spreading, lightly veined, sometimes whitish near the base, 15–33 mm long by 15–25 mm wide, often with a shallow apical notch; **stamens** numerous, in a few whorls; **filaments** pale yellow, 4–9 mm long, spreading to ascending; **anthers** yellow, *c.* 1 mm long; **pistils** included; **hypanthia** (ripe hips) bright red, persistent overwinter, glabrous, shiny, 9–16 mm long by 8–15 mm wide, the neck constricted or not. **FRUIT an achene**, 1-seeded, 6–25 per hip, golden tan, plump, 2.6–4.5 mm long by 1.8–3.5 mm wide by 1.5–3 mm thick, with silvery straight hairs 1–1.5 mm long to the outside and as an apical tuft.

■ **LEAVES** Alternate, compound (odd-pinnate), sometimes glandular; **leaf blades** 3–13 cm long by 3.5–8 cm wide; **leaflets** usually five or seven (nine), teeth mostly gland-tipped, apices rounded, 1–7 cm long by 6–40 mm wide, smallest near the base of the leaf, glabrous above, slightly hairy below and on the margins, often folded along the midrib; **petioles** downy, 10–25 mm long; **stipules** 10–22 mm long by 4–8 mm wide, wavy, attached to petiole except for the free tips, hairy below and on margins, glabrous above, with gland-tipped teeth; **lateral petiolules** 1–2 mm long; **terminal petiolule** 9–16 mm long; **rachis** downy.

■ **STEM** Erect, reddish brown, woody, usually one, persisting overwinter; 3–5 mm thick near the base; **prickles** tan, 2–9 mm long, thin, numerous, mostly straight with little basal thickening.

Rose—*Rosaceae* **Shrub** Pink Petals 5

LOW PRAIRIE ROSE *Rosa arkansana* Porter

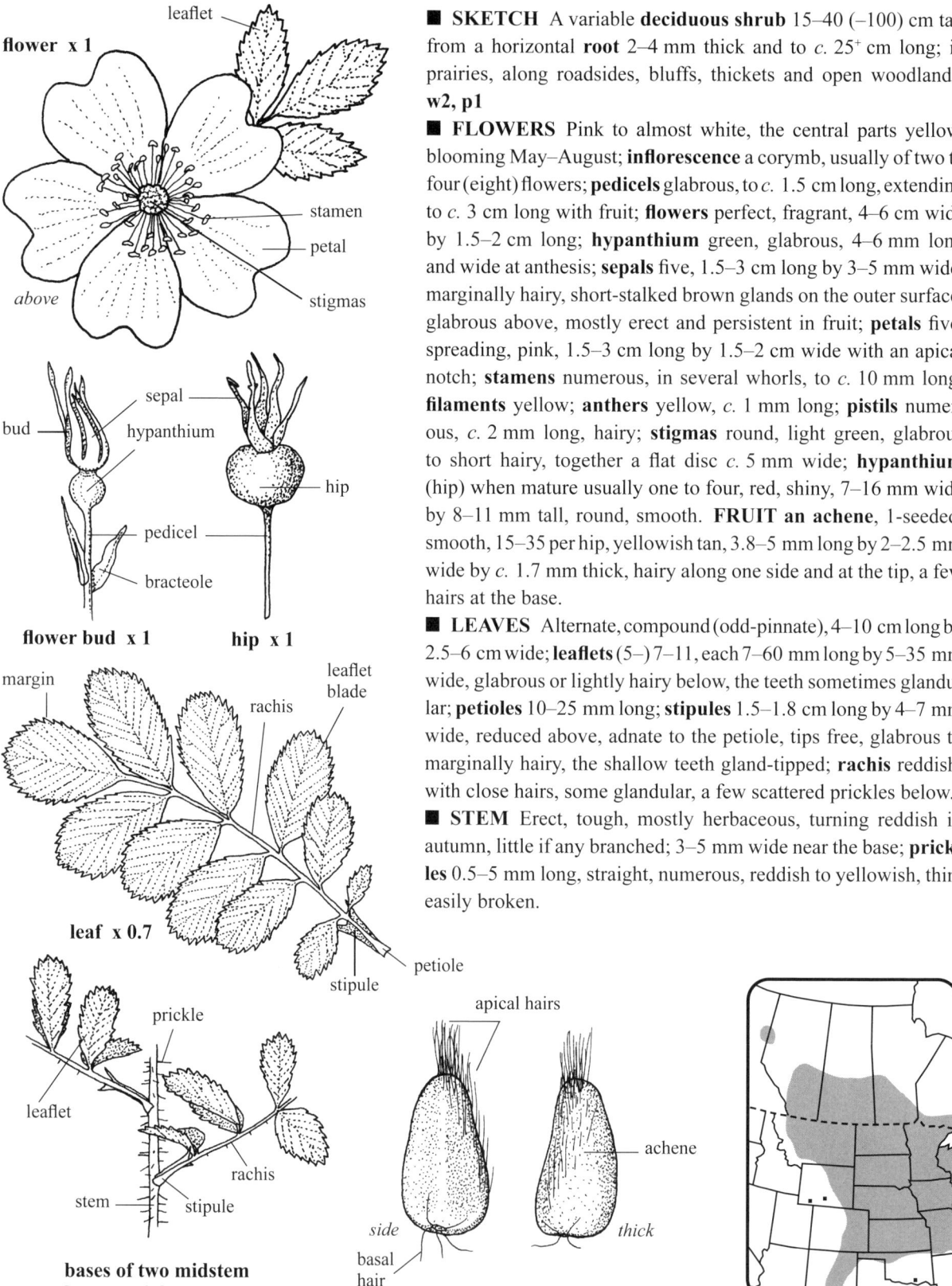

- **SKETCH** A variable **deciduous shrub** 15–40 (–100) cm tall from a horizontal **root** 2–4 mm thick and to *c.* 25+ cm long; in prairies, along roadsides, bluffs, thickets and open woodlands. **w2, p1**
- **FLOWERS** Pink to almost white, the central parts yellow, blooming May–August; **inflorescence** a corymb, usually of two to four (eight) flowers; **pedicels** glabrous, to *c.* 1.5 cm long, extending to *c.* 3 cm long with fruit; **flowers** perfect, fragrant, 4–6 cm wide by 1.5–2 cm long; **hypanthium** green, glabrous, 4–6 mm long and wide at anthesis; **sepals** five, 1.5–3 cm long by 3–5 mm wide, marginally hairy, short-stalked brown glands on the outer surface, glabrous above, mostly erect and persistent in fruit; **petals** five, spreading, pink, 1.5–3 cm long by 1.5–2 cm wide with an apical notch; **stamens** numerous, in several whorls, to *c.* 10 mm long; **filaments** yellow; **anthers** yellow, *c.* 1 mm long; **pistils** numerous, *c.* 2 mm long, hairy; **stigmas** round, light green, glabrous to short hairy, together a flat disc *c.* 5 mm wide; **hypanthium** (hip) when mature usually one to four, red, shiny, 7–16 mm wide by 8–11 mm tall, round, smooth. **FRUIT an achene**, 1-seeded, smooth, 15–35 per hip, yellowish tan, 3.8–5 mm long by 2–2.5 mm wide by *c.* 1.7 mm thick, hairy along one side and at the tip, a few hairs at the base.
- **LEAVES** Alternate, compound (odd-pinnate), 4–10 cm long by 2.5–6 cm wide; **leaflets** (5–) 7–11, each 7–60 mm long by 5–35 mm wide, glabrous or lightly hairy below, the teeth sometimes glandular; **petioles** 10–25 mm long; **stipules** 1.5–1.8 cm long by 4–7 mm wide, reduced above, adnate to the petiole, tips free, glabrous to marginally hairy, the shallow teeth gland-tipped; **rachis** reddish, with close hairs, some glandular, a few scattered prickles below.
- **STEM** Erect, tough, mostly herbaceous, turning reddish in autumn, little if any branched; 3–5 mm wide near the base; **prickles** 0.5–5 mm long, straight, numerous, reddish to yellowish, thin, easily broken.

Sunshine Rose, Prairie Wild Rose, Prairie Rose

Rose—*Rosaceae*   **Shrub** Pink Petals 5

Wood's Rose   *Rosa woodsii* Lindl.

■ **SKETCH** A variable, armed **deciduous shrub** 20–200 cm tall, solitary or in thickets; in stream valleys, ravines, marshes, along railways, roadsides, in open woodlands and sandhills. **w1, p1**

■ **FLOWERS** Pink, blooming May–August; **inflorescence** a corymbiform cyme, terminal on new growth; **pedicels** 5–20 mm long in fruit; **flowers** perfect, fragrant, 2–5 cm wide by 10–12 mm tall; **hypanthium** glabrous, green, 4–5 mm wide by 3–4 mm tall at anthesis, enlarging with fruit; **sepals** five, united at base, persistent, 0.8–2 cm long by 1.5–4 mm wide, entire, hairy below, sometimes hairs reddish gland-tipped, apices narrow and mostly glabrous; **petals** five, pink, flat, 2–2.5 cm long and wide; **stamens** numerous, each 3–8 mm long, the central three or four whorls together *c.* 16 mm wide; **filaments** yellow, filiform, 2–6 mm long; **anthers** yellow, 1.2–2 mm long; **pistils** numerous, a central cluster 3–4 mm wide; **ovaries** green, 1.5–1.8 mm long, hairy along one side; **styles** hairy, 1.5–3.5 mm long; **stigmas** round, 0.8–1 mm wide; **hypanthium** (hip) reddish, round, neckless, overwintering, 9–18 mm wide and long, wall 0.5–1 mm thick. **FRUIT an achene**, 1-seeded, 20–50 per hip, tan, smooth, 3.5–5 mm long by 2.2–2.8 mm wide by 1.6–1.8 mm thick, hairy along one side and with an apical tuft.

■ **LEAVES** Alternate, compound (odd-pinnate), toothed; **blades** 4–14 cm long by 3.5–11 cm wide; **leaflets** usually five to nine, mostly sessile, larger near the apices, each 1.5–5 cm long by 7–35 mm wide, teeth often gland-tipped at least along one margin, glabrous to hairy above, lighter green and hairy on raised veins below, lateral veins obvious; **petioles** 1–3 cm long; **stipules** paired, 1.5–2.5 cm long, each half 2–5 mm wide, glabrous to hairy, attached to the petiole for most of their length, free for 3–7 mm at the tips, finely toothed near the pointed apices, teeth usually glandular; **terminal petiolules** 8–12 mm long, hairy; **lateral petiolules** less than 1 mm long; **rachis** hairy, usually unarmed, reddish.

■ **STEM** Erect, armed, persistent; 8–20 mm thick near the base; **prickles** tan, 4–10 mm long, spreading to slightly descending especially below, expanded at the base, and some slightly curved, usually a **spreading pair** 6–7 mm long with a base 3–4 mm long at the base of a stipule (node), prickles above fewer and usually reduced; **branches** reddish brown, mostly hairless, alternate, glaucous, slightly shiny.

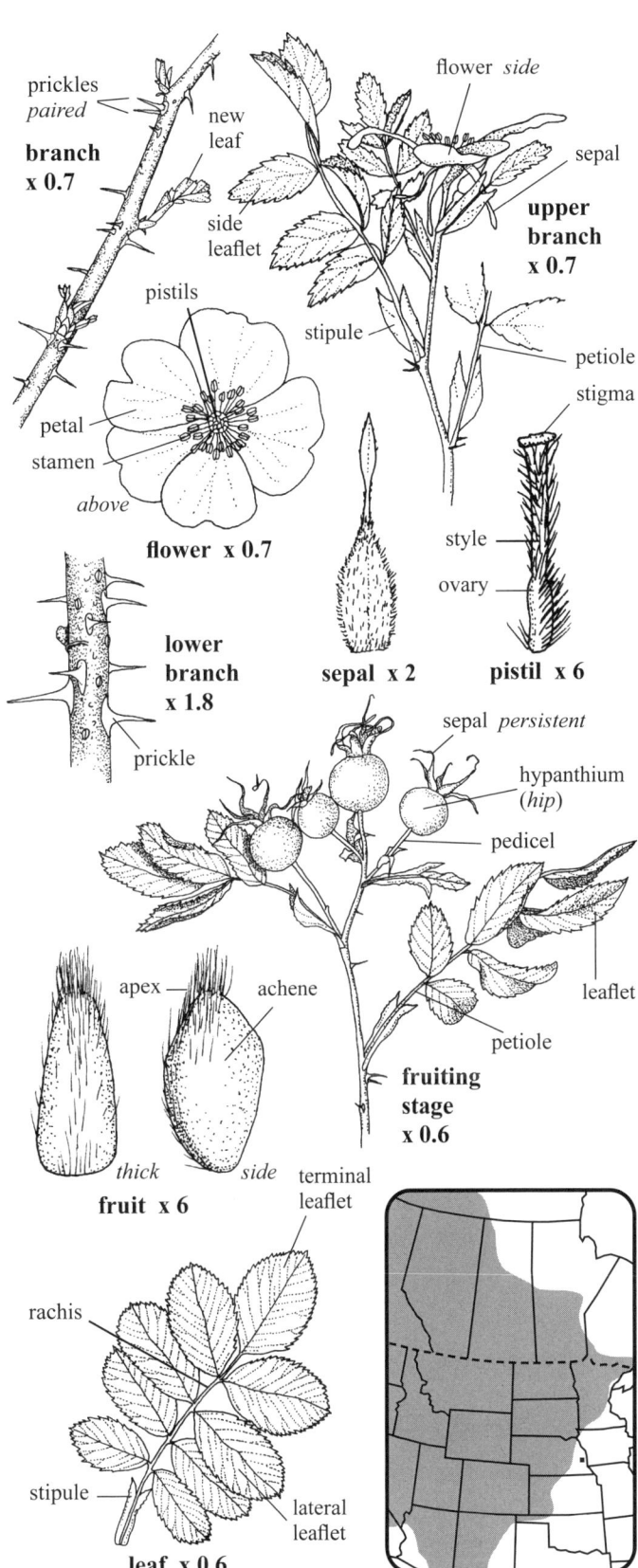

Wood's Wild Rose, Western Wild Rose, Common Wild Rose

Rose—*Rosaceae*   **Wildflower** Pink Petals 5 (6)

Stemless Raspberry *Rubus arcticus* L.

■ **SKETCH** An unarmed **perennial herb** 5–20 cm tall in small open colonies from dark reddish brown **rhizomes** 5–15+ cm long by 0.5–0.7 mm thick with **fibrous roots** 1–5 cm long; **stolons** absent; in wet fields, woodland edges, meadows, muskeg and tundra. **w1**

■ **FLOWERS** Pink to reddish pink, blooming June–July; **inflorescence** a terminal flower, at or slightly below the top leaves; **pedicels** erect, green, 3–30 mm long, white hairs descending; **flowers** perfect, fragrant, 1.5–2.5 cm wide by 1–1.5 cm tall, erect; **hypanthium** green, *c.* 4 mm wide by *c.* 2 mm tall, hairy, 12-ribbed; **sepals** five (6), ascending to spreading from top of hypanthium, 5.5–8 mm long by 1.5–2.5 mm wide, hairy on both sides and margins, veiny, reddish green below, reflexed with fruit; **petals** five (6), 10–15 mm long by 5–6 mm wide, veiny, a few gland-tipped hairs mostly on the claw; **stamens** glabrous, *c.* 50, 1.8–3.5 mm long, arranged in 4 or 5 rows (short ones close to pistil) around the pistils; **filaments** reddish pink to pale yellow, 1–3 mm long by *c.* 0.6 mm wide, narrowing at apices to a filiform stalk *c.* 0.2 mm long; **anthers** 2-lobed, *c.* 0.5 mm long, pale yellow; **pistils** numerous, together *c.* 2.5 mm wide and tall, green and glabrous; **ovary** *c.* 0.8 mm long; **styles** *c.* 1.7 mm long, curved inward, persistent and *c.* 1.3 mm long on fruit; **stigmas** rough, slightly wider than the styles. **FRUIT a drupelet**, red, juicy, shiny, 20–30 lightly united forming an erect cluster 10–13 mm wide by 8–10 mm tall; **drupelets** 3.5–4.5 mm long by *c.* 3 mm wide, 1-seeded, slightly transparent; **seeds** tan, smooth, dull, wedged-shaped, 2–2.2 mm long by 1.3–1.5 mm wide by 1–1.2 mm thick, with a thin pink line along the rounded side opposite the hilum.

■ **LEAVES** Alternate, compound, two to five per stem; **blades** 1.3–3.8 cm long by 1.1–5 cm wide; **leaflets** three, 5–35 mm long by 5–22 mm wide, toothed, slightly shiny and green above, reddish near the hairy margins, dull and green below with veins raised and hairy, less hairy above; **petioles** reddish green, 0.8–4.5 cm long, hairy; **stipules** entire, paired, veiny, 4–6.5 mm long by 1.5–3 mm wide, erect to spreading, margins hairy; **petiolules** reddish, 0.5–3 mm long, terminal one the longest.

■ **STEM** Erect, simple, leafy, pinkish green, hairy, round and solid; 1–1.3 mm wide near the base, covered with dark brown bracts below ground.

■ **SYN.** *Rubus acaulis* Michx. = *Rubus arcticus* ssp. *acaulis* (Michx.) Focke and *Cylactis arctica* ssp. *acaulis* (Michx.) W.A. Weber = *Rubus arcticus* ssp. *acaulis* (Michx.) Focke.

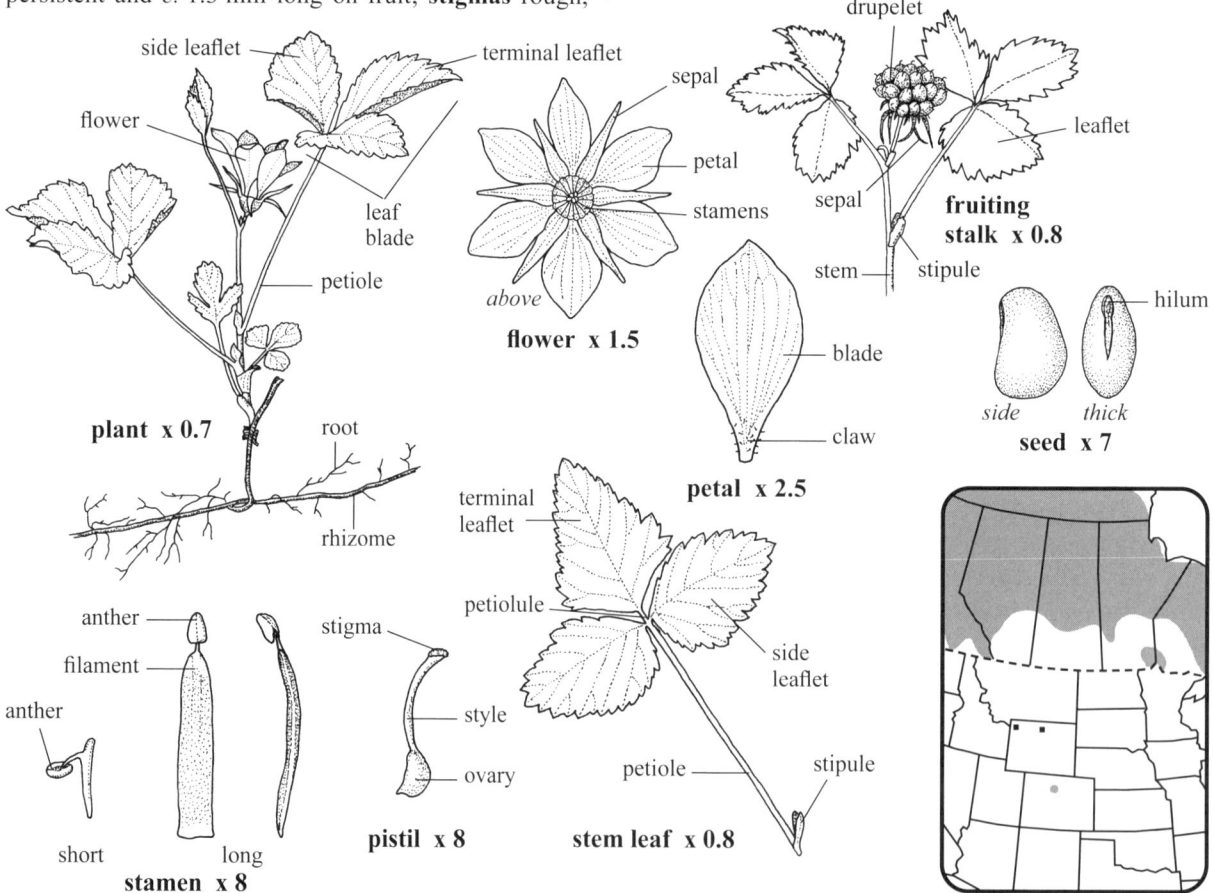

Dwarf Raspberry, Arctic Raspberry, Dwarf Nagoonberry, Arctic Blackberry, Northern Blackberry

Rose—*Rosaceae*   **Shrub**   White   Petals 5

WILD RED RASPBERRY   *Rubus idaeus* L.

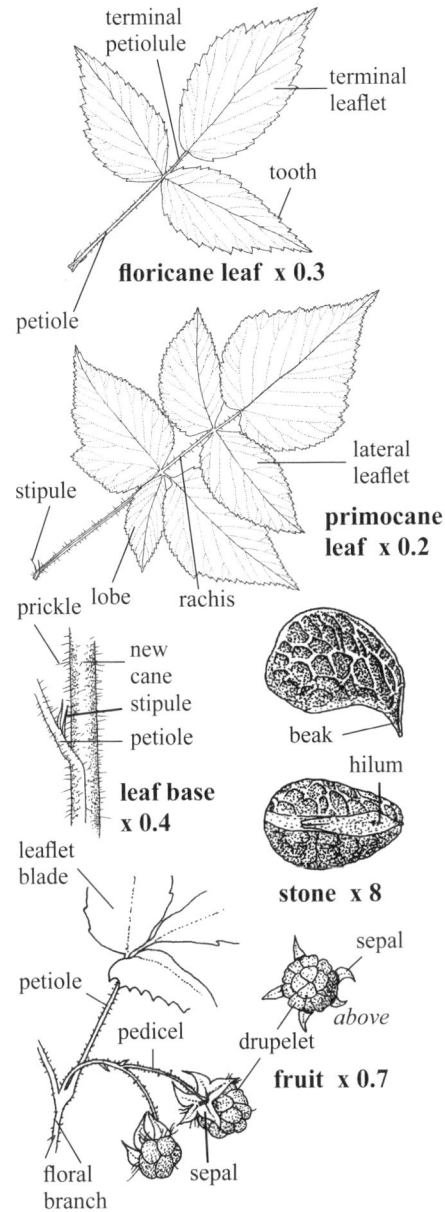

■ **SKETCH** A weakly armed **deciduous shrub** 1–2 (–3) m tall or long with **rootstocks** and **stolons**; in thickets, on wooded hillsides, streambanks, ravines, rocky sites, and clearings in coniferous forests. **w1**

■ **FLOWERS** White, blooming May–August; **inflorescence** a raceme with two to five flowers at the ends of new leafy, alternate branchlets, arising from leaf axils; **pedicels** 6–25 mm long, hairs sometimes glandular, spreading; **flowers** perfect, often drooping, 1–1.5 cm wide by 8–10 mm high; **sepals** five, pointed, reflexed, green, 6–7 mm long by 3–4.2 mm wide, hairy on both sides, glandular hairy below, persistent; **petals** five, quickly deciduous, erect, 5–6 mm long by *c.* 3 mm wide, tapered at base, between the sepals, glabrous, blunt; **stamens** numerous, slanting inward, *c.* 4 mm long; **filaments** white, glabrous; **anthers** pale yellow, *c.* 0.5 mm long, attached in the middle; **pistils** clustered, white, 4–4.2 mm long by 2.5–3 mm wide, about equal to the stamens; **ovary** hairy, *c.* 0.8 mm long; **style** filiform, *c.* 3 mm long; **stigma** obscure, *c.* 0.2 mm long. **FRUIT a drupelet**, 1-stoned, in aggregates 8–18 mm wide and long, reddish, each drupelet 4–5 mm long by 3–4.3 mm wide by 3–4 mm thick, soft, minutely hairy, with a slight groove to the outside; **stones** 1-seeded, tannish pink, 2.2–2.6 mm long by 1.5–2 mm wide by 1–1.6 mm thick, reticulate, beaked.

■ **LEAVES** Alternate, compound, trifoliate on reproductive shoots (floricanes) or with five leaflets on new, vegetative shoots (primocanes), margins toothed to doubly so; **floricane blades** 3.5–17 cm long by 3.5–19 cm wide, glabrous above, closely hairy and lighter green below; **leaflets** three, pointed, 3–13 cm long by 1.5–8 cm wide, terminal leaflet the largest; **petiolules** hairy, 0–42 mm long; **primocane leaflets** five, light green, tomentose below; **petioles** glandular-hairy, 1.3–8 cm long; **stipules** filiform, hairy, 3–10 mm long, often caducous.

■ **STEM** A **cane**; first year vegetative (**primocane**) and second year floral (**floricane**), erect to arched, round, reddish brown, several in a cluster, glandular or not, bark peeling; 5–8 mm thick near the base; **prickles** weak, not enlarged at base, simple, straight to slightly descending, 1–4 mm long, attached to and removed with the bark on older stems.

■ **SYN.** *Rubus strigosus* Michx. = *Rubus idaeus* ssp. *strigosus* (Michx.) Focke.

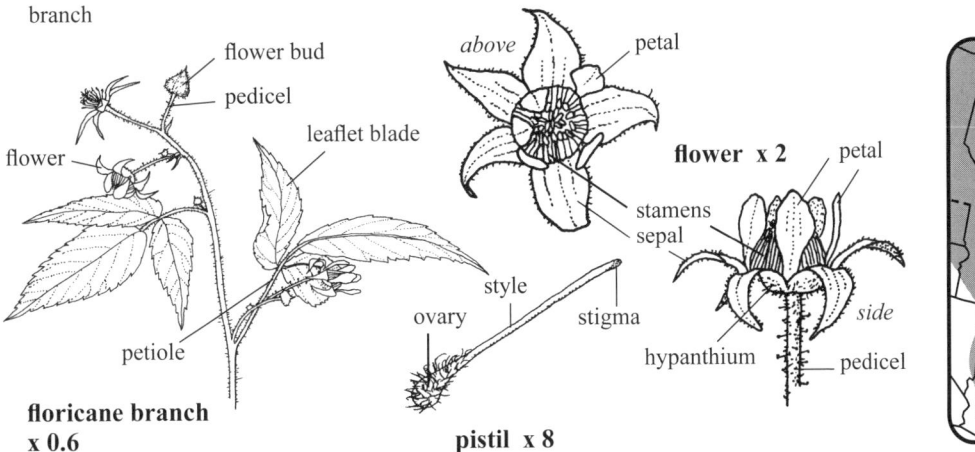

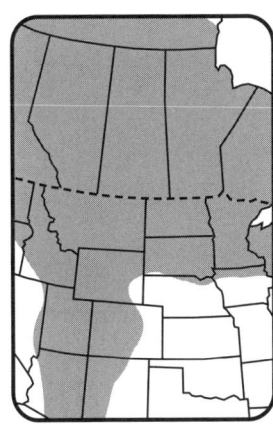

Red Raspberry, American Red Raspberry, Common Red Raspberry

Rose—*Rosaceae* **Wildflower** White, rarely pink Petals 5

Dewberry *Rubus pubescens* Raf.

■ **SKETCH** An unarmed, perennial **trailing herb** 10–50 cm tall from slightly hairy, reddish **stolons** 1–2 mm thick and to *c.* 100 cm long, rooting at nodes 2–15+ cm apart; in moist woods and bluffs, near stream banks and bogs. w1

■ **FLOWERS** White, rarely pink, blooming May–July; **inflorescence** a terminal cluster of 1–3 (–7) flowers; **pedicels** 1.5–2.5 cm long at anthesis, to *c.* 4.8 cm long with fruit, hairy; **flowers** perfect, erect, 8–12 mm wide by 12–13 mm long, in groups of two or three; **sepals** five (6), softly hairy, often glandular, 3–7 mm long by *c.* 2 mm wide, spreading then reflexed, persistent; **petals** five, white to light pink, usually erect, 4–10 mm long by 2–5 mm wide, veined, tapered at base; **stamens** numerous, 2–3 mm long before anthesis; **filaments** white, apices narrow and reflexed; **anthers** pale yellow, *c.* 0.6 mm long; **hypanthium** hairy, 3–4 mm wide by *c.* 2 mm tall; **pistils** *c.* 2 mm long; **ovary** green, *c.* 0.9 mm long, a few apical hairs; **stigma** rough, grooved, pale green. **FRUIT** a drupe, 4–10 mm long by 5–14 mm wide, reddish purple, 1–3 per stem; **druplets** 1-stoned, 2–22 per drupe, slowly separating from the receptacle, smooth, shiny, each 5–6 mm long by 4–6 mm wide with a narrow faint central groove; **stones** 1-seeded, pinkish white, 2.6–3 mm long by 1.8–2 mm wide by 1–1.2 mm thick, smooth or reticulate, ridges white, low and irregular, a white central ridge along one side, beak short, rounded at the other end.

■ **LEAVES** Alternate, compound, 2–5 on erect stems; **leaf blades** 5–10 cm long and 5–12 cm wide; **leaflets** three (5), toothed; **blades** dull, slightly lighter green below, 2–7.5 cm long by 2–4.5 cm wide, margins hairy, hairs ascending, slightly hairy above, hairy below on blade and along the raised veins; **petioles** hairy, 2.5–7.5 cm long, reddish at fruiting time, hairs deflexed or arched downward and of various lengths; **petiolules** 0–7 mm long, hairy, the terminal one the longest; **stipules** entire, 5–12 mm long by 2–8 mm wide, margins hairy with a few scattered hairs on the blades, sometimes lobed.

■ **STEM** Reproductive stems erect from nodes along stolons, hairy; 1.5–2 mm wide near the reddish base.

■ **SYN.** *Cylactis pubescens* (Raf.) W.A. Weber = *Rubus pubescens* var. *pubescens* Raf.

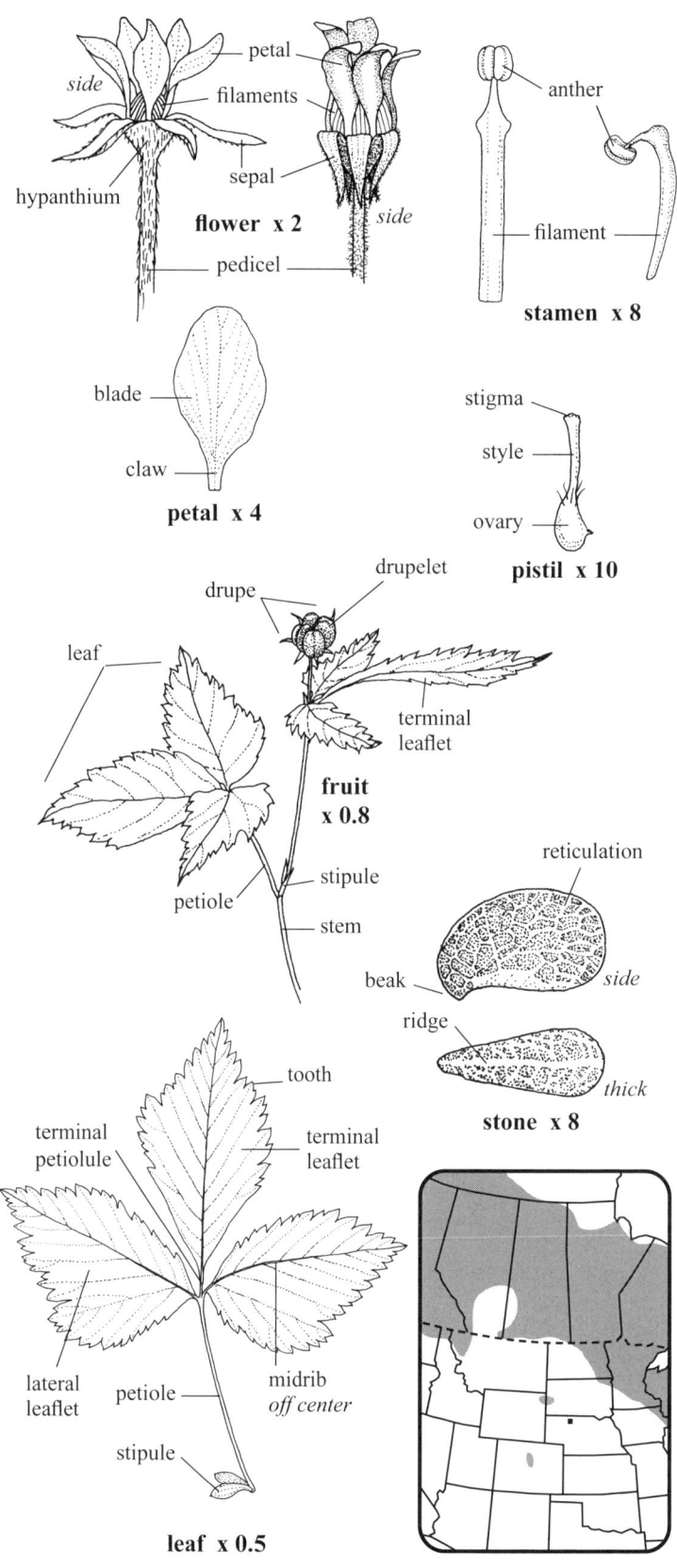

Dwarf Red Raspberry, Trailing Raspberry, Running Raspberry, Creeping Blackberry, Dwarf Blackberry

Three-toothed Cinquefoil *Sibbaldiopsis tridentata* (Ait.) Rydb.

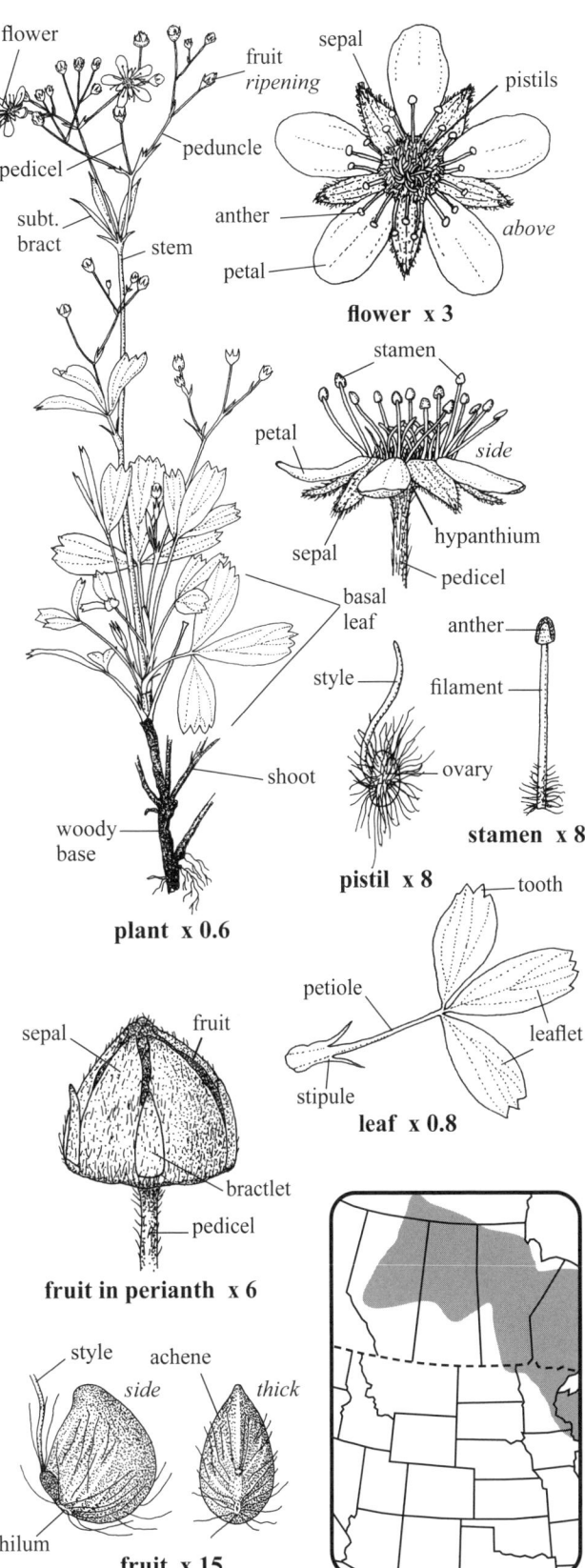

flower x 3

stamen x 8

pistil x 8

plant x 0.6

leaf x 0.8

fruit in perianth x 6

fruit x 15

■ **SKETCH** A hairy, tufted **evergreen herb** 5–30 cm tall with dark brown **rhizomes** to *c.* 10 cm long by 2–3 mm thick; in dry sandy sites, on rocky outcrops, prairie hillsides, cliffs, ledges and lakeshores.

■ **FLOWERS** White, blooming June–August; **inflorescence** a cyme, stiff, 1–4 per stem, together 3–8 cm long by 2–4 cm wide; **floral branches** 3–7 cm long, ascending; **peduncles** hairy, 1–3 cm long, ascending, stiff; **pedicels** stiff, green to reddish, ascending, 7–15 mm long, hairs white; **subtending bracts** (of pedicels) green, ascending, 2–3 mm long, pointed, entire, hairy; **flowers** white, 10–15 mm wide by 4–5 mm tall; **sepals** five, green, 3–4 mm long by 1.5–2 mm wide (including hairs), veined above, pointed, spreading, persistent, hairy below and on margins, marginal hairs tangled and reddish near tip; **bractlets** darker green than sepals, pointed, spreading, slightly hairy below and on margins, tips reddish, *c.* 2 mm long by *c.* 0.6 mm wide, between the sepals; **petals** five, white, blunt, spreading, 5–7 mm long by *c.* 3 mm wide, tapered slightly at base, attached at edge of hypanthium; **hypanthium** 3–4 mm wide, hairs ascending; **stamens** 20, three opposite each sepals, one opposite each petal, 3–3.5 mm long, ascending from near edge of hypanthium; **filaments** white, hairy near base; **anthers** *c.* 0.6 mm long and wide, white and orange; **pistils** *c.* 20, each 2.5–3 mm long; **ovary** green, very hairy, *c.* 1 mm long; **style** white, lateral, curved, hairless; **stigma** obscure; **fruiting heads** brown, 3.5–4 mm tall by *c.* 4 mm wide, erect. **FRUIT an achene**, 1-seeded, 8–18 per flower, 1–1.3 mm long by 1–1.1 mm wide by 0.7–0.8 mm thick, medium brown, smooth or with low ridges, persistent style attached to one side, long white hairs by hilum.

■ **LEAVES** Basal and stem, alternate, evergreen, compound, ascending, dark green, shiny and hairless above, pale green, dull and hairy below, hairs white and appressed, margins slightly hairy; **leaf blades** 1.3–2.5 cm long by 2–4 cm wide, turning red in the fall; **leaflets** three, firm, each 9–25 mm long by 4–10 mm wide, terminal leaflet usually the longest, 2–5 teeth at the blunt apices; **petioles** 1–4.5 cm long, slightly hairy; **petiolules** 0.5–1 mm long, terminal one the longest; **stipules** pale green, hairy, united to base of petioles as a sheath, TL 9–14 mm, the upper parts spreading and pointed, 4–6 mm long; **stem leaves** few, subtending floral branches, reduced and sessile above.

■ **STEM** Woody at base, persistent, stiff, branched, round, solid, white hairs appressed and ascending, new growth green and herbaceous above, base covered with sheaths of several leaves; 1–1.5 mm wide near base.

■ **SYN.** *Potentilla tridentata* Ait.

Shrubby Fivefingers

Rose—*Rosaceae* Shrub White Petals 5

Narrow-leaved Meadowsweet *Spiraea alba* Du Roi

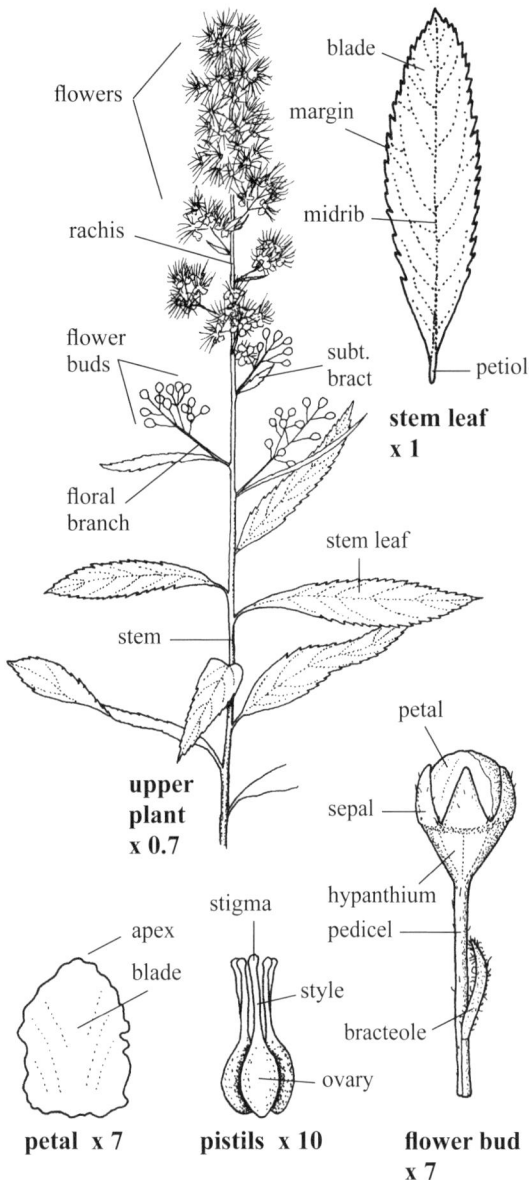

■ **SKETCH** A **deciduous shrub** 40–200 cm tall, singly or in colonies with a **woody root** 5–10 mm thick and to *c.* 10⁺ cm long; in moist open woodlands, on rocky outcrops, along shores, and in moist prairies and wetlands. **p2**

■ **FLOWERS** White, blooming June–August; **inflorescence** a panicle, 1.5–20 cm long by 2–7 cm wide with a fuzzy look when in bloom; **floral branches** ascending, reddish to yellowish brown, striated, glabrous to slightly hairy, long ones leafy, 1.5–30 cm long, reduced above, each subtended by a bract; **pedicels** 2–9 mm long, glabrous; **bracteoles** linear, margins hairy, *c.* 2 mm long, erect near middle of pedicel; **flower buds** *c.* 2 mm wide by *c.* 2.3 mm long; **flowers** perfect, fragrant, glabrous, 5–7 mm wide by 5–6 mm long, blooming sequence top to bottom; **rachis** light green; **hypanthium** 10-nerved, green and 1–1.8 mm long by 1.5–2 mm wide; **sepals** five, triangular, spreading, 1–1.5 mm long by 0.8–1 mm wide, slightly hairy on outside, more hairy inside with cobwebby hairs at apex; **petals** five, white, 2.3–3.5 mm long by 2.2–2.5 mm wide, spreading to quickly reflexed, flat, margins erose, soon deciduous; **stamens** 25–50, exserted to *c.* 5 mm, arising from outside of nectar ring which is tan and *c.* 2 mm wide; **filaments** bone white, 2–4 mm long, involute to spreading; **anthers** *c.* 0.5 mm long and wide; **pistils** often five, *c.* 2.2 mm long; **ovary** glabrous, *c.* 1 mm long; **styles** erect, *c.* 1.2 mm long, filiform; **stigma** slightly expanded. **FRUIT a follicle** (five), glabrous, 3-sided, 3–3.7 mm long (including beak) by *c.* 1 mm wide by *c.* 0.8 mm thick, opening lengthways along inner (ventral) suture and at apex of the outer suture; **seeds** tan, one or two per follicle, 2–2.3 mm long by 0.4–0.5 mm wide by *c.* 0.4 mm thick, 3-sided, glabrous, dull, pointed at base.

■ **LEAVES** Alternate, simple, spreading; **blades** glabrous or hairy below on veins, 1–7 cm long by 2–21 mm wide, tips sometimes with a partial twist, margins serrate and finely hairy; **petioles** 2–5 mm long, glabrous to slightly hairy; **stipules** absent.

■ **STEM** Woody, erect, yellowish brown, smooth, slightly angular and glaucous; 5–8 mm wide near the base.

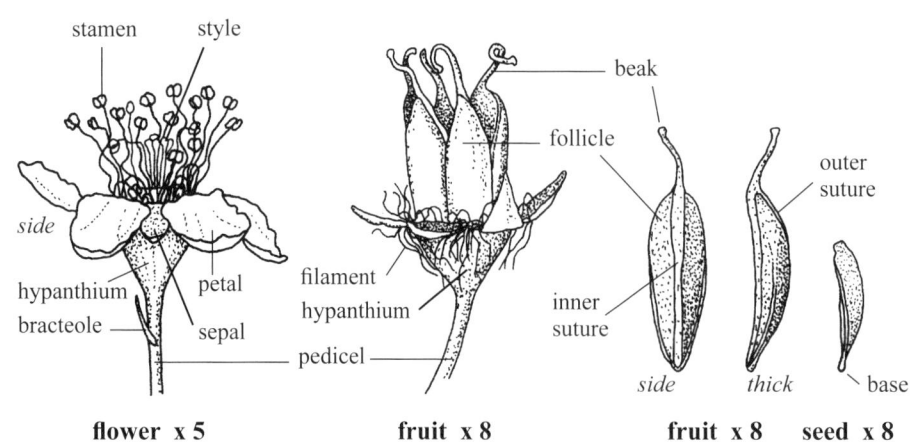

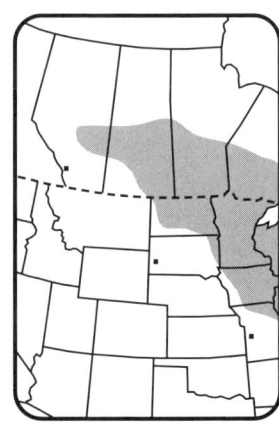

470 Meadowsweet, White Meadowsweet, Narrowleaf Spirea

Madder—*Rubiaceae* **Wildflower   White   Corolla 4-lobed**

Northern Bedstraw *Galium boreale* L.

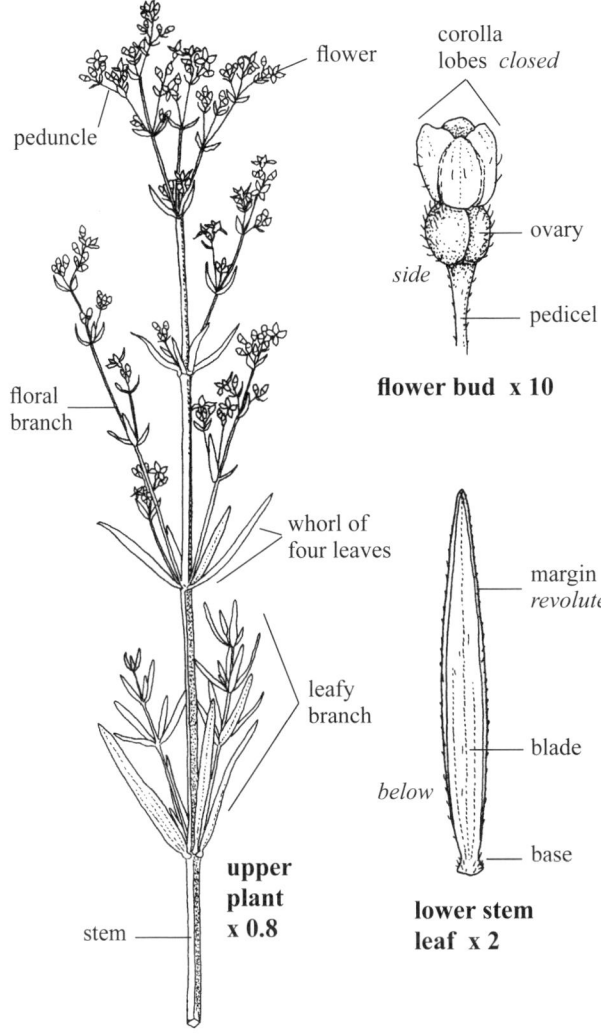

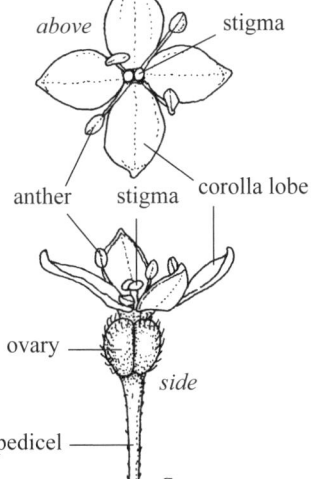

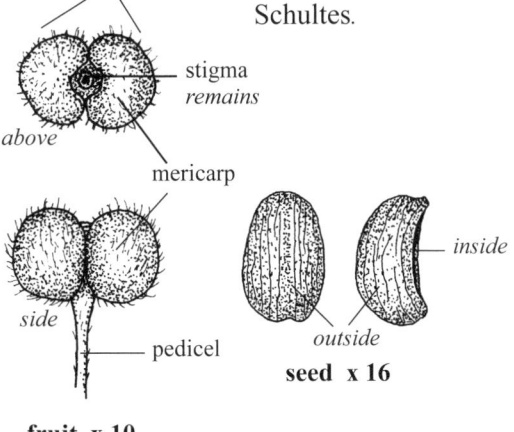

- **SKETCH** A **perennial herb** 20–90 cm tall from a reddish brown **rootstock** 1–1.5 mm thick; in moist prairies, along shores, rocky hillsides, roadsides, in valleys and open woodlands. **p1**
- **FLOWERS** White, blooming in May–August; **inflorescence** a cymose panicle, terminal and axillary, leafy, TL 5–40 cm by 2–12 cm wide; **floral branches** 2–26 cm long, reduced above; **peduncles** 3–6 mm long; **pedicels** green, 1–2.5 mm long, erect, hairs scattered; **subtending bracts** (of pedicels) green, leaflike, pointed, scabrous, 2–3 mm long by *c.* 1 mm wide; **flowers** perfect, fragrant, 3–6 mm wide by *c.* 2 mm tall, numerous; **calyx tube** tiny, ovoid, teeth absent; **corolla** 4-lobed, these *c.* 2 mm long by *c.* 1 mm wide, pointed, thick, a few scattered hairs below; **stamens** four, between the corolla lobes; **filaments** white, *c.* 0.6 mm long, attached to base of corolla; **anthers** yellow, *c.* 0.5 mm long; **ovary** 2-celled, light green, cleft, hairs stiff, ascending; **styles** two, included, *c.* 0.6 mm long, light green; **stigmas** slightly enlarged, round. **FRUIT** a **schizocarp**, 2-seeded, 1.3–1.5 mm tall by 1.8–2 mm wide by 1–1.2 mm thick, green ripening to dark brown, splitting in the middle, hairy to glabrous; **seeds** dark brown, 1–1.2 mm long by *c.* 0.6 mm thick by *c.* 0.7 mm wide, one per mericarp, concave on inside, convex on outside, long thin ribs with fine transverse lines between them.
- **LEAVES** Whorled, in fours, simple, entire, linear, sessile, ascending to spreading, some with leafy branches in axils; **blades** 3-veined, 0.5–6 cm long by 1–10 mm wide, firm, shiny to dull above, margins revolute, usually scabrous, base hairy.
- **STEM** Erect, stiff, green, branched above, square, each corner with a slightly raised ridge, scabrous from very short reflexed to spreading hairs; 1–3 mm wide near the base.
- **SYN.** *Galium septentrionale* Roemer & J.A. Schultes.

Madder—*Rubiaceae* **Wildflower   White   Corolla 3-lobed**

SMALL BEDSTRAW   *Galium trifidum* L.

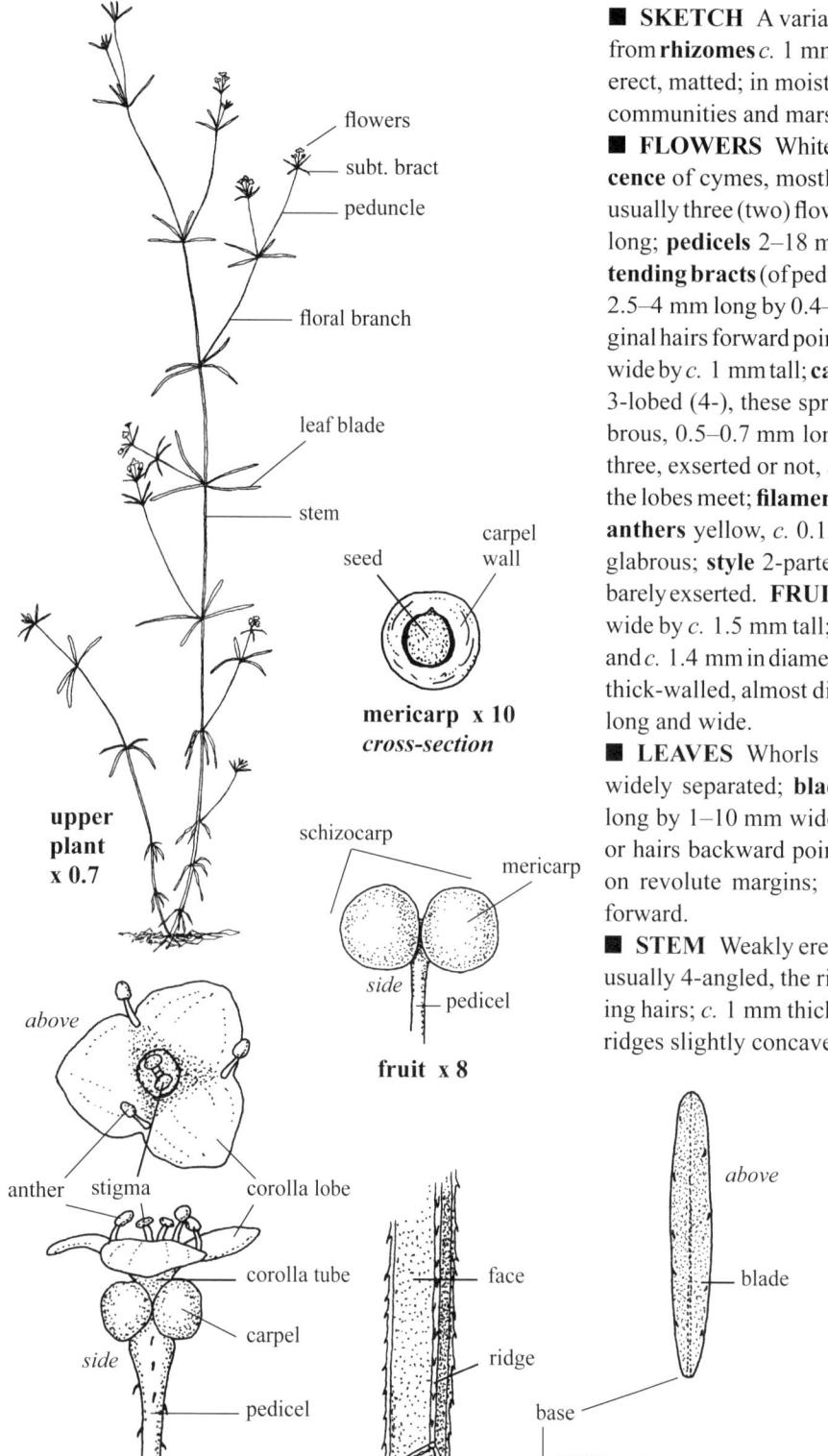

■ **SKETCH** A variable **perennial herb** 10–60 cm tall from **rhizomes** *c.* 1 mm thick and to *c.* 7⁺ cm long, weakly erect, matted; in moist woodlands, prairies, bogs, alpine communities and marshes.

■ **FLOWERS** White, blooming June–August; **inflorescence** of cymes, mostly axillary, at the end of branches, usually three (two) flowers; **branches** ascending, 1–5 cm long; **pedicels** 2–18 mm long, stiff hairs reflexed; **subtending bracts** (of pedicels) leaflike, three or four, sessile, 2.5–4 mm long by 0.4–1 mm wide, glabrous above, marginal hairs forward pointing; **flowers** perfect, 1.4–2.3 mm wide by *c.* 1 mm tall; **calyx** minute, teeth obsolete; **corolla** 3-lobed (4-), these spreading to slightly ascending, glabrous, 0.5–0.7 mm long by 0.7–0.8 mm wide; **stamens** three, exserted or not, attached to the corolla tube where the lobes meet; **filaments** *c.* 0.3 mm long, arched inward; **anthers** yellow, *c.* 0.1 mm long; **carpels** round, paired, glabrous; **style** 2-parted; **stigmas** capitate, pale yellow, barely exserted.   **FRUIT a schizocarp**, 2-seeded, *c.* 3 mm wide by *c.* 1.5 mm tall; **mericarps** paired, each 1-seeded and *c.* 1.4 mm in diameter, round, smooth, brown to black, thick-walled, almost distinct when ripe; **seeds** *c.* 0.7 mm long and wide.

■ **LEAVES** Whorls of four (5 or 6), simple, entire, widely separated; **blades** sessile, spreading, 5–26 mm long by 1–10 mm wide, 1-nerved, blunt tipped, smooth or hairs backward pointing below on raised midrib and on revolute margins; **hairs** above fewer and pointing forward.

■ **STEM** Weakly erect to decumbent, green, branched, usually 4-angled, the ridges with short downward pointing hairs; *c.* 1 mm thick near the base; **faces** between the ridges slightly concave, glabrous and green.

472   Threepetal Bedstraw

Madder—*Rubiaceae* Wildflower Greenish white Corolla 4-lobed

Sweet-scented Bedstraw *Galium triflorum* Michx.

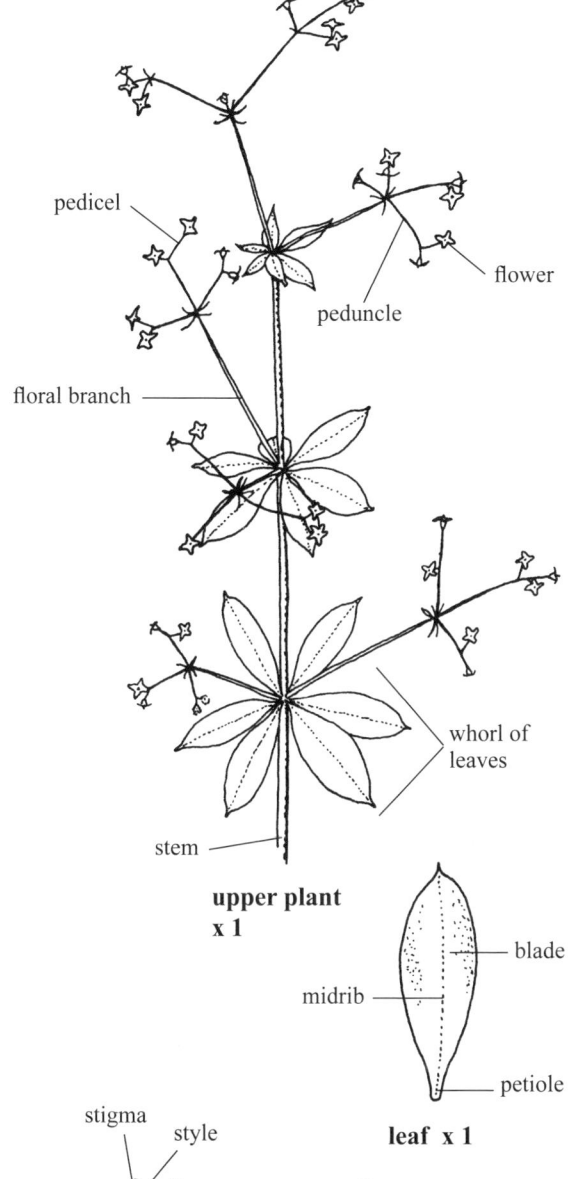

upper plant x 1

leaf x 1

■ **SKETCH** A **perennial herb** 20–120 cm long, usually scrambling, with slender **rhizomes**; in moist woodland edges along streams and rivers.

■ **FLOWERS** Greenish white, blooming May–September; **inflorescence** paniculate, axillary and terminal; **floral branches** usually one or two ascending from each leafy node; **peduncles** usually divided into two or three apical pedicels, ridged, mostly glabrous, with a few hairs near the base; **pedicels** mostly 1–15 mm long, glabrous; **subtending bracts** (of pedicels) one, *c.* 1 mm long, entire, pointed; **flowers** perfect, in twos or threes, each 2.2–3.3 mm wide by *c.* 2 mm long (including the ovary); **sepals** absent; **corolla** usually 4-lobed, each lobe 1–1.5 mm long by *c.* 0.6 mm wide, with a lighter green margin, pointed, thickish, spreading; **stamens** four, horizontal, between the corolla lobes; **filaments** attached to the corolla; **ovary** 2-celled, covered with white bristles, each with an apical hook; **styles** two, exserted; **stigmas** swollen. **FRUIT a schizocarp**, 1.5–2.2 mm in diameter by *c.* 1.5 mm tall (not including bristles); **mericarps** two per schizocarp, each 1-seeded, covered with hooked, hyaline bristles; **seeds** blackish brown, *c.* 1 mm long by *c.* 0.5 mm wide, glabrous.

■ **LEAVES** Whorls of six (four or five), simple, entire; **blades** flat, 0.5–6 cm long by 1–2 cm wide, with a short apical point, reduced above, usually glabrous above but with ascending hairs along the upper margins, glabrous below except for a few descending hairs along the raised midrib, lateral veins faint; **petioles** 0–2 mm long.

■ **STEM** Weak, leaning to scrambling, simple or branched, angled, the four ridges with deflexed hairs, and white spreading hairs near the leafy nodes, or hairless; 1–2 mm wide near the base.

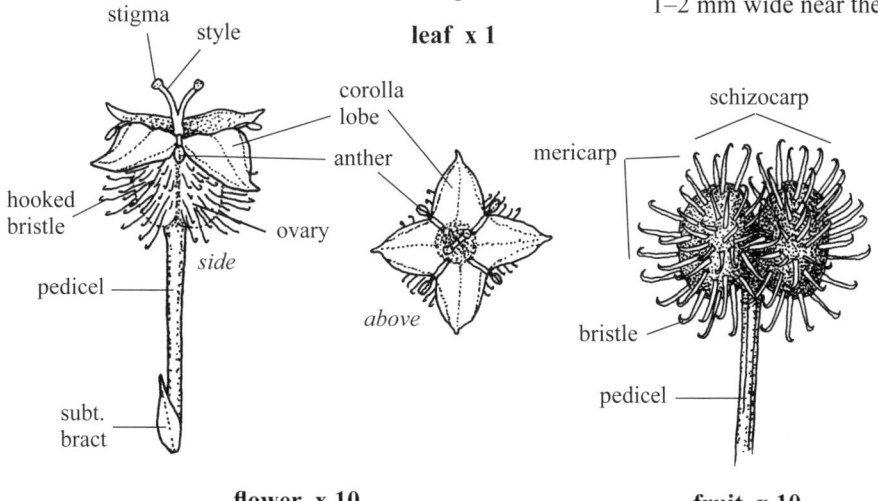

flower x 10   fruit x 10

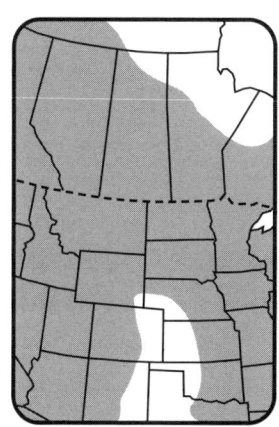

Fragrant Bedstraw   473

Madder—*Rubiaceae* **Wildflower** Whitish pink to pale purple Corolla tubular, 4-lobed

LONG-LEAVED BLUETS *Houstonia longifolia* Gaertn.

■ **SKETCH** A tufted **perennial herb** 9–25 cm tall from a reddish brown **rhizome** *c.* 1 mm thick; often on rocky outcrops, in forests, sandy prairies, open sandy woods and dunes.

■ **FLOWERS** Whitish pink to pale purple, blooming May–September; **inflorescence** a cyme, loose to crowded; **peduncles** 3–15 mm long, ascending, glabrous; **pedicels** 1–10 mm long, glabrous, ascending; **flowers** perfect, 7–8 mm wide by 6–9 mm long; **hypanthium** (ovary) green, glabrous, *c.* 1 mm long and wide; **calyx** tubular, 4-lobed, green, glabrous, lobes erect, pointed, 1.3–2.5 mm long by 0.7–1.2 mm wide, extending above the fruit; **corolla** tubular, 4-lobed, the lobes 2–3 mm long by 1.5–2 mm wide, pointed, minutely hairy, tube hairy inside, glabrous outside, pinkish at base of lobes and inside the tube, slightly veined; **stamens** four, situated at the top of the corolla tube between the corolla lobes; **filaments** *c.* 0.5 mm long, white, attached near center of the anther; **anthers** white, 1–1.2 mm long; **pistil** one, included; **ovary** (hypanthium) green, *c.* 1 mm long and wide; **style** one, filiform, erect, *c.* 3 mm long, glabrous; **stigma** 2-lobed, these *c.* 0.4 mm long, hairy. **FRUIT a capsule**, reddish green and turning dark brown when fully ripe, glabrous, slightly shiny, cleft in the middle, 2.5–3 mm long by 2–2.6 mm wide by *c.* 2 mm thick, opening across apex; **seeds** *c.* 25 per capsule, black, slightly shiny, of various shapes, 0.9–1.2 mm long by 0.6–0.8 mm wide by *c.* 0.2 mm thick, papillose (microscopic), one side convex, the opposite side concave with a low central ridge.

■ **LEAVES** Basal and stem, opposite, simple, entire, pointed, sessile, glabrous, ascending to spreading, pairs widely spaced; **basal leaves** usually absent by anthesis; **stem leaves** 6–30 mm long by 2–5 mm wide, reduced below, one-veined, dull, margins revolute, some ciliate; **stipules** paired, triangular, whitish to purplish, apices pointed and brown without a bristle, 1–2 mm long by 1–1.5 mm wide, margins entire or toothed, turning pink.

■ **STEM** Erect, simple to branched above, glabrous, squarish, 4-ridged, single or clustered, green above, pinkish purple near the 0.5–1 mm thick base; **nodes** glabrous or with short hairs.

■ **SYN.** *Hedyotis longifolia* (Gaertn.) Hook.

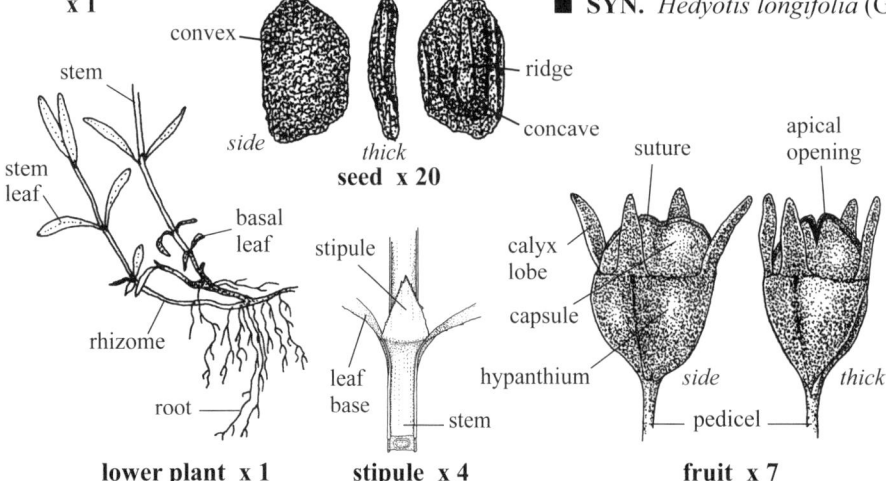
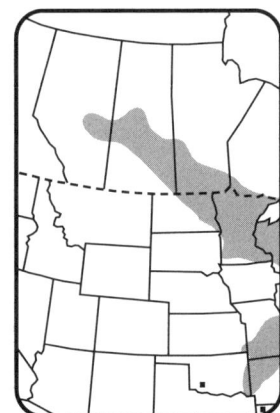

Slender-leaved Bluet, Longleaf Summer Bluet, Long-leaved Houstonia

Family Characteristics

Willow—Salicaceae — Subtending bract only

SKETCH Dicotyledonous **shrubs** or **trees**, deciduous. In northern and mountainous areas, often wet, but not aquatic. Only 2 genera in North America, *Salix* and *Populus*. Worldwide about 400 species.

FLOWERS Unisexual. Some flowers reach anthesis (bloom) before the leaves are fully expanded (mature). **Inflorescence** dioecious, an ascending (*Salix*), or hanging catkin or ament (*Populus*). **Perianth** (sepals and petals) absent. **Flowers** solitary, each subtended by a small entire or toothed bract (scale). **Stamens** 2 or more. **Pistil** 1. **Ovary** 1, with 1 locule; ovules numerous. **Style** 1 or obscure. **Stigmas** 2–4, often 2-lobed.

FRUIT A **capsule** with 2 or 4 valves, opening from apex, glabrous to hairy. **Seeds** many, each with long silky white hairs, carried aloft on spring winds.

LEAVES Alternate, simple, pointed, entire to toothed. **Stipules** present, often large on new summer growth.

STEM Shrubs woody, erect to leaning, green, gray to brown to reddish. **Tree trunks** to 1 m wide and to about 30 m tall, deeply furrowed. *Salix* (willow) buds pointed, pressed against twigs, 1-scaled. *Populus* (aspen) buds pointed, often aiming away from twigs, some sticky, several-scaled.

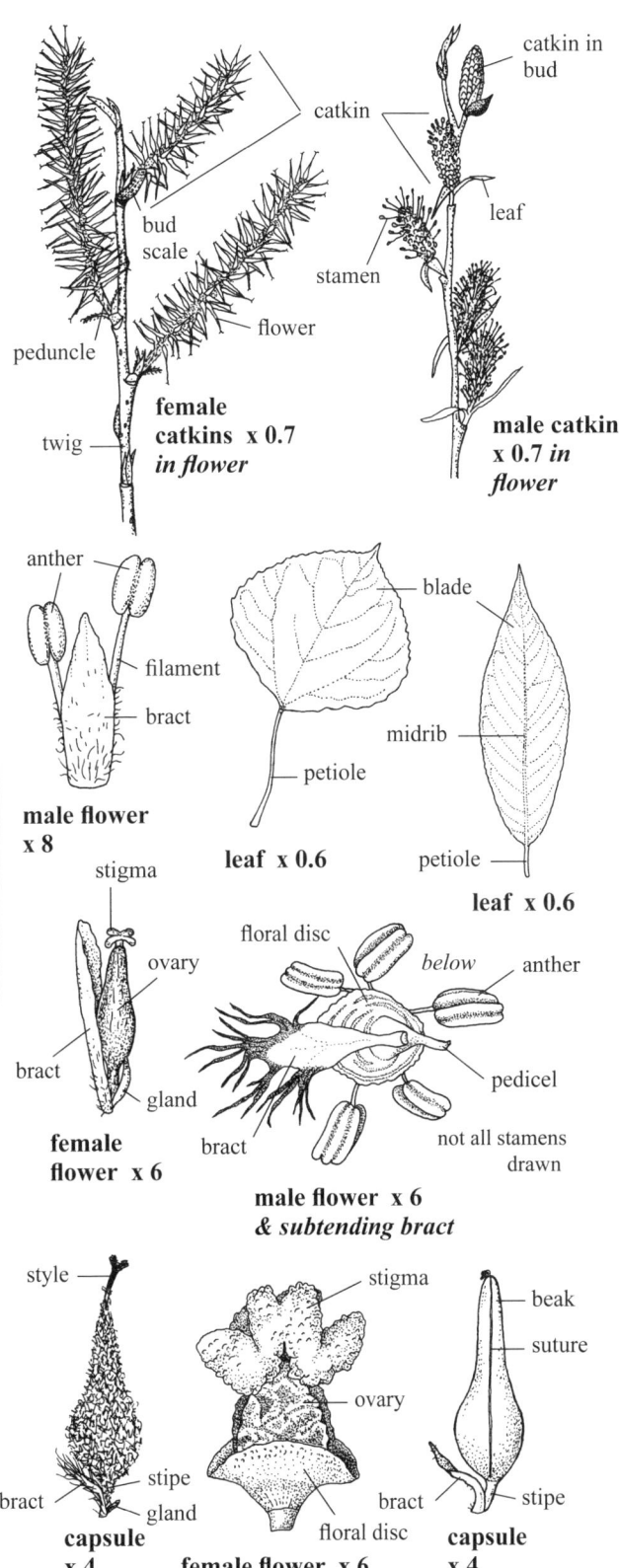

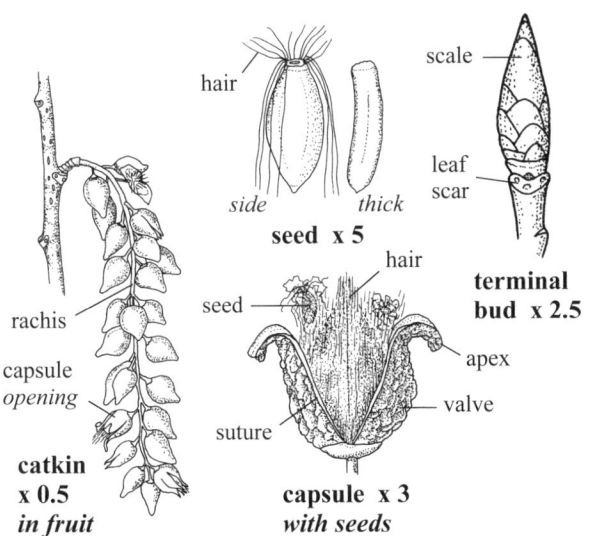

475

Willow—*Salicaceae* **Tree** Green and red Catkins

Balsam Poplar *Populus balsamifera* L.

■ **SKETCH** A large, variable **deciduous tree** 5–30 m tall in colonies; in forest and parklands, moist areas along rivers, streams and floodplains; **dioecious**. w1

■ **FLOWERS** Green and red (pink), blooming March–June; **inflorescence** of catkins, unisexual, numerous on upper branches, hanging and active before the leaves appear; <u>male</u> **catkins** reddish, loose; **rachis** with short white hairs; **pedicels** *c.* 0.8 mm long, pale yellow; **subtending bracts** (of pedicels) brown, *c.* 5 mm long and wide with dark filiform segments; **floral disc** pale green, 3–4 mm wide and long; **stamens** 12–60, exserted; **filaments** white, filiform, *c.* 1 mm long, attached on floral disc; **anthers** reddish, 1.3–1.5 mm long; <u>female</u> **catkins** green and orange, drooping, 6–15 cm long by 13–18 mm wide at anthesis; **pedicels** green, *c.* 0.5 (to 3 in fruit) mm long; **subtending bracts** (of pedicels) as above; **flowers** 4–5 mm wide by 5–6 mm long; **floral disc** green, glabrous, 3–4 mm wide by *c.* 2 mm tall by *c.* 3 mm thick; **ovary** green, ridged and furrowed; **stigmas** two, lobed, pinkish orange, angular, turning dark brown and persisting on apices of fruit; **fruiting catkins** 8–18 cm long by 20–22 mm wide from base of leaf clusters. **FRUIT a capsule**, (3-) 2-valved, 4–8 mm long by 4–5 mm wide, glabrous to hairy, opening from apex; **rachis** green, *c.* 2 mm thick, hairs white and spreading; **peduncle** 3–7 mm long, naked; **seeds** 26–44 per capsule, 1.6–2 mm long by 0.8–1 mm wide by *c.* 0.7 mm thick, tan, rounded at one end, silky white hairs *c.* 5 mm long at truncate end, these deciduous as a unit.

■ **LEAVES** In alternate clusters, simple, finely toothed to entire, fragrant, sticky when emerging; **spring blades** 6–12 (–15 in summer) cm long by 4–8 (–13 in summer) cm wide, slightly shiny green, lighter green to rusty and sticky below, margins ciliate (microscopic), midrib finely hairy below; **petioles** finely hairy, reddish green, 1–5 cm long; **stipules** brown, appressed, 7–10 mm long by 1–1.5 mm wide, entire, margins revolute and finely hairy near pointed apices, deciduous as the leaves expand.

■ **TRUNK** Single, branched, pale gray with horizontal streaks, becoming vertically furrowed near the base on older trees, ridges very rough; **dbh** 30–100+ cm; **branches** alternate, ascending above to descending below; **new twigs** alternate, smooth, shiny, bright reddish brown, grayish brown by the third year, glabrous to hairy; **terminal buds** 10–20 mm long; **side buds** 10–13 mm long by 3–4 mm wide, sticky, shiny, with several scales.

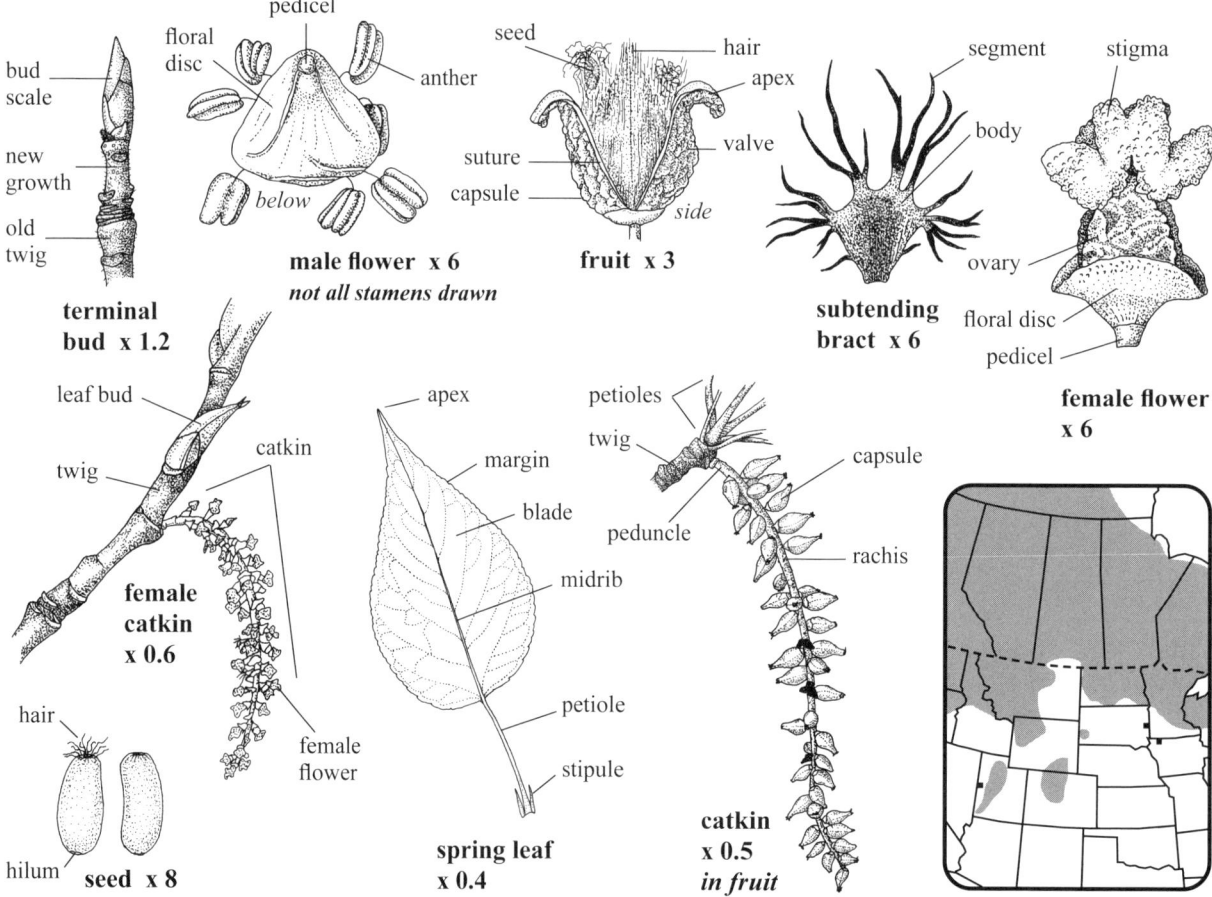

476   Black Poplar, Eastern Balsam Poplar

Willow—*Salicaceae*   Tree   Green and red   Catkins

Cottonwood   *Populus deltoides* Bartr. ex Marsh.

■ **SKETCH** A variable **deciduous tree** 15–30⁺ m tall; in moist sites along pond and lake edges, stream bottomlands and riverbanks; **dioecious. w2**

■ **FLOWERS** Green and red, blooming February–June; **inflorescence** of catkins, unisexual, numerous, mostly in the upper half of a tree, hanging before the leaves fully open; **male catkins** reddish, loose, 4–14 cm long by 1–2 cm wide, drooping; **rachis** glabrous, pale green, *c.* 1.5 mm thick (midway); **pedicels** 1–6 mm long, pale green, glabrous; **subtending bracts** (of pedicels) one, 5–8 mm long by 4–6 mm wide, tapered at tan base, blade divides into three or four main lobes, hairless and dark brown; **male florets** 40–70 per catkin; **floral disc** flat, *c.* 2.5 mm wide by *c.* 3 mm long, pale yellowish green, glabrous; **stamens** 35–80; **filaments** free, 0.5–1.5 mm long; **anthers** medium red, 1.5–1.9 mm long by *c.* 1 mm wide; **female catkins** 10–22 cm long by 2–3 cm wide in fruit; **pedicels** and **subtending bracts** as above; **female florets** glabrous, 20–40 per catkin, each *c.* 5 mm long and wide, green; **rachis** glabrous, *c.* 1.5 mm thick; **floral disc** green, glabrous, 3–5 mm long by *c.* 3 mm wide by *c.* 2 mm tall, thickish and irregular; **ovary** green, in disc, 3- or 4-valved; **stigmas** yellowish green, 3- or 4-lobed, one lobe *c.* 5 mm long by *c.* 3 mm wide. **FRUIT** a capsule, glabrous, 7–11 mm long by 6–8 mm wide, reddish, 3- or 4-valved, opening from a narrow apex; **seeds** 7–15 per capsule, tan, 3.2–4.2 mm long by 1–1.5 mm wide by 0.8–0.9 mm thick, dull, hairs soft, white to tawny, deciduous, 5–7 mm long.

■ **LEAVES** Alternate, simple, leathery, teeth blunt, glabrous; **blades** shiny above, lighter green below, veins obvious, 4–14 cm long and wide, turning yellowish tan in the fall, often with a few glands at the blade's base; **petioles** 3–13 cm long, glabrous, flattened near the blade.

■ **TRUNK** Erect, crown spreads little, dividing into large branches, usually taller than wide, bark greenish gray, longitudinally furrowed near the base, lighter gray and smooth above; **branches** and **twigs** mostly alternate, ascending; **new twigs** smooth, slightly shiny, light green to orangish tan, slightly zigzagging between the lateral buds; **lenticels** very narrow, 1–2 mm long, whitish, barely raised; **winter buds** tan, alternate, 2- or 3-scaled, 0.8–2.5 cm long by 4–5 mm wide, pointed, fragrant and resinous in the spring; **dbh** 20–150 cm.

■ **SYN.** *Populus sargentii* Dode = *Populus deltoides* ssp. *monilifera* (Ait.) Eckenwalder.

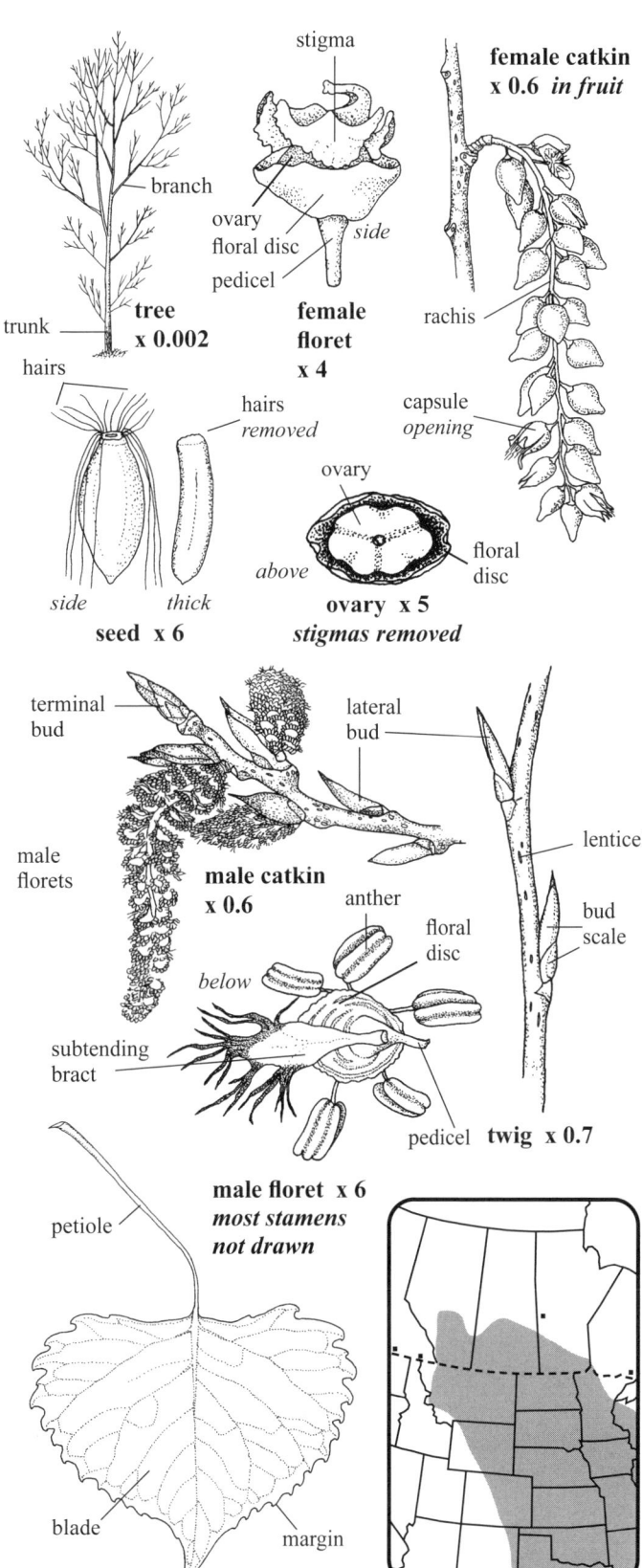

Eastern Cottonwood, Plains Cottonwood, Western Cottonwood

Willow—*Salicaceae* **Tree** Green to reddish purple Catkins

Aspen Poplar *Populus tremuloides* Michx.

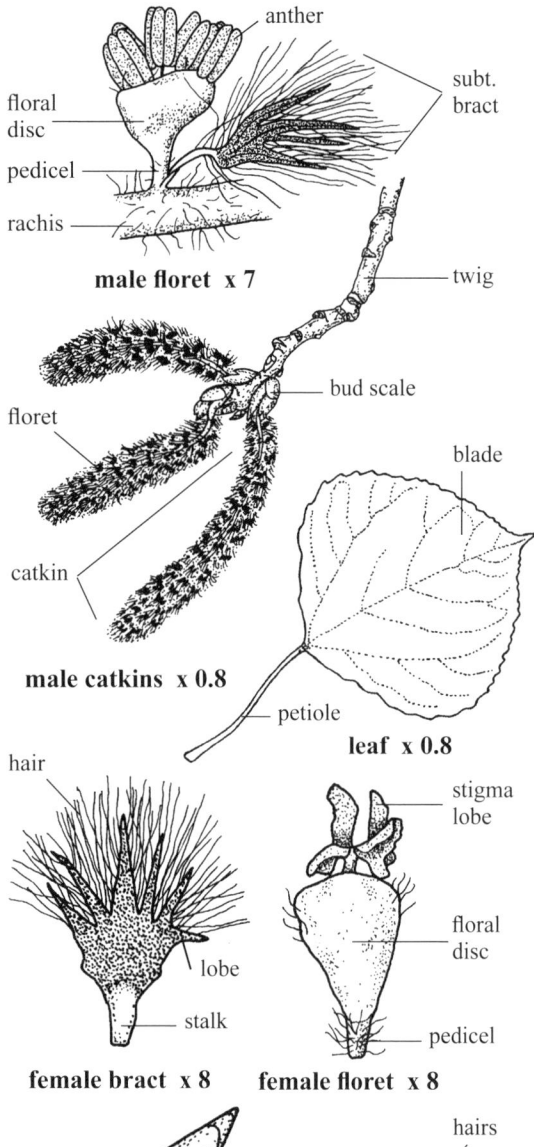

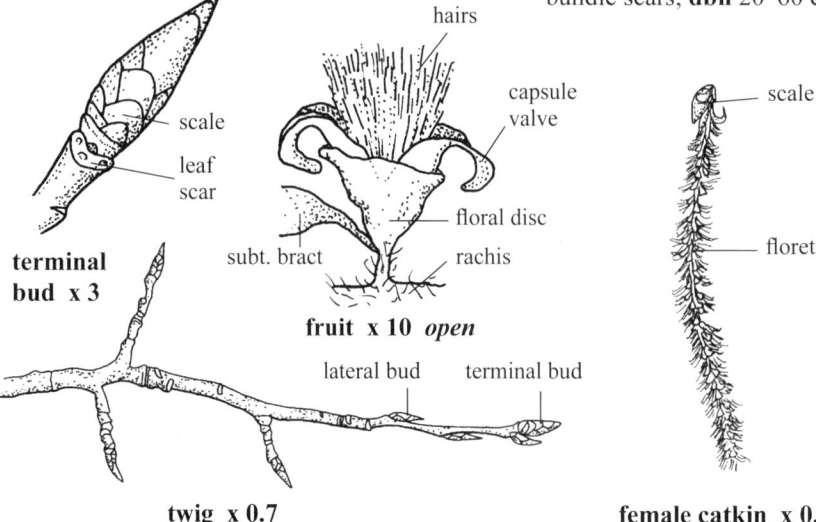

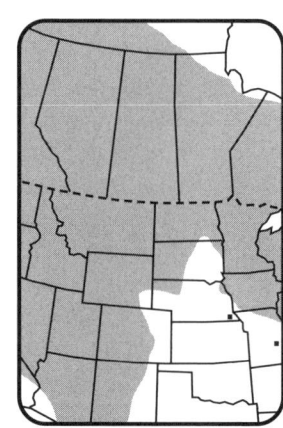

■ **SKETCH** A **deciduous tree** 5–30 m tall, often forming large patches of cloned trees from underground **suckers**; along prairie edges, in parklands, on mountain slopes, around ponds and streams; **dioecious**. w1

■ **FLOWERS** Green, reddish and fuzzy, blooming April–June before the leaves are fully developed; **inflorescence** of catkins, unisexual; **male catkins** 1.5–7 cm long by 8–11 mm wide, drooping; **rachis** *c.* 1 mm wide, hairy; **pedicels** slightly hairy, 0.5–1.5 mm long; **subtending bracts** (of pedicels) 3–4.5 mm long and wide (including the hairs), thin and flat, with three to six pointed lobes bearing marginal white hairs but hairless elsewhere, dark brown except for a light green stalk; <u>male florets</u> numerous, sparsely hairy; **floral disc** truncate, 1.5–2.1 mm across, slightly recessed above; **stamens** 5–12; **filaments** white, filiform, *c.* 1 mm long; **anthers** reddish purple, 1–1.2 mm long, 4-lobed; **female catkins** 4–8 (–11 with fruit) cm long by 7–9 (–12) mm wide, fuzzy, curled at first, becoming long and drooping at anthesis; **rachis** light green, *c.* 1 mm wide and very hairy; **pedicels** 0.5–1 mm long and hairy; **bracts** as above; <u>female florets</u> average *c.* 13 per cm of rachis; **floral disc** lightly hairy, 1.8–2.2 mm wide by 2.2–2.5 mm long by 1–1.2 mm thick; **style** short; **stigmas** reddish purple, irregularly lobed, exserted. **FRUIT** a capsule, 2–5 mm long by 3–3.5 mm wide when open, the two valves splitting open at the apex; **seeds** 5–12 per capsule, beige, 0.2–0.3 mm long, with silky white hairs 0.7–1.6 mm long.

■ **LEAVES** Alternate, simple, finely toothed, glabrous; **blades** 2–12 cm long by 2–10 cm wide, lighter green and slightly glaucous below, trembling, turning yellow in fall; **petioles** 1–6 cm long, light green to reddish, glabrous, flattened near the blade.

■ **TRUNK** Pale green to light gray and smooth above, shallow furrows near the base; **twigs** light greenish brown to reddish brown, glabrous, slightly shiny; **lenticels** tan, small, oval to elongated; **winter buds** usually 5–8 mm long, reddish brown, pointed, glabrous, with several scales; **leaf scars** oval, with three bundle scars; **dbh** 20–60 cm (–108).

478   Quaking Aspen, Trembling Aspen, Aspen

Willow—*Salicaceae* **Shrub** Pale green Catkins

Beaked Willow *Salix bebbiana* Sarg.

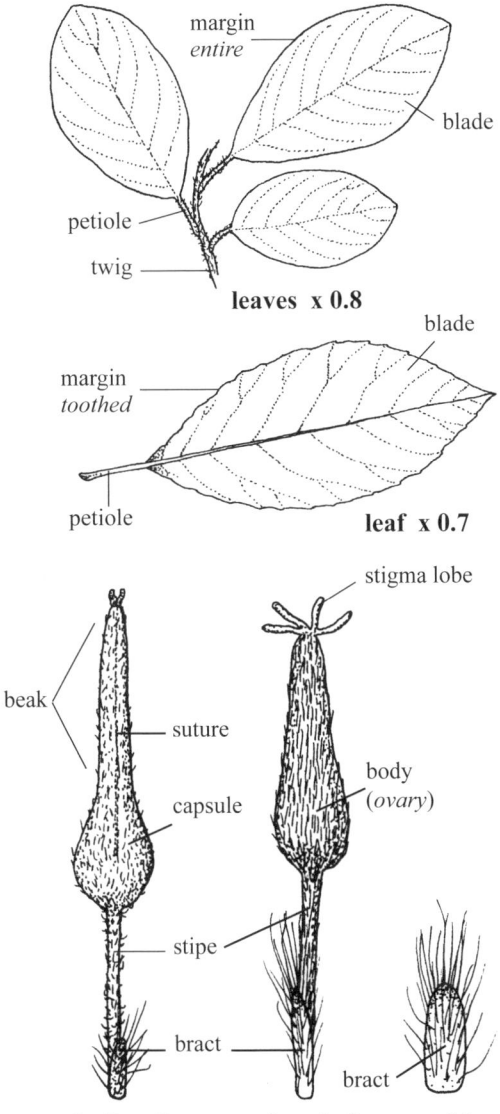

leaves x 0.8

leaf x 0.7

fruit x 5   female flower x 10

■ **SKETCH** A **deciduous shrub** 2–8 m tall; along streams, in wet meadows, swamps and woodland edges, sometimes forming thickets; **dioecious. w2**

■ **FLOWERS** Pale green, blooming April–June; **inflorescence** of catkins, unisexual, at the ends of new twigs, emerging and maturing with the leaves; **male catkins** 1–3 cm long; **stamens** two; **filaments** 5–6 mm long; **anthers** 0.5–0.8 mm long; **peduncles** 0.5–2 cm long, with two to four leaves 5–8 mm long; **female catkins** light green, 1.5–6 cm long by 6–12 mm wide; **rachis** hairy and light green; **female flowers** 4–6.5 mm long (including stipe) by 0.6–1 mm wide; **stipes** hairy, 2–3 mm long, in fruit to *c.* 6 mm long; **bract** usually much shorter than the stipe, light yellowish green, some with a reddish tip, the apex blunt, the body 1.5–1.7 mm long by *c.* 0.5 mm wide and slightly curved around the stipe, white hairs *c.* 1 mm long on the outside; **stigmas** 4-lobed, sessile; **catkins in fruit** 1.5–8 cm long to *c.* 2 cm wide. **FRUIT a capsule**, lightly hairy, 5–9 mm long, the beak curling open to release the seeds; **seeds** minute.

■ **LEAVES** Alternate, simple, entire to slightly toothed, emerging with the catkins; **blades** 2.3–8 cm long by 0.6–3 cm wide, widest above or at the middle, margins slightly wavy, hairy to glabrous above, glaucous and pubescent below (soft to touch), the veins raised and hairy; **petioles** 3–10 mm long, glandless, hairy; **stipules** deciduous, 2–6 mm long by 1–3 mm wide, shallowly dentate to almost entire, on vigorous summer shoots.

■ **STEM** Several; 3–15 cm wide near the base; **twigs** reddish brown to yellowish green, slightly pubescent among winter buds to glabrous and shiny, appearing jagged from jutting alternate leaf scars; **lenticels** oval, usually *c.* 0.5 mm long, slightly raised; **winter buds** golden brown, veined, 1-scaled, usually 3.8–5 mm long by 1–2 mm wide by 1–1.3 mm thick, slightly shiny, with short white hairs at the base and sometimes in the grooves; **leaf bundle scars** three.

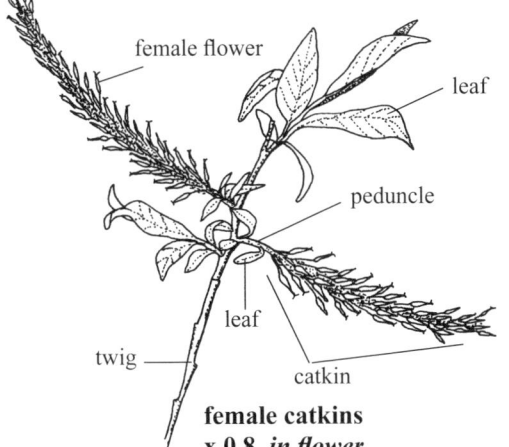

female catkins x 0.8 *in flower*

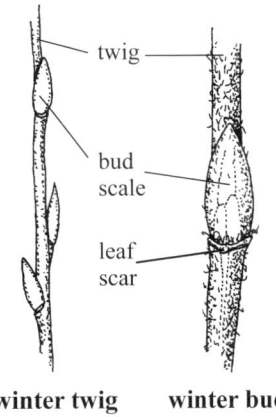

winter twig x 1.6   winter bud x 4

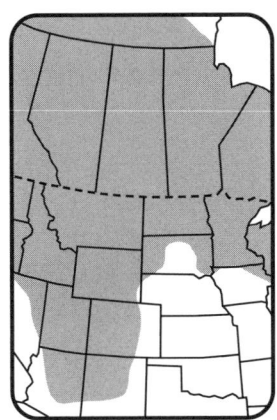

Bebb Willow, Diamond Willow, Gray Willow

Hoary Willow *Salix candida* Flucggé ex Willd.

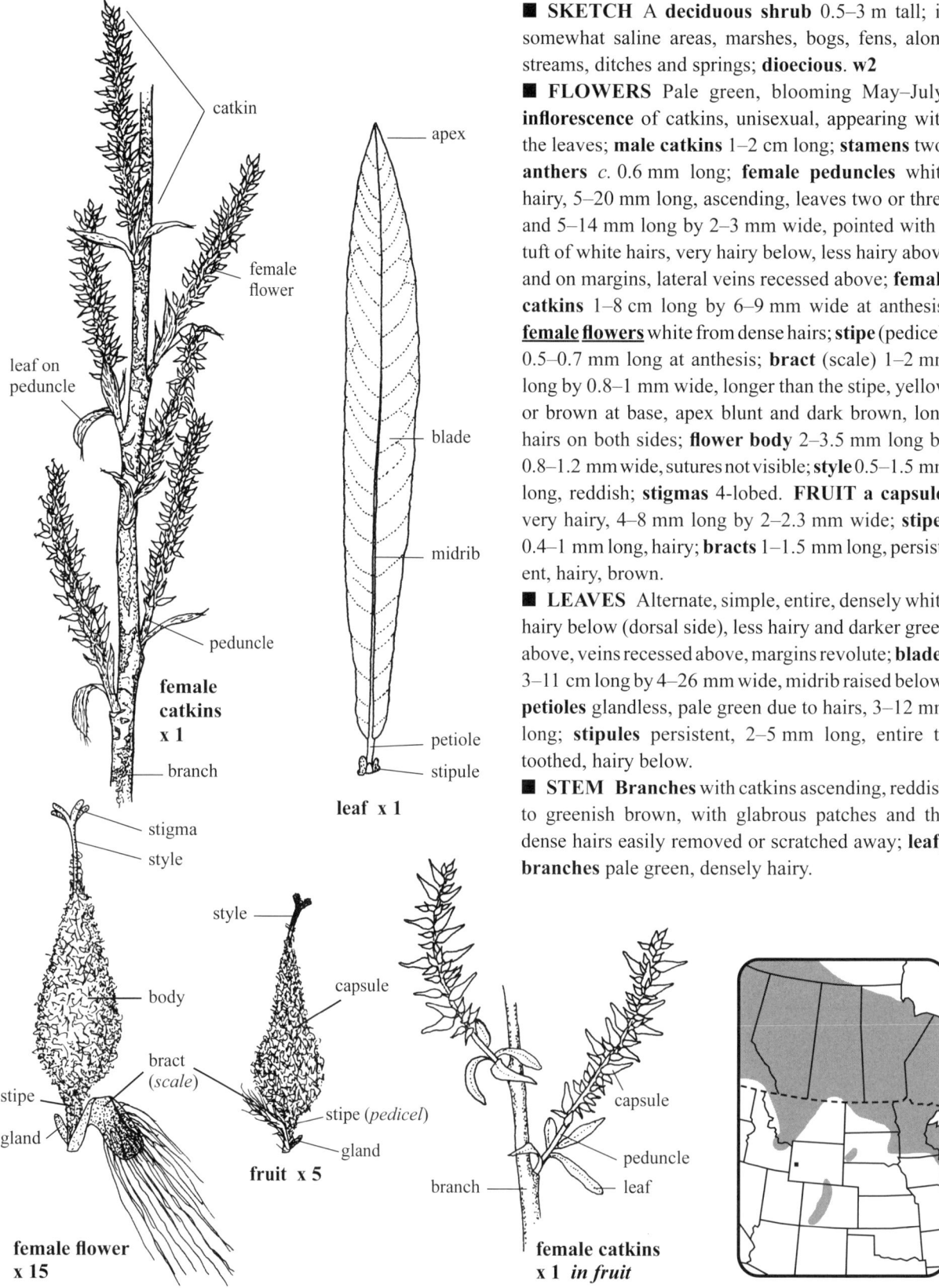

- **SKETCH** A **deciduous shrub** 0.5–3 m tall; in somewhat saline areas, marshes, bogs, fens, along streams, ditches and springs; **dioecious**. w2
- **FLOWERS** Pale green, blooming May–July; **inflorescence** of catkins, unisexual, appearing with the leaves; **male catkins** 1–2 cm long; **stamens** two; **anthers** *c.* 0.6 mm long; **female peduncles** white hairy, 5–20 mm long, ascending, leaves two or three and 5–14 mm long by 2–3 mm wide, pointed with a tuft of white hairs, very hairy below, less hairy above and on margins, lateral veins recessed above; **female catkins** 1–8 cm long by 6–9 mm wide at anthesis; **female flowers** white from dense hairs; **stipe** (pedicel) 0.5–0.7 mm long at anthesis; **bract** (scale) 1–2 mm long by 0.8–1 mm wide, longer than the stipe, yellow or brown at base, apex blunt and dark brown, long hairs on both sides; **flower body** 2–3.5 mm long by 0.8–1.2 mm wide, sutures not visible; **style** 0.5–1.5 mm long, reddish; **stigmas** 4-lobed. FRUIT a capsule, very hairy, 4–8 mm long by 2–2.3 mm wide; **stipes** 0.4–1 mm long, hairy; **bracts** 1–1.5 mm long, persistent, hairy, brown.
- **LEAVES** Alternate, simple, entire, densely white hairy below (dorsal side), less hairy and darker green above, veins recessed above, margins revolute; **blades** 3–11 cm long by 4–26 mm wide, midrib raised below; **petioles** glandless, pale green due to hairs, 3–12 mm long; **stipules** persistent, 2–5 mm long, entire to toothed, hairy below.
- **STEM Branches** with catkins ascending, reddish to greenish brown, with glabrous patches and the dense hairs easily removed or scratched away; **leafy branches** pale green, densely hairy.

Willow—*Salicaceae*

Shrub Green Catkins

Pussy Willow *Salix discolor* Muhl.

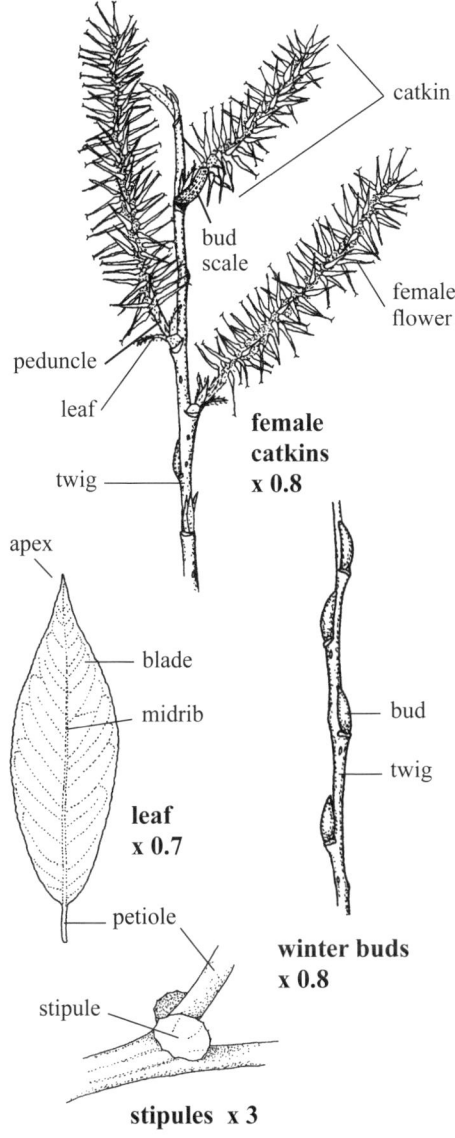

■ **SKETCH** A **deciduous shrub** 3–6 m tall, often with several stems in a clump; in moist areas around ponds, waterways and in swampy woods; **dioecious**. w2

■ **FLOWERS** Green, blooming March–June before the leaves are fully formed; **inflorescence** of catkins, unisexual; **male catkins** white and yellow, 2.5–4 cm long; **stamens** two; **anthers** yellow, 0.5–0.9 mm long; **peduncles** 4–12 mm long, hairy, with two or three leaves 4–5 mm long by *c.* 1.5 mm wide, these deciduous, white with hairs, or green and hairy below and on margins, hairless above; **female catkins** whitish green, somewhat cottony in appearance, 1.5–8 cm long by 11–15 mm wide (to 10 cm long in fruit), subsessile, alternate; **bud scales** often persistent at base of peduncle; **rachis** hairy; female flowers numerous, usually perpendicular to the rachis, TL including stipe 6–8 mm by 0.8–1.2 mm wide, hairy, narrowing toward apices; **stipe** hairy, 0.5–2 mm long, lengthening to 1–3 mm in fruit; **bract** 2–3 mm long by *c.* 0.8 mm wide, dark brown to black, tan at its base, with hairs *c.* 2 mm past the apex, hairy on both sides; **style** 0.5–0.8 mm long; **stigmas** green, 4-lobed, *c.* 1 mm wide, spreading. **FRUIT a capsule**, many-seeded, 6–12 mm long by *c.* 2 mm wide, beak long and tapered, finely hairy, splitting into two valves; **seeds** 1.5–1.8 mm long by 0.5–0.6 mm wide, grayish green; **hairs** white, 12–15 mm long.

■ **LEAVES** Alternate, simple, entire to finely irregularly toothed, emerging after anthesis; **early blades** 3–12 cm long by 0.5–3.5 cm wide, slightly glossy above, glabrous to lightly hairy and whitish glaucous below, teeth glandular, margins lightly hairy; **petioles** 3–28 mm long, not glandular, downy when young; **stipules** deciduous, 3–10 mm long and almost as wide, glabrous to lightly hairy, blunt to pointed, the shallow teeth with brown glands, mostly on new summer shoots.

■ **STEM** Erect, branching; **twigs** yellowish green to reddish brown, glabrous to slightly hairy; **new vigorous shoots** in summer are light green and downy; **previous year's twigs** hairless, darker greenish brown, rough from jutting leaf scars; **bark** grayish brown; **leaf scars** light tan, 2–3 mm long by *c.* 1 mm wide, crescent-shaped; **winter buds** alternate, 6–9 mm long by *c.* 2 mm wide by 2–3 mm thick, dark reddish brown, pointed, 1-scaled, minutely hairy at the base and inside at the twig.

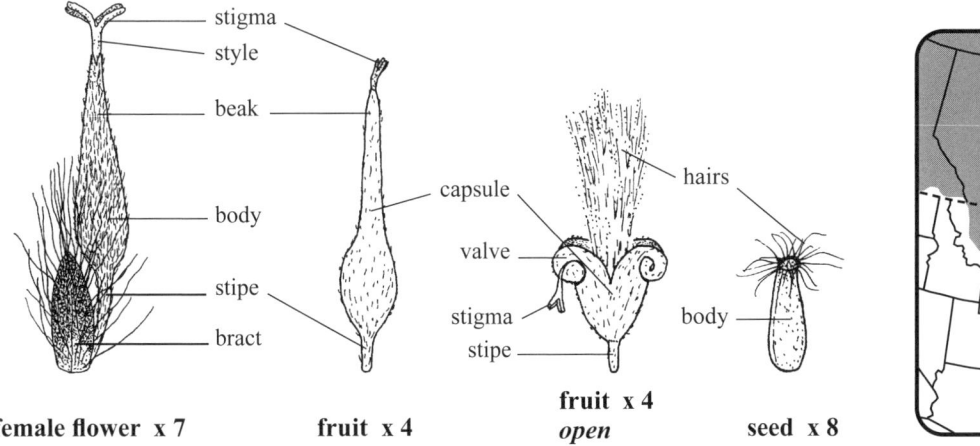

Willow—*Salicaceae* **Shrub** Green Catkins

Yellow Willow *Salix eriocephala* Michx.

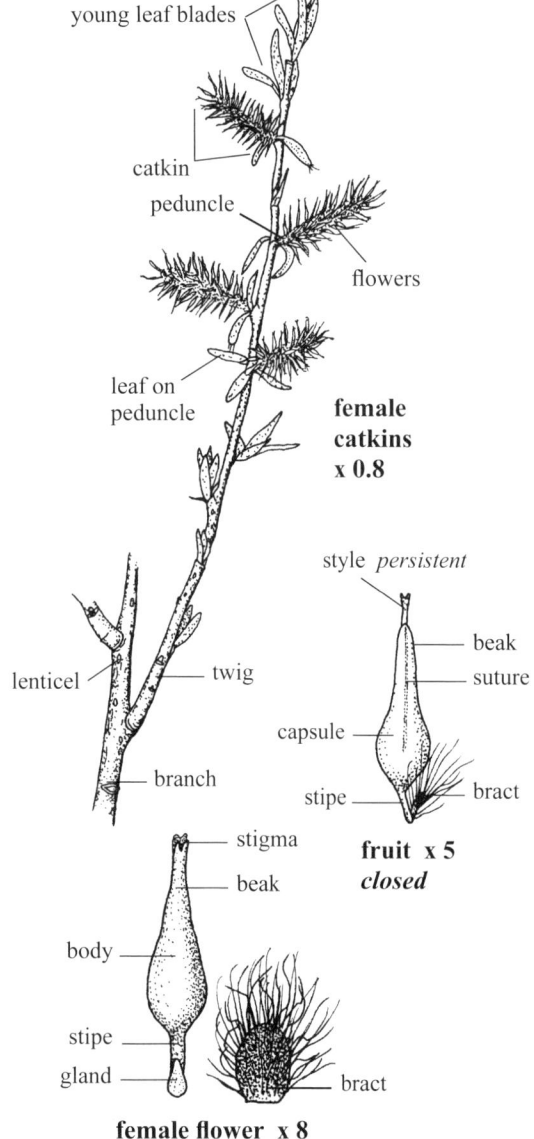

female catkins x 0.8

fruit x 5 *closed*

female flower x 8

■ **SKETCH** A variable **deciduous shrub** 2–7 m tall; along stream banks, in wet meadows and ditches; **dioecious. w1**

■ **FLOWERS** Green, blooming May–June, slightly before or as the leaves unfold; **inflorescence** of unisexual catkins; **male catkins** 2–5 cm long (not illustrated); **stamens** two; **anthers** yellow, 0.4–0.8 mm long; **peduncles** 1–14 mm long, reddish, hairs gray, leaves reflexed and one to four, green, 10–12 mm long by *c.* 3 mm wide, entire, glabrous but often with a tuft of white apical hairs; **female catkins** with flowers are 2–4 (–7) cm long by *c.* 1.2 cm wide, alternate, subsessile, interior of catkins fuzzy white from bract hairs, catkins (in fruit) 1.2–7 cm long; **rachis** not hairy; **female flowers** body 3–3.5 mm long (not including stipe), glabrous; **stipe** 0.7–3 mm long, glabrous; **bracts** *c.* 1.3 mm long by 0.8–1 mm wide, hairy, blunt, dark brown; **stigma** subsessile, 4-parted. FRUIT a capsule, pale green to reddish brown, glabrous, 4–8 mm long (including stipe) by 0.8–1.8 mm wide; **stipe** as above; **bract** dark brown, 2.3–3 mm long (including hairs) by 1–2 mm wide, as long as or slightly shorter than the stipe; **style** *c.* 1 mm long.

■ **LEAVES** Alternate, simple, finely toothed to entire, with an apical tuft of white hairs when young; **blades** 1.5–13 cm long by 0.5–3.5 cm wide, yellowish green above, slightly shiny, glabrous to glaucous or slightly hairy below; **teeth** glandless or glandular; **petioles** glandless, 4–26 mm long, hairless; **stipules** 8–18 mm long by 5–11 mm wide, teeth glandless, glaucous below, yellowish green above, on vigorous leafy summer shoots.

■ **STEM** Usually a few dozen branches in a clump; **bark** silvery gray on older branches, dull; **young branches** light yellowish green to yellowish gray, alternate, glabrous to short hairy and slightly shiny, with some short hairs above the leaf scars; **lenticels** darker than yellowish branches, raised, axis parallel to twig length.

■ **SYN.** *Salix cordata* Muhl., non Michx. and *Salix rigida* Muhl.

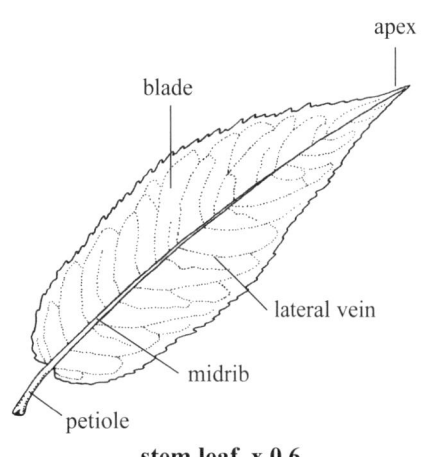

stem leaf x 0.6

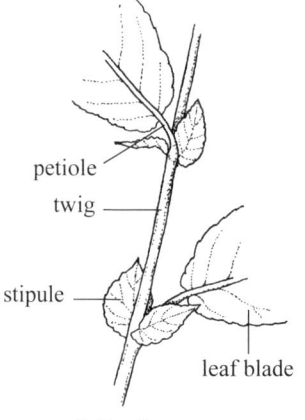

leaves x 0.7 *of summer*

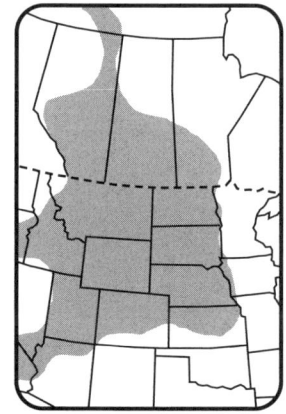

Sandbar Willow *Salix interior* Rowlee

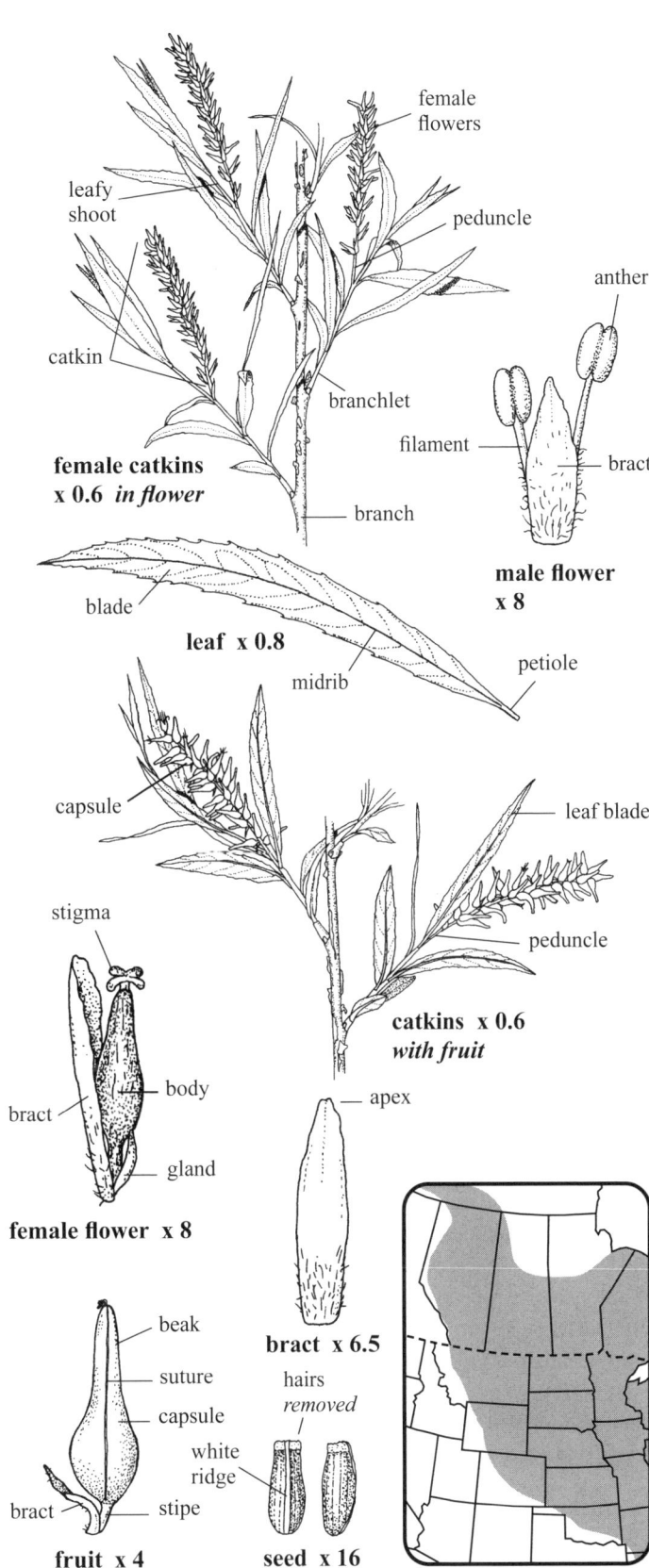

- **SKETCH** A **deciduous shrub** 0.5–6 m tall in thickets from creeping **rhizomes**; on islands and river sandbars, along rivers, sloughs, lakes and railway ditches; **dioecious**. w2
- **FLOWERS** Light green, blooming April–June; **inflorescence** of unisexual catkins on leafy branches, emerging with or after the leaves; **peduncles** slightly hairy to glabrous, 8–20 mm long; **male catkins** 2–4.5 cm long by 6–8 mm wide; **male flowers** 3–5 mm long; **bracts** 2.5–3 mm long by *c.* 1 mm wide, pale yellowish green, hairy on outside at the base, apical edges revolute and forming a point; **stamens** two, exserted; **filaments** pale green, 2–4 mm long, the lower half hairy; **anthers** yellow, *c.* 1 mm long; **female catkins** ascending, alternate, 1.5–8 cm long by 5–10 mm wide, in fruit to 15 mm wide, light yellowish green, glabrous; **female flowers** ascending, crowded above; TL 3–4 mm by *c.* 1 mm wide, green, slightly hairy on beak; **stipe** 0.5–1.3 mm long, slightly hairy; **gland** at base of stipe longer than or equal to stipe; **bracts** 4–4.5 mm long by *c.* 1 mm wide, pale yellowish green, revolute near apices, hairy near the base; **stigma** subsessile, 4-lobed. **FRUIT** a capsule, 5–9 mm long by 2–2.5 mm wide, glabrous or hairy; **stipes** 0.5–1.5 mm long; **bracts** pointed, yellowish tan, deciduous, 2.5–3.5 mm long, mostly glabrous to slightly hairy; **seeds** dark gray, 0.7–0.9 mm long by *c.* 0.3 mm wide and thick, one whitish ridge along one side, greenish near apices; **hairs** white, deciduous.
- **LEAVES** Alternate, simple, toothed to entire, yellowish green above, the same or slightly paler below, glabrous below when mature; **blades** tapered, 2–15 cm long by 2–14 mm wide, midrib slightly raised below, lateral veins faint; **petioles** glandless, 1–6 mm long, glabrous; **stipules** absent to minute.
- **STEM** Erect, in thickets covering several m² or more; **branches** pale brown to reddish brown, leafy, some shedding bark; **twigs** light yellow to orange, glabrous.
- **SYN.** *Salix exigua* ssp. *interior* (Rowlee) Cronq. and *Salix exigua* var. *exterior* (Fern.) C.F. Reed.

Long-leaf Willow

Willow—Salicaceae

Shrub Green Catkins

Velvet-fruited Willow *Salix maccalliana* Rowlee

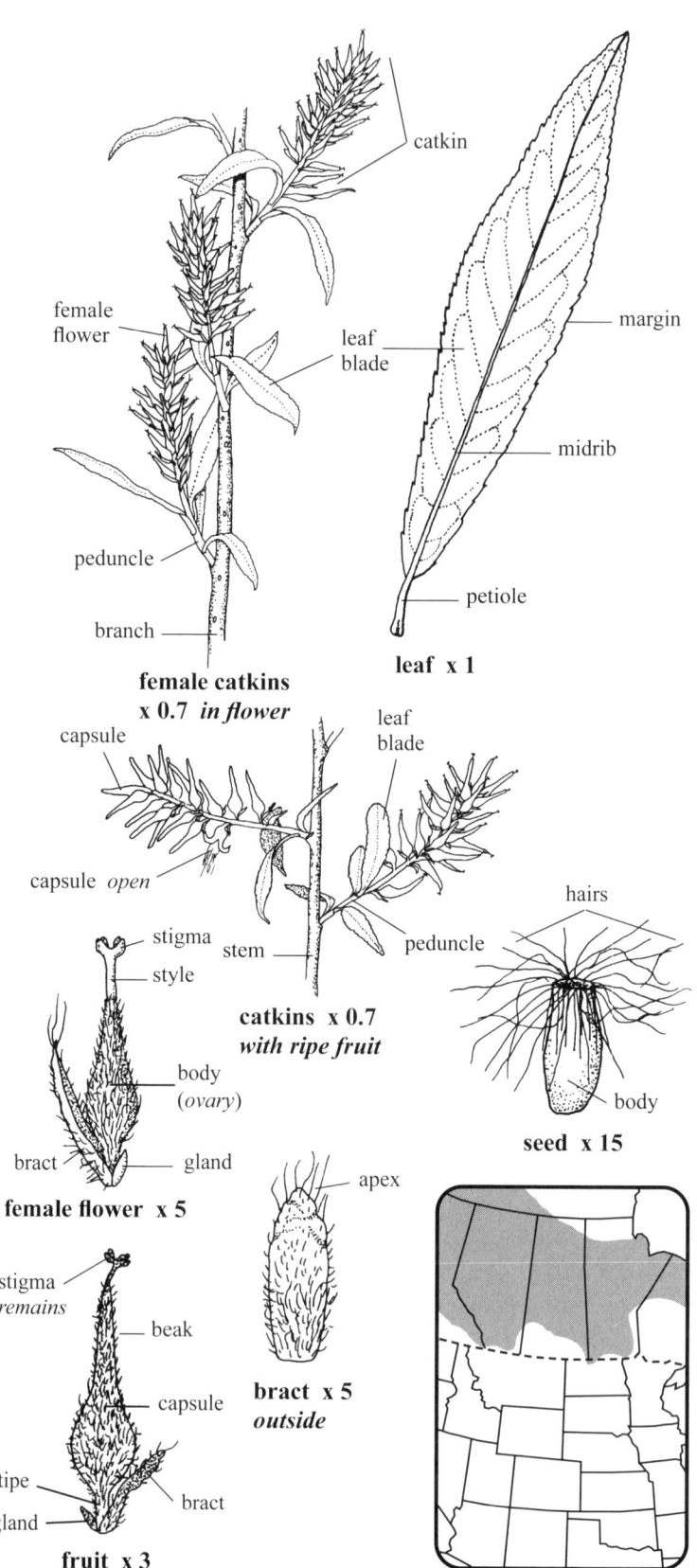

■ **SKETCH** A **deciduous shrub** 1–5 m tall in small clumps; along lakeshores, riverbanks, in thickets, bogs and muskeg; **dioecious**. w2

■ **FLOWERS** Pale green, blooming May–June; **inflorescence** a catkin, unisexual, emerging with the leaves; **peduncles** lightly hairy, 10–25 mm long, usually with two to four leaves 1–3 cm long by 5–9 mm wide, finely toothed, teeth glandular, glabrous, but with a few long hairs at apices when young; **male catkins** 2–3 cm long; <u>male flowers</u> with two stamens; **anthers** *c*. 1 mm long; **female catkins** 2–6 cm long by 1–1.5 cm wide; <u>female flowers</u> TL 6–7 mm, hairy; **stipe** (pedicel) hairy, 1–1.5 mm long; **bract** (scale) pale green, hairy on lower outside, less hairy near the blunt tip, 3–5 mm long by 1.5–1.8 mm wide, apex sometimes wrinkled, some turning dark brown in upper half when mature; **style** *c*. 1 mm long; **stigmas** 4-lobed. **FRUIT** a **capsule**, 6–11 mm long by 2–3 mm wide, hairy, long beaked; **stipe** hairy, 1–2 mm long; **bract** equal to longer than the stipe, 2–5 mm long, dark brown near apex or upper two-thirds, pale greenish below or throughout, more hairy near the base; **seeds** numerous, brown, 1.2–1.3 mm long by *c*. 0.5 mm wide by 0.3–0.4 mm thick; **hairs** white, 7–8 mm long and spreading.

■ **LEAVES** Alternate, simple, entire or with small teeth, these rounded and glandular; **blades** of young leaves reddish, glabrous but slightly hairy at pointed apices; **older blades** medium to dark green and shiny above, 4–9 cm long by 6–32 mm wide, flat, soft to thick and leathery, net-veined and lighter green below but not glaucous or hairy, midrib light green above, obvious, lateral veins faint, not recessed; **petioles** 3–10 mm long, slightly hairy above, glandular at base or not; **stipules** are glandular lobes 0–1.5 mm long.

■ **STEM** Several in a cluster, erect to ascending, 10–15 mm thick; **branches** alternate, reddish brown, shiny, glabrous; **twigs** yellowish gray, hairless.

484   McCalla's Willow

Willow—*Salicaceae* **Shrub** Green to yellow Catkins

Basket Willow   *Salix petiolaris* Sm.

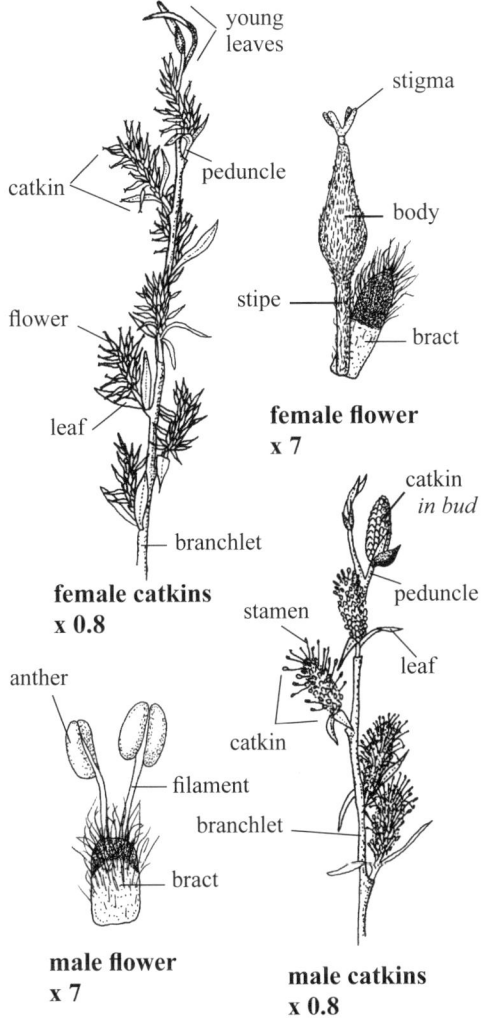

female catkins x 0.8

female flower x 7

male flower x 7

male catkins x 0.8

■ **SKETCH** A **deciduous shrub** 1–4 (–7) m tall; in clumps along shorelines, in ditches and wet meadows; **dioecious**. **w2**
■ **FLOWERS** Green to yellow, blooming May–June before the leaves are fully developed; **inflorescence** of catkins (aments), unisexual; **male peduncles** hairy, 3–5 mm long with two or three hairy leaves 10–13 mm long by 2–3 mm wide; **male catkins** 1–3.5 cm long by 8–11 mm wide, ascending; **stamens** two; **filaments** pale yellow, 3–4 mm long, glabrous; **anthers** yellow, 0.8–1 mm long; **bract** 1.8–2 mm long by *c.* 1 mm wide, the upper half to third dark brown, pale green below, apex blunt, very hairy especially on the outside; **female peduncles** spreading, 5–20 mm long with two to four leaves 1–2.5 cm long by 2.5–4.2 mm wide, entire, pointed, sessile, glabrous above, hairs on midrib below; **female catkins** 0.8–3.5 cm long by 1.5–2 cm wide in flower, to *c.* 6 cm long in fruit; <u>**female flowers**</u> hairy, erect to spreading, numerous, TL *c.* 5 mm; **body** *c.* 3 mm long (including the four-lobed stigmas); **stipe** 1.5–2 mm long, hairy; **bract** *c.* 2 mm long by *c.* 0.8 mm wide, apex dark brown, blunt, hairs mostly on the outside. **FRUIT a capsule**, light green to slightly reddish, 4–7 mm long (body) by 2–2.3 mm wide, reflexed at base of catkin, spreading to ascending at the catkin's apex, sparsely hairy throughout or more hairy near the base; **stipe** 1–5 mm long, hairy; **bract** persistent, dark brown above, slightly paler near the base, hairy on the outside, mostly hairless inside, wrinkled, about as long as the stipe; **seeds** *c.* 1.5 mm long by *c.* 0.4 mm wide; **hairs** white.
■ **LEAVES** Alternate, simple, serrate (some entire), teeth gland-tipped; **blades** 2–12 cm long by 5–25 mm wide, glaucous below, shiny and darker green above, slightly hairy above when young, glabrous with age; **petioles** 3–12 mm long, glands absent; **stipules** absent.
■ **STEM** Yellowish, glabrous; to *c.* 10 mm thick near the grayish base; **floral branches** numerous, reddish brown to dark brown, glabrous, not very shiny, erect.
■ **SYN.** *Salix gracilis* Anderss.

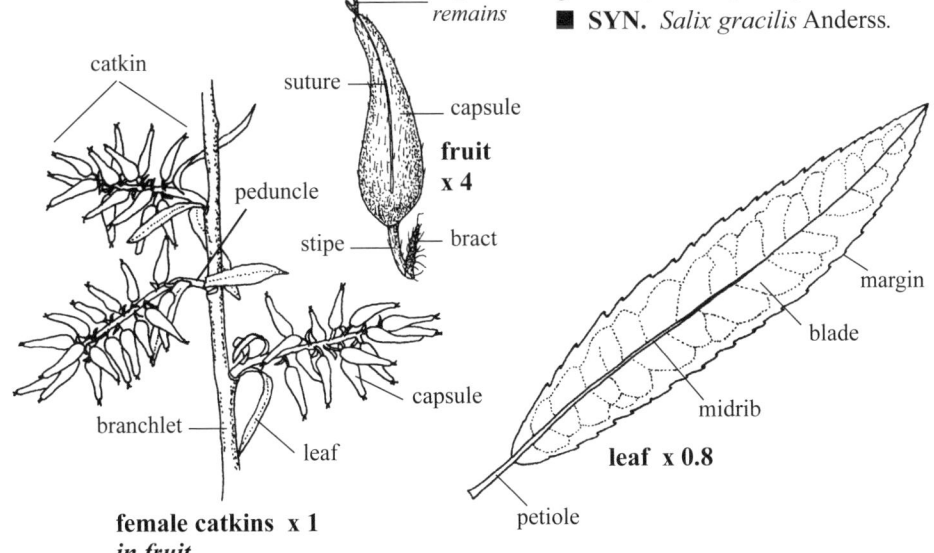

female catkins x 1 in fruit

fruit x 4

leaf x 0.8

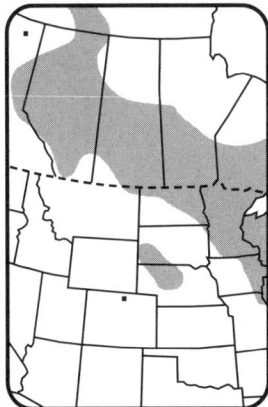

Meadow Willow, Slender Willow   485

Sandalwood—*Santalaceae* **Wildflower** White Sepals 5 (4)

BASTARD TOADFLAX *Comandra umbellata* (L.) Nutt.

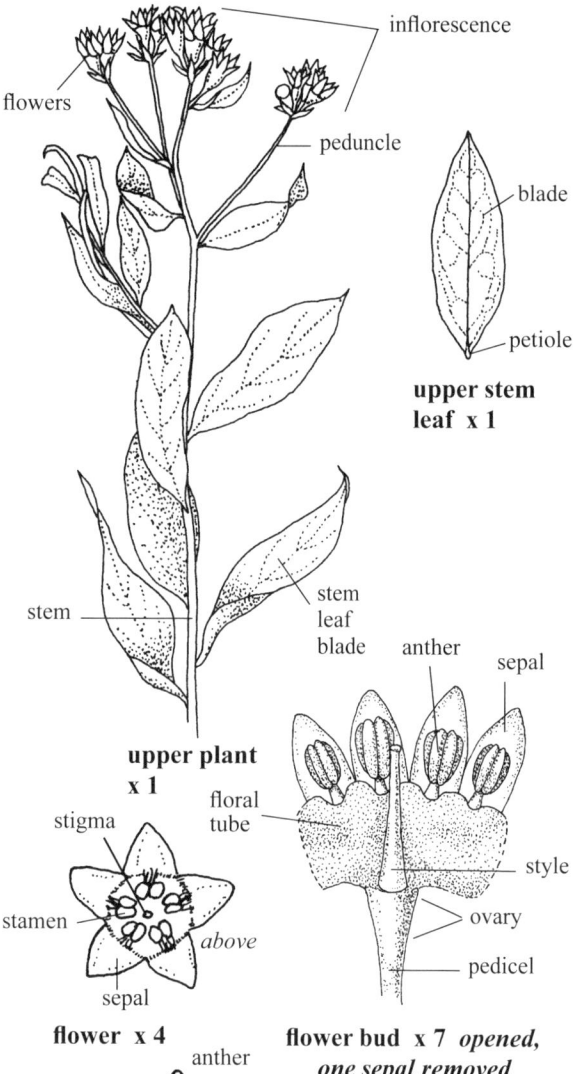

upper plant x 1
upper stem leaf x 1
flower x 4
flower bud x 7 *opened, one sepal removed*
fruit x 2

■ **SKETCH** A variable, glabrous **perennial herb** (often a root parasite) 8–50 cm tall, in small clusters of one to seven stems from a woody, horizontal **rhizome** 1–5 mm thick and to *c.* 15⁺ cm long; in meadows, moist to dry prairies, desert shrub and open wooded areas including sandy to rocky slopes. p2

■ **FLOWERS** White with some pink, blooming April–July; **inflorescence** paniculate to corymbose, 1–5 cm wide and tall, of 3- to 6-flowered cymules; **peduncles** 0.5–3.5 cm long, ascending from upper leaf axils, reduced above; **pedicels** 1–4 mm long; **subtending bracteoles** (of flowers) one, 2–4 mm long by 1.2–1.6 mm wide, pointed; **flowers** perfect, 5–7 mm wide by 4–5 mm long; **sepals** five (four), 2.5–4.5 mm long by 1.5–2 mm wide, attached to floral tube, petal-like, white, thick, hairy along inner basal margins, erect and persistent in fruit, then 3–4 mm long and reddish; **petals** absent; **stamens** five, opposite sepals; **filaments** 0.5–1 mm long, attached to the sepals at their base; **anthers** pinkish green, 0.4–1 mm long and wide; **style** one, tapered, filiform, 2–3 mm long; **stigma** round, green, level with the anthers. **FRUIT** drupe-like, dry, 1-seeded, thin-walled, falling while still green, erect to ascending, 4–8 mm long and wide (not including sepals), constricted at apex, suture circular and near the top, the top easily peeled away to reveal the seed inside; **seeds** one per fruit, about as big as the drupe, smooth.

■ **LEAVES** Alternate, simple, entire, ascending, imbricate, glabrous to glaucous, sessile to short petiolate, reduced above and below; **blades** 0.7–5.5 cm long by 2–15 mm wide, medium green to grayish green, dull, midrib faint, lateral veins obscure to obvious; **stipules** absent.

■ **STEM** Erect, simple to branched above, glabrous, rigid; 1.5–3 mm thick near the slightly woody base.

■ **SYN.** *Comandra pallida* A. DC. = *Comandra umbellata* ssp. *pallida* (A. DC.) Piehl and *Comandra richardsiana* Fern. = *Comandra umbellata* ssp. *umbellata* (L.) Nutt.

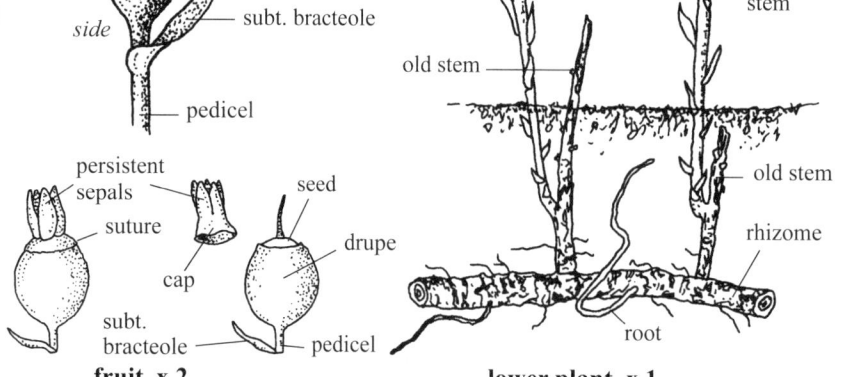

lower plant x 1

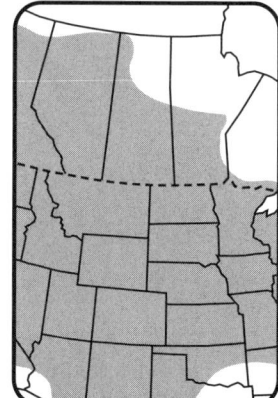

False Toad-flax, Pale Comandra, Star Toadflax

Saxifrage—*Saxifragaceae* **Wildflower** Yellowish green Sepals 4

Golden Saxifrage *Chrysosplenium tetrandrum* (Lund ex Malmgr.) Th. Fries

■ **SKETCH** A glabrous **perennial herb** 3–20 cm tall in colonies from white **stolons** 1–3 cm long by *c.* 0.5 mm thick with filiform **roots** 3–6 cm long; in wet mossy areas along woodland streams.

■ **FLOWERS** Yellowish green, blooming May–June; **inflorescence** a cyme, 3–4 mm tall by 8–14 mm wide, of several crowded flowers subtended by leaflike bracts at the stem's apex; **pedicels** 0–0.5 mm long; **flowers** perfect, 3–5.5 mm wide by 2.5–3.3 mm long, sessile or nearly so, squarish to rectangular; **lateral flowers** arise on apex of petiole; **hypanthium** *c.* 2 mm wide by *c.* 1.3 mm long; **sepals** four, *c.* 2 mm long by 2–3.5 mm wide, fleshy, erect to recurved, blunt, 3-veined, midrib obvious; **petals** absent; **stamens** two to eight, included; **filaments** green, *c.* 0.3 mm long; **anthers** yellow, 4-lobed, 0.5–0.8 mm long by *c.* 0.5 mm wide; **styles** two, *c.* 1 mm apart, *c.* 0.3 mm long, green, filiform; **stigmas** yellowish green, rough, obscure. **FRUIT a capsule**, green, indented in middle, *c.* 4 mm long by *c.* 2 mm wide by *c.* 3 mm deep (closed), suture along its upper length, enveloped by four sepals, opening across the apex; **seeds** usually 30–36 per capsule, glossy, reddish brown, 0.6–0.9 mm long by 0.5–0.6 mm wide by *c.* 0.5 mm thick, with a narrow ridge ending at the hilum, in two alternating rows the length of the capsule.

■ **LEAVES** Basal and stem, alternate, simple, toothed; **blades** 7–21 mm long by 5–17 mm wide, veins mostly obscure, dull, greenish yellow, slightly lighter below, glabrous, an opaque gland at the margin of each blunt tooth; **petioles** 0.3–4.5 cm long, reduced above, margins continue as ridges along stem; **stipules** absent.

■ **STEM** Erect, weak, glabrous, 3-sided, sides convex, usually branched above; 1–1.5 mm thick near the base.

■ **NOTE** Similar to *Chrysosplenium iowense*.

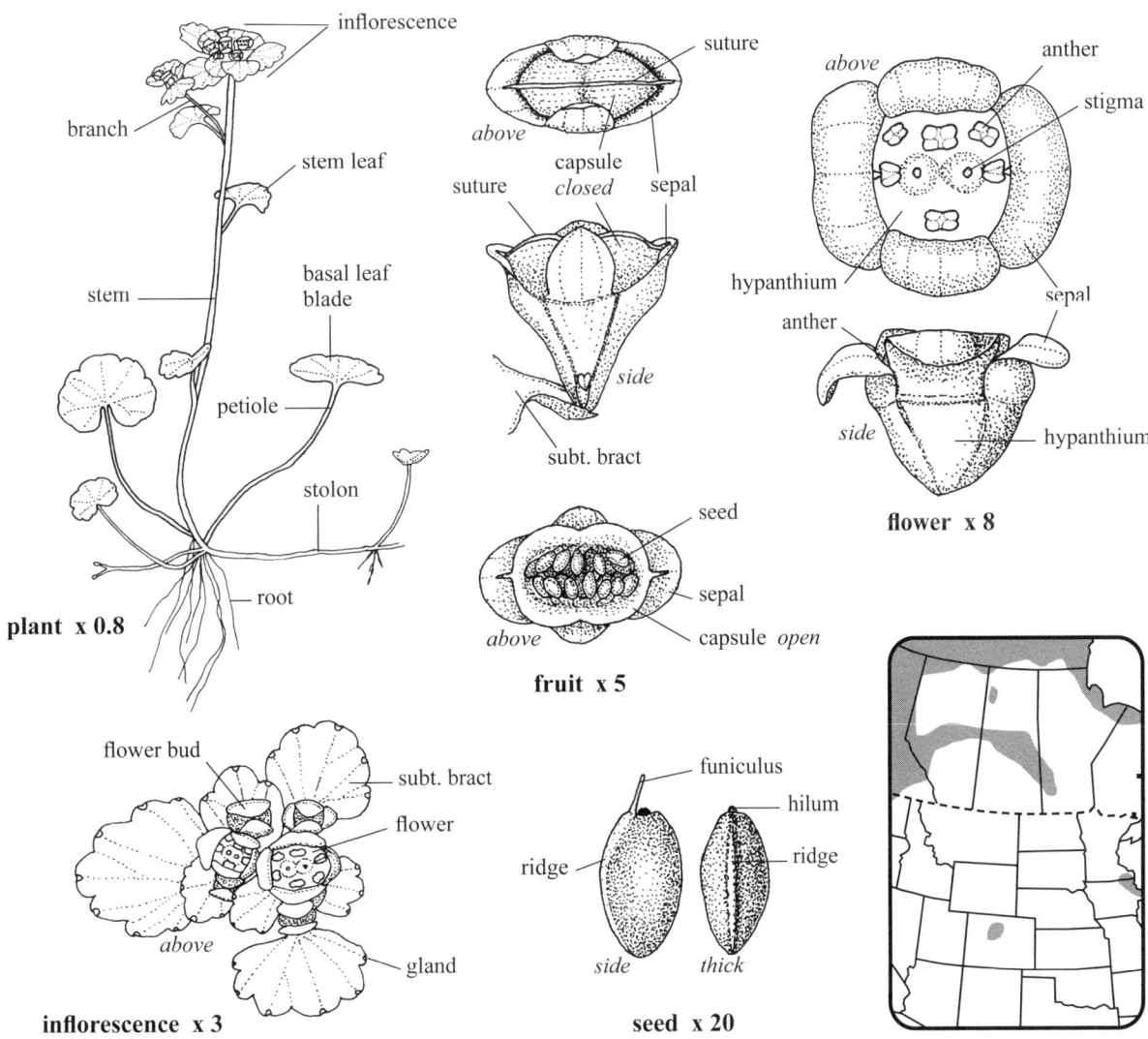

Northern Golden-carpet, Northern Golden-saxifrage, Northern Water-carpet, Green Saxifrage

Saxifrage—*Saxifragaceae*   **Wildflower**   Yellowish green to reddish green   Petals 5

ALUMROOT   *Heuchera richardsonii* R. Br.

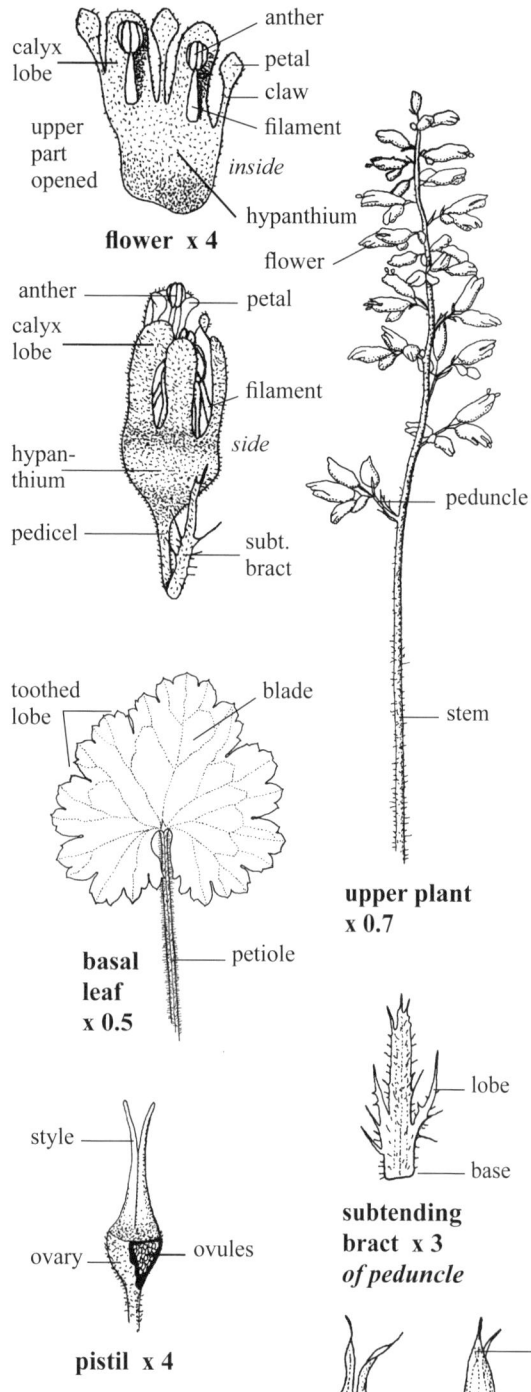

■ **SKETCH** A **perennial herb** 15–70 cm tall from a branched **caudex** 4–15 cm long by 1–1.5 cm wide; in moist to dry prairies, along shores, sandy hillsides and open woods. **p2**

■ **FLOWERS** Yellowish green to reddish green, blooming May–July; **inflorescence** a raceme or panicle, terminal, 3–23 cm long by 1.8–4 cm wide; **peduncles** 1–15+ mm long, hairs spreading and gland-tipped, glands often reddish; **subtending bracts** (of peduncles) leaflike, entire or with a few pointed lobes, 4–8 mm long by 1–4 mm wide, hairs gland-tipped, inner surface glabrous; **pedicels** 1–5 mm long, with gland-tipped hairs; **subtending bracts** (of pedicels) 2–4 mm long, glandular-hairy on outside; **flowers** perfect, crowded above, *c.* 4 mm wide by 8–10 mm long, asymmetrical, angled, closed at apices; **hypanthium** glandular-hairy, 4–5 mm long; **calyx** 5-lobed, green to reddish, lobes 2–3.8 mm long by 1.5–2 mm wide, flat, blunt, glandular-hairy on outside, some glands reddish; **petals** five, 2–3 mm long by 0.8–1 mm wide, clawed, green turning reddish; **stamens** five, opposite calyx lobes, barely exserted; **filaments** thick, light green, 1–2.8 mm long, attached near the base of calyx lobes; **anthers**, orange, 1–1.5 mm long; **ovary** with two carpels, each with numerous pink ovules; **styles** two, filiform, included, 1.5–3 mm long, round, spreading, widening at base. **FRUIT a capsule**, upright, tan, hairless, veiny, 10–12 mm long by 3–3.5 mm wide and thick (including beak), covered with persistent hypanthium, beak 2-parted, opening at apex; **seeds** blackish brown, numerous, 0.6–0.9 mm long by *c.* 0.4 mm wide and thick, covered with short, stout, pricklelike hairs.

■ **LEAVES** Basal, simple, lobed and coarsely toothed, 4–13; **blades** 1.5–7 cm long by 1.5–8 cm wide, lighter green and more hairy below, smooth above, palmately veined; **petioles** 3–17 cm long by *c.* 2 mm wide, hairy to almost glabrous, grooved above.

■ **STEM** Lime green, hollow, one to three per caudex, erect, hairs white, numerous, spreading and 1–3 mm long, gland-tipped or not, glands more numerous and reddish above near the base of the inflorescence; 2–4 mm thick near the base.

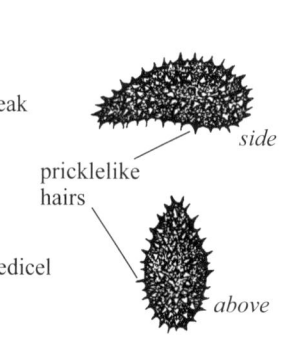

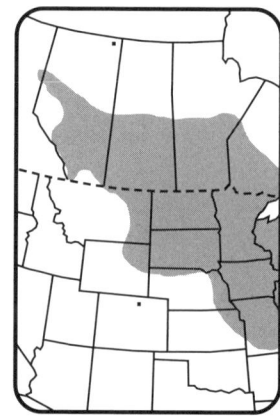

488   Richardson's Alumroot

Saxifrage—*Saxifragaceae* **Wildflower** Pale greenish yellow Petals 5, finely divided

Bishop's-cap *Mitella nuda* L.

■ **SKETCH** A glandular-hairy **perennial herb** 3–25 cm tall with a tan **rhizome** 5–15 cm long by *c.* 1 mm thick with dark scales or a bract from each node; in moist woodlands, bogs, swamps and along stream banks.

■ **FLOWERS** Pale greenish yellow, blooming April–June; **inflorescence** a raceme, terminal, one, 3–12 cm long by 10–13 mm wide of 2–10 flowers; **pedicels** pale green, 1–5 mm long, reduced above, glandular-hairy, ascending even with fruit; **flowers** perfect, 6–11 mm wide by *c.* 3 mm long; **rachis** *c.* 0.5 mm wide, glandular-hairy; **calyx** 5-lobed, TL *c.* 2 mm, lobes 1.5–1.8 mm long by 1.2–1.5 mm wide, spreading, tips sometimes descending, 3-veined above, glandular on margins; **petals** five, 3–5 mm long and wide, divisions filiform, often with four opposite divisions from the midvein, glabrous; **hypanthium** green, glandular-hairy, 1.5–2 mm wide and *c.* 1.5 mm long; **stamens** 10, erect from the disc, *c.* 1 mm long, included; **filaments** *c.* 0.5 mm long, pale greenish yellow; **anthers** yellow, *c.* 0.4 mm long; **pistil** green, *c.* 1.7 mm long by *c.* 1 mm wide, glandular; **ovary** *c.* 1.5 mm long by *c.* 1 mm wide, oval; **styles** two, spreading, 0.5–0.8 mm long, filiform, brown at first in bud, then green, not glandular. **FRUIT** a capsule, opening widely along ventral sutures to form the "miter" or bishop's cap-shaped structure, stiff, 5–7 mm long by 3–5 mm wide, apices pointed and deflexed, glandular on outside; **seeds** 5–31 per capsule, shiny, dark brown, 1–1.3 mm long by 0.8–0.9 mm wide and thick, reticulate beneath smooth coat, a low ridge along one side to a point at the hilum end.

■ **LEAVES** Basal and stem, alternate, simple, lobed (obscurely) and toothed, dull, slightly lighter green and shiny below; **basal blades** 1.5–5 cm long and wide, a few erect white hairs mostly along the veins on both sides; **petioles** pinkish green, 4–5.7 cm long, hairless below, more white hairs near the blade; **stem leaves** zero to two, similar to basal blades; **petioles** 0–12 mm long, ascending, glandular-hairy.

■ **STEM** Erect, not branched, glandular-hairy, more so above, stiff; 1–2 mm wide near the bare pinkish base.

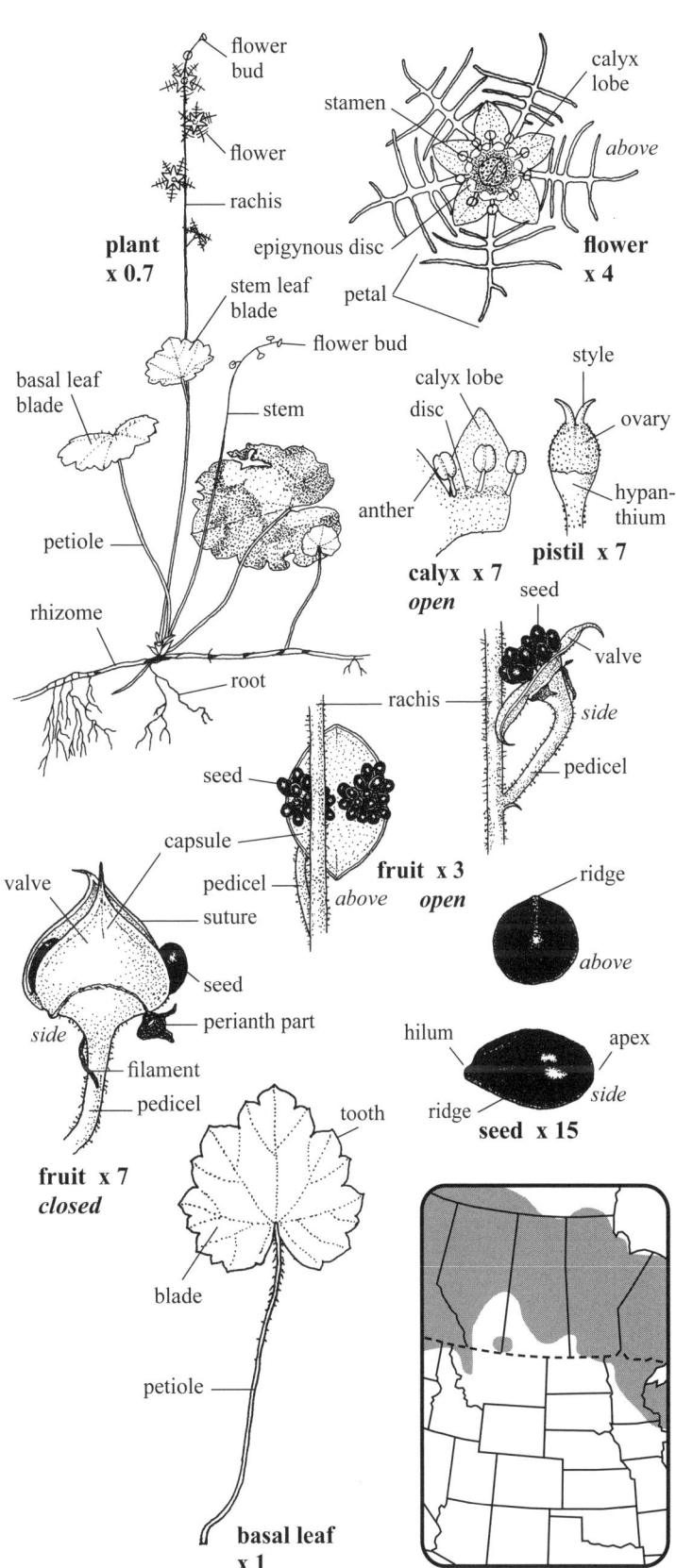

Bare-stem Bishop's-cap, Naked Bishop's-cap, Naked Miterwort

Saxifrage—*Saxifragaceae* — **Wildflower** White Petals 5

GLAUCOUS GRASS-OF-PARNASSUS *Parnassia glauca* Raf.

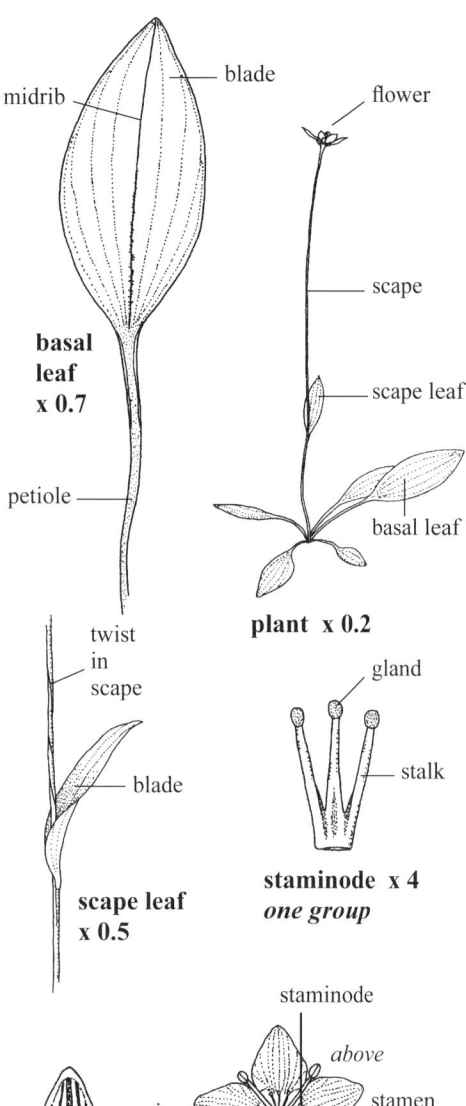

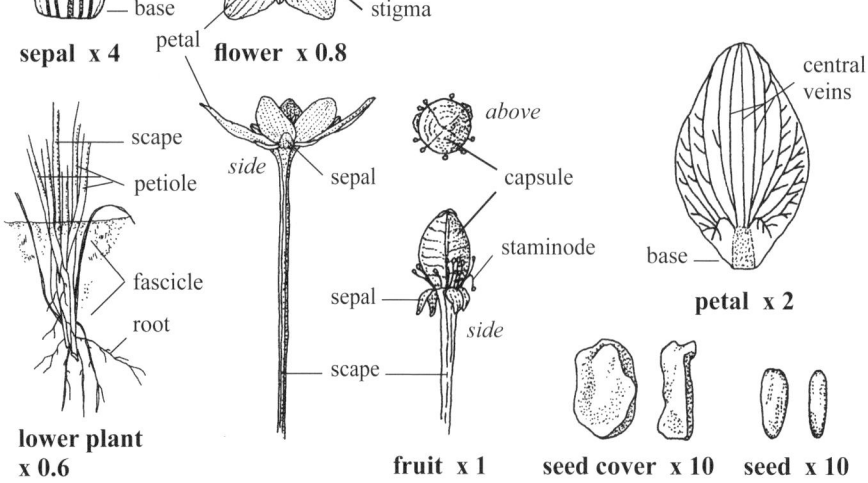

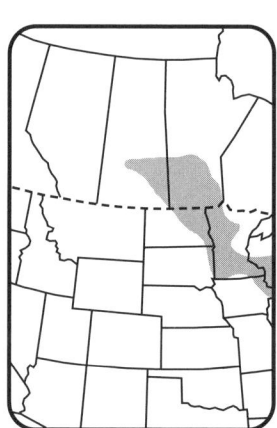

■ **SKETCH** A **perennial herb** 10–40 cm tall from **roots** and the **fascicled bases** of basal leaf petioles; in moist fields, swamps, bogs and along shores.

■ **FLOWERS** Creamy white, blooming July–October; **inflorescence** a single, terminal flower on each scape; **flowers** perfect, 15–30 mm wide by 7–9 mm tall; **sepals** five, 3–6 mm long by 2–2.5 mm wide, 3- or 5-veined, tips often dark brown, margins green hyaline, apices blunt, bases truncate, reflexed and persistent in fruit; **petals** five, blunt, glabrous, 7–18 mm long by 7–10 mm wide, 7- to 13-veined, veins green and obvious, five central veins usually not branched, lateral veins branched to the outside; **stamens** five, included, 7–9 mm long, horizontal between the petals; **filaments** white, 6–8 mm long; **anthers** tan, 1–2 mm long; **staminodia** 15, in five groups, each group of three sterile stamens opposite each petal, 3–7 mm long, stalks slightly widening and united near their bases, white, ascending, with a yellow apical gland, mostly erect with fruit; **pistil** one, 5–6 mm tall by 2–3 mm wide, glabrous, oval, slightly taller than the staminodia; **ovary** roundish in cross-section, 4-sutured, with transverse ridges; **styles** obscure; **stigmas** 4-lobed, *c.* 1.2 mm wide (together). **FRUIT** a capsule, 7–10 mm long by 7–8 mm wide, 4-valved, opening at the apex, tan, somewhat wrinkled transversely; **seed cover** angular, size and shape variable, tan, reticulate net-veined, 1–1.4 mm long by 0.5–0.9 mm wide by 0.4–0.6 mm thick; **seeds** 0.9–1 mm long by *c.* 0.4 mm wide by *c.* 0.3 mm thick, blunt at both ends.

■ **LEAVES** Basal and scape, simple, entire; **basal leaves** 4–13, glabrous, lighter green below, usually 7-veined, pointed; **blades** 1.5–6.5 cm long by 1–3.5 cm wide; **petioles** 1.2–14 cm long, glabrous, with a wide, shallow groove above; **scape leaf** (bractlike), one or absent, near the base, glabrous, 1–5 cm long by 5–18 mm wide, clasping the scape, ascending and sessile.

■ **STEM** A **scape**, erect but not stiff, angular from five longitudinal ridges, one to several per clump, hollow, hairless, glaucous, often twisted above the scape leaf; 1–1.5 mm thick near the base.

490   Fen Grass-of-parnassus, Grass of Parnassus

Saxifrage—*Saxifragaceae* **Wildflower** White Petals 5

Northern Grass-of-parnassus *Parnassia palustris* L.

■ **SKETCH** A variable **perennial herb** 6–35 cm tall with **fibrous roots**, solitary or tufted; in bogs, wet meadows, ditches, along streams and in aspen groves. **p2**

■ **FLOWERS** White, blooming July–September; **inflorescence** a terminal flower; **flowers** perfect, 14–25 mm wide by 6–8 mm tall; **sepals** five, entire, linear, 5–7 mm long by 2–3 mm wide, pliable, 3- or 5-veined, veins most obvious on inner surface, margins green, reflexed in fruit, a ridge between each runs down to the scape; **petals** five, each 8–12 mm long by 6–7 mm wide, glabrous, entire, with a tiny apical notch, with three or five main veins, the lateral ones branching, base tapered; **stamens** five, opposite the sepals, exserted or not; **filaments** ascending, white, glabrous, *c.* 7 mm long by *c.* 1 mm wide at base; **anthers** 1–2.2 mm long, pale yellow; **staminodia** in five groups, each group opposite a petal and between two stamen bases, TL of one group is 5–7 mm by 4–5 mm wide, with 9–15 sterile stamens 1–3.5 mm long and mostly erect, yellowish green, each stalk topped with a false gland 0.3–0.5 mm wide; **pistil** one, 7–7.5 mm long, glabrous, tapered gradually to the stigma; **ovary** 3.5–4 mm wide at the base; **stigmas** 4-lobed, *c.* 1.3 mm wide, pale green. **FRUIT a capsule**, 4-valved, tan, 7–10 mm long by 3.5–5 mm wide, opening by four apical teeth; **seeds** numerous, golden, the wing minutely net-veined and easily removed from seed, seed and wing TL 0.8–1 mm by 0.3–0.4 mm wide by *c.* 0.2 mm thick, seeds (wing removed), smooth, *c.* 0.5 mm long by *c.* 0.2 mm wide and thick.

■ **LEAVES** Basal and scape, simple, entire; **basal leaves** few to several, glabrous; **blades** slightly shiny above, paler green below, apices blunt, 7–30 mm long by 7–20 mm wide, faintly palmately veined; **petioles** 4–35 mm long; **scape leaf** one, with clasping auricles, sessile, glabrous, dull, 5-veined, *c.* 15 mm long by *c.* 11 mm wide.

■ **STEM** A scape, erect, 4- or 5-ridged, these narrow and obvious, unevenly spaced; *c.* 1 mm wide near the base.

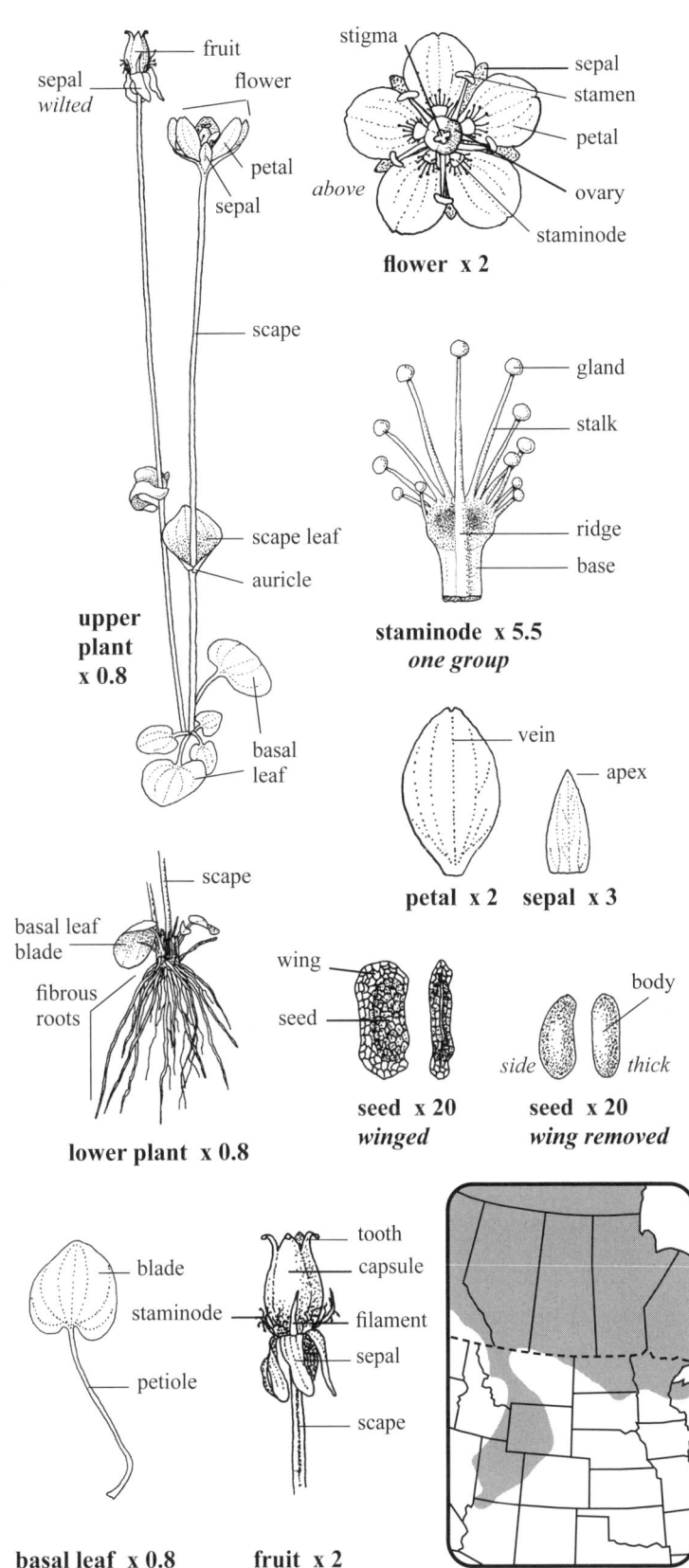

Mountain Grass-of-parnassus, Marsh Grass-of-parnassus

Early Saxifrage *Saxifraga virginiensis* Michx.

■ **SKETCH** A glandular-hairy **perennial herb** 4–25 cm tall with a tan, vertical **bulbous rootstock** 4–10 mm long and wide; in depressions on rocky outcrops and shady banks, covering several m² in the forest.

■ **FLOWERS** White, blooming April–May; **inflorescence** cymose-paniculate, 0.7–6 cm long by 0.7–7 cm wide, elongating, glands reddish brown at tips of hairs; **floral branches** 0.3–9 cm long, ascending, reduced above; **subtending bracts** (of floral branches) 5–40 mm long by 0.5–5 mm wide; **peduncles** 1–5 mm long; **subtending bracts** (of peduncles) 10–13 mm long by *c*. 1 mm wide, pointed; **pedicels** 1–6 mm long, green, hairy; **subtending bracts** (of pedicels) 5–7 mm long by *c*. 1 mm wide; **flowers** perfect, fragrant, 6–10 mm wide by 3–5 mm tall, congested; **hypanthium** 1–2 mm long by 3–4 mm wide; **calyx lobes** five, spreading to ascending, pale green, 2–5 mm long by 2–2.5 mm wide, persistent at base of fruit; **petals** five, white, pointed or slightly notched at apices, 4–5.5 mm long by 1.5–2.5 mm wide, ascending to spreading, veins obscure; **stamens** 10, 1.4–2 mm long, included; **filaments** white, 0.5–1 mm long, lengthening after anthesis; **anthers** yellow, 2-lobed, *c*. 0.9 mm long; **pistils** two, included, glabrous, *c*. 2 mm wide and long, each carpel with two locules and numerous ovules; **styles** two, tapered; **stigmas** pale green, flat and angled, slightly wider than styles. **FRUIT** a follicle, reddish, then ripening to tan, opening along an inner suture, each *c*. 4.5 mm long by *c*. 3 mm wide (across opening); **seeds** brown, numerous, 0.4–0.7 mm long by *c*. 0.2 mm wide, tapered at both ends, 10-ridged, the ridges papillose (microscopic).

■ **LEAVES** Basal, TL 2.5–10 cm (including petioles); **blades** 1.5–6 cm long by 1–3.5 cm wide, widely toothed to entire, simple, hairy on lower margins; **petioles** winged, 1–4 cm long by 2–7 mm wide, flat, ascending, hairy on margins with some hairs glandular.

■ **STEM** A **scape**, solitary, erect, oval, hollow, branched above, glandular-hairy, hairs spreading; 1–5 mm thick near the base.

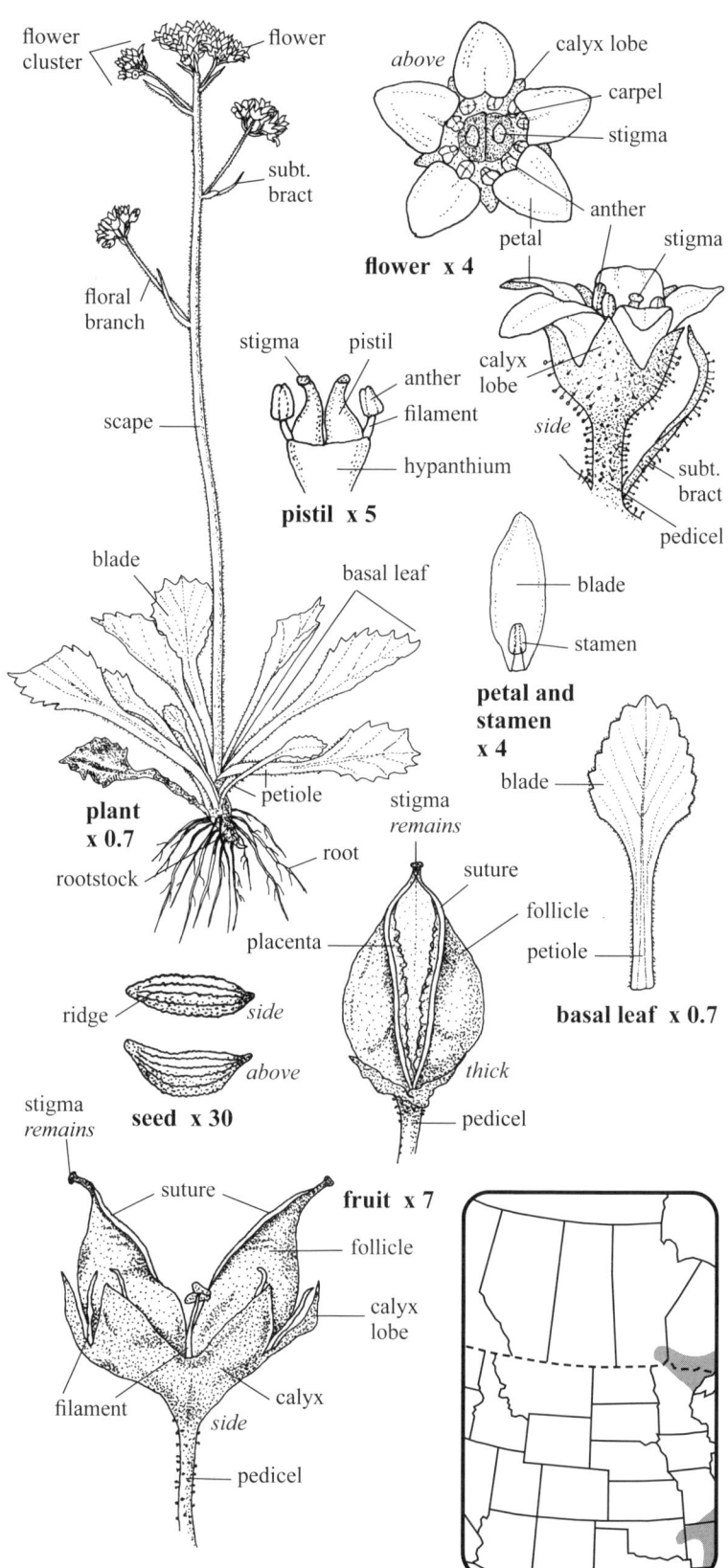

Family Characteristics

Figwort—Scrophulariaceae   Flower parts 5 (2 or 4)

SKETCH Dicotyledonous herbs (wildflowers), annual, biennial or perennial, sometimes shrubby, in dry to moist sites. Worldwide about 3,000 species, cosmopolitan, especially in the western United States.

FLOWERS Perfect, regular to bilabiate, (4-) 5-merous, sometimes with a spur. **Inflorescence** a spike, cyme, raceme or solitary in leaf axils, some with bractlets. **Calyx** tubular, 5-lobed (2- or 4-), persistent. **Corolla** tubular, 2-lipped, 5-lobed (4-), upper lip 3-lobed. **Stamens** often 4 (2 or 5), didynamous (2 long and 2 short), a fifth stamen is often sterile, inserted on the corolla tube between the lobes. **Pistil** 1, of 2 united locules. **Ovary** superior, ovules numerous. **Style** 1, erect. **Stigma** 1- or 2-parted.

FRUIT A capsule, 2-valved (4-), apical teeth variable, or a fleshy berry. **Seeds** numerous to few, some winged.

LEAVES Mostly opposite, alternate, or whorled, entire to toothed. **Stipules** often absent.

STEM Erect, solid, simple to branched.

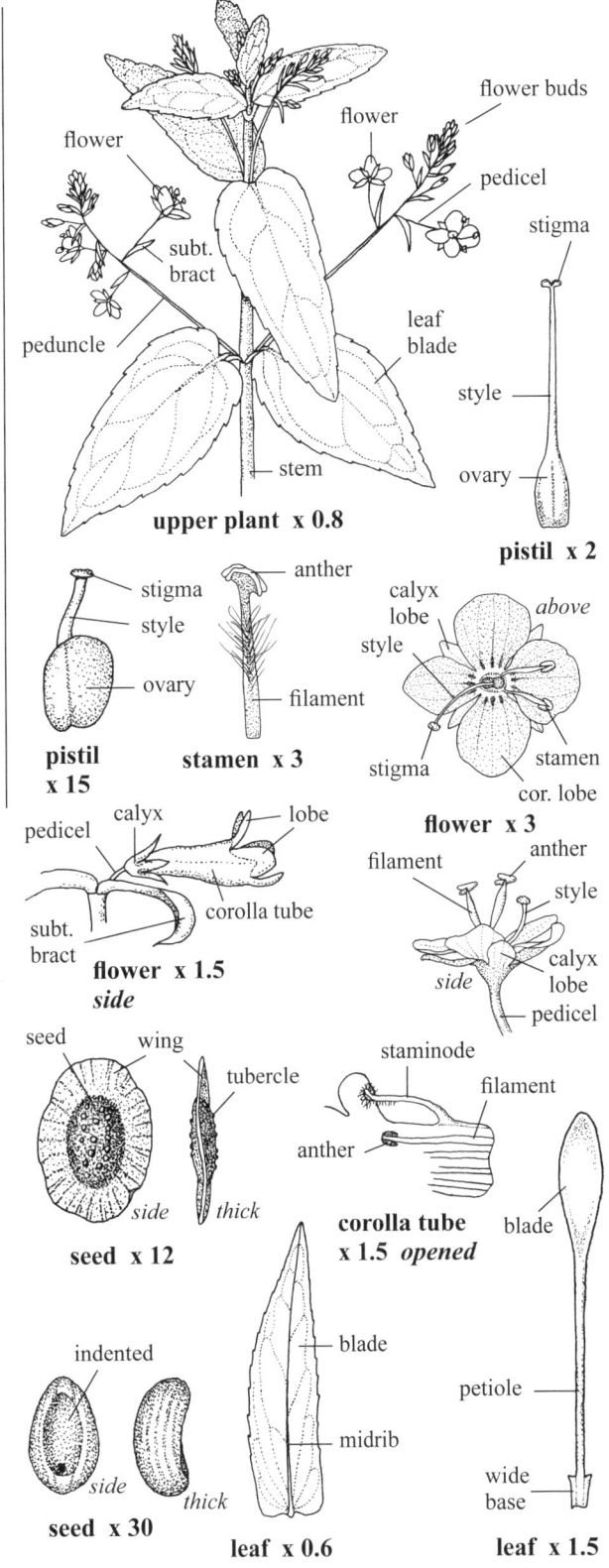

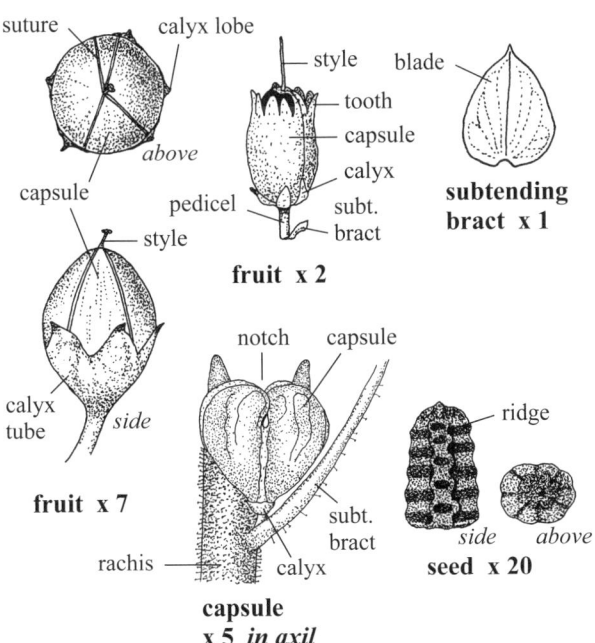

493

Figwort—*Scrophulariaceae* **Wildflower** Pink Corolla tubular, 5-lobed

SLENDER AGALINIS *Agalinis tenuifolia* (Vahl) Raf.

■ **SKETCH** A variable **annual herb** 10–60 cm tall with light tan **fibrous roots** 1–3 cm long; in dry woods, prairies and thickets to moist sites along lakeshores, stream banks and ditches.

■ **FLOWERS** Pink, blooming August–October; **inflorescence** a raceme of up to 15+ flowers along the main stem and floral branches; **pedicels** thin, 8–20 mm long, axillary; **flowers** 5-lobed, ascending, 8–12 mm wide by 10–15 mm long; **calyx** tubular, green, 5-ridged, hairless, 3.5–5.5 mm long, enclosing the fruit, 5-toothed, teeth 0.2–2 mm long, ascending; **corolla** tubular, 5-lobed, glabrous outside, 7–10 mm long by *c.* 1.3 mm wide at the base, pink spots in the throat against a white background; **corolla lobes** ciliate on margins, 4–5 mm long and wide; **stamens** four, included; **filaments** attached below to corolla tube, upper 3 mm free, outer two hairless, inner two hairy, light brown; **anthers** white, 1.3–2.2 mm long by *c.* 1 mm wide, very hairy on the side with sutures, tan on drying; **pistil** one, glabrous, 4–4.5 mm long; **ovary** *c.* 1.5 mm long by *c.* 1 mm wide; **style** one, filiform, exserted in fruit, bent at tip; **stigma** one, white, flattened, slightly wider than style. **FRUIT** a **capsule**, 3.8–6 mm long by 3–6 mm wide, smooth, dark brown, shiny, about level with the top of the calyx, blunt, slightly hairy at apex, 2-valved, splitting in half from the apex, style remains form a **beak** 0.1–0.3 mm long; **placenta** hairless, flat and central; **seeds** reticulate, dark brown, *c.* 120 per capsule, each 0.8–0.9 mm long by 0.3–0.5 mm wide by *c.* 0.3 mm thick, 4- or 5-sided.

■ **LEAVES** Opposite, entire, simple, sessile, linear, 2–7 cm long by 1–5 mm wide, pointed, ascending to spreading, subtending pedicels and floral branches, leaf bases continue as two low ridges along the stem; **basal leaves** absent.

■ **STEM** Erect, branching on tall plants, solid, green, round, slightly scabrous along the ridges, erect with fruit, turning blackish green on drying; 0.5–2 mm thick near the bare base.

■ **SYN.** *Gerardia tenuifolia* Vahl = *Agalinis tenuifolia* var. *tenuifolia* (Vahl) Raf.

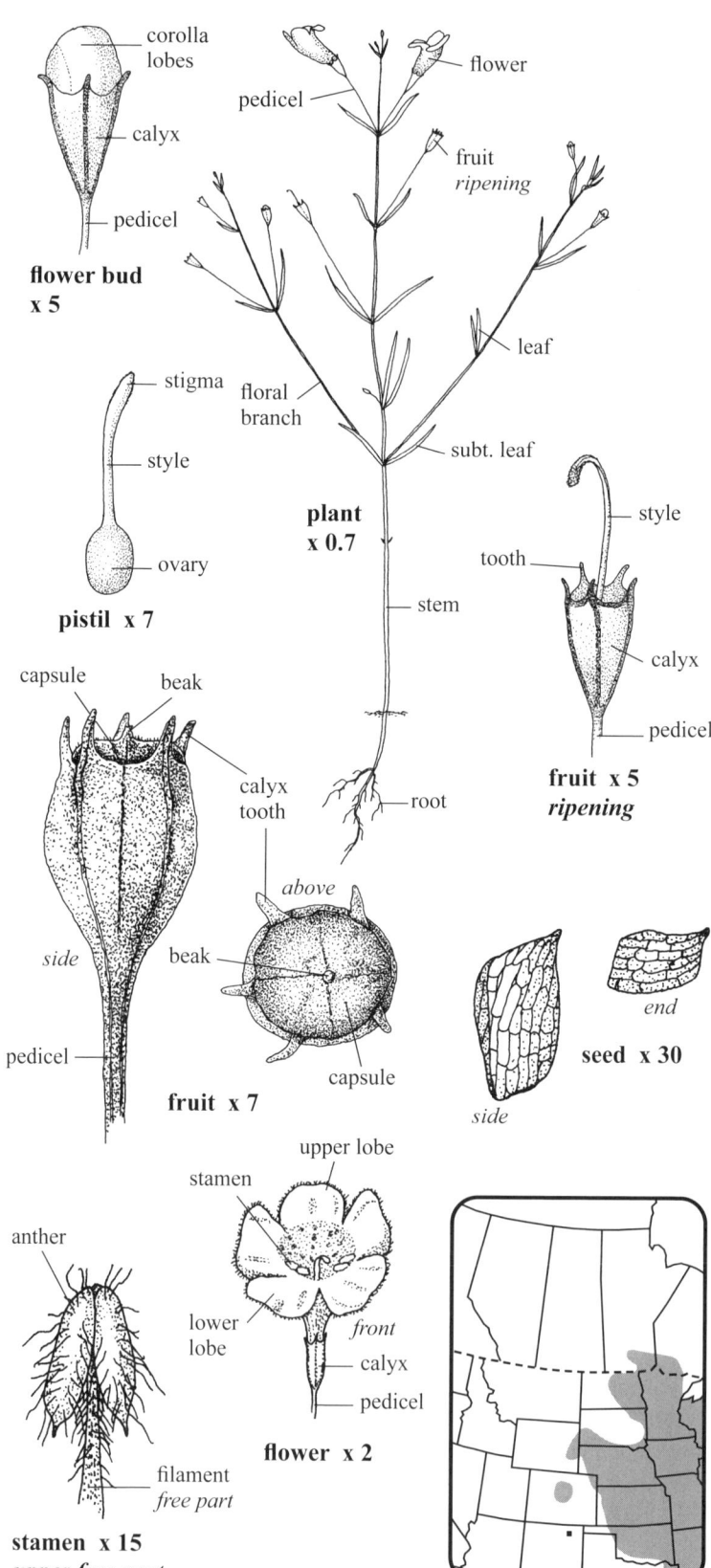

494 Slender Gerardia, Slender-leaved False Foxglove

Figwort—*Scrophulariaceae* — Wildflower   Pink and green   Corolla tubular, 2-lipped

Red Indian Paintbrush   *Castilleja miniata* Dougl. ex Hook.

■ **SKETCH** A variable **perennial herb** 25–100 cm tall from a woody **crown** or reddish brown **rhizome** 3–5 cm long by 1.5–2.5 mm thick with small curled bracts and fine **roots** 2–8 cm long; in moist meadows, thickets, streambanks and along woodland roadsides. p2

■ **FLOWERS** Pink and green, blooming June–September; **inflorescence** a spike, terminal and on upper floral branches, 2–15 cm long by 3–7.5 cm wide; **floral branches** ascending, 3–6 cm long; **upper subtending bracts** (of flowers) leaflike, reddish, white hairs often gland-tipped, 1.5–3 cm long by 8–15 mm wide, with two or four lobes or teeth per side, paper thin, reduced above; **flowers** perfect, sessile, 20–25 in terminal spike, each 3–4 mm wide by 25–44 mm long by *c.* 5 mm deep; **calyx** green at hairy base, reddish above, tubular, 17–30 mm long by *c.* 5 mm wide, 4-ridged, with numerous fine veins, 2-lobed, these 7–9 mm long, each with two erect teeth 3–9 mm long; **corolla** tubular, 25–44 mm long, hairy, nerved, exserted, 2-lipped; **upper lip** (galea) hooded, hairy below, pink and hairless on sides; **lower lip** green, 3-parted, curved, united most of their 2–2.5 mm length by a thin green membrane; **stamens** four, two long and two short (didynamous), included; **filaments** pale green, united to the corolla tube near the ovary; **anthers** pale orange, staggered along the filament, lower anther 2.3–2.5 mm long, upper anther 2.8–3 mm long and united in upper half to shorter anther, both with a few cottony hairs at base, located below the tip of the galea; **pistil** one, glabrous, exserted; **ovary** dark green, 4–5 mm long by *c.* 2 mm wide and deep, a groove leads to the apex; **style** filiform, 22–30 mm long, exserted, purplish at tip; **stigma** pinkish purple, *c.* 0.9 mm wide with a horizontal cleft; **fruiting stalks** erect, 2–8 cm long by 1.3–2 cm wide. **FRUIT a capsule**, enclosed in calyx, 8–13 mm long by 4–6 mm wide by 5–6 mm thick, slightly hairy, 2-valved, opening at apex; **seed wing** net-veined, loose and easily removed, 1.4–2.5 mm long by 0.6–1.1 mm wide and thick, tan; **seeds** 33–80 per capsule, light gray, 1–2 mm long by 0.4–0.8 mm wide and thick, slightly curved and grooved, wrinkled, short beaked.

■ **LEAVES** Alternate, simple, sessile, entire, ascending to spreading; **stem blades** 3–7 cm long by 6–15 mm wide, glabrous, green and dull.

■ **STEM** Erect, hollow below, ridged from leaf bases, branched, slightly hairy or not; 3–5 mm wide near the stiff base.

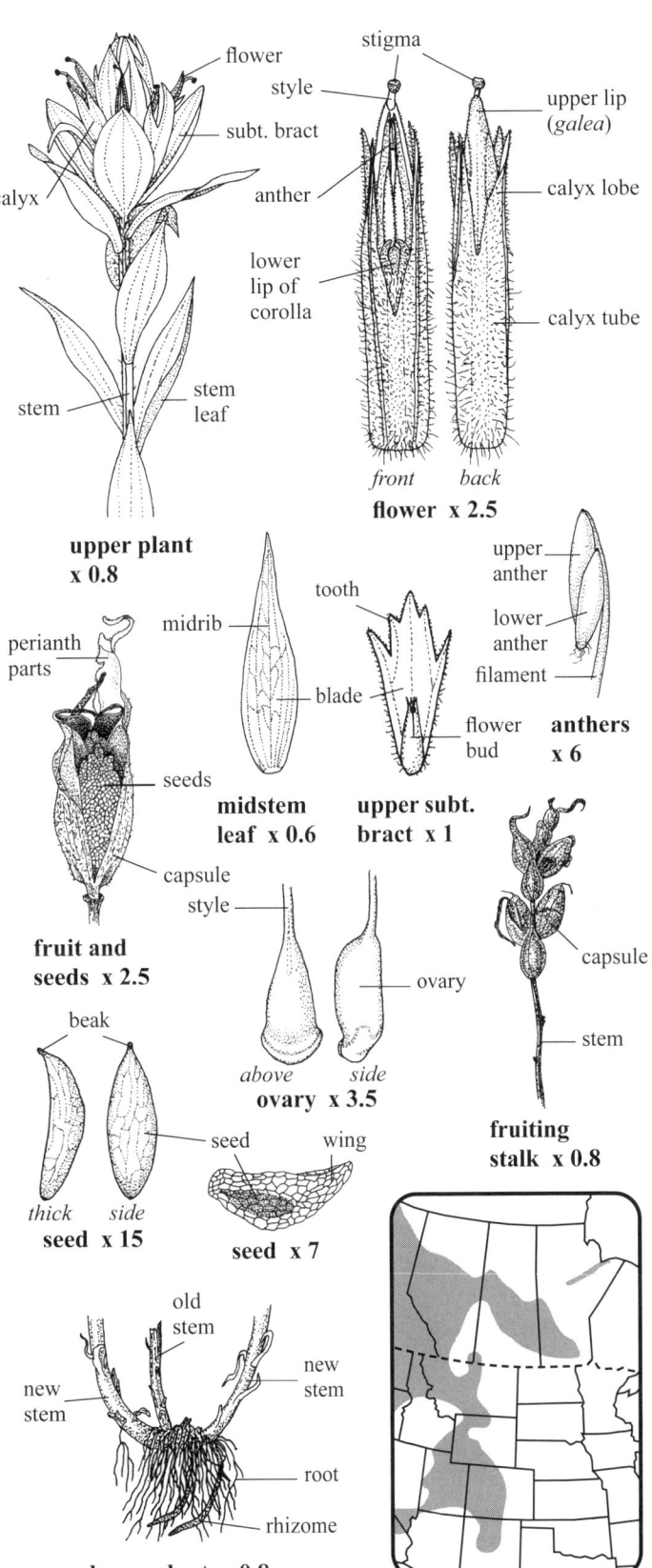

Common Red Paintbrush, Red Painted-cup, Giant Red Indian Paintbrush

Figwort—*Scrophulariaceae* **Wildflower** White with reddish blue stripes
Corolla 2-lipped, 5-lobed

SMALL-SNAPDRAGON   *Chaenorhinum minus* (L.) Lange

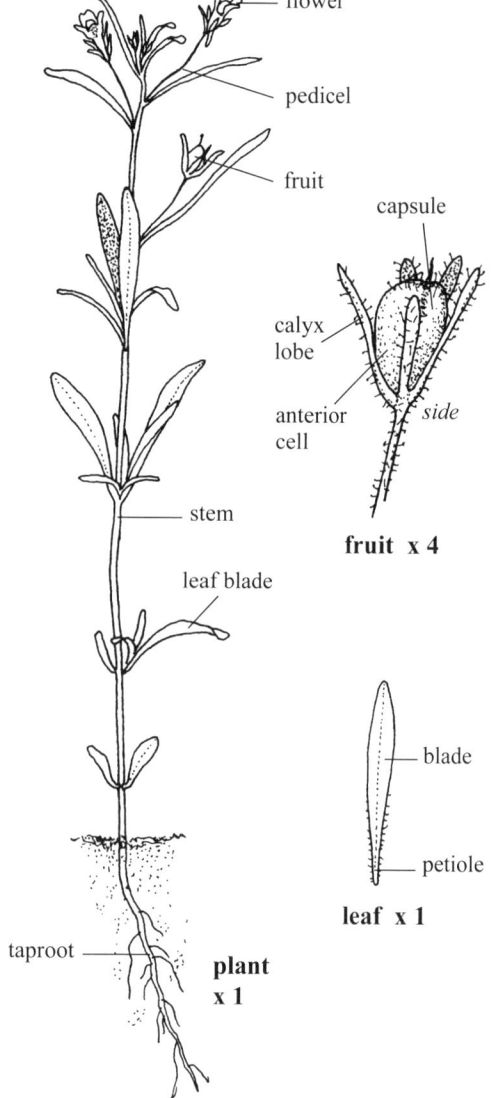

■ **SKETCH** An **annual herb** 5–40 cm tall from a white **taproot** 0.8–3 mm wide by 2–10⁺ cm long; in gravelly areas along railways and at loading docks. *Naturalized*

■ **FLOWERS** White with wide reddish purple to blue stripes, blooming May–August; **inflorescence** a raceme to *c.* 30 cm long by 2–3 cm wide; **pedicels** axillary, 4–13 mm long, to 18 mm long in fruit, densely covered with gland-tipped spreading hairs; **flowers** perfect, 3–5 mm wide by 7–9 mm long, including the glabrous **spur** 1.3–3 mm long; **calyx** deeply 5-lobed, 2–3.6 mm long, erect, lengthening to *c.* 5 mm in fruit, the lobes with glandular spreading hairs; **corolla** bilabiate; **upper lip** 2-lobed and erect, white with some glandular hairs on the back; **lower lip** 3-lobed, these mostly white except for two wide stripes entering along the recessed **palate** which does not completely close the yellow to pinkish purple throat; **stamens** four, included, two long and two short; **filaments** white, the long pair *c.* 3 mm, the short pair 1.6–1.8 mm long; **anthers** *c.* 0.8 mm long, light yellow with two reddish purple stripes; **ovary** *c.* 1 mm long and hairy; **style** 2–2.2 mm long. **FRUIT** a capsule, lightly hairy at the apex, asymmetrical with the anterior cell noticeably larger, 3–6 mm long by *c.* 3 mm wide, opening by pores, surrounded by the five calyx lobes; **seeds** about 60 per capsule, 0.5–0.8 mm long by *c.* 0.5 mm wide and thick, dark brown, with 12–15 ribs.

■ **LEAVES** Alternate above, opposite below, simple, entire, subtending pedicels above, smaller below; **blades** 0.5–3 cm long by 1–6 mm wide, slightly thickened and with some white glandular hairs along the margins near the tapered base; **petioles** 0–6 mm long.

■ **STEM** Erect, one to several, covered with spreading gland-tipped hairs, branching throughout or not; 1–3 mm thick near the base.

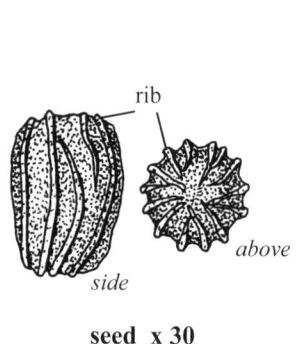

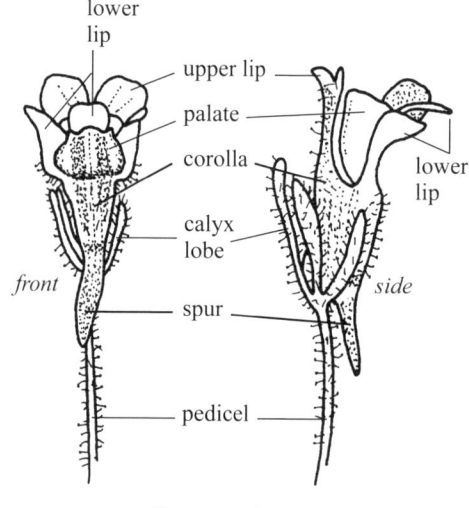

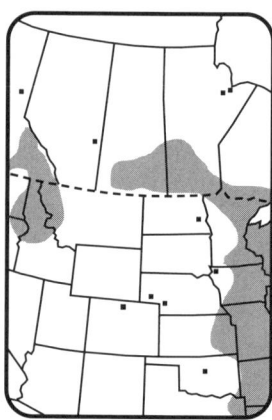

Dwarf Snapdragon

Figwort—*Scrophulariaceae* **Wildflower** White to pinkish Corolla tubular, 5-lobed

Mudwort *Limosella aquatica* L.

■ **SKETCH** A glabrous, tufted **annual herb** 3–10 cm tall or long from **fibrous roots** and reddish green **stolons** *c.* 1 mm wide; in mud at the water's edge (illustrated), or submerged in shallow pools to *c.* 15 cm deep.

■ **FLOWERS** White to pinkish, blooming June–October; **inflorescence** axillary, with one flower from each leaf axis; **peduncles** ascending, 1–10+ mm long (40 mm in fruit), usually shorter than the leaves; **flowers** perfect, apparently irregular, 2–3 mm wide by 2–2.3 mm tall; **calyx** erect, *c.* 2 mm long, 2.5–2.8 mm long in fruit; **calyx tube** 1–1.5 mm long, persistent around the fruit; **calyx lobes** five, pointed, 0.5–0.8 mm long, thick, apices spreading with tips often bent, a small reddish dot below were the lobes meet; **corolla** tubular; **corolla tube** greenish, *c.* 1.5 mm long; **corolla lobes** five, pointed to blunt, each 0.8–1 mm long by 0.4–0.7 mm wide, glabrous below with a few erect hairs scattered near the throat above, pinkish at the base of the lobes in the throat; **stamens** four (2 + 2), *c.* 0.5 mm long; **filaments** white, filiform, attached to the corolla tube below the lobes, two on one lobe opposite two on the other lobe, arching inward to the stigma; **anthers** purple, 0.2–0.3 mm long, slightly wider than the filaments; **pistil** 1.5–1.6 mm long; **ovary** green, roundish, asymmetrical, 0.8–0.9 mm long by *c.* 0.8 mm wide by *c.* 0.6 mm thick, glabrous; **style** and **stigma** *c.* 0.8 mm long; **style** one, white, filiform, offset, thicker than the filaments; **stigma** white, capitate, about twice as wide as the style. **FRUIT a capsule**, 4-valved, glabrous, 3–3.5 mm long (not including the 0.5 mm long style) by 2.3–3.3 mm wide; **seeds** brown, 0.5–0.7 mm long by *c.* 0.3 mm wide by *c.* 0.2 mm thick, numerous, several-ribbed with minute cross-reticulations.

■ **LEAVES** Appearing basal but congested in an alternate to opposite pattern on the short stem, simple, entire, glabrous, petiolate, ascending to spreading; **blades**, flat, slightly fleshy, 0.6–3 cm long by 1–12 mm wide, blunt, veins faint below, usually 3-veined, floating when growing in shallow water; **petioles** 1–12+ cm long depending on the water's depth, roundish, not grooved, broadening into a hyaline stipulelike base 1–2.5 mm long.

■ **STEM** Short, light tan, round; 1–1.3 mm wide.

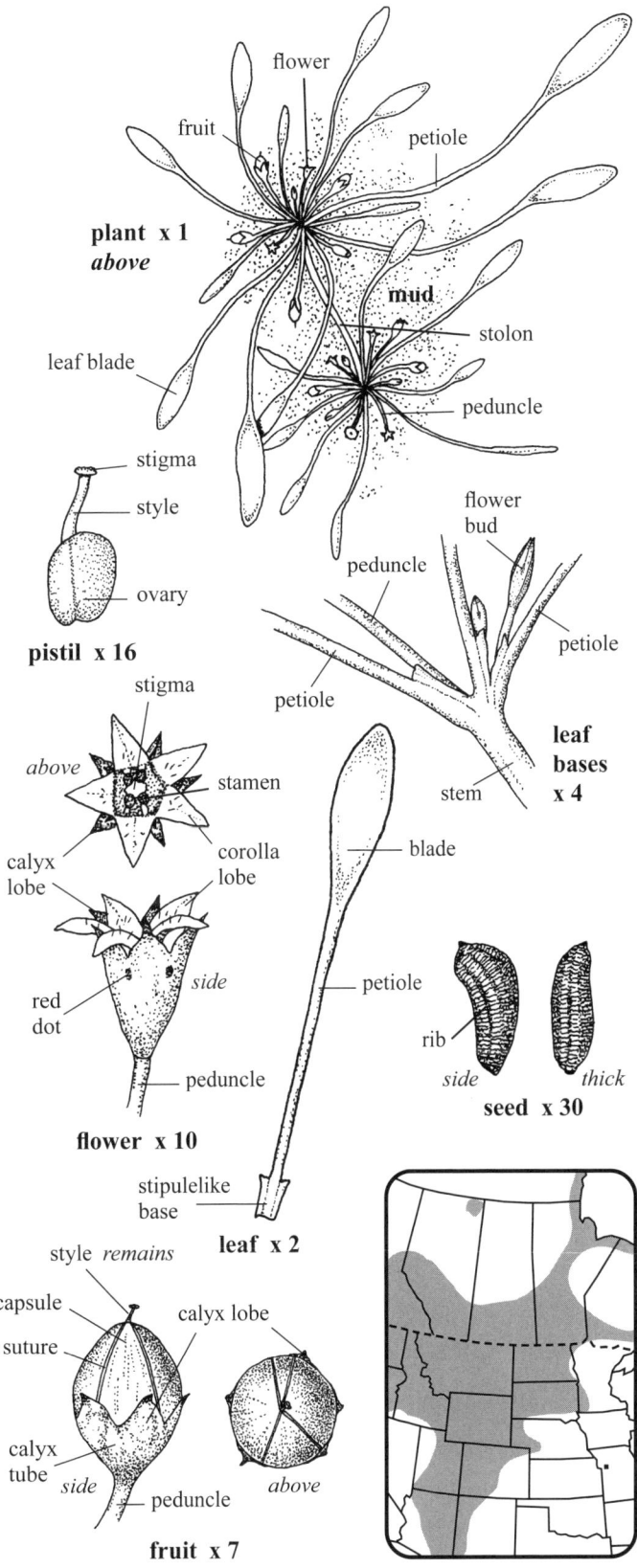

Awl-leaved Mudwort, Water Mudwort

Figwort—*Scrophulariaceae*

Wildflower Yellow, orange throat
Corolla tubular, 2-lipped, 5-lobed

BUTTER-AND-EGGS *Linaria vulgaris* P. Mill.

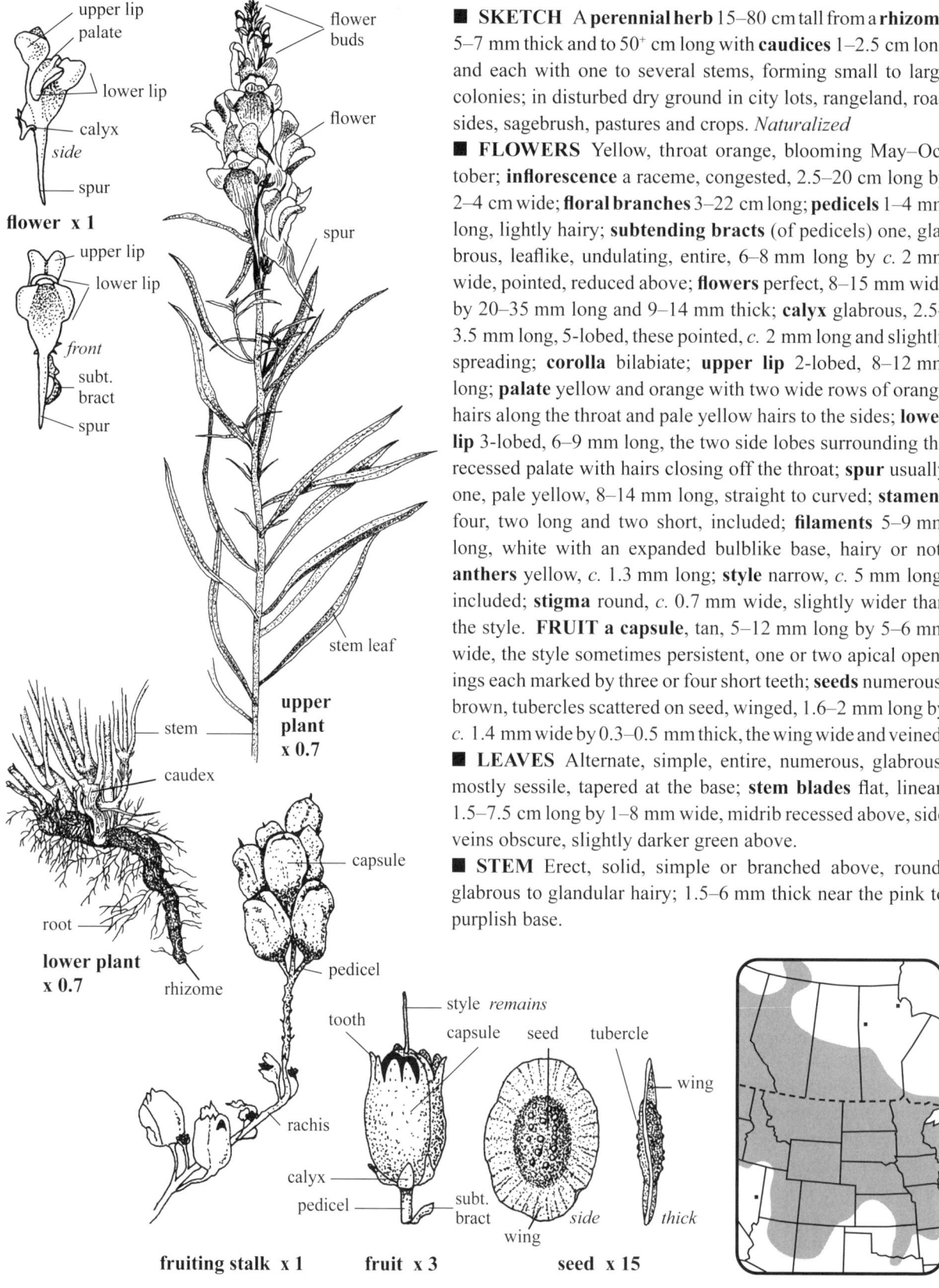

■ **SKETCH** A **perennial herb** 15–80 cm tall from a **rhizome** 5–7 mm thick and to 50+ cm long with **caudices** 1–2.5 cm long and each with one to several stems, forming small to large colonies; in disturbed dry ground in city lots, rangeland, road sides, sagebrush, pastures and crops. *Naturalized*

■ **FLOWERS** Yellow, throat orange, blooming May–October; **inflorescence** a raceme, congested, 2.5–20 cm long by 2–4 cm wide; **floral branches** 3–22 cm long; **pedicels** 1–4 mm long, lightly hairy; **subtending bracts** (of pedicels) one, glabrous, leaflike, undulating, entire, 6–8 mm long by *c.* 2 mm wide, pointed, reduced above; **flowers** perfect, 8–15 mm wide by 20–35 mm long and 9–14 mm thick; **calyx** glabrous, 2.5–3.5 mm long, 5-lobed, these pointed, *c.* 2 mm long and slightly spreading; **corolla** bilabiate; **upper lip** 2-lobed, 8–12 mm long; **palate** yellow and orange with two wide rows of orange hairs along the throat and pale yellow hairs to the sides; **lower lip** 3-lobed, 6–9 mm long, the two side lobes surrounding the recessed palate with hairs closing off the throat; **spur** usually one, pale yellow, 8–14 mm long, straight to curved; **stamens** four, two long and two short, included; **filaments** 5–9 mm long, white with an expanded bulblike base, hairy or not; **anthers** yellow, *c.* 1.3 mm long; **style** narrow, *c.* 5 mm long, included; **stigma** round, *c.* 0.7 mm wide, slightly wider than the style. **FRUIT** a capsule, tan, 5–12 mm long by 5–6 mm wide, the style sometimes persistent, one or two apical openings each marked by three or four short teeth; **seeds** numerous, brown, tubercles scattered on seed, winged, 1.6–2 mm long by *c.* 1.4 mm wide by 0.3–0.5 mm thick, the wing wide and veined.

■ **LEAVES** Alternate, simple, entire, numerous, glabrous, mostly sessile, tapered at the base; **stem blades** flat, linear, 1.5–7.5 cm long by 1–8 mm wide, midrib recessed above, side veins obscure, slightly darker green above.

■ **STEM** Erect, solid, simple or branched above, round, glabrous to glandular hairy; 1.5–6 mm thick near the pink to purplish base.

Yellow Toadflax, Toad-flax, Greater Butter-and-eggs

Figwort—*Scrophulariaceae* **Wildflower** Purple to blue, throat orange
Corolla tubular, 2-lipped

BLUE MONKEYFLOWER *Mimulus ringens* L.

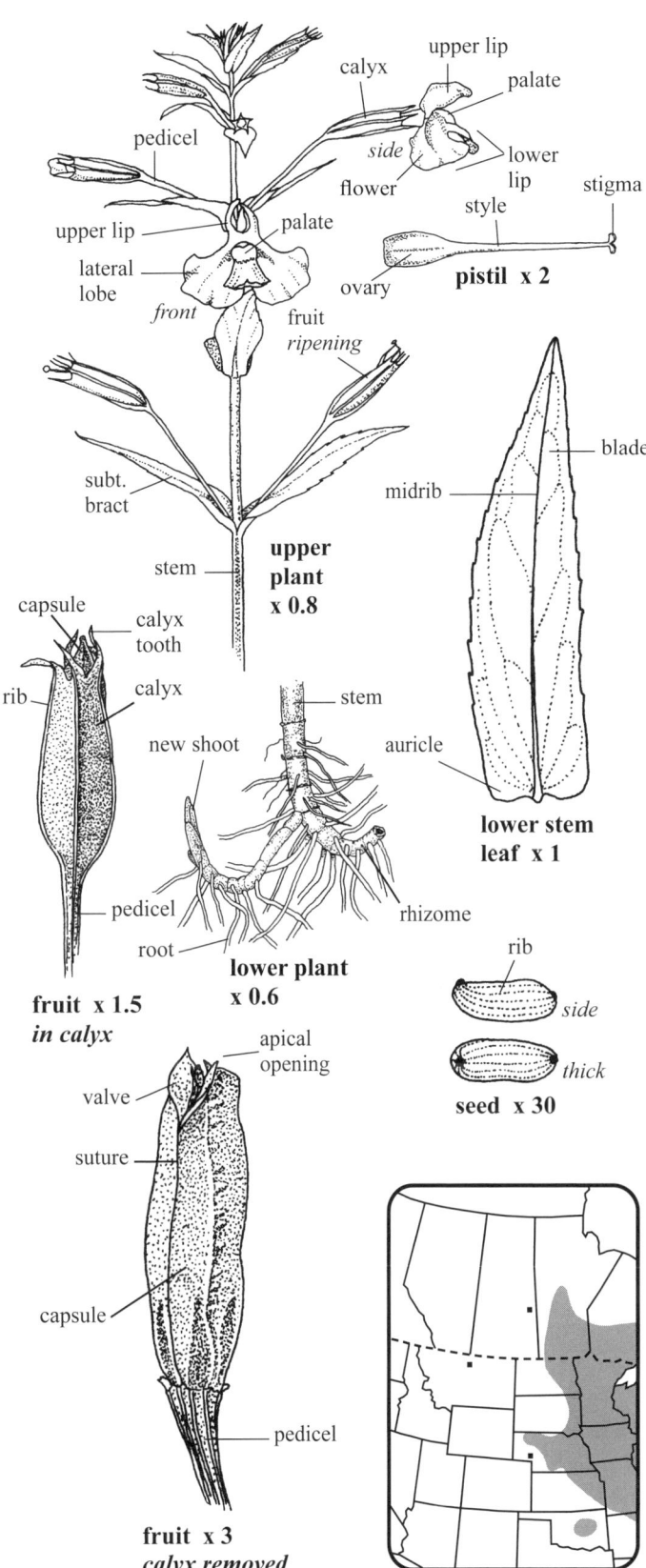

■ **SKETCH** A glabrous **perennial herb** 20–130 cm tall with white **rhizomes** 3–6 mm thick by 5–10 cm long and tan **fibrous roots** to *c.* 1 mm wide by *c.* 15 cm long from nodes; along pond and marsh edges, wet woods, sometimes in shallow water.

■ **FLOWERS** Purple to blue, throat orange, blooming June–September; **inflorescence** a raceme 10–15 cm long by 7–8 cm wide of 1–16 flowers; **pedicels** 2–4.5 cm long, ascending from axils of bracts; **flowers** perfect, opposite, last one day, 15–25 mm wide by 20–30 mm long; **calyx** tubular, 5-ribbed, green, TL 10–18 mm, hairless, sharply angled, 5-toothed, the teeth 2–6 mm long, scabrous on inside margin, short hairy inside especially near apices; **corolla** tubular, 2-lipped, glandular-hairy, with two hairy ridges in tube as a continuation of the orange palate; **corolla tube** *c.* 15 mm long, yellowish near the base; **upper lip** 2-lobed, these 8–11 mm long, ascending; **lower lip** 3-lobed, lateral lobes two, these 10–12 mm wide, middle lobe 8–10 mm wide, declining; **throat** closed, with two elevated ridges 3–5 mm wide forming a hairy palate; **stamens** four, often included, didynamous (2 long, 2 short), 15–17 mm long with the upper 5–7 mm of each filament free and hairless, lower filament hairy and attached to corolla tube; **anther sacs** divergent, *c.* 1.5 mm wide, wider than long, surrounding the stigma; **pistil** 16–20 mm long, glabrous; **ovary** green, *c.* 5 mm long by *c.* 2.5 mm wide, oval in cross-section; **style** filiform, white, finely hairy; **stigmas** 2-lobed, white, *c.* 1.6 mm wide (together), blunt. **FRUIT** a **capsule**, cylindrical, 10–14 mm long by *c.* 5 mm wide by 3–4.3 mm thick, 2-ribbed on opposite sides, veined, opening at apex along two sutures, tan, hairless, hidden by calyx; **seeds** hundreds per capsule, light brown, 0.3–0.5 mm long by *c.* 0.2 mm wide and thick, pointed at both dark brown ends, slightly shiny, *c.* 10 dark, papillose ribs.

■ **LEAVES** Opposite, simple, sessile, ascending to spreading, dull, lighter green below, hairless, weakly toothed, 2–10 cm long by 6–35 mm wide, reduced above and subtending the pedicels, auriculate, nerves faint, often with fascicled leaves in the axils; **lower blades** wilted by anthesis.

■ **STEM** Erect, mostly unbranched, stiff, hollow, smooth, squarish above with opposite wide grooves; round and pinkish near the 3–7 mm wide base.

Alleghany Monkeyflower, Square-stemmed Monkey-flower, Monkey Flower

Figwort—*Scrophulariaceae* — Wildflower Yellow Corolla tubular, 2-lipped

SWAMP LOUSEWORT *Pedicularis lanceolata* Michx.

■ **SKETCH** A hemiparasitic **perennial herb** 30–90 cm tall from tan **rhizomes** 3–4 mm thick; in moist prairies, bogs and marshes.

■ **FLOWERS** Pale yellow to creamy, blooming July–October; **inflorescence** a raceme (spikelike), 2–20 cm long by 3–4.5 cm wide, terminal and on lateral floral branches 5–8 cm long; **pedicels** 0–4 mm long; **subtending bracts** (of flowers) one, leaflike, longer to shorter than the flower, some with auricles; **flowers** perfect, loose below, crowded above, 15–25 mm long, laterally compressed, remain closed; **calyx** light green, 8–15 mm long including the two, toothed lobes at the apex, lobes 3–5 mm long and wide, spreading; **calyx tube** *c.* 8 mm long by 5–6 mm wide by *c.* 4 mm thick, glabrous; **corolla** bilabiate, TL 15–25 mm by 7–9 mm tall by 3–4 mm thick, glabrous, tubular below; **corolla tube** 9–11 mm long by 3–4 mm tall, flattened; **upper lip** (galea) 8–13 mm long, curved and united at the truncate apex, entire, and fitting slightly inside the lower lip; **lower lip** 7–12 mm long with three apical lobes, all lobes *c.* 3 mm long by *c.* 4 mm wide (flattened), inside the lip are two parallel, longitudinal ridges with white hairs on the lower half; **stamens** four, included, equal; **filaments** white, flattened, *c.* 2 cm long by *c.* 0.5 mm wide, winged, attached at base of corolla tube and hairy along the lower 4–5 mm; **anthers** yellow, 2.3–3 mm long; **pistil** TL 25–30 mm, included to exserted *c.* 1 mm past apex of the galea; **ovary** green, glabrous, 4–5 mm long, laterally compressed; **style** white, filiform, curved at apex to conform to top of upper lip, *c.* 25 mm long by *c.* 0.3 mm wide. **FRUIT a capsule**, ascending, sessile, dark brown, 9–13 mm long by 5–7 mm wide by 4–5 mm thick, veined, slightly flattened, 2-valved, opening at the pointed apex and along the inner suture facing the rachis, each half with a central partition for a total of four chambers, enclosed in persistent calyx except for the apex; **seeds** numerous, dark brown, 2.8–3.1 mm long by *c.* 1.5 mm wide by 0.8–1 mm thick, winged along three sides, hairless, striations slightly irregular.

■ **LEAVES** Opposite below, alternate above among the flowers, simple, toothed lobes to less than halfway to the midrib, ascending to spreading, sessile to short petiolate, scabrous; **blades** 2–12 cm long by 0.5–3 cm wide, lateral veins hairy above, less so below, midrib glabrous below.

■ **STEM** Erect, mostly round, sometimes decumbent at the base, one to five, smooth, simple to few branched, lower stem hairy in a wide band up from the base of the leaves, or glabrous; 1–6 mm thick near the base.

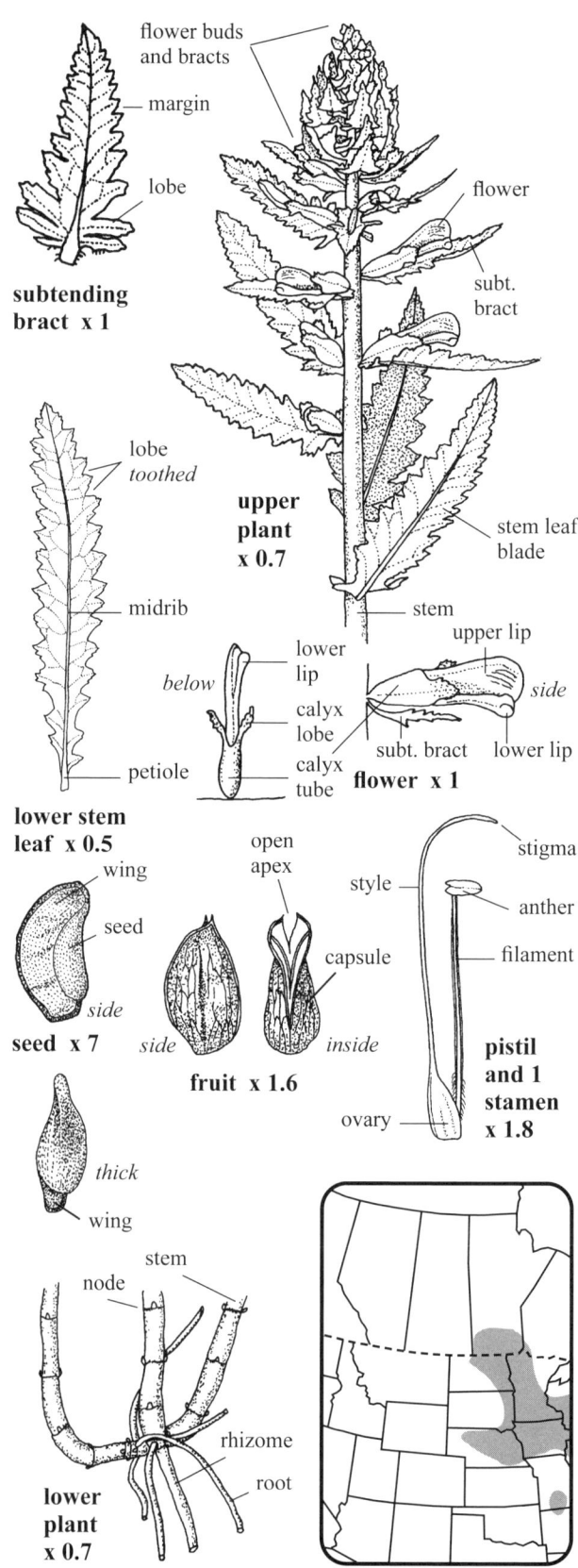

500

Figwort—*Scrophulariaceae* **Wildflower** Pale purple Corolla tubular, 2-lipped, 5-lobed

LILAC-FLOWERED BEARDTONGUE   *Penstemon gracilis* Nutt.

■ **SKETCH** A glandular-hairy **perennial herb** 15–60 cm tall with a **caudex** atop a **taproot**; in prairies, slough margins, open woods, sandhills, in valleys and dry sites in lower mountains. **p2**

■ **FLOWERS** Pale purple, whitish in throat and below, blooming May–August; **inflorescence** a thyrse, 4–20 cm long by 5–7 cm wide; **verticillasters** in the upper 2–5 nodes, 2- to 8-flowered; **subtending bracts** (of peduncles) 1.2–6 cm long, leaflike, reduced above; **peduncles** erect, 5–22 mm long, reduced above; **subtending bracts** (of pedicels) 5–10 mm long; **pedicels** 2–6 mm long, hairy; **flowers** perfect, 9–12 mm wide by 12–25 mm long, spreading; **calyx** tubular, 5-lobed, 6–8 mm long, glandular-hairy, 10-veined and persistent, lobes pointed, entire, 4–6 mm long by 1.5–2 mm wide, tips recurved; **corolla** tubular, bilabiate, glandular-hairy; **corolla tube** 2-grooved below with a central ridge; **upper lip** 2-lobed, those curled backward and 7–8 mm wide; **lower lip** 3-lobed, extended past upper lip; **throat** raised, white with several thin purple lines, white hairs *c.* 1 mm long; **stamens** four, didynamous (2 + 2), fertile, included, 12–15 mm long; **filaments** white, long ones attached along their 5–6 mm long wrinkled bases to the corolla tube, the two shorter filaments are free to their bases; **anthers** dark purple and white, *c.* 1.7 mm long by 1.5–2 mm wide with minute hairs along the sutures; **staminode** one, 14–16 mm long, slightly exserted, visible in throat, the upper 8–9 mm covered with dense yellow eglandular hairs on one side only; **pistil** glabrous, included, 12–15 mm long; **ovary** tapered, *c.* 3 mm long by *c.* 2 mm wide with four low ridges; **style** white, filiform, 9–12 mm long, tip bent; **stigma** white, round, *c.* 0.6 mm wide. **FRUIT a capsule**, glabrous, erect, brown, 2-valved, eventually opening at apex, 8–14 mm long by 6–8 mm wide, lower half enclosed in calyx, the six teeth recurved, style persistent but eventually breaking off; **seeds** numerous, black with a brown wing at one end, reticulate, 0.7–0.9 mm long by 0.3–0.5 mm wide by 0.2–0.3 mm thick, angular, usually 4-sided.

■ **LEAVES** Basal and stem, opposite, simple, toothed to entire, spreading, arched, pointed, dull, glabrous or with short hairs; **basal blades** 2.5–7 cm long by 4–15 mm wide; **petioles** 8–22 mm long; **stem blades** sessile, clasping, 3–9 cm long by 3–11 mm wide, reduced above, teeth short.

■ **STEM** Erect, unbranched, tough, one to several, green to reddish near the top, glabrous below, glandular-hairy above, hairs spreading and white; 2–3 mm wide at the curved to straight base.

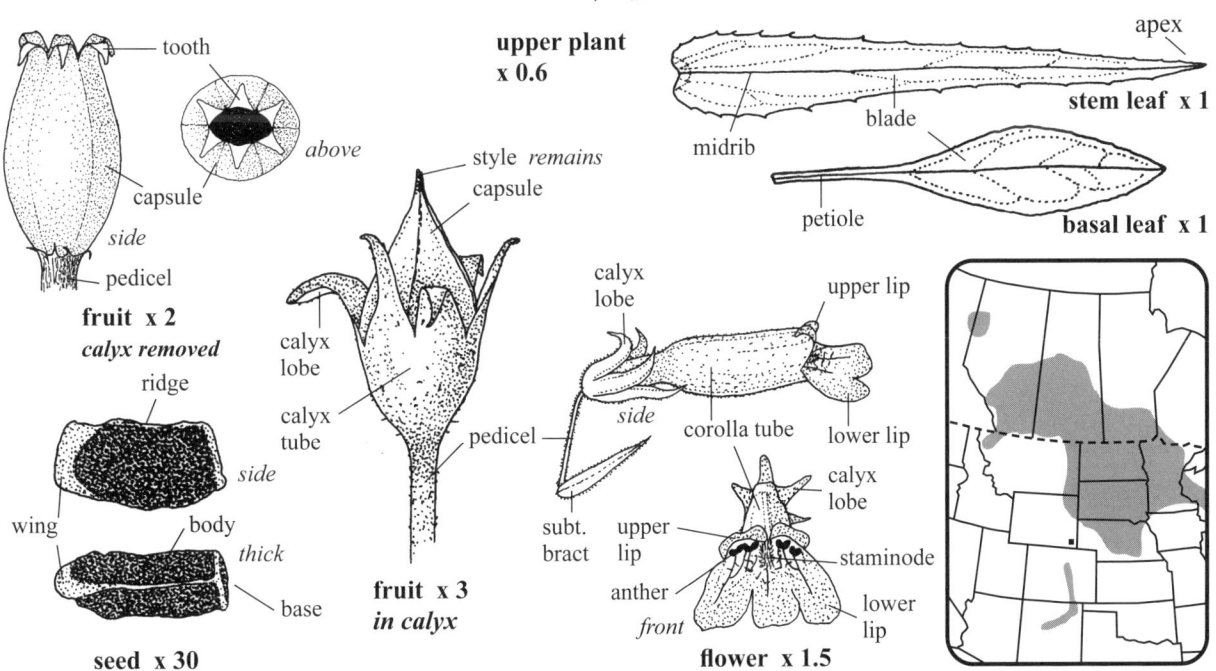

Slender Beardtongue, Lilac Penstemon, Lilac Beardtongue

Figwort—*Scrophulariaceae*

Wildflower Blue, rarely red to white
Corolla tubular, 2-lipped, 5-lobed

Smooth Blue Beardtongue   *Penstemon nitidus* Dougl. ex Benth.

- **SKETCH** A glabrous and glaucous, often tufted **perennial herb** 15–35 cm tall from a woody **caudex** atop a **taproot**, often forming small patches; along dry sandy hillsides and eroded river banks. p1
- **FLOWERS** Usually blue, rarely red to white, blooming May–June; **inflorescence** a thyrse, 4–17 cm long by 3–3.5 cm wide with verticillasters of 2–5 crowded flowers from axils of subtending bracts; **pedicels** green, erect to spreading at anthesis, 2–4 mm long, lengthening to 8–10 mm in fruit; **subtending bracts** (of pedicels) one, reduced to *c.* 5 mm long by *c.* 2 mm wide above, larger below; **flowers** perfect, 8–12 mm wide by 12–18 mm long by *c.* 10 mm tall at mouth; **calyx** 4–5 mm long, tubular, cut almost to the base, calyx teeth five, pointed, longer than the tube; **corolla** tubular, 2-lipped, tube 11–14 mm long, 5-veined, wrinkled on inner surface; **upper lip** is 2-lobed; **lower lip** 3-lobed; **stamens** four, fertile, 10–12 mm long, included, outer two attached at base, inner two attached *c.* 2 mm above base; **anthers** dark brown to purple, 0.7–1.2 mm long; **staminode** one, sterile, 1–2 mm longer than stamens, yellow hairs near the apex; **filaments** (of staminode) attached to base of corolla tube for about the lower 4 mm, not wrinkled, apex bluish purple, hairs long and short, dishevelled; **pistil** glabrous, included, 10–12 mm long; **ovary** 2–3 mm long by *c.* 1.2 mm wide, green, apex tapered; **style** filiform, *c.* 8 mm long by *c.* 0.2 mm wide; **stigma** white, *c.* 0.3 mm wide or slightly wider than the style. **FRUIT a capsule**, persistent, ascending, 4-valved, 8–13 mm long by 5–8 mm wide by 4–7 mm thick, opening from apex; **seeds** dull, *c.* 18 per capsule, 2.5–4.5 mm long by 2–3 mm wide and thick, dark brown, reticulate, angular, ridges thin and winglike; **septum** incomplete in each half.
- **LEAVES** Basal and stem, opposite, simple, entire, glabrous, glaucous; **basal leaves** TL 8–12 cm (including petiole) by 8–12 mm wide, spreading to ascending, faintly veined; **petioles** 3–6 cm long, tapered, slightly winged; **stem leaves** sessile, arched to erect, clasping, largest at midstem, mucronate, 2–5 cm long by 5–30 mm wide.
- **STEM** Erect, often branching, lighter green than the leaves, one to several stems from the caudex; 3–6 mm thick near the base.

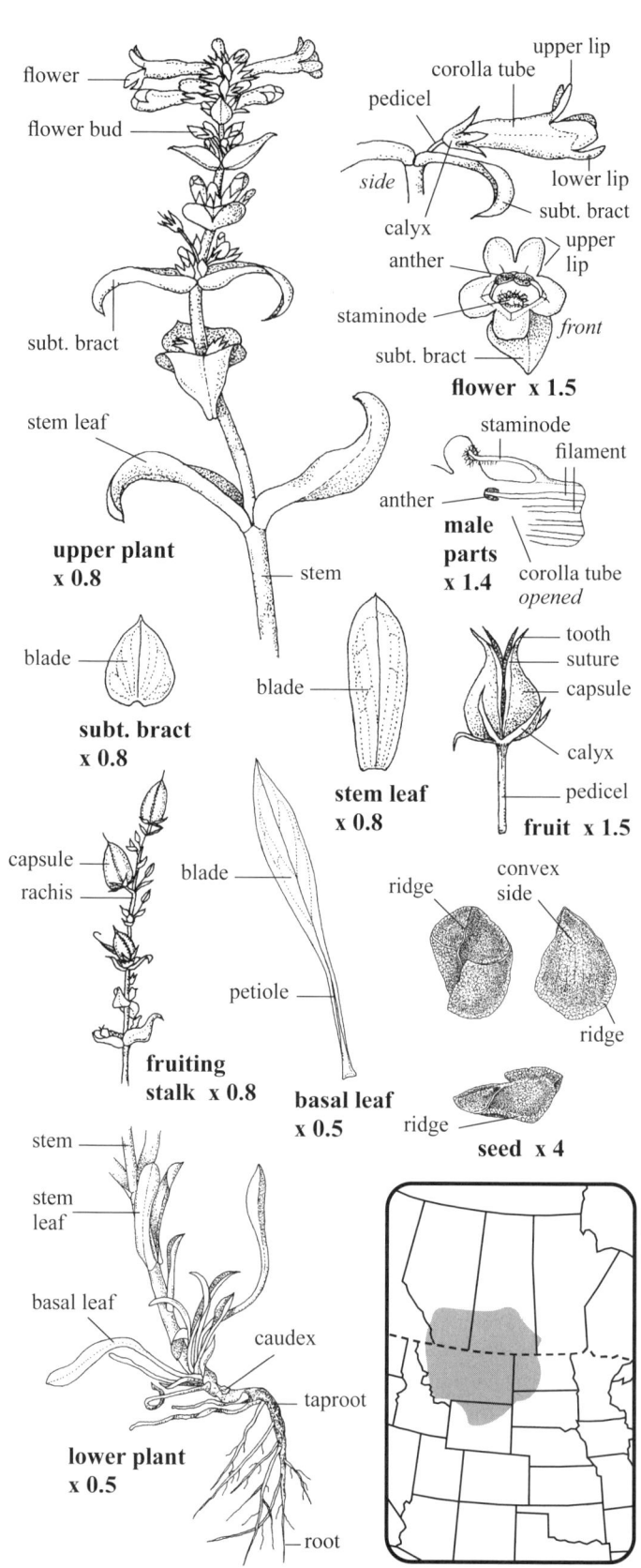

Figwort—*Scrophulariaceae* **Wildflower** Yellow Corolla tubular, 5-lobed

Common Mullein *Verbascum thapsus* L.

■ **SKETCH** An **annual, biennial** or **perennial herb** covered with multi-branched soft hairs throughout, 34–244 cm tall with a rough, brown **taproot** 12–30 cm long by 5–12 mm wide; in fields, edges of gravelly railways and roadsides. *Naturalized* **w2**

■ **FLOWERS** Yellow, blooming June–September; **inflorescence** a spike, compact, terminal, forming once, then the plant dies, erect, 6–113 cm long by 2–3.5 cm wide, sometimes with floral branches to 40 cm long at base; **flowers** perfect, sessile, yellow, 13–25 mm wide by 12–17 mm long, few to several blooming at one time; **subtending bracts** (of flowers) three, one large keeled bract and two smaller ones *c.* 9 mm long by *c.* 2 mm wide, reduced above, entire, pointed, very hairy; **calyx** tubular, 5-lobed, TL 6–10 mm, very hairy, lobes 4–5 mm long, pointed, erect around fruit; **corolla** tubular, 5-lobed, these ascending to spreading and 6–9 mm long by 4–9 mm wide, upper two lobes the smallest, hairless inside, hairy outside on lobes; **stamens** five; three upper **filaments** *c.* 6 mm long with yellow hairs and two lower ones *c.* 8 mm long are glabrous or slightly hairy, attached to base of corolla tube; **anthers** orange, hairless and curved; **pistil** one, 9–14 mm long; **ovary** very hairy, 3–5 mm long by *c.* 2.7 mm wide, pale green and tapered above to a green filiform style; **style** 6–8 mm long, hairy at base, glabrous above; **stigma** green, angled, *c.* 1 mm long; **fruiting stalks** erect, overwintering, brown. **FRUIT a capsule**, tan, 6–10.5 mm long by 5–6 mm wide and thick, 2-valved, apical opening, hairy, veined, in spirals on curved pedicels 2–4 mm long; **seeds** dark brown, numerous, 0.7–1 mm long by *c.* 0.6 mm wide and thick, ridged and angular.

■ **LEAVES** Basal and stem, hairs stalked and with several branches; **rosettes** 5–60 cm across of 5–55 leaves, often growing a few years; **basal leaves** twelve or more, spreading, TL 13–30 cm by 3–12 cm wide; **petioles** hairy, pale green, 3–7 cm long by 4–10 mm wide, not grooved; **stem leaves** alternate, numerous, imbricate, finely toothed, arched below to erect above, 2–56 cm long by 1–16 cm wide, reduced and sessile above, upper blades with wavy margins and twisted pointed tips, the blade edges continue along the stem forming wavy parallel wings.

■ **STEM** Erect, rarely branched, overwintering, mostly hidden by continuous leaf bases; 8–20 mm wide near the leafy hairy base.

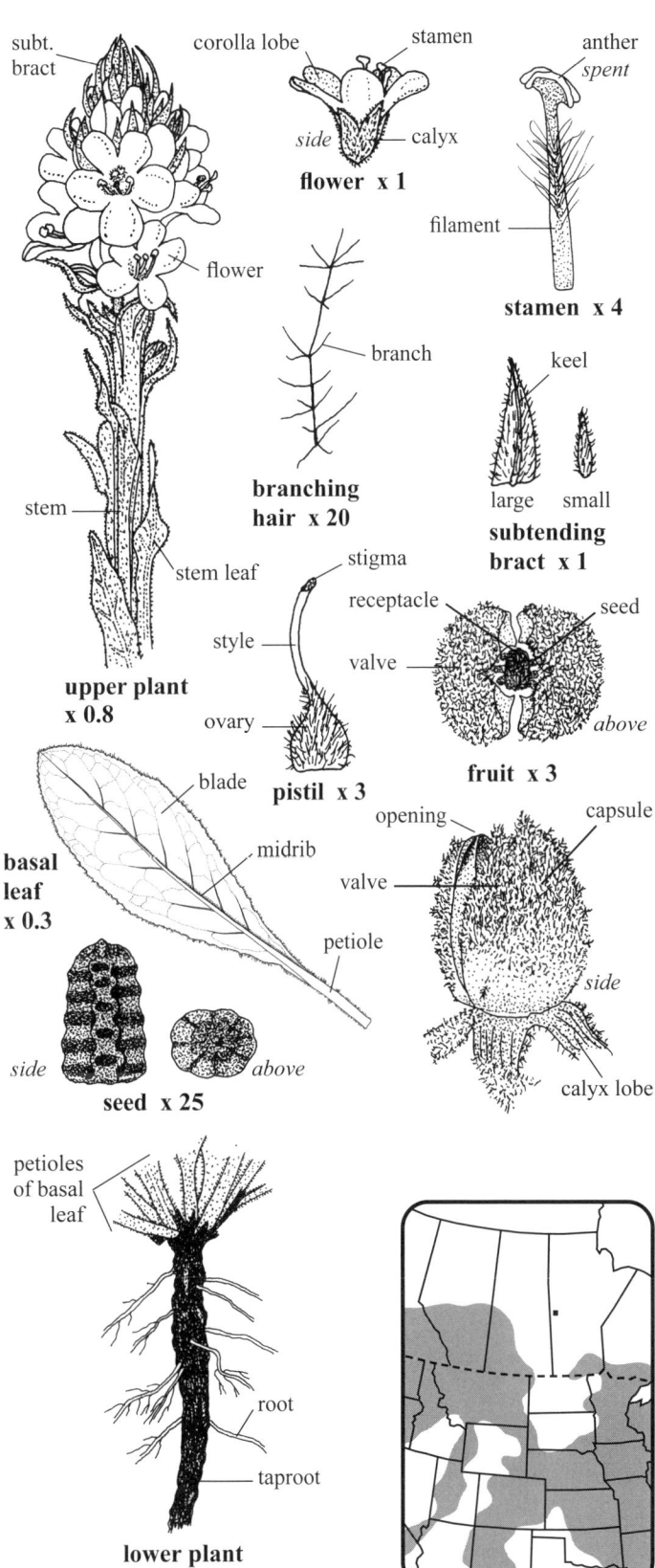

Great Mullein, Flannel Mullein, Flannel-plant, Woolly Mullein

Figwort—*Scrophulariaceae* **Wildflower** Light blue, center white Corolla tubular, 4-lobed

AMERICAN SPEEDWELL *Veronica americana* Schwein. ex Benth.

■ **SKETCH** A glabrous **perennial herb** 20–60 cm tall or long from a decumbent lower stem 5–20 cm long with **fibrous roots** 2–13 cm long from the underground nodes; in muddy soils of seeps, ditches, and along lakeshores, streams and wet meadows.

■ **FLOWERS** Light blue, centers white, blooming June–August; **inflorescence** a raceme, several from the upper leaf axils, each 1–14 cm long by 1–1.5 cm wide with 2–28 flowers per raceme; **peduncles** 1–8 cm long, ascending, green, naked; **pedicels** 3–13 (–18) mm long, spreading to ascending, longest with fruit; **subtending bracts** (of pedicels) green, 5–15 mm long by 0.7–2.3 mm wide, entire, pointed; **flowers** perfect, 5–9 mm wide by 4–5 mm long; **calyx** green, tubular, 4-lobed; **tube** *c.* 0.3 mm long; **lobes** 2.5–4.5 mm long by 1.2–1.6 mm wide, erect to spreading, 3-nerved; **corolla** tubular, 4-lobed; **tube** white, *c.* 0.8 mm long and slightly hairy on the inside; **lobes** 2–2.5 mm long by 1.2–3.5 mm wide, veins blue becoming reddish and wide at base; **stamens** two, exserted, 3.5–4 mm long, attached at base of widest lobe; **filaments** pale purple in middle, white at both tapered ends, *c.* 3 mm long; **anthers** 1–1.2 mm long, 2-lobed, pale pinkish purple; **pistil** one, exserted, 2.5–3.5 mm long; **ovary** green, 2-lobed, *c.* 0.5 mm long by 0.7–1 mm wide; **style** one, pinkish near apex in bud, persistent; **stigma** white, wider than style. **FRUIT a capsule**, 2.5–4 mm long by 2.5–4.3 mm wide by *c.* 2.3 mm thick, often slightly notched, somewhat transparent; **seeds** 30–45 per capsule, each 0.4–0.7 mm long by 0.4–0.6 mm wide by 0.2–0.3 mm thick, plano-convex, medium brown, smooth, marginal ridge faint, hilum on flat side; **receptacle** pale green, rough, white hairs stellate, two in the middle and each serving one side of the capsule.

■ **LEAVES** Opposite, simple, toothed, dull; **blades** 2–9.3 cm long by 7–30 mm wide; **petioles** 2–8 mm long, reduced above; **stipules** absent.

■ **STEM** Erect, rooting at lower nodes, smooth, pale green, solid, weak, sometimes branched, roundish; 3–6 mm thick near the decumbent base.

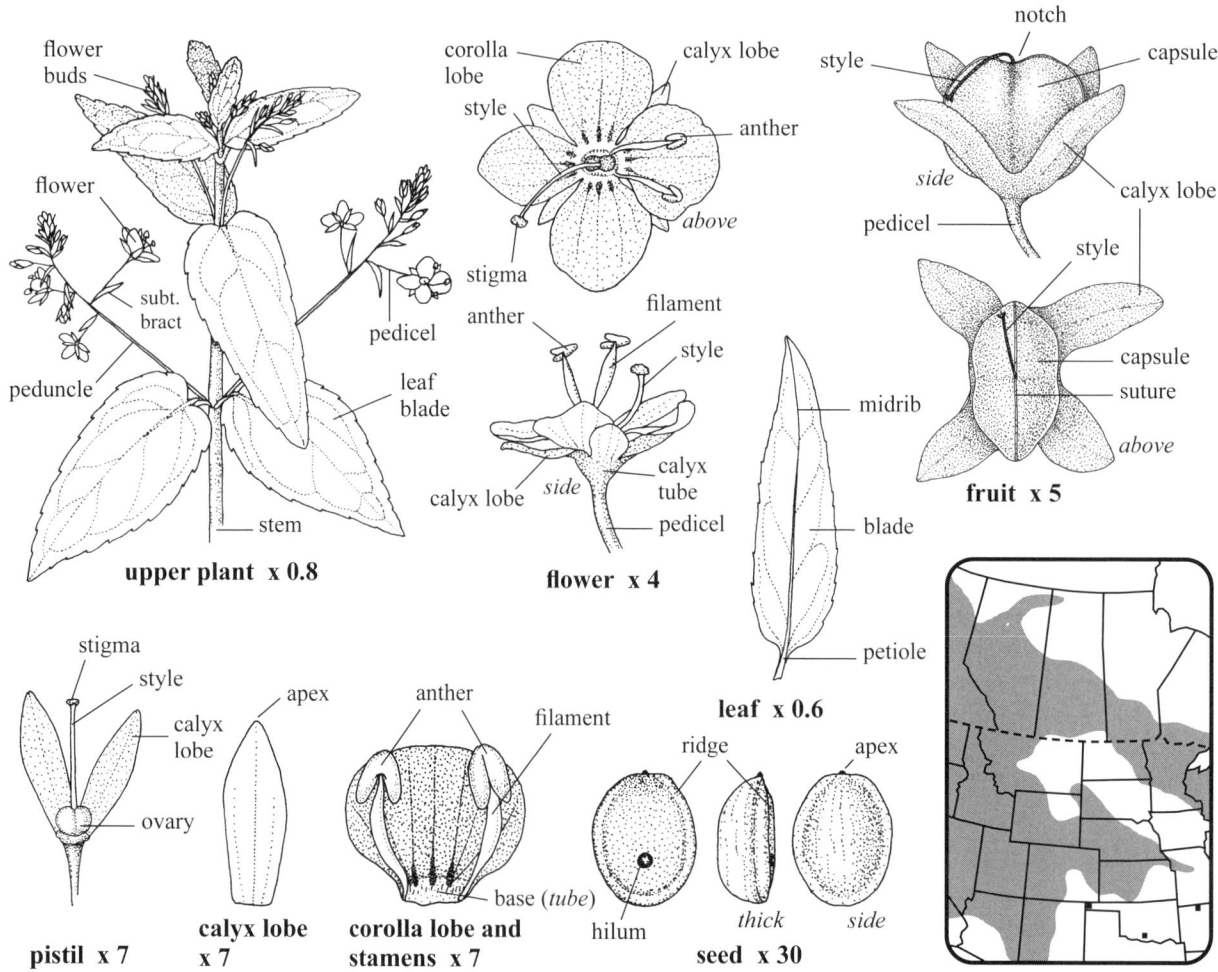

504 American Brooklime, Brooklime Speedwell

Figwort—*Scrophulariaceae* **Wildflower** White   Corolla tubular, 4-lobed

Hairy Speedwell *Veronica peregrina* L.

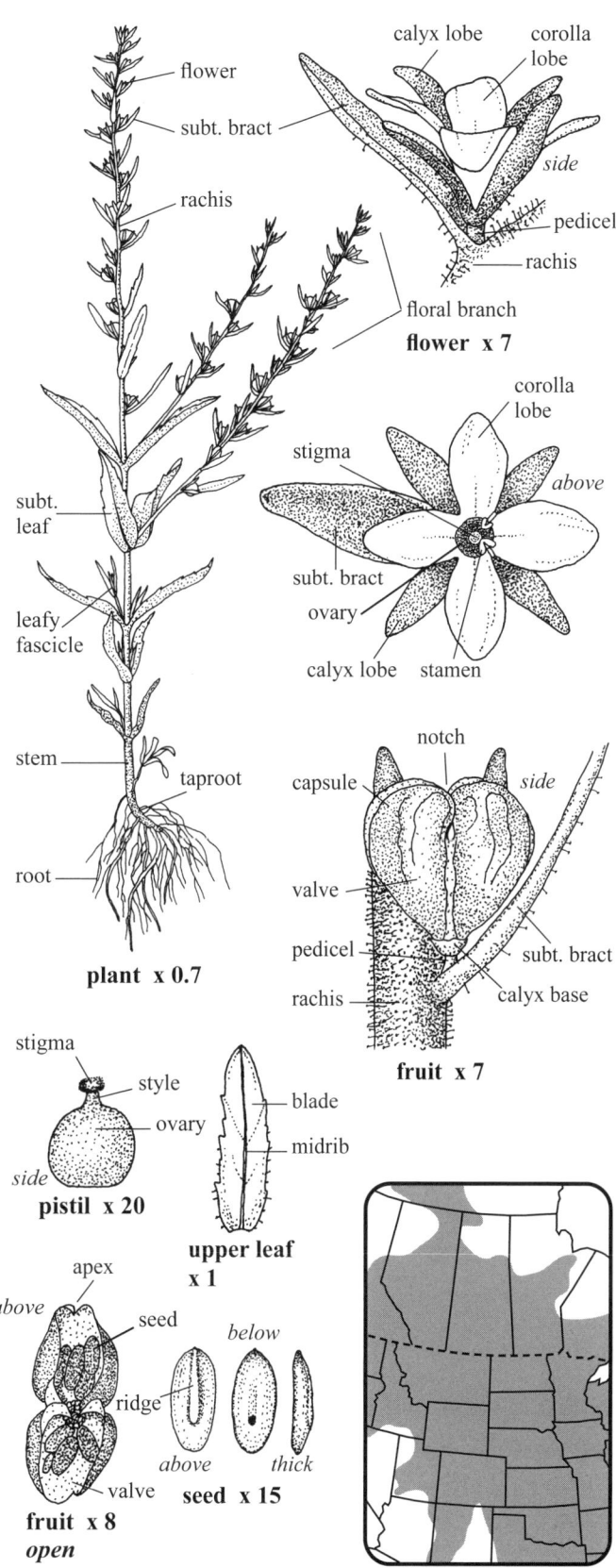

■ **SKETCH** A glandular-hairy to glabrous **annual herb** 5–40 cm tall from a reddish **taproot** 0.5–5 cm long by 1–4 mm wide with white **fibrous roots** 2–6 cm long; in wet meadows, shores of streams, marshes, lakes, rocky outcrops, lawns and gardens.

■ **FLOWERS** White, blooming April–August; **inflorescence** a raceme, 4–12 cm long by 1–2 cm wide, lengthening as fruit develops, terminal and axillary; **pedicels** 0.5–1.5 mm long, ascending, glabrous; **subtending bracts** (of pedicels) 4–15 mm long by 1–2.3 mm wide, green to reddish, slightly glandular-hairy on outside and margins; **flowers** white, one per bract axil, 3–6 mm wide by 2–3 mm long, ascending; **rachis** *c.* 1 mm thick, green to reddish, very glandular-hairy, hairs spreading; **calyx** 4-lobed, these 3.2–5.5 mm long by 0.8–1 mm wide, blunt, glandular-hairy on margins and outside, glabrous inside, united for *c.* 0.5 mm at base, longer than, reddish and erect around fruit; **corolla** tubular, 4-lobed, these 2–2.3 mm long by 0.9–1.5 mm wide, glabrous; **stamens** two, *c.* 0.6 mm long, attached at the inside base of one corolla lobe; **filaments** *c.* 0.4 mm long, hyaline; **anthers** yellow, *c.* 0.2 mm long; **pistil** one, *c.* 0.6 mm wide and long, green, glabrous; **ovary** *c.* 0.4 mm thick; **style** 0.1–0.4 mm long; **stigma** capitate, wider than style, white. **FRUIT a capsule**, reddish green, heart-shaped, notched in middle, becoming tan and hairless when fully ripe, 3–4.5 mm long by 3.5–5 mm wide by 1–2 mm thick when open, two-chambered, opening at apex; **seeds** golden yellow, 65–95 per capsule, flattened, oval, smooth, 0.7–1 mm long by *c.* 0.4 mm wide by *c.* 0.2 mm thick, flat on one side with a dark spot, a central ridge on the opposite side.

■ **LEAVES** Opposite below, simple, entire to shallowly toothed, sessile, often with leafy fascicles in axils or subtending short alternate floral branches; **blades** 10–40 mm long by 3–11 mm wide, ascending, with scattered marginal hairs on lower half, hairless on both sides, midrib obvious, lateral veins obscure, reddish green, lowest stem leaves wilted.

■ **STEM** Erect to prostrate, prostrate stems rooting at nodes, simple to branched, densely glandular-hairy, stiff, round and solid, branches all floral, ascending, opposite or alternate, some rebranching, lower branches to *c.* 15 cm long; 1.5–4 mm wide near the reddish base.

■ **NOTE** The subspecies *xalapensis* (shown) is glandular-hairy; the subspecies *peregrina* is glabrous.

Neckweed, Purslane Speedwell, Necklace-weed

Figwort—*Scrophulariaceae*   **Wildflower**   White to pinkish   Corolla tubular, 4-lobed

CULVER'S-ROOT   *Veronicastrum virginicum* (L.) Farw.

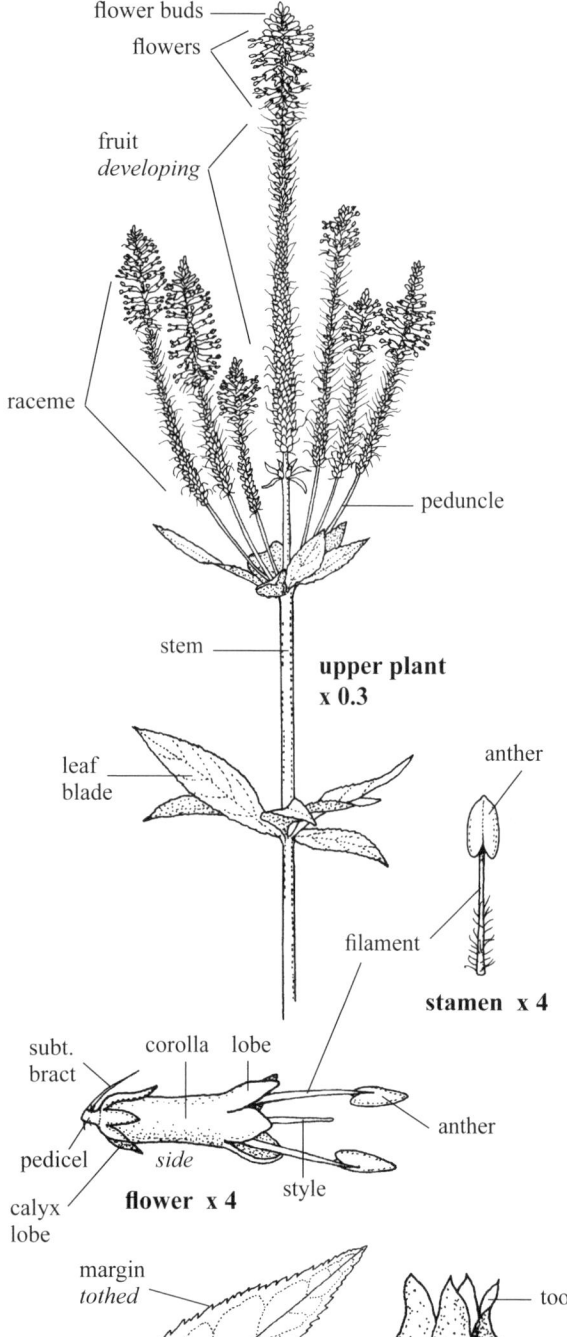

■ **SKETCH** A **perennial herb** 30–200 cm tall; along roadsides, in open woods, meadows and moist prairies. **p2**
■ **FLOWERS** White to pinkish, blooming June–September; **inflorescence** a raceme (spikelike), erect, one to several, each 8–20 cm long by *c.* 2 cm wide in bloom, the terminal one the longest with two to six shorter racemes rising from the uppermost whorl of leaves, the next lowest whorl of leaves may support up to four racemes; **peduncles** glabrous, 2–5 cm long, naked; **pedicels** 0.5–1 mm long; **subtending bracts** (of pedicels) one, glabrous, narrow, pointed, about as long as the calyx; **flowers** perfect, usually horizontal and *c.* 10 mm long; **calyx** 5-lobed (4-), these pointed, 1.5–2.5 mm long, glabrous with thin white scarious margins; **corolla** tubular, 4-lobed, white to pale pink, glabrous outside, 5–8 mm long with two lobes *c.* 2 mm long, and with two shorter lobes *c.* 1.5 mm long, these not spreading; **body** tubular, *c.* 1.8 mm wide, slightly hairy inside near the base; **stamens** two, exserted 4–5 mm past the corolla's lobes; **filaments** white, attached to the corolla, with white hairs along the lower half; **anthers** pinkish, 1–2 mm long; **style** one, white, filiform, exserted, persistent on green fruit; **stigma** obscure; **fruiting stalks** to *c.* 35 cm long by *c.* 13 mm wide, erect and fairly rigid. **FRUIT a capsule**, dark brown, slightly rough, hairless, 3–4.5 mm long by 2.5–3 mm wide, mostly round, not shiny, apical teeth four, each *c.* 1 mm long and slightly spreading; **seeds** golden to dark brown, 0.5–0.7 mm long by *c.* 0.4 mm wide by *c.* 0.2 mm thick, usually curved, indented along the concave surface, wrinkled to striated, 40–50 per capsule.
■ **LEAVES** Stem only, in whorls of six (two to seven per node), simple, toothed; **blades** 7–15 cm long by 1–3.5 cm wide, flat, usually horizontal, drooping near the base, glabrous and dark green above, lighter green below and light to very hairy or glabrous along the raised veins; **petioles** 0–8 mm long.
■ **STEM** Erect, glabrous, solid, round, green, branches few, slightly ridged; 5–10 mm thick near the base.

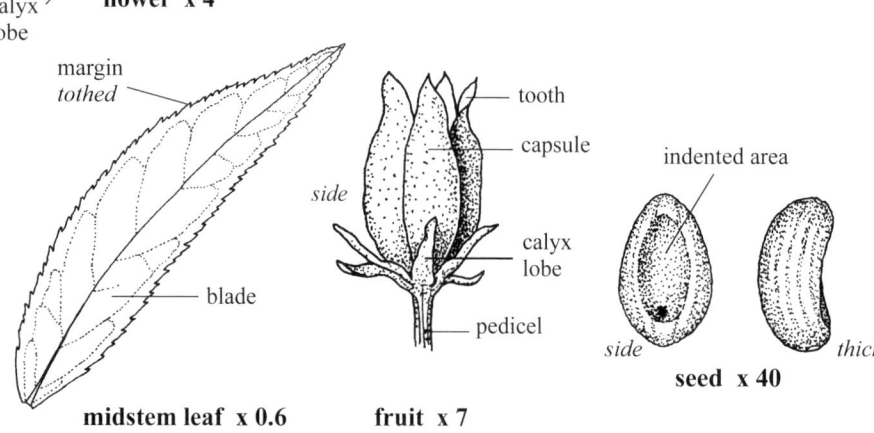

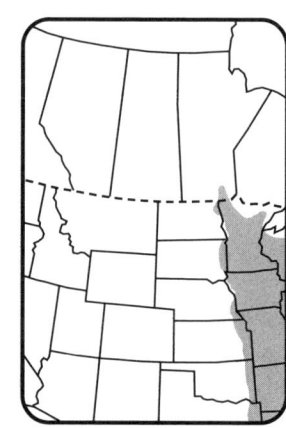

506

Nightshade—*Solanaceae*   **Wildflower**   Purple, center yellow   Corolla tubular, 5-lobed

BITTERSWEET *Solanum dulcamara* L.

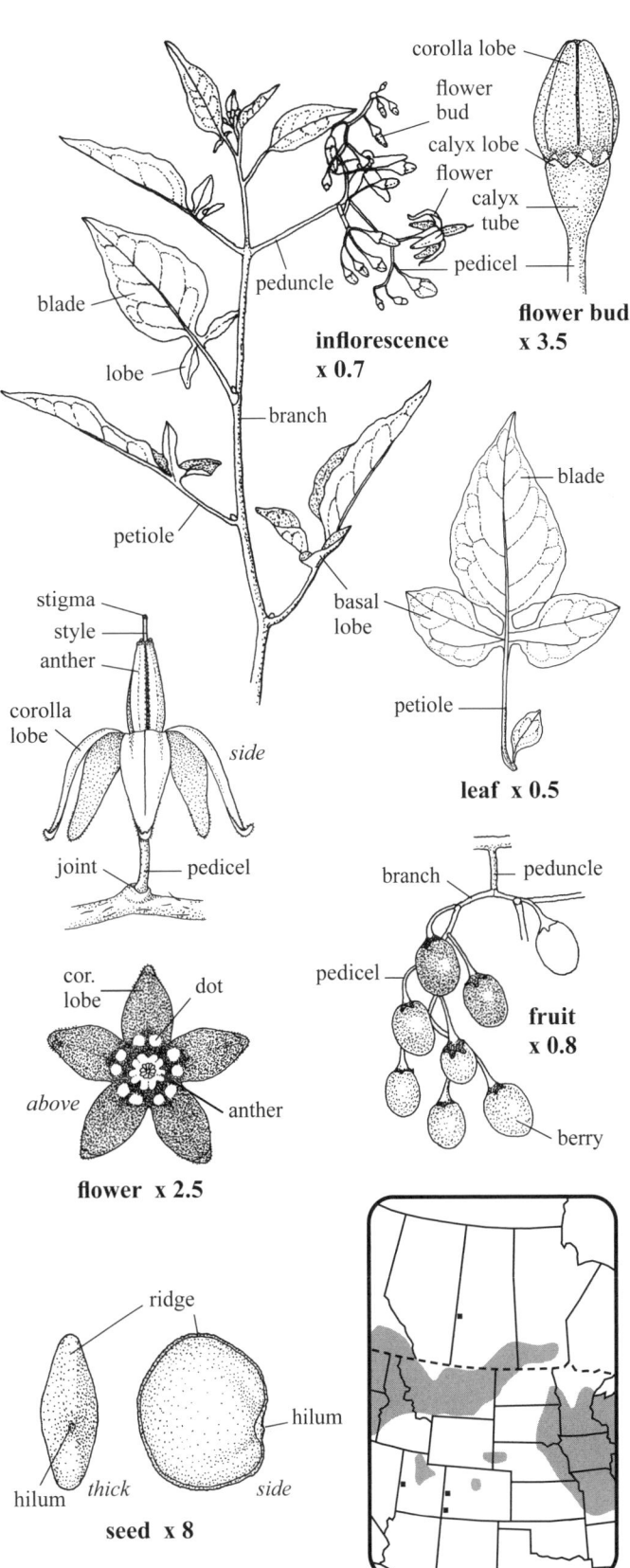

■ **SKETCH** An unarmed **perennial herb** 20–300 cm tall, climbing or clambering over vegetation; **stolons** woody, tan bark peeling, horizontal, 4–5 mm thick, and rooting at nodes with **fibrous roots** several cm long; in thickets, along fencerows, roadsides, ditches and canal banks. *Naturalized* w1

■ **FLOWERS** Purple, center (anthers) yellow, blooming May–September; **inflorescence** cymose or paniculate, 4–6 cm long and wide, hanging, branched; **peduncles** 1.0–4.5 cm long by *c.* 1 mm thick, some opposite a petiole; **pedicels** purple, 5–7 mm long, to *c.* 10 mm long with fruit, descending, *c.* 0.5 mm wide with a joint at the base, curved, hairs scattered; **flowers** *c.* 20 (7–30) per cluster, each 8–12 mm wide by *c.* 12 mm long; **calyx tube** dark brownish purple, *c.* 2.8 mm long by *c.* 2.5 mm wide, glabrous; **calyx lobes** five, *c.* 0.5 mm long by *c.* 1.5 mm wide, turning dark brown with fruit and tight against the berry; **corolla** tubular, 5-lobed, lobes 6–8 mm long by 2.5–3 mm wide, minutely hairy on the hooded tips, purple, eventually reflexed, central vein obvious, united at base for *c.* 1 mm into a tube; **two dots** pale green with a white margin at the base of each corolla lobe; **stamens** five; **anthers** deep yellow, connate, exserted, 4–6 mm long by 2–2.2 mm wide (together), forming a tube around the erect style; **ovary** light green, glabrous, *c.* 2 mm long by *c.* 1.5 mm wide, round, tapered above; **style** one, pale green, exserted *c.* 2 mm past the tips of the anthers, filiform, TL *c.* 6 mm; **stigma** capitate, light green, slightly wider than the style, unlobed. **FRUIT a berry**, red, shiny, glabrous, 5–10 mm long by 5.5–9 mm wide, oval, *c.* 20 per cluster, hanging; **seeds** pale tan, 3–32 per berry, flattened, 2.2–3 mm long by 2–2.6 mm wide by *c.* 1 mm thick, finely reticulate with a low ridge around the entire seed, slightly concave at the golden orange **hilum**.

■ **LEAVES** Alternate, simple, entire, often with two large basal lobes of varying size or absent; **main blades** dull, slightly lighter green below, 2.5–13 cm long by 1.5–9 cm wide, mostly hairy on veins on both sides, scattered hairs elsewhere, hairs mostly ascending and more numerous below (dorsal side); **basal lobes** 7–48 mm long by 4–30 mm wide; **petioles** 1.5–5.8 cm long, spreading to ascending, hairy especially in upper groove, hairs appressed and arched forward.

■ **STEM** Green, with scattered short, basifixed hairs, erect but weak and often leaning on or climbing over adjacent vegetation, tendrils absent; 3–5 mm wide near the woody base.

Nightshade, Climbing Nightshade, Bittersweet Nightshade, Deadly Nightshade

Linden—*Tiliaceae* **Tree** Pale yellow Petals 5

BASSWOOD *Tilia americana* L.

■ **SKETCH** A variable **deciduous tree** 20–35 m tall; along riverbanks and moist rich bottomlands, planted in cities; a soft, light **hardwood**. w2

■ **FLOWERS** Creamy yellow, blooming June–July; **inflorescence** of axillary cymes, with hundreds of cymes and thousands of flowers throughout a large tree; **peduncles** 5–8 cm long, the inner half attached to a linear, leaflike bract, the outer free part 1.5–6 cm long, hanging below the bract; **rays** two to four; **pedicels** 3–5 mm long; **bracts** entire, 4–14 cm long by 9–25 mm wide, thin, glabrous, and persistent into the winter; **bract stalk** 2–10 mm long; **flowers** perfect, fragrant, hanging, 9–14 per cyme, each 10–15 mm wide by 7–8 mm long; **sepals** five, 4–5 mm long by c. 3 mm wide, greenish yellow, convex on the outside, veins obscure; **petals** five, 5–10 mm long by c. 2 mm wide, tapered; **stamens** numerous but included; **staminodia** five, sterile, petal-like, tapered; **pistil** one, exerted; **style** c. 4 mm long, narrow; **stigma** white, flat, c. 1 mm wide. **FRUIT nutlike**, round, gray from fine hairs, indehiscent, 7–11 mm across, a rough wall c. 1 mm thick, overwintering on the tree; **seeds** one (two) per fruit, dark brown, smooth, flat on one side, c. 5 mm long by c. 4 mm wide by c. 3 mm thick, requires two years to germinate.

■ **LEAVES** Alternate, simple, toothed, pointed; **blades** glabrous above, with small tufts of white hairs along the midrib below, 5–20 cm long by 4.5–15 cm wide, base often uneven; **petioles** 1.3–8 cm long, glabrous.

■ **TRUNK** Young trees have smooth, light gray bark, breaking into long, flat, scaly ridges with age; **branches** alternate, ascending; **new twigs** reddish brown, hairless; **winter buds** alternate, red, 3–6 mm long, with two or three large scales, glabrous, somewhat lopsided; **pure stands** not formed, capable of regenerating from suckers around the base of stumps, also grows singly; **dbh** 15–50 cm.

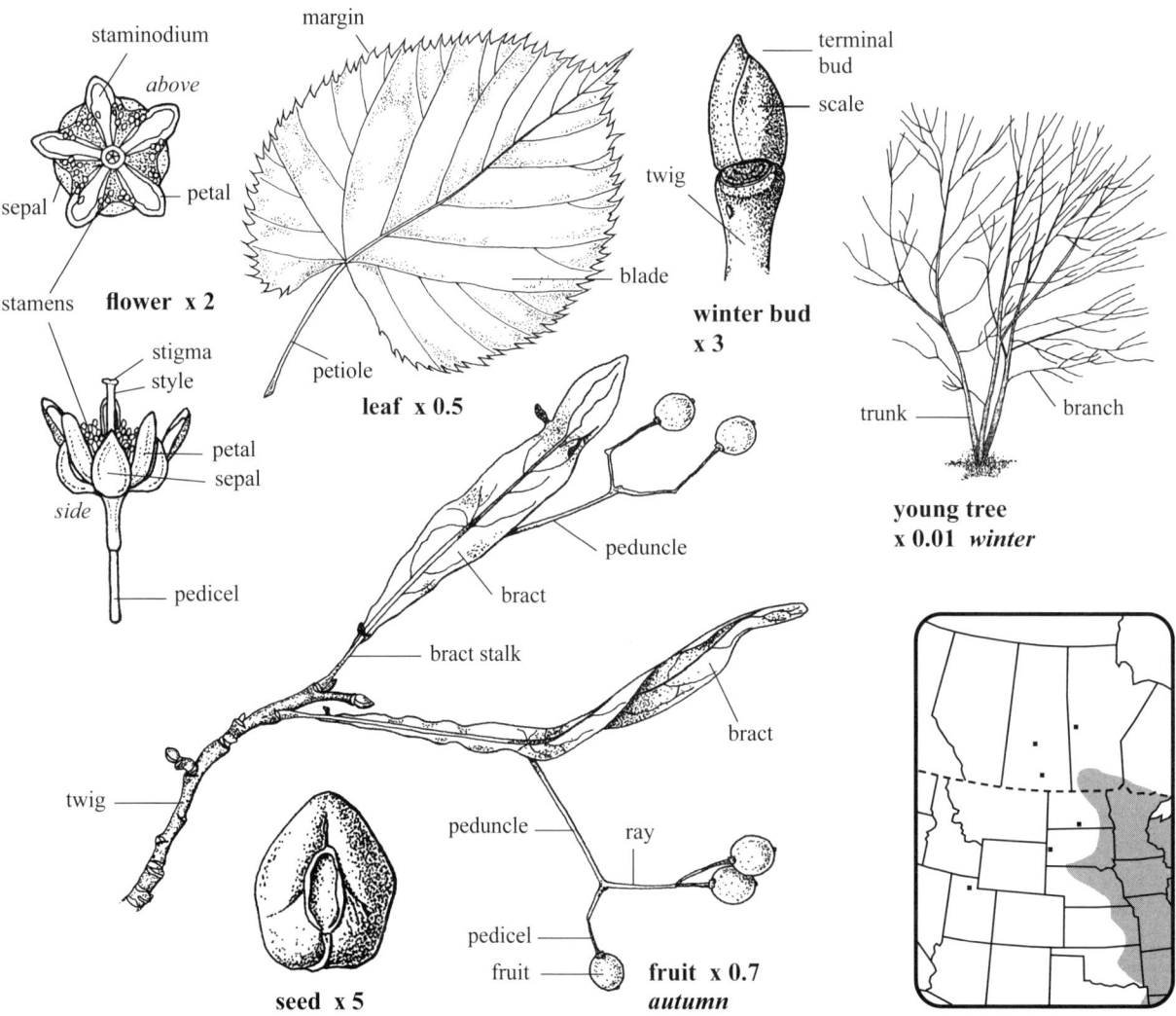

508 American Basswood, American Linden, Common Linden

Elm—*Ulmaceae*   **Tree**   Reddish green   Calyx tubular, 5- to 9-lobed

American Elm   *Ulmus americana* L.

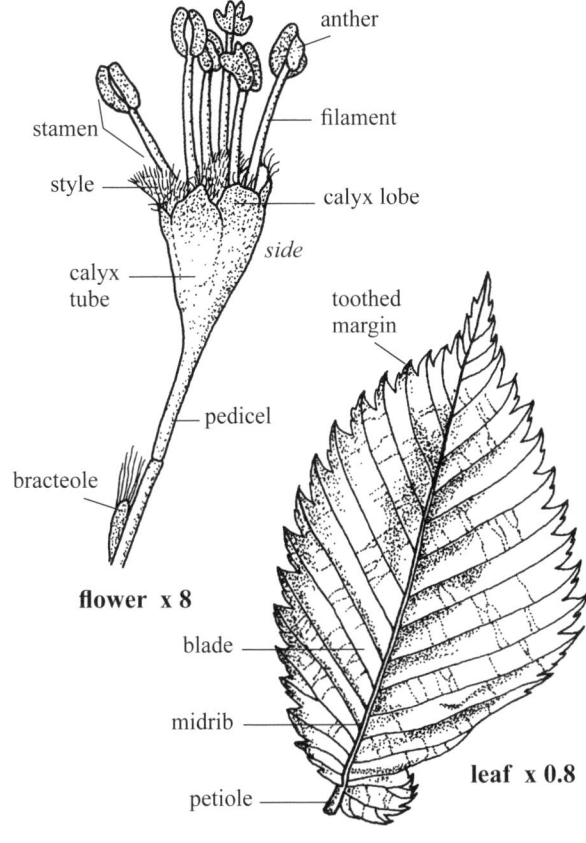

flower x 8

leaf x 0.8

fruit x 2

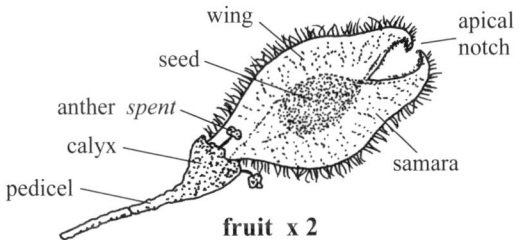

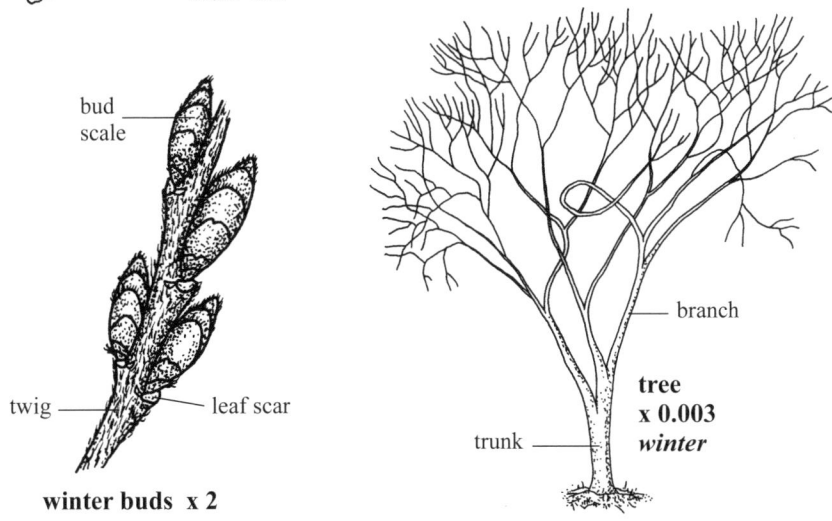

winter buds x 2

tree x 0.003 *winter*

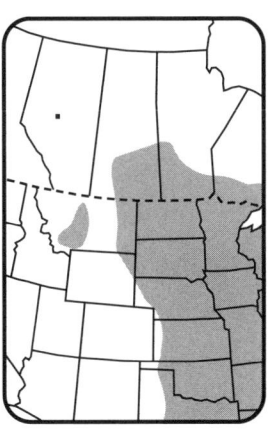

■ **SKETCH** A **deciduous tree** 10–35 m tall with spreading branches forming an open, round crown, with other trees or solitary; in the open, along rivers and streams, ravines and fencerows, in rich, well-drained soil; often planted. **w2**

■ **FLOWERS** Reddish green, blooming in February–May before the leaves appear; **inflorescence** of small fascicles of flowers; **pedicels** green, unequal, 3–18 mm long with fruit, glabrous; **bracteoles** one or two, reddish brown, 2–3 mm long including the hairy apices; **flowers** perfect, 10–20 per cluster; **calyx** tubular, 5- to 9-lobed, reddish green, 2–3 mm long by 2–4 mm wide, bell-shaped and with curly apical hair; **petals** absent; **stamens** five to nine (usually six), one for each calyx lobe, exserted; **filaments** stout, 3–4 mm long, whitish; **anthers** dark brownish purple, *c.* 1 mm long; **pollen** whitish green; **styles** 2-lobed, whitish green, *c.* 1.5 mm long, hairy, barely exserted, persistent as wings around the seed. **FRUIT a samara**, 1-seeded, numerous, winged, tan, some tinged with reddish purple, 9–16 mm long by 7–10 mm wide, flattened, 1–2 mm thick, the margins ciliate with white hairs, the apical notch 3–4 mm deep to the seed.

■ **LEAVES** Alternate, simple, margins doubly toothed, dark green; **blades** 3–15 cm long by 1.3–8 cm wide, pointed, glabrous to slightly scabrous above, less so below, base unequal; **petioles** 2–10 mm long, glabrous to lightly hairy; **stipules** linear, 5–7 mm long, hairy, deciduous.

■ **TRUNK** Bark medium gray, deeply furrowed with scaly flat ridges; **young twigs** lightly hairy to glabrous; **winter buds** alternate and often clustered, reddish brown, dull, 5–9 mm long by 3–3.5 mm wide by 2–3 mm thick, slightly flattened, 6- or 7-scaled, the scale margins ciliate with tan hairs; **lenticels** golden tan on twigs, round to oval, not raised, *c.* 0.8 mm long; **dbh** 40–100 cm.

White Elm

Nettle—*Urticaceae* **Wildflower** Light green Calyx 5-lobed

WOOD NETTLE   *Laportea canadensis* (L.) Weddell

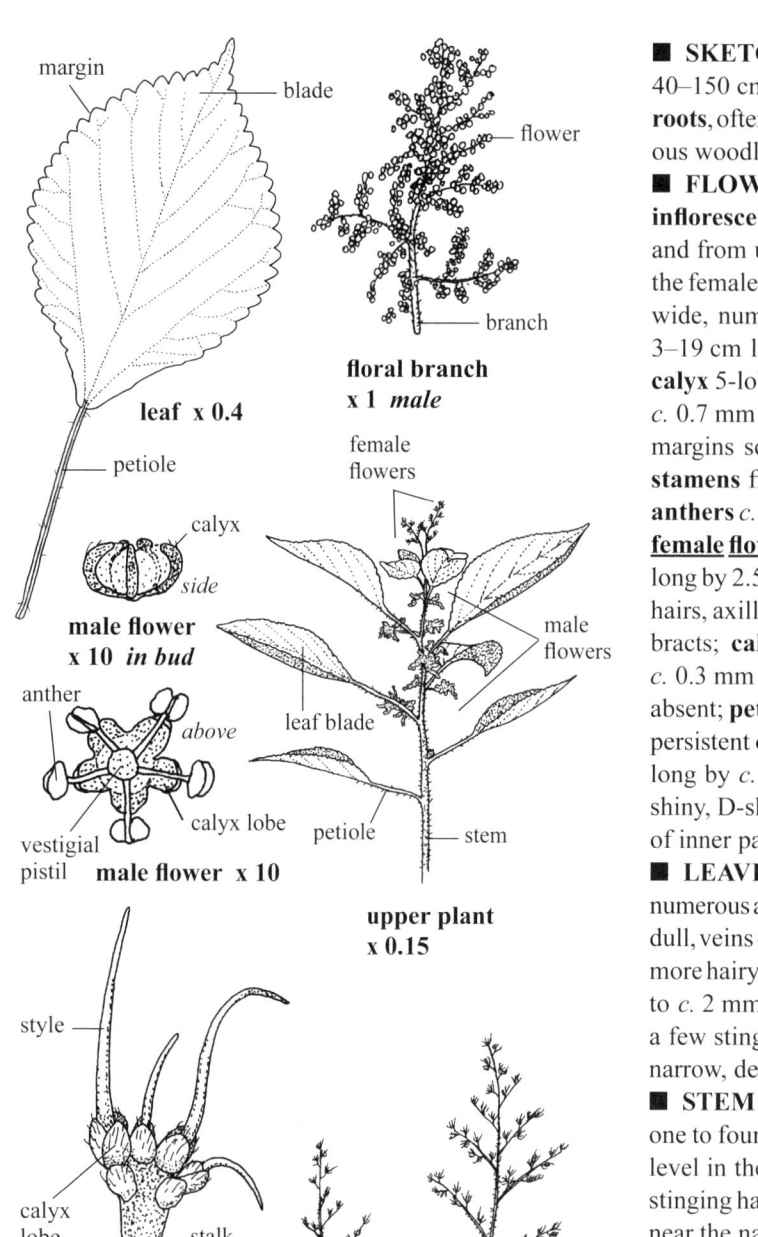

■ **SKETCH** A **perennial herb** with stinging hairs, 40–150 cm tall from a mass of 1–3 mm thick **tuberous roots**, often in colonies covering several m^2; in rich deciduous woodlands and along river valleys; **monoecious**.
■ **FLOWERS** Light green, blooming July–September; **inflorescence** paniculate, unisexual, branched, terminal and from upper leaf axils; **male flowers** axillary, below the female flowers, pale yellowish green, sessile, *c.* 2 mm wide, numerous along one or two horizontal branches 3–19 cm long and 3–8 cm wide, from upper leaf axils; **calyx** 5-lobed (2- to 4-), with a green rounded keel, each *c.* 0.7 mm long by *c.* 0.5 mm wide, lightly hairy and the margins scarious, spreading at anthesis; **petals** absent; **stamens** five, exserted; **filaments** green, *c.* 1 mm long; **anthers** *c.* 0.5 mm long; **pistil** round, vestigial, included; **female flowers** on two to five terminal branches 4–20 cm long by 2.5–13 cm wide with silvery, slightly descending hairs, axillary from the uppermost three or four (leaflike) bracts; **calyx** 2- to 4-lobed, hairy, *c.* 0.4 mm long by *c.* 0.3 mm wide, enlarging with fruit, outer ones may be absent; **petals** absent; **styles** thin, 1–3 mm long, hairless, persistent or not. **FRUIT an achene**, 1-seeded, 2–3 mm long by *c.* 2 mm wide by *c.* 0.7 mm thick, dark brown, shiny, D-shaped, slightly mottled or smooth, falling free of inner pair of enlarged calyx lobes.
■ **LEAVES** Alternate, simple, roundly toothed, more numerous above; **blades** 3–30 cm long by 1.5–18 cm wide, dull, veins obvious, glabrous or with scattered hairs above, more hairy below especially along the raised midrib, hairs to *c.* 2 mm long; **petioles** 1–15 cm long, grooved, with a few stinging hairs, reduced above; **stipules** brownish, narrow, deciduous.
■ **STEM** Erect but weak, solid, longitudinally grooved, one to four stems in a clump, easily broken off at ground level in the fall, unbranched, densely covered with stiff stinging hairs especially in the upper half; 2–10 mm thick near the naked base.

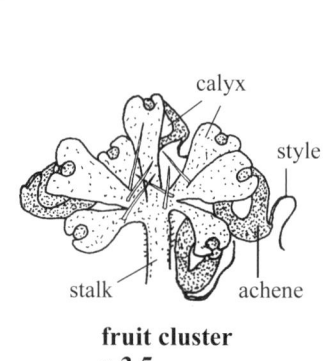

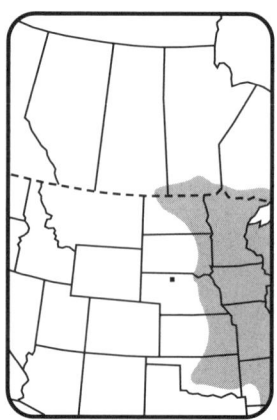

510   Canada Woodnettle, Canadian Wood-nettle

Stinging Nettle *Urtica dioica* L.

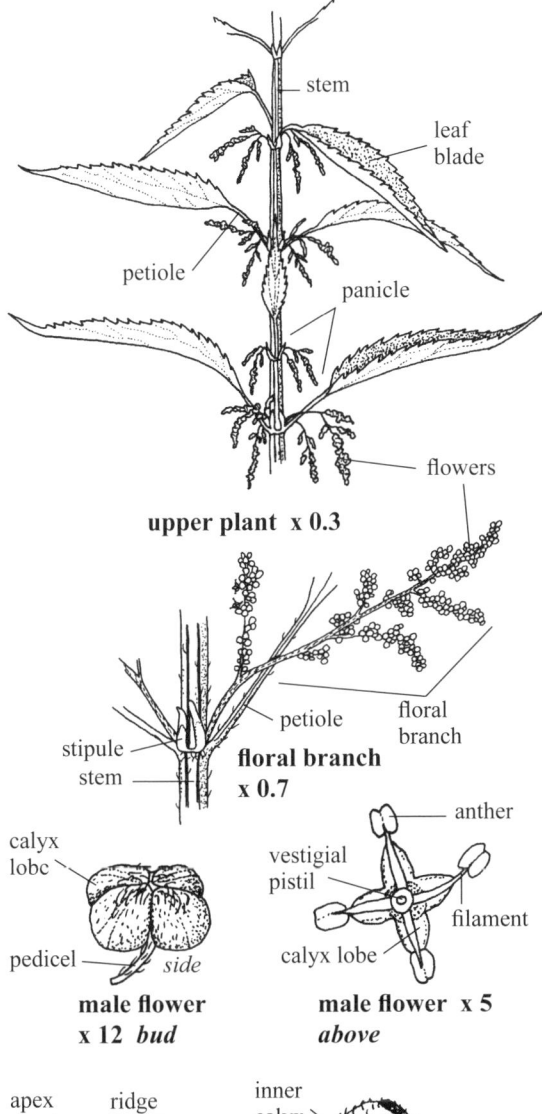

upper plant x 0.3

floral branch x 0.7

male flower x 12 *bud*

male flower x 5 *above*

fruit x 20

fruit x 10 *covered by calyx*

■ **SKETCH** A variable **perennial herb** 0.5–3 m tall, often in dense patches from **rhizomes** 2–5 mm thick and to *c.* 10⁺ cm long with pink offshoots; in woodland edges, old pastures, orchards and along streams, sloughs, slopes, river banks and ditches; **monoecious** but each plant mostly female or male; some **dioecious**.

■ **FLOWERS** Green, blooming May–October; **inflorescence** a panicle, each 1.5–8 cm long from the upper leaf axils; **pedicels** less than 1 mm long and lightly hairy; **subtending bracts** (of flower clusters) green, *c.* 1.5 mm long, hairy; **flower clusters** unisexual; **rachis** very hairy; **male flowers** 4–5 mm wide; **calyx** 4-lobed, *c.* 1.5 mm long by 0.7 mm wide, with a wide, green midvein, C-shaped, cupped in bud around the anthers, glabrous inside, hairy outside at the apex, tinged with red, spreading; **corolla** absent; **stamens** four (generally spent when the calyx opens), exserted, erect to spreading; **filaments** flat, slightly longer than the calyx lobes; **anthers** whitish yellow, *c.* 0.8 mm long and wide; **pistil** vestigial, round, included; **female flowers** 3–4.5 mm wide; **calyx** of four unequal parts: **1)** outer two parts inconspicuous; **2)** two inner parts 1–2 mm long and enclosing the fruit; **corolla** absent; **stigma** sessile and tufted. **FRUIT** an achene, 1-seeded, 1–1.5 mm long by 0.7–0.9 mm wide by *c.* 0.3 mm thick, smooth, with a tan, marginal ridge.

■ **LEAVES** Opposite, simple, coarsely toothed, upper ones often folded up along the midrib, lower ones less folded; **blades** 2.5–20 cm long by 0.6–13 cm wide, the veins raised below and with a few stinging hairs, dark green and mostly glabrous above; **petioles** 1–6 cm long, reduced above, grooved above and along the sides, with stinging hairs; **stipules** pale green, 5–15 mm long by 2–4 mm wide, free, paired, entire, pointed, erect, lightly hairy.

■ **STEM** Erect, hollow, fibrous and tough, simple to branched, bluntly square with four deep vertical grooves; 2–10 mm thick near the base; **stinging hairs** few to numerous and ascending.

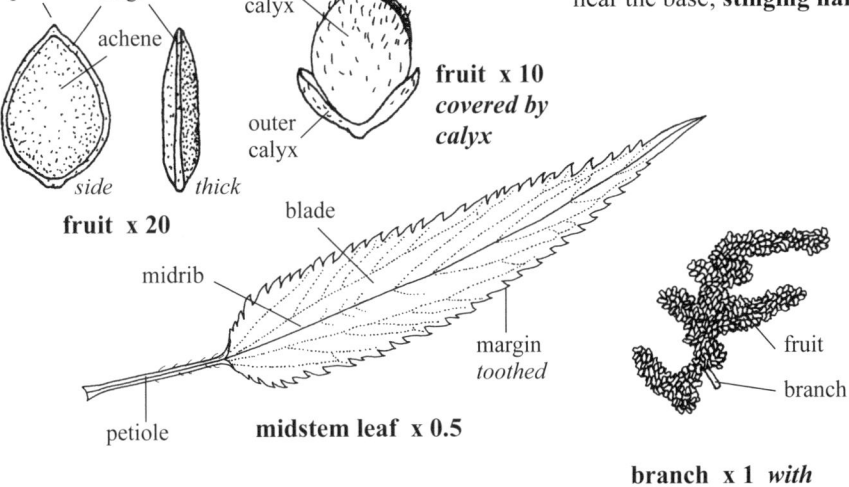

midstem leaf x 0.5

branch x 1 *with ripening fruit*

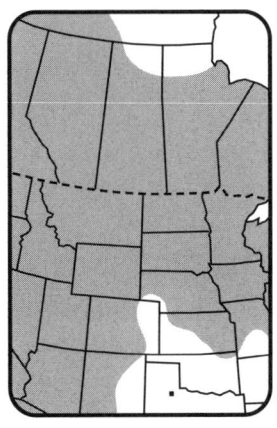

Valerian—*Valerianaceae*   **Wildflower**   White   Corolla tubular, 5-lobed

NORTHERN VALERIAN   *Valeriana dioica* L.

- **SKETCH** A **perennial herb** 10–70 cm tall with a branched, brown **rhizome** to *c.* 8 cm long by *c.* 5 mm thick with numerous whitish tan **fibrous roots** 1–10 cm long.
- **FLOWERS** White, blooming June–July; **inflorescence** a cyme 1–3 (–6) cm long by 1.5–4.5 cm wide, terminal, sometimes axillary from upper leaves, expanding through anthesis; **floral branches** 0.2–4.5 cm long (to 6 cm with fruit), ascending, one to three at nodes, reduced above; **flowers** perfect, or all female, numerous, sessile, 3–4 mm wide by 3.5–4.3 mm long; **subtending bracts** (of flowers) one, green, 1.5–6 mm long by 0.5–0.8 mm wide, entire; **rachis** green, grooved; **calyx** of 5–15 inrolled bristles in flower, these expanding and forming the pappus of the fruit; **corolla** tubular, 5-lobed; **tube** *c.* 2 mm long, lower tube, glabrous and ridged outside, hairy inside; **lobes** *c.* 1 mm long and wide, 1-nerved, glabrous, margins irregular; **stamens** three, white, exserted; **filaments** attached midway on corolla tube, *c.* 2.5 mm long, hairless; **anthers** 4-lobed, *c.* 0.8 mm long; **ovary** green, *c.* 1 mm long by *c.* 0.7 mm wide, flattened, ribbed, hairless; **style** one, erect, slightly exserted or not, *c.* 1.6 mm long, hyaline; **stigma** 3-lobed, microscopic; **fruiting heads** up to *c.* 15 cm long and *c.* 9 cm wide, open, diffuse. **FRUIT** an achene, 1-seeded, tan, 3–5 mm long by 1.5–1.8 mm wide by *c.* 0.5 mm thick, 6-ribbed, two side ribs and one midrib on one (inner side) and three central ribs on the outer side, hairless; **pappus** of 5–15 hairy bristles, each 4–6 mm long, together 8–12 mm wide and connected at their bases by a tan membrane *c.* 0.5 mm long, white hairs on bristles 0.5–1.5 mm long.
- **LEAVES** Basal and stem, opposite, simple; **basal leaves** usually four, TL 2–6 cm, lobed or entire; **blades** 7–22 (–80) mm long by 5–13 (–30) mm wide, slightly hairy along veins; **petioles** 1–2 cm long by *c.* 5 mm wide, flat, lower half marginally ciliate; **stem blades** dull, few, 2–7 cm long by 1.2–4 cm wide with three to six pairs of entire lobes, mostly ascending; **leaves** on new shoots several, blades glabrous.
- **STEM** Erect, weak, round, 4- to 6-ridged, hollow, glabrous, branched above; 2–4 mm wide near the green base.
- **SYN.** *Valeriana septentrionalis* Rydb. = *Valeriana dioica* var. *sylvatica* S. Wats.

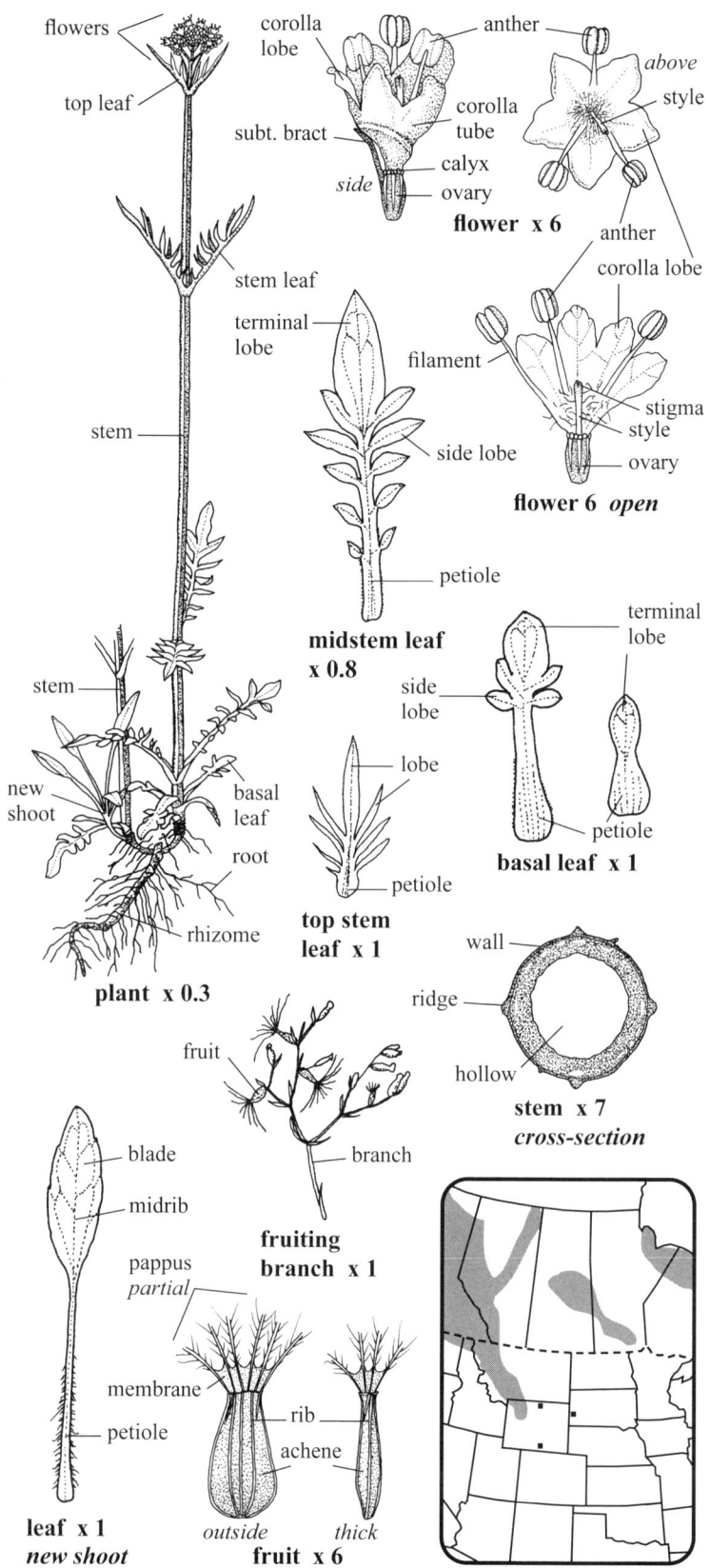

Vervain—*Verbenaceae* **Wildflower** Blue to purple Corolla tubular, 5-lobed

BRACTED VERVAIN *Verbena bracteata* Lag. & Rodr.

■ **SKETCH** An **annual** or short-lived, hairy **perennial herb**, decumbent to erect, 10–60 cm long or tall with a tan **taproot** 2–5 mm wide and to *c*. 15⁺ cm long, usually in small open colonies; in dry gravelly sites, roadsides, overgrazed prairies and mountain slopes.

■ **FLOWERS** Blue to purple, blooming April–October; **inflorescence** a spike, elongating, 1–20 cm long by 1–2 cm wide; **subtending bracts** (of flowers) one, 3–8 mm long by *c*. 1.3 mm wide (not including hairs), lengthening to *c*. 15 mm in fruit, green, tip reddish; **flowers** perfect, sessile, 3–4 mm wide by 6–7 mm long, throat white with white hairs; **calyx** 3–4 mm long by 0.9–1.1 mm wide, lobed; **calyx lobes** five, 0.5–0.6 mm wide, light green and hairy, one lobe less prominent than the others, erect around the fruit, tips distinct only for *c*. 0.4 mm, slightly reddish; **corolla** weakly bilabiate, 5.5–6 mm long, slightly hairy on outside, easily removed from calyx; **corolla tube** whitish near base, pinkish above, 4–5 mm long, 1.5–2 mm above base of corolla tube is an inner fringe of ascending white hairs *c*. 0.2 mm long; **corolla lobes** five, upper two close together, each *c*. 1.3 mm long by *c*. 1.1 mm wide and entire, the other lobes *c*. 1 mm long and weakly notched at apices; **stamens** four (5), mostly sessile in corolla tube at two levels; **filaments** shorter than the anthers; **anthers** pale yellow, *c*. 0.3 mm long; **ovary** glabrous, 4-lobed; **style** one, included. **FRUIT a nutlet**, up to four per calyx, separating when ripe, each 1-seeded, dark brown, 2–2.5 mm long by 0.7–0.8 mm wide (across rounded back) by *c*. 0.6 mm thick, blunt at both ends, four or five ribs on the lower half and net-veined above.

■ **LEAVES** Opposite, simple, toothed, often 3-lobed, the two lateral lobes much smaller than the large terminal lobe, flat, dull, slightly lighter green below, hairs on both surfaces spreading to ascending; **blades** 1–7 cm long by 0.5–3 cm wide, veins obvious, raised below, margins often revolute; **petioles** winged, hairy, 2–20 mm long.

■ **STEM** Simple to branched throughout, square, often with a groove, hairs simple, spreading; **branches** leafy or with flowers, one to several branches from the top of the taproot, short plants erect, taller plants decumbent; 1.5–4 mm thick near the hairy base.

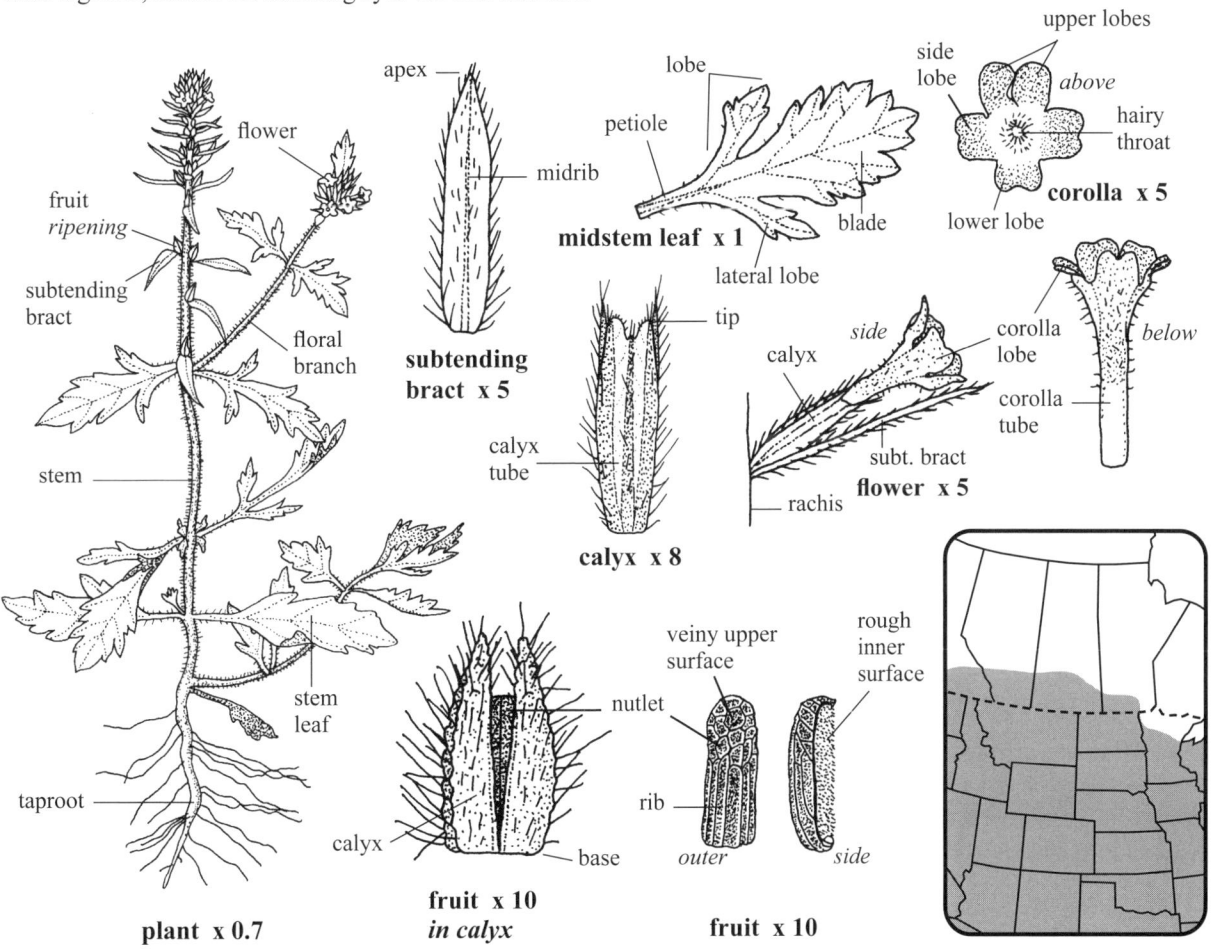

Bigbract Verbena, Carpet Vervain, Prostrate Vervain

Early Blue Violet *Viola adunca* Sm.

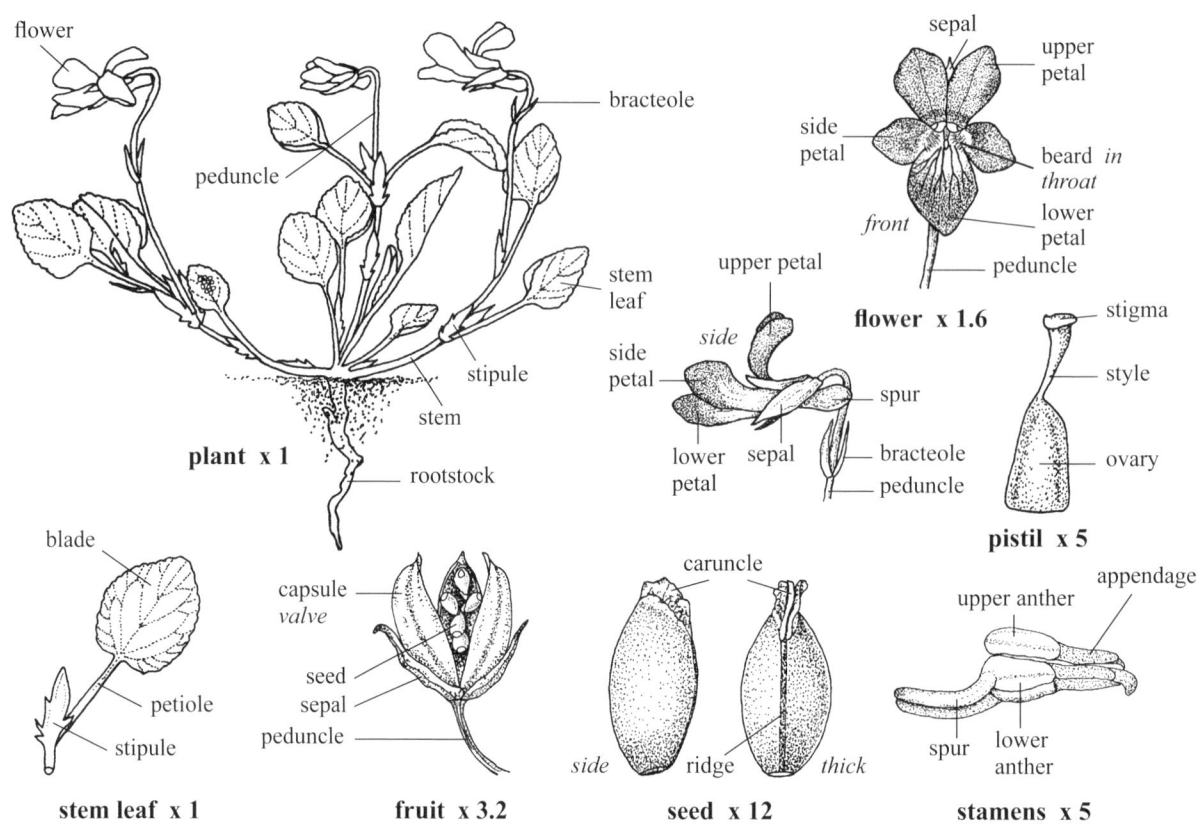

plant x 1
flower x 1.6
pistil x 5
stem leaf x 1
fruit x 3.2
seed x 12
stamens x 5

■ **SKETCH** A leafy **perennial herb** 2–15 cm tall in erect tufts from a woody **rootstock** 2–3 mm thick; sometimes with a **branched caudex** 2–4 cm long; in moist to dry prairies, rocky outcrops, mountain brush, woods, shaded mountain meadows and along banks of streams. p2

■ **FLOWERS** Blue to violet, blooming April–August; **inflorescence** a few+ flowers, each terminal on an erect peduncle; **peduncles** axillary, 2–12 cm long, strongly curved at spur, usually rising above the leaves; **bracteoles** paired, erect on peduncle; **flowers** perfect, 10–20 mm wide by 12–20 mm long; **late summer flowers** cleistogamous (not showy but seed producers); **sepals** five, green, entire and pointed, 5–8 mm long by 2–3 mm wide, with auricles 0.5–2 mm long; **petals** five, unequal, 8–18 mm long, side and lower petals with a white throat and dark purple veins; **lower petal** one, 5–7 mm wide with a blunt or sometimes hooked **spur** 4–6 mm long reaching to or beyond the peduncle; **side petals** two, each 4–7 mm wide, bearded with a tuft of white throat hairs; **upper petals** two, glabrous, each 4–8 mm wide, bent backward, sometimes crossed or strongly reflexed over the sepals and spur; **stamens** five, c. 4 mm long and included, the two lower stamens with a spur 2.5–3 mm long which extends into the lower petal's spur; **filaments** light green, broad; **anthers** green, 1.8–2 mm long; **appendage** apical, c. 1.8 mm long, turning orange; **pistil** green, glabrous, 4–5.5 mm long, included; **ovary** c. 3 mm long by c. 1.7 mm wide, enclosed by the five filaments; **style** slender, thicker above, c. 2.3 mm long; **stigma** c. 1 mm wide, slightly hairy to hairless, rounded. **FRUIT a capsule**, yellowish tan, 3-valved, 4–7 mm long, hairless, ejects seeds 1–3 m; **seeds** 1.5–2.3 mm long by c. 1 mm wide and thick; **caruncle** c. 0.6 mm long near the hilum, a dark low ridge runs from the base of the seed to the caruncle.

■ **LEAVES** Basal and stem, alternate, simple, finely toothed, glabrous to hairy; **blades** 1–3.5 cm long and wide, often curled near the base or folded upward along the midrib, withered leaves persist at the plant's base from the previous year; **petioles** 1–7.5 cm long, hairy, often slightly longer than the blade; **stipules** green, toothed to entire, 5–18 mm long.

■ **STEM** Prostrate or arched upward, several, 4–14 cm tall or long, leafy, glabrous to hairy.

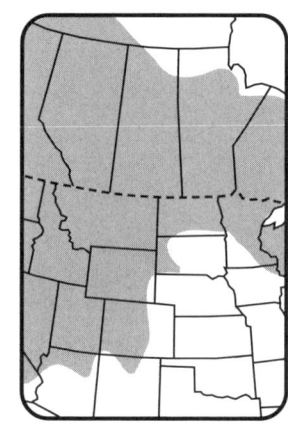

514 Hookedspur Violet

Violet—*Violaceae* **Wildflower** White with pink on back Petals 5

Western Canada Violet *Viola canadensis* L.

■ **SKETCH** A **perennial herb** 10–60 cm tall from tan **rhizomes** 1–4 mm wide and 6–20+ cm long; in damp meadows, mountain brush, thickets and open woods. **p2**

■ **FLOWERS** White with pink on back or front of petal tips, throat yellow, blooming April–September; **inflorescence** of single flowers from upper leaf axils; **peduncles** 1–8 cm long, glabrous to slightly hairy near the flowers, with a pair of bracteoles 3–4 mm long; **flowers** perfect, 12–17 mm wide by 8–15 mm long, cleistogamous flowers follow the earlier showy ones; **sepals** five, green, glabrous to ciliate, 4–9 mm long, pointed; **petals** five, a flush of pink to violet on the backs of at least the two upper petals, the other three petals have dark purple throat veins and the two side petals have small white beards in the inner yellow area; **upper petals** two, 4–7 mm wide; **side petals** two, 11–14 mm long by 5–7 mm wide; **lower petal** one, 12–14 mm long by 9–10 mm wide; **spur** *c.* 2 mm long, from the base of the lower petal, pale green, blunt, not reaching the peduncle; **stamens** five, hidden in throat, 2.5–3.2 mm long by 1–1.3 mm wide, surrounding the pistil; **filaments** pale green, *c.* 1.7 mm long with a winglike apical appendage *c.* 1 mm long which turns orange; **anthers** green, attached their full length to the wide filaments; **pistil** one, 4.2–4.5 mm long; **ovary** 2.1–2.3 mm long by 1.3–1.5 mm wide, glabrous to slightly hairy above; **style** and **stigma** together 2–2.2 mm long; **stigma** hairy along the sides, *c.* 0.7 mm wide. **FRUIT a capsule**, 3-valved, 5–10 mm long, glabrous or slightly hairy; **seeds** brown to purplish black, *c.* 15 per capsule, 1.6–2.2 mm long by *c.* 1.2 mm wide and thick, roundish with one low ridge, cottony hairs on body and ridge.

■ **LEAVES** Basal and stem, alternate, simple, toothed; **basal leaf blades** usually three to five, widely heart-shaped, margins shallowly toothed, 3–13 cm long by 2.5–12 cm wide, hairy below on the distinctive veins; **petioles** glabrous, 2–22 cm long; **stipules** paired, 1–2 cm long by 4–6 mm wide, entire, pointed, thin, glabrous; **stem leaves** few, similar, reduced above.

■ **STEM** Glabrous, erect, one to several; 2–5 mm thick near the base.

■ **SYN.** *Viola rugulosa* Greene = *Viola canadensis* var. *rugulosa* (Greene) C.L. Hitchc.

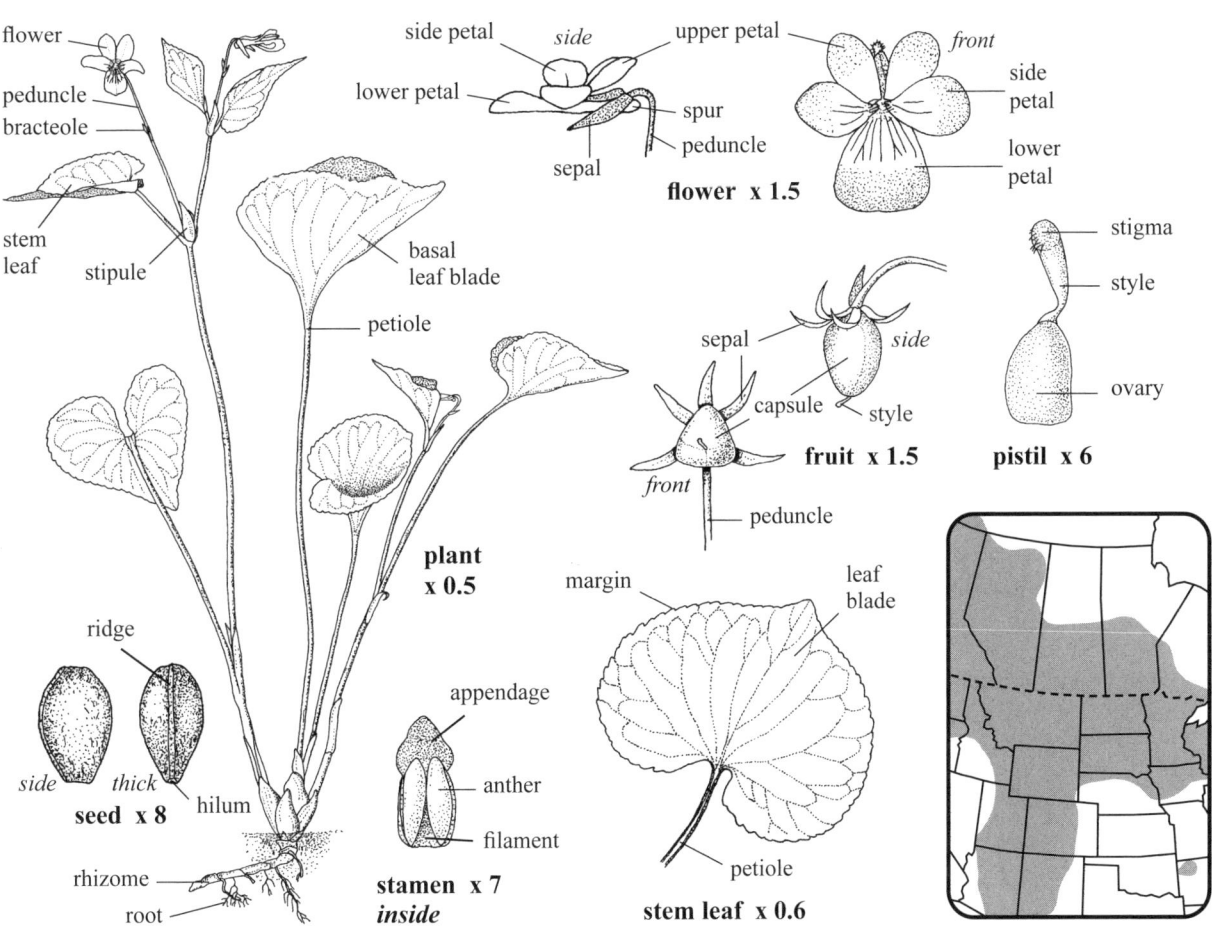

Tall White Violet, Canadian Western Violet, Canada Violet

Violet—*Violaceae* **Wildflower** Purple with white throat Petals 5

Northern Bog Violet *Viola nephrophylla* Greene

■ **SKETCH** A stemless **perennial herb** 5–12 cm tall from a light tan, rough **caudex** 2–3 cm long by 4–5 mm wide, usually curved and with pale **fibrous roots** 3–6 cm long from lower sections of caudex; in moist areas around sloughs and bogs, open woods, thickets and lake shores. **w2**

■ **FLOWERS** Pinkish purple with a white throat, blooming May–July; **inflorescence** a single flower, some cleistogamous; **peduncles** erect, glabrous, 6–11 cm long, recurved at apices; **flowers** perfect, 15–22 mm wide by 10–21 mm long by 10–11 mm high; **sepals** five, green, 6–9 mm long by 2–2.6 mm wide, margins glabrous or ciliate, pointed, persistent; **petals** five, flaring, purple; **upper petals** two, recurved, 10–16 mm long by *c.* 5 mm wide, whitish near the base, glabrous or bearded, lightly veined; **side petals** two, 12–19 mm long by 5–6 mm wide, lower half white with several scattered white hairs; **lower petal** one, hairy to glabrous, 15–21 mm long by *c.* 6 mm wide, veins dark purple; **spur** purple, 2–3 mm long, blunt; **stamens** five, included, 6–7 mm long (including spur and appendages); **appendages** orange, 1.4–1.7 mm long, flat, blunt, together forming a tube for the style and stigma; **anthers** pale green, *c.* 2 mm long; **spurs** green, *c.* 3 mm long, blunt, paired, one attached to each of the lower stamens, fitting into petal's spur; **pistil** 4.5–5 mm long, glabrous, included but *c.* 1 mm past the appendages; **ovary** green, bluntly triangular, *c.* 2 mm long by *c.* 1.5 mm wide at base, tapered; **style** *c.* 2 mm long, green, persistent; **stigma** *c.* 0.9 mm wide, with a pore and a ridge. **FRUIT a capsule**, ascending, 6–10 mm long by 4–6 mm wide, bluntly triangular, pale greenish yellow, 3-valved, valves *c.* 4 mm wide; **seeds** 45–57 per capsule, dark brown, 1.8–2.2 mm long (including aril) by *c.* 1 mm wide and thick, ejected 1–15 cm from the capsule when it opens; **aril** white, 0.9–1 mm long, a black ridge from the base of aril to seed's base.

■ **LEAVES** Erect, basal, arising from top of caudex, protruding from soil near base of peduncle; **blades** dull, toothed, simple, 1.5–5.5 cm long by 2.5–6 cm wide (flat), apices rounded to pointed, bases inrolled, hairy below especially on veins, hairless above; **petioles** 4–23 cm long, glabrous; **stipules** erect, whitish, entire, *c.* 10 mm long by 1.5–2 mm wide, pointed, usually below ground.

■ **STEM** Absent.

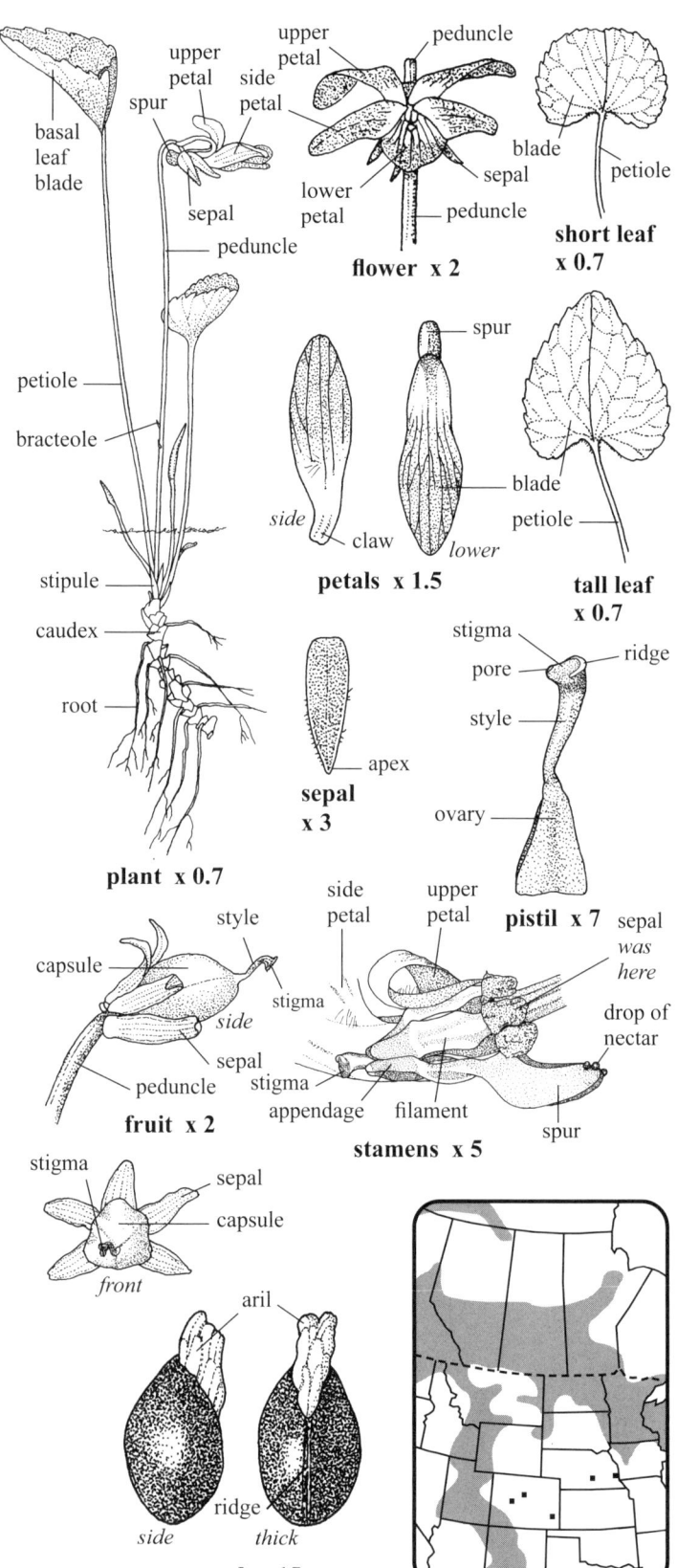

516   Bog Violet

Violet—*Violaceae* **Wildflower** Bluish purple, throat white Petals 5

CROWFOOT VIOLET *Viola pedatifida* G. Don

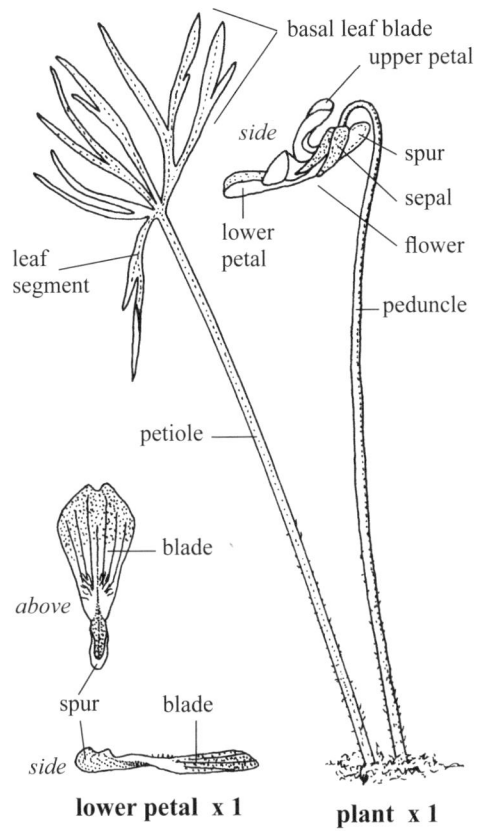

lower petal x 1

plant x 1

■ **SKETCH** A **perennial herb** 5–22 cm tall from a pale brown, vertical **caudex** 3–6 mm wide by 0.7–6+ cm long; in prairies, open woodlands and along exposed banks. p2

■ **FLOWERS** Bluish purple, throat white, blooming April–June; **inflorescence** a single flower at the end of erect peduncles; **peduncles** one to six, each 2.5–17 cm long, slightly hairy in the lower half, glabrous above; **cleistogamous flowers** on short peduncles may appear later in the season; **flowers** perfect, 10–30 mm wide by 18–30 mm long; **sepals** five, glabrous to slightly hairy, 6–14 mm long by 2–3.5 mm wide, pointed; **petals** five, all hairy in the throat of varying amounts; **lower petal** 6–10 mm wide and to *c.* 25 mm long, with several dark blue veins against the white throat; **spur** of lower petal blunt and 3–4 mm long; **side petals** two, *c.* 7 mm wide, very hairy with one or two dark blue veins; **upper petals** two, each *c.* 20 mm long by 7–8 mm wide, less hairy and white in their throats; **stamens** five, *c.* 5 mm long by 1.5 mm wide, surrounding the lower half of the pistil; **filaments** wide, the lower two with a green spur *c.* 4 mm long and each with a thin, apical orange appendage *c.* 2 mm long; **anthers** *c.* 3 mm long; **pistil** glabrous, 6–7 mm long; **style** *c.* 3 mm long; **stigma** flat, with an orifice to one side. **FRUIT a capsule**, 3-valved, tan, 8–14 mm long by 4–6 mm wide and thick, glabrous; **seeds** numerous, tan, slightly shiny, 1.7–2 mm long by 1–1.2 mm thick and wide with a pale line on one side; **aril** white, 0.6–0.9 mm long.

■ **LEAVES** Basal, compound, lobed, three to ten (only one shown); **early blades** deeply divided, 1.5–7 cm long by 4–10 cm wide, mostly 3-lobed, lobes divided into narrow segments; **segments** ciliate along the margins with the hairs more common near the tips; **late summer leaves** also deeply divided; **petioles** 2.5–20 cm long, slightly grooved, lightly hairy along the lower half, glabrous above; **stipules** short, below ground at bases of petioles.

■ **STEM** Absent.

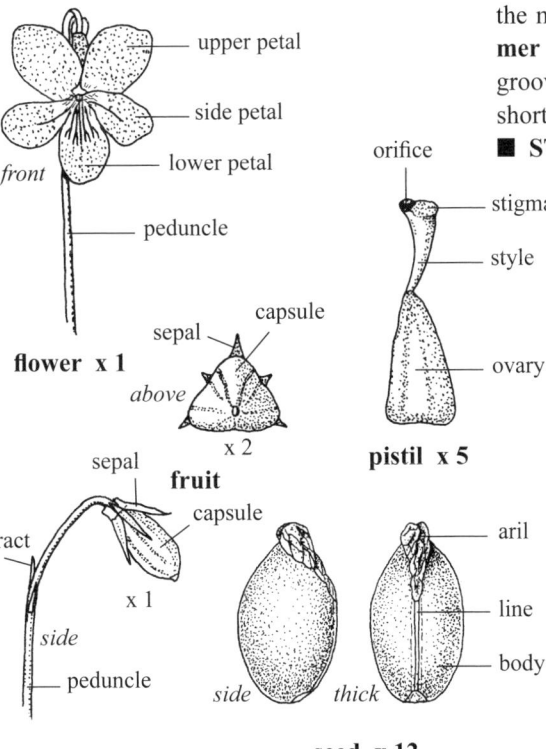

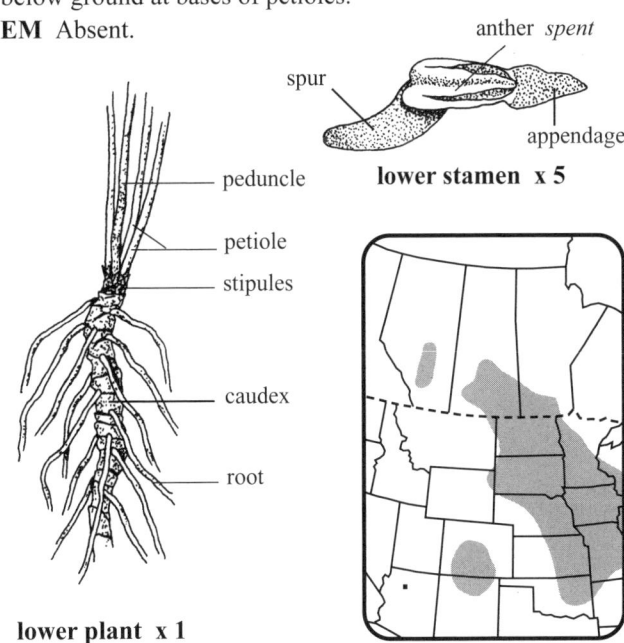

Prairie Violet, Larkspur-violet   517

Violet—*Violaceae*

Wildflower Yellow Petals 5

DOWNY YELLOW VIOLET *Viola pubescens* Ait.

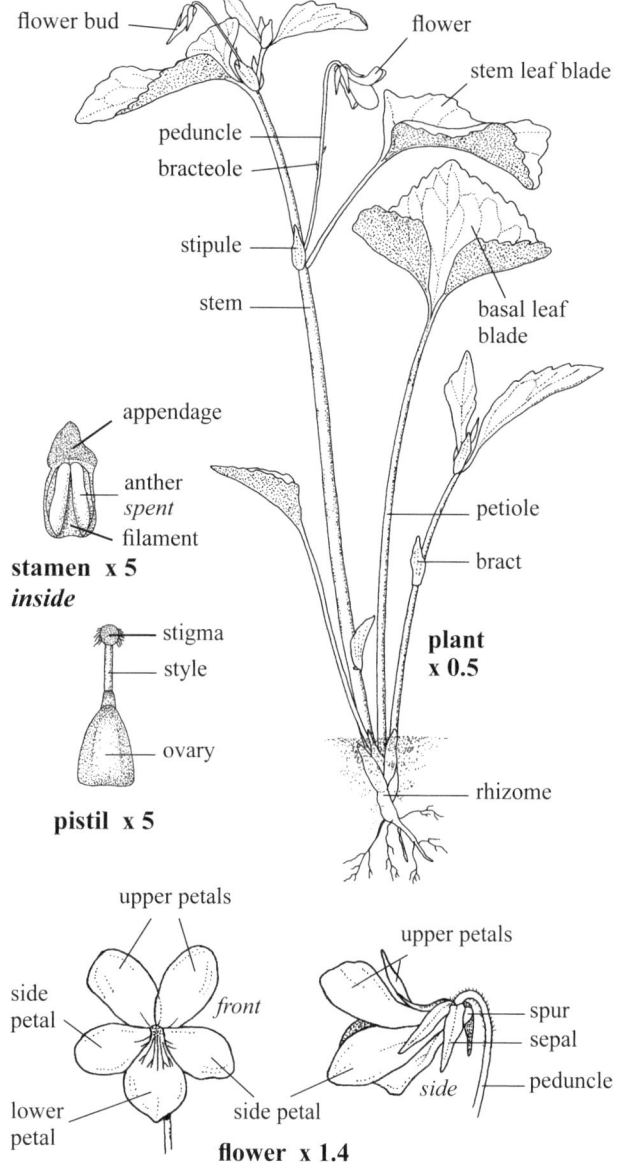

■ **SKETCH** A hairy to glabrous **perennial herb** 12–40 cm tall with a short, erect to horizontal **rhizome** 1.5–2.5 cm long by *c.* 5 mm wide; in patches in moist to dry woodlands, along banks of streams and in thickets. **w2, p2**

■ **FLOWERS** Yellow with brownish purple veins in the throat of all five petals but mainly on the bottom petal, blooming March–June; **inflorescence** axillary, flowers solitary; **peduncles** hairy, ascending, 2–8 cm long; **flowers** perfect, nodding, to *c.* 20 mm wide by *c.* 15 mm long, fragrant, cleistogamous flowers occur late in the season; **sepals** five, green, 6–10 mm long, glabrous, pointed, persistent; **corolla** of five petals; **upper petals** two, each 13–15 mm long by 6–7 mm wide; **side petals** two, each 14–16 mm long by 6–7 mm wide, with a tiny beard in the throat; **lower petal** one, *c.* 15 mm long by 7–8 mm wide, forming a blunt **spur** 2–3 mm long; **stamens** five, each *c.* 3 mm long by 1.3–1.5 mm wide; **filaments** green, wide with an apical appendage *c.* 1 mm long; **anthers** green, *c.* 1.8 mm long; **pistil** *c.* 4 mm long, surrounded by the stamens; **stigma** hairy on the sides. **FRUIT a capsule**, triangular, brown, 3-valved, each valve 8–14 mm long by 4–5 mm wide, woolly or glabrous, splitting open from apex; **seeds** white to light tan, 25–30 per capsule, each 2–3 mm long by 1.7–2 mm wide by *c.* 1.3 mm thick, smooth, slightly shiny, a whitish ridge to the seed's base; **caruncle** *c.* 1.3 mm long near the hilum.

■ **LEAVES** Basal and stem, alternate, simple, toothed, two to four near the top; **blades** wavy, widely heart-shaped, 3–10 cm long and wide, more downy below along the raised veins than above; **petioles** ascending, 1–19 cm long, shorter to longer than leaf blades; **stipules** paired, 8–20 mm long by 4–7 mm wide, entire or toothed, pointed, hairy or not, sometimes leafless below, as a bract.

■ **STEM** Round, smooth, stout, one to several, erect to decumbent, more downy in the upper half; 2–3 mm thick near the base.

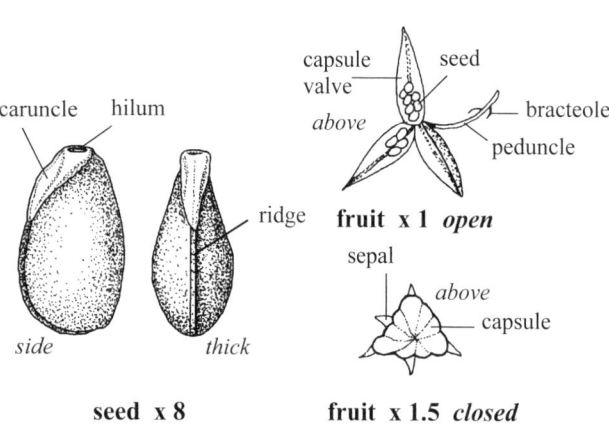

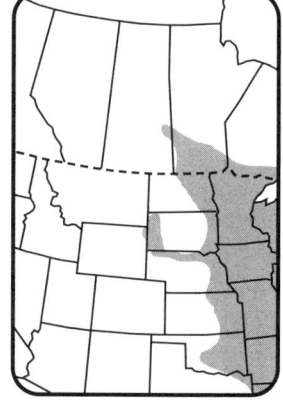

518 Smooth Yellow Violet

Grape—*Vitaceae*   **Vine**   Green   Petals 5

LARGE-TOOTHED VIRGINIA CREEPER   *Parthenocissus vitacea* (Knerr) A. S. Hitchc.

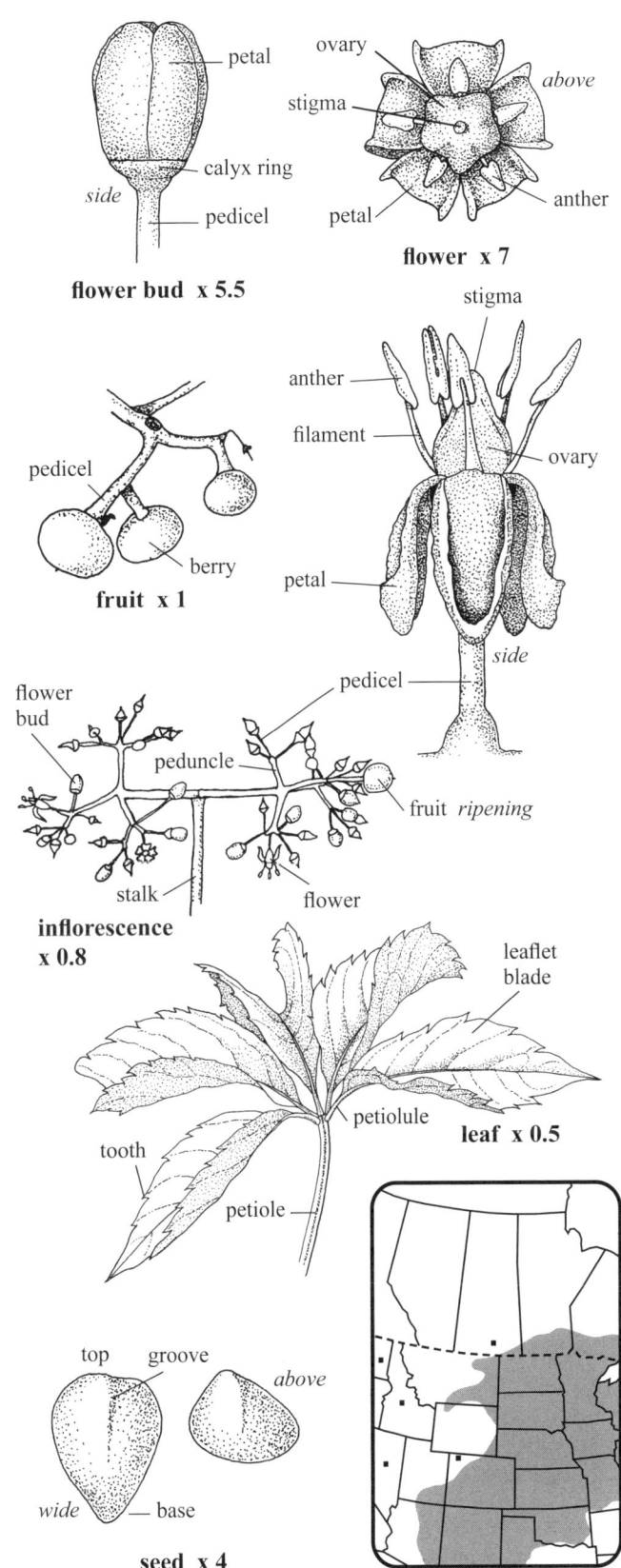

■ **SKETCH** A **woody vine** climbing 1 m or more high on trees or prostate along the ground for several meters, with **tendrils**; in woods, ravines, hillsides, rocky to non-rocky sites, in shade or sunlight, near seeps and springs. **w1, p2**

■ **FLOWERS** Green, blooming May–July; **inflorescence** a cyme, compound, the main axis dichotomously branched, 3.5–8 cm long by 3–6 cm wide; **inflorescence stalk** 5–11 cm long by 2–3 mm wide, green, glabrous, stiff; **pedicels** green, 4–5 mm long by *c*. 0.5 mm wide, stiff, glabrous, thickening and turning red with fruit; **flower buds** 3.5–4 mm long by 2.5–3 mm wide, glabrous; **flowers** perfect, 4–5 mm wide and tall, regular, 10–60 per inflorescence; **calyx** united in a continuous circular ring *c*. 1 mm long, the margin thin, light green and entire; **petals** five, green, reflexed, 3–3.3 mm long by 1.5–2 mm wide, hairless, thick, not persistent, margins raised, wavy, and light green, apices hooded; **stamens** five, exserted, *c*. 3 mm long; **filaments** pink to white, *c*. 2 mm long; **anthers** pale yellow, *c*. 1.5 mm long; **pistil** one, exserted, pinkish below, *c*. 2 mm tall by *c*. 1.5 mm wide, 5-lobed; **style** 0.8–1 mm long, pale yellow; **stigma** blunt, pale yellow, hairless. **FRUIT** a **berry**, bluish black with a slight bloom, dull, roundish, 7–12 mm wide, smooth, thin-walled; **seeds** 1–4 per berry, each 4.5–5.2 mm long by *c*. 4 mm wide by *c*. 3 mm thick, shiny, brown, base pointed, 3-sided, a shallow groove on outside.

■ **LEAVES** Alternate, compound, five leaflets in a palmate whorl, slightly shiny above, lighter green and duller below, reddish in fall; **petioles** 19–28 cm long by 3–5 mm wide, green, glabrous, dichotomous from short woody spur branches; **leaflet blades** 5–20 cm long by 3–13 cm wide, coarsely toothed, central blade the largest, two rear blades the smallest, hairy along the veins on both sides but more so below, or glabrous, young blades very shiny above and below, flat to V-shaped; **petiolules** 0.5–7 cm long, slightly winged near blade, green to reddish; **stipules** paired, green with hyaline margins, apices blunt, 7–9 mm long by 3–4 mm wide, spreading on new green shoots; **tendrils** alternate, light green, opposite leaf petioles near tips of new growth, stalk 3–6 cm long, dichotomously branching one to three times, tips usually without discs.

■ **STEM** Woody, bark light gray, slightly rough, not peeling or armed; 7–15 mm wide near the base.

Woodbine, Thicket Creeper

Monocotyledons

THIS FINAL LARGE section of the book deals with the second major group of vascular plants within the Angiosperms (flowering plants), the Monocotyledons. Within the monocots are some colorful prairie plants, many aquatics or semi-aquatics and the more distinguished, wind-pollinated grasses and sedges.

MONOCOTS ARE SEPARATED from the dicots by two main characteristics. The monocots have **1)** flower parts in 3s or 6s. The flowers may be quite large, colorful and showy, as in the lilies (Liliaceae) and orchids (Orchidaceae). Occasionally, this rule is broken. The pondweeds (Potamogetonaceae) have small, brown perianth parts, bracts (or sepals) in 4s; and **2)** blades of the leaves parallel-veined.

THE TWO LARGEST families in the monocots are the widespread grasses (Poaceae) and sedges (Cyperaceae). The *Carex* species of Cyperaceae are unique among the plants. Their flowers and leaves are not used in identification. Instead, the perigynium is used for identification. The perigynium, a modified bract or scale, encloses the ripe fruit, a 1-seeded achene.

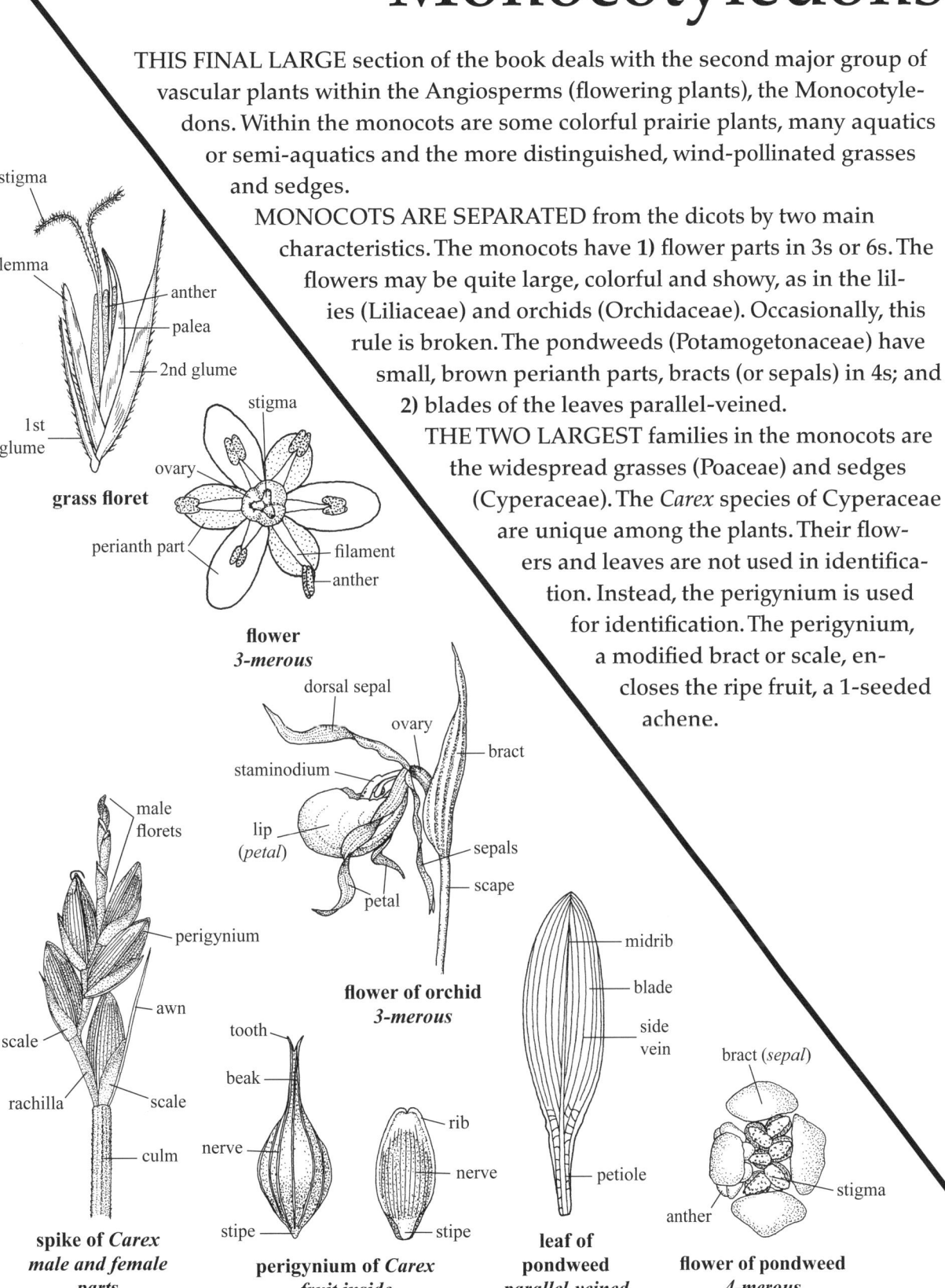

Calamus—*Acoraceae*   **Wildflower**   Pale greenish yellow   Perianth parts 6

SWEET FLAG   *Acorus americanus* (Raf.) Raf.

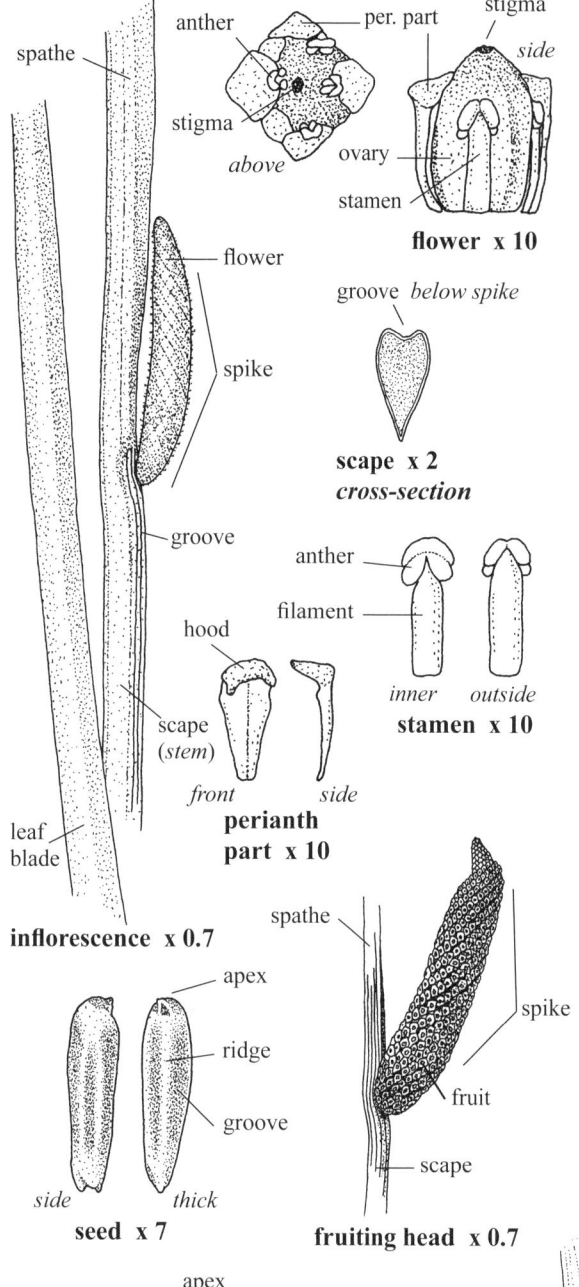

- **SKETCH** A glabrous **perennial herb** 60–150+ cm tall with a branching scented **rhizome** 10–13 mm thick by 10–20+ cm long with dark brown leaf bases 8–15 cm long by 8–13 mm wide; **roots** white, 1–2.5 mm wide by 5–15 cm long; in marshes, swamps and ditches. **w2**
- **FLOWERS** Pale greenish yellow, blooming May–August; **inflorescence** a spike, erect to ascending, 4–9 cm long by 8–12 mm wide, round in cross-section, slightly curved, sessile, appearing lateral; **flowers** perfect, diamond-shaped (4-sided), 2–2.5 mm wide by 2.5–3 mm long, spirally arranged and crowded on a thick inner spadix; **spathe** green, leaf-like, erect, 10–70 cm long by 8–17 mm wide by *c.* 3 mm thick, closed, flattened, pointed, easily bent; **perianth parts** six, 1.5–1.7 mm long by 0.6–1 mm wide, hooded, grayish hyaline, membranous, turning brown, fragmenting as fruit ripens, thicker near base; **stamens** six, opposite perianth parts, TL *c.* 1.8 mm (before anthers open); **filaments** wide, hairless; **anthers** pale yellow, lobed, 0.6–0.7 mm long by 0.7–0.8 mm wide, under perianth hood, eventually exserted; **pistil** green, hairless, 2–2.2 mm long by *c.* 1.5 mm wide and thick; **ovary** *c.* 1.5 mm long; **style** obscure, tapered; **stigma** one, flat, narrow; **fruiting spike** ascending, dark brown, 5–9 cm long by 10–15 mm wide. **FRUIT a berry**, light tan, not fleshy, 4–5 mm long by 1.5–3 mm wide by *c.* 2 mm thick, shiny, walls paper thin, angled, grooved and ridged; **seeds** 2–4 per berry, pale tan, angular, pitted, each 3.3–4 mm long by 0.7–1.1 mm wide and thick, dark brown at apices.
- **LEAVES** Simple, entire, flattened, crowded below, ascending, dull, open and clasping stem near the base, 90–180 cm long by 5–20 mm wide by 2–3 mm thick, linear, veins parallel, medium green, pinkish near the base.
- **STEM** A **scape**, flattened, stiff, green, erect, widely grooved on side below the spike, 3-sided, spongy inside; 3–5 mm thick near the base.
- **SYN.** *Acorus calamus* auct. non L.

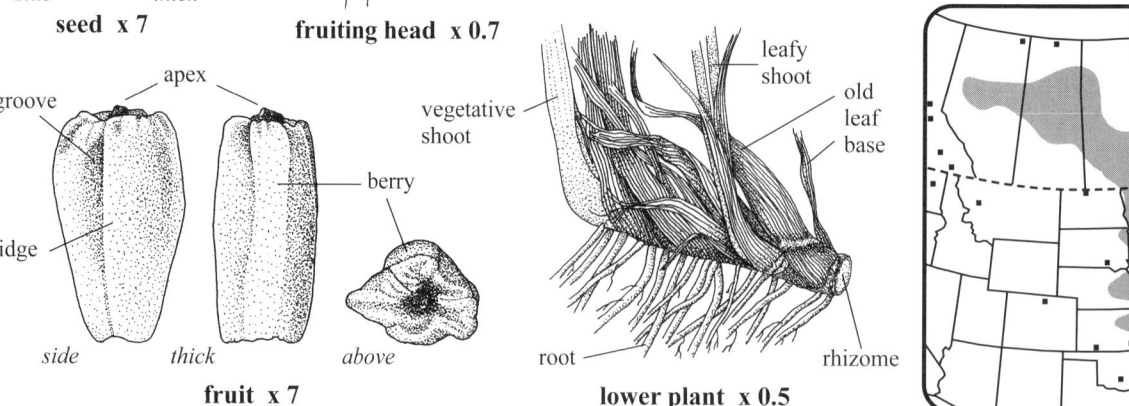

American Calamus, Ratroot, Calamus, Pepper-root, Sweet Cinnamon

Water-plantain—*Alismataceae* **Wildflower White Petals 3**

COMMON WATER-PLANTAIN *Alisma triviale* Pursh

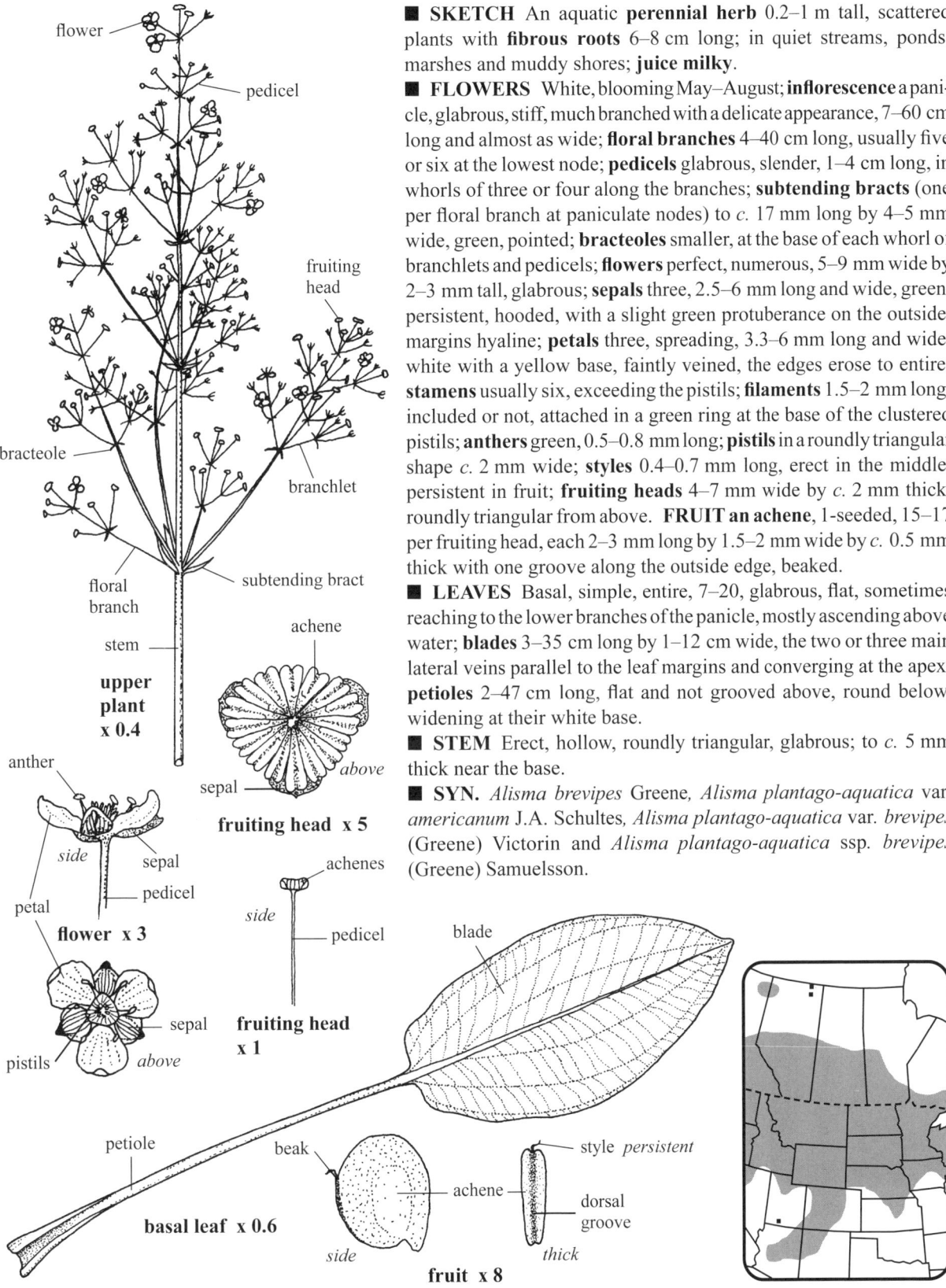

■ **SKETCH** An aquatic **perennial herb** 0.2–1 m tall, scattered plants with **fibrous roots** 6–8 cm long; in quiet streams, ponds, marshes and muddy shores; **juice milky**.

■ **FLOWERS** White, blooming May–August; **inflorescence** a panicle, glabrous, stiff, much branched with a delicate appearance, 7–60 cm long and almost as wide; **floral branches** 4–40 cm long, usually five or six at the lowest node; **pedicels** glabrous, slender, 1–4 cm long, in whorls of three or four along the branches; **subtending bracts** (one per floral branch at paniculate nodes) to *c.* 17 mm long by 4–5 mm wide, green, pointed; **bracteoles** smaller, at the base of each whorl of branchlets and pedicels; **flowers** perfect, numerous, 5–9 mm wide by 2–3 mm tall, glabrous; **sepals** three, 2.5–6 mm long and wide, green, persistent, hooded, with a slight green protuberance on the outside, margins hyaline; **petals** three, spreading, 3.3–6 mm long and wide, white with a yellow base, faintly veined, the edges erose to entire; **stamens** usually six, exceeding the pistils; **filaments** 1.5–2 mm long, included or not, attached in a green ring at the base of the clustered pistils; **anthers** green, 0.5–0.8 mm long; **pistils** in a roundly triangular shape *c.* 2 mm wide; **styles** 0.4–0.7 mm long, erect in the middle, persistent in fruit; **fruiting heads** 4–7 mm wide by *c.* 2 mm thick, roundly triangular from above. **FRUIT** an achene, 1-seeded, 15–17 per fruiting head, each 2–3 mm long by 1.5–2 mm wide by *c.* 0.5 mm thick with one groove along the outside edge, beaked.

■ **LEAVES** Basal, simple, entire, 7–20, glabrous, flat, sometimes reaching to the lower branches of the panicle, mostly ascending above water; **blades** 3–35 cm long by 1–12 cm wide, the two or three main lateral veins parallel to the leaf margins and converging at the apex; **petioles** 2–47 cm long, flat and not grooved above, round below, widening at their white base.

■ **STEM** Erect, hollow, roundly triangular, glabrous; to *c.* 5 mm thick near the base.

■ **SYN.** *Alisma brevipes* Greene, *Alisma plantago-aquatica* var. *americanum* J.A. Schultes, *Alisma plantago-aquatica* var. *brevipes* (Greene) Victorin and *Alisma plantago-aquatica* ssp. *brevipes* (Greene) Samuelsson.

Large Water Plantain, Northern Broad-leaved Water-plantain, Western Water-plantain

Water-plantain—*Alismataceae*   **Wildflower**   White   Petals 3

ARUM-LEAVED ARROWHEAD   *Sagittaria cuneata* Sheldon

■ **SKETCH** A variable, glabrous **perennial**, **aquatic herb** 20–110 cm tall from **rhizomes** (some prefer stolons) 1–3.5 mm thick by 3–15 cm long with **fibrous roots** and **corms** in shallow water to *c.* 50 cm deep; in marshes, ditches, along shores and streams, the base often submerged; **monoecious** or **dioecious**. w2

■ **FLOWERS** White, blooming June–September; **inflorescences** a raceme, one to three, each 1.5–25 cm long, by 2–10 cm wide, with 2–10 whorls, usually three flowers (two to eight) per whorl, sometimes branched at the lowest node; **pedicels** 5–50 mm long; **subtending bracts** (of pedicels) one, 5–40 mm long by 2–5 mm wide; **flowers** unisexual; **male flowers** above, 15–25 mm wide; **sepals** three, light green, 4–8 mm long by 2–6 mm wide, C-shaped; **petals** three, white, 8–15 mm long and wide; **stamens** 10–18, each *c.* 3 mm long; **filaments** light green, wide and fleshy, glabrous, sometimes two are united; **anthers** 1–1.3 mm long, pale yellow; **female flowers** below male ones; **pistils** green, numerous, in a globular head 5–8 mm wide by 4–5 mm long, more obvious once the petals fall; **fruiting heads** globular, 7–15 mm wide. **FRUIT** an achene, 1-seeded, beaked, 1.5–3 mm long by 1.3–2.5 mm wide by *c.* 0.4 mm thick, numerous, the wings pale brown but darker at the apex; **beak** erect, dark brown, 0.1–0.4 mm long.

■ **LEAVES** Basal, simple, entire, basally 2-lobed or not, submerged or not; **submerged leaves** not lobed, with no obvious petiole but gradually tapered to the base, ribbonlike to *c.* 50 cm long by 5–20 mm wide; **basal blades** aerial, erect, shorter to taller than the inflorescence, 1.5–30 cm long by 0.5–15 cm wide; **lobes** pointed and 4–12 cm long, or absent; **petioles** 2–50 cm long, angular, curved, flat above and round below, not grooved, spongy, 5–10 mm wide near the base; **submerged stems** may lack leaves, or develop floating leaves.

■ **STEM** A **scape**, round, smooth, spongy, erect, usually unbranched; 5–10 mm thick near the base.

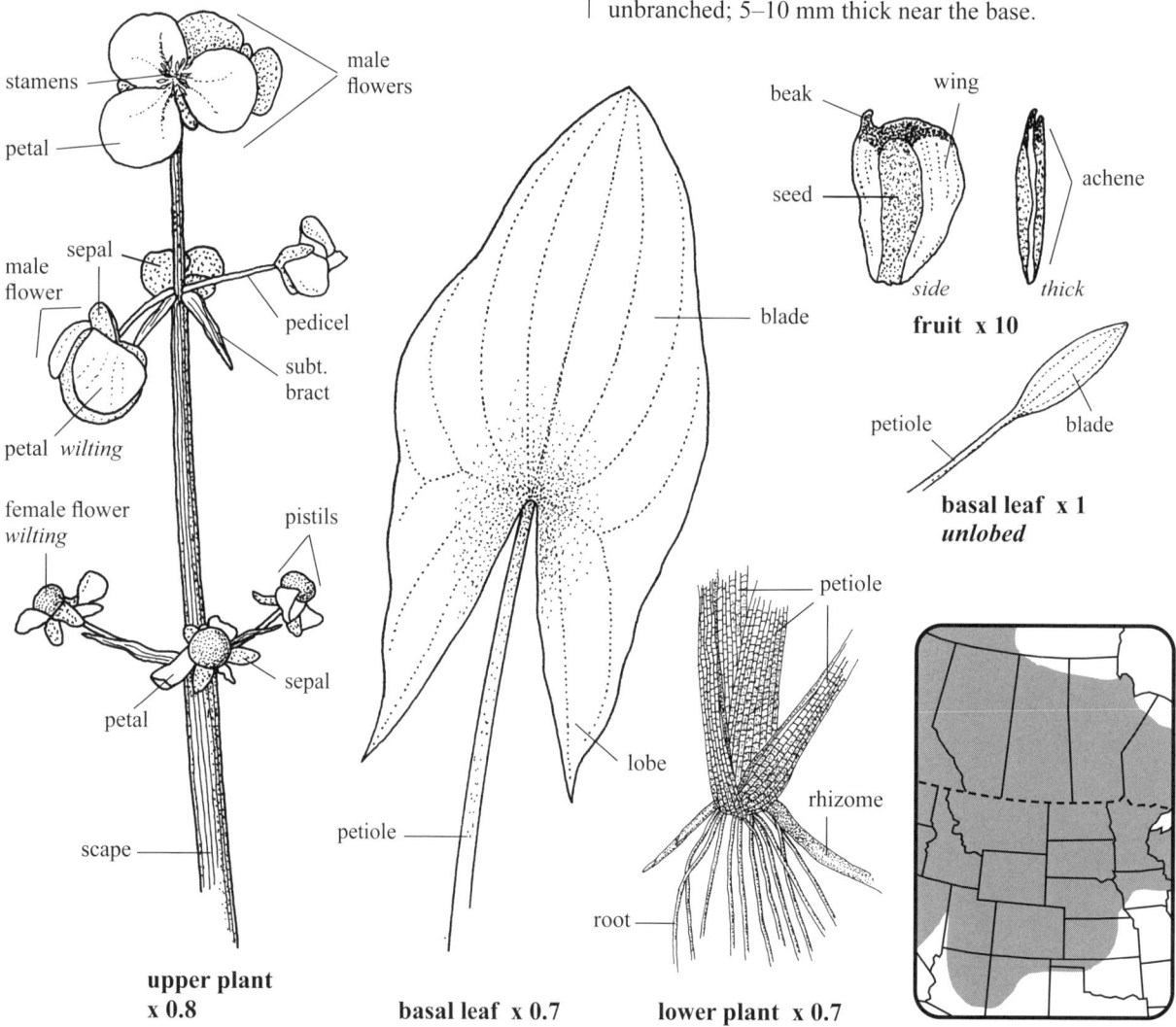

Water-plantain—*Alismataceae* **Wildflower** White, center yellow Petals 3

Broad-leaved Arrowhead *Sagittaria latifolia* Willd.

■ **SKETCH** A variable, aquatic, glabrous **perennial herb** 30–100 cm tall with orangish tan **rhizomes** 1.5–20+ cm long by 0.7–5.5 mm wide and autumnal, spotted **corms** 1–5 cm long and wide and orangish brown **roots** 2–20 cm long by *c.* 1 mm thick; in ditches and along muddy shores; **monoecious** to **dioecious**. w1

■ **FLOWERS** White, center yellow, blooming June–September; **inflorescence** a raceme, erect above water, one to several, male flowers above female ones, 3–9 whorls of three flowers each, more crowded above; **pedicels** green, ascending, 3–10 mm long (to 35 mm in fruit) by 1–1.3 mm thick, oval; **subtending bracts** (of whorls) 5–10 mm long by 6–7 mm wide, pale green, membranous, turning brown; **flowers** unisexual, lowest ones may be perfect, 18–35 mm wide by 6–7 mm long; **rachis** sharply triangular, twisted; **sepals** three, green, C-shaped, 4–7 mm long and wide, reflexed with fruit, veins numerous, united at base for *c.* 1 mm; **petals** three, white, 8–10 mm long and wide, claw lightly yellow, veins faint, quickly deciduous; **stamens** exserted, 25–40 in a central cluster 4–5 mm wide and *c.* 2.5 mm tall; **filaments** 0.6–0.9 mm long, greenish;
anthers 1–1.3 mm long, yellow; **pistils** numerous, greenish, together 5–7 mm wide by *c.* 4 mm tall, each *c.* 1.5 mm long by *c.* 1 mm wide by *c.* 0.2 mm thick with a transparent wing; **style** and **stigma** horizontal; **fruiting heads** 1–3 cm wide, roundish. **FRUIT an achene**, 1-seeded, winged or not, 2.5–3.5 mm long to *c.* 2 mm wide; **beak** 1–2 mm long, horizontal.

■ **LEAVES** Variable, several arising from the base of the scape; **blades** entire, erect, 3-lobed, the lobes 13–30(–50) cm long; **terminal lobe** 7–15 cm long by 9–20 mm (–12 cm) wide; **basal lobes** two, 2.3–15 cm long by 4–10+ mm wide, spreading or close together; **petioles** 6–60 cm long, roundly triangular, ascending.

■ **STEM** A **scape**, solid, length varies with depth of water, erect, usually unbranched, sharply triangular, soft but stiff with a low ridge on each face below the flowers; *c.* 5 mm wide near the base.

■ **SYN.** *Sagittaria engelmanniana* ssp. *longirostra* (Micheli) Bogin, *Sagittaria longirostra* (Micheli) J.G. Sm. and *Sagittaria variabilis* var. *obtusa* Engelm.

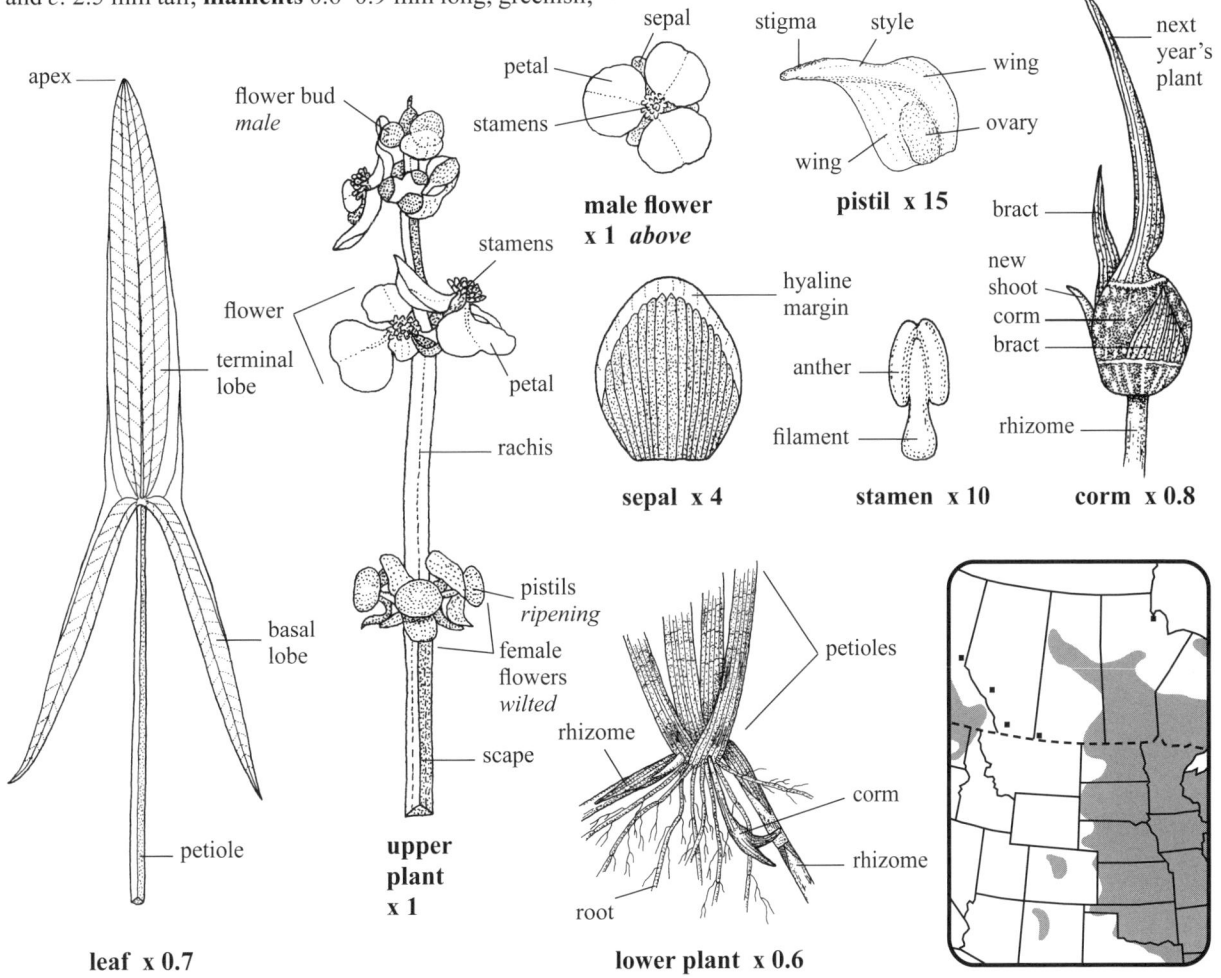

Arrowhead, Common Arrowhead, Duck-potato, Wapato

Arum—*Araceae* **Wildflower** **White** **Perianth absent**

WATER CALLA *Calla palustris* L.

■ **SKETCH** A colonial, aquatic **perennial herb** 30–50 cm tall with a shiny, smooth, greenish brown **rhizome** 10–30 cm long by 8–20 mm wide with pale green **roots** 2–25 cm long and up to *c.* 1 mm thick from **nodes** 0.7–5 cm apart with a ridge and dark brown veiny fragments of sheaths to *c.* 5 cm long; in water of marshes, ponds and wet boggy sites.

■ **FLOWERS** White, blooming May–June; **inflorescence** a spike 2–3 cm long by 9–13 mm wide, slightly curved, round in cross-section, with a blunt apex; **flowers** perfect, or the upper ones male, 2–3 mm wide by 3–3.5 mm long, sessile, crowded on a thick spadix; **perianth** absent; **spathe** entire, 3.5–6.5 cm long by 2–4 cm wide, white above, light green on underside, abruptly forming a fine point 4–16 mm long by *c.* 1 mm wide and curled; **stamens** six (9–12), white, 2–4 mm long, glabrous, several arising from the base of each ovary; **filaments** *c.* 0.5 mm wide; **anthers** 2-lobed, *c.* 0.5 mm long by *c.* 1 mm wide; **pistil** green, glabrous, whitish near base; **style** tapered, obscure; **stigma** *c.* 0.8 mm wide, white, quickly turning brown. **FRUIT a berry**, reddish, slightly shiny, 6–10 mm long by 4–12 mm wide, fleshy, crowded, grooved on sides, hairless, slightly tapered and whitish toward the base; **seeds** 2–8 per berry, 4–4.3 mm long by 1.8–2 mm wide by 1.7–1.9 mm thick, dark purplish brown with a blunt ridge on one side, the body with numerous low wide ribs below and pitted in the upper half, enclosed in clear gelatin.

■ **LEAVES** Basal, few, emergent, entire, simple, glabrous, heart-shaped, pointed, dull, turning yellow in late summer; **blades** 4–14 cm long by 4.5–10 cm wide, parallel veined, apical point 2–4 mm long and involute; **petioles** 13–35 cm long by *c.* 10 mm wide at base and tapered to *c.* 4 mm wide near the blade, flat above, round below, ascending, often curved; **basal sheaths** green, attached to base of petiole for 3–10 cm, open along one side; **ligules** 3–5 cm long, pointed, enveloping another leaf or scape.

■ **STEM** A **scape** 12–20 cm long (to *c.* 23 cm long with fruit) by 5–8 mm wide, arched upward, axillary, smooth and green.

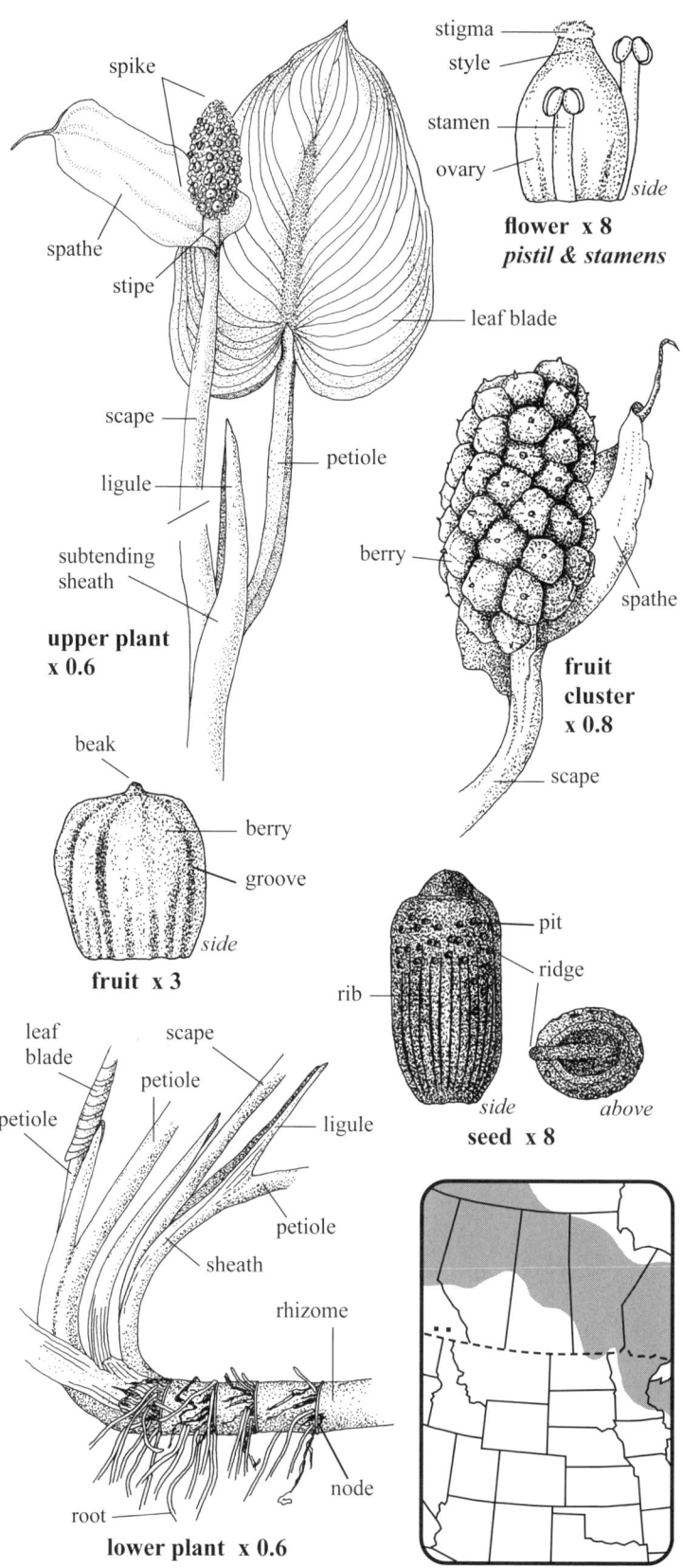

Marsh Calla, Wild Calla, Water Arum, Water Dragon

Flowering Rush—*Butomaceae*   **Wildflower**   Pink   Petals 3

FLOWERING RUSH   *Butomus umbellatus* L.

■ **SKETCH** A glabrous and glaucous **perennial herb** 0.6–1.5 m tall with white **fibrous roots** from a white short **rhizome** forming cloned groups; along shores of marshes, rivers, ponds and sloughs. *Naturalized*

■ **FLOWERS** Pink, blooming June–August, often sexually sterile due to cloning; **inflorescence** a compound cyme (umbel-like) of 8–48 flowers, terminal, one inflorescence per stem, each 8–20 cm wide by 7–14 cm tall; **subtending bracts** (of cyme) three, green turning brown at anthesis, ascending to spreading at anthesis, 20–26 mm long by 10–12 mm wide, with wide white scarious margins, entire, keeled in upper half, the keel green or brown, tapered to a narrow point; **subtending bracts** (of pedicels) membranous, 10–12 mm long, pointed; **pedicels** green, 4–12.2 cm long, ascending to arching upward, round, *c.* 1 mm thick; **flowers** perfect, regular, 25–32 mm wide by 7–12 mm tall; **sepals** three, spreading 10–12 mm long by 6–7 mm wide (flattened) with a wide brownish pink band below (dorsally) with pink margins, C-shaped, tips pointed to blunt, persistent and erect around fruit; **petals** three, pink, spreading to ascending, 14–16 mm long by 9–10 mm wide, midnerve dark pink below, and darker pink near the base, apices roundish and finely erose; **stamens** nine, in two whorls (6 + 3) from the base of pistils, slightly exserted, 10–11 mm long; **filaments** whitish pink, *c.* 1 mm wide and flattened at the base; **anthers** pink, 5–6 mm long before opening, *c.* 2 mm long when spent, basifixed, erect, opening from the base upwards; **pollen** deep yellow; **pistils** (carpels) six (5–7), wine red, included, 7–8 mm long and wide (together), forming a circle, inner surfaces touching, joined at bases; **ovaries** 6–7 mm wide (together) by *c.* 6 mm long, tapered to the style; **style** one, angled outward, 3–4 mm long, narrow; **stigma** 2-lobed, slightly wider than the style, *c.* 1 mm long, offset and pointed to the outside. **FRUIT a follicle**, body 5–6 mm long by 2–2.5 mm wide, dark brown with gray dots at the apex, usually four (6) per flower, beaked, the cluster 4–6 mm wide by 7–9 mm long (including styles), united along their inner surface; **seeds** brown, straight, hairless, 5- to 7-ribbed, 1–1.2 mm long by *c.* 0.3 mm wide, without endosperm, rarely formed.

■ **LEAVES** Congested near the base, simple, entire, linear, ascending, green, sharply triangular, usually 10–12, to *c.* 1 m long with the narrow grooved side 5–7 mm wide and the two slightly convex sides 7–9 mm wide; **blades** soft with hollow cells inside, gradually tapered and flattened toward the apex, not grooved near the tip, pale green hyaline margins form on the lower 15–20 cm on the grooved side, lower part of blades fit together with their sheaths enclosing the base of the stem.

■ **STEM** A **scape**, green, erect, one, smooth, solid, oval, slightly spongy; 3–5 mm wide below the inflorescence and 5–12 mm wide near the pinkish base.

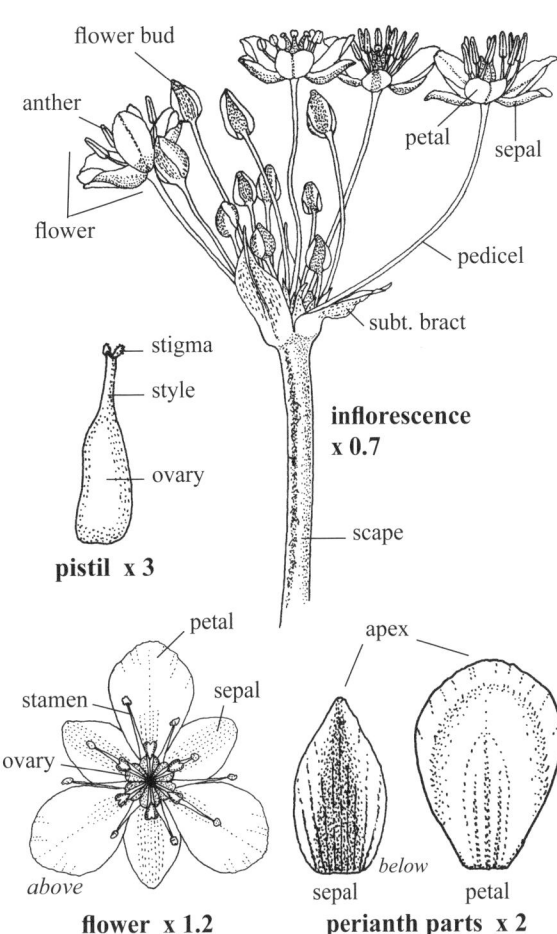

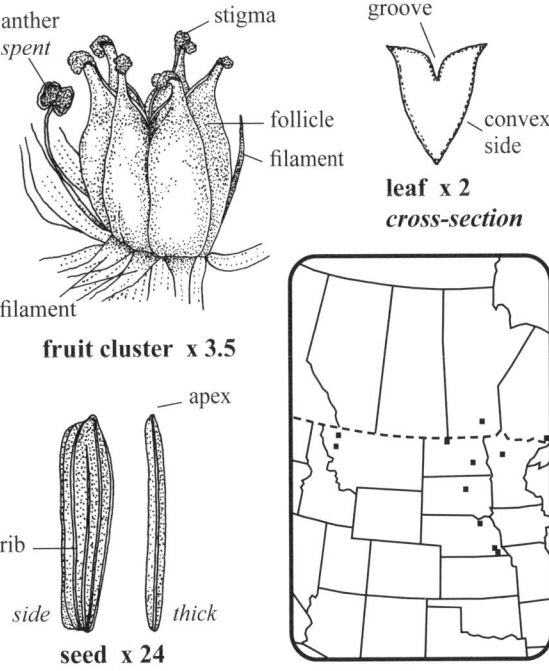

Grassy Rush

Sedge

Having turned to bow
as graciously
as water over stone

Sartwell's Sedge and Woolly Sedge
simply there
with tiny flowers
and a triangular stem
that the botanist draws
in his book

Humble as rain
when night is short
and night is long
and the nightwind's voice
is true

Family Characteristics

Sedge—Cyperaceae — Fruit under scales

(Non-*Carex* species)

SKETCH Monocotyledonous perennial herbs often forming extensive underground **rhizomes** in wet (some aquatic), cool regions including the arctic. Worldwide almost 4,000 species.

FLOWERS Perfect, small, crowded into spikelets, one under each fertile scale, generally not used in identification. **Inflorescence** a spikelet, 1 to many, solitary and terminal, or in tight to spreading, open clusters, or umbels often subtended by one to several involucral bracts. **Perianth** reduced to bristles or scales (often 6 per flower or fruit), or absent, arising from the base of the ovary/fruit. **Stamens** 3 or less, exserted, plants wind-pollinated. **Pistil** 1. **Ovary** 1-loculed with 1 ovule. **Style** 2- or 3-parted, sometimes enlarging and persistent as a tubercle atop the fruit. **Fertile scales** awned or not, variable.

FRUIT An achene, 1-seeded, lenticular if style is 2-parted; triangular if style is 3-parted, one under each fertile scale of a spikelet, often used in identification.

LEAVES Usually 3-ranked, long and grasslike with parallel venation and entire margins, sometimes reduced to a sheath and short blade or blade absent. **Ligules** often absent. **Sheaths** enclosing the stem, often closed, apices usually V-shaped or slightly concave.

STEM A **culm**, erect, green, solitary to tufted, round to triangular to oval, the corners often blunt, pithy inside, soft to firm.

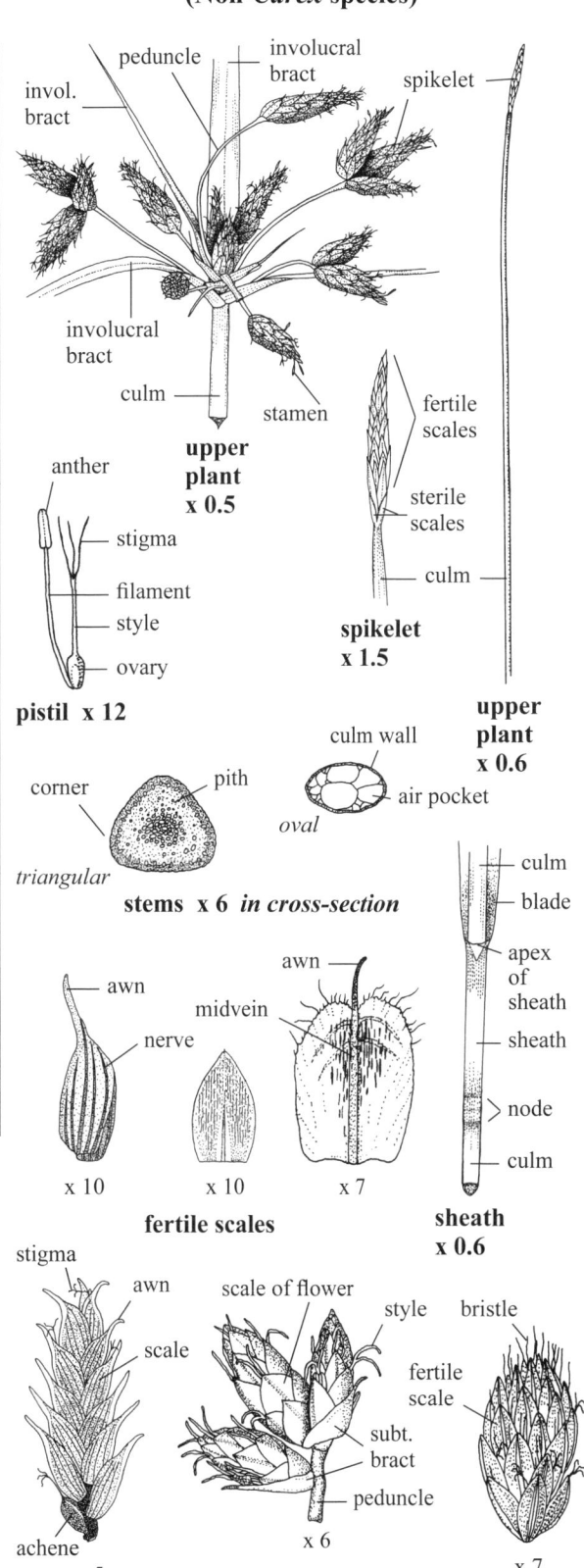

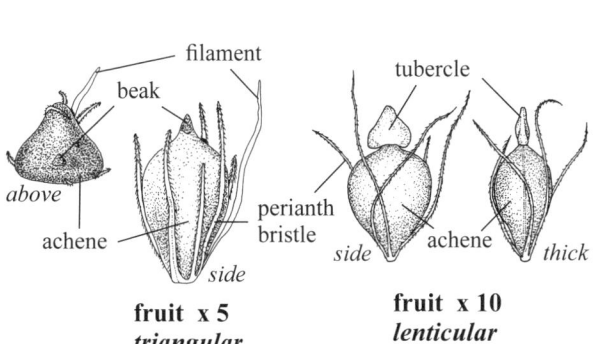

Family Characteristics

Sedge—Cyperaceae   Fruit in perigynium
(*Carex* species)

SKETCH Monocotyledonous perennial herbs with **fibrous roots** and often with **rhizomes**, often in temperate, alpine to arctic regions, wet habitats preferred, rare in tropics. Worldwide about 2,000 species.

FLOWERS Unisexual, greenish, small, one per scale, not used in identification. **Inflorescence** a spike of a few to many flowers, mostly monoecious, rarely dioecious. **Spikes** 1–many, mostly unisexual, male ones above or below female, or a spike androgynous or gynecandrous, each often subtended by a bract. **Perianth** absent. **Stamens** 3 (rarely 2), exserted. **Pistil** 1, enclosed in a modified bract (perigynium). **Ovary** 1-loculed with 1 ovule. **Style** 1, generally long, 2- or 3-parted (stigmas), exserted. **Perigynium** a modified bract enclosing the fruit, used in identification of species by its shape, size, nervation and beak. **Female scales** one per perigynium, longer to shorter and wider to narrower than perigynium, covering its dorsal side, usually not awned, also helpful in identification.

FRUIT An **achene**, 1-seeded, completely enclosed in perigynium, helpful in identification. Mostly brown, some shiny, one per perigynium. Lenticular if style is 2-parted; triangular if style is 3-parted.

LEAVES Three-ranked, simple, entire, narrow, green, flat, V-shaped or W-shaped. **Sheaths** closed, often V-shaped at apices. **Ligules** often unremarkable.

STEM A **culm**, erect, triangular, edges sharp to blunt, sometime scabrous, usually solid.

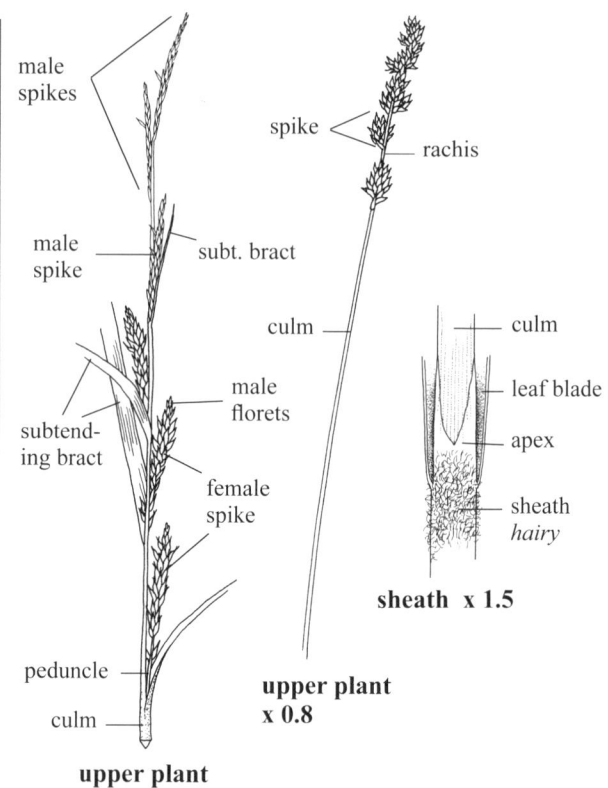

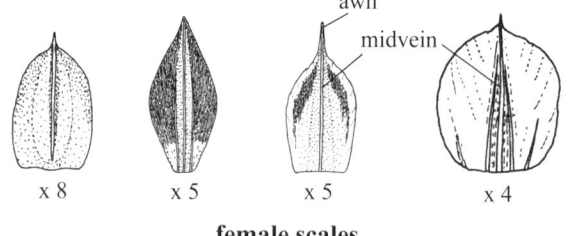

female scales

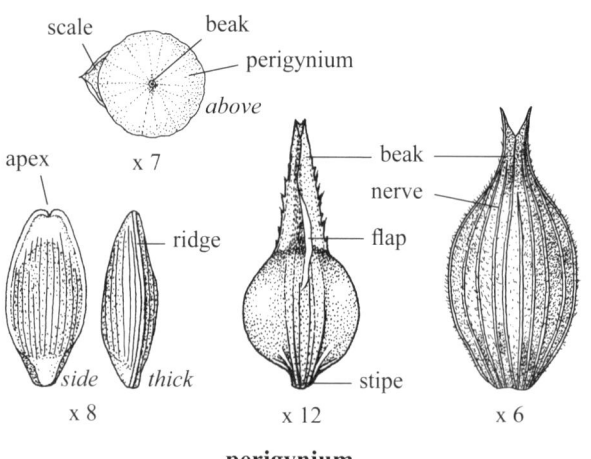

perigynium

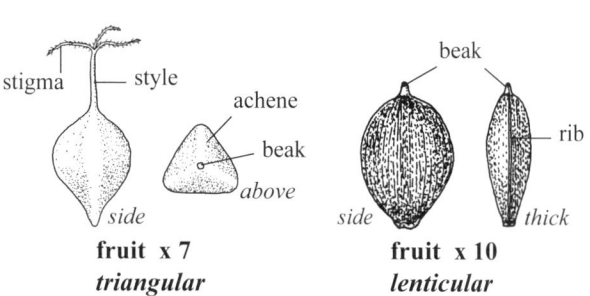

530

Sedge—*Cyperaceae* Sedge Perigynia hairless, 2–3.4 mm long

WATER SEDGE *Carex aquatilis* Wahlenb.

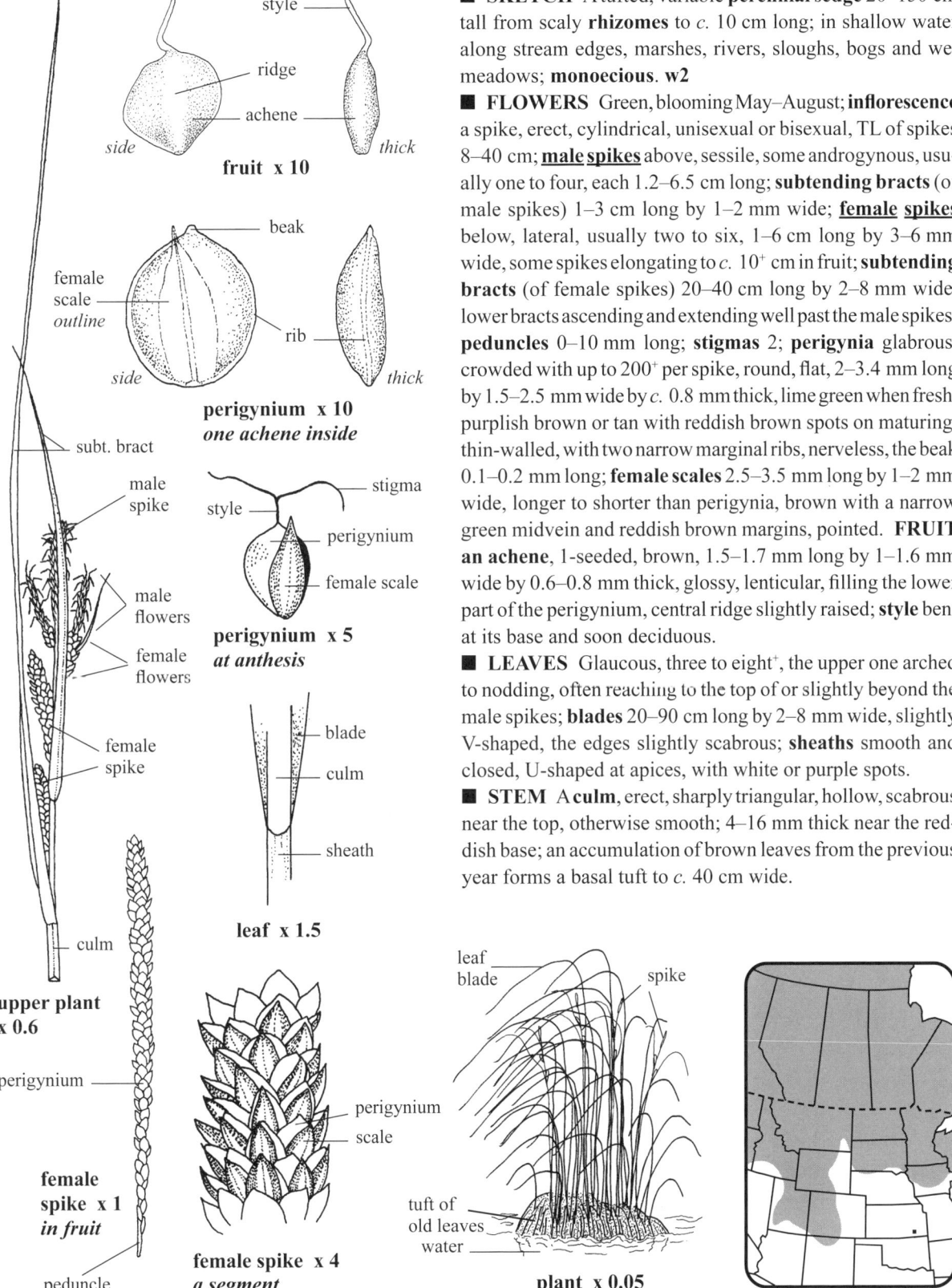

■ **SKETCH** A tufted, variable **perennial sedge** 20–150 cm tall from scaly **rhizomes** to *c.* 10 cm long; in shallow water along stream edges, marshes, rivers, sloughs, bogs and wet meadows; **monoecious**. w2

■ **FLOWERS** Green, blooming May–August; **inflorescence** a spike, erect, cylindrical, unisexual or bisexual, TL of spikes 8–40 cm; **male spikes** above, sessile, some androgynous, usually one to four, each 1.2–6.5 cm long; **subtending bracts** (of male spikes) 1–3 cm long by 1–2 mm wide; **female spikes** below, lateral, usually two to six, 1–6 cm long by 3–6 mm wide, some spikes elongating to *c.* 10^+ cm in fruit; **subtending bracts** (of female spikes) 20–40 cm long by 2–8 mm wide, lower bracts ascending and extending well past the male spikes; **peduncles** 0–10 mm long; **stigmas** 2; **perigynia** glabrous, crowded with up to 200^+ per spike, round, flat, 2–3.4 mm long by 1.5–2.5 mm wide by *c.* 0.8 mm thick, lime green when fresh, purplish brown or tan with reddish brown spots on maturing, thin-walled, with two narrow marginal ribs, nerveless, the beak 0.1–0.2 mm long; **female scales** 2.5–3.5 mm long by 1–2 mm wide, longer to shorter than perigynia, brown with a narrow green midvein and reddish brown margins, pointed. **FRUIT** an achene, 1-seeded, brown, 1.5–1.7 mm long by 1–1.6 mm wide by 0.6–0.8 mm thick, glossy, lenticular, filling the lower part of the perigynium, central ridge slightly raised; **style** bent at its base and soon deciduous.

■ **LEAVES** Glaucous, three to eight$^+$, the upper one arched to nodding, often reaching to the top of or slightly beyond the male spikes; **blades** 20–90 cm long by 2–8 mm wide, slightly V-shaped, the edges slightly scabrous; **sheaths** smooth and closed, U-shaped at apices, with white or purple spots.

■ **STEM** A **culm**, erect, sharply triangular, hollow, scabrous near the top, otherwise smooth; 4–16 mm thick near the reddish base; an accumulation of brown leaves from the previous year forms a basal tuft to *c.* 40 cm wide.

531

Sedge—*Cyperaceae* Sedge Perigynia hairless, 6–12 mm long

AWNED SEDGE *Carex atherodes* Spreng.

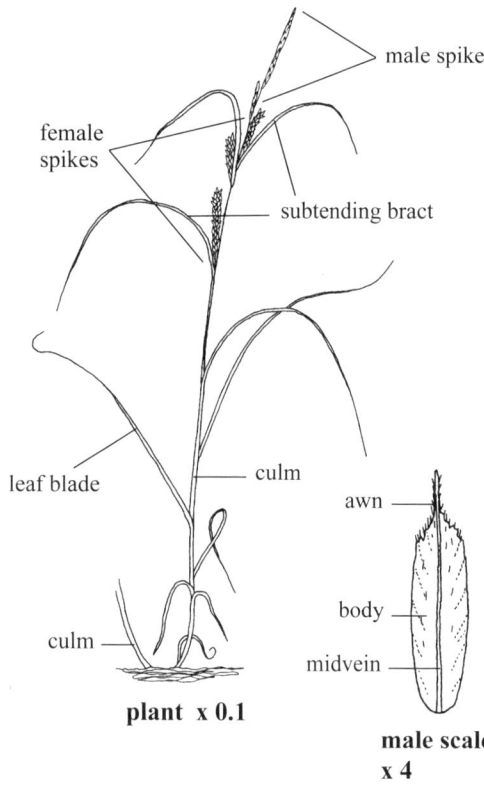

plant x 0.1

male scale x 4

■ **SKETCH** A loosely tufted **perennial sedge** 30–125 cm tall with long scaly **rhizomes**; in ditches, wet meadows and thickets, and wetland edges often covering several m^2; **monoecious**. w2

■ **FLOWERS** Green, blooming May–July; **inflorescence** a spike, unisexual, several, TL of spikes 12–60 cm; **male spikes** terminal, two to six, each 1.2–8 cm long by 3–5 mm wide in flower, overlapping; **subtending bracts** (of male spikes) to *c.* 15 cm long by *c.* 3 mm wide, reduced above, arched, scabrous on margins and below; **peduncles** 5–20 mm long, triangular, grooved; **male scales** imbricate, 7–10 mm long by 2–2.5 mm wide, apices erose, ciliate, V-shaped, edges hyaline, midvein yellowish green and extending as an awn 1–2 mm long with ascending hair; **female spikes** two to five, remote, each 30- to 90-flowered and 2.5–10 cm long by 8–12 mm wide in fruit, often topped with a few male florets; **subtending bracts** (of female spikes) leaflike, arched, 13–30 cm long by 3–9 mm wide, scabrous on edges and below, smooth above; **peduncles** 0–15 mm long; **stigmas** 3; **perigynia** brown, thin-walled, multi-nerved, 6–12 mm long by 2–3.8 mm wide by *c.* 1.8 mm thick, hairless; **beak** bidentate, 2–4 mm long, teeth 1.2–3 mm long, curved outward; **female scales** 6–8 mm long by 1.2–1.5 mm wide (flattened), mostly shorter than the perigynia, margins hyaline, awn 2–4 mm long. **FRUIT an achene**, 1-seeded, tiny compared to perigynium, 3-sided, ribbed, body 2.2–2.5 mm long by 1–1.2 mm wide, corners blunt, style straight and deciduous.

■ **LEAVES** Flat, arched, channelled, four to eight per culm, lower ones brown and curled by anthesis; **blades** 20–40 cm long by 3–12 mm wide, dull, smooth to scabrous on margins, often with white hairs below near blade's base, glabrous above; **ligules** membranous, *c.* 1 mm long, blunt, not hairy; **sheaths** very hairy at apices (appear fuzzy) to rarely glabrous, closed, apices V-shaped, lower ones laddering, brown and less hairy.

■ **STEM** A **culm**, erect, sharply triangular, glabrous; slightly arched from the 5–15 mm thick base.

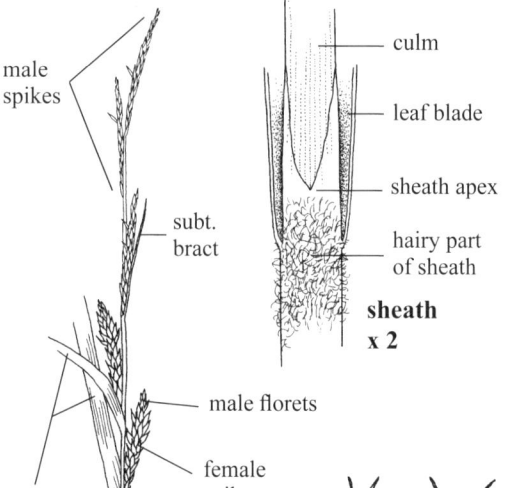

upper plant x 0.3

perigynium x 4

female scale x 4

fruit x 8

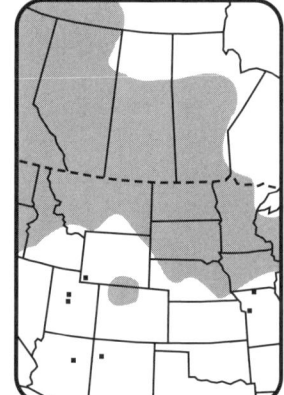

Wheat Sedge

Sedge—*Cyperaceae* Sedge Perigynia hairless, 2.4–3 mm long

GOLDEN SEDGE *Carex aurea* Nutt.

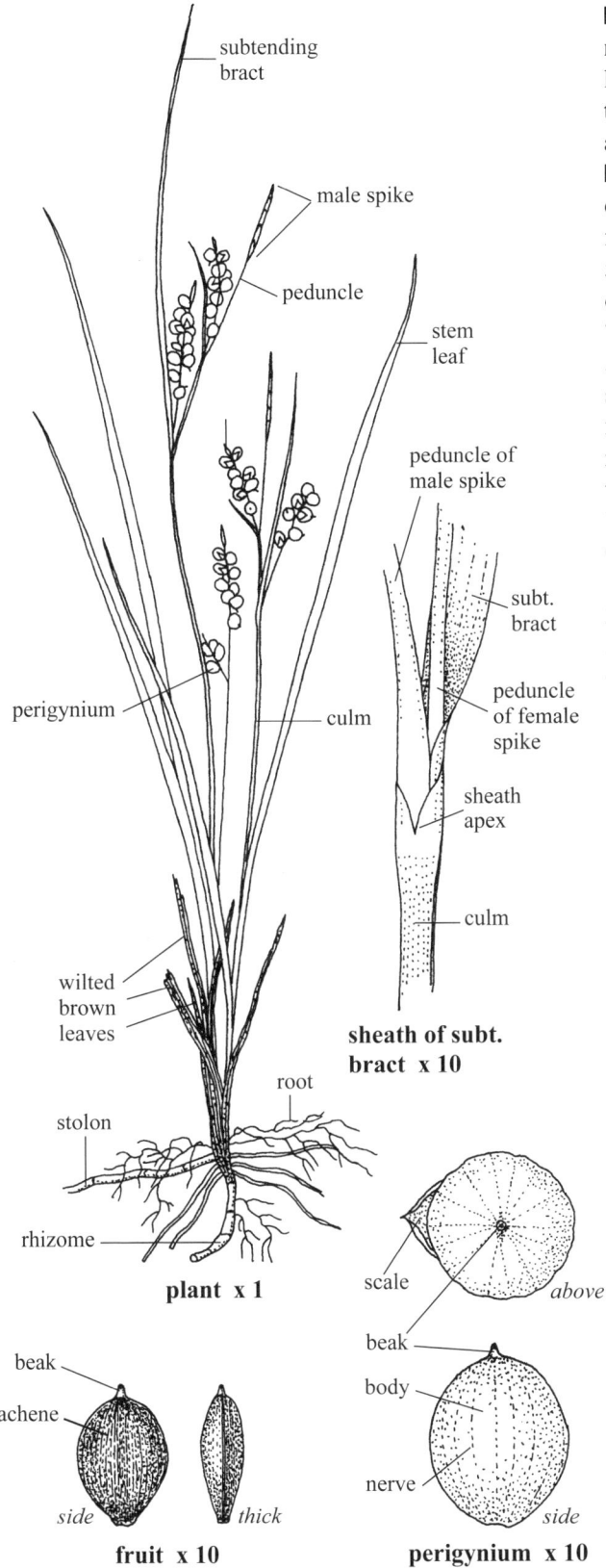

■ **SKETCH** A loosely tufted, yellowish green **perennial sedge** 8–40 cm tall with tan **rhizomes** 2–10⁺ cm long by *c.* 1 mm thick, and **stolons**, forming patches up to *c.* 1 m²; in moist fields, meadows, banks, damp woods and by springs; **monoecious. w2**

■ **FLOWERS** Green, blooming April–May; **inflorescence** a spike, mostly unisexual, separate, male above female, TL 2–10 cm, erect; **male spike** one, terminal, 5–15 mm long by *c.* 1 mm wide, persistent, sometimes one or more female florets at its apex; **peduncles** erect, 7–17 mm long; **female spikes** 2–5 per culm, each 5–25 mm long by *c.* 4 mm wide with 4–20 perigynia per spike, some with a male floret at their apex; **subtending bracts** (of female spikes) leaflike, 0.7–9 cm long, reduced above, the lowest one often extending beyond the inflorescence; **sheaths** (of subt. bracts) 1–8 mm long, reduced above, white, apices V-shaped; **peduncles** 0.7–6 cm long, ascending; **stigmas** 2; **perigynia** hairless, roundish, light orangish tan when ripe, smooth, 2.4–3 mm long by 1.5–2.2 mm wide, nerves numerous and coarse to faint, spongy at base below the fruit; **beak** 0–0.1 mm long, pointed, not bidentate; **female scales** light tan to reddish brown, 1.7–2 mm long by 1.2–1.5 mm wide, smaller than perigynia, awn 0.2–0.6 mm long, sides streaked with brown, midvein whitish tan or pale green. **FRUIT an achene**, 1-seeded, lenticular, dark brown, shiny, 1.5–2 mm long by 1.2–1.5 mm wide by *c.* 0.6 mm thick, above base of perigynium; **beak** 0.1–0.2 mm long.

■ **LEAVES** Usually a few fresh ones arising from the base among brown remnants; **blades** 5–25 cm long by 1.5–4 mm wide, tapered to a fine point, mostly flat to slightly V-shaped, often extending past the inflorescence; **sheaths** V-shaped at apices.

■ **STEM** A **culm**, erect, sharply triangular, smooth, green; 0.8–2 mm wide across one side near the base.

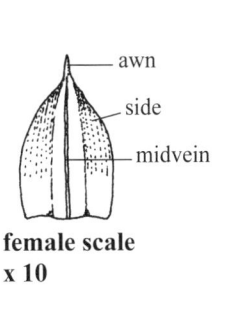

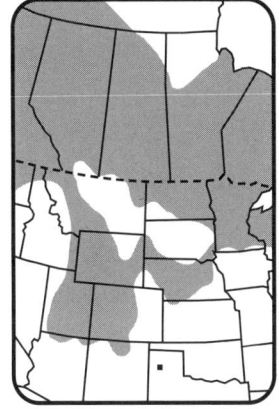

Golden-fruit Sedge 533

Sedge—*Cyperaceae*

BEBB'S SEDGE *Carex bebbii* Olney ex Fern.

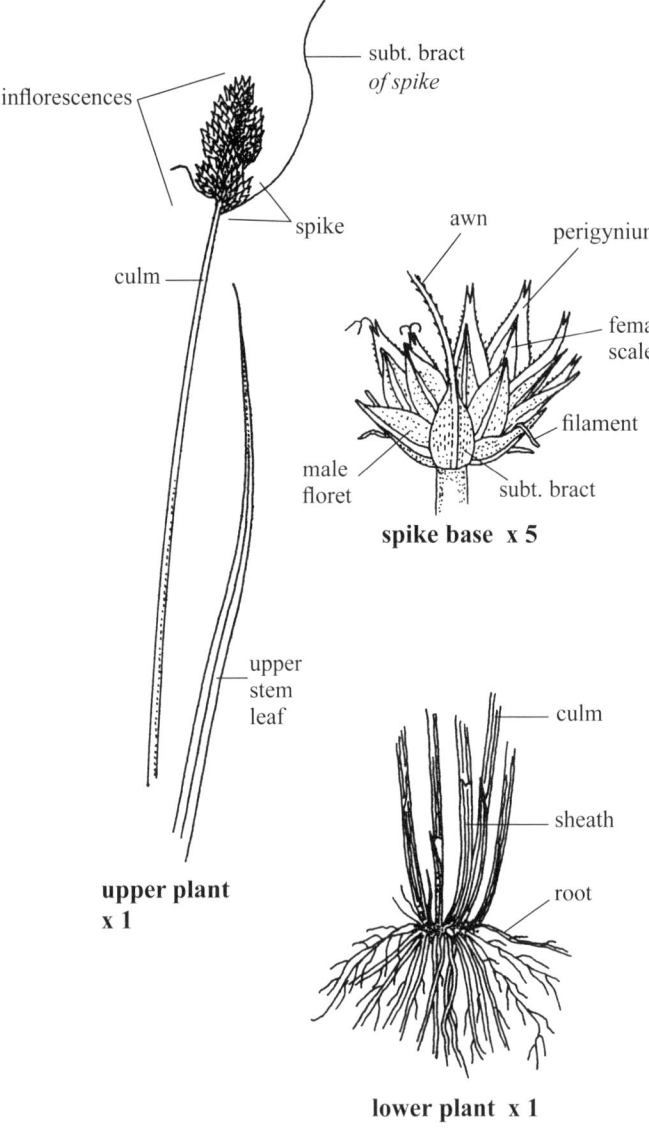

■ **SKETCH** A tufted **perennial sedge** 10–90 cm tall with short, dark brown **rhizomes** hidden by tan **fibrous roots**; in ditches, moist woods, swamps and meadows, along shorelines of marshy areas and beaver ponds; **monecious**. w2

■ **FLOWERS** Green, blooming April–May; **inflorescence** a spike, 3–12, crowded and imbricate, together 12–30 mm long by 8–11 mm wide; **spikes** bisexual, 5–10 mm long by 4–7 mm wide, sessile, gynecandrous (male florets at the base of each spike), the lowest spike sometimes slightly removed from the upper ones; **subtending bracts** of lowest two or three spikes often visible, tan, the lowest 1.2–5 cm long, sometimes exceeding the inflorescence, reduced above, filiform; **stigmas** 2; **perigynia** reddish brown, shiny, plano-convex, hairless, winged its full length, 2.5–3.8 mm long by 1.2–2 mm wide by *c.* 0.4 mm thick, 2- or 3-nerved dorsally and zero to few-nerved ventrally, tapered to a serrulate, bidentate, flat beak; **female scales** 2.3–3.5 mm long by 0.7–1.1 mm wide, upper half and point darker brown, midnerve tan and thin, base membranous and mostly transparent. **FRUIT an achene**, 1-seeded, yellowish brown, 1–1.3 mm long by 0.6–0.9 mm wide by *c.* 0.4 mm thick, ovate, smooth and shiny.

■ **LEAVES** Three or four per culm, 5–22 cm long by 2–4.5 mm wide, reduced below, upper leaves ascending and usually not reaching the inflorescence, blades flat and scabrous on margins; **sheaths** tight, V-shaped and usually torn at apices, white-hyaline ventrally.

■ **STEM** A **culm**, erect, hollow, sharply triangular, scabrous right beneath the inflorescence, smooth below; 2–3 mm thick near the brown, sheath-covered base.

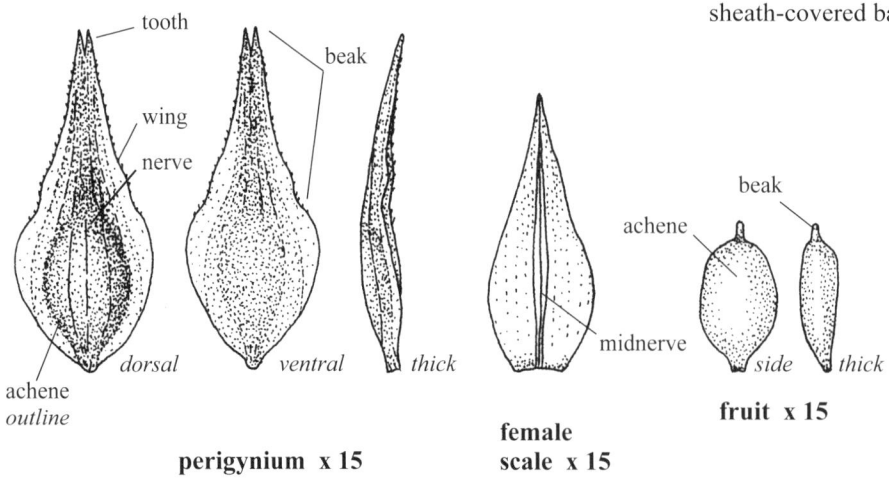

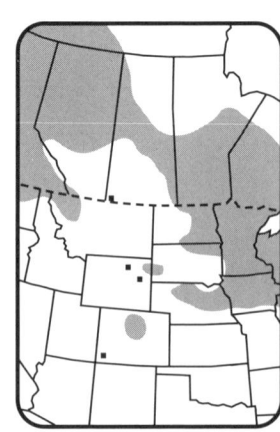

Sedge—*Cyperaceae* Sedge Perigynia hairless, 3–5 mm long

BROAD-FRUITED SEDGE *Carex brevior* (Dewey) Mackenzie

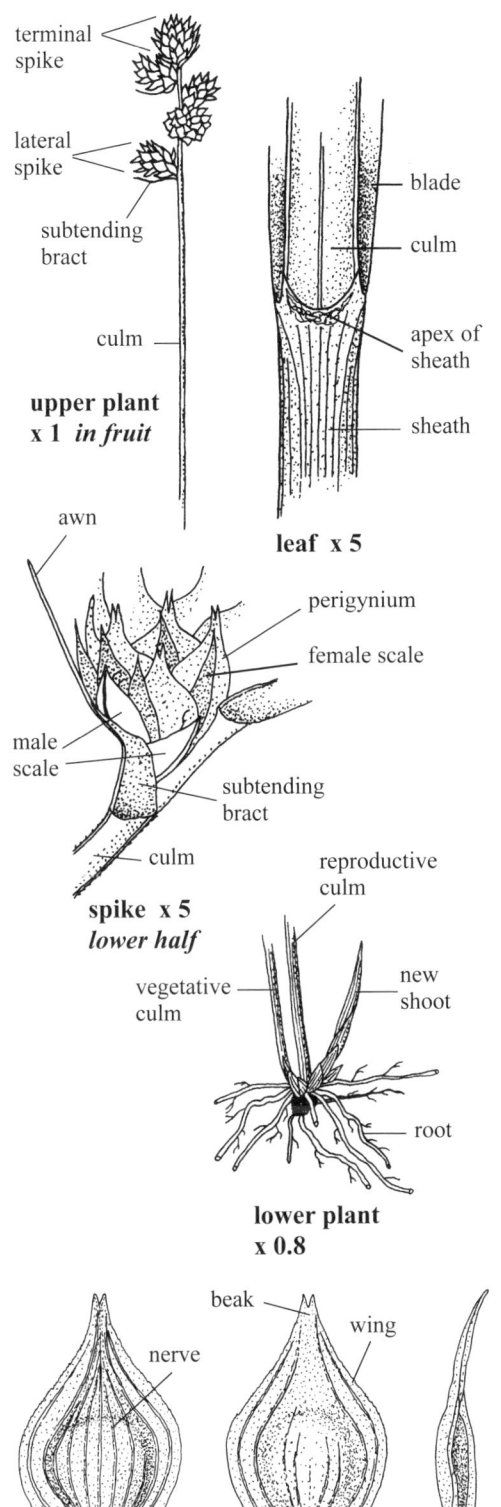

upper plant x 1 *in fruit*

leaf x 5

spike x 5 *lower half*

lower plant x 0.8

■ **SKETCH** A tufted **perennial sedge** 15–120 cm tall from short, **black rootstocks** with reddish tan roots *c.* 1 mm thick to 7+ cm long; in open moist woodlands, along marshes, dry prairies, mountains, in ditches and meadows; **monoecious. w2**
■ **FLOWERS** Green, blooming June–July; **inflorescence** a spike, together 1.5–6 cm long by 8–16 mm wide, situated well above the top leaf; **spikes** bisexual, sessile, usually four to seven, each 6–20 mm long by 4–12 mm wide, congested to remote, gynecandrous with only a few male flowers at the base of each spike and the female flowers above comprising most of the spike; **subtending bracts** (of spikes) one, usually not extending past the spike, body to *c.* 4 mm long by *c.* 2 mm wide, reduced above, base V-shaped and enveloping the base of the spike, the awn 3–4 mm long, reduced above; **male scales** usually one to five at the base of each spike, 2.5–3 mm long by *c.* 1 mm wide, similar in shape to the female scale; **stigmas** 2; **perigynia** tan, glabrous, 15–40 per spike, each 3–5 mm long by 2–3 mm wide by 0.5–1 mm thick, winged to base, thin-walled, not spongy, several nerves on the outside (dorsal), fewer nerves below (ventral); **beak** 0.8–1 mm long, bidentate, the teeth not spreading; **female scales** light brown, midvein and a few lateral veins visible, pointed, tip and upper margins hyaline, 2.6–4.3 mm long by *c.* 1 mm wide, narrower and slightly shorter than the perigynia. **FRUIT an achene**, 1-seeded, glabrous, shiny, filling the lower two-thirds of the perigynium, TL 1.7–2.2 mm by 1.2–1.8 mm wide by *c.* 0.5 mm thick, lenticular, with a short beak.
■ **LEAVES** Usually three to six per culm, ascending; **blades** 3–30 cm long by 2–4 mm wide, mostly at the lower half of the culm, scabrous on margins and below, glabrous above, tapered to a fine point; **sheaths** glabrous, slightly keeled, closed, concave at apices, lighter green below apices from whitish longitudinal stripes between the thin green veins for a few cm or more.
■ **STEM** A **culm**, erect to leaning in fruit, hollow, sharply to bluntly triangular, slightly scabrous below the inflorescence, a secondary ridge between each corner; smooth near the 1–2 mm thick base and for most of its length.

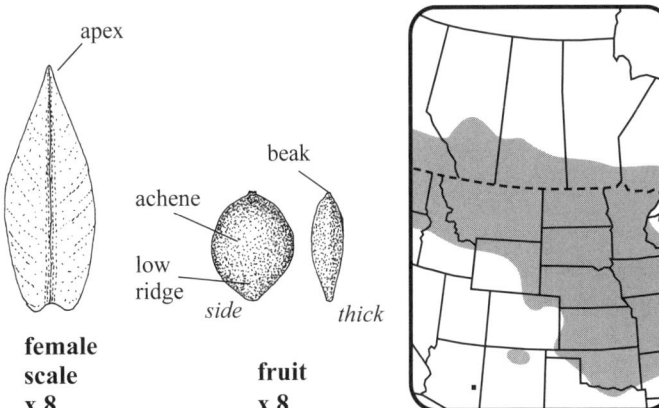

perigynium x 8 — dorsal, ventral, thick

female scale x 8

fruit x 8 — side, thick

Short-beaked Sedge 535

Sedge—*Cyperaceae* Sedge Perigynia hairless, 2.3–3.8 mm long

Hair-like Sedge *Carex capillaris* L.

■ **SKETCH** A tufted **perennial sedge**, yellowish green, 10–60 cm tall with **rhizomes** *c.* 1 mm long, and tangled brown **fibrous roots** to *c.* 8 cm long; in wet meadows, ditches, marshes, bogs, open forests and pond edges; **monoecious. w2**

■ **FLOWERS** Light green, blooming May–June; **inflorescence** a unisexual spike, TL of spikes 10–20 cm by 1.5–3 cm wide; **peduncles** 10–17 mm long by *c.* 0.1 mm thick; **male spike** one, terminal, erect, tan, 5–10 mm long by *c.* 1 mm wide; **peduncles** arched, 10–40 mm (from nodes) long by *c.* 0.1 mm thick; **female spikes** lateral, usually three (2 or 4), hanging, each 6- to 20-flowered and 5–20 mm long by 2–3 mm wide; **subtending bracts** (of female spikes) leaflike, ascending; **blades** 3–60 mm long by 1–1.7 mm wide, reduced above; **sheath** of lowest bract, 13–18 mm long, closed, tight, apex flat or shallowly V-shaped and whitish; **stigmas** 3; **perigynia** hairless, side ribs two, nerveless, greenish brown, shiny, 2.3–3.8 mm long by 0.8–1.5 mm wide by *c.* 1 mm thick, thin-walled, tapered to a short stipe; **beak** 0.6–1 mm long, teeth minute, straight, whitish; **female scales** generally cover the lower half of the perigynia, each 1.5–2.8 mm long by 1–1.5 mm wide (flat), midvein greenish brown with brown streaks near the base, margins wide and hyaline, apices blunt to pointed, white. **FRUIT an achene**, 1-seeded, triangular, corners rounded, TL 1.2–1.7 mm by 0.7–1 mm wide by *c.* 1 mm thick, filling the perigynium; **beak** straight, *c.* 0.1 mm long.

■ **LEAVES** Yellowish green, usually three or four near the base of the culm and one about midway which almost reaches the inflorescence; **blades** 2–20 cm long by 1–4 mm wide, flat, inrolled on drying, slightly scabrous margins near sheaths; **ligules** unremarkable; **sheaths** tight, apices flat to slightly V-shaped.

■ **STEM** A **culm**, smooth to slightly scabrous, ridged, bluntly triangular, sides slightly convex, erect to leaning with fruit, barely nodding; 2–2.5 mm wide at the base covered with dark brown sheaths and old brown leaves 2–4 cm long.

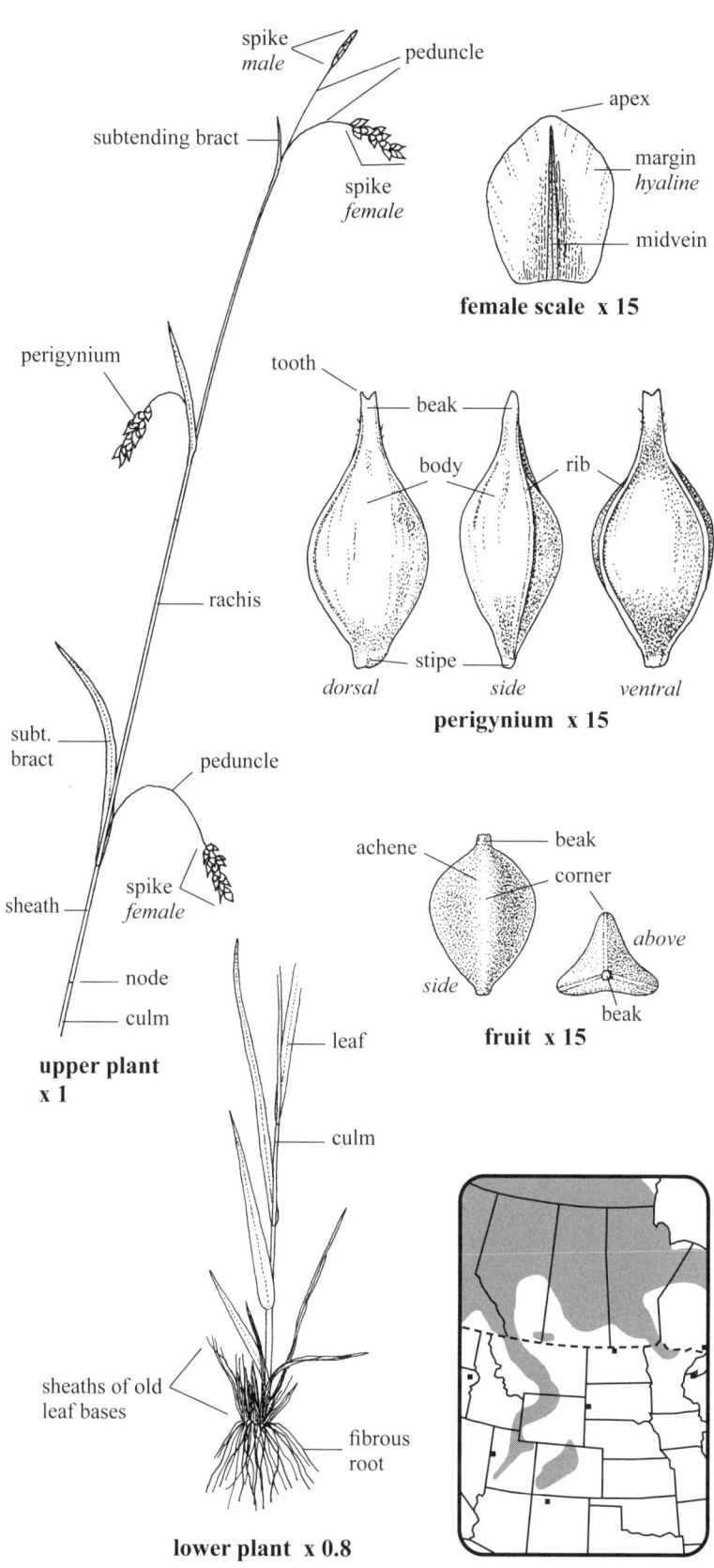

536

Sedge—*Cyperaceae* **Sedge** Perygnia hairless, 4–5.3 mm long

Dewey's Sedge *Carex deweyana* Schwein.

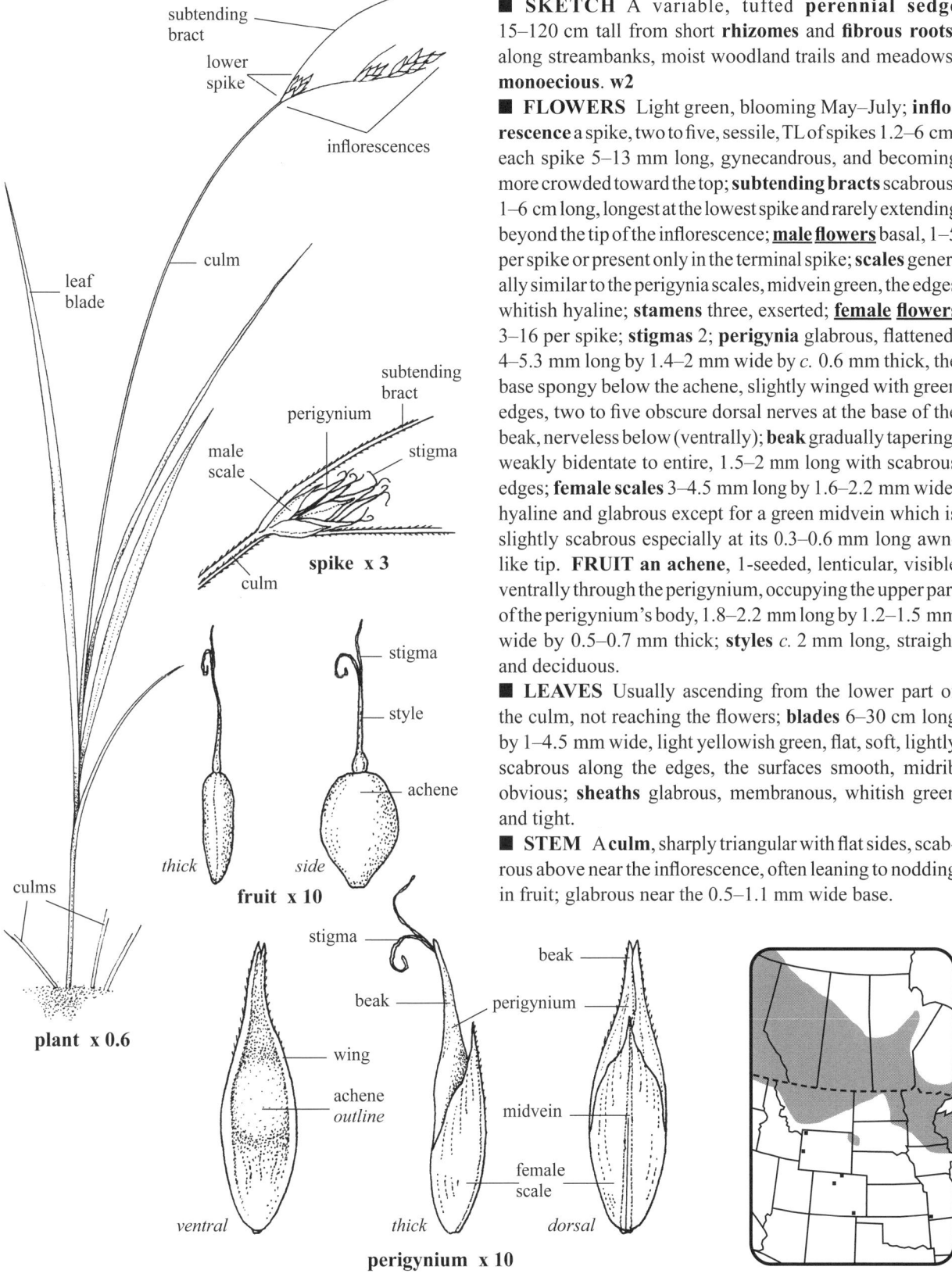

plant x 0.6
spike x 3
fruit x 10
perigynium x 10

■ **SKETCH** A variable, tufted **perennial sedge** 15–120 cm tall from short **rhizomes** and **fibrous roots**; along streambanks, moist woodland trails and meadows; **monoecious**. w2

■ **FLOWERS** Light green, blooming May–July; **inflorescence** a spike, two to five, sessile, TL of spikes 1.2–6 cm, each spike 5–13 mm long, gynecandrous, and becoming more crowded toward the top; **subtending bracts** scabrous, 1–6 cm long, longest at the lowest spike and rarely extending beyond the tip of the inflorescence; **male flowers** basal, 1–5 per spike or present only in the terminal spike; **scales** generally similar to the perigynia scales, midvein green, the edges whitish hyaline; **stamens** three, exserted; **female flowers** 3–16 per spike; **stigmas** 2; **perigynia** glabrous, flattened, 4–5.3 mm long by 1.4–2 mm wide by *c.* 0.6 mm thick, the base spongy below the achene, slightly winged with green edges, two to five obscure dorsal nerves at the base of the beak, nerveless below (ventrally); **beak** gradually tapering, weakly bidentate to entire, 1.5–2 mm long with scabrous edges; **female scales** 3–4.5 mm long by 1.6–2.2 mm wide, hyaline and glabrous except for a green midvein which is slightly scabrous especially at its 0.3–0.6 mm long awn-like tip. **FRUIT an achene**, 1-seeded, lenticular, visible ventrally through the perigynium, occupying the upper part of the perigynium's body, 1.8–2.2 mm long by 1.2–1.5 mm wide by 0.5–0.7 mm thick; **styles** *c.* 2 mm long, straight and deciduous.

■ **LEAVES** Usually ascending from the lower part of the culm, not reaching the flowers; **blades** 6–30 cm long by 1–4.5 mm wide, light yellowish green, flat, soft, lightly scabrous along the edges, the surfaces smooth, midrib obvious; **sheaths** glabrous, membranous, whitish green and tight.

■ **STEM** A **culm**, sharply triangular with flat sides, scabrous above near the inflorescence, often leaning to nodding in fruit; glabrous near the 0.5–1.1 mm wide base.

Round-fruit Short-scale Sedge

Sedge—*Cyperaceae*   **Sedge**   Perigynia shiny, dark brown, 2–3 mm long

TWO-STAMENED SEDGE   *Carex diandra* Schrank

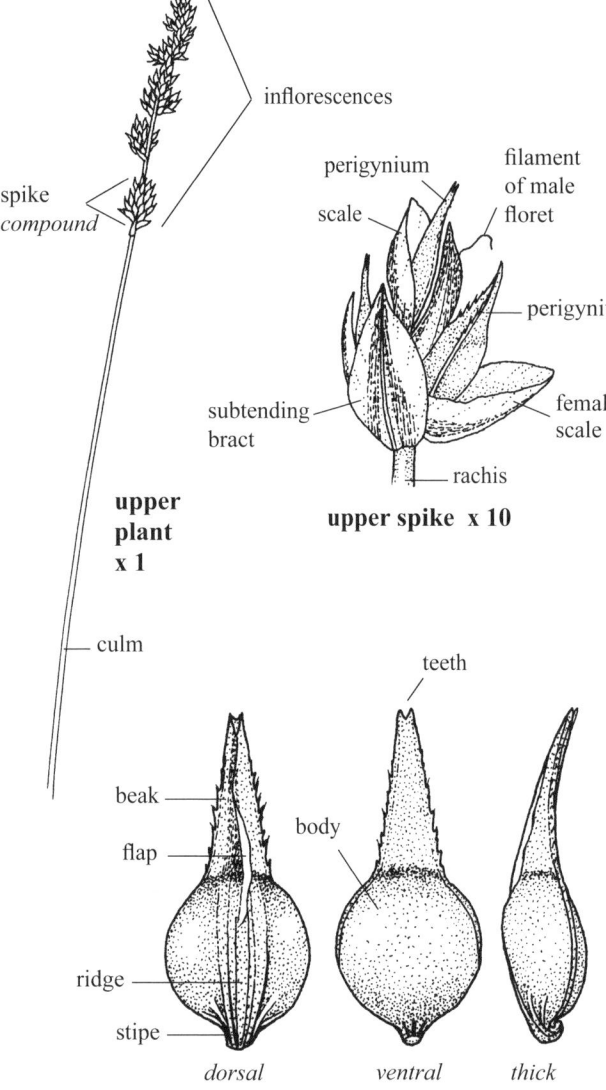

■ **SKETCH** A tufted **perennial sedge** 40–80 cm tall from short black **rhizomes**; in swamps, wet fields, fens and lakeshores of the Boreal forest; **monoecious. w2**

■ **FLOWERS** Green, blooming May–June; **inflorescence** a spike, TL 3–5 cm by 7–9 mm wide; **spikes** bisexual, androgynous, 6–11, each 3–8 mm long by 3–6 mm wide, reduced and more crowded above, lowest one compound with two or three branches, often remote; **subtending bracts** (of spikes) 3–15 mm long, reduced above, pointed, usually shorter than its spike; **male scales** tan, 2–5 mm long, erect at apex of each spike; **stigmas** 2; **perigynia** dark brown, shiny, thin, 2–3 mm long by 1–1.4 mm wide by 0.7–0.8 mm thick, 3–10 per spike, convex on both sides, not winged, the base spongy but dry and enveloping the elongate base of the achene; nerveless below, two main diverging nerves above (dorsal) with a slightly depressed area lighter or darker than the sides, usually with a thick stipe; **beak** tan, bidentate, margins strongly scabrous, a narrow hyaline flap extending to body; **female scales** *c.* 2.5 mm long by *c.* 1.7 mm wide (flattened), pointed, tan around midvein with reddish brown striations on both sides and across the base. **FRUIT an achene**, 1-seeded, lenticular, 1.1–1.4 mm long by 0.9–1 mm wide by 0.5–0.6 mm thick, slightly shiny to dull; **beak** short; **stipe** long.

■ **LEAVES** Two or three, ascending and arched, 4–36 cm long by 1–3.5 mm wide, reduced above, situated near base of culm, not reaching to inflorescence; **sheaths** membranous, scabrous, often with scattered microscopic dark red dots.

■ **STEM** A **culm**, sharply triangular, scabrous above along ridges, faces flat, smooth below with faces slightly convex, stiff but hollow, leaning with fruit.

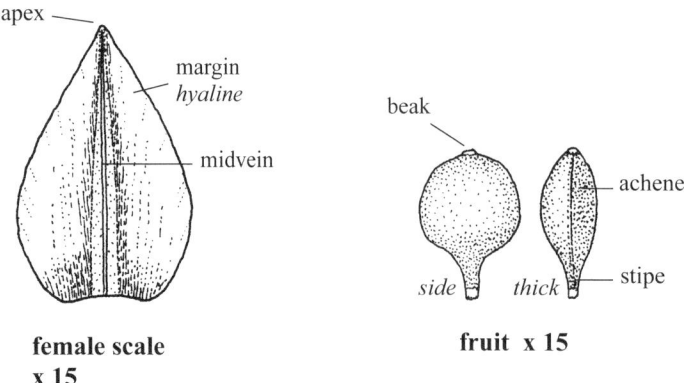

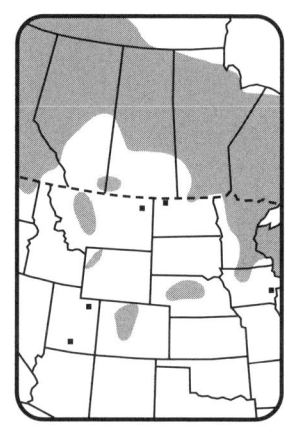

Lesser Tussock Sedge, Lesser Panicled Sedge

Two-seeded Sedge *Carex disperma* Dewey

■ **SKETCH** A loosely tufted **perennial sedge** 10–60 cm tall with brown, veiny bracteal **rhizomes** to *c.* 7 cm long by *c.* 0.8 mm thick with nodal **fibrous roots** 2–4 cm long; in bogs, springy sites, coniferous woods, swamps and ravines; **monoecious. w2**

■ **FLOWERS** Green, blooming March–June; **inflorescence** a spike, TL of spikes 1.2–4 cm by 3–6 mm wide; **spikes** androgynous, two to five, 3–6 mm long, 1–15 mm apart, crowded above, sessile, each with 1–6 female flowers and one (2) male flowers above; **subtending bracts** (of spikes) 2–5 (–20) mm long, reduced above, sides hyaline tan, midrib raised and light tan, pointed, extending as a short awn in lower bracts; **stigmas** 2; **perigynia** slightly shiny, 2–3 mm long by 1.3–1.7 mm wide by 1–1.1 mm thick, hairless, greenish yellow to dark brown, ascending to spreading, slightly flattened, sides convex, ventral side slightly flatter, nerves 20–24, some incomplete, two prominent ventral ribs, body thin-walled at apices, thick and spongy on sides and base; **beak** 0.1–0.2 mm long, entire; **female scales** 1.5–2 mm long by 1–1.2 mm wide, with wide hyaline margins and a pale tan midvein, shorter than perigynium. **FRUIT** an achene, 1-seeded, 1.5–1.8 mm long (including beak) by 1–1.2 mm wide by *c.* 0.6 mm thick, lenticular, shiny, brown to black; **beak** *c.* 0.2 mm long.

■ **LEAVES** Three to five, arising from near the base; **blades** flat, 2–30 cm long by 1–2 mm wide; **ligules** unremarkable; **sheaths** V-shaped at apices, tight, membranous.

■ **STEM** A **culm**, thin, weak, leaning with fruit, scabrous and sharply triangular below the inflorescence; smooth to slightly blunt near the 0.5–0.6 mm wide base; **veiny brown bracts** (three or four) covering the culm below the leaves, these 3–13 mm long, reduced below, tips rounded.

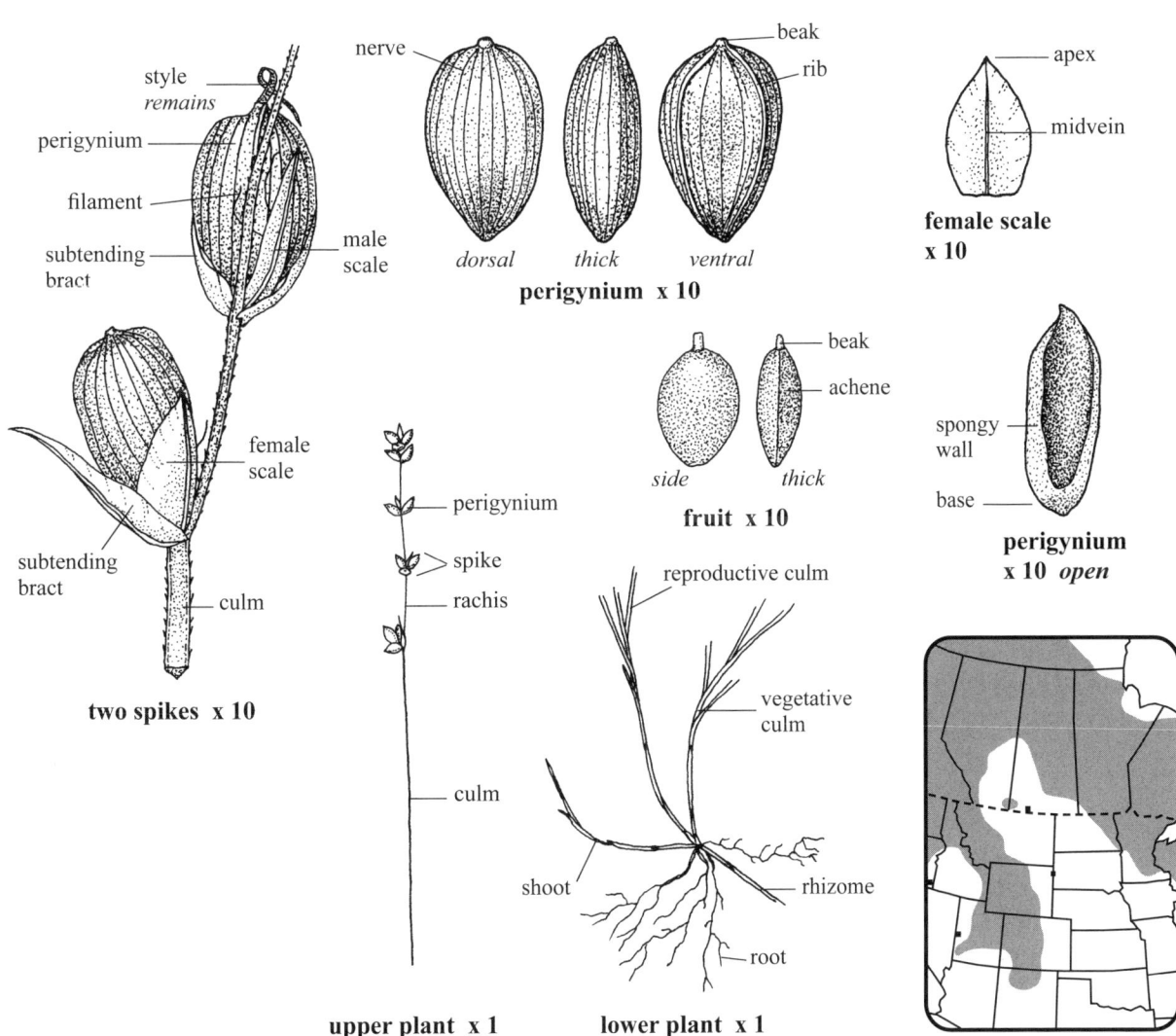

Thread-leaved Sedge *Carex filifolia* Nutt.

- **SKETCH** A tufted **perennial sedge** 5–35 cm tall from a dark brown **rootstock** less than 10 mm long; in dry prairies, on eroded slopes and ridges, in dry sandy soils; **monoecious**. w2, p1
- **FLOWERS** Brown, often blooming April–June; **inflorescence** a spike 1–2.5 cm long by 2–5 mm wide, terminal, androgynous; **male flowers** above, 3–20, together 6–10 mm long by 1–2 mm wide, tapered to a point; **male scales** 3–4 mm long by 1.5–2 mm wide (flattened), glabrous, imbricate, partly reddish brown, apices blunt and margins whitish hyaline; **female flowers** below, 2–18; **stigmas** 3, exserted; **perigynia** brown, thin-walled, 2–4.8 mm long by 1.3–2.3 mm wide, laterally 2-ribbed, body hairless, slightly hairy along the lateral ribs; **beak** 0.2–0.8 mm long, hairless to finely hairy near its base, not strongly bidentate; **female scales** 3–5 mm long by 2–4.5 mm wide, covering the perigynia, glabrous, margin easily torn, body mostly hyaline, midvein light tan with reddish brown sides, apices blunt with a mucro. **FRUIT an achene**, 1-seeded, 1.8–3.3 mm long by 1.1–1.9 mm wide, triangular, medium brown to black, slightly shiny, filling the perigynium, sides slightly concave or with one side slightly convex, corners blunt; **beak** straight, 0.2–0.4 mm long.
- **LEAVES** Basal, V-shaped, filiform, two or three emerging from the brown sheaths at the base of the culms; **blades** 3–25 cm long by 0.3–0.4 mm wide (inrolled), c. 0.8 mm wide (flattened), ascending, longer to shorter than culms, slightly scabrous; **new sheaths** closed, tubular, flat at apices, hyaline on inside, ridged longitudinally on outside; **old basal sheaths** extending 2–3 cm from the base.
- **STEM** A **culm**, erect, slightly triangular, stiff; c. 1 mm thick near the base (not including old sheaths).

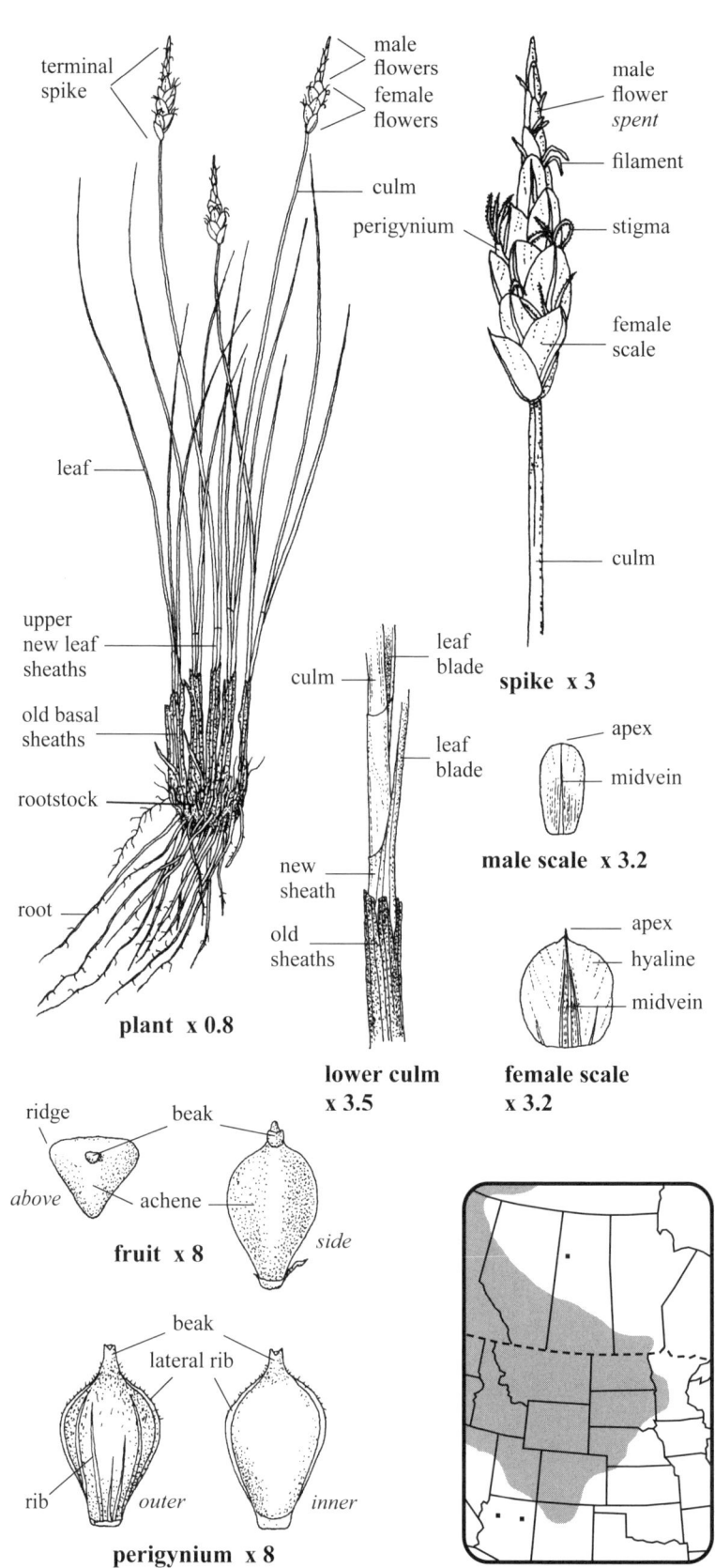

Hay Sedge *Carex foenea* var. *foenea* Willd.

■ **SKETCH** A variable **perennial sedge** 15–100 cm tall in open colonies from a **rhizome** to *c.* 1 m long by 1–1.5 mm thick covered with reddish brown veined scales 8–15 mm long, these pointed, imbricate, the tips usually fibrillose as the veins separate, the scales continuous along the base of the culms and becoming leafy; on rocky outcrops, open pine woods, sandy sites and alpine meadows; **monoecious. w2**

■ **FLOWERS** Green, blooming April–May; **inflorescence** a spike, 2–12, TL 1.2–5 cm by 3–6 mm wide, crowded above, the lowest two spikes 4–6 mm apart; **subtending bracts** (of spikes) pointed, 4–8 mm long, reduced above, shorter to longer than a spike, lowest ones with awns 0–4 mm long, terminal spike bractless; **spikes** sessile, erect, each 5.5–10 mm long, variable in their sexuality, male or female, some androgenous, others gynaecandrous; **stigmas** 2; **perigynia** beaked, hairless, variable, winged in upper half, 4–6.5 mm long by 1.8–2.4 mm wide by *c.* 1 mm thick, nerveless to 12-nerved dorsally and *c.* 6-nerved to nerveless below; **beak** bidentate, serrulate on edges, 1.2–3 mm long; **teeth** slightly incurved, *c.* 0.5 mm long; **female scales** pointed, 3–6 mm long by *c.* 2 mm wide, midvein green, margins and apices hyaline. **FRUIT an achene**, 1-seeded, brown, lenticular, 2–2.2 mm long by 1.5–1.8 mm wide by *c.* 0.7 mm thick, shiny; **beak** tiny.

■ **LEAVES** Three to five per culm, shorter than the culm, 1.5–30 cm long by 1–3 mm wide, reduced below, ascending, straight, stiff, not arched, two or three brown leaves near the base; **ligules** unremarkable; **sheaths** overlapping, closed on upper leaves, apices V-shaped.

■ **STEM** A **culm**, erect, triangular, 2–3 cm apart along the rhizome, slightly scabrous below the inflorescence, smooth with edges blunt near the 0.5–0.8 mm wide sheath-covered base.

■ **SYN.** *Carex siccata* Dewey.

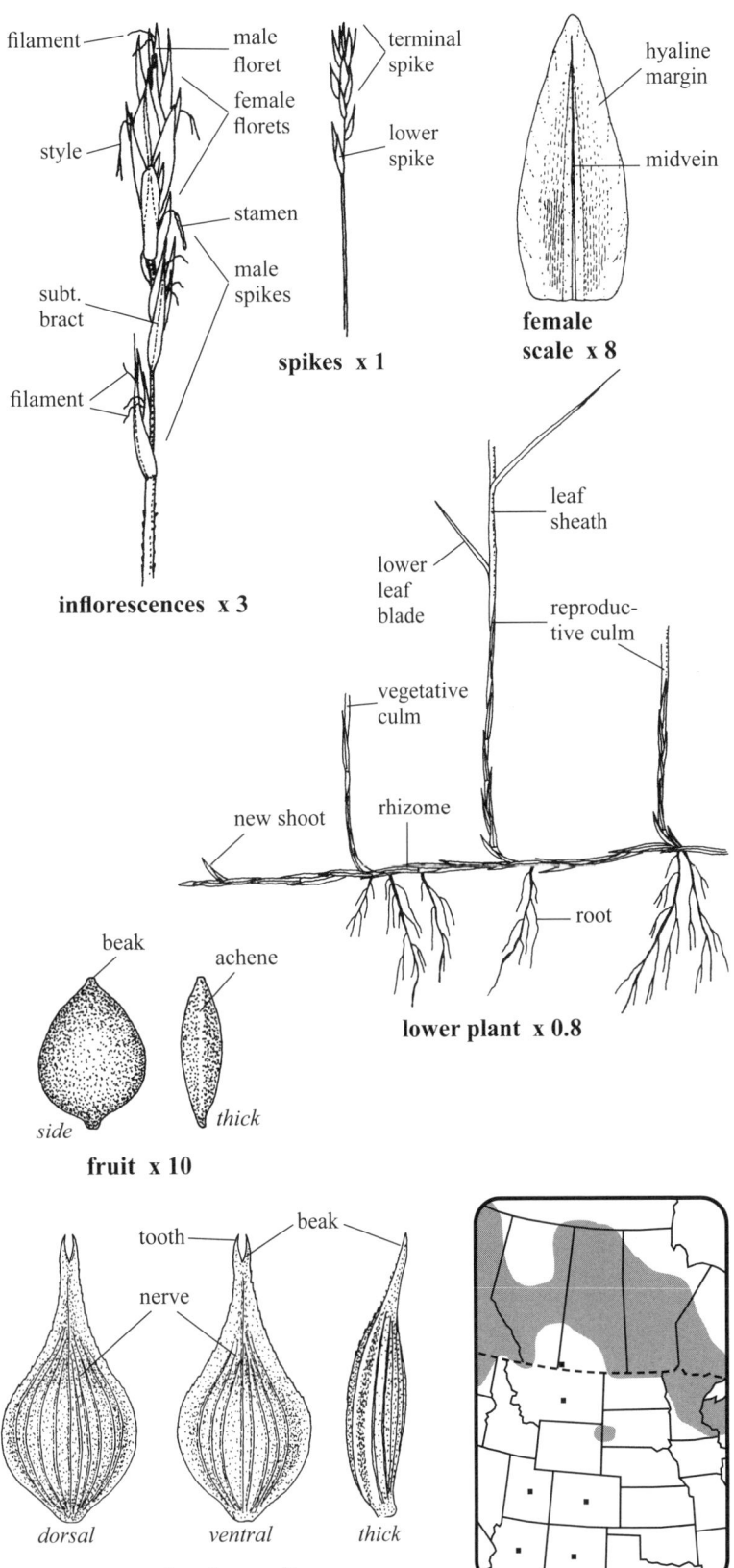

Dry-spike Sedge, Hillside Sedge

Sedge—*Cyperaceae* **Sedge** Perigynia beakless, hairless, 2.4–3.7 mm long

SLENDER SEDGE *Carex gracillima* Schwein.

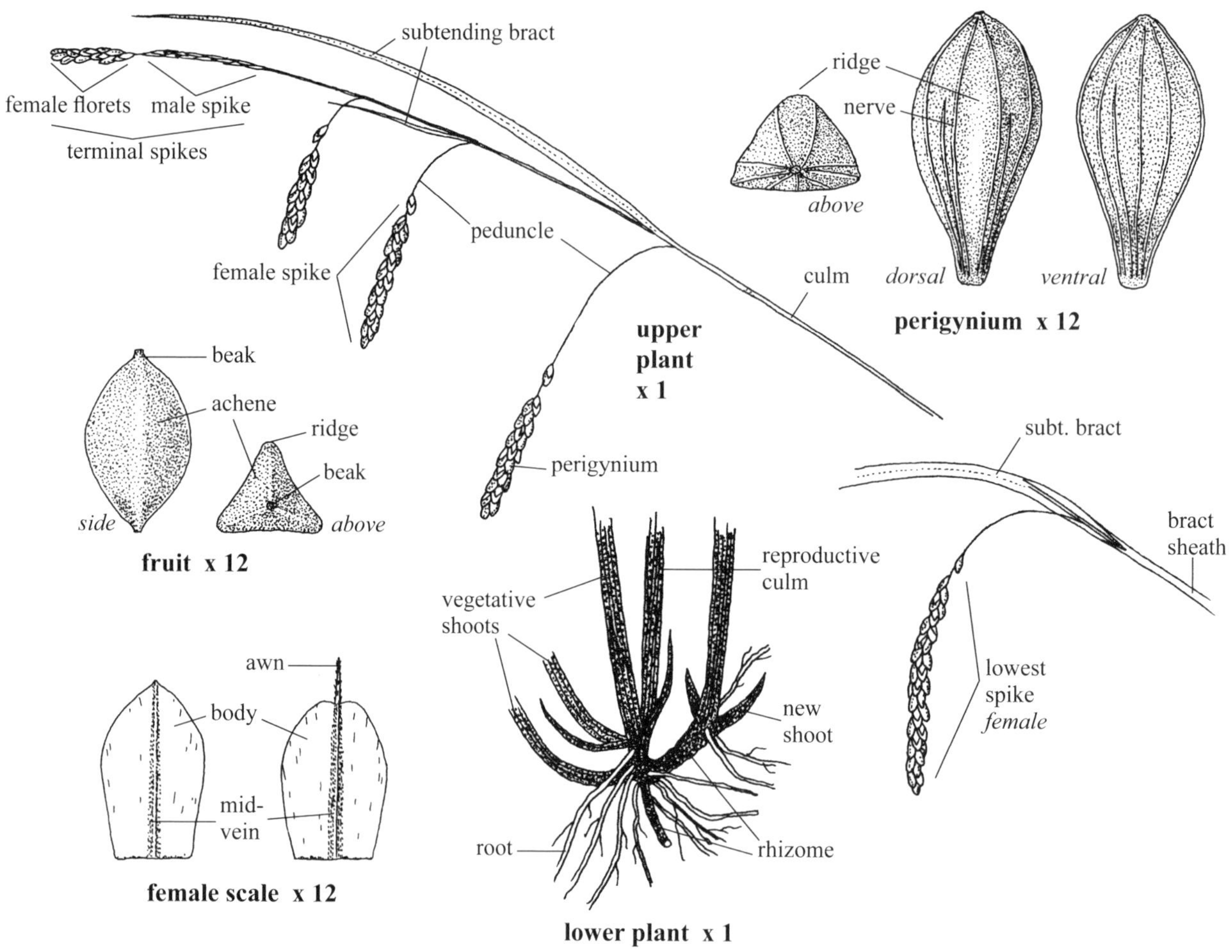

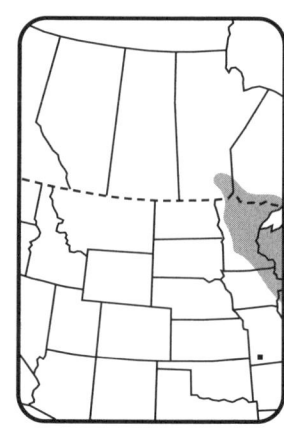

- **SKETCH** A tufted, variable **perennial sedge** 20–95 cm tall with dark **rhizomes** 2–4 cm long by 1–3 mm thick; in coniferous swamps, thickets, roadsides, moist to dry woods of parklands and forests; **monoecious**. w2
- **FLOWERS** Green, blooming May–June; **inflorescence** a spike, TL 8–40 cm, four or five spikes drooping on widely separated peduncles; **male spike** one, terminal, light tan, pointed, 15–33 mm long by *c*. 1 mm wide, sometimes with female florets (one to several) at its tip; **female spikes** 2–5, hanging, 10–60 mm long by *c*. 3 mm wide, the lowest two spikes up to *c*. 28 cm apart; **peduncles** 0.6–8 cm long, reduced above, emerging from subtending bracts; **subtending bracts** (of female spikes) 0.2–15 cm long by 0.5–4.5 mm wide, yellowish green, margins scabrous, ascending to drooping; **stigmas** 3; **perigynia** 2.4–3.7 mm long by 1.4–1.8 mm wide, green ripening to dark brown, beakless, triangular, nerved dorsally, less nerved or nerveless on the flat ventral side; **female scales** half to almost as long as the perigynia, ripening to tan with a fine scabrous point (awnlike) on the lower scales of a spike, upper scales awnless, body 1.8–2.2 mm long by 1.1–1.3 mm wide (flattened), midvein tan to slightly darker brown. **FRUIT an achene**, 1-seeded, light brown, *c*. 2 mm long by 1–1.2 mm wide by *c*. 1 mm thick, slightly shiny, triangular, sides concave, beak (style) straight.
- **LEAVES** Yellowish green, flat, scabrous on margins, 4 or 5 near the base, lowest ones short and wilted; **blades** ascending, 5–40 cm long by 3–10 mm wide, usually not reaching the inflorescence on tall plants; **sheaths** tight.
- **STEM** A **culm**, triangular, smooth, erect at first, leaning as fruit ripens; reddish brown near the 1.5–2.5 mm wide base.

542 Graceful Sedge

Sedge—*Cyperaceae* Sedge Perigynia hairless, 2–4 mm long

GRANULAR SEDGE *Carex granularis* Muhl. ex Willd.

- **SKETCH** A variable, tufted **perennial sedge** 20–100 cm tall from short **rhizomes**; in wet fields, ditches, swamps, bogs and along river valleys; **monoecious**. w2
- **FLOWERS** Green, blooming March–June; **inflorescence** a unisexual spike, together 3.5–35 cm long; <u>male spike</u> one, terminal, 0.6–4 cm long, partially hidden among the upper female spikes; **peduncles** 0–4 mm long, erect; **male scales** brown when dry, apices pointed, not awned, imbricate, slightly V-shaped, 2.5–3.3 mm long by 1–1.3 mm wide (flattened); <u>female spikes</u> two to four, erect, each 6–32 mm long by 3–6 mm wide; **subtending bracts** (of peduncles) leaflike, yellowish green, margins scabrous, 2–17 cm long by 1–7 mm wide, reduced above, ascending to spreading, the upper ones reaching beyond the top spike; **peduncles** 4–50 mm long, reduced above; **stigmas** 3, exserted; **perigynia** glabrous when ripe, 2–4 mm long by 1–2 mm wide, olive to brownish, slightly inflated, not spongy at the base, ascending to spreading at maturity, three or four main nerves with many finer nerves throughout, nerves more obvious when perigynia are lighter green, more obscure as it ripens to a darker brown, slightly 2-ribbed on the sides near the beak; **beak** straight or bent, 0.1–0.3 mm long, tubular, not bidentate; **female scales** 1.5–2.9 mm long by 0.8–1.4 mm wide, tan, V-shaped, acute to barely awned, usually shorter and narrower than perigynia, often reddish-dotted, midvein yellowish when mature. **FRUIT an achene**, 1-seeded, dark brown, 1.7–2.3 mm long by 1–1.4 mm wide, triangular, 3-ribbed, body finely striate, sides convex to concave, style persistent and bent.
- **LEAVES** Ascending to arched, glabrous, 5–35 cm long by 1–5 mm wide, flat, leaves of vegetative plants the widest, with two parallel ridges raised above and one raised midrib below; **sheaths** closed, to *c.* 2.5 cm long, sometimes red dotted ventrally.
- **STEM** A **culm**, sharply triangular below the female spikes, bluntly triangular among the lower leaves, smooth; 1–1.5 mm thick near the base.

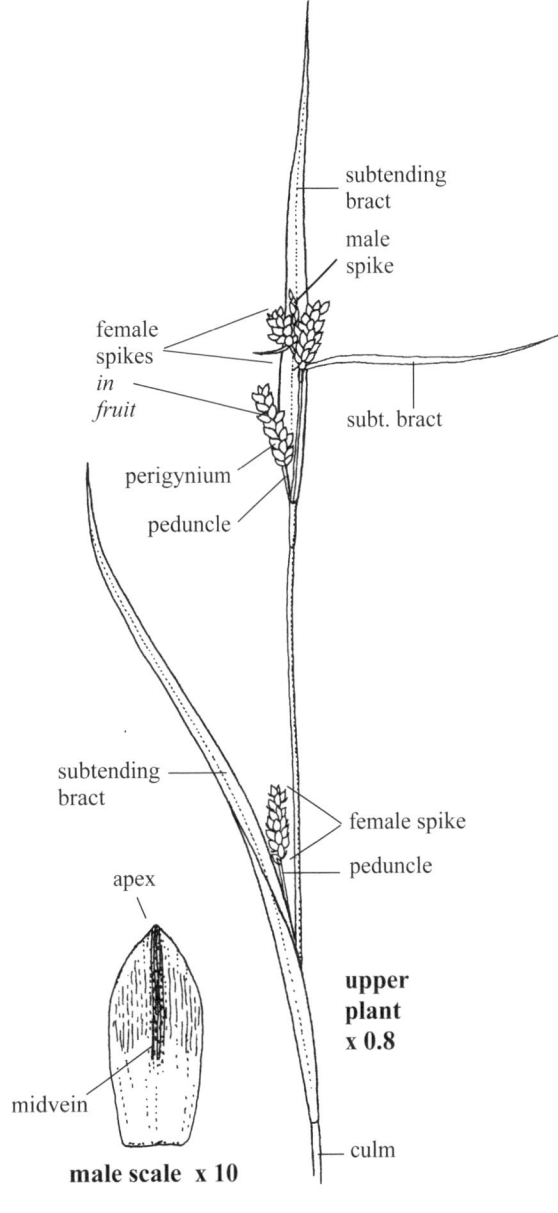

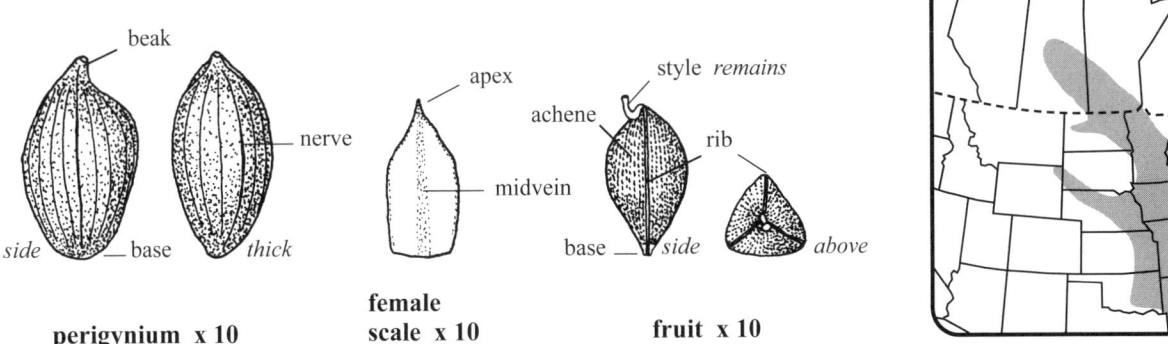

Limestone Meadow Sedge

Sedge—*Cyperaceae* Sedge Perigynia hairless, 3–3.5 mm long

NORTHERN BOG SEDGE *Carex gynocrates* Wormsk. ex Drej.

■ **SKETCH** A perennial sedge 5–30 cm tall from tan **rhizomes** 2–5 cm long by 0.4–0.6 mm thick, covered with veiny **bracts** 5–16 mm long with blunt tips; in sphagnum bogs, marshes and muskeg; **dioecious** or **monoecious**. w2

■ **FLOWERS** Green, blooming May–June; **inflorescence** a spike, one, variable, all female, androgynous, or all male; male spike one, 8–16 mm long (not shown); female spike of 5–18 perigynia, in fruit 6–15 mm long by 5–6 mm wide; **subtending bracts** absent; **stigmas** 2; **perigynia** slightly shiny, 20- to 24-nerved, hairless, brown to very dark brown at maturity, 3–3.5 mm long by 1.4–1.6 mm wide by 1.1–1.3 mm thick, crowded above, mostly spreading to slightly descending, two side ribs obvious or not, walls spongy and thickest near the flat base, thinner above; **beak** angled slightly upward, 0.5–0.8 mm long, minutely bidentate to entire; **female scales** brown, usually slightly shorter than perigynia and about as wide, 2–3.5 mm long by 1.4–2 mm wide (flat), midvein raised and tan, sometimes extending as an awn to *c*. 1 mm long on lower scales, margins light tan near apices, darker brown toward the base. **FRUIT an achene**, 1-seeded, medium brown, lenticular, 1.5–1.7 mm long (including beak) by *c*. 1.1 mm wide by 0.6–0.7 mm thick, glossy, 2-ribbed, raised off base of the perigynia by spongy matter; **beak** *c*. 0.3 mm long.

■ **LEAVES** Situated near the base, 2–6 lower leaves, usually brown and not reaching the fruit; **blades** 2–15 cm long by 0.4–0.7 mm wide, slightly involute; **sheaths** V-shaped at apices, closed, tight and membranous.

■ **STEM** A **culm**, roundly triangular throughout, green, smooth, usually two to four arising from one node, some vegetative; 0.4–0.6 mm wide near the tan base.

■ **SYN.** *Carex dioica* var. *gynocrates* (Wormsk. ex Drej.).

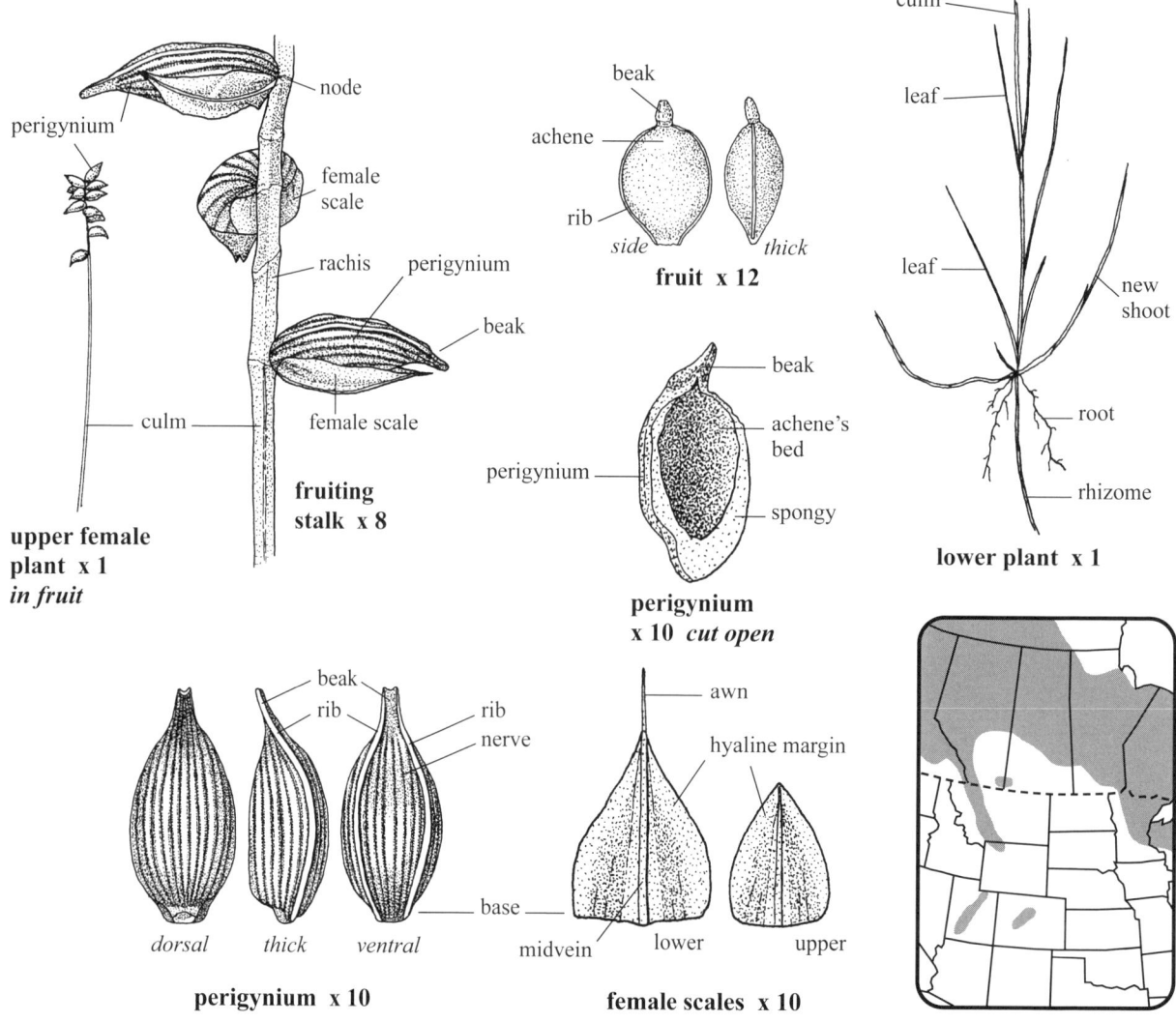

upper female plant x 1 *in fruit*

fruiting stalk x 8

fruit x 12

perigynium x 10 *cut open*

lower plant x 1

perigynium x 10

female scales x 10

Yellow Bog Sedge

Sedge—*Cyperaceae* **Sedge** Perigynia hairy, 4.5–6.7 mm long

Sand Sedge *Carex houghtoniana* Torr. ex Dewey

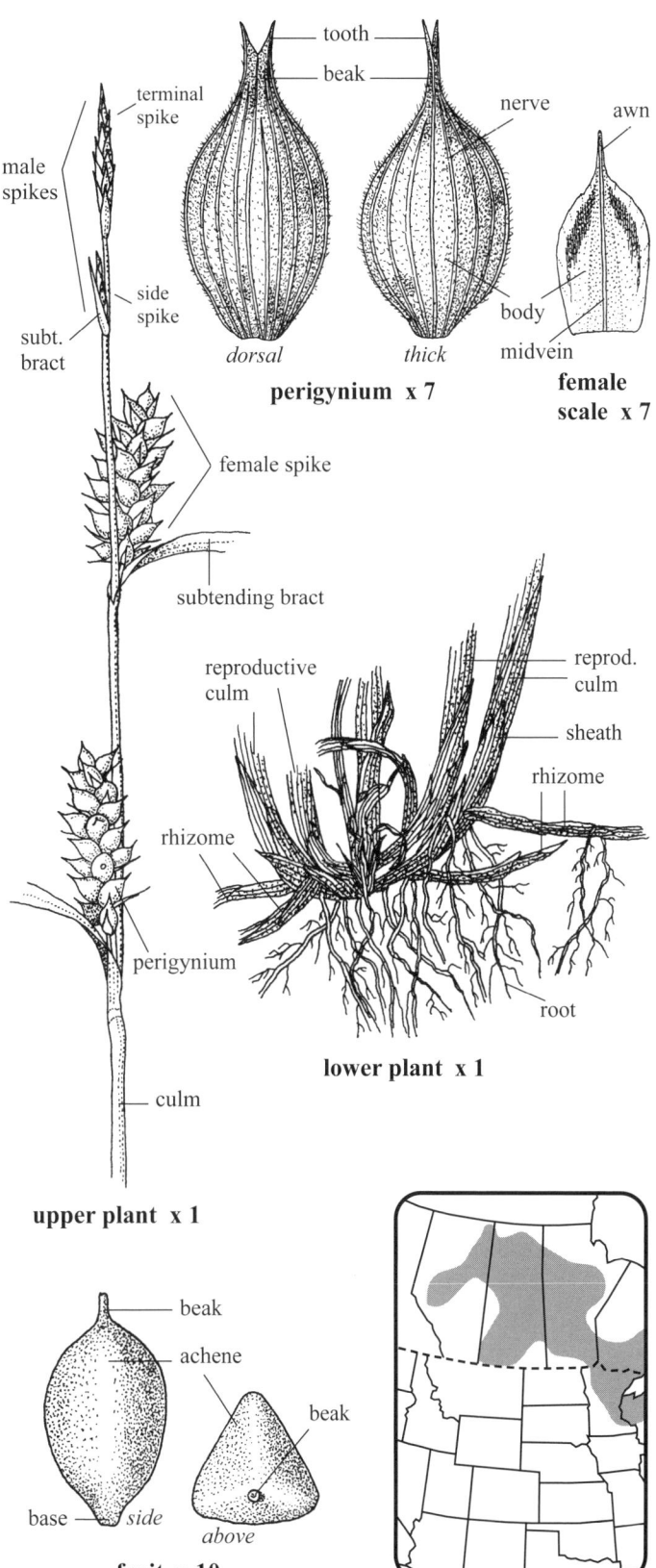

perigynium x 7

female scale x 7

lower plant x 1

upper plant x 1

fruit x 10

■ **SKETCH** A loosely tufted **perennial sedge** growing 20–80 (–100) cm tall with scaly **rhizomes** 3–15 cm long by 2–3 mm thick, tan to dark purplish brown with pointed white tips and pointed scales 5–15 mm long with tan, **fibrous roots** 1–4 cm long; in sandy to gravelly soils, open woods, on ledges and disturbed sites; **monoecious**. w2

■ **FLOWERS** Green, blooming May–June; **inflorescence** a spike, unisexual; **male spikes** one to three, TL 1.3–3.5 cm by 2–3 mm wide, erect, brown, side spike remote from the terminal one; **subtending bracts** (of side spikes) erect, green, *c.* 1 cm long; **peduncles** (of terminal spikes) scabrous, 14–33 mm long, lateral spike subsessile; **female spikes** one to four, ascending, remote, each 2–3 cm long by 10–13 mm wide; **subtending bracts** (of female spikes) leaflike, spreading, 5–21 cm long by 3–6 mm wide, reduced above, margins scabrous, tapered to a fine point; **sheaths** 1–4 mm long, V-shaped at apices; **peduncles** (of female spikes) 3–10 mm long; **stigmas** 3; **perigynia** ascending to spreading, in six columns, green to dark brown when ripe, not spongy, finely hairy, 4.5–6.7 mm long by 2–3 mm wide, body 17- to 22-nerved, the two side nerves ending on beak margins the most obvious; **beak** 1.2–2 mm long, scabrous on margins, bidentate, teeth erect to slightly spreading, 0.5–0.8 mm long; **female scales** 4–6 mm long by 1.8–2 mm wide, tan, upper margins hyaline, apices tapered to an **awn** up to *c.* 3 mm long, midvein raised with wide green to tan margins, dark brown streaks near the apices, scales persistent after perigynia fall. **FRUIT an achene**, 1-seeded, triangular, corners rounded, body 2.7–3.1 mm long by 1.8–2.1 mm wide, slightly shiny, brown, not ribbed; **beak** straight, *c.* 0.4 mm long.

■ **LEAVES** Flat, three to five green ones on lower half of culm, arched, margins scabrous; **blades** 4–27 cm long by 3–8 mm wide, reduced below, upper leaf the longest and barely reaching the lowest female spike when held erect, lowest two or three leaves bractlike, pointed, appressed; **blades** of vegetative plants 20–50 cm long by *c.* 7 mm wide, arched; **ligules** membranous, 1.5–14 mm long; **sheaths** tight, glabrous, basal ones dark reddish purple, V-shaped at apices, up to *c.* 5 cm long above.

■ **STEM** A **culm**, erect to leaning, stiff, sharply triangular with edges scabrous below inflorescence, faces of culm slightly convex, hollow above, solid below, edges more blunt below; the base 3–5 mm thick (including sheaths).

■ **SYN.** *Carex houghtonii* Torr.

Houghton's Sedge

Sedge—*Cyperaceae* Sedge Perigynia hairless, 10–15 mm long

SWOLLEN SEDGE *Carex intumescens* Rudge

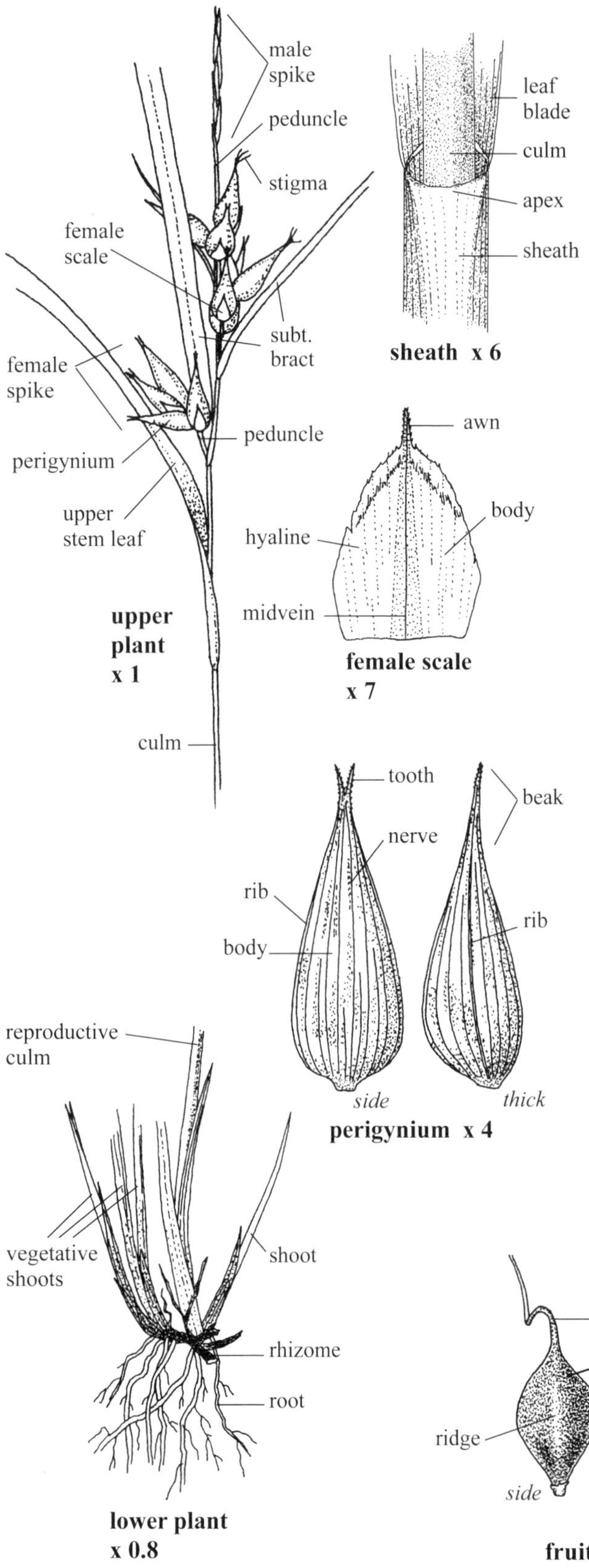

■ **SKETCH** A tufted, yellowish green **perennial sedge** 30–80 cm tall with woody **rhizomes** *c.* 5 mm long by 1.5–2 mm thick and tan **roots** to *c.* 12 cm long; in damp open woods, swamps and bogs; **monoecious. w2**

■ **FLOWERS** Green, blooming May–July; **inflorescence** a spike, unisexual, male above female, TL of spikes 2.5–10 cm; **male spikes** 1–3, erect, terminal, TL 13–26 mm by 1–1.5 mm wide, tan; **peduncles** (of male spikes) 5–40 mm long, slightly scabrous; **male scales** pale green to tan, 5–7 mm long by 0.8–1 mm wide, pointed, midnerves green, scabrous near apices; **female spikes** 1–4, each with 1–12 perigynia, erect from upper axils, 13–27 mm long by 14–25 mm wide; **peduncles** (of female spikes) 3–15 mm long; **subtending bracts** (of female spikes) leaflike, 2.7–15 cm long by 1–5 mm wide, flat, scabrous, extending well past the spikes, ascending to spreading; **stigmas** 3; **perigynia** green, shiny, roundish in cross-section, each 10–15 mm long by 4–8 mm wide, slightly transparent; **body** *c.* 20-nerved, not spongy or winged, hairless, thin-walled with two ribs; **beak** 3–5 mm long, bidentate, the teeth *c.* 1 mm long and slightly spreading; **female scales** pale green, 4–4.6 mm long by 2.7–3 mm wide (flattened), midvein pale green, often forming a short awn, faint striations near apical margins, sides membranous, hyaline to pale green. **FRUIT an achene**, 1-seeded, 4.5–5.5 mm long by 3–3.8 mm wide, triangular, edges blunt, sides concave near the base, brown, shiny, situated near the base of a perigynium; **style** persistent, straight to bent.

■ **LEAVES** Five to twelve, top leaf near the inflorescence and extending well past it; **stem blades** arched to ascending, 10–40 cm long by 3–9 mm wide, flat, yellowish green, scabrous on margins, tan leaves numerous at the base of entire clump; **sheaths** smooth.

■ **STEM** A **culm**, leaning with fruit, triangular, scabrous above on sharp corners, hairs ascending, sides slightly convex, corners smooth and blunt below; 2–3 mm thick near the pinkish base.

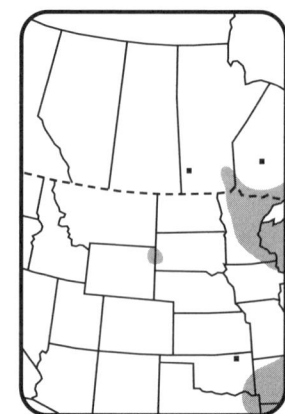

546 Greater Bladder Sedge, Bladder Sedge

Sedge—*Cyperaceae* **Sedge** Perigynia hairless, beakless, 2.5–5 mm long

BRISTLE-STALKED SEDGE *Carex leptalea* Wahlenb.

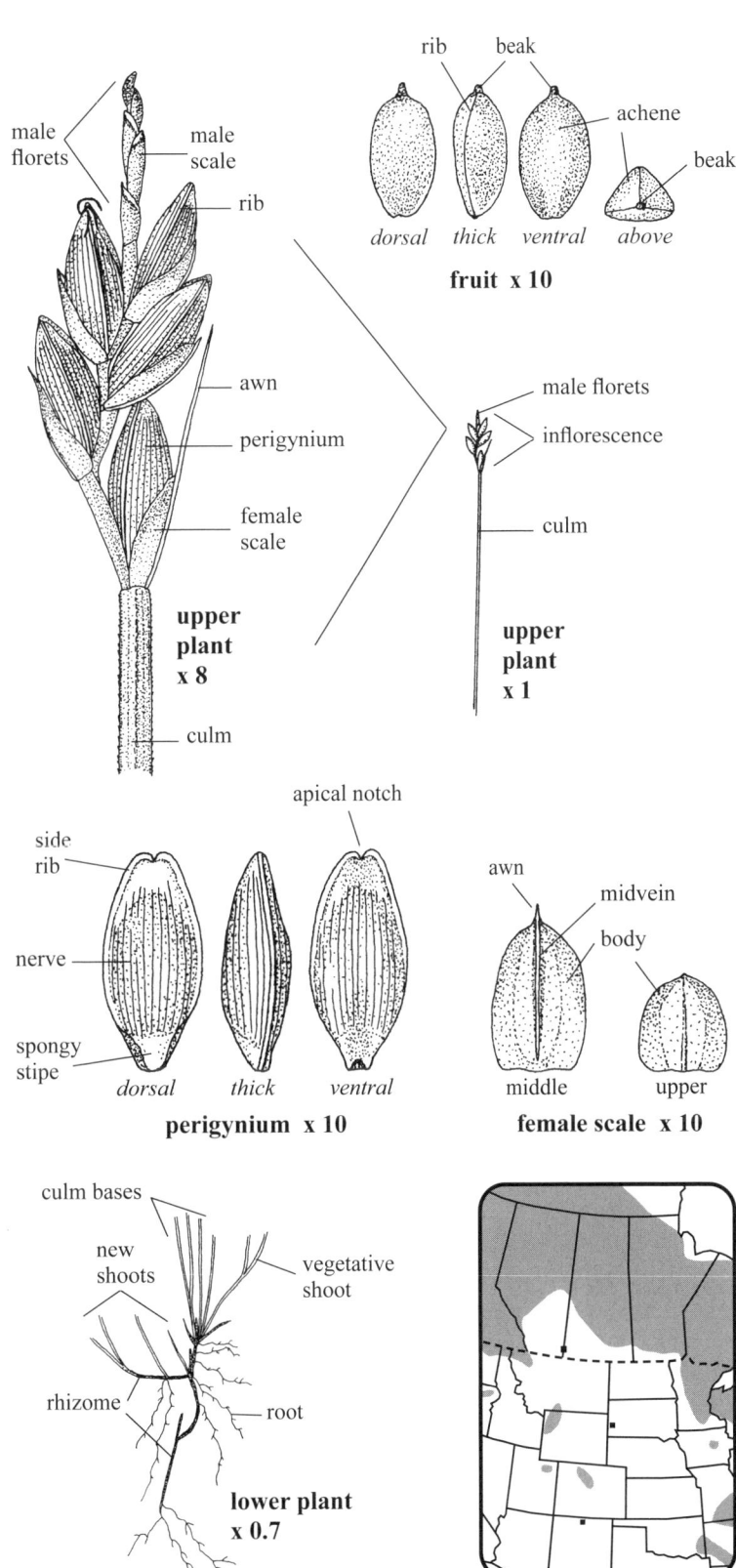

■ **SKETCH** A variable, tufted **perennial sedge** 10–70 cm tall with dark brown **rhizomes** 0.5–2 cm long by 0.6–1.2 mm thick, covered with veiny blunt bracts; **fibrous roots** filiform, 1–18 cm long; in wet woods, sphagnum bogs, ditches, shores and marshes. **w2**

■ **FLOWERS** Greenish, blooming March–June; **inflorescence** a spike, solitary, 4–16 mm long by 2.2–3.5 mm wide, female flowers below the male flowers; **male florets** erect, together 3–4 mm long; **male scales** darker brown at apices, blunt to pointed, not awned; **female florets** 1–10 (–13); **stigmas** 3; **perigynia** ascending, greenish, each 2.5–5 mm long by 1–1.6 mm wide by *c.* 1 mm thick, larger in the south, tapered at base to a spongy wide stipe, beakless, notched at rounded apex, *c.* 24-nerved, these thin and faint, two side ribs green, raised and ending at apex, slightly shiny, thin-walled, slightly transparent; **female scales** 1.4–4.6 mm long (including awn) by 1.3–1.4 mm wide, the body 1.8–2 mm long, awn of lowest scale up to *c.* 2.6 mm long, brown and scabrous, upper scales shorter and awnless, early deciduous, midvein brown, margins tannish orange. **FRUIT an achene**, 1-seeded, filling most of the perigynium, triangular, brown and shiny, 1.3–2 mm long by 0.9–1.2 mm wide by 0.7–0.9 mm thick, sides concave near the base, convex above; **beak** straight, 0.1–0.2 mm long, dark.

■ **LEAVES** One to three near the base, flat, lax, 1.5–25 cm long by 0.5–1.3 mm wide, reduced below, upper leaf the longest, midrib rounded and raised below, margins smooth, brown to green above; **ligules** unremarkable; **sheaths** tight, smooth, margins hyaline.

■ **STEM** A **culm**, green, slightly scabrous below the inflorescence, sharply triangular above with the sides slightly concave, sharply triangular and smooth below, weak; base 0.4–0.6 mm wide and covered in brown leafy sheaths.

Threadstem Sedge

Sedge—*Cyperaceae*

Sedge Perigynia hairless, beakless, 2.7–3.5 mm long

BOG SEDGE *Carex magellanica* Lam.

■ **SKETCH** A tufted **perennial sedge** 10–80 cm tall with stiff **rhizomes** 2–3 mm wide with veiny pointed **bracts** to *c*. 2 cm long and **roots** 3–18 cm long by 0.7–1 mm wide, soft, yellowish orange with root hairs to *c*. 1 mm long; in ditches, bogs, fens and muskeg in Boreal forest; **monoecious. w2**

■ **FLOWERS** Green, blooming May–June; **inflorescence** a spike, unisexual to bisexual, TL of spikes 3–5.5 cm; <u>**male spikes**</u> one, terminal, 5–12 mm long by *c*. 1 mm wide, erect, some with female florets; **male scales** linear and 5–7 mm long; **subtending bracts** (of male spikes) 12–15 mm long; **peduncles** (of male spikes) 10–18 mm long, erect, scabrous; <u>**female spikes**</u> two to four, hanging, each 7–20 mm long by 6–7 mm wide (including scale tips), often with a few male florets at their base; **subtending bracts** (of female spikes) 1.2–15 cm long by 1–2 mm wide, flat, leaflike, margins scabrous, reduced above, the lowest bract extending well past the inflorescence, its **sheath** 4–5 mm long, closed, tight, membranous, apices often torn; **peduncles** (of female spikes) arched, 1–4 cm long, very slender; **stigmas** 3; **perigynia** glaucous, pale green to brown, hairless, 2.7–3.5 mm long by 1.8–2.5 mm wide by 1–1.2 mm thick, side ribs two, nerves faint and paler green than the light green body, two to four nerves on each convex side; **beak** 0–0.2 mm long; **stipe** 0.3–0.5 mm long; **female scales** early deciduous, 2.8–7 mm long by 1–2 mm wide (flat), reduced above, 3-nerved, tapered to a point or an awn to *c*. 3 mm long, midvein green, sides brown, or all brown, margins often revolute. **FRUIT an achene**, 1-seeded, in the lower part of the perigynium atop the stipe, triangular, sides slightly concave, 1.5–2 mm long by 1.2–1.3 mm wide, pale olive to yellowish green; **beak** obscure, straight.

■ **LEAVES** One to five per culm in the lower half, arched, flat; **blades** 10–22 cm long by 2–4 mm wide, (to 70 cm long on vegetative culms); **ligules** membranous, unremarkable, a continuation from the hyaline margins of the sheath; **sheaths** glabrous, closed, hyaline, enveloping stem at base.

■ **STEM** A **culm**, triangular, scabrous to smooth and hollow above, smooth and solid below, faces flat to slightly convex.

■ **SYN.** *Carex paupercula* Michx. = *Carex magellanica* ssp. *irrigua* (Wahlenb.) Hultén.

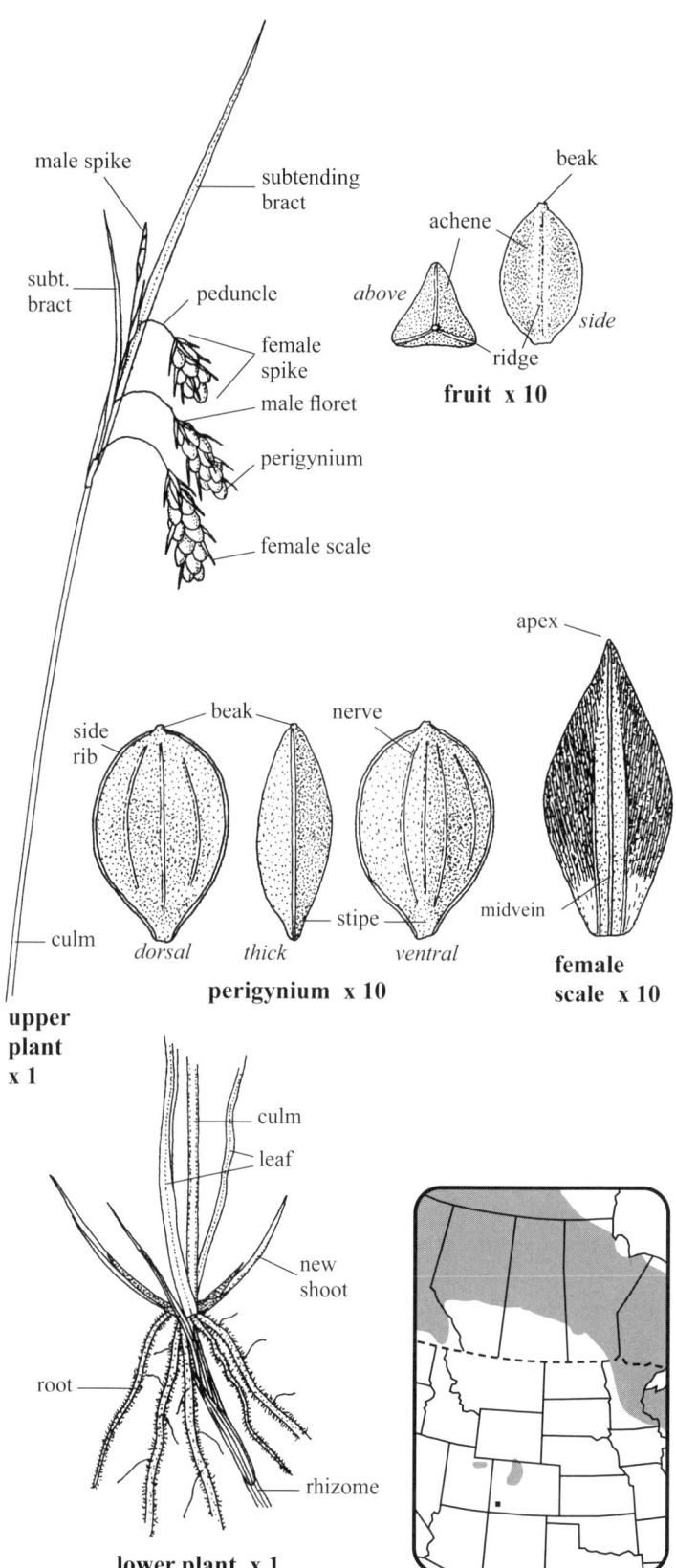

Boreal Bog Sedge

Sedge—*Cyperaceae* Sedge   Perigynia hairless, 1.8–3.5 mm long

Parry's Sedge   *Carex parryana* Dewey

■ **SKETCH** A loosely tufted **perennial sedge** 15–40 cm tall with long **rhizomes** *c.* 1 mm thick; in wet grassland, slough margins, ditches, swampy areas in prairies and parklands; **monoecious**. **w2**

■ **FLOWERS** Green, blooming April–May; **inflorescence** a spike, 1–6, erect to ascending, overlapping TL 3–8 cm; **terminal spike** the longest at 1.2–3 cm by 3–5 mm wide, with all male or all female florets or a mixture; **lateral spikes** 2–4, each 0.7–2.4 cm long, female or male; **male scales** *c.* 3 mm long, margins widely hyaline, central vein tan with dark brown striations; **peduncles** 1–15 mm long, ascending, green; **subtending bracts** (of peduncles), 1.8–6 cm long by 5–10 mm wide, usually broken off by the time fruit is ripe, dark brown at their bases, shorter to longer than spikes; **stigmas** 3 (2); **perigynia** 1.8–3.5 mm long by 1.3–2 mm wide by 0.8–0.9 mm thick, 5- or 6-nerved above (dorsally), and brown with spots near the apices or these absent, ventrally usually 3- or 5-nerved, with reddish brown dots near the apices, body flattened and prominently winged, hairless except for microscopic hairs on the two lateral ridges near the beak, body widest at the middle or slightly above, tapered to the short stipe, lateral ridges along the wings, convex above, flat ventrally, achene barely visible inside; **beak** 0.2–0.4 mm long, apex bidentate but teeth very short; **female scales** *c.* 2 mm long, awn (when present) tan, usually less than 1 mm long, body about as long and wide as the perigynium, with reddish brown striations, margins widely hyaline, as is the tip. **FRUIT an achene**, 1-seeded, 1.4–2.2 mm long by *c.* 1 mm wide by *c.* 0.7 mm thick, medium brown, triangular, three ribs light brown; **beak** 0.1–0.2 mm long.

■ **LEAVES** Ascending, margins scabrous, 8–18 cm long by 2–4 mm wide; **blades** linear, widely spaced above, upper leaf the longest, but not reaching the base of the inflorescence, usually two green leaves per culm with 2 or 3 brown leaves ascending from the base; **sheaths** whitish hyaline ventrally, glabrous, concave at apices.

■ **STEM** A **culm**, stiff, triangular, erect, slightly scabrous and hollow near the inflorescence, sides slightly convex and corners sharp above, culm and lower leaves surrounded by dark brown old leaf sheaths at the base; smooth below near the 1–1.5 mm wide base.

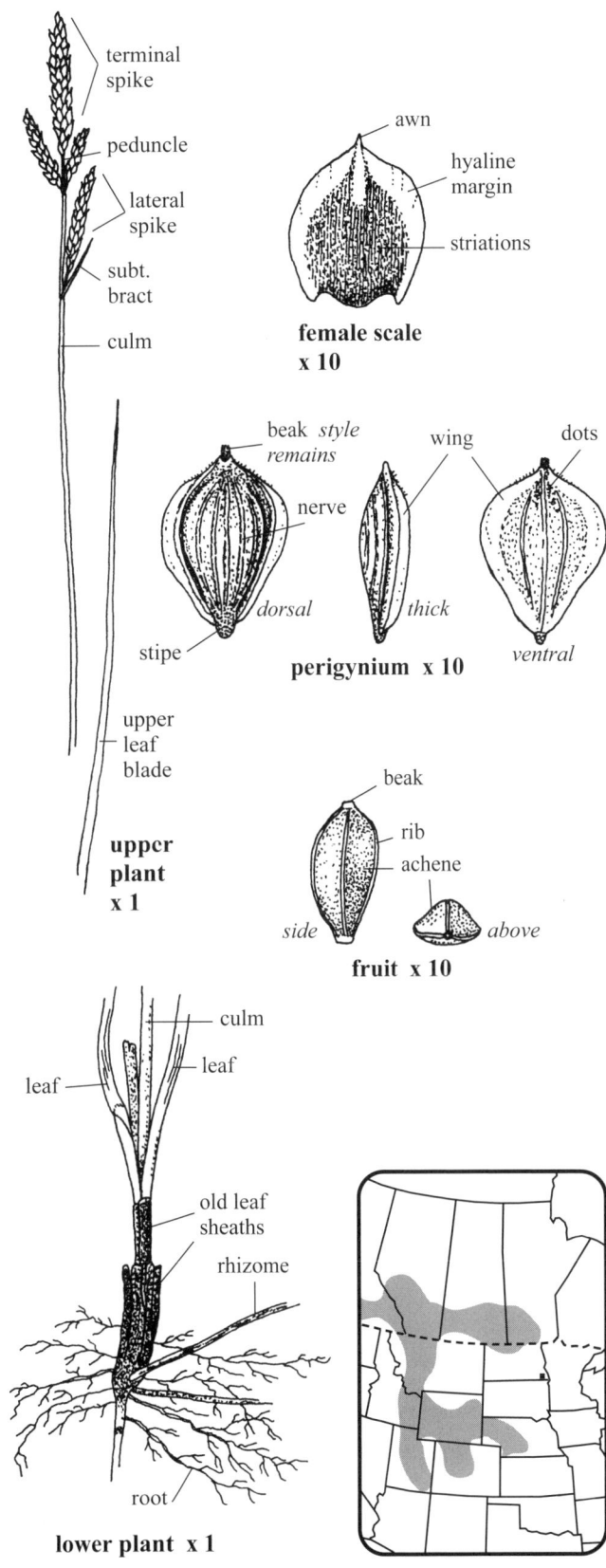

549

Sedge—*Cyperaceae* Sedge Perigynia hairy, 3–4.2 mm long

PECK'S SEDGE *Carex peckii* Howe

■ **SKETCH** A loosely tufted variable **perennial sedge** 10–55 cm high with reddish brown **rhizomes** 1–20 mm long by 1–2 mm thick; in open, rich, moist woodlands, outcrops and along riverbanks in Boreal forest. **w2**

■ **FLOWERS** Green, blooming April–May; **inflorescence** a spike, unisexual, TL of spikes 8–45 mm by 5–10 mm wide; **male spikes** one, terminal, 5–9 mm long, narrow, stalked; **female spikes** one to three, lateral, each 4–7 mm long, usually sessile, the lowest spike with 3–10 flowers, some spikes with apical male flowers; **subtending bracts** (of female spikes) 5–30 mm long, sometimes exceeding the inflorescence, reduced above; **stigmas** 3; **perigynia** slightly hairy, yellowish green, nervless, 3–4.2 mm long by 1–1.4 mm wide by 0.8–1.2 mm thick, with two lateral ribs along the outer, more convex surface, the stipe narrow and slightly spongy; **beak** bidentate, 0.5–1 mm long, the teeth 0.2–0.4 mm long, not spreading; **female scales** variable, 2–3.3 mm long by 1.2–1.8 mm wide (flattened), shorter than perigynia, pale green with reddish brown streaks, wide margins hyaline, beige near the base. **FRUIT an achene**, 1-seeded, dull, 2–2.4 mm long by 1–1.3 mm wide and 0.8–0.9 mm thick, medium reddish brown, triangular, the sides slightly convex, the ribs light and distinct; **stipe** obvious.

■ **LEAVES** Usually four blades from the lower half of the culm, 2–18 cm long by 1–3.3 mm wide, lightly scabrous, flat; **ligules** membranous, *c.* 0.3 mm long, apices flat; **sheaths** glabrous, tight, laddering absent.

■ **STEM** A **culm**, mostly erect, sharply triangular; 0.7–0.8 mm across one side near the reddish purple non-fibrous base.

■ **SYN.** *Carex nigromarginata* var. *elliptica* (Boott) Gleason and *Carex nigromarginata* var. *minor* (Boott) Gleason.

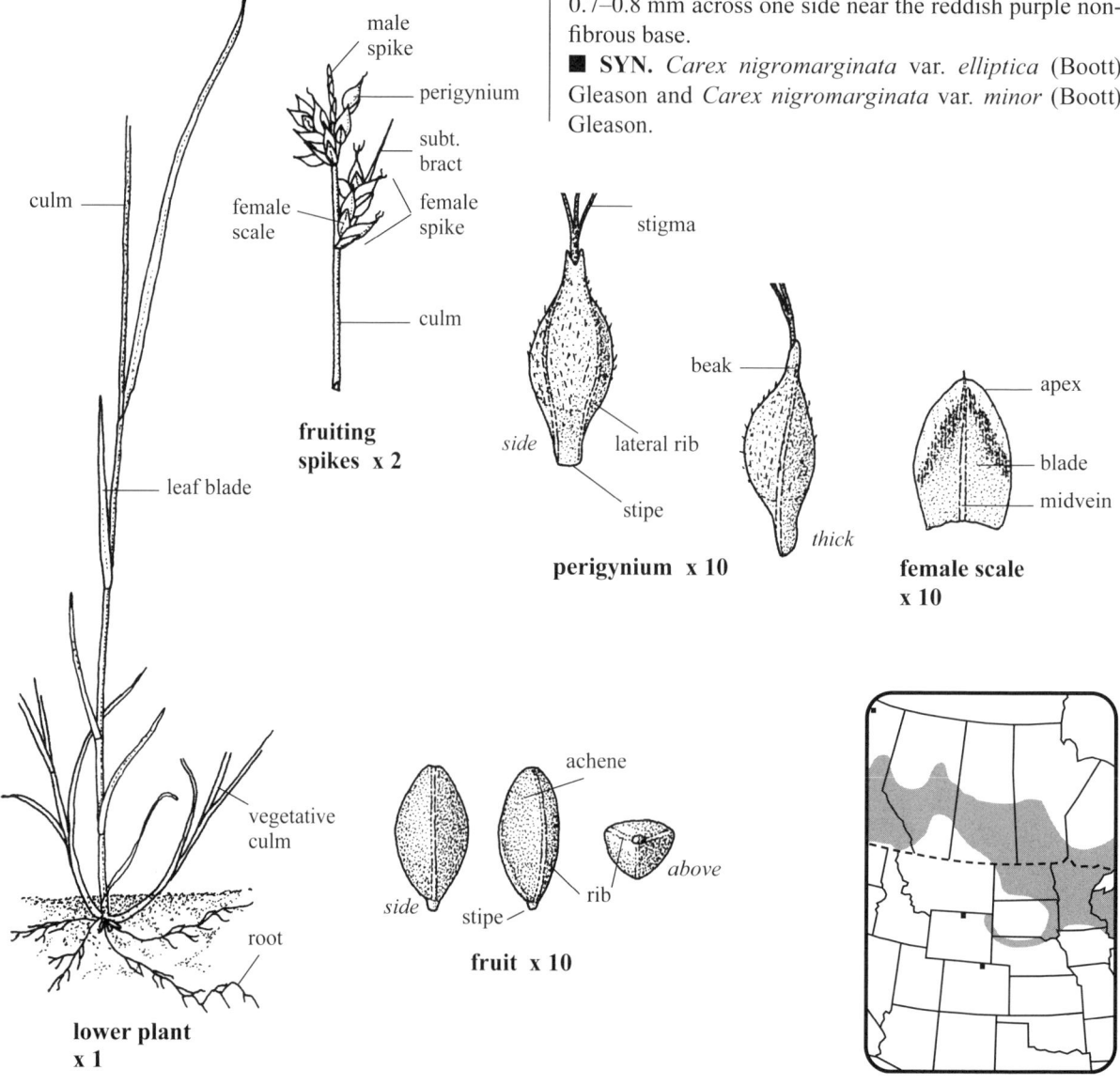

WOOLLY SEDGE   *Carex pellita* Muhl ex Willd.

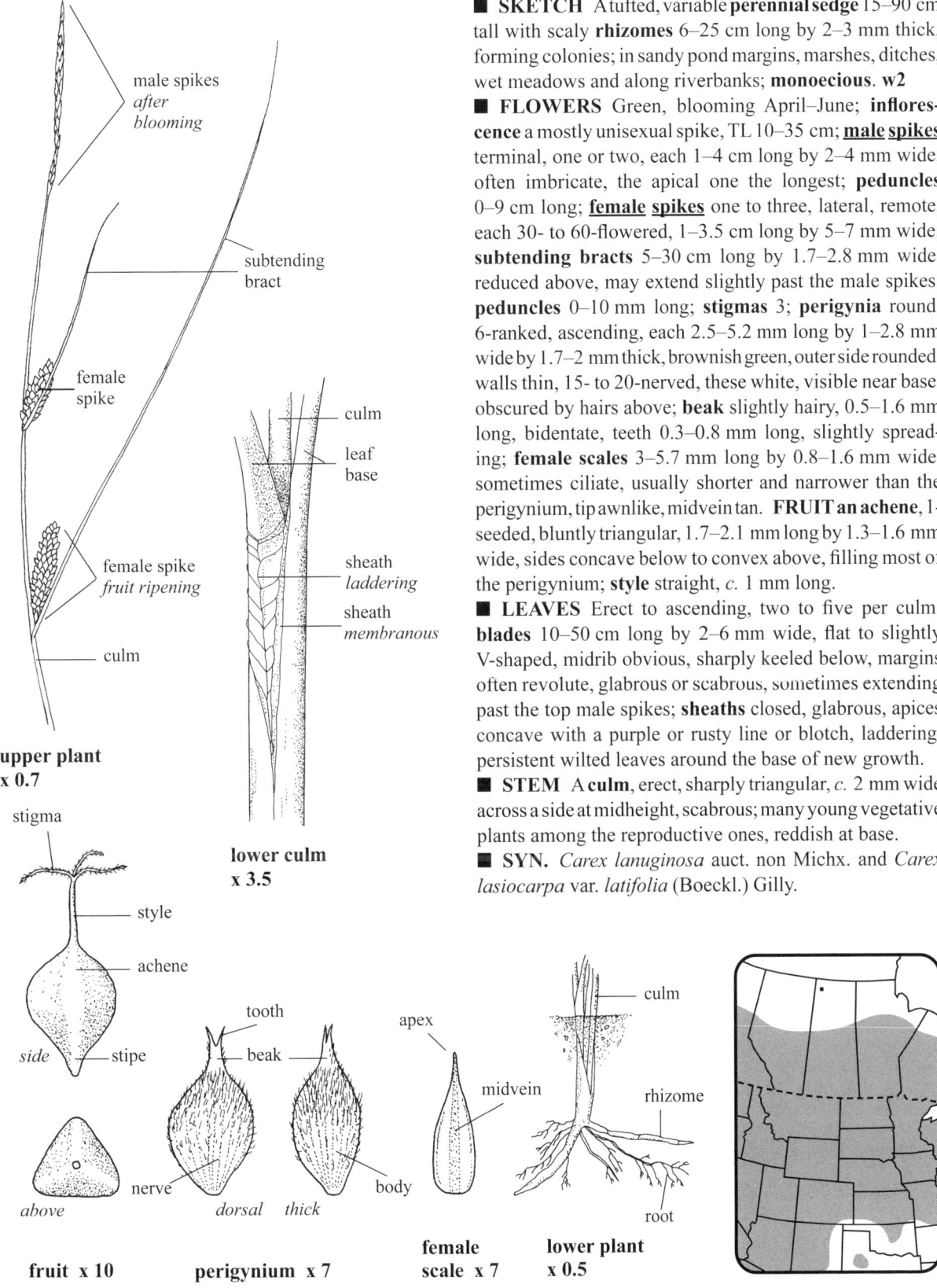

fruit x 10   perigynium x 7   female scale x 7   lower plant x 0.5

■ **SKETCH** A tufted, variable **perennial sedge** 15–90 cm tall with scaly **rhizomes** 6–25 cm long by 2–3 mm thick, forming colonies; in sandy pond margins, marshes, ditches, wet meadows and along riverbanks; **monoecious**. w2

■ **FLOWERS** Green, blooming April–June; **inflorescence** a mostly unisexual spike, TL 10–35 cm; **male spikes** terminal, one or two, each 1–4 cm long by 2–4 mm wide, often imbricate, the apical one the longest; **peduncles** 0–9 cm long; **female spikes** one to three, lateral, remote, each 30- to 60-flowered, 1–3.5 cm long by 5–7 mm wide; **subtending bracts** 5–30 cm long by 1.7–2.8 mm wide, reduced above, may extend slightly past the male spikes; **peduncles** 0–10 mm long; **stigmas** 3; **perigynia** round, 6-ranked, ascending, each 2.5–5.2 mm long by 1–2.8 mm wide by 1.7–2 mm thick, brownish green, outer side rounded, walls thin, 15- to 20-nerved, these white, visible near base, obscured by hairs above; **beak** slightly hairy, 0.5–1.6 mm long, bidentate, teeth 0.3–0.8 mm long, slightly spreading; **female scales** 3–5.7 mm long by 0.8–1.6 mm wide, sometimes ciliate, usually shorter and narrower than the perigynium, tip awnlike, midvein tan. **FRUIT an achene**, 1-seeded, bluntly triangular, 1.7–2.1 mm long by 1.3–1.6 mm wide, sides concave below to convex above, filling most of the perigynium; **style** straight, *c.* 1 mm long.

■ **LEAVES** Erect to ascending, two to five per culm; **blades** 10–50 cm long by 2–6 mm wide, flat to slightly V-shaped, midrib obvious, sharply keeled below, margins often revolute, glabrous or scabrous, sometimes extending past the top male spikes; **sheaths** closed, glabrous, apices concave with a purple or rusty line or blotch, laddering, persistent wilted leaves around the base of new growth.

■ **STEM** A **culm**, erect, sharply triangular, *c.* 2 mm wide across a side at midheight, scabrous; many young vegetative plants among the reproductive ones, reddish at base.

■ **SYN.** *Carex lanuginosa* auct. non Michx. and *Carex lasiocarpa* var. *latifolia* (Boeckl.) Gilly.

Sedge—*Cyperaceae* **Sedge** Perigynia hairy above, 2.5–3.5 mm long

Sun-loving Sedge   *Carex pensylvanica* Lam.

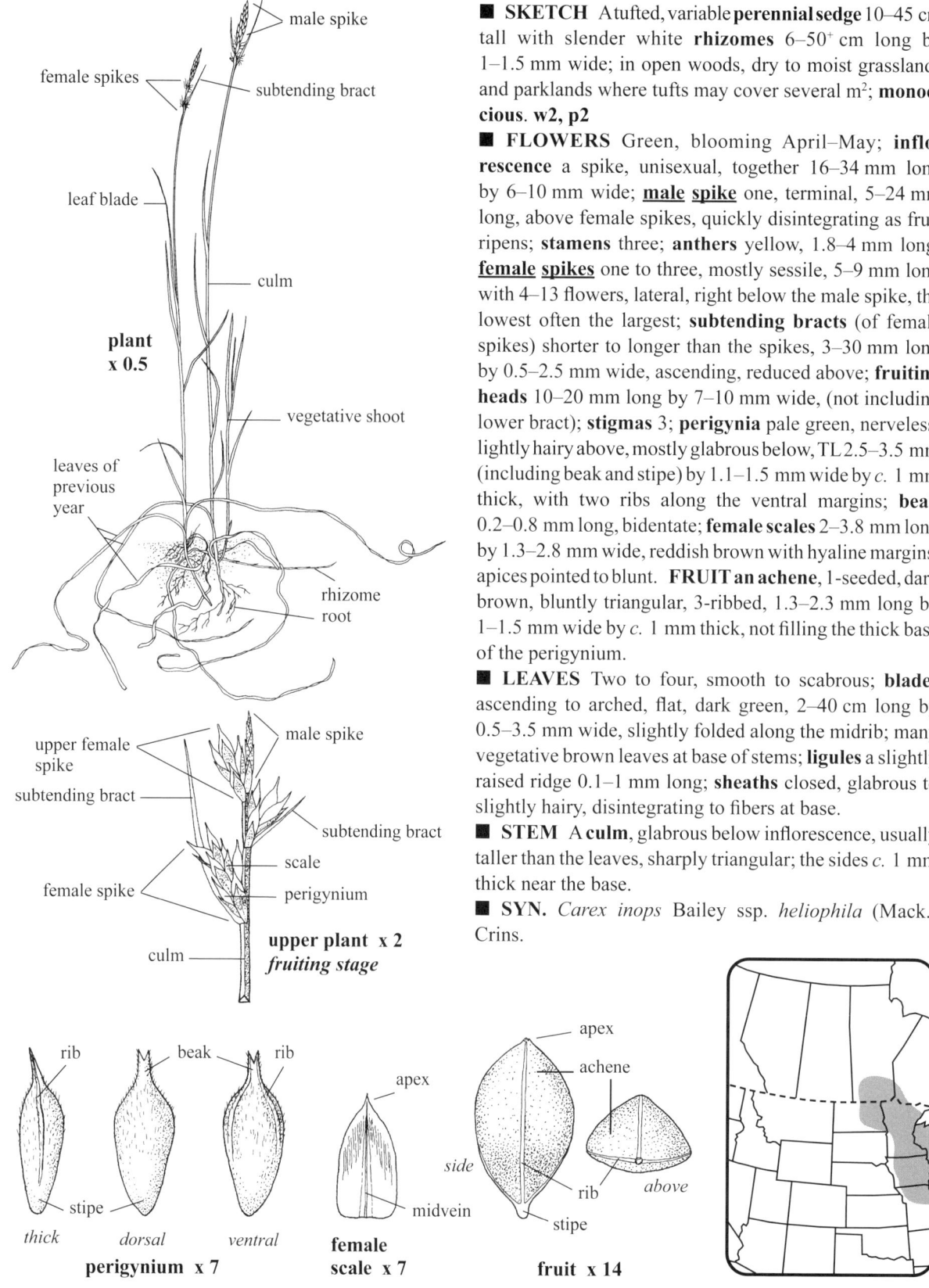

- **SKETCH** A tufted, variable **perennial sedge** 10–45 cm tall with slender white **rhizomes** 6–50+ cm long by 1–1.5 mm wide; in open woods, dry to moist grasslands and parklands where tufts may cover several m^2; **monoecious**. w2, p2
- **FLOWERS** Green, blooming April–May; **inflorescence** a spike, unisexual, together 16–34 mm long by 6–10 mm wide; **male spike** one, terminal, 5–24 mm long, above female spikes, quickly disintegrating as fruit ripens; **stamens** three; **anthers** yellow, 1.8–4 mm long; **female spikes** one to three, mostly sessile, 5–9 mm long with 4–13 flowers, lateral, right below the male spike, the lowest often the largest; **subtending bracts** (of female spikes) shorter to longer than the spikes, 3–30 mm long by 0.5–2.5 mm wide, ascending, reduced above; **fruiting heads** 10–20 mm long by 7–10 mm wide, (not including lower bract); **stigmas** 3; **perigynia** pale green, nerveless, lightly hairy above, mostly glabrous below, TL 2.5–3.5 mm (including beak and stipe) by 1.1–1.5 mm wide by *c*. 1 mm thick, with two ribs along the ventral margins; **beak** 0.2–0.8 mm long, bidentate; **female scales** 2–3.8 mm long by 1.3–2.8 mm wide, reddish brown with hyaline margins, apices pointed to blunt. **FRUIT an achene**, 1-seeded, dark brown, bluntly triangular, 3-ribbed, 1.3–2.3 mm long by 1–1.5 mm wide by *c*. 1 mm thick, not filling the thick base of the perigynium.
- **LEAVES** Two to four, smooth to scabrous; **blades** ascending to arched, flat, dark green, 2–40 cm long by 0.5–3.5 mm wide, slightly folded along the midrib; many vegetative brown leaves at base of stems; **ligules** a slightly raised ridge 0.1–1 mm long; **sheaths** closed, glabrous to slightly hairy, disintegrating to fibers at base.
- **STEM** A **culm**, glabrous below inflorescence, usually taller than the leaves, sharply triangular; the sides *c*. 1 mm thick near the base.
- **SYN.** *Carex inops* Bailey ssp. *heliophila* (Mack.) Crins.

Sedge—*Cyperaceae* **Sedge** Perigynia hairless, 2.5–3.8 mm long

GRACEFUL SEDGE *Carex praegracilis* W. Boott

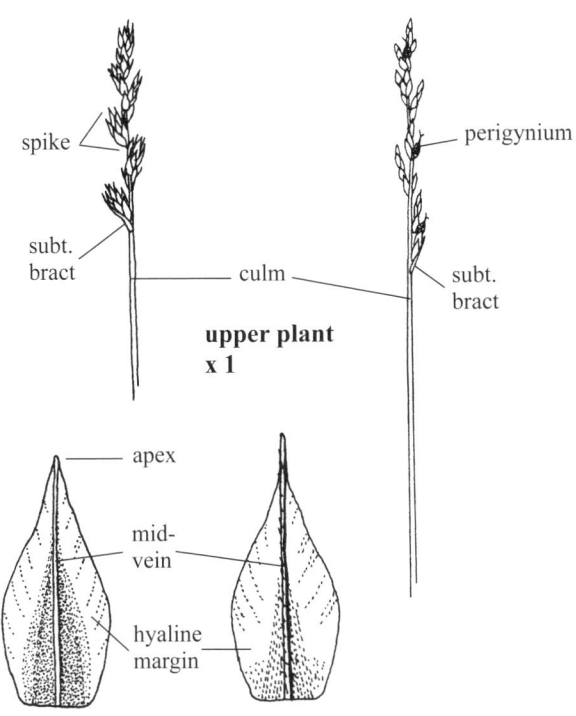

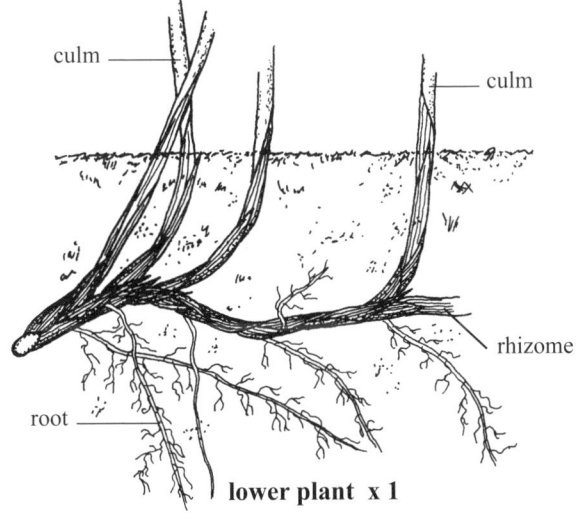

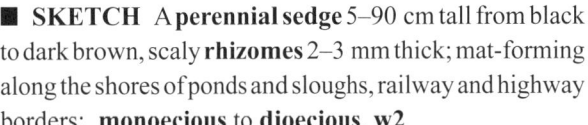

- **SKETCH** A **perennial sedge** 5–90 cm tall from black to dark brown, scaly **rhizomes** 2–3 mm thick; mat-forming along the shores of ponds and sloughs, railway and highway borders; **monoecious** to **dioecious**. **w2**
- **FLOWERS** Green, blooming May–June; **inflorescence** a spike, TL of spikes 2–6 cm by 3–10 mm wide; **spikes** often unisexual or androgynous, usually 5–18 (–25), these more or less sessile, each 3.5–10 mm long, overlapping above, the lower ones separate; <u>**male flowers**</u> above on androgynous spikes; **subtending bracts** (of spikes) one, brown to tan with a green midrib, rarely exceeding a spike, the lowest bract the largest at 4–25 mm long; **stigmas** 2; **perigynia** 4–11 per spike, flattened, plano-convex, body hairless, dull, light brown to brownish black, 2.5–3.8 mm long by 1.2–1.9 mm wide by *c.* 0.7 mm thick, 3- or 5-nerved only on the dorsal round side, ventral side nerveless, base spongy, stipe short; **beak** 0.7–1.2 mm long or up to about half the length of the perigynium's body, obliquely cut, bidentate, scabrous along the sides; **female scales** pointed, slightly longer or shorter and as wide as the perigynia, the midvein tan with a medium brown border, margins hyaline. **FRUIT an achene**, 1-seeded, medium brown, smooth, lenticular, body 1.4–1.8 mm long by 1–1.3 mm wide by 0.5–0.6 mm thick, two low ribs along the sides; **style** base straight and persistent.
- **LEAVES** Usually three or four blades near the base of the culm, 10–45 cm long by 1.2–3 mm wide, flattened, arching and tapered to a fine point; **ligules** typically short and membranous; **sheaths** dark brown to black, scabrous, apices truncate to V-shaped, closed, no laddering.
- **STEM** A **culm**, erect, stiff, sharply triangular and scabrous right below the inflorescence, smooth and less sharply triangular near the base; 1–2 mm thick across one side near the base.

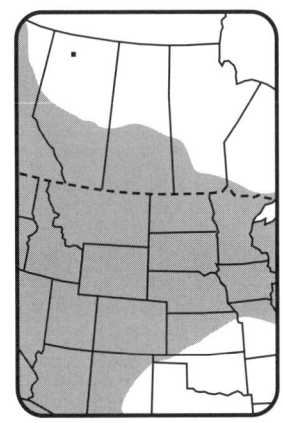

Field Sedge, Clustered Field Sedge

Sedge—Cyperaceae

Sedge Perigynia hairless, 3.5–6.2 mm long

CYPERUS-LIKE SEDGE *Carex pseudocyperus* L.

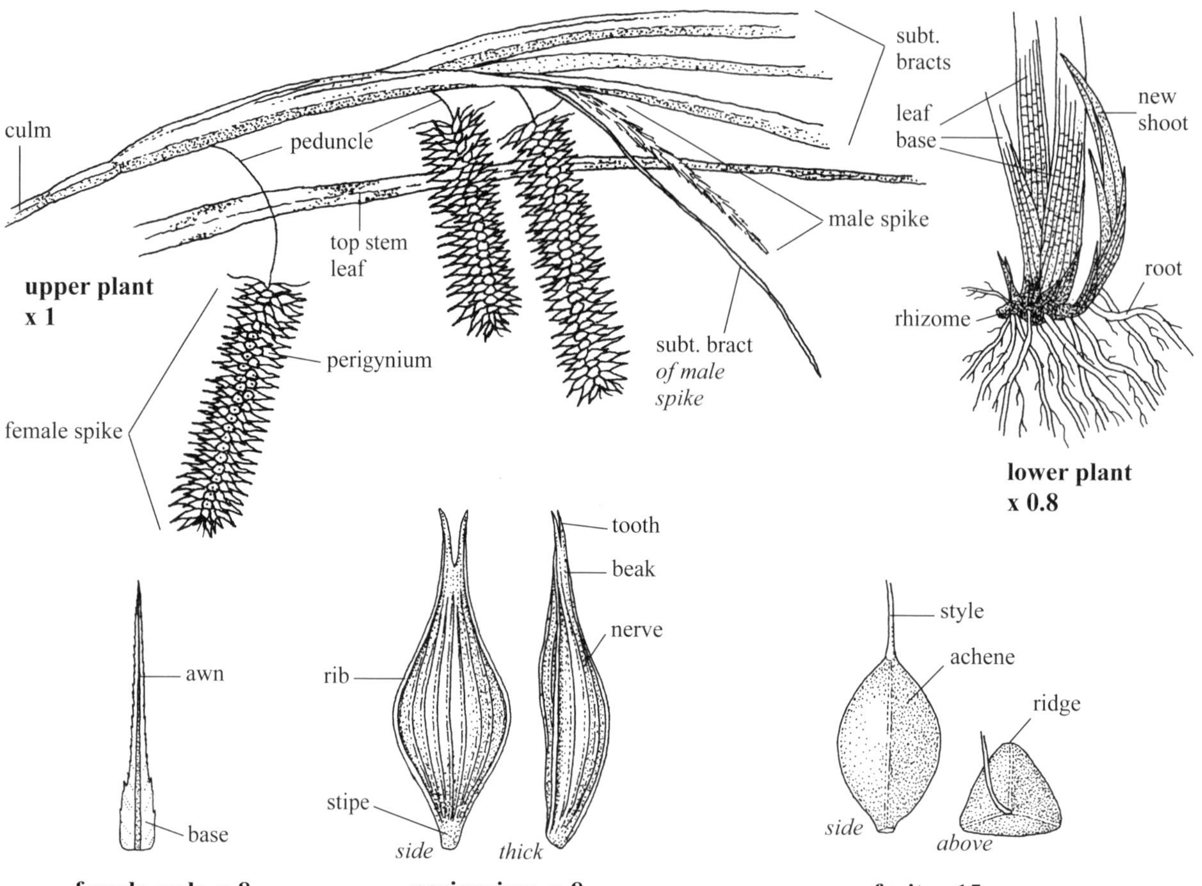

- **SKETCH** A tufted **perennial sedge** 30–100 cm tall with **rhizomes** usually *c.* 2 mm long and thick with tan **roots** *c.* 1 mm thick; along marsh edges, bogs, swamps and ditches in the forest; **monoecious**. w2
- **FLOWERS** Green, blooming May–June; **inflorescence** a spike, unisexual; **male spike** one, terminal, 3–7 cm long by *c.* 3 mm wide; **male scales** numerous, imbricate, *c.* 7 mm long; **subtending bracts** (of male spikes) 0–6 cm long; **peduncles** 5–13 mm long; **female spikes** 2–6, lateral, cylindrical, hanging, yellowish green ripening to tan, 2.5–7 cm long by 9–10 mm wide, some with male florets at apices; **subtending bracts** 4.3–55 cm long by 1–9 mm wide, reduced above, margins and blades scabrous, the lower ones extending beyond the spikes, sheaths white ventrally, V-shaped and the two margins sometimes overlapping; **peduncles** thin, 0.6–11 cm long, recurved; **stigmas** 3; **perigynia** in 11–13 columns per spike, mostly spreading, 12- to 20-nerved, 3.5–6.2 mm long by 1.5–2.2 mm wide by 1–1.2 mm thick, plano-convex, thin-walled, shiny, hairless, 3-sided with a short stipe, nerves same color as body; **beak** 1–2 mm long, teeth straight to slightly spreading, *c.* 1 mm long; **female scales** awnlike, 3–8 mm long by *c.* 0.6 mm wide, scabrous, flared at base. **FRUIT** an achene, 1-seeded, 1.5–1.7 mm long by 0.9–1 mm wide by 0.8–0.9 mm thick, golden brown, triangular, sides slightly convex, sometimes one side concave, situated at base of perigynium; **style** tan, persistent.
- **LEAVES** Three to five per culm, green, tan near the base; **blades** linear, 30–56 cm long by 7–15 mm wide, scabrous above, margins of lower blades smooth; **ligules** longer than wide, obvious; **sheaths** V-shaped, brown at apices, tight, green.
- **STEM** A **culm**, erect to leaning with fruit, sharply triangular, scabrous above and smooth at the base, solid, sides concave; 9–12 mm wide near the brown base (including leaf sheaths).

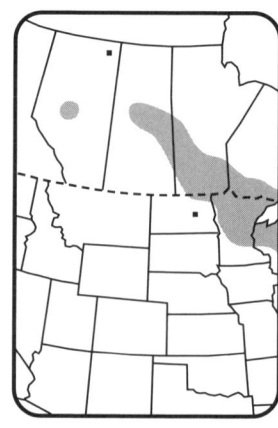

554 Cypress-like Sedge

Sedge—*Cyperaceae* **Sedge** Perigynia hairless, 2.5–3.8 mm long

Eastern Star Sedge *Carex radiata* (Wahlenb.) Small

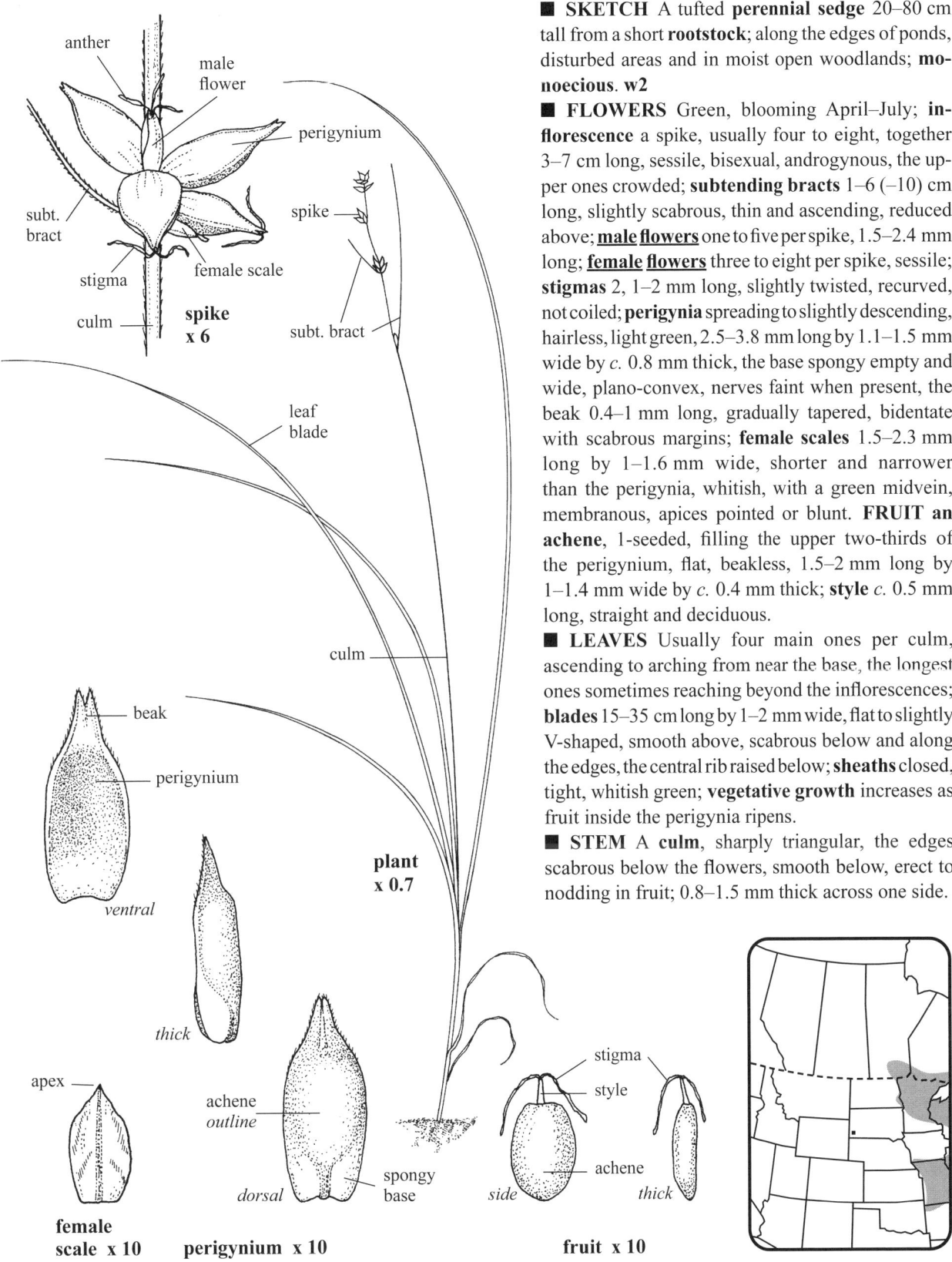

■ **SKETCH** A tufted **perennial sedge** 20–80 cm tall from a short **rootstock**; along the edges of ponds, disturbed areas and in moist open woodlands; **monoecious. w2**

■ **FLOWERS** Green, blooming April–July; **inflorescence** a spike, usually four to eight, together 3–7 cm long, sessile, bisexual, androgynous, the upper ones crowded; **subtending bracts** 1–6 (–10) cm long, slightly scabrous, thin and ascending, reduced above; male flowers one to five per spike, 1.5–2.4 mm long; female flowers three to eight per spike, sessile; **stigmas** 2, 1–2 mm long, slightly twisted, recurved, not coiled; **perigynia** spreading to slightly descending, hairless, light green, 2.5–3.8 mm long by 1.1–1.5 mm wide by *c.* 0.8 mm thick, the base spongy empty and wide, plano-convex, nerves faint when present, the beak 0.4–1 mm long, gradually tapered, bidentate with scabrous margins; **female scales** 1.5–2.3 mm long by 1–1.6 mm wide, shorter and narrower than the perigynia, whitish, with a green midvein, membranous, apices pointed or blunt. **FRUIT an achene**, 1-seeded, filling the upper two-thirds of the perigynium, flat, beakless, 1.5–2 mm long by 1–1.4 mm wide by *c.* 0.4 mm thick; **style** *c.* 0.5 mm long, straight and deciduous.

■ **LEAVES** Usually four main ones per culm, ascending to arching from near the base, the longest ones sometimes reaching beyond the inflorescences; **blades** 15–35 cm long by 1–2 mm wide, flat to slightly V-shaped, smooth above, scabrous below and along the edges, the central rib raised below; **sheaths** closed, tight, whitish green; **vegetative growth** increases as fruit inside the perigynia ripens.

■ **STEM** A **culm**, sharply triangular, the edges scabrous below the flowers, smooth below, erect to nodding in fruit; 0.8–1.5 mm thick across one side.

555

Sedge—*Cyperaceae* — Sedge Perigynia hairless, 7–10 mm long

Turned Sedge *Carex retrorsa* Schwein.

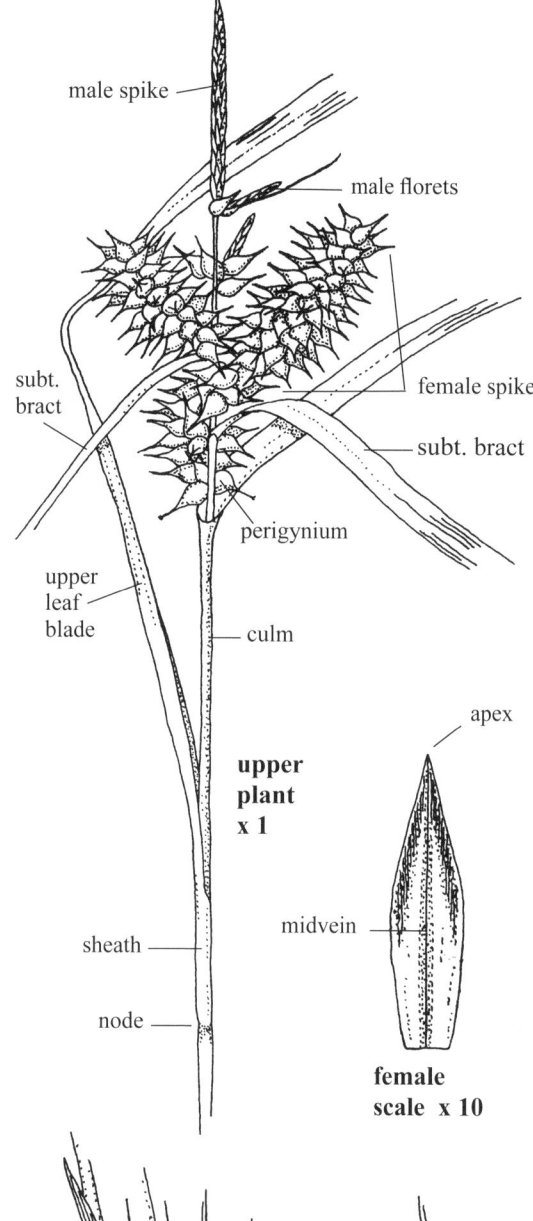

upper plant x 1

female scale x 10

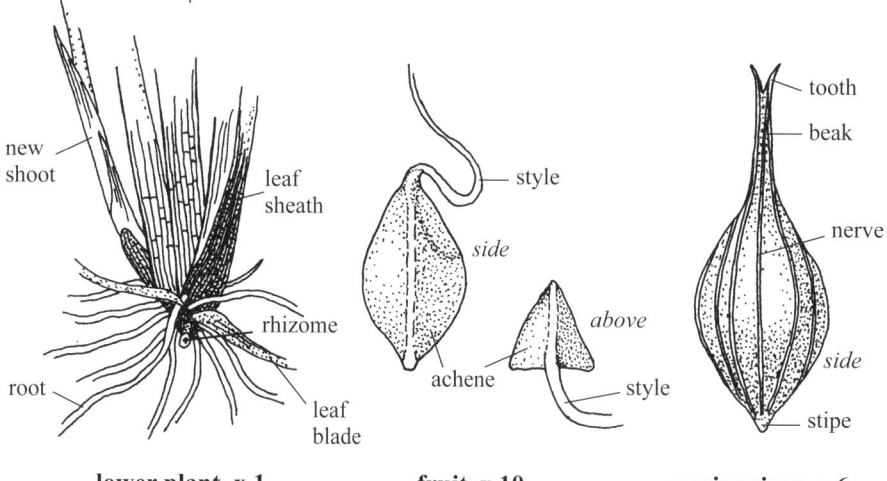

lower plant x 1   fruit x 10   perigynium x 6

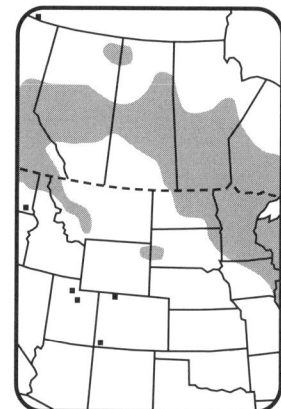

■ **SKETCH** A tufted **perennial sedge** 20–100 cm tall with hard, dark brown **rhizomes** *c.* 5 mm long by *c.* 2 mm thick with tan **roots** 2–10 cm long and *c.* 1 mm wide; in ditches, sloughs, marshes, sometimes in shallow quiet water, along rivers and reservoirs; **monoecious**. w2

■ **FLOWERS** Yellowish green, blooming April–May; **inflorescence** a spike, together 5–12 cm long by 3–4 cm wide; **male spikes** 1–3, above the female spikes, erect, 0.8–3.5 cm long by 1.5–2.5 mm wide; **male scales** *c.* 5 mm long by *c.* 1.5 mm wide, pointed, central area green and narrow, margins hyaline and tan, widest about midpoint, tapered to base, imbricate; **peduncles** 2–12 mm long, reduced above; **female spikes** 3–8, each 2.5–6 cm long by 15–20 mm wide, ascending, crowded, upper ones often with some apical male flowers; **subtending bracts** (of spikes) 0.5–80 cm long by 2–6 mm wide, foliaceous, reduced above, tapered to a fine point, green, arched to spreading, margins scabrous and slightly scabrous below along raised midrib, smooth above, much longer than inflorescence; **stigmas** 3; **perigynia** crowded, hairless, shiny, spreading to slightly descending, 7–10 mm long by 2–3.4 mm wide, light yellowish green, several nerved; **stipe** *c.* 0.5 mm long; **beak** bidentate, 2–4.5 mm long, narrow and nerved, teeth 0.5–1 mm long, curved outward; **body** inflated, not winged, roundish in cross-section, easily crushed; **female scales** pointed, 2.5–4.5 mm long by 1–1.8 mm wide, midvein light green, margins wide, hyaline and streaked with brown, shorter and more narrow than the perigynia. **FRUIT an achene**, 1-seeded, 3-sided, 1.5–2.5 mm long by 1.2–1.5 mm wide, situated at the base of the perigynium, sides slightly concave or convex, corners rounded, style curved or twisted.

■ **LEAVES** Four or five main blades, arched; **blades** 15–40 cm long by 3–10 mm wide, flat, grooved above, midrib slightly scabrous below, margins scabrous; **sheaths** loose, glabrous.

■ **STEM** A **culm**, triangular, the corners sharp above, faces round to convex; almost oval near the 1.5–4 mm thick base.

Sedge—*Cyperaceae* Sedge Perigynia hairless, 2.5–4.6 mm long

SARTWELL'S SEDGE   *Carex sartwellii* Dew.

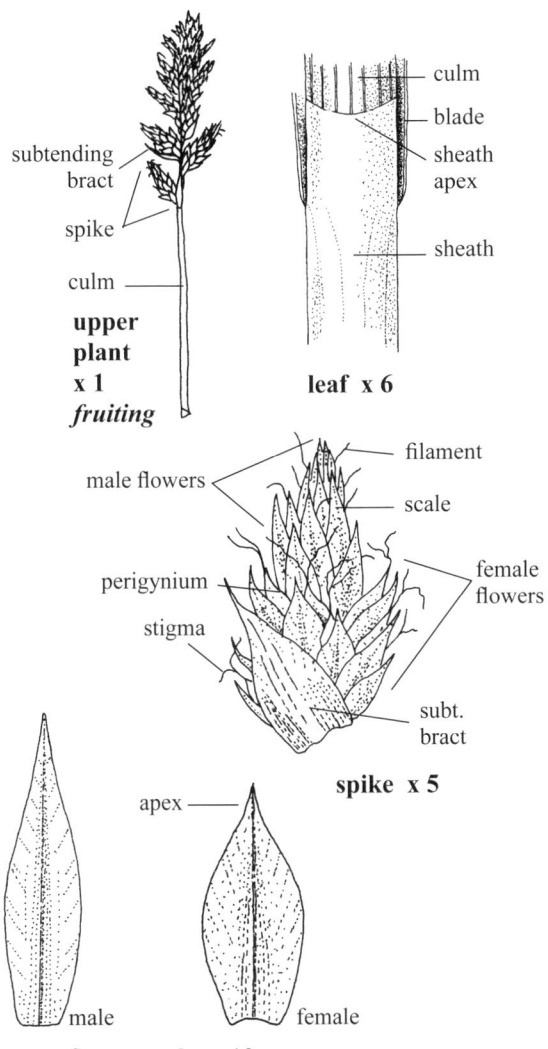

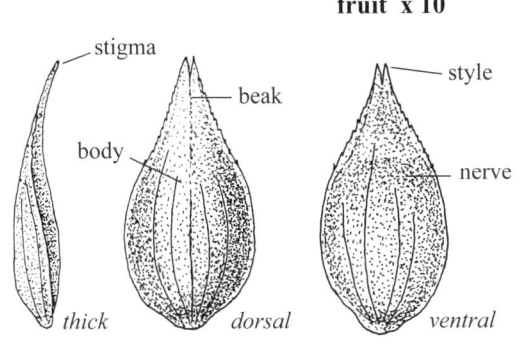

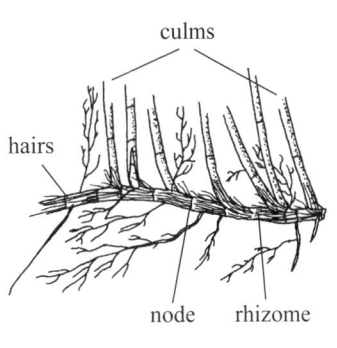

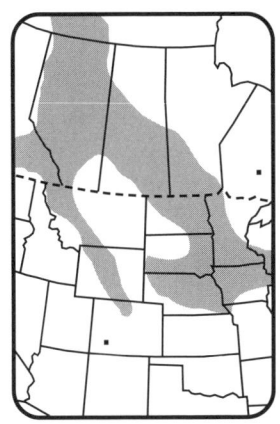

■ **SKETCH** A **perennial sedge** 30–120 cm tall with brown, fibrillose **rhizomes** 2–3 mm thick by 0.5–8+ cm long with fine black hairs *c.* 7 mm long from the nodes; in wet meadows, woodland edges, shallow marshes and bogs; **monoecious**. w2

■ **FLOWERS** Green to brown, blooming May–June; **inflorescence** a spike, 6–20, together 3–7 cm long by 7–13 mm wide; **spikes** 4–11 mm long by 1.5–7 mm wide (in fruit), sessile, androgynous, crowded above, ascending, middle to upper spikes may be entirely male in some inflorescences, or with fewer perigynia than the lower spikes; **subtending bracts** (of spikes) green, 0.4–2 cm long, V-shaped, widening near their bases, reduced above; **male scales** tan, soft with a dark brown pointed tip, 3.7–4 mm long by 0.8–1 mm wide, thin, 1- or 3-veined; **stigmas** 2; **perigynia** glabrous, 2.5–4.6 mm long by 1.3–2 mm wide by *c.* 0.5 mm thick, plano-convex, with several dark brown, thin nerves on each surface, upper margins scarious, not spongy at base; **beak** scabrous or not, 0.4–1.2 mm long, longer beaks tapered, teeth tiny; **female scales** soft, 2.5–4.3 mm long by 1–2 mm wide, generally covering most of a perigynium's body, flat, tan, pointed to blunt, hyaline along apical margins. **FRUIT an achene**, 1-seeded, golden tan, smooth, body 1–2.3 mm long by 0.7–0.9 mm wide by *c.* 0.4 mm thick, lenticular, not filling the perigynium, slightly convex on both sides and somewhat shiny; **beak** tiny, straight.

■ **LEAVES** Two to six per culm, well-spaced; **blades** 3–36 cm long by 2–4.5 mm wide, slightly scabrous below, glabrous above, margins entire and slightly involute, ascending to arched, not reaching the spikes, tapered to a fine point, lowest one or two leaves wilted by fruiting time; **vegetative shoots** with leaves to *c.* 40 cm long; **ligules** tubular, 2.2–8 mm long; **sheaths** concave at apices, closed, inside silvery light green, striate, easily torn.

■ **STEM** A **culm**, hollow, sharply triangular and scabrous below inflorescence; 2–4 mm thick near the base; **vegetative culms** scattered along the rhizomes.

557

Sedge—*Cyperaceae*　　　　　　　　　**Sedge**　Perigynia hairless, 4.5–7.3 mm long

SPRENGEL'S SEDGE　*Carex sprengelii* Dewey ex Spreng.

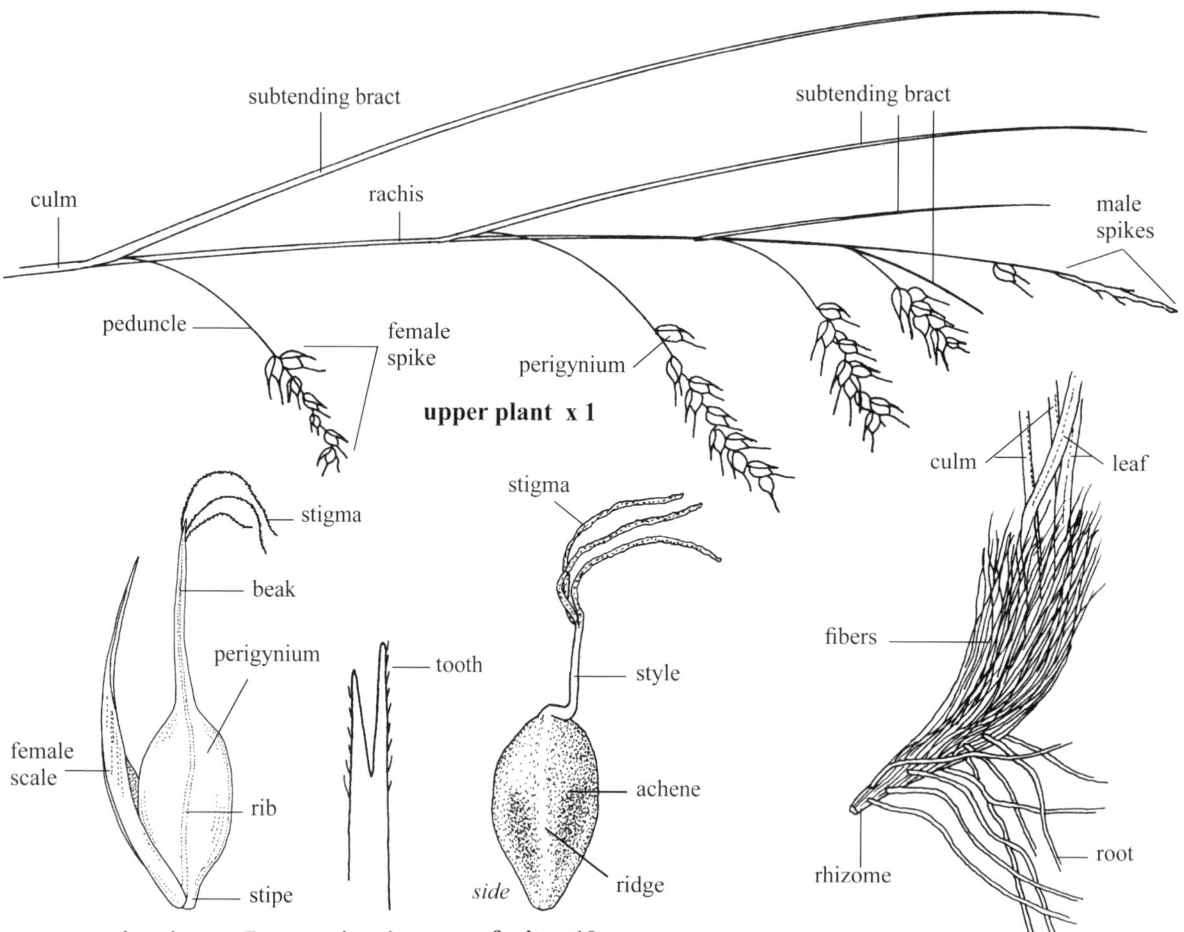

■ **SKETCH** A tufted **perennial sedge** 20–90 cm high from a **rhizome** 0.7–3 mm wide and to 15⁺ cm long, covered with fine, dark brown fibers 2–4 cm long from old leaves; in woodlands to moist semiopen prairies, sandy areas, thickets and moist canyons; **monoecious**. w2

■ **FLOWERS** Green, blooming May–June; **inflorescence** a spike, together 10–20 cm long; **male spikes** one to three, terminal, tan, together 1.5–3.8 cm long, each 1–2 cm long, disintegrating; **female spikes** two to five, 7- to 40-flowered, separate, each 1–3.8 cm long; **peduncles** 0.7–11 cm long by *c.* 0.2 mm thick, drooping to hanging; **subtending bracts** (of peduncles) one, leaflike, 1.5–20 cm long by 1–3 mm wide and scabrous, reduced above; **stigmas** 3; **perigynia** 4.5–7.3 mm long (including beak) by 1.3–2 mm wide, green to straw-colored, mostly round, glabrous and shiny, with two low ribs; **beak** 2.2–4.5 mm long, thin, with an apex slightly hairy and bidentate, the teeth not spreading; **female scales** as wide as, and shorter to as long as the perigynia, membranous, pale hyaline with a fine long scabrous point and green midvein. **FRUIT an achene**, 1-seeded, triangular, 2–3.2 mm long by 1.5–1.8 mm wide, glabrous, generally filling the upper body of the perigynium; **style** bent at the base and deciduous.

■ **LEAVES** Lime green, flat, four to eight per culm, scabrous; **blades** 6–35 cm long by 2–4 mm wide; **sheaths** closed, concave to flat at apices, whitish green, smooth.

■ **STEM** A **culm**, triangular with the edges blunt, rough above, leaning when in fruit; 2–3 mm wide near the brown base.

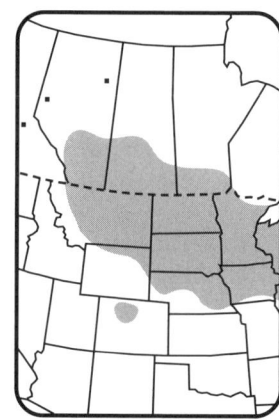

558　　Long-beaked Sedge

AWL-FRUITED SEDGE *Carex stipata* Muhl. ex Willd.

■ **SKETCH** A variable, tufted **perennial sedge** 10–120 cm tall from short **rhizomes** *c*. 5 mm thick with dark fibers and tan roots *c*. 1 mm thick; in swamps, bogs, thickets, moist woods and meadows of parklands; **monoecious**. w2

■ **FLOWERS** Yellowish green, blooming May–June; **inflorescence** a spike, 5–15, bisexual, branched at lower nodes, forming an elongated head 2–15 cm by 10–30 mm wide; **spikes** androgynous, 12–20 mm long by 10–13 mm wide, ascending, crowded above, more distant below; <u>male</u> <u>spikes</u> usually of one sessile floret 3–4 mm long, erect; **subtending bracts** (of spikes) 1–6.5 cm long by *c*. 1 mm wide, ascending to descending; **stigmas** 2; **perigynia** descending to ascending, plano-convex, 4–5.5 mm long by 1.6–2 mm wide by *c*. 1.3 mm thick, spongy at the flat tan base, shiny, with *c*. 21 reddish brown nerves in total, eight nerves below and the rest above on the convex dorsal side, 3–4 reaching into the beak; **beak** scabrous on margins, 2.5–3.5 mm long, sometimes dark brown, bidentate; **teeth** erect and *c*. 0.5 mm long; **female scales** shorter and not as wide as perigynia, 3–3.5 mm long by 1–1.2 mm wide (flattened), midvein green to tan, sides tan with dark markings, apices slightly darker brown, margins hyaline, sides revolute near apices. **FRUIT an achene**, 1-seeded, broadly lenticular, TL 1.5–2 mm by 1.2–1.7 mm wide by *c*. 0.9 mm thick, positioned below the beak on the thick spongy base, light brown, shiny; **beak** (style remains) light tan and cylindric.

■ **LEAVES** Two or three, yellowish green; **blades** 13–100 cm long by 3–15 mm wide, flat, usually not reaching to base of spikes, situated on lower third of stem, edges scabrous; **sheaths** of lower leaves often wrinkled at the V-shaped apices, thin, membranous, easily torn.

■ **STEM** A **culm**, erect, soft, leaning with fruit, sharply triangular throughout, sides concave above and slightly winged, sides slightly convex below; 3–5 mm wide near the fibrillose base.

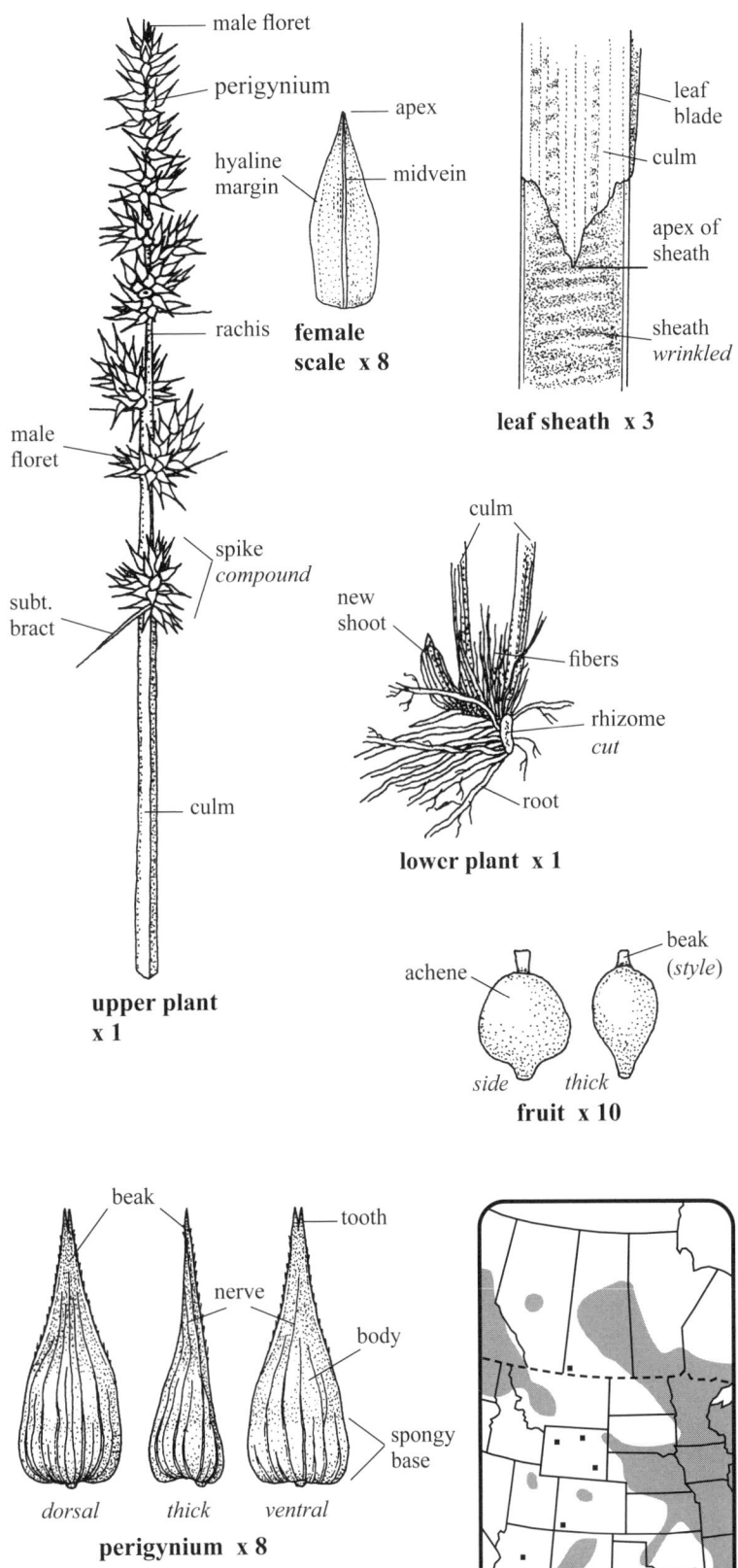

Sedge—*Cyperaceae* Sedge   Perigynia hairless, 3–4.5 mm long

QUILL SEDGE   *Carex tenera* Dewey

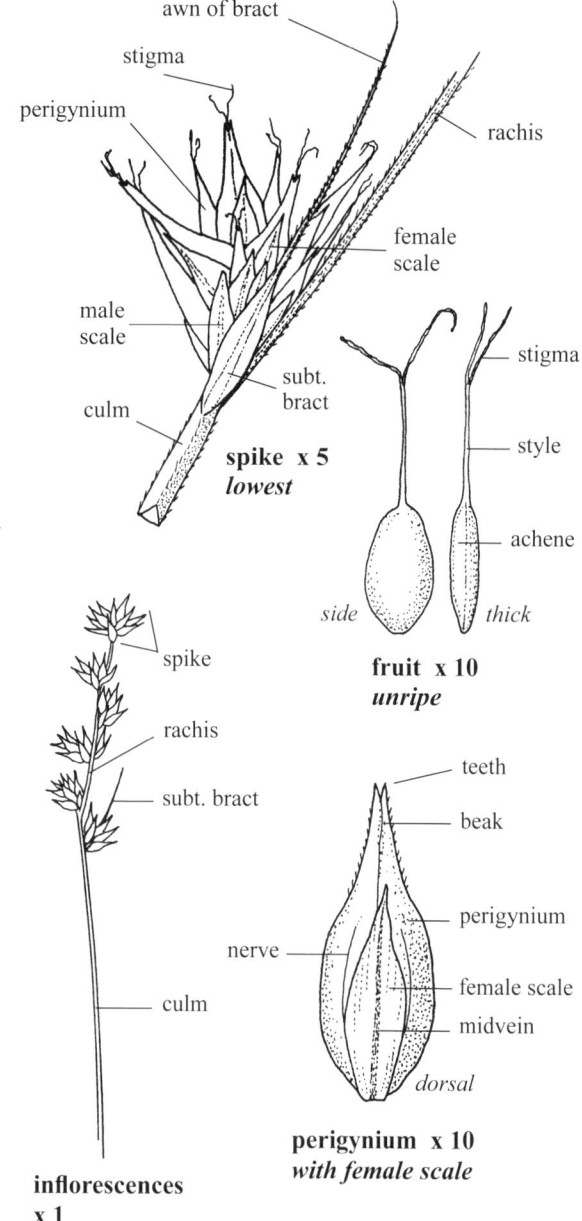

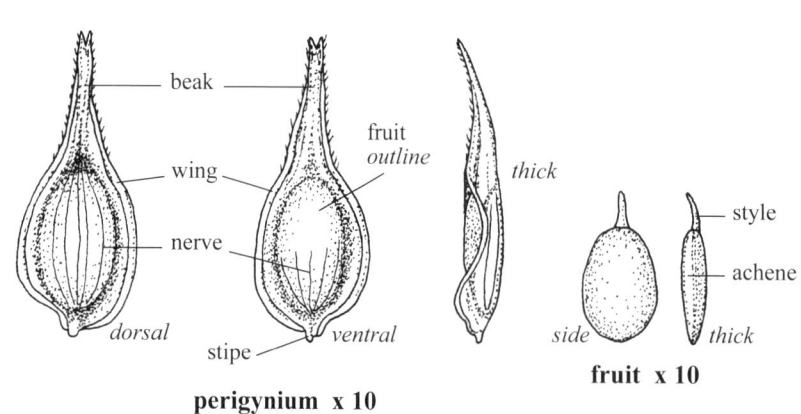

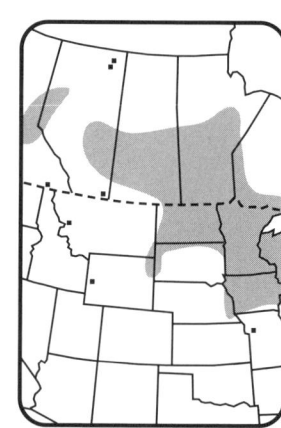

- **SKETCH** A tufted **perennial sedge** 30–90 cm tall from a short black **rootstock**; in openings of woodlands and in moist uplands; **monoecious**. w2
- **FLOWERS** Green, blooming May–June; **inflorescence** a spike, together 2–5 cm long by 6–9 mm wide, sometimes bent or nodding; **spikes** usually three to eight, gynaecandrous, sessile, 5–10 mm long by 4–5 mm wide, separate but close; **subtending bracts** (of spikes) scabrous, to *c.* 1 (–4) cm long and usually only from the lowest spike; **male flowers** two to four at base of spikes; **scales** 1.5–2.5 mm long; **female flowers** usually 3–14 per spike, mostly ascending, above male flowers; **stigmas** 2; **perigynia** glabrous, green to straw-colored, 3–4.5 mm long by 1.5–2 mm wide by *c.* 0.5 mm thick, 3- to 7-nerved dorsally, faintly 3- or 5-nerved ventrally, plano-convex, distinctly winged to the base; **beak** gradually tapered, bidentate, flat, with scabrous edges and about as long as the body; **female scales** pointed, 2.3–3.3 mm long, about half as wide as the perigynia, whitish hyaline with a green midvein. **FRUIT** an achene, 1-seeded, lenticular, 1.2–1.7 mm long by *c.* 1 mm wide by *c.* 0.3 mm thick, outline visible in perigynium; **style** straight and deciduous.
- **LEAVES** Three to six along the culm, not exceeding the flowers when upright; **blades** 10–35 cm long by 1–3 mm wide, flat, scabrous below, smooth above, horizontal to arching; **sheaths** closed, smooth to slightly wrinkled, apices U-shaped.
- **STEM** A **culm**, erect to almost horizontal in fruit, sharply triangular but with a slightly raised central ridge along each of the three sides giving it a rounded blunt feel, scabrous near the flowers, smooth below; *c.* 2 mm thick near the base.
- **NOTE** This species is in the complex and difficult section Ovales, the largest section of *Carex* in North America ("Flora of North America," www.efloras.org).

THIN-FLOWERED SEDGE *Carex tenuiflora* Wahl.

■ **SKETCH** A loosely tufted **perennial sedge** 10–60 cm tall or long with tan **rhizomes** 0.5–0.7 mm wide and covered with blunt, tan, veiny **bracts** 5–20 mm long which form filiform fibers; **fibrous roots** 1.5–21 cm long; in sphagnum bogs and wet woods; **monoecious. w2**

■ **FLOWERS** Green, blooming May–July; **inflorescence** a spike, 2–4, gynaecandrous, usually close together forming a silvery head 7–15 mm long by 6–9 mm wide in fruit; **subtending bracts** (of spikes) 2.5–5.5 mm long, reduced above, lower one usually awned, the **awn** tan or green, 1–3.8 (–15) mm long; **spikes** 4–9 mm long, mostly ascending to spreading with fruit; <u>male florets</u> 2–4 at base of each spike; **male scales** white hyaline, 2–3 mm long, pointed, midnerve green to brown; <u>female florets</u> above male ones, imbricate; **stigmas** 2; **perigynia** dull, hairless, 3–15 per spike, green, 3–3.5 mm long by 1.5–1.7 mm wide by 0.8–0.9 mm thick, flatter on ventral side, walls thickish (not transparent), slightly spongy at narrow base; 8 or 9 red nerves dorsally on green convex side, darker green below the beak, some nerves incomplete and not reaching the beak, 8 or 9 red nerves on ventral side, two side ribs obvious near beak; **beak** white, 0–0.2 mm long, flat to slightly notched; **female scales** persistent after perigynia fall, 2.8–3.5 mm long by 1.5–1.7 mm wide (flat), midvein with narrow green margins, most of scale white hyaline. **FRUIT an achene**, 1-seeded, lenticular, brown, slightly shiny, 1.5–2 mm long (including beak) by 1.1–1.3 mm wide by 0.6–0.7 mm thick, on spongy stipe; **beak** straight, *c.* 0.1 mm long or absent.

■ **LEAVES** Four or five situated near the base; **blades** flat, 8–22 cm long by 1–2 mm wide, brown below, top leaf usually green; **sheaths** tight, closed, V-shaped at apices.

■ **STEM** A **culm**, sharply triangular and scabrous on corner ridges above, smooth below, faces flat, erect at first, then usually horizontal with fruit; **base** *c.* 2 mm wide and covered with tan leaf sheaths, *c.* 0.5 mm wide with sheaths removed.

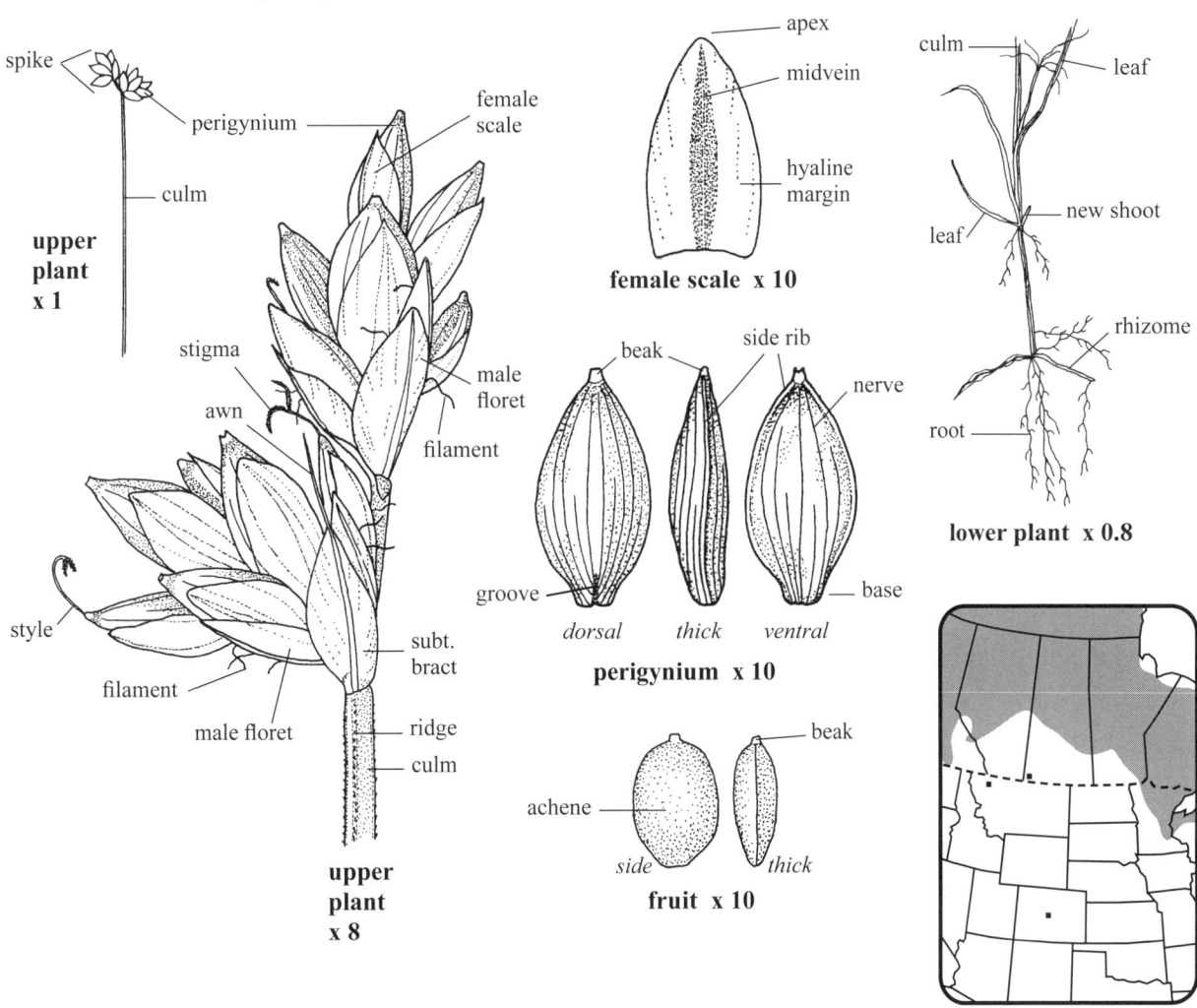

Sedge—*Cyperaceae* Sedge Perigynia hairless, 3.4–8.5 mm long

NORTHERN BEAKED SEDGE *Carex utriculata* Boott

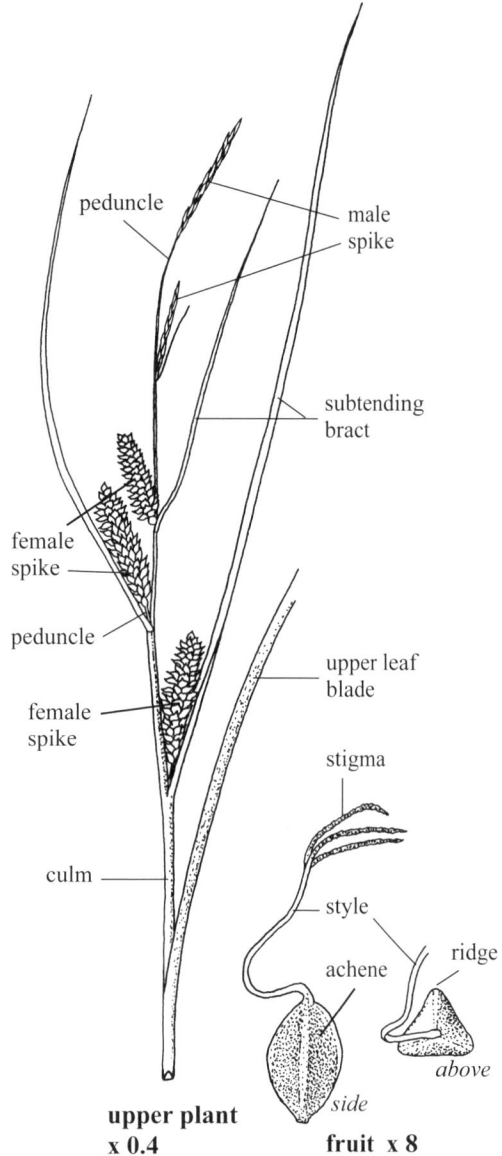

upper plant x 0.4
fruit x 8

■ **SKETCH** A tufted, glabrous **perennial sedge** 30–100 cm tall from woody **rhizomes** 2–4 mm thick and to *c.* 15⁺ cm long, forming colonies several m² in bogs, lakeshores, sloughs, marshes, fens and riverbanks; **monoecious**. w2

■ **FLOWERS** Green, blooming May–June; **inflorescence** a spike, unisexual, together 11–50 cm long; **male spikes** terminal, two to five, each 1.7–5 cm long by 1.5–3 mm wide (post anthesis), the lowest one remote; **subtending bracts** filiform, erect, 12–70 mm long, shorter to longer than the inflorescence, reduced above; **peduncles** 0–4 cm long, glabrous; **female spikes** two to five, below the male spikes, lateral, erect, 2.5–10 cm long by 8–15 mm wide; **subtending bracts** (of female spikes) one, leaflike, lower bracts may extend past the terminal male spike, ascending, scabrous, usually 2–4 mm wide, flat to slightly V-shaped; **peduncles** 0–10 mm long, erect; **stigmas** 3; **perigynia** inflated but not spongy at base, glabrous, tan to straw-colored, some purple tinged, 3.4–8.5 mm long by 1.7–3 mm wide, spreading to ascending in 8–10 columns, numerous, two lateral ribs are a continuation of the beak margins, roundish on back, 8- to 15-nerved; **beak** bidentate, smooth, 1–2.7 mm long, teeth usually not spreading but often bent and 0.3–1.2 mm long; **female scales** shorter to slightly longer than perigynia, tapered to a fine point or rarely awned, point sometimes scabrous or not, 1- or 3-veined, 2.6–7.5 mm long by 1–2 mm wide, not hiding the perigynium, tip dark brown and often involute, base tan. **FRUIT an achene**, 1-seeded, tan, triangular, 1.8–2 mm long by *c.* 1.2 mm wide, ridges rounded, not a tight fit at the base of the perigynium; **style** persistent, quickly bent above the achene.

■ **LEAVES** Three to six per culm, ascending to arched, yellowish green, flat to broadly V-shaped, scabrous or not, upper ones usually reaching past the top spike; **blades** 30–60 cm long by 3–15 mm wide, tapered to a fine point; **sheaths** not laddering, pale green, cross-veined on the inside, these visible when fresh.

■ **STEM** A **culm**, sharply triangular below the inflorescence to bluntly triangular near the base; 3–7 mm thick near the soft base.

■ **SYN.** *Carex rostrata* var. *utriculata* (Boott) Bailey.

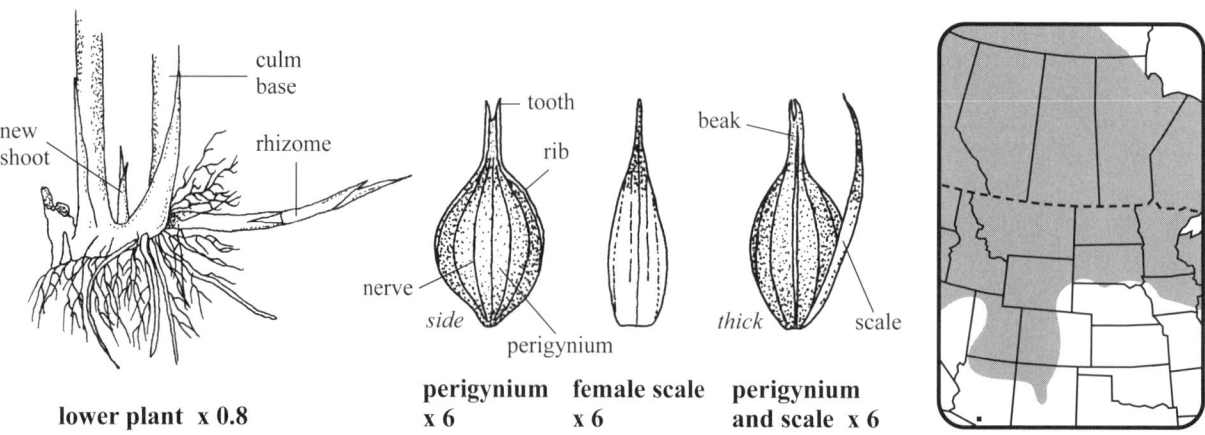

lower plant x 0.8
perigynium x 6
female scale x 6
perigynium and scale x 6

562 Northwest Territory Sedge

Sedge—*Cyperaceae* — Sedge Perigynia hairless, 2–4.2 mm long

GREEN SEDGE *Carex viridula* Michx.

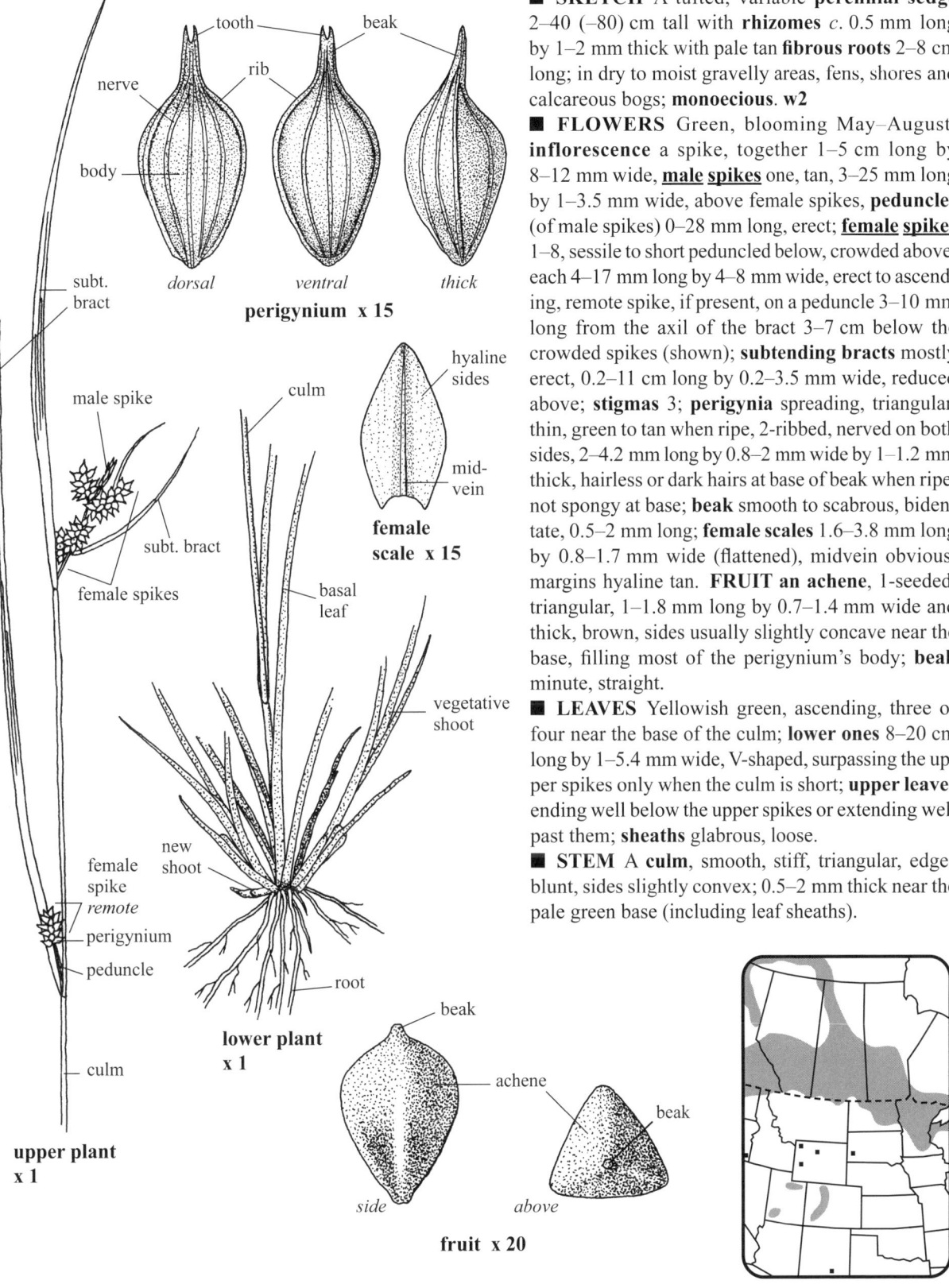

- **SKETCH** A tufted, variable **perennial sedge** 2–40 (–80) cm tall with **rhizomes** *c*. 0.5 mm long by 1–2 mm thick with pale tan **fibrous roots** 2–8 cm long; in dry to moist gravelly areas, fens, shores and calcareous bogs; **monoecious**. **w2**
- **FLOWERS** Green, blooming May–August; **inflorescence** a spike, together 1–5 cm long by 8–12 mm wide, **male spikes** one, tan, 3–25 mm long by 1–3.5 mm wide, above female spikes, **peduncles** (of male spikes) 0–28 mm long, erect; **female spikes** 1–8, sessile to short peduncled below, crowded above, each 4–17 mm long by 4–8 mm wide, erect to ascending, remote spike, if present, on a peduncle 3–10 mm long from the axil of the bract 3–7 cm below the crowded spikes (shown); **subtending bracts** mostly erect, 0.2–11 cm long by 0.2–3.5 mm wide, reduced above; **stigmas** 3; **perigynia** spreading, triangular, thin, green to tan when ripe, 2-ribbed, nerved on both sides, 2–4.2 mm long by 0.8–2 mm wide by 1–1.2 mm thick, hairless or dark hairs at base of beak when ripe, not spongy at base; **beak** smooth to scabrous, bidentate, 0.5–2 mm long; **female scales** 1.6–3.8 mm long by 0.8–1.7 mm wide (flattened), midvein obvious, margins hyaline tan. **FRUIT an achene**, 1-seeded, triangular, 1–1.8 mm long by 0.7–1.4 mm wide and thick, brown, sides usually slightly concave near the base, filling most of the perigynium's body; **beak** minute, straight.
- **LEAVES** Yellowish green, ascending, three or four near the base of the culm; **lower ones** 8–20 cm long by 1–5.4 mm wide, V-shaped, surpassing the upper spikes only when the culm is short; **upper leaves** ending well below the upper spikes or extending well past them; **sheaths** glabrous, loose.
- **STEM** A **culm**, smooth, stiff, triangular, edges blunt, sides slightly convex; 0.5–2 mm thick near the pale green base (including leaf sheaths).

Little Green Sedge

WHITE-SCALED SEDGE *Carex xerantica* Bailey

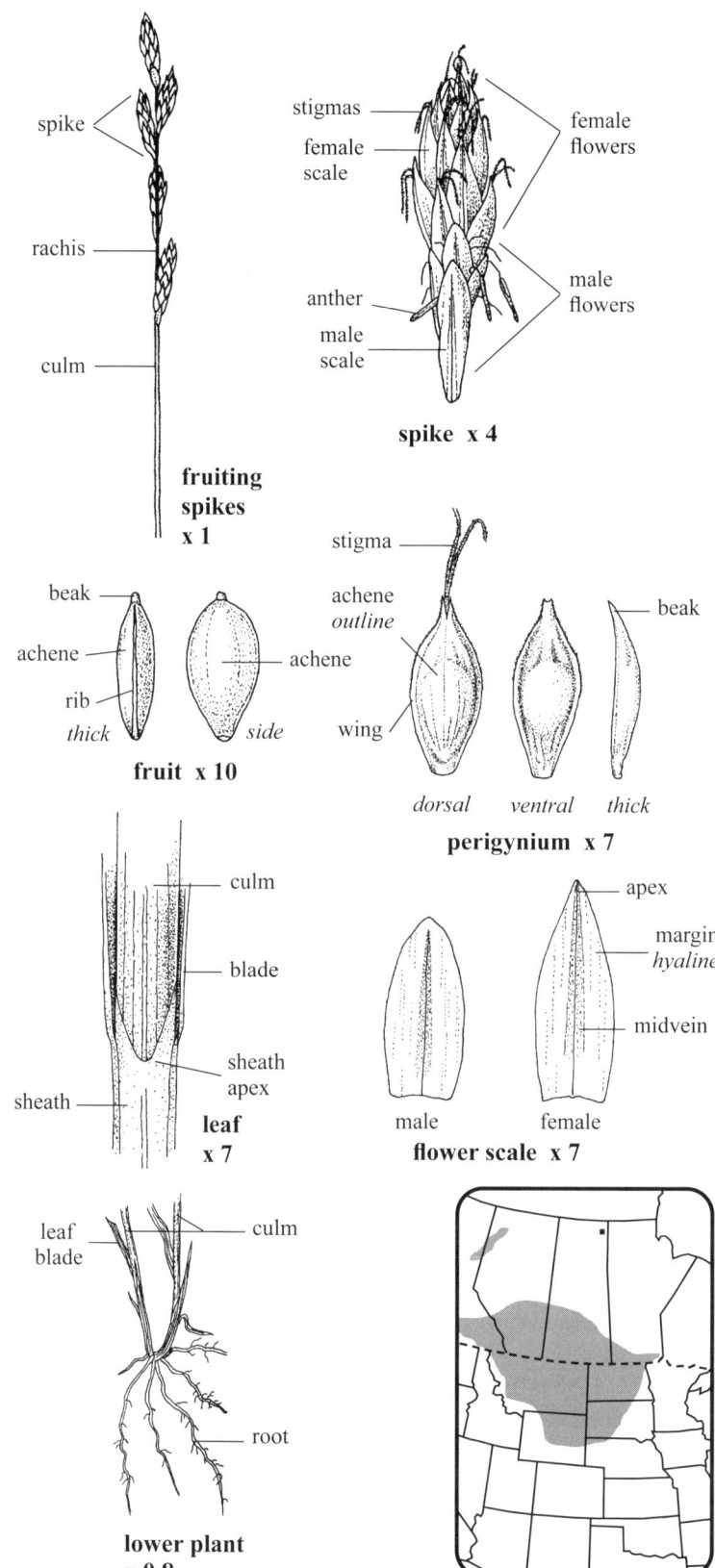

■ **SKETCH** A tufted **perennial sedge** 15–70 cm tall from a short thick **rhizome**; in open woods, rocky outcrops, talus slopes, meadows and prairies; **monoecious**. w2, p1

■ **FLOWERS** Green to brown, blooming May–August; **inflorescence** a spike, 3–6, terminal, stiff, together 2.5–5 cm long by 5–8 mm wide; **spikes** bisexual, gynaecandrous, sessile, 8–14 mm long; **male flowers** few to several at the narrow base; **male scales**, when fruit is ripe, pale whitish brown, body hyaline, midvein pale tan with a thin slightly darker ridge not extending to the narrow to broadly pointed apices; **stamens** 3; **anthers** 1.3–2 mm long; **female flowers** above male ones and form the bulk of each spike; **stigmas** 2; **perigynia** glabrous, whitish tan to medium tan across the lower half, 3.4–6 mm long by 1.4–2.6 mm wide by 0.4–0.6 mm thick, the base narrow and flat, several faint nerves from base on dorsal (convex) side, the ventral or concave side nerveless or with a few short nerves at the base below the achene's outline, narrowly winged to the base, upper margins slightly serrate to scabrous; **beak** slightly bidentate; **female scales** whitish to yellowish, 4–6 mm long by 1.5–2.4 mm wide, covering the perigynia, midvein extending from apex to near base. **FRUIT an achene**, 1-seeded, brown, lenticular, 1.7–3 mm long by 1–1.6 mm wide by 0.5–0.6 mm thick.

■ **LEAVES** Usually three to five per culm, ascending, not reaching the inflorescence; **blades** 2–20 cm long by 1–3 mm wide, reduced below, flat to involute on drying, tapered to a fine point, slightly scabrous near the apices, with a wrinkled appearance on the inside (ventral side) when fruit are ripe, two or three deep furrows on the inside; **ligules** unremarkable; **sheaths** tight, closed, V-shaped at sharp or blunt apices, not laddering, slightly keeled.

■ **STEM** A **culm**, green, smooth, erect, bluntly triangular, the sides slightly convex; 1–2 mm thick near the almost round base.

■ **NOTE** Resembles *Carex petasata*, which has larger perigynia that are distinctly veined on the ventral side ("Flora of North America," www.efloras.org).

AWNED NUT-GRASS *Cyperus squarrosus* L.

■ **SKETCH** A tufted **annual herb** 2–20 cm tall with filiform, reddish tan **fibrous roots** 2–6 cm long by *c.* 0.3 mm thick; along slough margins, in depressions of rocky outcrops, shorelines, seeps and sandy soils of prairies and parklands.

■ **FLOWERS** Pale green, blooming July–September; **inflorescence** a spike of sessile spikelets; **spikes** 1–4 (–6) per plant, TL 2–6 cm, each 6–20 mm long by 6–21 mm wide, usually one spike sessile, the others on ascending peduncles (rays), occasionally one ray and spike descending; **peduncles** (rays) 4–40 mm long, triangular, mostly ascending; **involucral bracts** (of spikes) 1.2–15 cm long by 0.5–3 mm wide across one side (not flattened), ascending to spreading below, main one erect, V-shaped, yellowish green; **spikelets** flat, pale green, ascending, ripening to reddish brown with a tan awn, sessile, 6–40 per spike, each 3–20 mm long by 2.5–3.3 mm wide by *c.* 1 mm thick; **rachilla** wingless, 2–7 mm long, hairless, dark reddish brown with fruit; **scales** 1.5–2.5 mm long at anthesis by *c.* 1 mm wide (flattened), nerves obscure then, 4- or 5-nerved on each side, these reddish brown and easily seen when fruit is ripe and scales brown, 5–34 per spikelet, 2-ranked, body V-shaped, transparent, fruit partially visible inside; **awn** 0.6–1.5 mm long, smooth, curved; **flowers** perfect, one per scale; **perianth** absent; **stamens** one per flower (scale), 1–1.5 mm long at anthesis, included to eventually exserted; **filaments** hyaline, filiform; **anthers** yellow, *c.* 0.5 mm long; **pistil** one, green, glabrous, 1.5–2 mm long; **ovary** 0.3–0.4 mm long; **style** one, exserted, hyaline, *c.* 1 mm long, turning brown; **stigmas** 3. **FRUIT** an achene, 1-seeded, triangular, dark brown, slightly shiny, 0.7–1.1 mm long by 0.4–0.5 mm wide and thick, at base of scale, smooth, the sides slightly convex to sometimes concave, ribs three, papillose.

■ **LEAVES** One to four near base of culm; **blades** 2–15 cm long by 1–3 mm wide (flat), one often reaching to the top spike, upper leaf the longest, V-shaped, clasping the culm; **sheaths** loose, apices truncate to slightly concave.

■ **STEM** A **culm**, sharply triangular, sides slightly concave, edges smooth, ascending to spreading, green turning yellowish as fruit ripens; 1.5–2.5 mm wide near the reddish brown base (including leaf sheaths).

■ **SYN.** *Cyperus aristatus* Rottb. and *Cyperus inflexus* Muhl.

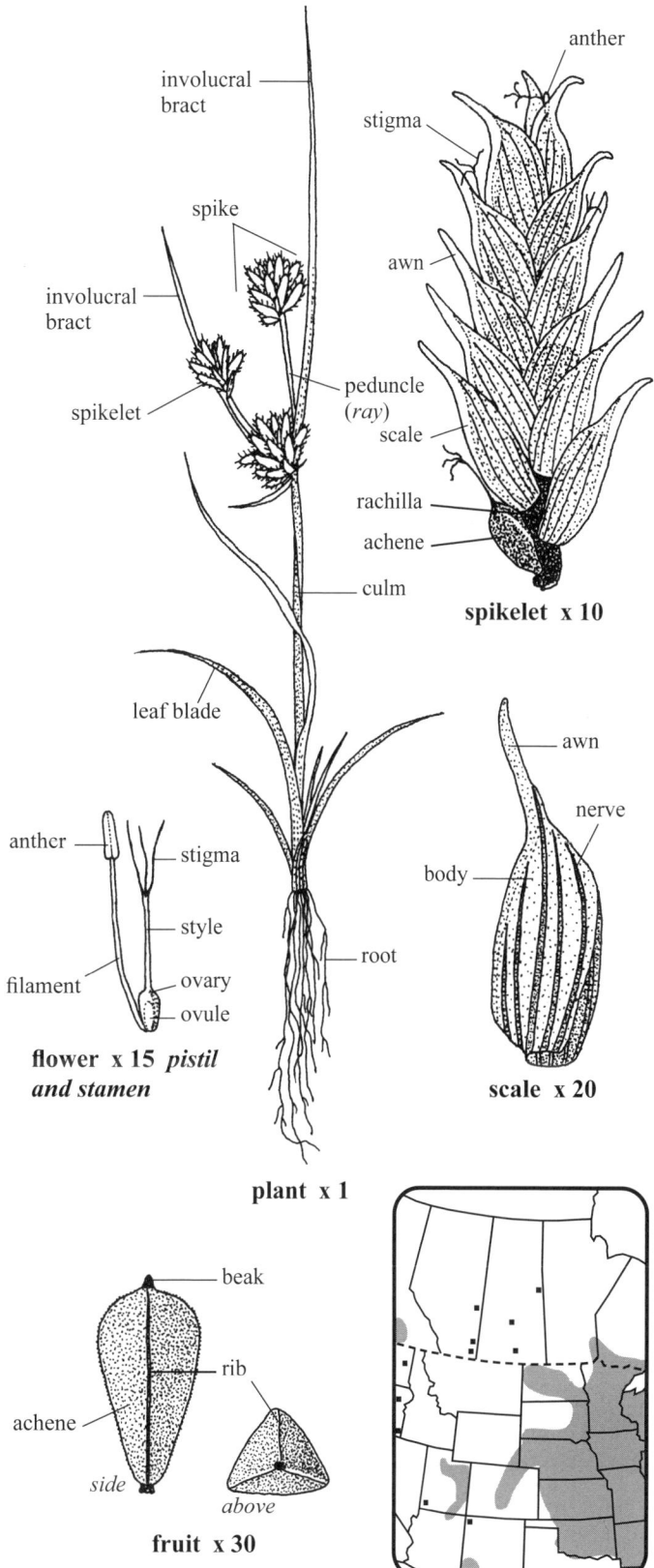

Nut-grass Scale 1.5–2.5 mm long, awned

Needle Spike-rush *Eleocharis acicularis* (L.) Roemer & J.A. Schultes

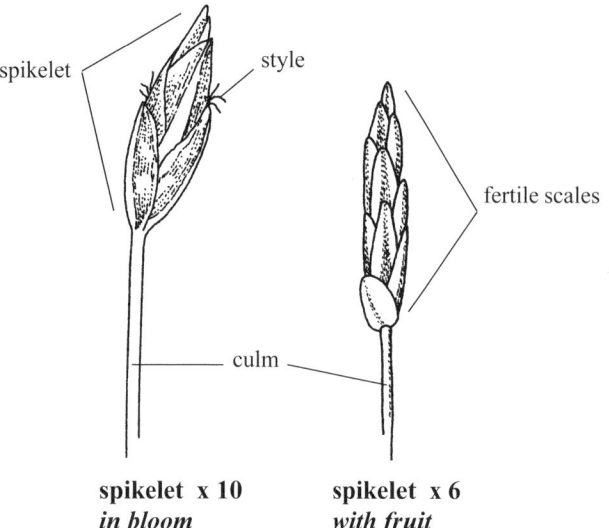

spikelet × 10
in bloom

spikelet × 6
with fruit

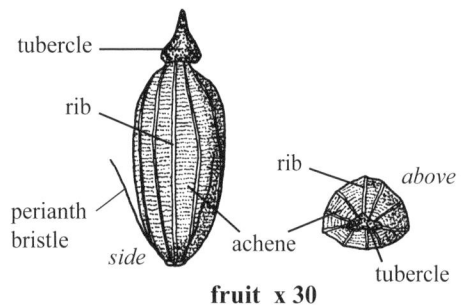

fruit × 30

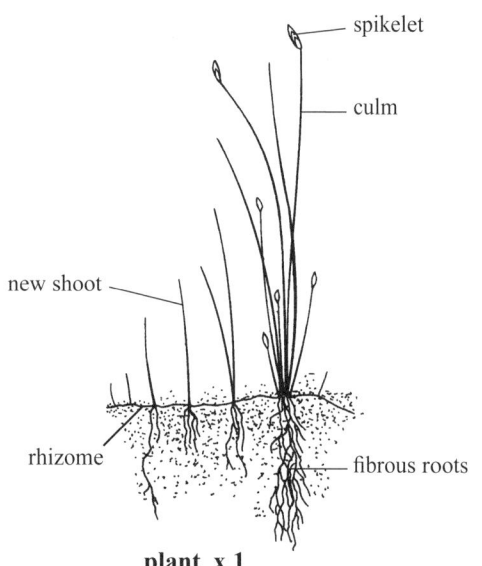

plant × 1

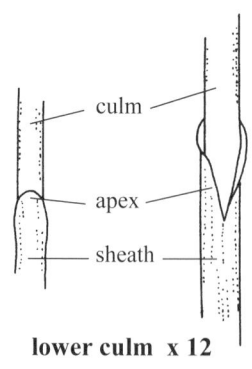

lower culm × 12

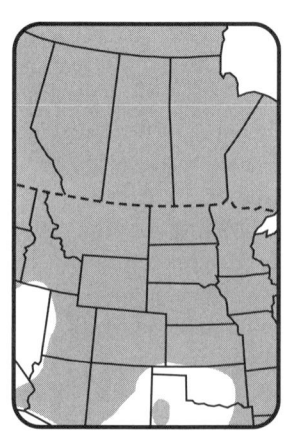

- **SKETCH** A variable, tufted **perennial herb** 2–20 (–60) cm tall from **rhizomes** 0.2–1.5 mm thick and to *c.* 12⁺ cm long, with **fibrous roots**; forming low green turf often covering several m²; in shallow water, along muddy shores, in dried ponds, roadside ditches, boggy meadows in mountains and marsh edges.

- **FLOWERS** Brown and green, blooming May–August; **inflorescence** a single spikelet, terminal, 2–8 mm long by 1–2 mm wide, and 3- to 15-flowered; **scales** all fertile, 4–25, each 1.2–2.5 (–3.5) mm long by 0.5–1 mm wide (flattened), the tip not sharply pointed, in fruit the scale midvein is green and wide, the edges white hyaline or streaked reddish brown; **flowers** perfect, one per scale; **stamens** three; **anthers** 0.7–1.4 mm long; **style** 3-parted. **FRUIT an achene**, 1-seeded, white to light gray, with 9–15 vertical ribs and numerous transverse lines between the ribs, one achene under and completely covered by each scale; **achene's body** 0.5–0.9 mm long by 0.4–0.6 mm wide by *c.* 0.3 mm thick, angled; **tubercles** (persistent style bases) pointed, dark green, 0.1–0.25 mm long by *c.* 0.2 mm wide, constricted at base; **perianth bristles** two to four or absent, as long as to usually shorter than the achene.

- **LEAVES** Reduced to a thin, basal, bladeless **sheath** usually 3–12 mm long, loose, reddish, the tip hyaline, rounded to pointed and membranous.

- **STEM** A **culm**, soft, green, simple, erect, round to angled and ridged, some compressed, glabrous; 0.2–0.6 mm thick near the base.

CREEPING SPIKE-RUSH *Eleocharis palustris* (L.) Roemer & J.A. Schultes

■ **SKETCH** A tufted variable **perennial herb** 20–115 cm high from a reddish, scaly **rhizome** 1–4 mm thick and 8–40+ mm long with **roots** to *c.* 15 cm long; in shallow, quiet water along shores of marshes and ponds, in wet meadows, mudflats and ditches, often covering several m^2.

■ **FLOWERS** Brown and green, blooming May–August; **inflorescence** a spikelet 5–22 mm long by 1.8–7 mm wide, terminal; **flowers** perfect, one per fertile scale; **sterile scales** two or three at the base of the spikelet but not encircling it, short; **fertile scales** imbricate, numerous, 2.3–5 mm long by 1.5–2.5 mm wide, completely covering the flower, tapered to a point, with two purple bands and a green midvein, turning tan in the autumn, the margins widely hyaline; **stamens** three; **style** (3-)2-parted. **FRUIT an achene**, 1-seeded, hard, shiny, golden to dark brown, slightly pitted, 1.1–2 mm long (not including tubercle) by 1–1.5 mm wide by 0.7–0.9 mm thick, lens-shaped; **tubercle** (enlarged base of persistent style) light tan, 0.3–0.7 mm long and wide, flattened, and slightly rough, persistent, the tip blunt; **perianth bristles** three to six, or absent, light tan, shorter to longer than the achene, with downward pointing hairs.

■ **LEAVES** Reduced to a bladeless **basal sheath** 1.5–17 cm long, tight fitting, closed, veins faint, apices pointed to flat and dark reddish brown.

■ **STEM** A **culm**, erect, green, simple, glabrous, round to oval, firm to soft, filled with several large air chambers; 1–5 mm thick at the reddish base.

■ **SYN.** *Eleocharis smallii* Britt.

■ **NOTE** *Eleocharis palustris* is the most widespread and common species of the extremely difficult circumboreal "*E. palustris* complex." The complex includes several species, including *E. macrostachya* ("Flora of North America," www.efloras.org).

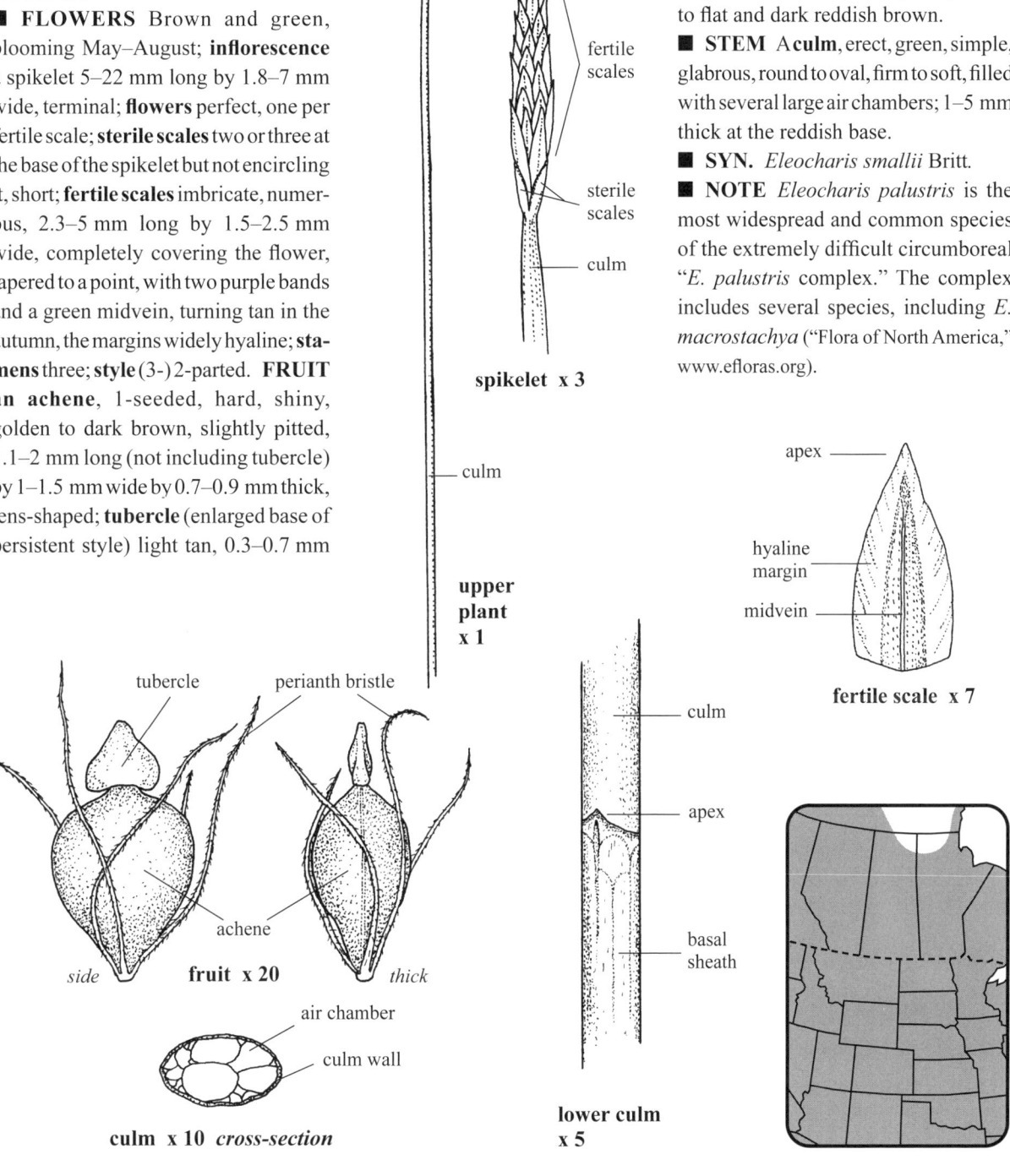

Sedge—*Cyperaceae* **Sedge** Spikelets 2–5, each 5–10 mm long

SLENDER COTTON-GRASS *Eriophorum gracile* W.D.J. Koch

■ **SKETCH** A **perennial sedge** 20–60 cm tall with a dark brown **rhizome** several cm long by *c.* 2 mm thick covered with blunt **scales** *c.* 1 cm long; **fibrous roots** thin, *c.* 6 cm long; in ditches, swamps, bogs, muskeg and wet meadows.

■ **FLOWERS** Brown, blooming May–June; **inflorescence** an umbel-like cyme of 2–5 spikelets, each 5–10 mm long (not including bristles); **peduncles** 5–30 mm long, tan to green, thicker at apices, with microscopic ascending white hairs; **subtending bracts** (of peduncles) two: **1)** outer one leaflike, 6–12 mm long, tan, tip tapered and blunt, and **2)** an inner loose bract, enveloping the base of peduncle, open, 4–5 mm long with a narrow blunt beak *c.* 0.8 mm long or truncate; **involucral bracts** one, erect, tan, channeled, apices narrowed and blunt, 0.5–2 cm long, shorter or longer than inflorescence; **rachilla** green, with deep sockets; **fertile scales** 3–4.2 mm long by 1–2.4 mm wide, 1–3 veined, midvein tan and narrow, ending below the scarious, blunt white apices, upper third of body and upper margins dark gray, tan and paler near the base; **anthers** *c.* 2 mm long; **pistil** one under each scale; **style** one, *c.* 1.5 mm long; **stigmas** 3, *c.* 2 mm long. **FRUIT an achene**, 1-seeded, beaked, 2.5–3.5 mm long by 0.8–1 mm wide by *c.* 0.5 mm thick, grayish brown, shiny, triangular, sides slightly convex, hairless, 3-ribbed, attached to rachilla at beak; **beak** tan, 0.3–0.5 mm long; **perianth bristles** 1.5–2 cm long, tan by achene, whitish at tips, thin, flattened and smooth.

■ **LEAVES** Basal and stem, entire, simple; **basal blades** four or five, crowded, 4–30 cm long by 1–2 mm wide, edges sharp, one side concave, the other two more or less flat, forming a short triangular tip; **stem leaves** usually three or four, sharply triangular, channeled below, becoming flatter but still triangular near the blunt apices; **top blade** well below the inflorescence, the longest at 1–4 cm; **sheaths** longer than blades, tight, 3–5.5 cm long, closed, apices brown, hyaline and V-shaped.

■ **STEM** A **culm**, erect to leaning, weak, bluntly triangular above to roundish below, hollow, channeled above; 1–2 mm thick at base.

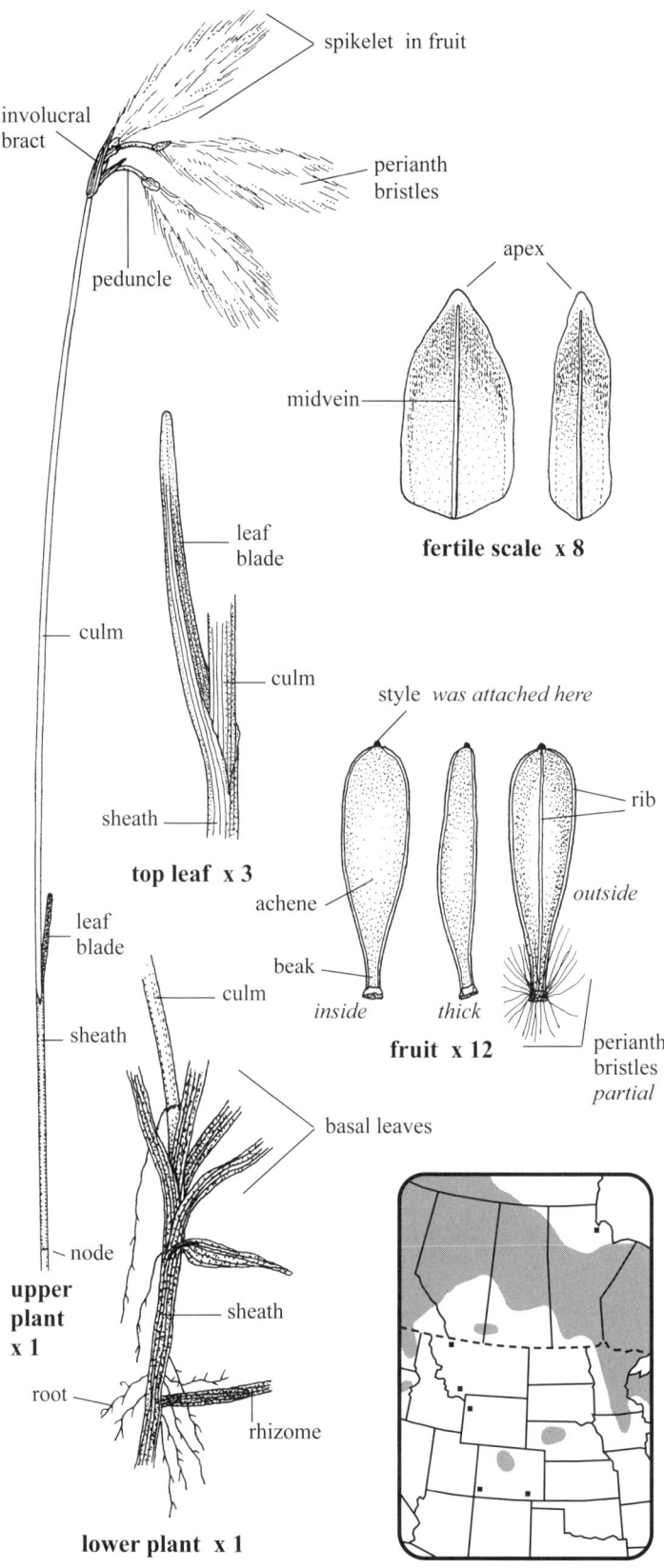

568

VISCID GREAT BULRUSH *Schoenoplectus acutus* (Muhl. ex Bigelow) A. & D. Löve

■ **SKETCH** A **perennial herb** 0.3–4 m tall with **rhizomes** 5–15 mm thick, light brown to brownish black, long; along borders of sloughs, ponds, lakes and in roadside ditches, often in rows from rhizomes. **w2**

■ **FLOWERS** Brown, blooming May–July; **inflorescence** a panicle, terminal, usually 2–5 cm long by 1.5–5 cm wide; **involucre bracts** one (2 or 3), erect, tan, terminal on culm, 1–8 cm long by 2–5 mm wide, shorter than to as tall as the inflorescence; **peduncles** (rays) green, 1–80 mm long, flat on one side, rounded on the other side, erect to spreading; **spikelets** 3–190, each 20- to 40-flowered, 5–24 mm long by 2.5–4 mm wide, in sessile clusters (glomerules) of mostly 2–10 (1–15) at the apices of peduncles; **flowers** perfect; **fertile scales** imbricate, 3–8 mm long by 2–3 mm wide, covering the fruit, with reddish brown striations, slightly sticky above, apices pointed or notched, hyaline margins hairy, midvein green and extending as a dark brown, often curved awn 0.5–1 mm long; **stamens** three, 2.6–3 mm long, yellow, exserted after the female parts; **filaments** white, oval, *c.* 3 mm long at anthesis, often persisting from the base of the achenes; **anthers** *c.* 2 mm long, apices hyaline and ragged to short hairy; **styles** (3-) 2-parted. **FRUIT an achene**, 1-seeded, plano-convex to lenticular to triangular, one per fertile scale, 1.8–3 mm long by 1.3–2 mm wide, brown to a lustrous black, beak 0.2–0.4 mm long; **perianth bristles** six (4–8), shorter than or as long as achene, hairs descending and base not hairy.

■ **LEAVES** Alternate, simple, entire, two to five at base of culm, sheaths only, or **blades** 2–23 cm long by 5–10 mm wide, ascending, pointed, darker green on outside (dorsal) and whitish on inner (ventral) side; **sheaths** closed, tight, sometimes laddering, whitish, apices slightly concave or V-shaped.

■ **STEM** A **culm**, dark green, smooth, erect, firm, oval below the inflorescence, dull, pith white; round near the 5–20 mm thick base.

■ **SYN.** *Scirpus acutus* Muhl. ex Bigelow = *Schoenoplectus acutus* var. *acutus* (Muhl. ex Bigelow) A. & D. Löve.

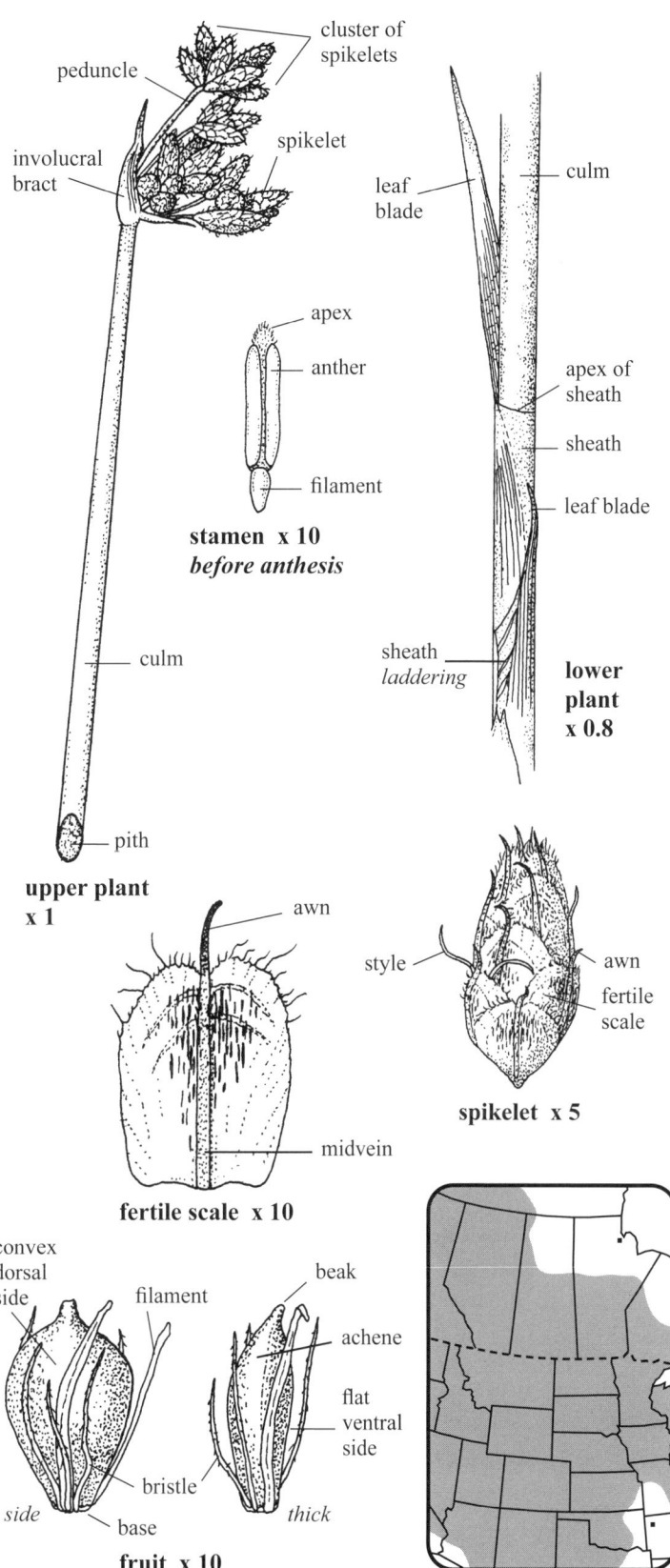

River Bulrush *Schoenoplectus fluviatilis* (Torr.) M.T. Strong

- **SKETCH** A glabrous **perennial herb** 50–200 cm tall with **rhizomes** 2–5 mm thick and **cormlike** thickenings 13–23 mm wide at the base of the stem below ground; in shallow freshwater marshes, along shores of rivers, ponds and lakes. **w1**
- **FLOWERS** Brown, blooming June–July; **inflorescence** umbel-like, 3.5–15 cm wide by 4–8 cm tall; **peduncles** flattened, several, 0.5–10 cm long, spreading to ascending; **involucral bracts** three to five, the three main ones V-shaped at their bases, yellowish green with brown spots or streaks; **upper bract** one, erect, 20–30 cm long by 4–15 mm wide, flat, slightly scabrous on margins, smooth on side facing flowers; **lateral involucral bracts** two, 9–22 cm long by 4–8 mm wide, ascending to descending; **other bracts** shorter, subtending each peduncle; **spikelets** 10–40, peduncled to sessile, 10–30 mm long by 4–10 mm wide, pointed, 1–4 (–8) at the ends of peduncles; **flowers** perfect; **fertile scales** imbricate, body 8–10 mm long by 3–6 mm wide (flattened), apices bidentate to ragged, awn 2–3.5 mm long, scabrous, hairs ascending; **stamens** two or three, exserted; **filaments** white, filiform; **anthers** pale yellow, *c.* 4 mm long; **style** 3-parted, filiform, 5–6 mm long, exserted. **FRUIT an achene**, 1-seeded, triangular, dark brown, shiny, tapered, 4–5.5 mm long by 2–2.9 mm wide; **perianth bristles** six (five), persistent, shorter to longer than the achene, hairs reflexed; **beak** tapered and to *c.* 0.8 mm long.
- **LEAVES** Alternate, simple, entire, four or five per culm, ascending to arched; **blades** 16–50 cm long by 3–20 mm wide, tapered to a fine point, usually glabrous, midrib recessed above, raised below; **sheaths** closed, yellowish green near the base, apices truncate to slightly concave with age and becoming brownish, lower sheaths obviously cross-veined; **nodes** a thin reddish brown line.
- **STEM** A **culm**, solid, erect, glabrous, solitary or several together, sharply triangular, the sides concave below the inflorescence, becoming flat to slightly convex below; 6–15 mm thick on one side near the base.
- **SYN.** *Bolboschoenus fluviatilis* (Torr.) Soják and *Scirpus fluviatilis* (Torr.) Gray.

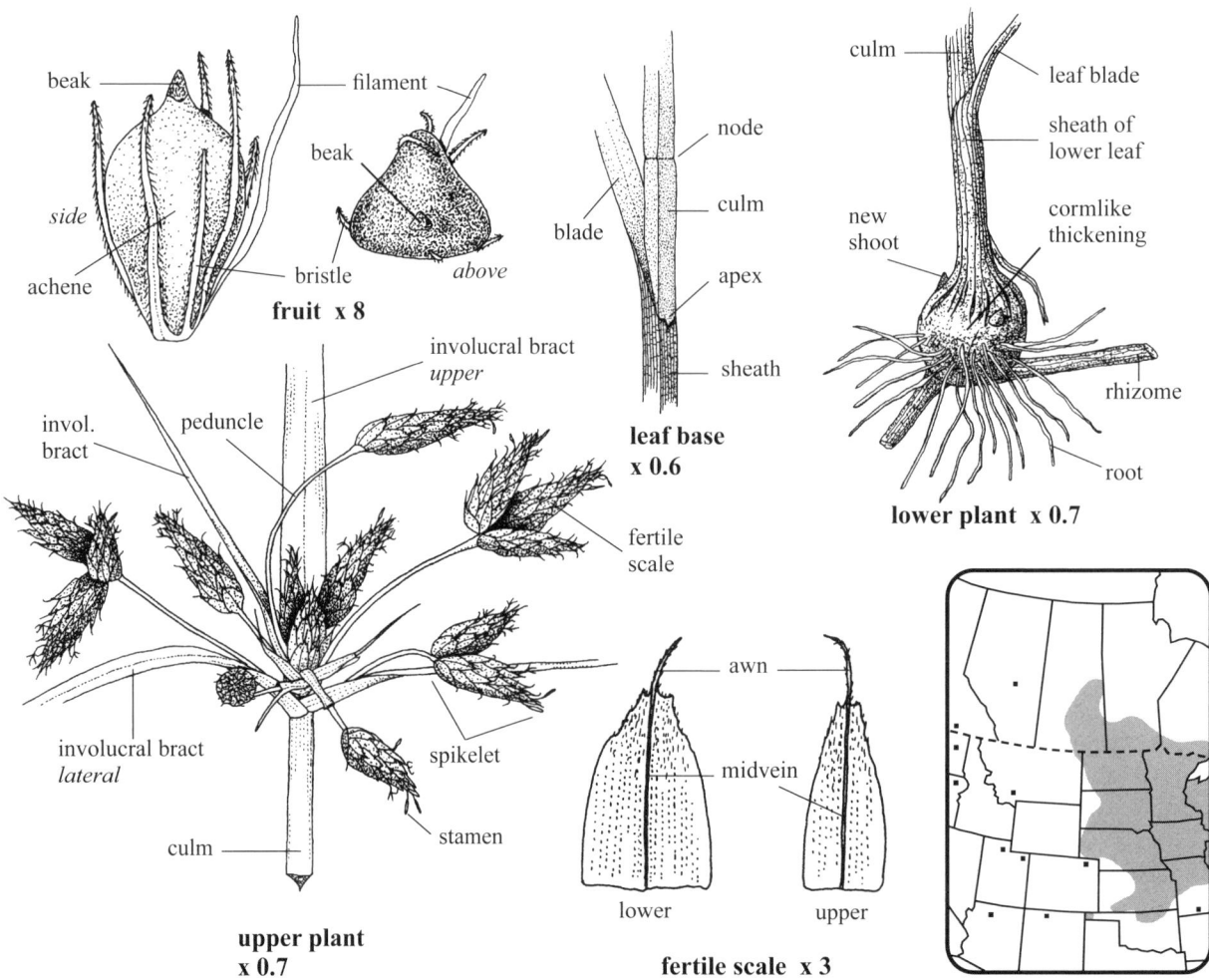

Sedge—*Cyperaceae* **Sedge** Spikelets 7–25 mm long

Prairie Bulrush *Schoenoplectus maritimus* (L.) Lye

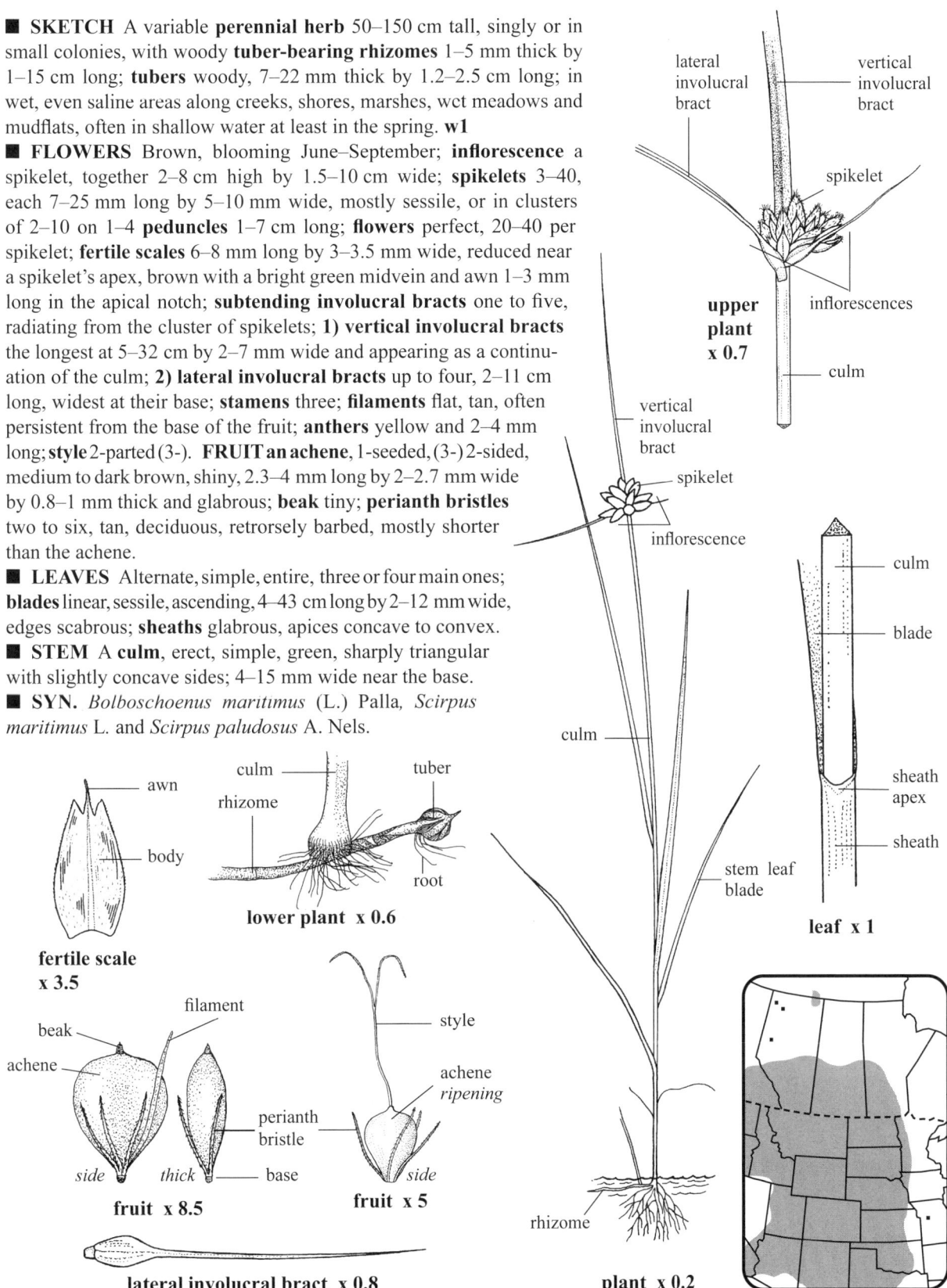

- **SKETCH** A variable **perennial herb** 50–150 cm tall, singly or in small colonies, with woody **tuber-bearing rhizomes** 1–5 mm thick by 1–15 cm long; **tubers** woody, 7–22 mm thick by 1.2–2.5 cm long; in wet, even saline areas along creeks, shores, marshes, wet meadows and mudflats, often in shallow water at least in the spring. **w1**
- **FLOWERS** Brown, blooming June–September; **inflorescence** a spikelet, together 2–8 cm high by 1.5–10 cm wide; **spikelets** 3–40, each 7–25 mm long by 5–10 mm wide, mostly sessile, or in clusters of 2–10 on 1–4 **peduncles** 1–7 cm long; **flowers** perfect, 20–40 per spikelet; **fertile scales** 6–8 mm long by 3–3.5 mm wide, reduced near a spikelet's apex, brown with a bright green midvein and awn 1–3 mm long in the apical notch; **subtending involucral bracts** one to five, radiating from the cluster of spikelets; **1) vertical involucral bracts** the longest at 5–32 cm by 2–7 mm wide and appearing as a continuation of the culm; **2) lateral involucral bracts** up to four, 2–11 cm long, widest at their base; **stamens** three; **filaments** flat, tan, often persistent from the base of the fruit; **anthers** yellow and 2–4 mm long; **style** 2-parted (3-). **FRUIT an achene**, 1-seeded, (3-) 2-sided, medium to dark brown, shiny, 2.3–4 mm long by 2–2.7 mm wide by 0.8–1 mm thick and glabrous; **beak** tiny; **perianth bristles** two to six, tan, deciduous, retrorsely barbed, mostly shorter than the achene.
- **LEAVES** Alternate, simple, entire, three or four main ones; **blades** linear, sessile, ascending, 4–43 cm long by 2–12 mm wide, edges scabrous; **sheaths** glabrous, apices concave to convex.
- **STEM** A **culm**, erect, simple, green, sharply triangular with slightly concave sides; 4–15 mm wide near the base.
- **SYN.** *Bolboschoenus maritimus* (L.) Palla, *Scirpus maritimus* L. and *Scirpus paludosus* A. Nels.

Cosmopolitan Bulrush, Saltmarsh Club-rush, Alkali Bulrush

Three-square Bulrush *Schoenoplectus pungens* (Vahl) Palla

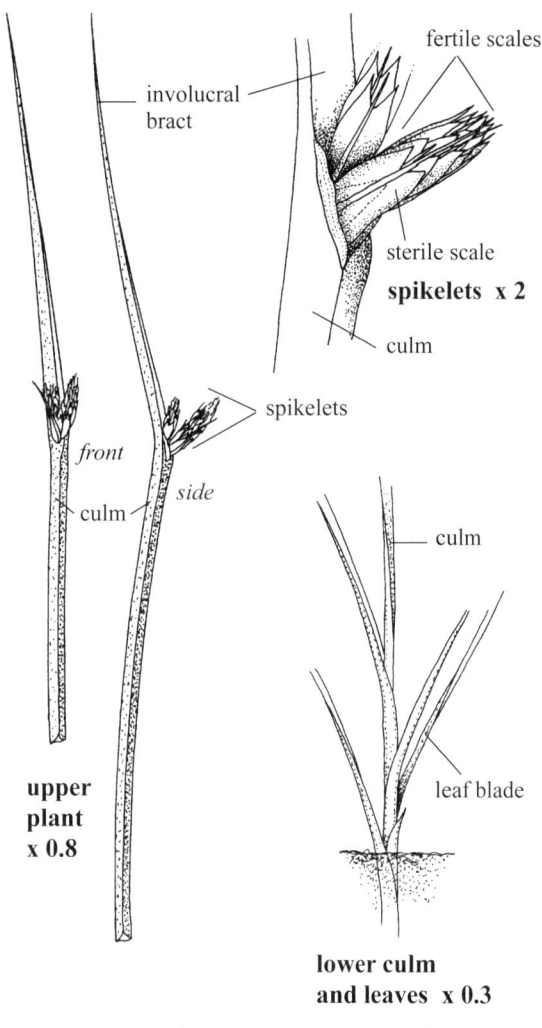

- **SKETCH** A variable **perennial herb** 20–200 cm tall in open colonies from brownish red **rhizomes** 1.5–6 mm wide by 1–12+ cm long; in damp areas along lakeshores, pond edges, river banks and marshes, even in shallow water. w1
- **FLOWERS** Brownish red, blooming May–August; **inflorescence** a spikelet, one to several in a cluster 0.6–3 cm long by 3–15 mm wide from a side eruption near the top; **involucral bracts** one, green, glabrous, 1.5–20 cm long, erect, appearing as a continuation of the culm, one side concave and whitish green; **spikelets** sessile, 7–23 mm long by 2.5–4 mm wide; **flowers** perfect; **lowest scale** sometimes sterile, with an awn-like tip to *c*. 5 mm long; **fertile scales** numerous, each covering one flower, overlapping, 3–6 mm long by 1.5–3 mm wide, orangish to brown with brownish red streaks, the bidentate tip with teeth 0.5–1 mm long, the midvein extended as an awn 1–1.5 mm long; **stamens** three; **filaments** tan, flat, glabrous, often persistent; **anthers** 2–3 mm long, pale yellow; **style** 2- or 3-parted. **FRUIT an achene**, 1-seeded, smooth, dark brown, lenticular to triangular, 2.5–3.4 mm long by 1.5–2.3 mm wide by *c*. 1 mm thick; **perianth bristles** usually six (2–8) and mostly shorter than the achene; **beak** 0.1–0.5 mm long.
- **LEAVES** Alternate, simple, entire, one to six per culm, deeply V-shaped, near the base of the culm, upper surface whitish green, the outer dorsal surface darker green like the culm, glabrous; **blades** sessile, mostly ascending, 2–70 cm long by 3–10 mm wide, linear, firm and pointed; **sheaths** closed, smooth, some with a reddish brown line at the concave apices.
- **STEM** A **culm**, glabrous, erect to slightly leaning, with air chambers, unbranched, sharply triangular, usually with two concave sides, the third side flat to slightly convex; 2–10 mm wide across one side near the base.
- **SYN.** *Scirpus americanus* auct. non Pers. = *Schoenoplectus pungens* var. *pungens* (Vahl) Palla and *Scirpus pungens* Vahl = *Schoenoplectus pungens* var. *pungens* (Vahl) Palla.

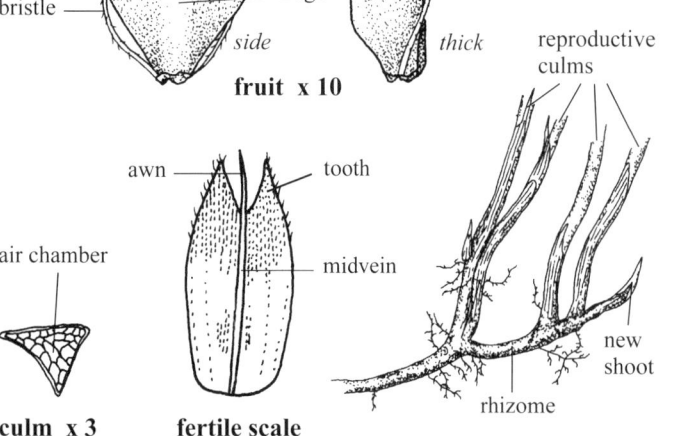

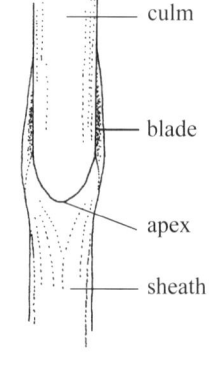

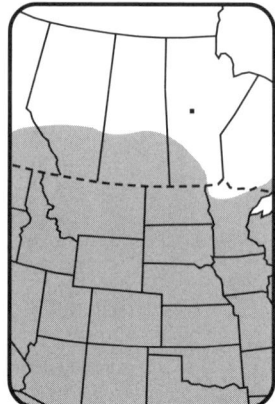

Common Threesquare, Three-square, Three-square Rush

Sedge—*Cyperaceae* Sedge Spikelets 4–15 mm long

GREAT BULRUSH *Schoenoplectus tabernaemontani* (K.C. Gmel.) Palla

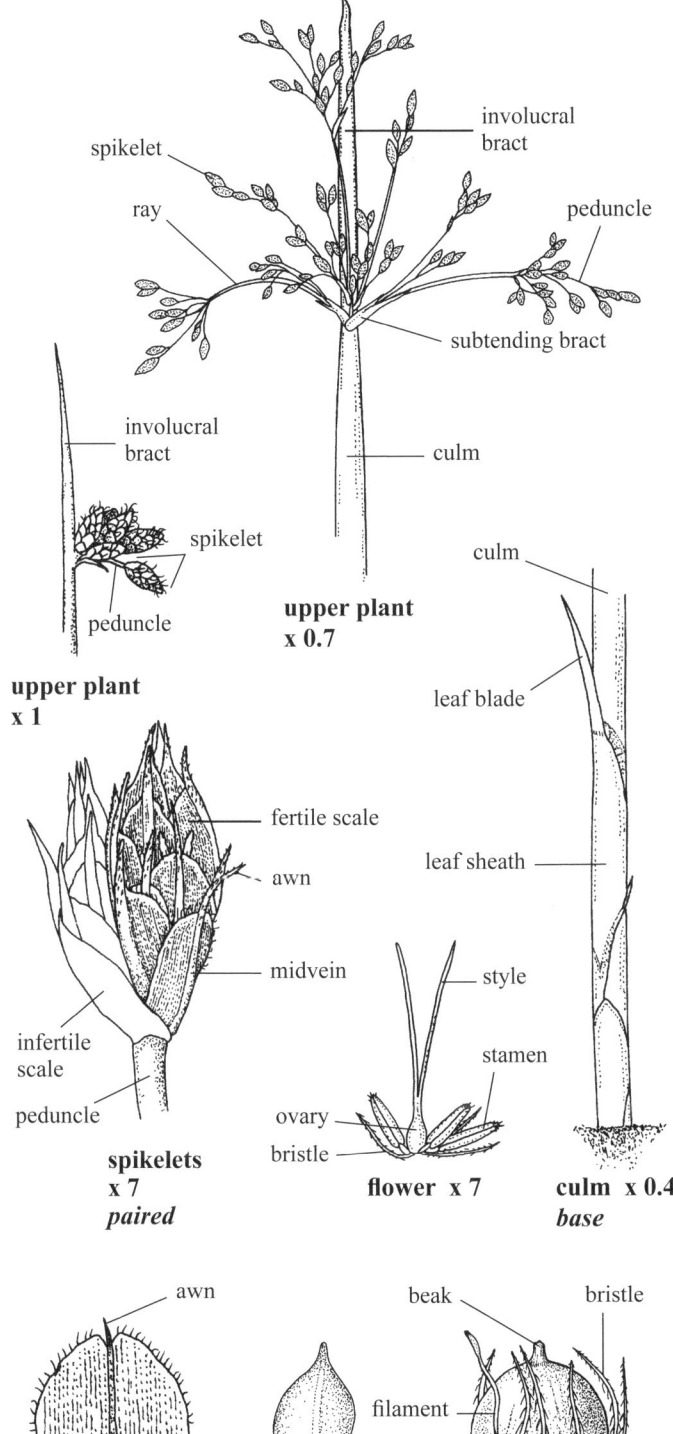

- **SKETCH** A variable, aquatic **perennial herb** 1–4 m tall with reddish brown **rhizomes** 5–12 mm thick and to 15+ cm long; in muddy sites, or ditch water, ponds, marshes and lakes, forming extensive borders often into the water. **w1**
- **FLOWERS** Brown, blooming June–August; **inflorescence** a panicle, one, terminal, 1–20 cm high and wide; **rays** ascending to drooping, 0.5–15 cm long and branching; **peduncles** 0–10 mm long; **involucral bracts** solitary, vertical, mostly 1–9 cm long; **spikelets** 15–200 per inflorescence, all solitary or some paired (rarely 3 or 4) at the end of a peduncle, each 4–15 mm long by 2–4 mm wide; **flowers** perfect, one under each fertile scale; **infertile scales** one, at bases of spikelets; **fertile scales** soft, thin, upper half reddish brown, midvein green, notched, 2–3.8 mm long by 1.5–3 mm wide, covering the fruit, upper margins ciliate, awn 0.2–1 mm long, scabrous; **stamens** included; **pistil** glabrous, *c.* 4 mm long; **styles** 2-parted, exserted. **FRUIT an achene**, 1-seeded, 1.5–2.8 mm long by 1.3–1.7 mm wide by 0.8–1 mm thick, shiny, dark brown, beak 0.2–0.3 mm long; **perianth bristles** six, equal to or slightly longer than the achene, tan, with downward hairs.
- **LEAVES** Alternate, simple, entire, absent or ascending near the base; **blades** linear, 0–18 cm long by 5–10 mm wide; **sheaths** closed, glabrous, whitish green on inside, some bladeless.
- **STEM** A **culm**, round, light green, glabrous, erect, soft from large air chambers, unbranched, often several growing in a line from a rhizome; 1–2 cm wide near the base.
- **SYN.** *Schoenoplectus validus* (Vahl) A. & D. Löve, *Scirpus tabernaemontani* K.C. Gmel. and *Scirpus validus* Vahl.

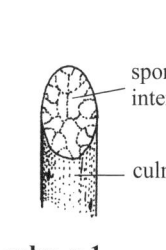

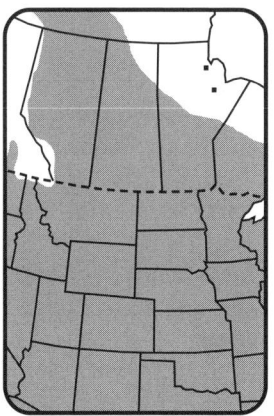

Softstem Bulrush, Softstem Club-rush 573

Sedge—*Cyperaceae*

WOOL-GRASS *Scirpus cyperinus* (L.) Kunth

■ **SKETCH** A variable, tufted, glabrous **perennial sedge** 40–200 cm tall with stiff woody **rhizomes** 1–2 mm long by 1–3 mm wide and tan **roots** 0.5–1 mm thick; in wet sites along marshes, meadows, swamps, bogs, lakeshores and rocky outcrops. **w1**

■ **FLOWERS** Greenish brown, blooming June–September; **inflorescence** an umbel, compound, with numerous spikelets, together 2–12 cm long by 3–8 cm wide, congested at top of culm or spreading widely; **involucral bracts** several but usually two or three main drooping ones of different lengths, 2–32 cm long by 1–4 mm wide, black to reddish brown at base, margins scabrous; **spikelets** in glomerules of 2–15, usually sessile, each 3–8 mm long by 2–3.5 mm wide; **flowers** perfect, numerous, one under each scale; **fertile scales** 1.1–2.2 mm long by 0.8–1 mm wide (flattened), midvein green to tan at anthesis with dark brown striations in upper half often reaching into hyaline margins, apices awnless but sometimes with a tiny mucro; **perianth bristles** six, *c.* 2 mm long at anthesis, smooth, exserted at apex of spikelet; **stamens** one or two per flower, *c.* 1 mm long; **filaments** white *c.* 0.5 mm long; **anthers** yellow, 0.5–0.7 mm long, eventually exserted; **pistil** one, green, 1.7–2 mm long; **ovary** 0.5–0.7 mm long; **style** straight, filiform; **stigmas** 3; **fruiting heads** fuzzy from tangled perianth bristles, drooping or not. **FRUIT an achene**, 1-seeded, white to slightly tan, smooth, 0.8–1 mm long by 0.3–0.5 mm wide by 0.2–0.3 mm thick, 3-sided, with three low ribs; **beak** short; **perianth bristles** twisted and smooth.

■ **LEAVES** Basal and stem, alternate, simple, entire, yellowish green, several along the culm, ascending to drooping, widely spaced, 4–75 cm long by 2–10 mm wide, reduced below, pointed, scabrous on margins; **basal leaves** numerous in the tuft of reproductive stems; **blades** V-shaped, ascending and arched; **lower stem leaves** ascending, arched, lowest ones brown by anthesis; **upper blade** may extend past inflorescence but drooping; **sheaths** closed, V-shaped at apices, 3–6 cm long.

■ **STEM** A **culm**, erect, stiff, simple, bluntly triangular, smooth, hollow, yellowish green, vegetative shoots from base; 3–5 mm thick near the base; **nodes** slightly darker green, not reddish, smooth.

■ **NOTE** *Scirpus atrocinctus* is now a separate species (Great Plains Flora Association, 1986).

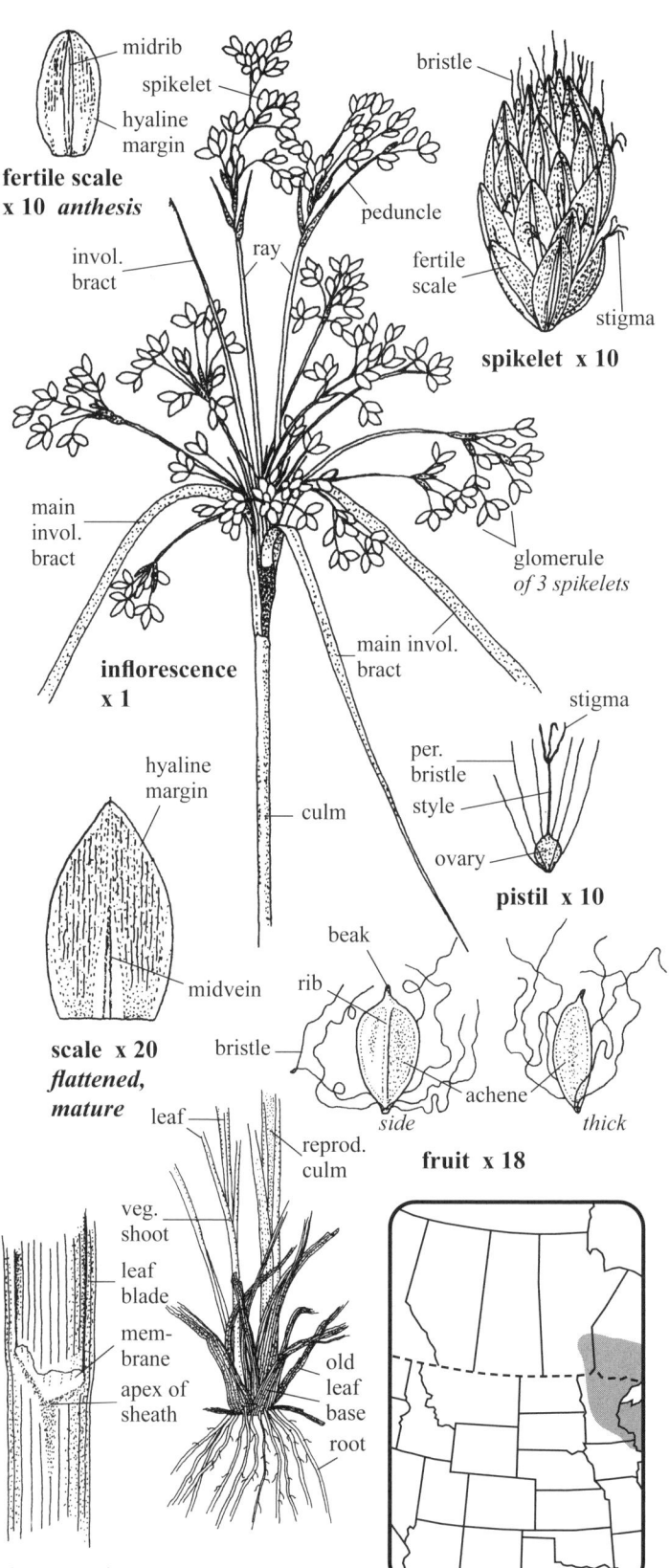

Sedge—*Cyperaceae* Sedge Spikelets 2.5–8 mm long

SMALL-FRUITED BULRUSH *Scirpus microcarpus* J. & K. Presl

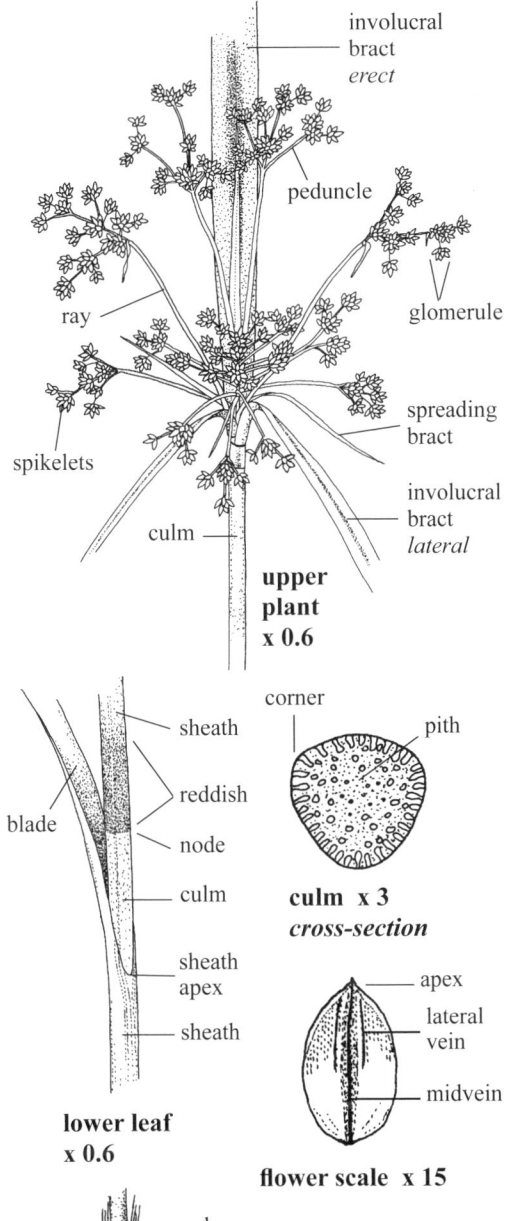

upper plant x 0.6

lower leaf x 0.6

culm x 3 cross-section

flower scale x 15

■ **SKETCH** A **perennial herb** 30–150 cm tall from a reddish brown **rhizome** 3–7 mm thick and to *c.* 20⁺ cm long; in moist sites of marshes, ditches, willow thicket openings, meadows and along streams, sometimes in shallow water. **w1**

■ **FLOWERS** Green, blooming May–August; **inflorescence** a compound umbel of spikelets in glomerules at the ends of rays, overall 5–20 cm long by 6–18 cm wide; **rays** 1–12 cm long by 1.5–2 mm wide, glabrous, erect to spreading, flat on one side; **involucral bracts** several, of various sizes; **main bracts** three, one erect and two lateral; **erect involucral bract** 15–30 cm long by 1–1.5 cm wide, channelled, glabrous inside, outside and margins scabrous; **lateral involucral bracts** (of rays) 0.6–5 cm long by 1–3 mm wide; **peduncles** 5–12⁺ mm long; **spikelets** 3–17 per glomerule, usually sessile, each spikelet 2.5–8 mm long by 1.2–3.5 mm wide, round in cross-section; **subtending bracts** (of spikelets) one, *c.* 1.5 mm long, whitish, membranous, appressed; **flowers** perfect, one per flower scale; **flower scales** 3-nerved, imbricate, 1.2–2 mm long by 0.8–1 mm wide (flattened), base hyaline and membranous, midvein green, apices sometimes mucronate, upper half darkly streaked; **stamens** three; **filaments** 0.4–0.5 mm long; **anthers** yellow, 0.8–1.2 mm long, emerging after the style; **pistil** 2–3 mm long; **ovary** glabrous, *c.* 0.5 mm long; **style** (3-) 2-parted, exserted, *c.* 2 mm long. **FRUIT** an achene, 1-seeded, glabrous, slightly shiny, light tan to whitish, 0.7–1.6 mm long by 0.6–1 mm wide by 0.3–0.4 mm thick, plano-convex to 3-angled; **beak** short and sharp; **perianth bristles** usually three or four (–6), shorter to longer than the achene, hairs descending.

■ **LEAVES** Basal and stem, alternate, simple, entire, linear; **basal blades** several, brown by anthesis, to *c.* 1 m long by 10–15 mm wide, spreading on the ground; **stem blades** green, 20–75 cm long by 3–15 (–20) mm wide, ascending to descending, scabrous below and along the margins, glabrous above, midrib channelled above; **sheaths** closed, apices concave to deeply V-shaped; **lower sheath bases** usually reddish for 1–4 cm above the nodal line.

■ **STEM** A **culm**, green, erect, not hollow; roundly triangular below inflorescence to almost round near the 5–15 mm thick base.

■ **SYN.** *Scirpus rubrotinctus* Fern.

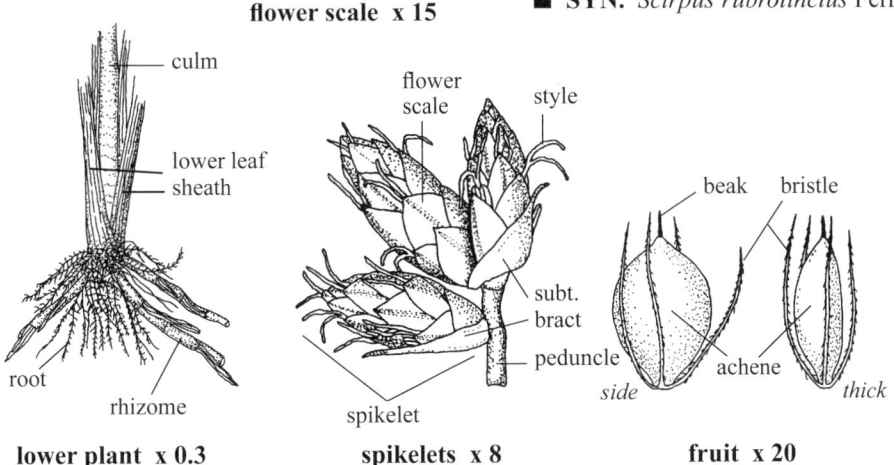

lower plant x 0.3   spikelets x 8   fruit x 20

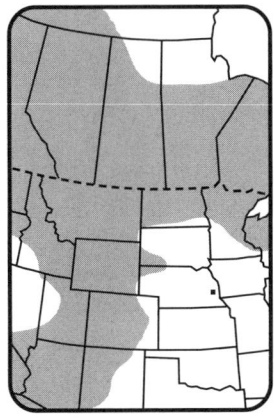

Panicled Bulrush, Red-fringe Bulrush, Small-flowered Bulrush

Pale-green Bulrush *Scirpus pallidus* (Britt.) Fern.

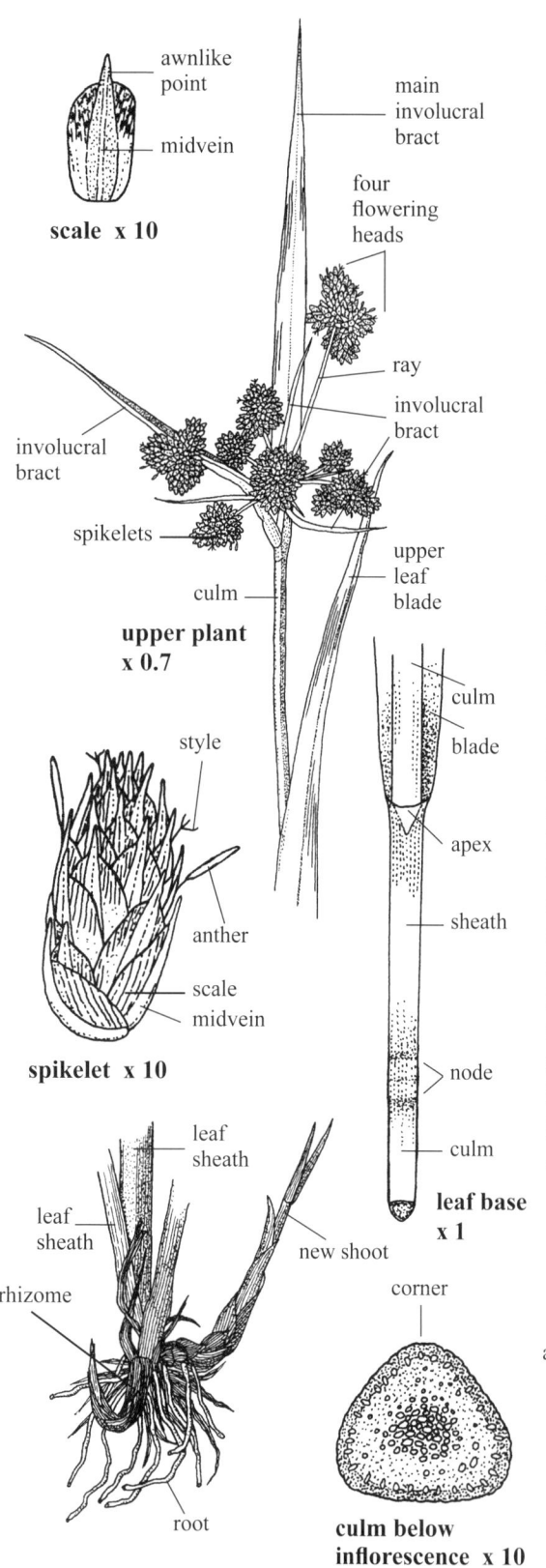

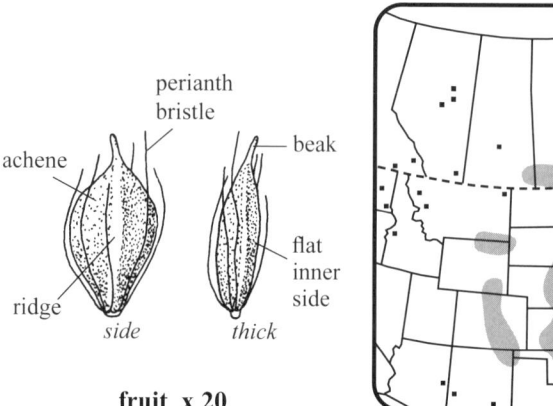

- **SKETCH** A tufted variable **perennial herb** 30–150 cm tall with tough **rhizomes** 6–10 mm thick by 1–2.5 cm long and covered with dark brown scales; along marsh shores, wet meadows and stream banks. **w1**
- **FLOWERS** Brown, blooming June–July; **inflorescence** umbel-like, with numerous spikelets in round heads, together 4–6 cm wide by 2.5–15 cm long; **rays** bluntly triangular, 2–70 mm long, erect to spreading; **flowering heads** round, 5–13 mm across, each of *c.* 100+ spikelets; **involucral bracts** several, leaflike, of various sizes, tapered to a fine point, erect to spreading; **main involucral bracts** erect, flat, 6–12 cm long by 3–6 mm wide, pointed, scabrous on margins and outside; **other involucral bracts** several, reduced in size; **spikelets** oval, sessile, 2–5 mm long by 1–2.5 mm wide; **scales** covering flowers, TL 1.8–2.8 mm by 0.8–1.3 mm wide (flattened), completely covering the fruit, blackish striate in the upper half as fruit ripens, midvein wide, green, awnlike point 0.4–0.6 (–1.2) mm long; **anthers** 0.8–1 mm long, exserted or included; **pistil** *c.* 1 mm long; **styles** (2-) 3-parted, exserted. **FRUIT** an achene, 1-seeded, 2- or 3-sided, light tan, 1–1.3 mm long (including beak) by 0.4–0.6 mm wide by 0.3–0.5 mm thick, ridged, widely triangular to plano-convex; **beak** *c.* 0.2 mm long; **perianth bristles** six (five), shorter to slightly longer than the achene, with tiny hairs pointing downward.
- **LEAVES** Alternate, simple, entire, 5–10 per culm, ascending to arching, closer together near the base; **blades** 8–55 cm long by 3–18 mm wide, largest about midstem, flat, slightly scabrous, smooth above, upper leaf often reaching into the inflorescence; **sheaths** closed, glabrous, base yellowish green, apices brown, rough, slightly concave, often torn and ragged.
- **STEM** A **culm**, erect, green, bluntly triangular below inflorescence and glabrous; base 4–8 mm thick; **nodes** slightly swollen and pale yellowish green for *c.* 5 mm.
- **SYN.** *Scirpus atrovirens* var. *pallidus* Britt.
- **NOTE** Similar to *Scirpus atrovirens* (Great Plains Flora Association, 1986).

Sedge—*Cyperaceae* **Bulrush** Spikelet one, 5–8 mm long

Alpine Cotton-grass *Trichophorum alpinum* (L.) Pers.

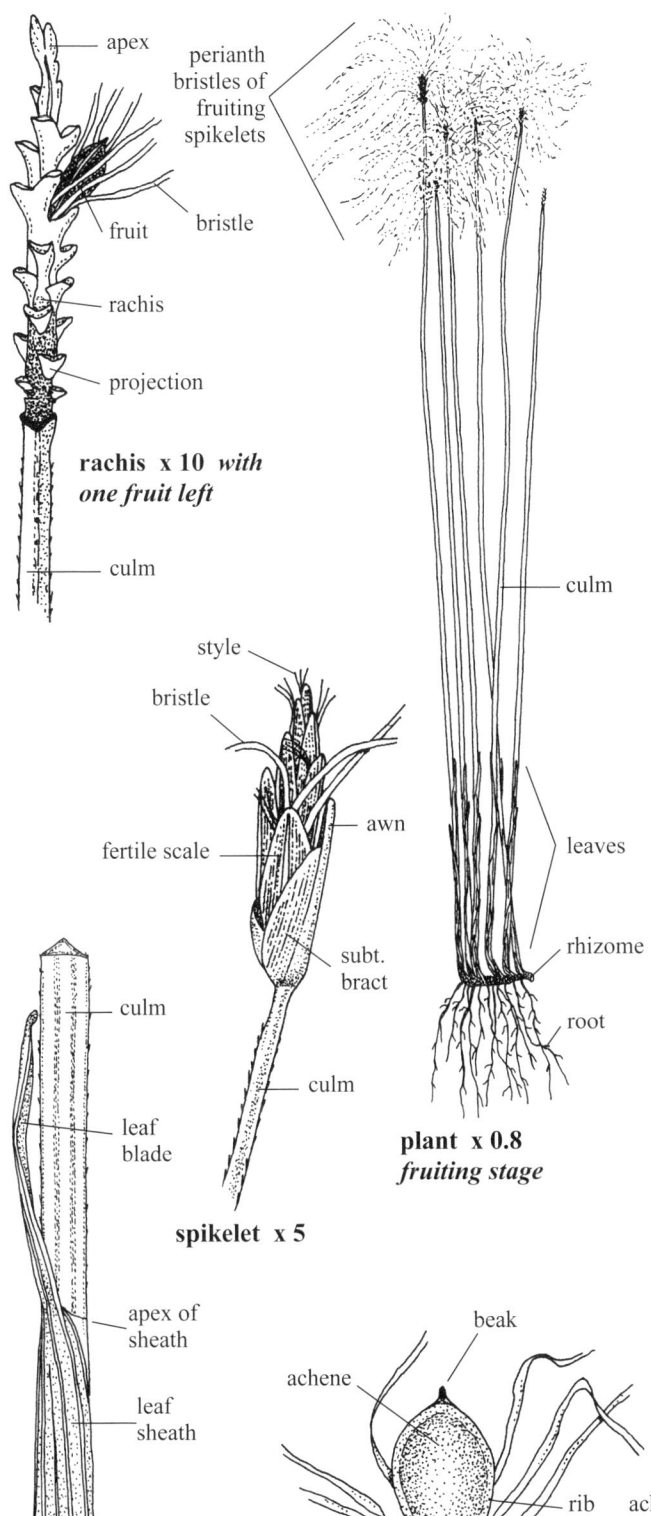

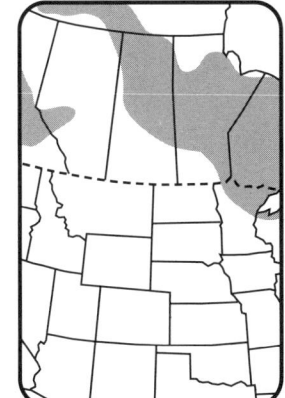

- **SKETCH** A tufted **perennial sedge** 6–45 cm tall with short brown **rhizomes** 1–2 mm thick and tan **fibrous roots** to *c*. 20 cm long; in fens, bogs, springs and wet gravelly areas.
- **FLOWERS** Green to brown, blooming May–June; **inflorescence** a spikelet, one, 5–8 mm long by 1–2.5 mm wide, terminal; **flowers** perfect, 15–20, only one per fertile scale; **subtending bracts** 4–7.8 mm long (including awn), awn blunt; **rachis** mostly tan, hairless, *c*. 1 mm wide, with projections in a spiral and each holding one fruit; **fertile scales** several, awnless, apices rounded, tan with hyaline margins, 3–4 mm long by 1–1.5 mm wide; **stamens** three; **pistil** one under each fertile scale; **styles** 3-parted, these brown, filiform and exserted at anthesis. **FRUIT** an achene, 1-seeded, 3-ribbed, medium to dark brown, 10–25 per spikelet, plano-convex, sides slightly convex, apex pointed, smooth and shiny, base flat, 1.3–1.5 mm long by 0.5–0.7 mm wide; **beak** 0.1–0.2 mm long, dark brown, somewhat pointed; **perianth bristles** six, white and flat, exserted at anthesis, smooth, 2–2.5 cm long, persistent over the summer; **seeds** smooth, yellowish tan, base pointed, apices rounded, 0.9–1 mm long by *c*. 0.6 mm wide and thick.
- **LEAVES** Alternate, six to eight, erect and appressed, imbricate over the lower 2–4 cm of a culm's base, tan by fruiting time; **blades** 0–20+ mm long, reduced below, apices blunt, grooved on inside; **sheaths** transparent on ventral side, making up most of the leaf, strongly veined and reddish brown near the base, apices slightly concave.
- **STEM** A **culm**, erect, sharply triangular and scabrous throughout, ascending hairs short and along the three ridges, 0.5–1 mm wide across one side below the inflorescence, reddish brown bases crowded along the rhizome.
- **SYN.** *Eriophorum alpinum* L. and *Scirpus hudsonianus* (Michx.) Fern.

Hudson Bay Bulrush, Alpine Bulrush, Alpine Leafless-bulrush

Iris—Iridaceae **Wildflower Yellow Petals 3**

Water Flag *Iris pseudacorus* L.

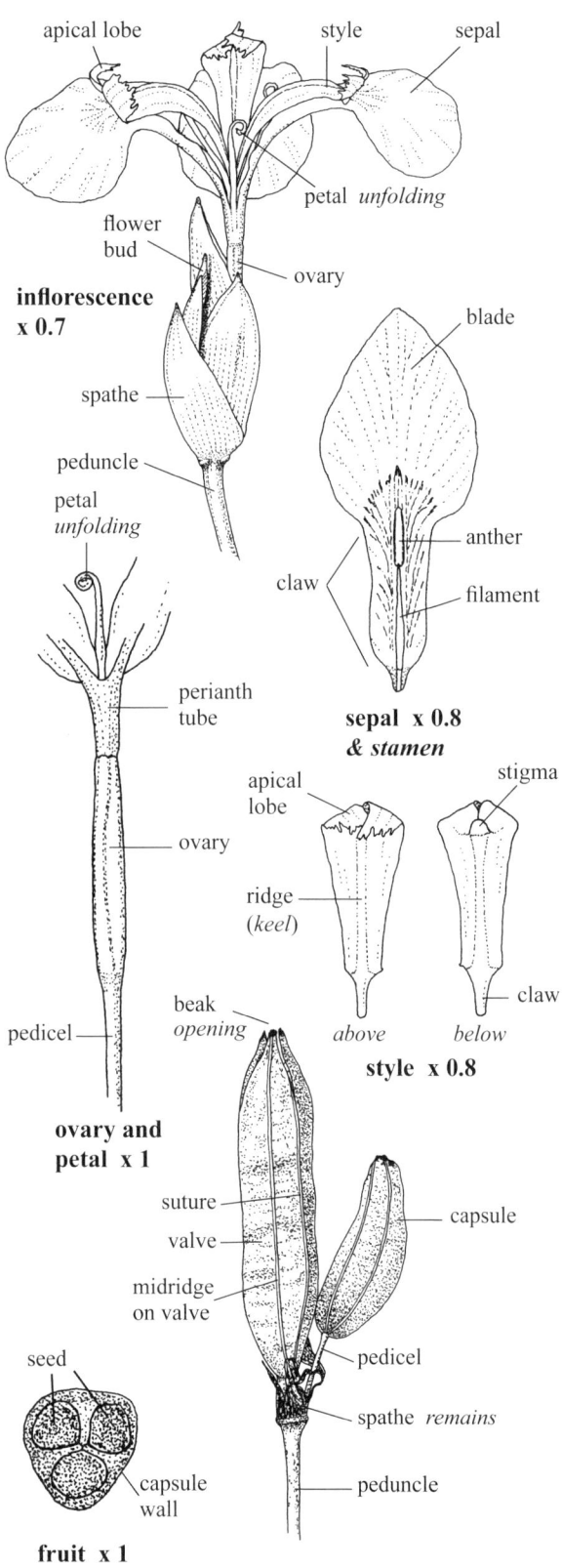
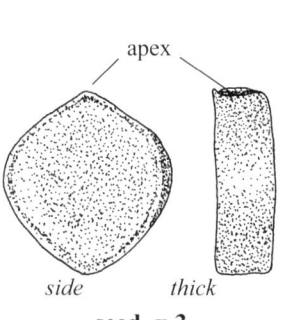
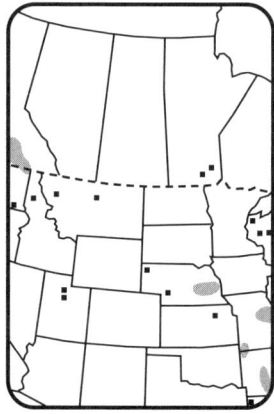

■ **SKETCH** A **perennial herb** 50–100+ cm tall in large clumps with pinkish brown **rhizomes** 10–30 mm wide and several cm long with veiny leafy base fragments and whitish tan wrinkled **roots** 1–3 mm wide by 4–15 cm long; in irrigation canals, along the edges of marshes, lakes and streams, sometimes in shallow water. *Naturalized*

■ **FLOWERS** Yellow, blooming April–July; **inflorescence** a raceme of 3–12 flowers per stalk; TL 20–30 cm; **pedicels** smooth, 7–20 mm long by *c.* 2 mm wide, pale green at anthesis, yellowish green with fruit; **spathes** of two bracts, 2–7 cm long, erect, outer one keeled, pointed; **flowers** perfect, 6–9 cm wide by 4–7 cm long; **sepals** three, 5.5–8 cm long by 2.5–4 cm wide, arched, yellow with brown veins along the claw, pointed; **claw** 2.5–3 cm long by 10–12 mm wide (flat); **petals** three, 2–4 cm long and 1–2 cm wide, ascending between the styles; **stamens** three, yellow; **filaments** pale yellow, 15–18 mm long by *c.* 2 mm wide, flat, lower 5–7 mm adnate to claw of sepal; **anthers** yellow and purple, *c.* 10 mm long by *c.* 2 mm wide; **ovary** green, glabrous, 2–3 cm long by *c.* 5 mm wide, triangular with blunt edges; **styles** ridged along the middle, 2–4 cm long by *c.* 1.3 cm wide, pale yellow, apical lobes two, ascending and recurved. **FRUIT a capsule**, buoyant, smooth, yellowish green, erect to drooping, 3-angled, 3-valved, slightly shiny, 3–8 cm long by 14–18 mm wide and thick, opening along three sutures at apex; **beak** thick and tapered, tip blackish, 5–10 mm long; **seeds** 14–77 per capsule, smooth, golden brown, 7–8 mm long by 6.5–7.8 mm wide by 2–4 mm thick, slightly shiny, slightly concave, stacked like chips.

■ **LEAVES** Basal (mostly) and stem, simple, entire, erect, sessile, smooth, glaucous, pointed, a few along the stem, 20–100 cm long by 13–30 mm wide by 1–2 mm thick.

■ **STEM** Erect, smooth, zigzagging in inflorescence, green to yellowish green with fruit, leafy, oval with blunt edges, often falling over before fruit is ripe; *c.* 5 mm thick near the base.

Pale-yellow Iris, Yellow Iris, Yellow Water Iris, Yellow-flag

Iris—*Iridaceae* **Wildflower** Purple with yellow center Petals 3

BLUE FLAG *Iris versicolor* L.

■ **SKETCH** A glabrous **perennial herb** 40–95 cm tall in clusters with a woody, reddish brown **rhizome** 8–10 mm thick and 5–10+ cm long, fibrillose; **roots** white, wrinkled, 1–3 mm wide and to *c.* 30 cm long; along shores, marshes and in swampy sites of open forests. p2

■ **FLOWERS** Purple with yellow center; blooming May–July; **inflorescence** a raceme of a few flowers, terminal; **pedicels** 1–2 cm long by 2–3 mm wide, green; **flowers** perfect, 8–10 cm wide by 6–7.5 cm tall (including ovary); **sepals** three, arched, 4.5–7 cm long by 2.5–4 cm wide, purple veined, the claw *c.* 3 cm long, yellow with purple veins; **petals** three, ascending, 3–5 cm long by 1–2 cm wide, veined; **spathes** of two bracts, green, entire, pointed, clasping, the outer bract 3–4 cm long, inner (upper) one 4–5.5 cm long; **stamens** three, glabrous, TL 3–3.5 cm, included, from base of sepals; **filaments** curved outward, lower green part 5–7 mm long, upper free purplish pink part 10–13 mm long by *c.* 2 mm wide, flat, tapered above; **anthers** purple, 0.8–1.7 cm long by *c.* 2 mm wide, 2-lobed, apices flat; **pistil** one, 6–7 cm long; **ovary** inferior, green, 1.5–2.5 cm long by 4–5 mm wide, 3-sided, edges blunt, sides slightly concave; **styles** three, blue, petal-like, 3–4 cm long by 10–14 mm wide, arching, 2-lobed at apices, each lobe 10–15 mm long by 5–7 mm wide, toothed and light purple; **stigma** pale purple, 5–6 mm wide by 3–4 mm long, entire, below style lobes, hairy above (against style), glabrous below. **FRUIT a capsule**, greenish brown, 2.5–6 cm long by 13–17 mm wide, erect to drooping, 3-valved, 3-sided, opening at apex, buoyant; **seeds** 60–90 per capsule, in three columns, stacked like chips, D-shaped, reddish brown, smooth, shiny, 5–7 mm long by 3–5 mm wide by 2–3 mm thick, slides slightly concave.

■ **LEAVES** Basal (mostly), and stem, alternate, entire, simple, sessile, sides folded together along the length of the midrib, pinkish purple near base; **basal blades** 5–90 cm long by 1–3 cm wide by 1–2 mm thick, erect, open near base to clasp stem; **stem leaves** one or two, 20–65 cm long by 8–25 mm wide, reduced above, erect, a dark brown line 1–2 mm wide at base of each stem blade.

■ **STEM** Erect, smooth, green, branched above, stiff, solid, oval, mostly erect with fruit; 5–10 mm thick near the base.

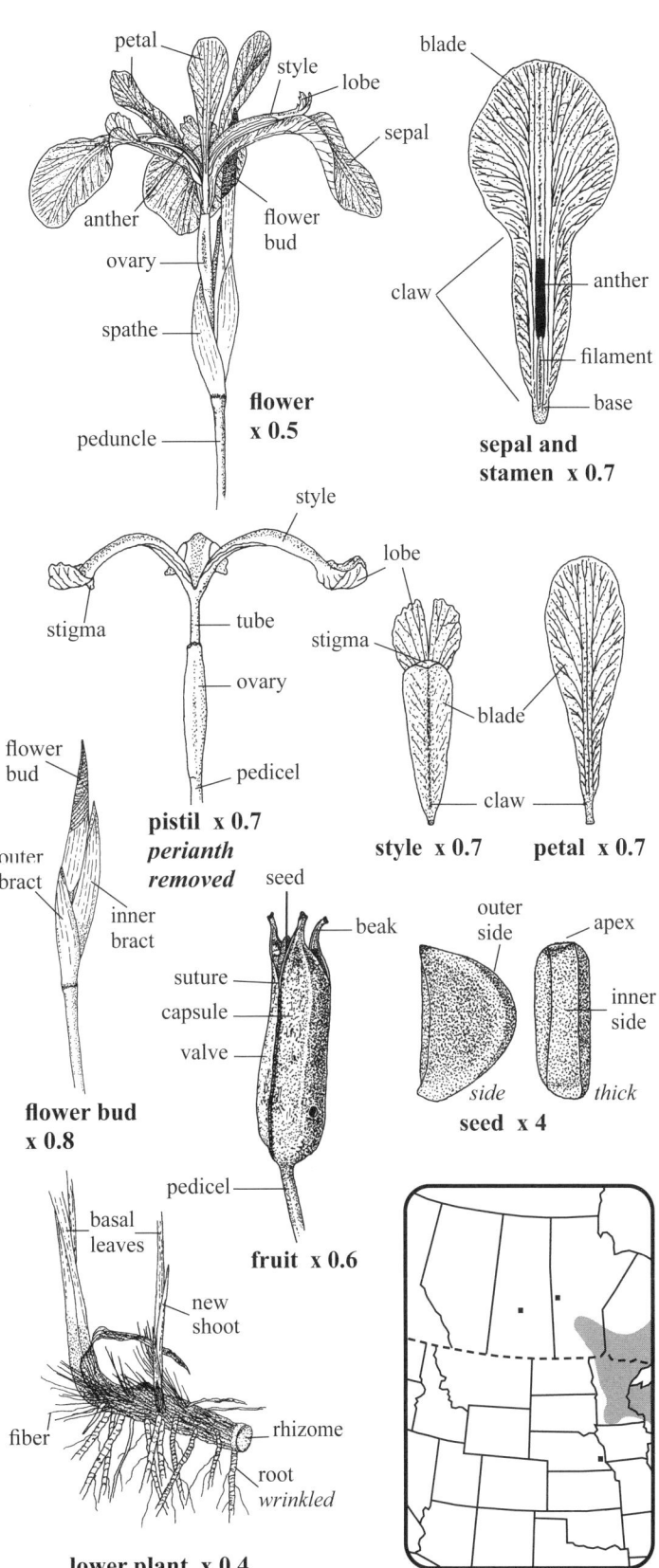

Harlequin Blue Flag, Larger Blue Flag, Northern Blue Flag

Iris—*Iridaceae* **Wildflower** Dark blue, center yellow Perianth parts 6

COMMON BLUE-EYED GRASS *Sisyrinchium montanum* Greene

■ **SKETCH** A tufted **perennial herb** 8–45 cm tall with obscure **rhizomes** and **fibrous roots** 4–10 cm long; in moist prairies, meadows, gravely flood plains, ditches and open woodlands. **p1**

■ **FLOWERS** Dark blue with a yellow center, blooming May–July; **inflorescence** a terminal umbel-like cluster; **pedicels** erect to slightly lax, 10–28 mm long, originate at the base of the spathe; **spathe** of two unequal, sessile or stalked, green, pointed bracts; **1) outer bracts** 3.5–7.5 cm long (averaging more than 4 cm long) by 1–2.2 mm wide, lower margins united for 2–5.5 mm, widens above the tip of the inner bract; **2) inner bracts** 1.5–3.5 cm long, exerted from near the base of the outer bract, margins narrowly hyaline; **flowers** perfect, one to six, 1.7–2.5 cm wide, emerge near the tip of the inner bract; **perianth parts** six, alike and equal, each 9–14 mm long by *c.* 4 mm wide, apices aristate, base yellow; **stamens** three; **filaments** 2–5 mm long; **anthers** 1.5–1.9 mm long, yellow; **ovary** glandular-hairy, 1.3–2.5 mm long; **style** 3-parted. **FRUIT a capsule**, round, green with six purple veins when fresh, ripening to a pale, papery brown, 4–6 mm long and wide; **seeds** black, *c.* 25 per capsule, each 1–2 mm long by *c.* 1.8 mm wide by *c.* 1.3 mm thick, glabrous, pointed at one end.

■ **LEAVES** Basal, simple, entire, linear, glabrous, erect, grasslike, 4–18 cm long by 1–4 mm wide, 2-ranked, arising from the base and usually not reaching past the midpoint of the stem.

■ **STEM** Erect, glaucous, winged, simple, entire to minutely toothed above, solitary or in tufts of up to 28+ reproductive stems, the wing slightly narrower than the stem; 2–3.5 mm wide near the base.

■ **NOTE** Similar to *Sisyrinchium mucronatum* on the northern Great Plains in which the stem wings are very narrow or absent (Great Plains Flora Association, 1986).

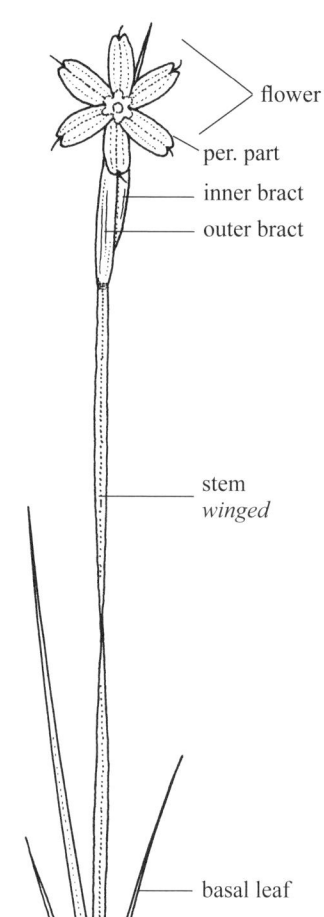

plant x 1

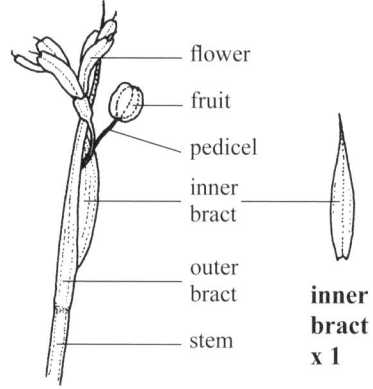

upper plant x 1

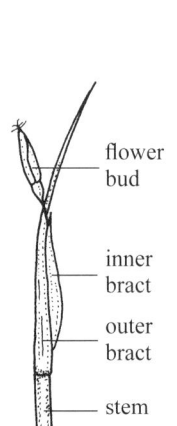

upper plant x 1

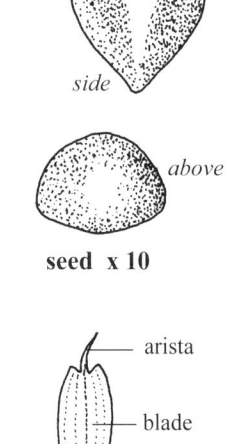

seed x 10

perianth x 2

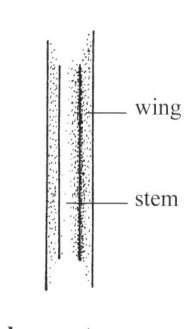

lower stem x 3

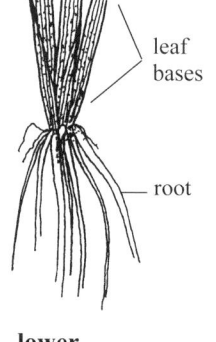

lower plant x 1

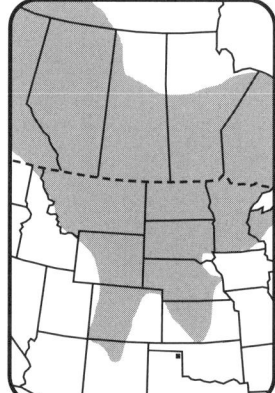

580 Strict Blue-eyed-grass, Blue-eyed Grass

Alpine Rush *Juncus alpinoarticulatus* Chaix

■ **SKETCH** A tufted, variable **perennial herb** 5–50 cm tall with a short, creeping **rhizome** 1–3 mm thick, without tubers; in ditches, wet prairies, sand bars and meadows, along stream shores and bogs.

■ **FLOWERS** Greenish brown, blooming June–September; **inflorescence** a panicle, ascending, 2.5–18 cm long by 0.5–6 cm wide of 5–25 glomerules; **floral branches** stiff, ascending, 0.5–1 mm thick; **peduncles** 0.2–3.5 cm long, smooth, ascending; **glomerules** 2–4 mm wide by 2–3.5 mm long, remote, each with 3–12 flowers; **involucral bracts** 2–10 cm long, leaflike, with septum inside; **bracteoles** (of glomerules) two, pointed, as long as to slightly shorter than the flowers; **flowers** perfect, 1–2 mm wide at anthesis by 2–3 mm long, sessile or some pedicellate; **perianth parts** six, in two whorls, from 1.6–3 mm long by 0.5–0.6 mm wide (flattened), persistent, green in middle, margins and apices reddish brown; 1) **outer perianth parts** three, slightly longer than inner parts, pointed; 2) **inner perianth parts** three, apices rounded, apical margins folded inward making it appear pointed; **stamens** six, included, opposite and about half the length of a perianth part; **filaments** hyaline, filiform, 0.7–0.8 mm long; **anthers** yellow, *c.* 0.4 mm long; **ovary** green, 3-lobed, *c.* 1 mm long by *c.* 0.7 wide, roundly triangular; **style** one, erect, filiform, *c.* 0.5 mm long, shorter than the stigmas; **stigmas** three, curled, exserted. **FRUIT a capsule**, erect, dark to pale brown, shiny, 3-valved, sides slightly concave, 2–3.4 mm long by 1–1.2 mm wide and thick, often exserted; **seeds** dark brown, numerous, pointed at both ends, ribbed longitudinally, shiny, 0.3–0.6 mm long by *c.* 0.2 mm wide and thick, a ridge along one side.

■ **LEAVES** Alternate, simple, entire, not basal, one to three per culm; **blades** ascending, rarely reaching the flowers, 4–18 cm long by 1–2 mm wide, glabrous, hollow, a wide central groove on the inner side extends 4–5 cm above the auricles before diminishing, upper part of blade oval to round in cross-section; **septa** complete and every 5–10 mm (seen inside the backlit leaves); **auricles** roundly blunt, membranous, 0.3–1.2 mm long, slightly imbricate; **sheaths** open, slightly keeled.

■ **STEM** A **culm**, hollow, stiff, one to several, erect, slightly oval; 2–3 mm thick near the base.

■ **SYN.** *Juncus alpinus* auct. non Vill. = *Juncus alpinoarticulatus* ssp. *nodulosus* (Wahlenb.) Hämet-Ahti.

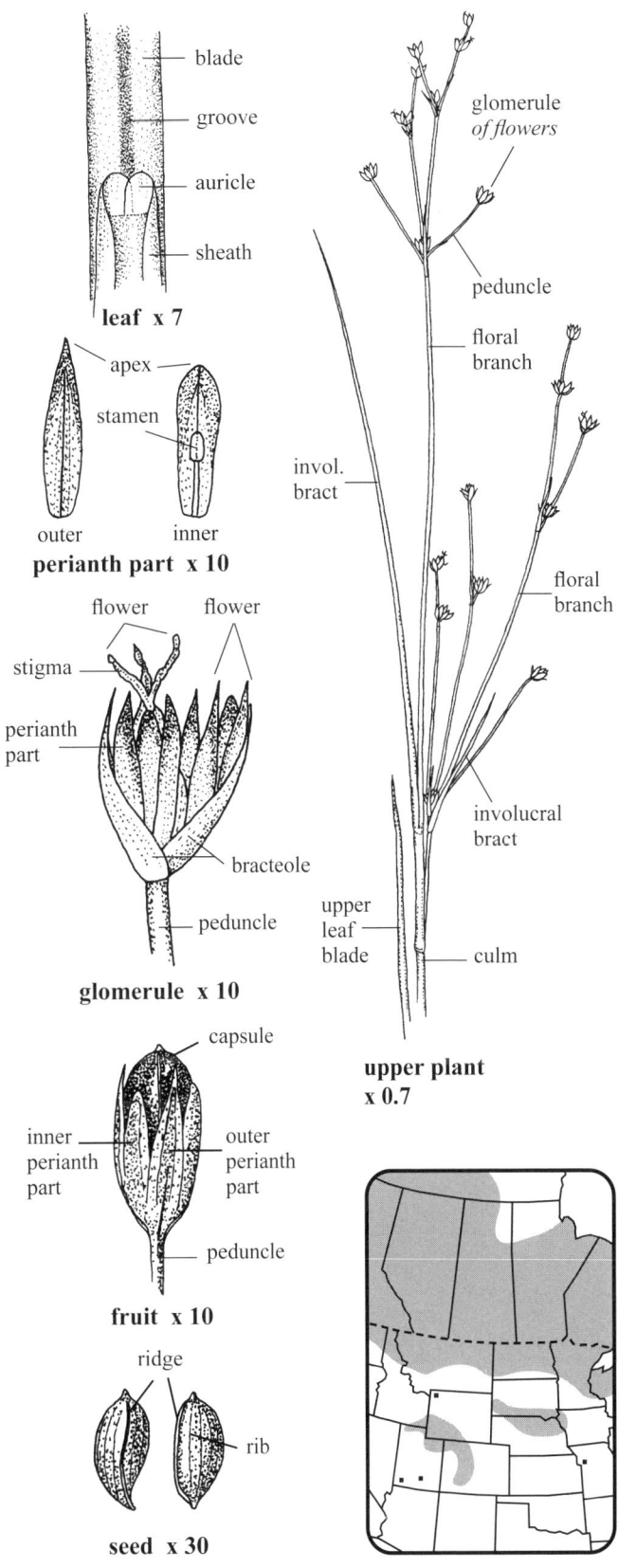

Rush—*Juncaceae* Rush Perianth parts 6, from 3.3–6 mm long

Baltic Rush *Juncus balticus* Willd.

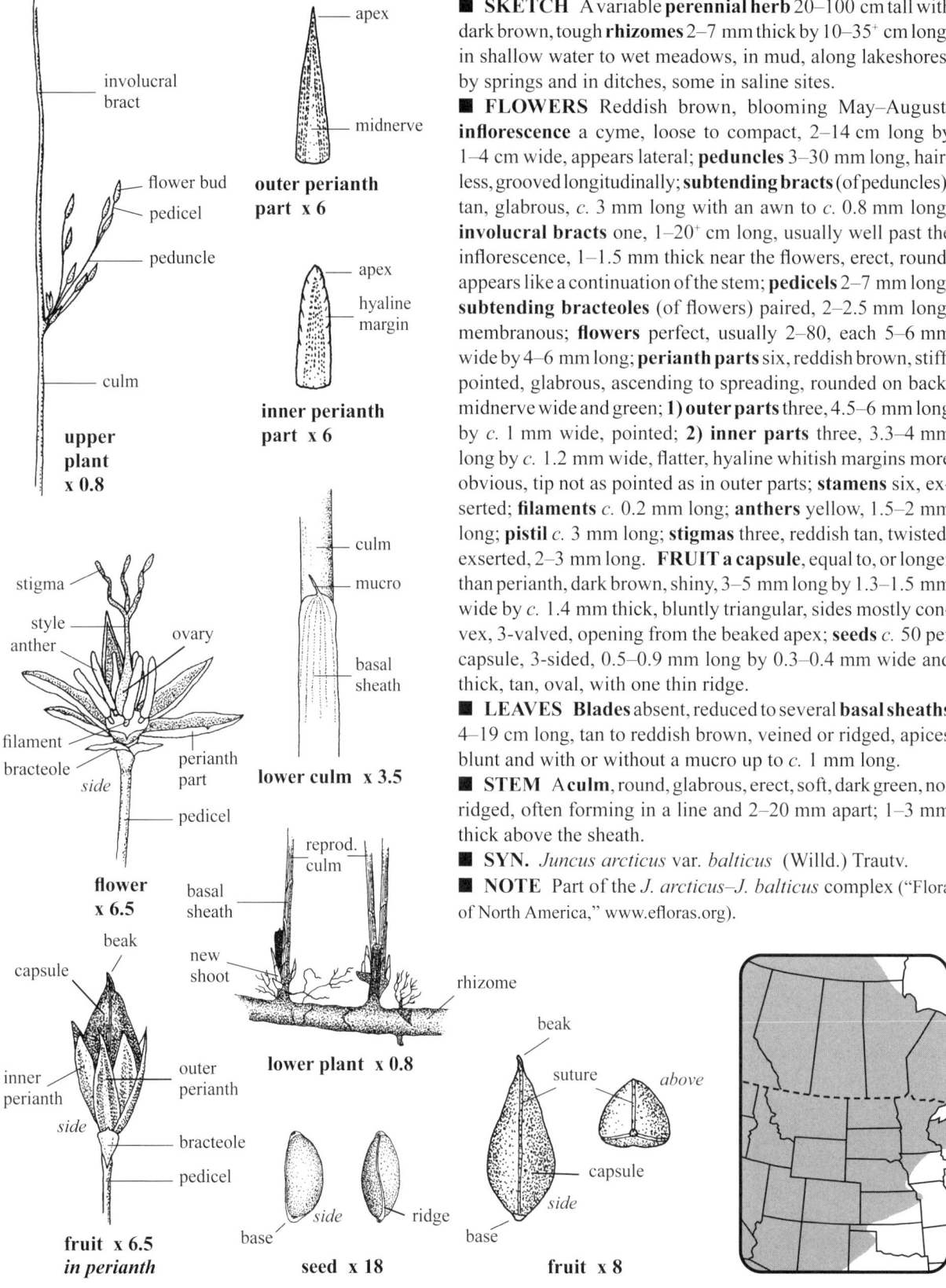

- **SKETCH** A variable **perennial herb** 20–100 cm tall with dark brown, tough **rhizomes** 2–7 mm thick by 10–35+ cm long; in shallow water to wet meadows, in mud, along lakeshores, by springs and in ditches, some in saline sites.
- **FLOWERS** Reddish brown, blooming May–August; **inflorescence** a cyme, loose to compact, 2–14 cm long by 1–4 cm wide, appears lateral; **peduncles** 3–30 mm long, hairless, grooved longitudinally; **subtending bracts** (of peduncles), tan, glabrous, *c*. 3 mm long with an awn to *c*. 0.8 mm long; **involucral bracts** one, 1–20+ cm long, usually well past the inflorescence, 1–1.5 mm thick near the flowers, erect, round, appears like a continuation of the stem; **pedicels** 2–7 mm long; **subtending bracteoles** (of flowers) paired, 2–2.5 mm long, membranous; **flowers** perfect, usually 2–80, each 5–6 mm wide by 4–6 mm long; **perianth parts** six, reddish brown, stiff, pointed, glabrous, ascending to spreading, rounded on back, midnerve wide and green; **1) outer parts** three, 4.5–6 mm long by *c*. 1 mm wide, pointed; **2) inner parts** three, 3.3–4 mm long by *c*. 1.2 mm wide, flatter, hyaline whitish margins more obvious, tip not as pointed as in outer parts; **stamens** six, exserted; **filaments** *c*. 0.2 mm long; **anthers** yellow, 1.5–2 mm long; **pistil** *c*. 3 mm long; **stigmas** three, reddish tan, twisted, exserted, 2–3 mm long. **FRUIT a capsule**, equal to, or longer than perianth, dark brown, shiny, 3–5 mm long by 1.3–1.5 mm wide by *c*. 1.4 mm thick, bluntly triangular, sides mostly convex, 3-valved, opening from the beaked apex; **seeds** *c*. 50 per capsule, 3-sided, 0.5–0.9 mm long by 0.3–0.4 mm wide and thick, tan, oval, with one thin ridge.
- **LEAVES** Blades absent, reduced to several **basal sheaths** 4–19 cm long, tan to reddish brown, veined or ridged, apices blunt and with or without a mucro up to *c*. 1 mm long.
- **STEM** A **culm**, round, glabrous, erect, soft, dark green, not ridged, often forming in a line and 2–20 mm apart; 1–3 mm thick above the sheath.
- **SYN.** *Juncus arcticus* var. *balticus* (Willd.) Trautv.
- **NOTE** Part of the *J. arcticus–J. balticus* complex ("Flora of North America," www.efloras.org).

Wire Rush

Rush—*Juncaceae*   **Rush**   Perianth parts 6, from 3–7 mm long

Toad Rush   *Juncus bufonius* L.

■ **SKETCH** A tufted, glabrous **annual herb** 5–40 cm tall with **fibrous roots**; in moist areas along lakes, ponds, rivers, marshes, seeps, springs and sand bars, often in brackish sites.

■ **FLOWERS** Green, blooming May–August; **inflorescence** a cyme with one to several flowers at the ends of usually two main glabrous branches; **involucral bracts** green, leaflike, longer to shorter than the inflorescence, 0.5–6 cm long by 0.5–0.8 mm wide, with a short membranous sheath at the base and sometimes with short auricles; **bracteoles** two, membranous, white hyaline, pointed, 1.5–2.2 mm long, enveloping the base of each flower; **flowers** perfect, mostly single, rarely in small heads, each 1.5–7 mm wide (perianth parts closed or spreading) by 3–7 mm long; **perianth parts** six, midnerve green, margins white hyaline; 1) **outer perianth parts** 3–7 mm long by 0.8–1.1 mm wide, pointed, margins hyaline, midnerve slightly raised; 2) **inner parts** 3–5 mm long by *c.* 1 mm wide (flattened), apices rounded; **stamens** three or six, 1–2.5 mm long, glabrous, one attached to base of each perianth part; **filaments** white, 0.8–1.8 mm long, filiform; **anthers** yellow, *c.* 0.5 mm long; **pistil** 1.7–2 mm long, glabrous; **ovary** green, *c.* 1.4 mm long by *c.* 0.7 mm wide by *c.* 0.5 mm thick, tapered at both ends; **style** white, *c.* 0.1 mm long; **stigmas** three, *c.* 0.5 mm long, not ascending, twisted, brown. **FRUIT a capsule**, triangular, brown, sides convex, slightly shiny, apex blunt, 3–4.5 mm long by 1.2–1.8 mm wide, opening along three sutures from apex, shorter to longer than perianth parts; **seeds** golden brown, 40–60 per capsule, 0.3–0.5 mm long by 0.2–0.3 mm wide and thick, striate, oval, dark brown at one end.

■ **LEAVES** Alternate, simple, entire, two or three per culm, ascending; **blades** 2–12 cm long by 0.5–1.5 mm wide, reduced below, involute, pliable, concave on inside, convex outside, no inner septa (tiny membranous divisions), solid but soft, slightly fleshy, usually not extending past inflorescence; **auricles** absent; **sheaths** light green, 1–1.5 cm long, membranous, open and imbricate at apices, tapering into margins of leaf blade, not hairy.

■ **STEM** A **culm**, erect, round to slightly oval, green, solid, pliable; 0.5–1.5 mm thick near the base.

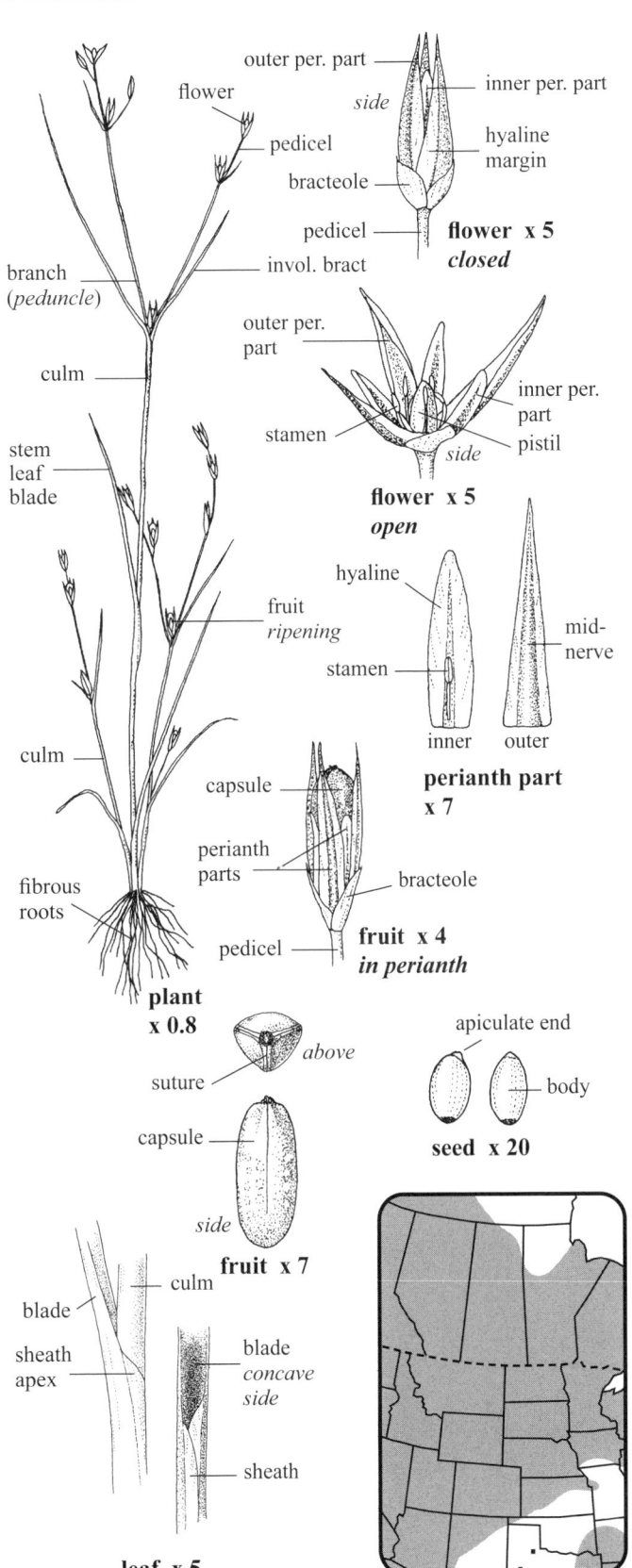

Seaside Rush

Rush—*Juncaceae* **Rush** Perianth parts 6, from 1.8–2.5 mm long

FLATTENED RUSH *Juncus compressus* Jacq.

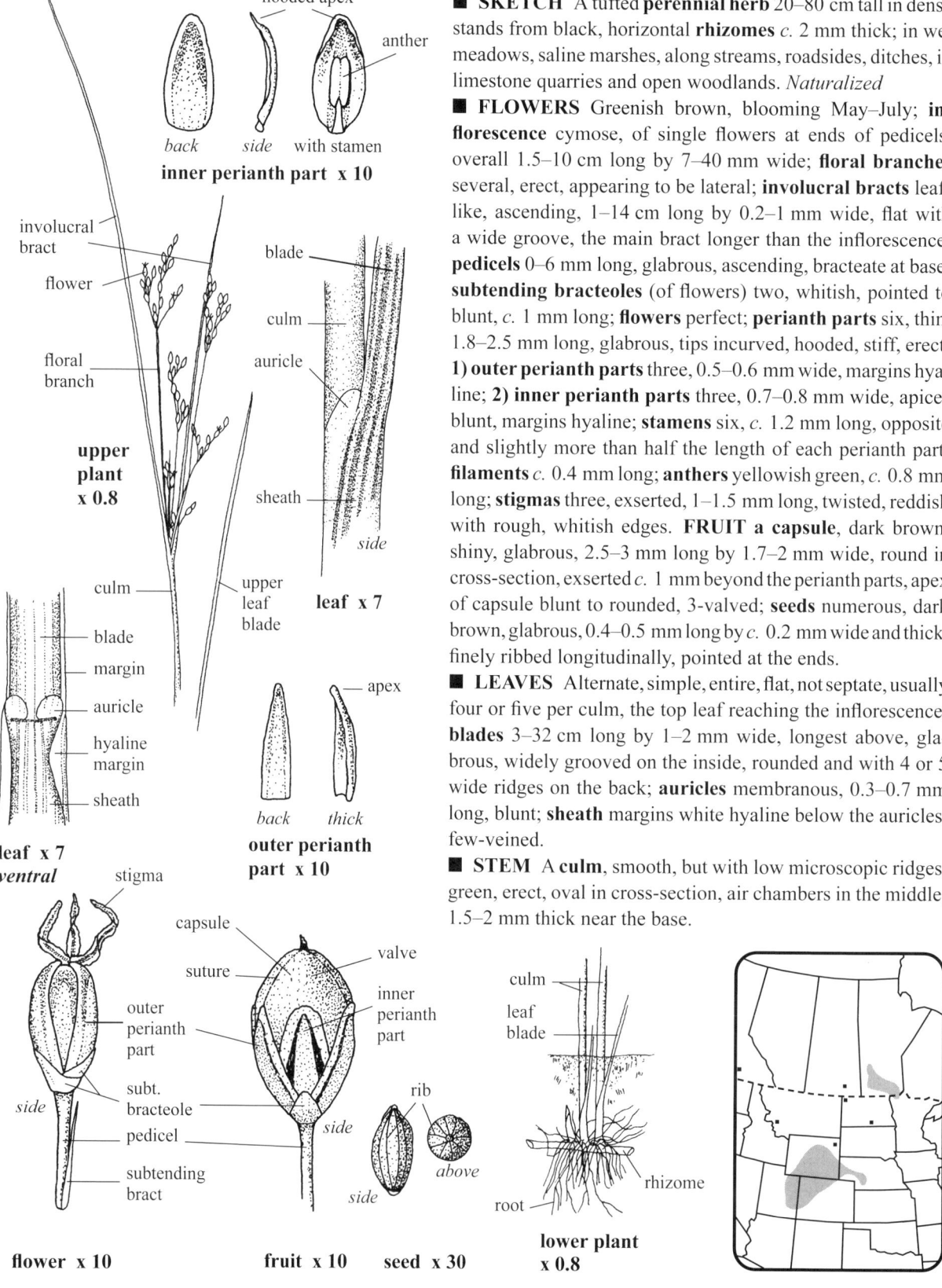

- **SKETCH** A tufted **perennial herb** 20–80 cm tall in dense stands from black, horizontal **rhizomes** *c.* 2 mm thick; in wet meadows, saline marshes, along streams, roadsides, ditches, in limestone quarries and open woodlands. *Naturalized*
- **FLOWERS** Greenish brown, blooming May–July; **inflorescence** cymose, of single flowers at ends of pedicels, overall 1.5–10 cm long by 7–40 mm wide; **floral branches** several, erect, appearing to be lateral; **involucral bracts** leaf-like, ascending, 1–14 cm long by 0.2–1 mm wide, flat with a wide groove, the main bract longer than the inflorescence; **pedicels** 0–6 mm long, glabrous, ascending, bracteate at base; **subtending bracteoles** (of flowers) two, whitish, pointed to blunt, *c.* 1 mm long; **flowers** perfect; **perianth parts** six, thin, 1.8–2.5 mm long, glabrous, tips incurved, hooded, stiff, erect; 1) **outer perianth parts** three, 0.5–0.6 mm wide, margins hyaline; 2) **inner perianth parts** three, 0.7–0.8 mm wide, apices blunt, margins hyaline; **stamens** six, *c.* 1.2 mm long, opposite and slightly more than half the length of each perianth part; **filaments** *c.* 0.4 mm long; **anthers** yellowish green, *c.* 0.8 mm long; **stigmas** three, exserted, 1–1.5 mm long, twisted, reddish with rough, whitish edges. **FRUIT a capsule**, dark brown, shiny, glabrous, 2.5–3 mm long by 1.7–2 mm wide, round in cross-section, exserted *c.* 1 mm beyond the perianth parts, apex of capsule blunt to rounded, 3-valved; **seeds** numerous, dark brown, glabrous, 0.4–0.5 mm long by *c.* 0.2 mm wide and thick, finely ribbed longitudinally, pointed at the ends.
- **LEAVES** Alternate, simple, entire, flat, not septate, usually four or five per culm, the top leaf reaching the inflorescence; **blades** 3–32 cm long by 1–2 mm wide, longest above, glabrous, widely grooved on the inside, rounded and with 4 or 5 wide ridges on the back; **auricles** membranous, 0.3–0.7 mm long, blunt; **sheath** margins white hyaline below the auricles, few-veined.
- **STEM** A **culm**, smooth, but with low microscopic ridges, green, erect, oval in cross-section, air chambers in the middle; 1.5–2 mm thick near the base.

LONG-STYLED RUSH *Juncus longistylis* Torr.

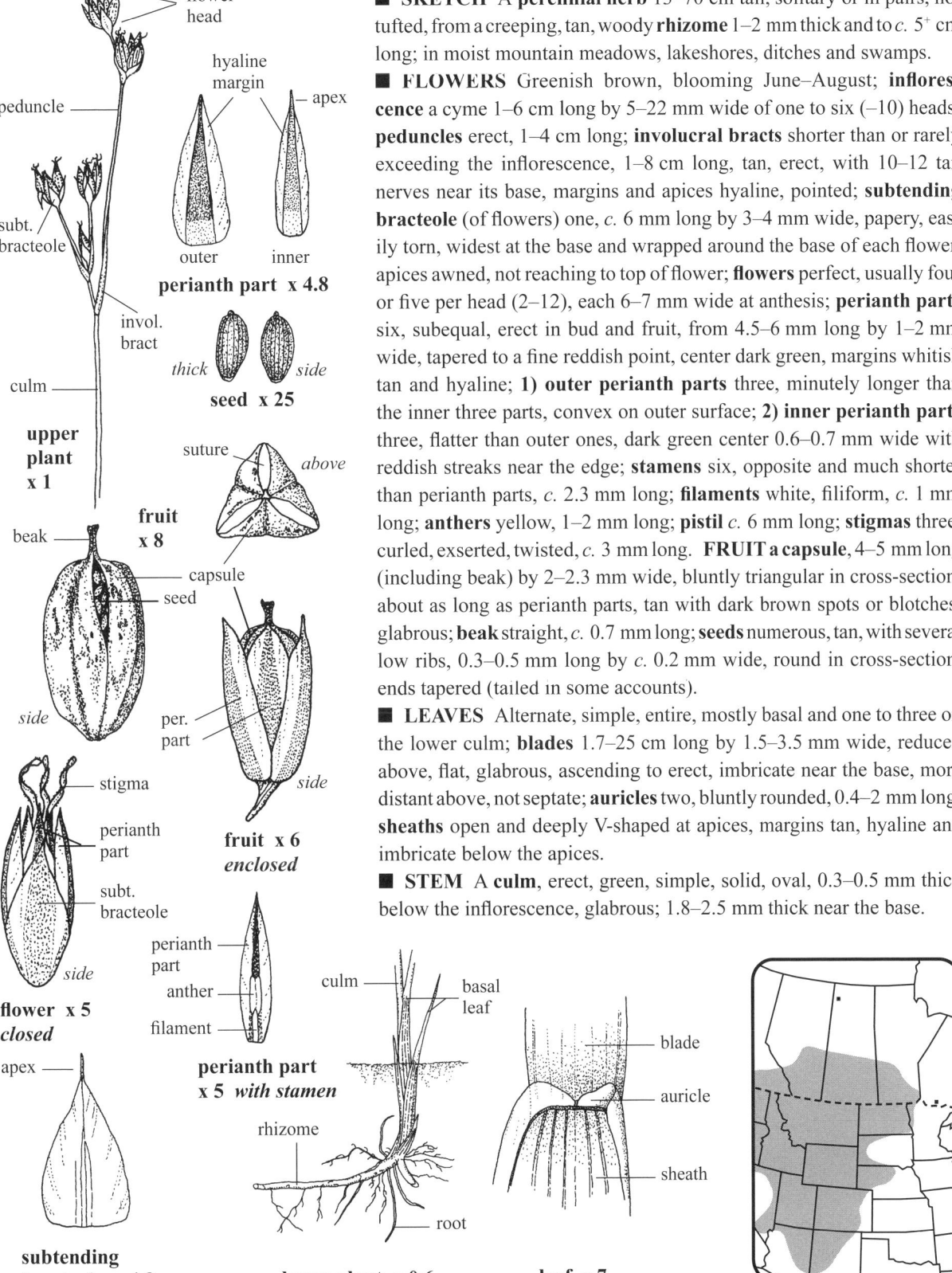

■ **SKETCH** A **perennial herb** 13–70 cm tall, solitary or in pairs, not tufted, from a creeping, tan, woody **rhizome** 1–2 mm thick and to *c.* 5⁺ cm long; in moist mountain meadows, lakeshores, ditches and swamps.

■ **FLOWERS** Greenish brown, blooming June–August; **inflorescence** a cyme 1–6 cm long by 5–22 mm wide of one to six (–10) heads; **peduncles** erect, 1–4 cm long; **involucral bracts** shorter than or rarely exceeding the inflorescence, 1–8 cm long, tan, erect, with 10–12 tan nerves near its base, margins and apices hyaline, pointed; **subtending bracteole** (of flowers) one, *c.* 6 mm long by 3–4 mm wide, papery, easily torn, widest at the base and wrapped around the base of each flower, apices awned, not reaching to top of flower; **flowers** perfect, usually four or five per head (2–12), each 6–7 mm wide at anthesis; **perianth parts** six, subequal, erect in bud and fruit, from 4.5–6 mm long by 1–2 mm wide, tapered to a fine reddish point, center dark green, margins whitish tan and hyaline; 1) **outer perianth parts** three, minutely longer than the inner three parts, convex on outer surface; 2) **inner perianth parts** three, flatter than outer ones, dark green center 0.6–0.7 mm wide with reddish streaks near the edge; **stamens** six, opposite and much shorter than perianth parts, *c.* 2.3 mm long; **filaments** white, filiform, *c.* 1 mm long; **anthers** yellow, 1–2 mm long; **pistil** *c.* 6 mm long; **stigmas** three, curled, exserted, twisted, *c.* 3 mm long. **FRUIT a capsule**, 4–5 mm long (including beak) by 2–2.3 mm wide, bluntly triangular in cross-section, about as long as perianth parts, tan with dark brown spots or blotches, glabrous; **beak** straight, *c.* 0.7 mm long; **seeds** numerous, tan, with several low ribs, 0.3–0.5 mm long by *c.* 0.2 mm wide, round in cross-section, ends tapered (tailed in some accounts).

■ **LEAVES** Alternate, simple, entire, mostly basal and one to three on the lower culm; **blades** 1.7–25 cm long by 1.5–3.5 mm wide, reduced above, flat, glabrous, ascending to erect, imbricate near the base, more distant above, not septate; **auricles** two, bluntly rounded, 0.4–2 mm long; **sheaths** open and deeply V-shaped at apices, margins tan, hyaline and imbricate below the apices.

■ **STEM** A **culm**, erect, green, simple, solid, oval, 0.3–0.5 mm thick below the inflorescence, glabrous; 1.8–2.5 mm thick near the base.

Rush—*Juncaceae* Rush Perianth parts 6, from 2.5–4 mm long

KNOTTED RUSH *Juncus nodosus* L.

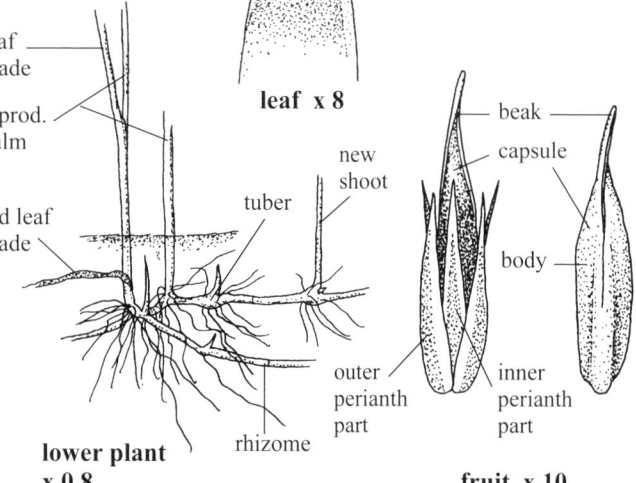
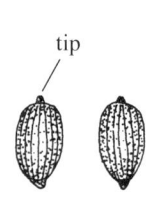
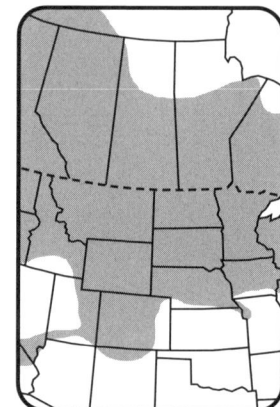

- **SKETCH** A glabrous **perennial herb** 20–70 cm tall from tan, smooth, creeping **rhizomes** 1–2 mm thick with **tubers** (or swollen nodes); in ditches, muddy shores of lakes and streams, swamps, fens, wet prairies and open woods, at times covering several m^2.
- **FLOWERS** Reddish green, blooming June–September; **inflorescence** a raceme, terminal, of 1–15 heads, TL 0.8–7 cm by 8–35 mm wide; **peduncles** green, ascending, 0–40 mm long, much narrower than the culm; **subtending bracts** (of peduncles) 2–15 mm long, tan to greenish; **flower heads** roundish, 6- to 30-flowered, each 5–12 mm wide, sessile to peduncled; **involucral bracts** leaflike, 1.5–13 cm long, usually exceeding the inflorescence; **subtending bracteoles** (of flowers) one, tan, 1-nerved, 2.5–3 mm long by *c.* 1 mm wide, tapered to a fine point; **flowers** perfect, 3–5 mm wide by 3–4 mm tall; **perianth parts** six, 2.5–4 mm long by 0.6–0.8 mm wide, equal or inner parts slightly longer, pointed, flat or with margins involute; **1) outer parts** three, slightly keeled, V-shaped especially in bud; **2) inner parts** three, flatter, margins hyaline; **stamens** six (3), opposite and shorter than each perianth part, 1.3–1.6 mm long; **filaments** filiform, 0.8–0.9 mm long; **anthers** 0.5–0.7 mm long, yellow; **pistil** *c.* 3.5 mm long; **ovary** *c.* 1.2 mm long by 0.7–0.8 mm wide, glabrous, tapered above; **style** *c.* 1 mm long; **stigmas** three, twisted, each part *c.* 1.2 mm long. **FRUIT** a capsule, shiny, dark brown, glabrous, 3.2–5 mm long by 0.7–0.8 mm wide, tapered to a beak, exserted; **seeds** numerous, 0.4–0.5 mm long by 0.2–0.3 mm wide, round in cross-section, ribs numerous.
- **LEAVES** Alternate, simple, entire, two to four per culm, oval, margins blunt, glabrous, green, tapered to a fine point; **blades** 5–20 (–30) cm long by 0.5–3 mm wide, ascending to erect, top leaf reaching to or past the inflorescence, shallowly grooved at auricles and 2–5 cm above; **septa** complete and usually 5–10 mm apart inside the leaves (visible when backlit); **auricles** two, tan, blunt, 0.3–1 mm long, not frayed; **sheaths** open above, margins imbricate below, glabrous.
- **STEM** A **culm**, erect, green, round, unbranched, glabrous; 1–3 mm thick near the base.

Rush

If the Knotted Rush could dream

it would dream rivers

summers spent leaning

into the wind

so go ahead and say it

the pure grace of green

the lush curve of a stem

O what do we really know

though Nature can instruct us

if we open our hearts to it

Torrey's Rush *Juncus torreyi* Coville

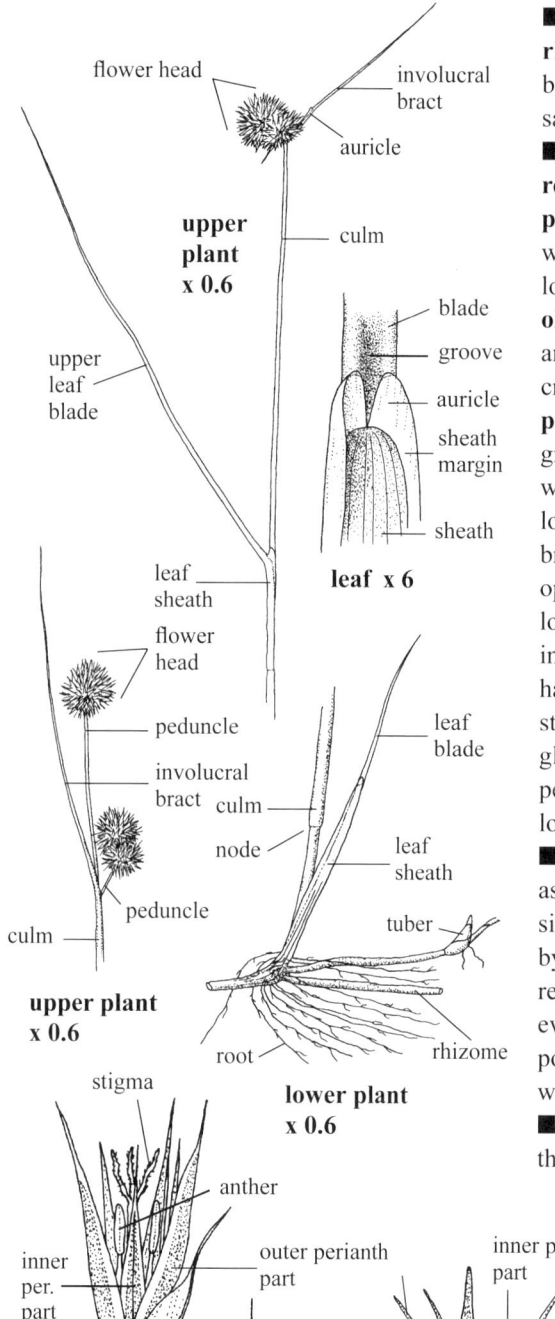

- **SKETCH** A **perennial herb** 15–90 cm tall from creeping, tan **rhizomes** 1–2 mm wide by 1–7+ cm long with **tubers** 3–4 mm wide by 7–10 mm long, forming open colonies; in ditches, marshes, on sand bars and in wet prairies.
- **FLOWERS** Greenish brown, blooming June–October; **inflorescence** 1–6 cm long of 1–13 round heads each 7–17 mm across; **peduncles** green, 5–40 mm long; **involucral bracts** one, leaflike with a sheath, short auricles and grooved to roundish, 1.5–16 cm long, usually longer than the inflorescence, erect to angled; **bracteoles** one per flower, 3.5–4.5 mm long by *c.* 1 mm wide, apices long and tapered, 1- or 3-nerved, shorter than perianth; **flowers** perfect, crowded, 25–100 per head; **perianth parts** six, unequal; **1) outer perianth parts** three, V-shaped, stiff, margins hyaline, center green, apices awnlike and brownish, 4–5.5 mm long by *c.* 0.8 mm wide (flattened); **2) inner perianth parts** three, smaller, 3–4.5 mm long by 0.4–0.5 mm wide, margins hyaline, middle green, apices brownish, thin; **stamens** six, 2–2.5 mm long, erect, included, one opposite each perianth part; **filaments** white hyaline, 1–1.5 mm long; **anthers** pale yellow, 0.8–1 mm long; **pistil** 3–3.3 mm long, included; **ovary** *c.* 1 mm long; **style** *c.* 1 mm long; **stigmas** three, hairy, curled, 1–1.3 mm long. **FRUIT a capsule**, dark brown to straw-colored, 4–5.7 mm long by 0.8–1 mm wide and thick, 3-angled, glabrous, 3-valved, apex equal to or slightly longer than the perianth; **seeds** *c.* 30 per capsule, brown, apiculate, 0.3–0.5 mm long by *c.* 0.2 mm wide and thick, with fine longitudinal ribs.
- **LEAVES** Alternate, simple, entire, two to five per culm, ascending, each with one wide groove along the inner or stem side; **blades** pointed, green to pinkish, roundish, 3–30 cm long by 1.2–5 mm wide, upper blade the longest, slightly longer or not reaching the base of the inflorescence; **septa** pale green, complete, every 2–5 mm apart inside the leaves; **auricles** two, blunt to slightly pointed, 1–3.5 mm long, membranous; **sheaths** 3–6 cm long, with wide, hyaline whitish, overlapping margins.
- **STEM** A **culm**, erect, stiff, oval, smooth, hollow; 2–5 mm thick near the base.

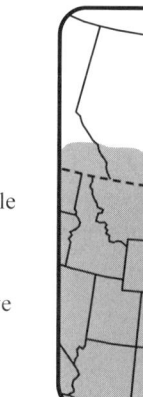

Rush—*Juncaceae* **Wood-rush** Perianth parts 6, from 2.5–4.3 mm long

Hairy Wood-rush   *Luzula acuminata* Raf.

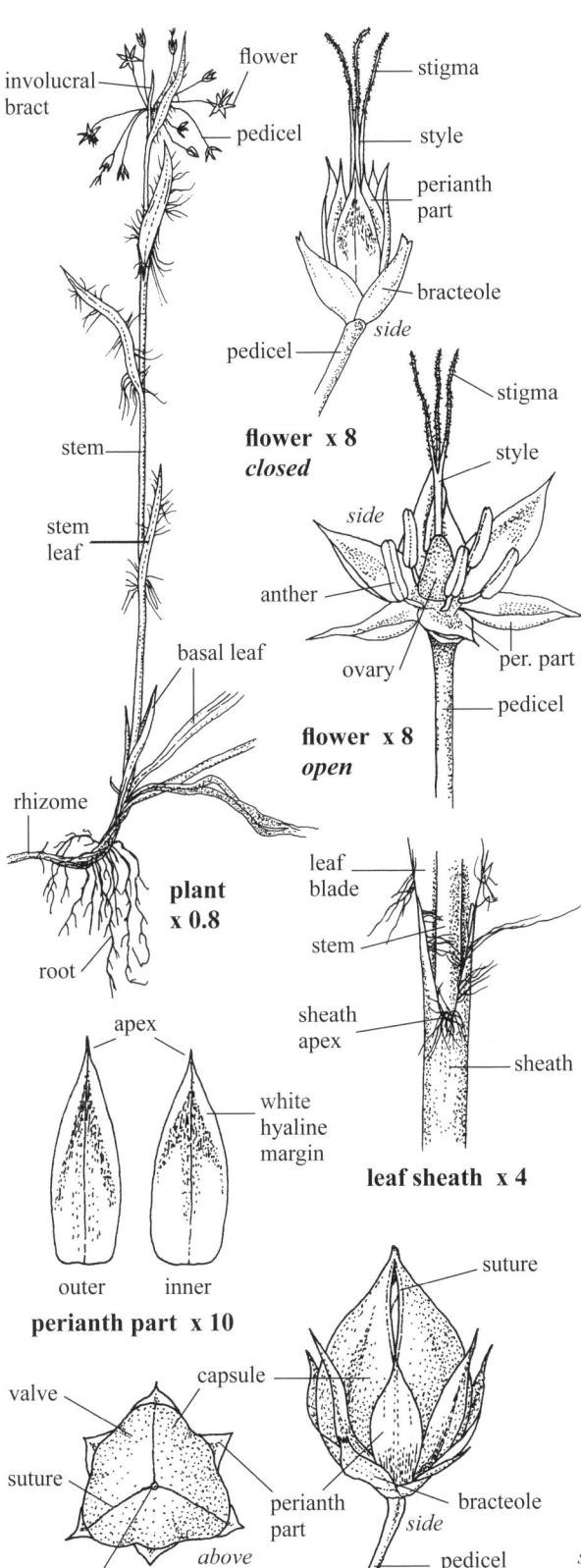

- **SKETCH** A variable, hairy **perennial herb** 10–40 cm tall, singly or in small open colonies, with a brown **rhizome** 5–10⁺ cm long by 1–2 mm thick; in open woods, ditches, meadows and rocky outcrops.
- **FLOWERS** Greenish brown, blooming May–June; **inflorescence** corymbose (umbel-like), short, terminal, 1.5–8 cm long and wide; **pedicels** green, glabrous, filiform, 0.5–4 cm long, extending to 6 cm with fruit, erect to declining, some branching; **subtending bracts** (of pedicels) leaflike, hairy or not, membranous, 2–5 mm long, wrapped around the lower part of the pedicel; **flowers** perfect, 10–35 per stem, each 4–7 mm wide and long; **involucral bracts** one, 2–15 mm long, white membranous near apex; **bracteoles** two, subtending each flower, 1.5–2 mm long by *c.* 1.2 mm wide, white, membranous, apices blunt and erose or with an awn to *c.* 0.5 mm long from a bidentate tip; **rachis** green, slightly hairy; **perianth parts** six, pointed, 2.5–4.3 mm long by 1–1.5 mm wide, spreading, margins whitish hyaline, centers reddish brown; **stamens** six, exserted; **filaments** white, *c.* 0.3 mm long; **anthers** yellow, 1–1.2 mm long, twisted when spent; **pistil** one, glabrous; **ovary** dark green, *c.* 1 mm long, bluntly triangular, tapered; **style** lighter green, *c.* 1 mm long; **stigmas** three, hairy, *c.* 1.5 mm long, green, filiform. **FRUIT a capsule**, 3–4.5 mm long by 2.5–3 mm wide, brown, roundly triangular, glabrous, longer than perianth parts, dull, pointed, opening at apex by three sutures, often pointing down; **seeds** three per capsule, dark reddish brown, shiny, roundish, 1.2–1.5 mm long by 1–1.2 mm wide and thick; **elaiosomes** white, the three intertwined at their apices, each *c.* 1.2 mm long (curled) and tapered.
- **LEAVES** Basal and stem, alternate, simple, entire, sessile; **basal blades** hairless, few, 5–30 cm long by 2–12 mm wide, spreading; **stem blades** ascending, 1–4 cm long by 1–5 mm wide, reduced above, two to four per stem, hairy on margins, hairs to *c.* 4 mm long, tangled; **sheaths** V-shaped at hairy apices, closed, body hairless.
- **STEM** Erect, green, simple; 1–2 mm thick at base.
- **SYN.** *Luzula pilosa* (L.) Willd. = *Luzula acuminata* var. *acuminata* Raf.

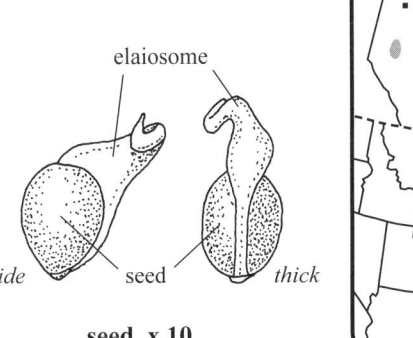

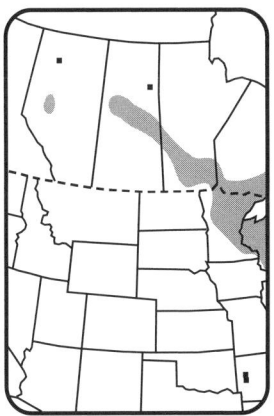

589

Rush—*Juncaceae*   **Wood-rush**   Perianth parts 6, from 1.5–2.5 mm long

SMALL-FLOWERED WOOD-RUSH   *Luzula parviflora* (Ehrh.) Desv.

■ **SKETCH** A loosely tufted **perennial herb** 20–80 cm tall with light tan **rhizomes** to *c.* 5 cm long by *c.* 1 mm wide with dark, narrow bracts 1–2 cm long from nodes, and brown **fibrous roots** 1–3 cm long; in moist open woods, mountain forests, thickets, shores, meadows and wetlands.

■ **FLOWERS** Greenish brown, blooming June–August; **inflorescence** a cyme, 5–15 cm long and wide; **involucral bracts** (of inflorescence) one, leaflike, erect, 1.5–5 cm long; **floral branches** green, 2–15 cm long, thin and drooping; **subtending bracts** (of floral branches) two, 3–33 mm long, the lower one open and pointed, often with a few scattered white hairs, the upper (inner) one enveloping the branch, pointed to flat and toothed; **pedicels** green, 1–12 mm long; **subtending bracts** (of pedicels) two, 1–2 mm long, as above; **flowers** perfect, 2.5–3 mm wide by *c.* 3 mm long; **subtending bracteoles** (of flowers) two, the lower one *c.* 0.9 mm long and wide, upper bract *c.* 1 mm long and wide, entire or toothed; **perianth parts** six, 1.5–2.5 mm long by 0.7–1 mm wide, imbricate at base, persistent and shorter than capsule, outer three with brownish apices, inner three pale green; **stamens** *c.* 1.2 mm long; **filament** slightly shorter than anther, *c.* 0.5 mm long; **anther** pale yellow, *c.* 0.7 mm long; **pistil** one, glabrous, 2–2.5 mm long; **ovary** 3-lobed, *c.* 0.7 mm wide; **style** white, erect; **stigmas** three, ascending, white, slightly longer than style. **FRUIT a capsule**, reddish brown to black, triangular, shiny, *c.* 2.5 mm long (including beak) by 1.3–1.5 mm wide, opening at apex; **beak** *c.* 0.2 mm long (persistent style); **seeds** three per capsule, 1.1–1.5 mm long by *c.* 0.8 mm wide, base angled and with cottony hairs (the funiculus) *c.* 1 mm long, rounded outside, inner sides flat, shiny, smooth, finely reticulate, brown with transparent ends.

■ **LEAVES** Basal and stem, alternate, grasslike, entire, dull to shiny, glabrous except for a tuft of white hairs 2–7 mm long at the base of the blade, the blade's tip brownish; **basal leaves** few, spreading, 6–17 cm long, imbricate; **stem blades** erect to ascending, three to nine along the stem, 3–30 cm long by 3–15 mm wide, reduced above and below; **sheaths** 2–3 cm long, tight, closed, glabrous.

■ **STEM** Erect, green, round, glabrous, solid; **nodes** not obvious; **central internodes** often reddish; 3–6 mm wide at the reddish brown base.

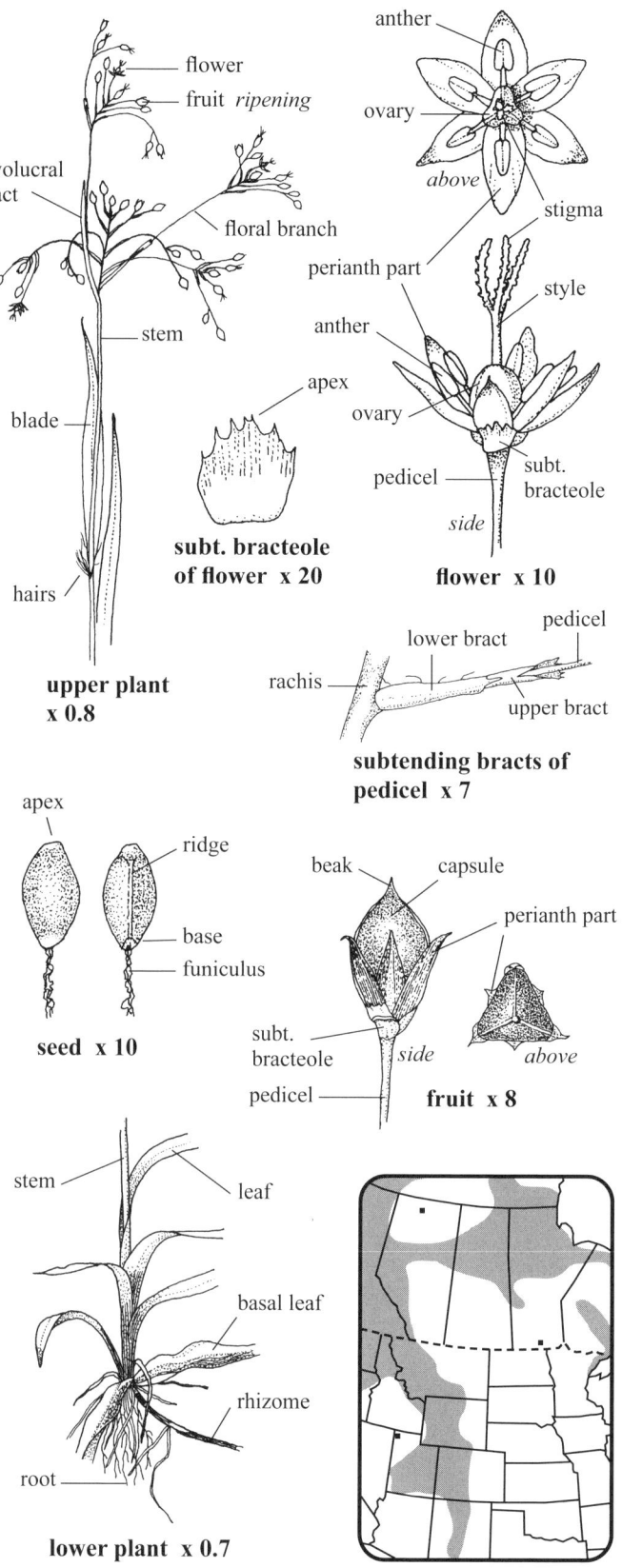

590

Arrow-grass—*Juncaginaceae*

Wildflower Green Perianth parts 6

SEASIDE ARROW-GRASS *Triglochin maritima* L.

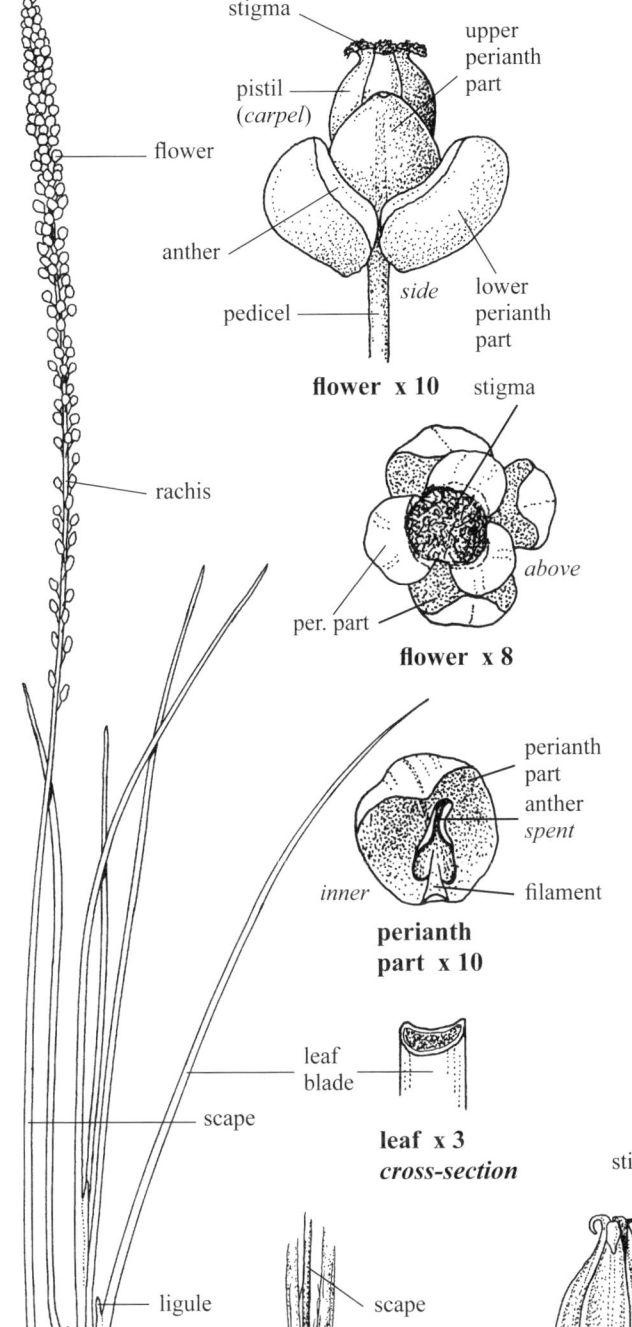

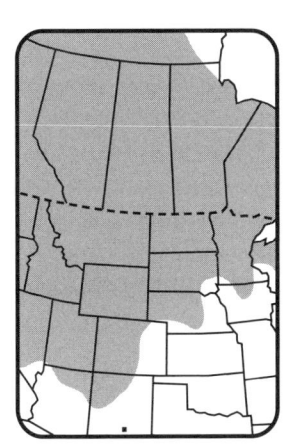

■ **SKETCH** A glabrous **perennial herb** 20–70 cm tall in clumps from a short, stout **rhizome** to *c.* 10 mm thick; along ditches, stream banks, bogs and beaver ponds, often in coastal saline areas. **w2**

■ **FLOWERS** Green, blooming April–August; **inflorescence** a raceme (spikelike), erect, 6–50 cm long by 5–13 mm wide, exceeding the leaves, about equal in length to the scape below, blooming sequence bottom to top; **pedicels** ascending, 1.5–4 mm long; **subtending bracts** absent; **flowers** perfect, numerous *c.* 3 mm wide and long, crowded above; **perianth parts** six, in two series, each 1.5–2 mm long, membranous, curved and ascending; **stamens** six, included, lower stamens in each flower releasing pollen first; **filaments** short, stout, green, one attached to the base of each perianth part and falling with it; **anthers** yellowish green, *c.* 1.8 mm long by *c.* 1.5 mm wide; **pistils** six, fertile, *c.* 2 mm long by *c.* 1.3 mm wide, composed of six carpels united to a central vertical carpophore; **styles** obscure; **stigmas** six, plumose, together *c.* 1 mm wide. **FRUIT a schizocarp**, brown, *c.* 2 mm wide; **mericarps** six, 1-seeded, each 2.5–5.5 mm long, base rounded, concave on the outside, falling separately; **beak** erect to recurved; **seeds** 2.5–3 mm long by 0.5–0.7 mm wide, with faint grooves, triangular.

■ **LEAVES** Basal, simple, entire, linear, fleshy and soft, half round, erect, not distinctively veined, 5–50+ cm long by 2–5 mm wide by *c.* 1 mm thick; **ligules** lobed or not, 0–5 mm long at the apex of a sheath; **basal sheaths** 0.7–2.5 cm long with overlapping membranous margins; **leaf bases** fibrous and persistent around the rhizome.

■ **STEM** A **scape**, round, erect, glabrous, stiff when in fruit; 3–8 mm thick near the green or purple base.

Arrow-grass—*Juncaginaceae* **Wildflower** Yellowish green Perianth parts 6

Marsh Arrow-grass   *Triglochin palustre* L.

- **SKETCH** A tufted, glabrous **perennial herb** 15–40 cm tall from white slender **rhizomes**; in brackish marshes, wet meadows, bogs, along shores and river banks. w2
- **FLOWERS** Yellowish green, blooming June–September; **inflorescence** a raceme 5–21 cm long by 3–5 mm wide, erect; **pedicels** erect to ascending, alternate, 1–5 mm long by *c.* 0.3 mm thick; **flowers** perfect, 1.8–2.5 mm wide by 2–2.5 mm tall; **perianth parts** six, in two series, green, entire, 1.1–1.6 mm long by 0.7–1 mm wide, C-shaped, with a short point, imbricate, the three lower parts falling early; **stamens** six, included, subsessile, one under each perianth part, the lower anthers spent early and deciduous, the upper anthers opening later; **filaments** obscure; **anthers** sessile, pale yellowish green, *c.* 1 mm long and wide; **pistils** six, green, glabrous, but with only three fertile carpels, *c.* 1 mm long by *c.* 0.8 mm wide; **style** obscure; **stigmas** three, hairs transparent and shiny, together *c.* 1 mm wide; **fruiting stalk** erect and to *c.* 23 cm long by 4–6 mm wide. **FRUIT a schizocarp**, brown; **mericarps** three, 1-seeded, separating from central carpophore, 5–8 mm long by *c.* 0.6 mm wide by *c.* 0.4 mm thick, triangular in cross-section, bases parting and tapered to a fine point; **carpophore** deeply 3-angled, tan, 6–7 mm long by *c.* 0.5 mm wide, the fruit fitting in the longitudinal depressions with the suture towards the middle; **seeds** 1–3 per flower, each 3–4 mm long by *c.* 0.4 mm wide by *c.* 0.3 mm thick, with a narrow low ridge or line on one side and shallow, longitudinal grooves on the other side.
- **LEAVES** Basal, simple, entire, glabrous, few to several, fleshy, attached at the scape's base by a membranous sheath; **blades** green, tapered to a fine point, 6–30 cm long by 1–2.5 mm wide, some reaching the inflorescence, oval in cross-section, erect to ascending, with a shallow groove; **ligules** membranous, of two blunt lobes 0.5–1 mm long; **sheaths** membranous, 5–10 cm long, apices open, enveloping the scape below.
- **STEM** A **scape**, green, hollow, stiff, erect, round, glabrous, often with tillers; 1–2 mm thick near the base.

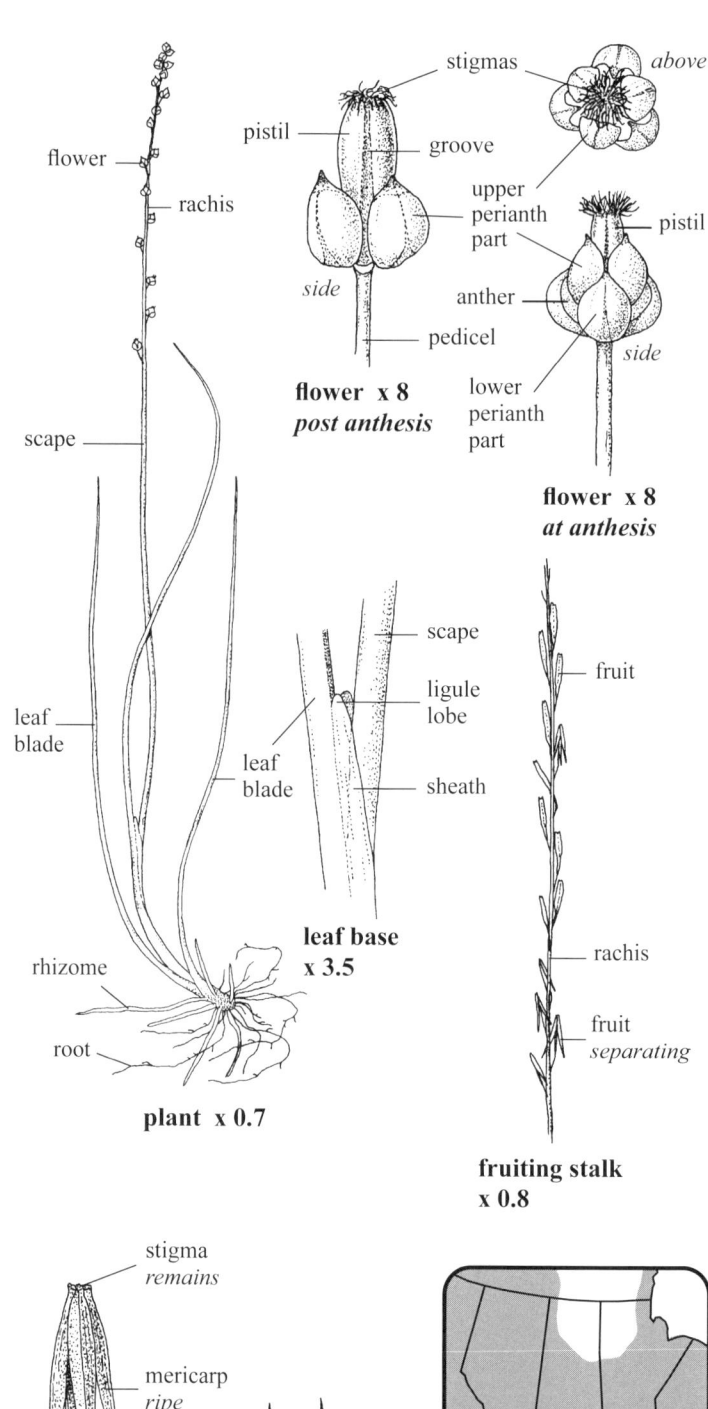

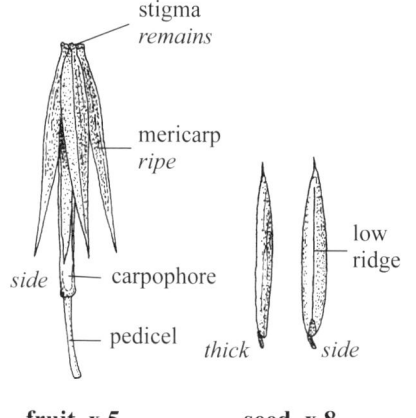

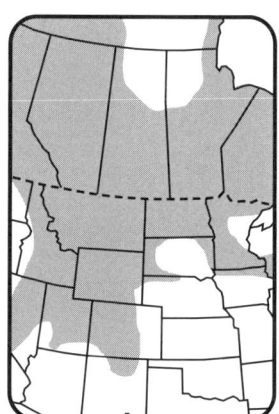

Duckweed—*Lemnaceae* **Wildflower** Green Perianth absent, flowers rarely seen

IVY-LEAVED DUCKWEED *Lemna trisulca* L.

■ **SKETCH** An aquatic **perennial herb** submerged to floating in long branched chains of seemingly tangled colonies from near the bottom (to 12 m deep) to across the water's surface (when flowering), not rooted in bottom; in clear, quiet water of ponds and sloughs.

■ **FLOWERS** Green, unisexual, rare; **perianth** absent; <u>male **flowers**</u> one stamen *c.* 1 mm long; **filament** much longer than anther; <u>female **flowers**</u> with one ovule; (flowers not seen by author). **FRUIT a utricle**, 1-seeded, 0.6–0.9 mm long, winged near apex; **seeds** 12–18 ribbed, adherent to fruit wall when ripe.

■ **LEAVES** and **STEM** A **thallus** (or frond), green, 6–10 mm long by 2–4 mm wide, 3–50 joined together, paper-thin, apices often curled, appears in triads, three veins obscure, with a slightly darker nodal spot above where the root and lateral vegetative buds originate, semi-transparent to dark green, narrowed to a stipe at the base; **stipes** (stalks) 5–15 mm long, wider near the thallus, flattened, edges blunt, *c.* 0.3 mm thick by 0.5–0.8 mm wide, often with one twist; **lateral vegetative thallus buds** are almost transparent, growing between the thin upper and lower layers of the parent thallus, easily removed with tweezers, new bud veins obscure; **roots** absent (early deciduous), or one per thallus, 5–25 mm long by 0.2 mm thick, attached to middle of thallus node near the stipe end; **root sheath** (cap) 1.8–2 mm long *c.* 0.3 mm wide, pointed, not winged, slightly darker green than the root.

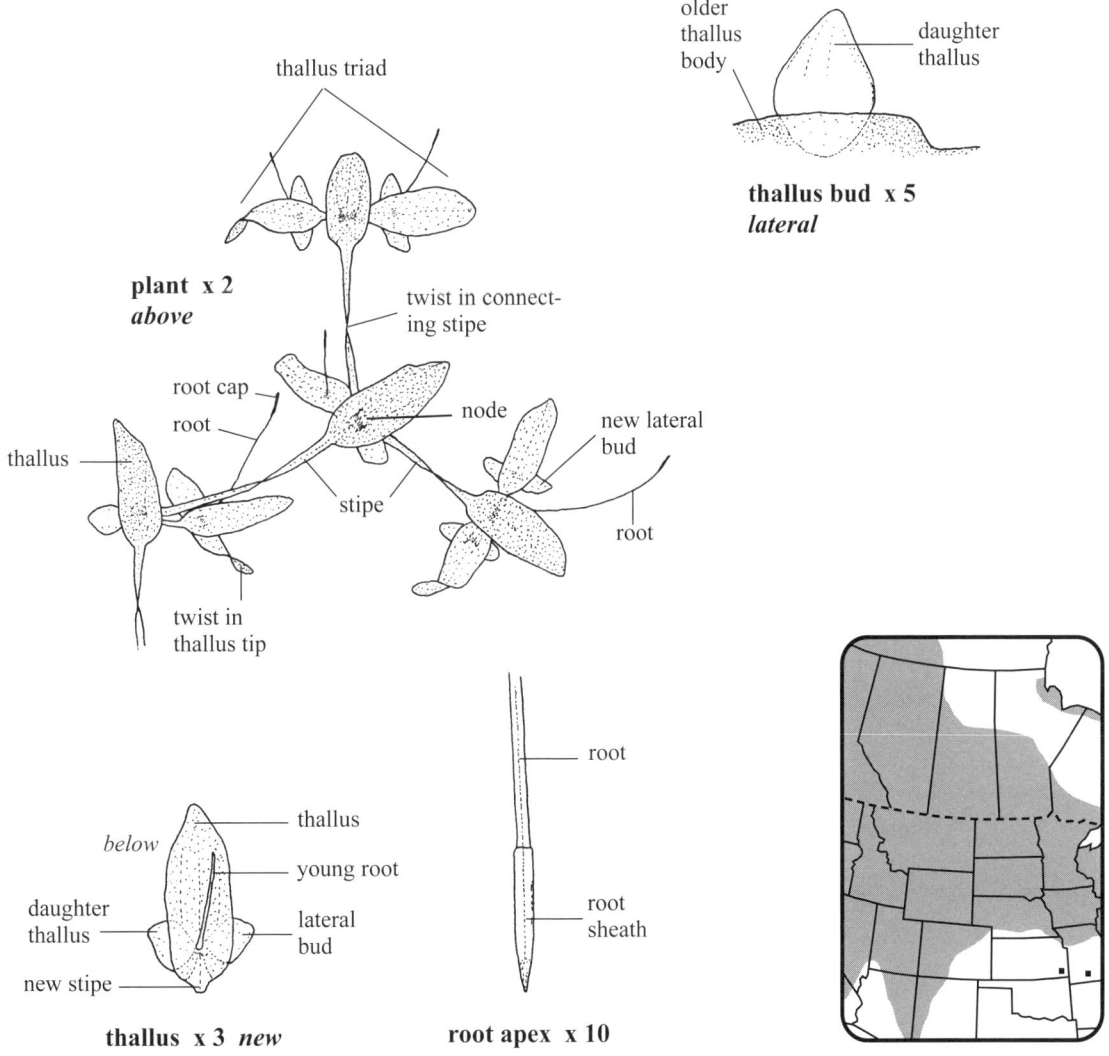

Star Duckweed

Duckweed—*Lemnaceae* **Wildflower** Green Perianth absent, flowers rarely seen

Lesser Duckweed *Lemna turionifera* Landolt

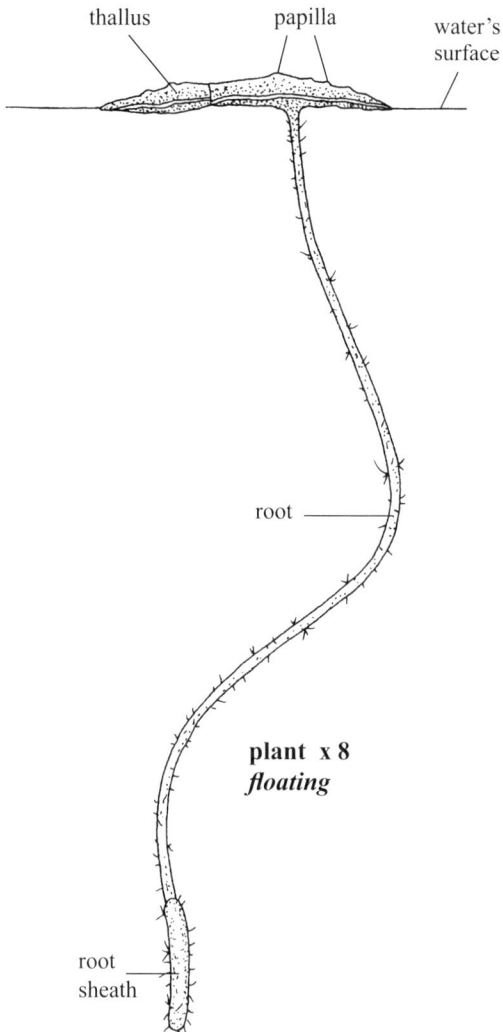

■ **SKETCH** An aquatic **perennial herb** floating freely on the water's surface in quiet ponds, sloughs, stream edges and ditches, often forming large mats covering the surface of ponds by late summer; **turions** (bulbets of compact fronds) 0.8–1.6 mm wide, rootless, overwinter by sinking to the bottom and then rising to the surface in the spring to resume growth; **monoecious. w1**

■ **FLOWERS** Green, rarely seen, blooming June–September, obscure, lacking a perianth and with minimal reproductive parts; **plants** reproduce mainly by vegetative or asexual budding from lateral pouches; <u>**male flowers**</u> with one stamen; <u>**female flowers**</u> usually one, of one pistil; **flower pouches** (when present) two, near the thallus base. **FRUIT a utricle**, 1-seeded, winged near apex; **seeds** with 30–60 obscure ribs, persistent in fruit.

■ **LEAVES** and **STEM** Undifferentiated, consisting of roundish, bright green, floating **thalli** (fronds) obscurely 3-nerved, 2–6 mm long by 2–2.5 mm wide by *c.* 0.5 mm thick, entire, mostly symmetrical, smooth and slightly convex above with a low central ridge with a few **papillae**, hairless, flat and usually reddish below, often with one or more small marginal thallus buds; **roots** a single, pale green, filiform strand usually from near the middle of the thallus opposite the nodal point, 2–14 cm long by *c.* 0.2 mm thick; **root sheath** (cap) wingless, 2–2.2 mm long by *c.* 0.3 mm wide, darker green than root, blunt, with microscopic brownish hairs.

■ **NOTE** Map includes the wide-ranging *Lemna turionifera* and the more restricted *Lemna minor*.

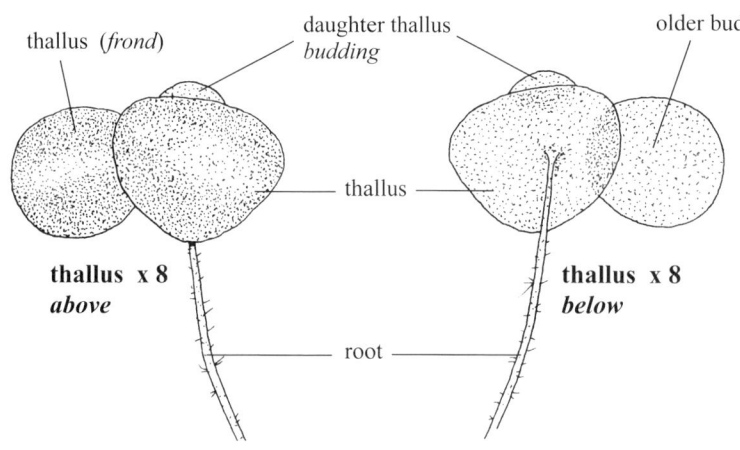

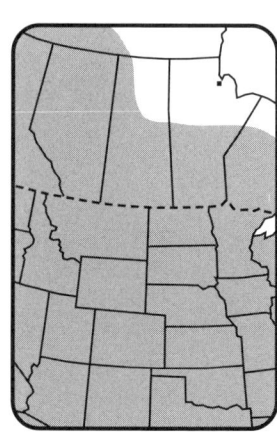

Duckweed—*Lemnaceae* **Wildflower** Green Flowers rarely seen Roots numerous

LARGER DUCKWEED *Spirodela polyrrhiza* (L.) Schleid

■ **SKETCH** An **aquatic perennial herb** in groups of two to five thalli, each group with two or more clusters of roots; floating on quiet water of marshes, ponds and sloughs among other duckweeds or in pure colonies; overwinters as minute turions; **monoecious. w1**

■ **FLOWERS** Rarely produced or observed (not illustrated), tiny, when present usually in two lateral **pouches**; **inflorescence** unisexual, appearing as a perfect flower without perianth parts; <u>**male flowers**</u> one or two, each consisting of one stamen; <u>**female flowers**</u> one, ovules 1 or 2. **FRUIT a utricle**, small-winged, 1–1.5 mm long, 2-seeded; **seeds** tiny, with 12–20 distinct ribs.

■ **LEAVES** and **STEM** Green and red, not distinctive; **thalli** (fronds) 3–10 mm long and wide, 3- to 21-nerved, oval, green above with one to a few microscopic papillae along the central nerve, usually reddish below, flat, thin, margin entire and sometimes with a thin, red line extending from below; **nerves** several and obscure, slightly recessed below, radiating from a slight, off-center depression (nodal point) which may have a reddish spot above, the depression resulting from where the roots meet a thallus; **stipes** (not shown) uniting thalli into a group, white and similar to the roots in appearance and thickness, best viewed from below, sometimes seen from above where a short gap exists between the thalli; **roots** 2–21, 0.2–3 cm long by *c.* 0.1 mm thick, easily broken from the thallus, not branched but clustered from a single nodal point, each with a hyaline margin; **sheath** (cap) of roots 0.5–1.3 mm long, pointed, slightly wider and slightly darker green than the root.

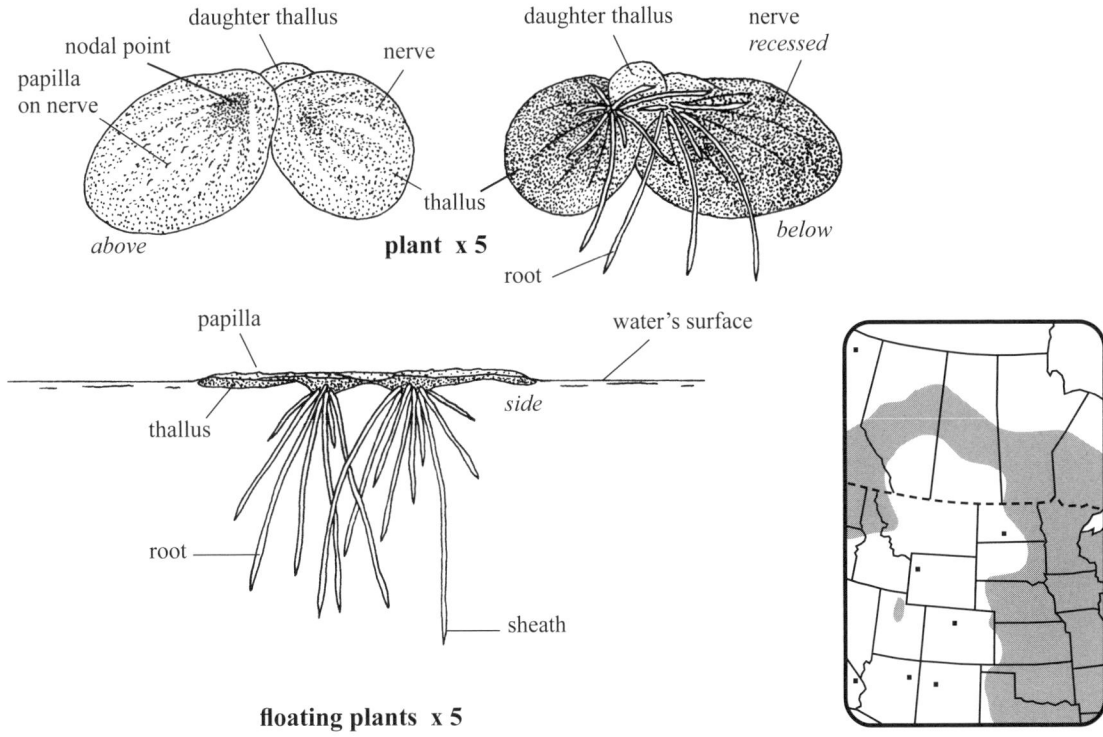

plant x 5

floating plants x 5

Common Duckmeat

Family Characteristics

Lily—Liliaceae — Flower parts 3, 4 or 6

SKETCH Monocotyledonous perennial herbs (wildflowers), sometimes woody, with rhizomes, bulbs and fleshy roots. Worldwide about 4,000 species.

FLOWERS Perfect, regular, often showy. **Inflorescence** a raceme, or rarely an umbel, solitary or several flowers. **Perianth** of 6 (4) parts, usually separate or united into a tube, the outer whorl (3 sepals) and inner whorl (3 petals) often similar in shape and color. **Stamens** usually 6 (4, 3), opposite the perianth parts, free or attached to perianth. **Pistil** 1. **Ovary** often superior, 3-lobed with 3 united locules; ovules numerous. **Style** 1 or 3. **Stigmas** 1 or 3, trilobed or 3-crested.

FRUIT A dry capsule or fleshy berry. **Seeds** one to over 100, flat to oval.

LEAVES Alternate or in whorls, rarely opposite, simple, mostly with parallel veins, entire, basal and stem.

STEM Erect, green, solid, usually simple, some branched.

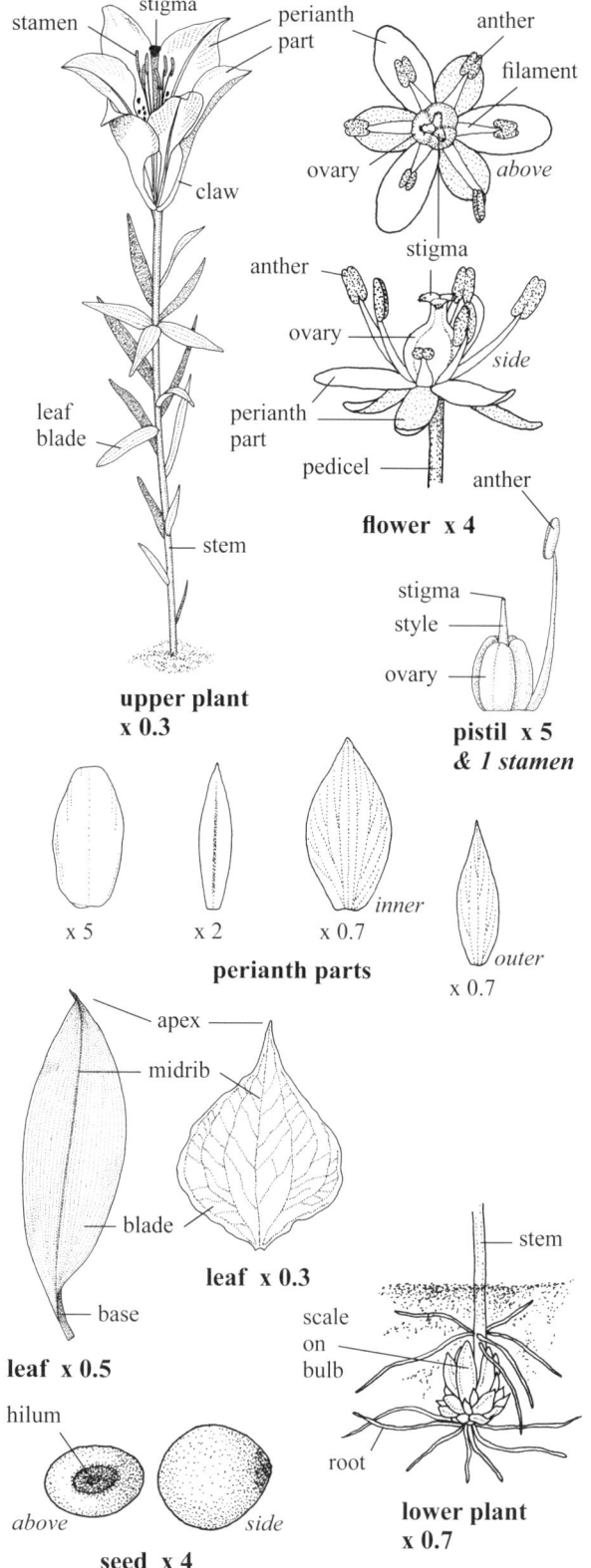

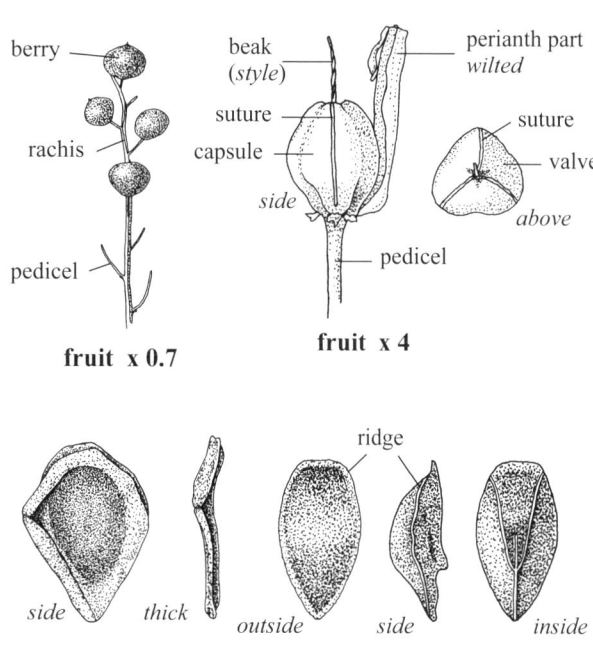

596

Lily—Liliaceae   **Wildflower** Pink Perianth parts 6

Wild Chives *Allium schoenoprasum* L.

■ **SKETCH** A tufted **perennial herb**, glabrous and glaucous, 20–70 cm tall from an oval, smooth white **bulb** *c.* 10 mm long by 4–5 mm wide, these often clumped together and covered with a membrane or sheath, onion-scented when bruised; **fibrous roots** white, 4–5 cm long by 0.5–0.8 mm thick; on rocky outcrops, banks, shores and moist meadows. Native and *naturalized*

■ **FLOWERS** Pink, blooming June–July; **inflorescence** an umbel, terminal, 2–3 cm long by 2–5 cm wide, 30- to 50-flowered; **spathe** (bract) dry and membranous, 9–12 mm long, wrapped entirely around the base of flowers, splitting into two halves, light purplish green, veined, pointed; **pedicels** glabrous and glaucous, green, 4–8 mm long, ascending to erect even with fruit; **flowers** perfect, 10–15 mm wide by 8–10 mm tall; **perianth parts** six, similar, 8–14 mm long by 1.8–2.2 mm wide, with a greenish brown midvein; **stamens** six, included, *c.* 5.5 mm long, opposite each perianth part; **filaments** white, *c.* 4.5 mm long; **anthers** light pink, *c.* 1 mm long; **pistil** one, *c.* 3 mm long, included; **ovary** 3-lobed, green, *c.* 2 mm long by *c.* 1.8 mm wide; **style** white, filiform, tapered, *c.* 1.2 mm long; **stigma** white, obscure, not as wide as style. **FRUIT a capsule**, reticulate, 3.5–4 mm long by *c.* 3.2 mm wide, tan, 3-valved, covered by perianth parts, concave at base, sutures three, opening from apex; **beak** (style) *c.* 2 mm long, erect; **seeds** 3 or 4 per capsule, ridged, black, 2.5–3.2 mm long by *c.* 1.3 mm wide by 1–1.4 mm thick.

■ **LEAVES** Basal and stem, alternate, simple, sessile, entire, thickish, hollow, glabrous, easily crushed; **stem leaves** one to three, ascending but usually not reaching the inflorescence, 9–25 cm long by 2–5 mm wide, tapered to a point, usually brown and wilting by anthesis, lower leaf 6–11 cm long, reduced in size compared to the upper leaf; **sheaths** closed and clasping the stem, extending to base of stem, V-shaped at hyaline apices; **ligules** unremarkable.

■ **STEM** A **scape**, erect, unbranched, solid below, slightly hollow above, oval; glabrous near the 2–4 mm wide base.

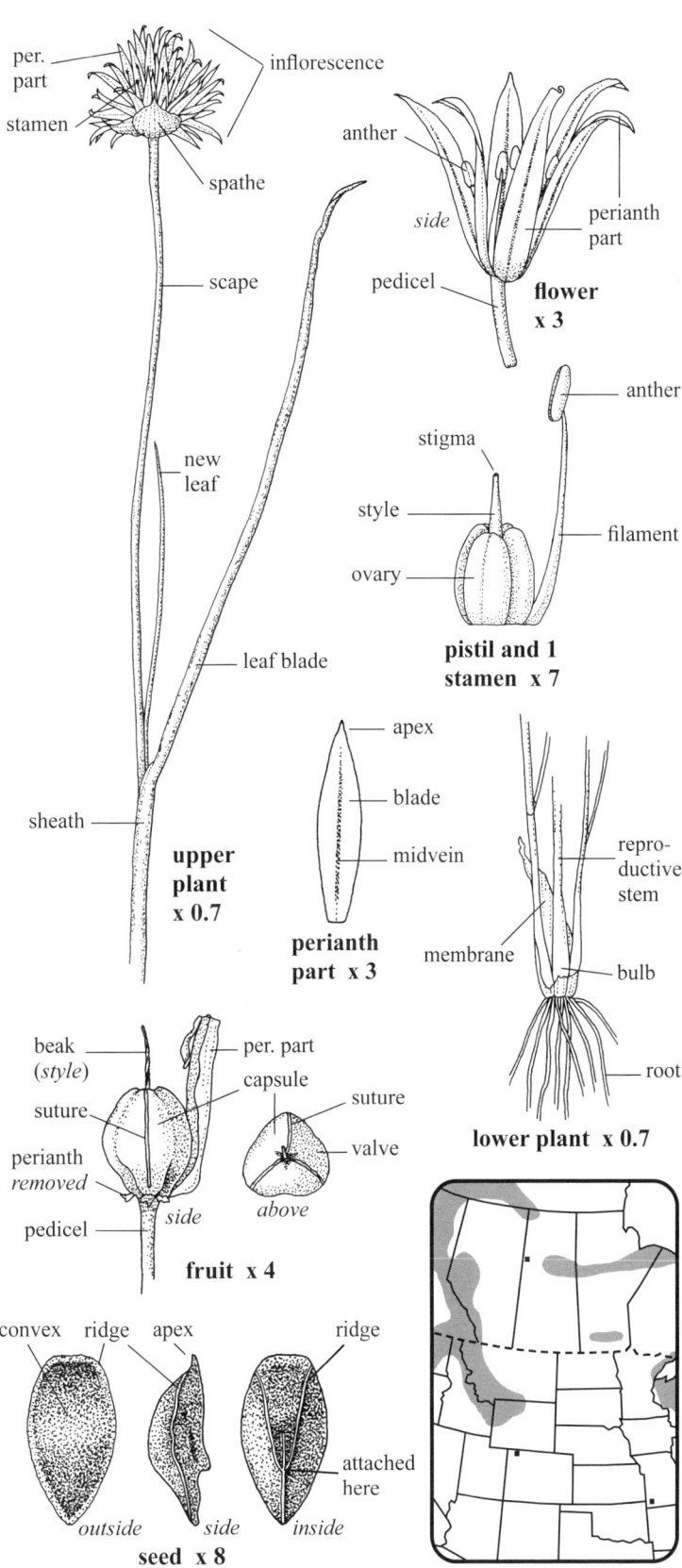

Siberian Wild Chives, Chives

Lily—*Liliaceae* **Wildflower** Pink to purple Perianth parts 6

PINK-FLOWERED ONION *Allium stellatum* Nutt. ex Ker-Gawl.

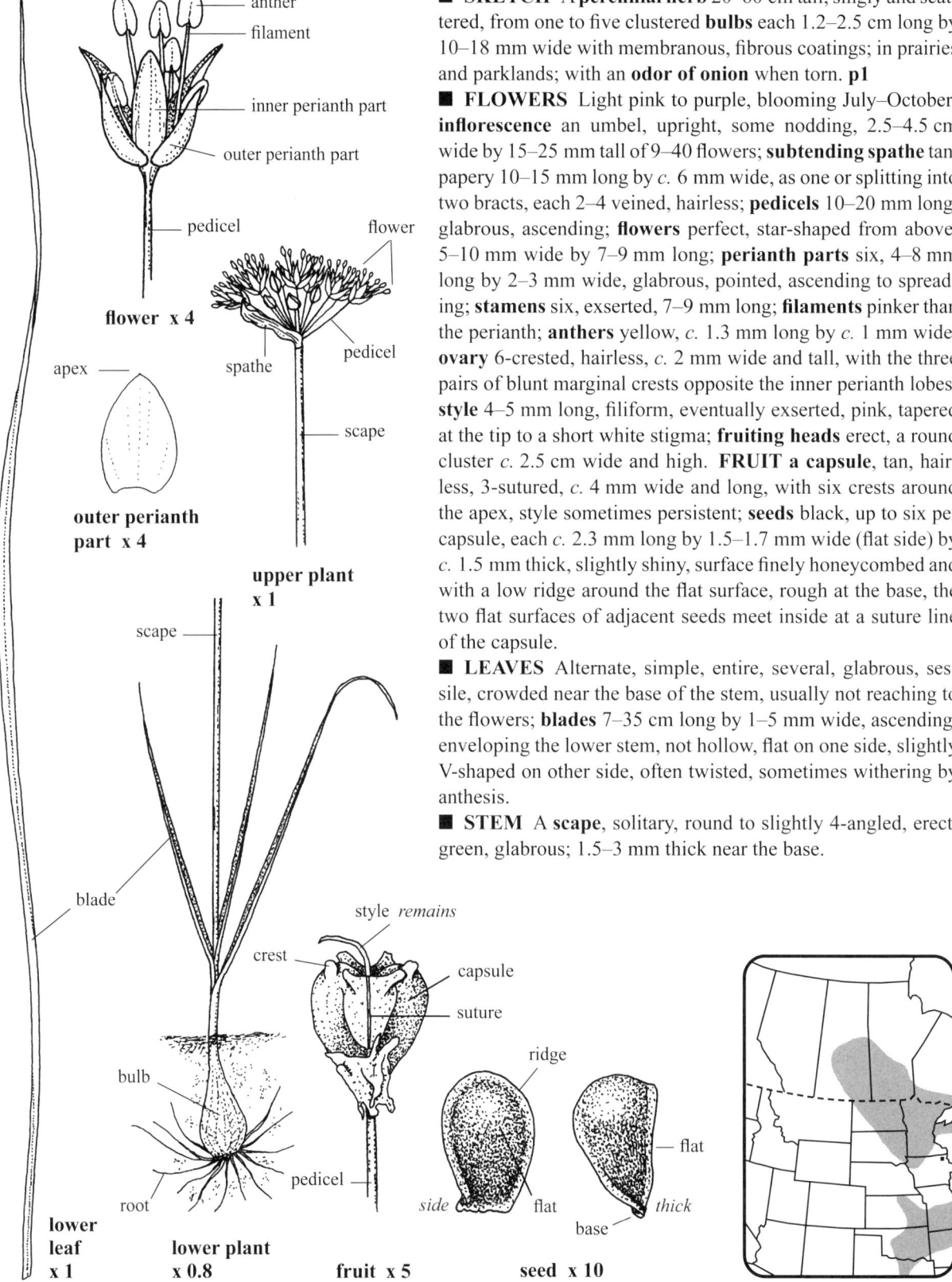

- **SKETCH** A **perennial herb** 20–60 cm tall, singly and scattered, from one to five clustered **bulbs** each 1.2–2.5 cm long by 10–18 mm wide with membranous, fibrous coatings; in prairies and parklands; with an **odor of onion** when torn. **p1**
- **FLOWERS** Light pink to purple, blooming July–October; **inflorescence** an umbel, upright, some nodding, 2.5–4.5 cm wide by 15–25 mm tall of 9–40 flowers; **subtending spathe** tan, papery 10–15 mm long by *c.* 6 mm wide, as one or splitting into two bracts, each 2–4 veined, hairless; **pedicels** 10–20 mm long, glabrous, ascending; **flowers** perfect, star-shaped from above, 5–10 mm wide by 7–9 mm long; **perianth parts** six, 4–8 mm long by 2–3 mm wide, glabrous, pointed, ascending to spreading; **stamens** six, exserted, 7–9 mm long; **filaments** pinker than the perianth; **anthers** yellow, *c.* 1.3 mm long by *c.* 1 mm wide; **ovary** 6-crested, hairless, *c.* 2 mm wide and tall, with the three pairs of blunt marginal crests opposite the inner perianth lobes; **style** 4–5 mm long, filiform, eventually exserted, pink, tapered at the tip to a short white stigma; **fruiting heads** erect, a round cluster *c.* 2.5 cm wide and high. **FRUIT a capsule**, tan, hairless, 3-sutured, *c.* 4 mm wide and long, with six crests around the apex, style sometimes persistent; **seeds** black, up to six per capsule, each *c.* 2.3 mm long by 1.5–1.7 mm wide (flat side) by *c.* 1.5 mm thick, slightly shiny, surface finely honeycombed and with a low ridge around the flat surface, rough at the base, the two flat surfaces of adjacent seeds meet inside at a suture line of the capsule.
- **LEAVES** Alternate, simple, entire, several, glabrous, sessile, crowded near the base of the stem, usually not reaching to the flowers; **blades** 7–35 cm long by 1–5 mm wide, ascending, enveloping the lower stem, not hollow, flat on one side, slightly V-shaped on other side, often twisted, sometimes withering by anthesis.
- **STEM** A **scape**, solitary, round to slightly 4-angled, erect, green, glabrous; 1.5–3 mm thick near the base.

Autumn Onion, Prairie Onion, Pink Wild Onion, Wild Onion

Lily—*Liliaceae* **Wildflower** Greenish yellow Perianth parts 6

Asparagus *Asparagus officinalis* L.

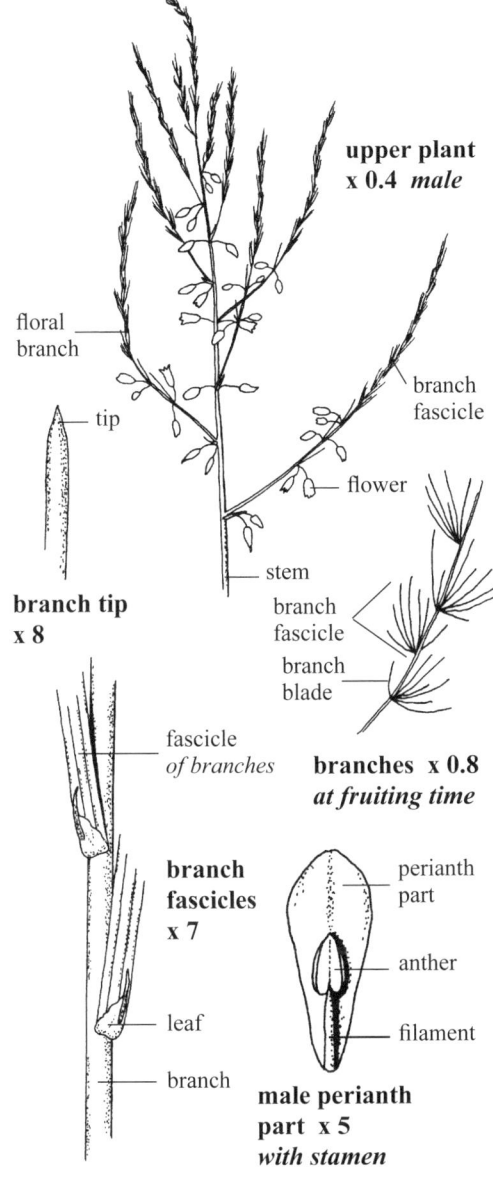

branch tip x 8

upper plant x 0.4 *male*

branches x 0.8 *at fruiting time*

branch fascicles x 7

male perianth part x 5 *with stamen*

■ **SKETCH** A glabrous **perennial herb** 0.7–2.5 m tall with a succulent **rhizome**; along woodland edges, fencerows, disturbed sites, ditches, roadsides and stream banks; **dioecious**. *Naturalized*
■ **FLOWERS** Greenish yellow, blooming May–September; **inflorescence** of racemes, these numerous with paired, often unisexual flowers along floral branches; **floral branches** to *c.* 100 cm long, rebranched; **male pedicels** 6–10 mm long, glabrous, curved, joint swollen and 2.5–3 mm from the base of the flower; **subtending bracts** (of pedicels) one, *c.* 2 mm long, bases wide, apices pointed; <u>male flowers</u> 2.5–4 mm wide by 4–6 mm long; **perianth parts** six, ascending, glabrous, persistent, blunt, imbricate; **outer perianth parts** three, 1.5–1.9 mm wide; **inner perianth parts** three, 2–2.5 mm wide (flattened), widest above the middle; **stamens** six, attached to base of perianth, 3.5–4 mm long, included; **filaments** 2–2.2 mm long, glabrous; **anthers** yellow, 1.5–1.8 mm long; **pistil** vestigial, *c.* 1 mm long, round, glabrous; **female pedicels** reflexed, 17–20 mm long, the joint 5–6 mm from the fruit; <u>female flowers</u> similar in size and shape to male flowers; **stamens** six, vestigial, white, 1–1.5 mm long, attached to base of perianth parts; **anthers** narrowly arrow-shaped; **pistil** glabrous 3–3.3 mm long, included; **ovary** *c.* 2 mm long by *c.* 1.3 mm wide, 3-grooved, not hairy; **style** *c.* 0.8 mm long; **stigma** 3-lobed. **FRUIT** a berry, red, fleshy, round, glabrous, slightly shiny, 6–10 mm wide and long, often hanging in pairs; **seeds** three to six per berry, black, 3.5–4 mm long by 2.8–3.3 mm wide by 2–2.5 mm thick, concave on one side.
■ **LEAVES** Alternate, simple, entire, subtending (branch fascicles), 1–3 mm long, tan, slightly erose, sessile, tip awnlike, base expanded and hardened.
■ **STEM** Erect, glabrous, round, green, turning yellow in the autumn as does most of the plant; 8–20+ mm thick near the base; **branches** in fascicles, appear leaflike; **branch blades** glabrous, pointed, 5–30 mm long by 0.3–0.4 mm wide, reduced towards the tips of floral branches, low ridged, roundly angular, 4–15 (–25) blades per fascicle, spreading at fruiting time, turning yellow in early autumn.

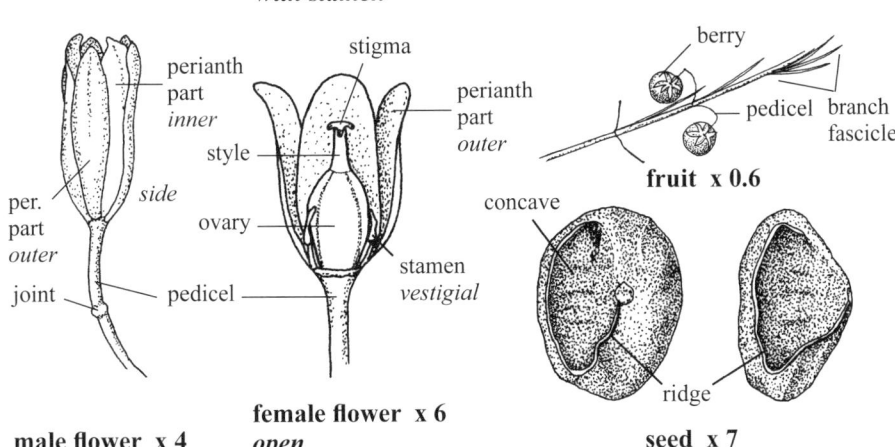

male flower x 4

female flower x 6 *open*

fruit x 0.6

seed x 7

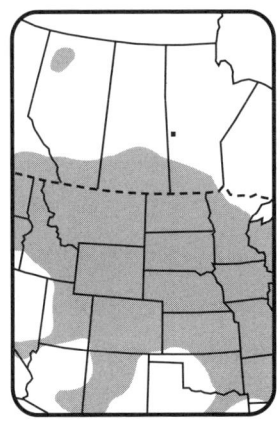

Garden Asparagus, Common Asparagus

Lily—*Liliaceae* **Wildflower** Yellow Perianth parts 6

YELLOW STAR-GRASS *Hypoxis hirsuta* (L.) Coville

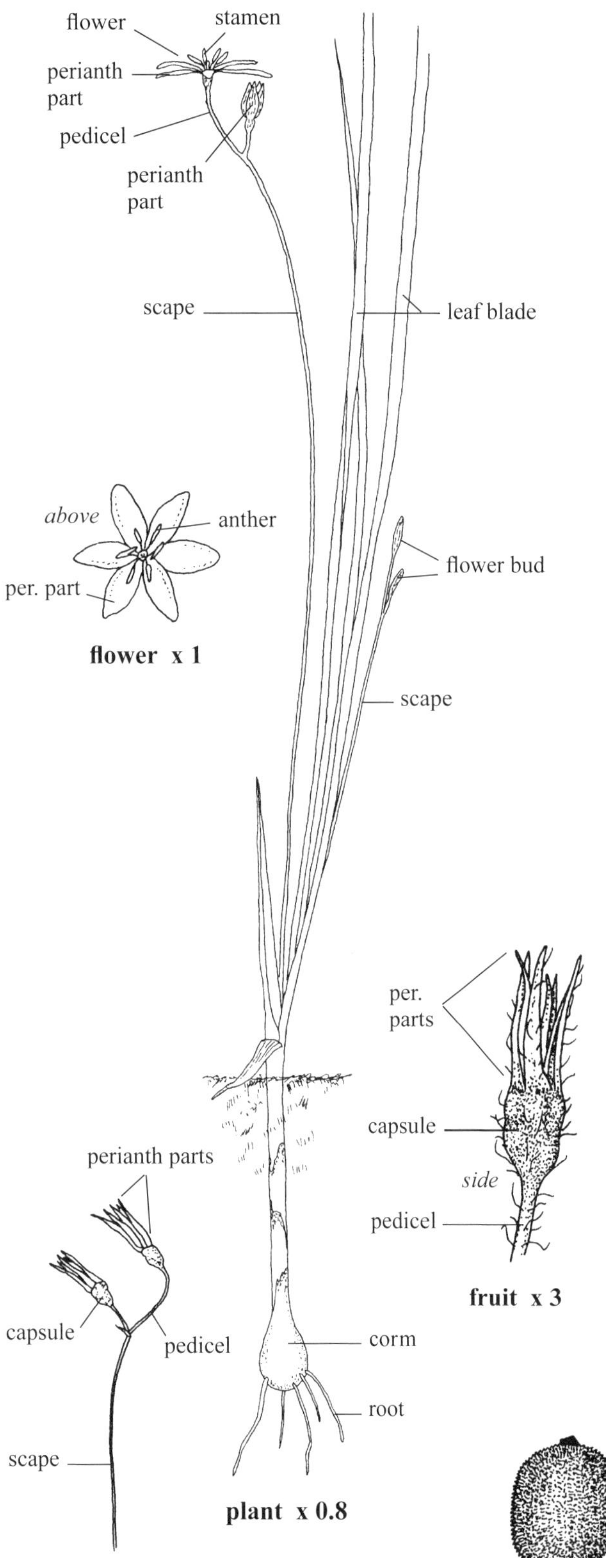

■ **SKETCH** A glabrous to hairy **perennial herb** 10–30 cm tall with a membranous-coated, white **corm** 7–25 mm long by 8–20 mm wide; in moist to dry prairies, open woods and thickets. **p2**

■ **FLOWERS** Yellow, blooming April–July; **inflorescence** an umbel of two to six (eight) flowers, terminal; **pedicels** green, 2–30 mm long, hairs thin and ascending; **flower buds** *c.* 8 mm long by *c.* 4 mm wide, pointed; **flowers** perfect, 15–25 mm wide by 5–8 mm long; **perianth parts** six, mostly similar with tips blunt to pointed, spreading, 8–14 mm long by 2.5–6 mm wide, glabrous and yellow above, erect and persistent on fruit; **three parts** green below with long thin ascending white hairs; **three parts** yellow below with a wide central green stripe, hairs white and mostly near the base; **stamens** six, exserted; **filaments** 2–3 mm long, yellow, bases wide and attached to middle of perianth bases; **anthers** yellow, 2–4 mm long; **style** erect, exserted, *c.* 3 mm long, thick, persistent on fruit; **stigma** triangular, with three sticky ridges. **FRUIT a capsule**, 3–5 mm long by *c.* 2.5 mm wide by *c.* 2 mm thick, green, hairy; **seeds** five to eight per capsule, dark brown to black, shiny, 1–1.5 mm long by 1–1.2 mm wide and thick, covered with numerous rows of short, stiff points.

■ **LEAVES** Basal, simple, entire, usually three to six; **blades** grasslike, erect, sessile, 5–25 cm long by 1–15 mm wide, flat to folded, whitish near the base, 5- to 9-nerved, tapered to a fine point, inner surface glabrous, outer (lower) surface smooth with fine scattered hairs along the raised keel.

■ **STEM** A **scape**, one to several per plant, erect but weak, filiform, 4–20 cm long by 0.6–0.8 mm wide, more hairy near the top, hairs thin, long, scattered and disheveled.

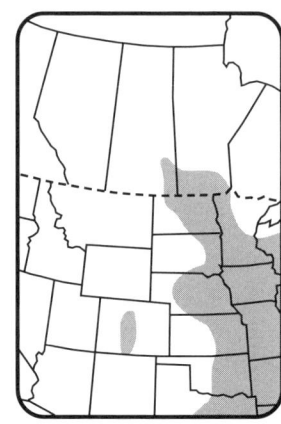

Stargrass, Common Goldstar, Eastern Yellow Star-grass

Lily—*Liliaceae* **Wildflower** Orange (red or yellow) Perianth parts 6

Wood Lily *Lilium philadelphicum* L.

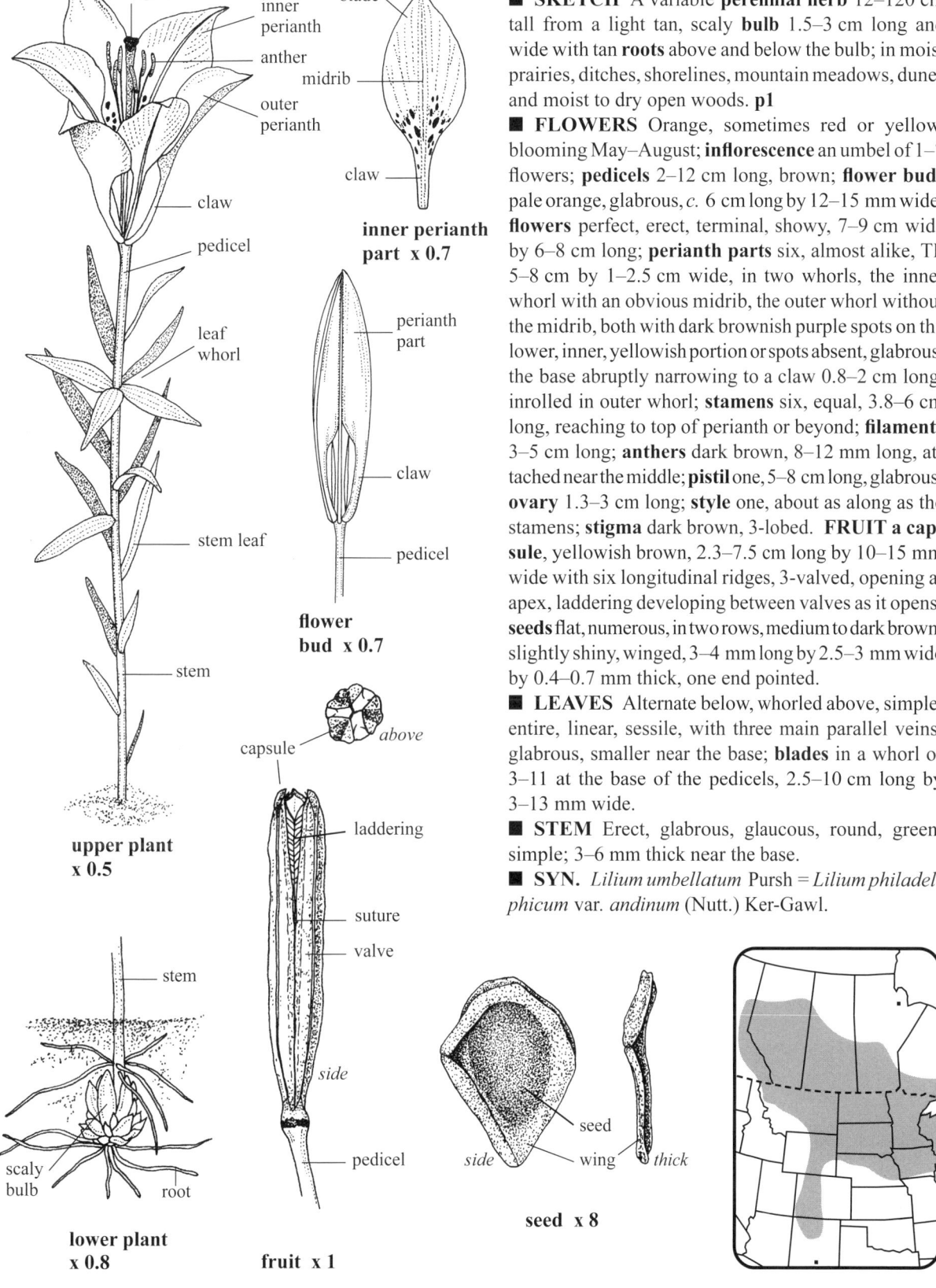

■ **SKETCH** A variable **perennial herb** 12–120 cm tall from a light tan, scaly **bulb** 1.5–3 cm long and wide with tan **roots** above and below the bulb; in moist prairies, ditches, shorelines, mountain meadows, dunes and moist to dry open woods. p1

■ **FLOWERS** Orange, sometimes red or yellow, blooming May–August; **inflorescence** an umbel of 1–7 flowers; **pedicels** 2–12 cm long, brown; **flower buds** pale orange, glabrous, *c.* 6 cm long by 12–15 mm wide; **flowers** perfect, erect, terminal, showy, 7–9 cm wide by 6–8 cm long; **perianth parts** six, almost alike, TL 5–8 cm by 1–2.5 cm wide, in two whorls, the inner whorl with an obvious midrib, the outer whorl without the midrib, both with dark brownish purple spots on the lower, inner, yellowish portion or spots absent, glabrous, the base abruptly narrowing to a claw 0.8–2 cm long, inrolled in outer whorl; **stamens** six, equal, 3.8–6 cm long, reaching to top of perianth or beyond; **filaments** 3–5 cm long; **anthers** dark brown, 8–12 mm long, attached near the middle; **pistil** one, 5–8 cm long, glabrous; **ovary** 1.3–3 cm long; **style** one, about as along as the stamens; **stigma** dark brown, 3-lobed. **FRUIT a capsule**, yellowish brown, 2.3–7.5 cm long by 10–15 mm wide with six longitudinal ridges, 3-valved, opening at apex, laddering developing between valves as it opens; **seeds** flat, numerous, in two rows, medium to dark brown, slightly shiny, winged, 3–4 mm long by 2.5–3 mm wide by 0.4–0.7 mm thick, one end pointed.

■ **LEAVES** Alternate below, whorled above, simple, entire, linear, sessile, with three main parallel veins, glabrous, smaller near the base; **blades** in a whorl of 3–11 at the base of the pedicels, 2.5–10 cm long by 3–13 mm wide.

■ **STEM** Erect, glabrous, glaucous, round, green, simple; 3–6 mm thick near the base.

■ **SYN.** *Lilium umbellatum* Pursh = *Lilium philadelphicum* var. *andinum* (Nutt.) Ker-Gawl.

Lily—Liliaceae **Wildflower** White Perianth parts 4

TWO-LEAVED SOLOMON'S-SEAL *Maianthemum canadense* Desf.

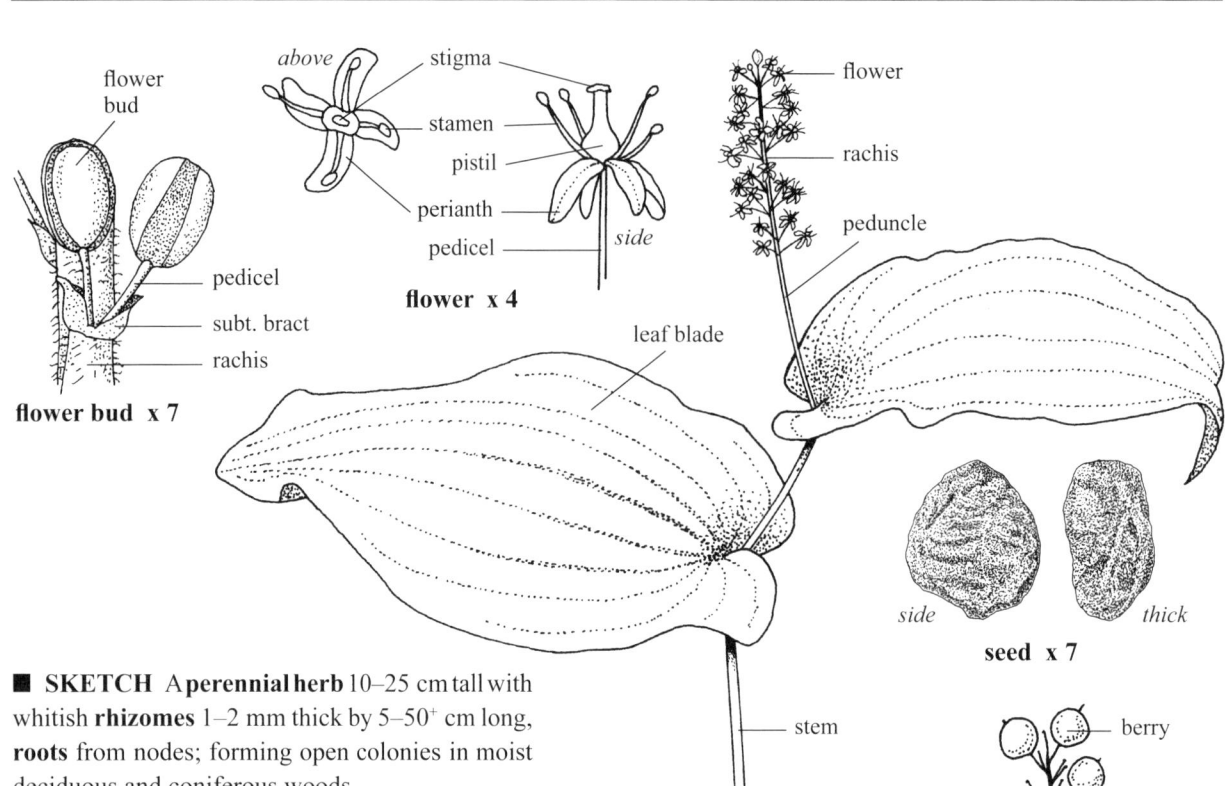

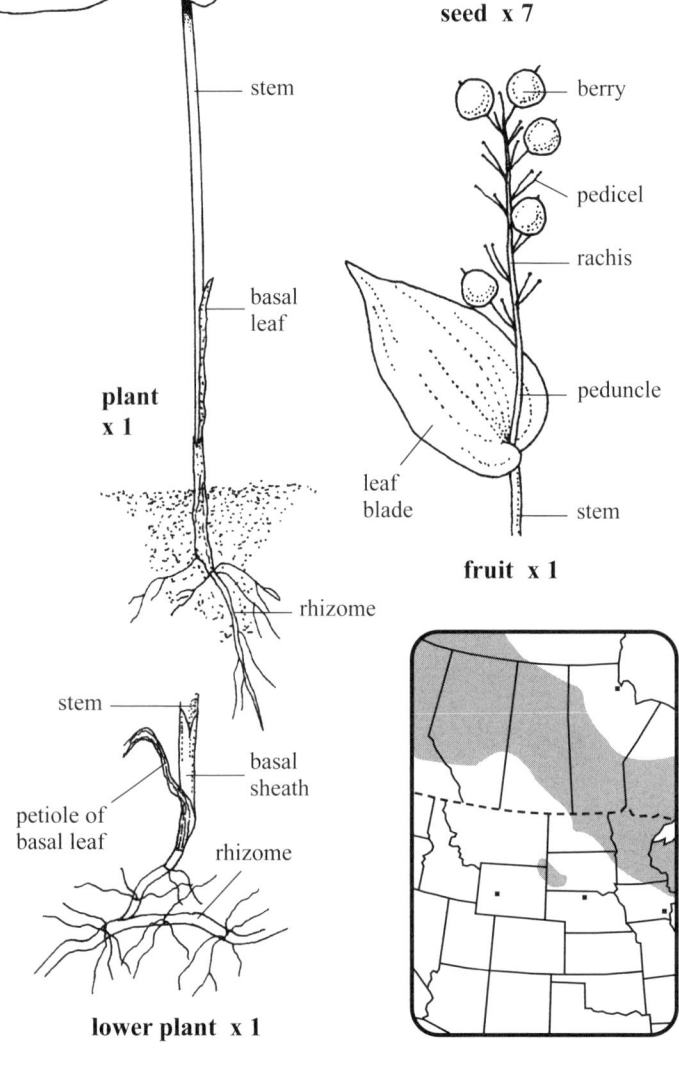

- **SKETCH** A perennial herb 10–25 cm tall with whitish **rhizomes** 1–2 mm thick by 5–50+ cm long, **roots** from nodes; forming open colonies in moist deciduous and coniferous woods.
- **FLOWERS** White, blooming May–July; **inflorescence** a raceme 1–6 cm long by 10–20 mm wide with 8–32 flowers; **peduncles** erect, lightly hairy, grooved, 1–5 cm long; **pedicels** 2–17 mm long, ascending, glabrous, usually in pairs; **subtending bracts** (of pedicels) *c.* 1 mm long and glabrous; **flowers** fragrant, mostly paired, 3–5 mm wide by 4–5 mm long; **rachis** hairy; **perianth parts** four, 2–3 mm long by *c.* 1 mm wide, deflexed; **stamens** four, white, exserted; **filaments** filiform, about as long as the pistil; **anthers** 0.2–0.4 mm long; **pistil** white and *c.* 1.2 mm wide; **stigma** 2-lobed. **FRUIT** a berry, red, sometimes speckled, 3–6 mm wide; **seeds** dark brown, dull, low ridged, *c.* 3 mm long by 2.6 mm wide by *c.* 2 mm thick, one or two per berry.
- **LEAVES** Basal and stem, alternate, simple, entire; **basal leaf** usually wilted by anthesis; **stem leaves** usually two (three), often clasping; **blades** pointed, margins slightly wavy, 1.5–10.5 cm long by 0.8–8 cm wide, the lower leaf the largest, main veins parallel, transverse veins faint, glabrous above, hairs on margins and veins below; **petioles** hairy, 0–3 mm long.
- **STEM** Erect, round and grooved, slightly hairy, unbranched; 1–2 mm thick near the base.

False Lily-of-the-valley, Canada Mayflower, Wild Lily-of-the-valley

Lily—*Liliaceae* **Wildflower** White Perianth parts 6

Star-flowered Solomon's-seal *Maianthemum stellatum* (L.) Link

■ **SKETCH** A **perennial herb** 20–65 cm high with a whitish **rhizome** 2–5 mm thick and 10–60 cm long; in moist woods, high slopes, desert sagebrush, meadows and along shores of rivers, lakes and streams. p2

■ **FLOWERS** White, blooming April–August; **inflorescence** a raceme (occasionally two), terminal, 1.5–6.5 cm long by 12–20 mm wide, almost sessile; **pedicels** 3–12 mm long, ascending; **flowers** perfect, 3–16, each 8–10 mm wide; **perianth parts** six, equal, similar, white, 3–7 mm long by 1–2 mm wide; **stamens** six, 2–5 mm long, opposite perianth parts; **filaments** white, attached to bases of perianth parts with the upper 1.5–2 mm free; **anthers** yellow, *c.* 0.8 mm long; **pistil** *c.* 4 mm long; **stigma** 3-lobed. **FRUIT a berry**, round, at first forming six reddish stripes, then ripening to an overall dark red, 7–11 mm wide by *c.* 7 mm long; **seeds** one to six per berry, light brown, roundly triangular, 2.5–3.7 mm long by 2.5–3.2 mm wide by *c.* 2.5 mm thick, smooth.

■ **LEAVES** Alternate, simple, entire, 2-ranked, sessile, 5–15 per plant; **blades** pointed, clasping, slightly folded to flat, 3–16 cm long by 0.7–4.5 cm wide, glabrous above, slightly hairy to glabrous below, veins parallel.

■ **STEM** Erect when young, nodding when in flower and fruit, unbranched, glabrous to slightly hairy, round, sometimes reddish at the 3–5 mm thick base; **lower sheaths** one or two, together extending for 3–8 cm.

■ **SYN.** *Smilacina stellata* (L.) Desf.

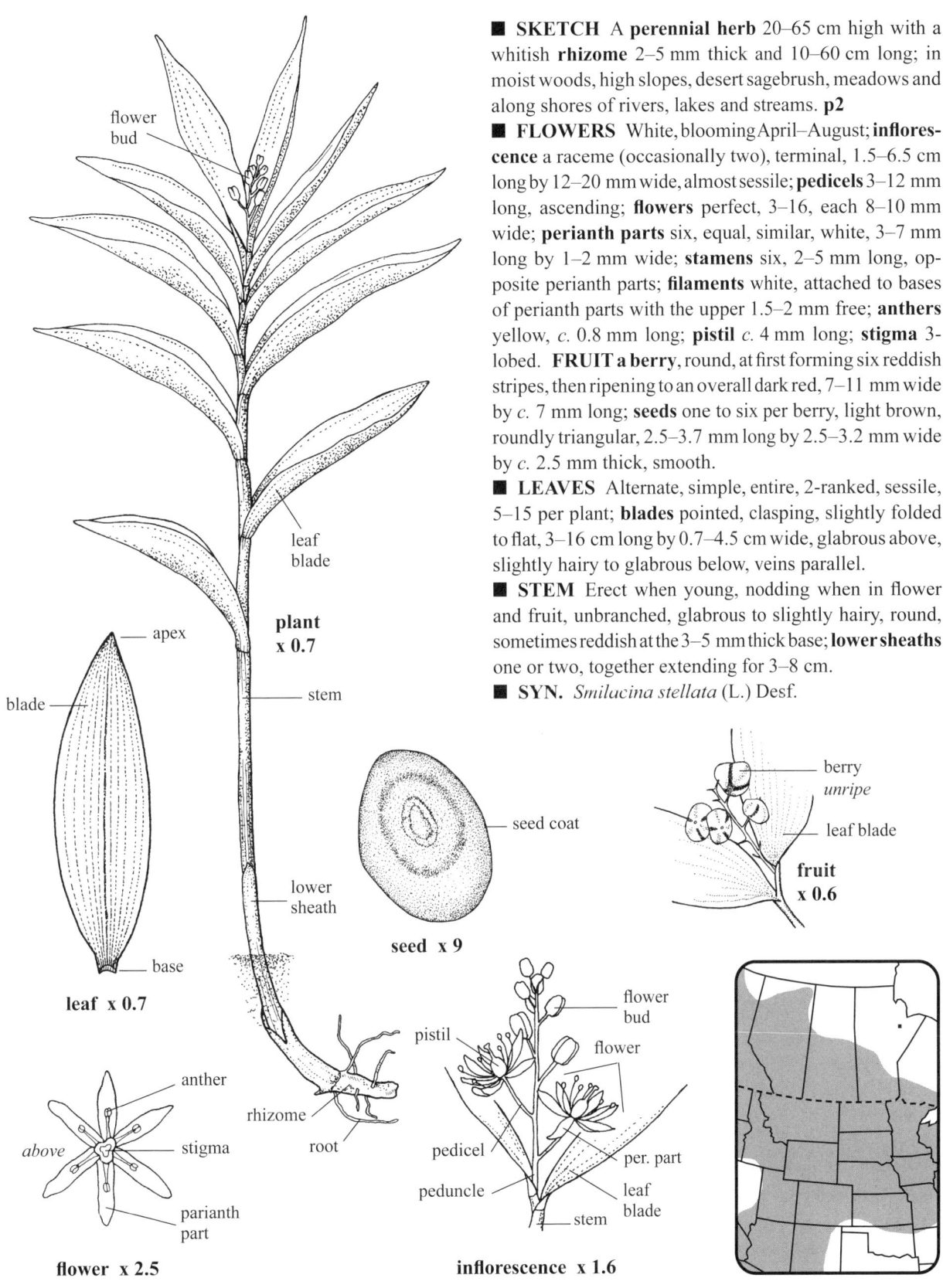

Starry Solomon's-seal, Starry False Lily of the Valley, Starry False Solomon's-seal

Lily—*Liliaceae* **Wildflower** White Perianth parts 6

THREE-LEAVED SOLOMON'S-SEAL *Maianthemum trifolium* (L.) Sloboda

■ **SKETCH** A glabrous **perennial herb** 5–40 cm tall in open colonies from a deep, whitish tan **rhizome** 2–30 cm long by *c.* 1.5 mm thick; in wet woods, bogs, swamps, fens and trail openings.

■ **FLOWERS** White, blooming May–July; **inflorescence** a raceme, 2–9 cm long by 15–30 mm wide of 3–15 flowers; **pedicels** 3–15 mm long (to 20 mm with fruit), curved slightly upward, green with purple streaks; **subtending bracts** (of pedicels) usually two (one), erect, one long and one short, 0.5–4.2 mm long, blunt, reduced above; **flowers** perfect, fragrant, 6–9 mm wide by 4.5–7 mm long; **rachis** green, slightly zigzagging, 3-sided; **perianth parts** six, similar, white, spreading to reflexed, blunt, 2.5–4 mm long by 1.5–2 mm wide, margins entire and revolute, veins obscure; **stamens** six, exserted, *c.* 3.5 mm long; **filaments** white, 2.5–3 mm long, narrow near the anther; **anthers** white, spotted with purple, 2-lobed, *c.* 0.8 mm long; **pistil** white, 2.5–2.8 mm long, exserted; **ovary** white, 3-lobed, *c.* 1.5 mm long by *c.* 1.8 mm wide; **styles** three, *c.* 1 mm long, erect, united except at spreading apices; **stigmas** three, rough, *c.* 1 mm wide. **FRUIT** a **berry**, juicy, dark red, shiny, 5–8 mm long by 4–9 mm wide, roundly 3-sided, slightly transparent; **seeds** 1–3 per berry, whitish pink, 3.5–4 mm long by 2.8–3.8 mm wide by 2–2.6 mm thick, smooth, dull, hilum oval.

■ **LEAVES** Alternate, simple, entire, arched, sessile, usually three (2 or 4) per stem, pointed; **blades** dull, 1–15 cm long by 1.2–5 cm wide, middle leaf the largest, much reduced above, veins parallel with about 12–16 on each side of midrib, glabrous, tapered at base and enveloping the stem; **stipules** absent.

■ **STEM** Erect, unbranched, green; base 1.5–2.5 mm thick; **bracts** below leaves, 3–5, membranous, each 8–12 mm long, often reaching to base of second leaf, tips blunt, lower ones dark brown and often torn.

■ **SYN.** *Smilacina trifolia* (L.) Desf.

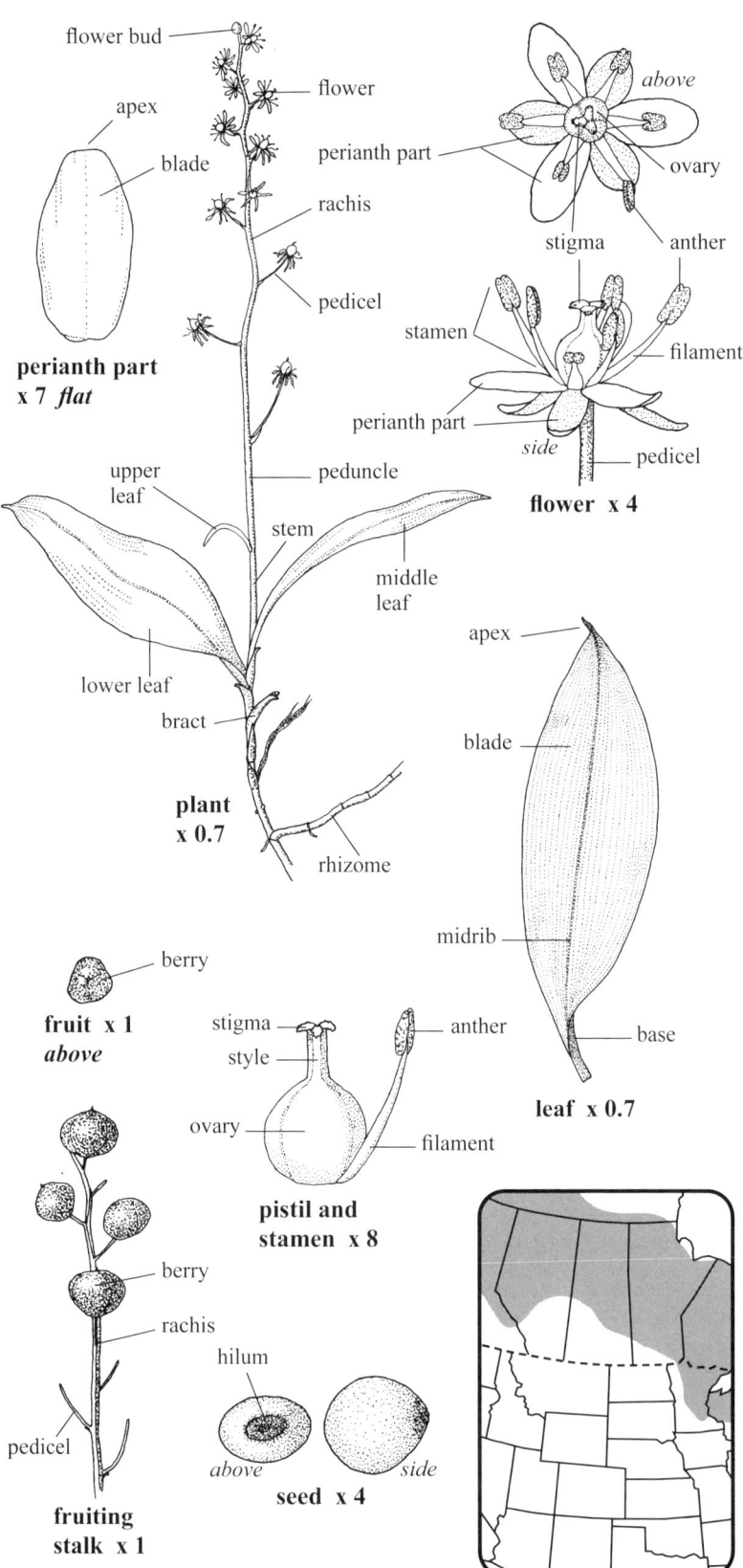

604 False Solomon's-seal, Threeleaf False Lily of the Valley, Three-leaved False Solomon's-seal

Lily—*Liliaceae* **Wildflower** Greenish white to creamy yellow Perianth tubular, 6-lobed

Common Solomon's-seal *Polygonatum biflorum* (Walt.) Ell.

■ **SKETCH** A variable, glabrous **perennial herb** 50–200 cm long, arching, with a knotty **rhizome** 7–30 mm wide and to 7+ cm long; in thickets, fields, and moist deciduous woodlands. **p2**

■ **FLOWERS** Greenish white to creamy yellow, blooming April–July; **inflorescence** of axillary clusters of one to ten flowers hanging from each peduncle of the middle nodes; **peduncles** 0.6–6 cm (–9 in fruit) long; **pedicels** subequal, 2–15 mm long, extending to 1–4 cm in fruit, usually with one fine bracteole near the base; **flowers** perfect, 5–30 per plant, hanging, tubular, thick, crisp, *c.* 10 mm wide by 12–22 mm long; **perianth** tubular, 6-lobed, these equal and spreading, each 3–6 mm long and pointed; **stamens** six, included; **filaments** inserted near the top of the perianth; **anthers** facing inward; **pistil** included; **style** one, filiform; **stigma** obscurely 3-lobed. **FRUIT a berry**, bluish black, 9–15 mm long and wide, smooth; **seeds** brown, smooth, 8–18 per berry, 2.9–4 mm long by 2.5–3.8 mm wide and thick.

■ **LEAVES** Alternate, simple, entire, 9–20+ per plant, sessile; **blades** 5.8–25 cm long by 2.2–13 cm wide, smaller near the top, glaucous below, with distinctive parallel veins, disintegrating as the berries ripen.

■ **STEM** Solitary, unbranched, glabrous and glaucous, round, erect to arched; 3–8 mm wide near the naked base.

■ **SYN.** *Polygonatum canaliculatum* auct. non (Muhl. ex Willd.) Pursh = *Polygonatum biflorum* var. *commutatum* (J.A. & J.H. Schultes) Morong and *Polygonatum commutatum* (J.A. & J.H. Schultes) A. Dietr. = *Polygonatum biflorum* var. *commutatum* (J.A. & J.H. Schultes) Morong.

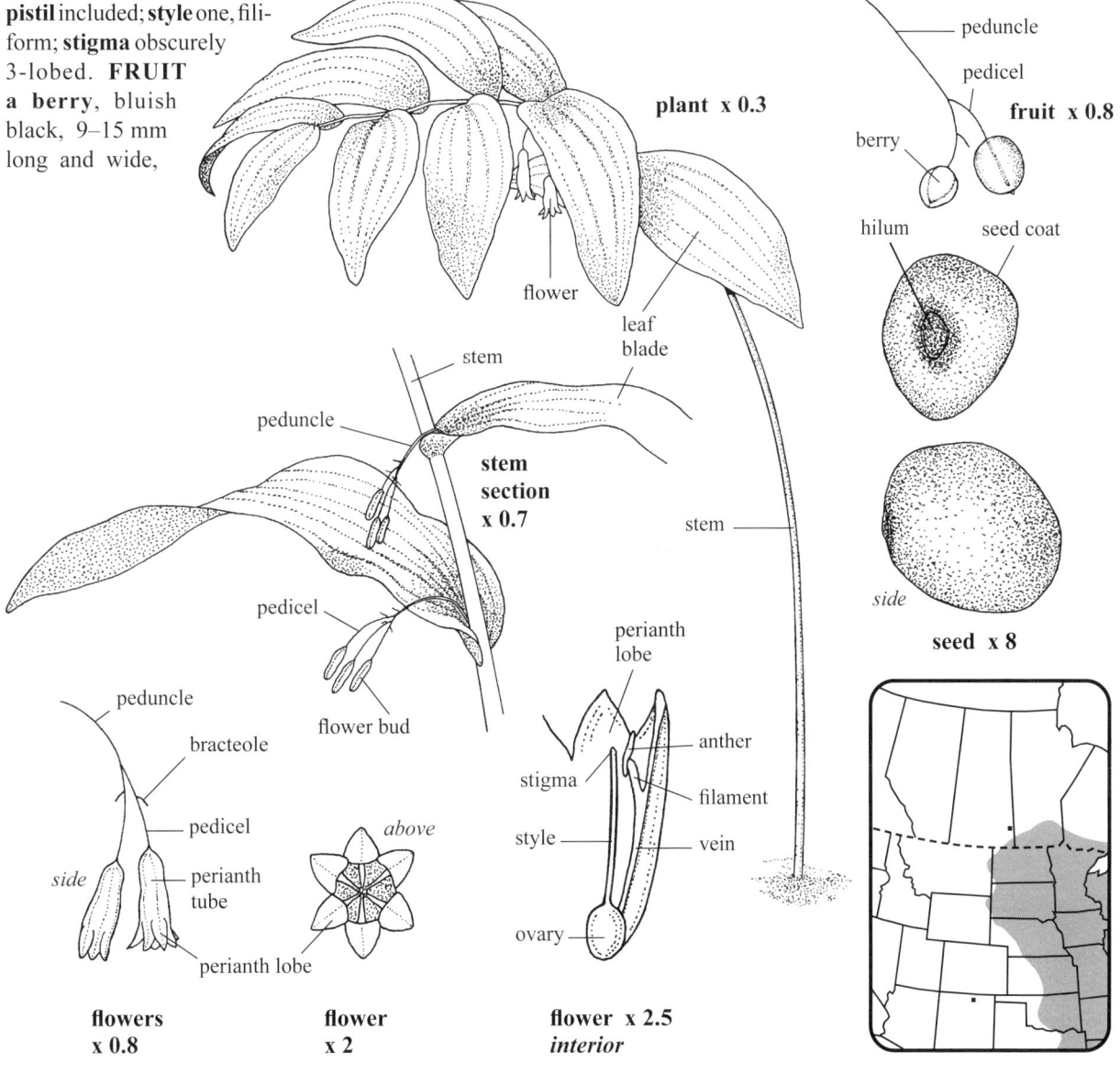

King Solomon's-seal, Smooth Solomon's-seal, Great Solomon's-seal

Lily—*Liliaceae* **Wildflower** White Petals 3

NODDING WAKEROBIN *Trillium cernuum* L.

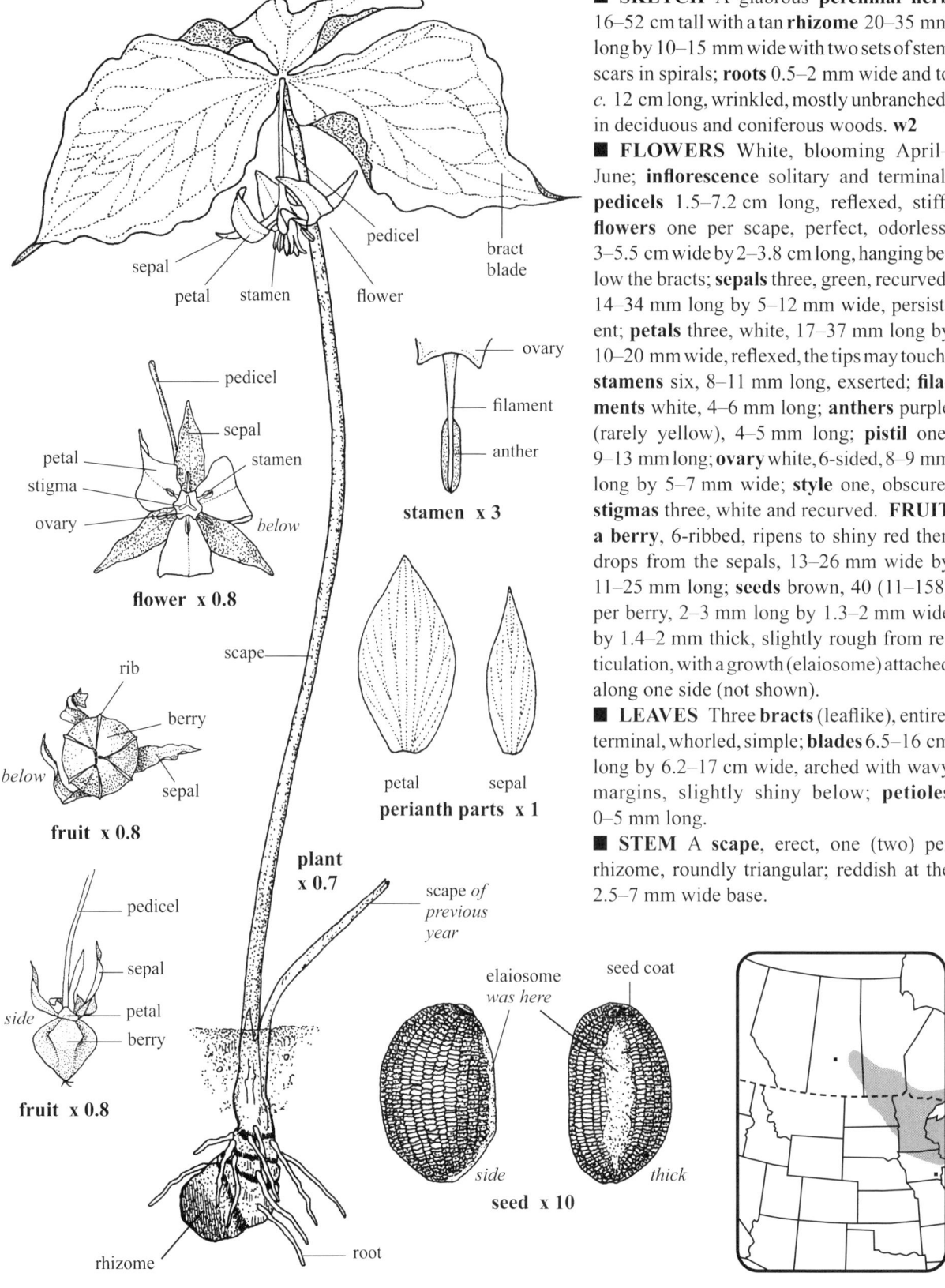

- **SKETCH** A glabrous **perennial herb** 16–52 cm tall with a tan **rhizome** 20–35 mm long by 10–15 mm wide with two sets of stem scars in spirals; **roots** 0.5–2 mm wide and to *c.* 12 cm long, wrinkled, mostly unbranched; in deciduous and coniferous woods. **w2**
- **FLOWERS** White, blooming April–June; **inflorescence** solitary and terminal; **pedicels** 1.5–7.2 cm long, reflexed, stiff; **flowers** one per scape, perfect, odorless, 3–5.5 cm wide by 2–3.8 cm long, hanging below the bracts; **sepals** three, green, recurved, 14–34 mm long by 5–12 mm wide, persistent; **petals** three, white, 17–37 mm long by 10–20 mm wide, reflexed, the tips may touch; **stamens** six, 8–11 mm long, exserted; **filaments** white, 4–6 mm long; **anthers** purple (rarely yellow), 4–5 mm long; **pistil** one, 9–13 mm long; **ovary** white, 6-sided, 8–9 mm long by 5–7 mm wide; **style** one, obscure; **stigmas** three, white and recurved. **FRUIT a berry**, 6-ribbed, ripens to shiny red then drops from the sepals, 13–26 mm wide by 11–25 mm long; **seeds** brown, 40 (11–158) per berry, 2–3 mm long by 1.3–2 mm wide by 1.4–2 mm thick, slightly rough from reticulation, with a growth (elaiosome) attached along one side (not shown).
- **LEAVES** Three **bracts** (leaflike), entire, terminal, whorled, simple; **blades** 6.5–16 cm long by 6.2–17 cm wide, arched with wavy margins, slightly shiny below; **petioles** 0–5 mm long.
- **STEM** A **scape**, erect, one (two) per rhizome, roundly triangular; reddish at the 2.5–7 mm wide base.

Lily—Liliaceae **Wildflower** Greenish white Perianth parts 6

Smooth Camas *Zigadenus elegans* Pursh

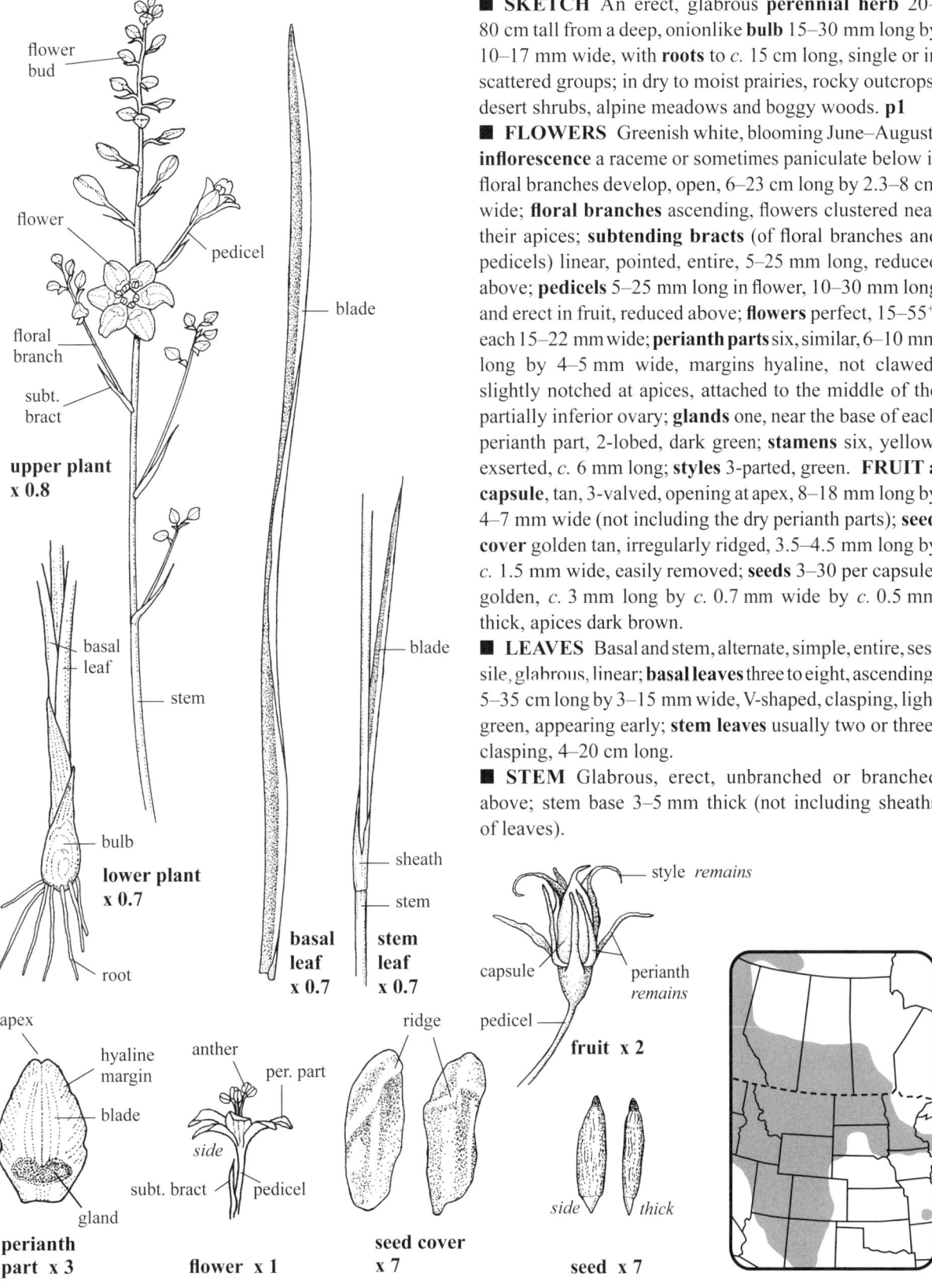

■ **SKETCH** An erect, glabrous **perennial herb** 20–80 cm tall from a deep, onionlike **bulb** 15–30 mm long by 10–17 mm wide, with **roots** to *c.* 15 cm long, single or in scattered groups; in dry to moist prairies, rocky outcrops, desert shrubs, alpine meadows and boggy woods. **p1**

■ **FLOWERS** Greenish white, blooming June–August; **inflorescence** a raceme or sometimes paniculate below if floral branches develop, open, 6–23 cm long by 2.3–8 cm wide; **floral branches** ascending, flowers clustered near their apices; **subtending bracts** (of floral branches and pedicels) linear, pointed, entire, 5–25 mm long, reduced above; **pedicels** 5–25 mm long in flower, 10–30 mm long and erect in fruit, reduced above; **flowers** perfect, 15–55[+], each 15–22 mm wide; **perianth parts** six, similar, 6–10 mm long by 4–5 mm wide, margins hyaline, not clawed, slightly notched at apices, attached to the middle of the partially inferior ovary; **glands** one, near the base of each perianth part, 2-lobed, dark green; **stamens** six, yellow, exserted, *c.* 6 mm long; **styles** 3-parted, green. **FRUIT** a **capsule**, tan, 3-valved, opening at apex, 8–18 mm long by 4–7 mm wide (not including the dry perianth parts); **seed cover** golden tan, irregularly ridged, 3.5–4.5 mm long by *c.* 1.5 mm wide, easily removed; **seeds** 3–30 per capsule, golden, *c.* 3 mm long by *c.* 0.7 mm wide by *c.* 0.5 mm thick, apices dark brown.

■ **LEAVES** Basal and stem, alternate, simple, entire, sessile, glabrous, linear; **basal leaves** three to eight, ascending, 5–35 cm long by 3–15 mm wide, V-shaped, clasping, light green, appearing early; **stem leaves** usually two or three, clasping, 4–20 cm long.

■ **STEM** Glabrous, erect, unbranched or branched above; stem base 3–5 mm thick (not including sheaths of leaves).

White Camas, Mountain Deathcamas

Family Characteristics

Orchid—Orchidaceae — Flower parts 3

SKETCH Monocotyledonous perennial herbs (wildflowers), an epiphyte or saprophyte, with bulbs, rhizomes, and fleshy or coral-like roots. Worldwide about 20,000 species, mainly in tropical forests to temperate regions.

FLOWERS Irregular, perfect, 3-merous in a wide range of colors, showy to quite small. **Inflorescence** a raceme or spike, or a solitary flower, each subtended by a bract. **Sepals** 3, mostly entire, like or unlike the petals, the lower two sometimes fused with a bifid tip. **Petals** 3, variously spotted or streaked, the lower one (lip) is unlike the side two, often with a spur and small lobes, (rarely fringed). **Stamens** 2 (1), fused to the style and three stigmas to form the column; anthers 1 (rarely 2), a staminodium (3rd anther) sometimes present. **Pistil** 1, of 1 or 3 united locules; ovules numerous. **Ovary** 1, inferior, thick, ribbed and may be twisted, often sessile or supported by a short pedicel usually visible with fruit. **Style** 1, often ending in a beak. **Stigmas** 3, sticky, functional or with one sterile.

FRUIT A **capsule**, erect to hanging, persistent, opening along 3 sutures. **Seeds** tiny, numerous and winged.

LEAVES Alternate, sometimes opposite, simple, entire, sheaths clasping the stem, sessile, one to several blades with parallel venation, ascending, arched to spreading. **Stipules** absent.

STEM Erect, solid, simple, green, round, glabrous to hairy, usually not branched.

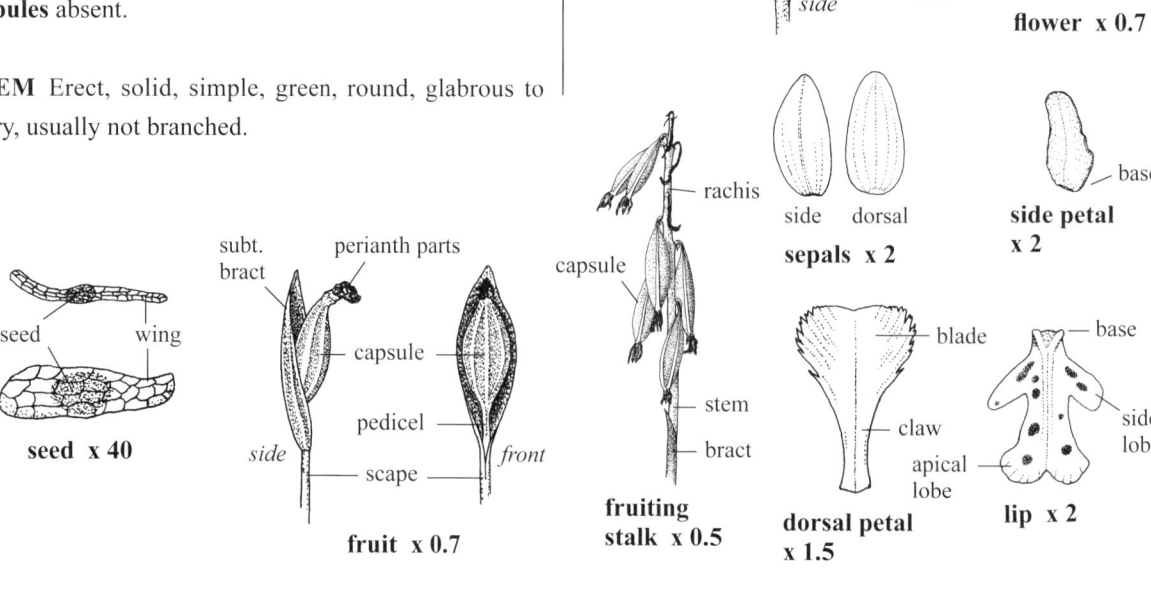

608

Orchid—*Orchidaceae* **Wildflower** White Petals 3, lip with purple spots

Round-leaved Orchid *Amerorchis rotundifolia* (Banks ex Pursh) Hultén

■ **SKETCH** A glabrous **perennial herb** 10–33 cm tall, in open colonies of up to 12 plants, with white fleshy **rhizomes** several cm long by 1–2 mm thick; in moist woods, swamps and open treed fens in sphagnum and along streambanks.

■ **FLOWERS** White with purplish tinges and spots, blooming June–August; **inflorescence** a raceme, 2–6 cm long by 1.7–3 cm wide of 2–17 flowers; **pedicels** green, 1–2 mm long; **subtending bracts** (of pedicels) green, 7–15 mm long by 2–5 mm wide, ascending; **flowers** perfect, 15–17 mm wide by 12–16 mm long by c. 10 mm tall; **sepals** three; **dorsal sepal** white, 6–7 mm long by 3.5–4.5 mm wide, slightly purplish on back, covering the two side petals; **side sepals** two, spreading, 7–9 mm long by 3–4 mm wide, white to slightly purply; **petals** three, side two purple or purplish along margin, erose along one side, sometimes with a purple spot on lower side, 5–6.5 mm long by 3–3.3 mm wide; **lip** (middle petal) white, 6–10 mm long by 5–8 mm wide, side lobes 2–3 mm long, blunt, apical lobe slightly wrinkled and notched, purple spots several, light yellow in throat; **spur** pale purple to white, 5–6 mm long by 0.8–1 mm wide and thick with two side projections at tip, curved away from ovary; **column** c. 2 mm long; **ovary** 12–14 mm long by c. 3 mm wide by c. 2.5 mm thick, slightly twisted, generally purplish green, tapered at base, with three wide rounded smooth ridges and two narrow, irregular purple ribs between the ridges. **FRUIT a capsule**, brown, erect, 12–15 mm long by 3–4 mm wide and thick, 3-ribbed, glabrous, dried perianth parts at apex; **seeds** numerous, 0.5–0.7 mm long by c. 0.1 mm wide and thick, tan, wings transparent, shiny, net-veined, straight to slightly curved.

■ **LEAVES** One at stem base, entire, pointed, slightly shiny; **blade** 3–11 cm long by 2–7.5 cm wide with six or seven veins on each side of midrib, slightly lighter green below, tip recurved, glabrous, ascending to spreading, base tapered forming a whitish sheath enclosing the stem.

■ **STEM** Erect, pale green, roundly angular, 4- or 5-sided, hollow; 1–1.5 mm wide above the leaf, wider below due to sheaths; **sheaths** two, alternate, white, closed, apices roundly pointed, V-shaped on opposite side of apices.

■ **SYN.** *Orchis rotundifolia* Banks ex Pursh.

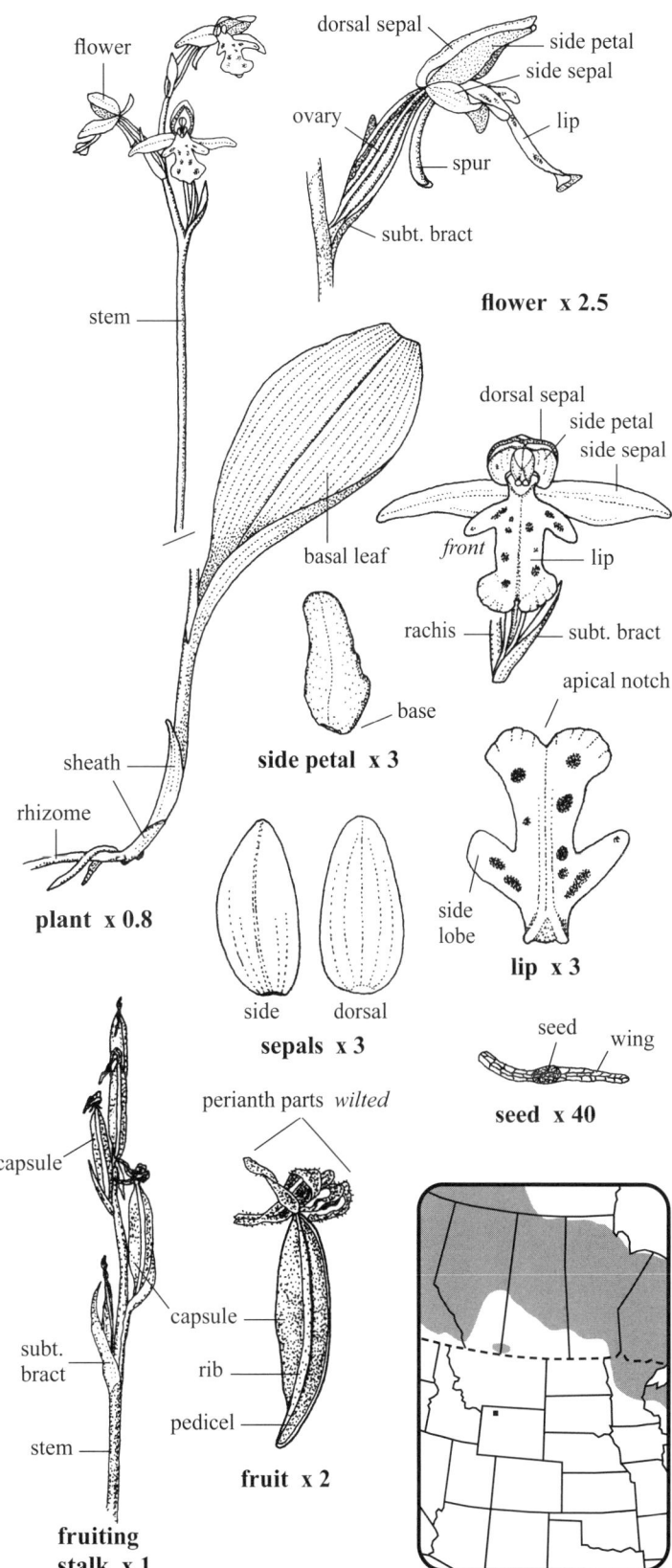

Small Round-leaved Orchis

Orchid—*Orchidaceae* **Wildflower** Yellowish green Petals 3

LONG-BRACTED ORCHID   *Coeloglossum viride* (L.) Hartman

■ **SKETCH** A variable, glabrous **perennial herb** 10–70 cm tall from paired, fleshy, white bisected **tubers** 1–2 cm long by 1.3–2.5 cm wide by 5–8 mm thick; **roots** spreading to ascending, white, 2–7 cm long by *c*. 2 mm wide from the tubers; in moist open mixed woods, dry to wet meadows, ditches, marshes, bogs and streamsides.

■ **FLOWERS** Yellowish green, blooming May–August; **inflorescence** a spike, 4.5–20 cm long by 2.5–3.3 cm wide (to *c*. 6 cm wide including bracts); **flowers** 4–60 per stem, glabrous, each 5–7 mm wide by 8–14 mm long; **subtending bracts** (of flowers) green, leaflike, pointed, ascending, sessile, 1–4.7 cm long by 1.5–9 mm wide, reduced above, covering the base of the ovary; **sepals** three; **dorsal sepal** one, 4–7 mm long by 2–3 mm wide; **side sepals** two, each 4–7 mm long by 2–4 mm wide, asymmetrical; **petals** three, two side petals mostly hidden by the dorsal sepal, these linear, curved inward, 3–5 mm long by *c*. 1 mm wide; **lip** (one petal) reflexed, 5–10 mm long by 2–3 mm wide, 2- or 3-lobed, the middle lobe smaller or absent, side lobes *c*. 1 mm long; **spur** from base of lip, 2–3 mm long by 1.8–2 mm wide, blunt; **anthers** two, pinkish, *c*. 1.5 mm long; **ovary** 2- or 3-ridged, twisted 180°, 6–7 mm long by 1–1.5 mm wide, partially hidden by bract. **FRUIT a capsule**, erect, 7–13 mm long by 2.5–4 mm wide and thick, the three flat ribs separating from above to release seeds; **seeds** numerous, winged, brown, 0.4–0.6 mm long by 0.1–0.2 mm wide and thick, shiny, wings net-veined, transparent, straight to slightly bent.

■ **LEAVES** Alternate, entire, three to six, sessile, simple, ascending, bases widening around stem; **blades** thin, soft, 5–18 cm long by 7–70 mm wide, reduced above and below, yellowish green, wilting with fruit.

■ **STEM** Green, erect, weak; 2–7 mm wide near the base.

■ **SYN.** *Coeloglossum bracteatum* (Muhl. ex Willd.) Parl. = *Coeloglossum viride* var. *virescens* (Muhl. ex Willd.) Luer and *Habenaria viridis* (L.) R. Br. ex Ait. f. = *Coeloglossum viride* var. *viride* (L.) Hartman.

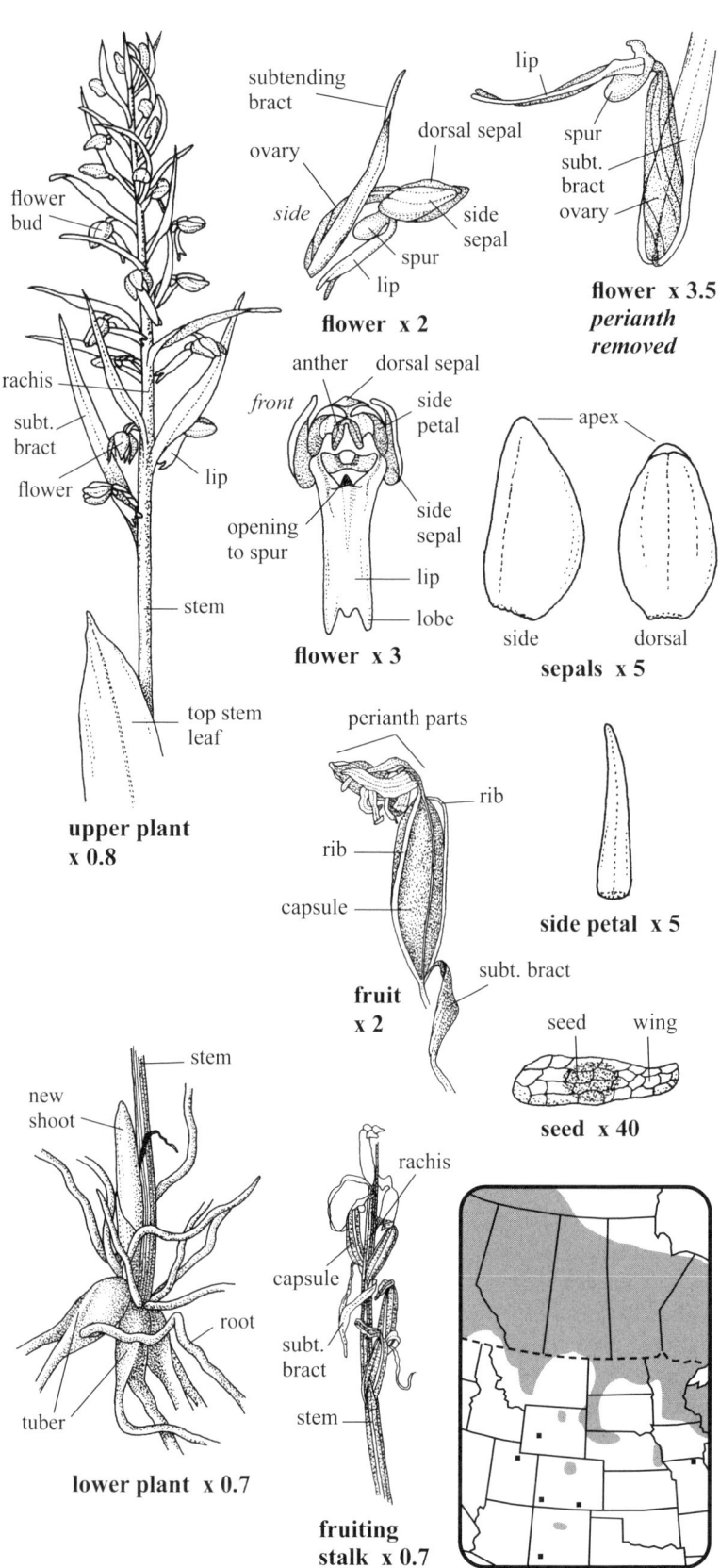

Bracted Orchid, Frog Orchid, Bracted Bog-orchid, Longbract Frog Orchid

Orchid—*Orchidaceae* **Wildflower** Pinkish brown Petals 3, lip 3-lobed, white and spotted

SPOTTED CORALROOT *Corallorhiza maculata* (Raf.) Raf.

■ **SKETCH** A variable, glabrous, saprophytic **perennial herb** 15–60 cm tall in small clusters from branched **rhizomes** 2–4 cm long by 4–5 mm thick with coral-like fingers 5–15 mm long by 4–5 mm wide, these pale yellow and easily broken off; in moist decaying litter in mixed woodlands, slopes and coniferous woods.

■ **FLOWERS** Pinkish brown, blooming June–August; **inflorescence** a raceme 5–20 cm long by 3–4 cm wide, loosely to densely flowered; **pedicels** 2–5 mm long, pinkish tan, ridged, twisted, reflexed with fruit; **subtending bracts** (of pedicels) 1.5–5 mm long by 1–2.2 mm wide, 3-lobed, side lobes thin and often curled; **flowers** fragrant, 6–40 per raceme, 15–18 mm wide by 10–16 mm long; **rachis** angled, *c.* 2 mm thick; **sepals** three, 3-veined, 5–15 mm long by 1.5–2.5 mm wide, usually not spotted; **side petals** two, 3-nerved, 5–11 mm long by 2–6 mm wide, spotted or not; **lip** 3-lobed, 5–8.5 mm long by 3–6 mm wide, white, reflexed, with several, reddish purple spots or not, apex of main lobe somewhat erose, side lobes raised, 2–2.5 mm long, thin, blunt, two central ridges in throat raised and *c.* 3 mm long; **claw** *c.* 1 mm long, yellow; **ovary** 5–14 mm long by 2–3 mm wide, 3-ridged; **column** yellow, arched, 4–7.5 mm long by *c.* 2 mm wide, may be spotted. **FRUIT a capsule**, hanging, hairless, brown, 1–2.3 cm long by 4–8 mm wide by 3–6 mm thick, persistent on old erect stems, ridges three; **seeds** pale yellow, 0.8–1.2 mm long by *c.* 0.2 mm wide (including wing), numerous, tapered at one or both ends, reticulate.

■ **LEAVES** Alternate, reduced to long enveloping sheaths (bracts) enclosing the stem for most of their length, usually four sheaths, the upper two partially above ground, 8–9 cm between the upper two tips, apices pointed and V-shaped, sometimes outlined with dark brown, veined; **lowest sheath** entirely underground, *c.* 2 cm long, apex ragged and brown.

■ **STEM** Erect, tan to whitish pink, solid, round in cross-section; 4–6 mm wide near the sheath-enclosed base.

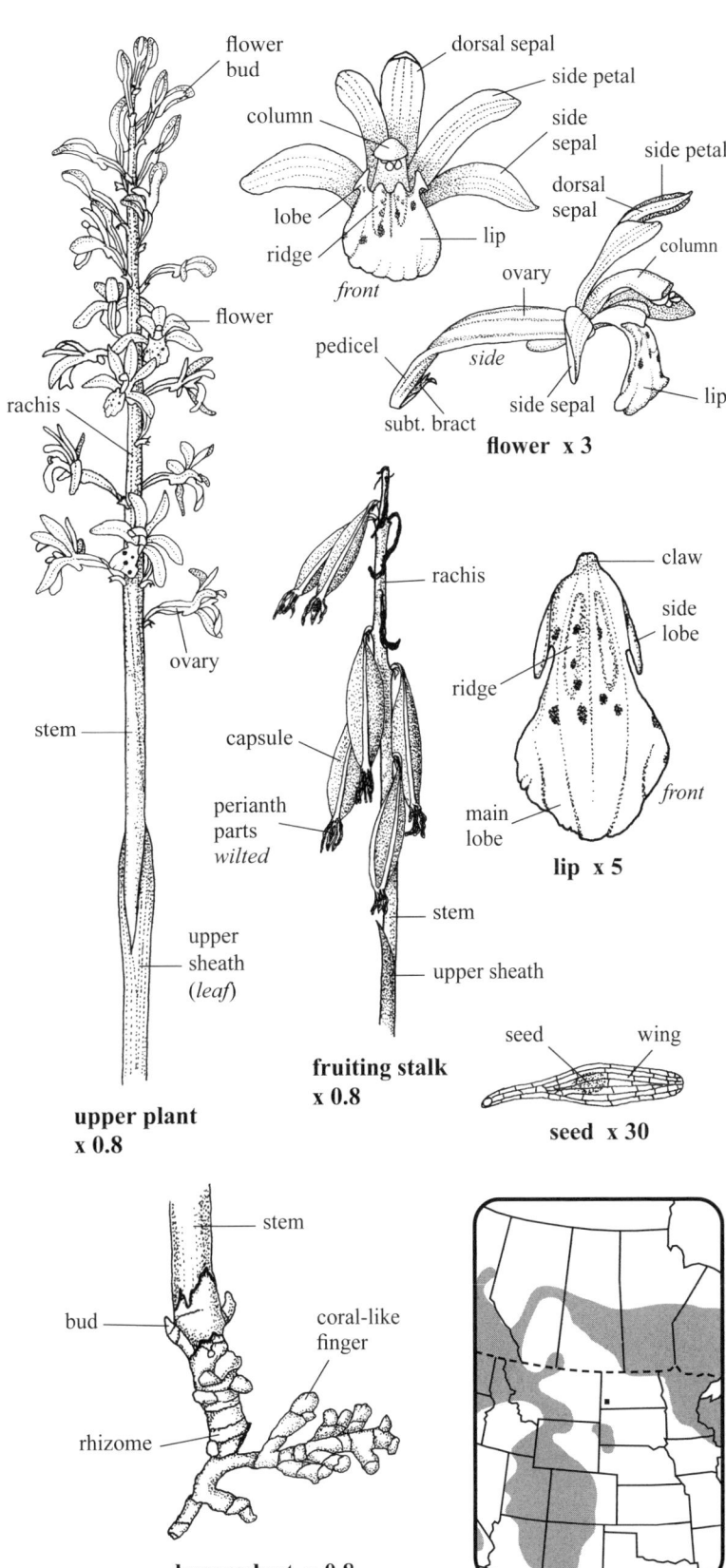

Summer Coralroot, Large Coralroot

Orchid—*Orchidaceae* **Wildflower** Pinkish purple with purple stripes   Petals 3, lip not lobed

STRIPED CORALROOT *Corallorhiza striata* Lindl.

■ **SKETCH** A glabrous, saprophytic **perennial herb** 10–50 cm tall in clumps from branching **rhizomes** with coral-like fingers 2–3 mm wide; in decaying forest litter.

■ **FLOWERS** Pinkish purple with purple stripes, blooming May–August; **inflorescence** a raceme 4–20 cm tall by 3–4 cm wide; **pedicels** 1–3 mm long, spreading, arched with fruit; **subtending bracts** (of pedicels) 2–5 mm long by *c.* 3 mm wide, glabrous, pinkish with a pale yellow base, partially covering the ovary base; **flowers** perfect, 4–35 per raceme, 15–22 mm wide by 11–15 mm long (not including ovary); **rachis** triangular, pinkish purple, sides flat, edges sharp, hollow with fruit; **sepals** three, 5.5–18 mm long by 2.5–5 mm wide, with three to five purple veins or unmarked, bases yellow; **side petals** two, similar to sepals; **lip** reddish purple to yellow, entire, pointed, basally reflexed, fleshy, margins involute, 3–16 mm long by 3–8 mm wide, upper side papillose (microscopic), deciduous; **lamella** fleshy, bilobed, *c.* 3.3 mm long, raised and shiny red; **column** slender, 3–6 mm long by *c.* 2 mm wide, curved, yellow with red spots; **pollinia** four, together *c.* 1 mm wide, yellow; **ovary** slightly twisted, 5–8 mm long, 3-ridged, ridges pink and wide. **FRUIT** a **capsule**, brown, hanging, usually overwintering on the erect stem, bluntly 3-sided, 3-ridged, 1.2–3 cm long by 5–9 mm wide and thick; **seeds** numerous, pale yellow, 0.8–1.5 mm long by *c.* 0.2 mm wide and thick, endosperm lacking, wing narrow, striated, clear, straight or curved.

■ **LEAVES** Alternate, reduced to five or six membranous sheaths (bracts) around the stem, apices V-shaped, dark brownish at the point, thinly veined, light tannish yellow; **upper sheaths** three, each 3–6 cm long; **three shorter sheaths** with apices ragged and below ground, 1–2 cm long.

■ **STEM** Erect, simple, solid, fleshy, pinkish purple, round, with low ridges, overwintering; base of dry stem (without sheaths) 2–5 mm thick, fresh stem with sheaths 4–10 mm thick.

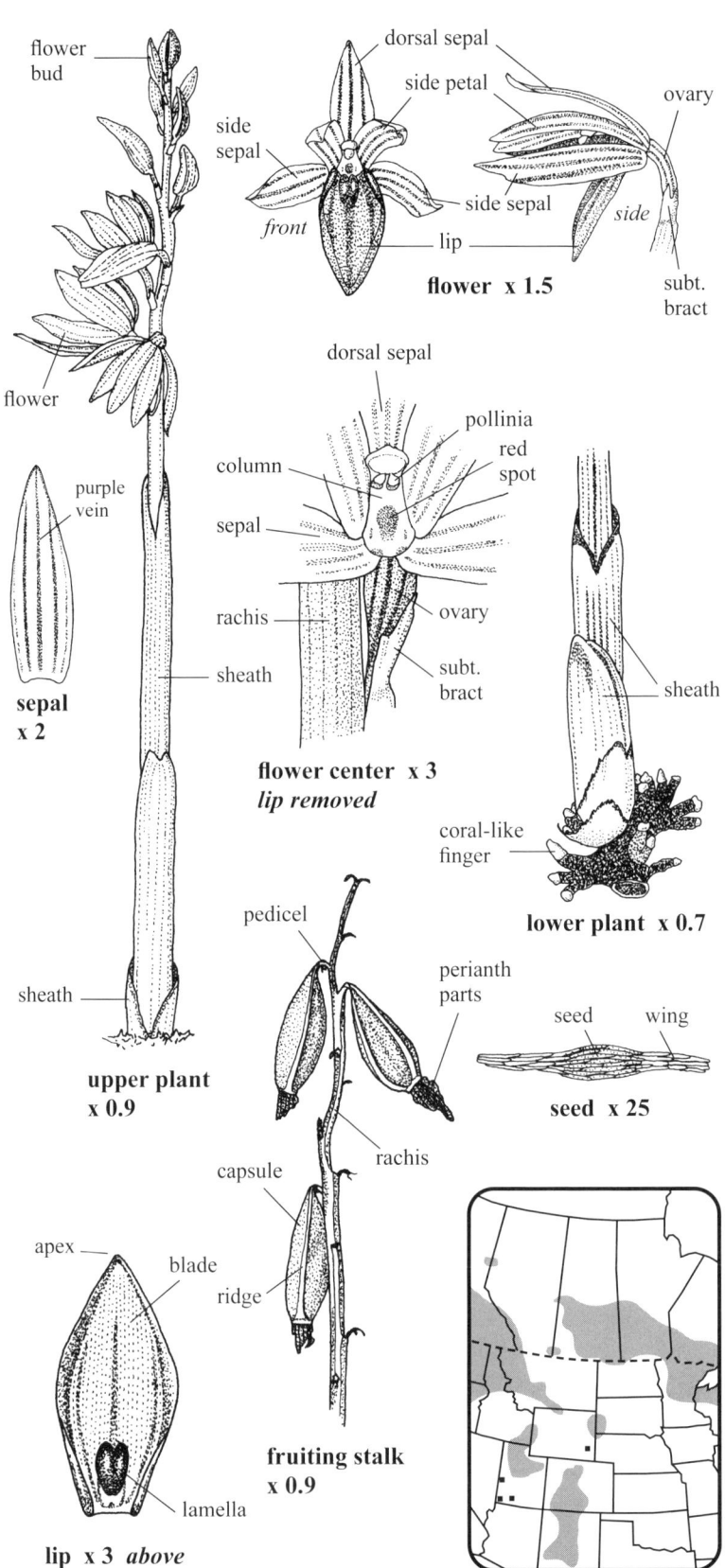

Orchid—*Orchidaceae*   **Wildflower**   Greenish yellow   Petals 3, lip 3-lobed

EARLY CORALROOT *Corallorhiza trifida* Châtelain

■ **SKETCH** A glabrous, saprophytic **perennial herb** 4–34 cm tall from branched tan **rhizomes** with coral-like fingers 1–3 cm long by 2–3 mm wide, these stiff and smooth; near marshes, in moist to dry woodlands, thickets, uplands and bogs.

■ **FLOWERS** Greenish yellow, blooming May–August; **inflorescence** a raceme 2–12 cm long by *c.* 2 cm wide; **pedicels** 1–1.5 mm long, green, twisted; **subtending bracts** (of pedicels) 0.8–3 mm long by 0.8–1 mm wide, pointed to blunt, with two side lobes; **flowers** perfect, spreading, 3–20 per raceme, each 5–7 mm wide by 8–10 mm long; **sepals** three, 1-nerved; **side sepals** 4–7 mm long by 1–1.2 mm wide; **dorsal sepal** entire, margins revolute; **side petals** two, greenish, 3–5 mm long by 1.2–2 mm wide, some with a reddish spot on the inside, membranous; **lip** white, reflexed from base, 3.5–5 mm long by 2–3 mm wide, reddish purple spots in throat, with two small side lobes and two grooves along the sides, apex blunt; **ovary** 2–6.5 mm long by 1.7–2 mm wide, ridged; **column** green with red spots, 2–5 mm long by *c.* 1 mm wide, wider near apex, curved forward, forming a green **mentum** on the ovary. **FRUIT a capsule**, green turning dark brown, 8–14 mm long by 5–6 mm wide, 3-ridged, these separating in the fall, persistent over winter on a recurved pedicel *c.* 2 mm long; **perianth** remains wilted at apex, 2–4 mm long, dark brown; **seeds** numerous, light tan, 0.7–0.9 mm long by 0.1–0.2 mm wide and thick, wing reticulate, straight to curved, one end narrow.

■ **LEAVES** Alternate, reduced to closed veiny sheaths 1–7 cm long, apices pointed and V-shaped on opposite side, pale green, the tops of two sheaths visible above ground.

■ **STEM** Erect, pale green, hollow above, solid below, angular below capsules; 2–4 mm wide at pale yellow, stiff base, *c.* 1.5 mm wide with fruit.

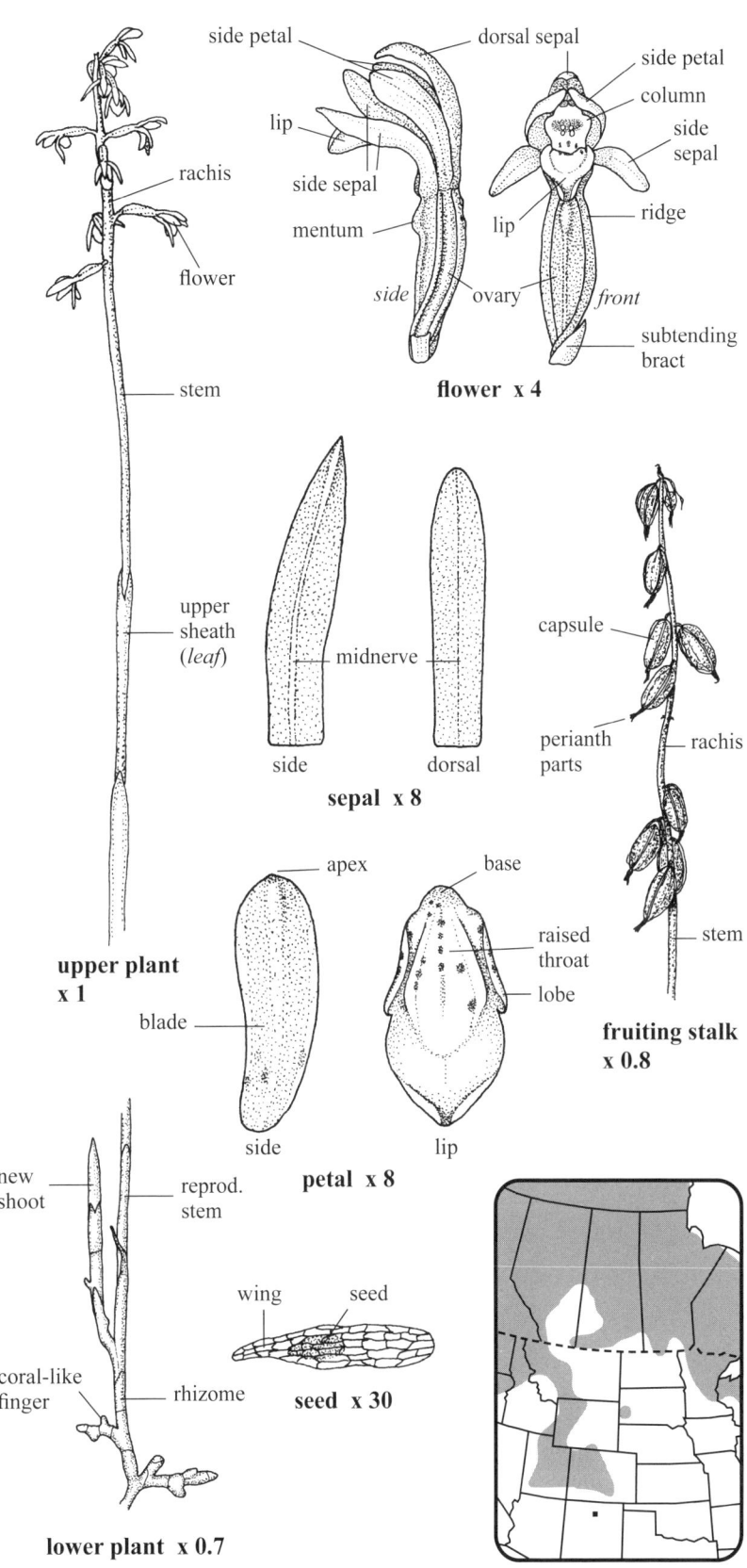

Pale Coralroot, Yellow Coralroot, Northern Coralroot

Orchid—*Orchidaceae* Wildflower White Petals 3

SMALL WHITE LADY'S-SLIPPER *Cypripedium candidum* Muhl. ex Willd.

■ **SKETCH** A **perennial herb** 10–40 cm tall in scattered clusters from short **rhizomes** and several thick **roots**; in moist prairies, pastures and open marshy shores. **p2**

■ **FLOWERS** White, blooming April–June; **inflorescence** a terminal flower; **pedicel** green, hairy, *c.* 1 cm long and erect in fruit; **subtending bracts** one, leaflike, 4–8 cm long by 12–18 mm wide (flattened); **flowers** perfect, fragrant, one (two) per stem, 2–4 cm wide by 2.5–3 cm long by 3.5–4.5 cm tall; **sepals** three, glandular-hairy on backs; **lower sepals** two, united and descending behind the lower lip, 1.3–3.5 cm long by 8–15 mm wide, undulate, green with brown veins, apex bidentate; **dorsal sepal** ascending, 1.5–3.5 cm long by 7–13 mm wide, green with several long reddish veins, glabrous inside (facing the flower); **petals** three; **side petals** two, linear, undulate, entire, 2.3–4.6 cm long by 3–6 mm wide, green with about five brownish to purple veins, descending, with one or two incomplete twists, hairy outside on midvein and inside near the base, some outside hairs glandular; **lower petal** (lip) forming a pouch, white with faint pinkish purple stripes outside, 16–27 mm long by 10–15 mm wide and high, glabrous outside, spotted red along internal lower veins, inside white hairy at base below the staminode, lip margins turned inward along opening; **stamens** two, one on either side of the staminode, fertile, yellow, 5–6 mm long by *c.* 1.3 mm wide, distinct part stiff, 2.5–3 mm long, smooth; **staminodia** yellow, 9–12 mm long by *c.* 3 mm wide, glabrous, stalked, blunt, extending into lower lip, reddish blotches on apical half; **ovary** green, hairy, *c.* 2 cm long by *c.* 4 mm wide, 6-ridged, not twisted, hidden by subtending bract; **stigmas** three, below the staminode, *c.* 4 mm wide and long, yellowish green. **FRUIT a capsule**, erect, 3-valved, 1.8–3 cm long by 1–1.5 cm wide, 6-ridged, hairs short; **seeds** numerous, *c.* 1 mm long, microscopic, wing net-veined and transparent.

■ **LEAVES** Alternate, simple, entire, three to six per plant, hairy on both sides, distinctly parallel veined, sessile, clasping, pointed, V-shaped, green, not striped, erect to ascending; **blades** 6–20 cm long by 2–5.3 cm wide (flattened).

■ **STEM** Erect, hairs short, nodding with the flower; 2–3 mm thick near the base (sheaths removed).

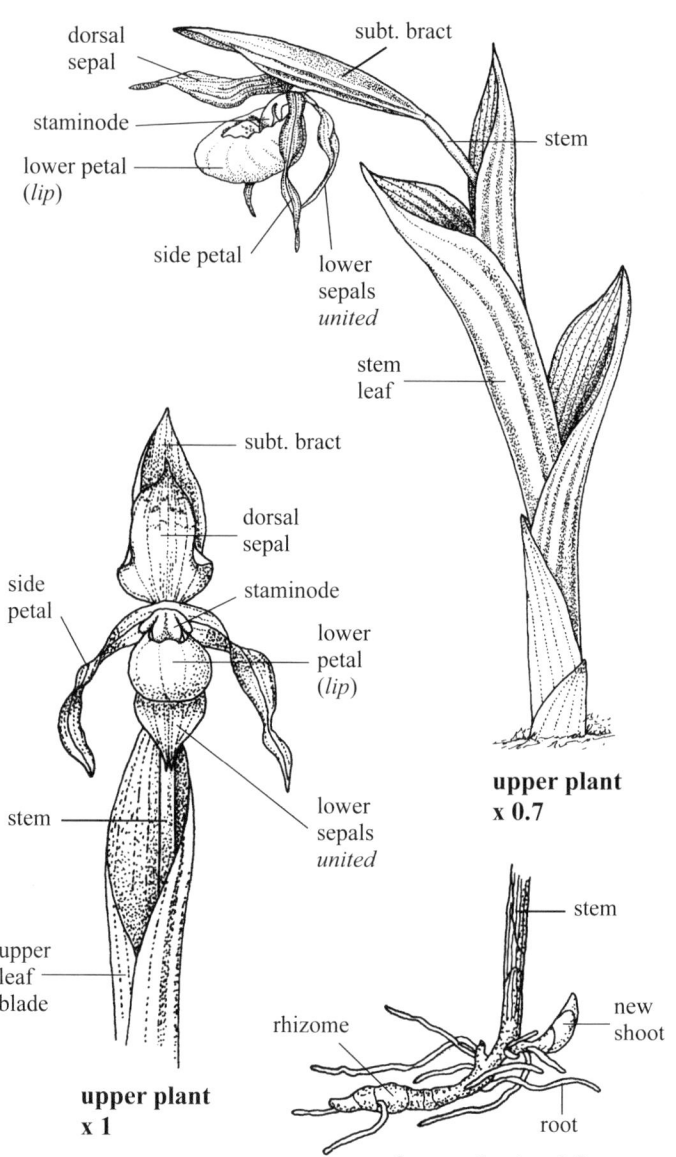
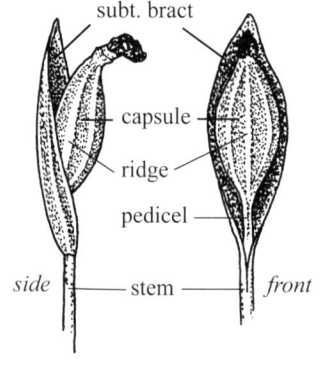
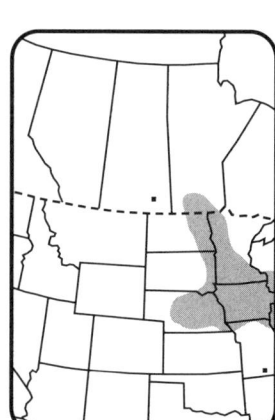

614 White Lady's Slipper

Orchid—*Orchidaceae*   **Wildflower**   Yellow   Petals 3

YELLOW LADY'S-SLIPPER   *Cypripedium parviflorum* Salisb.

■ **SKETCH** A variable **perennial herb** 15–57 cm tall with a **rhizome** 3–4 mm wide with tan **roots** *c.* 1 mm thick; solitary or in tight clusters to *c.* 50; in prairies, along woodland edges, in ditches, stream sides and alder swamps.

■ **FLOWERS** Yellow, blooming April–July; **inflorescence** terminal, mostly 1-flowered (2-), showy; **subtending bracts** (of pedicels) one, 3.5–11 cm long by 1–5 cm wide, upright, entire, green, hairs gland-tipped; **pedicels** 5–10⁺ mm long; **flowers** perfect, irregular, scented; **sepals** three, green with reddish veins, glandular-hairy; **dorsal sepal** 2–8 cm long by 1–4 cm wide, apex pointed; **side sepals** two, fused, 2–7.5 cm long by 0.5–2.8 cm wide, appearing as one except for two tiny apical teeth; **petals** three; **side petals** two, green with reddish veins, linear, twisted one to three times, rarely flat, tips pointed, 2.5–9.5 cm long by 5–12 mm wide, hairy at the base; **lower petal** (lip) yellow, hollow, pouch-shaped, mostly glabrous except for a patch of long, white, erect hairs at its base below the stigma, often with purple spots or veins on the inside, 1.5–5.5 cm long by 1.5–3 cm wide, horizontal to declining; **staminode** one, on a thick, fleshy stalk, yellow with purple marginal spots near its tip, *c.* 1 cm long by 5–6 mm wide, glabrous, covering the stigma and stamens below; **stamens** two, fertile, *c.* 5 mm long, yellow; **anthers** 2–3 mm long; **ovary** green, 20–25 mm long by 3–5 mm wide and thick, few ribbed, glandular-hairy; **stigmas** three, united, yellow, oval, *c.* 6 mm long by *c.* 5 mm wide. **FRUIT a capsule**, 2–3.7 cm long by 1–2 cm wide, 5- or 6-ribbed, slightly hairy, erect; **seeds** *c.* 1,500 per capsule, 1–1.2 mm long by *c.* 0.2 mm wide, winged.

■ **LEAVES** Alternate, simple, entire, pointed, three to six, about 10 parallel veins, smooth, clasping; **blades** 3–20 cm long by 1.5–11 cm wide, larger above, more hairy to the outside, some hairs gland-tipped.

■ **STEM** Erect, unbranched, glabrous to sparsely glandular-hairy; 3–9 mm thick near the sheath-covered base.

■ **SYN.** *Cypripedium calceolus* ssp. *parviflorum* (Salisb.) Hultén and *Cypripedium calceolus* var. *parviflorum* (Salisb.) Fern.

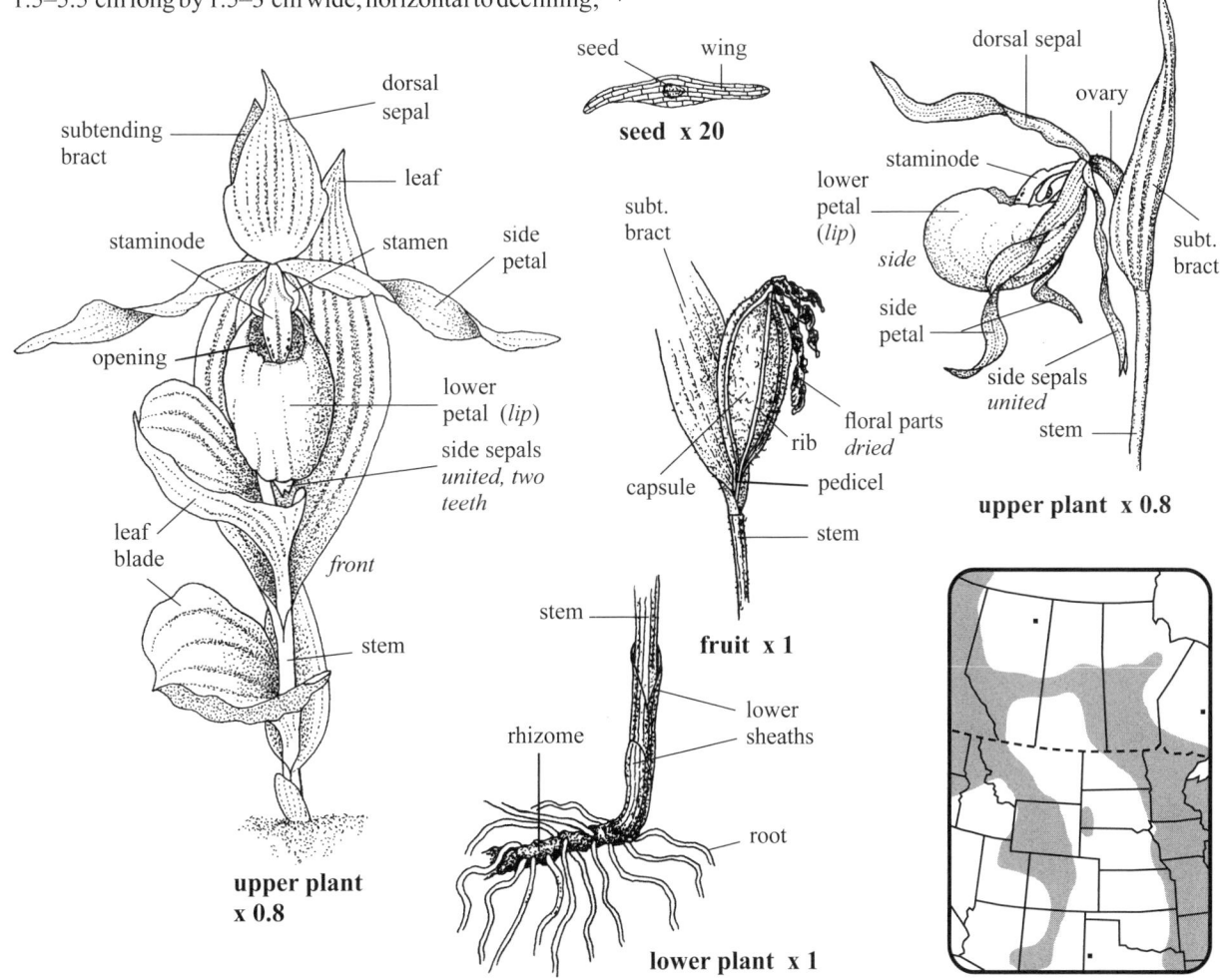

Yellow Moccasin Flower

Orchid—*Orchidaceae* Wildflower White Petals 3

LESSER RATTLESNAKE-PLANTAIN *Goodyera repens* (L.) R. Br. ex Ait. f.

■ **SKETCH** A glandular-hairy **perennial herb** 7–20 (–30) cm tall in open colonies from tan, branching **rhizomes** 2–7 cm long by 1–2 mm thick with short, fine **fibrous roots**; in woods and bogs, often in mossy sites.

■ **FLOWERS** White, blooming July–September; **inflorescence** a raceme (spikelike), one, erect, 7- to 36-flowered, 2–7 cm long by 8–11 mm wide; **pedicels** green, *c.* 1 mm long, hidden; **subtending bracts** (of pedicels) glandular-hairy, 5–12 mm long by 1.5–3 mm wide, reduced above, 1- or 3-nerved; **flowers** perfect, mostly one-sided or in a spiral, 3–6 mm wide by 8–10 mm long (including ovary) by 4–5 mm tall, ascending, crowded, hairy; **sepals** three, hairy on outside only; **dorsal sepal** weakly attached to two dorsal petals, 3.5–4.2 mm long by 2–2.9 mm wide (flat); **side sepals** 3–5 mm long by 2–3.5 mm wide (flat), midnerve pale green; **petals** three, white, not hairy; **dorsal petals** two, membranous, 3–5 mm long by 1.8–2 mm wide, midvein faint, entire to erose on outside margin near apices; **lip** 2–5 mm long by 1.5–3.2 mm wide, apex recurved, the sac 2–2.5 mm long by *c.* 2 mm wide, blunt, hidden by the side sepals; **ovary** green, glandular-hairy, 5–7 mm long by *c.* 2 mm wide by 1.5–1.7 mm thick, with three low wide ridges, the dorsal ridge the widest; **anthers** *c.* 1 mm long; **pollen** yellow. FRUIT **a capsule**, light brown, ascending to descending, 4–7 mm long by 2–4 mm wide and thick, very hairy with dry perianth parts at the apex; **seeds** numerous, brown, 0.5–0.7 mm long by 0.1–0.2 mm wide and thick, wing shiny, reticulate and transparent.

■ **LEAVES** Alternate, simple, entire, usually two to eight near the base; **lower blades** 1–4.5 cm long by 6–18 mm wide, spreading, side veins green to white, reticulate; **petioles** wide, 1–13 mm long, forming a closed sheath 5–12 mm long around the stem, sheaths imbricate; **upper leaves** (sheaths) two or three, TL 13–21 mm, erect, blunt, the apical 2–5 mm ascending, top leaf slightly hairy.

■ **STEM** Erect, pale green, hairless below, glandular-hairy above, round, solid, stiff; 1.5–2.5 mm thick near the base.

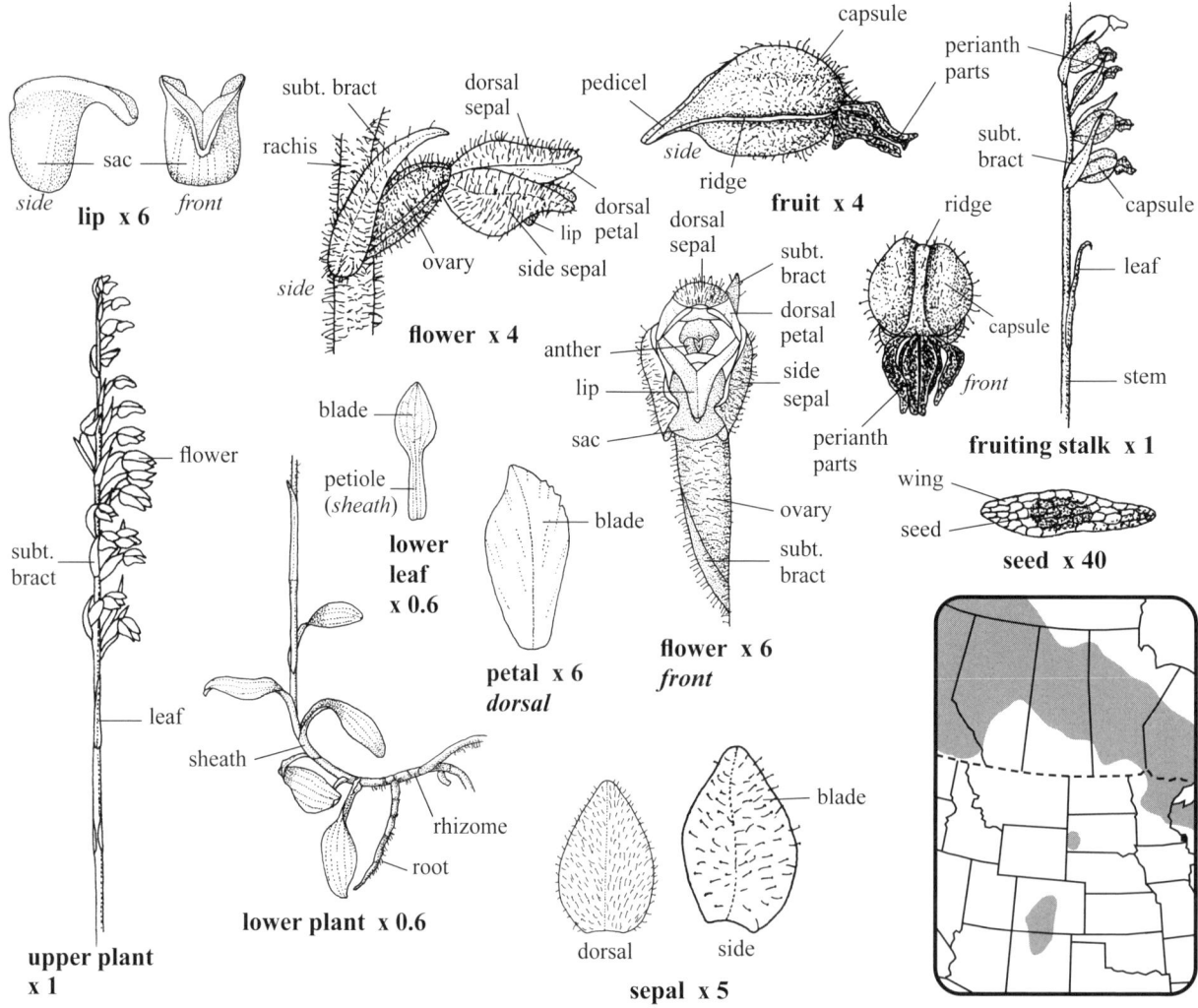

616 Dwarf Rattlesnake Plantain, Lesser Rattlesnake-orchid

Orchid—*Orchidaceae*   **Wildflower**   Reddish green to yellowish green   Petals 3

Heart-leaved Twayblade   *Listera cordata* (L.) R. Br. ex Ait. f.

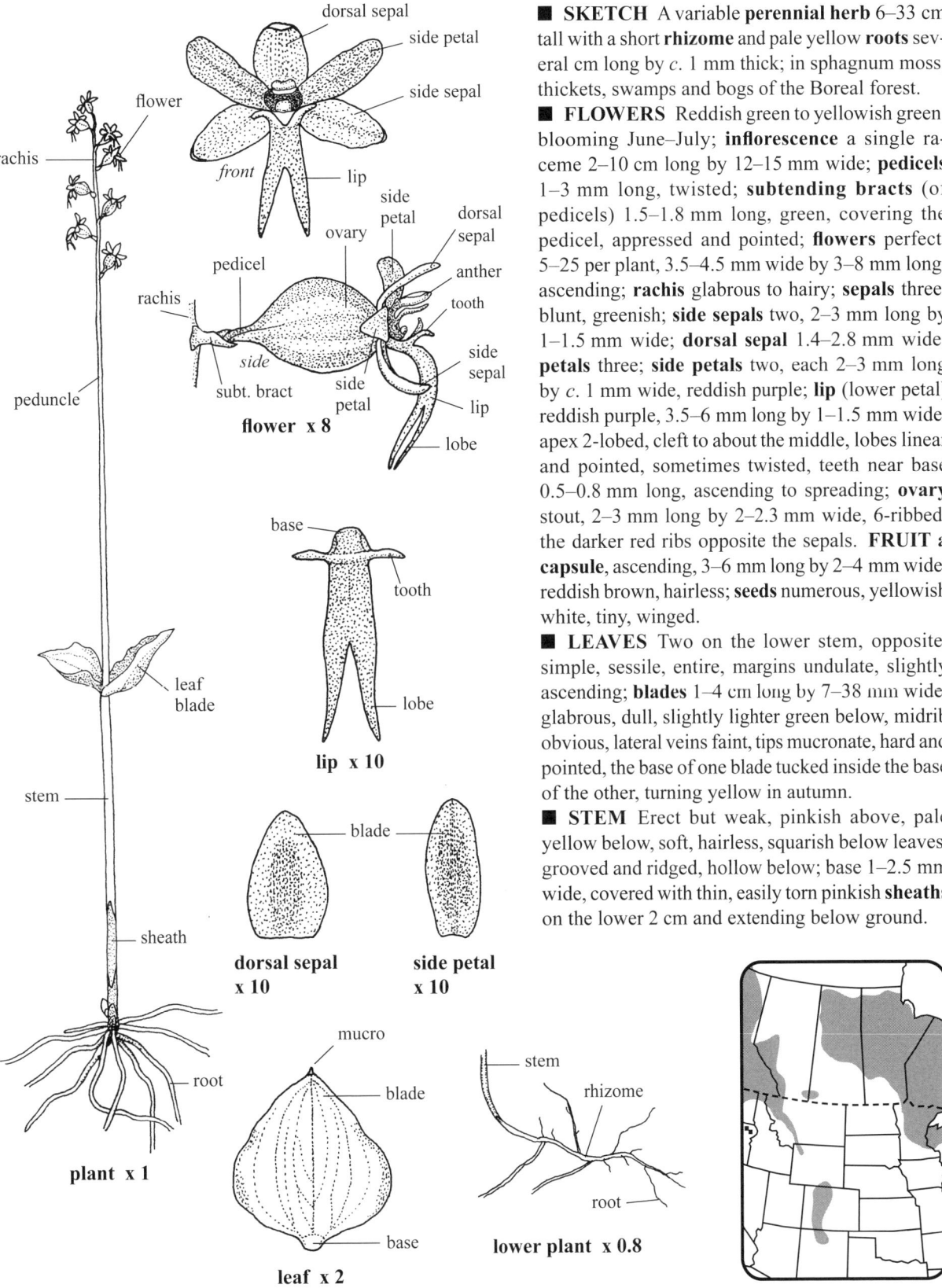

■ **SKETCH** A variable **perennial herb** 6–33 cm tall with a short **rhizome** and pale yellow **roots** several cm long by *c.* 1 mm thick; in sphagnum moss, thickets, swamps and bogs of the Boreal forest.

■ **FLOWERS** Reddish green to yellowish green, blooming June–July; **inflorescence** a single raceme 2–10 cm long by 12–15 mm wide; **pedicels** 1–3 mm long, twisted; **subtending bracts** (of pedicels) 1.5–1.8 mm long, green, covering the pedicel, appressed and pointed; **flowers** perfect, 5–25 per plant, 3.5–4.5 mm wide by 3–8 mm long, ascending; **rachis** glabrous to hairy; **sepals** three, blunt, greenish; **side sepals** two, 2–3 mm long by 1–1.5 mm wide; **dorsal sepal** 1.4–2.8 mm wide; **petals** three; **side petals** two, each 2–3 mm long by *c.* 1 mm wide, reddish purple; **lip** (lower petal) reddish purple, 3.5–6 mm long by 1–1.5 mm wide, apex 2-lobed, cleft to about the middle, lobes linear and pointed, sometimes twisted, teeth near base 0.5–0.8 mm long, ascending to spreading; **ovary** stout, 2–3 mm long by 2–2.3 mm wide, 6-ribbed, the darker red ribs opposite the sepals. **FRUIT a capsule**, ascending, 3–6 mm long by 2–4 mm wide, reddish brown, hairless; **seeds** numerous, yellowish white, tiny, winged.

■ **LEAVES** Two on the lower stem, opposite, simple, sessile, entire, margins undulate, slightly ascending; **blades** 1–4 cm long by 7–38 mm wide, glabrous, dull, slightly lighter green below, midrib obvious, lateral veins faint, tips mucronate, hard and pointed, the base of one blade tucked inside the base of the other, turning yellow in autumn.

■ **STEM** Erect but weak, pinkish above, pale yellow below, soft, hairless, squarish below leaves, grooved and ridged, hollow below; base 1–2.5 mm wide, covered with thin, easily torn pinkish **sheaths** on the lower 2 cm and extending below ground.

Orchid—*Orchidaceae* **Wildflower** Yellowish green Petals 3

GREEN-FLOWERED BOG ORCHID *Platanthera aquilonis* Sheviak

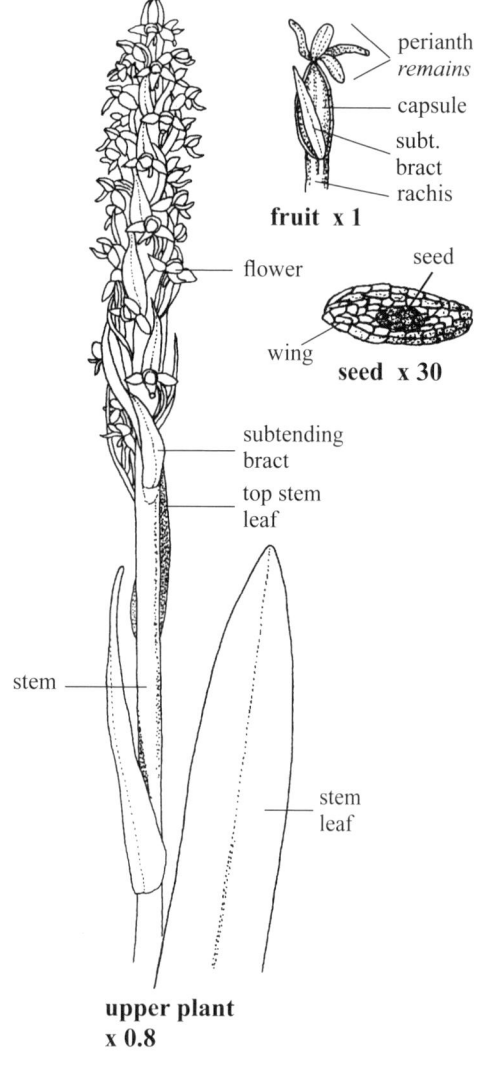

fruit x 1

seed x 30

upper plant x 0.8

■ **SKETCH** A glabrous **perennial herb**, solitary and scattered, 15–80 cm tall with several fleshy **roots** 1–2 mm thick and to *c.* 15+ cm long; in bogs and fens, roadside ditches, cool marshes, tundra, banks of streams, thickets and wet woodland edges to open damp meadows.

■ **FLOWERS** Yellowish green, blooming May–August; **inflorescence** a spike 5–25 cm long by 15–20 mm wide, densely to sparsely flowered; **subtending bracts** (of flowers) one, 8–40 mm long by 3–6 mm wide, reduced above, green, entire, pointed, glabrous, the lower bracts exceeding the flowers; **flowers** perfect, sessile, 7–12 mm wide by 14–16 mm long, throat glabrous, generally not spotted, more crowded above; **sepals** three; **dorsal sepal** oval, 3–7 mm long by 1.3–4 mm wide, apex rounded, green, partially covering the two upper petals; **side sepals** two, spreading to reflexed, entire, flat, 3–8 mm long by 1–3.5 mm wide; **petals** three, green, 3–9 mm long by 1–3 mm wide, somewhat fleshy, glabrous, two side ones not united above, together they form a hood over the column and stamens; **lower petal** (lip) spreading to ascending, gradually wider toward the base, entire, fleshy, blunt, 4–9 mm long by 1.5–2.5 mm wide; **spur** below the lip, 3–6 mm long by 1–1.3 mm wide, descending, cylindrical; **ovary** stout, 5–13 mm long. **FRUIT a capsule**, erect, sessile, 1–1.5 cm long by 4–7 mm wide; **seeds** numerous, golden, *c.* 0.2 mm long by *c.* 0.1 mm wide, wing net-veined, *c.* 0.7 mm long by *c.* 0.3 mm wide and long.

■ **LEAVES** Alternate, simple, entire, four to seven per stem; **blades** flat, veins obscure and generally parallel to margins, 2–25 cm long by 0.5–4 cm wide, reduced above, slightly fleshy, pointed, sessile, glabrous, clasping the stem, ascending, green; **basal sheaths** 2–4.5 cm long.

■ **STEM** Erect, glabrous, slightly ridged due to the continuation of leaf bases along the stem; 2–7 mm thick near the base.

■ **SYN.** *Habenaria hyperborea* (L.) R. Br. ex Ait. f. = *Platanthera hyperborea* var. *hyperborea* (L.) Lindl. and *Platanthera hyperborea* auct. non (L.) Lindl.

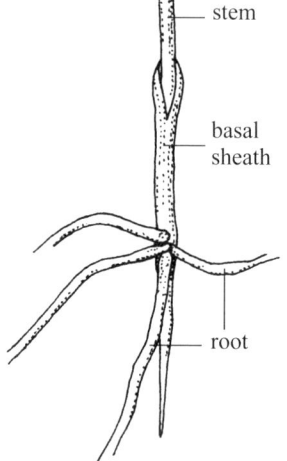

lower plant x 1

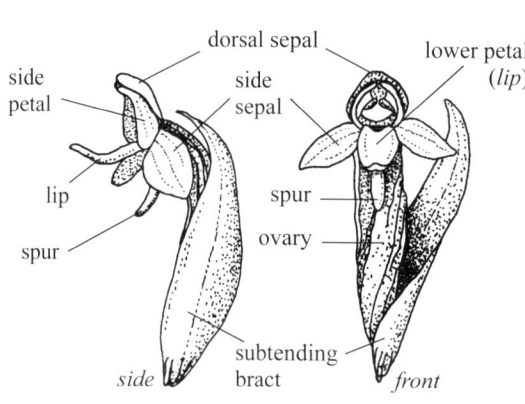

flower x 3

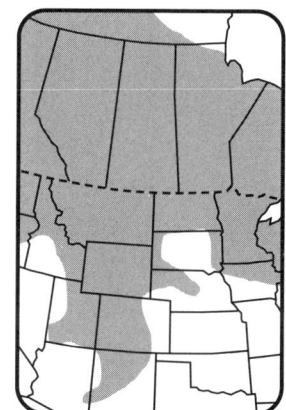

Orchid—*Orchidaceae* **Wildflower** Creamy white Petals 3

GREAT PLAINS WHITE FRINGED ORCHID *Platanthera praeclara* Sheviak & Bowles

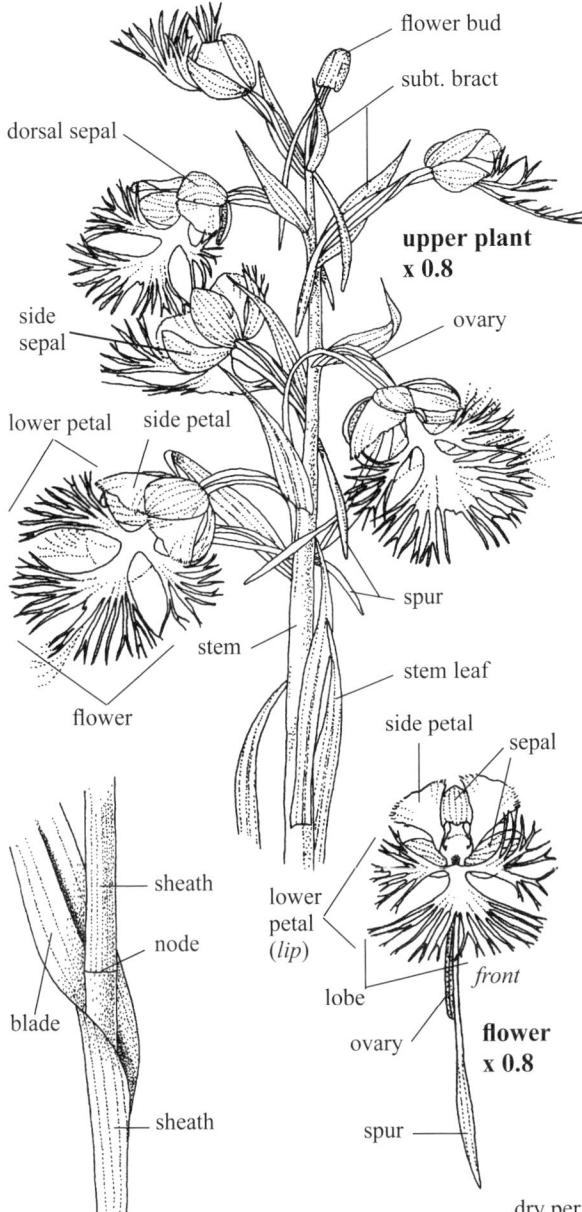

■ **SKETCH** A glabrous **perennial herb** 30–110 cm tall from coarse, fleshy **roots** and **tubers**, in scattered open colonies or singly; in very moist prairies, hay fields, open bogs, marshy sites and roadside ditches.

■ **FLOWERS** Creamy white, blooming May–August; **inflorescence** a spike, open, altogether 5–20 cm long by 3.5–11 cm wide; **subtending bracts** (of flowers) one, green, pointed, leaflike, 2.5–5.5 cm long by 7–10 mm wide, reduced above, veiny; **flowers** perfect, sessile, 1–25, each 2.5–4 cm wide by *c.* 3.3 cm long (not including spur), fragrant at night when pollination occurs; **sepals** three, glabrous, oval, C-shaped; **side sepals** 7–14 mm long by 5–10 mm wide, entire, pointed, 3- to 6-veined, creamy with green veins; **dorsal sepal** 7–12 mm long by 5–8 mm wide; **petals** three, fringed or toothed; **lower petal** (lip) 3-lobed, each lobe deeply fringed, 2–3.2 cm long by 2–4 cm wide, its **spur** deflexed, greenish yellow, 2.5–5.5 cm long by 1–3 mm wide, widest near its distant tip; **side petals** two, each 9–17 mm long by 7–14 mm wide, slightly toothed; **column** of fused stamens, style and stigma; **anther sacs** separate; **ovary** slightly twisted, 2–3 cm long by 2.5–3 mm wide, glabrous, 3-ridged, grooved. **FRUIT a capsule**, 2–3 cm long by 6–9 mm wide with three flat ridges *c.* 2 mm wide, persistent; **seeds** numerous, microscopic, lack endosperm, winged, released through splits.

■ **LEAVES** Alternate, simple, entire, linear, usually two to eight, sessile, glabrous, ascending, thick, clasping the stem, the midrib distinctive and raised, five or six thin lateral veins on either side of the midrib; **blades** whitish green on outside (dorsal) and along sheath, yellowish green on ventral side facing the stem, 7–25 cm long by 1–5 cm wide, pointed, reduced above, lower blades more spreading.

■ **STEM** Erect, glabrous, slightly ridged from leaf midrib extensions, not branched, fleshy; 5–12 mm wide near the base.

■ **SYN.** *Habenaria leucophaea* var. *praeclara* (Sheviak & Bowles) Cronq.

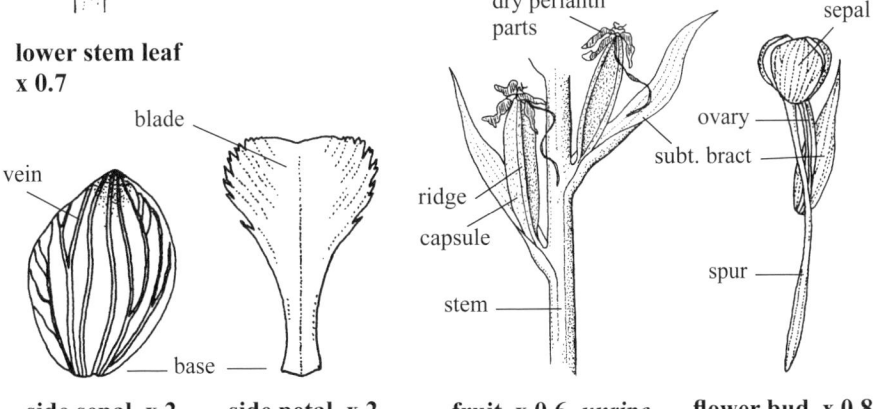

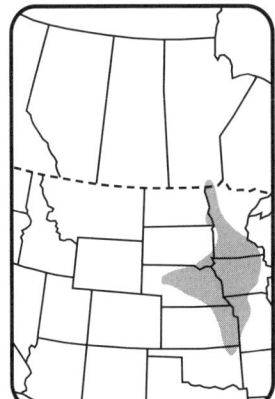

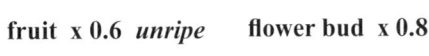

Western Prairie Fringed Orchid 619

Wild Grass

It might have been the day
the universe expanded
when the earth opened up
to the sky
and the wild grass bowed down
to the wind

That small round place
in the wild grass
where I slept like a deer
the sound of all the crickets
waking me
to the light of the moon on my hands

Later there was the cradle I wove
out of the blades of grass
and the gray Heron
I startled into the air

Family Characteristics

Grass—Poaceae — Flowers in spikelets

SKETCH Monocotyledonous herbs, annual or perennial, solitary, tufted or in colonies, with **fibrous roots** and often **rhizomes**. About 9,000 species worldwide.

FLOWERS Mostly perfect, male or female, or sterile, small, often greenish, wind-pollinated. **Spikelets** of 1–many florets (flowers) attached along the **rachilla** (axis of the spikelet). **Inflorescence** a spike, raceme or panicle with a central main axis or **rachis**. **Spikelets** subtended by two protective **glumes** (sterile bracts), the first glume slightly lower than the second one. **Florets** (flowers) consisting of an outer **lemma** and inner 2-nerved **palea** enclosing the stamens and pistil, or these sometimes sterile and reduced. Glumes and lemmas (nerved or not) may or may not have an awn. **Stamens** mostly 3 (2), often yellow, exserted or not. **Pistil** 1. **Ovary** 1-loculed with 1 ovule. **Styles** 2 (1), long to short. **Stigmas** 2, feathery, usually exserted.

FRUIT A **grain** (caryopsis), 1-seeded, sometimes enclosed by a hardened lemma and palea. **Pericarp** of fruit usually attached to the seed.

LEAVES Alternate, 2-ranked, basal or stem, each of 3 parts: **1)** the lower **sheath** arising from the node, enveloping the stem, open with margins overlapping, or united **2)** the **ligule** (a thin membrane or fringe of hairs) at the junction of the sheath and blade is visible when the blade is gently pulled away from the stem, and **3)** the **blade** is simple with parallel venation, ascending to arched, mostly flat, narrow, usually scabrous along the entire margins or inner ridges. **Auricles** (paired narrow projections of sheath margins) sometimes present at the collar near the ligule.

STEM A culm, round to oval, often hollow, enclosed by the sheath of the leaf, exposed below the floral branches; **nodes** solid, slightly enlarged, glabrous to hairy.

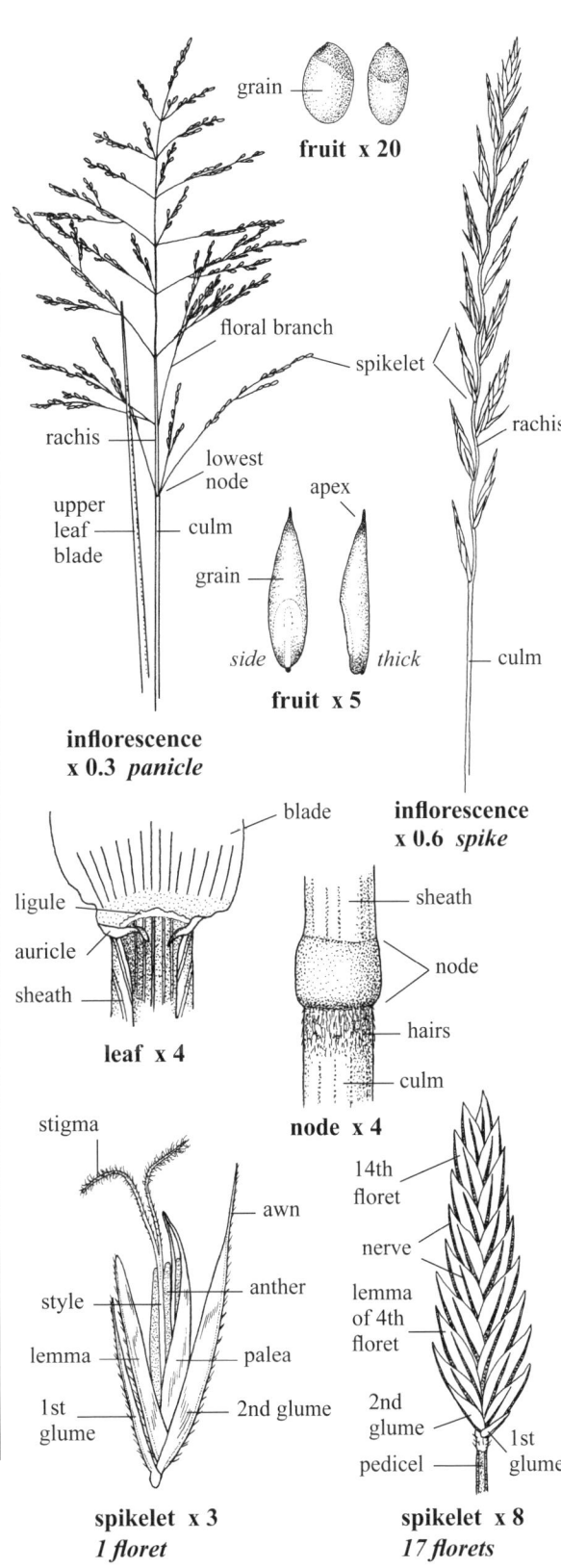

CRESTED WHEATGRASS *Agropyron cristatum* (L.) Gaertn.

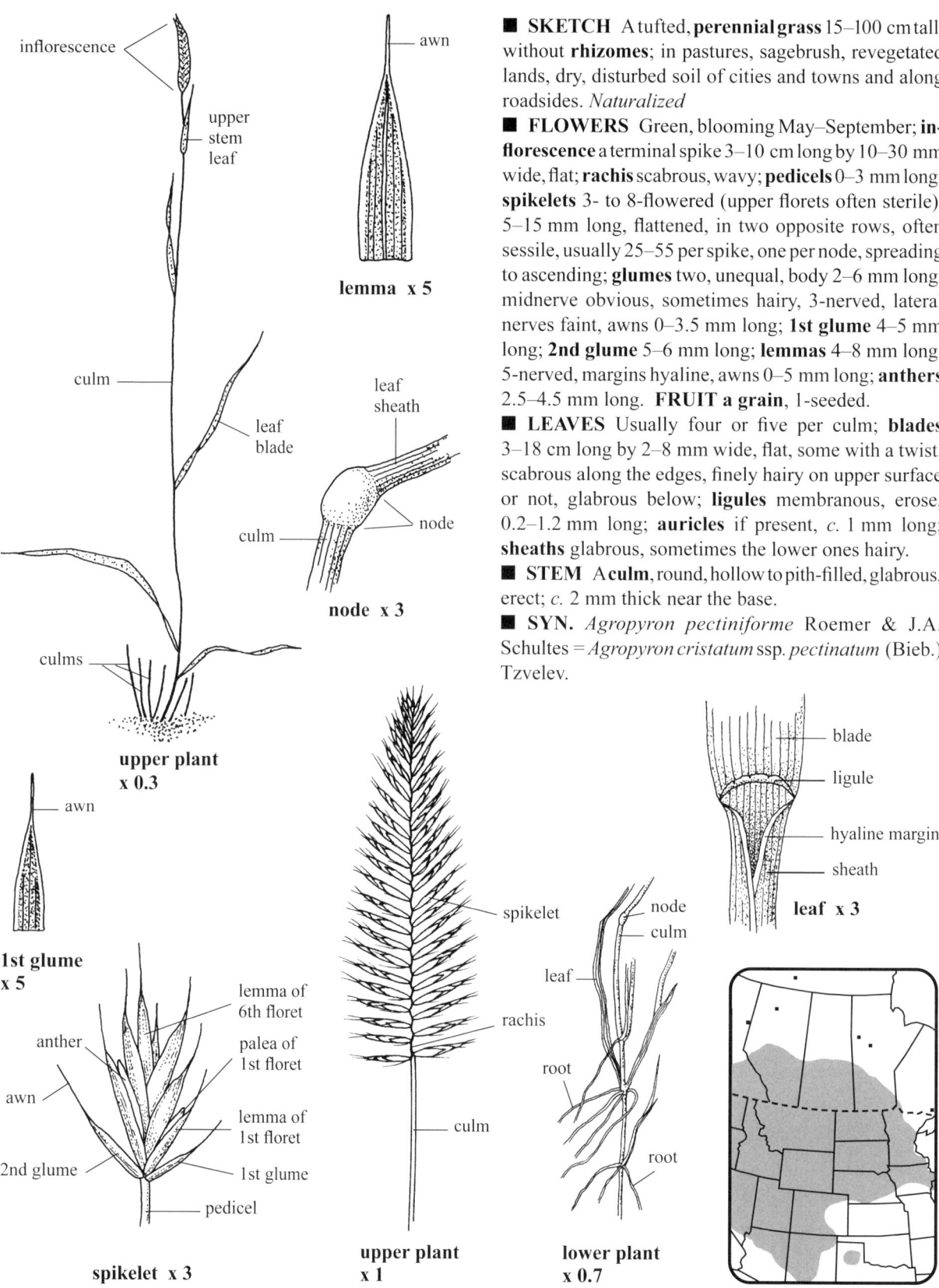

- **SKETCH** A tufted, **perennial grass** 15–100 cm tall, without **rhizomes**; in pastures, sagebrush, revegetated lands, dry, disturbed soil of cities and towns and along roadsides. *Naturalized*
- **FLOWERS** Green, blooming May–September; **inflorescence** a terminal spike 3–10 cm long by 10–30 mm wide, flat; **rachis** scabrous, wavy; **pedicels** 0–3 mm long; **spikelets** 3- to 8-flowered (upper florets often sterile), 5–15 mm long, flattened, in two opposite rows, often sessile, usually 25–55 per spike, one per node, spreading to ascending; **glumes** two, unequal, body 2–6 mm long, midnerve obvious, sometimes hairy, 3-nerved, lateral nerves faint, awns 0–3.5 mm long; **1st glume** 4–5 mm long; **2nd glume** 5–6 mm long; **lemmas** 4–8 mm long, 5-nerved, margins hyaline, awns 0–5 mm long; **anthers** 2.5–4.5 mm long. **FRUIT a grain**, 1-seeded.
- **LEAVES** Usually four or five per culm; **blades** 3–18 cm long by 2–8 mm wide, flat, some with a twist, scabrous along the edges, finely hairy on upper surface or not, glabrous below; **ligules** membranous, erose, 0.2–1.2 mm long; **auricles** if present, *c.* 1 mm long; **sheaths** glabrous, sometimes the lower ones hairy.
- **STEM** A **culm**, round, hollow to pith-filled, glabrous, erect; *c.* 2 mm thick near the base.
- **SYN.** *Agropyron pectiniforme* Roemer & J.A. Schultes = *Agropyron cristatum* ssp. *pectinatum* (Bieb.) Tzvelev.

Grass—*Poaceae* Grass Spikelets 1-flowered

REDTOP *Agrostis stolonifera* L.

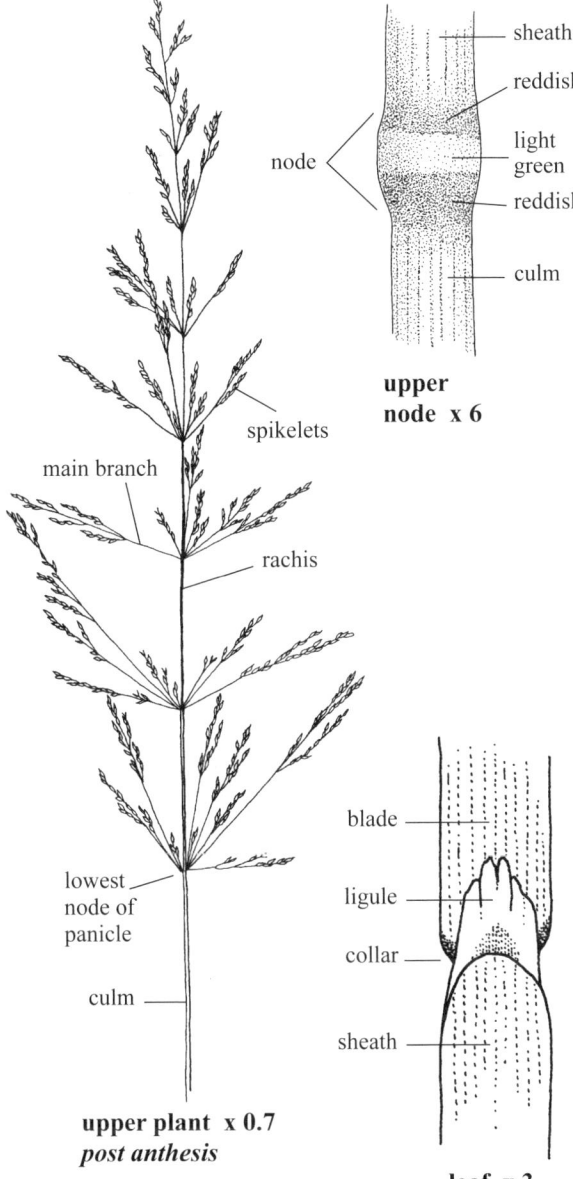

upper plant x 0.7
post anthesis

upper node x 6

leaf x 3

■ **SKETCH** A variable, reddish **perennial grass** 30–120 cm tall, in colonies from **stolons** or **rhizomes**; in moist areas of prairies to mountains, swampy ground and occasionally in shallow water at pond edges. *Naturalized*

■ **FLOWERS** Reddish purple, blooming June–September; **inflorescence** an open to compact panicle 5–30 cm long by 2–13 cm wide; **main branches** reddish, 1–10 cm long, strongly ascending to spreading, of various lengths, slightly scabrous, lower half to third of longer branches bare; **rachis** reddish purple, glabrous below, more scabrous above; **spikelets** 1-flowered, 1.5–3.5 mm long by *c.* 0.5 mm wide; **glumes** two, almost equal, 1.5–3.5 mm long, 1- or 3-nerved, lateral nerves obscure, midnerve often scabrous at least near its apex, awnless, pointed, reddish purple, a narrow green band at the base, rounded on the back below, more laterally compressed near the apex; **lemmas** 1.2–2.5 mm long by *c.* 0.5 mm wide, rounded dorsally, hyaline and softly membranous, more keeled near the awnless, pointed apices, 3- or 5-nerved, with thin red lines near apices; **paleae** 0.5–1.5 mm long, shorter than the lemmas; **stamens** three; **anthers** pale yellow, 0.5–1.3 mm long. **FRUIT a grain**, 1-seeded.

■ **LEAVES** Usually five or six per culm, lower ones wilted, upper two or three green and ascending; **blades** flat, scabrous to smooth, 4–26 cm long by 2–10 mm wide, top blades usually well below the panicle; **ligules** membranous, 1.5–7 mm long, tip flat to pointed, usually torn, not hairy, reddish purple near the base; **auricles** absent; **collars** often reddish purple, glabrous; **sheaths** open, smooth, round, margins glabrous, often reddish.

■ **STEM** A **culm**, erect, hollow, glabrous below the panicle; 1.5–3 mm thick near the decumbent base; **nodes** reddish brown on both sides of the light green area.

■ **SYN.** *Agrostis alba* var. *palustris* (Huds.) Pers., *Agrostis alba* var. *stolonifera* (L.) Sm. and *Agrostis palustris* Huds.

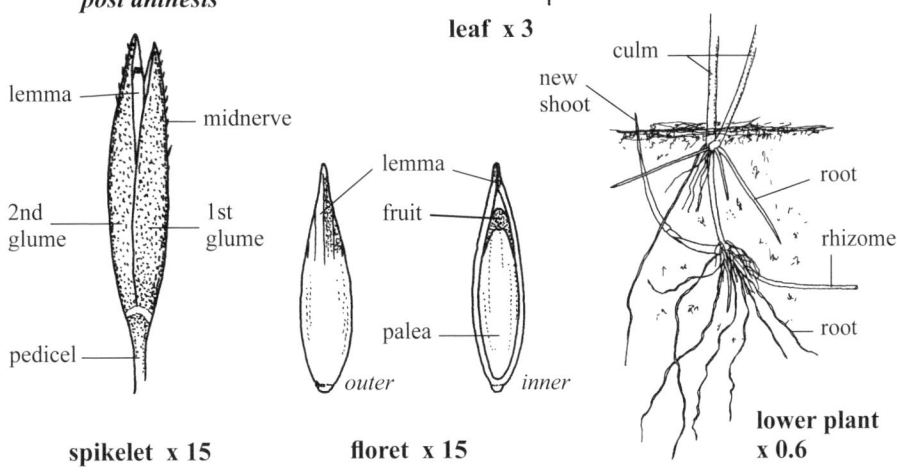

spikelet x 15   floret x 15   lower plant x 0.6

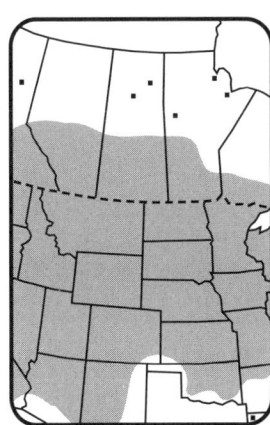

Spreading Bent, Creeping Bentgrass

Grass—*Poaceae*　　　　　　　　　　　　　　　　　　　　　　　　　　Grass　Spikelets 1-flowered

SHORT-AWNED FOXTAIL　*Alopecurus aequalis* Sobol.

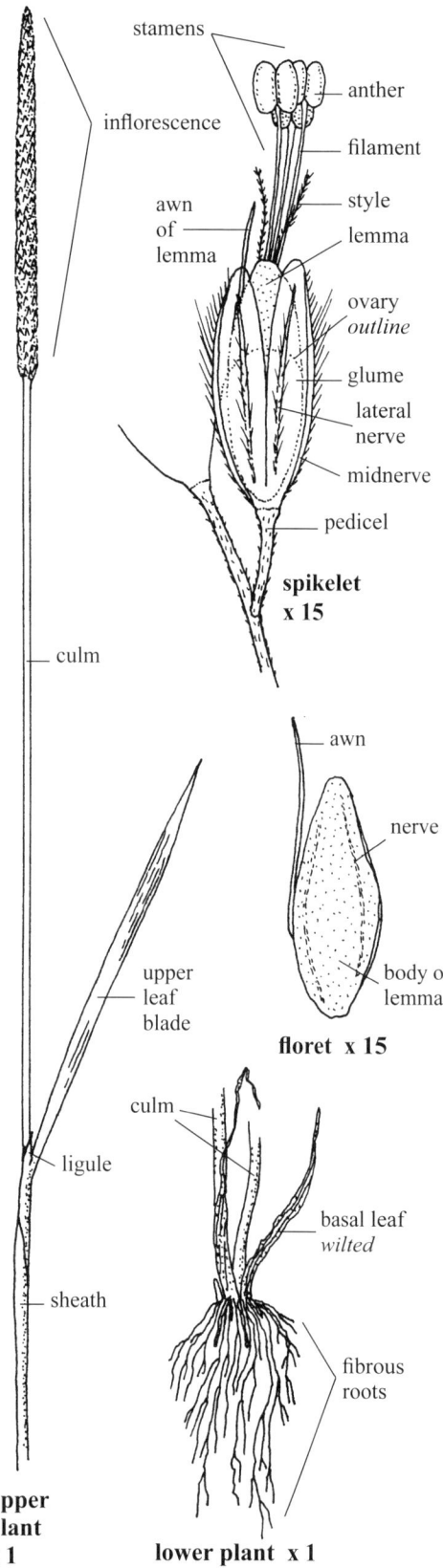

upper plant x 1

lower plant x 1

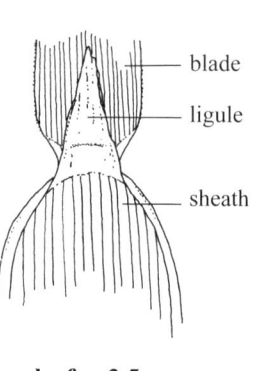

leaf x 3.5

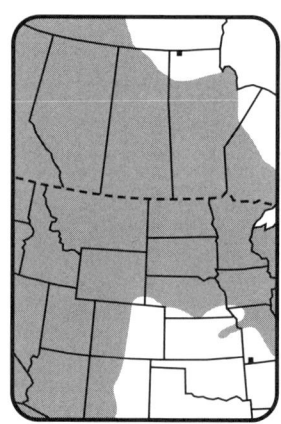

■ **SKETCH** A tufted **perennial grass** 15–65 cm tall with **fibrous roots**; in moist soils around sloughs, lake shores, ditches and river edges, sometimes in shallow water or dry ponds and moist saline meadows.

■ **FLOWERS** Pale green, blooming May–September, sometimes in the first year; **inflorescence** a panicle (spikelike), compact, cylindrical, 3–8 cm long by 3–6 mm wide, erect; **spikelets** 1-flowered, 1.6–2.8 mm long by 0.8–1 mm wide (not including hairs), laterally compressed, dropping early and easily with disarticulation below the glumes; **glumes** two, 1.6–2.5 mm long, united at base, nerves three hairy and green, awnless, body hyaline and transparent, apices blunt and scarious; **lemmas** green, 1.6–2.8 mm long, visible inside glumes, enclosing the fruit, soft, glabrous, faintly 3- or 5-nerved, body of lemma as long as glumes, open along side opposite its awn; **awn** barely exserted, *c.* 1 mm longer than the lemma, attached right below the lemma's middle, the free part *c.* 2 mm long, straight to slightly curved; **palea** absent; **stamens** three, exserted; **filaments** white, hyaline, glabrous, filiform; **anthers** pale yellow, 0.5–0.8 mm long, drying to orange; **styles** two, *c.* 3 mm long, hairy, exserted. **FRUIT a grain**, 1 seeded.

■ **LEAVES** Few per culm, upper ones ascending, 1–3 wilted leaves below; **blades** dull, hairless, medium green, margins lightly scabrous, 5–15 cm long by 1–6 mm wide, ribbed on inner surface, flaccid when floating; **sheaths** smooth, slightly inflated to accommodate fascicles or secondary culms and panicles emerging from the lower leaf axils, margins hyaline, hairless; **ligules** clasping the culm, membranous, tips sometimes folded back and down, tan, pointed, 3–7 mm long, tapered, edges slightly rolled or curved away from the culm, with dark brown marks on the body or edges.

■ **STEM** A **culm**, erect to decumbent below and sometimes rooting at the lower nodes, smooth, light green above, hollow; 1–2 mm wide near the brown base; **nodes** light yellowish green, hairless, smooth, slightly swollen, 2–3 mm long.

Short-awned Meadow Foxtail, Water Foxtail, Little Meadow-foxtail

Big Bluestem *Andropogon gerardii* Vitman

■ **SKETCH** A **perennial grass** 0.6–3 m tall, singly or in large mats from short **rhizomes**; in prairies, dry sandy areas and along roadsides. **p2**

■ **FLOWERS** Reddish green, blooming June–October; **inflorescence** a raceme (spikelike), two to seven per main branch, each 4–10 cm long; **main branches** one to six, each 8–22 cm long from the top and upper leaf axils, support the racemes; **spikelets** two at each rachis node; **1) sessile spikelet** 2-flowered with the lower floret vestigial, the upper floret perfect; **glumes** two, equal, 6–11 mm long by *c.* 1.2 mm wide; **1st glume** dorsally compressed with a tiny bidentate tip, two marginal scabrous nerves; **2nd glume** V-shaped, hyaline edges hairy, tip pointed and scabrous; **fertile lemmas** with a bidentate tip, margins hairy in the upper half, the body membranous and *c.* 5 mm long by *c.* 1 mm wide, its awn twisted (or absent), 8–25 mm long, hairs ascending; **2) pedicellate** (stalked) **spikelet** 1-flowered; **pedicel** flat, hairy, 4–6 mm long; **glumes** two, about equal and 3.5–12 mm long, each with a scabrous midnerve and hairy upper margin, pointed; **1st glume** with five or seven green nerves; **2nd glume** 5-nerved, fitting inside the 1st glume; **floret** male, or rarely perfect; **lemmas** and **paleae** hyaline, hairy at their tips, the awnless lemma slightly shorter than the 3-nerved palea; **anthers** brownish orange, 2.5–4.5 mm long. **FRUIT a grain**, 1-seeded, tan, dark brown at both ends, edges rounded, 4–4.5 mm long by *c.* 1 mm wide by *c.* 0.8 mm thick, hairless.

■ **LEAVES** Arching, green with longitudinal red stripes; **lower leaves** with scattered white hairs to *c.* 3.5 mm long on the sheaths and base of the blade's upper surface, those above hairless; **blades** flat, 5–50 cm long by 2–10 mm wide, glabrous to scabrous below, edges scabrous, tapered to a long fine point, the midrib wide, green or red; **ligules** creamy pink, a short fringed membrane 0.6–2.5 mm long, flat to pointed, sometimes torn; **auricles** absent; **sheaths** purplish near base, hairy below, glabrous above.

■ **STEM** A **culm**, erect, oval, solid, with a longitudinal groove; the base 2–4 mm thick; **nodes** often dark red, this color extending below the nodes for several cm.

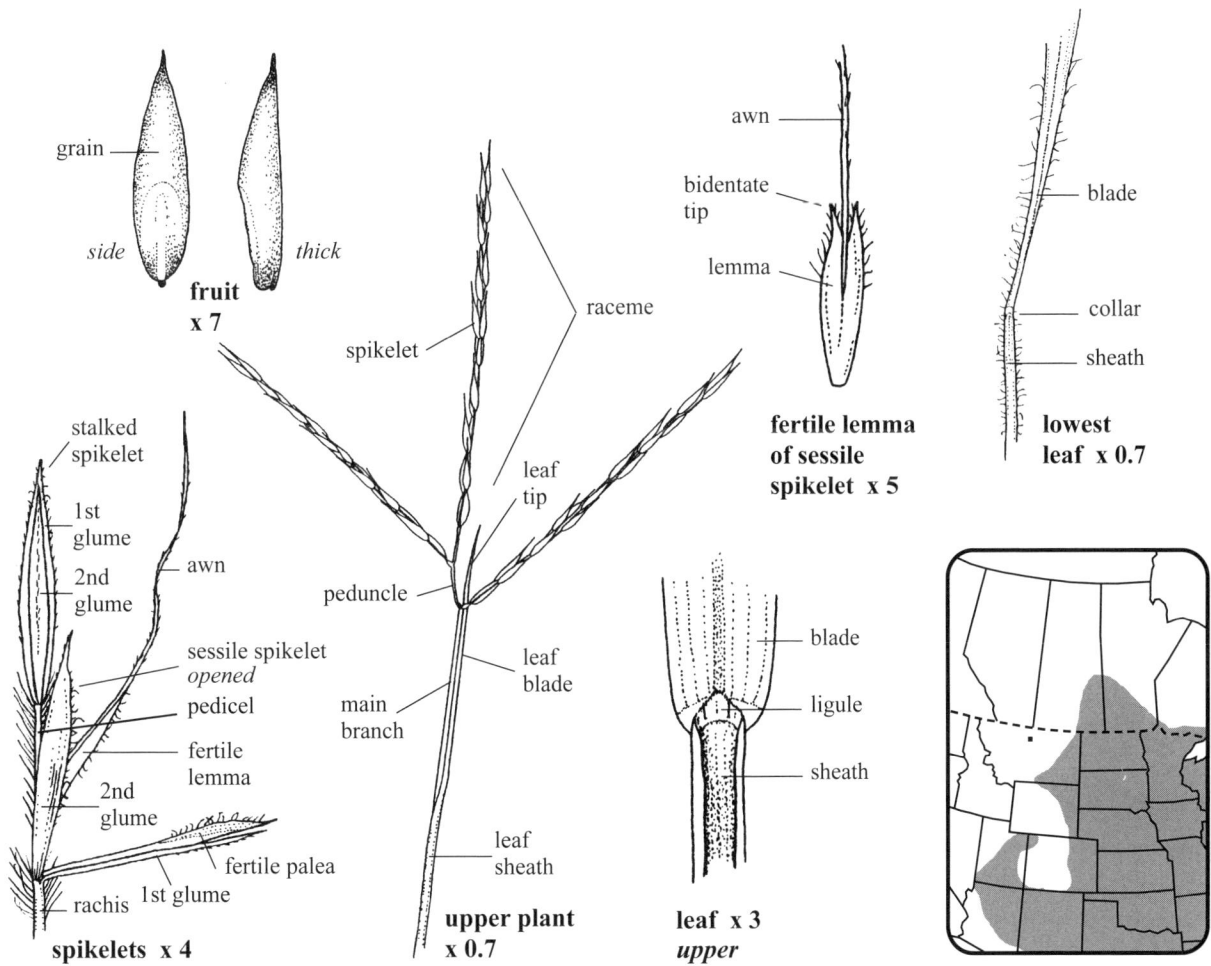

Grass—*Poaceae* Grass Spikelets 2- to 4-flowered

WILD OAT *Avena fatua* L.

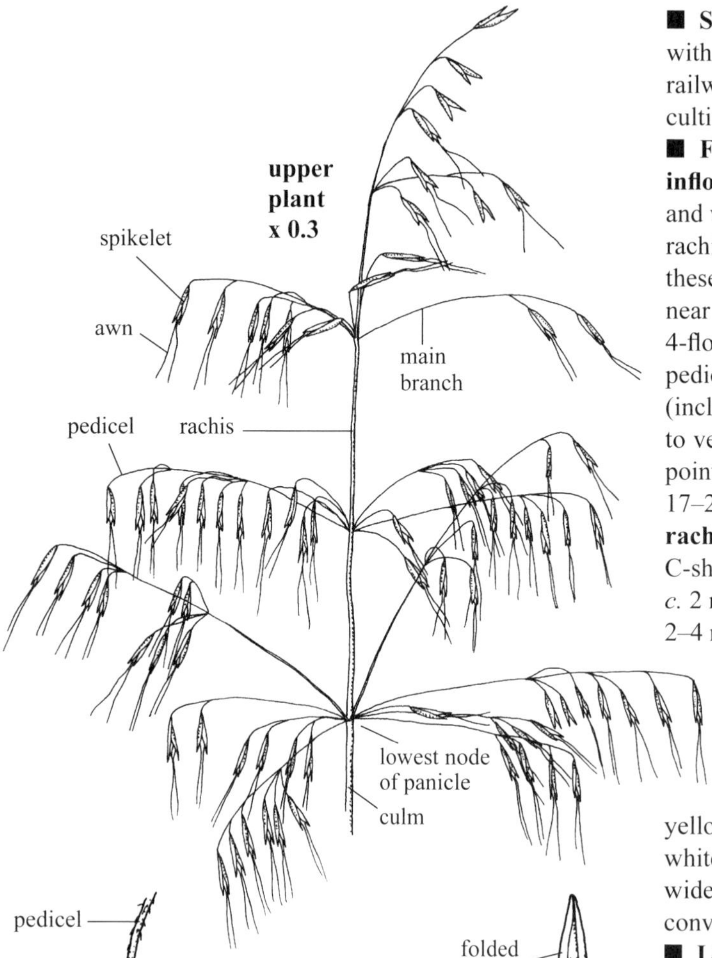

- **SKETCH** A tufted **annual grass** 30–120 cm tall with **fibrous roots**; in disturbed sites, city lots, along railways, valleys, foothills, roadsides, in gardens and cultivated fields. *Naturalized* **w1**
- **FLOWERS** Green, blooming March–October; **inflorescence** a panicle, several, open, 10–36 cm long and wide, with several widely spaced nodes along the rachis; **main branches** several from the lower nodes, these mostly spreading to ascending, glabrous, naked near the rachis and reduced above; **spikelets** 2- to 4-flowered, 30–70⁺ per panicle, hanging on curved pedicels near the ends of the branches, 3.5–6 cm long (including awns), the third or fourth floret is fertile to vestigial and *c.* 2 mm long; **glumes** two, subequal, pointed, persistent, not awned; **1st glume** 7-nerved, 17–23 mm long; **2nd glume** 9-nerved, 19–25 mm long; **rachilla** partially hairy; **lemmas** awned, 7-nerved, C-shaped, apices bidentate, body 14–20 mm long by *c.* 2 mm wide, hairy to glabrous, basal hairs stiff, white, 2–4 mm long; **awns** 1.8–4 cm long, brown near the base, attached 5–7 mm from the lemma's base, bent when dry; **paleae** 2-nerved, green, membranous, ciliate on upper margins, edges folded inward, *c.* 2 mm wide, slightly shorter and fitting inside the lemmas, apices bidentate; **anthers** three, yellow, 1.5–3 mm long. **FRUIT a grain**, 1-seeded, white to dark brown, hairy, 9–10 mm long by *c.* 2 mm wide by *c.* 1.5 mm thick, flat with a groove on one side, convex on the other side.
- **LEAVES** Ascending, three or four per culm; **blades** flat, slightly scabrous above, margins glabrous to hairy, 10–30 cm long by 4–14 mm wide; **ligules** 1.5–5 mm long, a pale blunt membrane; **auricles** absent; **sheaths** open, glabrous or with spreading hairs.
- **STEM** A **culm**, erect, hollow, round, green, glabrous, with tillers; 2–5 mm thick near the base; **nodes** green to brown, glabrous.

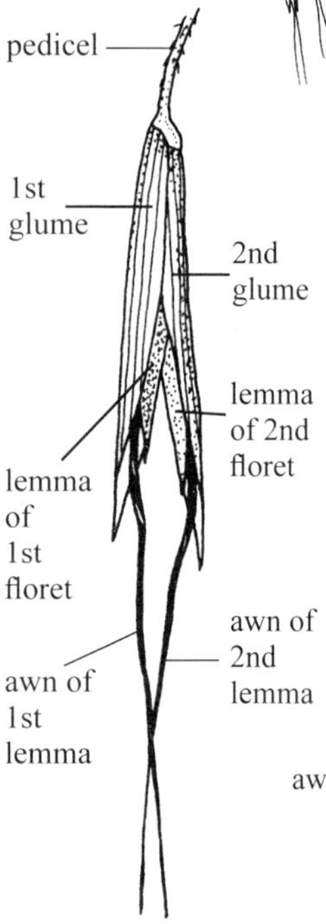

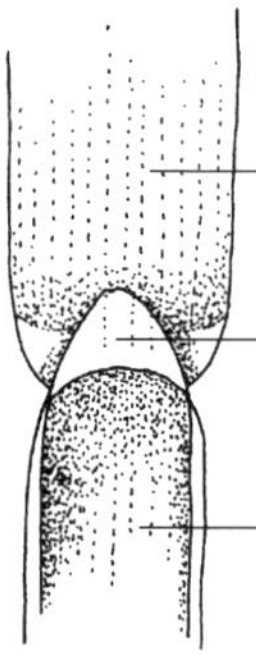

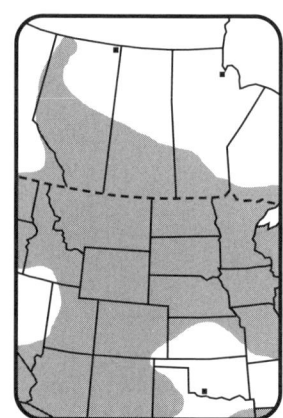

626

Grass—*Poaceae*   Grass   Spikelets 1-flowered

SLOUGH GRASS   *Beckmannia syzigachne* (Steud.) Fern.

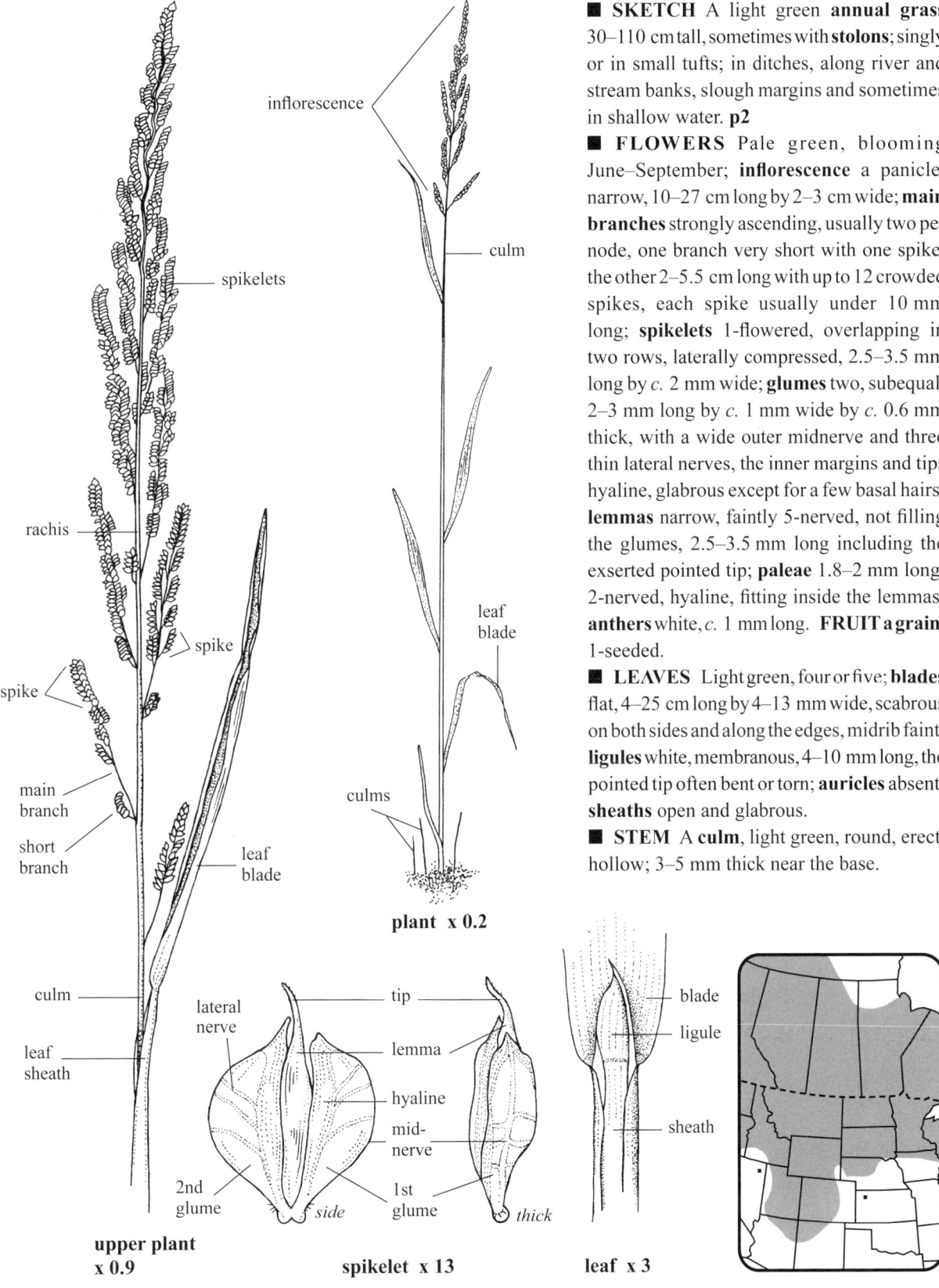

upper plant x 0.9   spikelet x 13   leaf x 3   plant x 0.2

■ **SKETCH** A light green **annual grass** 30–110 cm tall, sometimes with **stolons**; singly or in small tufts; in ditches, along river and stream banks, slough margins and sometimes in shallow water. **p2**

■ **FLOWERS** Pale green, blooming June–September; **inflorescence** a panicle, narrow, 10–27 cm long by 2–3 cm wide; **main branches** strongly ascending, usually two per node, one branch very short with one spike, the other 2–5.5 cm long with up to 12 crowded spikes, each spike usually under 10 mm long; **spikelets** 1-flowered, overlapping in two rows, laterally compressed, 2.5–3.5 mm long by *c.* 2 mm wide; **glumes** two, subequal, 2–3 mm long by *c.* 1 mm wide by *c.* 0.6 mm thick, with a wide outer midnerve and three thin lateral nerves, the inner margins and tips hyaline, glabrous except for a few basal hairs; **lemmas** narrow, faintly 5-nerved, not filling the glumes, 2.5–3.5 mm long including the exserted pointed tip; **paleae** 1.8–2 mm long, 2-nerved, hyaline, fitting inside the lemmas; **anthers** white, *c.* 1 mm long. **FRUIT** a grain, 1-seeded.

■ **LEAVES** Light green, four or five; **blades** flat, 4–25 cm long by 4–13 mm wide, scabrous on both sides and along the edges, midrib faint; **ligules** white, membranous, 4–10 mm long, the pointed tip often bent or torn; **auricles** absent; **sheaths** open and glabrous.

■ **STEM** A **culm**, light green, round, erect, hollow; 3–5 mm thick near the base.

American Sloughgrass

Side-oats Grama *Bouteloua curtipendula* (Michx.) Torr.

■ **SKETCH** A variable **perennial grass** 10–100 cm tall, singly or in tufts from tan **rhizomes** *c.* 2 mm wide and to *c.* 3⁺ cm long; in prairies, on bluffs along rivers, rocky slopes, in sagebrush and parklands. **p1**

■ **FLOWERS** Purple and green, blooming June–November; **inflorescence** a panicle with short branches, stiff, TL 6–25 cm by 8–10 mm wide; **spikes** 12–80, reflexed mostly along one side of the central rachis, crowded above, each 5–15 mm long; **branch rachis** dark gray, scabrous, pointed, bearing the spikelets; **spikelets** two to eight (–15) per spike, each 2- or 3-flowered and 4–8 mm long by *c.* 1 mm wide, reduced above, sessile; **glumes** two, unequal, 1-nerved, slightly purple, scabrous along the midnerve; **1st glume** awnlike, 2.5–6 mm long by *c.* 0.3 mm wide; **2nd glume** 4–8 mm long by 0.7–1 mm wide, hairy and mostly purple; **florets** of two types: **1) lower floret** one, perfect; **fertile lemmas** 3–6.5 mm long by 1–1.4 mm wide, the body creamy white, rolled at the edges around the palea inside, 3-nerved, each nerve ending in a scabrous, green awn 0.3–0.6 mm long; **fertile paleae** about as long as the lemmas, 2-nerved, ending in a short bidentate scabrous tip; **anthers** 2–3 mm long, red to orange; **2) upper florets** one or two; **sterile lemma's** body 0.5–3 mm long or reduced to usually three scabrous awns, the central awn 2–7 mm long; **palea** absent. **FRUIT a grain**, 1-seeded, 3–3.7 mm long by *c.* 1 mm wide by *c.* 0.5 mm thick, slightly darker brown near the base and apex, a small dark brown oval spot *c.* 0.5 mm long about midway on the flattest side.

■ **LEAVES** Flat, three to five, longer below; **blades** 3–30 cm long by 2–6 mm wide with a long tapered point, glabrous below to scabrous on both surfaces and the edges, midrib raised below; **ligules** usually a fringe of white hairs 0.3–0.7 mm long or sometimes only an erose membrane; **auricles** absent; **sheaths** open, glabrous above, the lower ones hairy with long hairs near the collar.

■ **STEM** A **culm**, erect, sometimes geniculate at the base, round, solid and white inside; *c.* 2 mm thick near the base; **nodes** reddish.

upper plant x 0.9

spike with 5 spikelets x 7

fertile lemma x 8

sterile lemma x 8

lower leaf x 1

fruit x 8

leaf x 8

spikelet x 8

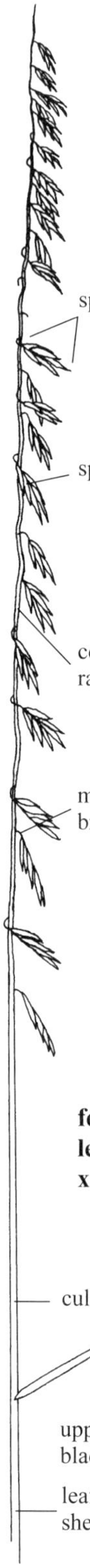

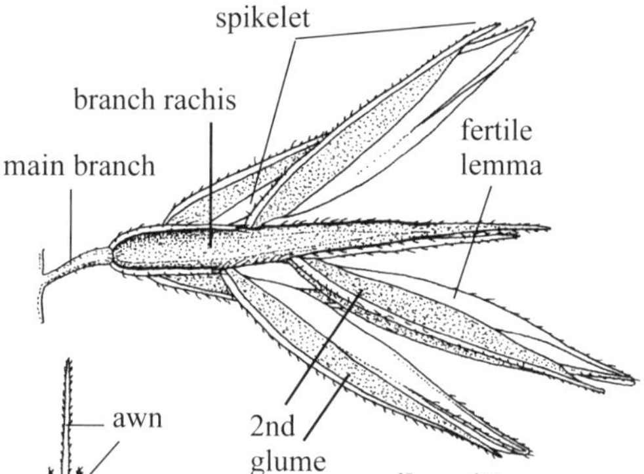

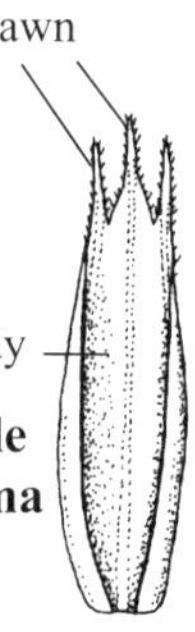

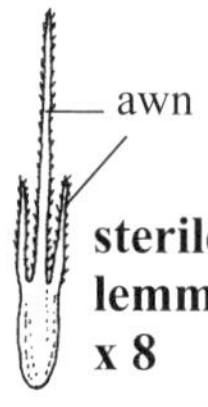

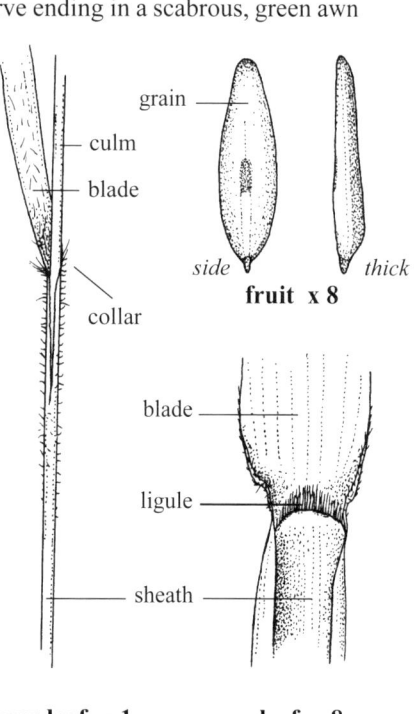

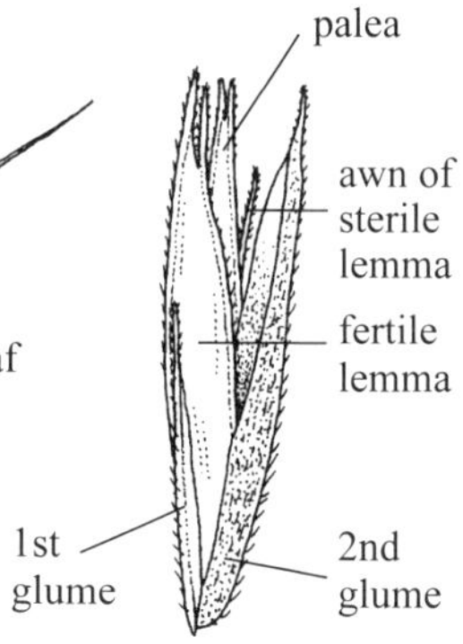

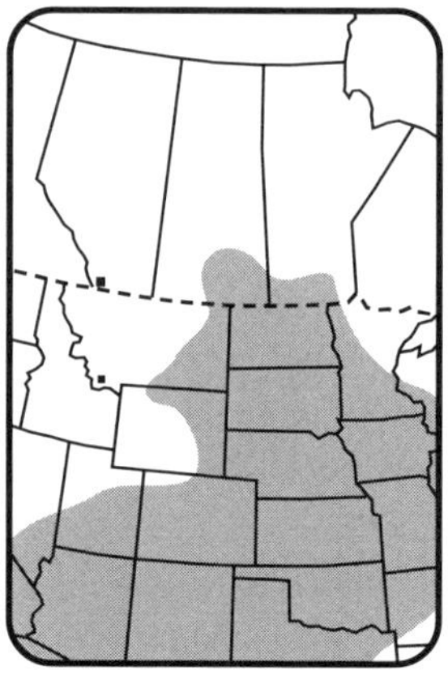

Grass—*Poaceae* Grass Spikelets 1- to 3-flowered

BLUE GRAMA *Bouteloua gracilis* (Willd. ex Kunth) Lag. ex Griffiths

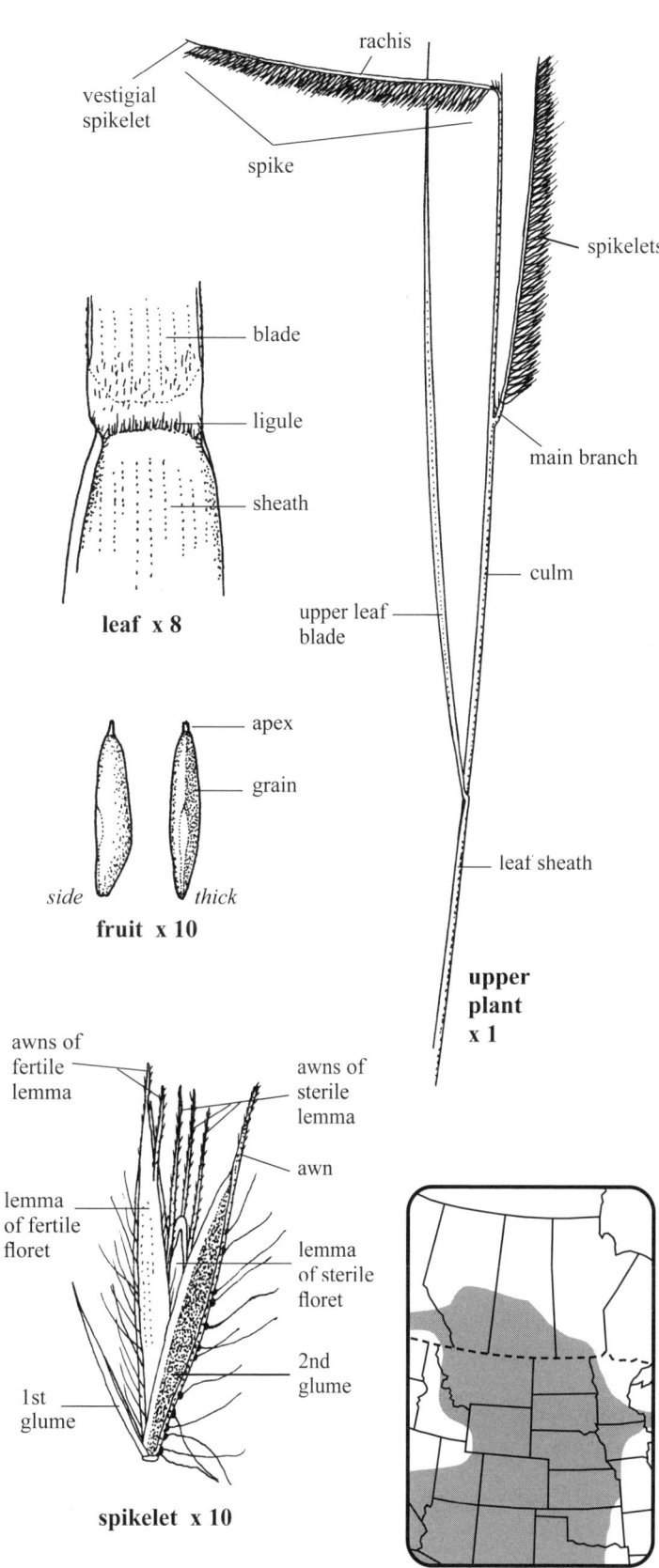

■ **SKETCH** A densely tufted **perennial grass** 15–70 cm tall from a short scaly **rhizome**; in dry prairies, mesas, foothills, rocky slopes, mountain meadows and parklands. **p1**

■ **FLOWERS** Purple and green, blooming May–October; **inflorescence** of one to three (six) spikes on straight to curved branches, each 1.2–7.5 cm long by 3–5 mm wide by 2–3 mm thick, horizontal to ascending, with a row of spikelets on either side of the purple rachis, and a vestigial spikelet at the apex; **spikelets** 1- to 3-flowered, 40–130 per spike, each 4–6 mm long with florets of two types: **1) lower floret** perfect with one or two sterile florets above; **glumes** two, unequal, 1-nerved; **1st glume** narrow, 1.5–3.5 mm long, slightly scabrous along its midnerve; **2nd glume** 3.5–5.5 mm long, purple with a white margin, white hairs to 1.5 mm long from raised bases on the midnerve, one awn to *c.* 1.7 mm long; **fertile lemmas** green, 3-nerved, 3.5–5.5 mm long (including three awns), hairy on midnerve and margins, the two lateral awns set back from the tip, awns 1–3 mm long; **paleae** to *c.* 4 mm long, 2-nerved, fitting tightly into lemmas; **stamens** three; **anthers** 2–3 mm long, pale yellow or purple, exserted; **2) upper florets** one or two, sterile, a tuft of white hairs to *c.* 2 mm long from the base; **lemmas** green, 1–3 mm long, with one to three scabrous awns 1–4 mm long, almost reaching to the tip of the spikelet. **FRUIT** a grain, 1-seeded, *c.* 2.5 mm long by *c.* 0.5 mm wide.

■ **LEAVES** Mostly basal, four or five per culm, scabrous above and on the margins, less so beneath; **blades** 2–22 cm long by 1–3.5 mm wide, flat to V-shaped, slightly involute and the margins slightly thickened; **ligules** a fringe of white hairs 0.2–0.5 mm long; **sheaths** not keeled, mostly glabrous but finely ridged longitudinally, the top edges sometimes with short to long white hairs.

■ **STEM** A **culm**, erect to ascending, solid, glabrous, geniculate at the 1–2 mm thick base; **nodes** glabrous to slightly hairy.

Eyelash Grass, Blue Grama Grass

Grass—*Poaceae*

Grass Spikelets 4- to 10-flowered, hairy

FRINGED BROME *Bromus ciliatus* L.

■ **SKETCH** A **perennial grass** 50–150 cm tall with a dark rough horizontal **caudex** 10–20 mm long by 5–8 mm thick; in moist meadows, open woods and thickets, bogs and open slopes. **w2, p2**

■ **FLOWERS** Light green, blooming July–September; **inflorescence** a panicle, loose, 7–30+ cm long by 7–10 cm wide, nodding mostly to one side, with 1–6 lax branches per node; **rachis** glabrous; **spikelets** 20–32+ per inflorescence, drooping, 4- to 10-flowered, each 14–28 mm long by 3–5 mm wide and thick (including hairs), pale green with obvious white hairs protruding along the lower three-quarters; **pedicels** green, hairs microscopic and ascending; **1st glume** 4.5–9.5 mm long by *c.* 1 mm wide (flattened), usually 1-nerved (or 3-nerved), pointed, central nerve scabrous, margins hyaline, middle light green; **2nd glume** 5.5–12 (–14) mm long by 1.5–2.2 mm wide (flattened), 3-nerved, midnerve scabrous, apex rounded, nerves brown at apex, glabrous, awn short or absent; **1st lemma** 8–15 mm long by 2–2.5 mm wide (flattened), 5- or 7-nerved, lower sides with ascending white hairs *c.* 1 mm long, nerves faint dorsally, glabrous to slightly hairy on back, midnerve scabrous; **awn** scabrous, 1.5–13 mm long, attached right below the brownish hyaline apex, the nerves slightly raised near the apex; **palea** usually *c.* 2 mm shorter than lemma body into which it fits, margins involute, two nerves green, margins with ascending hairs, apices slightly ragged and blunt, body pale green, *c.* 1 mm wide, sometimes with an awn *c.* 0.3 mm long; **anthers** variable in length, 0.7–3 (–4.5) mm long, included. **FRUIT** a grain, 1-seeded, 6–7 mm long (including apical hairs) by *c.* 1.5 mm wide by *c.* 0.6 mm thick, flattened, reddish brown with a tuft of white apical hairs *c.* 1 mm long, body hairless, tapered at base with a thin groove on the concave side.

■ **LEAVES** About eight per culm, ascending to erect above, the lowest few brown and lax; **blades** flat, margins scabrous, sometimes hairy on upper surface, glabrous below, 10–28 cm long by 4–14 mm wide; **sheaths** glabrous to hairy, especially above, less hairy below, not open to base (node), hairs mostly descending to spreading; **ligules** 0.5–1.5 mm long, ragged; **auricles** absent.

■ **STEM** A **culm**, erect, hollow, stiff, light green, solitary or in a loose bunch of 2–4; *c.* 2 mm thick near the brown base; **nodes** slightly swollen, 1.5–2.5 mm wide by 0.5–1 mm tall, glabrous or most often with descending white hairs at least from the lower part of the node, a dark green band *c.* 2 mm long right below the greenish brown node.

■ **SYN.** *Bromopsis canadensis* (Michx.) Holub = *Bromus ciliatus* var. *ciliatus* L.

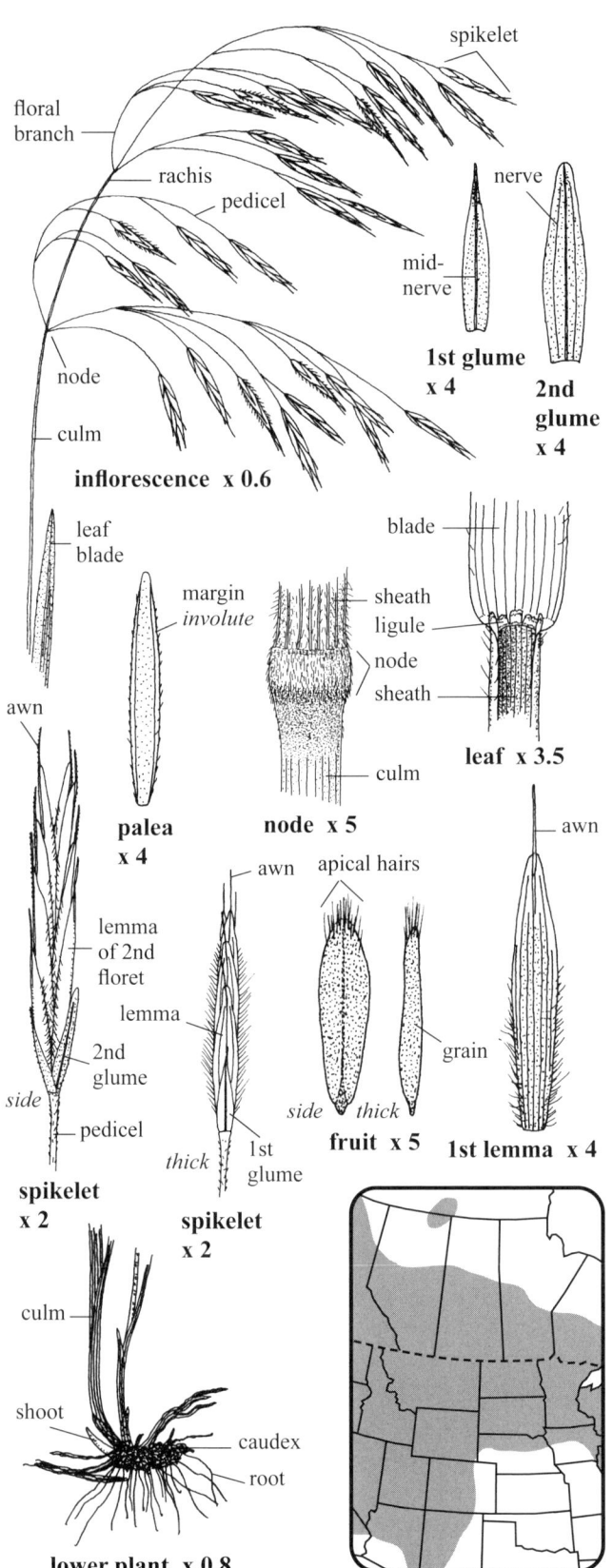

Grass—*Poaceae* Grass Spikelets 4- to 13-flowered

Smooth Brome *Bromus inermis* Leyss.

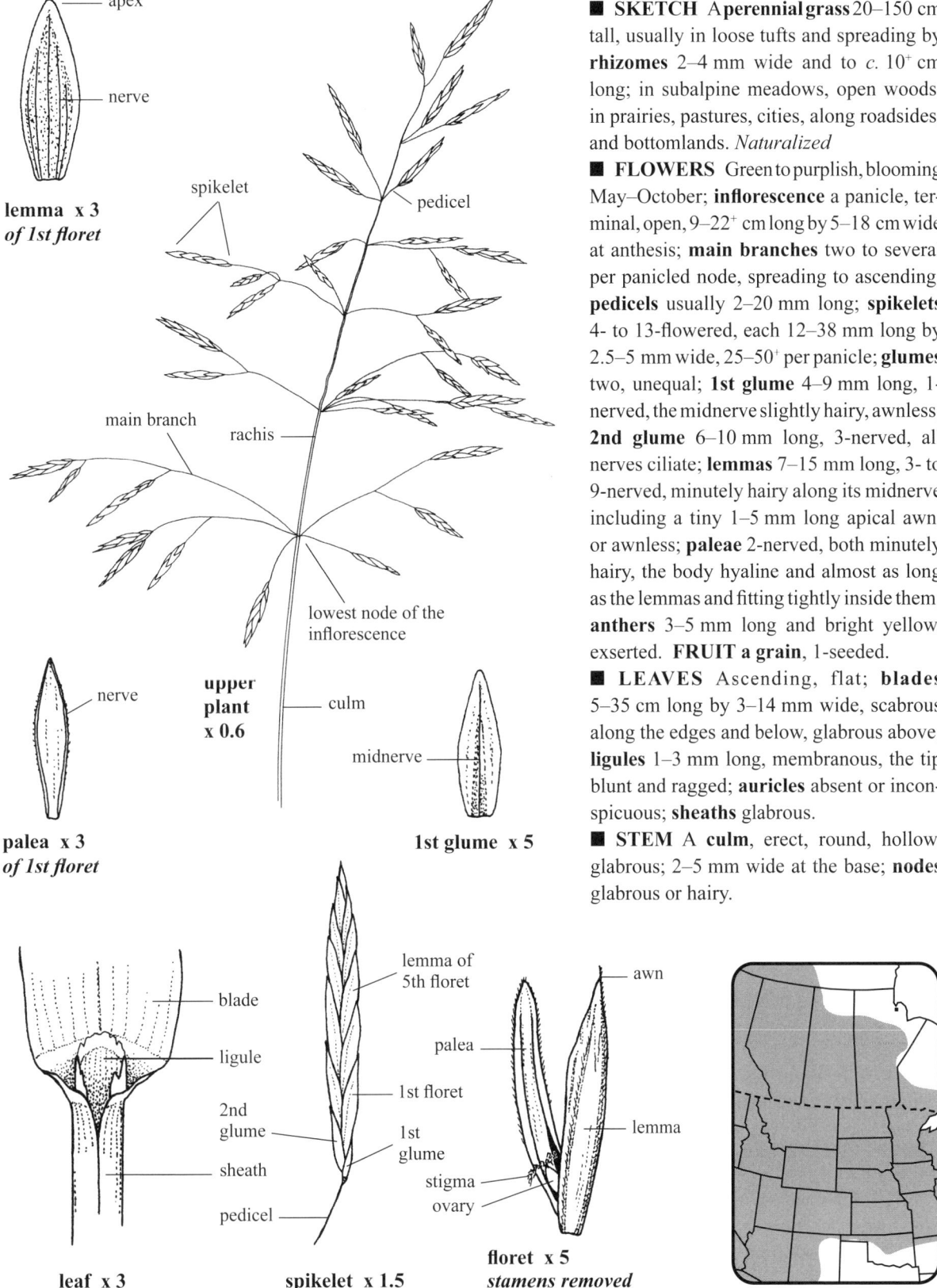

■ **SKETCH** A perennial grass 20–150 cm tall, usually in loose tufts and spreading by **rhizomes** 2–4 mm wide and to *c.* 10⁺ cm long; in subalpine meadows, open woods, in prairies, pastures, cities, along roadsides, and bottomlands. *Naturalized*

■ **FLOWERS** Green to purplish, blooming May–October; **inflorescence** a panicle, terminal, open, 9–22⁺ cm long by 5–18 cm wide at anthesis; **main branches** two to several per panicled node, spreading to ascending; **pedicels** usually 2–20 mm long; **spikelets** 4- to 13-flowered, each 12–38 mm long by 2.5–5 mm wide, 25–50⁺ per panicle; **glumes** two, unequal; **1st glume** 4–9 mm long, 1-nerved, the midnerve slightly hairy, awnless; **2nd glume** 6–10 mm long, 3-nerved, all nerves ciliate; **lemmas** 7–15 mm long, 3- to 9-nerved, minutely hairy along its midnerve including a tiny 1–5 mm long apical awn, or awnless; **paleae** 2-nerved, both minutely hairy, the body hyaline and almost as long as the lemmas and fitting tightly inside them; **anthers** 3–5 mm long and bright yellow, exserted. **FRUIT a grain**, 1-seeded.

■ **LEAVES** Ascending, flat; **blades** 5–35 cm long by 3–14 mm wide, scabrous along the edges and below, glabrous above; **ligules** 1–3 mm long, membranous, the tip blunt and ragged; **auricles** absent or inconspicuous; **sheaths** glabrous.

■ **STEM** A **culm**, erect, round, hollow, glabrous; 2–5 mm wide at the base; **nodes** glabrous or hairy.

631

Grass—*Poaceae* Grass Spikelets 5- to 10-flowered

NODDING BROME   *Bromus porteri* (Coult.) Nash

■ **SKETCH** A **perennial grass**, loosely tufted, 30–100 cm tall from **fibrous roots**, rhizomes absent; in meadows, thickets, sagebrush and woodland openings, often near water.

■ **FLOWERS** Green to tan, blooming June–August; **inflorescence** a panicle, loose, open, 5–20 cm long by 6–10 cm wide, nodding; **main branches** usually three at the lowest node in panicle, hairy, arched and drooping with one or two spikelets at the ends; **rachis** hairs ascending and thick around the rachis nodes; **pedicels** with white hairs ascending; **spikelets** 5- to 10-flowered, 17–30 mm long by 4–8.5 mm wide, awned, slightly flattened; **glumes** two, hairy, awnless, unequal, involute, soft, apices brownish, hairs ascending on outside, bluntly pointed; **1st glume** 2- or 3-nerved, lateral nerves faint and seen more easily from inner surface, sometimes only one lateral nerve, body 5–7.5 mm long by 1–1.3 mm wide (flattened), hairy except on margins and tip; **2nd glume** 3-nerved, 6.8–8.5 mm long by 1.8–2.5 mm wide (flattened), nerves distinct, not converging at apex; **rachilla** hairy on the convex surface only, glabrous on inner concave surface; **lemmas** obscurely 5-nerved, lowest one 8–15 mm long, slightly reduced above, hairy throughout, tips usually brownish and awned, awns 1.5–6 mm long, attached *c.* 0.5 mm back from the tip, scabrous; **paleae** 1–2.5 mm shorter than lemmas, apices blunt, 2-nerved, nerves ciliate, margins green; **anthers** pale yellow, 1–3 mm long. **FRUIT a grain**, 1-seeded, 6–7 mm long.

■ **LEAVES** Six to eight per culm, mostly ascending, upper leaf usually erect and reaching into the panicle; **blades** 10–28 cm long by 1–6 mm wide, flat or twisted, slightly bluish green, lowest blades with widely scattered long white hairs on both surfaces, upper leaves less hairy but more scabrous, especially on outside (dorsal); **ligules** membranous, 0.5–2.5 mm long, erose at blunt apices, streaked with brown, tip sometimes folded and hidden; **auricles** absent or only on lower leaves; **sheaths** slightly hairy to glabrous, often hairy below, slightly keeled, margins narrowly hyaline and hairless.

■ **STEM** A **culm**, green, erect to leaning, loosely clumped, round, hollow; *c.* 2 mm thick near the base; **nodes** slightly swollen, dark, glabrous to hairy, hairs appressed and descending from lower margin of the dark band.

■ **SYN.** *Bromus anomalus* auct. non Rupr. ex Fourn.

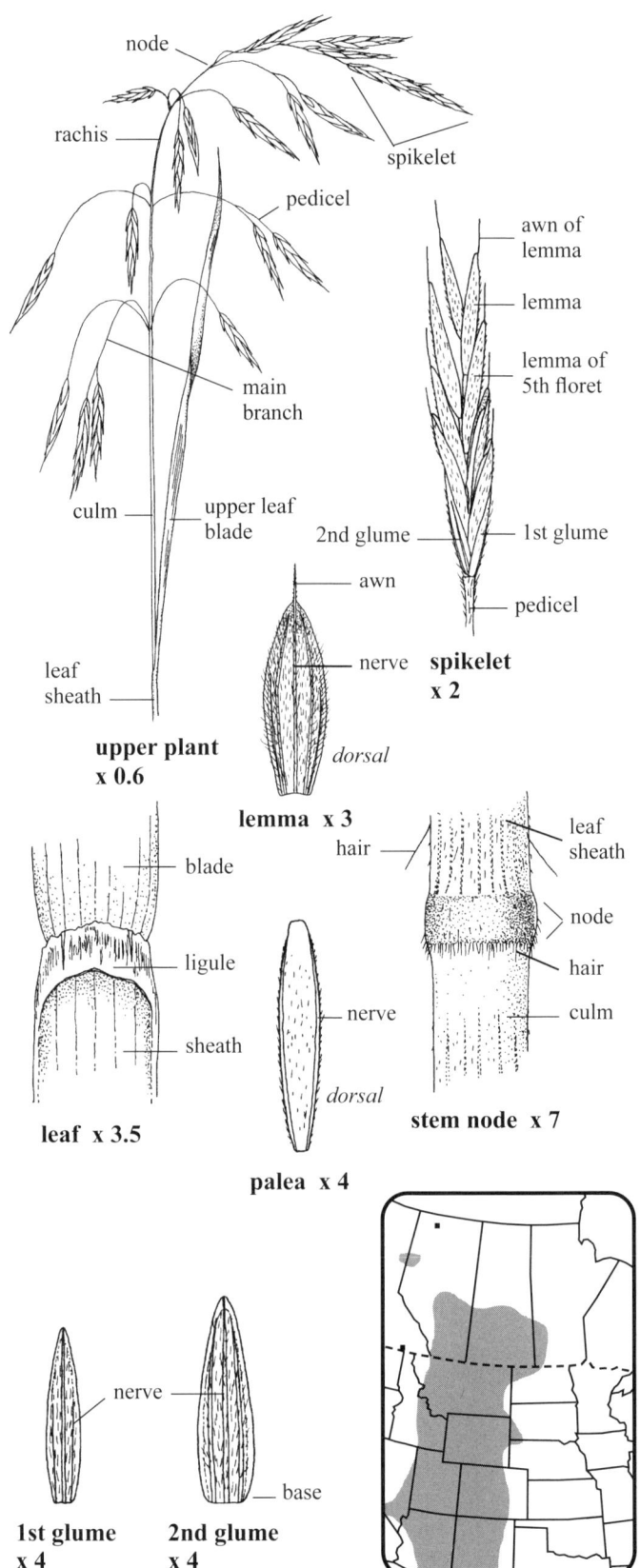

Porter's Brome, Anomalus Brome

Grass—*Poaceae*   **Grass** Spikelets 1-flowered

Marsh Reed Grass   *Calamagrostis canadensis* (Michx.) Beauv.

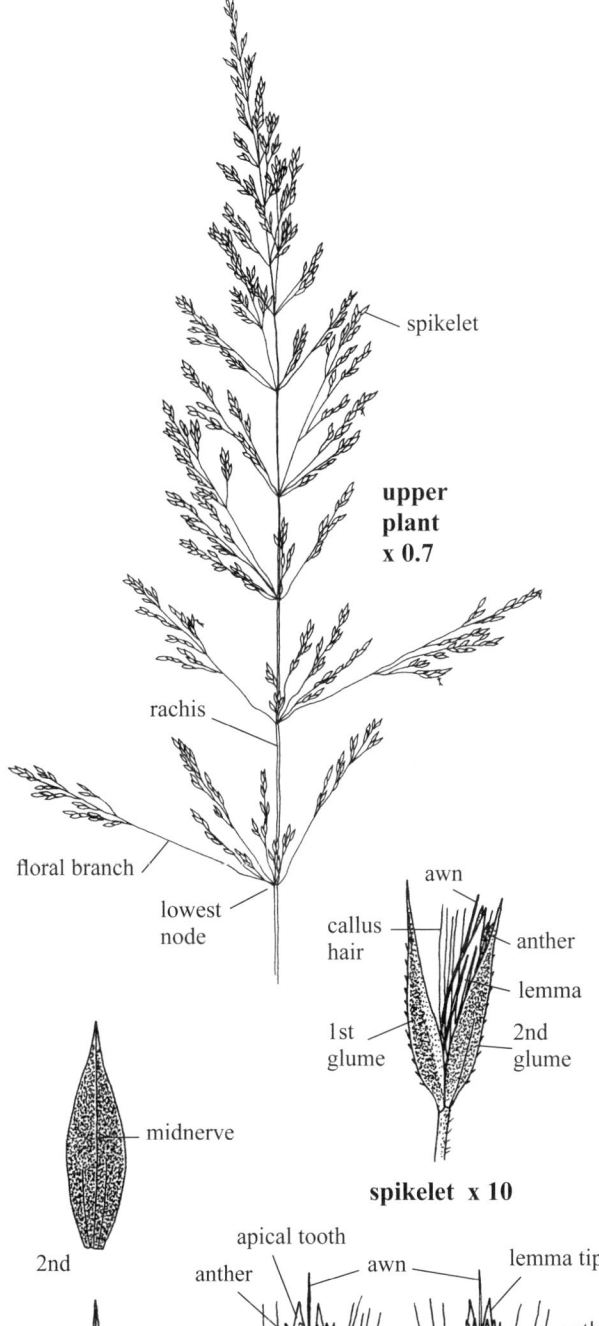

■ **SKETCH** A variable **perennial grass** 60–150 cm tall in open colonies; **rhizomes** brown, several cm long by *c.* 1 mm thick, covered with several tan bracts 10–20 mm long, often hidden in a mass of tan **fibrous roots** to *c.* 7 cm long; in marshy sites, sloughs and ravine bottoms. **w2**

■ **FLOWERS** Purplish green, blooming June–August; **inflorescence** a panicle, 7–23 cm long by 1–14 cm wide, open to slightly contracted, some nodding; **floral branches** spreading to ascending, more ascending above, usually six at the lower paniculate nodes, 1–7 cm long, reduced above, bases naked; **spikelets** 1-flowered, 1.5–6 mm long (depending on length of glumes); **glumes** two, purplish green, slightly scabrous, 1.5–6 mm long; **1st** (lowest) **glume** the longest, 1-nerved; **2nd glume** 3-nerved, the side nerves faint; **lemmas** 1.5–3.5 mm long by *c.* 1 mm wide, 5-nerved, nerves reddish purple, apices bidentate, membranous, transparent, easily torn, awned; **awns** straight, 0.2–2.2 mm long, extending slightly beyond the tip of the lemma, usually attached right below the middle of the lemma's body; **callus hairs** numerous, about as long as lemma; **paleae** membranous, transparent, *c.* 2 mm long, bidentate and ragged at the apices, with some purplish streaks, shorter than lemma and enclosed inside its base; **stamens** three; **anthers** purplish yellow, 1–1.7 mm long. **FRUIT a grain**, 1-seeded.

■ **LEAVES** Several, ascending, often arched; **blades** scabrous on margins, flat, 6–38 cm long by 1.5–8 mm wide, reduced above, ridges and furrows shallow, top leaf usually not reaching the base of the panicle; **bottom blades** short and brown by anthesis; **ligules** tan, membranous, 3–7 mm long, often torn; **sheaths** glabrous or with short white hairs near the collar, open to nodes, margins membranous, not hairy, some purplish; **auricles** absent.

■ **STEM** A **culm**, erect, hollow, single or a few in a tuft; some with vegetative shoots from midstem nodes; bases green to tan, 2–3 mm thick, covered with sheaths.

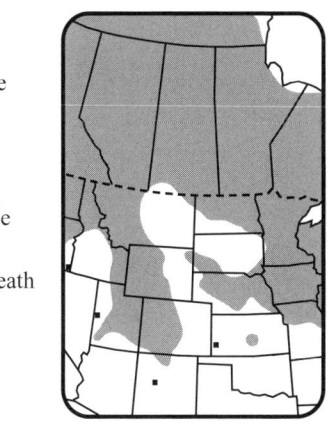

Bluejoint   633

Grass—*Poaceae*   **Grass**   Spikelets 1-flowered

NORTHERN REED GRASS   *Calamagrostis stricta* (Timm) Koel.

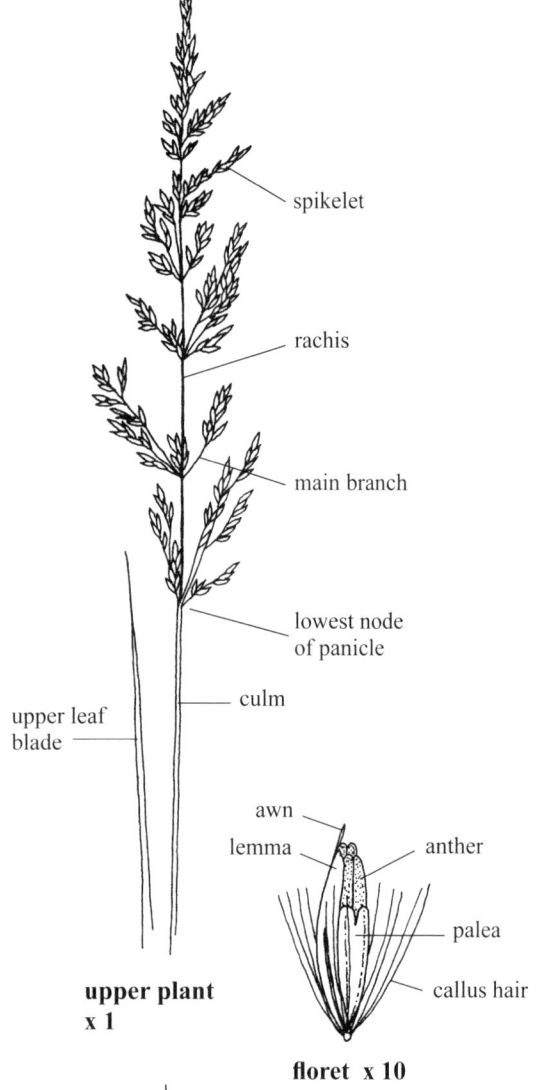

upper plant x 1

floret x 10

■ **SKETCH** A **perennial grass** 40–100 cm tall from white or brown **rhizomes** 1–2 mm thick and to *c.* 22⁺ cm long, not tufted, forming open patches; in moist fields, along marsh edges and streams, in open meadows among conifers in the mountains. **p2**

■ **FLOWERS** Greenish tan, blooming June–September; **inflorescence** a panicle 4–20 cm long by 0.8–4 cm wide, tan and narrow, usually five branches at the lowest node; **main branches** 1–6 cm long, shorter above, ascending and mostly obscured by the numerous small spikelets except at the bare bases of the longer branches; **pedicels** with hairs ascending; **spikelets** 1-flowered, perfect, 2–5 mm long; **glumes** two, tan, stiff, not awned, almost equal, 2–5 mm long, scabrous along the upper half of the midnerves; **1st glume** 1-nerved; **2nd glume** 3-nerved, the nerves light green; **lemmas** 5-nerved, 1.5–4 mm long, apices bidentate, slightly ragged, its 1–3 mm long awn attached about one-third above the base; **paleae** 2-nerved, hyaline, 1.8–3.2 mm long, apices bidentate, teeth blunt; **anthers** *c.* 2 mm long, tips reddish; **callus hairs** 2–3.5 mm long, white. **FRUIT a grain**, 1-seeded.

■ **LEAVES** Five or six per culm, upper three ascending, the top leaf may extend into the lower part of the panicle or not; **2nd blade** from the top usually the longest; **blades** 4–35 cm long by 1–5 mm wide, fairly stiff, involute to flat, not arched, tips pointed upward, scabrous along the margins and *c.* 12 scabrous longitudinal ridges on upper surface separated by deep, narrow furrows; **ligules** whitish tan, 1–6 mm long, tips slightly ragged; **auricles** absent; **collar** not hairy; **sheaths** open, glabrous.

■ **STEM** A **culm**, erect, hollow, slightly scabrous below the panicle; 2–3 mm thick near the base.

■ **SYN.** *Calamagrostis inexpansa* Gray = *Calamagrostis stricta* ssp. *inexpansa* (Gray) C.W. Greene, *Calamagrostis neglecta* (Ehrh.) P.G. Gaertn., B. Mey. & Scherb. = *Calamagrostis stricta* ssp. *stricta* (Timm) Koel. and *Calamagrostis crassiglumis* Thurb. = *Calamagrostis stricta* ssp. *inexpansa* (Gray) C.W. Greene.

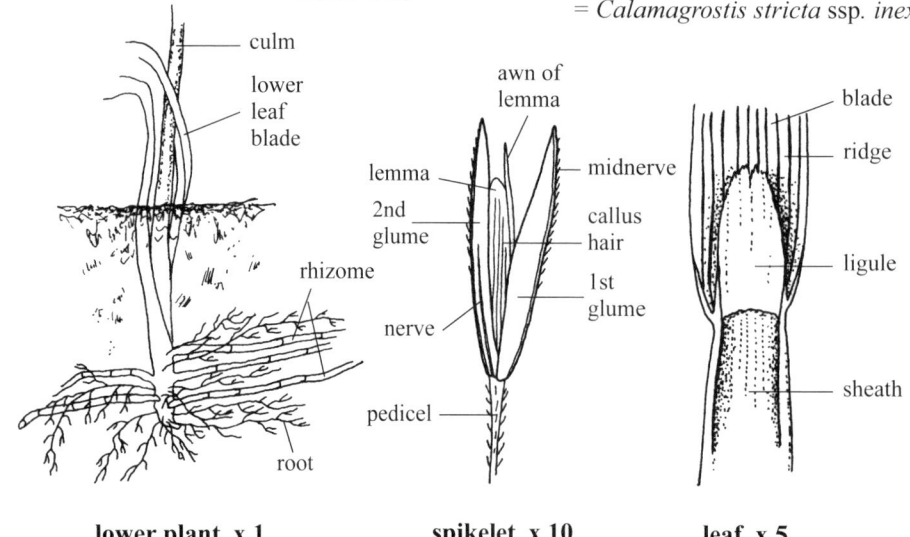

lower plant x 1   spikelet x 10   leaf x 5

634   Narrow Reed-grass, Slimstem Reedgrass

Grass—*Poaceae* Grass Spikelets 1-flowered

Sand Grass *Calamovilfa longifolia* (Hook.) Scribn.

■ **SKETCH** A **perennial grass**, often glaucous, 50–240 cm tall, singly or in loose colonies from long, scaly, tan **rhizomes** 2–3 mm thick; in sandy prairies, sand dunes, hillsides and open woods. **p2**

■ **FLOWERS** Green, blooming July–September; **inflorescence** a panicle 12–75 cm long by 1–23 cm wide, narrow to open; **main branches** erect to usually spreading at anthesis, 1–33 cm long, often congested within a few millimeters of a node, bare in lower quarter; **rachis** ribbed, not hairy; **pedicels** glabrous; **spikelets** 1-flowered, awnless, 4–9.5 mm long by 1.2–1.5 mm wide (before anthesis); **glumes** two, unequal, laterally compressed, pointed, 1-nerved, tip of midnerve slightly scabrous and darker green, sides scarious and whitish; **1st glume** 4–8 mm long; **2nd glume** 5–9.5 mm long; **lemmas** slightly longer than first glume and shorter than second glume, 1-nerved, pointed, slightly scabrous along midnerve near apices, laterally compressed near apices; **callus** hairs 2–4 mm long; **paleae** scabrous, firm, apices minutely bidentate, 2-nerved, slightly shorter or equal to the lemmas; **anthers** three, 3–5.5 mm long, exserted, reddish brown and pale green. **FRUIT a grain**, 1-seeded.

■ **LEAVES** Ascending, eight or nine, scabrous inside (ventral side), glabrous outside; **blades** with a fine point, 8–80 cm long by 2–12 mm wide, reduced above, some with a twist and arched, margins with a thin white line ventrally, top leaf often reaching into the lower part of the panicle, crowded at base; **ligules** a fringe of white hairs 0.3–2.3 mm long; **auricles** absent; **collar** pale green, with white hairs 2–5 mm long, often in a tuft; **sheaths** open, round, glabrous to hairy (lower ones), margins scarious, slightly reddish to white, imbricate toward the stem's base.

■ **STEM** A **culm**, round, erect, 3–5 mm thick near the base due to overlapping leaf sheaths, smooth below the panicle, hollow; **nodes** glabrous, slightly swollen, narrow, mostly hidden by the overlapping leaf sheaths.

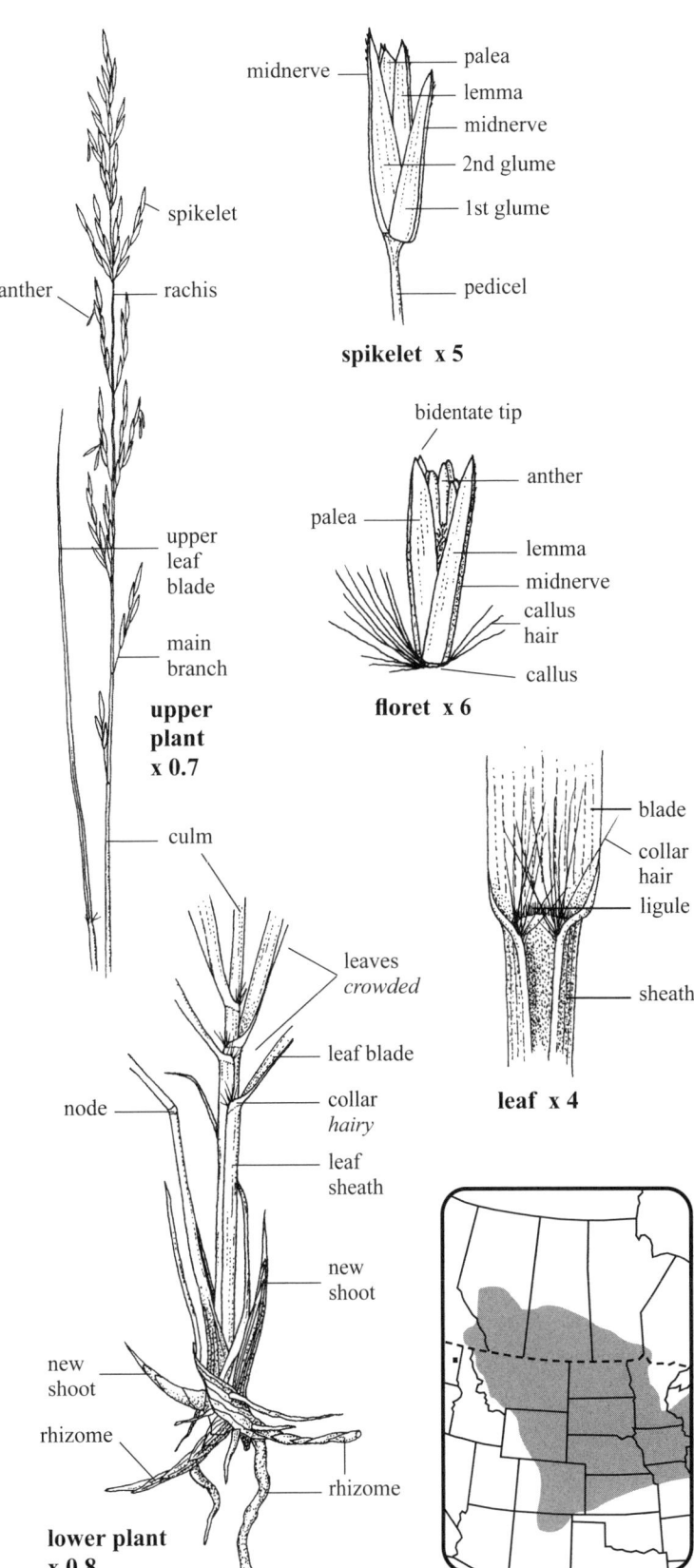

Prairie Sandreed, Sand Reed Grass

Grass—*Poaceae*   **Grass** Spikelets 1-flowered Stamen one

SLENDER WOOD GRASS *Cinna latifolia* (Trev. ex Göpp.) Griseb.

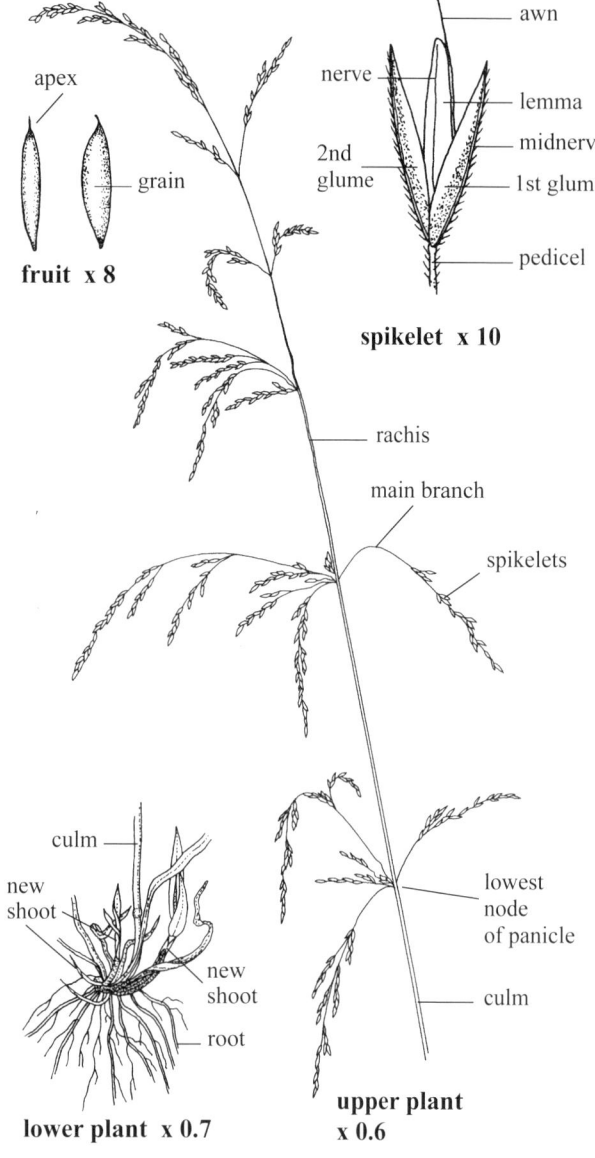

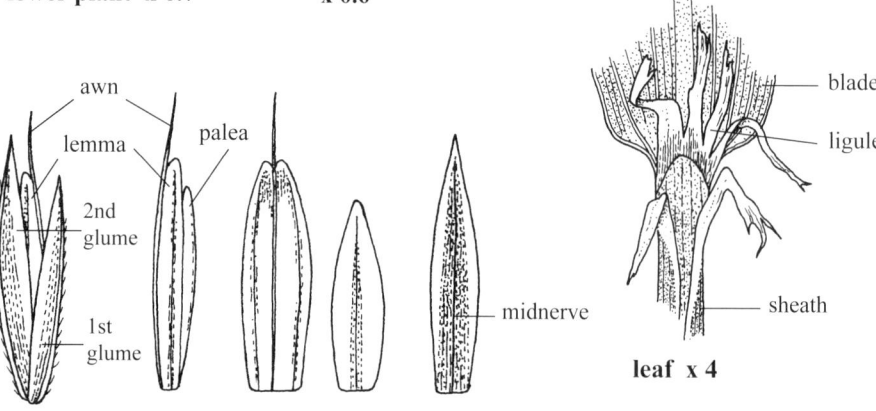

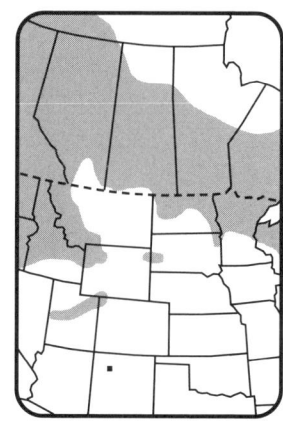

■ **SKETCH** A loosely tufted **perennial grass** 50–140 cm tall with short **rhizomes**; in damp woods, rocky outcrops and along wooded river and stream banks.
■ **FLOWERS** Green, blooming in June–September; **inflorescence** a panicle, open, 15–30+ cm long by 5–10 cm wide, nodding; **main branches** spreading to recurved, 1–12 cm long, four or five from the lower nodes, usually bare of spikelets in the lower quarter, lowest ones sometimes included in the upper leaf sheath; **rachis** glabrous to slightly scabrous above; **spikelets** 1-flowered, fertile, not stiff, TL 3–4.3 mm (including awn); **glumes** two, subequal, pointed, laterally compressed, turning purplish in fall, 1-nerved, often with short hairs on the midnerve, awnless, margins whitish; **1st glume** 1.9–3.5 mm long; **2nd glume** slightly longer at 2.3–4 mm; **lemmas** 3-nerved, thin, glabrous, TL 2.3–3.8 mm by *c.* 1.1 mm wide (flattened), body *c.* 2.5 mm long, apices bidentate; **awn** white, 0.3–1.5 mm long, free less than 0.5 mm back from the lemma's tip; **paleae** glabrous, 1-nerved, 2–3 mm long, shorter than the lemma, membranous, awnless; **anther** one, pale yellow, 0.5–0.8 mm long. **FRUIT** a **grain**, 1-seeded, 1.8–2 mm long by *c.* 0.5 mm wide by *c.* 0.3 mm thick.
■ **LEAVES** Four to eight per culm, slightly scabrous on both sides and along the margins; **blades** 10–25 cm long by 6–15 mm wide, flat, soft, thin, spreading to lax; **ligules** 1.5–7 mm long, white to tinged with purple, torn or not, apices irregular, some with a lateral projection; **auricles** absent; **sheaths** open, glabrous, round on upper leaves, slightly keeled on lower leaves.
■ **STEM** A **culm**, erect to leaning, hollow, glabrous; sometimes decumbent at the 2–3 mm thick base; **nodes** glabrous, slightly swollen.

636   Drooping Woodreed, Slender Wood-reed, Wood Reedgrass

Orchard Grass *Dactylis glomerata* L.

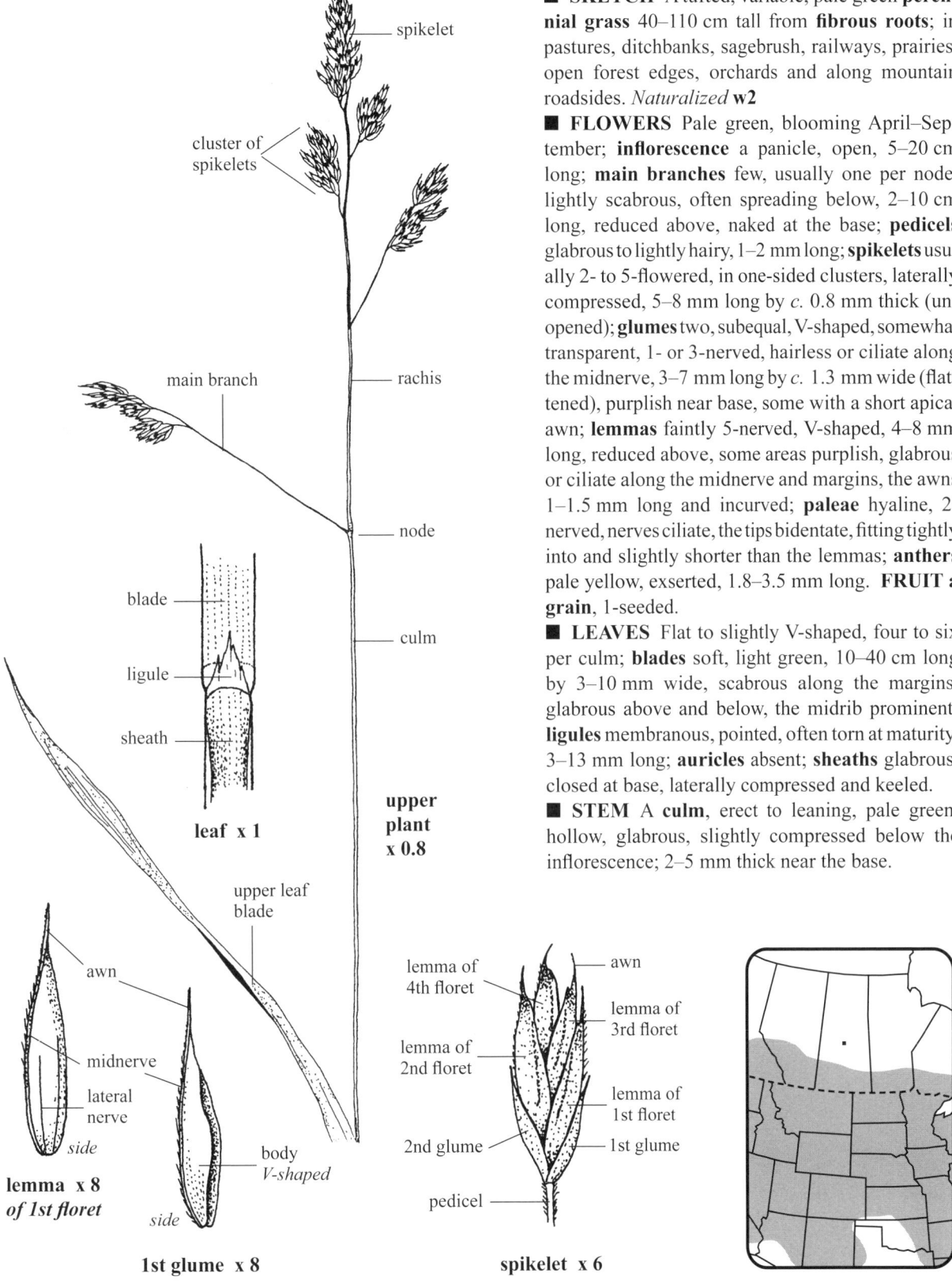

■ **SKETCH** A tufted, variable, pale green **perennial grass** 40–110 cm tall from **fibrous roots**; in pastures, ditchbanks, sagebrush, railways, prairies, open forest edges, orchards and along mountain roadsides. *Naturalized* **w2**

■ **FLOWERS** Pale green, blooming April–September; **inflorescence** a panicle, open, 5–20 cm long; **main branches** few, usually one per node, lightly scabrous, often spreading below, 2–10 cm long, reduced above, naked at the base; **pedicels** glabrous to lightly hairy, 1–2 mm long; **spikelets** usually 2- to 5-flowered, in one-sided clusters, laterally compressed, 5–8 mm long by *c.* 0.8 mm thick (unopened); **glumes** two, subequal, V-shaped, somewhat transparent, 1- or 3-nerved, hairless or ciliate along the midnerve, 3–7 mm long by *c.* 1.3 mm wide (flattened), purplish near base, some with a short apical awn; **lemmas** faintly 5-nerved, V-shaped, 4–8 mm long, reduced above, some areas purplish, glabrous or ciliate along the midnerve and margins, the awns 1–1.5 mm long and incurved; **paleae** hyaline, 2-nerved, nerves ciliate, the tips bidentate, fitting tightly into and slightly shorter than the lemmas; **anthers** pale yellow, exserted, 1.8–3.5 mm long. **FRUIT a grain**, 1-seeded.

■ **LEAVES** Flat to slightly V-shaped, four to six per culm; **blades** soft, light green, 10–40 cm long by 3–10 mm wide, scabrous along the margins, glabrous above and below, the midrib prominent; **ligules** membranous, pointed, often torn at maturity, 3–13 mm long; **auricles** absent; **sheaths** glabrous, closed at base, laterally compressed and keeled.

■ **STEM** A **culm**, erect to leaning, pale green, hollow, glabrous, slightly compressed below the inflorescence; 2–5 mm thick near the base.

Tufted Hair Grass *Deschampsia caespitosa* (L.) Beauv.

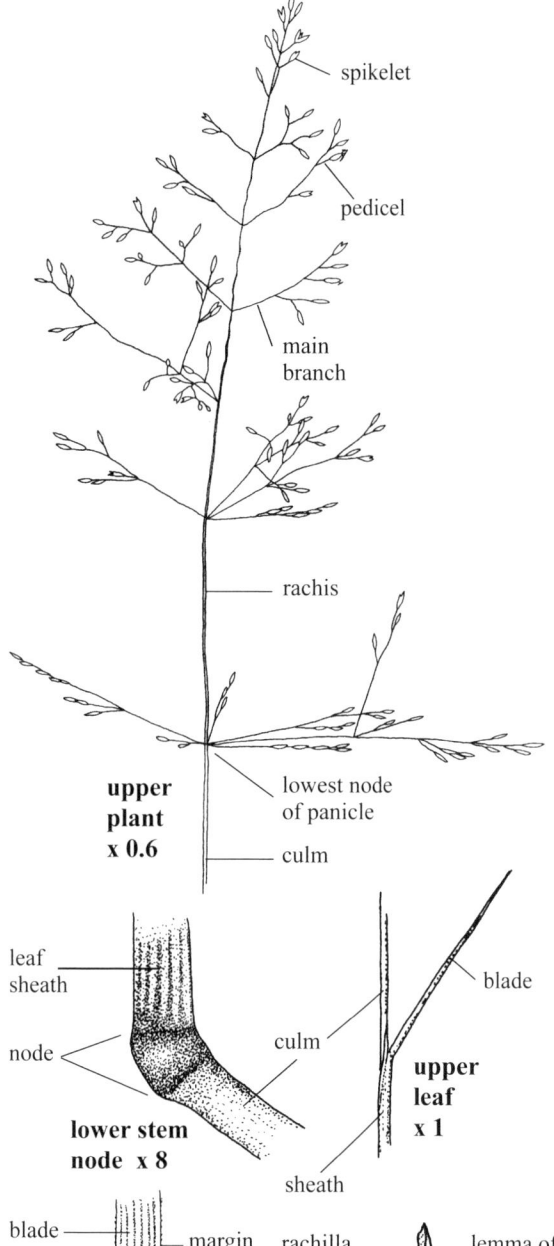

■ **SKETCH** A tufted **perennial grass** 20–120 cm tall; along ditch and stream banks, river bars, wet meadows, in forest clearings, marshes and dry to moist prairies. **p2**

■ **FLOWERS** Purple and tan, blooming June–September; **inflorescence** a panicle, open, 8–28 cm long by 7–15 cm wide, apex nodding, lowest paniculate node usually with five unequal main branches, the number reduce above; **main branches** scabrous, to *c.* 12 cm long, reduced above, ascending above to spreading below; **rachis** reddish purple, scabrous; **spikelets** 2- or rarely 3-flowered, 2.5–5.5 mm long; **glumes** two, streaked reddish purple, margins and apices whitish tan, apices pointed and slightly frayed; **1st glume** 1-nerved, 2–5 mm long, upper part tan; **2nd glume** 3-nerved, 3–5 mm long, slightly longer than 2nd floret; **rachilla** white, hairy, *c.* 1.8 mm long, extends alongside upper floret; **florets** perfect, slightly laterally compressed, *c.* 1 mm apart; **lemmas** membranous, nerveless to faintly 5-nerved, soft, laterally compressed, with several white hairs at the callus, apices erose, the lower one 2–3.5 mm long by *c.* 1 mm wide (flattened); **awns** reddish, 0.7–5 mm long, shorter to longer than the lemma, attached near the lemma's base and free most of its length; **paleae** membranous, slightly shorter than the lemmas, 2-nerved, tip tridentate; **anthers** 1–2 mm long. **FRUIT a grain**, 1-seeded.

■ **LEAVES** About six, mostly near the base, scabrous on margins and above to glabrous below, ridges and deep furrows on the upper surface; **blades** 2–33 cm long by 1–3.5 mm wide, reduced above, stiff, ascending, folded to involute, rarely flat, tips usually tan, top blade well below the inflorescence; **ligules** membranous, stiff, tan, pointed, apices often torn, 2–12 mm long, nerved; **auricles** absent; **collars** hairless; **sheaths** open, glabrous to slightly scabrous, keel very low.

■ **STEM** A **culm**, smooth below the panicle, hollow, erect to somewhat bent to geniculate below; 1–3 mm thick near the base; **nodes** enlarged, glabrous, darker than the culm.

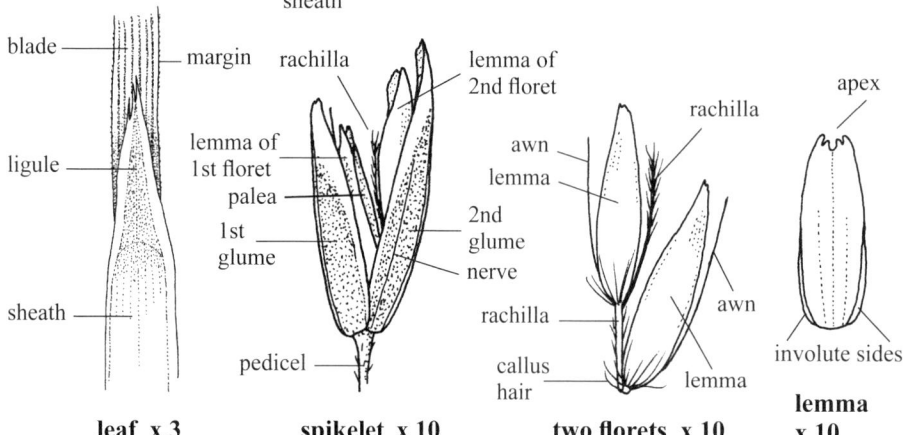

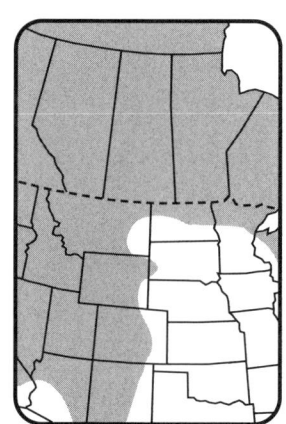

Grass—*Poaceae* **Grass** Spikelets 2-flowered

LEIBERG'S ROSETTE GRASS   *Dichanthelium leibergii* (Vasey) Freckmann

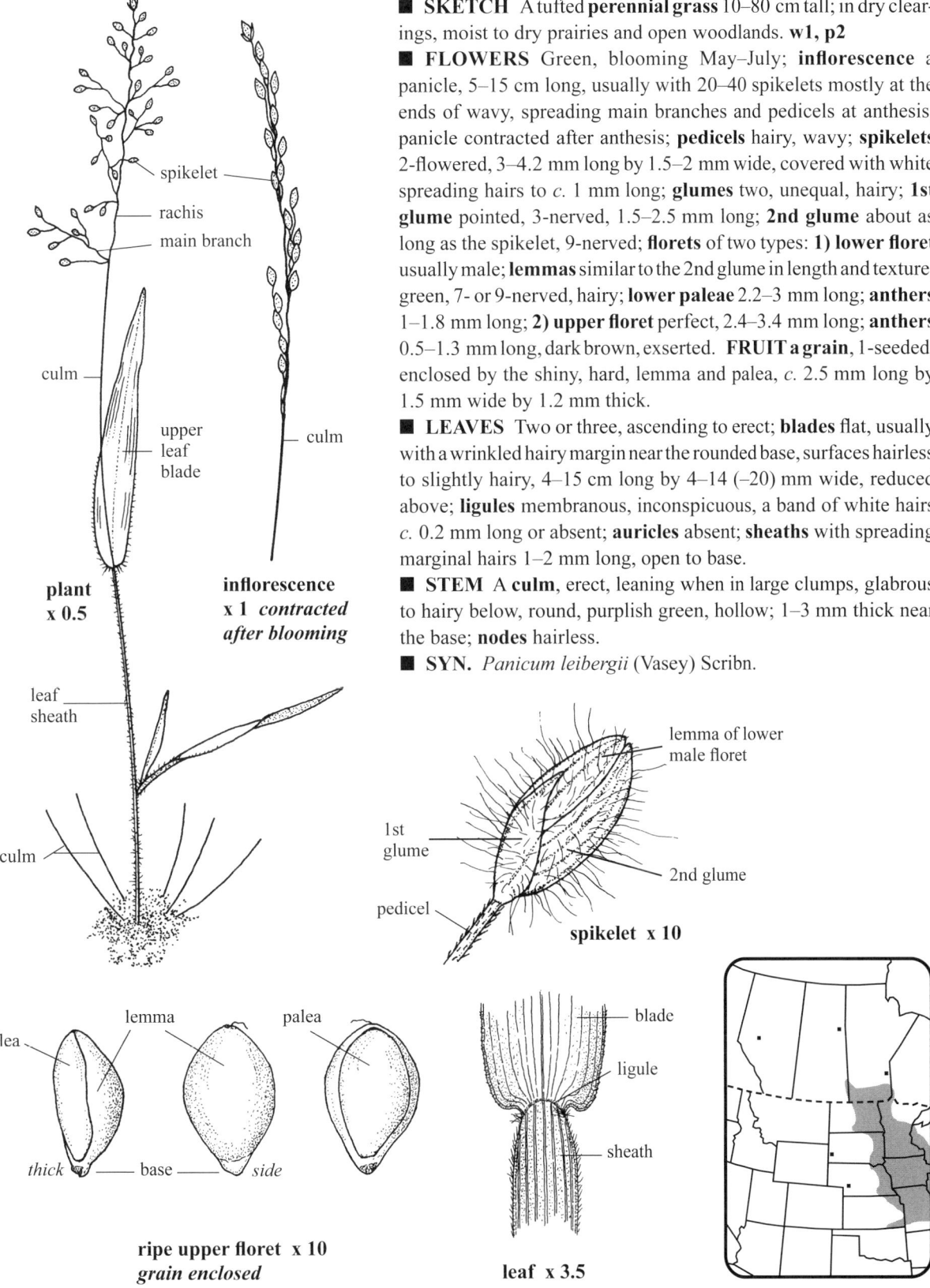

- **SKETCH** A tufted **perennial grass** 10–80 cm tall; in dry clearings, moist to dry prairies and open woodlands. **w1, p2**
- **FLOWERS** Green, blooming May–July; **inflorescence** a panicle, 5–15 cm long, usually with 20–40 spikelets mostly at the ends of wavy, spreading main branches and pedicels at anthesis, panicle contracted after anthesis; **pedicels** hairy, wavy; **spikelets** 2-flowered, 3–4.2 mm long by 1.5–2 mm wide, covered with white spreading hairs to *c.* 1 mm long; **glumes** two, unequal, hairy; **1st glume** pointed, 3-nerved, 1.5–2.5 mm long; **2nd glume** about as long as the spikelet, 9-nerved; **florets** of two types: **1) lower floret** usually male; **lemmas** similar to the 2nd glume in length and texture, green, 7- or 9-nerved, hairy; **lower paleae** 2.2–3 mm long; **anthers** 1–1.8 mm long; **2) upper floret** perfect, 2.4–3.4 mm long; **anthers** 0.5–1.3 mm long, dark brown, exserted. **FRUIT a grain**, 1-seeded, enclosed by the shiny, hard, lemma and palea, *c.* 2.5 mm long by 1.5 mm wide by 1.2 mm thick.
- **LEAVES** Two or three, ascending to erect; **blades** flat, usually with a wrinkled hairy margin near the rounded base, surfaces hairless to slightly hairy, 4–15 cm long by 4–14 (–20) mm wide, reduced above; **ligules** membranous, inconspicuous, a band of white hairs *c.* 0.2 mm long or absent; **auricles** absent; **sheaths** with spreading marginal hairs 1–2 mm long, open to base.
- **STEM** A **culm**, erect, leaning when in large clumps, glabrous to hairy below, round, purplish green, hollow; 1–3 mm thick near the base; **nodes** hairless.
- **SYN.** *Panicum leibergii* (Vasey) Scribn.

Leiberg Dichanthelium, Leiberg's Panicgrass

Grass—*Poaceae* Grass Spikelets 2-flowered

Barnyard Grass *Echinochloa crus-galli* (L.) Beauv.

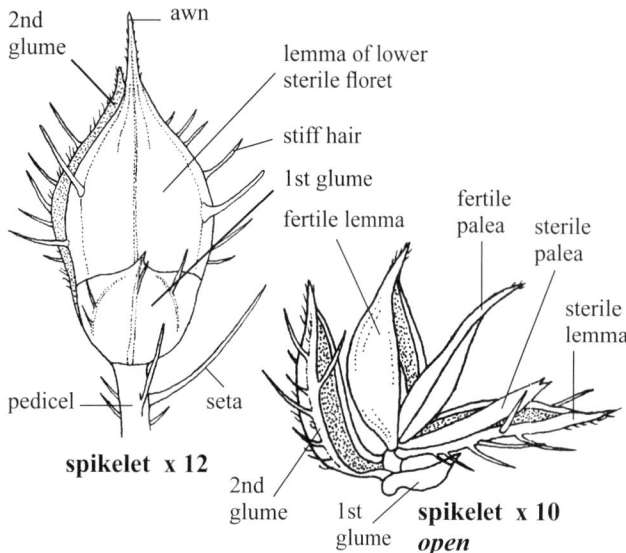

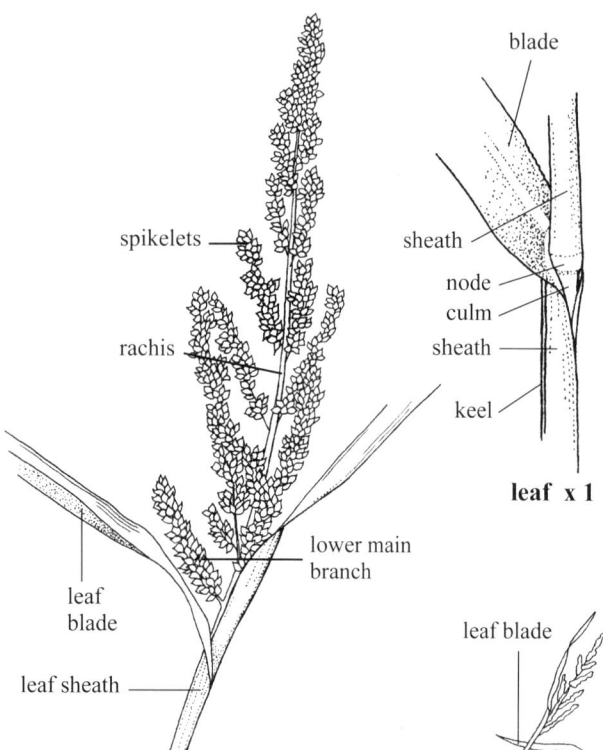

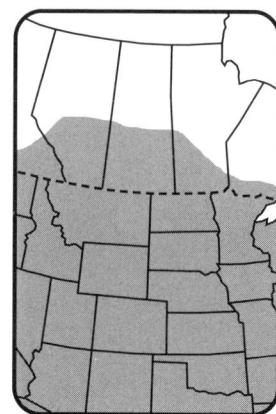

■ **SKETCH** A tufted, variable **annual grass** 30–200 cm tall from **fibrous roots** in moist to dry sites; in farmyards, cultivated fields, along ditches, reservoirs, river banks and sidewalks. *Naturalized*

■ **FLOWERS** Brownish red to green, blooming in July–November; **inflorescence** a panicle 10–25 cm long by 3–8 cm wide, terminal and from the lower nodes; **lower main branches** the longest and to *c.* 10 cm, usually more spreading than those above; **spikelets** 2-flowered, crowded, numerous, 2.5–4 mm long by 1–2.3 mm wide; **glumes** two, unequal; **1st glume** 1–2 mm long, 3-nerved with a very short tip and a few hyaline stiff hairs (setae); **2nd glume** 2.5–4 mm long, 5-nerved, with a scabrous tip not over 1 mm long, long stiff hairs on the nerves; **1) lower floret** sterile; **lemmas** 5-nerved, 2–4 mm long with a few stiff hairs on its nerves, with a scabrous **awn** 1–55 mm long; **paleae** hyaline, 2–3.5 mm long by *c.* 1.5 mm wide, hairless with a bidentate tip; **2) upper floret** fertile; **lemmas** glabrous, firm, nerves obscure, 2.5–4 mm long by *c.* 2 mm wide, tips short and hairy; **paleae** glabrous, almost as long, tips slightly hairy, fitting tightly inside the lemmas; **anthers** three, 0.5–1 mm long; **stigmas** two, feathery, reddish purple when exserted. **FRUIT** a grain, 1-seeded, orangish yellow, 1.7–3.2 mm long by 1–2 mm wide.

■ **LEAVES** Flat to V-shaped, four or five; **blades** glabrous on the surfaces, somewhat scabrous along the edges, 15–65 cm long by 8–25 mm wide, midrib wide, reddish green, raised below and continuing as a prominent keel along the sheath; **ligules** and **auricles** absent; **sheaths** glabrous, open, flattened and keeled.

■ **STEM** A **culm**, glabrous, prostrate to ascending to erect, stout, usually several in a tuft, some with tillers, these with smaller panicles; the base reddish, 5–8 mm wide and 3–4 mm thick.

Large Barnyard Grass

Grass—*Poaceae* **Grass** Spikelets 2–4 per node, each 3- to 7-flowered

Canada Wild Rye *Elymus canadensis* L.

■ **SKETCH** A variable, tufted **perennial grass** 80–220 cm tall with **fibrous roots** or rarely with **rhizomes** 1–2 mm thick by 1–4 cm long; in sandy to rocky sites, streambanks, woodland trails, ditches and disturbed sites. **p1**

■ **FLOWERS** Green, blooming March–August; **inflorescence** a spike, usually arched to erect, well exserted from upper leaf sheath, terminal, 6–30 cm long by 1–2 cm wide (including ascending awns); **rachis** flattened, slightly scabrous; **spikelets** 2–4 per rachis node, sessile, mostly appressed, usually 3- to 7-flowered; **glumes** two, subequal, stiff, awned, 11–30 mm long (including awn) by 0.8–1.5 mm wide, green turning tan with fruit, 2- or 5-nerved, nerves scabrous or not, margins scabrous to smooth, awn scabrous from ascending hairs; **lemmas** 3- or 5-nerved, bodies 8–13 mm long, pale green near the base, hairy to glabrous or with margins ciliate; **awns** (of lemmas) 12–50 mm long, spreading to recurved at maturity; **paleae** slightly shorter than the lemmas, 8–12 mm long, scabrous along the margins, apices usually bifid, outer surface hairy; **stamens** three, mostly included; **anthers** pale yellow, 2–3.3 mm long; **ovary** green, with a dense tuft of white apical hairs *c.* 1 mm long. FRUIT a grain, 1-seeded, 6–6.5 mm long (including apical hairs) by *c.* 1.5 mm wide by *c.* 1 mm thick, tan, concave on inside, convex on outside, tuft of white apical hairs 1–1.5 mm long.

■ **LEAVES** Usually four or five per culm, ascending to spreading; **blades** flat to involute in drier sites, sometimes glaucous, 10–40 cm long by 4–20 mm wide, glabrous below to scabrous above and on margins; **ligules** flat, membranous, entire to erose, 0.5–1.5 mm long; **auricles** 1.5–4 mm long, clasping; **sheaths** usually glabrous or slightly scabrous to ciliate.

■ **STEM** A **culm**, erect, arched, round, glabrous, hollow, green turning tan with fruit, not branched; 2–4 mm thick near the brown, often decumbent base; **nodes** glabrous, dark brown, obvious.

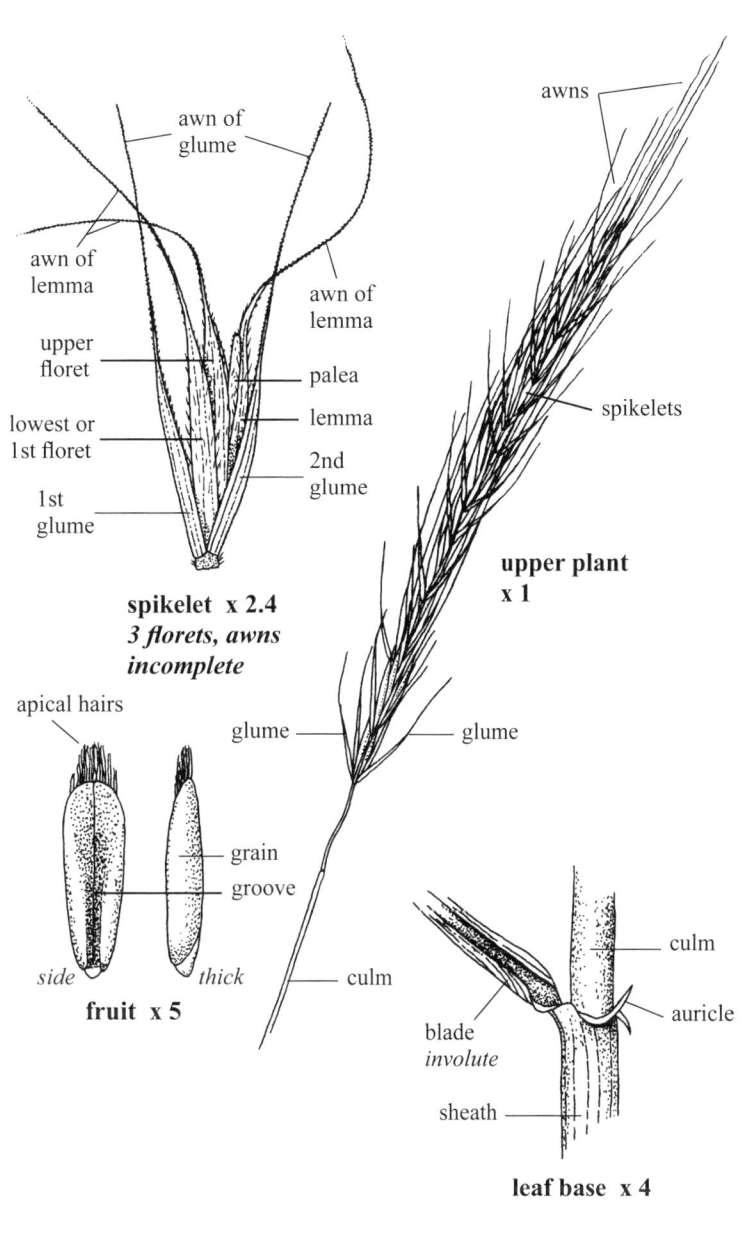

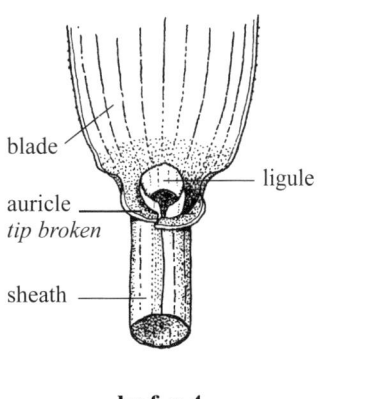

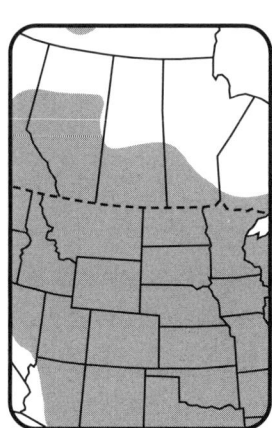

Nodding Wild Rye

Grass — Poaceae | Grass Spikelets 3- to 8-flowered

Quack Grass *Elymus repens* (L.) Gould

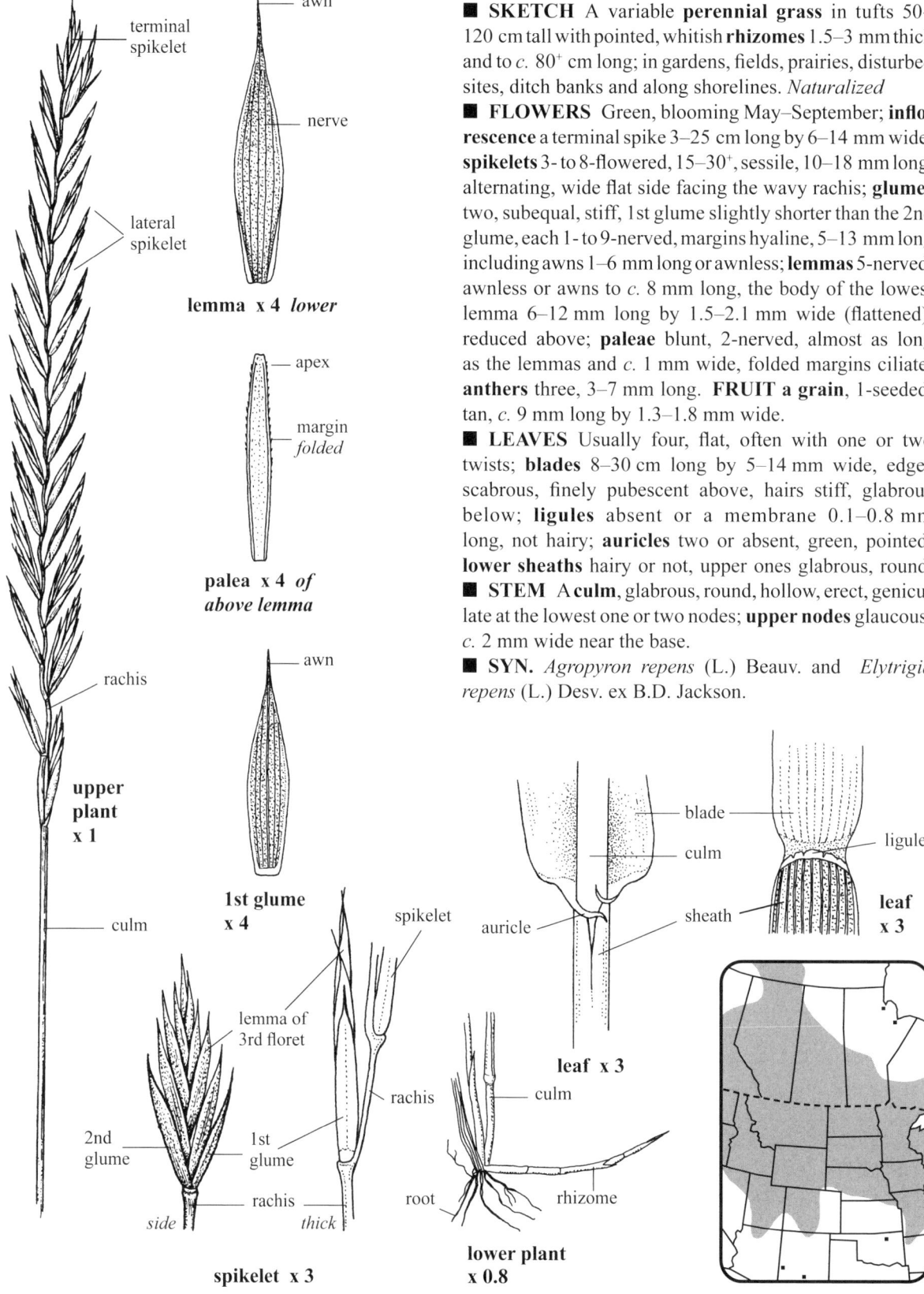

- **SKETCH** A variable **perennial grass** in tufts 50–120 cm tall with pointed, whitish **rhizomes** 1.5–3 mm thick and to *c.* 80+ cm long; in gardens, fields, prairies, disturbed sites, ditch banks and along shorelines. *Naturalized*
- **FLOWERS** Green, blooming May–September; **inflorescence** a terminal spike 3–25 cm long by 6–14 mm wide; **spikelets** 3- to 8-flowered, 15–30+, sessile, 10–18 mm long, alternating, wide flat side facing the wavy rachis; **glumes** two, subequal, stiff, 1st glume slightly shorter than the 2nd glume, each 1- to 9-nerved, margins hyaline, 5–13 mm long including awns 1–6 mm long or awnless; **lemmas** 5-nerved, awnless or awns to *c.* 8 mm long, the body of the lowest lemma 6–12 mm long by 1.5–2.1 mm wide (flattened), reduced above; **paleae** blunt, 2-nerved, almost as long as the lemmas and *c.* 1 mm wide, folded margins ciliate; **anthers** three, 3–7 mm long. **FRUIT** a grain, 1-seeded, tan, *c.* 9 mm long by 1.3–1.8 mm wide.
- **LEAVES** Usually four, flat, often with one or two twists; **blades** 8–30 cm long by 5–14 mm wide, edges scabrous, finely pubescent above, hairs stiff, glabrous below; **ligules** absent or a membrane 0.1–0.8 mm long, not hairy; **auricles** two or absent, green, pointed; **lower sheaths** hairy or not, upper ones glabrous, round.
- **STEM** A **culm**, glabrous, round, hollow, erect, geniculate at the lowest one or two nodes; **upper nodes** glaucous; *c.* 2 mm wide near the base.
- **SYN.** *Agropyron repens* (L.) Beauv. and *Elytrigia repens* (L.) Desv. ex B.D. Jackson.

Couch Grass, Creeping Wild Rye

Grass—*Poaceae* Grass   Spikelets 2 (3) per node, each 2- to 6-flowered

VIRGINIA WILD RYE   *Elymus virginicus* L.

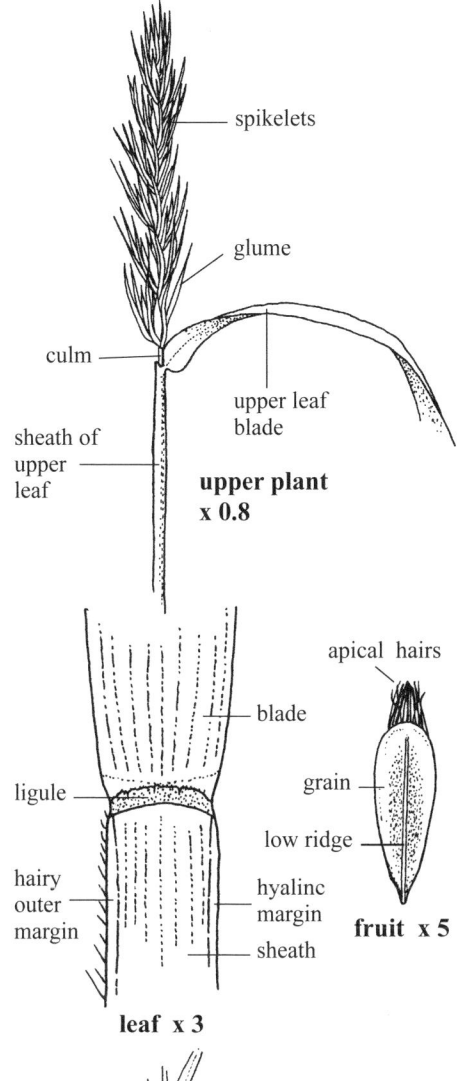

■ **SKETCH** A variable **perennial grass** 30–150 cm tall from **fibrous roots** in open tufts of one to several culms; along river bottomlands, shorelines, in cultivated sites, thickets, prairies and open woods.

■ **FLOWERS** Green, blooming April–September; **inflorescence** a spike, mostly erect, 5–17 cm long by 10–15 mm wide, partially enclosed in to exserted 0.2–15 cm past the sheath of the upper leaf; **rachis** ciliate; **spikelets** 2- to 6-flowered, 1.3–4.5 cm long (including awns), usually two (three) per rachis node, quickly deciduous; **glumes** two per spikelet, equal to subequal, stiff, 1–4 cm long (including awns when present) by 1–3 mm wide, longer to rarely shorter than the lemmas, slightly scabrous near the tapered tip, flat above but roundish, lighter green and hard near the base, with two major raised nerves and usually one or three narrow lesser nerves obscure at the base; **florets** perfect, pale green, first two florets sometimes awned; **lemmas** of lowest (1st) floret 5-nerved near apices, glabrous at base to scabrous at tips, stiff, body 7–12 mm long by *c.* 1.6 mm wide (not flattened), lateral nerves the full length, rounded on back; **awn** 0–4 cm long; **paleae** slightly shorter than the lemmas, stiff, apices bluntly pointed, the two lateral nerves scabrous near the apices; **3rd floret** if present often vestigial, TL *c.* 4 mm including the 0.3–4 mm long awn; **stamens** three, exserted; **anthers** pale yellow, 2–3 mm long. **FRUIT a grain**, 1-seeded, *c.* 6 mm long (including white hairs) by *c.* 1.7 mm wide, with a low ridge, apical hairs *c.* 1 mm long.

■ **LEAVES** Usually six or seven per culm, lowest one often brown by anthesis; **blades** 5–30 cm long by 3–18 mm wide, largest about the middle of the culm, slightly scabrous to glabrous, arching, dull, often with a twist near the culm, midrib obvious below, not above; **top leaf** often near the base of the inflorescence; **ligules** membranous, flat, tan, 0.3–1 mm long, apices ciliate and erose (microscopic); **auricles** absent or minute; **sheaths** smooth, round, outer margin with hairs *c.* 0.6 mm long, these more numerous near the sheath's base.

■ **STEM** A **culm**, erect, stiff, hollow, round, tillering from base; 1–3 mm thick near the base.

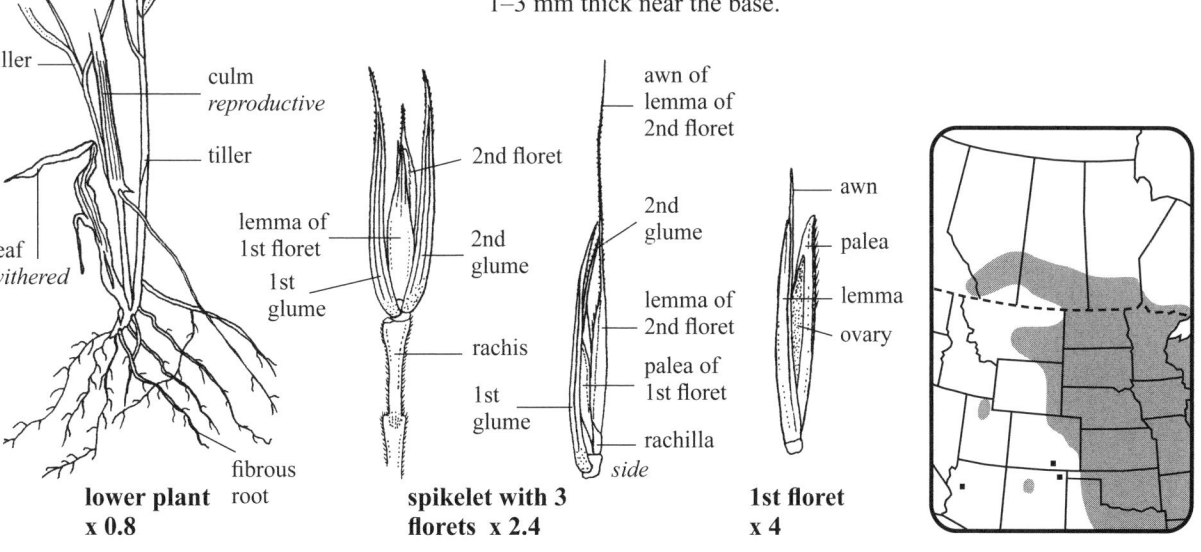

Stinkgrass *Eragrostis cilianensis* (All.) Vign. ex Janchen

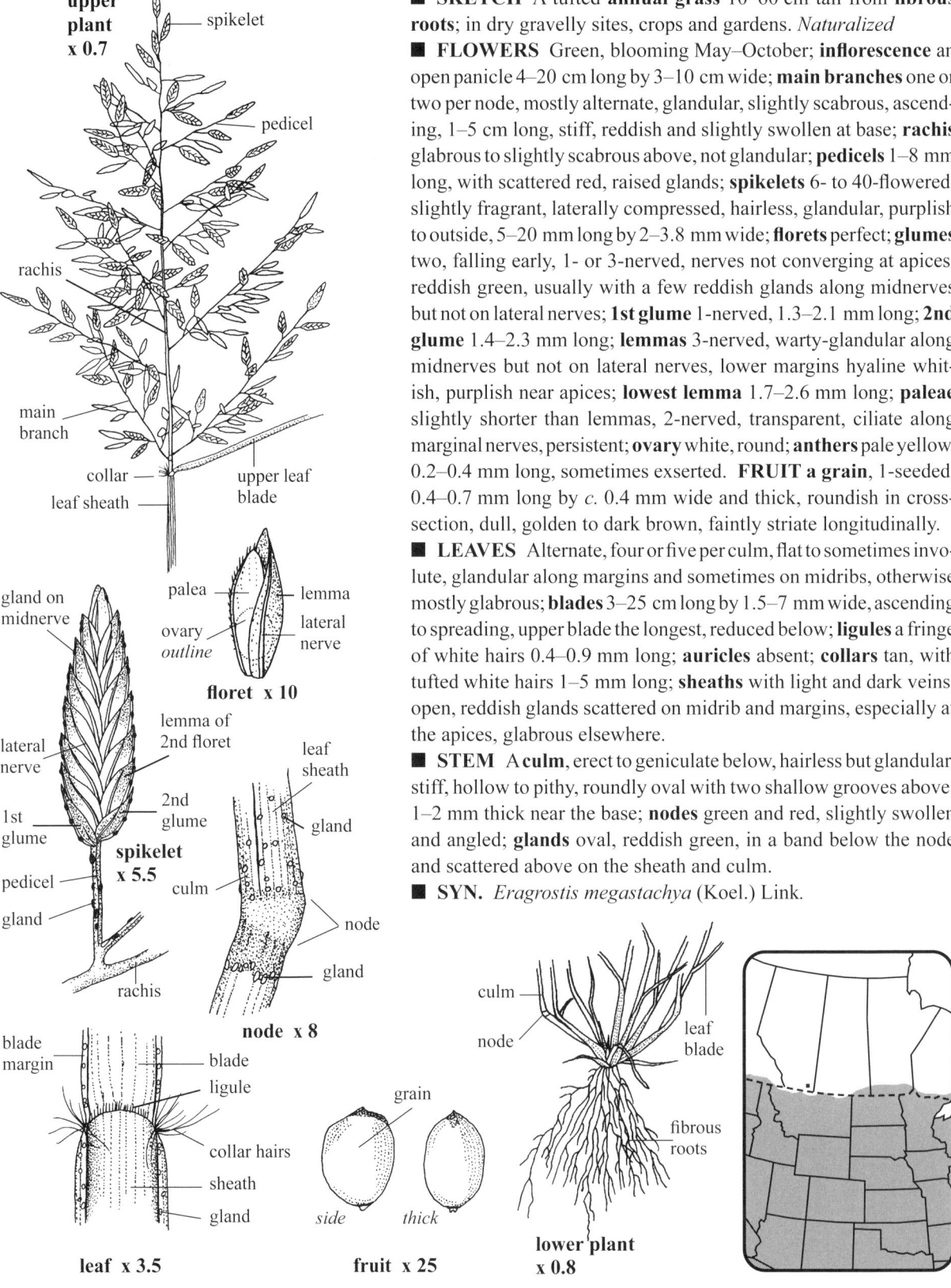

- **SKETCH** A tufted **annual grass** 10–60 cm tall from **fibrous roots**; in dry gravelly sites, crops and gardens. *Naturalized*
- **FLOWERS** Green, blooming May–October; **inflorescence** an open panicle 4–20 cm long by 3–10 cm wide; **main branches** one or two per node, mostly alternate, glandular, slightly scabrous, ascending, 1–5 cm long, stiff, reddish and slightly swollen at base; **rachis** glabrous to slightly scabrous above, not glandular; **pedicels** 1–8 mm long, with scattered red, raised glands; **spikelets** 6- to 40-flowered, slightly fragrant, laterally compressed, hairless, glandular, purplish to outside, 5–20 mm long by 2–3.8 mm wide; **florets** perfect; **glumes** two, falling early, 1- or 3-nerved, nerves not converging at apices, reddish green, usually with a few reddish glands along midnerves but not on lateral nerves; **1st glume** 1-nerved, 1.3–2.1 mm long; **2nd glume** 1.4–2.3 mm long; **lemmas** 3-nerved, warty-glandular along midnerves but not on lateral nerves, lower margins hyaline whitish, purplish near apices; **lowest lemma** 1.7–2.6 mm long; **paleae** slightly shorter than lemmas, 2-nerved, transparent, ciliate along marginal nerves, persistent; **ovary** white, round; **anthers** pale yellow, 0.2–0.4 mm long, sometimes exserted. **FRUIT a grain**, 1-seeded, 0.4–0.7 mm long by *c.* 0.4 mm wide and thick, roundish in cross-section, dull, golden to dark brown, faintly striate longitudinally.
- **LEAVES** Alternate, four or five per culm, flat to sometimes involute, glandular along margins and sometimes on midribs, otherwise mostly glabrous; **blades** 3–25 cm long by 1.5–7 mm wide, ascending to spreading, upper blade the longest, reduced below; **ligules** a fringe of white hairs 0.4–0.9 mm long; **auricles** absent; **collars** tan, with tufted white hairs 1–5 mm long; **sheaths** with light and dark veins, open, reddish glands scattered on midrib and margins, especially at the apices, glabrous elsewhere.
- **STEM** A **culm**, erect to geniculate below, hairless but glandular, stiff, hollow to pithy, roundly oval with two shallow grooves above; 1–2 mm thick near the base; **nodes** green and red, slightly swollen and angled; **glands** oval, reddish green, in a band below the node and scattered above on the sheath and culm.
- **SYN.** *Eragrostis megastachya* (Koel.) Link.

Grass—*Poaceae* Grass Spikelets 5- to 44-flowered

CREEPING LOVE GRASS   *Eragrostis hypnoides* (Lam.) B. S. P.

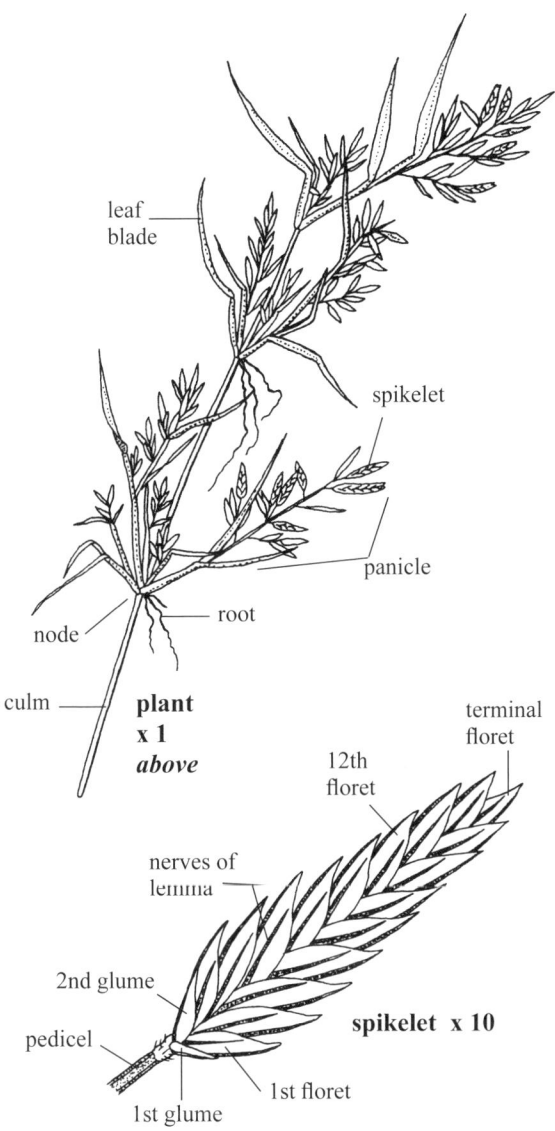

■ **SKETCH** A mat-forming **annual grass** 20–30 cm wide from prostrate branched stems with **fibrous roots** from the lower nodes; in sandy to muddy ground along river banks, near streams, ponds and lakes.

■ **FLOWERS** Light green, blooming June–October; **inflorescence** a panicle, each 1–7 cm long by 1–3 cm wide, numerous, compact to open; **floral branches** and **rachis** hairless, with longitudinal ridges; **spikelets** 5- to 44-flowered, laterally compressed, glandless, sessile or not, 3–16 mm long by 1.5–3 mm wide, usually minutely hairy at the base; **florets** perfect; **glumes** two, unequal, acute, hyaline, 1-nerved, the nerve green, obvious, scabrous or not; **1st glume** 0.4–1.1 mm long; **2nd glume** 0.7–1.5 mm long; **lemmas** 3-nerved, white hyaline, nerves green and obvious, not converging at the apices, midnerves minutely scabrous or not, lowest lemma 1.5–2 mm long by 0.5–0.6 mm wide (across one side); **paleae** minutely ciliate along the two green nerves, lower palea *c.* 1 mm long and shorter than its lemma, apex blunt to slightly ragged; **stamens** three, included to exserted; **anthers** two, pale yellow, *c.* 0.3 mm long. **FRUIT a grain**, 1-seeded, tan, 0.4–0.6 mm long by *c.* 0.4 mm wide by *c.* 0.3 mm thick, slightly flattened, dark brown at one end.

■ **LEAVES** Imbricate among the panicles; **blades** flat, 5–50 mm long by 1–3 mm wide, short hairy above along the ridges, hairs white and simple, glabrous below; **ligules** a fringe of white hairs 0.3–0.9 mm long; **sheaths** open, slightly keeled, glabrous to slightly hairy along both margins at least near the collar.

■ **STEM** A **culm**, decumbent, hollow, branched, tough, pinkish, glabrous, oval with four or five low, longitudinal ridges unevenly spaced; 1–1.5 mm thick near the base; **nodes** swollen, hairy at their bases, lower ones often with fibrous roots.

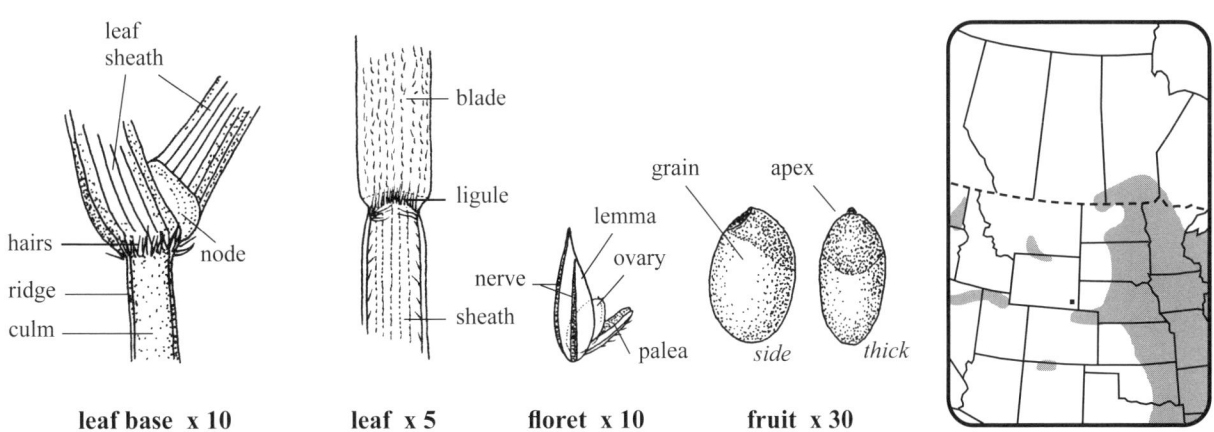

Smooth Love Grass, Teal Lovegrass

Grass—*Poaceae* Grass Spikelets 2- to 8-flowered

TALL MANNA GRASS *Glyceria grandis* S. Wats.

■ **SKETCH** A **perennial grass** 60–200 cm tall from **rhizomes** 2–3 mm wide and to *c.* 10⁺ cm long; in wet meadows, at the edges of ponds, sloughs, ditches and lakes, even in shallow standing water. **w1**

■ **FLOWERS** Green to reddish, blooming June–August; **inflorescence** an open panicle 18–40 cm long by 15–25 cm wide, slightly nodding; **main branches** ascending to arched, lower ones to *c.* 19 cm long, reduced above, usually three or four per node, redividing; **pedicels** glabrous; **spikelets** 2- to 8-flowered, each 4–7 mm long by 1–2 mm wide and *c.* 0.8 mm thick; **glumes** two, papery, subequal, 1-nerved, whitish, bluntly pointed, not reaching the apex of the lowest lemma; **1st glume** 1–2 mm long; **2nd glume** 1.5–2.4 mm long; **lemmas** with firm blunt apices, 7-nerved, the lowest the longest at 2.1–2.9 mm, reddish purple near the tip or throughout; **paleae** green, almost as long as the lemmas, curved, 2-nerved, fitting into the lemmas; **stamens** three; **anthers** 0.7–1.1 mm long. **FRUIT** a **grain**, 1-seeded.

■ **LEAVES** Usually four or five per culm, the top blade often reaching into the base of the panicle, erect; **blades** flat, greenish yellow, the edges slightly scabrous, 15–45 cm long by 7–14 mm wide, soft, generally glabrous on both sides, ascending; **ligules** 2–5 mm long, membranous, some forming a central point; **sheaths** greenish yellow, slightly keeled or not, glabrous, closed to open.

■ **STEM** A **culm**, stout, erect, hollow; 3–8 mm thick near the base.

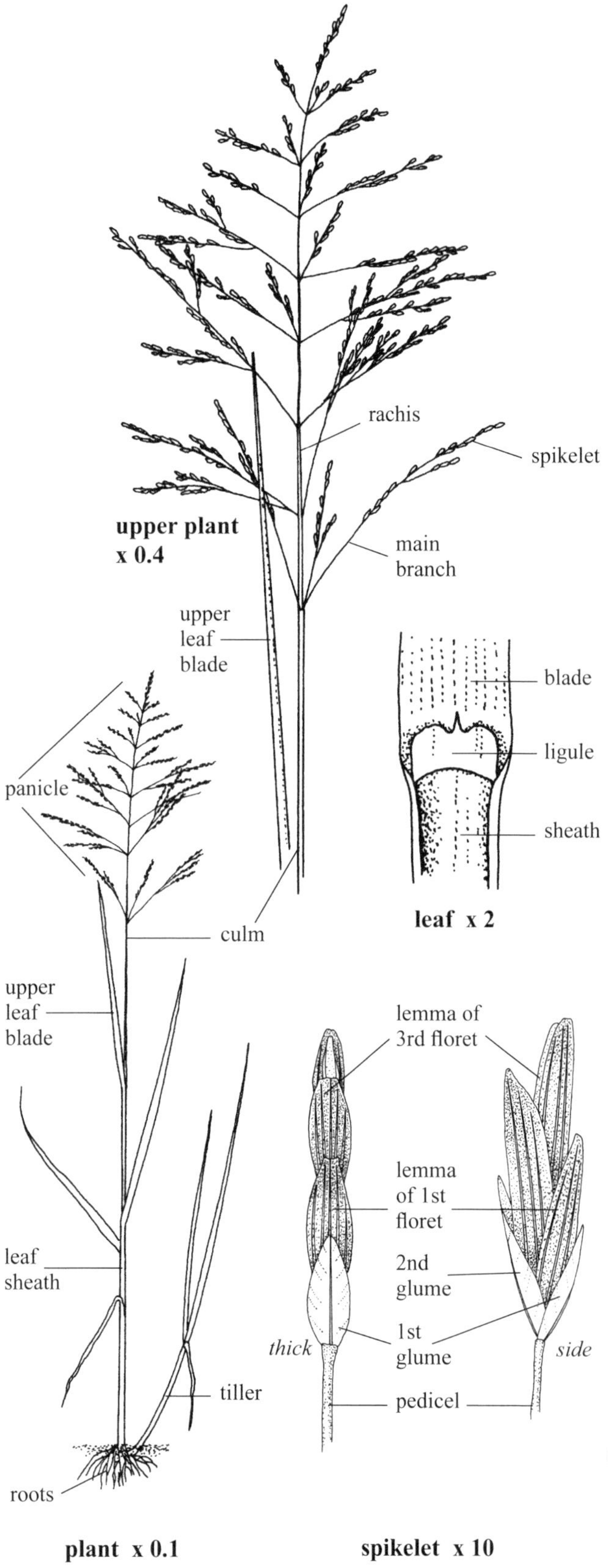

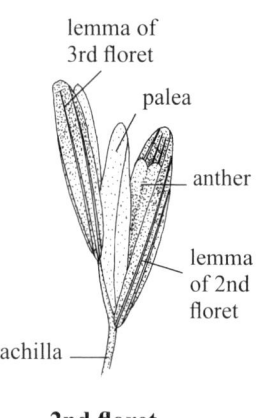

2nd floret x 10 *open*

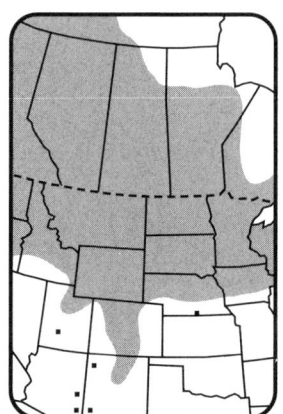

American Manna Grass

Grass—*Poaceae* Grass Spikelets 3- to 8-flowered

Fowl Manna Grass *Glyceria striata* (Lam.) A. S. Hitchc.

■ **SKETCH** A tufted **perennial grass** 30–140 cm tall from short, pink **rhizomes** 1–2 mm thick; in ditches, boggy meadows, along lakes, ponds, sloughs and wet woods, often in shallow water.

■ **FLOWERS** Green to brownish purple, blooming May–August; **inflorescence** a panicle 4–25 cm long by 10–18 cm wide, usually nodding; **main branches** spreading, arched, usually two to four at lower nodes, unequal in length, lowest branch to *c.* 15 cm long, slightly scabrous with spikelets on distant half; **rachis** scabrous, green; **pedicels** glabrous; **spikelets** 3- to 8-flowered, 2.5–4.5 mm long by 1.3–2.4 mm wide by *c.* 1 mm thick; **glumes** two, subequal, pointed, 1-nerved, shorter than the lowest lemma; **1st glume** 0.5–1 mm long; **2nd glume** 0.7–1.4 mm long; **lemmas** blunt, distinctly 7-nerved, the nerves not strongly converging at the scarious tip, lowest lemma 1.3–2 mm long; **paleae** 2-nerved, greenish, stiff, equal to or slightly longer than the lemmas; **stamens** two; **filaments** white, very thin; **anthers** light purplish, 0.3–0.5 mm long, quickly deciduous. **FRUIT a grain**, 1-seeded.

■ **LEAVES** Usually five to seven per culm, ascending to erect, upper leaf blade often reaching into lower part of the panicle; **blades** 10–40 cm long by 2–6 mm wide, slightly scabrous below and on margins, smooth above; **ligules** membranous, 1–3 mm long, not hairy, apices slightly ragged, easily torn, sometimes reddish on the sides by the collar; **auricles** absent; **collars** not hairy; **sheaths** tight, closed until near the collar, smooth to scabrous, very slightly keeled with a thin, low, ridge.

■ **STEM** A **culm**, erect, green, hollow, smooth below the panicle, sometimes decumbent at the base and rooting at the lowest nodes; 2–4 mm thick near the reddish base.

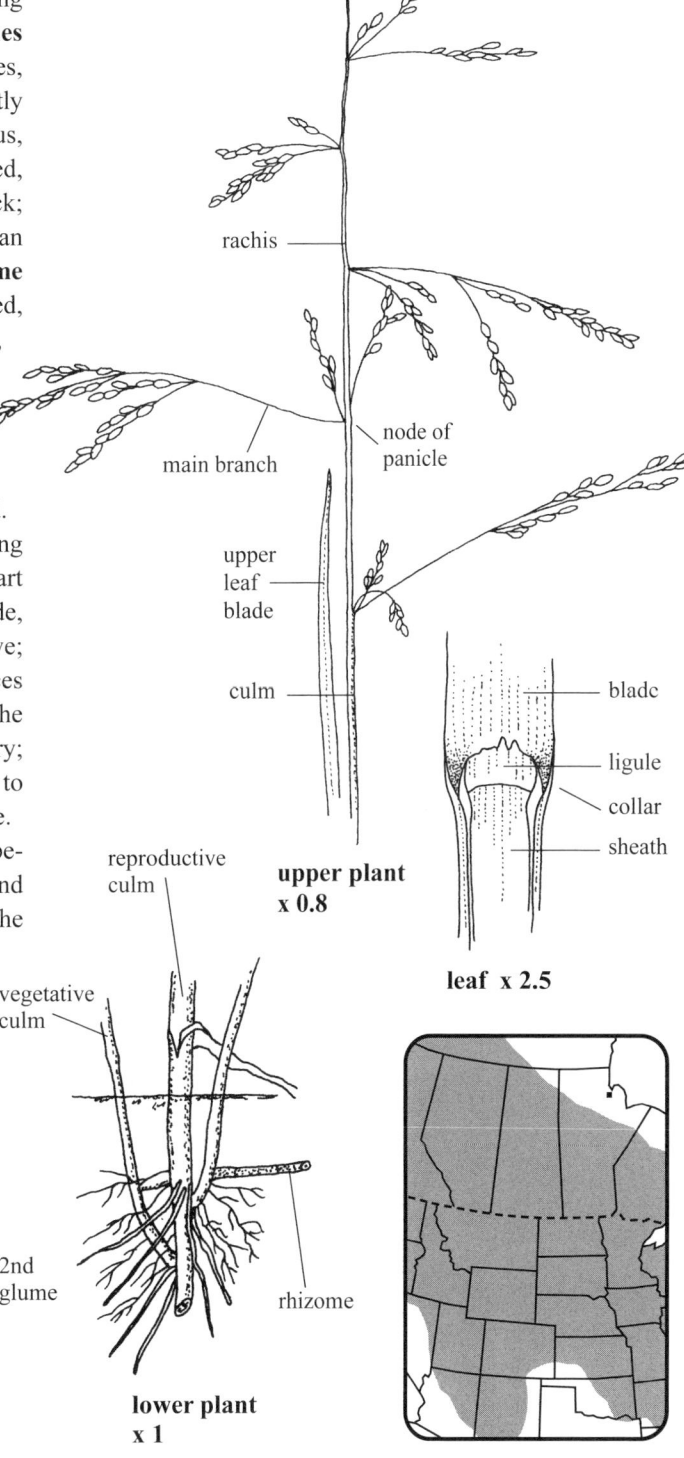

Grass—*Poaceae* Grass Spikelets 1-flowered

SPEAR GRASS *Hesperostipa comata* (Trin. & Rupr.) Barkworth

■ **SKETCH** A tufted **perennial grass** 35–100+ cm tall from **fibrous roots**; in dry prairies and pastures, deserts, in sandy to rocky soils. **p1**

■ **FLOWERS** Green to tan, blooming May–September; **inflorescence** a panicle, contracted to ascending, 10–35 cm long by 5–15 cm wide, the lower part of the panicle sometimes enclosed in the upper leaf's sheath; **spikelets** 1-flowered, 6–20 per panicle, two to four spikelets per lower node, to one per upper node; **main branches** and **rachis** slightly scabrous; **glumes** two, subequal, 1.8–3 cm long, bases wrapped around the lemma base and appressed to it, glabrous, hyaline, transparent except for five green nerves, the lateral nerves incomplete, C-shaped, tapered to an awnlike point, persistent and tan; **1st glume** 1.8–2.2 mm wide (flattened), more membranous and pliable than the lemmas; **2nd glume** 1.5–2.8 cm long, slightly shorter and wider than the first; **lemmas** awned, the body 8–13 mm long (including the 3 mm long hairy callus) by *c.* 1 mm thick, light brown when ripe, shiny, 5-nerved with ascending white hairs in the lower half or throughout, C-shaped, stiff, tightly wrapped around the paleae, margins hairy, a fringe of hairs 0.3–1 mm long at the lemma's apex; **awns** 6–20 cm long by *c.* 0.4 mm thick near the first bend, one or two times bent on drying, tightly coiled in the lower half, hairy near the first bend, less so above; **paleae** soft, glabrous, 6–7 mm long by *c.* 0.8 mm wide, 2-nerved, pointed, thin, light green; **stamens** three, included; **anthers** pale yellow, 0.1–0.6 mm long when floret is cleistogamous, 3.5–5.5 mm long normally. **FRUIT a grain**, 1-seeded, 0.6–1.5 cm long by *c.* 1 mm wide.

■ **LEAVES** Three or four per culm, turning brown early; **blades** ascending, 7–38 cm long by 1–4 mm wide, upper blade usually reaching into the lower part of the panicle, flat to involute near the tip, tapered to a fine point, ventral ridges and furrows obvious; **ligules** pointed to blunt, sometimes torn, brownish near base, apices ciliate, pointed to blunt, 0.7–10 mm long, reduced below; **auricles** absent; **sheaths** glabrous to hairy, open to base, margins hyaline and slightly scabrous.

■ **STEM** A **culm**, solid, green, oval, glabrous, stiff; 1.5–2.5 mm thick near the base; **nodes** slightly swollen, glabrous to hairy, *c.* 0.5 mm long, dark greenish brown, obscure.

■ **SYN.** *Stipa comata* Trin. & Rupr. = *Hesperostipa comata* ssp. *comata* (Trin. & Rupr.) Barkworth.

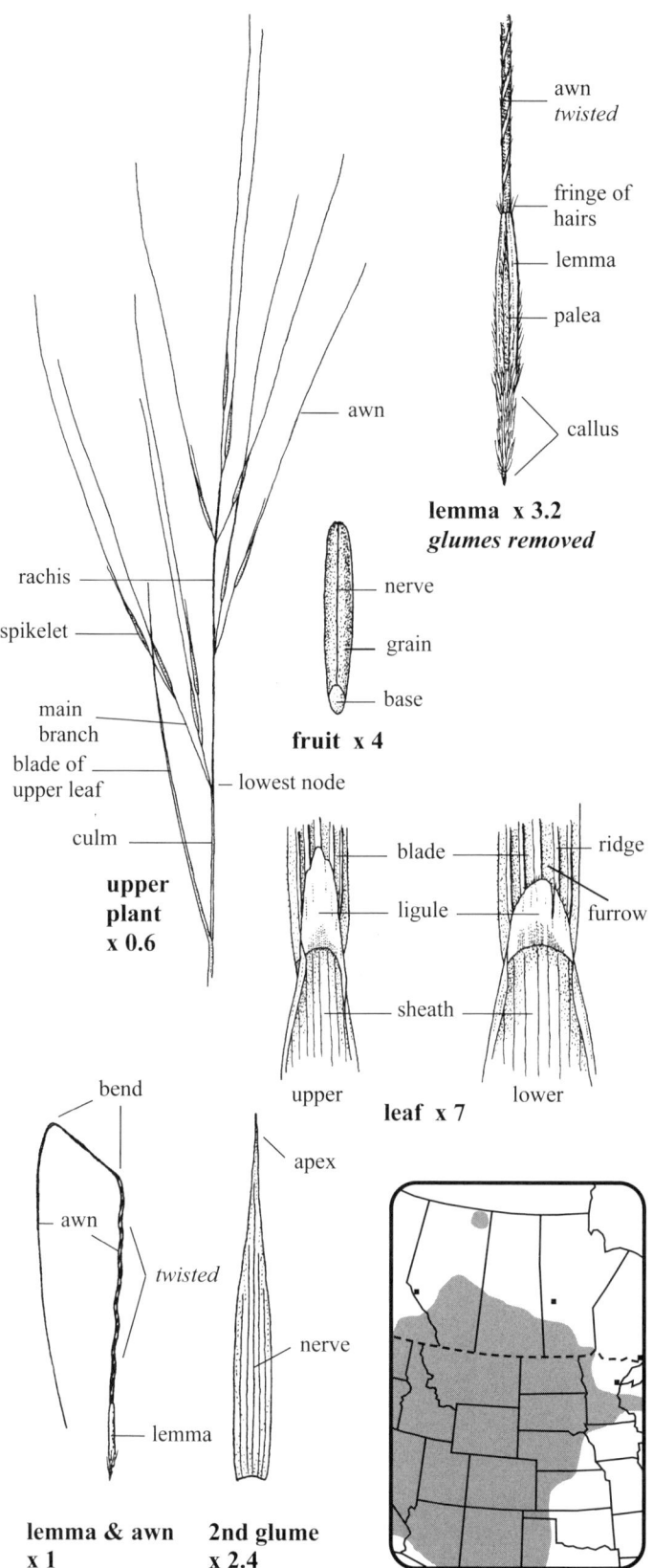

Needle and Thread

Grass—*Poaceae* Grass Spikelets 1-flowered

PORCUPINE GRASS   *Hesperostipa spartea* (Trin.) Barkworth

■ **SKETCH** A tufted **perennial grass** growing 50–140 cm tall; in moist to dry prairies, near bluffs and in parkland openings. **p2**

■ **FLOWERS** Green, blooming June–July; **inflorescence** a panicle 9–27 cm long, often nodding, usually free from upper leaf sheath; **spikelets** 1-flowered, 8–12 per inflorescence; **glumes** two, transparent, 5- or 7-nerved, about equal in length, 2–5 cm long including the awnlike tip; **lemma body** transparent, faintly 5-nerved, hairy near the base and along the margins, 10–25 mm long, C-shaped; **awn** 6–25 cm long, with several twists in the lower half, two bends about midlength as it dries, a tuft of short hairs identifies the awn/lemma junction; **callus** stiff and sharp, hairy and persistent at the base of the lemma; **paleae** glabrous, faintly nerved, pointed, fitting tightly into and filling the lemma's body; **lemma** and **palea** (a floret) break off above the tan glumes which persist on the plant; **stamens** three; **anthers** yellow, 7–11 mm long by *c.* 1 mm wide and easily seen through the transparent glumes and lemma; **cleistogamous florets** have anthers 0.3–1.5 mm long. **FRUIT a grain**, 1-seeded, dark brown, enclosed in lemma, body 1–2.2 mm long.

■ **LEAVES** Narrow, usually three per culm, upper one sometimes reaching into the panicle; **blades** 15–60 cm long by 1–5 mm wide, flat, ascending, with scattered white hairs and strongly ridged on the upper surface, glabrous and not strongly ridged below (dorsal); **ligules** white, pointed above, less so below, ragged, ciliate or not, membranous, 1–8 mm long, reduced below; **sheaths** glabrous, but some hairy along the margins.

■ **STEM** A **culm**, round, hollow, erect to leaning, sometimes hairy at lower nodes.

■ **SYN.** *Stipa spartea* Trin.

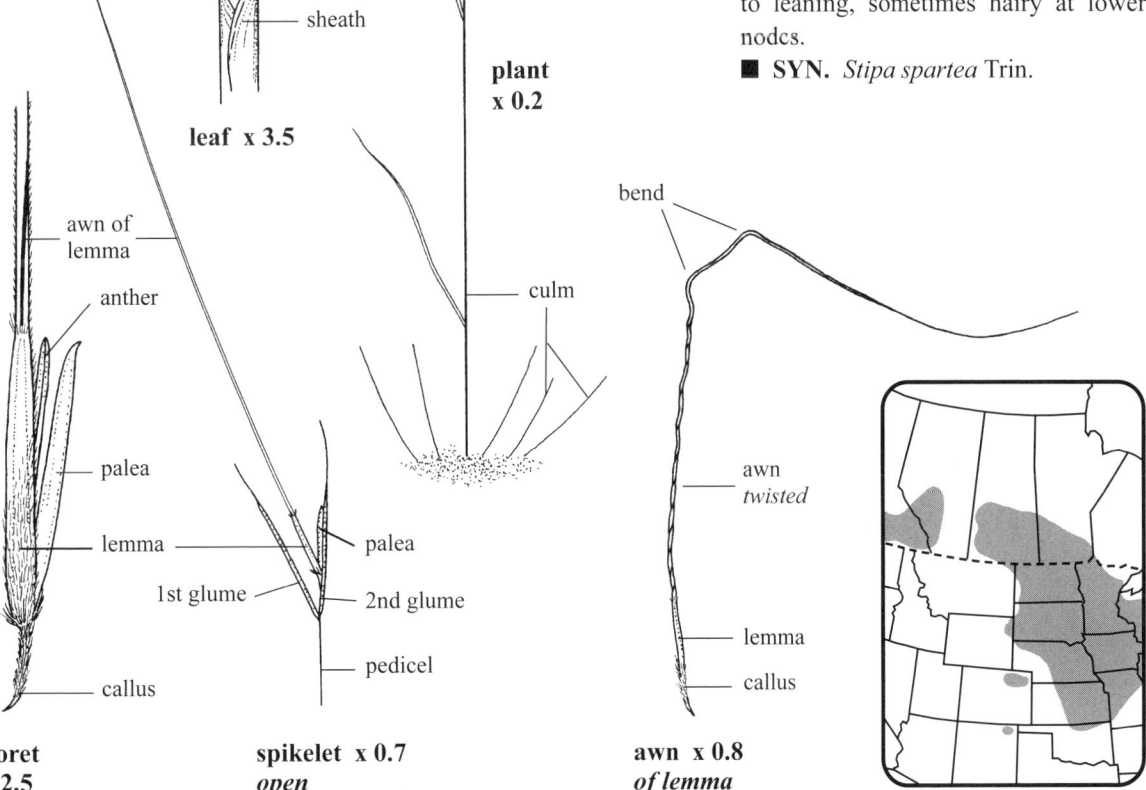

649

Grass—*Poaceae* — **Grass** Spikelets 3-flowered

Sweet Grass *Hierochloë odorata* (L.) Beauv.

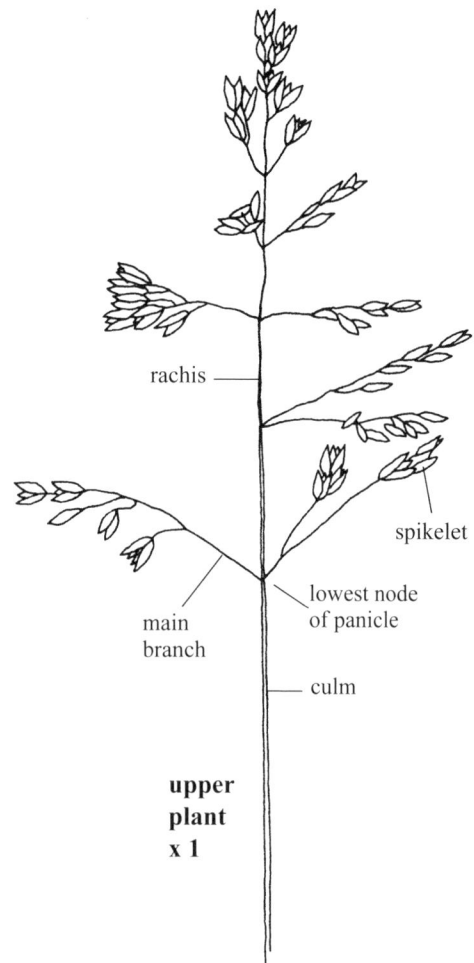

upper plant x 1

- **SKETCH** A **perennial grass** 20–70 cm tall from whitish **rhizomes** $c.$ 10⁺ cm long; in wet fields, willow thicket openings, along riverbanks, roadsides and wetlands. **p1**
- **FLOWERS** Pale green, shiny, fragrant, blooming May–August; **inflorescence** a panicle, open, 5–15 cm long by 5–9 cm wide, usually with two main branches from most nodes; **main branches** shiny, reddish green, 2–6 cm long, reduced above, often with a slight swelling near the rachis, bare near the base; **pedicels** smooth, shiny, red and green; **spikelets** 3-flowered, awnless, 4.5–6 mm long by 3–4 mm wide by 1.5–1.8 mm thick, shiny; **glumes** two, subequal, pale green, glabrous, membranous, rounded on the back, some with reddish streaks near the base; **1st glume** 3.5–5.5 mm long, 1- or 3-nerved; **2nd glume** 3.8–6 mm long, 3-nerved, wider than the 1st glume; **florets** of two types: **1) male florets** two, one above each glume; **lemmas** 5-nerved, 3.5–5 mm long, hairy near the apices and margins; **paleae** about as long as the lemmas, bidentate, 2-nerved; **stamens** three; **anthers** pale yellow, $c.$ 2 mm long; **2) perfect floret** one, terminal; **lemmas** 3- to 7-nerved, nerves faint, pale green, 2.5–4 mm long by $c.$ 1 mm wide, shiny, rounded at back, hairy at apices; **fertile paleae** as long as lemmas; **anthers** pale yellow and 1–1.5 mm long; **stigmas** two, feathery. **FRUIT a grain**, 1-seeded, 3–3.2 mm long by $c.$ 1 mm wide and thick.
- **LEAVES** Usually two or three per culm; **stem blades** 3–5 cm long by 2–6 mm wide, reduced above, upper blades not reaching the panicle, flat, glabrous, often tan and quickly fading; **vegetative plant blades** 10–30 cm long; **ligules** membranous, 1–5 mm long, reduced below, the apices slightly ragged and bluntly pointed to truncate; **auricles** absent; **sheaths** open, glabrous or slightly hairy near the collar.
- **STEM** A **culm**, glabrous, hollow, erect; 2–3 mm thick near the purplish base.
- **SYN.** *Hierochloë hirta* (Schrank) Borbs ssp. *arctica* (J. Presl) Weim. and *Anthoxanthium hirtum* (Schrank) Schout. & Veldk. ssp. *arcticum* (J. Presl) G. Tucker.

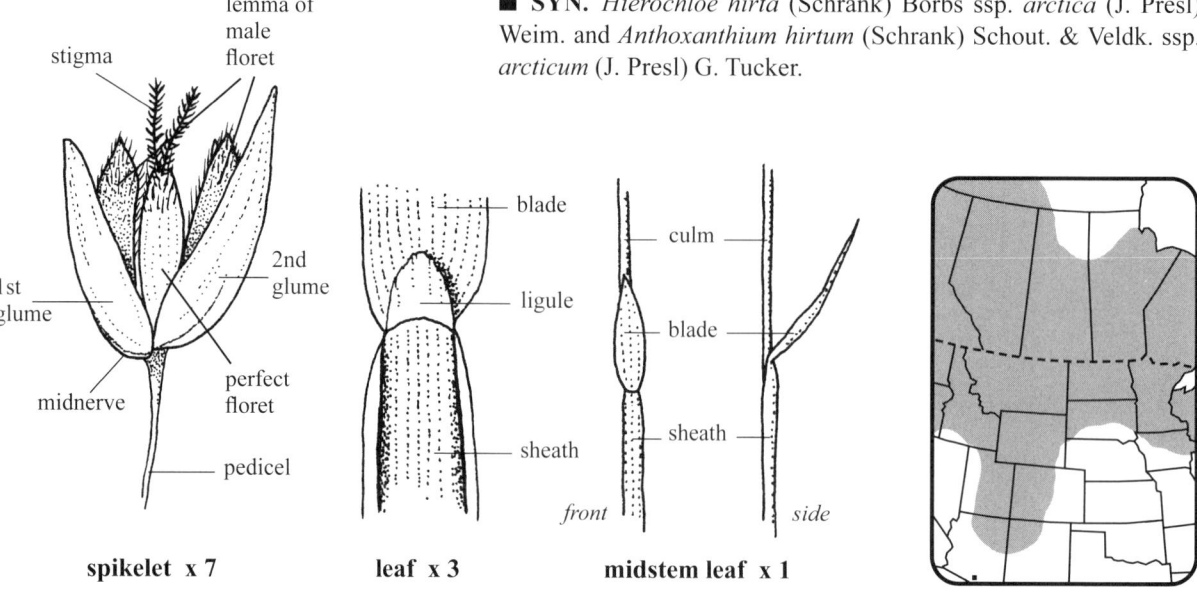

spikelet x 7 — leaf x 3 — midstem leaf x 1

Grass—*Poaceae* **Grass** Spikelets 3 per node, each 1-flowered, some sterile

Wild Barley *Hordeum jubatum* L.

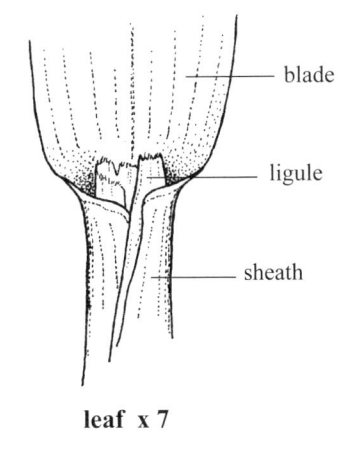

■ **SKETCH** A variable, tufted **annual** or **perennial grass** 30–100 cm tall with **fibrous roots**; along roadsides, alleys, clearings, in wet areas now dry, pastures, banks of streams and drying ditches, including saline areas.

■ **FLOWERS** Green, blooming May–September; **inflorescence** spikelike, terminal, nodding, 5–14 cm long by 5–8 cm wide (including awns); **spikelets** three per node with six awnlike glumes plus a central awn; **1) central spikelet** 1-flowered, perfect, sessile; **glumes** two, 1–8 cm long, awnlike, scabrous; **lemmas** obscurely 5-nerved, glabrous to scabrous, 5–8 mm long by *c*. 1 mm wide and tapered to a scabrous awn 0.5–9 cm long; **paleae** awnless, the body almost as long as the lemmas; **anthers** pale yellow, 0.8–1.9 mm long; **2) lateral spikelets** two, each 1-flowered and sterile, on curved pedicels 1–1.5 mm long; **glumes** two per spikelet, awnlike, scabrous, 0.5–7 cm long, similar to the glumes of the central fertile lemma; **lemma body** and its vestigial awn wavy, 4–10 mm long; **palea** absent. **FRUIT a grain**, 1-seeded.

■ **LEAVES** Usually five main ones, ascending, the edges scabrous, slightly scabrous above and below or smooth below; **blades** 3–15 cm long by 2–9 mm wide, flat, often with a twist; **ligules** 0.2–1 mm long, ragged, with tiny hairs along its flat top; **auricles** absent or minute; **lowest sheaths** hairy to smooth.

■ **STEM** A **culm**, round, hollow, glabrous, erect but nodding at the top, geniculate at the lowest, dark nodes; 1–2 mm wide near the base.

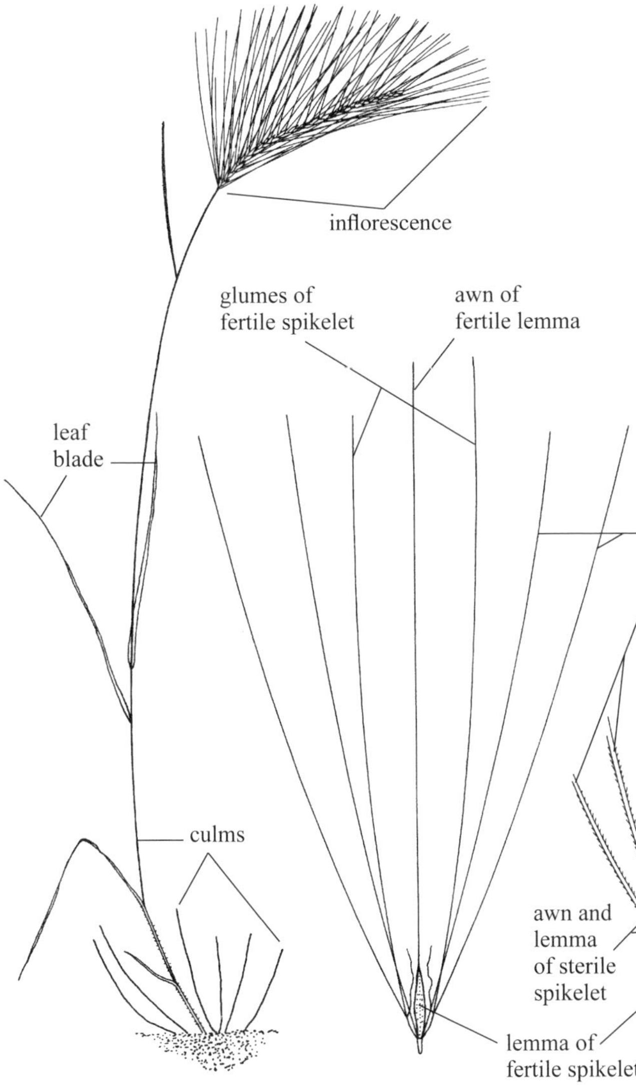

Squirreltail, Foxtail Barley

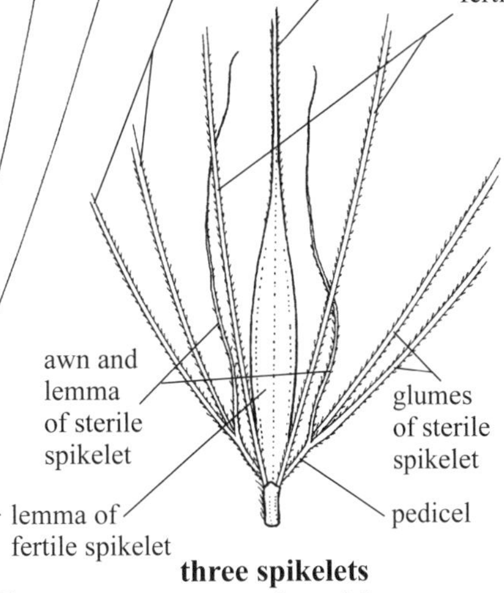

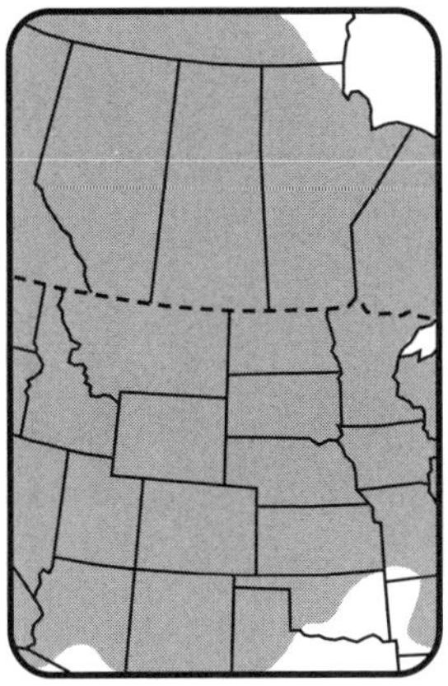

Grass—*Poaceae*

JUNE GRASS *Koeleria macrantha* (Ledeb.) J.A. Schultes

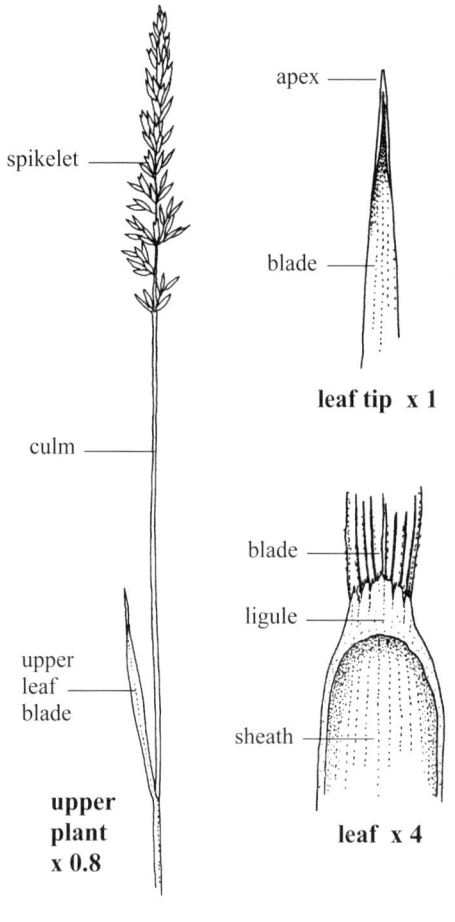

leaf tip x 1

leaf x 4

upper plant x 0.8

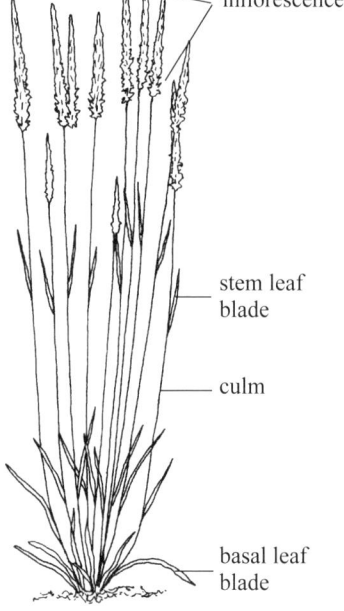

plant x 0.3

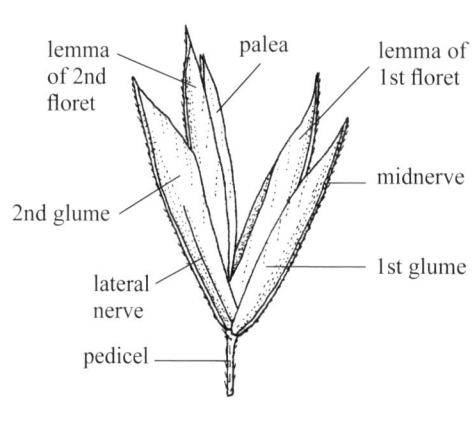

spikelet x 8

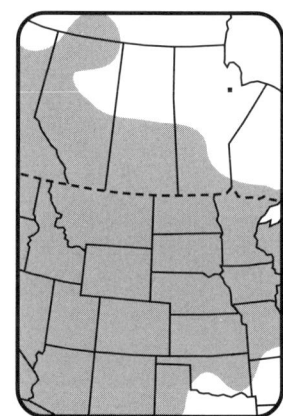

- **SKETCH** A tufted, variable **perennial grass** 10–60 cm tall; in dry prairies, open woodlands, sagebrush, rocky alpine slopes and clearings. **p1**
- **FLOWERS** Light green, blooming in May–August; **inflorescence** a panicle (spikelike) 3–20 cm long by 8–18 mm wide, often interrupted below; **main branches** ascending, slightly hairy; **rachis** green, hairs abundant, short, ascending; **pedicels** to *c.* 1 mm long, green to reddish, hairs ascending; **spikelets** usually 2-flowered (to 4-, but 4th floret vestigial), each 4–5.5 mm long by *c.* 1 mm thick, flattened laterally, breaking off above glumes; **glumes** two, unequal, awnless but pointed; **1st glume** 2–4.8 mm long, 1-nerved, scabrous along the green midnerve, hyaline, membranous; **2nd glume** 2.5–5.5 mm long, 1- or 3-nerved, hyaline, midnerve scabrous; **lemmas** 2.5–5.2 mm long, 5-nerved, lateral nerves faint, pointed and awnlike, midnerve slightly scabrous, margins hyaline; **paleae** hyaline, shiny, minutely shorter than the lemmas, bidentate; **stamens** three; **anthers** 1.2–2 mm long, yellow. **FRUIT a grain**, 1-seeded, yellowish brown, *c.* 0.5 mm long by *c.* 0.3 mm wide.
- **LEAVES Basal blades** 4–20 cm long by 2–3.5 mm wide, flat, tip keeled, lowest ones often tan and wilting by anthesis; **stem blades** usually two along the culm, 3–6 cm long by 1–3.7 mm wide, flat to involute to folded, tips prow-shaped, light bluish green, glabrous to hairy below, broad raised ribs and deep furrows above ligule; **ligules** white, 0.5–2 mm long, membranous, tips ragged and ciliate; **auricles** absent; **collars** glabrous to long hairy; **sheaths** open to the base, round to slightly keeled, smooth or with short to long stiff hairs especially near the top of the sheath.
- **STEM** A **culm**, hollow, round, glabrous or scabrous below panicle, green, erect; 1–2 mm thick near the base; **nodes** with descending microscopic hairs.
- **SYN.** *Koeleria cristata* auct. p.p. non Pers., *Koeleria gracilis* Pers., *Koeleria nitida* Nutt. and *Koeleria pyramidata* auct. p.p. non (Lam.) Beauv.

Grass—*Poaceae* **Grass** Glumes absent Spikelets 1-flowered

Rice Cut Grass *Leersia oryzoides* (L.) Sw.

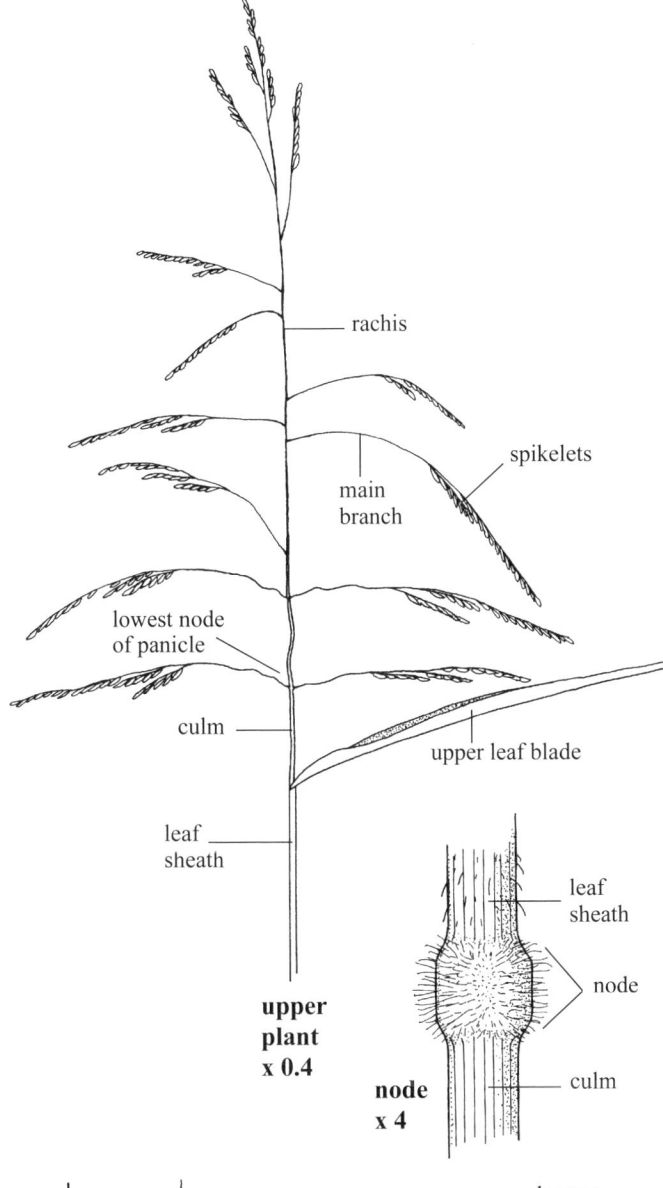

upper plant x 0.4

node x 4

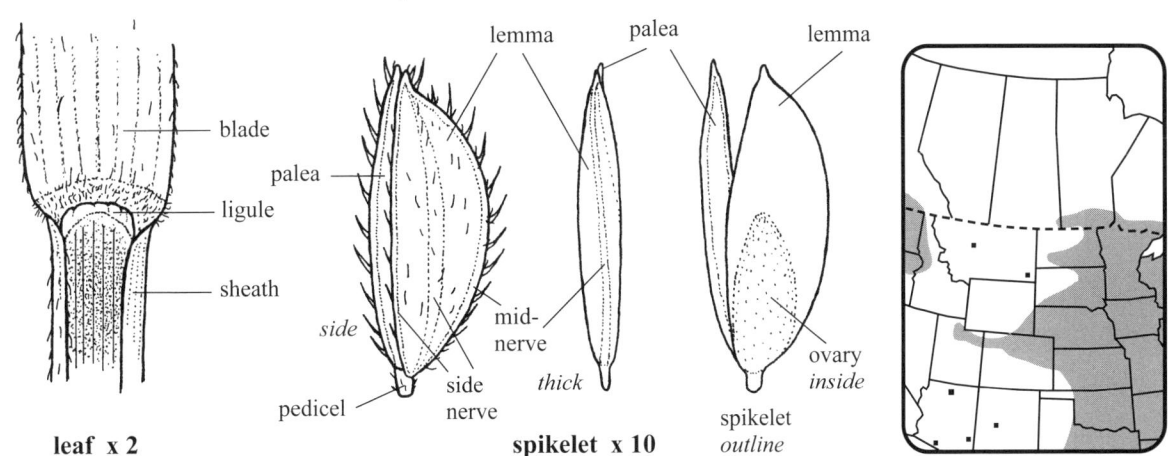

leaf x 2 spikelet x 10 spikelet *outline*

- **SKETCH** A tufted **perennial grass** 40–160 cm tall with long slender **rhizomes**; in ditches and moist meadows, along banks of rivers and streams, sometimes in colonies.
- **FLOWERS** Green, blooming from June–November; **inflorescence** a panicle, open, 10–30 cm long by 12–18+ cm wide, sometimes the lower branches enclosed in the upper leaf sheath; **main branches** 3–12 cm long, usually one or two per node or whorled, spreading to drooping; **spikelets** 1-flowered, 3.5–5.5 mm long by 1.4–1.8 mm wide by *c.* 0.5 mm thick, subsessile, laterally compressed, asymmetrical, imbricate along the outer half of the branches; **glumes** absent; **lemmas** 5-nerved, *c.* 4 mm long with rigid hairs ascending along the mid- and side nerves; **paleae** 3-nerved, hispid especially along the outer nerve, minutely longer than and fitting into the lemmas; **stamens** three (one or two); **anthers** 0.2–2.4 mm long (variable), pale yellow. **FRUIT a grain**, 1-seeded, reddish brown to tan, 2.5–3 mm long.
- **LEAVES** Usually several; **blades** flat, 7–30 cm long by 4–15 mm wide, reduced below and wilting by anthesis, glabrous to scabrous above, scabrous below and especially along the margins from downward pointing hairs; **ligules** 0.5–2 mm long, tan, membranous, flat-topped and often split; **auricles** absent; **sheaths** scabrous, slightly keeled and hairy at collar.
- **STEM** A **culm**, erect to leaning, weak, round, hollow, glabrous; 2–3 mm wide at the sometimes decumbent base; **nodes** hairy; **tillers** one or two, each 30–60 cm long from the lower nodes.

Cut Grass 653

Grass—*Poaceae* **Grass** Spikelets fuzzy, 3- to 5-flowered

Hairy Wild Rye *Leymus innovatus* (Beal) Pilger

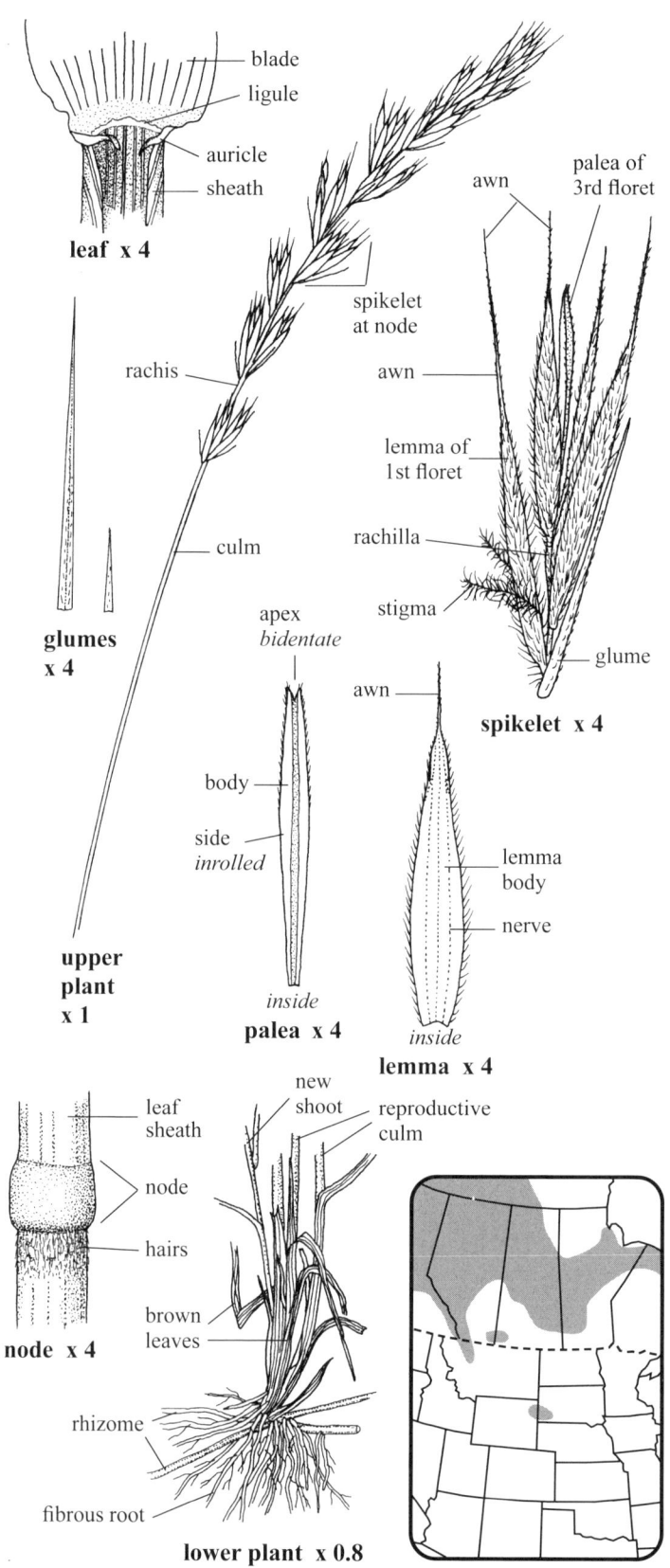

■ **SKETCH** A tufted or solitary **perennial grass** 40–135 cm tall from whitish tan **rhizomes** to *c.* 10 cm long by 1–2.5 mm thick, covered with dark brown scales; **fibrous roots** 1–6 cm long; in clearings, along roadsides and woodland openings.

■ **FLOWERS** Grayish green, blooming June–July; **inflorescence** a spike, nodding to erect at anthesis, 5–12 cm long by 10–15 mm wide, fuzzy; **spikelets** 3- to 5-flowered, spreading at anthesis, more erect with fruit, imbricate, sessile, usually two (1–3) per node, each 10–19 mm long (including awns) by 4–5 mm wide, lowest one often remote and smaller; **glumes** hairy, awnlike, one or two (sometimes absent), green, turning purple, one of the pair much longer, 0.2–11 mm long by 0.2–0.6 mm wide, stiff, 1-nerved, tapered to a very fine point, margins hyaline and revolute, very hairy to almost hairless; **rachilla** pale green and hairy; **lemmas** green to purple, stiff, outside covered with long white hairs to base of awn, hairless inside, 3- or 5-nerved, nerves mostly hidden by outside hairs, body 7–10 mm long by *c.* 1 mm wide, to *c.* 2 mm wide (unrolled and flat), hyaline near apices, two purple lines may run into teeth at the apices; **awns** 0.5–11 mm long, reduced or absent in the upper florets; **paleae** about as long as body of lemmas, white, margins involute, 2-nerved, the nerves green and hairy near the bidentate apices; **stamens** three, exserted; **anthers** yellow and purplish red, 3.5–6 mm long; **stigmas** two, feathery, *c.* 3 mm long, exserted. **FRUIT a grain**, 1-seeded, apex hairy.

■ **LEAVES** Two or three culm leaves, flat, 1.5–7 cm long by 3–12 mm wide, ascending, margins scabrous, top blade well below the inflorescence, lower blades congested and brown; **blades** of new vegetative shoots to *c.* 30 cm long and to *c.* 6 mm wide; **ligules** a hairless hyaline membrane, 0.1–0.3 mm long or to *c.* 2 mm long and ciliate, slightly ragged; **auricles** 1–1.5 mm long, curved, on young plants and lower leaves; **sheaths** 7–18 cm long, reduced below, open to base.

■ **STEM** A culm, glaucous, smooth, hollow, stiff; 1.8–3 mm wide near the base; **nodes** dark green, glabrous, lower nodes swollen to *c.* 4 mm wide, white hairs below nodes appressed and tangled for *c.* 2 mm.

■ **SYN.** *Elymus innovatus* Beal.

Downy Ryegrass, Downy Lyme Grass, Fuzzyspike Wild Rye

Grass—*Poaceae* Grass Spikelets 4- to 15-flowered

PERENNIAL RYE GRASS *Lolium perenne* L.

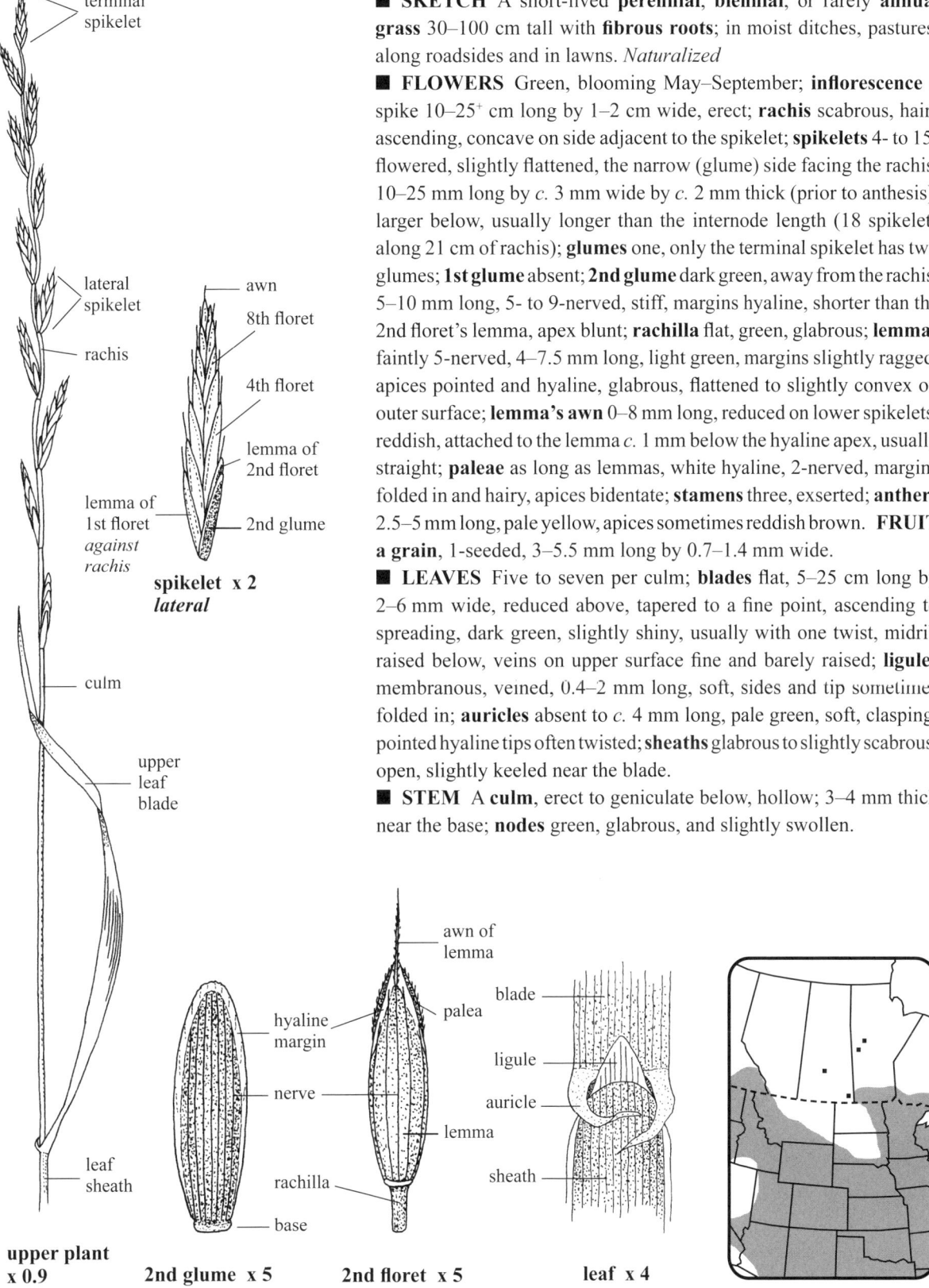

■ **SKETCH** A short-lived **perennial**, **biennial**, or rarely **annual grass** 30–100 cm tall with **fibrous roots**; in moist ditches, pastures, along roadsides and in lawns. *Naturalized*

■ **FLOWERS** Green, blooming May–September; **inflorescence** a spike 10–25+ cm long by 1–2 cm wide, erect; **rachis** scabrous, hairs ascending, concave on side adjacent to the spikelet; **spikelets** 4- to 15-flowered, slightly flattened, the narrow (glume) side facing the rachis, 10–25 mm long by *c.* 3 mm wide by *c.* 2 mm thick (prior to anthesis), larger below, usually longer than the internode length (18 spikelets along 21 cm of rachis); **glumes** one, only the terminal spikelet has two glumes; **1st glume** absent; **2nd glume** dark green, away from the rachis, 5–10 mm long, 5- to 9-nerved, stiff, margins hyaline, shorter than the 2nd floret's lemma, apex blunt; **rachilla** flat, green, glabrous; **lemmas** faintly 5-nerved, 4–7.5 mm long, light green, margins slightly ragged, apices pointed and hyaline, glabrous, flattened to slightly convex on outer surface; **lemma's awn** 0–8 mm long, reduced on lower spikelets, reddish, attached to the lemma *c.* 1 mm below the hyaline apex, usually straight; **paleae** as long as lemmas, white hyaline, 2-nerved, margins folded in and hairy, apices bidentate; **stamens** three, exserted; **anthers** 2.5–5 mm long, pale yellow, apices sometimes reddish brown. **FRUIT** a grain, 1-seeded, 3–5.5 mm long by 0.7–1.4 mm wide.

■ **LEAVES** Five to seven per culm; **blades** flat, 5–25 cm long by 2–6 mm wide, reduced above, tapered to a fine point, ascending to spreading, dark green, slightly shiny, usually with one twist, midrib raised below, veins on upper surface fine and barely raised; **ligules** membranous, veined, 0.4–2 mm long, soft, sides and tip sometimes folded in; **auricles** absent to *c.* 4 mm long, pale green, soft, clasping, pointed hyaline tips often twisted; **sheaths** glabrous to slightly scabrous, open, slightly keeled near the blade.

■ **STEM** A **culm**, erect to geniculate below, hollow; 3–4 mm thick near the base; **nodes** green, glabrous, and slightly swollen.

English Rye Grass

Grass—*Poaceae* **Grass** Spikelets 1-flowered (2-)

SCRATCH GRASS *Muhlenbergia asperifolia* (Nees & Meyen ex Trin.) Parodi

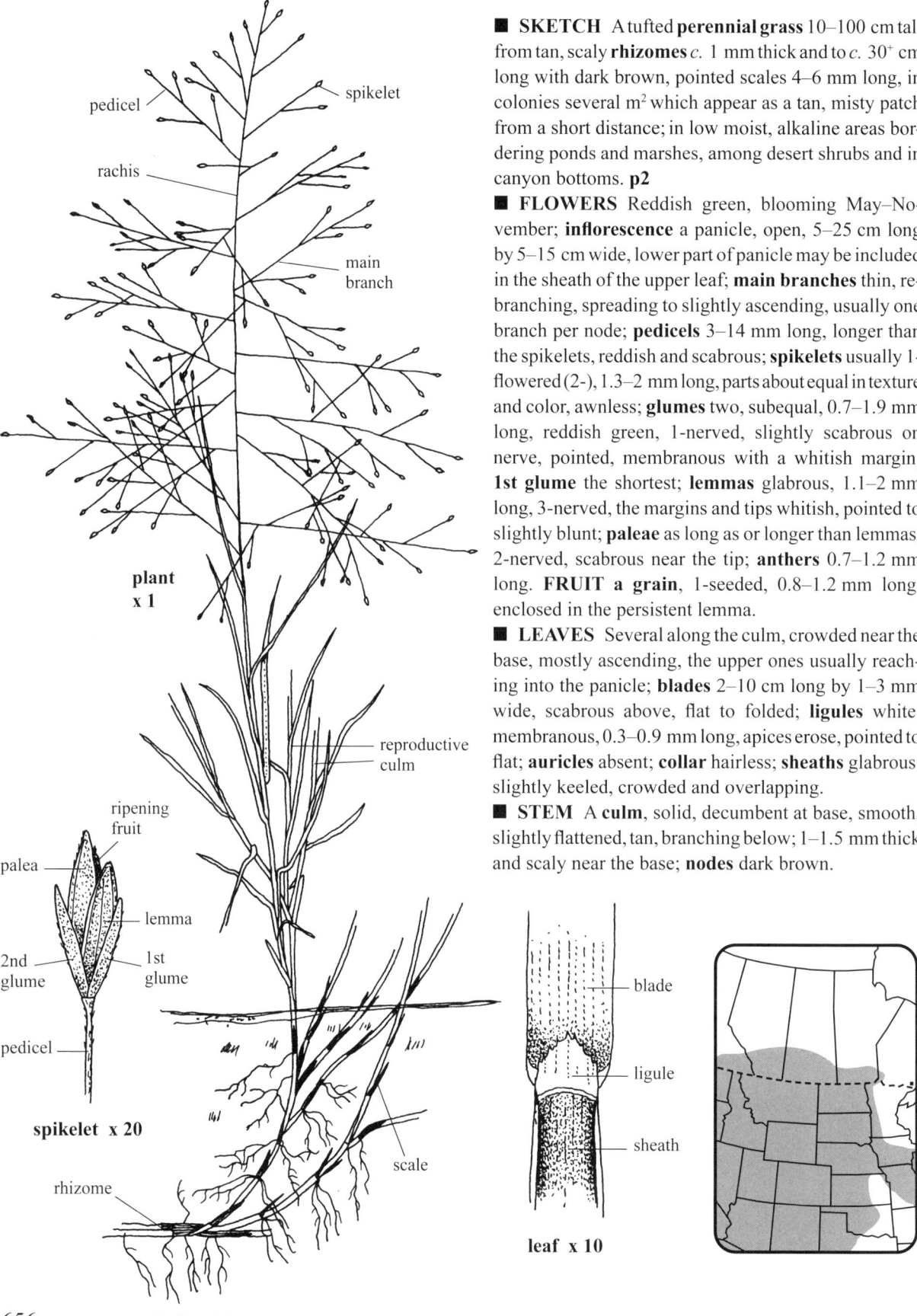

■ **SKETCH** A tufted **perennial grass** 10–100 cm tall from tan, scaly **rhizomes** *c.* 1 mm thick and to *c.* 30+ cm long with dark brown, pointed scales 4–6 mm long, in colonies several m² which appear as a tan, misty patch from a short distance; in low moist, alkaline areas bordering ponds and marshes, among desert shrubs and in canyon bottoms. **p2**

■ **FLOWERS** Reddish green, blooming May–November; **inflorescence** a panicle, open, 5–25 cm long by 5–15 cm wide, lower part of panicle may be included in the sheath of the upper leaf; **main branches** thin, rebranching, spreading to slightly ascending, usually one branch per node; **pedicels** 3–14 mm long, longer than the spikelets, reddish and scabrous; **spikelets** usually 1-flowered (2-), 1.3–2 mm long, parts about equal in texture and color, awnless; **glumes** two, subequal, 0.7–1.9 mm long, reddish green, 1-nerved, slightly scabrous on nerve, pointed, membranous with a whitish margin; **1st glume** the shortest; **lemmas** glabrous, 1.1–2 mm long, 3-nerved, the margins and tips whitish, pointed to slightly blunt; **paleae** as long as or longer than lemmas, 2-nerved, scabrous near the tip; **anthers** 0.7–1.2 mm long. FRUIT a grain, 1-seeded, 0.8–1.2 mm long, enclosed in the persistent lemma.

■ **LEAVES** Several along the culm, crowded near the base, mostly ascending, the upper ones usually reaching into the panicle; **blades** 2–10 cm long by 1–3 mm wide, scabrous above, flat to folded; **ligules** white, membranous, 0.3–0.9 mm long, apices erose, pointed to flat; **auricles** absent; **collar** hairless; **sheaths** glabrous, slightly keeled, crowded and overlapping.

■ **STEM** A **culm**, solid, decumbent at base, smooth, slightly flattened, tan, branching below; 1–1.5 mm thick and scaly near the base; **nodes** dark brown.

Alkali Muhly

Grass—*Poaceae* Grass Spikelets 1-flowered

Prairie Muhly *Muhlenbergia cuspidata* (Torr. ex Hook.) Rydb.

■ **SKETCH** A tufted **perennial grass** 20–70 cm tall from thickened, dark brown, hard culm bases and tan **roots** along a dark brown, horizontal **rhizome** 2–4 mm thick; in dry to moist prairies, often on upper slopes, and crests of river and ravine banks. p2

■ **FLOWERS** Green and purplish, blooming June–October; **inflorescence** a panicle, contracted, 5–15 cm long by 2–3 mm wide; **main branches** usually one at each node of the panicle, erect, reduced above; **rachis** ridged longitudinally, minutely hairy; **spikelets** 1-flowered, 2.5–4 mm long by *c.* 0.8 mm wide (before anthesis), slightly compressed laterally; **glumes** two, 1-nerved, light tan, subequal to equal, pointed, slightly scabrous on midnerve or not, 1.2–3.2 mm long, shorter than to less than half of lemma's length; **lemmas** purplish, 3-nerved, pointed to short-awned, 2.5–4 mm long by *c.* 1 mm wide (flattened), margins hyaline, pliable, glabrous to slightly hairy at the base and along lower midnerve or even on upper body; **paleae** slightly shorter than lemmas, 2-nerved, bidentate, body transparent, soft, hairless; **stamens** three; **anthers** 1.2–2 mm long, pale greenish yellow; **stigmas** two, reddish purple; **galls** (from insects in grain) greenish yellow, shiny, glabrous, 2–2.5 mm long by 1–1.5 mm wide. **FRUIT a grain**, 1-seeded, 1.5–2.3 mm long, brown.

■ **LEAVES** Ascending, several per culm, glabrous, smooth, flat to V-shaped; **blades** 2.5–23 cm long by 1–3 mm wide, reduced above and below, upper leaf blade usually not reaching the panicle, microscopically hairy and scabrous on inside ridges and margins, glabrous on outside (dorsally); **ligules** white, soft, membranous, 0.2–0.7 mm long, apices truncate and ragged; **auricles** absent; **sheaths** open, flattened around the oval culm, slightly keeled, glabrous with margins scarious.

■ **STEM** A **culm**, green, erect, solid, oval below the inflorescence, up to four reproductive culms from the lower leaf axil, slightly scabrous below the panicle, hairs white and ascending, longitudinal ridges numerous; 1–1.5 mm thick near the base; **nodes** mostly hidden by lower sheaths, upper nodes pale green and slightly swollen.

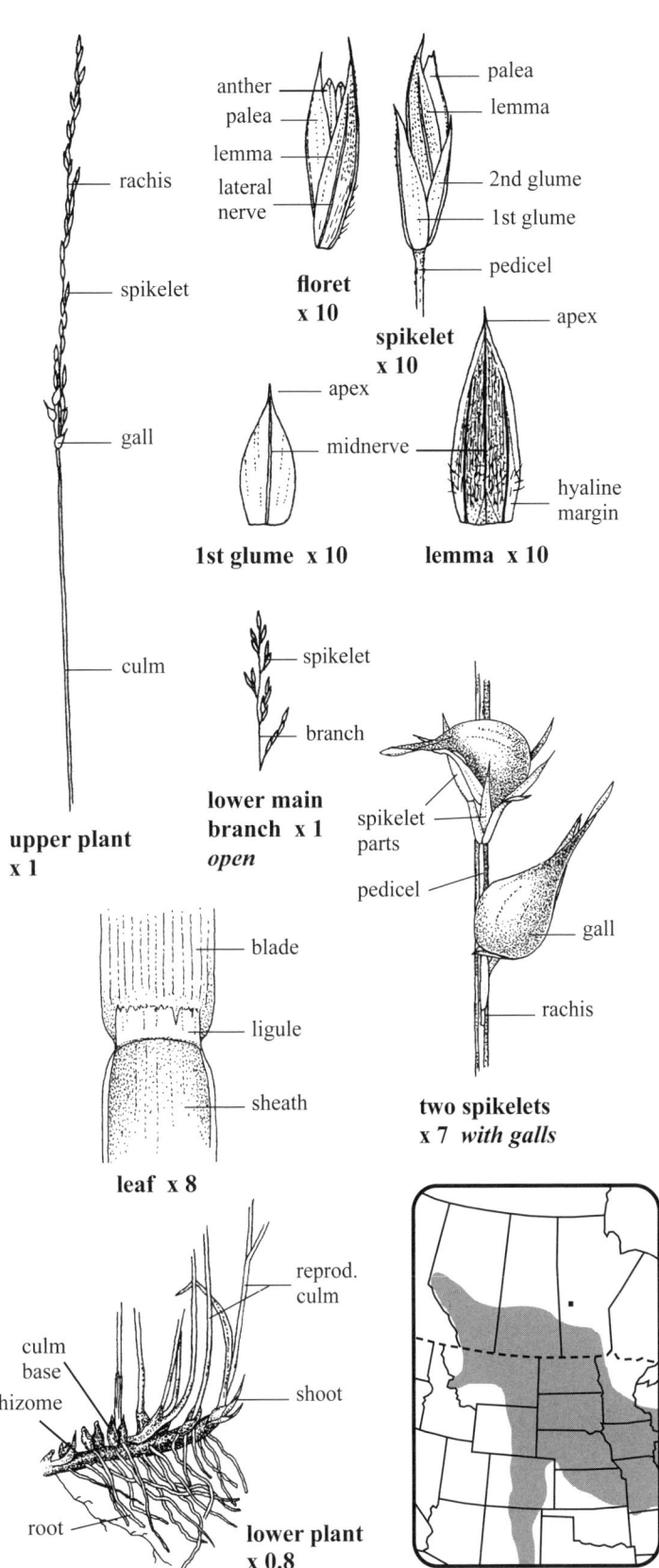

Stony-hills Muhly, Plains Muhly

Grass—*Poaceae* Grass Spikelets 1-flowered (2-)

MAT MUHLY *Muhlenbergia richardsonis* (Trin.) Rydb.

■ **SKETCH** A tufted **perennial grass** 5–70 cm tall from tan, smooth to scaly, stiff **rhizomes** 0.6–1 mm thick; in most to dry prairies, parklands, talus slopes, shores of rivers and saline sites. **p2**

■ **FLOWERS** Tan and gray, blooming July–September; **inflorescence** a panicle 2–15 cm long by 2–4 mm wide at anthesis; **main branches** erect, glabrous, 0.5–5 cm long, one or two per node with one branch usually much longer than the other branch; **rachis** glabrous; **pedicels** glabrous, 0.5–1.5 mm long; **spikelets** 1-flowered (rarely 2-flowered), each 2–3 mm long; **florets** perfect; **glumes** two, subequal, 1-nerved, 0.8–1.8 mm long, slightly scabrous along midnerve, stiff; **lemmas** acute, slightly scabrous, 2–3 mm long, tan with gray patches on sides and at apices, 3- or 5-nerved, lateral nerves faint and not merging at the apex, the midnerve sometimes extended into an apical **awn** 0.2–0.3 mm long; **paleae** 1.2–2.9 mm long, usually reaching to the base of the lemma's awn, apices slightly ragged, not obviously bidentate; **anthers** 1–1.6 mm long, yellow to purple; **stigmas** feathery, deep reddish purple. **FRUIT a grain**, 1-seeded, 1–1.5 mm long by *c.* 0.5 mm wide, brown.

■ **LEAVES** Five to seven per culm; **blades** ascending, 1–6.5 cm long by 0.5–4 mm wide, not overlapping, usually brown near the apices, the lower blades tan by anthesis, flat to involute, upper surface short hairy, glabrous below; **ligules** membranous, whitish tan, dry, apices pointed to flat and slightly ragged, hairless, 1–3 mm long, continuous with the sheath's hyaline margins; **collars** tan, not hairy; **sheaths** not keeled, stiff, open, white hyaline along both hairless margins, glabrous to minutely rough.

■ **STEM** A **culm**, solid, erect to decumbent, wiry, green, roundish below to slightly flattened above, minute ascending hairs near the base and above; minutely rough especially near the 0.5–1 mm thick base.

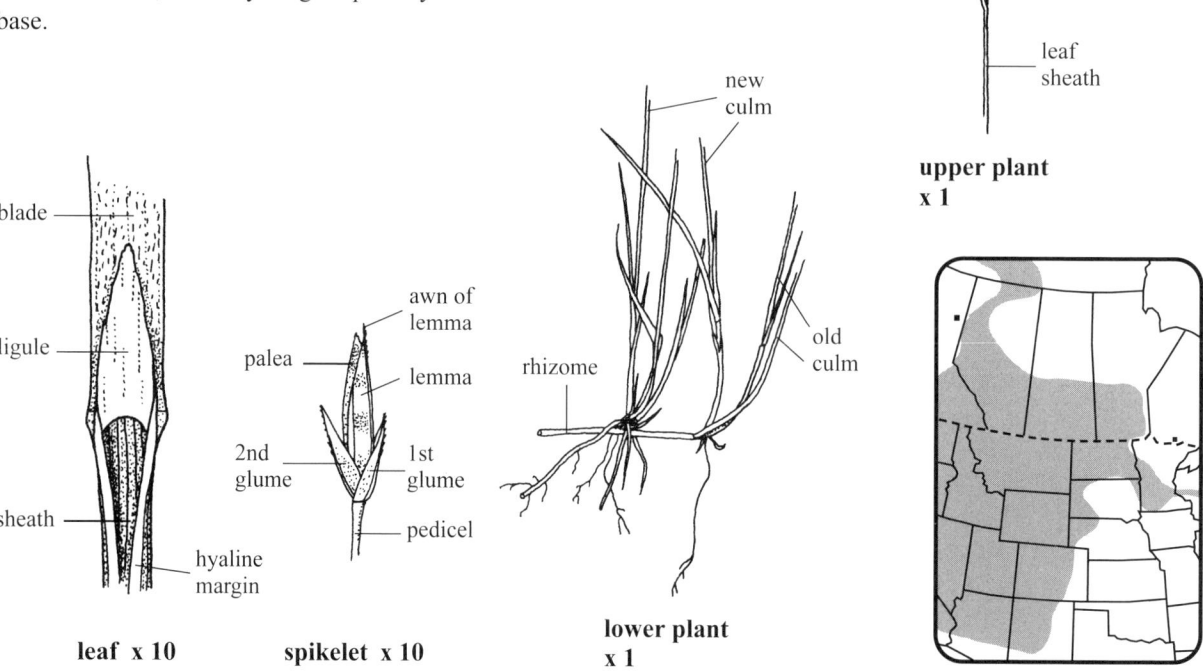

Matted Muhly

Green Needle Grass *Nassella viridula* (Trin.) Barkworth

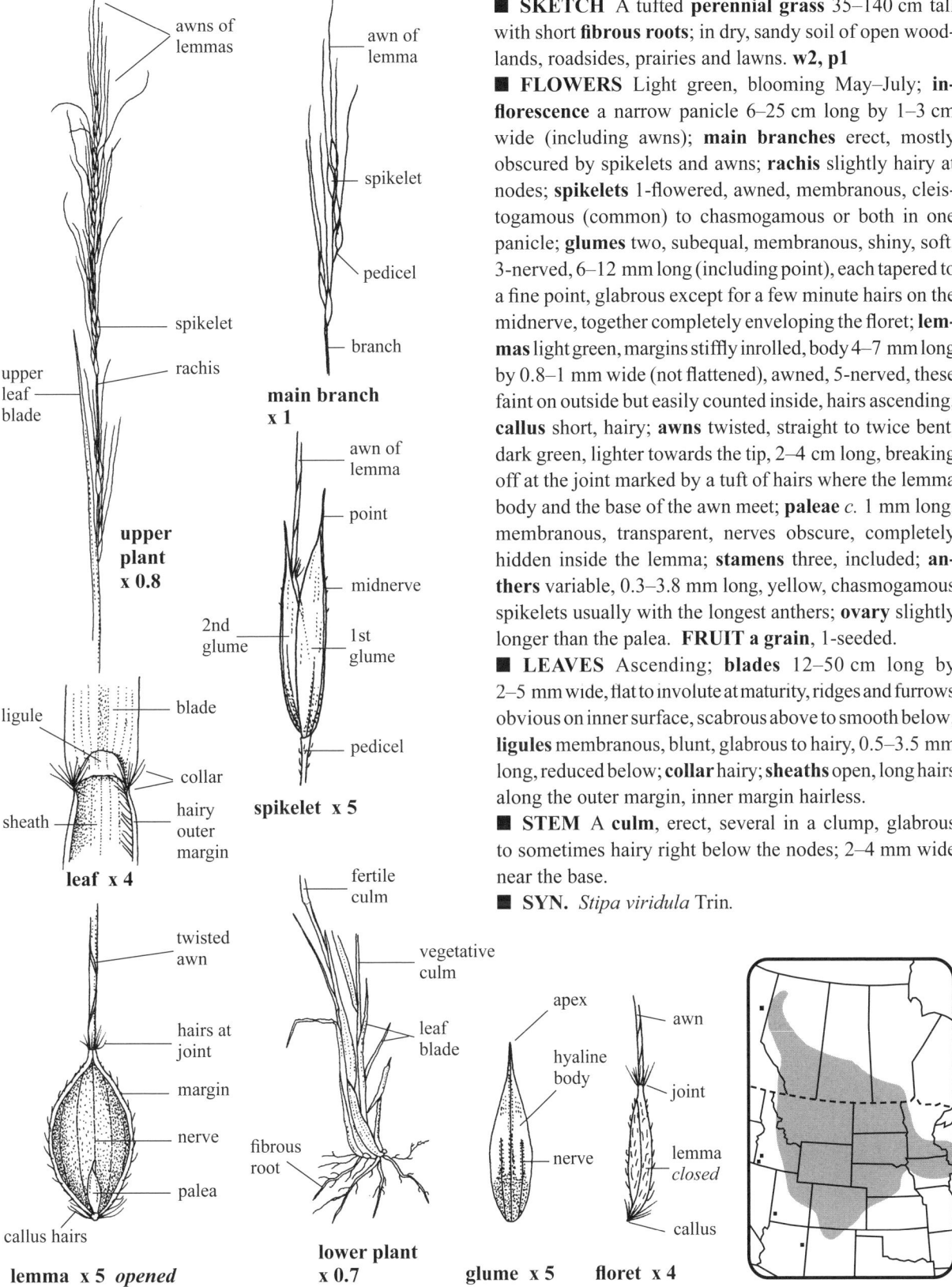

■ **SKETCH** A tufted **perennial grass** 35–140 cm tall with short **fibrous roots**; in dry, sandy soil of open woodlands, roadsides, prairies and lawns. **w2, p1**

■ **FLOWERS** Light green, blooming May–July; **inflorescence** a narrow panicle 6–25 cm long by 1–3 cm wide (including awns); **main branches** erect, mostly obscured by spikelets and awns; **rachis** slightly hairy at nodes; **spikelets** 1-flowered, awned, membranous, cleistogamous (common) to chasmogamous or both in one panicle; **glumes** two, subequal, membranous, shiny, soft, 3-nerved, 6–12 mm long (including point), each tapered to a fine point, glabrous except for a few minute hairs on the midnerve, together completely enveloping the floret; **lemmas** light green, margins stiffly inrolled, body 4–7 mm long by 0.8–1 mm wide (not flattened), awned, 5-nerved, these faint on outside but easily counted inside, hairs ascending; **callus** short, hairy; **awns** twisted, straight to twice bent, dark green, lighter towards the tip, 2–4 cm long, breaking off at the joint marked by a tuft of hairs where the lemma body and the base of the awn meet; **paleae** *c.* 1 mm long, membranous, transparent, nerves obscure, completely hidden inside the lemma; **stamens** three, included; **anthers** variable, 0.3–3.8 mm long, yellow, chasmogamous spikelets usually with the longest anthers; **ovary** slightly longer than the palea. **FRUIT a grain**, 1-seeded.

■ **LEAVES** Ascending; **blades** 12–50 cm long by 2–5 mm wide, flat to involute at maturity, ridges and furrows obvious on inner surface, scabrous above to smooth below; **ligules** membranous, blunt, glabrous to hairy, 0.5–3.5 mm long, reduced below; **collar** hairy; **sheaths** open, long hairs along the outer margin, inner margin hairless.

■ **STEM** A culm, erect, several in a clump, glabrous to sometimes hairy right below the nodes; 2–4 mm wide near the base.

■ **SYN.** *Stipa viridula* Trin.

Green Tussock Grass

WHITE-GRAINED MOUNTAIN RICE GRASS *Oryzopsis asperifolia* Michx.

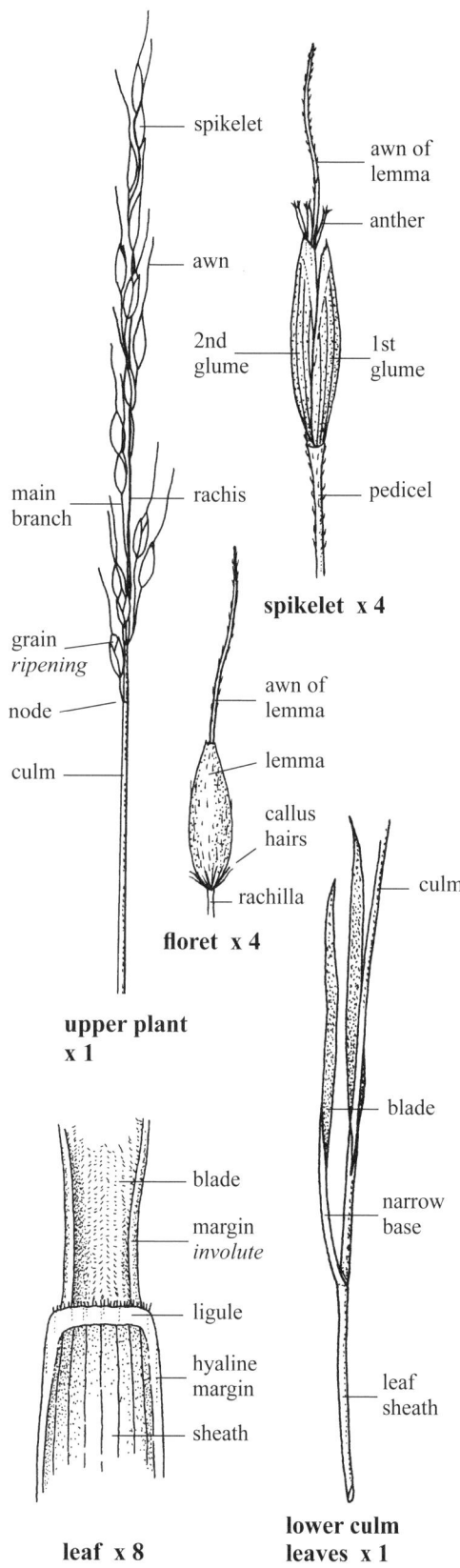

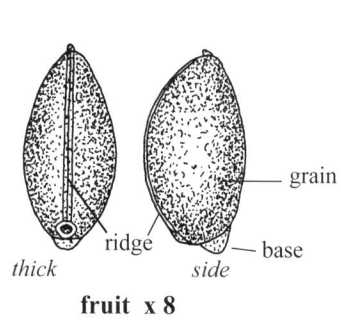

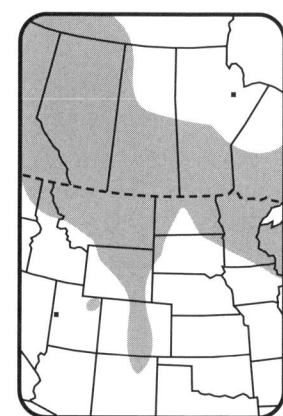

■ **SKETCH** A tufted **perennial grass** 25–100 cm tall from short **rhizomes**; in dry to moist deciduous to coniferous woodlands and parklands. **p2**

■ **FLOWERS** Green, blooming May–July; **inflorescence** a raceme or panicle 3.5–10 cm long by 5–10 mm wide, erect; **main branches** usually one or two per node, erect, appressed, or sometimes spreading; **pedicels** 5–11 mm long; **spikelets** 1-flowered, 10–20, each 5–9 mm long (not including awn) by 1.4–2 mm wide, one to three at the end of each main branch; **glumes** two, subequal, 5.5–8 mm long, C-shaped, enveloping the floret, not awned, soft, somewhat laterally compressed, margins and apices hyaline, apices pointed but slightly ragged, 7- to 11-nerved, the nerves not converging at the apices but ending at the hyaline margins; **lemmas** awned, pale green, the 7 or 9 nerves more visible on the inside and converging to the base of the awn, soft, 5–8 mm long by *c.* 1.5 mm wide, enveloping the paleae, persistent around fruit, hairs appressed but not abundant, callus hairs white and in a tuft *c.* 1 mm long; **awn of lemmas** 5–13 mm long, slightly curved but ascending, hairs microscopic and ascending; **paleae** 2-nerved, pale green, 5.3–6 mm long by *c.* 0.5 mm wide, slightly shorter than the lemmas, not flattened, rolled inward, tips blunt; **anthers** 3–4 mm long. **FRUIT a grain**, 1-seeded, slightly rough, 3–3.5 mm long by 1.5–1.8 mm wide and thick, round in cross-section, one long ridge on one side.

■ **LEAVES** On stem and from tillers, bluish green, evergreen; **blades** with numerous strong, short, hairy ridges above, smooth below, often involute at the narrow tapered base, flat to somewhat revolute towards the middle of the longer blades, with a stiff appearance, glaucous above, lower blades scabrous, 0.5–40 cm long by 2–10 mm wide, reduced above, erect to appressed; **ligules** white, membranes ciliate and slightly irregular, 0.1–0.4 mm long; **auricles** absent; **sheaths** open, smooth, margins pale green and hyaline, stiff, not hairy.

■ **STEM** A **culm**, smooth to slightly scabrous, hollow, erect to leaning; 1–4 mm thick near the purplish base; **nodes** slightly swollen, light green, glabrous.

Grass—*Poaceae*   Grass   Spikelets 2-flowered

WITCH GRASS   *Panicum capillare* L.

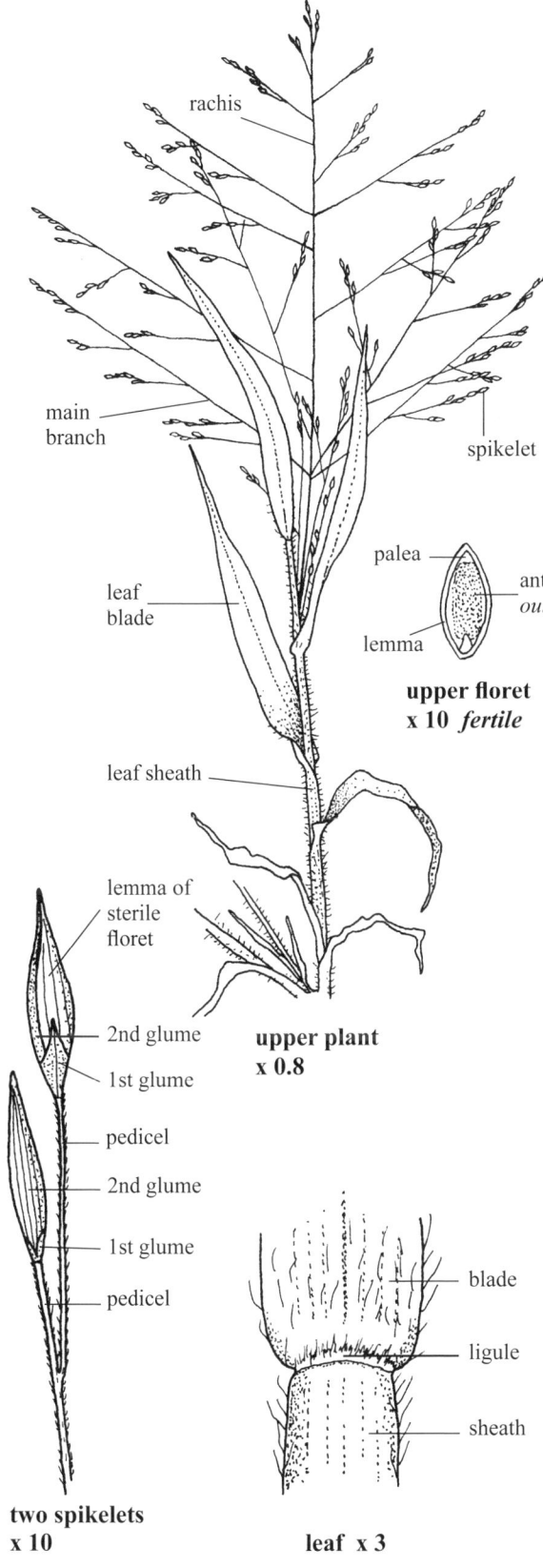

two spikelets
x 10

leaf x 3

main branch
base x 10

■ **SKETCH** A tufted **annual grass** 20–130 cm tall from **fibrous roots**; in disturbed dry sites, canyons, pastures, sandy prairies and along sidewalks. **w1, p1**

■ **FLOWERS** Green and purple, blooming May–November; **inflorescence** a panicle, delicate, open, 4–50 cm long and about as wide with some lower branches erect and included in the upper leaf sheath, breaking off at maturity; **main branches** scabrous, ascending to spreading, to *c.* 18+ cm long, bases of most swollen and hairy at the rachis; **pedicels** 1–30 mm long, scabrous; **spikelets** 2-flowered, glabrous, 2–3.8 mm long by 0.8–1 mm wide before anthesis; **glumes** two, very unequal; **1st glume** pointed, 0.8–1.5 mm long, 3- to 7-nerved, the central green nerve distinct, side nerves faint, often half purple; **2nd glume** 2–3.3 mm long by *c.* 1 mm wide, 7- or 9-nerved, often purplish, the central three nerves well developed; **1) lower floret** sterile, its lemma about as long as the 2nd glume, pointed, with 5- or 7-nerves, glabrous; **paleae** usually absent; **2) upper floret** fertile; **lemmas** smooth, shiny, light green, 1.3–2.1 mm long by 0.6–0.9 mm wide; **paleae** 2-nerved, inside the lemma; **anthers** golden brown, 0.7–1.1 mm long. **FRUIT a grain**, 1-seeded, enclosed in the golden, shiny, hard lemma and palea, 1.5–2 mm long by *c.* 0.7 mm wide.

■ **LEAVES** Alternate, yellowish green; **blades** mostly flat and ascending, 3–40 cm long by 2–18 mm wide, often covered with fine soft hairs above and below, more so near the base, scabrous along the margins; **lower blades** dried and withered; **ligules** 0.5–2 mm long with hairs from a membranous base; **sheaths** flattened and weakly keeled, with soft hairs spreading to slightly ascending and to *c.* 3 mm long.

■ **STEM** A **culm**, decumbent to erect, hollow, slightly flattened, often branched from the base, glabrous except slightly hairy below the inflorescence; 2–5 mm wide near the base; **nodes** hairy.

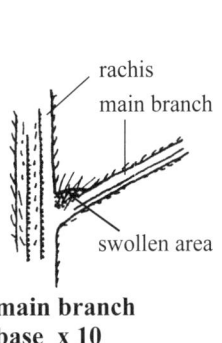
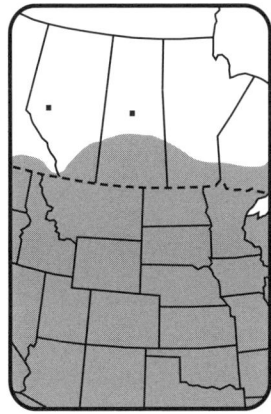

Common Panic Grass, Common Witchgrass

Grass—*Poaceae*

Grass Spikelets 2-flowered

BROOMCORN MILLET *Panicum miliaceum* L.

■ **SKETCH** A tufted **annual grass** 20–120 cm tall from **fibrous roots**; in dry disturbed sites, cultivated fields, along sidewalks and roadsides. *Naturalized* **w1**

■ **FLOWERS** Green, blooming in June–November; **inflorescence** a panicle, dense to open, 10–40 cm long by 5–20 cm wide; **main branches** scabrous and to *c.* 12 cm long, contracted to ascending, rebranched, the lowest ones often partially included in the upper leaf sheath; **spikelets** 2-flowered, glabrous, 4–5.5 mm long to *c.* 2.5 mm wide by *c.* 2 mm thick, somewhat dorsally flattened; **glumes** two, unequal; **1st glume** 2.5–3.5 mm long with a short purple tip, 5- or 7-nerved; **2nd glume** 4–5.5 mm long, 9- to 13-nerved; **florets** of two types: **1) lower floret** sterile; **lemmas** yellowish, minutely shorter than the 2nd glume, membranous, purple tipped, 9- to 15-nerved; **paleae** whitish with a bidentate tip, nerves obscure, *c.* 2 mm long; **2) upper floret** perfect; **lemma** fertile, shiny, plump, 2.7–3.7 mm long by *c.* 2 mm wide, enveloping the smooth palea, these hardening around the fruit; **anthers** 2–3 mm long; **stigmas** *c.* 1 mm long, reddish purple. **FRUIT a grain**, 1-seeded, *c.* 3 mm long by *c.* 2 mm wide, straw-colored to reddish brown to nearly black.

■ **LEAVES** Flat, mostly ascending; **blades** 6–40 cm long by 5–25 mm wide, edges scabrous, glabrous to hairy, the white hairs 2–5 mm long and spreading, these more common below and along the lower margins; **ligules** 1–3 mm long, white hairs on a short base; **auricles** absent; **sheaths** thinly to densely hairy, keeled, open, somewhat flattened, reddish green, the veins distinctive.

■ **STEM** A **culm**, prostrate to ascending, glabrous to hairy, stout, hollow, a dozen or more stalks and tillers per tuft; base 6–8 mm wide by 3–8 mm thick and reddish.

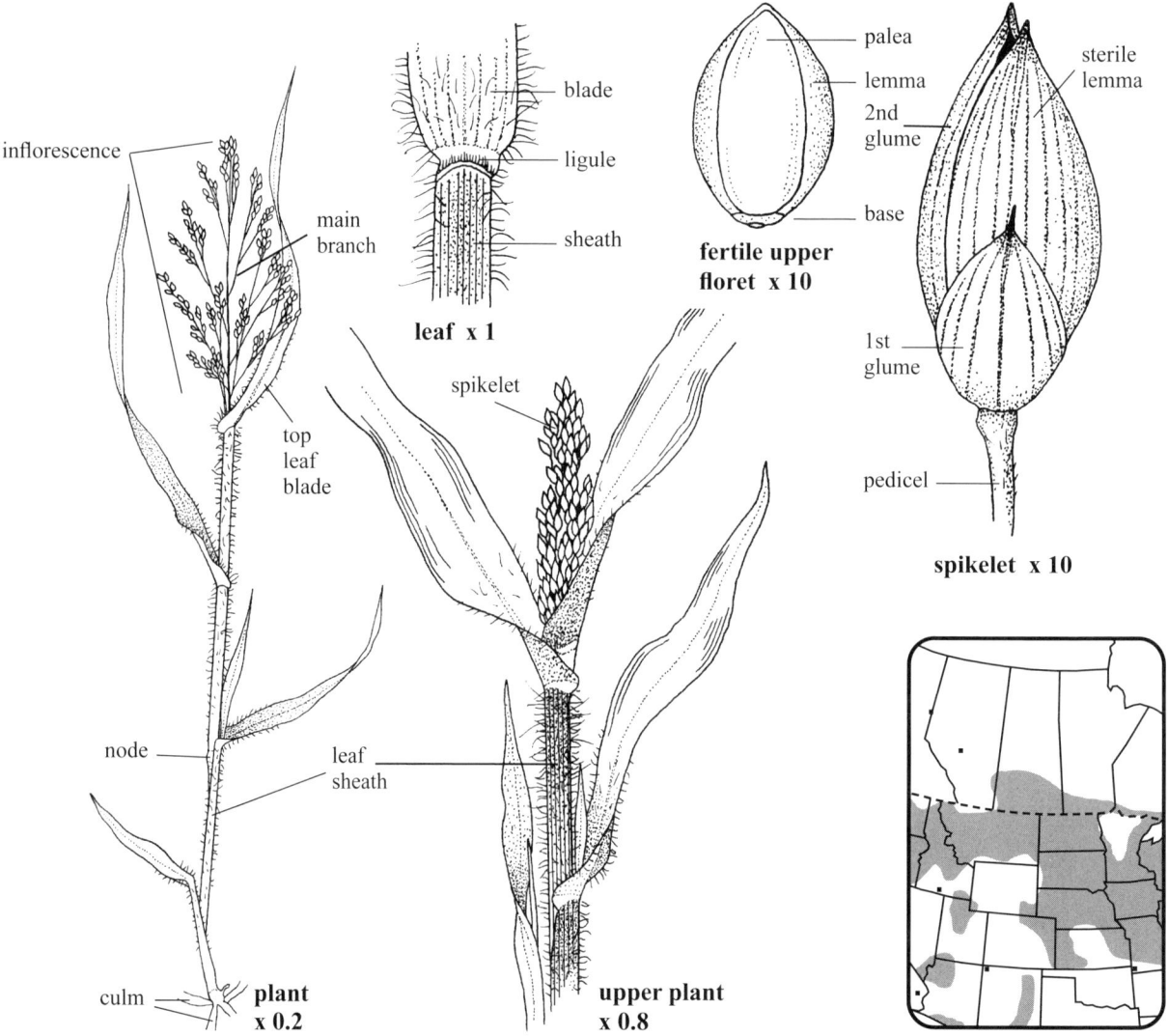

Hog Millet, Proso Millet, Panic Millet

Grass—*Poaceae* | Grass Spikelets 2-flowered

Switch Grass   *Panicum virgatum* L.

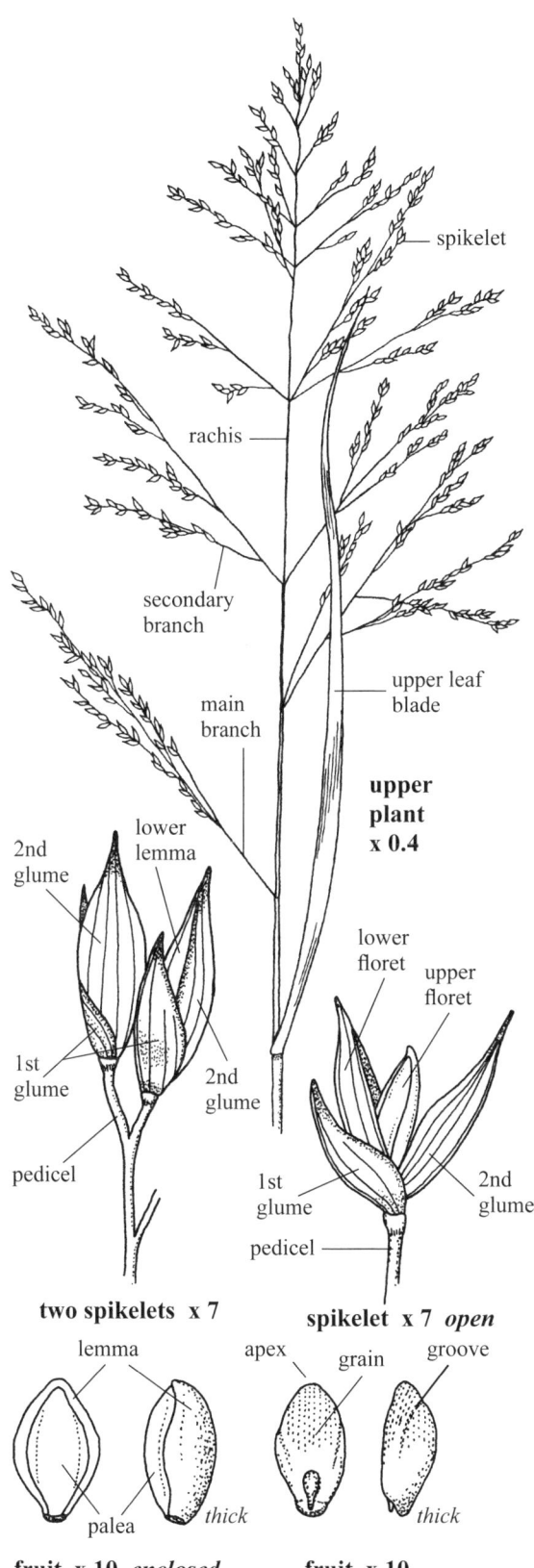

two spikelets x 7

spikelet x 7 *open*

fruit x 10 *enclosed*

fruit x 10

■ **SKETCH** A variable, solitary or tufted **perennial grass** 60–200⁺ cm tall with scaly **rhizomes** 2–3 mm wide and 1–5⁺ cm long; in dry to moist prairies, dunes, rocky stream beds, brackish marshes and open woods. **w2, p2**

■ **FLOWERS** Green and red, blooming June–October; **inflorescence** an open panicle 10–55 cm long by 10–30 cm wide; **main branches** spreading to ascending, the lower branches 7–20 cm long, one to four per node, rebranching, bare of spikelets near base; **rachis** slightly scabrous above; **spikelets** 2-flowered, numerous, appressed, overlapping, glabrous, 3–6 mm long by *c.* 1.5 mm wide and thick; **glumes** two, unequal, membranous, distinctly nerved, reddish purple at apices; **1st glume** 2.5–4.5 mm long by *c.* 2 mm wide (flattened), 3- or 5-nerved; **2nd glume** 7- or 9-nerved, 3–6 mm long by *c.* 2.5 mm wide; **florets** of two types: **1) lower floret** male or sterile; **lemmas** 7-nerved, similar in texture to the glumes, minutely shorter and fitting inside the 2nd glume; **paleae** 2.5–4 mm long, slightly shorter than lemmas and fitting inside, hyaline, 2-nerved, bidentate, margins folded in; **stamens** three, usually present; **anthers** golden brown, 1.5–2.3 mm long; **2) upper floret** perfect, hyaline, glabrous, 2.5–3.5 mm long by 1.2–1.5 mm wide; **lemmas** 3- or 5-nerved, shiny; **paleae** slightly shorter and fitting inside lemmas, 2-nerved, pale green; **anthers** golden tan, 1.5–2 mm long; **stigmas** feathery, turning deep purple at anthesis. **FRUIT a grain**, 1-seeded, enclosed in hard palea and lemma, 1.8–2.8 mm long by 1–1.5 mm wide by *c.* 0.8 mm thick, grooved, apex gray.

■ **LEAVES** Four to six per culm, flat, smooth, slightly scabrous along the margins, arched, the top leaf the shortest and often ascending into the panicle; **blades** 15–60 cm long by 2–15 mm wide, upper surface often with scattered white hairs, especially near the ligule; **ligules** of white hairs 2–6 mm long; **auricles** absent; **collars** white to reddish; **sheaths** mostly glabrous, open, round, often reddish near base or along the hairy margins.

■ **STEM** A **culm**, erect to leaning, firm, hollow; 2–5 mm thick near the base; **nodes** reddish or green with a red area above and below, glabrous to slightly hairy.

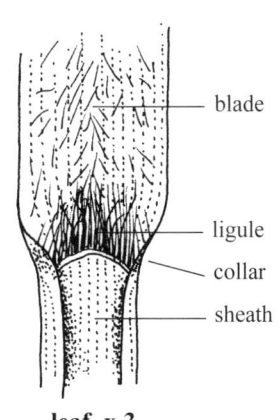

leaf x 3

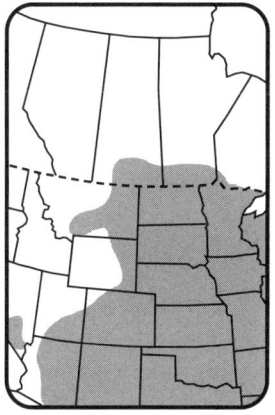

Grass—*Poaceae*   Grass   Spikelets 4- to 11-flowered (16-)

WESTERN WHEATGRASS   *Pascopyrum smithii* (Rydb.) A. Löve

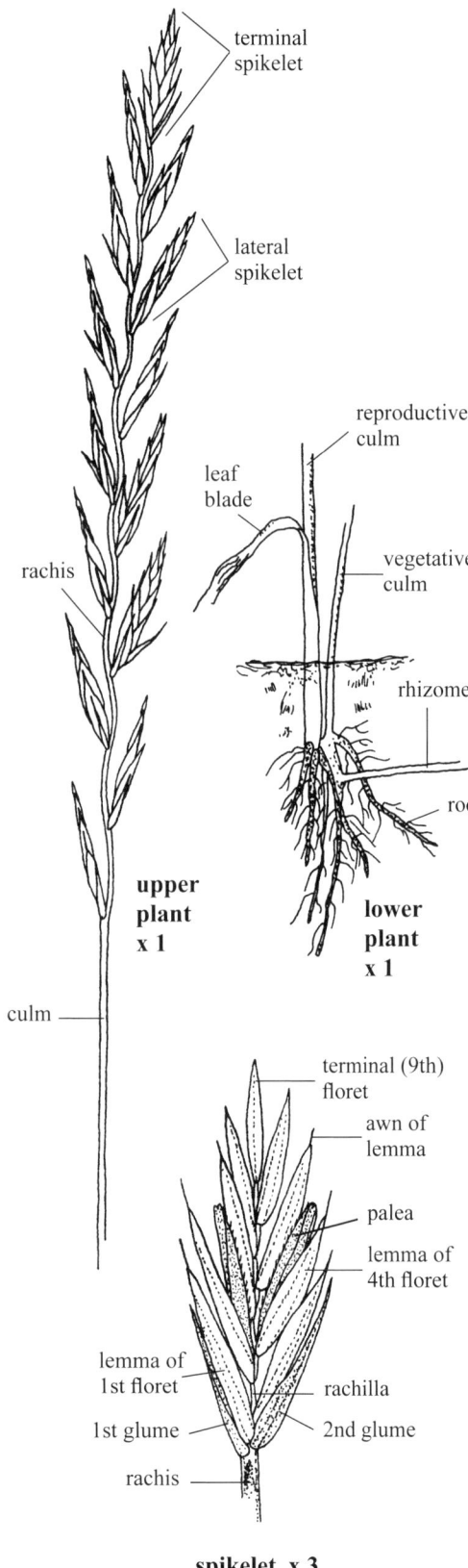

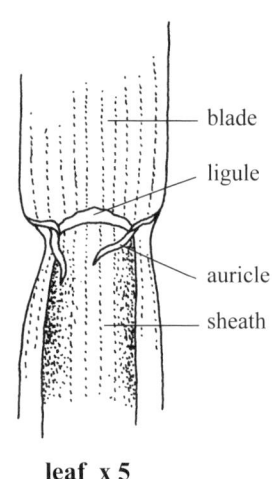

spikelet x 3

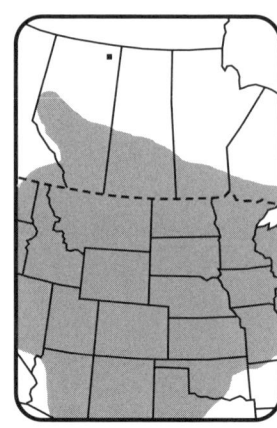

leaf x 5

■ **SKETCH** A glaucous, often glabrous, tufted or solitary **perennial grass** 30–100 cm tall from stiff **rhizomes** *c.* 1 mm thick and to *c.* 20⁺ cm long; in dry to moist prairies, alkaline areas, sagebrush deserts, sandhills, roadsides and along railways. **pl**

■ **FLOWERS** Green, blooming May–September; **inflorescence** a terminal, erect spike 5–24 cm long by 10–16 mm wide; **spikelets** 4- to 11-flowered (16-), sessile, 10–25 mm long, 6–17 spikelets per inflorescence, usually one per node (sometimes paired on lower nodes); **glumes** two, subequal, 3- to 7-nerved, nerves often obscure, strongly C-shaped, stiff, glabrous to hairy, broadest below the middle, edges hyaline, tapered to a short awn or not; **1st glume** 7–12 mm long, glabrous, 3- or 5-nerved, with a fine scabrous point; **2nd glume** 8–15 mm long, 7-nerved, tapered to a fine point; **rachilla** light green, not hairy; **lemmas** 5-nerved, stiff, glabrous to lightly hairy, C-shaped, 5–14 mm long (including awn), reduced above in spikelet, awns 0–4 mm long; **paleae** stiff, white, two nerves green, margins brown and scabrous, apices bidentate, slightly shorter than lemmas, awnless; **stamens** three; **anthers** pale yellow, 3–5 mm long. **FRUIT a grain**, 1-seeded, 4–5 mm long, apex hairy, enclosed in lemma and palea.

■ **LEAVES** Pale bluish green, stiff, four or five per culm, glaucous, ascending, upper leaf usually not reaching the inflorescence; **blades** 3.5–30 cm long by 2–7 mm wide, middle blade the longest and tapered to a long slender tip, margins involute especially in dry weather, ridges separated by deep furrows above, scabrous on the margins and above, glabrous below; **ligules** a narrow, tan membrane 0.2–0.8 mm long, hairless to ciliate, usually not torn; **auricles** tan to purple, 1–2 mm long, straight to curled, sometimes absent; **sheaths** round, stiff, open, usually glabrous or lower ones slightly hairy on margins.

■ **STEM** A **culm**, erect, stiff, hollow, round; 2–3 mm thick near the base; **nodes** brown to purple and glabrous.

■ **SYN.** *Agropyron smithii* Rydb., *Elymus smithii* (Rydb.) Gould and *Elytrigia smithii* (Rydb.) Nevski. Similar to *A. dasystachyum*.

Grass—*Poaceae* Grass Spikelets 3-flowered

REED CANARY GRASS *Phalaris arundinacea* L.

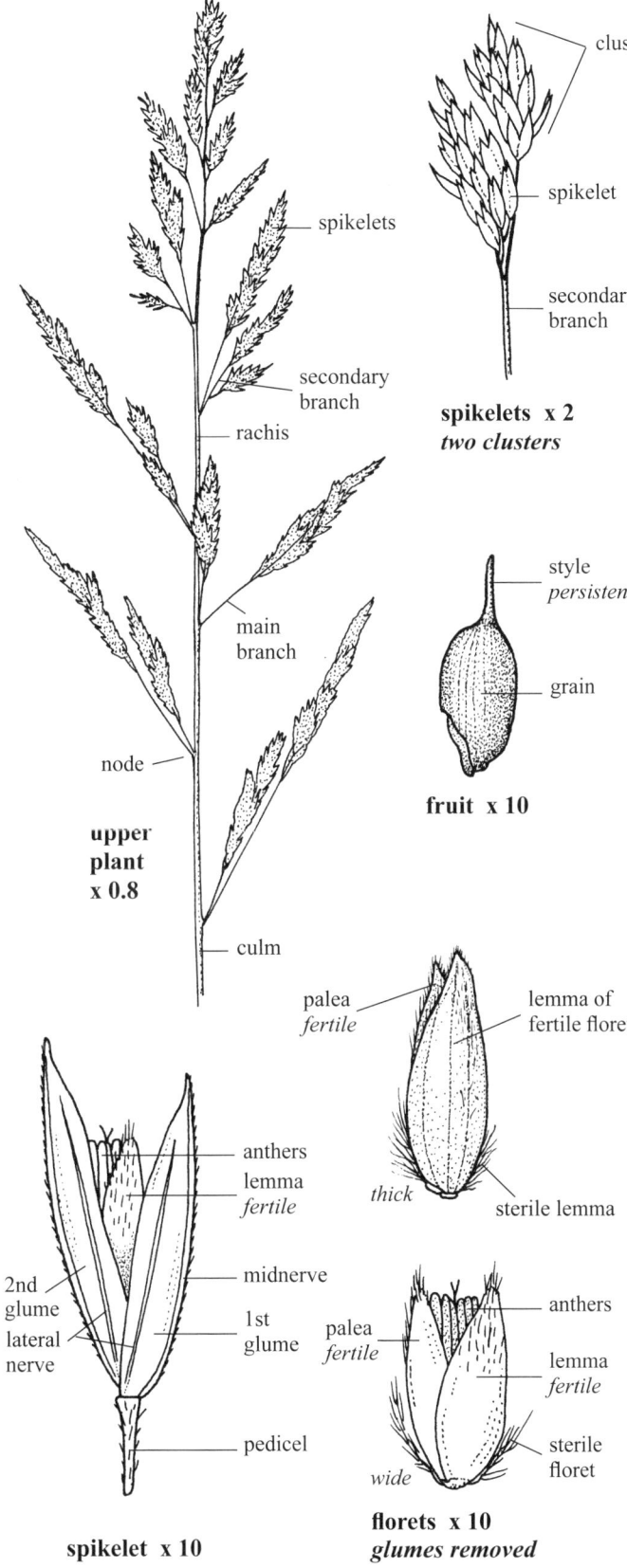
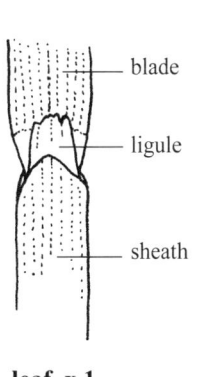
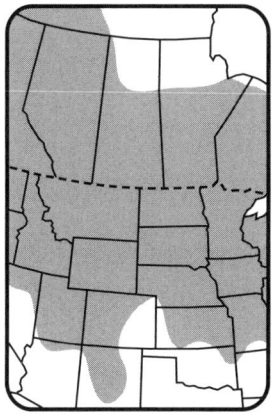

■ **SKETCH** A tufted **perennial grass** 60–210 cm tall from scaly, pinkish **rhizomes** 1–4 mm wide and to 5+ cm long; in conspicuous clumps *c.* 40 cm wide and at times almost continuous along river and lake shores, marshes, meadows, ditches, and boggy areas.

■ **FLOWERS** Greenish red, blooming May–October; **inflorescence** a panicle 5–20 (–30) cm long by 6–8 cm wide at anthesis, turning light tan and contracted after anthesis; **main branches** usually paired, one long and one short at each node, naked at their bases, the longest branch to *c.* 8 cm; **pedicels** hairy, 0.5–1 mm long; **spikelets** 3-flowered, 3.5–6 mm long, numerous, laterally compressed, forming lobes or clusters along the branches; **glumes** two, subequal, awnless, 3–6 mm long, 3-nerved, the midnerve scabrous; **florets** of two types: **1) sterile florets** two, each reduced to a thin hairy lemma 1–1.8 mm long on opposite sides at the base of the fertile floret; **2) fertile floret** one, 2.5–4 mm long; **lemmas** awnless, 2.8–4 mm long by *c.* 1.3 mm wide, faintly 3- or 5-nerved, slightly hairy above; **paleae** slightly shorter than the lemmas, hairy at apices; **anthers** 2–3.4 mm long. **FRUIT** a **grain**, 1-seeded, shiny, *c.* 2 mm long by *c.* 1 mm wide, brown, style persistent.

■ **LEAVES** Usually five to ten, flat, with one twist, largest at midstem; **blades** smooth with scabrous edges, 6–40 cm long by 5–20 mm wide; **ligules** membranous, apices rough, glabrous, 3–7 mm long; **auricles** if any, minute; **sheaths** open, margins glabrous, outer margin hyaline.

■ **STEM** A **culm**, stiff, erect to geniculate, glabrous, hollow; 2–5 mm thick near the base.

Grass—*Poaceae*

Grass Spikelets 3-flowered

Canary Grass *Phalaris canariensis* L.

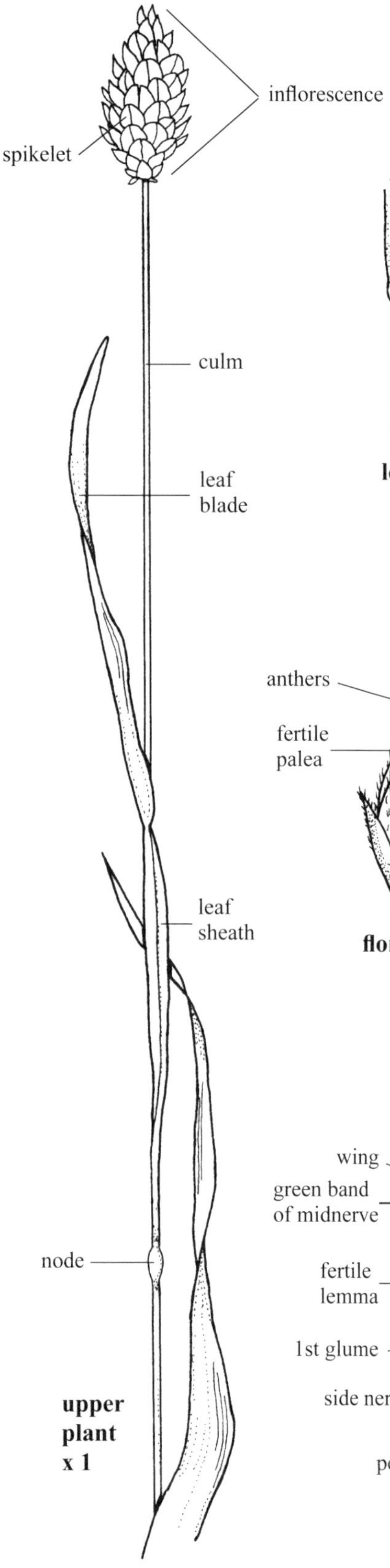

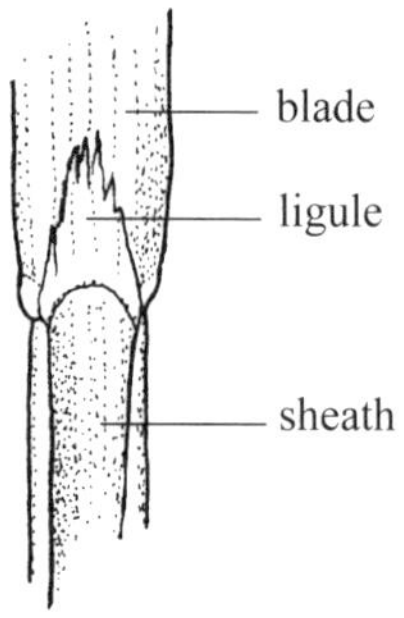

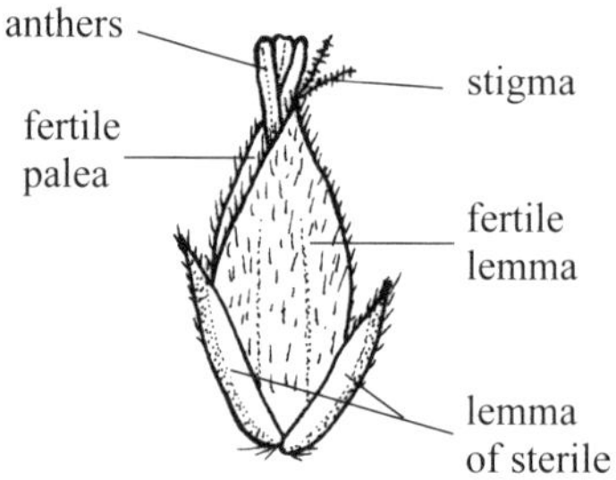

■ **SKETCH** A solitary or tufted **annual grass** 20–100 cm tall with **fibrous roots**; in gardens, dumps, yards, along roadsides and sidewalks; introduced from bird seed. *Naturalized*

■ **FLOWERS** Light green, blooming March–August; **inflorescence** a panicle (spikelike), dense, oval, 1.6–4 cm long by 10–16 mm wide; **pedicels** glabrous, *c.* 1 mm long; **spikelets** 3-flowered, laterally compressed; **glumes** two, subequal, winged, pale green, 6–9 mm long by 2–2.5 mm wide (one side), flattened, slightly hairy, stiff, papery, 3- or 5-nerved, midnerve is the wide green band on the sides, narrowly winged on the outside; **florets** of two types: **1) infertile florets** two, subequal, 2–3.8 mm long, attached at the base of the fertile floret, each reduced to a sterile lemma with a wide green midnerve and a hairy pointed tip; **2) fertile floret** one, perfect; **fertile lemma** hairy, 3.5–5.5 mm long, 2- to 5-nerved, these faint; **fertile palea** 3–5 mm long, slightly hairy, flat and fitting inside the lemma; **anthers** yellow, 2.7–3.5 mm long; **stigmas** white and feathery. **FRUIT a grain**, 1-seeded, 3.9–4.2 mm long.

■ **LEAVES** Ascending, three to eight main ones, often with one or two twists; **blades** mostly flat, 4–25 cm long by 4–10 mm wide, glabrous to scabrous, margins scabrous; **ligules** white, membranous, 2–7 mm long, truncate to pointed and often torn; **auricles** absent to minute; **sheaths** scabrous or not, upper ones loosely open around culm.

■ **STEM** A **culm**, round, hollow, erect, hairless, single or tufted; geniculate or not at the 1–2 mm wide base; **tillers** from the lowest nodes.

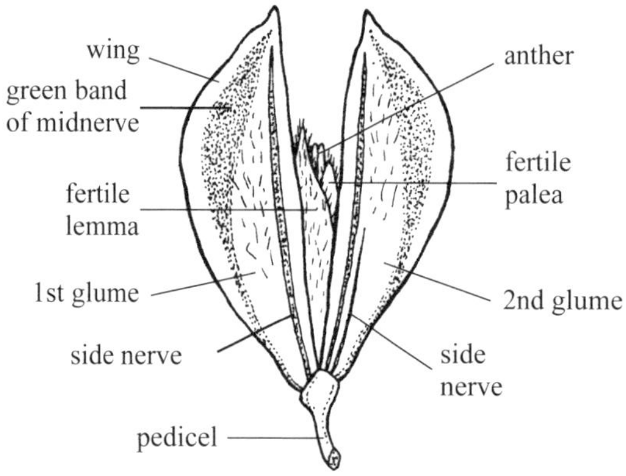

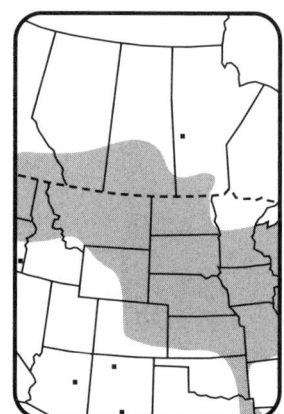

Annual Canary Grass, Common Canary Grass

Grass—*Poaceae*

TIMOTHY   *Phleum pratense* L.

■ **SKETCH** A perennial grass singly or in large tufts 30–130 cm tall with a brown, swollen, underground base; in pastures, ditches, rangeland, mountain meadows and lawns. *Naturalized*

■ **FLOWERS** Green, blooming April–September; **inflorescence** a panicle (spikelike), cylindrical, terminal, 2–12 (–23) cm long by 5–10 mm thick; **spikelets** 1-flowered, crowded, 2.3–5 mm long by *c.* 1 mm wide (not including hairs), laterally compressed; **glumes** two, subequal, 3-nerved, 2.3–5 mm long (including the 0.6–2 mm long, thick scabrous awn), stiff hairs to *c.* 1 mm long along the midnerve, the two side nerves also slightly scabrous, margins hyaline; **2nd glume's** inner margin with loose cottony hairs; **lemmas** and **paleae** membranous, delicate, much shorter than glumes; **lemmas** 1.5–2.4 mm long, 5- or 7-nerved (these faint), apices erose, slightly hairy along the midnerve leading up to its short scabrous awn, or awnless; **paleae** slightly shorter than the lemmas, hyaline; **anthers** 1.5–2 mm long. **FRUIT a grain**, 1-seeded.

■ **LEAVES** Four to six main ones, ascending; **blades** flat, 2.5–28 cm long by 3–12 mm wide, reduced above, often with a twist, scabrous on margins and below, midrib raised below, glabrous above; **ligules** 1.5–6 mm long, pointed to blunt, often ragged, membranous, glabrous, usually surrounding the stem; **auricles** absent; **sheaths** glabrous, rounded, open.

■ **STEM** A **culm**, erect, round, hollow, unbranched, mostly glabrous, one or two **bulblike nodes** below ground; sometimes geniculate near the 2–4 mm thick base; **tillers** may form.

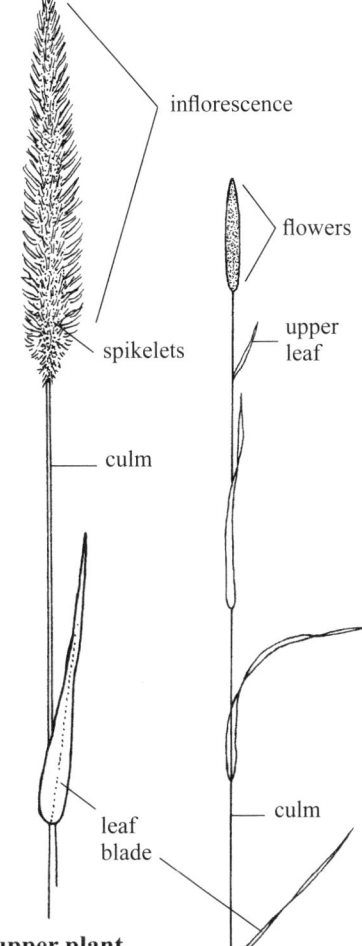

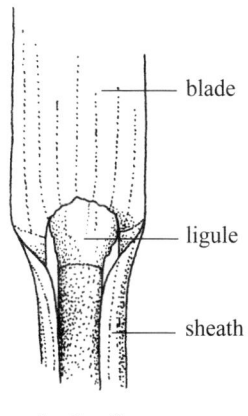

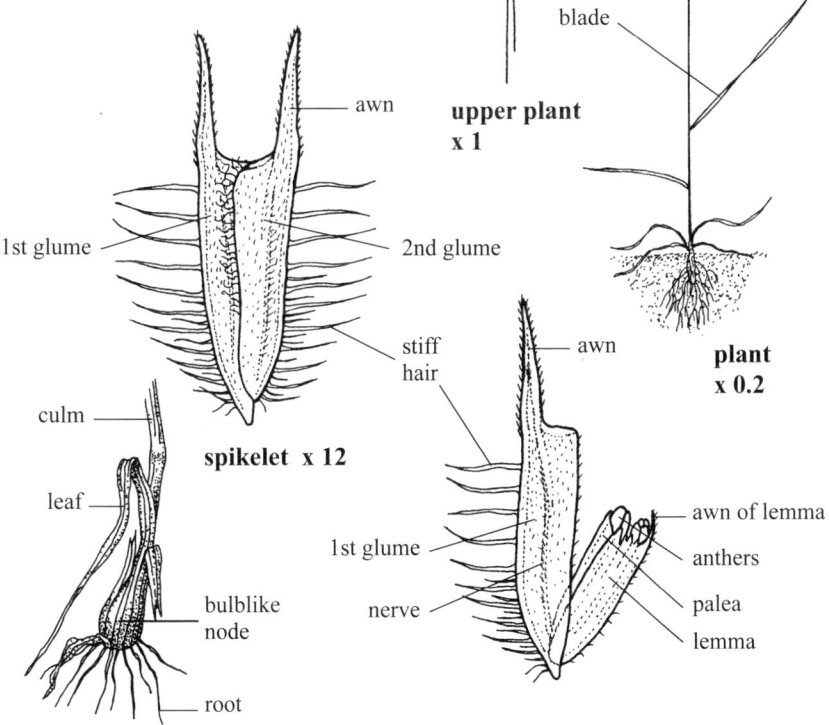

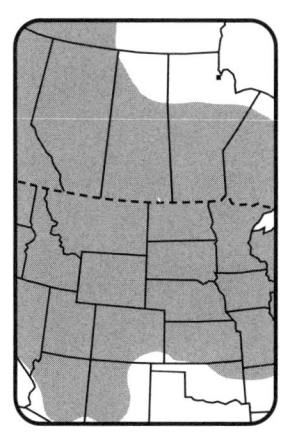

Common Timothy

Grass—*Poaceae* **Grass** Spikelets 3- to 8-flowered

Common Reed Grass *Phragmites australis* (Cav.) Trin. ex Steud.

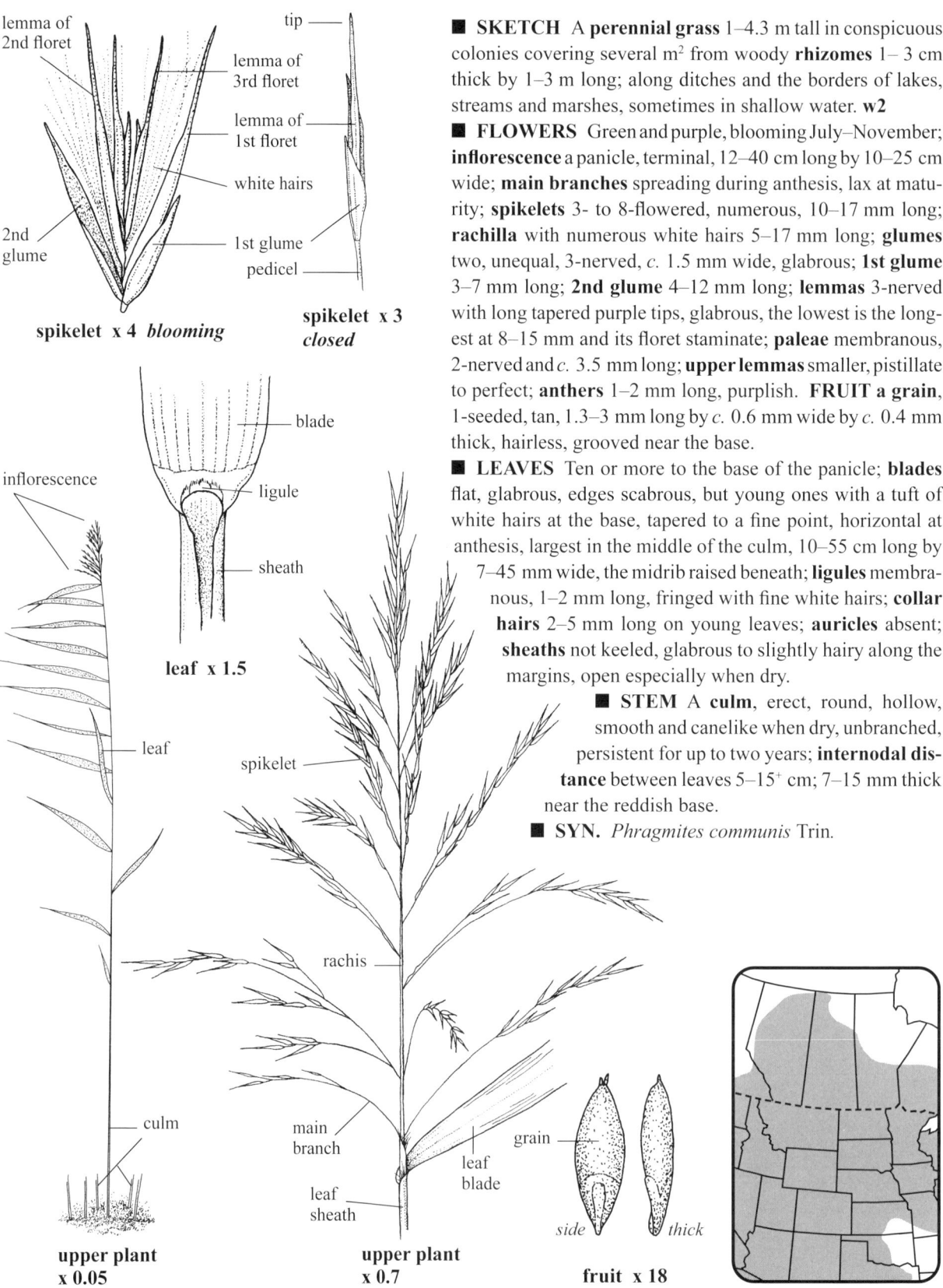

■ **SKETCH** A **perennial grass** 1–4.3 m tall in conspicuous colonies covering several m² from woody **rhizomes** 1– 3 cm thick by 1–3 m long; along ditches and the borders of lakes, streams and marshes, sometimes in shallow water. **w2**

■ **FLOWERS** Green and purple, blooming July–November; **inflorescence** a panicle, terminal, 12–40 cm long by 10–25 cm wide; **main branches** spreading during anthesis, lax at maturity; **spikelets** 3- to 8-flowered, numerous, 10–17 mm long; **rachilla** with numerous white hairs 5–17 mm long; **glumes** two, unequal, 3-nerved, *c.* 1.5 mm wide, glabrous; **1st glume** 3–7 mm long; **2nd glume** 4–12 mm long; **lemmas** 3-nerved with long tapered purple tips, glabrous, the lowest is the longest at 8–15 mm and its floret staminate; **paleae** membranous, 2-nerved and *c.* 3.5 mm long; **upper lemmas** smaller, pistillate to perfect; **anthers** 1–2 mm long, purplish. **FRUIT a grain**, 1-seeded, tan, 1.3–3 mm long by *c.* 0.6 mm wide by *c.* 0.4 mm thick, hairless, grooved near the base.

■ **LEAVES** Ten or more to the base of the panicle; **blades** flat, glabrous, edges scabrous, but young ones with a tuft of white hairs at the base, tapered to a fine point, horizontal at anthesis, largest in the middle of the culm, 10–55 cm long by 7–45 mm wide, the midrib raised beneath; **ligules** membranous, 1–2 mm long, fringed with fine white hairs; **collar hairs** 2–5 mm long on young leaves; **auricles** absent; **sheaths** not keeled, glabrous to slightly hairy along the margins, open especially when dry.

■ **STEM** A **culm**, erect, round, hollow, smooth and canelike when dry, unbranched, persistent for up to two years; **internodal distance** between leaves 5–15⁺ cm; 7–15 mm thick near the reddish base.

■ **SYN.** *Phragmites communis* Trin.

Grass—*Poaceae* Grass Spikelets 2- to 6-flowered

CANADA BLUE GRASS *Poa compressa* L.

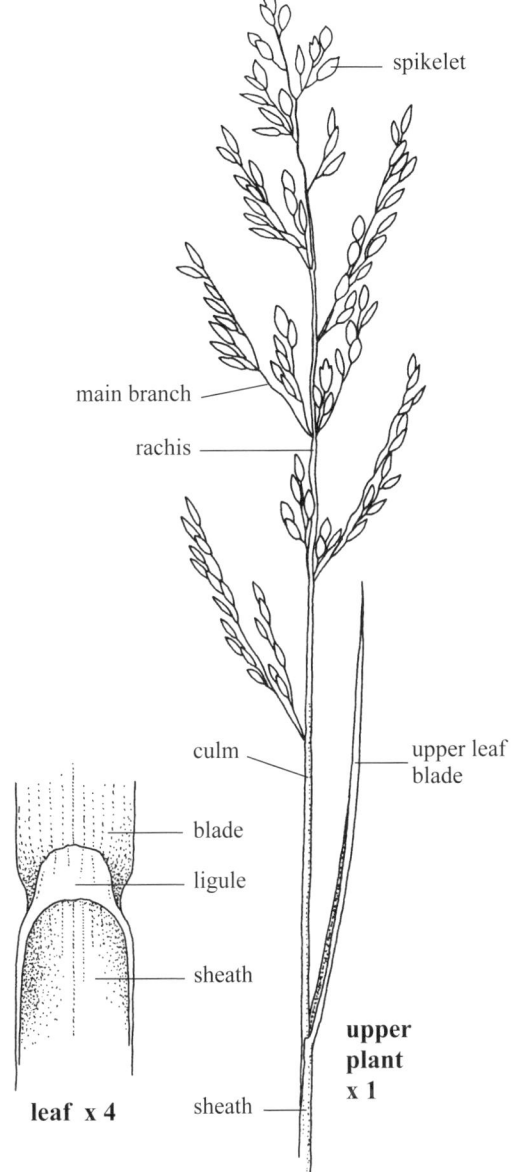

■ **SKETCH** A glabrous **perennial grass** 15–70 cm tall from tan, stiff, long, woody **rhizomes** 1–1.5 mm thick; in moist to dry prairies, wet meadows, sagebrush, along rocky sites and roadsides. *Naturalized* **w2**

■ **FLOWERS** Green, blooming June–September; **inflorescence** a panicle, compact to open, 1.5–10 cm long by 1.5–3.5 cm wide; **main branches** ascending, unequal, scabrous, one to three per node (usually two at the lowest node), 0.6–3.5 cm long, reduced above, with spikelets beginning close to their bases; **spikelets** 2- to 6-flowered, laterally flattened, 2.5–6 mm long by 1–2 mm wide by 0.7–0.9 mm thick prior to anthesis; **glumes** two, subequal, margins white hyaline, midnerves scabrous in upper half, tips pointed and some with purplish red to brown streaks; **1st glume** 1- or 3-nerved, 1.5–3 mm long; **2nd glume** 3-nerved, 1.8–3.3 mm long; **florets** perfect; **lemmas** 5-nerved, 2–3 mm long, hairy on lower part of midnerve and marginal nerves, cottony hairs at base, margins white hyaline, tips pointed, sometimes streaked with brown and white; **paleae** 2-nerved, these green and scabrous from ascending hairs, bidentate and slightly shorter than the lemmas, body hyaline membranous and stiff; **stamens** three; **anthers** yellow, 1–1.8 mm long. **FRUIT a grain**, 1-seeded.

■ **LEAVES** Usually five per culm, flat to folded along the midrib; **blades** 2–15 cm long by 1.2–4.5 mm wide, ascending, scabrous throughout, veins not deeply furrowed, lower blades usually brown by anthesis; **ligules** blunt, membranous, 0.7–3 mm long, entire to ragged; **auricles** absent; **collars** glabrous; **sheaths** compressed, very slightly scabrous, open above and closed near the base, keel is a continuation of the raised midrib of the leaf blade.

■ **STEM** A **culm**, erect to ascending, tough, hollow, oval especially along exposed culm below the inflorescence; slightly decumbent near the 1.5–2.5 mm wide base; **nodes** flattened.

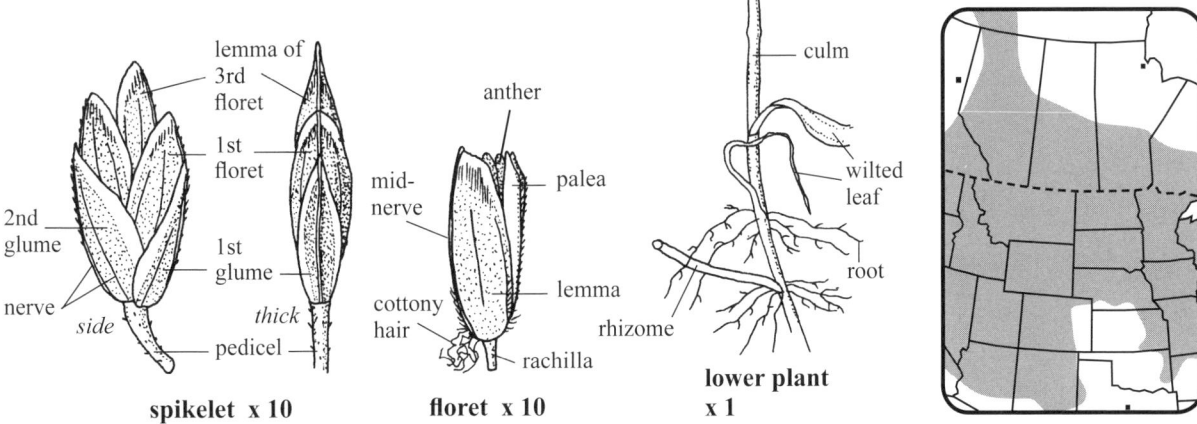

Flat-stem Blue Grass

Grass—*Poaceae* **Grass** Spikelets 2- to 5-flowered

Fowl Blue Grass   *Poa palustris* L.

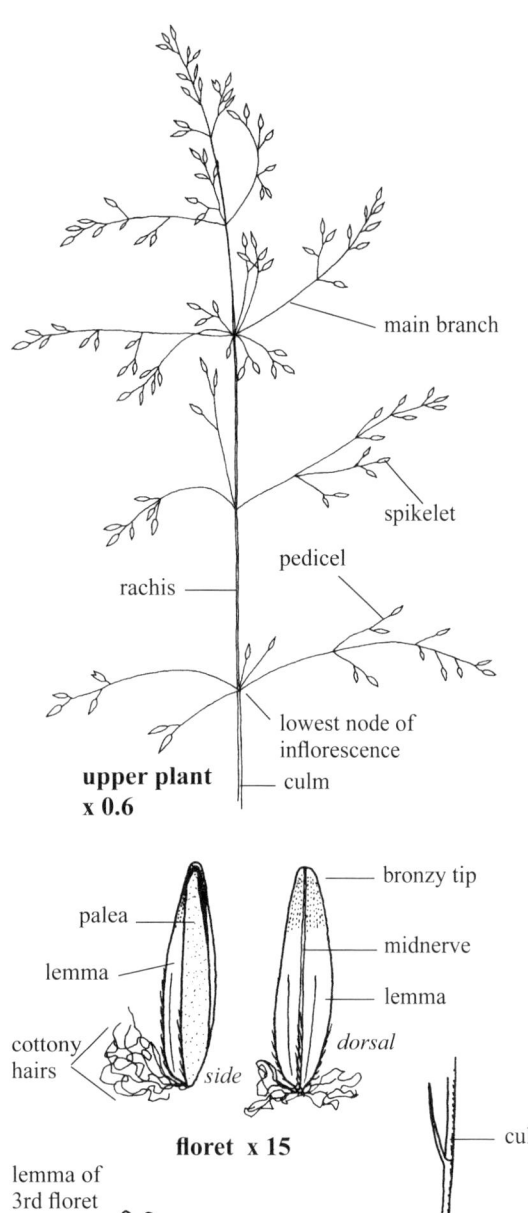

- **SKETCH** A **perennial grass** 30–140 cm tall in loose tufts or short lines, often rooting at lowest nodes, **stoloniferous** with **fibrous roots**; in wet fields, shallow water, along grassy shores of lakes and rivers. *Naturalized* **w1**
- **FLOWERS** Green with bronzy tips, blooming June–August; **inflorescence** a panicle, open, nodding, 7–30 cm long by 12–15 cm wide; **main branches** usually three to six at lower nodes, swollen at base, glabrous, spreading, some rebranched, to *c.* 8 cm long below; **spikelets** 2- to 5-flowered, 3–4.5 mm long, in the upper third to half of a branch; **glumes** two, subequal, purplish green, not awned, 1- or 3-nerved, 1.5–3 mm long, margins hyaline, tips brownish, weakly keeled, scabrous along midnerve at apices; **2nd glume** usually longer than the first; **lemmas** 2–3 mm long by 0.5–0.6 mm wide (not flattened), greenish purple, awnless, keeled, apices blunt to pointed and bronzy, 5-nerved, these obscure, short hairs along lower margins and midnerve, cottony hairs at base; **paleae** 2-nerved, about as long as lemmas, tips blunt, body hyaline except for green nerves; **stamens** three; **anthers** 0.8–1.4 mm long, pale yellow with reddish purple streaks. **FRUIT a grain**, 1-seeded.
- **LEAVES** Usually five along the culm; **blades** mostly flat, 5–25 cm long by 2–5 mm wide, ascending to arched, slightly scabrous, tips brown and slightly prow-shaped, upper stem leaf not reaching the base of the panicle; **ligules** whitish tan to white, finely hairy on outside, tips torn or entire, 2–6 mm long; **auricles** absent; **collars** hairless; **sheaths** glabrous to slightly scabrous and slightly keeled.
- **STEM** A **culm**, hollow, glabrous, round, weak, often leaning on other vegetation; 1–2 mm thick near the purplish decumbent base; **nodes** swollen, purplish to dark gray; **tillers** from lower nodes; **adventitious roots** and shoots at ground level amidst the tangle of fibrous roots.

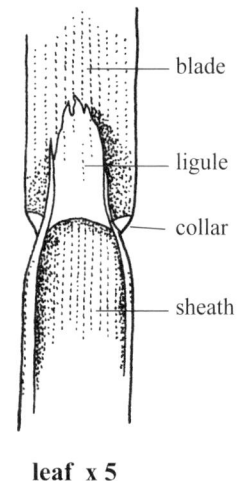

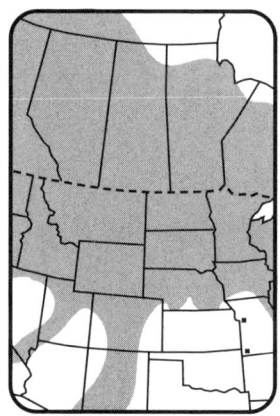

Fowl Meadow Grass

Kentucky Blue Grass *Poa pratensis* L.

upper plant x 1

spikelet x 10

spikelet x 10

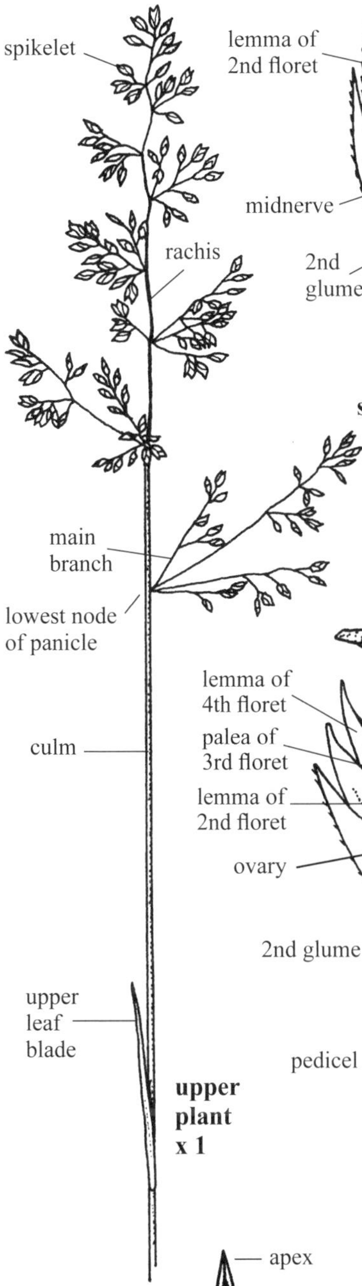

■ **SKETCH** A tufted, or sod-forming **perennial grass** 10–100 cm tall in open colonies from long, white, **rhizomes**; in lawns, dry to moist hillsides, pastures, disturbed sites, forest openings to subalpine sites. Native and *naturalized* **w2**

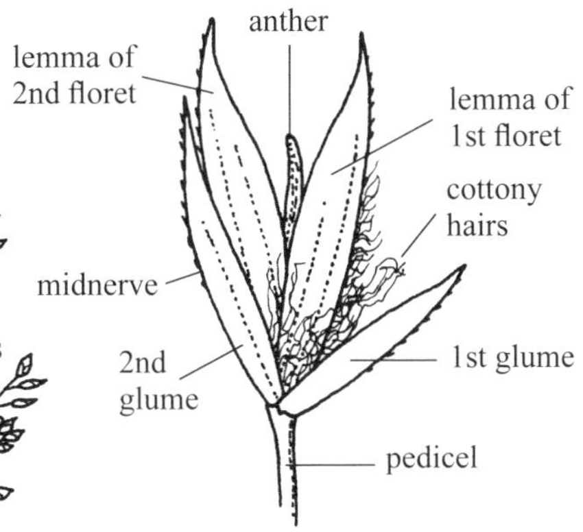

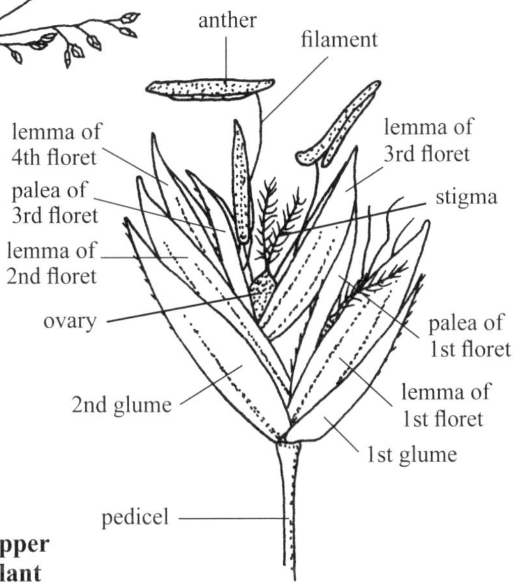

■ **FLOWERS** Green, some tinged with purple, blooming April–September; **inflorescence** an open panicle 4–17 cm long by 3–10 cm wide; **main branches** at the lowest node three to six, variable in length, to *c.* 12 cm long, mostly ascending and reduced above, naked near the rachis; **pedicels** glabrous, 0.5–3 mm long; **spikelets** 2- to 6-flowered, compressed, 3–6 mm long; **glumes** two, subequal, midnerves slightly scabrous; **1st glume** 1-nerved, 1.7–3.3 mm long; **2nd glume** 3-nerved, 2–3.8 mm long; **lemmas** keeled, 5-nerved, the lowest lemma the longest at 2.5–4 mm, long cottony hairs near the bases of the scabrous midnerves and on marginal nerves; **paleae** almost as long as the lemmas, scabrous along the two nerves; **anthers** light greenish yellow, 1–2 mm long, exserted; **stigmas** 2-parted, feathery, silvery white, *c.* 1 mm long, exserted. **FRUIT a grain**, 1-seeded.

■ **LEAVES** Usually three to five per culm, strongly ascending; **blades** flat to folded, prow-shaped at the tip, usually 1–40 cm long by 1–5 mm wide, glabrous above to lightly scabrous below; **ligules** 0.5–2.7 mm long, beige to white, membranous, the apices flat and finely hairy; **auricles** absent; **sheaths** glabrous to scabrous, slightly keeled, open in the upper half.

■ **STEM** A **culm**, erect, glabrous, round to slightly flattened, hollow; 1–2 mm thick near the base.

leaf blade tip x 10

leaf x 5

lower plant x 1

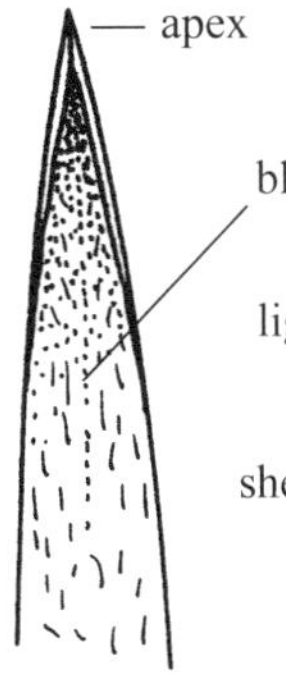

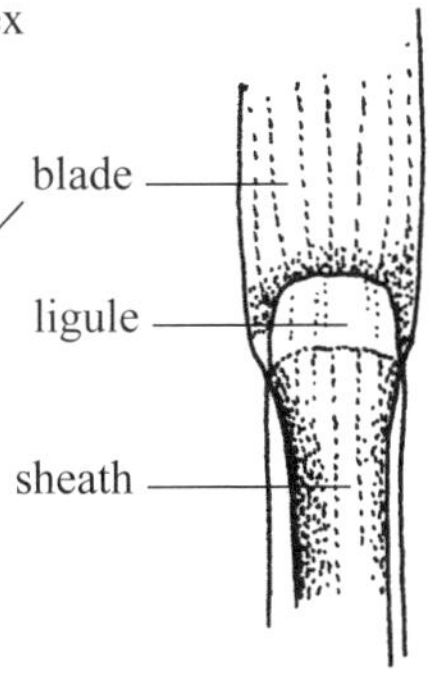

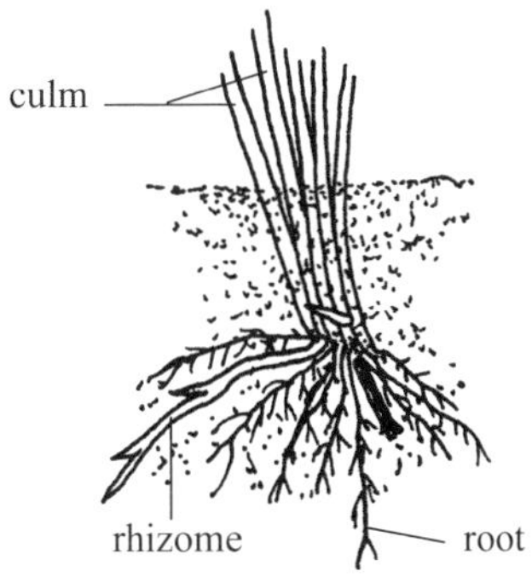

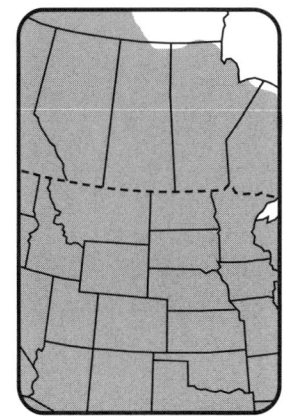

Grass—*Poaceae* **Grass** Spikelets (2) 3 per node, each 1- to 3-flowered

Russian Wild Rye *Psathyrostachys juncea* (Fisch.) Nevski

■ **SKETCH** A tufted **perennial grass** 30–100 cm tall with deep **fibrous roots** *c.* 1 mm thick, with stem bases enveloped by the fibrous remains of several old leaf sheaths; along roadsides, sidewalks, coulees in badlands and prairie pastures. *Naturalized*

■ **FLOWERS** Green, blooming May–July; **inflorescence** a terminal spike 6–12 cm long by 7–13 mm wide, erect, well above the upper leaf sheath; **rachis** green, hairy on margins, especially near the apex of each segment, disarticulating between the spikelets when mature; **spikelets** 1- to 3-flowered, with three (two) stiff spikelets per node, each 6–11 mm long; **rachilla** 2–5 mm long, white, hairy, often with a vestigial floret or lemma at its apex; **3rd floret** vestigial if 2nd floret is present, 2nd floret sometimes vestigial; **2nd floret** shorter than the 1st floret (if three florets present); **glumes** two, subequal, awnlike, up to six per node, each 3–9 mm long by 0.2–0.4 mm wide, scabrous, usually shorter than lemmas; **lemmas** 3- or 5-nerved, stiff, green with hyaline margins, C-shaped, 5–9 mm long, short hairy, some slightly reddish near the base, awns 0.5–2.3 mm long, scabrous; **paleae** hyaline, 2-nerved, green and ciliate on nerves, apices bidentate, awnless, body about as long as lemma's body; **stamens** three, exserted; **anthers** 3.5–5.5 mm long. **FRUIT a grain**, 1-seeded, 3–3.8 mm long.

■ **LEAVES** About five per culm near the base, ascending, glabrous; **blades** 1–30 cm long by 1–6 mm wide, ridged above (ventral side); **ligules** 0.2–1 mm long, flat-topped, often torn, apices ragged; **auricles** two, 1–1.5 mm long or absent, curled, whitish, stiff, easily broken; **sheaths** glaucous, glabrous, not keeled, open, upper margins white hyaline, one margin with a wide flat appendage 1–2 mm long.

■ **STEM** A **culm**, hollow, glabrous to scabrous above, round, base of culm 2–3 mm wide; **nodes** glabrous, with a reddish band 1.6–2 mm wide and a light green band *c.* 1 mm wide above and below the reddish band.

■ **SYN.** *Elymus junceus* Fisch.

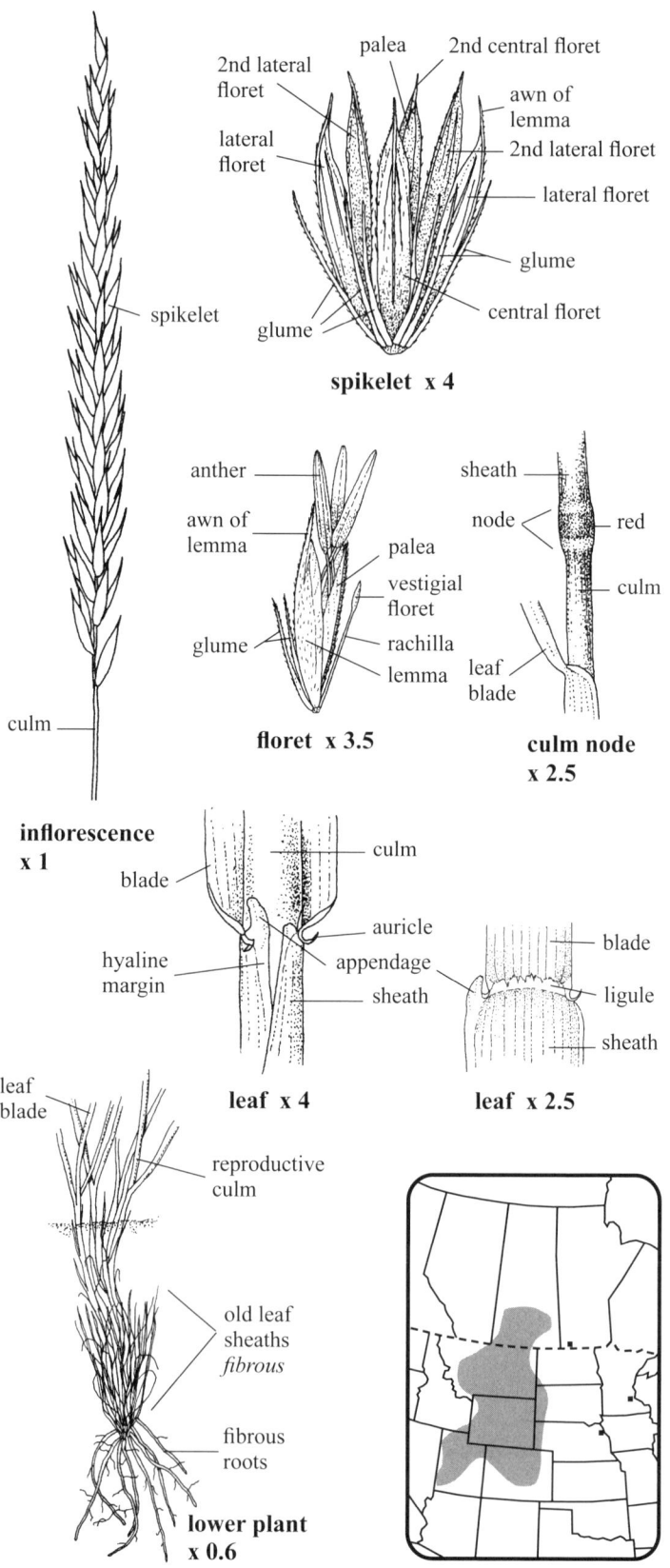

Grass—*Poaceae*   **Grass** Spikelets 3- to 7-flowered

SLENDER SALT-MEADOW GRASS *Puccinellia distans* (Jacq.) Parl.

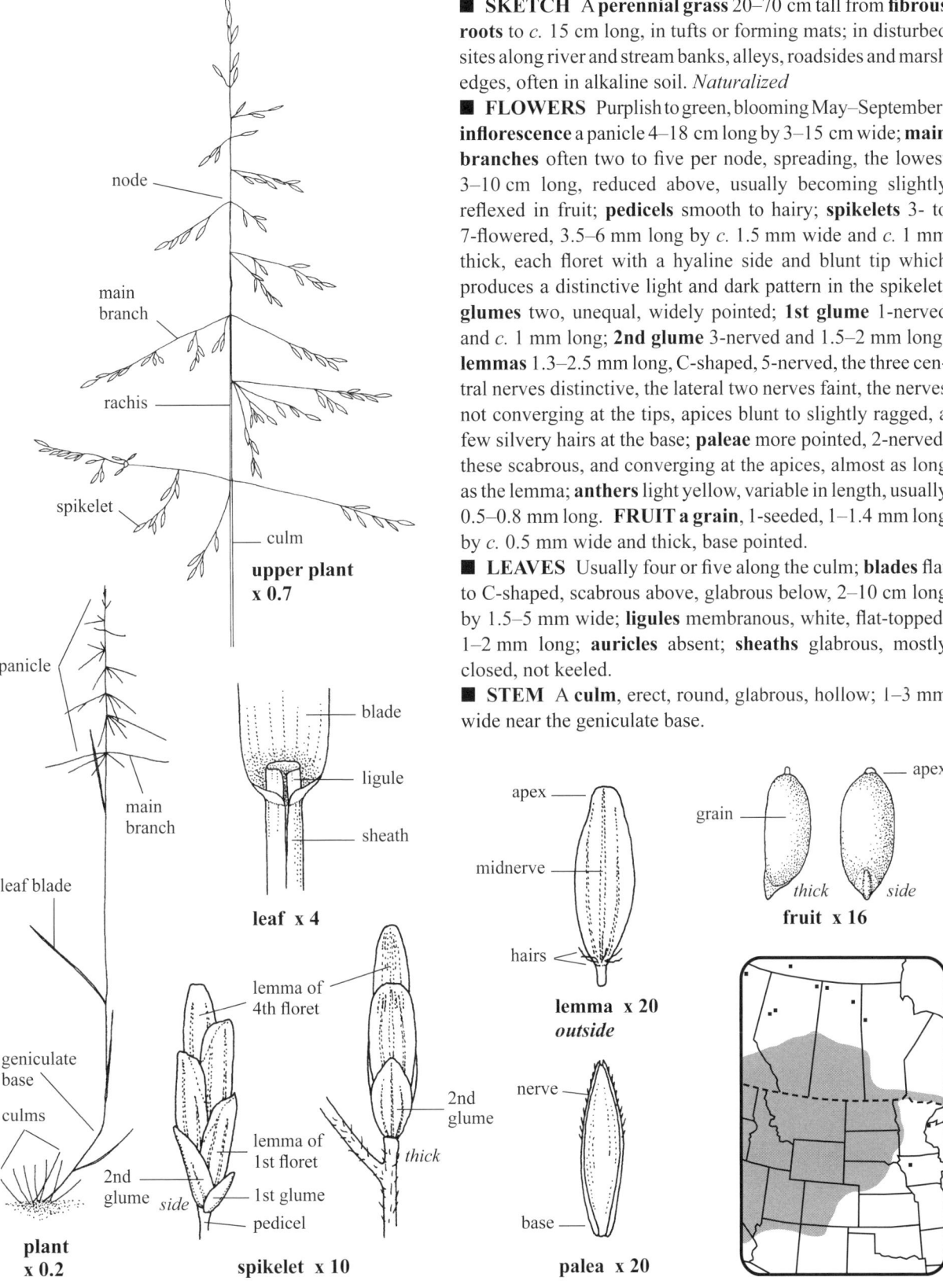

- **SKETCH** A **perennial grass** 20–70 cm tall from **fibrous roots** to *c.* 15 cm long, in tufts or forming mats; in disturbed sites along river and stream banks, alleys, roadsides and marsh edges, often in alkaline soil. *Naturalized*
- **FLOWERS** Purplish to green, blooming May–September; **inflorescence** a panicle 4–18 cm long by 3–15 cm wide; **main branches** often two to five per node, spreading, the lowest 3–10 cm long, reduced above, usually becoming slightly reflexed in fruit; **pedicels** smooth to hairy; **spikelets** 3- to 7-flowered, 3.5–6 mm long by *c.* 1.5 mm wide and *c.* 1 mm thick, each floret with a hyaline side and blunt tip which produces a distinctive light and dark pattern in the spikelet; **glumes** two, unequal, widely pointed; **1st glume** 1-nerved and *c.* 1 mm long; **2nd glume** 3-nerved and 1.5–2 mm long; **lemmas** 1.3–2.5 mm long, C-shaped, 5-nerved, the three central nerves distinctive, the lateral two nerves faint, the nerves not converging at the tips, apices blunt to slightly ragged, a few silvery hairs at the base; **paleae** more pointed, 2-nerved, these scabrous, and converging at the apices, almost as long as the lemma; **anthers** light yellow, variable in length, usually 0.5–0.8 mm long. **FRUIT a grain**, 1-seeded, 1–1.4 mm long by *c.* 0.5 mm wide and thick, base pointed.
- **LEAVES** Usually four or five along the culm; **blades** flat to C-shaped, scabrous above, glabrous below, 2–10 cm long by 1.5–5 mm wide; **ligules** membranous, white, flat-topped, 1–2 mm long; **auricles** absent; **sheaths** glabrous, mostly closed, not keeled.
- **STEM** A **culm**, erect, round, glabrous, hollow; 1–3 mm wide near the geniculate base.

Spreading Alkali-grass, Weeping Alkaligrass

Grass—*Poaceae* Grass Spikelets 3- to 8-flowered

Nuttall's Salt-meadow Grass *Puccinellia nuttalliana* (J.A. Schultes) A.S. Hitchc.

■ **SKETCH** A tufted **perennial grass** 25–80 cm tall; in pastures, alleys, along railways, and moist to sandy, often alkaline sites.

■ **FLOWERS** Green, blooming in June–August; **inflorescence** a panicle, delicate, open, 7–32 cm long by 5–25+ cm wide; **main branches** usually four to six at lowest node in panicle, these spreading to descending, 4–20 cm long, reduced above, scabrous and naked near the rachis; **pedicels** scabrous from ascending stiff hairs; **spikelets** green, 3- to 8-flowered, each 3.5–7 mm long by 1–1.6 mm wide; **glumes** two, unequal, margins hyaline, C-shaped, pointed, awnless, thin; **1st glume** 0.8–1.7 mm long, 1-nerved; **2nd glume** 1.1–2.2 mm long, 3-nerved, lateral nerves faint and incomplete; **rachilla** green, glabrous to slightly hairy; **florets** perfect; **lemmas** green, midnerve scabrous, apices slightly ragged, the lowest 1.8–2.8 mm long by *c.* 1 mm wide (flattened), reduced above, 5-nerved, lateral nerves faint and incomplete, C-shaped, not keeled, awnless, glabrous; **paleae** 2-nerved, apices slightly ragged, nerves green, hairy near apices, minutely longer to slightly shorter than the lemmas; **stamens** three, included to exserted; **anthers** light tan, 0.2–1.2 mm long. **FRUIT** a grain, 1-seeded, slightly shiny, 1–1.5 mm long by *c.* 0.5 mm wide by *c.* 0.4 mm thick, medium brown, bluntly pointed, feathery style parts often persistent at apex, flattened on one side, convex on opposite side, plump, not grooved.

■ **LEAVES** Ascending, three or four per culm, slightly scabrous on margins, flat to involute, 1–15 cm long by 0.8–2.7 mm wide, ribbed and cottony hairy on inner surface (ventral side); **ligules** membranous, slightly pointed, 1.5–3 mm long; **auricles** absent; **sheaths** ribbed, tight but open above, margins hyaline, not hairy.

■ **STEM** A **culm**, erect to leaning, narrowly hollow; 1–2 mm thick near the green base; **nodes** green, glabrous, slightly swollen.

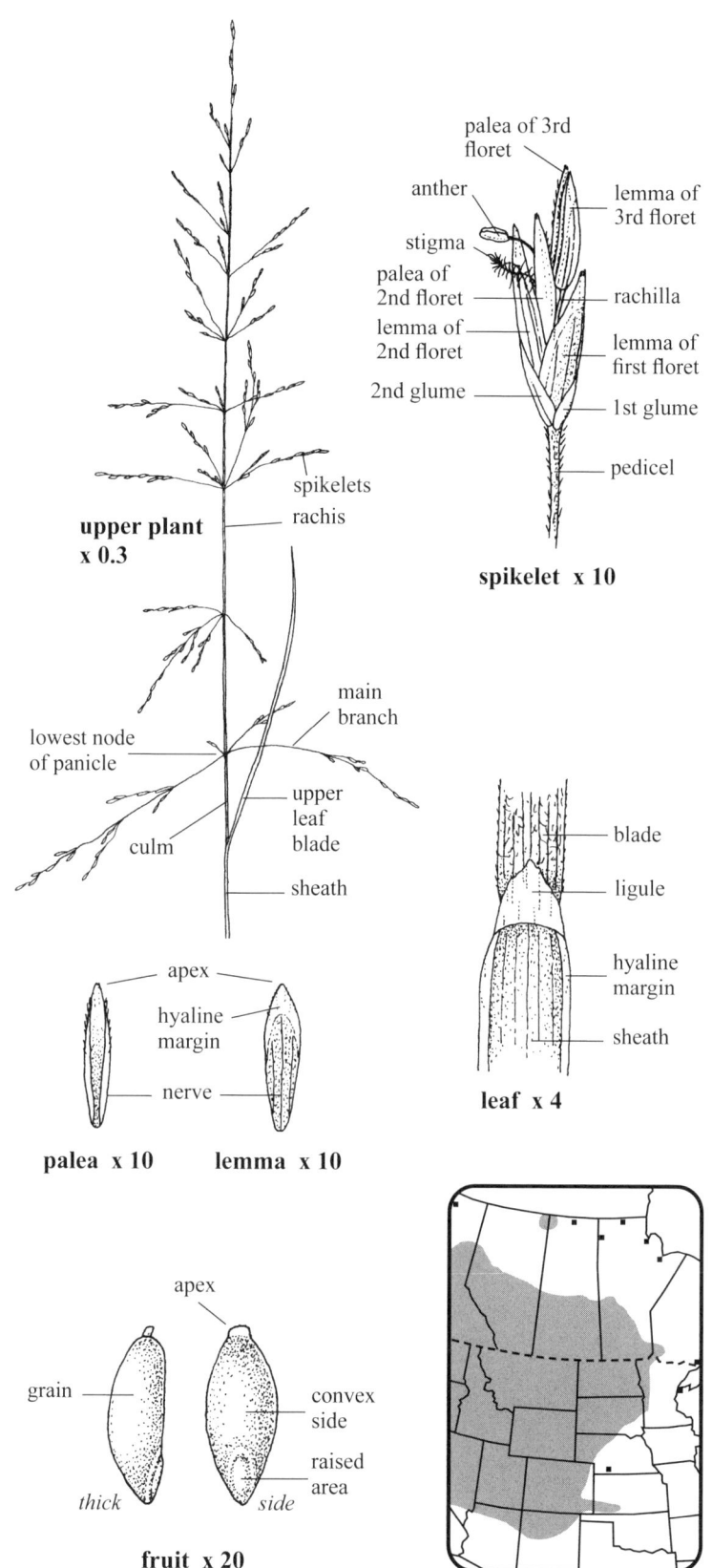

Nuttall's Alkaligrass

Purple Oat Grass *Schizachne purpurascens* (Torr.) Swallen

■ **SKETCH** A loosely tufted **perennial grass** 30–100+ cm tall with **rhizomes** 1–20 mm long by *c.* 1 mm thick; in moist to drier open woods, at forest edges and in northern prairies. **p2**

■ **FLOWERS** Green with purple bases, blooming May–August; **inflorescence** a panicle (or raceme) 5–15 cm long by 1–3 cm wide, weak and nodding; **main branches** ascending to spreading, 1–4 cm long, usually three (two) at the lowest node of panicle; **spikelets** 3- to 6-flowered, 10–20+ per inflorescence with two to five on each main branch at the lowest node, usually one per upper node, each 9–17 mm long (not including awns), disarticulating above the glumes; **rachis** triangular, scabrous; **glumes** two, unequal, purplish except for hyaline apices and upper margins, pointed, glabrous, lateral nerves straight, ending at hyaline margins; **1st glume** 3- or 5-nerved, 4–6.5 mm long; **2nd glume** 5-nerved, 5.5–8.5 mm long, lateral nerves incomplete; **lemmas** awned, apices hyaline and bidentate, teeth 1.2–2.4 mm long, margins hyaline, body 8–10 mm long, C-shaped, 5- to 9-nerved, nerves converging at apices; **awns** purplish, straight to bent on drying, 8–15 mm long by *c.* 0.1 mm thick, attached 2–3 mm below the lemma's apex, hairs ascending; **paleae** hyaline, 4–6 mm long by 0.8–1.1 mm wide (unflattened), 2-nerved, hairy on margins, stiff; **callus hairs** at base of floret numerous, 1–2 mm long and white; **stamens** three; **anthers** pale green, 1.2–1.7 mm long. **FRUIT a grain**, 1-seeded, *c.* 3.5 mm long, shiny.

■ **LEAVES** Three or four per culm, upper ones erect to ascending but usually not reaching the inflorescence; **blades** 5–35 cm long by 1.5–6 mm wide, glabrous, flat, many-ribbed; **ligules** tubular, purplish at bases, 0.5–2 mm long, membranous, hyaline, truncate and often torn, some with two lateral lobes; **auricles** absent; **sheaths** closed, margins hyaline, ribbed, not keeled, slightly scabrous.

■ **STEM** A **culm**, solid above, green, scabrous below the inflorescence and usually not branched, several leaves in a tuft form a base of tillers; 1–2 mm thick near the hollow decumbent base.

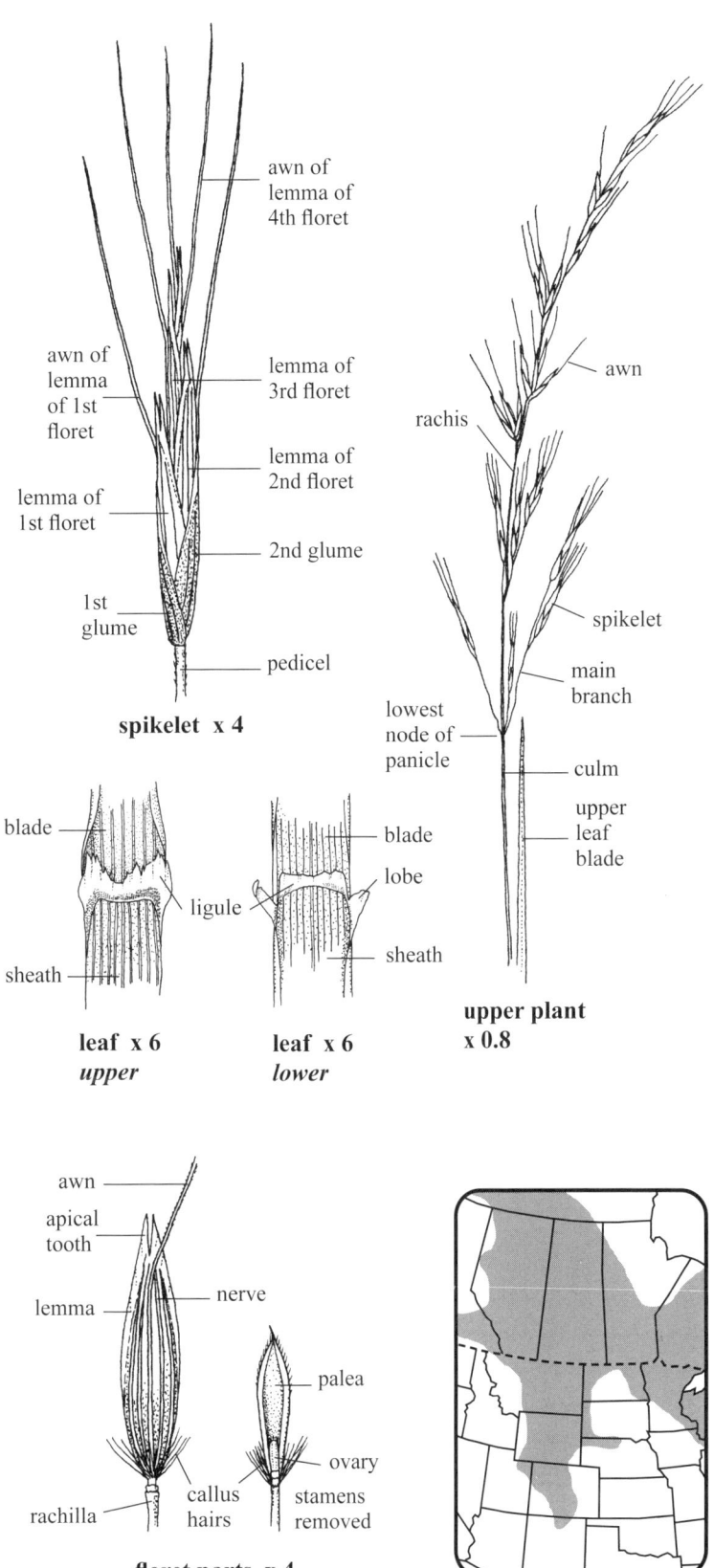

False Melic Grass, False Melic

Grass—*Poaceae* **Grass** Spikelets paired, sessile one perfect and 1-flowered

LITTLE BLUESTEM *Schizachyrium scoparium* (Michx.) Nash

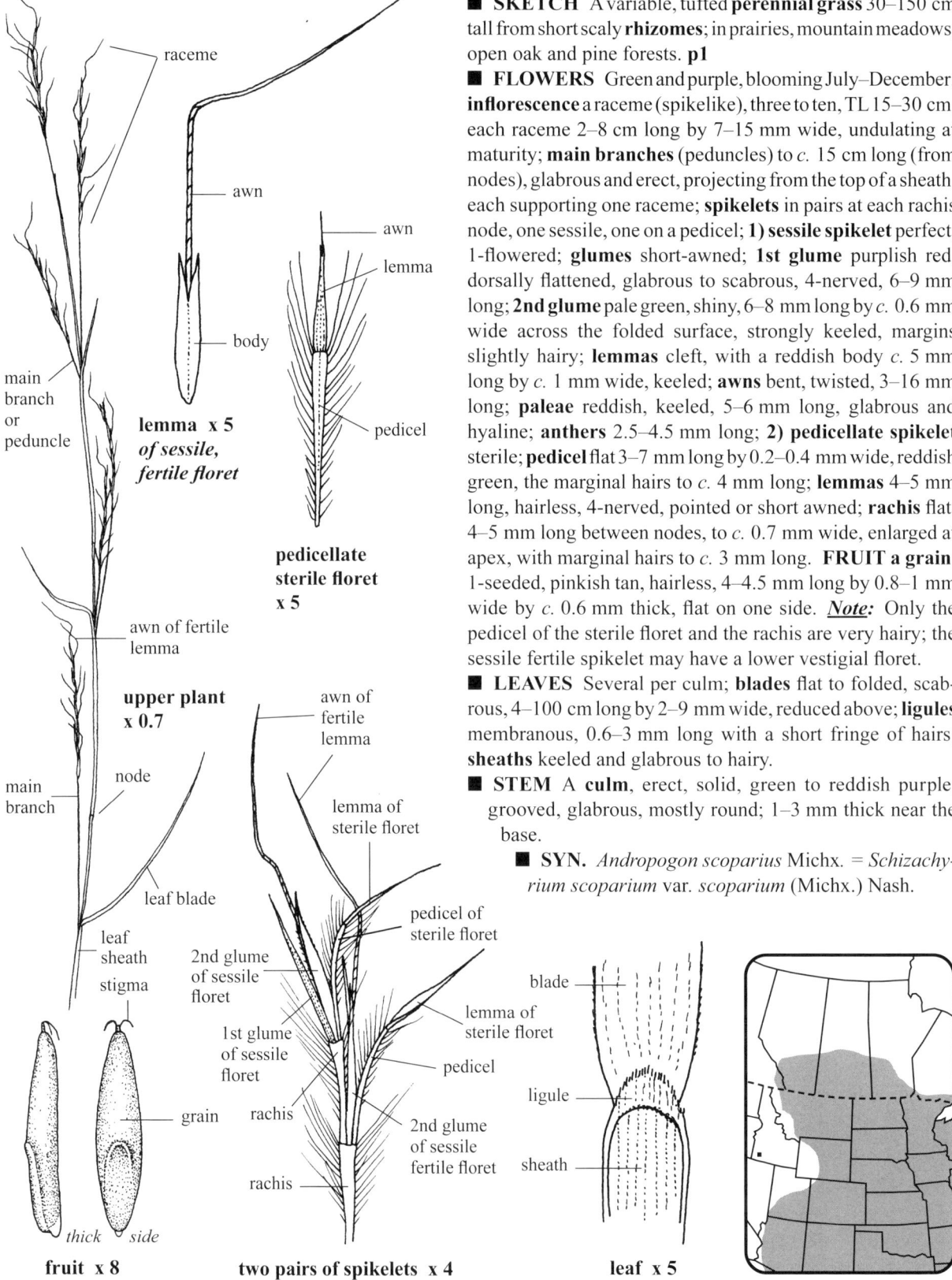

- **SKETCH** A variable, tufted **perennial grass** 30–150 cm tall from short scaly **rhizomes**; in prairies, mountain meadows, open oak and pine forests. **p1**
- **FLOWERS** Green and purple, blooming July–December; **inflorescence** a raceme (spikelike), three to ten, TL 15–30 cm, each raceme 2–8 cm long by 7–15 mm wide, undulating at maturity; **main branches** (peduncles) to *c.* 15 cm long (from nodes), glabrous and erect, projecting from the top of a sheath, each supporting one raceme; **spikelets** in pairs at each rachis node, one sessile, one on a pedicel; **1) sessile spikelet** perfect, 1-flowered; **glumes** short-awned; **1st glume** purplish red, dorsally flattened, glabrous to scabrous, 4-nerved, 6–9 mm long; **2nd glume** pale green, shiny, 6–8 mm long by *c.* 0.6 mm wide across the folded surface, strongly keeled, margins slightly hairy; **lemmas** cleft, with a reddish body *c.* 5 mm long by *c.* 1 mm wide, keeled; **awns** bent, twisted, 3–16 mm long; **paleae** reddish, keeled, 5–6 mm long, glabrous and hyaline; **anthers** 2.5–4.5 mm long; **2) pedicellate spikelet** sterile; **pedicel** flat 3–7 mm long by 0.2–0.4 mm wide, reddish green, the marginal hairs to *c.* 4 mm long; **lemmas** 4–5 mm long, hairless, 4-nerved, pointed or short awned; **rachis** flat, 4–5 mm long between nodes, to *c.* 0.7 mm wide, enlarged at apex, with marginal hairs to *c.* 3 mm long. **FRUIT** a grain, 1-seeded, pinkish tan, hairless, 4–4.5 mm long by 0.8–1 mm wide by *c.* 0.6 mm thick, flat on one side. <u>*Note:*</u> Only the pedicel of the sterile floret and the rachis are very hairy; the sessile fertile spikelet may have a lower vestigial floret.
- **LEAVES** Several per culm; **blades** flat to folded, scabrous, 4–100 cm long by 2–9 mm wide, reduced above; **ligules** membranous, 0.6–3 mm long with a short fringe of hairs; **sheaths** keeled and glabrous to hairy.
- **STEM** A **culm**, erect, solid, green to reddish purple, grooved, glabrous, mostly round; 1–3 mm thick near the base.
  - **SYN.** *Andropogon scoparius* Michx. = *Schizachyrium scoparium* var. *scoparium* (Michx.) Nash.

Small Bluestem, Little False Bluestem

Grass—*Poaceae* Grass Spikelets 2- to 4-flowered

SPRANGLETOP *Scolochloa festucacea* (Willd.) Link

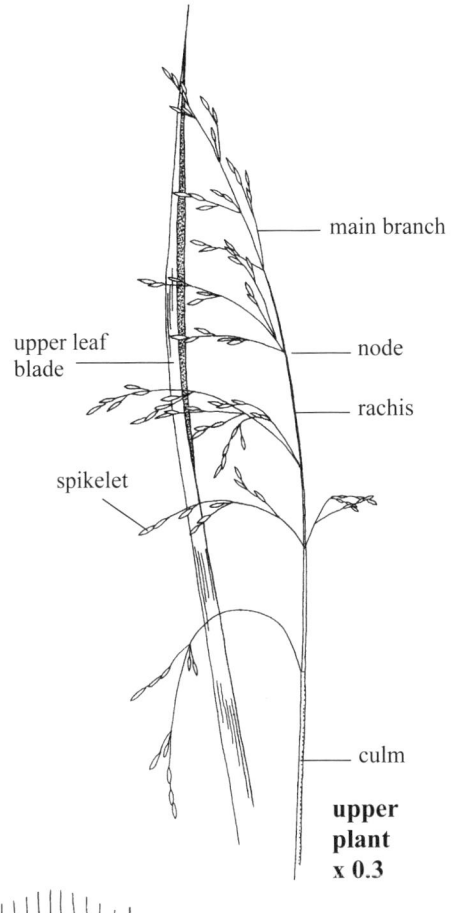

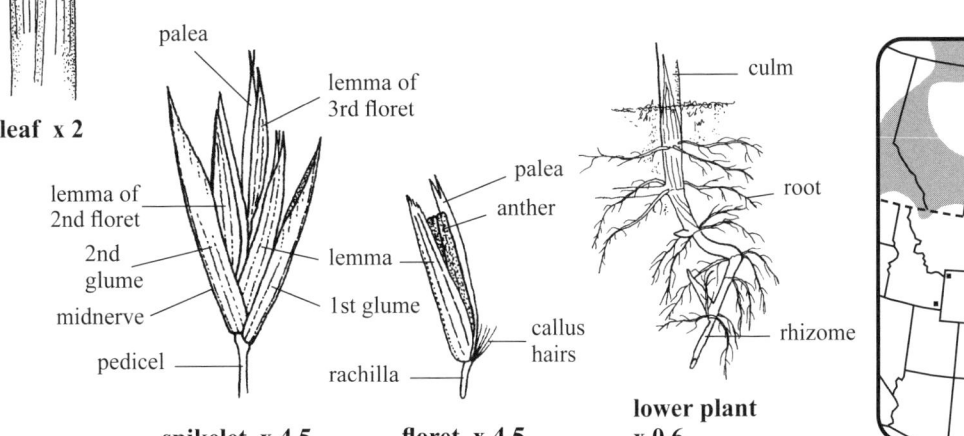

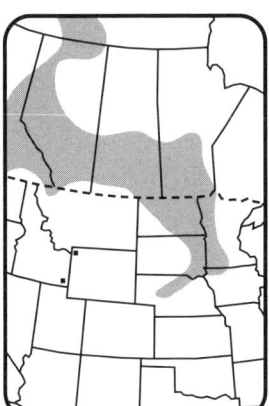

■ **SKETCH** A **perennial grass** 80–180 cm tall from long white **rhizomes** 3–5 mm thick; scattered in ditches, marshes, bordering sloughs and wet meadows, sometimes in quiet shallow water.

■ **FLOWERS** Green with purple markings, blooming in June–July; **inflorescence** a panicle, open, tip often nodding, 8–35 cm long by 12–18 cm wide; **main branches** to *c.* 18 cm long, reduced above, nodding to ascending, one to five per node, these often rebranched, slightly scabrous; **rachis** scabrous above; **pedicels** slightly scabrous to glabrous, hairs green, ascending; **spikelets** 2- to 4-flowered, perfect, 6–11 mm long by *c.* 1.3 mm thick, slightly compressed laterally; **glumes** two, unequal, stiff, pointed, membranous; **1st glume** 5–7.5 mm long, midnerve slightly scabrous, 3- or 5-nerved, nerves not obvious, awnless but midnerve may appear as a short awn less than 0.5 mm long; **2nd glume** 5.6–9 mm long, usually 5-nerved, these obscure, slightly longer than the 2nd floret, midnerve slightly scabrous, not awned or as above; **lemmas** of 1st floret rounded on back, 5–7.5 mm long, awnless, tips ragged, usually 5- or 7-nerved, these obscure below but more obvious at apices; **paleae** as long as lemmas, apices bidentate, 2-nerved, these scabrous; **stamens** three; **anthers** 2–3.3 mm long; **callus** with a tuft of white hairs *c.* 1 mm long. **FRUIT a grain**, 1-seeded.

■ **LEAVES** Flat, six to eight per culm, ascending, some with a twist; **blades** 12–50 cm long by 3–12 mm wide, upper leaf reaching into or past the panicle, glabrous to slightly scabrous on margins and above; **ligules** torn, ragged, membranous, 4–10 mm long, edges tan and a little thicker, hairless; **auricles** absent; **collars** hairless; **sheaths** round, stiff, flat, open to the base, glabrous, cross-veins visible in some.

■ **STEM** A **culm**, hollow, round, slightly scabrous below the panicle; 4–7 mm thick near the base, often with one or two vegetative stems (tillers) from the base.

Common Reed-grass, Common Rivergrass, Thatch Grass

Grass—*Poaceae*   Grass   Spikelets 2-flowered

Yellow Foxtail   *Setaria pumila* (Poir.) Roemer & J.A. Schultes

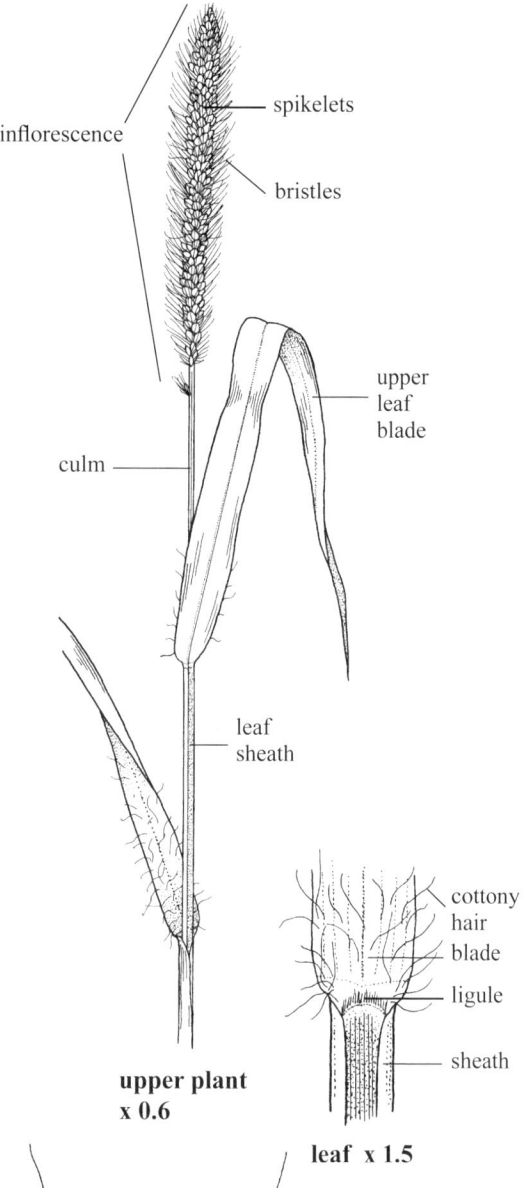

upper plant x 0.6

leaf x 1.5

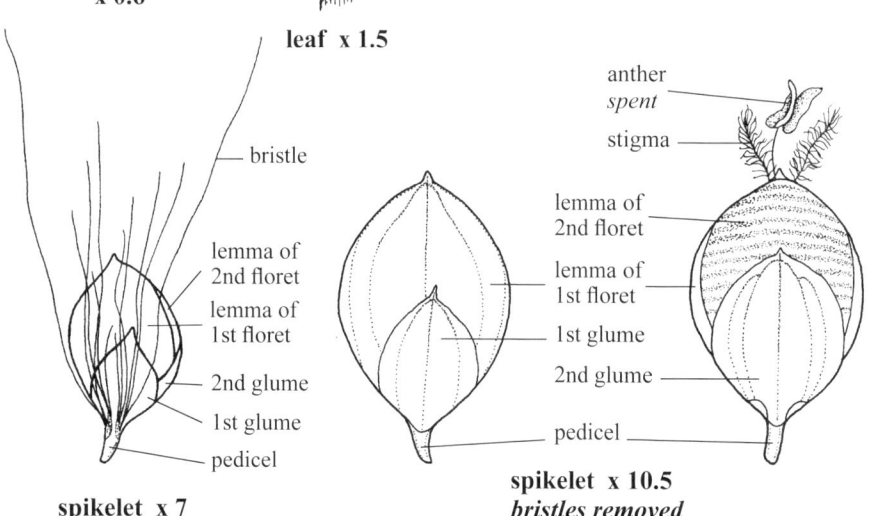

spikelet x 7

spikelet x 10.5
bristles removed

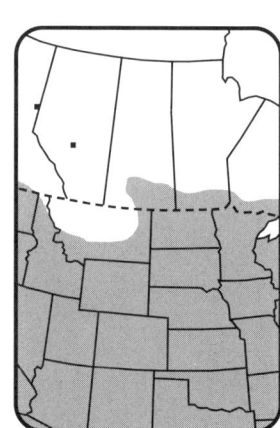

■ **SKETCH** A tufted **annual grass** 30–130 cm tall from **fibrous roots**; in gardens, fields and disturbed drier sites. *Naturalized* **w1**

■ **FLOWERS** Green, blooming June–October; **inflorescence** a panicle (spikelike), erect to slightly nodding, cylindrical, compact, 3–15 cm long by 12–20 mm wide with numerous ascending bristles; **bristles** pale brown to yellowish, stiff, 2–10 mm long, attached to pedicels below the 1st glume, usually 5–20 per spikelet, the hairs short and ascending; **spikelets** 2-flowered, 2.5–3.5 mm long by *c.* 2.2 mm wide, awnless, plump; **glumes** two, unequal and glabrous; **1st glume** 1–1.8 mm long, 3- or 5-nerved with a pointed tip; **2nd glume** 1.5–2.4 mm long, 5- or 7-nerved; **1) lower (1st) floret** staminate or rarely sterile, 2–3 mm long; **lemmas** 5- or 7-nerved, 2.2–3 mm long, pointed, glabrous, slightly transparent revealing the anthers; **paleae** 2-nerved, membranous, 2–3 mm long; **2) upper (2nd) floret** perfect, 2–3 mm long; **lemmas** unnerved, shiny, thick, corrugated, short tip bent into the tip of the 1st lemma into which it fits; **paleae** unnerved, thick, slightly shorter and less corrugated than the upper lemma into which they tightly fit; **anthers** medium brown, 1–2 mm long. **FRUIT a grain**, 1-seeded, 2.5–3.3 mm long by 1.5–2.2 mm wide by 1–1.5 mm thick, enclosed by the hard lemma and palea.

■ **LEAVES** Usually four to six main ones; **blades** ascending then nodding, flat, often with a twist, glabrous to scabrous on both surfaces and edges, 10–32 cm long by 4–12 mm wide, often with cottony hairs 5–10 mm long near the bases of the blades, the hairs sometimes forming white tufts near the ligule; **ligules** of fine white erect hairs 0.5–2.5 mm long from a minute membranous base; **auricles** absent; **sheaths** glabrous, closed, keeled, flattened, reddish near the lower nodes.

■ **STEM** A **culm**, finely grooved below the inflorescence, round, erect, pith-filled to hollow, some with **tillers** and adventitious roots at lower nodes; the geniculate base 3–5 mm wide.

Yellow Bristlegrass, Cattail Grass

Grass—*Poaceae* **Grass** Spikelets 2-flowered

Green Foxtail *Setaria viridis* (L.) Beauv.

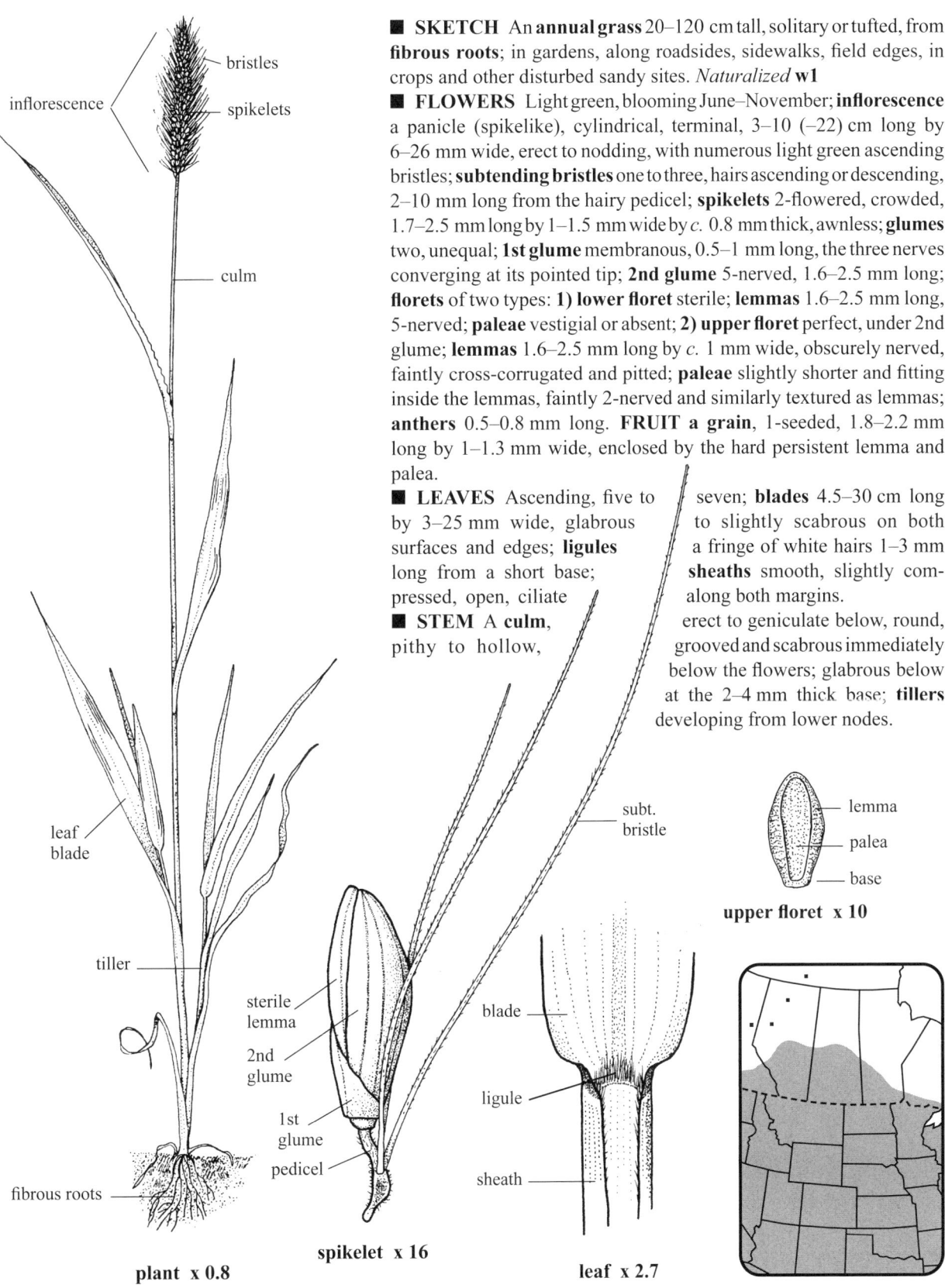

■ **SKETCH** An **annual grass** 20–120 cm tall, solitary or tufted, from **fibrous roots**; in gardens, along roadsides, sidewalks, field edges, in crops and other disturbed sandy sites. *Naturalized* **w1**

■ **FLOWERS** Light green, blooming June–November; **inflorescence** a panicle (spikelike), cylindrical, terminal, 3–10 (–22) cm long by 6–26 mm wide, erect to nodding, with numerous light green ascending bristles; **subtending bristles** one to three, hairs ascending or descending, 2–10 mm long from the hairy pedicel; **spikelets** 2-flowered, crowded, 1.7–2.5 mm long by 1–1.5 mm wide by *c.* 0.8 mm thick, awnless; **glumes** two, unequal; **1st glume** membranous, 0.5–1 mm long, the three nerves converging at its pointed tip; **2nd glume** 5-nerved, 1.6–2.5 mm long; **florets** of two types: **1) lower floret** sterile; **lemmas** 1.6–2.5 mm long, 5-nerved; **paleae** vestigial or absent; **2) upper floret** perfect, under 2nd glume; **lemmas** 1.6–2.5 mm long by *c.* 1 mm wide, obscurely nerved, faintly cross-corrugated and pitted; **paleae** slightly shorter and fitting inside the lemmas, faintly 2-nerved and similarly textured as lemmas; **anthers** 0.5–0.8 mm long. **FRUIT a grain**, 1-seeded, 1.8–2.2 mm long by 1–1.3 mm wide, enclosed by the hard persistent lemma and palea.

■ **LEAVES** Ascending, five to seven; **blades** 4.5–30 cm long by 3–25 mm wide, glabrous to slightly scabrous on both surfaces and edges; **ligules** a fringe of white hairs 1–3 mm long from a short base; **sheaths** smooth, slightly compressed, open, ciliate along both margins.

■ **STEM** A **culm**, pithy to hollow, erect to geniculate below, round, grooved and scabrous immediately below the flowers; glabrous below at the 2–4 mm thick base; **tillers** developing from lower nodes.

Green Foxtail Grass, Green Bristlegrass

Grass—*Poaceae*

INDIAN GRASS *Sorghastrum nutans* (L.) Nash

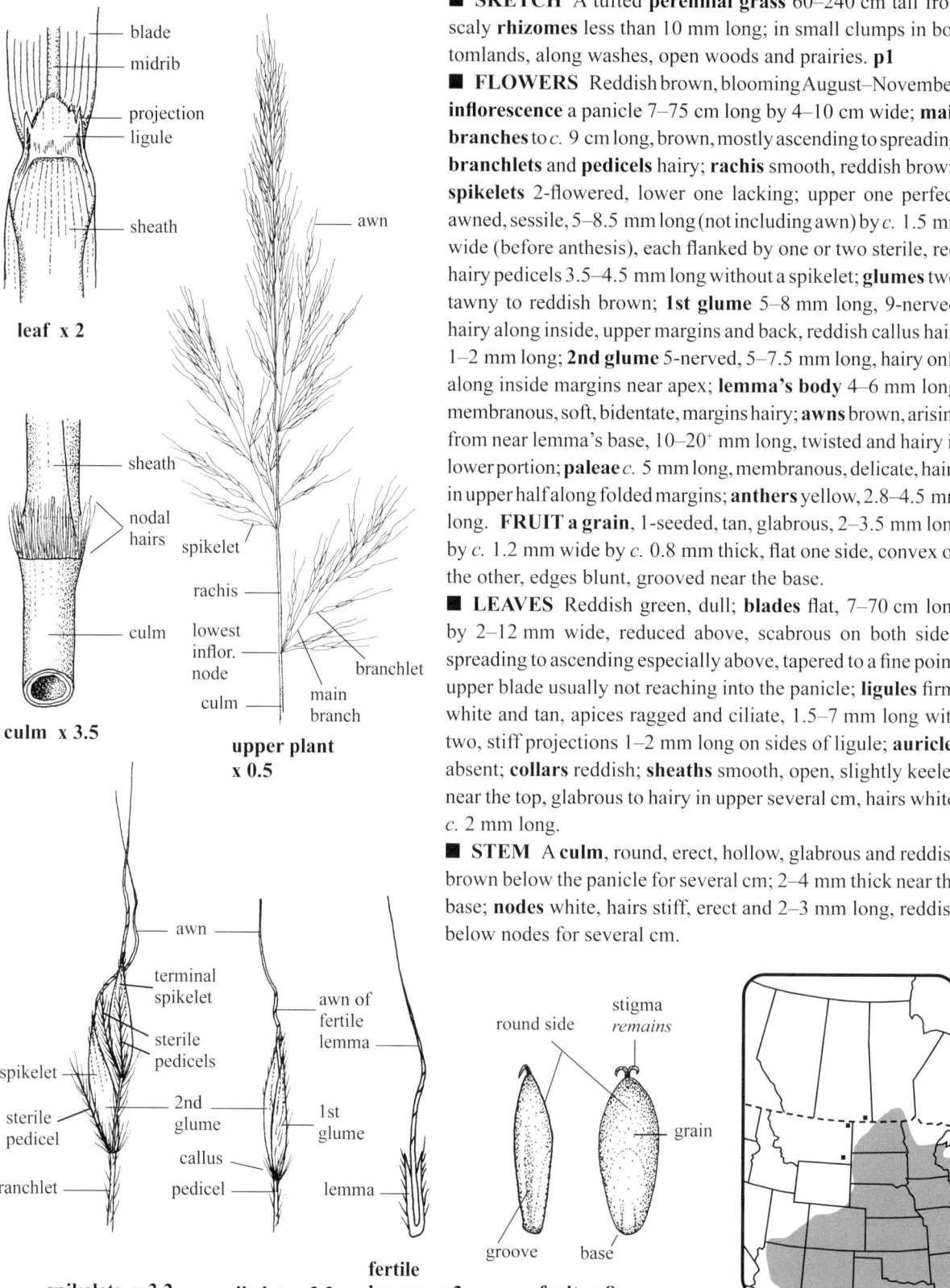

leaf x 2

culm x 3.5

upper plant x 0.5

spikelets x 3.2   spikelet x 3.2   fertile lemma x 3   fruit x 8

■ **SKETCH** A tufted **perennial grass** 60–240 cm tall from scaly **rhizomes** less than 10 mm long; in small clumps in bottomlands, along washes, open woods and prairies. **p1**

■ **FLOWERS** Reddish brown, blooming August–November; **inflorescence** a panicle 7–75 cm long by 4–10 cm wide; **main branches** to *c.* 9 cm long, brown, mostly ascending to spreading; **branchlets** and **pedicels** hairy; **rachis** smooth, reddish brown; **spikelets** 2-flowered, lower one lacking; upper one perfect, awned, sessile, 5–8.5 mm long (not including awn) by *c.* 1.5 mm wide (before anthesis), each flanked by one or two sterile, red, hairy pedicels 3.5–4.5 mm long without a spikelet; **glumes** two, tawny to reddish brown; **1st glume** 5–8 mm long, 9-nerved, hairy along inside, upper margins and back, reddish callus hairs 1–2 mm long; **2nd glume** 5-nerved, 5–7.5 mm long, hairy only along inside margins near apex; **lemma's body** 4–6 mm long, membranous, soft, bidentate, margins hairy; **awns** brown, arising from near lemma's base, 10–20$^+$ mm long, twisted and hairy in lower portion; **paleae** *c.* 5 mm long, membranous, delicate, hairy in upper half along folded margins; **anthers** yellow, 2.8–4.5 mm long. **FRUIT a grain**, 1-seeded, tan, glabrous, 2–3.5 mm long by *c.* 1.2 mm wide by *c.* 0.8 mm thick, flat one side, convex on the other, edges blunt, grooved near the base.

■ **LEAVES** Reddish green, dull; **blades** flat, 7–70 cm long by 2–12 mm wide, reduced above, scabrous on both sides, spreading to ascending especially above, tapered to a fine point, upper blade usually not reaching into the panicle; **ligules** firm, white and tan, apices ragged and ciliate, 1.5–7 mm long with two, stiff projections 1–2 mm long on sides of ligule; **auricles** absent; **collars** reddish; **sheaths** smooth, open, slightly keeled near the top, glabrous to hairy in upper several cm, hairs white, *c.* 2 mm long.

■ **STEM** A **culm**, round, erect, hollow, glabrous and reddish brown below the panicle for several cm; 2–4 mm thick near the base; **nodes** white, hairs stiff, erect and 2–3 mm long, reddish below nodes for several cm.

Yellow Indian Grass

Grass—*Poaceae*

ALKALI CORD GRASS *Spartina gracilis* Trin.

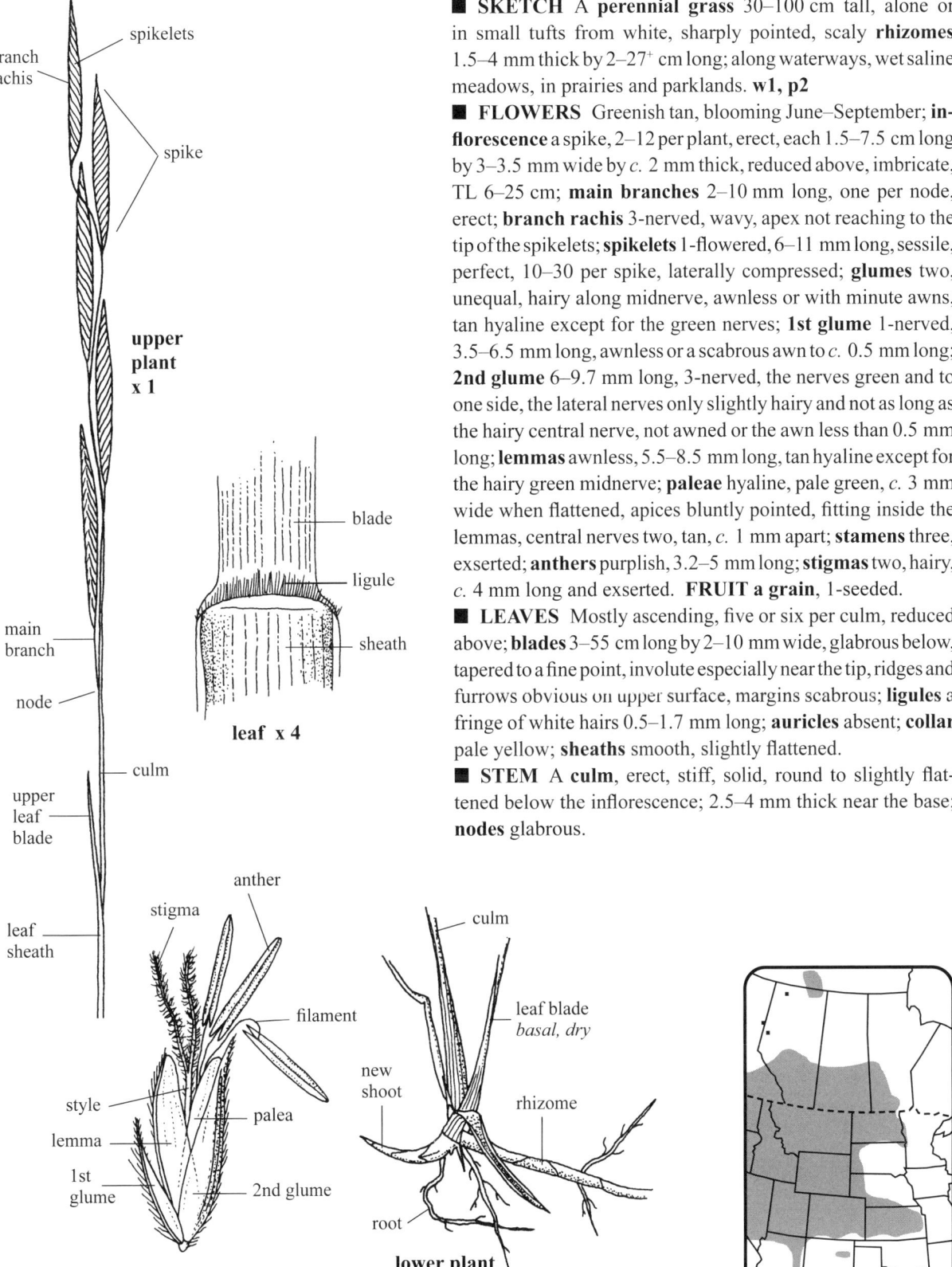

upper plant x 1

leaf x 4

spikelet x 5

lower plant x 1

- **SKETCH** A **perennial grass** 30–100 cm tall, alone or in small tufts from white, sharply pointed, scaly **rhizomes** 1.5–4 mm thick by 2–27⁺ cm long; along waterways, wet saline meadows, in prairies and parklands. **w1, p2**
- **FLOWERS** Greenish tan, blooming June–September; **inflorescence** a spike, 2–12 per plant, erect, each 1.5–7.5 cm long by 3–3.5 mm wide by *c.* 2 mm thick, reduced above, imbricate, TL 6–25 cm; **main branches** 2–10 mm long, one per node, erect; **branch rachis** 3-nerved, wavy, apex not reaching to the tip of the spikelets; **spikelets** 1-flowered, 6–11 mm long, sessile, perfect, 10–30 per spike, laterally compressed; **glumes** two, unequal, hairy along midnerve, awnless or with minute awns, tan hyaline except for the green nerves; **1st glume** 1-nerved, 3.5–6.5 mm long, awnless or a scabrous awn to *c.* 0.5 mm long; **2nd glume** 6–9.7 mm long, 3-nerved, the nerves green and to one side, the lateral nerves only slightly hairy and not as long as the hairy central nerve, not awned or the awn less than 0.5 mm long; **lemmas** awnless, 5.5–8.5 mm long, tan hyaline except for the hairy green midnerve; **paleae** hyaline, pale green, *c.* 3 mm wide when flattened, apices bluntly pointed, fitting inside the lemmas, central nerves two, tan, *c.* 1 mm apart; **stamens** three, exserted; **anthers** purplish, 3.2–5 mm long; **stigmas** two, hairy, *c.* 4 mm long and exserted. **FRUIT a grain**, 1-seeded.
- **LEAVES** Mostly ascending, five or six per culm, reduced above; **blades** 3–55 cm long by 2–10 mm wide, glabrous below, tapered to a fine point, involute especially near the tip, ridges and furrows obvious on upper surface, margins scabrous; **ligules** a fringe of white hairs 0.5–1.7 mm long; **auricles** absent; **collar** pale yellow; **sheaths** smooth, slightly flattened.
- **STEM** A **culm**, erect, stiff, solid, round to slightly flattened below the inflorescence; 2.5–4 mm thick near the base; **nodes** glabrous.

Grass—*Poaceae*

Grass Spikelets 1-flowered

PRAIRIE CORD GRASS *Spartina pectinata* Bosc ex Link

■ **SKETCH** A **perennial grass** 0.8–2.3 m tall, singly or in tufts from sharply pointed, scaly **rhizomes** 4–36+ cm long by 2–8 mm thick; along river and stream banks, in swamps, ditches and moist prairies. **w2**, **p2**

■ **FLOWERS** Green, blooming June–September; **inflorescence** a panicle of spikes, TL 14–50 cm by 10–20 cm wide; **floral branches** naked at base, scabrous, 2–40 mm long, reduced above, ascending, each with a spike; **spikes** 5–50, each 2.5–12 cm long by *c.* 5 mm wide, reduced above; **spikelets** 1-flowered, sessile, 10–80 per spike, each 10–25 mm long (including awns), laterally compressed, crowded in two rows along one side of the triangular, light purple rachis of spike; **glumes** two, unequal, scabrous along the midnerves and awns; **1st glume** 1-nerved, the body 4–8 mm long, narrow, with an awn 1–4 mm long; **2nd glume** 3-nerved, the body 9–14 mm long with an awn 3–10 mm long; **lemmas** 1-nerved, awnless, 6–10 mm long, the midnerve scabrous; **paleae** 2-nerved near the center, glabrous, as long as or longer than the lemmas; **stamens** three; **anthers** 4.5–7 mm long; **styles** 2-parted; **stigmas** spreading, hairy and exserted. **FRUIT a grain**, 1-seeded.

■ **LEAVES** Mostly erect, five or six per culm, the uppermost blade usually extending through the inflorescence; **blades** flat, becoming tightly involute on drying in the autumn, 20–120 cm long by 4–15 mm wide, the edges scabrous, strongly ridged above, glabrous below, tapered to a long fine point; **ligules** a fringe of hairs 2–4 mm long; **auricles** absent; **sheaths** glabrous.

■ **STEM** A **culm**, erect, round, hollow, persistent over winter; 3–10 mm thick near the reddish base.

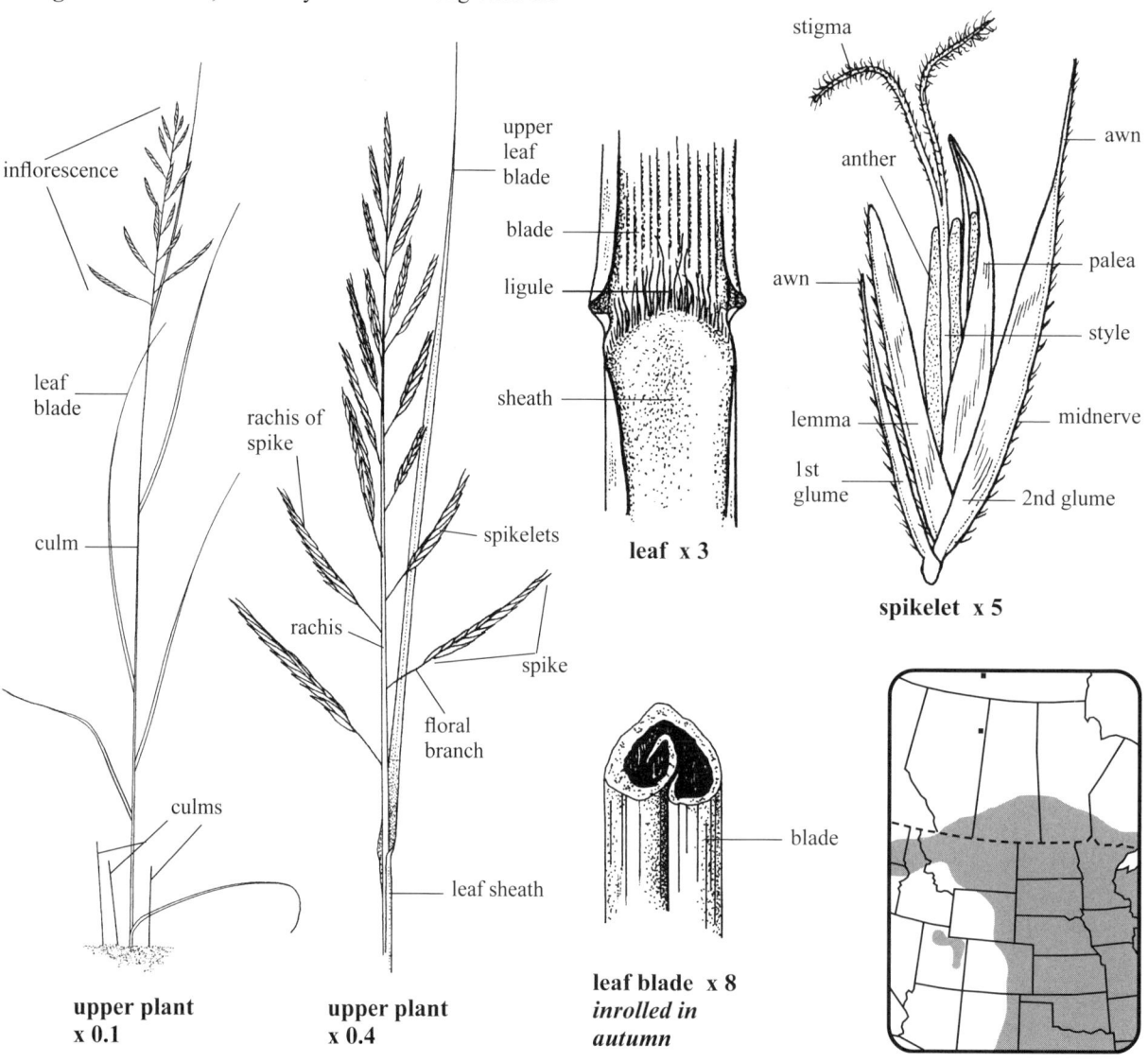

Slough Grass, Freshwater Cord Grass

Grass—*Poaceae* Grass Spikelets 1-flowered

Rough Dropseed *Sporobolus compositus* (Poir.) Merr.

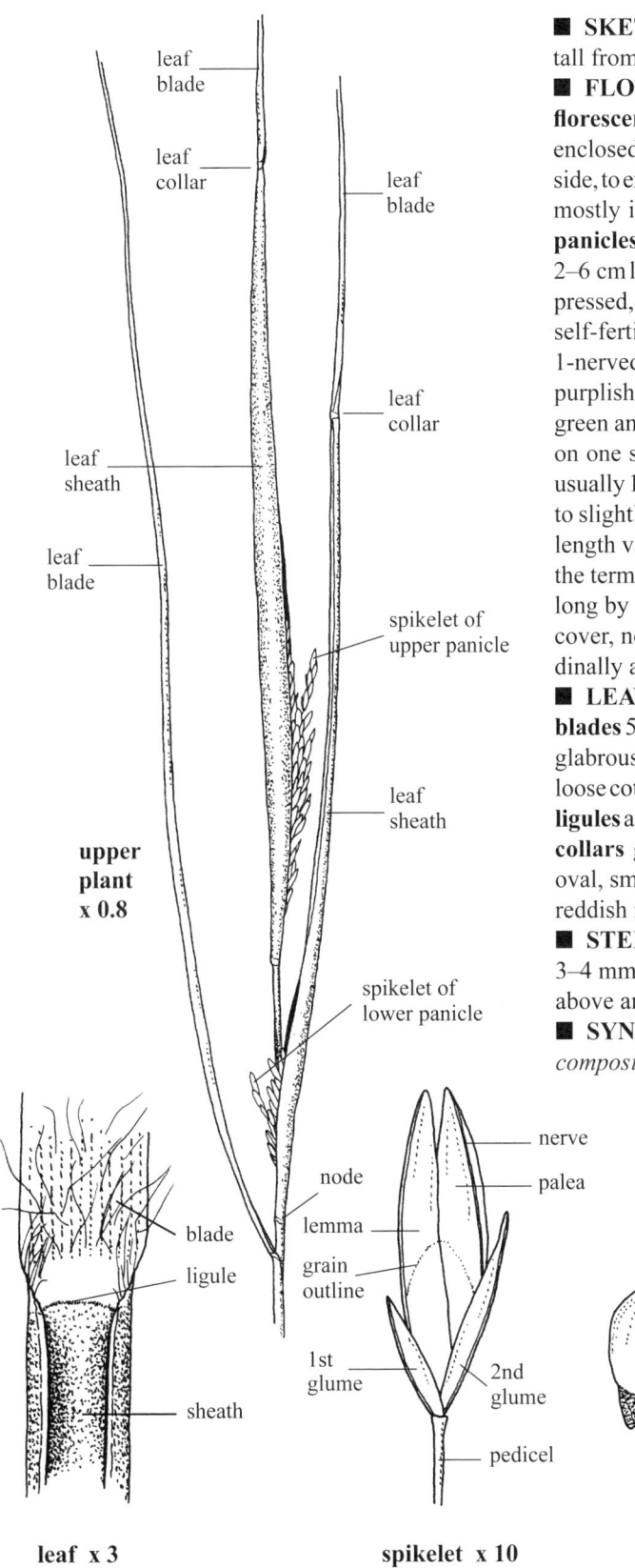

upper plant x 0.8

leaf x 3

spikelet x 10

■ **SKETCH** A variable, tufted **perennial grass** 20–150 cm tall from short **rhizomes**; in prairies and roadsides. **w2, p2**
■ **FLOWERS** Pale green, blooming July–November; **inflorescence** a panicle 5–30 cm long, one to three, contracted, enclosed in the upper leaf sheaths which are open along one side, to exserted; **upper panicle** 10–18 cm long by *c.* 1 cm wide, mostly included at anthesis, slightly exserted in fruit; **lower panicles** reduced, 2–4 cm long, also included; **main branches** 2–6 cm long, erect; **spikelets** 1-flowered, slightly laterally compressed, 4–6.5 (–10) mm long, opaque, awnless, some florets self-fertilized; **glumes** two, unequal, often slightly scabrous, 1-nerved, folded along midnerve, membranous, whitish to purplish; **1st glume** 1.5–4.5 mm long, midnerve tan to bright green and raised; **2nd glume** 2–5.5 mm long, *c.* 0.5 mm wide on one side; **lemmas** membranous, opaque, 3–6.5 mm long, usually 1-nerved, V-shaped, glabrous; **paleae** 2-nerved, shorter to slightly longer than the lemmas; **anthers** 0.2–3.2 mm long, length variable by location, longer ones often exserted and in the terminal panicle. **FRUIT a grain**, 1-seeded, 1.3–2.8 mm long by *c.* 1 mm wide by 0.9–1 mm thick, with a thin papery cover, not gummy when moistened, minutely striate longitudinally and slightly shiny, dark green at base.
■ **LEAVES** Up to ten per culm, ascending to spreading; **blades** 5–70 cm long by 2–10 mm wide, tapered to a fine point, glabrous to hairy below, flat, long ridges and furrows above, loose cottony hairs 1–4 mm long near the bases of lower blades; **ligules** a fringe of white hairs 0.1–0.4 mm long; **auricles** absent; **collars** glabrous above, lower ones hairy; **sheaths** round to oval, smooth, upper ones inflated and enclosing the panicles, reddish near the base, stiff.
■ **STEM** A **culm**, erect, not often branched, solid, smooth; 3–4 mm thick near the base; **nodes** green, glabrous, reddish above and below.
■ **SYN.** *Sporobolus asper* (Michx.) Kunth = *Sporobolus compositus* var. *compositus* (Poir.) Merr.

fruit x 10

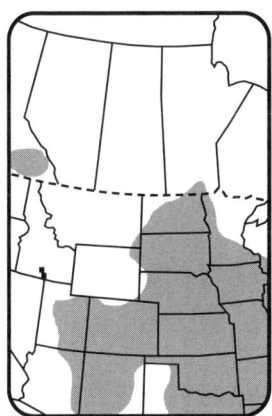

Head-like Dropseed, Composite Dropseed

Grass—*Poaceae*　　　　　　　　　　　　　　　　　　　　　　　　Grass　Spikelets 1-flowered

SAND DROPSEED　*Sporobolus cryptandrus* (Torr.) Gray

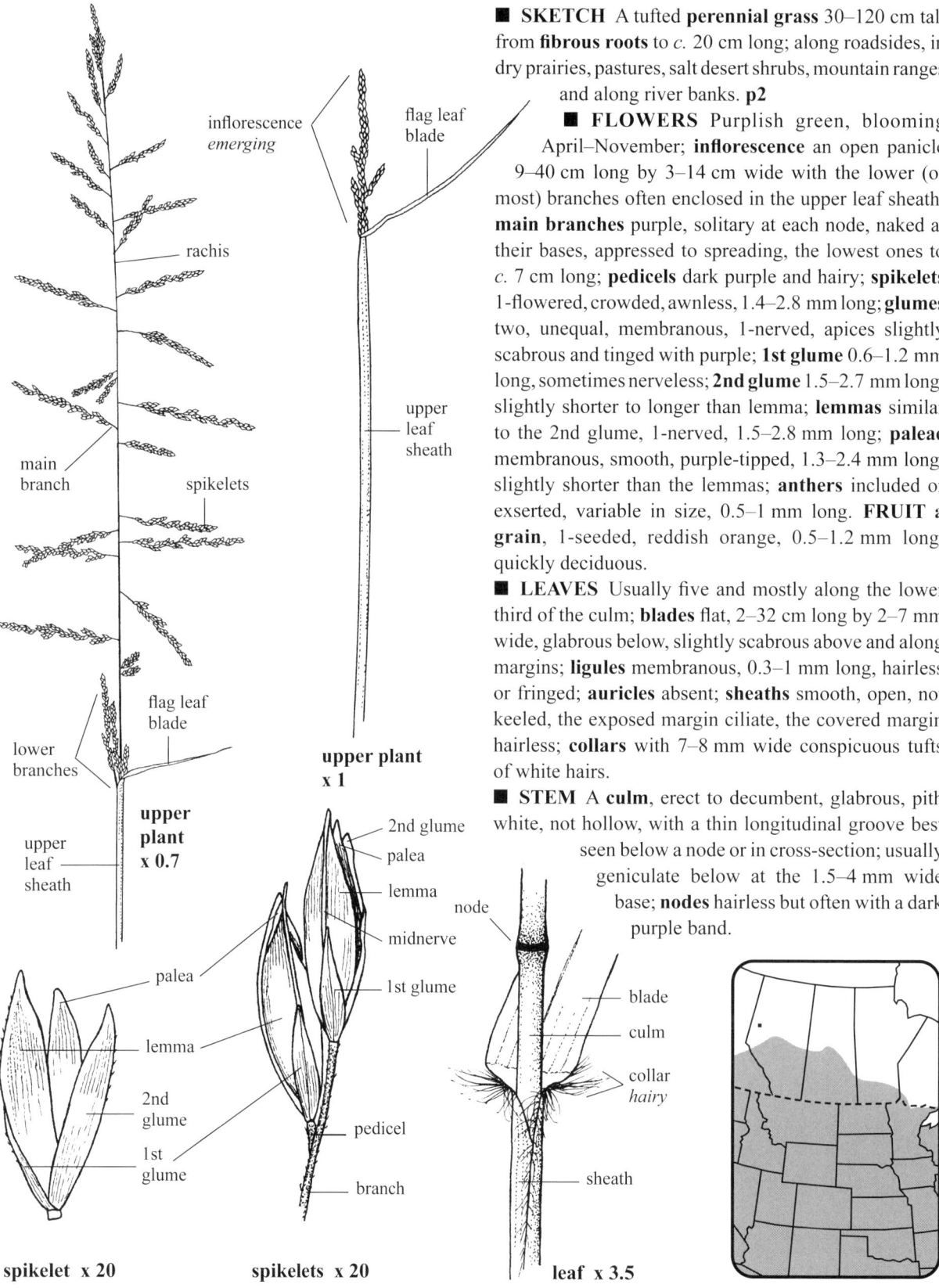

■ **SKETCH** A tufted **perennial grass** 30–120 cm tall from **fibrous roots** to *c.* 20 cm long; along roadsides, in dry prairies, pastures, salt desert shrubs, mountain ranges and along river banks. **p2**

■ **FLOWERS** Purplish green, blooming April–November; **inflorescence** an open panicle 9–40 cm long by 3–14 cm wide with the lower (or most) branches often enclosed in the upper leaf sheath; **main branches** purple, solitary at each node, naked at their bases, appressed to spreading, the lowest ones to *c.* 7 cm long; **pedicels** dark purple and hairy; **spikelets** 1-flowered, crowded, awnless, 1.4–2.8 mm long; **glumes** two, unequal, membranous, 1-nerved, apices slightly scabrous and tinged with purple; **1st glume** 0.6–1.2 mm long, sometimes nerveless; **2nd glume** 1.5–2.7 mm long, slightly shorter to longer than lemma; **lemmas** similar to the 2nd glume, 1-nerved, 1.5–2.8 mm long; **paleae** membranous, smooth, purple-tipped, 1.3–2.4 mm long, slightly shorter than the lemmas; **anthers** included or exserted, variable in size, 0.5–1 mm long. **FRUIT a grain**, 1-seeded, reddish orange, 0.5–1.2 mm long, quickly deciduous.

■ **LEAVES** Usually five and mostly along the lower third of the culm; **blades** flat, 2–32 cm long by 2–7 mm wide, glabrous below, slightly scabrous above and along margins; **ligules** membranous, 0.3–1 mm long, hairless or fringed; **auricles** absent; **sheaths** smooth, open, not keeled, the exposed margin ciliate, the covered margin hairless; **collars** with 7–8 mm wide conspicuous tufts of white hairs.

■ **STEM** A **culm**, erect to decumbent, glabrous, pith white, not hollow, with a thin longitudinal groove best seen below a node or in cross-section; usually geniculate below at the 1.5–4 mm wide base; **nodes** hairless but often with a dark purple band.

Grass—*Poaceae*

Grass Spikelets 1-flowered

Prairie Dropseed *Sporobolus heterolepis* (Gray) Gray

- **SKETCH** A tufted **perennial grass** 30–95 cm tall; in open woodlands, sandhills and moist to dry prairies. **p2**
- **FLOWERS** Light grayish tan, blooming July–November; **inflorescence** an open panicle 8–22 cm long by 2–11 cm wide, most branches ascending to spreading; **main branches** one to three per lower node, glabrous, mostly naked in lower third to half, shiny, lightly reddish green, lowest branch 1–11 cm long; **pedicels** usually appressed, 1–7 mm long; **spikelets** mostly 1-flowered (2-), 4–6 mm long, unawned, all floral parts about equal in color (grayish) and texture (membranous to papery); **glumes** two, unequal, slightly scabrous along the midnerve and pointed tips; **1st glume** 1-nerved or not, 1.5–4.5 mm long by *c.* 0.5 mm wide; **2nd glume** 1- or 3-nerved, 3–6 mm long by 1.3–1.6 mm wide, usually the length of the spikelet; **lemmas** pointed to blunt, 1-nerved, 3.2–4.2 mm long by *c.* 1.5 mm wide; **paleae** 2-nerved, slightly longer to shorter than the lemmas, 3.5–5.2 mm long by 1–1.2 mm wide, mostly fitting inside the lemmas; **anthers** three, dark reddish brown, 1.5–3 mm long. **FRUIT a grain**, 1-seeded, opaque, shiny, round, 1.5–2 mm long and wide.
- **LEAVES** Three or four per culm; **blades** ascending to arched, flat to folded, glabrous to scabrous along the edges, somewhat keeled, usually 7–45 cm long by 1.4–2.5 mm wide, reduced above; **ligules** a white fringed membrane 0.1–0.3 mm long; **sheaths** slightly keeled or not, glabrous above to hairy at lower nodes with some of the hairs 1–4 mm long from the sides and the collar.
- **STEM** A **culm**, solid, light green, erect to leaning, glabrous, in scattered clumps; 1–2 mm thick near the base.

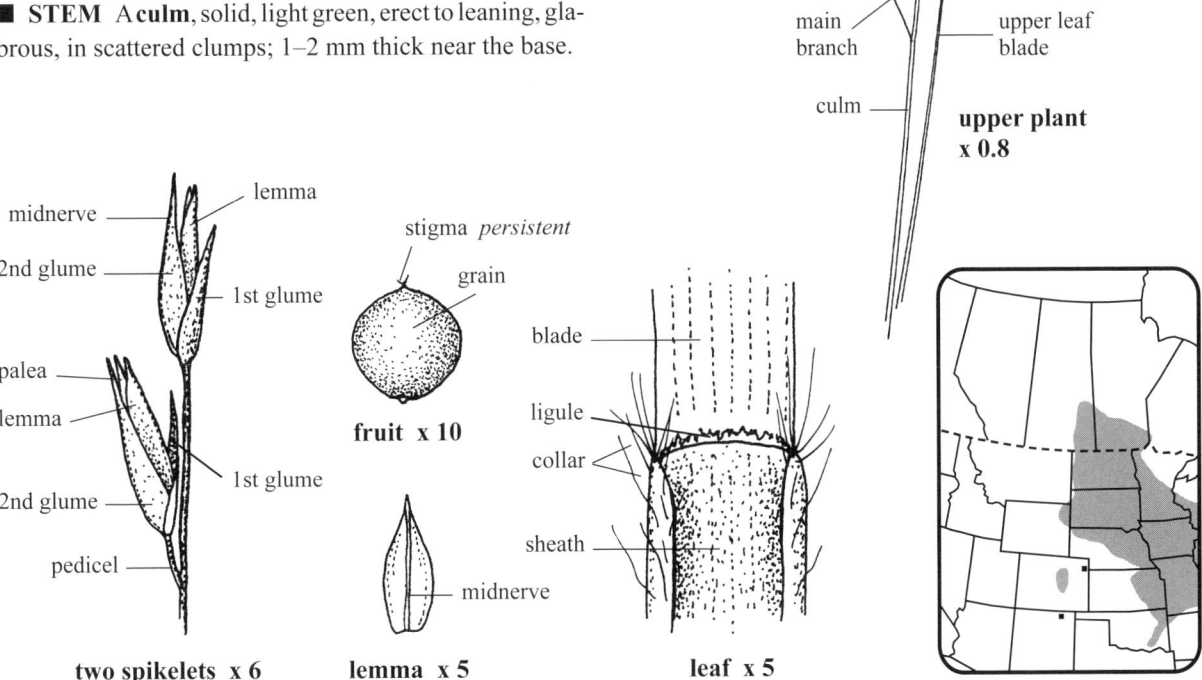

685

Grass—*Poaceae* Grass Spikelets 1-flowered

ANNUAL DROPSEED *Sporobolus neglectus* Nash

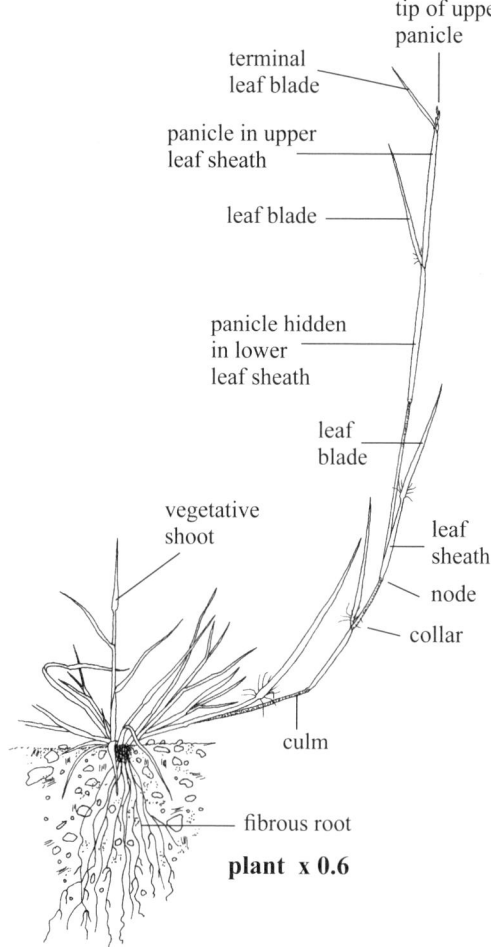
plant x 0.6

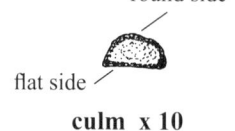
culm x 10
cross-section

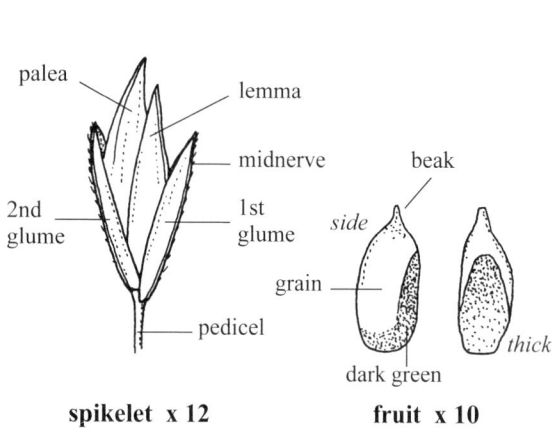
spikelet x 12

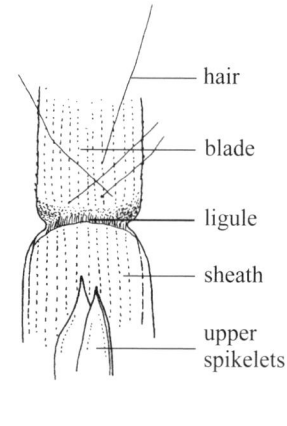
fruit x 10

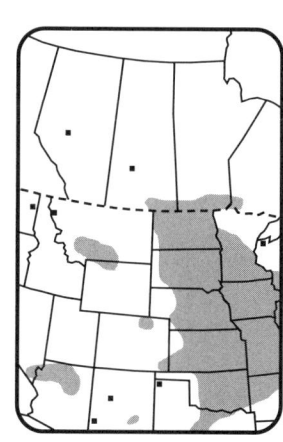
leaf x 7

■ **SKETCH** A tufted **annual grass** 10–45 cm tall from **fibrous roots** 5–10 cm long; in disturbed sandy or gravelly sites along railways, in sandy fields and parking lots. **w2**

■ **FLOWERS** Pale green, turning reddish purple when exposed to light, blooming August–November; **inflorescence** a panicle, contracted, 3–6 cm long by 2–5 mm wide, mostly enclosed in a leaf sheath, terminal and lateral; **main culms** two to five, each with one to three panicles sometimes partially exserted from their sheaths; **spikelets** 1-flowered, 1.8–2.8 mm long, pale green, membranous, slightly shiny, laterally flattened; **glumes** two, 1-nerved, subequal, shorter than floret; **1st glume** 1.2–2.5 mm long, scabrous on midnerve or not; **2nd glume** 1.5–2.6 mm long, scabrous on midnerve or smooth; **lemmas** 1-nerved, glabrous, awnless, pointed, 1.7–2.8 mm long, some purple-tinged; **paleae** 2-nerved, 1.5–3 mm long, shorter than to slightly longer than lemmas, smooth, pointed; **anthers** three, enclosed, 0.1–1.8 mm long, short and apparently nonfunctional below, terminal panicles with the longer, fertile anthers. **FRUIT a grain**, 1-seeded, brownish orange, 0.8–2 mm long by 0.8–0.9 mm wide by *c.* 0.6 mm thick with a dark green area along one side and the base, short-beaked, falling quickly from the spikelet.

■ **LEAVES** Several per culm; **blades** tapered to a fine point, 1–12 cm long by 1–2 mm wide, reduced above, ascending, scabrous along the margins, scattered white hairs 1–5 mm long, some papillose-based on upper or both sides at the base of blades, especially the lower blades; **lowest blades** brown by anthesis; **ligules** a dense fringe of white hairs 0.1–0.3 mm long; **auricles** absent; **collars** reddish purple; **sheaths** glabrous, slightly keeled, open and inflated to hold a panicle.

■ **STEM** A **culm**, decumbent to geniculate below, solid, glabrous to scabrous, wiry, flat on one side, other side round with several low ridges or longitudinal veins; 1–1.5 mm thick near the base; **nodes** slightly swollen, glabrous, reddish; **tillers** usually present.

■ **SYN.** *Sporobolus vaginiflorus* var. *neglectus* (Nash) Scribn.

Small Dropseed, Puffsheath Dropseed, Poverty Grass

ANNUAL WILD-RICE *Zizania aquatica* L.

■ **SKETCH** A variable aquatic **annual grass** 40–260 cm tall from orangish tan **fibrous roots** 1–15 cm long by 1–2 mm wide; in shallow slow moving water along shores of lakes, streams and ponds; **monoecious**. Cultivated in lakes. *Naturalized* **w1, p1**

■ **FLOWERS** Green to pale pink, blooming July–September; **inflorescence** a panicle 15–60 cm long; **flowers** unisexual, female above the male, all floral branches swollen and often hairy at base; **male branches** ascending to drooping, usually 2–4 per node, 7–15 cm long; **female branches** *c*. 20, erect to ascending, 5–12 cm long, reduced above; **rachis** glabrous; **male spikelets** numerous, 1-flowered, hanging; **glumes** absent; **lemmas** 5-nerved, midnerve hairy, pointed, 5–11 mm long with a hairy awn 0–3 mm long; **paleae** 3-nerved, margins hyaline, apices pointed, midnerve hairy or scabrous; **stamens** six; **filaments** clear, filiform, 1–2 mm long; **anthers** greenish yellow, 4–7.5 mm long; **female spikelets** ascending, 1-flowered, together 2–3 cm wide at anthesis by 12–33 cm long, some to *c*. 16 cm wide as fruit ripen; **glumes** absent; **lemmas** translucent, 10–12 mm long by *c*. 4 mm wide (flattened), with five rounded nerves, midnerve hairy at its base and apex, lateral four nerves hairy, nerves converging at hairy apices; **awns** erect, 1.5–6.5 cm long, hairs ascending; **paleae** 3-nerved, 12–15 mm long by *c*. 2 mm wide (flattened), margins involute, hairy apices pointed and fitting tightly inside the lemmas. **FRUIT a grain**, 1-seeded, 180–275, dark brown, slightly shiny, 8–15 mm long by 1.3–2 mm wide and thick, deeply grooved on lemma side, shallow groove on palea side, enclosed and dropping within firm palea and lemma.

■ **LEAVES** Usually three per culm, arching to spreading, margins scabrous with stiff hairs ascending; **blades** dull, mostly 5–100 cm long by 8–40 mm wide, midrib raised below; **ligules** membranous, 7–23 mm long, pointed, sometimes torn, white with brown streaks near margins; **auricles** absent; **sheaths** open almost to base, loose, cross-septate on inner side.

■ **STEM** A **culm**, erect, stiff, unbranched, hollow, smooth and light yellowish green; 1–2 cm wide near the base; often with several **vegetative shoots** clustered at the base; **nodes** darker green, hairs microscopic and ascending.

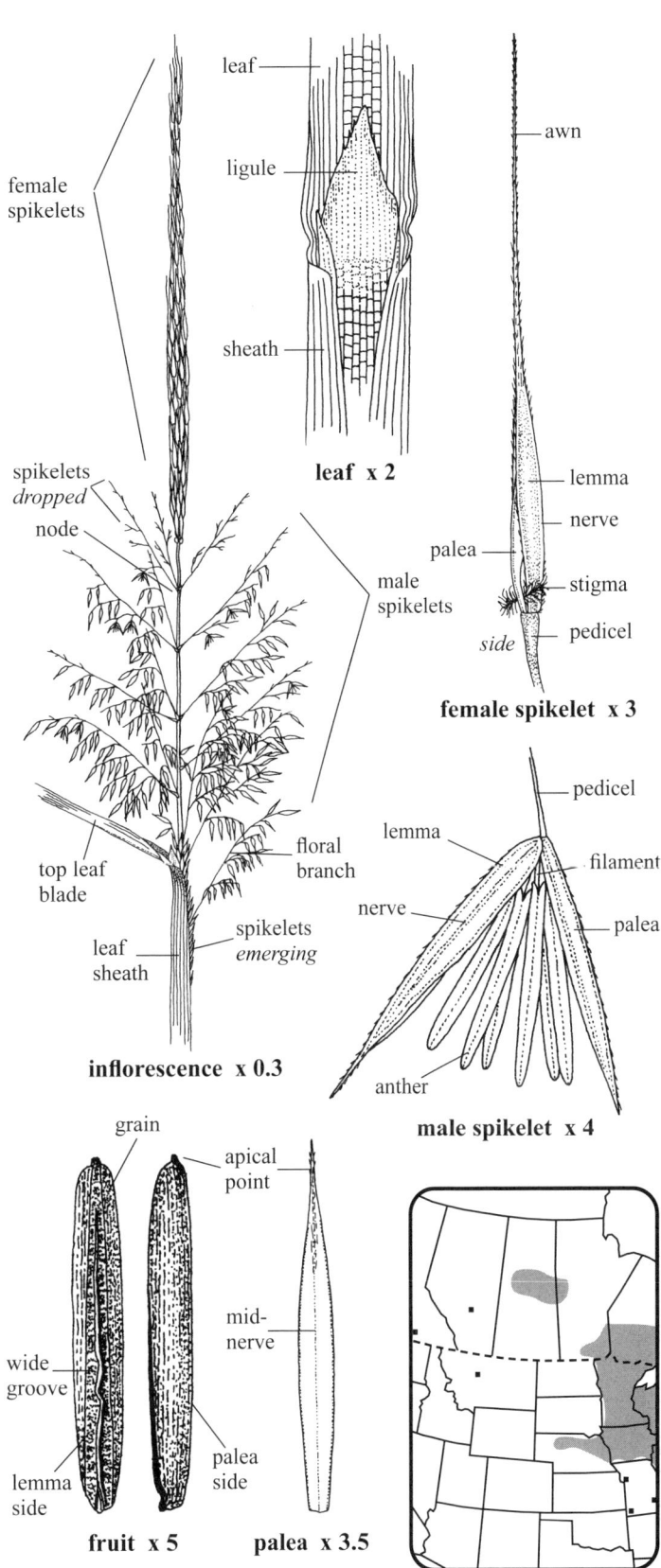

Pondweed—*Potamogetonaceae* | **Wildflower** Green Sepals (bracts) 4

NORTHERN PONDWEED *Potamogeton alpinus* Balbis

- **SKETCH** An aquatic, glabrous **perennial herb** up to 2 m in length depending on water's depth, with a reddish green **rhizome** 1–1.5 mm thick with **fibrous roots** 1–10 cm long at **nodes** 3–7 cm apart; in quiet lakes and ponds. w1
- **FLOWERS** Green, blooming July–September; **inflorescence** spikelike, 1–4 per stem, each 1.1–3.5 cm long by 4–6 mm wide at anthesis, mostly above water; **peduncles** reddish green, 2.5–15 cm long by 2–3 mm thick, often curved; **flowers** perfect, 24–50 per spike, spirally arranged, congested, each 2.5–3 mm wide by 1.5–2 mm tall, short-stalked; **sepals** (bracts) four, green, 1–1.4 mm long by 1.2–1.8 mm wide, C-shaped; **petals** absent; **stamens** four, one included under each sepal; **filaments** obscure; **anthers** white, 1–1.5 mm long and wide, 2-lobed; **pistils** four (5–8), each *c.* 1.4 mm tall by *c.* 0.7 mm wide by *c.* 0.5 mm thick, widely ridged on dorsal (outer) side; **ovary** green, slightly flattened; **style** obscure; **stigma** curved, pointing outward, *c.* 0.5 mm long, apex free and turning brown after anthesis; **fruiting heads** cylindrical, 1.5–3 cm long by 9–11 mm wide, usually horizontal on or below the surface. **FRUIT** drupelike, light olive brown, 2.5–3.8 mm long by 1.8–2.8 mm wide by 1.1–1.9 mm thick, smooth or dorsal keel prominent, slightly shiny; **beak** 0.2–1 mm long, curved outward.
- **LEAVES** Floating (opposite) and submerged (alternate); **1) floating leaves** at end of stem or absent, green; **blades** thin, entire, 4–6 (–10) cm long by 1–2.5 cm wide, roundly pointed, tapered to a petiole; **petioles** reddish green, often twisted once, 1–2 cm long, winged, wrinkled near blade's base; **stipules** membranous, transparent, pointed, 1–4 cm long, open along one side, apices often torn, veined, usually ascending and angled away from stem and petiole; **2) submerged leaves** 3–18 cm long by 5–15 (–20) mm wide, sessile, tapered at base, apices widely pointed, blades often twisted, 6- or 7- (11-) nerved, midrib wide near base with obvious net-veining on both sides, brownish green to reddish.
- **STEM** Round, submerged or at the surface, brown, simple or rarely branched, solid, smooth; 2–3 mm thick.

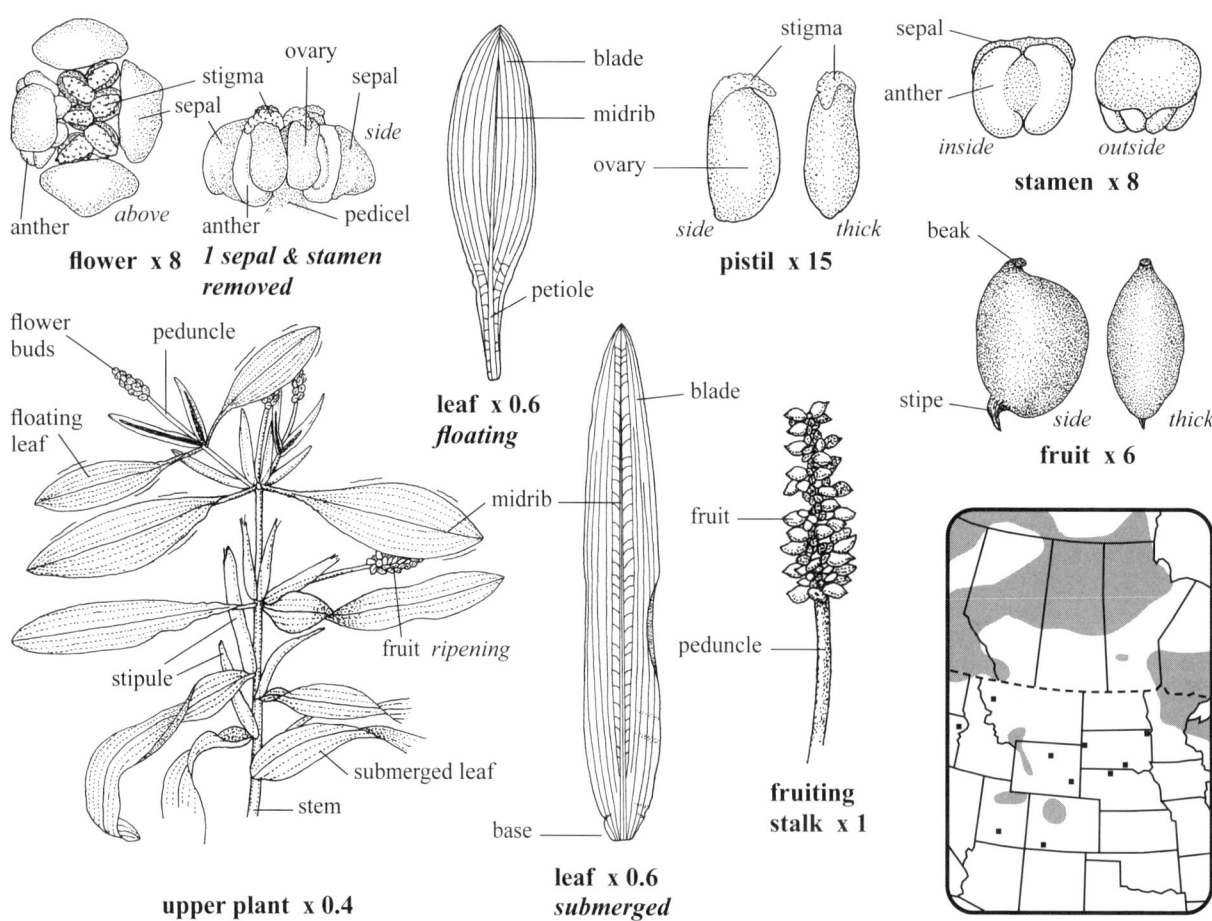

flower x 8   *1 sepal & stamen removed*   pistil x 15   stamen x 8

leaf x 0.6 *floating*   fruit x 6

upper plant x 0.4   leaf x 0.6 *submerged*   fruiting stalk x 1

Reddish Pondweed, Alpine Pondweed

Pondweed—*Potamogetonaceae* — Wildflower — Brown and white — Sepals (bracts) 4

LEAFY PONDWEED *Potamogeton foliosus* Raf.

■ **SKETCH** An aquatic, hairless **perennial herb** 30–80 cm long from **rhizomes**; in shallow water of ponds, lakes, streams and rivers; **turions** uncommon. w1

■ **FLOWERS** Brown and white, blooming June–August; **inflorescence** a spike, several per plant, each 1.5–7 mm long by *c.* 1.5 mm wide, often at ends of branches; **peduncles** 3–10 mm long, from leaf axils, light green to tan, slightly compressed at base and slightly thicker below the spike; **flowers** perfect, sessile, 2–8 per spike, in one or two whorls, crowded, 1.3–1.4 mm wide by *c.* 0.7 mm tall, staggered at the top of the peduncle; **sepals (bracts)** four, brown, transparent, covering the bases of the anthers, *c.* 0.8 mm long by 0.5–0.7 mm wide, tapered to a claw; **stamens** four, *c.* 0.5 mm long, exserted; **filaments** hidden, minute; **anthers** white, 2-lobed, each with a faint suture to the outside; **pollen** white; **pistils** four per flower, each 0.6–0.8 mm long; **ovary** green, tapered above to an obscure style; **stigmas** four, pinkish brown, rough, the four together *c.* 0.8 mm wide; **fruiting heads** ripen underwater, each 4–5 mm wide by 2–3 mm tall. **FRUIT** drupelike, 1-seeded, 1–4 per flower, 1.5–2.7 mm long by 1.3–2.2 mm wide by 0.9–1.3 mm thick, slightly shiny, spreading, olive to greenish brown, hairless, sides mostly rounded (convex), the dorsal keel obvious and often undulate, more prominent near the apex; **beak** 0.2–0.4 mm long, angled outward towards the dorsal keel; **embryo** one per fruit, white, curved, as in other *Potamogeton*.

■ **LEAVES** Alternate, sessile, entire, submerged, simple, brownish; **blades** 1- or 3-nerved, tapered and thin, darker brown near the pointed apices, 1.3–8.2 cm long by 0.3–2.3 mm wide, side nerves obscure, midrib light brown and most obvious in lower half of blade; **stipules** brownish, transparent, open and free to base, apices bluntly pointed, often torn and ragged, bases truncate, 0.4–2 cm long by *c.* 0.9 mm wide (flat), tubular, erect around the stem or angled away from it.

■ **STEM** Submerged, compressed, solid, light tan, branching and rebranching; 0.5–1 mm wide; **branches** alternate or opposite above.

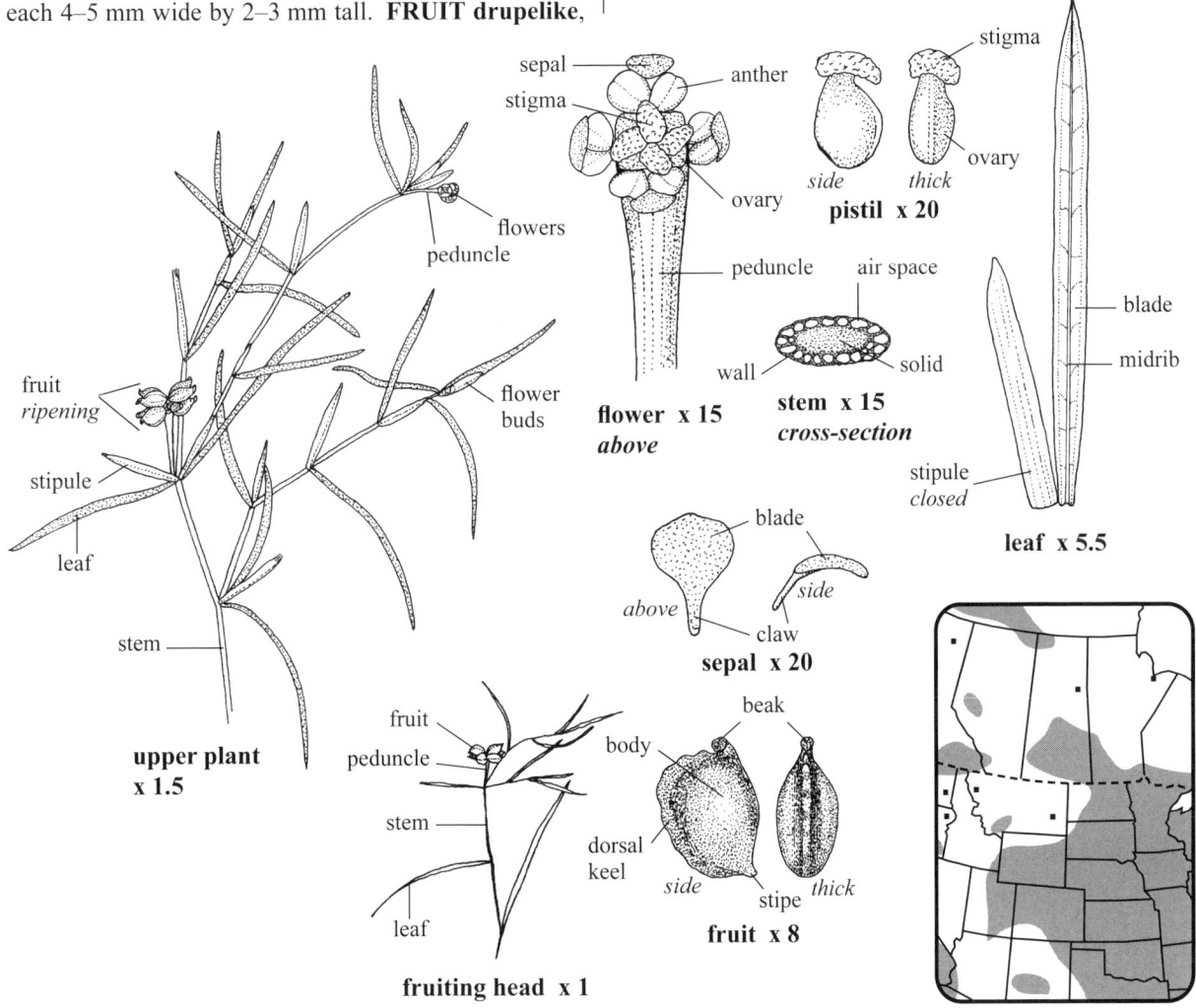

689

Pondweed—*Potamogetonaceae* **Wildflower** Whitish green Sepals (bracts) 4

FLOATING-LEAF PONDWEED *Potamogeton natans* L.

■ **SKETCH** A glabrous, aquatic **perennial herb** 0.3–1⁺ m long with whitish tan **rhizomes** several cm long and 2–3 mm thick with a few **fibrous roots** 2–3 cm long; in quiet ponds and lakes, shallow to deep water. w1

■ **FLOWERS** Whitish green, blooming July–August; **inflorescence** a spike, terminal, cylindrical, 1.7–5 cm long by 7–10 mm wide at anthesis, usually 2–5 cm above the water's surface; **peduncles** brown, 3–13 cm long by 3–4 mm wide, from leaf axils; **flowers** perfect, sessile, congested, each 3–4 mm wide by $c.$ 3 mm tall, spirally arranged; **sepals** (bracts) four, $c.$ 2 mm long by $c.$ 2.8 mm wide (flat), green with brownish apices; **stamens** four, included, one under each sepal; **filaments** green, 0.5–0.7 mm long; **anthers** white, 2-lobed, 1.6–1.8 mm long by $c.$ 1.8 mm wide; **pistils** four, green, squarish, each 2–2.2 mm long by 0.8–1 mm wide by 0.7–0.9 mm thick, the sides blunt; **ovary** smooth, green, 3- or 4-sided, rectangular; **style** obscure; **stigmas** light pink, each 1.3 mm long, arched; **fruiting heads** lying on water's surface, 3–5 cm long by 11–13 mm wide. **FRUIT** drupelike, 1-seeded, greenish to reddish brown, smooth and shiny, 3–5.8 mm long by 2–3.3 mm wide by 2.5–3 mm thick, in groups of four, dorsal keel round and obscure to slightly raised and narrow, sides convex.

■ **LEAVES** Floating and submerged, alternate, simple, entire; **1) floating**, several on the surface, flat; **blades** 3–11 cm long by 1–6 cm wide, mostly 19- to 35-veined, brownish green, shiny on both surfaces; **petioles** brown and pale green near bent apices, 3–30 cm long by 2–3 mm wide, reduced above; **stipules** membranous, stiff, brown, pointed, transparent, 4–10 cm long by 5–7 mm wide near base (unopened), free to base and ascending away from the stem, tubular; **2) submerged leaves** brown, sessile, rigid, linear, 10–40 cm long by 2–3 mm wide, concave on one side, convex on the other side, nerves few, obscure.

■ **STEM** Brown, round, hairless, grooved and ridged, solid but soft; 2–3 mm wide.

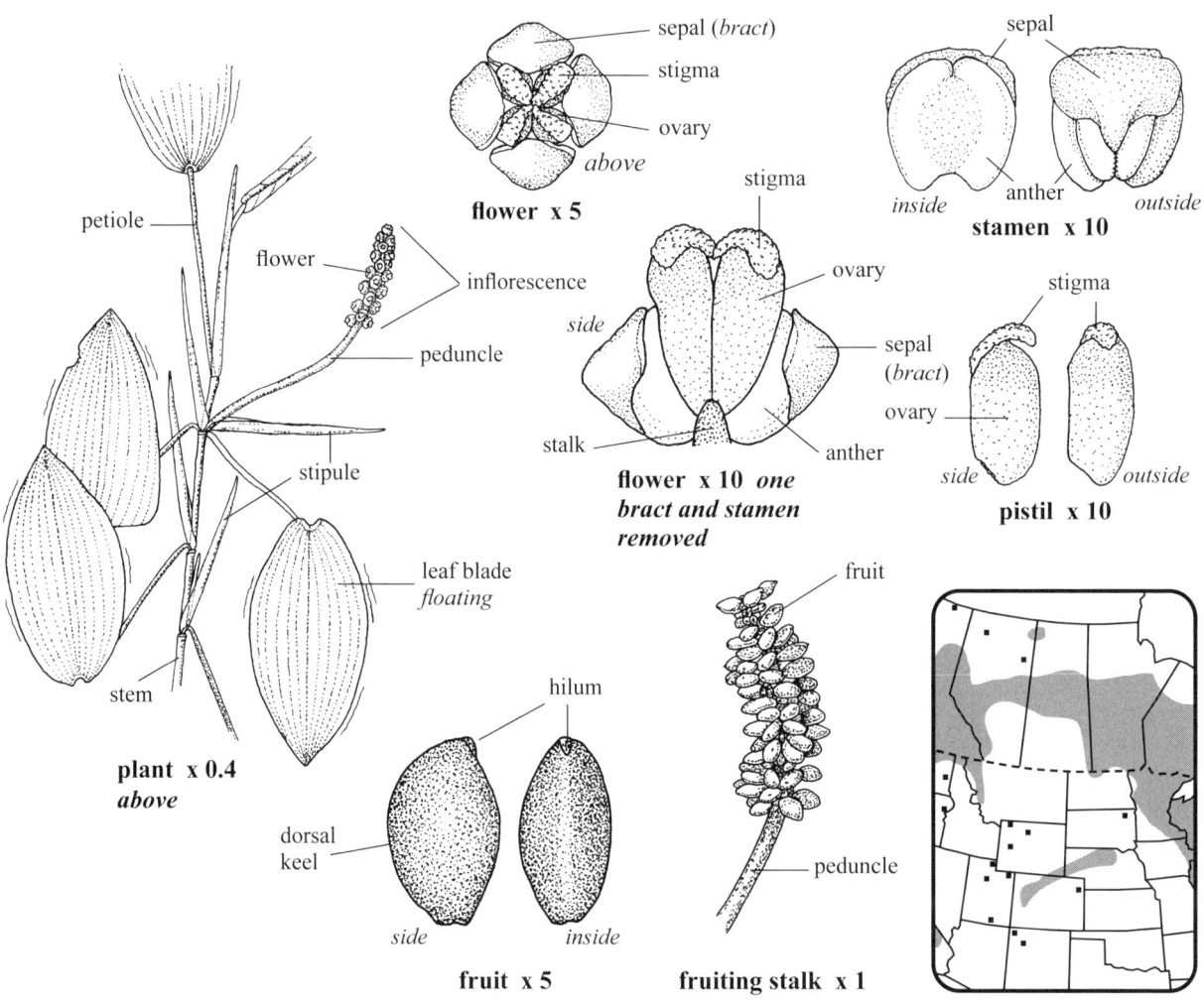

690 Floating Pondweed

Pondweed—*Potamogetonaceae* **Wildflower** Reddish green Sepals (bracts) 4

RICHARDSON'S PONDWEED *Potamogeton richardsonii* (Benn.) Rydb.

■ **SKETCH** A glabrous, aquatic **perennial herb**, mostly submerged, 30–100 cm tall with white to pink, unspotted **rhizomes** 1–2 mm wide by 5–50 mm long with **nodal roots** to *c.* 11 cm long, in scattered colonies; **turions** absent; in shallow to deep (4 m) freshwater of ponds, sloughs, lakes and streams. **w1**

■ **FLOWERS** Reddish green, blooming April–August; **inflorescence** a raceme (spikelike), terminal and axillary, one to four, each 10–40 mm long by 5–7 mm wide, with 4–12 whorls of flowers; **peduncles** slightly flattened, solid, not ridged, 1.5–20 cm long by 2–2.2 mm wide, some curved with fruit; **pedicels** 0.2–1.5 mm long, terminal one the longest; **flowers** perfect, 10–48 per raceme, each 3.5–4 mm wide by 2–3 mm tall; **rachis** green, *c.* 1 mm thick, round; **sepals** (bracts) four, green, *c.* 1.5 mm long by 1.5–2 mm wide, clawed, entire, glabrous, apices and margins hyaline; **stamens** four, included; **filaments** obscure; **anthers** 1.1–1.7 mm long by 1.5–1.7 mm wide, white; **pistils** four, each *c.* 1.8 mm long by *c.* 0.8 mm wide and distinct; **ovary** green, 1–1.3 mm long, glabrous, rounded, ridged along inner face; **style** collarlike; **stigmas** dark reddish green, *c.* 0.6 mm wide and tall, rough; **fruiting heads** each 15–22 mm long by 7–10 mm wide. **FRUIT** drupelike, 1-seeded, greenish brown, one to four per flower, each 2.5–4 mm long (including beak) by 1.7–3 mm wide by 1.7–1.9 mm thick, smooth, rarely keeled; **beak** 0.5–0.9 mm long, flat, angled.

■ **LEAVES** Alternate, simple, entire, submerged, spreading, glabrous, undulate, multi-nerved but three most obvious, pointed; **blades** sessile with auricles clasping, 1.5–13 cm long by 4–28 mm wide; **stipules** whitish, 0.8–2 cm long, eventually fragmenting.

■ **STEM** Flexous, leafy, simple to branched, roundish, not ridged, light green to pinkish; 1–2 mm thick.

■ **SYN.** *Potamogeton perfoliatus* var. *richardsonii* Benn.

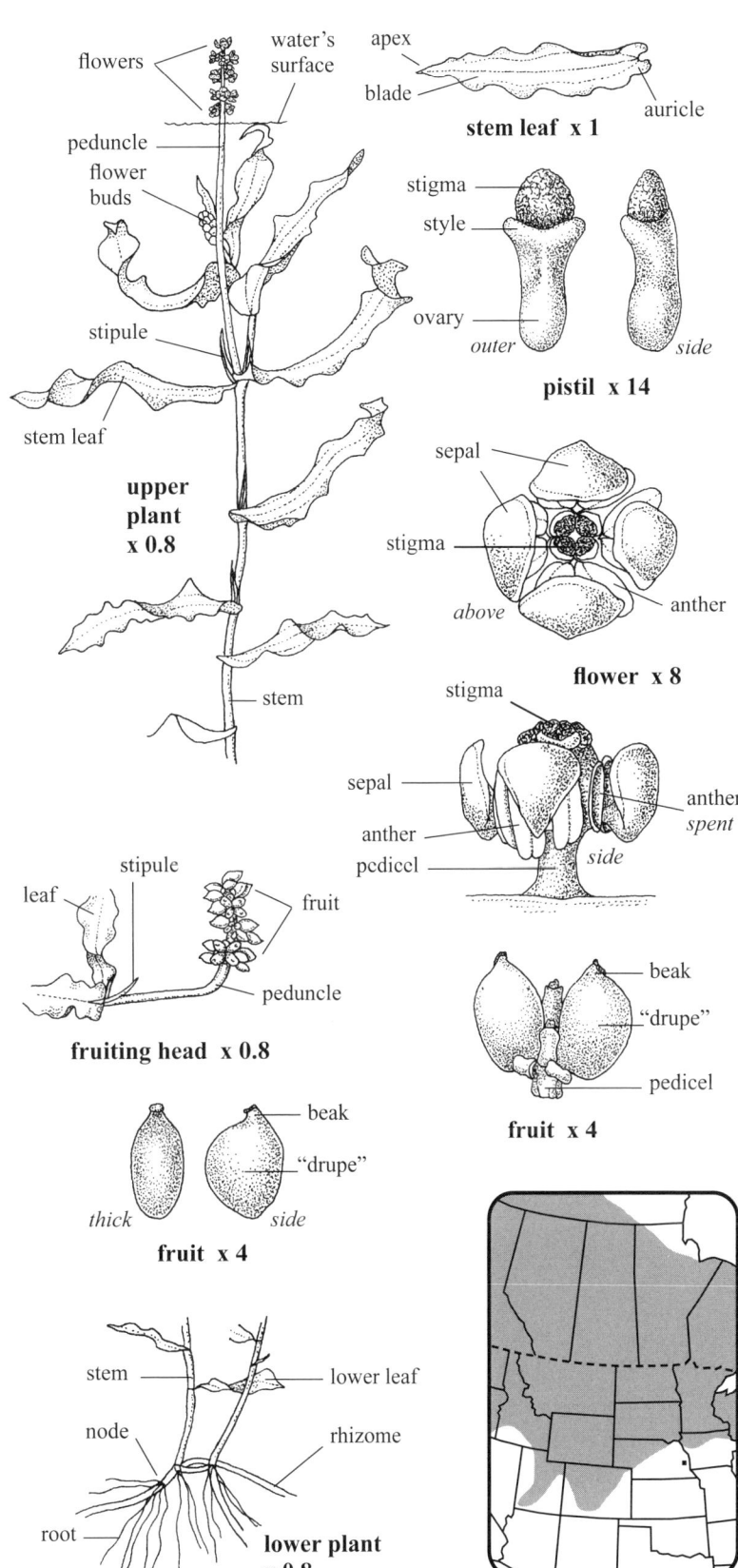

Red-headed Pondweed, Claspingleaf Pondweed

Pondweed—*Potamogetonaceae* — Wildflower Reddish green Sepals (bracts) 4

Slender Pondweed *Stuckenia filiformis* (Pers.) Börner

■ **SKETCH** A variable, glabrous **perennial aquatic herb** 10–100 cm long from white, smooth **rhizomes** *c.* 2 mm thick with a few **roots** at each node; in shallow, standing to gently flowing water of ditches and saline ponds, often a congested mass of thin leaves and stems covering several square meters. **w1**

■ **FLOWERS** Reddish green, blooming July–August; **inflorescence** a spike 1–5 cm long by 2–3 mm wide, with two to six whorls of flowers, the lower often remote; **peduncles** 2–15 cm long by 0.5–0.7 mm thick, pinkish, glabrous; **flowers** perfect, 2–2.5 mm wide by 1.5–2 mm tall, sessile, more congested above, some above the water's surface, terminal and lateral, usually two or three on opposite sides of a whorl; **sepals** (bracts) four per flower, each dark green, C-shaped, *c.* 1 mm wide, glabrous, clawed; **stamens** four, included; **filaments** obscure; **anthers** white, *c.* 0.5 mm long, the two lobes at the base of the bracts; **pollen** white; **pistil** brown, glabrous, *c.* 1 mm long by 0.7 mm wide; **style** obscure; **stigmas** rough, pinkish brown, the four together *c.* 1 mm across, each *c.* 0.8 mm long, pointed at the outside. **FRUIT** drupelike, 1-seeded, glabrous, brown, 2–3 mm long by 1.8–2.3 mm wide by 1.1–1.6 mm thick, round on back, keels and lateral ridges obscure, not transparent; **beak** obscure, flat, rough; **embryo** one per fruit, curved, tapered, 1.5–1.8 mm long by 1.2–1.3 mm wide by 0.5–0.7 mm thick, thickest at the base, finely striated, bone white to brown, round in cross-section.

■ **LEAVES** Alternate, simple, entire, mostly submerged or some at the surface, flat, apices tapered but not sharply pointed, midrib visible, lateral veins obscure; **blades** narrowly linear, 5–15 cm long by 0.2–3.2 mm wide, 1- or 3-nerved, often with one small shoot from the sheaths of the larger leaves; **ligules** a free, membranous elongation 5–20 mm long of the sheath, loose, single, apices blunt; **stipules** a membranous sheath 1–4 cm long, united to base of leaf blades for 5–10 mm, wrapped tightly around the stem but open to its base at the node.

■ **STEM** Erect to horizontal, floating, slightly angled, with marginal air pockets, glabrous, branched from lower nodes, less branched above; 1–2 mm thick near the base; **sheaths** tan, membranous, 10–15 mm long at each underground node.

■ **SYN.** *Potamogeton filiformis* Pers. = *Stuckenia filiformis* ssp. *filiformis* (Pers.) Börner.

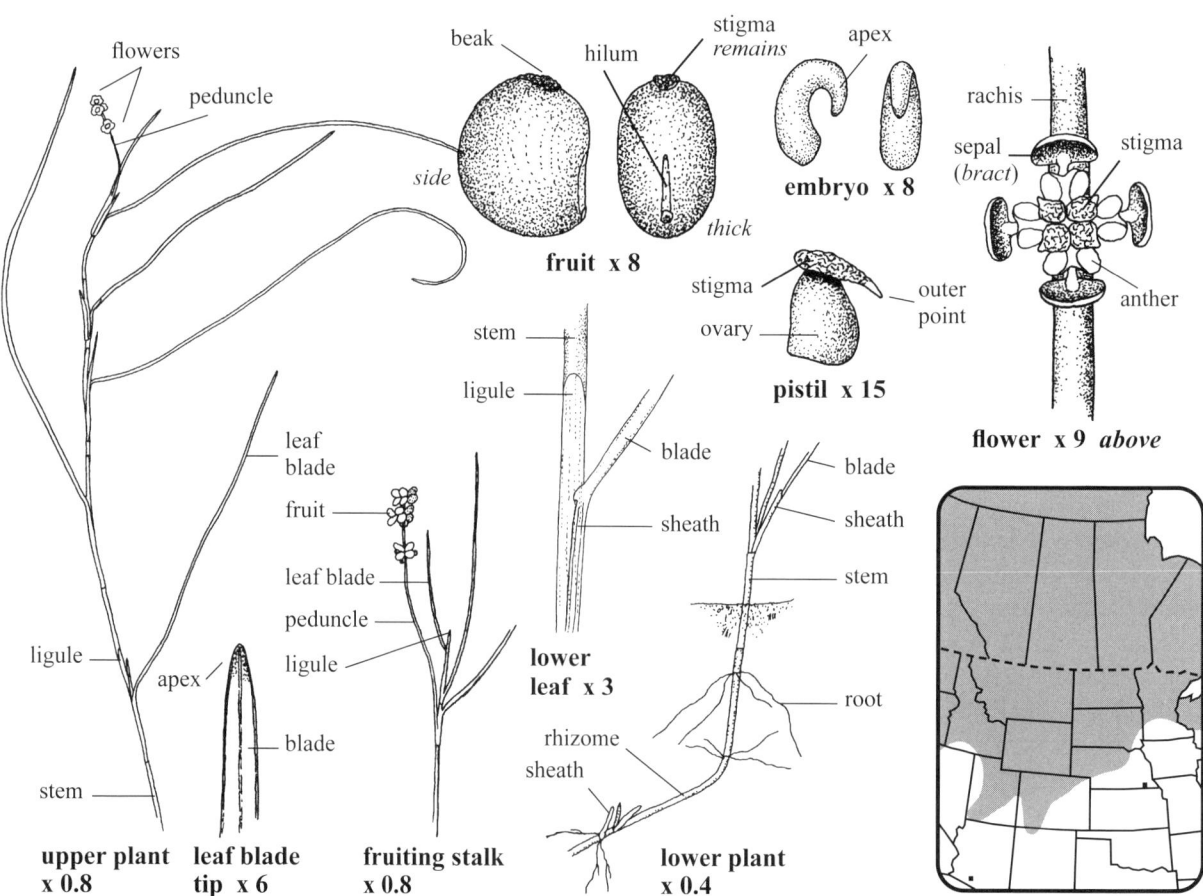

692 Thread-leaved Pondweed, Fineleaf Pondweed, Slender-leaf False Pondweed

Pondweed—*Potamogetonaceae* — **Wildflower** Brown Sepals (bracts) 4

Sago Pondweed *Stuckenia pectinata* (L.) Börner

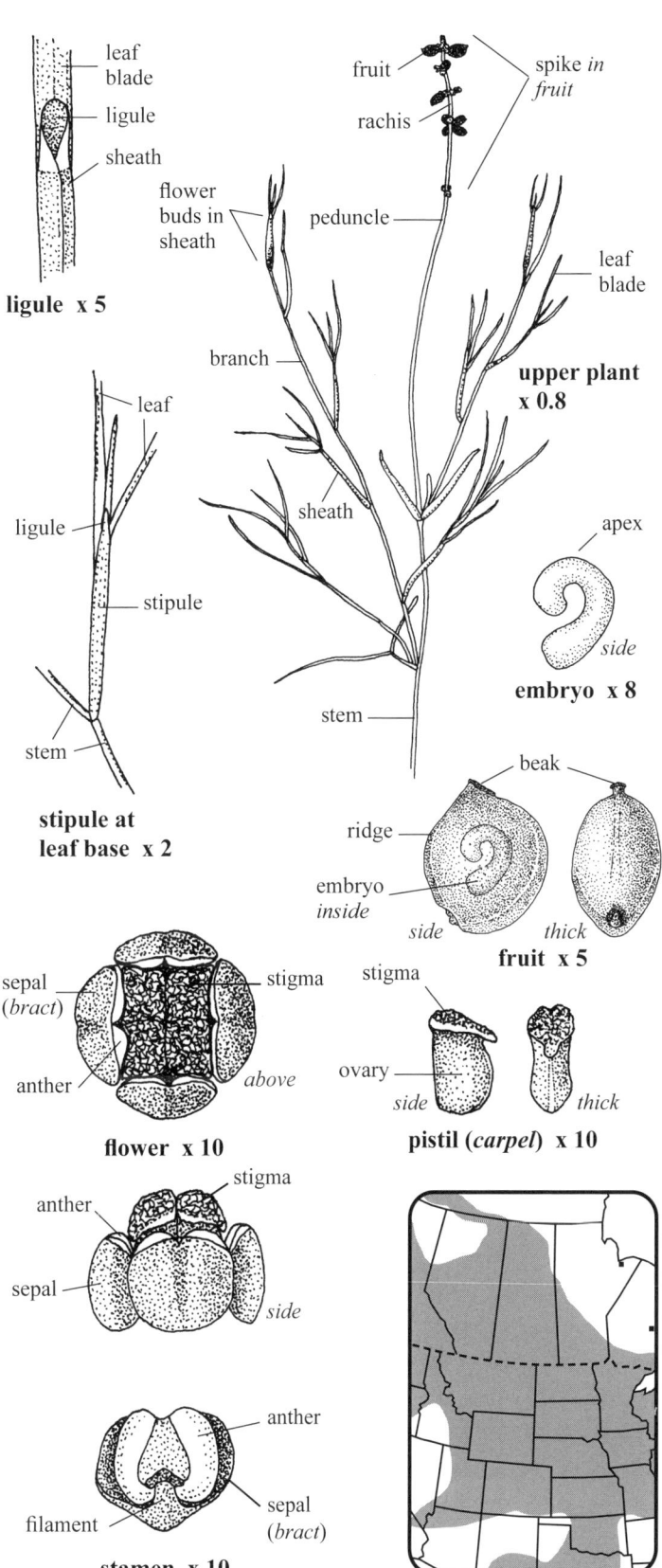

■ **SKETCH** A glabrous **aquatic perennial herb** 30–150 cm long from creeping **rhizomes** *c.* 1 mm thick with **tubers**, forming colonies several m^2; in shallow freshwater lakes, ponds and streams. **w1**

■ **FLOWERS** Brown, blooming June–September; **inflorescence** a spike, terminal and axillary, each 1–5 cm long by *c.* 5 mm wide, to *c.* 9 mm wide in fruit, near or at the water's surface; **flower buds** 2.5–3 mm wide by 2–2.5 mm tall; **peduncles** tan, not stiff, 3–15 cm long by *c.* 0.8 mm thick, not erect with fruit, often prostrate at or near the water's surface; **flowers** perfect, sessile, in 2–6 whorls, more crowded above; **sepals (bracts)** four, brown, membranous, C-shaped, 1.5–2 mm long and wide; **petals** absent; **stamens** four, under and inserted at the bases of the sepals; **filaments** brown, thick *c.* 1 mm long; **anthers** white, 2-lobed, *c.* 1 mm long by *c.* 1.3 mm wide; **pollen** white; **style** very short; **carpels** four, brown, apices touching; **stigmas** together *c.* 2 mm wide (corner to corner), outer end pointed, rough, each *c.* 1.5 mm long by *c.* 1 mm wide (in flower buds). **FRUIT** drupelike, brown, shiny, 2.5–4.3 mm long by 2.5–3.8 mm wide by 2–2.8 mm thick, 2-keeled, these wide and not distinctive, coat has pale reticulations below the smooth surface when fresh and wet; **beak** 0.2–1 mm long, angled; **embryos** white, soft, *c.* 2 mm long by *c.* 1.5 mm wide, centered in the fruit.

■ **LEAVES** Alternate, green, submerged, tapered to a fine point or blunt, glabrous, oval in cross-section, becoming flatter near apices, 3–15 cm long by 0.4–1 mm wide by 0.5–0.7 mm thick, margins entire and rounded, nerves three but not obvious on fresh leaves, appearing mostly 1-nerved; **stipules** forming a sheath 1.1–5 cm long, usually enclosing a fascicle of three or four new leaves or a single leaf blade to its base, open to the base with the sides overlapping, margins membranous and whitish when dry; **ligules** 1–2 mm long, erect, membranous, entire to somewhat ragged and torn, blunt.

■ **STEM** Lax, filiform, tan, not hollow, round, ridged lengthways (microscopic), branched throughout; 0.7–1 mm thick.

■ **SYN.** *Potamogeton pectinatus* L.

Fennel-leaved Pondweed, Slender Pondweed, Bushy Pondweed, Sago False Pondweed

Greenbrier—*Smilacaceae*

Vine   Green   Perianth parts 6

CARRIONFLOWER   *Smilax lasioneura* Hook.

■ **SKETCH** A variable **perennial herbaceous vine** 1–2.5 m tall, unarmed, with tendrils and a **rhizome** 7–20 mm thick and to *c.* 10+ cm long and knobby; in moist woods, thickets and sandy prairies; **dioecious**.

■ **FLOWERS** Green, blooming in April–July, lightly carrion-scented if at all; **inflorescence** an umbel 2–3 cm across, each with 30–40 unisexual flowers; **peduncles** 10–15+ per plant, axillary, horizontal, glabrous, 2–12 cm long (usually not past the leaf tip); **pedicels** 5–20 mm long; **perianth parts** six, green, similar at 3–6 mm long and 1–1.6 mm wide, reflexed; <u>male flowers</u> 6–9 mm wide; **stamens** six, exserted, *c.* 2 mm long; **filaments** flat, white, *c.* 2 mm long; **anthers** pale yellow, curved; <u>female flowers</u> 4–5 mm wide by 6–7 mm long, perianth as above; **staminodia** one to six, each *c.* 1 mm long, hairlike, erect; **ovary** glabrous, *c.* 3 mm long; **stigma** 3-lobed, *c.* 3 mm wide, subsessile. **FRUIT a berry**, round, bluish black with a bloom, 5–15 mm long by 6–11 mm wide, forming a tight cluster 2.5–4.5 cm wide and long, persisting into early winter; **seeds** one to six per berry, smooth and shiny, bluntly 3- or 4-sided, brownish red, 4–5 mm long by 3.5–4.5 mm wide by *c.* 3 mm thick.

■ **LEAVES** Alternate, variable, simple, entire, with several main parallel veins and delicate faint net veins between them; **blades** 4–12.5 cm long by 2.5–10 cm wide, pointed, lightly hairy to glabrous below; **petioles** 2–9 cm long; **tendrils** (modified stipules) paired, from most upper petiole bases, 5–10 cm long, unbranched.

■ **STEM** Erect, lax to nodding, grooved, glabrous, usually attached by tendrils to vegetation; stiff at the 3–5 mm thick, naked base; **internodal** length 8–55 mm, reduced above; **branches** axillary at the lower leaves, ascending and to *c.* 50 cm long.

■ **SYN.** *Smilax herbacea* var. *lasioneura* (Hook.) A. DC.

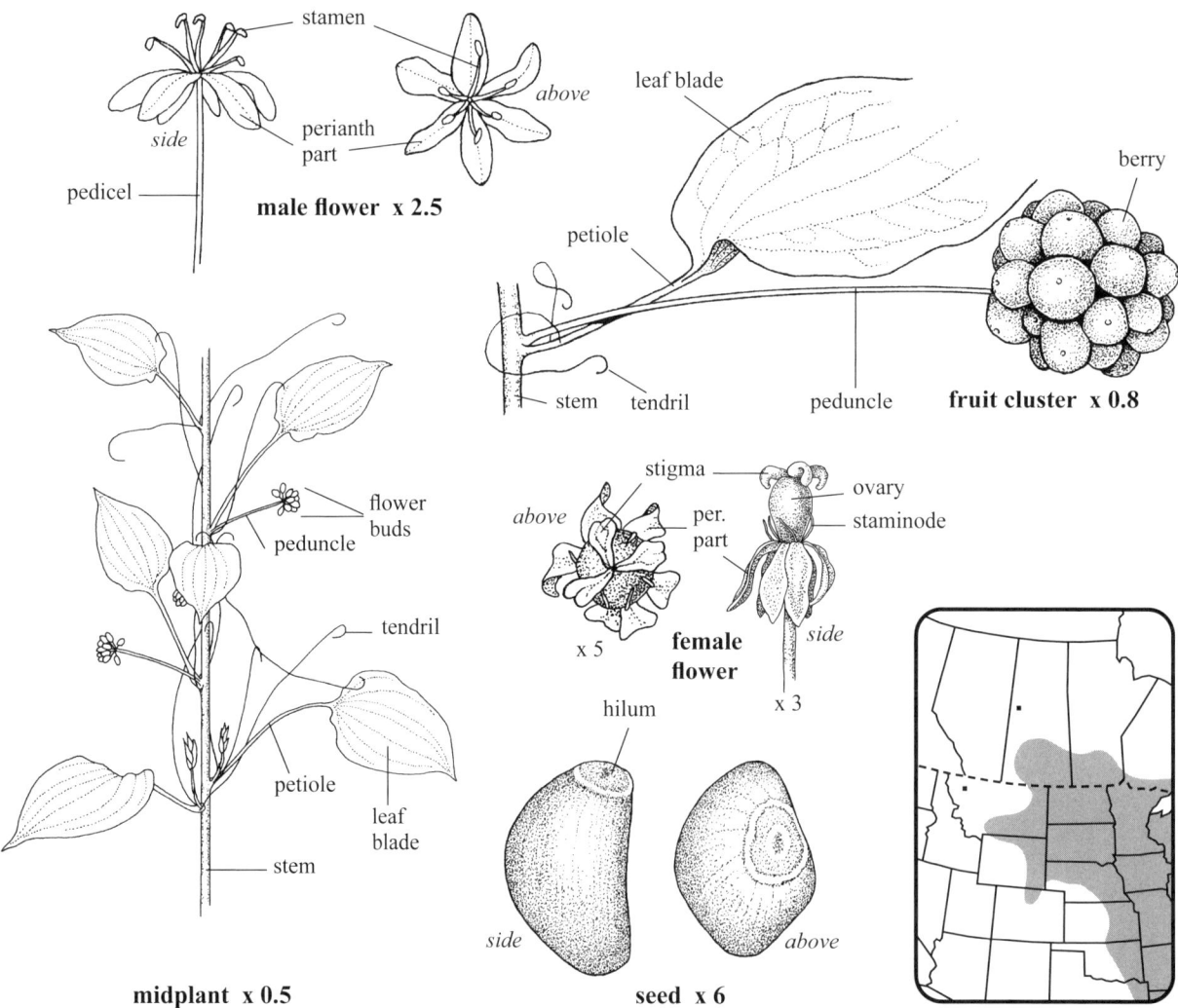

midplant x 0.5

seed x 6

694   Blue Ridge Carrion-flower

Bur-reed—*Sparganiaceae* **Wildflower** Green and white Perianth parts (scales) 3–6

STEMLESS BUR-REED *Sparganium emersum* Rehmann

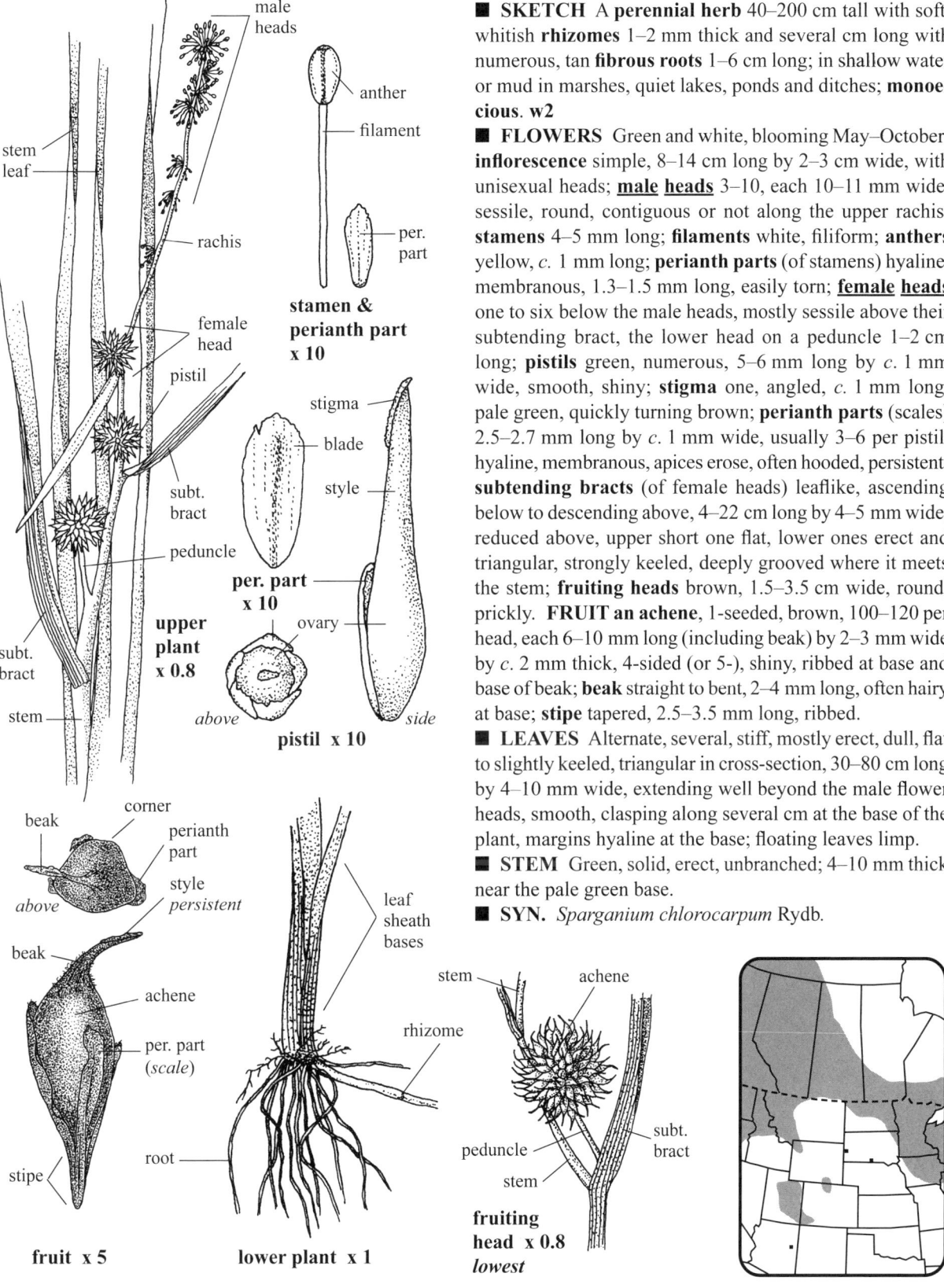

■ **SKETCH** A **perennial herb** 40–200 cm tall with soft, whitish **rhizomes** 1–2 mm thick and several cm long with numerous, tan **fibrous roots** 1–6 cm long; in shallow water or mud in marshes, quiet lakes, ponds and ditches; **monoecious**. w2

■ **FLOWERS** Green and white, blooming May–October; **inflorescence** simple, 8–14 cm long by 2–3 cm wide, with unisexual heads; male heads 3–10, each 10–11 mm wide, sessile, round, contiguous or not along the upper rachis; **stamens** 4–5 mm long; **filaments** white, filiform; **anthers** yellow, *c*. 1 mm long; **perianth parts** (of stamens) hyaline, membranous, 1.3–1.5 mm long, easily torn; female heads one to six below the male heads, mostly sessile above their subtending bract, the lower head on a peduncle 1–2 cm long; **pistils** green, numerous, 5–6 mm long by *c*. 1 mm wide, smooth, shiny; **stigma** one, angled, *c*. 1 mm long, pale green, quickly turning brown; **perianth parts** (scales) 2.5–2.7 mm long by *c*. 1 mm wide, usually 3–6 per pistil, hyaline, membranous, apices erose, often hooded, persistent; **subtending bracts** (of female heads) leaflike, ascending below to descending above, 4–22 cm long by 4–5 mm wide, reduced above, upper short one flat, lower ones erect and triangular, strongly keeled, deeply grooved where it meets the stem; **fruiting heads** brown, 1.5–3.5 cm wide, round, prickly. **FRUIT an achene**, 1-seeded, brown, 100–120 per head, each 6–10 mm long (including beak) by 2–3 mm wide by *c*. 2 mm thick, 4-sided (or 5-), shiny, ribbed at base and base of beak; **beak** straight to bent, 2–4 mm long, often hairy at base; **stipe** tapered, 2.5–3.5 mm long, ribbed.

■ **LEAVES** Alternate, several, stiff, mostly erect, dull, flat to slightly keeled, triangular in cross-section, 30–80 cm long by 4–10 mm wide, extending well beyond the male flower heads, smooth, clasping along several cm at the base of the plant, margins hyaline at the base; floating leaves limp.

■ **STEM** Green, solid, erect, unbranched; 4–10 mm thick near the pale green base.

■ **SYN.** *Sparganium chlorocarpum* Rydb.

Green-fruited Bur-reed, Simple-stem Bur-reed, Narrow-leaf Bur-reed

Bur-reed—*Sparganiaceae* **Wildflower** Greenish Perianth parts (scales) 5 (3–6)

BROAD-FRUITED BUR-REED *Sparganium eurycarpum* Engelm. ex Gray

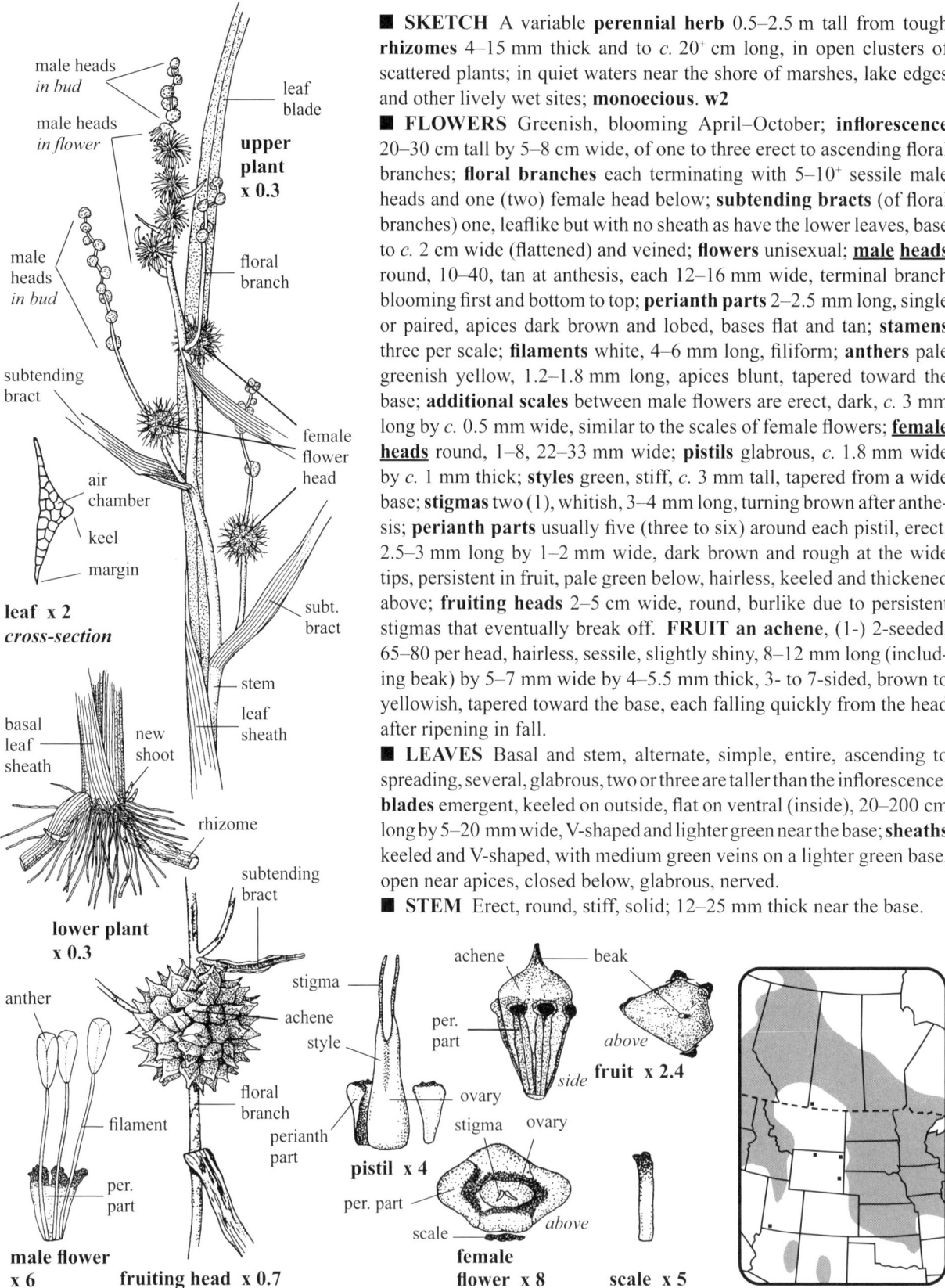

■ **SKETCH** A variable **perennial herb** 0.5–2.5 m tall from tough **rhizomes** 4–15 mm thick and to *c.* 20⁺ cm long, in open clusters of scattered plants; in quiet waters near the shore of marshes, lake edges and other lively wet sites; **monoecious. w2**

■ **FLOWERS** Greenish, blooming April–October; **inflorescence** 20–30 cm tall by 5–8 cm wide, of one to three erect to ascending floral branches; **floral branches** each terminating with 5–10⁺ sessile male heads and one (two) female head below; **subtending bracts** (of floral branches) one, leaflike but with no sheath as have the lower leaves, base to *c.* 2 cm wide (flattened) and veined; **flowers** unisexual; **male heads** round, 10–40, tan at anthesis, each 12–16 mm wide, terminal branch blooming first and bottom to top; **perianth parts** 2–2.5 mm long, single or paired, apices dark brown and lobed, bases flat and tan; **stamens** three per scale; **filaments** white, 4–6 mm long, filiform; **anthers** pale greenish yellow, 1.2–1.8 mm long, apices blunt, tapered toward the base; **additional scales** between male flowers are erect, dark, *c.* 3 mm long by *c.* 0.5 mm wide, similar to the scales of female flowers; **female heads** round, 1–8, 22–33 mm wide; **pistils** glabrous, *c.* 1.8 mm wide by *c.* 1 mm thick; **styles** green, stiff, *c.* 3 mm tall, tapered from a wide base; **stigmas** two (1), whitish, 3–4 mm long, turning brown after anthesis; **perianth parts** usually five (three to six) around each pistil, erect, 2.5–3 mm long by 1–2 mm wide, dark brown and rough at the wide tips, persistent in fruit, pale green below, hairless, keeled and thickened above; **fruiting heads** 2–5 cm wide, round, burlike due to persistent stigmas that eventually break off. **FRUIT an achene**, (1-) 2-seeded, 65–80 per head, hairless, sessile, slightly shiny, 8–12 mm long (including beak) by 5–7 mm wide by 4–5.5 mm thick, 3- to 7-sided, brown to yellowish, tapered toward the base, each falling quickly from the head after ripening in fall.

■ **LEAVES** Basal and stem, alternate, simple, entire, ascending to spreading, several, glabrous, two or three are taller than the inflorescence; **blades** emergent, keeled on outside, flat on ventral (inside), 20–200 cm long by 5–20 mm wide, V-shaped and lighter green near the base; **sheaths** keeled and V-shaped, with medium green veins on a lighter green base, open near apices, closed below, glabrous, nerved.

■ **STEM** Erect, round, stiff, solid; 12–25 mm thick near the base.

696   Giant Bur-reed

Bur-reed—*Sparganiaceae* **Wildflower**   Green   Perianth parts (scales) 3 or 4

CLUSTERED BUR-REED   *Sparganium glomeratum* (Laestad.) L. Neum.

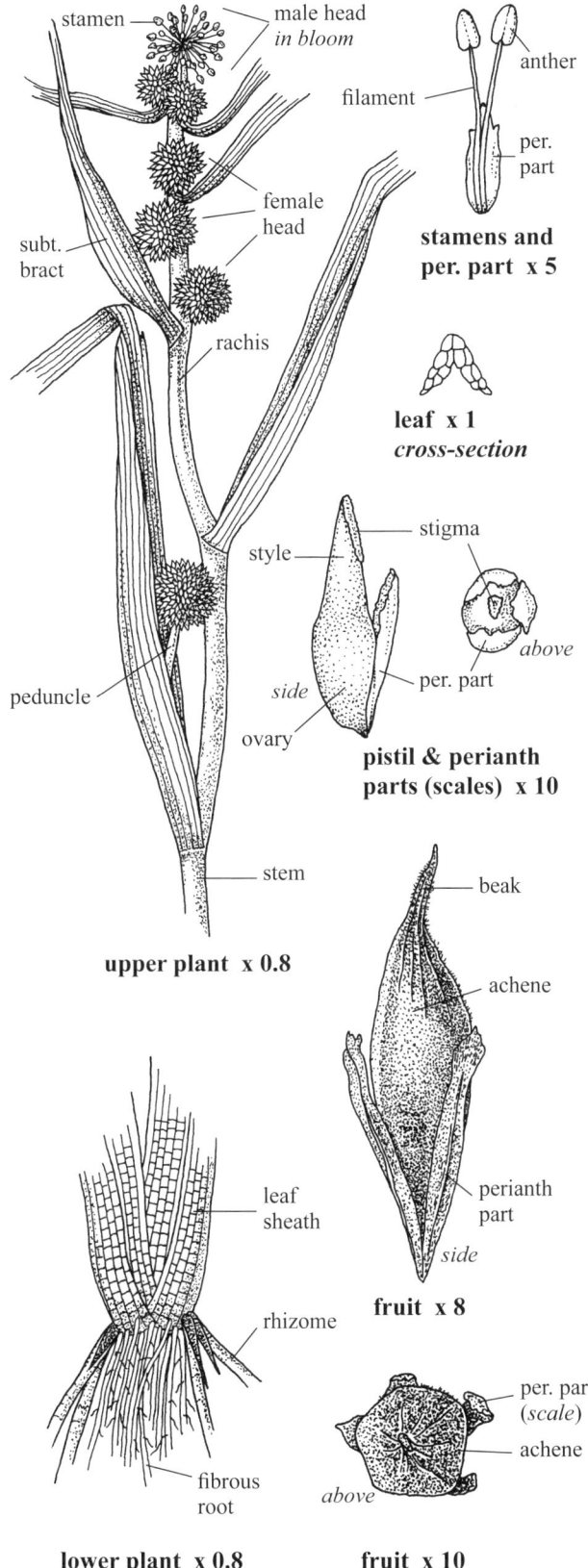

upper plant x 0.8

lower plant x 0.8

fruit x 10

fruit x 8

fruiting head x 1.5

■ **SKETCH** A **perennial herb** 20–60 cm tall with tan, soft but solid **rhizomes** 2–3 mm wide and to *c.* 30⁺ cm long with dark pointed sheaths 1.5–3.5 cm long and tan **fibrous roots** 5–10 cm long by *c.* 1 mm wide; along marshes and swamp edges, beaver ponds and ditches in the Boreal forest; **monoecious. w2**

■ **FLOWERS** Green, blooming June–August; **inflorescence** mostly erect, 4–11 cm long by 1.5–2.5 cm wide, with unisexual heads; male heads one (2), terminal, 10–12 mm wide, sessile, above female heads; **perianth parts** (scales) 3 or 4, transparent, 2–3 mm long by 0.6–0.9 mm wide (flattened), entire or toothed at apices; **stamens** 5–6 mm long, 1–4 attached to base of perianth parts; **filaments** 4–5 mm long, tan; **anthers** tan, *c.* 1 mm long; **rachis** smooth, oval, *c.* 5 mm wide, unbranched; female heads two to six per stem, each 9–12 mm wide, some supra-axillary, upper heads sessile; **peduncles** of lowest heads ascending, 2.5–4.5 cm long by 2–3 mm wide; **subtending bracts** (of female heads) 10–40 cm long by 5–7 mm wide, reduced and flat above, more V-shaped below, ascending to spreading; **style** one, *c.* 1 mm long; **stigma** one, angled; **fruiting heads** 12–20 mm wide, round, green turning slightly brown before the achenes fall. FRUIT an achene, 1-seeded, green to brown, shiny, beak and upper body glabrous to slightly hairy, ridged, TL 5–8 mm by 1.5–3 mm wide and thick, 3- to 5-sided; **beak** slightly bent, 1–2 mm long; **stipe** covered by a few perianth parts.

■ **LEAVES** Basal and stem, alternate, erect, entire, sessile, V-shaped, triangular, soft, two edges sharp, inner side concave, outer two sides slightly convex; **basal blades** four or five, 30–60 cm long by 6–11 mm wide (one side), dull, spongy, wilted by anthesis; **stem leaves** erect, usually two, 40–50 cm long by *c.* 1 cm wide, reduced above, triangular.

■ **STEM** Erect to floating, simple, smooth, solid and stiff, roundly triangular; *c.* 20 mm wide when bases of leaves are included, *c.* 6 mm thick when the base is naked.

697

Bur-reed—*Sparganiaceae*   **Wildflower**   Green   Perianth parts (scales) 3 or 4

SMALL BUR-REED   *Sparganium natans* L.

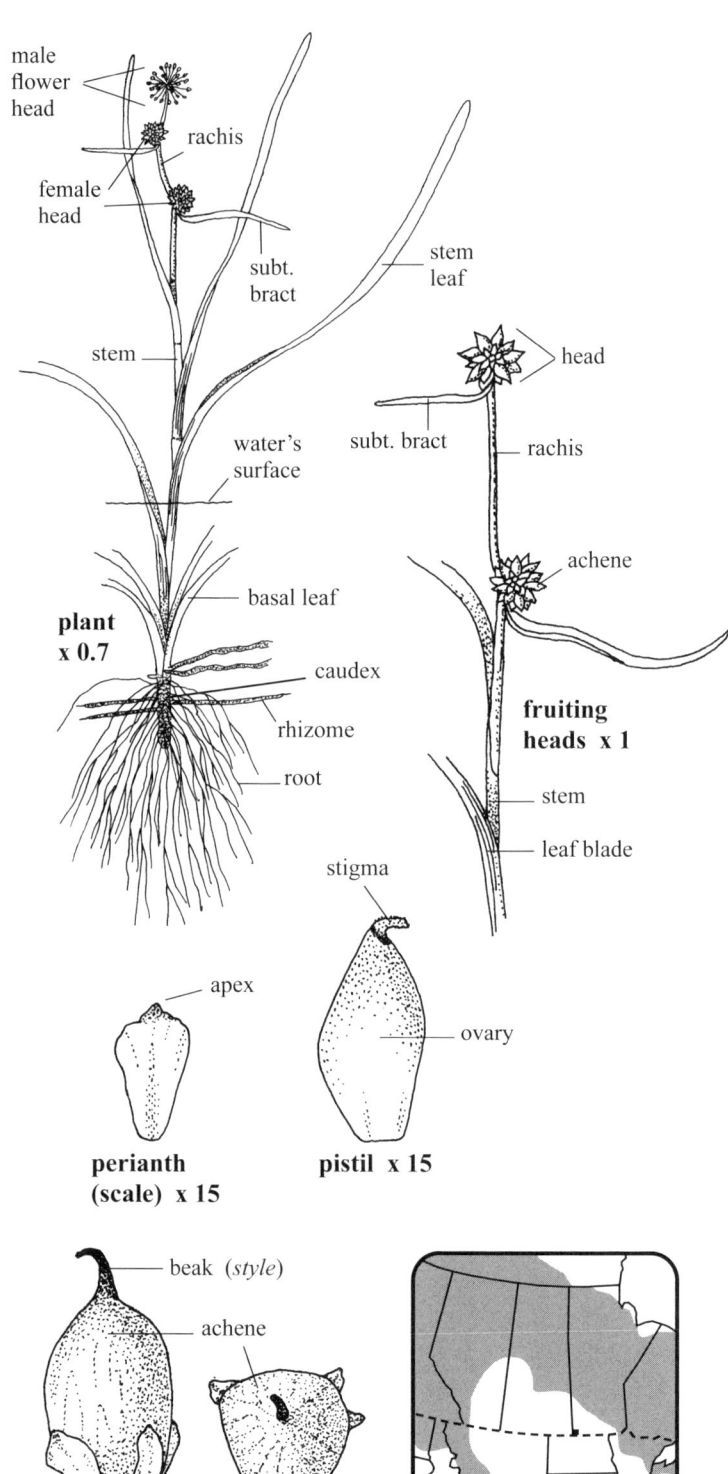

■ **SKETCH** An aquatic **perennial herb** 10–30 (–100) cm tall or long from a brown, erect **caudex** 10–20 mm long by 3–4 mm wide with brown **rhizomes** 2–5 cm long by *c.* 0.7 mm wide and **roots** 5–12 cm long by *c.* 0.5 mm wide, light tan, numerous; in shallow quiet water; **monoecious**. w2

■ **FLOWERS** Green, blooming June–September; **inflorescence** 2–5 cm long by *c.* 10 mm wide of 2–4 dense and well-spaced, round, unisexual heads; **subtending bracts** (of male heads) absent; **male head** one (2), terminating the inflorescence, *c.* 8 mm wide; **subtending bracts** (of stamens) brown, membranous, *c.* 1 mm long by *c.* 0.8 mm wide; **stamens** numerous, the heads whitish, 5–10 mm above the top female head, disintegrating quickly after anthesis; **filaments** filiform, white, 3–4 mm long; **anthers** pale yellow, 0.5–0.8 mm long; **subtending bracts** (of female heads) 10–35 mm long by 2–3 mm wide, leaflike, spreading, reduced above, tips rounded; **female heads** two (1–4), often sessile, axillary, 5–10 mm wide with *c.* 50 pistils, the upper two heads 12–30 mm apart; **rachis** green, erect, oval, curved to straight; **pistils** numerous, green, *c.* 2 mm long by *c.* 1 mm wide; **style** one, short, brown; **stigma** one, light gray, 0.5–1 mm long; **perianth parts** (scales) 3 or 4, at base of each pistil or fruit, each part 1–1.4 mm long by 0.7–1 mm wide, the tips erose and light brown at anthesis, persistent with fruit; **fruiting heads** brown, 7–12 mm wide, round, usually two per plant (stem). **FRUIT an achene**, 1-seeded, brown, smooth, 2–4 mm long by 1.3–2.5 mm wide, lightly veined, in perianth parts; **stipe** less than 1 mm long; **beak** (persistent style) 0.7–1.5 mm long, often bent.

■ **LEAVES** Basal and stem, alternate, simple, flat, entire, sessile, glabrous, grasslike, several per stem, spreading to ascending, erect or floating, upper ones reaching to top or slightly beyond the male flower head; **basal blades** 4–6, similar to stem leaves in shape, elongated to *c.* 60 cm when floating; **sheaths** overlapping and crowded, V-shaped, the hyaline margins tan, bases white; **stem blades** dull, 4–10 cm long by 2–7 mm wide, 2–4, reduced above.

■ **STEM** Erect to decumbent, sometimes floating; base 1–2 mm thick (with leaf sheaths removed); **nodes** slightly swollen, marked by a dark green line.

■ **SYN.** *Sparganium minimum* (Hartman) Wallr.

Cattail—*Typhaceae* **Wildflower** Green Perianth absent

Narrow-leaved Cattail *Typha angustifolia* L.

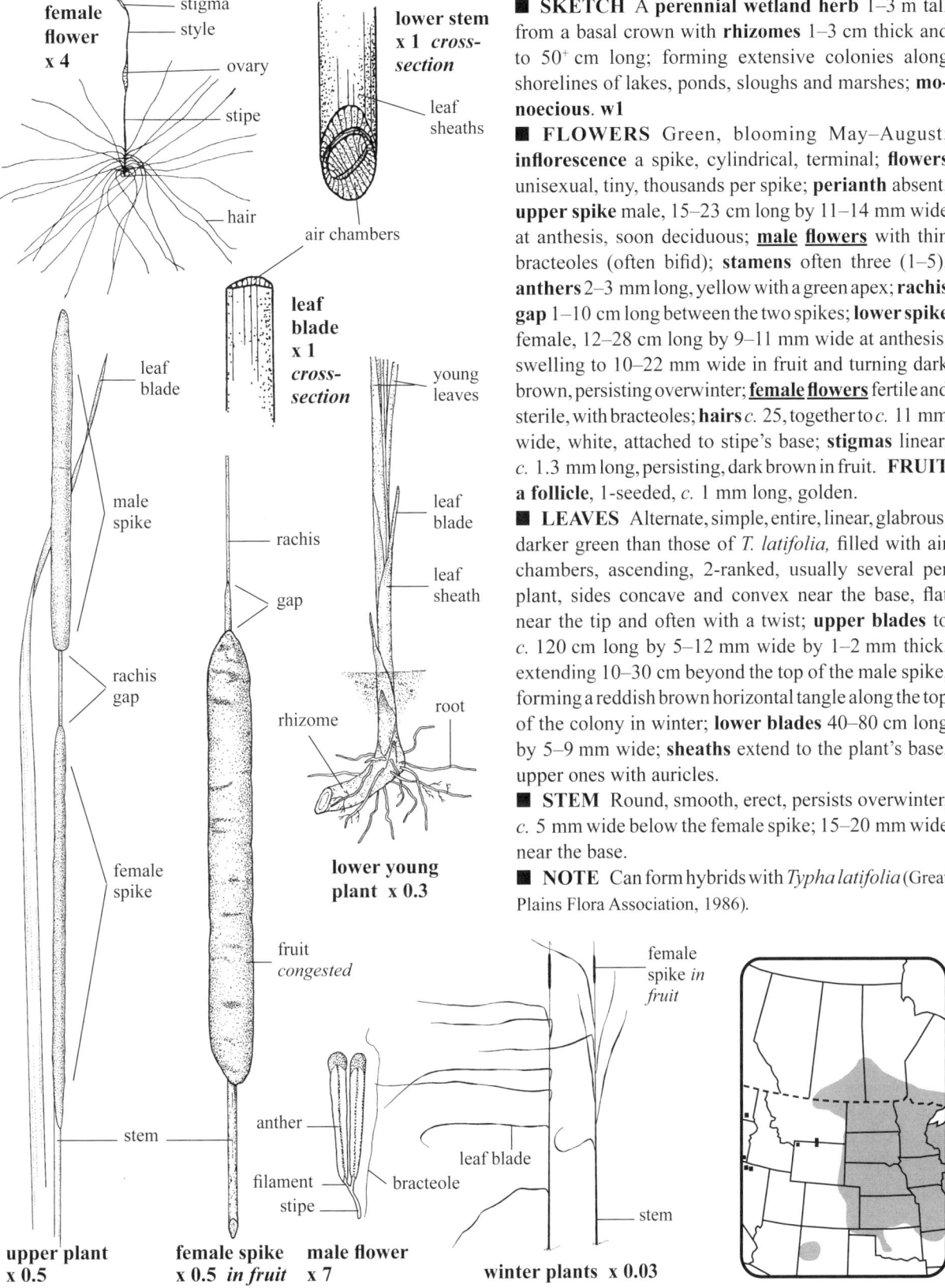

- **SKETCH** A perennial wetland herb 1–3 m tall from a basal crown with **rhizomes** 1–3 cm thick and to 50+ cm long; forming extensive colonies along shorelines of lakes, ponds, sloughs and marshes; **monoecious. w1**
- **FLOWERS** Green, blooming May–August; **inflorescence** a spike, cylindrical, terminal; **flowers** unisexual, tiny, thousands per spike; **perianth** absent; **upper spike** male, 15–23 cm long by 11–14 mm wide at anthesis, soon deciduous; <u>male flowers</u> with thin bracteoles (often bifid); **stamens** often three (1–5); **anthers** 2–3 mm long, yellow with a green apex; **rachis gap** 1–10 cm long between the two spikes; **lower spike** female, 12–28 cm long by 9–11 mm wide at anthesis, swelling to 10–22 mm wide in fruit and turning dark brown, persisting overwinter; <u>female flowers</u> fertile and sterile, with bracteoles; **hairs** c. 25, together to c. 11 mm wide, white, attached to stipe's base; **stigmas** linear, c. 1.3 mm long, persisting, dark brown in fruit. **FRUIT** a follicle, 1-seeded, c. 1 mm long, golden.
- **LEAVES** Alternate, simple, entire, linear, glabrous, darker green than those of *T. latifolia*, filled with air chambers, ascending, 2-ranked, usually several per plant, sides concave and convex near the base, flat near the tip and often with a twist; **upper blades** to c. 120 cm long by 5–12 mm wide by 1–2 mm thick, extending 10–30 cm beyond the top of the male spike, forming a reddish brown horizontal tangle along the top of the colony in winter; **lower blades** 40–80 cm long by 5–9 mm wide; **sheaths** extend to the plant's base, upper ones with auricles.
- **STEM** Round, smooth, erect, persists overwinter, c. 5 mm wide below the female spike; 15–20 mm wide near the base.
- **NOTE** Can form hybrids with *Typha latifolia* (Great Plains Flora Association, 1986).

Narrow-leaf Cat-tail   699

Cattail—*Typhaceae* — **Wildflower** Green Perianth absent

COMMON CATTAIL   *Typha latifolia* L.

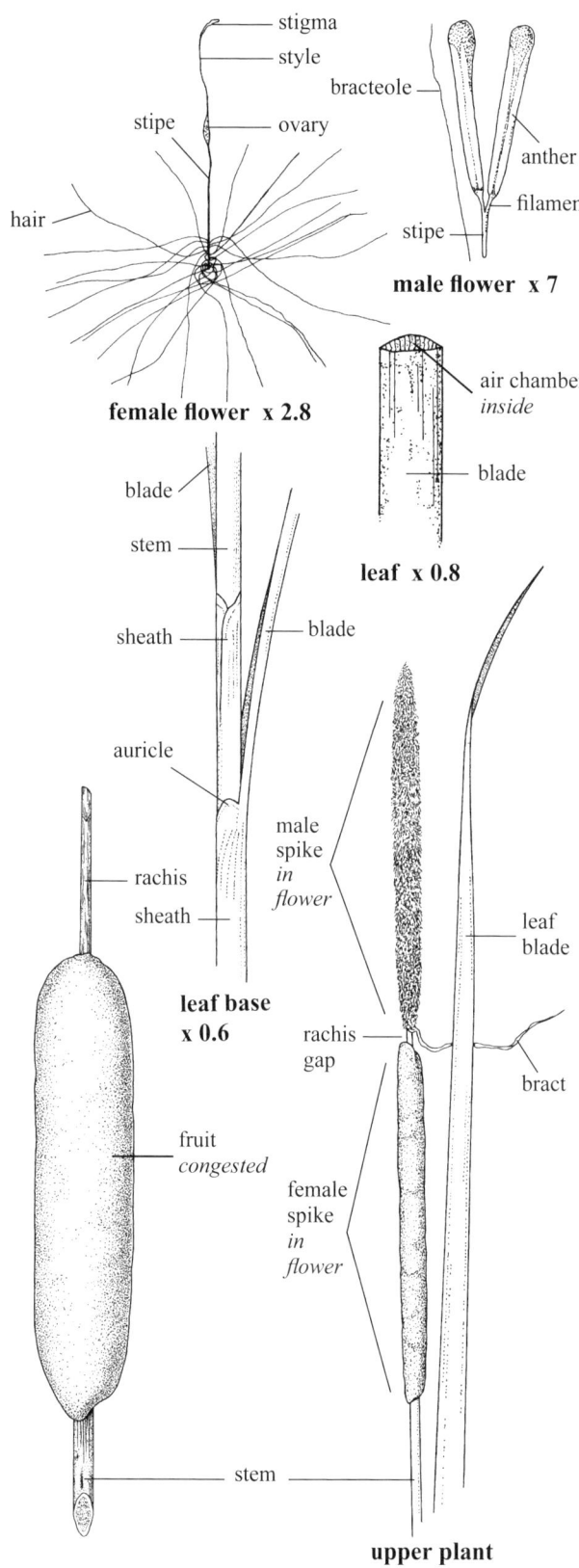

female flower x 2.8

male flower x 7

leaf x 0.8

leaf base x 0.6

fruiting spike x 0.4

upper plant x 0.3

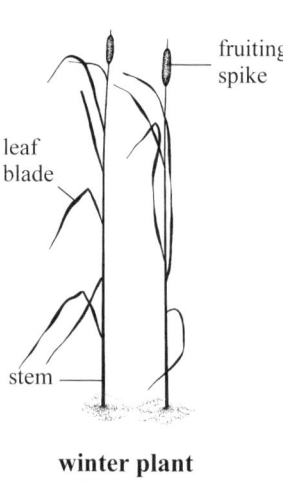

winter plant x 0.03

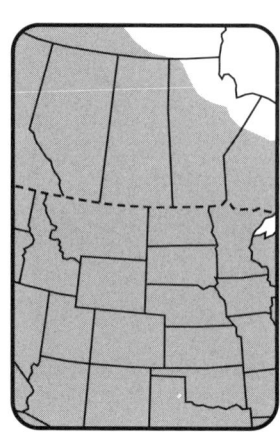

■ **SKETCH** A perennial wetland herb 1.5–3 m tall from a basal crown with **rhizomes** 1–3 cm thick and to 60+ cm long with aerial shoots forming large colonies; in shallow quiet freshwater bordering marshes, ponds and lakes; **monoecious. w1**

■ **FLOWERS** Green, blooming May–August; **inflorescence** a spike, terminal and cylindrical; **flowers** unisexual, tiny, thousands per spike; **perianth** lacking; **upper spike** male, 14–18 cm long by 10–20 mm wide at anthesis; male flowers soon deciduous, with hairlike bracteoles; **stamens** three (1–7); **anthers** 3–3.5 mm long; **rachis gap** 0–35 mm long between female and male spikes; **lower spike** female, fragrant, 10–25 cm long by 8–20 mm wide at anthesis, swelling to 2–3.5 cm wide in fruit, turning dark brown and persisting overwinter; female flowers without bracteoles, fertile and sterile; **hairs** c. 25, white, from stipe's base; **stipes** elongating in fruit; **stigma** oval, c. 0.9 mm long, persistent and brown in fruit. **FRUIT a follicle**, 1-seeded, numerous, golden, smooth, 0.6–1 mm long.

■ **LEAVES** Alternate, simple, entire, light green, 8–11, linear, sessile, glabrous, ascending with a twist near the flat pointed tip; **upper leaves** equal to or extending beyond the tip of the male spike by up to c. 25 cm; **blades** 20–100 cm long by 10–25 mm wide by 1–2 mm thick, shortest near the base, filled with air chambers, slightly concave and convex on opposite sides near the base; **sheaths** short auricled and reaching to the base of the plant.

■ **STEM** Smooth, round, tough, erect, c. 5 mm wide below the female spike; the base of the plant c. 2 cm wide due to the layering of leaf sheaths; stem and leaves overwinter.

■ **NOTE** Forms hybrids with *Typha angustifolia* (Great Plains Flora Association, 1986).

Glossary

Once, I was a visitor in a country where leaves are the common currency. A pocketful of silvery-backed poplar could buy you a meal or a clear glass of water. Maple leaves, wide as open hands, catalpa big as platters, would get you more: a shank of land, a twist of river.

— Terry Griggs

Accessory A "fruit" with fleshy (sometimes edible) parts arising from the developed receptacle and not derived from the pistil (e.g., a strawberry; *see also under* fruit).

Achene An indehiscent, dry, hard, one-seeded fruit with its wall adherent to the seed coat inside at only one point (e.g., fruit of a buttercup; *see also under* fruit). The fruit of Asteraceae—Aster, although described as an achene in this book, may be called a cypsela (pl. cypselae).

Acorn The fruit of an oak, a one-seeded, thick-walled nut with the fruit's cap developed from the united involucre (*see also under* fruit).

Acuminate Gradually tapering to a sharp, rather elongated point or tip (e.g., a tip of a leaf's blade).

Acute Not ending abruptly in a point, but gradually forming a point.

Adnate Different organs fused or grown together (e.g., a filament to a corolla's tube; *see also* connate).

Adventitious A plant unexpectedly (to us) developing organs in an unusual position (e.g., a root forming at the lower node of a stem).

Aerial Those parts of a plant which are usually visible above ground (e.g., stem, leaves, branches, tendrils, and flowers).

Aggregate A fruit formed by the joining of two or more ripe pistils that are originally separate in the flower (e.g., raspberry; *see also under* fruit).

Alternate Leaves or branches placed singly at each node along the stem; stamens and sepals spirally arranged and alongside each other in a flower (*see also* opposite, and whorl).

Ament Another name for a catkin (*see also* catkin).

Androecium All of the stamens (as a group) in one flower, sometimes with the filaments united into a tube.

Androgynous A plant's style of inflorescence, often a spike, with the female (pistillate) flowers situated below the male (staminate) flowers (e.g., in some sedges *Carex* spp.; *see also* gynaecandrous).

Angiosperm A large group of flowering plants, each of which typically has its ovules enclosed within an ovary (e.g., Red Clover, p. 323; *see also* gymnosperm); includes dicots and monocots.

Annual A plant with a lifestyle lasting one year, which runs the complete cycle from seed through flowers to fruit, then dies within a growing season (*see also* biennial, and perennial).

Anther The pollen-producing, upper, saclike part of a stamen, attached in various ways to the top of the filament (*see also* filament, and stamen).

Anther Head In Milkweed flowers the thick upper portion below the horns (e.g., Whorled Milkweed, p. 97).

Anthesis The time during a plant's reproductive period when its flowers (open or closed) allow pollination to occur.

Glossary

Anthocarp A fruit with some flower part persisting along with the pericarp (e.g., the calyx tube in *Mirabilis hirsuta*, p. 373; *see also* fruit).

Apex (pl. apices) The tip or far end of an organ (e.g., a leaf blade's tip).

Apical Concerned with the tip of a plant's organ.

Apiculate Usually refers to a leaf blade that abruptly forms a small, sometimes broad, and usually sharp flexible point at its tip.

Apiculum A tip, short and slender and soft (e.g., on a leaf's blade, or a stamen's anther).

Appendage A secondary, and often smaller part of an organ that is attached to the main structure (e.g., a small lobe on the lower part of a flower's petal).

Appressed One organ closely pressed (flattened) against a different organ (e.g., hairs on a leaf).

Aquatic A plant adapted to growing mostly submerged in water and not simply in a damp meadow or on shore alongside a lake (e.g., *Potamogeton richardsonii*, p. 691).

Aril An outgrowth, often fleshy or hairy, from the funiculus on a seed and which may cover the hilum; possibly assisting in seed dispersal (e.g., Seneca Snakeroot, p. 387; *see also* caruncle, and strophiole).

Arillate A seed with an aril (*see above*).

Arista The narrow, elongated point at the tip of a flower's petal (e.g., *Sisyrinchium montanum*, p. 580).

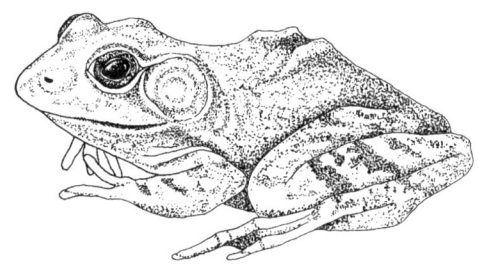

Auricle A thin or wide, fingerlike, characteristic appendage at some leaf blade bases (e.g., in some grasses).

Awn An apical, bristlelike appendage, usually round in cross-section, often rough or twisted (e.g., on some grass spikelet glumes or lemmas; forming the pappus of some Asteraceae—Aster; *see also* scale).

Axil The inner angle or area between two organs (e.g., between a leaf's petiole and the stem).

Axillary Concerned with a leaf's axil (e.g., flowers, branches or buds arising in the inner angle where a leaf grows from the stem).

Axis (pl. axes) The main, usually vertical, support upon which a plant's parts are arranged (e.g., leaves along the axis of a stem).

Banner
The upper, and usually largest, single petal of the five comprising the corolla in Fabaceae—Pea (also called the standard).

Barbed With bristles having sharply reflexed points along most of their length.

Basifixed Attached by the base (e.g., anthers onto their filaments, or hairs to a leaf; *see also* dolabriform).

Beak A narrow to thick, short to long, apical, firm point on a fruit or other structure (e.g., on a legume, an achene, and even a perigynium).

Bearded Bearing long or stiff hairs in a line, ring or tuft (e.g., on some stigmas or styles or the petals of some violets).

Berry A juicy, fleshy, indehiscent fruit with two to many seeds, maturing from the single ovary in a flower's pistil (e.g., *Lonicera tatarica*, p. 240; *see also* drupe, and *under* fruit).

Bidentate Having two teeth (e.g., some grass palea, or the beak of some perigynium in sedges; *see also* tridentate).

Biennial A plant with a lifestyle lasting for two years, which germinates, produces a rosette of leaves the first year, usually waiting to produce flowers and fruit in the second year before dying (*see also* annual, and perennial).

Glossary

Bifid Deeply two-cleft or two-lobed, usually from the tip or apex (e.g., a flower's petal).

Bilabiate Two-lipped (e.g., a flower's corolla in Lamiaceae—Mint).

Binomial The regulated, italicized, scientific, generic and specific name (in Latin) of an organism (e.g., *Acer negundo*; *see also* common name).

Bipinnate Twice compound, when the main and secondary divisions of a compound leaf are pinnate (*see also* pinnate).

Bisexual A flower with both functional female and male parts (e.g., a perfect flower).

Biternate A compound leaf with three leaflets, and with each leaflet divided again into smaller subleaflets or segments.

Black or White The color of most plant hairs, these sometimes adjacent.

Blade The wide, usually flat, expanded, thin portion of a leaf or petal, also called lamina (*see also* petiole, and claw).

Bloom A white to pale blue, fine, waxy, powdery covering of some fruit, leaves and stems, which is easily removed by stroking with a finger (*see also* glaucous).

Blossom A general term for a vascular plant's flower, especially those on shrubs or fruit trees.

Boreal Pertaining to the north (e.g., plants in the Boreal forest interacting with a Boreal Chickadee).

Bract A small leaf in an inflorescence, often subtending a flower or its stalk, or a modified leaf in a flower (e.g., a chaffy bract in some Asteraceae—Aster).

Bracteate Having bracts (e.g., in some styles of inflorescence such as a catkin, or in Bracted Vervain, p. 513).

Bracteole A smaller bract, often growing along a secondary axis such as on a flower's pedicel (e.g., Kalm's Lobelia, p. 364).

Branch A division of a stem from a node and often from a leaf axil; a large, rather thick, early division in the trunk of a tree, itself supporting flowers or redividing into smaller branches.

Branchlet The final end division of a branch, usually with buds.

Bristle A rather stiff and slender hair.

Bud An enclosed, very young, undeveloped leaf or flower, usually covered with one to several bud scales (specialized leaves).

Bulb An underground, vertical, modified bud covered with layers of fleshy, overlapping scales (e.g., *Zigadenus* sp., p. 607; *see also* corm).

Bur (or burr) A "fruit" with a rough or prickly covering formed from its pericarp, a persistent calyx or an involucre enclosing the fruit (e.g., *Xanthium strumarium*, p. 198; *see also under* fruit).

Butte A solitary mountain or hill with very steep sides and usually distinguished from a mesa by its smaller summit area.

Caducous
Plant parts which fall off quickly after full development (e.g., sepals, or stipules; *see also* deciduous).

Calcareous (soils) Plants growing on limestone (soil composed or calcium carbonate from organic remains of shells or coral).

Callus A hard, sometimes sharp protuberance (e.g., at the base of a lemma in Porcupine Grass, p. 649).

Calyx (pl. calyces) The outer whorl of a flower's perianth, composed individually of sepals, which may be separate or united into a tube with teeth or lobes (*see also* corolla).

Campanulate The bell-shaped perianth parts (e.g., a calyx, or more often the tubular corolla of some bluebells).

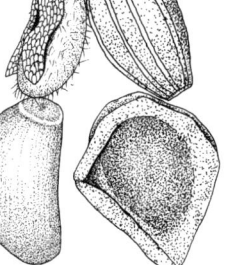

Glossary

Capillary Hairlike, or very slender (e.g., the bristles of a pappus on an achene).

Capitate Formed into a headlike, compact cluster (e.g., sessile flowers in Asteraceae—Aster, or some *Juncus* spp., or describing a wide stigma on a narrow style; *see also* head).

Capsule A dry, dehiscent fruit from fused carpels, splitting open when ripe along its sutures to release two or more seeds (*see also under* fruit).

Carpel The ovule-enclosing, fertile, modified leaf in a simple or compound pistil of a flower.

Carpophore A slender stalk from the receptacle which supports ripe carpels (e.g., in Apiaceae—Carrot); or a thick central stalk (e.g., Marsh Arrowgrass, p. 592).

Caruncle A corky textured appendage, believed to aid in seed dispersal, growing near and possibly covering the hilum of some seeds (e.g., *Euphorbia esula,* p. 292, and some violets; *see also* aril, and strophiole).

Catkin In a monoecious or a dioecious plant, a spikelike ascending or hanging inflorescence composed of many small, unisexual, bracteate flowers on an elongating axis, falling as a unit (e.g., in willows or poplars; *see also* ament).

Caudex (pl. caudices) The tough, sometimes branched and persistent, often woody, underground part between the base of an herbaceous stem and the upper part of its taproot or rhizome; from the caudex new annual, aerial herbaceous stems grow each year.

Cauline Associated with or on the stem; not basal (e.g., stem leaves).

Cespitose (or caespitose) Plants growing in tufts, mats or clumps; not solitary (e.g., some grasses, or other dryland species; *see also* colonial and solitary).

Chaffy Bracts Small, usually stiff, membranous, dry, vertical parts or scales (e.g., subtending, protecting, and separating disc florets on the receptacle in the flower heads of some Asteraceae—Aster).

Chalazal Collar A raised collar (necklike) at the apex of a seed where the coma hairs are attached.

Channelled Plant parts with one or more longitudinal grooves (e.g., on the inner, ventral surface of some grass leaf blades; *see also* furrowed).

Chasmogamous A perfect or unisexual flower open during pollination (most flowers of herbaceous plants; *see also* cleistogamous).

Chevron The pale green, inverted V-shaped pattern visible on the upper (ventral surface) of each of the three leaflets of some clover species.

Cilia Hairs along a margin, usually short to microscopic (e.g., along the margins of leaf blades, or the apex of some ligules of grasses).

Ciliate Fringed with tiny, crowded, marginal hairs (e.g., some leaves, or sepals).

Circumscissile Opening along a horizontal, circular line (suture) around a fruit or anther, the upper valve (top) usually coming off like a lid or cap (e.g., the fruit of some Plantain species, pp. 384–385).

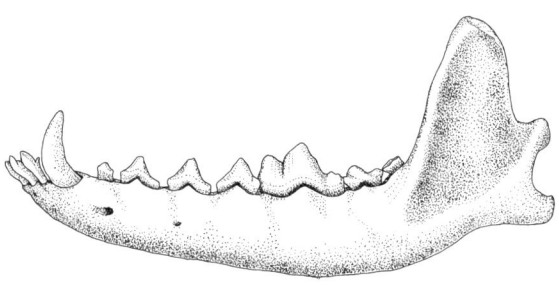

Claw The long or short, stalklike and narrow base of a petal or sepal (*see also* blade, and petiole).

Cleistogamous A flower which remains closed, is self-pollinated, and does produce seed (e.g., *Amphicarpaea bracteata,* p. 296, some violets, and Water-pepper, p. 396; *see also* chasmogamous).

Clone Individual plants with the same genetic makeup (genotype), the result of vegetative propagation from stolons or connecting underground parts (e.g., a grove of Aspen Poplar).

Glossary

Collar The outside zone where a leaf's blade meets the leaf's sheath (e.g., in grass).

Colonial Used chiefly for plants employing asexual reproduction through the growth of underground connectors such as rhizomes between aerial stems (*see also* solitary, and cespitose).

Column The united filaments or stamens, or when the filaments and style are united (e.g., in milkweeds, and orchids).

Coma A tuft of smooth, apical hairs that aids a seed in its dispersal by wind (e.g., in milkweeds; *see also* pappus).

Commissure The surface or inner face where organs (usually carpels) touch but later become separate when ripe (e.g., the schizocarps composed of two mericarps in Apiaceae—Carrot).

Common Name A locally or regionally unregulated variable plant name in a country's language; named by the people for the people (e.g., Manitoba Maple, p. 72; *see also* binomial).

Compound A portion of a plant made up of two or more parts (e.g., a leaf with two or more leaflets, or a pistil with two or more carpels; *see also* simple).

Concave A curved surface like the inside of a capsule (*see also* convex).

Cone A modified reproductive structure atop a reproductive stem of non-flowering plants (e.g., the cone (or strobilus) of Equisetaceae—Horsetail); the dense and roundish to somewhat elongated arrangement of crowded ovule-bearing scales and imbricate bracts along a central axis, often woody when mature (e.g., a cone from a spruce tree, p. 67).

Coniferous A plant, usually monoecious, producing or bearing cones (e.g., pine trees and some shrubs, often referred to as conifers or evergreens, and usually with simple leaves).

Connate Similar structures which are united (e.g., filaments, or petals forming a tube; *see also* adnate).

Convex A rounded surface like the outside of a capsule (*see also* concave).

Cordate A heart-shaped pattern; picture a stylized, inverted heart with rounded lobes and a sinus (space) between the basal lobes (e.g., the shape of some leaves).

Corm A fleshy, dense, underground vertical stem thickened for storage, often renewed each year, covered and protected with dry, papery leaf bases or scales (e.g., *see also* bulb).

Corolla The inner whorl of a flower's perianth, consisting of petals which may be united into a tube or separate, tiny to very showy (*see also* calyx); in disc florets of Asteraceae—Aster, the corolla may be composed of the lower, narrow corolla tube and an upper wider corolla limb with the corolla teeth or lobes at the apex.

Corona Petal-like appendages lying between a corolla and the stamens which, in milkweeds, consists of a wide hood and an ascending, incurved, narrow horn.

Corymb An inflorescence which is flat-topped or somewhat rounded, and gradually flowering from the outside inward, with the outer pedicels or peduncles the longest (*see also* cyme and *under* inflorescence).

Cotyledon The embryonic leaf within a seed; a plant's original leaf.

Creeping Plants that grow along or just below the ground and produce roots from nodes of the stem.

Crenate Margins with shallow, rounded teeth; scalloped (e.g., some leaf margins; *see also* dentate, and serrate).

Cucullate Somewhat hooded or hood-shaped (e.g., the tips of some flowers' sepals).

Culm The aerial stem of a grass or a sedge, rounded and often hollow in grasses, but triangular and solid in sedges (*Carex* spp.).

Glossary

Cup A cover formed from a united involucre (e.g., the cup of an acorn in *Quercus macrocarpa*, p. 327).

Cyathium In the genus *Euphorbia* (Spurges, p. 290–292), a cuplike perianth, appearing to house a perfect flower, but actually supports crowded, unisexual naked flowers comprised of one pistil and several male flowers each of which has one stamen but together appear in different stages of development.

Cylindrical Elongated and circular in cross-section (e.g., the shape of some clustered flowers' pistils and then their seed heads in Ranunculaceae—Buttercup).

Cyme A wide, flat to somewhat rounded style of inflorescence, with the flower's blooming sequence from the center to the outside or from the top downward (*see also* corymb, and *under* inflorescence).

Cymule A small cyme in an inflorescence (e.g., *Mentha arvensis*, p. 353).

Deciduous
Plant parts which are not persistent (e.g., leaves wilting and then falling at the end of the growing season, or flower parts falling after anthesis; *see also* caducous, and persistent).

Decumbent Reclining or lying along the ground, but with the end or tip ascending (e.g., a stem).

Deflexed A part abruptly bent downward (e.g., a fruit's pedicel; *see also* reflexed, and inflexed).

Dehiscent Ripe fruit, or an anther which opens in an orderly style by means of sutures, slits, valves, etc. (e.g., a capsule; *see also* indehiscent).

Dentate Usually describing leaves with sharp, spreading teeth perpendicular to a blade's margin (*see also* serrate, and crenate).

Determinate The terminal flower in an inflorescence that opens and matures before the flowers below it, thereby suppressing the continued lengthening of the stem's apex (*see also* indeterminate).

Diadelphous A flower's stamens arranged in two sets due to the union of the filaments (e.g., ten stamens as 9 + 1 in many Fabaceae—Pea).

Dicotyledons Flowering plants having seeds with two embryonic cotyledons (leaves), broad leaves with net venation, and flower parts usually arranged in fours or fives or in multiples thereof (*see also* monocotyledons, and cotyledon).

Didynamous A flower with its four stamens arranged in two sets of different lengths (e.g., 2 long and 2 short in Lamiaceae—Mint).

Dilated The lateral expansion of a part (e.g., the flat, lower lip of some orchids).

Dimorphic A part developing two distinct forms (e.g., some achenes in Chenopodiaceae—Goosefoot).

Dioecious A species having its staminate (male) and pistillate (female) flowers on separate plants (e.g., a plant's flowers are either all male or all female as in most willows; *see also* monoecious, and unisexual).

Disc (disk) The central portion of a flowering head made up of tiny disc florets (e.g., Asteraceae—Aster).

Disc Floret A small, mostly vertical flower, often perfect, or sometimes male, with a tubular corolla, situated and often crowded in the central disc portion of a flower head of Asteraceae—Aster (*see also* ligulate floret, and ray floret).

Discoid Head A flower head in Asteraceae—Aster composed entirely of disc florets, the ray florets absent (e.g., Pineappleweed, p. 161).

Dissected A plant part divided, sometimes repeatedly, into narrow segments or parts (e.g., a leaf; *see also* entire, and toothed).

Distal Away from the center and approaching or placed at the end (e.g., a bud or a flower at or near the end of a branch).

Distant Similar parts not growing close together (e.g., leaves widely separated; *see also* remote).

Distinct Separate, not united with similar plant parts (e.g., flower petals not united into a tube; *see also* free).

Dolabriform A part attached near or at its middle, rather than at its base (e.g., hairs, or anthers; *see also* basifixed).

Dorsal Refers to the back or outer surface of an organ (e.g., abaxial or the surface of a horizontal leaf closest to the ground and away from the stem of the plant (*see also* ventral).

Dorsally Compressed A part flattened from the back or outer surface, which allows all nerves or veins to be observed at a glance (e.g., a lemma, or glume; *see also* laterally compressed).

Dot Map A map with scattered dots which indicate where plants, now residing in herbaria, were collected; a common type of map generally used in journals and some books (*see also* range map).

Drupe An indehiscent fruit with a fleshy exocarp and one seed enclosed in a stony endocarp called a pit or stone (e.g., Red-fruited Choke Cherry, p. 462; *see also* berry, and *under* fruit).

Drupelet One part of a compound fruit made up of many loosely joined parts (e.g., many drupelets form a whole raspberry fruit).

Elaiosome
A lipid-rich growth on the seeds of *Trillium* spp.

Elater Microscopic, club-shaped bands attached to a spore and which unravel on drying to aid in its dispersal (*Equisetum* spp.).

Embryo A young plant enclosed in a seed that requires suitable conditions to germinate and grow to maturity.

Emergent Plants rooted underwater but with aerial reproductive or vegetative shoots (stems) above the water's surface (e.g., Cattails).

Endocarp The inner single layer of the pericarp or fruit wall (*see also* exocarp, mesocarp, and pericarp).

Entire A plant part which is whole and without divisions (e.g., a leaf blade's margin lacking teeth, lobes, or spines, etc.; *see also* dissected, and toothed).

Epidermis The outer, surface layer of cells on a leaf, flower, petal, or root.

Epigynous Parts growing on the apex of the inferior ovary, in fact or only in appearance (e.g., a flower's stamens, or perianth parts).

Epiphyte An independent plant growing upon another plant and not connected to the ground (e.g., some orchids).

Erose An irregularly shaped margin looking as though it has been chewed by an insect (e.g., the sepal of *Corydalis aurea*, p. 328).

Even-pinnate A compound leaf divided into an even number of leaflets with the terminal leaflet missing and sometimes replaced by a tendril (e.g., a leaf with six leaflets; *see also* odd-pinnate).

Exocarp The outer, often thin, single layer of the pericarp or fruit wall (*see also* endocarp, mesocarp, and pericarp).

Exserted Plant parts projecting beyond the envelope of other parts (e.g., stamens, or styles extending past the top of the corolla or petals; *see also* included).

Farinose
Covered with microscopic, irregularly shaped, mealy flakes (e.g., some leaves and flowers in Chenopodiaceae—Goosefoot; *see also* scurfy, and mealy).

Fascicle A tight bundle or cluster of leaves, flowers, stems or roots.

Female Describes a flower with only female reproductive parts (e.g., pistils in a pistillate flower); a female tree.

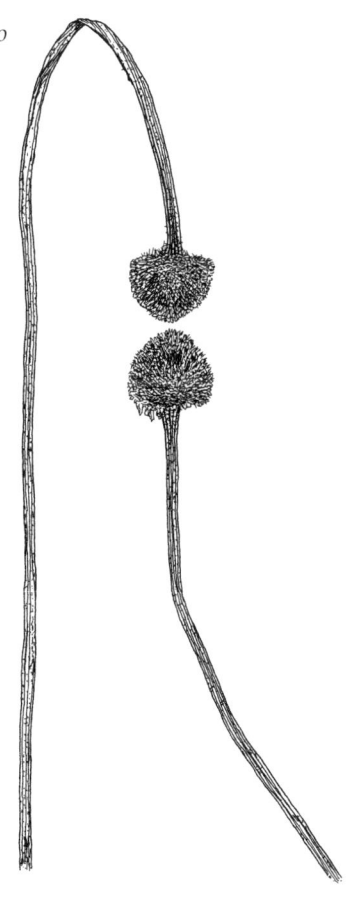

Glossary

Fen A wet, peaty area usually alkaline in nature due to its water contacting calcareous rocks below.

Fibrillose A dense covering of fine, quite long fibers (e.g., on rhizomes of some sedges such as *Carex sprengelii*, p. 558).

Fibrous Roots A root system with fine, branched, adventitious roots having the main divisions quite similar in thickness, often forming a dense cluster (e.g., grass roots; *see also* rhizome, and taproot).

Filament A stamen's stalk which gives support to the apical anther; thick or threadlike and longer to shorter than the anther (*see also* anther, and stamen).

Filiform Long and very slender, like a thread (e.g., the filament of a stamen).

Fleshy Succulent, thick and juicy (e.g., leaves, stems, or fruit of some plants).

Floret An inconspicuous to showy flower making up part of a flower head of the Asteraceae—Aster (*see also* disc, ligulate, and ray florets); in a grass spikelet, the palea, lemma, and reproductive parts together, with one to several florets in a spikelet (*see also* glume, palea, and lemma).

Floricane A flower-bearing, second-year stem of the genus *Rubus* (e.g., raspberry, p. 467; *see also* primocane).

Flower The reproductive parts and enveloping perianth parts necessary for the production of fruit and seeds in angiosperms.

Flower Bud An unexpanded flower, overwintering on twigs of trees and shrubs or below ground and enclosed in one or more protective scales (*see also* leaf bud).

Flower Head A dense cluster of sessile or almost sessile florets of one or two types arranged on a receptacle supported by a peduncle (e.g., in Asteraceae—Aster).

Foliaceous Leaflike in appearance (e.g., some sepals, calyx lobes, or bracts that resemble small leaves in texture, shape and color).

Follicle A dry, dehiscent, many-seeded fruit maturing from a simple ovary and opening, when ripe, along only one, ventral suture (e.g., in milkweeds; *see also* legume, loment, and *under* fruit).

Fornix (pl. fornices) Tiny, interior appendages in the upper throat of a flower's corolla (e.g., in Hoary Puccoon, p. 211).

Free One plant's organ not joined to a different organ (e.g., a stamen free from a petal; *see also* distinct).

Fruit The seed container developing from a mature ovary wall and any external persistent parts on a flowering plant. Twenty-three types are listed below:

    accessory A "fruit" with fleshy (sometimes edible) parts developed from the receptacle and not derived from the pistil (e.g., Smooth Wild Strawberry, p. 451).

    achene An indehiscent, dry, one-seeded fruit with its ovary wall united to the seed coat at one point (e.g., fruit of a buttercup).

    acorn The fruit of an oak, a one-seeded nut, the fruit's cap formed from the united involucre.

    aggregate A fruit formed by the joining of two or more ripe pistils that are separate in the flower (e.g., raspberry, p. 466).

    anthocarp A fruit with some flower part persisting along with the pericarp (e.g., the calyx tube in *Mirabilis hirsuta*, p. 373).

    berry A fleshy, indehiscent fruit with two to many seeds, maturing from a flower's single pistil (e.g., *Lonicera* sp.; *see also* drupe).

    bur (burr) A "fruit" with a rough or prickly covering formed from its pericarp, or by a persistent calyx, or an involucre enclosing the fruit (e.g., *Xanthium strumarium*, p. 198).

capsule A dry, dehiscent fruit, of fused carpels, splitting open when ripe along its sutures to release two or more seeds.

drupe A fleshy, indehiscent fruit with one seed enclosed in a stony endocarp called a pit or stone (*see also* berry).

follicle A dry, dehiscent, many-seeded fruit maturing from a single carpel and opening, when ripe, along only one suture (e.g., in milkweeds; *see also* legume, and loment).

grain A common term for the fruit of a grass; a dry, indehiscent, one-seeded fruit having the seed fused to the pericarp (fruit wall); known botanically as a caryopsis.

legume The fruit of Fabaceae—Pea that is bilaterally symmetrical, from one carpel, dehiscent (opening when ripe) along both sutures into two valves with seeds attached along the ventral suture (common name is a pod; *see also* loment, and follicle).

loment A fruit of the Fabaceae—Pea that is slightly constricted between the enclosed seeds and with each flat one-seeded segment breaking off at its joints when mature but not opening (*see also* legume, and follicle).

nut An indehiscent, usually one-seeded, thick-walled, hard fruit usually enclosed partially or wholly in an involucre or husk (e.g., a nut in an acorn; *see also* nutlet, and achene).

nutlet A small nut (*see also* nut).

pepo A fleshy, usually indehiscent fruit of the Cucurbitaceae—Cucumber, with a hard rind and pulpy interior filled with seeds.

pod A common name for a dry, dehiscent fruit with several seeds (e.g., a legume of Fabaceae—Pea).

pome A fleshy, indehiscent fruit developing from an inferior ovary with several locules (e.g., Hawthorn).

pyxis A capsule with a horizontal suture allowing the upper portion to lift off like a lid (e.g., Plantain).

samara A winged fruit, indehiscent, and one-seeded (e.g., Green Ash, p. 376).

schizocarp A dry, indehiscent fruit which splits into two, one-seeded, closed mericarps (e.g., Water-hemlock, p. 82 in Apiaceae—Carrot).

silique A dry, dehiscent fruit with a partition (septum) separating the two valves of the fruit (e.g., in Brassicaceae—Mustard; it *includes* the term silicle in this book).

utricle A bladderlike fruit, 1-seeded, indehiscent, may be very thin-walled (e.g., in *Amaranthus blitoides*, p. 75).

Funiculus In a flower's ovary, the stalk that connects the ovule to the ovary's wall or internal placenta, and then the ripe seed in the fruit.

Funnelform The tubular shape of a flower's corolla when its parts are united (e.g., Field Bindweed, p. 276).

Furrowed Having longitudinal channels or grooves (e.g., best observed near the collar on the inside of some leaf blades of grass; *see also* channelled).

Galea

The hoodlike, upper lip of a flower's two-lipped corolla (e.g., *Pedicularis lanceolata*, p. 500).

Geniculate Bent at or near the base of a plant's stem, or bent along the long awns of some grasses (e.g., *Hesperostipa spartea*, p. 649).

Genotype The genetic makeup of an individual plant with its genes interacting with the environment to produce variations in its external appearance (*see also* phenotype).

Glossary

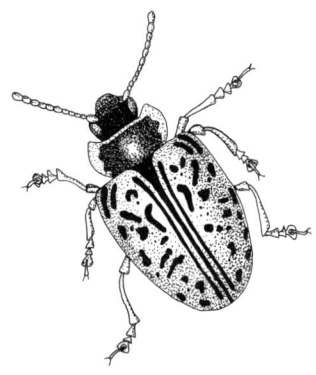

Genus (pl. genera) The smallest, natural taxonomic group placed below the family level and which includes one to many species within it (e.g., the genus *Carex*; *see also* species).

Gibbous A part swollen on one side giving it a slight asymmetry (e.g., Twining Honeysuckle, at the base of the flower's tubular corolla, p. 239).

Glabrous Smooth, not hairy and without glands (e.g., an entire plant or only some of its parts, such as a stem or seed; *see also* scabrous, and smooth).

Gland A secretory structure raised or in a slight depression (e.g., a gland-tipped hair, or glands in flowers or on a leaf's blade that secrete a sticky substance; *see also* punctate).

Glaucous Smooth parts coated with a thin, whitish to pale bluish, waxy bloom that rubs off easily with a finger (e.g., a stem, or succulent fruit; *see also* bloom).

Globose A part which is roundish (e.g., a capsule ripe with seeds).

Glomerule A small, compact, irregularly shaped cluster of tiny flowers (e.g., in some Chenopodiaceae—Goosefoot).

Glume Sterile bracts, usually paired, at the base of a typical grass spikelet, the bases of the two bracts usually slightly unequal in position (*see also* floret, lemma, and palea).

Grain A common term for the fruit of a grass; a dry, indehiscent, one-seeded fruit having the seed coat fused to the pericarp (fruit wall), sometimes enclosed by the hardened palea and lemma (known botanically as a caryopsis; *see also under* fruit).

Gymnosperm A group of plants with its members not having their ovules enclosed in an ovary (e.g., a pine tree; *see also* angiosperm).

Gynaecandrous A monoecious species with its spikelike inflorescence composed of female (pistillate) flowers situated above the male (staminate) flowers (e.g., in some *Carex* spp.; *see also* androgynous).

Gynobase An enlarged receptacle that bears the ovary and later the mature fruit (e.g., *Lappula squarrosa*, p. 210).

Gynodioecious A plant which is dioecious, but has some of its flowers perfect (hermaphroditic) on one plant and other flowers female (pistillate) on another plant.

Gynomonoecious Female (pistillate) flowers and perfect flowers situated in different parts of the plant's inflorescence.

Habitat
The place a plant occupies or where it grows, reproduces and dies (e.g., a desert, ditch, marsh, mountain, meadow, prairie, ravine, slough, and sidewalk crack, etc.).

Head Describes the cluster of numerous, crowded, sessile or subsessile flowers, of one or two types, on a compound receptacle supported by a peduncle in the Asteraceae—Aster family; a roundish cluster of crowded flowers in some *Juncus* spp. (*see also* capitate).

Herb A plant with non-woody aerial parts with flowers, the whole of which dies back each year to ground level; an annual, biennial, or perennial plant.

Herbarium A room with cabinets for the safe storage of dried, pressed, mounted and mapped collections of identified wild plant specimens found in an area, region, or country(ies); usually associated with a Museum or University, but may also be private (*see also* specimen, and type specimen).

Heterostylous A species having styles of two or three different lengths in the population, but all flowers on one plant will have styles of a similar length (e.g., the tristylous Purple Loosestrife, p. 366).

Glossary

Hilum (pl. hila) A scar on a seed that indicates where the ovule's stalk (funiculus) was attached to the seed coat; hila have eight shapes and range from apical to marginal positions on a seed.

Hip An enlarged, mature hypanthium, persistent and which encloses numerous achenes (e.g., roses).

Hispid A part that has long, rough, bristly or rigid hairs (e.g., Rice Cut Grass, p. 653; *see also* strigose).

Hood A small, petaloid blade, part of the corona, blunt and ascending (e.g., near the top of milkweed flowers).

Horn A petal-like, narrow part of the corona from the hood's base and arched over the anther head in the milkweeds (e.g., *Asclepias speciosa*, p. 95).

Hyaline Thin, translucent, and generally colorless (e.g., flower parts or their margins; *see also* membranous).

Hypanthium In flowers with petals, an enlarged receptacle below the calyx joined with the inferior ovary and formed from the fusion of the flower's lower perianth parts and stamens (e.g., Rosaceae—Rose).

Imbricate
Plant parts which overlap like shingles on a roof (e.g., the margins of most involucral bracts in the flower heads of Asteraceae—Aster).

Imperfect Those flowers that have either functional pistils or stamens, but not both; perianth parts may be present or absent (*see also* perfect, and unisexual).

Included Those parts that do not visibly protrude (e.g., stamens not reaching beyond the top of the corolla envelope; *see also* exserted).

Indehiscent Mature fruit that remains closed and does not open in an orderly style by means of valves, sutures or pores, etc., to release its seeds (e.g., a berry; *see also* dehiscent).

Indeterminate When the main axis of the inflorescence continues to elongate because the lower or outer flowers bloom first in the sequence (e.g., Common Mullein, p. 503; *see also* determinate).

Indigenous Native to a country, continent, a large region or an island; plants not introduced (*see also* introduced).

Inferior Those parts placed below other parts (e.g., a flower's ovary is inferior when the other flower parts are attached to its apex or to its hypanthium's apex with which it is fused; *see also* superior).

Inflexed Turned or bent inward (e.g., a hair; *see also* deflexed).

INFLORESCENCE A plant's flowers, including the axis (rachis), stalks and bracts, and how they are arranged, from one to many flowers per plant or per inflorescence; a plant may have one to many inflorescences such as several racemes. Six types are listed below:

    corymb An inflorescence that is flat-topped or somewhat rounded, and gradually flowering from the outside inward, with the outer pedicels or peduncles the longest (*see also* cyme).

    cyme A wide, flat to somewhat rounded inflorescence, with the flower's blooming sequence from the center to the outside or from the top downward (e.g., *Potentilla gracilis*, p. 458; *see also* corymb).

    panicle An inflorescence, irregularly compound, with flowers on pedicels or peduncles; a branched raceme.

    raceme An unbranched, elongated inflorescence with its alternate flowers on pedicels or peduncles (*see also* panicle).

    spike An unbranched, elongated inflorescence with sessile flowers along a central rachis (e.g., Quack Grass, p. 642).

    umbel A flat-topped, or rounded inflorescence in which the pedicels, of about equal length and each bearing a flower, ascend from a common point (e.g., Apiaceae—Carrot).

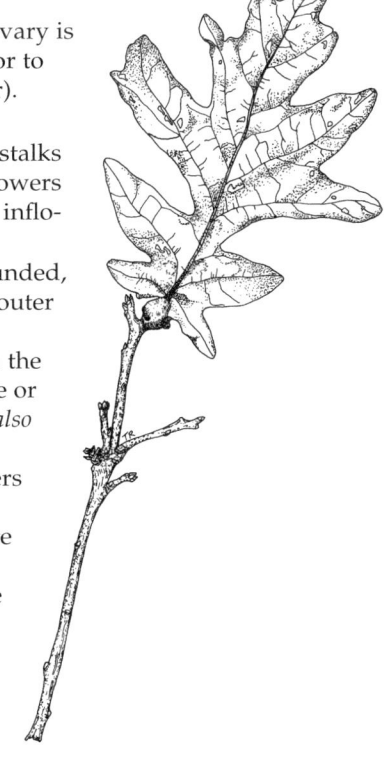

Glossary

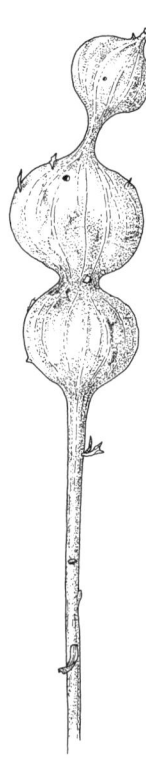

Internode The variable length of a stem located between two adjacent nodes (e.g., distance between two alternate leaves).

Introduced Plants brought in by various agents from another region or country; the first step in a species becoming naturalized where the habitat is suitable (*see also* indigenous).

Introrse Facing inward toward the axis (e.g., an anther facing the center of a flower).

Involucel A secondary and often smaller involucre (e.g., bracts subtending an umbellet in a compound umbel).

Involucral Bract One bract (of an involucre) in a whorl subtending a flower head in Asteraceae—Aster (e.g., Chicory, p. 120); one to several leaflike bracts subtending but often surpassing the inflorescence (e.g., River Bulrush, p. 570; also called phyllary).

Involucre A whorl or a few bracts under a flower, a flower cluster, a flower head, or the entire inflorescence, similar in role to the calyx of a solitary flower (e.g., in Asteraceae—Aster; also know as phyllaries).

Involute A plant part rolled inward or upward toward the upper side (e.g., the margins of a leaf's blade; *see also* revolute).

Joint
An articulation at a node in grasses, or a weak zone along a line where the structure will easily break off (e.g., positioned at the stem's base in tumbleweeds, or elsewhere along a flower's pedicel, or a branch).

Keel
A dorsal, central, raised ridge or midrib (e.g., along a grass leaf's sheath); the two, often partially united, lower petals of a flower with five petals (e.g., Fabaceae—Pea).

Key A short, textual description in a book or manual, often arranged in dichotomous pairs that present a choice of contrasting characters of a genus or any other taxonomic group (species) that may eventually lead to the identity of a plant and its name.

Laddering
When an old basal sheath disintegrates into a central vein and side veins resembling a rope ladder (e.g., Woolly Sedge, p. 551).

Lamella In some orchids, a fleshy lobed structure at the base of the lip (e.g., Striped Coralroot, p. 612).

Lanate Curly, intertwined hairs (e.g., leaf blades with a woolly appearance; *see also* woolly and tomentose).

Lateral Along or to the side (e.g., a lateral flower bud along a lateral floral branch).

Laterally Compressed Flattened from the sides, usually along the length of the midnerve or midrib, producing a V-shaped appearance (e.g., a whole grass spikelet, or a grass floret's lemma showing only half of the nerves present; *see also* dorsally compressed).

Lax Not rigid (e.g., a drooping inflorescence, or a leaf's blade).

Leaf (pl. leaves) A lateral or basal plant part usually borne on a stem at a node and dorsi-ventrally flattened into a blade (lamina) and often attached by a narrow petiole, which is usually subtended by a pair of stipules; a green organ of a plant where most of the photosynthesis and transpiration take place (e.g., a simple or compound leaf, often shed at the end of a growing season).

Leaf Bud On twigs, a bud covered with one or more scales, which opens in spring, revealing a plant's young leaves (e.g., Manitoba Maple; *see also* flower bud).

Leaf, Compound A leaf with two or more individual blades, each called a leaflet (*see also* leaf, simple).

Leaf, Simple A leaf comprised of a single blade, entire, toothed, or lobed and with or without a petiole; not compound (*see also* leaf, compound).

Glossary

Leaflet One part (blade and usually a petiolule) of a compound leaf attached to the central, supporting rachis; may be simple to compound, entire or toothed, sessile or stalked, and is usually without stipules.

Leaf Scar The mark left on a stem or branch from where the base of the leaf's petiole was attached during the growing season before its fall.

Leaf Sheath The lower part of the leaf below the blade that widens and partially or fully envelopes the stem or culm (e.g., most notable in Poaceae—Grass).

Legume The fruit of Fabaceae—Pea that is bilaterally symmetrical, with one carpel, usually dehiscent (opening when ripe) along both sutures, two-valved with seeds attached along the ventral suture (common name is a pod; *see also* loment, follicle, and *under* fruit).

Lemma The outer and lower of the two fertile bracts that enclose the pistil and stamens in a grass's floret (*see also* glume, palea, and floret).

Lenticel A corky spot for gas exchange, horizontal or vertical, slightly raised and often tan colored against the contrasting bark of a tree or shrub.

Lenticular Shaped like a lens which is convex on opposite sides, as are some seeds or one-seeded achenes (e.g., in some *Polygonum* spp.; *see also* terete).

Ligulate Floret A small but sometimes showy, perfect flower with its corolla expanded into a bladelike structure, which together form a ligulate head (e.g., Dandelion, p. 195, of the Asteraceae—Aster; *see also* ray floret, and disc floret).

Ligule A flattened, elongated, bladelike corolla, often with five minute apical teeth, in a ligulate or ray floret (often showy) of Asteraceae—Aster; the membranous or hairy projection (not showy) at the junction of a leaf's blade and sheath in Poaceae—Grass, or Cyperaceae—Sedge, which may help identify a species.

Limb The obviously upper, wider tubular part of the corolla above the corolla tube in disc florets of Asteraceae—Aster; also the expanded, usually lobelike, parts of a united corolla above the throat.

Linear Long and narrow in shape with the sides and veins more or less parallel (e.g., a grass blade).

Lip Each part of the two divisions of a bilabiate corolla or calyx (e.g., a united corolla cleft into an upper and lower portion); the lower pouchlike or flattened and often lobed petal of some orchids.

Lobe The partial division of a plant's leaf, tubular corolla, sepal, bract or a petal, especially when the apex is rounded and the sinus does not reach to the midrib (e.g., a Bur Oak leaf, p. 327).

Locule A compartment or cavity of an organ (e.g., in an anther, or within an ovary or fruit where the ovules are protected and the seeds ripen).

Loment A flat, specialized fruit (legume) of the Fabaceae—Pea that is slightly constricted between the enclosed seeds and with each one-seeded segment breaking off at its joints when mature but not opening (*see also* legume, follicle, and *under* fruit).

Lustrous Glossy or shiny (e.g., a leaf's blade or a fruit's seed).

Male
Describes a flower with only the male reproductive parts (stamens); a staminate flower; a tree with only male flowers.

Marginal Involved with the edges (e.g., hairs, or teeth along the edge of a leaf's blade).

Glossary

Mealy A part covered with small, irregular flakes or granules that can be easily scraped off (e.g., *Chenopodium* spp.; *see also* farinose and scurfy).

Membranous Thin, pliable, and sometimes translucent (e.g., some paleae or ligules in a grass; *see also* hyaline).

Mentum In some orchids, a bulge on the front of the ovary from a short extension from the column (e.g., Early Coralroot, p. 613).

Mericarp The mature one-seeded carpel of a longitudinally dehiscent fruit (e.g., one of the two segments of a schizocarp in Apiaceae—Carrot).

Mesa A flat-topped, mostly isolated, natural elevation generally larger than a butte (*see also* butte).

Mesocarp The middle, multilayered part of the fruit wall or pericarp that often becomes the fleshy part of a succulent fruit (*see also* endocarp, exocarp, and pericarp).

Midrib The central and often the most obvious vein of a leaf or a leaflike bract (*see also* nerve, vein, and rib).

Minute Very small, microscopic, inconspicuous (e.g., some tiny hairs, or glands).

Monocotyledon A large group of flowering plants having one cotyledon (embryonic seed leaf), parallel venation on the mostly narrow leaves, and with flower parts usually in threes or sixes, and with a fibrous, adventitious root system (e.g., grass; *see also* dicotyledons and cotyledons).

Monoecious A plant with pistillate (female) and staminate (male) unisexual flowers situated in different locations on the same plant (e.g., in most *Carex* and in Cattails; *see also* dioecious, and unisexual).

Montane Growing in mountainous country.

Mucro A pointed, short, abrupt and very fine projection (e.g., at the tip of a leaflet, petal or a bract).

Mucronate A part that has a mucro at its tip (*see above*).

Muricate Rough, with hard, short points or projections; pebbly (e.g., the coat of some seeds).

Naturalized
A plant (species) established for at least 10 years in a region to which it is not native; an introduced plant with a reproducing, thriving, local or expanding population.

Nectar Gland A secreting organ (nectary) in a flower that produces nectar to reward a visiting insect, thereby facilitating cross-pollination.

Needle A common name for a stiff, linear leaf of conifers (e.g., the leaf of a spruce tree).

Nerve A simple, rather straight, longitudinal vein (e.g., on a leaf, or on a glume, or lemma of Poaceae—Grass; *see also* midrib, and vein).

Net-veined A leaf's venation pattern, not parallel but forming a network throughout a blade.

Nocturnal Plants reproductively active at night (e.g., a night-blooming flower).

Nodding That part of a plant which faces or arches downward (e.g., the tip of an inflorescence, or a flower bud).

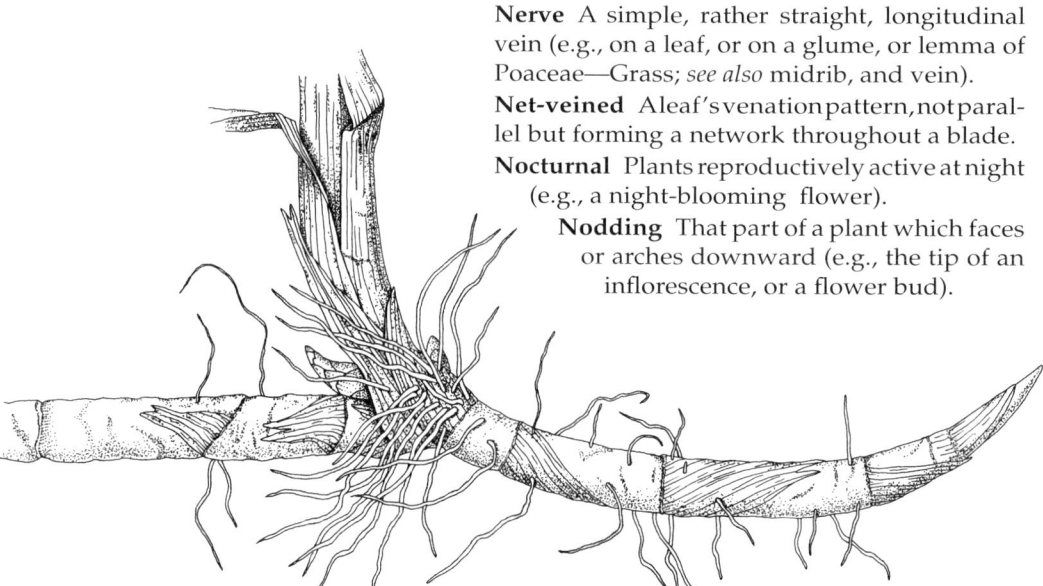

Glossary

Node That place on the stem where a leaf's petiole or sheath or a branch is or has been attached (e.g., sometimes slightly enlarged, or even hairy, as in Poaceae—Grass).

Nodulose A plant part with microscopic, knobby projections or knots (e.g., along some stems or leaves of *Equisetum arvense*, p. 49).

Nomenclature In botanical taxonomy, the systematic naming of taxa according to the International Code of Botanical Nomenclature; the most recently accepted taxonomic scientific name of a plant (often with one to many previous names as synonyms).

Nut A dry, indehiscent, usually one-seeded, thick-walled, hard fruit, often enclosed partially or wholly in an involucre or husk (e.g., a nut in an acorn; *see also* nutlet and achene; *see under* fruit).

Nutlet A small nut, but not an achene (*see also* nut; *see under* fruit).

Obscure
Not very obvious or distinctive, often due to the smallness of the part in our selective vision (e.g., marginal veins on a leaf).

Ocrea (pl. ocreae) A tubular, membranous sheath formed by the fusion of leaf stipules above the stem's nodes (e.g., *Polygonum* spp.).

Ocreola (pl. ocreolae) Membranous sheathing at the base of flowers, often covering the pedicels (e.g., *Polygonum* spp.).

Odd-pinnate A compound leaf which bears a terminal leaflet, resulting in an odd-number of leaflets (e.g., seven; *see also* even-pinnate).

Opposite Branches or leaves situated two at each node and across from each other along the stem (*see also* alternate, and whorl); one plant part before another part (e.g., a stamen in front of a petal rather than between two petals).

Ovary In the large group of plants known as angiosperms, the lower and usually expanded part of a flower's pistil that contains the ovules (*see also* pistil).

Ovule A tiny body in the flower's ovary that is usually fertilized and develops into a seed.

Ovuliferous Scale The ovule- and then seed-bearing, lateral woody structure of a female cone in the group of plants known as gymnosperms (e.g., White Spruce, p. 67).

Palea
(pl. paleae) The small, upper and inner fertile bract (along with the outer, usually longer, lemma) that subtends and mostly encloses the reproductive parts in a grass floret (*see also* lemma, glume, and floret).

Palmate Divided in a handlike style with the lobes, leaflets, or veins radiating like fingers from a common point near the petiole (e.g., a five-lobed leaf).

Palustrine A habitat best described as a wet site such as a marsh, swamp, ditch, bog or shoreline of a river or lake.

Panicle A style of inflorescence, irregularly compound, with flowers on pedicels or peduncles; a branched raceme generally longer than wide (*see also* raceme, and *under* inflorescence).

Papery A part that is thin and has a dry texture like that of paper (e.g., the glume, or lemma of a grass floret).

Papilla (pl. papillae) A tiny, microscopic, nipplelike, blunt projection (e.g., on some seeds; *see also* tubercle).

Papillose A surface covered with papillae giving it a microscopically rough surface (e.g., some seeds).

Pappus A modified calyx, situated atop the ovary, which forms a crown of bristles, awns, scales, etc., visible on fruit (e.g., on some achenes of Asteraceae—Aster; *see also* coma).

Glossary

Parallel-veined With straight, lateral veins parallel to the midrib (e.g., in a grass leaf).

Parietal Ovules attached to the inside of the external wall of the ovary (e.g., parietal placentation within one cavity of a flower's ovary).

Pedicel In an inflorescence, the stalk of one flower; in a grass, the stalk of a spikelet (*see also* peduncle).

Peduncle The stalk of an inflorescence (cluster of flowers) or a flower head in Asteraceae—Aster, or of a flower when it is the only one of an inflorescence (*see also* pedicel).

Pendulous Hanging down (e.g., catkins in Aspen Poplar, p. 478).

Pepo A fleshy, usually indehiscent, fruit of the Cucurbitaceae—Cucumber, with a hard rind and pulpy interior filled with seeds (*see under* fruit).

Perennial An herbaceous plant with a growth habit lasting three or more years (*see also* annual, and biennial).

Perfect A flower with functional stamens and pistils, and with perianth parts present or absent; a bisexual flower (*see also* unisexual, and imperfect).

Perianth A flower's protective envelope consisting of the corolla (petals) and the calyx (sepals) arranged in two whorls, with the sepals placed to the outside, or one whorl (often the petals) sometimes absent; generally used when petals and sepals are similar in appearance.

Pericarp A fruit's wall which is usually three-layered and developed from the ovary wall (*see also* endocarp, exocarp, and mesocarp).

Perigynium (pl. perigynia) The thin, modified bract (used for identification) surrounding a unisexual flower's pistil and then the single fruit (a one-seeded achene) in *Carex* spp. (sedges).

Persistent Plant parts that remain attached even when wilted and dry (e.g., the style at the tip of a legume; *see also* deciduous).

Petal One part of a flower's corolla, often colored and obvious, adapted to attract and guide insect pollinators.

Petaloid Similar to, or resembling a flower's petal, sometimes reduced.

Petiole The narrow stalk of a leaf that attaches the wide, flat blade to the stem (when absent the leaf is sessile; *see also* claw and blade).

Petiolule The stalk of a leaflet in a compound leaf, attached to the central rachis and often shorter than the leaf blade's petiole.

Phenotype The physical characteristics of a plant or species due to the environmental interactions with its genes (*see also* genotype).

Pinnate A compound leaf with several leaflets arranged along both sides of the supporting rachis; the leaf is called odd-pinnate if the terminal leaflet is present, and even-pinnate if the terminal leaflet is absent or replaced by a tendril (*see also* bipinnate, and tripinnate).

Pinnatifid A compound leaf divided in a pinnate style (*see above*).

Glossary

Pistil The central, ovule-bearing organ of a flower, each divided into stigma, style and ovary; eventually ripening into the fruit with enclosed seeds; one to many pistils per flower (*see also* ovary).

Pistillate A unisexual, female flower with a pistil or pistils, but no stamens, or at least not functional stamens (*see also* staminate).

Pit A small hollow or depression (e.g., in a seed's coat, or on a fruit).

Pith The central tissue of a stem or root, may be soft and spongy (e.g., Great Bulrush, p. 573).

Pitted Marked with small microscopic depressions; punctate (e.g., on a leaf's surface).

Placenta The tissue inside the pistil's ovary where the ovule or ovules are attached (there are several arrangements of placentation).

Plano-convex Refers to the shape of a fruit with one side flat and its opposite side convex (e.g., Viscid Great Bulrush, p. 569).

Plant A vegetable organism, usually photosynthetic and sedentary acquiring the soil's liquid-bearing minerals and gasses, with a few species catching and digesting insects or aquatic invertebrates.

Plumose With fine, elongated hairs; hairy in a feathery way (e.g., a persistent style on the fruit of *Geum triflorum*, p. 455).

Pod A common name for a dry, dehiscent fruit with several seeds (e.g., a legume of Fabaceae—Pea; *see also* fruit).

Pollen The dustlike grains containing male gametes produced and liberated from the anthers of flowers at anthesis, and which fertilize the female flower of the species.

Pollinium (pl. pollinia) A coherent mass of pollen grains suitable for pollination by insects (e.g., in orchids; *see* Striped Coralroot).

Polygamodioecious A dioecious species having some perfect (bisexual) flowers in addition to its unisexual flowers.

Polygamomonoecious A monoecious plant species that also bears some perfect (bisexual) flowers.

Polygamous A species with bisexual (perfect) and unisexual (female or male) flowers occurring on the same plant.

Pome A fleshy, indehiscent fruit that develops from an inferior ovary in an hypanthium with several locules (e.g., Hawthorn; *see also under* fruit).

Prickle A sharp outgrowth of a stem or twig's surface that is nonvascular, weak or stiff, and removable with the bark (e.g., on a rose stem; *see also* spine).

Primocane The first year's vegetative stem (e.g., the Wild Red Raspberry, p. 467; *see also* floricane).

Procumbent A prostrate stem growing along the ground, but without developing nodular roots; trailing (*see also* prostrate).

Prostrate A plant's part which lies flat on the ground (e.g., a stem; *see also* procumbent).

Pubescent Covered with short, soft, downy hairs; a general term describing any kind of a plant's hairiness.

Punctate With colored or translucent dots, or pits, usually of a glandular nature (e.g., on a calyx, or leaf blade; *see also* gland).

Pustule A microscopic, blisterlike elevation or projection (e.g., at the base of a hair).

Pyxis A capsule (fruit) with a horizontal suture allowing the upper portion to lift off like a lid (e.g., Plantain spp., pp. 384–385; *see under* fruit).

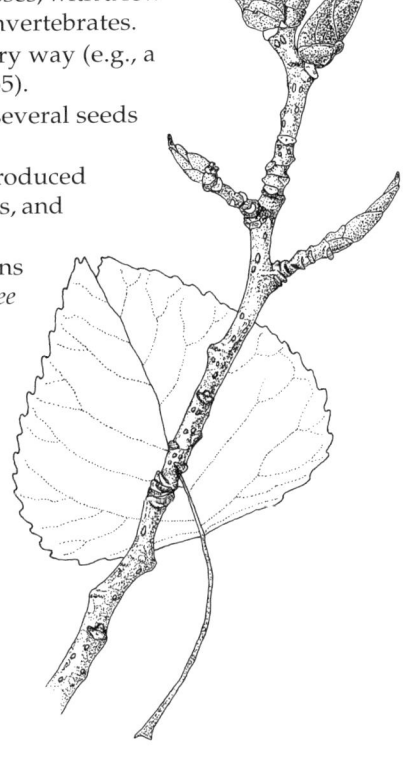

Glossary

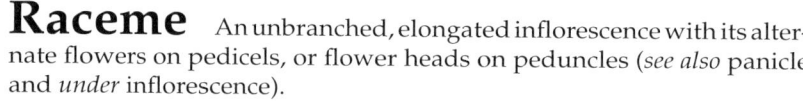

Raceme An unbranched, elongated inflorescence with its alternate flowers on pedicels, or flower heads on peduncles (*see also* panicle and *under* inflorescence).

Rachilla A secondary, usually hidden axis (e.g., in grasses, the central axis within a multiflowered spikelet that supports each floret; in multicompound leaves, a branch of the rachis).

Rachis (pl. rachises or rachides) The visible main axis of a compound leaf supporting the leaflets; or in an inflorescence supporting the main nodal branches or the sessile spikelets.

Radical Involved with the root; arising from, or near, the root (e.g., basal leaves).

Radicle Within a seed, the embryonic root situated below the seed leaf and consisting of internal vascular tissue and a root tip that becomes the plant's primary root.

Range Map A solid, usually gray area on a map denoting where a species generally grows and might be found (*see also* dot map).

Rank A row, especially a vertical one such as the two-ranked leaves along the culm of a grass; a taxonomic rank such as a species under the rank of genus.

Ray The stalk supporting an umbellet of a compound umbel, or branches in other complex inflorescences (e.g., some bulrushes, *Scirpus* spp.); another name of a ligule of a floret in Asteraceae—Aster.

Ray Floret A peripheral flower in a flowering head of Asteraceae—Aster, with the corolla's lower part tubular and the upper part expanded into a flat ligule or blade, usually with apical teeth; ray florets may be female, perfect or sterile (*see also* disc floret, and ligulate floret).

Receptacle The expanded end of the peduncle or pedicel which bears flowers or a flower (e.g., flower heads of Asteraceae—Aster).

Recurved Gradually arched backward or downward (e.g., a flower's pedicel).

Reflexed Abruptly bent or turned downward or backward (e.g., stem hairs; *see also* deflexed, and inflexed).

Regular Externally uniform or symmetrical in shape, orientation, size, or structure (e.g., a flower and its parts).

Remote Parts widely spaced or scattered (e.g., Kalm's Lobelia's flowers are widely separated along the rachis; *see also* distant).

Reticulation A network pattern revealing the regular crossing of the threadlike veins of a leaf's blade.

Retrorse Directed downward or even backward (e.g., stem hairs).

Revolute Rolled backward (e.g., a leaf blade's margin rolled toward the lower or outer dorsal side; *see also* involute).

Rhizome The underground portion of a plant's stem, creeping horizontally or with a vertical orientation, sometimes rooting at the nodes and vegetatively forming new stems and leaves (*see also* taproot, runner, stolon, and fibrous roots).

Rib The primary or prominently raised ridges on an achene or seed (*see also* midrib, nerve, and vein).

Rigid Stiff, inflexible (e.g., some plant hairs, or branches).

Root The descending, nongreen portion of a plant, mostly developing underground, absorbing moisture and nutrients from the surrounding soil, and anchoring the aerial parts.

Rootlet A very fine branch of a root, secondary in nature.

Rootstock A general term for a plant's vertical rhizome or any of its underground parts.

Rosette A group of organs radiating from the center in a circular whorl at ground level (e.g., leaves of the Dandelion, p. 195).

Rotate Shaped like a wheel (e.g., the spreading petals of a flower).

Rudimentary An imperfectly formed plant part that stops growing at an early stage (e.g., a rudimentary stamen; *see also* vestigial).

Rugose Wrinkled in appearance (e.g., a leaf blade's surface due to sunken veins).

Runner A common name for a stolon; a long, above ground, horizontal, narrow, lateral shoot, rooting at intervals from its nodes and eventually developing new individual plants of similar genotype (e.g., Smooth Wild Strawberry, p. 451; *see also* stolon, and rhizome).

Sagittate
A part shaped somewhat like an arrowhead (e.g., pointed leaf blades with their basal lobes pointing downward).

Saline Concerned with salt; those plants that can tolerate saline soil conditions (e.g., Saline Goosefoot, p. 268).

Samara A winged fruit, not fleshy, indehiscent, one-seeded (e.g., Green Ash, p. 376; *see also under* fruit).

Sap The juice of a plant containing sugars, mineral salts, and water, usually clear but sometimes milky in appearance.

Saprophyte A plant without green color and which obtains its food from dead organic matter (e.g., Spotted Coralroot, p. 611).

Scabrous Rough to the touch from short, stout hairs or papillae (e.g., a leaf blade's margins; *see also* glabrous and smooth).

Scale A thin, dry, flat, appressed organ; a wide pappus element atop an achene or ovary in Asteraceae—Aster, often narrowing into an apical awn or bristle (e.g., a fertile bract; *see also* awn); a modified bract subtending each perigynium in a sedge (*Carex* spp.).

Scape A flowering stem, a special peduncle, usually bare, arising from the ground amidst a rosette of basal leaves (e.g., Dandelion, p. 195).

Scar A mark left on a stem or branch by the petiole where the leaf was attached, or on a seed at the point where it was attached in the fruit.

Scarious Describes parts that are not green but are thin, dry and membranous (e.g., the margins of a leaf or bract).

Schizocarp A dry, indehiscent fruit that splits into two or more, one-seeded, closed mericarps each attached to a carpophore (e.g., Water-parsnip, p. 87, in Apiaceae—Carrot; *see also* fruit).

Scurfy Parts covered with small, branlike, irregularly shaped scales (e.g., on the surface of leaves, or flower parts in some Chenopodiaceae—Goosefoot; *see also* farinose, and mealy).

Scutellum A small, shield-shaped part of an organ (e.g., on the calyx of Blue Skullcap, p. 357, in Lamiaceae—Mint).

Secund Parts or organs directed mostly to one side (e.g., flower heads situated along the floral branches of some Goldenrod species).

Seed A fertilized and ripened ovule, often inside a fruit, which is the condensed version of a species.

Seed Coat The outer covering of the seed, developed from the integuments of the ovule (also called testa).

Seed Leaf A cotyledon, embryonic in nature and appearance.

Segment One narrow division of a leaf, petal, calyx, or flower's perianth.

Sepal One of the individual parts of a flower's calyx, which is usually green, situated in a whorl outside the whorl of petals, and often smaller and less showy than the petals, unless the petals are missing.

Septum (pl. septa) A partition (e.g., inside an ovary separating the chambers; the often complete, minute membrane across the inside of some leaf blades of *Juncus* spp.).

Serrate With large, sharp teeth pointing forward (e.g., on a leaf blade's margin; *see also* crenate and dentate).

Glossary

Serrulate Serrate as above, only with smaller teeth.

Sessile Without any type of a supporting stalk (e.g., a leaf blade or flower attached directly by its base).

Seta (pl. setae) A short bristle, sometimes in association with longer bristles (e.g., in a pappus of some achenes in Asteraceae—Aster).

Sheath A tubular structure surrounding part or all of an organ (e.g., the portion of a grass leaf below the blade which surrounds the culm).

Shield A tiny, outer visible cover, usually numerous, in the apical reproductive cone of Horsetails, *Equisetum* spp., pp. 49–55.

Shoot A young, rapidly growing, ascending stem, branch or twig which usually arises near the plant's base or even from a rhizome.

Shrub A low, usually several-stemmed woody plant, deciduous or coniferous; part of a forest's understory (e.g., American Hazelnut, p. 207).

Silique A dry, dehiscent fruit with a persistent partition (septum) separating the two walls and cavities of the fruit (e.g., in Brassicaceae—Mustard it *includes* silicle in this book; *see under* fruit).

Silky A texture due to a soft, fine, appressed covering of smooth, straight hairs (e.g., on some leaves).

Simple Parts that are whole and not compound or branched (e.g., a leaf with only one blade; *see also* compound).

Simple Leaf A leaf with but one blade, it being entire, toothed or lobed but not divided into leaflets (*see also* leaf, compound).

Simple Pistil A flower's pistil consisting of a single carpel with one or more ovules.

Sinuate With a deep, wavy, but untoothed margin (e.g., a leaf's margin).

Sinus The space between two lobes (e.g., in a leaf of Bur Oak, p. 327).

Smooth Parts that are not rough or hairy (e.g., a seed's coat; *see also* glabrous, and scabrous).

Solitary A one-flowered inflorescence; a single plant growing in a certain location; species with its individuals relatively isolated (*see also* colonial, and cespitose).

Spadix A spike with a thickened or fleshy central axis covered with crowded small flowers.

Spathe A membranous or showy bract, nerved, sometimes torn, beneath but once enclosing the inflorescence (onion species such as *Allium stellatum*, p. 598, or Water Calla, p. 526).

Species (pl. species) Plants capable of or actually interbreeding and reproducing in a wild population that is reproductively isolated; the taxonomic distinction below a genus (*see also* genus).

Specimen A plant, or a portion of it, which is dried, mounted, mapped, and identified for botanical study (e.g., a herbarium specimen; *see also* herbarium, and type specimen).

Spike An unbranched, elongated and narrow inflorescence with sessile flowers or spikelets situated along a central rachis (e.g., Virginia Wild Rye, p. 643; *see also under* inflorescence).

Spikelet A small and secondary spike, each of which contains one to many florets; the main unit of inflorescence in the grasses.

Spine A sharp, rigid, outgrowth from the wood of a stem or leaf blade's edge but without vascular tissue; often a modified branch or sometimes a petiole, stipule or other part (*see also* prickle).

Spiny A plant bearing spines along its stem, branches or leaves (e.g., *Cirsium* spp.).

Sporadic Plants growing here and there in a large area, without a continuous range, making them difficult to locate in the field and map.

Sporangiophore A modified branch or stalk, with several sporangia which bear the spores (e.g., *Equisetum arvense*, p. 49).

Glossary

Sporangium (pl. sporangia) A body bearing spores, borne on a stalk (sporangiophore) and covered with a shield (e.g., in the cones of Horsetail, *Equisetum* spp., pp. 49–55).

Spore A reproductive, single-celled, asexual body capable of developing directly into a new individual (e.g., Horsetail, *Equisetum* spp.).

Sporophore A spore-bearing branch or organ.

Spur A hollow, tubular projection, usually from the base of a flower's petal or sepal, often with nectar to attract insects; a short compact branchlet, sometimes ending in a point.

Stalk A general term for the supporting part of any organ (e.g., a petiole, peduncle, pedicel, filament or stipe).

Stamen The pollen-bearing male reproductive organ of a flower composed of two parts: a lower filament, and the apical anther (*see also* anther, and filament).

Staminate A unisexual, male flower having stamens and a vestigial pistil, in monoecious and dioecious plants (*see also* pistillate).

Staminode (pl. staminodia) A sterile stamen or a structure resembling one in a flower; some may be petal-like and showy, or rudimentary.

Standard In Fabaceae—Pea flowers, the upright, large, wide, single petal (called a banner in this book).

Stellate Radiating like the points of a star (e.g., leaf or stem hairs with three to several microscopic branches).

Stem The central axis of a herbaceous plant that bears the branches, leaves, flowers and fruit; usually aerial but can also develop underground as a horizontal, creeping rhizome necessary for a plant's vegetative reproduction.

Sterile A flower without functional pistils or stamens; the reproductive parts (all or some) being infertile or missing.

Stigma (pl. stigmas, stigmata) The receptive, usually sticky or hairy, tip or region of a pistil above the style to which the pollen adheres and germinates to begin the pollination process (*see also* style).

Stigmatic Joint In thistles, *Cirsium* spp., it marks the microscopic zone where the style and stigma meet.

Stinging Hair A hollow hair bearing a gland that secretes an acidic substance (e.g., Stinging Nettle, *Urtica dioica*, p. 511).

Stipe The linear, sterile stalk of a pistil, gland, or fruit (e.g., in female cattail, or willow flowers).

Stipitate A plant's part positioned and supported on a stipe.

Stipule A showy to tiny bractlike appendage at the base of a leaf's petiole, often in pairs, sometimes partially united, or attached to the petiole or stem, or modified as spines, or tendrils.

Stolon A horizontal stem, mostly above ground, which roots at its nodes or tip and vegetatively produces a new plant (e.g., Smooth Wild Strawberry, p. 451; *see also* runner and sucker).

Stomate (pl. stomata) The microscopic pores in the leaf's epidermis with two guard cells that open and close to regulate the exchange of gasses and moisture.

Stone The hard endocarp which encloses the one seed within a drupe.

Striate Marked with fine longitudinal lines or ridges, or streaks of color (e.g., on the fertile scales of *Scirpus cyperinus*, p. 574.

Strigose A part bearing sharp, stiff, straight, appressed hairs that point forward and often have a swollen base (*see also* hispid).

Strobilus (pl. strobili) Whorls of sporangiophores forming a conelike structure comprised of several sporangia covered with a shield (e.g., in *Equisetum* spp., pp. 49–55).

Strophiole An appendage at the hilum of some seeds (e.g., in *Moehringia lateriflora*, p. 252; *see also* caruncle, and aril).

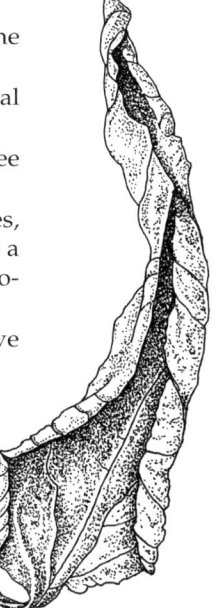

Glossary

Style In flowers, the usually lengthened narrow part of a pistil between the ovary and the stigma(s), at times quite obscure (*see also* stigma).

Submerged A plant part beneath the water's surface.

Subspecies The rank below species in taxonomy; abbreviated to ssp.

Subtend Located below or close to (e.g., a bract below a flower's stalk, or a leaf below a floral branch).

Succulent Juicy to fleshy (e.g., a fleshy, thickish stem, or leaf).

Sucker A vegetative shoot originating from below the ground (e.g., around the trunk's base in Basswood, p. 508; *see also* stolon).

Sulcate Grooved or furrowed lengthwise (e.g., a stem).

Sulcus A groove or furrow (e.g., on a seed).

Superior Placed above (e.g., an ovary inserted above and free from the flower's perianth or other parts around its base; the upper lip of a flower's corolla; *see also* inferior).

Suture A long line of dehiscence in a ripe fruit; a narrow groove that marks the seam of union or of separation of the valves (e.g., in a pod of Fabaceae—Pea).

Synonym In taxonomy the one or more older rejected scientific name(s) for a taxon compared to the one correct and current binomial.

Taproot
The main descending root with a direct continuation from the root's embryonic radicle (*see also* rhizome, and fibrous roots).

Tawny Dull, brownish yellow (e.g., the color of some pappus bristles).

Taxon (pl. taxa) A general term for any morphological unit or group, such as family, genus, species, etc.; the name of a botanical journal.

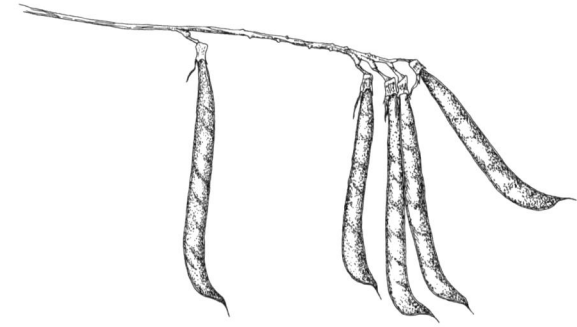

Taxonomy The systematic classification and description of organisms by a regulatory committee or a researcher.

Tendril A slender, twisting organ a plant uses to cling to a support; an organ modified from a stem, leaf, leaflet, or stipule.

Terete Circular in cross-section (e.g., a stem; *see also* lenticular).

Terminal At or belonging to the end or apex (e.g., a terminal bud at the end of a twig).

Thallus The green, vegetative combination of leaf and stem in Lemnaceae—Duckweed, also known as a frond, p. 593–595.

Thorn A pointed, hard appendage, branched or simple, more deeply seated at its base than a prickle and having vascular tissue; a dangerous spine (e.g., Round-leaved Hawthorn, p. 449).

Throat A flower's apical opening below the lobes into its tubular corolla or perianth.

Thyrse An elongated but compact, narrow inflorescence with opposite, lateral cymes (e.g., *Penstemon nitidus*, p. 502).

Tiller A sucker or branch from the base of the stem or culm (e.g., in some grasses).

Tomentose Densely hairy with short, matted, crooked, woolly hairs (e.g., leaves of *Salix candida*, p. 480; *see also* woolly, and lanate).

Glossary

Toothed With a dentate margin; a general term describing a leaf blade's margin when it has teeth (*see also* entire, and dissected).

Trailing A plant that grows in a prostrate, more or less horizontal, manner, often over adjacent vegetation, but not rooting at stem nodes (e.g., Wild Cucumber, p. 279).

Tree A woody, tall, long-lived plant with one main trunk and branches that form a generally distinct and elevated crown (e.g., American Elm, p. 509).

Triangular The cross-sectional shape of a culm (stem) of *Carex* spp.

Tridentate Three-toothed (*see also* bidentate).

Trifoliate A division with three leaves or three leaflets (e.g., Alsike Clover, p. 322).

Tripinnate Three times pinnate (e.g., *Thalictrum* spp., pp. 440–441; *see also* pinnate, and bipinnate).

Tristylous A population of flowering plants having flower styles in three different lengths but these not all in one plant's flowers (e.g., Purple Loosestrife, p. 366).

Truncate Ending abruptly; the base or apex mostly straight across or flat (e.g., some flower petals, seeds, and grass ligules).

Tuber A thick, short branch, usually an underground storage organ, attached to a narrower rhizome (e.g., *Helianthus tuberosus*, p. 148).

Tubercle A small tuber or tuberlike body; a brown oval growth on the valves of some *Rumex* spp., pp. 402–406, most obvious at fruiting stage; the apical structure on an achene of *Eleocharis acicularis*, p. 566.

Tufted Plants in a clustered, clumped or cespitose manner (e.g., many dryland grasses).

Turgid Swollen or expanded from the intake of water and not from air.

Turion A scaly, swollen offshoot of an underground rhizome (e.g., some aquatic plants produce and shed a turion that overwinters on the pond's bottom then germinates the next spring).

Tussock A tuft of grass or grasslike plants (e.g., Water Sedge, p. 531).

Twig A small shoot, or final branch of a tree or shrub with lateral buds and a terminal bud.

Type Specimen The original, illustrated or pressed specimen in a herbarium from which a plant species was described in text and given a published taxonomic name (*see also* herbarium, and specimen).

Umbel
A flat-topped or rounded inflorescence in which the pedicels, of about equal length, ascend from a common point (e.g., in Apiaceae—Carrot; *see also under* inflorescence).

Umbel, Compound A divided umbel in which each of the rays support an umbellet (e.g., Cow-parsnip, p. 83; *see also* umbellet).

Umbellet A secondary umbel of small flowers, each flower on a pedicel, with the whole umbellet supported by a ray (*see above*).

Unarmed Without prickles, thorns, or other sharp points (e.g., most of the plants described in this book).

Undulate Unevenly wavy up and down on the surface or margin (e.g., a leaf blade's margin in Curled Dock, p. 404).

Unisexual Flowers of one sex; female flowers with a pistil, or male flowers with stamens (*see also* perfect, imperfect, monoecious, and dioecious).

Utricle A bladderlike fruit, usually thin-walled, 1-seeded and indehiscent (e.g., Prostrate Amaranth, p. 75; *see also under* fruit).

Valve
A separable part of a fruit's wall that opens to release the seeds; one of the sections into which a mature capsule or legume splits (e.g., Golden-bean, p. 320); an enlarged sepal covering the fruit in *Rumex* spp., pp. 402–406.

Glossary

Variety A rank below that of a species but above the category of form in the taxonomic ordering and naming of plants.

Vascular Specialized plant tissue for transporting water and nutrients; in leaves, the vascular bundles in veins also help to support the narrow petiole and flat blade.

Vein A thread of vascular tissue often visible in a leaf's blade (*see also* nerve, and midrib).

Venation The pattern of veins visible on either side of a leaf's blade.

Ventral Referring to the front, adaxial (toward the axis) or inner surface of an organ; the upper surface of a horizontal leaf (*see also* dorsal).

Verticillaster A false whorl composed of pairs of opposite, clustered, condensed cymes in opposite leaf or bract axils (e.g., in Lamiaceae—Mint such as *Lycopus asper*, p. 351).

Vestigial Poorly developed organs that generally do not function (e.g., the tiny and not robust stamens in some pistillate [female] flowers; *see also* rudimentary).

Villose Having long, soft hairs, which are not matted or curved.

Vine An herbaceous or woody plant with a long, flexible stem that is not strong enough to support itself; usually attaches to nearby plants or fences with tendrils or the twisting of the stem itself.

Whorl
Arranged in a circle, as three or more leaves or flower parts or pedicels at a single node on the stem (*see also* opposite, and alternate).

Wing The lateral two petals in flowers of Fabaceae—Pea; a thin, dry, membranous, flat, broad extension of an organ, probably aiding in dispersal (e.g., part of a samara of Green Ash, p. 376).

Woody With a hard texture, often with smooth or rough bark.

Woolly Having curly, soft hairs, usually matted (*see also* lanate, and tomentose).

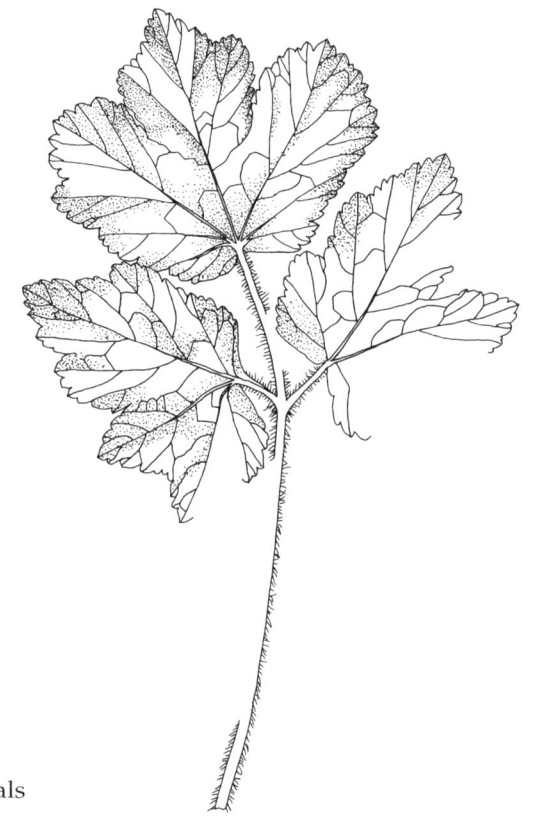

Abbreviations

& And (ampersand)
B or W Black or white
c. *Circa* (about)
dbh Diameter breast height
inflor. Inflorescence
invol. Involucre
p Native Prairie Restoration Species Priority (**p1**=high, **p2**=secondary)
p. Page
pl. Plural
pp. Pages
per. Perianth
reprod. Reproductive
sp. Species (singular)
spp. Species (plural)
ssp. Subspecies
subt. Subtending
TL Total Length
TW Total Width
var. Variety
w Wildlife Use of plants by birds and mammals (**w1**=high, **w2**=moderate)
x In a scientific name indicates a hybrid

References

Humans eat participants, not interactions; being relatively incompetent until quite recently, humans have by and large not generated cultural rules for the maintenance of interactions per se, but rather for the preservation of the participants.

— Daniel Janzen

Aarssen, L. W., I. V. Hall, and **K. I. N. Jensen.** 1986. The biology of Canadian weeds. 76. *Vicia angustifolia* L., *V. cracca* L., *V. sativa* L., *V. tetrasperma* (L.) Schreb., and *V. villosa* Roth. *Can. J. Plant Sci.* 66: 711–737.

Aiken, S. G. 1981. A conspectus of *Myriophyllum* (Haloragaceae) in North America. *Brittonia* 33: 57–69.

Aiken, S. G., and **S. J. Darbyshire.** 1983. *Grass Genera of Western Canadian Cattle Rangelands.* Monograph No. 29. Ottawa: Biosystematics Research Institute, Research Branch, Agriculture Canada.

Aiken, S. G., P. R. Newroth, and **I. Wile.** 1979. The biology of Canadian weeds. 34. *Myriophyllum spicatum* L. *Can. J. Plant Sci.* 59: 201–215.

Albee, B., L. Shultz, and **S. Goodrich.** 1988. *Atlas of the Vascular Plants of Utah.* Occasional Publication No. 7. Salt Lake City: Utah Museum of Natural History.

Allaby, M., ed. 1992. *The Concise Oxford Dictionary of Botany.* Oxford: Oxford University Press.

Alley, H. P., and **G. A. Lee.** 1969. *Weeds of Wyoming.* Bulletin 498. Laramie: Agricultural Experiment Station, University of Wyoming.

Allred, K. W. 1990. New Mexico grass types and a selected bibliography of New Mexico grass taxonomy. *Great Basin Naturalist* 50: 73–82.

Anderson, M. G. 1995. Interactions between *Lythrum salicaria* and native organisms: a critical review. *Environ. Manage.* 19: 225–231.

Applequist, W. L. 2002. A reassessment of the nomenclature of *Matricaria* L. and *Tripleurospermum* Sch. Bip. (Asteraceae). *Taxon* 51: 757–761.

Archibold, O. W., D. Brooks, and **L. Delanoy.** 1997. An investigation of the invasive shrub European Buckthorn, *Rhamnus cathartica* L., near Saskatoon, Saskatchewan. *Canadian Field-Naturalist* 111: 617–621.

Argus, G. W. 1973. The genus *Salix* in Alaska and the Yukon. Publications in Botany No. 2. Ottawa: National Museum of Natural Sciences, National Museums of Canada.

Argus, G. W. 1980. The typification and identity of *Salix eriocephala* Michx. (Salicaceae). *Brittonia* 32: 170–177.

Argus, G. W. 1997. Notes on the taxonomy and distribution of California *Salix*. *Madroño* 44: 115–136.

Argus, G. W., and **K. M. Pryer.** 1990. *Rare Vascular Plants In Canada: Our Natural Heritage.* Ottawa: Canadian Museum of Nature.

Argus, G. W., and **D. J. White.** 1977. *The Rare Vascular Plants of Ontario.* Syllogeus No. 14. Ottawa: National Museum of Natural Sciences.

Argus, G. W., and **D. J. White.** 1978. *The Rare Vascular Plants of Alberta.* Syllogeus No. 17. Ottawa: National Museum of Natural Sciences.

Argus, G. W., K. M. Pryer, D. J. White, and **C. J. Keddy,** eds. 1982–1987. *Atlas of the Rare Vascular Plants of Ontario.* 4 parts. Ottawa: National Museum of Natural Sciences, National Museums of Canada.

Armstrong, K. C. 1982. Hybrids between the tetraploids of *Bromus inermis* and *B. pumpellianus*. *Can. J. Botany* 60: 476–482.

Arnold, T. W., and **K. F. Higgins.** 1986. Effects of shrub coverages on birds of North Dakota mixed-grass prairies. *Canadian Field-Naturalist* 100: 10–14.

Austin, D. F. 1997. *Calystegia* (Convolvulaceae) in Texas. *Sida* 17: 837–839.

Avers, C. 1953. Biosystematic studies in *Aster*. 2. Isolating mechanisms and some phylogenetic considerations. *Evolution* 7: 317–327.

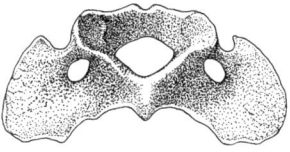

Badger, K. S., and **I. A. Ungar.** 1991. Life history and population dynamics of *Hordeum jubatum* along a soil salinity gradient. *Can. J. Botany* 69: 384–393.

Bahn, P. G., and **J. Vertut.** 1988. *Images of the Ice Age.* New York: Facts on File.

Bakshi, T. S., and **R. G. Holmberg.** 1986. *A Preliminary Biological Survey of Athabasca University Lands.* Athabasca University Internal Report, Alberta.

References

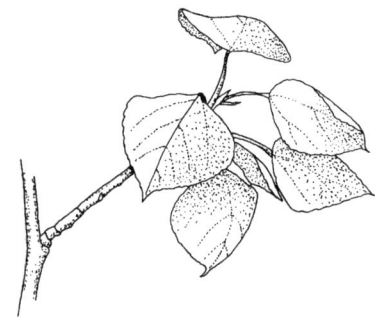

Balgooyen, C. P., and D. M. Waller. 1995. The use of *Clintonia borealis* and other indicators to gauge impacts of White-tailed Deer on plant communities in northern Wisconsin, USA. *Nat. Areas J.* 15: 308–318.

Barclay-Estrup, P., T. E. Duralia, and A. G. Harris. 1991. Flowering sequence of the orchid genus *Goodyera* in Thunder Bay district, Ontario. *Rhodora* 93: 141–147.

Bare, J. E. 1979. *Wildflowers and Weeds of Kansas*. Lawrence: Regents Press of Kansas.

Barkley, T. M. 1980. Taxonomic notes on *Senecio tomentosus* and its allies (Asteraceae). *Brittonia* 32: 291–308.

Barkley, T. M. 1983. *Field Guide to the Common Weeds of Kansas*. Lawrence: University Press of Kansas.

Barkworth, M. E. 1978. A taxonomic study of the large-glumed species of *Stipa* (Gramineae) occurring in Canada. *Can. J. Botany* 56: 606–625.

Barkworth, M. E., and M. A. Torres. 2001. Distribution and diagnostic characters of *Nassella* (Poaceae: Stipeae). *Taxon* 50: 439–468.

Barneby, R. C. 1964. *Atlas of the North American Astragalus*. Vol. 13, 2 parts. Bronx: Mem. New York Botanical Garden.

Barneby, R. C. 1989. *Intermountain Flora: Vascular Plants of the Intermountain West, USA. Vol. 3B: Fabales*. Bronx: New York Botanical Garden.

Barrett, S. C. H., and B. F. Wilson. 1981. Colonizing ability in the *Echinochloa crus-galli* complex (barnyard grass). 1. Variation in life history. *Can. J. Botany* 59: 1844–1860.

Bassett, I. J. 1967. Taxonomy of *Plantago* L. in North America: sections *Holopsyllium* Pilger, *Palaeopsyllium* Pilger, and *Lamprosantha* Decne. *Can. J. Botany* 45: 565–577.

Bassett, I. J. 1973. *The Plantains of Canada*. Monograph No. 7. Ottawa: Research Branch, Canada Department of Agriculture.

Bassett, I. J., and C. W. Crompton. 1973. The genus *Atriplex* (Chenopodiaceae) in Canada and Alaska. 3. Three hexaploid annuals: *A. subspicata*, *A. gmelinii*, and *A. alaskensis*. *Can. J. Botany* 51: 1715–1723.

Bassett, I. J., and C. W. Crompton. 1975. The biology of Canadian weeds. 11. *Ambrosia artemisiifolia* L. and *A. psilostachya* DC. *Can. J. Plant Sci.* 55: 463–476.

Bassett, I. J., and C. W. Crompton. 1978a. The biology of Canadian weeds. 32. *Chenopodium album* L. *Can. J. Plant Sci.* 58: 1061–1072.

Bassett, I. J., and C. W. Crompton. 1978b. The genus *Suaeda* (Chenopodiaceae) in Canada. *Can. J. Botany* 56: 581–591.

Bassett, I. J., and C. W. Crompton. 1982a. The biology of Canadian weeds. 55. *Ambrosia trifida* L. *Can. J. Plant Sci.* 62: 1003–1010.

Bassett, I. J., and C. W. Crompton. 1982b. The genus *Chenopodium* in Canada. *Can. J. Botany* 60: 586–610.

Bassett, I. J., and D. B. Munro. 1987. The biology of Canadian weeds. 79. *Atriplex patula* L., *A. prostrata* Boucher ex DC., and *A. rosea* L. *Can. J. Plant Sci.* 67: 1069–1082.

Bassett, I. J., and J. Terasmae. 1962. Ragweeds, *Ambrosia* species, in Canada and their history in postglacial time. *Can. J. Botany* 40: 141–150.

Bassett, I. J., C. W. Crompton, and D. W. Woodland. 1974. The family Urticaceae in Canada. *Can. J. Botany* 52: 503–516.

Bassett, I. J., C. W. Crompton, and D. W. Woodland. 1977. The biology of Canadian weeds. 21. *Urtica dioica* L. *Can. J. Plant Sci.* 57: 491–498.

Bassett, I. J., C. W. Crompton, J. McNeill, and P. M. Taschereau. 1983. The genus *Atriplex* (Chenopodiaceae) in Canada. Monograph No. 13. Ottawa: Communications Branch, Agriculture Canada.

Batson, W. 1975. *A Guide to the Genera of Native and Commonly Introduced Ferns and Seed Plants of Eastern North America*. Published by the author at the University of South Carolina.

Baum, B. R. 1968. On some relationships between *Avena sativa* and *A. fatua* (Gramineae) as studied from Canadian material. *Can. J. Botany* 46: 1013–1024.

Baum, B. R., and Bailey, L. G. 1990. Key and synopsis of North American *Hordeum* species. *Can. J. Botany* 68: 2433–2442.

Beardsley, P. M., and R. G. Olmstead. 2002. Redefining Phrymaceae: the placement of *Mimulus*, tribe Mimuleae, and *Phryma*. *Amer. J. Botany* 89: 1093–1102.

Beatley, J. C. 1973. Russian-thistle (*Salsola*) species in western United States. *J. Range Management* 26: 225–226.

Beatley, J. C. 1976. *Vascular Plants of the Nevada Test Site and Central-Southern Nevada: Ecological and Geographic Distribution*. TID–26881. Technical Information Center, Office of Technical Information, Energy Research and Development Administration.

Bégin, C., and L. Filion. 1999. Black Spruce (*Picea mariana*) architecture. *Can. J. Botany* 77: 664–672.

Bergeron, J. M., and L. Jodoin. 1987. Defining "high quality" food resources of herbivores: the case for Meadow Voles (*Microtus pennsylvanicus*). *Oecologia* 71: 510–517.

Best, K. F. 1977. The biology of Canadian weeds. 22. *Descurainia sophia* (L.) Webb. *Can. J. Plant Sci.* 57: 499–507.

Best, K. F., and **G. I. McIntyre.** 1975. The biology of Canadian weeds. 9. *Thlaspi arvense* L. *Can. J. Plant Sci.* 55: 279–292.

Best, K. F., J. Looman, and **J. B. Campbell.** 1971. *Prairie Grasses Identified and Described by Vegetative Characters.* Publication 1413. Ottawa: Canada Department of Agriculture.

Best, K. F., J. D. Banting, and **G. G. Bowes.** 1978. The biology of Canadian weeds. 31. *Hordeum jubatum* L. *Can. J. Plant Sci.* 58: 699–708.

Best, K. F., G. G. Bowes, A. G. Thomas, and **M. G. Maw.** 1980. The biology of Canadian weeds. 39. *Euphorbia esula* L. *Can. J. Plant Sci.* 60: 651–663.

Bhowmik, P. C., and **J. D. Bandeen.** 1976. The biology of Canadian Weeds. 19. *Asclepias syriaca* L. *Can. J. Plant Sci.* 56: 579–589.

Biddy, C. J. 1981. Food supply and diet of the Bearded Tit. *Bird Study* 28: 201–210.

Blunt, W., and **W. T. Stearn.** 1994. (new edition). *The Art of Botanical Illustration.* Woodbridge, Suffolk: Antique Collectors' Club in association with The Royal Botanic Gardens, Kew.

Boe, A., and **K. Fluharty.** 1993. Reproductive biology of a Canada Milk-vetch population from eastern South Dakota. *Prairie Naturalist* 25: 65–72.

Boileau, F., M. Crête, and **J. Huot.** 1994. Food habits of the Black Bear, *Ursus americanus,* and habitat use in Gaspésie Park, eastern Québec. *Canadian Field-Naturalist* 108: 162–169.

Booth, W. E., and **J. C. Wright.** 1962. *Flora of Montana. Part 2 Dicotyledons.* Bozeman: Montana State College.

Boraiah, G., and **M. Heimburger.** 1964. Cytotaxonomic studies on New World *Anemone* (Section *Eriocephalus*) with woody rootstocks. *Can. J. Botany* 42: 891–922.

Bough, M., J. C. Colosi, and **P. B. Cavers.** 1986. The major weedy biotypes of Proso Millet (*Panicum miliaceum*) in Canada. *Can. J. Botany* 64: 1188–1198.

Bowden, W. M. 1962. Cytotaxonomy of the native and adventive species of *Hordeum, Eremopyrum, Secale, Sitanion,* and *Triticum* in Canada. *Can. J. Botany* 40: 1675–1711.

Bowden, W. M. 1964. Cytotaxonomy of the species and interspecific hybrids of the genus *Elymus* in Canada and neighboring areas. *Can. J. Botany* 42: 547–601.

Bowden, W. M. 1965. Cytotaxonomy of the species and interspecific hybrids of the genus *Agropyron* in Canada and neighboring areas. *Can. J. Botany* 43: 1421–1448.

Bowers, J. E. 1981. Local floras of Arizona: an annotated bibliography. *Madroño* 28: 193–209.

Bradley, C., and **M. Fairbarns.** 1984. Vegetation, flora and special features of the candidate Goose Mountain Ecological Reserve. ENR Technical Report T/61. Natural Areas Technical Report No. 13. Edmonton: Alberta Energy and Natural Resources.

Brako, L., A. Y. Rossman, and **D. F. Farr.** 1995. *Scientific and Common Names of 7,000 Vascular Plants in the United States.* St. Paul, MN: The American Phytopathological Society.

Brayshaw, T. C. 1976. *Catkin-Bearing Plants of British Columbia.* Occasional Paper No. 18. Victoria: Royal British Columbia Musuem.

Brayshaw, T. C. 1989. *Buttercups, Waterlilies, and Their Relatives: (The order Ranales) in British Columbia.* Royal British Columbia Museum Memoir No. 1. Victoria: Royal British Columbia Museum.

Brayshaw, T. C. 1996. *Trees and Shrubs of British Columbia.* Victoria: Royal British Columbia Museum/Vancouver: UBC Press.

Brotherson, J. D. 1983. Species composition, distribution, and phytosociology of Kalsow Prairie, a mesic tall-grass prairie in Iowa. *Great Basin Naturalist* 43: 137–167.

Brouillet, L., et **J. C. Semple.** 1981. A propos du status taxonomique de *Solidago ptarmicoides.* *Can. J. Botany* 59: 17–21.

Brummitt, R. K. 1971. Relationship of *Heracleum lanatum* Michx. of North America to *H. sphondylium* of Europe. *Rhodora* 73: 578–584.

Brunsfeld, S. J., and **F. D. Johnson.** 1985. *Field Guide to the Willows of East-central Idaho.* Moscow: Forest, Wildlife and Range Experiment Station, University of Idaho.

Budd, A., and **K. Best.** 1969. *Wild Plants of the Canadian Prairies.* Publication 983. Ottawa: Canada Department of Agriculture.

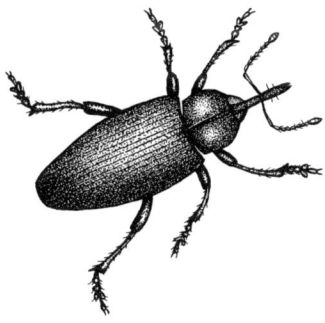

Burbridge, J. 1989. *Wildflowers of the Southern Interior of British Columbia and Adjacent Parts of Washington, Idaho and Montana.* Vancouver: University of British Columbia Press.

Butters, F. K. 1926. Notes on the range of *Maianthemum canadense* and its variety *interius.* *Rhodora* 28: 9–11.

References

Callow, J. M., H. A. Kantrud, and K. F. Higgins. 1992. First flowering dates and flowering periods of prairie plants at Woodworth, North Dakota. *Prairie Naturalist* 24: 57–64.

Campbell, J. B., K. F. Best, and A. C. Budd. 1973. *Ninety-nine Range Forage Plants of the Canadian Prairies*. Publication 964. Ottawa: Canada Department of Agriculture.

Campbell, J. J. N. 1996. (1209) Proposal to conserve the name *Elymus virginicus* (Poaceae) with a conserved type. *Taxon* 45: 128–129.

Canne, J. M. 1977. A revision of the genus *Galinsoga* (Compositae: *Heliantheae*). *Rhodora* 79: 319–389.

Cantino, P. D. 1981. Change of status for *Physostegia virginiana* var. *ledinghamii* (Labiatae) and evidence for a hybrid origin. *Rhodora* 83: 111–118.

Case, R. L., and J. B. Kauffman. 1997. Wild ungulate influences on the recovery of willows, Black Cottonwood and Thin-leaf Alder following cessation of cattle-grazing in northeastern Oregon. *Northwest Sci.* 71: 115–156.

Catling, P. M., A. A. Reznicek, and B. S. Brookes. 1988. The separation of *Carex disticha* and *Carex sartwellii* and the status of *Carex disticha* in North America. *Can. J. Botany* 66: 2323–2330.

Catling, P. M., and Spicer, K. W. 1987. The perennial *Juncus* of section *Poiophylli* in the Canadian prairie provinces. *Can. J. Botany* 65: 750–760.

Ceska, A., and M. A. M. Bell. 1973. *Utricularia* (Lentibulariaceae) in the Pacific Northwest. *Madroño* 22: 74–84.

Chen, C. J., M. G. Mendenhall, and B. L. Turner. 1994. Taxonomy of *Thermopsis* (Fabaceae) in North America. *Annals Missouri Botanical Garden* 81: 714–742.

Cherniawsky, D. M., and R. J. Bayer. 1998. Systematics of North American *Petasites* (Asteraceae: *Senecioneae*). 3. A taxonomic revision. *Can. J. Botany* 76: 2061–2075.

Chinnappa, C. C., and J. K. Morton. 1976. Studies on the *Stellaria longipes* Goldie complex—variation in wild populations. *Rhodora* 78: 488–502.

Chinnappa, C. C., and J. K. Morton. 1991. Studies on the *Stellaria longipes* complex (Caryophyllaceae)—Taxonomy. *Rhodora* 93: 129–135.

Chmielewski, J. G., and J. C. Semple. 2001. The biology of Canadian weeds. 113. *Symphyotrichum lanceolatum* (Willd.) Nesom [*Aster lanceolatus* Willd.] and *S. lateriflorum* (L.) Löve & Löve [*Aster lateriflorus* (L.) Britt.]. *Can. J. Plant Sci.* 81: 829–849.

Cholewa, A. F., and D. M. Henderson. 1984. Biosystematics of *Sisyrinchium* section *Bermudiana* (Iridaceae) of the Rocky Mountains. *Brittonia* 36: 342–363.

Christiansen, P., and M. Müller. 1999. *An Illustrated Guide to Iowa Prairie Plants*. Iowa City: University of Iowa Press.

Clarkson, R. W., and J. C. deVos. 1986. The Bullfrog, *Rana catesbeiana shaw*, in the lower Colorado River, Arizona/California. *J. Herpetol.* 20: 42–49.

Clevenger, S., and C. B. Heiser, Jr. 1963. *Helianthus laetiflorus* and *Helianthus rigidus*—Hybrids or species? *Rhodora* 65: 121–133.

Clokey, I. W. 1951. *Flora of the Charleston Mountains, Clark County, Nevada*. Berkeley: University of California Press.

Cochrane, T. S., and H. H. Iltis. 2000. *Atlas of the Wisconsin Prairie and Savanna Flora*. Madison: Technical Bulletin 191. Department of Natural Resources, Wisconsin.

Cody, W. J. 1996a. Additions and range extensions to the vascular plant flora of the Northwest Territories, Canada. *Canadian Field-Naturalist* 110: 260–270.

Cody, W. J. 1996b. *Flora of the Yukon Territory*. Ottawa: National Research Council, Research Press.

Cody, W. J., and D. M. Britton. 1989. *Ferns and Fern Allies of Canada*. Publication 1829/E. Ottawa: Agriculture Canada.

Cody, W. J., and K. Shaw. 1973. Canada Plum in southwestern Alberta. *Blue Jay* 31: 217–219.

Cody, W. J., and V. Wagner. 1981. The biology of Canadian weeds. 49. *Equisetum arvense* L. *Can. J. Plant Sci.* 61: 123–133.

Coffey, V. J., and S. B. Jones, Jr. 1980. Biosystematics of *Lysimachia* section *Seleucia* (Primulaceae). *Brittonia* 32: 309–322.

Cole, K. C. 1998. *The Universe and The Teacup: The Mathematics of Truth and Beauty*. New York: Harcourt Brace & Company.

Colorado Native Plant Society. 1989. *Rare Plants of Colorado*. Estes Park, CO: Published in cooperation with the Rocky Mountain Nature Association.

Confer, J. L., and P. A. Werner. 1985. Downy Woodpeckers at goldenrod galls. *J. Field Ornithology* 56: 56–64.

Correll, D. S., and H. B. Correll. 1972. *Aquatic and Wetland Plants of Southwestern United States*. Vol. 1., reissued in 1975 in 2 vols. Stanford: Stanford University Press.

Crawford, D. J. 1973. Morphology, flavonoid chemistry, and chromosome number of the *Chenopodium neomexicanum* complex. *Madroño* 22: 185–195.

Cresswell, S. E., and A. W. Robertson. 1994. Discrimination by pollen-collecting bumblebees among differentially rewarding flowers of an alpine wildflower, *Campanula rotundifolia*. *Oikos* 69: 304–308.

Crins, W. J., and P. W. Ball. 1983. The taxonomy of the *Carex pensylvanica* complex (Cyperaceae) in North America. *Can. J. Botany* 61: 1692–1717.

Croat, T. 1972. *Solidago canadensis* complex of the Great Plains. *Brittonia* 24: 317–326.

Croizat, L. 1945. "*Euphorbia esula*" in North America. *American Midland Naturalist* 33: 231–243.

Crompton, C. W., and I. J. Bassett. 1985. The biology of Canadian weeds. 65. *Salsola pestifer* A. Nels. *Can. J. Plant Sci.* 65: 379–388.

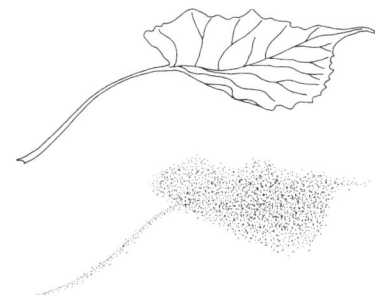

Crompton, C. W., A. E. Stahevitch, and W. A. Wojtas. 1990. Morphometric studies of the *Euphorbia esula* group (Euphorbiaceae) in North America. *Can. J. Botany* 68: 1978–1988.

Cronquist, A., A. H. Holmgren, N. H. Holmgren, J. L. Reveal, and P. K. Holmgren. 1972–1994. *Intermountain Flora: Vascular Plants of the Intermountain West, USA*. Vols. 1, 4, 5, and 6. New York: Hafner Publishing Co., The New York Botanical Garden: Columbia University Press.

Cross, A. F. 1991. Vegetation of two southeastern Arizona desert marshes. *Madroño* 38: 185–194.

Curtis, J. T. 1959. *The Vegetation of Wisconsin*. Madison: University of Wisconsin Press.

Darwent, A. L. 1975. The biology of Canadian weeds. 14. *Gypsophila paniculata* L. *Can. J. Plant Sci.* 55: 1049–1058.

Daubenmire, R. 1974. Taxonomic and ecologic relationships between *Picea glauca* and *Picea engelmannii*. *Can. J. Botany* 52: 1545–1560.

Davis, A. N., and T. L. Briggs. 1986. Dispersion pattern of aerial shoots of the Common Marsh Reed *Phragmites australis* (Poaceae). *Rhodora* 88: 325–330.

Davis, R. J. 1952. *Flora of Idaho*. Provo: Brigham Young University Press.

Delorit, R. J., and C. R. Gunn. 1986. *Seeds of Continental United States Legumes (Fabaceae)*. River Falls, WI: Agronomy Publications.

Desrochers, A. M., J. F. Bain, and S. I. Warwick. 1988a. The biology of Canadian weeds. 89. *Carduus nutans* L. and *Carduus acanthoides* L. *Can. J. Plant Sci.* 68: 1053–1068.

Desrochers, A. M., J. F. Bain, and S. I. Warwick. 1988b. A biosystematic study of the *Carduus nutans* complex in Canada. *Can. J. Botany* 66: 1621–1631.

De Vries, B. 1973. Nodding Trilliums in eastern Saskatchewan. *Blue Jay* 31: 214–217.

DeWet, J. M. J. 1981. Grasses and the culture history of man. *Annals Missouri Botanical Garden* 68: 87–104.

Dewey, D. R., and C. Hsiao. 1983. A cytogenetic basis for transferring Russian Wildrye from *Elymus* to *Psathyrostachys*. *Crop Science* 23: 123–126.

Dickenson, V. 1998. *Drawn From Life: Science and Art in the Portrayal of the New World*. Toronto: University of Toronto Press.

Dickinson, T., D. Metsger, J. Bull, and R. Dickinson. 2004. *The ROM Field Guide to Wildflowers of Ontario*. Toronto: Royal Ontario Museum and McClelland and Stewart.

Dick-Peddie, W. A. 1993. *A New Mexico Vegetation: Past, Present, and Future*. University of New Mexico Press.

Diggs, G. M. Jr., C. E. S. Taylor, and R. J. Taylor. 1997. *Chaenorhinum minus* (Scrophulariaceae) new to Texas. *Sida*: 17: 631.

Dix, R. L., and F. E. Smeins. 1967. The prairie, meadow, and marsh vegetation of Nelson County, North Dakota. *Can. J. Botany* 45: 21–58.

Dore, W. G., and J. McNeill. 1980. *Grasses of Ontario*. Monograph 26. Ottawa: Research Branch, Agriculture Canada.

Dorn, R. D. 1975. A systematic study of *Salix* section *Cordatae* in North America. *Can. J. Botany* 53: 1491–1522.

Dorn, R. D. 1976. A synopsis of American *Salix*. *Can. J. Botany* 54: 2769–2789.

Dorn, R. D. 1977a. *Manual of the Vascular Plants of Wyoming*. 2 vols. New York: Garland Publishing.

Dorn, R. D. 1977b. Willows of the Rocky Mountain States. *Rhodora* 79: 390–429.

Dorn, R. D. 1984. *Vascular Plants of Montana*. Cheyenne: Mountain West Publishing.

Dorn, R. D. 1992. *Vascular Plants of Wyoming*. 2nd ed. Cheyenne: Mountain West Publishing.

Dorn, R. D. 1995. A taxonomic study of *Salix* section *Cordatae* subsection *Luteae* (Salicaceae). *Brittonia* 47: 160–174.

Dorn, R. D. 2002. Noteworthy collections: Colorado and New Mexico. *Salix discolor*. *Madroño* 49: 54–58.

Douglas, B. J., A. G. Thomas, I. N. Morrison, and M. G. Maw. 1985. The biology of Canadian weeds. 70. *Setaria viridis* (L.) Beauv. *Can. J. Plant Sci.* 65: 669–690.

Douglas, G. W. 1995. *The Sunflower Family (Asteraceae) of British Columbia*. Vol. 2. Victoria: Royal British Columbia Museum.

Douglas, G. W., G. W. Argus, H. L. Dickson, and D. F. Brunton. 1981. *The Rare Vascular Plants of the Yukon*.

References

Syllogeus No. 28. Ottawa: The National Museum of Natural Sciences.

Douglas, G. W., G. B. Straley, and **D. V. Meidinger.** 1998. *Rare Native Vascular Plants of British Columbia.* Victoria: British Columbia Ministry of Environment.

Douglas, G. W., D. Meidinger, and **J. Pojar,** eds. 2002. *Illustrated Flora of British Columbia.* Volume 8, General Summary, Maps and Keys. Victoria: Ministry of Sustainable Resource Management and the British Columbia Ministry of Forests.

Doust, L. L., A. MacKinnon, and **J. L. Doust.** 1985. Biology of Canadian Weeds. 71. *Oxalis stricta* L., *O. corniculata* L., *O. dillenii* Jacq. ssp. *dillenii* and *O. dillenii* Jacq. ssp. *filipes* (Small) Eiten. *Can. J. Plant Sci.* 65: 691–709.

Dressler, R. 1981. *The Orchids: Natural History and Classification.* Massachusetts: Harvard University Press.

Drew, W. B. 1936. The North American representatives of *Ranunculus*, Batrachium. *Rhodora* 38: 1–47.

Dugle, J. R. 1966. A taxonomic study of western Canadian species in the genus *Betula. Can. J. Botany* 44: 929–1007.

Dunn, P. H. 1976. Distribution of *Carduus nutans, C. acanthoides, C. pycnocephalus,* and *C. crispus,* in the United States. *Weed Science* 24: 518–524.

Dunn, P. H. 1979. The distribution of Leafy Spurge (*Euphorbia esula*) and other weedy *Euphorbia* spp. in the United States. *Weed Science* 27: 509–516.

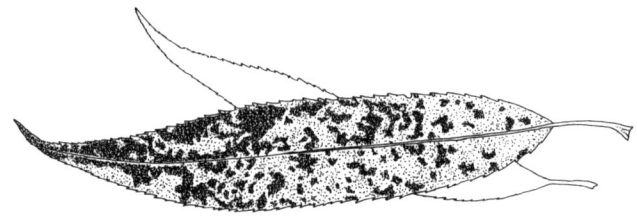

Ebinger, J. E. 1962. The varieties of *Luzula acuminata. Rhodora* 64: 74–83.

Eckert, C. G., B. Massonnet, and **J. J. Thomas.** 2000. Variation in sexual and clonal reproduction among introduced populations of Flowering Rush, *Butomus umbellatus* (Butomaceae). *Can. J. Botany* 78: 437–446.

Editorial Subcommittee of the Regional Technical Committee of Project NC-10. 1968. *Weeds of the North Central States.* Publication No. 36, Circular 718. Urbana: University of Illinois Agricultural Experiment Station.

Edwards, J. 1985. Effects of herbivory by moose on flower and fruit production of *Aralia nudicaulis. J. Ecol.* 73: 861–868.

Eilers, L. J., and **D. M. Roosa.** 1994. *The Vascular Plants of Iowa: An Annotated Checklist and Natural History.* Iowa City: University of Iowa Press.

Eiseley, L. 1998. *The Invisible Pyramid.* Lincoln: University of Nebraska Press.

Elias, T. S. 1989. *Field Guide to North American Trees.* Danbury, CT: Grolier Book Clubs.

Elisens, W. J., and **J. G. Packer.** 1980. A contribution to the taxonomy of the *Oxytropis campestris* complex in northwestern North America. *Can. J. Botany* 58: 1820–1831.

Elliott, C., and **J. T. Flinders.** 1985. Food habits of the Columbian Ground Squirrel, *Spermophilus columbianus,* in southcentral Idaho. *Canadian Field-Naturalist* 99: 327–330.

Epple, A. O., and **L. E. Epple.** 1995. *A Field Guide to the Plants of Arizona.* Helena, MT: Falcon Press.

Farrar, J. L. 1995. *Trees In Canada.* Ottawa: Fitzhenry & Whiteside, and the Canadian Forest Service, Natural Resources Canada, in cooperation with the Canada Communication Group–Publishing, Supply and Services Canada.

Fassett, N. C. 1939. *The Leguminous Plants of Wisconsin: The Taxonomy, Ecology, and Distribution of the Leguminosae Growing in the State Without Cultivation.* Madison: The University of Wisconsin Press.

Fassett, N. C. 1951. *Grasses of Wisconsin.* Madison: The University of Wisconsin Press.

Fassett, N. C. 1969. *A Manual of Aquatic Plants.* Madison: The University of Wisconsin Press.

Fassett, N. C. 1976. *Spring Flora of Wisconsin.* 4th ed., rev. by O. S. Thomson. Madison: The University of Wisconsin Press.

Fernald, M. L. 1933. Recent discoveries in the Newfoundland flora. *Rhodora* 35: 1–16, 47–63, 80–107, 120–140, 161–185, 203–223, 230–247, 265–283, 298–315, 327–346, 364–403.

Fernald, M. L. 1934. *Draba* in temperate northeastern America. *Rhodora* 36: 241–261, 285–305, 314–344, 353–371, 392–404.

Fernald, M. L. 1935. Critical plants of the upper Great Lakes region of Ontario and Michigan. *Rhodora* 37: 197–222, 238–262, 272–301, 324–341.

Fernald, M. L. 1946. The North American representatives of *Alisma plantago-aquatica. Rhodora* 48: 86–88.

Fernald, M. L. 1970. *Gray's Manual of Botany.* 8th (Centennial) ed. New York: Van Nostrand Reinhold.

Fernald, M. L., and **A. E. Brackett.** 1929. The representatives of *Eleocharis palustris* in North America. *Rhodora* 31: 57–77.

Fernald, M. L., and **C. A. Weatherby.** 1916. The genus *Puccinellia* in eastern North America. *Rhodora* 18: 1–23.

References

Fielding, D. J., M. A. Brusven, and **L. P. Kish.** 1996. Consumption of Diffuse Knapweed by two species of Polyphagous grasshoppers (*Orthoptera*: Acrididae) in southern Idaho. *Great Basin Naturalist* 56: 22–27.

Flora of North America Editorial Committee, ed. 1993–. *Flora of North America North of Mexico.* vols. 1–5, 19–27, several more in progress. New York: Oxford University Press.

Forestry Branch, Saskatchewan Tourism and Renewable Resources. 1980. *Guide to Forest Understory Vegetation in Saskatchewan.* Technical Bulletin No. 9/1980.

Frankton, C., and **I. J. Bassett.** 1968. The genus *Atriplex* (Chenopodiaceae) in Canada. I. Three introduced species: *A. heterosperma, A. oblongifolia,* and *A. hortensis. Can. J. Botany* 46: 1309–1313.

Frankton, C., and **R. J. Moore.** 1961. Cytotaxonomy, phylogeny, and Canadian distribution of *Cirsium undulatum* and *Cirsium flodmanii. Can. J. Botany* 39: 21–33.

Freckmann, R. W. 1972. *Grasses of Central Wisconsin.* Report No. 6. Madison: University of Wisconsin Press.

Freeman, C. C., R. L. McGregor, and **C. A. Morse.** 1998. Vascular plants new to Kansas. *Sida* 18: 593–604.

Frick, B. 1984. The biology of Canadian weeds. 62. *Lappula squarrosa* (Retz.) Dumort. *Can. J. Plant Sci.* 64: 375–386.

Fryxell, P. A. 1988. *Malvaceae of Mexico.* Systematic Botany Monographs. Vol. 25. The American Society of Plant Taxonomists.

Furlow, J. J. 1979. The systematics of the American species of *Alnus* (Betulaceae). *Rhodora* 81: 1–121, 151–248.

Gaiser, L. O. 1946. The genus *Liatris. Rhodora* 48: 165–183, 216–263, 273–326, 331–382, 393–412.

Gandhi, K. N., J. H. Wiersema, and **R. J. Soreng.** 2001. (1479) Proposal to conserve the name *Bouteloua gracilis* (Kunth) Griffiths against *B. gracilis* Vasey (Poaceae). *Taxon* 50: 573–575.

Gardner, R. C. 1974. Systematics of *Cirsium* (Compositae) in Wyoming. *Madroño* 22: 239–265.

Gauthier, G., and **J. Bédard.** 1991. Experimental tests of the palatability of forage plants in Greater Snow Geese. *J. Appl. Ecol.* 28: 491–500.

Gillett, J. M. 1963. *The Gentians of Canada, Alaska and Greenland.* Publication 1180. Ottawa: Research Branch, Canada Department of Agriculture.

Gillett, J. M., and **H. A. Senn.** 1960. Cytotaxonomy and infraspecific variation of *Agropyron smithii* Rydb. *Can. J. Botany* 38: 747–760.

Gillis, W. 1971. The systematics and ecology of Poison-Ivy and the Poison-Oaks (*Toxicodendron*, Anacardiaceae). *Rhodora* 73: 72–159, 161–237, 370–443, 465–539.

Giroux, J-F., Y. Bédard, and **J. Bédard.** 1984. Habitat use by Greater Snow Geese during the brood-rearing period. *Arctic* 37: 155–160.

Gleason, H. 1968. *The New Britton and Brown Illustrated Flora of the Northeastern United States and Adjacent Canada.* 3 vols. New York: The New York Botanical Garden with Hafner Publishing Company.

Glimn-Lacy, J., and **P. B. Kaufman.** 1984. *Botany Illustrated: Introduction To Plants, Major Groups, Flowering Plant Families.* New York: Van Nostrand Reinhold.

Glück, E. 1986. Flock size and habitat-dependent food and energy intake of foraging goldfinches. *Oecologia* 71: 149–155.

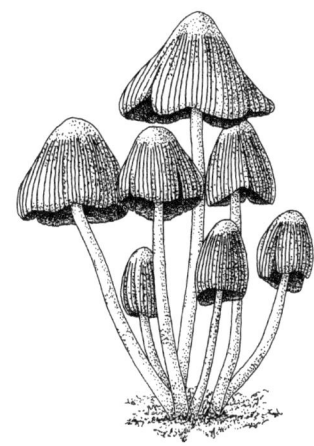

Godfrey, R. K., and **J. W. Wooten.** 1981. *Aquatic and Wetland Plants of Southeastern United States: Dicotyledons.* Athens: University of Georgia Press.

Gould, F. W. 1967. The grass genus *Andropogon* in the United States. *Brittonia* 19: 70–76.

Gould, F. W. 1975. *The Grasses of Texas.* College Station: Texas A&M University Press.

Gould, F. W. 1979. The genus *Bouteloua* (Poaceae). *Annals Missouri Botanical Garden* 66: 348–416.

Gould, F. W., and **C. A. Clark.** 1978. *Dichanthelium* (Poaceae) in the United States and Canada. *Annals Missouri Botanical Garden* 65: 1088–1132.

Gould, F. W., and **Z. J. Kapadia.** 1962. Biosystematic studies in the *Bouteloua curtipendula* complex. I. The aneuploid rhizomatous *B. curtipendula* of Texas. *Amer. J. Botany* 49: 887–891.

Gould, F. W., and **Z. J. Kapadia.** 1964. Biosystematic studies in the *Bouteloua curtipendula* complex 2. Taxonomy. *Brittonia* 16: 182–207.

Gould, F. W., M. A. Ali, and **D. E. Fairbrothers.** 1972. A revision of *Echinochloa* in the United States. *American Midland Naturalist* 87: 36–59.

Govaerts, R. 2001. How many species of seed plants are there? *Taxon* 50: 1085–1090.

References

Grace, J. B., and J. S. Harrison. 1986. The biology of Canadian weeds. 73. *Typha latifolia* L., *Typha angustifolia* L., and *Typha* xglauca Godr. *Can. J. Plant Sci.* 66: 361–379.

Great Plains Flora Association. 1977. *Atlas of the Flora of the Great Plains.* Edited by T. M. Barkley. Ames: Iowa State University Press.

Great Plains Flora Association. 1986, 1991. *Flora of the Great Plains.* Edited by T. M. Barkley. Lawrence: University Press of Kansas.

Griggs, T. 1992. *Quickening.* Erin, ON: The Porcupine's Quill.

Groh, H. 1949. Plants of clearing and trail between Peace River and Fort Vermilion, Alberta. *Canadian Field-Naturalist* 63: 119–134.

Gross, R. S., P. A. Werner, and W. R. Hawthorn. 1980. The biology of Canadian weeds. 38. *Arctium minus* (Hill) Bernh. and *A. lappa* L. *Can. J. Plant Sci.* 60: 621–634.

Grossman, J., ed. 1993. *The Chicago Manual of Style.* 14th ed. Chicago: University of Chicago Press.

Guglielmo, C. G., and W. H. Karasov. 1995. Nutrition quality of winter browse for Ruffed Grouse. *J. Wildlife Management* 59: 427–436.

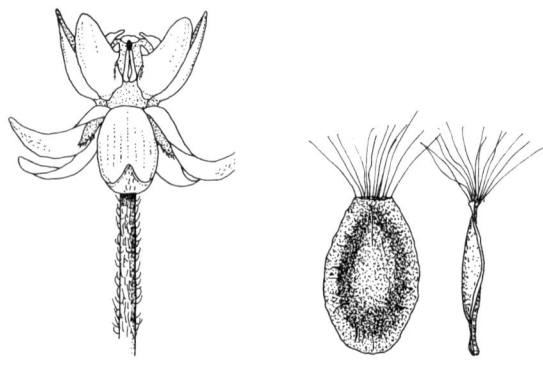

Haber, E. 1988. Hybridization of *Pyrola chlorantha* (Ericaceae) in North America. *Can. J. Botany* 66: 1993–2000.

Hall, I. V., and J. D. Sibley. 1976. The Biology of Canadian Weeds. 20. *Cornus canadensis* L. *Can. J. Plant Sci.* 56: 885–892.

Hall, I. V., E. Steiner, P. Threadgill, and R. W. Jones. 1988. The biology of Canadian weeds. 84. *Oenothera biennis* L. *Can. J. Plant Sci.* 68: 163–173.

Hallsten, G. P., Q. D. Skinner, and A. A. Beetle. 1987. *Grasses of Wyoming.* 3rd ed. Research Journal 202, Agricultural Experiment Station, University of Wyoming.

Hallworth, B., and M. Mychajluk. 1983. Nodding Thistle, *Carduus nutans*: an addition to the vascular flora of Alberta. *Canadian Field-Naturalist* 97: 328.

Hallworth, B., C. C. Chinnappa, S. Orser, and R. Dickinson. 1997. *Plants of Kananaskis Country.* Edmonton and Calgary, AB: The University of Alberta Press and The University of Calgary Press.

Hancock, J. F., S. Serçe, C. M. Portman, P. W. Callow, and J. J. Luby. 2004. Taxonomic variation among North and South American subspecies of *Fragaria virginiana* Miller and *Fragaria chiloensis* (L.) Miller. *Can. J. Botany* 82: 1632–1644.

Harms, L. J. 1968. Cytotaxonomic studies in *Eleocharis* subseries Palustres: central United States taxa. *Amer. J. Botany* 55: 966–974.

Harms, V. L. 1973. Taxonomic studies of North America *Sparganium*. I. *S. hyperboreum* and *S. minimum. Can J. Botany* 51: 1629–1641.

Harms, V. L. 2003. *Checklist of the Vascular Plants of Saskatchewan and the Provincially and Nationally Rare Native Plants in Saskatchewan.* Saskatoon: University Extension Press, University of Saskatchewan.

Harms, V. L., and G. F. Ledingham. 1986. The Narrow-leaved Cat-tail, *Typha angustifolia*, and the hybrid cat-tail, *T.* xglauca, newly reported from Saskatchewan. *Canadian Field-Naturalist* 100: 107–110.

Harms, V. L., J. H. Hudson, and J. Heilman-Ternier. 1980. Contributions to the vascular flora of boreal Saskatchewan, Canada. *Rhodora* 82: 239–280.

Harrington, H. D. 1954. *Manual of the Plants of Colorado.* 2nd ed. Denver: Sage Books.

Harrington, H. D., and L. W. Durrell. 1979. *How To Identify Plants.* First Swallow Press/Ohio University Press.

Harris, J. G., and M. W. Harris. 1994. *Plant Identification Terminology: An Illustrated Glossary.* Spring Lake, UT: Spring Lake Publishing.

Hartman, E. L., and M. L. Rottman. 1987. Alpine vascular flora of the Ruby Range, West Elk Mountains, Colorado. *Great Basin Naturalist* 47: 152–160.

Hatch, S. L., and J. Pluhar. 1993. *Texas Range Plants.* College Station, TX: A&M University Press.

Hatch, S. L., K. N. Gandhi, and L. E. Brown. 1990. *Checklist of the Vascular Plants of Texas.* Texas Agricultural Experimental Station. College Station, TX: A&M University System.

Hawthorn, W. R. 1974. The biology of Canadian weeds. 4. *Plantago major* and *P. rugelii. Can. J. Plant Sci.* 54: 383–396.

Headstrom, R., and B. Angell. 1984. *Suburban Wildflowers: An Introduction to the Common Wildflowers of Your Back Yard and Local Park.* Englewood Cliffs: Prentice-Hall.

Heard, S. B., and J. C. Semple. 1988. The *Solidago rigida* complex (Compositae: Astereae): a multivariate morphometric analysis and chromosome numbers. *Can. J. Botany* 66: 1800–1807.

Heiser, C. B., Jr. 1954. Variation and subspeciation in the Common Sunflower, *Helianthus annuus*. *American Midland Naturalist* 51: 287–305.

Henderson, D. M. 1976. A biosystematic study of Pacific northwestern Blue-eyed grasses (*Sisyrinchium*, Iridaceae). *Brittonia* 28: 149–176.

Henderson, N. C. 1962. A taxonomic revision of the genus *Lycopus* (Labiatae). *American Midland Naturalist* 68: 95–138.

Hickey, M., and C. King. 1981. *100 Families of Flowering Plants*. Cambridge: Cambridge University Press.

Hildahl, V., and M. Benum. 1987. *Heritage Trees of Manitoba*. Winnipeg: Manitoba Forestry Association.

Hington, T. M., and W. R. Clark. 1984. Impact of small mammals on the vegetation of reclaimed land in the northern Great Plains. *J. Range Management* 37: 438–441.

Hitchcock, A. S. 1971. *Manual of Grasses of the United States*. 2 vols. rev. New York: Dover Publications.

Holmgren, P., N. Holmgren, and L. Barnett, eds. 1990. *Index Herbariorum Part 1: The Herbaria of the World*. 8th ed. Bronx: New York Botanical Garden.

Hopkins, M. 1937. *Arabis* in eastern and central North America. *Rhodora* 39: 63–98, 106–148, 155–186.

Hotchkiss, N. 1972. *Common Marsh, Underwater and Floating-leaved Plants of the United States and Canada*. New York: Dover Publications.

Hotchkiss, N., and H. L. Dozier. 1949. Taxonomy and distribution of North American cat-tails. *American Midland Naturalist* 41: 237–254.

House, M., and S. Munro. 1979. *Plantae Occidentalis: 200 Years of Botanical Art in British Columbia*. Technical Bulletin No. 11. Vancouver: The Botanical Garden, University of British Columbia.

Hubbard, F. T. 1917. *Andropogon scoparius* in the United States and Canada. *Rhodora* 19: 100–105.

Hudson, J. 1977. *Carex In Saskatchewan*. Available from the author. Saskatoon, Saskatchewan.

Hultén, E. 1968. *Flora of Alaska and Neighboring Territories: A Manual of the Vascular Plants*. Stanford: Stanford University Press.

Hultén, E., and M. Fries. 1986. *Atlas of Northern European Vascular Plants North of the Tropic of Cancer*. 3 vols. Germany: Koeltz Scientific Books.

Hume, L., J. Martinez, and K. Best. 1983. The biology of Canadian weeds. 60. *Polygonum convolvulus* L. *Can. J. Plant Sci.* 63: 959–971.

Humphrey, R. R. 1970. (rev.) *Arizona Range Grasses*. Tucson: The University of Arizona Press.

Hurd, E. G., N. L. Shaw, J. Mastrogiuseppe, L. C. Smithman, and S. Goodrich. 1998. *Field Guide to Intermountain Sedges*. USDA, Forest Service. Fort Collins: Rocky Mountain Research Station.

Hutchinson, I., J. Colosi, and R. A. Lewin. 1984. The biology of Canadian weeds. 63. *Sonchus asper* (L.) Hill and *S. oleraceus* L. *Can. J. Plant Sci.* 64: 731–744.

Hyatt, P. E. 1998. Arkansas *Carex* (Cyperaceae): a briefly annotated list. *Sida* 18: 535–554.

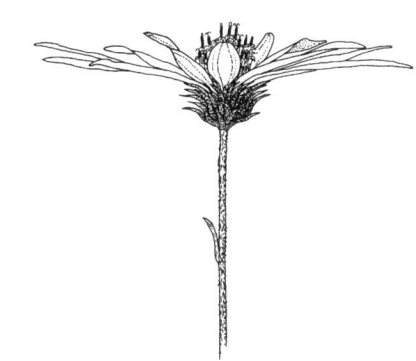

Isely, D., and S. L. Welsh. 1960. *Petalostemon candidum* and *P. occidentale* (Leguminosae). *Brittonia* 12: 114–118.

Jacobsen, T. D. 1979. Numerical analysis of variation between *Allium cernuum* and *Allium stellatum* (Liliaceae). *Taxon* 28: 517–523.

Janzen, D. H. 1977. Promising directions of study in tropical animal-plant interactions. *Annals Missouri Botanical Garden* 64: 706–736.

Jasieniuk, M. A., and E. A. Johnson. 1979. A vascular flora of the Caribou Range, Northwest Territories, Canada. *Rhodora* 81: 249–274.

Johnson, B. L. 1945. Cyto-taxonomic studies in *Oryzopsis*. *Bot. Gaz.* 107: 1–32.

Johnson, D., L. Kershaw, A. MacKinnon, and J. Pojar. 1995. *Plants of the Western Boreal Forest & Aspen Parkland*. Edmonton: Lone Pine Publishing.

Johnson, F. L., and T. H. Milby. 1989. *Oklahoma Botanical Literature*, Oklahoma Museum of Natural History Publication Series, University of Oklahoma Press.

Johnson, K., L. Fairfield, and R. Taylor. 1987. *Wildflowers of Churchill and the Hudson Bay Region*. Winnipeg: Manitoba Museum of Man and Nature.

Jones, A. G. 1978a. The taxonomy of *Aster* section *Multiflori* (Asteraceae) 1. Nomenclatural review and formal presentation of taxa. *Rhodora* 80: 319–357.

Jones, A. G. 1978b. The taxonomy of *Aster* section *Multiflori* (Asteraceae) 2. Biosystematic investigations. *Rhodora* 80: 453–490.

Jones, A. G. 1992. *Aster* and *Brachyactis* (Asteraceae) in Oklahoma. *Sida*, Bot. Misc. No. 8.

Jones, E. K., and N. C. Fassett. 1950. Subspecific variation in *Sporobolus cryptandrus*. *Rhodora* 52: 125–126.

References

Joyal, E. 1990. New variety of *Oxytropis campestris* (Fabaceae) from the Columbia Basin, Washington. *Great Basin Naturalist* 50: 373–377.

Kartesz, J. 1994. *A Synonymized Checklist of the Vascular Flora of the United States, Canada, and Greenland*. 2nd ed., Vol. 1. Oregon: Timber Press.

Kass, R. J. 1988. A checklist of the vascular plants of the House Range, Utah. *Great Basin Naturalist* 48: 102–116.

Kawano, S. 1963. Cytogeography and evolution of the *Deschampsia caespitosa* complex. *Can. J. Botany* 41: 719–742.

Kearney, T. H., R. H. Peebles, and **collaborators.** 1964. *Arizona Flora*. 2nd ed., with supplement by J. T. Howell and E. McClintock and collaborators. Berkeley: University of California Press.

Kelso, S. 1991. Taxonomy of *Primula* sects. *Aleuritia* and *Armerina* in North America. *Rhodora* 93: 67–99.

Kershaw, L., J. Gould, D. Johnson, and **J. Lancaster.** 2001. *Rare Vascular Plants of Alberta*. Edmonton: University of Alberta Press and the Canadian Forest Service, Alberta.

Kiviat, E. 1996. American Goldfinch nests in Purple Loosestrife. *Wilson Bulletin* 108: 182–186.

Kohli, B., and **J. G. Packer.** 1976. A contribution to the taxonomy of the *Potentilla pensylvanica* complex in North America. *Can. J. Botany* 54: 706–719.

Komarck, S. 1994. *Flora of the San Juans: A Field Guide to the Mountain Plants of Southwestern Colorado*. Kivaki Press.

Kotanen, P. M., J. Bergelson, and **D. L. Hazlett.** 1998. Habitats of native and exotic plants in Colorado shortgrass steppe: a comparative approach. *Can. J. Botany* 76: 664–672.

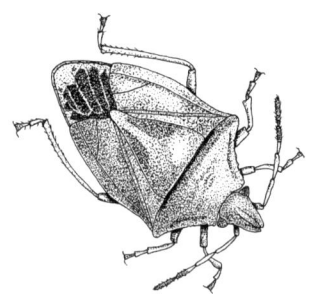

Koyama, T. 1963. The genus *Scirpus* Linn. Critical species of the section *Pterolepis*. *Can. J. Botany* 41: 1107–1131.

Koyama, T., and **S. Kawano.** 1964. Critical taxa of grasses with North American and Eastern Asiatic distribution. *Can. J. Botany* 42: 859–884.

Krajina, V. J., K. Klinka, and **J. Worrall.** 1982. *Distribution and Ecological Characteristics of Trees and Some Shrubs of British Columbia*. Vancouver: University of British Columbia, Faculty of Forestry.

Kuchel, S. D., and **L. P. Bruederle.** 2000. Allozyme data support a Eurasian origin for *Carex viridula* subsp. *viridula* var. *viridula* (Cyperaceae). *Madroño* 47: 147–158.

Kuehn, M. M., and **B. N. White.** 1999. Morphological analysis of genetically identified cattails *Typha latifolia*, *Typha angustifolia*, and *Typha xglauca*. *Can. J. Botany* 77: 906–912.

Lackschewitz, K. 1986. *Plants of West-Central Montana—Identification and Ecology: An Annotated Checklist*. General Technical Report INT-217. USDA, Forest Service, Intermountain Research Station.

Lamont, S. 1980. *Trees and Shrubs of the Qu'Appelle Valley*. Saskatchewan Culture and Youth/Museum of Natural History.

Landolt, E., ed. 1980. *Biosystematic investigations in the family of duckweeds* (Lemnaceae) vol. 1. Zürich: Veröffentlichungen des Geobotanischen Institutes der ETH.

Larrison, E. J., G. W. Patrick, W. H. Baker, and **J. A. Yaich.** 1977. *Washington Wildflowers*. Seattle: The Seattle Audubon Society.

Larson, B. M. H., and **S. C. H. Barrett.** 1998. Reproductive biology of island and mainland populations of *Primula mistassinica* (Primulaceae) on Lake Huron shorelines. *Can. J. Botany* 76: 1819–1827.

Larson, G. E. 1993. *Aquatic and Wetland Vascular Plants of the Northern Great Plains*. General Technical Report RM-238. Fort Collins, CO: USDA Forestry Service.

Lavin, M., and **C. Seibert.** 2005. *Grasses of Montana*. Bozeman, MT: Herbarium, Department of Plant Sciences and Plant Pathology, Montana State University.

Lawrence, D. L., and **J. T. Romo.** 1994. Tree and shrub communities of wooded draws near the Matador Research Station in southern Saskatchewan. *Canadian Field-Naturalist* 108: 397–412.

Lawrence, W. E. 1945. Some ecotypic relations of *Deschampsia caespitosa*. *Amer. J. Botany* 32: 298–314.

Leake, D. V., J. B. Leake, and **M. L. Roeder.** 1993. *Desert and Mountain Plants of the Southwest*. Norman: University of Oklahoma Press.

Lehr, J. H. 1978. *A Catalogue of the Flora of Arizona*. Phoenix: Desert Botanical Garden.

Leighton, A. 1998. Milkweed: the Monarch's prairie host. *Blue Jay* 56: 46–54.

Lemna, W. K., and **C. G. Messersmith.** 1990. The biology of Canadian weeds. 94. *Sonchus arvensis* L. *Can. J. Plant Sci.* 70: 509–532.

Leonard, R. I. 1993. *Guide to Grasses of the Lower Rio Grande Valley, Texas*. University Texas Pan. American Press.

Lesica, P., and S. Miles. 1999. Russian Olive invasion into cottonwood forests along a regulated river in north-central Montana. *Can. J. Botany* 77: 1077–1083.

Lesica, P., and S. Miles. 2001. Natural history and invasion of Russian Olive along eastern Montana rivers. *Western North American Naturalist* 61: 1–10.

Lewis, W. H. 1958. Minor forms of North American species of *Rosa*. *Rhodora* 60: 237–243.

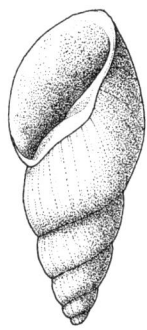

Lewis, W. H. 1959. A monograph of the genus *Rosa* in North America. 1. *R. acicularis*. *Brittonia* 11: 1–24.

Lichvar, R. W., R. D. Dorn, and E. F. Evert. 1983. New records for the vascular flora of Wyoming and Montana. *Great Basin Naturalist* 43: 739–740.

Lichvar, R. W., E. I. Collins, and D. H. Knight. 1985. Checklist of vascular plants for the Bighorn Canyon National Recreation Area, Wyoming and Montana. *Great Basin Naturalist* 45: 734–746.

Lindgren, C. J. 2003. A brief history of Purple Loosestrife, *Lythrum salicaria*, in Manitoba and its status in 2001. *Canadian Field-Naturalist* 117: 100–109.

Lindsay, M., ed. 1992. *The Visual Dictionary of Plants*. Toronto: Stoddart Publishing Company.

Lommasson, R. C. 1973. *Nebraska Wild Flowers*. Lincoln: University of Nebraska Press.

Long, R. W. 1966. Biosystematics of the *Helianthus nuttallii* complex (Compositae). *Brittonia* 18: 64–79.

Looman, J. 1976. Biological flora of the Canadian Prairie Provinces. 4. *Triglochin* L., the genus. *Can. J. Plant Sci.* 56: 725–732.

Looman, J. 1978. Biological flora of the Canadian Prairie Provinces. 5. *Koeleria gracilis* Pres. *Can. J. Plant Sci.* 58: 459–466.

Looman, J. 1983. *111 Range and Forage Plants of the Canadian Prairies*. Publication 1751. Ottawa: Research Branch, Agriculture Canada.

Looman, J. 1984. The Biological Flora of Canada. 4. *Shepherdia argentea* (Pursh) Nutt. Buffaloberry. *Canadian Field-Naturalist* 98: 231–244.

Looman, J., and K. F. Best, eds. 1994. *Budd's Flora of the Canadian Prairie Provinces*. Publication 1662. Ottawa: Ministry of Supply and Services Canada.

Löve, D., and P. Dansereau. 1959. Biosystematic studies on *Xanthium*: taxonomic appraisal and ecological status. *Can. J. Botany* 37: 173–208.

Lowcock, L. A., and R. W. Murphy. 1990. Seed dispersal via amphibian vectors: passive transport of Bur-marigold, *Bidens cernua*, achenes by migrating salamanders, genus *Ambystoma*. *Canadian Field-Naturalist* 104: 298–300.

Luer, C. A. 1975. *The Native Orchids of the United States and Canada, Excluding Florida*. Bronx: The New York Botanical Garden.

Luken, J. O., J. W. Thieret, and J. R. Kartesz. 1993. *Erucastrum gallicum* (Brassicaceae): invasion and spread in North America. *Sida* 15: 569–582.

MacRoberts, D. T. 1988. *A Documented Checklist and Atlas of the Vascular Flora of Louisiana*. 3 parts. Shreveport: Louisiana State University.

Maher, R. V., G. W. Argus, V. L. Harms, and J. H. Hudson. 1979. *The Rare Vascular Plants of Saskatchewan*. Syllogeus No. 20. Ottawa: National Museum of Natural Sciences.

Makowski, R. M. D., and I. N. Morrison. 1989. The biology of Canadian weeds. 91. *Malva pusilla* Sm. (= *M. rotundifolia* L.) *Can. J. Plant Sci.* 69: 861–879.

Mal, T. K., J. Lovett-Doust, L. Lovett-Doust, and G. A. Mulligan. 1992. The biology of Canadian weeds. 100. *Lythrum salicaria*. *Can. J. Plant Sci.* 72: 1305–1330.

Mangaly, J. K. 1968. A cytotaxonomic study of the herbaceous species of *Smilax*: section *Coprosmanthus*. *Rhodora* 70: 55–82.

Mariani, C. L., C. G. Earley, and C. McKinnon. 1993. Early nesting by the American Goldfinch, *Carduelis tristis* and subsequent parasitism by the Brown-headed Cowbird, *Molothrus ater*, in Ontario. *Canadian Field-Naturalist* 107: 349–350.

Martin, A. C., H. S. Zim, and A. L. Nelson. 1951. *American Wildlife & Plants: A Guide to Wildlife Food Habits*. New York: Dover Publications.

Martin, A., and W. Barkley. 1961. *Seed Identification Manual*. Berkeley: University of California Press.

Martin, W. C., and C. R. Hutchins. 1981. *A Flora of New Mexico*. 2 vols. Germany: Strauss and Cramer GmbH.

Matsumura, Y., and H. D. Harrington. 1955. *The True Aquatic Plants of Colorado*. Technical Bulletin 57. Fort Collins: Colorado Agricultural Experiment Station, Colorado Agricultural and Mechanical College.

Matthews, J. F., D. W. Ketron, and S. F. Zane. 1993. The biology and taxonomy of the *Portulaca oleracea* L. (Portulacaceae) complex in North America. *Rhodora* 95: 166–183.

References

Maun, M. A., and S. C. H. Barrett. 1986. The biology of Canadian weeds. 77. *Echinochloa crus-galli* (L.) Beauv. *Can. J. Plant Sci.* 66: 739–759.

Maw, M. G., A. G. Thomas, and A. Stahevitch. 1985. The biology of Canadian weeds. 66. *Artemisia absinthium* L. *Can. J. Plant Sci.* 65: 389–400.

Maze, J. 1968. Past hybridization between *Quercus macrocarpa* and *Quercus gambelii*. *Brittonia* 20: 321–333.

McClintock, K. A., and M. J. Waterway. 1994. Genetic differentiation between *Carex lasiocarpa* and *C. pellita* (Cyperaceae) in North America. *Amer. J. Botany* 81: 224–231.

McGregor, R. L. 1976. *History of Naturalized Kansas Plants*. Report No. 7. State Biological Survey of Kansas.

McJannet, C. L., G. W. Argus, S. A. Edlund, and J. Cayouette. 1993. *Rare vascular plants in the Canadian Arctic*. Syllogeus No. 72. Ottawa: Canadian Museum of Nature.

McJannet, C. L., G. W. Argus, and W. J. Cody. 1995. *Rare Vascular Plants in the Northwest Territories*. Syllogeus No. 73. Ottawa: Canadian Museum of Nature.

McLellan, B. N., and F. W. Hovey. 1995. The diet of Grizzly Bears in the Flatland River drainage of south-eastern British Columbia. *Can. J. Zoology* 73: 704–712.

McMillan, C. 1964. Ecotypic differentiation within four North American prairie grasses. I. Morphological variation within transplanted community fractions. *Amer. J. Botany* 51: 1119–1128.

McMillan, C., and J. Weiler. 1959. Cytogeography of *Panicum virgatum* in central North America. *Amer. J. Botany* 46: 590–593.

McNeill, J. 1977. The biology of Canadian weeds. 25. *Silene alba* (Miller) E. H. L. Krause. *Can. J. Plant Sci.* 57: 1103–1114.

McNeill, J. 1980. The biology of Canadian weeds. 46. *Silene noctiflora* L. *Can J. Plant Sci.* 60: 1243–1253.

McNeill, J., I. J. Bassett, and C. W. Crompton. 1977. *Suaeda calceoliformis*, the correct name for *Suaeda depressa* Auct. *Rhodora* 79: 133–137.

McNeill, J., I. J. Bassett, C. W. Crompton, and P. M. Taschereau. 1983. Taxonomic and nomenclatural notes on *Atriplex* L. (Chenopodiaceae). *Taxon* 32: 549–556.

McPherson, G. D., and J. G. Packer. 1974. A contribution to the taxonomy of *Viola adunca*. *Can. J. Botany* 52: 895–902.

McVaugh, R. 1936. Studies in the taxonomy and distribution of the eastern North American species of *Lobelia*. *Rhodora* 38: 241–263, 276–298, 305–329, 346–362.

Mertens, T. R., and P. H. Raven. 1965. Taxonomy of *Polygonium*, section *Polygonum* (Avicularia) in North America. *Madroño* 18: 85–92.

Michaels, A. 1997. *The Weight of Oranges: Miner's Pond*. Toronto: McClelland & Stewart.

Mitchell, R. J., J. D. Karron, K. G. Holmquist, and J. M. Bell. 2005. Patterns of multiple paternity in fruits of *Mimulus ringens* (Phrymaceae). *Amer. J. Botany* 92: 885–890.

Mitchell, R. S. 1968. Variation in the *Polygonum amphibium* complex and its taxonomic significance. University of California Publications in Botany, Vol. 45. Los Angeles: University of California Press.

Mitchell, R. S. 1978. *Rumex maritimus* L. versus *R. persicarioides* L. (Polygonaceae) in the western hemisphere. *Brittonia* 30: 293–296.

Miyanishi, K., and P. B. Cavers. 1980. The biology of Canadian weeds. 40. *Portulaca oleracea* L. *Can. J. Plant Sci.* 60: 953–963.

Miyanishi, K., O. Eriksson, and R. W. Wein. 1991. The biology of Canadian weeds. 98. *Potentilla anserina* L. *Can. J. Plant Sci.* 71: 791–801.

Mohlenbrock, R. H. 1970. *Flowering Plants – Flowering Rush to Rushes*. Vol. 1, and *Flowering Plants – Lilies to Orchids*, Vol. 2, Carbondale: Southern Illinois University Press.

Mohlenbrock, R. H., ed. 1972. *The Illustrated Flora of Illinois, Grasses – Bromus to Paspalum*. Carbondale: Southern Illinois University Press.

Mohlenbrock, R. H. 1973. *The Illustrated Flora of Illinois, Grasses – Panicum to Danthonia*. Carbondale: Southern Illinois University Press.

Mohlenbrock, R. H., ed. 1976. *The Illustrated Flora of Illinois, Sedges – Cyperus to Scleria*. Carbondale: Southern Illinois University Press.

Mohlenbrock, R. H. 1982. Illinois Convolvulaceae in the Missouri Botanical Garden herbarium. *Annals Missouri Botanical Garden* 69: 393–401.

Mohlenbrock, R. H., ed. 1990. *The Illustrated Flora of Illinois, Flowering Plants – Nightshades to Mistletoe*. Carbondale: Southern Illinois University Press.

Mohlenbrock, R. H., and D. M. Ladd. 1978. *Distribution of Illinois Vascular Plants*. Carbondale: Southern Illinois University Press.

Mohlenbrock, R. H., and M. Mohlenbrock. 1983. *Where Have All the Wildflowers Gone? A Region-by-Region Guide to Threatened or Endangered U. S. Wildflowers*. New York: MacMillan Publishing Company.

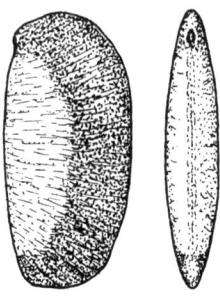

References

Moore, A. W. 1964. Note on non-leguminous nitrogen-fixing plants in Alberta. *Can. J. Botany* 42: 952–955.

Moore, R. J. 1972. Distribution of native and introduced knapweeds (Centaurea) in Canada and the United States. *Rhodora* 74: 331–346.

Moore, R. J. 1975. The biology of Canadian weeds. 13. *Cirsium arvense* (L.) Scop. *Can. J. Plant Sci.* 55: 1033–1048.

Moore, R. J., and C. Frankton. 1969. Cytotaxonomy of some *Cirsium* species of the eastern United States with a key to eastern species. *Can. J. Botany* 47: 1257–1275.

Moore, R. J., and C. Frankton. 1974. *The Thistles of Canada.* Monograph No. 10. Ottawa: Canadian Department of Agriculture, Research Branch.

Morden, C. W., and S. L. Hatch. 1996. Morphological variation and synopsis of the *Muhlenbergia repens* complex (Poaceae). *Sida* 17: 349–365.

Morgan, J. P., D. R. Collicutt, and J. D. Thompson. 1995. *Restoring Canada's Native Prairies: A Practical Manual.* Argyle, MB: Prairie Habitats.

Mosquin, T. 1966. A new taxonomy for *Epilobium angustifolium* L. (Onagraceae). *Brittonia* 18: 167–188.

Mosquin, T. 1971. Biosystematic studies in the North American species of *Linum*, section *Adenolinum* (Linaceae). *Can. J. Botany* 49: 1379–1388.

Moss, E. H. 1956. Ragweed in southeastern Alberta. *Can. J. Botany* 34: 763–767.

Moss, E. H. 1983. *Flora of Alberta.* 2nd ed., rev. by J. G. Packer. Toronto: University of Toronto Press.

Moyle, J., and N. Hotchkiss. 1945. *The Aquatic and Marsh Vegetation of Minnesota and its Value to Waterfowl.* Technical Bulletin No. 3. Minnesota Department of Conservation.

Mühlenbach, V. 1957. Adventitious and escaped plants new to Missouri. *Rhodora* 59: 27–31.

Mühlenbach, V. 1960. Adventive plants new to the Missouri flora. *American Midland Naturalist* 64: 161–168.

Mühlenbach, V. 1979. Contributions to the synanthropic (adventive) flora of the railroads in St. Louis, Missouri, U.S.A. *Annals Missouri Botanical Garden* 66: 1–108.

Mulligan, G. A. 1961. The genus *Lepidium* in Canada. *Madroño* 16: 77–90.

Mulligan, G. A. 1980. The genus *Cicuta* in North America. *Can. J. Botany* 58: 1755–1767.

Mulligan, G. A., and L. G. Bailey. 1975. The biology of Canadian weeds. 8. *Sinapis arvensis* L. *Can. J. Plant Sci.* 55: 171–183.

Mulligan, G. A., and I. J. Bassett. 1959. *Achillea millefolium* complex in Canada and portions of the United States. *Can. J. Botany* 37: 73–79.

Mulligan, G. A., and B. E. Junkins. 1977. The biology of Canadian weeds. 23. *Rhus radicans* L. *Can. J. Plant Sci.* 57: 515–523.

Mulligan, G. A., and P. G. Kevan. 1973. Color, brightness, and other floral characteristics attracting insects to the blossoms of some Canadian weeds. *Can. J. Botany* 51: 1939–1952.

Mulligan, G. A., and D. B. Munro. 1981a. The biology of Canadian weeds. 48. *Cicuta maculata* L., *C. douglasii* (DC.) Coult. & Rose and *C. virosa* L. *Can. J. Plant Sci.* 61: 93–105.

Mulligan, G. A., and D. B. Munro. 1981b. The biology of Canadian weeds. 51. *Prunus virginiana* L. and *P. serotina* Ehrh. *Can. J. Plant Sci.* 61: 977–992.

Mulligan, G. A., D. B. Munro, and J. McNeill. 1983. The status of *Stachys palustris* (Labiatae) in North America. *Can. J. Botany* 61: 679–682.

Munz, P. A. 1962. *California Desert Wildflowers.* Berkeley: University of California Press.

Munz, P. A. 1974. *A Flora of Southern California.* Berkeley: University of California Press.

Murley, M. R. 1951. Seeds of Crucifera in northeastern North America. *American Midland Naturalist* 46: 1–81.

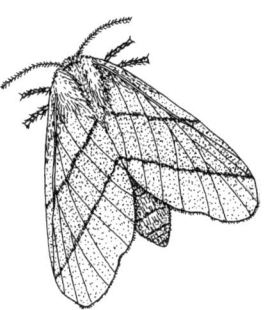

Najda, H. G., A. L. Darwent, and G. Hamilton. 1982. The biology of Canadian weeds. 54. *Crepis tectorum* L. *Can. J. Plant Sci.* 62: 473–481.

Nelson, B. E. 1984. rev. ed. *Vascular Plants of the Medicine Bow Mountains, Wyoming.* Jelm Mt. Press.

Nelson, R., and R. Williams. 1992. *Handbook of Rocky Mountain Plants.* R. Rinehart.

Newcomb, L. 1977. *Newcomb's Wildflower Guide: An Ingenious New Key System for Quick, Positive Field Identification of the Wildflowers, Flowering Shrubs and Vines of Northeastern and Northcentral North America.* Illustrated by G. Morrison. New York: Little, Brown and Company.

Norment, C. J., and M. E. Fuller. 1997. Breeding season frugivory by Harris' Sparrows (*Zonotrichia querula*) and White-crowned Sparrows (*Zonotrichia leucophrys*) in a low arctic ecosystem. *Can. J. Zoology* 75: 670–679.

O'Donovan, J. T., and M. P. Sharma. 1987. The biology of Canadian weeds. 80. *Galeopsis tetrahit* L. *Can. J. Plant Sci.* 67: 787–796.

References

Ogden, E. C. 1943. The broad-leaved species of *Potamogeton* of North America north of Mexico. *Rhodora* 45: 57–105, 119–163, 171–214.

Olesen, J. M. 1987. Heterostyly, homostyly, and long-distance dispersal of *Menyanthes trifoliata* to Greenland. *Can. J. Botany* 65: 1509–1513.

Oleskevich, C., S. F. Shamoun, and Z. K. Punja. 1996. The biology of Canadian weeds. 106. *Rubus strigosus* Michx., *Rubus parviflorus* Nutt., and *Rubus spectabilis* Pursh. *Can. J. Plant Sci.* 76: 187–201.

Olmstead, R. G., C. W. dePamphilis, A. D. Wolfe, N. D. Young, W. J. Elisons, and P. A. Reeves. 2001. Disintegration of the Scrophulariaceae. *Amer. J. Botany* 88: 348–361.

Ottenbreit, K. A., and R. J. Staniforth. 1994. Crossability of naturalized and cultivated *Lythrum* taxa. *Can. J. Botany* 72: 337–341.

Owens, J. N., and M. Molder. 1979. Sexual reproduction of White Spruce (*Picea glauca*). *Can. J. Botany* 57: 152–169.

Owensby, C. E. 1980. *Kansas Prairie Wildflowers*. Ames: Iowa State University Press.

Ownbey, G. B., and T. Morley. 1991. *Vascular Plants of Minnesota, A Checklist and Atlas*. Minneapolis: University of Minnesota Press.

Ownbey, G. B., and W. R. Smith. 1988. New and noteworthy plant records for Minnesota. *Rhodora* 90: 369–377.

Ownbey, M. 1950. Natural hybridization and amphiploidy in the genus *Tragopogon*. *Amer. J. Botany* 37: 487–498.

Packer, J. G. 1963. The taxonomy of some North American species of *Chrysosplenium* L., section *Alternifolia* Franchet. *Can. J. Botany* 41: 85–103.

Packer J. G., and C. E. Bradley. 1984. *A Checklist of the Rare Vascular Plants in Alberta*. Natural History Occasional Paper No. 5. Edmonton: Provincial Museum of Alberta.

Packer, J. G., and K. E. Denford. 1974. A contribution to the taxonomy of *Arctostaphylos uva-ursi*. *Can. J. Botany* 52: 743–753.

Parker, K. F., and L. B. Hamilton. 1972. *An Illustrated Guide to Arizona Weeds*. Tucson: The University of Arizona Press.

Parmelee, J. A. 1971. The genus *Gymnosporangium* in Western Canada. *Can. J. Botany* 49: 903–926.

Patten, B. C., Jr. 1954. The status of some American species of *Myriophyllum* as revealed by the discovery of intergrade material between *M. exalbescens* Fern. and *M. spicatum* L. in New Jersey. *Rhodora* 56: 213–225.

Patterson, P. A., K. E. Neiman, and J. R. Tonn. 1985. *Field Guide to Forest Plants of Northern Idaho*. General Technical Report INT–180. Ogden, UT: USDA, Forest Service, Intermountain Research Station.

Pavlick, L. E. 1995. *Bromus L. of North America*. Victoria: Royal British Columbia Museum.

Payne, W. W. 1978. A glossary of plant hair terminology. *Brittonia* 30: 239–255.

Peck, M. E. 1961. *A Manual of the Higher Plants of Oregon*. 2nd ed. Binfords and Mort Publishers.

Penner, R., G. E. E. Moodie, and R. J. Staniforth. 1999. The dispersal of fruits and seeds of Poison-ivy, *Toxicodendron radicans*, by Ruffed Grouse, *Bonasa umbellus*, and squirrels, *Tamiasciurus hudsonicus* and *Sciurus carolinensis*. *Canadian Field-Naturalist* 113: 616–620.

Peoples, A. D., R. L. Lochmiller, D. M. Leslie, Jr., J. C. Boren, and D. M. Engle. 1994. Essential amino acids in Northern Bobwhite foods. *J. Wildlife Management* 58: 167–175.

Perry, G., and J. McNeill. 1986. The nomenclature of *Eragrostis cilianensis* (Poaceae) and the contribution of Bellardi to Allioni's *Flora Pedemontana*. *Taxon* 35: 696–701.

Phillips, W. L., and R. L. Stuckey, comps. 1976. *Index to Plant Distribution Maps in North American Periodicals Through 1972*. Boston, MA: G. K. Hall & Company.

Phipps, J. B. 1998. Introduction to the red-fruited hawthorns (*Crataegus*, Rosaceae) of western North America. *Can. J. Botany* 76: 1863–1899.

Phipps, J. B., and M. Muniyamma. 1980. A taxonomic revision of *Crataegus* (Rosaceae) in Ontario. *Can. J. Botany* 58: 1621–1699.

Pip, E., and K. Simmons. 1986. Aquatic angiosperms at unusual depths in Shoal Lake, Manitoba–Ontario. *Canadian Field-Naturalist* 100: 354–358.

Pogan, E. 1963. Taxonomical value of *Alisma triviale* Pursh and *Alisma subcordatum* Rafin. *Can. J. Botany* 41: 1011–1013.

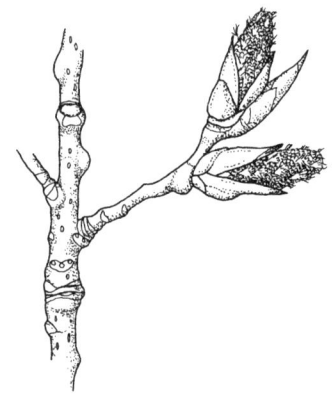

Pohl, R. W. 1966. The Grasses of Iowa. *Iowa State J. Science* 40: 341–566. Ames: Iowa State University Press.

Pohl, R. W. 1968. *How to Know the Grasses*. Dubuque, IA: Wm. C. Brown Company.

Pojar, J., and A. MacKinnon, comps. & eds. 1994. *Plants of Coastal British Columbia, Including Washington, Oregon & Alaska*. Vancouver: Lone Pine Publishing.

References

Porsild, A. E., and **W. J. Cody.** 1980. *Vascular Plants of Continental Northwest Territories, Canada.* Ottawa: National Museum of Natural Sciences, National Museums of Canada.

Porsild, A. E., and **D. T. Lid.** 1957. *Illustrated Flora of the Canadian Arctic Archipelago.* (rev. 1964). Bulletin No. 146, Biological Series No. 50. Ottawa: National Museum of Canada.

Porter, C. L. 1962. *A Flora of Wyoming.* Bulletin 402. Agriculture Experiment Station, University of Wyoming.

Pringle, J. S. 1967. Taxonomy of *Gentiana*, section *Pneumonanthae*, in eastern North America. *Brittonia* 19: 1–32.

Puff, C. 1976. The *Galium trifidum* group (*Galium* sect. *Aparinoides*, Rubiaceae). *Can. J. Botany* 54: 1911–1925.

Punter, E. 1994. *Inventory and Annotated Checklist of the Vascular Plants of the Manitoba Model Forest.* Pine Falls, MB: Manitoba Model Forest.

Punter, E., comp. 1995. *Manitoba's Vascular Plants.* Winnipeg, MB: Manitoba Conservation Data Center.

Pyšek, P., D. M. Richardson, Marcel Rejmánek, G. L. Webster, M. Williamson, and **J. Kirschner.** 2004. Alien plants in checklists and floras: towards better communication between taxonomists and ecologists. *Taxon* 53: 131–143.

Raju, **M. V. S.** 1990. *The Wild Oat Inflorescence and Seed: Anatomy, Development and Morphology.* Regina: Canadian Plains Research Center.

Rare and Endangered Plant Technical Committee of the Idaho Natural Areas Council: Vascular Plant Species of Concern in Idaho. 1981. Forestry, Wildlife and Range Experiment Station, Bulletin 34. Moscow, ID: University of Idaho.

Raven, P. H., and **D. P. Gregory.** 1972. Observations of meiotic chromosomes in *Gaura* (Onagraceae). *Brittonia* 24: 71–86.

Reaume, T. 2003. The biology of *Trillium cernuum* (Liliaceae). *Blue Jay* 61: 143–167.

Reaume, T., and **K. Szwaluk.** 2006. Clustered Bur-reed, *Sparganium glomeratum,* in Manitoba. *Blue Jay* 64: 220–221.

Reddoch, J. M., and **A. H. Reddoch.** 1997. The orchids in the Ottawa district: Floristics, phytogeography, population studies and historical review. *Canadian Field-Naturalist* 111: 1–185.

Reeder, J. R. 1951. *Setaria lutescens* an untenable name. *Rhodora* 53: 27–30.

Reinking, M. 1981. *Juncus* x *stuckeyi* (Juncaceae), a natural hybrid from northern Ohio. *Brittonia* 33: 170–178.

Reznicek, A. A. 1985. What is *Carex rostrata* Stokes? *Amer. J. Botany* 72: 966.

Reznicek, A. A. 1990. Evolution in sedges (*Carex*, Cyperaceae). *Can. J. Botany* 68: 1409–1432.

Reznicek, A. A., and **P. M. Catling.** 1987. *Carex praegracilis* (Cyperaceae) in eastern North America: a remarkable case of rapid invasion. *Rhodora* 89: 205–216.

Richards, E. L. 1968. A monograph of the genus *Ratibida*. *Rhodora* 70: 348–393.

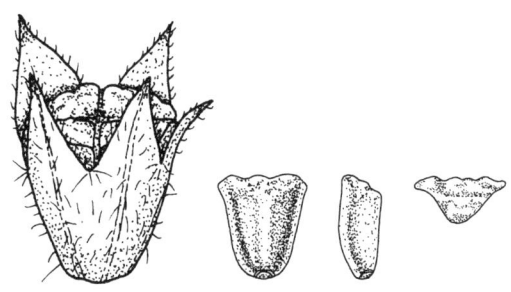

Rickett, H. W. 1944. *Cornus stolonifera* and *Cornus occidentalis*. *Brittonia* 5: 149–159.

Riley, J. L. 2003. *Flora of the Hudson Bay Lowland and its Postglacial Origins.* Ottawa: National Research Council Research Press, Canada.

Ringelman, J. K., and **J. R. Longcore.** 1982. Movements and wetland selection by brood-rearing Black Ducks. *J. Wildlife Management* 46: 615–621.

Ringius, G. S., R. A. Sims, and **S. J. Meades.** 1997. *Indicator Plant Species in Canadian Forests.* Ottawa: Canadian Forest Service, Natural Resources Canada.

Robison, H. W., and **A. T. Robert.** 1995. *Only in Arkansas: A Study of Endemic Plants and Animals of the State.* University of Arkansas Press.

Rogers, C. M. 1963. Yellow-flowered species of *Linum* in eastern North America. *Brittonia* 15: 97–122.

Rogers, C. M. 1968. Yellow-flowered species of *Linum* in Central America and western North America. *Brittonia* 20: 107–135.

Rolfsmeier, S. B. 1995. Keys and distributional maps for Nebraska Cyperaceae, part 1: *Bulbostylis, Cyperus, Dulichium, Eleocharis, Eriophorum, Fimbristylis, Fuirena, Lipocarpha,* and *Scirpus. Trans. Nebraska Acad. of Sciences* 22: 27–42.

Rolfsmeier, S. B., and **B. Wilson.** 1997. Keys and distributional maps for Nebraska Cyperaceae, part 2: *Carex* and *Scleria. Trans. Nebraska Acad. of Sciences* 24: 5–26.

Rollins, R. C. 1940. Studies in the genus *Hedysarum* in North America. *Rhodora* 42: 216–239.

Rollins, R. C. 1941. A monographic study of *Arabis* in western North America. *Rhodora* 43: 289–325, 348–411, 425–481.

Rollins, R. C. 1993. *The Cruciferae of Continental North America.* Stanford: Stanford University Press.

References

Ross, F. H., and J. A. Quinn. 1977. Phenology and reproductive allocation in *Andropogon scoparius* (Gramineae) populations in communities of different successional stages. *Amer. J. Botany* 64: 535–540.

Rossbach, G. B. 1939. Aquatic *Utricularis*. *Rhodora* 41: 113–128.

Rothrock, P. E., and A. A. Reznicek. 1998. Documented chromosome numbers 1998: 1. Chromosome numbers in *Carex* section *ovales* (Cyperaceae): additions, variations, and corrections. *Sida* 18: 587–592.

Rumble, M. A., and L. D. Flake. 1983. Management considerations to enhance use of stock ponds by waterfowl broods. *J. Range Management* 36: 691–694.

Rydberg, P. A. 1965. *Flora of the Prairies and Plains of Central North America*. New York: The New York Botanical Garden with the Hafner Publishing Company.

Salamun, P. J. 1952. A Black Hills variety of *Osmorhiza longistylis*. *American Midland Naturalist* 47: 251–253.

Saner, M. A., D. R. Clements, M. R. Hall, D. J. Doohan, and C. W. Crompton. 1995. The biology of Canadian weeds. 105. *Linaria vulgaris* Mill. *Can. J. Plant Sci.* 75: 525–537.

Sarkar, N. M. 1958. Cytotaxonomic studies on *Rumex* section *Axillares*. *Can. J. Botany* 36: 947–996.

Schnee, B. K., and D. M. Waller. 1986. Reproductive behavior of *Amphicarpaea bracteata* (Leguminosae), an amphicarpic annual. *Amer. J. Botany* 73: 376–386.

Schuyler, A. E. 1974. Typification and application of the names *Scirpus americanus* Pers., *S. olneyi* Gray, and *S. pungens* Vahl. *Rhodora* 76: 51–52.

Scoggan, H. J. 1957. *Flora of Manitoba*. Bulletin No. 140, Biological Series No. 47. Ottawa: National Museum of Canada.

Scoggan, H. J. 1978. *The Flora of Canada*. 4 parts. Ottawa: National Museum of Natural Sciences.

Scora, R. W. 1967. Interspecific relationships in the genus *Monarda* (Labiatae). University of California Publications in Botany Vol. 41. Los Angeles: University of California Press.

Semple, J. C. 1979. The cytogeography of *Aster lanceolatus* (synonyms *A. simplex* and *A. paniculatus*) in Ontario with additional counts from populations in the United States. *Can. J. Botany* 57: 397–402.

Semple, J. C. 1996. A revision of *Heterotheca* sect. *Phyllotheca* (Nutt.) Harms (Compositae: Astereae): The prairie and montane goldenasters of North America. Biology Series No. 37. Waterloo, ON: Department of Biology, University of Waterloo.

Semple, J. C., and G. R. Ringius. 1992. *The Goldenrods of Ontario: Solidago L. and Euthamia Nutt*. No. 36 (revised by J. Semple). Waterloo, ON: Department of Biology, University of Waterloo.

Semple, J. C., V. C. Blok, and P. Heiman. 1980. Morphological, anatomical, habit, and habitat differences among the goldenaster genera *Chrysopsis*, *Heterotheca*, and *Pityopsis* (Compositae–Astereae). *Can. J. Botany* 58: 147–163.

Semple, J. C., J. G. Chmielewski, K. S. Rao, and G. A. Allen. 1983. The cytogeography of *Aster lanceolatus*. II. A preliminary survey of the range including *A. hesperius*. *Can. J. Botany* 61: 434–441.

Semple, J. C., J. G. Chmielewski, and R. A. Brammall. 1990. A multivariate morphometric study of *Solidago nemoralis* (Compositae: Astereae) and comparison with *S. californica* and *S. sparsiflora*. *Can. J. Botany* 68: 2070–2082.

Semple, J. C., S. B. Heard, and C. S. Xiang. 1996. The Asters of Ontario (Compositae: Astereae): *Diplactis* Raf., *Oclemena* E.L. Green, *Doellingeria* Nees and *Aster* L. (including *Canadanthus* Nesom, *Symphyotrichum* Nees and *Virgulus* Raf.) No. 38. Waterloo, ON: Department of Biology, University of Waterloo.

Shan, R. H., and L. Constance. 1951. The genus *Sanicula* (Umbelliferae) in the Old World and the New. University of California Publications in Botany Vol. 25. Los Angeles: University of California Press.

Shapiro, A. M. 1982. A recondite breeding site for the Monarch (*Danaus plexippus*, Danaidae) in the Mountane Sierra Nevada. *J. Res. Lepid.* 20: 50–57.

Sharma, M. P., and W. H. Vanden Born. 1978. The biology of Canadian weeds. 27. *Avena fatua* L. *Can. J. Plant Sci.* 58: 141–157.

Sharma, N., P. Koul, and A. K. Koul. 1993. Pollination biology of some species of the genus *Plantago* L. *Bot. J. Linn. Soc.* 111: 129–138.

Sharpe, P. B., and B. Van Horne. 1998. Influence of habitat on behavior of Townsend's Ground Squirrels (*Sepermophilus townsendii*). *J. of Mammalogy* 79: 906–918.

Shaw, J., and R. L. Small. 2005. Chloroplast DNA phylogeny and phylogeography of the North American plums (*Prunus* subgenus *Prunus* Section *Prunocerasus*, Rosaceae). *Amer. J. Botany* 92: 2011–2030.

Shaw, R. J. 1989. *Vascular Plants of Northern Utah: An Identification Manual*. Utah State University Press.

Shaw, R. K. 1976. A taxonomic and ecologic study of the riverbottom forest on St. Mary River, Lee Creek, and

References

Belly River in southwestern Alberta, Canada. *Great Basin Naturalist* 36: 243–271.

Shay, J. M. 1999. *Annotated vascular plant species list for the Delta Marsh, Manitoba and surrounding area.* Occasional Publication No. 2. Winnipeg: University of Manitoba Field Station (Delta Marsh).

Shay, J. M., A. J. Macaulay, and K. A. Frego. 1988. A morphological comparison of *Scirpus acutus* and *S. validus* in southern Manitoba. *Can. J. Botany* 66: 2331–2337.

Shay, J. M., M. Herring, and B. S. Dyck. 2000. Dune colonization by mixed prairie in the Bald Head Hills, southwestern Manitoba. *Canadian Field-Naturalist* 114: 612–627.

Sherman, L. J. 1984. The effects of patchy food availability on nest-site selection and movement patterns of reproductively active female Meadow Voles, *Microtus pennsylvanicus*. *Holarct. Ecol.* 1: 249–299.

Shetler, S. G. 1963. A checklist and key to the species of *Campanula* native or commonly naturalized in North America. *Rhodora* 65: 319–337.

Sheviak, C. J., and M. L. Bowles. 1986. The prairie fringed orchids: a pollinator-isolated species pair. *Rhodora* 88: 267–290.

Shultz, L. M., E. E. Neely, and J. S. Tuhy. 1987. Flora of the Orange Cliffs of Utah. *Great Basin Naturalist* 47: 287–298.

Sieren, D. J. 1981. The taxonomy of the genus *Euthamia*. *Rhodora* 83: 551–579.

Simpson, M. G. 2006. *Plant Systematics.* Boston: Elsevier Academic Press, MA.

Singh, V., and R. Sattler. 1974. Floral development of *Butomus umbellatus*. *Can. J. Botany* 52: 223–230.

Sinnott, Q. P. 1985. A revision of *Ribes* L. subg. *Grossularia* (Mill.) Pers. sect. *Grossularia* (Mill.) Nutt. (Grossulariaceae) in North America. *Rhodora* 87: 189–286.

Smith, A. L. 1973. Life cycle of the marsh grass *Scolochloa festucacea*. *Can. J. Botany* 51: 1661–1668.

Smith, E. B. 1988. 2nd ed. *An Atlas and Annotated List of the Vascular Plants of Arkansas.* Fayetteville: Published by the author; out of print.

Smith, E. B., and H. M. Parker. 1971. A biosystematic study of *Coreopsis tinctoria* and *C. cardaminefolia* (Compositae). *Brittonia* 23: 161–170.

Smith, S. G. 1967. Experimental and natural hybrids in North American *Typha* (Typhaceae). *American Midland Naturalist* 78: 257–287.

Smith, S. G., and G. Yatskievych. 1996. Notes on the genus *Scirpus* sensu lato in Missouri. *Rhodora* 98: 168–179.

Smreciu, E. A., J. Hobden, and R. Hermesh. 1992. *A Checklist of the Vascular Flora in the Vicinity of the Oldman River Dam.* Alberta Environmental Centre and Wild Rose Consulting. Vegreville: Alberta, AECV92–R2.

Solbrig, O. T. 1965. The California species of *Gutierrezia* (Compositae–Astereae). *Madroño* 18: 75–83.

Soper, J. H., and M. L. Heimburger. 1982. *Shrubs of Ontario.* A Life Sciences Miscellaneous Publication. Toronto: Royal Ontario Museum.

Soreng, R. J. 1985. *Poa* L. in New Mexico, with a key to middle and southern Rocky Mountain species (Poaceae). *Great Basin Naturalist* 45: 395–422.

Soulliere, G. J. 1988. Density of suitable Wood Duck nest cavities in a northern hardwood forest. *J. Wildlife Management* 52: 86–89.

Spellenberg, R. 1979. *The Audubon Society Field Guide to North American Wildflowers – Western Region.* New York: Alfred A. Knopf.

Spence, J. R. 2005. Notes on significant collections and additions to the flora of Glen Canyon National Recreation Area, Utah and Arizona, between 1992 and 2004. *Western North American Naturalist* 65: 103–111.

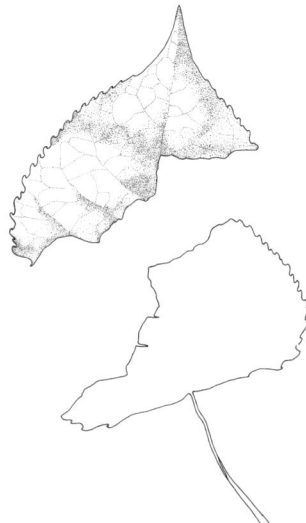

Spence, J. R., and H. Buchanan. 1993. 1993 update, checklist of the vascular plants of Bryce Canyon National Park, Utah. *Great Basin Naturalist* 53: 207–221.

Standley, L. A. 1985. *Systematic Botany Monographs.* Vol. 7. Systematics of the *Acutae* Group of *Carex* (Cyperaceae) in the Pacific Northwest. *The American Society of Plant Taxonomists* pp. 53–57.

Staniforth, R. J., and L. M. Bergeron. 1990. Annual smartweeds in the Prairie Provinces. *Canadian Field-Naturalist* 104: 526–533.

Staniforth, R. J., and P. B. Cavers. 1979. Distribution and habitats of four annual smartweeds in Ontario. *Canadian Field-Naturalist* 93: 378–385.

Staniforth, R. J., and P. A. Scott. 1991. Dynamics of weed populations in a northern subarctic community. *Can. J. Botany* 69: 814–821.

References

Stebbins, G. L., Jr. 1930. A revision of some North American species of *Calamagrostis*. *Rhodora* 32: 34–57.

Steel, M. G., P. B. Cavers, and **S. M. Lee.** 1983. The biology of Canadian weeds. 59. *Setaria glauca* (L.) Beauv. and *S. verticillata* (L.) Beauv. *Can. J. Plant Sci.* 63: 711–725.

Steeves, T. A., M. W. Steeves, and **A. R. Olson.** 1991. Flower development in *Amelanchier alnifolia* (Maloideae). *Can. J. Botany* 69: 844–857.

Stephens, H. A. 1969. *Trees, Shrubs, and Woody Vines in Kansas.* Lawrence: University Press of Kansas.

Stephens, H. A. 1973. *Woody Plants of the North Central Plains.* Lawrence: University Press of Kansas.

Stevens, O. A. 1932. The number and weight of seeds produced by weeds. *Amer. J. Botany* 19: 784–794.

Stevens, O. A. 1963. *Handbook of North Dakota Plants.* Fargo: North Dakota Institute for Regional Studies.

Stewart-Wade, S. M., S. Neumann, L. L. Collins, and **G. J. Boland.** 2002. The biology of Canadian weeds. 117. *Taraxacum officinale* G. H. Weber ex Wiggers. *Can. J. Plant Sci.* 82: 825–853.

Steyermark, J. A. 1963. *Flora of Missouri.* Vols. 1 & 2. Ames: Iowa State University Press.

St. John, H. 1963. *Flora of Southeastern Washington and Adjacent Idaho.* 3rd ed. Esondido, CA: Outdoor Pictures.

Stocking, K. 1955. Some considerations of the genera *Echinocystis* and *Echinopepon* in the United States and northern Mexico. *Madroño* 13: 84–100.

St. Pierre, R. G., and **T. A. Steeves.** 1990. Observations on shoot morphology, anthesis, flower number and seed production in the Saskatoon, *Amelanchier alnifolia* (Rosaceae). *Canadian Field-Naturalist* 104: 379–386.

Straley, G. B. 1986. Wild Bergamot, *Monarda fistulosa* (Lamiaceae), new to the Northwest Territories. *Canadian Field-Naturalist* 100: 380–381.

Straley, G. B., R. L. Taylor, and **G. W. Douglas.** 1985. *The Rare Vascular Plants of British Columbia.* Syllogeus No. 59. Ottawa: National Museum of Natural Sciences, National Museums of Canada.

Stubbendieck, J., E. C. Conard, and **B. P. Jansen.** 1989. *Common Legumes of the Great Plains: An Illustrated Guide.* Lincoln: University of Nebraska Press.

Stubbendieck, J., S. L. Hatch, and **C. H. Butterfield.** 1997. 5th ed. *North American Range Plants.* Lincoln: University of Nebraska Press.

Stuckey, R. L. 1981. Distributional history of *Juncus compressus* (Juncaceae) in North America. *Canadian Field-Naturalist* 95: 167–171.

Stutz, H. C., and **S. C. Sanderson.** 1998. Taxonomic clarification of *Atriplex nuttallii* (Chenopodiaceae) and its near relatives. *Sida* 18: 193–212.

Svenson, H. K. 1939. Monographic studies in the genus *Eleocharis* 5. *Rhodora* 41: 1–19, 43–77, 90–111.

Swanton, C. J., P. B. Cavers, D. R. Clements, and **M. J. Moore.** 1992. The biology of Canadian weeds. 101. *Helianthus tuberosus* L. *Can. J. Plant Sci.* 72: 1367–1382.

Szczawinski, A. F. 1962. *The Heather Family (Ericaceae) of British Columbia.* Handbook No. 19. Victoria: British Columbia Provincial Museum.

Szczawinski, A. F. 1975. *Orchids of British Columbia.* Handbook No. 16. Victoria: British Columbia Provincial Museum.

Taylor, R. J., and **C. E. S. Taylor.** 1994. *An Annotated List of the Ferns, Fern Allies, Gymnosperms and Flowering Plants of Oklahoma.* 3rd ed. Durant: Biology Department, Southeastern Oklahoma State University.

Taylor, R. L., and **B. MacBryde.** 1977. *Vascular Plants of British Columbia: A descriptive resource inventory.* Technical Bulletin No. 4. Vancouver: The University of British Columbia Press.

Taylor, T. M. C. 1970. *Pacific Northwest Ferns and Their Allies.* Toronto: University of Toronto Press.

Taylor, T. M. C. 1973a. *The Ferns and Fern-allies of British Columbia.* Handbook No. 12. Victoria: British Columbia Provincial Museum.

Taylor, T. M. C. 1973b. *The Rose Family of British Columbia.* Handbook No. 30. Victoria: British Columbia Provincial Museum.

Taylor, T. M. C. 1974a. *The Figwort Family (Scrophulariaceae) of British Columbia.* Handbook No. 33. Victoria: British Columbia Provincial Museum.

Taylor, T. M. C. 1974b. *The Lily Family of British Columbia.* Handbook No. 25. Victoria: British Columbia Provincial Museum.

Taylor, T. M. C. 1974c. *The Pea Family (Leguminosae) of British Columbia.* Handbook No. 32. Victoria: British Columbia Provincial Museum.

Taylor, T. M. C. 1983. *The Sedge Family (Cyperaceae).* Handbook No. 43. Victoria: British Columbia Provincial Museum.

Terrell, E. E. 1976. The correct name for Pearl Millet and Yellow Foxtail. *Taxon* 25: 297–304.

Thomas, V.G., and J.P. Prevett. 1980. The nutritional value of Arrow-grasses to geese at James Bay. *J. Wildlife Management* 44: 830–836.

Thompson, D.J., and D.G. Stout. 1991. Duration of the juvenile period in Diffuse Knapweed (*Centaurea diffusa*). *Can. J. Botany* 69: 368–371.

Thompson, F.L., L.A. Hermanutz, and D.J. Innes. 1998. The reproductive ecology of island populations of distylous *Menyanthes trifoliata* (Menyanthaceae). *Can. J. Botany* 76: 818–828.

Thoreau, H.D. 1993. *Faith in a Seed: The Dispersion of Seeds and Other Late Natural History Writings*. Edited by B.P. Dean. Washington, DC: Island Press.

Thoreau, H.D. 2000. *Wild Fruits*. Edited by B.P. Dean. New York: W.W. Norton & Company.

Timoney, K., and A. Robinson. 1998. *A floristic and landscape survey of the Ft. Assiniboine Sandhills Wildland Park*. Edson: Alberta Environmental Protection.

Tisch, E.L. 2001. *Corallorhiza maculata* var. *ozettensis* (Orchidaceae), a new Coral-root from coastal Washington. *Madroño* 48: 40–42.

Towner, H.F. 1977. The biosystematics of *Calylophus* (Onagraceae). *Annals Missouri Botanical Garden* 64: 48–120.

Trelease, W. 1967. *Winter Botany*. 3rd ed., New York: Dover Publications.

Turkington, R.A., and J.J. Burdon. 1983. The biology of Canadian weeds. 54. *Trifolium repens* L. *Can. J. Plant Sci.* 63: 243–266.

Turkington, R.A., and P.B. Cavers. 1979. The biology of Canadian weeds. 33. *Medicago lupulina* L. *Can. J. Plant Sci.* 59: 99–110.

Turkington, R.A., and G.D. Franko. 1980. The biology of Canadian weeds. 41. *Lotus corniculatus* L. *Can. J. Plant Sci.* 60: 965–979.

Turkington, R.A., P.B. Cavers, and E. Rempel. 1978. The biology of Canadian weeds. 29. *Melilotus alba* Desr. and *M. officinalis* (L.) Lam. *Can. J. Plant Sci.* 58: 523–537.

Turkington, R.A., N.C. Kenkel, and G.D. Franko. 1980. The biology of Canadian weeds. 42. *Stellaria media* (L.) Vill. *Can. J. Plant Sci.* 60: 981–992.

Turland, N. 1996. (1270) Proposal to reject the name *Malva rotundifolia* (Malvaceae). *Taxon* 45: 707–708.

Turner, B.L. 1959. *The Legumes of Texas*. Austin: University of Texas Press.

Turner, G.H. 1949. Plants of the Edmonton district of the province of Alberta. *Canadian Field-Naturalist* 63: 1–28.

Tyrl, R.J. 1975. Origin and distribution of polyploid *Achillea* (Compositae) in western North America. *Brittonia* 27: 187–196.

Tyrl, R.J., and U.T. Waterfall. 1998. *Identification of Oklahoma Plants: A Taxonomic Treatment Comprising Keys and Descriptions For the Vascular Plants of Oklahoma*. Stillwater, OK: Flora Oklahoma Incorporated.

Tyrl, R.J., T.G. Bidwell, and R.E. Masters. 2002. *Field Guide to Oklahoma Plants: Commonly Encountered Prairie, Shrubland, and Forest Species*. Illustrations by Bellamy P. Jansen. Stillwater, OK: Oklahoma State University.

Ueng, R., and I.E. Lindauer. 1993. Morphology and growth variations of *Agropyron smithii* Rydb. (Western Wheatgrass) at different salinity levels. *Great Basin Naturalist* 53: 367–372.

U.S. Department of Agriculture Staff. 1970. *Common Weeds of the United States*. New York: Dover Publications.

Van Bruggen, T. 1985. *The Vascular Plants of South Dakota*. 2nd ed. Ames: Iowa State University Press.

Vance, F.R., J.R. Jowsey, and J.S. McLean. 1984. *Wildflowers Across The Prairies*. Vancouver: Douglas & McIntyre.

Venning, F., and M. Saito. 1984. *A Guide to Field Identification of Wildflowers of North America*. New York: Golden Press.

Vines, R.A. 1960. *Trees, Shrubs and Woody Vines of the Southwest*. Austin: University of Texas Press.

Voss, E.G. 1961. Which side is up? A look at the leaves of *Oryzopsis*. *Rhodora* 63: 285–287.

Voss, E.G. 1966. Nomenclatural notes on monocots. *Rhodora* 68: 435–463.

Voss, E.G. 1972. *Michigan Flora*. part 1. Gymnosperms and Monocots. Bulletin 55. Ann Arbor: Cranbrook Institute of Science and the University of Michigan Herbarium.

Voss, E.G. 1985. *Michigan Flora*. part 2. Dicots (Saururaceae–Cornaceae). Bulletin 59. Ann Arbor: Cranbrook Institute of Science and the University of Michigan Herbarium.

Voss, E.G. 1996. *Michigan Flora*. part 3. Dicots (Pyrolaceae–Compositae). Bulletin 61. Ann Arbor: Cranbrook Institute of Science and the University of Michigan Herbarium.

Wagner, W.H., Jr., and T.F. Beals. 1958. Perennial ragweeds (*Ambrosia*) in Michigan, with the description of a new, intermediate taxon. *Rhodora* 60: 177–204.

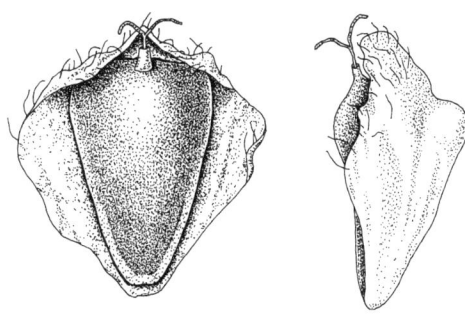

References

Wagnon, H. K. 1952. A revision of the genus *Bromus*, section *Bromopsis*, of North America. *Brittonia* 7: 415–480.

Wagstaff, S. J., and R. J. Taylor. 1988. Genecology of *Cerastium arvense* and *C. beeringianum* (Caryophyllaceae) in northwest Washington. *Madroño* 35: 266–277.

Walker, B. H., and R. T. Coupland. 1970. Herbaceous wetland vegetation in the aspen grove and grassland regions of Saskatchewan. *Can. J. Botany* 48: 1861–1878.

Warwick, S. I., and L. Black. 1982. The biology of Canadian weeds. 52. *Achillea millefolium* L. s. l. *Can. J. Plant Sci.* 62: 163–182.

Warwick, S. I., and R. D. Sweet. 1983. The biology of Canadian weeds. 58. *Galinsoga parviflora* and *G. quadriradiata* (= *G. ciliata*). *Can. J. Plant Sci.* 63: 695–709.

Warwick, S. I., and D. A. Wall. 1998. The biology of Canadian weeds. 108. *Erucastrum gallicum* (Willd.) O. E. Schulz. *Can. J. Plant Sci.* 78: 155–165.

Warwick, S. I., H. J. Beckie, A. G. Thomas, and T. McDonald. 2000. The biology of Canadian weeds. 8. *Sinapis arvensis* L. (updated). *Can. J. Plant Sci.* 80: 939–961.

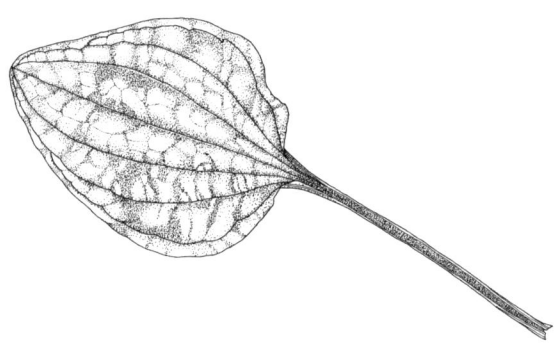

Warwick, S. I., A. Francis, and D. J. Susko. 2002. The biology of Canadian weeds. 9. *Thlaspi arvense* L. (updated). *Can. J. Plant Sci.* 82: 803–823.

Watson, A. K., and A. J. Renney. 1974. The biology of Canadian weeds. 6. *Centaurea diffusa* and *C. maculosa*. *Can. J. Plant Sci.* 54: 687–701.

Watts, B. D. 1990. Cover use and predator-related mortality in Song and Savannah Sparrows. *Auk* 107: 775–778.

Weatherby, C. A. 1926. On *Solidago rigida* L., and the application of old botanical names. *Rhodora* 28: 138–145.

Weaver, S. E. 2001. The biology of Canadian weeds. 115. *Conyza canadensis*. *Can. J. Plant Sci.* 81: 867–875.

Weaver, S. E., and M. J. Lechowicz. 1983. The biology of Canadian weeds. 56. *Xanthium strumarium* L. *Can. J. Plant Sci.* 63: 211–225.

Weaver, S. E., and E. L. McWilliams. 1980. The biology of Canadian weeds. 44. *Amaranthus retroflexus* L., *A. powellii* S. Wats. and *A. hybridus* L. *Can. J. Plant Sci.* 60: 1215–1234.

Weaver, S. E., and W. R. Riley. 1982. The biology of Canadian weeds. 53. *Convolvulus arvensis* L. *Can. J. Plant Sci.* 62: 461–472.

Webber, J. M., and P. W. Ball. 1984. The taxonomy of the *Carex rosea* (section *Phaestoglochin*) in Canada. *Can. J. Botany* 62: 2058–2073.

Weber, W. A. 1972. *Rocky Mountain Flora*. Boulder: Colorado Associated University Press.

Weber, W. A., and R. C. Wittmann. 1992. *Catalogue of the Colorado Flora: A Biodiversity Baseline*. Niwot, CO: University of Colorado Press.

Weber, W. A., and R. C. Wittmann. 1996a. *Colorado Flora: Eastern Slope*. 2nd ed. Niwot, CO: University Press of Colorado.

Weber, W. A., and R. C. Wittmann. 1996b. *Colorado Flora: Western Slope*. 2nd ed. Niwot, CO: University Press of Colorado.

Weber, W. A., B. C. Johnston, and R. C. Wittmann. 1981. Additions to the flora of Colorado. 7. *Brittonia* 33: 325–331.

Welsh, S. L. 1983. Utah flora: Compositae (Asteraceae). *Great Basin Naturalist* 43: 179–357.

Welsh, S. L. 1991. *Oxytropis* DC.—Names, basionyms, types, and synonyms—Flora North America Project. *Great Basin Naturalist* 51: 377–396.

Welsh, S. L. 1993. New taxa and new nomenclatural combinations in the Utah flora. *Rhodora* 95: 392–421.

Welsh, S. L. 1995. Names and types of *Hedysarum* L. (Fabaceae) in North America. *Great Basin Naturalist* 55: 66–73.

Welsh, S. L., and C. Crompton. 1995. Names and types of perennial *Atriplex* Linnaeus (Chenopodiaceae) in North America selectively exclusive of Mexico. *Great Basin Naturalist* 55: 322–334.

Welsh, S. L., N. D. Atwood, S. Goodrich, and L. C. Higgins, eds. 1987. *A Utah Flora. Great Basin Naturalist Memoirs*. No. 9. Provo, UT: Brigham Young University.

Wemple, D. K., and N. R. Lersten. 1966. An interpretation of the flower of *Petalostemon* (Leguminosae). *Brittonia* 18: 117–126.

Werner, P. A., and R. Rioux. 1977. The biology of Canadian weeds. 24. *Agropyron repens* (L.) Beauv. *Can. J. Plant Sci.* 57: 905–919.

Werner, P. A., and J. D. Soule. 1976. The biology of Canadian weeds. 18. *Potentilla recta* L., *P. norvegica* L., and *P. argentea* L. *Can. J. Plant Sci.* 56: 591–603.

Werner, P. A., I. K. Bradbury, and R. S. Gross. 1980. The biology of Canadian weeds. 45. *Solidago canadensis* L. *Can. J. Plant Sci.* 60: 1393–1409.

Wetmore, R. H., and A. L. Delisle. 1939. Studies in the genetics and cytology of two species in the genus *Aster* and their polymorphy in nature. *Amer. J. Botany* 26: 1–12.

References

Wheeler, L.C. 1941. *Euphorbia* subgenus *Chamaesyce* in Canada and the United States exclusive of southern Florida. *Rhodora* 43: 96–154, 168–205, 223–286.

Whitcomb, R.F. 1994. North American grasslands as insect habitat: A photographic essay. *Amer. Entomologist* 40: 92–101.

White, D.J., and K.L. Johnson. 1980. *The Rare Vascular Plants of Manitoba*. Syllogeus No. 27. Ottawa: National Museum of Natural Sciences.

White, D.J., E. Haber, and C. Keddy. 1993. *Invasive Plants of Natural Habitats in Canada*. Ottawa: Canadian Wildlife Service in cooperation with the Canadian Museum of Nature.

Widén, M., and B. Widén. 1999. Sex expression in the clonal gynodioecious herb *Glechoma hederacea* (Lamiaceae). *Can. J. Botany* 77: 1689–1698.

Widrlechner, M.P. 1998. The genus *Rubus* L. in Iowa. *Castanea* 63: 415–465.

Wilbur, H.M. 1976. Life history evolution in seven milkweeds of the genus *Asclepias*. *J. of Ecol.* 64: 223–240.

Wilbur, R.L. 1975. A revision of the North American genus *Amorpha* (Leguminosae—Psoraleae). *Rhodora* 77: 337–409.

Willson, M.F., and B.J. Rathcke. 1974. Adaptive design of the floral display in *Ascelpias syriaca* L. *American Midland Naturalist* 92: 47–57.

Wilson, B.L., R. Brainerd, M. Huso, K. Kuykendall, D. Lytjen, B. Newhouse, N. Otting, S. Sundberg, and P. Zika. 1999. *Atlas of Oregon Carex*. Occasional Paper No. 1. Corvallis: The Native Plant Society of Oregon.

Wilson, H.D. 2001. Informatics: new media and paths of data flow. *Taxon* 50: 381–387.

Wilson, T., and U. Posluszny. 2003. Complex tendril branching in two species of *Parthenocissus*: implications for the vitaceous shoot architecture. *Can. J. Botany* 81: 587–597.

Withycombe, C.L. 1924. On the function of bladders in *Utricularia vulgaris* L. *J. Linn. Soc. Botany* 46: 401–413.

Wittmann, R.C., and W.A. Weber. 1991. *A Catalogue of the Flora of Colorado*. Niwot: University Press of Colorado.

Wolf, S.J., and J. McNeill. 1986. Synopsis and achene morphology of *Polygonum* section *Polygonum* (Polygonaceae) in Canada. *Rhodora* 88: 457–479.

Wolf, S.J., and J. McNeill. 1987. Cytotaxonomic studies on *Polygonum* section *Polygonum* in eastern Canada and the adjacent United States. *Can. J. Botany* 65: 647–652.

Woo, S.L., A.G. Thomas, D.P. Peschken, G.G. Bowes, D.W. Douglas, V.L. Harms, and A.S. McClay. 1991. The biology of Canadian weeds. 99. *Matricaria perforata* Mérat (Asteraceae). *Can. J. Plant Sci.* 71: 1101–1119.

Woodland, D.W. 1997. *Contemporary Plant Systematics*. 2nd ed. Berrien Springs, Michigan.

Woodson, R.E., Jr. 1954. The North American species of *Asclepias*. L. *Annals Missouri Botanical Garden* 41: 1–211.

Wooten, E.O., and P.C. Standley. 1971. *Flora of New Mexico*. Lubrecht & Cramer.

Wyatt, R. 1996. More on the southward spread of Common Milkweed, *Asclepias syriaca* L. *Bull. Torrey Bot. Club* 123: 68–69.

Wydeven, A.P., and R.B. Dahlgren. 1983. Food habits of elk in the northern Great Plains. *J. Wildlife Management* 47: 916–923.

Yatskievych, G. 1999. *Steyermark's Flora of Missouri*. Vol. 1, rev. Jefferson City, St. Louis: The Missouri Department of Conservation in cooperation with The Missouri Botanical Garden Press.

Yeo, R.R. 1964. Life history of the Common Cattail. *Weeds* 12: 284–288.

Zandstra, I.I., and W.F. Grant. 1968. The biosystematics of the genus *Lotus* (Leguminosae) in Canada. 1. Cytotaxonomy. *Can. J. Botany* 46: 557–583.

Zentz, W.R., and W.R. Jacobi. 1989. Ecology of *Commandra umbellata* (Santalaceae) in western Wyoming. *Great Basin Naturalist* 49: 650–655.

Zika, P.F. 2003. Noteworthy collections: Washington. *Rhamnus cathartica*. *Madroño* 50: 313–314.

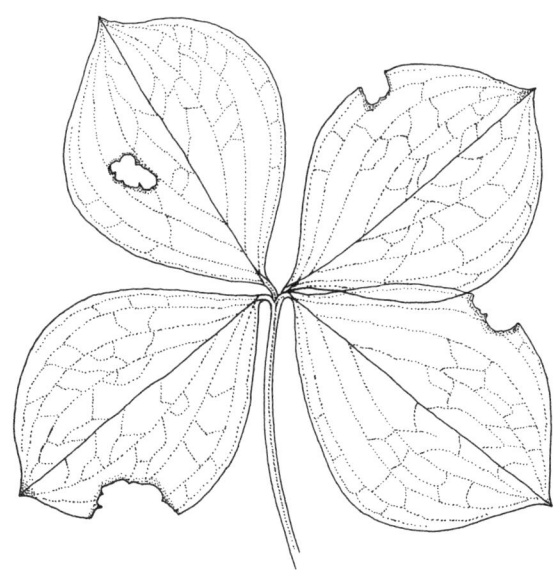

References

Websites

http://invader.dbs.umt.edu/
County distribution maps of weeds in Washington, Oregon, Idaho, Montana and Wyoming.

http://ucjeps.herb.berkeley.edu/jeps-list.html
Vascular plants of California, distribution maps in color by ecological or vegetative regions.

www.bonap.org
Biota of North America Program: distribution listed by state (Canada excluded); one common and one scientific name for vascular plants in North America.

www.csdl.tamu.edu/FLORA/arkansas/
A digital version of the 1988 book, *An Atlas and Annotated Checklist of the Vascular Plants of Arkansas*, by Edwin Smith with distribution maps by county.

www.csdl.tamu.edu/FLORA/taes/tracy/
A checklist of vascular plants of Texas. The state is divided into ten vegetative regions. Plants are followed by the numbers of the regions where they are found.

www.efloras.org or www.fna.org
Flora of North America: range maps, descriptive text, and scientific names of plants of North America north of Mexico. Information goes online as the volumes are published.

www.ipni.org/index.html
International plant name index.

www.itis.gov
Integrated Taxonomic Information System (ITIS).

www.nearctica.com/nathist/vascular/regcheck.htm
Links to regional and state or provincial checklists and distribution maps.

www.npwrc.usgs.gov
Northern Prairie Wildlife Research Center located in North Dakota, gives text, drawings, and distribution maps on 300 western wetland vascular species.

www.nr.usu.edu/Geography-Department/utgeog/utvatlas/ut-vascatlas.html
A digital version of the 1988 book, *Atlas of the Vascular Plants of Utah*, by Beverly Albee et al.

www.oregonflora.org
The Oregon plant atlas. Dot distributional maps for all the vascular species in the state.

www.rmh.uwyo.edu
Atlas of the vascular flora of Wyoming from the collection of the Rocky Mountain Herbarium (RMH); dots in counties.

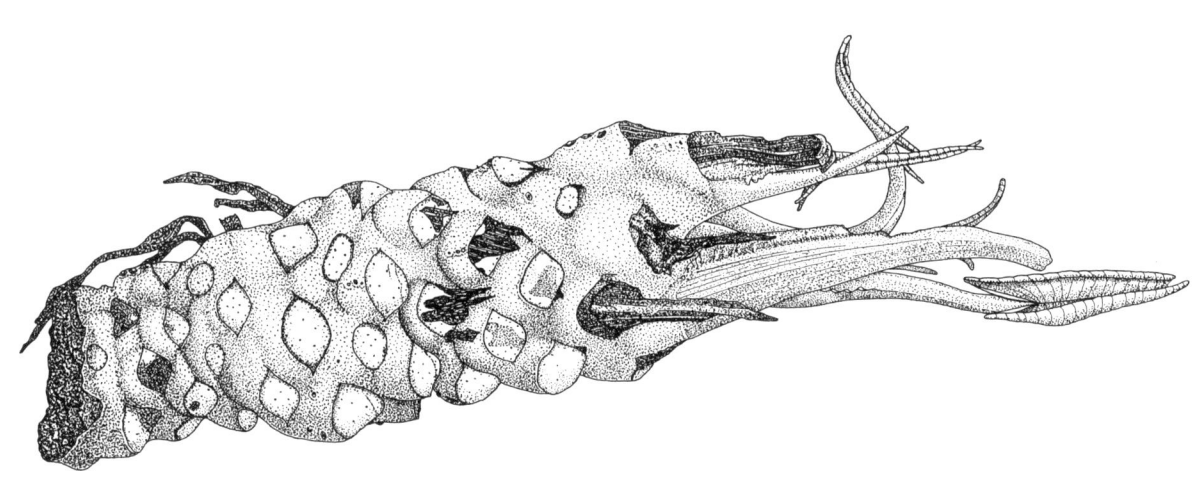

Index

Names are presented in four styles: **common name**; alternate common name; *Latin name*; and *Latin synonym*.

A

Abies balsamea 65
Absinthe 111
Absinthe Wormwood 111
Absinthium 111
Acer
 negundo 72
 saccharinum 73
 spicatum 74
Aceraceae—Maple 72–74
Achillea
 lanulosa 100
 millefolium 100
 millefolium var. *occidentalis* 100
 sibirica 101
Acoraceae—Calamus 522
Acorus
 americanus 522
 calamus 522
Actaea rubra 420
Agalinis, Slender 494
Agalinis
 tenuifolia 494
 tenuifolia var. *tenuifolia* 494

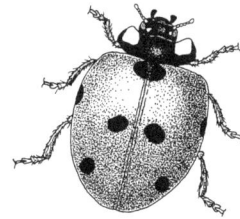

Agastache foeniculum 345
Agoseris, Pale 102
Agoseris glauca 102

Agrimonia gryposepala 444
Agrimony
 Hooked 444
 Tall Hairy 444
Agropyron
 cristatum 622
 cristatum ssp. *pectinatum* 622
 pectiniforme 622
 repens 642
 smithii 664
Agrostis
 alba var. *palustris* 623
 alba var. *stolonifera* 623
 palustris 623
 stolonifera 623
Alder
 Gray 201
 Green 202
 Mountain 202
 Sitka 202
 Speckled 201
Alderleaf Juneberry 445
Alexanders
 Golden 89
 Heart-leaf 88
 Heart-leaved 88
Alfalfa 314
Alfilaria 335
Alisma
 brevipes 523
 plantago-aquatica ssp. *brevipes* 523
 plantago-aquatica var. *americanum* 523
 plantago-aquatica var. *brevipes* 523
 triviale 523
Alismataceae—Water-Plantain 523–525
Alkali American-aster 184
Alkali Blite 269
Alkali Bulrush 571
Alkali Buttercup 435
Alkali Cord Grass 681
Alkali-grass (Alkaligrass)
 Nuttall's 674
 Spreading 673
 Weeping 673
Alkali Muhly 656

Alkali Plantain 384
Alleghany Monkeyflower 499
Allium
 schoenoprasum 597
 stellatum 598
Alnus
 crispa 202
 incana 201
 incana ssp. *rugosa* 201
 incana ssp. *tenuifolia* 201
 rugosa 201
 tenuifolia 201
 viridis 202
 viridus ssp. *crispa* 202
Alopecurus aequalis 624
Alpine Bulrush 577
Alpine Cotton-grass 577
Alpine Golden Wild Buckwheat 389
Alpine Hedysarum 308
Alpine Leafless-bulrush 577
Alpine Pondweed 688
Alpine Rush 581
Alpine Sweetvetch 308
Alsike Clover 322
Alumroot 488
 Richardson's 488
Amaranth
 Common 76
 Green 76
 Mat 75
 Prostrate 75
 Redroot 76
Amaranthaceae—Amaranth 75–76
Amaranth—Amaranthaceae 75–76
Amaranthus
 blitoides 75
 graecizans 75
 retroflexus 76
Ambrosia
 artemisiifolia 103
 artemisiifolia var. *elatior* 103
 elatior 103
 psilostachya 104
 trifida 105
Amelanchier alnifolia 445

747

Index

American-aster
  Alkali 184
  Boreal 183
  Fringed 185
  Lindley's 185
  New England 191
  Rough White Prairie 187
  Smooth Blue 188
  White Heath 186
  White Panicled 189
American Basswood 508
American Black Current 337
American Brooklime 504
American Bugleweed 350
American Bush-cranberry 246
American Calamus 522
American Cow-parsnip 83
American Dragonhead 346
American Elm 509
American Filbert 207
American Hazelnut 207
American Hedysarum 308
American Hog-peanut 296
American Larch 66
American Licorice 307
American Linden 508
American Manna Grass 646
American Milfoil 342
American Pasqueflower 431
American Purple Vetch 325
American Red Raspberry 467
American Silverberry 281
American Sloughgrass 627
American Speedwell 504
American Starflower 414
American Twinflower 238
American Vetch 325
American Water-horehound 350
American White Birch 205
American Wild Mint 353
American Wood Anemone 424
Amerorchis rotundifolia **609**
Amorpha, Dwarf Indigo-bush 295
Amorpha nana **295**
Amphicarpaea bracteata **296**
Anacardiaceae—Sumac 77–78
Anaphalis margaritacea **106**

Andropogon
  gerardii **625**
  scoparius 676
Androsace
  occidentalis **408**
  septentrionalis **409**
Anemone
  American Wood 424
  Canada 421
  Candle 422
  Crocus 431
  Cut-leaved 423
  Long-fruited 422
  Long-headed 422
  Meadow 421
  Pacific 423
  Red 423
  Wood 424
Anemone
  canadensis **421**
  cylindrica **422**
  multifida **423**
  patens 431
  quinquefolia **424**

Anis 80
Anise-root 84
Annual Canary Grass 666
Annual Dropseed 686
Annual Hawk's-beard 128
Annual Ragweed 103
Annual Sow-thistle 182
Annual Sunflower 144
Annual Wild-rice 687
Anomalus Brome 632
Antennaria
  athabascensis 108
  campestris 108
  howellii var. *athabascensis* 108
  howellii var. *campestris* 108
  microphylla **107**
  neglecta **108**
  nitida 107
  parvifolia 107
  rosea var. *nitida* 107

Anthoxanthium hirtum ssp. *arcticum* 650
Apiaceae—Carrot 79–89
Apocynaceae—Dogbane 90–91
Apocynum
  androsaemifolium **90**
  cannabinum **91**
  sibiricum 91
Apple
  Balsam 279
  Wild Balsam 279
Aquilegia
  brevistyla **425**
  canadensis **426**
Arabis
  confinis var. *interposita* 214
  X *divaricarpa* **214**
  holboellii **215**
  holboellii var. *retrofracta* 215
  retrofracta 215
Araceae—Arum 526
Araliaceae—Ginseng 92
Aralia nudicaulis **92**
Arborvitae 63
  Eastern 63
Arctic Blackberry 466
Arctic Bur-reed 698
Arctic Raspberry 466
Arctium
  minus **109**
  tomentosum **110**
Arctostaphylos uva-ursi **284**
Arenaria lateriflora 252
Argentina anserina **446**
Arrow-grass 591
  Marsh 592
  Seaside 591
  Slender 592
Arrow-grass—Juncaginaceae 591–592
Arrowhead 525
  Arum-leaved 524
  Broad-leaved 525
  Common 525
  Northern 524
Arrowleaf Sweet Coltsfoot 167
Arrow-leaf Tear-thumb 400
Arrow-leaved Colt's-foot 167
Arrowwood
  Downy 247
  Shortstalk 247

Index

Artemisia
  absinthium 111
  biennis 112
  campestris 113
  campestris ssp. *caudata* 113
  campestris var. *borealis* 113
  campestris var. *scouleriana* 113
  camporum 113
  canadensis 113
  caudata 113
  frigida 114
  ludoviciana 115

Artichoke, Jerusalem 148

Arum, Water 526

Arum—Araceae 526

Arum-leaved Arrowhead 524

Asclepiadaceae—Milkweed 93–97

Asclepias
  incarnata 93
  ovalifolia 94
  speciosa 95
  syriaca 96
  verticillata 97

Ash
  Black 375
  Green 376
  Red 376
  Swamp 375

Ashleaf Maple 72

Asparagus 599
  Common 599
  Garden 599

Asparagus officinalis 599

Aspen 478
  Quaking 478
  Trembling 478

Aspen Poplar 478

Aster
  Bog 192
  Boreal 183
  Bristly 192
  Calico 190
  Eastern Willow 189
  Flat-top 129
  Flat-topped White 129
  Fringed 185
  Heath 186, 187
  Lindley's 185
  Lindley's Blue 185
  Many-flowered 186
  Marsh 183
  New England 191
  Northern Bog 183
  One-sided 190
  Panicled 189

[Aster]
  Purple-stemmed 192
  Rayless 184
  Rayless Alkali 184
  Rush 183
  Small Blue 189
  Smooth 188
  Smooth Blue 188
  Sneezewort 162
  Swamp 192
  Tall Flat-top White 129
  Tufted White Prairie 186
  Umbellate 129
  White 186
  White Heath 186
  White Panicle 189
  White Prairie 187
  White Upland 162
  White Woodland 190
  Wood 190

Aster
  borealis 183
  brachyactis 184
  ciliolatus 185
  commutatus 187
  ericoides 186
  falcatus 187
  hesperius 189
  junceus 183
  junciformis 183
  laevis 188
  lanccolatus 189
  lateriflorus 190
  novae-angliae 191
  paniculatus 189
  pansus 186
  ptarmicoides 162
  pubentior 129
  puniceus 192
  simplex 189
  umbellatus 129

Asteraceae—Aster 98–198

Aster—Asteraceae 98–198

Astragalus
  agrestis 297
  bisulcatus 298
  canadensis 299
  caryocarpus 300
  crassicarpus 300
  crassicarpus var. *crassicarpus* 300
  danicus var. *dasyglottis* 297
  dasyglottis 297
  goniatus 297
  hypoglottis 297
  missouriensis 301
  pectinatus 302

Atriplex
  Garden 261
  Nuttall's 260

Atriplex
  gardneri 260
  hortensis 261
  nuttallii ssp. *gardneri* 260
  patula ssp. *hastata* 262
  patula var. *hastata* 262
  patula var. *triangularis* 262
  prostrata 262
  triangularis 262

Autumn Dwarf Gentian 332

Autumn Onion 598

Avena fatua 626

Avens
  Cut-leaf 453
  Large-leaved 453
  Long-plumed Purple 455
  Purple 454
  Three-flowered 455
  Water 454
  Yellow 452

Awl-fruited Sedge 559

Awl-leaved Mudwort 497

Awned Cyperus 565

Awned Flat Sedge 565

Awned Nut-grass 565

Awned Sedge 532

Awned Umbrella-sedge 565

Awn-fruited Sedge 559

Axseed 304

Axyris amaranthoides 263

B

Baby's-breath 251
  Tall 251

Babysbreath Gypsophila 251

Balkan Catchfly 254

Index

Balsam, Canada 65
Balsam Apple 279
  Wild 279
Balsam Fir 65
Balsaminaceae—Touch-me-not 199
Balsam Poplar 476
  Eastern 476
Baltic Rush 582
Baneberry 420
  Red 420
  Red and White 420
Banksian Pine 69
Bare-stem Bishop's-cap 489
Barley
  Foxtail 651
  Wild 651
Barnyard Grass 640
  Large 640
Basket Willow 485
Bassia scoparia 271
Basswood 508
  American 508
Bastard Toadflax 486
Beaked Hazelnut 208
Beaked Willow 479
Bean, Prairie Buck 320
Bearberry 284
  Common 284
  Red 284
Bearberry Manzanita 284
Bearded Flatsedge 565
Beardtongue
  Lilac 501
  Lilac-flowered 501
  Slender 501
  Smooth Blue 502
  Waxleaf 502
Beautiful Sunflower 147
Bebb's Sedge 534
Bebb Willow 479
Beckmannia syzigachne 627
Bedstraw
  Fragrant 473
  Northern 471
  Small 472
  Sweet-scented 473
  Threepetal 472
Beeblossom, Scarlet 380
Beech—Fagaceae 327

Bee Nettle 347
Beggarticks
  Common 117
  Devil's 117
  Nodding 116
  Smooth 116
Bellflower
  Bluebell 234
  Creeping 233
  Marsh 232
  Rover 233
Bellflower—Campanulaceae 232–234
Bent, Spreading 623
Bentgrass, Creeping 623
Bergamot
  Western Wild 354
  Wild 354
Bet, Bouncing 253
Betony, Marsh 358
Betula
  fontinalis 204
  glandulifera 206
  glandulosa 203
  glandulosa var. *glandulifera* 206
  nana 203
  occidentalis 204
  papyrifera 205
  pumila 206
  pumila var. *glandulifera* 206
Betulaceae—Birch 201–208
Bicknell's Crane's-bill 336
Bicknell's Geranium 336
Bicknell's Wild Geranium 336

Bidens
  cernua **116**
  frondosa 117
Biennial Campion 254
Biennial Sagewort 112
Biennial Wormwood 112

Big Bluestem 625
Bigbract Verbena 513
Big-seed Goosefoot 270
Bindweed 394
  Black 395
  Common 276
  Field 276
  Fringed Black 394
  Hedge 275
  Hedge False 275
Bine 235
Birch
  American White 205
  Black 204
  Bog 206
  Canoe 205
  Dwarf 206
  Glandular 203
  Mountain 204
  Paper 205
  Resin 203
  River 204
  Scrub 203
  Swamp 206
  Tundra Dwarf 203
  Water 204
  White 205
  Yellow 206
Birch—Betulaceae 201–208
Bird Cherry 461
Birdfoot Deervetch 312
Bird's-foot Trefoil 312
  Garden 312
Bird Vetch 326
Bishop's-cap 489
  Bare-stem 489
  Naked 489
Bitter Cress 217
  (Bitter-cress and Bittercress)
  Pennsylvania 217
  Quaker 217
Bitter Fleabane 131
Bittersweet 507
Bittersweet Nightshade 507
Black Ash 375
Blackberry
  Arctic 466
  Creeping 468
  Dwarf 468
  Northern 466
Black Bindweed 395
  Fringed 394
Black Birch 204

Index

Black Current
American 337
Canadian 339
Northern 339
Wild 337, 339

Black-eyed Susan 171

Black Hills Spruce 67

Black Medick 313

Black Poplar 476

Blacksamson 130

Black Snakeroot 86

Black Spruce 68

Bladder Campion 255

Bladder Sedge 546

Bladderwort
Common 360
Flat-leaved 359
Greater 360
Lesser 361
Small 361

**Bladderwort—
Lentibulariaceae 359–361**

Blanket Flower 139 (Blanketflower)
Great 139

Blazingstar (Blazing Star 157)
Dotted 158
Meadow 157
Rocky Mountain 157

Blite
Alkali 269
Coast 269
Low Sea 274
Sea 274
Strawberry 266
Western 274

Blite Goosefoot 266

Bluebell 234
Creeping 233

Bluebell Bellflower 234

Bluebell-of-Scotland 234

Bluebells
Northern 212
Tall 212

Blueberry 287
Late Low 287
Lowbush 287
Low Sweet 287
Sour-top 288
Velvet-leaved 288

Bluebur 210

Blue Columbine 425

Blue-eyed Grass 580
(Blue-eyed-grass)
Common 580
Strict 580

Blue Flag 579
Harlequin 579
Larger 579
Northern 579

Blue Flax 362
Wild 362

Blue Fly Honeysuckle 241

Blue Giant Hyssop 345

Blue Grama 629

Blue Grama Grass 629

Blue Grass
Canada 669
Flat-stem 669
Fowl 670
Kentucky 671

Bluejoint 633

Blue Lettuce 155
Common 155
Russian 155

Blue Monkeyflower 499

Blue Ridge Carrion-flower 694

Blue Skullcap 357

Bluestem
Big 625
Little 676
Little False 676
Small 676

Blue Stickseed 210

Bluets (Bluet)
Longleaf Summer 474
Long-leaved 474
Slender-leaved 474

Bluntleaf Grove Sandwort 252

Blunt-leaved Sandwort 252

Bob 274

Bog Aster 192
Northern 183

Bogbean 371

Bog Birch 206

Bog Labrador Tea 286

Bog Marsh Cress 226

Bog Orchid (Bog-orchid)
Bracted 610
Green-flowered 618
Northern Green 618

Bog Sedge 548

Bog Spruce 68

Bog Violet 516

Bog Wintergreen 417

Bog Yellowcress 226

Bolboschoenus
fluviatilis 570
maritimus 571

Boneset
Purple 137
Tall 137

Borage—Boraginaceae 209–212

Boraginaceae—Borage 209–212

Boreal American-aster 183

Boreal Aster 183

Boreal Bog Sedge 548

Bottle Gentian 331

Bouncing Bet 253

Bouteloua
curtipendula 628
gracilis 629

Box Elder 72

Bracted Bog-orchid 610

Bracted Orchid 610

Bracted Vervain 513

Branched Pepper-grass 225

Branched Pepperwort 225

Brassicaceae—Mustard 213–230

Brassica kaber 227

Breadroot
Indian 319
Large Indian 319
Silverleaf Indian 318

Bristlegrass
Green 679
Yellow 678

Bristle-stalked Sedge 547

Bristly Aster 192

Bristly Buttercup 437

Bristly Club-moss 57

Bristly Crowfoot 437

Bristly Sheepbur 210

751

Index

Brittle-stem Hemp-nettle 347
Broad-fruited Bur-reed 696
Broad-fruited Sedge 535
Broad-leaved Arrowhead 525
Broad-leaved Cat-tail 700
Broad-leaved Pussytoes 108
Brome
    Anomalus 632
    Fringed 630
    Nodding 632
    Porter's 632
    Smooth 631
Bromopsis canadensis 630
Bromus
    anomalus 632
    ciliatus 630
    ciliatus var. *ciliatus* 630
    inermis 631
    porteri 632
Brooklime, American 504
Brooklime Speedwell 504
Brook Lobelia 364
Broomcorn Millet 662
Broom Snakeweed 142
Broomweed 142
    Common 142
Bruner's Trumpetweed 137
Buck Bean, Prairie 320
Buck-bean 371
Buck-bean—Menyanthaceae 371
Buckbrush 243
Buckthorn 442
    Common 442
    European 442
Buckthorn—Rhamnaceae 442
Buckwheat 390
    Alpine Golden Wild 389
    Climbing 395
    Climbing False 401
    False 401
    Garden 390
    Wild 395
    Yellow Wild 389
Buckwheat—Polygonaceae 388–406
Buffalo-bean 300
Buffaloberry 282
    Canada 283
    Russet 283
    Silver 282
    Thorny 282

Bugleweed 352
    American 350
    Northern 352
    Rough 351
    Slender 352
Bulb-bearing Water-hemlock 81
Bulblet-bearing Water-hemlock 81
Bulbous Water Hemlock 81
Bullhead-lily 374
Bull Thistle 124
Bulrush
    Alkali 571
    Alpine 577
    Cosmopolitan 571
    Cottongrass 574
    Great 573
    Hard-stem 569
    Hudson Bay 577
    Pale 576
    Pale-green 576
    Panicled 575
    Prairie 571
    Red-fringe 575
    River 570
    Small-flowered 575
    Small-fruited 575
    Softstem 573
    Three-square 572
    Viscid Great 569
    Wool Grass 574
Bunchberry 277
    Canadian 277
Bunchberry Dogwood 277
Burdock
    Common 109
    Cotton 110
    Lesser 109
    Woolly 110

Bur Oak 327
Bur-reed
    Arctic 698
    Broad-fruited 696
    Clustered 697
    Giant 696
    Green-fruited 695
    Narrow-leaf 695
    Simple-stem 695
    Slender 698
    Small 698
    Stemless 695
Bur-reed—Sparganiaceae 695–698
Burr-marigold, Nodding 116
Bush Cinquefoil 450
Bush-cranberry
    American 246
    High 246
    Low 244
Bush-honeysuckle 237
    Northern 237
Bushy Knotweed 399
Bushy Peppergrass 225
Bushy Pondweed 693
Bushy Vetchling 311
Butomaceae—Flowering Rush 527
Butomus umbellatus 527
Butter-and-eggs 498
    Greater 498
Buttercup
    Alkali 435
    Bristly 437
    Celery-leaved 439
    Common 433
    Creeping 436
    Cursed 439
    Early Wood 432
    Kidney-leaved 432
    Labrador 438
    Littleleaf 432
    Meadow 433
    Pennsylvania 437
    Prairie 438
    Seaside 435
    Shore 435
    Small-flowered 432
    Smooth-leaved 432
    Tall 433
    Thread-leaved 434
    White Water 434
Buttercup—Ranunculaceae 419–441
Butterflyweed, Scarlet 380

Index

C

Calamagrostis
   canadensis **633**
   crassiglumis 634
   inexpansa 634
   neglecta 634
   stricta **634**
   stricta ssp. *inexpansa* 634
   stricta ssp. *stricta* 634

Calamovilfa longifolia **635**

Calamus 522
   American 522

Calamus—Acoraceae 522

Calico Aster 190

Calla
   Marsh 526
   Water 526
   Wild 526

Calla palustris **526**

Callitrichaceae—Water-starwort 231

Callitriche
   palustris **231**
   verna 231

Caltha palustris **427**

Calylophus serrulatus **377**

Calystegia
   sepium **275**
   sepium ssp. *sepium* 275

Camas
   Smooth 607
   White 607

Campanula
   aparinoides **232**
   rapunculoides **233**
   rotundifolia **234**

Campanulaceae—Bellflower 232–234

Campion
   Biennial 254
   Bladder 255
   White 255

Canada Anemone 421

Canada Balsam 65

Canada Blue Grass 669

Canada Buffaloberry 283

Canada Fleabane 125

Canada Goldenrod 175

Canada Gooseberry 340

Canada Hawkweed 151

Canada Mayflower 602

Canada Plum 460

Canada Thistle 121

Canada Violet 515

Canada Wild Rye 641

Canada Woodnettle 510

Canadian Black Currant 339

Canadian Bunchberry 277

Canadian Columbine 426

Canadian Goldenrod 175

Canadian Gooseberry 340

Canadian Horseweed 125

Canadian Milk-vetch 299

Canadian Moonseed 370

Canadian Plum 460

Canadian Thistle 121

Canadian Western Violet 515

Canadian Wood-nettle 510

Canary Grass 666
   Annual 666
   Common 666
   Reed 665

Candle Anemone 422

Canoe Birch 205

Caprifoliaceae—Honeysuckle 236–247

Capsella bursa-pastoris **216**

Caragana arborescens **303**

Caragana, Common 303

Caraway 80

Cardamine pensylvanica **217**

Carduus nutans **118**

Careless-weed 153

Carex
   aquatilis **531**
   atherodes **532**
   aurea **533**
   bebbii **534**
   brevior **535**
   capillaris **536**
   deweyana **537**
   diandra **538**
   dioica var. *gynocrates* 544
   disperma **539**
   filifolia **540**
   foenea **var.** *foenea* **541**
   gracillima **542**
   granularis **543**
   gynocrates **544**

[*Carex*]
   houghtoniana **545**
   houghtonii 545
   inops ssp. *heliophila* 552
   intumescens **546**
   lanuginosa 551
   lasiocarpa var. *latifolia* 551
   leptalea **547**
   magellanica **548**
   magellanica ssp. *irrigua* 548
   nigromarginata var. *elliptica* 550
   nigromarginata var. *minor* 550
   parryana **549**
   paupercula 548
   peckii **550**
   pellita **551**
   pensylvanica **552**
   praegracilis **553**
   pseudocyperus **554**
   radiata **555**
   retrorsa **556**
   rostrata var. *utriculata* 562
   sartwellii **557**
   siccata 541
   sprengelii **558**
   stipata **559**
   tenera **560**
   tenuiflora **561**
   utriculata **562**
   viridula **563**
   xerantica **564**

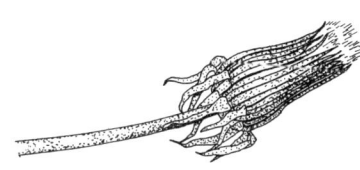

Carpet Phlox 386

Carpet Vervain 513

Carrionflower 694
   (Carrion-flower)
   Blue Ridge 694

Carrot—Apiaceae 79–89

Carum carvi 80

Caryophyllaceae—Pink 248–258

Castilleja miniata **495**

Catchfly
   Balkan 254
   Night-flowering 256
   Smooth 254

Cattail (Cat-tail)
   Broad-leaved 700
   Common 700
   Narrow-leaf 699
   Narrow-leaved 699

Index

Cattail Grass 678

Cattail—Typhaceae 699–700

Cedar
　Creeping 62
　Eastern White 63
　Northern White 63
　White 63

Celery-leaved Buttercup 439

Centaurea diffusa 119

Cerastium
　arvense 249
　fontanum 250
　fontanum ssp. *vulgare* 250
　holosteoides 250
　vulgatum 250

Ceratoides lanata 272

Chaenorhinum minus 496

Chamaerhodos 447

Chamaerhodos
　erecta 447
　erecta ssp. *nuttallii* 447
　nuttallii 447

Chamaesyce
　glyptosperma 290
　serpyllifolia 291
　serpyllifolia ssp. *serpyllifolia* 291

Chamerion
　angustifolium 378
　angustifolium ssp. *angustifolium* 378

Chamomile
　Scentless 197
　Wild 197

Charlie, Creeping 348

Charlock 227

Charlock Mustard 227

Cheeseweed Mallow 367

Chenopodiaceae—Goosefoot 259–274

Chenopodium
　album 264
　berlandieri 265
　capitatum 266
　fremontii 267
　gigantospermum 270
　glaucum 268
　hybridum 270
　hybridum ssp. *gigantospermum* 270
　hybridum var. *gigantospermum* 270
　rubrum 269
　simplex 270

Cherry
　Bird 461
　Choke 462
　Fire 461
　Pin 461
　Red-fruited Choke 462
　Wild Red 461

Chickweed 258
　Common 258
　Common Mouse-ear 250
　Field 249
　Mouse-ear 250
　Prairie 249
　Prairie Mouse-ear 249

Chicory 120

Chives 597
　Siberian Wild 597
　Wild 597

Chocolate-root 454

Choke Cherry 462
　(Chokecherry)
　Common 462
　Red-fruited 462

Christmas-green 56

Chrysanthemum
　leucanthemum 156

Chrysopsis
　hispida 150
　villosa 150

Chrysosplenium tetrandrum 487

Cicely, Smooth Sweet 84

Cichorium intybus 120

Cicuta
　bulbifera 81
　maculata 82

Cinna latifolia 636

Cinnamon, Sweet 522

Cinquefoil
　Bush 450
　Graceful 458
　Marsh 448
　Norwegian 459
　Rough 459
　Shrubby 450
　Silverweed 446
　Slender 458
　Tall 457
　Three-toothed 469
　White 457

Cirsium
　arvense 121
　flodmanii 122
　muticum 123
　vulgare 124

Clammy Cockle 256

Claspingleaf Pondweed 691

Clematis
　ligusticifolia 428
　tangutica 429

Clematis 429
　Western 428
　Western White 428

Climbing Buckwheat 395

Climbing False Buckwheat 401

Climbing Nightshade 507

Closed Bottle Gentian 331

Closed Gentian 331

Clover
　Alsike 322
　Dutch 324
　Field 321
　Hop 321
　Ladino 324
　Lesser Hop 321
　Low Hop 321
　Red 323
　Smaller Hop 321
　White 324
　Yellow Field 321

Club-moss
　Bristly 57
　Prickly-tree 58
　Stiff 57
　Trailing 56
　Tree 58

Club-moss—Lycopodiaceae 56–58

Club-rush
　Hard-stem 569
　River 570
　Saltmarsh 571
　Softstem 573

Index

Clustered Bur-reed 697
Clustered Field Sedge 553
Clustered Marsh Ragwort 173
Clustered Sunflower 146
Coast Blite 269
Cockle
  Clammy 256
  Sticky 256
  White 255
Cocklebur 198
  Rough 198
Cock's-head 297
Coeloglossum
  bracteatum 610
  viride 610
  viride var. *virescens* 610
  viride var. *viride* 610
Colorado Rubber-plant 152
Colorado Rubberweed 152
Colt's-foot (Coltsfoot)
  Arrowleaf Sweet 167
  Arrow-leaved 167
  Palmate 166
  Palmate-leaved 166
  Sweet 167
Columbine
  Blue 425
  Canadian 426
  Red 426
  Small-flowered 425
  Wild 426
Comandra, Pale 486
Comandra
  pallida 486
  richardsiana 486
  umbellata 486
  umbellata ssp. *pallida* 486
  umbellata ssp. *umbellata* 486
Comarum palustre 448
Common Amaranth 76
Common Arrowhead 525
Common Asparagus 599
Common Bearberry 284
Common Beggarticks 117
Common Bindweed 276
Common Bladderwort 360
Common Blue-eyed Grass 580
Common Blue Lettuce 155
Common Broomweed 142
Common Buckthorn 442

Common Burdock 109
Common Buttercup 433
Common Canary Grass 666
Common Caragana 303
Common Cattail 700
Common Chickweed 258
Common Chokecherry 462
Common Cowparsnip 83
Common Dandelion 195
Common Dogmustard 221
Common Duckmeat 595
Common Duckweed 594
Common Evening Primrose 381
Common Fleabane 135
Common Gaillardia 139
Common Goat's-beard 196
Common Goldstar 600
Common Groundsel 174
Common Harebell 234
Common Hemp-nettle 347
Common Hop 235
Common Horsetail 49
Common Juniper 61
Common Knotweed 393
Common Labrador Tea 286
Common Linden 508
Common Mare's-tail 343
Common Milkweed 96
Common Moonseed 370
Common Motherwort 349
Common Mouse-ear Chickweed 250
Common Mullein 503
Common Nettle 511
Common Ninebark 456
Common Panic Grass 661
Common Pepper-grass 224
Common Pink Wintergreen 417
Common Plantain 385
Common Poison-ivy 78
Common Purslane 407
Common Ragweed 103
Common Red Paintbrush 495
Common Red Raspberry 467
Common Reed 668

Common Reed Grass 668
Common Reed-grass 677
Common Rivergrass 677
Common Scouring-rush 51
Common Silverweed 446

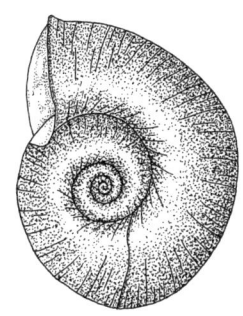

Common Skullcap 356
Common Smartweed 396
Common Sneezeweed 143
Common Solomon's-seal 605
Common Sow-thistle 182
Common Spike-rush 567
Common Starwort 258
Common Sunflower 144
Common Sweet Grass 650
Common Tansy 193
Common Threesquare 572
Common Tickseed 126
Common Timothy 667
Common Water Hemlock 82
Common Water-parsnip 87
Common Water-plantain 523
Common Water-starwort 231
Common Wild Rose 465
Common Witchgrass 661
Common Yarrow 100
Common Yellow Wood-sorrel 383
Composite Dropseed 683
Coneflower
  Cutleaf 172
  Green-headed 172
  Long-headed 170
  Prairie 170
  Purple 130
  Tall 172
  Upright Prairie 170
Convolvulaceae—Morning-glory 275–276

755

Index

Convolvulus
   arvensis 276
   sepium 275

Conyza
   canadensis 125
   canadensis var. *canadensis* 125

Coptis
   groenlandica 430
   trifolia 430

Corallorhiza
   maculata 611
   striata 612
   trifida 613

Coralroot
   Early 613
   Hooded 612
   Large 611
   Northern 613
   Pale 613
   Spotted 611
   Striped 612
   Summer 611
   Yellow 613

Cord Grass
   Alkali 681
   Freshwater 682
   Prairie 682

Coreopsis
   Garden 126
   Plains 126

Coreopsis tinctoria **126**

Cornaceae—Dogwood 277–278

Cornel, Dwarf 277

Corn-mustard 227

Cornus
   alba 278
   canadensis 277
   sericea 278
   sericea ssp. *sericea* 278
   stolonifera 278

Coronilla varia 304

Corydalis
   Golden 328
   Pale 329
   Pink 329

Corydalis
   aurea 328
   sempervirens 329

Corylus
   americana 207
   cornuta **208**

Cosmopolitan Bulrush 571

Cotton Burdock 110

Cotton-grass
   Alpine 577
   Slender 568

Cottongrass Bulrush 574

Cottonwood 477
   Eastern 477
   Plains 477
   Western 477

Couch Grass 642

Cowbane, Spotted 82

Cowberry 289

Cow Lily 374

Cow-parsnip 83
   (Cowparsnip)
   American 83
   Common 83

Cowslip 427

Cow Vetch 326

Cranberry
   Dry-ground 289
   Hog 284

Cranberrybush, European 246

Crane's-bill
   Bicknell's 336
   Northern 336

Crataegus
   chrysocarpa **449**
   columbiana var. *chrysocarpa* 449
   rotundifolia 449

Cream-colored Vetchling 309

Cream Pea 309

Creamy Peavine 309

Creeper
   Large-toothed Virginia 519
   Thicket 519

Creeping Bellflower 233

Creeping Bentgrass 623

Creeping Blackberry 468

Creeping Bluebell 233

Creeping Buttercup 436

Creeping Cedar 62

Creeping Charlie 348

Creeping Juniper 62

Creeping Love Grass 645

Creeping Savin 62

Creeping Snowberry 285

Creeping Spearwort 436

Creeping Spike-rush 567

Creeping Wild Rye 642

Crepis
   runcinata **127**
   tectorum **128**

Cress
   Bitter 217
   Bog Marsh 226
   Marsh 226
   Marsh Yellow 226
   Purple Rock 214
   Reflexed Rock 215
   Yellow 226

Crested Wheatgrass 622

Crocus, Prairie 431

Crocus Anemone 431

Crowfoot
   Bristly 437
   Cursed 439
   Seaside 435
   Whitewater 434

Crowfoot Violet 517

Crown-vetch
   (Crownvetch, Crown Vetch 304)
   Field 304
   Purple 304

Crunch-weed 227

Cucumber, Wild 279

Cucumber—Cucurbitaceae 279

Cucurbitaceae—Cucumber 279

Culver's-root 506

Cuman Ragweed 104

Cupressaceae—Cypress 61–63

Curled Dock 404

Curly-cup 141

Curly-cup Gumweed 141

Curly Dock 404

Curly-top Gumweed 141

Curlytop Knotweed 397

Index

Currant
  American Black 337
  Canadian Black 339
  Hudson Bay 339
  Northern Black 339
  Red 341
  Red Swamp 341
  Skunk 338
  Swamp 341
  Swamp Red 341
  Wild Black 337, 339
  Wild Red 341
Currant—Grossulariaceae 337–341
Cursed Buttercup 439
Cursed Crowfoot 439
Cut Grass 653
  Rice 653
Cut-leaf Avens 453
Cutleaf Coneflower 172
Cutleaf Ironplant 160
Cut-leaf Water-horehound 350
Cut-leaved Anemone 423
Cylactis
  arctica ssp. *acaulis* 466
  pubescens 468
Cyperaceae—Sedge 529–577
Cyperus
  aristatus 565
  inflexus 565
  squarrosus 565
Cyperus, Awned 565
Cyperus-like Sedge 554
Cypress—Cupressaceae 61–63
Cypress-like Sedge 554
Cypripedium
  calceolus ssp. *parviflorum* 615
  calceolus var. *parviflorum* 615
  candidum 614
  parviflorum 615

D

Dactylis glomerata 637
Daisy, Ox-eye 156
Daisy Fleabane 136
Dalea
  candida 305
  candida var. *candida* 305
  purpurea 306
  purpurea var. *purpurea* 306
Dame's-Rocket 223

Dame's Violet 223
Dandelion 195
  Common 195
  False 102
  Large-flowered False 102
  Red-seeded 194
Dasiphora
  floribunda 450
  fruticosa 450
Deadly Nightshade 507
Deathcamas, Mountain 607
Deciduous Traveler's-joy 428
Deervetch, Birdfoot 312
Delight, Single 415
Deschampsia caespitosa 638
Descurainia
  incana 218
  incana ssp. *incana* 218
  richardsonii 218
  sophia 219
Devil's Beggarticks 117
Devil's Pitchfork 117
Dewberry 468
Dewey's Sedge 537
Diamond Willow 479, 482
Dichanthelium, Leiberg 639
Dichanthelium leibergii 639
Diervilla lonicera 237
Diffuse Knapweed 119
Diphasiastrum complanatum 56
Disc Mayweed 161
Dock
  Curled 404
  Curly 404
  Golden 405
  Sour 402
  Triangular-valved 406
  Western 403
  White 406
  Willow 406
  Yellow 404
Dockleaf Smartweed 397
Doellingeria
  umbellata 129
  umbellata var. *pubens* 129
  umbellata var. *umbellata* 129
Dogbane
  Indian Hemp 91
  Prairie 91
  Spreading 90
Dogbane—Apocynaceae 90–91

Dog Mustard 221
Dogmustard, Common 221
Dog Nettle 347
Dogwood
  Bunchberry 277
  Dwarf 277
  Red-osier 278
Dogwood—Cornaceae 277–278
Doorweed 393
Dotted Blazingstar 158
Dotted Gayfeather 158

Downy Arrowwood 247
Downy Lyme Grass 654
Downy Ryegrass 654
Downy Yellow Violet 518
Draba, Woodland 220
Draba nemorosa 220
Dracocephalum
  parviflorum 346
  virginianum 355
Dragon, Water 526
Dragonhead 346
  American 346
  False 355
Drooping Woodreed 636
Dropseed
  Annual 686
  Composite 683
  Head-like 683
  Prairie 685
  Puffsheath 686
  Rough 683
  Sand 684
  Small 686
Dry-ground Cranberry 289
Dry-spike Sedge 541
Duckmeat, Common 595
Duck-potato 525

Index

Duckweed
  Common 594
  Ivy-leaved 593
  Larger 595
  Lesser 594
  Star 593
Duckweed—Lemnaceae 593–595
Dutch Clover 324
Dwarf Birch 206
  Tundra 203
Dwarf Blackberry 468
Dwarf Canadian Primrose 413
Dwarf Cornel 277
Dwarf Dogwood 277
Dwarf False Indigo 295
Dwarf Indigo-bush Amorpha 295
Dwarf Juniper 61
Dwarf Milkweed 94
Dwarf Nagoonberry 466
Dwarf Primrose 413
Dwarf Raspberry 466
Dwarf Rattlesnake Plantain 616
Dwarf Red Raspberry 468
Dwarf Scouring-rush 53
Dwarf Snapdragon 496
Dwarf Wild Indigo 295

E

Early Blue Violet 514
Early Coralroot 613
Early Meadow-rue 441
Early Saxifrage 492
Early Wood Buttercup 432
Early Yellow Locoweed 317
Eastern Arborvitae 63
Eastern Balsam Poplar 476
Eastern Cottonwood 477
Eastern Star Sedge 555
Eastern White Cedar 63
Eastern Willow Aster 189
Eastern Yellow Star-grass 600
Echinacea
  angustifolia **130**
    angustifolia var. *angustifolia* 130
    pallida var. *angustifolia* 130
***Echinochloa crus-galli* 640**

Echinocystis lobata 279
Eggs, Scrambled 328
Elaeagnaceae—Oleaster 280–283
Elaeagnus
  angustifolia 280
  commutata 281
Elder
  Box 72
  Red 242
  Red-berried 242
  Scarlet 242
Elderberry
  Red 242
  Stinking 242
Eleocharis
  acicularis **566**
  palustris **567**
  smallii 567
Elm
  American 509
  White 509
Elm—Ulmaceae 509
Elymus
  canadensis **641**
  innovatus 654
  junceus 672
  repens **642**
  smithii 664
  virginicus **643**
Elytrigia
  repens 642
  smithii 664
English Rye Grass 655
Epilobium
  adenocaulon 379
  angustifolium 378
  ciliatum **379**
    ciliatum ssp. *ciliatum* 379
    ciliatum ssp. *glandulosum* 379
    glandulosum 379
Equisetaceae—Horsetail 49–55
Equisetum
  arvense **49**
  fluviatile **50**
  hyemale **51**
  laevigatum **52**
  scirpoides **53**
  sylvaticum **54**
  variegatum **55**
    variegatum var. *variegatum* 55
Eragrostis
  cilianensis **644**
  hypnoides **645**
  megastachya 644

Ericaceae—Heath 284–289
Erigeron
  acris **131**
    acris ssp. *debilis* 131
    acris ssp. *politus* 131
    annuus ssp. *strigosus* 136
  caespitosus **132**
    canadensis 125
  glabellus **133**
  lonchophyllus **134**
  philadelphicus **135**
    ramosus 136
  strigosus **136**
    strigosus var. *strigosus* 136
Eriogonum flavum 389
Eriophorum
  alpinum **577**
  gracile **568**
Erodium cicutarium 335
Erucastrum gallicum 221
Erysimum cheiranthoides 222
Eupatorium maculatum 137
Euphorbia
  esula **292**
    esula var. *uralensis* 292
  glyptosperma **290**
  serpyllifolia **291**
    virgata 292
Euphorbiaceae—Spurge 290–292
European Buckthorn 442
European Cranberrybush 246
European Hop 235
European Stickseed 210
Eurotia lanata 272
Euthamia
  graminifolia **138**
    graminifolia var. *graminifolia* 138

758

Index

Evening-primrose
(Evening Primrose 381)
Common 381
Nuttall's 382
Shrubby 377
White 382
White-stemmed 382
Yellow 377, 381

Evening-primrose—
Onagraceae 377–382

Everlasting
Pearly 106
Small-leaved 107
Western Pearly 106

Eyelash Grass 629

F

Fabaceae—Pea 294–326

Fagaceae—Beech 327

Fagopyrum
esculentum **390**
sagittatum 390

Fallopia
cilinodis 394
convolvulus 395
scandens 401

Fall Sneezeweed 143

False Buckwheat 401
Climbing 401

False Dandelion 102

False Dragonhead 355

False Indigo
Dwarf 295
Fragrant 295

False Lily-of-the-valley 602

False London Rocket 229

False Melic 675

False Melic Grass 675

False Ragweed 153

False Solomon's-seal 604

False Sunflower 149
Rough 149

False Toadflax 486

Farewell Summer 190

Fat-hen 262

Felwort 332
Marsh 334

Fen Grass-of-parnassus 490

Fennel-leaved Pondweed 693

Feverfew 444

Fiddleleaf Hawksbeard 127

Field Bindweed 276

Field Chickweed 249

Field Clover 321
Yellow 321

Field Crown-vetch 304

Field Horsetail 49

Field Locoweed 317

Field Milk-vetch 297

Field Mint 353

Field Pennycress 230

Field Pussytoes 108

Field Sagewort 113

Field Sedge 553

Field Sow Thistle 181

Figwort—Scrophulariaceae 493–506

Filaria 335

Filbert, American 207

Fineleaf Pondweed 692

Fir, Balsam 65

Fire-berry Hawthorn 449

Fire Cherry 461

Fireweed 271, 378
Mexican 271
Narrow-leaf 378

Five-finger, Marsh 448

Fivefingers, Shrubby 469

Flag
Blue 579
Harlequin Blue 579
Larger Blue 579
Northern Blue 579
Sweet 522
Water 578

Flannel Mullein 503

Flannel-plant 503

Flat-leaved Bladderwort 359

Flatsedge, Bearded 565

Flat-stem Blue Grass 669

Flattened Rush 584

Flat-top Aster 129

Flat-top-goldenrod, Prairie 162

Flat-top Goldentop 138

Flat-topped Goldenrod 138

Flat-topped White Aster 129

Flax
Blue 362
Large-flowered Yellow 363
Lewis Wild 362
Prairie 362
Stiff 363
Stiffstem 363
Wild Blue 362
Yellow 363

Flax—Linaceae 362–363

Fleabane
Bitter 131
Canada 125
Common 135
Daisy 136
Hirsute 134
Marsh 173
Northern Daisy 131
Philadelphia 135
Prairie 136
Short-ray 134
Smooth 133
Spear-leaved 134
Streamside 133
Tufted 132

Flixweed 219

Floating-leaf Pondweed 690

Floating Pondweed 690

Flodman's Thistle 122

Flower
Blanket 139
Monkey 499
Pasque 431
Torch 455
Yellow Moccasin 615

Flowering Rush 527

Flowering Rush—Butomaceae 527

Fly Honeysuckle 241

Four-o'clock, Hairy 373

Four-o'clock—Nyctaginaceae 373

Fowl Blue Grass 670

Fowl Manna Grass 647

759

Index

Fowl Meadow Grass 670
Foxglove, Slender-leaved False 494
Foxtail
    Green 679
    Short-awned 624
    Short-awned Meadow 624
    Water 624
    Yellow 678
Foxtail Barley 651
Fragaria
    glauca 451
    ***virginiana* 451**
    virginiana ssp. *glauca* 451
Fragrant Bedstraw 473
Fragrant False Indigo 295
Fraxinus
    nigra 375
    pennsylvanica 376
Fremont's Goosefoot 267
Freshwater Cord Grass 682
Fringed American-aster 185
Fringed Aster 185
Fringed Black Bindweed 394
Fringed Brome 630
Fringed Gentian 333
    Lesser 333
    Narrow-leaved 333
    Small 333
Fringed Loosestrife 411
Fringed Orchid
    Great Plains White 619
    Western Prairie 619
Fringed Quickweed 140
Fringed Willowherb 379
Fringed Yellow Loosestrife 411
Frog Orchid 610
Fumariaceae—Fumitory 328–329
Fumitory—Fumariaceae 328–329
Fuzzyspike Wild Rye 654

G

Gaillardia 139
    Common 139
    Great-Flowered 139
Gaillardia aristata 139
Galeopsis
    bifida 347
    tetrahit var. *bifida* 347
Galinsoga 140

Galinsoga
    ciliata 140
    ***quadriradiata* 140**
Galium
    ***boreale* 471**
    septentrionale 471
    ***trifidum* 472**
    ***triflorum* 473**
Garden Asparagus 599
Garden Atriplex 261
Garden Bird's-foot Trefoil 312
Garden Buckwheat 390
Garden Coreopsis 126
Garden Orache 261
Garden Sorrel 402
Gardner's Saltbush 260
***Gaultheria hispidula* 285**
Gaura, Scarlet 380
Gaura coccinea 380
Gayfeather
    Dotted 158
    Strap-style 157
Gentian
    Autumn Dwarf 332
    Bottle 331
    Closed 331
    Closed Bottle 331
    Fringed 333
    Lesser Fringed 333
    Narrow-leaved Fringed 333
    Northern 330, 332
    Oblong-leaved 330
    Pleated 330
    Prairie 330
    Small Fringed 333
Gentiana
    acuta 332
    ***affinis* 330**
    amarella 332
    ***andrewsii* 331**
    crinita var. *browniana* 333
    procera 333
Gentianaceae—Gentian 330–334
Gentianella
    amarella 332
    amarella ssp. *acuta* 332
    crinita ssp. *procera* 333
Gentian—Gentianaceae 330–334
Gentianopsis
    procera 333
    virgata 333
Geraniaceae—Geranium 335–336

Geranium
    Bicknell's 336
    Bicknell's Wild 336
Geranium bicknellii 336
Geranium—Geraniaceae 335–336
Gerardia, Slender 494
Gerardia tenuifolia 494
Geum
    ***aleppicum* 452**
    ***macrophyllum* 453**
    macrophyllum var. *perincisum* 453
    perincisum 453
    ***rivale* 454**
    ***triflorum* 455**
Giant Bur-reed 696
Giant Goldenrod 176
Giant-hyssop 345
    (Giant Hyssop)
    Blue 345
Giant Ragweed 105
Giant Red Indian Paintbrush 495
Giant-seed Goosefoot 270
Giant Sumpweed 153

Gill-over-the-ground 348
Ginseng—Araliaceae 92
Glandular Birch 203
Glaucous Grass-of-parnassus 490
Glaucous Honeysuckle 239
Glaucous White Lettuce 169
Glaux maritima 410
Glechoma hederacea 348
Globemallow, Scarlet 369
Glow, Golden 172
Glyceria
    grandis 646
    striata 647

Index

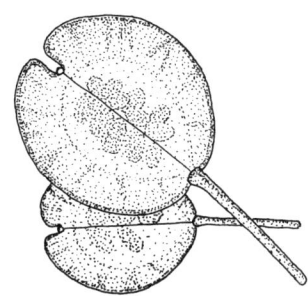

Glycyrrhiza lepidota 307

Goat-chicory, Pale 102

Goat's-beard (Goat's Beard 196)
    Common 196
    Yellow 196

Golden Alexanders 89

Golden-aster (Goldenaster)
    Hairy 150
    Hairy False 150

Golden-banner, Prairie 320

Golden-bean 320

Golden-carpet, Northern 487

Golden Corydalis 328

Golden Dock 405

Golden-fruit Sedge 533

Golden Glow 172

Golden Groundsel 164

Golden-hardhack 450

Golden Ragwort 164

Goldenrod
    Canada 175
    Canadian 175
    Flat-topped 138
    Giant 176
    Graceful 175
    Gray 180
    Hardleaf Flattop 163
    Lance-leaved 138
    Late 176
    Low 177
    Missouri 177
    Prairie 177
    Rigid 163
    Showy 180
    Soft 179
    Stiff 163
    Upland White 162
    Velvety 179

Golden Saxifrage 487
    (Golden-saxifrage)
    Northern 487

Golden Sedge 533

Golden Tickseed 126

Golden Tickweed 126

Goldentop, Flat-top 138

Golden Zizia 89

Goldstar, Common 600

Goldthread 430
    Three-leaf 430

Goodyera repens 616

Gooseberry
    Canada 340
    Canadian 340
    Northern 340
    Wild 340

Goosefoot
    Big-seed 270
    Blite 266
    Fremont's 267
    Giant-seed 270
    Maple-leaved 270
    Oak-leaf 268
    Pit-seed 265
    Red 269
    Saline 268
    Spear-leaved 273

Goosefoot—Chenopodiaceae 259–274

Graceful Cinquefoil 458

Graceful Goldenrod 175

Graceful Sedge 542, 553

Grama
    Blue 629
    Side-oats 628

Granular Sedge 543

Grape—Vitaceae 519

Grass
    Alkali Cord 681
    American Manna 646
    Annual Canary 666
    Barnyard 640
    Blue-eyed 580
    Blue Grama 629
    Canada Blue 669
    Canary 666
    Cattail 678
    Common Blue-eyed 580
    Common Canary 666
    Common Panic 661
    Common Reed 668
    Common Sweet 650
    Couch 642
    Creeping Love 645
    Cut 653
    Downy Lyme 654
    English Rye 655

[Grass]
    Eyelash 629
    False Melic 675
    Flat-stem Blue 669
    Fowl Blue 670
    Fowl Manna 647
    Fowl Meadow 670
    Freshwater Cord 682
    Green Foxtail 679
    Green Needle 659
    Green Tussock 659
    Indian 680
    June 652
    Kentucky Blue 671
    Large Barnyard 640
    Leiberg's Rosette 639
    Marsh Reed 633
    Northern Reed 634
    Northern Sweet 650
    Nuttall's Salt-meadow 674
    Orchard 637
    Perennial Rye 655
    Porcupine 649
    Poverty 686
    Prairie Cord 682
    Prairie June 652
    Prairie Koeler's 652
    Purple Oat 675
    Quack 642
    Reed Canary 665
    Rice Cut 653
    Sand 635
    Sand Reed 635
    Scratch 656
    Slender Salt-meadow 673
    Slender Wood 636
    Slough 627, 682
    Smooth Love 645
    Spear 648
    Sweet 650
    Switch 663
    Tall Manna 646
    Thatch 677
    Tufted Hair 638
    Vanilla 650
    Wand Panic 663
    White-grained Mountain Rice 660
    Witch 661
    Yellow Indian 680

Grass-of-parnassus (Grass of Parnassus 490)
    Fen 490
    Glaucous 490
    Marsh 491
    Mountain 491
    Northern 491

Grass—Poaceae 621–687

Index

Grassy Rush 527
Gray Alder 201
Gray Goldenrod 180
Gray Pine 69
Gray Ragwort 165
Gray Tansy Mustard 218
Gray Willow 479
Great Blanketflower 139
Great Bulrush 573
Greater Bladder Sedge 546
Greater Bladderwort 360
Greater Butter-and-eggs 498
Greater Creeping Spearwort 436
Great-flowered Gaillardia 139
Great Mullein 503
Great Plains White Fringed Orchid 619
Great Ragweed 105
Great Solomon's-seal 605
Great Willow-herb 378
Green Alder 202

Green Amaranth 76
Green Ash 376
Greenbrier—Smilacaceae 694
Green Bristlegrass 679
Green-flowered Bog Orchid 618
Greenflowered Wintergreen 418
Green Foxtail 679
Green Foxtail Grass 679
Green-fruited Bur-reed 695
Green-headed Coneflower 172
Greenish-flowered Pyrola 418
Greenish-flowered Wintergreen 418
Greenland Labrador-tea 286

Green Needle Grass 659
Green Saxifrage 487
Green Sedge 563
Green Sorrel 402
Green Tussock Grass 659
Green Wintergreen 418
Grindelia squarrosa 141
Groovebur, Tall Hairy 444
Grossulariaceae—Currant 337–341
Ground-cedar (Groundcedar) 56, 58
Ground-ivy 348
Ground Juniper 61
Ground Milkvetch 300
Ground-pine 58
  (Groundpine)
  Stiff 57
  Trailing 56
  Tree 58
Ground-plum 300
Groundsel 174
  Common 174
  Golden 164
  Silver-woolly 165
  Silvery 165
Grove Sandwort 252
Gumweed 141
  Curly-cup 141
  Curly-top 141
Gutierrezia
  diversifolia 142
  sarothrae 142
Gypsophila, Babysbreath 251
Gypsophila paniculata 251

H

Habenaria
  hyperborea 618
  leucophaea var. *praeclara* 619
  viridis 610
Hackelia
  americana 209
  ***deflexa* 209**
  deflexa var. *americana* 209
Hackmatack 66
Hair Grass, Tufted 638
Hair-like Sedge 536
Hairy False Goldenaster 150
Hairy Four-o'clock 373

Hairy Golden-aster 150
Hairy Hedgenettle 358
Hairy Speedwell 505
Hairy Umbrellawort 373
Hairy Wild Rye 654
Hairy Wood-rush 589
Halberd-leaved Saltbush 262
Haloragaceae—Water-Milfoil 342
Haplopappus spinulosus 160
Hardleaf Flattop Goldenrod 163
Hard Maple 73
Hard-stem Bulrush 569
Hard-stem Club-rush 569
Harebell 234
  Common 234
Harlequin Blue Flag 579
Harlequin-flower 329
Hawk's-beard (Hawksbeard)
  Annual 128
  Fiddleleaf 127
  Narrow-leaved 128
  Scapose 127
  Smooth 127
Hawkweed
  Canada 151
  Narrow-leaved 151
Hawthorn
  Fire-berry 449
  Northern 449
  Round-leaved 449
Hay Sedge 541
Hazelnut 207
  American 207
  Beaked 208
Head-like Dropseed 683
Heart-leaf Alexanders 88
Heartleaf Twayblade 617
Heartleaf Willow 482
Heart-leaved Alexanders 88
Heart-leaved Twayblade 617
Heath Aster 186, 187
Heath—Ericaceae 284–289
Hedge Bindweed 275
Hedge False Bindweed 275
Hedge-nettle
  (Hedgenettle 358)
  Hairy 358
  Marsh 358
  Swamp 358

762

Index

Hedyotis longifolia 474
Hedysarum
    Alpine 308
    American 308
Hedysarum alpinum 308
Helenium autumnale 143
Helianthus
    annuus 144
    maximiliani 145
    nuttallii 146
    pauciflorus 147
    tuberosus 148
Heliopsis helianthoides 149
Hemlock Waterparsnip 87
Hemp-nettle 347
    Brittle-stem 347
    Common 347
    Split-lip 347
Heracleum
    lanatum 83
    maximum 83
    sphondylium ssp. *montanum* 83
    sphondylium var. *lanatum* 83
Herb Sophia 219
Hesperis matronalis 223
Hesperostipa
    comata 648
    comata ssp. *comata* 648
    spartea 649
Heterotheca
    villosa 150
    villosa var. *minor* 150
    villosa var. *villosa* 150
Heuchera richardsonii 488
Hieracium umbellatum 151
Hierochloë
    hirta ssp. *arctica* 650
    odorata 650
High Bush-cranberry 246
Hillside Sedge 541
Hippochaete variegata 55
Hippuridaceae—**Mare's-tail** 343
Hippuris vulgaris 343
Hirsute Fleabane 134
Hoary Puccoon 211
Hoary Willow 480
Hog Cranberry 284
Hog Millet 662
Hog-peanut 296
    American 296

Hogweed, Little 407
Holboell's Rockcress 215
Honeysuckle
    Blue Fly 241
    Fly 241
    Glaucous 239
    Limber 239
    Mountain 239
    Mountain Fly 241
    Northern 241
    Tatarian 240
    Twining 239
    Wild 239
Honeysuckle—Caprifoliaceae 236–247
Hooded Coralroot 612
Hooded Skullcap 356
Hood's Phlox 386
Hooked Agrimony 444
Hookedspur Violet 514
Hop 235
    Common 235
    European 235
Hop Clover 321
    Lesser 321
    Low 321
    Smaller 321
Hops 235
Hordeum jubatum 651
Horehound, One Flower 352
Horse-mint 354
Horsetail
    Common 49
    Field 49
    Scouringrush 51
    Smooth 52
    Swamp 50
    Sylvan's 54
    Variegated 55
    Water 50
    Wood 54
    Woodland 54
Horsetail—Equisetaceae 49–55
Horseweed 125
    Canadian 125
Houghton's Sedge 545
Houstonia, Long-leaved 474
Houstonia longifolia 474
Hudson Bay Bulrush 577
Hudson Bay Currant 339
Humulus lupulus 235

Hymenoxys richardsonii 152
Hypoxis hirsuta 600
Hyssop
    Blue Giant 345
    Lavendar 345

I

Impatiens
    biflora 199
    capensis 199
Indian Breadroot 319
    Large 319
    Silverleaf 318
Indian Grass 680
    Yellow 680
Indian-hemp 91
Indian Hemp Dogbane 91
Indianpaint 266
Indian Paintbrush
    Giant Red 495
    Red 495
Indian-pipe 372
    (Indianpipe)
    One-flower 372
Indian Pipe—Monotropaceae 372
Indian Wild Rice 687
Indigo
    Dwarf False 295
    Dwarf Wild 295
    Fragrant False 295
Iridaceae—Iris 578–580
Iris
    Pale-yellow 578
    Yellow 578
    Yellow Water 578
Iris
    pseudacorus 578
    versicolor 579
Iris—Iridaceae 578–580
Ironplant
    Cutleaf 160
    Spiny 160

Index

Iva xanthifolia 153
Ivy-leaved Duckweed 593

J

Jack Pine 69
Jasmine
  Northern Rock 409
  Western Rock 408
Jerusalem Artichoke 148
Jewelweed 199
Joe-pye Weed 137
  Spotted 137
Juncaceae—Rush 581–590
Juncaginaceae—Arrow-grass 591–592
Juncus
  alpinoarticulatus **581**
  alpinoarticulatus ssp.
    nodulosus 581
  alpinus 581
  arcticus var. *balticus* 582
  balticus **582**
  bufonius **583**
  compressus **584**
  longistylis **585**
  nodosus **586**
  torreyi **588**
Juneberry 445
  Alderleaf 445
June Grass 652
  Prairie 652
Juniper
  Common 61
  Creeping 62
  Dwarf 61
  Ground 61
  Low 61
  Mountain 61
  Trailing 62
Juniperus
  communis **61**
  horizontalis **62**

K

Kalm's Lobelia 364
Kentucky Blue Grass 671
Kidney-leaved Buttercup 432
Kindlingweed 142
King's-cureall 381
King Solomon's-seal 605

Kinnikinnick 284
Knapweed
  Diffuse 119
  White 119
Knotted Rush 586
Knotweed
  Bushy 399
  Common 393
  Curlytop 397
  Leathery 391
  Oval-leaf 393
  Striate 391
  Water 392
  Yellow-flowered 399
Kochia 271
Kochia scoparia **271**
Koeleria
  cristata 652
  gracilis 652
  macrantha **652**
  nitida 652
  pyramidata 652
Krascheninnikovia lanata **272**

L

Labrador Buttercup 438
Labrador-tea 286
  (Labrador Tea)
  Bog 286
  Common 286
  Greenland 286
  Rusty 286
Lactuca
  oblongifolia 155
  pulchella 155
  scariola 154
  serriola **154**
  tatarica **155**
  tatarica var. *pulchella* 155
Lacy Tansyaster 160
Ladino Clover 324

Lady's-slipper (Lady's Slipper)
  Small White 614
  White 614
  Yellow 615
Lady's-thumb 398
  (Ladysthumb)
  Spotted 398
Lake Mistassini Primrose 413
Lamb's-quarters 264
Lamiaceae—Mint 344–358
Lance-leaved Goldenrod 138
Laportea canadensis **510**
Lappula
  americana 209
  echinata 210
  myosotis 210
  squarrosa 210
Larch 66
  American 66
Large Barnyard Grass 640
Large Coralroot 611
Large-flowered False
  Dandelion 102
Large-flowered Yellow Flax 363
Large Indian Breadroot 319
Large-leaved Avens 453
Large-leaved Watercrowfoot 434
Larger Blue Flag 579
Larger Duckweed 595
**Large-toothed Virginia
  Creeper** 519
Large Water Plantain 523
Larix laricina **66**
Larkspur-violet 517
Late Goldenrod 176
Late Low Blueberry 287
Late Yellow Locoweed 317
Lathyrus
  ochroleucus 309
  palustris **310**
  venosus **311**
Lavender Hyssop 345
Leafless-bulrush, Alpine 577
Leafy Pondweed 689
Leafy Spurge 292
Leathery Knotweed 391
Ledum groenlandicum **286**
Leersia oryzoides **653**

Leiberg Dichanthelium 639
Leiberg's Panicgrass 639
Leiberg's Rosette Grass 639
Lemna
   trisulca 593
   turionifera 594
Lemnaceae—Duckweed 593–595
Lentibulariaceae—Bladderwort 359–361
Leonurus cardiaca 349
Lepidium
   densiflorum 224
   ramosissimum 225
Lesser Bladderwort 361
Lesser Burdock 109
Lesser Duckweed 594
Lesser Fringed Gentian 333
Lesser Hop Clover 321
Lesser Panicled Sedge 538
Lesser Rattlesnake-orchid 616
Lesser Rattlesnake-plantain 616
Lesser Spearwort 436
Lesser Tussock Sedge 538
Lettuce
   Blue 155
   Common Blue 155
   Glaucous White 169
   Lobed Prickly 154
   Prickly 154
   Russian Blue 155
   White 168
Leucanthemum vulgare 156
Lewis Wild Flax 362
Leymus innovatus 654
Liatris
   ligulistylis 157
   punctata 158
Licorice
   American 307
   Wild 307
Lilac Beardtongue 501
Lilac-flowered Beardtongue 501

Lilac Penstemon 501
Liliaceae—Lily 596–607
Lilium
   philadelphicum 601
   philadelphicum var. *andinum* 601
   umbellatum 601
Lily
   Cow 374
   Wild 601
   Wild Wood 601
   Wood 601
Lily—Liliaceae 596–607
Lily-of-the-valley (Lily of the Valley)
   False 602
   Starry False 603
   Threeleaf False 604
   Wild 602
Limber Honeysuckle 239
Limestone Meadow Sedge 543
Limosella aquatica 497
Linaceae—Flax 362–363
Linaria vulgaris 498
Linden
   American 508
   Common 508
Linden—Tiliaceae 508
Lindley's American-aster 185
Lindley's Aster 185
Lindley's Blue Aster 185
Lingonberry 289
Linnaea
   americana 238
   borealis 238
   borealis ssp. *americana* 238
Linum
   lewisii 362
   lewisii var. *lewisii* 362
   perenne var. *lewisii* 362
   rigidum 363
Lionsheart, Virginia 355
Listera cordata 617
Lithospermum canescens 211
Little Bluestem 676
Little False Bluestem 676
Little Green Sedge 563
Little Ground Rose 447
Little Hogweed 407
Littleleaf Buttercup 432
Littleleaf Pussytoes 107

Little Meadow-foxtail 624
Little Rose 447
Liverleaf Wintergreen 417
Lobed Prickly Lettuce 154
Lobelia
   Brook 364
   Kalm's 364
   Ontario 364
   Palespike 365
   Spiked 365
Lobelia
   kalmii 364
   spicata 365
Lobeliaceae—Lobelia 364–365
Lobelia—Lobeliaceae 364–365
Locoweed
   Early Yellow 317
   Field 317
   Late Yellow 317
   Northern Yellow 317
Lolium perenne 655
Lomatogonium rotatum 334
Long-beaked Sedge 558
Long-bracted Orchid 610
Longbract Frog Orchid 610
Long-fruited Anemone 422
Long-headed Anemone 422
Long-headed Coneflower 170
Long-headed Thimbleweed 422
Longleaf Starwort 257
Longleaf Summer Bluet 474
Long-leaf Willow 483
Long-leaved Bluets 474
Long-leaved Houstonia 474
Long-leaved Stitchwort 257
Long-plumed Purple Avens 455
Long-styled Rush 585
Lonicera
   caerulea var. *villosa* 241
   dioica 239
   tatarica 240
   villosa 241
   villosa var. *villosa* 241
Loosestrife
   Fringed 411
   Fringed Yellow 411
   Purple 366
   Tufted 412
Loosestrife—Lythraceae 366

Index

Lotus corniculatus 312
Lousewort, Swamp 500
Love Grass (Lovegrass)
  Creeping 645
  Smooth 645
  Teal 645
Lowbush Blueberry 287
Low Bush-cranberry 244
Low Goldenrod 177
Low Hop Clover 321
Low Juniper 61
Low Mallow 368
Low Prairie Rose 464
Low Sea Blite 274
Low Sweet Blueberry 287
Lucerne 314
Lungwort, Tall 212
Luzula
  acuminata 589
  acuminata var. *acuminata* 589
  parviflora 590
  pilosa 589
Lychnis alba 255
Lycopodium
  annotinum 57
  complanatum 56
  dendroideum 58
  obscurum var. *dendroideum* 58
Lycopus
  americanus 350
  asper 351
  uniflorus 352
  virginicus var. *pauciflorus* 352
Lygodesmia juncea 159
Lyme Grass, Downy 654
Lysimachia
  ciliata 411
  thyrsiflora 412
Lythraceae—Loosestrife 366
Lythrum salicaria 366

M

Machaeranthera
  pinnatifida 160
  pinnatifida var. *pinnatifida* 160
Madder—Rubiaceae 471–474
Mad-dog Skullcap 357
Maianthemum
  canadense 602
  stellatum 603
  trifolium 604
Maidenhair 455
Mallow
  Cheeseweed 367
  Low 368
  Red False 369
  Round-leaved 368
  Scarlet 369
  Small-flowered 367
  Small-fruited 367
  Small-whorl 367
Mallow—Malvaceae 367–369
Malva
  parviflora 367
  pusilla 368
  rotundifolia 368
Malvaceae—Mallow 367–369
Malvastrum coccineum 369
Manitoba Maple 72
Manna Grass
  American 646
  Fowl 647
  Tall 646
Mannikin Twayblade 617
Manybranched Pepperweed 225
Many-flowered Aster 186
Many-flowered Yarrow 101
Manzanita, Bearberry 284
Maple
  Ashleaf 72
  Hard 73
  Manitoba 72
  Mountain 74
  River 73
  Silver 73
  Soft 73
  Sugar 73
  White 74
Maple—Aceraceae 72–74
Maple-leaved Goosefoot 270
Mare's-tail 343
  Common 343

Mare's-tail—Hippuridaceae 343
Marsh Arrow-grass 592
Marsh Aster 183
Marsh Bellflower 232
Marsh Betony 358
Marsh Calla 526
Marsh Cinquefoil 448
Marsh Cress 226
  Bog 226
Marsh Felwort 334
Marsh Five-finger 448
Marsh Fleabane 173
Marsh Grass-of-parnassus 491
Marsh Hedge-nettle 358
Marshlocks, Purple 448
Marsh-marigold 427
  Yellow 427
Marsh Pea 310
Marsh-pepper Smartweed 396
Marsh Ragwort 173
  Clustered 173
Marsh Reed Grass 633
Marsh Skullcap 356
Marsh-trefoil 371
Marsh Valerian 512
Marsh Vetchling 310
Marsh Yellow Cress 226
Maryland Black Snakeroot 86
Maryland Sanicle 86
Mat Amaranth 75
Mat Muhly 658
Mat Pigweed 75
Matricaria
  discoidea 161
  maritima ssp. *inodora* 197
  maritima var. *agrestis* 197
  matricarioides 161
  perforata 197
Matted Muhly 658
Matweed 75
Maximillian's Sunflower 145
Mayflower, Canada 602
Maystar 414
Mayweed
  Disc 161
  Scentless False 197
McCalla's Willow 484

Index

Meadow Anemone 421
Meadow Blazingstar 157
Meadow Buttercup 433
Meadow-foxtail, Little 624
Meadow Grass, Fowl 670
Meadow-parsnip 88
Meadow-rue
  Early 441
  Tall 440
  Veiny 441
  Veiny-leaf 441
Meadowsweet 470
  Narrow-leaved 470
  White 470
Meadow Willow 485
Meadow Zizia 88
Medicago
  falcata 314
  lupulina 313
  sativa 314
  sativa ssp. *falcata* 314
Medick, Black 313
Melic, False 675
Melilotus
  alba 315
  officinalis 316

Menispermaceae—Moonseed 370
Menispermum canadense 370
Mentha arvensis 353
Menyanthaceae—Buck-bean 371
Menyanthes trifoliata 371
Mertensia, Tall 212
Mertensia paniculata 212
Mexican Fireweed 271
Mexican-hat, Red-spike 170
Michaelmas-daisy 145
Mild Water-pepper 396
Milfoil 100
  American 342

Milk-vetch 299 (Milkvetch)
  Canadian 299
  Field 297
  Ground 300
  Missouri 301
  Narrow-leaved 302
  Purple 297
  Tine-leaved 302
  Two-grooved 298
Milkweed
  Common 96
  Dwarf 94
  Oval-leaved 94
  Showy 95
  Silky 96
  Swamp 93
  Whorled 97
Milkweed—Asclepiadaceae 93–97
Milkwort—Polygalaceae 387
Millet
  Broomcorn 662
  Hog 662
  Panic 662
  Proso 662
Mimulus ringens 499
Miner's Peppergrass 224
Mint
  American Wild 353
  Field 353
  Wild 353
Mint—Lamiaceae 344–358
Mirabilis hirsuta 373
Missouri Goldenrod 177
Missouri Milk-vetch 301
Missouri Willow 482
Mistassini Primrose 413
Mitella nuda 489
Miterwort, Naked 489
Moccasin Flower, Yellow 615
Moehringia lateriflora 252
Moldavica parviflora 346
Monarda fistulosa 354
Moneses uniflora 415
Monkeyflower
  (Monkey-flower, Monkey Flower 499)
  Alleghany 499
  Blue 499
  Square-stemmed 499
Monolepis nuttalliana 273
Monotropaceae—Indian Pipe 372

Monotropa uniflora 372
Moonseed 370
  Canadian 370
  Common 370
Moonseed—Menispermaceae 370
Mooseberry 244
Morning-glory, Wild 275
Morning-glory—Convolvulaceae 275–276
Moss Phlox 386
Mossycup Oak 327
Mother-of-the-evening 223
Motherwort 349
  Common 349
Moundscale 260
Mountain Alder 202
Mountain Birch 204
Mountain-cranberry, Northern 289
Mountain Deathcamas 607
Mountain Fly Honeysuckle 241
Mountain Grass-of-parnassus 491
Mountain Honeysuckle 239
Mountain Juniper 61
Mountain Maple 74
Mountain Rice Grass, White-grained 660
Mountain Sneezeweed 143
Mountain Tansymustard 218
Mouse-ear Chickweed 250
Mudwort 497
  Awl-leaved 497
  Water 497
Mugwort, Western 115
Muhlenbergia
  asperifolia 656
  cuspidata 657
  richardsonis 658
Muhly
  Alkali 656
  Mat 658
  Matted 658
  Plains 657
  Prairie 657
  Stony-hills 657
Mullein
  Common 503
  Flannel 503
  Great 503
  Woolly 503

Index

Musk Thistle 118
Mustard
  Charlock 227
  Dog 221
  Gray Tansy 218
  Small Tumbleweed 229
  Tall Hedge 229
  Tansy 218, 219
  Treacle 222
  Tumbleweed 228
  Tumbling 228
  Wild 227
  Wormseed 222
Mustard—Brassicaceae 213–230
Myriophyllum
  exalbescens 342
  ***sibiricum* 342**
  spicatum var. *capillaceum* 342
  spicatum ssp. *exalbescens* 342
  spicatum var. *exalbescens* 342
  spicatum ssp. *squamosum* 342
  spicatum var. *squamosum* 342

N

Nagoonberry, Dwarf 466
Naked Bishop's-cap 489
Naked Miterwort 489
Nannyberry 245
Narrow-leaf Bur-reed 695
Narrow-leaf Cat-tail 699
Narrow-leaf Fireweed 378
Narrowleaf Spirea 470
Narrow-leaved Cattail 699
Narrow-leaved Fringed Gentian 333
Narrow-leaved Hawk's-beard 128
Narrow-leaved Hawkweed 151
Narrow-leaved Meadowsweet 470
Narrow-leaved Milk-vetch 302
Narrow-leaved Poison-vetch 302
Narrow-leaved Sunflower 145
Narrow Reed-grass 634
Nassella viridula **659**
Naumburgia thyrsiflora 412
Neat Pussytoes 107
Necklace-weed 505
Neckweed 505
Needle and Thread 648
Needle Spike-rush 566

Nettle
  Bee 347
  Common 511
  Dog 347
  Stinging 511
  Wood 510
Nettle—Urticaceae 510–511
New England American-aster 191
New England Aster 191
Nightcaps 424
Night-flowering Catchfly 256
Nightflowering Silene 256

Nightshade 507
  Bittersweet 507
  Climbing 507
  Deadly 507
Nightshade—Solanaceae 507
Ninebark 456
  Common 456
Nodding Beggarticks 116
Nodding Brome 632
Nodding Burr-marigold 116
Nodding Plumeless-thistle 118
Nodding Smartweed 397
Nodding Stickseed 209
Nodding Thistle 118
Nodding Trillium 606
Nodding Wakerobin 606
Nodding Wild Rye 641
Northern Arrowhead 524
Northern Beaked Sedge 562
Northern Bedstraw 471
Northern Blackberry 466
Northern Black Currant 339
Northern Bluebells 212
Northern Blue Flag 579
Northern Bog Aster 183
Northern Bog Sedge 544

Northern Bog Violet 516
Northern Broad-leaved Water-plantain 523
Northern Bugleweed 352
Northern Bush-honeysuckle 237
Northern Coralroot 613
Northern Crane's-bill 336
Northern Daisy Fleabane 131
Northern Gentian 330, 332
Northern Golden-carpet 487
Northern Golden-saxifrage 487
Northern Gooseberry 340
Northern Grass-of-parnassus 491
Northern Green Bog-orchid 618
Northern Green Orchid 618
Northern Green Rush 581
Northern Hawthorn 449
Northern Honeysuckle 241
Northern Mountain-cranberry 289
Northern Pondweed 688
Northern Pygmy-flower 409
Northern Reed Grass 634
Northern Rock Jasmine 409
Northern Running-pine 56
Northern Scouring-rush 55
Northern Starflower 414
Northern Sweet Grass 650
Northern Valerian 512
Northern Water-carpet 487
Northern Water-horehound 352
Northern White Cedar 63
Northern Wild Rice 687
Northern Willowherb 379
Northern Yellow Locoweed 317
Northwest Territory Sedge 562
Norwegian Cinquefoil 459
Nuphar
  lutea ssp. *variegata* 374
  variegata 374
Nut-grass 565
  Awned 565
Nuttall's Alkaligrass 674
Nuttall's Atriplex 260
Nuttall's Evening-primrose 382
Nuttall's Povertyweed 273

Index

Nuttall's Salt-meadow Grass 674
Nuttall's Sunflower 146
Nyctaginaceae—Four-o'clock 373
Nymphaeaceae—Water-lily 374

O

Oak
    Bur 327
    Mossycup 327
Oak-leaf Goosefoot 268
Oat, Wild 626
Oblong-leaved Gentian 330
Oenothera
    biennis 381
    nuttallii 382
    serrulata 377
Oldman 111
Old-man-in-the-spring 174
Old Man's Whiskers 455
Oleaceae—Olive 375–376
Oleaster 280
Oleaster—Elaeagnaceae 280–283
Oligoneuron
    album 162
    rigidum 163
Olive, Russian 280
Olive—Oleaceae 375–376
Onagraceae—Evening-primrose 377–382
One-flowered Pyrola 415
One-flowered Wintergreen 415
One Flower Horehound 352
One-flower Indianpipe 372
One-sided Aster 190
One-sided Pyrola 416
One-sided Wintergreen 416
Onion
    Autumn 598
    Pink-flowered 598
    Pink Wild 598
    Prairie 598
    Wild 598
Ontario Lobelia 364
Orach, Thinleaf 262
Orache 262
    Garden 261
Orange Touch-me-not 199

Orchard Grass 637
Orchid
    Bracted 610
    Frog 610
    Great Plains White Fringed 619
    Green-flowered Bog 618
    Long-bracted 610
    Longbract Frog 610
    Northern Green 618
    Round-leaved 609
    Western Prairie Fringed 619
Orchidaceae—Orchid 608–619
Orchid—Orchidaceae 608–619
Orchis, Small Round-leaved 609
Orchis rotundifolia 609
Oriental Virgin's-bower 429
Orthilia secunda 416
Oryzopsis asperifolia 660
Osmorhiza
    aristata var. *longistylis* 84
    longistylis 84
Oswego-tea 354
Oval-leaf Knotweed 393
Oval-leaved Milkweed 94
Oxalidaceae—Wood-sorrel 383
Oxalis
    europaea 383
    stricta 383
Ox-eye 149 (Oxeye)
    Rough 149
    Smooth 149
    Sweet 149
Ox-eye Daisy 156
Oxytropis
    campestris 317
    campestris var. *varians* 317
    varians 317

P

Pacific Anemone 423
Pacific Wormwood 113
Packera
    aurea 164
    cana 165
Paintbrush
    Common Red 495
    Giant Red Indian 495
    Red Indian 495
Painted-cup, Red 495
Paiuteweed 274

Pale Agoseris 102
Pale Bulrush 576
Pale Comandra 486
Pale Coralroot 613
Pale Corydalis 329
Pale Goat-chicory 102
Pale-green Bulrush 576
Pale Persicaria 397
Pale Smartweed 397
Palespike Lobelia 365
Pale Spike-rush 567
Pale Vetchling 309
Pale-yellow Iris 578
Palmate Coltsfoot 166
Palmate-leaved Colt's-foot 166
Panicgrass (Panic Grass)
    Common 661
    Leiberg's 639
    Wand 663
Panicled Aster 189
Panicled Bulrush 575
Panic Millet 662
Panicum
    capillare 661
    leibergii 639
    miliaceum 662
    virgatum 663
Paper Birch 205
Parasol Whitetop 129
Parilla, Yellow 370
Parnassia
    glauca 490
    palustris 491
Parnassus, Grass of 490
Parry's Sedge 549
Parsnip, Wild 85

Index

Parthenocissus vitacea 519
Pascopyrum smithii 664
Pasque Flower 431 (Pasqueflower)
   American 431
Pastinaca sativa 85
Pasture Sage 114
Pasture Sagewort 114
Pea
   Cream 309
   Marsh 310
   Veiny 311
   Yellow 320
Pea—Fabaceae 294–326
Pearly Everlasting 106
Peashrub, Siberian 303
Peavine
   Creamy 309
   Purple 311
   Wild 311
Peck's Sedge 550
Pedicularis lanceolata 500
Pediomelum
   argophyllum 318
   esculentum 319
Pennsylvania Bittercress 217
Pennsylvania Buttercup 437
Pennsylvania Sedge 552
Pennycress, Field 230
Penstemon
   Lilac 501
   Waxleaf 502
Penstemon
   gracilis **501**
   nitidus **502**
Pentaphylloides floribunda 450

Pepper-grass (Peppergrass, Pepper Grass 224)
   Branched 225
   Bushy 225
   Common 224
   Miner's 224
   Prairie 224
Pepper-root 522
Pepperweed, Manybranched 225
Pepperwort, Branched 225
Perennial Ragweed 104
Perennial Rye Grass 655
Perennial Sow-thistle 181
Persicaria
   Pale 397
   Swamp 392
Persicaria
   amphibia 392
   hydropiper 396
   lapathifolia 397
   maculata 398
   sagittata 400
Petalostemon
   candidum 305
   candidus 305
   purpureum 306
   purpureus 306
Petalostemum (see *Petalostemon*)
Petasites
   frigidus **var.** *palmatus* **166**
   frigidus **var.** *sagittatus* **167**
   palmatus 166
   sagittatus 167
Phalaris
   arundinacea **665**
   canariensis 666
Philadelphia Fleabane 135
Phleum pratense **667**
Phlox
   Carpet 386
   Hood's 386
   Moss 386
   Spiny 386
Phlox hoodii **386**
Phlox—Polemoniaceae 386
Phragmites
   australis **668**
   communis 668
Physocarpus
   intermedius 456
   opulifolius **456**
   opulifolius var. *intermedius* 456

Physostegia
   formosior 355
   virginiana **355**
   virginiana ssp. *virginiana* 355
Picea
   glauca **67**
   mariana **68**
Pigweed 264
   Mat 75
   Prostrate 75
   Red 269
   Red-root 76
   Rough 76
   Russian 263
Pinaceae—Pine 65–69
Pin Cherry 461
Pine
   Banksian 69
   Gray 69
   Jack 69
   Scrub 69
Pineappleweed 161
Pine—Pinaceae 65–69
Pink—Caryophyllaceae 248–258
Pink Corydalis 329
Pink-flowered Onion 598
Pink Pyrola 417
Pink Wild Onion 598
Pink Wintergreen 417
   Common 417
Pinus banksiana **69**
Pitchfork, Devil's 117
Pit-seed Goosefoot 265
Plains Coreopsis 126
Plains Cottonwood 477
Plains Muhly 657
Plains Wormwood 113
Plains Yellow Primrose 377
Plantaginaceae—Plantain 384–385
Plantago
   eriopoda **384**
   major **385**
Plantain
   Alkali 384
   Common 385
   Dwarf Rattlesnake 616
   Large Water 523
   Redwool 384
   Saline 384
Plantain—Plantaginaceae 384–385

Index

Platanthera
  aquilonis 618
  hyperborea 618
  hyperborea var. *hyperborea* 618
  praeclara 619

Pleated Gentian 330

Plum
  Canada 460
  Canadian 460

Plumeless-thistle, Nodding 118

Pneumonanthe affinis 330

Poa
  compressa 669
  palustris 670
  pratensis 671

Poaceae—Grass 621–687

Poetry
  "Horsetail" 48
  "Rush" 587
  "Sedge" 528
  "The Shrub" 200
  "Tree" 64
  "Wild Flower" 178
  "Wild Grass" 620
  "Wild Vine" 293

Poison-ivy 78
  Common 78
  Western 78

Poison-vetch, Narrow-leaved 302

Polemoniaceae—Phlox 386

Polygalaceae—Milkwort 387

Polygala senega 387

Polygonaceae—Buckwheat 388–406

Polygonatum
  biflorum 605
  biflorum var. *commutatum* 605
  canaliculatum 605
  commutatum 605

Polygonum
  achoreum 391
  amphibium var. **emersum** 392
  arenastrum 393
  aviculare 393
  aviculare var. *arenastrum* 393
  cilinode 394
  coccineum 392
  convolvulus var. **convolvulus** 395
  erectum ssp. *achoreum* 391
  exsertum 399
  fagopyrum 390
  hydropiper 396
  lapathifolium 397
  persicaria 398
  prolificum 399
  ramosissimum 399
  ramosissimum var. *prolificum* 399
  ramosissimum var. *ramosissimum* 399
  sagittatum 400
  scandens var. **scandens** 401

Pond-lily
  Small Yellow 374
  Yellow 374

Pondweed
  Alpine 688
  Bushy 693
  Claspingleaf 691
  Fennel-leaved 693
  Fineleaf 692
  Floating 690
  Floating-leaf 690
  Leafy 689
  Northern 688
  Reddish 688
  Red-headed 691
  Richardson's 691
  Sago 693
  Sago False 693
  Slender 692, 693
  Slender-leaf False 692
  Thread-leaved 692

Pondweed—Potamogetonaceae 688–693

Poplar
  Aspen 478
  Balsam 476
  Black 476
  Eastern Balsam 476

Populus
  balsamifera 476
  deltoides 477
  deltoides ssp. *monilifera* 477
  sargentii 477
  tremuloides 478

Porcupine Grass 649

Porter's Brome 632

Portulacaceae—Purslane 407

Portulaca oleracea 407

Potamogeton
  alpinus 688
  filiformis 692
  foliosus 689
  natans 690
  pectinatus 693
  perfoliatus var. *richardsonii* 691
  richardsonii 691

Potamogetonaceae—Pondweed 688–693

Potentilla
  anserina 446
  arguta 457
  floribunda 450
  fruticosa 450
  gracilis 458
  norvegica 459
  palustris 448
  tridentata 469

Poverty Grass 686

Poverty Weed 273
  (Povertyweed) Nuttall's 273

Prairie Buck Bean 320

Prairie Bulrush 571

Prairie Buttercup 438

Prairie Chickweed 249

Prairie-clover
  Purple 306
  Violet 306
  White 305

Prairie Coneflower 170
  Upright 170

Prairie Cord Grass 682

Prairie Crocus 431

Prairie Dogbane 91

Prairie Dropseed 685

Prairie Flat-top-goldenrod 162

Prairie Flax 362

Prairie Fleabane 136

Prairie Gentian 330

Prairie Golden-banner 320

Prairie Goldenrod 177

Prairie June Grass 652

Prairie Koeler's Grass 652

Prairie Mouse-ear Chickweed 249

Prairie Muhly 657

Index

Prairie Onion 598
Prairie Peppergrass 224
Prairie Rose 464
Prairie Sage 115
Prairie Sagebrush 114
Prairie Sagewort 114
Prairie Sandreed 635
Prairie Smoke 455
Prairie Thermopsis 320
Prairie Turnip 319
Prairie Violet 517
Prairie Wild Rose 464
Prenanthes
  alba **168**
  racemosa **169**
Prickly Lettuce 154
Prickly Rose 463
Prickly-tree Club-moss 58
Prickly Wild Rose 463
Primrose
  Common Evening 381
  Dwarf 413
  Dwarf Canadian 413
  Evening 381
  Lake Mistassini 413
  Mistassini 413
  Plains Yellow 377
  White-stemmed Evening 382
Primrose—Primulaceae 408–414
Primulaceae—Primrose 408–414
Primula mistassinica **413**
Princess-pine 58
Proso Millet 662
Prostrate Amaranth 75
Prostrate Pigweed 75
Prostrate Vervain 513
Prunus
  nigra **460**
  pensylvanica **461**
  virginiana **462**
Psathyrostachys juncea **672**
Psoralea, Silverleaf 318
Psoralea
  argophylla 318
  esculenta 319
Puccinellia
  distans **673**
  nuttalliana **674**

Puccoon, Hoary 211
Puddingberry 277
Puffsheath Dropseed 686
Pulsatilla patens **431**
Purple Avens 454
  Long-plumed 455
Purple Boneset 137
Purple Coneflower 130
Purple Crownvetch 304
Purple-leaved Willowherb 379
Purple Loosestrife 366
Purple Marshlocks 448
Purple Milk-vetch 297
Purple Oat Grass 675
Purple Peavine 311
Purple Prairie-clover 306
Purple Rattlesnakeroot 169
Purple Rock Cress 214
Purple-stemmed Aster 192
Purple Vetch 325
  American 325
Purple-white Tufted Vetch 326

Pursh Seepweed 274
Purslane 407
  Common 407
Purslane—Portulaceae 407
Purslane Speedwell 505
Pusley 407
Pussytoes
  Broad-leaved 108
  Field 108
  Littleleaf 107
  Neat 107
  Small-leaf 107

Pussy Willow 481
Pygmyflower 409
  (Pygmy-flower)
  Northern 409
  Western 408
Pygmyflower Rockjasmine 409
Pyrola
  Greenish-flowered 418
  One-flowered 415
  One-sided 416
  Pink 417
Pyrola
  asarifolia **417**
  asarifolia ssp. *asarifolia* 417
  chlorantha **418**
  secunda 416
  uliginosa 417
  uniflora 415
  virens **418**
Pyrolaceae—Wintergreen 415–418

Q

Quack Grass 642
Quaker Bitter-cress 217
Quaking Aspen 478
Quercus macrocarpa **327**
Quickweed 140
  Fringed 140
Quill Sedge 560

R

Rabbitberry 283
Ragweed
  Annual 103
  Common 103
  Cuman 104
  False 153
  Giant 105
  Great 105
  Perennial 104
  Short 103
  Western 104
Ragwort
  Clustered Marsh 173
  Golden 164
  Gray 165
  Marsh 173
  Swamp 173
Ranunculaceae—Buttercup 419–441

Index

Ranunculus
  abortivus 432
  acris 433
  aquatilis 434
  cymbalaria 435
  flammula 436
  flammula var. *filiformis* 436
  pensylvanicus 437
  reptans 436
  rhomboideus 438
  sceleratus 439

Raspberry
  American Red 467
  Arctic 466
  Common Red 467
  Dwarf 466
  Dwarf Red 468
  Red 467
  Running 468
  Stemless 466
  Trailing 468
  Wild Red 467

Ratibida columnifera 170

Ratroot 522

Rattlesnake-orchid, Lesser 616

Rattlesnake-plantain
  (Rattlesnake Plantain)
  Dwarf 616
  Lesser 616

Rattlesnakeroot
  Purple 169
  White 168

Rayless Alkali Aster 184

Rayless Aster 184

Red and White Baneberry 420

Red Anemone 423

Red Ash 376

Red Baneberry 420

Red Bearberry 284

Red-berried Elder 242

Red Clover 323

Red Columbine 426

Red Currant 341
  Swamp 341
  Wild 341

Reddish Pondweed 688

Red Elder 242

Red Elderberry 242

Red False Mallow 369

Red-fringe Bulrush 575

Red-fruited Choke Cherry 462

Red Goosefoot 269

Red-headed Pondweed 691

Red Indian Paintbrush 495

Redleg 398

Redosier 278

Red-osier Dogwood 278

Red Painted-cup 495

Red Pigweed 269

Red Raspberry 467
  American 467
  Common 467
  Dwarf 468
  Wild 467

Redroot Amaranth 76

Red-root Pigweed 76

Red-seeded Dandelion 194

Redshank 398

Red-spike Mexican-hat 170

Redstem Stork's-bill 335

Red Swamp Currant 341

Redtop 623

Red Willow 278

Red Windflower 423

Redwool Plantain 384

Reed, Common 668

Reed Canary Grass 665

Reed Grass (Reed-grass, Reedgrass)
  Common 668, 677
  Marsh 633
  Narrow 634
  Northern 634
  Sand 635
  Slimstem 634
  Wood 636

Reflexed Rock Cress 215

Resin Birch 203

Retrorse Sedge 556

Rhamnaceae—Buckthorn 442

Rhamnus cathartica 442

Rhododendron groenlandicum 286

Rhombic-leaved Sunflower 147

Rhus
  glabra 77
  radicans var. *rydbergii* 78
  radicans var. *vulgaris* 78

Ribes
  americanum 337
  glandulosum 338
  hudsonianum 339
  oxyacanthoides 340
  oxyacanthoides ssp. *setosum* 340
  rubrum var. *propinquum* 341
  setosum 340
  triste 341

Rib-seed Sandmat 290

Rice (see Wild Rice)

Rice Cut Grass 653

Rice Grass (Ricegrass)
  Roughleaf 660
  White-grained Mountain 660

Richardson's Alumroot 488

Richardson's Pondweed 691

Ridge-seeded Spurge 290

Rigid Goldenrod 163

River Birch 204

River Bulrush 570

River Club-rush 570

Rivergrass, Common 677

River Maple 73

Rock Cress (Rockcress)
  Holboell's 215
  Purple 214
  Reflexed 215

Rocket
  False London 229
  Sweet 223

Rock-harlequin 329

Rockjasmine (Rock Jasmine)
  Northern 409
  Pygmyflower 409
  Western 408

Rocky Mountain Blazing Star 157

Root, Seneca 387

Rorippa
  hispida 226
  palustris 226
  palustris ssp. *hispida* 226

Index

Rosa
  acicularis 463
  arkansana 464
  woodsii 465
Rosaceae—Rose 443–470
Rose
  Common Wild 465
  Little 447
  Little Ground 447
  Low Prairie 464
  Prairie 464
  Prairie Wild 464
  Prickly 463
  Prickly Wild 463
  Sunshine 464
  Western Wild 465
  Wood's 465
  Wood's Wild 465
Rose—Rosaceae 443–470
Rough Bugleweed 351
Rough Cinquefoil 459
Rough Cocklebur 198
Rough Dropseed 683
Rough False Sunflower 149
Roughleaf Ricegrass 660
Rough Oxeye 149
Rough Pigweed 76
Rough Water-horehound 351
Rough White Prairie American-aster 187
Round-fruited Rush 584
Round-fruit Short-scale Sedge 537
Round-leaf Thimbleweed 421
Round-leaved Hawthorn 449
Round-leaved Mallow 368
Round-leaved Orchid 609
Rover Bellflower 233
Rubber-plant, Colorado 152
Rubberweed, Colorado 152

Rubiaceae—Madder 471–474
Rubus
  acaulis 466
  arcticus 466
  arcticus ssp. *acaulis* 466
  idaeus 467
  idaeus ssp. *strigosus* 467
  pubescens 468
  pubescens var. *pubescens* 468
  strigosus 467
Rudbeckia
  ampla 172
  hirta 171
  hirta var. *pulcherrima* 171
  laciniata 172
  laciniata var. *ampla* 172
  serotina 171
Rumex
  acetosa 402
  ***aquaticus* var. *fenestratus* 403**
  ***crispus* 404**
  fueginus 405
  ***maritimus* 405**
  occidentalis 403
  ***salicifolius* var. *mexicanus* 406**
  triangulivalvis 406
Running-pine, Northern 56
Running Raspberry 468
Rush
  Alpine 581
  Baltic 582
  Flattened 584
  Flowering 527
  Grassy 527
  Knotted 586
  Long-styled 585
  Northern Green 581
  Round-fruited 584
  Seaside 583
  Three-square 572
  Toad 583
  Torrey's 588
  Wire 582
Rush Aster 183
Rush—Juncaceae 581–590
Rush Skeletonplant 159
Rush Skeleton-weed 159
Russet Buffaloberry 283
Russian Blue Lettuce 155
Russian Olive 280
Russian Pigweed 263
Russian Wild Rye 672
Rusty Labrador-tea 286

Rye
  Canada Wild 641
  Creeping Wild 642
  Fuzzyspike Wild 654
  Hairy Wild 654
  Nodding Wild 641
  Russian Wild 672
  Virginia Wild 643
Rye Grass (Ryegrass)
  Downy 654
  English 655
  Perennial 655

S

Sage
  Pasture 114
  Prairie 115
  White 115, 272
Sagebrush
  Prairie 114
  White 115
Sage-leafed Willow 480
Sagewort
  Biennial 112
  Field 113
  Pasture 114
  Prairie 114
  Western 113
Sagittaria
  cuneata 524
  engelmanniana ssp. *longirostra* 525
  ***latifolia* 525**
  longirostra 525
  variabilis var. *obtusa* 525
Sago False Pondweed 693
Sago Pondweed 693
Salicaceae—Willow 475–485
Saline Goosefoot 268
Saline Plantain 384
Salix
  ***bebbiana* 479**
  ***candida* 480**
  cordata 482
  ***discolor* 481**
  ***eriocephala* 482**
  exigua ssp. *interior* 483
  exigua var. *exterior* 483
  gracilis 485
  ***interior* 483**
  ***maccalliana* 484**
  ***petiolaris* 485**
  rigida 482
Salsify, Western 196

Index

Saltbush
    Gardner's 260
    Halberd-leaved 262

Saltmarsh Club-rush 571

Salt-meadow Grass
    Nuttall's 674
    Slender 673

Sambucus
    pubens 242
    ***racemosa* 242**
    racemosa var. *racemosa* 242

Sandalwood—Santalaceae 486

Sandbar Willow 483

Sand Dropseed 684

Sand Grass 635

Sandmat (Sand-mat)
    Rib-seed 290
    Thyme-leaved 291

Sandreed, Prairie 635

Sand Reed Grass 635

Sand Sedge 545

Sandwort
    Bluntleaf Grove 252
    Blunt-leaved 252
    Grove 252

Sanicle 86
    Maryland 86

Sanicula marilandica 86

Santalaceae—Sandalwood 486

Saponaria officinalis 253

Sarsaparilla, Wild 92

Sartwell's Sedge 557

Saskatoon 445

Saskatoon Serviceberry 445

Savin, Creeping 62

Saxifragaceae—Saxifrage 487–492

Saxifraga virginiensis 492

Saxifrage
    Early 492
    Golden 487
    Green 487

Saxifrage—Saxifragaceae 487–492

Scapose Hawk's-beard 127

Scarlet Beeblossom 380

Scarlet Butterflyweed 380

Scarlet Elder 242

Scarlet Gaura 380

Scarlet Globemallow 369

Scarlet Mallow 369

Scarlet Sumac 77

Scentless Chamomile 197

Scentless False Mayweed 197

Schizachne purpurascens **675**

Schizachyrium
    ***scoparium* 676**
    scoparium var. *scoparium* 676

Schoenoplectus
    ***acutus* 569**
    acutus var. *acutus* 569
    ***fluviatilis* 570**
    ***maritimus* 571**
    ***pungens* 572**
    pungens var. *pungens* 572
    ***tabernaemontani* 573**
    validus 573

Scirpus
    acutus 569
    americanus 572
    atrovirens var. *pallidus* 576
    ***cyperinus* 574**
    fluviatilis 570
    hudsonianus 577
    maritimus 571
    ***microcarpus* 575**
    ***pallidus* 576**
    paludosus 571
    pungens 572
    rubrotinctus 575
    tabernaemontani 573
    validus 573

Scolochloa festucacea **677**

Scotch Thistle 124

Scouring-rush
    Common 51
    Dwarf 53
    Northern 55
    Smooth 52
    Tall 51
    Variegated 55

Scouringrush Horsetail 51

Scrambled Eggs 328

Scratch Grass 656

Scrophulariaceae—Figwort 493–506

Scrub Birch 203

Scrub Pine 69

Scurf-pea, Silver-leaf 318

Scutellaria
    epilobiifolia 356
    galericulata 356
    lateriflora 357

Sea-blite (Sea Blite 274)
    Low 274
    Western 274

Sea-milkwort 410

Seaside Arrow-grass 591

Seaside Buttercup 435

Seaside Crowfoot 435

Seaside Rush 583

Securigera varia 304

Sedge
    Awl-fruited 559
    Awned 532
    Awned Flat 565
    Awn-fruited 559
    Bebb's 534
    Bladder 546
    Bog 548
    Boreal Bog 548
    Bristle-stalked 547
    Broad-fruited 535
    Clustered Field 553
    Cyperus-like 554
    Cypress-like 554
    Dewey's 537
    Dry-spike 541
    Eastern Star 555
    Field 553
    Golden 533
    Golden-fruit 533
    Graceful 542, 553
    Granular 543
    Greater Bladder 546
    Green 563
    Hair-like 536
    Hay 541
    Hillside 541
    Houghton's 545
    Lesser Panicled 538
    Lesser Tussock 538
    Limestone Meadow 543
    Little Green 563
    Long-beaked 558
    Northern Beaked 562
    Northern Bog 544
    Northwest Territory 562
    Parry's 549
    Peck's 550

Index

[Sedge]
    Pennsylvania 552
    Quill 560
    Retrorse 556
    Round-fruit Short-scale 537
    Sand 545
    Sartwell's 557
    Short-beaked 535
    Slender 542
    Soft-leaved 539
    Sparse-flowered 561
    Sprengel's 558
    Stalk-grained 559
    Sun-loving 552
    Swollen 546
    Thin-flowered 561
    Thread-leaved 540
    Threadstem 547
    Turned 556
    Two-seeded 539
    Two-stamened 538
    Water 531
    Wheat 532
    White-scaled 564
    Woolly 551
    Yellow Bog 544

Sedge—Cyperaceae 529–577

Seepweed, Pursh 274

Seneca Root 387

Seneca Snakeroot 387

Senecio
    aureus 164
    canus 165
    congestus **173**
    palustris 173
    vulgaris **174**

Serviceberry, Saskatoon 445

Setaria
    pumila **678**
    viridis **679**

Shaggy-soldier 140

Sheepberry 245

Sheepbur, Bristly 210

Shepherdia
    argentea **282**
    canadensis 283

Shepherd's-purse 216

Shore Buttercup 435

Short-awned Foxtail 624

Short-awned Meadow Foxtail 624

Short-beaked Sedge 535

Short Ragweed 103

Short-ray Fleabane 134

Shortspike Watermilfoil 342

Shortstalk Arrowwood 247

Showy Goldenrod 180

Showy Milkweed 95

Showy Sunflower 144

Shrubby Cinquefoil 450

Shrubby Evening-primrose 377

Shrubby Fivefingers 469

Sibbaldiopsis tridentata **469**

Siberian Peashrub 303

Siberian Water-milfoil 342

Siberian Wild Chives 597

Siberian Yarrow 101

Sidebells 416

Sidebells Wintergreen 416

Side-oats Grama 628

Silene, Nightflowering 256

Silene
    alba 255
    csereii **254**
    latifolia **255**
    latifolia ssp. *alba* 255
    noctiflora **256**
    pratensis 255

Silky Milkweed 96

Silverberry 281
    American 281

Silver Buffaloberry 282

Silverleaf Indian Breadroot 318

Silverleaf Psoralea 318

Silver-leaf Scurf-pea 318

Silver Maple 73

Silverweed 446
    Common 446

Silverweed Cinquefoil 446

Silver-woolly Groundsel 165

Silvery Groundsel 165

Simple-stem Bur-reed 695

Sinapis arvensis **227**

Single Delight 415

Sisymbrium
    altissimum **228**
    loeselii **229**
    sophia 219

Sisyrinchium montanum **580**

Sitka Alder 202

Sium suave **87**

Skeletonplant, Rush 159

Skeletonweed 159
    (Skeleton-weed)
    Rush 159

Skullcap
    Blue 357
    Common 356
    Hooded 356
    Mad-dog 357
    Marsh 356

Skunkberry 338

Skunk Currant 338

Slender Agalinis 494

Slender Arrow-grass 592

Slender Beardtongue 501

Slender Bugleweed 352

Slender Bur-reed 698

Slender Cinquefoil 458

Slender Cotton-grass 568

Slender Gerardia 494

Slender-leaf False Pondweed 692

Slender-leaved Bluet 474

Slender-leaved False Foxglove 494

Slender Pondweed 692, 693

Slender Salt-meadow Grass 673

Slender Sedge 542

Slender Willow 485

Slender Wood Grass 636

Slender Wood-reed 636

Slimstem Reedgrass 634

Slipper, White Lady's 614

Slough Grass 627, 682
    (Sloughgrass)
    American 627

Index

Small Bedstraw 472
Small Bladderwort 361
Small Blue Aster 189
Small Bluestem 676
Small Bur-reed 698
Small Dropseed 686
Smaller Hop Clover 321
Small-flowered Bulrush 575
Small-flowered Buttercup 432
Small-flowered Columbine 425
Small-flowered Mallow 367
Small-flowered Wood-rush 590
Small Fringed Gentian 333
Small-fruited Bulrush 575
Small-fruited Mallow 367
Small-leaf Pussytoes 107
Small-leaved Everlasting 107
Small Round-leaved Orchis 609
Small-snapdragon 496
Small Tumbleweed Mustard 229
Small White Lady's-slipper 614
Small-whorl Mallow 367
Small Yellow Pond-lily 374
Smartweed
   Common 396
   Dockleaf 397
   Marsh-pepper 396
   Nodding 397
   Pale 397
   Water 392
Smilacaceae—Greenbrier 694
Smilacina
   stellata 603
   trifolia 604
Smilax
   herbacea var. *lasioneura* 694
   lasioneura 694
Smoke, Prairie 455
Smooth Aster 188
Smooth Beggarticks 116
Smooth Blue American-aster 188
Smooth Blue Aster 188
Smooth Blue Beardtongue 502
Smooth Brome 631
Smooth Camas 607
Smooth Catchfly 254
Smooth Fleabane 133

Smooth Hawk's-beard 127
Smooth Horsetail 52
Smooth-leaved Buttercup 432
Smooth Love Grass 645
Smooth Oxeye 149
Smooth Scouring-rush 52
Smooth Solomon's-seal 605
Smooth Sumac 77
Smooth Sweet Cicely 84
Smooth Wild Strawberry 451
Smooth Yellow Violet 518
Snakeroot 86
   Black 86
   Maryland Black 86
   Seneca 387
Snakeweed 142
   Broom 142
Snapdragon, Dwarf 496
Snapweed 199
   Spotted 199
Sneezeweed 143
   Common 143
   Fall 143
   Mountain 143
Sneezewort Aster 162
Snowberry
   Creeping 285
   Western 243
Soapberry 283
Soapwort 253
Soft Goldenrod 179
Soft-leaved Sedge 539
Soft Maple 73
Softstem Bulrush 573
Softstem Club-rush 573
Solanaceae—Nightshade 507

Solanum dulcamara 507
Solidago
   canadensis 175
   gigantea 176
   graminifolia 138
   missouriensis 177
   mollis 179
   nemoralis 180
   ptarmicoides 162
   rigida 163
Solomon's-seal
   Common 605
   False 604
   Great 605
   King 605
   Smooth 605
   Star-flowered 603
   Starry 603
   Starry False 603
   Three-leaved 604
   Three-leaved False 604
   Two-leaved 602
Sonchus
   arvensis 181
   arvensis ssp. *uliginosus* 181
   oleraceus 182
   uliginosus 181
Sophia, Herb 219
Sorghastrum nutans 680
Sorrel
   Garden 402
   Green 402
Sour Dock 402
Sour-top Blueberry 288
Sowbane 270
Sow-thistle (Sow Thistle)
   Annual 182
   Common 182
   Field 181
   Perennial 181
Sparganiaceae—Bur-reed 695–698
Sparganium
   emersum 695
   eurycarpum 696
   chlorocarpum 695
   glomeratum 697
   minimum 698
   natans 698
Sparse-flowered Sedge 561
Spartina
   gracilis 681
   pectinata 682
Spatterdock 374
Spear Grass 648

Index

Spear-leaved Fleabane 134
Spear-leaved Goosefoot 273
Spearwort 436
   Creeping 436
   Greater Creeping 436
   Lesser 436
Speckled Alder 201
Speedwell
   American 504
   Brooklime 504
   Hairy 505
   Purslane 505
Sphaeralcea
   coccinea 369
   coccinea ssp. *coccinea* 369
Spiked Lobelia 365
Spiked Water-milfoil 342
Spike-rush
   Common 567
   Creeping 567
   Needle 566
   Pale 567
Spiny Ironplant 160
Spiny Phlox 386
Spiraea alba 470
Spirea, Narrowleaf 470
Spirodela polyrrhiza 595
Split-lip Hemp-nettle 347
Sporobolus
   asper 683
   compositus 683
   compositus var. *compositus* 683
   cryptandrus 684
   heterolepis 685
   neglectus 686
   vaginiflorus var. *neglectus* 686
Spotted Coralroot 611
Spotted Cowbane 82
Spotted Joe-pye Weed 137
Spotted Ladysthumb 398
Spotted Snapweed 199
Spotted Touch-me-not 199
Spotted Water-hemlock 82
Sprangletop 677
Spreading Alkali-grass 673
Spreading Bent 623
Spreading Dogbane 90
Sprengel's Sedge 558
Spring Water-starwort 231

Spruce
   Black 68
   Black Hills 67
   Bog 68
   Swamp 68
   White 67
Spurge
   Leafy 292
   Ridge-seeded 290
   Thyme-leaved 291
Spurge—Euphorbiaceae 290–292
Square-stemmed Monkey-flower 499
Squashberry 244
Squirreltail 651
Stachys
   palustris var. *pilosa* 358
   pilosa 358
   pilosa var. *pilosa* 358
Stalk-grained Sedge 559
Star Duckweed 593
Starflower 414
   American 414
   Northern 414
Star-flowered Solomon's-seal 603
Star-grass (Stargrass 600)
   Eastern Yellow 600
   Yellow 600
Starry False Lily of the Valley 603
Starry False Solomon's-seal 603
Starry Solomon's-seal 603
Star Toadflax 486
Starwort
   Common 258
   Longleaf 257
Steironema ciliatum 411
Stellaria
   longifolia 257
   media 258
Stemless Bur-reed 695
Stemless Raspberry 466
Stickseed
   Blue 210
   European 210
   Nodding 209
Sticky Cockle 256
Stiff Club-moss 57
Stiff Flax 363
Stiff Goldenrod 163
Stiff Ground-pine 57

Stiffstem Flax 363
Stiff Sunflower 147
Stinging Nettle 511
Stinkgrass 644
Stinking Elderberry 242
Stinkweed 230
Stipa
   comata 648
   spartea 649
   viridula 659
Stitchwort, Long-leaved 257
Stony-hills Muhly 657

Stork's-bill 335
   Redstem 335
Strap-style Gayfeather 157
Strawberry
   Smooth Wild 451
   Virginia 451
   Wild 451
Strawberry Blite 266
Streamside Fleabane 133
Striate Knotweed 391
Strict Blue-eyed-grass 580
Striped Coralroot 612
Stuckenia
   filiformis 692
   filiformis ssp. *filiformis* 692
   pectinata 693

Index

Suaeda
   calceoliformis **274**
   depressa 274
   maritima var. *americana* 274

Sugar Maple 73

Sumac
   Scarlet 77
   Smooth 77

Sumac—Anacardiaceae 77–78

Summer, Farewell 190

Summer Coralroot 611

Summer-cypress 271

Sumpweed, Giant 153

Sundrops, Yellow 377

Sunflower
   Annual 144
   Beautiful 147
   Clustered 146
   Common 144
   False 149
   Maximillian's 145
   Narrow-leaved 145
   Nuttall's 146
   Rhombic-leaved 147
   Rough False 149
   Showy 144
   Stiff 147
   Tuberous-rooted 146

Sun-loving Sedge 552

Sunshine Rose 464

Susan, Black-eyed 171

Swamp Ash 375

Swamp Aster 192

Swamp Birch 206

Swamp Currant 341
   Red 341

Swamp Hedgenettle 358

Swamp Horsetail 50

Swamp Lousewort 500

Swamp Milkweed 93

Swamp Persicaria 392

Swamp Ragwort 173

Swamp Red Currant 341

Swamp Spruce 68

Swamp Thistle 123

Sweet Cicely, Smooth 84

Sweet Cinnamon 522

Sweet-clover
   White 315
   Yellow 316

Sweet Colt's-foot 167

Sweet Flag 522

Sweet Grass 650
   Common 650
   Northern 650

Sweet Oxeye 149

Sweet Rocket 223

Sweet-scented Bedstraw 473

Sweetvetch, Alpine 308

Switch Grass 663

Swollen Sedge 546

Sylvan's Horsetail 54

Symphoricarpos occidentalis 243

Symphyotrichum
   boreale **183**
   ciliatum **184**
   ciliolatum **185**
   ericoides **186**
   ericoides var. *ericoides* 186
   ericoides var. *pansum* 186
   falcatum **187**
   falcatum var. *commutatum* 187
   falcatum var. *falcatum* 187
   laeve **188**
   laeve var. *laeve* 188
   lanceolatum **189**
   lanceolatum var. *hesperium* 189
   lanceolatum var. *lanceolatum* 189
   lateriflorum **190**
   lateriflorum var. *lateriflorum* 190
   novae-angliae **191**
   puniceum **192**
   puniceum var. *puniceum* 192

T

Tall Baby's-breath 251

Tall Bluebells 212

Tall Boneset 137

Tall Buttercup 433

Tall Cinquefoil 457

Tall Coneflower 172

Tall Flat-top White Aster 129

Tall Hairy Agrimony 444

Tall Hairy Groovebur 444

Tall Hedge Mustard 229

Tall Lungwort 212

Tall Manna Grass 646

Tall Meadow-rue 440

Tall Mertensia 212

Tall Scouring-rush 51

Tall White Violet 515

Tall Wormwood 113

Tamarack 66

Tanacetum vulgare **193**

Tansy 193
   Common 193

Tansyaster, Lacy 160

Tansy Mustard 218, 219
   (Tansymustard)
   Gray 218
   Mountain 218

Taraxacum
   erythrospermum 194
   laevigatum **194**
   officinale **195**

Tatarian Honeysuckle 240

Teal Lovegrass 645

Tear-thumb 400
   Arrow-leaf 400

Thalictrum
   confine 441
   dasycarpum **440**
   venulosum **441**

Thatch Grass 677

Thermopsis, Prairie 320

Thermopsis rhombifolia **320**

Thicket Creeper 519

Thimbleweed
   Long-headed 422
   Round-leaf 421

Thin-flowered Sedge 561

Thinleaf Orach 262

Index

Thistle
    Bull 124
    Canada 121
    Canadian 121
    Field Sow 181
    Flodman's 122
    Musk 118
    Nodding 118
    Scotch 124
    Swamp 123
Thlaspi arvense 230
Thorny Buffaloberry 282
Thread-leaved Buttercup 434
Thread-leaved Pondweed 692
Thread-leaved Sedge 540
Threadstem Sedge 547
Three-flowered Avens 455
Threeleaf False Lily of the Valley 604
Three-leaf Goldthread 430
Three-leaved False Solomon's-seal 604
Three-leaved Solomon's-seal 604
Threepetal Bedstraw 472
Three-square 572 (Threesquare)
    Common 572
Three-square Bulrush 572
Three-square Rush 572
Three-toothed Cinquefoil 469

Thuja occidentalis 63
Thyme-leaved Sand-mat 291
Thyme-leaved Spurge 291
Tickseed
    Common 126
    Golden 126
Tickweed, Golden 126
Tilia americana 508
Tiliaceae—Linden 508
Timothy 667
    Common 667
Tine-leaved Milk-vetch 302
Toadflax (Toad-flax 498)
    Bastard 486
    False 486
    Star 486
    Yellow 498
Toad Rush 583
Torch Flower 455
Torrey's Rush 588
Touch-me-not
    Orange 199
    Spotted 199
Touch-me-not—Balsaminaceae 199
Toxicodendron
    radicans var. *rydbergii* 78
    rydbergii 78
Tragopogon dubius 196
Trailing Club-moss 56
Trailing Ground-pine 56
Trailing Juniper 62
Trailing Raspberry 468
Traveler's-joy, Deciduous 428
Treacle Mustard 222
Tree Club-moss 58
Tree Groundpine 58
Trefoil
    Bird's-foot 312
    Garden Bird's-foot 312
Trembling Aspen 478
Triangular-valved Dock 406
Trichophorum alpinum 577
Trientalis borealis 414
Trifolium
    campestre 321
    hybridum 322
    pratense 323
    procumbens 321
    repens 324

Triglochin
    maritima 591
    palustre 592
Trillium, Nodding 606
Trillium cernuum 606
Trimorpha
    acris 131
    lonchophylla 134
Tripleurospermum perforata 197
Trumpetweed, Bruner's 137
Tuberous-rooted Sunflower 146
Tufted Fleabane 132
Tufted Hair Grass 638
Tufted Loosestrife 412
Tufted Vetch 326
    Purple-white 326
Tufted White Prairie Aster 186
Tufted Yellow-loosestrife 412
Tumbleweed 75
Tumbleweed Mustard 228
Tumbling Mustard 228
Tundra Dwarf Birch 203
Turned Sedge 556
Turnip, Prairie 319
Twayblade
    Heartleaf 617
    Heart-leaved 617
    Mannikin 617
Twinflower 238
    American 238
Twining Honeysuckle 239
Twinsisters 240
Two-grooved Milk-vetch 298
Two-grooved Vetch 298
Two-leaved Solomon's-seal 602
Two-seeded Sedge 539
Two-stamened Sedge 538
Typha
    angustifolia 699
    latifolia 700
Typhaceae—Cattail 699–700

U

Ulmaceae—Elm 509
Ulmus americana 509
Umbellate Aster 129
Umbrellaplant, Yellow 389

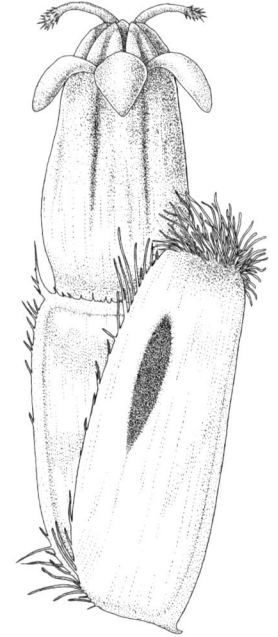

Index

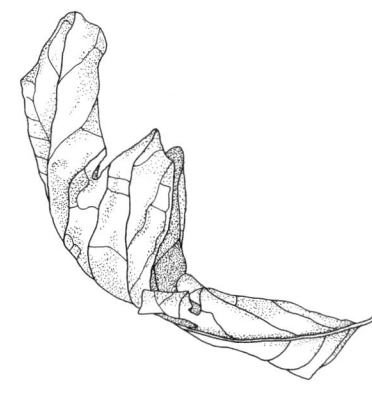

Umbrella-sedge, Awned 565
Umbrellawort, Hairy 373
Upland White Goldenrod 162
Upright Prairie Coneflower 170
Upright Yellow Wood-sorrel 383
Urticaceae—Nettle 510–511
Urtica dioica 511
Utricularia
  intermedia 359
  macrorhiza 360
  minor 361
  vulgaris 360

Vaccinium
  angustifolium 287
  myrtilloides 288
  vitis-idaea 289
Valerian
  Marsh 512
  Northern 512
Valerianaceae—Valerian 512
Valerian
  dioica 512
  dioica var. *sylvatica* 512
  septentrionalis 512
Valerian—Valerianaceae 512
Vanilla Grass 650
Variegated Horsetail 55
Variegated Scouring-rush 55
Veiny-leaf Meadow-rue 441
Veiny Meadow-rue 441
Veiny Pea 311
Veiny Vetchling 311
Velvet-fruited Willow 484

Velvet-leaved Blueberry 288
Velvety Goldenrod 179
Verbascum thapsus 503
Verbena, Bigbract 513
Verbena bracteata 513
Verbenaceae—Vervain 513
Vernal Water-starwort 231
Veronica
  americana 504
  peregrina 505
Veronicastrum virginicum 506
Vervain
  Bracted 513
  Carpet 513
  Prostrate 513
Vervain—Verbenaceae 513
Vetch
  American 325
  American Purple 325
  Bird 326
  Cow 326
  Crown 304
  Purple 325
  Purple-white Tufted 326
  Tufted 326
  Two-grooved 298
  Wild 325
Vetchling 310
  Bushy 311
  Cream-colored 309
  Marsh 310
  Pale 309
  Veiny 311
  Yellow 309
Viburnum
  edule 244
  lentago 245
  opulus 246
  opulus var. *americanum* 246
  rafinesquianum 247
  trilobum 246
Vicia
  americana 325
  cracca 326
Viola
  adunca 514
  canadensis 515
  canadensis var. *regulosa* 515
  nephrophylla 516
  pedatifida 517
  pubescens 518
  regulosa 515
Violaceae—Violet 514–518

Violet
  Bog 516
  Canada 515
  Canadian Western 515
  Crowfoot 517
  Dame's 223
  Downy Yellow 518
  Early Blue 514
  Hookedspur 514
  Northern Bog 516
  Prairie 517
  Smooth Yellow 518
  Tall White 515
  Western Canada 515
Violet Prairie-clover 306
Violet—Violaceae 514–518
Virginia Creeper, Large-toothed 519
Virginia Lionsheart 355
Virginia Strawberry 451
Virginia Wild Rye 643
Virgin's-bower
  Oriental 429
  Western 428
Viscid Great Bulrush 569
Vitaceae—Grape 519

Wakerobin, Nodding 606
Wallflower, Wormseed 222
Wand Panic Grass 663
Wapato 525
Water Arum 526
Water Avens 454
Waterberry 241
Water Birch 204
Water Calla 526
Water-carpet, Northern 487
Watercrowfoot, Large-leaved 434
Water Dragon 526
Water Flag 578
Water Foxtail 624
Water-hemlock 82
  (Water Hemlock)
  Bulb-bearing 81
  Bulblet-bearing 81
  Bulbous 81
  Common 82
  Spotted 82

Index

Water-horehound 350
  American 350
  Cut-leaf 350
  Northern 352
  Rough 351
  Western 351
Water Horsetail 50
Water Knotweed 392
Water-lily—Nymphaeaceae 374
Water-milfoil (Watermilfoil)
  Siberian 342
  Shortspike 342
  Spiked 342
Water-milfoil—Haloragaceae 342
Water Mudwort 497
Water-parsnip 87 (Waterparsnip)
  Common 87
  Hemlock 87
Water-pepper 396
  Mild 396
Water-plantain (Water Plantain)
  Common 523
  Large 523
  Northern Broad-leaved 523
  Western 523
Water-plantain—
  Alismataceae 523–525
Water Sedge 531
Water Smartweed 392
Water-starwort 231
  Common 231
  Spring 231
  Vernal 231
Water-starwort—
  Callitrichaceae 231
Waxleaf Beardtongue 502
Waxleaf Penstemon 502
Weed
  Joe-pye 137
  Poverty 273
  Spotted Joe-pye 137
Weeping Alkaligrass 673
Western Blite 274
Western Canada Violet 515
Western Clematis 428
Western Cottonwood 477
Western Dock 403
Western Mugwort 115
Western Pearly Everlasting 106
Western Poison-ivy 78

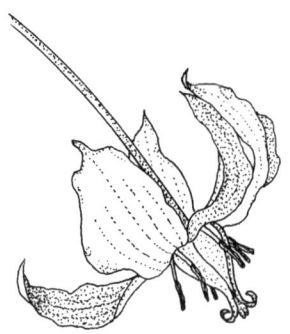

Western Prairie Fringed Orchid 619
Western Pygmyflower 408
Western Ragweed 104
Western Rock Jasmine 408
Western Sagewort 113
Western Salsify 196
Western Sea-blite 274
Western Snowberry 243
Western Violet, Canadian 515
Western Virgin's-bower 428
Western Water-horehound 351
Western Water-plantain 523
Western Wheatgrass 664
Western White Clematis 428
Western Wild Bergamot 354
Western Wild Rose 465
Western Yarrow 100
Wheatgrass
  Crested 622
  Western 664
Wheat Sedge 532
Whip-poor-will-flower 606
Whiskers, Old Man's 455
White Aster 186
  Flat-topped 129
  Tall Flat-top 129
White Birch 205
  American 205
White Camas 607
White Campion 255
White Cedar 63
  Eastern 63
  Northern 63
White Cinquefoil 457
White Clover 324

White Cockle 255
White Dock 406
White Elm 509
White Evening-primrose 382
White-grained Mountain Rice
  Grass 660
White Heath American-aster 186
White Heath Aster 186
White Knapweed 119
White Lady's Slipper 614
White Lettuce 168
  Glaucous 169
White Maple 74
White Meadowsweet 470
White Panicle Aster 189
White Panicled American-aster 189
White Prairie Aster 187
White Prairie-clover 305
White Rattlesnakeroot 168
White Sage 115, 272
White Sagebrush 115
White-scaled Sedge 564
White Spruce 67
White-stemmed Evening
  Primrose 382
White Sweet-clover 315
Whitetop 136
  Parasol 129
White Upland Aster 162
White Water Buttercup 434
Whitewater Crowfoot 434
White Woodland Aster 190
Whitlow-grass
  Woodland 220
  Yellow 220
Whorled Milkweed 97
Wild Balsam Apple 279
Wild Barley 651
Wild Bergamot 354
  Western 354
Wild Black Currant 337, 339
Wild Blue Flax 362
Wild Buckwheat 395
  Alpine Golden 389
  Yellow 389
Wild Calla 526
Wild Chamomile 197

Index

Wild Chives 597
Wild Columbine 426
Wild Cucumber 279
Wild Gooseberry 340
Wild Honeysuckle 239
Wild Licorice 307
Wild Lily 601
Wild Lily-of-the-valley 602
Wild Mint 353
   American 353
Wild Morning-glory 275
Wild Mustard 227
Wild Oat 626
Wild Onion 598
Wild Parsnip 85
Wild Peavine 311
Wild Red Cherry 461
Wild Red Currant 341

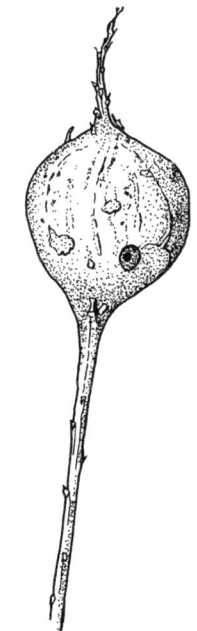

Wild Red Raspberry 467
Wild-rice (Wild Rice 687)
   Annual 687
   Indian 687
   Northern 687
Wild Rose
   Common 465
   Prairie 464
   Prickly 463
   Western 465
   Wood's 465

Wild Rye
   Canada 641
   Creeping 642
   Fuzzyspike 654
   Hairy 654
   Nodding 641
   Russian 672
   Virginia 643
Wild Sarsaparilla 92
Wild Strawberry 451
   Smooth 451
Wild Vetch 325
Wild Wood Lily 601
Willow
   Basket 485
   Beaked 479
   Bebb 479
   Diamond 479, 482
   Gray 479
   Heartleaf 482
   Hoary 480
   Long-leaf 483
   McCalla's 484
   Meadow 485
   Missouri 482
   Pussy 481
   Red 278
   Sage-leafed 480
   Sandbar 483
   Slender 485
   Velvet-fruited 484
   Yellow 482
Willow Dock 406
Willowherb (Willow-herb)
   Fringed 379
   Great 378
   Northern 379
   Purple-leaved 379
Willow—Salicaceae 475–485
Windflower 424
   Red 423
Winterfat 272
Wintergreen
   Bog 417
   Common Pink 417
   Green 418
   Greenflowered 418
   Greenish-flowered 418
   Liverleaf 417
   One-flowered 415
   One-sided 416
   Pink 417
   Sidebells 416
Wintergreen—Pyrolaceae 415–418
Wire Rush 582

Witch Grass 661 (Witchgrass)
   Common 661
Wolfberry 243
Wolf's-milk 292
Wolf-willow 281
Wood Anemone 424
   American 424
Wood Aster 190
Woodbine 519
Wood Grass, Slender 636
Wood Horsetail 54
Woodland Draba 220
Woodland Horsetail 54
Woodland Whitlow-grass 220
Wood Lily 601
Wood Nettle 510
   (Woodnettle, Wood-nettle)
   Canada 510
   Canadian 510
Wood-nymph 415
Wood-reed (Woodreed)
   Drooping 636
   Slender 636
Wood Reedgrass 636
Wood-rush
   Hairy 589
   Small-flowered 590
Wood-sorrel
   Common Yellow 383
   Upright Yellow 383
   Yellow 383
Wood-sorrel—Oxalidaceae 383
Wood's Rose 465
Wood's Wild Rose 465
Wool-grass 574
Wool Grass Bulrush 574
Woolly Burdock 110
Woolly Mullein 503
Woolly Sedge 551
Wormseed Mustard 222
Wormseed Wallflower 222
Wormwood
   Absinthe 111
   Biennial 112
   Pacific 113
   Plains 113
   Tall 113
Woundwort 358

Index

X

Xanthium strumarium 198

Y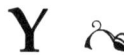

Yarrow 100
  Common 100
  Many-flowered 101
  Siberian 101
  Western 100
Yellow Avens 452
Yellow Birch 206
Yellow Bog Sedge 544
Yellow Bristlegrass 678
Yellow Coralroot 613
Yellow Cress 226 (Yellowcress)
  Bog 226
  Marsh 226
Yellow Dock 404
Yellow Evening-primrose 377, **381**
Yellow Field Clover 321
Yellow-flag 578
Yellow Flax 363
  Large-flowered 363
Yellow-flowered Knotweed 399

Yellow Foxtail 678
Yellow Goat's-beard 196
Yellow Indian Grass 680
Yellow Iris 578
Yellow Lady's-slipper 615
Yellow-loosestrife (Yellow Loosestrife)
  Fringed 411
  Tufted 412
Yellow Marsh-marigold 427
Yellow Moccasin Flower 615
Yellow Parilla 370
Yellow Pea 320
Yellow Pond-lily 374
  Small 374
Yellow Star-grass 600
  Eastern 600
Yellow Sundrops 377
Yellow Sweet-clover 316
Yellow Toadflax 498
Yellow Umbrellaplant 389
Yellow Vetchling 309

Yellow Violet
  Downy 518
  Smooth 518
Yellow Water Iris 578
Yellow Whitlow-grass 220
Yellow Wild Buckwheat 389
Yellow Willow 482
Yellow Wood-sorrel 383
  Common 383
  Upright 383

Z

Zigadenus elegans 607
Zizania aquatica 687
Zizia
  Golden 89
  Meadow 88
Zizia
  aptera 88
  aurea 89

(incomplete) Short Index—Common

Absinthe 111
Alder 201
Alexanders 88
Amaranth 75
Anemone 421, 431
Arrow-grass 591
Arrowhead 524
Arrowwood 247
Ash 375
Asparagus 599
Aster 129, 183
Atriplex 260
Avens 452
Barley 651
Basswood 508
Bearberry 284
Beardtongue 501
Bedstraw 471
Beggarticks 116
Bellflower 232
Bindweed 275, 394
Birch 203
Bishop's-cap 489
Bittersweet 507
Black-eyed Susan 171
Bladderwort 359
Blazingstar 157
Bluebur 210
Blue Grass 669
Blueberry 287
Blue-eyed Grass 580
Bluestem 625, 676
Brome 630
Broomweed 142
Buckwheat 390, 395, 401
Buffaloberry 282
Bulrush 569, 575
Bunchberry 277
Burdock 109
Bur-reed 695
Bush-cranberry 244
Buttercup 432
Calla 526
Canary Grass 665
Caraway 80
Carrionflower 694
Catchfly 254
Cattail 699
Cedar 63
Chamomile 197
Cherry 461
Chickweed 249, 258
Chicory 120
Cinquefoil 448, 450, 457
Clover 321
Club-moss 56
Cocklebur 198
Colt's-foot 166
Columbine 425
Coneflower 130, 170, 172
Coralroot 611
Cord Grass 681

Corydalis 328
Cotton-grass 568, 577
Cow-parsnip 83
Cucumber 279
Currant 337
Daisy 156
Dandelion 194
Dewberry 468
Dock 402
Dogbane 90
Dogwood 278
Dropseed 683
Duckweed 593
Elm 509
Evening-primrose 377, 381
False Dandelion 102
False Ragweed 153
Fir 65
Fireweed 378
Flag 522, 578
Flax 362
Fleabane 125, 131
Foxtail 624, 678
Fringed Orchid 619
Gaillardia 139
Galinsoga 140
Gentian 330
Geranium 336
Giant-hyssop 345
Goat's-beard 196
Goldenrod 138, 162, 175
Goosefoot 265, 273
Grama 628
Grass-of-parnassus 490
Ground-ivy 348
Ground-pine 58
Groundsel 165, 174
Gumweed 141
Hawk's-beard 127
Hawkweed 151
Hawthorn 449
Hazelnut 207
Hedge-nettle 358
Hog-peanut 296
Honeysuckle 239
Hop 235
Horsetail 49
Indian Grass 680
Indian-hemp 91
Indian-pipe 372
Ironplant 160
Joe-pye Weed 137
June Grass 652
Juniper 61
Knapweed 119
Knotweed 391
Labrador-tea 286
Lady's-slipper 614
Lamb's-quarters 264
Lettuce 154, 168
Licorice 307
Lily 601

Lobelia 364
Locoweed 317
Loosestrife 366, 411
Mallow 367
Manna Grass 646
Maple 72
Mare's-tail 343
Marsh-marigold 427
Meadow-rue 440
Medick 313
Milk-vetch 297
Milkweed 93
Mint 353
Mudwort 497
Muhly 657
Mullein 503
Mustard 218, 221, 227
Nannyberry 245
Nettle 510
Nut-grass 565
Oak 327
Oat 626
Oat Grass 675
Olive 280
Onion 598
Orchard Grass 637
Orchid 609, 618
Paintbrush 495
Parilla 370
Parsnip 85
Pepper-grass 224
Pigweed 76, 263
Pine 69
Pineappleweed 161
Plantain 384
Plum 460
Poison-ivy 78
Pond-lily 374
Pondweed 688
Poplar 476
Prairie-clover 305
Psoralea 318
Puccoon 211
Purslane 407
Pussytoes 107
Pygmyflower 408
Quack Grass 642
Ragweed 103, 153
Ragwort 164, 173
Raspberry 466
Reed Grass 633, 668
Rock Cress 214
Rose 463
Rubberweed 152
Rush 527, 581
Sage 114
Salt-meadow Grass 673
Sarsaparilla 92
Saskatoon 445
Saxifrage 487, 492
Scouring-rush 51
Sea-blite 274

Sedge 531
Shepherd's-purse 216
Silverweed 446
Skeletonweed 159
Skullcap 356
Slough Grass 627
Snakeroot 86, 387
Sneezeweed 143
Solomon's-seal 602
Sow-thistle 181
Spear Grass 648
Speedwell 504
Spike-rush 566
Spruce 67
Spurge 291
Starflower 414
Stickseed 209
Stinkgrass 644
Stinkweed 230
Stitchwort 257
Strawberry 451
Sumac 77
Sunflower 144
Sweet-clover 315
Sweet Flag 522
Sweet Grass 650
Switch Grass 663
Tamarack 66
Tansy 193
Tansy Mustard 218
Thistle 118, 121
Touch-me-not 199
Trefoil 312
Twinflower 238
Umbrellawort 373
Valerian 512
Vervain 513
Vetch 325
Vetchling 309
Violet 514
Virginia Creeper 519
Virgin's-bower 428
Wakerobin 606
Water-hemlock 81
Water-horehound 350
Water-milfoil 342
Water-parsnip 87
Water-plantain 523
Water Sedge 531
Wheatgrass 622, 664
Wild Rye 641, 643, 654, 672
Wild-rice 687
Willow 479
Winterfat 272
Wintergreen 415
Witch Grass 661
Wolf-willow 281
Wood-rush 589
Wood-sorrel 383
Wool-grass 574
Wormwood 112
Yarrow 100